Handbook of Mellin Transforms

Advances in Applied Mathematics

Series Editor: Daniel Zwillinger

CRC Standard Curves and Surfaces with Mathematica®, Third Edition
David H. von Seggern

Handbook of Peridynamic Modeling
Floriin Bobaru, John T. Foster, Philippe H. Geubelle, and Stewart A. Silling

Advanced Engineering Mathematics with MATLAB, Fourth Edition
Dean G. Duffy

Linear and Complex Analysis for Applications
John P. D'Angelo

Quadratic Programming with Computer Programs
Michael J. Best

Green's Functions with Applications, Second Edition
Dean G. Duffy

Introduction to Radar Analysis, Second Edition
Bassem R. Mahafza

CRC Standard Mathematical Tables and Formulas, 33rd Edition
Dan Zwillinger

The Second-Order Adjoint Sensitivity Analysis Methodology
Dan Gabriel Cacuci

Handbook of Mellin Transforms
Yury A. Brychkov, Oleg I. Marichev, and Nikolay V. Savischenko

Handbook of Mellin Transforms

Yu. A. Brychkov
O. I. Marichev
N. V. Savischenko

CRC Press
Taylor & Francis Group
Boca Raton London New York

CRC Press is an imprint of the
Taylor & Francis Group, an **informa** business

First published in paperback 2024

First published 2019 by CRC Press
2385 NW Executive Center Drive, Suite 320, Boca Raton FL 33431

and by CRC Press
4 Park Square, Milton Park, Abingdon, Oxon, OX14 4RN

CRC Press is an imprint of Taylor & Francis Group, LLC

Publisher's Note
The publisher has gone to great lengths to ensure the quality of this reprint but points out that some imperfections in the original copies may be apparent.

ISBN: 978-1-138-35335-0 (hbk)
ISBN: 978-1-03-291840-2 (pbk)
ISBN: 978-0-429-43425-9 (ebk)

DOI: 10.1201/9780429434259

Visit the Taylor & Francis Web site at
http://www.taylorandfrancis.com

and the CRC Press Web site at
http://www.crcpress.com

Contents

Preface

The Mellin transformation was introduced by a Finnish mathematician Robert Hjalmar Mellin in his paper "Über die fundamentale Wichtigkeit des Satzes von Cauchy für die Theorien der Gamma– und der hypergeometrischen Funktionen. Acta Soc. Fennicae, 1896, 21, 1–115." At present, it is widely used in various problems of pure and applied mathematics, in particular, in the theory of differential and integral equations, and the theory of Dirichlet series. It found extensive applications in mathematical physics, number theory, mathematical statistics, theory of asymptotic expansions, and especially, in the theory of special functions and integral transformations. Using the Mellin transformation, many classical integral transforms can be represented as compositions of direct and inverse Laplace transforms.

This handbook contains tables of the direct Mellin transforms of the form

$$F(s) = \mathfrak{M}[f(x);\, s] = \int_0^\infty x^{s-1} f(x)\, dx, \qquad s = \sigma + i\tau.$$

Since the majority of integrals can be reduced to the form of the corresponding Mellin transforms with a specific choice of parameters, this book can also be considered as a handbook of definite and indefinite integrals. By changes of variables, the Mellin transform can be turned into the Fourier and Laplace transforms.

The inverse Mellin transform has the form

$$f(x) = \mathfrak{M}^{-1}[F(s);\, x] = \frac{1}{2\pi i} \int_{\sigma-i\infty}^{\sigma+i\infty} x^{-s} F(s)\, ds, \qquad \alpha < \sigma < \beta;$$

see Appendix I.

The main text is introduced by a fairly detailed list of contents, from which the required formulas can easily be found. The tables are arranged in two columns. The left-hand column of each page shows function $f(x)$ and the right-hand column gives the corresponding Mellin transform $F(s)$. For the sake of compactness, abbreviated notation is used. For example, the formula 3.14.9.1 (the formula 1 of the Subsection 3.14.9)

No.	$f(x)$	$F(s)$
1	$\begin{Bmatrix} S(ax) \\ C(ax) \end{Bmatrix} K_\nu(bx)$	$\frac{2^{s+\delta-1} a^{\delta+1/2}}{3^\delta \sqrt{\pi}\, b^{s+\delta+1/2}} \Gamma\left(\frac{2s-2\nu+2\delta+1}{4}\right) \Gamma\left(\frac{2s+2\nu+2\delta+1}{4}\right) \times {}_3F_2\left(\begin{matrix} \frac{2\delta+1}{4},\ \frac{2s-2\nu+2\delta+1}{4},\ \frac{2s+2\nu+2\delta+1}{4} \\ \frac{2\delta+1}{2},\ \frac{2\delta+5}{4};\ -\frac{a^2}{b^2} \end{matrix}\right)$ $[a,\ \operatorname{Re} b > 0;\ \operatorname{Re} s > \lvert\operatorname{Re}\nu\rvert - (2 \pm 1)/2]$

where $\delta = \begin{Bmatrix} 1 \\ 0 \end{Bmatrix}$, is a contraction of the two formulas

1	$S\,(ax)\,K_\nu\,(bx)$	$\dfrac{2^s\,a^{3/2}}{3\sqrt{\pi}\,b^{s+3/2}}\,\Gamma\left(\dfrac{2s-2\nu+3}{2}\right)\Gamma\left(\dfrac{2s+2\nu+3}{2}\right)$ $\times\,{}_3F_2\left(\begin{matrix}\frac{3}{4},\ \frac{2s-2\nu+3}{2},\ \frac{2s+2\nu+3}{2}\\ \frac{3}{2},\ \frac{7}{4};\ -\frac{a^2}{b^2}\end{matrix}\right)$ $[a,\ \mathrm{Re}\,b>0;\ \mathrm{Re}\,s>\lvert\mathrm{Re}\,\nu\rvert-3/2]$

(in which only the upper sign and the upper expression in the curly brackets are taken) and

2	$C\,(ax)\,K_\nu\,(bx)$	$\dfrac{2^{s-1}\,a^{1/2}}{\sqrt{\pi}\,b^{s+1/2}}\,\Gamma\left(\dfrac{2s-2\nu+1}{2}\right)\Gamma\left(\dfrac{2s+2\nu+1}{2}\right)$ $\times\,{}_3F_2\left(\begin{matrix}\frac{1}{4},\ \frac{2s-2\nu+1}{2},\ \frac{2s+2\nu+1}{2}\\ \frac{1}{2},\ \frac{5}{4};\ -\frac{a^2}{b^2}\end{matrix}\right)$ $[a,\ \mathrm{Re}\,b>0;\ \mathrm{Re}\,s>\lvert\mathrm{Re}\,\nu\rvert-1/2]$

(in which only the lower sign and the lower expression in the curly brackets are taken).

The formula $a,\ b<\mathrm{Re}\,s<c,\ d$ is an abbreviated form of the inequality

$$\max\,(a,\ b)<\mathrm{Re}\,s<\min\,(c,\ d)\,.$$

In all chapters, unless other restrictions are indicated, $k,\ l,\ m,\ n,\ p,\ q=0,\ 1,\ 2,\ldots$

Some integrals are considered in the sense of the principal value.

Various functional relations that will be useful for evaluation of Mellin transforms are given at the beginning of every section. More formulas can be found at `http://functions.wolfram.com`.

In the preparation of this handbook, use was made, above all, of the books of H. Bateman, A. Erdélyi, W. Magnus, F. Oberhettinger, and F. G. Tricomi [1], Yu. A. Brychkov [3], O. I. Marichev [14], I. S. Gradshteyn and I. M. Ryzhik [13], V. A. Ditkin and A. P. Prudnikov [10], F. Oberhettinger [15], and A. P. Prudnikov, Yu. A. Brychkov, and O. I. Marichev [18–23]. An appreciable part of the formulas were obtained by the authors.

Appendix I contains some properties of Mellin transforms and examples of their application.

Appendix II is devoted to conditions of convergences of integrals.

The bibliographic sources and notations are given at the end of the book.

This handbook is intended for researchers, engineers, post-graduate students, university students, and generally for anyone who uses mathematical methods.

Chapter 1

General Formulas

1.1. Transforms Containing Arbitrary Functions

1.1.1. Basic formulas

Notation: $F_1(s) = \mathfrak{M}[f_1(x); s]$, $F_2(s) = \mathfrak{M}[f_2(x); s]$.

No.	$f(x)$	$F(s)$
1	$\frac{1}{2\pi i}\int_{c-i\infty}^{c+i\infty} F(s)\, x^{-s}\, ds$	$F(s)$
2	$\int_0^\infty f_1\left(\frac{x}{t}\right) f_2(t)\, \frac{dt}{t}$	$F_1(s)\, F_2(s)$

1.1.2. $f(ax^r)$ and the power function

Condition: $\operatorname{Im}\beta = 0$, $\beta \neq 0$.

1	$f(ax)$	$a^{-s}F(s)$
2	$x^\alpha f(x)$	$F(s+\alpha)$
3	$f(x^\beta)$	$\frac{1}{\lvert\beta\rvert} F\left(\frac{s}{\beta}\right)$
4	$f(ax^\beta)$	$\frac{1}{\lvert\beta\rvert} a^{-s/\beta} F\left(\frac{s}{\beta}\right)$
5	$x^\alpha f(x^\beta)$	$\frac{1}{\lvert\beta\rvert} F\left(\frac{s+\alpha}{\beta}\right)$
6	$x^\alpha f(ax^\beta)$	$\frac{1}{\lvert\beta\rvert} a^{-(s+\alpha)/\beta} F\left(\frac{s+\alpha}{\beta}\right)$

1.1.3. $f(ax^r)$ and elementary functions

Condition: $\operatorname{Im}\beta = 0,\ \beta \neq 0$.

1	$\ln x\, f(x)$	$F'(s)$
2	$\ln^m x\, f(x)$	$F^{(m)}(s)$
3	$x^\alpha \ln^m x\, f(x)$	$F^{(m)}(s+\alpha)$
4	$\ln^m x\, f(x^\beta)$	$\frac{\operatorname{sgn}\beta}{\beta^{m+1}} F^{(m)}\left(\frac{s}{\beta}\right)$
5	$\ln x\, f(ax)$	$a^{-s}\left[F'(s) - \ln a\, F(s)\right]$
6	$\ln^m x\, f(ax)$	$(-1)^m a^{-s} \sum_{k=0}^{m} (-1)^k \binom{m}{k} \ln^{m-k} a\, F^{(k)}(s)$
7	$\ln^m x\, f(ax^\beta)$	$\frac{(-1)^m \operatorname{sgn}\beta}{\beta^{m+1}} a^{-s/\beta} \sum_{k=0}^{m} (-1)^k \binom{m}{k} \ln^{m-k} a\, F^{(k)}\left(\frac{s}{\beta}\right)$
8	$x^\alpha \ln^m x\, f(x^\beta)$	$\frac{\operatorname{sgn}\beta}{\beta^{m+1}} F^{(m)}\left(\frac{s+\alpha}{\beta}\right)$
9	$x^\alpha \ln x\, f(ax^\beta)$	$\frac{\operatorname{sgn}\beta}{\beta^2} a^{-(s+\alpha)/\beta}\left[-\ln a\, F\left(\frac{s+\alpha}{\beta}\right) + F'\left(\frac{s+\alpha}{\beta}\right)\right]$
10	$x^\alpha \ln^m x\, f(ax^\beta)$	$\frac{(-1)^m \operatorname{sgn}\beta}{\beta^{m+1}} a^{-(s+\alpha)/\beta} \sum_{k=0}^{m} (-1)^k \binom{m}{k} \ln^{m-k} a\, F^{(k)}\left(\frac{s+\alpha}{\beta}\right)$
11	$x^\alpha e^{bx} f(ax^\beta)$	$\frac{\operatorname{sgn}\beta}{\beta} a^{-(s+\alpha)/\beta} \sum_{n=0}^{\infty} \frac{\left(a^{-1/\beta} b\right)^n}{n!} F\left(\frac{s+n+\alpha}{\beta}\right)$

1.1.4. Derivatives of $f(x)$

1	$f'(x)$	$(1-s)\, F(s-1)$ $\quad \left[x^{s-1} f(x)\big\vert_{x=0} = x^{s-1} f(x)\big\vert_{x=\infty} = 0\right]$
2	$f^{(n)}(x)$	$(-1)^n\, \Gamma\left[\begin{matrix} s \\ s-n \end{matrix}\right] F(s-n) = \Gamma\left[\begin{matrix} n+1-s \\ 1-s \end{matrix}\right] F(s-n)$ $\quad \left[\begin{matrix} x^{s-k} f^{(n-k)}(x)\big\vert_{x=0} = x^{s-k} f^{(n-k)}(x)\big\vert_{x=\infty} = 0, \\ k = 1, 2, \ldots, n \end{matrix}\right]$

No.	$f(x)$	$F(s)$
3	$\left(x\dfrac{d}{dx}\right)^n f(x)$	$(-s)^n F(s)$ $\left[\begin{array}{l} x^s \left(x\dfrac{d}{dx}\right)^k f(x)\Big\vert_{x=0} = x^s \left(x\dfrac{d}{dx}\right)^k f(x)\Big\vert_{x=\infty} = 0, \\ k = 0, 1, \dots, n-1 \end{array}\right]$
4	$\left(\dfrac{d}{dx}x\right)^n f(x)$	$(1-s)^n F(s)$ $\left[\begin{array}{l} x^s \left(\dfrac{d}{dx}x\right)^k f(x)\Big\vert_{x=0} = x^s \left(\dfrac{d}{dx}x\right)^k f(x)\Big\vert_{x=\infty} = 0, \\ k = 0, 1, \dots, n-1 \end{array}\right]$
5	$\left(x^{1-\alpha}\dfrac{d}{dx}\right)^n f(x)$	$(-\alpha)^n \,\Gamma\!\left[\begin{array}{c} \frac{s}{\alpha} \\ \frac{s-n\alpha}{\alpha} \end{array}\right] F(s-n\alpha)$
6		$= \alpha^n \,\Gamma\!\left[\begin{array}{c} \frac{(n+1)\alpha-s}{\alpha} \\ \frac{\alpha-s}{\alpha} \end{array}\right] F(s-n\alpha) \qquad [\alpha \neq 0]$ $\left[\begin{array}{l} x^{s-k\alpha} \left(x^{1-\alpha}\dfrac{d}{dx}\right)^{n-k} f(x)\Big\vert_{x=0} \\ \qquad = x^{s-k\alpha} \left(x^{1-\alpha}\dfrac{d}{dx}\right)^{n-k} f(x)\Big\vert_{x=\infty} = 0, \\ k = 1, 2, \dots, n \end{array}\right]$
7	$\left(\dfrac{d}{dx}x^{1-\beta}\right)^n f(x)$	$\beta^n \,\Gamma\!\left[\begin{array}{c} \frac{1-s+n\beta}{\beta} \\ \frac{1-s}{\beta} \end{array}\right] F(s-n\beta) \qquad [\beta \neq 0]$ $\left[\begin{array}{l} x^{s-k\beta} \left(\dfrac{d}{dx}x^{1-\beta}\right)^{n-k} f(x)\Big\vert_{x=0} \\ \qquad = x^{s-k\beta} \left(\dfrac{d}{dx}x^{1-\beta}\right)^{n-k} f(x)\Big\vert_{x=\infty} = 0, \\ k = 1, 2, \dots, n \end{array}\right]$
8	$\left(x^{1-\alpha}\dfrac{d}{dx}x^{1-\beta}\right)^n f(x)$	$(\alpha+\beta-1)^n \,\Gamma\!\left[\begin{array}{c} \frac{n(\alpha+\beta-1)+\alpha-s}{\alpha+\beta-1} \\ \frac{\alpha-s}{\alpha+\beta-1} \end{array}\right] F(s-n\alpha-n\beta+n)$ $[\alpha+\beta-1 \neq 0]$
9		$= \prod\limits_{k=0}^{n-1} [\alpha - s + k(\alpha+\beta-1)]\; F(s-n\alpha-n\beta+n)$ $\left[\begin{array}{l} x^{s-k(\alpha+\beta-1)} \left(x^{1-\alpha}\dfrac{d}{dx}x^{1-\beta}\right)^{n-k} f(x)\Big\vert_{x=0} \\ \quad = x^{s-k(\alpha+\beta-1)} \left(x^{1-\alpha}\dfrac{d}{dx}x^{1-\beta}\right)^{n-k} f(x)\Big\vert_{x=\infty} = 0, \\ k = 1, 2, \dots, n \end{array}\right]$

No.	$f(x)$	$F(s)$
10	$\left(x^{1-\alpha}\dfrac{d}{dx}\,x^{\alpha}\right)^{n} f(x)$	$(\alpha-s)^{n} F(s)$ $\left[\begin{aligned} & x^{s}\left(x^{1-\alpha}\frac{d}{dx}\,x^{\alpha}\right)^{n-k} f(x)\Big\vert_{x=0} \\ & \qquad = x^{s}\left(x^{1-\alpha}\frac{d}{dx}\,x^{\alpha}\right)^{n-k} f(x)\Big\vert_{x=\infty} = 0, \\ & k=1,\,2,\,\dots,\,n \end{aligned}\right]$
11	$\dfrac{\partial}{\partial a} f(x,\,a)$	$\dfrac{\partial}{\partial a} F(s,\,a)$

1.1.5. Integrals containing $f(x)$

Notation: $F_1(s)=\mathfrak{M}[f_1(x)\,;\,s]$, $F_2(s)=\mathfrak{M}[f_2(x)\,;\,s]$.

1	$\int_0^\infty f_1(xt)\, f_2(t)\, dt$	$F_1(s)\, F_2(1-s)$
2	$\int_0^\infty t^{\alpha} f_1(xt)\, f_2(t)\, dt$	$F_1(s)\, F_2(1-s+\alpha)$
3	$x^{\alpha}\int_0^\infty f_1(xt)\, f_2(t)\, dt$	$F_1(s+\alpha)\, F_2(1-s-\alpha)$
4	$x^{\alpha}\int_0^\infty t^{\beta} f_1(xt)\, f_2(t)\, dt$	$F_1(s+\alpha)\, F_2(1-s-\alpha+\beta)$
5	$\int_0^\infty f_1\left(\dfrac{x}{t}\right) f_2(t)\, dt$	$F_1(s)\, F_2(s+1)$
6	$\int_0^\infty t^{\alpha} f_1\left(\dfrac{x}{t}\right) f_2(t)\, dt$	$F_1(s)\, F_2(s+\alpha+1)$
7	$x^{\alpha}\int_0^\infty f_1\left(\dfrac{x}{t}\right) f_2(t)\, dt$	$F_1(s+\alpha)\, F_2(s+\alpha+1)$
8	$x^{\alpha}\int_0^\infty t^{\beta} f_1\left(\dfrac{x}{t}\right) f_2(t)\, dt$	$F_1(s+\alpha)\, F_2(s+\alpha+\beta+1)$
9	$\int_0^\infty f_1\left(\dfrac{t}{x}\right) f_2(t)\, dt$	$F_1(-s)\, F_2(s+1)$
10	$\int_0^\infty f_1(x^{\alpha} t^{\beta})\, f_2(t^{\gamma})\, dt$	$\dfrac{1}{\lvert\alpha\rvert} F_1\left(\dfrac{s}{\alpha}\right) \dfrac{1}{\lvert\gamma\rvert} F_2\left(\dfrac{\alpha-\beta s}{\alpha\gamma}\right)$ $[\alpha,\,\beta,\,\gamma\neq 0]$

No.	$f(x)$	$F(s)$
11	$\int_0^x f(t)\,dt$	$-\frac{1}{s}F(s+1)$ [$\operatorname{Re} s < 0$]
12	$\int_0^x \ldots \int_0^x f(t)\,(dt)^n = \int_0^x \frac{(x-t)^{n-1}}{(n-1)!} f(t)\,dt$	$\frac{(-1)^n}{(s)_n} F(s+n)$ [$\operatorname{Re} s < 1-n$]
13	$\int_0^x \frac{(x-t)^{\alpha-1}}{\Gamma(\alpha)} f(t)\,dt \equiv \left(I_{0+}^{\alpha} f\right)(x)$	$\Gamma\left[\begin{matrix}1-s-\alpha\\ 1-s\end{matrix}\right] F(s+\alpha)$ [$\operatorname{Re}\alpha > 0;\ \operatorname{Re}(s+\alpha) < 1$]
14	$\int_x^\infty f(t)\,dt$	$\frac{1}{s}F(s+1)$ [$\operatorname{Re} s > 0$]
15	$\int_x^\infty \ldots \int_x^\infty f(t)\,(dt)^n = \int_x^\infty \frac{(t-x)^{n-1}}{(n-1)!} f(t)\,dt$	$\frac{1}{(s)_n} F(s+n)$ [$\operatorname{Re} s > 0$]
16	$\int_x^\infty \frac{(t-x)^{\alpha-1}}{\Gamma(\alpha)} f(t)\,dt \equiv \left(I_{-}^{\alpha} f\right)(x)$	$\Gamma\left[\begin{matrix}s\\ s+\alpha\end{matrix}\right] F(s+\alpha)$ [$\operatorname{Re}\alpha,\ \operatorname{Re} s > 0$]
17	$x^\gamma \left(I_{0+}^{\alpha} x^\beta f\right)(x)$	$\Gamma\left[\begin{matrix}1-s-\alpha-\gamma\\ 1-s-\gamma\end{matrix}\right] F(s+\alpha+\beta+\gamma)$ [$\operatorname{Re}\alpha > 0;\ \operatorname{Re}(s+\alpha+\gamma) < 1$]
18	$x^\gamma \left(I_{-}^{\alpha} x^\beta f\right)(x)$	$\Gamma\left[\begin{matrix}s+\gamma\\ s+\alpha+\gamma\end{matrix}\right] F(s+\alpha+\beta+\gamma)$ [$\operatorname{Re}\alpha,\ \operatorname{Re}(s+\gamma) > 0$]
19	$\int_0^\infty e^{-xt} f(t)\,dt$	$\Gamma(s)\,F(1-s)$ [$\operatorname{Re} s > 0$]
20	$x^\alpha \int_0^\infty t^\beta e^{-xt} f(t)\,dt$	$\Gamma(s+\alpha)\,F(1-s-\alpha+\beta)$ [$\operatorname{Re}(s+\alpha) > 0$]
21	$\int_0^\infty e^{-t/x} f(t)\,dt$	$\Gamma(-s)\,F(s+1)$ [$\operatorname{Re} s < 0$]
22	$x^\alpha \int_0^\infty t^\beta e^{-t/x} f(t)\,dt$	$\Gamma(-s-\alpha)\,F(s+\alpha+\beta+1)$ [$\operatorname{Re}(s+\alpha) < 0$]
23	$\int_0^\infty e^{-x/t} f(t)\,dt$	$\Gamma(s)\,F(s+1)$ [$\operatorname{Re} s > 0$]

No.	$f(x)$	$F(s)$
24	$x^{\alpha}\int_0^{\infty} t^{\beta} e^{-x/t} f(t)\, dt$	$\Gamma(s+\alpha)\ F(s+\alpha+\beta+1)$ $[\operatorname{Re}(s+\alpha)>0]$
25	$\int_0^{\infty} \cos(xt)\, f(t)\, dt$	$\cos\frac{s\pi}{2}\, \Gamma(s)\, \mathfrak{M}[f(x);\, 1-s]$ $[\operatorname{Re} s>0]$
26	$\int_0^{\infty} \sin(xt)\, f(t)\, dt$	$\sin\frac{s\pi}{2}\, \Gamma(s)\, \mathfrak{M}[f(x);\, 1-s]$ $[\operatorname{Re} s>0]$
27	$\int_0^{\infty} \sqrt{xt}\, J_{\nu}(xt)\, f(t)\, dt$	$2^{s-1/2}\, \Gamma\begin{bmatrix} \frac{2s+2\nu+1}{4} \\ \frac{3-2s+2\nu}{4} \end{bmatrix} \mathfrak{M}[f(x);\, 1-s]$
28	$\int_0^{\infty} \sqrt{xt}\, K_{\nu}(xt)\, f(t)\, dt$	$2^{s-3/2}\, \Gamma\left(\frac{2s+2\nu+1}{4}\right) \Gamma\left(\frac{2s-2\nu+1}{4}\right) \mathfrak{M}[f(x);\, 1-s]$
29	$\int_0^{\infty} \sqrt{xt}\, Y_{\nu}(xt)\, f(t)\, dt$	$\frac{2^{s-1/2}}{\pi}\, \sin\frac{(2\nu-2s-3)\pi}{4}\, \Gamma\left(\frac{2s-2\nu+1}{4}\right)$ $\times\, \Gamma\left(\frac{2s+2\nu+1}{4}\right) \mathfrak{M}[f(x);\, 1-s]$
30	$\int_0^{\infty} \sqrt{xt}\, \mathbf{H}_{\nu}(xt)\, f(t)\, dt$	$2^{s-1/2}\, \tan\frac{(2s+2\nu+1)\pi}{4}\, \Gamma\begin{bmatrix} \frac{2s+2\nu+1}{4} \\ \frac{3-2s+2\nu}{4} \end{bmatrix} \mathfrak{M}[f(x);\, 1-s]$

Chapter 2

Elementary Functions

2.1. Algebraic Functions

More formulas can be obtained from the corresponding sections due to the relations

$$\frac{1}{\sqrt{z+1}+1}=\frac{1}{2}\,{}_2F_1\left(\begin{matrix}\frac{1}{2},\,1\\2;\,-z\end{matrix}\right),\quad \frac{1}{\sqrt{\sqrt{z+1}+1}}=\frac{1}{\sqrt{2}}\,{}_2F_1\left(\begin{matrix}\frac{1}{4},\,\frac{3}{4}\\\frac{3}{2};\,-z\end{matrix}\right),$$

$$\frac{1}{\sqrt{1-\sqrt{z}}}+\frac{1}{\sqrt{1+\sqrt{z}}}=2\,{}_2F_1\left(\begin{matrix}\frac{1}{4},\,\frac{3}{4}\\\frac{1}{2};\,z\end{matrix}\right),\quad \frac{1}{(1-\sqrt{z})^{3/2}}+\frac{1}{(1+\sqrt{z})^{3/2}}=2\,{}_2F_1\left(\begin{matrix}\frac{3}{4},\,\frac{5}{4}\\\frac{1}{2};\,z\end{matrix}\right),$$

$$(z+1)^a={}_1F_0\left(\begin{matrix}-a\\-z\end{matrix}\right)={}_2F_1\left(\begin{matrix}-a,\,b\\b;\,-z\end{matrix}\right),\quad (z+1)^a=\frac{1}{\Gamma(-a)}\,G_{11}^{11}\left(z\,\middle|\,\begin{matrix}a+1\\0\end{matrix}\right),$$

$$\frac{1}{1-z}=\pi\,G_{22}^{11}\left(z\,\middle|\,\begin{matrix}0,\,1/2\\0,\,1/2\end{matrix}\right),\quad (1-x)_+^{\alpha-1}=\Gamma(\alpha)\,G_{11}^{10}\left(x\,\middle|\,\begin{matrix}\alpha\\0\end{matrix}\right),\quad (x-1)_+^{\alpha-1}=\Gamma(\alpha)\,G_{11}^{01}\left(x\,\middle|\,\begin{matrix}\alpha\\0\end{matrix}\right).$$

2.1.1. $(a^r-x^r)_+^\alpha$ and $(x^r-a^r)_+^\alpha$

No.	$f(x)$	$F(s)$	
1	$\theta(a-x)$	$\frac{a^s}{s}$	$[a,\ \operatorname{Re} s>0]$
2	$\theta(x-a)$	$-\frac{a^s}{s}$	$[a>0;\ \operatorname{Re} s<0]$
3	$\theta(x-a)-\theta(x-b)$	$\frac{b^s-a^s}{s}$	$[0<a<b;\ \operatorname{Re} s>0]$
4	$\theta(a-x)\,x^\alpha$	$\frac{a^{s+\alpha}}{s+\alpha}$	$[a,\ \operatorname{Re}(s+\alpha)>0]$
5	$\theta(x-a)\,x^\alpha$	$-\frac{a^{s+\alpha}}{s+\alpha}$	$[a>0;\ \operatorname{Re}(s+\alpha)<0]$

No.	$f(x)$	$F(s)$
6	$(a-x)_+^{\alpha-1}$	$a^{s+\alpha-1}\,\mathrm{B}(\alpha,\,s)$ $\qquad [a,\ \mathrm{Re}\,\alpha,\ \mathrm{Re}\,s>0]$
7	$(x-a)_+^{\alpha-1}$	$a^{s+\alpha-1}\,\mathrm{B}(\alpha,\,1-\alpha-s)$ $\qquad [a,\ \mathrm{Re}\,\alpha>0;\ \mathrm{Re}\,(\alpha+s)<1]$
8	$(a^r-x^r)_+^{\alpha-1}$	$\dfrac{a^{s+(\alpha-1)r}}{r}\,\mathrm{B}\left(\dfrac{s}{r},\,\alpha\right)$ $\qquad [a,\ r,\ \mathrm{Re}\,\alpha,\ \mathrm{Re}\,s>0]$
9	$(x^r-a^r)_+^{\alpha-1}$	$\dfrac{a^{s+(\alpha-1)r}}{r}\,\mathrm{B}\left(\alpha,\,1-\alpha-\dfrac{s}{r}\right)$ $\qquad [a,\ r,\ \mathrm{Re}\,\alpha>0;\ \mathrm{Re}\,s<r\,(1-\mathrm{Re}\,\alpha)]$
10	$x^\alpha\,(a-x)_+^{\beta-1}$	$a^{s+\alpha+\beta-1}\,\mathrm{B}(s+\alpha,\,\beta)$ $\qquad [a,\ \mathrm{Re}\,\beta,\ \mathrm{Re}\,(s+\alpha)>0]$
11	$x^\alpha\,(x-a)_+^{\beta-1}$	$a^{s+\alpha+\beta-1}\,\mathrm{B}(1-s-\alpha-\beta,\,\beta)$ $\qquad [a,\ \mathrm{Re}\,\beta,\ \mathrm{Re}\,(s+\alpha+\beta)<1]$

2.1.2. $(ax+b)^\rho$ and $|x-a|^\rho$

1	$\dfrac{1}{a-x}$	$\pi a^{s-1}\cot(s\pi)$ $\qquad [a>0;\ 0<\mathrm{Re}\,s<1]$
2	$\dfrac{a}{a-x}-\displaystyle\sum_{k=0}^{n}\left(\dfrac{x}{a}\right)^k$	$\pi a^s\cot(s\pi)$ $\qquad [a>0;\ -n-1<\mathrm{Re}\,s<-n]$
3	$\dfrac{1}{(ax+b)^\rho}$	$\dfrac{b^{s-\rho}}{a^s}\,\mathrm{B}(s,\,\rho-s)$ $\qquad [0<\mathrm{Re}\,s<\mathrm{Re}\,\rho;\ \lvert\arg a\rvert,\ \lvert\arg b\rvert<\pi]$
4	$\dfrac{1}{(a-x)^n}$	$-\dfrac{\pi\,(-a)^{s-n}}{(n-1)!\sin(s\pi)}\displaystyle\prod_{k=1}^{n-1}(s-k)$ $\qquad [0<\mathrm{Re}\,s<n;\ n=1,\,2,\ldots;\ \lvert\arg(-a)\rvert<\pi]$
5	$\dfrac{1}{(x+a)^\rho}-\dfrac{1}{x^\rho}$	$a^{s-\rho}\,\mathrm{B}(s,\,\rho-s)$ $\qquad [-1<\mathrm{Re}\,s<0,\ \mathrm{Re}\,\rho;\ \lvert\arg a\rvert<\pi]$
6	$\dfrac{a^\rho}{(x+a)^\rho}+\dfrac{\rho x}{a}-1$	$a^s\,\mathrm{B}(s,\,\rho-s)$ $\qquad [-2<\mathrm{Re}\,s<-1,\ \mathrm{Re}\,\rho;\ \lvert\arg a\rvert<\pi]$
7	$\dfrac{a^\rho}{(x+a)^\rho}-\displaystyle\sum_{k=0}^{n}\binom{-\rho}{k}\left(\dfrac{x}{a}\right)^k$	$a^s\,\mathrm{B}(s,\,\rho-s)$ $\qquad [-n-1<\mathrm{Re}\,s<-n,\ \mathrm{Re}\,\rho;\ \lvert\arg a\rvert<\pi]$

No.	$f(x)$	$F(s)$
8	$\dfrac{1}{\lvert x-a\rvert^{\rho}}$	$a^{s-\rho}\sec\dfrac{\rho\pi}{2}\cos\dfrac{(2s-\rho)\pi}{2}\,\mathrm{B}(s,\,\rho-s)$
9		$=\dfrac{\pi a^{s-\rho}}{\Gamma(\rho)}\sec\dfrac{\rho\pi}{2}\,\Gamma\left[\begin{matrix}s,\,\rho-s\\ \frac{2s-\rho+1}{2},\,\frac{1-2s+\rho}{2}\end{matrix}\right]$ $[a>0;\ 0<\operatorname{Re}s<\operatorname{Re}\rho<1]$
10	$\dfrac{\operatorname{sgn}(a-x)}{\lvert x-a\rvert^{\rho}}$	$\pi\,a^{s-\rho}\csc\dfrac{\rho\pi}{2}\,\Gamma\left[\begin{matrix}s,\,\rho-s\\ \rho,\,\frac{2s-\rho+2}{2},\,\frac{\rho-2s}{2}\end{matrix}\right]$ $[a>0;\ 0<\operatorname{Re}s<\operatorname{Re}\rho<1]$

2.1.3. $(ax+b)^{\rho}(cx+d)^{\sigma}$

1	$\dfrac{1}{(ax+b)(cx+d)}$	$\dfrac{\pi(ac)^{1-s}}{(bc-ad)\sin(s\pi)}\left[(ad)^{s-1}-(bc)^{s-1}\right]$ $[0<\operatorname{Re}s<2;\ \lvert\arg(b/a)\rvert,\ \lvert\arg(d/c)\rvert<\pi]$
2	$\dfrac{1}{(x+a)(b-x)}$	$\dfrac{\pi}{a+b}\left[\dfrac{a^{s-1}}{\sin(s\pi)}+b^{s-1}\cot(s\pi)\right]$ $[b>0;\ 0<\operatorname{Re}s<2;\ \lvert\arg a\rvert<\pi]$
3	$\dfrac{1}{(x-a)(x-b)}$	$\pi\cot(s\pi)\,\dfrac{a^{s-1}-b^{s-1}}{b-a}$ $[a>b>0;\ 0<\operatorname{Re}s<2]$
4	$\dfrac{1}{(x+a)^{\rho}(x-b)}$	$a^{-\rho}(-b)^{s-1}\,\mathrm{B}(s,\,\rho-s+1)\,{}_2F_1\left(\begin{matrix}\rho,\,s;\,\frac{a+b}{a}\\ \rho+1\end{matrix}\right)$ $[a\neq0;\ 0<\operatorname{Re}s<\operatorname{Re}\rho+1;\ \lvert\arg a\rvert<\pi,\ \lvert\arg(-b)\rvert<\pi]$
5	$\dfrac{1}{(x+a)^{\rho}(x-b)}$	$-\dfrac{\pi b^{s-1}}{(a+b)^{\rho}}\cot[(s-\rho)\pi]-\dfrac{a^{s-\rho}}{a+b}\,\mathrm{B}(s,\,\rho-s)\,{}_2F_1\left(\begin{matrix}1,\,1-\rho;\,\frac{a}{a+b}\\ s-\rho+1\end{matrix}\right)$ $[a\neq0;\ b>0;\ 0<\operatorname{Re}s<\operatorname{Re}\rho+1]$
6	$\dfrac{1}{(ax+b)^{\rho}(cx+d)^{\sigma}}$	$\dfrac{d^{s-\sigma}}{b^{\rho}c^{s}}\,\mathrm{B}(s,\,\rho+\sigma-s)\,{}_2F_1\left(\begin{matrix}\rho,\,s;\,\frac{bc-ad}{bc}\\ \rho+\sigma\end{matrix}\right)$ $[0<\operatorname{Re}s<\operatorname{Re}(\rho+\sigma);\ \lvert\arg(b/a)\rvert,\ \lvert\arg(d/c)\rvert<\pi]$

2.1.4. $(a-x)_+^{\rho}(bx+c)^{\sigma}$ and $(x-a)_+^{\rho}(bx+c)^{\sigma}$

1	$\dfrac{\theta(a-x)}{x+a}$	$\dfrac{a^{s-1}}{2}\left[\psi\left(\dfrac{s+1}{2}\right)-\psi\left(\dfrac{s}{2}\right)\right]$ $[a,\ \operatorname{Re}s>0]$
2	$\dfrac{\theta(a-x)}{(bx+c)^{\rho}}$	$\dfrac{a^{s}}{sc^{\rho}}\,{}_2F_1\left(\begin{matrix}\rho,\,s;\,-\frac{ab}{c}\\ s+1\end{matrix}\right)$ $\left[\begin{matrix}a,\ \operatorname{Re}s>0;\\ \lvert\arg(bx+c)\rvert<\pi\text{ for }0\le x\le a\end{matrix}\right]$

No.	$f(x)$	$F(s)$
3	$\dfrac{\theta(x-a)}{(bx+c)^{\rho}}$	$\dfrac{a^{s-\rho}b^{-\rho}}{\rho-s}\, {}_2F_1\left(\begin{matrix}\rho,\ \rho-s;\ -\frac{c}{ab}\\ 1-s+\rho\end{matrix}\right)$ $[a>0;\ b\neq 0;\ \operatorname{Re}s<\operatorname{Re}\rho;\ \lvert\arg(bx+c)\rvert<\pi \text{ for } x\geq a]$
4	$(a-x)_+^{\rho}\,(bx+c)^{\rho}$	$\left(\dfrac{ac}{b}\right)^{(s+\rho)/2}(ab+c)^{\rho}\,\Gamma(\rho+1)\,\Gamma(s)\,\mathrm{P}_{\rho}^{-s-\rho}\left(\dfrac{c-ab}{c+ab}\right)$ $[a,\ \operatorname{Re}s>0;\ \operatorname{Re}\rho>-1;\ \lvert\arg(bx+c)\rvert<\pi \text{ for } 0\leq x\leq a]$
5	$(a-x)_+^{\rho}\,(bx+c)^{\sigma}$	$a^{s+\rho}c^{\sigma}\,\mathrm{B}(\rho+1,\ s)\ {}_2F_1\left(\begin{matrix}-\sigma,\ s;\ -\frac{ab}{c}\\ s+\rho+1\end{matrix}\right)$ $[a,\ \operatorname{Re}s>0;\ \operatorname{Re}\rho>-1;\ \lvert\arg(bx+c)\rvert<\pi \text{ for } 0\leq x\leq a]$
6	$(x-a)_+^{\rho}\,(bx+c)^{\rho}$	$\left(\dfrac{ac}{b}\right)^{(s+\rho)/2}(ab+c)^{\rho}\,\Gamma(\rho+1)\,\Gamma(-s-2\rho)\,\mathrm{P}_{\rho}^{s+\rho}\left(\dfrac{ab-c}{ab+c}\right)$ $\left[\begin{matrix}a>0;\ \operatorname{Re}\rho>-1;\ \operatorname{Re}s<-2\operatorname{Re}\rho;\\ \lvert\arg(bx+c)\rvert<\pi \text{ for } x\geq a\end{matrix}\right]$
7	$(x-a)_+^{\rho}\,(bx+c)^{\sigma}$	$a^{s+\rho+\sigma}\,b^{\sigma}\,\mathrm{B}(\rho+1,\ -s-\rho-\sigma)\ {}_2F_1\left(\begin{matrix}-\sigma,\ -s-\rho-\sigma\\ 1-s-\sigma;\ -\frac{c}{ab}\end{matrix}\right)$ $\left[\begin{matrix}a>0;\ \operatorname{Re}\rho>-1;\ \operatorname{Re}s<-\operatorname{Re}(\rho+\sigma)\\ \lvert\arg(bx+c)\rvert<\pi \text{ for } x\geq a\end{matrix}\right]$
8	$\dfrac{(a-x)_+^{\rho}}{(bx+c)^{\rho+1/2}}$	$\dfrac{a^{s+\rho}}{c^{\rho+1/2}}\,\mathrm{B}(s,\ \rho+1)\ {}_2F_1\left(\begin{matrix}\frac{2\rho+1}{2},\ s;\ -\frac{ab}{c}\\ s+\rho+1\end{matrix}\right)$ $[a,\ \operatorname{Re}s>0;\ \operatorname{Re}\rho>-1;\ \lvert\arg(bx+c)\rvert<\pi \text{ for } 0\leq x\leq a]$
9	$\dfrac{(a-x)_+^{\rho}}{(bx+c)^{\rho+3/2}}$	$\dfrac{a^{s+\rho}}{c^{\rho+3/2}}\,\mathrm{B}(s,\ \rho+1)\ {}_2F_1\left(\begin{matrix}\frac{2\rho+3}{2},\ s;\ -\frac{ab}{c}\\ s+\rho+1\end{matrix}\right)$ $[a,\ \operatorname{Re}s>0;\ \operatorname{Re}\rho>-1;\ \lvert\arg(bx+c)\rvert<\pi \text{ for } 0\leq x\leq a]$
10	$\dfrac{(x-a)_+^{\rho}}{(bx+c)^{\rho+1/2}}$	$\dfrac{a^{s-1/2}}{b^{\rho+1/2}}\,\mathrm{B}\left(\dfrac{1-2s}{2},\ \rho+1\right)\ {}_2F_1\left(\begin{matrix}\frac{2\rho+1}{2},\ \frac{1-2s}{2}\\ \frac{3-2s+2\rho}{2};\ -\frac{c}{ab}\end{matrix}\right)$ $\left[\begin{matrix}a>0;\ \operatorname{Re}\rho>-1;\ \operatorname{Re}s<1/2;\\ \lvert\arg(bx+c)\rvert<\pi \text{ for } x\geq a\end{matrix}\right]$
11	$\dfrac{(x-a)_+^{\rho}}{(bx+c)^{\rho+3/2}}$	$\dfrac{a^{s-3/2}}{b^{\rho+3/2}}\,\mathrm{B}\left(\dfrac{3-2s}{2},\ \rho+1\right)\ {}_2F_1\left(\begin{matrix}\frac{2\rho+3}{2},\ \frac{3-2s}{2}\\ \frac{5-2s+2\rho}{2};\ -\frac{c}{ab}\end{matrix}\right)$ $\left[\begin{matrix}a>0;\ \operatorname{Re}\rho>-1;\ \operatorname{Re}s<3/2\\ \lvert\arg(bx+c)\rvert<\pi \text{ for } x\geq a\end{matrix}\right]$

2.1.5. $(ax^\mu + b)^\rho (cx^\nu + d)^\sigma$

1	$\dfrac{1}{(ax^\mu+1)^\rho (bx^\mu+1)^r}$	$\dfrac{a^{-s/\mu}}{\mu}\,\mathrm{B}\left(\dfrac{s}{\mu},\,\rho+r-\dfrac{s}{\mu}\right)\,{}_2F_1\left(\begin{matrix} r,\,\frac{s}{\mu};\,\frac{a-b}{a} \\ \rho+r\end{matrix}\right)$ $[\mu>0;\,0<\mathrm{Re}\,s<\mu\,\mathrm{Re}\,(\rho+r);\,\lvert\arg a\rvert,\,\lvert\arg b\rvert<\pi]$
2	$\dfrac{1}{(x+a)(x^2+b^2)}$	$\dfrac{\pi}{2(a^2+b^2)}\left[\dfrac{ab^{s-2}}{\sin(s\pi/2)}-\dfrac{b^{s-1}}{\cos(s\pi/2)}+\dfrac{2a^{s-1}}{\sin(s\pi)}\right]$ $[\mathrm{Re}\,b>0;\,0<\mathrm{Re}\,s<3;\,\lvert\arg a\rvert<\pi]$
3	$\dfrac{1}{(x^2+a^2)(b^2-x^2)}$	$\dfrac{\pi}{2(a^2+b^2)}\left(a^{s-2}\csc\dfrac{s\pi}{2}+b^{s-2}\cot\dfrac{s\pi}{2}\right)$ $[a^2+b^2\neq 0;\,0<\mathrm{Re}\,s<4]$
4	$\dfrac{1}{(x^2+a)(x^2+b)}$	$\dfrac{\pi}{2(a-b)}\csc\dfrac{s\pi}{2}\left(b^{s/2-1}-a^{s/2-1}\right)$ $[0<\mathrm{Re}\,s<4;\,\lvert\arg a\rvert<\pi;\,\lvert\arg b\rvert<\pi]$
5	$\dfrac{1}{\left(x^{1/n}+a^{1/n}\right)^\rho}$	$na^{s-\rho/n}\,\mathrm{B}\,(ns,\,\rho-ns)$ $\quad [a>0;\,0<n\,\mathrm{Re}\,s<\mathrm{Re}\,\rho]$
6	$\dfrac{(x/a)^\alpha-(x/a)^\beta}{x-a}$	$\pi\,a^{s-1}\dfrac{\sin[(\alpha-\beta)\pi]}{\sin[(s+\alpha)\pi]\sin[(s+\beta)\pi]}$ $[a>0;\,-\mathrm{Re}\,\alpha,\,-\mathrm{Re}\,\beta<\mathrm{Re}\,s<1-\mathrm{Re}\,\alpha,\,1-\mathrm{Re}\,\beta]$
7	$\dfrac{x^\mu-1}{x^\nu-1}$	$\dfrac{\pi}{\nu}\sin\dfrac{\mu\pi}{\nu}\csc\dfrac{s\pi}{\nu}\csc\dfrac{(s+\mu)\pi}{\nu}$ $\quad [0<\mathrm{Re}\,s<\nu-\mu]$
8	$\dfrac{x^\mu-1}{x^{\mu n}-1}$	$\dfrac{\pi}{\mu n}\sin\dfrac{\pi}{n}\csc\dfrac{s\pi}{\mu n}\csc\dfrac{(s+\mu)\pi}{\mu n}$ $\quad [0<\mathrm{Re}\,s<(n-1)\,\mu;\,n\geq 2]$
9	$\dfrac{x-1}{x^n-1}$	$\dfrac{\pi}{n}\sin\dfrac{\pi}{n}\csc\dfrac{s\pi}{n}\csc\dfrac{(s+1)\pi}{n}$ $\quad [0<\mathrm{Re}\,s<n-1;\,n\geq 2]$
10	$\dfrac{x^\mu-a^\mu}{x-a}$	$\pi a^{s+\mu-1}\sin(\mu\pi)\csc(s\pi)\csc[(s+\mu)\pi]$ $[a>0;\,0<\mathrm{Re}\,s<1;\,0<\mathrm{Re}\,(s+\mu)<1]$
11	$\dfrac{x^\mu-x^{-\mu}}{x^\nu-x^{-\nu}}$	$\dfrac{\pi\sin(\mu\pi/\nu)}{\nu[\cos(\mu\pi/\nu)+\cos(s\pi/\nu)]}$ $[-\mathrm{Re}\,(\mu+\nu),\,\mathrm{Re}\,(\mu-\nu)<\mathrm{Re}\,s<\mathrm{Re}\,(\mu+\nu),\,\mathrm{Re}\,(\nu-\mu)]$

2.1.6. $(a-x)_+^{\alpha-1}(x^n+b^n)^r$ and $(x-a)_+^{\alpha-1}(x^n+b^n)^r$

1	$(a-x)_+^{\alpha-1}(x^n+b^n)^r$	$a^{s+\alpha-1}b^{nr}\,\mathrm{B}(s,\,\alpha)\;{}_{n+1}F_n\left(\begin{matrix}-r,\,\frac{s}{n},\,\frac{s+1}{n},\ldots,\frac{s+n-1}{n};\,-\left(\frac{a}{b}\right)^n\\ \frac{s+\alpha}{n},\,\frac{s+\alpha+1}{n},\ldots,\frac{s+\alpha+n-1}{n}\end{matrix}\right)$ $[a,\ \mathrm{Re}\,\alpha>0;\ b\neq0;\ \mathrm{Re}\,s>0;\ n=1,2,\ldots]$
2	$(x-a)_+^{\alpha-1}(x^n+b^n)^r$	$a^{s+nr+\alpha-1}\,\mathrm{B}(1-s-nr-\alpha,\,\alpha)$ $\times\;{}_{n+1}F_n\left(\begin{matrix}-r,\,-\frac{s+nr+\alpha-1}{n},\,-\frac{s+nr+\alpha-2}{n},\ldots,-\frac{s+nr+\alpha-n}{n}\\ -\frac{s+nr-1}{n},\,-\frac{s+nr-2}{n},\ldots,-\frac{s+nr-n}{n};\,-\left(\frac{b}{a}\right)^n\end{matrix}\right)$ $[a,\ \mathrm{Re}\,\alpha>0;\ b\neq0;\ \mathrm{Re}\,s<1-nr-\alpha;\ n=1,2,\ldots]$

2.1.7. $(ax^2+bx+c)^\rho\,(dx+e)$

1	$\dfrac{1}{ax^2+bx+c}$	$-\dfrac{\pi}{\sqrt{b^2-4ac}}\left[\csc(s\pi)\left(\dfrac{\sqrt{b^2-4ac}+b}{2a}\right)^{s-1}+\cot(s\pi)\left(\dfrac{\sqrt{b^2-4ac}-b}{2a}\right)^{s-1}\right]$ $\left[\begin{matrix}a,b,c\text{ are real; }a>0;\ b^2-4ac>0;\\ -\sqrt{b^2-4ac}-b<0<\sqrt{b^2-4ac}-b;\ 0<\mathrm{Re}\,s<2\end{matrix}\right]$
2		$=\dfrac{\pi\cot(s\pi)}{\sqrt{b^2-4ac}}\left[\left(\dfrac{-\sqrt{b^2-4ac}-b}{2a}\right)^{s-1}-\left(\dfrac{\sqrt{b^2-4ac}-b}{2a}\right)^{s-1}\right]$ $\left[\begin{matrix}a,b,c\text{ are real; }a>0;\ b^2-4ac>0;\\ \sqrt{b^2-4ac}+b<0;\ 0<\mathrm{Re}\,s<2\end{matrix}\right]$
3		$=\dfrac{\pi\csc(s\pi)}{\sqrt{b^2-4ac}}\left[\left(\dfrac{b-\sqrt{b^2-4ac}}{2a}\right)^{s-1}-\left(\dfrac{\sqrt{b^2-4ac}+b}{2a}\right)^{s-1}\right]$ $\left[\begin{matrix}(\lvert\mathrm{Im}\,a\rvert+\lvert\mathrm{Im}\,b\rvert+\lvert\mathrm{Im}\,c\rvert\neq0)\text{ or }(a,b,c\text{ are real; }a>0;\\ b^2-4ac>0;\ \sqrt{b^2-4ac}-b<0;\ 0<\mathrm{Re}\,s<2)\end{matrix}\right]$
4	$\dfrac{1}{ax^2+bx+a}$	$\dfrac{2\pi\cot(s\pi)}{\sqrt{b^2-4a^2}}\sinh\left[(s-1)\ln\dfrac{-\sqrt{b^2-4a^2}-b}{2a}\right]$ $\left[\begin{matrix}a,b\text{ are real; }a>0;\ b^2-4a^2>0;\\ \sqrt{b^2-4a^2}+b<0;\end{matrix}\right]$
5		$=\dfrac{2\pi\csc(s\pi)}{\sqrt{b^2-4a^2}}\sinh\left[(s-1)\ln\dfrac{b-\sqrt{b^2-4a^2}}{2a}\right]$ $[\lvert\mathrm{Im}\,a\rvert+\lvert\mathrm{Im}\,b\rvert\neq0;\ 0<\mathrm{Re}\,s<2]$
6	$\dfrac{1}{x^2+2x\cos(\beta\pi)+1}$	$-\dfrac{\pi}{\sin(\beta\pi)}\,\Gamma\left[\begin{matrix}s,\,1-s\\ \beta s-\beta,\,1-\beta s+\beta\end{matrix}\right]$ $[\lvert\beta\rvert<1;\ 0<\mathrm{Re}\,s<2]$

No.	$f(x)$	$F(s)$
7	$\dfrac{x+a}{(x+b)(x+c)}$	$\dfrac{\pi}{\sin(s\pi)}\left[\dfrac{b-a}{b-c}\,b^{s-1}+\dfrac{c-a}{c-b}\,c^{s-1}\right]$ $\left[\begin{matrix}0<\operatorname{Re}s<1;\\ \lvert\arg b\rvert,\ \lvert\arg c\rvert<\pi\end{matrix}\right]$
8	$\dfrac{x+a}{(x+a)^2+b^2}$	$\dfrac{\pi}{\sin(s\pi)}\left(a^2+b^2\right)^{s/2-1/2}\cos\left[(1-s)\arctan\dfrac{b}{a}\right]$ $[ab\neq 0;\ 0<\operatorname{Re}s<1]$
9	$\dfrac{1}{(ax^2+2bx+c)^{\rho}}$	$a^{-s/2}c^{s/2-\rho}\,\mathrm{B}(s,\ 2\rho-s)\ {}_2F_1\left(\begin{matrix}\frac{s}{2},\ \frac{2\rho-s}{2}\\ \frac{2\rho+1}{2};\ 1-\frac{b^2}{ac}\end{matrix}\right)$ $[a>0;\ b^2<ac;\ 0<\operatorname{Re}s<2\operatorname{Re}\rho]$

2.1.8. Algebraic functions of $\sqrt{ax+b}$

1	$\dfrac{1}{\left(\sqrt{x+a}\pm\sqrt{a}\right)^{\rho}}$	$\pm\rho\,(4a)^{s-\rho/2}\,\Gamma\left[\begin{matrix}\frac{2s-(1\mp1)\rho}{2},\ \rho-2s\\ \frac{2-2s+(1\pm1)\rho}{2}\end{matrix}\right]$ $[(1\mp1)\operatorname{Re}\rho/2<\operatorname{Re}s<\operatorname{Re}\rho/2;\ \lvert\arg a\rvert<\pi]$
2	$\dfrac{1}{\left(\sqrt{x+a}\pm\sqrt{x}\right)^{\rho}}$	$\pm\rho\,2^{-2s}a^{s-\rho/2}\,\Gamma\left[\begin{matrix}2s,\ \frac{\pm\rho-2s}{2}\\ \frac{2s\pm\rho+2}{2}\end{matrix}\right]$ $[0<\operatorname{Re}s<\pm\operatorname{Re}\rho/2;\ \lvert\arg a\rvert<\pi]$
3	$\dfrac{1}{\sqrt{x+a}\left(\sqrt{x+a}\pm\sqrt{a}\right)^{\rho}}$	$2^{2s-\rho}a^{s-(\rho+1)/2}\,\mathrm{B}\left(\dfrac{2s-(1\mp1)\rho}{2},\ 1-2s+\rho\right)$ $[(1\mp1)\operatorname{Re}\rho/2<\operatorname{Re}s<(\operatorname{Re}\rho+1)/2;\ \lvert\arg a\rvert<\pi]$
4	$\dfrac{1}{\sqrt{x+a}\left(\sqrt{x+a}\pm\sqrt{x}\right)^{\rho}}$	$2^{1-2s}a^{s+(\rho-1)/2}\,\mathrm{B}\left(2s,\ \dfrac{1-2s\mp\rho}{2}\right)$ $[0<\operatorname{Re}s<(1\mp\operatorname{Re}\rho)/2;\ \lvert\arg a\rvert<\pi]$
5	$\dfrac{1}{\left(\sqrt{x}+\sqrt{a}\right)^{\rho}}+\dfrac{1}{\lvert\sqrt{x}-\sqrt{a}\rvert^{\rho}}$	$2\sqrt{\pi}\,a^{s-\rho/2}\,\Gamma\left[\begin{matrix}\frac{1-\rho}{2},\ \frac{\rho-2s}{2},\ s\\ \frac{\rho}{2},\ \frac{2s-\rho+1}{2},\ \frac{1-2s}{2}\end{matrix}\right]$ $[a>0;\ 0<\operatorname{Re}s<\operatorname{Re}\rho/2<1/2]$
6	$\dfrac{1}{\left(x+a+\sqrt{a(2x+a)}\right)^{\rho}}$	$\rho 2^{s-\rho+1}a^{s-\rho}\,\Gamma\left[\begin{matrix}2\rho-2s,\ s\\ 1-s+2\rho\end{matrix}\right]$ $[0<\operatorname{Re}s<\operatorname{Re}\rho;\ \lvert\arg a\rvert<\pi]$

No.	$f(x)$	$F(s)$
7	$\dfrac{(2x+a)^{-1/2}}{\left[x+a+\sqrt{a(2x+a)}\right]^{\rho}}$	$2^{s-\rho}\,a^{s-\rho-1/2}\,\mathrm{B}\,(1-2s+2\rho,\,s)$ $[0<\operatorname{Re}s<\operatorname{Re}\rho+1/2;\ \lvert\arg a\rvert<\pi]$
8	$\dfrac{(x+a)^{-1/2}}{\left[x+a+b+2\sqrt{b\,(x+a)}\,\right]^{\rho}}$	$2^{2s-2\rho}a^{s-\rho-1}\sqrt{b}\,\mathrm{B}\,(1-2s+2\rho,\,s)\ {}_2F_1\!\left(\begin{matrix}\frac{2\rho+1}{2},\,1-s+\rho\\ 1-s+2\rho;\,\frac{a-b}{a}\end{matrix}\right)$ $[0<\operatorname{Re}s<\operatorname{Re}\rho+1/2;\ \lvert\arg a\rvert,\,\lvert\arg b\rvert,\,\lvert\arg(b/a)\rvert<\pi]$

2.1.9. Algebraic functions of $\sqrt{ax^2+bx+c}$

1	$\dfrac{1}{\sqrt{x^2+2x\cos\beta+1}}$	$\dfrac{\pi}{\sin(s\pi)}\,P_{s-1}(\cos\beta)$ $[\lvert\beta\rvert<\pi;\ 0<\operatorname{Re}s<1]$
2	$\dfrac{1}{\left(\sqrt{x^2+a^2}\pm a\right)^{\rho}}$	$\pm 2^{s-\rho-1}\rho a^{s-\rho}\,\Gamma\!\left[\begin{matrix}\frac{s-(1\mp1)\rho}{2},\,\rho-s\\ \frac{2-s+(1\pm1)\rho}{2}\end{matrix}\right]$ $[\operatorname{Re}a>0;\ (1\mp1)\,\rho<\operatorname{Re}s<\operatorname{Re}\rho]$
3	$\dfrac{1}{\left(\sqrt{x^2+a^2}\pm x\right)^{\rho}}$	$\pm\dfrac{\rho a^{s-\rho}}{2^{s+1}}\,\Gamma\!\left[\begin{matrix}s,\,\frac{\pm\rho-s}{2}\\ \frac{s\pm\rho+2}{2}\end{matrix}\right]$ $[\operatorname{Re}a>0;\ 0<\operatorname{Re}s<\pm\operatorname{Re}\rho]$
4	$\dfrac{1}{\sqrt{x^2+1}\left(\sqrt{x^2+1}+a\right)^{\rho}}$	$\dfrac{a^{-\rho}}{2}\,\mathrm{B}\!\left(\dfrac{1-s}{2},\,\dfrac{s}{2}\right)\,{}_2F_1\!\left(\begin{matrix}\frac{\rho}{2},\,\frac{\rho+1}{2}\\ \frac{s+1}{2};\,\frac{1}{a^2}\end{matrix}\right)$ $+\,a^{s-\rho-1}\,\mathrm{B}\,(s-1,\,1-s+\rho)\ {}_2F_1\!\left(\begin{matrix}\frac{1-s+\rho}{2},\,\frac{2-s+\rho}{2}\\ \frac{3-s}{2};\,\frac{1}{a^2}\end{matrix}\right)$ $[\operatorname{Re}a>-1;\ 0<\operatorname{Re}s<\operatorname{Re}\rho+1]$
5	$\dfrac{\left(x^2+1\right)^{-1/2}}{\left(\cos\beta\pm i\sin\beta\sqrt{x^2+1}\right)^{\rho}}$	$\left(\dfrac{\sin\beta}{2}\right)^{(1-s)/2}\Gamma\!\left[\begin{matrix}\frac{s}{2},\,1-s+\rho\\ \rho\end{matrix}\right]$ $\times\left[\dfrac{1}{\sqrt{\pi}}\,\mathrm{Q}_{\rho-(s+1)/2}^{(s-1)/2}(\cos\beta)\mp\dfrac{i\sqrt{\pi}}{2}\,\mathrm{P}_{\rho-(s+1)/2}^{(s-1)/2}(\cos\beta)\right]$ $[0<\operatorname{Re}s<\operatorname{Re}\rho+1]$
6	$\dfrac{\left(x^2+1\right)^{-1/2}}{\left(\sqrt{(a^2-1)(x^2+1)}+a\right)^{\rho}}$	$\dfrac{\left(a^2-1\right)^{-\rho/2}}{2}\,\mathrm{B}\!\left(\dfrac{s}{2},\,\dfrac{1-s+\rho}{2}\right)\,{}_2F_1\!\left(\begin{matrix}\frac{\rho}{2},\,\frac{1-s+\rho}{2}\\ \frac{1}{2};\,\frac{a^2}{a^2-1}\end{matrix}\right)$ $-\,\dfrac{\left(a^2-1\right)^{-(\rho+1)/2}}{2a\,(1-s+\rho)}\,\mathrm{B}\!\left(\dfrac{s}{2},\,\dfrac{2-s+\rho}{2}\right)\left[{}_2F_1\!\left(\begin{matrix}\frac{\rho+1}{2},\,\frac{2-s+\rho}{2}\\ -\frac{1}{2};\,\frac{a^2}{a^2-1}\end{matrix}\right)\right.$ $\left.-\,\left(1+a^2(2-s+2\rho)\right)\,{}_2F_1\!\left(\begin{matrix}\frac{\rho+1}{2},\,\frac{2-s+\rho}{2}\\ \frac{1}{2};\,\frac{a^2}{a^2-1}\end{matrix}\right)\right]$ $[\operatorname{Re}a>1;\ \operatorname{Re}\rho>0;\ \operatorname{Re}s<\operatorname{Re}\rho+1]$

No.	$f(x)$	$F(s)$
7	$\dfrac{1}{\sqrt{x^2+a^2}\left(\sqrt{x^2+a^2}+a\right)^{\rho}}$	$(2a)^{s-\rho-1}\,\mathrm{B}\left(\frac{s}{2},\,1-s+\rho\right)$ $[\operatorname{Re}a>0;\;0<\operatorname{Re}s<\operatorname{Re}\rho+1]$
8	$\dfrac{1}{\sqrt{x^2+a^2}\left(\sqrt{x^2+a^2}+b\right)^{\rho}}$	$(2a)^{s-\rho-1}\,\mathrm{B}\left(\frac{s}{2},\,1-s+\rho\right)\,{}_2F_1\left(\begin{matrix}1-s+\rho,\;\rho\\ \frac{2-s+2\rho}{2};\;\frac{a-b}{2a}\end{matrix}\right)$ $\left[\begin{matrix}\operatorname{Re}a>0;\;0<\operatorname{Re}s<\operatorname{Re}\rho+1;\\ \lvert\arg(b/a+1)\rvert<\pi\end{matrix}\right]$
9	$\dfrac{1}{\sqrt{x^2+a^2}\left(\sqrt{x^2+a^2}-a\right)^{\rho}}$	$(2a)^{s-\rho-1}\,\mathrm{B}\left(\frac{s}{2}-\rho,\,1-s+\rho\right)$ $[\operatorname{Re}a>0;\;2\operatorname{Re}\rho<\operatorname{Re}s<\operatorname{Re}\rho+1]$
10	$\dfrac{1}{\sqrt{x^2+a^2}\left(\sqrt{x^2+a^2}\pm x\right)^{\rho}}$	$2^{-s}a^{s-\rho-1}\,\mathrm{B}\left(s,\,\frac{1-s\pm\rho}{2}\right)$ $[\operatorname{Re}a>0;\;0<\operatorname{Re}s<1\pm\operatorname{Re}\rho]$
11	$\dfrac{1}{\sqrt{x^2+a^2}\left(\sqrt{x^2+a^2}+bx\right)^{\rho}}$	$2^{-s}a^{s-\rho-1}\,\mathrm{B}\left(s,\,\frac{1-s+\rho}{2}\right)\,{}_2F_1\left(\begin{matrix}\rho,\;s;\;\frac{1-b}{2}\\ \frac{s+\rho+1}{2}\end{matrix}\right)$ $[\operatorname{Re}a>0;\;0<\operatorname{Re}s<\operatorname{Re}\rho+1;\;\lvert\arg(b+1)\rvert<\pi]$
12	$\dfrac{1}{\left(x+a+\sqrt{(x+a)^2-a^2}\right)^{\rho}}$	$2^{1-s}\rho a^{s-\rho}\,\Gamma\left[\begin{matrix}2s,\;\rho-s\\ s+\rho+1\end{matrix}\right]$ $[0<\operatorname{Re}s<\operatorname{Re}\rho;\;\lvert\arg a\rvert<\pi]$
13	$\dfrac{1}{\left(x+a+\sqrt{(x+a)^2-b^2}\right)^{\rho}}$	$2^{-\rho}a^{s-\rho}\,\mathrm{B}(s,\,\rho-s)\,{}_2F_1\left(\begin{matrix}\frac{\rho-s}{2},\;\frac{\rho-s+1}{2}\\ \rho+1;\;\frac{b^2}{a^2}\end{matrix}\right)$ $[\lvert b\rvert\le\lvert a\rvert;\;0<\operatorname{Re}s<\operatorname{Re}\rho;\;\lvert\arg a\rvert<\pi]$
14		$=\dfrac{\rho\left(a^2-b^2\right)^{s/2}}{(ib)^{\rho}}\,\Gamma(s)\,\Gamma(\rho-s)\,\mathrm{P}_s^{-\rho}\left(\dfrac{a}{\sqrt{a^2-b^2}}\right)$ $[0<b<a;\;0<\operatorname{Re}s<\operatorname{Re}\rho]$
15	$\dfrac{1}{\left(x+a+\sqrt{(x+a)^2-b^2x^2}\right)^{\rho}}$	$2^{-\rho}a^{s-\rho}\,\mathrm{B}(s,\,\rho-s)\,{}_2F_1\left(\begin{matrix}\frac{s}{2},\;\frac{s+1}{2}\\ \rho+1;\;b^2\end{matrix}\right)$ $[\lvert b\rvert\le 1;\;0<\operatorname{Re}s<\operatorname{Re}\rho;\;\lvert\arg a\rvert<\pi]$
16	$\dfrac{(x+2a)^{-1/2}}{\left(x+a+\sqrt{x^2+2ax}\right)^{\rho}}$	$\dfrac{a^{s-\rho-1/2}}{2^{s-1/2}}\,\mathrm{B}\left(2s,\,\frac{1-2s+2\rho}{2}\right)$ $[0<\operatorname{Re}s<\operatorname{Re}\rho+1/2;\;\lvert\arg a\rvert<\pi]$
17	$\dfrac{\left(x^2+2ax\right)^{-1/2}}{\left(x+a+\sqrt{x^2+2ax}\right)^{\rho}}$	$\dfrac{2^{s-1}a^{s-\rho-1}}{\sqrt{\pi}}\,\Gamma\left[\begin{matrix}s,\;1-s+\rho,\;\frac{2s-1}{2}\\ s+\rho\end{matrix}\right]$ $[1/2<\operatorname{Re}s<\operatorname{Re}\rho+1;\;\lvert\arg a\rvert<\pi]$

No.	$f(x)$	$F(s)$
18	$\dfrac{\left[(x+a)^2-b^2\right]^{-1/2}}{\left[x+a+\sqrt{(x+a)^2-b^2}\right]^{\rho}}$	$2^{-\rho}a^{s-\rho-1}\,\mathrm{B}\,(s,\,1-s+\rho)\;{}_2F_1\left(\begin{matrix}\frac{1-s+\rho}{2},\,\frac{2-s+\rho}{2}\\ \rho+1;\,\frac{b^2}{a^2}\end{matrix}\right)$ $[\lvert b\rvert<\lvert a\rvert;\;0<\operatorname{Re}s<\operatorname{Re}\rho+1;\;\lvert\arg a\rvert<\pi]$
19	$\dfrac{\left[(x+a)^2-b^2x^2\right]^{-1/2}}{\left[x+a+\sqrt{(x+a)^2-b^2x^2}\right]^{\rho}}$	$2^{-\rho}a^{s-\rho-1}\,\mathrm{B}(s,\,\rho-s+1)\;{}_2F_1\left(\begin{matrix}\frac{s}{2},\,\frac{s+1}{2}\\ \rho+1;\,b^2\end{matrix}\right)$ $[\lvert b\rvert<1;\;0<\operatorname{Re}s<\operatorname{Re}\rho+1;\;\lvert\arg a\rvert<\pi]$
20	$\dfrac{1}{\left(\sqrt{x^2+a^2}+\sqrt{b^2x^2+a^2}\right)^{\rho}}$	$2^{-\rho-1}a^{s-\rho}\,\mathrm{B}\left(\dfrac{s}{2},\,\dfrac{\rho-s}{2}\right)\;{}_2F_1\left(\begin{matrix}\frac{\rho+1}{2},\,\frac{s}{2}\\ \rho+1;\,1-b^2\end{matrix}\right)$ $[\operatorname{Re}a,\;\operatorname{Re}b>0;\;0<\operatorname{Re}s<\operatorname{Re}\rho]$
21	$\dfrac{1}{\left(\sqrt{x^2+a^2}+\sqrt{x^2+b^2}\right)^{\rho}}$	$2^{-\rho-1}a^{s-\rho}\,\mathrm{B}\left(\dfrac{s}{2},\,\dfrac{\rho-s}{2}\right)\;{}_2F_1\left(\begin{matrix}\frac{\rho-s}{2},\,\frac{\rho+1}{2}\\ \rho+1;\,\frac{a^2-b^2}{a^2}\end{matrix}\right)$ $[\operatorname{Re}a,\;\operatorname{Re}b>0;\;0<\operatorname{Re}s<\operatorname{Re}\rho]$
22	$\dfrac{(x+a)^{-1/2}}{\left(x+bx+a+2\sqrt{bx\,(x+a)}\right)^{\rho}}$	$\dfrac{a^{s-\rho-1/2}\sqrt{b}}{2^{2s-1}}\,\mathrm{B}\left(2s,\,\dfrac{1-2s+2\rho}{2}\right)\;{}_2F_1\left(\begin{matrix}\frac{2\rho+1}{2},\,\frac{2s+1}{2}\\ \frac{2s+2\rho+1}{2};\,1-b\end{matrix}\right)$ $[0<\operatorname{Re}s<\operatorname{Re}\rho+1/2;\;\lvert\arg a\rvert,\,\lvert\arg b\rvert<\pi]$
23	$\dfrac{(x^2+a^2)^{-1/2}(x^2+b^2)^{-1/2}}{\left(\sqrt{x^2+a^2}+\sqrt{x^2+b^2}\right)^{\rho}}$	$\dfrac{a^{s-\rho-2}}{2^{\rho+1}}\,\mathrm{B}\left(\dfrac{s}{2},\,\dfrac{2-s+\rho}{2}\right)\;{}_2F_1\left(\begin{matrix}\frac{\rho+1}{2},\,\frac{2-s+\rho}{2}\\ \rho+1;\,\frac{a^2-b^2}{a^2}\end{matrix}\right)$ $[\operatorname{Re}a,\;\operatorname{Re}b>0;\;0<\operatorname{Re}s<\operatorname{Re}\rho+2]$
24	$\dfrac{(x^2+a^2)^{-1/2}(b^2x^2+a^2)^{-1/2}}{\left(\sqrt{x^2+a^2}+\sqrt{b^2x^2+a^2}\right)^{\rho}}$	$\dfrac{a^{s-\rho-2}}{2^{\rho+1}}\,\mathrm{B}\left(\dfrac{s}{2},\,\dfrac{2-s+\rho}{2}\right)\;{}_2F_1\left(\begin{matrix}\frac{\rho+1}{2},\,\frac{s}{2}\\ \rho+1;\,1-b^2\end{matrix}\right)$ $[\operatorname{Re}a,\;\operatorname{Re}b>0;\;0<\operatorname{Re}s<\operatorname{Re}\rho+2]$

2.1.10. Various algebraic functions

1	$(a-x)_+^{-\alpha}+\dfrac{\sin\left[(c-\alpha)\,\pi\right]}{\sin(c\pi)}\,(x-a)_+^{-\alpha}$	$\dfrac{\pi a^{s-\alpha}}{\sin(c\pi)\;\Gamma(\alpha)}\,\Gamma\left[\begin{matrix}s,\,\alpha-s\\ s-c+1,\,c-s\end{matrix}\right]$ $[a>0;\;0<\operatorname{Re}s<\operatorname{Re}\alpha<1]$
2	$\dfrac{\sin\left[(c-\alpha)\,\pi\right]}{\sin(c\pi)}\,(a-x)_+^{-\alpha}+(x-a)_+^{-\alpha}$	$\dfrac{\pi a^{s-\alpha}}{\sin(c\pi)\;\Gamma(\alpha)}\,\Gamma\left[\begin{matrix}s,\,\alpha-s\\ s+c-\alpha,\,1-s-c+\alpha\end{matrix}\right]$ $[a>0;\;0<\operatorname{Re}s<\operatorname{Re}\alpha<1]$

No.	$f(x)$	$F(s)$
3	$\theta(a-x)\left(\sqrt{a-x}+\sqrt{a}\right)^{-1/2}$ $+\theta(x-a)\,x^{-1/2}\left(\sqrt{x}+\sqrt{x-a}\right)^{1/2}$	$\dfrac{\sqrt{\pi}\,a^{s-1/4}}{2^{3/2}}\Gamma\left[\begin{matrix}\frac{1-4s}{4},\ s\\ \frac{4s+1}{4},\ \frac{3-2s}{2}\end{matrix}\right]$ $[a>0;\ 0<\operatorname{Re}s<1/4]$
4	$\theta(a-x)\left(\sqrt{a-x}+\sqrt{a}\right)^{1/2}$ $+\sqrt{a}\,\theta(x-a)\left(\sqrt{x}+\sqrt{x-a}\right)^{-1/2}$	$\dfrac{\sqrt{\pi}\,a^{s+1/4}}{2^{3/2}}\Gamma\left[\begin{matrix}\frac{1-4s}{4},\ s\\ \frac{4s+5}{4},\ \frac{1-2s}{2}\end{matrix}\right]$ $[a>0;\ 0<\operatorname{Re}s<1/4]$
5	$(a-x)_+^{-1/2}\sqrt{\sqrt{a-x}+\sqrt{a}}$ $+(x-a)_+^{-1/2}\sqrt{\sqrt{x}+\sqrt{x-a}}$	$\sqrt{2\pi}\,a^{s-1/4}\,\Gamma\left[\begin{matrix}\frac{1-4s}{4},\ s\\ \frac{4s+1}{4},\ \frac{1-2s}{2}\end{matrix}\right]$ $[a>0;\ 0<\operatorname{Re}s<1/4]$
6	$(a-x)_+^{-1/2}\sqrt{\sqrt{a}-\sqrt{a-x}}$ $-(x-a)_+^{-1/2}\sqrt{\sqrt{x}+\sqrt{x-a}}$	$-\sqrt{2\pi}\,a^{s-1/4}\,\Gamma\left[\begin{matrix}\frac{1-4s}{4},\ \frac{2s+1}{2}\\ \frac{4s+1}{4},\ 1-s\end{matrix}\right]$ $[a>0;\ -1/2<\operatorname{Re}s<1/4]$
7	$(a-x)_+^{-1/2}\sqrt{\sqrt{a-x}+\sqrt{a}}$ $-(x-a)_+^{-1/2}\sqrt{\sqrt{x}-\sqrt{x-a}}$	$\sqrt{2\pi}\,a^{s-1/4}\,\Gamma\left[\begin{matrix}\frac{3-4s}{4},\ s\\ \frac{4s+3}{4},\ \frac{1-2s}{2}\end{matrix}\right]$ $[a>0;\ 0<\operatorname{Re}s<3/4]$
8	$(a-x)_+^{-1/2}\left[\left(\sqrt{a}+\sqrt{a-x}\right)^{\rho}\right.$ $\left.+\left(\sqrt{a}-\sqrt{a-x}\right)^{\rho}\right]$	$2\sqrt{\pi}\,a^{s+(\rho-1)/2}\,\Gamma\left[\begin{matrix}s,\ s+\rho\\ \frac{2s+\rho}{2},\ \frac{2s+\rho+1}{2}\end{matrix}\right]$ $[a,\ \operatorname{Re}s>0,\ -\operatorname{Re}\rho]$
9	$(x-a)_+^{-1/2}\left[\left(\sqrt{x}+\sqrt{x-a}\right)^{\rho}\right.$ $\left.+\left(\sqrt{x}-\sqrt{x-a}\right)^{\rho}\right]$	$2\sqrt{\pi}\,a^{s+(\rho-1)/2}\,\Gamma\left[\begin{matrix}\frac{1-2s-\rho}{2},\ \frac{1-2s+\rho}{2}\\ 1-s,\ \frac{1-2s}{2}\end{matrix}\right]$ $[a>0;\ \operatorname{Re}s<(1-\lvert\operatorname{Re}\rho\rvert)/2]$
10	$\left(x^2-a^2\right)_+^{-1/2}\left[\left(x+\sqrt{x^2-a^2}\right)^{\rho}\right.$ $\left.+\left(x-\sqrt{x^2-a^2}\right)^{\rho}\right]$	$2^{-s}a^{s+\rho-1}\,\mathrm{B}\left(\dfrac{1-s-\rho}{2},\ \dfrac{1-s+\rho}{2}\right)$ $[\operatorname{Re}s<\lvert\operatorname{Re}\rho\rvert+1]$
11	$\left(a^2-x^2\right)_+^{-1/2}\left[\left(a+\sqrt{a^2-x^2}\right)^{\rho}\right.$ $\left.+\left(a-\sqrt{a^2-x^2}\right)^{\rho}\right]$	$(2a)^{s+\rho-1}\,\mathrm{B}\left(\dfrac{s}{2},\ \dfrac{s+2\rho}{2}\right)$ $[\operatorname{Re}s>0,\ -2\operatorname{Re}\rho]$

No.	$f(x)$	$F(s)$
12	$\dfrac{\left(\sqrt{a}+\sqrt{a-x}\right)^{\rho}-\left(\sqrt{a}-\sqrt{a-x}\right)^{\rho}}{\sqrt{a-x}}$	$2^{2s+\rho}\,\dfrac{\sin(\rho\pi)}{\pi}\,a^{s+(\rho-1)/2}\,\Gamma(1-2s-\rho)\,\Gamma(s) \times\Gamma(s+\rho)\quad \left[\begin{matrix}0,\ -\operatorname{Re}\rho<\operatorname{Re}s<(1-\operatorname{Re}\rho)/2;\\ -\pi<\arg a\le\pi\end{matrix}\right]$
13	$\dfrac{\left(\sqrt{x}+\sqrt{x-a}\right)^{\rho}-\left(\sqrt{x}-\sqrt{x-a}\right)^{\rho}}{\sqrt{x-a}}$	$2^{1-2s}\,\dfrac{\sin(\rho\pi)}{\pi}\,a^{s+(\rho-1)/2}\,\Gamma(2s)\,\Gamma\left(\dfrac{1-2s-\rho}{2}\right) \times\Gamma\left(\dfrac{1-2s+\rho}{2}\right)\quad \left[\begin{matrix}0<\operatorname{Re}s<(1-\lvert\operatorname{Re}\rho\rvert)/2;\\ -\pi<\arg a\le\pi\end{matrix}\right]$
14	$\dfrac{\left(x+\sqrt{x^2-a^2}\right)^{\rho}-\left(x-\sqrt{x^2-a^2}\right)^{\rho}}{\sqrt{x^2-a^2}}$	$2^{-s}\,\dfrac{\sin(\rho\pi)}{\pi}\,a^{s+\rho-1}\,\Gamma\left(\dfrac{1-s-\rho}{2}\right)\Gamma\left(\dfrac{1-s+\rho}{2}\right) \times\Gamma(s)\quad \left[\begin{matrix}0<\operatorname{Re}s<1-\lvert\operatorname{Re}\rho\rvert;\\ -\pi/2<\arg a\le\pi/2\end{matrix}\right]$
15	$\theta(a-x)\left[\left(\sqrt{a}+\sqrt{a-x}\right)^{\rho} - \left(\sqrt{a}-\sqrt{a-x}\right)^{\rho}\right]$	$2^{2s+\rho}\rho\,a^{s+\rho/2}\,\Gamma\left[\begin{matrix}s,\ s+\rho\\ 2s+\rho+1\end{matrix}\right]\quad [a>0;\ \operatorname{Re}s>0,\ -\operatorname{Re}\rho]$
16	$\theta(a-x)\left[\left(\sqrt{a+x}+\sqrt{a-x}\right)^{\rho} - \left(\sqrt{a+x}-\sqrt{a-x}\right)^{\rho}\right]$	$2^{s+\rho-2}\rho\,a^{s+\rho/2}\,\Gamma\left[\begin{matrix}\frac{s}{2},\ \frac{s+\rho}{2}\\ \frac{2s+\rho+2}{2}\end{matrix}\right]\quad [a>0;\ \operatorname{Re}s>0,\ -\operatorname{Re}\rho]$
17	$\theta(a-x)\left[\left(\sqrt{\sqrt{a}+\sqrt{x}}-\sqrt{\sqrt{a}-\sqrt{x}}\right)^{\rho} - a^{-\rho/4}\left(\sqrt{\sqrt{a}+\sqrt{x}}+\sqrt{\sqrt{a}-\sqrt{x}}\right)^{\rho}\right]$	$-2^{2s+\rho-1}\rho\,a^{s+\rho/4}\,\Gamma\left[\begin{matrix}s,\ \frac{2s+\rho}{2}\\ \frac{4s+\rho+2}{2}\end{matrix}\right]\quad [a>0;\ \operatorname{Re}s>0,\ -\operatorname{Re}\rho/2]$
18	$\dfrac{\theta(a-x)}{\sqrt{a-x}}\left[\left(\sqrt{\sqrt{a}+\sqrt{x}}-\sqrt{\sqrt{a}-\sqrt{x}}\right)^{\rho} + \left(\sqrt{\sqrt{a}+\sqrt{x}}+\sqrt{\sqrt{a}-\sqrt{x}}\right)^{\rho}\right]$	$2^{2s+\rho}a^{s+(\rho-2)/4}\,\mathrm{B}\left(s,\ \dfrac{2s+\rho}{2}\right)\quad [a>0;\ \operatorname{Re}s>0,\ -\operatorname{Re}\rho/2]$
19	$\dfrac{\theta(a-x)}{\sqrt{a-x}}\left[\left(\sqrt{a}-\sqrt{a-x}\right)^{\rho} + \left(\sqrt{a}+\sqrt{a+x}\right)^{\rho}\right]$	$2^{2s+\rho}a^{s+(\rho-1)/2}\,\mathrm{B}(s,\ s+\rho)\quad [a>0;\ \operatorname{Re}s>0,\ -\operatorname{Re}\rho]$
20	$\dfrac{\theta(a-x)}{\sqrt{a^2-x^2}}\left[\left(\sqrt{a+x}-\sqrt{a-x}\right)^{\rho} + \left(\sqrt{a+x}+\sqrt{a-x}\right)^{\rho}\right]$	$2^{s+\rho-1}a^{s+(\rho-2)/2}\,\mathrm{B}\left(\dfrac{s}{2},\ \dfrac{s+\rho}{2}\right)\quad [a>0;\ \operatorname{Re}s>0,\ -\operatorname{Re}\rho]$
21	$\theta(a-x)\left[\left(a-\sqrt{a^2-x^2}\right)^{\rho} - \left(a+\sqrt{a^2-x^2}\right)^{\rho}\right]$	$-2^{s+\rho-1}\rho\,a^{s+\rho}\,\Gamma\left[\begin{matrix}\frac{s}{2},\ \frac{s+2\rho}{2}\\ s+\rho+1\end{matrix}\right]\quad [a>0;\ \operatorname{Re}s>0,\ -2\operatorname{Re}\rho]$

No.	$f(x)$	$F(s)$
22	$\dfrac{\theta(a-x)}{\sqrt{a^2-x^2}}\left[\left(a-\sqrt{a^2-x^2}\right)^{\rho}+\left(a+\sqrt{a^2-x^2}\right)^{\rho}\right]$	$(2a)^{s+\rho-1}\,\mathrm{B}\left(\dfrac{s}{2},\,\dfrac{s+2\rho}{2}\right)$ $[a>0;\ \operatorname{Re}s>0,\ -2\operatorname{Re}\rho]$
23	$\theta(x-a)\left[\left(\sqrt{x}-\sqrt{x-a}\right)^{\rho}-\left(\sqrt{x}+\sqrt{x-a}\right)^{\rho}\right]$	$-2^{-2s}\rho\,a^{s+\rho/2}\,\Gamma\left[\begin{matrix}\frac{-2s-\rho}{2},\ \frac{-2s+\rho}{2}\\ 1-2s\end{matrix}\right]$ $[a>0;\ \operatorname{Re}s<-\lvert\operatorname{Re}\rho\rvert/2]$
24	$\theta(x-a)\left[\left(\sqrt{x+a}-\sqrt{x-a}\right)^{\rho}-\left(\sqrt{x+a}+\sqrt{x-a}\right)^{\rho}\right]$	$-2^{-s+\rho/2-2}\rho\,a^{s+\rho/2}\,\Gamma\left[\begin{matrix}\frac{-2s-\rho}{4},\ \frac{-2s+\rho}{4}\\ 1-s\end{matrix}\right]$ $[a>0;\ \operatorname{Re}s<-\lvert\operatorname{Re}\rho\rvert/2]$
25	$\theta(x-a)\left[\left(\sqrt{\sqrt{x}+\sqrt{a}}+\sqrt{\sqrt{x}-\sqrt{a}}\right)^{\rho}-\left(\sqrt{\sqrt{x}+\sqrt{a}}-\sqrt{\sqrt{x}-\sqrt{a}}\right)^{\rho}\right]$	$2^{-2s+\rho/2-1}\rho\,a^{s+\rho/4}\,\Gamma\left[\begin{matrix}\frac{-4s-\rho}{4},\ \frac{-4s+\rho}{4}\\ 1-2s\end{matrix}\right]$ $[a>0;\ \operatorname{Re}s<-\lvert\operatorname{Re}\rho\rvert/4]$
26	$\dfrac{\theta(x-a)}{\sqrt{x-a}}\left[\left(\sqrt{\sqrt{x}+\sqrt{a}}-\sqrt{\sqrt{x}-\sqrt{a}}\right)^{\rho}+\left(\sqrt{\sqrt{x}+\sqrt{a}}+\sqrt{\sqrt{x}-\sqrt{a}}\right)^{\rho}\right]$	$2^{-2s+\rho/2+1}a^{s+(\rho-2)/4}\,\mathrm{B}\left(\dfrac{2-4s-\rho}{4},\,\dfrac{2-4s+\rho}{4}\right)$ $[a>0;\ \operatorname{Re}s<(2-\lvert\operatorname{Re}\rho\rvert)/4]$
27	$\dfrac{\theta(x-a)}{\sqrt{x-a}}\left[\left(\sqrt{x}-\sqrt{x-a}\right)^{\rho}+\left(\sqrt{x}+\sqrt{x-a}\right)^{\rho}\right]$	$2^{1-2s}a^{s+(\rho-1)/2}\,\mathrm{B}\left(\dfrac{1-2s-\rho}{2},\,\dfrac{1-2s+\rho}{2}\right)$ $[a>0;\ \operatorname{Re}s<(1-\lvert\operatorname{Re}\rho\rvert)/2]$
28	$\dfrac{\theta(x-a)}{\sqrt{x^2-a^2}}\left[\left(\sqrt{x+a}-\sqrt{x-a}\right)^{\rho}+\left(\sqrt{x+a}+\sqrt{x-a}\right)^{\rho}\right]$	$2^{-s+\rho/2}a^{s+(\rho-2)/2}\,\mathrm{B}\left(\dfrac{2-2s-\rho}{4},\,\dfrac{2-2s+\rho}{4}\right)$ $[a>0;\ \operatorname{Re}s<(2-\lvert\operatorname{Re}\rho\rvert)/2]$
29	$\theta(x-a)\left[\left(x-\sqrt{x^2-a^2}\right)^{\rho}-\left(x+\sqrt{x^2-a^2}\right)^{\rho}\right]$	$-2^{-s-1}\rho\,a^{s+\rho}\,\Gamma\left[\begin{matrix}\frac{-s-\rho}{2},\ \frac{-s+\rho}{2}\\ 1-s\end{matrix}\right]$ $[a>0;\ \operatorname{Re}s<-\lvert\operatorname{Re}\rho\rvert]$
30	$\dfrac{\theta(x-a)}{\sqrt{x^2-a^2}}\left[\left(x-\sqrt{x^2-a^2}\right)^{\rho}+\left(x+\sqrt{x^2-a^2}\right)^{\rho}\right]$	$2^{-s}a^{s+\rho-1}\,\mathrm{B}\left(\dfrac{1-s-\rho}{2},\,\dfrac{1-s+\rho}{2}\right)$ $[a>0;\ \operatorname{Re}s<1-\lvert\operatorname{Re}\rho\rvert]$
31	$\left[a^2x^2+(bx-x-1)^2-2ax\,(bx+x+1)\right]^{-1/2}$	$\pi\csc(s\pi)\,F_4(1,\,s;\,1,\,1;\,a,\,b)$ $[0<\operatorname{Re}s<1]$

2.2. The Exponential Function

More formulas can be obtained from the corresponding sections due to the relations

$$a^z = e^{z \ln a}, \quad e^z = {}_0F_0(z) = {}_1F_1(a;\, a;\, z), \quad e^{-z} = G_{01}^{10}\left(z \,\middle|\, \begin{matrix} \cdot \\ 0 \end{matrix}\right).$$

2.2.1. $e^{-ax^r - bx^p}$

No.	$f(x)$	$F(s)$
1	e^{-ax}	$\frac{\Gamma(s)}{a^s}$ [$\operatorname{Re} a,\ \operatorname{Re} s > 0$ or $(\operatorname{Re} a = 0;\ 0 < \operatorname{Re} s < 1)$]
2	$e^{-ax} - \sum_{k=0}^{n-1} \frac{(-ax)^k}{k!}$	$\frac{\Gamma(s)}{a^s}$ [$\operatorname{Re} a \geq 0;\ -n < \operatorname{Re} s < 1-n;\ n = 1, 2, \ldots$]
3	e^{iax}	$\frac{2^{s-1}\sqrt{\pi}}{a^s}\left(\Gamma\left[\begin{matrix}\frac{s}{2}\\ \frac{1-s}{2}\end{matrix}\right] + i\,\Gamma\left[\begin{matrix}\frac{s+1}{2}\\ \frac{2-s}{2}\end{matrix}\right]\right)$ [$a > 0;\ 0 < \operatorname{Re} s < 1$]
4	$e^{-(a+ib)x}$	$\frac{\Gamma(s)}{(a^2+b^2)^{s/2}} \exp\left(-is \arctan \frac{b}{a}\right)$ [$a,\ \operatorname{Re} s > 0$ or $(a > 0;\ 0 < \operatorname{Re} s < 1)$]
5	$\left\{\begin{matrix}\theta(a-x)\\ \theta(x-a)\end{matrix}\right\} e^{-bx}$	$b^{-s}\left\{\begin{matrix}\gamma(s,\, ab)\\ \Gamma(s,\, ab)\end{matrix}\right\}$ $\left[a > 0;\ \left\{\begin{matrix}\operatorname{Re} s > 0\\ \operatorname{Re} b > 0\end{matrix}\right\}\right]$
6	$e^{-ax^2 - bx}$	$\frac{\Gamma(s)}{(2a)^{s/2}} e^{b^2/(8a)} D_{-s}\left(\frac{b}{\sqrt{2a}}\right)$ $\left[\begin{matrix}(\operatorname{Re} a, \operatorname{Re} s > 0) \text{ or } (\operatorname{Re} b, \operatorname{Re} s > 0;\ \operatorname{Re} a = 0) \text{ or}\\ (0 < \operatorname{Re} s < 2;\ \operatorname{Re} a = \operatorname{Re} b = 0;\ \operatorname{Im} a \neq 0)\end{matrix}\right]$
7	$e^{ax - bx^n}$	$\frac{b^{-s/n}}{n} \sum_{k=0}^{n-1} \frac{a^k b^{-k/n}}{k!} \Gamma\left(\frac{s+k}{n}\right) {}_2F_n\left(\begin{matrix}1,\ \frac{s+k}{n};\ \frac{a^n}{bn^n}\\ \Delta(n,\, k+1)\end{matrix}\right)$ [$\operatorname{Re} b > 0;\ n \geq 2$]
8	$e^{-ax - b/x}$	$2\left(\frac{b}{a}\right)^{s/2} K_s\left(2\sqrt{ab}\right)$ [$\operatorname{Re} a,\ \operatorname{Re} b > 0$]
9	$e^{-ax - b/x^2}$	$\frac{b^{s/2}}{2}\Gamma\left(-\frac{s}{2}\right) {}_0F_2\left(\begin{matrix}-\frac{a^2 b}{4}\\ \frac{1}{2},\ \frac{s+2}{2}\end{matrix}\right) - \frac{ab^{(s+1)/2}}{2}\Gamma\left(-\frac{s+1}{2}\right) {}_0F_2\left(\begin{matrix}-\frac{a^2 b}{4}\\ \frac{3}{2},\ \frac{s+3}{2}\end{matrix}\right) + a^{-s}\Gamma(s)\, {}_0F_2\left(\begin{matrix}-\frac{a^2 b}{4}\\ \frac{1-s}{2},\ \frac{2-s}{2}\end{matrix}\right)$ [$\operatorname{Re} a,\ \operatorname{Re} b > 0$]
10	$e^{ia(x+b/x)/2}$	$i\pi b^{s/2} e^{-is\pi/2} H_{-s}^{(1)}(a\sqrt{b})$ [$\operatorname{Im} a > 0;\ \operatorname{Im}(ab) > 0$]

No.	$f(x)$	$F(s)$
11	$\begin{Bmatrix} \theta(a-x) \\ \theta(x-a) \end{Bmatrix} e^{-b/x^{\mu}}$	$\frac{b^{s/\mu}}{\mu} \begin{Bmatrix} \Gamma(-s/\mu,\, b/a^{\mu}) \\ \gamma(-s/\mu,\, b/a^{\mu}) \end{Bmatrix}$ $[a,\ \operatorname{Re} b,\ \operatorname{Re}\mu > 0;\ \operatorname{Re} s < 0]$
12	$e^{-ax^{\mu}}$	$\frac{a^{-s/\mu}}{\mu} \Gamma\left(\frac{s}{\mu}\right)$ $[\mu,\ \operatorname{Re} a,\ \operatorname{Re} s > 0]$
13	$1 - e^{-ax^{\pm\mu}}$	$-\frac{a^{\mp s/\mu}}{\mu} \Gamma\left(\pm\frac{s}{\mu}\right)$ $\left[\begin{matrix} \mu,\ \operatorname{Re} a > 0; \\ -(1\pm1)\mu/2 < \operatorname{Re} s < (1\mp1)\mu/2 \end{matrix}\right]$

2.2.2. $e^{bx^m(a-x)^n}$ and algebraic functions

No.	$f(x)$	$F(s)$
1	$\begin{Bmatrix} \theta(a-x) \\ \theta(x-a) \end{Bmatrix} x^{\alpha} e^{-bx}$	$b^{-s-\alpha} \begin{Bmatrix} \gamma(s+\alpha,\, ab) \\ \Gamma(s+\alpha,\, ab) \end{Bmatrix}$ $[a,\ \operatorname{Re} b,\ \operatorname{Re}(s+\alpha) > 0]$
2	$(a-x)_+^{\alpha-1} e^{bx}$	$a^{s+\alpha-1}\, \mathrm{B}(s,\,\alpha)\ {}_1F_1\left(\begin{matrix} s;\ ab \\ s+\alpha \end{matrix}\right)$ $[\operatorname{Re}\alpha,\ \operatorname{Re} s > 0]$
3	$(x-a)_+^{\alpha-1} e^{-bx}$	$a^{s+\alpha-1} e^{-ab}\, \Gamma(\alpha)\ \Psi(\alpha,\, s+\alpha;\, ab)$ $[\operatorname{Re} b,\ \operatorname{Re} s > 0]$
4	$\left(a^2-x^2\right)_+^{\alpha-1} e^{-bx}$	$\frac{a^{s+2\alpha-2}}{2} \mathrm{B}\left(\alpha,\, \frac{s}{2}\right) {}_1F_2\left(\begin{matrix} \frac{s}{2};\ \frac{a^2b^2}{4} \\ \frac{1}{2},\ \frac{s}{2}+\alpha \end{matrix}\right) - \frac{a^{s+2\alpha-1}b}{2} \mathrm{B}\left(\alpha,\, \frac{s+1}{2}\right) {}_1F_2\left(\begin{matrix} \frac{s+1}{2};\ \frac{a^2b^2}{4} \\ \frac{3}{2},\ \frac{s+1}{2}+\alpha \end{matrix}\right)$ $[a,\ \operatorname{Re} s,\ \operatorname{Re}\alpha > 0]$
5	$\left(x^2-a^2\right)_+^{\alpha-1} e^{-bx}$	$\frac{a^{s+2\alpha-2}}{2} \mathrm{B}\left(\alpha,\, 1-\alpha-\frac{s}{2}\right) {}_1F_2\left(\begin{matrix} \frac{s}{2};\ \frac{a^2b^2}{4} \\ \frac{1}{2},\ \alpha+\frac{s}{2} \end{matrix}\right) - \frac{a^{s+2\alpha-1}b}{2} \mathrm{B}\left(\alpha,\, \frac{1-s}{2}-\alpha\right) {}_1F_2\left(\begin{matrix} \frac{s+1}{2};\ \frac{a^2b^2}{4} \\ \frac{3}{2},\ \alpha+\frac{s+1}{2} \end{matrix}\right) + b^{-s-2\alpha+2}\, \Gamma(s+2\alpha-2)\ {}_1F_2\left(\begin{matrix} 1-\alpha;\ \frac{a^2b^2}{4} \\ \frac{3-s-2\alpha}{2},\ \frac{4-s-2\alpha}{2} \end{matrix}\right)$
6		$= \frac{\Gamma(\alpha)}{\sqrt{\pi}} \left(\frac{2a}{b}\right)^{\alpha-1/2} K_{\alpha-1/2}(ab)$ $[a,\ \operatorname{Re} b,\ \operatorname{Re}\alpha > 0]$
7	$\frac{e^{-bx}}{(x+a)^{\rho}}$	$a^{s-\rho}\, \Gamma(s)\ \Psi(s,\, s-\rho+1;\, ab)$
8		$= \frac{a^{(s-\rho-1)/2}}{b^{(s-\rho+1)/2}} e^{ab/2}\, \Gamma(s)\ W_{(1-\rho-s)/2,\,(s-\rho)/2}(ab)$ $[\operatorname{Re} b,\ \operatorname{Re} s > 0;\ \lvert\arg a\rvert < \pi]$

No.	$f(x)$	$F(s)$
9	$\dfrac{e^{-bx}}{x+a}$	$a^{s-1}\,e^{ab}\,\Gamma(s)\,\Gamma(1-s,\,ab)$ $\left[\begin{array}{l}(\operatorname{Re}b,\ \operatorname{Re}s>0)\ \text{or}\\ (\operatorname{Re}b=0;\ 0<\operatorname{Re}s<1);\ \lvert\arg a\rvert<\pi\end{array}\right]$
10	$\dfrac{e^{-bx}}{x-a}$	$\dfrac{\pi e^{-ab}\csc(s\pi)}{b^{s-1}\Gamma(1-s)}\,E_s(-ab)+i\pi e^{-ab}a^{s-1}$ $[a,\ \operatorname{Re}b,\ \operatorname{Re}s>0;\ s\neq 1]$
11	$\dfrac{e^{-bx}}{(x^2+a^2)^{\rho}}$	$\dfrac{\Gamma(s-2\rho)}{b^{s-2\rho}}\,{}_1F_2\left(\begin{matrix}\rho;\ \frac{2\rho-s+1}{2}\\ \frac{2\rho-s+2}{2};\ -\frac{a^2b^2}{4}\end{matrix}\right)$ $+\dfrac{a^{s-2\rho}}{2}\,\mathrm{B}\left(\dfrac{s}{2},\,\dfrac{2\rho-s}{2}\right)\,{}_1F_2\left(\begin{matrix}\frac{s}{2};\ -\frac{a^2b^2}{4}\\ \frac{1}{2},\ \frac{s-2\rho+2}{2}\end{matrix}\right)$ $-\dfrac{a^{s-2\rho+1}b}{2}\,\mathrm{B}\left(\dfrac{s+1}{2},\,\dfrac{2\rho-s-1}{2}\right)\,{}_1F_2\left(\begin{matrix}\frac{s+1}{2};\ -\frac{a^2b^2}{4}\\ \frac{3}{2},\ \frac{s-2\rho+3}{2}\end{matrix}\right)$ $\left[\begin{array}{l}\operatorname{Re}a,\ \operatorname{Re}b,\ \operatorname{Re}s>0\ \text{or}\\ (\operatorname{Re}b=0;\ \operatorname{Re}(s-2\rho)<1)\end{array}\right]$
12	$\dfrac{e^{-bx}}{(x^2+a^2)^{n}}$	$\dfrac{(-1)^n\,\Gamma(s-1)}{2\,(n-1)!}\,D_t^{n-1}\Big[t^{(s-2)/2}\big(e^{ib\sqrt{t}+i\pi s/2}\,\Gamma\big(2-s,\,ib\sqrt{t}\big)$ $+\,e^{-ib\sqrt{t}-i\pi s/2}\,\Gamma\big(2-s,\,-ib\sqrt{t}\big)\big)\Big]\Big\rvert_{t=a^2}$ $[\operatorname{Re}a,\ \operatorname{Re}b,\ \operatorname{Re}s>0;\ n=1,\,2,\ldots]$
13	$\dfrac{e^{-bx}}{x^2-a^2}$	$\dfrac{\Gamma(s-2)}{b^{s-2}}\,{}_1F_2\left(\begin{matrix}1;\ \frac{a^2b^2}{4}\\ \frac{3-s}{2},\ \frac{4-s}{2}\end{matrix}\right)-\dfrac{\pi a^{s-2}}{2\sin(s\pi)}\left[e^{ab}+e^{-ab}\cos(s\pi)\right]$ $[a,\ \operatorname{Re}b,\ \operatorname{Re}s>0]$
14	$(a-x)_+^{\alpha-1}\,(b-x)^{-\alpha}\,e^{cx}$	$a^{s+\alpha-1}b^{-\alpha}\,\mathrm{B}(s,\,\alpha)\,\Phi_1\left(s,\,\alpha;\ s+\alpha;\ \dfrac{a}{b},\,ac\right)$ $[0<a<\lvert b\rvert;\ \operatorname{Re}s,\ \operatorname{Re}\alpha>0]$
15	$\left(\sqrt{x+a}+\sqrt{a}\right)^{\rho}e^{-bx}$	$\dfrac{\rho\sqrt{a}}{b^{s+(\rho-1)/2}}\,\Gamma\left(\dfrac{2s+\rho-1}{2}\right)\,{}_2F_2\left(\begin{matrix}\frac{1+\rho}{2},\ \frac{1-\rho}{2};\ ab\\ \frac{3}{2},\ \frac{3-2s-\rho}{2}\end{matrix}\right)$ $+\,b^{-s-\rho/2}\,\Gamma\left(\dfrac{2s+\rho}{2}\right)\,{}_2F_2\left(\begin{matrix}\frac{\rho}{2},\ -\frac{\rho}{2};\ ab\\ \frac{1}{2},\ \frac{2-2s-\rho}{2}\end{matrix}\right)$ $-\,2^{2s+\rho}\rho a^{s+\rho/2}\,\Gamma\left[\begin{matrix}s,\ -2s-\rho\\ 1-s-\rho\end{matrix}\right]\,{}_2F_2\left(\begin{matrix}s,\ s+\rho;\ ab\\ \frac{2s+\rho+1}{2},\ \frac{2s+\rho+2}{2}\end{matrix}\right)$ $[\operatorname{Re}b,\ \operatorname{Re}s>0;\ \lvert\arg a\rvert<\pi]$

No.	$f(x)$	$F(s)$
16	$\left(\sqrt{x+a}-\sqrt{a}\right)^{\rho} e^{-bx}$	$-\dfrac{\rho\sqrt{a}}{b^{s+(\rho-1)/2}}\Gamma\left(\dfrac{2s+\rho-1}{2}\right) {}_2F_2\left(\begin{matrix}\frac{1+\rho}{2}, \frac{1-\rho}{2}; ab\\ \frac{3}{2}, \frac{3-2s-\rho}{2}\end{matrix}\right)$ $+ b^{-s-\rho/2}\Gamma\left(\dfrac{2s+\rho}{2}\right) {}_2F_2\left(\begin{matrix}\frac{\rho}{2}, -\frac{\rho}{2}; ab\\ \frac{1}{2}, \frac{2-2s-\rho}{2}\end{matrix}\right)$ $+ 2^{2s+\rho}\rho a^{s+\rho/2}\Gamma\left[\begin{matrix}s+\rho, -2s-\rho\\ 1-s\end{matrix}\right] {}_2F_2\left(\begin{matrix}s, s+\rho; ab\\ \frac{2s+\rho+1}{2}, \frac{2s+\rho+2}{2}\end{matrix}\right)$ $[\operatorname{Re} b,\ \operatorname{Re}(s+\rho)>0;\ \lvert\arg a\rvert<\pi]$
17	$\dfrac{\left(\sqrt{x+a}-\sqrt{a}\right)^{\rho}}{\sqrt{x+a}} e^{-bx}$	$b^{(1-\rho)/2-s}\Gamma\left(\dfrac{2s+\rho-1}{2}\right) {}_2F_2\left(\begin{matrix}\frac{1+\rho}{2}, \frac{1-\rho}{2}; ab\\ \frac{1}{2}, \frac{3-2s-\rho}{2}\end{matrix}\right)$ $-\dfrac{\rho\sqrt{a}}{b^{s+\rho/2-1}}\Gamma\left(\dfrac{2s+\rho-2}{2}\right) {}_2F_2\left(\begin{matrix}\frac{2+\rho}{2}, \frac{2-\rho}{2}; ab\\ \frac{3}{2}, \frac{4-2s-\rho}{2}\end{matrix}\right)$ $+ 2^{2s+\rho} a^{s+(\rho-1)/2} \mathrm{B}(1-2s-\rho, s+\rho) {}_2F_2\left(\begin{matrix}s, s+\rho; ab\\ \frac{2s+\rho}{2}, \frac{2s+\rho+1}{2}\end{matrix}\right)$ $[\operatorname{Re} b,\ \operatorname{Re}(s+\rho)>0;\ \lvert\arg a\rvert<\pi]$
18	$\left(\sqrt{x+a}\pm\sqrt{x}\right)^{\rho} e^{-bx}$	$\mp\dfrac{\rho a^{s+\rho/2}}{2^{2s}}\Gamma\left[\begin{matrix}2s, \frac{-2s\mp\rho}{2}\\ \frac{2s\mp\rho+2}{2}\end{matrix}\right] {}_2F_2\left(\begin{matrix}s, \frac{2s+1}{2}; ab\\ \frac{2s-\rho+2}{2}, \frac{2s+\rho+2}{2}\end{matrix}\right)$ $+\dfrac{2^{\pm\rho}a^{(\rho\mp\rho)/2}}{b^{s\pm\rho/2}}\Gamma\left(s\pm\dfrac{\rho}{2}\right) {}_2F_2\left(\begin{matrix}\mp\frac{\rho}{2}, \frac{1\mp\rho}{2}; ab\\ 1\mp\rho, \frac{2-2s\mp\rho}{2}\end{matrix}\right)$ $[\operatorname{Re} b,\ \operatorname{Re} s>0;\ \lvert\arg a\rvert<\pi]$
19	$\dfrac{\left(\sqrt{x+a}+\sqrt{x}\right)^{\rho}}{\sqrt{x+a}} e^{-bx}$	$-\dfrac{\pi a^{s+(\rho-1)/2}}{2^{2s-1}}\csc\dfrac{(2s+\rho-1)\pi}{2}\Gamma\left[\begin{matrix}2s\\ \frac{2s-\rho+1}{2}, \frac{2s+\rho+1}{2}\end{matrix}\right]$ $\times {}_2F_2\left(\begin{matrix}s, \frac{2s+1}{2}; ab\\ \frac{2s-\rho+1}{2}, \frac{2s+\rho+1}{2}\end{matrix}\right) - \dfrac{2^{\rho}\pi}{b^{s+(\rho-1)/2}}\sec\dfrac{(2s+\rho)\pi}{2}$ $\times\left[\Gamma\left(\dfrac{3-2s-\rho}{2}\right)\right]^{-1} {}_2F_2\left(\begin{matrix}\frac{1-\rho}{2}, \frac{2-\rho}{2}; ab\\ 1-\rho, \frac{3-2s-\rho}{2}\end{matrix}\right)$ $[\operatorname{Re} b,\ \operatorname{Re} s>0;\ \lvert\arg a\rvert<\pi]$
20	$\dfrac{\left(\sqrt{x+a}-\sqrt{x}\right)^{\rho}}{\sqrt{x+a}} e^{-bx}$	$-\dfrac{\pi a^{s+(\rho-1)/2}}{2^{2s-1}}\csc\dfrac{(2s-\rho-1)\pi}{2}\Gamma\left[\begin{matrix}2s\\ \frac{2s-\rho+1}{2}, \frac{2s+\rho+1}{2}\end{matrix}\right]$ $\times {}_2F_2\left(\begin{matrix}s, \frac{2s+1}{2}; ab\\ \frac{2s-\rho+1}{2}, \frac{2s+\rho+1}{2}\end{matrix}\right) - \dfrac{2^{-\rho}\pi a^{\rho}}{b^{s-(\rho+1)/2}}\sec\dfrac{(2s-\rho)\pi}{2}$ $\times\left[\Gamma\left(\dfrac{3-2s+\rho}{2}\right)\right]^{-1} {}_2F_2\left(\begin{matrix}\frac{1+\rho}{2}, \frac{2+\rho}{2}; ab\\ 1+\rho, \frac{3-2s+\rho}{2}\end{matrix}\right)$ $[\operatorname{Re} b,\ \operatorname{Re} s>0;\ \lvert\arg a\rvert<\pi]$

No.	$f(x)$	$F(s)$
21	$\left(\sqrt{x^2+a^2}+a\right)^\rho e^{-bx}$	$-2^{s+\rho-1}\rho a^{s+\rho}\,\Gamma\left[\begin{matrix}\frac{s}{2},\ -s-\rho\\ \frac{2-s-2\rho}{2}\end{matrix}\right]{}_2F_3\left(\begin{matrix}\frac{s}{2},\ \frac{s+2\rho}{2};\ -\frac{a^2b^2}{4}\\ \frac{1}{2},\ \frac{s+\rho+1}{2},\ \frac{s+\rho+2}{2}\end{matrix}\right)$ $+\,2^{s+\rho}\rho a^{s+\rho+1}b\,\Gamma\left[\begin{matrix}\frac{s+1}{2},\ -s-\rho-1\\ \frac{1-s-2\rho}{2}\end{matrix}\right]{}_2F_3\left(\begin{matrix}\frac{s+1}{2},\ \frac{s+2\rho+1}{2};\ -\frac{a^2b^2}{4}\\ \frac{3}{2},\ \frac{s+\rho+2}{2},\ \frac{s+\rho+3}{2}\end{matrix}\right)$ $+\,\frac{\rho a\,\Gamma(s+\rho-1)}{b^{s+\rho-1}}\,{}_2F_3\left(\begin{matrix}\frac{1+\rho}{2},\ \frac{1-\rho}{2};\ -\frac{a^2b^2}{4}\\ \frac{3}{2},\ \frac{2-s-\rho}{2},\ \frac{3-s-\rho}{2}\end{matrix}\right)$ $+\,\frac{\Gamma(s+\rho)}{b^{s+\rho}}\,{}_2F_3\left(\begin{matrix}\frac{\rho}{2},\ -\frac{\rho}{2};\ -\frac{a^2b^2}{4}\\ \frac{1}{2},\ \frac{1-s-\rho}{2},\ \frac{2-s-\rho}{2}\end{matrix}\right)$ $[\operatorname{Re} a,\ \operatorname{Re} b,\ \operatorname{Re} s>0]$
22	$\left(\sqrt{x^2+a^2}-a\right)^\rho e^{-bx}$	$2^{s+\rho-1}\rho a^{s+\rho}\,\Gamma\left[\begin{matrix}\frac{s+2\rho}{2},\ -s-\rho\\ \frac{2-s}{2}\end{matrix}\right]{}_2F_3\left(\begin{matrix}\frac{s}{2},\ \frac{s+2\rho}{2};\ -\frac{a^2b^2}{4}\\ \frac{1}{2},\ \frac{s+\rho+1}{2},\ \frac{s+\rho+2}{2}\end{matrix}\right)$ $-\,2^{s+\rho}\rho\, a^{s+\rho+1}b\,\Gamma\left[\begin{matrix}\frac{s+2\rho+1}{2},\ -s-\rho-1\\ \frac{1-s}{2}\end{matrix}\right]{}_2F_3\left(\begin{matrix}\frac{s+1}{2},\ \frac{s+2\rho+1}{2};\ -\frac{a^2b^2}{4}\\ \frac{3}{2},\ \frac{s+\rho+2}{2},\ \frac{s+\rho+3}{2}\end{matrix}\right)$ $-\,\frac{\rho a}{b^{s+\rho-1}}\,\Gamma(s+\rho-1)\,{}_2F_3\left(\begin{matrix}\frac{1+\rho}{2},\ \frac{1-\rho}{2};\ -\frac{a^2b^2}{4}\\ \frac{3}{2},\ \frac{2-s-\rho}{2},\ \frac{3-s-\rho}{2}\end{matrix}\right)$ $+\,\frac{\Gamma(s+\rho)}{b^{s+\rho}}\,{}_2F_3\left(\begin{matrix}\frac{\rho}{2},\ -\frac{\rho}{2};\ -\frac{a^2b^2}{4}\\ \frac{1}{2},\ \frac{1-s-\rho}{2},\ \frac{2-s-\rho}{2}\end{matrix}\right)$ $[\operatorname{Re} a,\ \operatorname{Re} b,\ \operatorname{Re}(s+2\rho)>0]$
23	$\frac{\left(\sqrt{x^2+a^2}+a\right)^\rho}{\sqrt{x^2+a^2}}\,e^{-bx}$	$(2a)^{s+\rho-1}\,\Gamma\left[\begin{matrix}\frac{s}{2},\ 1-s-\rho\\ \frac{2-s-2\rho}{2}\end{matrix}\right]{}_2F_3\left(\begin{matrix}\frac{s}{2},\ \frac{s+2\rho}{2};\ -\frac{a^2b^2}{4}\\ \frac{1}{2},\ \frac{s+\rho}{2},\ \frac{s+\rho+1}{2}\end{matrix}\right)$ $-\,(2a)^{s+\rho}\,b\,\Gamma\left[\begin{matrix}\frac{s+1}{2},\ -s-\rho\\ \frac{1-s-2\rho}{2}\end{matrix}\right]{}_2F_3\left(\begin{matrix}\frac{s+1}{2},\ \frac{s+2\rho+1}{2};\ -\frac{a^2b^2}{4}\\ \frac{3}{2},\ \frac{s+\rho+1}{2},\ \frac{s+\rho+2}{2}\end{matrix}\right)$ $+\,\frac{\Gamma(\rho+s-1)}{b^{s+\rho-1}}\,{}_2F_3\left(\begin{matrix}\frac{1+\rho}{2},\ \frac{1-\rho}{2};\ -\frac{a^2b^2}{4}\\ \frac{1}{2},\ \frac{2-s-\rho}{2},\ \frac{3-s-\rho}{2}\end{matrix}\right)$ $+\,\frac{\rho a}{b^{s+\rho-2}}\,\Gamma(s+\rho-2)\,{}_2F_3\left(\begin{matrix}\frac{2+\rho}{2},\ \frac{2-\rho}{2};\ -\frac{a^2b^2}{4}\\ \frac{3}{2},\ \frac{3-s-\rho}{2},\ \frac{4-s-\rho}{2}\end{matrix}\right)$ $[\operatorname{Re} a,\ \operatorname{Re} b,\ \operatorname{Re} s>0]$
24	$\frac{\left(\sqrt{x^2+a^2}-a\right)^\rho}{\sqrt{x^2+a^2}}\,e^{-bx}$	$(2a)^{s+\rho-1}\,\mathrm{B}\left(\frac{s+2\rho}{2},\ 1-s-\rho\right){}_2F_3\left(\begin{matrix}\frac{s}{2},\ \frac{s+2\rho}{2};\ -\frac{a^2b^2}{4}\\ \frac{1}{2},\ \frac{s+\rho}{2},\ \frac{s+\rho+1}{2}\end{matrix}\right)$ $-\,(2a)^{s+\rho}\,b\,\mathrm{B}\left(\frac{s+2\rho+1}{2},\ -s-\rho\right){}_2F_3\left(\begin{matrix}\frac{s+1}{2},\ \frac{s+2\rho+1}{2};\ -\frac{a^2b^2}{4}\\ \frac{3}{2},\ \frac{s+\rho+1}{2},\ \frac{s+\rho+2}{2}\end{matrix}\right)$ $+\,\frac{\Gamma(s+\rho-1)}{b^{s+\rho-1}}\,{}_2F_3\left(\begin{matrix}\frac{1+\rho}{2},\ \frac{1-\rho}{2};\ -\frac{a^2b^2}{4}\\ \frac{1}{2},\ \frac{2-s-\rho}{2},\ \frac{3-s-\rho}{2}\end{matrix}\right)$ $-\,\frac{\rho\, a}{b^{s+\rho-2}}\,\Gamma(s+\rho-2)\,{}_2F_3\left(\begin{matrix}\frac{2+\rho}{2},\ \frac{2-\rho}{2};\ -\frac{a^2b^2}{4}\\ \frac{3}{2},\ \frac{3-s-\rho}{2},\ \frac{4-s-\rho}{2}\end{matrix}\right)$ $[\operatorname{Re} a,\ \operatorname{Re} b,\ \operatorname{Re}(s+2\rho)>0]$

No.	$f(x)$	$F(s)$
25	$\left(\sqrt{x^2+a^2}+x\right)^{\rho} e^{-bx}$	$-\frac{\rho\, a^{s+\rho}}{2^{s+1}}\,\Gamma\left[\begin{matrix} s,\ -\frac{s+\rho}{2} \\ \frac{s-\rho+2}{2}\end{matrix}\right] {}_2F_3\left(\begin{matrix}\frac{s}{2},\ \frac{s+1}{2};\ -\frac{a^2b^2}{4} \\ \frac{1}{2},\ \frac{s+\rho+2}{2},\ \frac{s-\rho+2}{2}\end{matrix}\right) + \frac{\rho\, a^{s+\rho+1}\, b}{2^{s+2}}\,\Gamma\left[\begin{matrix} s+1,\ -\frac{s+\rho+1}{2} \\ \frac{s-\rho+3}{2}\end{matrix}\right] {}_2F_3\left(\begin{matrix}\frac{s+1}{2},\ \frac{s+2}{2};\ -\frac{a^2b^2}{4} \\ \frac{3}{2},\ \frac{s+\rho+3}{2},\ \frac{s-\rho+3}{2}\end{matrix}\right) + \frac{2^{\rho}\Gamma(s+\rho)}{b^{s+\rho}}\, {}_2F_3\left(\begin{matrix}-\frac{\rho}{2},\ \frac{1-\rho}{2};\ -\frac{a^2b^2}{4} \\ 1-\rho,\ \frac{1-s-\rho}{2},\ \frac{2-s-\rho}{2}\end{matrix}\right)$ [Re a, Re b, Re $s>0$]
26	$\left(\sqrt{x^2+a^2}-x\right)^{\rho} e^{-bx}$	$\frac{\rho\, a^{s+\rho}}{2^{s+1}}\,\Gamma\left[\begin{matrix} s,\ \frac{\rho-s}{2} \\ \frac{s+\rho+2}{2}\end{matrix}\right] {}_2F_3\left(\begin{matrix}\frac{s}{2},\ \frac{s+1}{2};\ -\frac{a^2b^2}{4} \\ \frac{1}{2},\ \frac{s-\rho+2}{2},\ \frac{s+\rho+2}{2}\end{matrix}\right) - \frac{\rho a^{s+\rho+1}\, b}{2^{s+2}}\,\Gamma\left[\begin{matrix} s+1,\ \frac{\rho-s-1}{2} \\ \frac{s+\rho+3}{2}\end{matrix}\right] {}_2F_3\left(\begin{matrix}\frac{s+1}{2},\ \frac{s+2}{2};\ -\frac{a^2b^2}{4} \\ \frac{3}{2},\ \frac{s-\rho+3}{2},\ \frac{s+\rho+3}{2}\end{matrix}\right) + \frac{a^{2\rho}\Gamma(s-\rho)}{2^{\rho} b^{s-\rho}}\, {}_2F_3\left(\begin{matrix}\frac{\rho}{2},\ \frac{\rho+1}{2};\ -\frac{a^2b^2}{4} \\ \rho+1,\ \frac{1-s+\rho}{2},\ \frac{2-s+\rho}{2}\end{matrix}\right)$ [Re a, Re b, Re $s>0$]
27	$\frac{\left(\sqrt{x^2+a^2}+x\right)^{\rho}}{\sqrt{x^2+a^2}}\, e^{-bx}$	$\frac{a^{s+\rho-1}}{2^{s}}\,\Gamma\left[\begin{matrix} s,\ \frac{1-s-\rho}{2} \\ \frac{s-\rho+1}{2}\end{matrix}\right] {}_2F_3\left(\begin{matrix}\frac{s}{2},\ \frac{s+1}{2};\ -\frac{a^2b^2}{4} \\ \frac{1}{2},\ \frac{s+\rho+1}{2},\ \frac{s-\rho+1}{2}\end{matrix}\right) - \frac{a^{s+\rho} b}{2^{s+1}}\,\Gamma\left[\begin{matrix} s+1,\ -\frac{s+\rho}{2} \\ \frac{s-\rho+2}{2}\end{matrix}\right] {}_2F_3\left(\begin{matrix}\frac{s+1}{2},\ \frac{s+2}{2};\ -\frac{a^2b^2}{4} \\ \frac{3}{2},\ \frac{s+\rho+2}{2},\ \frac{s-\rho+2}{2}\end{matrix}\right) + \frac{2^{\rho}\Gamma(s+\rho-1)}{b^{s+\rho-1}}\, {}_2F_3\left(\begin{matrix}\frac{1-\rho}{2},\ \frac{2-\rho}{2};\ -\frac{a^2b^2}{4} \\ 1-\rho,\ \frac{2-s-\rho}{2},\ \frac{3-s-\rho}{2}\end{matrix}\right)$ [Re a, Re b, Re $s>0$]
28	$\frac{\left(\sqrt{x^2+a^2}-x\right)^{\rho}}{\sqrt{x^2+a^2}}\, e^{-bx}$	$\frac{a^{s+\rho-1}}{2^{s}}\,\mathrm{B}\left(s,\ \frac{1-s+\rho}{2}\right) {}_2F_3\left(\begin{matrix}\frac{s}{2},\ \frac{s+1}{2};\ -\frac{a^2b^2}{4} \\ \frac{1}{2},\ \frac{s-\rho+1}{2},\ \frac{s+\rho+1}{2}\end{matrix}\right) - \frac{a^{s+\rho}\, b}{2^{s+1}}\,\mathrm{B}\left(s+1,\ \frac{\rho-s}{2}\right) {}_2F_3\left(\begin{matrix}\frac{s+1}{2},\ \frac{s+2}{2};\ -\frac{a^2b^2}{4} \\ \frac{3}{2},\ \frac{s-\rho+2}{2},\ \frac{s+\rho+2}{2}\end{matrix}\right) + \frac{2^{-\rho}a^{2\rho}}{b^{s-\rho-1}}\,\Gamma(s-\rho-1)\, {}_2F_3\left(\begin{matrix}\frac{\rho+1}{2},\ \frac{\rho+2}{2};\ -\frac{a^2b^2}{4} \\ \rho+1,\ \frac{2-s+\rho}{2},\ \frac{3-s+\rho}{2}\end{matrix}\right)$ [Re a, Re b, Re $s>0$]
29	$(a-x)_{+}^{\alpha-1}\, e^{bx^2}$	$a^{s+\alpha-1}\,\mathrm{B}(s,\ \alpha)\ {}_2F_2\left(\begin{matrix}\frac{s}{2},\ \frac{s+1}{2};\ a^2b \\ \frac{s+\alpha}{2},\ \frac{s+\alpha+1}{2}\end{matrix}\right)$ [a, Re α, Re $s>0$]
30	$(a-x)_{+}^{\alpha-1}\, e^{bx^n}$	$a^{s+\alpha-1}\,\mathrm{B}(\alpha,\ s)\ {}_nF_n\left(\begin{matrix}\Delta(n,\ s);\ a^n b \\ \Delta(n,\ s+\alpha)\end{matrix}\right)$ [a, Re α, Re $s>0$; $n=1,2,\ldots$]

No.	$f(x)$	$F(s)$
31	$(a-x)_+^{\alpha-1}\, e^{bx(a-x)}$	$a^{s+\alpha-1}\,\mathrm{B}(s,\,\alpha)\;{}_2F_2\left(\begin{matrix}s,\,\alpha;\;\frac{a^2b}{4}\\ \frac{s+\alpha}{2},\;\frac{s+\alpha+1}{2}\end{matrix}\right)$ [$a,$ $\operatorname{Re}\alpha,$ $\operatorname{Re}s>0$]
32	$(a-x)_+^{\alpha-1}\, e^{bx^2(a-x)^2}$	$a^{s+\alpha-1}\,\mathrm{B}(s,\,\alpha)\;{}_4F_4\left(\begin{matrix}\frac{s}{2},\;\frac{s+1}{2},\;\frac{\alpha}{2},\;\frac{\alpha+1}{2};\;\frac{a^4b}{16}\\ \frac{s+\alpha}{4},\;\frac{s+\alpha+1}{4},\;\frac{s+\alpha+2}{4},\;\frac{s+\alpha+3}{4}\end{matrix}\right)$ [$a,$ $\operatorname{Re}s>0$]
33	$(a-x)_+^{\alpha-1}\, e^{b(a-x)^n}$	$a^{s+\alpha-1}\;\mathrm{B}(\alpha,\,s)\;{}_nF_n\left(\begin{matrix}\Delta(n,\,\alpha)\,;\;a^nb\\ \Delta(n,\,s+\alpha)\end{matrix}\right)$ [$a,$ $\operatorname{Re}\alpha,$ $\operatorname{Re}s>0;$ $n=1,2,\ldots$]

2.2.3. $e^{\varphi(x)}$ and algebraic functions

1	$(x-a)_+^{\alpha-1}\, e^{b/x}$	$a^{s+\alpha-1}\,\mathrm{B}(1-s-\alpha,\,\alpha)\;{}_1F_1\left(\begin{matrix}1-s-\alpha\\ 1-s;\;\frac{b}{a}\end{matrix}\right)$ [$a>0;$ $\operatorname{Re}(s+\alpha)<0$]
2	$\dfrac{e^{-b/x}}{(x+a)^{\rho}}$	$a^{s-\rho}\,\Gamma(\rho-s)\;\Psi\left(\rho-s;\;\dfrac{b}{a};\;1-s\right)$ [$\operatorname{Re}b>0;$ $\operatorname{Re}\rho>\operatorname{Re}s>0;$ $\lvert\arg a\rvert<\pi$]
3	$\dfrac{e^{-b/x}}{x+a}$	$a^{s-1}\,e^{b/a}\,\Gamma(1-s)\,\Gamma\left(s,\;\dfrac{b}{a}\right)$ [$\operatorname{Re}b>0;$ $\operatorname{Re}s<1;$ $\lvert\arg a\rvert<\pi$]
4	$(x-a)_+^{\alpha-1}\, e^{b/x^2}$	$a^{s+\alpha-1}\,\mathrm{B}(1-s-\alpha,\,\alpha)\;{}_2F_2\left(\begin{matrix}\frac{1-s-\alpha}{2},\;\frac{2-s-\alpha}{2}\\ \frac{1-s}{2},\;\frac{2-s}{2};\;\frac{b}{a^2}\end{matrix}\right)$ [$a>0;$ $\operatorname{Re}(s+\alpha)<0$]
5	$(1-x)_+^{\alpha}\, e^{-a/(1-x)}$	$e^{-a}\,\Gamma(s)\,\Psi(s;\;a;\;-\alpha)$ [$\operatorname{Re}a,$ $\operatorname{Re}s>0$]
6	$(1-x)_+^{-1/2}\, e^{-a/(1-x)}$	$2^s e^{-a/2}\,\Gamma(s)\,D_{-2s}\left(\sqrt{2a}\right)$ [$\operatorname{Re}a,$ $\operatorname{Re}s>0$]
7	$\left(1-x^2\right)_+^{-1/2}\, e^{-a/(1-x)}$	$e^{-a/2}\,\Gamma(s)\;D_{-s}^2\left(\sqrt{a}\right)$ [$\operatorname{Re}a,$ $\operatorname{Re}s>0$]
8	$(x-1)_+^{\alpha}\, e^{-a/(x-1)}$	$\Gamma(-s-\alpha)\,\Psi(-s-\alpha;\;-\alpha;\;a)$ [$\operatorname{Re}a>0;$ $\operatorname{Re}s<-\operatorname{Re}\alpha$]
9	$(x-1)_+^{-1/2}\, e^{-a/(x-1)}$	$\dfrac{e^{a/2}}{2^{s-1/2}}\,\Gamma\left(\dfrac{1}{2}-s\right)D_{2s-1}\left(\sqrt{2a}\right)$ [$\operatorname{Re}a>0;$ $\operatorname{Re}s<1/2$]
10	$\left(x^2-1\right)_+^{-1/2}\, e^{-a/(x-1)}$	$e^{a/2}\,\Gamma(1-s)\;D_{s-1}^2\left(\sqrt{a}\right)$ [$\operatorname{Re}a>0,$ $\operatorname{Re}s<1$]

No.	$f(x)$	$F(s)$
11	$\dfrac{e^{b/(x+a)}}{(x+a)^{\rho}}$	$a^{s-\rho}\,\mathrm{B}(s,\,\rho-s)\;{}_1F_1\left(\begin{matrix}\rho-s\\ \rho;\;\frac{b}{a}\end{matrix}\right)$ $[0<\operatorname{Re}s<\operatorname{Re}\rho;\;\lvert\arg a\rvert<\pi]$
12	$\dfrac{e^{bx/(x+a)}}{(x+a)^{\rho}}$	$a^{s-\rho}\,\mathrm{B}(s,\,\rho-s)\;{}_1F_1\left(\begin{matrix}s\\ \rho;\;b\end{matrix}\right)$ $[0<\operatorname{Re}s<\operatorname{Re}\rho;\;\lvert\arg a\rvert<\pi]$
13	$(a-x)_+^{\alpha-1}\,(b-x)^{-\alpha}$ $\times\, e^{c/(b-x)}$	$a^{s+\alpha-1}b^{-\alpha}e^{c/(b-a)}\,\mathrm{B}(\alpha,\,s)\,\Phi_1\left(\alpha,\,s;\,s+\alpha;\,\dfrac{a}{b},\,\dfrac{ac}{b(a-b)}\right)$ $[0<a<b;\;\operatorname{Re}s,\;\operatorname{Re}\alpha>0]$
14	$\left(x^2+1\right)^{\alpha}e^{-a/(x^2+1)}$	$\dfrac{e^{-a}}{2}\,\mathrm{B}\left(\dfrac{s}{2},\,-\dfrac{s}{2}-\alpha\right)\;{}_1F_1\left(\begin{matrix}\frac{s}{2};\;a\\ -\alpha\end{matrix}\right)$ $[0<\operatorname{Re}s<-2\operatorname{Re}\alpha]$
15	$\left(1-x^2\right)_+^{-1/2}e^{-ax/(1-x)}$	$e^{a/2}\,\Gamma(s)\;D_{-s}^2\left(\sqrt{a}\right)$ $[\operatorname{Re}a,\;\operatorname{Re}s>0]$
16	$\left(1-x^2\right)_+^{-1/2}$ $\times\, e^{-a(1+x)/(1-x)}$	$\Gamma(s)\;D_{-s}^2\left(\sqrt{2a}\right)$ $[\operatorname{Re}a,\;\operatorname{Re}s>0]$
17	$\left(x^2-1\right)_+^{-1/2}$ $\times\, e^{-a(x+1)/(x-1)}$	$\Gamma(1-s)\;D_{s-1}^2\left(\sqrt{2a}\right)$ $[\operatorname{Re}s<1]$
18	$\left(1-x^2\right)_+^{-1/2}e^{a(x-x^{-1})}$	$\dfrac{\sqrt{\pi^3a}}{2\sqrt{2}}\left[J_{s/2}(a)\,Y_{(s-1)/2}(a)-J_{(s-1)/2}(a)\,Y_{s/2}(a)\right]$ $[\operatorname{Re}a>0]$
19	$\left(x^2-1\right)_+^{-1/2}e^{a(x^{-1}-x)}$	$\dfrac{\sqrt{\pi^3a}}{2\sqrt{2}}\left[J_{(1-s)/2}(a)\,Y_{-s/2}(a)-J_{-s/2}(a)\,Y_{(1-s)/2}(a)\right]$ $[\operatorname{Re}a>0]$
20	$\left(1-x^2\right)_+^{-1/2}$ $\times\, e^{-(ax+b)/(1-x^2)}$	$e^{-b/2}\,\Gamma(s)\;D_{-s}\left(\sqrt{b+\sqrt{b^2-a^2}}\right)D_{-s}\left(\sqrt{b-\sqrt{b^2-a^2}}\right)$ $[\operatorname{Re}(a+b),\;\operatorname{Re}s>0]$
21	$\left(x^2-1\right)_+^{-1/2}$ $\times\, e^{-(ax+b)/(x^2-1)}$	$e^{b/2}\,\Gamma(1-s)\;D_{s-1}\left(\sqrt{b+\sqrt{b^2-a^2}}\right)D_{s-1}\left(\sqrt{b-\sqrt{b^2-a^2}}\right)$ $[\operatorname{Re}(a+b)>0;\;\operatorname{Re}s<1]$
22	$\left(1+x^2\right)^{\alpha}$ $\times\, e^{-a(1-x^2)/(1+x^2)}$	$\dfrac{1}{2}\,(2a)^{\alpha/2}\,\mathrm{B}\left(\dfrac{s}{2},\,-\dfrac{s+2\alpha}{2}\right)M_{-(s+\alpha)/2,\,-(\alpha+1)/2}(2a)$ $[\operatorname{Re}a>0;\;0<\operatorname{Re}s<-2\operatorname{Re}\alpha]$

No.	$f(x)$	$F(s)$
23	$\left(1-x^2\right)_+^{\alpha} e^{-ax^2/(1-x^2)}$	$\dfrac{a^{\alpha/2}}{2} e^{a/2}\,\Gamma\left(\dfrac{s}{2}\right) W_{-(s+\alpha)/2,\,-(\alpha+1)/2}(a)$ [$\operatorname{Re} a,\ \operatorname{Re} s>0$]
24	$\left(1-x^2\right)_+^{-1/2}$ $\times e^{-(ax+b)^2/(1-x^2)}$	$e^{(a^2-b^2)/2}\,\Gamma(s)\, D_{-s}\left(\sqrt{2}\,a\right) D_{-s}\left(\sqrt{2}\,b\right)$ [$\operatorname{Re}(a+b),\ \operatorname{Re} s>0$]
25	$\left(1-x^2\right)_+^{-1/2}$ $\times e^{-(bx^2+ax+b)/(1-x^2)}$	$\Gamma(s)\, D_{-s}\left(\sqrt{2b+\sqrt{4b^2-a^2}}\right) D_{-s}\left(\sqrt{2b-\sqrt{4b^2-a^2}}\right)$ [$\operatorname{Re}(a+2b),\ \operatorname{Re} s>0$]
26	$\left(1-x^2\right)_+^{-1/2}$ $\times e^{-(bx^2+ax)/(1-x^2)}$	$e^{b/2}\Gamma(s)\, D_{-s}\left(\sqrt{b+\sqrt{b^2-a^2}}\right) D_{-s}\left(\sqrt{b-\sqrt{b^2-a^2}}\right)$ [$\operatorname{Re}(a+b),\ \operatorname{Re} s>0$]
27	$\left(x^2-1\right)_+^{-1/2}$ $\times e^{-(bx^2+ax+b)/(x^2-1)}$	$\Gamma(1-s)\, D_{s-1}\left(\sqrt{2b+\sqrt{4b^2-a^2}}\right) D_{s-1}\left(\sqrt{2b-\sqrt{4b^2-a^2}}\right)$ [$\operatorname{Re}(a+2b)>0;\ \operatorname{Re} s<1$]
28	$e^{-b\sqrt{x+a}}$	$\dfrac{b}{\sqrt{\pi}}\left(\dfrac{2\sqrt{a}}{b}\right)^{s+1/2}\Gamma(s)\, K_{s+1/2}\left(\sqrt{a}\,b\right)$ [$\operatorname{Re} a,\ \operatorname{Re} b,\ \operatorname{Re} s>0$]
29	$\dfrac{e^{-b\sqrt{x+a}}}{\sqrt{x+a}}$	$\dfrac{2}{\sqrt{\pi}}\left(\dfrac{2\sqrt{a}}{b}\right)^{s-1/2}\Gamma(s)\, K_{s-1/2}\left(\sqrt{a}\,b\right)$ [$a,\ \operatorname{Re} b,\ \operatorname{Re} s>0$]
30	$\dfrac{e^{ia\sqrt{x^2+1}}}{\sqrt{x^2+1}}$	$\dfrac{i\sqrt{\pi}}{2}\left(\dfrac{a}{2}\right)^{(1-s)/2}\Gamma\left(\dfrac{s}{2}\right) H_{(1-s)/2}^{(1)}(a)$ [$\operatorname{Im} a,\ \operatorname{Re} s>0$]
31	$\theta(x-a)\, e^{-b\sqrt{x^2-a^2}}$	$\dfrac{a^{(s+1)/2}}{b^{(s-1)/2}} S_{(s-3)/2,\,(s+1)/2}(ab)$ [$a,\ \operatorname{Re} b>0$]
32	$\left(a^2-x^2\right)_+^{-1/2} e^{-b\sqrt{a^2-x^2}}$	$\dfrac{\sqrt{\pi}}{2}\left(\dfrac{2a}{b}\right)^{(s-1)/2}\Gamma\left(\dfrac{s}{2}\right)\left[I_{(s-1)/2}(ab)-\mathbf{L}_{(s-1)/2}(ab)\right]$ [$a,\ \operatorname{Re} b,\ \operatorname{Re} s>0$]
33	$\left(x^2-a^2\right)_+^{-1/2} e^{-b\sqrt{x^2-a^2}}$	$\dfrac{\sqrt{\pi}}{2}\left(\dfrac{2a}{b}\right)^{(s-1)/2}\Gamma\left(\dfrac{s}{2}\right)\left[\mathbf{H}_{(s-1)/2}(ab)-Y_{(s-1)/2}(ab)\right]$ [$a,\ \operatorname{Re} b>0$]

2.2.4. $(e^{ax} \pm c)^{\rho}\, e^{-bx}$

1	$\dfrac{1}{e^{ax}+1}$	$\dfrac{1-2^{1-s}}{a^s}\Gamma(s)\,\zeta(s)$ [Re a, Re $s>0$]
2	$\dfrac{1}{e^{ax}-1}$	$\dfrac{\Gamma(s)}{a^s}\zeta(s)$ [Re $a>0$; Re $s>1$]
3	$\dfrac{1}{e^{ax}-c}$	$\dfrac{\Gamma(s)}{a^s c}\,\mathrm{Li}_s(c)$ [Re $s>1$; $\lvert\arg(1-c)\rvert<\pi$]
4	$\dfrac{e^{-bx}}{e^{ax}+1}$	$\dfrac{\Gamma(s)}{(2a)^s}\left[\zeta\left(s,\,\dfrac{a+b}{2a}\right)-\zeta\left(s,\,\dfrac{2a+b}{2a}\right)\right]$ [Re a, Re $(a+b)$, Re $s>0$]
5	$\dfrac{e^{-bx}}{e^{ax}-1}$	$\dfrac{\Gamma(s)}{a^s}\zeta\left(s,\,\dfrac{a+b}{a}\right)$ [Re a, Re $(a+b)>0$; Re $s>1$]
6	$\left(\dfrac{1}{e^x-1}-\dfrac{1}{x}+\dfrac{1}{2}\right)e^{-ax}$	$\Gamma(s)\left[\zeta(s,\,a)-\dfrac{a^{-s}}{2}+\dfrac{a^{1-s}}{1-s}\right]$ [Re $a>0$; Re $s>-1$]
7	$\left(\dfrac{1}{e^x-1}-\dfrac{1}{x}\right)e^{-ax}$	$\Gamma(s)\left[\zeta(s,\,a)-a^{-s}+\dfrac{a^{1-s}}{1-s}\right]$ [Re a, Re $s>0$]
8	$\dfrac{e^{-bx}}{e^{ax}-c}$	$\dfrac{\Gamma(s)}{a^s}\Phi\left(c,\,s,\,\dfrac{a+b}{a}\right)$ [(Re a, Re $(a+b)>0$; Re $s>0$; $\lvert\arg(1-c)\rvert<\pi$) or ($\lvert c\rvert\le 1$; $c\neq 1$; Re $s>0$) or ($c=1$; Re $s>1$)]
9	$\dfrac{1}{(e^{ax}-1)^2}$	$\dfrac{\Gamma(s)}{a^s}\left[\zeta(s-1)-\zeta(s)\right]$ [Re $a>0$; Re $s>2$]
10	$\dfrac{e^{-bx}}{(e^{ax}-1)^2}$	$\dfrac{\Gamma(s)}{a^{s+1}}\left[a\,\zeta\left(s-1,\,\dfrac{a+b}{a}\right)-(a+b)\,\zeta\left(s,\,\dfrac{a+b}{a}\right)\right]$ [Re a, Re $(a+b)>0$; Re $s>2$]
11	$\dfrac{e^{-bx}}{(e^{ax}-c)^2}$	$\dfrac{\Gamma(s)}{a^{s+1}c}\left[a\Phi\left(c,\,s-1,\,\dfrac{a+b}{a}\right)-(a+b)\,\Phi\left(c,\,s,\,\dfrac{a+b}{a}\right)\right]$ [(Re a, Re $(a+b)>0$; Re $s>0$; $\lvert\arg(1-c)\rvert<\pi$) or ($\lvert c\rvert<1$; $c\neq 1$; Re $s>0$) or ($c=1$; Re $s>1$)]
12	$(e^{bx}+c)^n\,e^{-ax}$	$c^n\,\Gamma(s)\sum_{k=0}^{n}\binom{n}{k}\dfrac{(a-bk)^{-s}}{c^k}$ [Re $s>0$; Re $a>n$ Re b]

2.3. Hyperbolic Functions

More formulas can be obtained from the corresponding sections due to the relations

$$\sinh z = -\sinh(-z) = \frac{e^z - e^{-z}}{2} = -i\sin(iz), \quad \cosh z = \cosh(-z) = \frac{e^z + e^{-z}}{2} = \cos(iz),$$

$$\sinh z = z\,{}_0F_1\left(\frac{3}{2};\, \frac{z^2}{4}\right), \quad \cosh z = {}_0F_1\left(\frac{1}{2};\, \frac{z^2}{4}\right),$$

$$\sinh z = \frac{\sqrt{\pi}\, z}{2} G_{02}^{10}\left(-\frac{z^2}{4}\,\middle|\, \begin{matrix} \cdot \\ 0,\, -1/2 \end{matrix}\right), \quad \cosh z = \sqrt{\pi}\, G_{02}^{10}\left(-\frac{z^2}{4}\,\middle|\, \begin{matrix} \cdot \\ 0,\, 1/2 \end{matrix}\right).$$

2.3.1. Rational functions of $\sinh x$ and $\cosh x$

No.	$f(x)$	$F(s)$	
1	$\sinh(ax)$	$i(-ia)^{-s}\sin\frac{s\pi}{2}\,\Gamma(s)$	$[\operatorname{Re} a = 0;\ \lvert\operatorname{Re} s\rvert < 1]$
2	$\cosh(ax)$	$(ia)^{-s}\cos\frac{s\pi}{2}\,\Gamma(s)$	$[\operatorname{Re} a = 0;\ 0 < \operatorname{Re} s < 1]$
3	$\sinh(ax) - ax$	$i(-ia)^{-s}\sin\frac{s\pi}{2}\,\Gamma(s)$	$[\operatorname{Re} a = 0;\ -3 < \operatorname{Re} s < -1]$
4	$\cosh(ax) - 1$	$(ia)^{-s}\cos\frac{s\pi}{2}\,\Gamma(s)$	$[\operatorname{Re} a = 0;\ -2 < \operatorname{Re} s < 0]$
5	$\cosh(ax) - \frac{a^2x^2}{2} - 1$	$(ia)^{-s}\cos\frac{s\pi}{2}\,\Gamma(s)$	$[\operatorname{Re} a = 0;\ -4 < \operatorname{Re} s < -2]$
6	$\sinh(ax) - \sum_{k=0}^{n}\frac{(ax)^{2k+1}}{(2k+1)!}$	$i(-ia)^{-s}\sin\frac{s\pi}{2}\,\Gamma(s)$	$[\operatorname{Re} a = 0;\ -2n-3 < \operatorname{Re} s < -2n-1]$
7	$\cosh(ax) - \sum_{k=0}^{n}\frac{(ax)^{2k}}{(2k)!}$	$(ia)^{-s}\cos\frac{s\pi}{2}\,\Gamma(s)$	$[\operatorname{Re} a = 0;\ -2(n+1) < \operatorname{Re} s < -2n]$
8	$\operatorname{sech}(ax)$	$\frac{2^{1-2s}}{a^s}\Gamma(s)\left[\zeta\left(s,\, \frac{1}{4}\right) - \zeta\left(s,\, \frac{3}{4}\right)\right]$	$[\operatorname{Re} a,\ \operatorname{Re} s > 0]$
9	$\operatorname{csch}(ax)$	$\frac{2^s - 1}{2^{s-1}a^s}\Gamma(s)\,\zeta(s)$	$[\operatorname{Re} a > 0;\ \operatorname{Re} s > 1]$
10	$\operatorname{csch}(ax) - \frac{1}{ax}$	$2(1-2^{-s})\,a^{-s}\,\Gamma(s)\,\zeta(s)$	$[\operatorname{Re} a > 0;\ \lvert\operatorname{Re} s\rvert < 1]$
11	$\operatorname{sech}^2(ax)$	$\frac{4}{(2a)^s}(1-2^{2-s})\,\Gamma(s)\,\zeta(s-1)$	$[\operatorname{Re} a,\ \operatorname{Re} s > 0]$

No.	$f(x)$	$F(s)$
12	$\operatorname{csch}^2(ax)$	$\dfrac{2^{2-s}}{a^s}\Gamma(s)\zeta(s-1)$ [Re a, Re $s > 2$]
13	$\dfrac{\sinh(ax)}{\sinh(bx)}$	$\dfrac{\Gamma(s)}{(2b)^s}\left[\zeta\left(s, \dfrac{b-a}{2b}\right) - \zeta\left(s, \dfrac{b+a}{2b}\right)\right]$ [Re $b >$ |Re a|; Re $s > 0$]
14	$\dfrac{\cosh(ax)}{\sinh(bx)}$	$\dfrac{\Gamma(s)}{(2b)^s}\left[\zeta\left(s, \dfrac{b-a}{2b}\right) + \zeta\left(s, \dfrac{b+a}{2b}\right)\right]$ [Re $b >$ |Re a|; Re $s > 1$]
15	$\dfrac{\sinh(ax)}{\cosh(bx)}$	$\dfrac{\Gamma(s)}{(4b)^s}\left[\zeta\left(s, \dfrac{b-a}{4b}\right) - \zeta\left(s, \dfrac{b+a}{4b}\right) + \zeta\left(s, \dfrac{3b+a}{4b}\right) - \zeta\left(s, \dfrac{3b-a}{4b}\right)\right]$ [Re $b >$ |Re a|; Re $s > -1$]
16	$\dfrac{\cosh(ax)}{\cosh(bx)}$	$\dfrac{\Gamma(s)}{(4b)^s}\left[\zeta\left(s, \dfrac{b-a}{4b}\right) + \zeta\left(s, \dfrac{b+a}{4b}\right) - \zeta\left(s, \dfrac{3b+a}{4b}\right) - \zeta\left(s, \dfrac{3b-a}{4b}\right)\right]$ [Re $b >$ |Re a|; Re $s > 0$]
17	$\dfrac{\sinh(ax)}{\cosh(bx)}$	$\dfrac{\Gamma(s)}{(2b)^s}\left[\Phi\left(-1, s, \dfrac{b-a}{2b}\right) - \Phi\left(-1, s, \dfrac{b+a}{2b}\right)\right]$ [Re $b >$ |Re a|; Re $s > 1$]
18	$\dfrac{\sinh(ax)}{\cosh^2(ax)}$	$\dfrac{2^{3-2s}\Gamma(s)}{a^s}\left[\zeta\left(s-1, \dfrac{1}{4}\right) - \zeta\left(s-1, \dfrac{3}{4}\right)\right]$ [Re $a > 0$; Re $s > 1$]
19	$\dfrac{\cosh(ax)}{\sinh^2(ax)}$	$\dfrac{2}{a^s}\left(1-2^{1-s}\right)\Gamma(s)\zeta(s-1)$ [Re $a > 0$; Re $s > 2$]
20	$\dfrac{1}{\cosh x + \cos\theta}$	$2^{s-1}\pi^s \csc\theta \csc\dfrac{s\pi}{2}\left[\zeta\left(1-s, \dfrac{\pi-\theta}{2\pi}\right) - \zeta\left(1-s, \dfrac{\pi+\theta}{2\pi}\right)\right]$ [|θ| $< \pi$; Re $s > 0$]
21	$\dfrac{\sinh(x/2)}{\cosh x + \cos\theta}$	$2^{2s-3}\pi^s \csc\dfrac{\theta}{2} \sec\dfrac{s\pi}{2}\left[\zeta\left(1-s, \dfrac{\pi-\theta}{4\pi}\right) - \zeta\left(1-s, \dfrac{\pi+\theta}{4\pi}\right) + \zeta\left(1-s, \dfrac{3\pi+\theta}{4\pi}\right) - \zeta\left(1-s, \dfrac{3\pi-\theta}{4\pi}\right)\right]$ [|θ| $< \pi$; Re $s > 0$]
22	$\dfrac{\cosh(x/2)}{\cosh x + \cos\theta}$	$2^{2s-3}\pi^s \sec\dfrac{\theta}{2} \csc\dfrac{s\pi}{2}\left[\zeta\left(1-s, \dfrac{\pi-\theta}{4\pi}\right) + \zeta\left(1-s, \dfrac{\pi+\theta}{4\pi}\right) - \zeta\left(1-s, \dfrac{3\pi+\theta}{4\pi}\right) - \zeta\left(1-s, \dfrac{3\pi+\theta}{4\pi}\right)\right]$ [|θ| $< \pi$; Re $s > 0$]

No.	$f(x)$	$F(s)$
23	$\dfrac{\sinh(ax)\sinh(bx)}{\cosh(2ax)+\cosh(2bx)}$	$\dfrac{2^{-1-2s}\Gamma(s)}{(a^2-b^2)^{s/2}}\left[\left(\dfrac{a+b}{a-b}\right)^{s/2}-\left(\dfrac{a+b}{a-b}\right)^{-s/2}\right] \times\left[\zeta\left(s,\,\dfrac{1}{4}\right)-\zeta\left(s,\,\dfrac{3}{4}\right)\right]$ $[\operatorname{Re}a,\ \operatorname{Re}b>0;\ \operatorname{Re}s>-2]$
24	$\tanh(ax)$	$\dfrac{2^{1-s}-1}{2^{s-1}a^s}\,\Gamma(s)\,\zeta(s)$ $[\operatorname{Re}a>0;\ -1<\operatorname{Re}s<0]$
25	$\tanh(ax)-1$	$\dfrac{2^{1-s}-1}{2^{s-1}a^s}\,\Gamma(s)\,\zeta(s)$ $[a,\ \operatorname{Re}s>0]$
26	$\coth(ax)-1$	$\dfrac{\Gamma(s)}{2^{s-1}a^s}\,\zeta(s)$ $[a>0;\ \operatorname{Re}s>1]$

2.3.2. Hyperbolic and algebraic functions

Notation: $\delta=\begin{Bmatrix}1\\0\end{Bmatrix}$.

1	$(a-x)_+^{\alpha-1}\begin{Bmatrix}\sinh(bx)\\ \cosh(bx)\end{Bmatrix}$	$\dfrac{a^{s+\alpha-1}}{2}\,\mathrm{B}(\alpha,s)\left[{}_1F_1\left(\begin{matrix}s;\ ab\\ s+\alpha\end{matrix}\right)\mp{}_1F_1\left(\begin{matrix}s;\ -ab\\ s+\alpha\end{matrix}\right)\right]$ $[a,\ \operatorname{Re}\alpha>0;\ \operatorname{Re}s>-(1\pm1)/2]$
2	$(a^2-x^2)_+^{\alpha-1}\times\begin{Bmatrix}\sinh(bx)\\ \cosh(bx)\end{Bmatrix}$	$\dfrac{a^{s+2\alpha+\delta-2}b^{\delta}}{2}\,\mathrm{B}\left(\alpha,\,\dfrac{s+\delta}{2}\right)\,{}_1F_2\left(\begin{matrix}\frac{s+\delta}{2};\ \frac{a^2b^2}{4}\\ \frac{2\delta+1}{2},\ \frac{s+2\alpha+\delta}{2}\end{matrix}\right)$ $[a,\ \operatorname{Re}\alpha>0;\ \operatorname{Re}s>-\delta]$
3	$\dfrac{1}{(x+a)^{\rho}}\sinh\dfrac{b}{x+a}$	$a^{s-\rho-1}b\,\mathrm{B}(s,\,\rho-s+1)\ {}_2F_3\left(\begin{matrix}\frac{\rho-s+1}{2},\ \frac{\rho-s+2}{2}\\ \frac{3}{2},\ \frac{\rho+1}{2},\ \frac{\rho+2}{2};\ \frac{b^2}{4a^2}\end{matrix}\right)$ $[0<\operatorname{Re}s<\operatorname{Re}\rho+1;\ \lvert\arg a\rvert<\pi]$
4	$\dfrac{1}{(x+a)^{\rho}}\cosh\dfrac{b}{x+a}$	$a^{s-\rho}\,\mathrm{B}(s,\,\rho-s)\ {}_2F_3\left(\begin{matrix}\frac{\rho-s}{2},\ \frac{\rho-s+1}{2}\\ \frac{1}{2},\ \frac{\rho}{2},\ \frac{\rho+1}{2};\ \frac{b^2}{4a^2}\end{matrix}\right)$ $[0<\operatorname{Re}s<\operatorname{Re}\rho;\ \lvert\arg a\rvert<\pi]$
5	$\dfrac{1}{(x+a)^{\rho}}\sinh\dfrac{bx}{x+a}$	$a^{s-\rho}b\,\mathrm{B}(s+1,\,\rho-s)\ {}_2F_3\left(\begin{matrix}\frac{s+1}{2},\ \frac{s+2}{2};\ \frac{b^2}{4}\\ \frac{3}{2},\ \frac{\rho+1}{2},\ \frac{\rho+2}{2}\end{matrix}\right)$ $[-1<\operatorname{Re}s<\operatorname{Re}\rho;\ \lvert\arg a\rvert<\pi]$
6	$\dfrac{1}{(x+a)^{\rho}}\cosh\dfrac{bx}{x+a}$	$a^{s-\rho}\,\mathrm{B}(s,\,\rho-s)\ {}_2F_3\left(\begin{matrix}\frac{s}{2},\ \frac{s+1}{2};\ \frac{b^2}{4}\\ \frac{1}{2},\ \frac{\rho}{2},\ \frac{\rho+1}{2}\end{matrix}\right)$ $[0<\operatorname{Re}s<\operatorname{Re}\rho;\ \lvert\arg a\rvert<\pi]$

No.	$f(x)$	$F(s)$
7	$\dfrac{1}{(x^2+a^2)^\rho}\sinh\dfrac{bx}{x^2+a^2}$	$\dfrac{a^{s-2\rho-1}b}{2}\,\mathrm{B}\left(\dfrac{s+1}{2},\,\dfrac{1-s+2\rho}{2}\right)\,{}_2F_3\left(\begin{matrix}\frac{s+1}{2},\,\frac{1-s+2\rho}{2};\,-\frac{b^2}{16a^2}\\ \frac{3}{2},\,\frac{\rho+1}{2},\,\frac{\rho+2}{2}\end{matrix}\right)$ [$\operatorname{Re}a>0;\ -1<\operatorname{Re}s<2\operatorname{Re}\rho+1$]
8	$\dfrac{1}{(x^2+a^2)^\rho}\cosh\dfrac{bx}{x^2+a^2}$	$\dfrac{a^{s-2\rho}}{2}\,\mathrm{B}\left(\dfrac{s}{2},\,\dfrac{2\rho-s}{2}\right)\,{}_2F_3\left(\begin{matrix}\frac{s}{2},\,\frac{2\rho-s}{2};\,\frac{b^2}{16a^2}\\ \frac{1}{2},\,\frac{\rho}{2},\,\frac{\rho+1}{2}\end{matrix}\right)$ [$\operatorname{Re}a>0;\ 0<\operatorname{Re}s<2\operatorname{Re}\rho$]
9	$(a-x)_+^{(\delta-1)/2}(bx+1)^\alpha \times\left\{\begin{matrix}\sinh\left(c\sqrt{a-x}\right)\\ \cosh\left(c\sqrt{a-x}\right)\end{matrix}\right\}$	$\dfrac{\sqrt{\pi}\,a^{s+\delta-1/2}c^\delta}{\delta+1}\,\Gamma\left[\begin{matrix}s\\ \frac{2s+2\delta+1}{2}\end{matrix}\right] \times\Xi_2\left(-\alpha,\,s;\,\dfrac{2s+2\delta+1}{2};\,-ab,\,\dfrac{ac^2}{4}\right)$ [$a,\ \operatorname{Re}s>0;\ \lvert\arg(ab+1)\rvert<\pi$]
10	$(x-a)_+^{(\delta-1)/2}(1-x+a)_+^\alpha \times\left\{\begin{matrix}\sinh\left(b\sqrt{x-a}\right)\\ \cosh\left(b\sqrt{x-a}\right)\end{matrix}\right\}$	$\dfrac{\sqrt{\pi}\,(a+1)^{s-1}b^\delta}{\delta+1}\,\Gamma\left[\begin{matrix}\alpha+1\\ \frac{2\alpha+2\delta+3}{2}\end{matrix}\right] \times\Xi_2\left(1-s,\,\alpha+1;\,\dfrac{2\alpha+2\delta+3}{2};\,\dfrac{1}{a+1},\,\dfrac{b^2}{4}\right)$ [$a,\ \operatorname{Re}s>0$]
11	$(a-x)_+^{\alpha-1} \times\left\{\begin{matrix}\sinh\left(b\sqrt{x(a-x)}\right)\\ \cosh\left(b\sqrt{x(a-x)}\right)\end{matrix}\right\}$	$a^{s+\alpha+\delta-1}b^\delta\,\mathrm{B}\left(\dfrac{2\alpha+\delta}{2},\,\dfrac{2s+\delta}{2}\right)\,{}_2F_3\left(\begin{matrix}\frac{2\alpha+\delta}{2},\,\frac{2s+\delta}{2};\,\frac{a^2b^2}{16}\\ \frac{2\delta+1}{2},\,\frac{s+\alpha+\delta}{2},\,\frac{s+\alpha+\delta+1}{2}\end{matrix}\right)$ [$a>0;\ \operatorname{Re}\alpha,\ \operatorname{Re}s>-\delta/2$]

2.3.3. Hyperbolic functions and e^{ax}

1	$e^{-ax}\sinh(ax)$	$-\dfrac{a^{-s}}{2^{s+1}}\,\Gamma(s)$ [$-1<\operatorname{Re}s<0;\ \lvert\arg a\rvert\le\pi/2$]
2	$e^{-ax}\left\{\begin{matrix}\sinh(bx)\\ \cosh(bx)\end{matrix}\right\}$	$\dfrac{\Gamma(s)}{2}\left[(a-b)^{-s}\mp(a+b)^{-s}\right]$ $\left[\begin{matrix}(\operatorname{Re}a>\lvert\operatorname{Re}b\rvert;\ \operatorname{Re}s>-(1\pm1)/2)\ \text{or}\\ (\operatorname{Re}a+\lvert\operatorname{Re}b\rvert=0;\ \operatorname{Re}s<1)\end{matrix}\right]$
3	$e^{-ax}\left\{\begin{matrix}\sinh(bx)\\ \cosh(bx)\end{matrix}\right\}^n$	$\dfrac{\Gamma(s)}{2^n}\sum_{k=0}^{n}(\mp1)^{n-k}\binom{n}{k}\left[a+(n-2k)b\right]^{-s}$ [$\operatorname{Re}a>n\lvert\operatorname{Re}b\rvert;\ \operatorname{Re}s>-(1\pm1)/2$]

No.	$f(x)$	$F(s)$
4	$e^{-ax}\sinh^{2n}(bx)$	$\dfrac{(-1)^n\Gamma(s)}{2^{2n}a^s}\dbinom{2n}{n}+\dfrac{\Gamma(s)}{2^{2n}}\sum_{k=0}^{n-1}(-1)^k\dbinom{2n}{k}$ $\times\left[(a-(2n-2k)b)^{-s}+(a+(2n-2k)b)^{-s}\right]$ $[\operatorname{Re}(a-2nb)>0;\ \operatorname{Re}s>-2n]$
5	$e^{-ax}\sinh^{2n+1}(bx)$	$\dfrac{\Gamma(s)}{2^{2n+1}}\sum_{k=0}^{n}(-1)^k\dbinom{2n+1}{k}\Big[(a-(2n-2k+1)b)^{-s}$ $-(a+(2n-2k+1)b)^{-s}\Big]$ $[\operatorname{Re}(a-(2n+1)b)>0;\ \operatorname{Re}s>-2n-1]$
6	$e^{-ax}\cosh^{n}(bx)$	$\dfrac{(1+(-1)^n)\Gamma(s)}{2^{n+1}a^s}\dbinom{n}{n/2}+\dfrac{\Gamma(s)}{2^n}\sum_{k=0}^{[(n-1)/2]}\dbinom{n}{k}$ $\times\left[(a-(n-2k)b)^{-s}+(a+(n-2k)b)^{-s}\right]$ $[\operatorname{Re}(a-nb),\ \operatorname{Re}s>0]$
7	$\dfrac{e^{-ax}}{\sinh(bx)}$	$\dfrac{2^{1-s}}{b^s}\Gamma(s)\,\zeta\left(s,\,\dfrac{a+b}{2b}\right)$ $[\operatorname{Re}a>-\lvert\operatorname{Re}b\rvert;\ \operatorname{Re}s>1]$
8	$\dfrac{e^{-ax}}{\cosh(bx)}$	$\dfrac{2^{1-2s}}{b^s}\Gamma(s)\left[\zeta\left(s,\,\dfrac{b+a}{4b}\right)-\zeta\left(s,\,\dfrac{3b+a}{4b}\right)\right]$ $[\operatorname{Re}a>-\lvert\operatorname{Re}b\rvert;\ \operatorname{Re}s>0]$
9	$\dfrac{e^{-ax}}{\cosh(ax)}$	$\dfrac{2^{1-s}}{a^s}\left(1-2^{1-s}\right)\Gamma(s)\,\zeta(s)$ $[\operatorname{Re}a,\ \operatorname{Re}s>0;\ s\neq1]$
10	$e^{-ax}\tanh(bx)$	$\dfrac{\Gamma(s)}{2^{2s-1}b^s}\left[\zeta\left(s,\,\dfrac{a}{4b}\right)-\zeta\left(s,\,\dfrac{a+2b}{4b}\right)\right]-\dfrac{\Gamma(s)}{a^s}$ $[\operatorname{Re}a>0;\ \operatorname{Re}s>-1]$
11	$e^{-ax}\coth(bx)$	$\dfrac{2^{1-s}}{b^s}\Gamma(s)\,\zeta\left(s,\,\dfrac{a}{2b}\right)-\dfrac{\Gamma(s)}{a^s}$ $[b,\ \operatorname{Re}a>0;\ \operatorname{Re}s>1]$
12	$\dfrac{1}{e^{ax}+1}\begin{Bmatrix}\sinh(bx)\\ \cosh(bx)\end{Bmatrix}$	$\dfrac{\Gamma(s)}{2(2a)^s}\left[\zeta\left(s,\,\dfrac{a-b}{2a}\right)\mp\zeta\left(s,\,\dfrac{a+b}{2a}\right)-\zeta\left(s,\,\dfrac{2a-b}{2a}\right)\right.$ $\left.\pm\zeta\left(s,\,\dfrac{2a+b}{2a}\right)\right]$ $[\operatorname{Re}a>\lvert\operatorname{Re}b\rvert;\ \operatorname{Re}s>-(1\pm1)/2]$
13	$\dfrac{1}{e^{ax}-1}\begin{Bmatrix}\sinh(bx)\\ \cosh(bx)\end{Bmatrix}$	$\dfrac{\Gamma(s)}{2a^s}\left[\zeta\left(s,\,\dfrac{a-b}{a}\right)\mp\zeta\left(s,\,\dfrac{a+b}{a}\right)\right]$ $[\operatorname{Re}a>\lvert\operatorname{Re}b\rvert;\ \operatorname{Re}s>(1\mp1)/2]$

No.	$f(x)$	$F(s)$
14	$\dfrac{e^{-ax}}{\cosh(ax)+\cos\theta}$	$\left(\dfrac{2\pi}{a}\right)^{s}\csc\theta\csc(s\pi)\left[\cos\dfrac{2\theta+s\pi}{2}\,\zeta\left(1-s,\,\dfrac{\pi+\theta}{2\pi}\right)-\cos\dfrac{2\theta-s\pi}{2}\,\zeta\left(1-s,\,\dfrac{\pi-\theta}{2\pi}\right)\right]$ $[\lvert\theta\rvert<\pi;\ \operatorname{Re}a,\ \operatorname{Re}s>0]$
15	$\dfrac{e^{-ax}}{\cosh(ax)-\cos\theta}$	$\dfrac{i\Gamma(s)}{a^{s}\sin\theta}\left[e^{i\theta}\operatorname{Li}_s\left(e^{-i\theta}\right)-e^{-i\theta}\operatorname{Li}_s\left(e^{i\theta}\right)\right]$ $[\lvert\theta\rvert\neq 2\pi n;\ \operatorname{Re}a,\ \operatorname{Re}s>0]$
16	$\theta(a-x)\,e^{bx}\sinh(a-x)$	$e^{-a}a^{s+1}\,\Gamma\begin{bmatrix}s\\ s+2\end{bmatrix}\Phi_2(s,\,1;\,s+2;\,ab+a,\,2a)$ $[a,\ \operatorname{Re}s>0]$
17	$(a-x)_+^{-1/2}\,e^{cx}\times\begin{Bmatrix}\sinh\left(b\sqrt{a-x}\right)\\ \cosh\left(b\sqrt{a-x}\right)\end{Bmatrix}$	$\dfrac{\sqrt{\pi}}{2^{\delta}}\,a^{s+\delta-1/2}b^{\delta}\,\Gamma\begin{bmatrix}s\\ \frac{2s+2\delta+1}{2}\end{bmatrix}\Phi_3\left(s;\,\dfrac{2s+2\delta+1}{2};\,ac,\,\dfrac{ab^2}{4}\right)$ $[a,\ \operatorname{Re}s>0]$

2.3.4. Hyperbolic functions and $e^{\varphi(x)}$

Notation: $\delta=\begin{Bmatrix}1\\0\end{Bmatrix}$.

1	$e^{-ax^2}\begin{Bmatrix}\sinh(bx)\\ \cosh(bx)\end{Bmatrix}$	$\dfrac{e^{b^2/(8a)}}{2^{s/2+1}a^{s/2}}\,\Gamma(s)\left[D_{-s}\left(-\dfrac{b}{\sqrt{2a}}\right)\mp D_{-s}\left(\dfrac{b}{\sqrt{2a}}\right)\right]$ $[\operatorname{Re}a>0;\ \operatorname{Re}s>-(1\pm1)/2]$
2	$e^{-ax^2-bx}\begin{Bmatrix}\sinh(cx)\\ \cosh(cx)\end{Bmatrix}$	$\dfrac{e^{(b^2+c^2)/(8a)}}{2^{s/2+1}a^{s/2}}\,\Gamma(s)\left[e^{-bc/(4a)}D_{-s}\left(\dfrac{b-c}{\sqrt{2a}}\right)\mp e^{bc/(4a)}\times D_{-s}\left(\dfrac{b+c}{\sqrt{2a}}\right)\right]$ $[\operatorname{Re}a>0;\ \operatorname{Re}s>-(1\pm1)/2]$
3	$e^{-ax-b/x}\begin{Bmatrix}\sinh(cx)\\ \cosh(cx)\end{Bmatrix}$	$\left(\dfrac{b}{a-c}\right)^{s/2}K_s\left(2\sqrt{ab-bc}\right)\mp\left(\dfrac{b}{a+c}\right)^{s/2}K_s\left(2\sqrt{ab+bc}\right)$ $[\operatorname{Re}a>\lvert\operatorname{Re}c\rvert;\ \operatorname{Re}b>0]$
4	$\left(a^2-x^2\right)_+^{-1/2}e^{-b/(a^2-x^2)}\times\begin{Bmatrix}\sinh\left[cx/\left(a^2-x^2\right)\right]\\ \cosh\left[cx/\left(a^2-x^2\right)\right]\end{Bmatrix}$	$\dfrac{2^{(2s-3)/4+\delta}a^{s-1}}{\sqrt{c}}\,e^{-b/(2a^2)}\left(b+\sqrt{b^2-a^2c^2}\right)^{1/4}\Gamma\left(\dfrac{s+\delta}{2}\right)\times D_{-s}\left(\dfrac{\sqrt{b+\sqrt{b^2-a^2c^2}}}{a}\right)M_{(1-2s)/4,\,\pm1/4}\left(\dfrac{b-\sqrt{b^2-a^2c^2}}{2a^2}\right)$ $[a>0;\ b>ac>0;\ \operatorname{Re}s>-\delta]$

No.	$f(x)$	$F(s)$
5	$(x^2-a^2)_+^{-1/2}\, e^{-b/(x^2-a^2)}$ $\times \left\{ \begin{matrix} \sinh\left[cx/(x^2-a^2)\right] \\ \cosh\left[cx/(x^2-a^2)\right] \end{matrix} \right\}$	$\dfrac{a^{s-1}}{2^{(2s+1)/4-\delta}\sqrt{c}}\, e^{b/(2a^2)} \left(b+\sqrt{b^2-a^2c^2}\right)^{1/4} \Gamma\left(\dfrac{1-s+\delta}{2}\right)$ $\times D_{s-1}\left(\dfrac{\sqrt{b+\sqrt{b^2-a^2c^2}}}{a}\right) M_{(2s-1)/4,\,\pm 1/4}\left(\dfrac{b-\sqrt{b^2-a^2c^2}}{2a^2}\right)$ $[a>0;\ b>ac>0;\ \operatorname{Re} s<\delta+1]$
6	$(a^2-x^2)_+^{-1/2}$ $\times e^{-b(a^2+x^2)/(a^2-x^2)}$ $\times \left\{ \begin{matrix} \sinh\left[cx/(a^2-x^2)\right] \\ \cosh\left[cx/(a^2-x^2)\right] \end{matrix} \right\}$	$\dfrac{2^{(2s-3)/4+\delta} a^{s-3/4}}{\sqrt{c}} \left(2ab+\sqrt{4a^2b^2-c^2}\right)^{1/4}$ $\times \Gamma\left(\dfrac{s+\delta}{2}\right) D_{-s}\left(\dfrac{\sqrt{2ab+\sqrt{4a^2b^2-c^2}}}{\sqrt{a}}\right)$ $\times M_{(1-2s)/4,\,\pm 1/4}\left(\dfrac{2ab-\sqrt{4a^2b^2-c^2}}{2a}\right)$ $[a>0;\ 2ab>c>0;\ \operatorname{Re} s>-\delta]$
7	$(x^2-a^2)_+^{-1/2}$ $\times e^{-b(x^2+a^2)/(x^2-a^2)}$ $\times \left\{ \begin{matrix} \sinh\left[cx/(x^2-a^2)\right] \\ \cosh\left[cx/(x^2-a^2)\right] \end{matrix} \right\}$	$\dfrac{a^{s-3/4}}{2^{(2s+1)/4-\delta}\sqrt{c}} \left(2ab+\sqrt{4a^2b^2-c^2}\right)^{1/4}$ $\times \Gamma\left(\dfrac{1-s+\delta}{2}\right) D_{s-1}\left(\dfrac{\sqrt{2ab+\sqrt{4a^2b^2-c^2}}}{\sqrt{a}}\right)$ $\times M_{(2s-1)/4,\,\pm 1/4}\left(\dfrac{2ab-\sqrt{4a^2b^2-c^2}}{2a}\right)$ $[a>0;\ 2ab>c>0;\ \operatorname{Re} s<\delta+1]$
8	$\dfrac{1}{\sqrt{x^2+a^2}}\, e^{-b/(x^2+a^2)}$ $\times \left\{ \begin{matrix} \sinh\left[cx/(x^2+a^2)\right] \\ \cosh\left[cx/(x^2+a^2)\right] \end{matrix} \right\}$	$\dfrac{2^{\delta-1/2} a^{s-1/2}}{\sqrt{c}}\, e^{-b/(2a^2)}\, \mathrm{B}\left(\dfrac{1-s+\delta}{2},\, \dfrac{s+\delta}{2}\right)$ $\times M_{(2s-1)/4,\,\pm 1/4}\left(\dfrac{\sqrt{b^2+a^2c^2}-b}{2a^2}\right)$ $\times M_{(1-2s)/4,\,\pm 1/4}\left(\dfrac{\sqrt{b^2+a^2c^2}+b}{2a^2}\right)$ $[\operatorname{Re} a,\ b,\ c>0;\ -\delta<\operatorname{Re} s<\delta+1]$
9	$\dfrac{1}{\sqrt{x^2+a^2}}\, e^{-b(a^2-x^2)/(a^2+x^2)}$ $\times \left\{ \begin{matrix} \sinh\left[cx/(x^2+a^2)\right] \\ \cosh\left[cx/(x^2+a^2)\right] \end{matrix} \right\}$	$\dfrac{2^{\delta-1/2} a^{s-1/2}}{\sqrt{c}}\, \mathrm{B}\left(\dfrac{1-s+\delta}{2},\, \dfrac{s+\delta}{2}\right)$ $\times M_{(2s-1)/4,\,\pm 1/4}\left(\dfrac{\sqrt{4a^2b^2+c^2}-2ab}{2a}\right)$ $\times M_{(1-2s)/4,\,\pm 1/4}\left(\dfrac{\sqrt{4a^2b^2+c^2}+2ab}{2a}\right)$ $[\operatorname{Re} a,\ b,\ c>0;\ -\delta<\operatorname{Re} s<\delta+1]$

2.4. Trigonometric Functions

More formulas can be obtained from the corresponding sections due to the relations

$$\sin z = -\sin(-z) = \cos\left(\frac{\pi}{2} - z\right) = -\cos\left(z + \frac{\pi}{2}\right) = \frac{e^{iz} - e^{-iz}}{2i} = -i\sinh(iz),$$

$$\cos z = \cos(-z) = \sin\left(\frac{\pi}{2} - z\right) = \sin\left(z + \frac{\pi}{2}\right) = \frac{e^{iz} + e^{-iz}}{2} = \cosh(iz),$$

$$\begin{Bmatrix} \sin z \\ \cos z \end{Bmatrix} = \sqrt{\frac{\pi z}{2}}\, J_{\pm 1/2}(z), \quad \sin z = z\, {}_0F_1\left(\frac{3}{2}; -\frac{z^2}{4}\right), \quad \cos z = {}_0F_1\left(\frac{1}{2}; -\frac{z^2}{4}\right),$$

$$\sin z = \frac{\sqrt{\pi}\, z}{\sqrt{z^2}} G_{02}^{10}\left(\frac{z^2}{4} \middle| \begin{matrix} \cdot \\ 1/2,\, 0 \end{matrix}\right), \quad \cos z = \sqrt{\pi}\, G_{02}^{10}\left(\frac{z^2}{4} \middle| \begin{matrix} \cdot \\ 0,\, 1/2 \end{matrix}\right).$$

2.4.1. $\sin(ax+b)$ and $\cos(ax+b)$

Notation: $\delta = \begin{Bmatrix} 1 \\ 0 \end{Bmatrix}$.

No.	$f(x)$	$F(s)$	
1	$\begin{Bmatrix} \sin(ax) \\ \cos(ax) \end{Bmatrix}$	$a^{-s} \begin{Bmatrix} \sin(s\pi/2) \\ \cos(s\pi/2) \end{Bmatrix} \Gamma(s)$	$[a > 0;\ -\delta < \operatorname{Re} s < 1]$
2	$\begin{Bmatrix} \sin(ax) \\ \cos(ax) \end{Bmatrix}$	$a^{-s} \begin{Bmatrix} \sin(s\pi/2) \\ \cos(s\pi/2) \end{Bmatrix} \Gamma(s)$	$[a > 0;\ -\delta < \operatorname{Re} s < 1]$
3	$\sin(ax) - ax$	$a^{-s} \sin\frac{s\pi}{2} \Gamma(s)$	$[a > 0;\ -3 < \operatorname{Re} s < -1]$
4	$\cos(ax) - 1$	$a^{-s} \cos\frac{s\pi}{2} \Gamma(s)$	$[a > 0;\ -2 < \operatorname{Re} s < 0]$
5	$\cos(ax) + \frac{a^2x^2}{2} - 1$	$a^{-s} \cos\frac{s\pi}{2} \Gamma(s)$	$[a > 0;\ -4 < \operatorname{Re} s < -2]$
6	$\sin(ax) - \sum_{k=0}^{n} \frac{(-1)^k (ax)^{2k+1}}{(2k+1)!}$	$a^{-s} \sin\frac{s\pi}{2} \Gamma(s)$	$[a > 0;\ -2n-3 < \operatorname{Re} s < -2n-1]$
7	$\cos(ax) - \sum_{k=0}^{n} \frac{(-1)^k (ax)^{2k}}{(2k)!}$	$a^{-s} \cos\frac{s\pi}{2} \Gamma(s)$	$[a > 0;\ -2(n+1) < \operatorname{Re} s < -2n]$
8	$\theta(a-x) \begin{Bmatrix} \sin(bx) \\ \cos(bx) \end{Bmatrix}$	$\frac{i^{(1\pm1)/2}}{2} b^{-s} \left[e^{-is\pi/2}\gamma(s,\, iab) \mp e^{is\pi/2}\gamma(s,\, -iab)\right]$	$[a > 0;\ \operatorname{Re} s > -(1\pm1)/2;\ \lvert\arg b\rvert < \pi]$

No.	$f(x)$	$F(s)$
9	$\begin{Bmatrix} \sin(ax+b) \\ \cos(ax+b) \end{Bmatrix}$	$\dfrac{\Gamma(s)}{a^s} \begin{Bmatrix} \sin(s\pi/2+b) \\ \cos(s\pi/2+b) \end{Bmatrix}$ $[a>0;\ 0<\operatorname{Re}s<1]$
10	$\begin{Bmatrix} \sin(ax+\theta\pi) \\ \cos(ax+\theta\pi) \end{Bmatrix}$	$\dfrac{\sqrt{\pi}}{2}\left(\dfrac{2}{a}\right)^s \Gamma\left[\begin{matrix} \frac{s}{2},\ \frac{s+1}{2} \\ \frac{s+2\theta-\delta+1}{2},\ \frac{-s-2\theta+\delta+1}{2} \end{matrix}\right]$ $[a>0;\ 0<\operatorname{Re}s<1]$
11	$b\sin(ax)+c\cos(ax)$	$b\sqrt{\dfrac{b^2+c^2}{b^2}}\,\dfrac{\Gamma(s)}{a^s}\,\sin\left(\dfrac{s\pi}{2}+\arctan\dfrac{c}{b}\right)$ $[a>0;\ 0<\operatorname{Re}s<1]$

2.4.2. Trigonometric and algebraic functions

Notation: $\delta = \begin{Bmatrix} 1 \\ 0 \end{Bmatrix}$.

1	$(a-x)_+^{\alpha-1}\begin{Bmatrix} \sin(bx) \\ \cos(bx) \end{Bmatrix}$	$\dfrac{i^{(1\pm1)/2}}{2}\,a^{s+\alpha-1}\,\mathrm{B}(\alpha,\ s)\left[{}_1F_1\left(\begin{matrix} s;\ -i\,ab \\ s+\alpha \end{matrix}\right) \mp {}_1F_1\left(\begin{matrix} s;\ i\,ab \\ s+\alpha \end{matrix}\right)\right]$ $[a,\ \operatorname{Re}\alpha>0,\ \operatorname{Re}s>-(1\pm1)/2]$
2	$(a^2-x^2)_+^{\alpha-1}\begin{Bmatrix} \sin(bx) \\ \cos(bx) \end{Bmatrix}$	$\dfrac{a^{s+2\alpha+\delta-2}b^\delta}{2}\,\mathrm{B}\left(\alpha,\ \dfrac{s+\delta}{2}\right)\,{}_1F_2\left(\begin{matrix} \frac{s+\delta}{2};\ -\frac{a^2b^2}{4} \\ \frac{2\delta+1}{2},\ \frac{s+2\alpha+\delta}{2} \end{matrix}\right)$ $[a,\ \operatorname{Re}\alpha>0;\ \operatorname{Re}s>-\delta]$
3	$\dfrac{\sin(bx)}{(x+a)^\rho}$	$a^{s-\rho+1}b\,\mathrm{B}(s+1,\ \rho-s-1)\ {}_2F_3\left(\begin{matrix} \frac{s+1}{2},\ \frac{s+2}{2};\ -\frac{a^2b^2}{4} \\ \frac{3}{2},\ \frac{s-\rho+2}{2},\ \frac{s-\rho+3}{2} \end{matrix}\right)$ $+\,b^{\rho-s}\sin\dfrac{(s-\rho)\pi}{2}\,\Gamma(s-\rho)\ {}_2F_3\left(\begin{matrix} \frac{\rho}{2},\ \frac{\rho+1}{2};\ -\frac{a^2b^2}{4} \\ \frac{1}{2},\ \frac{1-s+\rho}{2},\ \frac{2-s+\rho}{2} \end{matrix}\right)$ $+\,\dfrac{\rho a}{b^{s-\rho-1}}\,\Gamma(s-\rho-1)\cos\dfrac{(\rho-s)\pi}{2}\ {}_2F_3\left(\begin{matrix} \frac{\rho+1}{2},\ \frac{\rho+2}{2};\ -\frac{a^2b^2}{4} \\ \frac{3}{2},\ \frac{2-s+\rho}{2},\ \frac{3-s+\rho}{2} \end{matrix}\right)$ $[b>0;\ -1<\operatorname{Re}s<\operatorname{Re}\rho+1;\ \lvert\arg a\rvert<\pi]$
4	$\dfrac{\cos(bx)}{(x+a)^\rho}$	$a^{s-\rho}\,\mathrm{B}(s,\ \rho-s)\ {}_2F_3\left(\begin{matrix} \frac{s}{2},\ \frac{s+1}{2};\ -\frac{a^2b^2}{4} \\ \frac{1}{2},\ \frac{s-\rho+1}{2},\ \frac{s-\rho+2}{2} \end{matrix}\right)$ $+\,b^{\rho-s}\Gamma(s-\rho)\cos\dfrac{(s-\rho)\pi}{2}\ {}_2F_3\left(\begin{matrix} \frac{\rho}{2},\ \frac{\rho+1}{2};\ -\frac{a^2b^2}{4} \\ \frac{1}{2},\ \frac{1-s+\rho}{2},\ \frac{2-s+\rho}{2} \end{matrix}\right)$ $+\,\dfrac{\rho a}{b^{s-\rho-1}}\Gamma(s-\rho-1)\sin\dfrac{(\rho-s)\pi}{2}\ {}_2F_3\left(\begin{matrix} \frac{\rho+1}{2},\ \frac{\rho+2}{2};\ -\frac{a^2b^2}{4} \\ \frac{3}{2},\ \frac{2-s+\rho}{2},\ \frac{3-s+\rho}{2} \end{matrix}\right)$ $[b>0;\ 0<\operatorname{Re}s<\operatorname{Re}\rho+1;\ \lvert\arg a\rvert<\pi]$
5	$\dfrac{1}{x+a}\begin{Bmatrix} \sin(bx) \\ \cos(bx) \end{Bmatrix}$	$\dfrac{i^{(1\pm1)/2}}{2}\,a^{s-1}\Gamma(s)\left[e^{iab}\,\Gamma(1-s,\ iab) \mp e^{-iab}\,\Gamma(1-s,\ -iab)\right]$ $[b>0;\ -(1\pm1)/2<\operatorname{Re}s<2;\ \lvert\arg a\rvert<\pi]$

No.	$f(x)$	$F(s)$
6	$\frac{1}{x-a}\begin{Bmatrix}\sin(bx)\\ \cos(bx)\end{Bmatrix}$	$-\frac{a}{b^{s-2}}\Gamma(s-2)\begin{Bmatrix}\sin(s\pi/2)\\ \cos(s\pi/2)\end{Bmatrix}{}_1F_2\left(\begin{matrix}1;\, -\frac{a^2b^2}{4}\\ \frac{3-s}{2},\, \frac{4-s}{2}\end{matrix}\right)$ $\mp\frac{\Gamma(s-1)}{b^{s-1}}\begin{Bmatrix}\cos(s\pi/2)\\ \sin(s\pi/2)\end{Bmatrix}{}_1F_2\left(\begin{matrix}1;\, -\frac{a^2b^2}{4}\\ \frac{2-s}{2},\, \frac{3-s}{2}\end{matrix}\right)$ $-\pi a^{s-1}\cot(s\pi)\begin{Bmatrix}\sin(ab)\\ \cos(ab)\end{Bmatrix}$ $[a,\, b>0;\, -\delta<\operatorname{Re}s<2]$
7	$\frac{1}{(x^2+a^2)^\rho}\begin{Bmatrix}\sin(bx)\\ \cos(bx)\end{Bmatrix}$	$\frac{a^{s-2\rho+\delta}b^\delta}{2}\,\mathrm{B}\left(\frac{s+\delta}{2},\, \frac{2\rho-s-\delta}{2}\right){}_1F_2\left(\begin{matrix}\frac{s+\delta}{2};\, \frac{a^2b^2}{4}\\ \frac{2\delta+1}{2},\, \frac{s-2\rho+\delta+2}{2}\end{matrix}\right)$ $+b^{2\rho-s}\begin{Bmatrix}\sin[(s-2\rho)\pi/2]\\ \cos[(s-2\rho)\pi/2]\end{Bmatrix}\Gamma(s-2\rho)\;{}_1F_2\left(\begin{matrix}\rho;\, \frac{a^2b^2}{4}\\ \frac{2\rho-s+1}{2},\, \frac{2\rho-s+2}{2}\end{matrix}\right)$ $[b,\, \operatorname{Re}a>0;\, -\delta<\operatorname{Re}s<2\operatorname{Re}\rho+1]$
8	$\frac{1}{x^2+a^2}\begin{Bmatrix}\sin(bx)\\ \cos(bx)\end{Bmatrix}$	$\frac{\pi a^{s-2}}{2}\begin{Bmatrix}\sinh(ab)\sec(s\pi/2)\\ \cosh(ab)\csc(s\pi/2)\end{Bmatrix}$ $-\frac{\Gamma(s-2)}{b^{s-2}}\begin{Bmatrix}\sin(s\pi/2)\\ \cos(s\pi/2)\end{Bmatrix}{}_1F_2\left(\begin{matrix}1;\, \frac{a^2b^2}{4}\\ \frac{3-s}{2},\, \frac{4-s}{2}\end{matrix}\right)$ $[b,\, \operatorname{Re}a>0;\, -(1\pm1)/2<\operatorname{Re}s<3]$
9	$\frac{1}{x^2-a^2}\begin{Bmatrix}\sin(bx)\\ \cos(bx)\end{Bmatrix}$	$\pm\frac{\pi a^{s-2}}{2}\begin{Bmatrix}\sin(ab)\tan(s\pi/2)\\ \cos(ab)\cot(s\pi/2)\end{Bmatrix}$ $-\frac{\Gamma(s-2)}{b^{s-2}}\begin{Bmatrix}\sin(s\pi/2)\\ \cos(s\pi/2)\end{Bmatrix}{}_1F_2\left(\begin{matrix}1;\, -\frac{a^2b^2}{4}\\ \frac{3-s}{2},\, \frac{4-s}{2}\end{matrix}\right)$ $[a,\, b>0;\, -(1\pm1)/2<\operatorname{Re}s<3]$
10	$\frac{1}{x^4+a^4}\sin(bx)$	$b^{4-s}\sin\frac{s\pi}{2}\,\Gamma(s-4)\;{}_1F_4\left(\begin{matrix}1;\, -\frac{a^4b^4}{256}\\ \frac{5-s}{4},\, \frac{6-s}{4},\, \frac{7-s}{4},\, \frac{8-s}{4}\end{matrix}\right)+\frac{\pi a^{s-4}}{2}\sec\frac{s\pi}{2}$ $\times\left(\cos\frac{s\pi}{4}\sinh\frac{ab}{\sqrt2}\cos\frac{ab}{\sqrt2}-\sin\frac{s\pi}{4}\cosh\frac{ab}{\sqrt2}\sin\frac{ab}{\sqrt2}\right)$ $[b>0;\, -1<\operatorname{Re}s<5;\, \lvert\arg a\rvert<\pi/4]$
11	$\frac{1}{x^4+a^4}\cos(bx)$	$b^{4-s}\cos\frac{s\pi}{2}\,\Gamma(s-4)\;{}_1F_4\left(\begin{matrix}1;\, -\frac{a^4b^4}{256}\\ \frac{5-s}{4},\, \frac{6-s}{4},\, \frac{7-s}{4},\, \frac{8-s}{4}\end{matrix}\right)$ $+\frac{\pi a^{s-4}}{4}\left(\csc\frac{s\pi}{4}\cosh\frac{ab}{\sqrt2}\cos\frac{ab}{\sqrt2}-\sec\frac{s\pi}{4}\sinh\frac{ab}{\sqrt2}\sin\frac{ab}{\sqrt2}\right)$ $[b>0;\, 0<\operatorname{Re}s<5;\, \lvert\arg a\rvert<\pi/4]$

No.	$f(x)$	$F(s)$
12	$\left(\sqrt{x^2+a^2}+a\right)^{\nu}\sin(bx)$	$-2^{s+\nu}\nu\pi a^{s+\nu+1}b\csc[(s+\nu)\pi]$ $\times\Gamma\left[\begin{matrix}\frac{s+1}{2}\\ \frac{1-s-2\nu}{2},\ s+\nu+2\end{matrix}\right]{}_2F_3\left(\begin{matrix}\frac{s+1}{2},\ \frac{s+2\nu+1}{2};\ \frac{a^2b^2}{4}\\ \frac{3}{2},\ \frac{s+\nu+2}{2},\ \frac{s+\nu+3}{2}\end{matrix}\right)$ $+\frac{\nu\pi ab^{-s-\nu+1}}{2\Gamma(2-s-\nu)}\csc\frac{(s+\nu)\pi}{2}\,{}_2F_3\left(\begin{matrix}\frac{1-\nu}{2},\ \frac{1+\nu}{2};\ \frac{a^2b^2}{4}\\ \frac{3}{2},\ \frac{2-s-\nu}{2},\ \frac{3-s-\nu}{2}\end{matrix}\right)$ $+\frac{\pi b^{-s-\nu}}{2\Gamma(1-s-\nu)}\sec\frac{(s+\nu)\pi}{2}\,{}_2F_3\left(\begin{matrix}-\frac{\nu}{2},\ \frac{\nu}{2};\ \frac{a^2b^2}{4}\\ \frac{1}{2},\ \frac{1-s-\nu}{2},\ \frac{2-s-\nu}{2}\end{matrix}\right)$ $[b,\ \operatorname{Re}a>0;\ -1<\operatorname{Re}s<1-\operatorname{Re}\nu]$
13	$\left(\sqrt{x^2+a^2}+a\right)^{\nu}\cos(bx)$	$2^{s+\nu-1}\nu\pi a^{s+\nu}\csc[(s+\nu)\pi]$ $\times\Gamma\left[\begin{matrix}\frac{s}{2}\\ \frac{2-s-2\nu}{2},\ s+\nu+1\end{matrix}\right]{}_2F_3\left(\begin{matrix}\frac{s}{2},\ \frac{s+2\nu}{2};\ \frac{a^2b^2}{4}\\ \frac{1}{2},\ \frac{s+\nu+1}{2},\ \frac{s+\nu+2}{2}\end{matrix}\right)$ $+\frac{\pi b^{-s-\nu}}{2\Gamma(1-s-\nu)}\csc\frac{(s+\nu)\pi}{2}\,{}_2F_3\left(\begin{matrix}-\frac{\nu}{2},\ \frac{\nu}{2};\ \frac{a^2b^2}{4}\\ \frac{1}{2},\ \frac{1-s-\nu}{2},\ \frac{2-s-\nu}{2}\end{matrix}\right)$ $-\frac{\nu\pi ab^{-s-\nu+1}}{2\Gamma(2-s-\nu)}\sec\frac{(s+\nu)\pi}{2}\,{}_2F_3\left(\begin{matrix}\frac{1-\nu}{2},\ \frac{1+\nu}{2};\ \frac{a^2b^2}{4}\\ \frac{3}{2},\ \frac{2-s-\nu}{2},\ \frac{3-s-\nu}{2}\end{matrix}\right)$ $[b,\ \operatorname{Re}a>0;\ 0<\operatorname{Re}s<1-\operatorname{Re}\nu]$
14	$\left(\sqrt{x^2+a^2}+x\right)^{\nu}\sin(bx)$	$2^{-s-2}\nu\pi a^{s+\nu+1}b\sec\frac{(s+\nu)\pi}{2}$ $\times\Gamma\left[\begin{matrix}s+1\\ \frac{s-\nu+3}{2},\ \frac{s+\nu+3}{2}\end{matrix}\right]{}_2F_3\left(\begin{matrix}\frac{s+1}{2},\ \frac{s+2}{2};\ \frac{a^2b^2}{4}\\ \frac{3}{2},\ \frac{s-\nu+3}{2},\ \frac{s+\nu+3}{2}\end{matrix}\right)$ $+\frac{2^{\nu-1}\pi b^{-s-\nu}}{\Gamma(1-s-\nu)}\sec\frac{(s+\nu)\pi}{2}\,{}_2F_3\left(\begin{matrix}\frac{1-\nu}{2},\ -\frac{\nu}{2};\ \frac{a^2b^2}{4}\\ 1-\nu,\ \frac{1-s-\nu}{2},\ \frac{2-s-\nu}{2}\end{matrix}\right)$ $[b,\ \operatorname{Re}a>0;\ -1<\operatorname{Re}s<1-\operatorname{Re}\nu]$
15	$\left(\sqrt{x^2+a^2}+x\right)^{\nu}\cos(bx)$	$2^{-s-1}\nu\pi a^{s+\nu}\csc\frac{(s+\nu)\pi}{2}$ $\times\Gamma\left[\begin{matrix}s\\ \frac{s-\nu+2}{2},\ \frac{s+\nu+2}{2}\end{matrix}\right]{}_2F_3\left(\begin{matrix}\frac{s}{2},\ \frac{s+1}{2};\ \frac{a^2b^2}{4}\\ \frac{1}{2},\ \frac{s-\nu+2}{2},\ \frac{s+\nu+2}{2}\end{matrix}\right)$ $+\frac{2^{\nu-1}\pi b^{-s-\nu}}{\Gamma(1-s-\nu)}\csc\frac{(s+\nu)\pi}{2}\,{}_2F_3\left(\begin{matrix}\frac{1-\nu}{2},\ -\frac{\nu}{2};\ \frac{a^2b^2}{4}\\ 1-\nu,\ \frac{1-s-\nu}{2},\ \frac{2-s-\nu}{2}\end{matrix}\right)$ $[b,\ \operatorname{Re}a>0;\ 0<\operatorname{Re}s<1-\operatorname{Re}\nu]$

No.	$f(x)$	$F(s)$
16	$\dfrac{\left(\sqrt{x^2+a^2}+a\right)^{\nu}}{\sqrt{x^2+a^2}}\sin(bx)$	$-2^{s+\nu}\pi a^{s+\nu}b\csc[(s+\nu)\pi]\times\Gamma\begin{bmatrix}\frac{s+1}{2}\\ \frac{1-s-2\nu}{2},\ s+\nu+1\end{bmatrix}{}_2F_3\left(\begin{matrix}\frac{s+1}{2},\ \frac{s+2\nu+1}{2};\ \frac{a^2b^2}{4}\\ \frac{3}{2},\ \frac{s+\nu+1}{2},\ \frac{s+\nu+2}{2}\end{matrix}\right)+\dfrac{\pi b^{-s-\nu+1}}{2\Gamma(2-s-\nu)}\csc\dfrac{(s+\nu)\pi}{2}\,{}_2F_3\left(\begin{matrix}\frac{1-\nu}{2},\ \frac{1+\nu}{2};\ \frac{a^2b^2}{4}\\ \frac{1}{2},\ \frac{2-s-\nu}{2},\ \frac{3-s-\nu}{2}\end{matrix}\right)-\dfrac{\nu\pi ab^{-s-\nu+2}}{2\Gamma(3-s-\nu)}\sec\dfrac{(s+\nu)\pi}{2}\,{}_2F_3\left(\begin{matrix}\frac{2-\nu}{2},\ \frac{2+\nu}{2};\ \frac{a^2b^2}{4}\\ \frac{3}{2},\ \frac{3-s-\nu}{2},\ \frac{4-s-\nu}{2}\end{matrix}\right)$ $[b,\ \operatorname{Re}a>0;\ -1<\operatorname{Re}s<2-\operatorname{Re}\nu]$
17	$\dfrac{\left(\sqrt{x^2+a^2}+a\right)^{\nu}}{\sqrt{x^2+a^2}}\cos(bx)$	$-2^{s+\nu-1}\pi a^{s+\nu-1}\csc[(s+\nu-1)\pi]\times\Gamma\begin{bmatrix}\frac{s}{2}\\ \frac{2-s-2\nu}{2},\ s+\nu\end{bmatrix}{}_2F_3\left(\begin{matrix}\frac{s}{2},\ \frac{s+2\nu}{2};\ \frac{a^2b^2}{4}\\ \frac{1}{2},\ \frac{s+\nu}{2},\ \frac{s+\nu+1}{2}\end{matrix}\right)-\dfrac{\nu\pi ab^{-s-\nu+2}}{2\Gamma(3-s-\nu)}\csc\dfrac{(s+\nu)\pi}{2}\,{}_2F_3\left(\begin{matrix}\frac{2-\nu}{2},\ \frac{2+\nu}{2};\ \frac{a^2b^2}{4}\\ \frac{3}{2},\ \frac{3-s-\nu}{2},\ \frac{4-s-\nu}{2}\end{matrix}\right)-\dfrac{\pi b^{-s-\nu+1}}{2\Gamma(2-s-\nu)}\sec\dfrac{(s+\nu)\pi}{2}\,{}_2F_3\left(\begin{matrix}\frac{1-\nu}{2},\ \frac{1+\nu}{2};\ \frac{a^2b^2}{4}\\ \frac{1}{2},\ \frac{2-s-\nu}{2},\ \frac{3-s-\nu}{2}\end{matrix}\right)$ $[b,\ \operatorname{Re}a>0;\ 0<\operatorname{Re}s<2-\operatorname{Re}\nu]$
18	$\dfrac{\left(\sqrt{x^2+a^2}+x\right)^{\nu}}{\sqrt{x^2+a^2}}\sin(bx)$	$-2^{-s-1}\pi a^{s+\nu}b\csc\dfrac{(s+\nu)\pi}{2}\times\Gamma\begin{bmatrix}s+1\\ \frac{s-\nu+2}{2},\ \frac{s+\nu+2}{2}\end{bmatrix}{}_2F_3\left(\begin{matrix}\frac{s+1}{2},\ \frac{s+2}{2};\ \frac{a^2b^2}{4}\\ \frac{3}{2},\ \frac{s-\nu+2}{2},\ \frac{s+\nu+2}{2}\end{matrix}\right)+\dfrac{2^{\nu-1}\pi b^{-s-\nu+1}}{\Gamma(2-s-\nu)}\csc\dfrac{(s+\nu)\pi}{2}\,{}_2F_3\left(\begin{matrix}\frac{1-\nu}{2},\ \frac{2-\nu}{2};\ \frac{a^2b^2}{4}\\ 1-\nu,\ \frac{2-s-\nu}{2},\ \frac{3-s-\nu}{2}\end{matrix}\right)$ $[b,\ \operatorname{Re}a>0;\ -1<\operatorname{Re}s<2-\operatorname{Re}\nu]$
19	$\dfrac{\left(\sqrt{x^2+a^2}+x\right)^{\nu}}{\sqrt{x^2+a^2}}\cos(bx)$	$-2^{-s}\pi a^{s+\nu-1}\csc\dfrac{(s+\nu-1)\pi}{2}\times\Gamma\begin{bmatrix}s\\ \frac{s-\nu+1}{2},\ \frac{s+\nu+1}{2}\end{bmatrix}{}_2F_3\left(\begin{matrix}\frac{s}{2},\ \frac{s+1}{2};\ \frac{a^2b^2}{4}\\ \frac{1}{2},\ \frac{s-\nu+1}{2},\ \frac{s+\nu+1}{2}\end{matrix}\right)-\dfrac{2^{\nu-1}\pi b^{-s-\nu+1}}{\Gamma(2-s-\nu)}\sec\dfrac{(s+\nu)\pi}{2}\,{}_2F_3\left(\begin{matrix}\frac{1-\nu}{2},\ \frac{2-\nu}{2};\ \frac{a^2b^2}{4}\\ 1-\nu,\ \frac{2-s-\nu}{2},\ \frac{3-s-\nu}{2}\end{matrix}\right)$ $[b,\ \operatorname{Re}a>0;\ 0<\operatorname{Re}s<2-\operatorname{Re}\nu]$

No.	$f(x)$	$F(s)$
20	$(x^2-a^2)_+^{-1/2} \times \left[\left(x+\sqrt{x^2-a^2}\right)^\nu + \left(x-\sqrt{x^2-a^2}\right)^\nu\right] \times \sin(bx)$	$\frac{a^{s+\nu}b}{2^{s+1}}\Gamma\left[\begin{matrix}-\frac{s+\nu}{2}, \frac{\nu-s}{2} \\ -s\end{matrix}\right] {}_2F_3\left(\begin{matrix}\frac{s+1}{2}, \frac{s+2}{2}; -\frac{a^2b^2}{4} \\ \frac{3}{2}, \frac{s-\nu+2}{2}, \frac{s+\nu+2}{2}\end{matrix}\right) + \frac{2^{\nu-1}\pi b^{-s-\nu+1}}{\Gamma(2-s-\nu)} \times \csc\frac{(s+\nu)\pi}{2}\, {}_2F_3\left(\begin{matrix}\frac{1-\nu}{2}, \frac{2-\nu}{2}; -\frac{a^2b^2}{4} \\ 1-\nu, \frac{2-s-\nu}{2}, \frac{3-s-\nu}{2}\end{matrix}\right) + \frac{2^{-\nu-1}\pi a^{2\nu}b^{-s+\nu+1}}{\Gamma(2-s+\nu)}\csc\frac{(s-\nu)\pi}{2} \times {}_2F_3\left(\begin{matrix}\frac{\nu+1}{2}, \frac{\nu+2}{2}; -\frac{a^2b^2}{4} \\ 1+\nu, \frac{2-s+\nu}{2}, \frac{3-s+\nu}{2}\end{matrix}\right)$ $[a,\, b>0;\ \operatorname{Re}s<2-\lvert\operatorname{Re}\nu\rvert]$
21	$(x^2-a^2)_+^{-1/2} \times \left[\left(x+\sqrt{x^2-a^2}\right)^\nu + \left(x-\sqrt{x^2-a^2}\right)^\nu\right] \times \cos(bx)$	$\frac{a^{s+\nu-1}}{2^s}\Gamma\left[\begin{matrix}\frac{1-s-\nu}{2}, \frac{1-s+\nu}{2} \\ 1-s\end{matrix}\right] {}_2F_3\left(\begin{matrix}\frac{s}{2}, \frac{s+1}{2}; -\frac{a^2b^2}{4} \\ \frac{1}{2}, \frac{s-\nu+1}{2}, \frac{s+\nu+1}{2}\end{matrix}\right) - \frac{2^{\nu-1}\pi b^{-s-\nu+1}}{\Gamma(2-s-\nu)} \times \sec\frac{(s+\nu)\pi}{2}\, {}_2F_3\left(\begin{matrix}\frac{1-\nu}{2}, \frac{2-\nu}{2}; -\frac{a^2b^2}{4} \\ 1-\nu, \frac{2-s-\nu}{2}, \frac{3-s-\nu}{2}\end{matrix}\right) - \frac{2^{-\nu-1}\pi a^{2\nu}b^{-s+\nu+1}}{\Gamma(2-s+\nu)}\sec\frac{(s-\nu)\pi}{2}\, {}_2F_3\left(\begin{matrix}\frac{\nu+1}{2}, \frac{\nu+2}{2}; -\frac{a^2b^2}{4} \\ 1+\nu, \frac{2-s+\nu}{2}, \frac{3-s+\nu}{2}\end{matrix}\right)$ $[a,\, b>0;\ \operatorname{Re}s<2-\lvert\operatorname{Re}\nu\rvert]$
22	$\frac{1}{(x+a)^\rho}\left\{\begin{matrix}\sin[b/(x+a)] \\ \cos[b/(x+a)]\end{matrix}\right\}$	$a^{s-\rho-\delta}b^\delta\,\mathrm{B}(s,\, \rho-s+\delta)\, {}_2F_3\left(\begin{matrix}\frac{\rho-s+\delta}{2}, \frac{\rho-s+\delta+1}{2}; -\frac{b^2}{4a^2} \\ \frac{2\delta+1}{2}, \frac{\rho+\delta}{2}, \frac{\rho+\delta+1}{2}\end{matrix}\right)$ $[0<\operatorname{Re}s<\operatorname{Re}\rho+\delta;\ \lvert\arg a\rvert<\pi]$
23	$\frac{1}{(x+a)^\rho} \times \left\{\begin{matrix}\sin[bx/(x+a)] \\ \cos[bx/(x+a)]\end{matrix}\right\}$	$a^{s-\rho}b^\delta\,\mathrm{B}(s+\delta,\, \rho-s)\, {}_2F_3\left(\begin{matrix}\frac{s+\delta}{2}, \frac{s+\delta+1}{2}; -\frac{b^2}{4} \\ \frac{2\delta+1}{2}, \frac{\rho+\delta}{2}, \frac{\rho+\delta+1}{2}\end{matrix}\right)$ $[-\delta<\operatorname{Re}s<\operatorname{Re}\rho;\ \lvert\arg a\rvert<\pi]$
24	$(1-x^2)_+^{-1/2} \times \left\{\begin{matrix}\sin(ax-a/x) \\ \cos(ax-a/x)\end{matrix}\right\}$	$+\sqrt{\frac{\pi a}{2}}\left\{\begin{matrix}I_{s/2}(a)\, K_{(s-1)/2}(a) \\ I_{(s-1)/2}(a)\, K_{s/2}(a)\end{matrix}\right\}$ $[a>0;\ \operatorname{Re}s>\ \ 1]$
25	$(x^2-1)_+^{-1/2} \times \left\{\begin{matrix}\sin(ax-a/x) \\ \cos(ax-a/x)\end{matrix}\right\}$	$\sqrt{\frac{\pi a}{2}}\left\{\begin{matrix}I_{(1-s)/2}(a)\, K_{s/2}(a) \\ I_{-s/2}(a)\, K_{(s-1)/2}(a)\end{matrix}\right\}$ $[a>0;\ \operatorname{Re}s<2]$

No.	$f(x)$	$F(s)$
26	$(a-x)_+^{(\delta-1)/2}(bx+1)^\alpha \times \begin{Bmatrix} \sin\left(c\sqrt{a-x}\right) \\ \cos\left(c\sqrt{a-x}\right) \end{Bmatrix}$	$\dfrac{\sqrt{\pi}\, a^{s+\delta-1/2} c^\delta}{\delta+1} \Gamma\left[\begin{matrix} s \\ \frac{2s+2\delta+1}{2} \end{matrix}\right] \times \Xi_2\left(-\alpha,\, s;\, \dfrac{2s+2\delta+1}{2};\, -ab,\, -\dfrac{ac^2}{4}\right)$ $[a,\ \operatorname{Re} s > 0;\ \lvert\arg(ab+1)\rvert < \pi]$
27	$(x-a)_+^{(\delta-1)/2}(1-x+a)_+^\alpha \times \begin{Bmatrix} \sin\left(b\sqrt{x-a}\right) \\ \cos\left(b\sqrt{x-a}\right) \end{Bmatrix}$	$\dfrac{\sqrt{\pi}\,(a+1)^{s-1} b^\delta}{\delta+1} \Gamma\left[\begin{matrix} \alpha+1 \\ \frac{2\alpha+2\delta+3}{2} \end{matrix}\right] \times \Xi_2\left(1-s,\, \alpha+1;\, \dfrac{2\alpha+2\delta+3}{2};\, \dfrac{1}{a+1},\, -\dfrac{b^2}{4}\right)$ $[a,\ \operatorname{Re} s > 0]$
28	$(a-x)_+^{\alpha-1} \times \begin{Bmatrix} \sin\left(b\sqrt{x(a-x)}\right) \\ \cos\left(b\sqrt{x(a-x)}\right) \end{Bmatrix}$	$a^{s+\alpha+\delta-1} b^\delta \operatorname{B}\left(\dfrac{2\alpha+\delta}{2},\, \dfrac{2s+\delta}{2}\right) {}_2F_3\left(\begin{matrix} \frac{2\alpha+\delta}{2},\, \frac{2s+\delta}{2};\, -\frac{a^2b^2}{16} \\ \frac{2\delta+1}{2},\, \frac{s+\alpha+\delta}{2},\, \frac{s+\alpha+\delta+1}{2} \end{matrix}\right)$ $[a > 0;\ \operatorname{Re} s > -\delta/2]$
29	$\begin{Bmatrix} \sin\left(b\sqrt{x^2+a^2}\right) \\ \cos\left(b\sqrt{x^2+a^2}\right) \end{Bmatrix}$	$\pm\dfrac{2^{(s-3)/2}\sqrt{\pi}\, a^{(s+1)/2}}{b^{(s-1)/2}} \Gamma\left(\dfrac{s}{2}\right)\left[\begin{Bmatrix} \cos(s\pi/2) \\ \sin(s\pi/2) \end{Bmatrix} J_{(s+1)/2}(ab) \mp \begin{Bmatrix} \sin(s\pi/2) \\ \cos(s\pi/2) \end{Bmatrix} Y_{(s+1)/2}(ab)\right]$ $[\operatorname{Re} a,\ b > 0;\ 0 < \operatorname{Re} s < 1]$
30	$\dfrac{1}{\sqrt{x^2+a^2}} \times \begin{Bmatrix} \sin\left(b\sqrt{x^2+a^2}\right) \\ \cos\left(b\sqrt{x^2+a^2}\right) \end{Bmatrix}$	$\pm 2^{(s-3)/2}\sqrt{\pi}\left(\dfrac{a}{b}\right)^{(s-1)/2} \Gamma\left(\dfrac{s}{2}\right) \begin{Bmatrix} J_{(1-s)/2}(ab) \\ Y_{(1-s)/2}(ab) \end{Bmatrix}$ $[\operatorname{Re} a,\ b > 0;\ 0 < \operatorname{Re} s < 2]$
31	$\theta(a-x) \times \begin{Bmatrix} \sin\left(b\sqrt{a^2-x^2}\right) \\ \cos\left(b\sqrt{a^2-x^2}\right) \end{Bmatrix}$	$\pm\dfrac{a^{(s+1)/2}}{b^{(s-1)/2}} \begin{Bmatrix} 2^{(s-3)/2}\sqrt{\pi}\, J_{(s+1)/2}(ab) \\ s_{(s-3)/2,\,(s+1)/2}(ab) \end{Bmatrix}$ $[a,\ \operatorname{Re} s > 0]$
32	$\left(a^2-x^2\right)_+^{-1/2} \times \sin\left(b\sqrt{a^2-x^2}\right)$	$\dfrac{\sqrt{\pi}}{2}\left(\dfrac{2a}{b}\right)^{(s-1)/2} \Gamma\left(\dfrac{s}{2}\right) \mathbf{H}_{(s-1)/2}(ab)$ $[a,\ \operatorname{Re} s > 0]$
33	$\left(a^2-x^2\right)_+^{-1/2} \times \cos\left(b\sqrt{a^2-x^2}\right)$	$2^{(s-3)/2}\sqrt{\pi}\left(\dfrac{a}{b}\right)^{(s-1)/2} \Gamma\left(\dfrac{s}{2}\right) J_{(s-1)/2}(ab)$ $[a,\ \operatorname{Re} s > 0]$

No.	$f(x)$	$F(s)$
34	$\theta(x-a)\sin\left(b\sqrt{x^2-a^2}\right)$	$\dfrac{2^{(s-1)/2}a^{(s+1)/2}}{\sqrt{\pi}\,b^{(s-1)/2}}\sin\dfrac{s\pi}{2}\,\Gamma\left(\dfrac{s}{2}\right)K_{(s+1)/2}(ab)$ $[a,\,b>0;\ \operatorname{Re}s<1]$
35	$\theta(x-a)\cos\left(b\sqrt{x^2-a^2}\right)$	$\dfrac{2^{(s-3)/2}\sqrt{\pi}\,a^{(s+1)/2}}{b^{(s-1)/2}}\Gamma\left(\dfrac{s}{2}\right)\left[I_{-(s+1)/2}(ab)-\mathbf{L}_{(s+1)/2}(ab)-\dfrac{(ab)^{(s-1)/2}}{2^{(s-3)/2}\sqrt{\pi}\,s\,\Gamma\left(\frac{s}{2}\right)}\right]$ $[a,\,b>0;\ \operatorname{Re}s<1]$
36	$\left(x^2-a^2\right)_+^{-1/2}\times\sin\left(b\sqrt{x^2-a^2}\right)$	$\dfrac{\sqrt{\pi}}{2}\left(\dfrac{2a}{b}\right)^{(s-1)/2}\Gamma\left(\dfrac{s}{2}\right)\left[I_{(1-s)/2}(ab)-\mathbf{L}_{(s-1)/2}(ab)\right]$ $[a,\,b>0;\ \operatorname{Re}s<2]$
37	$\left(x^2-a^2\right)_+^{-1/2}\times\cos\left(b\sqrt{x^2-a^2}\right)$	$\dfrac{1}{\sqrt{\pi}}\left(\dfrac{2a}{b}\right)^{(s-1)/2}\sin\dfrac{s\pi}{2}\,\Gamma\left(\dfrac{s}{2}\right)K_{(s-1)/2}(ab)$ $[a>0;\ \operatorname{Re}s<2]$
38	$\dfrac{1}{(x^2+a^2)^\rho}\cos\dfrac{bx}{x^2+a^2}$	$\dfrac{a^{s-2\rho}}{2}\mathrm{B}\left(\dfrac{s}{2},\,\rho-\dfrac{s}{2}\right){}_2F_3\left(\begin{matrix}\frac{s}{2},\,\rho-\frac{s}{2};\,-\frac{b^2}{16a^2}\\ \frac{1}{2},\,\frac{\rho}{2},\,\frac{\rho+1}{2}\end{matrix}\right)$ $[\operatorname{Re}a>0;\ 0<\operatorname{Re}s<2\operatorname{Re}\rho]$
39	$\dfrac{1}{(x^2+a^2)^\rho}\sin\dfrac{bx}{x^2+a^2}$	$\dfrac{a^{s-2\rho-1}b}{2}\mathrm{B}\left(\dfrac{s+1}{2},\,\dfrac{1-s+2\rho}{2}\right){}_2F_3\left(\begin{matrix}\frac{s+1}{2},\,\frac{1-s+2\rho}{2};\,-\frac{b^2}{16a^2}\\ \frac{3}{2},\,\frac{\rho+1}{2},\,\frac{\rho+2}{2}\end{matrix}\right)$ $[\operatorname{Re}a>0;\ -1<\operatorname{Re}s<2\operatorname{Re}\rho+1]$

2.4.3. Trigonometric and the exponential functions

Notation: $\delta=\begin{Bmatrix}1\\0\end{Bmatrix}$.

1	$e^{-ax}\begin{Bmatrix}\sin(ax)\\ \cos(ax)\end{Bmatrix}$	$2^{-s/2}a^{-s}\begin{Bmatrix}\sin(s\pi/4)\\ \cos(s\pi/4)\end{Bmatrix}\Gamma(s)$ $\left[\begin{matrix}(\operatorname{Re}s>-\delta;\ \lvert\arg a\rvert<\pi/4)\ \text{or}\\ (-\delta<\operatorname{Re}s<1;\ \lvert\arg a\rvert=\pi/4)\end{matrix}\right]$
2	$e^{-ax/\sqrt{3}}\begin{Bmatrix}\sin(ax)\\ \cos(ax)\end{Bmatrix}$	$2^{-s}3^{s/2}a^{-s}\begin{Bmatrix}\sin(s\pi/3)\\ \cos(s\pi/3)\end{Bmatrix}\Gamma(s)$ $\left[\begin{matrix}(\operatorname{Re}s>-\delta;\ \lvert\arg a\rvert<\pi/6)\ \text{or}\\ (-\delta<\operatorname{Re}s<1;\ \lvert\arg a\rvert=\pi/6)\end{matrix}\right]$
3	$e^{-ax}\begin{Bmatrix}\sin(bx)\\ \cos(bx)\end{Bmatrix}$	$\dfrac{\Gamma(s)}{(a^2+b^2)^{s/2}}\begin{Bmatrix}\sin[s\arctan(b/a)]\\ \cos[s\arctan(b/a)]\end{Bmatrix}$ $\left[\begin{matrix}(\operatorname{Re}a>\lvert\operatorname{Im}b\rvert;\ \operatorname{Re}s>-(1\pm1)/2)\ \text{or}\\ (\operatorname{Re}a+\lvert\operatorname{Im}b\rvert=0;\ \operatorname{Re}s<1)\end{matrix}\right]$

No.	$f(x)$	$F(s)$
4	$e^{-\sqrt{3}\,ax}\begin{Bmatrix}\sin(ax)\\ \cos(ax)\end{Bmatrix}$	$(2a)^{-s}\begin{Bmatrix}\sin(s\pi/6)\\ \cos(s\pi/6)\end{Bmatrix}\Gamma(s)$ $\left[\begin{matrix}(\operatorname{Re}s>-\delta;\ \lvert\arg a\rvert<\pi/3)\ \text{or}\\ (-\delta<\operatorname{Re}s<1;\ \lvert\arg a\rvert=\pi/3)\end{matrix}\right]$
5	$e^{-(\sqrt{2}+1)ax}\begin{Bmatrix}\sin(ax)\\ \cos(ax)\end{Bmatrix}$	$2^{-s}\left(1+\frac{1}{\sqrt{2}}\right)^{-s/2}a^{-s}\begin{Bmatrix}\sin(s\pi/8)\\ \cos(s\pi/8)\end{Bmatrix}\Gamma(s)$ $\left[\begin{matrix}(\operatorname{Re}s>-\delta;\ \lvert\arg a\rvert<3\pi/8)\ \text{or}\\ (-\delta<\operatorname{Re}s<1;\ \lvert\arg a\rvert=3\pi/8)\end{matrix}\right]$
6	$e^{-\sqrt{1+2/\sqrt{5}}\,ax}\begin{Bmatrix}\sin(ax)\\ \cos(ax)\end{Bmatrix}$	$\left(2+\frac{2}{\sqrt{5}}\right)^{-s/2}a^{-s}\begin{Bmatrix}\sin(s\pi/5)\\ \cos(s\pi/5)\end{Bmatrix}\Gamma(s)$ $\left[\begin{matrix}(\operatorname{Re}s>-\delta;\ \lvert\arg a\rvert<3\pi/10)\ \text{or}\\ (-\delta<\operatorname{Re}s<1;\ \lvert\arg a\rvert=3\pi/10)\end{matrix}\right]$
7	$e^{-ax\cos\theta}\begin{Bmatrix}\sin(ax\sin\theta)\\ \cos(ax\sin\theta)\end{Bmatrix}$	$\frac{\Gamma(s)}{a^s}\begin{Bmatrix}\sin(s\theta)\\ \cos(s\theta)\end{Bmatrix}$ $[a>0;\ \lvert\theta\rvert<\pi/2;\ \operatorname{Re}s>-(1\pm1)/2]$
8	$e^{-x\cos(\theta\pi)}\times\begin{Bmatrix}\sin[x\sin(\theta\pi)]\\ \cos[x\sin(\theta\pi)]\end{Bmatrix}$	$\pi\,\Gamma\left[\begin{matrix}s\\ \frac{1-\delta}{2}+\theta s,\ \frac{1+\delta}{2}-\theta s\end{matrix}\right]$ $[\lvert\theta\rvert<1/2;\ \operatorname{Re}s>-\delta]$
9	$\theta(a-x)\,e^{-bx}\begin{Bmatrix}\sin(cx)\\ \cos(cx)\end{Bmatrix}$	$\frac{i^{(1\pm1)/2}}{2}\left[(b+ic)^{-s}\gamma(s,\,ab+iac)\mp(b-ic)^{-s}\gamma(s,\,ab-iac)\right]$ $[a>0;\ \operatorname{Re}s>-(1\pm1)/2]$
10	$\theta(x-a)\,e^{-bx}\begin{Bmatrix}\sin(cx)\\ \cos(cx)\end{Bmatrix}$	$\frac{i^{(1\pm1)/2}}{2}\left[(b+ic)^{-s}\Gamma(s,\,ab+iac)\mp(b-ic)^{-s}\Gamma(s,\,ab-iac)\right]$ $[a,\ \operatorname{Re}b>0]$
11	$e^{-ax^2}\begin{Bmatrix}\sin(bx)\\ \cos(bx)\end{Bmatrix}$	$\frac{b^\delta}{2a^{(s+\delta)/2}}\Gamma\left(\frac{s+\delta}{2}\right){}_1F_1\left(\begin{matrix}\frac{s+\delta}{2}\\ \frac{2\delta+1}{2};\ -\frac{b^2}{4a}\end{matrix}\right)$ $[\operatorname{Re}a>0;\ \operatorname{Re}s>-\delta]$
12	$e^{-ax^2-bx}\begin{Bmatrix}\sin(cx)\\ \cos(cx)\end{Bmatrix}$	$\frac{i^{(1\pm1)/2}\Gamma(s)}{2(2a)^{s/2}}e^{(b^2-c^2)/(8a)}\left[e^{ibc/(4a)}D_{-s}\left(\frac{b+ic}{\sqrt{2a}}\right)\mp e^{-ibc/(4a)}D_{-s}\left(\frac{b-ic}{\sqrt{2a}}\right)\right]$ $[\operatorname{Re}a>0;\ \operatorname{Re}s>-(1\pm1)/2]$
13	$e^{-c/x}\begin{Bmatrix}\sin(bx)\\ \cos(bx)\end{Bmatrix}$	$i^{(1\pm1)/2}\left(\frac{c}{b}\right)^{s/2}\left[e^{-is\pi/4}K_s\left(2e^{i\pi/4}\sqrt{bc}\right)\mp e^{is\pi/4}K_s\left(2e^{-i\pi/4}\sqrt{bc}\right)\right]$ $[b,\ \operatorname{Re}c>0,\ \operatorname{Re}s<1]$
14	$e^{-ax-c/x}\begin{Bmatrix}\sin(bx)\\ \cos(bx)\end{Bmatrix}$	$i^{(1\pm1)/2}c^{s/2}\left[(a+ib)^{-s/2}K_s\left(2\sqrt{ac+ibc}\right)\mp(a-ib)^{-s/2}K_s\left(2\sqrt{ac-ibc}\right)\right]$ $[\operatorname{Re}a>\lvert\operatorname{Im}b\rvert;\ \operatorname{Re}c>0]$

No.	$f(x)$	$F(s)$
15	$e^{-a/x^2}\begin{Bmatrix}\sin(bx)\\ \cos(bx)\end{Bmatrix}$	$\frac{\Gamma(s)}{b^s}\begin{Bmatrix}\sin(s\pi/2)\\ \cos(s\pi/2)\end{Bmatrix}{}_0F_2\left(\begin{matrix}-;\ \frac{ab^2}{4}\\ \frac{1-s}{2},\ \frac{2-s}{2}\end{matrix}\right)$ $+\frac{a^{(s+\delta)/2}b^\delta}{2}\Gamma\left(-\frac{s+\delta}{2}\right){}_0F_2\left(\begin{matrix}-;\ \frac{ab^2}{4}\\ \frac{2\delta+1}{2},\ \frac{s+\delta+2}{2}\end{matrix}\right)$ $[b,\ \operatorname{Re}a>0;\ \operatorname{Re}s<1]$
16	$e^{-a\sqrt{x}}\begin{Bmatrix}\sin(bx)\\ \cos(bx)\end{Bmatrix}$	$i^{(1\pm1)/2}\frac{\Gamma(2s)}{(2b)^s}\left[e^{-i\left(a^2+4bs\pi\right)/(8b)}D_{-2s}\left(\frac{ae^{-\pi i/4}}{\sqrt{2b}}\right)\right.$ $\left.\mp e^{i\left(a^2+4bs\pi\right)/(8b)}D_{-2s}\left(\frac{ae^{\pi i/4}}{\sqrt{2b}}\right)\right]$ $[b,\ \operatorname{Re}a>0;\ \operatorname{Re}s>-(1\pm1)/2]$
17	$\frac{1}{e^{ax}-1}\begin{Bmatrix}\sin(bx)\\ \cos(bx)\end{Bmatrix}$	$\frac{i^{(1\pm1)/2}}{2a^s}\Gamma(s)\left[\zeta\left(s,\ \frac{a+ib}{a}\right)\mp\zeta\left(s,\ \frac{a-ib}{a}\right)\right]$ $[\operatorname{Re}a>\lvert\operatorname{Im}b\rvert;\ \operatorname{Re}s>(1\mp1)/2]$
18	$\frac{1}{e^{ax}+1}\begin{Bmatrix}\sin(bx)\\ \cos(bx)\end{Bmatrix}$	$\frac{i^{(1\pm1)/2}}{2^{s+1}a^s}\Gamma(s)\left[\zeta\left(s,\ \frac{a+ib}{2a}\right)\mp\zeta\left(s,\ \frac{a-ib}{2a}\right)\right.$ $\left.-\zeta\left(s,\ \frac{2a+ib}{2a}\right)\pm\zeta\left(s,\ \frac{2a-ib}{2a}\right)\right]$ $[\operatorname{Re}a>\lvert\operatorname{Im}b\rvert;\ \operatorname{Re}s>-(1\pm1)/2]$
19	$(a-x)_+^{(\delta-1)/2}e^{cx}$ $\times\begin{Bmatrix}\sin\left(b\sqrt{a-x}\right)\\ \cos\left(b\sqrt{a-x}\right)\end{Bmatrix}$	$\frac{\sqrt{\pi}}{2^\delta}a^{s+\delta-1/2}b^\delta\,\Gamma\left[\begin{matrix}s\\ \frac{2s+2\delta+1}{2}\end{matrix}\right]\Phi_3\left(s;\ \frac{2s+2\delta+1}{2};\ ac,\ -\frac{ab^2}{4}\right)$ $[a,\ \operatorname{Re}s>0]$
20	$e^{-ax}\begin{Bmatrix}\sin\left(bx^2+ax\right)\\ \cos\left(bx^2+ax\right)\end{Bmatrix}$	$\frac{\Gamma(s)}{(2b)^{s/2}}e^{a^2/(4b)}\begin{Bmatrix}\sin(s\pi/4)\\ \cos(s\pi/4)\end{Bmatrix}D_{-s}\left(\frac{a}{\sqrt{b}}\right)$ $[b>0;\ \operatorname{Re}s>-(1\pm1)/2;\ \lvert\arg a\rvert<\pi/4]$
21	$\left(a^2-x^2\right)_+^{-1/2}e^{-b/(a^2-x^2)}$ $\times\begin{Bmatrix}\sin\left[cx/\left(a^2-x^2\right)\right]\\ \cos\left[cx/\left(a^2-x^2\right)\right]\end{Bmatrix}$	$\frac{2^{(2s-3)/4+\delta}a^{s-1}}{\sqrt{c}}e^{-b/(2a^2)}\left(\sqrt{b^2+a^2c^2}+b\right)^{1/4}\Gamma\left(\frac{s+\delta}{2}\right)$ $\times D_{-s}\left(\frac{\sqrt{\sqrt{b^2+a^2c^2}+b}}{a}\right)M_{(2s-1)/4,\,\pm1/4}\left(\frac{\sqrt{b^2+a^2c^2}-b}{2a^2}\right)$ $[a,\ b,\ c>0;\ \operatorname{Re}s>-\delta]$

No.	$f(x)$	$F(s)$
22	$(x^2-a^2)_+^{-1/2} \times e^{-b/(x^2-a^2)} \times \begin{Bmatrix} \sin[cx/(x^2-a^2)] \\ \cos[cx/(x^2-a^2)] \end{Bmatrix}$	$\frac{a^{s-1}}{2^{(2s+1)/4-\delta}\sqrt{c}} e^{b/(2a^2)} \left(\sqrt{b^2+a^2c^2}+b\right)^{1/4} \times \Gamma\left(\frac{1-s+\delta}{2}\right) D_{s-1}\left(\frac{\sqrt{\sqrt{b^2+a^2c^2}+b}}{a}\right) \times M_{(1-2s)/4,\,\pm 1/4}\left(\frac{\sqrt{b^2+a^2c^2}-b}{2a^2}\right)$ $[a,\, b,\, c>0;\ \operatorname{Re} s<\delta+1]$
23	$(a^2-x^2)_+^{-1/2} \times e^{-b(a^2+x^2)/(a^2-x^2)} \times \begin{Bmatrix} \sin[cx/(a^2-x^2)] \\ \cos[cx/(a^2-x^2)] \end{Bmatrix}$	$\frac{2^{(2s-3)/4+\delta}a^{s-3/4}}{\sqrt{c}} \left(\sqrt{4a^2b^2+c^2}+2ab\right)^{1/4} \times \Gamma\left(\frac{s+\delta}{2}\right) D_{-s}\left(\frac{\sqrt{\sqrt{4a^2b^2+c^2}+2ab}}{\sqrt{a}}\right) \times M_{(2s-1)/4,\,\pm 1/4}\left(\frac{\sqrt{4a^2b^2+c^2}-2ab}{2a}\right)$ $[a,\, b,\, c>0;\ \operatorname{Re} s>-\delta]$
24	$(x^2-a^2)_+^{-1/2} \times e^{-b(x^2+a^2)/(x^2-a^2)} \times \begin{Bmatrix} \sin[cx/(x^2-a^2)] \\ \cos[cx/(x^2-a^2)] \end{Bmatrix}$	$\frac{a^{s-3/4}}{2^{(2s+1)/4-\delta}\sqrt{c}} \left(\sqrt{4a^2b^2+c^2}+2ab\right)^{1/4} \times \Gamma\left(\frac{1-s+\delta}{2}\right) D_{s-1}\left(\frac{\sqrt{\sqrt{4a^2b^2+c^2}+2ab}}{\sqrt{a}}\right) \times M_{(1-2s)/4,\,\pm 1/4}\left(\frac{\sqrt{4a^2b^2+c^2}-2ab}{2a}\right)$ $[a,\, b,\, c>0;\ \operatorname{Re} s<\delta+1]$
25	$\frac{1}{\sqrt{x^2+a^2}} e^{-b/(x^2+a^2)} \times \begin{Bmatrix} \sin[cx/(x^2+a^2)] \\ \cos[cx/(x^2+a^2)] \end{Bmatrix}$	$\frac{2^{\delta-1/2}a^{s-1/2}}{\sqrt{c}} e^{-b/(2a^2)} \mathrm{B}\left(\frac{1-s+\delta}{2}, \frac{s+\delta}{2}\right) \times M_{(1-2s)/4,\,\pm 1/4}\left(\frac{b-\sqrt{b^2-a^2c^2}}{2a^2}\right) \times M_{(1-2s)/4,\,\pm 1/4}\left(\frac{b+\sqrt{b^2-a^2c^2}}{2a^2}\right)$ $[\operatorname{Re} a,\, b,\, c>0;\ -\delta<\operatorname{Re} s<\delta+1]$
26	$\frac{1}{\sqrt{x^2+a^2}} \times e^{-b(a^2-x^2)/(a^2+x^2)} \times \begin{Bmatrix} \sin[cx/(x^2+a^2)] \\ \cos[cx/(x^2+a^2)] \end{Bmatrix}$	$\frac{2^{\delta-1/2}a^{s-1/2}}{\sqrt{c}} \mathrm{B}\left(\frac{1-s+\delta}{2}, \frac{s+\delta}{2}\right) \times M_{(1-2s)/4,\,\pm 1/4}\left(\frac{2ab-\sqrt{4a^2b^2-c^2}}{2a}\right) \times M_{(1-2s)/4,\,\pm 1/4}\left(\frac{2ab+\sqrt{4a^2b^2-c^2}}{2a}\right)$ $[\operatorname{Re} a,\, b,\, c>0;\ -\delta<\operatorname{Re} s<\delta+1]$

2.4.4. Trigonometric and hyperbolic functions

Notation: $\delta = \begin{Bmatrix} 1 \\ 0 \end{Bmatrix}$.

1	$(a-x)_+^{\alpha-1} \times \begin{Bmatrix} \sinh(bx)\sin(bx) \\ \cosh(bx)\cos(bx) \end{Bmatrix}$	$a^{s+\alpha+2\delta-1} b^{2\delta}\, \mathrm{B}(\alpha,\, s+2\delta) \times {}_4F_7\left(\begin{matrix} \Delta(4,\, s+2\delta);\ \pm\frac{a^4b^4}{64} \\ \frac{2\delta+1}{4},\ \frac{2\delta+3}{4},\ \frac{2\delta+1}{2},\ \Delta(4,\, s+\alpha+2\delta) \end{matrix} \right)$ $[a,\ \operatorname{Re}\alpha > 0;\ \operatorname{Re} s > -2\delta]$
2	$(a-x)_+^{\alpha-1} \times \begin{Bmatrix} \cosh(bx)\sin(bx) \\ \sinh(bx)\cos(bx) \end{Bmatrix}$	$a^{s+\alpha} b\, \mathrm{B}(\alpha,\, s+1)\ {}_4F_7\left(\begin{matrix} \Delta(4,\, s+1);\ -\frac{a^4b^4}{64} \\ \frac{1}{2},\ \frac{3}{4},\ \frac{5}{4},\ \Delta(4,\, s+\alpha+1) \end{matrix} \right) \pm \frac{a^{s+\alpha+2}b^3}{3}\, \mathrm{B}(\alpha,\, s+3)\ {}_4F_7\left(\begin{matrix} \Delta(4,\, s+3);\ -\frac{a^4b^4}{64} \\ \frac{1}{4},\ \frac{1}{2},\ \frac{3}{4},\ \Delta(4,\, s+\alpha+3) \end{matrix} \right)$ $[a,\ \operatorname{Re}\alpha > 0;\ \operatorname{Re} s > -1]$
3	$(a^2-x^2)_+^{\alpha-1} \times \begin{Bmatrix} \sinh(bx)\sin(bx) \\ \cosh(bx)\cos(bx) \end{Bmatrix}$	$\frac{a^{s+2\alpha+2\delta-2} b^{2\delta}}{2}\, \mathrm{B}\left(\alpha,\, \frac{s+2\delta}{2}\right) \times {}_2F_5\left(\begin{matrix} \frac{s+2\delta}{4},\ \frac{s+2\delta+2}{4};\ -\frac{a^4b^4}{64} \\ \frac{2\delta+1}{4},\ \frac{2\delta+3}{4},\ \frac{2\delta+1}{2},\ \frac{s+2\alpha+2\delta}{4},\ \frac{s+2\alpha+2\delta+2}{4} \end{matrix} \right)$ $[a,\ \operatorname{Re}\alpha > 0;\ \operatorname{Re} s > -2\delta]$
4	$(a^2-x^2)_+^{\alpha-1} \times \begin{Bmatrix} \cosh(bx)\sin(bx) \\ \sinh(bx)\cos(bx) \end{Bmatrix}$	$\frac{a^{s+2\alpha-1} b}{2}\, \mathrm{B}\left(\alpha,\, \frac{s+1}{2}\right)\ {}_2F_5\left(\begin{matrix} \frac{s+1}{4},\ \frac{s+3}{4};\ -\frac{a^4b^4}{64} \\ \frac{1}{2},\ \frac{3}{4},\ \frac{5}{4},\ \frac{s+2\alpha+1}{4},\ \frac{s+2\alpha+3}{4} \end{matrix} \right) \pm \frac{a^{s+2\alpha+1}b^3}{6}\, \mathrm{B}\left(\alpha,\, \frac{s+3}{2}\right)\ {}_2F_5\left(\begin{matrix} \frac{s+3}{4},\ \frac{s+5}{4};\ -\frac{a^4b^4}{64} \\ \frac{5}{4},\ \frac{3}{2},\ \frac{7}{4},\ \frac{s+2\alpha+3}{4},\ \frac{s+2\alpha+5}{4} \end{matrix} \right)$ $[a,\ \operatorname{Re}\alpha > 0;\ \operatorname{Re} s > -1]$
5	$e^{-ax} \begin{Bmatrix} \sinh(bx)\sin(bx) \\ \cosh(bx)\cos(bx) \end{Bmatrix}$	$a^{-s-2\delta} b^{2\delta}\, \Gamma(s+2\delta)\ {}_4F_3\left(\begin{matrix} \Delta(4,\, s+2\delta);\ -\frac{4b^4}{a^4} \\ \frac{2\delta+1}{4},\ \frac{2\delta+3}{4},\ \frac{2\delta+1}{2} \end{matrix} \right)$ $[\operatorname{Re} a > \lvert\operatorname{Re} b\rvert + \lvert\operatorname{Im} b\rvert;\ \operatorname{Re} s > -2\delta]$
6	$e^{-ax} \begin{Bmatrix} \cosh(bx)\sin(bx) \\ \sinh(bx)\cos(bx) \end{Bmatrix}$	$a^{-s-1} b\, \Gamma(s+1)\ {}_4F_3\left(\begin{matrix} \Delta(4,\, s+1) \\ \frac{1}{2},\ \frac{3}{4},\ \frac{5}{4};\ -\frac{4b^4}{a^4} \end{matrix} \right) \pm \frac{a^{-s-3}b^3}{3}\, \Gamma(s+3)\ {}_4F_3\left(\begin{matrix} \Delta(4,\, s+3) \\ \frac{5}{4},\ \frac{3}{2},\ \frac{7}{4};\ -\frac{4b^4}{a^4} \end{matrix} \right)$ $[\operatorname{Re} a > \lvert\operatorname{Re} b\rvert + \lvert\operatorname{Im} b\rvert;\ \operatorname{Re} s > -1]$
7	$e^{-ax^2} \begin{Bmatrix} \sinh(bx)\sin(bx) \\ \cosh(bx)\cos(bx) \end{Bmatrix}$	$\frac{a^{-(s+2\delta)/2} b^{2\delta}}{2}\, \Gamma\left(\frac{s+2\delta}{2}\right)\ {}_2F_3\left(\begin{matrix} \frac{s+2\delta}{4},\ \frac{s+2\delta+2}{4};\ -\frac{b^4}{16a^2} \\ \frac{2\delta+1}{4},\ \frac{2\delta+3}{4},\ \frac{2\delta+1}{2} \end{matrix} \right)$ $[\operatorname{Re} a > 0;\ \operatorname{Re} s > -2\delta]$

No.	$f(x)$	$F(s)$
8	$e^{-ax^2}\begin{Bmatrix}\cosh(bx)\sin(bx)\\ \sinh(bx)\cos(bx)\end{Bmatrix}$	$\frac{a^{-(s+1)/2}b}{2}\Gamma\left(\frac{s+1}{2}\right){}_2F_3\left(\begin{matrix}\frac{s+1}{4},\frac{s+3}{4}\\ \frac{1}{2},\frac{3}{4},\frac{5}{4};-\frac{b^4}{16a^2}\end{matrix}\right) \pm\frac{a^{-(s+3)/2}b^3}{6}\Gamma\left(\frac{s+3}{2}\right){}_2F_3\left(\begin{matrix}\frac{s+3}{4},\frac{s+5}{4}\\ \frac{5}{4},\frac{3}{2},\frac{7}{4};-\frac{b^4}{16a^2}\end{matrix}\right)$ $[\operatorname{Re} a>0;\ \operatorname{Re} s>-1]$

2.4.5. Products of trigonometric functions

Notation: $\lambda_n=\dfrac{1+(-1)^n}{2}$, $\mu_n=\dfrac{(-1)^m+(-1)^n}{2}$.

1	$\sin^2(ax)$	$-\frac{a^{-s}}{2^{s+1}}\cos\frac{s\pi}{2}\Gamma(s)$ $[a>0;\ -2<\operatorname{Re} s<0]$
2	$\sin^2(ax)-\frac{1}{2}$	$-\frac{a^{-s}}{2^{s+1}}\cos\frac{s\pi}{2}\Gamma(s)$ $[a>0;\ 0<\operatorname{Re} s<1]$
3	$\cos^2(ax)-\frac{1}{2}$	$\frac{a^{-s}}{2^{s+1}}\cos\frac{s\pi}{2}\Gamma(s)$ $[a>0;\ 0<\operatorname{Re} s<1]$
4	$\cos^2(ax)-1$	$\frac{a^{-s}}{2^{s+1}}\cos\frac{s\pi}{2}\Gamma(s)$ $[a>0;\ -2<\operatorname{Re} s<0]$
5	$\sin^2(ax)-a^2x^2$	$-\frac{a^{-s}}{2^{s+1}}\cos\frac{s\pi}{2}\Gamma(s)$ $[a>0;\ -4<\operatorname{Re} s<-2]$
6	$\cos^2(ax)+a^2x^2-1$	$\frac{a^{-s}}{2^{s+1}}\cos\frac{s\pi}{2}\Gamma(s)$ $[a>0;\ -4<\operatorname{Re} s<-2]$
7	$\sin^3(ax)$	$\frac{3^{s+1}-1}{4}(3a)^{-s}\sin\frac{s\pi}{2}\Gamma(s)$ $[a>0;\ -2<\operatorname{Re} s<0]$
8	$\cos^3(ax)$	$\frac{3^{s+1}+1}{4}(3a)^{-s}\cos\frac{s\pi}{2}\Gamma(s)$ $[a>0;\ 0<\operatorname{Re} s<1]$
9	$\cos^3(ax)+\frac{3}{2}a^2x^2-1$	$\frac{3^{s+1}+1}{4}(3a)^{-s}\cos\frac{s\pi}{2}\Gamma(s)$ $[a>0;\ -4<\operatorname{Re} s<-2]$
10	$\sin^n(ax)$	$2^{s-n}\sqrt{\pi}\,\lvert a\rvert^{-s}\operatorname{sgn}^n a\,\Gamma\left[\begin{matrix}\frac{s+2\lambda}{2}\\ \frac{2\lambda-s+1}{2}\end{matrix}\right] \times\sum_{j=0}^{\lfloor(n-1)/2\rfloor}(-1)^{[n/2]+j}\frac{n!\,(n-2j)^{-s}}{j!\,(n-j)!}$ $\left[\begin{matrix}\lambda=(1-(-1)^n)/4;\ s\neq-2(\lambda+k);\\ \operatorname{Im} a=0,\ a\neq0;\ -n<\operatorname{Re} s<2\lambda;\ n\geq1\end{matrix}\right]$

No.	$f(x)$	$F(s)$
11	$\sin^{2n}(ax)$	$\frac{\sqrt{\pi}}{2^{2n}a^{s}}\Gamma\left[\begin{matrix}\frac{s}{2}\\ \frac{1-s}{2}\end{matrix}\right]\sum_{k=0}^{n-1}(-1)^{n+k}\frac{(2n)!}{k!\,(2n-k)!}(n-k)^{-s}$ $[a>0;\ -2<\operatorname{Re}s<0;\ n\geq 1]$
12	$\sin^{2n+1}(ax)$	$\frac{\sqrt{\pi}}{2^{2n-s+1}a^{s}}\Gamma\left[\begin{matrix}\frac{s+1}{2}\\ \frac{2-s}{2}\end{matrix}\right]\sum_{k=0}^{n}(-1)^{n+k}\frac{(2n+1)!}{k!\,(2n-k+1)!}$ $\times(2n-2k+1)^{-s}$ $[a>0;\ \lvert\operatorname{Re}s\rvert<1]$
13	$\cos^{2n+1}(ax)$	$2^{s-2n-1}\sqrt{\pi}\,(2n+1)!\,a^{-s}\,\Gamma\left[\begin{matrix}\frac{s}{2}\\ \frac{1-s}{2}\end{matrix}\right]$ $\times\sum_{k=0}^{n}\frac{(2n-2k+1)^{-s}}{k!\,(2n-k+1)!}$ $[a>0;\ 0<\operatorname{Re}s<1]$
14	$\cos^{n}(ax)-1$	$2^{1-n}a^{-s}\cos\frac{s\pi}{2}\,\Gamma(s)\sum_{k=0}^{\frac{n}{2}-1}\binom{n}{k}(n-2k)^{-s}$ $[a>0;\ -2<\operatorname{Re}s<0]$
15	$\cos^{2n}(ax)-1$	$2^{-2n}(2n)!\sqrt{\pi}\,a^{-s}\,\Gamma\left[\begin{matrix}\frac{s}{2}\\ \frac{1-s}{2}\end{matrix}\right]\sum_{k=0}^{n-1}\frac{(n-k)^{-s}}{k!\,(2n-k)!}$ $[a>0;\ -2<\operatorname{Re}s<0;\ n\geq 1]$
16	$\sin^{n}(ax)-\frac{(-1)^{n}+1}{2^{n+1}}\binom{n}{n/2}$	$\sqrt{\pi}\,\Gamma\left[\begin{matrix}\frac{2s+(-1)^{n+1}+1}{4}\\ \frac{-2s+(-1)^{n+1}+3}{4}\end{matrix}\right]$ $\times\sum_{k=0}^{[(n-1)/2]}(-1)^{[n/2]-k}\binom{n}{k}\frac{2^{s-n}}{[a(n-2k)]^{s}}$ $[a>0;\ ((-1)^{n}-1)/2<\operatorname{Re}s<1]$
17	$\cos^{n}(ax)-\frac{(-1)^{n}+1}{2^{n+1}}\binom{n}{n/2}$	$2^{1-n}a^{-s}\cos\frac{s\pi}{2}\,\Gamma(s)\sum_{k=0}^{[(n-1)/2]}\binom{n}{k}(n-2k)^{-s}$ $[a>0;\ 0<\operatorname{Re}s<1]$
18	$\sin^{m}(ax)$ $-2^{1-m}\sum_{j=0}^{n}\frac{(-1)^{j}(ax)^{m+2j}}{(m+2j)!}$ $\times\sum_{k=0}^{[(m-1)/2]}\binom{m}{k}\frac{(-1)^{k}}{(m-2k)^{-m-2j}}$	$\sqrt{\pi}\,\Gamma\left[\begin{matrix}-\frac{s+m+2n}{2},\ \frac{s+m+2n+2}{2}\\ \frac{1-s}{2},\ \frac{2-s}{2}\end{matrix}\right]$ $\times\sum_{k=0}^{[(m-1)/2]}(-1)^{n+k+1}\binom{m}{k}\frac{2^{s-m}}{[a(m-2k)]^{s}}$ $[a>0;\ -m-2n-2<\operatorname{Re}s<-m-2n]$

No.	$f(x)$	$F(s)$
19	$\cos^m(ax) - 2^{1-m}\sum_{j=1}^{n}\frac{(-1)^j(ax)^{2j}}{(2j)!} \times \sum_{k=0}^{[(m-1)/2]}\binom{m}{k}(m-2k)^{2j} - 1$	$2^{1-m}a^{-s}\cos\frac{s\pi}{2}\Gamma(s)\sum_{k=0}^{[(m-1)/2]}\binom{m}{k}(m-2k)^{-s}$ $[a>0;\ -2n-2<\operatorname{Re}s<-2n]$
20	$\left\{\begin{matrix}\sin(ax)\sin(bx)\\ \cos(ax)\cos(bx)\end{matrix}\right\}$	$\frac{1}{2}\cos\frac{s\pi}{2}\Gamma(s)\left[\lvert a-b\rvert^{-s}\mp(a+b)^{-s}\right]$ $[a,\,b>0;\ a\neq b;\ -(1\pm1)<\operatorname{Re}s<1]$
21	$\sin(ax)\cos(bx)$	$\frac{1}{2}\sin\frac{s\pi}{2}\Gamma(s)\left[(a+b)^{-s}+\lvert a-b\rvert^{-s}\operatorname{sgn}(a-b)\right]$ $[a,\,b>0;\ a\neq b;\ \lvert\operatorname{Re}s\rvert<1]$
22	$\sin^m(ax)\left[\sin^n(bx) - 2^{-n}\lambda_n\binom{n}{n/2}\right]$	$(-2)^{-m-n+1}\frac{(s-1)^{\lambda_{m+1}\lambda_{n+1}}}{s^{\lambda_m\lambda_n}}\sin\frac{(s-\mu_n)\pi}{2}\Gamma(s+\mu_n)$ $\times\left\{\sum_{k=0}^{[(m-1)/2]}(-1)^{[m/2]-k}\binom{m}{k}\sum_{j=0}^{[(n-1)/2]}(-1)^{[n/2]-j}\binom{n}{j}\right.$ $\times\left[[a(m-2k)+b(n-2j)]^{-s}\right.$ $\left.\left.+\frac{(-1)^{m+(n-m)\theta(a(m-2k)-b(n-2j))}}{\lvert a(m-2k)-b(n-2j)\rvert^{s}}\right]\right\}+\frac{(-1)^{n-1}}{2^{m+n-1}}\binom{m}{m/2}$ $\times\frac{\lambda_m}{b^s}\Gamma(s)\sin\frac{(s-\lambda_n)\pi}{2}\sum_{k=0}^{[(n-1)/2]}\binom{n}{k}\frac{(-1)^{[n/2]-k}}{(n-2k)^s}$ $[a,\,b,\,m,\,n>0;\ -m-n\lambda_{n+1}<\operatorname{Re}s<1]$
23	$\sin^m(ax)\left[\cos^n(bx) - 2^{-n}\lambda_n\binom{n}{n/2}\right]$	$2^{-m-n+1}\binom{m}{m/2}\frac{\lambda_m}{b^s}\cos\frac{s\pi}{2}\Gamma(s)\sum_{k=0}^{[(n-1)/2]}\binom{n}{k}(n-2k)^{-s}$ $-(-1)^m 2^{-m-n+1}\sin\frac{(s-\lambda_m)\pi}{2}$ $\times\Gamma(s)\sum_{k=0}^{[(m-1)/2]}(-1)^{[m/2]-k}\binom{m}{k}$ $\times\sum_{j=0}^{[(n-1)/2]}\binom{n}{j}\left\{[a(m-2k)+b(n-2j)]^{-s}\right.$ $\left.+\frac{(-1)^{m\theta(b(n-2j)-a(m-2k))}}{\lvert a(m-2k)-b(n-2j)\rvert^{s}}\right\}$ $[a,\,b,\,m,\,n>0;\ -m<\operatorname{Re}s<1]$

No.	$f(x)$	$F(s)$
24	$\cos^n(bx)\left[\sin^m(ax) - 2^{-m}\lambda_m\binom{m}{m/2}\right]$	$(-1)^{m-1}2^{-m-n+1}\binom{n}{n/2}\frac{\lambda_n}{a^s}\sin\frac{(s-\lambda_m)\pi}{2} \times\Gamma(s)\sum_{k=0}^{[(m-1)/2]}\binom{m}{k}\frac{(-1)^{[m/2]-k}}{(m-2k)^s} - (-1)^m 2^{-m-n+1}\sin\frac{(s-\lambda_m)\pi}{2} \times\Gamma(s)\sum_{k=0}^{[(m-1)/2]}(-1)^{[m/2]-k}\binom{m}{k}\sum_{j=0}^{[(n-1)/2]}\binom{n}{j} \times\left\{[b(n-2j)+a(m-2k)]^{-s} + \frac{(-1)^{m\theta(b(n-2j)-a(m-2k))}}{\lvert b(n-2j)-a(m-2k)\rvert^s}\right\}$ $[a,\, b,\, m,\, n>0;\ -m\lambda_{m+1}<\operatorname{Re}s<1]$
25	$\cos^m(ax)\left[\cos^n(bx) - 2^{-n}\lambda_n\binom{n}{n/2}\right]$	$2^{-m-n+1}\binom{m}{m/2}\frac{\lambda_m}{b^s}\cos\frac{s\pi}{2}\Gamma(s)\sum_{k=0}^{[(n-1)/2]}\binom{n}{k}(n-2k)^{-s} + 2^{-m-n+1}\cos\frac{s\pi}{2}\Gamma(s)\sum_{j=0}^{[(m-1)/2]}\binom{m}{j} \times\sum_{k=0}^{[(n-1)/2]}\binom{n}{k}\left\{[b(n-2j)+a(m-2k)]^{-s} + \lvert b(n-2j)-a(m-2k)\rvert^{-s}\right\}$ $[a,\, b,\, m,\, n>0;\ 0<\operatorname{Re}s<1]$
26	$\sin^m(ax)\sin^{2n}(bx)$	$(-1)^{n+[(m+1)/2]}2^{-m-2n+1}s^{-\lambda_m}\sin\frac{(\lambda_m-s)\pi}{2} \times\Gamma(\lambda_m+s)\sum_{k=0}^{[(m-1)/2]}(-1)^k\binom{m}{k}\sum_{j=0}^{n-1}(-1)^j\binom{2n}{j} \times\left\{[a(m-2k)+2b(n-j)]^{-s} + \frac{(-1)^{m+(2n-m)\theta(a(m-2k)-2b(n-j))}}{\lvert a(m-2k)-2b(n-j)\rvert^s}\right\} + 2^{-m-2n-s+1}\binom{m}{m/2}\frac{\lambda_m}{b^s}\cos\frac{s\pi}{2}\Gamma(s) \times\sum_{k=0}^{n-1}\binom{2n}{k}\frac{(-1)^{k+n}}{(n-k)^s} + (-1)^{[m/2]}2^{-m-2n+1}a^{-s} \times\binom{2n}{n}\sin\frac{(\lambda_m+s)\pi}{2}\Gamma(s)\sum_{k=0}^{[(m-1)/2]}\binom{m}{k}\frac{(-1)^k}{(m-2k)^s}$ $[a,\, b,\, m,\, n>0;\ -m-2n<\operatorname{Re}s<\lambda_{m+1}]$

No.	$f(x)$	$F(s)$
27	$\sin^m(ax)\cos^n(bx)$	$(-1)^{m+1}\,2^{-m-n+1}\binom{n}{n/2}\frac{\lambda_n}{a^s}\sin\frac{(s-\lambda_m)\pi}{2}\times\Gamma(s)\sum_{k=0}^{[(m-1)/2]}\binom{m}{k}\frac{(-1)^{k+[m/2]}}{(m-2k)^s}+2^{-m-n+1}\binom{m}{m/2}\frac{\lambda_m}{b^s}\cos\frac{s\pi}{2}\Gamma(s)\sum_{k=0}^{[(n-1)/2]}\binom{n}{k}\frac{1}{(n-2k)^s}-(-1)^m\,2^{-m-n+1}\sin\frac{(s-\lambda_m)\pi}{2}\Gamma(s)\sum_{k=0}^{[(m-1)/2]}(-1)^{[m/2]-k}\binom{m}{k}\times\sum_{j=0}^{[(n-1)/2]}\binom{n}{j}\left\{[b(n-2j)+a(m-2k)]^{-s}+\frac{(-1)^{m\theta(b(n-2j)-a(m-2k))}}{\lvert a(m-2k)-b(n-2j)\rvert^s}\right\}$ $[a,\,b,\,m,\,n>0;\ -m<\operatorname{Re}s<1-\delta_{(-1)^n+(-1)^m-2,\,0}]$
28	$\cos^m(ax)\cos^{2n-1}(bx)$	$2^{-m-2n+2}\binom{m}{m/2}\frac{\lambda_m}{b^s}\cos\frac{s\pi}{2}\Gamma(s)\sum_{k=0}^{n-1}\binom{2n-1}{k}\frac{1}{(2n-2k-1)^s}+2^{-m-2n+2}\cos\frac{s\pi}{2}\Gamma(s)\sum_{k=0}^{n-1}\binom{2n-1}{k}\times\sum_{j=0}^{[(m-1)/2]}\binom{m}{j}\left\{[a(m-2j)+b(2n-2k-1)]^{-s}+\lvert a(m-2j)-b(2n-2k-1)\rvert^{-s}\right\}$ $[a,\,b,\,m,\,n>0;\ 0<\operatorname{Re}s<1]$
29	$\begin{Bmatrix}\sin(ax^2)\\\cos(ax^2)\end{Bmatrix}\sin(bx)$	$\frac{a^{-(s+1)/2}b}{2}\Gamma\left(\frac{s+1}{2}\right)\begin{Bmatrix}\cos[(1-s)\pi/4]\\\sin[(1-s)\pi/4]\end{Bmatrix}\times{}_2F_3\left(\begin{matrix}\frac{s+1}{2},\,\frac{s+3}{2};\,-\frac{b^4}{64a^2}\\\frac{1}{2},\,\frac{3}{4},\,\frac{5}{4}\end{matrix}\right)\mp\frac{a^{-(s+3)/2}b^3}{12}\Gamma\left(\frac{s+3}{2}\right)\times\begin{Bmatrix}\cos[(s+1)\pi/4]\\\sin[(s+1)\pi/4]\end{Bmatrix}{}_2F_3\left(\begin{matrix}\frac{s+3}{4},\,\frac{s+5}{4};\,-\frac{b^4}{64a^2}\\\frac{5}{4},\,\frac{3}{2},\,\frac{7}{4}\end{matrix}\right)$ $[a,\,b>0;\ -1-(1\pm1)<\operatorname{Re}s<2]$
30	$\begin{Bmatrix}\sin(ax^2)\\\cos(ax^2)\end{Bmatrix}\cos(bx)$	$\frac{a^{-s/2}}{2}\Gamma\left(\frac{s}{2}\right)\begin{Bmatrix}\sin(s\pi/4)\\\cos(s\pi/4)\end{Bmatrix}{}_2F_3\left(\begin{matrix}\frac{s}{4},\,\frac{s+2}{4};\,-\frac{b^4}{64a^2}\\\frac{1}{4},\,\frac{1}{2},\,\frac{3}{4}\end{matrix}\right)\mp\frac{a^{-s/2-1}b^2}{4}\Gamma\left(\frac{s+2}{2}\right)\begin{Bmatrix}\cos(s\pi/4)\\\sin(s\pi/4)\end{Bmatrix}{}_2F_3\left(\begin{matrix}\frac{s+2}{4},\,\frac{s+4}{4};\,-\frac{b^4}{64a^2}\\\frac{3}{4},\,\frac{5}{4},\,\frac{3}{2}\end{matrix}\right)$ $[a,\,b>0;\ -(1\pm1)<\operatorname{Re}s<2]$

No.	$f(x)$	$F(s)$
31	$\begin{Bmatrix} \sin(ax)\sin(b/x) \\ \cos(ax)\cos(b/x) \end{Bmatrix}$	$\pm\frac{\pi}{4}\left(\frac{b}{a}\right)^{s/2} \csc\frac{s\pi}{2}\left[J_s\left(2\sqrt{ab}\right) - J_{-s}\left(2\sqrt{ab}\right) \pm \frac{2\sin(s\pi)}{\pi} K_s\left(2\sqrt{ab}\right)\right]$ $[a,\, b > 0;\ \lvert\operatorname{Re} s\rvert < (3 \pm 1)/2]$
32	$\begin{Bmatrix} \sin(ax)\cos(b/x) \\ \cos(ax)\sin(b/x) \end{Bmatrix}$	$\frac{\pi}{4}\left(\frac{b}{a}\right)^{s/2} \sec\frac{s\pi}{2}\left[J_s\left(2\sqrt{ab}\right) + J_{-s}\left(2\sqrt{ab}\right) \pm \frac{2\sin(s\pi)}{\pi} K_s\left(2\sqrt{ab}\right)\right]$ $[a,\, b > 0;\ \lvert\operatorname{Re} s\rvert < 1]$
33	$\sin(ax)\sin(bx)\sin(cx)$	$\frac{\Gamma(s)}{4} \sin\frac{s\pi}{2}\left[\frac{1}{(a+b-c)^s} - \frac{1}{(a+b+c)^s} + \frac{\operatorname{sgn}(a-b+c)}{\lvert a-b+c\rvert^s} - \frac{\operatorname{sgn}(a-b-c)}{\lvert a-b-c\rvert^s}\right]$ $\left[\begin{matrix} a > 0;\ \operatorname{Im} b = \operatorname{Im} c = 0;\ b > \lvert c\rvert; \\ a - b \neq \lvert c\rvert;\ -3 < \operatorname{Re} s < 1 \end{matrix}\right]$
34	$\sin(ax)\sin(bx)\cos(cx)$	$\frac{\Gamma(s)}{4} \cos\frac{s\pi}{2}\left[-\frac{1}{(a+b-c)^s} - \frac{1}{(a+b+c)^s} + \frac{1}{\lvert a-b+c\rvert^s} + \frac{1}{\lvert a-b-c\rvert^s}\right]$ $\left[\begin{matrix} a > 0;\ \operatorname{Im} b = \operatorname{Im} c = 0;\ b > \lvert c\rvert; \\ a - b \neq \lvert c\rvert;\ -2 < \operatorname{Re} s < 1 \end{matrix}\right]$
35	$\sin(ax)\cos(bx)\cos(cx)$	$\frac{\Gamma(s)}{4} \sin\frac{s\pi}{2}\left[\frac{1}{(a+b-c)^s} + \frac{1}{(a+b+c)^s} + \frac{\operatorname{sgn}(a-b+c)}{\lvert a-b+c\rvert^s} + \frac{\operatorname{sgn}(a-b-c)}{\lvert a-b-c\rvert^s}\right]$ $\left[\begin{matrix} a > 0;\ \operatorname{Im} b = \operatorname{Im} c = 0;\ b > \lvert c\rvert; \\ a - b \neq \lvert c\rvert;\ -1 < \operatorname{Re} s < 1 \end{matrix}\right]$
36	$\cos(ax)\cos(bx)\cos(cx)$	$\frac{\Gamma(s)}{4} \cos\frac{s\pi}{2}\left[\frac{1}{(a+b-c)^s} + \frac{1}{(a+b+c)^s} + \frac{1}{\lvert a-b+c\rvert^s} + \frac{1}{\lvert a-b-c\rvert^s}\right]$ $\left[\begin{matrix} a > 0;\ \operatorname{Im} b = \operatorname{Im} c = 0;\ b > \lvert c\rvert; \\ a - b \neq \lvert c\rvert;\ 0 < \operatorname{Re} s < 1 \end{matrix}\right]$
37	$e^{-ax}\begin{Bmatrix} \sin^n(bx) \\ \cos^n(bx) \end{Bmatrix}$	$\begin{Bmatrix} (-i)^n \\ 1 \end{Bmatrix}\frac{\Gamma(s)}{2^n}\sum_{k=0}^{n} (\mp 1)^{n-k}\binom{n}{k}\left[a + ib(n-2k)\right]^{-s}$ $[\operatorname{Re} a > n\lvert\operatorname{Im} b\rvert;\ \operatorname{Re} s > -(1 \pm 1)n/2]$

No.	$f(x)$	$F(s)$
38	$e^{-ax}\sin^{2n}(bx)$	$\frac{\Gamma(s)}{2^{2n}a^s}\binom{2n}{n}+\frac{(-1)^n\Gamma(s)}{2^{2n}}\sum_{k=0}^{n-1}(-1)^k\binom{2n}{k}$ $\times\left[(a-i(2n-2k)b)^{-s}+(a+i(2n-2k)b)^{-s}\right]$ $[\operatorname{Re}(a-2inb)>0;\ \operatorname{Re}s>-2n]$
39	$e^{-ax}\sin^{2n+1}(bx)$	$-\frac{(-1)^n i\Gamma(s)}{2^{2n+1}}\sum_{k=0}^{n}(-1)^k\binom{2n+1}{k}$ $\times\left[(a-i(2n-2k+1)b)^{-s}-(a+i(2n-2k+1)b)^{-s}\right]$ $[\operatorname{Re}(a-i(2n+1)b)>0;\ \operatorname{Re}s>-2n-1]$
40	$e^{-ax}\cos^{n}(bx)$	$\frac{[1+(-1)^n]\Gamma(s)}{2^{n+1}a^s}\binom{n}{n/2}+\frac{\Gamma(s)}{2^n}\sum_{k=0}^{[(n-1)/2]}\binom{n}{k}$ $\times\left[(a-i(n-2k)b)^{-s}+(a+i(n-2k)b)^{-s}\right]$ $[\operatorname{Re}(a-inb)>0;\ \operatorname{Re}s>0]$

2.4.6. $\operatorname{sinc}^n(bx)$ and elementary functions

1	$\operatorname{sinc}(ax)$	$\frac{2^{s-2}\sqrt{\pi}}{a^s}\Gamma\left[\begin{matrix}\frac{s}{2}\\ \frac{3-s}{2}\end{matrix}\right]$ $[a>0;\ 0<\operatorname{Re}s<2]$
2	$e^{-ax}\operatorname{sinc}(ax)$	$\frac{2^{3s/2-4}}{\sqrt{\pi}\,a^s}\Gamma\left[\begin{matrix}\frac{s}{4},\ \frac{s+1}{4},\ \frac{s+2}{4}\\ \frac{5-s}{4}\end{matrix}\right]$ $\left[\begin{matrix}(\lvert\arg a\rvert<\pi/4;\ \operatorname{Re}s>0)\ \text{or}\\ (\lvert\arg a\rvert=\pi/4;\ 0<\operatorname{Re}s<2)\end{matrix}\right]$
3	$e^{-ax}\operatorname{sinc}(bx)$	$\frac{\Gamma(s-1)}{b(a^2+b^2)^{(s-1)/2}}\sin\left[(s-1)\arctan\frac{b}{a}\right]$ $\left[\begin{matrix}(\operatorname{Re}a>\lvert\operatorname{Im}b\rvert;\ \operatorname{Re}s>0)\ \text{or}\\ (\operatorname{Re}a=\lvert\operatorname{Im}b\rvert;\ 0<\operatorname{Re}s<2)\end{matrix}\right]$
4	$e^{-ax^2}\operatorname{sinc}(bx)$	$\frac{a^{-s/2}}{2}\Gamma\left(\frac{s}{2}\right)\,{}_1F_1\left(\begin{matrix}\frac{s}{2}\\ \frac{3}{2};\ -\frac{b^2}{4a}\end{matrix}\right)$ $[\operatorname{Re}a,\ \operatorname{Re}s>0]$
5	$e^{-ax^2-bx}\operatorname{sinc}(cx)$	$\frac{i\Gamma(s-1)}{2c(2a)^{(s-1)/2}}e^{(b^2-c^2)/(8a)}\left[e^{ibc/(4a)}D_{1-s}\left(\frac{b+ic}{\sqrt{2a}}\right)\right.$ $\left.\mp e^{-ibc/(4a)}D_{1-s}\left(\frac{b-ic}{\sqrt{2a}}\right)\right]$ $[\operatorname{Re}a,\ \operatorname{Re}s>0]$

No.	$f(x)$	$F(s)$
6	$\sin(ax)\,\mathrm{sinc}(ax)$	$-(2a)^{-s}\sin\frac{s\pi}{2}\,\Gamma(s-1)$ $[a>0;\ \lvert\mathrm{Re}\,s\rvert<1]$
7	$\cos(ax)\,\mathrm{sinc}(ax)$	$-(2a)^{-s}\cos\frac{s\pi}{2}\,\Gamma(s-1)$ $[a>0;\ 0<\mathrm{Re}\,s<2]$
8	$\mathrm{sinc}^2(ax)$	$2^{1-s}a^{-s}\cos\frac{s\pi}{2}\,\Gamma(s-2)$ $[a>0;\ 0<\mathrm{Re}\,s<2]$
9	$\mathrm{sinc}^3(ax)$	$\frac{a^{-s}}{4}\left(3-3^{3-s}\right)\cos\frac{s\pi}{2}\,\Gamma(s-3)$ $[a>0;\ 0<\mathrm{Re}\,s<4]$
10	$\mathrm{sinc}^{2n}(ax)$	$\frac{\sqrt{\pi}}{2^{2n}a^s}\,\Gamma\left[\begin{matrix}\frac{s-2n}{2}\\ \frac{1-s+2n}{2}\end{matrix}\right]\sum_{k=0}^{n-1}(-1)^{n+k}\frac{(2n)!}{k!\,(2n-k)!}(n-k)^{2n-s}$ $[a>0;\ 0<\mathrm{Re}\,s<2n]$
11	$\mathrm{sinc}^{2n+1}(ax)$	$\frac{\sqrt{\pi}}{2^{4n-s+2}a^s}\,\Gamma\left[\begin{matrix}\frac{s-2n}{2}\\ \frac{3-s+2n}{2}\end{matrix}\right]\sum_{k=0}^{n}(-1)^{n+k}\frac{(2n+1)!}{k!\,(2n-k+1)!}$ $\times(2n-2k+1)^{2n-s+1}$ $[a>0;\ 0<\mathrm{Re}\,s<2n+2]$
12	$\mathrm{sinc}^n(ax)$ $-\frac{(-1)^n+1}{2^{n+1}}\frac{\binom{n}{n/2}}{(ax)^n}$	$\frac{\sqrt{\pi}}{a^s}\,\Gamma\left[\begin{matrix}\frac{2s-2n+(-1)^{n+1}+1}{4}\\ \frac{-2s+2n+(-1)^{n+1}+3}{4}\end{matrix}\right]$ $\times\sum_{k=0}^{[(n-1)/2]}(-1)^{[n/2]-k}\binom{n}{k}2^{s-2n}(n-2k)^{n-s}$ $[a>0;\ (2n+(-1)^n-1)/2<\mathrm{Re}\,s<n+1]$
13	$e^{-ax}\,\mathrm{sinc}^n(bx)$	$(-i)^n\frac{\Gamma(s-n)}{(2b)^n}\sum_{k=0}^{n}(-1)^{n-k}\binom{n}{k}\left[a+ib(n-2k)\right]^{n-s}$ $[\mathrm{Re}\,a>n\,\lvert\mathrm{Im}\,b\rvert;\ \mathrm{Re}\,s>0]$
14	$\mathrm{sinc}\left(b\sqrt{x^2+a^2}\right)$	$\frac{2^{(s-3)/2}\sqrt{\pi}\,a^{(s-1)/2}}{b^{(s+1)/2}}\Gamma\left(\frac{s}{2}\right)J_{(1-s)/2}(ab)$ $[\mathrm{Re}\,a>0;\ 0<\mathrm{Re}\,s<2]$
15	$\theta(a-x)$ $\times\,\mathrm{sinc}\left(b\sqrt{a^2-x^2}\right)$	$\frac{2^{(s-3)/2}\sqrt{\pi}\,a^{(s-1)/2}}{b^{(s+1)/2}}\Gamma\left(\frac{s}{2}\right)\mathbf{H}_{(s-1)/2}(ab)$ $[a,\ \mathrm{Re}\,s>0]$
16	$\theta(x-a)$ $\times\,\mathrm{sinc}\left(b\sqrt{x^2-a^2}\right)$	$\frac{2^{(s-3)/2}\sqrt{\pi}\,a^{(s-1)/2}}{b^{(s+1)/2}}\Gamma\left(\frac{s}{2}\right)\left[I_{(1-s)/2}(ab)-\mathbf{L}_{(s-1)/2}(ab)\right]$ $[a,\ b>0;\ \mathrm{Re}\,s<2]$

2.5. The Logarithmic Function

More formulas can be obtained from the corresponding sections due to the relations

$$\ln(z+1) = z\,{}_2F_1(1,\,1;\,2;\,-z), \quad \ln\left(\sqrt{z+1}+\sqrt{z}\right) = \sqrt{z}\,{}_2F_1\left(\frac{1}{2},\,\frac{1}{2};\,\frac{3}{2};\,-z\right),$$

$$\ln\frac{1+\sqrt{z}}{1-\sqrt{z}} = 2\sqrt{z}\,{}_2F_1\left(\frac{1}{2},\,1;\,\frac{3}{2};\,z\right), \quad \frac{\ln\left(\sqrt{z+1}+\sqrt{z}\right)}{\sqrt{z+1}} = \sqrt{z}\,{}_2F_1\left(1,\,1;\,\frac{3}{2};\,-z\right),$$

$$\ln^2\left(\sqrt{z+1}+\sqrt{z}\right) = z\,{}_3F_2\left(1,\,1,\,1;\,\frac{3}{2},\,2;\,-z\right),$$

$$\ln(z+1) = G_{22}^{12}\left(z\,\middle|\,\begin{matrix}1,\,1\\1,\,0\end{matrix}\right), \quad \ln\left(\sqrt{z+1}\pm\sqrt{z}\right) = \pm\frac{1}{2\sqrt{\pi}}\,G_{22}^{12}\left(z\,\middle|\,\begin{matrix}1,\,1\\1/2,\,0\end{matrix}\right),$$

$$\frac{\ln\left(\sqrt{z+1}+\sqrt{z}\right)}{\sqrt{z+1}} = \frac{\sqrt{\pi}}{2}\,G_{33}^{22}\left(z\,\middle|\,\begin{matrix}1/2,\,1/2,\\1/2,\,1/2,\,0\end{matrix}\right),$$

$$\ln^2\left(\sqrt{z+1}+\sqrt{z}\right) = \frac{\sqrt{\pi}}{2}\,G_{33}^{13}\left(z\,\middle|\,\begin{matrix}1,\,1,\,1\\1,\,0,\,1/2\end{matrix}\right).$$

2.5.1. $\ln(bx)$ and algebraic functions

No.	$f(x)$	$F(s)$
1	$\left\{\begin{matrix}\theta(a-x)\\\theta(x-a)\end{matrix}\right\}\ln\frac{x}{a}$	$\mp\frac{a^s}{s^2}$ $[a>0;\ \pm\operatorname{Re}s>0]$
2	$\left\{\begin{matrix}\theta(a-x)\\\theta(x-a)\end{matrix}\right\}\ln(bx)$	$\mp\frac{a^s\,[1-s\ln(ab)]}{s^2}$ $[a>0;\ \pm\operatorname{Re}s>0;\ \lvert\arg b\rvert<\pi]$
3	$(a-x)_+^{\alpha-1}\ln(bx)$	$a^{s+\alpha-1}\,\mathrm{B}(s,\,\alpha)\,[\psi(s)-\psi(s+\alpha)+\ln(ab)]$ $[a,\ \operatorname{Re}\alpha,\ \operatorname{Re}s>0;\ \lvert\arg b\rvert<\pi]$
4	$(x-a)_+^{\alpha-1}\ln(bx)$	$a^{s+\alpha-1}\,\mathrm{B}(1-s-\alpha,\,\alpha)\,[\psi(1-s)-\psi(1-s-\alpha)+\ln(ab)]$ $[a,\ \operatorname{Re}\alpha>0;\ \operatorname{Re}(s+\alpha)<1;\ \lvert\arg b\rvert<\pi]$
5	$\left(a^2-x^2\right)_+^{\alpha-1}\ln(bx)$	$\frac{a^{s+2\alpha-2}}{2}\,\mathrm{B}\left(\alpha,\,\frac{s}{2}\right)\left[\frac{1}{2}\psi\left(\frac{s}{2}\right)-\frac{1}{2}\psi\left(\frac{s}{2}+\alpha\right)+\ln(ab)\right]$ $[a,\ \operatorname{Re}\alpha,\ \operatorname{Re}s>0;\ \lvert\arg b\rvert<\pi]$
6	$\left(x^2-a^2\right)_+^{\alpha-1}\ln(bx)$	$\frac{a^{s+2\alpha-2}}{2}\,\mathrm{B}\left(\alpha,\,\frac{2-s-2\alpha}{2}\right)\left[\frac{1}{2}\psi\left(\frac{2-s}{2}\right)-\frac{1}{2}\psi\left(\frac{2-s-2\alpha}{2}\right)+\ln(ab)\right]$ $[a,\ \operatorname{Re}\alpha>0;\ \operatorname{Re}(s+2\alpha)<2;\ \lvert\arg b\rvert<\pi]$
7	$\theta(a-x)\,\frac{\ln x}{x+a}$	$\frac{a^{s-1}}{4}\left\{2\ln a\left[\psi\left(\frac{s+1}{2}\right)-\psi\left(\frac{s}{2}\right)\right]+\psi'\left(\frac{s+1}{2}\right)-\psi'\left(\frac{s}{2}\right)\right\}$ $[a,\ \operatorname{Re}s>0]$

No.	$f(x)$	$F(s)$
8	$\dfrac{\ln x}{x+a}$	$\pi a^{s-1}\csc(s\pi)\,[\ln a-\pi\cot(s\pi)]$ $\quad[0<\operatorname{Re}s<1;\ \lvert\arg a\rvert<\pi]$
9	$\dfrac{\ln x}{a-x}$	$\pi a^{s-1}\left[\ln a\cot(s\pi)-\dfrac{\pi}{\sin^2(s\pi)}\right]$ $\quad[a>0;\ 0<\operatorname{Re}s<1]$
10	$\dfrac{\ln x}{(x+a)(x-1)}$	$\dfrac{\pi\csc^2(s\pi)}{a+1}\left\{\pi-a^{s-1}\left[\sin(s\pi)\ln a-\pi\cos(s\pi)\right]\right\}$ $\quad[0<\operatorname{Re}s<2;\ s\neq1;\ \lvert\arg a\rvert<\pi]$
11	$\dfrac{\ln x}{(x+a)^2}$	$\dfrac{\pi(1-s)\,a^{s-2}}{\sin(s\pi)}\left[\ln a-\pi\cot(s\pi)+\dfrac{1}{s-1}\right]$ $\quad[0<\operatorname{Re}s<2;\ s\neq1;\ \lvert\arg a\rvert<\pi]$
12	$\dfrac{\ln x}{(x+a)(x+b)}$	$\dfrac{\pi\csc(s\pi)}{a-b}\left[b^{s-1}\ln b-a^{s-1}\ln a-\pi\left(b^{s-1}-a^{s-1}\right)\cot(s\pi)\right]$ $\quad[0<\operatorname{Re}s<2;\ s\neq1;\ \lvert\arg a\rvert,\ \lvert\arg b\rvert<\pi]$
13	$\dfrac{\ln(x/b)}{(x+a)(x+b)}$	$\dfrac{\pi}{(b-a)\sin(s\pi)}\left[a^{s-1}\ln\dfrac{a}{b}+\pi\left(b^{s-1}-a^{s-1}\right)\cot(s\pi)\right]$ $\quad[0<\operatorname{Re}s<2;\ \lvert\arg a\rvert,\ \lvert\arg b\rvert<\pi]$
14	$\dfrac{\ln x}{(x+a)(x+b)(x+c)}$	$\pi\csc(s\pi)\left[\dfrac{a^{s-1}(\pi\cot(s\pi)-\ln a)}{(a-b)(c-a)}+\dfrac{b^{s-1}(\pi\cot(s\pi)-\ln b)}{(a-b)(b-c)}+\dfrac{c^{s-1}(\pi\cot(s\pi)-\ln c)}{(a-c)(c-b)}\right]$ $\quad[0<\operatorname{Re}s<3;\ s\neq1;\ \lvert\arg a\rvert,\ \lvert\arg b\rvert,\ \lvert\arg c\rvert<\pi]$

2.5.2. $\ln(bx+c)$ and algebraic functions

1	$\theta(a-x)\ln(x+a)$	$\dfrac{a^s}{s}\left\{\ln(2a)-\dfrac{1}{2}\left[\psi\left(\dfrac{s+2}{2}\right)-\psi\left(\dfrac{s+1}{2}\right)\right]\right\}$ $\quad[a,\ \operatorname{Re}s>0]$
2	$\theta(a-x)\ln(bx+c)$	$\dfrac{a^s}{s}\left[\ln\left(\dfrac{ab}{c}+1\right)+\ln c-\dfrac{ab}{c}\,\Phi\left(-\dfrac{ab}{c},\ 1,\ s+1\right)\right]$ $\quad\left[\begin{array}{c}a,\ \operatorname{Re}c,\ \operatorname{Re}s>0;\ \operatorname{Re}(c/b)\geq0\text{ or}\\ \operatorname{Re}(c/b)\leq-1;\ \operatorname{Im}(c/b)\neq0\end{array}\right]$
3	$\begin{Bmatrix}\ln(ax+1)\\ \ln\lvert ax-1\rvert\end{Bmatrix}$	$\dfrac{\pi a^{-s}}{s}\begin{Bmatrix}\csc(s\pi)\\ \cot(s\pi)\end{Bmatrix}$ $\quad[-1<\operatorname{Re}s<0;\ \lvert\arg a\rvert<\pi]$
4	$\dfrac{\ln(x+a)}{(x+a)^{\rho}}$	$a^{s-\rho}\,\mathrm{B}(s,\ \rho-s)\,[\psi(\rho)-\psi(\rho-s)+\ln a]$ $\quad[0<\operatorname{Re}s<\operatorname{Re}\rho;\ \lvert\arg a\rvert<\pi]$

No.	$f(x)$	$F(s)$
5	$(a-x)_+^{\alpha-1}\ln(bx+c)$	$\frac{a^{s+\alpha}b}{c}\,\mathrm{B}(s+1,\,\alpha)\;{}_3F_2\left(\begin{matrix}1,\,1,\,s+1;\,-\frac{ab}{c}\\2,\,s+\alpha+1\end{matrix}\right)+a^{s+\alpha-1}\ln c\,\mathrm{B}(s,\,\alpha)$ $[a,\ \mathrm{Re}\,\alpha,\ \mathrm{Re}\,s>0;\ \lvert\arg(bx+c)\rvert<\pi \text{ at } 0<x<a]$
6	$(x-a)_+^{\alpha-1}\ln(bx+c)$	$\frac{a^{s+\alpha-2}c}{b}\,\mathrm{B}(\alpha,\,2-s-\alpha)\;{}_3F_2\left(\begin{matrix}1,\,1,\,2-s-\alpha\\2,\,2-s;\,-\frac{c}{ab}\end{matrix}\right)+a^{s+\alpha-1}$ $\times\,\mathrm{B}(\alpha,\,1-s-\alpha)\left[\psi(1-s)-\psi(1-s-\alpha)+\log\frac{ab}{c}+\log c\right]$ $[a,\ \mathrm{Re}\,\alpha>0;\ \mathrm{Re}(s+\alpha)<1;\ \lvert\arg(bx+c)\rvert<\pi \text{ at } x>a]$
7	$\left(a^2-x^2\right)_+^{\alpha-1}\ln(bx+c)$	$\frac{a^{s+2\alpha+1}b^3}{6c^3}\,\mathrm{B}\left(\alpha,\,\frac{s+3}{2}\right)\;{}_3F_2\left(\begin{matrix}1,\,\frac{3}{2},\,\frac{s+3}{2};\,\frac{a^2b^2}{c^2}\\\frac{5}{2},\,\frac{s+2\alpha+3}{2}\end{matrix}\right)$ $-\frac{a^{s+2\alpha}b^2}{4c^2}\,\mathrm{B}\left(\alpha,\,\frac{s+2}{2}\right)\;{}_3F_2\left(\begin{matrix}1,\,1,\,\frac{s+2}{2};\,\frac{a^2b^2}{c^2}\\2,\,\frac{s+2\alpha+2}{2}\end{matrix}\right)$ $+\frac{a^{s+2\alpha-1}b}{2c}\,\mathrm{B}\left(\alpha,\,\frac{s+1}{2}\right)+\frac{a^{s+2\alpha-2}\ln c}{2}\,\mathrm{B}\left(\alpha,\,\frac{s}{2}\right)$ $[a,\ \mathrm{Re}\,\alpha,\ \mathrm{Re}\,s>0;\ \lvert\arg(bx+c)\rvert<\pi \text{ at } 0<x<a]$
8	$\left(x^2-a^2\right)_+^{\alpha-1}\ln(bx+c)$	$\frac{a^{s+2\alpha-5}c^3}{6b^3}\,\mathrm{B}\left(\alpha,\,\frac{5-s-2\alpha}{2}\right)\;{}_3F_2\left(\begin{matrix}1,\,\frac{3}{2},\,\frac{5-s-2\alpha}{2}\\\frac{5}{2},\,\frac{5-s}{2};\,\frac{c^2}{a^2b^2}\end{matrix}\right)$ $-\frac{a^{s+2\alpha-4}c^2}{4b^2}\,\mathrm{B}\left(\alpha,\,\frac{4-s-2\alpha}{2}\right)\;{}_3F_2\left(\begin{matrix}1,\,1,\,\frac{4-s-2\alpha}{2}\\2,\,\frac{4-s}{2};\,\frac{c^2}{a^2b^2}\end{matrix}\right)$ $+\frac{a^{s+2\alpha-3}c}{2b}\,\mathrm{B}\left(\alpha,\,\frac{3-s-2\alpha}{2}\right)$ $+\frac{a^{s+2\alpha-2}}{2}\,\mathrm{B}\left(\alpha,\,\frac{2-s-2\alpha}{2}\right)\left[\frac{1}{2}\psi\left(\frac{2-s}{2}\right)\right.$ $\left.-\frac{1}{2}\psi\left(\frac{2-s-2\alpha}{2}\right)+\log\frac{ab}{c}+\log c\right]$ $[a,\ \mathrm{Re}\,\alpha>0;\ \mathrm{Re}(s+2\alpha)<2;\ \lvert\arg(bx+c)\rvert<\pi \text{ at } x>a]$
9	$(a-x)_+^{\alpha-1}$ $\times\ln[b(a-x)+1]$	$a^{s+\alpha}b\,\mathrm{B}(s,\,\alpha+1)\;{}_3F_2\left(\begin{matrix}1,\,1,\,s+1\\2,\,s+\alpha+1;\,-ab\end{matrix}\right)$ $[a,\ \mathrm{Re}\,s>0;\ \mathrm{Re}\,\alpha>-1]$
10	$\theta(a-x)(bx+1)^{\alpha}$ $\times\ln[c(a-x)+1]$	$\frac{a^{s+1}c}{s(s+1)}\,F_3(-\alpha,\,1,\,s,\,1;\,s+2;\,-ab,\,-ac)$ $[a,\ \mathrm{Re}\,s>0;\ \lvert\arg b\rvert,\ \lvert\arg(1+ac)\rvert<\pi]$

2.5.3. $\ln\dfrac{ax+b}{cx+d}$, $\ln\left|\dfrac{ax+b}{cx+d}\right|$ and algebraic functions

1	$\ln\dfrac{ax+b}{ax+c}$	$\dfrac{\pi a^{-s}}{s}\,(b^s-c^s)\csc(s\pi)$ $[0<\operatorname{Re}s<1;\ \lvert\arg a\rvert,\ \lvert\arg b\rvert,\ \lvert\arg c\rvert<\pi]$
2	$\ln\left\lvert\dfrac{ax+b}{ax-c}\right\rvert$	$\dfrac{\pi a^{-s}}{s}\csc(s\pi)\,[b^s-c^s\cos(s\pi)]$ $[a,\ b,\ c>0;\ 0<\operatorname{Re}s<1]$
3	$\ln\left\lvert\dfrac{x+a}{x-a}\right\rvert$	$\dfrac{\pi a^s}{s}\tan\dfrac{s\pi}{2}$ $[a>0;\ \lvert\operatorname{Re}s\rvert<1;\ s\neq0]$
4	$\theta(a-x)\ln\left[\dfrac{c(a-x)}{b-x}+1\right]$	$\dfrac{a^{s+1}c}{s(s+1)b}F_1\left(1,\ s,\ 1;\ s+2;\ \dfrac{a}{b},\ -\dfrac{ac}{b}\right)$ $[0<a<b;\ \operatorname{Re}s>0;\ \lvert\arg c\rvert<\pi]$
5	$\dfrac{1}{(x+a)^\rho}\ln\left(\dfrac{b}{x+a}+1\right)$	$a^{s-\rho-1}b\,\mathrm{B}(s,\ 1-s+\rho)\ {}_3F_2\left(\begin{matrix}1,\ 1,\ 1-s+\rho\\ 2,\ \rho+1;\ -\frac{b}{a}\end{matrix}\right)$ $[0<\operatorname{Re}s<\operatorname{Re}\rho+1;\ \lvert\arg a\rvert,\ \lvert\arg b\rvert<\pi]$
6	$\dfrac{1}{(x+a)^\rho}\ln\dfrac{x+a+b}{x+a-b}$	$a^{s-\rho-1}b\,\mathrm{B}(s,\ 1-s+\rho)\ {}_4F_3\left(\begin{matrix}\frac{1}{2},\ 1,\ \frac{1-s+\rho}{2},\ \frac{2-s+\rho}{2}\\ \frac{3}{2},\ \frac{\rho+1}{2},\ \frac{\rho+2}{2};\ \frac{b^2}{a^2}\end{matrix}\right)$ $[0<\operatorname{Re}s<\operatorname{Re}\rho+1;\ \lvert\arg a\rvert,\ \lvert\arg b\rvert<\pi]$
7	$\dfrac{1}{(x+a)^\rho}\ln\dfrac{(1+b)x+a}{(1-b)x+a}$	$2a^{s-\rho}b\,\mathrm{B}(s+1,\ \rho-s)\ {}_4F_3\left(\begin{matrix}\frac{1}{2},\ 1,\ \frac{s+1}{2},\ \frac{s+2}{2}\\ \frac{3}{2},\ \frac{\rho+1}{2},\ \frac{\rho+2}{2};\ b^2\end{matrix}\right)$ $[-1<\operatorname{Re}s<\operatorname{Re}\rho;\ \lvert\arg a\rvert,\ \lvert\arg b\rvert<\pi]$
8	$(a-x)_+^{\alpha-1}\ln\dfrac{1+bx}{1-bx}$	$a^{s+\alpha}b\,\mathrm{B}(s+1,\ \alpha)\ {}_4F_3\left(\begin{matrix}\frac{1}{2},\ \frac{1}{2},\ \frac{s+1}{2},\ \frac{s+2}{2};\ a^2b^2\\ \frac{3}{2},\ \frac{s+\alpha+1}{2},\ \frac{s+\alpha+2}{2}\end{matrix}\right)$ $[a,\ \operatorname{Re}\alpha>0;\ \operatorname{Re}s>-1]$
9	$(a-x)_+^{\alpha-1}\ln\dfrac{1+b(a-x)}{1-b(a-x)}$	$a^{s+\alpha}b\,\mathrm{B}(s,\ \alpha+1)\ {}_4F_3\left(\begin{matrix}\frac{1}{2},\ \frac{1}{2},\ \frac{\alpha+1}{2},\ \frac{\alpha+2}{2};\ a^2b^2\\ \frac{3}{2},\ \frac{s+\alpha+1}{2},\ \frac{s+\alpha+2}{2}\end{matrix}\right)$ $[a,\ \operatorname{Re}s>0;\ \operatorname{Re}\alpha>-1]$
10	$(a^2-x^2)_+^{\alpha-1}\ln\dfrac{1+bx}{1-bx}$	$\dfrac{a^{s+2\alpha-1}b}{2}\,\mathrm{B}\left(\dfrac{s+1}{2},\ \alpha\right)\ {}_3F_2\left(\begin{matrix}\frac{1}{2},\ \frac{1}{2},\ \frac{s+1}{2};\ a^2b^2\\ \frac{3}{2},\ \frac{s+2\alpha+1}{2}\end{matrix}\right)$ $[a,\ \operatorname{Re}\alpha>0;\ \operatorname{Re}s>-1]$

2.5.4. $\ln\left(ax^2+bx+c\right)$ and algebraic functions

1	$\ln\left(x^2+1\right)$	$\frac{\pi}{s}\csc\frac{s\pi}{2}$ $[-2<\operatorname{Re}s<0]$
2	$\ln\left[(x-1)^2\right]$	$\frac{2\pi}{s}\cot(s\pi)$ $[-1<\operatorname{Re}s<0]$
3	$\ln\left(x^2+2ax+1\right)$	$\frac{2\pi\cos(s\arccos a)}{s\sin(s\pi)}$ $[-1<a\le 1;\ -1<\operatorname{Re}s<0]$
4	$\frac{\ln\left(x^2+a^2\right)}{x+a}$	$\frac{\pi a^{s-1}}{2}\left\{\frac{2}{s}\csc\frac{s\pi}{2}-\frac{2}{s+1}\sec\frac{s\pi}{2}+\left[\ln\left(4a^4\right)-4\pi\cot(s\pi)\right]\csc(s\pi)+\sec\frac{s\pi}{2}\,\Phi\left(-1,\,1,\,\frac{s+3}{2}\right)-\csc\frac{s\pi}{2}\,\Phi\left(-1,\,1,\,\frac{s+2}{2}\right)\right\}$ $[0<\operatorname{Re}s<1;\ \operatorname{Re}a>0]$
5	$(a-x)_+^{\alpha-1}\ln\left(bx^2+1\right)$	$a^{s+\alpha+1}b\,\mathrm{B}(s+2,\,\alpha)\,{}_4F_3\left(\begin{matrix}1,\,1,\,\frac{s+2}{2},\,\frac{s+3}{2};\,-a^2b\\2,\,\frac{s+\alpha+2}{2},\,\frac{s+\alpha+3}{2}\end{matrix}\right)$ $[a,\ \operatorname{Re}\alpha>0;\ \operatorname{Re}s>-2;\ \lvert\arg b\rvert<\pi]$
6	$(a-x)_+^{\alpha-1}\times\ln\left[b(a-x)^2+1\right]$	$a^{s+\alpha+1}b\,\mathrm{B}(s,\,\alpha+2)\,{}_4F_3\left(\begin{matrix}1,\,1,\,\frac{\alpha+2}{2},\,\frac{\alpha+3}{2};\,-a^2b\\2,\,\frac{s+\alpha+2}{2},\,\frac{s+\alpha+3}{2}\end{matrix}\right)$ $[a,\ \operatorname{Re}s>0;\ \operatorname{Re}\alpha>-2;\ \lvert\arg b\rvert<\pi]$
7	$(a-x)_+^{\alpha-1}\times\ln(bx(a-x)+1)$	$a^{s+\alpha+1}b\,\mathrm{B}(s+1,\,\alpha+1)\,{}_4F_3\left(\begin{matrix}1,\,1,\,s+1,\,\alpha+1;\,-\frac{a^2b}{4}\\2,\,\frac{s+\alpha+2}{2},\,\frac{s+\alpha+3}{2}\end{matrix}\right)$ $\left[a>0;\ \operatorname{Re}s,\ \operatorname{Re}\alpha>-1;\ \left\lvert\arg\left(4+a^2b\right)\right\rvert<\pi\right]$

2.5.5. $\ln\frac{ax^2+bx+c}{dx^2+ex+f}$ and algebraic functions

1	$\ln\frac{x^2+2x\cos\theta+1}{x^2}$	$-2\pi\,\Gamma\left[\begin{matrix}s,\,-s\\\frac{\pi+2\theta s}{2\pi},\,\frac{\pi-2\theta s}{2\pi}\end{matrix}\right]$ $[\lvert\theta\rvert<\pi;\ 0<\operatorname{Re}s<1]$
2	$\ln\frac{(x+a)^2+c^2}{(x+b)^2+c^2}$	$\frac{2\pi}{s\sin(s\pi)}\left[\left(a^2+c^2\right)^{s/2}\cos\left(s\arctan\frac{c}{a}\right)-\left(b^2+c^2\right)^{s/2}\cos\left(s\arctan\frac{c}{b}\right)\right]$ $[a,\,b,\,c>0;\ 0<\operatorname{Re}s<1]$
3	$\ln\frac{x^2+2abx+a^2}{(x+a)^2}$	$\frac{2\pi a^s}{s}\csc(s\pi)\left[\cos(s\arccos b)-1\right]$ $[a>0;\ -1<b\le 1;\ \lvert\operatorname{Re}s\rvert<1]$

No.	$f(x)$	$F(s)$
4	$(a-x)_+^{\alpha-1} \times \ln \dfrac{1+bx(a-x)}{1-bx(a-x)}$	$2a^{s+\alpha+1} b\,\mathrm{B}(s+1,\,\alpha+1)\; {}_6F_5\left(\begin{matrix}\frac{1}{2},\,1,\,\Delta(2,\,\alpha+1),\,\Delta(2,\,s+1)\\ \frac{3}{2},\,\Delta(4,\,s+\alpha+2);\;\frac{a^4b^2}{16}\end{matrix}\right)$ $[a>0;\ \operatorname{Re}s,\ \operatorname{Re}\alpha>-1]$
5	$\dfrac{1}{(x+a)^\rho}\ln\left[\dfrac{b}{(x+a)^2}+1\right]$	$a^{s-\rho-2} b\,\mathrm{B}(s,\,\rho-s+2)\; {}_4F_3\left(\begin{matrix}1,\,1,\,\frac{\rho-s+2}{2},\,\frac{\rho-s+3}{2}\\ 2,\,\frac{\rho+2}{2},\,\frac{\rho+3}{2};\,-\frac{b}{a^2}\end{matrix}\right)$ $[0<\operatorname{Re}s<\operatorname{Re}\rho+2;\ \lvert\arg a\rvert,\ \lvert\arg b\rvert<\pi]$
6	$\dfrac{1}{(x+a)^\rho}\ln\left[\dfrac{bx^2}{(x+a)^2}+1\right]$	$a^{s-\rho} b\,\mathrm{B}(s+2,\,\rho-s)\; {}_4F_3\left(\begin{matrix}1,\,1,\,\frac{s+2}{2},\,\frac{s+3}{2}\\ 2,\,\frac{\rho+2}{2},\,\frac{\rho+3}{2};\,-b\end{matrix}\right)$ $[-2<\operatorname{Re}s<\operatorname{Re}\rho;\ \lvert\arg a\rvert,\ \lvert\arg b\rvert<\pi]$

2.5.6. $\ln(\varphi(x))$ and algebraic functions

1	$\ln\dfrac{\sqrt{x+a}\pm\sqrt{a}}{\sqrt{x}}$	$\pm\dfrac{a^s}{2\sqrt{\pi}s}\Gamma(s)\Gamma\left(\dfrac{1}{2}-s\right)$ $[0<\operatorname{Re}s<1/2;\ \lvert\arg a\rvert<\pi]$
2	$\ln\dfrac{\sqrt{x+a}\pm\sqrt{x}}{\sqrt{a}}$	$\mp\dfrac{a^s}{2\sqrt{\pi}s}\Gamma(-s)\Gamma\left(s+\dfrac{1}{2}\right)$ $[-1/2<\operatorname{Re}s<0;\ \lvert\arg a\rvert<\pi]$
3	$\theta(a-x)\ln\dfrac{\sqrt{a-x}+\sqrt{a}}{2\sqrt{a}}$	$\dfrac{\sqrt{\pi}a^s}{2s}\Gamma\left[\begin{matrix}s\\ \frac{2s+1}{2}\end{matrix}\right]-\dfrac{a^s}{s^2}\left(s\ln 2+\dfrac{1}{2}\right)$ $[a,\ \operatorname{Re}s>0]$
4	$\dfrac{1}{\sqrt{x+a}}\ln\dfrac{\sqrt{x+a}\pm\sqrt{x}}{\sqrt{a}}$	$\pm\dfrac{\pi^{3/2}a^{s-1/2}}{2}\sec(s\pi)\Gamma\left[\begin{matrix}\frac{1-2s}{2}\\ 1-s\end{matrix}\right]$ $[\lvert\operatorname{Re}s\rvert<1/2;\ \lvert\arg a\rvert<\pi]$
5	$\dfrac{1}{\sqrt{x+a}}\ln\dfrac{\sqrt{x+a}\pm\sqrt{a}}{\sqrt{x}}$	$\pm\dfrac{\pi^{3/2}a^{s-1/2}}{2}\csc(s\pi)\,\Gamma\left[\begin{matrix}s\\ \frac{2s+1}{2}\end{matrix}\right]$ $[0<\operatorname{Re}s<1;\ \lvert\arg a\rvert<\pi]$
6	$\dfrac{\ln\left(\sqrt{x+a}\pm\sqrt{a}\right)}{\sqrt{x+a}}$	$2^{2s-1}a^{s-1/2}\,\mathrm{B}(s,\,1-2s)\left[\psi(1-s)-\psi\left(\dfrac{1-2s}{2}\right)+\ln a-\left\{\begin{matrix}0\\ 2\pi\cot(s\pi)\end{matrix}\right\}\right]$ $[0<\operatorname{Re}s<1/2;\ \lvert\arg a\rvert<\pi]$
7	$\dfrac{\ln\left(\sqrt{x+a}\pm\sqrt{x}\right)}{\sqrt{x+a}}$	$2^{-2s}a^{s-1/2}\,\mathrm{B}\left(2s,\,\dfrac{1-2s}{2}\right)[\ln a\pm\pi\tan(s\pi)]$ $\left[\begin{matrix}(0<\operatorname{Re}s<1/2;\ \lvert\arg a\rvert<\pi;\ a\neq1)\ \text{or}\\ (\lvert\operatorname{Re}s\rvert<1/2\ \text{for}\ a=1)\end{matrix}\right]$
8	$\dfrac{1}{(x+a)^\rho}\ln\dfrac{\sqrt{x+a}+b}{\sqrt{x+a}-b}$	$2a^{s-\rho-1/2}b\,\mathrm{B}\left(s,\,\rho-s+\frac{1}{2}\right)\,{}_3F_2\left(\begin{matrix}\frac{1}{2},\,1,\,\rho-s+\frac{1}{2}\\ \frac{3}{2},\,\rho+\frac{1}{2};\,\frac{b^2}{a}\end{matrix}\right)$ $[0<\operatorname{Re}s<\operatorname{Re}\rho+1/2;\ \lvert\arg(1-b^2/a)\rvert<\pi]$

No.	$f(x)$	$F(s)$
9	$\left(\sqrt{x+a}\pm\sqrt{x}\right)^{\rho}$ $\times\ln\left(\sqrt{x+a}\pm\sqrt{x}\right)$	$\mp\frac{2^{-2s}\rho a^{s+\rho/2}}{2s\mp\rho}\,\mathrm{B}\left(2s,\,\frac{\mp\rho-2s}{2}\right)$ $\times\left[\ln a\mp\psi\left(\frac{\mp\rho-2s}{2}\right)\pm\psi\left(\frac{2s\mp\rho+2}{2}\right)+\frac{2}{\rho}\right]$ $\left[0<\mathrm{Re}\,s<\mp\mathrm{Re}\,\rho/2;\ (\mathrm{Re}\,s>-1/2\text{ for }a=1);\ \lvert\arg a\rvert<\pi\right]$
10	$\frac{\left(\sqrt{x+a}\pm\sqrt{x}\right)^{\rho}}{\sqrt{x+a}}$ $\times\ln\left(\sqrt{x+a}\pm\sqrt{x}\right)$	$2^{-2s}a^{s+(\rho-1)/2}\,\mathrm{B}\left(2s,\,\frac{1\mp\rho-2s}{2}\right)$ $\times\left[\ln a\mp\psi\left(\frac{1\mp\rho-2s}{2}\right)\pm\psi\left(\frac{1\mp\rho+2s}{2}\right)\right]$ $\left[0<\mathrm{Re}\,s<1\mp\mathrm{Re}\,\rho/2;\ (\mathrm{Re}\,s>-1\text{ for }a=1);\ \lvert\arg a\rvert<\pi\right]$
11	$\theta(x-a)\ln\frac{\sqrt{x-a}+\sqrt{x}}{\sqrt{a}}$	$-\frac{\sqrt{\pi}a^{s}}{2s}\,\Gamma\left[\begin{matrix}-s\\ \frac{1}{2}-s\end{matrix}\right]$ $[a>0;\ \mathrm{Re}\,s<0]$
12	$\theta(a-x)\ln\frac{\sqrt{a-x}+\sqrt{a}}{\sqrt{x}}$	$\frac{\sqrt{\pi}a^{s}}{2s}\,\Gamma\left[\begin{matrix}s\\ s+\frac{1}{2}\end{matrix}\right]$ $[a,\ \mathrm{Re}\,s>0]$
13	$\theta(x-a)\ln\frac{\sqrt{x-a}+\sqrt{x}}{\sqrt{x}}$	$-\frac{a^{s}}{2s^{2}}-\frac{\sqrt{\pi}\,a^{s}}{2s}\,\Gamma\left[\begin{matrix}-s\\ \frac{1}{2}-s\end{matrix}\right]$ $[a>0;\ \mathrm{Re}\,s<0]$
14	$\theta(a-x)\ln\frac{\sqrt{a}+\sqrt{a-x}}{\sqrt{a}-\sqrt{a-x}}$	$\frac{\sqrt{\pi}a^{s}}{s}\,\Gamma\left[\begin{matrix}s\\ s+\frac{1}{2}\end{matrix}\right]$ $[a,\ \mathrm{Re}\,s>0]$
15	$\theta(x-a)\ln\frac{\sqrt{x}+\sqrt{x-a}}{\sqrt{x}-\sqrt{x-a}}$	$-\frac{\sqrt{\pi}a^{s}}{s}\,\Gamma\left[\begin{matrix}-s\\ \frac{1}{2}-s\end{matrix}\right]$ $[a>0;\ \mathrm{Re}\,s<0]$
16	$(a-x)_{+}^{\alpha-1}\ln\frac{1+b\sqrt{a-x}}{1-b\sqrt{a-x}}$	$a^{s+\alpha-1/2}b\,\mathrm{B}\left(s,\,\frac{2\alpha+1}{2}\right)\,{}_{3}F_{2}\left(\begin{matrix}\frac{1}{2},\ \frac{1}{2},\ \frac{2\alpha+1}{2};\ ab^{2}\\ \frac{3}{2},\ \frac{2s+2\alpha+1}{2}\end{matrix}\right)$ $\left[a,\ \mathrm{Re}\,s>0;\ \mathrm{Re}\,\alpha>-1/2;\ \lvert\arg\left(1-ab^{2}\right)\rvert<\pi\right]$
17	$\theta(a-x)(bx+1)^{\nu}$ $\times\ln\frac{\sqrt{a}+\sqrt{a-x}}{\sqrt{x}}$	$\frac{\sqrt{\pi}\,a^{s}}{2s}\,\Gamma\left[\begin{matrix}s\\ \frac{2s+1}{2}\end{matrix}\right]\,{}_{3}F_{2}\left(\begin{matrix}-\nu,\ s,\ s;\ -ab\\ \frac{2s+1}{2},\ s+1\end{matrix}\right)$ $\left[a,\ \mathrm{Re}\,s>0;\ \lvert\arg(1+ab)\rvert<\pi\right]$
18	$\theta(a-x)(bx+1)^{\alpha}\ln\left(c\sqrt{a-x}\right.$ $\left.+\sqrt{c^{2}(a-x)+1}\right)$	$\frac{\sqrt{\pi}\,a^{s+1/2}c}{2}\,\Gamma\left[\begin{matrix}s\\ \frac{2s+3}{2}\end{matrix}\right]F_{3}\left(-\alpha,\ \frac{1}{2},\ s,\ \frac{1}{2};\ \frac{2s+3}{2};\ -ab,\ -ac^{2}\right)$ $\left[a,\ \mathrm{Re}\,s>0;\ \lvert\arg b\rvert,\ \lvert\arg\left(1+ac^{2}\right)\rvert<\pi\right]$

No.	$f(x)$	$F(s)$
19	$\theta(a-x)(bx+1)^{\alpha} \times \ln \frac{c+\sqrt{a-x}}{c-\sqrt{a-x}}$	$\frac{\sqrt{\pi}\, a^{s+1/2}}{c} \Gamma\left[\begin{matrix} s \\ \frac{2s+3}{2} \end{matrix}\right] F_3\left(-\alpha, \frac{1}{2}, s, 1; \frac{2s+3}{2}; -ab, \frac{a}{c^2}\right)$ $[a,\ \operatorname{Re} s>0;\ \lvert\arg b\rvert,\ \lvert\arg(1-a/c^2)\rvert<\pi]$
20	$\theta(a-x)\ln \frac{\sqrt{b-x}+c\sqrt{a-x}}{\sqrt{b-x}-c\sqrt{a-x}}$	$a^{s+1/2}\sqrt{\frac{\pi}{b}}\, c\,\Gamma\left[\begin{matrix} s \\ \frac{2s+3}{2} \end{matrix}\right] F_1\left(\frac{1}{2}, s, 1; \frac{2s+3}{2}; \frac{a}{b}, \frac{ac^2}{b}\right)$ $[a,\ \operatorname{Re} s>0;\ \lvert\arg(1-a/b)\rvert,\ \lvert\arg(1-ac^2/b)\rvert<\pi]$
21	$\theta(a-x)\ln\left(c\sqrt{\frac{a-x}{b-x}} + \sqrt{\frac{c^2(a-x)}{b-x}+1}\right)$	$\frac{1}{2}\, a^{s+1/2}\sqrt{\frac{\pi}{b}}\, c\,\Gamma\left[\begin{matrix} s \\ \frac{2s+3}{2} \end{matrix}\right] F_1\left(\frac{1}{2}, s, \frac{1}{2}; \frac{2s+3}{2}; \frac{a}{b}, -\frac{ac^2}{b}\right)$ $[a,\ \operatorname{Re} s>0;\ \lvert\arg(1-a/b)\rvert,\ \lvert\arg(1+ac^2/b)\rvert<\pi]$
22	$\theta(a-x)\ln\left(a+\sqrt{a^2-x^2}\right)$	$\frac{a^s}{s}\left[\frac{\sqrt{\pi}}{2}\Gamma\left[\begin{matrix} \frac{s}{2} \\ \frac{s+1}{2} \end{matrix}\right] - \frac{1}{s} + \ln a\right]$ $[a,\ \operatorname{Re} s>0]$
23	$\ln \frac{\sqrt{a^2+x^2}+a}{2a}$	$\frac{a^s}{2\sqrt{\pi}\, s}\Gamma\left(\frac{1-s}{2}\right)\Gamma\left(\frac{s}{2}\right)$ $[\operatorname{Re} a>0;\ -2<\operatorname{Re} s<0]$
24	$\ln \frac{\sqrt{x^2+a^2}+x}{a}$	$-\frac{a^s}{2s}\,\mathrm{B}\left(\frac{s+1}{2}, -\frac{s}{2}\right)$ $[\operatorname{Re} a>0;\ -1<\operatorname{Re} s<0]$
25	$\theta(x-a)\ln \frac{\sqrt{x}+\sqrt{x-a}}{\sqrt{x}-\sqrt{x-a}}$	$\sqrt{\pi}\, a^s\,\Gamma\left[\begin{matrix} -s, -s \\ \frac{1}{2}-s, 1-s \end{matrix}\right]$ $[a>0;\ \operatorname{Re} s<0]$
26	$\theta(a-x)\ln \frac{\sqrt{a^2-x^2}+a}{x}$	$\frac{\sqrt{\pi}\, a^s}{2s}\Gamma\left[\begin{matrix} \frac{s}{2} \\ \frac{s+1}{2} \end{matrix}\right]$ $[a,\ \operatorname{Re} s>0]$
27	$\theta(x-a)\ln \frac{\sqrt{x^2-a^2}+x}{a}$	$-\frac{\sqrt{\pi}\, a^s}{2s}\Gamma\left[\begin{matrix} -\frac{s}{2} \\ \frac{1-s}{2} \end{matrix}\right]$ $[a>0;\ \operatorname{Re} s<0]$
28	$\ln \frac{\sqrt{x^2+a^2}+x}{2x}$	$-\frac{a^s}{2\sqrt{\pi}\, s}\Gamma\left(-\frac{s}{2}\right)\Gamma\left(\frac{s+1}{2}\right)$ $[\operatorname{Re} a>0;\ 0<\operatorname{Re} s<2]$
29	$\frac{\ln\left(\sqrt{x^2+a^2}\pm x\right)}{\sqrt{x^2+a^2}}$	$2^{-s}a^{s-1}\,\mathrm{B}\left(s, \frac{1-s}{2}\right)\left(\ln a \pm \frac{\pi}{2}\tan\frac{s\pi}{2}\right)$ $[\operatorname{Re} a>0;\ 0<\operatorname{Re} s<1]$
30	$\left(\sqrt{x^2+a^2}+x\right)^{\alpha} \times \ln\left(\sqrt{x^2+a^2}+x\right)$	$\frac{2^{-s}a^{s+\alpha}}{s-\alpha}\,\mathrm{B}\left(s, -\frac{s+\alpha}{2}\right)\left[\frac{\alpha}{2}\psi\left(-\frac{s+\alpha}{2}\right) - \frac{\alpha}{2}\psi\left(\frac{s-\alpha+2}{2}\right) - \alpha\ln a - 1\right]$ $\left[\begin{matrix}\operatorname{Re} a>0;\ 0<\operatorname{Re} s<-\operatorname{Re}\alpha \\ (\operatorname{Re} s>-1 \text{ for } a=1)\end{matrix}\right]$

No.	$f(x)$	$F(s)$
31	$(a-x)_+^{\alpha-1}\ln\dfrac{b+\sqrt{x(a-x)}}{b-\sqrt{x(a-x)}}$	$\dfrac{2a^{s+\alpha}}{b}\,\mathrm{B}\left(\dfrac{2\alpha+1}{2},\,\dfrac{2s+1}{2}\right)\,{}_4F_3\left(\begin{matrix}\frac{1}{2},\,1,\,\frac{2\alpha+1}{2},\,\frac{2s+1}{2};\,\frac{a^2}{4b^2}\\ \frac{3}{2},\,\frac{s+\alpha+1}{2},\,\frac{s+\alpha+2}{2}\end{matrix}\right)$ $\left[a>0;\ \mathrm{Re}\,s,\ \mathrm{Re}\,\alpha>-1;\ \left|\arg\left(1-a^2/(4b^2)\right)\right|<\pi\right]$
32	$(a-x)_+^{\alpha-1}\ln\Big[bx(a-x)$ $+\sqrt{b^2x^2(a-x)^2+1}\Big]$	$a^{s+\alpha+1}b\,\mathrm{B}(s+1,\,\alpha+1)$ $\times\,{}_6F_5\left(\begin{matrix}\frac{1}{2},\,\frac{1}{2},\,\Delta(2,\,\alpha+1),\,\Delta(2,\,s+1)\\ \frac{3}{2},\,\Delta(4,\,s+\alpha+2);\,-\frac{a^4b^2}{16}\end{matrix}\right)$ $\left[a>0;\ \mathrm{Re}\,s,\ \mathrm{Re}\,\alpha>-1;\ \left|\arg\left(1+a^4b^2/16\right)\right|<\pi\right]$
33	$(a-x)_+^{\alpha-1}\ln\big(b\sqrt{a-x}$ $+\sqrt{b^2(a-x)+1}\big)$	$a^{\alpha+s-1/2}b\,\mathrm{B}\left(s,\,\dfrac{2\alpha+1}{2}\right)\,{}_3F_2\left(\begin{matrix}\frac{1}{2},\,\frac{1}{2},\,\frac{2\alpha+1}{2};\,-ab^2\\ \frac{3}{2},\,\frac{2s+2\alpha+1}{2}\end{matrix}\right)$ $\left[a,\ \mathrm{Re}\,s>0;\ \mathrm{Re}\,\alpha>-1/2;\ \left|\arg\left(1+ab^2\right)\right|<\pi\right]$
34	$(a-x)_+^{\alpha-1}$ $\times\ln\left(bx+\sqrt{b^2x^2+1}\right)$	$a^{\alpha+s}b\,\mathrm{B}(s+1,\,\alpha)\,{}_4F_3\left(\begin{matrix}\frac{1}{2},\,\frac{1}{2},\,\frac{s+1}{2},\,\frac{s+2}{2};\,-a^2b^2\\ \frac{3}{2},\,\frac{s+\alpha+1}{2},\,\frac{s+\alpha+2}{2}\end{matrix}\right)$ $\left[a,\ \mathrm{Re}\,\alpha>0;\ \mathrm{Re}\,s>-1;\ \left|\arg\left(1+a^2b^2\right)\right|<\pi\right]$
35	$(a-x)_+^{\alpha-1}\ln\big[b(a-x)$ $+\sqrt{b^2(a-x)^2+1}\big]$	$a^{\alpha+s}b\,\mathrm{B}(s,\,\alpha+1)\,{}_4F_3\left(\begin{matrix}\frac{1}{2},\,\frac{1}{2},\,\frac{\alpha+1}{2},\,\frac{\alpha+2}{2};\,-a^2b^2\\ \frac{3}{2},\,\frac{s+\alpha+1}{2},\,\frac{s+\alpha+2}{2}\end{matrix}\right)$ $\left[a,\ \mathrm{Re}\,s>0;\ \mathrm{Re}\,\alpha>-1;\ \left|\arg\left(1+a^2b^2\right)\right|<\pi\right]$
36	$\left(a^2-x^2\right)_+^{\alpha-1}$ $\times\ln\left(bx+\sqrt{b^2x^2+1}\right)$	$\dfrac{a^{2\alpha+s-1}b}{2}\,\mathrm{B}\left(\dfrac{s+1}{2},\,\alpha\right)\,{}_3F_2\left(\begin{matrix}\frac{1}{2},\,\frac{1}{2},\,\frac{s+1}{2};\,-a^2b^2\\ \frac{3}{2},\,\frac{s+2\alpha+1}{2}\end{matrix}\right)$ $\left[a,\ \mathrm{Re}\,\alpha>0;\ \mathrm{Re}\,s>-1;\ \left|\arg\left(1+a^2b^2\right)\right|<\pi\right]$
37	$\theta(x-a)$ $\times\ln\dfrac{cx+\sqrt{x^2+c^2x^2-b^2}}{\sqrt{x^2-b^2}}$	$-\dfrac{a^sc}{s}\,F_2\left(\dfrac{1}{2},\,\dfrac{1}{2},\,-\dfrac{s}{2};\,\dfrac{3}{2},\,\dfrac{2-s}{2};\,-c^2,\,\dfrac{b^2}{a^2}\right)$ $\left[\begin{matrix}a>0;\ \mathrm{Re}\,s<0;\\ \left|\arg\left(1-b^2/a^2\right)\right|,\ \left|\arg\left(1+c^2\right)\right|<\pi\end{matrix}\right]$
38	$\theta(a-x)\ln\dfrac{a+\sqrt{a^2-x^2}}{a-\sqrt{a^2-x^2}}$	$\dfrac{\sqrt{\pi}\,a^s}{s}\,\Gamma\left[\begin{matrix}\frac{s}{2}\\ \frac{s+1}{2}\end{matrix}\right]$ $\quad[a,\ \mathrm{Re}\,s>0]$
39	$\theta(x-a)\ln\dfrac{x+\sqrt{x^2-a^2}}{x-\sqrt{x^2-a^2}}$	$-\dfrac{\sqrt{\pi}}{s}\,a^s\,\Gamma\left[\begin{matrix}-\frac{s}{2}\\ \frac{1-s}{2}\end{matrix}\right]$ $\quad[a>0;\ \mathrm{Re}\,s<0]$
40	$\dfrac{1}{\sqrt{a+x}}\ln\dfrac{x-(b-c)^2x+a}{x-(b+c)^2x+a}$	$\dfrac{4bc}{a^{1/2-s}}\,\mathrm{B}\left(s+1,\,\dfrac{1}{2}-s\right)\,F_4\left(1,\,s+1;\,\dfrac{3}{2},\,\dfrac{3}{2};\,b^2,\,c^2\right)$ $\left[-1<\mathrm{Re}\,s<1/2;\ \left|\arg a\right|,\ \left|\arg b\right|,\ \left|\arg c\right|<\pi\right]$

2.5.7. $\ln(\varphi(x))$ and the exponential function

1	$e^{-ax}\ln x$	$a^{-s}\Gamma(s)[\psi(s)-\ln a]$ $[\operatorname{Re}a,\ \operatorname{Re}s>0]$
2	$\theta(a-x)\,e^{bx}\ln(1+c(a-x))$	$\dfrac{a^{s+1}c}{s(s+1)}\,\Xi_1(1,\,s,\,1;\,s+2;\,-ac,\,ab)$ $[a,\ \operatorname{Re}s>0;\ \lvert\arg(ac+1)\rvert<\pi]$
3	$\theta(a-x)\,e^{bx}\ln\dfrac{\sqrt{a}+\sqrt{a-x}}{\sqrt{x}}$	$\dfrac{a^s\sqrt{\pi}}{2s}\,\Gamma\left[\begin{matrix}s\\ \frac{2s+1}{2}\end{matrix}\right]\,{}_2F_2\left(\begin{matrix}s,\,s;\,ab\\ \frac{2s+1}{2},\,s+1\end{matrix}\right)$ $[a,\ \operatorname{Re}s>0]$
4	$\theta(a-x)\,e^{bx^2}\ln\dfrac{\sqrt{a}+\sqrt{a-x}}{\sqrt{x}}$	$\dfrac{\sqrt{\pi}\,a^s}{2s}\,\Gamma\left[\begin{matrix}s\\ \frac{2s+1}{2}\end{matrix}\right]\,{}_3F_3\left(\begin{matrix}\frac{s}{2},\,\frac{s}{2},\,\frac{s+1}{2};\,a^2b\\ \frac{2s+1}{4},\,\frac{2s+3}{4},\,\frac{s+2}{2}\end{matrix}\right)$ $[a,\ \operatorname{Re}s>0]$
5	$\theta(a-x)\,e^{bx}\ln\dfrac{1+c\sqrt{a-x}}{1-c\sqrt{a-x}}$	$\sqrt{\pi}a^{s+1/2}c\,\Gamma\left[\begin{matrix}s\\ \frac{2s+3}{2}\end{matrix}\right]\Xi_1\left(\frac{1}{2},\,s,\,1;\,s+\frac{3}{2};\,ac^2,\,ab\right)$ $[a,\ \operatorname{Re}s>0;\ \lvert ac^2\rvert<1]$
6	$\dfrac{\ln x}{e^x+1}$	$\Gamma(s)\left\{\left[2^{1-s}\ln 2+\left(1-2^{1-s}\right)\psi(s)\right]\zeta(s)+\left(1-2^{1-s}\right)\zeta'(s)\right\}$ $[\operatorname{Re}s>0]$
7	$\ln(1+e^{-ax})$	$\dfrac{1-2^{-s}}{a^s}\,\Gamma(s)\,\zeta(s+1)$ $[\operatorname{Re}a,\ \operatorname{Re}s>0]$
8	$\ln(1-e^{-ax})$	$-\dfrac{\Gamma(s)}{a^s}\,\zeta(s+1)$ $[\operatorname{Re}a,\ \operatorname{Re}s>0]$

2.5.8. The logarithmic and hyperbolic or trigonometric functions

Notation: $\delta=\left\{\begin{matrix}1\\0\end{matrix}\right\}$.

1	$\ln\tanh(ax)$	$\dfrac{2^{-s}-2}{(2a)^s}\,\Gamma(s)\,\zeta(s+1)$ $[a,\ \operatorname{Re}s>0]$
2	$\theta(1-x)\left\{\begin{matrix}\sin(ax)\\ \cos(ax)\end{matrix}\right\}\ln^n x$	$\dfrac{(-1)^n\,n!a^\delta}{(s+\delta)^{n+1}}\,{}_{n+1}F_{n+2}\left(\begin{matrix}\frac{s+\delta}{2},\,\frac{s+\delta}{2},\,\ldots,\,\frac{s+\delta}{2};\,-\frac{a^2}{4}\\ \delta+\frac{1}{2},\,\frac{s+\delta+2}{2},\,\frac{s+\delta+2}{2},\,\ldots,\,\frac{s+\delta+2}{2}\end{matrix}\right)$ $[a>0;\ \operatorname{Re}s>-\delta]$
3	$\left\{\begin{matrix}\sin(ax)\\ \cos(ax)\end{matrix}\right\}\ln x$	$\dfrac{\Gamma(s)}{a^s}\left\{\begin{matrix}\sin(s\pi/2)\\ \cos(s\pi/2)\end{matrix}\right\}\left[\psi(s)-\ln a\pm\dfrac{\pi}{2}\tan^{\mp1}\dfrac{s\pi}{2}\right]$ $[a>0;\ -(1\pm1)/2<\operatorname{Re}s<1]$

No.	$f(x)$	$F(s)$
4	$e^{-ax}\sin(bx)\ln x$	$\dfrac{\Gamma(s)}{(a^2+b^2)^{s/2}}\sin\left(s\arctan\dfrac{b}{a}\right)\left[\psi(s)-\dfrac{1}{2}\ln(a^2+b^2)+\arctan\dfrac{b}{a}\cot\left(s\arctan\dfrac{b}{a}\right)\right]$ $[\operatorname{Re}a>\lvert\operatorname{Im}b\rvert;\ \operatorname{Re}s>-1]$
5	$e^{-ax}\cos(bx)\ln x$	$\dfrac{\Gamma(s)}{(a^2+b^2)^{s/2}}\cos\left(s\arctan\dfrac{b}{a}\right)\left[\psi(s)-\dfrac{1}{2}\ln(a^2+b^2)-\arctan\dfrac{b}{a}\tan\left(s\arctan\dfrac{b}{a}\right)\right]$ $[\operatorname{Re}a>\lvert\operatorname{Im}b\rvert;\ \operatorname{Re}s>0]$
6	$\theta(a-x)\begin{Bmatrix}\sinh(bx)\\ \cosh(bx)\end{Bmatrix}\times\ln\dfrac{\sqrt{a}+\sqrt{a-x}}{\sqrt{x}}$	$\dfrac{\sqrt{\pi}\,a^{s+\delta}b^{\delta}}{2(s+\delta)}\Gamma\left[\begin{matrix}s+\delta\\ \frac{2s+2\delta+1}{2}\end{matrix}\right]{}_3F_4\left(\begin{matrix}\frac{s+\delta}{2},\ \frac{s+\delta}{2},\ \frac{s+\delta+1}{2};\ \frac{a^2b^2}{4}\\ \frac{2\delta+1}{2},\ \frac{2s+2\delta+1}{4},\ \frac{2s+2\delta+3}{4},\ \frac{s+\delta+2}{2}\end{matrix}\right)$ $[a>0;\ \operatorname{Re}s>-\delta]$
7	$\theta(a-x)\begin{Bmatrix}\sin(bx)\\ \cos(bx)\end{Bmatrix}\times\ln\dfrac{\sqrt{a}+\sqrt{a-x}}{\sqrt{x}}$	$\dfrac{\sqrt{\pi}\,a^{s+\delta}b^{\delta}}{2(s+\delta)}\Gamma\left[\begin{matrix}s+\delta\\ \frac{2s+2\delta+1}{2}\end{matrix}\right]{}_3F_4\left(\begin{matrix}\frac{s+\delta}{2},\ \frac{s+\delta}{2},\ \frac{s+\delta+1}{2};\ -\frac{a^2b^2}{4}\\ \frac{2\delta+1}{2},\ \frac{2s+2\delta+1}{4},\ \frac{2s+2\delta+3}{4},\ \frac{s+\delta+2}{2}\end{matrix}\right)$ $[a>0;\ \operatorname{Re}s>-\delta]$
8	$\theta(a-x)\begin{Bmatrix}\sin(bx)\\ \cos(bx)\end{Bmatrix}\times\ln\dfrac{a+\sqrt{a^2-x^2}}{x}$	$\dfrac{\sqrt{\pi}\,a^{s+\delta}b^{\delta}}{2(s+\delta)}\Gamma\left[\begin{matrix}\frac{s+\delta}{2}\\ \frac{s+\delta+1}{2}\end{matrix}\right]{}_2F_3\left(\begin{matrix}\frac{s+\delta}{2},\ \frac{s+\delta}{2};\ -\frac{a^2b^2}{4}\\ \frac{2\delta+1}{2},\ \frac{s+\delta+1}{2},\ \frac{s+\delta+2}{2}\end{matrix}\right)$ $[a>0;\ \operatorname{Re}s>-\delta]$
9	$\theta(a-x)\times\begin{Bmatrix}\sinh(bx)\sin(bx)\\ \cosh(bx)\cos(bx)\end{Bmatrix}\times\ln\dfrac{\sqrt{a}+\sqrt{a-x}}{\sqrt{x}}$	$\dfrac{\sqrt{\pi}\,a^{s+2\delta}b^{2\delta}}{2(s+2\delta)}\Gamma\left[\begin{matrix}s+2\delta\\ \frac{2s+4\delta+1}{2}\end{matrix}\right]\times{}_5F_8\left(\begin{matrix}\frac{s+2\delta}{4},\ \Delta(4,\ s+2\delta);\ -\frac{a^4b^4}{64}\\ \frac{2\delta+1}{4},\ \frac{2\delta+3}{4},\ \frac{2\delta+1}{2},\ \Delta\left(4,\ \frac{2s+4\delta+1}{2}\right),\ \frac{s+2\delta+4}{4}\end{matrix}\right)$ $[a>0;\ \operatorname{Re}s>-2\delta-1]$
10	$\theta(a-x)\times\begin{Bmatrix}\sinh(bx)\sin(bx)\\ \cosh(bx)\cos(bx)\end{Bmatrix}\times\ln\dfrac{a^2+\sqrt{a^4-x^4}}{x^2}$	$\dfrac{\sqrt{\pi}\,a^{s+2\delta}b^{2\delta}}{2(s+2\delta)}\Gamma\left[\begin{matrix}\frac{s+2\delta}{4}\\ \frac{s+2\delta+2}{4}\end{matrix}\right]\times{}_2F_5\left(\begin{matrix}\frac{s+2\delta}{4},\ \frac{s+2\delta}{4};\ -\frac{a^4b^4}{64}\\ \frac{2\delta+1}{4},\ \frac{2\delta+3}{4},\ \frac{2\delta+1}{2},\ \frac{s+2\delta+2}{4},\ \frac{s+2\delta+4}{4}\end{matrix}\right)$ $[a>0;\ \operatorname{Re}s>-2\delta]$

No.	$f(x)$	$F(s)$
11	$\theta(a-x) \times \begin{Bmatrix} \sinh(bx)\cos(bx) \\ \cosh(bx)\sin(bx) \end{Bmatrix} \times \ln\dfrac{\sqrt{a}+\sqrt{a-x}}{\sqrt{x}}$	$\dfrac{\sqrt{\pi}\,a^{s+1}b}{2(s+1)}\Gamma\begin{bmatrix} s+1 \\ \frac{2s+3}{2} \end{bmatrix} {}_5F_8\left(\begin{matrix} \frac{s+1}{4}, \frac{s+1}{4}, \frac{s+2}{4}, \frac{s+3}{4}, \frac{s+4}{4}; -\frac{a^4b^4}{64} \\ \frac{1}{2}, \frac{3}{4}, \frac{5}{4}, \frac{2s+3}{8}, \frac{2s+5}{8}, \frac{2s+7}{8}, \frac{2s+9}{8}, \frac{s+5}{4} \end{matrix}\right) \mp \dfrac{\sqrt{\pi}\,a^{s+3}b^3}{6(s+3)}\Gamma\begin{bmatrix} s+3 \\ \frac{2s+7}{2} \end{bmatrix} \times {}_5F_8\left(\begin{matrix} \frac{s+3}{4}, \frac{s+3}{4}, \frac{s+4}{4}, \frac{s+5}{4}, \frac{s+6}{4}; -\frac{a^4b^4}{64} \\ \frac{5}{4}, \frac{3}{2}, \frac{7}{4}, \frac{2s+7}{8}, \frac{2s+9}{8}, \frac{2s+11}{8}, \frac{2s+13}{8}, \frac{s+7}{4} \end{matrix}\right)$ $[a>0;\ \operatorname{Re} s > -1]$
12	$e^{-x}\sin(a\ln x)$	$-i\,\Gamma(s+ia)\sinh\ln\dfrac{\Gamma(s+ia)}{\lvert\Gamma(s+ia)\rvert}$ $[\operatorname{Re} s > \lvert\operatorname{Im} a\rvert]$
13	$e^{-x}\cos(a\ln x)$	$\Gamma(s+ia)\cosh\ln\dfrac{\Gamma(s+ia)}{\lvert\Gamma(s+ia)\rvert}$ $[\operatorname{Re} s > \lvert\operatorname{Im} a\rvert]$
14	$\theta(1-x)\begin{Bmatrix} \sin(a\ln x) \\ \cos(a\ln x) \end{Bmatrix}$	$\mp\dfrac{1}{s^2+a^2}\begin{Bmatrix} a \\ s \end{Bmatrix}$ $[\operatorname{Re} s > 0]$
15	$\theta(a-x)\sin\left(b\ln\dfrac{x}{a}\right)$	$-\dfrac{a^s b}{s^2+b^2}$ $[a>0;\ \operatorname{Re} s > \lvert\operatorname{Im} b\rvert]$
16	$\theta(x-a)\sin\left(b\ln\dfrac{x}{a}\right)$	$\dfrac{a^s b}{s^2+b^2}$ $[a>0;\ \operatorname{Re} s < -\lvert\operatorname{Im} b\rvert]$

2.5.9. Products of logarithms

1	$\ln x\ln(x^2+1)$	$-\dfrac{\pi}{2s^2}\csc\dfrac{s\pi}{2}\left(\pi s\cot\dfrac{s\pi}{2}+2\right)$ $[-2 < \operatorname{Re} s < 0]$
2	$\ln^2 x\ln(x^2+1)$	$\dfrac{\pi}{8s^3}\csc^3\dfrac{s\pi}{2}\left[3\pi^2s^2+\left(\pi^2s^2-8\right)\cos(s\pi)+4\pi s\sin(s\pi)+8\right]$ $[-2 < \operatorname{Re} s < 0]$
3	$\theta(a-x)\ln^2(a-x)$	$\dfrac{a^s}{s}\left\{[\psi(s+1)-\ln a+\mathbf{C}]^2-\psi'(s+1)+\dfrac{\pi^2}{6}\right\}$ $[a,\ \operatorname{Re} s > 0;\ \operatorname{Re} s > -2 \text{ for } a=1]$
4	$\theta(x-a)\ln^2(x-a)$	$-\dfrac{a^s}{s}\left\{[\psi(-s)-\ln a+\mathbf{C}]^2+\psi'(-s)+\dfrac{\pi^2}{6}\right\}$ $[a>0;\ \operatorname{Re} s < 0]$
5	$\theta(a-x)\ln^n(a-x)$	$a^s\dfrac{\partial^n}{\partial\beta^n}\left[a^\beta\,\mathrm{B}(\beta+1,\ s)\right]\Big\rvert_{\beta=0}$ $[a,\ \operatorname{Re} s > 0;\ \operatorname{Re} s > -n \text{ for } a=1]$

No.	$f(x)$	$F(s)$
6	$(a-x)_+^{\alpha-1}\ln^n(a-x)$	$a^{s-1}\dfrac{\partial^n}{\partial\alpha^n}\left[a^\alpha\,\mathrm{B}(\alpha,\,s)\right]$ $[a,\ \mathrm{Re}\,\alpha,\ \mathrm{Re}\,s>0;\ \mathrm{Re}\,s>-n\text{ for }a=1]$
7	$(x-a)_+^{\alpha-1}\ln^n(x-a)$	$a^{s-1}\dfrac{\partial^n}{\partial\alpha^n}\left[a^\alpha\,\mathrm{B}(\alpha,\,1-s-\alpha)\right]$ $[a,\ \mathrm{Re}\,\alpha>0;\ \mathrm{Re}\,s<1-\mathrm{Re}\,\alpha]$
8	$\dfrac{\theta(a-x)}{(bx+c)^\rho}\ln^n(a-x)$	$\dfrac{a^s}{c^\rho}\dfrac{\partial^n}{\partial\beta^n}\left[a^\beta\,\mathrm{B}(\beta+1,\,s)\;{}_2F_1\left(\begin{matrix}\rho,\,s;\,-\frac{ab}{c}\\ s+\beta+1\end{matrix}\right)\right]\Bigg\vert_{\beta=0}$ $[a,\ \mathrm{Re}\,s>0;\ \vert\arg(bx+c)\vert<\pi\text{ for }0\le x\le a]$
9	$\dfrac{\theta(x-a)}{(bx+c)^\rho}\ln^n(x-a)$	$\dfrac{a^{s-\rho}}{b^\rho}\dfrac{\partial^n}{\partial\beta^n}\left[a^\beta\,\mathrm{B}(\beta+1,\,\rho-s-\beta)\;{}_2F_1\left(\begin{matrix}\rho,\,\rho-s-\beta\\ 1-s+\rho;\,-\frac{c}{ab}\end{matrix}\right)\right]\Bigg\vert_{\beta=0}$ $[a>0;\ \mathrm{Re}\,s<\mathrm{Re}\,\rho;\ \vert\arg(bx+c)\vert<\pi\text{ for }a\le x<\infty]$
10	$(a-x)_+^{\alpha-1}\ln^n(bx+c)$	$a^{s+\alpha-1}\,\mathrm{B}(\alpha,\,s)\dfrac{\partial^n}{\partial\beta^n}\left[c^\beta\;{}_2F_1\left(\begin{matrix}-\beta,\,s;\,-\frac{ab}{c}\\ s+\alpha\end{matrix}\right)\right]\Bigg\vert_{\beta=0}$ $[a,\ \mathrm{Re}\,\alpha,\ \mathrm{Re}\,s>0;\ \vert\arg(bx+c)\vert<\pi\text{ for }0\le x\le a]$
11	$\dfrac{\theta(x-a)}{(bx+c)^\rho}\ln^n(bx+c)$	$(-1)^n\dfrac{\partial^n}{\partial\rho^n}\left[\dfrac{a^{s-\rho}b^{-\rho}}{\rho-s}\;{}_2F_1\left(\begin{matrix}\rho,\,\rho-s;\,-\frac{c}{ab}\\ 1-s+\rho\end{matrix}\right)\right]$ $[a>0;\ \mathrm{Re}\,s<\mathrm{Re}\,\rho;\ \vert\arg(bx+c)\vert<\pi\text{ for }a\le x<\infty]$
12	$\dfrac{1}{(bx+c)^\rho}\ln^n(bx+c)$	$(-1)^n\left(\dfrac{c}{b}\right)^s\dfrac{\partial^n}{\partial\rho^n}\left[c^{-\rho}\,\mathrm{B}(\rho-s,\,s)\right]$ $\left[\begin{matrix}a>0;\ 0<\mathrm{Re}\,s<\mathrm{Re}\,\rho;\\ \vert\arg(bx+c)\vert<\pi\text{ for }0\le x<\infty\end{matrix}\right]$
13	$\dfrac{\theta(a-x)\ln^\alpha(a/x)}{b^2x^2-2abx\cos\theta+a^2}$	$\dfrac{ia^{s-2}\Gamma(\alpha+1)}{2b\sin\theta}\left[\Phi\left(be^{-i\theta},\,\alpha+1,\,s-1\right)-\Phi\left(be^{i\theta},\,\alpha+1,\,s-1\right)\right]$ $[a,\,b,\ \mathrm{Re}\,\alpha>0;\ 0<\mathrm{Re}\,s<2;\ \vert\theta\vert<\pi]$
14	$\theta(a-x)\ln\dfrac{\sqrt{a-x}+\sqrt{a}}{\sqrt{x}}$ $\times\ln(bx+1)$	$\dfrac{\sqrt{\pi}\,a^{s+1}b\,\Gamma(s+1)}{2s(s+1)\,\Gamma\left(\frac{2s+3}{2}\right)}\left[(s+1)\;{}_3F_2\left(\begin{matrix}1,\,1,\,s+1\\ 2,\,\frac{2s+3}{2};\,-ab\end{matrix}\right)\right.$ $\left.-\;{}_3F_2\left(\begin{matrix}1,\,s+1,\,s+1\\ \frac{2s+3}{2},\,s+2;\,-ab\end{matrix}\right)\right]$ $[a>0;\ \mathrm{Re}\,s>-1;\ \vert\arg b\vert<\pi]$
15	$\theta(a-x)\ln\dfrac{\sqrt{a-x}+\sqrt{a}}{\sqrt{x}}$ $\times\ln\dfrac{1+bx}{1-bx}$	$\dfrac{\sqrt{\pi}\,a^{s+1}b\,\Gamma(s+1)}{s(s+1)\,\Gamma\left(\frac{2s+3}{2}\right)}\left[(s+1)\;{}_4F_3\left(\begin{matrix}\frac{1}{2},\,1,\,\frac{s+1}{2},\,\frac{s+2}{2};\,a^2b^2\\ \frac{3}{2},\,\frac{2s+3}{4},\,\frac{2s+5}{4}\end{matrix}\right)\right.$ $\left.-\;{}_4F_3\left(\begin{matrix}1,\,\frac{s+1}{2},\,\frac{s+1}{2},\,\frac{s+2}{2};\,a^2b^2\\ \frac{2s+3}{4},\,\frac{2s+5}{4},\,\frac{s+3}{2}\end{matrix}\right)\right]$ $[a>0;\ \mathrm{Re}\,s>-1;\ \vert\arg\left(1-a^2b^2\right)\vert<\pi]$

No.	$f(x)$	$F(s)$
16	$\theta(a-x)\ln\frac{\sqrt{a-x}+\sqrt{a}}{\sqrt{x}}$ $\times\ln\left(bx+\sqrt{b^2x^2+1}\right)$	$\frac{\sqrt{\pi}\,a^{s+1}b\,\Gamma(s+1)}{2(s+1)\,\Gamma\left(\frac{2s+3}{2}\right)}\,{}_5F_4\left(\begin{matrix}\frac{1}{2},\ \frac{1}{2},\ \frac{s+1}{2},\ \frac{s+1}{2},\ \frac{s+2}{2};\ -a^2b^2\\ 1,\ \frac{2s+3}{4},\ \frac{2s+5}{4},\ \frac{s+3}{2}\end{matrix}\right)$ $\left[a>0;\ \mathrm{Re}\,s>-1;\ \lvert\arg\left(1+a^2b^2\right)\rvert<\pi\right]$
17	$\frac{\theta(a-x)}{\sqrt{b^2x^2+1}}\ln\frac{\sqrt{a-x}+\sqrt{a}}{\sqrt{x}}$ $\times\ln\left(bx+\sqrt{b^2x^2+1}\right)$	$\frac{\sqrt{\pi}\,a^{s+1}b\,\Gamma(s+1)}{2(s+1)\,\Gamma\left(\frac{2s+3}{2}\right)}\,{}_5F_4\left(\begin{matrix}1,\ 1,\ \frac{s+1}{2},\ \frac{s+1}{2},\ \frac{s+2}{2};\ -a^2b^2\\ \frac{3}{2},\ \frac{2s+3}{4},\ \frac{2s+5}{4},\ \frac{s+3}{2}\end{matrix}\right)$ $\left[a>0;\ \mathrm{Re}\,s>-1;\ \lvert\arg\left(1+a^2b^2\right)\rvert<\pi\right]$
18	$\theta(a-x)\ln\frac{\sqrt{a-x}+\sqrt{a}}{\sqrt{x}}$ $\times\ln\left(bx^2+1\right)$	$\frac{\sqrt{\pi}\,a^{s+2}b\,\Gamma(s+2)}{2s(s+2)\,\Gamma\left(\frac{2s+5}{2}\right)}\left[(s+2)\,{}_4F_3\left(\begin{matrix}1,\ 1,\ \frac{s+2}{2},\ \frac{s+3}{2};\ -a^2b\\ 2,\ \frac{2s+5}{4},\ \frac{2s+7}{4}\end{matrix}\right)\right.$ $\left.-\,2\,{}_4F_3\left(\begin{matrix}1,\ \frac{s+2}{2},\ \frac{s+2}{2},\ \frac{s+3}{2};\ -a^2b\\ \frac{2s+5}{4},\ \frac{2s+7}{4},\ \frac{s+4}{2}\end{matrix}\right)\right]$ $\left[a>0;\ \mathrm{Re}\,s>-2;\ \lvert\arg\left(1-a^2b^2\right)\rvert<\pi\right]$
19	$(a-x)_+^{\alpha-1}\ln^2\left[bx(a-x)\right.$ $\left.+\sqrt{b^2x^2(a-x)^2+1}\right]$	$a^{s+\alpha+3}b^2\,\mathrm{B}(s+2,\ \alpha+2)$ $\times\,{}_7F_6\left(\begin{matrix}1,\ 1,\ 1,\ \Delta(2,\ \alpha+2),\ \Delta(2,\ s+2)\\ \frac{3}{2},\ 2,\ \Delta(4,\ s+\alpha+4);\ -\frac{a^4b^2}{16}\end{matrix}\right)$ $\left[\mathrm{Re}\,\alpha,\ \mathrm{Re}\,s>-2;\ \lvert\arg\left(16+a^4b^2\right)\rvert<\pi\right]$
20	$(a-x)_+^{\alpha-1}\ln^2\left[b(a-x)\right.$ $\left.+\sqrt{b^2(a-x)^2+1}\right]$	$a^{s+\alpha+1}b^2\,\mathrm{B}(s,\ \alpha+2)\ {}_4F_3\left(\begin{matrix}1,\ 1,\ \frac{\alpha+2}{2},\ \frac{\alpha+3}{2};\ -a^2b^2\\ \frac{3}{2},\ \frac{s+\alpha+2}{2},\ \frac{s+\alpha+3}{2}\end{matrix}\right)$ $\left[a,\ \mathrm{Re}\,s>0;\ \mathrm{Re}\,\alpha>-2;\ \lvert\arg\left(1+a^2b^2\right)\rvert<\pi\right]$
21	$(a-x)_+^{\alpha-1}\ln^2\left[b\sqrt{a-x}\right.$ $\left.+\sqrt{b^2(a-x)+1}\right]$	$a^{s+\alpha}b^2\,\mathrm{B}(s,\ \alpha+1)\ {}_4F_3\left(\begin{matrix}1,\ 1,\ 1,\ \alpha+1;\ -ab^2\\ \frac{3}{2},\ 2,\ s+\alpha+1\end{matrix}\right)$ $\left[a,\ \mathrm{Re}\,s>0;\ \mathrm{Re}\,\alpha>-1;\ \lvert\arg\left(1+ab^2\right)\rvert<\pi\right]$
22	$(a-x)_+^{\alpha-1}\ln^2\left[b\sqrt{x(a-x)}\right.$ $\left.+\sqrt{1+b^2x(a-x)}\right]$	$a^{s+\alpha+1}b^2\,\mathrm{B}(s+1,\ \alpha+1)\ {}_5F_4\left(\begin{matrix}1,\ 1,\ 1,\ \alpha+1,\ s+1;\ -\frac{a^2b^2}{4}\\ \frac{3}{2},\ 2,\ \frac{s+\alpha+2}{2},\ \frac{s+\alpha+3}{2}\end{matrix}\right)$ $\left[a>0;\ \mathrm{Re}\,\alpha,\ \mathrm{Re}\,s>-1;\ \lvert\arg\left(4+a^2b^2\right)\rvert<\pi\right]$
23	$(a-x)_+^{\alpha-1}$ $\times\ln^2\left(bx+\sqrt{b^2x^2+1}\right)$	$a^{s+\alpha+1}b^2\,\mathrm{B}(s+2,\ \alpha)\ {}_4F_3\left(\begin{matrix}1,\ 1,\ \frac{s+2}{2},\ \frac{s+3}{2};\ -a^2b^2\\ \frac{3}{2},\ \frac{s+\alpha+2}{2},\ \frac{s+\alpha+3}{2}\end{matrix}\right)$ $\left[a,\ \mathrm{Re}\,\alpha>0;\ \mathrm{Re}\,s>-2;\ \lvert\arg\left(1+a^2b^2\right)\rvert<\pi\right]$
24	$\left(a^2-x^2\right)_+^{\alpha-1}$ $\times\ln^2\left(bx+\sqrt{b^2x^2+1}\right)$	$\frac{a^{s+2\alpha}b^2}{2}\ \mathrm{B}\left(\frac{s+2}{2},\ \alpha\right)\ {}_4F_3\left(\begin{matrix}1,\ 1,\ 1,\ \frac{s+2}{2};\ -a^2b^2\\ \frac{3}{2},\ 2,\ \frac{s+2\alpha+2}{2}\end{matrix}\right)$ $\left[a,\ \mathrm{Re}\,\alpha>0;\ \mathrm{Re}\,s>-2;\ \lvert\arg\left(1+a^2b^2\right)\rvert<\pi\right]$

No.	$f(x)$	$F(s)$
25	$\theta(a-x)\ln\frac{a+\sqrt{a^2-x^2}}{x}\times\ln\frac{b+x}{b-x}$	$\frac{2\sqrt{\pi}\,a^{s+1}}{bs^2(s+1)}\Gamma\left[\begin{matrix}\frac{s+1}{2}\\ \frac{s}{2}\end{matrix}\right]\left[(s+1)\,{}_3F_2\left(\begin{matrix}\frac{1}{2},\,1,\,\frac{s+1}{2}\\ \frac{3}{2},\,\frac{s+2}{2};\,\frac{a^2}{b^2}\end{matrix}\right)-{}_3F_2\left(\begin{matrix}1,\,\frac{s+1}{2},\,\frac{s+1}{2}\\ \frac{s+2}{2},\,\frac{s+3}{2};\,\frac{a^2}{b^2}\end{matrix}\right)\right]$ $\left[a>0;\ \operatorname{Re}s>-1;\ \lvert\arg(1-a^2/b^2)\rvert<\pi\right]$
26	$\theta(a-x)\ln\frac{a+\sqrt{a^2-x^2}}{x}\times\ln\left(bx+\sqrt{b^2x^2+1}\right)$	$\frac{\sqrt{\pi}\,a^{s+1}b}{s^2(s+1)}\Gamma\left[\begin{matrix}\frac{s+1}{2}\\ \frac{s}{2}\end{matrix}\right]\left[(s+1)\,{}_3F_2\left(\begin{matrix}\frac{1}{2},\,\frac{1}{2},\,\frac{s+1}{2}\\ \frac{3}{2},\,\frac{s+2}{2};\,-a^2b^2\end{matrix}\right)-{}_3F_2\left(\begin{matrix}\frac{1}{2},\,\frac{s+1}{2},\,\frac{s+1}{2}\\ \frac{s+2}{2},\,\frac{s+3}{2};\,-a^2b^2\end{matrix}\right)\right]$ $\left[a>0;\ \operatorname{Re}s>-1;\ \lvert\arg(1+a^2b^2)\rvert<\pi\right]$
27	$\frac{\theta(a-x)}{\sqrt{b^2x^2+1}}\ln\frac{a+\sqrt{a^2-x^2}}{x}\times\ln\left(bx+\sqrt{b^2x^2+1}\right)$	$\frac{\sqrt{\pi}\,a^{s+1}b}{2(s+1)}\Gamma\left[\begin{matrix}\frac{s+1}{2}\\ \frac{s+2}{2}\end{matrix}\right]{}_4F_3\left(\begin{matrix}1,\,1,\,\frac{s+1}{2},\,\frac{s+1}{2}\\ \frac{3}{2},\,\frac{s+2}{2},\,\frac{s+3}{2};\,-a^2b^2\end{matrix}\right)$ $\left[a>0;\ \operatorname{Re}s>-1;\ \lvert\arg\left(1+a^2b^2\right)\rvert<\pi\right]$
28	$\theta(a-x)\ln\frac{a+\sqrt{a^2-x^2}}{a-\sqrt{a^2-x^2}}\times\ln\left(bx+\sqrt{b^2x^2+1}\right)$	$\frac{4\sqrt{\pi}\,a^{s+1}b}{s^2(s+1)^2}\Gamma\left[\begin{matrix}\frac{s+3}{2}\\ \frac{s}{2}\end{matrix}\right]\left[(s+1)\,{}_3F_2\left(\begin{matrix}\frac{1}{2},\,\frac{1}{2},\,\frac{s+1}{2}\\ \frac{3}{2},\,\frac{s+2}{2};\,-a^2b^2\end{matrix}\right)-{}_3F_2\left(\begin{matrix}\frac{1}{2},\,\frac{s+1}{2},\,\frac{s+1}{2}\\ \frac{s+2}{2},\,\frac{s+3}{2};\,-a^2b^2\end{matrix}\right)\right]$ $\left[a>0;\ \operatorname{Re}s>-1;\ \lvert\arg(1+a^2b^2)\rvert<\pi\right]$
29	$\theta(1-x)\,e^{ax}\ln^n x$	$\frac{(-1)^n\,n!}{s^{n+1}}\,{}_{n+1}F_{n+1}\left(\begin{matrix}s,\,s,\,\ldots,\,s;\,a\\ s+1,\,s+1,\,\ldots,\,s+1\end{matrix}\right)$ $[\operatorname{Re}s>0]$
30	$e^{-ax^\alpha}\ln^n x$	$\frac{1}{\alpha}\left(\frac{\partial}{\partial s}\right)^n\left[a^{-s/\alpha}\,\Gamma\left(\frac{s}{\alpha}\right)\right]$ $[\alpha,\ \operatorname{Re}a,\ \operatorname{Re}s>0]$
31	$e^{-ax}\ln^2 x$	$\frac{\Gamma(s)}{a^s}\left\{[\psi(s)-\ln a]^2+\psi'(s)\right\}$ $[\operatorname{Re}a,\ \operatorname{Re}s>0]$
32	$e^{-ax}\ln^3 x$	$\frac{\Gamma(s)}{a^s}\left\{[\psi(s)-\ln a]^3+3[\psi(s)-\ln a]\,\psi'(s)+\psi''(s)\right\}$ $[\operatorname{Re}a,\ \operatorname{Re}s>0]$
33	$\frac{e^{-ax}}{(bx+c)^\rho}\ln^n(bx+c)$	$(-1)^n\left(\frac{c}{b}\right)^s\Gamma(s)\,\frac{\partial^n}{\partial\rho^n}\left[c^{-\rho}\,\Psi\left(s,\,s-\rho+1;\,\frac{ac}{b}\right)\right]$ $\left[(\operatorname{Re}a,\ \operatorname{Re}s>0)\ \text{or}\ (\operatorname{Re}s>-n\ \text{for}\ c=1);\ \lvert\arg(bx+c)\rvert<\pi\ \text{for}\ x\ge 0\right]$
34	$e^{-ax^2-bx}\ln^n x$	$\frac{\partial^n}{\partial s^n}\left[(4a)^{-s/2}\,\Gamma(s)\,\Psi\left(\frac{s}{2},\,\frac{1}{2};\,\frac{b^2}{4a}\right)\right]$ $\left[\operatorname{Re}a,\ \operatorname{Re}s>0\ \text{or}\ (\operatorname{Re}a=0;\ \operatorname{Re}b,\ \operatorname{Re}s>0)\ \text{or}\ (\operatorname{Re}a=\operatorname{Re}b=0;\ \operatorname{Im}a\neq 0;\ 0<\operatorname{Re}s<2)\right]$

No.	$f(x)$	$F(s)$
35	$e^{-ax-b/x}\ln^n x$	$2\frac{\partial^n}{\partial s^n}\left[\left(\frac{b}{a}\right)^{s/2}K_s\left(2\sqrt{ab}\right)\right]$ $[\operatorname{Re}a,\ \operatorname{Re}b>0]$
36	$\theta(1-x)\begin{Bmatrix}\sinh(ax)\\ \cosh(ax)\end{Bmatrix}\ln^n x$	$\frac{(-1)^n n!a^\delta}{(s+\delta)^{n+1}}\,{}_{n+1}F_{n+2}\left(\begin{matrix}\frac{s+\delta}{2},\ \frac{s+\delta}{2},\ldots,\ \frac{s+\delta}{2};\ \frac{a^2}{4}\\ \frac{2\delta+1}{2},\ \frac{s+\delta+2}{2},\ \frac{s+\delta+2}{2},\ldots,\ \frac{s+\delta+2}{2}\end{matrix}\right)$ $[a>0;\ \operatorname{Re}s>-\delta]$
37	$\begin{Bmatrix}\sin(ax)\\ \cos(ax)\end{Bmatrix}\ln^2 x$	$\frac{\Gamma(s)}{a^s}\begin{Bmatrix}\sin(s\pi/2)\\ \cos(s\pi/2)\end{Bmatrix}\left[\left(\psi(s)-\ln a\pm\frac{\pi}{2}\tan^{\mp1}\frac{s\pi}{2}\right)^2+\psi'(s)-\frac{\pi^2}{4}\begin{Bmatrix}\csc(s\pi/2)\\ \sec(s\pi/2)\end{Bmatrix}^2\right]$ $[a>0;\ -(1\pm1)/2<\operatorname{Re}s<1]$
38	$\begin{Bmatrix}\sin(ax)\\ \cos(ax)\end{Bmatrix}\ln^n x$	$\frac{\partial^n}{\partial s^n}\left[\frac{\Gamma(s)}{a^s}\begin{Bmatrix}\sin(s\pi/2)\\ \cos(s\pi/2)\end{Bmatrix}\right]$ $[a>0;\ -(1\pm1)/2<\operatorname{Re}s<1]$
39	$\theta(a-x)\ln\frac{\sqrt{a}+\sqrt{a-x}}{\sqrt{x}}\times\ln^2\left(bx+\sqrt{b^2x^2+1}\right)$	$\frac{\sqrt{\pi}\,a^{s+2}b^2}{2s(s+2)}\Gamma\left[\begin{matrix}s+2\\ \frac{2s+5}{2}\end{matrix}\right]\left[(s+2)\,{}_5F_4\left(\begin{matrix}1,\ 1,\ 1,\ \frac{s+2}{2},\ \frac{s+3}{2};\ -a^2b^2\\ \frac{3}{2},\ 2,\ \frac{2s+5}{4},\ \frac{2s+7}{4}\end{matrix}\right)-2\,{}_5F_4\left(\begin{matrix}1,\ 1,\ \frac{s+2}{2},\ \frac{s+2}{2},\ \frac{s+3}{2};\ -a^2b^2\\ \frac{3}{2},\ \frac{2s+5}{4},\ \frac{2s+7}{4},\ \frac{s+4}{2}\end{matrix}\right)\right]$ $[a>0;\ \operatorname{Re}s>-2]$
40	$\theta(a-x)\ln\frac{a+\sqrt{a^2-x^2}}{x}\times\ln^2\left(bx+\sqrt{b^2x^2+1}\right)$	$\frac{\sqrt{\pi}\,a^{s+2}b^2}{2(s+2)}\Gamma\left[\begin{matrix}\frac{s+2}{2}\\ \frac{s+3}{2}\end{matrix}\right]{}_5F_4\left(\begin{matrix}1,\ 1,\ 1,\ \frac{s+2}{2},\ \frac{s+2}{2}\\ \frac{3}{2},\ 2,\ \frac{s+3}{2},\ \frac{s+4}{2};\ -a^2b^2\end{matrix}\right)$ $[a>0;\ \operatorname{Re}s>-2]$
41	$\frac{\theta(1-x)\ln^n x}{\ln^2 x+a^2}$	$\frac{1}{a}\frac{\partial^n}{\partial s^n}\left[\sin(as)\operatorname{ci}(as)-\cos(as)\operatorname{si}(as)\right]$ $[a,\ \operatorname{Re}s>0]$
42	$\frac{\theta(1-x)\ln x}{\ln^2 x+a^2}$	$\sin(as)\operatorname{si}(as)+\cos(as)\operatorname{ci}(as)$ $[a,\ \operatorname{Re}s>0]$
43	$\frac{\theta(1-x)}{(\ln x-a)^n}$	$\frac{1}{(n-1)!}\left[s^{n-1}e^{as}\operatorname{Ei}(-as)-\sum_{k=1}^{n-1}(n-k-1)!\frac{s^{k-1}}{(-a)^{n-k}}\right]$ $[a,\ \operatorname{Re}s>0]$
44	$\frac{\theta(1-x)}{\ln x\left[\ln^2(-\ln x)+\pi^2\right]}$	$\nu(s)-e^s$ $[\operatorname{Re}s>0]$
45	$\frac{\theta(1-x)\ln(-\ln x)}{\sqrt{-\ln x}\left[\ln^2(-\ln x)+\pi^2\right]}$	$\pi\left[\nu\left(s,\ -\frac{1}{2}\right)-e^s\right]$ $[\operatorname{Re}s>0]$

2.6. Inverse Trigonometric Functions

More formulas can be obtained from the corresponding sections due to the relations

$$\arcsin z = z\,{}_2F_1\left(\frac{1}{2}, \frac{1}{2}; \frac{3}{2}; z^2\right), \quad \arccos z = \frac{\pi}{2} - z\,{}_2F_1\left(\frac{1}{2}, \frac{1}{2}; \frac{3}{2}; z^2\right),$$

$$\frac{\arcsin z}{\sqrt{1-z^2}} = z\,{}_2F_1\left(1, 1; \frac{3}{2}; z^2\right), \quad \frac{\arccos z}{\sqrt{1-z^2}} = \frac{\pi}{2\sqrt{1-z^2}} - z\,{}_2F_1\left(1, 1; \frac{3}{2}; z^2\right),$$

$$\arcsin^2 z = z^2\,{}_3F_2\left(1, 1, 1; \frac{3}{2}, 2; z^2\right), \quad \arctan z = z\,{}_2F_1\left(1, \frac{1}{2}; \frac{3}{2}; -z^2\right),$$

$$\operatorname{arccot} z = \frac{\pi z}{2}\sqrt{\frac{1}{z^2}}\sqrt{\frac{1}{z^2+1}}\sqrt{z^2+1} - z\,{}_2F_1\left(\frac{1}{2}, 1; \frac{3}{2}; -z^2\right),$$

$$\operatorname{arccsc} z = \frac{1}{z}\,{}_2F_1\left(\frac{1}{2}, \frac{1}{2}; \frac{3}{2}; \frac{1}{z^2}\right), \quad \operatorname{arcsec} z = \frac{\pi}{2} - \frac{1}{z}\,{}_2F_1\left(\frac{1}{2}, \frac{1}{2}; \frac{3}{2}; \frac{1}{z^2}\right),$$

$$\arcsin z = -\frac{1}{2\sqrt{\pi}\,z} G_{22}^{12}\left(-z^2 \,\middle|\, \begin{matrix} 3/2, 3/2 \\ 1, 1/2 \end{matrix}\right), \quad \arcsin^2 z = -\frac{\sqrt{\pi}}{2} G_{33}^{13}\left(-z^2 \,\middle|\, \begin{matrix} 1, 1, 1 \\ 1, 0, 1/2 \end{matrix}\right),$$

$$\operatorname{arccsc} z = \frac{\sqrt{-z^2}}{2\sqrt{\pi}\,z} G_{22}^{21}\left(-z^2 \,\middle|\, \begin{matrix} 1/2, 1 \\ 0, 0 \end{matrix}\right), \quad \operatorname{arcsec} z = \frac{\pi}{2} - \frac{1}{2\sqrt{\pi}\,z} G_{22}^{12}\left(-\frac{1}{z^2} \,\middle|\, \begin{matrix} 1/2, 1/2 \\ 0, -1/2 \end{matrix}\right),$$

$$\arctan z = \frac{1}{2z} G_{22}^{12}\left(z^2 \,\middle|\, \begin{matrix} 1, 3/2 \\ 1, 1/2 \end{matrix}\right).$$

2.6.1. $\arcsin(\varphi(x))$, $\arccos(\varphi(x))$, and algebraic functions

No.	$f(x)$	$F(s)$
1	$\arcsin(ax)$	$\frac{i\,(ia)^{-s}}{2\sqrt{\pi}\,s}\Gamma\left(\frac{s+1}{2}\right)\Gamma\left(-\frac{s}{2}\right)$ $[-1 < \operatorname{Re} s < 0]$
2	$\arccos(ax) - \frac{\pi}{2}$	$\frac{(-a)^{-(s+1)/2}\,a^{(1-s)/2}}{2\sqrt{\pi}\,s}\Gamma\left(\frac{s+1}{2}\right)\Gamma\left(-\frac{s}{2}\right)$ $[-1 < \operatorname{Re} s < 0]$
3	$\arcsin(ax) - ax$	$-\frac{i\,(ia)^{-s}}{\sqrt{\pi}\,s^2}\Gamma\left(\frac{2-s}{2}\right)\Gamma\left(\frac{s+1}{2}\right)$ $[\operatorname{Re}(ia) > 0;\ -3 < \operatorname{Re} s < -1]$
4	$\arccos(ax) + ax - \frac{\pi}{2}$	$-\frac{i\,(ia)^{-s}}{2\sqrt{\pi}\,s}\Gamma\left(\frac{s+1}{2}\right)\Gamma\left(-\frac{s}{2}\right)$ $[\operatorname{Im} a < 0;\ -3 < \operatorname{Re} s < -1]$
5	$\arcsin(ax) - \sum_{k=0}^{n}\frac{(1/2)_k\,(ax)^{2k+1}}{(2k+1)\,k!}$	$-\frac{i\sqrt{\pi}\,(ia)^{-s}}{s^2}\sec\frac{s\pi}{2}\,\Gamma\left[\begin{matrix}\frac{2-s}{2}\\ \frac{1-s}{2}\end{matrix}\right]$ $[\operatorname{Re}(ia) > 0;\ -3-2n < \operatorname{Re} s < -1-2n]$
6	$\arccos(ax) - \frac{\pi}{2} + \sum_{k=0}^{n}\frac{(1/2)_k\,(ax)^{2k+1}}{(2k+1)\,k!}$	$\frac{i\sqrt{\pi}\,(ia)^{-s}}{2s}\sec\frac{s\pi}{2}\,\Gamma\left[\begin{matrix}-\frac{s}{2}\\ \frac{1-s}{2}\end{matrix}\right]$ $[\operatorname{Im} a < 0;\ -2n-3 < \operatorname{Re} s < -2n-1]$

No.	$f(x)$	$F(s)$
7	$\frac{1}{\sqrt{1-a^2x^2}}\arcsin(ax)$	$-\frac{i\pi^{3/2}(ia)^{-s}}{4}\sec\frac{s\pi}{2}\,\Gamma\left[\begin{matrix}\frac{1-s}{2}\\ \frac{2-s}{2}\end{matrix}\right]$ $[\operatorname{Re}(ia)>0;\ \lvert\operatorname{Re}s\rvert<1]$
8	$\theta(a-x)\left\{\begin{matrix}\arcsin(x/a)\\ \arccos(x/a)\end{matrix}\right\}$	$\frac{(\pi\pm\pi)a^s}{4s}\mp\frac{\sqrt{\pi}\,a^s}{s^2}\Gamma\left[\begin{matrix}\frac{s+1}{2}\\ \frac{s}{2}\end{matrix}\right]$ $[a>0;\ \operatorname{Re}s>-(1\pm1)/2]$
9	$(a-x)_+^{\alpha-1}\arcsin(bx)$	$a^{s+\alpha}b\,\mathrm{B}(s+1,\,\alpha)\ {}_4F_3\left(\begin{matrix}\frac{1}{2},\,\frac{1}{2},\,\frac{s+1}{2},\,\frac{s+2}{2};\,a^2b^2\\ \frac{3}{2},\,\frac{s+\alpha+1}{2},\,\frac{s+\alpha+2}{2}\end{matrix}\right)$ $[a,\ \operatorname{Re}\alpha>0;\ \operatorname{Re}s>-1]$
10	$(a^2-x^2)_+^{\alpha-1}\arcsin(bx)$	$\frac{a^{s+2\alpha-1}b}{2}\mathrm{B}\left(\frac{s+1}{2},\,\alpha\right)\ {}_3F_2\left(\begin{matrix}\frac{1}{2},\,\frac{1}{2},\,\frac{s+1}{2};\,ab^2\\ \frac{3}{2},\,\frac{s+2\alpha+1}{2}\end{matrix}\right)$ $[a,\ \operatorname{Re}\alpha>0;\ \operatorname{Re}s>-1]$
11	$\frac{\theta(a-x)}{(x^2+b^2)^{\rho}}\arccos\frac{x}{a}$	$\frac{\pi a^s}{2^{s+1}b^{2\rho}}\Gamma\left[\begin{matrix}s\\ \frac{s+2}{2},\,\frac{s+2}{2}\end{matrix}\right]\ {}_3F_2\left(\begin{matrix}\rho,\,\frac{s}{2},\,\frac{s+1}{2}\\ \frac{s+2}{2},\,\frac{s+2}{2};\,-\frac{a^2}{b^2}\end{matrix}\right)$ $[a,\ \operatorname{Re}b,\ \operatorname{Re}s>0]$
12	$(a-x)_+^{\alpha-1}\arcsin(b(a-x))$	$a^{s+\alpha}b\,\mathrm{B}(s,\,\alpha+1)\ {}_4F_3\left(\begin{matrix}\frac{1}{2},\,\frac{1}{2},\,\frac{\alpha+1}{2},\,\frac{\alpha+2}{2};\,a^2b^2\\ \frac{3}{2},\,\frac{s+\alpha+1}{2},\,\frac{s+\alpha+2}{2}\end{matrix}\right)$ $[a,\ \operatorname{Re}s>0;\ \operatorname{Re}\alpha>-1]$
13	$(a-x)_+^{\alpha-1}\arcsin(bx(a-x))$	$a^{s+\alpha+1}b\,\mathrm{B}(s+1,\,\alpha+1)$ $\times\ {}_6F_5\left(\begin{matrix}\frac{1}{2},\,\frac{1}{2},\,\Delta(2,\,s+1),\,\Delta(2,\,\alpha+1)\\ \frac{3}{2},\,\Delta(4,\,s+\alpha+2);\,\frac{a^4b^2}{16}\end{matrix}\right)$ $[\operatorname{Re}s,\ \operatorname{Re}\alpha>-1]$
14	$(a-x)_+^{\alpha-1}\arcsin\left(b\sqrt{a-x}\right)$	$a^{s+\alpha-1/2}b\,\mathrm{B}\left(s,\,\frac{2\alpha+1}{2}\right)\ {}_3F_2\left(\begin{matrix}\frac{1}{2},\,\frac{1}{2},\,\frac{2\alpha+1}{2};\,ab^2\\ \frac{3}{2},\,\frac{2s+2\alpha+1}{2}\end{matrix}\right)$ $[a,\ \operatorname{Re}s>0;\ \operatorname{Re}\alpha>-1/2]$
15	$\theta(a-x)(bx+1)^{\alpha}$ $\times\arcsin\left(c\sqrt{a-x}\right)$	$\frac{\sqrt{\pi}a^{s+1/2}c}{2}\Gamma\left[\begin{matrix}s\\ \frac{2s+3}{2}\end{matrix}\right]F_3\left(-\alpha,\,\frac{1}{2},\,s,\,\frac{1}{2};\,s+\frac{3}{2};\,-ab,\,ac^2\right)$ $[a,\ \operatorname{Re}s>0;\ a\lvert b\rvert,\ a\lvert c^2\rvert<1;\ \lvert\arg(ab+1)\rvert<\pi]$
16	$\theta(a-x)\frac{(bx+1)^{\alpha}}{\sqrt{1-c^2(a-x)}}$ $\times\arcsin\left(c\sqrt{a-x}\right)$	$\frac{\sqrt{\pi}a^{s+1/2}c}{2}\Gamma\left[\begin{matrix}s\\ \frac{2s+3}{2}\end{matrix}\right]F_3\left(-\alpha,\,1,\,s,\,1;\,s+\frac{3}{2};\,-ab,\,ac^2\right)$ $\left[\begin{matrix}a,\ \operatorname{Re}s>0;\ a\lvert b\rvert,\ a\lvert c^2\rvert<1;\\ \lvert\arg(ab+1)\rvert,\ \lvert\arg(ac^2+1)\rvert<\pi\end{matrix}\right]$

No.	$f(x)$	$F(s)$
17	$(a-x)_+^{\alpha-1} \times \arcsin\left(b\sqrt{x(a-x)}\right)$	$a^{s+\alpha} b \,\mathrm{B}\left(\frac{2s+1}{2}, \frac{2\alpha+1}{2}\right) {}_4F_3\left(\begin{matrix}\frac{1}{2}, \frac{1}{2}, \frac{2s+1}{2}, \frac{2\alpha+1}{2}; \frac{a^2b^2}{4} \\ \frac{3}{2}, \frac{s+\alpha+1}{2}, \frac{s+\alpha+2}{2}\end{matrix}\right)$ [a, $\operatorname{Re} s$, $\operatorname{Re}\alpha > -1/2$]
18	$\frac{1}{(x+a)^\rho} \arcsin \frac{b}{x+a}$	$a^{s-\rho-1} b \,\mathrm{B}(s, 1-s+\rho) \, {}_4F_3\left(\begin{matrix}\frac{1}{2}, \frac{1}{2}, \frac{1-s+\rho}{2}, \frac{2-s+\rho}{2} \\ \frac{3}{2}, \frac{\rho+1}{2}, \frac{\rho+2}{2}; \frac{b^2}{a^2}\end{matrix}\right)$ [$0 < \operatorname{Re} s < \operatorname{Re}\rho + 1$; $\lvert\arg a\rvert < \pi$]
19	$\frac{1}{\sqrt{(x+a)^2 - b^2}\,(x+a)^\rho} \times \arcsin \frac{b}{x+a}$	$a^{s-\rho-2} b \,\mathrm{B}(s, 2-s+\rho) \, {}_4F_3\left(\begin{matrix}1, 1, \frac{2-s+\rho}{2}, \frac{3-s+\rho}{2} \\ \frac{3}{2}, \frac{\rho+2}{2}, \frac{\rho+3}{2}; \frac{b^2}{a^2}\end{matrix}\right)$ [$0 < \operatorname{Re} s < \operatorname{Re}\rho + 2$; $\lvert\arg a\rvert < \pi$]
20	$\frac{1}{(x+a)^\rho} \arcsin \frac{bx}{x+a}$	$a^{s-\rho} b \,\mathrm{B}(s+1, \rho-s) \, {}_4F_3\left(\begin{matrix}\frac{1}{2}, \frac{1}{2}, \frac{s+1}{2}, \frac{s+2}{2} \\ \frac{3}{2}, \frac{\rho+1}{2}, \frac{\rho+2}{2}; b^2\end{matrix}\right)$ [$-1 < \operatorname{Re} s < \operatorname{Re}\rho$; $\lvert\arg a\rvert < \pi$]
21	$\frac{(x+a)^{-\rho}}{\sqrt{1 - \frac{b^2x^2}{(x+a)^2}}} \arcsin \frac{bx}{x+a}$	$a^{s-\rho} b \,\mathrm{B}(s+1, \rho-s) \, {}_4F_3\left(\begin{matrix}1, 1, \frac{s+1}{2}, \frac{s+2}{2} \\ \frac{3}{2}, \frac{\rho+1}{2}, \frac{\rho+2}{2}; b^2\end{matrix}\right)$ [$-1 < \operatorname{Re} s < \operatorname{Re}\rho$; $\lvert\arg a\rvert < \pi$]
22	$\frac{1}{(x^2+a^2)^\rho} \arcsin \frac{bx}{x^2+a^2}$	$\frac{a^{s-2\rho-1} b}{2} \,\mathrm{B}\left(\frac{s+1}{2}, \frac{1-s+2\rho}{2}\right) {}_4F_3\left(\begin{matrix}\frac{1}{2}, \frac{1}{2}, \frac{s+1}{2}, \frac{1-s+2\rho}{2} \\ \frac{3}{2}, \frac{\rho+1}{2}, \frac{\rho+2}{2}; \frac{b^2}{4a^2}\end{matrix}\right)$ [$\operatorname{Re} a > 0$; $-1 < \operatorname{Re} s < 2\operatorname{Re}\rho + 1$]
23	$\frac{(x^2+a^2)^{-\rho}}{\sqrt{1 - \frac{b^2x^2}{(x^2+a^2)^2}}} \times \arcsin \frac{bx}{x^2+a^2}$	$\frac{a^{s-2\rho-1} b}{2} \,\mathrm{B}\left(\frac{s+1}{2}, \frac{1-s+2\rho}{2}\right) {}_4F_3\left(\begin{matrix}1, 1, \frac{s+1}{2}, \frac{1-s+2\rho}{2} \\ \frac{3}{2}, \frac{\rho+1}{2}, \frac{\rho+2}{2}; \frac{b^2}{4a^2}\end{matrix}\right)$ [$\operatorname{Re} a > 0$; $-1 < \operatorname{Re} s < 2\operatorname{Re}\rho + 1$]
24	$\frac{(x+a)^{-\rho}}{\sqrt{a - b^2 + x}} \arcsin \frac{b}{\sqrt{x+a}}$	$a^{s-\rho-1} b \,\mathrm{B}(s, 1-s+\rho) \, {}_3F_2\left(\begin{matrix}1, 1, 1-s+\rho \\ \frac{3}{2}, \rho+1; \frac{b^2}{a}\end{matrix}\right)$ [$\lvert b^2\rvert < a$; $0 < \operatorname{Re} s < \operatorname{Re}\rho + 1$]
25	$\theta(a-x) \times \arcsin\left(c\sqrt{\frac{a-x}{b-x}}\right)$	$a^{s+1/2} \sqrt{\frac{\pi}{b}}\, c \frac{\Gamma(s)}{2\Gamma\left(s+\frac{3}{2}\right)} F_1\left(\frac{1}{2}, s, \frac{1}{2}; s+\frac{3}{2}; \frac{a}{b}, \frac{ac^2}{b}\right)$ [$a < \lvert b\rvert, \lvert b/c^2\rvert$; a, $\operatorname{Re} s > 0$]

No.	$f(x)$	$F(s)$
26	$\frac{\theta(a-x)}{\sqrt{c^2(x-a)+b-x}}$ $\times \arcsin\left(c\sqrt{\frac{a-x}{b-x}}\right)$	$\frac{\sqrt{\pi}a^{s+1/2}c}{2b}\frac{\Gamma(s)}{\Gamma\left(s+\frac{3}{2}\right)}F_1\left(1,\, s,\, 1;\, s+\frac{3}{2};\, \frac{a}{b},\, \frac{ac^2}{b}\right)$ $[a<\lvert b\rvert,\, \lvert b/c^2\rvert;\, a,\ \operatorname{Re}s>0]$
27	$\theta(x-a)\arcsin\frac{cx}{\sqrt{x^2-b^2}}$	$-\frac{a^s c}{s}F_2\left(\frac{1}{2},\, \frac{1}{2},\, -\frac{s}{2};\, \frac{3}{2},\, 1-\frac{s}{2};\, c^2,\, \frac{b^2}{a^2}\right)$ $[a>b>0;\ \operatorname{Re}s<0;\ \lvert\arg c\rvert<\pi]$
28	$\frac{\theta(x-a)}{\sqrt{x^2(1-c^2)-b^2}}$ $\times \arcsin\frac{cx}{\sqrt{x^2-b^2}}$	$\frac{a^{s-1}c}{1-s}F_2\left(1,\, 1,\, \frac{1-s}{2};\, \frac{3}{2},\, \frac{3-s}{2};\, c^2,\, \frac{b^2}{a^2}\right)$ $[a>b>0;\ \operatorname{Re}s<0;\ \lvert\arg c\rvert<\pi]$
29	$\theta(x-a)\arccos\frac{a}{x}$	$\frac{\sqrt{\pi}\,a^s}{s^2}\Gamma\left[\begin{matrix}\frac{1-s}{2}\\ -\frac{s}{2}\end{matrix}\right]$ $[a>0;\ \operatorname{Re}s<0]$
30	$\arccos\left(\sqrt{ax+1}-\sqrt{ax}\right)$	$\frac{a^{-s}}{\sqrt{\pi}\,s}\sin(s\pi)\,\Gamma(-2s)\,\Gamma\left(\frac{4s+1}{2}\right)$ $[-1/4<\operatorname{Re}s<0;\ \lvert\arg a\rvert<\pi]$
31	$\arccos\frac{\sqrt{ax}}{\sqrt{ax+1}+1}$	$\frac{a^{-s}}{\sqrt{\pi}\,s}\sin(s\pi)\,\Gamma\left(\frac{1}{2}-2s\right)\Gamma(2s)$ $[0<\operatorname{Re}s<1/4;\ \lvert\arg a\rvert<\pi]$
32	$\arccos\frac{\sqrt{ax+1}-1}{\sqrt{ax}}$	$\frac{a^{-s}}{\sqrt{\pi}\,s}\sin(s\pi)\,\Gamma\left(\frac{1}{2}-2s\right)\Gamma(2s)$ $[0<\operatorname{Re}s<1/4;\ \lvert\arg a\rvert<\pi]$
33	$\arccos\frac{1}{\sqrt{ax}+\sqrt{ax+1}}$	$\frac{a^{-s}}{\sqrt{\pi}\,s}\sin(s\pi)\,\Gamma(-2s)\,\Gamma\left(\frac{4s+1}{2}\right)$ $[-1/4<\operatorname{Re}s<0;\ \lvert\arg a\rvert<\pi]$

2.6.2. $\arcsin(\varphi(x))$, $\arccos(\varphi(x))$, and the exponential function

1	$\theta(a-x)\,e^{bx}$ $\times \arcsin\left(c\sqrt{a-x}\right)$	$\sqrt{\pi}a^{s+1/2}c\frac{\Gamma(s)}{2\Gamma\left(s+\frac{3}{2}\right)}\Xi_1\left(\frac{1}{2},\, s,\, \frac{1}{2};\, s+\frac{3}{2};\, ac^2,\, ab\right)$ $[a,\ \operatorname{Re}s>0]$
2	$\frac{\theta(a-x)}{\sqrt{1-c^2(a-x)}}e^{bx}$ $\times \arcsin\left(c\sqrt{a-x}\right)$	$\sqrt{\pi}a^{s+1/2}c\frac{\Gamma(s)}{2\Gamma\left(s+\frac{3}{2}\right)}\Xi_1\left(1,\, s,\, 1;\, s+\frac{3}{2};\, ac^2,\, ab\right)$ $[a,\ \operatorname{Re}s>0]$

No.	$f(x)$	$F(s)$
3	$\theta(a-x)\, e^{bx} \arccos\frac{x}{a}$	$\frac{(2a)^s}{s+1}\left[ab\,\Gamma\left[\begin{matrix}\frac{s+2}{2},\ \frac{s+2}{2}\\ s+2\end{matrix}\right] {}_2F_3\left(\begin{matrix}\frac{s+1}{2},\ \frac{s+2}{2};\ \frac{a^2b^2}{4}\\ \frac{3}{2},\ \frac{s+3}{2},\ \frac{s+3}{2}\end{matrix}\right) + \frac{2}{s}\Gamma\left[\begin{matrix}\frac{s+3}{2},\ \frac{s+3}{2}\\ s+2\end{matrix}\right] {}_2F_3\left(\begin{matrix}\frac{s}{2},\ \frac{s+1}{2};\ \frac{a^2b^2}{4}\\ \frac{1}{2},\ \frac{s+2}{2},\ \frac{s+2}{2}\end{matrix}\right)\right]$ $[a,\ \mathrm{Re}\, s>0]$
4	$\theta(a-x)\, e^{bx^2} \arccos\frac{x}{a}$	$\frac{\sqrt{\pi}a^s}{2s}\Gamma\left[\begin{matrix}\frac{s+1}{2}\\ \frac{s+2}{2}\end{matrix}\right] {}_2F_2\left(\begin{matrix}\frac{s}{2},\ \frac{s+1}{2};\ a^2b\\ \frac{s+2}{2},\ \frac{s+2}{2}\end{matrix}\right)$ $[a,\ \mathrm{Re}\, s>0]$

2.6.3. arccos (bx) and hyperbolic or trigonometric functions

Notation: $\delta=\begin{Bmatrix}1\\0\end{Bmatrix}$.

1	$\theta(a-x)\begin{Bmatrix}\sinh(bx)\\ \sin(bx)\end{Bmatrix}\arccos\frac{x}{a}$	$\frac{\sqrt{\pi}\,a^{s+1}b}{(s+1)^2}\Gamma\left[\begin{matrix}\frac{s+2}{2}\\ \frac{s+1}{2}\end{matrix}\right] {}_2F_3\left(\begin{matrix}\frac{s+1}{2},\ \frac{s+2}{2};\ \pm\frac{a^2b^2}{4}\\ \frac{3}{2},\ \frac{s+3}{2},\ \frac{s+3}{2}\end{matrix}\right)$ $[a>0;\ \mathrm{Re}\, s>-1]$
2	$\theta(a-x)\begin{Bmatrix}\cosh(bx)\\ \cos(bx)\end{Bmatrix}\arccos\frac{x}{a}$	$\frac{\sqrt{\pi}\,a^s}{s^2}\Gamma\left[\begin{matrix}\frac{s+1}{2}\\ \frac{s}{2}\end{matrix}\right] {}_2F_3\left(\begin{matrix}\frac{s}{2},\ \frac{s+1}{2};\ \pm\frac{a^2b^2}{4}\\ \frac{1}{2},\ \frac{s+2}{2},\ \frac{s+2}{2}\end{matrix}\right)$ $[a,\ \mathrm{Re}\, s>0]$
3	$\theta(a-x)\begin{Bmatrix}\sinh(bx^2)\\ \cosh(bx^2)\end{Bmatrix}\arccos\frac{x}{a}$	$\frac{\sqrt{\pi}\,a^{s+2\delta}b^{2\delta}}{(s+2\delta)^2}\Gamma\left[\begin{matrix}\frac{s+2\delta+1}{2}\\ \frac{s+2\delta}{2}\end{matrix}\right] \times {}_3F_4\left(\begin{matrix}\frac{s+2\delta}{4},\ \frac{s+2\delta+1}{4},\ \frac{s+2\delta+3}{4};\ \frac{a^4b^2}{4}\\ \frac{2\delta+1}{2},\ \frac{s+2\delta+2}{4},\ \frac{s+2\delta+4}{4},\ \frac{s+2\delta+4}{4}\end{matrix}\right)$ $[a>0;\ \mathrm{Re}\, s>-2\delta]$
4	$\theta(a-x)\begin{Bmatrix}\sin(bx)\sinh(bx)\\ \cos(bx)\cosh(bx)\end{Bmatrix} \times\arccos\frac{x}{a}$	$\frac{\sqrt{\pi}\,a^{s+2\delta}b^{2\delta}}{2(s+2\delta)}\Gamma\left[\begin{matrix}\frac{s+2\delta+1}{2}\\ \frac{s+2\delta+2}{2}\end{matrix}\right] \times {}_3F_6\left(\begin{matrix}\frac{s+2\delta}{4},\ \frac{s+2\delta+1}{4},\ \frac{s+2\delta+3}{4};\ -\frac{a^4b^4}{64}\\ \frac{2\delta+1}{4},\ \frac{2\delta+3}{4},\ \frac{2\delta+1}{2},\ \frac{s+2\delta+2}{4},\ \frac{s+2\delta+4}{4},\ \frac{s+2\delta+4}{4}\end{matrix}\right)$ $[a>0;\ \mathrm{Re}\, s>-(2\delta+1)]$
5	$\theta(a-x)\begin{Bmatrix}\cosh(bx)\sin(bx)\\ \sinh(bx)\cos(bx)\end{Bmatrix} \times\arccos\frac{x}{a}$	$\frac{\sqrt{\pi}\,a^{s+1}b}{2(s+1)}\Gamma\left[\begin{matrix}\frac{s+2}{2}\\ \frac{s+3}{2}\end{matrix}\right] {}_3F_6\left(\begin{matrix}\frac{s+1}{4},\ \frac{s+2}{4},\ \frac{s+4}{4};\ -\frac{a^4b^4}{64}\\ \frac{1}{2},\ \frac{3}{4},\ \frac{5}{4},\ \frac{s+3}{4},\ \frac{s+5}{4},\ \frac{s+5}{4}\end{matrix}\right) \pm\frac{\sqrt{\pi}\,a^{s+3}b^3}{6(s+3)}\Gamma\left[\begin{matrix}\frac{s+4}{2}\\ \frac{s+5}{2}\end{matrix}\right] {}_3F_6\left(\begin{matrix}\frac{s+3}{4},\ \frac{s+4}{4},\ \frac{s+6}{4};\ -\frac{a^4b^4}{64}\\ \frac{5}{4},\ \frac{3}{2},\ \frac{7}{4},\ \frac{s+5}{4},\ \frac{s+7}{4},\ \frac{s+7}{4}\end{matrix}\right)$ $[a>0;\ \mathrm{Re}\, s>-1]$

No.	$f(x)$	$F(s)$
6	$\theta(a-x)\left\{\begin{matrix}\sinh(b\sqrt{x})\sin(b\sqrt{x})\\ \cosh(b\sqrt{x})\cos(b\sqrt{x})\end{matrix}\right\}$ $\times\arccos\frac{x}{a}$	$\frac{\sqrt{\pi}\,a^{s+\delta}b^{2\delta}}{2(s+\delta)}\Gamma\left[\begin{matrix}\frac{s+\delta+1}{2}\\ \frac{s+\delta+2}{2}\end{matrix}\right]$ $\times{}_2F_5\left(\begin{matrix}\frac{s+\delta}{2},\ \frac{s+\delta+1}{2};\ -\frac{a^2b^4}{64}\\ \frac{2\delta+1}{4},\ \frac{2\delta+3}{4},\ \frac{2\delta+1}{2},\ \frac{s+\delta+2}{2},\ \frac{s+\delta+2}{2}\end{matrix}\right)$ $[a>0;\ \operatorname{Re}s>-\delta]$
7	$\theta(a-x)\left\{\begin{matrix}\cosh(b\sqrt{x})\sin(b\sqrt{x})\\ \sinh(b\sqrt{x})\cos(b\sqrt{x})\end{matrix}\right\}$ $\times\arccos\frac{x}{a}$	$\frac{\sqrt{\pi}\,a^{s+1/2}b}{2s+1}\Gamma\left[\begin{matrix}\frac{2s+3}{4}\\ \frac{2s+5}{4}\end{matrix}\right]{}_2F_5\left(\begin{matrix}\frac{2s+1}{4},\ \frac{2s+3}{4};\ -\frac{a^2b^4}{64}\\ \frac{1}{2},\ \frac{3}{4},\ \frac{5}{4},\ \frac{2s+5}{4},\ \frac{2s+5}{4}\end{matrix}\right)$ $\pm\frac{\sqrt{\pi}\,a^{s+3/2}b^3}{3(2s+3)}\Gamma\left[\begin{matrix}\frac{2s+5}{4}\\ \frac{2s+7}{4}\end{matrix}\right]{}_2F_5\left(\begin{matrix}\frac{2s+3}{4},\ \frac{2s+5}{4};\ -\frac{a^2b^4}{64}\\ \frac{5}{4},\ \frac{3}{2},\ \frac{7}{4},\ \frac{2s+7}{4},\ \frac{2s+7}{4}\end{matrix}\right)$ $[a>0;\ \operatorname{Re}s>-1/2]$

2.6.4. Trigonometric functions of inverse trigonometric functions

1	$\theta(a-x)\sin\left(\nu\arccos\frac{x}{a}\right)$	$\frac{\nu\pi a^s}{2^{s+1}}\Gamma\left[\begin{matrix}s\\ \frac{s-\nu+1}{2},\ \frac{s+\nu+1}{2}\end{matrix}\right]$ $[a,\ \operatorname{Re}s>0]$
2	$\frac{1}{\sqrt{a^2-x^2}}\sin\left(\nu\arccos\frac{x}{a}\right)$	$\frac{a^{s-1}}{2^{s+1}\pi}\sin(\nu\pi)\Gamma\left[s,\ \frac{1-s-\nu}{2},\ \frac{1-s+\nu}{2}\right]$ $[0<\operatorname{Re}s<1-\lvert\operatorname{Re}\nu\rvert;\ \lvert\arg a\rvert<\pi]$
3	$(a^2-x^2)_+^{-1/2}\cos\left(\nu\arccos\frac{x}{a}\right)$	$\frac{\pi a^{s-1}}{2^s}\Gamma\left[\begin{matrix}s\\ \frac{s+\nu+1}{2},\ \frac{s-\nu+1}{2}\end{matrix}\right]$ $[a,\ \operatorname{Re}s>0]$
4	$(x^2-a^2)_+^{-1/2}\cos\left(\nu\operatorname{arcsec}\frac{x}{a}\right)$	$\pi(2a)^{s-1}\Gamma\left[\begin{matrix}1-s\\ \frac{2-s-\nu}{2},\ \frac{2-s+\nu}{2}\end{matrix}\right]$ $[a>0;\ \operatorname{Re}s<1]$
5	$\theta(a-x)\sin\left(\nu\arcsin\sqrt{1-\frac{x}{a}}\right)$	$\frac{\nu\sqrt{\pi}\,a^s}{2}\Gamma\left[\begin{matrix}s,\ \frac{2s+1}{2}\\ \frac{2s+\nu+2}{2},\ \frac{2s-\nu+2}{2}\end{matrix}\right]$ $[a,\ \operatorname{Re}s>0]$
6	$(a^2-x^2)_+^{-1/2}\cos\left(\nu\arccos\frac{x^2-a^2}{a^2}\right)$	$\frac{\pi a^{s-1}}{2^s}\Gamma\left[\begin{matrix}s\\ \frac{s+2\nu+1}{2},\ \frac{s-2\nu+1}{2}\end{matrix}\right]$ $[a,\ \operatorname{Re}s>0]$
7	$(x^2-a^2)_+^{-1/2}\cos\left(\nu\arccos\frac{a}{x}\right)$	$\pi(2a)^{s-1}\Gamma\left[\begin{matrix}1-s\\ \frac{2-s-\nu}{2},\ \frac{2-s+\nu}{2}\end{matrix}\right]$ $[a>0;\ \operatorname{Re}s<1]$
8	$(1-x)_+^{-1/2}\left(1+\sqrt{1-x}\right)^\nu\cos\frac{\pi\nu}{2}$ $-x^{\nu/2}(x-1)_+^{-1/2}\sin\left(\nu\arcsin\frac{1}{\sqrt{x}}\right)$	$\sqrt{\pi}\,\Gamma\left[\begin{matrix}s,\ \frac{2-2s-\nu}{2}\\ \frac{2s+\nu+1}{2},\ 1-s-\nu\end{matrix}\right]$ $[0<\operatorname{Re}s<1-\operatorname{Re}\nu/2]$

No.	$f(x)$	$F(s)$
9	$(1-x)_+^{-1/2}\left(1+\sqrt{1-x}\right)^\nu \sin\frac{\pi\nu}{2}$ $+ x^{\nu/2}(x-1)_+^{-1/2}\cos\left(\nu\arcsin\frac{1}{\sqrt{x}}\right)$	$\sqrt{\pi}\,\Gamma\left[\begin{matrix} s, \frac{1-2s-\nu}{2} \\ \frac{2s+\nu}{2}, 1-s-\nu \end{matrix}\right]$ $[0<\operatorname{Re}s<(1-\operatorname{Re}\nu)/2]$
10	$(1-x)_+^{-1/2}\sin\left(\nu\arcsin\sqrt{x}\right)$ $-\cos\frac{\nu\pi}{2}(x-1)_+^{-1/2}\left(\sqrt{x}+\sqrt{x-1}\right)^\nu$	$-\sqrt{\pi}\,\Gamma\left[\begin{matrix} \frac{2s+1}{2}, \frac{1-2s-\nu}{2} \\ \frac{2s-\nu+1}{2}, 1-s \end{matrix}\right]$ $[-1/2<\operatorname{Re}s<(1-\operatorname{Re}\nu)/2]$
11	$(1-x)_+^{-1/2}\cos\left(\nu\arcsin\sqrt{x}\right)$ $+\sin\frac{\nu\pi}{2}(x-1)_+^{-1/2}\left(\sqrt{x}+\sqrt{x-1}\right)^\nu$	$\sqrt{\pi}\,\Gamma\left[\begin{matrix} s, \frac{1-2s-\nu}{2} \\ \frac{2s-\nu+1}{2}, \frac{1-2s}{2} \end{matrix}\right]$ $[0<\operatorname{Re}s<(1-\operatorname{Re}\nu)/2]$
12	$\theta(1-x)\sin\left(\nu\arcsin\sqrt{x}\right)$ $+\sin\frac{\pi\nu}{2}\theta(x-1)\left(\sqrt{x}+\sqrt{x-1}\right)^\nu$	$\frac{\nu\sqrt{\pi}}{2}\Gamma\left[\begin{matrix} \frac{2s+1}{2}, -\frac{2s+\nu}{2} \\ \frac{2s-\nu+2}{2}, 1-s \end{matrix}\right]$ $[-1/2<\operatorname{Re}s<-\operatorname{Re}\nu/2]$
13	$\theta(1-x)\left(1+\sqrt{1-x}\right)^\nu\sin\frac{\nu\pi}{2}$ $+\theta(x-1)\,x^{\nu/2}\sin\left(\nu\arcsin\frac{1}{\sqrt{x}}\right)$	$\frac{\nu\sqrt{\pi}}{2}\Gamma\left[\begin{matrix} s, \frac{1-2s-\nu}{2} \\ \frac{2s+\nu+2}{2}, 1-s-\nu \end{matrix}\right]$ $[0<\operatorname{Re}s<(1-\operatorname{Re}\nu)/2]$
14	$\theta(1-x)\cos\left(\nu\arcsin\sqrt{x}\right)$ $+\cos\frac{\nu\pi}{2}\theta(x-1)\left(\sqrt{x}+\sqrt{x-1}\right)^\nu$	$-\frac{\nu\sqrt{\pi}}{2}\Gamma\left[\begin{matrix} s, -\frac{2s+\nu}{2} \\ \frac{2s-\nu+2}{2}, \frac{1-2s}{2} \end{matrix}\right]$ $[0<\operatorname{Re}s<-\operatorname{Re}\nu/2]$
15	$\theta(1-x)\left(1+\sqrt{1-x}\right)^\nu\cos\frac{\nu\pi}{2}$ $+\theta(x-1)\,x^{\nu/2}\cos\left(\nu\arcsin\frac{1}{\sqrt{x}}\right)$	$-\frac{\nu\sqrt{\pi}}{2}\Gamma\left[\begin{matrix} s, -\frac{2s+\nu}{2} \\ \frac{2s+\nu+1}{2}, 1-\nu-s \end{matrix}\right]$ $[0<\operatorname{Re}s<-\operatorname{Re}\nu/2]$
16	$(1-x)_+^{-1/2}\sin\left(\nu\arccos\sqrt{x}\right)$ $+(x-1)_+^{-1/2}\sinh\left(\nu\operatorname{arccosh}\sqrt{x}\right)$	$\frac{\sin(\nu\pi)}{2\pi^{3/2}}\Gamma\left[s, \frac{2s+1}{2}, \frac{1-2s-\nu}{2}, \frac{1-2s+\nu}{2}\right]$ $[\lvert\operatorname{Re}\nu\rvert<1;\ 0<\operatorname{Re}s<(1-\lvert\operatorname{Re}\nu\rvert)/2]$
17	$(1-x)_+^{-1/2}\sinh\left(\nu\operatorname{arccosh}\frac{1}{\sqrt{x}}\right)$ $+(x-1)_+^{-1/2}\sin\left(\nu\arccos\frac{1}{\sqrt{x}}\right)$	$\frac{\sin(\nu\pi)}{2\pi^{3/2}}\Gamma\left[\frac{2s+\nu}{2}, \frac{2s-\nu}{2}, \frac{1-2s}{2}, 1-s\right]$ $[\lvert\operatorname{Re}\nu\rvert<1;\ \lvert\operatorname{Re}\nu\rvert/2<\operatorname{Re}s<1/2]$

2.6.5. $\arcsin(\varphi(x))$, $\arccos(\varphi(x))$, and the logarithmic function

1	$\theta(a-x)\ln(bx+1)\arccos\frac{x}{a}$	$\frac{\sqrt{\pi}\,a^{s+1}b}{2(s+1)}\Gamma\left[\begin{matrix}\frac{s+2}{2}\\ \frac{s+3}{2}\end{matrix}\right]{}_6F_5\left(\begin{matrix}\frac{1}{2},\frac{1}{2},1,1,\frac{s+1}{2},\frac{s+2}{2}\\ \frac{1}{2},1,\frac{3}{2},\frac{s+3}{2},\frac{s+3}{2}\end{matrix};a^2b^2\right) - \frac{\sqrt{\pi}\,a^{s+2}b^2}{4(s+2)}\Gamma\left[\begin{matrix}\frac{s+3}{2}\\ \frac{s+4}{2}\end{matrix}\right]{}_6F_5\left(\begin{matrix}1,1,\frac{3}{2},\frac{3}{2},\frac{s+2}{2},\frac{s+3}{2}\\ \frac{3}{2},\frac{3}{2},2,\frac{s+4}{2},\frac{s+4}{2}\end{matrix};a^2b^2\right)$ $[a>0;\ \operatorname{Re}s>-1;\ \lvert\arg b\rvert<\pi]$
2	$\theta(a-x)\ln\frac{\sqrt{a}+\sqrt{a-x}}{\sqrt{x}}\times\arcsin(bx)$	$\frac{\sqrt{\pi}\,a^{s+1}b}{2(s+1)}\Gamma\left[\begin{matrix}s+1\\ \frac{2s+3}{2}\end{matrix}\right]{}_5F_4\left(\begin{matrix}\frac{1}{2},\frac{1}{2},\frac{s+1}{2},\frac{s+1}{2},\frac{s+2}{2}\\ \frac{3}{2},\frac{2s+3}{4},\frac{2s+5}{4},\frac{s+3}{2}\end{matrix};a^2b^2\right)$ $[a>0;\ \operatorname{Re}s>-1]$
3	$\theta(a-x)\ln\frac{\sqrt{a}+\sqrt{a-x}}{\sqrt{x}}\times\frac{\arcsin(bx)}{\sqrt{1-b^2x^2}}$	$\frac{\sqrt{\pi}\,a^{s+1}b}{2(s+1)}\Gamma\left[\begin{matrix}s+1\\ \frac{2s+3}{2}\end{matrix}\right]{}_5F_4\left(\begin{matrix}1,1,\frac{s+1}{2},\frac{s+1}{2},\frac{s+2}{2}\\ \frac{3}{2},\frac{2s+3}{4},\frac{2s+5}{4},\frac{s+3}{2}\end{matrix};a^2b^2\right)$ $[a>0;\ \operatorname{Re}s>-1]$
4	$\theta(a-x)\ln\frac{a+\sqrt{a^2-x^2}}{x}\times\arcsin(bx)$	$\frac{\sqrt{\pi}\,a^{s+1}b}{s^2(s+1)}\left[\begin{matrix}\frac{s+1}{2}\\ \frac{s}{2}\end{matrix}\right]\left[(s+1)\,{}_3F_2\left(\begin{matrix}\frac{1}{2},\frac{1}{2},\frac{s+1}{2}\\ \frac{3}{2},\frac{s+2}{2}\end{matrix};a^2b^2\right) - {}_3F_2\left(\begin{matrix}\frac{1}{2},\frac{s+1}{2},\frac{s+1}{2}\\ \frac{s+2}{2},\frac{s+3}{2}\end{matrix};a^2b^2\right)\right]$ $[a>0;\ \operatorname{Re}s>-1;\ \lvert\arg(1+a^2/b^2)\rvert<\pi]$
5	$\theta(a-x)\ln\frac{a+\sqrt{a^2-x^2}}{x}\times\frac{\arcsin(bx)}{\sqrt{1-b^2x^2}}$	$\frac{\sqrt{\pi}\,a^{s+1}b}{2(s+1)}\Gamma\left[\begin{matrix}\frac{s+1}{2}\\ \frac{s+2}{2}\end{matrix}\right]{}_4F_3\left(\begin{matrix}1,1,\frac{s+1}{2},\frac{s+1}{2}\\ \frac{3}{2},\frac{s+2}{2},\frac{s+3}{2}\end{matrix};a^2b^2\right)$ $[a>0;\ \operatorname{Re}s>-1;\ \lvert\arg(1+a^2b^2)\rvert<\pi]$
6	$(a-x)_+^{-1/2}\arcsin\sqrt{\frac{a-x}{a}} + (x-a)_+^{-1/2}\times\ln\frac{\sqrt{x}+\sqrt{x-a}}{\sqrt{a}}$	$\frac{a^{s-1/2}}{2\sqrt{\pi}}\Gamma\left[s,\frac{2s+1}{2},\frac{1-2s}{2},\frac{1-2s}{2}\right]$ $[a>0;\ 0<\operatorname{Re}s<1/2]$
7	$(x-a)_+^{-1/2}\arcsin\sqrt{\frac{x-a}{x}} + (a-x)_+^{-1/2}\times\ln\frac{\sqrt{a}+\sqrt{a-x}}{\sqrt{x}}$	$\frac{a^{s-1/2}}{2\sqrt{\pi}}\Gamma\left[s,s,1-s,\frac{1-2s}{2}\right]$ $[a>0;\ 0<\operatorname{Re}s<1/2]$

No.	$f(x)$	$F(s)$
8	$\theta(a-x)\ln(bx^2+1)$ $\times \arccos\frac{x}{a}$	$\frac{\sqrt{\pi}\,a^{s+2}b}{s(s+2)^2}\Gamma\left[\begin{matrix}\frac{s+3}{2}\\ \frac{s+2}{2}\end{matrix}\right]\left[(s+2)\,{}_3F_2\left(\begin{matrix}1,\,1,\,\frac{s+3}{2}\\ 2,\,\frac{s+4}{2};\,-a^2b\end{matrix}\right)\right.$ $\left.-2\,{}_3F_2\left(\begin{matrix}1,\,\frac{s+2}{2},\,\frac{s+3}{2}\\ \frac{s+4}{2},\,\frac{s+4}{2};\,-a^2b\end{matrix}\right)\right]$ $[\mathrm{Re}\,s>-2;\ \lvert\arg(1+a^2b)\rvert<\pi]$
9	$\theta(a-x)\ln\frac{b+x}{b-x}\arccos\frac{x}{a}$	$\frac{\sqrt{\pi}\,a^{s+1}}{2b(s+1)}\Gamma\left[\begin{matrix}\frac{s}{2}\\ \frac{s+3}{2}\end{matrix}\right]\left[(s+1)\,{}_3F_2\left(\begin{matrix}\frac{1}{2},\,1,\,\frac{s+2}{2};\,\frac{a^2}{b^2}\\ \frac{3}{2},\,\frac{s+3}{2}\end{matrix}\right)\right.$ $\left.-{}_3F_2\left(\begin{matrix}1,\,\frac{s+1}{2},\,\frac{s+2}{2};\,\frac{a^2}{b^2}\\ \frac{s+3}{2},\,\frac{s+3}{2}\end{matrix}\right)\right]$ $[a>0;\ \mathrm{Re}\,s>-1;\ \lvert\arg(b^2-a^2)\rvert<\pi]$
10	$\theta(a-x)\arccos\frac{x}{a}$ $\times\ln\left(bx+\sqrt{1+b^2x^2}\right)$	$\frac{\sqrt{\pi}\,a^{s+1}b}{4(s+1)}\Gamma\left[\begin{matrix}\frac{s}{2}\\ \frac{s+3}{2}\end{matrix}\right]\left[(s+1)\,{}_3F_2\left(\begin{matrix}\frac{1}{2},\,\frac{1}{2},\,\frac{s+2}{2}\\ \frac{3}{2},\,\frac{s+3}{2};\,-a^2b^2\end{matrix}\right)\right.$ $\left.-{}_3F_2\left(\begin{matrix}\frac{1}{2},\,\frac{s+1}{2},\,\frac{s+2}{2}\\ \frac{s+3}{2},\,\frac{s+3}{2};\,-a^2b^2\end{matrix}\right)\right]$ $[a>0;\ \mathrm{Re}\,s>-1;\ \lvert\arg(1+a^2b^2)\rvert<\pi]$
11	$\theta(a-x)\arccos\frac{x}{a}$ $\times\ln^2\left(bx+\sqrt{1+b^2x^2}\right)$	$\frac{\sqrt{\pi}\,a^{s+2}b^2}{2s(s+2)}\Gamma\left[\begin{matrix}\frac{s+3}{2}\\ \frac{s+4}{2}\end{matrix}\right]\left[(s+2)\,{}_4F_3\left(\begin{matrix}1,\,1,\,1,\,\frac{s+3}{2};\,-a^2b^2\\ \frac{3}{2},\,2,\,\frac{s+4}{2}\end{matrix}\right)\right.$ $\left.-2\,{}_4F_3\left(\begin{matrix}1,\,1,\,\frac{s+2}{2},\,\frac{s+3}{2};\,-a^2b^2\\ \frac{3}{2},\,\frac{s+4}{2},\,\frac{s+4}{2}\end{matrix}\right)\right]$ $[a>0;\ \mathrm{Re}\,s>-2;\ \lvert\arg(1+a^2b^2)\rvert<\pi]$
12	$\frac{\theta(a-x)}{\sqrt{1+b^2x^2}}\arccos\frac{x}{a}$ $\times\ln\left(bx+\sqrt{1+b^2x^2}\right)$	$\frac{\sqrt{\pi}\,a^{s+1}b}{2(s+1)}\Gamma\left[\begin{matrix}\frac{s+2}{2}\\ \frac{s+3}{2}\end{matrix}\right]{}_4F_3\left(\begin{matrix}1,\,1,\,\frac{s+1}{2},\,\frac{s+2}{2};\,-a^2b^2\\ \frac{3}{2},\,\frac{s+3}{2},\,\frac{s+3}{2}\end{matrix}\right)$ $[a>0;\ \mathrm{Re}\,s>-1;\ \lvert\arg(1+a^2b^2)\rvert<\pi]$

2.6.6. $\arctan(\varphi(x))$ and $\operatorname{arccot}(bx)$

1	$\left\{\begin{matrix}\arctan(ax)\\ \operatorname{arccot}(ax)\end{matrix}\right\}$	$\mp\frac{\pi a^{-s}}{2s}\sec\frac{s\pi}{2}$ $[\mathrm{Re}\,a>0;\ 0<\mp\mathrm{Re}\,s<1]$
2	$\arctan(ax)-ax$	$-\frac{\pi a^{-s}}{2s}\sec\frac{s\pi}{2}$ $[\mathrm{Re}\,a>0;\ -3<\mathrm{Re}\,s<-1]$

No.	$f(x)$	$F(s)$
3	$\operatorname{arccot}(ax)+ax-\frac{\pi}{2}\,ax\sqrt{\frac{1}{a^2x^2}}$	$\frac{\pi a^{-s}}{2s}\sec\frac{s\pi}{2}$ [Re $a>0$; $-3<$ Re $s<-1$]
4	$\arctan(ax)-\sum\limits_{k=0}^{n}(-1)^k\frac{(ax)^{2k+1}}{2k+1}$	$-\frac{\pi a^{-s}}{2s}\sec\frac{s\pi}{2}$ [Re $a>0$; $-2n-3<$ Re $s<2n-1$]
5	$\operatorname{arccot}(ax)-\frac{\pi}{2}\,ax\sqrt{\frac{1}{a^2x^2}}+\sum\limits_{k=0}^{n}(-1)^k\frac{(ax)^{2k+1}}{2k+1}$	$\frac{\pi a^{-s}}{2s}\sec\frac{s\pi}{2}$ [Re $a>0$; $-2n-3<$ Re $s<-2n-1$]
6	$\theta(a-x)\begin{Bmatrix}\arctan(x/a)\\ \operatorname{arccot}(x/a)\end{Bmatrix}$	$\frac{a^s}{4s}\left[\pi\pm\psi\left(\frac{s+1}{4}\right)\mp\psi\left(\frac{s+3}{4}\right)\right]$ [$a>0$; Re $s>-(1\pm1)/2$]
7	$(a-x)_+^{\alpha-1}\arctan(bx)$	$a^{s+\alpha}b\,\mathrm{B}(s+1,\,\alpha)\;{}_4F_3\left(\begin{matrix}\frac{1}{2},\,1,\,\frac{s+1}{2},\,\frac{s+2}{2};\,-a^2b^2\\ \frac{3}{2},\,\frac{s+\alpha+1}{2},\,\frac{s+\alpha+2}{2}\end{matrix}\right)$ [a, Re $\alpha>0$; Re $s>-1$]
8	$(a^2-x^2)_+^{\alpha-1}\arctan(bx)$	$\frac{a^{s+2\alpha-1}b}{2}\,\mathrm{B}\left(\frac{s+1}{2},\,\alpha\right)\;{}_3F_2\left(\begin{matrix}\frac{1}{2},\,1,\,\frac{s+1}{2};\,-a^2b^2\\ \frac{3}{2},\,\frac{s+2\alpha+1}{2}\end{matrix}\right)$ [a, Re $\alpha>0$; Re $s>-1$]
9	$(a-x)_+^{\alpha-1}\arctan[b(a-x)]$	$a^{s+\alpha}b\,\mathrm{B}(s,\,\alpha+1)\;{}_4F_3\left(\begin{matrix}\frac{1}{2},\,1,\,\frac{\alpha+1}{2},\,\frac{\alpha+2}{2};\,-a^2b^2\\ \frac{3}{2},\,\frac{s+\alpha+1}{2},\,\frac{s+\alpha+2}{2}\end{matrix}\right)$ [a, Re $s>0$; Re $\alpha>-1$]
10	$(a-x)_+^{\alpha-1}\times\arctan\left(b\sqrt{a-x}\right)$	$a^{s+\alpha-1/2}b\,\mathrm{B}\left(s,\,\alpha+\frac{1}{2}\right)\;{}_3F_2\left(\begin{matrix}\frac{1}{2},\,1,\,\frac{2\alpha+1}{2};\,-ab^2\\ \frac{3}{2},\,\frac{2s+2\alpha+1}{2}\end{matrix}\right)$ [a, Re $s>0$; Re $\alpha>-1/2$]
11	$\theta(a-x)(bx+1)^\alpha\times\arctan\left(c\sqrt{a-x}\right)$	$\frac{\sqrt{\pi}a^{s+1/2}c}{2}\,\Gamma\left[\begin{matrix}s\\ \frac{2s+3}{2}\end{matrix}\right]F_3\left(-\alpha,\,\frac{1}{2},\,s,\,1;\,\frac{2s+3}{2};\,-ab,\,-ac^2\right)$ [a, Re $s>0$; $\lvert\arg b\rvert<\pi$]
12	$(a-x)_+^{\alpha-1}\times\arctan\left(b\sqrt{x(a-x)}\right)$	$a^{s+\alpha}b\,\mathrm{B}\left(s+\frac{1}{2},\,\alpha+\frac{1}{2}\right)\;{}_4F_3\left(\begin{matrix}\frac{1}{2},\,1,\,\frac{2s+1}{2},\,\frac{2\alpha+1}{2};\,-\frac{a^2b^2}{4}\\ \frac{3}{2},\,\frac{s+\alpha+1}{2},\,\frac{s+\alpha+2}{2}\end{matrix}\right)$ [$a>0$; Re s, Re $\alpha>-1/2$]

No.	$f(x)$	$F(s)$
13	$\frac{1}{(x+a)^{\rho}} \arctan \frac{b}{x+a}$	$a^{s-\rho-1} b \, \mathrm{B}(s, 1-s+\rho) \, {}_4F_3\left(\begin{matrix} \frac{1}{2}, 1, \frac{1-s+\rho}{2}, \frac{2-s+\rho}{2} \\ \frac{3}{2}, \frac{\rho+1}{2}, \frac{\rho+2}{2}; -\frac{b^2}{a^2} \end{matrix}\right)$ $[0 < \operatorname{Re} s < \operatorname{Re} \rho + 1; \ \lvert\arg a\rvert < \pi]$
14	$\frac{1}{(x+a)^{\rho}} \arctan \frac{b}{\sqrt{x+a}}$	$a^{s-\rho-1/2} b \, \mathrm{B}\left(s, \frac{1}{2} - s + \rho\right) {}_3F_2\left(\begin{matrix} \frac{1}{2}, 1, \frac{1-2s+2\rho}{2} \\ \frac{3}{2}, \frac{2\rho+1}{2}; -\frac{b^2}{a} \end{matrix}\right)$ $[0 < \operatorname{Re} s < \operatorname{Re} \rho + 1/2; \ \lvert\arg a\rvert < \pi]$
15	$\frac{1}{(x+a)^{\rho}} \arctan \frac{bx}{x+a}$	$a^{s-\rho} b \, \mathrm{B}(s+1, \rho-s) \, {}_4F_3\left(\begin{matrix} \frac{1}{2}, 1, \frac{s+1}{2}, \frac{s+2}{2} \\ \frac{3}{2}, \frac{\rho+1}{2}, \frac{\rho+2}{2}; -b^2 \end{matrix}\right)$ $[-1 < \operatorname{Re} s < \operatorname{Re} \rho; \ \lvert\arg a\rvert < \pi]$
16	$\theta(x-a) \arctan \frac{bx}{\sqrt{x^2-c^2}}$	$-\frac{a^s b}{s} F_2\left(\frac{1}{2}, 1, -\frac{s}{2}; \frac{3}{2}, \frac{2-s}{2}; -b^2, \frac{c^2}{a^2}\right)$ $[a > c > 0; \ \operatorname{Re} s < 0; \ \lvert\arg b\rvert < \pi]$
17	$\theta(a-x)$ $\times \arctan\left(c\sqrt{\frac{a-x}{b-x}}\right)$	$\frac{\sqrt{\pi}\, a^{s+1/2} c}{2\sqrt{b}} \Gamma\left[\begin{matrix} s \\ \frac{2s+3}{2} \end{matrix}\right] F_1\left(\frac{1}{2}, s, 1; s+\frac{3}{2}; \frac{a}{b}, -\frac{ac^2}{b}\right)$ $[b > a > 0; \ \operatorname{Re} s > 0]$
18	$\frac{1}{\sqrt{a+x}} \arctan\left[2bcx \right.$ $\left. \times \frac{1}{x - i(b^2+c^2)x + a}\right]$	$2a^{s-1/2} bc \, \mathrm{B}\left(s+1, \frac{1}{2} - s\right) F_4\left(1, s+1; \frac{3}{2}, \frac{3}{2}; ib^2, ic^2\right)$ $[-1 < \operatorname{Re} s < 1/2; \ \lvert\arg a\rvert < \pi]$
19	$\theta(1-x) \arctan \frac{\ln(-\ln x)}{\pi}$	$\frac{\pi}{s}\left[e^s - \nu(s) - \frac{1}{2}\right]$ $[\operatorname{Re} s > 0]$
20	$\theta(1-x) \arctan \frac{\pi}{\ln(-\ln x)}$	$\frac{\pi}{s}[\nu(s) - 2\sinh s]$ $[\operatorname{Re} s > 0]$

2.6.7. $\arctan(\varphi(x))$ and the exponential function

1	$e^{-ax} \arctan(bx)$	$\frac{\pi}{2a^s}\Gamma(s) - \frac{a^{1-s}}{b}\Gamma(s-1) \, {}_2F_3\left(\begin{matrix} \frac{1}{2}, 1; -\frac{a^2}{4b^2} \\ \frac{3}{2}, \frac{2-s}{2}, \frac{3-s}{2} \end{matrix}\right) - \frac{\pi a b^{-s-1}}{2(s+1)} \csc \frac{s\pi}{2}$ $\times \, {}_1F_2\left(\begin{matrix} \frac{s+1}{2}; -\frac{a^2}{4b^2} \\ \frac{3}{2}, \frac{s+3}{2} \end{matrix}\right) - \frac{\pi b^{-s}}{2s} \sec \frac{s\pi}{2} \, {}_1F_2\left(\begin{matrix} \frac{s}{2}; -\frac{a^2}{4b^2} \\ \frac{1}{2}, \frac{s+2}{2} \end{matrix}\right)$ $[b, \operatorname{Re} a > 0; \ \operatorname{Re} s > -1]$

No.	$f(x)$	$F(s)$
2	$e^{-ax^2}\arctan(bx)$	$\frac{\pi a^{-s/2}}{4}\Gamma\left(\frac{s}{2}\right)-\frac{\pi b^{-s}}{2s}\sec\frac{s\pi}{2}\,{}_1F_1\left(\begin{matrix}\frac{s}{2};\ \frac{a}{b^2}\\ \frac{s+2}{2}\end{matrix}\right)$ $-\frac{a^{(1-s)/2}}{2b}\Gamma\left(\frac{s-1}{2}\right)\,{}_2F_2\left(\begin{matrix}\frac{1}{2},\ 1;\ \frac{a}{b^2}\\ \frac{3}{2},\ \frac{3-s}{2}\end{matrix}\right)$ $[b,\ \operatorname{Re}a>0;\ \operatorname{Re}s>-1]$
3	$\theta(a-x)\,e^{bx}$ $\times\arctan\left(c\sqrt{a-x}\right)$	$\frac{\sqrt{\pi}\,a^{s+1/2}c}{2}\Gamma\left[\begin{matrix}s\\ s+\frac{3}{2}\end{matrix}\right]\Xi_1\left(\frac{1}{2},\ s,\ 1;\ s+\frac{3}{2};\ -ac^2,\ ab\right)$ $[a,\ \operatorname{Re}s>0]$
4	$\arctan\left(ae^{-x}\right)$	$\frac{a}{2^{s+1}}\Gamma(s)\,\Phi\left(-a^2,\ s+1,\ \frac{1}{2}\right)$ $[a,\ \operatorname{Re}s>0]$

2.6.8. $\arctan(\varphi(x))$ and trigonometric functions

1	$\sin(ax)\arctan(bx)$	$\frac{\pi}{2a^s}\sin\frac{s\pi}{2}\Gamma(s)+\frac{\pi ab^{-s-1}}{2(s+1)}\csc\frac{s\pi}{2}\,{}_1F_2\left(\begin{matrix}\frac{s+1}{2};\ \frac{a^2}{4b^2}\\ \frac{3}{2},\ \frac{s+3}{2}\end{matrix}\right)$ $+\frac{a^{1-s}}{b}\cos\frac{s\pi}{2}\Gamma(s-1)\,{}_2F_3\left(\begin{matrix}\frac{1}{2},\ 1;\ \frac{a^2}{4b^2}\\ \frac{3}{2},\ \frac{3-s}{2},\ 1-\frac{s}{2}\end{matrix}\right)$ $[a,\ b>0;\ \lvert\operatorname{Re}s\rvert<1]$
2	$\cos(ax)\arctan(bx)$	$\frac{\pi}{2a^s}\cos\frac{s\pi}{2}\Gamma(s)-\frac{\pi b^{-s}}{2s}\sec\frac{s\pi}{2}\,{}_1F_2\left(\begin{matrix}\frac{s}{2};\ \frac{a^2}{4b^2}\\ \frac{1}{2},\ \frac{s+2}{2}\end{matrix}\right)$ $-\frac{a^{1-s}}{b}\sin\frac{s\pi}{2}\Gamma(s-1)\,{}_2F_3\left(\begin{matrix}\frac{1}{2},\ 1;\ \frac{a^2}{4b^2}\\ \frac{3}{2},\ \frac{2-s}{2},\ \frac{3-s}{2}\end{matrix}\right)$ $[a,\ b>0;\ 0<\operatorname{Re}s<1]$
3	$\sin(ax)\arctan\frac{b}{x}$	$-\frac{\pi ab^{s+1}}{2(s+1)}\csc\frac{s\pi}{2}\,{}_1F_2\left(\begin{matrix}\frac{s+1}{2};\ \frac{a^2b^2}{4}\\ \frac{3}{2},\ \frac{s+3}{2}\end{matrix}\right)$ $-\frac{b}{a^{s-1}}\cos\frac{s\pi}{2}\Gamma(s-1)\,{}_2F_3\left(\begin{matrix}\frac{1}{2},\ 1;\ \frac{a^2b^2}{4}\\ \frac{3}{2},\ \frac{2-s}{2},\ \frac{3-s}{2}\end{matrix}\right)$ $[a,\ b>0;\ -1<\operatorname{Re}s<2]$
4	$\cos(ax)\arctan\frac{b}{x}$	$\frac{\pi b^s}{2s}\sec\frac{s\pi}{2}\,{}_1F_2\left(\begin{matrix}\frac{s}{2};\ \frac{a^2b^2}{4}\\ \frac{1}{2},\ \frac{s+2}{2}\end{matrix}\right)$ $+\frac{b}{a^{s-1}}\sin\frac{s\pi}{2}\Gamma(s-1)\,{}_2F_3\left(\begin{matrix}\frac{1}{2},\ 1;\ \frac{a^2b^2}{4}\\ \frac{3}{2},\ \frac{2-s}{2},\ \frac{3-s}{2}\end{matrix}\right)$ $[a,\ b>0;\ 0<\operatorname{Re}s<2]$

No.	$f(x)$	$F(s)$
5	$\frac{1}{(x^2+a^2)^{\nu/2}} \times \left\{\begin{matrix}\sin(\nu\arctan(x/a))\\ \cos(\nu\arctan(x/a))\end{matrix}\right\}$	$a^{s-\nu}\left\{\begin{matrix}\sin(s\pi/2)\\ \cos(s\pi/2)\end{matrix}\right\}\mathrm{B}(s,\,\nu-s)$ $[\mathrm{Re}\,a>0;\; -(1\pm1)/2<\mathrm{Re}\,s<\mathrm{Re}\,\nu]$
6	$\frac{1}{(x^2+a^2)^{\nu/2}} \times \left\{\begin{matrix}\sin[\nu\,\mathrm{arccot}\,(x/a)]\\ \cos[\nu\,\mathrm{arccot}\,(x/a)]\end{matrix}\right\}$	$a^{s-\nu}\left\{\begin{matrix}\sin[(\nu-s)\pi/2]\\ \cos[(\nu-s)\pi/2]\end{matrix}\right\}\mathrm{B}(s,\,\nu-s)$ $[\mathrm{Re}\,a>0;\, 0<\mathrm{Re}\,s<(1\pm1)/2+\mathrm{Re}\,\nu]$
7	$(1-x)_+^{-1/2}\left[\left(1+\sqrt{1-x}\right)^{\nu} - \left(1-\sqrt{1-x}\right)^{\nu}\right] + 2x^{\nu/2}(x-1)_+^{-1/2} \times \sin\left(\nu\arctan\sqrt{x-1}\right)$	$\frac{\sin(\nu\pi)}{\pi^{3/2}}\,\Gamma\left[s,\, s+\nu,\, \frac{1-2s-\nu}{2},\, \frac{2-2s-\nu}{2}\right]$ $[\lvert\mathrm{Re}\,\nu\rvert<1;\, 0, -\mathrm{Re}\,\nu<\mathrm{Re}\,s<(1-\mathrm{Re}\,\nu)/2]$
8	$(x-1)_+^{-1/2}\left[\left(\sqrt{x}+\sqrt{x-1}\right)^{\nu} - \left(\sqrt{x}-\sqrt{x-1}\right)^{\nu}\right] + 2(1-x)_+^{-1/2} \times \sin\left(\nu\arctan\sqrt{\frac{1-x}{x}}\right)$	$\frac{\sin(\nu\pi)}{\pi^{3/2}}\,\Gamma\left[s,\, \frac{2s+1}{2},\, \frac{1-2s-\nu}{2},\, \frac{1-2s+\nu}{2}\right]$ $[\lvert\mathrm{Re}\,\nu\rvert<1;\, 0<\mathrm{Re}\,s<(1-\lvert\mathrm{Re}\,\nu\rvert)/2]$
9	$\frac{1}{(x^2+2ax\cos\varphi+a^2)^{\rho}} \times \left\{\begin{matrix}\sin u\\ \cos u\end{matrix}\right\}$ $u=2\rho\arctan\frac{a\sin\varphi}{x+a\cos\varphi}$	$a^{s-2\rho}\left\{\begin{matrix}\sin[(2\rho-s)\varphi]\\ \cos[(2\rho-s)\varphi]\end{matrix}\right\}\mathrm{B}(s,\,2\rho-s)$ $[a>0;\, 0\le\varphi<\pi;\, 0<\mathrm{Re}\,s<2\,\mathrm{Re}\,\rho]$

2.6.9. $\arctan(\varphi(x))$ and the logarithmic function

1	$\theta(a-x)\ln\frac{\sqrt{a}+\sqrt{a-x}}{\sqrt{x}} \times \arctan(bx)$	$\frac{\sqrt{\pi}\,a^{s+1}b}{2s(s+1)}\,\Gamma\begin{bmatrix}s+1\\ \frac{2s+3}{2}\end{bmatrix}\left[(s+1)\;{}_4F_3\left(\begin{matrix}\frac{1}{2},\,1,\,\frac{s+1}{2},\,\frac{s+2}{2}\\ \frac{3}{2},\,\frac{2s+3}{4},\,\frac{2s+5}{4};\,-a^2b^2\end{matrix}\right) - {}_4F_3\left(\begin{matrix}1,\,\frac{s+1}{2},\,\frac{s+1}{2},\,\frac{s+2}{2}\\ \frac{2s+3}{4},\,\frac{2s+5}{4},\,\frac{s+3}{2};\,-a^2b^2\end{matrix}\right)\right]$ $[a>0;\; \mathrm{Re}\,s>-1;\, \lvert\arg(1+a^2b^2)\rvert<\pi]$

No.	$f(x)$	$F(s)$
2	$\theta(a-x)\ln\dfrac{a+\sqrt{a^2-x^2}}{x} \times \arctan(bx)$	$\dfrac{\sqrt{\pi}\,a^{s+1}b}{s(s+1)}\Gamma\left[\begin{matrix}\frac{s+1}{2}\\ \frac{s}{2}\end{matrix}\right] {}_4F_3\left(\begin{matrix}\frac{1}{2},\ 1,\ \frac{s+1}{2},\ \frac{s+1}{2}\\ \frac{3}{2},\ \frac{s+2}{2},\ \frac{s+3}{2};\ -a^2b^2\end{matrix}\right)$ $[a>0;\ \operatorname{Re}s>-1;\ \lvert\arg(1+a^2b^2)\rvert<\pi]$
3	$\theta(a-x)\ln\dfrac{a+\sqrt{a^2-x^2}}{a-\sqrt{a^2-x^2}} \times \arctan(bx)$	$\dfrac{2\sqrt{\pi}\,a^{s+1}b}{s^2(s+1)}\Gamma\left[\begin{matrix}\frac{s+1}{2}\\ \frac{s}{2}\end{matrix}\right]\left[(s+1)\,{}_3F_2\left(\begin{matrix}\frac{1}{2},\ 1,\ \frac{s+1}{2}\\ \frac{3}{2},\ \frac{s+2}{2};\ -a^2b^2\end{matrix}\right) - {}_3F_2\left(\begin{matrix}1,\ \frac{s+1}{2},\ \frac{s+1}{2}\\ \frac{s+2}{2},\ \frac{s+3}{2};\ -a^2b^2\end{matrix}\right)\right]$ $[a>0;\ \operatorname{Re}s>-1;\ \lvert\arg(1+a^2b^2)\rvert<\pi]$

2.6.10. $\operatorname{arccsc}(\varphi(x))$ and algebraic functions

1	$\operatorname{arccsc}(ax)$	$\dfrac{i(ia)^{-s}}{2\sqrt{\pi}\,s}\Gamma\left(\dfrac{s}{2}\right)\Gamma\left(\dfrac{1-s}{2}\right)$	$[\operatorname{Im}a<0;\ 0<\operatorname{Re}s<1]$
2	$\theta(x-a)\operatorname{arccsc}\dfrac{x}{a}$	$-\dfrac{\sqrt{\pi}\,a^s}{s^2}\Gamma\left[\begin{matrix}\frac{1-s}{2}\\ -\frac{s}{2}\end{matrix}\right]-\dfrac{\pi a^s}{2s}$	$[a>0;\ \operatorname{Re}s<0]$
3	$\dfrac{\operatorname{arccsc}(ax)}{\sqrt{a^2x^2-1}}$	$-\dfrac{\pi^{3/2}(ia)^{-s}}{4}\csc\dfrac{s\pi}{2}\Gamma\left[\begin{matrix}\frac{s}{2}\\ \frac{s+1}{2}\end{matrix}\right]$	$[\operatorname{Im}a<0;\ 0<\operatorname{Re}s<2]$
4	$\theta(a-x)\operatorname{arccsc}\dfrac{a}{x}$	$-\dfrac{\sqrt{\pi}\,a^s}{s^2}\Gamma\left[\begin{matrix}\frac{s+1}{2}\\ \frac{s}{2}\end{matrix}\right]+\dfrac{\pi a^s}{2s}$	$[a,\ \operatorname{Re}s>0]$
5	$\operatorname{arccsc}^2(ax)$	$-\dfrac{\pi^{3/2}(ia)^{-s}}{2s}\csc\dfrac{s\pi}{2}\Gamma\left[\begin{matrix}\frac{s}{2}\\ \frac{s+1}{2}\end{matrix}\right]$	$[\operatorname{Im}a<0;\ 0<\operatorname{Re}s<2]$

2.6.11. $\operatorname{arcsec}(bx)$ and algebraic functions

1	$\theta(x-a)\operatorname{arcsec}\dfrac{x}{a}$	$\dfrac{\sqrt{\pi}\,a^s}{s^2}\Gamma\left[\begin{matrix}\frac{1-s}{2}\\ -\frac{s}{2}\end{matrix}\right]$	$[a>0;\ \operatorname{Re}s<0]$
2	$\operatorname{arcsec}(ax)-\dfrac{\pi}{2}$	$\dfrac{i}{2\sqrt{\pi}\,s}\left(-\dfrac{1}{a^2}\right)^{s/2}\Gamma\left(\dfrac{1-s}{2}\right)\Gamma\left(\dfrac{s}{2}\right)$	$[\operatorname{Re}a>0;\ 0<\operatorname{Re}s<1]$
3	$\operatorname{arcsec}^2(ax)-\dfrac{\pi^2}{4}$	$-\dfrac{\pi^{3/2}(ia)^{-s}}{s}e^{is\pi/2}\csc(s\pi)\Gamma\left[\begin{matrix}\frac{s}{2}\\ \frac{s+1}{2}\end{matrix}\right]$	$[\operatorname{Im}a<0;\ 0<\operatorname{Re}s<1]$

2.6.12. Products of inverse trigonometric functions

1	$\theta(a-x)\arcsin^2(bx)$	$\frac{a^s\arcsin^2(ab)}{s}-\frac{2a^{s+2}b^2}{s(s+2)}\,{}_3F_2\left(\begin{matrix}1,\,1,\,\frac{s+2}{2}\\ \frac{3}{2},\,\frac{s+4}{2};\,a^2b^2\end{matrix}\right)$ $[a>0;\ \operatorname{Re}s>-2;\ \lvert\arg(1-a^2b^2)\rvert<\pi]$
2	$\arcsin^2(ax)$	$-\frac{\pi^{3/2}(ia)^{-s}}{2s}\csc\frac{s\pi}{2}\,\Gamma\left[\begin{matrix}-\frac{s}{2}\\ \frac{1-s}{2}\end{matrix}\right]$ $[\operatorname{Im}a<0;\ -2<\operatorname{Re}s<0]$
3	$\arccos^2(ax)-\frac{\pi^2}{4}$	$-\frac{\pi^{3/2}(ia)^{-s}}{2s}e^{is\pi/2}\csc(s\pi)\,\Gamma\left[\begin{matrix}-\frac{s}{2}\\ \frac{1-s}{2}\end{matrix}\right]$ $[\operatorname{Im}a<0;\ -1<\operatorname{Re}s<0]$
4	$\theta(a-x)\arcsin(bx)$ $\times\arccos\frac{x}{a}$	$\frac{\sqrt{\pi}\,a^{s+1}b}{4(s+1)}\,\Gamma\left[\begin{matrix}\frac{s}{2}\\ \frac{s+3}{2}\end{matrix}\right]\left[(s+1)\,{}_3F_2\left(\begin{matrix}\frac{1}{2},\,\frac{1}{2},\,\frac{s+2}{2}\\ \frac{3}{2},\,\frac{s+3}{2};\,a^2b^2\end{matrix}\right)-{}_3F_2\left(\begin{matrix}\frac{1}{2},\,\frac{s+1}{2},\,\frac{s+2}{2}\\ \frac{s+3}{2},\,\frac{s+3}{2};\,a^2b^2\end{matrix}\right)\right]$ $[a>0;\ \operatorname{Re}s>-1;\ \lvert\arg(1-a^2b^2)\rvert<\pi]$
5	$\frac{\theta(a-x)}{\sqrt{1-b^2x^2}}\arcsin(bx)$ $\times\arccos\frac{x}{a}$	$\frac{\sqrt{\pi}\,a^{s+1}b}{2(s+1)}\,\Gamma\left[\begin{matrix}\frac{s+2}{2}\\ \frac{s+3}{2}\end{matrix}\right]{}_4F_3\left(\begin{matrix}1,\,1,\,\frac{s+1}{2},\,\frac{s+2}{2}\\ \frac{3}{2},\,\frac{s+3}{2},\,\frac{s+3}{2};\,a^2b^2\end{matrix}\right)$ $[a>0;\ \operatorname{Re}s>-1;\ \lvert\arg(1-a^2b^2)\rvert<\pi]$
6	$\theta(a-x)\arctan(bx)$ $\times\arccos\frac{x}{a}$	$\frac{\sqrt{\pi}\,a^{s+1}b}{4(s+1)}\,\Gamma\left[\begin{matrix}\frac{s}{2}\\ \frac{s+3}{2}\end{matrix}\right]\left[(s+1)\,{}_3F_2\left(\begin{matrix}\frac{1}{2},\,1,\,\frac{s+2}{2}\\ \frac{3}{2},\,\frac{s+3}{2};\,-a^2b^2\end{matrix}\right)-{}_3F_2\left(\begin{matrix}1,\,\frac{s+1}{2},\,\frac{s+2}{2}\\ \frac{s+3}{2},\,\frac{s+3}{2};\,-a^2b^2\end{matrix}\right)\right]$ $[a>0;\ \operatorname{Re}s>-1;\ \lvert\arg(1+a^2b^2)\rvert<\pi]$
7	$(a-x)_+^{\alpha-1}\arcsin^2(bx)$	$a^{s+\alpha+1}b^2\,\mathrm{B}(s+2,\,\alpha)\,{}_5F_4\left(\begin{matrix}1,\,1,\,1,\,\frac{s+2}{2},\,\frac{s+3}{2};\,a^2b^2\\ \frac{3}{2},\,2,\,\frac{s+\alpha+2}{2},\,\frac{s+\alpha+3}{2}\end{matrix}\right)$ $[a,\ \operatorname{Re}\alpha>0;\ \operatorname{Re}s>-2;\ \lvert\arg(1-a^2b^2)\rvert<\pi]$
8	$(a^2-x^2)_+^{\alpha-1}\arcsin^2(bx)$	$\frac{a^{s+2\alpha}b^2}{2}\,\mathrm{B}\left(\frac{s+2}{2},\,\alpha\right){}_4F_3\left(\begin{matrix}1,\,1,\,1,\,\frac{s+2}{2};\,a^2b^2\\ \frac{3}{2},\,2,\,\frac{s+2\alpha+2}{2}\end{matrix}\right)$ $[a,\ \operatorname{Re}\alpha>0;\ \operatorname{Re}s>-2;\ \lvert\arg(1-a^2b^2)\rvert<\pi]$
9	$\theta(a-x)\ln\frac{\sqrt{a}+\sqrt{a-x}}{\sqrt{x}}$ $\times\arcsin^2(bx)$	$\frac{\sqrt{\pi}\,a^{s+2}b^2}{2s}\,\Gamma\left[\begin{matrix}s+2\\ \frac{2s+5}{2}\end{matrix}\right]\left[{}_5F_4\left(\begin{matrix}1,\,1,\,1,\,\frac{s+2}{2},\,\frac{s+3}{2}\\ \frac{3}{2},\,2,\,\frac{2s+5}{4},\,\frac{2s+7}{4};\,a^2b^2\end{matrix}\right)-\frac{2}{s+2}\,{}_5F_4\left(\begin{matrix}1,\,1,\,\frac{s+2}{2},\,\frac{s+2}{2},\,\frac{s+3}{2}\\ \frac{3}{2},\,\frac{2s+5}{4},\,\frac{2s+7}{4},\,\frac{s+4}{2};\,a^2b^2\end{matrix}\right)\right]$ $[a>0;\ \operatorname{Re}s>-2;\ \lvert\arg(1-a^2b^2)\rvert<\pi]$

No.	$f(x)$	$F(s)$
10	$\theta(a-x)\ln\frac{a+\sqrt{a^2-x^2}}{x}$ $\times\arcsin^2(bx)$	$\frac{\sqrt{\pi}\,a^{s+2}b^2}{2(s+2)}\Gamma\left[\begin{matrix}\frac{s+2}{2}\\ \frac{s+3}{2}\end{matrix}\right]{}_5F_4\left(\begin{matrix}1,\,1,\,1,\,\frac{s+2}{2},\,\frac{s+2}{2};\,a^2b^2\\ \frac{3}{2},\,2,\,\frac{s+3}{2},\,\frac{s+4}{2}\end{matrix}\right)$ $[a>0;\ \operatorname{Re}s>-2;\ \lvert\arg(1-a^2b^2)\rvert<\pi]$
11	$(a-x)_+^{\alpha-1}$ $\times\arcsin^2(b(a-x))$	$a^{s+\alpha+1}b^2\,\mathrm{B}(s,\,\alpha+2)\,{}_5F_4\left(\begin{matrix}1,\,1,\,1,\,\frac{\alpha+2}{2},\,\frac{\alpha+3}{2};\,a^2b^2\\ \frac{3}{2},\,2,\,\frac{s+\alpha+2}{2},\,\frac{s+\alpha+3}{2}\end{matrix}\right)$ $[a,\ \operatorname{Re}s>0;\ \operatorname{Re}\alpha>-2;\ \lvert\arg(1-a^2b^2)\rvert<\pi]$
12	$(a-x)_+^{\alpha-1}$ $\times\arcsin^2(bx(a-x))$	$a^{s+\alpha+3}b^2\,\mathrm{B}(s+2,\,\alpha+2)$ $\times\,{}_7F_6\left(\begin{matrix}1,\,1,\,1,\,\Delta(2,\,\alpha+2),\,\Delta(2,\,s+2)\\ \frac{3}{2},\,2,\,\Delta(4,\,s+\alpha+4);\,\frac{a^4b^2}{16}\end{matrix}\right)$ $[a>0;\ \operatorname{Re}s,\ \operatorname{Re}\alpha>-2;\ \lvert\arg(16-a^4b^2)\rvert<\pi]$
13	$(a-x)_+^{\alpha-1}$ $\times\arcsin^2(b\sqrt{a-x})$	$a^{s+\alpha}b^2\,\mathrm{B}(s,\,\alpha+1)\,{}_4F_3\left(\begin{matrix}1,\,1,\,1,\,\alpha+1;\,ab^2\\ \frac{3}{2},\,2,\,s+\alpha+1\end{matrix}\right)$ $[a,\ \operatorname{Re}s>0;\ \operatorname{Re}\alpha>-1;\ \lvert\arg(1-ab^2)\rvert<\pi]$
14	$(a-x)_+^{\alpha-1}$ $\times\arcsin^2\left(b\sqrt{x(a-x)}\right)$	$a^{s+\alpha+1}b^2\,\mathrm{B}(s+1,\,\alpha+1)\,{}_5F_4\left(\begin{matrix}1,\,1,\,1,\,\alpha+1,\,s+1;\,\frac{a^2b^2}{4}\\ \frac{3}{2},\,2,\,\frac{s+\alpha+2}{2},\,\frac{s+\alpha+3}{2}\end{matrix}\right)$ $[a>0;\ \operatorname{Re}s,\ \operatorname{Re}\alpha>-1;\ \lvert\arg(4-a^2b^2)\rvert<\pi]$
15	$(a-x)_+^{\alpha-1}$ $\times\arcsin^2\left(bx\sqrt{a-x}\right)$	$a^{s+\alpha+2}b^2\,\mathrm{B}(s+2,\,\alpha+1)\,{}_6F_5\left(\begin{matrix}1,\,1,\,1,\,\alpha+1,\,\frac{s+2}{2},\,\frac{s+3}{2};\,\frac{4a^3b^2}{27}\\ \frac{3}{2},\,2,\,\frac{s+\alpha+3}{3},\,\frac{s+\alpha+4}{3},\,\frac{s+\alpha+5}{3}\end{matrix}\right)$ $[a>0;\ \operatorname{Re}\alpha>-1;\ \operatorname{Re}s>-2;\ \lvert\arg(27-4a^3b^2)\rvert<\pi]$
16	$\frac{1}{(x+a)^\rho}\arcsin^2\frac{b}{x+a}$	$a^{s-\rho-2}b^2\,\mathrm{B}(s,\,2-s+\rho)\,{}_5F_4\left(\begin{matrix}1,\,1,\,1,\,\frac{2-s+\rho}{2},\,\frac{3-s+\rho}{2}\\ \frac{3}{2},\,2,\,\frac{\rho+2}{2},\,\frac{\rho+3}{2};\,\frac{b^2}{a^2}\end{matrix}\right)$ $[0<\operatorname{Re}s<\operatorname{Re}\rho+2;\ \lvert\arg a\rvert<\pi]$
17	$\frac{1}{(x+a)^\rho}\arcsin^2\frac{bx}{x+a}$	$a^{s-\rho}b^2\,\mathrm{B}(s+2,\,\rho-s)\,{}_5F_4\left(\begin{matrix}1,\,1,\,1,\,\frac{s+2}{2},\,\frac{s+3}{2}\\ \frac{3}{2},\,2,\,\frac{\rho+2}{2},\,\frac{\rho+3}{2};\,b^2\end{matrix}\right)$ $[-2<\operatorname{Re}s<\operatorname{Re}\rho;\ \lvert\arg a\rvert<\pi]$
18	$\theta(a-x)\arcsin^2(bx)$ $\times\arccos\frac{x}{a}$	$\frac{\sqrt{\pi}\,a^{s+2}b^2}{2s(s+2)}\Gamma\left[\begin{matrix}\frac{s+3}{2}\\ \frac{s+4}{2}\end{matrix}\right]\left[(s+2)\,{}_4F_3\left(\begin{matrix}1,\,1,\,1,\,\frac{s+3}{2}\\ \frac{3}{2},\,2,\,\frac{s+4}{2};\,a^2b^2\end{matrix}\right)\right.$ $\left.-\,2\,{}_4F_3\left(\begin{matrix}1,\,1,\,\frac{s+2}{2},\,\frac{s+3}{2}\\ \frac{3}{2},\,\frac{s+4}{2},\,\frac{s+4}{2};\,a^2b^2\end{matrix}\right)\right]$ $[a>0;\ \operatorname{Re}s>-2;\ \lvert\arg(1-a^2b^2)\rvert<\pi]$

2.7. Inverse Hyperbolic Functions

More formulas can be obtained from the corresponding sections due to the relations

$$\operatorname{arcsinh} z = \ln\left(z + \sqrt{z^2+1}\right); \ \operatorname{arccosh} z = \ln\left(z + \sqrt{z^2-1}\right), \ -\pi/2 < \arg z \leq \pi/2;$$

$$\operatorname{arcsinh} z = z\,{}_2F_1\left(\frac{1}{2}, \frac{1}{2}; \frac{3}{2}; -z^2\right), \quad \operatorname{arccosh} z = \frac{\sqrt{z-1}}{\sqrt{1-z}}\left[\frac{\pi}{2} - z\,{}_2F_1\left(\frac{1}{2}, \frac{1}{2}; \frac{3}{2}; z^2\right)\right],$$

$$\frac{\operatorname{arcsinh} z}{\sqrt{z^2+1}} = z\,{}_2F_1\left(1, 1; \frac{3}{2}; -z^2\right), \quad \operatorname{arcsinh}^2 z = z^2\,{}_3F_2\left(1, 1, 1; \frac{3}{2}, 2; -z^2\right),$$

$$\operatorname{arctanh} z = \frac{1}{2}\left[\ln(1+z) - \ln(1-z)\right], \quad \operatorname{arctanh} z = z\,{}_2F_1\left(1, \frac{1}{2}; \frac{3}{2}; z^2\right),$$

$$\operatorname{arccoth} z = \frac{1}{2}\left[\ln\frac{z+1}{z} - \ln\frac{z-1}{z}\right],$$

$$\operatorname{arccoth} z = -\frac{\pi z}{2}\sqrt{-\frac{1}{z^2}}\sqrt{\frac{1}{1-z^2}}\sqrt{1-z^2} + z\,{}_2F_1\left(\frac{1}{2}, 1; \frac{3}{2}; z^2\right),$$

$$\operatorname{arccsch} z = \frac{1}{z}\,{}_2F_1\left(\frac{1}{2}, \frac{1}{2}; \frac{3}{2}; -\frac{1}{z^2}\right), \quad \operatorname{arcsech} z = \frac{\sqrt{z^{-1}-1}}{\sqrt{1-z^{-1}}}\left[\frac{\pi}{2} - \frac{1}{z}\,{}_2F_1\left(\frac{1}{2}, \frac{1}{2}; \frac{3}{2}; \frac{1}{z^2}\right)\right],$$

$$\operatorname{arcsinh} z = \frac{1}{2\sqrt{\pi}\,z}\,G_{22}^{12}\left(z^2 \,\middle|\, \begin{matrix} 3/2,\, 3/2 \\ 1,\, 1/2 \end{matrix}\right),$$

$$\operatorname{arccosh} z = \frac{\sqrt{z-1}}{\sqrt{1-z}}\left[\frac{\pi}{2} - \frac{z}{2\sqrt{\pi}}\,G_{22}^{12}\left(-z^2 \,\middle|\, \begin{matrix} 1/2,\, 1/2 \\ 0,\, -1/2 \end{matrix}\right)\right],$$

$$\operatorname{arctanh} z = -\frac{1}{2z}\,G_{22}^{12}\left(-z^2 \,\middle|\, \begin{matrix} 1,\, 3/2 \\ 1,\, 1/2 \end{matrix}\right), \quad \operatorname{arccoth} z = \frac{1}{2z}\,G_{22}^{12}\left(-\frac{1}{z^2} \,\middle|\, \begin{matrix} 0,\, 1/2 \\ 0,\, -1/2 \end{matrix}\right).$$

2.7.1. $\operatorname{arcsinh}^n(\varphi(x))$ and elementary functions

No.	$f(x)$	$F(s)$
1	$\operatorname{arcsinh}(ax)$	$\frac{a^{-s}}{2s^2}\,\mathrm{B}\left(\frac{s+1}{2}, \frac{2-s}{2}\right)$ [$\operatorname{Re} a > 0$; $-1 < \operatorname{Re} s < 0$]
2	$\operatorname{arcsinh}(ax) - ax$	$\frac{a^{-s}}{2s^2}\,\mathrm{B}\left(\frac{s+1}{2}, \frac{2-s}{2}\right)$ [$\operatorname{Re} a > 0$; $-3 < \operatorname{Re} s < -1$]
3	$\operatorname{arcsinh}(ax) - \sum_{k=0}^{n} (-1)^k \frac{(1/2)_k}{k!\,(2k+1)} (ax)^{2k+1}$	$\frac{a^{-s}}{2s^2}\,\mathrm{B}\left(\frac{s+1}{2}, \frac{2-s}{2}\right)$ [$\operatorname{Re} a > 0$; $-2n-3 < \operatorname{Re} s < -2n-1$]
4	$\operatorname{arcsinh}(ax) - \ln(2ax) + \frac{1}{2}\sum_{k=1}^{n} (-1)^k \frac{(1/2)_k}{k!\,k} (ax)^{-2k}$	$\frac{a^{-s}}{2s^2}\,\mathrm{B}\left(\frac{s+1}{2}, \frac{2-s}{2}\right)$ [$\operatorname{Re} a > 0$; $2n < \operatorname{Re} s < 2n+2$]

No.	$f(x)$	$F(s)$
5	$\theta(a-x)\operatorname{arcsinh}(bx)$	$-\frac{a^{s+1}b}{s(s+1)}\,{}_2F_1\left(\begin{matrix}\frac{1}{2},\ \frac{s+1}{2}\\ \frac{s+3}{2};\ -a^2b^2\end{matrix}\right)+\operatorname{arcsinh}(ab)\,\frac{a^s}{s}$ $[a>0;\ \operatorname{Re}s>-1]$
6	$(a-x)_+^{\alpha-1}\operatorname{arcsinh}(bx)$	$a^{s+\alpha}b\,\mathrm{B}(\alpha,\ s+1)\ {}_4F_3\left(\begin{matrix}\frac{1}{2},\ \frac{1}{2},\ \frac{s+1}{2},\ \frac{s+2}{2};\ -a^2b^2\\ \frac{3}{2},\ \frac{s+\alpha+1}{2},\ \frac{s+\alpha+2}{2}\end{matrix}\right)$ $[a,\ \operatorname{Re}\alpha>0;\ \operatorname{Re}s>-1]$
7	$\frac{1}{\sqrt{a^2x^2+1}}\operatorname{arcsinh}(ax)$	$\frac{\pi^{3/2}a^{-s}}{4}\sec\frac{s\pi}{2}\,\Gamma\left[\begin{matrix}\frac{1-s}{2}\\ \frac{2-s}{2}\end{matrix}\right]$ $[\operatorname{Re}a>0;\ -1<\operatorname{Re}s<1]$
8	$\theta(a-x)\arccos\frac{x}{a}\operatorname{arcsinh}(bx)$	$\frac{\sqrt{\pi}a^{s+1}b}{2(s+1)}\,\Gamma\left[\begin{matrix}\frac{s+2}{2}\\ \frac{s+3}{2}\end{matrix}\right]\,{}_4F_3\left(\begin{matrix}\frac{1}{2},\ \frac{1}{2},\ \frac{s+1}{2},\ \frac{s+2}{2}\\ \frac{3}{2},\ \frac{s+3}{2},\ \frac{s+3}{2};\ -a^2b^2\end{matrix}\right)$ $[a>0;\ \operatorname{Re}s>-1]$
9	$\frac{1}{\sqrt{a^2x^2+1}}\operatorname{arcsinh}\frac{1}{ax}$	$\frac{\pi^{3/2}a^{-s}}{4}\csc\frac{s\pi}{2}\,\Gamma\left[\begin{matrix}\frac{s}{2}\\ \frac{s+1}{2}\end{matrix}\right]$ $[\operatorname{Re}a>0;\ 0<\operatorname{Re}s<2]$
10	$\operatorname{arcsinh}\sqrt{\frac{\sqrt{ax+1}-1}{2}}$	$-\frac{a^{-s}}{4s}\,\mathrm{B}\left(s+\frac{1}{2},\ -s\right)$ $[-1/2<\operatorname{Re}s<0;\ \lvert\arg a\rvert<\pi]$
11	$\operatorname{arcsinh}\sqrt{\frac{\sqrt{ax+1}-\sqrt{ax}}{2\sqrt{ax}}}$	$\frac{a^{-s}}{4s}\,\mathrm{B}\left(s,\ \frac{1}{2}-s\right)$ $[0<\operatorname{Re}s<1/2;\ \lvert\arg a\rvert<\pi]$
12	$\operatorname{arcsinh}^2(ax)$	$-\frac{\pi^{3/2}a^{-s}}{s^2}\csc\frac{s\pi}{2}\,\Gamma\left[\begin{matrix}\frac{2-s}{2}\\ \frac{1-s}{2}\end{matrix}\right]$ $[\operatorname{Re}a>0;\ -2<\operatorname{Re}s<0]$
13	$(a-x)_+^{\alpha-1}\operatorname{arcsinh}^2(bx)$	$a^{s+\alpha+1}b^2\,\mathrm{B}(\alpha,\ s+2)\ {}_5F_4\left(\begin{matrix}1,\ 1,\ 1,\ \frac{s+2}{2},\ \frac{s+3}{2};\ -a^2b^2\\ \frac{3}{2},\ 2,\ \frac{s+\alpha+2}{2},\ \frac{s+\alpha+3}{2}\end{matrix}\right)$ $[a,\ \operatorname{Re}\alpha>0;\ \operatorname{Re}s>-2]$
14	$\theta(a-x)\arccos\frac{x}{a}\operatorname{arcsinh}^2(bx)$	$\frac{\sqrt{\pi}a^{s+2}b^2}{2(s+2)}\,\Gamma\left[\begin{matrix}\frac{s+3}{2}\\ \frac{s+4}{2}\end{matrix}\right]\,{}_5F_4\left(\begin{matrix}1,\ 1,\ 1,\ \frac{s+2}{2},\ \frac{s+3}{2}\\ \frac{3}{2},\ 2,\ \frac{s+4}{2},\ \frac{s+4}{2};\ -a^2b^2\end{matrix}\right)$ $[a>0;\ \operatorname{Re}s>-2]$
15	$\operatorname{arcsinh}^2\sqrt{\frac{\sqrt{ax+1}-1}{2}}$	$\frac{\pi^{3/2}a^{-s}}{8s}\csc(s\pi)\,\Gamma\left[\begin{matrix}-s\\ \frac{1-2s}{2}\end{matrix}\right]$ $[-1<\operatorname{Re}s<0;\ \lvert\arg a\rvert<\pi]$
16	$\operatorname{arcsinh}^2\sqrt{\frac{\sqrt{ax+1}-\sqrt{ax}}{2\sqrt{ax}}}$	$\frac{\pi^{3/2}a^{-s}}{8s}\csc(s\pi)\,\Gamma\left[\begin{matrix}s\\ \frac{2s+1}{2}\end{matrix}\right]$ $[0<\operatorname{Re}s<1;\ \lvert\arg a\rvert<\pi]$

2.7.2. $\operatorname{arccosh}^n(\varphi(x))$ and elementary functions

1	$\operatorname{arccosh}(ax)+\frac{i\pi}{2}$	$\frac{(ia)^{-s}}{2s^2}\,\mathrm{B}\left(\frac{s+1}{2},\,1-s\right)$ $[\operatorname{Im}a<0;\ -1<\operatorname{Re}s<0]$
2	$\operatorname{arccosh}(ax)-\frac{\pi}{2}\frac{\sqrt{ax-1}}{\sqrt{1-ax}}+\frac{\sqrt{ax-1}}{\sqrt{1-ax}}\sum_{k=0}^{n}\frac{(1/2)_k}{k!\,(2k+1)}(ax)^{2k+1}$	$\frac{i\sqrt{a-1}}{2\sqrt{\pi}\sqrt{1-a}\,s^2}(ia)^{-s}\,\mathrm{B}\left(\frac{s+1}{2},\,1-s\right)$ $[\operatorname{Im}a<0;\,-2n-3<\operatorname{Re}s<-2n-1]$
3	$\operatorname{arccosh}(ax)-\frac{1}{2}\ln\left(-4a^2x^2\right)+\frac{\pi\sqrt{-a^2}}{2a}+\frac{1}{2}\sum_{k=1}^{n}\frac{(1/2)_k}{k\,k!}(ax)^{-2k}$	$\frac{a^{-s}e^{is\pi/2}}{2s^2}\,\mathrm{B}\left(\frac{s+1}{2},\,1-s\right)$ $[\operatorname{Im}a>0;\ 2n<\operatorname{Re}s<2n+2]$
4	$\operatorname{arccosh}\left(\sqrt{ax}+\sqrt{ax+1}\right)$	$-\frac{a^{-s}}{s}\cos(s\pi)\,\mathrm{B}\left(-2s,\,2s+\frac{1}{2}\right)$ $[-1/4<\operatorname{Re}s<0;\ \lvert\arg a\rvert<\pi]$
5	$\operatorname{arccosh}\frac{1}{\sqrt{ax+1}-\sqrt{ax}}$	$-\frac{a^{-s}}{s}\cos(s\pi)\,\mathrm{B}\left(-2s,\,2s+\frac{1}{2}\right)$ $[-1/4<\operatorname{Re}s<0;\ \lvert\arg a\rvert<\pi]$
6	$\operatorname{arccosh}\frac{\sqrt{ax}}{\sqrt{ax+1}-1}$	$\frac{a^{-s}}{s}\cos(s\pi)\,\mathrm{B}\left(2s,\,\frac{1}{2}-2s\right)$ $[0<\operatorname{Re}s<1/4;\ \lvert\arg a\rvert<\pi]$
7	$\operatorname{arccosh}\frac{\sqrt{ax+1}+1}{\sqrt{ax}}$	$\frac{a^{-s}}{s}\cos(s\pi)\,\mathrm{B}\left(2s,\,\frac{1}{2}-2s\right)$ $[0<\operatorname{Re}s<1/4;\ \lvert\arg a\rvert<\pi]$
8	$\operatorname{arccosh}^2(ax)+\frac{\pi^2}{4}$	$\frac{\pi^{3/2}}{s}(ia)^{-s}e^{is\pi/2}\csc(s\pi)\,\Gamma\left[\begin{matrix}-\frac{s}{2}\\ \frac{1-s}{2}\end{matrix}\right]$ $[\operatorname{Im}a<0;\ -1<\operatorname{Re}s<0]$

2.7.3. $\operatorname{arctanh}(ax)$ and elementary functions

1	$\operatorname{arctanh}(ax)$	$\frac{i\pi}{2s}(ia)^{-s}\sec\frac{s\pi}{2}$ $[\operatorname{Im}a<0;\ -1<\operatorname{Re}s<0]$
2	$\operatorname{arctanh}(ax)-ax$	$\frac{i\pi}{2s}(ia)^{-s}\sec\frac{s\pi}{2}$ $[\operatorname{Im}a<0;\ -3<\operatorname{Re}s<-1]$

No.	$f(x)$	$F(s)$
3	$\operatorname{arctanh}(ax) - \sum_{k=0}^{n} \frac{(ax)^{2k+1}}{2k+1}$	$\frac{i\pi}{2s}(ia)^{-s}\sec\frac{s\pi}{2}$ $[\operatorname{Im} a < 0;\ -2n-3 < \operatorname{Re} s < -2n-1]$
4	$\operatorname{arctanh}(ax) - \frac{\pi i}{2} - \sum_{k=0}^{n} \frac{(ax)^{-2k-1}}{2k+1}$	$\frac{i\pi}{2s}\left(\frac{i}{a}\right)^{s}\sec\frac{s\pi}{2}$ $[\operatorname{Im} a > 0;\ 2n+1 < \operatorname{Re} s < 2n+3]$
5	$(a-x)_{+}^{\alpha-1}\operatorname{arctanh}(bx)$	$a^{s+\alpha} b \operatorname{B}(\alpha,\ s+1)\ {}_4F_3\left(\begin{matrix}\frac{1}{2},\ 1,\ \frac{s+1}{2},\ \frac{s+2}{2};\ a^2b^2\\ \frac{3}{2},\ \frac{s+\alpha+1}{2},\ \frac{s+\alpha+2}{2}\end{matrix}\right)$ $[a,\ \operatorname{Re}\alpha > 0;\ \operatorname{Re} s > -1]$
6	$\theta(a-x)\ln\frac{\sqrt{a-x}+\sqrt{a}}{\sqrt{x}}$ $\times \operatorname{arctanh}(bx)$	$\frac{\sqrt{\pi}a^{s+1}b}{2s(s+1)}\Gamma\left[\begin{matrix}s+1\\ \frac{2s+3}{2}\end{matrix}\right]\left[(s+1)\ {}_4F_3\left(\begin{matrix}\frac{1}{2},\ 1,\ \frac{s+1}{2},\ \frac{s+2}{2}\\ \frac{3}{2},\ \frac{2s+3}{4},\ \frac{2s+5}{4};\ a^2b^2\end{matrix}\right) - {}_4F_3\left(\begin{matrix}1,\ \frac{s+1}{2},\ \frac{s+1}{2},\ \frac{s+2}{2}\\ \frac{2s+3}{4},\ \frac{2s+5}{4},\ \frac{s+3}{2};\ a^2b^2\end{matrix}\right)\right]$ $[a > 0;\ \operatorname{Re} s > -1]$
7	$\theta(a-x)\arccos\frac{x}{a}\operatorname{arctanh}(bx)$	$\frac{\sqrt{\pi}a^{s+1}b}{2(s+1)}\Gamma\left[\begin{matrix}\frac{s+2}{2}\\ \frac{s+3}{2}\end{matrix}\right]{}_4F_3\left(\begin{matrix}\frac{1}{2},\ 1,\ \frac{s+1}{2},\ \frac{s+2}{2}\\ \frac{3}{2},\ \frac{s+3}{2},\ \frac{s+3}{2};\ a^2b^2\end{matrix}\right)$ $[a > 0;\ \operatorname{Re} s > -1]$

2.7.4. arccoth (ax) and algebraic functions

1	$\operatorname{arccoth}(ax)$	$-\frac{i\pi}{2s}(-ia)^{-s}\sec\frac{s\pi}{2}$ $[\operatorname{Im} a > 0;\ 0 < \operatorname{Re} s < 1]$
2	$\operatorname{arccoth}(ax) - \frac{\pi i}{2} - ax$	$\frac{i\pi}{2s}(ia)^{-s}\sec\frac{s\pi}{2}$ $[\operatorname{Im} a < 0;\ -3 < \operatorname{Re} s < -1]$
3	$\operatorname{arccoth}(ax) - \frac{\pi i}{2} - \sum_{k=0}^{n}\frac{(ax)^{2k+1}}{2k+1}$	$\frac{i\pi}{2s}(ia)^{-s}\sec\frac{s\pi}{2}$ $[\operatorname{Im} a < 0;\ -2n-3 < \operatorname{Re} s < -2n-1]$
4	$\operatorname{arccoth}(ax) - \sum_{k=0}^{n}\frac{(ax)^{-2k-1}}{2k+1}$	$-\frac{i\pi}{2s}\left(-\frac{i}{a}\right)^{s}\sec\frac{s\pi}{2}$ $[\operatorname{Im} a < 0;\ 2n+1 < \operatorname{Re} s < 2n+3]$

2.7.5. $\operatorname{arcsech}^n(\varphi(x))$ and elementary functions

1	$\operatorname{arcsech}(ax) + \frac{i\pi}{2}$	$\frac{a^{-s}}{2s^2} e^{is\pi/2} \mathrm{B}\left(\frac{s+2}{2}, \frac{1-s}{2}\right)$ $[\operatorname{Im} a > 0;\ 0 < \operatorname{Re} s < 1]$
2	$\operatorname{arcsech}(ax) - \frac{1}{2}\ln\left(-\frac{4}{a^2x^2}\right)$ $+ \frac{\pi a}{2}\sqrt{-\frac{1}{a^2x^2}} + \frac{1}{2}\sum_{k=1}^{n}\frac{(1/2)_k}{k!\,k}(ax)^{2k}$	$\frac{(ia)^{-s}}{s^2} \mathrm{B}\left(\frac{s+2}{2}, \frac{1-s}{2}\right)$ $[\operatorname{Im} a < 0;\ -2n-2 < \operatorname{Re} s < -2n]$
3	$\operatorname{arcsech}\left(\sqrt{ax+1} - \sqrt{ax}\right)$	$-\frac{a^{-s}}{s}\cos(s\pi)\,\mathrm{B}\left(2s+\frac{1}{2}, -2s\right)$ $[-1/4 < \operatorname{Re} s < 0;\ \lvert\arg a\rvert < \pi]$
4	$\operatorname{arcsech}\frac{1}{\sqrt{ax} + \sqrt{ax+1}}$	$-\frac{a^{-s}}{s}\cos(s\pi)\,\mathrm{B}\left(2s+\frac{1}{2}, -2s\right)$ $[-1/4 < \operatorname{Re} s < 0;\ \lvert\arg a\rvert < \pi]$
5	$\operatorname{arcsech}\frac{\sqrt{ax}}{\sqrt{ax+1}+1}$	$\frac{a^{-s}}{s}\cos(s\pi)\,\mathrm{B}\left(2s, \frac{1}{2}-2s\right)$ $[0 < \operatorname{Re} s < 1/4;\ \lvert\arg a\rvert < \pi]$
6	$\operatorname{arcsech}\frac{\sqrt{ax+1}-1}{\sqrt{ax}}$	$\frac{a^{-s}}{s}\cos(s\pi)\,\mathrm{B}\left(2s, \frac{1}{2}-2s\right)$ $[0 < \operatorname{Re} s < 1/4;\ \lvert\arg a\rvert < \pi]$
7	$\operatorname{arcsech}\left(\sqrt{a^2x^2+1} - ax\right)$	$-\frac{a^{-s}}{s}\cos\frac{s\pi}{2}\,\mathrm{B}\left(s+\frac{1}{2}, -s\right)$ $[\operatorname{Re} a > 0;\ -1/2 < \operatorname{Re} s < 0]$
8	$\operatorname{arcsech}\frac{1}{ax + \sqrt{a^2x^2+1}}$	$-\frac{a^{-s}}{s}\cos\frac{s\pi}{2}\,\mathrm{B}\left(s+\frac{1}{2}, -s\right)$ $[\operatorname{Re} a > 0;\ -1/2 < \operatorname{Re} s < 0]$
9	$\operatorname{arcsech}\frac{\sqrt{a^2x^2+1}-1}{ax}$	$\frac{a^{-s}}{s}\cos\frac{s\pi}{2}\,\mathrm{B}\left(s, \frac{1}{2}-s\right)$ $[\operatorname{Re} a > 0;\ 0 < \operatorname{Re} s < 1/2]$
10	$\operatorname{arcsech}\frac{ax}{\sqrt{a^2x^2+1}+1}$	$\frac{a^{-s}}{s}\cos\frac{s\pi}{2}\,\mathrm{B}\left(s, \frac{1}{2}-s\right)$ $[\operatorname{Re} a > 0;\ 0 < \operatorname{Re} s < 1/2]$
11	$\operatorname{arcsech}^2(ax) + \frac{\pi^2}{4}$	$\frac{\pi^{3/2}a^{-s}}{s}\csc(s\pi)\,\Gamma\left[\begin{matrix}\frac{s}{2}\\ \frac{s+1}{2}\end{matrix}\right]$ $[0 < \operatorname{Re} s < 1;\ \lvert\arg a\rvert < \pi]$

2.7.6. $\operatorname{arccsch}^n(\varphi(x))$ and elementary functions

1	$\operatorname{arccsch}(ax)$	$\frac{a^{-s}}{2s^2}\mathrm{B}\left(\frac{s+2}{2}, \frac{1-s}{2}\right)$ [$\operatorname{Re} a > 0;\ 0 < \operatorname{Re} s < 1$]
2	$\operatorname{arccsch}(ax) - \sum_{k=0}^{n}(-1)^k \frac{(1/2)_k}{(2k+1)\,k!}(ax)^{-2k-1}$	$\frac{a^{-s}}{2s^2}\mathrm{B}\left(\frac{s+2}{2}, \frac{1-s}{2}\right)$ [$2n+1 < \operatorname{Re} s < 2n+3;\ -\pi/2 \le \arg a < \pi/2$]
3	$\operatorname{arccsch}(ax) - \frac{1}{2ax}\left(\frac{1}{a^2x^2}\right)^{-1/2}\ln\frac{4}{a^2x^2}$	$\frac{a^{-s-1}}{2s^2}\left(\frac{1}{a^2}\right)^{1/2}\mathrm{B}\left(\frac{s+2}{2}, \frac{1-s}{2}\right)$ [$-2 < \operatorname{Re} s < 0;\ -\pi/2 \le \arg a < \pi/2$]
4	$\operatorname{arccsch}(ax) - \frac{1}{2ax}\left(\frac{1}{a^2x^2}\right)^{-1/2}\left[\ln\frac{4}{a^2x^2} - \sum_{k=1}^{n}(-1)^k\frac{(1/2)_k}{k!\,k}(ax)^{2k}\right]$	$\frac{a^{-s-1}}{2s^2}\left(\frac{1}{a^2}\right)^{1/2}\mathrm{B}\left(\frac{s+2}{2}, \frac{1-s}{2}\right)$ [$-2n-2 < \operatorname{Re} s < -2n;\ -\pi/2 \le \arg a < \pi/2$]
5	$\frac{1}{\sqrt{a^2x^2+1}}\operatorname{arccsch}(ax)$	$\frac{\pi^{3/2}\left(a^2\right)^{(1-s)/2}}{4a}\csc\frac{s\pi}{2}\,\Gamma\left[\begin{matrix}\frac{s}{2}\\ \frac{s+1}{2}\end{matrix}\right]$ [$\operatorname{Re} a \ne 0;\ 0 < \operatorname{Re} s < 2$]
6	$\frac{1}{\sqrt{a^2x^2+1}}\operatorname{arccsch}\frac{1}{ax}$	$\frac{\pi^{3/2}a^{-s}}{4}\sec\frac{s\pi}{2}\,\Gamma\left[\begin{matrix}\frac{1-s}{2}\\ \frac{2-s}{2}\end{matrix}\right]$ [$\operatorname{Re} a > 0;\ -1 < \operatorname{Re} s < 1$]
7	$\operatorname{arccsch}\sqrt{\frac{2}{\sqrt{ax+1}-1}}$	$-\frac{a^{-s}}{4s}\mathrm{B}\left(s+\frac{1}{2}, -s\right)$ [$-1/2 < \operatorname{Re} s < 0;\ \lvert\arg a\rvert < \pi$]
8	$\operatorname{arccsch}\sqrt{\frac{2\sqrt{ax}}{\sqrt{ax+1}-\sqrt{ax}}}$	$\frac{a^{-s}}{4s}\mathrm{B}\left(s, \frac{1}{2}-s\right)$ [$0 < \operatorname{Re} s < 1/2;\ \lvert\arg a\rvert < \pi$]
9	$\operatorname{arccsch}^2(ax)$	$\frac{\pi^{3/2}a^{-s}}{2s}\csc\frac{s\pi}{2}\,\Gamma\left[\begin{matrix}\frac{s}{2}\\ \frac{s+1}{2}\end{matrix}\right]$ [$\operatorname{Re} a > 0;\ 0 < \operatorname{Re} s < 2$]
10	$\operatorname{arccsch}^2\sqrt{\frac{2}{\sqrt{ax+1}-1}}$	$\frac{\pi^{3/2}a^{-s}}{8s}\csc(s\pi)\,\Gamma\left[\begin{matrix}-s\\ \frac{1-2s}{2}\end{matrix}\right]$ [$-1 < \operatorname{Re} s < 0;\ \lvert\arg a\rvert < \pi$]
11	$\operatorname{arccsch}^2\sqrt{\frac{2\sqrt{ax}}{\sqrt{ax+1}-\sqrt{ax}}}$	$\frac{\pi^{3/2}a^{-s}}{8s}\csc(s\pi)\,\Gamma\left[\begin{matrix}s\\ \frac{2s+1}{2}\end{matrix}\right]$ [$0 < \operatorname{Re} s < 1;\ \lvert\arg a\rvert < \pi$]

2.7.7. Hypebolic functions of inverse hyperbolic functions

1	$\sinh\left(\nu \operatorname{arcsinh} \frac{x}{a}\right)$	$\frac{\nu a^s}{4\sqrt{\pi}} \cos \frac{\nu\pi}{2}\, \Gamma\left[\begin{matrix} \frac{s+1}{2}, -\frac{s+\nu}{2}, \frac{\nu-s}{2} \\ \frac{2-s}{2} \end{matrix}\right]$ $[\operatorname{Re} a > 0;\ -1 < \operatorname{Re} s < -\lvert\operatorname{Re} \nu\rvert]$
2	$\sinh\left(\nu \operatorname{arccsch} \frac{x}{a}\right)$	$\frac{\nu a^s}{4\sqrt{\pi}} \cos \frac{\nu\pi}{2}\, \Gamma\left[\begin{matrix} \frac{1-s}{2}, \frac{s-\nu}{2}, \frac{s+\nu}{2} \\ \frac{s+2}{2} \end{matrix}\right]$ $[\operatorname{Re} a > 0;\ \lvert\operatorname{Re} \nu\rvert < \operatorname{Re} s < 1]$
3	$\frac{1}{\sqrt{x^2+a^2}} \sinh\left(\nu \operatorname{arcsinh} \frac{x}{a}\right)$	$\frac{a^{s-1}}{2\sqrt{\pi}} \sin \frac{\nu\pi}{2}\, \Gamma\left[\begin{matrix} \frac{s+1}{2}, \frac{1-\nu-s}{2}, \frac{\nu-s+1}{2} \\ \frac{2-s}{2} \end{matrix}\right]$ $[\operatorname{Re} a > 0;\ -1 < \operatorname{Re} s < 1 - \lvert\operatorname{Re} \nu\rvert]$
4	$\frac{1}{\sqrt{x^2+a^2}} \sinh\left(\nu \operatorname{arccsch} \frac{x}{a}\right)$	$\frac{a^{s-1}}{2\sqrt{\pi}} \sin \frac{\nu\pi}{2}\, \Gamma\left[\begin{matrix} \frac{2-s}{2}, \frac{s-\nu}{2}, \frac{s+\nu}{2} \\ \frac{s+1}{2} \end{matrix}\right]$ $[\operatorname{Re} a > 0;\ \lvert\operatorname{Re} \nu\rvert < \operatorname{Re} s < 2]$
5	$\lvert a-x\rvert^\nu \sinh\left(\nu \operatorname{arctanh} \frac{2\sqrt{ax}}{a+x}\right)$	$-\sqrt{\pi}\, a^{s+\nu}\, \Gamma\left[\begin{matrix} \frac{2\nu+1}{2} \\ -\nu \end{matrix}\right] \Gamma\left[\begin{matrix} \frac{2s+1}{2}, \frac{1-2\nu-2s}{2} \\ 1-s,\ s+\nu+1 \end{matrix}\right]$ $\left[\begin{matrix} a > 0;\ \operatorname{Re} \nu > -1/2; \\ -1/2 < \operatorname{Re} s < 1/2 - \operatorname{Re} \nu \end{matrix}\right]$
6	$\lvert a-x\rvert^\nu \sinh\left(\nu \operatorname{arccoth} \frac{a+x}{2\sqrt{ax}}\right)$	$-\sqrt{\pi}\, a^{s+\nu}\, \Gamma\left[\begin{matrix} \frac{2\nu+1}{2} \\ -\nu \end{matrix}\right] \Gamma\left[\begin{matrix} \frac{2s+1}{2}, \frac{1-2\nu-2s}{2} \\ 1-s,\ s+\nu+1 \end{matrix}\right]$ $\left[\begin{matrix} a > 0;\ \operatorname{Re} \nu > -1/2; \\ -1/2 < \operatorname{Re} s < 1/2 - \operatorname{Re} \nu \end{matrix}\right]$
7	$\theta(a-x) \sinh\left(\nu \operatorname{arcsech} \frac{x}{a}\right)$	$\frac{\nu\sqrt{\pi}\, a^s}{4}\, \Gamma\left[\begin{matrix} \frac{s-\nu}{2}, \frac{s+\nu}{2} \\ \frac{s+1}{2}, \frac{s+2}{2} \end{matrix}\right]$ $[a > 0;\ \operatorname{Re} s > \lvert\operatorname{Re} \nu\rvert]$
8	$\theta(x-a) \sinh\left(\nu \operatorname{arccosh} \frac{x}{a}\right)$	$\frac{\nu\sqrt{\pi}\, a^s}{4}\, \Gamma\left[\begin{matrix} \frac{-s-\nu}{2}, \frac{-s+\nu}{2} \\ \frac{1-s}{2}, \frac{2-s}{2} \end{matrix}\right]$ $[a > 0;\ \operatorname{Re} s < -\lvert\operatorname{Re} \nu\rvert]$
9	$\theta(a-x) \sinh\left(\nu \operatorname{arctanh} \sqrt{1-\frac{x}{a}}\right)$	$\frac{\nu\sqrt{\pi}\, a^s}{2}\, \Gamma\left[\begin{matrix} \frac{2s-\nu}{2}, \frac{2s+\nu}{2} \\ \frac{2s+1}{2},\ s+1 \end{matrix}\right]$ $[a > 0;\ \operatorname{Re} s > \lvert\operatorname{Re} \nu\rvert/2]$
10	$\theta(x-a) \sinh\left(\nu \operatorname{arctanh} \sqrt{1-\frac{a}{x}}\right)$	$\frac{\nu\sqrt{\pi}\, a^s}{2}\, \Gamma\left[\begin{matrix} \frac{-2s-\nu}{2}, \frac{-2s+\nu}{2} \\ \frac{1-2s}{2},\ 1-s \end{matrix}\right]$ $[a > 0;\ \operatorname{Re} s < -\lvert\operatorname{Re} \nu\rvert/2]$
11	$\frac{1}{\sqrt{x^2+a^2}} \cosh\left(\nu \operatorname{arcsinh} \frac{x}{a}\right)$	$\frac{a^{s-1}}{2\sqrt{\pi}} \cos \frac{\nu\pi}{2}\, \Gamma\left[\begin{matrix} \frac{s}{2}, \frac{1-s-\nu}{2}, \frac{1-s+\nu}{2} \\ \frac{1-s}{2} \end{matrix}\right]$ $[\operatorname{Re} a > 0;\ 0 < \operatorname{Re} s < 1 - \lvert\operatorname{Re} \nu\rvert]$

No.	$f(x)$	$F(s)$
12	$\frac{1}{\sqrt{x^2+a^2}} \cosh\left(\nu \operatorname{arccsch} \frac{x}{a}\right)$	$\frac{a^{s-1}}{2\sqrt{\pi}} \cos \frac{\nu\pi}{2} \, \Gamma\left[\begin{matrix} \frac{1-s}{2}, \frac{s-\nu}{2}, \frac{s+\nu}{2} \\ \frac{s}{2} \end{matrix}\right]$ $[\operatorname{Re} a > 0;\ \lvert\operatorname{Re}\nu\rvert < \operatorname{Re} s < 1]$
13	$\lvert a-x\rvert^{\nu} \cosh\left(\nu \operatorname{arctanh}\left(\frac{2\sqrt{ax}}{x+a}\right)\right)$	$\sqrt{\pi}\, a^{s+\nu}\, \Gamma\left[\begin{matrix} \frac{2\nu+1}{2}, -s-\nu, s \\ -\nu, \frac{1-2s}{2}, \frac{2s+2\nu+1}{2} \end{matrix}\right]$ $[a > 0;\ \operatorname{Re}\nu > -1/2;\ 0 < \operatorname{Re} s < -\operatorname{Re}\nu]$
14	$\lvert a-x\rvert^{\nu} \cosh\left(\nu \operatorname{arccoth}\left(\frac{x+a}{2\sqrt{ax}}\right)\right)$	$\sqrt{\pi}\, a^{s+\nu}\, \Gamma\left[\begin{matrix} \frac{2\nu+1}{2}, -s-\nu, s \\ -\nu, \frac{1-2s}{2}, \frac{2s+2\nu+1}{2} \end{matrix}\right]$ $[a > 0;\ \operatorname{Re}\nu > -1/2;\ 0 < \operatorname{Re} s < -\operatorname{Re}\nu]$
15	$\frac{\theta(a-x)}{\sqrt{a^2-x^2}} \cosh\left(\nu \operatorname{arcsech} \frac{x}{a}\right)$	$\frac{\sqrt{\pi}\, a^{s-1}}{2} \Gamma\left[\begin{matrix} \frac{s-\nu}{2}, \frac{s+\nu}{2} \\ \frac{s}{2}, \frac{s+1}{2} \end{matrix}\right]$ $[a > 0;\ \operatorname{Re} s > \lvert\operatorname{Re}\nu\rvert]$
16	$\frac{\theta(a-x)}{\sqrt{a-x}} \cosh\left(\nu \operatorname{arctanh} \sqrt{1-\frac{x}{a}}\right)$	$\sqrt{\pi}\, a^{s-1}\, \Gamma\left[\begin{matrix} \frac{2s-\nu}{2}, \frac{2s+\nu}{2} \\ s, \frac{2s+1}{2} \end{matrix}\right]$ $[a > 0;\ \operatorname{Re} s > \lvert\operatorname{Re}\nu\rvert/2]$
17	$\frac{\theta(x-a)}{\sqrt{x^2-a^2}} \cosh\left(\nu \operatorname{arccosh} \frac{x}{a}\right)$	$\frac{\sqrt{\pi}\, a^{s-1}}{2} \Gamma\left[\begin{matrix} \frac{1-s-\nu}{2}, \frac{1-s+\nu}{2} \\ \frac{1-s}{2}, \frac{2-s}{2} \end{matrix}\right]$ $[a > 0;\ \operatorname{Re} s < 1 - \lvert\operatorname{Re}\nu\rvert]$
18	$\frac{\theta(x-a)}{\sqrt{x-a}} \cosh\left(\nu \operatorname{arctanh} \sqrt{\frac{x}{a}-1}\right)$	$\sqrt{\pi}\, a^{s-1/2}\, \Gamma\left[\begin{matrix} \frac{1-2s-\nu}{2}, \frac{1-2s+\nu}{2} \\ \frac{1-2s}{2}, 1-s \end{matrix}\right]$ $[a > 0;\ \operatorname{Re} s < (1 - \lvert\operatorname{Re}\nu\rvert)/2]$

Chapter 3

Special Functions

3.1. The Gamma $\Gamma(z)$, Psi $\psi(z)$, and Zeta $\zeta(z)$ Functions

More formulas can be obtained from the corresponding sections due to the relations

$$\Gamma(z) = \lim_{w \to \infty} \frac{w^z}{z} \, {}_1F_1(z;\, z+1;\, -w),$$

$$\Gamma(1-z)\,\Gamma(1+z) = \frac{z\pi}{\sin(z\pi)}, \quad \Gamma\left(z+\frac{1}{2}\right)\Gamma\left(\frac{1}{2}-z\right) = \frac{\pi}{\cos(z\pi)},$$

$$\psi(z) = (z-1)\, {}_3F_2(1,\, 1,\, 2-z;\, 2,\, 2;\, 1) - \mathbf{C}, \quad \psi(-z) = \frac{1}{z} + \pi\cot(z\pi) + \psi(z),$$

$$\psi^{(n)}(z) = (-1)^{n+1}\, n!\, z^{-n-1}\, {}_{n+2}F_{n+1}(1,\, z,\, z, \ldots, z;\, z+1,\, z+1, \ldots, z+1;\, 1),$$

$$\psi^{(n)}(z \pm m) = \psi^{(n)}(z) \pm (-1)^n\, n! \sum_{k=(1\mp 1)/2}^{m-(1\pm 1)/2} \frac{1}{(z \pm k)^{n+1}},$$

$$\zeta(s) = \operatorname{Li}_s(1), \quad \operatorname{Re} s > 1; \quad \zeta(s,\, a+n) = \zeta(s,\, a) - \sum_{k=0}^{n-1} \frac{1}{\left((a+k)^2\right)^{s/2}},$$

$$\zeta(s,\, a-n) = \zeta(s,\, a) + \sum_{k=0}^{n-1} \frac{1}{\left((a+k-n)^2\right)^{s/2}}.$$

3.1.1. $\Gamma(\varphi(x))$

No.	$f(x)$	$F(s)$	
1	$\frac{a^x}{\Gamma(x+b)}$	$a^{1-b}\,\mu(a,\, s-1,\, b-1)$	$[a,\ \operatorname{Re} b,\ \operatorname{Re} s > 0]$
2	$\ln \frac{\sqrt{x}\,\Gamma(x)}{\Gamma\left(x+\frac{1}{2}\right)}$	$\frac{\sec(s\pi/2)}{(2\pi)^s}\left(1-2^{-s-1}\right)\Gamma(s)\,\zeta(s+1)$	$[0 < \operatorname{Re} s < 1]$
3	$\frac{x^c a^x}{\Gamma(x+b+1)}$	$a^{-b}\,\Gamma(s+c)\,\mu(a,\, s+c-1,\, b)$	$[\operatorname{Re}(s+c) > 0]$

No.	$f(x)$	$F(s)$	
4	$\dfrac{\theta(1-x)}{\Gamma(1-\ln x)}$	$\nu(e^{-s})$	
5	$\dfrac{\theta(1-x)}{\Gamma(b-\ln x+1)}$	$e^{bs}\nu(e^{-s},\, b)$	
6	$\dfrac{\theta(1-x)(-\ln x)^{c}}{\Gamma(b-\ln x+1)}$	$\Gamma(c+1)\, e^{bs}\mu(e^{-s},\, c,\, b)$	

3.1.2. $\psi(ax+b)$

1	$\psi(x+1)+\mathbf{C}$	$-\dfrac{\pi}{\sin(s\pi)}\,\zeta(1-s)$	$[-1<\operatorname{Re} s<0]$
2	$\psi(x+a)-\psi(x+b)$	$\dfrac{\pi}{\sin(s\pi)}[\zeta(1-s,\, b)-\zeta(1-s,\, a)]$	$[a,\, b>0;\ 0<\operatorname{Re} s<1]$
3	$\ln x-\psi(x+1)$	$\dfrac{\pi}{\sin(s\pi)}\,\zeta(1-s)$	$[0<\operatorname{Re} s<1]$
4	$\ln x-\psi\left(x+\dfrac{1}{2}\right)$	$\dfrac{2^{1-s}-1}{\sin(s\pi)}\,\zeta(s)$	$[0<\operatorname{Re} s<1]$
5	$\ln(x+1)-\psi(x+1)$	$\dfrac{\pi}{\sin(s\pi)}\left[\zeta(1-s)+\dfrac{1}{s}\right]$	$[0<\operatorname{Re} s<1]$

3.1.3. $\psi^{(n)}(ax+b)$

1	$\dfrac{1}{x}-\psi'(x+1)$	$\dfrac{\pi(s-1)}{\sin(s\pi)}\,\zeta(2-s)$	$[1<\operatorname{Re} s<2]$
2	$\dfrac{1}{x+1}-\psi'(x+1)$	$\dfrac{\pi(s-1)}{\sin(s\pi)}\left[\zeta(2-s)+\dfrac{1}{s-1}\right]$	$[0<\operatorname{Re} s<2]$
3	$\psi^{(n)}(x+1)$	$\dfrac{(-1)^{n-1}\pi}{\sin(s\pi)}\,(1-s)_n\,\zeta(1-s+n)$	$[0<\operatorname{Re} s<n]$

3.1.4. $\zeta(\nu,\, ax+b)$

1	$\zeta(\nu,\, ax+b)$	$a^{-s}\,\mathrm{B}(s,\, \nu-s)\,\zeta(\nu-s,\, b)$	$[\operatorname{Re}\nu,\ \operatorname{Re} b>0;\ 0<\operatorname{Re} s<\operatorname{Re}\nu-1]$
2	$\zeta(\nu,\, x)-\dfrac{1}{x^{\nu}}$	$\mathrm{B}(s,\, \nu-s)\,\zeta(\nu-s)$	$[0<\operatorname{Re} s<\operatorname{Re}\nu-1]$

3.2. The Polylogarithm $\mathrm{Li}_n(z)$

More formulas can be obtained from the corresponding sections due to the relations

$$\mathrm{Li}_n(z) = z\,{}_{n+1}F_n(1, 1, \ldots, 1; 2, 2, \ldots, 2; z),$$
$$\mathrm{Li}_n(-z) = -G_{n+1,\,n+1}^{1,\,n+1}\left(z\,\middle|\,\begin{matrix}1, 1, \ldots, 1\\ 1, 0, \ldots, 0\end{matrix}\right).$$

3.2.1. $\mathrm{Li}_n(bx)$ and algebraic functions

No.	$f(x)$	$F(s)$
1	$\theta(a-x)\,\mathrm{Li}_2\left(\frac{x}{a}\right)$	$\frac{a^s}{s^2}\left[\frac{\pi^2 s}{6} - \psi(s+1) - \mathbf{C}\right]$ $[a>0;\ \mathrm{Re}\,s>-1]$
2	$\mathrm{Li}_n(-ax)$	$(-1)^n \frac{\pi\csc(s\pi)}{a^s s^n}$ $[-1<\mathrm{Re}\,s<0;\ \lvert\arg a\rvert<\pi]$
3	$\theta(a-x)\,\mathrm{Li}_n(-bx)$	$\frac{a^{s+1}b}{s(s+1)}\,{}_{n+1}F_n\left(\begin{matrix}1, 1, \ldots, 1, s+1\\ 2, \ldots, 2, s+2;\ -ab\end{matrix}\right) - \frac{a^{s+1}b}{s}\,{}_{n+1}F_n\left(\begin{matrix}1, 1, \ldots, 1\\ 2, \ldots, 2;\ -ab\end{matrix}\right)$ $[a>0;\ \mathrm{Re}\,s>-1;\ \lvert\arg b\rvert<\pi]$
4	$(a-x)_+^{\alpha-1}\,\mathrm{Li}_n(-bx)$	$-a^{s+\alpha}b\,\mathrm{B}(\alpha, s+1)\,{}_{n+2}F_{n+1}\left(\begin{matrix}1, 1, \ldots, 1, s+1;\ -ab\\ 2, \ldots, 2, s+\alpha+1\end{matrix}\right)$ $[a,\ \mathrm{Re}\,\alpha>0;\ \mathrm{Re}\,s>-1;\ \lvert\arg b\rvert<\pi]$
5	$(x-a)_+^{\alpha-1}\,\mathrm{Li}_n(-bx)$	$-a^{s+\alpha}b\,\mathrm{B}(\alpha, -s-\alpha)\,{}_{n+2}F_{n+1}\left(\begin{matrix}1, 1, \ldots, 1, s+1;\ -ab\\ 2, \ldots, 2, s+\alpha+1\end{matrix}\right) + (-1)^{n+1}\frac{\pi\csc[(s+\alpha)\pi]}{b^{s+\alpha-1}(s+\alpha-1)^n} \times {}_{n+1}F_n\left(\begin{matrix}1-\alpha, 1-s-\alpha, \ldots, 1-s-\alpha\\ 2-s-\alpha, \ldots, 2-s-\alpha;\ -ab\end{matrix}\right)$ $[a,\ \mathrm{Re}\,\alpha>0;\ \mathrm{Re}\,(s+\alpha)<1;\ \lvert\arg b\rvert<\pi]$
6	$\frac{1}{(x+a)^\rho}\,\mathrm{Li}_n(-bx)$	$-a^{s-\rho+1}b\,\mathrm{B}(s+1, \rho-s-1) \times {}_{n+2}F_{n+1}\left(\begin{matrix}1, 1, \ldots, 1, s+1;\ ab\\ 2, \ldots, 2, s-\rho+2\end{matrix}\right) + \frac{\pi b^{\rho-s}}{(\rho-s)^n} \times \csc[(s-\rho)\pi]\,{}_{n+1}F_n\left(\begin{matrix}\rho, \rho-s, \ldots, \rho-s;\ ab\\ \rho-s+1, \ldots, \rho-s+1\end{matrix}\right)$ $[-1<\mathrm{Re}\,s<\mathrm{Re}\,\rho;\ \lvert\arg a\rvert, \lvert\arg b\rvert<\pi]$

No.	$f(x)$	$F(s)$
7	$\frac{1}{x-a}\operatorname{Li}_n(-bx)$	$\pi a^s b\cot(s\pi)\,{}_{n+1}F_n\left(\begin{matrix}1,\,1,\,\ldots,1;\,-ab\\2,\,2,\,\ldots,2\end{matrix}\right) - \frac{\pi\, b^{1-s}}{(1-s)^n}\csc(s\pi)\,{}_{n+1}F_n\left(\begin{matrix}1,\,1-s,\,\ldots,\,1-s;\,-ab\\2-s,\,\ldots,\,2-s\end{matrix}\right)$ $[a>0;\ \lvert\operatorname{Re}s\rvert<1;\ \lvert\arg b\rvert<\pi]$
8	$(a-x)_+^{\alpha-1}\operatorname{Li}_2(-bx^2)$	$-a^{s+\alpha+1}b\,\mathrm{B}(\alpha,\,s+2)\,{}_5F_4\left(\begin{matrix}1,\,1,\,1,\,\frac{s+2}{2},\,\frac{s+3}{2};\,-a^2b\\2,\,2,\,\frac{s+\alpha+2}{2},\,\frac{s+\alpha+3}{2}\end{matrix}\right)$ $[a,\ \operatorname{Re}\alpha>0;\ \operatorname{Re}s>-2]$
9	$\frac{1}{(x+a)^\rho}\operatorname{Li}_2\left(\frac{b}{x+a}\right)$	$a^{s-\rho-1}b\,\mathrm{B}(s,\,1-s+\rho)\,{}_4F_3\left(\begin{matrix}1,1,1,\,1-s+\rho\\2,2,\,\rho+1;\,\frac{b}{a}\end{matrix}\right)$ $[0<\operatorname{Re}s<\operatorname{Re}\rho+1;\ \lvert\arg a\rvert<\pi]$
10	$\frac{1}{(x+a)^\rho}\operatorname{Li}_2\left(\frac{bx}{x+a}\right)$	$a^{s-\rho}b\,\mathrm{B}(s+1,\,\rho-s)\,{}_4F_3\left(\begin{matrix}1,1,1,\,s+1\\2,2,\,\rho+1;\,b\end{matrix}\right)$ $[-1<\operatorname{Re}s<\operatorname{Re}\rho;\ \lvert\arg a\rvert<\pi]$
11	$(a-x)_+^{\alpha-1}\operatorname{Li}_2(bx(a-x))$	$a^{s+\alpha+1}\,b\,\mathrm{B}(s+1,\,\alpha+1)\,{}_5F_4\left(\begin{matrix}1,\,1,\,1,\,s+1,\,\alpha+1;\,\frac{a^2b}{4}\\2,\,2,\,\frac{s+\alpha+2}{2},\,\frac{s+\alpha+3}{2}\end{matrix}\right)$ $[a>0;\ \operatorname{Re}s,\ \operatorname{Re}\alpha>-1;\ \lvert\arg(4-a^2b)\rvert<\pi]$

3.2.2. $\operatorname{Li}_n(bx)$ and the logarithmic or inverse trigonometric functions

1	$\theta(a-x)\ln\frac{\sqrt{a}+\sqrt{a-x}}{\sqrt{x}} \times\operatorname{Li}_2(bx)$	$\frac{\sqrt{\pi}\,a^{s+1}b}{2s(s+1)}\Gamma\left[\begin{matrix}s\\\frac{2s+3}{2}\end{matrix}\right]\left[{}_3F_2\left(\begin{matrix}1,\,s+1,\,s+1\\\frac{2s+3}{2},\,s+2;\,ab\end{matrix}\right) - (s+1)\,{}_3F_2\left(\begin{matrix}1,\,1,\,s+1\\2,\,\frac{2s+3}{2};\,ab\end{matrix}\right)+s(s+1)\,{}_4F_3\left(\begin{matrix}1,\,1,\,1,\,s+1\\2,\,2,\,\frac{2s+3}{2};\,ab\end{matrix}\right)\right]$ $[a>0;\ \operatorname{Re}s>-2;\ \lvert\arg(1-ab)\rvert<\pi]$
2	$\theta(a-x)\arccos\sqrt{\frac{x}{a}} \times\operatorname{Li}_2(bx)$	$\frac{\sqrt{\pi}\,a^{s+1}b}{2s^2(s+1)}\Gamma\left[\begin{matrix}\frac{2s+3}{2}\\s+2\end{matrix}\right]\left[{}_3F_2\left(\begin{matrix}1,\,s+1,\,\frac{2s+3}{2}\\s+2,\,s+2;\,ab\end{matrix}\right) - (s+1)\,{}_3F_2\left(\begin{matrix}1,\,1,\,s+1\\2,\,\frac{2s+3}{2};\,ab\end{matrix}\right)+s(s+1)\,{}_4F_3\left(\begin{matrix}1,\,1,\,1,\,\frac{2s+3}{2}\\2,\,2,\,s+2;\,ab\end{matrix}\right)\right]$ $[a>0;\ \operatorname{Re}s>-1;\ \lvert\arg(1-ab)\rvert<\pi]$

3.3. The Exponential Integral Ei (z)

More formulas can be obtained from the corresponding sections due to the relations

$$\operatorname{Ei}(z) = -e^{z}\,\Psi(1;\,1;\,-z) + \frac{1}{2}\left(\ln z - \ln\frac{1}{z}\right) - \ln(-z),$$

$$\operatorname{Ei}(z) = z\,{}_2F_2(1,\,1;\,2,\,2;\,z) + \frac{1}{2}\left(\ln z - \ln\frac{1}{z}\right) + \mathbf{C},$$

$$\operatorname{Ei}(-z) = -G_{12}^{20}\left(z\,\middle|\,\begin{matrix}1\\0,\,0\end{matrix}\right), \quad \operatorname{Ei}(-z) = -e^{-z}G_{12}^{21}\left(z\,\middle|\,\begin{matrix}0\\0,\,0\end{matrix}\right).$$

3.3.1. Ei $(\varphi(x))$ and algebraic functions

No.	$f(x)$	$F(s)$
1	$\operatorname{Ei}(-ax)$	$-\frac{a^{-s}}{s}\Gamma(s)$ $[a,\ \operatorname{Re}s>0]$
2	$\operatorname{Ei}(-ax-b)$	$-\left(\frac{b}{a}\right)^{s}\Gamma(s)\Gamma(-s,\,b)$ $[a,\ \operatorname{Re}s>0;\ \lvert\arg b\rvert<\pi]$
3	$(a-x)_{+}^{\alpha-1}\operatorname{Ei}(-bx)$	$-a^{s+\alpha}b\,\mathrm{B}(s+1,\,\alpha)\,{}_3F_3\left(\begin{matrix}s+1,\,1,\,1;\,-ab\\s+\alpha+1,\,2,\,2\end{matrix}\right)$ $+\,a^{s+\alpha-1}\,\mathrm{B}(s,\,\alpha)\,[\psi(s)-\psi(s+\alpha)+\ln(ab)+\mathbf{C}]$ $[a,\ \operatorname{Re}\alpha,\ \operatorname{Re}s>0;\ \lvert\arg b\rvert<\pi]$
4	$(x-a)_{+}^{\alpha-1}\operatorname{Ei}(-bx)$	$-a^{s+\alpha}b\,\mathrm{B}(\alpha,\,-s-\alpha)\,{}_3F_3\left(\begin{matrix}1,\,1,\,s+1;\,-ab\\2,\,2,\,s+\alpha+1\end{matrix}\right)$ $-\,b^{-s-\alpha+1}\frac{\Gamma(s+\alpha-1)}{s+\alpha-1}\,{}_2F_2\left(\begin{matrix}1-\alpha,\,1-s-\alpha;\,-ab\\2-s-\alpha,\,2-s-\alpha\end{matrix}\right)$ $+\,a^{s+\alpha-1}\,\mathrm{B}(\alpha,\,1-s-\alpha)\,[\psi(1-s)-\psi(1-s-\alpha)+\ln(ab)+\mathbf{C}]$ $[a,\ \operatorname{Re}b,\ \operatorname{Re}\alpha>0;\ \operatorname{Re}(s+\alpha)<1]$
5	$\frac{1}{(x+a)^{\rho}}\operatorname{Ei}(-bx)$	$-a^{s-\rho+1}\,b\,\mathrm{B}(s+1,\,\rho-s-1)\,{}_3F_3\left(\begin{matrix}1,\,1,\,s+1;\,ab\\2,\,2,\,s-\rho+2\end{matrix}\right)$ $+\,\frac{b^{\rho-s}\,\Gamma(s-\rho)}{\rho-s}\,{}_2F_2\left(\begin{matrix}\rho,\,\rho-s;\,ab\\\rho-s+1,\,\rho-s+1\end{matrix}\right)$ $+\,a^{s-\rho}\,\mathrm{B}(s,\,\rho-s)\,[\psi(s)-\psi(\rho-s)+\ln(ab)+\mathbf{C}]$ $[\operatorname{Re}b>0;\ 0<\operatorname{Re}s<\rho;\ \lvert\arg a\rvert<\pi]$
6	$\frac{1}{x+a}\operatorname{Ei}(-bx)$	$-\frac{b^{1-s}\,\Gamma(s-1)}{s-1}\,{}_2F_2\left(\begin{matrix}1,\,1-s;\,ab\\2-s,\,2-s\end{matrix}\right)$ $-\,\pi a^{s-1}\csc(s\pi)\left[\pi\cot(s\pi)+\Gamma(0,\,-ab)+\ln\frac{1}{a}+\ln(-a)\right]$ $[\operatorname{Re}b>0;\ 0<\operatorname{Re}s<1;\ \lvert\arg a\rvert<\pi]$

No.	$f(x)$	$F(s)$
7	$\dfrac{1}{x-a}\operatorname{Ei}(-bx)$	$\pi a^{s-1}\cot(s\pi)\left[2\pi\csc(2s\pi)-\operatorname{Ei}(-ab)\right]+\dfrac{b^{1-s}}{1-s}\Gamma(s-1)$ $\times\,{}_2F_2\left(\begin{matrix}1,\,1-s;\,-ab\\2-s,\,2-s\end{matrix}\right)$ $[a,\ \operatorname{Re}b>0;\ 0<\operatorname{Re}s<1]$
8	$\left(a^2-x^2\right)_+^{\alpha-1}\operatorname{Ei}(-bx)$	$\dfrac{a^{s+2\alpha}b^2}{8}\operatorname{B}\left(\alpha,\,\dfrac{s+2}{2}\right){}_3F_4\left(\begin{matrix}1,\,1,\,\frac{s+2}{2};\,\frac{a^2b^2}{4}\\\frac{3}{2},\,2,\,2,\,\frac{s+2\alpha+2}{2}\end{matrix}\right)$ $-\dfrac{a^{s+2\alpha-1}b}{2}\operatorname{B}\left(\alpha,\,\dfrac{s+1}{2}\right){}_2F_3\left(\begin{matrix}\frac{1}{2},\,\frac{s+1}{2};\,\frac{a^2b^2}{4}\\\frac{3}{2},\,\frac{3}{2},\,\frac{s+2\alpha+1}{2}\end{matrix}\right)$ $+\dfrac{a^{s+2\alpha-2}}{2}\operatorname{B}\left(\alpha,\,\dfrac{s}{2}\right)\left[\dfrac{1}{2}\psi\left(\dfrac{s}{2}\right)-\dfrac{1}{2}\psi\left(\dfrac{s+2\alpha}{2}\right)\right.$ $\left.+\ln(ab)+\mathbf{C}\right]$ $[a,\ \operatorname{Re}\alpha,\ \operatorname{Re}s>0;\ \lvert\arg b\rvert<\pi]$
9	$\left(x^2-a^2\right)_+^{\alpha-1}\operatorname{Ei}(-bx)$	$\dfrac{a^{s+2\alpha}b^2}{8}\operatorname{B}\left(\alpha,\,-\dfrac{s+2\alpha}{2}\right){}_3F_4\left(\begin{matrix}1,\,1,\,\frac{s+2}{2};\,\frac{a^2b^2}{4}\\\frac{3}{2},\,2,\,2,\,\frac{s+2\alpha+2}{2}\end{matrix}\right)$ $-\dfrac{a^{s+2\alpha-1}b}{2}\operatorname{B}\left(\alpha,\,-\dfrac{s+2\alpha-1}{2}\right){}_2F_3\left(\begin{matrix}\frac{1}{2},\,\frac{s+1}{2};\,\frac{a^2b^2}{4}\\\frac{3}{2},\,\frac{3}{2},\,\frac{s+2\alpha+1}{2}\end{matrix}\right)$ $-\dfrac{\Gamma(s+2\alpha-2)}{s+2\alpha-2}b^{-s-2\alpha+2}{}_2F_3\left(\begin{matrix}1-\alpha,\,-\frac{s+2\alpha-2}{2};\,\frac{a^2b^2}{4}\\-\frac{s+2\alpha-3}{2},\,-\frac{s+2\alpha-4}{2},\,-\frac{s+2\alpha-4}{2}\end{matrix}\right)$ $+\dfrac{a^{s+2\alpha-2}}{2}\operatorname{B}\left(\alpha,\,-\dfrac{s+2\alpha-2}{2}\right)\left[-\dfrac{1}{2}\psi\left(-\dfrac{s+2\alpha-2}{2}\right)+\ln(ab)\right.$ $\left.+\dfrac{1}{2}\psi\left(-\dfrac{s-2}{2}\right)+\mathbf{C}\right]$ $[a,\ \operatorname{Re}b,\ \operatorname{Re}\alpha>0;\ \operatorname{Re}(s+2\alpha)<2]$

3.3.2. $\operatorname{Ei}(\varphi(x))$ and the exponential function

1	$e^{\pm ax}\operatorname{Ei}(\mp ax)$	$-\dfrac{\pi}{a^s}\left\{\begin{matrix}\csc(s\pi)\\\cot(s\pi)\end{matrix}\right\}\Gamma(s)$ $[a>0;\ 0<\operatorname{Re}s<1]$
2	$e^{-ax}\operatorname{Ei}(-bx)$	$-\dfrac{\Gamma(s)}{s(a+b)^s}{}_2F_1\left(\begin{matrix}1,\,s;\,\frac{a}{a+b}\\s+1\end{matrix}\right)$ $[\operatorname{Re}(a+b),\ \operatorname{Re}s>0;\ \lvert\arg b\rvert<\pi]$
3	$e^{-ax}\operatorname{Ei}(bx)$	$-\dfrac{\pi}{a^s}\cot(s\pi)\Gamma(s)+\dfrac{\Gamma(s-1)}{b(a-b)^{s-1}}{}_2F_1\left(\begin{matrix}1,\,1;\,\frac{b-a}{b}\\2-s\end{matrix}\right)$ $[\operatorname{Re}a>b>0;\ \operatorname{Re}s>0]$
4	$e^{-a/x}\operatorname{Ei}(-bx)$	$a^s\Gamma(-s)\left[\dfrac{ab}{s+1}{}_2F_3\left(\begin{matrix}1,\,1;\,ab\\2,\,2,\,s+2\end{matrix}\right)-\psi(-s)+\ln(ab)+\mathbf{C}\right]$ $-\dfrac{b^{-s}}{s}\Gamma(s)\,{}_1F_2\left(\begin{matrix}-s;\,ab\\1-s,\,1-s\end{matrix}\right)$ $[\operatorname{Re}a,\ \operatorname{Re}b>0]$

No.	$f(x)$	$F(s)$
5	$e^{-a\sqrt{x}}\operatorname{Ei}(-bx)$	$\frac{2a}{(2s+1)\,b^{s+1/2}}\Gamma\left(\frac{2s+1}{2}\right)\,{}_2F_2\left(\begin{matrix}\frac{2s+1}{2},\ \frac{2s+1}{2}\\ \frac{3}{2},\ \frac{2s+3}{2};\ \frac{a^2}{4b}\end{matrix}\right) - \frac{\Gamma(s)}{sb^s}\,{}_2F_2\left(\begin{matrix}s,\ s;\ \frac{a^2}{4b}\\ \frac{1}{2},\ s+1\end{matrix}\right)$ $\left[\left(\operatorname{Re}b,\ \operatorname{Re}s>0\right) \text{ or } \left(\operatorname{Re}b=0;\ \operatorname{Re}a,\ \operatorname{Re}s>0\right) \text{ or } \left(\operatorname{Re}b=\operatorname{Re}a=0;\ 0<\operatorname{Re}s<2\right);\ \left(\operatorname{Im}b=0 \text{ or } \left(\operatorname{Im}b\neq0;\ \operatorname{Re}a>0\right) \text{ or } \left(\operatorname{Im}b\neq0;\ \operatorname{Re}a=0;\ 2\operatorname{Re}s<1\right)\right)\right]$
6	$e^{ax}\operatorname{Ei}(-ax-b)$	$-\frac{\pi a^{-s}}{\sin(s\pi)}\Gamma(s,\,b)$ $[0<\operatorname{Re}s<1]$
7	$e^{ax}\left[\operatorname{Ei}(-2ax) - \operatorname{Ei}(-ax)\right]$	$\frac{a^{-s}}{2}\Gamma(s)\left[\psi\left(\frac{2-s}{2}\right)-\psi\left(\frac{1-s}{2}\right)\right]$ $[0<\operatorname{Re}s<1;\ \lvert\arg a\rvert<\pi]$
8	$e^{bx}\operatorname{Ei}(-u_+) + e^{-bx}\operatorname{Ei}(u_-)$ $u_\pm=b\left(\sqrt{x^2+a^2}\pm a\right)$	$-\sqrt{\pi}\,a^{(s+1)/2}\left(\frac{b}{2}\right)^{(1-s)/2}\cot\frac{s\pi}{2}\,\Gamma\left(\frac{s}{2}\right)K_{(s+1)/2}(ab)$ $[b,\ \operatorname{Re}a>0;\ 0<\operatorname{Re}s<1]$

3.3.3. Ei (bx) and hyperbolic or trigonometric functions

Notation: $\delta=\left\{\begin{matrix}1\\0\end{matrix}\right\}$.

1	$\left\{\begin{matrix}\sin(ax)\\ \cos(ax)\end{matrix}\right\}\operatorname{Ei}(-bx)$	$-\frac{a^\delta}{(s+\delta)\,b^{s+\delta}}\Gamma(s+\delta)\,{}_3F_2\left(\begin{matrix}\frac{s+\delta}{2},\ \frac{s+\delta}{2},\ \frac{s+\delta+1}{2}\\ \frac{2\delta+1}{2},\ \frac{s+\delta+2}{2};\ -\frac{a^2}{b^2}\end{matrix}\right)$ $[a,\ b>0;\ \operatorname{Re}s>-\delta]$
2	$\left\{\begin{matrix}\sin(a\sqrt{x})\\ \cos(a\sqrt{x})\end{matrix}\right\}\operatorname{Ei}(-bx)$	$-\frac{2a^\delta}{(2s+\delta)\,b^{s+\delta/2}}\Gamma\left(\frac{2s+\delta}{2}\right)\,{}_2F_2\left(\begin{matrix}\frac{2s+\delta}{2},\ \frac{2s+\delta}{2};\ -\frac{a^2}{4b}\\ \frac{2\delta+1}{2},\ \frac{2s+\delta+2}{2}\end{matrix}\right)$ $[\operatorname{Re}a,\ \operatorname{Re}(a+b),\ \operatorname{Re}s>0]$
3	$e^{bx}\sin(ax)\operatorname{Ei}(-bx)$	$\frac{a^{1-s}}{b}\Gamma(s-1)\cos\frac{s\pi}{2}\,{}_3F_2\left(\begin{matrix}\frac{1}{2},\ 1,\ 1;\ -\frac{a^2}{b^2}\\ \frac{2-s}{2},\ \frac{3-s}{2}\end{matrix}\right) - \frac{a^{2-s}}{b^2}\Gamma(s-2)\sin\frac{s\pi}{2}\,{}_3F_2\left(\begin{matrix}1,\ 1,\ \frac{3}{2};\ -\frac{a^2}{b^2}\\ \frac{3-s}{2},\ \frac{4-s}{2}\end{matrix}\right) + \frac{\pi\csc(s\pi)}{(a^2+b^2)^{s/2}}\Gamma(s)\sin\left(s\arctan\frac{a}{b}\right)$ $[a>0;\ -1<\operatorname{Re}s<2;\ \lvert\arg b\rvert<\pi]$

No.	$f(x)$	$F(s)$
4	$e^{-bx}\sin(ax)\operatorname{Ei}(bx)$	$\frac{\pi a^{1-s}}{2b\,\Gamma(2-s)}\csc\frac{s\pi}{2}\,{}_3F_2\left(\begin{matrix}\frac{1}{2},\,1,\,1;\,-\frac{a^2}{b^2}\\ \frac{2-s}{2},\,\frac{3-s}{2}\end{matrix}\right) - \frac{\pi a^{2-s}}{2b^2\,\Gamma(3-s)}\sec\frac{s\pi}{2}\,{}_3F_2\left(\begin{matrix}1,\,1,\,\frac{3}{2};\,-\frac{a^2}{b^2}\\ \frac{3-s}{2},\,\frac{4-s}{2}\end{matrix}\right) - \frac{\pi\cot(s\pi)}{(a^2+b^2)^{s/2}}\Gamma(s)\sin\left(s\arctan\frac{a}{b}\right)$ $[a,\,b>0;\;-1<\operatorname{Re}s<2]$
5	$e^{bx}\cos(ax)\operatorname{Ei}(-bx)$	$-\frac{a^{1-s}}{b}\Gamma(s-1)\sin\frac{s\pi}{2}\,{}_3F_2\left(\begin{matrix}\frac{1}{2},\,1,\,1;\,-\frac{a^2}{b^2}\\ \frac{2-s}{2},\,\frac{3-s}{2}\end{matrix}\right) - \frac{a^{2-s}}{b^2}\Gamma(s-2)\cos\frac{s\pi}{2}\,{}_3F_2\left(\begin{matrix}1,\,1,\,\frac{3}{2};\,-\frac{a^2}{b^2}\\ \frac{3-s}{2},\,\frac{4-s}{2}\end{matrix}\right) - \frac{\pi\csc(s\pi)}{(a^2+b^2)^{s/2}}\Gamma(s)\cos\left(s\arctan\frac{a}{b}\right)$ $[a>0;\;0<\operatorname{Re}s<2;\;\lvert\arg b\rvert<\pi]$
6	$e^{-bx}\cos(ax)\operatorname{Ei}(bx)$	$-\frac{\pi a^{1-s}}{2b\,\Gamma(2-s)}\sec\frac{s\pi}{2}\,{}_3F_2\left(\begin{matrix}\frac{1}{2},\,1,\,1;\,-\frac{a^2}{b^2}\\ \frac{2-s}{2},\,\frac{3-s}{2}\end{matrix}\right) - \frac{\pi a^{2-s}}{2b^2\,\Gamma(3-s)}\csc\frac{s\pi}{2}\,{}_3F_2\left(\begin{matrix}1,\,1,\,\frac{3}{2};\,-\frac{a^2}{b^2}\\ \frac{3-s}{2},\,\frac{4-s}{2}\end{matrix}\right) - \frac{\pi\cot(s\pi)}{(a^2+b^2)^{s/2}}\Gamma(s)\cos\left(s\arctan\frac{a}{b}\right)$ $[a,\,b>0;\;0<\operatorname{Re}s<2]$
7	$\left\{\begin{matrix}\sin(ax)\sinh(ax)\\ \cos(ax)\cosh(ax)\end{matrix}\right\}\times\operatorname{Ei}(-bx)$	$-\frac{a^{2\delta}}{b^{s+2\delta}(s+2\delta)}\Gamma(s+2\delta)\,{}_5F_4\left(\begin{matrix}\frac{s+2\delta}{4},\,\Delta(4,\,s+2\delta);\,-\frac{4a^4}{b^4}\\ \frac{2\delta+1}{4},\,\frac{2\delta+3}{4},\,\frac{2\delta+1}{2},\,\frac{s+2\delta+4}{4}\end{matrix}\right)$ $[a,\,b>0;\;\operatorname{Re}s>-2\delta]$
8	$\left\{\begin{matrix}\cos(ax)\sinh(ax)\\ \sin(ax)\cosh(ax)\end{matrix}\right\}\times\operatorname{Ei}(-bx)$	$\pm\frac{a^3b^{-s-3}}{3(s+3)}\Gamma(s+3)\,{}_5F_4\left(\begin{matrix}\frac{s+3}{4},\,\Delta(4,\,s+3)\\ \frac{5}{4},\,\frac{3}{2},\,\frac{7}{4},\,\frac{s+7}{4};\,-\frac{4a^4}{b^4}\end{matrix}\right) - \frac{ab^{-s-1}}{s+1}\Gamma(s+1)\,{}_5F_4\left(\begin{matrix}\frac{s+1}{4},\,\Delta(4,\,s+1)\\ \frac{1}{2},\,\frac{3}{4},\,\frac{5}{4},\,\frac{s+5}{4};\,-\frac{4a^4}{b^4}\end{matrix}\right)$ $[a,\,b>0;\;\operatorname{Re}s>-1]$

3.3.4. $e^{ax}\ln^n x\operatorname{Ei}(bx)$

1	$\ln(ax)\operatorname{Ei}(-bx)$	$\frac{b^{-s}}{s}\Gamma(s)\left[\ln\frac{b}{a}-\psi(s)+\frac{1}{s}\right]$ $[\operatorname{Re}a,\;\operatorname{Re}b,\;\operatorname{Re}s>0]$

No.	$f(x)$	$F(s)$
2	$\ln^n x\,\mathrm{Ei}(-ax)$	$-\dfrac{d^n}{ds^n}\left[\dfrac{\Gamma(s)}{a^s s}\right]$ [Re a, Re $s>0$]
3	$e^{ax}\ln x\,\mathrm{Ei}(-ax)$	$\dfrac{\pi\Gamma(s)}{a^s\sin(s\pi)}\left[\pi\cot(s\pi)-\psi(s)+\ln a\right]$ [$0<\mathrm{Re}\,s<1$; $\lvert\arg a\rvert<\pi$]
4	$e^{-ax}\ln x\,\mathrm{Ei}(-bx)$	$\dfrac{\Gamma(s)}{(a+b)^s}\left[\left[\ln(a+b)-\psi(s)\right]\Phi\left(\dfrac{a}{a+b},\,1,\,s\right)+\Phi\left(\dfrac{a}{a+b},\,2,\,s\right)\right]$ [Re $(a+b)$, Re $s>0$; $\lvert\arg b\rvert<\pi$]
5	$e^{-ax}\ln^n x\,\mathrm{Ei}(-bx)$	$-\dfrac{d^n}{ds^n}\left[\dfrac{\Gamma(s)}{(a+b)^s}\Phi\left(\dfrac{a}{a+b},\,1,\,s\right)\right]$ [Re $(a+b)$, Re $s>0$; $\lvert\arg b\rvert<\pi$]
6	$e^{\pm ax}\ln^n x\,\mathrm{Ei}(\mp ax)$	$-\pi\dfrac{d^n}{ds^n}\left[\dfrac{\Gamma(s)}{a^s}\begin{Bmatrix}\csc(s\pi)\\ \cot(s\pi)\end{Bmatrix}\right]$ $\left[0<\mathrm{Re}\,s<1;\ \begin{Bmatrix}\lvert\arg a\rvert<\pi\\ a>0\end{Bmatrix}\right]$

3.3.5. Products of Ei (ax)

No.	$f(x)$	$F(s)$
1	$\mathrm{Ei}^2(-ax)$	$\dfrac{a^{-s}\Gamma(s)}{2^{s-1}s}\Phi\left(\dfrac{1}{2},\,1,\,s\right)$ [a, Re $s>0$]
2	$\mathrm{Ei}(-ax)\,\mathrm{Ei}(-bx)$	$\dfrac{\Gamma(s)}{a^s}\left[\dfrac{bs}{a(s+1)}\,{}_4F_3\left(\begin{matrix}1,\,1,\,s+1,\,s+1\\ 2,\,2,\,s+2;\,-\frac{b}{a}\end{matrix}\right)+\dfrac{1}{s}\left(\dfrac{1}{s}-\psi(s)-\mathbf{C}+\ln\dfrac{a}{b}\right)\right]$ [$a+b$, Re $s>0$]
3	$e^{ax}\,\mathrm{Ei}^2(-ax)$	$\dfrac{\Gamma(s)}{2a^s}\left[\dfrac{4\pi^2\cos(s\pi)}{\sin^2(s\pi)}+\psi'\left(\dfrac{1-s}{2}\right)-\psi'\left(\dfrac{2-s}{2}\right)\right]$ [a, Re $s>0$]
4	$e^{-ax}\,\mathrm{Ei}(-bx)\,\mathrm{Ei}(bx)$	$\dfrac{\pi}{sb^s}\cot\dfrac{s\pi}{2}\Gamma(s)\,{}_3F_2\left(\begin{matrix}\frac{s}{2},\,\frac{s}{2},\,\frac{s+1}{2}\\ \frac{1}{2},\,\frac{s+2}{2};\,\frac{a^2}{b^2}\end{matrix}\right)+\dfrac{\pi a}{(s+1)\,b^{s+1}}\tan\dfrac{s\pi}{2}\Gamma(s+1)\times{}_2F_1\left(\begin{matrix}\frac{s+1}{2},\,\frac{s+1}{2},\,\frac{s+2}{2}\\ \frac{3}{2},\,\frac{s+3}{2};\,\frac{a^2}{b^2}\end{matrix}\right)-\dfrac{a^{2-s}\Gamma(s-2)}{b^2}\,{}_4F_3\left(\begin{matrix}1,\,1,\,1,\,\frac{3}{2};\,\frac{a^2}{b^2}\\ 2,\,\frac{3-s}{2},\,\frac{4-s}{2}\end{matrix}\right)$ [b, Re a, Re $s>0$]
5	$\ln(ax)\,\mathrm{Ei}^2(-bx)$	$\dfrac{2^{1-s}b^{-s}}{s}\Gamma(s)\left\{\dfrac{2}{s}\left[\psi(s)-\dfrac{1}{s}-\ln 2\right]{}_2F_1\left(\begin{matrix}1,\,1;\,-1\\ s+1\end{matrix}\right)+\ln\dfrac{a}{b}\Phi\left(\dfrac{1}{2},\,1,\,s\right)-\Phi\left(\dfrac{1}{2},\,2,\,s\right)\right\}$ [b, Re a, Re $s>0$]

3.4. The Sine $\operatorname{si}(z)$, $\operatorname{Si}(z)$, and Cosine $\operatorname{ci}(z)$ Integrals

More formulas can be obtained from the corresponding sections due to the relations

$$\operatorname{si}(z) = \operatorname{Si}(z) - \frac{\pi}{2}; \quad \operatorname{ci}(z) = \frac{1}{2}\left[\operatorname{Ei}(-iz) + \operatorname{Ei}(iz)\right], \ [\operatorname{Re} z > 0];$$

$$\operatorname{si}(z) = -\frac{\pi}{2}\left(\frac{\sqrt{z^2}}{z} + 1\right) + \frac{i}{2}\left[\operatorname{Ei}(-iz) - \operatorname{Ei}(iz)\right], \ [\operatorname{Re} z \neq 0];$$

$$\operatorname{Si}(z) = z\,{}_1F_2\left(\frac{1}{2}; \frac{3}{2}, \frac{3}{2}; -\frac{z^2}{4}\right),$$

$$\operatorname{ci}(z) = -\frac{z^2}{4}\,{}_2F_3\left(1, 1; 2, 2, \frac{3}{2}; -\frac{z^2}{4}\right) + \ln z + \mathbf{C},$$

$$\operatorname{ci}(z) = -\frac{\sqrt{\pi}}{2}\,G_{13}^{20}\left(\frac{z^2}{4}\,\middle|\,\begin{matrix}1\\0,\,0,\,1/2\end{matrix}\right) - \frac{\ln z^2}{2} + \ln z,$$

$$\operatorname{Si}(z) = \frac{\sqrt{\pi z^2}}{2z}\left[\sqrt{\pi} - G_{13}^{20}\left(\frac{z^2}{4}\,\middle|\,\begin{matrix}1\\0,\,1/2,\,0\end{matrix}\right)\right], \quad \operatorname{Si}(z) = \frac{\sqrt{\pi z^2}}{2z}\,G_{13}^{11}\left(\frac{z^2}{4}\,\middle|\,\begin{matrix}1\\1/2,\,0,\,0\end{matrix}\right),$$

$$\operatorname{Si}(z) = \frac{\sqrt{\pi}\,z}{4}\,G_{13}^{11}\left(\frac{z^2}{4}\,\middle|\,\begin{matrix}1/2\\0,\,-1/2,\,-1/2\end{matrix}\right).$$

3.4.1. $\operatorname{si}(ax)$, $\operatorname{Si}(ax)$, and $\operatorname{ci}(ax)$

No.	$f(x)$	$F(s)$
1	$\operatorname{si}(ax)$	$-\frac{\Gamma(s)}{a^s s}\sin\frac{s\pi}{2}$ $[a > 0;\ 0 < \operatorname{Re} s < 2]$
2	$\operatorname{ci}(ax)$	$-\frac{\Gamma(s)}{a^s s}\cos\frac{s\pi}{2}$ $[a > 0;\ 0 < \operatorname{Re} s < 2]$
3	$\operatorname{Si}(ax)$	$-\frac{\Gamma(s)}{a^s s}\sin\frac{s\pi}{2}$ $[a > 0;\ -1 < \operatorname{Re} s < 0]$

3.4.2. $\operatorname{si}(bx)$, $\operatorname{ci}(bx)$, and algebraic functions

1	$(a-x)_+^{\alpha-1}\operatorname{si}(bx)$	$a^{s+\alpha}b\,\mathrm{B}(\alpha, s+1)\,{}_3F_4\left(\begin{matrix}\frac{1}{2}, \frac{s+1}{2}, \frac{s+2}{2}; -\frac{a^2b^2}{4}\\ \frac{3}{2}, \frac{3}{2}, \frac{s+\alpha+1}{2}, \frac{s+\alpha+2}{2}\end{matrix}\right) - \frac{\pi}{2}a^{s+\alpha-1}\,\mathrm{B}(\alpha, s)$ $[a, b, \operatorname{Re}\alpha, \operatorname{Re} s > 0]$
2	$(a-x)_+^{\alpha-1}\operatorname{ci}(bx)$	$-\frac{a^{s+\alpha+1}b^2}{4}\,\mathrm{B}(\alpha, s+2)\,{}_4F_5\left(\begin{matrix}1, 1, \frac{s+2}{2}, \frac{s+3}{2}; -\frac{a^2b^2}{4}\\ \frac{3}{2}, 2, 2, \frac{s+\alpha+2}{2}, \frac{s+\alpha+3}{2}\end{matrix}\right)$ $+ a^{s+\alpha-1}\,\mathrm{B}(\alpha, s)\left[\psi(s) - \psi(s+\alpha) + \log(ab) + \mathbf{C}\right]$ $[a, b, \operatorname{Re}\alpha, \operatorname{Re} s > 0]$

No.	$f(x)$	$F(s)$
3	$\left(a^2-x^2\right)_+^{\alpha-1}\operatorname{si}(bx)$	$\frac{a^{s+2\alpha-1}b}{2}\,\mathrm{B}\left(\alpha,\,\frac{s+1}{2}\right)\,{}_2F_3\left(\begin{matrix}\frac{1}{2},\,\frac{s+1}{2};\,-\frac{a^2b^2}{4}\\ \frac{3}{2},\,\frac{3}{2},\,\frac{s+2\alpha+1}{2}\end{matrix}\right)-\frac{\pi a^{s+2\alpha-2}}{4}\,\mathrm{B}\left(\alpha,\,\frac{s}{2}\right)$ $[a,\,b,\,\operatorname{Re}\alpha,\,\operatorname{Re}s>0]$
4	$\left(a^2-x^2\right)_+^{\alpha-1}\operatorname{ci}(bx)$	$-\frac{a^{s+2\alpha}b^2}{8}\,\mathrm{B}\left(\alpha,\,\frac{s+2}{2}\right)\,{}_3F_4\left(\begin{matrix}1,\,1,\,\frac{s+2}{2};\,-\frac{a^2b^2}{4}\\ \frac{3}{2},\,2,\,2,\,\frac{s+2\alpha+2}{2}\end{matrix}\right)$ $+\frac{a^{s+2\alpha-2}}{2}\,\mathrm{B}\left(\alpha,\,\frac{s}{2}\right)\left[\frac{1}{2}\psi\left(\frac{s}{2}\right)-\frac{1}{2}\psi\left(\frac{s+2\alpha}{2}\right)+\ln(ab)+\mathbf{C}\right]$ $[a,\,b,\,\operatorname{Re}\alpha,\,\operatorname{Re}s>0]$
5	$\frac{1}{(x^2+a^2)^\rho}\operatorname{si}(bx)$	$-\frac{a^{s-2\rho+3}b^3}{36}\,\mathrm{B}\left(\frac{s+3}{2},\,\frac{2\rho-s-3}{2}\right)\,{}_3F_4\left(\begin{matrix}1,\,\frac{3}{2},\,\frac{s+3}{2};\,\frac{a^2b^2}{4}\\ 2,\,\frac{5}{2},\,\frac{5}{2},\,\frac{s-2\rho+5}{2}\end{matrix}\right)$ $+\frac{a^{s-2\rho+1}b}{2}\,\mathrm{B}\left(\frac{s+1}{2},\,\frac{2\rho-s-1}{2}\right)$ $-\frac{\pi a^{s-2\rho}}{4}\,\mathrm{B}\left(\frac{s}{2},\,\frac{2\rho-s}{2}\right)+\frac{b^{2\rho-s}}{2\rho-s}\,\Gamma(s-2\rho)$ $\times\sin\frac{(s-2\rho)\pi}{2}\,{}_2F_3\left(\begin{matrix}\rho,\,\frac{2\rho-s}{2};\,\frac{a^2b^2}{4}\\ \frac{2\rho-s+1}{2},\,\frac{2\rho-s+2}{2},\,\frac{2\rho-s+2}{2}\end{matrix}\right)$ $[b,\,\operatorname{Re}a>0;\,0<\operatorname{Re}s<2\operatorname{Re}\rho+2]$
6	$\frac{1}{(x^2+a^2)^\rho}\operatorname{ci}(bx)$	$-\frac{a^{s-2\rho+2}b^2}{8}\,\mathrm{B}\left(\frac{s+2}{2},\,\frac{2\rho-s-2}{2}\right)\,{}_3F_4\left(\begin{matrix}1,\,1,\,\frac{s+2}{2};\,\frac{a^2b^2}{4}\\ \frac{3}{2},\,2,\,2,\,\frac{s-2\rho+4}{2}\end{matrix}\right)$ $+\frac{a^{s-2\rho}}{2}\,\mathrm{B}\left(\frac{s}{2},\,\frac{2\rho-s}{2}\right)\left[\frac{1}{2}\psi\left(\frac{s}{2}\right)-\frac{1}{2}\psi\left(\frac{2\rho-s}{2}\right)+\ln(ab)+\mathbf{C}\right]$ $+\frac{b^{2\rho-s}}{2\rho-s}\,\Gamma(s-2\rho)\cos\frac{(s-2\rho)\pi}{2}\,{}_2F_3\left(\begin{matrix}\rho,\,\frac{2\rho-s}{2};\,\frac{a^2b^2}{4}\\ \frac{2\rho-s+1}{2},\,\frac{2\rho-s+2}{2},\,\frac{2\rho-s+2}{2}\end{matrix}\right)$ $[b,\,\operatorname{Re}a>0;\,0<\operatorname{Re}s<2\operatorname{Re}\rho+2]$
7	$\frac{1}{x^2-a^2}\operatorname{si}(bx)$	$-\frac{\pi b^{2-s}}{2(2-s)\Gamma(3-s)}\sec\frac{s\pi}{2}\,{}_2F_3\left(\begin{matrix}1,\,\frac{2-s}{2};\,-\frac{a^2b^2}{4}\\ \frac{3-s}{2},\,\frac{4-s}{2},\,\frac{4-s}{2}\end{matrix}\right)$ $+\frac{\pi a^{s-2}}{2}\tan\frac{s\pi}{2}\operatorname{Si}(ab)+\frac{\pi^2a^{s-2}}{4}\cot\frac{s\pi}{2}$ $[a,\,b>0;\,0<\operatorname{Re}s<4]$
8	$\frac{1}{x^2-a^2}\operatorname{ci}(bx)$	$\frac{\pi b^{2-s}}{2(2-s)\Gamma(3-s)}\csc\frac{s\pi}{2}\,{}_2F_3\left(\begin{matrix}1,\,\frac{2-s}{2};\,-\frac{a^2b^2}{4}\\ \frac{3-s}{2},\,\frac{4-s}{2},\,\frac{4-s}{2}\end{matrix}\right)$ $-\frac{\pi a^{s-2}}{2}\cot\frac{s\pi}{2}\operatorname{ci}(ab)+\frac{\pi^2a^{s-2}}{4}\csc^2\frac{s\pi}{2}$ $[a,\,b>0;\,0<\operatorname{Re}s<4]$

No.	$f(x)$	$F(s)$
9	$\dfrac{1}{(x+a)^{\rho}}\operatorname{Si}\left(\dfrac{b}{x+a}\right)$	$a^{s-\rho-1}b\,\mathrm{B}(s,\,1-s+\rho)\;{}_3F_4\left(\begin{matrix}\frac{1}{2},\,\frac{1-s+\rho}{2},\,\frac{2-s+\rho}{2};\,-\frac{b^2}{4a^2}\\ \frac{3}{2},\,\frac{3}{2},\,\frac{\rho+1}{2},\,\frac{\rho+2}{2}\end{matrix}\right)$ [$0<\operatorname{Re}s<\operatorname{Re}\rho+1;\ \lvert\arg a\rvert<\pi$]
10	$\dfrac{1}{(x+a)^{\rho}}\operatorname{Si}\left(\dfrac{bx}{x+a}\right)$	$a^{s-\rho}b\,\mathrm{B}(s+1,\,\rho-s)\;{}_3F_4\left(\begin{matrix}\frac{1}{2},\,\frac{s+1}{2},\,\frac{s+2}{2};\,-\frac{b^2}{4}\\ \frac{3}{2},\,\frac{3}{2},\,\frac{\rho+1}{2},\,\frac{\rho+2}{2}\end{matrix}\right)$ [$-1<\operatorname{Re}s<\operatorname{Re}\rho;\ \lvert\arg a\rvert<\pi$]
11	$\dfrac{1}{(x^2+a^2)^{\rho}}$ $\times\operatorname{Si}\left(\dfrac{bx}{x^2+a^2}\right)$	$\dfrac{a^{s-2\rho-1}b}{2}\,\mathrm{B}\left(\dfrac{s+1}{2},\,\dfrac{1-s+2\rho}{2}\right)\;{}_3F_4\left(\begin{matrix}\frac{1}{2},\,\frac{s+1}{2},\,\frac{1-s+2\rho}{2};\,-\frac{b^2}{16a^2}\\ \frac{3}{2},\,\frac{3}{2},\,\frac{\rho+1}{2},\,\frac{\rho+2}{2}\end{matrix}\right)$ [$\operatorname{Re}a>0;\ -1<\operatorname{Re}s<2\operatorname{Re}\rho+1$]

3.4.3. si (bx), ci (bx), and the exponential function

1	$e^{-ax}\begin{Bmatrix}\operatorname{si}(bx)\\ \operatorname{ci}(bx)\end{Bmatrix}$	$\pm\dfrac{a\,\Gamma(s+1)}{b^{s+1}(s+1)}\begin{Bmatrix}\cos(s\pi/2)\\ \sin(s\pi/2)\end{Bmatrix}{}_3F_2\left(\begin{matrix}\frac{s+1}{2},\,\frac{s+1}{2},\,\frac{s+2}{2}\\ \frac{3}{2},\,\frac{s+3}{2};\,-\frac{a^2}{b^2}\end{matrix}\right)$ $-\dfrac{\Gamma(s)}{b^s s}\begin{Bmatrix}\sin(s\pi/2)\\ \cos(s\pi/2)\end{Bmatrix}{}_3F_2\left(\begin{matrix}\frac{s}{2},\,\frac{s}{2},\,\frac{s+1}{2}\\ \frac{1}{2},\,\frac{s+2}{2};\,-\frac{a^2}{b^2}\end{matrix}\right)$ [$b,\ \operatorname{Re}a,\ \operatorname{Re}s>0$]
2	$e^{-ax^2}\operatorname{si}(bx)$	$-\dfrac{b^3}{36a^{(s+3)/2}}\,\Gamma\left(\dfrac{s+3}{2}\right){}_3F_3\left(\begin{matrix}1,\,\frac{3}{2},\,\frac{s+3}{2}\\ 2,\,\frac{5}{2},\,\frac{5}{2};\,-\frac{b^2}{4a}\end{matrix}\right)$ $+\dfrac{b}{2a^{(s+1)/2}}\,\Gamma\left(\dfrac{s+1}{2}\right)-\dfrac{\pi}{4a^{s/2}}\,\Gamma\left(\dfrac{s}{2}\right)$ [$b,\ \operatorname{Re}a,\ \operatorname{Re}s>0$]
3	$e^{-ax^2}\operatorname{ci}(bx)$	$-\dfrac{b^2}{8a^{(s+2)/2}}\,\Gamma\left(\dfrac{s+2}{2}\right){}_3F_3\left(\begin{matrix}1,\,1,\,\frac{s+2}{2}\\ \frac{3}{2},\,2,\,2;\,-\frac{b^2}{4a}\end{matrix}\right)$ $+\dfrac{\Gamma(s/2)}{4a^{s/2}}\left[\psi\left(\dfrac{s}{2}\right)+\ln\dfrac{b^2}{a}+2\mathbf{C}\right]$ [$b,\ \operatorname{Re}a,\ \operatorname{Re}s>0$]

3.4.4. si (bx), ci (bx), and trigonometric functions

1	$\sin(ax)\operatorname{si}(bx)$	$\dfrac{b\,\Gamma(s+1)}{a^{s+1}}\,\cos\dfrac{s\pi}{2}\,{}_3F_2\left(\begin{matrix}\frac{1}{2},\,\frac{s+1}{2},\,\frac{s+2}{2}\\ \frac{3}{2},\,\frac{3}{2};\,\frac{b^2}{a^2}\end{matrix}\right)-\dfrac{\pi\Gamma(s)}{2a^s}\,\sin\dfrac{s\pi}{2}$ [$0<b<a;\ -1<\operatorname{Re}s<2$]
2	$\sin(ax)\operatorname{si}(bx)$	$-\dfrac{a\Gamma(s+1)}{b^{s+1}(s+1)}\,\cos\dfrac{s\pi}{2}\,{}_3F_2\left(\begin{matrix}\frac{s+1}{2},\,\frac{s+1}{2},\,\frac{s+2}{2}\\ \frac{3}{2},\,\frac{s+3}{2};\,\frac{a^2}{b^2}\end{matrix}\right)$ [$0<a<b;\ -1<\operatorname{Re}s<2$]

No.	$f(x)$	$F(s)$
3	$\sin(ax)\operatorname{ci}(bx)$	$\frac{b^2\Gamma(s+2)}{4a^{s+2}} \sin\frac{s\pi}{2}\, {}_4F_3\left(\begin{matrix}1,\, 1,\, \frac{s+2}{2},\, \frac{s+3}{2} \\ \frac{3}{2},\, 2,\, 2;\, \frac{b^2}{a^2}\end{matrix}\right) + \frac{\Gamma(s)}{a^s} \sin\frac{s\pi}{2} \left[\mathbf{C} + \psi(s) + \frac{\pi}{2} \cot\frac{s\pi}{2} + \ln\frac{b}{a}\right]$ $[0 < b < a;\ -1 < \operatorname{Re} s < 2]$
4	$\sin(ax)\operatorname{ci}(bx)$	$\frac{a\Gamma(s+1)}{b^{s+1}(s+1)} \sin\frac{s\pi}{2}\, {}_3F_2\left(\begin{matrix}\frac{s+1}{2},\, \frac{s+1}{2},\, \frac{s+2}{2} \\ \frac{3}{2},\, \frac{s+3}{2};\, \frac{a^2}{b^2}\end{matrix}\right)$ $[0 < a < b;\ -1 < \operatorname{Re} s < 2]$
5	$\cos(ax)\operatorname{si}(bx)$	$-\frac{b\,\Gamma(s+1)}{a^{s+1}} \sin\frac{s\pi}{2}\, {}_3F_2\left(\begin{matrix}\frac{1}{2},\, \frac{s+1}{2},\, \frac{s+2}{2} \\ \frac{3}{2},\, \frac{3}{2};\, \frac{b^2}{a^2}\end{matrix}\right) - \frac{\pi\Gamma(s)}{2a^s} \cos\frac{s\pi}{2}$ $[0 < b < a;\ 0 < \operatorname{Re} s < 2]$
6	$\cos(ax)\operatorname{si}(bx)$	$-\frac{\Gamma(s)}{b^s s} \sin\frac{s\pi}{2}\, {}_3F_2\left(\begin{matrix}\frac{s}{2},\, \frac{s}{2},\, \frac{s+1}{2} \\ \frac{1}{2},\, \frac{s+2}{2};\, \frac{a^2}{b^2}\end{matrix}\right)$ $[0 < a < b;\ 0 < \operatorname{Re} s < 2]$
7	$\cos(ax)\operatorname{ci}(bx)$	$\frac{b^2\Gamma(s+2)}{4a^{s+2}} \cos\frac{s\pi}{2}\, {}_4F_3\left(\begin{matrix}1,\, 1,\, \frac{s+2}{2},\, \frac{s+3}{2} \\ \frac{3}{2},\, 2,\, 2;\, \frac{b^2}{a^2}\end{matrix}\right) + \frac{\Gamma(s)}{a^s} \cos\frac{s\pi}{2} \left[\mathbf{C} + \psi(s) - \frac{\pi}{2} \tan\frac{s\pi}{2} + \ln\frac{b}{a}\right]$ $[0 < b < a;\ 0 < \operatorname{Re} s < 2]$
8	$\cos(ax)\operatorname{ci}(bx)$	$-\frac{\Gamma(s)}{b^s s} \cos\frac{s\pi}{2}\, {}_3F_2\left(\begin{matrix}\frac{s}{2},\, \frac{s}{2},\, \frac{s+1}{2} \\ \frac{1}{2},\, \frac{s+2}{2};\, \frac{a^2}{b^2}\end{matrix}\right)$ $[0 < a < b;\ 0 < \operatorname{Re} s < 2]$
9	$\sin(ax)\operatorname{ci}(ax) - \cos(ax)\operatorname{si}(ax)$	$\frac{\pi}{2a^s}\Gamma(s)\sec\frac{s\pi}{2}$ $[a > 0;\ 0 < \operatorname{Re} s < 1]$
10	$\cos(ax)\operatorname{ci}(ax) + \sin(ax)\operatorname{si}(ax)$	$-\frac{\pi}{2a^s}\Gamma(s)\csc\frac{s\pi}{2}$ $[a > 0;\ 0 < \operatorname{Re} s < 2]$
11	$\cos(ax)\operatorname{ci}(ax) + \sin(ax)\operatorname{Si}(ax)$	$-\frac{\pi}{2a^s} \cos\frac{s\pi}{2} \cot\frac{s\pi}{2}\,\Gamma(s)$ $[a > 0;\ 0 < \operatorname{Re} s < 1]$
12	$\sin(ax)\operatorname{ci}(ax) - \cos(ax)\operatorname{Si}(ax)$	$\frac{\pi}{2a^s} \sin\frac{s\pi}{2} \tan\frac{s\pi}{2}\,\Gamma(s)$ $[a > 0;\ -1 < \operatorname{Re} s < 1]$

No.	$f(x)$	$F(s)$
13	$\sin\left(b\sqrt{x^2+a^2}\right)$ $\times \operatorname{si}\left(b\sqrt{x^2+a^2}\right)$ $+\cos\left(b\sqrt{x^2+a^2}\right)$ $\times \operatorname{ci}\left(b\sqrt{x^2+a^2}\right)$	$-\frac{\pi a^{(s+1)/2}}{2^{(s+3)/2}b^{(s-1)/2}}\csc\frac{s\pi}{2}\,\Gamma(s)\,\Gamma\left(\frac{1-s}{2}\right)J_{-(s+1)/2}(ab)$ $-\frac{2^{(s-5)/2}\pi^{3/2}a^{(s+1)/2}}{b^{(s-1)/2}}\Gamma\left(\frac{s}{2}\right)\left[\sec\frac{s\pi}{2}J_{(s+1)/2}(ab)\right.$ $\left.+\csc\frac{s\pi}{2}\mathbf{H}_{(s+1)/2}(ab)\right]+\frac{\pi a^s}{2s}\csc\frac{s\pi}{2}$ $[a,\,b>0;\;0<\operatorname{Re}s<2]$
14	$e^{-ax}[\sin(bx)\operatorname{ci}(bx)$ $-\cos(bx)\operatorname{si}(bx)]$	$\frac{\pi\,\Gamma(s)}{2b^s}\sec\frac{s\pi}{2}\,{}_2F_1\left(\begin{matrix}\frac{s}{2},\,\frac{s+1}{2}\\ \frac{1}{2};\,-\frac{a^2}{b^2}\end{matrix}\right)+\frac{\pi a\,\Gamma(s+1)}{2b^{s+1}}\csc\frac{s\pi}{2}$ $\times\,{}_2F_1\left(\begin{matrix}\frac{s+1}{2},\,\frac{s+2}{2}\\ \frac{3}{2};\,-\frac{a^2}{b^2}\end{matrix}\right)+\frac{\Gamma(s-1)}{a^{s-1}b}\,{}_3F_2\left(\begin{matrix}\frac{1}{2},\,1,\,1\\ \frac{2-s}{2},\,\frac{3-s}{2};\,-\frac{a^2}{b^2}\end{matrix}\right)$ $[b,\;\operatorname{Re}a,\;\operatorname{Re}s>0]$
15	$e^{-ax}[\cos(bx)\operatorname{ci}(bx)$ $+\sin(bx)\operatorname{si}(bx)]$	$\frac{\pi a\,\Gamma(s+1)}{2b^{s+1}}\sec\frac{s\pi}{2}\,{}_2F_1\left(\begin{matrix}\frac{s+1}{2},\,\frac{s+2}{2}\\ \frac{3}{2};\,-\frac{a^2}{b^2}\end{matrix}\right)-\frac{\pi\,\Gamma(s)}{2b^s}\csc\frac{s\pi}{2}$ $\times\,{}_2F_1\left(\begin{matrix}\frac{s}{2},\,\frac{s+1}{2}\\ \frac{1}{2};\,-\frac{a^2}{b^2}\end{matrix}\right)-\frac{\Gamma(s-2)}{a^{s-2}b^2}\,{}_3F_2\left(\begin{matrix}\frac{3}{2},\,1,\,1\\ \frac{3-s}{2},\,\frac{4-s}{2};\,-\frac{a^2}{b^2}\end{matrix}\right)$ $[b,\;\operatorname{Re}a,\;\operatorname{Re}s>0]$

3.4.5. Si (bx) and the logarithmic or inverse trigonometric functions

1	$\theta(a-x)\ln\frac{\sqrt{a-x}+\sqrt{a}}{\sqrt{x}}$ $\times\operatorname{Si}(bx)$	$\frac{\sqrt{\pi}\,a^{s+1}b}{2s}\Gamma\left[\begin{matrix}s+1\\ \frac{2s+3}{2}\end{matrix}\right]\left[{}_3F_4\left(\begin{matrix}\frac{1}{2},\,\frac{s+1}{2},\,\frac{s+2}{2};\,-\frac{a^2b^2}{4}\\ \frac{3}{2},\,\frac{3}{2},\,\frac{2s+3}{4},\,\frac{2s+5}{4}\end{matrix}\right)\right.$ $\left.-\frac{1}{s+1}\,{}_3F_4\left(\begin{matrix}\frac{s+1}{2},\,\frac{s+1}{2},\,\frac{s+2}{2};\,-\frac{a^2b^2}{4}\\ \frac{3}{2},\,\frac{2s+3}{4},\,\frac{2s+5}{4},\,\frac{s+3}{2}\end{matrix}\right)\right]$ $[a>0;\;\operatorname{Re}s>-1]$
2	$\theta(a-x)\ln\frac{a+\sqrt{a^2-x^2}}{x}$ $\times\operatorname{Si}(bx)$	$\frac{\sqrt{\pi}\,a^{s+1}b}{2s(s+1)}\Gamma\left[\begin{matrix}\frac{s+1}{2}\\ \frac{s+2}{2}\end{matrix}\right]\left[(s+1)\,{}_2F_3\left(\begin{matrix}\frac{1}{2},\,\frac{s+1}{2};\,-\frac{a^2b^2}{4}\\ \frac{3}{2},\,\frac{3}{2},\,\frac{s+2}{2}\end{matrix}\right)\right.$ $\left.-\,{}_2F_3\left(\begin{matrix}\frac{s+1}{2},\,\frac{s+1}{2};\,-\frac{a^2b^2}{4}\\ \frac{3}{2},\,\frac{s+2}{2},\,\frac{s+3}{2}\end{matrix}\right)\right]$ $[a>0;\;\operatorname{Re}s>-1]$
3	$\theta(a-x)\arccos\frac{x}{a}\operatorname{Si}(bx)$	$\frac{\sqrt{\pi}\,a^{s+1}b}{2(s+1)}\Gamma\left[\begin{matrix}\frac{s+2}{2}\\ \frac{s+3}{2}\end{matrix}\right]{}_3F_4\left(\begin{matrix}\frac{1}{2},\,\frac{s+1}{2},\,\frac{s+2}{2};\,-\frac{a^2b^2}{4}\\ \frac{3}{2},\,\frac{3}{2},\,\frac{s+3}{2},\,\frac{s+3}{2}\end{matrix}\right)$ $[a>0;\;\operatorname{Re}s>-1]$

3.4.6. Si (bx), si (bx), ci (bx), and Ei $(-ax^r)$

1	$\operatorname{Ei}(-ax)\operatorname{Si}(bx)$	$-\frac{b\,\Gamma(s)}{a^{s+1}}\left[{}_3F_2\left(\begin{matrix}\frac{1}{2},\ \frac{s+1}{2},\ \frac{s+2}{2}\\ \frac{3}{2},\ \frac{3}{2};\ -\frac{b^2}{a^2}\end{matrix}\right)-\frac{1}{s+1}\,{}_3F_2\left(\begin{matrix}\frac{s+1}{2},\ \frac{s+1}{2},\ \frac{s+2}{2}\\ \frac{3}{2},\ \frac{s+3}{2};\ -\frac{b^2}{a^2}\end{matrix}\right)\right]$ $[a,\ b>0;\ \operatorname{Re}s>-1]$
2	$\operatorname{Ei}(-ax)\operatorname{si}(bx)$	$\frac{b^3\Gamma(s+3)}{18a^{s+3}(s+3)}\,{}_5F_4\left(\begin{matrix}1,\ \frac{3}{2},\ \frac{s+3}{2},\ \frac{s+3}{2},\ \frac{s+4}{2}\\ 2,\ \frac{5}{2},\ \frac{5}{2},\ \frac{s+5}{2};\ -\frac{b^2}{a^2}\end{matrix}\right)-\frac{b\,\Gamma(s+1)}{a^{s+1}(s+1)}+\frac{\pi\Gamma(s)}{2a^s s}$ $[b,\ \operatorname{Re}a,\ \operatorname{Re}s>0]$
3	$\operatorname{Ei}(-ax)\operatorname{ci}(bx)$	$\frac{b^2\Gamma(s+2)}{4\,a^{s+2}(s+2)}\,{}_5F_4\left(\begin{matrix}1,\ 1,\ \frac{s+2}{2},\ \frac{s+2}{2},\ \frac{s+3}{2}\\ 2,\ 2,\ \frac{3}{2},\ \frac{s+4}{2};\ -\frac{b^2}{a^2}\end{matrix}\right)$ $-\frac{\Gamma(s)}{a^s s}\left[\psi(s)-\frac{1}{s}+\ln\frac{b}{a}+\mathbf{C}\right]$ $[b,\ \operatorname{Re}a,\ \operatorname{Re}s>0]$
4	$\operatorname{Ei}\left(-ax^2\right)\operatorname{si}(bx)$	$\frac{b^3}{18a^{(s+3)/2}(s+3)}\Gamma\left(\frac{s+3}{2}\right)\,{}_4F_4\left(\begin{matrix}1,\ \frac{3}{2},\ \frac{s+3}{2},\ \frac{s+3}{2}\\ 2,\ \frac{5}{2},\ \frac{5}{2},\ \frac{s+5}{2};\ -\frac{b^2}{4a}\end{matrix}\right)$ $-\frac{b}{a^{(s+1)/2}(s+1)}\Gamma\left(\frac{s+1}{2}\right)+\frac{\pi}{2a^{s/2}s}\Gamma\left(\frac{s}{2}\right)$ $[a,\ \operatorname{Re}b,\ \operatorname{Re}s>0]$
5	$\operatorname{Ei}\left(-ax^2\right)\operatorname{ci}(bx)$	$\frac{b^2}{4a^{s/2+1}(s+2)}\Gamma\left(\frac{s+2}{2}\right)\,{}_4F_4\left(\begin{matrix}1,\ 1,\ \frac{s+2}{2},\ \frac{s+2}{2}\\ 2,\ 2,\ \frac{3}{2},\ \frac{s+4}{2};\ -\frac{b^2}{4a}\end{matrix}\right)$ $-\frac{\Gamma(s/2)}{a^{s/2}s}\left[\frac{1}{2}\psi\left(\frac{s}{2}\right)-\frac{1}{s}+\ln\frac{b}{\sqrt{a}}+\mathbf{C}\right]$ $[b,\ \operatorname{Re}a,\ \operatorname{Re}s>0]$

3.4.7. $\operatorname{si}^2(bx)+\operatorname{ci}^2(bx)$ and trigonometric functions

1	$\operatorname{si}^2(ax)+\operatorname{ci}^2(ax)$	$\frac{\pi\Gamma(s)}{a^s s}\csc\frac{s\pi}{2}$ $[a>0;\ 0<\operatorname{Re}s<2]$
2	$\sin(ax)\left[\operatorname{si}^2(bx)+\operatorname{ci}^2(bx)\right]$	$-\frac{a^{2-s}\Gamma(s-2)}{b^2}\sin\frac{s\pi}{2}\,{}_4F_3\left(\begin{matrix}1,\ 1,\ 1,\ \frac{3}{2};\ \frac{a^2}{b^2}\\ 2,\ \frac{3-s}{2},\ \frac{4-s}{2}\end{matrix}\right)$ $+\frac{\pi a\,\Gamma(s+1)}{b^{s+1}(s+1)}\sec\frac{s\pi}{2}\,{}_3F_2\left(\begin{matrix}\frac{s+1}{2},\ \frac{s+1}{2},\ \frac{s+2}{2}\\ \frac{3}{2},\ \frac{s+3}{2};\ \frac{a^2}{b^2}\end{matrix}\right)$ $[a,\ b>0;\ -1<\operatorname{Re}s<2]$
3	$\cos(ax)\left[\operatorname{si}^2(bx)+\operatorname{ci}^2(bx)\right]$	$-\frac{a^{2-s}\Gamma(s-2)}{b^2}\cos\frac{s\pi}{2}\,{}_4F_3\left(\begin{matrix}1,\ 1,\ 1,\ \frac{3}{2};\ \frac{a^2}{b^2}\\ 2,\ \frac{3-s}{2},\ \frac{4-s}{2}\end{matrix}\right)$ $+\frac{\pi\,\Gamma(s)}{b^s s}\csc\frac{s\pi}{2}\,{}_3F_2\left(\begin{matrix}\frac{s}{2},\ \frac{s+1}{2},\ \frac{s}{2}\\ \frac{1}{2},\ \frac{s+2}{2};\ \frac{a^2}{b^2}\end{matrix}\right)$ $[a,\ b>0;\ 0<\operatorname{Re}s<2]$

3.4.8. Products of $\mathrm{si}\,(bx)$ and $\mathrm{ci}\,(bx)$

1	$\mathrm{si}\,(ax)\,\mathrm{si}\,(bx)$	$-\frac{a^{-s-1}b}{s+1}\cos\frac{s\pi}{2}\,\Gamma\,(s+1)\;{}_4F_3\left(\begin{matrix}\frac{1}{2},\,\frac{s+1}{2},\,\frac{s+1}{2},\,\frac{s+2}{2}\\ \frac{3}{2},\,\frac{3}{2},\,\frac{s+3}{2};\,\frac{b^2}{a^2}\end{matrix}\right) + \frac{\pi}{2a^s s}\sin\frac{s\pi}{2}\,\Gamma\,(s)$ $[0<b<a;\;0<\mathrm{Re}\,s<2]$
2	$\mathrm{si}\,(ax)\,\mathrm{ci}\,(bx)$	$-\frac{a^{-s-2}b^2}{4\,(s+2)}\sin\frac{s\pi}{2}\,\Gamma\,(s+2)\;{}_5F_4\left(\begin{matrix}1,\,1,\,\frac{s+2}{2},\,\frac{s+2}{2},\,\frac{s+3}{2}\\ \frac{3}{2},\,2,\,2,\,\frac{s+4}{2};\,\frac{b^2}{a^2}\end{matrix}\right) - \frac{\Gamma\,(s)}{a^s s}\sin\frac{s\pi}{2}\left[\psi\,(s)+\frac{\pi}{2}\cot\frac{s\pi}{2}-\frac{1}{s}+\ln\frac{b}{a}+\mathbf{C}\right]$ $[0<b<a;\;0<\mathrm{Re}\,s<2]$
3	$\mathrm{si}\,(ax)\,\mathrm{ci}\,(bx)$	$\frac{a^3b^{-s-3}}{18\,(s+3)}\sin\frac{s\pi}{2}\,\Gamma\,(s+3)\;{}_5F_4\left(\begin{matrix}1,\,\frac{3}{2},\,\frac{s+1}{2},\,\frac{s+1}{2},\,\frac{s+4}{2}\\ 2,\,\frac{5}{2},\,\frac{5}{2},\,\frac{s+5}{2};\,\frac{a^2}{b^2}\end{matrix}\right) + \frac{a}{b^{s+1}\,(s+1)}\sin\frac{s\pi}{2}\,\Gamma\,(s+1)+\frac{\pi}{2b^s s}\cos\frac{s\pi}{2}\,\Gamma\,(s)$ $[0<a<b;\;0<\mathrm{Re}\,s<2]$
4	$\mathrm{ci}\,(ax)\,\mathrm{ci}\,(bx)$	$-\frac{a^{-s-2}b^2}{4\,(s+2)}\cos\frac{s\pi}{2}\,\Gamma\,(s+2)\;{}_5F_4\left(\begin{matrix}1,\,1,\,\frac{s+2}{2},\,\frac{s+2}{2},\,\frac{s+3}{2}\\ \frac{3}{2},\,2,\,2,\,\frac{s+4}{2};\,\frac{b^2}{a^2}\end{matrix}\right) - \frac{\Gamma\,(s)}{a^s s}\cos\frac{s\pi}{2}\left[\psi\,(s)-\frac{\pi}{2}\tan\frac{s\pi}{2}-\frac{1}{s}+\ln\frac{b}{a}+\mathbf{C}\right]$ $[0<b<a;\;0<\mathrm{Re}\,s<2]$
5	$\left[\sin\,(x)\,\mathrm{ci}\,(2x) - \cos\,(x)\,\mathrm{Si}\,(2x)\right]^2$	$\frac{2^{-s-4}}{s}\,\Gamma\,(s)\left\{\pi^2 s\,[3-\cos\,(s\pi)]\sec\frac{s\pi}{2}+4\pi\,[1+\cos\,(s\pi)] \times\csc\frac{s\pi}{2}+4s\cos\frac{s\pi}{2}\left[\psi'\left(\frac{s+1}{2}\right)-\psi'\left(\frac{s}{2}\right)\right]\right\}$ $[-2<\mathrm{Re}\,s<0]$
6	$\left[\sin\,(x)\,\mathrm{ci}\,(2x) - \cos\,(x)\,\mathrm{Si}\,(2x)\right] \times\left[\cos\,(x)\,\mathrm{ci}\,(2x) + \sin\,(x)\,\mathrm{Si}\,(2x)\right]$	$2^{-s-3}\Gamma\,(s)\left\{\frac{\pi^2}{2}\,[\cos\,(s\pi)+3]\csc\frac{s\pi}{2} +\sin\frac{s\pi}{2}\left[3\psi'\left(\frac{s+1}{2}\right)-4\psi'\,(s)-\psi'\left(\frac{s}{2}\right)\right]\right\}$ $[-1<\mathrm{Re}\,s<1]$

3.5. Hyperbolic Sine shi (z) and Cosine chi (z) Integrals

More formulas can be obtained from the corresponding sections due to the relations

$$\operatorname{shi}(z) = -i\operatorname{Si}(iz), \quad \operatorname{shi}(z) = z\,{}_1F_2\left(\frac{1}{2}; \frac{3}{2}, \frac{3}{2}; \frac{z^2}{4}\right),$$

$$\operatorname{chi}(z) = \operatorname{ci}(iz) - \frac{\pi i}{2}, \quad \operatorname{chi}(z) = \frac{z^2}{4}\,{}_2F_3\left(1, 1; 2, 2, \frac{3}{2}; \frac{z^2}{4}\right) + \ln z + \mathbf{C},$$

$$\operatorname{chi}(z) = -\frac{\sqrt{\pi}}{2}\,G_{13}^{20}\left(-\frac{z^2}{4}\,\middle|\,\begin{matrix}1\\0,\,0,\,1/2\end{matrix}\right) + \frac{1}{2}\Big[\ln z - \ln(-z)\Big].$$

3.5.1. shi (bx), chi (bx), and algebraic functions

No.	$f(x)$	$F(s)$
1	$(a-x)_+^{\alpha-1}\operatorname{shi}(bx)$	$a^{s+\alpha}b\,\mathrm{B}(\alpha, s+1)\,{}_3F_4\left(\begin{matrix}\frac{1}{2}, \frac{s+1}{2}, \frac{s+2}{2}; \frac{a^2b^2}{4}\\ \frac{3}{2}, \frac{3}{2}, \frac{s+\alpha+1}{2}, \frac{s+\alpha+2}{2}\end{matrix}\right)$ $[a,\ \operatorname{Re}\alpha,\ \operatorname{Re}s > 0]$
2	$(a-x)_+^{\alpha-1}\operatorname{chi}(bx)$	$\frac{a^{s+\alpha+1}b^2}{4}\,\mathrm{B}(\alpha, s+2)\,{}_4F_5\left(\begin{matrix}1, 1, \frac{s+2}{2}, \frac{s+3}{2}; \frac{a^2b^2}{4}\\ \frac{3}{2}, 2, 2, \frac{s+\alpha+2}{2}, \frac{s+\alpha+3}{2}\end{matrix}\right)$ $+\,a^{s+\alpha-1}\,\mathrm{B}(\alpha, s)\Big[\psi(s) - \psi(s+\alpha) + \log(ab) + \mathbf{C}\Big]$ $[a,\ \operatorname{Re}\alpha,\ \operatorname{Re}s > 0]$
3	$(a^2-x^2)_+^{\alpha-1}\operatorname{shi}(bx)$	$\frac{a^{s+2\alpha-1}b}{2}\,\mathrm{B}\left(\alpha, \frac{s+1}{2}\right)\,{}_2F_3\left(\begin{matrix}\frac{1}{2}, \frac{s+1}{2}; \frac{a^2b^2}{4}\\ \frac{3}{2}, \frac{3}{2}, \frac{s+2\alpha+1}{2}\end{matrix}\right)$ $[a,\ \operatorname{Re}\alpha,\ \operatorname{Re}s > 0]$
4	$(a^2-x^2)_+^{\alpha-1}\operatorname{chi}(bx)$	$\frac{a^{s+2\alpha}b^2}{8}\,\mathrm{B}\left(\alpha, \frac{s+2}{2}\right)\,{}_3F_4\left(\begin{matrix}1, 1, \frac{s+2}{2}; \frac{a^2b^2}{4}\\ \frac{3}{2}, 2, 2, \frac{s+2\alpha+2}{2}\end{matrix}\right)$ $+\,\frac{a^{s+2\alpha-2}}{2}\,\mathrm{B}\left(\alpha, \frac{s}{2}\right)\left[\frac{1}{2}\psi\left(\frac{s}{2}\right) - \frac{1}{2}\psi\left(\frac{s+2\alpha}{2}\right) + \ln(ab) + \mathbf{C}\right]$ $[a,\ \operatorname{Re}\alpha,\ \operatorname{Re}s > 0]$
5	$\frac{1}{(x+a)^\rho}\operatorname{shi}\left(\frac{b}{x+a}\right)$	$a^{s-\rho-1}b\,\mathrm{B}(s, 1-s+\rho)\,{}_3F_4\left(\begin{matrix}\frac{1}{2}, \frac{1-s+\rho}{2}, \frac{2-s+\rho}{2}; \frac{b^2}{4a^2}\\ \frac{3}{2}, \frac{3}{2}, \frac{\rho+1}{2}, \frac{\rho+2}{2}\end{matrix}\right)$ $[0 < \operatorname{Re}s < \operatorname{Re}\rho + 1;\ \lvert\arg a\rvert < \pi]$
6	$\frac{1}{(x+a)^\rho}\operatorname{shi}\left(\frac{bx}{x+a}\right)$	$a^{s-\rho}b\,\mathrm{B}(s+1, \rho-s)\,{}_3F_4\left(\begin{matrix}\frac{1}{2}, \frac{s+1}{2}, \frac{s+2}{2}; \frac{b^2}{4}\\ \frac{3}{2}, \frac{3}{2}, \frac{\rho+1}{2}, \frac{\rho+2}{2}\end{matrix}\right)$ $[-1 < \operatorname{Re}s < \operatorname{Re}\rho;\ \lvert\arg a\rvert < \pi]$

No.	$f(x)$	$F(s)$
7	$\dfrac{1}{(x^2+a^2)^\rho}\,\mathrm{shi}\left(\dfrac{bx}{x^2+a^2}\right)$	$\dfrac{a^{s-2\rho-1}b}{2}\,\mathrm{B}\left(\dfrac{s+1}{2},\,\dfrac{1-s+2\rho}{2}\right)\,{}_3F_4\left(\begin{matrix}\frac{1}{2},\,\frac{s+1}{2},\,\frac{1-s+2\rho}{2};\,\frac{b^2}{16a^2}\\ \frac{3}{2},\,\frac{3}{2},\,\frac{\rho+1}{2},\,\frac{\rho+2}{2}\end{matrix}\right)$ $[\mathrm{Re}\,a>0;\ -1<\mathrm{Re}\,s<2\,\mathrm{Re}\,\rho+1]$

3.5.2. shi (bx), chi (bx), and the exponential function

1	$e^{-ax}\,\mathrm{shi}\,(bx)$	$\dfrac{b^3}{18\,a^{s+3}}\,\Gamma(s+3)\,{}_4F_3\left(\begin{matrix}1,\,\frac{3}{2},\,\frac{s+3}{2},\,\frac{s+4}{2}\\ 2,\,\frac{5}{2},\,\frac{5}{2};\,\frac{b^2}{a^2}\end{matrix}\right)+\dfrac{b}{a^{s+1}}\,\Gamma(s+1)$ $[\mathrm{Re}\,a>\lvert\mathrm{Re}\,b\rvert;\ \mathrm{Re}\,s>0]$
2	$e^{-ax}\,\mathrm{chi}\,(bx)$	$\dfrac{b^2}{4a^{s+2}}\,\Gamma(s+2)\,{}_4F_3\left(\begin{matrix}1,\,1,\,\frac{s+2}{2},\,\frac{s+3}{2}\\ \frac{3}{2},\,2,\,2;\,\frac{b^2}{a^2}\end{matrix}\right)+\dfrac{\Gamma(s)}{a^s}\left[\psi(s)+\ln\dfrac{b}{a}+\mathbf{C}\right]$ $[\mathrm{Re}\,a>\lvert\mathrm{Re}\,b\rvert;\ \mathrm{Re}\,s>0]$
3	$e^{-ax^2}\,\mathrm{shi}\,(bx)$	$\dfrac{b^3}{36a^{(s+3)/2}}\,\Gamma\left(\dfrac{s+3}{2}\right)\,{}_3F_3\left(\begin{matrix}1,\,\frac{3}{2},\,\frac{s+3}{2}\\ 2,\,\frac{5}{2},\,\frac{5}{2};\,\frac{b^2}{4a}\end{matrix}\right)+\dfrac{b}{2a^{(s+1)/2}}\,\Gamma\left(\dfrac{s+1}{2}\right)$ $[\mathrm{Re}\,a,\ \mathrm{Re}\,s>0;\ \lvert\arg b\rvert<\pi]$
4	$e^{-ax^2}\,\mathrm{chi}\,(bx)$	$\dfrac{b^2}{8a^{s/2+1}}\,\Gamma\left(\dfrac{s+2}{2}\right)\,{}_3F_3\left(\begin{matrix}1,\,1,\,\frac{s+2}{2}\\ \frac{3}{2},\,2,\,2;\,\frac{b^2}{4a}\end{matrix}\right)$ $+\dfrac{1}{2a^{s/2}}\,\Gamma\left(\dfrac{s}{2}\right)\left[\dfrac{1}{2}\,\psi\left(\dfrac{s}{2}\right)+\ln\dfrac{b}{\sqrt{a}}+\mathbf{C}\right]$ $[\mathrm{Re}\,a,\ \mathrm{Re}\,s>0;\ \lvert\arg b\rvert<\pi]$

3.5.3. shi (bx) and the logarithmic or inverse trigonometric functions

1	$\theta(a-x)\ln\dfrac{\sqrt{a-x}+\sqrt{a}}{\sqrt{x}}$ $\times\,\mathrm{shi}\,(bx)$	$\dfrac{\sqrt{\pi}\,a^{s+1}b}{2s}\,\Gamma\left[\begin{matrix}s+1\\ \frac{2s+3}{2}\end{matrix}\right]\left[{}_3F_4\left(\begin{matrix}\frac{1}{2},\,\frac{s+1}{2},\,\frac{s+2}{2};\,\frac{a^2b^2}{4}\\ \frac{3}{2},\,\frac{3}{2},\,\frac{2s+3}{4},\,\frac{2s+5}{4}\end{matrix}\right)\right.$ $\left.-\dfrac{1}{s+1}\,{}_3F_4\left(\begin{matrix}\frac{s+1}{2},\,\frac{s+1}{2},\,\frac{s+2}{2};\,\frac{a^2b^2}{4}\\ \frac{3}{2},\,\frac{2s+3}{4},\,\frac{2s+5}{4},\,\frac{s+3}{2}\end{matrix}\right)\right]$ $[a>0;\ \mathrm{Re}\,s>-1]$
2	$\theta(a-x)\arccos\dfrac{x}{a}\,\mathrm{shi}\,(bx)$	$\dfrac{\sqrt{\pi}\,a^{s+1}b}{2(s+1)}\,\Gamma\left[\begin{matrix}\frac{s+2}{2}\\ \frac{s+3}{2}\end{matrix}\right]\,{}_3F_4\left(\begin{matrix}\frac{1}{2},\,\frac{s+1}{2},\,\frac{s+2}{2};\,\frac{a^2b^2}{4}\\ \frac{3}{2},\,\frac{3}{2},\,\frac{s+3}{2},\,\frac{s+3}{2}\end{matrix}\right)$ $[a>0;\ \mathrm{Re}\,s>-1]$

3.6. erf (z), erfc (z), and erfi (z)

More formulas can be obtained from the corresponding sections due to the relations

$$\begin{Bmatrix} \operatorname{erf}(z) \\ \operatorname{erfc}(z) \end{Bmatrix} = \frac{1}{\sqrt{\pi}} \begin{Bmatrix} \gamma(1/2,\, z^2) \\ \Gamma(1/2,\, z^2) \end{Bmatrix}, \quad \begin{Bmatrix} \operatorname{erf}(z) \\ \operatorname{erfi}(z) \end{Bmatrix} = \frac{2z}{\sqrt{\pi}}\, {}_1F_1\left(\frac{1}{2};\, \frac{3}{2};\, \mp z^2\right),$$

$$\begin{Bmatrix} \operatorname{erf}(z) \\ \operatorname{erfi}(z) \end{Bmatrix} = \frac{z}{\sqrt{\pm z^2}}\left[1 - \frac{e^{-z^2}}{\sqrt{\pi}}\, \Psi\left(\frac{1}{2};\, \frac{1}{2};\, \pm z^2\right)\right], \quad \operatorname{erf}(z) = -i \operatorname{erfi}(iz) = 1 - \operatorname{erfc}(z),$$

$$\operatorname{erfc}(z) = \frac{z}{\sqrt{z^2}}\left[\frac{e^{-z^2}}{\sqrt{\pi}}\, \Psi\left(\frac{1}{2};\, \frac{1}{2};\, z^2\right) - 1\right] + 1, \quad \operatorname{erfc}(z) = 1 - \frac{2z}{\sqrt{\pi}}\, {}_1F_1\left(\frac{1}{2};\, \frac{3}{2};\, -z^2\right),$$

$$\operatorname{erf}(z) = \frac{\sqrt{2}\, z}{\sqrt{-iz^2}}\left[C(-iz^2) - iS(-iz^2)\right],$$

$$\operatorname{erf}(z) = \frac{z}{\sqrt{\pi z^2}}\, G_{12}^{11}\left(z^2 \,\middle|\, \begin{matrix} 1 \\ 1/2,\, 0 \end{matrix}\right), \quad \operatorname{erfc}(\sqrt{z}) = \frac{1}{\sqrt{\pi}}\, G_{12}^{20}\left(z \,\middle|\, \begin{matrix} 1 \\ 0,\, 1/2 \end{matrix}\right),$$

$$\operatorname{erfi}(z) = \frac{z}{\sqrt{-\pi z^2}}\, G_{12}^{11}\left(-z^2 \,\middle|\, \begin{matrix} 1 \\ 1/2,\, 0 \end{matrix}\right).$$

3.6.1. erf $(ax+b)$, erfc $\left(ax+bx^{-1}\right)$

No.	$f(x)$	$F(s)$
1	$\operatorname{erf}(ax+b) - \operatorname{erf}(cx+b)$	$\dfrac{e^{-b^2}\left(c^{-s} - a^{-s}\right)}{2^s\sqrt{\pi}}\, \Gamma(s)\, \Psi\left(\begin{matrix} \frac{s+1}{2} \\ \frac{1}{2};\, b^2 \end{matrix}\right)$ $[\operatorname{Re} s > 0;\ \lvert\arg a\rvert,\ \lvert\arg c\rvert < \pi/4]$
2	$\operatorname{erf}(ax+b) - \operatorname{erf}(cx+d)$	$\dfrac{\Gamma(s)}{2^{(s-1)/2}\sqrt{\pi}}\left[c^{-s}e^{-d^2/2}D_{-s-1}\left(\sqrt{2}\, d\right) - a^{-s}e^{-b^2/2}D_{-s-1}\left(\sqrt{2}\, b\right)\right]$ $[\operatorname{Re} s > 0;\ \lvert\arg a\rvert,\ \lvert\arg c\rvert < \pi/4]$
3	$\operatorname{erfc}\left(ax \pm \dfrac{b}{x}\right)$	$\dfrac{2b}{\sqrt{\pi}\, s}\left(\dfrac{b}{a}\right)^{(s-1)/2} e^{\mp 2ab}\left[K_{(s+1)/2}(2ab) \mp K_{(s-1)/2}(2ab)\right]$ $[b > 0;\ \lvert\arg a\rvert < \pi/4]$

3.6.2. erf (bx), erfc (bx), and algebraic functions

1	$\begin{Bmatrix} \operatorname{erf}(ax) \\ \operatorname{erfc}(ax) \end{Bmatrix}$	$\mp\dfrac{a^{-s}}{\sqrt{\pi}\, s}\, \Gamma\left(\dfrac{s+1}{2}\right)$ $\left[\begin{Bmatrix} -1 < \operatorname{Re} s < 0 \\ \operatorname{Re} s > 0 \end{Bmatrix};\ \lvert\arg a\rvert < \pi/4\right]$
2	$(a-x)_+^{\alpha-1} \begin{Bmatrix} \operatorname{erf}(bx) \\ \operatorname{erfc}(bx) \end{Bmatrix}$	$\pm\dfrac{2a^{s+\alpha}b}{\sqrt{\pi}}\, \mathrm{B}(s+1,\, \alpha)\, {}_3F_3\left(\begin{matrix} \frac{1}{2},\, \frac{s+1}{2},\, \frac{s+2}{2};\, -a^2b^2 \\ \frac{3}{2},\, \frac{s+\alpha+2}{2},\, \frac{s+\alpha+1}{2} \end{matrix}\right) + \dfrac{1 \mp 1}{2}\, a^{s+\alpha-1}\, \mathrm{B}(s,\, \alpha)$ $[a,\ \operatorname{Re}\alpha > 0;\ \operatorname{Re} s > -(1 \pm 1)/2]$

No.	$f(x)$	$F(s)$
3	$(x-a)_+^{\alpha-1}\begin{Bmatrix}\operatorname{erf}(bx)\\ \operatorname{erfc}(bx)\end{Bmatrix}$	$\pm\frac{2a^{s+\alpha}b}{\sqrt{\pi}}\,\mathrm{B}(\alpha,\,-s-\alpha)\;{}_3F_3\left(\begin{matrix}\frac{1}{2},\,\frac{s+1}{2},\,\frac{s+2}{2};\,-a^2b^2\\ \frac{3}{2},\,\frac{s+\alpha+2}{2},\,\frac{s+\alpha+1}{2}\end{matrix}\right)$ $\pm\frac{\Gamma\left(\frac{s+\alpha}{2}\right)}{\sqrt{\pi}\,b^{s+\alpha-1}\,(1-s-\alpha)}\;{}_3F_3\left(\begin{matrix}\frac{1-\alpha}{2},\,\frac{2-\alpha}{2},\,\frac{1-s-\alpha}{2};\,-a^2b^2\\ \frac{1}{2},\,\frac{2-s-\alpha}{2},\,\frac{3-s-\alpha}{2}\end{matrix}\right)$ $\pm\frac{a\,(1-\alpha)\,\Gamma\left(\frac{s+\alpha-1}{2}\right)}{\sqrt{\pi}\,b^{s+\alpha-2}\,(2-s-\alpha)}\;{}_3F_3\left(\begin{matrix}\frac{2-\alpha}{2},\,\frac{3-\alpha}{2},\,\frac{2-s-\alpha}{2};\,-a^2b^2\\ \frac{3}{2},\,\frac{3-s-\alpha}{2},\,\frac{4-s-\alpha}{2}\end{matrix}\right)$ $+\frac{1\mp 1}{2}\,a^{s+\alpha-1}\,\mathrm{B}(\alpha,\,1-\alpha-s)$ $\left[\operatorname{Re}\alpha>0,\,\begin{Bmatrix}a>0;\;\operatorname{Re}(s+\alpha)<1\\ \operatorname{Re}a>0\end{Bmatrix};\,\lvert\arg b\rvert<\pi/4\right]$
4	$(a^2-x^2)_+^{\alpha-1}\begin{Bmatrix}\operatorname{erf}(bx)\\ \operatorname{erfc}(bx)\end{Bmatrix}$	$\pm\frac{a^{s+2\alpha-1}b}{\sqrt{\pi}}\,\mathrm{B}\left(\frac{s+1}{2},\,\alpha\right)\;{}_2F_2\left(\begin{matrix}\frac{1}{2},\,\frac{s+1}{2};\,-a^2b^2\\ \frac{3}{2},\,\frac{s+2\alpha+1}{2}\end{matrix}\right)$ $+\frac{1\mp 1}{4}\,a^{s+2\alpha-2}\,\mathrm{B}\left(\frac{s}{2},\,\alpha\right)$ $[a,\;\operatorname{Re}\alpha>0;\;\operatorname{Re}s>-(1\pm 1)/2]$
5	$(x^2-a^2)_+^{\alpha-1}\begin{Bmatrix}\operatorname{erf}(bx)\\ \operatorname{erfc}(bx)\end{Bmatrix}$	$\pm\frac{a^{s+2\alpha-1}b}{\sqrt{\pi}}\,\mathrm{B}\left(\frac{1-s-2\alpha}{2},\,\alpha\right)\;{}_2F_2\left(\begin{matrix}\frac{1}{2},\,\frac{s+1}{2};\,-a^2b^2\\ \frac{3}{2},\,\frac{s+2\alpha+1}{2}\end{matrix}\right)$ $\pm\frac{b^{2-s-2\alpha}}{\sqrt{\pi}\,(2-s-2\alpha)}\,\Gamma\left(\frac{s+2\alpha-1}{2}\right)$ $\times\;{}_2F_2\left(\begin{matrix}1-\alpha,\,\frac{2-s-2\alpha}{2};\,-a^2b^2\\ \frac{3-s-2\alpha}{2},\,\frac{4-s-2\alpha}{2}\end{matrix}\right)$ $+\frac{1\mp 1}{4}\,a^{s+2\alpha-2}\,\mathrm{B}\left(\frac{2-s-2\alpha}{2},\,\alpha\right)$ $\left[\operatorname{Re}\alpha>0,\,\begin{Bmatrix}a>0;\;\operatorname{Re}(s+2\alpha)<2\\ a>0\end{Bmatrix};\,\lvert\arg b\rvert<\pi/4\right]$
6	$\frac{1}{(x+a)^{\rho}}\begin{Bmatrix}\operatorname{erf}(bx)\\ \operatorname{erfc}(bx)\end{Bmatrix}$	$\pm\frac{2a^{s-\rho+1}b}{\sqrt{\pi}}\,\mathrm{B}(s+1,\,\rho-s-1)\;{}_3F_3\left(\begin{matrix}\frac{1}{2},\,\frac{s+1}{2},\,\frac{s+2}{2};\,-a^2b^2\\ \frac{3}{2},\,\frac{s-\rho+2}{2},\,\frac{s-\rho+3}{2}\end{matrix}\right)$ $\pm\frac{1}{\sqrt{\pi}\,b^{s-\rho}\,(\rho-s)}\,\Gamma\left(\frac{s-\rho+1}{2}\right)\;{}_3F_3\left(\begin{matrix}\frac{\rho}{2},\,\frac{\rho+1}{2},\,\frac{\rho-s}{2};\,-a^2b^2\\ \frac{1}{2},\,\frac{\rho-s+1}{2},\,\frac{\rho-s+2}{2}\end{matrix}\right)$ $\mp\frac{\rho a}{\sqrt{\pi}\,b^{s-\rho-1}\,(\rho-s+1)}\,\Gamma\left(\frac{s-\rho}{2}\right)$ $\times\;{}_3F_3\left(\begin{matrix}\frac{\rho+1}{2},\,\frac{\rho+2}{2},\,\frac{\rho-s+1}{2};\,-a^2b^2\\ \frac{3}{2},\,\frac{\rho-s+2}{2},\,\frac{\rho-s+3}{2}\end{matrix}\right)+\frac{1\mp 1}{2}\,a^{s-\rho}\,\mathrm{B}(s,\,\rho-s)$ $\left[\begin{Bmatrix}-1<\operatorname{Re}s<\operatorname{Re}\rho\\ \operatorname{Re}s>0\end{Bmatrix};\,\lvert\arg a\rvert,\,4\lvert\arg b\rvert<\pi\right]$

No.	$f(x)$	$F(s)$
7	$\frac{1}{x-a}\left\{\begin{matrix}\operatorname{erf}(bx)\\ \operatorname{erfc}(bx)\end{matrix}\right\}$	$\mp\frac{\pi a^{s-1}}{b}\cot(s\pi)\operatorname{erf}(ab)\pm\frac{\Gamma\left(\frac{s}{2}\right)}{\sqrt{\pi}\,b^{s-1}(1-s)}\,{}_2F_2\left(\begin{matrix}1,\ \frac{1-s}{2};\ -a^2b^2\\ \frac{2-s}{2},\ \frac{3-s}{2}\end{matrix}\right)$ $\pm\frac{a\Gamma\left(\frac{s-1}{2}\right)}{\sqrt{\pi}\,b^{s-2}(2-s)}\,{}_2F_2\left(\begin{matrix}1,\ \frac{2-s}{2};\ -a^2b^2\\ \frac{3-s}{2},\ \frac{4-s}{2}\end{matrix}\right)-\frac{\pi\mp\pi}{2}a^{s-1}\cot(s\pi)$ $[a>0;\ \lvert\operatorname{Re}s\rvert<1;\ \lvert\arg b\rvert<\pi/4]$
8	$\frac{1}{(x^2+a^2)^{\rho}}\left\{\begin{matrix}\operatorname{erf}(bx)\\ \operatorname{erfc}(bx)\end{matrix}\right\}$	$\pm\frac{a^{s-2\rho+1}b}{\sqrt{\pi}}\operatorname{B}\left(\frac{s+1}{2},\ \frac{2\rho-s-1}{2}\right)\,{}_2F_2\left(\begin{matrix}\frac{1}{2},\ \frac{s+1}{2};\ a^2b^2\\ \frac{3}{2},\ \frac{s-2\rho+3}{2}\end{matrix}\right)$ $\pm\frac{b^{2\rho-s}}{\sqrt{\pi}(2\rho-s)}\Gamma\left(\frac{s-2\rho+1}{2}\right)\,{}_2F_2\left(\begin{matrix}\rho,\ \frac{2\rho-s}{2};\ a^2b^2\\ \frac{2\rho-s+1}{2},\ \frac{2\rho-s+2}{2}\end{matrix}\right)$ $+\frac{(1\mp1)}{4}a^{s-2\rho}\operatorname{B}\left(\frac{s}{2},\ \frac{2\rho-s}{2}\right)$ $\left[\operatorname{Re}a>0;\ \left\{\begin{matrix}-1<\operatorname{Re}s<2\operatorname{Re}\rho\\ \operatorname{Re}s>0\end{matrix}\right\};\ \lvert\arg b\rvert<\pi/4\right]$
9	$\frac{1}{x^2-a^2}\left\{\begin{matrix}\operatorname{erf}(bx)\\ \operatorname{erfc}(bx)\end{matrix}\right\}$	$\pm\frac{\pi a^{s-2}}{2}\tan\frac{s\pi}{2}\operatorname{erf}(ab)\pm\frac{b^{2-s}}{\sqrt{\pi}(2-s)}\Gamma\left(\frac{s-1}{2}\right)$ $\times\,{}_2F_2\left(\begin{matrix}1,\ \frac{2-s}{2};\ -a^2b^2\\ \frac{3-s}{2},\ \frac{4-s}{2}\end{matrix}\right)-\frac{(1\mp1)\,\pi a^{s-2}}{4}\cot\frac{s\pi}{2}$ $\left[a>0;\ \left\{\begin{matrix}-1<\operatorname{Re}s<2\\ \operatorname{Re}s>0\end{matrix}\right\};\ \lvert\arg b\rvert<\pi/4\right]$
10	$(ax^2+b)^n\operatorname{erfc}(cx)$	$\frac{b^n}{\sqrt{\pi}\,c^s s}\Gamma\left(\frac{s+1}{2}\right)\,{}_3F_1\left(\begin{matrix}-n,\ \frac{s}{2},\ \frac{s+1}{2}\\ \frac{s+2}{2};\ -\frac{a}{bc^2}\end{matrix}\right)$ $[\operatorname{Re}s>0;\ \lvert\arg c\rvert<\pi/4]$

3.6.3. erf (bx), erfc (bx), and the exponential function

1	$e^{-ax}\left\{\begin{matrix}\operatorname{erf}(bx)\\ \operatorname{erfc}(bx)\end{matrix}\right\}$	$\mp\frac{1}{\sqrt{\pi}\,b^s s}\Gamma\left(\frac{s+1}{2}\right)\,{}_2F_2\left(\begin{matrix}\frac{s}{2},\ \frac{s+1}{2}\\ \frac{1}{2},\ \frac{s+2}{2};\ \frac{a^2}{4b^2}\end{matrix}\right)\pm\frac{a}{\sqrt{\pi}\,b^{s+1}(s+1)}$ $\times\Gamma\left(\frac{s+2}{2}\right)\,{}_2F_2\left(\begin{matrix}\frac{s+1}{2},\ \frac{s+2}{2}\\ \frac{3}{2},\ \frac{s+3}{2};\ \frac{a^2}{4b^2}\end{matrix}\right)+\frac{1\pm1}{2a^s}\Gamma(s)$ $\left[\left\{\begin{matrix}\operatorname{Re}a>0,\ \operatorname{Re}s>-1\\ \operatorname{Re}s>0\end{matrix}\right\};\ \lvert\arg b\rvert<\pi/4\right]$
2	$e^{-ax^2}\left\{\begin{matrix}\operatorname{erf}(bx)\\ \operatorname{erfc}(bx)\end{matrix}\right\}$	$\pm\frac{b}{\sqrt{\pi}\,a^{(s+1)/2}}\Gamma\left(\frac{s+1}{2}\right)\,{}_2F_1\left(\begin{matrix}\frac{1}{2},\ \frac{s+1}{2}\\ \frac{3}{2};\ -\frac{b^2}{a}\end{matrix}\right)+\frac{1\mp1}{4a^{s/2}}\Gamma\left(\frac{s}{2}\right)$ $[\operatorname{Re}a>0;\ \operatorname{Re}s>-(1\pm1)/2;\ \lvert\arg b\rvert<\pi/4]$
3	$e^{a^2x^2}\operatorname{erfc}(ax)$	$\frac{a^{-s}}{2}\Gamma\left(\frac{s}{2}\right)\sec\frac{s\pi}{2}$ $[0<\operatorname{Re}s<1;\ \lvert\arg a\rvert<\pi/4]$

No.	$f(x)$	$F(s)$
4	$e^{-a^2x^2}\,\mathrm{erfi}\,(ax)$	$\dfrac{\pi}{2a^s\Gamma\left(\frac{2-s}{2}\right)}\sec\dfrac{s\pi}{2}$ $[\lvert\mathrm{Re}\,s\rvert<1;\ \lvert\arg a\rvert<\pi/4]$
5	$e^{ax^2}\,\mathrm{erfc}\,(bx)$	$\dfrac{b^{-s}}{\sqrt{\pi}\,s}\Gamma\left(\dfrac{s+1}{2}\right){}_2F_1\left(\begin{matrix}\frac{s}{2},\ \frac{s+1}{2}\\ \frac{s+2}{2};\ \frac{a}{b^2}\end{matrix}\right)$ $[\mathrm{Re}\,(b^2-a),\ \mathrm{Re}\,s>0]$
6	$e^{-a/x}\left\{\begin{matrix}\mathrm{erf}\,(bx)\\ \mathrm{erfc}\,(bx)\end{matrix}\right\}$	$\dfrac{1\mp1}{2}a^s\,\Gamma(-s)\pm\dfrac{2a^{s+1}b}{\sqrt{\pi}}\Gamma(-s-1)\,{}_1F_3\left(\begin{matrix}\frac{1}{2};\ -\frac{a^2b^2}{4}\\ \frac{3}{2},\ \frac{s+2}{2},\ \frac{s+3}{2}\end{matrix}\right)$ $\mp\dfrac{1}{\sqrt{\pi}\,b^s s}\Gamma\left(\dfrac{s+1}{2}\right){}_1F_3\left(\begin{matrix}-\frac{s}{2};\ -\frac{a^2b^2}{4}\\ \frac{1}{2},\ \frac{1-s}{2},\ \frac{2-s}{2}\end{matrix}\right)$ $\pm\dfrac{a}{\sqrt{\pi}\,b^{s-1}(s-1)}\Gamma\left(\dfrac{s}{2}\right){}_1F_3\left(\begin{matrix}\frac{1-s}{2};\ -\frac{a^2b^2}{4}\\ \frac{3}{2},\ \frac{2-s}{2},\ \frac{3-s}{2}\end{matrix}\right)$ $\left[\left\{\begin{matrix}\mathrm{Re}\,a>0;\ \mathrm{Re}\,s<0\\ \mathrm{Re}\,a>0\end{matrix}\right\};\ \lvert\arg b\rvert<\pi/4\right]$
7	$e^{-a/x^2}\left\{\begin{matrix}\mathrm{erf}\,(bx)\\ \mathrm{erfc}\,(bx)\end{matrix}\right\}$	$\dfrac{1\mp1}{4}a^{s/2}\,\Gamma\left(-\dfrac{s}{2}\right)\pm\dfrac{a^{(s+1)/2}b}{\sqrt{\pi}}\Gamma\left(-\dfrac{s+1}{2}\right){}_1F_2\left(\begin{matrix}\frac{1}{2};\ ab^2\\ \frac{3}{2},\ \frac{s+3}{2}\end{matrix}\right)$ $\mp\dfrac{1}{\sqrt{\pi}\,b^s s}\Gamma\left(\dfrac{s+1}{2}\right){}_1F_2\left(\begin{matrix}-\frac{s}{2};\ ab^2\\ \frac{1-s}{2},\ \frac{2-s}{2}\end{matrix}\right)$ $\left[\left\{\begin{matrix}\mathrm{Re}\,a>0;\ \mathrm{Re}\,s<0\\ \mathrm{Re}\,a>0\end{matrix}\right\};\ \lvert\arg b\rvert<\pi/4\right]$
8	$e^{-ax-b^2x^2}\,\mathrm{erfi}\,(bx)$	$\dfrac{\Gamma(s-1)}{\sqrt{\pi}\,a^{s-1}b}\,{}_2F_2\left(\begin{matrix}\frac{1}{2},\ 1;\ \frac{a^2}{4b^2}\\ \frac{2-s}{2},\ \frac{3-s}{2}\end{matrix}\right)+\dfrac{\Gamma(s/2)}{2b^s}\tan\dfrac{s\pi}{2}\,{}_1F_1\left(\begin{matrix}\frac{s}{2};\ \frac{a^2}{4b^2}\\ \frac{1}{2}\end{matrix}\right)$ $+\dfrac{a}{2b^{s+1}}\Gamma\left(\dfrac{s+1}{2}\right)\cot\dfrac{s\pi}{2}\,{}_1F_1\left(\begin{matrix}\frac{s+1}{2};\ \frac{a^2}{4b^2}\\ \frac{3}{2}\end{matrix}\right)$ $[\mathrm{Re}\,a>0;\ \mathrm{Re}\,s>-1;\ \lvert\arg b\rvert<\pi/4]$
9	$e^{-ax+b^2x^2}\,\mathrm{erfc}\,(bx)$	$\dfrac{\Gamma(s-1)}{\sqrt{\pi}\,a^{s-1}b}\,{}_2F_2\left(\begin{matrix}\frac{1}{2},\ 1;\ -\frac{a^2}{4b^2}\\ \frac{2-s}{2},\ \frac{3-s}{2}\end{matrix}\right)+\dfrac{\Gamma\left(\frac{s}{2}\right)}{2b^s}\sec\dfrac{s\pi}{2}\,{}_1F_1\left(\begin{matrix}\frac{s}{2};\ -\frac{a^2}{4b^2}\\ \frac{1}{2}\end{matrix}\right)$ $+\dfrac{a}{2b^{s+1}}\Gamma\left(\dfrac{s+1}{2}\right)\csc\dfrac{s\pi}{2}\,{}_1F_1\left(\begin{matrix}\frac{s+1}{2};\ -\frac{a^2}{4b^2}\\ \frac{3}{2}\end{matrix}\right)$ $[\mathrm{Re}\,a,\ \mathrm{Re}\,s>0;\ \lvert\arg b\rvert<\pi/4]$
10	$e^{-ax-bx^2}\,\mathrm{erf}\,(cx)$	$\dfrac{c}{\sqrt{\pi}\,b^{(s+1)/2}}\Gamma\left(\dfrac{s+1}{2}\right)\Psi_1\left(\dfrac{s+1}{2},\ \dfrac{1}{2},\ \dfrac{3}{2},\ \dfrac{1}{2};\ -\dfrac{c^2}{b};\ \dfrac{a^2}{4b}\right)$ $-\dfrac{ac}{\sqrt{\pi}\,b^{(s+2)/2}}\Gamma\left(\dfrac{s+2}{2}\right)\Psi_1\left(\dfrac{s+2}{2},\ \dfrac{1}{2},\ \dfrac{3}{2},\ \dfrac{3}{2};\ -\dfrac{c^2}{b};\ \dfrac{a^2}{4b}\right)$ $[\mathrm{Re}\,b,\ \mathrm{Re}\,(b+c^2)>0;\ \mathrm{Re}\,s>-1]$

No.	$f(x)$	$F(s)$
11	$e^{-b^2x^2-a/x^2}\,\mathrm{erfi}\,(bx)$	$-\dfrac{\pi a^{s/4}}{2b^{s/2}}\sec\dfrac{s\pi}{2}\left[\mathbf{L}_{s/2}\,(2b\sqrt{a})-I_{-s/2}\,(2b\sqrt{a})\right]$ $[\mathrm{Re}\,a>0;\ \mathrm{Re}\,s<1;\ s\neq-1,-3,\ldots;\ \lvert\arg b\rvert<\pi/4]$
12	$e^{b^2x^2-a/x^2}\,\mathrm{erfc}\,(bx)$	$\dfrac{\pi a^{s/4}}{2b^{s/2}}\sec\dfrac{s\pi}{2}\left[\mathbf{H}_{s/2}\,(2b\sqrt{a})-Y_{s/2}\,(2b\sqrt{a})\right]$ $[\mathrm{Re}\,a>0;\ \mathrm{Re}\,s<1;\ s\neq-1,-3,\ldots;\ \lvert\arg b\rvert<\pi/4]$
13	$e^{a^2x^2}\,\mathrm{erfc}\,(ax+b)$	$\dfrac{\Gamma(s)}{\sqrt{\pi}\,(2a)^s}\,\Gamma\left(\dfrac{1-s}{2},\,b^2\right)$ $\quad[0<\mathrm{Re}\,s<1;\ \lvert\arg a\rvert<\pi/4]$
14	$e^{-a^2x}\,\mathrm{erfi}\,(a\sqrt{x})$	$a^{-2s}\,\Gamma\begin{bmatrix}\frac{1-2s}{2},\ \frac{2s+1}{2}\\ 1-s\end{bmatrix}$ $\quad[0<\lvert\mathrm{Re}\,s\rvert<1/2;\ \lvert\arg a\rvert<\pi/4]$
15	$\theta(a-x)\,e^{bx}\,\mathrm{erf}\,(c\sqrt{a-x})$	$a^{s+1/2}c\,\Gamma\begin{bmatrix}s\\ \frac{2s+3}{2}\end{bmatrix}\Phi_2\left(s,\,\dfrac{1}{2};\,\dfrac{2s+3}{2};\,ab,\,-ac^2\right)$ $\quad[a,\ \mathrm{Re}\,s>0]$

3.6.4. erf (bx), erfc (bx), erfi (bx), and algebraic or the exponential functions

1	$(a-x)_+^{\alpha-1}\,e^{b^2x^2}$ $\times\begin{Bmatrix}\mathrm{erf}\,(bx)\\ \mathrm{erfc}\,(bx)\end{Bmatrix}$	$\pm\dfrac{2a^{s+\alpha}b}{\sqrt{\pi}}\,\mathrm{B}\,(s+1,\,\alpha)\ {}_3F_3\left(\begin{matrix}1,\ \frac{s+1}{2},\ \frac{s+2}{2};\ a^2b^2\\ \frac{3}{2},\ \frac{s+\alpha+1}{2},\ \frac{s+\alpha+2}{2}\end{matrix}\right)$ $+\dfrac{1\mp1}{2}\,a^{s+\alpha-1}\,\mathrm{B}\,(s,\,\alpha)\ {}_2F_2\left(\begin{matrix}\frac{s}{2},\ \frac{s+1}{2};\ a^2b^2\\ \frac{s+\alpha}{2},\ \frac{s+\alpha+1}{2}\end{matrix}\right)$ $[a,\ \mathrm{Re}\,\alpha>0;\ \mathrm{Re}\,s>-(1\pm1)/2]$
2	$\left(a^2-x^2\right)_+^{\alpha-1}\,e^{b^2x^2}$ $\times\begin{Bmatrix}\mathrm{erf}\,(bx)\\ \mathrm{erfc}\,(bx)\end{Bmatrix}$	$\pm\dfrac{a^{s+2\alpha-1}b}{\sqrt{\pi}}\,\mathrm{B}\left(\dfrac{s+1}{2},\,\alpha\right)\ {}_2F_2\left(\begin{matrix}1,\ \frac{s+1}{2};\ a^2b^2\\ \frac{3}{2},\ \frac{s+2\alpha+1}{2}\end{matrix}\right)$ $+\dfrac{1\mp1}{4}\,a^{s+2\alpha-2}\,\mathrm{B}\left(\dfrac{s}{2},\,\alpha\right)\ {}_1F_1\left(\begin{matrix}\frac{s}{2};\ a^2b^2\\ \frac{s+2\alpha}{2};\end{matrix}\right)$ $[a,\ \mathrm{Re}\,\alpha>0;\ \mathrm{Re}\,s>-(1\pm1)/2]$
3	$\left(x^2-a^2\right)_+^{\alpha-1}\,e^{\mp b^2x^2}$ $\times\begin{Bmatrix}\mathrm{erfi}\,(bx)\\ \mathrm{erfc}\,(bx)\end{Bmatrix}$	$\pm\dfrac{a^{s+2\alpha-1}b}{\sqrt{\pi}}\,\mathrm{B}\left(\dfrac{1-s-2\alpha}{2},\,\alpha\right)\ {}_2F_2\left(\begin{matrix}1,\ \frac{s+1}{2};\ \mp a^2b^2\\ \frac{3}{2},\ \frac{s+2\alpha+1}{2}\end{matrix}\right)$ $+\dfrac{1\mp1}{4}\,a^{s+2\alpha-2}\,\mathrm{B}\left(\dfrac{2-s-2\alpha}{2},\,\alpha\right)\ {}_1F_1\left(\begin{matrix}\frac{s}{2};\ a^2b^2\\ \frac{s+2\alpha}{2};\end{matrix}\right)$ $\pm\dfrac{b^{2-s-2\alpha}}{2}\begin{Bmatrix}\tan\left[(s+2\alpha)\,\pi/2\right]\\ \sec\left[(s+2\alpha)\,\pi/2\right]\end{Bmatrix}$ $\times\Gamma\left(\dfrac{s+2\alpha-2}{2}\right)\ {}_1F_1\left(\begin{matrix}1-\alpha;\ \mp a^2b^2\\ \frac{4-s-2\alpha}{2}\end{matrix}\right)$ $[a,\ \mathrm{Re}\,\alpha>0;\ \mathrm{Re}\,(s+2\alpha)<3;\ \lvert\arg b\rvert<\pi/4]$

No.	$f(x)$	$F(s)$
4	$\dfrac{e^{\mp b^2x^2}}{(x+a)^\rho}\begin{Bmatrix}\operatorname{erfi}(bx)\\ \operatorname{erfc}(bx)\end{Bmatrix}$	$\pm\dfrac{2a^{s-\rho+1}b}{\sqrt{\pi}}\,\mathrm{B}(s+1,\,\rho-s-1)\;{}_3F_3\left(\begin{matrix}1,\,\frac{s+1}{2},\,\frac{s+2}{2};\;\mp a^2b^2\\ \frac{3}{2},\,\frac{s-\rho+2}{2},\,\frac{s-\rho+3}{2}\end{matrix}\right)$ $\mp\dfrac{b^{\rho-s}}{2}\begin{Bmatrix}\tan[(\rho-s)\pi/2]\\ \sec[(\rho-s)\pi/2]\end{Bmatrix}\Gamma\left(\dfrac{s-\rho}{2}\right)$ $\times\,{}_2F_2\left(\begin{matrix}\frac{\rho}{2},\,\frac{\rho+1}{2};\,\mp a^2b^2\\ \frac{1}{2},\,\frac{\rho-s+2}{2}\end{matrix}\right)\pm\dfrac{\rho ab^{\rho-s+1}}{2}\begin{Bmatrix}\cot[(s-\rho)\pi/2]\\ \csc[(s-\rho)\pi/2]\end{Bmatrix}$ $\times\,\Gamma\left(\dfrac{s-\rho-1}{2}\right)\,{}_2F_2\left(\begin{matrix}\frac{\rho+1}{2},\,\frac{\rho+2}{2};\,\mp a^2b^2\\ \frac{3}{2},\,\frac{3-s+\rho}{2}\end{matrix}\right)$ $+\dfrac{1\mp 1}{2}\,a^{s-\rho}\,\mathrm{B}(s,\,\rho-s)\;{}_2F_2\left(\begin{matrix}\frac{s}{2},\,\frac{s+1}{2};\,a^2b^2\\ \frac{s-\rho+1}{2},\,\frac{s-\rho+2}{2}\end{matrix}\right)$ $[-(1\pm 1)/2<\operatorname{Re}s<\operatorname{Re}\rho+1;\;\lvert\arg a\rvert,\,4\lvert\arg b\rvert<\pi]$
5	$\dfrac{e^{-b^2x^2}}{x+a}\operatorname{erfi}(bx)$	$\dfrac{a^{s-1}}{2}\,e^{-a^2b^2}\Big[i^{s-1}\cot\dfrac{s\pi}{2}\,\Gamma\left(\dfrac{s+1}{2}\right)\gamma\left(\dfrac{1-s}{2},-a^2b^2\right)$ $-\,i^s\tan\dfrac{s\pi}{2}\,\Gamma\left(\dfrac{s}{2}\right)\gamma\left(\dfrac{2-s}{2},-a^2b^2\right)-\dfrac{2\pi}{\sin(s\pi)}\operatorname{erfi}(ab)\Big]$ $[-1<\operatorname{Re}s<2;\;\lvert\arg a\rvert,\,4\lvert\arg b\rvert<\pi]$
6	$\dfrac{e^{-b^2x^2}}{x-a}\operatorname{erfi}(bx)$	$-\pi a^{s-1}\,e^{-a^2b^2}\cot(s\pi)\operatorname{erfi}(ab)$ $-\dfrac{b^{1-s}}{2}\cot\dfrac{s\pi}{2}\,\Gamma\left(\dfrac{s-1}{2}\right)\,{}_1F_1\left(\begin{matrix}1;\,-a^2b^2\\ \frac{3-s}{2}\end{matrix}\right)$ $+\dfrac{ab^{2-s}}{2}\tan\dfrac{s\pi}{2}\,\Gamma\left(\dfrac{s-2}{2}\right)\,{}_1F_1\left(\begin{matrix}1;\,-a^2b^2\\ \frac{4-s}{2}\end{matrix}\right)$ $[a>0;\;-1<\operatorname{Re}s<2;\;\lvert\arg b\rvert<\pi/4]$
7	$\dfrac{e^{\mp b^2x^2}}{(x^2+a^2)^\rho}\begin{Bmatrix}\operatorname{erfi}(bx)\\ \operatorname{erfc}(bx)\end{Bmatrix}$	$\pm\dfrac{a^{s-2\rho+1}b}{\sqrt{\pi}}\,\mathrm{B}\left(\dfrac{s+1}{2},\,\dfrac{2\rho-s-1}{2}\right)\,{}_2F_2\left(\begin{matrix}1,\,\frac{s+1}{2};\,\pm a^2b^2\\ \frac{3}{2},\,\frac{s-2\rho+3}{2}\end{matrix}\right)$ $\mp\dfrac{b^{2\rho-s}}{2}\begin{Bmatrix}\tan[(2\rho-s)\pi/2]\\ \sec[(2\rho-s)\pi/2]\end{Bmatrix}\Gamma\left(\dfrac{s-2\rho}{2}\right)\,{}_1F_1\left(\begin{matrix}\rho;\,\pm a^2b^2\\ \frac{2-s+2\rho}{2}\end{matrix}\right)$ $+\dfrac{1\mp 1}{4}\,a^{s-2\rho}\,\mathrm{B}\left(\dfrac{s}{2},\,\dfrac{2\rho-s}{2}\right)\,{}_1F_1\left(\begin{matrix}\frac{s}{2};\,-a^2b^2\\ \frac{s-2\rho+2}{2};\end{matrix}\right)$ $[\operatorname{Re}a>0;\;-(1\pm 1)/2<\operatorname{Re}s<2\operatorname{Re}\rho+1;\;\lvert\arg b\rvert<\pi/4]$
8	$\dfrac{e^{b^2x^2}}{x^2+a^2}\operatorname{erfc}(bx)$	$\dfrac{\pi a^{s-2}}{2}\,e^{-a^2b^2}\sec\dfrac{s\pi}{2}\Big[\cot\dfrac{s\pi}{2}-\operatorname{erfi}(ab)$ $+\dfrac{i^{s-2}}{\pi}\,\Gamma\left(\dfrac{s}{2}\right)\gamma\left(\dfrac{2-s}{2},\,-a^2b^2\right)\Big]$ $[\operatorname{Re}a>0;\;0<\operatorname{Re}s<3;\;\lvert\arg b\rvert<\pi/4]$

No.	$f(x)$	$F(s)$
9	$\dfrac{e^{-b^2x^2}}{x^2+a^2}\operatorname{erfi}(bx)$	$\dfrac{\pi a^{s-2}}{2}e^{a^2b^2}\sec\dfrac{s\pi}{2}\left[\operatorname{erf}(ab)-\dfrac{1}{\Gamma\left(\frac{2-s}{2}\right)}\gamma\left(\dfrac{2-s}{2},\,a^2b^2\right)\right]$ [$\operatorname{Re}a>0;\ -1<\operatorname{Re}s<3;\ \lvert\arg b\rvert<\pi/4$]
10	$\dfrac{e^{b^2x^2}}{x^2-a^2}\operatorname{erfc}(bx)$	$\dfrac{\pi a^{s-2}}{2}e^{a^2b^2}\tan\dfrac{s\pi}{2}\operatorname{erfc}(ab)-\dfrac{\pi a^{s-2}}{\sin(s\pi)}e^{a^2b^2}$ $-\dfrac{b^{2-s}}{2}\sec\dfrac{s\pi}{2}\Gamma\left(\dfrac{s-2}{2}\right){}_1F_1\left(\begin{matrix}1;\ a^2b^2\\ \frac{4-s}{2}\end{matrix}\right)$ [$a>0;\ 0<\operatorname{Re}s<3;\ \lvert\arg b\rvert<\pi/4$]
11	$\dfrac{e^{-b^2x^2}}{x^2-a^2}\operatorname{erfi}(bx)$	$\dfrac{\pi a^{s-2}}{2}e^{-a^2b^2}\tan\dfrac{s\pi}{2}\operatorname{erfi}(ab)$ $+\dfrac{b^{2-s}}{2}\tan\dfrac{s\pi}{2}\Gamma\left(\dfrac{s-2}{2}\right){}_1F_1\left(\begin{matrix}1;\ -a^2b^2\\ \frac{4-s}{2}\end{matrix}\right)$ [$a>0;\ -1<\operatorname{Re}s<3;\ \lvert\arg b\rvert<\pi/4$]

3.6.5. erf $(\varphi(x))$, erfc $(\varphi(x))$, and algebraic functions

No.	$f(x)$	$F(s)$
1	$(a-x)_+^{\alpha-1}$ $\times\operatorname{erf}\left(b\sqrt{x(a-x)}\right)$	$\dfrac{2}{\sqrt{\pi}}a^{s+\alpha}b\,\mathrm{B}\left(\dfrac{2\alpha+1}{2},\,\dfrac{2s+1}{2}\right){}_3F_3\left(\begin{matrix}\frac{1}{2},\ \frac{2\alpha+1}{2},\ \frac{2s+1}{2};\ -\frac{a^2b^2}{4}\\ \frac{3}{2},\ \frac{s+\alpha+1}{2},\ \frac{s+\alpha+2}{2}\end{matrix}\right)$ [$a>0;\ \operatorname{Re}\alpha,\ \operatorname{Re}s>-1/2$]
2	$(a-x)_+^{\alpha-1}\operatorname{erf}(bx(a-x))$	$\dfrac{2}{\sqrt{\pi}}a^{s+\alpha+1}b\,\mathrm{B}(s+1,\,\alpha+1)\,{}_5F_5\left(\begin{matrix}\frac{1}{2},\ \Delta(2,\,s+1),\ \Delta(2,\,\alpha+1)\\ \frac{3}{2},\ \Delta(4,\,s+\alpha+2);\ -\frac{a^4b^2}{16}\end{matrix}\right)$ [$a>0;\ \operatorname{Re}\alpha,\ \operatorname{Re}s>-1$]
3	$\theta(1-x)\operatorname{erfc}\left(\dfrac{ax+b}{\sqrt{1-x^2}}\right)$	$\sqrt{\dfrac{2}{\pi}}e^{(a^2-b^2)/2}\Gamma(s)\,D_{-s}\left(\sqrt{2}\,a\right)D_{-s-1}\left(\sqrt{2}\,b\right)$ [$\operatorname{Re}s,\ \operatorname{Re}b>0$]
4	$\theta(x-a)\operatorname{erf}\left(\dfrac{bx}{\sqrt{x^2-c^2}}\right)$	$-\dfrac{2a^sb}{\sqrt{\pi}\,s}\Psi_1\left(\dfrac{1}{2},\,-\dfrac{s}{2};\,\dfrac{2-s}{2},\,\dfrac{3}{2};\,\dfrac{c^2}{a^2},\,-b^2\right)$ [$a>0;\ \operatorname{Re}s<0;\ \lvert c\rvert<a$]
5	$\dfrac{1}{(x+a)^\rho}\operatorname{erf}\left(\dfrac{bx}{x+a}\right)$	$\dfrac{2a^{s-\rho}b}{\sqrt{\pi}}\mathrm{B}(s+1,\,\rho-s)\,{}_3F_3\left(\begin{matrix}\frac{1}{2},\ \frac{s+1}{2},\ \frac{s+2}{2}\\ \frac{3}{2},\ \frac{\rho+1}{2},\ \frac{\rho+2}{2};\ -b^2\end{matrix}\right)$ [$-1<\operatorname{Re}s<\operatorname{Re}\rho;\ \lvert\arg a\rvert<\pi$]

No.	$f(x)$	$F(s)$
6	$\dfrac{1}{(x^2+a^2)^\rho}\operatorname{erf}\left(\dfrac{bx}{x^2+a^2}\right)$	$\dfrac{a^{s-2\rho-1}b}{\sqrt{\pi}}\,\mathrm{B}\left(\dfrac{s+1}{2},\,\dfrac{1-s+2\rho}{2}\right)\,{}_3F_3\left(\begin{matrix}\frac{1}{2},\,\frac{s+1}{2},\,\frac{1-s+2\rho}{2}\\ \frac{3}{2},\,\frac{\rho+1}{2},\,\frac{\rho+2}{2};\,-\frac{b^2}{4a^2}\end{matrix}\right)$ [$\operatorname{Re}a>0;\ -1<\operatorname{Re}s<2\operatorname{Re}\rho+1$]

3.6.6. erf $(\varphi(x))$, erfc $(\varphi(x))$, and the exponential function

1	$(a-x)_+^{\alpha-1}\,e^{b^2x(a-x)}$ $\times\operatorname{erf}\left(b\sqrt{x(a-x)}\right)$	$\dfrac{2}{\sqrt{\pi}}\,a^{s+\alpha}b\,\mathrm{B}\left(s+\dfrac{1}{2},\,\alpha+\dfrac{1}{2}\right)\,{}_3F_3\left(\begin{matrix}1,\,\frac{2s+1}{2},\,\frac{2\alpha+1}{2};\,\frac{a^2b^2}{4}\\ \frac{3}{2},\,\frac{s+\alpha+1}{2},\,\frac{s+\alpha+2}{2}\end{matrix}\right)$ [$a>0;\ \operatorname{Re}\alpha,\ \operatorname{Re}s>-1/2$]
2	$(a-x)_+^{\alpha-1}\,e^{b^2x^2(a-x)^2}$ $\times\operatorname{erf}(bx(a-x))$	$\dfrac{2}{\sqrt{\pi}}\,a^{s+\alpha+1}b\,\mathrm{B}(s+1,\,\alpha+1)\,{}_6F_5\left(\begin{matrix}1,\,\Delta(2,\,s+1),\,\Delta(2,\,\alpha+1)\\ \frac{3}{2},\,\Delta(4,\,s+\alpha+2);\,\frac{a^4b^2}{16}\end{matrix}\right)$ [$a>0;\ \operatorname{Re}s,\ \operatorname{Re}\alpha>-1$]
3	$\dfrac{\theta(x-a)}{\sqrt{x^2-b^2}}\,e^{a^2x^2/(x^2-b^2)}$ $\times\operatorname{erf}\left(\dfrac{cx}{\sqrt{x^2-c^2}}\right)$	$\dfrac{2a^{s-1}c}{\sqrt{\pi}\,(1-s)}\,\Psi_1\left(1,\,\dfrac{1-s}{2};\,\dfrac{3-s}{2},\,\dfrac{3}{2};\,\dfrac{c^2}{a^2},\,-b^2\right)$ [$a>0;\ \operatorname{Re}s<0;\ \lvert c\rvert<a$]
4	$\dfrac{1}{(x+a)^\rho}\,e^{b^2x^2/(x+a)^2}$ $\times\operatorname{erf}\left(\dfrac{bx}{x+a}\right)$	$\dfrac{2a^{s-\rho}b}{\sqrt{\pi}}\,\mathrm{B}(s+1,\,\rho-s)\,{}_3F_3\left(\begin{matrix}1,\,\frac{s+1}{2},\,\frac{s+2}{2}\\ \frac{3}{2},\,\frac{\rho+1}{2},\,\frac{\rho+2}{2};\,b^2\end{matrix}\right)$ [$-1<\operatorname{Re}s<\operatorname{Re}\rho;\ \lvert\arg a\rvert<\pi$]
5	$\dfrac{1}{(x^2+a^2)^\rho}\,e^{b^2x^2/(x^2+a^2)^2}$ $\times\operatorname{erf}\left(\dfrac{bx}{x^2+a^2}\right)$	$\dfrac{a^{s-2\rho-1}b}{2\sqrt{\pi}}\,\mathrm{B}\left(\dfrac{s+1}{2},\,\dfrac{1-s+2\rho}{2}\right)\,{}_3F_3\left(\begin{matrix}1,\,\frac{s+1}{2},\,\frac{1-s+2\rho}{2}\\ \frac{3}{2},\,\frac{\rho+1}{2},\,\frac{\rho+2}{2};\,\frac{b^2}{4a^2}\end{matrix}\right)$ [$\operatorname{Re}a>0;\ -1<\operatorname{Re}s<2\operatorname{Re}\rho+1$]

3.6.7. erf (bx), erfc (bx), and trigonometric functions

Notation: $\delta=\begin{Bmatrix}1\\0\end{Bmatrix}$.

1	$\begin{Bmatrix}\sin(ax)\\ \cos(ax)\end{Bmatrix}\operatorname{erf}(bx)$	$-\dfrac{a^\delta b^{-s-\delta}}{\sqrt{\pi}\,(s+\delta)}\,\Gamma\left(\dfrac{s+\delta+1}{2}\right)\,{}_2F_2\left(\begin{matrix}\frac{s+\delta}{2},\,\frac{s+\delta+1}{2};\,-\frac{a^2}{4b^2}\\ \frac{2\delta+1}{2},\,\frac{s+\delta+2}{2}\end{matrix}\right)$ $+\dfrac{\Gamma(s)}{a^s}\begin{Bmatrix}\sin(s\pi/2)\\ \cos(s\pi/2)\end{Bmatrix}$ [$a>0;\ -\delta-1<\operatorname{Re}s<1;\ \lvert\arg b\rvert<\pi/4$]

No.	$f(x)$	$F(s)$
2	$\begin{Bmatrix} \sin(ax^2) \\ \cos(ax^2) \end{Bmatrix} \operatorname{erf}(bx)$	$-\frac{a^{\delta} b^{-s-2\delta}}{\sqrt{\pi}(s+2\delta)} \Gamma\left(\frac{s+2\delta+1}{2}\right) {}_3F_2\left(\begin{matrix} \frac{s+2\delta}{4}, \frac{s+2\delta+1}{4}, \frac{s+2\delta+3}{4} \\ \frac{2\delta+1}{2}, \frac{s+2\delta+4}{4}; -\frac{a^2}{b^4} \end{matrix}\right) + \frac{a^{-s/2}}{2} \Gamma\left(\frac{s}{2}\right) \begin{Bmatrix} \sin(s\pi/4) \\ \cos(s\pi/4) \end{Bmatrix}$ $[a > 0;\ -2\delta - 1 < \operatorname{Re} s < 2;\ \lvert\arg b\rvert < \pi/4]$
3	$\begin{Bmatrix} \sin(ax^2) \\ \cos(ax^2) \end{Bmatrix} \operatorname{erfc}(bx)$	$\frac{a^{\delta} b^{-s-2\delta}}{\sqrt{\pi}(s+2\delta)} \Gamma\left(\frac{s+2\delta+1}{2}\right) {}_3F_2\left(\begin{matrix} \frac{s+2\delta}{4}, \frac{s+2\delta+1}{4}, \frac{s+2\delta+3}{4} \\ \frac{2\delta+1}{2}, \frac{s+2\delta+4}{4}; -\frac{a^2}{b^4} \end{matrix}\right)$ $[\operatorname{Re} s > -2\delta;\ \operatorname{Re} b^2 > \lvert\operatorname{Im} a\rvert]$
4	$\sin(a\sqrt{x}) \operatorname{erfc}(bx)$	$\frac{2ab^{-s-1/2}}{\sqrt{\pi}(2s+1)} \Gamma\left(\frac{2s+3}{4}\right) {}_2F_4\left(\begin{matrix} \frac{2s+1}{4}, \frac{2s+3}{4}; \frac{a^4}{256b^2} \\ \frac{1}{2}, \frac{3}{4}, \frac{5}{4}, \frac{2s+5}{4} \end{matrix}\right) - \frac{a^3 b^{-s-3/2}}{3\sqrt{\pi}(2s+3)} \Gamma\left(\frac{2s+5}{4}\right) {}_2F_4\left(\begin{matrix} \frac{2s+3}{4}, \frac{2s+5}{4}; \frac{a^4}{256b^2} \\ \frac{5}{4}, \frac{3}{2}, \frac{7}{4}, \frac{2s+7}{4} \end{matrix}\right)$ $[\operatorname{Re} s > -1/2;\ \lvert\arg b\rvert < \pi/4]$
5	$\cos(a\sqrt{x}) \operatorname{erfc}(bx)$	$\frac{b^{-s}}{\sqrt{\pi}\, s} \Gamma\left(\frac{s+1}{2}\right) {}_2F_4\left(\begin{matrix} \frac{s}{2}, \frac{s+1}{2}; \frac{a^4}{256b^2} \\ \frac{1}{4}, \frac{1}{2}, \frac{3}{4}, \frac{s+2}{2} \end{matrix}\right) - \frac{a^2 b^{-s-1}}{2\sqrt{\pi}(s+1)} \Gamma\left(\frac{s+2}{2}\right) {}_2F_4\left(\begin{matrix} \frac{s+1}{2}, \frac{s+2}{2}; \frac{a^4}{256b^2} \\ \frac{3}{4}, \frac{5}{4}, \frac{3}{2}, \frac{s+3}{2} \end{matrix}\right)$ $[\operatorname{Re} s > 0;\ \lvert\arg b\rvert < \pi/4]$
6	$\begin{Bmatrix} \sin^{2n}(ax) \\ \cos^{2n}(ax) \end{Bmatrix} \operatorname{erfc}(bx)$	$\frac{2^{-2n} b^{-s}}{\sqrt{\pi}\, s} \Gamma\left(\frac{s+1}{2}\right) \left[2 \sum_{k=0}^{n-1} (\mp 1)^{n-k} \binom{2n}{k} \times {}_2F_2\left(\begin{matrix} \frac{s}{2}, \frac{s+1}{2}; -(n-k)^2 \frac{a^2}{b^2} \\ \frac{1}{2}, \frac{s+2}{2} \end{matrix}\right) + \binom{2n}{n}\right]$ $[a > 0;\ \operatorname{Re} s > -2n\delta;\ \lvert\arg b\rvert < \pi/4;\ n \geq 1]$
7	$\begin{Bmatrix} \sin^{2n+1}(ax) \\ \cos^{2n+1}(ax) \end{Bmatrix} \operatorname{erfc}(bx)$	$\frac{2^{-2n} a^{\delta} b^{-s-\delta}}{\sqrt{\pi}(s+\delta)} \Gamma\left(\frac{s+\delta+1}{2}\right) \sum_{k=0}^{n} (\mp 1)^{n-k} (2n-2k+1)^{\delta} \times \binom{2n+1}{k} {}_2F_2\left(\begin{matrix} \frac{s+\delta}{2}, \frac{s+\delta+1}{2}; -\left(n-k+\frac{1}{2}\right)^2 \frac{a^2}{b^2} \\ \frac{2\delta+1}{2}, \frac{s+\delta+2}{2} \end{matrix}\right)$ $[a > 0;\ \operatorname{Re} s > -(2n+3)\delta;\ \lvert\arg b\rvert < \pi/4]$
8	$\begin{Bmatrix} \sinh(ax)\sin(ax) \\ \cosh(ax)\cos(ax) \end{Bmatrix} \times \operatorname{erfc}(bx)$	$\frac{a^{2\delta} b^{-s-2\delta}}{\sqrt{\pi}(s+2\delta)} \Gamma\left(\frac{s+2\delta+1}{2}\right) \times {}_3F_4\left(\begin{matrix} \frac{s+2\delta}{4}, \frac{s+2\delta+1}{4}, \frac{s+2\delta+3}{4} \\ \frac{2\delta+1}{4}, \frac{2\delta+3}{4}, \frac{2\delta+1}{2}, \frac{s+2\delta+4}{4}; -\frac{a^4}{16b^4} \end{matrix}\right)$ $[a > 0;\ \operatorname{Re} s > -2\delta;\ \lvert\arg b\rvert < \pi/4]$

No.	$f(x)$	$F(s)$
9	$\left\{\begin{matrix}\sinh(ax)\cos(ax)\\ \cosh(ax)\sin(ax)\end{matrix}\right\} \times \operatorname{erfc}(bx)$	$\frac{ab^{-s-1}}{\sqrt{\pi}(s+1)}\Gamma\left(\frac{s+2}{2}\right){}_3F_4\left(\begin{matrix}\frac{s+1}{4}, \frac{s+2}{4}, \frac{s+4}{4}\\ \frac{1}{2}, \frac{3}{4}, \frac{5}{4}, \frac{s+5}{4}; -\frac{a^4}{16b^4}\end{matrix}\right) \mp \frac{a^3b^{-s-1}}{3\sqrt{\pi}(s+3)}\Gamma\left(\frac{s+4}{2}\right){}_3F_4\left(\begin{matrix}\frac{s+3}{4}, \frac{s+4}{4}, \frac{s+6}{4}\\ \frac{5}{4}, \frac{3}{2}, \frac{7}{4}, \frac{s+7}{4}; -\frac{a^4}{16b^4}\end{matrix}\right)$ $[a>0;\ \operatorname{Re}s>-1;\ \lvert\arg b\rvert<\pi/4]$

3.6.8. erfc (bx), erfi (bx), and the exponential or trigonometric functions

Notation: $\delta = \left\{\begin{matrix}1\\0\end{matrix}\right\}$.

1	$e^{-b^2x^2}\left\{\begin{matrix}\sin(ax)\\ \cos(ax)\end{matrix}\right\} \times \operatorname{erfi}(bx)$	$\mp\frac{\Gamma(s-1)}{\sqrt{\pi}\,a^{s-1}b}\left\{\begin{matrix}\cos(s\pi/2)\\ \sin(s\pi/2)\end{matrix}\right\}{}_2F_2\left(\begin{matrix}\frac{1}{2}, 1; -\frac{a^2}{4b^2}\\ \frac{2-s}{2}, \frac{3-s}{2}\end{matrix}\right) \mp \frac{a^\delta}{2b^{s+\delta}}\left\{\begin{matrix}\cot(s\pi/2)\\ \tan(s\pi/2)\end{matrix}\right\}\Gamma\left(\frac{s+\delta}{2}\right){}_1F_1\left(\begin{matrix}\frac{s+\delta}{2}; -\frac{a^2}{4b^2}\\ \frac{2\delta+1}{2}\end{matrix}\right)$ $[a>0;\ -\delta-1<\operatorname{Re}s<2;\ \lvert\arg b\rvert<\pi/4]$
2	$e^{b^2x^2}\left\{\begin{matrix}\sin(ax)\\ \cos(ax)\end{matrix}\right\} \times \operatorname{erfc}(bx)$	$\mp\frac{\Gamma(s-1)}{\sqrt{\pi}\,a^{s-1}b}\left\{\begin{matrix}\cos(s\pi/2)\\ \sin(s\pi/2)\end{matrix}\right\}{}_2F_2\left(\begin{matrix}\frac{1}{2}, 1; \frac{a^2}{4b^2}\\ \frac{2-s}{2}, \frac{3-s}{2}\end{matrix}\right) \mp \frac{a^\delta}{2b^{s+\delta}}\left\{\begin{matrix}\csc(s\pi/2)\\ \sec(s\pi/2)\end{matrix}\right\}\Gamma\left(\frac{s+\delta}{2}\right){}_1F_1\left(\begin{matrix}\frac{s+\delta}{2}; \frac{a^2}{4b^2}\\ \frac{2\delta+1}{2}\end{matrix}\right)$ $[a>0;\ -\delta<\operatorname{Re}s<2;\ \lvert\arg b\rvert<\pi/4]$
3	$e^{-b^2x^2}\left\{\begin{matrix}\sin(ax^2)\\ \cos(ax^2)\end{matrix}\right\} \times \operatorname{erfi}(bx)$	$\frac{1}{2\sqrt{\pi}\,a^{(s-1)/2}b}\left\{\begin{matrix}\sin[(s-1)\pi/4]\\ \cos[(s-1)\pi/4]\end{matrix}\right\}\Gamma\left(\frac{s-1}{2}\right){}_3F_2\left(\begin{matrix}\frac{1}{4}, \frac{3}{4}, 1; -\frac{a^2}{b^4}\\ \frac{3-s}{4}, \frac{5-s}{4}\end{matrix}\right) - \frac{1}{4\sqrt{\pi}\,a^{(s-3)/2}b^3}\left\{\begin{matrix}\sin[(s+1)\pi/4]\\ \cos[(s+1)\pi/4]\end{matrix}\right\} \times\Gamma\left(\frac{s-3}{2}\right){}_3F_2\left(\begin{matrix}\frac{3}{4}, 1, \frac{5}{4}; -\frac{a^2}{b^4}\\ \frac{5-s}{4}, \frac{7-s}{4}\end{matrix}\right) + \frac{a^\delta}{2b^{s+2\delta}}\tan\frac{s\pi}{2}\,\Gamma\left(\frac{s+2\delta}{2}\right){}_2F_1\left(\begin{matrix}\frac{s+2\delta}{4}, \frac{s+2\delta+2}{4}\\ \frac{2\delta+1}{2}; -\frac{a^2}{b^4}\end{matrix}\right)$ $[a>0;\ -2\delta-1<\operatorname{Re}s<3;\ \lvert\arg b\rvert<\pi/4]$
4	$e^{b^2x^2}\sin(ax^2)\operatorname{erfc}(bx)$	$-\frac{a^{(1-s)/2}}{4\sqrt{\pi}\,b}\cos\frac{s\pi}{2}\csc\frac{(s+1)\pi}{4}\Gamma\left(\frac{s-1}{2}\right){}_3F_2\left(\begin{matrix}\frac{1}{4}, \frac{3}{4}, 1; -\frac{a^2}{b^4}\\ \frac{3-s}{4}, \frac{5-s}{4}\end{matrix}\right) + \frac{a^{(3-s)/2}}{8\sqrt{\pi}\,b^3}\cos\frac{s\pi}{2}\sec\frac{(s+1)\pi}{4}\Gamma\left(\frac{s-3}{2}\right){}_3F_2\left(\begin{matrix}1, \frac{3}{4}, \frac{5}{4}; -\frac{a^2}{b^4}\\ \frac{5-s}{4}, \frac{7-s}{4}\end{matrix}\right) - \frac{1}{2}\left(a^2+b^4\right)^{-s/4}\sec\frac{s\pi}{2}\sin\left(\frac{s}{2}\arctan\frac{a}{b^2}\right)\Gamma\left(\frac{s}{2}\right)$ $[a>0;\ -2<\operatorname{Re}s<3;\ \lvert\arg b\rvert<\pi/4]$

No.	$f(x)$	$F(s)$
5	$e^{b^2x^2}\cos(ax^2)\operatorname{erfc}(bx)$	$-\frac{a^{(1-s)/2}}{4\sqrt{\pi}\,b}\cos\frac{s\pi}{2}\csc\frac{(s-1)\pi}{4}\Gamma\left(\frac{s-1}{2}\right){}_3F_2\left(\begin{matrix}\frac{1}{4},\frac{3}{4},1;-\frac{a^2}{b^4}\\ \frac{3-s}{4},\frac{5-s}{4}\end{matrix}\right)$ $+\frac{a^{(3-s)/2}}{8\sqrt{\pi}\,b^3}\cos\frac{s\pi}{2}\csc\frac{(s+1)\pi}{4}\Gamma\left(\frac{s-3}{2}\right){}_3F_2\left(\begin{matrix}1,\frac{3}{4},\frac{5}{4};-\frac{a^2}{b^4}\\ \frac{5-s}{4},\frac{7-s}{4}\end{matrix}\right)$ $+\frac{1}{2}(a^2+b^4)^{-(s+2)/4}\sec\frac{s\pi}{2}\Gamma\left(\frac{s}{2}\right)$ $\times\left[a\sin\left(\frac{s+2}{2}\arctan\frac{a}{b^2}\right)+b^2\cos\left(\frac{s+2}{2}\arctan\frac{a}{b^2}\right)\right]$ $[a>0;\ 0<\operatorname{Re}s<3;\ \lvert\arg b\rvert<\pi/4]$

3.6.9. erf (bx), erfc (bx), and the logarithmic function

1	$\ln x\operatorname{erf}(ax)$	$\frac{a^{-s}}{\sqrt{\pi}\,s}\Gamma\left(\frac{s+1}{2}\right)\left[\ln a+\frac{1}{s}-\frac{1}{2}\psi\left(\frac{s+1}{2}\right)\right]$ $[-1<\operatorname{Re}s<0;\ \lvert\arg a\rvert<\pi/4]$
2	$\ln(x^2+a^2)\left\{\begin{matrix}\operatorname{erf}(bx)\\ \operatorname{erfc}(bx)\end{matrix}\right\}$	$\mp\frac{a^2b^{2-s}}{\sqrt{\pi}\,s}\Gamma\left(\frac{s-1}{2}\right){}_2F_2\left(\begin{matrix}1,1;a^2b^2\\ 2,\frac{3-s}{2}\end{matrix}\right)$ $\mp\frac{2a^2b^{2-s}}{\sqrt{\pi}\,s(s-2)}\Gamma\left(\frac{s-1}{2}\right){}_2F_2\left(\begin{matrix}1,\frac{2-s}{2};a^2b^2\\ \frac{3-s}{2},\frac{4-s}{2}\end{matrix}\right)$ $\pm\frac{b^{-s}}{\sqrt{\pi}\,s}\Gamma\left(\frac{s+1}{2}\right)\left[\frac{2}{s}-\psi\left(\frac{s+1}{2}\right)+2\ln b\right]$ $\pm\left[\frac{\pi a^s}{s}\operatorname{erfi}(ab)+\frac{\sqrt{\pi}\,i^{1-s}}{s\,b^s}\gamma\left(\frac{s+1}{2},-a^2b^2\right)\right]\sec\frac{s\pi}{2}$ $+\left\{\begin{matrix}0\\1\end{matrix}\right\}\frac{\pi a^s}{s}\csc\frac{s\pi}{2}$ $\left[\operatorname{Re}a>0;\ \left\{\begin{matrix}-1<\operatorname{Re}s<0\\ \operatorname{Re}s>0\end{matrix}\right\};\ \lvert\arg b\rvert<\pi/4\right]$
3	$\ln\lvert x^2-a^2\rvert\left\{\begin{matrix}\operatorname{erf}(bx)\\ \operatorname{erfc}(bx)\end{matrix}\right\}$	$\pm\frac{a^2b^{2-s}}{\sqrt{\pi}\,s}\Gamma\left(\frac{s-1}{2}\right){}_2F_2\left(\begin{matrix}1,1;-a^2b^2\\ 2,\frac{3-s}{2}\end{matrix}\right)$ $\pm\frac{2a^2b^{2-s}}{\sqrt{\pi}\,s(s-2)}\Gamma\left(\frac{s-1}{2}\right){}_2F_2\left(\begin{matrix}1,\frac{2-s}{2};-a^2b^2\\ \frac{3-s}{2},\frac{4-s}{2}\end{matrix}\right)$ $\pm\frac{b^{-s}}{\sqrt{\pi}\,s}\Gamma\left(\frac{s+1}{2}\right)\left[\frac{2}{s}-\psi\left(\frac{s+1}{2}\right)+\pi\tan\frac{s\pi}{2}+2\ln b\right]$ $\mp\left[\frac{\pi a^s}{s}\operatorname{erf}(ab)+\frac{\sqrt{\pi}\,b^{-s}}{s}\Gamma\left(\frac{s+1}{2},a^2b^2\right)\right]\tan\frac{s\pi}{2}$ $+\left\{\begin{matrix}0\\1\end{matrix}\right\}\frac{\pi a^s}{s}\cot\frac{s\pi}{2}$ $\left[a>0;\ \left\{\begin{matrix}-1<\operatorname{Re}s<0\\ \operatorname{Re}s>0\end{matrix}\right\};\ \lvert\arg b\rvert<\pi/4\right]$

No.	$f(x)$	$F(s)$
4	$\theta(a-x)\ln\frac{\sqrt{a}+\sqrt{a-x}}{\sqrt{x}}$ $\times\operatorname{erf}(bx)$	$\frac{a^{s+1}b}{s}\Gamma\left[\begin{matrix}s+1\\ \frac{2s+3}{2}\end{matrix}\right]{}_3F_3\left(\begin{matrix}\frac{1}{2},\,\frac{s+1}{2},\,\frac{s+2}{2};\,-a^2b^2\\ \frac{3}{2},\,\frac{2s+3}{4},\,\frac{2s+5}{4}\end{matrix}\right)$ $-\frac{a^{s+1}b}{s(s+1)}\Gamma\left[\begin{matrix}s+1\\ \frac{2s+3}{2}\end{matrix}\right]{}_3F_3\left(\begin{matrix}\frac{s+1}{2},\,\frac{s+1}{2},\,\frac{s+2}{2};\,-a^2b^2\\ \frac{2s+3}{4},\,\frac{2s+5}{4},\,\frac{s+3}{2}\end{matrix}\right)$ $[a>0;\ \operatorname{Re}s>-1]$
5	$\theta(a-x)\ln\frac{\sqrt{a}+\sqrt{a-x}}{\sqrt{x}}$ $\times e^{b^2x^2}\operatorname{erf}(bx)$	$\frac{a^{s+1}b}{s+1}\Gamma\left[\begin{matrix}s+1\\ \frac{2s+3}{2}\end{matrix}\right]{}_4F_4\left(\begin{matrix}1,\,\frac{s+1}{2},\,\frac{s+1}{2},\,\frac{s+2}{2};\,a^2b^2\\ \frac{3}{2},\,\frac{2s+3}{4},\,\frac{2s+5}{4},\,\frac{s+3}{2}\end{matrix}\right)$ $[a>0;\ \operatorname{Re}s>-1]$
6	$\theta(a-x)\ln\frac{a+\sqrt{a^2-x^2}}{x}$ $\times\operatorname{erf}(bx)$	$\frac{2a^{s+1}b}{s(s+1)}\Gamma\left[\begin{matrix}\frac{s+1}{2}\\ \frac{s}{2}\end{matrix}\right]{}_3F_3\left(\begin{matrix}\frac{1}{2},\,\frac{s+1}{2},\,\frac{s+1}{2};\,-a^2b^2\\ \frac{3}{2},\,\frac{s+2}{2},\,\frac{s+3}{2}\end{matrix}\right)$ $[a>0;\ \operatorname{Re}s>-1]$
7	$\ln^n x\operatorname{erf}(ax)$	$-\frac{1}{\sqrt{\pi}}\frac{\partial^n}{\partial s^n}\left[\frac{a^{-s}}{s}\Gamma\left(\frac{s+1}{2}\right)\right]$ $[-1<\operatorname{Re}s<0;\ \lvert\arg a\rvert<\pi/4]$
8	$\theta(a-x)\ln^n\frac{x}{a}\operatorname{erf}(bx)$	$\frac{2(-1)^n\,n!\,a^{s+1}b}{\sqrt{\pi}(s+1)^{n+1}}\,{}_{n+2}F_{n+2}\left(\begin{matrix}\frac{1}{2},\,\frac{s+1}{2},\ldots,\frac{s+1}{2}\\ \frac{3}{2},\,\frac{s+3}{2},\ldots,\frac{s+3}{2};\,-a^2b^2\end{matrix}\right)$ $[a>0;\ \operatorname{Re}s>0]$
9	$\theta(a-x)e^{b^2x^2}\ln^n\frac{x}{a}$ $\times\operatorname{erf}(bx)$	$\frac{2(-1)^n\,n!\,a^{s+1}b}{\sqrt{\pi}(s+1)^{n+1}}\,{}_{n+2}F_{n+2}\left(\begin{matrix}1,\,\frac{s+1}{2},\ldots,\frac{s+1}{2}\\ \frac{3}{2},\,\frac{s+3}{2},\ldots,\frac{s+3}{2};\,a^2b^2\end{matrix}\right)$ $[a>0;\ \operatorname{Re}s>-1]$

3.6.10. erf (ax) and inverse trigonometric functions

1	$\theta(1-x)\left\{\begin{matrix}\arcsin x\\ \arccos x\end{matrix}\right\}$ $\times\operatorname{erf}(ax)$	$\frac{(1\pm1)\sqrt{\pi}}{4s}\left[\sqrt{\pi}\operatorname{erf}(a)-a^{-s}\gamma\left(\frac{s+1}{2},\,a^2\right)\right]$ $\mp\frac{a}{2(s+1)}\Gamma\left[\begin{matrix}\frac{s}{2}\\ \frac{s+3}{2}\end{matrix}\right]\left[(s+1)\,{}_2F_2\left(\begin{matrix}\frac{1}{2},\,\frac{s+2}{2};\,-a^2\\ \frac{3}{2},\,\frac{s+3}{2}\end{matrix}\right)\right.$ $\left.-\,{}_2F_2\left(\begin{matrix}\frac{s+1}{2},\,\frac{s+2}{2};\,-a^2\\ \frac{s+3}{2},\,\frac{s+3}{2}\end{matrix}\right)\right]$ $[\operatorname{Re}s>0]$
2	$\theta(a-x)\arccos\frac{x}{a}$ $\times\operatorname{erf}(bx)$	$\frac{a^{s+1}b}{2}\Gamma\left[\begin{matrix}\frac{s}{2}\\ \frac{s+3}{2}\end{matrix}\right]\left[{}_2F_2\left(\begin{matrix}\frac{1}{2},\,\frac{s+2}{2}\\ \frac{3}{2},\,\frac{s+3}{2};\,-a^2b^2\end{matrix}\right)\right.$ $\left.-\frac{1}{s+1}\,{}_2F_2\left(\begin{matrix}\frac{s+1}{2},\,\frac{s+2}{2}\\ \frac{s+3}{2},\,\frac{s+3}{2};\,-a^2b^2\end{matrix}\right)\right]$ $[a>0;\ \operatorname{Re}s>-1]$

No.	$f(x)$	$F(s)$
3	$\theta(a-x)\, e^{b^2x^2} \arccos\frac{x}{a} \times \operatorname{erf}(bx)$	$\frac{a^{s+1}b}{s+1}\Gamma\left[\begin{matrix}\frac{s+2}{2}\\ \frac{s+3}{2}\end{matrix}\right] {}_3F_3\left(\begin{matrix}1,\ \frac{s+1}{2},\ \frac{s+2}{2}\\ \frac{3}{2},\ \frac{s+3}{2},\ \frac{s+3}{2};\ a^2b^2\end{matrix}\right)$ $[a>0;\ \operatorname{Re}s>-1]$
4	$\arctan x \operatorname{erf}(ax)$	$\frac{a^{1-s}}{\sqrt{\pi}\, s}\Gamma\left(\frac{s}{2}\right)\left[\frac{1}{s-1}\, {}_2F_2\left(\begin{matrix}1,\ \frac{1-s}{2};\ a^2\\ \frac{2-s}{2},\ \frac{3-s}{2}\end{matrix}\right)+{}_2F_2\left(\begin{matrix}\frac{1}{2},\ 1;\ a^2\\ \frac{3}{2},\ \frac{2-s}{2}\end{matrix}\right)\right]$ $+\frac{\pi}{2s}\csc\frac{s\pi}{2}\operatorname{erfi}(a)+\frac{\sqrt{\pi}\left(-a^2\right)^{(1-s)/2}}{2as}\csc\frac{s\pi}{2}$ $\times\gamma\left(\frac{s+1}{2},-a^2\right)-\frac{\sqrt{\pi}\, a^{-s}}{2s}\Gamma\left(\frac{s+1}{2}\right)$ $[-2<\operatorname{Re}s<0;\ \lvert\arg a\rvert<\pi/4]$

3.6.11. erf (bx) and Ei $\left(-ax^2\right)$

1	$\operatorname{Ei}\left(-ax^2\right)\operatorname{erf}(bx)$	$-\frac{2a^{-(s+1)/2}b}{\sqrt{\pi}\,(s+1)}\Gamma\left(\frac{s+1}{2}\right) {}_3F_2\left(\begin{matrix}\frac{1}{2},\ \frac{s+1}{2},\ \frac{s+1}{2}\\ \frac{3}{2},\ \frac{s+3}{2};\ -\frac{b^2}{a}\end{matrix}\right)$ $[\operatorname{Re}a>0;\ \operatorname{Re}s>-1;\ \lvert\arg b\rvert<\pi/4]$
2	$e^{b^2x^2}\operatorname{Ei}\left(-ax^2\right)\operatorname{erf}(bx)$	$-\frac{2a^{-(s+1)/2}b}{\sqrt{\pi}\,(s+1)}\Gamma\left(\frac{s+1}{2}\right) {}_3F_2\left(\begin{matrix}1,\ \frac{s+1}{2},\ \frac{s+1}{2}\\ \frac{3}{2},\ \frac{s+3}{2};\ \frac{b^2}{a}\end{matrix}\right)$ $\left[\operatorname{Re}\left(a-b^2\right)>0;\ \operatorname{Re}s>-1;\ \lvert\arg b\rvert<\pi/4\right]$

3.6.12. erf (bx), erfc (bx), and si (ax), ci (ax), Si (ax)

1	$\operatorname{si}(ax)\operatorname{erf}(bx)$	$\frac{a^3b^{-s-3}}{18\sqrt{\pi}\,(s+3)}\Gamma\left(\frac{s+4}{2}\right) {}_4F_4\left(\begin{matrix}1,\ \frac{3}{2},\ \frac{s+3}{2},\ \frac{s+4}{2}\\ 2,\ \frac{5}{2},\ \frac{5}{2},\ \frac{s+5}{2};\ -\frac{a^2}{4b^2}\end{matrix}\right)$ $-\frac{ab^{-s-1}}{\sqrt{\pi}\,(s+1)}\Gamma\left(\frac{s+2}{2}\right)-\frac{a^{-s}}{s}\sin\frac{s\pi}{2}\Gamma(s)+\frac{\sqrt{\pi}}{2b^s s}\Gamma\left(\frac{s+1}{2}\right)$ $[a>0;\ -1<\operatorname{Re}s<2;\ \lvert\arg b\rvert<\pi/4]$
2	$\operatorname{ci}(ax)\operatorname{erf}(bx)$	$\frac{a^2b^{-s-2}}{4\sqrt{\pi}\,(s+2)}\Gamma\left(\frac{s+3}{2}\right) {}_4F_4\left(\begin{matrix}1,\ 1,\ \frac{s+2}{2},\ \frac{s+3}{2}\\ \frac{3}{2},\ 2,\ 2,\ \frac{s+4}{2};\ -\frac{a^2}{4b^2}\end{matrix}\right)$ $+\frac{b^{-s}}{\sqrt{\pi}\, s}\Gamma\left(\frac{s+1}{2}\right)\left[\frac{1}{s}-\frac{1}{2}\psi\left(\frac{s+1}{2}\right)+\ln\frac{b}{a}-\mathbf{C}\right]$ $-\frac{a^{-s}}{s}\Gamma(s)\cos\frac{s\pi}{2}$ $[a>0;\ -1<\operatorname{Re}s<2;\ \lvert\arg b\rvert<\pi/4]$

No.	$f(x)$	$F(s)$
3	$\operatorname{Si}(ax)\operatorname{erfc}(bx)$	$\frac{a\,\Gamma(s/2)}{2\sqrt{\pi}\,b^{s+1}}\left[{}_2F_2\left(\begin{matrix}\frac{1}{2},\ \frac{s+2}{2}\\ \frac{3}{2},\ \frac{3}{2};\ -\frac{a^2}{4b^2}\end{matrix}\right)-\frac{1}{s+1}\,{}_2F_2\left(\begin{matrix}\frac{s+1}{2},\ \frac{s+2}{2}\\ \frac{3}{2},\ \frac{s+3}{2};\ -\frac{a^2}{4b^2}\end{matrix}\right)\right]$ $[a>0;\ \operatorname{Re}s>-1;\ \lvert\arg b\rvert<\pi/4]$

3.6.13. Products of erf (ax), erfc (bx), erfi (cx)

1	$\left\{\begin{matrix}\operatorname{erf}(ax)\operatorname{erf}(bx)\\ \operatorname{erfc}(ax)\operatorname{erfc}(bx)\end{matrix}\right\}$	$-\frac{2b}{\pi a^{s+1}(s+1)}\,\Gamma\left(\frac{s+2}{2}\right)\,{}_3F_2\left(\begin{matrix}\frac{1}{2},\ \frac{s+1}{2},\ \frac{s+2}{2}\\ \frac{3}{2},\ \frac{s+3}{2};\ -\frac{b^2}{a^2}\end{matrix}\right)\mp\frac{1}{\sqrt{\pi}\,s}\left\{\begin{matrix}b^{-s}\\ a^{-s}\end{matrix}\right\}\Gamma\left(\frac{s+1}{2}\right)$ $\left[\left\{\begin{matrix}-2<\operatorname{Re}s<0\\ \operatorname{Re}s>0\end{matrix}\right\};\ \lvert\arg a\rvert,\ \lvert\arg b\rvert<\pi/4\right]$
2	$\operatorname{erfi}(ax)\operatorname{erfc}(ax)$	$\frac{a^{-s}}{\sqrt{\pi}\,s}\tan\frac{s\pi}{4}\,\Gamma\left(\frac{s+1}{2}\right)$ $[-1<\operatorname{Re}s<2;\ \lvert\arg a\rvert<\pi/4]$
3	$\operatorname{erf}(ax)\operatorname{erfc}(bx)$	$\frac{2b}{\pi a^{s+1}(s+1)}\,\Gamma\left(\frac{s+2}{2}\right)\,{}_3F_2\left(\begin{matrix}\frac{1}{2},\ \frac{s+1}{2},\ \frac{s+2}{2}\\ \frac{3}{2},\ \frac{s+3}{2};\ -\frac{b^2}{a^2}\end{matrix}\right)+\frac{1}{\sqrt{\pi}\,s}\left(b^{-s}-a^{-s}\right)\Gamma\left(\frac{s+1}{2}\right)$ $[\operatorname{Re}s>-1;\ \lvert\arg a\rvert,\ \lvert\arg b\rvert<\pi/4]$
4	$1-\operatorname{erf}^2(ax)$	$\frac{2}{\pi a^s}\,\Gamma\left(\frac{s}{2}\right)\,{}_2F_1\left(\begin{matrix}\frac{1}{2},\ \frac{s+2}{2}\\ \frac{3}{2};\ -1\end{matrix}\right)$ $[\operatorname{Re}s>0;\ \lvert\arg a\rvert<\pi/4]$
5	$\operatorname{erf}^2(ax)$	$\frac{2}{\pi(1+s)a^s}\,\Gamma\left(\frac{s+2}{2}\right)\,{}_3F_2\left(\begin{matrix}\frac{1}{2},\ \frac{s+1}{2},\ \frac{s+2}{2}\\ \frac{3}{2},\ \frac{s+3}{2};\ -1\end{matrix}\right)-\frac{a^{-s}}{\sqrt{\pi}\,s}\,\Gamma\left(\frac{s+1}{2}\right)$ $[-2<\operatorname{Re}s<0;\ \lvert\arg a\rvert<\pi/4]$
6	$(a-x)_+^{\alpha-1}$ $\times\operatorname{erf}\left(b\sqrt[4]{x(a-x)}\right)$ $\times\operatorname{erfi}\left(b\sqrt[4]{x(a-x)}\right)$	$\frac{4}{\pi}\,a^{s+\alpha}b^2\,\mathrm{B}\left(\frac{2\alpha+1}{2},\ \frac{2s+1}{2}\right)\,{}_4F_5\left(\begin{matrix}\frac{1}{2},\ 1,\ \frac{2\alpha+1}{2},\ \frac{2s+1}{2};\ \frac{a^2b^2}{16}\\ \frac{3}{4},\ \frac{3}{2},\ \frac{5}{4},\ \frac{s+\alpha+1}{2},\ \frac{s+\alpha+2}{2}\end{matrix}\right)$ $[a>0;\ \operatorname{Re}s,\ \operatorname{Re}\alpha>-1/2]$
7	$\operatorname{erfi}(ax)\operatorname{erf}(ax)\operatorname{erfc}(bx)$	$\frac{4a^2b^{-s-2}}{\pi^{3/2}(s+2)}\,\Gamma\left(\frac{s+3}{2}\right)\,{}_5F_4\left(\begin{matrix}\frac{1}{2},\ 1,\ \frac{s+2}{4},\ \frac{s+3}{4},\ \frac{s+4}{4}\\ \frac{3}{4},\ \frac{5}{4},\ \frac{3}{2},\ \frac{s+6}{4};\ \frac{a^4}{4b^4}\end{matrix}\right)$ $\left[\operatorname{Re}\left(b^2-a^2\right)>0;\ \operatorname{Re}s>-2;\ \lvert\arg a\rvert,\ \lvert\arg b\rvert<\pi/4\right]$

3.6.14. Products of erf (ax), erfc (bx), erfi (cx), and algebraic functions

1	$(a-x)_+^{\alpha-1}$ $\times\operatorname{erf}(bx)\operatorname{erfi}(bx)$	$\frac{4a^{s+\alpha+1}b^2}{\pi}\,\mathrm{B}(\alpha,\,s+2)\;{}_6F_7\left(\begin{matrix}\frac{1}{2},\,1,\,\Delta(4,\,s+2);\,\frac{a^4b^4}{4}\\ \frac{3}{4},\,\frac{5}{4},\,\frac{3}{2},\,\Delta(4,\,s+\alpha+2)\end{matrix}\right)$ $[a,\ \operatorname{Re}\alpha>0;\ \operatorname{Re}s>-2]$
2	$(a^2-x^2)_+^{\alpha-1}$ $\times\operatorname{erf}(bx)\operatorname{erfi}(bx)$	$\frac{2a^{s+2\alpha}b^2}{\pi}\,\mathrm{B}\left(\alpha,\,\frac{s+2}{2}\right)\;{}_4F_5\left(\begin{matrix}\frac{1}{2},\,1,\,\frac{s+2}{4},\,\frac{s+4}{4};\,\frac{a^4b^4}{4}\\ \frac{3}{4},\,\frac{5}{4},\,\frac{3}{2},\,\frac{s+2\alpha+2}{4},\,\frac{s+2\alpha+4}{4}\end{matrix}\right)$ $[a,\ \operatorname{Re}\alpha>0;\ \operatorname{Re}s>-2]$

3.6.15. Products of erf (ax), erfc (bx), erfi (cx), and the exponential function

1	$e^{-ax^2}\operatorname{erfi}(bx)\operatorname{erf}(bx)$	$\frac{2b^2}{\pi a^{s/2+1}}\,\Gamma\left(\frac{s+2}{2}\right)\;{}_4F_3\left(\begin{matrix}\frac{1}{2},\,1,\,\frac{s+2}{4},\,\frac{s+4}{4}\\ \frac{3}{2},\,\frac{3}{4},\,\frac{5}{4};\,\frac{b^4}{a^2}\end{matrix}\right)$ $[\operatorname{Re}a>\operatorname{Re}b^2;\ \operatorname{Re}s>-2;\ \lvert\arg b\rvert<\pi/4]$
2	$e^{-a^2x^2}\operatorname{erfi}(ax)\operatorname{erf}(bx)$	$-\frac{2a}{\pi b^{s+1}(s+1)}\,\Gamma\left(\frac{s+2}{2}\right)\;{}_3F_2\left(\begin{matrix}1,\,\frac{s+1}{2},\,\frac{s+2}{2}\\ \frac{3}{2},\,\frac{s+3}{2};\,-\frac{a^2}{b^2}\end{matrix}\right)$ $+\frac{a^{-s}}{2}\,\Gamma\left(\frac{s}{2}\right)\tan\frac{s\pi}{2}$ $[-2<\operatorname{Re}s<1;\ \lvert\arg a\rvert,\ \lvert\arg b\rvert<\pi/4]$
3	$e^{-(a^2+b^2)x^2}\operatorname{erfi}(ax)$ $\times\operatorname{erfi}(bx)$	$-\frac{b}{\sqrt{\pi}\,a^{s+1}}\cot\frac{s\pi}{2}\,\Gamma\left(\frac{s+1}{2}\right)\;{}_2F_1\left(\begin{matrix}1,\,\frac{s+1}{2}\\ \frac{3}{2};\,-\frac{b^2}{a^2}\end{matrix}\right)$ $-\frac{b^{1-s}}{2\sqrt{\pi}\,a}\cot\frac{s\pi}{2}\,\Gamma\left(\frac{s-1}{2}\right)\;{}_2F_1\left(\begin{matrix}\frac{1}{2},\,1\\ \frac{3-s}{2};\,-\frac{b^2}{a^2}\end{matrix}\right)$ $[\lvert\operatorname{Re}s\rvert<2;\ \lvert\arg a\rvert,\ \lvert\arg b\rvert<\pi/4]$
4	$e^{b^2x^2}\operatorname{erfc}(ax)\operatorname{erfc}(bx)$	$-\frac{2b}{\pi a^{s+1}(s+1)}\,\Gamma\left(\frac{s+2}{2}\right)\;{}_3F_2\left(\begin{matrix}1,\,\frac{s+1}{2},\,\frac{s+2}{2}\\ \frac{3}{2},\,\frac{s+3}{2};\,\frac{b^2}{a^2}\end{matrix}\right)$ $+\frac{a^{-s}}{s\sqrt{\pi}}\,\Gamma\left(\frac{s+1}{2}\right)\;{}_2F_1\left(\begin{matrix}\frac{s}{2},\,\frac{s+1}{2}\\ \frac{s+2}{2};\,\frac{b^2}{a^2}\end{matrix}\right)$ $[\operatorname{Re}s>0;\ \lvert\arg a\rvert,\ \lvert\arg b\rvert<\pi/4]$
5	$e^{a^2x^2}\operatorname{erf}(ax)\operatorname{erfc}(bx)$	$\frac{2a}{\pi b^{s+1}(s+1)}\,\Gamma\left(\frac{s+2}{2}\right)\;{}_3F_2\left(\begin{matrix}1,\,\frac{s+1}{2},\,\frac{s+2}{2}\\ \frac{3}{2},\,\frac{s+3}{2};\,\frac{a^2}{b^2}\end{matrix}\right)$ $[\operatorname{Re}(b^2-a^2)>0;\ \operatorname{Re}s>-1;\ \lvert\arg b\rvert<\pi/4]$

No.	$f(x)$	$F(s)$
6	$e^{b^2x^2}\operatorname{erf}(ax)\operatorname{erfc}(bx)$	$\frac{2b}{\pi a^{s+1}(s+1)}\Gamma\left(\frac{s+2}{2}\right){}_3F_2\left(\begin{matrix}1,\, \frac{s+1}{2},\, \frac{s+2}{2}\\ \frac{3}{2},\, \frac{s+3}{2};\, \frac{b^2}{a^2}\end{matrix}\right)$ $-\frac{1}{\sqrt{\pi}\,a^s s}\Gamma\left(\frac{s+1}{2}\right){}_2F_1\left(\begin{matrix}\frac{s}{2},\, \frac{s+1}{2}\\ \frac{s+2}{2};\, \frac{b^2}{a^2}\end{matrix}\right)+\frac{b^{-s}}{2}\Gamma\left(\frac{s}{2}\right)\sec\frac{s\pi}{2}$ $[\lvert\operatorname{Re} s\rvert<1;\ \lvert\arg a\rvert,\ \lvert\arg b\rvert<\pi/4]$
7	$e^{-ax^4}\operatorname{erf}(bx)\operatorname{erfi}(bx)$	$\frac{b^2}{\pi a^{(s+2)/4}}\Gamma\left(\frac{s+2}{4}\right){}_3F_3\left(\begin{matrix}\frac{1}{2},\, 1,\, \frac{s+2}{4};\, \frac{b^4}{4a}\\ \frac{3}{4},\, \frac{5}{4},\, \frac{3}{2}\end{matrix}\right)$ $[\operatorname{Re} a>0;\ \operatorname{Re} s>-2;\ \lvert\arg b\rvert<\pi/4]$

3.6.16. Products of erf (ax), erfc (bx), erfi (cx), and the logarithmic function

1	$\theta(a-x)\ln\frac{\sqrt{a-x}+\sqrt{a}}{x}$ $\times\operatorname{erf}(bx)\operatorname{erfi}(bx)$	$\frac{2a^{s+2}b^2}{\sqrt{\pi}(s+2)}\Gamma\left[\begin{matrix}s+2\\ \frac{2s+5}{2}\end{matrix}\right]{}_7F_8\left(\begin{matrix}\frac{1}{2},\, 1,\, \frac{s+2}{4},\, \Delta\left(4,\, \frac{s+4}{2}\right);\, \frac{a^4b^4}{4}\\ \frac{3}{4},\, \frac{5}{4},\, \frac{3}{2},\, \frac{s+6}{4},\, \Delta\left(4,\, \frac{s+2}{2}\right)\end{matrix}\right)$ $[a>0;\ \operatorname{Re} s>-2]$
2	$\theta(a-x)\ln\frac{\sqrt{a^2-x^2}+a}{x}$ $\times\operatorname{erf}(bx)\operatorname{erfi}(bx)$	$\frac{a^{s+2}b^2}{\sqrt{\pi}}\Gamma\left[\begin{matrix}\frac{s}{2}\\ \frac{s+3}{2}\end{matrix}\right]\left[{}_4F_3\left(\begin{matrix}\frac{1}{2},\, 1,\, \frac{s+2}{4},\, \frac{s+4}{4};\, \frac{a^2b^4}{4}\\ \frac{3}{4},\, \frac{5}{4},\, \frac{3}{2},\, \frac{s+3}{4},\, \frac{s+5}{4}\end{matrix}\right)\right.$ $\left.-\frac{2}{s+2}\,{}_4F_3\left(\begin{matrix}1,\, \frac{s+2}{4},\, \frac{s+2}{4},\, \frac{s+4}{4};\, \frac{a^2b^4}{4}\\ \frac{3}{4},\, \frac{5}{4},\, \frac{s+3}{4},\, \frac{s+5}{4},\, \frac{s+6}{4}\end{matrix}\right)\right]$ $[a>0;\ \operatorname{Re} s>-2]$
3	$\theta(a-x)\ln^n\frac{x}{a}$ $\times\operatorname{erf}(bx)\operatorname{erfi}(bx)$	$\frac{4(-1)^n n!\,a^{s+2}b^2}{\pi(s+2)^{n+1}}\,{}_{n+3}F_{n+4}\left(\begin{matrix}\frac{1}{2},\, 1,\, \frac{s+2}{4},\, \ldots,\, \frac{s+2}{4};\, \frac{a^4b^4}{4}\\ \frac{3}{4},\, \frac{5}{4},\, \frac{3}{2},\, \frac{s+8}{4},\, \ldots,\, \frac{s+8}{4}\end{matrix}\right)$ $[a>0;\ \operatorname{Re} s>-2]$

3.6.17. Products of erf (ax), erfc (bx), erfi (cx), and inverse trigonometric functions

1	$\theta(a-x)\arccos\frac{x}{a}$ $\times\operatorname{erf}(bx)\operatorname{erfi}(bx)$	$\frac{2a^{s+2}b^2}{\sqrt{\pi}\,s}\Gamma\left[\begin{matrix}\frac{s+3}{2}\\ \frac{s+4}{2}\end{matrix}\right]\left[{}_4F_5\left(\begin{matrix}\frac{1}{2},\, 1,\, \frac{s+3}{4},\, \frac{s+5}{4};\, \frac{a^4b^4}{4}\\ \frac{3}{4},\, \frac{5}{4},\, \frac{3}{2},\, \frac{s+4}{4},\, \frac{s+6}{4}\end{matrix}\right)\right.$ $\left.-\frac{2}{s+2}\,{}_4F_5\left(\begin{matrix}1,\, \frac{s+2}{4},\, \frac{s+3}{4},\, \frac{s+5}{4};\, \frac{a^4b^4}{4}\\ \frac{3}{4},\, \frac{5}{4},\, \frac{s+4}{4},\, \frac{s+6}{4},\, \frac{s+6}{4}\end{matrix}\right)\right]$ $[a>0;\ \operatorname{Re} s>-2]$

3.7. The Fresnel Integrals $S(z)$ and $C(z)$

More formulas can be obtained from the corresponding sections due to the relations

$$\begin{Bmatrix} S(z) \\ C(z) \end{Bmatrix} = \frac{1 \pm i}{4}\left[\operatorname{erf}\left(\frac{(1+i)\sqrt{z}}{\sqrt{2}}\right) \mp \operatorname{erfi}\left(\frac{(1+i)\sqrt{z}}{\sqrt{2}}\right)\right],$$

$$\begin{Bmatrix} S(z) \\ C(z) \end{Bmatrix} = \begin{Bmatrix} i \\ 1 \end{Bmatrix}\sqrt{z}\left\{\frac{1}{2\sqrt{2iz}}\left[1 - \frac{e^{-iz}}{\sqrt{\pi}}\Psi\left(\frac{1}{2}, \frac{1}{2}; iz\right)\right] \mp \frac{1}{2\sqrt{-2iz}}\left[1 - \frac{e^{iz}}{\sqrt{\pi}}\Psi\left(\frac{1}{2}, \frac{1}{2}; -iz\right)\right]\right\},$$

$$S(z) = \frac{1}{3}\sqrt{\frac{2z^3}{\pi}}\,{}_1F_2\left(\frac{3}{4}; \frac{3}{2}, \frac{7}{4}; -\frac{z^2}{4}\right), \quad C(z) = \sqrt{\frac{2z}{\pi}}\,{}_1F_2\left(\frac{1}{4}; \frac{1}{2}, \frac{5}{4}; -\frac{z^2}{4}\right),$$

$$S(z) = \frac{\pi z^{3/8}}{\sqrt{2}(-\sqrt{z})^{3/4}} G_{13}^{10}\left(-\frac{z^2}{4}\,\middle|\, \begin{matrix} 1 \\ 3/4, 1/4, 0 \end{matrix}\right), \quad C(z) = \frac{\pi z^{1/8}}{\sqrt{2}(-\sqrt{z})^{1/4}} G_{13}^{10}\left(-\frac{z^2}{4}\,\middle|\, \begin{matrix} 1 \\ 1/4, 3/4, 0 \end{matrix}\right),$$

$$S(\sqrt{z^2}) = \frac{1}{2} - \frac{1}{2} G_{13}^{20}\left(\frac{z^2}{4}\,\middle|\, \begin{matrix} 1 \\ 0, 3/4, 1/4 \end{matrix}\right), \quad C(\sqrt{z^2}) = \frac{1}{2} - \frac{1}{2} G_{13}^{20}\left(\frac{z^2}{4}\,\middle|\, \begin{matrix} 1 \\ 0, 1/4, 3/4 \end{matrix}\right),$$

$$S^2(\sqrt{z^2}) + C^2(\sqrt{z^2}) = \frac{1}{\sqrt{2}} G_{24}^{12}\left(\frac{z^2}{4}\,\middle|\, \begin{matrix} 1/2, 1 \\ 1/2, 3/4, 1/4, 0 \end{matrix}\right).$$

3.7.1. $S(\varphi(x))$, $C(\varphi(x))$, and algebraic functions

Notation: $\delta = \begin{Bmatrix} 1 \\ 0 \end{Bmatrix}$.

No.	$f(x)$	$F(s)$
1	$\begin{Bmatrix} S(ax) \\ C(ax) \end{Bmatrix}$	$-\frac{a^{-s}}{\sqrt{2\pi}\,s}\Gamma\left(\frac{2s+1}{2}\right)\begin{Bmatrix} \sin[(2s+1)\pi/4] \\ \cos[(2s+1)\pi/4] \end{Bmatrix}$ $[a > 0;\ -1 \mp 1/2 < \operatorname{Re} s < 0]$
2	$\frac{1}{2} - \begin{Bmatrix} S(ax) \\ C(ax) \end{Bmatrix}$	$\frac{a^{-s}}{\sqrt{2\pi}\,s}\Gamma\left(\frac{2s+1}{2}\right)\begin{Bmatrix} \sin[(2s+1)\pi/4] \\ \cos[(2s+1)\pi/4] \end{Bmatrix}$ $[a > 0;\ 0 < \operatorname{Re} s < 3/2]$
3	$(a-x)_+^{\alpha-1}\begin{Bmatrix} S(bx) \\ C(bx) \end{Bmatrix}$	$\sqrt{\frac{2}{\pi}}\,\frac{a^{s+\alpha+\delta-1/2}b^{\delta+1/2}}{2\delta+1}\,\mathrm{B}\left(s+\delta+\frac{1}{2}, \alpha\right) \times {}_3F_4\left(\begin{matrix} \frac{2\delta+1}{4}, \frac{2s+3}{4}, \frac{2s+4\delta+1}{4}; -\frac{a^2b^2}{4} \\ \frac{2\delta+1}{2}, \frac{2\delta+5}{4}, \frac{2s+2\alpha+3}{4}, \frac{2s+2\alpha+4\delta+1}{4} \end{matrix}\right)$ $[a,\ b,\ \operatorname{Re}\alpha > 0;\ \operatorname{Re} s > -\delta - 1/2]$
4	$(a^2-x^2)_+^{\alpha-1}\begin{Bmatrix} S(bx) \\ C(bx) \end{Bmatrix}$	$\frac{a^{s+2\alpha+\delta-3/2}b^{\delta+1/2}}{(2\delta+1)\sqrt{2\pi}}\,\mathrm{B}\left(\frac{2s+2\delta+1}{4}, \alpha\right) \times {}_2F_3\left(\begin{matrix} \frac{2s+2\delta+1}{4}, \frac{2\delta+1}{4}; -\frac{a^2b^2}{4} \\ \frac{2\delta+1}{2}, \frac{2\delta+5}{4}, \frac{2s+4\alpha+2\delta+1}{4} \end{matrix}\right)$ $[a,\ b,\ \operatorname{Re}\alpha > 0;\ \operatorname{Re} s > -\delta - 1/2]$

No.	$f(x)$	$F(s)$
5	$\dfrac{1}{(x^2+a^2)^{\rho}}\begin{Bmatrix} S(bx) \\ C(bx) \end{Bmatrix}$	$\dfrac{a^{s-2\rho+\delta+1/2}b^{\delta+1/2}}{(2\delta+1)\sqrt{2\pi}}\,\mathrm{B}\left(\dfrac{4\rho-2s-2\delta-1}{4},\,\dfrac{2s+2\delta+1}{4}\right)$ $\times\,{}_2F_3\left(\begin{matrix} \frac{2\delta+1}{4},\,\frac{2s+2\delta+1}{4};\,\frac{a^2b^2}{4} \\ \frac{2\delta+5}{4},\,\frac{2\delta+1}{2},\,\frac{2s-4\rho+2\delta+5}{4} \end{matrix}\right)$ $+\,\dfrac{b^{2\rho-s}}{\sqrt{2\pi}\,(2\rho-s)}\begin{Bmatrix} \sin[(2s-4\rho+1)\pi/4] \\ \cos[(2s-4\rho+1)\pi/4] \end{Bmatrix}$ $\times\,\Gamma\left(s-2\rho+\dfrac{1}{2}\right)\,{}_2F_3\left(\begin{matrix} \rho,\,\frac{2\rho-s}{2};\,\frac{a^2b^2}{4} \\ \frac{2-s+2\rho}{2},\,\frac{1-2s+4\rho}{4},\,\frac{3-2s+4\rho}{4} \end{matrix}\right)$ [b, $\operatorname{Re} a>0$; $-\delta-1/2<\operatorname{Re} s<2\operatorname{Re}\rho$]
6	$\dfrac{1}{x^2-a^2}\begin{Bmatrix} S(bx) \\ C(bx) \end{Bmatrix}$	$\dfrac{b^{2-s}}{\sqrt{2\pi}\,(s-2)}\,\Gamma\left(\dfrac{2s-3}{2}\right)\begin{Bmatrix} \sin[(2s+1)\pi/4] \\ \cos[(2s+1)\pi/4] \end{Bmatrix}$ $\times\,{}_2F_3\left(\begin{matrix} 1,\,\frac{2-s}{2};\,-\frac{a^2b^2}{4} \\ \frac{4-s}{2},\,\frac{5-2s}{4},\,\frac{7-2s}{4} \end{matrix}\right)\pm\sqrt{\dfrac{\pi}{2}}\,\dfrac{a^{s+\delta-3/2}b^{\delta+1/2}}{2\delta+1}$ $\times\left(\tan\dfrac{(2s+1)\pi}{4}\right)^{\pm 1}\,{}_1F_2\left(\begin{matrix} \frac{2\delta+1}{4};\,-\frac{a^2b^2}{4} \\ \frac{2\delta+1}{2},\,\frac{2\delta+5}{4} \end{matrix}\right)$ [a, $b>0$; $-\delta-1/2<\operatorname{Re} s<2$]
7	$(a-x)_+^{\alpha-1}$ $\times\begin{Bmatrix} S(bx(a-x)) \\ C(bx(a-x)) \end{Bmatrix}$	$\sqrt{\dfrac{2}{\pi}}\,\dfrac{a^{s+\alpha+2\delta}b^{\delta+1/2}}{2\delta+1}\,\mathrm{B}\left(s+\delta+\dfrac{1}{2},\,\alpha+\delta+\dfrac{1}{2}\right)$ $\times\,{}_5F_6\left(\begin{matrix} \Delta\left(2,\,\frac{2s+2\delta+1}{2}\right),\,\Delta\left(2,\,\frac{2\alpha+2\delta+1}{2}\right),\,\frac{2\delta+1}{4} \\ \frac{2\delta+1}{2},\,\frac{2\delta+5}{4},\,\Delta(4,\,s+\alpha+2\delta+1);\,-\frac{a^2b^2}{64} \end{matrix}\right)$ [a, $b>0$; $\operatorname{Re}\alpha$, $\operatorname{Re} s>-\delta-1/2$]
8	$(a-x)_+^{\alpha-1}$ $\times\begin{Bmatrix} S\left(b\sqrt{x(a-x)}\right) \\ C\left(b\sqrt{x(a-x)}\right) \end{Bmatrix}$	$\sqrt{\dfrac{2}{\pi}}\,\dfrac{a^{s+\alpha+\delta-1/2}b^{\delta+1/2}}{2\delta+1}\,\mathrm{B}\left(\dfrac{4s+2\delta+1}{4},\,\dfrac{4\alpha+2\delta+1}{2}\right)$ $\times\,{}_5F_6\left(\begin{matrix} \frac{4s+2\delta+1}{4},\,\frac{4\alpha+2\delta+1}{2},\,\frac{2\delta+1}{4};\,-\frac{a^2b^2}{16} \\ \frac{2\delta+1}{2},\,\frac{2\delta+5}{4},\,\frac{2s+2\alpha+3}{4},\,\frac{2s+2\alpha+4\delta+1}{4} \end{matrix}\right)$ [a, $b>0$; $\operatorname{Re}\alpha$, $\operatorname{Re} s>-(2\delta+1)/4$]

3.7.2. $S(bx)$, $C(bx)$, and the exponential function

Notation: $\delta=\begin{Bmatrix} 1 \\ 0 \end{Bmatrix}$.

1	$e^{-ax}\begin{Bmatrix} S(bx) \\ C(bx) \end{Bmatrix}$	$\sqrt{\dfrac{2}{\pi}}\,\dfrac{b^{\delta+1/2}}{(2\delta+1)\,a^{s+\delta+1/2}}\,\Gamma\left(\dfrac{2s+2\delta+1}{2}\right)\,{}_3F_2\left(\begin{matrix} \frac{2\delta+1}{4},\,\frac{2s+3}{4},\,\frac{2s+4\delta+1}{4} \\ \frac{2\delta+1}{2},\,\frac{2\delta+5}{4};\,-\frac{b^2}{a^2} \end{matrix}\right)$ [b, $\operatorname{Re} a>0$; $\operatorname{Re} s>-\delta-1/2$]

No.	$f(x)$	$F(s)$
2	$e^{-ax^2}\begin{Bmatrix} S(bx) \\ C(bx) \end{Bmatrix}$	$\dfrac{b^{\delta+1/2}}{\sqrt{2\pi}\,(2\delta+1)\,a^{(2s+2\delta+1)/4}}\,\Gamma\left(\dfrac{2s+2\delta+1}{4}\right) \times {}_2F_2\left(\begin{matrix}\frac{2\delta+1}{4}, \frac{2s+2\delta+1}{4} \\ \frac{2\delta+1}{2}, \frac{2\delta+5}{4};\, -\frac{b^2}{4a}\end{matrix}\right)$ $[b,\ \operatorname{Re} a > 0;\ \operatorname{Re} s > -\delta - 1/2]$

3.7.3. $S(\varphi(x))$, $C(\varphi(x))$, and trigonometric functions

Notation: $\delta = \begin{Bmatrix} 1 \\ 0 \end{Bmatrix}$.

No.	$f(x)$	$F(s)$
1	$\begin{Bmatrix} \sin(ax) \\ \cos(ax) \end{Bmatrix} S(bx)$	$\dfrac{1}{3}\sqrt{\dfrac{2}{\pi}}\,\dfrac{b^{3/2}}{a^{s+3/2}}\,\sin\dfrac{(\pm 1 - 2s)\pi}{4}\,\Gamma\left(s+\dfrac{3}{2}\right)\,{}_3F_2\left(\begin{matrix}\frac{3}{4}, \frac{2s+3}{4}, \frac{2s+5}{4} \\ \frac{3}{2}, \frac{7}{4};\, \frac{b^2}{a^2}\end{matrix}\right)$ $[a > b > 0;\ -(3\pm 2)/2 < \operatorname{Re} s < 1]$
2	$\begin{Bmatrix} \sin(ax) \\ \cos(ax) \end{Bmatrix} C(bx)$	$\sqrt{\dfrac{2}{\pi}}\,\dfrac{b^{1/2}}{a^{s+1/2}}\,\cos\dfrac{(\pm 1 - 2s)\pi}{4}\,\Gamma\left(s+\dfrac{1}{2}\right)\,{}_3F_2\left(\begin{matrix}\frac{1}{4}, \frac{2s+1}{4}, \frac{2s+3}{4} \\ \frac{1}{2}, \frac{5}{4};\, \frac{b^2}{a^2}\end{matrix}\right)$ $[a > b > 0;\ -(2\pm 1)/2 < \operatorname{Re} s < 1/2]$
3	$\begin{Bmatrix} \sin(ax) \\ \cos(ax) \end{Bmatrix} S(bx)$	$\dfrac{a^{-s}}{2}\cos\dfrac{(s-\delta)\pi}{2}\,\Gamma(s) - \dfrac{a^{\delta} b^{-s-\delta}}{\sqrt{2\pi}\,(s+\delta)}\cos\dfrac{(2s+2\delta-1)\pi}{4} \times \Gamma\left(\dfrac{2s+2\delta+1}{2}\right)\,{}_3F_2\left(\begin{matrix}\frac{2s+\delta+1}{4}, \frac{2s+3}{4}, \frac{2s+5\delta}{4} \\ \frac{2\delta+1}{2}, \frac{s+\delta+2}{2};\, \frac{a^2}{b^2}\end{matrix}\right)$ $[b > a > 0;\ -(2\delta+3)/2 < \operatorname{Re} s < 1]$
4	$\begin{Bmatrix} \sin(ax) \\ \cos(ax) \end{Bmatrix} C(bx)$	$\dfrac{a^{-s}}{2}\cos\dfrac{(s-\delta)\pi}{2}\,\Gamma(s) + \dfrac{a^{\delta} b^{-s-\delta}}{\sqrt{2\pi}\,(s+\delta)}\sin\dfrac{(2s+2\delta-1)\pi}{4} \times \Gamma\left(\dfrac{2s+2\delta+1}{2}\right)\,{}_3F_2\left(\begin{matrix}\frac{2s+\delta+1}{4}, \frac{2s+3}{4}, \frac{2s+5\delta}{4} \\ \frac{2\delta+1}{2}, \frac{s+\delta+2}{2};\, \frac{a^2}{b^2}\end{matrix}\right)$ $[b > a > 0;\ -(2\delta+1)/2 < \operatorname{Re} s < 1]$
5	$\begin{Bmatrix} \sin(ax) \\ \cos(ax) \end{Bmatrix} C(ax) \mp \begin{Bmatrix} \cos(ax) \\ \sin(ax) \end{Bmatrix} S(ax)$	$\dfrac{\pi a^{-s}}{2\sqrt{2}\,\Gamma(1-s)}\csc\left(\dfrac{\pi}{4} \mp \dfrac{s\pi}{2}\right)$ $[a > 0;\ -(2\pm 1)/2 < \operatorname{Re} s < (3\mp 1)/4]$

No.	$f(x)$	$F(s)$
6	$\begin{Bmatrix} \sin(ax) \\ \cos(ax) \end{Bmatrix} \left[\frac{1}{2} - S(ax)\right] \pm \begin{Bmatrix} \cos(ax) \\ \sin(ax) \end{Bmatrix} \left[\frac{1}{2} - C(ax)\right]$	$\frac{\Gamma(s)}{2\sqrt{2}\,a^s} \csc\left(\frac{\pi}{4} \pm \frac{s\pi}{2}\right)$ $[a > 0;\ 0 < \operatorname{Re} s < (2 \pm 1)/2]$
7	$\cos u \left[\frac{1}{2} - C(u)\right] + \sin u \left[\frac{1}{2} - S(u)\right]$, $u = b\sqrt{x^2 + a^2}$	$\frac{a^{(s+1)/2}}{4\sqrt{2\pi}\, b^{(s-1)/2}} \Gamma\left[\begin{matrix} \frac{s}{2},\ \frac{3-2s}{4} \\ \frac{3}{4} \end{matrix}\right] S_{s/2-1,\,(s+1)/2}(ab)$ $[a,\, b > 0;\ 0 < \operatorname{Re} s < 3/2]$
8	$\cos u \left[\frac{1}{2} - S(u)\right] - \sin u \left[\frac{1}{2} - C(u)\right]$, $u = b\sqrt{x^2 + a^2}$	$\frac{a^{(s+1)/2}}{2\sqrt{2\pi}\, b^{(s-1)/2}} \Gamma\left[\begin{matrix} \frac{s}{2},\ \frac{1-2s}{4} \\ \frac{1}{4} \end{matrix}\right] S_{s/2,\,(s+1)/2}(ab)$ $[a,\, b > 0;\ 0 < \operatorname{Re} s < 1/2]$
9	$\frac{\cos u}{u} \left[\frac{1}{2} - C(u)\right] + \frac{\sin u}{u} \left[\frac{1}{2} - S(u)\right]$, $u = b\sqrt{x^2 + a^2}$	$-\frac{2\sqrt{2\pi}}{3} a^{s-1/2} b^{-1/2} \csc\frac{(2s-1)\pi}{4} \Gamma\left[\begin{matrix} \frac{s}{2} \\ -\frac{3}{4},\ \frac{2s+3}{4} \end{matrix}\right] \times {}_1F_2\left(\begin{matrix} 1;\ -\frac{a^2b^2}{4} \\ \frac{5}{4},\ \frac{2s+3}{4} \end{matrix}\right) + 2^{-(s+2)/2} \left(\frac{a}{b}\right)^{(s-1)/2} b^{-1} \times \left[\csc\frac{(2s-1)\pi}{4} \Gamma(s-1)\, \Gamma\left(\frac{3-s}{2}\right) J_{(1-s)/2}(ab) + \sqrt{\pi}\, 2^{s-3/2} \sec\frac{s\pi}{2}\, \Gamma\left(\frac{s}{2}\right) J_{(s-1)/2}(ab)\right]$ $[a,\, b > 0;\ 0 < \operatorname{Re} s < 5/2]$
10	$\frac{\cos u}{u} \left[\frac{1}{2} - S(u)\right] - \frac{\sin u}{u} \left[\frac{1}{2} - C(u)\right]$, $u = b\sqrt{x^2 + a^2}$	$\frac{\sqrt{2\pi}}{12} a^{s+1/2} b^{1/2} \csc\frac{(2s+1)\pi}{4} \Gamma\left[\begin{matrix} \frac{s}{2} \\ \frac{3}{4},\ \frac{2s+5}{4} \end{matrix}\right] \times {}_1F_2\left(\begin{matrix} 1;\ -\frac{a^2b^2}{4} \\ \frac{7}{4},\ \frac{2s+5}{4} \end{matrix}\right) + 2^{-(s+2)/2} \left(\frac{a}{b}\right)^{(s-1)/2} b^{-1} \times \left[\csc\frac{(2s+1)\pi}{4} \Gamma(s-1)\, \Gamma\left(\frac{3-s}{2}\right) J_{(1-s)/2}(ab) + \sqrt{\pi}\, 2^{s-3/2} \sec\frac{s\pi}{2}\, \Gamma\left(\frac{s}{2}\right) J_{(s-1)/2}(ab)\right]$ $[a,\, b > 0;\ 0 < \operatorname{Re} s < 3/2]$

3.7.4. $S(bx)$, $C(bx)$, and the logarithmic function

Notation: $\delta = \begin{Bmatrix} 1 \\ 0 \end{Bmatrix}$.

1	$\theta(a-x)\ln\dfrac{\sqrt{a-x}+\sqrt{a}}{\sqrt{x}} \times \begin{Bmatrix} S(bx) \\ C(bx) \end{Bmatrix}$	$\dfrac{a^{s+\delta+1/2}b^{\delta+1/2}}{(2\delta+1)\sqrt{2}\,s}\Gamma\left[\begin{matrix}\frac{2s+2\delta+1}{2} \\ s+\delta+1\end{matrix}\right] \times \left[{}_3F_4\left(\begin{matrix}\frac{2\delta+1}{4}, \frac{2s+2\delta+1}{4}, \frac{2s+2\delta+3}{4}; -\frac{a^2b^2}{4} \\ \frac{2\delta+1}{2}, \frac{2\delta+5}{4}, \frac{s+\delta+1}{2}, \frac{s+\delta+2}{2}\end{matrix}\right) - \dfrac{2\delta+1}{2s+2\delta+1}\,{}_3F_4\left(\begin{matrix}\frac{2s+2\delta+1}{4}, \frac{2s+2\delta+1}{4}, \frac{2s+2\delta+3}{4}; -\frac{a^2b^2}{4} \\ \frac{2\delta+1}{2}, \frac{s+\delta+1}{2}, \frac{s+\delta+2}{2}, \frac{2s+2\delta+5}{4}\end{matrix}\right)\right]$ $[a>0;\ \mathrm{Re}\,s > -(2\delta+1)/2]$
2	$\theta(a-x)\ln\dfrac{\sqrt{a^2-x^2}+a}{x} \times \begin{Bmatrix} S(bx) \\ C(bx) \end{Bmatrix}$	$\dfrac{a^{s+\delta+1/2}b^{\delta+1/2}}{(2\delta+1)\sqrt{2}\,s}\Gamma\left[\begin{matrix}\frac{2s+2\delta+1}{4} \\ \frac{2s+2\delta+3}{4}\end{matrix}\right] \times \left[{}_2F_3\left(\begin{matrix}\frac{2\delta+1}{4}, \frac{2s+2\delta+1}{4}; -\frac{a^2b^2}{4} \\ \frac{2\delta+1}{2}, \frac{2\delta+5}{4}, \frac{2s+2\delta+3}{2}\end{matrix}\right) - \dfrac{2\delta+1}{2s+2\delta+1}\,{}_3F_4\left(\begin{matrix}\frac{2s+2\delta+1}{4}, \frac{2s+2\delta+1}{4}; -\frac{a^2b^2}{4} \\ \frac{2\delta+1}{2}, \frac{2s+2\delta+3}{4}, \frac{2s+2\delta+5}{4}\end{matrix}\right)\right]$ $[a>0;\ \mathrm{Re}\,s > -(2\delta+1)/2]$

3.7.5. $S(bx)$, $C(bx)$, and $\mathrm{si}(ax)$, $\mathrm{ci}(ax)$

1	$\mathrm{si}(ax)\,S(bx)$	$\dfrac{(2b)^{3/2}}{3\sqrt{\pi}\,a^{s+3/2}(2s+3)}\sin\dfrac{(2s-1)\pi}{4}\,\Gamma\left(\dfrac{2s+3}{2}\right) \times {}_4F_3\left(\begin{matrix}\frac{3}{4}, \frac{2s+3}{4}, \frac{2s+3}{4}, \frac{2s+5}{4} \\ \frac{3}{2}, \frac{7}{4}, \frac{2s+7}{4}; \frac{b^2}{a^2}\end{matrix}\right)$ $\left[\begin{matrix}(0<b<a;\ -3/2<\mathrm{Re}\,s<2) \text{ or} \\ (b=a>0;\ -3/2<\mathrm{Re}\,s<1)\end{matrix}\right]$
2	$\mathrm{si}(ax)\,S(bx)$	$\dfrac{a^3}{18\sqrt{2\pi}\,b^{s+3}(s+3)}\cos\dfrac{(2s+1)\pi}{4}\,\Gamma\left(\dfrac{2s+7}{2}\right) \times {}_5F_4\left(\begin{matrix}1, \frac{3}{2}, \frac{s+3}{2}, \frac{2s+7}{4}, \frac{2s+9}{4} \\ 2, \frac{5}{2}, \frac{5}{2}, \frac{s+5}{2}; \frac{a^2}{b^2}\end{matrix}\right) + {} + \dfrac{a}{\sqrt{2\pi}\,b^{s+1}(s+1)}\cos\dfrac{(2s+1)\pi}{4}\,\Gamma\left(\dfrac{2s+3}{2}\right) + \dfrac{\Gamma\left(\frac{2s+1}{2}\right)}{\sqrt{2\pi}\,b^s s}\cos\dfrac{(1-2s)\pi}{4} + \dfrac{\Gamma(s)}{2a^s s}\cos\dfrac{(s-1)\pi}{2}$ $[b>a>0;\ -3/2<\mathrm{Re}\,s<2]$

No.	$f(x)$	$F(s)$
3	$\operatorname{si}(ax)\, C(bx)$	$\dfrac{2\sqrt{2b}}{\sqrt{\pi}\, a^{s+1/2}(2s+1)} \cos\dfrac{(2s-1)\pi}{4}\, \Gamma\left(\dfrac{2s+1}{2}\right) \times {}_4F_3\left(\begin{matrix}\frac{1}{4}, \frac{2s+1}{4}, \frac{2s+1}{4}, \frac{2s+3}{4} \\ \frac{1}{2}, \frac{5}{4}, \frac{2s+5}{4};\ \frac{b^2}{a^2}\end{matrix}\right)$ $\left[\begin{matrix}(0 < b < a;\ -1/2 < \operatorname{Re} s < 2) \text{ or} \\ (b = a > 0;\ -1/2 < \operatorname{Re} s < 1)\end{matrix}\right]$
4	$\operatorname{ci}(ax)\, S(bx)$	$-\dfrac{(2b)^{3/2}}{3\sqrt{\pi}\, a^{s+3/2}(2s+3)} \sin\dfrac{(2s+1)\pi}{4}\, \Gamma\left(\dfrac{2s+3}{2}\right) \times {}_4F_3\left(\begin{matrix}\frac{3}{4}, \frac{2s+3}{4}, \frac{2s+3}{4}, \frac{2s+5}{4} \\ \frac{3}{2}, \frac{7}{4}, \frac{2s+7}{4};\ \frac{b^2}{a^2}\end{matrix}\right)$ $\left[\begin{matrix}(0 < b < a;\ -3/2 < \operatorname{Re} s < 2) \text{ or} \\ (b = a > 0;\ -3/2 < \operatorname{Re} s < 1)\end{matrix}\right]$
5	$\operatorname{ci}(ax)\, C(bx)$	$\dfrac{2\sqrt{2b}\, a^{-s-1/2}}{\sqrt{\pi}(2s+1)} \cos\dfrac{(2s+1)\pi}{4}\, \Gamma\left(\dfrac{2s+1}{2}\right) \times {}_4F_3\left(\begin{matrix}\frac{1}{4}, \frac{2s+1}{4}, \frac{2s+1}{4}, \frac{2s+3}{4} \\ \frac{1}{2}, \frac{5}{4}, \frac{2s+5}{4};\ \frac{b^2}{a^2}\end{matrix}\right)$ $\left[\begin{matrix}(0 < b < a;\ -1/2 < \operatorname{Re} s < 2) \text{ or} \\ (b = a > 0;\ -1/2 < \operatorname{Re} s < 1)\end{matrix}\right]$

3.7.6. $S(bx)$, $C(bx)$, and $\operatorname{erf}(a\sqrt{x})$, $\operatorname{erfc}(a\sqrt{x})$

Notation: $\delta = \begin{Bmatrix}1\\0\end{Bmatrix}$.

1	$\operatorname{erf}(a\sqrt{x}) \begin{Bmatrix} S(bx) \\ C(bx) \end{Bmatrix}$	$-\dfrac{2^{3/2} b^{\delta+1/2}}{\pi(2\delta+1)(2s+2\delta+1)\, a^{2s+2\delta+1}}\, \Gamma(s+\delta+1) \times {}_4F_3\left(\begin{matrix}\frac{2\delta+1}{4}, \frac{s+\delta+1}{2}, \frac{s+\delta+2}{2}, \frac{2s+2\delta+1}{4} \\ \frac{2\delta+1}{2}, \frac{2\delta+5}{4}, \frac{2s+2\delta+5}{4};\ -\frac{b^2}{a^4}\end{matrix}\right) - \dfrac{1}{\sqrt{2\pi}\, b^s s}\, \Gamma\left(\dfrac{2s+1}{2}\right) \begin{Bmatrix} \sin[(2s+1)\pi/4] \\ \cos[(2s+1)\pi/4] \end{Bmatrix}$ $[b > 0;\ -1-\delta < \operatorname{Re} s < 0;\ \lvert\arg a\rvert < \pi/4]$
2	$\operatorname{erfc}(a\sqrt{x}) \begin{Bmatrix} S(bx) \\ C(bx) \end{Bmatrix}$	$\dfrac{2^{3/2} b^{\delta+1/2}}{\pi(2\delta+1)(2s+2\delta+1)\, a^{2s+2\delta+1}}\, \Gamma(s+\delta+1) \times {}_4F_3\left(\begin{matrix}\frac{2\delta+1}{4}, \frac{s+\delta+1}{2}, \frac{s+\delta+2}{2}, \frac{2s+2\delta+1}{4} \\ \frac{2\delta+1}{2}, \frac{2\delta+5}{4}, \frac{2s+2\delta+5}{4};\ -\frac{b^2}{a^4}\end{matrix}\right)$ $[b > 0;\ \operatorname{Re} s > -(2\delta+1)/2;\ \lvert\arg a\rvert < \pi/4]$

3.7.7. Products of $S(bx)$ and $C(bx)$

1	$S(ax)\,S(bx)$	$\frac{b^{3/2}}{3\pi s a^{s+3/2}} \sin\frac{s\pi}{2}\,\Gamma(s+2)\,{}_3F_2\left(\begin{matrix}\frac{3}{4},\,\frac{s+2}{2},\,\frac{s+3}{2}\\ \frac{3}{2},\,\frac{7}{4};\,\frac{b^2}{a^2}\end{matrix}\right) - \frac{a^{-s-3/2}b^{3/2}}{\pi s(2s+3)} \sin\frac{s\pi}{2}\,\Gamma(s+2)\,{}_3F_2\left(\begin{matrix}\frac{s+2}{2},\,\frac{s+3}{2},\,\frac{2s+3}{4}\\ \frac{3}{2},\,\frac{2s+7}{4};\,\frac{b^2}{a^2}\end{matrix}\right) - \frac{\sqrt{\pi}\,b^{-s}}{2^{5/2}s\,\Gamma\left(\frac{1-2s}{2}\right)} \csc\frac{(2s+3)\pi}{4}$ $[a,\,b>0;\ -3<\operatorname{Re}s<0]$
2	$S(ax)\,C(bx)$	$\frac{b^{1/2}}{\pi a^{s+1/2}} \cos\frac{s\pi}{2}\,\Gamma(s)\,{}_3F_2\left(\begin{matrix}\frac{1}{4},\,\frac{s+1}{2},\,\frac{s+2}{2}\\ \frac{1}{2},\,\frac{5}{4};\,\frac{b^2}{a^2}\end{matrix}\right) + \frac{b^{1/2}}{\pi(2s+1)a^{s+1/2}} \cos\frac{s\pi}{2}\,\Gamma(s)\,{}_3F_2\left(\begin{matrix}\frac{s+1}{2},\,\frac{s+2}{2},\,\frac{2s+1}{4}\\ \frac{1}{2},\,\frac{2s+5}{4};\,\frac{b^2}{a^2}\end{matrix}\right) - \frac{\sqrt{\pi}\,b^{-s}}{2^{5/2}s\,\Gamma\left(\frac{1-2s}{2}\right)} \csc\frac{(2s+1)\pi}{4}$ $[a,\,b>0;\ -2<\operatorname{Re}s<0]$
3	$C(ax)\,C(bx)$	$\frac{b^{1/2}}{\pi a^{s+1/2}} \sin\frac{s\pi}{2}\,\Gamma(s)\,{}_3F_2\left(\begin{matrix}\frac{1}{4},\,\frac{s+1}{2},\,\frac{s+2}{2}\\ \frac{1}{2},\,\frac{5}{4};\,\frac{b^2}{a^2}\end{matrix}\right) - \frac{b^{1/2}}{\pi(2s+1)a^{s+1/2}} \sin\frac{s\pi}{2}\,\Gamma(s)\,{}_3F_2\left(\begin{matrix}\frac{s+1}{2},\,\frac{s+2}{2},\,\frac{2s+1}{4}\\ \frac{1}{2},\,\frac{2s+5}{4};\,\frac{b^2}{a^2}\end{matrix}\right) - \frac{\sqrt{\pi}\,b^{-s}}{2^{5/2}s\,\Gamma\left(\frac{1-2s}{2}\right)} \csc\frac{(2s+1)\pi}{4}$ $[a,\,b>0;\ -1<\operatorname{Re}s<0]$
4	$C^2(ax)-S^2(ax)$	$\frac{2}{\pi a^s} \sin\frac{s\pi}{2}\,\Gamma(s)\,{}_2F_1\left(\begin{matrix}\frac{1}{2},\,s+1\\ \frac{3}{2};\,-1\end{matrix}\right)$ $[a>0;\ 0<\operatorname{Re}s<1]$
5	$C^2(ax)+S^2(ax)$	$-\frac{\sqrt{\pi}}{2sa^s\,\Gamma\left(\frac{1-2s}{2}\right)} \sec\frac{s\pi}{2}$ $[a>0;\ -1<\operatorname{Re}s<0]$
6	$\left[\frac{1}{2}-C(ax)\right]^2 + \left[\frac{1}{2}-S(ax)\right]^2$	$\frac{a^{-s}}{2\sqrt{\pi}\,s} \sec\frac{s\pi}{2}\,\Gamma\left(\frac{2s+1}{2}\right)$ $[a>0;\ 0<\operatorname{Re}s<1]$

3.8. The Incomplete Gamma Function $\Gamma(\nu, z)$ and $\gamma(\nu, z)$

More formulas can be obtained from the corresponding sections due to the relations

$$\Gamma(-1, z) = \mathrm{Ei}(-z) + \frac{e^{-z}}{z} + \frac{1}{2}\left[\ln\left(-\frac{1}{z}\right) - \ln(-z)\right] + \ln z,$$

$$\Gamma\left(-\frac{1}{2}, z\right) = \frac{2e^{-z}}{\sqrt{z}} - 2\sqrt{\pi}\,\mathrm{erfc}\left(\sqrt{z}\right),$$

$$\Gamma(0, z) = -\mathrm{Ei}(-z) + \frac{1}{2}\left[\ln(-z) - \ln\left(-\frac{1}{z}\right)\right] - \ln z, \quad \Gamma\left(\frac{1}{2}, z\right) = \sqrt{\pi}\,\mathrm{erfc}\left(\sqrt{z}\right),$$

$$\Gamma(1, z) = e^{-z}, \quad \Gamma(n, z) = (n-1)!\, e^{-z} \sum_{k=0}^{n-1} \frac{z^k}{k!},$$

$$\Gamma(\nu, z) = \Gamma(\nu) - \frac{z^\nu e^{-z}}{\nu}\, {}_1F_1(1; \nu+1; z), \quad \gamma(\nu, z) = \Gamma(\nu) - e^{-z}\,\Psi(1-\nu; 1-\nu; z),$$

$$\Gamma(\nu, z) = e^{-z}\,\Psi(1-\nu; 1-\nu; z), \quad \gamma(\nu, z) = \frac{z^\nu}{\nu}\, {}_1F_1(\nu; \nu+1; -z),$$

$$\Gamma(\nu, z) = G_{12}^{20}\left(z \,\middle|\, \begin{matrix} 1 \\ 0, \nu \end{matrix}\right), \quad \gamma(\nu, z) = G_{12}^{11}\left(z \,\middle|\, \begin{matrix} 1 \\ \nu, 0 \end{matrix}\right).$$

3.8.1. $\Gamma(\nu, ax)$, $\gamma(\nu, ax)$, and algebraic functions

No.	$f(x)$	$F(s)$
1	$\left\{\begin{matrix} \Gamma(\nu, ax) \\ \gamma(\nu, ax) \end{matrix}\right\}$	$\pm\frac{a^{-s}}{s}\,\Gamma(s+\nu)$ $\quad [\mathrm{Re}\, a,\ \pm\mathrm{Re}\, s,\ \mathrm{Re}(s+\nu) > 0]$
2	$(a-x)_+^{\alpha-1}\left\{\begin{matrix} \Gamma(\nu, bx) \\ \gamma(\nu, bx) \end{matrix}\right\}$	$\mp\frac{a^{s+\alpha+\nu-1}b^\nu}{\nu}\,\mathrm{B}(\alpha, s+\nu)\, {}_2F_2\left(\begin{matrix} \nu, s+\nu; -ab \\ \nu+1, s+\alpha+\nu \end{matrix}\right) + \frac{1\pm1}{2}\, a^{s+\alpha-1}\,\Gamma\left[\begin{matrix} s, \alpha, \nu \\ s+\alpha \end{matrix}\right]$ $\quad \left[a,\ \mathrm{Re}\,\alpha,\ \mathrm{Re}(s+\nu) > 0,\ \left\{\begin{matrix} \mathrm{Re}\, s > 0 \\ \mathrm{Re}\,\nu > 0 \end{matrix}\right\}\right]$
3	$(x-a)_+^{\alpha-1}\left\{\begin{matrix} \Gamma(\nu, bx) \\ \gamma(\nu, bx) \end{matrix}\right\}$	$\mp\frac{a^{s+\alpha+\nu-1}b^\nu}{\nu}\,\mathrm{B}(\alpha, 1-s-\alpha-\nu)\, {}_2F_2\left(\begin{matrix} \nu, s+\nu; -ab \\ \nu+1, s+\alpha+\nu \end{matrix}\right) \mp \frac{b^{1-s-\alpha}}{1-s-\alpha}\,\Gamma(s+\alpha+\nu-1) \times {}_2F_2\left(\begin{matrix} 1-\alpha, 1-s-\alpha; -ab \\ 2-s-\alpha-\nu, 2-s-\alpha \end{matrix}\right) + \left\{\begin{matrix} 1 \\ 0 \end{matrix}\right\} a^{s+\alpha-1}\Gamma(\nu)\,\mathrm{B}(\alpha, 1-s-\alpha)$ $\quad \left[a,\ \mathrm{Re}\,\alpha > 0;\ \left\{\begin{matrix} \mathrm{Re}\, b > 0 \\ \mathrm{Re}\, b,\ \mathrm{Re}\,\nu > 0;\ \mathrm{Re}(s+\alpha) < 1 \end{matrix}\right\}\right]$

No.	$f(x)$	$F(s)$
4	$\frac{1}{(x+a)^{\rho}} \begin{Bmatrix} \Gamma(\nu, bx) \\ \gamma(\nu, bx) \end{Bmatrix}$	$\mp \frac{a^{s+\nu-\rho} b^{\nu}}{\nu} \mathrm{B}(s+\nu, -s-\nu+\rho) \times {}_2F_2\left(\begin{matrix} \nu, s+\nu; ab \\ \nu+1, s+\nu-\rho+1 \end{matrix}\right) \pm \frac{b^{-s+\rho}}{s-\rho} \Gamma(s+\nu-\rho) \times {}_2F_2\left(\begin{matrix} \rho, -s+\rho; ab \\ 1-s+\rho, 1-s-\nu+\rho \end{matrix}\right) + \frac{1 \pm 1}{2} a^{s-\rho} \Gamma(\nu) \mathrm{B}(s, -s+\rho)$ $\left[\begin{matrix} \mathrm{Re}(s+\nu) > 0; \lvert\arg a\rvert < \pi; \\ \begin{Bmatrix} \mathrm{Re}\, b, \mathrm{Re}\, s > 0 \\ \mathrm{Re}\, b, \mathrm{Re}\, \nu > 0; 0 < \mathrm{Re}\, s < \mathrm{Re}\, \rho \end{Bmatrix} \end{matrix}\right]$
5	$\frac{1}{x-a} \begin{Bmatrix} \Gamma(\nu, bx) \\ \gamma(\nu, bx) \end{Bmatrix}$	$\pm \pi a^{s-1} \cot[(s+\nu)\pi] \gamma(\nu, ab) \mp \frac{b^{1-s}}{1-s} \Gamma(s+\nu-1) \times {}_2F_2\left(\begin{matrix} 1, 1-s; -ab \\ 2-s, 2-s-\nu \end{matrix}\right) - \frac{\pi \pm \pi}{2} a^{s-1} \cot(s\pi) \Gamma(\nu)$ $\left[a, \mathrm{Re}(s+\nu) > 0; \begin{Bmatrix} \mathrm{Re}\, b, \mathrm{Re}\, s > 0 \\ \mathrm{Re}\, b, \mathrm{Re}\, \nu > 0; \mathrm{Re}\, s < 1 \end{Bmatrix}\right]$
6	$(a-x)_{+}^{\alpha-1} \times \begin{Bmatrix} \Gamma(\nu, bx(a-x)) \\ \gamma(\nu, bx(a-x)) \end{Bmatrix}$	$\mp \frac{a^{s+\alpha+2\nu-1} b^{\nu}}{\nu} \mathrm{B}(s+\nu, \alpha+\nu) \times {}_3F_3\left(\begin{matrix} \nu, \alpha+\nu, s+\nu; -\frac{a^2 b}{4} \\ \nu+1, \frac{s+\alpha+2\nu}{2}, \frac{s+\alpha+2\nu+1}{2} \end{matrix}\right) + \frac{1 \pm 1}{2} a^{s+\alpha-1} \Gamma(\nu) \mathrm{B}(s, \alpha)$ $[a, \mathrm{Re}\, \nu, \mathrm{Re}(\alpha+\nu), \mathrm{Re}(s+\nu) > 0]$
7	$\theta(x-a) \gamma\left(\nu, \frac{cx}{x-b}\right)$	$-\frac{a^s c^{\nu}}{\nu s} \Psi_1\left(\nu, -s; 1-s, \nu+1; \frac{b}{a}, -c\right)$ $[a > 0; \lvert b\rvert < a; \mathrm{Re}(s+\nu) < -1]$

3.8.2. $\Gamma(\nu, ax)$, $\gamma(\nu, ax)$, and the exponential function

1	$e^{ax} \Gamma(\nu, ax)$	$\frac{\pi \csc[(s+\nu)\pi]}{a^s} \Gamma\left[\begin{matrix} s \\ 1-\nu \end{matrix}\right]$ $[\mathrm{Re}\, a, \mathrm{Re}\, s > 0; 0 < \mathrm{Re}(s+\nu) < 1]$
2	$e^{-ax} \begin{Bmatrix} \Gamma(\nu, bx) \\ \gamma(\nu, bx) \end{Bmatrix}$	$\frac{1 \pm 1}{2a^s} \Gamma(\nu) \Gamma(s) \mp \frac{b^{\nu}}{\nu a^{s+\nu}} \Gamma(s+\nu) {}_2F_1\left(\begin{matrix} \nu, s+\nu \\ \nu+1; -\frac{b}{a} \end{matrix}\right)$ $\left[\mathrm{Re}\, a, \mathrm{Re}\, b, \mathrm{Re}(s+\nu), \begin{Bmatrix} \mathrm{Re}\, s \\ \mathrm{Re}\, \nu \end{Bmatrix} > 0\right]$

No.	$f(x)$	$F(s)$
3	$(a-x)_+^{\alpha-1}\, e^{bx}\,\Gamma(\nu, bx)$	$a^{s+\alpha-1}\Gamma(\nu)\,\mathrm{B}(\alpha, s)\;{}_1F_1\left(\begin{matrix} s;\, ab \\ s+\alpha \end{matrix}\right) - \frac{a^{s+\alpha+\nu-1}b^{\nu}}{\nu}\,\mathrm{B}(\alpha, s+\nu)\;{}_2F_2\left(\begin{matrix} 1,\, s+\nu;\, ab \\ \nu+1,\, s+\alpha+\nu \end{matrix}\right)$ $[a,\ \operatorname{Re}\alpha>0;\ \operatorname{Re}s>0,\ -\operatorname{Re}\nu;\ \lvert\arg b\rvert<\pi]$
4	$(x-a)_+^{\alpha-1}\, e^{bx}\,\Gamma(\nu, bx)$	$-\frac{a^{s+\alpha+\nu-1}b^{\nu}}{\nu}\,\mathrm{B}(\alpha, 1-s-\alpha-\nu)\;{}_2F_2\left(\begin{matrix} 1,\, s+\nu;\, ab \\ \nu+1,\, s+\alpha+\nu \end{matrix}\right) + a^{s+\alpha-1}\Gamma(\nu)\,\mathrm{B}(\alpha, 1-s-\alpha)\;{}_1F_1\left(\begin{matrix} s;\, ab \\ s+\alpha \end{matrix}\right) - \frac{\pi b^{1-s-\alpha}}{\sin[(s+\alpha+\nu)\pi]}\,\Gamma\left[\begin{matrix} s+\alpha-1 \\ 1-\nu \end{matrix}\right]\;{}_1F_1\left(\begin{matrix} 1-\alpha;\, ab \\ 2-s-\alpha \end{matrix}\right)$ $[a,\ \operatorname{Re}\alpha>0;\ \operatorname{Re}(s+\alpha+\nu)<2;\ \lvert\arg b\rvert<\pi]$
5	$\frac{e^{bx}}{(x+a)^{\rho}}\,\Gamma(\nu, bx)$	$a^{s-\rho}\,\Gamma(\nu)\,\mathrm{B}(s, \rho-s)\;{}_1F_1\left(\begin{matrix} s;\, -ab \\ s-\rho+1 \end{matrix}\right) - \frac{a^{s+\nu-\rho}b^{\nu}}{\nu} \times \mathrm{B}(s+\nu, \rho-\nu-s)\;{}_2F_2\left(\begin{matrix} 1,\, s+\nu;\, -ab \\ \nu+1,\, s+\nu-\rho+1 \end{matrix}\right) + \frac{\pi b^{\rho-s}}{\sin[(s+\nu-\rho)\pi]}\,\Gamma\left[\begin{matrix} s-\rho \\ 1-\nu \end{matrix}\right]\;{}_1F_1\left(\begin{matrix} \rho;\, -ab \\ \rho-s+1 \end{matrix}\right)$ $[\operatorname{Re}s>0;\ 0<\operatorname{Re}(s+\nu)<\operatorname{Re}\rho+1;\ \lvert\arg a\rvert,\ \lvert\arg b\rvert<\pi]$
6	$\frac{e^{bx}}{x-a}\,\Gamma(\nu, bx)$	$-\pi a^{s-1}e^{ab}\,\Gamma(\nu)\cot(s\pi) + \pi a^{s-1}e^{-ab}\cot[(s+\nu)\pi]\;\gamma(\nu, ba) - \frac{\pi a^{s-1}e^{ab}}{\sin[(s+\nu)\pi]}\,\Gamma\left[\begin{matrix} s \\ 1-\nu \end{matrix}\right]\gamma(1-s, ab)$ $[a,\ \operatorname{Re}s>0;\ 0<\operatorname{Re}(s+\nu)<1;\ \lvert\arg b\rvert<\pi]$
7	$e^{-ax^2}\left\{\begin{matrix} \Gamma(\nu, bx) \\ \gamma(\nu, bx) \end{matrix}\right\}$	$\frac{1\pm 1}{4a^{s/2}}\,\Gamma(\nu)\,\Gamma\left(\frac{s}{2}\right) \mp \frac{a^{-(s+\nu)/2}b^{\nu}}{2\nu}\,\Gamma\left(\frac{s+\nu}{2}\right)\;{}_2F_2\left(\begin{matrix} \frac{\nu}{2},\, \frac{s+\nu}{2} \\ \frac{1}{2},\, \frac{\nu+2}{2};\, \frac{b^2}{4a} \end{matrix}\right) \pm \frac{a^{-(s+\nu+1)/2}b^{\nu+1}}{2(\nu+1)}\,\Gamma\left(\frac{s+\nu+1}{2}\right)\;{}_2F_2\left(\begin{matrix} \frac{\nu+1}{2},\, \frac{s+\nu+1}{2} \\ \frac{3}{2},\, \frac{\nu+3}{2};\, \frac{b^2}{4a} \end{matrix}\right)$ $\left[\operatorname{Re}a,\ \operatorname{Re}b,\ \operatorname{Re}(s+\nu),\ \left\{\begin{matrix} \operatorname{Re}s \\ \operatorname{Re}\nu \end{matrix}\right\}>0\right]$
8	$e^{-a/x}\left\{\begin{matrix} \Gamma(\nu, bx) \\ \gamma(\nu, bx) \end{matrix}\right\}$	$\frac{1\pm 1}{2}\,a^{s}\,\Gamma(\nu)\,\Gamma(-s) \mp \frac{a^{s+\nu}b^{\nu}}{\nu}\;{}_1F_2\left(\begin{matrix} \nu;\, ab \\ \nu+1,\, s+\nu+1 \end{matrix}\right) \times \Gamma(-s-\nu) \pm \frac{\Gamma(s+\nu)}{b^{s}s}\;{}_1F_2\left(\begin{matrix} -s;\, ab \\ 1-s,\, 1-s-\nu \end{matrix}\right)$ $\left[\operatorname{Re}a,\ \left\{\begin{matrix} \operatorname{Re}b \\ \operatorname{Re}\nu,\ \operatorname{Re}b,\ \operatorname{Re}(-s) \end{matrix}\right\}>0\right]$

No.	$f(x)$	$F(s)$
9	$e^{ax-b/x}\,\Gamma(\nu, ax)$	$2^{2-s-2\nu}\left(\frac{b}{a}\right)^{s/2}\Gamma(1-s-\nu)\,S_{s+2\nu-1,\,-s}(2\sqrt{ab})$ $[\operatorname{Re} b>0;\ \operatorname{Re}(s+\nu)<1;\ \lvert\arg a\rvert<\pi]$
10	$\theta(a-x)\,e^{bx}$ $\times\,\gamma(\nu, c(a-x))$	$a^{s+\nu}c^{\nu}\,\Gamma\begin{bmatrix} s,\ \nu \\ s+\nu+1 \end{bmatrix}\Phi_2(s,\ \nu;\ s+\nu+1;\ ab,\ -ac)$ $[a,\ \operatorname{Re} s>0;\ \operatorname{Re}\nu>-1]$
11	$(a-x)_+^{-\nu}\,e^{b(a-x)}$ $\times\,\gamma(\nu, b(a-x))$	$\frac{a^s b^{\nu}}{\nu s}\,{}_2F_2\left(\begin{matrix} 1,\ 1;\ ab \\ \nu+1,\ s+1 \end{matrix}\right)$ $[a,\ \operatorname{Re} s>0;\ 0<\operatorname{Re}\nu<1]$
12	$(a-x)_+^{\alpha-1}\,e^{bx(a-x)}$ $\times\left\{\begin{matrix}\Gamma(\nu, bx(a-x)) \\ \gamma(\nu, bx(a-x))\end{matrix}\right\}$	$\mp\frac{a^{s+\alpha+2\nu-1}b^{\nu}}{\nu}\,\mathrm{B}(s+\nu,\ \alpha+\nu)\,{}_3F_3\left(\begin{matrix} 1,\ \alpha+\nu,\ s+\nu;\ \frac{a^2b}{4} \\ \nu+1,\ \frac{s+\alpha+2\nu}{2},\ \frac{s+\alpha+2\nu+1}{2} \end{matrix}\right)$ $+\frac{1\pm1}{2}\,a^{s+\alpha-1}\,\Gamma(\nu)\,\mathrm{B}(s,\ \alpha)\,{}_2F_2\left(\begin{matrix} s,\ \alpha;\ \frac{a^2b}{4} \\ \frac{s+\alpha}{2},\ \frac{s+\alpha+1}{2} \end{matrix}\right)$ $[a,\ \operatorname{Re}\nu,\ \operatorname{Re}(s+\nu),\ \operatorname{Re}(\alpha+\nu)>0]$
13	$\frac{e^{b/(x+a)}}{(x+a)^{\rho}}\,\gamma\left(\nu,\ \frac{b}{x+a}\right)$	$\frac{a^{s-\nu-\rho}b^{\nu}}{\nu}\,\mathrm{B}(s,\ \nu+\rho-s)\,{}_2F_2\left(\begin{matrix} 1,\ \nu+\rho-s \\ \nu+1,\ \nu+\rho;\ \frac{b}{a} \end{matrix}\right)$ $[0<\operatorname{Re} s<\operatorname{Re}(\nu+\rho);\ \lvert\arg a\rvert<\pi]$
14	$\frac{e^{bx/(x+a)}}{(x+a)^{\rho}}\,\gamma\left(\nu,\ \frac{bx}{x+a}\right)$	$\frac{a^{s-\rho}b^{\nu}}{\nu}\,\mathrm{B}(s+\nu,\ \rho-s)\,{}_2F_2\left(\begin{matrix} 1,\ s+\nu;\ b \\ \nu+1,\ \nu+\rho \end{matrix}\right)$ $[-\operatorname{Re}\nu<\operatorname{Re} s<\operatorname{Re}\rho;\ \lvert\arg a\rvert<\pi]$
15	$\theta(x-a)\,(x-b)^{\nu-1}$ $\times\,e^{cx/(x-b)}\,\gamma\left(\nu,\ \frac{cx}{x-b}\right)$	$\frac{a^{s+\nu-1}c^{\nu}}{\nu(1-s-\nu)}\,\Psi_1\left(1,\ 1-s-\nu;\ 2-s-\nu,\ \nu+1;\ \frac{b}{a},\ c\right)$ $[a>0;\ \lvert b\rvert<a;\ \operatorname{Re}(s+\nu)<-1]$

3.8.3. $\Gamma(\nu, ax)$, $\gamma(\nu, ax)$, and trigonometric functions

Notation: $\delta=\left\{\begin{matrix}1\\0\end{matrix}\right\}$.

1	$\sin(ax)\left\{\begin{matrix}\Gamma(\nu, bx) \\ \gamma(\nu, bx)\end{matrix}\right\}$	$\pm\frac{a\,\Gamma(s+\nu+1)}{b^{s+1}(s+1)}\,{}_3F_2\left(\begin{matrix} \frac{s+1}{2},\ \frac{s+\nu+1}{2},\ \frac{s+\nu+2}{2} \\ \frac{3}{2},\ \frac{s+3}{2};\ -\frac{a^2}{b^2} \end{matrix}\right)+\frac{1\mp1}{2a^s}\sin\frac{s\pi}{2}\,\Gamma(s)\,\Gamma(\nu)$ $\left[a,\ \operatorname{Re} b>0;\ \left\{\begin{matrix}\operatorname{Re} s>-1;\ \operatorname{Re}(s+\nu)>-1 \\ \operatorname{Re}\nu>0,\ -\operatorname{Re}\nu-1<\operatorname{Re} s<1\end{matrix}\right\}\right]$

No.	$f(x)$	$F(s)$
2	$\cos(ax)\begin{Bmatrix}\Gamma(\nu,\, bx)\\ \gamma(\nu,\, bx)\end{Bmatrix}$	$\pm\dfrac{\Gamma(s+\nu)}{b^s s}\, {}_3F_2\left(\begin{matrix}\frac{s}{2},\, \frac{s+\nu}{2},\, \frac{s+\nu+1}{2}\\ \frac{1}{2},\, \frac{s+2}{2};\, -\frac{a^2}{b^2}\end{matrix}\right) + \dfrac{1\mp 1}{2a^s}\cos\dfrac{s\pi}{2}\,\Gamma(s)\,\Gamma(\nu)$ $\left[a,\ \operatorname{Re} b>0;\ \begin{Bmatrix}\operatorname{Re} s,\ \operatorname{Re}(s+\nu)>0\\ \operatorname{Re}\nu>0;\ -\operatorname{Re}\nu<\operatorname{Re} s<1\end{Bmatrix}\right]$
3	$\sin(a\sqrt{x})\begin{Bmatrix}\Gamma(\nu,\, bx)\\ \gamma(\nu,\, bx)\end{Bmatrix}$	$\pm\dfrac{2a}{(2s+1)\,b^{s+1/2}}\,\Gamma\left(\dfrac{2s+2\nu+1}{2}\right)\, {}_2F_2\left(\begin{matrix}\frac{2s+1}{2},\, \frac{2s+2\nu+1}{2}\\ \frac{3}{2},\, \frac{2s+3}{2};\, -\frac{a^2}{4b}\end{matrix}\right)$ $+\dfrac{1\mp 1}{a^{2s}}\sin(s\pi)\,\Gamma(\nu)\,\Gamma(2s)$ $\left[a,\ \operatorname{Re} b>0;\ \begin{Bmatrix}\operatorname{Re} s,\ \operatorname{Re}(s+\nu)>-1/2\\ \operatorname{Re}\nu>0;\ -\operatorname{Re}\nu-1/2<\operatorname{Re} s<1/2\end{Bmatrix}\right]$
4	$\cos(a\sqrt{x})\begin{Bmatrix}\Gamma(\nu,\, bx)\\ \gamma(\nu,\, bx)\end{Bmatrix}$	$\pm\dfrac{\Gamma(s+\nu)}{b^s s}\, {}_2F_2\left(\begin{matrix}s,\, s+\nu;\, -\frac{a^2}{4b}\\ \frac{1}{2},\, s+1\end{matrix}\right) + \dfrac{1\mp 1}{a^{2s}}\cos(s\pi)\,\Gamma(\nu)\,\Gamma(2s)$ $\left[a,\ \operatorname{Re} b>0;\ \begin{Bmatrix}\operatorname{Re} s,\ \operatorname{Re}(s+\nu)>0\\ \operatorname{Re}\nu>0;\ -\operatorname{Re}\nu<\operatorname{Re} s<1/2\end{Bmatrix}\right]$
5	$e^{bx}\begin{Bmatrix}\sin(a\sqrt{x})\\ \cos(a\sqrt{x})\end{Bmatrix}$ $\times\,\Gamma(\nu,\, bx)$	$\dfrac{\pi a^{\delta}}{b^{s+\delta/2}}\csc\dfrac{(2s+2\nu+\delta)\,\pi}{2}\,\Gamma\left[\begin{matrix}\frac{2s+\delta}{2}\\ 1-\nu\end{matrix}\right]\, {}_1F_1\left(\begin{matrix}\frac{2s+\delta}{2};\, \frac{a^2}{4b}\\ \frac{2\delta+1}{2}\end{matrix}\right)$ $-\dfrac{2b^{\nu-1}}{a^{2s+2\nu-2}}\,\Gamma(2s+2\nu-2)\begin{Bmatrix}\sin[(s+\nu)\,\pi]\\ \cos[(s+\nu)\,\pi]\end{Bmatrix}$ $\times\, {}_2F_2\left(\begin{matrix}1,\, 1-\nu;\, \frac{a^2}{4b}\\ \frac{3-2s-2\nu}{2},\, 2-s-\nu\end{matrix}\right)$ $\left[\begin{matrix}a>0;\ \operatorname{Re} s>-\delta/2;\\ -\delta/2<\operatorname{Re}(s+\nu)<3/2;\ \lvert\arg b\rvert<\pi\end{matrix}\right]$

3.8.4. $\Gamma(\nu,\, ax)$, $\gamma(\nu,\, ax)$, and the logarithmic function

1	$\theta(a-x)\ln\dfrac{\sqrt{a}+\sqrt{a-x}}{\sqrt{x}}$ $\times\begin{Bmatrix}\Gamma(\nu,\, bx)\\ \gamma(\nu,\, bx)\end{Bmatrix}$	$\mp\dfrac{\sqrt{\pi}\, a^{s+\nu} b^{\nu}}{2\nu\,(s+\nu)}\,\Gamma\left[\begin{matrix}s+\nu\\ \frac{2s+2\nu+1}{2}\end{matrix}\right]\, {}_3F_3\left(\begin{matrix}\nu,\, s+\nu,\, s+\nu;\, -ab\\ \nu+1,\, \frac{2s+2\nu+1}{2},\, s+\nu+1\end{matrix}\right)$ $+\dfrac{1\pm 1}{4s}\sqrt{\pi}\, a^s\,\Gamma\left[\begin{matrix}s\\ \frac{2s+1}{2}\end{matrix}\right]$ $[a,\ \operatorname{Re}\nu,\ \operatorname{Re}(s+\nu+1)>0]$
2	$\theta(a-x)\ln\dfrac{\sqrt{a}+\sqrt{a-x}}{\sqrt{x}}$ $\times\, e^{bx}\begin{Bmatrix}\Gamma(\nu,\, bx)\\ \gamma(\nu,\, bx)\end{Bmatrix}$	$\mp\dfrac{\sqrt{\pi}\, a^{s+\nu} b^{\nu}}{2\nu\,(s+\nu)}\,\Gamma\left[\begin{matrix}s+\nu\\ \frac{2s+2\nu+1}{2}\end{matrix}\right]\, {}_3F_3\left(\begin{matrix}1,\, s+\nu,\, s+\nu;\, ab\\ \nu+1,\, \frac{2s+2\nu+1}{2},\, s+\nu+1\end{matrix}\right)$ $+\dfrac{1\pm 1}{4s}\sqrt{\pi}\, a^s\,\Gamma\left[\begin{matrix}s\\ \frac{2s+1}{2}\end{matrix}\right]\, {}_2F_2\left(\begin{matrix}s,\, s;\, ab\\ \frac{2s+1}{2},\, s+1\end{matrix}\right)$ $[a,\ \operatorname{Re}\nu,\ \operatorname{Re}(s+\nu+1)>0]$

3.8.5. $\gamma(\nu, ax)$ and inverse trigonometric functions

1	$\theta(a-x)\arccos\frac{x}{a}$ $\times\gamma(\nu, bx)$	$\frac{(2a)^{s+\nu}b^{\nu}}{\nu(s+\nu)}\Gamma\left[\begin{matrix}\frac{s+\nu+1}{2}, \frac{s+\nu+3}{2}\\ s+\nu+2\end{matrix}\right]{}_3F_4\left(\begin{matrix}\frac{\nu}{2}, \frac{s+\nu}{2}, \frac{s+\nu+1}{2}; \frac{a^2b^2}{4}\\ \frac{1}{2}, \frac{\nu+2}{2}, \frac{s+\nu+2}{2}, \frac{s+\nu+2}{2}\end{matrix}\right) - \frac{2^{s+\nu}a^{s+\nu+1}b^{\nu+1}}{(\nu+1)(s+\nu+1)}\Gamma\left[\begin{matrix}\frac{s+\nu+2}{2}, \frac{s+\nu+2}{2}\\ s+\nu+2\end{matrix}\right] \times{}_3F_4\left(\begin{matrix}\frac{\nu+1}{2}, \frac{s+\nu+1}{2}, \frac{s+\nu+2}{2}; \frac{a^2b^2}{4}\\ \frac{3}{2}, \frac{\nu+3}{2}, \frac{s+\nu+3}{2}, \frac{s+\nu+3}{2}\end{matrix}\right)$ $[a,\ \mathrm{Re}(s+\nu+1)>0]$
2	$\theta(a-x)e^{bx}\arccos\frac{x}{a}$ $\times\gamma(\nu, bx)$	$\frac{2^{s+\nu}a^{s+\nu+1}b^{\nu+1}}{\nu(\nu+1)(s+\nu+1)}\Gamma\left[\begin{matrix}\frac{s+\nu+2}{2}, \frac{s+\nu+2}{2}\\ s+\nu+2\end{matrix}\right] \times{}_3F_4\left(\begin{matrix}1, \frac{s+\nu+1}{2}, \frac{s+\nu+2}{2}; \frac{a^2b^2}{4}\\ \frac{\nu+2}{2}, \frac{\nu+3}{2}, \frac{s+\nu+3}{2}, \frac{s+\nu+3}{2}\end{matrix}\right)+\frac{2^{s+\nu+1}a^{s+\nu}b^{\nu}}{\nu(s+\nu)(s+\nu+1)} \times\Gamma\left[\begin{matrix}\frac{s+\nu+3}{2}, \frac{s+\nu+3}{2}\\ s+\nu+2\end{matrix}\right]{}_3F_4\left(\begin{matrix}1, \frac{s+\nu+1}{2}, \frac{s+\nu}{2}; \frac{a^2b^2}{4}\\ \frac{\nu+1}{2}, \frac{\nu+2}{2}, \frac{s+\nu+2}{2}, \frac{s+\nu+2}{2}\end{matrix}\right)$ $[a,\ \mathrm{Re}(s+\nu+1)>0]$

3.8.6. $\Gamma(\nu, ax)$, $\gamma(\nu, ax)$, and $\mathrm{Ei}(bx)$

1	$\mathrm{Ei}(-ax)\left\{\begin{matrix}\Gamma(\nu, bx)\\ \gamma(\nu, bx)\end{matrix}\right\}$	$\pm\frac{b^{\nu}\Gamma(s+\nu)}{\nu(s+\nu)a^{s+\nu}}{}_3F_2\left(\begin{matrix}\nu, s+\nu, s+\nu; -\frac{b}{a}\\ \nu+1, s+\nu+1\end{matrix}\right)-\frac{1\pm1}{2a^s s}\Gamma(\nu)\Gamma(s)$ $\left[\mathrm{Re}\,a,\ \mathrm{Re}\,b,\ \mathrm{Re}(s+\nu)>0;\ \left\{\begin{matrix}\mathrm{Re}\,s\\ \mathrm{Re}\,\nu\end{matrix}\right\}>0\right]$

3.8.7. $\Gamma(\nu, ax)$, $\gamma(\nu, ax)$, and $\mathrm{erf}(bx^r)$, $\mathrm{erfc}(bx^r)$, $\mathrm{erfi}(bx^r)$

1	$\mathrm{erfc}(ax)\left\{\begin{matrix}\Gamma(\nu, bx)\\ \gamma(\nu, bx)\end{matrix}\right\}$	$\mp\frac{a^{-s-\nu}b^{\nu}}{\sqrt{\pi}\,\nu(s+\nu)}\Gamma\left(\frac{s+\nu+1}{2}\right){}_3F_3\left(\begin{matrix}\frac{\nu}{2}, \frac{s+\nu}{2}, \frac{s+\nu+1}{2}\\ \frac{1}{2}, \frac{\nu+2}{2}, \frac{s+\nu+2}{2}; \frac{b^2}{4a^2}\end{matrix}\right) \pm\frac{a^{-s-\nu-1}b^{\nu+1}}{\sqrt{\pi}(\nu+1)(s+\nu+1)}\Gamma\left(\frac{s+\nu+2}{2}\right) \times{}_3F_3\left(\begin{matrix}\frac{\nu+1}{2}, \frac{s+\nu+1}{2}, \frac{s+\nu+2}{2}\\ \frac{3}{2}, \frac{\nu+3}{2}, \frac{s+\nu+3}{2}; \frac{b^2}{4a^2}\end{matrix}\right)+\frac{1\pm1}{2}\frac{a^{-s}}{\sqrt{\pi}\,s}\Gamma(\nu)\Gamma\left(\frac{s+1}{2}\right)$ $[\mathrm{Re}\,b,\ \mathrm{Re}(a^2+b)>0;\ \mathrm{Re}\,s>-\mathrm{Re}\,\nu,\ 0]$
2	$\mathrm{erf}(a\sqrt{x})\left\{\begin{matrix}\Gamma(\nu, bx)\\ \gamma(\nu, bx)\end{matrix}\right\}$	$\pm\frac{b^{\nu}}{\sqrt{\pi}\,\nu(s+\nu)a^{2s+2\nu}}\Gamma\left(s+\nu+\frac{1}{2}\right){}_3F_2\left(\begin{matrix}\nu, s+\nu, s+\nu+\frac{1}{2}\\ \nu+1, s+\nu+1; -\frac{b}{a^2}\end{matrix}\right) -\frac{1\pm1}{2\sqrt{\pi}\,a^{2s}s}\Gamma(\nu)\Gamma\left(\frac{2s+1}{2}\right)\pm\frac{\Gamma(s+\nu)}{b^s s}$ $\left[\mathrm{Re}\,b>0;\ \mathrm{Re}(s+\nu)>-1/2;\ \left\{\begin{matrix}\mathrm{Re}\,s\\ \mathrm{Re}\,\nu\end{matrix}\right\}>0;\ \lvert\arg a\rvert<\pi/4\right]$

No.	$f(x)$	$F(s)$
3	$\operatorname{erfc}(a\sqrt{x})\begin{Bmatrix}\Gamma(\nu, bx)\\ \gamma(\nu, bx)\end{Bmatrix}$	$\mp\frac{b^{\nu}}{\sqrt{\pi}\,\nu\,(s+\nu)\,a^{2s+2\nu}}\Gamma\left(s+\nu+\frac{1}{2}\right)\,{}_3F_2\left(\begin{matrix}\nu,\, s+\nu,\, s+\nu+\frac{1}{2}\\ \nu+1,\, s+\nu+1;\, -\frac{b}{a^2}\end{matrix}\right)$ $+\frac{1\pm 1}{2\sqrt{\pi}\,a^{2s}s}\Gamma(\nu)\,\Gamma\left(\frac{2s+1}{2}\right)$ $\left[\operatorname{Re} b,\ \operatorname{Re}(s+\nu),\ \begin{Bmatrix}\operatorname{Re} s\\ \operatorname{Re}\nu\end{Bmatrix} > 0;\ \lvert\arg a\rvert < \pi/4\right]$
4	$e^{bx}\operatorname{erfc}(a\sqrt{x})\,\gamma(\nu, bx)$	$\frac{b^{\nu}}{\sqrt{\pi}\,\nu\,a^{2s+2\nu}\,(s+\nu)}\Gamma\left(s+\nu+\frac{1}{2}\right)\,{}_3F_2\left(\begin{matrix}1,\, s+\nu,\, s+\nu+\frac{1}{2}\\ \nu+1,\, s+\nu+1;\, \frac{b}{a^2}\end{matrix}\right)$ $\left[\operatorname{Re}(a^2-b),\ \operatorname{Re}\nu,\ \operatorname{Re}(s+\nu) > 0;\ \lvert\arg b\rvert < \pi/4\right]$
5	$e^{a^2x}\operatorname{erf}(a\sqrt{x})\,\Gamma(\nu, bx)$	$\frac{4a}{\sqrt{\pi}\,b^{s+1/2}\,(2s+1)}\Gamma\left(s+\nu+\frac{1}{2}\right)\,{}_3F_2\left(\begin{matrix}1,\, s+\frac{1}{2},\, s+\nu+\frac{1}{2}\\ \frac{3}{2},\, s+\frac{3}{2};\, \frac{a^2}{b}\end{matrix}\right)$ $\left[\operatorname{Re}(b-a^2) > 0;\ \operatorname{Re} s,\ \operatorname{Re}(s+\nu) > -1/2; \lvert\arg a\rvert < \pi/4\right]$
6	$\operatorname{erfi}(a\sqrt{x})\operatorname{erf}(a\sqrt{x})$ $\times\,\Gamma(\nu, bx)$	$\frac{4a^2}{\pi\,b^{s+1}\,(s+1)}\Gamma(s+\nu+1)\,{}_5F_4\left(\begin{matrix}\frac{1}{2},\, 1,\, \frac{s+1}{2},\, \frac{s+\nu+1}{2},\, \frac{s+\nu+2}{2}\\ \frac{3}{4},\, \frac{5}{4},\, \frac{3}{2},\, \frac{s+3}{2};\, \frac{a^4}{b^2}\end{matrix}\right)$ $\left[a,\ \operatorname{Re} b > 0;\ \operatorname{Re} s,\ \operatorname{Re}(s+\nu) > -1\right]$

3.8.8. Products of $\Gamma(\mu, ax)$ and $\gamma(\nu, ax)$

1	$\Gamma(\nu, -ax)\,\Gamma(\nu, ax)$	$\frac{\pi(-a)^{-s/2}a^{-s/2}}{s}\csc\frac{(s+2\nu)\pi}{2}\,\Gamma\left[\begin{matrix}s+\nu\\ 1-\nu\end{matrix}\right]$ $\left[\operatorname{Re} a > 0;\ 0,\ -2\operatorname{Re}\nu < \operatorname{Re} s < 2-2\operatorname{Re}\nu\right]$
2	$\Gamma(\mu, ax)\begin{Bmatrix}\Gamma(\nu, bx)\\ \gamma(\nu, bx)\end{Bmatrix}$	$\mp\frac{b^{\nu}\,\Gamma(s+\mu+\nu)}{\nu\,(s+\nu)\,a^{s+\nu}}\,{}_3F_2\left(\begin{matrix}\nu,\, s+\nu,\, s+\mu+\nu\\ \nu+1,\, s+\nu+1;\, -\frac{b}{a}\end{matrix}\right)+\frac{1\pm 1}{2a^s s}\Gamma(\nu)\,\Gamma(s+\mu)$ $\left[\begin{matrix}\operatorname{Re}(s+\mu),\ \operatorname{Re}(s+\nu),\ \operatorname{Re}(s+\nu+\mu) > 0;\\ \begin{Bmatrix}\operatorname{Re}(a+b) > 0\\ \operatorname{Re} b,\ \operatorname{Re}(a+b) > 0\end{Bmatrix}\end{matrix}\right]$
3	$\gamma(\mu, ax)\,\gamma(\nu, bx)$	$-\frac{b^{\nu}\,\Gamma(s+\mu+\nu)}{\nu\,(s+\nu)\,a^{s+\nu}}\,{}_3F_2\left(\begin{matrix}\nu,\, s+\nu,\, s+\mu+\nu\\ \nu+1,\, s+\nu+1;\, -\frac{b}{a}\end{matrix}\right)$ $-\frac{\Gamma(\mu)\,\Gamma(s+\nu)}{b^s s}$ $\left[\begin{matrix}\operatorname{Re} a,\ \operatorname{Re} b,\ \operatorname{Re}\mu,\ \operatorname{Re}\nu,\\ \operatorname{Re}(s+\mu+\nu) > 0;\ \operatorname{Re} s < 0\end{matrix}\right]$
4	$e^{-ax}\gamma(\mu, bx)\,\gamma(\nu, cx)$	$\frac{b^{\mu}c^{\nu}\,\Gamma(s+\mu+\nu)}{\mu\nu a^{s+\mu+\nu}}F_2\left(s+\mu+\nu,\, \mu,\, \nu;\, \mu+1,\, \nu+1;\, -\frac{b}{a},\, -\frac{c}{a}\right)$ $\left[\begin{matrix}\operatorname{Re} a,\ \operatorname{Re}(a+b),\ \operatorname{Re}(a+c),\\ \operatorname{Re}(a+b+c),\ \operatorname{Re}(s+\mu+\nu) > 0\end{matrix}\right]$

3.9. The Parabolic Cylinder Function $D_\nu(z)$

More formulas can be obtained from the corresponding sections due to the relations

$$D_{-1/2}(z)=\frac{\sqrt{\pi}}{2}\left[\sqrt[4]{z^2}\,I_{-1/4}\left(\frac{z^2}{4}\right)-\frac{z}{\sqrt[4]{z^2}}I_{1/4}\left(\frac{z^2}{4}\right)\right],\quad D_0(z)=e^{-z^2/4},$$

$$D_{1/2}(z)=\frac{\sqrt{\pi}}{4}\left[z\sqrt[4]{z^2}\,I_{-1/4}\left(\frac{z^2}{4}\right)+\left(z^2\right)^{3/4}I_{-3/4}\left(\frac{z^2}{4}\right)\right.$$
$$\left.-\left(z^2\right)^{3/4}I_{1/4}\left(\frac{z^2}{4}\right)-z\sqrt[4]{z^2}\,I_{3/4}\left(\frac{z^2}{4}\right)\right],\quad D_1(z)=ze^{-z^2/4},$$

$$D_{2n+\varepsilon}(z)=(-1)^n\,2^n n!\,z^{\varepsilon}e^{-z^2/4}L_n^{(2\varepsilon-1)/2}\left(\frac{z^2}{2}\right),\quad [\varepsilon=0\text{ or }1;\ n=0,\,1,\,2,\ldots];$$

$$D_\nu(z)=2^{\nu/2}e^{-z^2/4}\Psi\left(-\frac{\nu}{2};\,\frac{1}{2};\,\frac{z^2}{2}\right)=2^{(\nu-1)/2}ze^{-z^2/4}\Psi\left(\frac{1-\nu}{2};\,\frac{3}{2};\,\frac{z^2}{2}\right),$$

$$D_\nu(z)=2^{\nu/2}e^{-z^2/4}\left[\frac{1}{\Gamma((1-\nu)/2)}\,{}_1F_1\left(-\frac{\nu}{2};\,\frac{1}{2};\,\frac{z^2}{2}\right)-\frac{\sqrt{2\pi}\,z}{\Gamma(-\nu/2)}\,z\,{}_1F_1\left(\frac{1-\nu}{2};\,\frac{3}{2};\,\frac{z^2}{2}\right)\right],$$

$$D_\nu(z)=2^{-\nu/2}e^{-z^2/4}H_\nu\left(\frac{z}{\sqrt{2}}\right),\quad D_\nu(\sqrt{2z})=2^{\nu/2}e^{z/2}\,G_{12}^{20}\left(z\,\middle|\,\begin{matrix}(1-\nu)/2\\0,\,1/2\end{matrix}\right).$$

3.9.1. $D_\nu(bx)$ and elementary functions

Notation: $\delta=\left\{\begin{matrix}1\\0\end{matrix}\right\}$.

No.	$f(x)$	$F(s)$
1	$D_\nu(ax)$	$\frac{2^{(\nu-s)/2}\sqrt{\pi}}{a^s}\Gamma\left[\begin{matrix}s\\\frac{s-\nu+1}{2}\end{matrix}\right]{}_2F_1\left(\begin{matrix}\frac{s}{2},\,\frac{s+1}{2}\\\frac{s-\nu+1}{2};\,\frac{1}{2}\end{matrix}\right)$ $[\operatorname{Re}s>0;\ \lvert\arg a\rvert<\pi/4]$
2	$e^{a^2x^2/4}D_\nu(ax)$	$\frac{a^{-s}}{2^{(s+\nu)/2+1}}\Gamma\left[\begin{matrix}s,\,-\frac{s+\nu}{2}\\-\nu\end{matrix}\right]$ $\left[\begin{matrix}0<\operatorname{Re}s<-\operatorname{Re}\nu;\\\lvert\arg a\rvert<3\pi/4\end{matrix}\right]$
3	$e^{-a^2x^2/4}D_\nu(ax)$	$\frac{2^{(\nu-s)/2}\sqrt{\pi}}{a^s}\Gamma\left[\begin{matrix}s\\\frac{s-\nu+1}{2}\end{matrix}\right]$ $[\operatorname{Re}s>0;\ \lvert\arg a\rvert<\pi/4]$
4	$e^{-a^2x^2/4}D_\nu(-ax)$	$\frac{2^{(\nu-s)/2}}{\sqrt{\pi}\,a^s}\cos\frac{(s+\nu)\pi}{2}\,\Gamma\left(\frac{1-s+\nu}{2}\right)\Gamma(s)$ $[0<\operatorname{Re}s<\operatorname{Re}\nu+1;\ \lvert\arg a\rvert<\pi/4]$
5	$e^{-ax^2}D_\nu(bx)$	$\frac{2^{(\nu-s)/2}\sqrt{\pi}}{b^s}\Gamma\left[\begin{matrix}s\\\frac{s-\nu+1}{2}\end{matrix}\right]{}_2F_1\left(\begin{matrix}\frac{s}{2},\,\frac{s+1}{2}\\\frac{s-\nu+1}{2};\,\frac{b^2-4a}{2b^2}\end{matrix}\right)$ $\left[\begin{matrix}(\operatorname{Re}s,\ \operatorname{Re}(4a+b^2)>0)\text{ or}\\(\operatorname{Re}(4a+b^2)=0;\,0<\operatorname{Re}s<-\operatorname{Re}\nu)\end{matrix}\right]$

No.	$f(x)$	$F(s)$
6	$\left(a^2-x^2\right)_+^{\alpha-1} e^{\pm b^2x^2/4}$ $\times D_\nu(bx)$	$\frac{2^{\nu/2-1}\sqrt{\pi}\,a^{s+2\alpha-2}}{\Gamma\left(\frac{1-\nu}{2}\right)} \mathrm{B}\left(\alpha,\,\frac{s}{2}\right) {}_2F_2\left(\begin{matrix}\frac{\mp2\nu+1\mp1}{4},\,\frac{s}{2}\\ \frac{1}{2},\,\frac{s+2\alpha}{2};\,\frac{a^2b^2}{2}\end{matrix}\right)$ $-\frac{2^{(\nu-1)/2}\sqrt{\pi}\,a^{s+2\alpha-1}b}{\Gamma\left(-\frac{\nu}{2}\right)} \mathrm{B}\left(\alpha,\,\frac{s+1}{2}\right) {}_2F_2\left(\begin{matrix}\frac{\mp2\nu+3\mp1}{4},\,\frac{s+1}{2}\\ \frac{3}{2},\,\frac{s+2\alpha+1}{2};\,\frac{a^2b^2}{2}\end{matrix}\right)$ $[a,\ \mathrm{Re}\,\alpha,\ \mathrm{Re}\,s>0]$
7	$\frac{e^{\pm b^2x^2/4}}{\left(x^2+a^2\right)^\rho} D_\nu(bx)$	$\frac{2^{\nu/2-1}\sqrt{\pi}\,a^{s-2\rho}}{\Gamma\left(\frac{1-\nu}{2}\right)} \mathrm{B}\left(\frac{s}{2},\,\frac{2\rho-s}{2}\right) {}_2F_2\left(\begin{matrix}\frac{1\mp1\mp2\nu}{4},\,\frac{s}{2}\\ \frac{1}{2},\,\frac{s-2\rho+2}{2};\,\mp\frac{a^2b^2}{2}\end{matrix}\right)$ $-\frac{2^{(\nu-1)/2}\sqrt{\pi}}{\Gamma\left(-\frac{\nu}{2}\right)} a^{s-2\rho+1}\,b\,\mathrm{B}\left(\frac{s+1}{2},\,\frac{2\rho-s-1}{2}\right)$ $\times\,{}_2F_2\left(\begin{matrix}\frac{3\mp1\mp2\nu}{4},\,\frac{s+1}{2}\\ \frac{3}{2},\,\frac{s-2\rho+3}{2};\,\mp\frac{a^2b^2}{2}\end{matrix}\right) + \frac{2^{\rho-(s\pm\nu)/2}\sqrt{\pi}}{\left[2\sqrt{\pi}\,\Gamma(-\nu)\right]^{(1\pm1)/2}}$ $\times\frac{b^{2\rho-s}\Gamma(s-2\rho)}{\Gamma^{\mp1}\left(\frac{1\mp1-2\nu\pm4\rho\mp2s}{4}\right)}\,{}_2F_2\left(\begin{matrix}\rho,\,\frac{4\rho-2s\mp2\nu+1\mp1}{4}\\ \frac{2\rho-s+2}{2},\,\frac{2\rho-s+1}{2};\,\mp\frac{a^2b^2}{2}\end{matrix}\right)$ $\left[\mathrm{Re}\,a>0;\ \left\{\begin{matrix}0<\mathrm{Re}\,s<\mathrm{Re}\,(2\rho-\nu)\\ \mathrm{Re}\,s>0\end{matrix}\right\};\ \lvert\arg b\rvert<(2\pm1)\,\pi/4\right]$
8	$e^{-ax+b^2x^2/4}D_\nu(bx)$	$-\frac{\pi}{2^{(s+\nu+2)/2}b^s}\csc\frac{(s+\nu)\,\pi}{2}\,\Gamma\left[\begin{matrix}s\\ -\nu,\ \frac{s+\nu+2}{2}\end{matrix}\right]$ $\times\,{}_2F_2\left(\begin{matrix}\frac{s}{2},\,\frac{s+1}{2};\,-\frac{a^2}{2b^2}\\ \frac{1}{2},\,\frac{s+\nu+2}{2}\end{matrix}\right)+\frac{\pi a}{2^{(s+\nu+3)/2}b^{s+1}}$ $\times\sec\frac{(s+\nu)\,\pi}{2}\,\Gamma\left[\begin{matrix}s+1\\ -\nu,\ \frac{s+\nu+3}{2}\end{matrix}\right]{}_2F_2\left(\begin{matrix}\frac{s+1}{2},\,\frac{s+2}{2};\,-\frac{a^2}{2b^2}\\ \frac{3}{2},\,\frac{s+\nu+3}{2}\end{matrix}\right)$ $+\frac{\pi b^\nu\csc\left[(s+\nu)\,\pi\right]}{a^{\nu+s}\Gamma(1-s-\nu)}\,{}_2F_2\left(\begin{matrix}-\frac{\nu}{2},\,\frac{1-\nu}{2};\,-\frac{a^2}{2b^2}\\ \frac{1-s-\nu}{2},\,\frac{2-s-\nu}{2}\end{matrix}\right)$ $[\mathrm{Re}\,a,\ \mathrm{Re}\,s>0;\ \lvert\arg b\rvert<3\pi/4]$
9	$e^{-a/x+b^2x^2/4}D_\nu(bx)$	$\frac{b^{-s}}{2^{(s+\nu+2)/2}}\Gamma\left[\begin{matrix}s,\ -\frac{s+\nu}{2}\\ -\nu\end{matrix}\right]{}_1F_3\left(\begin{matrix}-\frac{s+\nu}{2};\,\frac{a^2b^2}{8}\\ \frac{1}{2},\,\frac{1-s}{2},\,\frac{2-s}{2}\end{matrix}\right)$ $-\frac{b^{-s}}{2^{(s+\nu+1)/2}}\Gamma\left[\begin{matrix}s-1,\ -\frac{s+\nu-1}{2}\\ -\nu\end{matrix}\right]{}_1F_3\left(\begin{matrix}-\frac{s+\nu-1}{2};\,\frac{a^2b^2}{8}\\ \frac{3}{2},\,\frac{2-s}{2},\,\frac{3-s}{2}\end{matrix}\right)$ $-\sqrt{\pi}\,2^{(\nu+1)/2}a^{s+1}b\,\Gamma\left[\begin{matrix}-s-1\\ -\frac{\nu}{2}\end{matrix}\right]{}_1F_3\left(\begin{matrix}\frac{1-\nu}{2};\,\frac{a^2b^2}{8}\\ \frac{3}{2},\,\frac{s+2}{2},\,\frac{s+3}{2}\end{matrix}\right)$ $+\sqrt{\pi}\,2^{\nu/2}a^s\,\Gamma\left[\begin{matrix}-s\\ \frac{1-\nu}{2}\end{matrix}\right]{}_1F_3\left(\begin{matrix}-\frac{\nu}{2};\,\frac{a^2b^2}{8}\\ \frac{1}{2},\,\frac{s+1}{2},\,\frac{s+2}{2}\end{matrix}\right)$ $[\mathrm{Re}\,a>0;\ \mathrm{Re}\,(s+\nu)<0;\ \lvert\arg b\rvert<3\pi/4]$

No.	$f(x)$	$F(s)$
10	$e^{-a/x-b^2x^2/4}D_\nu(bx)$	$\frac{\sqrt{\pi}\,b^{-s}}{2^{(s-\nu)/2}}\Gamma\left[\begin{matrix}s\\ \frac{s-\nu+1}{2}\end{matrix}\right]{}_1F_3\left(\begin{matrix}-\frac{s-\nu-1}{2};\,-\frac{a^2b^2}{8}\\ \frac{1}{2},\,\frac{1-s}{2},\,\frac{2-s}{2}\end{matrix}\right)$ $-\frac{\sqrt{\pi}\,ab^{1-s}}{2^{(s-\nu-1)/2}}\Gamma\left[\begin{matrix}s-1\\ \frac{s-\nu}{2}\end{matrix}\right]{}_1F_3\left(\begin{matrix}-\frac{s-\nu-2}{2};\,-\frac{a^2b^2}{8}\\ \frac{3}{2},\,\frac{2-s}{2},\,\frac{3-s}{2}\end{matrix}\right)$ $-\sqrt{\pi}\,2^{(\nu+1)/2}a^{s+1}b\,\Gamma\left[\begin{matrix}-s-1\\ -\frac{\nu}{2}\end{matrix}\right]{}_1F_3\left(\begin{matrix}\frac{\nu+2}{2};\,-\frac{a^2b^2}{8}\\ \frac{3}{2},\,\frac{s+2}{2},\,\frac{s+3}{2}\end{matrix}\right)$ $+\sqrt{\pi}\,2^{\nu/2}a^s\,\Gamma\left[\begin{matrix}-s\\ \frac{1-\nu}{2}\end{matrix}\right]{}_1F_3\left(\begin{matrix}\frac{\nu+1}{2};\,-\frac{a^2b^2}{8}\\ \frac{1}{2},\,\frac{s+1}{2},\,\frac{s+2}{2}\end{matrix}\right)$ $[\operatorname{Re}a>0;\ \lvert\arg b\rvert<\pi/4]$
11	$e^{-a/x^2\pm b^2x^2/4}D_\nu(bx)$	$2^{\nu/2-1}\sqrt{\pi}\,a^{s/2}\,\Gamma\left[\begin{matrix}-\frac{s}{2}\\ \frac{1-\nu}{2}\end{matrix}\right]{}_1F_2\left(\begin{matrix}\frac{\mp 2\nu+1\mp 1}{4}\\ \frac{1}{2},\,\frac{s+2}{2};\,\mp\frac{ab^2}{2}\end{matrix}\right)$ $-2^{(\nu-1)/2}\sqrt{\pi}\,a^{(s+1)/2}b\,\Gamma\left[\begin{matrix}-\frac{s+1}{2}\\ -\frac{\nu}{2}\end{matrix}\right]$ $\times{}_1F_2\left(\begin{matrix}\mp\frac{\nu}{2}+\frac{3\mp 1}{4}\\ \frac{3}{2},\,\frac{s+3}{2};\,\mp\frac{ab^2}{2}\end{matrix}\right)+\frac{2^{-(s\pm\nu)/2}\sqrt{\pi}\,b^{-s}}{\left[2\sqrt{\pi}\,\Gamma(-\nu)\right]^{(1\pm 1)/2}}$ $\times\Gamma(s)\,\Gamma^{\pm 1}\left(\frac{\mp 2s-2\nu+1\mp 1}{4}\right){}_1F_2\left(\begin{matrix}\frac{\mp 2\nu-2s+1\mp 1}{4}\\ \frac{1-s}{2},\,\frac{2-s}{2};\,\mp\frac{ab^2}{2}\end{matrix}\right)$ $\left[\begin{Bmatrix}\operatorname{Re}a>0;\ \operatorname{Re}(s+\nu)<0\\ \operatorname{Re}a>0\end{Bmatrix};\ \lvert\arg b\rvert<(2\pm 1)\,\pi/4\right]$
12	$e^{\pm b^2x^2/4}\sin(ax)\,D_\nu(bx)$	$\frac{2^{(\nu-s-1)/2}a\sqrt{\pi}}{b^{s+1}}\left[2^{\nu+1}\sqrt{\pi}\,\Gamma(-\nu)\right]^{-(1\pm 1)/2}\Gamma(s+1)$ $\times\Gamma^{\pm 1}\left(\frac{-2\nu\mp 2s\mp 3+1}{4}\right){}_2F_2\left(\begin{matrix}\frac{s+1}{2},\,\frac{s+2}{2};\,\pm\frac{a^2}{2b^2}\\ \frac{3}{2},\,\frac{2s\pm 2\nu+5\pm 1}{4}\end{matrix}\right)$ $+\frac{1\pm 1}{2a^{s+\nu}}b^\nu\sin\frac{(s+\nu)\pi}{2}\Gamma(s+\nu)\,{}_2F_2\left(\begin{matrix}-\frac{\nu}{2},\,\frac{1-\nu}{2};\,\frac{a^2}{2b^2}\\ \frac{1-s-\nu}{2},\,\frac{2-\nu-s}{2}\end{matrix}\right)$ $\left[a>0;\ \begin{Bmatrix}-1<\operatorname{Re}s<1-\operatorname{Re}\nu\\ \operatorname{Re}s>-1\end{Bmatrix};\ \lvert\arg b\rvert<(2\pm 1)\,\pi/4\right]$
13	$e^{\pm b^2x^2/4}\cos(ax)$ $\times D_\nu(bx)$	$\frac{2^{(\nu-s)/2}\sqrt{\pi}}{b^s}\left[2^{\nu+1}\sqrt{\pi}\,\Gamma(-\nu)\right]^{-(1\pm 1)/2}\Gamma(s)$ $\times\Gamma^{\pm 1}\left(\frac{-2\nu\mp 2s\mp 1+1}{4}\right){}_2F_2\left(\begin{matrix}\frac{s}{2},\,\frac{s+1}{2};\,\pm\frac{a^2}{2b^2}\\ \frac{1}{2},\,\frac{2s\pm 2\nu+3\pm 1}{4}\end{matrix}\right)$ $+\frac{1\pm 1}{2a^{s+\nu}}b^\nu\cos\frac{(s+\nu)\pi}{2}\Gamma(s+\nu)\,{}_2F_2\left(\begin{matrix}-\frac{\nu}{2},\,\frac{1-\nu}{2};\,\frac{a^2}{2b^2}\\ \frac{1-s-\nu}{2},\,\frac{2-\nu-s}{2}\end{matrix}\right)$ $\left[a>0;\ \begin{Bmatrix}0<\operatorname{Re}s<1-\operatorname{Re}\nu\\ \operatorname{Re}s>0\end{Bmatrix};\ \lvert\arg b\rvert<(2\pm 1)\,\pi/4\right]$

No.	$f(x)$	$F(s)$
14	$e^{b^2x^2/4}\sin\left(ax^2\right)D_\nu(bx)$	$\frac{\pi a}{2^{(s+\nu+4)/2}b^{s+2}}\csc\frac{(s+\nu)\pi}{2}\,\Gamma\left[\begin{matrix}s+2\\ -\nu,\ \frac{s+\nu+4}{2}\end{matrix}\right]$ $\times\,{}_4F_3\left(\begin{matrix}\frac{s+2}{4},\ \frac{s+3}{4},\ \frac{s+4}{4},\ \frac{s+5}{4}\\ \frac{3}{2},\ \frac{s+\nu+4}{4},\ \frac{s+\nu+6}{4};\ -\frac{4a^2}{b^4}\end{matrix}\right)+\frac{(\nu-1)\,\nu\,b^{\nu-2}}{4a^{(s+\nu-2)/2}}\cos\frac{(s+\nu)\pi}{4}$ $\times\,\Gamma\left(\frac{s+\nu-2}{2}\right){}_4F_3\left(\begin{matrix}\frac{2-\nu}{4},\ \frac{3-\nu}{4},\ \frac{4-\nu}{4},\ \frac{5-\nu}{4}\\ \frac{3}{2},\ -\frac{s+\nu-4}{4},\ -\frac{s+\nu-6}{4};\ -\frac{4a^2}{b^4}\end{matrix}\right)$ $+\,\frac{\pi a^{-(s+\nu)/2}b^\nu}{4\,\Gamma\left(-\frac{s+\nu-2}{2}\right)}\sec\frac{(s+\nu)\pi}{4}\,{}_4F_3\left(\begin{matrix}-\frac{\nu}{4},\ \frac{1-\nu}{4},\ \frac{2-\nu}{4},\ \frac{3-\nu}{4}\\ \frac{1}{2},\ -\frac{s+\nu-2}{4},\ -\frac{s+\nu-4}{4};\ -\frac{4a^2}{b^4}\end{matrix}\right)$ $[a>0;\ -2<\operatorname{Re}s<2-\operatorname{Re}\nu;\ \lvert\arg b\rvert<3\pi/4]$
15	$e^{b^2x^2/4}\cos\left(ax^2\right)D_\nu(bx)$	$-\frac{\pi}{2^{(s+\nu+2)/2}b^{s}}\csc\frac{(s+\nu)\pi}{2}\,\Gamma\left[\begin{matrix}s\\ -\nu,\ \frac{s+\nu+2}{2}\end{matrix}\right]$ $\times\,{}_4F_3\left(\begin{matrix}\frac{s}{4},\ \frac{s+1}{4},\ \frac{s+2}{4},\ \frac{s+3}{4}\\ \frac{1}{2},\ \frac{s+\nu+2}{4},\ \frac{s+\nu+4}{4};\ -\frac{4a^2}{b^4}\end{matrix}\right)+\frac{(\nu-1)\,\nu\pi b^{\nu-2}}{8a^{(s+\nu-2)/2}\,\Gamma\left(-\frac{s+\nu-4}{2}\right)}$ $\times\sec\frac{(s+\nu)\pi}{4}\,{}_4F_3\left(\begin{matrix}\frac{2-\nu}{4},\ \frac{3-\nu}{4},\ \frac{4-\nu}{4},\ \frac{5-\nu}{4}\\ \frac{3}{2},\ -\frac{s+\nu-4}{4},\ -\frac{s+\nu-6}{4};\ -\frac{4a^2}{b^4}\end{matrix}\right)$ $+\,\frac{b^\nu}{2a^{(s+\nu)/2}}\cos\frac{(s+\nu)\pi}{4}\,\Gamma\left(\frac{s+\nu}{2}\right)$ $\times\,{}_4F_3\left(\begin{matrix}-\frac{\nu}{4},\ \frac{1-\nu}{4},\ \frac{2-\nu}{4},\ \frac{3-\nu}{4}\\ \frac{1}{2},\ -\frac{s+\nu-2}{4},\ -\frac{s+\nu-4}{4};\ -\frac{4a^2}{b^4}\end{matrix}\right)$ $[a>0;\ 0<\operatorname{Re}s<2-\operatorname{Re}\nu;\ \lvert\arg b\rvert<3\pi/4]$
16	$e^{-b^2x^2/4}\left\{\begin{matrix}\sin\left(ax^2\right)\\ \cos\left(ax^2\right)\end{matrix}\right\}$ $\times\,D_\nu(bx)$	$\frac{\sqrt{\pi}\,a^\delta b^{-s-2\delta}}{2^{(s-\nu+2\delta)/2}}\,\Gamma\left[\begin{matrix}s+2\delta\\ \frac{s-\nu+2\delta+1}{2}\end{matrix}\right]{}_4F_3\left(\begin{matrix}\frac{s+2\delta}{4},\ \frac{s+2\delta+1}{4},\ \frac{s+2\delta+2}{4},\ \frac{s+2\delta+3}{4}\\ \frac{2\delta+1}{2},\ \frac{s-\nu+2\delta+1}{4},\ \frac{s-\nu+2\delta+3}{4};\ -\frac{4a^2}{b^4}\end{matrix}\right)$ $[a>0;\ \operatorname{Re}s>-2\delta;\ \lvert\arg b\rvert<\pi/4]$

3.9.2. $D_\nu(bx)$ and $\operatorname{erf}(ax)$, $\operatorname{erfc}(ax)$

1	$e^{\pm b^2x^2/4}\operatorname{erf}(ax)\,D_\nu(bx)$	$\frac{2^{(\nu-s+1)/2}a}{b^{s+1}\left[2^{\nu+1}\sqrt{\pi}\,\Gamma(-\nu)\right]^{(1\pm1)/2}}\,\Gamma(s+1)$ $\times\,\Gamma^{\pm1}\left(\frac{1\mp2s-2\nu\mp3}{4}\right){}_3F_2\left(\begin{matrix}\frac{1}{2},\ \frac{s+1}{2},\ \frac{s+2}{2};\ \pm\frac{2a^2}{b^2}\\ \frac{3}{2},\ \frac{2s\pm2\nu+5\pm1}{4}\end{matrix}\right)$ $-\,\frac{(1\pm1)\,a^{-s-\nu}b^\nu}{2\sqrt{\pi}\,(s+\nu)}\,\Gamma\left(\frac{s+\nu+1}{2}\right){}_3F_2\left(\begin{matrix}-\frac{\nu}{2},\ \frac{1-\nu}{2},\ -\frac{s+\nu}{2}\\ \frac{2-s-\nu}{2},\ \frac{1-s-\nu}{2};\ \frac{2a^2}{b^2}\end{matrix}\right)$ $\left[\operatorname{Re}a>0;\ \left\{\begin{matrix}-1<\operatorname{Re}s<-\operatorname{Re}\nu\\ \operatorname{Re}s>-1\end{matrix}\right\};\ \lvert\arg b\rvert<(2\pm1)\,\pi/4\right]$

No.	$f(x)$	$F(s)$
2	$e^{\pm b^2x^2/4}\operatorname{erfc}(ax)$ $\times D_\nu(bx)$	$\frac{2^{\nu/2}a^{-s}}{s}\Gamma\left[\begin{matrix}\frac{s+1}{2}\\ \frac{1-\nu}{2}\end{matrix}\right]{}_3F_2\left(\begin{matrix}\frac{1\mp 1\mp 2\nu}{4},\ \frac{s}{2},\ \frac{s+1}{2}\\ \frac{1}{2},\ \frac{s+2}{2};\ \pm\frac{b^2}{2a^2}\end{matrix}\right)$ $-\frac{2^{(\nu+1)/2}a^{-s-1}b}{s+1}\Gamma\left[\begin{matrix}\frac{s+2}{2}\\ -\frac{\nu}{2}\end{matrix}\right]{}_3F_2\left(\begin{matrix}\frac{3\mp 1\mp 2\nu}{4},\ \frac{s+1}{2},\ \frac{s+2}{2}\\ \frac{3}{2},\ \frac{s+3}{2};\ \pm\frac{b^2}{2a^2}\end{matrix}\right)$ $[\operatorname{Re}s>0;\ \lvert\arg a\rvert<\pi/4;\ \lvert\arg b\rvert<(2\pm 1)\pi/4]$
3	$e^{-b^2x^2/4}\operatorname{erfc}(ax)$ $\times[D_\nu(-bx)-D_\nu(bx)]$	$-\frac{2^{(\nu+3)/2}b}{\pi a^{s+1}(s+1)}\sin\frac{\nu\pi}{2}\Gamma\left(\frac{\nu+2}{2}\right)\Gamma\left(\frac{s+2}{2}\right){}_3F_2\left(\begin{matrix}\frac{\nu+2}{2},\ \frac{s+1}{2},\ \frac{s+2}{2}\\ \frac{3}{2},\ \frac{s+3}{2};\ -\frac{b^2}{2a^2}\end{matrix}\right)$ $[\operatorname{Re}s,\ \operatorname{Re}(2a^2+b^2)>0;\ \lvert\arg a\rvert<\pi/4]$

3.9.3. Products of $D_\mu(bx^r)$

1	$D_{-\nu-1}(ax)\,D_\nu(ax)$	$\frac{\pi}{2^{s+1/2}a^s}\Gamma\left[\begin{matrix}s\\ \frac{s-2\nu+2}{4},\ \frac{s+2\nu+4}{4}\end{matrix}\right]$ $[\operatorname{Re}s>0;\ \lvert\arg a\rvert<\pi/4]$
2	$D_\nu\left(e^{\pi i/4}ax\right)D_\nu\left(\frac{ax}{e^{\pi i/4}}\right)$	$\frac{\sqrt{\pi}}{2^{s+1}a^s}\Gamma\left[\begin{matrix}s,\ -\frac{s+2\nu}{4}\\ -\nu,\ \frac{s-2\nu+2}{4}\end{matrix}\right]$ $[\operatorname{Re}a>0;\ 0<\operatorname{Re}s<-2\operatorname{Re}\nu]$
3	$e^{(a^2+b^2)x^2/4}D_\mu(ax)$ $\times D_\nu(bx)$	$\frac{\sqrt{\pi}}{2^{(s+\mu-\nu)/2+1}a^s}\Gamma\left[\begin{matrix}s,\ -\frac{s+\mu}{2}\\ -\mu,\ \frac{1-\nu}{2}\end{matrix}\right]{}_3F_2\left(\begin{matrix}-\frac{\nu}{2},\ \frac{s}{2},\ \frac{s+1}{2}\\ \frac{1}{2},\ \frac{s+\mu+2}{2};\ -\frac{b^2}{a^2}\end{matrix}\right)$ $-\frac{\sqrt{\pi}\,b}{2^{(s+\mu-\nu)/2+1}a^{s+1}}\Gamma\left[\begin{matrix}s+1,\ -\frac{s+\mu+1}{2}\\ -\mu,\ -\frac{\nu}{2}\end{matrix}\right]{}_3F_2\left(\begin{matrix}\frac{1-\nu}{2},\ \frac{s+1}{2},\ \frac{s+2}{2}\\ \frac{3}{2},\ \frac{s+\mu+3}{2};\ -\frac{b^2}{a^2}\end{matrix}\right)$ $+\frac{a^\mu}{2^{(s+\mu+\nu)/2+1}b^{s+\mu}}\Gamma\left[\begin{matrix}s+\mu,\ -\frac{s+\mu+\nu}{2}\\ -\nu\end{matrix}\right]{}_3F_2\left(\begin{matrix}-\frac{\mu}{2},\ \frac{1-\mu}{2},\ -\frac{s+\mu+\nu}{2}\\ \frac{1-s-\mu}{2},\ \frac{2-s-\mu}{2};\ -\frac{b^2}{a^2}\end{matrix}\right)$ $[0<\operatorname{Re}s<-\operatorname{Re}(\mu+\nu);\ \lvert\arg a\rvert,\ \lvert\arg b\rvert<3\pi/4]$
4	$e^{(a^2-b^2)x^2/4}D_\mu(ax)$ $\times D_\nu(bx)$	$\frac{\pi}{2^{(s-\mu-\nu)/2}b^s}\Gamma\left[\begin{matrix}s\\ \frac{1-\mu}{2},\ \frac{s-\nu+1}{2}\end{matrix}\right]{}_3F_2\left(\begin{matrix}-\frac{\mu}{2},\ \frac{s}{2},\ \frac{s+1}{2}\\ \frac{1}{2},\ \frac{s-\nu+1}{2};\ \frac{a^2}{b^2}\end{matrix}\right)$ $-\frac{\pi\,a}{2^{(s-\mu-\nu)/2}b^{s+1}}\Gamma\left[\begin{matrix}s+1\\ -\frac{\mu}{2},\ \frac{s-\nu+2}{2}\end{matrix}\right]{}_3F_2\left(\begin{matrix}\frac{1-\mu}{2},\ \frac{s+1}{2},\ \frac{s+2}{2}\\ \frac{3}{2},\ \frac{s-\nu+2}{2};\ \frac{a^2}{b^2}\end{matrix}\right)$ $[\operatorname{Re}s>0;\ \lvert\arg a\rvert<3\pi/4,\ \lvert\arg b\rvert<\pi/4]$
5	$e^{-(a^2+b^2)x^2/4}D_\mu(ax)$ $\times D_\nu(bx)$	$\frac{\pi}{2^{(s-\mu-\nu)/2}a^s}\Gamma\left[\begin{matrix}s\\ \frac{1-\nu}{2},\ \frac{s-\mu+1}{2}\end{matrix}\right]{}_3F_2\left(\begin{matrix}\frac{\nu+1}{2},\ \frac{s}{2},\ \frac{s+1}{2}\\ \frac{1}{2},\ \frac{s-\mu+1}{2};\ -\frac{b^2}{a^2}\end{matrix}\right)$ $-\frac{\pi b}{2^{(s-\mu-\nu)/2}a^{s+1}}\Gamma\left[\begin{matrix}s+1\\ -\frac{\nu}{2},\ \frac{s-\mu+2}{2}\end{matrix}\right]{}_3F_2\left(\begin{matrix}\frac{\nu+2}{2},\ \frac{s+1}{2},\ \frac{s+2}{2}\\ \frac{3}{2},\ \frac{s-\mu+2}{2};\ -\frac{b^2}{a^2}\end{matrix}\right)$ $[\operatorname{Re}s>0;\ \lvert\arg a\rvert,\ \lvert\arg b\rvert<\pi/4]$

3.10. The Bessel Function $J_\nu(z)$

More formulas can be obtained from the corresponding sections due to the relations

$$J_{\pm 1/2}(z) = \sqrt{\frac{2}{\pi}}\frac{1}{\sqrt{z}}\begin{Bmatrix}\sin z\\ \cos z\end{Bmatrix}, \quad J_{\pm 3/2}(z) = \sqrt{\frac{2}{\pi}}\frac{1}{z^{3/2}}\left[\pm\begin{Bmatrix}\sin z\\ \cos z\end{Bmatrix} - z\begin{Bmatrix}\cos z\\ \sin z\end{Bmatrix}\right];$$

$$J_\nu(z) = \frac{1}{\sin(\nu\pi)}\left[Y_{-\nu}(z) - Y_\nu(z)\cos(\nu\pi)\right], \quad [\nu \neq 0,\ \pm 1,\ \pm 2, \ldots];$$

$$J_\nu(z) = \frac{1}{2}\left[H_\nu^{(1)}(z) + H_\nu^{(2)}(z)\right], \quad J_\nu(z) = \frac{z^\nu}{(iz)^\nu} I_\nu(iz),$$

$$J_\nu(z) = \frac{(z/2)^\nu}{\Gamma(\nu+1)}\, {}_0F_1\left(\nu+1;\ -\frac{z^2}{4}\right), \quad J_\nu(z) = z^\nu \left(z^2\right)^{-\nu/2} G_{02}^{10}\left(\frac{z^2}{4}\,\middle|\, \begin{matrix}\cdot\\ \nu/2,\ -\nu/2\end{matrix}\right),$$

$$J_\nu(z) = \pi\left(\frac{z}{2}\right)^\nu G_{13}^{10}\left(-\frac{z^2}{4}\,\middle|\, \begin{matrix}1/2\\ 0,\ -\nu,\ 1/2\end{matrix}\right),$$

$$J_\nu(z) = \pi z^\nu \left(-z^2\right)^{-\nu/2} G_{13}^{10}\left(-\frac{z^2}{4}\,\middle|\, \begin{matrix}(\nu+1)/2\\ \nu/2,\ -\nu/2,\ (\nu+1)/2\end{matrix}\right).$$

3.10.1. $J_\nu(bx)$ and algebraic functions

No.	$f(x)$	$F(s)$
1	$1 - J_0(ax)$	$-\frac{2^{s-1}}{a^s}\Gamma\begin{bmatrix}\frac{s+2}{2}\\ \frac{2-s}{2}\end{bmatrix}$ $[a > 0;\ -2 < \operatorname{Re} s < 0]$
2	$J_\nu(ax)$	$\frac{2^{s-1}}{a^s}\Gamma\begin{bmatrix}\frac{s+\nu}{2}\\ \frac{2-s+\nu}{2}\end{bmatrix}$ $[a > 0;\ -\operatorname{Re}\nu < \operatorname{Re} s < 3/2]$
3	$J_\nu(ax) - \frac{2^{-\nu}(ax)^\nu}{\Gamma(\nu+1)}$	$-\frac{2^{s-1}}{a^s}\Gamma\begin{bmatrix}-\frac{s+\nu}{2},\ \frac{s+\nu+2}{2}\\ \frac{2-s-\nu}{2},\ \frac{2-s+\nu}{2}\end{bmatrix}$ $[a > 0;\ -\operatorname{Re}\nu - 2 < \operatorname{Re} s < 3/2,\ -\operatorname{Re}\nu]$
4	$J_\nu(ax) \pm J_{-\nu}(ax)$	$\pm\frac{1}{\pi}\left(\frac{2}{a}\right)^s \begin{Bmatrix}\cos(\nu\pi/2)\sin(s\pi/2)\\ \sin(\nu\pi/2)\cos(s\pi/2)\end{Bmatrix}\Gamma\left(\frac{s-\nu}{2}\right)\Gamma\left(\frac{s+\nu}{2}\right)$ $[a > 0;\ \lvert\operatorname{Re}\nu\rvert < \operatorname{Re} s < 3/2]$
5	$(a-x)_+^{\alpha-1} J_\nu(bx)$	$a^{s+\alpha+\nu-1}\left(\frac{b}{2}\right)^\nu \Gamma\begin{bmatrix}\alpha,\ s+\nu\\ \nu+1,\ s+\alpha+\nu\end{bmatrix}$ $\times\, {}_2F_3\left(\begin{matrix}\frac{s+\nu}{2},\ \frac{s+\nu+1}{2};\ -\frac{ab^2}{4}\\ \nu+1,\ \frac{s+\alpha+\nu}{2},\ \frac{s+\alpha+\nu+1}{2}\end{matrix}\right)$ $[a,\ \operatorname{Re}\alpha,\ \operatorname{Re}(s+\nu) > 0]$
6	$\left(a^2 - x^2\right)_+^{\alpha-1} J_\nu(bx)$	$\frac{a^{s+2\alpha+\nu-2} b^\nu}{2^{\nu+1}}\Gamma\begin{bmatrix}\alpha,\ \frac{s+\nu}{2}\\ \nu+1,\ \frac{s+2\alpha+\nu}{2}\end{bmatrix} {}_1F_2\left(\begin{matrix}\frac{s+\nu}{2};\ -\frac{a^2b^2}{4}\\ \nu+1,\ \frac{s+2\alpha+\nu}{2}\end{matrix}\right)$ $[a,\ \operatorname{Re}\alpha,\ \operatorname{Re}(s+\nu) > 0]$

No.	$f(x)$	$F(s)$
7	$\frac{1}{(x+a)^\rho} J_\nu(bx)$	$a^{s+\nu-\rho}\left(\frac{b}{2}\right)^\nu \Gamma\left[\begin{matrix} s+\nu,\, \rho-\nu-s \\ \nu+1,\, \rho \end{matrix}\right] {}_2F_3\left(\begin{matrix} \frac{s+\nu}{2},\, \frac{s+\nu+1}{2};\, -\frac{a^2b^2}{4} \\ \nu+1,\, \frac{s+\nu-\rho+1}{2},\, \frac{s+\nu-\rho+2}{2} \end{matrix}\right)$ $+\frac{2^{s-\rho-1}}{b^{s-\rho}} \Gamma\left[\begin{matrix} \frac{s-\rho+\nu}{2} \\ \frac{2-s+\nu+\rho}{2} \end{matrix}\right] {}_2F_3\left(\begin{matrix} \frac{\rho}{2},\, \frac{\rho+1}{2};\, -\frac{a^2b^2}{4} \\ \frac{1}{2},\, \frac{2-s-\nu+\rho}{2},\, \frac{2-s+\nu+\rho}{2} \end{matrix}\right)$ $-\frac{\rho\, a 2^{s-\rho-2}}{b^{s-\rho-1}} \Gamma\left[\begin{matrix} \frac{s+\nu-\rho-1}{2} \\ \frac{3-s+\nu+\rho}{2} \end{matrix}\right] {}_2F_3\left(\begin{matrix} \frac{\rho+1}{2},\, \frac{\rho+2}{2};\, -\frac{a^2b^2}{4} \\ \frac{3}{2},\, \frac{3-s-\nu+\rho}{2},\, \frac{3-s+\nu+\rho}{2} \end{matrix}\right)$ $[b>0;\, -\operatorname{Re}\nu < \operatorname{Re}s < \operatorname{Re}\rho + 3/2;\, \lvert\arg a\rvert < \pi]$
8	$\frac{1}{x+a} J_\nu(bx)$	$2^{s-2}b^{-s+1} \Gamma\left[\begin{matrix} \frac{s+\nu-1}{2} \\ -\frac{s-\nu-3}{2} \end{matrix}\right] {}_1F_2\left(\begin{matrix} 1;\, -\frac{a^2b^2}{4} \\ -\frac{s-\nu-3}{2},\, -\frac{s+\nu-3}{2} \end{matrix}\right)$ $-2^{s-3}ab^{2-s} \Gamma\left[\begin{matrix} \frac{s+\nu-2}{2} \\ -\frac{s-\nu-4}{2} \end{matrix}\right] {}_1F_2\left(\begin{matrix} 1;\, -\frac{a^2b^2}{4} \\ -\frac{s-\nu-4}{2},\, -\frac{s+\nu-4}{2} \end{matrix}\right)$ $+\pi a^{s-1} \csc[(s+\nu)\pi]\, J_\nu(ab)$ $[b>0;\, -\operatorname{Re}\nu < \operatorname{Re}s < 5/2;\, \lvert\arg a\rvert < \pi]$
9	$\frac{1}{x-a} J_\nu(bx)$	$\frac{2^{s-2}}{b^{s-1}} \Gamma\left[\begin{matrix} \frac{s+\nu-1}{2} \\ \frac{3-s+\nu}{2} \end{matrix}\right] {}_1F_2\left(\begin{matrix} 1;\, -\frac{a^2b^2}{4} \\ \frac{3-s-\nu}{2},\, \frac{3-s+\nu}{2} \end{matrix}\right)$ $+\frac{2^{s-3}a}{b^{s-2}} \Gamma\left[\begin{matrix} \frac{s+\nu-2}{2} \\ \frac{4-s+\nu}{2} \end{matrix}\right] {}_1F_2\left(\begin{matrix} 1;\, -\frac{a^2b^2}{4} \\ \frac{4-s-\nu}{2},\, \frac{4-s+\nu}{2} \end{matrix}\right)$ $-\pi a^{s-1} \cot[(s+\nu)\pi]\, J_\nu(ab)$ $[a,\, b>0;\, -\operatorname{Re}\nu < \operatorname{Re}s < 5/2]$
10	$\frac{1}{(x^2+a^2)^\rho} J_\nu(bx)$	$\frac{a^{s+\nu-2\rho}b^\nu}{2^{\nu+1}} \Gamma\left[\begin{matrix} \frac{s+\nu}{2},\, \frac{2\rho-\nu-s}{2} \\ \nu+1,\, \rho \end{matrix}\right] {}_1F_2\left(\begin{matrix} \frac{s+\nu}{2};\, \frac{a^2b^2}{4} \\ \nu+1,\, \frac{s+\nu-2\rho+2}{2} \end{matrix}\right)$ $+\frac{2^{s-2\rho-1}}{b^{s-2\rho}} \Gamma\left[\begin{matrix} \frac{s+\nu-2\rho}{2} \\ \frac{2-s+\nu+2\rho}{2} \end{matrix}\right] {}_1F_2\left(\begin{matrix} \rho;\, \frac{a^2b^2}{4} \\ \frac{2-s-\nu+2\rho}{2},\, \frac{2-s+\nu+2\rho}{2} \end{matrix}\right)$ $[\operatorname{Re}a,\, b>0;\, -\operatorname{Re}\nu < \operatorname{Re}s < 2\operatorname{Re}\rho + 3/2]$
11	$\frac{1}{x^2+a^2} J_\nu(bx)$	$2^{s-3}b^{2-s} \Gamma\left[\begin{matrix} \frac{s+\nu-2}{2} \\ \frac{4-s+\nu}{2} \end{matrix}\right] {}_1F_2\left(\begin{matrix} 1;\, \frac{a^2b^2}{4} \\ \frac{4-s-\nu}{2},\, \frac{4-s+\nu}{2} \end{matrix}\right)$ $+\frac{\pi a^{s-2}}{2} \csc\frac{(s+\nu)\pi}{2}\, I_\nu(ab)$ $[\operatorname{Re}a,\, b>0;\, -\operatorname{Re}\nu < \operatorname{Re}s < 7/2]$
12	$\frac{1}{x^2-a^2} J_\nu(bx)$	$\frac{2^{s-3}}{b^{s-2}} \Gamma\left[\begin{matrix} \frac{s+\nu-2}{2} \\ \frac{4-s+\nu}{2} \end{matrix}\right] {}_1F_2\left(\begin{matrix} 1;\, -\frac{a^2c^2}{4} \\ \frac{4-s-\nu}{2},\, \frac{4-s+\nu}{2} \end{matrix}\right) - \frac{\pi a^{s-2}}{2} \cot\frac{(s+\nu)\pi}{2}\, J_\nu(ab)$ $[a,\, b>0;\, -\operatorname{Re}\nu < \operatorname{Re}s < 7/2]$

No.	$f(x)$	$F(s)$
13	$\frac{1}{(x^4+a^4)^\rho} J_\nu(bx)$	$\frac{2^{s-4\rho-1}}{b^{s-4\rho}} \Gamma\begin{bmatrix} \frac{s+\nu-4\rho}{2} \\ \frac{2-s+\nu+4\rho}{2} \end{bmatrix} \times {}_1F_4\left(\begin{matrix} \rho;\, -\frac{a^4b^4}{256} \\ \frac{4-s-\nu+4\rho}{4},\, -\frac{s+\nu-4\rho-1}{4},\, \frac{2-s+\nu+4\rho}{4},\, \frac{4-s+\nu+4\rho}{4} \end{matrix}\right)$ $- \frac{a^{s+\nu-4\rho+2} b^{\nu+2}}{2^{\nu+4}} \Gamma\begin{bmatrix} \frac{s+\nu+2}{4},\, \rho-\frac{s+\nu+2}{4} \\ \nu+2,\, \rho \end{bmatrix} \times {}_1F_4\left(\begin{matrix} \frac{s+\nu+2}{4};\, -\frac{a^4b^4}{256} \\ \frac{3}{2},\, \frac{\nu+2}{2},\, \frac{\nu+3}{2},\, \frac{s+\nu-4\rho+3}{4} \end{matrix}\right)$ $+ \frac{a^{s+\nu-4\rho} b^{\nu}}{2^{\nu+2}} \Gamma\begin{bmatrix} \frac{s+\nu}{4},\, \rho-\frac{s+\nu}{4} \\ \nu+1,\, \rho \end{bmatrix} {}_1F_4\left(\begin{matrix} \frac{s+\nu}{4};\, -\frac{a^4b^4}{256} \\ \frac{1}{2},\, \frac{\nu+1}{2},\, \frac{\nu+2}{2},\, \frac{s+\nu-4\rho+4}{4} \end{matrix}\right)$ $[b>0;\ -\operatorname{Re}\nu < \operatorname{Re}s < 4\operatorname{Re}\rho + 3/2;\ \lvert\arg a\rvert < \pi/4]$
14	$\frac{1}{x^4-a^4} J_\nu(bx)$	$\frac{2^{s-5}}{b^{s-4}} \Gamma\begin{bmatrix} \frac{s+\nu-4}{2} \\ \frac{6-s+\nu}{2} \end{bmatrix} {}_1F_4\left(\begin{matrix} 1;\, \frac{a^4b^4}{256} \\ \frac{8-s-\nu}{4},\, \frac{8-s+\nu}{4},\, \frac{6-s-\nu}{4},\, \frac{6-s+\nu}{4} \end{matrix}\right)$ $- \frac{\pi a^{s+\nu-2} b^{\nu+2}}{2^{\nu+4}\Gamma(\nu+2)} \tan\frac{(s+\nu)\pi}{4}\, {}_0F_3\left(\begin{matrix} \frac{a^4b^4}{256} \\ \frac{3}{2},\, \frac{\nu+2}{2},\, \frac{\nu+3}{2} \end{matrix}\right)$ $- \frac{\pi a^{s+\nu-4} b^{\nu}}{2^{\nu+2}\Gamma(\nu+1)} \cot\frac{(s+\nu)\pi}{4}\, {}_0F_3\left(\begin{matrix} \frac{a^4b^4}{256} \\ \frac{1}{2},\, \frac{\nu+1}{2},\, \frac{\nu+2}{2} \end{matrix}\right)$ $[a,\ b>0;\ -\operatorname{Re}\nu < \operatorname{Re}s < 11/2]$
15	$\left(\sqrt{x^2+a^2}+a\right)^\rho J_\nu(bx)$	$\frac{2^{s+\rho-1}}{b^{s+\rho}} \Gamma\begin{bmatrix} \frac{s+\rho+\nu}{2} \\ \frac{2-s+\nu-\rho}{2} \end{bmatrix} {}_2F_3\left(\begin{matrix} -\frac{\rho}{2},\, \frac{\rho}{2};\, \frac{a^2b^2}{4} \\ \frac{1}{2},\, \frac{2-s-\nu-\rho}{2},\, \frac{2-s+\nu-\rho}{2} \end{matrix}\right)$ $+ \frac{2^{s+\rho-2}\rho a}{b^{s+\rho-1}} \Gamma\begin{bmatrix} \frac{s+\rho+\nu-1}{2} \\ \frac{3-s+\nu-\rho}{2} \end{bmatrix} {}_2F_3\left(\begin{matrix} \frac{1-\rho}{2},\, \frac{1+\rho}{2};\, \frac{a^2b^2}{4} \\ \frac{3}{2},\, \frac{3-s-\nu-\rho}{2},\, \frac{3-s+\nu-\rho}{2} \end{matrix}\right)$ $- 2^{s+\rho-1}\rho a^{s+\rho+\nu} b^{\nu} \Gamma\begin{bmatrix} -s-\rho-\nu,\, \frac{s+\nu}{2} \\ \nu+1,\, 1-\frac{s+2\rho+\nu}{2} \end{bmatrix}$ $\times {}_2F_3\left(\begin{matrix} \frac{s+\nu}{2},\, \frac{s+2\rho+\nu}{2};\, \frac{a^2b^2}{4} \\ \nu+1,\, \frac{s+\rho+\nu+1}{2},\, \frac{s+\rho+\nu+2}{2} \end{matrix}\right)$ $[b,\ \operatorname{Re}a>0;\ -\operatorname{Re}\nu < \operatorname{Re}s < -\operatorname{Re}\rho + 3/2]$
16	$\frac{\left(\sqrt{x^2+a^2}+a\right)^\rho}{\sqrt{x^2+a^2}} J_\nu(bx)$	$\frac{2^{s+\rho-2}}{b^{s+\rho-1}} \Gamma\begin{bmatrix} \frac{s+\nu+\rho-1}{2} \\ \frac{3-s+\nu-\rho}{2} \end{bmatrix} {}_2F_3\left(\begin{matrix} \frac{1-\rho}{2},\, \frac{1+\rho}{2};\, \frac{a^2b^2}{4} \\ \frac{1}{2},\, \frac{3-s-\nu-\rho}{2},\, \frac{3-s+\nu-\rho}{2} \end{matrix}\right)$ $+ \frac{2^{s+\rho-3}\rho a}{b^{s+\rho-2}} \Gamma\begin{bmatrix} \frac{s+\nu+\rho-2}{2} \\ \frac{4-s+\nu-\rho}{2} \end{bmatrix} {}_2F_3\left(\begin{matrix} \frac{2-\rho}{2},\, \frac{2+\rho}{2};\, \frac{a^2b^2}{4} \\ \frac{3}{2},\, \frac{4-s-\nu-\rho}{2},\, \frac{4-s+\nu-\rho}{2} \end{matrix}\right)$ $+ 2^{s+\rho-1} a^{s+\rho+\nu-1} b^{\nu} \Gamma\begin{bmatrix} 1-s-\nu-\rho,\, \frac{s+\nu}{2} \\ \nu+1,\, 1-\frac{s+\nu+2\rho}{2} \end{bmatrix}$ $\times {}_2F_3\left(\begin{matrix} \frac{s+\nu}{2},\, \frac{s+2\rho+\nu}{2};\, \frac{a^2b^2}{4} \\ \nu+1,\, \frac{s+\nu+\rho}{2},\, \frac{s+\nu+\rho+1}{2} \end{matrix}\right)$ $[b,\ \operatorname{Re}a>0;\ -\operatorname{Re}\nu < \operatorname{Re}s < 5/2 - \operatorname{Re}\rho]$

No.	$f(x)$	$F(s)$
17	$\left(\sqrt{x^2+a^2}\pm x\right)^\rho J_\nu(bx)$	$\frac{2^{s\pm 2\rho-1}a^{\rho\mp\rho}}{b^{s\pm\rho}}\Gamma\left[\begin{matrix}\frac{s+\nu\pm\rho}{2}\\ \frac{2-s+\nu\mp\rho}{2}\end{matrix}\right] {}_2F_3\left(\begin{matrix}\mp\frac{\rho}{2},\ \frac{1\mp\rho}{2};\ \frac{a^2b^2}{4}\\ 1\mp\rho,\ \frac{2-s-\nu\mp\rho}{2},\ \frac{2-s+\nu\mp\rho}{2}\end{matrix}\right)$ $\mp\frac{\rho a^{s+\rho+\nu}b^\nu}{2^{s+2\nu+1}}\Gamma\left[\begin{matrix}s+\nu,\ -\frac{s+\nu\pm\rho}{2}\\ \nu+1,\ \frac{s+\nu\mp\rho+2}{2}\end{matrix}\right] {}_2F_3\left(\begin{matrix}\frac{s+\nu}{2},\ \frac{s+\nu+1}{2};\ \frac{a^2b^2}{4}\\ \nu+1,\ \frac{s+\nu+\rho+2}{2},\ \frac{s+\nu-\rho+2}{2}\end{matrix}\right)$ $[b,\ \operatorname{Re} a>0;\ -\operatorname{Re}\nu<\operatorname{Re} s<3/2\mp\operatorname{Re}\rho]$
18	$\frac{\left(\sqrt{x^2+a^2}\pm x\right)^\rho}{\sqrt{x^2+a^2}} J_\nu(bx)$	$\frac{2^{s\pm 2\rho-2}a^{\rho\mp\rho}}{b^{s\pm\rho-1}}\Gamma\left[\begin{matrix}\frac{s+\nu\pm\rho-1}{2}\\ \frac{3-s+\nu\mp\rho}{2}\end{matrix}\right] {}_2F_3\left(\begin{matrix}\frac{1\mp\rho}{2},\ \frac{2\mp\rho}{2};\ \frac{a^2b^2}{4}\\ 1\mp\rho,\ \frac{3-s-\nu\mp\rho}{2},\ \frac{3-s+\nu\mp\rho}{2}\end{matrix}\right)$ $+\frac{a^{s+\rho+\nu-1}b^\nu}{2^{s+2\nu}}\Gamma\left[\begin{matrix}s+\nu,\ -\frac{s+\nu\pm\rho-1}{2}\\ \nu+1,\ \frac{s+\nu\mp\rho+1}{2}\end{matrix}\right]$ $\times {}_2F_3\left(\begin{matrix}\frac{s+\nu}{2},\ \frac{s+\nu+1}{2};\ \frac{a^2b^2}{4}\\ \nu+1,\ \frac{s+\nu+\rho+1}{2},\ \frac{s+\nu-\rho+1}{2}\end{matrix}\right)$ $[b,\ \operatorname{Re} a>0;\ -\operatorname{Re}\nu<\operatorname{Re} s<5/2\mp\operatorname{Re}\rho]$

3.10.2. $J_\nu(\varphi(x))$ and algebraic functions

1	$\theta(1-x)\,J_\nu\left(\frac{a}{x}-ax\right)$	$I_{(\nu+s)/2}(a)\,K_{(\nu-s)/2}(a)$ $[a>0;\ \operatorname{Re}(s+\nu)<3/2]$
2	$\theta(x-1)\,J_\nu\left(ax-\frac{a}{x}\right)$	$I_{(\nu-s)/2}(a)\,K_{(\nu+s)/2}(a)$ $[a>0;\ \operatorname{Re}\nu>-1;\ \operatorname{Re} s<3/2]$
3	$J_\nu\left(a\left\vert x-\frac{1}{x}\right\vert\right)$	$I_{(\nu-s)/2}(a)\ K_{(\nu+s)/2}(a)+I_{(\nu+s)/2}(a)\ K_{(\nu-s)/2}(a)$ $[a>0;\ \operatorname{Re}\nu>-1;\ \vert\operatorname{Re} s\vert<3/2]$
4	$J_\nu\left(ax+\frac{a}{x}\right)$	$-\frac{\pi}{2}\left[J_{(\nu-s)/2}(a)\ Y_{(\nu+s)/2}(a)+J_{(\nu+s)/2}(a)\ Y_{(\nu-s)/2}(a)\right]$ $[a>0;\ \vert\operatorname{Re} s\vert<3/2]$
5	$(a-x)_+^{\alpha-1}$ $\times J_\nu\big(bx(a-x)\big)$	$a^{s+\alpha+2\nu-1}\left(\frac{b}{2}\right)^\nu\Gamma\left[\begin{matrix}s+\nu,\ \alpha+\nu\\ \nu+1,\ s+\alpha+2\nu\end{matrix}\right]$ $\times {}_4F_5\left(\begin{matrix}\Delta(2,\ \alpha+\nu),\ \Delta(2,\ s+\nu)\\ \nu+1,\ \Delta(4,\ s+\alpha+2\nu);\ -\frac{a^4b^2}{64}\end{matrix}\right)$ $[a,\ \operatorname{Re}(\alpha+\nu),\ \operatorname{Re}(s+\nu)>0]$
6	$(a^2-x^2)_+^{\nu/2}$ $\times J_\nu\left(b\sqrt{a^2-x^2}\right)$	$\frac{2^{s/2-1}a^{s/2+\nu}}{b^{s/2}}\Gamma\left(\frac{s}{2}\right)J_{s/2+\nu}(ab)$ $[a,\ \operatorname{Re} s>0;\ \operatorname{Re}\nu>-1]$

No.	$f(x)$	$F(s)$
7	$\left(x^2-a^2\right)_+^{\nu/2}$ $\times J_\nu\left(b\sqrt{x^2-a^2}\right)$	$\dfrac{a^{s/2+\nu}}{\Gamma\left(\frac{2-s}{2}\right)}\left(\dfrac{2}{b}\right)^{s/2}K_{s/2+\nu}(ab)$ $[a,\, b>0;\ \operatorname{Re}\nu>-1;\ \operatorname{Re}s<3/2-\operatorname{Re}\nu]$
8	$\left(x^2+a^2\right)^\rho$ $\times J_\nu\left(b\sqrt{x^2+a^2}\right)$	$\dfrac{2^{s+2\rho-1}\pi\, b^{-s-2\rho}}{\Gamma\left(\frac{2-s+\nu-2\rho}{2}\right)\Gamma\left(\frac{2-s-\nu-2\rho}{2}\right)}\csc\dfrac{(s+\nu+2\rho)\pi}{2}$ $\times\,{}_1F_2\left(\begin{matrix}\frac{2-s}{2};\ -\frac{a^2b^2}{4}\\ \frac{2-s+\nu-2\rho}{2},\ \frac{2-s-\nu-2\rho}{2}\end{matrix}\right)$ $-\dfrac{\pi a^{s+\nu+2\rho}b^\nu}{2^{\nu+1}}\csc\dfrac{(s+\nu+2\rho)\pi}{2}$ $\times\Gamma\left[\begin{matrix}\frac{s}{2}\\ \nu+1,\ -\frac{\nu+2\rho}{2},\ \frac{s+\nu+2\rho+2}{2}\end{matrix}\right]{}_1F_2\left(\begin{matrix}\frac{\nu+2\rho+2}{2};\ -\frac{a^2b^2}{4}\\ \nu+1,\ \frac{s+\nu+2\rho+2}{2}\end{matrix}\right)$ $[\operatorname{Re}a,\, b>0;\ 0<\operatorname{Re}s<-2\operatorname{Re}\rho+1/2]$
9	$\left(x^2+a^2\right)^{\nu/2}$ $\times J_\nu\left(b\sqrt{x^2+a^2}\right)$	$\dfrac{2^{s/2-1}a^{s/2+\nu}}{b^{s/2}}\Gamma\left(\dfrac{s}{2}\right)\left[J_{s/2+\nu}(ab)\cos\dfrac{s\pi}{2}-Y_{s/2+\nu}(ab)\sin\dfrac{s\pi}{2}\right]$ $[\operatorname{Re}a,\, b>0;\ 0<\operatorname{Re}s<3/2-\operatorname{Re}\nu]$
10	$\left(x^2+a^2\right)^{-\nu/2}$ $\times J_\nu\left(b\sqrt{x^2+a^2}\right)$	$\dfrac{2^{s/2-1}a^{s/2-\nu}}{b^{s/2}}\Gamma\left(\dfrac{s}{2}\right)J_{\nu-s/2}(ab)$ $[\operatorname{Re}a,\, b>0;\ 0<\operatorname{Re}s<\operatorname{Re}\nu+3/2]$
11	$\dfrac{1}{(x+a)^\rho}J_\nu\left(\dfrac{b}{x+a}\right)$	$\dfrac{a^{s-\nu-\rho}}{\Gamma(\nu+1)}\left(\dfrac{b}{2}\right)^\nu \mathrm{B}(s,\,\nu+\rho-s)\;{}_2F_3\left(\begin{matrix}\frac{\nu+\rho-s}{2},\ \frac{\nu+\rho-s+1}{2};\ -\frac{b^2}{4a^2}\\ \nu+1,\ \frac{\nu+\rho}{2},\ \frac{\nu+\rho+1}{2}\end{matrix}\right)$ $[0<\operatorname{Re}s<\operatorname{Re}(\nu+\rho);\ \lvert\arg a\rvert<\pi]$
12	$\dfrac{1}{(x+a)^\rho}J_\nu\left(\dfrac{bx}{x+a}\right)$	$a^{s-\rho}\left(\dfrac{b}{2}\right)^\nu\dfrac{\mathrm{B}(s+\nu,\,\rho-s)}{\Gamma(\nu+1)}\;{}_2F_3\left(\begin{matrix}\frac{s+\nu}{2},\ \frac{s+\nu+1}{2};\ -\frac{b^2}{4}\\ \nu+1,\ \frac{\nu+\rho}{2},\ \frac{\nu+\rho+1}{2}\end{matrix}\right)$ $[-\operatorname{Re}\nu<\operatorname{Re}s<\operatorname{Re}\rho;\ \lvert\arg a\rvert<\pi]$
13	$\dfrac{1}{\left(x^2+a^2\right)^\rho}$ $\times J_\nu\left(\dfrac{bx}{x^2+a^2}\right)$	$\dfrac{2^{-\nu-1}a^{s-\nu-2\rho}b^\nu}{\Gamma(\nu+1)}\mathrm{B}\left(\dfrac{s+\nu}{2},\,\dfrac{\nu+2\rho-s}{2}\right)$ $\times\,{}_2F_3\left(\begin{matrix}\frac{s+\nu}{2},\ \frac{\nu+2\rho-s}{2};\ -\frac{b^2}{16a^2}\\ \nu+1,\ \frac{\nu+\rho}{2},\ \frac{\nu+\rho+1}{2}\end{matrix}\right)$ $[\operatorname{Re}a,\, b>0;\ -\operatorname{Re}\nu<\operatorname{Re}s<\operatorname{Re}(\nu+2\rho)]$

3.10.3. $J_\nu(\varphi(x))$ and the exponential function

1	$e^{-ax}J_\nu(bx)$	$\frac{b^\nu}{2^\nu a^{s+\nu}}\Gamma\begin{bmatrix} s+\nu \\ \nu+1 \end{bmatrix} {}_2F_1\left(\begin{matrix} \frac{s+\nu}{2},\ \frac{s+\nu+1}{2} \\ \nu+1;\ -\frac{b^2}{a^2} \end{matrix}\right)$ $[\mathrm{Re}(s+\nu)>0;\ \mathrm{Re}\,a>\lvert\mathrm{Im}\,b\rvert]$
2	$e^{-ax^2}J_\nu(bx)$	$\frac{2^{-\nu-1}b^\nu}{a^{(s+\nu)/2}}\Gamma\begin{bmatrix} \frac{s+\nu}{2} \\ \nu+1 \end{bmatrix} {}_1F_1\left(\begin{matrix} \frac{s+\nu}{2};\ -\frac{b^2}{4a} \\ \nu+1 \end{matrix}\right)$ $[\mathrm{Re}\,a>0;\ \mathrm{Re}(s+\nu)>0;\ \lvert\arg b\rvert<\pi]$
3	$e^{-a\sqrt{x}}J_\nu(bx)$	$\frac{2^{s-1}}{b^s}\Gamma\begin{bmatrix} \frac{s+\nu}{2} \\ \frac{2-s+\nu}{2} \end{bmatrix} {}_2F_3\left(\begin{matrix} \frac{s-\nu}{2},\ \frac{s+\nu}{2} \\ \frac{1}{4},\ \frac{1}{2},\ \frac{3}{4};\ -\frac{a^4}{64b^2} \end{matrix}\right) - \frac{2^{s-1/2}a}{b^{s+1/2}}\Gamma\begin{bmatrix} \frac{2s+2\nu+1}{4} \\ \frac{3-2s+2\nu}{4} \end{bmatrix} {}_2F_3\left(\begin{matrix} \frac{2s-2\nu+1}{4},\ \frac{2s+2\nu+1}{4} \\ \frac{1}{2},\ \frac{3}{4},\ \frac{5}{4};\ -\frac{a^4}{64b^2} \end{matrix}\right) + \frac{2^{s-1}a^2}{b^{s+1}}\Gamma\begin{bmatrix} \frac{s+\nu+1}{2} \\ \frac{1-s+\nu}{2} \end{bmatrix} {}_2F_3\left(\begin{matrix} \frac{s-\nu+1}{2},\ \frac{s+\nu+1}{2} \\ \frac{3}{4},\ \frac{5}{4},\ \frac{3}{2};\ -\frac{a^4}{64b^2} \end{matrix}\right) - \frac{2^{s-1/2}a^3}{3b^{s+3/2}}\Gamma\begin{bmatrix} \frac{2s+2\nu+3}{4} \\ \frac{1-2s+2\nu}{4} \end{bmatrix} {}_2F_3\left(\begin{matrix} \frac{2s-2\nu+3}{4},\ \frac{2s+2\nu+3}{4} \\ \frac{5}{4},\ \frac{3}{2},\ \frac{7}{4};\ -\frac{a^4}{64b^2} \end{matrix}\right)$ $[b,\ \mathrm{Re}\,a,\ \mathrm{Re}(s+\nu)>0]$
4	$e^{-a/x}J_\nu(bx)$	$\frac{2^{s-1}}{b^s}\Gamma\begin{bmatrix} \frac{s+\nu}{2} \\ \frac{2-s+\nu}{2} \end{bmatrix} {}_0F_3\left(\begin{matrix} -\frac{a^2b^2}{16} \\ \frac{1}{2},\ \frac{2-s-\nu}{2},\ \frac{2-s+\nu}{2} \end{matrix}\right) - \frac{2^{s-2}a}{b^{s-1}}\Gamma\begin{bmatrix} \frac{s+\nu-1}{2} \\ \frac{3-s+\nu}{2} \end{bmatrix} {}_0F_3\left(\begin{matrix} -\frac{a^2b^2}{16} \\ \frac{3}{2},\ \frac{3-s-\nu}{2},\ \frac{3-s+\nu}{2} \end{matrix}\right) + \frac{a^{s+\nu}b^\nu}{2^\nu}\Gamma\begin{bmatrix} -s-\nu \\ \nu+1 \end{bmatrix} {}_0F_3\left(\begin{matrix} -\frac{a^2b^2}{16} \\ \nu+1,\ \frac{s+\nu+1}{2},\ \frac{s+\nu+2}{2} \end{matrix}\right)$ $[b,\ \mathrm{Re}\,a>0;\ \mathrm{Re}\,s<3/2]$
5	$e^{-a/x^2}J_\nu(bx)$	$\frac{2^{s-1}}{b^s}\Gamma\begin{bmatrix} \frac{s+\nu}{2} \\ \frac{2-s+\nu}{2} \end{bmatrix} {}_0F_2\left(\begin{matrix} \frac{ab^2}{4} \\ \frac{2-s-\nu}{2},\ \frac{2-s+\nu}{2} \end{matrix}\right) + \frac{a^{(s+\nu)/2}b^\nu}{2^{\nu+1}}\Gamma\begin{bmatrix} -\frac{s+\nu}{2} \\ \nu+1 \end{bmatrix} {}_0F_2\left(\begin{matrix} \frac{ab^2}{4} \\ \nu+1,\ \frac{s+\nu+2}{2} \end{matrix}\right)$ $[b,\ \mathrm{Re}\,a>0;\ \mathrm{Re}\,s<3/2]$
6	$(a-x)_+^{\alpha-1}\,e^{\pm ibx}J_\nu(bx)$	$a^{s+\alpha+\nu-1}\left(\frac{b}{2}\right)^\nu\Gamma\begin{bmatrix} \alpha,\ s+\nu \\ \nu+1,\ s+\alpha+\nu \end{bmatrix} {}_2F_2\left(\begin{matrix} \frac{2\nu+1}{2},\ s+\nu;\ \pm 2iab \\ 2\nu+1,\ s+\alpha+\nu \end{matrix}\right)$ $[a,\ b,\ \mathrm{Re}\,\alpha,\ \mathrm{Re}(s+\nu)>0]$
7	$(a-x)_+^{\nu/2}\,e^{bx} \times J_\nu\left(c\sqrt{a-x}\right)$	$a^{s+\nu}\left(\frac{c}{2}\right)^\nu\Gamma\begin{bmatrix} s \\ s+\nu+1 \end{bmatrix}\Phi_3\left(s;\ s+\nu+1;\ ab,\ -\frac{ac^2}{4}\right)$ $[a,\ \mathrm{Re}\,s>0;\ \mathrm{Re}\,\nu>-1]$

No.	$f(x)$	$F(s)$
8	$\left(a^2-x^2\right)_+^{-1} e^{-b/(a^2-x^2)}$ $\times J_\nu\left(\frac{cx}{a^2-x^2}\right)$	$\frac{a^{s-1}}{c} e^{-b/(2a^2)}\,\Gamma\left[\begin{matrix}\frac{s+\nu}{2}\\ \nu+1\end{matrix}\right] M_{(s-1)/2,\,\nu/2}\left(\frac{\sqrt{b^2+a^2c^2}-b}{2a^2}\right)$ $\times W_{(1-s)/2,\,\nu/2}\left(\frac{\sqrt{b^2+a^2c^2}+b}{2a^2}\right)$ $[a,\ b,\ c,\ \mathrm{Re}\,(s+\nu)>0]$
9	$\left(x^2-a^2\right)_+^{-1} e^{-b/(x^2-a^2)}$ $\times J_\nu\left(\frac{cx}{x^2-a^2}\right)$	$-\frac{a^{s-1}}{c} e^{b/(2a^2)}\,\Gamma\left[\begin{matrix}\frac{2-s+\nu}{2}\\ \nu+1\end{matrix}\right] M_{(1-s)/2,\,\nu/2}\left(\frac{\sqrt{b^2+a^2c^2}-b}{2a^2}\right)$ $\times W_{(s-1)/2,\,\nu/2}\left(\frac{\sqrt{b^2+a^2c^2}+b}{2a^2}\right)$ $[a,\ b,\ c>0;\ \mathrm{Re}\,s<\mathrm{Re}\,\nu]$
10	$\frac{e^{\pm 2a^2b/(x^2+a^2)}}{x^2+a^2}$ $\times J_\nu\left(\frac{2cx}{x^2+a^2}\right)$	$\frac{a^{s-1}}{2c} e^{\pm b}\,\Gamma\left[\begin{matrix}\frac{2-s+\nu}{2},\ \frac{s+\nu}{2}\\ \nu+1,\ \nu+1\end{matrix}\right] M_{\mp(1-s)/2,\,\nu/2}\left(\frac{ab-\sqrt{a^2b^2-c^2}}{a}\right)$ $\times M_{\mp(1-s)/2,\,\nu/2}\left(\frac{ab+\sqrt{a^2b^2-c^2}}{a}\right)$ $[\mathrm{Re}\,a,\ b,\ c>0;\ -\mathrm{Re}\,\nu<\mathrm{Re}\,s<\mathrm{Re}\,\nu+2]$

3.10.4. $J_\nu(bx)$ and trigonometric functions

Notation: $\delta=\left\{\begin{matrix}1\\0\end{matrix}\right\}$.

1	$\left\{\begin{matrix}\sin(ax)\\ \cos(ax)\end{matrix}\right\} J_\nu(ax)$	$\frac{(2a)^{-s}}{\sqrt{\pi}}\left\{\begin{matrix}\sin[(s+\nu)\pi/2]\\ \cos[(s+\nu)\pi/2]\end{matrix}\right\}\Gamma\left[\begin{matrix}\frac{1-2s}{2},\ s+\nu\\ 1-s+\nu\end{matrix}\right]$ $[a>0;\ -\mathrm{Re}\,\nu-\delta<\mathrm{Re}\,s<1/2]$
2	$\left\{\begin{matrix}\sin(ax+b)\\ \cos(ax+b)\end{matrix}\right\} J_\nu(ax)$	$\frac{(2a)^{-s}}{\sqrt{\pi}}\left\{\begin{matrix}\sin[(s+\nu)\pi/2+b]\\ \cos[(s+\nu)\pi/2+b]\end{matrix}\right\}\Gamma\left[\begin{matrix}\frac{1-2s}{2},\ s+\nu\\ 1-s+\nu\end{matrix}\right]$ $[a>0;\ -\mathrm{Re}\,\nu<\mathrm{Re}\,s<1/2]$
3	$\left\{\begin{matrix}\sin(ax)\\ \cos(ax)\end{matrix}\right\} J_\nu(bx)$	$\frac{2^{s+\delta-1}a^\delta}{b^{s+\delta}}\,\Gamma\left[\begin{matrix}\frac{s+\nu+\delta}{2}\\ \frac{2-s+\nu-\delta}{2}\end{matrix}\right] {}_2F_1\left(\begin{matrix}\frac{s-\nu+\delta}{2},\ \frac{s+\nu+\delta}{2}\\ \frac{2\delta+1}{2};\ \frac{a^2}{b^2}\end{matrix}\right)$ $[0<a<b;\ -\mathrm{Re}\,\nu-\delta<\mathrm{Re}\,s<3/2]$
4	$\left\{\begin{matrix}\sin(ax)\\ \cos(ax)\end{matrix}\right\} J_\nu(bx)$	$\frac{b^\nu}{2^\nu a^{s+\nu}}\left\{\begin{matrix}\sin[(s+\nu)\pi/2]\\ \cos[(s+\nu)\pi/2]\end{matrix}\right\}\Gamma\left[\begin{matrix}s+\nu\\ \nu+1\end{matrix}\right] {}_2F_1\left(\begin{matrix}\frac{s+\nu}{2},\ \frac{s+\nu+1}{2}\\ \nu+1;\ \frac{b^2}{a^2}\end{matrix}\right)$ $[0<b<a;\ -\mathrm{Re}\,\nu-\delta<\mathrm{Re}\,s<3/2]$

No.	$f(x)$	$F(s)$
5	$\begin{Bmatrix} \sin(ax^2) \\ \cos(ax^2) \end{Bmatrix} J_\nu(bx)$	$\frac{b^\nu}{2^{\nu+1}a^{(s+\nu)/2}} \begin{Bmatrix} \sin[(s+\nu)\pi/4] \\ \cos[(s+\nu)\pi/4] \end{Bmatrix} \Gamma\left[\begin{matrix} \frac{s+\nu}{2} \\ \nu+1 \end{matrix}\right] \times {}_2F_3\left(\begin{matrix} \frac{s+\nu}{4}, \frac{s+\nu+2}{4}; -\frac{b^4}{64a^2} \\ \frac{1}{2}, \frac{\nu+1}{2}, \frac{\nu+2}{2} \end{matrix}\right) \mp \frac{b^{\nu+2}}{2^{\nu+3}a^{(s+\nu)/2+1}} \begin{Bmatrix} \cos[(s+\nu)\pi/4] \\ \sin[(s+\nu)\pi/4] \end{Bmatrix} \Gamma\left[\begin{matrix} \frac{s+\nu+2}{2} \\ \nu+2 \end{matrix}\right] \times {}_2F_3\left(\begin{matrix} \frac{s+\nu+2}{4}, \frac{s+\nu+4}{4}; -\frac{b^4}{64a^2} \\ \frac{3}{2}, \frac{\nu+2}{2}, \frac{\nu+3}{2} \end{matrix}\right)$ $[a, b > 0;\ -\operatorname{Re}\nu - 2\delta < \operatorname{Re} s < 5/2]$
6	$\sin(a\sqrt{x})\, J_\nu(bx)$	$\frac{2^{s-1/2}a}{b^{s+1/2}} \Gamma\left[\begin{matrix} \frac{2s+2\nu+1}{4} \\ \frac{3-2s+2\nu}{4} \end{matrix}\right] {}_2F_3\left(\begin{matrix} \frac{2s-2\nu+1}{4}, \frac{2s+2\nu+1}{4} \\ \frac{1}{2}, \frac{3}{4}, \frac{5}{4}; -\frac{a^4}{64b^2} \end{matrix}\right) - \frac{2^{s-1/2}a^3}{3b^{s+3/2}} \Gamma\left[\begin{matrix} \frac{2s+2\nu+3}{4} \\ \frac{1-2s+2\nu}{4} \end{matrix}\right] {}_2F_3\left(\begin{matrix} \frac{2s-2\nu+3}{4}, \frac{2s+2\nu+3}{4} \\ \frac{5}{4}, \frac{3}{2}, \frac{7}{4}; -\frac{a^4}{64b^2} \end{matrix}\right)$ $[a, b > 0;\ -\operatorname{Re}\nu - 1/2 < \operatorname{Re} s < 3/2]$
7	$\cos(a\sqrt{x})\, J_\nu(bx)$	$\frac{2^{s-1}}{b^s} \Gamma\left[\begin{matrix} \frac{s+\nu}{2} \\ \frac{2-s+\nu}{2} \end{matrix}\right] {}_2F_3\left(\begin{matrix} \frac{s-\nu}{2}, \frac{s+\nu}{2} \\ \frac{1}{4}, \frac{1}{2}, \frac{3}{4}; -\frac{a^4}{64b^2} \end{matrix}\right) - \frac{2^{s-1}a^2}{b^{s+1}} \Gamma\left[\begin{matrix} \frac{s+\nu+1}{2} \\ \frac{1-s+\nu}{2} \end{matrix}\right] {}_2F_3\left(\begin{matrix} \frac{s-\nu+1}{2}, \frac{s+\nu+1}{2} \\ \frac{3}{4}, \frac{5}{4}, \frac{3}{2}; -\frac{a^4}{64b^2} \end{matrix}\right)$ $[a, b > 0;\ -\operatorname{Re}\nu < \operatorname{Re} s < 3/2]$
8	$\sin\frac{a}{x}\, J_\nu(bx)$	$\frac{2^{s-2}a}{b^{s-1}} \Gamma\left[\begin{matrix} \frac{s+\nu-1}{2} \\ \frac{3-s+\nu}{2} \end{matrix}\right] {}_0F_3\left(\begin{matrix} \frac{a^2b^2}{16} \\ \frac{3}{2}, \frac{3-s-\nu}{2}, \frac{3-s+\nu}{2} \end{matrix}\right) - a^{s+\nu}\left(\frac{b}{2}\right)^\nu \sin\frac{(s+\nu)\pi}{2}\, \Gamma\left[\begin{matrix} -s-\nu \\ \nu+1 \end{matrix}\right] \times {}_0F_3\left(\begin{matrix} \frac{a^2b^2}{16} \\ \nu+1, \frac{s+\nu+1}{2}, \frac{s+\nu+2}{2} \end{matrix}\right)$ $[a, b > 0;\ -\operatorname{Re}\nu - 1 < \operatorname{Re} s < 5/2]$
9	$\cos\frac{a}{x}\, J_\nu(bx)$	$\frac{2^{s-1}}{b^s} \Gamma\left[\begin{matrix} \frac{s+\nu}{2} \\ \frac{2-s+\nu}{2} \end{matrix}\right] {}_0F_3\left(\begin{matrix} \frac{a^2b^2}{16} \\ \frac{1}{2}, \frac{2-s-\nu}{2}, \frac{2-s+\nu}{2} \end{matrix}\right) + a^{s+\nu}\left(\frac{b}{2}\right)^\nu \cos\frac{(s+\nu)\pi}{2}\, \Gamma\left[\begin{matrix} -s-\nu \\ \nu+1 \end{matrix}\right] \times {}_0F_3\left(\begin{matrix} \frac{a^2b^2}{16} \\ \nu+1, \frac{s+\nu+1}{2}, \frac{s+\nu+2}{2} \end{matrix}\right)$ $[a, b > 0;\ -\operatorname{Re}\nu - 1 < \operatorname{Re} s < 3/2]$

No.	$f(x)$	$F(s)$
10	$\sin(ax)\,J_\nu(ax)$ $\pm\cos(ax)\,J_{-\nu}(ax)$	$-\dfrac{2^{2-s}a^{-s}}{\pi^{3/2}}\sin\dfrac{(2\nu\mp1)\pi}{4}\cos\dfrac{(2s\mp1)\pi}{4}$ $\times\cos\dfrac{(s-\nu)\pi}{2}\sin\dfrac{(s+\nu)\pi}{2}\,\Gamma\left[\dfrac{1}{2}-s,\ s-\nu,\ s+\nu\right]$ $[a>0;\ -\operatorname{Re}\nu-1,\ \operatorname{Re}\nu<\operatorname{Re}s<(2\mp1)/2]$
11	$\cos(ax)\,J_\nu(ax)$ $\pm\sin(ax)\,J_{-\nu}(ax)$	$\pm\dfrac{2^{2-s}a^{-s}}{\pi^{3/2}}\sin\dfrac{(2\nu\pm1)\pi}{4}\cos\dfrac{(2s\mp1)\pi}{4}$ $\times\sin\dfrac{(s-\nu)\pi}{2}\cos\dfrac{(s+\nu)\pi}{2}\,\Gamma\left[\dfrac{1}{2}-s,\ s-\nu,\ s+\nu\right]$ $[a>0;\ \operatorname{Re}\nu-1,\ -\operatorname{Re}\nu<\operatorname{Re}s<(2\mp1)/2]$
12	$e^{-ax}\begin{Bmatrix}\sin(bx)\\ \cos(bx)\end{Bmatrix}J_\nu(bx)$	$\dfrac{(2b)^{\nu+\delta}}{\sqrt{\pi}\,a^{s+\nu+\delta}}\,\Gamma\begin{bmatrix}\frac{2\nu+2\delta+1}{2},\ s+\nu+\delta\\ 2\nu+\delta+1\end{bmatrix}$ $\times{}_4F_3\left(\begin{matrix}\frac{2\nu+3}{4},\ \frac{2\nu+4\delta+1}{4},\ \frac{s+\nu+1}{2},\ \frac{s+\nu+2\delta}{2}\\ \frac{2\delta+1}{2},\ \nu+1,\ \frac{2\nu+2\delta+1}{2};\ -\frac{4b^2}{a^2}\end{matrix}\right)$ $[b,\ \operatorname{Re}a>0;\ \operatorname{Re}(s+\nu)>-\delta]$
13	$e^{-a\sqrt{x}}\begin{Bmatrix}\sin(a\sqrt{x})\\ \cos(a\sqrt{x})\end{Bmatrix}$ $\times J_\nu(bx)$	$\dfrac{2^{s+1/2}a^3}{3b^{s+3/2}}\,\Gamma\begin{bmatrix}\frac{2s+2\nu+3}{4}\\ \frac{1-2s+2\nu}{4}\end{bmatrix}{}_2F_3\left(\begin{matrix}\frac{2s+2\nu+3}{4},\ \frac{2s-2\nu+3}{4}\\ \frac{3}{2},\ \frac{5}{4},\ \frac{7}{4};\ \frac{a^4}{16b^2}\end{matrix}\right)$ $\pm\dfrac{2^{s-1/2}a}{b^{s+1/2}}\,\Gamma\begin{bmatrix}\frac{2s+2\nu+1}{4}\\ \frac{3-2s+2\nu}{4}\end{bmatrix}{}_2F_3\left(\begin{matrix}\frac{2s+2\nu+1}{4},\ \frac{2s-2\nu+1}{4}\\ \frac{1}{2},\ \frac{3}{4},\ \frac{5}{4};\ \frac{a^4}{16b^2}\end{matrix}\right)$ $\mp\dfrac{2^{s+\delta-1}a^{2\delta}}{b^{s+\delta}}\,\Gamma\begin{bmatrix}\frac{s+\nu+\delta}{2}\\ \frac{2-s+\nu-\delta}{2}\end{bmatrix}{}_2F_3\left(\begin{matrix}\frac{s+\nu+\delta}{2},\ \frac{s-\nu+\delta}{2}\\ \frac{3}{4},\ \frac{2\delta+1}{2},\ \frac{4\delta+1}{4};\ \frac{a^4}{16b^2}\end{matrix}\right)$ $[b>0;\ \operatorname{Re}(s+\nu)>-\delta/2;\ \lvert\arg a\rvert<\pi/4]$

3.10.5. $J_\nu(bx)$ and the logarithmic function

1	$\theta(a-x)\ln\dfrac{\sqrt{a}+\sqrt{a-x}}{\sqrt{x}}$ $\times J_\nu(bx)$	$\dfrac{\sqrt{\pi}\,a^{s+\nu}}{2(s+\nu)}\left(\dfrac{b}{2}\right)^{\nu}\Gamma\begin{bmatrix}s+\nu\\ \nu+1,\ \frac{2s+2\nu+1}{2}\end{bmatrix}$ $\times{}_3F_4\left(\begin{matrix}\frac{s+\nu}{2},\ \frac{s+\nu}{2},\ \frac{s+\nu+1}{2};\ -\frac{a^2b^2}{4}\\ \nu+1,\ \frac{2s+2\nu+1}{4},\ \frac{2s+2\nu+3}{4},\ \frac{s+\nu+2}{2}\end{matrix}\right)$ $[a,\ \operatorname{Re}(s+\nu)>0]$
2	$\theta(a-x)\ln\dfrac{a+\sqrt{a^2-x^2}}{x}$ $\times J_\nu(bx)$	$\dfrac{\sqrt{\pi}\,a^{s+\nu}}{2(s+\nu)}\left(\dfrac{b}{2}\right)^{\nu}\Gamma\begin{bmatrix}\frac{s+\nu+1}{2}\\ \nu+1,\ \frac{s+\nu}{2}\end{bmatrix}{}_2F_3\left(\begin{matrix}\frac{s+\nu}{2},\ \frac{s+\nu}{2};\ -\frac{a^2b^2}{4}\\ \nu+1,\ \frac{s+\nu+1}{2},\ \frac{s+\nu+2}{2}\end{matrix}\right)$ $[a,\ \operatorname{Re}(s+\nu)>0]$

3.10.6. $J_\nu(bx)$ and inverse trigonometric functions

Notation: $\delta = \begin{Bmatrix} 1 \\ 0 \end{Bmatrix}$.

1	$\theta(a-x)\arccos\frac{x}{a}\,J_\nu(bx)$	$\frac{\sqrt{\pi}\,a^{s+\nu}}{(s+\nu)^2}\left(\frac{b}{2}\right)^\nu \Gamma\left[\begin{matrix}\frac{s+\nu+1}{2}\\ \nu+1,\ \frac{s+\nu}{2}\end{matrix}\right] {}_2F_3\left(\begin{matrix}\frac{s+\nu}{2},\ \frac{s+\nu+1}{2};\ -\frac{a^2b^2}{4}\\ \nu+1,\ \frac{s+\nu+2}{2},\ \frac{s+\nu+2}{2}\end{matrix}\right)$ $[a,\ \operatorname{Re}(s+\nu)>0]$
2	$\theta(a-x)\begin{Bmatrix}\sin(bx)\\ \cos(bx)\end{Bmatrix}$ $\times\arccos\frac{x}{a}\,J_\nu(bx)$	$\frac{\sqrt{\pi}\,a^{s+\nu+\delta}b^{\nu+\delta}}{2^{\nu+1}(s+\nu+\delta)}\Gamma\left[\begin{matrix}\frac{s+\nu+\delta+1}{2}\\ \nu+1,\ \frac{s+\nu+\delta+2}{2}\end{matrix}\right]$ $\times\,{}_4F_5\left(\begin{matrix}\frac{2\nu+2\delta+1}{2},\ \frac{2\nu+2\delta+3}{2},\ \frac{s+\nu+\delta}{2},\ \frac{s+\nu+\delta+1}{2};\ -a^2b^2\\ \frac{2\delta+1}{2},\ \frac{2\nu+\delta+1}{2},\ \frac{2\nu+\delta+2}{2},\ \frac{s+\nu+\delta+2}{2},\ \frac{s+\nu+\delta+2}{2}\end{matrix}\right)$ $[a>0;\ \operatorname{Re}(s+\nu)>-\delta]$

3.10.7. $J_\nu(bx)$ and $\operatorname{Ei}(ax^r)$

1	$\operatorname{Ei}(-ax)\,J_\nu(bx)$	$-\frac{a^{-s-\nu}}{s+\nu}\left(\frac{b}{2}\right)^\nu \Gamma\left[\begin{matrix}s+\nu\\ \nu+1\end{matrix}\right] {}_3F_2\left(\begin{matrix}\frac{s+\nu}{2},\ \frac{s+\nu}{2},\ \frac{s+\nu+1}{2}\\ \nu+1,\ \frac{s+\nu+2}{2};\ -\frac{b^2}{a^2}\end{matrix}\right)$ $\left[\begin{matrix}(\operatorname{Re}a>\lvert\operatorname{Im}b\rvert;\operatorname{Im}a=0;\operatorname{Re}(s+\nu)>0)\ \text{or}\\ (\operatorname{Re}a+\lvert\operatorname{Im}b\rvert=0;\operatorname{Im}a=0;-\operatorname{Re}\nu<\operatorname{Re}s<5/2)\ \text{or}\\ (\operatorname{Re}a\ge 0;\operatorname{Im}a\ne 0;\operatorname{Im}b=0;-\operatorname{Re}\nu<\operatorname{Re}s<3/2)\end{matrix}\right]$
2	$\operatorname{Ei}(-ax^2)\,J_\nu(bx)$	$-\frac{a^{-(s+\nu)/2}}{s+\nu}\left(\frac{b}{2}\right)^\nu \Gamma\left[\begin{matrix}\frac{s+\nu}{2}\\ \nu+1\end{matrix}\right] {}_2F_2\left(\begin{matrix}\frac{s+\nu}{2},\ \frac{s+\nu}{2};\ -\frac{b^2}{4a}\\ \nu+1,\ \frac{s+\nu+2}{2}\end{matrix}\right)$ $[\operatorname{Re}a,\ \operatorname{Re}(s+\nu)>0]$
3	$e^{\pm ax}\operatorname{Ei}(\mp ax)\,J_\nu(bx)$	$-\frac{\pi}{a^{s+\nu}}\left(\frac{b}{2}\right)^\nu \begin{Bmatrix}\csc[(s+\nu)\pi]\\ \cot[(s+\nu)\pi]\end{Bmatrix}\Gamma\left[\begin{matrix}s+\nu\\ \nu+1\end{matrix}\right] {}_2F_1\left(\begin{matrix}\frac{s+\nu}{2},\ \frac{s+\nu+1}{2}\\ \nu+1;\ -\frac{b^2}{a^2}\end{matrix}\right)$ $\mp\frac{2^{s-2}}{ab^{s-1}}\Gamma\left[\begin{matrix}\frac{s+\nu-1}{2}\\ \frac{3-s+\nu}{2}\end{matrix}\right] {}_3F_2\left(\begin{matrix}\frac{1}{2},\ 1,\ 1;\ -\frac{b^2}{a^2}\\ \frac{3-s-\nu}{2},\ \frac{3-s+\nu}{2}\end{matrix}\right)$ $+\frac{2^{s-3}}{a^2b^{s-2}}\Gamma\left[\begin{matrix}\frac{s+\nu-2}{2}\\ \frac{4-s+\nu}{2}\end{matrix}\right] {}_3F_2\left(\begin{matrix}1,\ 1,\ \frac{3}{2};\ -\frac{b^2}{a^2}\\ \frac{4-s-\nu}{2},\ \frac{4-s+\nu}{2}\end{matrix}\right)$ $[b,\ \operatorname{Re}a>0;\ -\operatorname{Re}\nu<\operatorname{Re}s<5/2]$
4	$e^{\pm ax^2}\operatorname{Ei}(\mp ax^2)\,J_\nu(bx)$	$-\frac{\pi b^\nu}{2^{\nu+1}a^{(s+\nu)/2}}\begin{Bmatrix}\csc[(s+\nu)\pi/2]\\ \cot[(s+\nu)\pi/2]\end{Bmatrix}\Gamma\left[\begin{matrix}\frac{s+\nu}{2}\\ \nu+1\end{matrix}\right] {}_1F_1\left(\begin{matrix}\frac{s+\nu}{2};\ \pm\frac{b^2}{4a}\\ \nu+1\end{matrix}\right)$ $\mp\frac{2^{s-3}}{ab^{s-2}}\Gamma\left[\begin{matrix}\frac{s+\nu-2}{2}\\ \frac{4-s+\nu}{2}\end{matrix}\right] {}_2F_2\left(\begin{matrix}1,\ 1;\ \pm\frac{b^2}{4a}\\ \frac{4-s-\nu}{2},\ \frac{4-s+\nu}{2}\end{matrix}\right)$ $[b,\ \operatorname{Re}a>0;\ -\operatorname{Re}\nu<\operatorname{Re}s<5/2]$

3.10.8. $J_\nu(bx)$ and $\operatorname{si}(ax^r)$, $\operatorname{Si}(ax)$, or $\operatorname{ci}(ax^r)$

Notation: $\delta = \begin{Bmatrix} 1 \\ 0 \end{Bmatrix}$.

1	$\begin{Bmatrix} \operatorname{si}(ax) \\ \operatorname{ci}(ax) \end{Bmatrix} J_\nu(bx)$	$-\frac{a^{-s-\nu}}{s+\nu}\left(\frac{b}{2}\right)^\nu \begin{Bmatrix} \sin[(s+\nu)\pi/2] \\ \cos[(s+\nu)\pi/2] \end{Bmatrix} \Gamma\begin{bmatrix} s+\nu \\ \nu+1 \end{bmatrix} \times {}_3F_2\left(\begin{matrix} \frac{s+\nu}{2}, \frac{s+\nu}{2}, \frac{s+\nu+1}{2} \\ \nu+1, \frac{s+\nu+2}{2}; \frac{b^2}{a^2} \end{matrix}\right)$ $[0 < b < a;\ -\operatorname{Re}\nu < \operatorname{Re}s < 5/2]$
2	$\operatorname{si}(ax) J_\nu(bx)$	$\frac{2^s a}{b^{s+1}} \Gamma\begin{bmatrix} \frac{s+\nu+1}{2} \\ \frac{1-s+\nu}{2} \end{bmatrix} {}_3F_2\left(\begin{matrix} \frac{1}{2}, \frac{s-\nu+1}{2}, \frac{s+\nu+1}{2} \\ \frac{3}{2}, \frac{3}{2}; \frac{a^2}{b^2} \end{matrix}\right) - \frac{\pi 2^{s-2}}{b^s} \Gamma\begin{bmatrix} \frac{s+\nu}{2} \\ \frac{2-s+\nu}{2} \end{bmatrix}$ $[0 < a < b;\ -\operatorname{Re}\nu < \operatorname{Re}s < 5/2]$
3	$\operatorname{ci}(ax) J_\nu(bx)$	$\frac{2^{s-2}}{b^s} \Gamma\begin{bmatrix} \frac{s+\nu}{2} \\ \frac{2-s+\nu}{2} \end{bmatrix} \left[\frac{a^2(s^2-\nu^2)}{2b^2} {}_4F_3\left(\begin{matrix} 1, 1, \frac{s-\nu+2}{2}, \frac{s+\nu+2}{2} \\ \frac{3}{2}, 2, 2; \frac{a^2}{b^2} \end{matrix}\right) + \psi\left(\frac{s+\nu}{2}\right) + \psi\left(\frac{2-s+\nu}{2}\right) + 2\ln\frac{2a}{b} + 2\mathbf{C}\right]$ $[0 < a < b;\ -\operatorname{Re}\nu < \operatorname{Re}s < 5/2]$
4	$\begin{Bmatrix} \operatorname{si}(ax^2) \\ \operatorname{ci}(ax^2) \end{Bmatrix} J_\nu(bx)$	$-\frac{(b/2)^\nu}{a^{(s+\nu)/2}(s+\nu)} \begin{Bmatrix} \sin[(s+\nu)\pi/4] \\ \cos[(s+\nu)\pi/4] \end{Bmatrix} \Gamma\begin{bmatrix} \frac{s+\nu}{2} \\ \nu+1 \end{bmatrix} \times {}_3F_4\left(\begin{matrix} \frac{s+\nu}{4}, \frac{s+\nu}{4}, \frac{s+\nu+2}{4}; -\frac{b^4}{64a^2} \\ \frac{1}{2}, \frac{\nu+1}{2}, \frac{\nu+2}{2}, \frac{s+\nu+4}{4} \end{matrix}\right)$ $\pm \frac{(b/2)^{\nu+2}}{a^{(s+\nu)/2+1}(s+\nu+2)} \begin{Bmatrix} \cos[(s+\nu)\pi/4] \\ \sin[(s+\nu)\pi/4] \end{Bmatrix} \Gamma\begin{bmatrix} \frac{s+\nu+2}{2} \\ \nu+2 \end{bmatrix} \times {}_3F_4\left(\begin{matrix} \frac{s+\nu+2}{4}, \frac{s+\nu+2}{4}, \frac{s+\nu+4}{4}; -\frac{b^4}{64a^2} \\ \frac{3}{2}, \frac{\nu+2}{2}, \frac{\nu+3}{2}, \frac{s+\nu+6}{4} \end{matrix}\right)$ $[a, b > 0;\ -\operatorname{Re}\nu < \operatorname{Re}s < 5/2]$
5	$\left[\begin{Bmatrix} \sin x \\ \cos x \end{Bmatrix} \operatorname{ci}(2x) \mp \begin{Bmatrix} \cos x \\ \sin x \end{Bmatrix} \operatorname{Si}(2x)\right] \times J_\nu(x)$	$-\frac{2^{-s-1}}{\sqrt{\pi}} \begin{Bmatrix} \sin[(s+\nu)\pi/2] \\ \cos[(s+\nu)\pi/2] \end{Bmatrix} \Gamma\begin{bmatrix} s+\nu, \frac{1-2s}{2} \\ 1-s+\nu \end{bmatrix} \times \left[\psi\left(\frac{1-s-\nu+\delta}{2}\right) \mp \psi\left(\frac{1-s+\nu}{2}\right) \pm \psi\left(\frac{2-s+\nu}{2}\right) - \psi\left(\frac{s+\nu+\delta}{2}\right)\right]$ $[-\delta/2 < \operatorname{Re}(s+\nu) < 3/2]$

3.10.9. $J_\nu(bx)$ and $\operatorname{erf}(ax^r)$, $\operatorname{erfc}(ax^r)$, or $\operatorname{erfi}(ax^r)$

Notation: $\delta = \begin{Bmatrix} 1 \\ 0 \end{Bmatrix}$.

1	$\begin{Bmatrix} \operatorname{erf}(ax) \\ \operatorname{erfc}(ax) \end{Bmatrix} J_\nu(bx)$	$\mp \frac{a^{-s-\nu}}{\sqrt{\pi}(s+\nu)} \left(\frac{b}{2}\right)^\nu \Gamma\left[\begin{matrix} \frac{s+\nu+1}{2} \\ \nu+1 \end{matrix}\right] {}_2F_2\left(\begin{matrix} \frac{s+\nu}{2}, \frac{s+\nu+1}{2}; -\frac{b^2}{4a^2} \\ \nu+1, \frac{s+\nu+2}{2} \end{matrix}\right) + 2^{s-2}\frac{1 \pm 1}{b^s} \Gamma\left[\begin{matrix} \frac{s+\nu}{2} \\ \frac{2-s+\nu}{2} \end{matrix}\right]$ $\left[\begin{Bmatrix} b>0;\ -1-\operatorname{Re}\nu < \operatorname{Re}s < 3/2 \\ \operatorname{Re}(s+\nu) > 0 \end{Bmatrix};\ \lvert\arg a\rvert < \pi/4\right]$
2	$\begin{Bmatrix} \operatorname{erf}(a\sqrt{x}) \\ \operatorname{erfc}(a\sqrt{x}) \end{Bmatrix} J_\nu(bx)$	$\mp \frac{a^{-2(s+\nu)}}{\sqrt{\pi}(s+\nu)} \left(\frac{b}{2}\right)^\nu \Gamma\left[\begin{matrix} \frac{2s+2\nu+1}{2} \\ \nu+1 \end{matrix}\right] {}_3F_2\left(\begin{matrix} \frac{s+\nu}{2}, \frac{2s+2\nu+1}{4}, \frac{2s+2\nu+3}{4} \\ \nu+1, \frac{s+\nu+2}{2}; -\frac{b^2}{a^4} \end{matrix}\right) + 2^{s-2}\frac{1 \pm 1}{b^s} \Gamma\left[\begin{matrix} \frac{s+\nu}{2} \\ \frac{2-s+\nu}{2} \end{matrix}\right]$ $\left[b>0; \begin{Bmatrix} -\operatorname{Re}\nu - 1/2 < \operatorname{Re}s < 3/2 \\ \operatorname{Re}(s+\nu) > 0 \end{Bmatrix};\ \lvert\arg a\rvert < \pi/4\right]$
3	$\operatorname{erf}\left(\frac{a}{x}\right) J_\nu(bx)$	$\frac{a^{s+\nu}}{\sqrt{\pi}(s+\nu)} \left(\frac{b}{2}\right)^\nu \Gamma\left[\begin{matrix} \frac{1-s-\nu}{2} \\ \nu+1 \end{matrix}\right] {}_1F_3\left(\begin{matrix} \frac{s+\nu}{2}; \frac{a^2b^2}{4} \\ \nu+1, \frac{s+\nu+1}{2}, \frac{s+\nu+2}{2} \end{matrix}\right) + \frac{a}{\sqrt{\pi}} \left(\frac{2}{b}\right)^{s-1} \Gamma\left[\begin{matrix} \frac{s+\nu-1}{2} \\ \frac{3-s+\nu}{2} \end{matrix}\right] {}_1F_3\left(\begin{matrix} \frac{1}{2}; \frac{a^2b^2}{4} \\ \frac{3}{2}, \frac{3-s-\nu}{2}, \frac{3-s+\nu}{2} \end{matrix}\right)$ $[b>0;\ -\operatorname{Re}\nu < \operatorname{Re}s < 5/2;\ \lvert\arg a\rvert < \pi/4]$
4	$e^{\mp a^2x^2} \begin{Bmatrix} \operatorname{erfi}(ax) \\ \operatorname{erfc}(ax) \end{Bmatrix} \times J_\nu(bx)$	$\frac{b^\nu}{2^{\nu+1} a^{s+\nu}} \Gamma\left[\begin{matrix} \frac{s+\nu}{2} \\ \nu+1 \end{matrix}\right] \begin{Bmatrix} \tan[(s+\nu)\pi/2] \\ \sec[(s+\nu)\pi/2] \end{Bmatrix} {}_1F_1\left(\begin{matrix} \frac{s+\nu}{2}; \mp\frac{b^2}{4a^2} \\ \nu+1 \end{matrix}\right) + \frac{2^{s-2}b^{1-s}}{\sqrt{\pi}\,a} \Gamma\left[\begin{matrix} \frac{s+\nu-1}{2} \\ \frac{3-s+\nu}{2} \end{matrix}\right] {}_2F_2\left(\begin{matrix} \frac{1}{2}, 1; \mp\frac{b^2}{4a^2} \\ \frac{3-s-\nu}{2}, \frac{3-s+\nu}{2} \end{matrix}\right)$ $[b>0;\ -\operatorname{Re}\nu - (1 \pm 1)/2 < \operatorname{Re}s < 5/2;\ \lvert\arg a\rvert < \pi/4]$
5	$e^{-a^2x} \operatorname{erfi}(a\sqrt{x}) J_\nu(bx)$	$\frac{2^{s+1/2}ab^{-s-1/2}}{\sqrt{\pi}} \Gamma\left[\begin{matrix} \frac{2s+2\nu+1}{4} \\ \frac{3-2s+2\nu}{4} \end{matrix}\right] {}_3F_2\left(\begin{matrix} 1, \frac{2s-2\nu+1}{4}, \frac{2s+2\nu+1}{4} \\ \frac{3}{4}, \frac{5}{4}; -\frac{a^4}{b^2} \end{matrix}\right) - \frac{2^{s+5/2}a^3b^{-s-3/2}}{3\sqrt{\pi}} \Gamma\left[\begin{matrix} \frac{2s+2\nu+3}{4} \\ \frac{1-2s+2\nu}{4} \end{matrix}\right] {}_3F_2\left(\begin{matrix} 1, \frac{2s-2\nu+3}{4}, \frac{2s+2\nu+3}{4} \\ \frac{5}{4}, \frac{7}{4}; -\frac{a^4}{b^2} \end{matrix}\right)$ $[b>0;\ -\operatorname{Re}\nu - 1/2 < \operatorname{Re}s < 2;\ \lvert\arg a\rvert < \pi/4]$
6	$e^{a^2x} \operatorname{erfc}(a\sqrt{x}) J_\nu(bx)$	$\frac{2^{1-2s-3\nu}a^{-2s-2\nu}b^\nu}{\sqrt{\pi}} \Gamma\left[\begin{matrix} \frac{1-2s-2\nu}{2}, 2s+2\nu \\ \nu+1 \end{matrix}\right] {}_2F_1\left(\begin{matrix} \frac{s+\nu}{2}, \frac{s+\nu+1}{2} \\ \nu+1; -\frac{b^2}{a^4} \end{matrix}\right) + \frac{2^{s-3/2}b^{1/2-s}}{a\sqrt{\pi}} \Gamma\left[\begin{matrix} \frac{2s+2\nu-1}{4} \\ \frac{5-2s+2\nu}{4} \end{matrix}\right] {}_3F_2\left(\begin{matrix} \frac{1}{4}, \frac{3}{4}, 1; -\frac{b^2}{a^4} \\ \frac{5-2s-2\nu}{4}, \frac{5-2s+2\nu}{4} \end{matrix}\right) - \frac{2^{s-7/2}b^{3/2-s}}{a^3\sqrt{\pi}} \Gamma\left[\begin{matrix} \frac{2s+2\nu-3}{4} \\ \frac{7-2s+2\nu}{4} \end{matrix}\right] {}_3F_2\left(\begin{matrix} \frac{3}{4}, 1, \frac{5}{4}; -\frac{b^2}{a^4} \\ \frac{7-2s-2\nu}{4}, \frac{7-2s+2\nu}{4} \end{matrix}\right)$ $[b>0;\ -\operatorname{Re}\nu < \operatorname{Re}s < 2;\ \lvert\arg a\rvert < \pi/4]$

No.	$f(x)$	$F(s)$
7	$\begin{Bmatrix} \sin(bx) \\ \cos(bx) \end{Bmatrix} \operatorname{erfc}(ax) \times J_\nu(bx)$	$\frac{(2b)^{\nu+\delta}}{\pi a^{s+\nu+\delta}(s+\nu+\delta)} \Gamma\left[\begin{matrix} \frac{2\nu+2\delta+1}{2}, \frac{s+\nu+\delta+1}{2} \\ 2\nu+\delta+1 \end{matrix}\right] \times {}_4F_4\left(\begin{matrix} \frac{2\nu+3}{4}, \frac{2\nu+4\delta+1}{4}, \frac{s+\nu+1}{2}, \frac{s+\nu+2\delta}{2}; -\frac{b^2}{a^2} \\ \frac{2\delta+1}{2}, \nu+1, \frac{2\nu+2\delta+1}{2}, \frac{s+\nu+\delta+2}{2} \end{matrix}\right)$ $[\operatorname{Re}(s+\nu) > -(1\pm1)/2;\ \lvert\arg a\rvert < \pi/4]$

3.10.10. $J_\nu(bx)$ and $S(ax^r)$, $C(ax^r)$

Notation: $\delta = \begin{Bmatrix} 1 \\ 0 \end{Bmatrix}$.

1	$\begin{Bmatrix} S(ax) \\ C(ax) \end{Bmatrix} J_\nu(bx)$	$-\frac{2^{-\nu-1/2} b^\nu}{\sqrt{\pi}\, a^{s+\nu}(s+\nu)} \Gamma\left[\begin{matrix} \frac{2s+2\nu+1}{2} \\ \nu+1 \end{matrix}\right] \times \begin{Bmatrix} \cos[(2s+2\nu+1)\pi/4] \\ \sin[(2s+2\nu+1)\pi/4] \end{Bmatrix} \times {}_3F_2\left(\begin{matrix} \frac{s+\nu}{2}, \frac{2s+2\nu+1}{4}, \frac{2s+2\nu+3}{4} \\ \nu+1, \frac{s+\nu+2}{2}; \frac{b^2}{a^2} \end{matrix}\right) + \frac{2^{s-2}}{b^s} \Gamma\left[\begin{matrix} \frac{s+\nu}{2} \\ \frac{2-s+\nu}{2} \end{matrix}\right]$ $\left[-(2\pm1)/2 - \operatorname{Re}\nu < \operatorname{Re}s < \begin{Bmatrix} 3/2 \text{ for } 0<b<a \\ 1 \text{ for } 0<b=a \end{Bmatrix}\right]$
2	$\begin{Bmatrix} S(ax) \\ C(ax) \end{Bmatrix} J_\nu(bx)$	$\frac{2^{s+\delta} a^{\delta+1/2}}{3^\delta \sqrt{\pi}\, b^{s+\delta+1/2}} \Gamma\left[\begin{matrix} \frac{2\nu+2s+2\delta+1}{4} \\ \frac{2\nu-2s-2\delta+3}{4} \end{matrix}\right] \times {}_3F_2\left(\begin{matrix} \frac{2\delta+1}{4}, \frac{2s+2\delta+2\nu+1}{4}, \frac{2s+2\delta-2\nu+1}{4} \\ \frac{2\delta+5}{4}, \frac{2\delta+1}{2}; \frac{a^2}{b^2} \end{matrix}\right)$ $[0<a<b;\ -(2\pm1)/2 - \operatorname{Re}\nu < \operatorname{Re}s < 3/2]$
3	$\begin{Bmatrix} S(ax^2) \\ C(ax^2) \end{Bmatrix} J_\nu(bx)$	$-\frac{a^{-(s+\nu)/2} b^\nu}{2^{\nu+1/2}\sqrt{\pi}(s+\nu)} \Gamma\left[\begin{matrix} \frac{s+\nu+1}{2} \\ \nu+1 \end{matrix}\right] \begin{Bmatrix} \sin[(s+\nu+1)\pi/4] \\ \cos[(s+\nu+1)\pi/4] \end{Bmatrix} \times {}_3F_4\left(\begin{matrix} \frac{s+\nu}{4}, \frac{s+\nu+1}{4}, \frac{s+\nu+3}{4}; -\frac{b^4}{64a^2} \\ \frac{1}{2}, \frac{\nu+1}{2}, \frac{\nu+2}{2}, \frac{s+\nu+4}{4} \end{matrix}\right) \pm \frac{a^{-(s+\nu)/2-1} b^{\nu+2}}{2^{\nu+5/2}\sqrt{\pi}(s+\nu+2)} \Gamma\left[\begin{matrix} \frac{s+\nu+3}{2} \\ \nu+2 \end{matrix}\right] \times \begin{Bmatrix} \cos[(s+\nu+1)\pi/4] \\ \sin[(s+\nu+1)\pi/4] \end{Bmatrix} \times {}_3F_4\left(\begin{matrix} \frac{s+\nu+2}{4}, \frac{s+\nu+3}{4}, \frac{s+\nu+5}{4}; -\frac{b^4}{64a^2} \\ \frac{3}{2}, \frac{\nu+2}{2}, \frac{\nu+3}{2}, \frac{s+\nu+6}{4} \end{matrix}\right) + \frac{2^{s-2}}{b^s} \Gamma\left[\begin{matrix} \frac{s+\nu}{2} \\ \frac{2-s+\nu}{2} \end{matrix}\right]$ $[a, b>0;\ -2\mp1 - \operatorname{Re}\nu < \operatorname{Re}s < 3/2]$

3.10.11. $J_\nu(bx)$ and $\Gamma(\mu, ax^r)$, $\gamma(\mu, ax^r)$

Notation: $\delta = \begin{Bmatrix} 1 \\ 0 \end{Bmatrix}$.

1	$\begin{Bmatrix} \gamma(\mu, ax) \\ \Gamma(\mu, ax) \end{Bmatrix} J_\nu(bx)$	$\mp \dfrac{a^{-s-\nu}}{s+\nu} \left(\dfrac{b}{2}\right)^\nu \Gamma\begin{bmatrix} s+\mu+\nu \\ \nu+1 \end{bmatrix} {}_3F_2\left(\begin{matrix} \frac{s+\nu}{2}, \frac{s+\mu+\nu}{2}, \frac{s+\mu+\nu+1}{2} \\ \nu+1, \frac{s+\nu+2}{2}; -\frac{b^2}{a^2} \end{matrix}\right) + \dfrac{2^{s-1}\delta}{b^s} \Gamma\begin{bmatrix} \mu, \frac{s+\nu}{2} \\ \frac{2-s+\nu}{2} \end{bmatrix}$ $\left[b, \operatorname{Re} a, \operatorname{Re}(s+\mu+\nu) > 0; \begin{Bmatrix} \operatorname{Re}\mu > 0; \operatorname{Re} s < 3/2 \\ \operatorname{Re}(s+\nu) > 0 \end{Bmatrix}\right]$
2	$\begin{Bmatrix} \gamma(\mu, ax^2) \\ \Gamma(\mu, ax^2) \end{Bmatrix} J_\nu(bx)$	$\mp \dfrac{a^{-(s+\nu)/2}}{s+\nu} \left(\dfrac{b}{2}\right)^\nu \Gamma\begin{bmatrix} \frac{s+2\mu+\nu}{2} \\ \nu+1 \end{bmatrix} {}_2F_2\left(\begin{matrix} \frac{s+\nu}{2}, \frac{s+2\mu+\nu}{2}; -\frac{b^2}{4a} \\ \nu+1, \frac{s+\nu+2}{2} \end{matrix}\right) + \dfrac{2^{s-1}\delta}{b^s} \Gamma\begin{bmatrix} \mu, \frac{s+\nu}{2} \\ \frac{2-s+\nu}{2} \end{bmatrix}$ $\left[b, \operatorname{Re} a, \operatorname{Re}(s+2\mu+\nu) > 0; \begin{Bmatrix} \operatorname{Re}\mu > 0; \operatorname{Re} s < 3/2 \\ \operatorname{Re}(s+\nu) > 0 \end{Bmatrix}\right]$

3.10.12. $J_\nu(bx)$ and $D_\nu(ax^r)$

Notation: $\delta = \begin{Bmatrix} 1 \\ 0 \end{Bmatrix}$.

1	$e^{-a^2x^2/4} D_\mu(ax)\, J_\nu(bx)$	$\dfrac{2^{(s+\mu-\nu)/2-1} b^\nu}{a^{s+\nu}} \Gamma\begin{bmatrix} \frac{s+\nu}{2}, \frac{s+\nu+1}{2} \\ \nu+1, \frac{s+\nu-\mu+1}{2} \end{bmatrix} {}_2F_2\left(\begin{matrix} \frac{s+\nu}{2}, \frac{s+\nu+1}{2}; -\frac{b^2}{2a^2} \\ \nu+1, \frac{s+\nu-\mu+1}{2} \end{matrix}\right)$ $[\operatorname{Re}(s+\nu) > 0; \lvert\arg a\rvert < \pi/4]$
2	$e^{a^2x^2/4} D_\mu(ax)\, J_\nu(bx)$	$\dfrac{b^\nu}{2^{(s+\mu+3\nu)/2+1} a^{s+\nu}} \Gamma\begin{bmatrix} s+\nu, -\frac{s+\nu+\mu}{2} \\ \nu+1, -\mu \end{bmatrix} {}_2F_2\left(\begin{matrix} \frac{s+\nu}{2}, \frac{s+\nu+1}{2}; \frac{b^2}{2a^2} \\ \nu+1, \frac{s+\nu+\mu+2}{2} \end{matrix}\right) + \dfrac{2^{s+\mu-1} a^\mu}{b^{s+\mu}} \Gamma\begin{bmatrix} \frac{s+\nu+\mu}{2} \\ -\frac{s-\nu+\mu-2}{2} \end{bmatrix} {}_2F_2\left(\begin{matrix} \frac{1-\mu}{2}, -\frac{\mu}{2}; \frac{b^2}{2a^2} \\ -\frac{s+\nu+\mu-2}{2}, -\frac{s-\nu+\mu-2}{2} \end{matrix}\right)$ $[b > 0; -\operatorname{Re}\nu < \operatorname{Re} s < 3/2 - \operatorname{Re}\mu; \lvert\arg a\rvert < 3\pi/4]$
3	$e^{-a^2/(4x^2)} D_\mu\left(\dfrac{a}{x}\right) \times J_\nu(bx)$	$\dfrac{\sqrt{\pi}\, 2^{(2s+\mu-2)/2}}{b^s} \Gamma\begin{bmatrix} \frac{s+\nu}{2} \\ \frac{1-\mu}{2}, \frac{2-s+\nu}{2} \end{bmatrix} {}_1F_3\left(\begin{matrix} \frac{\mu+1}{2}; \frac{a^2b^2}{8} \\ \frac{1}{2}, \frac{2-s-\nu}{2}, \frac{2-s+\nu}{2} \end{matrix}\right) - \dfrac{\sqrt{\pi}\, 2^{(2s+\mu-3)/2} a}{b^{s-1}} \Gamma\begin{bmatrix} \frac{s+\nu-1}{2} \\ -\frac{\mu}{2}, \frac{3-s+\nu}{2} \end{bmatrix} {}_1F_3\left(\begin{matrix} \frac{\mu+2}{2}; \frac{a^2b^2}{8} \\ \frac{3}{2}, \frac{3-s-\nu}{2}, \frac{3-s+\nu}{2} \end{matrix}\right) + \sqrt{\pi}\, 2^{(s+\mu-\nu)/2} a^{s+\nu} b^\nu \Gamma\begin{bmatrix} -\nu \\ s+\nu+1, \frac{1-s-\mu-\nu}{2} \end{bmatrix} \times \sin(\nu\pi) \csc[(s+\nu)\pi]\, {}_1F_3\left(\begin{matrix} \frac{s+\mu+\nu+1}{2}; \frac{a^2b^2}{8} \\ \nu+1, \frac{s+\nu+1}{2}, \frac{s+\nu+2}{2} \end{matrix}\right)$ $[b > 0; \operatorname{Re}(\mu-\nu) < \operatorname{Re} s < 3/2; \lvert\arg a\rvert < \pi/4]$

No.	$f(x)$	$F(s)$
4	$D_{-\mu-1}(a\sqrt{x})\,D_\mu(a\sqrt{x})$ $\times J_\nu(bx)$	$\frac{2^{1/2-2s-3\nu}\pi b^\nu}{a^{2s+2\nu}}\Gamma\left[\begin{matrix}2s+2\nu\\ \nu+1,\ \frac{s+\mu+\nu+2}{2},\ \frac{s-\mu+\nu+1}{2}\end{matrix}\right]$ $\times {}_4F_3\left(\begin{matrix}\frac{s+\nu}{2},\ \frac{s+\nu+1}{2},\ \frac{2s+2\nu+1}{4},\ \frac{2s+2\nu+3}{4}\\ \nu+1,\ \frac{s+\mu+\nu+2}{2},\ \frac{s-\mu+\nu+1}{2};\ -\frac{4b^2}{a^4}\end{matrix}\right)$ $[b,\ \mathrm{Re}(s+\nu)>0;\ \lvert\arg a\rvert<\pi/4]$

3.10.13. Products of $J_\mu(ax)$

1	$J_\nu^2(ax)$	$\frac{a^{-s}}{2\sqrt{\pi}}\Gamma\left[\begin{matrix}\frac{1-s}{2},\ \frac{s+2\nu}{2}\\ \frac{2-s}{2},\ \frac{2-s+2\nu}{2}\end{matrix}\right]$ $[a>0;\ -2\,\mathrm{Re}\,\nu<\mathrm{Re}\,s<1]$
2	$J_{\nu-1}(ax)\,J_\nu(ax)$	$\frac{a^{-s}}{2\sqrt{\pi}}\Gamma\left[\begin{matrix}\frac{2-s}{2},\ \frac{s+2\nu-1}{2}\\ \frac{3-s}{2},\ \frac{1-s+2\nu}{2}\end{matrix}\right]$ $[a>0;\ 1-2\,\mathrm{Re}\,\nu<\mathrm{Re}\,s<2]$
3	$J_{-\nu}(ax)\,J_\nu(ax)$	$\frac{a^{-s}}{2\sqrt{\pi}}\Gamma\left[\begin{matrix}\frac{s}{2},\ \frac{1-s}{2}\\ \frac{2-s-2\nu}{2},\ \frac{2-s+2\nu}{2}\end{matrix}\right]$ $[a>0;\ 0<\mathrm{Re}\,s<1]$
4	$J_{-\nu-1}(ax)\,J_\nu(ax)$ $+\frac{2\sin(\pi\nu)}{\pi ax}$	$-\frac{a^{-s}}{2\sqrt{\pi}}\Gamma\left[\begin{matrix}\frac{1-s}{2},\ \frac{2-s}{2},\ \frac{s+1}{2}\\ \frac{3-s}{2},\ \frac{1-s-2\nu}{2},\ \frac{3-s+2\nu}{2}\end{matrix}\right]$ $[a>0;\ \lvert\mathrm{Re}\,s\rvert<1]$
5	$J_{-\nu-2}(ax)\,J_\nu(ax)$ $-\frac{4(\nu+1)\sin(\pi\nu)}{\pi a^2x^2}$	$\frac{a^{-s}}{2\sqrt{\pi}}\Gamma\left[\begin{matrix}\frac{1-s}{2},\ \frac{2-s}{2},\ \frac{s+2}{2}\\ \frac{4-s}{2},\ -\frac{s+2\nu}{2},\ \frac{4-s+2\nu}{2}\end{matrix}\right]$ $[a>0;\ -2<\mathrm{Re}\,s<1]$
6	$J_{-n-\nu-1}(ax)\,J_\nu(ax)$ $-\frac{2}{\sqrt{\pi}}\sum_{k=0}^{[n/2]}\frac{(-1)^{[(n+1)/2]+k}}{k!}$ $\times\frac{([n/2]-k+1)_{n-[n/2]}}{(ax)^{n-2k+1}}$ $\times\Gamma\left[\begin{matrix}k-n+\left[\frac{n}{2}\right]+\frac{1}{2}\\ k+\nu+1,\ k-n-\nu\end{matrix}\right]$	$\frac{(-1)^{n+1}a^{-s}}{2\sqrt{\pi}}\Gamma\left[\begin{matrix}\frac{1-s}{2},\ \frac{2-s}{2},\ \frac{s+n+1}{2}\\ \frac{3-s+n}{2},\ -\frac{s+n+2\nu-1}{2},\ \frac{3-s+n+2\nu}{2}\end{matrix}\right]$ $[a>0;\ -n-1<\mathrm{Re}\,s<1]$
7	$J_\mu(ax)\,J_\nu(ax)$	$\frac{2^{s-1}}{a^s}\Gamma\left[\begin{matrix}1-s,\ \frac{s+\mu+\nu}{2}\\ \frac{2-s+\mu-\nu}{2},\ \frac{2-s-\mu+\nu}{2},\ \frac{2-s+\mu+\nu}{2}\end{matrix}\right]$ $[a>0;\ -\mathrm{Re}(\mu+\nu)<\mathrm{Re}\,s<1]$

No.	$f(x)$	$F(s)$
8	$J_\mu(ax)\,J_\nu(ax) - \dfrac{2^{-\mu-\nu}(ax)^{\mu+\nu}}{\Gamma(\mu+1)\,\Gamma(\nu+1)}$	$\dfrac{2^{s-1}}{a^s}\,\Gamma\left[\begin{matrix}1-s,\ \frac{s+\mu+\nu}{2}\\ \frac{2-s+\mu-\nu}{2},\ \frac{2-s-\mu+\nu}{2},\ \frac{2-s+\mu+\nu}{2}\end{matrix}\right]$ $[a>0;\ -\operatorname{Re}(\mu+\nu)-2<\operatorname{Re}s<-\operatorname{Re}(\mu+\nu),\ 1]$
9	$J_\nu(ax)\,J_\nu(bx)$	$\dfrac{2^{s-1}(ab)^\nu}{(a+b)^{s+2\nu}}\,\Gamma\left[\begin{matrix}\frac{s+2\nu}{2}\\ \nu+1,\ \frac{2-s}{2}\end{matrix}\right]\,{}_2F_1\left(\begin{matrix}\frac{2\nu+1}{2},\ \frac{s+2\nu}{2}\\ 2\nu+1;\ \frac{4ab}{(a+b)^2}\end{matrix}\right)$ $[a,\ b>0;\ a\neq b;\ -2\operatorname{Re}\nu<\operatorname{Re}s<2]$
10	$J_\mu(ax)\,J_\nu(bx)$	$\dfrac{2^{s-1}b^\nu}{a^{s+\nu}}\,\Gamma\left[\begin{matrix}\frac{s+\mu+\nu}{2}\\ \nu+1,\ \frac{2-s+\mu-\nu}{2}\end{matrix}\right]\,{}_2F_1\left(\begin{matrix}\frac{s+\mu+\nu}{2},\ \frac{s-\mu+\nu}{2}\\ \nu+1;\ \frac{b^2}{a^2}\end{matrix}\right)$ $[0<b<a;\ -(\mu+\nu)<\operatorname{Re}s<2]$
11	$J_\mu(a\sqrt{x})\,J_\nu(bx)$	$\dfrac{2^{s-\mu/2-1}a^\mu}{b^{s+\mu/2}}\,\Gamma\left[\begin{matrix}\frac{2s+\mu+2\nu}{4}\\ \mu+1,\ \frac{4-2s-\mu+2\nu}{4}\end{matrix}\right]\,{}_2F_3\left(\begin{matrix}\frac{2s+\mu-2\nu}{4},\ \frac{2s+\mu+2\nu}{4}\\ \frac{1}{2},\ \frac{\mu+1}{2},\ \frac{\mu+2}{2};\ -\frac{a^4}{64b^2}\end{matrix}\right)$ $-\dfrac{2^{s-\mu/2-2}a^{\mu+2}}{b^{s+\mu/2+1}}\,\Gamma\left[\begin{matrix}\frac{2s+\mu+2\nu+2}{4}\\ \mu+2,\ \frac{2-2s-\mu+2\nu}{4}\end{matrix}\right]$ $\times\,{}_2F_3\left(\begin{matrix}\frac{2s+\mu-2\nu+2}{4},\ \frac{2s+\mu+2\nu+2}{4}\\ \frac{3}{2},\ \frac{\mu+2}{2},\ \frac{\mu+3}{2};\ -\frac{a^4}{64b^2}\end{matrix}\right)$ $[a,\ b>0;\ -\operatorname{Re}(\nu+\mu/2)<\operatorname{Re}s<7/4]$
12	$J_\mu\left(\dfrac{a}{x}\right)J_\nu(bx)$	$\dfrac{a^\mu b^{\mu-s}}{2^{2\mu-s+1}}\,\Gamma\left[\begin{matrix}\frac{s-\mu+\nu}{2}\\ \mu+1,\ \frac{2-s+\mu+\nu}{2}\end{matrix}\right]\,{}_0F_3\left(\begin{matrix}\frac{a^2b^2}{16}\\ \mu+1,\ \frac{2-s+\mu-\nu}{2},\ \frac{2-s+\mu+\nu}{2}\end{matrix}\right)$ $+\dfrac{a^{s+\nu}b^\nu}{2^{s+2\nu+1}}\,\Gamma\left[\begin{matrix}\frac{\mu-\nu-s}{2}\\ \nu+1,\ \frac{s+\mu+\nu+2}{2}\end{matrix}\right]$ $\times\,{}_0F_3\left(\begin{matrix}\frac{a^2b^2}{16}\\ \nu+1,\ \frac{s-\mu+\nu+2}{2},\ \frac{s+\mu+\nu+2}{2}\end{matrix}\right)$ $[a,\ b>0;\ -\operatorname{Re}\nu-3/2<\operatorname{Re}s<\operatorname{Re}\mu+3/2]$
13	$J_\nu^2(ax)\pm J_{-\nu}^2(ax)$	$\pm\dfrac{a^{-s}}{\pi^{3/2}}\left\{\begin{matrix}\cos(\nu\pi)\sin(s\pi/2)\\ \sin(\nu\pi)\cos(s\pi/2)\end{matrix}\right\}\Gamma\left[\begin{matrix}\frac{1-s}{2},\ \frac{s}{2}-\nu,\ \frac{s}{2}+\nu\\ \frac{2-s}{2}\end{matrix}\right]$ $[a>0;\ 2\lvert\operatorname{Re}\nu\rvert<\operatorname{Re}s<(3+1)/2],$
14	$J_\mu(ax)\,J_\nu(ax) \pm J_{-\mu}(ax)\,J_{-\nu}(ax)$	$\pm\dfrac{1}{\pi}\left(\dfrac{2}{a}\right)^s\left\{\begin{matrix}\cos[(\mu+\nu)\pi/2]\sin(s\pi/2)\\ \sin[(\mu+\nu)\pi/2]\cos(s\pi/2)\end{matrix}\right\}$ $\times\,\Gamma\left[\begin{matrix}1-s,\ \frac{s-\mu-\nu}{2},\ \frac{s+\mu+\nu}{2}\\ \frac{2-s+\mu-\nu}{2},\ \frac{2-s-\mu+\nu}{2}\end{matrix}\right]$ $[a>0;\ \lvert\operatorname{Re}(\mu+\nu)\rvert<\operatorname{Re}s<(3\mp1)/2]$

3.10.14. $J_\mu(bx)\,J_\nu(cx)$ and the exponential or trigonometric functions

Notation: $\delta = \begin{Bmatrix}1\\0\end{Bmatrix}$.

1	$e^{-ax}J_\mu(bx)\,J_\nu(bx)$	$a^{-s-\mu-\nu}\left(\frac{b}{2}\right)^{\mu+\nu}\Gamma\left[\begin{matrix}s+\mu+\nu\\ \mu+1,\ \nu+1\end{matrix}\right]$ $\times\,{}_4F_3\left(\begin{matrix}\frac{\mu+\nu+1}{2},\ \frac{\mu+\nu+2}{2},\ \frac{s+\mu+\nu}{2},\ \frac{s+\mu+\nu+1}{2}\\ \mu+1,\ \nu+1,\ \mu+\nu+1;\ -\frac{4b^2}{a^2}\end{matrix}\right)$ $[\operatorname{Re}a>2\lvert\operatorname{Im}b\rvert];\ \operatorname{Re}(s+\mu+\nu)>0$
2	$e^{-ax}J_\mu(bx)\,J_\nu(cx)$	$\frac{b^\mu c^\nu}{2^{\mu+\nu}a^{s+\mu+\nu}}\Gamma\left[\begin{matrix}s+\mu+\nu\\ \mu+1,\ \nu+1\end{matrix}\right]$ $\times F_4\left(\frac{s+\mu+\nu}{2},\ \frac{s+\mu+\nu+1}{2};\ \mu+1,\ \nu+1;\ -\frac{b^2}{a^2},\ -\frac{c^2}{a^2}\right)$ $[\operatorname{Re}a>\lvert\operatorname{Im}b\rvert+\lvert\operatorname{Im}c\rvert;\ \operatorname{Re}(s+\mu+\nu)>0]$
3	$e^{-ax^2}J_\mu(bx)\,J_\nu(bx)$	$\frac{b^{\mu+\nu}}{2^{\mu+\nu+1}a^{(s+\mu+\nu)/2}}\Gamma\left[\begin{matrix}\frac{s+\mu+\nu}{2}\\ \mu+1,\ \nu+1\end{matrix}\right]$ $\times\,{}_3F_3\left(\begin{matrix}\frac{s+\mu+\nu}{2},\ \frac{\mu+\nu+1}{2},\ \frac{\mu+\nu+2}{2};\ -\frac{b^2}{a}\\ \mu+1,\ \nu+1,\ \mu+\nu+1\end{matrix}\right)$ $[\operatorname{Re}a,\ \operatorname{Re}(s+\mu+\nu)>0]$
4	$e^{-ax^2}J_\mu(bx)\,J_\nu(cx)$	$\frac{b^\mu c^\nu}{2^{\mu+\nu+1}a^{(s+\mu+\nu)/2}}\Gamma\left[\begin{matrix}\frac{s+\mu+\nu}{2}\\ \mu+1,\ \nu+1\end{matrix}\right]$ $\times\Psi_2\left(\frac{s+\mu+\nu}{2};\ \mu+1,\ \nu+1;\ -\frac{b^2}{4a},\ -\frac{c^2}{4a}\right)$ $[\operatorname{Re}a,\ \operatorname{Re}(s+\mu+\nu)>0]$
5	$\begin{Bmatrix}\sin(ax)\\ \cos(ax)\end{Bmatrix}$ $\times J_\mu(bx)\,J_\nu(bx)$	$\frac{(b/2)^{\mu+\nu}}{a^{s+\mu+\nu}}\Gamma\left[\begin{matrix}s+\mu+\nu\\ \mu+1,\ \nu+1\end{matrix}\right]\begin{Bmatrix}\sin[(s+\mu+\nu)\pi/2]\\ \cos[(s+\mu+\nu)\pi/2]\end{Bmatrix}$ $\times\,{}_4F_3\left(\begin{matrix}\frac{\mu+\nu+1}{2},\ \frac{\mu+\nu+2}{2},\ \frac{s+\mu+\nu}{2},\ \frac{s+\mu+\nu+1}{2}\\ \mu+1,\ \nu+1,\ \mu+\nu+1;\ \frac{4b^2}{a^2}\end{matrix}\right)$ $\left[\begin{matrix}(0<2b<a;\ \operatorname{Re}s<2;\ \operatorname{Re}(s+\mu+\nu)>-(1\pm1)/2)\\ \text{or }(a=2b>0;\ \operatorname{Re}s<1)\end{matrix}\right]$
6	$\begin{Bmatrix}\sin(ax)\\ \cos(ax)\end{Bmatrix}$ $\times J_\mu(bx)\,J_\nu(bx)$	$\frac{2^{s+\delta-1}a^\delta}{b^{s+\delta}}\Gamma\left[\begin{matrix}1-s-\delta,\ \frac{s+\mu+\nu+\delta}{2}\\ \frac{2-s+\mu-\nu-\delta}{2},\ \frac{2-s-\mu+\nu-\delta}{2},\ \frac{2-s+\mu+\nu-\delta}{2}\end{matrix}\right]$ $\times\,{}_4F_3\left(\begin{matrix}\frac{s-\mu-\nu+\delta}{2},\ \frac{s+\mu-\nu+\delta}{2},\ \frac{s-\mu+\nu+\delta}{2},\ \frac{s+\mu+\nu+\delta}{2}\\ \frac{2\delta+1}{2},\ \frac{s+\delta}{2},\ \frac{s+\delta+1}{2};\ \frac{a^2}{4b^2}\end{matrix}\right)$ $-\frac{(\mu^2-\nu^2)\,a^{2-s}}{2\pi b^2}\sin\frac{(\mu-\nu)\pi}{2}\begin{Bmatrix}\sin(s\pi/2)\\ \cos(s\pi/2)\end{Bmatrix}$ $\times\Gamma(s-2)\,{}_4F_3\left(\begin{matrix}\frac{2-\mu-\nu}{2},\ \frac{2-\mu+\nu}{2},\ \frac{\mu-\nu+2}{2},\ \frac{\mu+\nu+2}{2}\\ \frac{3}{2},\ \frac{3-s}{2},\ \frac{4-s}{2};\ \frac{a^2}{4b^2}\end{matrix}\right)\mp$

No.	$f(x)$	$F(s)$
		$\mp \dfrac{a^{1-s}}{\pi b} \cos \dfrac{(\mu-\nu)\pi}{2} \begin{Bmatrix} \cos(s\pi/2) \\ \sin(s\pi/2) \end{Bmatrix}$ $\times \Gamma(s-1)\, {}_4F_3\left(\begin{matrix} \frac{1-\mu-\nu}{2}, \frac{1-\mu+\nu}{2}, \frac{\mu-\nu+1}{2}, \frac{\mu+\nu+1}{2} \\ \frac{1}{2}, \frac{2-s}{2}, \frac{3-s}{2}; \frac{a^2}{4b^2} \end{matrix} \right)$ $[0 < a < 2b;\ \operatorname{Re} s < 2;\ \operatorname{Re}(s+\mu+\nu) > -\delta]$

3.10.15. $J_\mu(bx)\, J_\nu(bx)$ and the logarithmic function

No.	$f(x)$	$F(s)$
1	$\theta(a-x)$ $\times \ln \dfrac{\sqrt{a-x}+\sqrt{a}}{\sqrt{x}}$ $\times J_\mu(bx)\, J_\nu(bx)$	$\dfrac{\sqrt{\pi}\, a^{s+\mu+\nu} b^{\mu+\nu}}{2^{\mu+\nu+1}(s+\mu+\nu)} \Gamma\left[\begin{matrix} s+\mu+\nu \\ \mu+1,\ \nu+1,\ s+\mu+\nu+\frac{1}{2} \end{matrix} \right]$ $\times {}_5F_6\left(\begin{matrix} \frac{\mu+\nu+1}{2}, \frac{\mu+\nu+2}{2}, \frac{s+\mu+\nu}{2}, \frac{s+\mu+\nu}{2}, \frac{s+\mu+\nu+1}{2}; -a^2b^2 \\ \mu+1,\ \nu+1,\ \mu+\nu+1,\ \Delta\left(2, \frac{2s+2\mu+2\nu+1}{2}\right), \frac{s+\mu+\nu+2}{2} \end{matrix} \right)$ $[a > 0;\ \operatorname{Re}(s+\mu+\nu) > 0]$
2	$\theta(a-x)$ $\times \ln \dfrac{\sqrt{a^2-x^2}+a}{x}$ $\times J_\mu(bx)\, J_\nu(bx)$	$\dfrac{\sqrt{\pi}\, a^{s+\mu+\nu} b^{\mu+\nu}}{2^{\mu+\nu+1}(s+\mu+\nu)} \Gamma\left[\begin{matrix} \frac{s+\mu+\nu}{2} \\ \mu+1,\ \nu+1,\ \frac{s+\mu+\nu+1}{2} \end{matrix} \right]$ $\times {}_4F_5\left(\begin{matrix} \frac{\mu+\nu+1}{2}, \frac{\mu+\nu+2}{2}, \frac{s+\mu+\nu}{2}, \frac{s+\mu+\nu}{2}; -a^2b^2 \\ \mu+1,\ \nu+1,\ \mu+\nu+1,\ \frac{s+\mu+\nu+1}{2}, \frac{s+\mu+\nu+2}{2} \end{matrix} \right)$ $[a > 0;\ \operatorname{Re}(s+(\mu+\nu)/2) > 0]$

3.10.16. $J_\mu(bx)\, J_\nu(bx)$ and inverse trigonometric functions

No.	$f(x)$	$F(s)$
1	$\theta(a-x) \arccos \dfrac{x}{a}$ $\times J_\mu(bx)\, J_\nu(bx)$	$\dfrac{2^{-\mu-\nu-1}\sqrt{\pi}\, a^{\mu+\nu+s} b^{\mu+\nu}}{\Gamma(\mu+1)\Gamma(\nu+1)(s+\mu+\nu)} \Gamma\left[\begin{matrix} \frac{s+\mu+\nu+1}{2} \\ \frac{s+\mu+\nu+2}{2} \end{matrix} \right]$ $\times {}_4F_5\left(\begin{matrix} \frac{\mu+\nu+1}{2}, \frac{\mu+\nu+2}{2}, \frac{s+\mu+\nu}{2}, \frac{s+\mu+\nu+1}{2}; -a^2b^2 \\ \mu+1,\ \nu+1,\ \mu+\nu+1,\ \frac{s+\mu+\nu+2}{2}, \frac{s+\mu+\nu+2}{2} \end{matrix} \right)$ $[a,\ \operatorname{Re}(s+\mu+\nu) > 0]$

3.10.17. $J_\mu(bx)\, J_\nu(bx)$ and $\operatorname{Ei}(-ax^r)$

No.	$f(x)$	$F(s)$
1	$\operatorname{Ei}(\ ax)\, J_\mu(bx)\, J_\nu(bx)$	$\dfrac{2^{-\mu-\nu} a^{-s-\mu-\nu} b^{\mu+\nu}}{s+\mu+\nu} \Gamma\left[\begin{matrix} s+\mu+\nu \\ \mu+1,\ \nu+1 \end{matrix} \right]$ $\times {}_5F_4\left(\begin{matrix} \frac{\mu+\nu+1}{2}, \frac{\mu+\nu+2}{2}, \frac{s+\mu+\nu}{2}, \frac{s+\mu+\nu}{2}, \frac{s+\mu+\nu+1}{2} \\ \mu+1,\ \nu+1,\ \mu+\nu+1,\ \frac{s+\mu+\nu+2}{2}; -\frac{4b^2}{a^2} \end{matrix} \right)$ $[\operatorname{Re}(s+\mu+\nu) > 0;\ \operatorname{Re} s > \lvert \operatorname{Im} b \rvert]$

No.	$f(x)$	$F(s)$
2	$\operatorname{Ei}(-ax^2)\,J_\mu(bx)\,J_\nu(bx)$	$-\dfrac{a^{-(s+\mu+\nu)/2}}{s+\mu+\nu}\left(\dfrac{b}{2}\right)^{\mu+\nu}\Gamma\left[\begin{matrix}\frac{s+\mu+\nu}{2}\\ \mu+1,\,\nu+1\end{matrix}\right] \times {}_4F_4\left(\begin{matrix}\frac{\mu+\nu+1}{2},\,\frac{\mu+\nu+2}{2},\,\frac{s+\mu+\nu}{2},\,\frac{s+\mu+\nu}{2};\,-\frac{b^2}{a}\\ \mu+1,\,\nu+1,\,\mu+\nu+1,\,\frac{s+\mu+\nu+2}{2}\end{matrix}\right)$ $[\operatorname{Re}a,\ \operatorname{Re}(s+\mu+\nu)>0]$

3.10.18. $J_\mu(bx)\,J_\nu(bx)$ and $\operatorname{erfc}(ax)$, $\operatorname{erf}(a/x)$, $\Gamma(\lambda,\,ax)$

1	$\operatorname{erfc}(ax)\,J_\mu(bx)\,J_\nu(bx)$	$\dfrac{a^{-s-\mu-\nu}}{\sqrt{\pi}\,(s+\mu+\nu)}\left(\dfrac{b}{2}\right)^{\mu+\nu}\Gamma\left[\begin{matrix}\frac{s+\mu+\nu+1}{2}\\ \mu+1,\,\nu+1\end{matrix}\right] \times {}_4F_4\left(\begin{matrix}\frac{\mu+\nu+1}{2},\,\frac{\mu+\nu+2}{2},\,\frac{s+\mu+\nu}{2},\,\frac{s+\mu+\nu+1}{2};\,-\frac{b^2}{a^2}\\ \mu+1,\,\nu+1,\,\mu+\nu+1,\,\frac{s+\mu+\nu+2}{2}\end{matrix}\right)$ $[\operatorname{Re}(s+\mu+\nu)>0;\ \lvert\arg a\rvert<\pi/4]$
2	$\operatorname{erf}\left(\dfrac{a}{x}\right)J_\mu(bx)\,J_\nu(bx)$	$\dfrac{2^{s-1}ab^{1-s}}{\sqrt{\pi}}\Gamma\left[\begin{matrix}2-s,\,\frac{s+\mu+\nu-1}{2}\\ \frac{3-s+\mu-\nu}{2},\,\frac{3-s-\mu+\nu}{2},\,\frac{3-s+\mu+\nu}{2}\end{matrix}\right] \times {}_3F_5\left(\begin{matrix}\frac{1}{2},\,\frac{2-s}{2},\,\frac{3-s}{2};\,a^2b^2\\ \frac{3}{2},\,\frac{3-s-\mu-\nu}{2},\,\frac{3-s+\mu-\nu}{2},\,\frac{3-s-\mu+\nu}{2},\,\frac{3-s+\mu+\nu}{2}\end{matrix}\right)$ $+\,2^{s-1}a^{s+\mu+\nu}b^{\mu+\nu}\sec\dfrac{(s+\mu+\nu)\pi}{2}$ $\times\,\Gamma\left[\begin{matrix}\frac{s+\mu+\nu}{2}\\ \mu+1,\,\nu+1,\,s+\mu+\nu+1\end{matrix}\right] \times {}_3F_5\left(\begin{matrix}\frac{\mu+\nu+1}{2},\,\frac{\mu+\nu+2}{2},\,\frac{s+\mu+\nu}{2};\,a^2b^2\\ \mu+1,\,\nu+1,\,\mu+\nu+1,\,\frac{s+\mu+\nu+1}{2},\,\frac{s+\mu+\nu+2}{2}\end{matrix}\right)$ $[b>0;\ -\operatorname{Re}(\mu+\nu)<\operatorname{Re}s<2;\ \lvert\arg a\rvert<\pi/4]$
3	$\Gamma(\lambda,\,ax)\,J_\mu(bx)\,J_\nu(bx)$	$\dfrac{a^{-(s+\mu+\nu)}\,(b/2)^{\mu+\nu}}{s+\mu+\nu}\Gamma\left[\begin{matrix}s+\lambda+\mu+\nu\\ \mu+1,\,\nu+1\end{matrix}\right] \times {}_5F_4\left(\begin{matrix}\frac{\mu+\nu+1}{2},\,\frac{\mu+\nu+2}{2},\,\frac{s+\mu+\nu}{2},\,\frac{s+\lambda+\mu+\nu}{2},\,\frac{s+\lambda+\mu+\nu+1}{2}\\ \mu+1,\,\nu+1,\,\mu+\nu+1,\,\frac{s+\mu+\nu+2}{2};\,-\frac{4b^2}{a^2}\end{matrix}\right)$ $\left[\begin{matrix}(\operatorname{Re}a>0;\ \operatorname{Re}a>2\lvert\operatorname{Im}b\rvert;\ -\operatorname{Re}(\lambda+\mu+\nu),\\ -\operatorname{Re}(\mu+\nu)<\operatorname{Re}s)\ \text{or}\\ (\operatorname{Re}a>0;\ \operatorname{Re}a+2\lvert\operatorname{Im}b\rvert=0;\ -\operatorname{Re}(\lambda+\mu+\nu),\\ -\operatorname{Re}(\mu+\nu)<\operatorname{Re}s<3-\operatorname{Re}\nu)\ \text{or}\\ (\operatorname{Re}a=0;\ b>0;\ -\operatorname{Re}(\lambda+\mu+\nu),\\ -\operatorname{Re}(\mu+\nu)<\operatorname{Re}s<3-\operatorname{Re}\nu)\end{matrix}\right]$

No.	$f(x)$	$F(s)$
4	$\Gamma(\lambda,\, ax^2)\, J_\mu(bx)$ $\times J_\nu(bx)$	$\dfrac{a^{-(s+\mu+\nu)/2}\,(b/2)^{\mu+\nu}}{s+\mu+\nu}\,\Gamma\left[\begin{matrix}\frac{s+2\lambda+\mu+\nu}{2}\\ \mu+1,\ \nu+1\end{matrix}\right]$ $\times\, {}_4F_4\left(\begin{matrix}\frac{\mu+\nu+1}{2},\ \frac{\mu+\nu+2}{2},\ \frac{s+\mu+\nu}{2},\ \frac{s+2\lambda+\mu+\nu}{2}\\ \mu+1,\ \nu+1,\ \mu+\nu+1,\ \frac{s+\mu+\nu+2}{2};\ -\frac{b^2}{a}\end{matrix}\right)$ $\left[\begin{matrix}(\operatorname{Re} a>0;\ -\operatorname{Re}(2\lambda+\mu+\nu),\ -\operatorname{Re}(\mu+\nu)<\operatorname{Re} s)\ \text{or}\\ (\operatorname{Re} a=\operatorname{Im} b=0;\ -\operatorname{Re}(2\lambda+\mu+\nu),\ -\operatorname{Re}(\mu+\nu)<\operatorname{Re} s<5-2\operatorname{Re}\nu)\end{matrix}\right]$

3.10.19. $J_\mu(\varphi(x))\, J_\nu(\psi(x))$

1	$J_{\pm\nu}(u_+)\, J_\nu(u_-)$ $u_\pm = b\left(\sqrt{x^2+a^2}\pm a\right)$	$\dfrac{1}{2\sqrt{\pi}}\left(\dfrac{a}{b}\right)^{s/2}\Gamma\left[\begin{matrix}\frac{s+2\nu}{2},\ \frac{1-s}{2}\\ \frac{2-s+2\nu}{2}\end{matrix}\right]$ $\times\left\{\begin{matrix}J_{-s/2}(2ab)\\ \cos(\nu\pi)\, J_{-s/2}(2ab)-\sin(\nu\pi)\, Y_{-s/2}(2ab)\end{matrix}\right\}$ $[b,\ \operatorname{Re} a>0;\ -2\operatorname{Re}\nu<\operatorname{Re} s<1]$

3.10.20. $J_\mu(\varphi(x))\, J_\nu(\psi(x))$ and algebraic functions

1	$(a^2-x^2)_+^{\alpha-1}$ $\times J_\nu(bx)\, J_\mu(bx)$	$\dfrac{1}{2}\, a^{s+2\alpha+\mu+\nu-2}\left(\dfrac{b}{2}\right)^{\mu+\nu}\Gamma\left[\begin{matrix}\alpha,\ \frac{s+\mu+\nu}{2}\\ \mu+1,\ \nu+1,\ \frac{s+2\alpha+\mu+\nu}{2}\end{matrix}\right]$ $\times\, {}_3F_4\left(\begin{matrix}\frac{\mu+\nu+1}{2},\ \frac{\mu+\nu+2}{2},\ \frac{s+\mu+\nu}{2};\ -a^2b^2\\ \mu+1,\ \nu+1,\ \mu+\nu+1,\ \frac{s+2\alpha+\mu+\nu}{2}\end{matrix}\right)$ $[a,\ \operatorname{Re}\alpha,\ \operatorname{Re}(s+\mu+\nu)>0]$
2	$\dfrac{1}{(x^2+a^2)^\rho}$ $\times J_\mu(bx)\, J_\nu(bx)$	$\dfrac{1}{2}\left(\dfrac{b}{2}\right)^{2\rho-s}\Gamma\left[\begin{matrix}1-s+2\rho,\ \frac{s+\mu+\nu-2\rho}{2}\\ \frac{2-s+\mu-\nu+2\rho}{2},\ \frac{2-s+\mu+\nu+2\rho}{2},\ \frac{2-s-\mu+\nu+2\rho}{2}\end{matrix}\right]$ $\times\, {}_3F_4\left(\begin{matrix}\rho,\ \frac{1-s+2\rho}{2},\ \frac{2-s+2\rho}{2};\ a^2b^2\\ \frac{2-s-\mu-\nu+2\rho}{2},\ \frac{2-s+\mu-\nu+2\rho}{2},\ \frac{2-s+\mu+\nu+2\rho}{2},\ \frac{2-s+\nu-\mu+2\rho}{2}\end{matrix}\right)$ $+\dfrac{a^{s-2\rho}}{2}\left(\dfrac{ab}{2}\right)^{\mu+\nu}\Gamma\left[\begin{matrix}\frac{2\rho-\mu-\nu-s}{2},\ \frac{s+\mu+\nu}{2}\\ \mu+1,\ \nu+1,\ \rho\end{matrix}\right]$ $\times\, {}_3F_4\left(\begin{matrix}\frac{\mu+\nu+1}{2},\ \frac{\mu+\nu+2}{2},\ \frac{s+\mu+\nu}{2};\ a^2b^2\\ \mu+1,\ \nu+1,\ \mu+\nu+1,\ \frac{s+\mu+\nu-2\rho+2}{2}\end{matrix}\right)$ $[b,\ \operatorname{Re} a,\ \operatorname{Re}(s+\mu+\nu)>0;\ \operatorname{Re}(s-2\rho)<1]$
3	$(a-x)_+^{\alpha-1}$ $\times J_\mu(bx(a-x))$ $\times J_\nu(bx(a-x))$	$a^{s+\alpha+2\mu+2\nu-1}\left(\dfrac{b}{2}\right)^{\mu+\nu}\Gamma\left[\begin{matrix}s+\mu+\nu,\ \alpha+\mu+\nu\\ \mu+1,\ \nu+1,\ s+\alpha+2\mu+2\nu\end{matrix}\right]$ $\times\, {}_6F_7\left(\begin{matrix}\Delta(2,\ \mu+\nu+1),\ \Delta(2,\ s+\mu+\nu),\ \Delta(2,\ \alpha+\mu+\nu)\\ \mu+1,\ \nu+1,\ \mu+\nu+1,\ \Delta(4,\ s+\alpha+2\mu+2\nu);\ -\frac{a^4b^2}{16}\end{matrix}\right)$ $[a>0;\ \operatorname{Re} s,\ \operatorname{Re}\alpha>-\operatorname{Re}(\mu+\nu)]$

No.	$f(x)$	$F(s)$
4	$\dfrac{1}{(x+a)^{\rho}} \times J_{\mu}\left(\dfrac{bx}{x+a}\right) \times J_{\nu}\left(\dfrac{bx}{x+a}\right)$	$a^{s-\rho}\left(\dfrac{b}{2}\right)^{\mu+\nu}\dfrac{\mathrm{B}(\rho-s,\,s+\mu+\nu)}{\Gamma(\mu+1)\Gamma(\nu+1)} \times {}_4F_5\left(\begin{matrix}\frac{\mu+\nu+1}{2},\,\frac{\mu+\nu+2}{2},\,\frac{s+\mu+\nu}{2},\,\frac{s+\mu+\nu+1}{2};\,-b^2\\ \mu+1,\,\nu+1,\,\mu+\nu+1,\,\frac{\mu+\nu+\rho}{2},\,\frac{\mu+\nu+\rho+1}{2}\end{matrix}\right)$ $[\mathrm{Re}(\mu+\nu)<\mathrm{Re}\,s<\mathrm{Re}\,\rho;\ \lvert\arg a\rvert<\pi]$
5	$\dfrac{1}{(x^2+a^2)^{\rho}} \times J_{\mu}\left(\dfrac{bx}{x^2+a^2}\right) \times J_{\nu}\left(\dfrac{bx}{x^2+a^2}\right)$	$\dfrac{a^{s-\mu-\nu-2\rho}b^{\mu+\nu}}{2^{\mu+\nu+1}\Gamma(\mu+1)\Gamma(\nu+1)}\mathrm{B}\left(\dfrac{s+\mu+\nu}{2},\,\dfrac{-s+\mu+\nu+2\rho}{2}\right) \times {}_4F_5\left(\begin{matrix}\frac{\mu+\nu+1}{2},\,\frac{\mu+\nu+2}{2},\,\frac{s+\mu+\nu}{2},\,\frac{-s+\mu+\nu+2\rho}{2}\\ \mu+1,\,\nu+1,\,\mu+\nu+1,\,\frac{\mu+\nu+\rho}{2},\,\frac{\mu+\nu+\rho+1}{2};\,-\frac{b^2}{4a^2}\end{matrix}\right)$ $[\mathrm{Re}\,a>0;\ -\mathrm{Re}(\mu+\nu)<\mathrm{Re}\,s<\mathrm{Re}(\mu+\nu+2\rho)]$

3.10.21. $J_{\lambda}(ax^r)\,J_{\mu}(bx^r)\,J_{\nu}(cx)$

1	$J_{\lambda}(ax)\,J_{\mu}(ax)\,J_{\nu}(bx)$	$\dfrac{2^{s-1}a^{\mu+\lambda}}{b^{s+\mu+\lambda}}\Gamma\left[\begin{matrix}\frac{s+\lambda+\mu+\nu}{2}\\ \mu+1,\,\lambda+1,\,\frac{\nu-\mu-\lambda-s+2}{2}\end{matrix}\right] \times {}_4F_3\left(\begin{matrix}\frac{\lambda+\mu+1}{2},\,\frac{\lambda+\mu+2}{2},\,\frac{s+\lambda+\mu-\nu}{2},\,\frac{s+\lambda+\mu+\nu}{2}\\ \lambda+1,\,\mu+1,\,\lambda+\mu+1;\,\frac{4a^2}{b^2}\end{matrix}\right)$ $[0<2a<b;\ -\mathrm{Re}(\lambda+\mu+\nu)<\mathrm{Re}\,s<5/2]$
2	$J_{\lambda}(ax)\,J_{\mu}(ax)\,J_{\nu}(bx)$	$\dfrac{2^{s-2}}{\pi ab^{s-1}}\cos\dfrac{(\lambda-\mu)\pi}{2}\,\Gamma\left[\begin{matrix}\frac{s+\nu-1}{2}\\ \frac{3-s+\nu}{2}\end{matrix}\right] \times {}_4F_3\left(\begin{matrix}\frac{1-\lambda-\mu}{2},\,\frac{\lambda-\mu+1}{2},\,\frac{1-\lambda+\mu}{2},\,\frac{\lambda+\mu+1}{2}\\ \frac{1}{2},\,\frac{3-s-\nu}{2},\,\frac{3-s+\nu}{2};\,\frac{b^2}{4a^2}\end{matrix}\right)$ $+\dfrac{2^{s-1}b^{\nu}}{a^{s+\nu}}\Gamma\left[\begin{matrix}\frac{s+\lambda+\mu+\nu}{2},\,1-s-\nu\\ \nu+1,\,\frac{2-s+\lambda+\mu-\nu}{2},\,\frac{2-s+\lambda-\mu-\nu}{2},\,\frac{2-s+\mu-\lambda-\nu}{2}\end{matrix}\right] \times {}_4F_3\left(\begin{matrix}\frac{s-\lambda-\mu+\nu}{2},\,\frac{s+\lambda-\mu+\nu}{2},\,\frac{s-\lambda+\mu+\nu}{2},\,\frac{s+\lambda+\mu+\nu}{2}\\ \nu+1,\,\frac{s+\nu}{2},\,\frac{s+\nu+1}{2};\,\frac{b^2}{4a^2}\end{matrix}\right)$ $+\dfrac{2^{s-4}(\lambda^2-\mu^2)}{\pi a^2b^{s-2}}\sin\dfrac{(\lambda-\mu)\pi}{2}\,\Gamma\left[\begin{matrix}\frac{s+\nu-2}{2}\\ \frac{4-s+\nu}{2}\end{matrix}\right] \times {}_4F_3\left(\begin{matrix}\frac{2-\lambda-\mu}{2},\,\frac{\lambda-\mu+2}{2},\,\frac{2-\lambda+\mu}{2},\,\frac{\lambda+\mu+2}{2}\\ \frac{3}{2},\,\frac{4-s-\nu}{2},\,\frac{4-s+\nu}{2};\,\frac{b^2}{4a^2}\end{matrix}\right)$ $[0<b<2a;\ -\mathrm{Re}(\lambda+\mu+\nu)<\mathrm{Re}\,s<5/2]$

No.	$f(x)$	$F(s)$
3	$J_\lambda(ax)\,J_\mu(bx)\,J_\nu(cx)$	$\frac{2^{s-1}a^\lambda b^\mu}{c^{s+\lambda+\mu}}\Gamma\left[\begin{matrix}\frac{s+\lambda+\mu+\nu}{2}\\ \lambda+1,\ \mu+1,\ \frac{2-s-\lambda-\mu+\nu}{2}\end{matrix}\right]$ $\times F_4\left(\frac{s+\lambda+\mu-\nu}{2},\ \frac{s+\lambda+\mu+\nu}{2};\ \lambda+1,\ \mu+1;\ \frac{a^2}{c^2},\ \frac{b^2}{c^2}\right)$ $[a,\ b,\ \mathrm{Re}\,(s+\lambda+\mu+\nu)>0;\ c>a+b;\ \mathrm{Re}\,s<5/2]$
4	$J_\lambda\left(\frac{a}{x}\right)J_\mu\left(\frac{a}{x}\right)$ $\times J_\nu(bx)$	$\frac{a^{\lambda+\mu}b^{\lambda+\mu-s}}{2^{2\lambda+2\mu-s+1}}\Gamma\left[\begin{matrix}\frac{s-\lambda-\mu+\nu}{2}\\ \lambda+1,\ \mu+1,\ \frac{2-s+\lambda+\mu+\nu}{2}\end{matrix}\right]$ $\times {}_2F_5\left(\begin{matrix}\frac{\lambda+\mu+1}{2},\ \frac{\lambda+\mu+2}{2};\ \frac{a^2b^2}{4}\\ \lambda+1,\ \mu+1,\ \lambda+\mu+1,\ \frac{2-s+\lambda+\mu-\nu}{2},\ \frac{2-s+\lambda+\mu+\nu}{2}\end{matrix}\right)$ $+\frac{a^{s+\nu}b^\nu}{2^{s+2\nu+1}}\Gamma\left[\begin{matrix}s+\nu+1,\ \frac{\lambda+\mu-\nu-s}{2}\\ \nu+1,\ \frac{s+\lambda+\mu+\nu+2}{2},\ \frac{s-\lambda+\mu+\nu+2}{2},\ \frac{s+\lambda-\mu+\nu+2}{2}\end{matrix}\right]$ $\times {}_2F_5\left(\begin{matrix}\frac{s+\nu+1}{2},\ \frac{s+\nu+2}{2};\ \frac{a^2b^2}{4}\\ \nu+1,\ \frac{s+\lambda+\mu+\nu+2}{2},\ \frac{s-\lambda-\mu+\nu+2}{2},\ \frac{s-\lambda+\mu+\nu+2}{2},\ \frac{s+\lambda-\mu+\nu+2}{2}\end{matrix}\right)$ $[a,\ b>0;\ -\mathrm{Re}\,\nu-1<\mathrm{Re}\,s<\mathrm{Re}\,(\lambda+\mu)+3/2]$
5	$e^{-ax}\prod_{k=1}^{n}J_{\nu_k}(b_kx)$	$\frac{\prod_{k=1}^{n}(b_k/2)^{\nu_k}}{\left(a+i\sum_{k=1}^{n}b_k\right)^{s+\nu}}\Gamma\left[\begin{matrix}s+\nu\\ \nu_1+1,\ \nu_2+1,\ldots,\ \nu_n+1\end{matrix}\right]$ $\times F_A^{(n)}\left(s+\nu,\ (\nu_n)+\frac{1}{2};\ 2(\nu_n)+1;\ \frac{2i(b_n)}{a+i\sum_{k=1}^{n}b_k}\right)$ $\left[\nu=\sum_{k=1}^{n}\nu_k;\ \mathrm{Re}\,a>\sum_{k=1}^{n}\mathrm{Im}\,b_k;\ \mathrm{Re}\,(s+\nu)>0\right]$
6		$=\frac{\prod_{k=1}^{n}(b_k/2)^{\nu_k}}{a^{s+\nu}}\Gamma\left[\begin{matrix}s+\nu\\ \nu_1+1,\ \nu_2+1,\ldots,\ \nu_n+1\end{matrix}\right]$ $\times F_C^{(n)}\left(\frac{s+\nu}{2},\ \frac{s+\nu+1}{2};\ (\nu_n)+1;\ -\frac{(b_n^2)}{a^2}\right)$ $\left[\nu=\sum_{k=1}^{n}\nu_k;\ \mathrm{Re}\,a,\ \mathrm{Re}\,(s+\nu)>0\right]$
7	$e^{-ax}\prod_{k=1}^{m}\sin(b_kx)$ $\times\prod_{k=1}^{n}\cos(c_kx)$ $\times\prod_{k=1}^{p}J_{\nu_k}(d_kx)$	$\frac{\prod_{k=1}^{m}b_k\prod_{k=1}^{p}(d_k/2)^{\nu_k}}{a^{s+m+\nu}}\Gamma\left[\begin{matrix}s+m+\nu\\ \nu_1+1,\ \nu_2+1,\ldots,\ \nu_p+1\end{matrix}\right]$ $\times F_C^{(m+n+p)}\left(\frac{s+m+\nu}{2},\ \frac{s+m+\nu+1}{2};\ \underbrace{\frac{3}{2},\ldots,\frac{3}{2}}_{m},\ \underbrace{\frac{1}{2},\ldots,\frac{1}{2}}_{n},\right.$ $\left.(\nu_p)+1;\ -\frac{(b_m^2)}{a^2},\ -\frac{(c_n^2)}{a^2},\ -\frac{(d_p^2)}{a^2}\right)$ $\left[\nu=\sum_{k=1}^{p}\nu_k;\ \mathrm{Re}\,a,\ \mathrm{Re}\,(s+\nu)>0\right]$

3.11. The Bessel Function $Y_\nu(z)$

More formulas can be obtained from the corresponding sections due to the relations

$$Y_{\pm 1/2}(z) = \mp\sqrt{\frac{2}{\pi}}\frac{1}{\sqrt{z}}\begin{Bmatrix}\cos z\\ \sin z\end{Bmatrix}, \quad Y_{\pm 3/2}(z) = \sqrt{\frac{2}{\pi}}\frac{1}{z^{3/2}}\left[\mp z\begin{Bmatrix}\sin z\\ \cos z\end{Bmatrix} - \begin{Bmatrix}\cos z\\ \sin z\end{Bmatrix}\right],$$

$$Y_\nu(z) = \csc(\nu\pi)\left[J_\nu(z)\cos(\nu\pi) - J_{-\nu}(z)\right], \quad [\nu \neq 0, \pm 1, \pm 2, \ldots];$$

$$Y_n(z) = \lim_{\nu\to n} Y_\nu(z), \quad [n = 0, \pm 1, \pm 2, \ldots];$$

$$Y_\nu(z) = \frac{1}{2i}\left[H_\nu^{(1)}(z) - H_\nu^{(2)}(z)\right],$$

$$Y_\nu(z) = -\frac{2}{\pi}\left\{i^\nu K_\nu(iz) + \left[\ln(iz) - \ln z\right] J_\nu(z)\right\}, \quad [\nu \neq 0, \pm 1, \pm 2, \ldots];$$

$$Y_\nu(z) = \frac{(iz)^{-\nu} z^{-\nu}}{\pi}\left\{\pi\csc(\nu\pi)\left[\cos(\nu\pi) z^{2\nu} - (iz)^{2\nu}\right] I_\nu(iz) - 2(iz)^{2\nu} K_\nu(iz)\right\},$$
$$[\nu \neq 0, \pm 1, \pm 2, \ldots];$$

$$Y_\nu(z) = -\frac{\cos(\nu\pi)\Gamma(-\nu)}{\pi}\left(\frac{z}{2}\right)^\nu {}_0F_1\left(1+\nu; -\frac{z^2}{4}\right) - \frac{\Gamma(\nu)}{\pi}\left(\frac{2}{z}\right)^\nu {}_0F_1\left(1-\nu; -\frac{z^2}{4}\right),$$

$$Y_\nu\left(\sqrt{z^2}\right) = G_{13}^{20}\left(\frac{z^2}{4}\,\middle|\,\begin{matrix} -(\nu+1)/2 \\ \nu/2, \ -\nu/2, \ -(\nu+1)/2\end{matrix}\right).$$

3.11.1. $Y_\nu(bx)$ and algebraic functions

No.	$f(x)$	$F(s)$
1	$Y_\nu(ax)$	$-\frac{2^{s-1}}{\pi a^s}\cos\frac{(s-\nu)\pi}{2}\Gamma\left(\frac{s-\nu}{2}\right)\Gamma\left(\frac{s+\nu}{2}\right)$ $[a>0;\ \lvert\operatorname{Re}\nu\rvert < \operatorname{Re}s < 3/2]$
2	$(a-x)_+^{\alpha-1} Y_\nu(bx)$	$-\frac{2^\nu a^{s+\alpha-\nu-1}b^{-\nu}}{\pi}\Gamma(\nu)\mathrm{B}(\alpha, s-\nu)\, {}_2F_3\left(\begin{matrix}\frac{s-\nu}{2}, \frac{s-\nu+1}{2}; -\frac{a^2b^2}{4}\\ 1-\nu, \frac{s+\alpha-\nu}{2}, \frac{s+\alpha-\nu+1}{2}\end{matrix}\right)$ $-\frac{2^{-\nu}a^{s+\alpha+\nu-1}b^\nu}{\pi}\cos(\pi\nu)\Gamma(-\nu)\mathrm{B}(\alpha, s+\nu)$ $\times\, {}_2F_3\left(\begin{matrix}\frac{s+\nu}{2}, \frac{s+\nu+1}{2}; -\frac{a^2b^2}{4}\\ 1+\nu, \frac{s+\alpha+\nu}{2}, \frac{s+\alpha+\nu+1}{2}\end{matrix}\right)$ $[a,\ \operatorname{Re}\alpha>0;\ \operatorname{Re}s > \lvert\operatorname{Re}\nu\rvert]$
3	$\frac{1}{x-a}Y_\nu(bx)$	$\frac{(2a)^{s-1}}{\pi}\left[\cos\frac{(s-\nu)\pi}{2}\Gamma\left(\frac{s-\nu}{2}\right)\Gamma\left(\frac{s+\nu}{2}\right)S_{1-s,\nu}(ab)\right.$ $\left.-2\sin\frac{(s-\nu)\pi}{2}\Gamma\left(\frac{s-\nu+1}{2}\right)\Gamma\left(\frac{s+\nu+1}{2}\right)S_{-s,\nu}(ab)\right]$ $+\pi a^{s-1}J_\nu(ab) \quad [a, b>0;\ \lvert\operatorname{Re}\nu\rvert < \operatorname{Re}s < 5/2]$

No.	$f(x)$	$F(s)$
4	$\frac{1}{(x+a)^{\rho}} Y_\nu(bx)$	$-\frac{a^{s+\nu-\rho}}{\pi}\left(\frac{b}{2}\right)^{\nu}\cos(\nu\pi)\,\Gamma\left[\begin{matrix}-\nu,\ s+\nu,\ -\nu+\rho-s\\ \rho\end{matrix}\right]$ $\times\, {}_2F_3\left(\begin{matrix}\frac{s+\nu}{2},\ \frac{s+\nu+1}{2};\ -\frac{a^2b^2}{4}\\ \nu+1,\ \frac{s+\nu-\rho+1}{2},\ \frac{s+\nu-\rho+2}{2}\end{matrix}\right) - \frac{a^{s-\nu-\rho}}{\pi}\left(\frac{2}{b}\right)^{\nu}$ $\times\,\Gamma\left[\begin{matrix}\nu,\ s-\nu,\ \nu+\rho-s\\ \rho\end{matrix}\right] {}_2F_3\left(\begin{matrix}\frac{s-\nu}{2},\ \frac{s-\nu+1}{2};\ -\frac{a^2b^2}{4}\\ 1-\nu,\ \frac{s-\nu-\rho+1}{2},\ \frac{s-\nu-\rho+2}{2}\end{matrix}\right)$ $-\frac{1}{2\pi}\left(\frac{b}{2}\right)^{\rho-s}\cos\frac{(s-\nu-\rho)\pi}{2}\,\Gamma\left(\frac{s+\nu-\rho}{2}\right)$ $\times\,\Gamma\left(\frac{s-\nu-\rho}{2}\right) {}_2F_3\left(\begin{matrix}\frac{\rho}{2},\ \frac{\rho+1}{2};\ -\frac{a^2b^2}{4}\\ \frac{1}{2},\ \frac{\rho-\nu-s+2}{2},\ \frac{\rho+\nu-s+2}{2}\end{matrix}\right)$ $-\frac{\rho a}{2\pi}\left(\frac{b}{2}\right)^{\rho-s+1}\sin\frac{(\nu+\rho-s)\pi}{2}\,\Gamma\left(\frac{s+\nu-\rho-1}{2}\right)$ $\times\,\Gamma\left(\frac{s-\nu-\rho-1}{2}\right) {}_2F_3\left(\begin{matrix}\frac{\rho+1}{2},\ \frac{\rho+2}{2};\ -\frac{a^2b^2}{4}\\ \frac{3}{2},\ \frac{\rho-\nu-s+3}{2},\ \frac{\rho+\nu-s+3}{2}\end{matrix}\right)$ $[b>0;\ \lvert\operatorname{Re}\nu\rvert < \operatorname{Re}s < \operatorname{Re}\rho + 3/2;\ \lvert\arg a\rvert < \pi]$
5	$\frac{1}{x+a} Y_\nu(bx)$	$-\frac{(2a)^{s-1}}{\pi}\left[\cos\frac{(s-\nu)\pi}{2}\,\Gamma\left(\frac{s+\nu}{2}\right)\Gamma\left(\frac{s-\nu}{2}\right) S_{1-s,\,\nu}(ab)\right.$ $\left.+\,2\sin\frac{(s-\nu)\pi}{2}\,\Gamma\left(\frac{s-\nu+1}{2}\right)\Gamma\left(\frac{s+\nu+1}{2}\right) S_{-s,\,\nu}(ab)\right]$ $[b>0;\ \lvert\operatorname{Re}\nu\rvert < \operatorname{Re}s < 5/2;\ \lvert\arg a\rvert < \pi]$
6	$(a^2-x^2)_+^{\alpha-1} Y_\nu(bx)$	$-\frac{2^{\nu-1}a^{s+2\alpha-\nu-2}}{\pi b^{\nu}}\Gamma(\nu)\,\mathrm{B}\left(\alpha,\ \frac{s-\nu}{2}\right) {}_1F_2\left(\begin{matrix}\frac{s-\nu}{2};\ -\frac{a^2b^2}{4}\\ 1-\nu,\ \frac{s+2\alpha-\nu}{2}\end{matrix}\right)$ $-\frac{a^{s+2\alpha+\nu-2}b^{\nu}}{\pi 2^{\nu+1}}\cos(\pi\nu)\,\Gamma(-\nu)\,\mathrm{B}\left(\alpha,\ \frac{s+\nu}{2}\right)$ $\times\, {}_1F_2\left(\begin{matrix}\frac{s+\nu}{2};\ -\frac{a^2b^2}{4}\\ 1+\nu,\ \frac{s+2\alpha+\nu}{2}\end{matrix}\right)$ $[a,\ \operatorname{Re}\alpha>0;\ \operatorname{Re}s > \lvert\operatorname{Re}\nu\rvert]$
7	$(x^2-a^2)_+^{\alpha-1} Y_\nu(bx)$	$\frac{2^{s+2\alpha-3}}{\pi b^{s+2\alpha-2}}\cos\frac{(s+2\alpha-\nu)\pi}{2}\,\Gamma\left(\frac{s+2\alpha-\nu-2}{2}\right)$ $\times\,\Gamma\left(\frac{s+2\alpha+\nu-2}{2}\right) {}_1F_2\left(\begin{matrix}1-\alpha;\ -\frac{a^2b^2}{4}\\ \frac{4-s-2\alpha-\nu}{2},\ \frac{4-s-2\alpha+\nu}{2}\end{matrix}\right)$ $-\frac{2^{\nu-1}a^{s+2\alpha-\nu-2}}{\pi b^{\nu}}\Gamma(\nu)\,\mathrm{B}\left(\alpha,\ \frac{-s-2\alpha+\nu+2}{2}\right)$ $\times\, {}_1F_2\left(\begin{matrix}\frac{s-\nu}{2};\ -\frac{a^2b^2}{4}\\ 1-\nu,\ \frac{s+2\alpha-\nu}{2}\end{matrix}\right) - \frac{2^{-\nu-1}a^{s+2\alpha+\nu-2}b^{\nu}}{\pi}\cos(\pi\nu)$ $\times\,\Gamma(-\nu)\,\mathrm{B}\left(\alpha,\ \frac{-s-2\alpha-\nu+2}{2}\right) {}_1F_2\left(\begin{matrix}\frac{s+\nu}{2};\ -\frac{a^2b^2}{4}\\ 1+\nu,\ \frac{s+2\alpha+\nu}{2}\end{matrix}\right)$ $[a,\ b,\ \operatorname{Re}\alpha>0;\ \operatorname{Re}(s+2\alpha) < 7/2]$

No.	$f(x)$	$F(s)$
8	$\dfrac{1}{(x^2+a^2)^\rho} Y_\nu(bx)$	$\dfrac{a^{s+\nu-2\rho}}{2}\left(\dfrac{b}{2}\right)^\nu \cot(\nu\pi)\,\Gamma\left[\begin{matrix}\frac{s+\nu}{2},\ \frac{-s-\nu+2\rho}{2}\\ \nu+1,\ \rho\end{matrix}\right] {}_1F_2\left(\begin{matrix}\frac{s+\nu}{2};\ \frac{a^2b^2}{4}\\ \nu+1,\ \frac{s+\nu-2\rho+2}{2}\end{matrix}\right)$ $-\dfrac{a^{s-\nu-2\rho}}{2\sin(\nu\pi)}\left(\dfrac{2}{b}\right)^\nu \Gamma\left[\begin{matrix}\frac{s-\nu}{2},\ \frac{\nu+2\rho-s}{2}\\ 1-\nu,\ \rho\end{matrix}\right] {}_1F_2\left(\begin{matrix}\frac{s-\nu}{2};\ \frac{a^2b^2}{4}\\ 1-\nu,\ \frac{s-\nu-2\rho+2}{2}\end{matrix}\right)$ $-\dfrac{1}{2\pi}\left(\dfrac{b}{2}\right)^{2\rho-s} \cos\dfrac{(\nu+2\rho-s)\pi}{2}\,\Gamma\left(\dfrac{s+\nu-2\rho}{2}\right)$ $\times\Gamma\left(\dfrac{s-\nu-2\rho}{2}\right) {}_1F_2\left(\begin{matrix}\rho;\ \frac{a^2b^2}{4}\\ \frac{2-s-\nu+2\rho}{2},\ \frac{2-s+\nu+2\rho}{2}\end{matrix}\right)$ $[\operatorname{Re} a,\ b>0;\ \lvert\operatorname{Re}\nu\rvert<\operatorname{Re} s<2\operatorname{Re}\rho+3/2]$
9	$\dfrac{1}{x^2+a^2} Y_\nu(bx)$	$\dfrac{1}{2\pi}\left(\dfrac{b}{2}\right)^{2-s}\cos\dfrac{(s-\nu)\pi}{2}\,\Gamma\left(\dfrac{s-\nu-2}{2}\right)\Gamma\left(\dfrac{s+\nu-2}{2}\right)$ $\times {}_1F_2\left(\begin{matrix}1;\ \frac{a^2b^2}{4}\\ \frac{4-s-\nu}{2},\ \frac{4-s+\nu}{2}\end{matrix}\right)+\dfrac{\pi a^{s-2}}{2}\csc(\nu\pi)\csc\dfrac{(\nu-s)\pi}{2}$ $\times I_{-\nu}(ab)+\dfrac{\pi a^{s-2}}{2}\cot(\nu\pi)\csc\dfrac{(s+\nu)\pi}{2} I_\nu(ab)$ $[\operatorname{Re} a,\ b>0;\ \lvert\operatorname{Re}\nu\rvert<\operatorname{Re} s<7/2]$
10	$\dfrac{1}{x^2-a^2} Y_\nu(bx)$	$\dfrac{1}{2\pi}\left(\dfrac{b}{2}\right)^{2-s}\cos\dfrac{(s-\nu)\pi}{2}\,\Gamma\left(\dfrac{s-\nu-2}{2}\right)\Gamma\left(\dfrac{s+\nu-2}{2}\right)$ $\times {}_1F_2\left(\begin{matrix}1;\ -\frac{a^2b^2}{4}\\ \frac{4-s-\nu}{2},\ \frac{4-s+\nu}{2}\end{matrix}\right)-\dfrac{\pi a^{s-2}}{2}\cot(\nu\pi)\cot\dfrac{(s+\nu)\pi}{2}$ $\times J_\nu(ab)+\dfrac{\pi a^{s-2}}{2}\csc(\nu\pi)\cot\dfrac{(s-\nu)\pi}{2} J_{-\nu}(ab)$ $[a,\ b>0;\ \lvert\operatorname{Re}\nu\rvert<\operatorname{Re} s<7/2]$
11	$\left(\sqrt{x^2+a^2}\pm x\right)^\rho Y_\nu(bx)$	$-\dfrac{2^{s\pm2\rho-1}a^{\rho\mp\rho}}{\pi b^{s\pm\rho}}\cos\dfrac{(\nu\mp\rho-s)\pi}{2}\,\Gamma\left(\dfrac{s\pm\rho+\nu}{2}\right)$ $\times\Gamma\left(\dfrac{s\pm\rho-\nu}{2}\right) {}_2F_3\left(\begin{matrix}\mp\frac{\rho}{2},\ \frac{1\mp\rho}{2};\ \frac{a^2b^2}{4}\\ 1\mp\rho,\ \frac{2\mp\rho-\nu-s}{2},\ \frac{2\mp\rho+\nu-s}{2}\end{matrix}\right)$ $-\dfrac{\rho a^{s+\rho+\nu}b^\nu}{2^{s+2\nu+1}}\cos(\nu\pi)\csc\dfrac{(\mp\rho-\nu-s)\pi}{2}$ $\times\Gamma\left[\begin{matrix}-\nu,\ s+\nu\\ \frac{s\mp\rho+\nu+2}{2},\ \frac{s\pm\rho+\nu+2}{2}\end{matrix}\right] {}_2F_3\left(\begin{matrix}\frac{s+\nu}{2},\ \frac{s+\nu+1}{2};\ \frac{a^2b^2}{4}\\ 1+\nu,\ \frac{s\pm\rho+\nu+2}{2},\ \frac{s\mp\rho+\nu+2}{2}\end{matrix}\right)$ $-\dfrac{\rho a^{s+\rho-\nu}}{2^{s-2\nu+1}b^\nu}\csc\dfrac{(\nu\mp\rho-s)\pi}{2}\,\Gamma\left[\begin{matrix}\nu,\ s-\nu\\ \frac{s\mp\rho-\nu+2}{2},\ \frac{s\pm\rho-\nu+2}{2}\end{matrix}\right]$ $\times {}_2F_3\left(\begin{matrix}\frac{s-\nu}{2},\ \frac{s-\nu+1}{2};\ \frac{a^2b^2}{4}\\ 1-\nu,\ \frac{s\pm\rho-\nu+2}{2},\ \frac{s\mp\rho-\nu+2}{2}\end{matrix}\right)$ $[\operatorname{Re} a,\ b>0;\ \lvert\operatorname{Re}\nu\rvert<\operatorname{Re} s<3/2\mp\operatorname{Re}\rho]$

No.	$f(x)$	$F(s)$
12	$\frac{\left(\sqrt{x^2+a^2}\pm x\right)^\rho}{\sqrt{x^2+a^2}} Y_\nu(bx)$	$-\frac{2^{s\pm 2\rho-2}a^{\rho\mp\rho}}{\pi b^{s\pm\rho-1}} \cos\frac{(\nu\mp\rho-s+1)\pi}{2} \Gamma\left(\frac{s\pm\rho+\nu-1}{2}\right)$ $\times \Gamma\left(\frac{s\pm\rho-\nu-1}{2}\right) {}_2F_3\left(\begin{matrix}\frac{1\mp\rho}{2}, \frac{2\mp\rho}{2}; \frac{a^2b^2}{4}\\ 1\mp\rho, \frac{3-s\mp\rho-\nu}{2}, \frac{3-s\mp\rho+\nu}{2}\end{matrix}\right)$ $-\frac{a^{s+\rho+\nu-1}b^\nu}{2^{s+2\nu}} \cos(\nu\pi)\csc\frac{(1\mp\rho-\nu-s)\pi}{2}$ $\times \Gamma\left[\begin{matrix}-\nu, s+\nu\\ \frac{s\mp\rho+\nu+1}{2}, \frac{s\pm\rho+\nu+1}{2}\end{matrix}\right] {}_2F_3\left(\begin{matrix}\frac{s+\nu}{2}, \frac{s+\nu+1}{2}; \frac{a^2b^2}{4}\\ 1+\nu, \frac{s\pm\rho+\nu+1}{2}, \frac{s\mp\rho+\nu+1}{2}\end{matrix}\right)$ $-\frac{a^{s+\rho-\nu-1}}{2^{s-2\nu}b^\nu} \csc\frac{(\nu\mp\rho-s+1)\pi}{2} \Gamma\left[\begin{matrix}\nu, s-\nu\\ \frac{s\mp\rho-\nu+1}{2}, \frac{s\pm\rho-\nu+1}{2}\end{matrix}\right]$ $\times {}_2F_3\left(\begin{matrix}\frac{s-\nu}{2}, \frac{s-\nu+1}{2}; \frac{a^2b^2}{4}\\ 1-\nu, \frac{s\pm\rho-\nu+1}{2}, \frac{s\mp\rho-\nu+1}{2}\end{matrix}\right)$ $[\operatorname{Re} a, b>0; \lvert\operatorname{Re}\nu\rvert<\operatorname{Re} s<5/2\mp\operatorname{Re}\rho]$

3.11.2. $Y_\nu(\varphi(x))$ and algebraic functions

1	$\left(x^2+a^2\right)^{\nu/2}$ $\times Y_\nu\left(b\sqrt{x^2+a^2}\right)$	$\frac{2^{s/2-1}a^{s/2+\nu}}{b^{s/2}} \Gamma\left(\frac{s}{2}\right)\left[Y_{s/2+\nu}(ab)\cos\frac{s\pi}{2} + J_{s/2+\nu}(ab)\sin\frac{s\pi}{2}\right]$ $[a, b>0; 0<\operatorname{Re} s<3/2-\operatorname{Re}\nu]$
2	$\left(x^2+a^2\right)^{-\nu/2}$ $\times Y_\nu\left(b\sqrt{x^2+a^2}\right)$	$\frac{2^{s/2-1}a^{s/2-\nu}}{b^{s/2}} \Gamma\left(\frac{s}{2}\right) Y_{\nu-s/2}(ab)$ $[a, b>0; 0<\operatorname{Re} s<\operatorname{Re}\nu+3/2]$
3	$\left(a^2-x^2\right)_+^{\nu/2}$ $\times Y_\nu\left(b\sqrt{a^2-x^2}\right)$	$2^{s/2-1}a^{s/2+\nu}b^{-s/2}\cot(\nu\pi)\,\Gamma\left(\frac{s}{2}\right) J_{s/2+\nu}(ab)$ $-\frac{2^\nu a^s b^{-\nu}}{s\pi}\Gamma(\nu)\, {}_1F_2\left(\begin{matrix}1; -\frac{a^2b^2}{4}\\ 1-\nu, \frac{s+2}{2}\end{matrix}\right)$ $[a, b, \operatorname{Re} s>0;\ \operatorname{Re}\nu>-1]$
4	$\left(a^2-x^2\right)_+^{-\nu/2}$ $\times Y_\nu\left(b\sqrt{a^2-x^2}\right)$	$-2^{s/2-1}a^{s/2-\nu}b^{-s/2}\csc(\nu\pi)\,\Gamma\left(\frac{s}{2}\right) J_{s/2-\nu}(ab)$ $-\frac{2^{-\nu}a^s b^\nu}{s\pi}\cos(\nu\pi)\,\Gamma(-\nu)\, {}_1F_2\left(\begin{matrix}1; -\frac{a^2b^2}{4}\\ 1+\nu, \frac{s+2}{2}\end{matrix}\right)$ $[a, b, \operatorname{Re} s>0;\ \operatorname{Re}\nu<1]$

No.	$f(x)$	$F(s)$
5	$Y_\nu\left(ax+\frac{a}{x}\right)$	$\frac{\pi}{2}\left[J_{(\nu-s)/2}(a)\,J_{(\nu+s)/2}(a)-Y_{(\nu-s)/2}(a)\,Y_{(\nu+s)/2}(a)\right]$ $[a>0;\ \lvert\operatorname{Re}s\rvert<3/2]$

3.11.3. $Y_\nu(bx)$ and the exponential function

1	$e^{-ax}\,Y_\nu(bx)$	$-\left(\frac{b}{2}\right)^{\nu}\frac{\cos(\nu\pi)}{\pi a^{s+\nu}}\Gamma(-\nu)\,\Gamma(s+\nu)\ {}_2F_1\left(\begin{matrix}\frac{s+\nu}{2},\ \frac{s+\nu+1}{2}\\ 1+\nu;\ -\frac{b^2}{a^2}\end{matrix}\right)$ $-\frac{a^{\nu-s}}{\pi}\left(\frac{2}{b}\right)^{\nu}\Gamma(\nu)\,\Gamma(s-\nu)\ {}_2F_1\left(\begin{matrix}\frac{s-\nu}{2},\ \frac{s-\nu+1}{2}\\ 1-\nu;\ -\frac{b^2}{a^2}\end{matrix}\right)$ $[\operatorname{Re}a>\lvert\operatorname{Im}b\rvert;\ \operatorname{Re}s>\lvert\operatorname{Re}\nu\rvert]$
2	$e^{\pm iax}\,Y_\nu(ax)$	$-\frac{e^{\pm(s+\nu)\pi i/2}}{\pi^{3/2}(2a)^s}\Gamma\left[s+\nu,\ s-\nu,\ \frac{1-2s}{2}\right]$ $\times\left[2\cos(\nu\pi)\cos(s\pi)\mp i\sin[(s+\nu)\pi]\right]$ $[a>0;\ \lvert\operatorname{Re}\nu\rvert<\operatorname{Re}s<1/2]$
3	$e^{-ax^2}\,Y_\nu(bx)$	$-\frac{2^{-\nu-1}b^{\nu}}{\pi a^{(s+\nu)/2}}\cos(\nu\pi)\,\Gamma(-\nu)\,\Gamma\left(\frac{s+\nu}{2}\right)\,{}_1F_1\left(\begin{matrix}\frac{s+\nu}{2};\ -\frac{b^2}{4a}\\ 1+\nu\end{matrix}\right)$ $-\frac{2^{\nu-1}b^{-\nu}}{\pi a^{(s-\nu)/2}}\Gamma(\nu)\,\Gamma\left(\frac{s-\nu}{2}\right)\,{}_1F_1\left(\begin{matrix}\frac{s-\nu}{2};\ -\frac{b^2}{4a}\\ 1-\nu\end{matrix}\right)$ $[\operatorname{Re}a,\ b>0;\ \operatorname{Re}s>\lvert\operatorname{Re}\nu\rvert]$

3.11.4. $Y_\nu(bx)$ and trigonometric functions

Notation: $\delta=\begin{Bmatrix}1\\0\end{Bmatrix}$.

1	$\begin{Bmatrix}\sin(ax)\\ \cos(ax)\end{Bmatrix}Y_\nu(ax)$	$\pm\frac{2^{1-s}a^{-s}}{\pi^{3/2}}\begin{Bmatrix}\sin^2[(s-\nu)\pi/2]\sin[(s+\nu)\pi/2]\\ \cos^2[(s-\nu)\pi/2]\cos[(s+\nu)\pi/2]\end{Bmatrix}$ $\times\Gamma\left[\frac{1-2s}{2},\ s-\nu,\ s+\nu\right]$ $[a>0;\ \lvert\operatorname{Re}\nu\rvert-\delta<\operatorname{Re}s<1/2]$
2	$\begin{Bmatrix}\sin(ax)\\ \cos(ax)\end{Bmatrix}Y_\nu(bx)$	$\pm\frac{2^{s+\delta-1}a^{\delta}}{\pi b^{s+\delta}}\begin{Bmatrix}\sin[(s-\nu)\pi/2]\\ \cos[(s-\nu)\pi/2]\end{Bmatrix}\Gamma\left(\frac{s-\nu+\delta}{2}\right)$ $\times\Gamma\left(\frac{s+\nu+\delta}{2}\right)\,{}_2F_1\left(\begin{matrix}\frac{s-\nu+\delta}{2},\ \frac{s+\nu+\delta}{2}\\ \frac{2\delta+1}{2};\ \frac{a^2}{b^2}\end{matrix}\right)$ $[0<a<b;\ \lvert\operatorname{Re}\nu\rvert-\delta<\operatorname{Re}s<3/2]$

No.	$f(x)$	$F(s)$
3	$\begin{Bmatrix} \sin(ax) \\ \cos(ax) \end{Bmatrix} Y_\nu(bx)$	$-\frac{b^\nu \cos(\nu\pi)}{2^{\nu+1} a^{s+\nu}} \begin{Bmatrix} \sec[(s+\nu)\pi/2] \\ \csc[(s+\nu)\pi/2] \end{Bmatrix} \Gamma\begin{bmatrix} -\nu \\ 1-s-\nu \end{bmatrix} \times {}_2F_1\left(\begin{matrix} \frac{s+\nu}{2}, \frac{s+\nu+1}{2} \\ 1+\nu;\ \frac{b^2}{a^2} \end{matrix}\right) - \frac{2^{\nu-1} a^{\nu-s}}{b^\nu} \begin{Bmatrix} \sec[(s-\nu)\pi/2] \\ \csc[(s-\nu)\pi/2] \end{Bmatrix} \times \Gamma\begin{bmatrix} \nu \\ 1-s+\nu \end{bmatrix} {}_2F_1\left(\begin{matrix} \frac{s-\nu}{2}, \frac{s-\nu+1}{2} \\ 1-\nu;\ \frac{b^2}{a^2} \end{matrix}\right)$ $[0 < b < a;\ \lvert\operatorname{Re}\nu\rvert - \delta < \operatorname{Re}s < 3/2]$
4	$\begin{Bmatrix} \sin(ax+b) \\ \cos(ax+b) \end{Bmatrix} Y_\nu(ax)$	$\mp\frac{(2a)^{-s}}{\sqrt{\pi}} \begin{Bmatrix} \cos[(s+\nu)\pi/2+b] \\ \sin[(s+\nu)\pi/2+b] \end{Bmatrix} \Gamma\begin{bmatrix} \frac{1-2s}{2}, s+\nu \\ 1-s+\nu \end{bmatrix} - \frac{2^{1-s} a^{-s}}{\pi^{3/2}} \cos(\nu\pi) \begin{Bmatrix} \sin[(\nu-s)\pi/2+b] \\ \cos[(\nu-s)\pi/2+b] \end{Bmatrix} \times \Gamma\left(\frac{1-2s}{2}\right) \Gamma(s-\nu)\,\Gamma(s+\nu)$ $[a > 0;\ \operatorname{Re}\nu < \operatorname{Re}s < 1/2]$

3.11.5. $Y_\nu(bx)$ and the logarithmic function

No.	$f(x)$	$F(s)$
1	$\ln x\, Y_\nu(ax)$	$\frac{2^{s-2}}{a^s} \Gamma\left(\frac{s+\nu}{2}\right) \Gamma\left(\frac{s-\nu}{2}\right) \left\{ \sin\frac{(s-\nu)\pi}{2} - \frac{1}{\pi} \cos\frac{(s-\nu)\pi}{2} \left[\psi\left(\frac{s+\nu}{2}\right) + \psi\left(\frac{s-\nu}{2}\right) - 2\ln\frac{a}{2} \right] \right\}$ $[a > 0;\ \lvert\operatorname{Re}\nu\rvert < \operatorname{Re}s < 3/2]$

3.11.6. $Y_\nu(bx)$ and $\operatorname{Ei}(ax^r)$

No.	$f(x)$	$F(s)$
1	$\operatorname{Ei}(-ax)\, Y_\nu(bx)$	$\frac{\cos(\nu\pi)\, a^{-s-\nu} (b/2)^\nu}{\pi(s+\nu)} \Gamma(-\nu)\,\Gamma(s+\nu)\, {}_3F_2\left(\begin{matrix} \frac{s+\nu}{2}, \frac{s+\nu}{2}, \frac{s+\nu+1}{2} \\ \nu+1, \frac{s+\nu+2}{2};\ -\frac{b^2}{a^2} \end{matrix}\right) + \frac{a^{\nu-s} (b/2)^{-\nu}}{\pi(s-\nu)} \Gamma(\nu)\,\Gamma(s-\nu)\, {}_3F_2\left(\begin{matrix} \frac{s-\nu}{2}, \frac{s-\nu}{2}, \frac{s-\nu+1}{2} \\ 1-\nu, \frac{s-\nu+2}{2};\ -\frac{b^2}{a^2} \end{matrix}\right)$ $[b,\ \operatorname{Re}a > 0;\ \operatorname{Re}s > \lvert\operatorname{Re}\nu\rvert]$
2	$\operatorname{Ei}(-ax^2)\, Y_\nu(bx)$	$\frac{2^\nu a^{(\nu-s)/2}}{\pi b^\nu (s-\nu)} \Gamma(\nu)\,\Gamma\left(\frac{s-\nu}{2}\right) {}_2F_2\left(\begin{matrix} \frac{s-\nu}{2}, \frac{s-\nu}{2};\ -\frac{b^2}{4a} \\ 1-\nu, \frac{s-\nu+2}{2} \end{matrix}\right) + \frac{b^\nu \cos(\nu\pi)}{2^\nu (s+\nu)\pi a^{(s+\nu)/2}} \Gamma(-\nu)\,\Gamma\left(\frac{s+\nu}{2}\right) {}_2F_2\left(\begin{matrix} \frac{s+\nu}{2}, \frac{s+\nu}{2};\ -\frac{b^2}{4a} \\ 1+\nu, \frac{s+\nu+2}{2} \end{matrix}\right)$ $[a,\ \operatorname{Re}b > 0;\ \operatorname{Re}s > \lvert\operatorname{Re}\nu\rvert]$

No.	$f(x)$	$F(s)$
3	$e^{\pm ax}\operatorname{Ei}(\mp ax)\,Y_\nu(bx)$	$-\dfrac{2^\nu\Gamma(\nu)\,\Gamma(s-\nu)}{a^{s-\nu}b^\nu}\begin{Bmatrix}\csc[(\nu-s)\pi]\\ \cot[(\nu-s)\pi]\end{Bmatrix}{}_2F_1\left(\begin{matrix}\frac{s-\nu}{2},\ \frac{s-\nu+1}{2}\\ 1-\nu;\ -\frac{b^2}{a^2}\end{matrix}\right) + \dfrac{\cos(\nu\pi)}{2^\nu a^{s+\nu}b^{-\nu}}\Gamma(-\nu)\,\Gamma(s+\nu) \times\begin{Bmatrix}\csc[(s+\nu)\pi]\\ \cot[(s+\nu)\pi]\end{Bmatrix}{}_2F_1\left(\begin{matrix}\frac{s+\nu}{2},\ \frac{s+\nu+1}{2}\\ 1+\nu;\ -\frac{b^2}{a^2}\end{matrix}\right) \pm\dfrac{2^{s-2}}{\pi ab^{s-1}}\sin\dfrac{(s-\nu)\pi}{2}\,\Gamma\left(\dfrac{s+\nu-1}{2}\right)\Gamma\left(\dfrac{s-\nu-1}{2}\right) \times{}_3F_2\left(\begin{matrix}\frac{1}{2},\ 1,\ 1;\ -\frac{b^2}{a^2}\\ \frac{3-s-\nu}{2},\ \frac{3-s+\nu}{2}\end{matrix}\right) + \dfrac{2^{s-3}}{\pi a^2b^{s-2}}\cos\dfrac{(s-\nu)\pi}{2} \times\Gamma\left(\dfrac{s+\nu-2}{2}\right)\Gamma\left(\dfrac{s-\nu-2}{2}\right){}_3F_2\left(\begin{matrix}1,\ 1,\ \frac{3}{2};\ -\frac{b^2}{a^2}\\ \frac{4-s-\nu}{2},\ \frac{4-s+\nu}{2}\end{matrix}\right)$ $[b,\ \operatorname{Re}a>0;\ \lvert\operatorname{Re}\nu\rvert<\operatorname{Re}s<5/2]$

3.11.7. $Y_\nu(bx)$ and $\operatorname{si}(ax)$, $\operatorname{ci}(ax)$

1	$\begin{Bmatrix}\operatorname{si}(ax)\\ \operatorname{ci}(ax)\end{Bmatrix}Y_\nu(bx)$	$\dfrac{2^\nu a^{\nu-s}}{\pi b^\nu(s-\nu)}\Gamma(\nu)\,\Gamma(s-\nu)\begin{Bmatrix}\sin[(s-\nu)\pi/2]\\ \cos[(s-\nu)\pi/2]\end{Bmatrix} \times{}_3F_2\left(\begin{matrix}\frac{s-\nu}{2},\ \frac{s-\nu}{2},\ \frac{s-\nu+1}{2}\\ 1-\nu,\ \frac{s-\nu+2}{2};\ \frac{b^2}{a^2}\end{matrix}\right) + \dfrac{b^\nu\Gamma(-\nu)\,\Gamma(s+\nu)}{2^\nu\pi a^{s+\nu}(s+\nu)} \times\cos(\nu\pi)\begin{Bmatrix}\sin[(s+\nu)\pi/2]\\ \cos[(s+\nu)\pi/2]\end{Bmatrix}{}_3F_2\left(\begin{matrix}\frac{s+\nu}{2},\ \frac{s+\nu}{2},\ \frac{s+\nu+1}{2}\\ 1+\nu,\ \frac{s+\nu+2}{2};\ \frac{b^2}{a^2}\end{matrix}\right)$ $\left[\begin{matrix}0<b\le a;\ \lvert\operatorname{Re}\nu\rvert<\operatorname{Re}s<5/2\text{ for }b<a;\\ \lvert\operatorname{Re}\nu\rvert<\operatorname{Re}s<3/2\text{ for }b=a\end{matrix}\right]$

3.11.8. $Y_\nu(bx)$ and $\operatorname{erf}(ax)$, $\operatorname{erfc}(ax)$, $\operatorname{erfi}(ax)$

1	$\begin{Bmatrix}\operatorname{erf}(ax)\\ \operatorname{erfc}(ax)\end{Bmatrix}Y_\nu(bx)$	$\mp\dfrac{2^\nu a^{\nu-s}}{\pi^{3/2}b^\nu(\nu-s)}\Gamma(\nu)\,\Gamma\left(\dfrac{s-\nu+1}{2}\right) \times{}_2F_2\left(\begin{matrix}\frac{s-\nu}{2},\ \frac{s-\nu+1}{2};\ -\frac{b^2}{4a^2}\\ 1-\nu,\ \frac{s-\nu+2}{2}\end{matrix}\right)\pm\dfrac{b^\nu\cos(\nu\pi)}{2^\nu\pi^{3/2}a^{s+\nu}(s+\nu)} \times\Gamma(-\nu)\,\Gamma\left(\dfrac{s+\nu+1}{2}\right){}_2F_2\left(\begin{matrix}\frac{s+\nu}{2},\ \frac{s+\nu+1}{2};\ -\frac{b^2}{4a^2}\\ 1+\nu,\ \frac{s+\nu+2}{2}\end{matrix}\right) -\dfrac{1\pm1}{\pi b^s}2^{s-2}\cos\dfrac{(s-\nu)\pi}{2}\,\Gamma\left(\dfrac{s+\nu}{2}\right)\Gamma\left(\dfrac{s-\nu}{2}\right)$ $\left[\begin{matrix}b>0;\ \operatorname{Re}s>\lvert\operatorname{Re}\nu\rvert-(1\pm1)/2;\\ \lvert\arg a\rvert<\pi/2;\ \operatorname{Re}s<3/2\text{ for erf}\end{matrix}\right]$

No.	$f(x)$	$F(s)$
2	$\left\{\begin{matrix} \operatorname{erf}(a\sqrt{x}) \\ \operatorname{erfc}(a\sqrt{x}) \end{matrix}\right\} Y_\nu(bx)$	$\mp \frac{2^{s+1/2}a}{\pi^{3/2}b^{s+1/2}} \Gamma\left(\frac{2s+2\nu+1}{4}\right) \Gamma\left(\frac{2s-2\nu+1}{4}\right) \times \cos\frac{(2s-2\nu+1)\pi}{4} \, {}_3F_2\left(\begin{matrix} \frac{1}{4}, \frac{2s+2\nu+1}{4}, \frac{2s-2\nu+1}{4} \\ \frac{1}{2}, \frac{5}{4}; -\frac{a^4}{b^2} \end{matrix}\right) \mp \frac{2^{s+3/2}a^3}{3\pi^{3/2}b^{s+3/2}} \Gamma\left(\frac{2s+2\nu+3}{4}\right) \Gamma\left(\frac{2s-2\nu+3}{4}\right) \times \sin\frac{(2s-2\nu+1)\pi}{4} \, {}_3F_2\left(\begin{matrix} \frac{3}{4}, \frac{2s+2\nu+3}{4}, \frac{2s-2\nu+3}{4} \\ \frac{3}{2}, \frac{7}{4}; -\frac{a^4}{b^2} \end{matrix}\right) - \frac{1\mp 1}{\pi b^s} 2^{s-2} \cos\frac{(s-\nu)\pi}{2} \Gamma\left(\frac{s+\nu}{2}\right) \Gamma\left(\frac{s-\nu}{2}\right)$ $[b>0;\ \lvert\operatorname{Re}\nu\rvert < \operatorname{Re}s + (1\pm 1)/4;\ \lvert\arg a\rvert < \pi/4;\ \operatorname{Re}s < 3/2 \text{ for erf}]$
3	$e^{\mp a^2x^2} \left\{\begin{matrix} \operatorname{erfi}(ax) \\ \operatorname{erfc}(ax) \end{matrix}\right\} \times Y_\nu(bx)$	$-\frac{b^\nu \cos(\nu\pi)}{2^{\nu+(1\pm 1)/2}a^{s+\nu}} \left\{\begin{matrix} \sec[(s+\nu)\pi/2] \\ \csc[(s+\nu)\pi] \end{matrix}\right\} \times \Gamma\left[\begin{matrix} -\nu \\ \frac{2-s-\nu}{2} \end{matrix}\right] {}_1F_1\left(\begin{matrix} \frac{s+\nu}{2}; \mp\frac{b^2}{4a^2} \\ 1+\nu \end{matrix}\right) - \frac{2^{\nu-(1\pm 1)/2}}{a^{s-\nu}b^\nu} \left\{\begin{matrix} \sec[(s-\nu)\pi/2] \\ \csc[(s-\nu)\pi] \end{matrix}\right\} \times \Gamma\left[\begin{matrix} \nu \\ \frac{2-s+\nu}{2} \end{matrix}\right] {}_1F_1\left(\begin{matrix} \frac{s-\nu}{2}; \mp\frac{b^2}{4a^2} \\ 1-\nu \end{matrix}\right) - \frac{2^{s-2}b^{1-s}}{\pi^{3/2}a} \Gamma\left(\frac{s+\nu-1}{2}\right) \times \Gamma\left(\frac{s-\nu-1}{2}\right) \sin\frac{(s-\nu)\pi}{2} \, {}_2F_2\left(\begin{matrix} \frac{1}{2}, 1; \mp\frac{b^2}{4a^2} \\ \frac{3-s-\nu}{2}, \frac{3-s+\nu}{2} \end{matrix}\right)$ $[b>0;\ \lvert\operatorname{Re}\nu\rvert - (1\pm 1)/2 < \operatorname{Re}s < 5/2;\ \lvert\arg a\rvert < (2\mp 1)\pi/4]$

3.11.9. $Y_\nu(bx)$ and $S(ax)$, $C(ax)$

Notation: $\delta = \left\{\begin{matrix} 1 \\ 0 \end{matrix}\right\}$.

1	$\left\{\begin{matrix} S(ax) \\ C(ax) \end{matrix}\right\} Y_\nu(bx)$	$-\frac{2^{s+\delta}a^{1/2+\delta}}{3^\delta \pi^{3/2} b^{s+\delta+1/2}} \cos\frac{(2s-2\nu+2\delta+1)\pi}{4} \times \Gamma\left(\frac{2s-2\nu+2\delta+1}{4}\right) \Gamma\left(\frac{2s+2\nu+2\delta+1}{4}\right) \times {}_3F_2\left(\begin{matrix} \frac{2\delta+1}{4}, \frac{2s-2\nu+2\delta+1}{4}, \frac{2s+2\nu+2\delta+1}{4} \\ \frac{2\delta+5}{4}, \frac{2\delta+1}{2}; \frac{a^2}{b^2} \end{matrix}\right)$ $[a, b>0;\ \lvert\operatorname{Re}\nu\rvert - (2\pm 1)/2 < \operatorname{Re}s < 3/2]$

3.11.10. $Y_\nu(bx)$ and $\gamma(\mu, ax)$, $\Gamma(\mu, ax)$

1	$\begin{Bmatrix} \gamma(\mu, ax) \\ \Gamma(\mu, ax) \end{Bmatrix} Y_\nu(bx)$	$\mp \frac{2^{s+\mu-1} a^\mu}{\mu \pi b^{s+\mu}} \Gamma\left(\frac{s+\mu-\nu}{2}\right) \Gamma\left(\frac{s+\mu+\nu}{2}\right) \times \cos\frac{(s+\mu-\nu)\pi}{2} \, {}_3F_2\left(\begin{matrix} \frac{\mu}{2}, \frac{s+\mu-\nu}{2}, \frac{s+\mu+\nu}{2} \\ \frac{1}{2}, \frac{\mu+2}{2}; -\frac{a^2}{b^2} \end{matrix}\right) \mp \frac{2^{s+\mu} a^{\mu+1}}{(\mu+1)\pi b^{s+\mu+1}} \Gamma\left(\frac{s+\mu-\nu+1}{2}\right) \Gamma\left(\frac{s+\mu+\nu+1}{2}\right) \times \sin\frac{(s+\mu-\nu)\pi}{2} \, {}_3F_2\left(\begin{matrix} \frac{\mu+1}{2}, \frac{s+\mu+\nu+1}{2}, \frac{s+\mu-\nu+1}{2} \\ \frac{3}{2}, \frac{\mu+3}{2}; -\frac{a^2}{b^2} \end{matrix}\right) - \frac{1 \mp 1}{\pi b^s} 2^{s-1} \cos\frac{(s-\nu)\pi}{2} \Gamma\left[\mu, \frac{s-\nu}{2}, \frac{s+\nu}{2}\right]$ $\left[\begin{matrix} b, \operatorname{Re} a > 0; \operatorname{Re}(s+\mu) > \lvert\operatorname{Re}\nu\rvert; \\ \begin{Bmatrix} \operatorname{Re}\mu > 0; \operatorname{Re} s < 3/2 \\ \operatorname{Re} s > \lvert\operatorname{Re}\nu\rvert \end{Bmatrix} \end{matrix}\right]$
2	$\begin{Bmatrix} \gamma(\mu, ax^2) \\ \Gamma(\mu, ax^2) \end{Bmatrix} Y_\nu(bx)$	$\mp \frac{2^\nu a^{(\nu-s)/2}}{\pi b^\nu (\nu - s)} \Gamma(\nu) \Gamma\left(\frac{s+2\mu-\nu}{2}\right) \times {}_2F_2\left(\begin{matrix} \frac{s-\nu}{2}, \frac{s+2\mu-\nu}{2}; -\frac{b^2}{4a} \\ 1-\nu, \frac{s-\nu+2}{2} \end{matrix}\right) \pm \frac{b^\nu \cos(\nu\pi)}{2^\nu \pi a^{(s+\nu)/2} (s+\nu)} \times \Gamma(-\nu) \Gamma\left(\frac{s+2\mu+\nu}{2}\right) {}_2F_2\left(\begin{matrix} \frac{s+\nu}{2}, \frac{s+2\mu+\nu}{2}; -\frac{b^2}{4a} \\ 1+\nu, \frac{s+\nu+2}{2} \end{matrix}\right) - \frac{1 \pm 1}{\pi b^s} 2^{s-1} \cos\frac{(s-\nu)\pi}{2} \Gamma\left[\mu, \frac{s+\nu}{2}, \frac{s-\nu}{2}\right]$ $\left[\begin{matrix} b, \operatorname{Re} a > 0; \operatorname{Re}(s+2\mu) > \lvert\operatorname{Re}\nu\rvert; \\ \begin{Bmatrix} \operatorname{Re}\mu > 0; \operatorname{Re} s < 3/2 \\ \operatorname{Re} s > \lvert\operatorname{Re}\nu\rvert \end{Bmatrix} \end{matrix}\right]$
3	$e^{ax^2} \Gamma(\mu, ax^2) Y_\nu(bx)$	$-\frac{2^{\nu-1} a^{(\nu-s)/2}}{b^\nu} \Gamma\left[\begin{matrix} \nu, \frac{s-\nu}{2} \\ 1-\mu \end{matrix}\right] \csc\frac{(s+2\mu-\nu)\pi}{2} \times {}_1F_1\left(\begin{matrix} \frac{s-\nu}{2}; \frac{b^2}{4a} \\ 1-\nu \end{matrix}\right) - \frac{b^\nu}{2^{\nu+1} a^{(s+\nu)/2}} \Gamma\left[\begin{matrix} -\nu, \frac{s+\nu}{2} \\ 1-\mu \end{matrix}\right] \times \cos(\nu\pi) \csc\frac{(s+2\mu+\nu)\pi}{2} \, {}_1F_1\left(\begin{matrix} \frac{s+\nu}{2}; \frac{b^2}{4a} \\ 1+\nu \end{matrix}\right) + \frac{2^{s+2\mu-3} a^{\mu-1}}{\pi b^{s+2\mu-2}} \Gamma\left(\frac{s+2\mu+\nu-2}{2}\right) \Gamma\left(\frac{s+2\mu-\nu-2}{2}\right) \times \cos\frac{(s-\nu+2\mu)\pi}{2} \, {}_2F_2\left(\begin{matrix} 1, 1-\mu; \frac{b^2}{4a} \\ \frac{4-s-2\mu+\nu}{2}, \frac{4-s-2\mu-\nu}{2} \end{matrix}\right)$ $\left[\begin{matrix} b > 0; \operatorname{Re} s > \lvert\operatorname{Re}\nu\rvert; \lvert\arg a\rvert < \pi; \\ \lvert\operatorname{Re}\nu\rvert < \operatorname{Re}(s+2\mu) < 7/2 \end{matrix}\right]$

3.11.11. $Y_\nu(bx)$ and $D_\mu(ax^r)$

1	$e^{a^2x/4}D_\mu(a\sqrt{x})\,Y_\nu(bx)$	$\frac{2^{2s-\mu-7/2}a^{-2s}}{\pi^2\,\Gamma(-\mu)}\,G_{55}^{44}\left(\frac{4b^2}{a^4}\,\middle\vert\,\begin{matrix}-\frac{\nu+1}{2},\ \frac{1-2s}{4},\ \frac{3-2s}{4},\ \frac{1-s}{2},\ \frac{2-s}{2}\\ -\frac{\nu+1}{2},\ -\frac{\nu}{2},\ \frac{\nu}{2},\ -\frac{2s+\mu}{4},\ -\frac{2s+\mu-2}{4}\end{matrix}\right)$ $[b>0;\ \mathrm{Re}\,(2s+\mu)<3,\ \mathrm{Re}\,s>\lvert\mathrm{Re}\,\nu\rvert;\ \lvert\arg a\rvert<3\pi/4]$
2	$e^{-a^2x/4}D_\mu(a\sqrt{x})\,Y_\nu(bx)$	$-\frac{2^{\mu/2+2\nu-s+1}a^{2\nu-2s}b^{-\nu}}{\sqrt{\pi}}\,\Gamma\left[\begin{matrix}\nu,\ 2s-2\nu\\ \frac{2s-\mu-2\nu+1}{2}\end{matrix}\right]$ $\times\,{}_4F_3\left(\begin{matrix}\frac{s-\nu}{2},\ \frac{s-\nu+1}{2},\ \frac{2s-2\nu+1}{4},\ \frac{2s-2\nu+3}{4}\\ 1-\nu,\ \frac{2s-\mu-2\nu+1}{4},\ \frac{2s-\mu-2\nu+3}{4};\ -\frac{4b^2}{a^4}\end{matrix}\right)$ $-\frac{2^{\mu/2-2\nu-s+1}a^{-2\nu-2s}b^{\nu}}{\sqrt{\pi}}\cos(\nu\pi)\,\Gamma\left[\begin{matrix}-\nu,\ 2s+2\nu\\ \frac{2s-\mu+2\nu+1}{2}\end{matrix}\right]$ $\times\,{}_4F_3\left(\begin{matrix}\frac{s+\nu}{2},\ \frac{s+\nu+1}{2},\ \frac{2s+2\nu+1}{4},\ \frac{2s+2\nu+3}{4}\\ \nu+1,\ \frac{2s-\mu+2\nu+1}{4},\ \frac{2s-\mu+2\nu+3}{4};\ -\frac{4b^2}{a^4}\end{matrix}\right)$ $[b>0;\ \mathrm{Re}\,s>\lvert\mathrm{Re}\,\nu\rvert;\ \lvert\arg a\rvert<\pi/4]$

3.11.12. $Y_\nu(\varphi(x))$ and $J_\mu(\psi(x))$

1	$\cos(ax)\,J_\nu(ax)$ $\pm\sin(ax)\,Y_\nu(ax)$	$\frac{2^{s-1}}{a^s}\,\Gamma\left[\begin{matrix}\frac{s-\nu}{2},\ \frac{s+\nu}{2}\\ \frac{s\mp 2a-\nu}{2},\ \frac{2-s\pm 2a+\nu}{2}\end{matrix}\right]$ $[a>0;\ \lvert\mathrm{Re}\,\nu\rvert<\mathrm{Re}\,s<3/2]$
2	$\begin{Bmatrix}\sin(ax+b)\\ \cos(ax+b)\end{Bmatrix}J_\nu(ax)$ $\mp\begin{Bmatrix}\cos(ax+b)\\ \sin(ax+b)\end{Bmatrix}Y_\nu(ax)$	$\pm\frac{2^{1-s}a^{-s}}{\pi^{3/2}}\cos(\nu\pi)\begin{Bmatrix}\cos[(\nu-s)\pi/2+b]\\ \sin[(\nu-s)\pi/2+b]\end{Bmatrix}$ $\times\,\Gamma\left(\frac{1-2s}{2}\right)\Gamma(s-\nu)\,\Gamma(s+\nu)$ $[a>0;\ \lvert\mathrm{Re}\,\nu\rvert<\mathrm{Re}\,s<1/2]$
3	$\begin{Bmatrix}\sin(ax+b)\\ \cos(ax+b)\end{Bmatrix}J_\nu(ax)$ $\pm\begin{Bmatrix}\cos(ax+b)\\ \sin(ax+b)\end{Bmatrix}Y_\nu(ax)$	$\frac{2^{1-s}a^{-s}}{\sqrt{\pi}}\begin{Bmatrix}\sin[(s+\nu)\pi/2+b]\\ \cos[(s+\nu)\pi/2+b]\end{Bmatrix}\Gamma\left[\begin{matrix}\frac{1-2s}{2},\ s+\nu\\ 1-s+\nu\end{matrix}\right]$ $\mp\frac{2^{1-s}a^{-s}}{\pi^{3/2}}\cos(\nu\pi)\begin{Bmatrix}\cos[(\nu-s)\pi/2+b]\\ \sin[(\nu-s)\pi/2+b]\end{Bmatrix}$ $\times\,\Gamma\left(\frac{1-2s}{2}\right)\Gamma(s-\nu)\,\Gamma(s+\nu)$ $[a>0;\ \lvert\mathrm{Re}\,\nu\rvert<\mathrm{Re}\,s<1/2]$
4	$J_\nu(ax)\,Y_\nu(ax)$	$-\frac{a^{-s}}{2\sqrt{\pi}}\,\Gamma\left[\begin{matrix}\frac{s}{2},\ \frac{s+2\nu}{2}\\ \frac{s+1}{2},\ \frac{2-s+2\nu}{2}\end{matrix}\right]$ $[a>0;\ 0,\ -2\,\mathrm{Re}\,\nu<\mathrm{Re}\,s<2]$

No.	$f(x)$	$F(s)$
5	$J_{-\nu}(ax)\,Y_\nu(ax)$	$-\dfrac{a^{-s}}{2\pi^{3/2}}\cos\dfrac{(s-2\nu)\pi}{2}\,\Gamma\left[\begin{matrix}\frac{s}{2},\ \frac{1-s}{2},\ \frac{s-2\nu}{2}\\ \frac{2-s-2\nu}{2}\end{matrix}\right]$ $[a>0;\ 0,\ 2\operatorname{Re}\nu<\operatorname{Re}s<1]$
6	$J_\mu(ax)\,Y_\nu(ax)$	$-\dfrac{2^{s-1}}{\pi a^s}\cos\dfrac{(s+\mu-\nu)\pi}{2}\,\Gamma\left[\begin{matrix}1-s,\ \frac{s+\mu-\nu}{2},\ \frac{s+\mu+\nu}{2}\\ \frac{2-s+\mu-\nu}{2},\ \frac{2-s+\mu+\nu}{2}\end{matrix}\right]$ $[a>0;\ \lvert\operatorname{Re}\nu\rvert-\operatorname{Re}\mu<\operatorname{Re}s<1]$
7	$J_\mu(ax)\,Y_\nu(bx)$	$-\dfrac{2^{s-1}a^\mu b^{-s-\mu}}{\pi}\cos\dfrac{(s+\mu-\nu)\pi}{2}\,\Gamma\left[\begin{matrix}\frac{s+\mu-\nu}{2},\ \frac{s+\mu+\nu}{2}\\ \mu+1\end{matrix}\right]$ $\times\,{}_2F_1\left(\begin{matrix}\frac{s+\mu-\nu}{2},\ \frac{s+\mu+\nu}{2}\\ \mu+1;\ \frac{a^2}{b^2}\end{matrix}\right)$ $[0<a<b;\ \lvert\operatorname{Re}\nu\rvert-\operatorname{Re}\mu<\operatorname{Re}s<2]$
8	$J_\nu(ax)\,Y_\nu(bx)$	$-\dfrac{2^{s-1}}{\pi}\left(a^2-b^2\right)^{-s/2}\Gamma\left(\dfrac{s}{2}\right)\Gamma\left(\dfrac{s+2\nu}{2}\right)$ $\times\left[\cos\dfrac{s\pi}{2}\,P_{-s/2}^{-\nu}\left(\dfrac{a^2+b^2}{a^2-b^2}\right)\right.$ $\left.+\dfrac{2e^{-i\nu\pi}}{\Gamma\left(\frac{2-s+2\nu}{2}\right)\Gamma\left(\frac{s+2\nu}{2}\right)}\,Q_{-s/2}^{\nu}\left(\dfrac{a^2+b^2}{a^2-b^2}\right)\right]$ $[0<b<a;\ 0,\ -2\operatorname{Re}\nu<\operatorname{Re}s<2]$
9	$J_\mu(ax)\,Y_\nu(bx)$	$-\dfrac{2^{s-1}b^\nu}{\pi a^{s+\nu}}\cos(\nu\pi)\,\Gamma\left[\begin{matrix}-\nu,\ \frac{s+\mu+\nu}{2}\\ \frac{2-s+\mu-\nu}{2}\end{matrix}\right]{}_2F_1\left(\begin{matrix}\frac{s-\mu+\nu}{2},\ \frac{s+\mu+\nu}{2}\\ 1+\nu;\ \frac{b^2}{a^2}\end{matrix}\right)$ $-\dfrac{2^{s-1}a^{\nu-s}}{\pi b^\nu}\Gamma\left[\begin{matrix}\nu,\ \frac{s+\mu-\nu}{2}\\ \frac{2-s+\mu+\nu}{2}\end{matrix}\right]{}_2F_1\left(\begin{matrix}\frac{s-\mu-\nu}{2},\ \frac{s+\mu-\nu}{2}\\ 1-\nu;\ \frac{b^2}{a^2}\end{matrix}\right)$ $[0<b<a;\ \lvert\operatorname{Re}\nu\rvert-\operatorname{Re}\mu<\operatorname{Re}s<2]$
10	$J_\nu(ax)\,Y_{-\nu}(bx)$	$-\dfrac{2^{s-1}}{\pi}\left(b^2-a^2\right)^{-s/2}\cos\dfrac{(s+2\nu)\pi}{2}$ $\times\,\Gamma\left(\dfrac{s}{2}\right)\Gamma\left(\dfrac{s+2\nu}{2}\right)P_{-s/2}^{-\nu}\left(\dfrac{b^2+a^2}{b^2-a^2}\right)$ $[0<a<b;\ 0,\ -2\operatorname{Re}\nu<\operatorname{Re}s<2]$
11	$J_\nu(ax)\,Y_{-\nu}(bx)$	$-\dfrac{2^{s-1}}{\pi}\left(a^2-b^2\right)^{-s/2}\Gamma\left(\dfrac{s}{2}\right)\Gamma\left(\dfrac{s+2\nu}{2}\right)$ $\times\left[\cos\dfrac{(s+2\nu)\pi}{2}\,P_{-s/2}^{-\nu}\left(\dfrac{a^2+b^2}{a^2-b^2}\right)\right.$ $\left.+\dfrac{2e^{-i\nu\pi}\cos(\nu\pi)}{\Gamma\left(\frac{2-s+2\nu}{2}\right)\Gamma\left(\frac{s+2\nu}{2}\right)}\,Q_{-s/2}^{\nu}\left(\dfrac{a^2+b^2}{a^2-b^2}\right)\right]$ $[0<b<a;\ 0,\ -2\operatorname{Re}\nu<\operatorname{Re}s<2]$

No.	$f(x)$	$F(s)$
12	$J_\mu(ax)\,Y_\nu(ax)$ $+\,J_\nu(ax)\,Y_\mu(ax)$	$-\frac{2^{s-1}}{\pi a^s}\,\Gamma\left[\begin{matrix}\frac{s+\mu-\nu}{2},\ \frac{s-\mu+\nu}{2},\ \frac{s+\mu+\nu}{2}\\ s,\ \frac{2-s+\mu+\nu}{2}\end{matrix}\right]$ $[a>0;\ -\operatorname{Re}(\mu+\nu),\ \lvert\operatorname{Re}(\mu-\nu)\rvert<\operatorname{Re}s<2]$
13	$J_\mu(ax)\,Y_\nu(ax)$ $-\,J_\nu(ax)\,Y_\mu(ax)$	$\frac{2^{s-1}}{\pi^2 a^s}\sin[(\mu-\nu)\pi]\,\Gamma\left[\begin{matrix}1-s,\ \frac{s-\mu+\nu}{2},\ \frac{s+\mu-\nu}{2},\ \frac{s+\mu+\nu}{2}\\ \frac{2-s+\mu+\nu}{2}\end{matrix}\right]$ $[a>0;\ \lvert\operatorname{Re}\nu\rvert-\operatorname{Re}\mu<\operatorname{Re}s<1]$
14	$J_\nu(ax)\,Y_{-\nu}(ax)$ $+\,J_{-\nu}(ax)\,Y_\nu(ax)$	$-\frac{a^{-s}}{\sqrt{\pi}}\,\Gamma\left[\begin{matrix}\frac{s-2\nu}{2},\ \frac{s+2\nu}{2}\\ \frac{2-s}{2},\ \frac{s+1}{2}\end{matrix}\right]$ $[a>0;\ 2\lvert\operatorname{Re}\nu\rvert<\operatorname{Re}s<2]$
15	$J_\nu(ax)\,Y_{-\nu}(ax)$ $-\,J_{-\nu}(ax)\,Y_\nu(ax)$	$\frac{a^{-s}}{2\pi^{5/2}}\sin(2\nu\pi)\,\Gamma\left[\frac{s}{2},\ \frac{1-s}{2},\ \frac{s-2\nu}{2},\ \frac{s+2\nu}{2}\right]$ $[a>0;\ 2\lvert\operatorname{Re}\nu\rvert<\operatorname{Re}s<1]$
16	$J_\nu(ax)\,Y_{-\nu}(bx)$ $+\,J_{-\nu}(ax)\,Y_\nu(bx)$	$-\frac{2^{s-1}}{a^{s-\nu}b^\nu}\left[\cos(\nu\pi)\csc\frac{s\pi}{2}+\csc\frac{(s-2\nu)\pi}{2}\right]$ $\times\Gamma\left[\begin{matrix}\nu\\ \frac{2-s}{2},\ \frac{2-s+2\nu}{2}\end{matrix}\right]\,{}_2F_1\left(\begin{matrix}\frac{s}{2},\ \frac{s-2\nu}{2}\\ 1-\nu;\ \frac{b^2}{a^2}\end{matrix}\right)$ $-\frac{2^{s-1}}{a^{s+\nu}b^{-\nu}}\left[\cos(\nu\pi)\csc\frac{s\pi}{2}+\csc\frac{(s+2\nu)\pi}{2}\right]$ $\times\Gamma\left[\begin{matrix}-\nu\\ \frac{2-s}{2},\ \frac{2-s-2\nu}{2}\end{matrix}\right]\,{}_2F_1\left(\begin{matrix}\frac{s}{2},\ \frac{s+2\nu}{2}\\ 1+\nu;\ \frac{b^2}{a^2}\end{matrix}\right)$ $[a>b>0;\ 2\lvert\operatorname{Re}\nu\rvert<\operatorname{Re}s<2]$
17	$J_\nu(u_-)\,Y_\nu(u_+)$ $u_\pm=b\left(\sqrt{x^2+a^2}\pm a\right)$	$\frac{1}{2\sqrt{\pi}}\left(\frac{a}{b}\right)^{s/2}\Gamma\left[\begin{matrix}\frac{s+2\nu}{2},\ \frac{1-s}{2}\\ \frac{2-s+2\nu}{2}\end{matrix}\right]Y_{-s/2}(2ab)$ $[a,\,b>0;\ -2\operatorname{Re}\nu<\operatorname{Re}s<1]$
18	$J_\nu(u_-)\,Y_{-\nu}(u_+)$ $u_\pm=b\left(\sqrt{x^2+a^2}\pm a\right)$	$\frac{1}{2\sqrt{\pi}}\left(\frac{a}{b}\right)^{s/2}\Gamma\left[\begin{matrix}\frac{s+2\nu}{2},\ \frac{1-s}{2}\\ \frac{2-s+2\nu}{2}\end{matrix}\right]\left[\sin(\nu\pi)\,J_{-s/2}(2ab)\right.$ $\left.+\cos(\nu\pi)\,Y_{-s/2}(2ab)\right]$ $[a,\,b>0;\ -2\operatorname{Re}\nu<\operatorname{Re}s<1]$
19	$J_\nu(u_-)\,Y_\nu(u_+)$ $-\,J_\nu(u_+)\,Y_\nu(u_-)$ $u_\pm=b\left(\sqrt{x^2+a^2}\pm a\right)$	$\frac{\cos(\nu\pi)}{\pi^{3/2}}\left(\frac{a}{b}\right)^{s/2}\Gamma\left(\frac{1-s}{2}\right)\Gamma\left(\frac{s}{2}-\nu\right)\Gamma\left(\frac{s}{2}+\nu\right)$ $\times J_{s/2}(2ab)$ $[a,\,b>0;\ 2\lvert\operatorname{Re}\nu\rvert<\operatorname{Re}s<1]$

3.11.13. $Y_\nu(bx)$, $J_\nu(bx)$, and trigonometric functions

Notation: $\delta = \begin{Bmatrix} 1 \\ 0 \end{Bmatrix}$.

1	$\begin{Bmatrix} \sin(ax) \\ \cos(ax) \end{Bmatrix} \times J_\nu(bx)\, Y_\nu(bx)$	$-\frac{a^\delta b^{-s-\delta}}{2\sqrt{\pi}} \Gamma\begin{bmatrix} \frac{s+\delta}{2}, \frac{s+2\nu+\delta}{2} \\ \frac{s+\delta+1}{2}, \frac{2-s+2\nu-\delta}{2} \end{bmatrix} {}_3F_2\left(\begin{matrix} \frac{s+\delta}{2}, \frac{s+2\nu+\delta}{2}, \frac{s-2\nu+\delta}{2} \\ \frac{2\delta+1}{2}, \frac{s+\delta+1}{2}; \frac{a^2}{4b^2} \end{matrix}\right)$ $[0 < a < 2b;\ -\delta,\ -2\operatorname{Re}\nu - \delta < \operatorname{Re} s < 2]$
2	$\begin{Bmatrix} \sin(ax) \\ \cos(ax) \end{Bmatrix} \times J_\nu(bx)\, Y_\nu(bx)$	$-\frac{\Gamma(s)}{\nu\pi a^s} \begin{Bmatrix} \sin(s\pi/2) \\ \cos(s\pi/2) \end{Bmatrix} {}_3F_2\left(\begin{matrix} \frac{1}{2}, \frac{s}{2}, \frac{s+1}{2} \\ 1-\nu, 1+\nu; \frac{4b^2}{a^2} \end{matrix}\right) - \frac{\cos(\nu\pi)}{\pi a^{s+2\nu}} \left(\frac{b}{2}\right)^{2\nu} \begin{Bmatrix} \sin[(s+2\nu)\pi/2] \\ \cos[(s+2\nu)\pi/2] \end{Bmatrix} \times \Gamma\begin{bmatrix} -\nu, s+2\nu \\ \nu+1 \end{bmatrix} {}_3F_2\left(\begin{matrix} \frac{2\nu+1}{2}, \frac{s+2\nu}{2}, \frac{s+2\nu+1}{2} \\ \nu+1, 2\nu+1; \frac{4b^2}{a^2} \end{matrix}\right)$ $[0 < 2b < a;\ -\delta,\ -2\operatorname{Re}\nu - \delta < \operatorname{Re} s < 2]$

3.11.14. $Y_\nu(bx)$, $J_\nu(bx)$, and $S(ax)$, $C(ax)$

Notation: $\delta = \begin{Bmatrix} 1 \\ 0 \end{Bmatrix}$.

1	$\begin{Bmatrix} S(ax) \\ C(ax) \end{Bmatrix} J_\nu(bx)\, Y_\nu(bx)$	$-\frac{a^{\delta+1/2} b^{-s-\delta-1/2}}{(2\delta+1)\sqrt{2}\,\pi} \Gamma\begin{bmatrix} \frac{2s+2\delta+1}{4}, \frac{2s+4\nu+2\delta+1}{4} \\ \frac{2s+2\delta+3}{4}, \frac{3-2s+4\nu-2\delta}{2} \end{bmatrix} \times {}_4F_3\left(\begin{matrix} \frac{2\delta+1}{4}, \frac{2s+2\delta+1}{4}, \frac{2s-4\nu+2\delta+1}{4}, \frac{2s+4\nu+2\delta+1}{4} \\ \frac{2\delta+1}{2}, \frac{2\delta+5}{4}, \frac{2s+2\delta+3}{4}; \frac{a^2}{4b^2} \end{matrix}\right)$ $[0 < a < 2b;\ -2\nu - \delta - 1/2,\ -\delta - 1/2 < \operatorname{Re} s < 1]$
2	$\begin{Bmatrix} S(ax) \\ C(ax) \end{Bmatrix} J_\nu(bx)\, Y_\nu(bx)$	$\frac{a^{-s-2\nu} b^{2\nu}}{\sqrt{2}\,\pi(s+2\nu)} \begin{Bmatrix} \sin[(2s+4\nu+1)\pi/4] \\ \cos[(2s+4\nu+1)\pi/4] \end{Bmatrix} \Gamma\begin{bmatrix} -\nu, \frac{2s+4\nu+1}{2} \\ \frac{1-2\nu}{2}, 2\nu+1 \end{bmatrix} \times {}_4F_3\left(\begin{matrix} \frac{2\nu+1}{2}, \frac{s+2\nu}{2}, \frac{2s+4\nu+1}{4}, \frac{2s+4\nu+3}{4} \\ \nu+1, 2\nu+1, \frac{s+2\nu+2}{2}; \frac{4b^2}{a^2} \end{matrix}\right) + \frac{a^{-s}}{\sqrt{2}\,\pi^{3/2}\nu s} \begin{Bmatrix} \sin[(2s+1)\pi/4] \\ \cos[(2s+1)\pi/4] \end{Bmatrix} \Gamma\left(s+\frac{1}{2}\right) \times {}_4F_3\left(\begin{matrix} \frac{1}{2}, \frac{s}{2}, \frac{2s+1}{4}, \frac{2s+3}{4} \\ 1-\nu, 1+\nu, \frac{s+2}{2}; \frac{4b^2}{a^2} \end{matrix}\right) - \frac{b^{-s}}{4\sqrt{\pi}} \Gamma\begin{bmatrix} \frac{s}{2}, \frac{s+2\nu}{2} \\ \frac{s+1}{2}, \frac{2-s+2\nu}{2} \end{bmatrix}$ $[0 < 2b < a;\ -2\nu - \delta - 1/2,\ -\delta - 1/2 < \operatorname{Re} s < 1]$

3.11.15. $Y_\nu(ax)$ and $J_\lambda(bx)\, J_\mu(cx)$

Notation: $\delta = \left\{\begin{matrix} 1 \\ 0 \end{matrix}\right\}.$

1	$J_\lambda(ax)\, J_\mu(ax)\, Y_\nu(bx)$	$-\frac{2^{s-1}a^{\lambda+\mu}}{\pi b^{s+\lambda+\mu}} \cos\frac{(s+\lambda+\mu-\nu)\pi}{2}\, \Gamma\left[\begin{matrix}\frac{s+\lambda+\mu+\nu}{2},\ \frac{s+\lambda+\mu-\nu}{2} \\ \lambda+1,\ \mu+1\end{matrix}\right] \times {}_4F_3\left(\begin{matrix}\frac{\lambda+\mu+1}{2},\ \frac{\lambda+\mu+2}{2},\ \frac{s+\lambda+\mu+\nu}{2},\ \frac{s+\lambda+\mu-\nu}{2} \\ \lambda+1,\ \mu+1,\ \lambda+\mu+1;\ \frac{4a^2}{b^2}\end{matrix}\right)$ $[0 < 2a < b;\ \lvert\operatorname{Re}\nu\rvert - \operatorname{Re}(\lambda+\mu) < \operatorname{Re} s < 5/2]$
2	$J_\lambda(ax)\, J_\mu(bx)\, Y_\nu(bx)$	$-\frac{2^{s-1}a^{\nu-\mu-s}}{\pi b^{\nu-\mu}}\, \Gamma\left[\begin{matrix}\nu,\ \frac{s+\lambda+\mu-\nu}{2} \\ \mu+1,\ \frac{2-s+\lambda+\nu-\mu}{2}\end{matrix}\right] \times {}_4F_3\left(\begin{matrix}\frac{\mu-\nu+1}{2},\ \frac{\mu-\nu+2}{2},\ \frac{s-\lambda+\mu-\nu}{2},\ \frac{s+\lambda+\mu-\nu}{2} \\ \mu+1,\ 1-\nu,\ \mu-\nu+1;\ \frac{4b^2}{a^2}\end{matrix}\right) - \frac{2^{s-1}b^{\mu+\nu}}{\pi a^{s+\mu+\nu}} \cos(\nu\pi)\, \Gamma\left[\begin{matrix}-\nu,\ \frac{s+\lambda+\mu+\nu}{2} \\ \mu+1,\ \frac{2-s+\lambda-\mu-\nu}{2}\end{matrix}\right] \times {}_4F_3\left(\begin{matrix}\frac{\mu+\nu+1}{2},\ \frac{\mu+\nu+2}{2},\ \frac{s+\mu+\nu-\lambda}{2},\ \frac{s+\lambda+\mu+\nu}{2} \\ \mu+1,\ \nu+1,\ \mu+\nu+1;\ \frac{4b^2}{a^2}\end{matrix}\right)$ $[0 < 2b < a;\ \lvert\operatorname{Re}\nu\rvert - \operatorname{Re}(\lambda+\mu) < \operatorname{Re} s < 5/2]$
3	$J_\lambda(ax)\, J_\mu(bx)\, Y_\nu(bx)$	$\frac{2^{s-2}}{\pi a^{s-1}b} \sin\frac{(\mu-\nu)\pi}{2}\, \Gamma\left[\begin{matrix}\frac{s+\lambda-1}{2} \\ \frac{3-s+\lambda}{2}\end{matrix}\right] \times {}_4F_3\left(\begin{matrix}\frac{\mu+\nu+1}{2},\ \frac{\mu-\nu+1}{2},\ \frac{\nu-\mu+1}{2},\ \frac{1-\mu-\nu}{2} \\ \frac{1}{2},\ \frac{3-s-\lambda}{2},\ \frac{3-s+\lambda}{2};\ \frac{a^2}{4b^2}\end{matrix}\right) + 2^{s-4}\frac{\nu^2-\mu^2}{\pi a^{s-2}b^4} \cos\frac{(\mu-\nu)\pi}{2}\, \Gamma\left[\begin{matrix}\frac{s+\lambda-2}{2} \\ \frac{4-s+\lambda}{2}\end{matrix}\right] \times {}_4F_3\left(\begin{matrix}\frac{\mu+\nu+2}{2},\ \frac{\mu-\nu+2}{2},\ \frac{\nu-\mu+2}{2},\ \frac{2-\mu-\nu}{2} \\ \frac{3}{2},\ \frac{4-s-\lambda}{2},\ \frac{4-s+\lambda}{2};\ \frac{a^2}{4b^2}\end{matrix}\right) - \frac{2^{s-1}a^{\lambda}}{\pi b^{s+\lambda}} \cos\frac{(s+\lambda+\mu-\nu)\pi}{2} \times \Gamma\left[\begin{matrix}\frac{s+\lambda+\mu+\nu}{2},\ \frac{s+\lambda+\mu-\nu}{2},\ 1-s-\lambda \\ \lambda+1,\ \frac{2-s-\lambda+\mu+\nu}{2},\ \frac{2-s-\lambda+\mu-\nu}{2}\end{matrix}\right] \times {}_4F_3\left(\begin{matrix}\frac{s+\lambda+\mu+\nu}{2},\ \frac{s+\lambda-\mu+\nu}{2},\ \frac{s+\lambda+\mu-\nu}{2},\ \frac{s+\lambda-\mu-\nu}{2} \\ \lambda+1,\ \frac{s+\lambda}{2},\ \frac{s+\lambda+1}{2};\ \frac{a^2}{4b^2}\end{matrix}\right)$ $[0 < a < 2b;\ \lvert\operatorname{Re}\nu\rvert - \operatorname{Re}(\lambda+\mu) < \operatorname{Re} s < 5/2]$

3.11.16. Products of $Y_\nu(\varphi(x))$

1	$Y_\nu^2(ax)$	$\frac{a^{-s}}{\sqrt{\pi}}\Gamma\left[\begin{matrix}\frac{s}{2}, \frac{s-2\nu}{2}, \frac{s+2\nu}{2} \\ \frac{s+1}{2}, \frac{s-2\nu+1}{2}, \frac{1-s+2\nu}{2}\end{matrix}\right] + \frac{a^{-s}}{2\sqrt{\pi}}\Gamma\left[\begin{matrix}\frac{1-s}{2}, \frac{s+2\nu}{2} \\ \frac{2-s}{2}, \frac{2-s+2\nu}{2}\end{matrix}\right]$ $[a>0;\ 2\lvert\operatorname{Re}\nu\rvert < \operatorname{Re}s < 1]$
2	$Y_\mu(ax)\,Y_\nu(bx)$	$\frac{2^{s-1}a^{\mu}}{\pi^2 b^{s+\mu}}\cos(\mu\pi)\cos\frac{(s+\mu-\nu)\pi}{2} \times\Gamma\left[-\mu, \frac{s+\mu-\nu}{2}, \frac{s+\mu+\nu}{2}\right] \times {}_2F_1\left(\begin{matrix}\frac{s+\mu-\nu}{2}, \frac{s+\mu+\nu}{2} \\ 1+\mu;\ \frac{a^2}{b^2}\end{matrix}\right) + \frac{2^{s-1}b^{\mu-s}}{\pi^2 a^{\mu}}\cos\frac{(s-\mu-\nu)\pi}{2} \times\Gamma\left[\mu, \frac{s-\mu-\nu}{2}, \frac{s-\mu+\nu}{2}\right] {}_2F_1\left(\begin{matrix}\frac{s-\mu-\nu}{2}, \frac{s-\mu+\nu}{2} \\ 1-\mu;\ \frac{a^2}{b^2}\end{matrix}\right)$ $[0<a<b;\ \lvert\operatorname{Re}\mu\rvert + \lvert\operatorname{Re}\nu\rvert < \operatorname{Re}s < 2]$
3	$J_\nu^2(ax) - Y_\nu^2(ax)$	$-\frac{a^{-s}}{\sqrt{\pi}}\Gamma\left[\begin{matrix}\frac{s}{2}, \frac{s-2\nu}{2}, \frac{s+2\nu}{2} \\ \frac{s+1}{2}, \frac{s-2\nu+1}{2}, \frac{1-s+2\nu}{2}\end{matrix}\right]$ $[a>0;\ 2\lvert\operatorname{Re}\nu\rvert < \operatorname{Re}s < 2]$
4	$J_\nu^2(ax) + Y_\nu^2(ax)$	$\frac{a^{-s}}{\pi^{5/2}}\cos(\nu\pi)\,\Gamma\left[\frac{s}{2}, \frac{1-s}{2}, \frac{s-2\nu}{2}, \frac{s+2\nu}{2}\right]$ $[a>0;\ 2\lvert\operatorname{Re}\nu\rvert < \operatorname{Re}s < 1]$
5	$Y_\nu^2(ax) \pm Y_{-\nu}^2(ax)$	$\frac{a^{-s}}{\sqrt{\pi}}\left\{\begin{matrix}\cos(\nu\pi)\left[\cot(s\pi) + 3\csc(s\pi)\right] \\ \sin(\nu\pi)\end{matrix}\right\}\Gamma\left[\begin{matrix}\frac{s-2\nu}{2}, \frac{s+2\nu}{2} \\ \frac{2-s}{2}, \frac{s+1}{2}\end{matrix}\right]$ $[a>0;\ 2\lvert\operatorname{Re}\nu\rvert < \operatorname{Re}s < (3\mp 1)/2]$
6	$J_\mu(ax)\,J_\nu(ax) - Y_\mu(ax)\,Y_\nu(ax)$	$-\frac{1}{\sqrt{\pi}\,a^s}\Gamma\left[\begin{matrix}\frac{s+\mu+\nu}{2}, \frac{s+\mu-\nu}{2}, \frac{s-\mu+\nu}{2}, \frac{s-\mu-\nu}{2} \\ \frac{s}{2}, \frac{s+1}{2}, \frac{s-\mu-\nu+1}{2}, \frac{1-s+\mu+\nu}{2}\end{matrix}\right]$ $[a>0;\ (\lvert\operatorname{Re}\mu\rvert + \lvert\operatorname{Re}\nu\rvert) < \operatorname{Re}s < 2]$
7	$J_\mu(ax)\,J_\nu(ax) + Y_\mu(ax)\,Y_\nu(ax)$	$\frac{2^{s-1}a^{-s}}{\pi^2}\cos(\mu\pi)\,\Gamma\left[\begin{matrix}1-s, \frac{s-\mu-\nu}{2}, \frac{s+\mu-\nu}{2}, \frac{s+\mu+\nu}{2} \\ \frac{2-s+\mu-\nu}{2}\end{matrix}\right] + \frac{2^{s-1}a^{-s}}{\pi^2}\cos(\nu\pi)\,\Gamma\left[\begin{matrix}1-s, \frac{s-\mu-\nu}{2}, \frac{s-\mu+\nu}{2}, \frac{s+\mu+\nu}{2} \\ \frac{2-s-\mu+\nu}{2}\end{matrix}\right]$ $[\lvert\operatorname{Re}(\mu-\nu)\rvert, \lvert\operatorname{Re}(\mu+\nu)\rvert < \operatorname{Re}s < 1,\ \mu+\nu \neq 0, \pm 1, \ldots]$

No.	$f(x)$	$F(s)$
8	$J_{-\nu}(ax)J_\nu(ax)$ $+Y_{-\nu}(ax)Y_\nu(ax)$	$\frac{a^{-s}}{\pi^{3/2}}\cos^2(\nu\pi)\sec\frac{s\pi}{2}\,\Gamma\left[\begin{matrix}\frac{s}{2},\ \frac{s-2\nu}{2},\ \frac{s+2\nu}{2}\\ \frac{s+1}{2}\end{matrix}\right]$ $[a>0;\ 2\lvert\operatorname{Re}\nu\rvert<\operatorname{Re}s<1]$
9	$J_{-\nu}(ax)J_\nu(ax)$ $-Y_{-\nu}(ax)Y_\nu(ax)$	$-\frac{a^{-s}}{\pi^{3/2}}\cos\frac{s\pi}{2}\,\Gamma\left[\begin{matrix}\frac{s}{2},\ \frac{s-2\nu}{2},\ \frac{s+2\nu}{2}\\ \frac{s+1}{2}\end{matrix}\right]$ $[a>0;\ 2\lvert\operatorname{Re}\nu\rvert<\operatorname{Re}s<2]$
10	$Y_\mu(ax)Y_\nu(ax)$ $-Y_{-\mu}(ax)Y_{-\nu}(ax)$	$\frac{1}{\pi^2}\left(\frac{2}{a}\right)^s\sin\frac{s\pi}{2}\sin\frac{(\mu+\nu)\pi}{2}$ $\times\Gamma\left[\begin{matrix}\frac{s-\mu-\nu}{2},\ \frac{s+\mu-\nu}{2},\ \frac{s-\mu+\nu}{2},\ \frac{s+\mu+\nu}{2}\\ s\end{matrix}\right]$ $-\frac{1}{\pi}\left(\frac{2}{a}\right)^s\cos\frac{s\pi}{2}\sin\frac{(\mu+\nu)\pi}{2}\,\Gamma\left[\begin{matrix}1-s,\ \frac{s-\mu-\nu}{2},\ \frac{s+\mu+\nu}{2}\\ \frac{2-s+\mu-\nu}{2},\ \frac{2-s-\mu+\nu}{2}\end{matrix}\right]$ $[a>0;\ \lvert\operatorname{Re}(\mu-\nu)\rvert,\ \lvert\operatorname{Re}(\mu+\nu)\rvert<\operatorname{Re}s<1]$
11	$Y_\mu(ax)Y_\nu(ax)$ $+Y_{-\mu}(ax)Y_{-\nu}(ax)$	$\frac{1}{\pi^2}\left(\frac{2}{a}\right)^s\cos\frac{s\pi}{2}\cos\frac{(\mu+\nu)\pi}{2}$ $\times\Gamma\left[\begin{matrix}\frac{s-\mu-\nu}{2},\ \frac{s+\mu-\nu}{2},\ \frac{s-\mu+\nu}{2},\ \frac{s+\mu+\nu}{2}\\ s\end{matrix}\right]$ $+\frac{1}{\pi}\left(\frac{2}{a}\right)^s\sin\frac{s\pi}{2}\cos\frac{(\mu+\nu)\pi}{2}\,\Gamma\left[\begin{matrix}1-s,\ \frac{s-\mu-\nu}{2},\ \frac{s+\mu+\nu}{2}\\ \frac{2-s+\mu-\nu}{2},\ \frac{2-s-\mu+\nu}{2}\end{matrix}\right]$ $[a>0;\ \lvert\operatorname{Re}(\mu-\nu)\rvert,\ \lvert\operatorname{Re}(\mu+\nu)\rvert<\operatorname{Re}s<1]$
12	$Y_\nu\left(b\sqrt{x^2+a^2}+ab\right)$ $\times Y_\nu\left(b\sqrt{x^2+a^2}-ab\right)$	$-\frac{1}{2\pi^{3/2}}\left(\frac{a}{b}\right)^{s/2}\Gamma\left[\frac{1-s}{2},\ \frac{s-2\nu}{2},\ \frac{s+2\nu}{2}\right]$ $\times\left[\sin\frac{(s-2\nu)\pi}{2}J_{-s/2}(2ab)+2\cos(\nu\pi)Y_{s/2}(2ab)\right]$ $[b,\ \operatorname{Re}a>0;\ 2\lvert\operatorname{Re}\nu\rvert<\operatorname{Re}s<1]$
13	$J_\nu(u_+)J_\nu(u_-)$ $+Y_\nu(u_+)Y_\nu(u_-)$ $u_\pm=b\left(\sqrt{x^2+a^2}\pm a\right)$	$\frac{\cos(\nu\pi)}{\pi^{3/2}}\left(\frac{a}{b}\right)^{s/2}\Gamma\left[\frac{1-s}{2},\ \frac{s-2\nu}{2},\ \frac{s+2\nu}{2}\right]$ $\times\left[\sin\frac{s\pi}{2}J_{-s/2}(2ab)-\cos\frac{s\pi}{2}Y_{-s/2}(2ab)\right]$ $[a,\ b>0;\ 2\lvert\operatorname{Re}\nu\rvert<\operatorname{Re}s<1]$

3.12. The Hankel Functions $H_\nu^{(1)}(z)$ and $H_\nu^{(2)}(z)$

More formulas can be obtained from the corresponding sections due to the relations ($j = 1, 2$)

$$H_{\pm 1/2}^{(j)}(z) = \left((-1)^j i\right)^{(1\pm 1)/2} \sqrt{\frac{2}{\pi}} \frac{1}{\sqrt{z}} e^{-(-1)^j iz},$$

$$H_{n-1/2}^{(j)}(z) = \sqrt{\frac{2}{\pi}} \frac{1}{\sqrt{z}} e^{i(-1)^j(n\pi/2 - z)} \sum_{k=0}^{n-1} (-1)^{kj} \frac{(n+k-1)!}{k!\,(n-k-1)!} (2iz)^{-k},$$

$$H_\nu^{(j)}(z) = J_\nu(z) - (-1)^j i\, Y_\nu(z).$$

3.12.1. $H_\nu^{(1)}(ax)$, $H_\nu^{(2)}(ax)$

No.	$f(x)$	$F(s)$
1	$H_\nu^{(j)}(ax)$ $j = 1, 2$	$\frac{2^{s-1} a^{-s}}{\pi} e^{(-1)^{j+1}(s-\nu-1)\pi i/2} \Gamma\left(\frac{s-\nu}{2}\right) \Gamma\left(\frac{s+\nu}{2}\right)$ $[a > 0;\ -\operatorname{Re}\nu < \operatorname{Re}(\nu + s) < 3/2]$

3.12.2. $H_\nu^{(1)}(bx)$, $H_\nu^{(2)}(bx)$, and the exponential function

1	$e^{-ax^2} H_\nu^{(1)}(bx)$	$-\frac{i 2^{\nu-1} a^{(\nu-s)/2} b^{-\nu}}{\pi} \Gamma(\nu) \Gamma\left(\frac{s-\nu}{2}\right) {}_1F_1\left(\begin{matrix}\frac{s-\nu}{2}\\ 1-\nu;\ -\frac{b^2}{4a}\end{matrix}\right)$ $-\frac{i e^{-i\pi\nu} a^{-(s+\nu)/2} b^\nu}{2^{\nu+1}\pi} \Gamma(-\nu) \Gamma\left(\frac{s+\nu}{2}\right) {}_1F_1\left(\begin{matrix}\frac{s+\nu}{2}\\ \nu+1;\ -\frac{b^2}{4a}\end{matrix}\right)$ $[\operatorname{Re} a > 0;\ \operatorname{Re} s > \lvert\operatorname{Re}\nu\rvert]$
2	$e^{-ax^2} H_\nu^{(2)}(bx)$	$\frac{a^{(1-s)/2}}{\pi b} e^{\nu\pi i/2 - b^2/(8a)} \Gamma\left(\frac{s-\nu}{2}\right) \Gamma\left(\frac{s+\nu}{2}\right)$ $\times W_{(1-s)/2,\,\nu/2}\left(-\frac{b^2}{4a}\right)$ $[\operatorname{Re} a > 0;\ \operatorname{Re} s > \lvert\operatorname{Re}\nu\rvert]$

3.12.3. $H_\nu^{(1)}(ax)$, $H_\nu^{(2)}(ax)$, and trigonometric functions

1	$\begin{Bmatrix}\sin(ax+b)\\ \cos(ax+b)\end{Bmatrix}$ $\times H_\nu^{(1)}(ax)$	$\mp \frac{i^{(1\pm 1)/2} (2a)^{-s} e^{i(b+(s+\nu)\pi/2)}}{\sqrt{\pi}} \Gamma\left[\begin{matrix}\frac{1}{2}-s,\ s+\nu\\ 1-s+\nu\end{matrix}\right]$ $-\frac{i\, 2^{1-s} a^{-s}}{\pi^{3/2}} \cos(\nu\pi) \begin{Bmatrix}\sin[b+(\nu-s)\pi/2]\\ \cos[b+(\nu-s)\pi/2]\end{Bmatrix}$ $\times \Gamma\left(\frac{1}{2}-s\right) \Gamma(s-\nu) \Gamma(s+\nu)$ $[a > 0;\ \lvert\operatorname{Re}\nu\rvert < \operatorname{Re} s < 1/2]$

No.	$f(x)$	$F(s)$
2	$\begin{Bmatrix} \sin(ax+b) \\ \cos(ax+b) \end{Bmatrix} \times H_\nu^{(2)}(ax)$	$\frac{i^{(1\pm1)/2}(2a)^{-s}e^{-i(b+(s+\nu)\pi/2)}}{\sqrt{\pi}} \Gamma\left[\begin{matrix} \frac{1}{2}-s,\ s+\nu \\ 1-s+\nu \end{matrix}\right] + \frac{i\,2^{1-s}a^{-s}}{\pi^{3/2}}\cos(\nu\pi) \begin{Bmatrix} \sin[b+(\nu-s)\pi/2] \\ \cos[b+(\nu-s)\pi/2] \end{Bmatrix} \times \Gamma\left(\frac{1}{2}-s\right)\Gamma(s-\nu)\Gamma(s+\nu)$ $[a>0;\ \lvert\mathrm{Re}\,\nu\rvert<\mathrm{Re}\,s<1/2]$

3.12.4. $H_\nu^{(1)}(bx)$, $H_\nu^{(2)}(bx)$, and $J_\mu(ax)$

No.	$f(x)$	$F(s)$
1	$J_\mu(ax)H_\nu^{(j)}(bx)$ $j=1,\,2$	$\frac{2^{s-1}a^\mu}{\pi b^{s+\mu}} e^{(-1)^{j+1}(s+\mu-\nu-1)\pi i/2}\Gamma\left[\begin{matrix} \frac{s+\mu+\nu}{2},\ \frac{s+\mu-\nu}{2} \\ \mu+1 \end{matrix}\right] \times {}_2F_1\left(\begin{matrix} \frac{s+\mu+\nu}{2},\ \frac{s+\mu-\nu}{2} \\ \mu+1;\ \frac{a^2}{b^2} \end{matrix}\right)$ $\left[\begin{matrix} \lvert a\rvert<\lvert b\rvert;\ \mathrm{Re}\big(ib+(-1)^j ia\big)>0 \text{ for } \mathrm{Re}(s+\mu\pm\nu)>0 \text{ and} \\ \mathrm{Re}\big(ib+(-1)^j ia\big)=0 \text{ for } \mathrm{Re}(s+\mu\pm\nu)>0;\ \mathrm{Re}\,s<2 \end{matrix}\right]$
2	$J_\mu(ax)H_\nu^{(j)}(ax)$ $j=1,\,2$	$\frac{2^{s-1}a^{-s}}{\pi} e^{(-1)^{j+1}(s+\mu-\nu-1)\pi i/2}\Gamma\left[\begin{matrix} \frac{s+\mu-\nu}{2},\ \frac{s+\mu+\nu}{2},\ 1-s \\ \frac{\mu-\nu-s+2}{2},\ \frac{\mu+\nu-s+2}{2} \end{matrix}\right]$ $[a>0;\ -\mathrm{Re}(\mu\pm\nu)<\mathrm{Re}\,s<1]$

3.12.5. Products of $H_\mu^{(1)}(ax)$ and $H_\nu^{(2)}(ax)$

No.	$f(x)$	$F(s)$
1	$H_\mu^{(j)}(ax)H_\nu^{(j)}(ax)$ $j=1,\,2$	$-\frac{2^{s-1}a^{-s}}{\pi^2} e^{(-1)^{j+1}(s-\mu-\nu)\pi i/2}\Gamma\left[\begin{matrix} \frac{s-\mu-\nu}{2},\ \frac{s-\mu+\nu}{2},\ \frac{s+\mu-\nu}{2},\ \frac{s+\mu+\nu}{2} \\ s \end{matrix}\right]$ $[a>0;\ \lvert\mathrm{Re}\,\mu\rvert+\lvert\mathrm{Re}\,\nu\rvert<\mathrm{Re}\,s<1]$
2	$H_\mu^{(j)}(ax)H_\nu^{(j)}(bx)$ $j=1,\,2$	$-\frac{2^{s-1}b^\nu}{\pi^2 a^{s+\nu}} e^{(-1)^{j+1}(s-\mu-\nu)\pi i/2}\Gamma\left[\begin{matrix} \frac{s-\nu-\mu}{2},\ \frac{s-\mu+\nu}{2},\ \frac{s+\mu-\nu}{2},\ \frac{s+\mu+\nu}{2} \\ s \end{matrix}\right] \times {}_2F_1\left(\begin{matrix} \frac{s-\mu+\nu}{2},\ \frac{s+\mu+\nu}{2} \\ s;\ \frac{a^2-b^2}{a^2} \end{matrix}\right)$ $[a,\,b>0;\ \lvert\mathrm{Re}\,\mu\rvert+\lvert\mathrm{Re}\,\nu\rvert<\mathrm{Re}\,s<1]$
3	$H_\nu^{(1)}(ax)H_\nu^{(2)}(ax)$	$\frac{a^{-s}}{\pi^{5/2}}\cos(\nu\pi)\Gamma\left[\frac{s}{2}-\nu,\ \frac{s}{2}+\nu,\ \frac{s}{2},\ \frac{1-s}{2}\right]$ $\left[\begin{matrix} a>0; \\ 2\lvert\mathrm{Re}\,\nu\rvert<\mathrm{Re}\,s<1 \end{matrix}\right]$
4	$H_\mu^{(1)}(ax)H_\nu^{(2)}(ax)$	$\frac{2^{s-1}}{\pi a^s}\Gamma\left[\begin{matrix} 1-s,\ \frac{s+\mu-\nu}{2},\ \frac{s-\mu+\nu}{2} \\ 1-\frac{s+\mu+\nu}{2},\ 1-\frac{s-\mu-\nu}{2} \end{matrix}\right] \times \left[\csc\frac{(s-\mu-\nu)\pi}{2} + e^{i\pi(\nu-\mu)}\csc\frac{(s+\mu+\nu)\pi}{2}\right]$ $[a>0;\ \max(\lvert\mathrm{Re}(\mu+\nu)\rvert,\ \lvert\mathrm{Re}(\mu-\nu)\rvert)<\mathrm{Re}\,s<1]$

3.13. The Modified Bessel Function $I_\nu(z)$

More formulas can be obtained from the corresponding sections due to the relations

$$I_{\pm 1/2}(z)=\sqrt{\frac{2}{\pi}}\frac{1}{\sqrt{z}}\begin{Bmatrix}\sinh z\\ \cosh z\end{Bmatrix},\quad I_{\pm 3/2}(z)=\sqrt{\frac{2}{\pi}}\frac{1}{z^{3/2}}\left[z\begin{Bmatrix}\cosh z\\ \sinh z\end{Bmatrix}-\begin{Bmatrix}\sinh z\\ \cosh z\end{Bmatrix}\right],$$

$$I_{\pm n\pm 1/2}(z)=\frac{1}{\sqrt{2\pi}}\frac{1}{\sqrt{z}}\sum_{k=0}^{n}\frac{(n+k)!}{k!\,(n-k)!}\frac{(-1)^k e^z\pm(-1)^{n+1}e^{-z}}{(2z)^k},\quad [n=0,\,1,\,2,\ldots];$$

$$I_\nu(z)=i^\nu J_\nu(iz),\quad I_\nu(z)=\frac{(z/2)^\nu}{\Gamma(\nu+1)}\,{}_0F_1\left(\nu+1;\,\frac{z^2}{4}\right)$$

$$I_\nu(z)=\frac{z^\nu e^{-z}}{2^\nu\,\Gamma(\nu+1)}\,{}_1F_1\left(\nu+\frac{1}{2};\,2\nu+1;\,2z\right),$$

$$I_\nu(z)=\pi\left(\frac{z}{2}\right)^\nu G_{13}^{10}\left(\frac{z^2}{4}\,\middle|\,\begin{matrix}1/2\\ 0,\,-\nu,\,1/2\end{matrix}\right),$$

$$I_\nu(z)=z^{\nu/2}(-z)^{-\nu/2}G_{02}^{10}\left(-\frac{z^2}{4}\,\middle|\,\begin{matrix}\cdot\\ \nu/2,\,-\nu/2\end{matrix}\right),$$

$$I_\nu(z)=\pi z^\nu\left(z^2\right)^{-\nu/2}G_{13}^{10}\left(\frac{z^2}{4}\,\middle|\,\begin{matrix}(\nu+1)/2\\ \nu/2,\,-\nu/2,\,(\nu+1)/2\end{matrix}\right).$$

3.13.1. $I_\nu(\varphi(x))$ and algebraic functions

No.	$f(x)$	$F(s)$
1	$(a-x)_+^{\alpha-1}I_\nu(bx)$	$\frac{a^{s+\alpha+\nu-1}}{\Gamma(\nu+1)}\left(\frac{b}{2}\right)^\nu \mathrm{B}(\alpha,\,s+\nu)\,{}_2F_3\left(\begin{matrix}\frac{s+\nu}{2},\,\frac{s+\nu+1}{2};\,\frac{a^2b^2}{4}\\ \nu+1,\,\frac{s+\alpha+\nu}{2},\,\frac{s+\alpha+\nu+1}{2}\end{matrix}\right)$ $[a,\ \mathrm{Re}\,\alpha,\ \mathrm{Re}\,(s+\nu)>0]$
2	$\left(a^2-x^2\right)_+^{\alpha-1}I_\nu(bx)$	$\frac{a^{s+2\alpha+\nu-2}b^\nu}{2^{\nu+1}}\Gamma\left[\begin{matrix}\alpha,\,\frac{s+\nu}{2}\\ \nu+1,\,\frac{s+2\alpha+\nu}{2}\end{matrix}\right]{}_1F_2\left(\begin{matrix}\frac{s+\nu}{2};\,\frac{a^2b^2}{4}\\ \nu+1,\,\frac{s+2\alpha+\nu}{2}\end{matrix}\right)$ $[a,\ \mathrm{Re}\,\alpha,\ \mathrm{Re}\,(s+\nu)>0]$
3	$(a-x)_+^{\alpha-1}I_\nu(b(a-x))$	$\frac{a^{s+\alpha+\nu-1}}{\Gamma(\nu+1)}\left(\frac{b}{2}\right)^\nu \mathrm{B}(\alpha+\nu,\,s)\,{}_2F_3\left(\begin{matrix}\frac{\alpha+\nu}{2},\,\frac{\alpha+\nu+1}{2};\,\frac{a^2b^2}{4}\\ \nu+1,\,\frac{s+\alpha+\nu}{2},\,\frac{s+\alpha+\nu+1}{2}\end{matrix}\right)$ $[a,\ \mathrm{Re}\,(\alpha+\nu),\ \mathrm{Re}\,s>0]$
4	$(a-x)_+^{\alpha-1}$ $\times I_\nu(bx(a-x))$	$a^{s+\alpha+2\nu-1}\left(\frac{b}{2}\right)^\nu\Gamma\left[\begin{matrix}\alpha+\nu,\,s+\nu\\ \nu+1,\,s+\alpha+2\nu\end{matrix}\right]$ $\times\,{}_4F_5\left(\begin{matrix}\Delta(2,\,\alpha+\nu),\,\Delta(2,\,s+\nu)\\ \nu+1,\,\Delta(4,\,s+\alpha+2\nu);\,\frac{a^4b^2}{64}\end{matrix}\right)$ $[a,\ \mathrm{Re}\,(\alpha+\nu),\ \mathrm{Re}\,(s+\nu)>0]$
5	$(a-x)_+^{\nu/2}I_\nu\left(b\sqrt{a-x}\right)$	$\frac{2^s a^{(s+\nu)/2}}{b^s}\Gamma(s)\,I_{s+\nu}\left(\sqrt{a}\,b\right)$ $[a,\ \mathrm{Re}\,s>0;\ \mathrm{Re}\,\nu>-1]$

No.	$f(x)$	$F(s)$
6	$(a-x)_+^{\alpha-1} I_\nu\left(b\sqrt{a-x}\right)$	$\frac{a^{s+\alpha+\nu/2-1}}{2}\left(\frac{b}{2}\right)^\nu \Gamma\left[\begin{matrix}\frac{2\alpha+\nu}{2}, s\\ \nu+1, \frac{2s+2\alpha+\nu}{2}\end{matrix}\right] {}_1F_2\left(\begin{matrix}\frac{2\alpha+\nu}{2}; \frac{ab^2}{4}\\ \nu+1, \frac{2s+2\alpha+\nu}{2}\end{matrix}\right)$ $[a,\ \mathrm{Re}(\alpha+\nu/2),\ \mathrm{Re}\, s>0]$
7	$(a-x)_+^{\nu/2}(bx+1)^\mu$ $\times I_\nu\left(c\sqrt{a-x}\right)$	$a^{s+\nu}\left(\frac{c}{2}\right)^\nu \Gamma\left[\begin{matrix}s\\ s+\nu+1\end{matrix}\right] \Xi_2\left(-\mu, s; s+\nu+1; -ab, \frac{ac^2}{4}\right)$ $[a,\ \mathrm{Re}\, s>0;\ \lvert\arg b\rvert<\pi]$
8	$(a-x)_+^{\alpha-1}$ $\times I_\nu\left(b\sqrt{x(a-x)}\right)$	$a^{s+\alpha+\nu-1}\left(\frac{b}{2}\right)^\nu \Gamma\left[\begin{matrix}\frac{2\alpha+\nu}{2}, \frac{2s+\nu}{2}\\ \nu+1, s+\alpha+\nu\end{matrix}\right]$ $\times\, {}_2F_3\left(\begin{matrix}\frac{2\alpha+\nu}{2}, \frac{2s+\nu}{2}; \frac{a^2b^2}{16}\\ \nu+1, \frac{s+\alpha+\nu}{2}, \frac{s+\alpha+\nu+1}{2}\end{matrix}\right)$ $[a,\ \mathrm{Re}(\alpha+\nu/2)>0;\ \mathrm{Re}(s+\nu/2)>-1]$
9	$\frac{1}{(x+a)^\rho} I_\nu\left(\frac{b}{x+a}\right)$	$\frac{a^{s-\nu-\rho}}{\Gamma(\nu+1)}\left(\frac{b}{2}\right)^\nu \mathrm{B}(s, \nu+\rho-s)\ {}_2F_3\left(\begin{matrix}\frac{\nu+\rho-s}{2}, \frac{\nu+\rho-s+1}{2}\\ \nu+1, \frac{\nu+\rho}{2}, \frac{\nu+\rho+1}{2}; \frac{b^2}{4a^2}\end{matrix}\right)$ $[0<\mathrm{Re}\, s<\mathrm{Re}(\nu+\rho);\ \lvert\arg a\rvert<\pi]$
10	$\frac{1}{(x+a)^\rho} I_\nu\left(\frac{bx}{x+a}\right)$	$a^{s-\rho}\left(\frac{b}{2}\right)^\nu \frac{\mathrm{B}(s+\nu, \rho-s)}{\Gamma(\nu+1)}\ {}_2F_3\left(\begin{matrix}\frac{s+\nu}{2}, \frac{s+\nu+1}{2}; \frac{b^2}{4}\\ \nu+1, \frac{\nu+\rho}{2}, \frac{\nu+\rho+1}{2}\end{matrix}\right)$ $[-\mathrm{Re}\,\nu<\mathrm{Re}\, s<\mathrm{Re}\,\rho;\ \lvert\arg a\rvert<\pi]$
11	$\frac{1}{(x^2+a^2)^\rho}$ $\times I_\nu\left(\frac{bx}{x^2+a^2}\right)$	$\frac{2^{-\nu-1}a^{s-\nu-2\rho}b^\nu}{\Gamma(\nu+1)} \mathrm{B}\left(\frac{s+\nu}{2}, \frac{\nu+2\rho-s}{2}\right)\ {}_2F_3\left(\begin{matrix}\frac{s+\nu}{2}, \frac{\nu+2\rho-s}{2}; \frac{b^2}{16a^2}\\ \nu+1, \frac{\nu+\rho}{2}, \frac{\nu+\rho+1}{2}\end{matrix}\right)$ $[\mathrm{Re}\, a>0;\ -\mathrm{Re}\,\nu<\mathrm{Re}\, s<\mathrm{Re}(\nu+2\rho)]$

3.13.2. $I_\nu(\varphi(x))$ and the exponential function

1	$e^{-ax} I_\nu(bx)$	$a^{-s-\nu}\left(\frac{b}{2}\right)^\nu \Gamma\left[\begin{matrix}s+\nu\\ \nu+1\end{matrix}\right] {}_2F_1\left(\begin{matrix}\frac{s+\nu}{2}, \frac{s+\nu+1}{2}\\ \nu+1; \frac{b^2}{a^2}\end{matrix}\right)$ $[\mathrm{Re}(s+\nu)>0;\ \mathrm{Re}\, a>\lvert\mathrm{Re}\, b\rvert]$
2	$e^{-ax} I_\nu(ax)$	$\frac{(2a)^{-s}}{\sqrt{\pi}}\Gamma\left[\begin{matrix}s+\nu, \frac{1-2s}{2}\\ 1-s+\nu\end{matrix}\right]$ $[\mathrm{Re}\, a>0;\ -\mathrm{Re}\,\nu<\mathrm{Re}\, s<1/2]$
3	$e^{-ax^2} I_\nu(bx)$	$\frac{2^{-\nu-1}b^\nu}{a^{(s+\nu)/2}}\Gamma\left[\begin{matrix}\frac{s+\nu}{2}\\ \nu+1\end{matrix}\right] {}_1F_1\left(\begin{matrix}\frac{s+\nu}{2}; \frac{b^2}{4a}\\ \nu+1\end{matrix}\right)$ $[\mathrm{Re}\, a,\ \mathrm{Re}(s+\nu)>0]$

No.	$f(x)$	$F(s)$
4	$e^{-ax-b\sqrt{x}} I_\nu(ax)$	$\sqrt{\frac{2}{\pi a}}\, b^{1-2s}\,\Gamma(2s-1)\; {}_2F_2\left(\begin{matrix}\frac{1-2\nu}{2},\ \frac{1+2\nu}{2}\\ 1-s,\ \frac{3-2s}{2};\ \frac{b^2}{8a}\end{matrix}\right) + \frac{(2a)^{-s}}{\sqrt{\pi}}\,\Gamma\left[\begin{matrix}\frac{1-2s}{2},\ s+\nu\\ 1-s+\nu\end{matrix}\right] {}_2F_2\left(\begin{matrix}s-\nu,\ s+\nu\\ \frac{1}{2},\ \frac{2s+1}{2};\ \frac{b^2}{8a}\end{matrix}\right) - \frac{b}{\sqrt{\pi}\,(2a)^{s+1/2}}\,\Gamma\left[\begin{matrix}-s,\ \frac{2s+2\nu+1}{2}\\ \frac{1-2s+2\nu}{2}\end{matrix}\right] {}_2F_2\left(\begin{matrix}\frac{2s-2\nu+1}{2},\ \frac{2s+2\nu+1}{2}\\ \frac{3}{2},\ s+1;\ \frac{b^2}{8a}\end{matrix}\right)$ $[\operatorname{Re} a,\ \operatorname{Re} b,\ \operatorname{Re}(s+\nu) > 0]$
5	$(a-x)_+^{\alpha-1} e^{\pm bx} I_\nu(bx)$	$a^{s+\alpha+\nu-1}\left(\frac{b}{2}\right)^\nu \Gamma\left[\begin{matrix}\alpha,\ s+\nu\\ \nu+1,\ s+\alpha+\nu\end{matrix}\right] {}_2F_2\left(\begin{matrix}\frac{2\nu+1}{2},\ s+\nu;\ \pm 2ab\\ 2\nu+1,\ s+\alpha+\nu\end{matrix}\right)$ $[a,\ \operatorname{Re}\alpha,\ \operatorname{Re}(s+\nu) > 0]$
6	$(x-a)_+^{\alpha-1} e^{-bx} I_\nu(bx)$	$\frac{(2b)^{1-s-\alpha}}{\sqrt{\pi}}\,\Gamma\left[\begin{matrix}s+\alpha+\nu-1,\ \frac{3-2s-2\alpha}{2}\\ 2-s-\alpha+\nu\end{matrix}\right] \times {}_2F_2\left(\begin{matrix}1-\alpha,\ \frac{3-2s-2\alpha}{2};\ -2ab\\ 2-s-\alpha-\nu,\ 2-s-\alpha+\nu\end{matrix}\right) + a^{s+\alpha+\nu-1}\left(\frac{b}{2}\right)^\nu \Gamma\left[\begin{matrix}\alpha,\ 1-s-\alpha-\nu\\ \nu+1,\ 1-s-\nu\end{matrix}\right] {}_2F_2\left(\begin{matrix}\frac{2\nu+1}{2},\ s+\nu;\ -2ab\\ 2\nu+1,\ s+\alpha+\nu\end{matrix}\right)$ $[a,\ \operatorname{Re}\alpha,\ \operatorname{Re} b > 0;\ \operatorname{Re}(s+\nu) < 3/2]$
7	$\frac{e^{-bx}}{(x+a)^\rho} I_\nu(bx)$	$\frac{(2b)^{\rho-s}}{\sqrt{\pi}}\,\Gamma\left[\begin{matrix}s+\nu-\rho,\ \frac{1-2s+2\rho}{2}\\ 1-s+\nu+\rho\end{matrix}\right] \times {}_2F_2\left(\begin{matrix}\rho,\ \frac{1-2s+2\rho}{2};\ 2ab\\ 1-s-\nu+\rho,\ 1-s+\nu+\rho\end{matrix}\right) + a^{s+\nu-\rho}\left(\frac{b}{2}\right)^\nu \Gamma\left[\begin{matrix}s+\nu,\ \rho-\nu-s\\ \nu+1,\ \rho\end{matrix}\right] {}_2F_2\left(\begin{matrix}\frac{2\nu+1}{2},\ s+\nu;\ 2ab\\ 2\nu+1,\ s+\nu-\rho+1\end{matrix}\right)$ $[\operatorname{Re} b > 0;\ -\operatorname{Re}\nu < \operatorname{Re} s < \operatorname{Re}\rho + 1/2;\ \lvert\arg a\rvert < \pi]$
8	$(a-x)_+^\nu e^{bx}$ $\times I_\nu(c(a-x))$	$\frac{a^{s+2\nu}(2c)^\nu e^{-ac}}{\sqrt{\pi}}\,\Gamma\left[\begin{matrix}s,\ \nu+\frac{1}{2}\\ s+2\nu+1\end{matrix}\right] \times \Phi_2\left(s,\ \nu+\frac{1}{2};\ s+2\nu+1;\ a(b+c),\ 2ac\right)$ $[a,\ \operatorname{Re} s > 0;\ \operatorname{Re}\nu > -1/2]$
9	$\frac{e^{-bx}}{(x+a)^\nu} I_\nu(bx+ab)$	$\frac{a^{(s-1)/2-\nu}}{\sqrt{\pi}\,(2b)^{(s+1)/2}}\,\Gamma\left[\begin{matrix}s,\ \frac{1}{2}-s+\nu\\ 1-s+2\nu\end{matrix}\right] M_{-s/2,\,\nu-s/2}(2ab)$ $[a,\ \operatorname{Re} b > 0;\ 0 < \operatorname{Re} s < \operatorname{Re}\nu + 1/2]$

No.	$f(x)$	$F(s)$
10	$\dfrac{e^{-bx}}{(x+a)^{\rho}} I_\nu(bx+ab)$	$a^{s+\nu-\rho}\left(\dfrac{b}{2}\right)^{\nu} e^{ab}\,\Gamma\left[\begin{matrix} s,\, \rho-\nu-s \\ \nu+1,\, \rho-\nu \end{matrix}\right] {}_2F_2\left(\begin{matrix} \frac{2\nu+1}{2},\, \nu-\rho+1;\, -2ab \\ 2\nu+1,\, s+\nu-\rho+1 \end{matrix}\right)$ $+\dfrac{(2b)^{\rho-s}e^{ab}}{\sqrt{\pi}}\,\Gamma\left[\begin{matrix} s+\nu-\rho,\, \frac{1}{2}-s+\rho \\ 1-s+\nu+\rho \end{matrix}\right]$ $\times {}_2F_2\left(\begin{matrix} 1-s,\, \frac{1}{2}-s+\rho;\, -2ab \\ 1-s-\nu+\rho,\, 1-s+\nu+\rho \end{matrix}\right)$ $[a,\ \operatorname{Re} b>0;\ 0<\operatorname{Re} s<\operatorname{Re}\rho+1/2]$
11	$(a-x)_+^{\alpha-1} e^{bx(a-x)}$ $\times I_\nu(bx(a-x))$	$a^{s+\alpha+2\nu-1}\left(\dfrac{b}{2}\right)^{\nu}\dfrac{\mathrm{B}(\alpha+\nu,\, s+\nu)}{\Gamma(\nu+1)}$ $\times {}_3F_3\left(\begin{matrix} \frac{2\nu+1}{2},\, \alpha+\nu,\, s+\nu;\, \frac{a^2b}{2} \\ 2\nu+1,\, \frac{s+\alpha+2\nu}{2},\, \frac{s+\alpha+2\nu+1}{2} \end{matrix}\right)$ $[a,\ \operatorname{Re}(s+\nu)>0;\ \operatorname{Re}(\alpha+\nu)>-1]$
12	$(a-x)_+^{\nu/2} e^{bx}$ $\times I_\nu(c\sqrt{a-x})$	$a^{s+\nu}\left(\dfrac{c}{2}\right)^{\nu}\Gamma\left[\begin{matrix} s \\ s+\nu+1 \end{matrix}\right]\Phi_3\left(s;\, s+\nu+1;\, ab,\, \dfrac{ac^2}{4}\right)$ $[a,\ \operatorname{Re} s>0;\ \operatorname{Re}\nu>-1]$
13	$\dfrac{e^{b/(x+a)}}{(x+a)^{\rho}} I_\nu\left(\dfrac{b}{x+a}\right)$	$a^{s-\nu-\rho}\left(\dfrac{b}{2}\right)^{\nu}\dfrac{\mathrm{B}(s,\, \nu+\rho-s)}{\Gamma(\nu+1)}\, {}_2F_2\left(\begin{matrix} \frac{2\nu+1}{2},\, \nu+\rho-s \\ 2\nu+1,\, \nu+\rho;\, \frac{2b}{a} \end{matrix}\right)$ $[0<\operatorname{Re} s<\operatorname{Re}(\nu+\rho);\ \lvert\arg a\rvert<\pi]$
14	$\dfrac{\theta(x-c)}{\sqrt{x-b}} e^{ax/(x-b)}$ $\times I_\nu\left(\dfrac{ax}{x-b}\right)$	$\dfrac{a^{\nu}c^{s-1/2}}{2^{\nu-1}(1-2s)\Gamma(\nu+1)}$ $\times \Psi_1\left(\dfrac{2\nu+1}{2},\, \dfrac{1-2s}{2};\, \dfrac{3-2s}{2},\, 2\nu+1;\, \dfrac{b}{c},\, 2a\right)$ $[a>0;\ c>b>0]$
15	$(a^2-x^2)_+^{-1} e^{-b/(a^2-x^2)}$ $\times I_\nu\left(\dfrac{cx}{a^2-x^2}\right)$	$\dfrac{a^{s-1}}{c} e^{-b/(2a^2)}\,\Gamma\left[\begin{matrix} \frac{s+\nu}{2} \\ \nu+1 \end{matrix}\right] M_{(1-s)/2,\, \nu/2}\left(\dfrac{b-\sqrt{b^2-a^2c^2}}{2a^2}\right)$ $\times W_{(1-s)/2,\, \nu/2}\left(\dfrac{b+\sqrt{b^2-a^2c^2}}{2a^2}\right)$ $[b>ac>0;\ a,\ \operatorname{Re}(s+\nu)>0]$
16	$(x^2-a^2)_+^{-1} e^{-b/(x^2-a^2)}$ $\times I_\nu\left(\dfrac{cx}{x^2-a^2}\right)$	$\dfrac{a^{s-1}}{c} e^{b/(2a^2)}\,\Gamma\left[\begin{matrix} \frac{2-s+\nu}{2} \\ \nu+1 \end{matrix}\right] M_{(s-1)/2,\, \nu/2}\left(\dfrac{b-\sqrt{b^2-a^2c^2}}{2a^2}\right)$ $\times W_{(s-1)/2,\, \nu/2}\left(\dfrac{b+\sqrt{b^2-a^2c^2}}{2a^2}\right)$ $[b>ac>0;\ a>0;\ \operatorname{Re}(s-\nu)<2]$

No.	$f(x)$	$F(s)$
17	$\frac{1}{x^2+a^2}\, e^{b/\left(x^2+a^2\right)} \times I_\nu\left(\frac{cx}{x^2+a^2}\right)$	$\frac{a^{s-1}}{c}\, e^{b/(2a^2)}\, \Gamma\left[\begin{matrix}\frac{s+\nu}{2},\, \frac{2-s+\nu}{2}\\ \nu+1,\, \nu+1\end{matrix}\right] M_{(s-1)/2,\,\nu/2}\left(\frac{\sqrt{b^2+a^2c^2}+b}{2a^2}\right) \times M_{(1-s)/2,\,\nu/2}\left(\frac{\sqrt{b^2+a^2c^2}-b}{2a^2}\right)$ $[b,\ \operatorname{Re} a>0;\ -\operatorname{Re}\nu<\operatorname{Re}s<\operatorname{Re}\nu+2]$

3.13.3. $I_\nu(ax)$ and trigonometric functions

Notation: $\delta = \left\{\begin{matrix}1\\0\end{matrix}\right\}$.

1	$e^{-ax}\left\{\begin{matrix}\sin(bx)\\ \cos(bx)\end{matrix}\right\} I_\nu(ax)$	$\left(\frac{a}{2}\right)^\nu b^{-s-\nu}\, \Gamma\left[\begin{matrix}s+\nu\\ \nu+1\end{matrix}\right] \left\{\begin{matrix}\sin[(s+\nu)\pi/2]\\ \cos[(s+\nu)\pi/2]\end{matrix}\right\} \times {}_4F_3\left(\begin{matrix}\frac{2\nu+1}{4},\, \frac{2\nu+3}{4},\, \frac{s+\nu}{2},\, \frac{s+\nu+1}{2}\\ \frac{1}{2},\, \frac{2\nu+1}{2},\, \nu+1;\, -\frac{4a^2}{b^2}\end{matrix}\right) + (-1)^\delta\, \frac{a^{\nu+1} b^{-s-\nu-1}}{2^\nu} \left\{\begin{matrix}\cos[(s+\nu)\pi/2]\\ \sin[(s+\nu)\pi/2]\end{matrix}\right\} \times \Gamma\left[\begin{matrix}s+\nu+1\\ \nu+1\end{matrix}\right] {}_4F_3\left(\begin{matrix}\frac{2\nu+3}{4},\, \frac{2\nu+5}{4},\, \frac{s+\nu+1}{2},\, \frac{s+\nu+2}{2}\\ \frac{3}{2},\, \nu+1,\, \frac{2\nu+3}{2};\, -\frac{4a^2}{b^2}\end{matrix}\right)$ $[0<2a<b;\ -\delta-\operatorname{Re}\nu<\operatorname{Re}s<3/2]$
2	$e^{-ax}\left\{\begin{matrix}\sin(bx)\\ \cos(bx)\end{matrix}\right\} I_\nu(ax)$	$\frac{(2a)^{-s-\delta}\, b^\delta}{\sqrt{\pi}}\, \Gamma\left[\begin{matrix}s+\nu+\delta,\, \frac{1-2s-2\delta}{2}\\ 1-s+\nu-\delta\end{matrix}\right] \times {}_4F_3\left(\begin{matrix}\frac{s-\nu+1}{2},\, \frac{s-\nu+2\delta}{2},\, \frac{s+\nu+1}{2},\, \frac{s+\nu+2\delta}{2}\\ \frac{2\delta+1}{2},\, \frac{2s+3}{4},\, \frac{2s+4\delta+1}{4};\, -\frac{b^2}{4a^2}\end{matrix}\right) - \frac{b^{1/2-s}}{2\sqrt{2\pi a}} \cos(s\pi) \times \csc\frac{(2s+2\delta-1)\pi}{4}\, \Gamma\left(\frac{2s-1}{2}\right) {}_4F_3\left(\begin{matrix}\frac{1+2\nu}{4},\, \frac{1-2\nu}{4},\, \frac{3+2\nu}{4},\, \frac{3-2\nu}{4}\\ \frac{1}{2},\, \frac{3-2s}{4},\, \frac{5-2s}{4};\, -\frac{b^2}{4a^2}\end{matrix}\right) - \frac{(4\nu^2-1)\, b^{3/2-s}}{16\sqrt{2\pi}\, a^{3/2}} \cos(s\pi) \csc\frac{(2s+2\delta-3)\pi}{4} \times \Gamma\left(\frac{2s-3}{2}\right) {}_4F_3\left(\begin{matrix}\frac{3+2\nu}{4},\, \frac{3-2\nu}{4},\, \frac{5+2\nu}{4},\, \frac{5-2\nu}{4}\\ \frac{3}{2},\, \frac{5-2s}{4},\, \frac{7-2s}{4};\, -\frac{b^2}{4a^2}\end{matrix}\right)$ $[0<b<2a;\ -\operatorname{Re}\nu-\delta<\operatorname{Re}s<3/2]$
3	$e^{-ax}\left\{\begin{matrix}\sin(b\sqrt{x})\\ \cos(b\sqrt{x})\end{matrix}\right\} \times I_\nu(ax)$	$\frac{(-1)^\delta}{b^{2s-1}} \sqrt{\frac{2}{\pi a}} \left\{\begin{matrix}\cos(s\pi)\\ \sin(s\pi)\end{matrix}\right\} \Gamma(2s-1)\, {}_2F_2\left(\begin{matrix}\frac{1-2\nu}{2},\, \frac{1+2\nu}{2}\\ \frac{3-2s}{2},\, 1-s;\, -\frac{b^2}{8a}\end{matrix}\right) + \frac{b^\delta}{(2a)^{s+\delta/2}\sqrt{\pi}}\, \Gamma\left[\begin{matrix}\frac{1-2s-\delta}{2},\, \frac{2s+2\nu+\delta}{2}\\ \frac{2-2s+2\nu+\delta}{2}\end{matrix}\right] {}_2F_2\left(\begin{matrix}\frac{2s-2\nu+\delta}{2},\, \frac{2s+2\nu+\delta}{2}\\ \frac{2\delta+1}{2},\, \frac{2s+\delta+1}{2};\, -\frac{b^2}{8a}\end{matrix}\right)$ $[b,\ \operatorname{Re} a>0;\ -\operatorname{Re}\nu-\delta/2<\operatorname{Re}s<1]$

3.13.4. $I_\nu(ax)$ and the logarithmic function

1	$\theta(a-x)\ln\dfrac{\sqrt{a}+\sqrt{a-x}}{\sqrt{x}} \times I_\nu(bx)$	$\dfrac{\sqrt{\pi}\,a^{s+\nu}}{2(s+\nu)}\left(\dfrac{b}{2}\right)^\nu \Gamma\left[\begin{matrix} s+\nu \\ \nu+1,\ \frac{2s+2\nu+1}{2}\end{matrix}\right] \times {}_3F_4\left(\begin{matrix}\frac{s+\nu}{2},\ \frac{s+\nu}{2},\ \frac{s+\nu+1}{2};\ \frac{a^2b^2}{4} \\ \nu+1,\ \frac{2s+2\nu+1}{4},\ \frac{2s+2\nu+3}{4},\ \frac{s+\nu+2}{2}\end{matrix}\right)$ $[a,\ \mathrm{Re}(s+\nu)>0]$
2	$\theta(a-x)\ln\dfrac{a+\sqrt{a^2-x^2}}{x} \times I_\nu(bx)$	$\dfrac{\sqrt{\pi}\,a^{s+\nu}}{2(s+\nu)}\left(\dfrac{b}{2}\right)^\nu \Gamma\left[\begin{matrix} \frac{s+\nu}{2} \\ \nu+1,\ \frac{s+\nu+1}{2}\end{matrix}\right] {}_2F_3\left(\begin{matrix}\frac{s+\nu}{2},\ \frac{s+\nu}{2};\ \frac{a^2b^2}{4} \\ \nu+1,\ \frac{s+\nu+1}{2},\ \frac{s+\nu+2}{2}\end{matrix}\right)$ $[a,\ \mathrm{Re}(s+\nu)>0]$
3	$\theta(a-x)\,e^{bx}\ln\dfrac{\sqrt{a}+\sqrt{a-x}}{\sqrt{x}} \times I_\nu(bx)$	$\dfrac{\sqrt{\pi}\,a^{s+\nu}b^\nu}{2^{\nu+1}(s+\nu)} \Gamma\left[\begin{matrix} s+\nu \\ \nu+1,\ \frac{2s+2\nu+1}{2}\end{matrix}\right] \times {}_3F_3\left(\begin{matrix}\frac{2\nu+1}{2},\ s+\nu,\ s+\nu;\ 2ab \\ 2\nu+1,\ \frac{2s+2\nu+1}{2},\ s+\nu+1\end{matrix}\right)$ $[a,\ \mathrm{Re}(s+\nu)>0]$

3.13.5. $I_\nu(ax)$ and inverse trigonometric functions

1	$\theta(a-x)\arccos\dfrac{x}{a}\,I_\nu(bx)$	$\dfrac{\sqrt{\pi}\,a^{s+\nu}}{(s+\nu)^2}\left(\dfrac{b}{2}\right)^\nu \Gamma\left[\begin{matrix} \frac{s+\nu+1}{2} \\ \nu+1,\ \frac{s+\nu}{2}\end{matrix}\right] {}_2F_3\left(\begin{matrix}\frac{s+\nu}{2},\ \frac{s+\nu+1}{2};\ \frac{a^2b^2}{4} \\ \nu+1,\ \frac{s+\nu+2}{2},\ \frac{s+\nu+2}{2}\end{matrix}\right)$ $[a,\ \mathrm{Re}(s+\nu)>0]$
2	$\theta(a-x)\,e^{bx}\arccos\dfrac{x}{a}\,I_\nu(bx)$	$\dfrac{2^{-\nu}\sqrt{\pi}\,a^{s+\nu+1}b^{\nu+1}}{(s+\nu+1)^2} \Gamma\left[\begin{matrix} \frac{s+\nu+2}{2} \\ \nu+1,\ \frac{s+\nu+1}{2}\end{matrix}\right] \times {}_4F_5\left(\begin{matrix}\frac{2\nu+3}{4},\ \frac{2\nu+5}{4},\ \frac{s+\nu+1}{2},\ \frac{s+\nu+2}{2} \\ \frac{3}{2},\ \nu+1,\ \frac{2\nu+3}{2},\ \frac{s+\nu+3}{2},\ \frac{s+\nu+3}{2};\ a^2b^2\end{matrix}\right) + \dfrac{2^{-\nu}\sqrt{\pi}\,a^{s+\nu}b^\nu}{(s+\nu)^2} \Gamma\left[\begin{matrix} \frac{s+\nu+1}{2} \\ \nu+1,\ \frac{s+\nu}{2}\end{matrix}\right] \times {}_4F_5\left(\begin{matrix}\frac{2\nu+1}{4},\ \frac{2\nu+3}{4},\ \frac{s+\nu}{2},\ \frac{s+\nu+1}{2};\ a^2b^2 \\ \frac{1}{2},\ \frac{2\nu+1}{2},\ \nu+1,\ \frac{s+\nu+2}{2},\ \frac{s+\nu+2}{2}\end{matrix}\right)$ $[a,\ \mathrm{Re}(s+\nu)>0]$

3.13.6. $I_\nu(ax)$ and $\mathrm{Ei}(bx^r)$

1	$\mathrm{Ei}(-ax)\,I_\nu(bx)$	$-\dfrac{b^\nu}{2^\nu a^{s+\nu}(s+\nu)} \Gamma\left[\begin{matrix} s+\nu \\ \nu+1\end{matrix}\right] {}_3F_2\left(\begin{matrix}\frac{s+\nu}{2},\ \frac{s+\nu}{2},\ \frac{s+\nu+1}{2} \\ \nu+1,\ \frac{s+\nu+2}{2};\ \frac{b^2}{a^2}\end{matrix}\right)$ $[\mathrm{Re}(a-b),\ \mathrm{Re}(s+\nu)>0]$

No.	$f(x)$	$F(s)$
2	$\operatorname{Ei}\left(-ax^2\right) I_\nu(bx)$	$-\frac{a^{-(s+\nu)/2}b^\nu}{2^\nu(s+\nu)}\Gamma\left[\begin{matrix}\frac{s+\nu}{2}\\ \nu+1\end{matrix}\right] {}_2F_2\left(\begin{matrix}\frac{s+\nu}{2},\ \frac{s+\nu}{2};\ \frac{b^2}{4a}\\ \nu+1,\ \frac{s+\nu+2}{2}\end{matrix}\right)$ $[\operatorname{Re} a,\ \operatorname{Re}(s+\nu)>0]$
3	$e^{-ax}\operatorname{Ei}(-bx) I_\nu(ax)$	$-\frac{b^{-s-\nu}}{s+\nu}\left(\frac{a}{2}\right)^\nu\Gamma\left[\begin{matrix}s+\nu\\ \nu+1\end{matrix}\right] {}_3F_2\left(\begin{matrix}\frac{2\nu+1}{2},\ s+\nu,\ s+\nu\\ 2\nu+1,\ s+\nu+1;\ -\frac{2a}{b}\end{matrix}\right)$ $[\operatorname{Re} b,\ \operatorname{Re} a,\ \operatorname{Re}(s+\nu)>0]$
4	$e^{(\pm b-a)x}\operatorname{Ei}(\mp bx) I_\nu(ax)$	$-\frac{\pi(a/2)^\nu}{b^{s+\nu}}\left\{\begin{matrix}\csc(s+\nu)\pi\\ \cot(s+\nu)\pi\end{matrix}\right\}\Gamma\left[\begin{matrix}s+\nu\\ \nu+1\end{matrix}\right] {}_2F_1\left(\begin{matrix}\frac{2\nu+1}{2},\ s+\nu\\ 2\nu+1;\ \pm\frac{2a}{b}\end{matrix}\right)$ $\mp\frac{(2a)^{1-s}}{\sqrt{\pi}b}\Gamma\left[\begin{matrix}s+\nu-1,\ \frac{3-2s}{2}\\ 2-s+\nu\end{matrix}\right] {}_3F_2\left(\begin{matrix}1,\ 1,\ \frac{3-2s}{2};\ \pm\frac{2a}{b}\\ 2-s-\nu,\ 2-s+\nu\end{matrix}\right)$ $[\operatorname{Re} a,\ \operatorname{Re} b>0;\ -\operatorname{Re}\nu<\operatorname{Re} s<3/2]$
5	$e^x\operatorname{Ei}(-2x) I_\nu(x)$	$-\frac{2^{-s}\sqrt{\pi}}{s+\nu}\sec(\nu\pi)\Gamma\left[\begin{matrix}s+\nu\\ \frac{1-2\nu}{2},\ 2\nu+1\end{matrix}\right] {}_3F_2\left(\begin{matrix}\frac{2\nu+1}{2},\ s+\nu,\ s+\nu\\ 2\nu+1,\ s+\nu+1;\ 1\end{matrix}\right)$ $[-\operatorname{Re}\nu<\operatorname{Re} s<3/2]$

3.13.7. $I_\nu(ax)$ and $\operatorname{si}(bx)$, $\operatorname{ci}(bx)$

1	$e^{-ax}\left\{\begin{matrix}\operatorname{si}(bx)\\ \operatorname{ci}(bx)\end{matrix}\right\} I_\nu(ax)$	$-\frac{b^{-s-\nu}}{s+\nu}\left(\frac{a}{2}\right)^\nu\Gamma\left[\begin{matrix}s+\nu\\ \nu+1\end{matrix}\right]\left\{\begin{matrix}\sin[(s+\nu)\pi/2]\\ \cos[(s+\nu)\pi/2]\end{matrix}\right\}$ $\times {}_5F_4\left(\begin{matrix}\frac{2\nu+1}{4},\ \frac{2\nu+3}{4},\ \frac{s+\nu}{2},\ \frac{s+\nu}{2},\ \frac{s+\nu+1}{2}\\ \frac{1}{2},\ \frac{2\nu+1}{2},\ \nu+1,\ \frac{s+\nu+2}{2};\ -\frac{4a^2}{b^2}\end{matrix}\right)$ $\pm\frac{a^{\nu+1}b^{-s-\nu-1}}{2^\nu(s+\nu+1)}\Gamma\left[\begin{matrix}s+\nu+1\\ \nu+1\end{matrix}\right]\left\{\begin{matrix}\cos[(s+\nu)\pi/2]\\ \sin[(s+\nu)\pi/2]\end{matrix}\right\}$ $\times {}_5F_4\left(\begin{matrix}\frac{2\nu+3}{4},\ \frac{2\nu+5}{4},\ \frac{s+\nu+1}{2},\ \frac{s+\nu+1}{2},\ \frac{s+\nu+2}{2}\\ \frac{3}{2},\ \nu+1,\ \frac{2\nu+3}{2},\ \frac{s+\nu+3}{2};\ -\frac{4a^2}{b^2}\end{matrix}\right)$ $[b,\ \operatorname{Re} a>0;\ -\operatorname{Re}\nu<\operatorname{Re} s<5/2]$

3.13.8. $I_\nu(ax)$ and $\operatorname{erf}(bx^r)$, $\operatorname{erfc}(bx^r)$

1	$\operatorname{erfc}(bx) I_\nu(ax)$	$\frac{a^\nu}{2^{s+2\nu}b^{s+\nu}}\Gamma\left[\begin{matrix}s+\nu\\ \nu+1,\ \frac{s+\nu+2}{2}\end{matrix}\right] {}_2F_2\left(\begin{matrix}\frac{s+\nu}{2},\ \frac{s+\nu+1}{2};\ \frac{a^2}{4b^2}\\ \nu+1,\ \frac{s+\nu+2}{2}\end{matrix}\right)$ $[\operatorname{Re} a,\ \operatorname{Re}(s+\nu)>0;\ \lvert\arg b\rvert<\pi/4]$

No.	$f(x)$	$F(s)$
2	$\operatorname{erfc}(b\sqrt{x})\, I_\nu(ax)$	$\frac{b^{-2(s+\nu)}}{\sqrt{\pi}(s+\nu)}\left(\frac{a}{2}\right)^\nu \Gamma\begin{bmatrix}\frac{2s+2\nu+1}{2}\\ \nu+1\end{bmatrix} {}_3F_2\left(\begin{matrix}\frac{s+\nu}{2},\ \frac{2s+2\nu+1}{4},\ \frac{2s+2\nu+3}{4}\\ \nu+1,\ \frac{s+\nu+2}{2};\ \frac{a^2}{b^4}\end{matrix}\right)$ $\left[\operatorname{Re}\left(b^2-a\right),\ \operatorname{Re}(s+\nu)>0;\ \lvert\arg b\rvert<\pi/4\right]$
3	$e^{-ax}\left\{\begin{matrix}\operatorname{erf}(bx)\\ \operatorname{erfc}(bx)\end{matrix}\right\} I_\nu(ax)$	$\mp\frac{(a/2)^\nu}{\sqrt{\pi}\, b^{s+\nu}(s+\nu)}\Gamma\begin{bmatrix}\frac{s+\nu+1}{2}\\ \nu+1\end{bmatrix} {}_4F_4\left(\begin{matrix}\frac{2\nu+1}{4},\ \frac{2\nu+3}{4},\ \frac{s+\nu}{2},\ \frac{s+\nu+1}{2}\\ \frac{1}{2},\ \frac{2\nu+1}{2},\ \nu+1,\ \frac{s+\nu+2}{2};\ \frac{a^2}{b^2}\end{matrix}\right)$ $\pm\frac{a^{\nu+1}}{2^\nu\sqrt{\pi}\, b^{s+\nu+1}(s+\nu+1)}\Gamma\begin{bmatrix}\frac{s+\nu+2}{2}\\ \nu+1\end{bmatrix}$ $\times {}_4F_4\left(\begin{matrix}\frac{2\nu+3}{4},\ \frac{2\nu+5}{4},\ \frac{s+\nu+1}{2},\ \frac{s+\nu+2}{2}\\ \frac{3}{2},\ \nu+1,\ \frac{2\nu+3}{2},\ \frac{s+\nu+3}{2};\ \frac{a^2}{b^2}\end{matrix}\right)+\frac{(1\pm1)}{2^{s+1}\sqrt{\pi}a^s}\Gamma\begin{bmatrix}s+\nu,\ \frac{1-2s}{2}\\ 1-s+\nu\end{bmatrix}$ $\left[\operatorname{Re} a>0;\ \left\{\begin{matrix}-\operatorname{Re}\nu-1<\operatorname{Re} s<1/2\\ \operatorname{Re}(s+\nu)>0\end{matrix}\right\};\ \lvert\arg b\rvert<\pi/4\right]$
4	$e^{-ax}\left\{\begin{matrix}\operatorname{erf}(b\sqrt{x})\\ \operatorname{erfc}(b\sqrt{x})\end{matrix}\right\}$ $\times I_\nu(ax)$	$\mp\frac{a^\nu}{2^\nu(s+\nu)\sqrt{\pi}\, b^{2(s+\nu)}}\Gamma\begin{bmatrix}\frac{2s+2\nu+1}{2}\\ \nu+1\end{bmatrix} {}_3F_2\left(\begin{matrix}\frac{2\nu+1}{2},\ \frac{2s+2\nu+1}{2},\ s+\nu\\ 2\nu+1,\ s+\nu+1;\ -\frac{2a}{b^2}\end{matrix}\right)$ $+\frac{(1\pm1)}{2^{s+1}\sqrt{\pi}a^s}\Gamma\begin{bmatrix}s+\nu,\ \frac{1-2s}{2}\\ 1-s+\nu\end{bmatrix}$ $\left[\operatorname{Re} a>0;\ \left\{\begin{matrix}-\operatorname{Re}\nu-1/2<\operatorname{Re} s<1/2\\ \operatorname{Re}(s+\nu)>0\end{matrix}\right\};\ \lvert\arg b\rvert<\pi/4\right]$
5	$\operatorname{erfc}\left(bx^2\right) J_\nu(ax)\, I_\nu(ax)$	$\frac{a^{2\nu}}{2^{2\nu}\sqrt{\pi}\, b^{s/2+\nu}(s+2\nu)}\Gamma\begin{bmatrix}\frac{s+2\nu+2}{4}\\ \nu+1,\ \nu+1\end{bmatrix}$ $\times {}_2F_4\left(\begin{matrix}\frac{s+2\nu}{4},\ \frac{s+2\nu+2}{4};\ -\frac{a^2}{64b^2}\\ \nu+1,\ \frac{\nu+1}{2},\ \frac{\nu+2}{2},\ \frac{s+2\nu+4}{4}\end{matrix}\right)$ $\left[\begin{matrix}\operatorname{Re}(s+2\nu)>0;\\ \lvert\arg b\rvert<\pi/4\end{matrix}\right]$

3.13.9. $I_\nu(ax)$ and $S(bx)$, $C(bx)$

1	$e^{-ax}\left\{\begin{matrix}S(bx)\\ C(bx)\end{matrix}\right\} I_\nu(ax)$	$-\frac{a^\nu b^{-s-\nu}}{2^{\nu+1/2}\sqrt{\pi}(s+\nu)}\Gamma\begin{bmatrix}\frac{2s+2\nu+1}{2}\\ \nu+1\end{bmatrix}\left\{\begin{matrix}\cos[(1-2s-2\nu)\pi/4]\\ \sin[(1-2s-2\nu)\pi/4]\end{matrix}\right\}$ $\times {}_5F_4\left(\begin{matrix}\frac{2\nu+1}{4},\ \frac{2\nu+3}{4},\ \frac{s+\nu}{2},\ \frac{2s+2\nu+1}{4},\ \frac{2s+2\nu+3}{4}\\ \frac{1}{2},\ \nu+1,\ \frac{2\nu+1}{2},\ \frac{s+\nu+2}{2};\ -\frac{4a^2}{b^2}\end{matrix}\right)$ $\pm\frac{a^{\nu+1}b^{-s-\nu-1}}{2^{\nu+1/2}\sqrt{\pi}(s+\nu+1)}\Gamma\begin{bmatrix}\frac{2s+2\nu+3}{2}\\ \nu+1\end{bmatrix}\left\{\begin{matrix}\sin[(1-2s-2\nu)\pi/4]\\ \cos[(1-2s-2\nu)\pi/4]\end{matrix}\right\}$ $\times {}_5F_4\left(\begin{matrix}\frac{2\nu+3}{4},\ \frac{2\nu+5}{4},\ \frac{s+\nu+1}{2},\ \frac{2s+2\nu+3}{4},\ \frac{2s+2\nu+5}{4}\\ \frac{3}{2},\ \nu+1,\ \frac{2\nu+3}{2},\ \frac{s+\nu+3}{2};\ -\frac{4a^2}{b^2}\end{matrix}\right)$ $+\frac{(2a)^{-s}}{2\sqrt{\pi}}\Gamma\begin{bmatrix}s+\nu,\ \frac{1-2s}{2}\\ 1-s+\nu\end{bmatrix}$ $[b,\ \operatorname{Re} a>0;\ -(2\pm1)/2-\operatorname{Re}\nu<\operatorname{Re} s<2]$

3.13.10. $I_\nu(ax)$ and $\gamma(\mu, bx)$, $\Gamma(\mu, bx^r)$

1	$\Gamma(\mu, bx)\, I_\nu(ax)$	$\frac{b^{-s-\nu}}{s+\nu}\left(\frac{a}{2}\right)^\nu \Gamma\left[\begin{matrix} s+\mu+\nu \\ \nu+1 \end{matrix}\right] {}_3F_2\left(\begin{matrix} \frac{s+\nu}{2}, \frac{s+\mu+\nu}{2}, \frac{s+\mu+\nu+1}{2} \\ \nu+1, \frac{s+\nu+2}{2}; \frac{a^2}{b^2} \end{matrix}\right)$ $[\operatorname{Re} b > \lvert\operatorname{Re} a\rvert;\ \operatorname{Re} s > -\operatorname{Re}\nu,\ -\operatorname{Re}(\mu+\nu)]$
2	$\Gamma(\mu, bx^2)\, I_\nu(ax)$	$\frac{b^{-(s+\nu)/2}}{s+\nu}\left(\frac{a}{2}\right)^\nu \Gamma\left[\begin{matrix} \frac{s+2\mu+\nu}{2} \\ \nu+1 \end{matrix}\right] {}_2F_2\left(\begin{matrix} \frac{s+\nu}{2}, \frac{s+2\mu+\nu}{2}; \frac{a^2}{4b} \\ \nu+1, \frac{s+\nu+2}{2} \end{matrix}\right)$ $[\operatorname{Re} b,\ \operatorname{Re}(s+\nu),\ \operatorname{Re}(s+2\mu+\nu) > 0]$
3	$e^{-ax}\left\{\begin{matrix} \gamma(\mu, bx) \\ \Gamma(\mu, bx) \end{matrix}\right\} I_\nu(ax)$	$\mp\frac{b^{-s-\nu}}{s+\nu}\left(\frac{a}{2}\right)^\nu \Gamma\left[\begin{matrix} s+\mu+\nu \\ \nu+1 \end{matrix}\right] {}_3F_2\left(\begin{matrix} \frac{2\nu+1}{2}, s+\nu, s+\mu+\nu \\ 2\nu+1, s+\nu+1; -\frac{2a}{b} \end{matrix}\right)$ $+\frac{1\pm 1}{2^{s+1}\sqrt{\pi}a^s}\Gamma\left[\begin{matrix} \mu, s+\nu, \frac{1-2s}{2} \\ 1-s+\nu \end{matrix}\right]$ $\left[\operatorname{Re} a,\ \operatorname{Re} b,\ \operatorname{Re}(s+\mu+\nu) > 0;\ \left\{\begin{matrix} \operatorname{Re}\mu > 0;\ \operatorname{Re} s < 1/2 \\ \operatorname{Re}(s+\nu) > 0 \end{matrix}\right\}\right]$

3.13.11. $I_\nu(ax)$ and $D_\mu(bx^r)$

1	$e^{-a^2x^2/4} D_\mu(ax)\, I_\nu(bx)$	$\frac{\sqrt{\pi}\, b^\nu}{2^{(s+3\nu-\mu)/2} a^{s+\nu}} \Gamma\left[\begin{matrix} s+\nu \\ \nu+1, \frac{s-\mu+\nu+1}{2} \end{matrix}\right] {}_2F_2\left(\begin{matrix} \frac{s+\nu}{2}, \frac{s+\nu+1}{2}; \frac{b^2}{2a^2} \\ \nu+1, \frac{s-\mu+\nu+1}{2} \end{matrix}\right)$ $[-\operatorname{Re}\nu < \operatorname{Re} s < 5/2 - \operatorname{Re}\mu;\ \lvert\arg a\rvert < \pi/4]$
2	$e^{-a^2x^2/4-bx} D_\mu(ax)$ $\times I_\nu(bx)$	$\frac{\sqrt{\pi}\, b^\nu}{2^{(s+3\nu-\mu)/2} a^{s+\nu}} \Gamma\left[\begin{matrix} s+\nu \\ \nu+1, \frac{s-\mu+\nu+1}{2} \end{matrix}\right]$ $\times {}_4F_4\left(\begin{matrix} \frac{2\nu+1}{4}, \frac{2\nu+3}{4}, \frac{s+\nu}{2}, \frac{s+\nu+1}{2}; \frac{2b^2}{a^2} \\ \frac{1}{2}, \frac{2\nu+1}{2}, \nu+1, \frac{s-\mu+\nu+1}{2} \end{matrix}\right) - \frac{\sqrt{\pi}\, b^{\nu+1}}{2^{(s-\mu+3\nu+1)/2} a^{s+\nu+1}}$ $\times \Gamma\left[\begin{matrix} s+\nu+1 \\ \nu+1, \frac{s-\mu+\nu+2}{2} \end{matrix}\right] {}_4F_4\left(\begin{matrix} \frac{2\nu+3}{4}, \frac{2\nu+5}{4}, \frac{s+\nu+1}{2}, \frac{s+\nu+2}{2}; \frac{2b^2}{a^2} \\ \frac{3}{2}, \nu+1, \frac{2\nu+3}{2}, \frac{s-\mu+\nu+2}{2} \end{matrix}\right)$ $[\operatorname{Re} b > 0;\ -\operatorname{Re}\nu < \operatorname{Re} s < 5/2 - \operatorname{Re}\mu;\ \lvert\arg a\rvert < \pi/4]$
3	$e^{(\pm a^2/4-b)x} D_\mu(a\sqrt{x})$ $\times I_\nu(bx)$	$\frac{2^{-s-2\nu\mp(\mu+1\mp1)/2} b^\nu}{a^{2(s+\nu)}} \Gamma^{\pm1}\left(\frac{1\mp1-2\mu\mp4\nu\mp4s}{4}\right) \Gamma\left[\begin{matrix} 2s+2\nu \\ \nu+1 \end{matrix}\right]$ $\times \left\{\begin{matrix} \Gamma^{-1}(-\mu) \\ \sqrt{\pi} \end{matrix}\right\} {}_3F_2\left(\begin{matrix} \frac{2\nu+1}{2}, s+\nu, \frac{2s+2\nu+1}{2} \\ 2\nu+1, \frac{4s\pm2\mu+4\nu+3\pm1}{4}; \pm\frac{4b}{a^2} \end{matrix}\right)$ $+\frac{(1\pm1)\, a^\mu}{2\sqrt{\pi}\,(2b)^{s+\mu/2}} \Gamma\left[\begin{matrix} \frac{2s+\mu+2\nu}{2}, \frac{1-2s-\mu}{2} \\ \frac{2-2s-\mu+2\nu}{2} \end{matrix}\right] {}_3F_2\left(\begin{matrix} -\frac{\mu}{2}, \frac{1-\mu}{2}, \frac{1-\mu-2s}{2}; \frac{4b}{a^2} \\ \frac{2-2s-\mu-2\nu}{2}, \frac{2-2s-\mu+2\nu}{2} \end{matrix}\right)$ $\left[\operatorname{Re} b > 0;\ \lvert\arg a\rvert < (2\pm1)\pi/4;\ \left\{\begin{matrix} -\operatorname{Re}\nu < \operatorname{Re} s < (1-\operatorname{Re}\mu)/2 \\ \operatorname{Re}(s+\nu) > 0 \end{matrix}\right\}\right]$

No.	$f(x)$	$F(s)$
4	$D_{-\mu-1}(a\sqrt{x})\,D_\mu(a\sqrt{x})$ $\times I_\nu(bx)$	$\frac{2^{1/2-2s-3\nu}\pi b^\nu}{a^{2(s+\nu)}}\Gamma\left[\begin{matrix}2s+2\nu\\ \nu+1,\ \frac{s-\mu+\nu+1}{2},\ \frac{s+\mu+\nu+2}{2}\end{matrix}\right] \times {}_4F_3\left(\begin{matrix}\frac{s+\nu}{2},\ \frac{s+\nu+1}{2},\ \frac{2s+2\nu+1}{4},\ \frac{2s+2\nu+3}{4}\\ \nu+1,\ \frac{s-\mu+\nu+1}{2},\ \frac{s+\mu+\nu+2}{2};\ \frac{4b^2}{a^4}\end{matrix}\right)$ $[\mathrm{Re}\,(a^2-2b),\ \mathrm{Re}\,(s+\nu)>0]$

3.13.12. $I_\nu(ax)$ and $J_\mu(bx^r)$, $Y_\mu(bx^r)$

No.	$f(x)$	$F(s)$
1	$e^{-ax}J_\mu(bx)\,I_\nu(ax)$	$\frac{2^{s-1}a^\nu}{b^{s+\nu}}\Gamma\left[\begin{matrix}\frac{s+\mu+\nu}{2}\\ \nu+1,\ \frac{2-s+\mu-\nu}{2}\end{matrix}\right]{}_4F_3\left(\begin{matrix}\frac{2\nu+1}{4},\ \frac{2\nu+3}{4},\ \frac{s+\nu-\mu}{2},\ \frac{s+\nu+\mu}{2}\\ \frac{1}{2},\ \frac{2\nu+1}{2},\ \nu+1;\ -\frac{4a^2}{b^2}\end{matrix}\right) - \frac{2^s a^{\nu+1}}{b^{s+\nu+1}}\Gamma\left[\begin{matrix}\frac{s+\mu+\nu+1}{2}\\ \nu+1,\ \frac{1-s+\mu-\nu}{2}\end{matrix}\right]{}_4F_3\left(\begin{matrix}\frac{2\nu+3}{4},\ \frac{2\nu+5}{4},\ \frac{s-\mu+\nu+1}{2},\ \frac{s+\mu+\nu+1}{2}\\ \frac{3}{2},\ \nu+1,\ \frac{2\nu+3}{2};\ -\frac{4a^2}{b^2}\end{matrix}\right)$ $[0<2a<b;\ \mathrm{Re}\,(s+\mu+\nu)>0;\ \mathrm{Re}\,s<2]$
2	$e^{-ax}J_\mu(bx)\,I_\nu(ax)$	$\frac{2^{s-2}b^{1/2-s}}{\sqrt{\pi a}}\Gamma\left[\begin{matrix}\frac{2s+2\mu-1}{4}\\ \frac{5-2s+2\mu}{4}\end{matrix}\right]{}_4F_3\left(\begin{matrix}\frac{1+2\nu}{4},\ \frac{1-2\nu}{4},\ \frac{3+2\nu}{4},\ \frac{3-2\nu}{4}\\ \frac{1}{2},\ \frac{5-2s-2\mu}{4},\ \frac{5-2s+2\mu}{4};\ -\frac{b^2}{4a^2}\end{matrix}\right) - \frac{2^{s-6}b^{3/2-s}}{\sqrt{\pi}a^{3/2}}(4\nu^2-1)\Gamma\left[\begin{matrix}\frac{2s+2\mu-3}{4}\\ \frac{7-2s+2\mu}{4}\end{matrix}\right] \times {}_4F_3\left(\begin{matrix}\frac{3+2\nu}{4},\ \frac{3-2\nu}{4},\ \frac{5+2\nu}{4},\ \frac{5-2\nu}{4}\\ \frac{3}{2},\ \frac{7-2s-2\mu}{4},\ \frac{7-2s+2\mu}{4};\ -\frac{b^2}{4a^2}\end{matrix}\right) + \frac{2^{-s-2\mu}b^\mu}{\sqrt{\pi}\,a^{s+\mu}}\Gamma\left[\begin{matrix}\frac{1-2s-2\mu}{2},\ s+\mu+\nu\\ \mu+1,\ 1-s-\mu+\nu\end{matrix}\right] \times {}_4F_3\left(\begin{matrix}\frac{s+\mu-\nu}{2},\ \frac{s+\mu-\nu+1}{2},\ \frac{s+\mu+\nu}{2},\ \frac{s+\mu+\nu+1}{2}\\ \mu+1,\ \frac{2s+2\mu+1}{4},\ \frac{2s+2\mu+3}{4};\ -\frac{b^2}{4a^2}\end{matrix}\right)$ $[0<b<2a;\ \mathrm{Re}\,(s+\mu+\nu)>0;\ \mathrm{Re}\,s<2]$
3	$e^{-ax^2}J_\nu(bx)\,I_\mu(ax^2)$	$\frac{2^{s-5/2}b^{1-s}}{\sqrt{\pi\,a}}\Gamma\left[\begin{matrix}\frac{s+\nu-1}{2}\\ \frac{3-s+\nu}{2}\end{matrix}\right]{}_2F_2\left(\begin{matrix}\frac{1-2\mu}{2},\ \frac{1+2\mu}{2};\ -\frac{b^2}{8a}\\ \frac{3-s-\nu}{2},\ \frac{3-s+\nu}{2}\end{matrix}\right) + \frac{2^{(s+3\nu)/2-1}b^\nu}{\sqrt{\pi}\,a^{(s+\nu)/2}}\Gamma\left[\begin{matrix}\frac{s+2\mu+\nu}{2},\ \frac{1-s-\nu}{2}\\ \nu+1,\ \frac{2-s+2\mu-\nu}{2}\end{matrix}\right]{}_2F_2\left(\begin{matrix}\frac{s-2\mu+\nu}{2},\ \frac{s+2\mu+\nu}{2}\\ \nu+1,\ \frac{s+\nu+1}{2};\ -\frac{b^2}{8a}\end{matrix}\right)$ $[b>0;\ -\mathrm{Re}\,(2\mu+\nu)<\mathrm{Re}\,s<5/2;\ \lvert\arg a\rvert<\pi/2]$
4	$e^{-ax}J_\mu(b\sqrt{x})\,I_\nu(ax)$	$\frac{a^{-s-\mu/2}b^\mu}{2^{s+3\mu/2}\sqrt{\pi}}\Gamma\left[\begin{matrix}\frac{1-2s-\mu}{2},\ \frac{2s+\mu+2\nu}{2}\\ \mu+1,\ \frac{2-2s-\mu+2\nu}{2}\end{matrix}\right]{}_2F_2\left(\begin{matrix}\frac{2s+\mu-2\nu}{2},\ \frac{2s+\mu+2\nu}{2}\\ \mu+1,\ \frac{2s+\mu+1}{2};\ -\frac{b^2}{8a}\end{matrix}\right) + \frac{2^{2s-3/2}b^{1-2s}}{\sqrt{\pi a}}\Gamma\left[\begin{matrix}\frac{2s+\mu-1}{2}\\ \frac{3-2s+\mu}{2}\end{matrix}\right]{}_2F_2\left(\begin{matrix}\frac{1-2\nu}{2},\ \frac{1+2\nu}{2};\ -\frac{b^2}{8a}\\ \frac{3-2s-\mu}{2},\ \frac{3-2s+\mu}{2}\end{matrix}\right)$ $[b,\ \mathrm{Re}\,a,\ \mathrm{Re}\,(s+\nu+\mu/2)>0;\ \mathrm{Re}\,s<5/2]$

No.	$f(x)$	$F(s)$
5	$\theta(a-x)\ln\dfrac{\sqrt{a-x}+\sqrt{a}}{\sqrt{x}}$ $\times J_\nu(bx)\,I_\nu(bx)$	$\dfrac{2^{-2\nu-1}\sqrt{\pi}\,a^{s+2\nu}b^{2\nu}}{s+2\nu}\,\Gamma\left[\begin{matrix}s+2\nu\\ \nu+1,\ \nu+1,\ \frac{2s+4\nu+1}{2}\end{matrix}\right]$ $\times{}_5F_8\left(\begin{matrix}\frac{s+2\nu}{4},\ \Delta(4,\ s+2\nu);\ -\frac{a^4b^4}{64}\\ \frac{\nu+1}{2},\ \frac{\nu+2}{2},\ \nu+1,\ \Delta\left(4,\ \frac{2s+4\nu+1}{2}\right),\ \frac{s+2\nu+4}{2}\end{matrix}\right)$ $[a,\ \mathrm{Re}\,(s+2\nu)>0]$
6	$\theta(a-x)\ln\dfrac{a^2+\sqrt{a^4-x^4}}{x^2}$ $\times J_\nu(bx)\,I_\nu(bx)$	$\dfrac{a^{s+2\nu}b^{2\nu}}{2^{-s/2+\nu+3}}\,\Gamma\left[\begin{matrix}\frac{s+2\nu}{4},\ \frac{s+2\nu}{4}\\ \frac{s+2\nu+2}{2},\ \nu+1,\ \nu+1\end{matrix}\right]$ $\times{}_2F_5\left(\begin{matrix}\frac{s+2\nu}{4},\ \frac{s+2\nu}{4};\ -\frac{a^4b^4}{64}\\ \frac{\nu+1}{2},\ \frac{\nu+2}{2},\ \nu+1,\ \frac{s+2\nu+2}{4},\ \frac{s+2\nu+2}{4}\end{matrix}\right)$ $[a,\ \mathrm{Re}\,(s+2\nu)>0]$
7	$\theta(a-x)\arccos\dfrac{x}{a}$ $\times J_\nu(bx)\,I_\nu(bx)$	$\dfrac{2^{-2\nu-1}\sqrt{\pi}\,a^{s+2\nu}b^{2\nu}}{s+2\nu}\,\Gamma\left[\begin{matrix}\frac{s+2\nu+1}{2}\\ \nu+1,\ \nu+1,\ \frac{s+2\nu+2}{2}\end{matrix}\right]$ $\times{}_3F_6\left(\begin{matrix}\frac{s+2\nu}{4},\ \frac{s+2\nu+1}{4},\ \frac{s+2\nu+3}{4};\ -\frac{a^4b^4}{64}\\ \frac{\nu+1}{2},\ \frac{\nu+2}{2},\ \nu+1,\ \frac{s+2\nu+2}{4},\ \frac{s+2\nu+4}{2},\ \frac{s+2\nu+4}{2}\end{matrix}\right)$ $[a,\ \mathrm{Re}\,(s+2\nu)>0]$
8	$\Gamma(\mu,\ ax)\,J_\nu(bx)\,I_\nu(bx)$	$\dfrac{a^{-s-2\nu}(b/2)^{2\nu}}{s+2\nu}\,\Gamma\left[\begin{matrix}s+\mu+2\nu\\ \nu+1,\ \nu+1\end{matrix}\right]$ $\times{}_5F_4\left(\begin{matrix}\frac{s+2\nu}{4},\ \Delta(4,\ s+\mu+2\nu);\ -\frac{4b^4}{a^4}\\ \frac{\nu+1}{2},\ \frac{\nu+2}{2},\ \nu+1,\ \frac{s+2\nu+4}{4}\end{matrix}\right)$ $[\mathrm{Re}\,a>\lvert\mathrm{Im}\,b\rvert+\lvert\mathrm{Re}\,b\rvert;\ \mathrm{Re}\,(s+2\nu)>-\mathrm{Re}\,\mu,\ 0]$
9	$\mathrm{erfc}\,(ax)\,J_\nu(bx)\,I_\nu(bx)$	$\dfrac{a^{-s-2\nu}b^{2\nu}}{2^{2\nu}\sqrt{\pi}\,(s+2\nu)}\,\Gamma\left[\begin{matrix}\frac{s+2\nu+1}{2}\\ \nu+1,\ \nu+1\end{matrix}\right]{}_3F_3\left(\begin{matrix}\frac{1}{2},\ \frac{s+2\nu}{4},\ \frac{s+2\nu+3}{4}\\ \frac{\nu+1}{2},\ \frac{\nu+2}{2},\ \nu+1;\ -\frac{b^4}{16a^4}\end{matrix}\right)$ $[\mathrm{Re}\,(s+2\nu)>0;\ \lvert\arg a\rvert<\pi/4]$
10	$e^{-ax}\,Y_\mu(bx)\ I_\nu(ax)$	$-\dfrac{2^{s-1}a^\nu}{\pi\,b^{s+\nu}\,\Gamma(\nu+1)}\cos\dfrac{(s-\mu+\nu)\pi}{2}\,\Gamma\left(\dfrac{s+\mu+\nu}{2}\right)$ $\times\Gamma\left(\dfrac{s-\mu+\nu}{2}\right){}_4F_3\left(\begin{matrix}\frac{2\nu+1}{4},\ \frac{2\nu+3}{4},\ \frac{s+\mu+\nu}{2},\ \frac{s-\mu+\nu}{2}\\ \frac{1}{2},\ \frac{2\nu+1}{2},\ \nu+1;\ -\frac{4a^2}{b^2}\end{matrix}\right)$ $-\dfrac{2^s a^{\nu+1}}{\pi\,b^{s+\nu+1}}\sin\dfrac{(s-\mu+\nu)\pi}{2}\,\Gamma\left[\begin{matrix}\frac{s+\mu+\nu+1}{2},\ \frac{s-\mu+\nu+1}{2}\\ \nu+1\end{matrix}\right]$ $\times{}_4F_3\left(\begin{matrix}\frac{2\nu+3}{4},\ \frac{2\nu+5}{4},\ \frac{s+\mu+\nu+1}{2},\ \frac{s-\mu+\nu+1}{2}\\ \frac{3}{2},\ \nu+1,\ \frac{2\nu+3}{2};\ -\frac{4a^2}{b^2}\end{matrix}\right)$ $[b,\ \mathrm{Re}\,a>0;\ \lvert\mathrm{Re}\,\mu\rvert-\mathrm{Re}\,\nu<\mathrm{Re}\,s<2]$

3.13.13. Products of $I_\nu(\varphi(x))$

1	$e^{-ax} I_\mu(bx)\, I_\nu(cx)$	$\frac{b^\mu c^\nu}{2^{\mu+\nu} a^{s+\mu+\nu}} \Gamma\left[\begin{matrix} s+\mu+\nu \\ \mu+1,\, \nu+1 \end{matrix}\right] \times F_4\left(\frac{s+\mu+\nu}{2},\, \frac{s+\mu+\nu+1}{2};\, \mu+1,\, \nu+1;\, \frac{b^2}{a^2},\, \frac{c^2}{a^2}\right)$ $[\operatorname{Re} a > \vert\operatorname{Re} b\vert + \vert\operatorname{Re} c\vert;\ \operatorname{Re}(s+\mu+\nu) > 0]$
2	$e^{-ax} I_\mu(bx)\, I_\nu(bx)$	$a^{-s-\mu-\nu} \left(\frac{b}{2}\right)^{\mu+\nu} \Gamma\left[\begin{matrix} s+\mu+\nu \\ \mu+1,\, \nu+1 \end{matrix}\right] \times {}_4F_3\left(\begin{matrix} \frac{\mu+\nu+1}{2},\, \frac{\mu+\nu+2}{2},\, \frac{s+\mu+\nu}{2},\, \frac{s+\mu+\nu+1}{2} \\ \mu+1,\, \nu+1,\, \mu+\nu+1;\, \frac{4b^2}{a^2} \end{matrix}\right)$ $[\operatorname{Re} a > 2\vert\operatorname{Re} b\vert;\ \operatorname{Re}(s+\mu+\nu) > 0]$
3	$e^{-(a+b)x} I_\mu(ax)\, I_\nu(bx)$	$\frac{a^{-s-\nu} b^\nu}{2^{s+2\nu}\sqrt{\pi}} \Gamma\left[\begin{matrix} \frac{1-2s-2\nu}{2},\, s+\mu+\nu \\ 1-s+\mu-\nu,\, \nu+1 \end{matrix}\right] \times {}_3F_2\left(\begin{matrix} \frac{2\nu+1}{2},\, s-\mu+\nu,\, s+\mu+\nu \\ 2\nu+1,\, \frac{2s+2\nu+1}{2};\, -\frac{b}{a} \end{matrix}\right) + \frac{b^{1/2-s}}{2^s \pi \sqrt{a}} \times \Gamma\left[\begin{matrix} \frac{2s+2\nu-1}{2},\, 1-s \\ \frac{3-2s+2\nu}{2} \end{matrix}\right] {}_3F_2\left(\begin{matrix} \frac{1-2\mu}{2},\, \frac{1+2\mu}{2},\, 1-s \\ \frac{3-2s-2\nu}{2},\, \frac{3-2s+2\nu}{2};\, -\frac{b}{a} \end{matrix}\right)$ $[\operatorname{Re}(a+b) > 0;\ -\operatorname{Re}(\mu+\nu) < \operatorname{Re} s < 1]$
4	$e^{-ax^2} I_\mu(bx)\, I_\nu(cx)$	$\frac{b^\mu c^\nu}{2^{\mu+\nu+1} a^{(s+\mu+\nu)/2}} \Gamma\left[\begin{matrix} \frac{s+\mu+\nu}{2} \\ \mu+1,\, \nu+1 \end{matrix}\right] \times \Psi_2\left(\frac{s+\mu+\nu}{2};\, \mu+1,\, \nu+1;\, \frac{b^2}{4a},\, \frac{c^2}{4a}\right)$ $[\operatorname{Re} a,\ \operatorname{Re}(s+\mu+\nu) > 0]$
5	$e^{-ax^2} I_\mu(bx)\, I_\nu(bx)$	$\frac{b^{\mu+\nu}}{2^{\mu+\nu+1} a^{(s+\mu+\nu)/2}} \Gamma\left[\begin{matrix} \frac{s+\mu+\nu}{2} \\ \mu+1,\, \nu+1 \end{matrix}\right] \times {}_3F_3\left(\begin{matrix} \frac{\mu+\nu+1}{2},\, \frac{\mu+\nu+2}{2},\, \frac{s+\mu+\nu}{2};\, \frac{b^2}{a} \\ \mu+1,\, \nu+1,\, \mu+\nu+1 \end{matrix}\right)$ $[\operatorname{Re} a,\ \operatorname{Re}(s+\mu+\nu) > 0]$
6	$\theta(a-x) \times \ln\frac{\sqrt{a-x}+\sqrt{a}}{\sqrt{x}} \times I_\mu(bx)\, I_\nu(bx)$	$\frac{\sqrt{\pi}\, a^{s+\mu+\nu} b^{\mu+\nu}}{2^{\mu+\nu+1}(s+\mu+\nu)} \Gamma\left[\begin{matrix} s+\mu+\nu \\ \mu+1,\, \nu+1,\, s+\mu+\nu+\frac{1}{2} \end{matrix}\right] \times {}_5F_6\left(\begin{matrix} \frac{\mu+\nu+1}{2},\, \frac{\mu+\nu+2}{2},\, \frac{s+\mu+\nu}{2},\, \frac{s+\mu+\nu}{2},\, \frac{s+\mu+\nu+1}{2};\, a^2b^2 \\ \mu+1,\, \nu+1,\, \mu+\nu+1,\, \Delta\left(2,\, \frac{2s+2\mu+2\nu+1}{2}\right),\, \frac{s+\mu+\nu+2}{2} \end{matrix}\right)$ $[a > 0;\ \operatorname{Re}(s+\mu+\nu) > 0]$

No.	$f(x)$	$F(s)$
7	$\theta(a-x)$ $\times \ln \frac{\sqrt{a^2-x^2}+a}{x}$ $\times I_\mu(bx) I_\nu(bx)$	$\frac{\sqrt{\pi}\, a^{s+\mu+\nu} b^{\mu+\nu}}{2^{\mu+\nu+1}(s+\mu+\nu)} \Gamma\left[\begin{matrix} \frac{s+\mu+\nu}{2} \\ \mu+1,\ \nu+1,\ \frac{s+\mu+\nu+1}{2} \end{matrix}\right] \times {}_4F_5\left(\begin{matrix} \frac{\mu+\nu+1}{2},\ \frac{\mu+\nu+2}{2},\ \frac{s+\mu+\nu}{2},\ \frac{s+\mu+\nu}{2};\ a^2b^2 \\ \mu+1,\ \nu+1,\ \mu+\nu+1,\ \frac{s+\mu+\nu+1}{2},\ \frac{s+\mu+\nu+2}{2} \end{matrix}\right)$ $[a>0;\ \operatorname{Re}(s+(\mu+\nu)/2)>0]$
8	$\theta(a-x) \arccos \frac{x}{a}$ $\times I_\mu(bx) I_\nu(bx)$	$\frac{\sqrt{\pi}\, a^{s+\mu+\nu} b^{\mu+\nu}}{2^{\mu+\nu+1}(s+\mu+\nu)} \Gamma\left[\begin{matrix} \frac{s+\mu+\nu+1}{2} \\ \mu+1,\ \nu+1,\ \frac{s+\mu+\nu+2}{2} \end{matrix}\right] \times {}_4F_5\left(\begin{matrix} \frac{\mu+\nu+1}{2},\ \frac{\mu+\nu+2}{2},\ \frac{s+\mu+\nu}{2},\ \frac{s+\mu+\nu+1}{2};\ a^2b^2 \\ \mu+1,\ \nu+1,\ \mu+\nu+1,\ \frac{s+\mu+\nu+2}{2},\ \frac{s+\mu+\nu+2}{2} \end{matrix}\right)$ $[a,\ \operatorname{Re}(s+\mu+\nu)>0]$
9	$\operatorname{Ei}(-ax) I_\mu(bx) I_\nu(bx)$	$-\frac{a^{-(s+\mu+\nu)}}{s+\mu+\nu} \left(\frac{b}{2}\right)^{\mu+\nu} \Gamma\left[\begin{matrix} s+\mu+\nu \\ \mu+1,\ \nu+1 \end{matrix}\right] \times {}_5F_4\left(\begin{matrix} \frac{\mu+\nu+1}{2},\ \frac{\mu+\nu+2}{2},\ \frac{s+\mu+\nu}{2},\ \frac{s+\mu+\nu}{2},\ \frac{s+\mu+\nu+1}{2} \\ \mu+1,\ \nu+1,\ \mu+\nu+1,\ \frac{s+\mu+\nu+2}{2};\ \frac{4b^2}{a^2} \end{matrix}\right)$ $[a>2\lvert\operatorname{Re} b\rvert;\ \operatorname{Re}(s+\mu+\nu)>0]$
10	$\operatorname{Ei}\left(-ax^2\right) I_\mu(bx) I_\nu(bx)$	$-\frac{a^{-(s+\mu+\nu)/2}}{s+\mu+\nu} \left(\frac{b}{2}\right)^{\mu+\nu} \Gamma\left[\begin{matrix} \frac{s+\mu+\nu}{2} \\ \mu+1,\ \nu+1 \end{matrix}\right] \times {}_4F_4\left(\begin{matrix} \frac{\mu+\nu+1}{2},\ \frac{\mu+\nu+2}{2},\ \frac{s+\mu+\nu}{2},\ \frac{s+\mu+\nu}{2};\ \frac{b^2}{a} \\ \mu+1,\ \nu+1,\ \mu+\nu+1,\ \frac{s+\mu+\nu+2}{2} \end{matrix}\right)$ $[\operatorname{Re} a,\ \operatorname{Re}(s+\mu+\nu)>0;\ \lvert\arg b\rvert<\pi]$
11	$\operatorname{erfc}(ax)\ I_\mu(bx) I_\nu(bx)$	$\frac{a^{-(s+\mu+\nu)}}{\sqrt{\pi}(s+\mu+\nu)} \left(\frac{b}{2}\right)^{\mu+\nu} \Gamma\left[\begin{matrix} \frac{s+\mu+\nu+1}{2} \\ \mu+1,\ \nu+1 \end{matrix}\right] \times {}_4F_4\left(\begin{matrix} \frac{\mu+\nu+1}{2},\ \frac{\mu+\nu+2}{2},\ \frac{s+\mu+\nu}{2},\ \frac{s+\mu+\nu+1}{2};\ \frac{b^2}{a^2} \\ \mu+1,\ \nu+1,\ \mu+\nu+1,\ \frac{s+\mu+\nu+2}{2} \end{matrix}\right)$ $[\operatorname{Re}(s+\mu+\nu)>0;\ \lvert\arg a\rvert<\pi/4]$
12	$\Gamma(\lambda,\ ax) I_\mu(bx) I_\nu(bx)$	$\frac{a^{-(s+\mu+\nu)}}{s+\mu+\nu} \left(\frac{b}{2}\right)^{\mu+\nu} \Gamma\left[\begin{matrix} s+\lambda+\mu+\nu \\ \mu+1,\ \nu+1 \end{matrix}\right] \times {}_5F_4\left(\begin{matrix} \frac{\mu+\nu+1}{2},\ \frac{\mu+\nu+2}{2},\ \frac{s+\mu+\nu}{2},\ \frac{s+\lambda+\mu+\nu}{2},\ \frac{s+\lambda+\mu+\nu+1}{2} \\ \mu+1,\ \nu+1,\ \mu+\nu+1,\ \frac{s+\mu+\nu+2}{2};\ \frac{4b^2}{a^2} \end{matrix}\right)$ $[\operatorname{Re}(a-2b)>0;\ \operatorname{Re}(s+\mu+\nu)>-\operatorname{Re}\lambda,\ 0]$
13	$I_\mu(ax) I_\nu(ax)$ $-I_{-\mu}(ax) I_{-\nu}(ax)$	$-\frac{\sin(\mu+\nu)\pi}{2\pi^{3/2}a^s} \Gamma\left[\begin{matrix} \frac{s+\mu+\nu}{2},\ \frac{s-\mu-\nu}{2},\ \frac{1-s}{2},\ \frac{2-s}{2} \\ \frac{2-s+\mu-\nu}{2},\ \frac{2-s-\mu+\nu}{2} \end{matrix}\right]$ $[\lvert\operatorname{Re}(\mu+\nu)\rvert<\operatorname{Re} s<1]$

No.	$f(x)$	$F(s)$
14	$(a-x)_+^{\alpha-1}$ $\times I_\mu(bx(a-x))$ $\times I_\nu(bx(a-x))$	$a^{s+\alpha+2\mu+2\nu-1}\left(\frac{b}{2}\right)^{\mu+\nu}\Gamma\left[\begin{matrix}\alpha+\mu+\nu,\ s+\mu+\nu\\ \mu+1,\ \nu+1,\ s+\alpha+2\mu+2\nu\end{matrix}\right]$ $\times\, {}_3F_2\left(\begin{matrix}\Delta(2,\ \mu+\nu+1),\ \Delta(2,\ \alpha+\mu+\nu),\ \Delta(2,\ s+\mu+\nu)\\ \mu+1,\ \nu+1,\ \mu+\nu+1,\ \Delta(4,\ s+\alpha+\mu+\nu);\ \frac{a^4b^2}{64}\end{matrix}\right)$ $[a>0;\ \operatorname{Re}s,\ \operatorname{Re}\alpha>-\operatorname{Re}(\mu+\nu)]$
15	$\frac{1}{(x+a)^\rho}I_\mu\left(\frac{bx}{x+a}\right)$ $\times I_\nu\left(\frac{bx}{x+a}\right)$	$a^{s-\rho}\left(\frac{b}{2}\right)^{\mu+\nu}\frac{\mathrm{B}(\rho-s,\ s+\mu+\nu)}{\Gamma(\mu+1)\Gamma(\nu+1)}$ $\times\, {}_4F_5\left(\begin{matrix}\frac{\mu+\nu+1}{2},\ \frac{\mu+\nu+2}{2},\ \frac{s+\mu+\nu}{2},\ \frac{s+\mu+\nu+1}{2}\\ \mu+1,\ \nu+1,\ \mu+\nu+1,\ \frac{\mu+\nu+\rho}{2},\ \frac{\mu+\nu+\rho+1}{2};\ b^2\end{matrix}\right)$ $[\operatorname{Re}(\mu+\nu)<\operatorname{Re}s<\operatorname{Re}\rho;\ \lvert\arg a\rvert<\pi]$
16	$\frac{1}{(x^2+a^2)^\rho}I_\mu\left(\frac{bx}{x^2+a^2}\right)$ $\times I_\nu\left(\frac{bx}{x^2+a^2}\right)$	$\frac{a^{s-\mu-\nu-2\rho}b^{\mu+\nu}}{2^{\mu+\nu+1}\Gamma(\mu+1)\Gamma(\nu+1)}\mathrm{B}\left(\frac{s+\mu+\nu}{2},\ \frac{\mu+\nu+2\rho-s}{2}\right)$ $\times\, {}_4F_5\left(\begin{matrix}\frac{\mu+\nu+1}{2},\ \frac{\mu+\nu+2}{2},\ \frac{s+\mu+\nu}{2},\ \frac{\mu+\nu+2\rho-s}{2}\\ \mu+1,\ \nu+1,\ \mu+\nu+1,\ \frac{\mu+\nu+\rho}{2},\ \frac{\mu+\nu+\rho+1}{2};\ \frac{b^2}{4a^2}\end{matrix}\right)$ $[\operatorname{Re}a>0;\ -\operatorname{Re}(\mu+\nu)<\operatorname{Re}s<\operatorname{Re}(\mu+\nu+2\rho)]$
17	$e^{-ax}\prod_{k=1}^{n}I_{\nu_k}(b_kx)$	$\frac{\Gamma(s+\nu)}{a^{s+\nu}}\prod_{k=1}^{n}\frac{(b_k/2)^{\nu_k}}{\Gamma(\nu_k+1)}F_C^{(n)}\left(\frac{s+\nu}{2},\ \frac{s+\nu+1}{2};\ (\nu_n)+1;\ \frac{(b_n^2)}{a^2}\right)$ $\left[\nu=\sum_{k=1}^{n}\nu_k;\ \operatorname{Re}a>\sum_{k=1}^{n}\lvert\operatorname{Re}b_k\rvert;\ \operatorname{Re}(s+\nu)>0\right]$
18	$e^{-ax}\prod_{k=1}^{m}\sin(b_kx)$ $\times\prod_{k=1}^{n}\cos(c_kx)$ $\times\prod_{k=1}^{p}I_{\nu_k}(d_kx)$	$\frac{\prod_{k=1}^{m}b_k\prod_{k=1}^{p}(d_k/2)^{\nu_k}}{a^{s+m+\nu}}\Gamma\left[\begin{matrix}s+m+\nu\\ \nu_1+1,\ \nu_2+1,\dots,\ \nu_p+1\end{matrix}\right]$ $\times F_C^{(m+n+p)}\Bigl(\frac{s+m+\nu}{2},\ \frac{s+m+\nu+1}{2};\ \underbrace{\frac{3}{2},\dots,\frac{3}{2}}_{m},\ \underbrace{\frac{1}{2},\dots,\frac{1}{2}}_{n},$ $(\nu_p)+1;\ -\frac{(b_m^2)}{a^2},\ -\frac{(c_n^2)}{a^2},\ \frac{(d_p^2)}{a^2}\Bigr)$ $\Bigl[\nu=\sum_{k=1}^{p}\nu_k;$ $\operatorname{Re}a>\sum_{k=1}^{m}\lvert\operatorname{Im}b_k\rvert+\sum_{k=1}^{n}\lvert\operatorname{Im}c_k\rvert+\sum_{k=1}^{p}\lvert\operatorname{Re}d_k\rvert;\ \operatorname{Re}(s+m+\nu)>0\Bigr]$
19	$e^{-ax}\prod_{k=1}^{m}J_{\mu_k}(b_kx)$ $\times\prod_{k=1}^{n}I_{\nu_k}(c_kx)$	$\frac{\prod_{k=1}^{m}(b_k/2)^{\mu_k}\prod_{k=1}^{n}(c_k/2)^{\nu_k}}{a^{s+\mu+\nu}}\Gamma\left[\begin{matrix}s+\mu+\nu\\ (\mu_m)+1,\ (\nu_n)+1\end{matrix}\right]$ $\times F_C^{(n+m)}\left(\Delta(2,\ s+\mu+\nu);\ (\mu_m)+1,\ (\nu_n)+1;\ -\frac{(b_m^2)}{a^2},\ \frac{(c_n^2)}{a^2}\right)$ $\Bigl[\mu=\sum_{k=1}^{m}\mu_k,\ \nu=\sum_{k=1}^{n}\nu_k;\ \operatorname{Re}a>\sum_{k=1}^{m}\lvert\operatorname{Im}b_k\rvert+\sum_{k=1}^{n}\lvert\operatorname{Re}c_k\rvert;$ $\operatorname{Re}(s+\mu+\nu)>0\Bigr]$

3.14. The Macdonald Function $K_\nu(z)$

More formulas can be obtained from the corresponding sections due to the relations

$$K_{\pm 1/2}(z) = \sqrt{\frac{2}{\pi}}\,\frac{1}{\sqrt{z}}\,e^{-z}, \quad K_{\pm 3/2}(z) = \sqrt{\frac{2}{\pi}}\,\frac{z+1}{z^{3/2}}\,e^{-z};$$

$$K_{(n+1)/2}(z) = \sqrt{\frac{2}{\pi}}\,\frac{1}{\sqrt{z}}\,e^{-z}\sum_{k=0}^{n}\frac{(n+k)!}{k!\,(n-k)!\,(2z)^k} = n!\sqrt{\frac{2}{\pi}}\,\frac{1}{\sqrt{z}}\,\frac{e^{-z}}{(-2z)^n}\,L_n^{-2n-1}(2z);$$

$$K_\nu(z) = \frac{\pi}{2\sin(\nu\pi)}\left[I_{-\nu}(z) - I_\nu(z)\right], \quad [\nu \neq 0, \pm 1, \pm 2, \ldots];$$

$$K_n(z) = \lim_{\nu\to n} K_\nu(z), \quad [n = 0, \pm 1, \pm 2, \ldots];$$

$$K_\nu(z) = 2^{-\nu-1}z^\nu\,\Gamma(-\nu)\;{}_0F_1\left(1+\nu;\,\frac{z^2}{4}\right) + 2^{\nu-1}z^{-\nu}\,\Gamma(\nu)\;{}_0F_1\left(1-\nu;\,\frac{z^2}{4}\right);$$

$$K_\nu(z) = \frac{1}{2}\,G_{02}^{20}\left(\frac{z^2}{4}\,\middle|\,\begin{matrix}\cdot\\ \nu/2,\,-\nu/2\end{matrix}\right), \ \operatorname{Re} z > 0; \quad K_\nu(z) = \sqrt{\pi}\,e^z\,G_{12}^{20}\left(2z\,\middle|\,\begin{matrix}1/2\\ -\nu,\,\nu\end{matrix}\right).$$

3.14.1. $K_\nu(ax^r)$ and algebraic functions

No.	$f(x)$	$F(s)$
1	$(a-x)_+^{\alpha-1}\,K_\nu(bx)$	$2^{\nu-1}a^{s+\alpha-\nu-1}b^{-\nu}\,\Gamma(\nu)\,\mathrm{B}(\alpha,\,s-\nu)\;{}_2F_3\left(\begin{matrix}\frac{s-\nu}{2},\,\frac{s-\nu+1}{2};\,\frac{a^2b^2}{4}\\ 1-\nu,\,\frac{s+\alpha-\nu}{2},\,\frac{s+\alpha-\nu+1}{2}\end{matrix}\right) + \frac{a^{s+\alpha+\nu-1}b^\nu}{2^{\nu+1}}\,\Gamma(-\nu)\,\mathrm{B}(\alpha,\,s+\nu)\;{}_2F_3\left(\begin{matrix}\frac{s+\nu}{2},\,\frac{s+\nu+1}{2};\,\frac{a^2b^2}{4}\\ 1+\nu,\,\frac{s+\alpha+\nu}{2},\,\frac{s+\alpha+\nu+1}{2}\end{matrix}\right)$ $[a,\ \operatorname{Re}\alpha > 0;\ \operatorname{Re} s > \lvert\operatorname{Re}\nu\rvert]$
2	$(x-a)_+^{\alpha-1}\,K_\nu(bx)$	$\frac{a^{s+\alpha-\nu-1}}{2^{1-\nu}b^\nu}\,\Gamma(\nu)\ \mathrm{B}(\alpha,\,1-s-\alpha+\nu)\;{}_2F_3\left(\begin{matrix}\frac{s-\nu}{2},\,\frac{s-\nu+1}{2};\,\frac{a^2b^2}{4}\\ 1-\nu,\,\frac{s+\alpha-\nu}{2},\,\frac{s+\alpha-\nu+1}{2}\end{matrix}\right) + \frac{a^{s+\alpha+\nu-1}}{2^{\nu+1}b^{-\nu}}\,\Gamma(-\nu)\,\mathrm{B}(\alpha,\,1-s-\alpha-\nu)\;{}_2F_3\left(\begin{matrix}\frac{s+\nu}{2},\,\frac{s+\nu+1}{2};\,\frac{a^2b^2}{4}\\ 1+\nu,\,\frac{s+\alpha+\nu}{2},\,\frac{s+\alpha+\nu+1}{2}\end{matrix}\right) + \frac{2^{s+\alpha-3}}{b^{s+\alpha-1}}\,\Gamma\left(\frac{s+\alpha-\nu-1}{2}\right)\Gamma\left(\frac{s+\alpha+\nu-1}{2}\right) \times {}_2F_3\left(\begin{matrix}\frac{1-\alpha}{2},\,\frac{2-\alpha}{2};\,\frac{a^2b^2}{4}\\ \frac{1}{2},\,\frac{3-s-\alpha-\nu}{2},\,\frac{3-s-\alpha+\nu}{2}\end{matrix}\right) - \frac{(\alpha-1)\,2^{s+\alpha-4}a}{b^{s+\alpha-2}}\,\Gamma\left(\frac{s+\alpha-\nu-2}{2}\right)\Gamma\left(\frac{s+\alpha+\nu-2}{2}\right) \times {}_2F_3\left(\begin{matrix}\frac{2-\alpha}{2},\,\frac{3-\alpha}{2};\,\frac{a^2b^2}{4}\\ \frac{3}{2},\,\frac{4-s-\alpha-\nu}{2},\,\frac{4-s-\alpha+\nu}{2}\end{matrix}\right)$ $[a,\ \operatorname{Re} b,\ \operatorname{Re}\alpha > 0]$
3	$K_\nu(ax)$	$\frac{2^{s-2}}{a^s}\,\Gamma\left(\frac{s-\nu}{2}\right)\Gamma\left(\frac{s+\nu}{2}\right)$ $[\operatorname{Re} a > 0;\ \operatorname{Re} s > \lvert\operatorname{Re}\nu\rvert]$

No.	$f(x)$	$F(s)$
4	$\frac{1}{x-a}K_\nu(bx)$	$\frac{2^{s-3}}{b^{s-1}}\Gamma\left(\frac{s+\nu-1}{2}\right)\Gamma\left(\frac{s-\nu-1}{2}\right){}_1F_2\left(\begin{matrix}1;\ \frac{a^2b^2}{4}\\ \frac{3-s-\nu}{2},\ \frac{3-s+\nu}{2}\end{matrix}\right) + \frac{2^{s-4}a}{b^{s-2}}\Gamma\left(\frac{s+\nu-2}{2}\right)\Gamma\left(\frac{s-\nu-2}{2}\right){}_1F_2\left(\begin{matrix}1;\ \frac{a^2b^2}{4}\\ \frac{4-s-\nu}{2},\ \frac{4-s+\nu}{2}\end{matrix}\right) + \frac{\pi^2a^{s-1}}{2\sin(\nu\pi)}\left[\cot[(s+\nu)\pi]\,I_\nu(ab) - \cot[(s-\nu)\pi]\,I_{-\nu}(ab)\right]$ $[a,\ \operatorname{Re}b>0;\ \operatorname{Re}s>\lvert\operatorname{Re}\nu\rvert]$
5	$\frac{1}{(x+a)^\rho}K_\nu(bx)$	$\frac{2^{s-\rho-2}}{b^{s-\rho}}\Gamma\left(\frac{s+\nu-\rho}{2}\right)\Gamma\left(\frac{s-\nu-\rho}{2}\right){}_2F_3\left(\begin{matrix}\frac{\rho}{2},\ \frac{\rho+1}{2};\ \frac{a^2b^2}{4}\\ \frac{1}{2},\ \frac{2-s-\nu+\rho}{2},\ \frac{2-s+\nu+\rho}{2}\end{matrix}\right) - \rho a\frac{2^{s-\rho-3}}{b^{s-\rho-1}}\Gamma\left(\frac{s+\nu-\rho-1}{2}\right)\Gamma\left(\frac{s-\nu-\rho-1}{2}\right) \times {}_2F_3\left(\begin{matrix}\frac{\rho+1}{2},\ \frac{\rho+2}{2};\ \frac{a^2b^2}{4}\\ \frac{3}{2},\ \frac{3-s-\nu+\rho}{2},\ \frac{3-s+\nu+\rho}{2}\end{matrix}\right) + \frac{2^{\nu-1}a^{s-\nu-\rho}}{b^\nu} \times \Gamma\left[\begin{matrix}\nu,\ s-\nu,\ \nu+\rho-s\\ \rho\end{matrix}\right]{}_2F_3\left(\begin{matrix}\frac{s-\nu}{2},\ \frac{s-\nu+1}{2};\ \frac{a^2b^2}{4}\\ 1-\nu,\ \frac{s-\nu-\rho+1}{2},\ \frac{s-\nu-\rho+2}{2}\end{matrix}\right) + \frac{a^{s+\nu-\rho}b^\nu}{2^{\nu+1}}\Gamma\left[\begin{matrix}-\nu,\ s+\nu,\ \rho-\nu-s\\ \rho\end{matrix}\right] \times {}_2F_3\left(\begin{matrix}\frac{s+\nu}{2},\ \frac{s+\nu+1}{2};\ \frac{a^2b^2}{4}\\ 1+\nu,\ \frac{s+\nu-\rho+1}{2},\ \frac{s+\nu-\rho+2}{2}\end{matrix}\right)$ $[\operatorname{Re}b>0;\ \operatorname{Re}s>\lvert\operatorname{Re}\nu\rvert;\ \lvert\arg a\rvert<\pi]$
6	$\frac{1}{x+a}K_\nu(bx)$	$\frac{2^{s-3}}{b^{s-1}}\Gamma\left(\frac{s-\nu-1}{2}\right)\Gamma\left(\frac{s+\nu-1}{2}\right){}_1F_2\left(\begin{matrix}1;\ \frac{a^2b^2}{4}\\ \frac{3-s-\nu}{2},\ \frac{3-s+\nu}{2}\end{matrix}\right) - \frac{2^{s-4}a}{b^{s}}\Gamma\left(\frac{s-\nu-2}{2}\right)\Gamma\left(\frac{s+\nu-2}{2}\right){}_1F_2\left(\begin{matrix}1;\ \frac{a^2b^2}{4}\\ \frac{4-s-\nu}{2},\ \frac{4-s+\nu}{2}\end{matrix}\right) + \frac{\pi a^{s-1}}{\sin[(s-\nu)\pi]}\left[K_\nu(ab) + \frac{\pi\cos(s\pi)}{\sin[(s+\nu)\pi]}I_\nu(ab)\right]$ $[\operatorname{Re}b>0;\ \operatorname{Re}s>\lvert\operatorname{Re}\nu\rvert;\ \lvert\arg a\rvert<\pi]$
7	$\frac{1}{(x^2+a^2)^\rho}K_\nu(bx)$	$\frac{a^{s-2\rho-\nu}}{2^{2-\nu}b^\nu}\Gamma\left[\begin{matrix}\nu,\ \frac{s-\nu}{2},\ \frac{\nu+2\rho-s}{2}\\ \rho\end{matrix}\right]{}_1F_2\left(\begin{matrix}\frac{s-\nu}{2};\ -\frac{a^2b^2}{4}\\ 1-\nu,\ \frac{s-\nu-2\rho+2}{2}\end{matrix}\right) + \frac{a^{s+\nu-2\rho}b^\nu}{2^{\nu+2}}\Gamma\left[\begin{matrix}-\nu,\ \frac{s+\nu}{2},\ \frac{2\rho-\nu-s}{2}\\ \rho\end{matrix}\right]{}_1F_2\left(\begin{matrix}\frac{s+\nu}{2};\ -\frac{a^2b^2}{4}\\ 1+\nu,\ \frac{s+\nu-2\rho+2}{2}\end{matrix}\right) + \frac{2^{s-2\rho-2}}{b^{s-2\rho}}\Gamma\left(\frac{s+\nu-2\rho}{2}\right)\Gamma\left(\frac{s-\nu-2\rho}{2}\right) \times {}_1F_2\left(\begin{matrix}\rho;\ -\frac{a^2b^2}{4}\\ \frac{2-s-\nu+2\rho}{2},\ \frac{2-s+\nu+2\rho}{2}\end{matrix}\right)$ $[\operatorname{Re}a,\ \operatorname{Re}b>0;\ \operatorname{Re}s>\lvert\operatorname{Re}\nu\rvert]$

No.	$f(x)$	$F(s)$
8	$\left(a^2-x^2\right)_+^{\alpha-1} K_\nu(bx)$	$\frac{a^{s+2\alpha-2}}{4}\Gamma(\alpha)\left[\left(\frac{ab}{2}\right)^{-\nu}\Gamma\left[\begin{matrix}\nu,\ \frac{s-\nu}{2}\\ \frac{s+2\alpha-\nu}{2}\end{matrix}\right]{}_1F_2\left(\begin{matrix}\frac{s-\nu}{2};\ \frac{a^2b^2}{4}\\ 1-\nu,\ \frac{s+2\alpha-\nu}{2}\end{matrix}\right) + \left(\frac{ab}{2}\right)^{\nu}\Gamma\left[\begin{matrix}-\nu,\ \frac{s+\nu}{2}\\ \frac{s+2\alpha+\nu}{2}\end{matrix}\right]{}_1F_2\left(\begin{matrix}\frac{s+\nu}{2};\ \frac{a^2b^2}{4}\\ 1+\nu,\ \frac{s+2\alpha+\nu}{2}\end{matrix}\right)\right]$ $[a,\ \operatorname{Re} b,\ \operatorname{Re}\alpha>0;\ \operatorname{Re} s>\lvert\operatorname{Re}\nu\rvert]$
9	$\left(x^2-a^2\right)_+^{\alpha-1} K_\nu(bx)$	$\frac{2^{\nu-2}a^{s+2\alpha-\nu-2}}{b^\nu}\Gamma(\nu)\,\mathrm{B}\left(\alpha,\ \frac{2-s-2\alpha+\nu}{2}\right) \times {}_1F_2\left(\begin{matrix}\frac{s-\nu}{2};\ \frac{a^2b^2}{4}\\ 1-\nu,\ \frac{s+2\alpha-\nu}{2}\end{matrix}\right) + \frac{a^{s+2\alpha+\nu-2}b^\nu}{2^{\nu+2}}\Gamma(-\nu) \times \mathrm{B}\left(\alpha,\ \frac{2-s-2\alpha-\nu}{2}\right){}_1F_2\left(\begin{matrix}\frac{s+\nu}{2};\ \frac{a^2b^2}{4}\\ 1+\nu,\ \frac{s+2\alpha+\nu}{2}\end{matrix}\right) + \frac{2^{s+2\alpha-4}}{b^{s+2\alpha-2}}\Gamma\left(\frac{s+2\alpha+\nu-2}{2}\right)\Gamma\left(\frac{s+2\alpha-\nu-2}{2}\right) \times {}_1F_2\left(\begin{matrix}1-\alpha;\ \frac{a^2b^2}{4}\\ \frac{4-s-2\alpha-\nu}{2},\ \frac{4-s-2\alpha+\nu}{2}\end{matrix}\right)$ $[a,\ \operatorname{Re} b,\ \operatorname{Re}\alpha>0]$
10	$\frac{1}{x^2-a^2}K_\nu(bx)$	$\frac{2^{s-4}}{b^{s-2}}\Gamma\left(\frac{s+\nu-2}{2}\right)\Gamma\left(\frac{s-\nu-2}{2}\right){}_1F_2\left(\begin{matrix}1;\ \frac{a^2b^2}{4}\\ \frac{4-s-\nu}{2},\ \frac{4-s+\nu}{2}\end{matrix}\right) + \frac{\pi^2a^{s-2}}{4\sin(\nu\pi)}\left[\cot\frac{(s+\nu)\pi}{2}I_\nu(ab)-\cot\frac{(s-\nu)\pi}{2}I_{-\nu}(ab)\right]$ $[a,\ \operatorname{Re} b>0;\ \operatorname{Re} s>\lvert\operatorname{Re}\nu\rvert]$
11	$\left(\sqrt{x^2+a^2}\pm a\right)^\rho$ $\times K_\nu(bx)$	$\pm\frac{2^{s+\rho-3}\rho a}{b^{s+\rho-1}}\Gamma\left(\frac{s+\rho+\nu-1}{2}\right)\Gamma\left(\frac{s+\rho-\nu-1}{2}\right) \times {}_2F_3\left(\begin{matrix}\frac{1+\rho}{2},\ \frac{1-\rho}{2};\ -\frac{a^2b^2}{4}\\ \frac{3}{2},\ \frac{3-s-\rho-\nu}{2},\ \frac{3-s-\rho+\nu}{2}\end{matrix}\right) + \frac{2^{s+\rho-2}}{b^{s+\rho}}\Gamma\left(\frac{s+\rho+\nu}{2}\right) \times \Gamma\left(\frac{s+\rho-\nu}{2}\right){}_2F_3\left(\begin{matrix}-\frac{\rho}{2},\ \frac{\rho}{2};\ -\frac{a^2b^2}{4}\\ \frac{1}{2},\ \frac{2-s-\rho-\nu}{2},\ \frac{2-s-\rho+\nu}{2}\end{matrix}\right) \mp\frac{2^{s+\rho-2}\rho a^{s+\rho-\nu}}{b^\nu}\Gamma\left[\begin{matrix}\nu,\ \frac{s+\rho\mp\rho-\nu}{2},\ -s-\rho+\nu\\ -\frac{s+\rho\pm\rho-\nu-2}{2}\end{matrix}\right] \times {}_2F_3\left(\begin{matrix}\frac{s-\nu}{2},\ \frac{s+2\rho-\nu}{2};\ -\frac{a^2b^2}{4}\\ 1-\nu,\ \frac{s+\rho-\nu+1}{2},\ \frac{s+\rho-\nu+2}{2}\end{matrix}\right) \mp 2^{s+\rho-2}\rho a^{s+\rho+\nu}b^\nu\,\Gamma\left[\begin{matrix}-\nu,\ \frac{s+\nu+\rho\mp\rho}{2},\ -s-\rho-\nu\\ -\frac{s+\rho\pm\rho+\nu-2}{2}\end{matrix}\right] \times {}_2F_3\left(\begin{matrix}\frac{s+\nu}{2},\ \frac{s+2\rho+\nu}{2};\ -\frac{a^2b^2}{4}\\ 1+\nu,\ \frac{s+\rho+\nu+1}{2},\ \frac{s+\rho+\nu+2}{2}\end{matrix}\right)$ $[\operatorname{Re} a,\ \operatorname{Re} b>0;\ \operatorname{Re}(s+\rho\mp\rho)>\lvert\operatorname{Re}\nu\rvert]$

No.	$f(x)$	$F(s)$
12	$\frac{\left(\sqrt{x^2+a^2}\pm a\right)^\rho}{\sqrt{x^2+a^2}} \times K_\nu(bx)$	$\frac{2^{s+\rho-3}}{b^{s+\rho-1}}\Gamma\left(\frac{s+\rho+\nu-1}{2}\right)\Gamma\left(\frac{s+\rho-\nu-1}{2}\right) \times {}_2F_3\left(\begin{matrix}\frac{1+\rho}{2}, \frac{1-\rho}{2}; -\frac{a^2b^2}{4}\\ \frac{1}{2}, \frac{3-s-\rho-\nu}{2}, \frac{3-s-\rho+\nu}{2}\end{matrix}\right) \pm \frac{2^{s+\rho-4}\rho a}{b^{s+\rho-2}}\Gamma\left(\frac{s+\rho+\nu-2}{2}\right) \times \Gamma\left(\frac{s+\rho-\nu-2}{2}\right) {}_2F_3\left(\begin{matrix}\frac{2+\rho}{2}, \frac{2-\rho}{2}; -\frac{a^2b^2}{4}\\ \frac{3}{2}, \frac{4-s-\rho-\nu}{2}, \frac{4-s-\rho+\nu}{2}\end{matrix}\right) + \frac{2^{s+\rho-2}a^{s+\rho-\nu-1}}{b^\nu}\Gamma\left[\begin{matrix}\nu, \frac{s+\rho\mp\rho-\nu}{2}, 1-s-\rho+\nu\\ -\frac{s+\rho\pm\rho-\nu-2}{2}\end{matrix}\right] \times {}_2F_3\left(\begin{matrix}\frac{s-\nu}{2}, \frac{s+2\rho-\nu}{2}; -\frac{a^2b^2}{4}\\ 1-\nu, \frac{s+\rho-\nu}{2}, \frac{s+\rho-\nu+1}{2}\end{matrix}\right) + 2^{s+\rho-2}a^{s+\rho+\nu-1}b^\nu\,\Gamma\left[\begin{matrix}-\nu, \frac{s+\rho\mp\rho+\nu}{2}, 1-s-\rho-\nu\\ -\frac{s+\rho\pm\rho+\nu-2}{2}\end{matrix}\right] \times {}_2F_3\left(\begin{matrix}\frac{s+\nu}{2}, \frac{s+2\rho+\nu}{2}; -\frac{a^2b^2}{4}\\ 1+\nu, \frac{s+\rho+\nu}{2}, \frac{s+\rho+\nu+1}{2}\end{matrix}\right)$ $[\operatorname{Re} a,\ \operatorname{Re} b>0;\ \operatorname{Re}(s+\rho\mp\rho)>\lvert\operatorname{Re}\nu\rvert]$
13	$\left(\sqrt{x^2+a^2}\pm x\right)^\rho \times K_\nu(bx)$	$\frac{2^{s\pm2\rho-2}a^{\rho\mp\rho}}{b^{s\pm\rho}}\Gamma\left(\frac{s\pm\rho+\nu}{2}\right)\Gamma\left(\frac{s\pm\rho-\nu}{2}\right) \times {}_2F_3\left(\begin{matrix}\mp\frac{\rho}{2}, \frac{1\mp\rho}{2}; -\frac{a^2b^2}{4}\\ 1\mp\rho, -\frac{s\pm\rho+\nu-2}{2}, -\frac{s\pm\rho-\nu-2}{2}\end{matrix}\right) \mp \frac{2^{2\nu-s-2}\rho a^{s+\rho-\nu}}{b^\nu} \times \Gamma\left[\begin{matrix}\nu, s-\nu, \frac{\nu\mp\rho-s}{2}\\ \frac{s\mp\rho-\nu+2}{2}\end{matrix}\right] {}_2F_3\left(\begin{matrix}\frac{s-\nu}{2}, \frac{s-\nu+1}{2}; -\frac{a^2b^2}{4}\\ 1-\nu, \frac{s\mp\rho-\nu+2}{2}, \frac{s\pm\rho-\nu+2}{2}\end{matrix}\right) \mp \frac{\rho a^{s+\rho+\nu}b^\nu}{2^{s+2\nu+2}}\Gamma\left[\begin{matrix}-\nu, s+\nu, -\frac{s\pm\rho+\nu}{2}\\ \frac{s\mp\rho+\nu+2}{2}\end{matrix}\right] \times {}_2F_3\left(\begin{matrix}\frac{s+\nu}{2}, \frac{s+\nu+1}{2}; -\frac{a^2b^2}{4}\\ 1+\nu, \frac{s+\nu-\rho+2}{2}, \frac{s+\rho+\nu+2}{2}\end{matrix}\right)$ $[\operatorname{Re} a,\ \operatorname{Re} b>0;\ \operatorname{Re} s>\lvert\operatorname{Re}\nu\rvert]$
14	$\frac{\left(\sqrt{x^2+a^2}\pm x\right)^\rho}{\sqrt{x^2+a^2}} \times K_\nu(bx)$	$\frac{2^{s\pm2\rho-3}a^{\rho\mp\rho}}{b^{s\pm\rho-1}}\Gamma\left(\frac{s\pm\rho+\nu-1}{2}\right)\Gamma\left(\frac{s\pm\rho-\nu-1}{2}\right) \times {}_2F_3\left(\begin{matrix}\frac{2\mp\rho}{2}, \frac{1\mp\rho}{2}; -\frac{a^2b^2}{4}\\ 1\mp\rho, -\frac{s\pm\rho+\nu-3}{2}, -\frac{s-\nu\pm\rho-3}{2}\end{matrix}\right) + \frac{a^{s+\rho-\nu-1}b^{-\nu}}{2^{s-2\nu+1}} \times \Gamma\left[\begin{matrix}\nu, \frac{\nu\mp\rho-s+1}{2}, s-\nu\\ \frac{s\mp\rho-\nu+1}{2}\end{matrix}\right] {}_2F_3\left(\begin{matrix}\frac{s-\nu}{2}, \frac{s-\nu+1}{2}; -\frac{a^2b^2}{4}\\ 1-\nu, \frac{s-\nu\mp\rho+1}{2}, \frac{s-\nu\pm\rho+1}{2}\end{matrix}\right) + \frac{a^{s+\rho+\nu-1}\,b^\nu}{2^{s+2\nu+1}}\Gamma\left[\begin{matrix}-\nu, \frac{1-s+\mp\rho-\nu}{2}, s+\nu\\ \frac{s\mp\rho+\nu+1}{2}\end{matrix}\right] \times {}_2F_3\left(\begin{matrix}\frac{s+\nu}{2}, \frac{s+\nu+1}{2}; -\frac{a^2b^2}{4}\\ 1+\nu, \frac{s-\rho+\nu+1}{2}, \frac{s+\rho+\nu+1}{2}\end{matrix}\right)$ $[\operatorname{Re} a,\ \operatorname{Re} b>0;\ \operatorname{Re} s>\lvert\operatorname{Re}\nu\rvert]$

3.14.2. $K_\nu(\varphi(x))$ and algebraic functions

1	$(x-a)_+^{\alpha-1} K_\nu\left(b\sqrt{x-a}\right)$	$\frac{2^{\nu-1}a^{s+\alpha-\nu/2-1}}{b^\nu}\Gamma\left[\begin{matrix}\nu,\ \frac{2\alpha-\nu}{2},\ \frac{2-2s-2\alpha+\nu}{2}\\ 1-s\end{matrix}\right]{}_1F_2\left(\begin{matrix}\frac{2\alpha-\nu}{2};\ -\frac{ab^2}{4}\\ 1-\nu,\ \frac{2s+2\alpha-\nu}{2}\end{matrix}\right)$ $+\frac{a^{s+\alpha+\nu/2-1}b^\nu}{2^{\nu+1}}\Gamma\left[\begin{matrix}-\nu,\ \frac{2\alpha+\nu}{2},\ \frac{2-2s-2\alpha-\nu}{2}\\ 1-s\end{matrix}\right]{}_1F_2\left(\begin{matrix}\frac{2\alpha+\nu}{2};\ -\frac{ab^2}{4}\\ 1+\nu,\ \frac{2s+2\alpha+\nu}{2}\end{matrix}\right)$ $+\frac{2^{2s+2\alpha-3}}{b^{2(s+\alpha-1)}}\Gamma\left(s+\alpha+\frac{\nu}{2}-1\right)\Gamma\left(s+\alpha-\frac{\nu}{2}-1\right)$ $\times{}_1F_2\left(\begin{matrix}1-s;\ -\frac{ab^2}{4}\\ \frac{4-2s-2\alpha-\nu}{2},\ \frac{4-2s-2\alpha+\nu}{2}\end{matrix}\right)$ $[a,\ \mathrm{Re}\,b>0;\ \mathrm{Re}\,\alpha+\lvert\mathrm{Re}\,\nu\rvert>0]$
2	$(x+a)^{\pm\nu/2} K_\nu\left(b\sqrt{x+a}\right)$	$a^{(s\pm\nu)/2}\left(\frac{2}{b}\right)^s\Gamma(s)\,K_{s\pm\nu}\left(\sqrt{a}\,b\right)$ $[\mathrm{Re}\,a,\ \mathrm{Re}\,b,\ \mathrm{Re}\,s>0]$
3	$(a-x)_+^{-\nu/2} K_\nu\left(b\sqrt{a-x}\right)$	$\frac{2^{-\nu-1}a^s b^\nu}{s}\Gamma(-\nu)\,{}_1F_2\left(\begin{matrix}1;\ \frac{ab^2}{4}\\ \nu+1,\ s+1\end{matrix}\right)$ $+\frac{2^{s-1}\pi\,a^{(s-\nu)/2}}{b^s}\csc(\nu\pi)\,\Gamma(s)\,I_{s-\nu}\left(\sqrt{a}\,b\right)$ $[a,\ \mathrm{Re}\,b,\ \mathrm{Re}\,s,\ \mathrm{Re}\,(s-\nu)>0]$
4	$\left(x^2-a^2\right)_+^{\nu/2}$ $\times K_\nu\left(b\sqrt{x^2-a^2}\right)$	$-\frac{2^{\nu-1}a^s b^{-\nu}}{s}\Gamma(\nu)\,{}_1F_2\left(\begin{matrix}1;\ -\frac{a^2b^2}{4}\\ 1-\nu,\ \frac{s+2}{2}\end{matrix}\right)$ $+\frac{2^{s/2-2}\pi^2 a^{s/2+\nu}b^{-s/2}}{\Gamma\left(\frac{2-s}{2}\right)}\csc\frac{(s+2\nu)\pi}{2}$ $\times\left[\csc\frac{s\pi}{2}\,J_{-s/2-\nu}(ab)+\csc(\nu\pi)\,J_{s/2+\nu}(ab)\right]$ $[a,\ \mathrm{Re}\,b>0;\ \mathrm{Re}\,\nu>-1]$
5	$\theta(1-x)\,K_\nu\left(ax-\frac{a}{x}\right)$	$\frac{\pi^2}{4}\csc(\nu\pi)\left[J_{(\nu+s)/2}(a)\,Y_{(s-\nu)/2}(a)-J_{(s-\nu)/2}(a)\,Y_{(\nu+s)/2}(a)\right]$ $[\mathrm{Re}\,a>0;\ \lvert\mathrm{Re}\,\nu\rvert<1]$
6	$\theta(x-1)\,K_\nu\left(ax-\frac{a}{x}\right)$	$\frac{\pi^2}{4}\csc(\nu\pi)\left[J_{(\nu-s)/2}(a)\,Y_{-(s+\nu)/2}(a)\right.$ $\left.-J_{-(s+\nu)/2}(a)\,Y_{(\nu-s)/2}(a)\right]$ $[\mathrm{Re}\,a>0;\ \lvert\mathrm{Re}\,\nu\rvert<1]$
7	$K_0\left(a\left\lvert x-\frac{1}{x}\right\rvert\right)$	$\frac{\pi^2}{4}\left[J_{s/2}^2(a)+Y_{s/2}^2(a)\right]$ $[\mathrm{Re}\,a>0]$
8	$K_\nu\left(ax+\frac{a}{x}\right)$	$K_{(s+\nu)/2}(a)\,K_{(\nu-s)/2}(a)$ $[\mathrm{Re}\,a>0]$

No.	$f(x)$	$F(s)$
9	$\left(\dfrac{bx+a}{ax+b}\right)^{\nu/2} K_\nu\left(\sqrt{u}\right)$ $u = ab\left(x+\dfrac{1}{x}\right)+a^2+b^2$	$2K_{s+\nu/2}(a)\, K_{s-\nu/2}(b)$ $\quad [a,\, b>0]$

3.14.3. $K_\nu(\varphi(x))$ and the exponential function

No.	$f(x)$	$F(s)$
1	$e^{-ax} K_\nu(ax)$	$\dfrac{\sqrt{\pi}}{(2a)^s}\Gamma\left[\begin{matrix} s-\nu,\, s+\nu \\ \frac{2s+1}{2}\end{matrix}\right]$ $\quad [\operatorname{Re} a>0;\ \operatorname{Re} s>\lvert\operatorname{Re}\nu\rvert]$
2	$e^{ax} K_\nu(ax)$	$\dfrac{\cos(\nu\pi)}{\sqrt{\pi}\,(2a)^s}\Gamma\left[s-\nu,\, s+\nu,\, \dfrac{1-2s}{2}\right]$ $\quad [\operatorname{Re} a>0;\ \lvert\operatorname{Re}\nu\rvert<\operatorname{Re} s<1/2]$
3	$e^{-ax} K_\nu(bx)$	$\dfrac{\sqrt{\pi}a^{\nu-s}}{2^s b^\nu}\Gamma\left[\begin{matrix} s-\nu,\, s+\nu \\ \frac{2s+1}{2}\end{matrix}\right] {}_2F_1\left(\begin{matrix}\frac{s-\nu}{2},\ \frac{s-\nu+1}{2} \\ \frac{2s+1}{2};\ \frac{a^2-b^2}{a^2}\end{matrix}\right)$
4		$= \dfrac{e^{-i\pi\nu}\,\Gamma(s-\nu)}{(a^2-b^2)^{s/2}} Q_{s-1}^{\nu}\left(\dfrac{a}{\sqrt{a^2-b^2}}\right)$
5		$= \sqrt{\dfrac{\pi}{2b}}\dfrac{\Gamma(s-\nu)\,\Gamma(s+\nu)}{(b^2-a^2)^{(2s-1)/4}} \mathrm{P}_{\nu-1/2}^{1/2-s}\left(\dfrac{a}{b}\right)$ $[\operatorname{Re}(a+b)>0;\ \operatorname{Re} s>\lvert\operatorname{Re}\nu\rvert]$
6	$(a-x)_+^{\alpha-1} e^{\pm bx} K_\nu(bx)$	$\dfrac{2^{\nu-1}a^{s+\alpha-\nu-1}}{b^\nu}\Gamma\left[\begin{matrix}\alpha,\ \nu,\ s-\nu \\ s+\alpha-\nu\end{matrix}\right] {}_2F_2\left(\begin{matrix}\frac{1-2\nu}{2},\ s-\nu;\ \pm 2ab \\ 1-2\nu,\ s+\alpha-\nu\end{matrix}\right)$ $+\dfrac{a^{s+\alpha+\nu-1}b^\nu}{2^{\nu+1}}\Gamma\left[\begin{matrix}\alpha,\ -\nu,\ s+\nu \\ s+\alpha+\nu\end{matrix}\right] {}_2F_2\left(\begin{matrix}\frac{1+2\nu}{2}, s+\nu;\ \pm 2ab \\ 1+2\nu,\ s+\alpha+\nu\end{matrix}\right)$ $[a,\ \operatorname{Re}\alpha>0;\ \operatorname{Re} s>\lvert\operatorname{Re}\nu\rvert]$
7	$(x-a)_+^{\alpha-1} e^{\pm bx} K_\nu(bx)$	$\dfrac{2^{\nu-1}a^{s+\alpha-\nu-1}}{b^\nu}\Gamma\left[\begin{matrix}\alpha,\ \nu,\ 1-s-\alpha+\nu \\ 1-s+\nu\end{matrix}\right] {}_2F_2\left(\begin{matrix}\frac{1-2\nu}{2},\ s-\nu;\ \pm 2ab \\ 1-2\nu,\ s+\alpha-\nu\end{matrix}\right)$ $+\dfrac{a^{s+\alpha+\nu-1}b^\nu}{2^{\nu+1}}\Gamma\left[\begin{matrix}\alpha,\ -\nu,\ 1-s-\alpha-\nu \\ 1-s-\nu\end{matrix}\right] {}_2F_2\left(\begin{matrix}\frac{1+2\nu}{2},\ s+\nu;\ \pm 2ab \\ 1+2\nu,\ s+\alpha+\nu\end{matrix}\right)$ $\mp\dfrac{\sqrt{\pi}}{(2b)^{s+\alpha-1}}\left[\cos(\nu\pi)\sec(s+\alpha)\pi\right]^{(1\pm 1)/2}$ $\times\Gamma\left[\begin{matrix}s+\alpha+\nu-1,\ s+\alpha-\nu-1 \\ \frac{2s+2\alpha-1}{2}\end{matrix}\right]$ $\times\, {}_2F_2\left(\begin{matrix}1-\alpha,\ \frac{3-2s-2\alpha}{2};\ \pm 2ab \\ 2-s-\alpha-\nu,\ 2-s-\alpha+\nu\end{matrix}\right)$ $\left[a,\ \operatorname{Re}\alpha>0;\ \left\{\begin{matrix}\operatorname{Re} b>0;\ \operatorname{Re}(s+\alpha)<3/2 \\ \operatorname{Re} b>0\end{matrix}\right\}\right]$

No.	$f(x)$	$F(s)$
8	$\dfrac{e^{\pm bx}}{(x+a)^{\rho}}K_{\nu}(bx)$	$\dfrac{\sqrt{\pi}}{(2b)^{s-\rho}}\left[\cos(\nu\pi)\sec(\rho-s)\pi\right]^{(1\pm1)/2}\Gamma\left[\begin{matrix}s+\nu-\rho,\ s-\nu-\rho\\ \frac{2s-2\rho+1}{2}\end{matrix}\right]$ $\times\,{}_2F_2\left(\begin{matrix}\rho,\ \frac{1-2s+2\rho}{2};\ \mp2ab\\ 1-s-\nu+\rho,\ 1-s+\nu+\rho\end{matrix}\right)+\dfrac{2^{\nu-1}a^{s-\nu-\rho}}{b^{\nu}}$ $\times\,\Gamma\left[\begin{matrix}\nu,\ s-\nu,\ \rho+\nu-s\\ \rho\end{matrix}\right]{}_2F_2\left(\begin{matrix}\frac{1-2\nu}{2},\ s-\nu;\ \mp2ab\\ 1-2\nu,\ s-\nu-\rho+1\end{matrix}\right)$ $+\dfrac{a^{s+\nu-\rho}b^{\nu}}{2^{\nu+1}}\Gamma\left[\begin{matrix}-\nu,\ s+\nu,\ \rho-\nu-s\\ \rho\end{matrix}\right]$ $\times\,{}_2F_2\left(\begin{matrix}\frac{1+2\nu}{2},\ s+\nu;\ \mp2ab\\ 1+2\nu,\ s+\nu-\rho+1\end{matrix}\right)$ $\left[\operatorname{Re}s>\lvert\operatorname{Re}\nu\rvert;\ \left\{\begin{matrix}\operatorname{Re}b>0;\ \operatorname{Re}(s-\rho)<1/2\\ \operatorname{Re}b>0\end{matrix}\right\};\ \lvert\arg a\rvert<\pi\right]$
9	$\dfrac{e^{\pm bx}}{x-a}K_{\nu}(bx)$	$\mp\dfrac{\sqrt{\pi}\left[\cos(\nu\pi)\sec(s\pi)\right]^{(1\pm1)/2}}{(2b)^{s-1}}\Gamma\left[\begin{matrix}s+\nu-1,\ s-\nu-1\\ \frac{2s-1}{2}\end{matrix}\right]$ $\times\,{}_2F_2\left(\begin{matrix}1,\ \frac{3-2s}{2};\ \pm2ab\\ 2-s-\nu,\ 2-s+\nu\end{matrix}\right)$ $-\dfrac{\pi a^{s-\nu-1}}{2^{1-\nu}b^{\nu}}\Gamma(\nu)\cot\left[(s-\nu)\pi\right]\,{}_1F_1\left(\begin{matrix}\frac{1-2\nu}{2};\ \pm2ab\\ 1-2\nu\end{matrix}\right)$ $-\dfrac{\pi a^{s+\nu-1}b^{\nu}}{2^{\nu+1}}\Gamma(-\nu)\cot\left[(s+\nu)\pi\right]\,{}_1F_1\left(\begin{matrix}\frac{2\nu+1}{2};\ \pm2ab\\ 2\nu+1\end{matrix}\right)$ $\left[\operatorname{Re}s>\lvert\operatorname{Re}\nu\rvert;\ \left\{\begin{matrix}a,\ \operatorname{Re}b>0;\ \operatorname{Re}s<3/2\\ a,\ \operatorname{Re}b>0\end{matrix}\right\}\right]$
10	$e^{-ax^2}K_{\nu}(bx)$	$\dfrac{a^{(1-s)/2}}{2b}e^{b^2/(8a)}\Gamma\left(\dfrac{s-\nu}{2}\right)\Gamma\left(\dfrac{s+\nu}{2}\right)W_{(1-s)/2,\,\nu/2}\left(\dfrac{b^2}{4a}\right)$ $[\operatorname{Re}a>0;\ \operatorname{Re}s>\lvert\operatorname{Re}\nu\rvert]$
11	$e^{-a\sqrt{x}}K_{\nu}(bx)$	$\dfrac{2^{s-2}}{b^{s}}\Gamma\left(\dfrac{s-\nu}{2}\right)\Gamma\left(\dfrac{s+\nu}{2}\right){}_2F_3\left(\begin{matrix}\frac{s-\nu}{2},\ \frac{s+\nu}{2}\\ \frac{1}{4},\ \frac{1}{2},\ \frac{3}{4};\ \frac{a^4}{64b^2}\end{matrix}\right)$ $-\dfrac{2^{s-3/2}}{b^{s+1/2}}\Gamma\left(\dfrac{2s-2\nu+1}{4}\right)\Gamma\left(\dfrac{2s+2\nu+1}{4}\right)$ $\times\,{}_2F_3\left(\begin{matrix}\frac{2s-2\nu+1}{4},\ \frac{2s+2\nu+1}{4}\\ \frac{1}{2},\ \frac{3}{4},\ \frac{5}{4};\ \frac{a^4}{64b^2}\end{matrix}\right)+\dfrac{2^{s-2}a^{2}}{b^{s+1}}\Gamma\left(\dfrac{s-\nu+1}{2}\right)$ $\times\,\Gamma\left(\dfrac{s+\nu+1}{2}\right){}_2F_3\left(\begin{matrix}\frac{s-\nu+1}{2},\ \frac{s+\nu+1}{2}\\ \frac{3}{4},\ \frac{5}{4},\ \frac{3}{2};\ \frac{a^4}{64b^2}\end{matrix}\right)$ $-\dfrac{2^{s-3/2}a^{3}}{3b^{s+3/2}}\Gamma\left(\dfrac{2s-2\nu+3}{4}\right)\Gamma\left(\dfrac{2s+2\nu+3}{4}\right)$ $\times\,{}_2F_3\left(\begin{matrix}\frac{2s-2\nu+3}{4},\ \frac{2s+2\nu+3}{4}\\ \frac{5}{4},\ \frac{3}{2},\ \frac{7}{4};\ \frac{a^4}{64b^2}\end{matrix}\right)$ $[\operatorname{Re}b>0;\ \operatorname{Re}s>\lvert\operatorname{Re}\nu\rvert]$

No.	$f(x)$	$F(s)$
12	$e^{-a/x}K_\nu(bx)$	$\frac{2^{s-2}}{b^s}\Gamma\left(\frac{s-\nu}{2}\right)\Gamma\left(\frac{s+\nu}{2}\right){}_0F_3\left(\frac{a^2b^2}{16} \Big\vert \frac{1}{2}, \frac{2-s-\nu}{2}, \frac{2-s+\nu}{2}\right)$ $-\frac{2^{s-3}a}{b^{s-1}}\Gamma\left(\frac{s-\nu-1}{2}\right)\Gamma\left(\frac{s+\nu-1}{2}\right){}_0F_3\left(\frac{a^2b^2}{16} \Big\vert \frac{3}{2}, \frac{3-s-\nu}{2}, \frac{3-s+\nu}{2}\right)$ $+\frac{a^{s+\nu}b^\nu}{2^{\nu+1}}\Gamma(-\nu)\Gamma(-s-\nu)\,{}_0F_3\left(\frac{a^2b^2}{16} \Big\vert 1+\nu, \frac{s+\nu+1}{2}, \frac{s+\nu+2}{2}\right)$ $+\frac{2^{\nu-1}a^{s-\nu}}{b^\nu}\Gamma(\nu)\Gamma(\nu-s)\,{}_0F_3\left(\frac{a^2b^2}{16} \Big\vert 1-\nu, \frac{s-\nu+1}{2}, \frac{s-\nu+2}{2}\right)$ $[\operatorname{Re} a,\ \operatorname{Re} b>0]$
13	$e^{-a/x^2}K_\nu(bx)$	$\frac{2^{s-2}}{b^s}\Gamma\left(\frac{s-\nu}{2}\right)\Gamma\left(\frac{s+\nu}{2}\right){}_0F_2\left(-\frac{ab^2}{4} \Big\vert \frac{2-s-\nu}{2}, \frac{2-s+\nu}{2}\right)$ $+\frac{a^{(s+\nu)/2}b^\nu}{2^{\nu+2}}\Gamma(-\nu)\Gamma\left(-\frac{s+\nu}{2}\right){}_0F_2\left(-\frac{ab^2}{4} \Big\vert 1+\nu, \frac{s+\nu+2}{2}\right)$ $+\frac{a^{(s-\nu)/2}}{2^{2-\nu}b^\nu}\Gamma(\nu)\Gamma\left(\frac{\nu-s}{2}\right){}_0F_2\left(-\frac{ab^2}{4} \Big\vert 1-\nu, \frac{s-\nu+2}{2}\right)$ $[\operatorname{Re} a,\ \operatorname{Re} b>0]$
14	$e^{\mp bx-a/x}K_\nu(bx)$	$\frac{a^{s+\nu}b^\nu}{2^{\nu+1}}\Gamma(-\nu)\Gamma(-\nu-s)\,{}_1F_2\left(\frac{1+2\nu}{2}; \pm 2ab \Big\vert 1+2\nu, s+\nu+1\right)$ $+\frac{a^{s-\nu}b^{-\nu}}{2^{1-\nu}}\Gamma(\nu)\Gamma(\nu-s)\,{}_1F_2\left(\frac{1-2\nu}{2}; \pm 2ab \Big\vert 1-2\nu, s-\nu+1\right)$ $+\frac{\sqrt{\pi}}{(2b)^s}\left(\frac{\cos(\nu\pi)}{\cos(s\pi)}\right)^{(1\mp 1)/2}\Gamma\left[\begin{matrix} s-\nu,\ s+\nu \\ \frac{2s+1}{2}\end{matrix}\right]$ $\times\,{}_1F_2\left(\frac{1-2s}{2}; \pm 2ab \Big\vert 1-s-\nu, 1-s+\nu\right)$ $[\operatorname{Re} a,\ \operatorname{Re} b>0]$
15	$(a^2-x^2)_+^{-1}$ $\times\exp\left(-b\frac{a^2+x^2}{a^2-x^2}\right)$ $\times K_\nu\left(\frac{cx}{a^2-x^2}\right)$	$\frac{a^{s-1}}{2c}\Gamma\left(\frac{s-\nu}{2}\right)\Gamma\left(\frac{s+\nu}{2}\right)W_{(1-s)/2,\,\nu/2}\left(\frac{2ab+\sqrt{4a^2b^2-c^2}}{2a}\right)$ $\times W_{(1-s)/2,\,\nu/2}\left(\frac{2ab-\sqrt{4a^2b^2-c^2}}{2a}\right)$ $[a,\ b,\ \operatorname{Re} c>0;\ \operatorname{Re} s>\lvert\operatorname{Re}\nu\rvert]$
16	$(x^2-a^2)_+^{-1}$ $\times\exp\left(-b\frac{a^2+x^2}{a^2-x^2}\right)$ $\times K_\nu\left(\frac{cx}{x^2-a^2}\right)$	$\frac{a^{s-1}}{2c}\Gamma\left(\frac{2-\nu-s}{2}\right)\Gamma\left(\frac{2+\nu-s}{2}\right)$ $\times W_{(s-1)/2,\nu/2}\left(\frac{2ab+\sqrt{4a^2b^2-c^2}}{2a}\right)$ $\times W_{(s-1)/2,\,\nu/2}\left(\frac{2ab-\sqrt{4a^2b^2-c^2}}{2a}\right)$ $[a,\ b,\ \operatorname{Re} c>0;\ \operatorname{Re} s<\lvert\operatorname{Re}\nu\rvert+2]$

3.14.4. $K_\nu(ax)$ and hyperbolic or trigonometric functions

Notation: $\delta = \begin{Bmatrix} 1 \\ 0 \end{Bmatrix}$.

1	$\begin{Bmatrix} \sinh(ax) \\ \sin(ax) \end{Bmatrix} K_\nu(bx)$	$\frac{2^{s-1}a}{b^{s+1}} \Gamma\left(\frac{s-\nu+1}{2}\right) \Gamma\left(\frac{s+\nu+1}{2}\right) {}_2F_1\left(\begin{matrix} \frac{s-\nu+1}{2}, \frac{s+\nu+1}{2} \\ \frac{3}{2}; \pm\frac{a^2}{b^2} \end{matrix}\right)$ $[\operatorname{Re} b > \lvert\operatorname{Re} a\rvert;\ \operatorname{Re} s > \lvert\operatorname{Re} \nu\rvert - 1]$
2	$\begin{Bmatrix} \cosh(ax) \\ \cos(ax) \end{Bmatrix} K_\nu(bx)$	$\frac{2^{s-2}}{b^s} \Gamma\left(\frac{s-\nu}{2}\right) \Gamma\left(\frac{s+\nu}{2}\right) {}_2F_1\left(\begin{matrix} \frac{s-\nu}{2}, \frac{s+\nu}{2} \\ \frac{1}{2}; \pm\frac{a^2}{b^2} \end{matrix}\right)$ $[\operatorname{Re} b > \lvert\operatorname{Re} a\rvert;\ \operatorname{Re} s > \lvert\operatorname{Re} \nu\rvert]$
3	$[1-\cos(ax)]\, K_\nu(bx)$	$\frac{2^{s-1}a^2}{b^{s+2}} \Gamma\left(\frac{s-\nu+2}{2}\right) \Gamma\left(\frac{s+\nu+2}{2}\right) {}_3F_2\left(\begin{matrix} 1, \frac{s-\nu+2}{2}, \frac{s+\nu+2}{2} \\ \frac{3}{2}, 2; -\frac{a^2}{b^2} \end{matrix}\right)$ $[\operatorname{Re} b > \lvert\operatorname{Im} a\rvert;\ \operatorname{Re} s > \lvert\operatorname{Re} \nu\rvert - 2]$
4	$\begin{Bmatrix} \sinh(ax+b) \\ \cosh(ax+b) \end{Bmatrix} K_\nu(ax)$	$\frac{2^{-s-1}a^{-s}e^b}{\sqrt{\pi}} \cos(\nu\pi)\, \Gamma\left(\frac{1-2s}{2}\right) \Gamma(s-\nu)\, \Gamma(s+\nu)$ $\mp 2^{-s-1}\sqrt{\pi}\, a^{-s} e^{-b}\, \Gamma\left[\begin{matrix} s-\nu, s+\nu \\ \frac{2s+1}{2} \end{matrix}\right]$ $[\operatorname{Re} a \geq 0;\ \lvert\operatorname{Re} \nu\rvert < \operatorname{Re} s < 1/2]$
5	$\begin{Bmatrix} \sin(ax^2) \\ \cos(ax^2) \end{Bmatrix} K_\nu(bx)$	$\frac{2^{-\nu-2}b^\nu}{a^{(s+\nu)/2}} \begin{Bmatrix} \sin[(s+\nu)\pi/4] \\ \cos[(s+\nu)\pi/4] \end{Bmatrix} \Gamma(-\nu)\, \Gamma\left(\frac{s+\nu}{2}\right)$ $\times {}_2F_3\left(\begin{matrix} \frac{s+\nu}{4}, \frac{s+\nu+2}{4}; -\frac{b^4}{64a^2} \\ \frac{1}{2}, \frac{\nu+1}{2}, \frac{\nu+2}{2} \end{matrix}\right) + \frac{2^{\nu-2}b^{-\nu}}{a^{(s-\nu)/2}}$ $\times \begin{Bmatrix} \sin[(s-\nu)\pi/4] \\ \cos[(s-\nu)\pi/4] \end{Bmatrix} \Gamma(\nu)\, \Gamma\left(\frac{s-\nu}{2}\right)$ $\times {}_2F_3\left(\begin{matrix} \frac{s-\nu}{4}, \frac{s-\nu+2}{4}; -\frac{b^4}{64a^2} \\ \frac{1}{2}, \frac{1-\nu}{2}, \frac{2-\nu}{2} \end{matrix}\right) \mp \frac{2^{-\nu-4}\, b^{\nu+2}}{a^{(s+\nu+2)/2}}$ $\times \begin{Bmatrix} \cos[(s+\nu)\pi/4] \\ \sin[(s+\nu)\pi/4] \end{Bmatrix} \Gamma(-\nu-1)\, \Gamma\left(\frac{s+\nu+2}{2}\right)$ $\times {}_2F_3\left(\begin{matrix} \frac{s+\nu+2}{4}, \frac{s+\nu+4}{4}; -\frac{b^4}{64a^2} \\ \frac{3}{2}, \frac{\nu+2}{2}, \frac{\nu+3}{2} \end{matrix}\right) \mp \frac{2^{\nu-4}b^{2-\nu}}{a^{(s-\nu+2)/2}}$ $\times \begin{Bmatrix} \cos[(s-\nu)\pi/4] \\ \sin[(s-\nu)\pi/4] \end{Bmatrix} \Gamma(\nu-1)\, \Gamma\left(\frac{s-\nu+2}{2}\right)$ $\times {}_2F_3\left(\begin{matrix} \frac{s-\nu+2}{4}, \frac{s-\nu+4}{4}; -\frac{b^4}{64a^2} \\ \frac{3}{2}, \frac{2-\nu}{2}, \frac{3-\nu}{2} \end{matrix}\right)$ $[a,\ \operatorname{Re} b > 0;\ \operatorname{Re} s > \lvert\operatorname{Re} \nu\rvert - 1 \mp 1]$

No.	$f(x)$	$F(s)$
6	$\left\{\begin{matrix}\sin(a\sqrt{x})\\ \cos(a\sqrt{x})\end{matrix}\right\} K_\nu(bx)$	$\frac{2^{s+\delta/2-2}\,a^\delta}{b^{s+\delta/2}}\,\Gamma\left(\frac{2s-2\nu+\delta}{4}\right)\Gamma\left(\frac{2s+2\nu+\delta}{4}\right)$ $\times\,{}_2F_3\left(\begin{matrix}\frac{2s-2\nu+\delta}{4},\ \frac{2s+2\nu+\delta}{4}\\ \frac{1}{2},\ \frac{3}{4},\ \frac{4\delta+1}{4};\ \frac{a^4}{64b}\end{matrix}\right) - \frac{2^{s+\delta/2-2}\,a^{\delta+2}}{3^\delta b^{s+\delta/2+1}}$ $\times\,\Gamma\left(\frac{2s-2\nu+\delta+2}{2}\right)\Gamma\left(\frac{2s+2\nu+\delta+2}{2}\right)$ $\times\,{}_2F_3\left(\begin{matrix}\frac{2s-2\nu+\delta+2}{4},\ \frac{2s+2\nu+\delta+2}{4}\\ \frac{5}{4},\ \frac{3}{2},\ \frac{4\delta+3}{4};\ \frac{a^4}{64b}\end{matrix}\right)$ $[a,\ \operatorname{Re} b>0;\ \lvert\operatorname{Re}\nu\rvert<\operatorname{Re} s+\delta/2]$
7	$\left\{\begin{matrix}\sin(a/x)\\ \cos(a/x)\end{matrix}\right\} K_\nu(bx)$	$2^{s-\delta-2}\,a^\delta b^{\delta-s}\,\Gamma\left(\frac{s-\nu-\delta}{2}\right)\Gamma\left(\frac{s+\nu-\delta}{2}\right)$ $\times\,{}_0F_3\left(\begin{matrix}-\frac{a^2b^2}{16}\\ \frac{2\delta+1}{2},\ \frac{2-s-\nu+\delta}{2},\ \frac{2-s+\nu+\delta}{2}\end{matrix}\right)$ $\mp\frac{a^{s+\nu}\,b^\nu}{2^{\nu+1}}\left\{\begin{matrix}\sin[(s+\nu)\pi/2]\\ \cos[(s+\nu)\pi/2]\end{matrix}\right\}$ $\times\,\Gamma(-\nu)\,\Gamma(-s-\nu)\,{}_0F_3\left(\begin{matrix}-\frac{a^2b^2}{16}\\ 1+\nu,\ \frac{s+\nu+1}{2},\ \frac{s+\nu+2}{2}\end{matrix}\right)$ $\mp\frac{2^{\nu-1}\,a^{s-\nu}}{b^\nu}\left\{\begin{matrix}\sin[(s-\nu)\pi/2]\\ \cos[(s-\nu)\pi/2]\end{matrix}\right\}$ $\times\,\Gamma(\nu)\,\Gamma(-s+\nu)\,{}_0F_3\left(\begin{matrix}-\frac{a^2b^2}{16}\\ 1-\nu,\ \frac{s-\nu+1}{2},\ \frac{s-\nu+2}{2}\end{matrix}\right)$ $[a,\ \operatorname{Re} b>0;\ \lvert\operatorname{Re}\nu\rvert<\operatorname{Re} s+1]$
8	$\left\{\begin{matrix}\sin(ax)\sinh(ax)\\ \cos(ax)\cosh(ax)\end{matrix}\right\}$ $\times K_\nu(bx)$	$2^{s+2\delta-2}a^{2\delta}b^{-s-2\delta}\,\Gamma\left(\frac{s-\nu+2\delta}{2}\right)\Gamma\left(\frac{s+\nu+2\delta}{2}\right)$ $\times\,{}_4F_3\left(\begin{matrix}\frac{s-\nu+2\delta}{4},\ \frac{s-\nu+2\delta+2}{4},\ \frac{s+\nu+2\delta}{4},\ \frac{s+\nu+2\delta+2}{4}\\ \frac{2\delta+1}{4},\ \frac{2\delta+3}{4},\ \frac{2\delta+1}{2};\ -\frac{4a^4}{b^4}\end{matrix}\right)$ $[\operatorname{Re} b>\lvert\operatorname{Re} a\rvert+\lvert\operatorname{Im} a\rvert;\ \operatorname{Re} s>\lvert\operatorname{Re}\nu\rvert-2\delta]$
9	$\left\{\begin{matrix}\sin(ax)\cosh(ax)\\ \cos(ax)\sinh(ax)\end{matrix}\right\}$ $\times K_\nu(bx)$	$\frac{2^{s-1}a}{b^{s+1}}\,\Gamma\left(\frac{s-\nu+1}{2}\right)\Gamma\left(\frac{s+\nu+1}{2}\right)$ $\times\,{}_4F_3\left(\begin{matrix}\frac{s-\nu+1}{4},\ \frac{s-\nu+3}{4},\ \frac{s+\nu+1}{4},\ \frac{s+\nu+3}{4}\\ \frac{1}{2},\ \frac{3}{4},\ \frac{5}{4};\ -\frac{4a^4}{b^4}\end{matrix}\right)$ $\pm\frac{2^{s+1}a^3}{3b^{s+3}}\,\Gamma\left(\frac{s-\nu+3}{2}\right)\Gamma\left(\frac{s+\nu+3}{2}\right)$ $\times\,{}_4F_3\left(\begin{matrix}\frac{s-\nu+3}{4},\ \frac{s-\nu+5}{4},\ \frac{s+\nu+3}{4},\ \frac{s+\nu+5}{4}\\ \frac{3}{4},\ \frac{5}{4},\ \frac{7}{2};\ -\frac{4a^4}{b^4}\end{matrix}\right)$ $[\operatorname{Re} b>\lvert\operatorname{Re} a\rvert+\lvert\operatorname{Im} a\rvert;\ \operatorname{Re} s>\lvert\operatorname{Re}\nu\rvert-1]$

No.	$f(x)$	$F(s)$
10	$e^{-bx}\left\{\begin{matrix}\sin(ax)\\ \cos(ax)\end{matrix}\right\}K_\nu(bx)$	$\frac{\sqrt{\pi}\,a^\delta}{(2b)^{s+\delta}}\Gamma\left[\begin{matrix}s-\nu+\delta,\ s+\nu+\delta\\ \frac{2s+2\delta+1}{2}\end{matrix}\right] \times {}_4F_3\left(\begin{matrix}\frac{s-\nu+1}{2},\ \frac{s+\nu+1}{2},\ \frac{s-\nu+2\delta}{2},\ \frac{s+\nu+2\delta}{2}\\ \frac{2\delta+1}{2},\ \frac{2s+3}{4},\ \frac{2s+4\delta+1}{4};\ -\frac{a^2}{4b^2}\end{matrix}\right)$ $[a,\ \operatorname{Re} b>0;\ \operatorname{Re} s>\lvert\operatorname{Re}\nu\rvert-\delta]$
11	$e^{-bx}\left\{\begin{matrix}\sin(a\sqrt{x})\\ \cos(a\sqrt{x})\end{matrix}\right\}K_\nu(bx)$	$\frac{\sqrt{\pi}\,a^\delta}{(2b)^{s+\delta/2}}\Gamma\left[\begin{matrix}\frac{2s-2\nu+\delta}{2},\ \frac{2s+2\nu+\delta}{2}\\ \frac{2s+\delta+1}{2}\end{matrix}\right] {}_2F_2\left(\begin{matrix}\frac{2s-2\nu+\delta}{2},\ \frac{2s+2\nu+\delta}{2}\\ \frac{2\delta+1}{2},\ \frac{2s+\delta+1}{2};\ -\frac{a^2}{8b}\end{matrix}\right)$ $[a,\ \operatorname{Re} b>0;\ \operatorname{Re} s>\lvert\operatorname{Re}\nu\rvert-\delta/2]$
12	$e^{-bx}\left\{\begin{matrix}\sin(ax)\cosh(ax)\\ \cos(ax)\sinh(ax)\end{matrix}\right\} \times K_\nu(bx)$	$\frac{\sqrt{\pi}\,a}{(2b)^{s+1}}\Gamma\left[\begin{matrix}s-\nu+1,\ s+\nu+1\\ \frac{2s+3}{2}\end{matrix}\right] \times {}_8F_7\left(\begin{matrix}\Delta(4,\ s-\nu+1),\ \Delta(4,\ s+\nu+1)\\ \frac{1}{2},\ \frac{3}{4},\ \frac{5}{4},\ \Delta\left(4,\ \frac{2s+3}{2}\right);\ -\frac{a^4}{4b^4}\end{matrix}\right) \pm \frac{\sqrt{\pi}\,a^3}{3(2b)^{s+3}}\Gamma\left[\begin{matrix}s-\nu+3,\ s+\nu+3\\ \frac{2s+7}{2}\end{matrix}\right] \times {}_8F_7\left(\begin{matrix}\Delta(4,\ s-\nu+3),\ \Delta(4,\ s+\nu+3)\\ \frac{5}{4},\ \frac{3}{2},\ \frac{7}{4},\ \Delta\left(4,\ \frac{2s+7}{2}\right);\ -\frac{a^4}{4b^4}\end{matrix}\right)$ $[\operatorname{Re} b>(\operatorname{Re} a+\operatorname{Im} a)/2;\ \operatorname{Re} s>\lvert\operatorname{Re}\mu\rvert+\lvert\operatorname{Re}\nu\rvert]$
13	$e^{-bx}K_\nu(bx) \times\left\{\begin{matrix}\sin(a\sqrt{x})\sinh(a\sqrt{x})\\ \cos(a\sqrt{x})\cosh(a\sqrt{x})\end{matrix}\right\}$	$\frac{\sqrt{\pi}\,a^{2\delta}}{(2b)^{s+\delta}}\Gamma\left[\begin{matrix}s-\nu+\delta,\ s+\nu+\delta\\ \frac{2s+2\delta+1}{2}\end{matrix}\right] \times {}_4F_5\left(\begin{matrix}\frac{s-\nu+\delta}{2},\ \frac{s-\nu+\delta+1}{2},\ \frac{s+\nu+\delta}{2},\ \frac{s+\nu+\delta+1}{2};\ -\frac{a^4}{64b^2}\\ \frac{2\delta+1}{4},\ \frac{2\delta+3}{4},\ \frac{2\delta+1}{2},\ \frac{2s+2\delta+1}{4},\ \frac{2s+2\delta+3}{4}\end{matrix}\right)$ $[\operatorname{Re} b>0;\ \operatorname{Re} s>\lvert\operatorname{Re}\nu\rvert-\delta]$
14	$e^{-bx}K_\nu(bx) \times\left\{\begin{matrix}\sin(a\sqrt{x})\cosh(a\sqrt{x})\\ \cos(a\sqrt{x})\sinh(a\sqrt{x})\end{matrix}\right\}$	$\frac{\sqrt{\pi}\,a}{(2b)^{s+1/2}}\Gamma\left[\begin{matrix}\frac{2s-2\nu+1}{2},\ \frac{2s+2\nu+1}{2}\\ s+1\end{matrix}\right] \times {}_4F_5\left(\begin{matrix}\frac{2s-2\nu+1}{4},\ \frac{2s-2\nu+3}{4},\ \frac{2s+2\nu+1}{4},\ \frac{2s+2\nu+3}{4}\\ \frac{1}{2},\ \frac{3}{4},\ \frac{5}{4},\ \frac{s+1}{2},\ \frac{s+2}{2};\ -\frac{a^4}{64b^2}\end{matrix}\right) \pm \frac{\sqrt{\pi}\,a^3}{3(2b)^{s+3/2}}\Gamma\left[\begin{matrix}\frac{2s-2\nu+3}{2},\ \frac{2s+2\nu+3}{2}\\ s+2\end{matrix}\right] \times {}_4F_5\left(\begin{matrix}\frac{2s-2\nu+3}{4},\ \frac{2s-2\nu+5}{4},\ \frac{2s+2\nu+3}{2},\ \frac{2s+2\nu+5}{4}\\ \frac{5}{4},\ \frac{3}{2},\ \frac{7}{4},\ \frac{s+2}{2},\ \frac{s+3}{2};\ -\frac{a^4}{64b^2}\end{matrix}\right)$ $[\operatorname{Re} b>0;\ \operatorname{Re} s>\lvert\operatorname{Re}\nu\rvert-1/2]$

3.14.5. $K_\nu(ax)$ and the logarithmic function

1	$\ln x\, K_\nu(ax)$	$\dfrac{2^{s-3}}{a^s}\Gamma\left(\dfrac{s-\nu}{2}\right)\Gamma\left(\dfrac{s+\nu}{2}\right)\left[\psi\left(\dfrac{s-\nu}{2}\right)+\psi\left(\dfrac{s+\nu}{2}\right)-2\ln\dfrac{a}{2}\right]$ [$\operatorname{Re} a>0;\ \lvert\operatorname{Re} s\rvert>\operatorname{Re}\nu$]
2	$\ln^n x\, K_\nu(ax)$	$\dfrac{\partial^n}{\partial s^n}\left[\dfrac{2^{s-2}}{a^s}\Gamma\left(\dfrac{s-\nu}{2}\right)\Gamma\left(\dfrac{s+\nu}{2}\right)\right]$ [$\operatorname{Re} a>0;\ \lvert\operatorname{Re} s\rvert>\operatorname{Re}\nu$]

3.14.6. $K_\nu(ax)$ and $\operatorname{Ei}(bx^r)$

1	$\operatorname{Ei}(-ax)\,K_\nu(bx)$	$\dfrac{2^{\nu-1}a^{\nu-s}b^{-\nu}}{\nu-s}\Gamma(\nu)\Gamma(s-\nu)\,{}_3F_2\left(\begin{matrix}\frac{s-\nu}{2},\ \frac{s-\nu}{2},\ \frac{s-\nu+1}{2}\\ 1-\nu,\ \frac{s-\nu+2}{2};\ \frac{b^2}{a^2}\end{matrix}\right) - \dfrac{2^{-\nu-1}a^{-s-\nu}b^{\nu}}{s+\nu}\Gamma(-\nu)\Gamma(s+\nu)\,{}_3F_2\left(\begin{matrix}\frac{s+\nu}{2},\ \frac{s+\nu}{2},\ \frac{s+\nu+1}{2}\\ 1+\nu,\ \frac{s+\nu+2}{2};\ \frac{b^2}{a^2}\end{matrix}\right)$ [$\operatorname{Re} a,\ \operatorname{Re} b>0;\ \operatorname{Re} s>\lvert\operatorname{Re}\nu\rvert$]
2	$\operatorname{Ei}(-ax^2)\,K_\nu(bx)$	$\dfrac{2^{\nu-1}a^{(\nu-s)/2}}{b^\nu(\nu-s)}\Gamma(\nu)\,\Gamma\left(\dfrac{s-\nu}{2}\right){}_2F_2\left(\begin{matrix}\frac{s-\nu}{2},\ \frac{s-\nu}{2};\ \frac{b^2}{4a}\\ 1-\nu,\ \frac{s-\nu+2}{2}\end{matrix}\right) - \dfrac{2^{-\nu-1}b^{\nu}}{a^{(s+\nu)/2}(s+\nu)}\Gamma(-\nu)\,\Gamma\left(\dfrac{s+\nu}{2}\right){}_2F_2\left(\begin{matrix}\frac{s+\nu}{2},\ \frac{s+\nu}{2};\ \frac{b^2}{4a}\\ 1+\nu,\ \frac{s+\nu+2}{2}\end{matrix}\right)$ [$\operatorname{Re} a,\ \operatorname{Re} b>0;\ \operatorname{Re} s>\lvert\operatorname{Re}\nu\rvert$]
3	$e^{\pm ax}\operatorname{Ei}(\mp ax)\,K_\nu(bx)$	$\dfrac{2^{\nu-1}\pi}{a^{s-\nu}b^\nu}\Gamma(\nu)\,\Gamma(s-\nu)\begin{Bmatrix}\csc[(\nu-s)\pi]\\ \cot[(\nu-s)\pi]\end{Bmatrix}{}_2F_1\left(\begin{matrix}\frac{s-\nu}{2},\ \frac{s-\nu+1}{2}\\ 1-\nu;\ \frac{b^2}{a^2}\end{matrix}\right) - \dfrac{\pi b^\nu}{2^{\nu+1}a^{s+\nu}}\Gamma(-\nu)\,\Gamma(s+\nu)\begin{Bmatrix}\csc[(s+\nu)\pi]\\ \cot[(s+\nu)\pi]\end{Bmatrix}{}_2F_1\left(\begin{matrix}\frac{s+\nu}{2},\ \frac{s+\nu+1}{2}\\ \nu+1;\ \frac{b^2}{a^2}\end{matrix}\right) \mp \dfrac{2^{s-3}}{ab^{s-1}}\Gamma\left(\dfrac{s-\nu-1}{2}\right)\Gamma\left(\dfrac{s+\nu-1}{2}\right){}_3F_2\left(\begin{matrix}\frac{1}{2},\ 1,\ 1;\ \frac{b^2}{a^2}\\ \frac{3-s-\nu}{2},\ \frac{3-s+\nu}{2}\end{matrix}\right) + \dfrac{2^{s-4}}{a^2b^{s-2}}\Gamma\left(\dfrac{s-\nu-2}{2}\right)\Gamma\left(\dfrac{s+\nu-2}{2}\right){}_3F_2\left(\begin{matrix}1,\ 1,\ \frac{3}{2};\ \frac{b^2}{a^2}\\ \frac{4-s-\nu}{2},\ \frac{4-s+\nu}{2}\end{matrix}\right)$ [$\operatorname{Re} a,\ \operatorname{Re} b>0;\ \operatorname{Re} s>\lvert\operatorname{Re}\nu\rvert$]
4	$e^{\pm bx}\operatorname{Ei}(-ax)\,K_\nu(bx)$	$\dfrac{2^{\nu-1}a^{\nu-s}}{b^\nu(\nu-s)}\Gamma(\nu)\,\Gamma(s-\nu)\,{}_3F_2\left(\begin{matrix}\frac{1-2\nu}{2},\ s-\nu,\ s-\nu\\ 1-2\nu,\ s-\nu+1;\ \pm\frac{2b}{a}\end{matrix}\right) - \dfrac{2^{-\nu-1}b^\nu}{a^{s+\nu}(s+\nu)}\Gamma(-\nu)\,\Gamma(s+\nu)\,{}_3F_2\left(\begin{matrix}\frac{1+2\nu}{2},\ s+\nu,\ s+\nu\\ 1+2\nu,\ s+\nu+1;\ \pm\frac{2b}{a}\end{matrix}\right)$ [$\operatorname{Re} a,\ \operatorname{Re} b>0;\ \operatorname{Re} s>\lvert\operatorname{Re}\nu\rvert$]

No.	$f(x)$	$F(s)$
5	$e^{(a\mp b)x}\,\mathrm{Ei}(-ax)\,K_\nu(bx)$	$-\frac{2^{\nu-1}\pi}{a^{s-\nu}b^\nu}\Gamma(\nu)\,\Gamma(s-\nu)\csc[(s-\nu)\pi]\;{}_2F_1\left(\begin{matrix}\frac{1-2\nu}{2},\; s-\nu\\ 1-2\nu;\; \pm\frac{2b}{a}\end{matrix}\right)$ $-\frac{\pi b^\nu}{2^{\nu+1}a^{s+\nu}}\Gamma(-\nu)\,\Gamma(s+\nu)\csc[(s+\nu)\pi]$ $\times\,{}_2F_1\left(\begin{matrix}\frac{1+2\nu}{2},\; s+\nu\\ 1+2\nu;\; \pm\frac{2b}{a}\end{matrix}\right)\mp\frac{\sqrt{\pi}}{a\,(2b)^{s-1}}\left[\frac{\cos(\nu\pi)}{\cos(s\pi)}\right]^{(1\mp1)/2}$ $\times\,\Gamma\left[\begin{matrix}s-\nu-1,\; s+\nu-1\\ \frac{2s-1}{2}\end{matrix}\right]{}_3F_2\left(\begin{matrix}1,\;1,\;\frac{3-2s}{2};\;\pm\frac{2b}{a}\\ 2-s-\nu,\;2-s+\nu\end{matrix}\right)$ $\left[\mathrm{Re}\,a>0;\;\mathrm{Re}\,s>\lvert\mathrm{Re}\,\nu\rvert;\;\left\{\begin{matrix}\mathrm{Re}\,b>0\\ \mathrm{Re}\,s<3/2;\;\lvert\arg b\rvert<\pi\end{matrix}\right\}\right]$
6	$e^{\pm(a+b)x}\,\mathrm{Ei}(ax)\,K_\nu(bx)$	$-\frac{2^{\nu-1}\pi}{a^{s-\nu}b^\nu}\Gamma(\nu)\,\Gamma(s-\nu)\cot[(s-\nu)\pi]\;{}_2F_1\left(\begin{matrix}\frac{1-2\nu}{2},\; s-\nu\\ 1-2\nu;\; \mp\frac{2b}{a}\end{matrix}\right)$ $-\frac{\pi b^\nu}{2^{\nu+1}a^{s+\nu}}\Gamma(-\nu)\,\Gamma(s+\nu)\cot[(s+\nu)\pi]$ $\times\,{}_2F_1\left(\begin{matrix}\frac{1+2\nu}{2},\; s+\nu\\ 1+2\nu;\; \mp\frac{2b}{a}\end{matrix}\right)\pm\frac{\sqrt{\pi}}{a\,(2b)^{s-1}}\left[\frac{\cos(\nu\pi)}{\cos(s\pi)}\right]^{(1\mp1)/2}$ $\times\,\Gamma\left[\begin{matrix}s-\nu-1,\; s+\nu-1\\ \frac{2s-1}{2}\end{matrix}\right]{}_3F_2\left(\begin{matrix}1,\;1,\;\frac{3-2s}{2};\;\mp\frac{2b}{a}\\ 2-s-\nu,\;2-s+\nu\end{matrix}\right)$ $\left[\mathrm{Re}\,a>0;\;\mathrm{Re}\,s>\lvert\mathrm{Re}\,\nu\rvert;\;\left\{\begin{matrix}\mathrm{Re}\,b>0\\ \mathrm{Re}\,s<3/2;\;\lvert\arg b\rvert<\pi\end{matrix}\right\}\right]$

3.14.7. $K_\nu(ax)$ and $\mathrm{Si}(bx)$, $\mathrm{si}(bx)$, $\mathrm{ci}(bx)$

1	$\mathrm{Si}(ax)\,K_\nu(bx)$	$\frac{2^{s-1}a}{b^{s+1}}\Gamma\left(\frac{s-\nu+1}{2}\right)\Gamma\left(\frac{s+\nu+1}{2}\right){}_3F_2\left(\begin{matrix}\frac{1}{2},\;\frac{s-\nu+1}{2},\;\frac{s+\nu+1}{2}\\ \frac{3}{2},\;\frac{3}{2};\;-\frac{a^2}{b^2}\end{matrix}\right)$ $[\mathrm{Re}\,s>\lvert\mathrm{Re}\,\nu\rvert-1;\;\mathrm{Re}\,b>\lvert\mathrm{Im}\,a\rvert]$
2	$\left\{\begin{matrix}\mathrm{si}(ax)\\ \mathrm{ci}(ax)\end{matrix}\right\}K_\nu(bx)$	$-\frac{2^{\nu-1}a^{\nu-s}}{b^s(s-\nu)}\Gamma(\nu)\,\Gamma(s-\nu)\left\{\begin{matrix}\sin[(s-\nu)\pi/2]\\ \cos[(s-\nu)\pi/2]\end{matrix}\right\}$ $\times\,{}_3F_2\left(\begin{matrix}\frac{s-\nu}{2},\;\frac{s-\nu}{2},\;\frac{s-\nu+1}{2}\\ 1-\nu,\;\frac{s-\nu+2}{2};\;-\frac{b^2}{a^2}\end{matrix}\right)-\frac{2^{-\nu-1}b^\nu}{a^{s+\nu}(s+\nu)}$ $\times\left\{\begin{matrix}\sin[(s+\nu)\pi/2]\\ \cos[(s+\nu)\pi/2]\end{matrix}\right\}{}_3F_2\left(\begin{matrix}\frac{s+\nu}{2},\;\frac{s+\nu}{2},\;\frac{s+\nu+1}{2}\\ 1+\nu,\;\frac{s+\nu+2}{2};\;-\frac{b^2}{a^2}\end{matrix}\right)$ $[a,\;\mathrm{Re}\,b>0;\;\mathrm{Re}\,s>\lvert\mathrm{Re}\,\nu\rvert]$

No.	$f(x)$	$F(s)$
3	$e^{-bx}\,\mathrm{si}(ax)\,K_\nu(bx)$	$\frac{\sqrt{\pi}\,a}{(2b)^{s+1}}\,\Gamma\left[\begin{matrix}s-\nu+1,\ s+\nu+1\\ \frac{2s+3}{2}\end{matrix}\right] \times {}_5F_4\left(\begin{matrix}\frac{1}{2},\ \frac{s-\nu+1}{2},\ \frac{s-\nu+2}{2},\ \frac{s+\nu+1}{2},\ \frac{s+\nu+2}{2}\\ \frac{3}{2},\ \frac{3}{2},\ \frac{s+3}{2},\ \frac{s+5}{2};\ -\frac{a^2}{4b^2}\end{matrix}\right) - \frac{\pi^{3/2}}{2^{s+1}b^s}\,\Gamma\left[\begin{matrix}s-\nu,\ s+\nu\\ \frac{2s+1}{2}\end{matrix}\right]$ $[a,\ \mathrm{Re}\,b>0;\ \mathrm{Re}\,s>\lvert\mathrm{Re}\,\nu\rvert]$
4	$e^{-bx}\,\mathrm{ci}(ax)\,K_\nu(bx)$	$\frac{\pi^{3/2}}{2^{s+1}b^s}\,\Gamma\left[\begin{matrix}s-\nu,\ s+\nu\\ \frac{2s+1}{2}\end{matrix}\right]\left[\psi(s-\nu)+\psi(s+\nu) - \psi\left(s+\frac{1}{2}\right)-\ln\frac{2b}{a}+\mathbf{C}\right] - \frac{\sqrt{\pi}\,a^2}{2^{s+4}b^{s+2}}\,\Gamma\left[\begin{matrix}s-\nu+2,\ s+\nu+2\\ \frac{2s+5}{2}\end{matrix}\right] \times {}_6F_5\left(\begin{matrix}1,\ 1,\ \frac{s-\nu+2}{2},\ \frac{s-\nu+3}{2},\ \frac{s+\nu+2}{2},\ \frac{s+\nu+3}{2}\\ \frac{3}{2},\ 2,\ 2,\ \frac{2s+5}{4},\ \frac{2s+7}{4};\ -\frac{a^2}{4b^2}\end{matrix}\right)$ $[a,\ \mathrm{Re}\,b>0;\ \mathrm{Re}\,s>\lvert\mathrm{Re}\,\nu\rvert]$
5	$e^{bx}\,\mathrm{si}(ax)\,K_\nu(bx)$	$\frac{a\cos(\pi\nu)}{2^{s+1}\sqrt{\pi}\,b^{s+1}}\,\Gamma\left[-s-\frac{1}{2},\ s-\nu+1,\ s+\nu+1\right] \times {}_5F_4\left(\begin{matrix}\frac{1}{2},\ \frac{s-\nu+1}{2},\ \frac{s-\nu+2}{2},\ \frac{s+\nu+1}{2},\ \frac{s+\nu+2}{2}\\ \frac{3}{2},\ \frac{3}{2},\ \frac{2s+3}{4},\ \frac{2s+5}{4};\ -\frac{a^2}{4b^2}\end{matrix}\right) + \frac{\sqrt{2\pi}}{(2s-1)\,a^{s-1/2}\sqrt{b}}\cos\frac{(2s+1)\pi}{4}\,\Gamma\left(s-\frac{1}{2}\right) \times {}_5F_4\left(\begin{matrix}\frac{1-2s}{4},\ \frac{1-2\nu}{4},\ \frac{3-2\nu}{4},\ \frac{2\nu+1}{4},\ \frac{2\nu+3}{4}\\ \frac{1}{2},\ \frac{3-2s}{4},\ \frac{5-2s}{4},\ \frac{5-2s}{4};\ -\frac{a^2}{4b^2}\end{matrix}\right) + \frac{\sqrt{2\pi}\,(4\nu^2-1)}{8\,(2s-3)\,a^{s-3/2}b^{3/2}}\sin\frac{(2s+1)\pi}{4}\,\Gamma\left(s-\frac{3}{2}\right) \times {}_5F_4\left(\begin{matrix}\frac{3-2s}{4},\ \frac{3-2\nu}{4},\ \frac{5-2\nu}{4},\ \frac{2\nu+3}{4},\ \frac{2\nu+5}{4}\\ \frac{3}{2},\ \frac{5-2s}{4},\ \frac{7-2s}{4},\ \frac{7-2s}{4};\ -\frac{a^2}{4b^2}\end{matrix}\right) - \frac{\sqrt{\pi}\,\cos(\pi\nu)}{2^{s+1}b^s}\,\Gamma\left[\frac{1}{2}-s,\ s-\nu,\ s+\nu\right]$ $[a,\ \mathrm{Re}\,b>0;\ \lvert\mathrm{Re}\,\nu\rvert<\mathrm{Re}\,s<3/2]$
6	$e^{bx}\,\mathrm{ci}(ax)\,K_\nu(bx)$	$-\frac{a^2\cos(\pi\nu)}{2^{s+4}\sqrt{\pi}\,b^{s+2}}\,\Gamma\left[-s-\frac{3}{2},\ s-\nu+2,\ s+\nu+2\right] \times {}_6F_5\left(\begin{matrix}1,\ 1,\ \frac{s-\nu+2}{2},\ \frac{s-\nu+3}{2},\ \frac{s+\nu+2}{2},\ \frac{s+\nu+3}{2}\\ \frac{3}{2},\ 2,\ 2,\ \frac{2s+5}{4},\ \frac{2s+7}{4};\ -\frac{a^2}{4b^2}\end{matrix}\right) -$

No.	$f(x)$	$F(s)$
		$-\frac{\sqrt{2\pi}}{(2s-1)\,a^{s-1/2}\sqrt{b}}\sin\frac{(2s+1)\,\pi}{4}\,\Gamma\left(s-\frac{1}{2}\right)$ $\times\,{}_5F_4\left(\begin{matrix}\frac{1-2s}{4},\ \frac{1-2\nu}{4},\ \frac{3-2\nu}{4},\ \frac{2\nu+1}{4},\ \frac{2\nu+3}{4}\\ \frac{1}{2},\ \frac{3-2s}{4},\ \frac{5-2s}{4},\ \frac{5-2s}{4};\ -\frac{a^2}{4b^2}\end{matrix}\right)$ $+\frac{\sqrt{\pi}\,(4\nu^2-1)}{2^{5/2}\,(2s-3)\,a^{s-3/2}b^{3/2}}\cos\frac{(2s+1)\,\pi}{4}\,\Gamma\left(s-\frac{3}{2}\right)\times$ $\times\,{}_5F_4\left(\begin{matrix}\frac{3-2s}{4},\ \frac{3-2\nu}{4},\ \frac{5-2\nu}{4},\ \frac{2\nu+3}{4},\ \frac{2\nu+5}{4}\\ \frac{3}{2},\ \frac{5-2s}{4},\ \frac{7-2s}{4},\ \frac{7-2s}{4};\ -\frac{a^2}{4b^2}\end{matrix}\right)$ $+\frac{\cos(\pi\nu)}{\sqrt{\pi}\,(2b)^s}\,\Gamma\left[\frac{1}{2}-s,\ s-\nu,\ s+\nu\right]$ $\times\left[\psi(s-\nu)+\psi(s+\nu)-\psi\left(\frac{1}{2}-s\right)+\ln\frac{a}{2b}+\mathbf{C}\right]$ $[a,\ \operatorname{Re}b>0;\ \lvert\operatorname{Re}\nu\rvert<\operatorname{Re}s<3/2]$

3.14.8. $K_\nu(ax)$ and $\operatorname{erf}(bx^r)$, $\operatorname{erfi}(bx^r)$, $\operatorname{erfc}(bx^r)$

1	$\left\{\begin{matrix}\operatorname{erf}(ax)\\ \operatorname{erfc}(ax)\end{matrix}\right\}K_\nu(bx)$	$\pm\frac{2^{\nu-1}a^{\nu-s}}{\sqrt{\pi}\,b^\nu\,(\nu-s)}\,\Gamma(\nu)\,\Gamma\left(\frac{s-\nu+1}{2}\right)$ $\times\,{}_2F_2\left(\begin{matrix}\frac{s-\nu}{2},\ \frac{s-\nu+1}{2};\ \frac{b^2}{4a^2}\\ 1-\nu,\ \frac{s-\nu+2}{2}\end{matrix}\right)\mp\frac{2^{-\nu-1}\,b^\nu}{\sqrt{\pi}\,a^{s+\nu}\,(\nu+s)}\,\Gamma(-\nu)$ $\times\,\Gamma\left(\frac{s+\nu+1}{2}\right)\,{}_2F_2\left(\begin{matrix}\frac{s+\nu}{2},\ \frac{s+\nu+1}{2};\ \frac{b^2}{4a^2}\\ 1+\nu,\ \frac{s+\nu+2}{2}\end{matrix}\right)$ $+\frac{(1\pm1)\,2^{s-3}}{b^s}\,\Gamma\left(\frac{s-\nu}{2}\right)\Gamma\left(\frac{s+\nu}{2}\right)$ $[\operatorname{Re}b>0;\ \operatorname{Re}s>\lvert\operatorname{Re}\nu\rvert-(1\pm1)/2;\ \lvert\arg a\rvert<\pi/4]$
2	$\left\{\begin{matrix}\operatorname{erf}(a\sqrt{x})\\ \operatorname{erfc}(a\sqrt{x})\end{matrix}\right\}K_\nu(bx)$	$\pm\frac{2^{s-1/2}a}{\sqrt{\pi}\,b^{s+1/2}}\,\Gamma\left(\frac{2s-2\nu+1}{4}\right)$ $\times\,\Gamma\left(\frac{2s+2\nu+1}{4}\right)\,{}_3F_2\left(\begin{matrix}\frac{1}{4},\ \frac{2s-2\nu+1}{4},\ \frac{2s+2\nu+1}{4}\\ \frac{1}{2},\ \frac{5}{4};\ \frac{a^4}{b^2}\end{matrix}\right)$ $\mp\frac{2^{s+1/2}a^3}{3\sqrt{\pi}\,b^{s+3/2}}\,\Gamma\left(\frac{2s-2\nu+3}{4}\right)$ $\times\,\Gamma\left(\frac{2s+2\nu+3}{4}\right)\,{}_3F_2\left(\begin{matrix}\frac{3}{4},\ \frac{2s-2\nu+3}{4},\ \frac{2s+2\nu+3}{4}\\ \frac{3}{2},\ \frac{7}{4};\ \frac{a^4}{b^2}\end{matrix}\right)$ $+\frac{(1\mp1)\,2^{s-3}}{b^s}\,\Gamma\left(\frac{s-\nu}{2}\right)\Gamma\left(\frac{s+\nu}{2}\right)$ $[\operatorname{Re}b>0;\ \operatorname{Re}s>\lvert\operatorname{Re}\nu\rvert-(1\pm1)/4;\ \lvert\arg a\rvert<\pi/4]$

No.	$f(x)$	$F(s)$
3	$e^{\pm bx}\operatorname{erf}(a\sqrt{x})\,K_\nu(bx)$	$\mp\frac{2a}{(2b)^{s+1/2}}\left(\frac{\cos(\nu\pi)}{\sin(s\pi)}\right)^{(1\pm1)/2}\Gamma\left[\begin{matrix}\frac{2s-2\nu+1}{2},\ \frac{2s+2\nu+1}{2}\\ s+1\end{matrix}\right]$ $\times\,{}_3F_2\left(\begin{matrix}\frac{1}{2},\ \frac{2s-2\nu+1}{2},\ \frac{2s+2\nu+1}{2}\\ \frac{3}{2},\ s+1;\ \pm\frac{a^2}{2b}\end{matrix}\right)$ $+\frac{(1\pm1)\,a^{1-2s}\,\Gamma(s)}{\sqrt{2b}\,(1-2s)}\,{}_3F_2\left(\begin{matrix}\frac{1-2\nu}{2},\ \frac{1+2\nu}{2},\ \frac{1-2s}{2}\\ 1-s,\ \frac{3-2s}{2};\ \frac{a^2}{2b}\end{matrix}\right)$ $\left[\operatorname{Re}b>0;\ \lvert\arg a\rvert<\frac{\pi}{4};\ \left\{\begin{matrix}\lvert\operatorname{Re}\nu\rvert-1/2<\operatorname{Re}s<1/2\\ \lvert\operatorname{Re}\nu\rvert-1/2<\operatorname{Re}s\end{matrix}\right\}\right]$
4	$e^{\pm bx}\operatorname{erfc}(a\sqrt{x})\,K_\nu(bx)$	$\pm\frac{2a}{(2b)^{s+1/2}}\left(\frac{\cos(\nu\pi)}{\sin(s\pi)}\right)^{(1\pm1)/2}\Gamma\left[\begin{matrix}\frac{2s-2\nu+1}{2},\ \frac{2s+2\nu+1}{2}\\ s+1\end{matrix}\right]$ $\times\,{}_3F_2\left(\begin{matrix}\frac{1}{2},\ \frac{2s-2\nu+1}{2},\ \frac{2s+2\nu+1}{2}\\ \frac{3}{2},\ s+1;\ \pm\frac{a^2}{2b}\end{matrix}\right)$ $-\frac{(1\pm1)\,a^{1-2s}\,\Gamma(s)}{\sqrt{2b}\,(1-2s)}\,{}_3F_2\left(\begin{matrix}\frac{1-2\nu}{2},\ \frac{1+2\nu}{2},\ \frac{1-2s}{2}\\ 1-s,\ \frac{3-2s}{2};\ \frac{a^2}{2b}\end{matrix}\right)$ $+\frac{\sqrt{\pi}}{(2b)^s}\,\Gamma\left[\begin{matrix}s-\nu,\ s+\nu\\ \frac{2s+1}{2}\end{matrix}\right]\left(\frac{\cos(\nu\pi)}{\cos(s\pi)}\right)^{(1\pm1)/2}$ $[\operatorname{Re}b>0;\ \operatorname{Re}s>\lvert\operatorname{Re}\nu\rvert;\ \lvert\arg a\rvert<\pi/4]$
5	$e^{a^2x}\operatorname{erf}(a\sqrt{x})\,K_\nu(bx)$	$\frac{2^{s-1/2}ab^{-s-1/2}}{\sqrt{\pi}}\,\Gamma\left(\frac{2s-2\nu+1}{4}\right)\Gamma\left(\frac{2s+2\nu+1}{4}\right)$ $\times\,{}_3F_2\left(\begin{matrix}1,\ \frac{2s-2\nu+1}{4},\ \frac{2s+2\nu+1}{4}\\ \frac{3}{4},\ \frac{5}{4};\ \frac{a^4}{b^2}\end{matrix}\right)$ $+\frac{2^{s+3/2}a^3b^{-s-3/2}}{3\sqrt{\pi}}\,\Gamma\left(\frac{2s-2\nu+3}{4}\right)$ $\times\Gamma\left(\frac{2s+2\nu+3}{4}\right)\,{}_3F_2\left(\begin{matrix}1,\ \frac{2s-2\nu+3}{4},\ \frac{2s+2\nu+3}{4}\\ \frac{5}{4},\ \frac{7}{4};\ \frac{a^4}{b^2}\end{matrix}\right)$ $\left[\operatorname{Re}b,\ \operatorname{Re}\left(b-a^2\right)>0;\ \operatorname{Re}s>\lvert\operatorname{Re}\nu\rvert-1/2\right]$
6	$e^{-(a^2+b)x}\operatorname{erfi}(a\sqrt{x})$ $\times K_\nu(bx)$	$\frac{a}{2^{s-1/2}b^{s+1/2}}\,\Gamma\left[\begin{matrix}\frac{2s-2\nu+1}{2},\ \frac{2s+2\nu+1}{2}\\ s+1\end{matrix}\right]{}_3F_2\left(\begin{matrix}1,\ \frac{2s-2\nu+1}{2},\ \frac{2s+2\nu+1}{2}\\ \frac{3}{2},\ s+1;\ -\frac{a^2}{2b}\end{matrix}\right)$ $[\operatorname{Re}b>0;\ \operatorname{Re}s>\lvert\operatorname{Re}\nu\rvert-1/2;\ \lvert\arg a\rvert<\pi/4]$
7	$e^{(a^2-b)x}\operatorname{erf}(a\sqrt{x})$ $\times K_\nu(bx)$	$\frac{ab^{-s-1/2}}{2^{s-1/2}}\,\Gamma\left[\begin{matrix}\frac{2s-2\nu+1}{2},\ \frac{2s+2\nu+1}{2}\\ s+1\end{matrix}\right]{}_3F_2\left(\begin{matrix}1,\ \frac{2s-2\nu+1}{2},\ \frac{2s+2\nu+1}{2}\\ \frac{3}{2},\ s+1;\ \frac{a^2}{2b}\end{matrix}\right)$ $[\operatorname{Re}b>0;\ \operatorname{Re}s>\lvert\operatorname{Re}\nu\rvert-1/2;\ \lvert\arg a\rvert<3\pi/4]$

No.	$f(x)$	$F(s)$
8	$e^{(a^2-b)x}\operatorname{erfc}(a\sqrt{x})$ $\times K_\nu(bx)$	$-\frac{a}{2^{s-1/2}b^{s+1/2}}\Gamma\left[\begin{matrix}\frac{2s-2\nu+1}{2},\ \frac{2s+2\nu+1}{2}\\ s+1\end{matrix}\right]{}_3F_2\left(\begin{matrix}1,\ \frac{2s-2\nu+1}{2},\ \frac{2s+2\nu+1}{2}\\ \frac{3}{2},\ s+1;\ \frac{a^2}{2b}\end{matrix}\right)$ $+\frac{\sqrt{\pi}}{(2b)^s}\Gamma\left[\begin{matrix}s-\nu,\ s+\nu\\ \frac{2s+1}{2}\end{matrix}\right]{}_2F_1\left(\begin{matrix}s-\nu,\ s+\nu\\ \frac{2s+1}{2};\ \frac{a^2}{2b}\end{matrix}\right)$ $[\operatorname{Re}b>0;\ \operatorname{Re}s>\lvert\operatorname{Re}\nu\rvert;\ \lvert\arg a\rvert<3\pi/4]$
9	$\left\{\begin{matrix}\operatorname{erfi}(a\sqrt{x})\\ \operatorname{erfc}(a\sqrt{x})\end{matrix}\right\}$ $\times e^{(\mp a^2+b)x}K_\nu(bx)$	$\frac{(2b)^{1/2-s}}{a}\frac{\cos(\nu\pi)}{\sin(s\pi)}\Gamma\left[\begin{matrix}\frac{2s-2\nu-1}{2},\ \frac{2s+2\nu-1}{2}\\ s\end{matrix}\right]$ $\times{}_3F_2\left(\begin{matrix}\frac{1}{2},\ 1,\ 1-s;\ \frac{2b}{a^2}\\ \frac{3-2s-2\nu}{2},\ \frac{3-2s+2\nu}{2}\end{matrix}\right)+\frac{\pi b^\nu}{2^{\nu+(1\pm1)/2}a^{2s+2\nu}}$ $\times\Gamma\left[\begin{matrix}-\nu\\ 1-s-\nu\end{matrix}\right]\left\{\begin{matrix}\sec[(s+\nu)\pi]\\ \csc[2(s+\nu)\pi]\end{matrix}\right\}{}_2F_1\left(\begin{matrix}\frac{1+2\nu}{2},\ s+\nu\\ 1+2\nu;\ \frac{2b}{a^2}\end{matrix}\right)$ $+\frac{2^{\nu-(1\pm1)/2}}{a^{2s-2\nu}b^\nu}\Gamma\left[\begin{matrix}\nu\\ 1-s+\nu\end{matrix}\right]\left\{\begin{matrix}\sec[(s-\nu)\pi]\\ \csc[2(s-\nu)\pi]\end{matrix}\right\}{}_2F_1\left(\begin{matrix}\frac{1-2\nu}{2},\ s-\nu\\ 1-2\nu;\ \frac{2b}{a^2}\end{matrix}\right)$ $\left[\begin{matrix}\operatorname{Re}b>0;\ \lvert\operatorname{Re}\nu\rvert-(1\pm1)/4<\operatorname{Re}s<1;\\ \lvert\arg a\rvert<(2\mp1)\pi/4\end{matrix}\right]$
10	$\operatorname{erf}(a\sqrt{x})\operatorname{erfi}(a\sqrt{x})$ $\times K_\nu(bx)$	$\frac{2^{s+1}a^2b^{-s-1}}{\pi}\Gamma\left(\frac{s-\nu+1}{2}\right)\Gamma\left(\frac{s+\nu+1}{2}\right)$ $\times{}_4F_3\left(\begin{matrix}\frac{1}{2},1,\ \frac{s-\nu+1}{2},\ \frac{s+\nu+1}{2}\\ \frac{3}{4},\ \frac{5}{4},\ \frac{3}{2};\ \frac{a^4}{b^2}\end{matrix}\right)$ $[\operatorname{Re}b,\ \operatorname{Re}(b-2a^2)>0;\ \operatorname{Re}s>\lvert\operatorname{Re}\nu\rvert-1]$

3.14.9. $K_\nu(ax)$ and $S(bx)$, $C(bx)$

Notation: $\delta=\left\{\begin{matrix}1\\0\end{matrix}\right\}$.

1	$\left\{\begin{matrix}S(ax)\\ C(ax)\end{matrix}\right\}K_\nu(bx)$	$\frac{2^{s+\delta-1}a^{\delta+1/2}}{3^\delta\sqrt{\pi}\,b^{s+\delta+1/2}}\Gamma\left(\frac{2s-2\nu+2\delta+1}{4}\right)\Gamma\left(\frac{2s+2\nu+2\delta+1}{4}\right)$ $\times{}_3F_2\left(\begin{matrix}\frac{2\delta+1}{4},\ \frac{2s-2\nu+2\delta+1}{4},\ \frac{2s+2\nu+2\delta+1}{4}\\ \frac{2\delta+1}{2},\ \frac{2\delta+5}{4};\ -\frac{a^2}{b^2}\end{matrix}\right)$ $[a,\ \operatorname{Re}b>0;\ \operatorname{Re}s>\lvert\operatorname{Re}\nu\rvert-(2\pm1)/2]$
2	$e^{-bx}\left\{\begin{matrix}S(ax)\\ C(ax)\end{matrix}\right\}K_\nu(bx)$	$\frac{\sqrt{2}\,a^{\delta+1/2}}{(2\delta+1)(2b)^{s+\delta+1/2}}\Gamma\left[\begin{matrix}\frac{2s-2\nu+2\delta+1}{2},\ \frac{2s+2\nu+2\delta+1}{2}\\ s+\delta+1\end{matrix}\right]$ $\times{}_5F_4\left(\begin{matrix}\frac{2\delta+1}{4},\ \frac{2s-2\nu+3}{4},\ \frac{2s+2\nu+3}{4},\ \frac{2s-2\nu+4\delta+1}{4},\ \frac{2s+2\nu+4\delta+1}{4}\\ \frac{2\delta+1}{2},\ \frac{2\delta+5}{4},\ \frac{s+2\delta+1}{2},\ \frac{s+2}{2};\ -\frac{a^2}{4b^2}\end{matrix}\right)$ $[a,\ \operatorname{Re}b>0;\ \operatorname{Re}s>\lvert\operatorname{Re}\nu\rvert-(2\pm1)/2]$

No.	$f(x)$	$F(s)$
3	$e^{bx}\begin{Bmatrix} S(ax) \\ C(ax) \end{Bmatrix} K_\nu(bx)$	$\pm\frac{2^{-s-\delta}a^{\delta+1/2}b^{-s-\delta-1/2}}{\pi^\delta(2\delta+1)}\frac{\cos(\nu\pi)}{\sin(s\pi)}\Gamma\left[\begin{matrix}\frac{2s-2\nu+2\delta+1}{2},\ \frac{2s+2\nu+2\delta+1}{2}\\ s+\delta+1\end{matrix}\right] \times {}_5F_4\left(\begin{matrix}\frac{2\delta+1}{4},\ \frac{2s-2\nu+3}{4},\ \frac{2s+2\nu+3}{4},\ \frac{2s-2\nu+4\delta+1}{4},\ \frac{2s+2\nu+4\delta+1}{4}\\ \frac{2\delta+1}{2},\ \frac{2\delta+5}{4},\ \frac{s+2\delta+1}{2},\ \frac{s+2}{2};\ -\frac{a^2}{4b^2}\end{matrix}\right) + \frac{a^{1/2-s}}{\sqrt{b}(1-2s)}\begin{Bmatrix}\sin(s\pi/2)\\ \cos(s\pi/2)\end{Bmatrix}\Gamma(s) \times {}_5F_4\left(\begin{matrix}\frac{1-2\nu}{4},\ \frac{3-2\nu}{4},\ \frac{2\nu+1}{4},\ \frac{2\nu+3}{4},\ \frac{1-2s}{4}\\ \frac{1}{2},\ \frac{1-s}{2},\ \frac{2-s}{2},\ \frac{5-2s}{4};\ -\frac{a^2}{4b^2}\end{matrix}\right) \pm\frac{(4\nu^2-1)\,a^{3/2-s}}{8b^{3/2}(2s-3)}\begin{Bmatrix}\cos(s\pi/2)\\ \sin(s\pi/2)\end{Bmatrix}\Gamma(s-1) \times {}_5F_4\left(\begin{matrix}\frac{3-2\nu}{4},\ \frac{5-2\nu}{4},\ \frac{2\nu+3}{4},\ \frac{2\nu+5}{4},\ \frac{3-2s}{4}\\ \frac{3}{2},\ \frac{2-s}{2},\ \frac{3-s}{2},\ \frac{7-2s}{4};\ -\frac{a^2}{4b^2}\end{matrix}\right)$ $[a>0;\ \lvert\operatorname{Re}\nu\rvert-(2\pm1)/2<\operatorname{Re}s<1/2;\ \lvert\arg b\rvert<\pi]$

3.14.10. $K_\nu(ax)$ and $\Gamma(\mu,\, bx)$, $\gamma(\mu,\, bx)$

1	$\begin{Bmatrix}\gamma(\mu,\,ax)\\ \Gamma(\mu,\,ax)\end{Bmatrix}K_\nu(bx)$	$\pm\frac{2^{s+\mu-2}a^\mu}{\mu\, b^{s+\mu}}\Gamma\left(\frac{s+\mu-\nu}{2}\right)\Gamma\left(\frac{s+\mu+\nu}{2}\right) \times {}_3F_2\left(\begin{matrix}\frac{\mu}{2},\ \frac{s+\mu-\nu}{2},\ \frac{s+\mu+\nu}{2}\\ \frac{1}{2},\ \frac{\mu+2}{2};\ \frac{a^2}{b^2}\end{matrix}\right) \mp\frac{2^{s+\mu-1}a^{\mu+1}}{(\mu+1)\,b^{s+\mu+1}} \times\Gamma\left(\frac{s+\mu-\nu+1}{2}\right)\Gamma\left(\frac{s+\mu+\nu+1}{2}\right) \times {}_3F_2\left(\begin{matrix}\frac{\mu+1}{2},\ \frac{s+\mu-\nu+1}{2},\ \frac{s+\mu+\nu+1}{2}\\ \frac{3}{2},\ \frac{\mu+3}{2};\ \frac{a^2}{b^2}\end{matrix}\right) + 2^{s-3}\frac{1\pm1}{b^s}\Gamma\left[\mu,\ \frac{s+\nu}{2},\ \frac{s-\nu}{2}\right]$ $\left[\operatorname{Re}a,\ \operatorname{Re}b>0;\ \operatorname{Re}(s+\mu)>\lvert\operatorname{Re}\nu\rvert;\ \begin{Bmatrix}\operatorname{Re}\mu>0\\ \operatorname{Re}s>\lvert\operatorname{Re}\nu\rvert\end{Bmatrix}\right]$
2	$\begin{Bmatrix}\gamma(\mu,\,ax^2)\\ \Gamma(\mu,\,ax^2)\end{Bmatrix}K_\nu(bx)$	$\pm\frac{2^{\nu-1}a^{(\nu-s)/2}}{(\nu-s)\,b^\nu}\Gamma(\nu)\,\Gamma\left(\frac{s+2\mu-\nu}{2}\right) \times {}_2F_2\left(\begin{matrix}\frac{s-\nu}{2},\ \frac{s+2\mu-\nu}{2};\ \frac{b^2}{4a}\\ 1-\nu,\ \frac{s-\nu+2}{2}\end{matrix}\right) \mp\frac{2^{-\nu-1}b^\nu}{(\nu+s)\,a^{(s+\nu)/2}}\Gamma(-\nu) \times\Gamma\left(\frac{s+2\mu+\nu}{2}\right){}_2F_2\left(\begin{matrix}\frac{s+\nu}{2},\ \frac{s+2\mu+\nu}{2};\ \frac{b^2}{4a}\\ 1+\nu,\ \frac{s+\nu+2}{2}\end{matrix}\right) + 2^{s-3}\frac{1\pm1}{b^s}\Gamma\left[\mu,\ \frac{s+\nu}{2},\ \frac{s-\nu}{2}\right]$ $\left[\operatorname{Re}a,\ \operatorname{Re}b>0;\ \operatorname{Re}(s+2\mu)>\lvert\operatorname{Re}\nu\rvert;\ \begin{Bmatrix}\operatorname{Re}\mu>0\\ \operatorname{Re}s>\lvert\operatorname{Re}\nu\rvert\end{Bmatrix}\right]$

No.	$f(x)$	$F(s)$
3	$e^{-bx}\left\{\begin{matrix}\gamma(\mu,\, ax)\\ \Gamma(\mu,\, ax)\end{matrix}\right\} \times K_\nu(bx)$	$\pm\frac{\sqrt{\pi}\, a^\mu}{\mu\,(2b)^{s+\mu}}\,\Gamma\left[\begin{matrix}s+\mu-\nu,\; s+\mu+\nu\\ \frac{2s+2\mu+1}{2}\end{matrix}\right] \times {}_3F_2\left(\begin{matrix}\mu,\; s+\mu-\nu,\; s+\mu+\nu\\ \mu+1,\; \frac{2s+2\mu+1}{2};\; -\frac{a}{2b}\end{matrix}\right) + \frac{(1\mp 1)\sqrt{\pi}}{2^{s+1}b^s}\,\Gamma\left[\begin{matrix}\mu,\; s+\nu,\; s-\nu\\ \frac{2s+1}{2}\end{matrix}\right]$ $\left[\operatorname{Re} a,\ \operatorname{Re} b>0;\ \operatorname{Re}(s+\mu)>\lvert\operatorname{Re}\nu\rvert;\ \left\{\begin{matrix}\operatorname{Re}\mu>0\\ \operatorname{Re} s>\lvert\operatorname{Re}\nu\rvert\end{matrix}\right\}\right]$
4	$e^{bx}\gamma(\mu,\, ax)\, K_\nu(bx)$	$\frac{a^\mu\cos(\nu\pi)}{\sqrt{\pi}\,\mu\,(2b)^{s+\mu}}\,\Gamma\left[-\frac{2s+2\mu-1}{2},\; s+\mu-\nu,\; s+\mu+\nu\right] \times {}_3F_2\left(\begin{matrix}\mu,\; s+\mu-\nu,\; s+\mu+\nu\\ \mu+1,\; \frac{2s+2\mu+1}{2};\; \frac{a}{2b}\end{matrix}\right) + \frac{a^{1/2-s}}{1-2s}\sqrt{\frac{2\pi}{b}} \times \Gamma\left(\frac{2s+2\mu-1}{2}\right)\, {}_3F_2\left(\begin{matrix}\frac{1+2\nu}{2},\; \frac{1-2\nu}{2},\; \frac{1-2s}{2}\\ \frac{3-2s}{2},\; \frac{3-2s-2\mu}{2};\; \frac{a}{2b}\end{matrix}\right)$ $[\operatorname{Re} a,\ \operatorname{Re} b,\ \operatorname{Re}\mu>0;\ \operatorname{Re}(s+\mu)>\lvert\operatorname{Re}\nu\rvert;\ \operatorname{Re} s<1/2]$
5	$e^{bx}\,\Gamma(\mu,\, ax)\, K_\nu(bx)$	$\frac{2^{\nu-1}a^{\nu-s}}{(s-\nu)\,b^\nu}\,\Gamma(\nu)\,\Gamma(s+\mu-\nu)\; {}_3F_2\left(\begin{matrix}\frac{1-2\nu}{2},\; s-\nu,\; s+\mu-\nu\\ 1-2\nu,\; s-\nu+1;\; \frac{2b}{a}\end{matrix}\right) + \frac{2^{-\nu-1}a^{-\nu-s}}{(s+\nu)\,b^{-\nu}}\,\Gamma(-\nu)\,\Gamma(s+\mu+\nu) \times {}_3F_2\left(\begin{matrix}\frac{1+2\nu}{2},\; s+\nu,\; s+\mu+\nu\\ 1+2\nu,\; s+\nu+1;\; \frac{2b}{a}\end{matrix}\right)$ $[\operatorname{Re} a,\ \operatorname{Re} b>0;\ \operatorname{Re}(s+\mu)>\lvert\operatorname{Re}\nu\rvert,\ \operatorname{Re} s>\lvert\operatorname{Re}\nu\rvert]$
6	$e^{(a\pm b)x}\,\Gamma(\mu,\, ax)\, K_\nu(bx)$	$\frac{\sqrt{\pi}}{(2b)^s}\,[\cos(\nu\pi)\sec(s\pi)]^{(1\pm 1)/2}\,\Gamma\left[\begin{matrix}\mu,\; s-\nu,\; s+\nu\\ \frac{2s+1}{2}\end{matrix}\right] \times {}_2F_1\left(\begin{matrix}s-\nu,\; s+\nu\\ \frac{2s+1}{2};\; \mp\frac{a}{2b}\end{matrix}\right) - \frac{\sqrt{\pi}\, a^\mu}{\mu\,(2b)^{s+\mu}} \times [\cos(\nu\pi)\sec[(s+\mu)\pi]]^{(1\pm 1)/2}\,\Gamma\left[\begin{matrix}s+\mu-\nu,\; s+\mu+\nu\\ \frac{2s+2\mu+1}{2}\end{matrix}\right] \times {}_3F_2\left(\begin{matrix}1,\; s+\mu-\nu,\; s+\mu+\nu\\ \mu+1,\; \frac{2s+2\mu+1}{2};\; \mp\frac{a}{2b}\end{matrix}\right) - \frac{(1\pm 1)\,\pi^{3/2}\,a^{1/2-s}}{2\sqrt{2b}\,\cos[(s+\mu)\pi]}\,\Gamma\left[\begin{matrix}\frac{2s-1}{2}\\ 1-\mu\end{matrix}\right]\, {}_2F_1\left(\begin{matrix}\frac{1-2\nu}{2},\; \frac{1+2\nu}{2}\\ \frac{3-2s}{2};\; -\frac{a}{2b}\end{matrix}\right)$ $\left[\begin{matrix}\operatorname{Re} a>0;\ \operatorname{Re} s,\ \operatorname{Re}(s+\mu)>\lvert\operatorname{Re}\nu\rvert;\\ \left\{\begin{matrix}\operatorname{Re}(s+\mu)<3/2;\ \lvert\arg b\rvert<\pi\\ \operatorname{Re} b>0\end{matrix}\right\}\end{matrix}\right]$

No.	$f(x)$	$F(s)$
7	$e^{a/x\pm bx}\Gamma\left(\mu,\,\frac{a}{x}\right)K_\nu(bx)$	$\frac{\sqrt{\pi}}{(2b)^s}[\cos(\nu\pi)\sec(s\pi)]^{(1\pm1)/2}\Gamma\left[\begin{matrix}\mu,\,s-\nu,\,s+\nu\\ \frac{2s+1}{2}\end{matrix}\right]$ $\times{}_1F_2\left(\begin{matrix}\frac{1-2s}{2};\,\pm2ab\\ 1-s-\nu,\,1-s+\nu\end{matrix}\right)-\frac{\sqrt{\pi}\,a^\mu}{\mu(2b)^{s-\mu}}$ $\times[\cos(\nu\pi)\sec[(s-\mu)\pi]]^{(1\pm1)/2}\Gamma\left[\begin{matrix}s-\mu-\nu,\,s-\mu+\nu\\ \frac{2s-2\mu+1}{2}\end{matrix}\right]$ $\times{}_2F_3\left(\begin{matrix}1,\,\frac{1-2s+2\mu}{2};\,\pm2ab\\ \mu+1,\,1-s+\mu-\nu,\,1-s+\mu+\nu\end{matrix}\right)$ $+\frac{\pi a^{s+\nu}b^\nu}{2^{\nu+1}\sin[(\mu-\nu-s)\pi]}\Gamma\left[\begin{matrix}-\nu,\,-s-\nu\\ 1-\mu\end{matrix}\right]{}_1F_2\left(\begin{matrix}\frac{1+2\nu}{2};\,\pm2ab\\ 1+2\nu,\,s+\nu+1\end{matrix}\right)$ $+\frac{2^{\nu-1}\pi a^{s-\nu}b^{-\nu}}{\sin[(\mu+\nu-s)\pi]}\Gamma\left[\begin{matrix}\nu,\,\nu-s\\ 1-\mu\end{matrix}\right]{}_1F_2\left(\begin{matrix}\frac{1-2\nu}{2};\,\pm2ab\\ 1-2\nu,\,s-\nu+1\end{matrix}\right)$ $\left[\begin{matrix}\mathrm{Re}\,a>0;\ \mathrm{Re}\,(s-\mu)>\lvert\mathrm{Re}\,\nu\rvert-1;\\ \left\{\begin{matrix}\lvert\arg b\rvert<\pi;\ \mathrm{Re}\,s,\ \mathrm{Re}\,(s-\mu)<1/2\\ \mathrm{Re}\,b>0\end{matrix}\right\}\end{matrix}\right]$

3.14.11. $K_\nu(ax)$ and $D_\mu(b\sqrt{x})$

Notation: $\delta=\left\{\begin{matrix}1\\0\end{matrix}\right\}.$

1	$e^{(\pm a^2/4-b)x}$ $\times D_\mu(a\sqrt{x})K_\nu(bx)$	$\frac{2^{(\mu-2s)/2}\pi}{b^s}\Gamma\left[\begin{matrix}s-\nu,\,s+\nu\\ \frac{1-\mu}{2},\,\frac{2s+1}{2}\end{matrix}\right]{}_3F_2\left(\begin{matrix}\frac{1\mp\mu-\delta}{2},\,s-\nu,\,s+\nu\\ \frac{1}{2},\,\frac{2s+1}{2};\,\pm\frac{a^2}{4b}\end{matrix}\right)$ $-\frac{2^{(\mu-2s)/2}\pi a}{b^{s+1/2}}\Gamma\left[\begin{matrix}\frac{2s-2\nu+1}{2},\,\frac{2s+2\nu+1}{2}\\ -\frac{\mu}{2},\,s+1\end{matrix}\right]$ $\times{}_3F_2\left(\begin{matrix}\frac{2-\delta\mp\mu}{2},\,\frac{2s-2\nu+1}{2},\,\frac{2s+2\nu+1}{2}\\ \frac{3}{2},\,s+1;\,\pm\frac{a^2}{4b}\end{matrix}\right)$ $[\mathrm{Re}\,b>0;\ \mathrm{Re}\,s>\lvert\mathrm{Re}\,\nu\rvert;\ \lvert\arg a\rvert<(2\pm1)\pi/4]$
2	$e^{(-a^2/4+b)x}$ $\times D_\mu(a\sqrt{x})K_\nu(bx)$	$\frac{2^{\mu/2+2\nu-s+1}\pi^{3/2}a^{2\nu-2s}}{2\sin(\nu\pi)\,b^\nu}\Gamma\left[\begin{matrix}2s-2\nu\\ 1-\nu,\,\frac{2s-\mu-2\nu+1}{2}\end{matrix}\right]$ $\times{}_3F_2\left(\begin{matrix}\frac{1-2\nu}{2},\,s-\nu,\,\frac{2s-2\nu+1}{2}\\ 1-2\nu,\,\frac{2s-\mu-2\nu+1}{2};\,\frac{4b}{a^2}\end{matrix}\right)$ $-\frac{2^{\mu/2-2\nu-s+1}\pi^{3/2}a^{-2s-2\nu}}{2\sin(\nu\pi)\,b^{-\nu}}\Gamma\left[\begin{matrix}2s+2\nu\\ 1+\nu,\,\frac{2s-\mu+2\nu+1}{2}\end{matrix}\right]$ $\times{}_3F_2\left(\begin{matrix}\frac{1+2\nu}{2},\,s+\nu,\,\frac{2s+2\nu+1}{2}\\ 1+2\nu,\,\frac{2s-\mu+2\nu+1}{2};\,\frac{4b}{a^2}\end{matrix}\right)$ $[\mathrm{Re}\,b>0;\ \mathrm{Re}\,s>\lvert\mathrm{Re}\,\nu\rvert;\ \lvert\arg a\rvert<\pi/4]$

3.14.12. $K_\nu(\varphi(x))$ and $J_\mu(\psi(x))$

Notation: $\delta = \left\{\begin{matrix} 1 \\ 0 \end{matrix}\right\}$.

1	$J_\mu(ax)\,K_\nu(bx)$	$\frac{2^{s-2}a^\mu}{b^{s+\mu}}\,\Gamma\left[\begin{matrix}\frac{s+\mu-\nu}{2},\ \frac{s+\mu+\nu}{2}\\ \mu+1\end{matrix}\right]{}_2F_1\left(\begin{matrix}\frac{s+\mu-\nu}{2},\ \frac{s+\mu+\nu}{2}\\ \mu+1;\ -\frac{a^2}{b^2}\end{matrix}\right)$ $[\operatorname{Re} b > \lvert\operatorname{Im} a\rvert;\ \operatorname{Re}(s+\mu) > \lvert\operatorname{Re}\nu\rvert]$
2	$J_\nu(ax)\,K_\nu(ax)$	$\frac{2^{s-3}}{a^s}\,\Gamma\left[\begin{matrix}\frac{s}{2},\ \frac{s+2\nu}{4}\\ \frac{4-s+2\nu}{4}\end{matrix}\right]$ $[\operatorname{Re} s,\ \operatorname{Re}(s+2\nu) > 0;\ \lvert\arg a\rvert < \pi/4]$
3	$[J_\nu(ax) - J_{-\nu}(ax)]\,K_\nu(ax)$	$-\frac{2^{3s/2-3}}{\sqrt{\pi}\,a^s}\sin\frac{\nu\pi}{2}\,\Gamma\left[\begin{matrix}s,\ \frac{s+2\nu}{4},\ \frac{s-2\nu}{4}\\ \frac{2-s}{4}\end{matrix}\right]$ $[\operatorname{Re} s > 2\lvert\operatorname{Re}\nu\rvert;\ \lvert\arg a\rvert < \pi/4]$
4	$[J_\nu(ax) + J_{-\nu}(ax)]\,K_\nu(ax)$	$\frac{2^{3s/2-3}}{\sqrt{\pi}\,a^s}\cos\frac{\nu\pi}{2}\,\Gamma\left[\begin{matrix}\frac{s+2}{4},\ \frac{s+2\nu}{4},\ \frac{s-2\nu}{4}\\ \frac{4-s}{4}\end{matrix}\right]$ $[\operatorname{Re} s > 2\lvert\operatorname{Re}\nu\rvert;\ \lvert\arg a\rvert < \pi/4]$
5	$J_\mu(ax^2)\,K_\nu(bx)$	$\frac{2^{(s-\nu)/2-3}b^\nu}{a^{(s+\nu)/2}}\,\Gamma\left[\begin{matrix}-\nu,\ \frac{s+2\mu+\nu}{4}\\ \frac{s+2\mu-\nu+4}{4}\end{matrix}\right]{}_2F_3\left(\begin{matrix}\frac{\nu-s-2\mu}{4},\ \frac{\nu+s+2\mu}{4}\\ \frac{1}{2},\ \frac{\nu+1}{2},\ \frac{\nu+2}{2};\ -\frac{b^4}{64a^2}\end{matrix}\right)$ $-\frac{2^{(s-\nu)/2-5}\,b^{\nu+2}}{a^{(s+\nu)/2+1}}\,\Gamma\left[\begin{matrix}-\nu-1,\ \frac{s+2\mu+\nu+2}{4}\\ \frac{s+2\mu-\nu+2}{4}\end{matrix}\right]$ $\times\,{}_2F_3\left(\begin{matrix}\frac{2-s-2\mu+\nu}{4},\ \frac{s+2\mu+\nu+2}{4}\\ \frac{3}{2},\ \frac{\nu+2}{2},\ \frac{\nu+3}{2};\ -\frac{b^4}{64a^2}\end{matrix}\right)$ $+\frac{2^{(s+\nu)/2-3}\,a^{(\nu-s)/2}}{b^\nu}\,\Gamma\left[\begin{matrix}\nu,\ \frac{s+2\mu-\nu}{4}\\ \frac{s+2\mu+\nu+4}{4}\end{matrix}\right]{}_2F_3\left(\begin{matrix}\frac{s+2\mu-\nu}{4},\ -\frac{s+2\mu+\nu}{4}\\ \frac{1}{2},\ \frac{1-\nu}{2},\ \frac{2-\nu}{2};\ -\frac{b^4}{64a^2}\end{matrix}\right)$ $-\frac{2^{(s+\nu)/2-5}\,a^{(\nu-s)/2-1}}{b^{\nu-2}}\,\Gamma\left[\begin{matrix}\nu+1,\ \frac{s+2\mu-\nu+2}{4}\\ \frac{s+2\mu+\nu+2}{4}\end{matrix}\right]$ $\times\,{}_2F_3\left(\begin{matrix}\frac{2-s-2\mu-\nu}{4},\ \frac{s+2\mu-\nu+2}{4}\\ \frac{3}{2},\ \frac{2-\nu}{2},\ \frac{3-\nu}{2};\ -\frac{b^4}{64a^2}\end{matrix}\right)$ $[a,\ \operatorname{Re} b > 0;\ \operatorname{Re} s > \lvert\operatorname{Re}(\nu-2\mu)\rvert]$
6	$J_\mu(a\sqrt{x})\,K_\nu(bx)$	$\frac{2^{s-\mu/2-2}\,a^\mu}{b^{s+\mu/2}}\,\Gamma\left[\begin{matrix}\frac{2s+\mu-2\nu}{4},\ \frac{2s+\mu+2\nu}{4}\\ \mu+1\end{matrix}\right]{}_2F_3\left(\begin{matrix}\frac{2s+\mu-2\nu}{4},\ \frac{2s+\mu+2\nu}{4}\\ \frac{1}{2},\ \frac{\mu+1}{2},\ \frac{\mu+2}{2};\ \frac{a^4}{64b^2}\end{matrix}\right)$ $-\frac{2^{s-\mu/2-3}\,a^{\mu+2}}{b^{s+\mu/2+1}}\,\Gamma\left[\begin{matrix}\frac{2s+\mu-2\nu+2}{4},\ \frac{2s+\mu+2\nu+2}{2}\\ \mu+2\end{matrix}\right]$ $\times\,{}_2F_3\left(\begin{matrix}\frac{2s+\mu-2\nu+2}{4},\ \frac{2s+\mu+2\nu+2}{4}\\ \frac{3}{2},\ \frac{\mu+2}{2},\ \frac{\mu+3}{2};\ \frac{a^4}{64b^2}\end{matrix}\right)$ $[a,\ \operatorname{Re} b > 0;\ \operatorname{Re} s > \lvert\operatorname{Re}\nu\rvert - \operatorname{Re}\mu/2]$

No.	$f(x)$	$F(s)$
7	$J_\mu\left(\frac{a}{x}\right) K_\nu(bx)$	$\frac{a^{s+\nu}b^\nu}{2^{s+2\nu+2}}\Gamma\left[\begin{matrix}-\nu,\ \frac{\mu-\nu-s}{2}\\ \frac{s+\mu+\nu+2}{2}\end{matrix}\right]{}_0F_3\left(\begin{matrix}-\frac{a^2b^2}{16}\\ 1+\nu,\ \frac{s-\mu+\nu+2}{2},\ \frac{s+\mu+\nu+2}{2}\end{matrix}\right)$ $+\frac{2^{2\nu-s-2}}{a^{\nu-s}b^\nu}\Gamma\left[\begin{matrix}\nu,\ \frac{\mu+\nu-s}{2}\\ \frac{s+\mu-\nu+2}{2}\end{matrix}\right]{}_0F_3\left(\begin{matrix}-\frac{a^2b^2}{16}\\ 1-\nu,\ \frac{s-\mu-\nu+2}{2},\ \frac{s+\mu-\nu+2}{2}\end{matrix}\right)$ $+\frac{2^{s-2\mu-2}a^\mu}{b^{s-\mu}}\Gamma\left[\begin{matrix}\frac{s-\mu+\nu}{2},\ \frac{s-\mu-\nu}{2}\\ \mu+1\end{matrix}\right]$ $\times{}_0F_3\left(\begin{matrix}-\frac{a^2b^2}{16}\\ \mu+1,\ \frac{2-s+\mu-\nu}{2},\ \frac{2-s+\mu+\nu}{2}\end{matrix}\right)$ $[a,\ \operatorname{Re} b>0;\ \operatorname{Re} s>\lvert\operatorname{Re}\nu\rvert-3/2]$
8	$\frac{1}{(x^4+a^4)^\rho}J_\nu(bx)K_\nu(bx)$	$\frac{a^{s-4\rho}}{8\nu}\mathrm{B}\left(\frac{s}{4},\ \frac{4\rho-s}{4}\right){}_1F_4\left(\begin{matrix}\frac{s}{4};\ \frac{a^4b^4}{64}\\ \frac{1}{2},\ \frac{2-\nu}{2},\ \frac{2+\nu}{2},\ \frac{s-2\rho+2}{2}\end{matrix}\right)$ $-\frac{a^{s-4\rho+2}b^2}{16(\nu^2-1)}\mathrm{B}\left(\frac{s+2}{4},\ \frac{4\rho-s-2}{4}\right)$ $\times{}_1F_4\left(\begin{matrix}\frac{s+2}{4};\ \frac{a^4b^4}{64}\\ \frac{3}{2},\ \frac{3-\nu}{2},\ \frac{3+\nu}{2},\ \frac{s-4\rho+6}{4}\end{matrix}\right)$ $+\frac{b^{2\nu}a^{s+2\nu-4\rho}}{2^{2\nu+3}}\Gamma\left[\begin{matrix}-\nu,\ \frac{s+2\nu}{4},\ \frac{4\rho-2\nu-s}{4}\\ \nu+1,\ \rho\end{matrix}\right]$ $\times{}_1F_4\left(\begin{matrix}\frac{s+2\nu}{4};\ \frac{a^4b^4}{64}\\ \nu+1,\ \frac{\nu+1}{2},\ \frac{\nu+2}{2},\ \frac{2\nu-4\rho+\delta+4}{4}\end{matrix}\right)$ $+\frac{2^{s-4\rho-3}}{b^{s-4\rho}}\Gamma\left[\begin{matrix}\frac{s-4\rho}{2},\ \frac{s+2\nu-4\rho}{4}\\ \frac{4-s+2\nu+4\rho}{4}\end{matrix}\right]$ $\times{}_1F_4\left(\begin{matrix}\rho;\ \frac{a^4b^4}{64}\\ \frac{2-s+4\rho}{4},\ \frac{4-s+4\rho}{4},\ \frac{4-s+4\rho-2\nu}{4},\ \frac{4-s+4\rho+2\nu}{4}\end{matrix}\right)$ $[\operatorname{Re} s,\ \operatorname{Re}(s-2\nu)>0;\ \lvert\arg a\rvert,\ \lvert\arg b\rvert<\pi/4]$
9	$e^{\pm bx}J_\mu(a\sqrt{x})K_\nu(bx)$	$\frac{\sqrt{\pi}\,a^\mu b^{-s-\mu/2}}{2^{s+3\mu/2}}\left[\frac{\cos(\nu\pi)}{\pi}\Gamma\left(\frac{1-2s-\mu}{2}\right)\right]^{(1\pm1)/2}$ $\times\Gamma\left[\begin{matrix}\frac{2s+\mu+2\nu}{2},\ \frac{2s+\mu-2\nu}{2}\\ \mu+1,\ \frac{2s+\mu+1}{2}\end{matrix}\right]{}_2F_2\left(\begin{matrix}\frac{2s+\mu+2\nu}{2},\ \frac{2s+\mu-2\nu}{2}\\ \mu+1,\ \frac{2s+\mu+1}{2};\ \pm\frac{a^2}{8b}\end{matrix}\right)$ $+2^{2s-5/2}\sqrt{\pi}\,\frac{1\pm1}{a^{2s-1}\sqrt{b}}\Gamma\left[\begin{matrix}\frac{2s+\mu-1}{2}\\ \frac{3-2s+\mu}{2}\end{matrix}\right]$ $\times{}_2F_2\left(\begin{matrix}\frac{1-2\nu}{2},\ \frac{1+2\nu}{2}\\ \frac{3-2s-\mu}{2},\ \frac{3-2s+\mu}{2};\ \pm\frac{a^2}{8b}\end{matrix}\right)$ $\left[\operatorname{Re}(2s+\mu)>2\lvert\operatorname{Re}\nu\rvert;\ \left\{\begin{matrix}\operatorname{Re} b>0\\ a,\ \operatorname{Re} b>0;\ \operatorname{Re} s<5/4\end{matrix}\right\}\right]$

No.	$f(x)$	$F(s)$
10	$\left\{\begin{matrix}\sin(ax)\\ \cos(ax)\end{matrix}\right\} J_\mu(ax)\, K_\nu(bx)$	$2^{s+\delta-2} a^{\mu+\delta} b^{-s-\mu-\delta}\, \Gamma\left[\begin{matrix}\frac{s+\mu-\nu+\delta}{2},\ \frac{s+\mu+\nu+\delta}{2}\\ \mu+1\end{matrix}\right]$ $\times\, {}_4F_3\left(\begin{matrix}\frac{2\mu+2\delta+1}{4},\ \frac{2\mu+2\delta+3}{4},\ \frac{s+\mu-\nu+\delta}{2},\ \frac{s+\mu+\nu+\delta}{2}\\ \frac{2\delta+1}{2},\ \frac{2\mu+2\delta+1}{2},\ \mu+1;\ -\frac{4a^2}{b^2}\end{matrix}\right)$ $[\operatorname{Re} b > 2\lvert\operatorname{Im} a\rvert;\ \operatorname{Re} s > \lvert\operatorname{Re}\nu\rvert - \operatorname{Re}\mu - \delta]$
11	$\left\{\begin{matrix}\sin(ax^2)\\ \cos(ax^2)\end{matrix}\right\} J_\nu(bx)\, K_\nu(bx)$	$\frac{a^{-s/2}}{4\nu}\, \Gamma\left(\frac{s}{2}\right) \left\{\begin{matrix}\sin(s\pi/4)\\ \cos(s\pi/4)\end{matrix}\right\} {}_2F_3\left(\begin{matrix}\frac{s}{2},\ \frac{s+2}{4};\ \frac{b^4}{16a^2}\\ \frac{1}{2},\ \frac{2-\nu}{2},\ \frac{2+\nu}{2}\end{matrix}\right)$ $\mp \frac{a^{-s/2-1} b^2}{8(\nu^2-1)}\, \Gamma\left(\frac{s+2}{2}\right) \left\{\begin{matrix}\cos(s\pi/4)\\ \sin(s\pi/4)\end{matrix}\right\}$ $\times\, {}_2F_3\left(\begin{matrix}\frac{s+2}{4},\ \frac{s+4}{4};\ \frac{b^4}{16a^2}\\ \frac{3}{2},\ \frac{3-\nu}{2},\ \frac{3+\nu}{2}\end{matrix}\right) + \frac{2^{2\nu-2} b^{2\nu}}{a^{s/2+\nu}}\, \Gamma\left[\begin{matrix}-\nu,\ \frac{s+2\nu}{2}\\ \nu+1\end{matrix}\right]$ $\times \left\{\begin{matrix}\sin[(2\nu+s)\pi/4]\\ \cos[(2\nu+s)\pi/4]\end{matrix}\right\} {}_2F_3\left(\begin{matrix}\frac{s+2\nu}{4},\ \frac{s+2\nu+2}{4};\ \frac{b^4}{16a^2}\\ \nu+1,\ \frac{\nu+1}{2},\ \frac{\nu+2}{2}\end{matrix}\right)$ $[a > 0;\ \operatorname{Re} s,\ \operatorname{Re}(s+2\nu) > -1 \mp 1;\ \lvert\arg b\rvert < \pi/2]$
12	$J_\lambda(ax)\, J_\mu(bx)\, K_\nu(cx)$	$\frac{2^{s-2} a^\lambda b^\mu}{c^{s+\lambda+\mu}}\, \Gamma\left[\begin{matrix}\frac{s+\lambda+\mu-\nu}{2},\ \frac{s+\lambda+\mu+\nu}{2}\\ \lambda+1,\ \mu+1\end{matrix}\right]$ $\times\, F_4\left(\frac{s+\lambda+\mu-\nu}{2},\ \frac{s+\lambda+\mu+\nu}{2};\ \lambda+1,\ \mu+1;\ -\frac{a^2}{c^2},\ -\frac{b^2}{c^2}\right)$ $\left[\begin{matrix}\lvert c\rvert > \lvert a\rvert + \lvert b\rvert;\ \operatorname{Re} c > \lvert\operatorname{Im} a\rvert + \lvert\operatorname{Im} b\rvert;\\ \operatorname{Re}(s+\lambda+\mu) > \lvert\operatorname{Re}\nu\rvert\end{matrix}\right]$
13	$J_\lambda(ax)\, J_\mu(ax)\, K_\nu(bx)$	$\frac{2^{s-2} a^{\lambda+\mu}}{b^{s+\lambda+\mu}}\, \Gamma\left[\begin{matrix}\frac{s+\lambda+\mu+\nu}{2},\ \frac{s+\lambda+\mu-\nu}{2}\\ \mu+1,\ \lambda+1\end{matrix}\right]$ $\times\, {}_4F_3\left(\begin{matrix}\frac{\lambda+\mu+1}{2},\ \frac{\lambda+\mu+2}{2},\ \frac{s+\lambda+\mu+\nu}{2},\ \frac{s+\lambda+\mu-\nu}{2}\\ \lambda+1,\ \mu+1,\ \lambda+\mu+1;\ -\frac{4a^2}{b^2}\end{matrix}\right)$ $[\lvert b\rvert > 2\lvert a\rvert;\ \operatorname{Re} c > 2\lvert\operatorname{Im} a\rvert;\ \operatorname{Re}(s+\lambda+\mu) > \lvert\operatorname{Re}\nu\rvert]$
14	$J_\mu(ax^2)\, J_\nu(bx)\, K_\nu(bx)$	$\frac{1}{8\nu}\left(\frac{2}{a}\right)^{s/2} \Gamma\left[\begin{matrix}\frac{2\mu+s}{4}\\ \frac{2\mu-s+4}{4}\end{matrix}\right] {}_2F_3\left(\begin{matrix}\frac{s-2\mu}{4},\ \frac{s+2\mu}{4};\ \frac{b^4}{16a^2}\\ \frac{1}{2},\ \frac{2-\nu}{2},\ \frac{2+\nu}{2}\end{matrix}\right)$ $-\frac{2^{s/2-3} b^2}{(\nu^2-1)\, a^{s/2+1}}\, \Gamma\left[\begin{matrix}\frac{s+2\mu+2}{4}\\ \frac{2-s+2\mu}{4}\end{matrix}\right] {}_2F_3\left(\begin{matrix}\frac{s-2\mu+2}{4},\ \frac{s+2\mu+2}{4};\ \frac{b^4}{16a^2}\\ \frac{3}{2},\ \frac{3-\nu}{2},\ \frac{3+\nu}{2}\end{matrix}\right)$ $+\frac{2^{s/2-\nu-3}}{a^{s/2+\nu} b^{-2\nu}}\, \Gamma\left[\begin{matrix}-\nu,\ \frac{s+2\mu+2\nu}{4}\\ \nu+1,\ \frac{4-s+2\mu-2\nu}{4}\end{matrix}\right] {}_2F_3\left(\begin{matrix}\frac{s-2\mu+2\nu}{4},\ \frac{s+2\mu+2\nu}{4}\\ \nu+1,\ \frac{\nu+1}{2},\ \frac{\nu+2}{2};\ \frac{b^4}{16a^2}\end{matrix}\right)$ $[a,\ \operatorname{Re} b > 0;\ \operatorname{Re}(s+2\mu),\ \operatorname{Re}(s+2\mu+2\nu) > 0]$

No.	$f(x)$	$F(s)$
15	$\prod_{j=1}^{n} J_{\mu_j}(a_j x) K_\nu(bx)$	$2^{s-2} b^{-s-\lambda} \Gamma\left(\frac{s+\lambda-\nu}{2}\right) \Gamma\left(\frac{s+\lambda+\nu}{2}\right) \times \prod_{j=1}^{n} \frac{a_j^{\mu_j}}{\Gamma(\mu_j+1)} F_C^{(n)}\left(\frac{s+\lambda-\nu}{2}, \frac{s+\lambda+\nu}{2}; \mu_1+1, \ldots, \mu_n+1; -\frac{a_1^2}{b^2}, \ldots, -\frac{a_n^2}{b^2}\right)$ $\left[\lambda = \sum_{j=1}^{n} \mu_j;\ \operatorname{Re} b > \sum_{j=1}^{n} \lvert\operatorname{Im} a_j\rvert;\ \operatorname{Re}(s+\lambda) > \lvert\operatorname{Re}\nu\rvert\right]$
16	$J_\nu\left(b\sqrt{\sqrt{x^2+a^2}-a}\right) \times K_\nu\left(b\sqrt{\sqrt{x^2+a^2}+a}\right)$	$2^{3s/2-1} \left(\frac{a}{b^2}\right)^{s/2} \Gamma\left[\begin{matrix}\frac{s+\nu}{2}\\ \frac{2-s+\nu}{2}\end{matrix}\right] K_s\left(\sqrt{2a}\, b\right)$ $[a,\, b > 0;\ \operatorname{Re} s > -\operatorname{Re}\nu]$

3.14.13. $K_\nu(\varphi(x))$ and $Y_\nu(\psi(x))$

1	$Y_\nu(ax) K_\nu(ax)$	$-\frac{2^{s-3}}{\pi a^s} \cos\frac{(s-2\nu)\pi}{4} \Gamma\left[\frac{s}{2}, \frac{s-2\nu}{4}, \frac{s+2\nu}{4}\right]$ $[\operatorname{Re} s > 2\lvert\operatorname{Re}\nu\rvert;\ \lvert\arg a\rvert < \pi/4]$
2	$Y_\mu(ax) K_\nu(bx)$	$-\frac{2^{s-2} b^{\mu-s}}{\pi a^\mu} \Gamma\left[\mu, \frac{s-\mu-\nu}{2}, \frac{s-\mu+\nu}{2}\right] \times {}_2F_1\left(\begin{matrix}\frac{s-\mu-\nu}{2}, \frac{s-\mu+\nu}{2}\\ 1-\mu;\ -\frac{a^2}{b^2}\end{matrix}\right) - \frac{2^{s-2} a^\mu}{\pi b^{s+\mu}} \cos(\mu\pi) \times \Gamma\left[-\mu, \frac{s+\mu-\nu}{2}, \frac{s+\mu+\nu}{2}\right] {}_2F_1\left(\begin{matrix}\frac{s+\mu-\nu}{2}, \frac{s+\mu+\nu}{2}\\ 1+\mu;\ -\frac{a^2}{b^2}\end{matrix}\right)$ $[\operatorname{Re} b > \lvert\operatorname{Im} a\rvert;\ \operatorname{Re} s > \lvert\operatorname{Re}\mu\rvert + \lvert\operatorname{Re}\nu\rvert]$
3	$Y_\mu\left(\frac{a}{x}\right) K_\nu(bx)$	$-\frac{2^{s-2\mu-2} a^\mu}{\pi b^{s-\mu}} \cos(\mu\pi)\, \Gamma\left[-\mu, \frac{s+\nu-\mu}{2}, \frac{s-\nu-\mu}{2}\right] \times {}_0F_3\left(\begin{matrix}-\frac{a^2 b^2}{16}\\ 1+\mu, \frac{2-s+\mu-\nu}{2}, \frac{2-s+\mu+\nu}{2}\end{matrix}\right) - \frac{2^{s+2\mu-2}}{\pi a^\mu b^{s+\mu}} \Gamma\left[\mu, \frac{s+\mu+\nu}{2}, \frac{s+\mu-\nu}{2}\right] \times {}_0F_3\left(\begin{matrix}-\frac{a^2 b^2}{16}\\ 1-\mu, \frac{2-s-\mu-\nu}{2}, \frac{2-s-\mu+\nu}{2}\end{matrix}\right) - \frac{a^{s+\nu} b^\nu}{2^{s+2\nu+2}\pi} \times \cos\frac{(s+\mu+\nu)\pi}{2} \Gamma\left[-\nu, \frac{\mu-\nu-s}{2}, -\frac{\mu+\nu+s}{2}\right] \times$

No.	$f(x)$	$F(s)$
		$\times\, {}_0F_3\left(\begin{matrix} -\frac{a^2b^2}{16} \\ 1+\nu,\ \frac{s-\mu+\nu+2}{2},\ \frac{s+\mu+\nu+2}{2}\end{matrix}\right) - \frac{a^{s-\nu}b^{-\nu}}{2^{s-2\nu+2}\pi}$ $\times \cos\frac{(s+\mu-\nu)\pi}{2}\,\Gamma\left[\nu,\ \frac{\mu+\nu-s}{2},\ \frac{\nu-\mu-s}{2}\right]$ $\times\, {}_0F_3\left(\begin{matrix} -\frac{a^2b^2}{16} \\ 1-\nu,\ \frac{s-\mu-\nu+2}{2},\ \frac{s+\mu-\nu+2}{2}\end{matrix}\right)$ $[a,\ \operatorname{Re} b>0;\ \operatorname{Re} s>\lvert\operatorname{Re}\nu\rvert-3/2]$
4	$Y_\nu\left(b\sqrt{\sqrt{x^2+a^2}-a}\right)$ $\times K_\nu\left(b\sqrt{\sqrt{x^2+a^2}+a}\right)$	$-2^{3s/2-1}\left(\frac{a}{b^2}\right)^{s/2}\Gamma\left[\begin{matrix}\frac{s-\nu}{2},\ \frac{s+\nu}{2}\\ \frac{s-\nu+1}{2},\ \frac{\nu-s+1}{2}\end{matrix}\right]K_s\left(\sqrt{2a}\,b\right)$ $[a,\ b>0;\ \operatorname{Re} s>\lvert\operatorname{Re}\nu\rvert]$

3.14.14. $K_\nu(ax)$ and $J_\nu(ax)$, $Y_\nu(ax)$

1	$\left[\cos\frac{\nu\pi}{2}J_\nu(ax)-\sin\frac{\nu\pi}{2}Y_\nu(ax)\right]$ $\times K_\nu(ax)$	$\frac{2^{s-3}}{a^s}\Gamma\left[\begin{matrix}\frac{s}{2},\ \frac{s-2\nu}{4},\ \frac{s+2\nu}{4}\\ \frac{s}{4},\ \frac{4-s}{4}\end{matrix}\right]$ $[a>0;\ \operatorname{Re} s>2\lvert\operatorname{Re}\nu\rvert]$
2	$\left[\sin\frac{\nu\pi}{2}J_\nu(ax)+\cos\frac{\nu\pi}{2}Y_\nu(ax)\right]$ $\times K_\nu(ax)$	$-\frac{2^{s-3}}{a^s}\Gamma\left[\begin{matrix}\frac{s}{2},\ \frac{s-2\nu}{4},\ \frac{s+2\nu}{4}\\ \frac{2-s}{4},\ \frac{2+s}{4}\end{matrix}\right]$ $[a>0;\ \operatorname{Re} s>2\lvert\operatorname{Re}\nu\rvert]$
3	$\frac{2}{\pi}K_0(ax)-Y_0(ax)$	$\frac{2^{2s-2}}{a^s}\Gamma\left[\begin{matrix}\frac{s}{4},\ \frac{s}{4}\\ \frac{2-s}{4},\ \frac{2-s}{4}\end{matrix}\right]$ $[a>0;\ 0<\operatorname{Re} s<3/4]$

3.14.15. $K_\nu(\varphi(x))$ and $I_\mu(\psi(x))$

1	$I_\mu(ax)K_\nu(bx)$	$\frac{2^{s-2}a^\mu}{b^{s+\mu}}\Gamma\left[\begin{matrix}\frac{s+\mu-\nu}{2},\ \frac{s+\mu+\nu}{2}\\ \mu+1\end{matrix}\right]{}_2F_1\left(\begin{matrix}\frac{s+\mu-\nu}{2},\ \frac{s+\mu+\nu}{2}\\ \mu+1;\ \frac{a^2}{b^2}\end{matrix}\right)$ $[\operatorname{Re} b>\lvert\operatorname{Re} a\rvert;\ \operatorname{Re}(s+\mu)>\lvert\operatorname{Re}\nu\rvert]$
2	$I_\mu(ax)K_\nu(ax)$	$\frac{2^{s-2}}{a^s}\Gamma\left[\begin{matrix}\frac{s+\mu-\nu}{2},\ \frac{s+\mu+\nu}{2},\ 1-s\\ \frac{2-s+\mu-\nu}{2},\ \frac{2-s+\mu+\nu}{2}\end{matrix}\right]$ $[\operatorname{Re} a>0;\ \lvert\operatorname{Re}\nu\rvert-\operatorname{Re}\mu<\operatorname{Re} s<1]$

No.	$f(x)$	$F(s)$
3	$I_\nu(a\sqrt{x})\,K_\nu(bx)$	$\frac{2^{s-\mu/2-2}\,a^\mu}{b^{s+\mu/2}}\,\Gamma\left[\begin{matrix}\frac{2s+\mu-2\nu}{4},\ \frac{2s+\mu+2\nu}{4}\\ \mu+1\end{matrix}\right]\,{}_2F_3\left(\begin{matrix}\frac{2s+\mu-2\nu}{4},\ \frac{2s+\mu+2\nu}{4}\\ \frac{1}{2},\ \frac{\mu+1}{2},\ \frac{\mu+2}{2};\ \frac{a^2}{64b^2}\end{matrix}\right)$ $+\frac{2^{s-\mu/2-3}\,a^{\mu+2}}{b^{s+\mu/2+1}}\,\Gamma\left[\begin{matrix}\frac{2s+\mu-2\nu+2}{4},\ \frac{2s+\mu+2\nu+2}{4}\\ \mu+2\end{matrix}\right]$ $\times\,{}_2F_3\left(\begin{matrix}\frac{2s+\mu-2\nu+2}{4},\ \frac{2s+\mu+2\nu+2}{4}\\ \frac{3}{2},\ \frac{\mu+2}{2},\ \frac{\mu+3}{2};\ \frac{a^4}{64b^2}\end{matrix}\right)$ $[\operatorname{Re} b>0;\ \operatorname{Re} s>\lvert\operatorname{Re}\nu\rvert-\operatorname{Re}\mu/2]$
4	$[I_\nu(ax)+I_{-\nu}(ax)]$ $\times K_\nu(ax)$	$\frac{\cos(\nu\pi)}{2\sqrt{\pi}\,a^s}\,\Gamma\left[\begin{matrix}\frac{s-2\nu}{2},\ \frac{s+2\nu}{2},\ \frac{1-s}{2}\\ \frac{2-s}{2}\end{matrix}\right]$ $[a>0;\ 2\lvert\operatorname{Re}\nu\rvert<\operatorname{Re} s<1]$
5	$I_\nu\left(b\sqrt{x^2+a^2}-ab\right)$ $\times K_\nu\left(b\sqrt{x^2+a^2}+ab\right)$	$\frac{1}{2\sqrt{\pi}}\left(\frac{a}{b}\right)^{s/2}\Gamma\left[\begin{matrix}\frac{s+2\nu}{2},\ \frac{1-s}{2}\\ \frac{2-s+2\nu}{2}\end{matrix}\right]K_{s/2}(2ab)$ $[a,\ \operatorname{Re} b>0;\ -2\operatorname{Re}\nu<\operatorname{Re} s<1]$
6	$I_\mu(ax)\,K_\nu(ax)$ $-\,I_\nu(ax)\,K_\mu(ax)$	$\frac{a^{-s}}{2\sqrt{\pi}}\,\sin\frac{(\nu-\mu)\pi}{2}\,\Gamma\left[\begin{matrix}\frac{s+\mu+\nu}{2},\ \frac{s-\mu+\nu}{2},\ \frac{s+\mu-\nu}{2},\ \frac{2-s}{2}\\ \frac{s+1}{2},\ \frac{2-s+\mu+\nu}{2}\end{matrix}\right]$ $[a>0;\ \operatorname{Re}(\nu-\mu),\ \lvert\operatorname{Re}\mu\rvert-\operatorname{Re}\nu<\operatorname{Re} s<1]$
7	$I_\mu(ax)\,K_\nu(ax)$ $+\,I_\nu(ax)\,K_\mu(ax)$	$\frac{a^{-s}}{2\sqrt{\pi}}\,\cos\frac{(\mu-\nu)\pi}{2}\,\Gamma\left[\begin{matrix}\frac{s+\mu+\nu}{2},\ \frac{s-\mu+\nu}{2},\ \frac{s+\mu-\nu}{2},\ \frac{1-s}{2}\\ \frac{s}{2},\ \frac{2-s+\mu+\nu}{2}\end{matrix}\right]$ $[a>0;\ -\operatorname{Re}(\mu-\nu),\ \lvert\operatorname{Re}\nu\rvert-\operatorname{Re}\mu<\operatorname{Re} s<1]$

3.14.16. $K_\nu(ax)$, $I_\mu(\varphi(x))$, and the exponential function

Notation: $\delta=\left\{\begin{matrix}1\\0\end{matrix}\right\}$.

1	$e^{\pm ax}I_\mu(ax)\,K_\nu(bx)$	$\frac{2^{s-2}a^\mu}{b^{s+\mu}}\,\Gamma\left[\begin{matrix}\frac{s+\mu-\nu}{2},\ \frac{s+\mu+\nu}{2}\\ \mu+1\end{matrix}\right]\,{}_4F_3\left(\begin{matrix}\frac{2\mu+1}{4},\ \frac{2\mu+3}{4},\ \frac{s+\mu-\nu}{2},\ \frac{s+\mu+\nu}{2}\\ \frac{1}{2},\ \frac{2\mu+1}{2},\ \mu+1;\ \frac{4a^2}{b^2}\end{matrix}\right)$ $\pm\frac{2^{s-1}a^{\mu+1}}{b^{s+\mu+1}}\,\Gamma\left[\begin{matrix}\frac{s+\mu-\nu+1}{2},\ \frac{s+\mu+\nu+1}{2}\\ \mu+1\end{matrix}\right]$ $\times\,{}_4F_3\left(\begin{matrix}\frac{2\mu+3}{4},\ \frac{2\mu+5}{4},\ \frac{s+\mu-\nu+1}{2},\ \frac{s+\mu+\nu+1}{2}\\ \frac{3}{2},\ \mu+1,\ \frac{2\mu+3}{2};\ \frac{4a^2}{b^2}\end{matrix}\right)$ $\left[\left\{\begin{matrix}\operatorname{Re} a,\ \operatorname{Re} b>0\\ \operatorname{Re} b>\lvert\operatorname{Re} a\rvert\end{matrix}\right\};\ \operatorname{Re}(s+\mu)>\lvert\operatorname{Re}\nu\rvert\right]$

No.	$f(x)$	$F(s)$
2	$e^{-ax} I_\mu(bx) K_\nu(bx)$	$\frac{b^{\mu+\nu}}{2^{\mu+\nu+1} a^{s+\mu+\nu}} \Gamma\left[\begin{matrix} -\nu,\ s+\mu+\nu \\ \mu+1 \end{matrix}\right] \times {}_4F_3\left(\begin{matrix} \frac{\mu+\nu+1}{2},\ \frac{\mu+\nu+2}{2},\ \frac{s+\mu+\nu}{2},\ \frac{s+\mu+\nu+1}{2} \\ \mu+1,\ \nu+1,\ \mu+\nu+1;\ \frac{4b^2}{a^2} \end{matrix}\right) + \frac{2^{\nu-\mu-1} b^{\mu-\nu}}{a^{s+\mu-\nu}} \Gamma\left[\begin{matrix} \nu,\ s+\mu-\nu \\ \mu+1 \end{matrix}\right] \times {}_4F_3\left(\begin{matrix} \frac{\mu-\nu+1}{2},\ \frac{\mu-\nu+2}{2},\ \frac{s+\mu-\nu}{2},\ \frac{s+\mu-\nu+1}{2} \\ \mu+1,\ 1-\nu,\ \mu-\nu+1;\ \frac{4b^2}{a^2} \end{matrix}\right)$ $[\operatorname{Re} a,\ \operatorname{Re} b > 0;\ \operatorname{Re}(s+\mu) > \lvert\operatorname{Re}\nu\rvert]$
3	$e^{-ax \pm bx} I_\mu(ax) K_\nu(bx)$	$\frac{\sqrt{\pi}\, a^\mu}{2^{s+2\mu}\, b^{s+\mu}} (\cos(\nu\pi) \sec[(s+\mu)\pi])^{(1\pm1)/2} \times \Gamma\left[\begin{matrix} s+\mu-\nu,\ s+\mu+\nu \\ \mu+1,\ \frac{2s+2\mu+1}{2} \end{matrix}\right] {}_3F_2\left(\begin{matrix} \frac{2\mu+1}{2},\ s+\mu-\nu,\ s+\mu+\nu \\ 2\mu+1,\ \frac{2s+2\mu+1}{2};\ \pm\frac{a}{b} \end{matrix}\right) - \frac{(1\pm1)\pi}{2^{s+1} a^{s-1/2} \sqrt{b}} \sec[(s+\mu)\pi]\, \Gamma\left[\begin{matrix} 1-s \\ \frac{3-2s-2\mu}{2},\ \frac{3-2s+2\mu}{2} \end{matrix}\right] \times {}_3F_2\left(\begin{matrix} \frac{1-2\nu}{2},\ \frac{1+2\nu}{2},\ 1-s \\ \frac{3-2s-2\mu}{2},\ \frac{3-2s+2\mu}{2};\ \pm\frac{a}{b} \end{matrix}\right)$ $\left[\operatorname{Re} a > 0;\ \operatorname{Re}(s+\mu) > \lvert\operatorname{Re}\nu\rvert;\ \left\{\begin{matrix} \operatorname{Re} s < 1;\ \lvert\arg b\rvert < \pi \\ \operatorname{Re} b > 0 \end{matrix}\right\}\right]$
4	$e^{-ax^2} I_\mu(bx) K_\nu(bx)$	$\frac{a^{-(s+\mu+\nu)/2}\, b^{\mu+\nu}}{2^{\mu+\nu+2}} \Gamma\left[\begin{matrix} -\nu,\ \frac{s+\mu+\nu}{2} \\ \mu+1 \end{matrix}\right] {}_3F_3\left(\begin{matrix} \frac{\mu+\nu+1}{2},\ \frac{\mu+\nu+2}{2},\ \frac{s+\mu+\nu}{2};\ \frac{b^2}{a} \\ \mu+1,\ \nu+1,\ \mu+\nu+1 \end{matrix}\right) + \frac{2^{\nu-\mu-2}\, b^{\mu-\nu}}{a^{(s+\mu-\nu)/2}} \Gamma\left[\begin{matrix} \nu,\ \frac{s+\mu-\nu}{2} \\ \mu+1 \end{matrix}\right] {}_3F_3\left(\begin{matrix} \frac{\mu-\nu+1}{2},\ \frac{\mu-\nu+2}{2},\ \frac{s+\mu-\nu}{2};\ \frac{b^2}{a} \\ \mu+1,\ 1-\nu,\ \mu-\nu+1 \end{matrix}\right)$ $[\operatorname{Re} a,\ \operatorname{Re} b > 0;\ \operatorname{Re}(s+\mu) > \lvert\operatorname{Re}\nu\rvert]$
5	$e^{-ax^2} I_\mu(ax^2) K_\nu(bx)$	$\frac{2^{-(s+3\nu)/2-2} b^\nu}{\sqrt{\pi}\, a^{(s+\nu)/2}} \Gamma\left[\begin{matrix} -\nu,\ \frac{1-s-\nu}{2},\ \frac{s+2\mu+\nu}{2} \\ \frac{1-s+2\mu-\nu}{2} \end{matrix}\right] {}_2F_2\left(\begin{matrix} \frac{s-2\mu+\nu}{2},\ \frac{s+2\mu+\nu}{2} \\ 1+\nu,\ \frac{s+\nu+1}{2};\ \frac{b^2}{8a} \end{matrix}\right) + \frac{2^{(3\nu-s)/2-2} b^\nu}{\sqrt{\pi}\, a^{(s-\nu)/2}} \Gamma\left[\begin{matrix} \nu,\ \frac{1-s+\nu}{2},\ \frac{s+2\mu-\nu}{2} \\ \frac{2-s+2\mu+\nu}{2} \end{matrix}\right] {}_2F_2\left(\begin{matrix} \frac{s-2\mu-\nu}{2},\ \frac{s+2\mu-\nu}{2} \\ 1-\nu,\ \frac{s-\nu+1}{2};\ \frac{b^2}{8a} \end{matrix}\right) + \frac{2^{s-7/2}\, b^{1-s}}{\sqrt{\pi a}} \Gamma\left(\frac{s-\nu-1}{2}\right) \Gamma\left(\frac{s+\nu-1}{2}\right) \times {}_2F_2\left(\begin{matrix} \frac{1-2\mu}{2},\ \frac{1+2\mu}{2} \\ \frac{3-s-\nu}{2},\ \frac{3-s+\nu}{2};\ \frac{b^2}{8a} \end{matrix}\right)$ $[\operatorname{Re} a,\ \operatorname{Re} b > 0;\ \operatorname{Re}(s+2\mu) > \lvert\operatorname{Re}\nu\rvert]$

No.	$f(x)$	$F(s)$
6	$e^{-ax} I_\mu(b\sqrt{x})\, K_\nu(ax)$	$\frac{\sqrt{\pi}\, b^\mu}{2^{s+3\mu/2} a^{s+\mu/2}} \Gamma\left[\begin{matrix} \frac{2s-2\nu+\mu}{2}, \frac{2s+2\nu+\mu}{2} \\ \mu+1, \frac{2s+\mu+1}{2} \end{matrix}\right] {}_2F_2\left(\begin{matrix} \frac{2s-2\nu+\mu}{2}, \frac{2s+2\nu+\mu}{2} \\ \mu+1, \frac{2s+\mu+1}{2}; \frac{b^2}{8a} \end{matrix}\right)$ $[\operatorname{Re} a,\ \operatorname{Re} b > 0;\ \operatorname{Re}(s+\mu/2) > \lvert\operatorname{Re}\nu\rvert]$
7	$\left\{\begin{matrix} \sinh(ax) \\ \cosh(ax) \end{matrix}\right\}$ $\times I_\mu(ax)\, K_\nu(bx)$	$2^{s+\delta-2} a^{\mu+\delta} b^{-s-\mu-\delta} \Gamma\left[\begin{matrix} \frac{s+\mu-\nu+\delta}{2}, \frac{s+\mu+\nu+\delta}{2} \\ \mu+1 \end{matrix}\right]$ $\times {}_4F_3\left(\begin{matrix} \frac{2\mu+2\delta+1}{4}, \frac{2\mu+2\delta+3}{4}, \frac{s+\mu-\nu+\delta}{2}, \frac{s+\mu+\nu+\delta}{2} \\ \frac{2\delta+1}{2}, \frac{2\mu+2\delta+1}{2}, \mu+1; \frac{4a^2}{b^2} \end{matrix}\right)$ $[\operatorname{Re} b > 2\lvert\operatorname{Re} a\rvert;\ \operatorname{Re} s > \lvert\operatorname{Re}\nu\rvert - \operatorname{Re}\mu - \delta]$
8	$\left\{\begin{matrix} \sin(ax) \\ \cos(ax) \end{matrix}\right\}$ $\times I_\mu(bx)\, K_\nu(bx)$	$\frac{2^{-\mu-\nu-1} b^{\mu+\nu}}{a^{s+\mu+\nu}} \left\{\begin{matrix} \sin[(s+\mu+\nu)\pi/2] \\ \cos[(s+\mu+\nu)\pi/2] \end{matrix}\right\} \Gamma\left[\begin{matrix} -\nu, s+\mu+\nu \\ \mu+1 \end{matrix}\right]$ $\times {}_4F_3\left(\begin{matrix} \frac{\mu+\nu+1}{2}, \frac{\mu+\nu+2}{2}, \frac{s+\mu+\nu}{2}, \frac{s+\mu+\nu+1}{2} \\ \mu+1, \nu+1, \mu+\nu+1; -\frac{4b^2}{a^2} \end{matrix}\right)$ $+ \frac{2^{\nu-\mu-1} b^{\mu-\nu}}{a^{s+\mu-\nu}} \left\{\begin{matrix} \sin[(s+\mu-\nu)\pi/2] \\ \cos[(s+\mu-\nu)\pi/2] \end{matrix}\right\} \Gamma\left[\begin{matrix} \nu, s+\mu-\nu \\ \mu+1 \end{matrix}\right]$ $\times {}_4F_3\left(\begin{matrix} \frac{\mu-\nu+1}{2}, \frac{\mu-\nu+2}{2}, \frac{s+\mu-\nu}{2}, \frac{s+\mu-\nu+1}{2} \\ \mu+1, 1-\nu, \mu-\nu+1; -\frac{4b^2}{a^2} \end{matrix}\right)$ $[a,\ \operatorname{Re} b > 0;\ \lvert\operatorname{Re}\nu\rvert - \operatorname{Re}\mu - (1\pm 1)/2 < \operatorname{Re} s < 2]$
9	$I_\lambda(ax)\, I_\mu(bx)\, K_\nu(cx)$	$\frac{2^{s-1} a^\lambda b^\mu}{c^{s+\lambda+\mu}} \Gamma\left[\begin{matrix} \frac{s+\lambda+\mu-\nu}{2}, \frac{s+\lambda+\mu+\nu}{2} \\ \lambda+1, \mu+1 \end{matrix}\right]$ $\times F_4\left(\frac{s+\lambda+\mu-\nu}{2}, \frac{s+\lambda+\mu+\nu}{2}; \lambda+1, \mu+1; \frac{a^2}{c^2}, \frac{b^2}{c^2}\right)$ $\left[\begin{matrix} \lvert c\rvert > \lvert a\rvert + \lvert b\rvert;\ \operatorname{Re} c > \lvert\operatorname{Re} a\rvert + \lvert\operatorname{Re} b\rvert; \\ \operatorname{Re}(s+\lambda+\mu) > \lvert\operatorname{Re}\nu\rvert \end{matrix}\right]$
10	$I_\lambda(ax)\, I_\mu(ax)\, K_\nu(bx)$	$\frac{2^{s-2} a^{\lambda+\mu}}{b^{s+\lambda+\mu}} \Gamma\left[\begin{matrix} \frac{s+\lambda+\mu-\nu}{2}, \frac{s+\lambda+\mu+\nu}{2} \\ \lambda+1, \mu+1 \end{matrix}\right]$ $\times {}_4F_3\left(\begin{matrix} \frac{\lambda+\mu+1}{2}, \frac{\lambda+\mu+2}{2}, \frac{s+\lambda+\mu-\nu}{2}, \frac{s+\lambda+\mu+\nu}{2} \\ \lambda+1, \mu+1, \lambda+\mu+1; \frac{4a^2}{b^2} \end{matrix}\right)$ $[\operatorname{Re} b > 2\lvert\operatorname{Re} a\rvert;\ \operatorname{Re}(s+\lambda+\mu) > \lvert\operatorname{Re}\nu\rvert]$

3.14.17. $K_\nu(ax)$ and $I_\mu(ax)$, $J_\lambda(bx)$

1	$J_\lambda(ax)\,I_\mu(bx)\,K_\nu(bx)$	$\frac{2^{s-2}\,b^{\mu+\nu}}{a^{s+\mu+\nu}}\,\Gamma\left[\begin{matrix}-\nu,\ \frac{s+\lambda+\mu+\nu}{2}\\ \mu+1,\ \frac{2-s+\lambda-\mu-\nu}{2}\end{matrix}\right] \times {}_4F_3\left(\begin{matrix}\frac{\mu+\nu+1}{2},\ \frac{\mu+\nu+2}{2},\ \frac{s-\lambda+\mu+\nu}{2},\ \frac{s+\lambda+\mu+\nu}{2}\\ \mu+1,\ \nu+1,\ \mu+\nu+1;\ -\frac{4b^2}{a^2}\end{matrix}\right) + \frac{2^{s-2}\,b^{\mu-\nu}}{a^{s+\mu-\nu}}\,\Gamma\left[\begin{matrix}\nu,\ \frac{s+\lambda+\mu-\nu}{2}\\ \mu+1,\ \frac{2-s+\lambda-\mu+\nu}{2}\end{matrix}\right] \times {}_4F_3\left(\begin{matrix}\frac{\mu-\nu+1}{2},\ \frac{\mu-\nu+2}{2},\ \frac{s-\lambda+\mu-\nu}{2},\ \frac{s+\lambda+\mu-\nu}{2}\\ \mu+1,\ 1-\nu,\ \mu-\nu+1;\ -\frac{4b^2}{a^2}\end{matrix}\right)$ $[a,\ \operatorname{Re} b>0;\ \lvert\operatorname{Re}\nu\rvert-\operatorname{Re}(\lambda+\mu)<\operatorname{Re} s<5/2]$
2	$J_\lambda(ax)\,I_\mu(bx)\,K_\nu(cx)$	$\frac{2^{s-2}\,a^\lambda b^\mu}{c^{s+\lambda+\mu}}\,\Gamma\left[\begin{matrix}\frac{s+\lambda+\mu-\nu}{2},\ \frac{s+\lambda+\mu+\nu}{2}\\ \lambda+1,\ \mu+1\end{matrix}\right] \times F_4\left(\frac{s+\lambda+\mu-\nu}{2},\ \frac{s+\lambda+\mu+\nu}{2};\ \lambda+1,\ \mu+1;\ -\frac{a^2}{c^2},\ \frac{b^2}{c^2}\right)$ $[\operatorname{Re} c>\lvert\operatorname{Im} a\rvert+\lvert\operatorname{Re} b\rvert;\ \operatorname{Re}(s+\lambda+\mu)>\lvert\operatorname{Re}\nu\rvert]$

3.14.18. Products of $K_\mu(\varphi(x))$

1	$K_\mu(ax)\,K_\nu(bx)$	$\frac{2^{s-3}a^{\nu-s}}{b^\nu}\,\Gamma\left[\nu,\ \frac{s-\mu-\nu}{2},\ \frac{s+\mu-\nu}{2}\right]\,{}_2F_1\left(\begin{matrix}\frac{s-\mu-\nu}{2},\ \frac{s+\mu-\nu}{2}\\ 1-\nu;\ \frac{b^2}{a^2}\end{matrix}\right) + \frac{2^{s-3}a^{-\nu-s}}{b^{-\nu}}\,\Gamma\left[-\nu,\ \frac{s-\mu+\nu}{2},\ \frac{s+\mu+\nu}{2}\right]\,{}_2F_1\left(\begin{matrix}\frac{s-\mu+\nu}{2},\ \frac{s+\mu+\nu}{2}\\ 1+\nu;\ \frac{b^2}{a^2}\end{matrix}\right)$ $[\operatorname{Re}(a+b)>0;\ \operatorname{Re} s>\lvert\operatorname{Re}\mu\rvert+\lvert\operatorname{Re}\nu\rvert]$
2	$K_\nu(ax)\,K_\nu(bx)$	$\sqrt{\frac{\pi}{ab}}\,\frac{2^{s-3}}{\lvert a^2-b^2\rvert^{(s-1)/2}}\,\Gamma\left[\frac{s}{2},\ \frac{s-2\nu}{2},\ \frac{s+2\nu}{2}\right]\,P_{\nu-1/2}^{(1-s)/2}\left(\frac{a^2+b^2}{2ab}\right)$
3		$=\frac{e^{i\nu\pi}2^{s-2}}{\lvert a^2-b^2\rvert^{s/2}}\,\Gamma\left(\frac{s}{2}\right)\,\Gamma\left(\frac{s+2\nu}{2}\right)\,Q_{s/2-1}^{-\nu}\left(\frac{a^2+b^2}{\lvert a^2-b^2\rvert}\right)$ $[\operatorname{Re}(a+b)>0;\ \operatorname{Re} s>2\lvert\operatorname{Re}\nu\rvert]$
4	$K_\mu(ax)\,K_\nu(ax)$	$\frac{2^{s-3}}{a^s\,\Gamma(s)}\,\Gamma\left[\frac{s+\mu+\nu}{2},\ \frac{s+\mu-\nu}{2},\ \frac{s+\nu-\mu}{2},\ \frac{s-\mu-\nu}{2}\right]$ $[\operatorname{Re} a>0;\ \operatorname{Re} s>\lvert\operatorname{Re}\mu\rvert+\lvert\operatorname{Re}\nu\rvert]$

No.	$f(x)$	$F(s)$
5	$K_\mu(a\sqrt{x})\,K_\nu(bx)$	$\frac{2^{s-\mu/2-3}a^\mu}{b^{s+\mu/2}}\Gamma\left[-\mu,\,\frac{2s+\mu-2\nu}{4},\,\frac{2s+\mu+2\nu}{4}\right]$ $\times\,{}_2F_3\left(\begin{matrix}\frac{2s+\mu-2\nu}{4},\,\frac{2s+\mu+2\nu}{4}\\ \frac{1}{2},\,\frac{1+\mu}{2},\,\frac{2+\mu}{2};\,\frac{a^4}{64b^2}\end{matrix}\right)$ $+\frac{2^{s+\mu/2-3}a^{-\mu}}{b^{s-\mu/2}}\Gamma\left[\mu,\,\frac{2s-\mu-2\nu}{4},\,\frac{2s-\mu+2\nu}{4}\right]$ $\times\,{}_2F_3\left(\begin{matrix}\frac{2s-\mu-2\nu}{4},\,\frac{2s-\mu+2\nu}{4}\\ \frac{1}{2},\,\frac{1-\mu}{2},\,\frac{2-\mu}{2};\,\frac{a^4}{64b^2}\end{matrix}\right)-\frac{2^{s-\mu/2-4}a^{\mu+2}}{b^{s+\mu/2+1}}$ $\times\,\Gamma\left[-\mu-1,\,\frac{2s+\mu-2\nu+2}{4},\,\frac{2s+\mu+2\nu+2}{4}\right]$ $\times\,{}_2F_3\left(\begin{matrix}\frac{2s+\mu-2\nu+2}{4},\,\frac{2s+\mu+2\nu+2}{4}\\ \frac{3}{2},\,\frac{2+\mu}{2},\,\frac{3+\mu}{2};\,\frac{a^4}{64b^2}\end{matrix}\right)-\frac{2^{s+\mu/2-4}a^{2-\mu}}{b^{s-\mu/2+1}}$ $\times\,\Gamma\left[\mu-1,\,\frac{2s-\mu-2\nu+2}{4},\,\frac{2s-\mu+2\nu+2}{4}\right]$ $\times\,{}_2F_3\left(\begin{matrix}\frac{2s-\mu-2\nu+2}{4},\,\frac{2s-\mu+2\nu+2}{4}\\ \frac{3}{2},\,\frac{2-\mu}{2},\,\frac{3-\mu}{2};\,\frac{a^4}{64b^2}\end{matrix}\right)$ $[\operatorname{Re}b>0;\ \operatorname{Re}s>\lvert\operatorname{Re}\mu\rvert/2+\lvert\operatorname{Re}\nu\rvert]$
6	$K_\mu\left(\frac{a}{x}\right)K_\nu(bx)$	$\frac{2^{s-2\mu-3}\,a^\mu}{b^{s-\mu}}\Gamma\left[-\mu,\,\frac{s-\mu-\nu}{2},\,\frac{s-\mu+\nu}{2}\right]$ $\times\,{}_0F_3\left(\begin{matrix}\frac{a^2b^2}{16}\\ 1+\mu,\,\frac{2-s+\mu-\nu}{2},\,\frac{2-s+\mu+\nu}{2}\end{matrix}\right)$ $+\frac{2^{s+2\mu-3}a^{-\mu}}{b^{s+\mu}}\Gamma\left[\mu,\,\frac{s+\mu-\nu}{2},\,\frac{s+\mu+\nu}{2}\right]$ $\times\,{}_0F_3\left(\begin{matrix}\frac{a^2b^2}{16}\\ 1-\mu,\,\frac{2-s-\mu-\nu}{2},\,\frac{2-s-\mu+\nu}{2}\end{matrix}\right)$ $+\frac{a^{s+\nu}\,b^\nu}{2^{s+2\nu+3}}\Gamma\left[-\nu,\,\frac{\mu-\nu-s}{2},\,-\frac{\mu+\nu+s}{2}\right]$ $\times\,{}_0F_3\left(\begin{matrix}\frac{a^2b^2}{16}\\ 1+\nu,\,\frac{s-\mu+\nu+2}{2},\,\frac{s+\mu+\nu+2}{2}\end{matrix}\right)+\frac{a^{s-\nu}b^{-\nu}}{2^{s-2\nu+3}}$ $\times\,\Gamma\left[\nu,\,\frac{\mu+\nu-s}{2},\,\frac{\nu-\mu-s}{2}\right]{}_0F_3\left(\begin{matrix}\frac{a^2b^2}{16}\\ 1-\nu,\,\frac{s-\mu-\nu+2}{2},\,\frac{s+\mu-\nu+2}{2}\end{matrix}\right)$ $[\operatorname{Re}a,\ \operatorname{Re}b>0]$
7	$K_\nu\left(b\sqrt{x^2+a^2}-ab\right)$ $\times\,K_\nu\left(b\sqrt{x^2+a^2}+ab\right)$	$\frac{\sqrt{\pi}}{2}\left(\frac{a}{b}\right)^{s/2}\Gamma\left[\begin{matrix}\frac{s-2\nu}{2},\,\frac{s+2\nu}{2}\\ \frac{s+1}{2}\end{matrix}\right]K_{s/2}(2ab)$ $[\operatorname{Re}a,\ \operatorname{Re}b>0;\ \operatorname{Re}s>2\lvert\operatorname{Re}\nu\rvert]$

3.14.19. Products of $K_{\mu}(ax^{r})$ and the exponential function

1	$e^{(a\pm b)x}K_{\mu}(ax)\,K_{\nu}(bx)$	$\frac{\sqrt{\pi}\,a^{\mu}}{2^{s+2\mu+1}\,b^{s+\mu}}\,\Gamma\left[\begin{matrix}-\mu,\ s+\mu-\nu,\ s+\mu+\nu\\ \frac{2s+2\mu+1}{2}\end{matrix}\right] \times[\cos(\nu\pi)\sec(s+\mu)\pi]^{(1\pm1)/2} \times{}_3F_2\left(\begin{matrix}\frac{1+2\mu}{2},\ s+\mu-\nu,\ s+\mu+\nu\\ 2\mu+1,\ \frac{2s+2\mu+1}{2};\ \mp\frac{a}{b}\end{matrix}\right)+\frac{\sqrt{\pi}\,a^{-\mu}}{2^{s-2\mu+1}b^{s-\mu}} \times\Gamma\left[\begin{matrix}\mu,\ s-\mu-\nu,\ s-\mu+\nu\\ \frac{2s-2\mu+1}{2}\end{matrix}\right][\cos(\nu\pi)\sec(s-\mu)\pi]^{(1\pm1)/2} \times{}_3F_2\left(\begin{matrix}\frac{1-2\mu}{2},\ s-\nu-\mu,\ s+\nu-\mu\\ 1-2\mu,\ \frac{2s-2\mu+1}{2};\ \mp\frac{a}{b}\end{matrix}\right) +\frac{(1\pm1)\cos(\mu\pi)}{2^{s+1}a^{s-1/2}\sqrt{b}}\,\Gamma\left[\frac{2s-2\mu-1}{2},\ \frac{2s+2\mu-1}{2},\ 1-s\right] \times{}_3F_2\left(\begin{matrix}\frac{1+2\nu}{2},\ \frac{1-2\nu}{2},\ 1-s;\ \mp\frac{a}{b}\\ \frac{3-2s-2\mu}{2},\ \frac{3-2s+2\mu}{2}\end{matrix}\right)$ $\left[\operatorname{Re}s>\lvert\operatorname{Re}\mu\rvert+\lvert\operatorname{Re}\nu\rvert;\ \begin{Bmatrix}\operatorname{Re}a,\ \operatorname{Re}b>0;\ \operatorname{Re}s<1\\ \operatorname{Re}b>0\end{Bmatrix}\right]$
2	$e^{-(a+b)x}K_{\mu}(ax)\,K_{\nu}(bx)$	$\frac{\sqrt{\pi}\,b^{\nu}}{2^{s+2\nu+1}a^{s+\nu}}\,\Gamma\left[\begin{matrix}-\nu,\ s+\mu+\nu,\ s-\mu+\nu\\ \frac{2s+2\nu+1}{2}\end{matrix}\right] \times{}_3F_2\left(\begin{matrix}\frac{1+2\nu}{2},\ s+\mu+\nu,\ s-\mu+\nu\\ 1+2\nu,\ \frac{2s+2\nu+1}{2};\ -\frac{b}{a}\end{matrix}\right) +\frac{\sqrt{\pi}\,a^{\nu-s}\,b^{-\nu}}{2^{s-2\nu+1}}\,\Gamma\left[\begin{matrix}\nu,\ s+\mu-\nu,\ s-\mu-\nu\\ \frac{2s-2\nu+1}{2}\end{matrix}\right] \times{}_3F_2\left(\begin{matrix}\frac{1-2\nu}{2},\ s+\mu-\nu,\ s-\mu-\nu\\ 1-2\nu,\ \frac{2s-2\nu+1}{2};\ -\frac{b}{a}\end{matrix}\right)$ $[\operatorname{Re}(a+b)>0;\ \operatorname{Re}s>\lvert\operatorname{Re}\mu\rvert+\lvert\operatorname{Re}\nu\rvert]$
3	$e^{\pm ax^{2}}K_{\mu}(ax^{2})\,K_{\nu}(bx)$	$\frac{\sqrt{\pi}\,b^{\nu}}{2^{(s+3\nu)/2+2}a^{(s+\nu)/2}}\left[\cos(\mu\pi)\sec\frac{(s+\nu)\pi}{2}\right]^{(1\pm1)/2} \times\Gamma\left[\begin{matrix}-\nu,\ \frac{s+2\mu+\nu}{2},\ \frac{s-2\mu+\nu}{2}\\ \frac{s+\nu+1}{2}\end{matrix}\right]{}_2F_2\left(\begin{matrix}\frac{s+2\mu+\nu}{2},\ \frac{s-2\mu+\nu}{2}\\ \nu+1,\ \frac{s+\nu+1}{2};\ \mp\frac{b^{2}}{8a}\end{matrix}\right) +\frac{\sqrt{\pi}\,a^{(\nu-s)/2}b^{-\nu}}{2^{(s-3\nu)/2+2}}\left[\cos(\mu\pi)\sec\frac{(s-\nu)\pi}{2}\right]^{(1\pm1)/2} \times\Gamma\left[\begin{matrix}\nu,\ \frac{s+2\mu-\nu}{2},\ \frac{s-2\mu-\nu}{2}\\ \frac{s-\nu+1}{2}\end{matrix}\right]{}_2F_2\left(\begin{matrix}\frac{s+2\mu-\nu}{2},\ \frac{s-2\mu-\nu}{2}\\ 1-\nu,\ \frac{s-\nu+1}{2};\ \mp\frac{b^{2}}{8a}\end{matrix}\right)+\frac{(1\pm1)}{2^{9/2-s}b^{s-1}} \times\sqrt{\frac{\pi}{a}}\,\Gamma\left(\frac{s-\nu-1}{2}\right)\Gamma\left(\frac{s+\nu-1}{2}\right){}_2F_2\left(\begin{matrix}\frac{1+2\mu}{2},\ \frac{1-2\mu}{2}\\ \frac{3-s-\nu}{2},\ \frac{3-s+\nu}{2};\ -\frac{b^{2}}{8a}\end{matrix}\right)$ $\left[\operatorname{Re}s>2\lvert\operatorname{Re}\mu\rvert+\lvert\operatorname{Re}\nu\rvert;\ \begin{Bmatrix}\operatorname{Re}a>0\\ \operatorname{Re}b>0\end{Bmatrix}\right]$

No.	$f(x)$	$F(s)$
4	$e^{\pm a/x^2} K_\mu\left(\frac{a}{x^2}\right) K_\nu(bx)$	$\frac{2^{s-3\mu-3}a^{\mu}}{b^{s-2\mu}}\Gamma\left[-\mu, \frac{s-2\mu+\nu}{2}, \frac{s-2\mu-\nu}{2}\right] \times {}_1F_3\left(\begin{matrix}\frac{1+2\mu}{2};\ \pm\frac{ab^2}{2}\\ 1+2\mu,\ \frac{1-s+2\mu-\nu}{2},\ \frac{1-s+2\mu+\nu}{2}\end{matrix}\right) + \frac{2^{s+3\mu-3}a^{-\mu}}{b^{s+2\mu}}\Gamma\left[\mu, \frac{s+2\mu-\nu}{2}, \frac{s+2\mu+\nu}{2}\right] \times {}_1F_3\left(\begin{matrix}\frac{1-2\mu}{2};\ \pm\frac{ab^2}{2}\\ 1-2\mu,\ \frac{2-s-2\mu-\nu}{2},\ \frac{2-s-2\mu+\nu}{2}\end{matrix}\right) + \frac{\sqrt{\pi}\,a^{(s+\nu)/2}b^{\nu}}{2^{(\nu-s)/2+2}}\Gamma\left[\begin{matrix}-\nu, \frac{2\mu-\nu-s}{2}, \frac{-2\mu-\nu-s}{2}\\ \frac{1-s-\nu}{2}\end{matrix}\right] \times\left[\cos(\mu\pi)\sec\frac{(s+\nu)\pi}{2}\right]^{(1\pm1)/2} \times {}_1F_3\left(\begin{matrix}\frac{s+\nu+1}{2};\ \pm\frac{ab^2}{2}\\ 1+\nu,\ \frac{s-2\mu+\nu+2}{2},\ \frac{s+2\mu+\nu+2}{2}\end{matrix}\right) + \frac{2^{(s+\nu)/2-2}\sqrt{\pi}}{a^{(\nu-s)/2}b^{\nu}}\Gamma\left[\begin{matrix}\nu, \frac{2\mu+\nu-s}{2}, \frac{-2\mu+\nu-s}{2}\\ \frac{1-s+\nu}{2}\end{matrix}\right]\times\left[\cos(\mu\pi)\sec\frac{(s-\nu)\pi}{2}\right]^{(1\pm1)/2}\times{}_1F_3\left(\begin{matrix}\frac{s-\nu+1}{2};\ \pm\frac{ab^2}{2}\\ 1-\nu,\ \frac{s-2\mu-\nu+2}{2},\ \frac{s+2\mu-\nu+2}{2}\end{matrix}\right)$ $\left[\operatorname{Re} b>0;\ \begin{Bmatrix}\operatorname{Re} s>\lvert\operatorname{Re}\nu\rvert-1\\ \operatorname{Re} a>0\end{Bmatrix}\right]$

3.14.20. Products of $K_\mu(ax^r)$ and trigonometric or hyperbolic functions

Notation: $\delta = \begin{Bmatrix}1\\0\end{Bmatrix}$.

1	$\sin(ax)\,K_\mu(bx)\,K_\nu(bx)$	$\frac{2^{s-2}a}{b^{s+1}}\Gamma\left[\begin{matrix}\frac{s+\mu+\nu+1}{2}, \frac{s+\mu-\nu+1}{2}, \frac{s-\mu+\nu+1}{2}, \frac{s-\mu-\nu+1}{2}\\ s+1\end{matrix}\right] \times {}_4F_3\left(\begin{matrix}\frac{s+\mu+\nu+1}{2}, \frac{s+\mu-\nu+1}{2}, \frac{s-\mu+\nu+1}{2}, \frac{s-\mu-\nu+1}{2}\\ \frac{3}{2}, \frac{s+1}{2}, \frac{s+2}{2};\ -\frac{a^2}{4b^2}\end{matrix}\right)$ $[2\operatorname{Re} b>\lvert\operatorname{Im} a\rvert;\ \operatorname{Re} s>\lvert\operatorname{Re}\mu\rvert+\lvert\operatorname{Re}\nu\rvert-1]$
2	$\cos(ax)\,K_\mu(bx)\,K_\nu(bx)$	$\frac{2^{s-3}}{b^{s}}\Gamma\left[\begin{matrix}\frac{s+\mu+\nu}{2}, \frac{s+\mu-\nu}{2}, \frac{s-\mu+\nu}{2}, \frac{s-\mu-\nu}{2}\\ s\end{matrix}\right] \times {}_4F_3\left(\begin{matrix}\frac{s+\mu+\nu}{2}, \frac{s+\mu-\nu}{2}, \frac{s-\mu+\nu}{2}, \frac{s-\mu-\nu}{2}\\ \frac{1}{2}, \frac{s}{2}, \frac{s+1}{2};\ -\frac{a^2}{4b^2}\end{matrix}\right)$ $[2\operatorname{Re} b>\lvert\operatorname{Im} a\rvert;\ \operatorname{Re} s>\lvert\operatorname{Re}\mu\rvert+\lvert\operatorname{Re}\nu\rvert]$

No.	$f(x)$	$F(s)$
3	$\begin{Bmatrix}\sin(ax)\sinh(ax)\\ \cos(ax)\cosh(ax)\end{Bmatrix}$ $\times K_\mu(bx)\,K_\nu(bx)$	$\frac{\sqrt{\pi}\,a^{2\delta}}{4b^{s+2\delta}}\,\Gamma\left[\begin{matrix}\frac{s-\mu-\nu+2\delta}{2}, \frac{s-\mu+\nu+2\delta}{2}, \frac{s+\mu-\nu+2\delta}{2}, \frac{s+\mu+\nu+2\delta}{2}\\ \frac{s+2\delta}{2}, \frac{s+2\delta+1}{2}\end{matrix}\right]$ $\times {}_8F_7\left(\begin{matrix}\Delta\left(2, \frac{s-\mu-\nu+2\delta}{2}\right), \Delta\left(2, \frac{s-\mu+\nu+2\delta}{2}\right), \Delta\left(2, \frac{s+\mu-\nu+2\delta}{2}\right), \Delta\left(2, \frac{s+\mu+\nu+2\delta}{2}\right)\\ \frac{2\delta+1}{4}, \frac{2\delta+3}{4}, \frac{2\delta+1}{2}, \Delta(4, s+2\delta)\end{matrix}; -\frac{a^4}{4b^4}\right)$ $[\operatorname{Re} b > (\operatorname{Re} a + \operatorname{Im} a)/2;\ \operatorname{Re} s > \lvert\operatorname{Re}\mu\rvert + \lvert\operatorname{Re}\nu\rvert]$
4	$\begin{Bmatrix}\sin(ax)\cosh(ax)\\ \cos(ax)\sinh(ax)\end{Bmatrix}$ $\times K_\mu(bx)\,K_\nu(bx)$	$\frac{\sqrt{\pi}\,a}{4b^{s+1}}\,\Gamma\left[\begin{matrix}\frac{s-\mu-\nu+1}{2}, \frac{s-\mu+\nu+1}{2}, \frac{s+\mu-\nu+1}{2}, \frac{s+\mu+\nu+1}{2}\\ \frac{s+1}{2}, \frac{s+2}{2}\end{matrix}\right]$ $\times {}_8F_7\left(\begin{matrix}\Delta\left(2, \frac{s-\mu-\nu+1}{2}\right), \Delta\left(2, \frac{s-\mu+\nu+1}{2}\right), \Delta\left(2, \frac{s+\mu-\nu+1}{2}\right), \Delta\left(2, \frac{s+\mu+\nu+1}{2}\right)\\ \frac{1}{2}, \frac{3}{4}, \frac{5}{4}, \Delta(4, s+1)\end{matrix}; -\frac{a^4}{4b^4}\right)$ $\pm \frac{\sqrt{\pi}\,a^3}{12b^{s+3}}\,\Gamma\left[\begin{matrix}\frac{s-\mu-\nu+3}{2}, \frac{s-\mu+\nu+3}{2}, \frac{s+\mu-\nu+3}{2}, \frac{s+\mu+\nu+3}{2}\\ \frac{s+3}{2}, \frac{s+4}{2}\end{matrix}\right]$ $\times {}_8F_7\left(\begin{matrix}\Delta\left(2, \frac{s-\mu-\nu+3}{2}\right), \Delta\left(2, \frac{s-\mu+\nu+3}{2}\right), \Delta\left(2, \frac{s+\mu-\nu+3}{2}\right), \Delta\left(2, \frac{s+\mu+\nu+3}{2}\right)\\ \frac{5}{4}, \frac{3}{2}, \frac{7}{4}, \Delta(4, s+3)\end{matrix}; -\frac{a^4}{4b^4}\right)$ $[\operatorname{Re} b > (\operatorname{Re} a + \operatorname{Im} a)/2;\ \operatorname{Re} s > \lvert\operatorname{Re}\mu\rvert + \lvert\operatorname{Re}\nu\rvert]$
5	$\begin{Bmatrix}\sin(a\sqrt{x})\sinh(a\sqrt{x})\\ \cos(a\sqrt{x})\cosh(a\sqrt{x})\end{Bmatrix}$ $\times K_\mu(bx)\,K_\nu(bx)$	$\frac{2^{s+\delta-3}a^{2\delta}}{b^{s+\delta}}\,\Gamma\left[\begin{matrix}\frac{s-\mu-\nu+\delta}{2}, \frac{s-\mu+\nu+\delta}{2}, \frac{s+\mu-\nu+\delta}{2}, \frac{s+\mu+\nu+\delta}{2}\\ s+\delta\end{matrix}\right]$ $\times {}_4F_5\left(\begin{matrix}\frac{s-\mu-\nu+\delta}{2}, \frac{s-\mu+\nu+\delta}{2}, \frac{s+\mu-\nu+\delta}{2}, \frac{s+\mu+\nu+\delta}{2}\\ \frac{2\delta+1}{4}, \frac{2\delta+3}{4}, \frac{2\delta+1}{2}, \frac{s+\delta}{2}, \frac{s+\delta+1}{2}\end{matrix}; -\frac{a^4}{64b^2}\right)$ $[\operatorname{Re} b > 0;\ \operatorname{Re} s > \lvert\operatorname{Re}\mu\rvert + \lvert\operatorname{Re}\nu\rvert - \delta]$
6	$\begin{Bmatrix}\sin(a\sqrt{x})\cosh(a\sqrt{x})\\ \cos(a\sqrt{x})\sinh(a\sqrt{x})\end{Bmatrix}$ $\times K_\mu(bx)\,K_\nu(bx)$	$\frac{2^{s-5/2}a}{b^{s+1/2}}\,\Gamma\left[\begin{matrix}\frac{2s-2\mu-2\nu+1}{4}, \frac{2s-2\mu+2\nu+1}{4}, \frac{2s+2\mu-2\nu+1}{4}, \frac{2s+2\mu+2\nu+1}{4}\\ \frac{2s+1}{2}\end{matrix}\right]$ $\times {}_4F_5\left(\begin{matrix}\frac{2s-2\mu-2\nu+1}{4}, \frac{2s-2\mu+2\nu+1}{4}, \frac{2s+2\mu-2\nu+1}{4}, \frac{2s+2\mu+2\nu+1}{4}\\ \frac{1}{2}, \frac{3}{4}, \frac{5}{4}, \frac{2s+1}{4}, \frac{2s+3}{4}\end{matrix}; -\frac{a^4}{64b^2}\right)$ $\pm \frac{2^{s-3/2}a^3}{3b^{s+3/2}}\,\Gamma\left[\begin{matrix}\frac{2s-2\mu-2\nu+3}{4}, \frac{2s-2\mu+2\nu+3}{4}, \frac{2s+2\mu-2\nu+3}{4}, \frac{2s+2\mu+2\nu+3}{4}\\ \frac{2s+3}{2}\end{matrix}\right]$ $\times {}_4F_5\left(\begin{matrix}\frac{2s-2\mu-2\nu+3}{4}, \frac{2s-2\mu+2\nu+3}{4}, \frac{2s+2\mu-2\nu+3}{4}, \frac{2s+2\mu+2\nu+3}{4}\\ \frac{5}{4}, \frac{3}{2}, \frac{7}{4}, \frac{2s+3}{4}, \frac{2s+5}{4}\end{matrix}; -\frac{a^4}{64b^2}\right)$ $[\operatorname{Re} b > 0;\ \operatorname{Re} s > \lvert\operatorname{Re}\mu\rvert + \lvert\operatorname{Re}\nu\rvert - 1/2]$

3.14.21. Products of $K_\nu(ax)$ and $\operatorname{erf}(b\sqrt{x})$, $\operatorname{erfi}(b\sqrt{x})$

Notation: $\delta = \begin{Bmatrix} 1 \\ 0 \end{Bmatrix}$.

1	$\operatorname{erf}(a\sqrt{x})\operatorname{erfi}(a\sqrt{x}) \times K_\mu(bx) K_\nu(bx)$	$\frac{2^s a^2 b^{-s-1}}{\pi} \Gamma\left[\begin{matrix} \frac{s-\mu-\nu+1}{2}, \frac{s-\mu+\nu+1}{2}, \frac{s+\mu-\nu+1}{2}, \frac{s+\mu+\nu+1}{2} \\ s+1 \end{matrix}\right] \times {}_6F_5\left(\begin{matrix} \frac{1}{2}, 1, \frac{s-\mu-\nu+1}{2}, \frac{s-\mu+\nu+1}{2}, \frac{s+\mu-\nu+1}{2}, \frac{s+\mu+\nu+1}{2} \\ \frac{3}{4}, \frac{5}{4}, \frac{3}{2}, \frac{s+1}{2}, \frac{s+2}{2}; \frac{a^4}{4b^2} \end{matrix}\right)$ $[\operatorname{Re} b > 0;\ \operatorname{Re} s > (\lvert\operatorname{Re}\mu\rvert + \lvert\operatorname{Re}\nu\rvert) - 1]$

3.14.22. Products of $K_\nu(ax)$ and $S(cx)$, $C(cx)$

Notation: $\delta = \begin{Bmatrix} 1 \\ 0 \end{Bmatrix}$.

1	$\begin{Bmatrix} S(ax) \\ C(ax) \end{Bmatrix} \times K_\mu(bx) K_\nu(bx)$	$\frac{a^{\delta+1/2} b^{-s-\delta-1/2}}{2\sqrt{2}(2\delta+1)} \Gamma\left[\begin{matrix} \frac{2s-2\mu-2\nu+2\delta+1}{4}, \frac{2s-2\mu+2\nu+2\delta+1}{4} \\ \frac{2s+2\delta+1}{4} \end{matrix}\right] \times \Gamma\left[\begin{matrix} \frac{2s+2\mu-2\nu+2\delta+1}{4}, \frac{2s+2\mu+2\nu+2\delta+1}{4} \\ \frac{2s+2\delta+3}{4} \end{matrix}\right] \times {}_5F_4\left(\begin{matrix} \frac{2\delta+1}{4}, \frac{2s-2\mu-2\nu+2\delta+1}{4}, \frac{2s-2\mu+2\nu+2\delta+1}{4}, \frac{2s+2\mu-2\nu+2\delta+1}{4}, \frac{2s+2\mu+2\nu+2\delta+1}{4} \\ \frac{2\delta+1}{2}, \frac{2\delta+5}{4}, \frac{2s+2\delta+1}{4}, \frac{2s+2\delta+3}{4}; -\frac{a^2}{4b^2} \end{matrix}\right)$ $[\operatorname{Re} b > \lvert\operatorname{Im} a\rvert;\ \operatorname{Re} s > \lvert\operatorname{Re}\mu\rvert + \lvert\operatorname{Re}\nu\rvert - \delta - 1/2]$

3.14.23. Products of $K_\nu(ax)$ and $J_\lambda(bx^r)$, $I_\mu(cx^r)$

1	$J_\lambda(ax) K_\mu(bx) K_\nu(cx)$	$\frac{2^{s-3} a^\lambda}{c^{s+\lambda}\Gamma(\lambda+1)} \left\{ \left(\frac{b}{c}\right)^\mu \Gamma\left[-\mu, \frac{s+\lambda+\mu-\nu}{2}, \frac{s+\lambda+\mu+\nu}{2}\right] \times F_4\left(\frac{s+\lambda+\mu-\nu}{2}, \frac{s+\lambda+\mu+\nu}{2}; \lambda+1, \mu+1; -\frac{a^2}{c^2}; \frac{b^2}{c^2}\right) + \left(\frac{b}{c}\right)^{-\mu} \Gamma\left[\mu, \frac{s+\lambda-\mu-\nu}{2}, \frac{s+\lambda-\mu+\nu}{2}\right] \times F_4\left(\frac{s+\lambda-\mu-\nu}{2}, \frac{s+\lambda-\mu+\nu}{2}; \lambda+1, 1-\mu; -\frac{a^2}{c^2}, \frac{b^2}{c^2}\right) \right\}$ $[\operatorname{Re}(b+c) > \lvert\operatorname{Im} a\rvert;\ \operatorname{Re}(s+\lambda) > \lvert\operatorname{Re}\mu\rvert + \lvert\operatorname{Re}\nu\rvert]$

No.	$f(x)$	$F(s)$
2	$\begin{Bmatrix} J_\lambda(ax) \\ I_\lambda(ax) \end{Bmatrix} K_\mu(bx)$ $\times K_\nu(bx)$	$\frac{2^{s-3}a^\lambda}{b^{s+\lambda}} \Gamma\left[\begin{matrix} \frac{s+\lambda+\mu+\nu}{2}, \frac{s+\lambda+\mu-\nu}{2}, \frac{s+\lambda-\mu+\nu}{2}, \frac{s+\lambda-\mu-\nu}{2} \\ \lambda+1, s+\lambda \end{matrix}\right]$ $\times {}_4F_3\left(\begin{matrix} \frac{s+\lambda+\mu+\nu}{2}, \frac{s+\lambda+\mu-\nu}{2}, \frac{s+\lambda-\mu+\nu}{2}, \frac{s+\lambda-\mu-\nu}{2} \\ \lambda+1, \frac{s+\lambda}{2}, \frac{s+\lambda+1}{2}; \mp\frac{a^2}{4b^2} \end{matrix}\right)$ $\left[2\operatorname{Re} b > \begin{Bmatrix} \lvert\operatorname{Im} a\rvert \\ \lvert\operatorname{Re} a\rvert \end{Bmatrix}; \operatorname{Re}(s+\lambda) > \lvert\operatorname{Re}\mu\rvert + \lvert\operatorname{Re}\nu\rvert\right]$
3	$J_\mu(a\sqrt{x})\, K_\mu(a\sqrt{x})$ $\times K_\nu(bx)$	$\frac{2^{s-3}}{\mu b^s}\Gamma\left(\frac{s-\nu}{2}\right)\Gamma\left(\frac{s+\nu}{2}\right) {}_2F_3\left(\begin{matrix}\frac{s-\nu}{2}, \frac{s+\nu}{2}; -\frac{a^4}{16b^2} \\ \frac{1}{2}, \frac{2-\mu}{2}, \frac{2+\mu}{2}\end{matrix}\right) - \frac{2^{s-3}a^2}{b^{s+1}(\mu^2-1)}$ $\times \Gamma\left(\frac{s-\nu+1}{2}\right)\Gamma\left(\frac{s+\nu+1}{2}\right) {}_2F_3\left(\begin{matrix}\frac{s-\nu+1}{2}, \frac{s+\nu+1}{2}; -\frac{a^4}{16b^2} \\ \frac{3}{2}, \frac{3-\mu}{2}, \frac{3+\mu}{2}\end{matrix}\right)$ $+ \frac{2^{s-\mu-3}a^{2\mu}}{b^{s+\mu}}\Gamma\left[\begin{matrix}-\mu, \frac{s+\mu-\nu}{2}, \frac{s+\mu+\nu}{2} \\ \mu+1\end{matrix}\right] {}_2F_3\left(\begin{matrix}\frac{s+\mu-\nu}{2}, \frac{s+\mu+\nu}{2}; -\frac{a^4}{16b^2} \\ \mu+1, \frac{\mu+1}{2}, \frac{\mu+2}{2}\end{matrix}\right)$ $[a,\ \operatorname{Re} b,\ \operatorname{Re}(s+\nu),\ \operatorname{Re}(s+\mu+\nu) > 0]$
4	$I_\lambda(ax)\, K_\mu(ax)\, K_\nu(bx)$	$\frac{2^{s-3}a^{\lambda+\mu}}{b^{s+\lambda+\mu}}\Gamma\left[\begin{matrix}-\mu, \frac{s+\lambda+\mu+\nu}{2}, \frac{s+\lambda+\mu-\nu}{2} \\ \lambda+1\end{matrix}\right]$ $\times {}_4F_3\left(\begin{matrix}\frac{\lambda+\mu+1}{2}, \frac{\lambda+\mu+2}{2}, \frac{s+\lambda+\mu+\nu}{2}, \frac{s+\lambda+\mu-\nu}{2} \\ \lambda+1, \mu+1, \lambda+\mu+1; \frac{4a^2}{b^2}\end{matrix}\right) + \frac{2^{s-3}a^{\lambda-\mu}}{b^{s+\lambda-\mu}}\Gamma(\mu)$ $\times \Gamma\left[\begin{matrix}\frac{s+\lambda-\mu+\nu}{2}, \frac{s+\lambda-\mu-\nu}{2} \\ \lambda+1\end{matrix}\right] {}_4F_3\left(\begin{matrix}\frac{\lambda-\mu+1}{2}, \frac{\lambda-\mu+2}{2}, \frac{s+\lambda-\mu+\nu}{2}, \frac{s+\lambda-\mu-\nu}{2} \\ 1-\mu, \lambda+1, \lambda-\mu+1; \frac{4a^2}{b^2}\end{matrix}\right)$ $[\operatorname{Re} a,\ \operatorname{Re} b > 0;\ \operatorname{Re}(s+\lambda) > \lvert\operatorname{Re}\mu\rvert + \lvert\operatorname{Re}\nu\rvert]$
5	$K_\lambda(ax)\, K_\mu(bx)\, K_\nu(cx)$	$\frac{2^{s-4}}{c^s}\,[A(\lambda, \mu) + A(\lambda, -\mu) + A(-\lambda, \mu) + A(-\lambda, -\mu)]$ $A(\lambda, \mu) = \left(\frac{a}{c}\right)^\lambda \left(\frac{b}{c}\right)^\mu \Gamma\left[-\lambda, -\mu, \frac{s+\lambda+\mu-\nu}{2}, \frac{s+\lambda+\mu+\nu}{2}\right]$ $\times F_4\left(\frac{s+\lambda+\mu-\nu}{2}, \frac{s+\lambda+\mu+\nu}{2}; \lambda+1, \mu+1; \frac{a^2}{c^2}, \frac{b^2}{c^2}\right)$ $[\operatorname{Re}(a+b+c) > 0;\ \operatorname{Re} s > \lvert\operatorname{Re}\lambda\rvert + \lvert\operatorname{Re}\mu\rvert + \lvert\operatorname{Re}\nu\rvert]$
6	$K_\lambda(ax)\, K_\mu(ax)\, K_\nu(bx)$	$\frac{2^{s-4}b^{-\nu}}{a^{s-\nu}}\Gamma\left[\begin{matrix}\nu, \frac{s+\lambda+\mu-\nu}{2}, \frac{s-\lambda+\mu-\nu}{2}, \frac{s+\lambda-\mu-\nu}{2}, \frac{s-\lambda-\mu-\nu}{2} \\ s-\nu\end{matrix}\right]$ $\times {}_4F_3\left(\begin{matrix}\frac{s+\lambda+\mu-\nu}{2}, \frac{s-\lambda+\mu-\nu}{2}, \frac{s+\lambda-\mu-\nu}{2}, \frac{s-\lambda-\mu-\nu}{2} \\ 1-\nu, \frac{s-\nu}{2}, \frac{s-\nu+1}{2}; \frac{b^2}{4a^2}\end{matrix}\right)$ $+ \frac{2^{s-4}b^\nu}{a^{s+\nu}}\Gamma\left[\begin{matrix}-\nu, \frac{s+\lambda+\mu+\nu}{2}, \frac{s-\lambda+\mu+\nu}{2}, \frac{s+\lambda-\mu+\nu}{2}, \frac{s-\lambda-\mu+\nu}{2} \\ s+\nu\end{matrix}\right]$ $\times {}_4F_3\left(\begin{matrix}\frac{s+\lambda+\mu+\nu}{2}, \frac{s-\lambda+\mu+\nu}{2}, \frac{s+\lambda-\mu+\nu}{2}, \frac{s-\lambda-\mu+\nu}{2} \\ 1+\nu, \frac{s+\nu}{2}, \frac{s+\nu+1}{2}; \frac{b^2}{4a^2}\end{matrix}\right)$ $[\operatorname{Re}(2a+b) > 0;\ \operatorname{Re} s > \lvert\operatorname{Re}\lambda\rvert + \lvert\operatorname{Re}\mu\rvert + \lvert\operatorname{Re}\nu\rvert]$

3.15. The Struve Functions $\mathbf{H}_\nu(z)$ and $\mathbf{L}_\nu(z)$

More formulas can be obtained from the corresponding sections due to the relations

$$\mathbf{H}_{\pm 1/2}(z) = \sqrt{\frac{2}{\pi}} \frac{1}{\sqrt{z}} \left\{ \begin{matrix} 1-\cos z \\ \sin z \end{matrix} \right\}, \quad \mathbf{L}_{\pm 1/2}(z) = \sqrt{\frac{2}{\pi}} \frac{1}{\sqrt{z}} \left\{ \begin{matrix} \cosh z - 1 \\ \sinh z \end{matrix} \right\},$$

$$\mathbf{H}_{-n-1/2}(z) = (-1)^n J_{n+1/2}(z), \quad \mathbf{L}_{-n-1/2}(z) = I_{n+1/2}(z),$$

$$\left\{ \begin{matrix} \mathbf{H}_\nu(z) \\ \mathbf{L}_\nu(z) \end{matrix} \right\} = \frac{2^{-\nu} z^{\nu+1}}{\sqrt{\pi}\,\Gamma(\nu+3/2)} \, {}_1F_2\left(1;\, \frac{3}{2},\, \nu+\frac{3}{2};\, \mp\frac{z^2}{4}\right),$$

$$\mathbf{H}_\nu(z) = z^{\nu+1}\left(z^2\right)^{-(\nu+1)/2} G_{13}^{11}\left(\frac{z^2}{4} \,\middle|\, \begin{matrix} (\nu+1)/2 \\ (\nu+1)/2,\, -\nu/2,\, \nu/2 \end{matrix}\right).$$

3.15.1. $\mathbf{H}_\nu(bx)$, $\mathbf{L}_\nu(bx)$, and algebraic functions

No.	$f(x)$	$F(s)$
1	$\mathbf{H}_\nu(ax)$	$\frac{2^{s-1}}{a^s} \tan\frac{(s+\nu)\pi}{2}\, \Gamma\left[\begin{matrix} \frac{s+\nu}{2} \\ \frac{2-s+\nu}{2} \end{matrix}\right]$ $[a>0;\ \operatorname{Re} s < 3/2;\ \lvert\operatorname{Re}(s+\nu)\rvert < 1]$
2	$(a-x)_+^{\alpha-1} \left\{ \begin{matrix} \mathbf{H}_\nu(bx) \\ \mathbf{L}_\nu(bx) \end{matrix} \right\}$	$\frac{a^{s+\alpha+\nu}\, b^{\nu+1}}{2^\nu \sqrt{\pi}\,\Gamma\left(\frac{2\nu+3}{2}\right)} \mathrm{B}(\alpha,\, s+\nu+1)\, {}_3F_4\left(\begin{matrix} 1,\, \frac{s+\nu+1}{2},\, \frac{s+\nu+2}{2};\, \mp\frac{a^2b^2}{4} \\ \frac{3}{2},\, \frac{2\nu+3}{2},\, \frac{s+\alpha+\nu+1}{2},\, \frac{s+\alpha+\nu+2}{2} \end{matrix}\right)$ $[a,\ \operatorname{Re}\alpha > 0;\ \operatorname{Re}(s+\nu) > -1]$
3	$(x-a)_+^{\alpha-1}\, \mathbf{H}_\nu(bx)$	$\frac{a^{s+\alpha+\nu}\, b^{\nu+1}}{2^\nu\sqrt{\pi}} \Gamma\left[\begin{matrix} \alpha,\, -s-\alpha-\nu \\ \frac{2\nu+3}{2},\, -s-\nu \end{matrix}\right] {}_3F_4\left(\begin{matrix} 1,\, \frac{s+\nu+1}{2},\, \frac{s+\nu+2}{2};\, -\frac{a^2b^2}{4} \\ \frac{3}{2},\, \frac{2\nu+3}{2},\, \frac{s+\alpha+\nu+1}{2},\, \frac{s+\alpha+\nu+2}{2} \end{matrix}\right)$ $+\frac{\pi}{2}\left(\frac{b}{2}\right)^{1-\alpha-s} \csc\frac{(s+\alpha+\nu)\pi}{2} \frac{1}{\Gamma\left(\frac{3-s-\alpha+\nu}{2}\right)\Gamma\left(\frac{3-s-\alpha-\nu}{2}\right)}$ $\times\, {}_2F_3\left(\begin{matrix} \frac{1-\alpha}{2},\, \frac{2-\alpha}{2};\, -\frac{a^2b^2}{4} \\ \frac{1}{2},\, \frac{3-s-\alpha-\nu}{2},\, \frac{3-s-\alpha+\nu}{2} \end{matrix}\right)$ $-\frac{\pi a}{2}\left(\frac{b}{2}\right)^{2-\alpha-s} \sec\frac{(s+\alpha+\nu)\pi}{2} \frac{1-\alpha}{\Gamma\left(\frac{4-s-\alpha-\nu}{2}\right)\Gamma\left(\frac{4-s-\alpha+\nu}{2}\right)}$ $\times\, {}_2F_3\left(\begin{matrix} \frac{2-\alpha}{2},\, \frac{3-\alpha}{2};\, -\frac{a^2b^2}{4} \\ \frac{3}{2},\, \frac{4-s-\alpha+\nu}{2},\, \frac{4-s-\alpha-\nu}{2} \end{matrix}\right)$ $[a,\ b,\ \operatorname{Re}\alpha > 0;\ \operatorname{Re}(s+\alpha) < 5/2,\ 2-\operatorname{Re}\nu]$
4	$\left(a^2-x^2\right)_+^{\alpha-1}$ $\times \left\{ \begin{matrix} \mathbf{H}_\nu(bx) \\ \mathbf{L}_\nu(bx) \end{matrix} \right\}$	$\frac{a^{s+2\alpha+\nu-1}\, b^{\nu+1}}{2^{\nu+1}\sqrt{\pi}\,\Gamma\left(\frac{2\nu+3}{2}\right)} \mathrm{B}\left(\alpha,\, \frac{s+\nu+1}{2}\right) {}_2F_3\left(\begin{matrix} 1,\, \frac{s+\nu+1}{2};\, \mp\frac{a^2b^2}{4} \\ \frac{3}{2},\, \frac{2\nu+3}{2},\, \frac{s+2\alpha+\nu+1}{2} \end{matrix}\right)$ $[a,\ \operatorname{Re}\alpha > 0;\ \operatorname{Re}(s+\nu) > -1]$

No.	$f(x)$	$F(s)$
5	$\left(x^2-a^2\right)_+^{\alpha-1}\mathbf{H}_\nu(bx)$	$\frac{a^{s+2\alpha+\nu-1}\,b^{\nu+1}}{2^{\nu+1}\sqrt{\pi}\,\Gamma\left(\frac{2\nu+3}{2}\right)}\mathrm{B}\left(\alpha,\,\frac{1-s-2\alpha-\nu}{2}\right){}_2F_3\left(\begin{matrix}1,\ \frac{s+\nu+1}{2};\ -\frac{a^2b^2}{4}\\ \frac{3}{2},\ \frac{2\nu+3}{2},\ \frac{s+2\alpha+\nu+1}{2}\end{matrix}\right) - \frac{\pi}{2}\left(\frac{b}{2}\right)^{2-2\alpha-s}\frac{\sec\frac{(s+2\alpha+\nu)\pi}{2}}{\Gamma\left(\frac{4-s-2\alpha-\nu}{2}\right)\Gamma\left(\frac{4-s-2\alpha+\nu}{2}\right)} \times{}_1F_2\left(\begin{matrix}1-\alpha;\ -\frac{a^2b^2}{4}\\ \frac{4-s-2\alpha+\nu}{2},\ \frac{4-s-2\alpha-\nu}{2}\end{matrix}\right)$ $[a,\ \operatorname{Re}\alpha>0;\ \operatorname{Re}(s+2\alpha)<7/2,\ 3-\operatorname{Re}\nu]$
6	$\frac{1}{(x+a)^\rho}\mathbf{H}_\nu(bx)$	$\frac{a^{s+\nu-\rho+1}b^{\nu+1}}{2^\nu\sqrt{\pi}\,\Gamma\left(\frac{2\nu+3}{2}\right)}\mathrm{B}(s+\nu+1,\ \rho-\nu-s-1) \times{}_3F_4\left(\begin{matrix}1,\ \frac{s+\nu+1}{2},\ \frac{s+\nu+2}{2};\ -\frac{a^2b^2}{4}\\ \frac{3}{2},\ \frac{2\nu+3}{2},\ \frac{s+\nu-\rho+2}{2},\ \frac{s+\nu-\rho+3}{2}\end{matrix}\right)+\frac{\pi}{2}\left(\frac{b}{2}\right)^{\rho-s} \times\frac{\sec\frac{(s+\nu-\rho)\pi}{2}}{\Gamma\left(\frac{2-s-\nu+\rho}{2}\right)\Gamma\left(\frac{2-s+\nu+\rho}{2}\right)}{}_2F_3\left(\begin{matrix}\frac{\rho}{2},\ \frac{\rho+1}{2};\ -\frac{a^2b^2}{4}\\ \frac{1}{2},\ \frac{2-s-\nu+\rho}{2},\ \frac{2-s+\nu+\rho}{2}\end{matrix}\right) - \frac{\pi a}{2}\left(\frac{b}{2}\right)^{\rho-s+1}\frac{\rho\csc\frac{(s+\nu-\rho)\pi}{2}}{\Gamma\left(\frac{3-s-\nu+\rho}{2}\right)\Gamma\left(\frac{3-s+\nu+\rho}{2}\right)} \times{}_2F_3\left(\begin{matrix}\frac{\rho+1}{2},\ \frac{\rho+2}{2};\ -\frac{a^2b^2}{4}\\ \frac{3}{2},\ \frac{3-s-\nu+\rho}{2},\ \frac{3-s+\nu+\rho}{2}\end{matrix}\right)$ $\left[\begin{matrix}b>0;\ \operatorname{Re}(s-\rho)<3/2;\\ -1<\operatorname{Re}(s+\nu)<\operatorname{Re}\rho+1;\ \lvert\arg a\rvert<\pi\end{matrix}\right]$
7	$\frac{1}{x-a}\mathbf{H}_\nu(bx)$	$-\pi\,a^{s-1}\cot[(s+\nu)\pi]\,\mathbf{H}_\nu(ab) + \frac{2^{s-2}\pi\csc\frac{(s+\nu)\pi}{2}}{b^{s-1}\Gamma\left(\frac{3-s-\nu}{2}\right)\Gamma\left(\frac{3-s+\nu}{2}\right)}{}_1F_2\left(\begin{matrix}1;\ -\frac{a^2b^2}{4}\\ \frac{3-s-\nu}{2},\ \frac{3-s+\nu}{2}\end{matrix}\right) - \frac{\pi}{2}\left(\frac{b}{2}\right)^{2-s}\frac{a\sec\frac{(s+\nu)\pi}{2}}{\Gamma\left(\frac{4-s-\nu}{2}\right)\Gamma\left(\frac{4-s+\nu}{2}\right)}{}_1F_2\left(\begin{matrix}1;\ -\frac{a^2b^2}{4}\\ \frac{4-s-\nu}{2},\ \frac{4-s+\nu}{2}\end{matrix}\right)$ $[a,\ b>0;\ -1<\operatorname{Re}(s+\nu)<3;\ \operatorname{Re}s<5/2]$
8	$\frac{1}{(x^2+a^2)^\rho}\mathbf{H}_\nu(bx)$	$\frac{a^{s+\nu-2\rho+1}\,b^{\nu+1}}{2^{\nu+1}\sqrt{\pi}\,\Gamma\left(\frac{2\nu+3}{2}\right)}\mathrm{B}\left(\frac{s+\nu+1}{2},\ \frac{2\rho-\nu-s-1}{2}\right) \times{}_2F_3\left(\begin{matrix}1,\ \frac{s+\nu+1}{2};\ \frac{a^2b^2}{4}\\ \frac{3}{2},\ \frac{2\nu+3}{2},\ \frac{s+\nu-2\rho+3}{2}\end{matrix}\right)+\frac{\pi\,(b/2)^{2\rho-s}}{2\Gamma\left(\frac{2-s-\nu+2\rho}{2}\right)\Gamma\left(\frac{2-s+\nu+2\rho}{2}\right)} \times\sec\frac{(s+\nu-2\rho)\pi}{2}\,{}_1F_2\left(\begin{matrix}\rho;\ \frac{a^2b^2}{4}\\ \frac{2-s-\nu+2\rho}{2},\ \frac{2-s+\nu+2\rho}{2}\end{matrix}\right)$ $\left[\begin{matrix}b,\ \operatorname{Re}a>0;\ \operatorname{Re}(s-2\rho)<3/2;\\ -1<\operatorname{Re}(s+\nu)<2\operatorname{Re}\rho+1\end{matrix}\right]$

No.	$f(x)$	$F(s)$
9	$\frac{1}{x^2-a^2}\mathbf{H}_\nu(bx)$	$\frac{\pi a^{s-2}}{2}\tan\frac{(s+\nu)\pi}{2}\mathbf{H}_\nu(ab)+2^{s-3}b^{2-s}\tan\frac{(s+\nu)\pi}{2}$ $\times\Gamma\left[\begin{matrix}\frac{s+\nu-2}{2}\\ \frac{4-s+\nu}{2}\end{matrix}\right]{}_1F_2\left(\begin{matrix}1;\,-\frac{a^2b^2}{4}\\ \frac{4-s-\nu}{2},\,\frac{4-s+\nu}{2}\end{matrix}\right)$ $[a,\,b>0;\ \operatorname{Re}s<7/2;\ -1<\operatorname{Re}(s+\nu)<3]$
10	$(a-x)_+^{\alpha-1}$ $\times\left\{\begin{matrix}\mathbf{H}_\nu(bx(a-x))\\ \mathbf{L}_\nu(bx(a-x))\end{matrix}\right\}$	$\frac{a^{s+\alpha+2\nu+1}b^{\nu+1}}{2^\nu\sqrt{\pi}}\Gamma\left[\begin{matrix}\alpha+\nu+1,\,s+\nu+1\\ \frac{2\nu+3}{2},\,s+\alpha+2\nu+2\end{matrix}\right]$ $\times{}_5F_6\left(\begin{matrix}1,\,\Delta(2,\,\alpha+\nu+1),\,\Delta(2,\,s+\nu+1)\\ \frac{3}{2},\,\frac{2\nu+3}{2},\,\Delta(4,\,s+\alpha+2\nu+2);\,\mp\frac{a^4b^2}{64}\end{matrix}\right)$ $[a>0;\ \operatorname{Re}(s+\nu),\ \operatorname{Re}(\alpha+\nu)>-1]$
11	$(a-x)_+^{\alpha-1}$ $\times\left\{\begin{matrix}\mathbf{H}_\nu(b\sqrt{x(a-x)})\\ \mathbf{L}_\nu(b\sqrt{x(a-x)})\end{matrix}\right\}$	$\frac{a^{s+\alpha+\nu}b^{\nu+1}}{2^\nu\sqrt{\pi}}\Gamma\left[\begin{matrix}\frac{2\alpha+\nu+1}{2},\,\frac{2s+\nu+1}{2}\\ \frac{2\nu+3}{2},\,s+\alpha+\nu+1\end{matrix}\right]$ $\times{}_3F_4\left(\begin{matrix}1,\,\frac{2\alpha+\nu+1}{2},\,\frac{2s+\nu+1}{2};\,\mp\frac{a^2b^2}{16}\\ \frac{3}{2},\,\frac{2\nu+3}{2},\,\frac{s+\alpha+\nu+1}{2},\,\frac{s+\alpha+\nu+2}{2}\end{matrix}\right)$ $[a>0;\ \operatorname{Re}s,\ \operatorname{Re}\alpha>-\operatorname{Re}(\nu+1)/2]$
12	$(x^2+a^2)^{-\nu/2}$ $\times\mathbf{H}_\nu(b\sqrt{x^2+a^2})$	$\frac{1}{(2a)^\nu}\left(\frac{a}{b}\right)^{s/2}\Gamma\left(\frac{s}{2}\right)\left[\frac{1}{\pi}\Gamma\left[\begin{matrix}\frac{1-s}{2}\\ \frac{2\nu+1}{2}\end{matrix}\right]S_{s/2+\nu,s/2-\nu}(ab)\right.$ $\left.+2^{\nu-s/2-1}Y_{\nu-s/2}(ab)\right]$ $[a,\,b>0;\,0<\operatorname{Re}s<1,\ \operatorname{Re}\nu+3/2]$
13	$(x^2+a^2)^{\nu/2}$ $\times\mathbf{H}_\nu(b\sqrt{x^2+a^2})$	$\frac{2^{s/2-1}a^{s/2+\nu}}{b^{s/2}}\Gamma\left(\frac{s}{2}\right)\sec\frac{(2\nu+s)\pi}{2}$ $\times\left[\cos(\nu\pi)\mathbf{H}_{s/2+\nu}(ab)+\sin(s\pi)J_{-s/2-\nu}(ab)\right]$ $[a,\,b>0;\,0<\operatorname{Re}s<1-2\operatorname{Re}\nu,\,3/2-\operatorname{Re}\nu]$
14	$(a^2-x^2)_+^{-\nu/2}$ $\times\mathbf{H}_\nu(b\sqrt{a^2-x^2})$	$\frac{a^{s+1}b^{\nu+1}}{2^{\nu+2}}\Gamma\left[\begin{matrix}\frac{s}{2}\\ \frac{2\nu+3}{2},\,\frac{s+3}{2}\end{matrix}\right]{}_1F_2\left(\begin{matrix}1;\,-\frac{a^2b^2}{4}\\ \frac{2\nu+3}{2},\,\frac{s+3}{2}\end{matrix}\right)$ $[a,\,b,\,\operatorname{Re}s>0]$
15	$(a^2-x^2)_+^{\nu/2}$ $\times\left\{\begin{matrix}\mathbf{H}_\nu(b\sqrt{a^2-x^2})\\ \mathbf{L}_\nu(b\sqrt{a^2-x^2})\end{matrix}\right\}$	$\frac{2^{s/2-1}a^{s/2+\nu}}{b^{s/2}}\Gamma\left(\frac{s}{2}\right)\left\{\begin{matrix}\mathbf{H}_{s/2+\nu}(ab)\\ \mathbf{L}_{s/2+\nu}(ab)\end{matrix}\right\}$ $[a,\,b,\,\operatorname{Re}s>0;\ \operatorname{Re}\nu>-3/2]$

No.	$f(x)$	$F(s)$
16	$\dfrac{1}{(x+a)^{\rho}} \times \begin{Bmatrix} \mathbf{H}_{\nu}(bx/(x+a)) \\ \mathbf{L}_{\nu}(bx/(x+a)) \end{Bmatrix}$	$\dfrac{a^{s-\rho}b^{\nu+1}}{2^{\nu}\sqrt{\pi}\,\Gamma\left(\frac{2\nu+3}{2}\right)} \mathrm{B}(s+\nu+1,\, \rho-s)\; {}_3F_4\left(\begin{matrix} 1,\, \frac{s+\nu+1}{2},\, \frac{s+\nu+2}{2};\, \mp\frac{b^2}{4} \\ \frac{3}{2},\, \frac{2\nu+3}{2},\, \frac{\nu+\rho+1}{2},\, \frac{\nu+\rho+2}{2} \end{matrix}\right)$ $[-\operatorname{Re}\nu-1<\operatorname{Re}s<\operatorname{Re}\rho;\ \lvert\arg a\rvert<\pi]$
17	$\dfrac{1}{(x^2+a^2)^{\rho}} \times \begin{Bmatrix} \mathbf{H}_{\nu}(bx/(x^2+a^2)) \\ \mathbf{L}_{\nu}(bx/(x^2+a^2)) \end{Bmatrix}$	$\dfrac{a^{s-\nu-2\rho-1}}{\sqrt{\pi}\,\Gamma\left(\frac{2\nu+3}{2}\right)} \left(\dfrac{b}{2}\right)^{\nu+1} \mathrm{B}\left(\dfrac{s+\nu+1}{2},\, \dfrac{1-s+\nu+2\rho}{2}\right)$ $\times\; {}_3F_4\left(\begin{matrix} 1,\, \frac{s+\nu+1}{2},\, \frac{1-s+\nu+2\rho}{2};\, \mp\frac{b^2}{16a^2} \\ \frac{3}{2},\, \frac{2\nu+3}{2},\, \frac{\nu+\rho+1}{2},\, \frac{\nu+\rho+2}{2} \end{matrix}\right)$ $[\operatorname{Re}a>0;\ -\operatorname{Re}\nu-1<\operatorname{Re}s<\operatorname{Re}(\nu+2\rho)+1]$

3.15.2. $\mathbf{H}_{\nu}(bx)$, $\mathbf{L}_{\nu}(bx)$, and the exponential function

1	$e^{-ax}\begin{Bmatrix} \mathbf{H}_{\nu}(bx) \\ \mathbf{L}_{\nu}(bx) \end{Bmatrix}$	$\dfrac{b^{\nu+1}}{2^{\nu}\sqrt{\pi}\,a^{s+\nu+1}} \Gamma\left[\begin{matrix} s+\nu+1 \\ \frac{2\nu+3}{2} \end{matrix}\right] {}_3F_2\left(\begin{matrix} 1,\, \frac{s+\nu+1}{2},\, \frac{s+\nu+2}{2} \\ \frac{3}{2},\, \frac{2\nu+3}{2};\, \mp\frac{b^2}{a^2} \end{matrix}\right)$ $[b,\ \operatorname{Re}a>0;\ \operatorname{Re}(s+\nu)>-1]$
2	$e^{-ax^2}\begin{Bmatrix} \mathbf{H}_{\nu}(bx) \\ \mathbf{L}_{\nu}(bx) \end{Bmatrix}$	$\dfrac{(b/2)^{\nu+1}}{\sqrt{\pi}a^{(s+\nu+1)/2}} \Gamma\left[\begin{matrix} \frac{s+\nu+1}{2} \\ \frac{2\nu+3}{2} \end{matrix}\right] {}_2F_2\left(\begin{matrix} 1,\, \frac{s+\nu+1}{2};\, \mp\frac{b^2}{4a} \\ \frac{3}{2},\, \frac{2\nu+3}{2} \end{matrix}\right)$ $[\operatorname{Re}a>0;\ \operatorname{Re}(s+\nu)>-1]$
3	$e^{-a/x^2}\,\mathbf{H}_{\nu}(bx)$	$\dfrac{a^{(s+\nu+1)/2}b^{\nu+1}}{2^{\nu+1}\sqrt{\pi}} \Gamma\left[\begin{matrix} -\frac{s+\nu+1}{2} \\ \frac{2\nu+3}{2} \end{matrix}\right] {}_1F_3\left(\begin{matrix} 1;\, \frac{ab^2}{4} \\ \frac{3}{2},\, \frac{2\nu+3}{2},\, \frac{s+\nu+3}{2} \end{matrix}\right)$ $+\dfrac{2^{s-1}\pi}{b^s\,\Gamma\left(\frac{2-s-\nu}{2}\right)\Gamma\left(\frac{2-s+\nu}{2}\right)} \sec\dfrac{(s+\nu)\pi}{2}\; {}_0F_2\left(\begin{matrix} \frac{ab^2}{4} \\ \frac{2-s-\nu}{2},\, \frac{2-s+\nu}{2} \end{matrix}\right)$ $[b,\ \operatorname{Re}a>0;\ \operatorname{Re}s<3/2, 1-\operatorname{Re}\nu]$

3.15.3. $\mathbf{H}_{\nu}(bx)$, $\mathbf{L}_{\nu}(bx)$, and trigonometric functions

Notation: $\delta=\begin{Bmatrix} 1 \\ 0 \end{Bmatrix}$.

1	$\begin{Bmatrix} \sin(ax) \\ \cos(ax) \end{Bmatrix} \mathbf{H}_{\nu}(bx)$	$\pm\dfrac{a^{-s-\nu-1}}{2^{\nu}\sqrt{\pi}} \Gamma\left[\begin{matrix} s+\nu+1 \\ \frac{2\nu+3}{2} \end{matrix}\right] \begin{Bmatrix} \cos[(s+\nu)\pi/2] \\ \sin[(s+\nu)\pi/2] \end{Bmatrix}$ $\times\; {}_3F_2\left(\begin{matrix} \frac{s+\nu+1}{2},\, \frac{s+\nu+2}{2},\, 1 \\ \frac{3}{2},\, \frac{2\nu+3}{2};\, \frac{b^2}{a^2} \end{matrix}\right)$ $[0<b<a;\ \operatorname{Re}s<3/2;\ -\delta-1<\operatorname{Re}(s+\nu)<2]$

No.	$f(x)$	$F(s)$
2	$\begin{Bmatrix} \sin(ax) \\ \cos(ax) \end{Bmatrix} \mathbf{H}_\nu(bx)$	$\mp \frac{(b/2)^{\nu-1}}{\sqrt{\pi}\, a^{s+\nu-1}} \Gamma\left[\begin{matrix} s+\nu-1 \\ \frac{2\nu+1}{2} \end{matrix}\right] \begin{Bmatrix} \cos[(s+\nu)\pi/2] \\ \sin[(s+\nu)\pi/2] \end{Bmatrix} \times {}_3F_2\left(\begin{matrix} \frac{1}{2}, 1, \frac{1-2\nu}{2};\ \frac{a^2}{b^2} \\ \frac{2-s-\nu}{2}, \frac{3-s-\nu}{2} \end{matrix}\right) + \frac{2^{s+\delta-1}\pi a^\delta \sec\frac{(s+\nu+\delta)\pi}{2}}{b^{s+\delta}\Gamma\left(\frac{2-s-\nu-\delta}{2}\right)\Gamma\left(\frac{2-s+\nu-\delta}{2}\right)} {}_2F_1\left(\begin{matrix} \frac{s-\nu+\delta}{2}, \frac{s+\nu+\delta}{2} \\ \frac{2\delta+1}{2};\ \frac{a^2}{b^2} \end{matrix}\right)$ $[0 < a < b;\ \operatorname{Re} s < 3/2;\ -\delta - 1 < \operatorname{Re}(s+\nu) < 2]$

3.15.4. $\mathbf{H}_\nu(bx)$, $\mathbf{L}_\nu(bx)$, and the logarithmic or inverse trigonometric functions

1	$\theta(a-x)\ln\frac{\sqrt{a}+\sqrt{a-x}}{\sqrt{x}} \times \begin{Bmatrix} \mathbf{H}_\nu(bx) \\ \mathbf{L}_\nu(bx) \end{Bmatrix}$	$\frac{a^{s+\nu+1}(b/2)^{\nu+1}}{s+\nu+1} \Gamma\left[\begin{matrix} s+\nu+1 \\ \frac{2\nu+3}{2}, \frac{2s+2\nu+3}{2} \end{matrix}\right] \times {}_4F_5\left(\begin{matrix} 1, \frac{s+\nu+1}{2}, \frac{s+\nu+1}{2}, \frac{s+\nu+2}{2};\ \mp\frac{a^2b^2}{4} \\ \frac{3}{2}, \frac{2\nu+3}{2}, \frac{2s+2\nu+3}{2}, \frac{2s+2\nu+5}{2}, \frac{s+\nu+3}{2} \end{matrix}\right)$ $[a > 0;\ \operatorname{Re}(s+\nu) > -1]$
2	$\theta(a-x)\ln\frac{a+\sqrt{a^2-x^2}}{x} \times \begin{Bmatrix} \mathbf{H}_\nu(bx) \\ \mathbf{L}_\nu(bx) \end{Bmatrix}$	$\frac{a^{s+\nu+1}(b/2)^{\nu+1}}{s+\nu+1} \Gamma\left[\begin{matrix} \frac{s+\nu+1}{2} \\ \frac{2\nu+3}{2}, \frac{s+\nu+2}{2} \end{matrix}\right] {}_3F_4\left(\begin{matrix} 1, \frac{s+\nu+1}{2}, \frac{s+\nu+1}{2};\ \mp\frac{a^2b^2}{4} \\ \frac{3}{2}, \frac{2\nu+3}{2}, \frac{s+\nu+2}{2}, \frac{s+\nu+3}{2} \end{matrix}\right)$ $[a > 0;\ \operatorname{Re}(s+\nu) > -1]$
3	$\theta(a-x)\arccos\frac{x}{a} \times \begin{Bmatrix} \mathbf{H}_\nu(bx) \\ \mathbf{L}_\nu(bx) \end{Bmatrix}$	$\frac{a^{s+\nu+1}b^{\nu+1}}{2^\nu(s+\nu+1)^2} \Gamma\left[\begin{matrix} \frac{s+\nu+2}{2} \\ \frac{2\nu+3}{2}, \frac{s+\nu+1}{2} \end{matrix}\right] {}_3F_4\left(\begin{matrix} 1, \frac{s+\nu+1}{2}, \frac{s+\nu+2}{2};\ \mp\frac{a^2b^2}{4} \\ \frac{3}{2}, \frac{2\nu+3}{2}, \frac{s+\nu+3}{2}, \frac{s+\nu+3}{2} \end{matrix}\right)$ $[a > 0;\ \operatorname{Re}(s+\nu) > -1]$

3.15.5. $\mathbf{H}_\nu(bx)$, $\mathbf{L}_\nu(bx)$, and $\Gamma(\mu, ax)$

1	$\Gamma(\mu, ax) \begin{Bmatrix} \mathbf{H}_\nu(bx) \\ \mathbf{L}_\nu(bx) \end{Bmatrix}$	$\frac{2^{-\nu}a^{-s-\nu-1}b^{\nu+1}}{\sqrt{\pi}(s+\nu+1)} \Gamma\left[\begin{matrix} s+\mu+\nu+1 \\ \frac{2\nu+3}{2} \end{matrix}\right] \times {}_4F_3\left(\begin{matrix} 1, \frac{s+\nu+1}{2}, \frac{s+\mu+\nu+1}{2}, \frac{s+\mu+\nu+2}{2} \\ \frac{3}{2}, \frac{2\nu+3}{2}, \frac{s+\nu+3}{2};\ \mp\frac{b^2}{a^2} \end{matrix}\right)$ $\left[\operatorname{Re} a > \begin{Bmatrix} \lvert\operatorname{Im} b\rvert \\ \lvert\operatorname{Re} b\rvert \end{Bmatrix};\ \operatorname{Re}(s+\nu+1) > -\operatorname{Re}\mu, 0\right]$

3.15.6. $\mathbf{H}_\nu(bx)$, $\mathbf{L}_\nu(bx)$, and $\operatorname{Ei}(-ax^2)$, $\operatorname{erfc}(ax^r)$, $D_\mu(ax)$

1	$\operatorname{Ei}(-ax^2)\begin{Bmatrix}\mathbf{H}_\nu(bx)\\ \mathbf{L}_\nu(bx)\end{Bmatrix}$	$-\dfrac{a^{-(s+\nu+1)/2}b^{\nu+1}}{2^\nu\sqrt{\pi}\,(s+\nu+1)}\Gamma\left[\begin{matrix}\frac{s+\nu+1}{2}\\ \frac{2\nu+3}{2}\end{matrix}\right]{}_3F_3\left(\begin{matrix}1,\ \frac{s+\nu+1}{2},\ \frac{s+\nu+1}{2};\ \mp\frac{b^2}{4a}\\ \frac{3}{2},\ \frac{2\nu+3}{2},\ \frac{s+\nu+3}{2}\end{matrix}\right)$ $[\operatorname{Re}a>0;\ \operatorname{Re}(s+\nu)>-1]$
2	$\operatorname{erfc}(ax)\begin{Bmatrix}\mathbf{H}_\nu(bx)\\ \mathbf{L}_\nu(bx)\end{Bmatrix}$	$\dfrac{a^{-s-\nu-1}b^{\nu+1}}{2^\nu\pi\,(s+\nu+1)}\Gamma\left[\begin{matrix}\frac{s+\nu+2}{2}\\ \frac{2\nu+3}{2}\end{matrix}\right]{}_3F_3\left(\begin{matrix}1,\ \frac{s+\nu+1}{2},\ \frac{s+\nu+2}{2}\\ \frac{3}{2},\ \frac{2\nu+3}{2},\ \frac{s+\nu+3}{2};\ \mp\frac{b^2}{4a^2}\end{matrix}\right)$ $[b>0;\ \operatorname{Re}(s+\nu)>-1;\ \lvert\arg a\rvert<\pi/4]$
3	$\operatorname{erfc}(a\sqrt{x})\begin{Bmatrix}\mathbf{H}_\nu(bx)\\ \mathbf{L}_\nu(bx)\end{Bmatrix}$	$\dfrac{a^{-2s-2\nu-2}b^{\nu+1}}{2^\nu\pi\,(s+\nu+1)}\Gamma\left[\begin{matrix}\frac{2s+2\nu+3}{2}\\ \frac{2\nu+3}{2}\end{matrix}\right]{}_4F_3\left(\begin{matrix}1,\ \frac{s+\nu+1}{2},\ \frac{2s+2\nu+3}{4},\ \frac{2s+2\nu+5}{4}\\ \frac{3}{2},\ \frac{2\nu+3}{2},\ \frac{s+\nu+3}{2};\ \mp\frac{b^2}{a^4}\end{matrix}\right)$ $\left[\operatorname{Re}a^2>\begin{Bmatrix}\lvert\operatorname{Im}b\rvert\\ \lvert\operatorname{Re}b\rvert\end{Bmatrix};\ \operatorname{Re}(s+\nu)>-1\right]$
4	$e^{a^2x^2/4}D_\mu(ax)$ $\times\begin{Bmatrix}\mathbf{H}_\nu(bx)\\ \mathbf{L}_\nu(bx)\end{Bmatrix}$	$\dfrac{a^{-s-\nu-1}b^{\nu+1}}{2^{(s+\mu+3\nu+3)/2}\sqrt{\pi}}\Gamma\left[\begin{matrix}s+\nu+1,\ -\frac{s+\mu+\nu+1}{2}\\ -\mu,\ \frac{2\nu+3}{2}\end{matrix}\right]$ $\times{}_3F_3\left(\begin{matrix}1,\ \frac{s+\nu+1}{2},\ \frac{s+\nu+2}{2}\\ \frac{3}{2},\ \frac{2\nu+3}{2},\ \frac{s+\mu+\nu+3}{2};\ \pm\frac{b^2}{2a^2}\end{matrix}\right)+\dfrac{2^{s+\mu-1}\pi\,a^\mu b^{-s-\mu}}{\Gamma\left(\frac{2-s-\mu-\nu}{2}\right)\Gamma\left(\frac{2-s-\mu+\nu}{2}\right)}$ $\times\sec\dfrac{(s+\mu+\nu)\pi}{2}\,{}_2F_2\left(\begin{matrix}-\frac{\mu}{2},\ \frac{1-\mu}{2};\ \pm\frac{b^2}{2a^2}\\ \frac{2-s-\mu-\nu}{2},\ \frac{2-s-\mu+\nu}{2}\end{matrix}\right)$ $\left[\begin{matrix}b>0;\ \operatorname{Re}(s+\mu)<1-\operatorname{Re}\nu;\ \operatorname{Re}(s+\mu)<3/2;\\ \operatorname{Re}(s+\nu)>-1;\ \lvert\arg a\rvert<3\pi/4\end{matrix}\right]$
5	$e^{-a^2x^2/4}D_\mu(ax)$ $\times\begin{Bmatrix}\mathbf{H}_\nu(bx)\\ \mathbf{L}_\nu(bx)\end{Bmatrix}$	$\dfrac{2^{(-s+\mu-3\nu-1)/2}b^{\nu+1}}{a^{s+\nu+1}}\Gamma\left[\begin{matrix}s+\nu+1\\ \frac{2\nu+3}{2},\ \frac{s-\mu+\nu+2}{2}\end{matrix}\right]$ $\times{}_3F_3\left(\begin{matrix}1,\ \frac{s+\nu+1}{2},\ \frac{s+\nu+2}{2};\ \mp\frac{b^2}{2a^2}\\ \frac{3}{2},\ \frac{2\nu+3}{2},\ \frac{s-\mu+\nu+2}{2}\end{matrix}\right)$ $[\operatorname{Re}(s+\nu)>-1;\ 4\lvert\arg a\rvert,\ \lvert\arg b\rvert<\pi]$

3.15.7. $\mathbf{H}_\nu(bx)$ and $J_\mu(ax)$

1	$J_\mu(ax)\,\mathbf{H}_\nu(bx)$	$\dfrac{2^s b^{\nu+1}}{\sqrt{\pi}\,a^{s+\nu+1}}\Gamma\left[\begin{matrix}\frac{s+\mu+\nu+1}{2}\\ \frac{2\nu+3}{2},\ \frac{1-s+\mu-\nu}{2}\end{matrix}\right]{}_3F_2\left(\begin{matrix}1,\ \frac{s-\mu+\nu+1}{2},\ \frac{s+\mu+\nu+1}{2}\\ \frac{3}{2},\ \frac{2\nu+3}{2};\ \frac{b^2}{a^2}\end{matrix}\right)$ $[0<b<a;\ \operatorname{Re}s<2;\ -\operatorname{Re}\mu-1<\operatorname{Re}(s+\nu)<5/2]$
2	$J_\mu(ax)\,\mathbf{H}_\nu(bx)$	$\dfrac{2^{s-1}b^{\nu-1}}{\sqrt{\pi}\,a^{s+\nu-1}}\Gamma\left[\begin{matrix}\frac{s+\mu+\nu-1}{2}\\ \frac{2\nu+1}{2},\ \frac{3-s+\mu-\nu}{2}\end{matrix}\right]{}_3F_2\left(\begin{matrix}\frac{1}{2},\ 1,\ \frac{1}{2}-\nu;\ \frac{a^2}{b^2}\\ \frac{3-s-\mu-\nu}{2},\ \frac{3-s+\mu-\nu}{2}\end{matrix}\right)$ $+\dfrac{2^{s-1}\pi\,a^\mu}{b^{s+\mu}}\dfrac{\sec\frac{(s+\mu+\nu)\pi}{2}}{\Gamma\left[\mu+1,\ \frac{2-s-\mu-\nu}{2},\ \frac{2-s-\mu+\nu}{2}\right]}{}_2F_1\left(\begin{matrix}\frac{s+\mu-\nu}{2},\ \frac{s+\mu+\nu}{2}\\ \mu+1;\ \frac{a^2}{b^2}\end{matrix}\right)$ $[0<a<b;\ \operatorname{Re}s<2;\ -\operatorname{Re}\mu-1<\operatorname{Re}(s+\nu)<5/2]$

No.	$f(x)$	$F(s)$
3	$J_\mu\left(\frac{a}{x}\right)\mathbf{H}_\nu(bx)$	$\frac{2^{s-2\mu-1}\pi a^\mu}{b^{s-\mu}}\frac{\sec\frac{(s-\mu+\nu)\pi}{2}}{\Gamma\left[\mu+1,\,\frac{2-s+\mu-\nu}{2},\,\frac{2-s+\mu+\nu}{2}\right]}$ $\times{}_0F_3\left(\begin{matrix}\frac{a^2b^2}{16}\\ \mu+1,\,\frac{2-s+\mu-\nu}{2},\,\frac{2-s+\mu+\nu}{2}\end{matrix}\right)+\frac{a^{s+\nu+1}b^{\nu+1}}{2^{s+2\nu+2}\sqrt{\pi}}$ $\times\Gamma\left[\begin{matrix}\frac{-s+\mu-\nu-1}{2}\\ \frac{2\nu+3}{2},\,\frac{s+\mu+\nu+3}{2}\end{matrix}\right]{}_1F_4\left(\begin{matrix}1;\,\frac{a^2b^2}{16}\\ \frac{3}{2},\,\frac{2\nu+3}{2},\,\frac{s-\mu+\nu+3}{2},\,\frac{s+\mu+\nu+3}{2}\end{matrix}\right)$ $[a,\,b>0;\ \operatorname{Re}(s-\mu)<3/2;\ -5/2<\operatorname{Re}(s+\nu)<\operatorname{Re}\mu+1]$

3.15.8. $\mathbf{H}(bx)$, $\mathbf{L}_\nu(bx)$, and $K_\mu(ax^r)$

1	$K_\mu(ax)\left\{\begin{matrix}\mathbf{H}_\nu(bx)\\ \mathbf{L}_\nu(bx)\end{matrix}\right\}$	$\frac{2^{s-1}b^{\nu+1}}{\sqrt{\pi}\,a^{s+\nu+1}}\Gamma\left[\begin{matrix}\frac{s-\mu+\nu+1}{2},\,\frac{s+\mu+\nu+1}{2}\\ \frac{2\nu+3}{2}\end{matrix}\right]{}_3F_2\left(\begin{matrix}1,\,\frac{s-\mu+\nu+1}{2},\,\frac{s+\mu+\nu+1}{2}\\ \frac{3}{2},\,\frac{2\nu+3}{2};\,\mp\frac{b^2}{a^2}\end{matrix}\right)$ $\left[\operatorname{Re}a>\left\{\begin{matrix}\lvert\operatorname{Im}b\rvert\\ \lvert\operatorname{Re}b\rvert\end{matrix}\right\};\ \operatorname{Re}(s\pm\mu+\nu)>-1\right]$
2	$e^{-ax}K_\mu(ax)\left\{\begin{matrix}\mathbf{H}_\nu(bx)\\ \mathbf{L}_\nu(bx)\end{matrix}\right\}$	$\frac{b^{\nu+1}}{2^{s+2\nu+1}a^{s+\nu+1}}\Gamma\left[\begin{matrix}s-\mu+\nu+1,s+\mu+\nu+1\\ \frac{2\nu+3}{2},\,\frac{2s+2\nu+3}{2}\end{matrix}\right]$ $\times{}_5F_4\left(\begin{matrix}1,\,\frac{s-\mu+\nu+1}{2},\,\frac{s-\mu+\nu+2}{2},\,\frac{s+\mu+\nu+1}{2},\,\frac{s+\mu+\nu+2}{2}\\ \frac{3}{2},\,\frac{2\nu+3}{2},\,\frac{2s+2\nu+3}{4},\,\frac{2s+2\nu+5}{4};\,\mp\frac{b^2}{4a^2}\end{matrix}\right)$ $\left[\left\{\begin{matrix}\operatorname{Re}a>\lvert\operatorname{Im}b\rvert\\ \operatorname{Re}a>\lvert\operatorname{Re}b\rvert\end{matrix}\right\};\ \operatorname{Re}(s+\nu)>\lvert\operatorname{Re}\mu\rvert-1\right]$
3	$e^{\mp ax^2}K_\mu(ax^2)\,\mathbf{H}_\nu(bx)$	$\pm\frac{2^{-(s+3\nu+3)/2}b^{\nu+1}}{a^{(s+\nu+1)/2}}\left[\cos(\mu\pi)\csc\frac{(s+\nu)\pi}{2}\right]^{(1\mp1)/2}$ $\times\Gamma\left[\begin{matrix}\frac{s-2\mu+\nu+1}{2},\,\frac{s+2\mu+\nu+1}{2}\\ \frac{2\nu+3}{2},\,\frac{s+\nu+2}{2}\end{matrix}\right]{}_3F_3\left(\begin{matrix}1,\,\frac{s-2\mu+\nu+1}{2},\,\frac{s+2\mu+\nu+1}{2}\\ \frac{3}{2},\,\frac{2\nu+3}{2},\,\frac{s+\nu+2}{2};\,\mp\frac{b^2}{8a}\end{matrix}\right)$ $+\frac{(1\mp1)\,2^{s-7/2}\pi^{3/2}\csc\frac{(s+\nu)\pi}{2}}{\sqrt{a}\,b^{s-1}\Gamma\left(\frac{3-s-\nu}{2}\right)\Gamma\left(\frac{3-s+\nu}{2}\right)}\,{}_2F_2\left(\begin{matrix}\frac{1-2\mu}{2},\,\frac{1+2\mu}{2};\,\frac{b^2}{8a}\\ \frac{3-s-\nu}{2},\,\frac{3-s+\nu}{2}\end{matrix}\right)$ $\left[\begin{matrix}\operatorname{Re}(s+\nu)>2\lvert\operatorname{Re}\mu\rvert-1;\ \lvert\arg a\rvert<(2\mp1)\pi/2;\\ \left\{\begin{matrix}\lvert\arg b\rvert<\pi\\ b>0;\ \operatorname{Re}(s+\nu)<2;\ \operatorname{Re}s<5/2\end{matrix}\right\}\end{matrix}\right]$
4	$e^{-ax^2}K_\mu(ax^2)\,\mathbf{L}_\nu(bx)$	$\frac{2^{-(s+3\nu+3)/2}b^{\nu+1}}{a^{(s+\nu+1)/2}}\Gamma\left[\begin{matrix}\frac{s-2\mu+\nu+1}{2},\,\frac{s+2\mu+\nu+1}{2}\\ \frac{2\nu+3}{2},\,\frac{s+\nu+2}{2}\end{matrix}\right]$ $\times{}_3F_3\left(\begin{matrix}1,\,\frac{s-2\mu+\nu+1}{2},\,\frac{s+2\mu+\nu+1}{2}\\ \frac{3}{2},\,\frac{2\nu+3}{2},\,\frac{s+\nu+2}{2};\,\frac{b^2}{8a}\end{matrix}\right)$ $[\operatorname{Re}a>0;\ \operatorname{Re}(s+\nu)>2\lvert\operatorname{Re}\mu\rvert-1;\ \lvert\arg b\rvert<\pi]$

No.	$f(x)$	$F(s)$
5	$K_\lambda(ax)\,K_\mu(ax)$ $\times\begin{Bmatrix}\mathbf{H}_\nu(bx)\\ \mathbf{L}_\nu(bx)\end{Bmatrix}$	$\dfrac{2^{s-2}b^{\nu+1}}{\sqrt{\pi}\,a^{s+\mu+1}}\,\Gamma\left[\begin{matrix}\frac{s+\lambda+\mu+\nu+1}{2},\ \frac{s+\lambda-\mu+\nu+1}{2},\ \frac{s-\lambda+\mu+\nu+1}{2},\ \frac{s-\lambda-\mu+\nu+1}{2}\\ \frac{2\nu+3}{2},\ s+\nu+1\end{matrix}\right]$ $\times\,{}_5F_4\left(\begin{matrix}1,\ \frac{s+\lambda+\mu+\nu+1}{2},\ \frac{s-\lambda+\mu+\nu+1}{2},\ \frac{s-\lambda-\mu+\nu+1}{2},\ \frac{s+\lambda-\mu+\nu+1}{2}\\ \frac{3}{2},\ \frac{2\nu+3}{2},\ \frac{s+\nu+1}{2},\ \frac{s+\nu+2}{2};\ \mp\frac{b^2}{4a^2}\end{matrix}\right)$ $\left[2\operatorname{Re}a>\begin{Bmatrix}\lvert\operatorname{Im}b\rvert\\ \lvert\operatorname{Re}b\rvert\end{Bmatrix};\ \operatorname{Re}(s+\nu)>\lvert\operatorname{Re}\lambda\rvert+\lvert\operatorname{Re}\mu\rvert-1\right]$

3.15.9. $\mathbf{H}_\nu(\varphi(x))-Y_\nu(\varphi(x))$, $I_{\pm\nu}(\varphi(x))-\mathbf{L}_\nu(\varphi(x))$

1	$\mathbf{H}_\nu(ax)-Y_\nu(ax)$	$\dfrac{2^{s-1}a^{-s}}{\pi}\cos(\nu\pi)\sec\dfrac{(s+\nu)\pi}{2}\,\Gamma\left(\dfrac{s-\nu}{2}\right)\Gamma\left(\dfrac{s+\nu}{2}\right)$ $[\operatorname{Re}a>0;\ \lvert\operatorname{Re}\nu\rvert<\operatorname{Re}s<1-\operatorname{Re}\nu]$
2	$I_\nu(ax)-\mathbf{L}_\nu(ax)$	$2^{s-1}a^{-s}\sec\dfrac{(s+\nu)\pi}{2}\,\Gamma\left[\begin{matrix}\frac{s+\nu}{2}\\ \frac{2-s+\nu}{2}\end{matrix}\right]$ $[\operatorname{Re}a>0;\ -\operatorname{Re}\nu<\operatorname{Re}s<1-\operatorname{Re}\nu]$
3	$I_{-\nu}(ax)-\mathbf{L}_\nu(ax)$	$2^{s-1}a^{-s}\cos(\nu\pi)\sec\dfrac{(s+\nu)\pi}{2}\,\Gamma\left[\begin{matrix}\frac{s-\nu}{2}\\ \frac{2-s-\nu}{2}\end{matrix}\right]$ $[\operatorname{Re}a>0;\ -\operatorname{Re}\nu<\operatorname{Re}s<1+\operatorname{Re}\nu]$
4	$(a^2-x^2)_+^{\alpha-1}$ $\times[I_{\pm\nu}(bx)-\mathbf{L}_\nu(bx)]$	$-\dfrac{a^{s+2\alpha+\nu-1}b^{\nu+1}}{2^{\nu+1}\sqrt{\pi}\,\Gamma\left(\frac{2\nu+3}{2}\right)}\,\mathrm{B}\left(\alpha,\ \dfrac{s+\nu+1}{2}\right){}_2F_3\left(\begin{matrix}1,\ \frac{s+\nu+1}{2};\ \frac{a^2b^2}{4}\\ \frac{3}{2},\ \frac{2\nu+3}{2},\ \frac{s+2\alpha+\nu+1}{2}\end{matrix}\right)$ $+\dfrac{a^{s+2\alpha\pm\nu-2}b^{\pm\nu}}{2^{1\pm\nu}\,\Gamma(1\pm\nu)}\,\mathrm{B}\left(\alpha,\ \dfrac{s\pm\nu}{2}\right){}_1F_2\left(\begin{matrix}\frac{s\pm\nu}{2};\ \frac{a^2b^2}{4}\\ 1\pm\nu,\ \frac{s+2\alpha\pm\nu}{2}\end{matrix}\right)$ $\left[a,\ \operatorname{Re}\alpha>0;\begin{Bmatrix}\operatorname{Re}(s+\nu)>0\\ -\operatorname{Re}s-1<\operatorname{Re}\nu<\operatorname{Re}s\end{Bmatrix}\right]$
5	$(x^2-a^2)_+^{\alpha-1}$ $\times[I_{\pm\nu}(bx)-\mathbf{L}_\nu(bx)]$	$-\dfrac{a^{s+2\alpha+\nu-1}}{\sqrt{\pi}\,\Gamma\left(\frac{2\nu+3}{2}\right)}\left(\dfrac{b}{2}\right)^{\nu+1}\mathrm{B}\left(\alpha,\ \dfrac{1-s-2\alpha-\nu}{2}\right)$ $\times\,{}_2F_3\left(\begin{matrix}1,\ \frac{s+\nu+1}{2};\ \frac{a^2b^2}{4}\\ \frac{3}{2},\ \frac{2\nu+3}{2},\ \frac{s+2\alpha+\nu+1}{2}\end{matrix}\right)-\dfrac{2^{s+2\alpha-3}}{b^{s+2\alpha-2}}\cos^{(1\mp1)/2}(\nu\pi)$ $\times\sec\dfrac{(s+2\alpha+\nu)\pi}{2}\,\Gamma\left[\begin{matrix}\frac{s+2\alpha\pm\nu-2}{2}\\ \frac{4-s-2\alpha\pm\nu}{2}\end{matrix}\right]{}_1F_2\left(\begin{matrix}1-\alpha;\ \frac{a^2b^2}{4}\\ \frac{4-s-2\alpha-\nu}{2},\ \frac{4-s-2\alpha+\nu}{2}\end{matrix}\right)$ $+\dfrac{a^{s+2\alpha\pm\nu-2}b^{\pm\nu}}{2^{1\pm\nu}\,\Gamma(1\pm\nu)}\,\mathrm{B}\left(\alpha,\ \dfrac{2-s-2\alpha\mp\nu}{2}\right){}_1F_2\left(\begin{matrix}\frac{s\pm\nu}{2};\ \frac{a^2b^2}{4}\\ 1\pm\nu,\ \frac{s+2\alpha\pm\nu}{2}\end{matrix}\right)$ $[a,\ \operatorname{Re}b,\ \operatorname{Re}\alpha>0;\ \operatorname{Re}(s+2\alpha+\nu)<3]$

No.	$f(x)$	$F(s)$
6	$\dfrac{1}{(x^2+a^2)^\rho}$ $\times\left[I_{\pm\nu}(bx)-\mathbf{L}_\nu(bx)\right]$	$-\dfrac{a^{s+\nu-2\rho+1}b^{\nu+1}}{2^{\nu+1}\sqrt{\pi}\,\Gamma\left(\frac{2\nu+3}{2}\right)}\,\mathrm{B}\left(\dfrac{s+\nu+1}{2},\,\dfrac{2\rho-\nu-s-1}{2}\right)$ $\times\,{}_2F_3\left(\begin{matrix}1,\,\frac{s+\nu+1}{2};\,-\frac{a^2b^2}{4}\\ \frac{3}{2},\,\frac{2\nu+3}{2},\,\frac{s+\nu-2\rho+3}{2}\end{matrix}\right)+\dfrac{2^{s-2\rho-1}}{b^{s-2\rho}}\cos^{(1\mp1)/2}(\nu\pi)$ $\times\sec\dfrac{(s+\nu-2\rho)\pi}{2}\,\Gamma\left[\begin{matrix}\frac{s\pm\nu-2\rho}{2}\\ \frac{\rho\pm\nu-s+2}{2}\end{matrix}\right]{}_1F_2\left(\begin{matrix}\rho;\,-\frac{a^2b^2}{4}\\ \frac{2-s-\nu+2\rho}{2},\,\frac{2-s+\nu+2\rho}{2}\end{matrix}\right)$ $+\dfrac{a^{s\pm\nu-2\rho}b^{\pm\nu}}{2^{1\pm\nu}\Gamma(1\pm\nu)}\,\mathrm{B}\left(\dfrac{s\pm\nu}{2},\,\dfrac{2\rho-s\mp\nu}{2}\right){}_1F_2\left(\begin{matrix}\frac{s\pm\nu}{2};\,-\frac{a^2b^2}{4}\\ 1\pm\nu,\,\frac{s\pm\nu-2\rho+2}{2}\end{matrix}\right)$ $\left[\mathrm{Re}\,a,\ \mathrm{Re}\,b>0;\ \mathrm{Re}\,(s+\nu-2\rho)<1;\ \left\{\begin{matrix}\mathrm{Re}\,(s+\nu)>0\\ -1-\mathrm{Re}\,s<\mathrm{Re}\,\nu<\mathrm{Re}\,s\end{matrix}\right\}\right]$
7	$\dfrac{1}{x^2-a^2}$ $\times\left[I_{\pm\nu}(bx)-\mathbf{L}_\nu(bx)\right]$	$-\dfrac{2^{s-3}}{b^{s-2}}\cos^{(1\mp1)/2}(\nu\pi)\sec\dfrac{(s+\nu)\pi}{2}\,\Gamma\left[\begin{matrix}\frac{s\pm\nu-2}{2}\\ \frac{4-s\pm\nu}{2}\end{matrix}\right]{}_1F_2\left(\begin{matrix}1;\,\frac{a^2b^2}{4}\\ \frac{4-s-\nu}{2},\,\frac{4-s+\nu}{2}\end{matrix}\right)$ $-\dfrac{\pi a^{s-2}}{2}\left[\tan\dfrac{(s+\nu)\pi}{2}\,\mathbf{L}_\nu(ab)+\cot\dfrac{(s\pm\nu)\pi}{2}\,I_{\pm\nu}(ab)\right]$ $\left[a,\ \mathrm{Re}\,b>0;\ \mathrm{Re}\,(s+\nu)<3;\ \left\{\begin{matrix}\mathrm{Re}\,(s+\nu)>0\\ -\mathrm{Re}\,s-1<\mathrm{Re}\,\nu<\mathrm{Re}\,s\end{matrix}\right\}\right]$
8	$e^{-ax}\left[I_{\pm\nu}(bx)-\mathbf{L}_\nu(bx)\right]$	$a^{-s\mp\nu}\left(\dfrac{b}{2}\right)^{\pm\nu}\Gamma\left[\begin{matrix}s\pm\nu\\ 1\pm\nu\end{matrix}\right]{}_2F_1\left(\begin{matrix}\frac{s\pm\nu}{2},\,\frac{s\pm\nu+1}{2}\\ 1\pm\nu;\,\frac{b^2}{a^2}\end{matrix}\right)$ $-\dfrac{a^{-s-\nu-1}b^{\nu+1}}{2^\nu\sqrt{\pi}}\,\Gamma\left[\begin{matrix}s+\nu+1\\ \frac{2\nu+3}{2}\end{matrix}\right]{}_3F_2\left(\begin{matrix}1,\,\frac{s+\nu+1}{2},\,\frac{s+\nu+2}{2}\\ \frac{3}{2},\,\frac{2\nu+3}{2};\,\frac{b^2}{a^2}\end{matrix}\right)$ $\left[\begin{matrix}(\mathrm{Re}\,a>\lvert\mathrm{Re}\,b\rvert;\ \mathrm{Re}\,s>-\mathrm{Re}\,\nu)\ \text{or}\\ (\mathrm{Re}\,a=\mathrm{Re}\,b=0;\ \mp\mathrm{Re}\,\nu<\mathrm{Re}\,s<3/2,\,2-\mathrm{Re}\,\nu)\end{matrix}\right]$
9	$e^{-ax}\left[Y_\nu(bx)-\mathbf{H}_\nu(bx)\right]$	$-\dfrac{a^{-\nu-s+1}(b/2)^{\nu-1}}{\pi^{3/2}}\cos(\nu\pi)\,\Gamma\left(\dfrac{1-2\nu}{2}\right)$ $\times\,\Gamma(s+\nu-1)\,{}_3F_2\left(\begin{matrix}\frac{1}{2},\,1,\,\frac{1-2\nu}{2};\,-\frac{a^2}{b^2}\\ \frac{2-s-\nu}{2},\,\frac{3-s-\nu}{2}\end{matrix}\right)$ $-\dfrac{2^s ab^{-s-1}}{\pi}\cos(\nu\pi)\csc\dfrac{(s+\nu)\pi}{2}\,\Gamma\left(\dfrac{s-\nu+1}{2}\right)$ $\times\,\Gamma\left(\dfrac{s+\nu+1}{2}\right){}_2F_1\left(\begin{matrix}\frac{s-\nu+1}{2},\,\frac{s+\nu+1}{2}\\ \frac{3}{2};\,-\frac{a^2}{b^2}\end{matrix}\right)-\dfrac{2^{s-1}b^{-s}}{\pi}\cos(\nu\pi)$ $\times\sec\dfrac{(s+\nu)\pi}{2}\,\Gamma\left(\dfrac{s-\nu}{2}\right)\Gamma\left(\dfrac{s+\nu}{2}\right){}_2F_1\left(\begin{matrix}\frac{s-\nu}{2},\,\frac{s+\nu}{2}\\ \frac{1}{2};\,-\frac{a^2}{b^2}\end{matrix}\right)$ $\left[\begin{matrix}(\mathrm{Re}\,a>\lvert\mathrm{Im}\,b\rvert;\ \mathrm{Re}\,s>\lvert\mathrm{Re}\,\nu\rvert)\ \text{or}\\ (\mathrm{Re}\,a=0,\,b>0;\ \lvert\mathrm{Re}\,\nu\rvert<\mathrm{Re}\,s<3/2,\,2-\mathrm{Re}\,\nu)\end{matrix}\right]$

No.	$f(x)$	$F(s)$
10	$\sin(ax)$ $\times\left[I_{\pm\nu}(bx)-\mathbf{L}_\nu(bx)\right]$	$-\frac{2^{-\nu}b^{\nu+1}}{\sqrt{\pi}\,a^{s+\nu+1}}\cos\frac{(s+\nu)\pi}{2}\,\Gamma\left[\begin{matrix}s+\nu+1\\ \frac{2\nu+3}{2}\end{matrix}\right]{}_3F_2\left(\begin{matrix}1,\ \frac{s+\nu+1}{2},\ \frac{s+\nu+2}{2}\\ \frac{3}{2},\ \frac{2\nu+3}{2};\ -\frac{b^2}{a^2}\end{matrix}\right)$ $+\frac{(b/2)^{\pm\nu}}{a^{s\pm\nu}}\sin\frac{(s\pm\nu)\pi}{2}\,\Gamma\left[\begin{matrix}s\pm\nu\\ 1\pm\nu\end{matrix}\right]{}_2F_1\left(\begin{matrix}\frac{s\pm\nu}{2},\ \frac{s\pm\nu+1}{2}\\ 1\pm\nu;\ -\frac{b^2}{a^2}\end{matrix}\right)$ $\left[a,\ \operatorname{Re}b>0;\ \left\{\begin{matrix}-1<\operatorname{Re}(s+\nu)<2\\ -2<\operatorname{Re}(s+\nu)<2,\ 2\operatorname{Re}s+1\end{matrix}\right\}\right]$
11	$\cos(ax)$ $\times\left[I_{\pm\nu}(bx)-\mathbf{L}_\nu(bx)\right]$	$\frac{2^{-\nu}b^{\nu+1}}{\sqrt{\pi}\,a^{s+\nu+1}}\sin\frac{(s+\nu)\pi}{2}\,\Gamma\left[\begin{matrix}s+\nu+1\\ \frac{2\nu+3}{2}\end{matrix}\right]{}_3F_2\left(\begin{matrix}1,\ \frac{s+\nu+1}{2},\ \frac{s+\nu+2}{2}\\ \frac{3}{2},\ \frac{2\nu+3}{2};\ -\frac{b^2}{a^2}\end{matrix}\right)$ $+\frac{(b/2)^{\pm\nu}}{a^{s\pm\nu}}\cos\frac{(s\pm\nu)\pi}{2}\,\Gamma\left[\begin{matrix}s\pm\nu\\ 1\pm\nu\end{matrix}\right]{}_2F_1\left(\begin{matrix}\frac{s\pm\nu}{2},\ \frac{s\pm\nu+1}{2}\\ 1\pm\nu;\ -\frac{b^2}{a^2}\end{matrix}\right)$ $\left[a,\ \operatorname{Re}b>0;\ \left\{\begin{matrix}0<\operatorname{Re}(s+\nu)<2\\ -1<\operatorname{Re}(s+\nu)<2,\ 2\operatorname{Re}s\end{matrix}\right\}\right]$
12	$J_\mu(ax)$ $\times\left[I_{\pm\nu}(bx)-\mathbf{L}_\nu(bx)\right]$	$-\frac{2^{s}b^{\nu+1}}{\sqrt{\pi}\,a^{s+\nu+1}}\Gamma\left[\begin{matrix}\frac{s+\mu+\nu+1}{2}\\ \frac{2\nu+3}{2},\ \frac{1-s+\mu-\nu}{2}\end{matrix}\right]{}_3F_2\left(\begin{matrix}1,\ \frac{s-\mu+\nu+1}{2},\ \frac{s+\mu+\nu+1}{2}\\ \frac{3}{2},\ \frac{2\nu+3}{2};\ -\frac{b^2}{a^2}\end{matrix}\right)$ $+\frac{2^{s-1}b^{\pm\nu}}{a^{s\pm\nu}}\Gamma\left[\begin{matrix}\frac{s+\mu\pm\nu}{2}\\ 1\pm\nu,\ \frac{2-s+\mu\mp\nu}{2}\end{matrix}\right]{}_2F_1\left(\begin{matrix}\frac{s-\mu\pm\nu}{2},\ \frac{s+\mu\pm\nu}{2}\\ 1\pm\nu;\ -\frac{b^2}{a^2}\end{matrix}\right)$ $\left[\begin{matrix}a,\ \operatorname{Re}b>0;\ \operatorname{Re}(s+\nu)<5/2;\\ \left\{\begin{matrix}\operatorname{Re}(s+\mu+\nu)>0\\ -1<\operatorname{Re}(s+\mu+\nu)<2\operatorname{Re}(s+\mu)\end{matrix}\right\}\end{matrix}\right]$
13	$\left(x^2+a^2\right)^{\nu/2}$ $\times\left[\mathbf{H}_\nu\left(b\sqrt{x^2+a^2}\right)\right.$ $\left.-Y_\nu\left(b\sqrt{x^2+a^2}\right)\right]$	$2^{(s-2)/2}a^{s/2+\nu}b^{-s/2}\cos(\nu\pi)\sec\frac{(s+2\nu)\pi}{2}\,\Gamma\left(\frac{s}{2}\right)$ $\times\left[\mathbf{H}_{s/2+\nu}(ab)-Y_{s/2+\nu}(ab)\right]$ $[a,\ b>0;\ 0<\operatorname{Re}s<1-2\operatorname{Re}\nu]$
14	$\left(x^2+a^2\right)^{-\nu/2}$ $\times\left[\mathbf{H}_\nu\left(b\sqrt{x^2+a^2}\right)\right.$ $\left.-Y_\nu\left(b\sqrt{x^2+a^2}\right)\right]$	$\frac{a^{s/2-\nu}}{2^\nu\pi b^{s/2}}\Gamma\left[\begin{matrix}\frac{s}{2},\ \frac{1-s}{2}\\ \frac{2\nu+1}{2}\end{matrix}\right]S_{s/2+\nu,\,s/2-\nu}(ab)$ $[a,\ b>0;\ 0<\operatorname{Re}s<1]$
15	$\left(x^2+a^2\right)^{\nu/2}$ $\times\left[I_{-\nu}\left(b\sqrt{x^2+a^2}\right)\right.$ $\left.-\mathbf{L}_\nu\left(b\sqrt{x^2+a^2}\right)\right]$	$\frac{2^{s/2-1}a^{s/2+\nu}}{b^{s/2}}\cos(\nu\pi)\sec\frac{(s+2\nu)\pi}{2}\,\Gamma\left(\frac{s}{2}\right)$ $\times\left[I_{-s/2-\nu}(ab)-\mathbf{L}_{s/2+\nu}(ab)\right]$ $[a,\ b,\ \operatorname{Re}s>0;\ \operatorname{Re}\nu<1/2]$

3.16. The Anger $\mathbf{J}_\nu(z)$ and Weber $\mathbf{E}_\nu(z)$ Functions

More formulas can be obtained from the corresponding sections due to the relations

$$\mathbf{E}_0(z) = -\mathbf{H}_0(z), \quad \mathbf{E}_1(z) = \frac{2}{\pi} - \mathbf{H}_1(z), \quad \mathbf{J}_{\pm n}(z) = J_{\pm n}(z),$$

$$\left\{ \begin{matrix} \mathbf{E}_\nu(z) \\ \mathbf{J}_\nu(z) \end{matrix} \right\} = \frac{1}{\nu\pi} \left\{ \begin{matrix} 1-\cos(\nu\pi) \\ \sin(\nu\pi) \end{matrix} \right\} {}_1F_2\left(1;\, 1-\frac{\nu}{2},\, 1+\frac{\nu}{2};\, -\frac{z^2}{4}\right)$$

$$\mp \frac{1}{(1-\nu^2)\pi} \left\{ \begin{matrix} 1+\cos(\nu\pi) \\ \sin(\nu\pi) \end{matrix} \right\} {}_1F_2\left(1;\, \frac{3-\nu}{2},\, \frac{3+\nu}{2};\, -\frac{z^2}{4}\right);$$

$$\left\{ \begin{matrix} \mathbf{E}_\nu(z) \\ \mathbf{J}_\nu(z) \end{matrix} \right\} = G_{35}^{22}\left(\frac{z^2}{4} \middle| \begin{matrix} 0,\, 1/2,\, (3-2\nu\pm 1)/4 \\ 0,\, 1/2,\, -\nu/2,\, \nu/2,\, (3-2\nu\pm 1)/4 \end{matrix} \right), \quad [-\pi/2 < \arg z \leq \pi/2].$$

3.16.1. $\mathbf{J}_\nu(\varphi(x))$, $\mathbf{E}_\nu(\varphi(x))$, and algebraic functions

No.	$f(x)$	$F(s)$
1	$\left\{ \begin{matrix} \mathbf{J}_\nu(ax) \\ \mathbf{E}_\nu(ax) \end{matrix} \right\}$	$\dfrac{2^s \pi\, a^{-s} \csc(s\pi)}{\Gamma\left(\frac{2-s-\nu}{2}\right)\Gamma\left(\frac{2-s+\nu}{2}\right)} \left\{ \begin{matrix} \cos[(\nu-s)\pi/2] \\ \sin[(\nu-s)\pi/2] \end{matrix} \right\}$ $[a>0;\ 0<\operatorname{Re} s<1]$
2	$\mathbf{J}_\nu(ax) \pm \mathbf{J}_{-\nu}(ax)$	$\dfrac{2^s \pi\, a^{-s}}{\Gamma\left(\frac{2-s-\nu}{2}\right)\Gamma\left(\frac{2-s+\nu}{2}\right)} \left\{ \begin{matrix} \cos(\nu\pi/2)\csc(s\pi/2) \\ \sin(\nu\pi/2)\sec(s\pi/2) \end{matrix} \right\}$ $[a>0;\ -(1\mp 1)/2<\operatorname{Re} s<(5\pm 1)/4]$
3	$(a-x)_+^{\alpha-1} \left\{ \begin{matrix} \mathbf{J}_\nu(bx) \\ \mathbf{E}_\nu(bx) \end{matrix} \right\}$	$\left\{ \begin{matrix} -\sin(\nu\pi) \\ 1+\cos(\nu\pi) \end{matrix} \right\} \dfrac{a^{s+\alpha} b}{(\nu^2-1)\pi} \mathrm{B}(s+1,\, \alpha) \times {}_3F_4\left(\begin{matrix} 1,\, \frac{s+1}{2},\, \frac{s+2}{2};\, -\frac{a^2b^2}{4} \\ \frac{3-\nu}{2},\, \frac{3+\nu}{2},\, \frac{s+\alpha+1}{2},\, \frac{s+\alpha+2}{2} \end{matrix} \right) + \left\{ \begin{matrix} \sin(\nu\pi) \\ 1+\cos(\nu\pi) \end{matrix} \right\} \dfrac{a^{s+\alpha-1}}{\nu\pi} \mathrm{B}(s,\, \alpha)\, {}_3F_4\left(\begin{matrix} 1,\, \frac{s}{2},\, \frac{s+1}{2};\, -\frac{a^2b^2}{4} \\ \frac{2-\nu}{2},\, \frac{2+\nu}{2},\, \frac{s+\alpha}{2},\, \frac{s+\alpha+1}{2} \end{matrix} \right)$ $[a,\ \operatorname{Re}\nu,\ \operatorname{Re} s>0]$
4	$(a^2-x^2)_+^{\alpha-1} \times \left\{ \begin{matrix} \mathbf{J}_\nu(bx) \\ \mathbf{E}_\nu(bx) \end{matrix} \right\}$	$\dfrac{a^{s+2\alpha-2}}{2\nu\pi} \mathrm{B}\left(\alpha,\, \frac{s}{2}\right) \left\{ \begin{matrix} \sin(\nu\pi) \\ 1-\cos(\nu\pi) \end{matrix} \right\} {}_2F_3\left(\begin{matrix} 1,\, \frac{s}{2};\, -\frac{a^2b^2}{4} \\ \frac{2-\nu}{2},\, \frac{2+\nu}{2},\, \frac{s+2\alpha}{2} \end{matrix} \right) \pm \dfrac{a^{s+2\alpha-1} b}{2(1\quad \nu^2)\pi} \mathrm{B}\left(\alpha,\, \frac{s+1}{2}\right) \times \left\{ \begin{matrix} \sin(\nu\pi) \\ 1+\cos(\nu\pi) \end{matrix} \right\} {}_2F_3\left(\begin{matrix} 1,\, \frac{s+1}{2};\, -\frac{a^2b^2}{4} \\ \frac{3-\nu}{2},\, \frac{3+\nu}{2},\, \frac{s+2\alpha+1}{2} \end{matrix} \right)$ $[a,\ \operatorname{Re}\nu,\ \operatorname{Re} s>0]$

No.	$f(x)$	$F(s)$
5	$\left(x^2-a^2\right)_+^{\alpha-1}\mathbf{J}_\nu(bx)$	$\frac{a^{s+2\alpha-2}}{2\nu\pi}\sin(\nu\pi)\,\Gamma\left[\begin{matrix}\alpha,\ -\frac{s+2\alpha-2}{2}\\ \frac{2-s}{2}\end{matrix}\right]{}_2F_3\left(\begin{matrix}1,\ \frac{s}{2};\ -\frac{a^2b^2}{4}\\ \frac{2-\nu}{2},\ \frac{2+\nu}{2},\ \frac{s+2\alpha}{2}\end{matrix}\right) + \frac{a^{s+2\alpha-1}b}{2\pi\left(1-\nu^2\right)}\sin(\nu\pi)\,\Gamma\left[\begin{matrix}\alpha,\ -\frac{s+2\alpha-1}{2}\\ \frac{1-s}{2}\end{matrix}\right]{}_2F_3\left(\begin{matrix}1,\ \frac{s+1}{2};\ -\frac{a^2b^2}{4}\\ \frac{3-\nu}{2},\ \frac{3+\nu}{2},\ \frac{s+2\alpha+1}{2}\end{matrix}\right) + \frac{2^\nu\pi^{3/2}b^{-s-2\alpha+2}}{\Gamma\left(-\frac{s+2\alpha+\nu-4}{2}\right)\Gamma\left(\frac{s+2\alpha-\nu-1}{2}\right)\Gamma(-s-2\alpha+\nu+3)} \times\csc\left[(s+2\alpha)\pi\right]{}_1F_2\left(\begin{matrix}1-\alpha;\ -\frac{a^2b^2}{4}\\ -\frac{s+2\alpha+\nu-4}{2},\ -\frac{s+2\alpha-\nu-4}{2}\end{matrix}\right)$ $[a,\ b,\ \operatorname{Re}\mu>0;\ \operatorname{Re}(s+2\mu)<3]$
6	$\left(x^2-a^2\right)_+^{\alpha-1}\mathbf{E}_\nu(bx)$	$\frac{a^{s+2\alpha-2}}{\nu\pi}\sin^2\frac{\nu\pi}{2}\,\Gamma\left[\begin{matrix}\alpha,\ -\frac{s+2\alpha-2}{2}\\ \frac{2-s}{2}\end{matrix}\right]{}_2F_3\left(\begin{matrix}1,\ \frac{s}{2};\ -\frac{a^2b^2}{4}\\ \frac{2-\nu}{2},\ \frac{2+\nu}{2},\ \frac{s+2\alpha}{2}\end{matrix}\right) - \frac{a^{s+2\alpha-1}b}{\left(1-\nu^2\right)\pi}\cos^2\frac{\nu\pi}{2}\,\Gamma\left[\begin{matrix}\alpha,\ -\frac{s+2\alpha-1}{2}\\ \frac{1-s}{2}\end{matrix}\right]{}_2F_3\left(\begin{matrix}1,\ \frac{s+1}{2};\ -\frac{a^2b^2}{4}\\ \frac{3-\nu}{2},\ \frac{3+\nu}{2},\ \frac{s+2\alpha+1}{2}\end{matrix}\right) + \frac{2^{s+2\alpha-2}\pi b^{-s-2\alpha+2}}{\Gamma\left(-\frac{s+2\alpha+\nu-4}{2}\right)\Gamma\left(-\frac{s+2\alpha-\nu-4}{2}\right)}\csc\left[(s+2\alpha)\pi\right] \times\sin\frac{(s+2\alpha-\nu)\pi}{2}\,{}_1F_2\left(\begin{matrix}1-\alpha;\ -\frac{a^2b^2}{4}\\ -\frac{s+2\alpha+\nu-4}{2},\ -\frac{s+2\alpha-\nu-4}{2}\end{matrix}\right)$ $[a,\ b,\ \operatorname{Re}\mu>0;\ \operatorname{Re}(s+2\mu)<3]$
7	$\frac{1}{\left(x^2+a^2\right)^\rho}\left\{\begin{matrix}\mathbf{J}_\nu(bx)\\ \mathbf{E}_\nu(bx)\end{matrix}\right\}$	$\frac{a^{s-2\rho}}{2\nu\pi}\mathrm{B}\left(\frac{s}{2},\ \frac{2\rho-s}{2}\right)\left\{\begin{matrix}\sin(\nu\pi)\\ 1-\cos(\nu\pi)\end{matrix}\right\}{}_2F_3\left(\begin{matrix}1,\ \frac{s}{2};\ \frac{a^2b^2}{4}\\ \frac{2-\nu}{2},\ \frac{2+\nu}{2},\ \frac{s-2\rho+2}{2}\end{matrix}\right) \pm\frac{a^{s-2\rho+1}b}{2\left(1-\nu^2\right)\pi}\mathrm{B}\left(\frac{s+1}{2},\ \rho-\frac{s+1}{2}\right)\left\{\begin{matrix}\sin(\nu\pi)\\ 1+\cos(\nu\pi)\end{matrix}\right\} \times{}_2F_3\left(\begin{matrix}1,\ \frac{s+1}{2};\ \frac{a^2b^2}{4}\\ \frac{3-\nu}{2},\ \frac{3+\nu}{2},\ \frac{s-2\rho+3}{2}\end{matrix}\right) - \left(\frac{b}{2}\right)^{2\rho-s}\frac{\pi\csc\left[(2\rho-s)\pi\right]}{\Gamma\left(\frac{2-s-\nu+2\rho}{2}\right)\Gamma\left(\frac{2-s+\nu+2\rho}{2}\right)} \times\left\{\begin{matrix}\cos\left[(\nu-s+2\rho)\pi/2\right]\\ \sin\left[(\nu-s+2\rho)\pi/2\right]\end{matrix}\right\}{}_1F_2\left(\begin{matrix}\rho;\ \frac{a^2b^2}{4}\\ \frac{2-s-\nu+2\rho}{2},\ \frac{2-s+\nu+2\rho}{2}\end{matrix}\right)$ $[b,\ \operatorname{Re}a,\ \operatorname{Re}s>0;\ \operatorname{Re}(s-2\rho)<1]$
8	$\frac{1}{x^2-a^2}\left\{\begin{matrix}\mathbf{J}_\nu(bx)\\ \mathbf{E}_\nu(bx)\end{matrix}\right\}$	$-\frac{\pi a^{s-2}}{2\nu\pi}\cot\frac{s\pi}{2}\left\{\begin{matrix}\sin(\nu\pi)\\ 1-\cos(\nu\pi)\end{matrix}\right\}{}_1F_2\left(\begin{matrix}1;\ -\frac{a^2b^2}{4}\\ \frac{2-\nu}{2},\ \frac{2+\nu}{2}\end{matrix}\right) \pm\frac{\pi a^{s-1}b}{2\left(1-\nu^2\right)\pi}\tan\frac{s\pi}{2}\left\{\begin{matrix}\sin(\nu\pi)\\ 1+\cos(\nu\pi)\end{matrix}\right\}{}_1F_2\left(\begin{matrix}1;\ -\frac{a^2b^2}{4}\\ \frac{3-\nu}{2},\ \frac{3+\nu}{2}\end{matrix}\right) - \left(\frac{b}{2}\right)^{2-s}\frac{\pi\csc(s\pi)}{\Gamma\left(\frac{4-s-\nu}{2}\right)\Gamma\left(\frac{4-s+\nu}{2}\right)}\left\{\begin{matrix}\cos\left[(\nu-s)\pi/2\right]\\ \sin\left[(\nu-s)\pi/2\right]\end{matrix}\right\} \times{}_1F_2\left(\begin{matrix}1;\ -\frac{a^2b^2}{4}\\ \frac{4-s-\nu}{2},\ \frac{4-s+\nu}{2}\end{matrix}\right)$ $[a,\ b>0;\ 0<\operatorname{Re}s<3]$

No.	$f(x)$	$F(s)$
9	$(x^2+a^2)^{\nu/2}$ $\times\left[\mathbf{J}_\nu\left(b\sqrt{x^2+a^2}\right)\right.$ $\left.-\mathbf{J}_{-\nu}\left(b\sqrt{x^2+a^2}\right)\right]$	$\frac{2^{s/2}\pi\, a^{s/2+\nu}b^{-s/2}}{\Gamma\left(\frac{2-s}{2}\right)}\sin\frac{\nu\pi}{2}\sec\frac{(s+\nu)\pi}{2}J_{-(s+2\nu)/2}(ab)$ $-\frac{a^{s+\nu+1}b}{4\pi}\sin(\nu\pi)\,\Gamma\left[\begin{matrix}\frac{s}{2},\ -\frac{s+\nu+1}{2}\\ \frac{3-\nu}{2}\end{matrix}\right]{}_1F_2\left(\begin{matrix}1;\ -\frac{a^2b^2}{4}\\ \frac{3-\nu}{2},\ \frac{s+\nu+3}{2}\end{matrix}\right)$ $[a,\, b>0;\ 0<\operatorname{Re}s<1-\operatorname{Re}\nu]$
10	$(x^2+a^2)^{\nu/2}$ $\times\left[\mathbf{J}_\nu\left(b\sqrt{x^2+a^2}\right)\right.$ $\left.+\mathbf{J}_{-\nu}\left(b\sqrt{x^2+a^2}\right)\right]$	$\frac{2^{s/2}\pi\, a^{s/2+\nu}b^{-s/2}}{\Gamma\left(\frac{2-s}{2}\right)}\cos\frac{\nu\pi}{2}\csc\frac{(s+\nu)\pi}{2}J_{-(s+2\nu)/2}(ab)$ $-\frac{a^{s+\nu}}{2\pi}\sin(\nu\pi)\,\Gamma\left[\begin{matrix}\frac{s}{2},\ -\frac{s+\nu}{2}\\ \frac{2-\nu}{2}\end{matrix}\right]{}_1F_2\left(\begin{matrix}1;\ -\frac{a^2b^2}{4}\\ \frac{2-\nu}{2},\ \frac{s+\nu+2}{2}\end{matrix}\right)$ $[a,\, b>0;\ 0<\operatorname{Re}s<3/2-\operatorname{Re}\nu]$

3.16.2. $\mathbf{J}_\nu(bx)$, $\mathbf{E}_\nu(bx)$, and the exponential or trigonometric functions

1	$e^{-ax}\begin{Bmatrix}\mathbf{J}_\nu(bx)\\ \mathbf{E}_\nu(bx)\end{Bmatrix}$	$\frac{1}{\nu\pi a^s}\begin{Bmatrix}\sin(\nu\pi)\\ 1-\cos(\nu\pi)\end{Bmatrix}\Gamma(s)\,{}_3F_2\left(\begin{matrix}1,\ \frac{s}{2},\ \frac{s+1}{2};\ -\frac{b^2}{a^2}\\ \frac{2-\nu}{2},\ \frac{2+\nu}{2}\end{matrix}\right)$ $\pm\frac{b}{(1-\nu^2)\pi a^{s+1}}\begin{Bmatrix}\sin(\nu\pi)\\ 1+\cos(\nu\pi)\end{Bmatrix}\Gamma(s+1)\,{}_3F_2\left(\begin{matrix}1,\ \frac{s+1}{2},\ \frac{s+2}{2}\\ \frac{3-\nu}{2},\ \frac{3+\nu}{2};\ -\frac{b^2}{a^2}\end{matrix}\right)$ $[\operatorname{Re}s>0;\ \operatorname{Re}a>\lvert\operatorname{Im}b\rvert]$
2	$e^{-ax^2}\begin{Bmatrix}\mathbf{J}_\nu(bx)\\ \mathbf{E}_\nu(bx)\end{Bmatrix}$	$\frac{a^{-s/2}}{2\nu\pi}\begin{Bmatrix}\sin(\nu\pi)\\ 1-\cos(\nu\pi)\end{Bmatrix}\Gamma\left(\frac{s}{2}\right){}_2F_2\left(\begin{matrix}1,\ \frac{s}{2};\ -\frac{b^2}{4a}\\ \frac{2-\nu}{2},\ \frac{2+\nu}{2}\end{matrix}\right)$ $-\frac{a^{-(s+1)/2}b}{2(1-\nu^2)\pi}\begin{Bmatrix}\sin(\nu\pi)\\ 1+\cos(\nu\pi)\end{Bmatrix}$ $\times\Gamma\left(\frac{s+1}{2}\right){}_2F_2\left(\begin{matrix}1,\ \frac{s+1}{2};\ -\frac{b^2}{4a}\\ \frac{3-\nu}{2},\ \frac{3+\nu}{2}\end{matrix}\right)$ $[b,\ \operatorname{Re}a,\ \operatorname{Re}s>0]$
3	$\sin(ax)\begin{Bmatrix}\mathbf{J}_\nu(bx)\\ \mathbf{E}_\nu(bx)\end{Bmatrix}$	$\frac{\Gamma(s)}{\nu\pi a^s}\sin\frac{s\pi}{2}\begin{Bmatrix}\sin(\nu\pi)\\ 1-\cos(\nu\pi)\end{Bmatrix}{}_3F_2\left(\begin{matrix}1,\ \frac{s}{2},\ \frac{s+1}{2}\\ \frac{2-\nu}{2},\ \frac{2+\nu}{2};\ \frac{b^2}{a^2}\end{matrix}\right)$ $\pm\frac{b\,\Gamma(s+1)}{(1-\nu^2)\pi\, a^{s+1}}\cos\frac{s\pi}{2}$ $\times\begin{Bmatrix}\sin(\nu\pi)\\ 1+\cos(\nu\pi)\end{Bmatrix}{}_3F_2\left(\begin{matrix}1,\ \frac{s+1}{2},\ \frac{s+2}{2}\\ \frac{3-\nu}{2},\ \frac{3+\nu}{2};\ \frac{b^2}{a^2}\end{matrix}\right)$ $\left[\begin{matrix}0<b\le a;\ -1<\operatorname{Re}s<3/2 \text{ for } b<a;\\ -1<\operatorname{Re}s<1/2 \text{ for } b=a\end{matrix}\right]$

No.	$f(x)$	$F(s)$
4	$\sin(ax)\begin{Bmatrix}\mathbf{J}_\nu(bx)\\ \mathbf{E}_\nu(bx)\end{Bmatrix}$	$\mp\frac{\Gamma(s-1)}{\nu\pi a^{s-1}b}\cos\frac{s\pi}{2}\begin{Bmatrix}\sin(\nu\pi)\\ 1+\cos(\nu\pi)\end{Bmatrix}{}_3F_2\left(\begin{matrix}1,\ \frac{1-\nu}{2},\ \frac{1+\nu}{2}\\ \frac{2-s}{2},\ \frac{3-s}{2};\ \frac{a^2}{b^2}\end{matrix}\right)$ $+\frac{\nu\Gamma(s-2)}{\pi a^{s-2}b^2}\sin\frac{s\pi}{2}\begin{Bmatrix}\sin(\nu\pi)\\ 1-\cos(\nu\pi)\end{Bmatrix}{}_3F_2\left(\begin{matrix}1,\ \frac{2-\nu}{2},\ \frac{2+\nu}{2}\\ \frac{3-s}{2},\ \frac{4-s}{2};\ \frac{a^2}{b^2}\end{matrix}\right)$ $-\pi a\left(\frac{2}{b}\right)^{s+1}\frac{\csc(s\pi)}{\Gamma\left(\frac{1-s-\nu}{2}\right)\Gamma\left(\frac{1-s+\nu}{2}\right)}\begin{Bmatrix}\cos[(\nu-s-1)\pi/2]\\ \sin[(\nu-s-1)\pi/2]\end{Bmatrix}$ $\times{}_2F_1\left(\begin{matrix}\frac{s-\nu+1}{2},\ \frac{s+\nu+1}{2}\\ \frac{3}{2};\ \frac{a^2}{b^2}\end{matrix}\right)$ $[0<a\le b;\ -1<\operatorname{Re}s<3/2 \text{ for } a<b;\ -1<\operatorname{Re}s<1/2 \text{ for } a=b]$
5	$\cos(ax)\begin{Bmatrix}\mathbf{J}_\nu(bx)\\ \mathbf{E}_\nu(bx)\end{Bmatrix}$	$\frac{\Gamma(s)}{\nu\pi a^s}\cos\frac{s\pi}{2}\begin{Bmatrix}\sin(\nu\pi)\\ 1-\cos(\nu\pi)\end{Bmatrix}{}_3F_2\left(\begin{matrix}1,\ \frac{s}{2},\ \frac{s+1}{2}\\ \frac{2-\nu}{2},\ \frac{2+\nu}{2};\ \frac{b^2}{a^2}\end{matrix}\right)$ $\mp\frac{b\Gamma(s+1)}{(1-\nu^2)\pi a^{s+1}}\sin\frac{s\pi}{2}\begin{Bmatrix}\sin(\nu\pi)\\ 1+\cos(\nu\pi)\end{Bmatrix}{}_3F_2\left(\begin{matrix}1,\ \frac{s+1}{2},\ \frac{s+2}{2}\\ \frac{3-\nu}{2},\ \frac{3+\nu}{2};\ \frac{b^2}{a^2}\end{matrix}\right)$ $[0<b\le a;\ 0<\operatorname{Re}s<3/2 \text{ for } b<a;\ 0<\operatorname{Re}s<1/2 \text{ for } b=a]$
6	$\cos(ax)\begin{Bmatrix}\mathbf{J}_\nu(bx)\\ \mathbf{E}_\nu(bx)\end{Bmatrix}$	$\pm\frac{a^{1-s}\Gamma(s-1)}{\nu\pi b}\sin\frac{s\pi}{2}\begin{Bmatrix}\sin(\nu\pi)\\ 1+\cos(\nu\pi)\end{Bmatrix}{}_3F_2\left(\begin{matrix}1,\ \frac{1-\nu}{2},\ \frac{1+\nu}{2}\\ \frac{2-s}{2},\ \frac{3-s}{2};\ \frac{a^2}{b^2}\end{matrix}\right)$ $+\frac{\nu a^{2-s}\Gamma(s-2)}{\pi b^2}\cos\frac{s\pi}{2}\begin{Bmatrix}\sin(\nu\pi)\\ 1-\cos(\nu\pi)\end{Bmatrix}$ $\times{}_3F_2\left(\begin{matrix}1,\ \frac{2+\nu}{2},\ \frac{2-\nu}{2}\\ \frac{3-s}{2},\ \frac{4-s}{2};\ \frac{a^2}{b^2}\end{matrix}\right)+\frac{\pi(2/b)^s\csc(s\pi)}{\Gamma\left(\frac{2-s-\nu}{2}\right)\Gamma\left(\frac{2-s+\nu}{2}\right)}$ $\times\begin{Bmatrix}\cos[(\nu-s)\pi/2]\\ \sin[(\nu-s)\pi/2]\end{Bmatrix}{}_2F_1\left(\begin{matrix}\frac{s-\nu}{2},\ \frac{s+\nu}{2}\\ \frac{1}{2};\ \frac{a^2}{b^2}\end{matrix}\right)$ $[0<a\le b;\ 0<\operatorname{Re}s<3/2 \text{ for } a<b;\ 0<\operatorname{Re}s<1/2 \text{ for } a=b]$

3.16.3. $\mathbf{J}_\nu(bx)$, $\mathbf{E}_\nu(bx)$, and $\operatorname{Ei}(-ax^2)$ or $\operatorname{erfc}(ax)$

1	$\operatorname{Ei}(-ax^2)\begin{Bmatrix}\mathbf{J}_\nu(bx)\\ \mathbf{E}_\nu(bx)\end{Bmatrix}$	$\mp\frac{a^{-s/2}}{\pi}\left[\frac{1}{\nu s}\begin{Bmatrix}\sin(\nu\pi)\\ \cos(\nu\pi)-1\end{Bmatrix}\Gamma\left(\frac{s}{2}\right){}_3F_3\left(\begin{matrix}1,\ \frac{s}{2},\ \frac{s}{2};\ -\frac{b^2}{4a}\\ \frac{2-\nu}{2},\ \frac{2+\nu}{2},\ \frac{s+2}{2}\end{matrix}\right)\right.$ $-\frac{a^{-1/2}b}{(\nu^2-1)(s+1)}\begin{Bmatrix}\sin(\nu\pi)\\ \cos(\nu\pi)+1\end{Bmatrix}$ $\left.\times\Gamma\left(\frac{s+1}{2}\right){}_3F_3\left(\begin{matrix}1,\ \frac{s+1}{2},\ \frac{s+1}{2};\ -\frac{b^2}{4a}\\ \frac{3-\nu}{2},\ \frac{3+\nu}{2},\ \frac{s+3}{2}\end{matrix}\right)\right]$ $[a,\ \operatorname{Re}s>0 \text{ or } (\operatorname{Re}a,\ b>0;\ \lvert\operatorname{Im}a\rvert\ne 0;\ 0<\operatorname{Re}s<1)]$

No.	$f(x)$	$F(s)$
2	$\operatorname{erfc}(ax)\begin{Bmatrix}\mathbf{J}_\nu(bx)\\ \mathbf{E}_\nu(bx)\end{Bmatrix}$	$\frac{a^{-s-1}}{\pi^{3/2}}\left[\frac{a}{\nu s}\begin{Bmatrix}\sin(\nu\pi)\\ 1-\cos(\nu\pi)\end{Bmatrix}\Gamma\left(\frac{s+1}{2}\right){}_3F_3\left(\begin{matrix}1, \frac{s}{2}, \frac{s+1}{2}; -\frac{b^2}{4a^2}\\ \frac{2-\nu}{2}, \frac{2+\nu}{2}, \frac{s+2}{2}\end{matrix}\right)\right.$ $\mp\frac{b}{(\nu^2-1)(s+1)}\begin{Bmatrix}\sin(\nu\pi)\\ \cos(\nu\pi)+1\end{Bmatrix}$ $\left.\times\Gamma\left(\frac{s+2}{2}\right){}_3F_3\left(\begin{matrix}1, \frac{s+1}{2}, \frac{s+2}{2}; -\frac{b^2}{4a^2}\\ \frac{3-\nu}{2}, \frac{3+\nu}{2}, \frac{s+3}{2}\end{matrix}\right)\right]$ $\left[\begin{matrix}(\operatorname{Re}s>0; \lvert\arg a\rvert<\pi/4)\text{ or}\\ (0<\operatorname{Re}s<7/2; \lvert\operatorname{Im}b\rvert\neq 0; \lvert\arg a\rvert<\pi/4)\end{matrix}\right]$

3.16.4. $\mathbf{J}_\nu(bx)$, $\mathbf{E}_\nu(bx)$, and $J_\mu(ax)$

No.	$f(x)$	$F(s)$
1	$J_\mu(ax)\begin{Bmatrix}\mathbf{J}_\nu(bx)\\ \mathbf{E}_\nu(bx)\end{Bmatrix}$	$\frac{2^{s-1}}{\nu\pi a^s}\begin{Bmatrix}\sin(\nu\pi)\\ 1-\cos(\nu\pi)\end{Bmatrix}\Gamma\left[\begin{matrix}\frac{s+\mu}{2}\\ \frac{2-s+\mu}{2}\end{matrix}\right]{}_3F_2\left(\begin{matrix}1, \frac{s-\mu}{2}, \frac{s+\mu}{2}\\ \frac{2-\nu}{2}, \frac{2+\nu}{2}; \frac{b^2}{a^2}\end{matrix}\right)$ $\pm\frac{2^s b}{(1-\nu^2)\pi a^{s+1}}\begin{Bmatrix}\sin(\nu\pi)\\ 1+\cos(\nu\pi)\end{Bmatrix}$ $\times\Gamma\left[\begin{matrix}\frac{s+\mu+1}{2}\\ \frac{1-s+\mu}{2}\end{matrix}\right]{}_3F_2\left(\begin{matrix}1, \frac{s-\mu+1}{2}, \frac{s+\mu+1}{2}\\ \frac{3-\nu}{2}, \frac{3+\nu}{2}; \frac{b^2}{a^2}\end{matrix}\right)$ $\left[\begin{matrix}0<b\le a; -\operatorname{Re}\mu<\operatorname{Re}s<2\text{ for }b<a;\\ -\operatorname{Re}\mu<\operatorname{Re}s<1\text{ for }b=a\end{matrix}\right]$
2	$J_\mu(ax)\begin{Bmatrix}\mathbf{J}_\nu(bx)\\ \mathbf{E}_\nu(bx)\end{Bmatrix}$	$\pm\frac{1}{2\pi b}\left(\frac{a}{2}\right)^{1-s}\begin{Bmatrix}\sin(\nu\pi)\\ 1+\cos(\nu\pi)\end{Bmatrix}\Gamma\left[\begin{matrix}\frac{s+\mu-1}{2}\\ \frac{3-s+\mu}{2}\end{matrix}\right]$ $\times{}_3F_2\left(\begin{matrix}1, \frac{1-\nu}{2}, \frac{1+\nu}{2}\\ \frac{3-s-\mu}{2}, \frac{3-s+\mu}{2}; \frac{a^2}{b^2}\end{matrix}\right)-\frac{\nu a^{2-s}}{2^{3-s}\pi b^2}\begin{Bmatrix}\sin(\nu\pi)\\ 1-\cos(\nu\pi)\end{Bmatrix}$ $\times\Gamma\left[\begin{matrix}\frac{s+\mu-2}{2}\\ \frac{4-s+\mu}{2}\end{matrix}\right]{}_3F_2\left(\begin{matrix}1, \frac{2-\nu}{2}, \frac{2+\nu}{2}\\ \frac{4-s-\mu}{2}, \frac{4-s+\mu}{2}; \frac{a^2}{b^2}\end{matrix}\right)$ $+\frac{2^s\pi a^\mu}{b^{s+\mu}}\frac{\csc[(s+\mu)\pi]}{\Gamma\left[\mu+1, \frac{2-s-\mu-\nu}{2}, \frac{2-s-\mu+\nu}{2}\right]}$ $\times\begin{Bmatrix}\cos[(\nu-s-\mu)\pi/2]\\ \sin[(\nu-s-\mu)\pi/2]\end{Bmatrix}{}_2F_1\left(\begin{matrix}\frac{s+\mu-\nu}{2}, \frac{s+\mu+\nu}{2}\\ \mu+1; \frac{a^2}{b^2}\end{matrix}\right)$ $\left[\begin{matrix}0<a\le b; -\operatorname{Re}\mu<\operatorname{Re}s<2\text{ for }a<b;\\ -\operatorname{Re}\mu<\operatorname{Re}s<1\text{ for }a=b\end{matrix}\right]$
3	$J_\nu(ax)-\mathbf{J}_\nu(ax)$	$-\frac{2^{s-1}\sin(\nu\pi)}{a^s\sin(s\pi)}\Gamma\left[\begin{matrix}\frac{s+\nu}{2}\\ \frac{2-s+\nu}{2}\end{matrix}\right]$ $[0, -\operatorname{Re}\nu<\operatorname{Re}s<1; \lvert\arg a\rvert<\pi]$

3.17. The Kelvin Functions $\operatorname{ber}_\nu(z)$, $\operatorname{bei}_\nu(z)$, and $\operatorname{ker}_\nu(z)$, $\operatorname{kei}_\nu(z)$

More formulas can be obtained from the corresponding sections due to the relations

$$\begin{Bmatrix}\operatorname{bei}_\nu(z)\\ \operatorname{ber}_\nu(z)\end{Bmatrix} = \frac{1}{\Gamma(\nu+2)}\begin{Bmatrix}\cos(3\pi\nu/4)\\ \sin(3\pi\nu/4)\end{Bmatrix}\left(\frac{z}{2}\right)^{\nu+2} {}_0F_3\left(\frac{3}{2}, \frac{\nu+2}{2}, \frac{\nu+3}{2}; -\frac{z^4}{256}\right)$$
$$+\frac{1}{\Gamma(\nu+1)}\begin{Bmatrix}\sin(3\pi\nu/4)\\ \cos(3\pi\nu/4)\end{Bmatrix}\left(\frac{z}{2}\right)^{\nu+2} {}_0F_3\left(\frac{1}{2}, \frac{\nu+1}{2}, \frac{\nu+2}{2}; -\frac{z^4}{256}\right),$$

$$\begin{Bmatrix}\operatorname{kei}_\nu(z)\\ \operatorname{ker}_\nu(z)\end{Bmatrix} = -2^{-\nu-3}\begin{Bmatrix}\cos(\nu\pi/4)\\ \sin(\nu\pi/4)\end{Bmatrix}\Gamma(-\nu-1)\, z^{\nu+2}\, {}_0F_3\left(\frac{3}{2}, \frac{\nu+2}{2}, \frac{\nu+3}{2}; -\frac{z^4}{256}\right)$$
$$\mp 2^{-\nu-1}\begin{Bmatrix}\sin(\nu\pi/4)\\ \cos(\nu\pi/4)\end{Bmatrix}\Gamma(-\nu)\, z^{\nu}\, {}_0F_3\left(\frac{1}{2}, \frac{\nu+1}{2}, \frac{\nu+2}{2}; -\frac{z^4}{256}\right)$$
$$-2^{\nu-3}\begin{Bmatrix}\cos(3\pi\nu/4)\\ \sin(3\pi\nu/4)\end{Bmatrix}\Gamma(\nu-1)\, z^{-\nu+2}\, {}_0F_3\left(\frac{3}{2}, \frac{2-\nu}{2}, \frac{3-\nu}{2}; -\frac{z^4}{256}\right)$$
$$\mp 2^{\nu-1}\begin{Bmatrix}\sin(3\nu\pi/4)\\ \cos(3\nu\pi/4)\end{Bmatrix}\Gamma(\nu)\, z^{-\nu}\, {}_0F_3\left(\frac{1}{2}, \frac{1-\nu}{2}, \frac{2-\nu}{2}; -\frac{z^4}{256}\right);$$

$$\begin{Bmatrix}\operatorname{ber}_\nu(z)\\ \operatorname{bei}_\nu(z)\end{Bmatrix} = \pi\, G_{15}^{20}\left(\frac{z^4}{256}\,\middle|\, \begin{matrix}\frac{4\nu+1\pm1}{4}\\ \frac{\nu}{4}, \frac{2+\nu}{4}, -\frac{\nu}{4}, \frac{2-\nu}{4}, \frac{4\nu+1\pm1}{4}\end{matrix}\right), \quad [-\pi/4 \le \arg z \le \pi/4];$$

$$\begin{Bmatrix}\operatorname{ker}_\nu(z)\\ \operatorname{kei}_\nu(z)\end{Bmatrix} = \pm\frac{1}{4}\, G_{15}^{40}\left(\frac{z^4}{256}\,\middle|\, \begin{matrix}\frac{2\nu+1\pm1}{4}\\ -\frac{\nu}{4}, \frac{\nu}{4}, \frac{2-\nu}{4}, \frac{\nu+2}{4}, \frac{2\nu+1\pm1}{4}\end{matrix}\right), \quad [-\pi/4 \le \arg z \le \pi/4].$$

3.17.1. $\operatorname{ber}_\nu(bx)$, $\operatorname{bei}_\nu(bx)$, $\operatorname{ker}_\nu(bx)$, $\operatorname{kei}_\nu(bx)$, and algebraic functions

No.	$f(x)$	$F(s)$
1	$(a-x)_+^{\alpha-1}\begin{Bmatrix}\operatorname{ber}_\nu(bx)\\ \operatorname{bei}_\nu(bx)\end{Bmatrix}$	$\frac{a^{s+\alpha+\nu-1}b^\nu}{2^\nu\,\Gamma(\nu+1)}\begin{Bmatrix}\cos(3\pi\nu/4)\\ \sin(3\pi\nu/4)\end{Bmatrix}\mathrm{B}(\alpha, s+\nu) \times {}_4F_7\left(\begin{matrix}\Delta(4, s+\nu);\ -\frac{a^4b^4}{256}\\ \frac{1}{2}, \frac{\nu+1}{2}, \frac{\nu+2}{2}, \Delta(4, s+\alpha+\nu)\end{matrix}\right) \mp \frac{a^{s+\alpha+\nu+1}b^{\nu+2}}{2^{\nu+2}\,\Gamma(\nu+2)}\begin{Bmatrix}\sin(3\pi\nu/4)\\ \cos(3\pi\nu/4)\end{Bmatrix}\mathrm{B}(\alpha, s+\nu+2) \times {}_4F_7\left(\begin{matrix}\Delta(4, s+\nu+2);\ -\frac{a^4b^4}{256}\\ \frac{3}{2}, \frac{\nu+2}{2}, \frac{\nu+3}{2}, \Delta(4, s+\alpha+\nu+2)\end{matrix}\right)$ $\left[\begin{matrix}a,\ \operatorname{Re}\alpha>0;\\ \operatorname{Re}(s+\nu)>0\end{matrix}\right]$
2	$(a^2-x^2)_+^{\alpha-1} \times \begin{Bmatrix}\operatorname{ber}_\nu(bx)\\ \operatorname{bei}_\nu(bx)\end{Bmatrix}$	$\frac{a^{s+\nu+2\alpha-2}b^\nu}{2^{\nu+1}\,\Gamma(\nu+1)}\begin{Bmatrix}\cos(3\pi\nu/4)\\ \sin(3\pi\nu/4)\end{Bmatrix}\mathrm{B}\left(\alpha, \frac{s+\nu}{2}\right) \times {}_2F_5\left(\begin{matrix}\Delta\left(2, \frac{s+\nu}{2}\right);\ -\frac{a^4b^4}{256}\\ \frac{1}{2}, \Delta(2, \nu+1), \Delta\left(2, \frac{s+2\alpha+\nu}{2}\right)\end{matrix}\right) \mp \frac{a^{s+\nu+2\alpha}b^{\nu+2}}{2^{\nu+3}\,\Gamma(\nu+2)}\begin{Bmatrix}\sin(3\pi\nu/4)\\ \cos(3\pi\nu/4)\end{Bmatrix}\mathrm{B}\left(\alpha, \frac{s+\nu+2}{2}\right) \times {}_2F_5\left(\begin{matrix}\Delta\left(2, \frac{s+\nu+2}{2}\right);\ -\frac{a^4b^4}{256}\\ \frac{3}{2}, \Delta(2, \nu+2), \Delta\left(2, \frac{s+2\alpha+\nu+2}{2}\right)\end{matrix}\right)$ $\left[\begin{matrix}a,\ \operatorname{Re}\alpha>0;\\ \operatorname{Re}(s+\nu)>0\end{matrix}\right]$

No.	$f(x)$	$F(s)$
3	$\begin{Bmatrix}\operatorname{ker}_\nu(ax)\\ \operatorname{kei}_\nu(ax)\end{Bmatrix}$	$\pm\frac{2^{s-2}}{a^s}\begin{Bmatrix}\cos[(s+2\nu)\pi/4]\\ \sin[(s+2\nu)\pi/4]\end{Bmatrix}\Gamma\left(\frac{s-\nu}{2}\right)\Gamma\left(\frac{s+\nu}{2}\right)$ $[\operatorname{Re}s>\lvert\operatorname{Re}\nu\rvert;\ \lvert\arg a\rvert<\pi/4]$
4	$(a-x)_+^{\alpha-1}\begin{Bmatrix}\operatorname{ker}_\nu(bx)\\ \operatorname{kei}_\nu(bx)\end{Bmatrix}$	$-\frac{a^{s+\alpha+\nu+1}b^{\nu+2}}{2^{\nu+3}}\begin{Bmatrix}\sin(\pi\nu/4)\\ \cos(\pi\nu/4)\end{Bmatrix}\Gamma(-\nu-1)\,\mathrm{B}(\alpha,\,s+\nu+2)$ $\times\,{}_4F_7\left(\begin{matrix}\Delta(4,\,s+\nu+2);\,-\frac{a^4b^4}{256}\\ \frac{3}{2},\,\frac{\nu+2}{2},\,\frac{\nu+3}{2},\,\Delta(4,\,s+\alpha+\nu+2)\end{matrix}\right)$ $\pm\frac{a^{s+\alpha+\nu-1}b^{\nu}}{2^{\nu+1}}\begin{Bmatrix}\cos(\pi\nu/4)\\ \sin(\pi\nu/4)\end{Bmatrix}\Gamma(-\nu)\,\mathrm{B}(\alpha,\,s+\nu)$ $\times\,{}_4F_7\left(\begin{matrix}\Delta(4,\,s+\nu);\,-\frac{a^4b^4}{256}\\ \frac{1}{2},\,\frac{\nu+1}{2},\,\frac{\nu+2}{2},\,\Delta(4,\,s+\alpha+\nu)\end{matrix}\right)$ $-\frac{2^{\nu-3}a^{s+\alpha-\nu+1}}{b^{\nu-2}}\begin{Bmatrix}\sin(3\pi\nu/4)\\ \cos(3\pi\nu/4)\end{Bmatrix}\Gamma(\nu-1)\,\mathrm{B}(\alpha,\,s-\nu+2)$ $\times\,{}_4F_7\left(\begin{matrix}\Delta(4,\,s-\nu+2);\,-\frac{a^4b^4}{256}\\ \frac{3}{2},\,\frac{2-\nu}{2},\,\frac{3-\nu}{2},\,\Delta(4,\,s+\alpha-\nu+2)\end{matrix}\right)$ $\pm\frac{2^{\nu-1}a^{s+\alpha-\nu-1}}{b^{\nu}}\begin{Bmatrix}\cos(3\pi\nu/4)\\ \sin(3\pi\nu/4)\end{Bmatrix}\Gamma(\nu)\,\mathrm{B}(\alpha,\,s-\nu)$ $\times\,{}_4F_7\left(\begin{matrix}\Delta(4,\,s-\nu);\,-\frac{a^4b^4}{256}\\ \frac{1}{2},\,\frac{1-\nu}{2},\,\frac{2-\nu}{2},\,\Delta(4,\,s+\alpha-\nu)\end{matrix}\right)$ $[a,\ \operatorname{Re}\alpha>0;\ \operatorname{Re}s>\lvert\operatorname{Re}\nu\rvert]$

3.17.2. $\operatorname{ber}_\nu(bx)$, $\operatorname{bei}_\nu(bx)$, $\operatorname{ker}_\nu(bx)$, $\operatorname{kei}_\nu(bx)$, and the exponential function

1	$e^{-ax}\begin{Bmatrix}\operatorname{ber}_\nu(bx)\\ \operatorname{bei}_\nu(bx)\end{Bmatrix}$	$a^{-s-\nu}\left(\frac{b}{2}\right)^{\nu}\begin{Bmatrix}\cos(3\nu\pi/4)\\ \sin(3\nu\pi/4)\end{Bmatrix}\Gamma\begin{bmatrix}s+\nu\\ \nu+1\end{bmatrix}{}_4F_3\left(\begin{matrix}\Delta(4,\,s+\nu);\,-\frac{b^4}{a^4}\\ \frac{1}{2},\,\Delta(2,\,\nu+1)\end{matrix}\right)$ $\mp a^{-s-\nu-2}\left(\frac{b}{2}\right)^{\nu+2}\begin{Bmatrix}\sin(3\nu\pi/4)\\ \cos(3\nu\pi/4)\end{Bmatrix}\Gamma\begin{bmatrix}s+\nu+2\\ \nu+2\end{bmatrix}$ $\times\,{}_4F_3\left(\begin{matrix}\Delta(4,\,s+\nu+2);\,-\frac{b^4}{a^4}\\ \frac{3}{2},\,\Delta(2,\,\nu+2)\end{matrix}\right)$ $[\sqrt{2}\operatorname{Re}a>\operatorname{Re}b+\lvert\operatorname{Im}b\rvert;\ \operatorname{Re}(s+\nu)>0]$
2	$e^{-ax^2}\begin{Bmatrix}\operatorname{ber}_\nu(bx)\\ \operatorname{bei}_\nu(bx)\end{Bmatrix}$	$\frac{b^{\nu}}{2^{\nu+1}a^{(s+\nu)/2}}\begin{Bmatrix}\cos(3\nu\pi/4)\\ \sin(3\nu\pi/4)\end{Bmatrix}\Gamma\begin{bmatrix}\frac{s+\nu}{2}\\ \nu+1\end{bmatrix}{}_2F_3\left(\begin{matrix}\Delta\left(2,\,\frac{s+\nu}{2}\right);\,-\frac{b^4}{64a^2}\\ \frac{1}{2},\,\Delta(2,\,\nu+1)\end{matrix}\right)$ $\mp\frac{b^{\nu+2}}{2^{\nu+3}a^{(s+\nu)/2+1}}\begin{Bmatrix}\sin(3\nu\pi/4)\\ \cos(3\nu\pi/4)\end{Bmatrix}\Gamma\begin{bmatrix}\frac{s+\nu+2}{2}\\ \nu+2\end{bmatrix}$ $\times\,{}_2F_3\left(\begin{matrix}\Delta\left(2,\,\frac{s+\nu+2}{2}\right);\,-\frac{b^4}{64a^2}\\ \frac{3}{2},\,\Delta(2,\,\nu+2)\end{matrix}\right)$ $[\operatorname{Re}a,\ \operatorname{Re}(s+\nu)>0]$

No.	$f(x)$	$F(s)$
3	$e^{-ax}\begin{Bmatrix}\operatorname{ker}_\nu(bx)\\ \operatorname{kei}_\nu(bx)\end{Bmatrix}$	$\pm\frac{2^{s-2}}{b^s}\begin{Bmatrix}\cos[(s+2\nu)\pi/4]\\ \sin[(s+2\nu)\pi/4]\end{Bmatrix}\Gamma\left(\frac{s+\nu}{2}\right)\Gamma\left(\frac{s-\nu}{2}\right)$ $\times\,{}_4F_3\left(\begin{matrix}\Delta\left(2,\frac{s+\nu}{2}\right),\ \Delta\left(2,\frac{s-\nu}{2}\right)\\ \frac{1}{4},\ \frac{1}{2},\ \frac{3}{4};\ -\frac{a^4}{b^4}\end{matrix}\right)$ $\mp\frac{2^{s-1}a}{b^{s+1}}\begin{Bmatrix}\cos[(s+2\nu+1)\pi/4]\\ \sin[(s+2\nu+1)\pi/4]\end{Bmatrix}\Gamma\left(\frac{s+\nu+1}{2}\right)$ $\times\,\Gamma\left(\frac{s-\nu+1}{2}\right){}_4F_3\left(\begin{matrix}\Delta\left(2,\frac{s+\nu+1}{2}\right),\ \Delta\left(2,\frac{s-\nu+1}{2}\right)\\ \frac{1}{2},\ \frac{3}{4},\ \frac{5}{4};\ -\frac{a^4}{b^4}\end{matrix}\right)$ $-\frac{2^{s-1}a^2}{b^{s+2}}\begin{Bmatrix}\sin[(s+2\nu)\pi/4]\\ \cos[(s+2\nu)\pi/4]\end{Bmatrix}\Gamma\left(\frac{s+\nu+2}{2}\right)$ $\times\,\Gamma\left(\frac{s-\nu+2}{2}\right){}_4F_3\left(\begin{matrix}\Delta\left(2,\frac{s+\nu+2}{2}\right),\ \Delta\left(2,\frac{s-\nu+2}{2}\right)\\ \frac{3}{4},\ \frac{5}{4},\ \frac{3}{2};\ -\frac{a^4}{b^4}\end{matrix}\right)$ $+\frac{2^s a^3}{3b^{s+3}}\begin{Bmatrix}\sin[(s+2\nu+1)\pi/4]\\ \cos[(s+2\nu+1)\pi/4]\end{Bmatrix}\Gamma\left(\frac{s+\nu+3}{2}\right)$ $\times\,\Gamma\left(\frac{s-\nu+3}{2}\right){}_4F_3\left(\begin{matrix}\Delta\left(2,\frac{s+\nu+3}{2}\right),\ \Delta\left(2,\frac{s-\nu+3}{2}\right)\\ \frac{5}{4},\ \frac{3}{2},\ \frac{7}{4};\ -\frac{a^4}{b^4}\end{matrix}\right)$ $\left[\operatorname{Re}\left(\sqrt{2}a+b\right)>\lvert\operatorname{Im}b\rvert;\ \operatorname{Re}s>\lvert\operatorname{Re}\nu\rvert\right]$
4	$e^{-ax^2}\begin{Bmatrix}\operatorname{ker}_\nu(bx)\\ \operatorname{kei}_\nu(bx)\end{Bmatrix}$	$\pm\frac{2^{\nu-2}}{a^{(s-\nu)/2}b^\nu}\begin{Bmatrix}\cos(3\pi\nu/4)\\ \sin(3\pi\nu/4)\end{Bmatrix}\Gamma(\nu)$ $\times\,\Gamma\left(\frac{s-\nu}{2}\right){}_2F_3\left(\begin{matrix}\frac{s-\nu}{4},\ \frac{s-\nu+2}{4}\\ \frac{1}{2},\ \frac{1-\nu}{2},\ \frac{2-\nu}{2};\ -\frac{b^4}{64a^2}\end{matrix}\right)$ $-\frac{2^{\nu-4}}{a^{(s-\nu+2)/2}b^{\nu-2}}\begin{Bmatrix}\sin(3\pi\nu/4)\\ \cos(3\pi\nu/4)\end{Bmatrix}\Gamma(\nu-1)$ $\times\,\Gamma\left(\frac{s-\nu+2}{2}\right){}_2F_3\left(\begin{matrix}\frac{s-\nu+2}{4},\ \frac{s-\nu+4}{4}\\ \frac{3}{2},\ \frac{2-\nu}{2},\ \frac{3-\nu}{2};\ -\frac{b^4}{64a^2}\end{matrix}\right)$ $\pm\frac{2^{-\nu-2}b^\nu}{a^{(s+\nu)/2}}\begin{Bmatrix}\cos(\pi\nu/4)\\ \sin(\pi\nu/4)\end{Bmatrix}\Gamma(-\nu)$ $\times\,\Gamma\left(\frac{s+\nu}{2}\right){}_2F_3\left(\begin{matrix}\frac{s+\nu}{4},\ \frac{s+\nu+2}{4}\\ \frac{1}{2},\ \frac{\nu+1}{2},\ \frac{\nu+2}{2};\ -\frac{b^4}{64a^2}\end{matrix}\right)$ $-\frac{2^{-\nu-4}b^{\nu+2}}{a^{(s+\nu+2)/2}}\begin{Bmatrix}\sin(\pi\nu/4)\\ \cos(\pi\nu/4)\end{Bmatrix}\Gamma(-\nu-1)$ $\times\,\Gamma\left(\frac{s+\nu+2}{2}\right){}_2F_3\left(\begin{matrix}\frac{s+\nu+2}{4},\ \frac{s+\nu+4}{4}\\ \frac{3}{2},\ \frac{\nu+2}{2},\ \frac{\nu+3}{2};\ -\frac{b^4}{64a^2}\end{matrix}\right)$ $[\operatorname{Re}a>0;\ \operatorname{Re}s>\lvert\operatorname{Re}\nu\rvert]$

3.17.3. $\mathrm{ker}_\nu(bx)$, $\mathrm{kei}_\nu(bx)$, and trigonometric functions

Notation: $\delta = \begin{Bmatrix} 1 \\ 0 \end{Bmatrix}$.

1	$\sin(ax)\begin{Bmatrix} \mathrm{ker}_\nu(bx) \\ \mathrm{kei}_\nu(bx) \end{Bmatrix}$	$U(1)$ $U(\delta) = \pm\dfrac{2^{s+\delta-2}a^\delta}{b^{s+\delta}}\begin{Bmatrix} \cos[(s+2\nu+\delta)\pi/4] \\ \sin[(s+2\nu+\delta)\pi/4] \end{Bmatrix}\Gamma\left(\dfrac{s-\nu+\delta}{2}\right)$ $\times\Gamma\left(\dfrac{s+\nu+\delta}{2}\right){}_4F_3\left(\begin{matrix}\Delta\left(2,\,\frac{s+\nu+\delta}{2}\right),\,\Delta\left(2,\,\frac{s-\nu+\delta}{2}\right)\\ \frac{1}{2},\,\frac{3}{4},\,\frac{4\delta+1}{4};\,-\frac{a^4}{b^4}\end{matrix}\right)$ $+\dfrac{2^{s+\delta-1}a^{\delta+2}}{3^\delta b^{s+\delta+2}}\begin{Bmatrix} \sin[(s+2\nu+\delta)\pi/4] \\ \cos[(s+2\nu+\delta)\pi/4] \end{Bmatrix}\Gamma\left(\dfrac{s-\nu+\delta+2}{2}\right)$ $\times\Gamma\left(\dfrac{s+\nu+\delta+2}{2}\right){}_4F_3\left(\begin{matrix}\Delta\left(2,\,\frac{s-\nu+\delta+2}{2}\right),\,\Delta\left(2,\,\frac{s+\nu+\delta+2}{2}\right)\\ \frac{5}{4},\,\frac{3}{2},\,\frac{4\delta+3}{4};\,-\frac{a^4}{b^4}\end{matrix}\right)$ $[a>0;\ \mathrm{Re}\,s > \lvert\mathrm{Re}\,\nu\rvert - 1;\ \lvert\arg b\rvert < \pi/4]$
2	$\cos(ax)\begin{Bmatrix} \mathrm{ker}_\nu(bx) \\ \mathrm{kei}_\nu(bx) \end{Bmatrix}$	$U(0)$ $\left[\begin{matrix} a>0;\ \mathrm{Re}\,s > \lvert\mathrm{Re}\,\nu\rvert; \\ \lvert\arg b\rvert < \pi/4;\ U(\delta):\text{ see 3.17.3.1}\end{matrix}\right]$

3.17.4. $\mathrm{ber}_\nu(bx)$, $\mathrm{bei}_\nu(bx)$, $\mathrm{ker}_\nu(bx)$, $\mathrm{kei}_\nu(bx)$, and $\mathrm{Ei}(-ax^r)$

1	$\mathrm{Ei}(-ax)\begin{Bmatrix} \mathrm{ber}_\nu(bx) \\ \mathrm{bei}_\nu(bx) \end{Bmatrix}$	$-\dfrac{a^{-s-\nu}b^\nu}{2^\nu(s+\nu)}\begin{Bmatrix} \cos(3\nu\pi/4) \\ \sin(3\nu\pi/4) \end{Bmatrix}\Gamma\begin{bmatrix} s+\nu \\ \nu+1 \end{bmatrix}$ $\times{}_5F_4\left(\begin{matrix}\frac{s+\nu}{4},\,\Delta(4,\,s+\nu)\\ \frac{1}{2},\,\Delta(2,\,\nu+1),\,\frac{s+\nu+4}{4};\,-\frac{b^4}{a^4}\end{matrix}\right)$ $\pm\dfrac{a^{-s-\nu-2}b^{\nu+2}}{2^{\nu+2}(s+\nu+2)}\begin{Bmatrix} \sin(3\nu\pi/4) \\ \cos(3\nu\pi/4) \end{Bmatrix}\Gamma\begin{bmatrix} s+\nu+2 \\ \nu+2 \end{bmatrix}$ $\times{}_5F_4\left(\begin{matrix}\frac{s+\nu+2}{4},\,\Delta(4,\,s+\nu+2)\\ \frac{3}{2},\,\Delta(2,\,\nu+2),\,\frac{s+\nu+6}{4};\,-\frac{b^4}{a^4}\end{matrix}\right)$ $[\mathrm{Re}(\sqrt{2}\,a-b) > \lvert\mathrm{Im}\,b\rvert;\ \mathrm{Re}(s+\nu)>0]$
2	$\mathrm{Ei}(-ax^2)\begin{Bmatrix} \mathrm{ber}_\nu(bx) \\ \mathrm{bei}_\nu(bx) \end{Bmatrix}$	$-\dfrac{a^{-(s+\nu)/2}b^\nu}{2^\nu(s+\nu)}\begin{Bmatrix} \cos(3\nu\pi/4) \\ \sin(3\nu\pi) \end{Bmatrix}\Gamma\begin{bmatrix} \frac{s+\nu}{2} \\ \nu+1 \end{bmatrix}$ $\times{}_3F_4\left(\begin{matrix}\frac{s+\nu}{4},\,\frac{s+\nu}{4},\,\frac{s+\nu+2}{4};\,-\frac{b^4}{64a^2}\\ \frac{1}{2},\,\Delta(2,\,\nu+1),\,\frac{s+\nu+4}{4}\end{matrix}\right)$ $\pm\dfrac{a^{-(s+\nu)/2-1}c^{\nu+2}}{2^{\nu+2}(s+\nu+2)}\begin{Bmatrix} \sin(3\nu\pi/4) \\ \cos(3\nu\pi/4) \end{Bmatrix}\Gamma\begin{bmatrix} \frac{s+\nu+2}{2} \\ \nu+2 \end{bmatrix}$ $\times{}_2F_5\left(\begin{matrix}\frac{s+\nu+2}{4},\,\frac{s+\nu+2}{4},\,\frac{s+\nu+4}{4};\,-\frac{c^4}{64a^2}\\ \frac{3}{2},\,\Delta(2,\,\nu+2),\,\frac{s+\nu+6}{4}\end{matrix}\right)$ $[\mathrm{Re}\,a,\ \mathrm{Re}(s+\nu)>0]$

3.17.5. $\mathrm{ber}_\nu(bx)$, $\mathrm{bei}_\nu(bx)$, $\mathrm{ker}_\nu(bx)$, $\mathrm{kei}_\nu(bx)$, and the Bessel functions

1	$J_\mu(ax)\begin{Bmatrix}\mathrm{ker}_\nu(bx)\\ \mathrm{kei}_\nu(bx)\end{Bmatrix}$	$\pm\frac{2^{s-2}a^\mu}{b^{s+\mu}}\begin{Bmatrix}\cos[(s+\mu+2\nu)\pi/4]\\ \sin[(s+\mu+2\nu)\pi/4]\end{Bmatrix}\Gamma\left[\begin{matrix}\frac{s+\mu+\nu}{2},\ \frac{s+\mu-\nu}{2}\\ \mu+1\end{matrix}\right] \times {}_4F_3\left(\begin{matrix}\Delta\left(2,\ \frac{s+\mu+\nu}{2}\right),\ \Delta\left(2,\ \frac{s+\mu-\nu}{2}\right)\\ \frac{1}{2},\ \Delta(2,\ \mu+1);\ -\frac{a^4}{b^4}\end{matrix}\right)+\frac{2^{s-2}a^{\mu+2}}{b^{s+\mu+2}} \times\begin{Bmatrix}\sin[(s+\mu+2\nu)\pi/4]\\ \cos[(s+\mu+2\nu)\pi/4]\end{Bmatrix}\Gamma\left[\begin{matrix}\frac{s+\mu+\nu+2}{2},\ \frac{s+\mu-\nu+2}{2}\\ \mu+2\end{matrix}\right] \times {}_4F_3\left(\begin{matrix}\Delta\left(2,\ \frac{s+\mu+\nu+2}{2}\right),\ \Delta\left(2,\ \frac{s+\mu-\nu+2}{2}\right)\\ \frac{3}{2},\ \Delta(2,\ \mu+2);\ -\frac{a^4}{b^4}\end{matrix}\right)$ $[a>0;\ \mathrm{Re}\,(s+\mu)>\lvert\mathrm{Re}\,\nu\rvert;\ \lvert\arg b\rvert<\pi]$
2	$K_\mu(ax)\begin{Bmatrix}\mathrm{ber}_\nu(bx)\\ \mathrm{bei}_\nu(bx)\end{Bmatrix}$	$\frac{2^{s-2}b^\nu}{a^{s+\nu}}\begin{Bmatrix}\cos(3\nu\pi/4)\\ \sin(3\nu\pi/4)\end{Bmatrix}\Gamma\left[\begin{matrix}\frac{s-\mu+\nu}{2},\ \frac{s+\mu+\nu}{2}\\ \nu+1\end{matrix}\right] \times {}_4F_3\left(\begin{matrix}\Delta\left(2,\ \frac{s-\mu+\nu}{2}\right),\ \Delta\left(2,\ \frac{s+\mu+\nu}{2}\right)\\ \frac{1}{2},\ \Delta(2,\ \nu+1);\ -\frac{b^4}{a^4}\end{matrix}\right) \mp\frac{2^{s-2}b^{\nu+2}}{a^{s+\nu+2}}\begin{Bmatrix}\sin(3\nu\pi/4)\\ \cos(3\nu\pi/4)\end{Bmatrix}\Gamma\left[\begin{matrix}\frac{s-\mu+\nu+2}{2},\ \frac{s+\mu+\nu+2}{2}\\ \nu+2\end{matrix}\right] \times {}_4F_3\left(\begin{matrix}\Delta\left(2,\ \frac{s-\mu+\nu+2}{2}\right),\ \Delta\left(2,\ \frac{s+\mu+\nu+2}{2}\right)\\ \frac{3}{2},\ \Delta(2,\ \nu+2);\ -\frac{b^4}{a^4}\end{matrix}\right)$ $[\mathrm{Re}\,(\sqrt{2}\,a-b)>\lvert\mathrm{Im}\,b\rvert;\ \mathrm{Re}\,(s+\nu)>\lvert\mathrm{Re}\,\mu\rvert]$
3	$K_\mu(ax^2)\begin{Bmatrix}\mathrm{ber}_\nu(bx)\\ \mathrm{bei}_\nu(bx)\end{Bmatrix}$	$\frac{2^{(s-\nu)/2-3}b^\nu}{a^{(s+\nu)/2}}\begin{Bmatrix}\cos(3\nu\pi/4)\\ \sin(3\nu\pi/4)\end{Bmatrix}\Gamma\left[\begin{matrix}\frac{s-2\mu+\nu}{4},\ \frac{s+2\mu+\nu}{4}\\ \nu+1\end{matrix}\right] \times {}_2F_3\left(\begin{matrix}\frac{s-2\mu+\nu}{4},\ \frac{s+2\mu+\nu}{4};\ -\frac{b^4}{64a^2}\\ \frac{1}{2},\ \frac{\nu+1}{2},\ \frac{\nu+2}{2}\end{matrix}\right)+\frac{2^{(s-\nu)/2-4}b^{\nu+2}}{a^{(s+\nu)/2+1}} \times\begin{Bmatrix}\sin(3\nu\pi/4)\\ \cos(3\nu\pi/4)\end{Bmatrix}\Gamma\left[\begin{matrix}\frac{s-2\mu+\nu+2}{4},\ \frac{s+2\mu+\nu+2}{4}\\ \nu+2\end{matrix}\right] \times {}_2F_3\left(\begin{matrix}\frac{s-2\mu+\nu+2}{4},\ \frac{s+2\mu+\nu+2}{4};\ -\frac{b^4}{64a^2}\\ \frac{3}{2},\ \frac{\nu+2}{2},\ \frac{\nu+3}{2}\end{matrix}\right)$ $\left[\begin{matrix}\mathrm{Re}\,a>0;\\ \mathrm{Re}\,(s+\nu)>2\lvert\mathrm{Re}\,\mu\rvert\end{matrix}\right]$

3.17.6. $\varphi(x)\left(\mathrm{ber}_\nu^2(bx)+\mathrm{bei}_\nu^2(bx)\right)$ and $\mathrm{ker}_\nu^2(bx)+\mathrm{kei}_\nu^2(bx)$

1	$e^{-ax}\left[\mathrm{ber}_\nu^2(bx)+\mathrm{bei}_\nu^2(bx)\right]$	$\frac{b^{2\nu}}{2^{2\nu}a^{s+2\nu}}\Gamma\left[\begin{matrix}s+2\nu\\ \nu+1,\ \nu+1\end{matrix}\right]{}_4F_3\left(\begin{matrix}\Delta(4,\ s+2\nu)\\ \Delta(2,\ \nu+1),\ \nu+1;\ \frac{4b^4}{a^4}\end{matrix}\right)$ $[\mathrm{Re}\,a>\sqrt{2}\,(\mathrm{Re}\,b+\lvert\mathrm{Im}\,b\rvert);\ \mathrm{Re}\,(s+2\nu)>0]$
2	$e^{-ax^2}\left[\mathrm{ber}_\nu^2(bx)+\mathrm{bei}_\nu^2(bx)\right]$	$\frac{b^{2\nu}}{2^{2\nu+1}a^{s/2+\nu}}\Gamma\left[\begin{matrix}\frac{s+2\nu}{2}\\ \nu+1,\ \nu+1\end{matrix}\right]{}_2F_3\left(\begin{matrix}\Delta\left(2,\ \frac{s+2\nu}{2}\right)\\ \Delta(2,\ \nu+1),\ \nu+1;\ \frac{b^4}{16a^4}\end{matrix}\right)$ $[\mathrm{Re}\,a,\ \mathrm{Re}\,(s+2\nu)>0]$

No.	$f(x)$	$F(s)$
3	$\theta(a-x)\ln\frac{\sqrt{a-x}+\sqrt{a}}{\sqrt{x}}$ $\times\left[\operatorname{ber}_\nu^2(bx)+\operatorname{bei}_\nu^2(bx)\right]$	$\frac{2^{-2\nu-1}\sqrt{\pi}\,a^{s+2\nu}b^{2\nu}}{s+2\nu}\Gamma\left[\begin{matrix}s+2\nu\\ \nu+1,\ \nu+1,\ \frac{2s+4\nu+1}{2}\end{matrix}\right]$ $\times{}_5F_8\left(\begin{matrix}\frac{s+2\nu}{4},\ \Delta(4,\ s+2\nu);\ \frac{a^4b^4}{64}\\ \frac{\nu+1}{2},\ \frac{\nu+2}{2},\ \nu+1,\ \Delta\left(4,\ s+2\nu+\frac{1}{2}\right),\ \frac{s+2\nu+4}{4}\end{matrix}\right)$ $[a,\ \operatorname{Re}(s+2\nu)>0]$
4	$\theta(a-x)\arccos\frac{x}{a}$ $\times\left[\operatorname{ber}_\nu^2(bx)+\operatorname{bei}_\nu^2(bx)\right]$	$\frac{2^{-2\nu-1}\sqrt{\pi}\,a^{s+2\nu}b^{2\nu}}{s+2\nu}\Gamma\left[\begin{matrix}\frac{s+2\nu+1}{2}\\ \nu+1,\ \nu+1,\ \frac{s+2\nu+2}{2}\end{matrix}\right]$ $\times{}_3F_6\left(\begin{matrix}\frac{s+2\nu}{4},\ \frac{s+2\nu+1}{4},\ \frac{s+2\nu+3}{4};\ \frac{a^4b^4}{64}\\ \frac{\nu+1}{2},\ \frac{\nu+2}{2},\ \nu+1,\ \frac{s+2\nu+2}{4},\ \frac{s+2\nu+4}{4},\ \frac{s+2\nu+4}{4}\end{matrix}\right)$ $[a>0;\ \operatorname{Re}(s+2\nu)>-1]$
5	$\Gamma(\mu,\ ax)$ $\times\left[\operatorname{ber}_\nu^2(bx)+\operatorname{bei}_\nu^2(bx)\right]$	$\frac{a^{-s-2\nu}(b/2)^{2\nu}}{s+2\nu}\Gamma\left[\begin{matrix}s+\mu+2\nu\\ \nu+1,\ \nu+1\end{matrix}\right]$ $\times{}_5F_4\left(\begin{matrix}\frac{s+2\nu}{4},\ \Delta(4,\ s+\mu+2\nu);\ \frac{4b^4}{a^4}\\ \frac{\nu+1}{2},\ \frac{\nu+2}{2},\ \nu+1,\ \frac{s+2\nu+4}{4}\end{matrix}\right)$ $[\operatorname{Re}(a-\sqrt{2}\,b)>0;\ \operatorname{Re}(s+2\nu)>-\operatorname{Re}\mu,\ 0]$
6	$\operatorname{erfc}(ax)$ $\times\left[\operatorname{ber}_\nu^2(bx)+\operatorname{bei}_\nu^2(bx)\right]$	$\frac{a^{-s-2\nu}b^{2\nu}}{2^{2\nu}\sqrt{\pi}\,(s+2\nu)}\Gamma\left[\begin{matrix}\frac{s+2\nu+1}{2}\\ \nu+1,\ \nu+1\end{matrix}\right]{}_3F_3\left(\begin{matrix}\frac{1}{2},\ \frac{s+2\nu}{4},\ \frac{s+2\nu+3}{4}\\ \frac{\nu+1}{2},\ \frac{\nu+2}{2},\ \nu+1;\ \frac{b^4}{16a^4}\end{matrix}\right)$ $[\operatorname{Re}(s+2\nu)>0;\ \lvert\arg a\rvert<\pi/4]$
7	$K_\mu(ax^2)$ $\times\left[\operatorname{ber}_\nu^2(bx)+\operatorname{bei}_\nu^2(bx)\right]$	$\frac{2^{s/2-\nu-3}b^{2\nu}}{a^{s/2+\nu}}\Gamma\left[\begin{matrix}\frac{s-2\mu+2\nu}{4},\ \frac{s+2\mu+2\nu}{4}\\ \nu+1,\ \nu+1\end{matrix}\right]$ $\times{}_2F_5\left(\begin{matrix}\frac{s-2\mu+2\nu}{4},\ \frac{s+2\mu+2\nu}{4};\ \frac{b^4}{16a^2}\\ \frac{\nu+1}{2},\ \frac{\nu+2}{2},\ \nu+1\end{matrix}\right)$ $[\operatorname{Re}a>0;\ \operatorname{Re}(s+2\nu)>2\lvert\operatorname{Re}\mu\rvert]$
8	$\operatorname{ker}_\nu^2(ax)+\operatorname{kei}_\nu^2(ax)$	$\frac{2^{s-4}}{a^s}\Gamma\left[\frac{s}{2},\ \frac{s-2\nu}{4},\ \frac{s+2\nu}{4}\right]$ $[\operatorname{Re}s>2\lvert\operatorname{Re}\nu\rvert;\ \lvert\arg a\rvert<\pi/4]$

3.17.7. Products of $\operatorname{ber}_\nu(bx)$, $\operatorname{bei}_\nu(bx)$, $\operatorname{ker}_\nu(bx)$, $\operatorname{kei}_\nu(bx)$

1	$\operatorname{ber}_\nu(ax)\left\{\begin{matrix}\operatorname{ker}_\nu(ax)\\ \operatorname{kei}_\nu(ax)\end{matrix}\right\}$	$\pm\frac{a^{-s}}{8\sqrt{\pi}}\left\{\begin{matrix}\cos(s\pi/4)\\ \sin(s\pi/4)\end{matrix}\right\}\Gamma\left[\begin{matrix}\frac{s}{2},\ \frac{1-s}{2},\ \frac{s+2\nu}{2}\\ \frac{2-s+2\nu}{2}\end{matrix}\right]$ $\pm\frac{2^{s-4}}{a^s}\left\{\begin{matrix}\cos[(s+6\nu)\pi/4]\\ \sin[(s+6\nu)\pi/4]\end{matrix}\right\}\Gamma\left[\begin{matrix}\frac{s}{2},\ \frac{s+2\nu}{4}\\ \frac{4-s+2\nu}{4}\end{matrix}\right]$ $[a>0;\ 0,\ -2\operatorname{Re}\nu<\operatorname{Re}s<2]$

No.	$f(x)$	$F(s)$
2	$\operatorname{ber}_{-\nu}(ax)\begin{Bmatrix}\operatorname{ker}_\nu(ax)\\ \operatorname{kei}_\nu(ax)\end{Bmatrix}$	$\pm\frac{a^{-s}}{8\sqrt{\pi}}\begin{Bmatrix}\cos[(s+4\nu)\pi/4]\\ \sin[(s+4\nu)\pi/4]\end{Bmatrix}\Gamma\left[\begin{matrix}\frac{s}{2},\frac{1-s}{2},\frac{s-2\nu}{2}\\ \frac{2-s-2\nu}{2}\end{matrix}\right]$ $\pm\frac{2^{s-4}}{a^s}\begin{Bmatrix}\cos([(s-2\nu)\pi/4]\\ \sin[(s-2\nu)\pi/4]\end{Bmatrix}\Gamma\left[\begin{matrix}\frac{s}{2},\frac{s-2\nu}{4}\\ \frac{4-s-2\nu}{4}\end{matrix}\right]$ $[a>0;\ 0,\ 2\operatorname{Re}\nu<\operatorname{Re}s<1]$
3	$\operatorname{bei}_\nu(ax)\begin{Bmatrix}\operatorname{ker}_\nu(ax)\\ \operatorname{kei}_\nu(ax)\end{Bmatrix}$	$\frac{2^{s-4}}{a^s}\begin{Bmatrix}\sin[(s+6\nu)\pi/4]\\ \cos[(s+6\nu)\pi/4]\end{Bmatrix}\Gamma\left[\begin{matrix}\frac{s}{2},\frac{s+2\nu}{4}\\ \frac{4-s+2\nu}{4}\end{matrix}\right]$ $-\frac{a^{-s}}{8\sqrt{\pi}}\begin{Bmatrix}\sin[s\pi/4]\\ \cos[s\pi/4]\end{Bmatrix}\Gamma\left[\begin{matrix}\frac{s}{2},\frac{1-s}{2},\frac{s+2\nu}{4}\\ \frac{2-s+2\nu}{2}\end{matrix}\right]$ $[a>0;\ 0,\ -2\operatorname{Re}\nu<\operatorname{Re}s<2]$
4	$\operatorname{bei}_{-\nu}(ax)\begin{Bmatrix}\operatorname{ker}_\nu(ax)\\ \operatorname{kei}_\nu(ax)\end{Bmatrix}$	$\frac{2^{s-4}}{a^s}\begin{Bmatrix}\sin[(s-2\nu)\pi/4]\\ \cos[(s-2\nu)\pi/4]\end{Bmatrix}\Gamma\left[\begin{matrix}\frac{s}{2},\frac{s-2\nu}{4}\\ \frac{4-s-2\nu}{4}\end{matrix}\right]$ $-\frac{a^{-s}}{8\sqrt{\pi}}\begin{Bmatrix}\sin[(s+4\nu)\pi/4]\\ \cos[(s+4\nu)\pi/4]\end{Bmatrix}\Gamma\left[\begin{matrix}\frac{s}{2},\frac{1-s}{2},\frac{s-2\nu}{2}\\ \frac{2-s-2\nu}{2}\end{matrix}\right]$ $[a>0;\ 0,\ 2\operatorname{Re}\nu<\operatorname{Re}s<1]$
5	$\begin{Bmatrix}\operatorname{ker}_\nu^2(ax)\\ \operatorname{kei}_\nu^2(ax)\end{Bmatrix}$	$\frac{2^{s-5}}{a^s}\Gamma\left[\frac{s}{2},\frac{s-2\nu}{4},\frac{s+2\nu}{4}\right]$ $\pm\frac{\sqrt{\pi}}{8a^s}\cos\left(\frac{s\pi}{4}+\nu\pi\right)\Gamma\left[\begin{matrix}\frac{s}{2},\frac{s-2\nu}{2},\frac{s+2\nu}{2}\\ \frac{s+1}{2}\end{matrix}\right]$ $[\lvert\operatorname{Re}\nu\rvert<\operatorname{Re}s<2;\ \lvert\arg a\rvert\le\pi/4]$
6	$\operatorname{kei}_\nu(ax)\operatorname{ker}_\nu(ax)$	$-\frac{\sqrt{\pi}}{8a^s}\sin\frac{(s+4\nu)\pi}{4}\Gamma\left[\begin{matrix}\frac{s}{2},\frac{s-2\nu}{2},\frac{s+2\nu}{2}\\ \frac{s+1}{2}\end{matrix}\right]$ $[\operatorname{Re}s>2\lvert\operatorname{Re}\nu\rvert;\ \lvert\arg a\rvert\le\pi/4]$
7	$\operatorname{kei}_{-\nu}(ax)\begin{Bmatrix}\operatorname{ker}_\nu(ax)\\ \operatorname{kei}_\nu(ax)\end{Bmatrix}$	$2^{s-5}a^{-s}\begin{Bmatrix}\sin(\pi\nu)\\ \cos(\pi\nu)\end{Bmatrix}\Gamma\left[\frac{s}{2},\frac{s-2\nu}{4},\frac{s+2\nu}{4}\right]$ $-\frac{\sqrt{\pi}a^{-s}}{8}\begin{Bmatrix}\sin(s\pi/4)\\ \cos(s\pi/4)\end{Bmatrix}\Gamma\left[\begin{matrix}\frac{s}{2},\frac{s-2\nu}{2},\frac{s+2\nu}{2}\\ \frac{s+1}{2}\end{matrix}\right]$ $[2\lvert\operatorname{Re}\nu\rvert<\operatorname{Re}s<2;\ \lvert\arg a\rvert\le\pi/4]$
8	$\begin{Bmatrix}\operatorname{ker}_{-\nu}(ax)\operatorname{ker}_\nu(ax)\\ \operatorname{kei}_{-\nu}(ax)\operatorname{kei}_\nu(ax)\end{Bmatrix}$	$\frac{2^{s-5}}{a^s}\cos(\nu\pi)\Gamma\left[\frac{s}{2},\frac{s-2\nu}{4},\frac{s+2\nu}{4}\right]$ $\pm\frac{\sqrt{\pi}}{8a^s}\cos\frac{s\pi}{4}\Gamma\left[\begin{matrix}\frac{s}{2},\frac{s-2\nu}{2},\frac{s+2\nu}{2}\\ \frac{s+1}{2}\end{matrix}\right]$ $[2\lvert\operatorname{Re}\nu\rvert<\operatorname{Re}s<2;\ \lvert\arg a\rvert\le\pi/4]$

3.18. The Airy Functions Ai (z) and Bi (z)

More formulas can be obtained from the corresponding sections due to the relations

$$\mathrm{Ai}\,(z) = \frac{1}{\pi}\sqrt{\frac{z}{3}}\,K_{1/3}\left(\frac{2}{3}\,z^{3/2}\right),$$

$$\mathrm{Bi}\,(z) = \sqrt{\frac{z}{3}}\left[I_{-1/3}\left(\frac{2}{3}\,z^{3/2}\right) + I_{1/3}\left(\frac{2}{3}\,z^{3/2}\right)\right],$$

$$\begin{Bmatrix}\mathrm{Ai}\,(z)\\ \mathrm{Bi}\,(z)\end{Bmatrix} = \frac{1}{3^{(5\pm3)/12}\,\Gamma\left(\frac{2}{3}\right)}\,{}_0F_1\left(\frac{2}{3};\,\frac{z^3}{9}\right) \mp \frac{z}{3^{(1\pm3)/12}\,\Gamma\left(\frac{1}{3}\right)}\,{}_0F_1\left(\frac{4}{3};\,\frac{z^3}{9}\right),$$

$$\begin{Bmatrix}\mathrm{Ai}'\,(z)\\ \mathrm{Bi}'\,(z)\end{Bmatrix} = \frac{z^2}{2\times 3^{(5\pm3)/12}\,\Gamma\left(\frac{2}{3}\right)}\,{}_0F_1\left(\frac{5}{3};\,\frac{z^3}{9}\right) \mp \frac{1}{3^{(1\pm3)/12}\,\Gamma\left(\frac{1}{3}\right)}\,{}_0F_1\left(\frac{1}{3};\,\frac{z^3}{9}\right);$$

$$\mathrm{Ai}\,(z) = \frac{1}{2\sqrt[6]{3}\,\pi}\,G_{02}^{20}\left(\frac{z^3}{9}\,\middle|\,\begin{matrix}\cdot\\ 0,\,1/3\end{matrix}\right),\quad [-\pi/3 < \arg z \le \pi/3];$$

$$\mathrm{Bi}\,(z) = \frac{2\pi}{\sqrt[6]{3}}\,G_{24}^{20}\left(\frac{z^3}{9}\,\middle|\,\begin{matrix}1/6,\,2/3\\ 0,\,1/3,\,1/6,\,2/3\end{matrix}\right),\quad [-\pi/3 < \arg z \le \pi/3];$$

$$\mathrm{Ai}'\,(z) = -\frac{\sqrt[6]{3}}{2\pi}\,G_{02}^{20}\left(\frac{z^3}{9}\,\middle|\,\begin{matrix}\cdot\\ 0,\,2/3\end{matrix}\right),\quad [-\pi/3 < \arg z \le \pi/3];$$

$$\mathrm{Bi}'\,(z) = -2\sqrt[6]{3}\,\pi\,G_{24}^{20}\left(\frac{z^3}{9}\,\middle|\,\begin{matrix}-1/6,\,1/3\\ 0,\,2/3,\,-1/6,\,1/3\end{matrix}\right),\quad [-\pi/3 < \arg z \le \pi/3].$$

3.18.1. Ai (bx), Ai′ (bx), Bi (bx), and algebraic functions

No.	$f(x)$	$F(s)$
1	$\mathrm{Ai}\,(ax)$	$\frac{3^{(4s-7)/6}}{2\pi a^s}\,\Gamma\left(\frac{s}{3}\right)\Gamma\left(\frac{s+1}{3}\right)$ $[\mathrm{Re}\,s > 0;\ \lvert\arg a\rvert < \pi/3]$
2	$(a-x)_+^{\alpha-1}\begin{Bmatrix}\mathrm{Ai}\,(bx)\\ \mathrm{Bi}\,(bx)\end{Bmatrix}$	$\frac{a^{s+\alpha-1}}{3^{(5\pm3)/12}\Gamma\,(2/3)}\,\mathrm{B}\,(\alpha,\,s)\,{}_3F_4\left(\begin{matrix}\frac{s}{3},\,\frac{s+1}{3},\,\frac{s+2}{3};\,\frac{a^3b^3}{9}\\ \frac{2}{3},\,\frac{s+\alpha}{3},\,\frac{s+\alpha+1}{3},\,\frac{s+\alpha+2}{3}\end{matrix}\right)$ $\mp\frac{a^{s+\alpha}b}{3^{(1\pm3)/12}\Gamma\,(1/3)}\,\mathrm{B}\,(\alpha,\,s+1)\,{}_3F_4\left(\begin{matrix}\frac{s+1}{3},\,\frac{s+2}{3},\,\frac{s+3}{3};\,\frac{a^3b^3}{9}\\ \frac{4}{3},\,\frac{s+\alpha+1}{3},\,\frac{s+\alpha+2}{3},\,\frac{s+\alpha+3}{3}\end{matrix}\right)$ $[a,\ \mathrm{Re}\,\alpha,\ \mathrm{Re}\,s > 0]$
3	$(a^3-x^3)_+^{\alpha-1}\begin{Bmatrix}\mathrm{Ai}\,(bx)\\ \mathrm{Bi}\,(bx)\end{Bmatrix}$	$\frac{a^{s+3\alpha-3}\,\Gamma\,(1/3)}{2\cdot 3^{(11+3)/12}\,\pi}\,\mathrm{B}\left(\alpha,\,\frac{s}{3}\right)\,{}_1F_2\left(\begin{matrix}\frac{s}{3};\,\frac{a^3b^3}{9}\\ \frac{2}{3},\,\frac{s+3\alpha}{3}\end{matrix}\right)$ $\mp\frac{a^{s+3\alpha-2}b\,\Gamma\,(2/3)}{2\cdot 3^{(7\pm3)/12}\,\pi}\,\mathrm{B}\left(\alpha,\,\frac{s+1}{3}\right)\,{}_1F_2\left(\begin{matrix}\frac{s+1}{3};\,\frac{a^3b^3}{9}\\ \frac{4}{3},\,\frac{s+3\alpha+1}{3}\end{matrix}\right)$ $[a,\ \mathrm{Re}\,\alpha,\ \mathrm{Re}\,s > 0]$

No.	$f(x)$	$F(s)$
4	$\left(x^3-a^3\right)_+^{\alpha-1}\operatorname{Ai}(bx)$	$-\dfrac{a^{s+3\alpha-2}b}{3^{4/3}\Gamma(1/3)}\operatorname{B}\left(\alpha,\,\dfrac{2-s-3\alpha}{3}\right){}_1F_2\left(\begin{matrix}\frac{s+1}{3};\ \frac{a^3b^3}{9}\\ \frac{4}{3},\ \frac{s+3\alpha+1}{3}\end{matrix}\right)$ $-\dfrac{a^{s+3\alpha-3}}{3^{2/3}\Gamma(-1/3)}\operatorname{B}\left(\alpha,\,\dfrac{3-s-3\alpha}{3}\right){}_1F_2\left(\begin{matrix}\frac{s}{3};\ \frac{a^3b^3}{9}\\ \frac{2}{3},\ \frac{s+3\alpha}{3}\end{matrix}\right)$ $-\dfrac{3^{2s/3+2\alpha-11/3}}{b^{s+3\alpha-3}}\left\{\sin\dfrac{(2s+6\alpha-1)\pi}{6}\,\Gamma\left[\begin{matrix}\frac{s+3\alpha-2}{3}\\ \frac{6-s-3\alpha}{3}\end{matrix}\right]\right.$ $\left.+\cos\dfrac{(s+3\alpha)\pi}{3}\,\Gamma\left[\begin{matrix}\frac{s+3\alpha-3}{3}\\ \frac{5-s-3\alpha}{3}\end{matrix}\right]\right\}{}_1F_2\left(\begin{matrix}1-\alpha;\ \frac{a^3b^3}{9}\\ \frac{5-s-3\alpha}{3},\ \frac{6-s-3\alpha}{3}\end{matrix}\right)$ $[a,\ \operatorname{Re}\alpha>0;\ \lvert\arg b\rvert<\pi/3]$
5	$\dfrac{1}{\left(x^3+a^3\right)^{\rho}}\operatorname{Ai}(bx)$	$-\dfrac{a^{s-3\rho+1}b}{3^{4/3}\,\Gamma(1/3)}\operatorname{B}\left(\dfrac{s+1}{3},\,-\dfrac{s-3\rho+1}{3}\right){}_1F_2\left(\begin{matrix}\frac{s+1}{3};\ -\frac{a^3b^3}{9}\\ \frac{4}{3},\ \frac{s-3\rho+4}{3}\end{matrix}\right)$ $-\dfrac{a^{s-3\rho}}{3^{2/3}\,\Gamma(-1/3)}\operatorname{B}\left(\dfrac{s}{3},\,-\dfrac{s-3\rho}{3}\right){}_1F_2\left(\begin{matrix}\frac{s}{3};\ -\frac{a^3b^3}{9}\\ \frac{2}{3},\ \frac{s-3\rho+3}{3}\end{matrix}\right)$ $+\dfrac{3^{2s/3-2\rho-5/3}}{b^{s-3\rho}}\left\{\cos\dfrac{(s-3\rho)\pi}{3}\,\Gamma\left[\begin{matrix}\frac{s-3\rho}{3}\\ -\frac{s-3\rho-2}{3}\end{matrix}\right]\right.$ $\left.+\sin\dfrac{(2s-6\rho-1)\pi}{6}\,\Gamma\left[\begin{matrix}\frac{s-3\rho+1}{3}\\ -\frac{s-3\rho-3}{3}\end{matrix}\right]\right\}{}_1F_2\left(\begin{matrix}\rho;\ -\frac{a^3b^3}{9}\\ -\frac{s-3\rho-2}{3},\ -\frac{s-3\rho-3}{3}\end{matrix}\right)$ $[\operatorname{Re}s>0;\ \lvert\arg a\rvert<\pi/3]$
6	$\operatorname{Ai}'(ax)$	$-\dfrac{3^{(4s-5)/6}}{2\pi}a^{-s}\,\Gamma\left(\dfrac{s}{3}\right)\Gamma\left(\dfrac{s+2}{3}\right)$ $[\operatorname{Re}s>0;\ \lvert\arg a\rvert<\pi/3]$

3.18.2. $\operatorname{Ai}(bx)$, $\operatorname{Ai}'(bx)$, $\operatorname{Bi}(bx)$, and the exponential function

1	$e^{-ax}\operatorname{Ai}(bx)$	$\dfrac{3^{-(s+1)/3}a\,b^{-s-2}}{4\Gamma(1-s)}\left[\dfrac{b^2}{3^{1/3}a}\csc\dfrac{s\pi}{3}\csc\dfrac{(s+1)\pi}{3}\right.$ $\times\Gamma\left(\dfrac{1-s}{3}\right){}_2F_2\left(\begin{matrix}\frac{s}{3},\ \frac{s+1}{3}\\ \frac{1}{3},\ \frac{2}{3};\ -\frac{a^3}{3b^3}\end{matrix}\right)-3^{1/3}b\sec\dfrac{(2s+1)\pi}{6}$ $\times\csc\dfrac{(s+1)\pi}{3}\,\Gamma\left(\dfrac{3-s}{3}\right){}_2F_2\left(\begin{matrix}\frac{s+1}{3},\ \frac{s+2}{3}\\ \frac{2}{3},\ \frac{4}{3};\ -\frac{a^3}{3b^3}\end{matrix}\right)$ $\left.+\dfrac{as}{2}\sec\dfrac{(2s+1)\pi}{6}\csc\dfrac{s\pi}{3}\,\Gamma\left(\dfrac{2-s}{3}\right){}_2F_2\left(\begin{matrix}\frac{s+2}{3},\ \frac{s+3}{3}\\ \frac{4}{3},\ \frac{5}{3};\ -\frac{a^3}{3b^3}\end{matrix}\right)\right]$ $[\operatorname{Re}a,\ \operatorname{Re}s>0;\ \lvert\arg b\rvert<\pi/3]$

No.	$f(x)$	$F(s)$
2	$e^{-ax^{3/2}}\begin{Bmatrix}\operatorname{Ai}(bx)\\ \operatorname{Bi}(bx)\end{Bmatrix}$	$\frac{3^{(-11\mp3)/12}}{\pi a^{2s/3}}\Gamma\left(\frac{1}{3}\right)\Gamma\left(\frac{2s}{3}\right){}_2F_1\left(\begin{matrix}\frac{s}{3},\ \frac{2s+3}{6}\\ \frac{2}{3};\ \frac{4b^3}{9a^2}\end{matrix}\right)$ $\mp\frac{3^{(-7\mp3)/12}\,b}{\pi a^{2(s+1)/3}}\Gamma\left(\frac{2}{3}\right)\Gamma\left(\frac{2s+2}{3}\right){}_2F_1\left(\begin{matrix}\frac{s+1}{3},\ \frac{2s+5}{6}\\ \frac{4}{3};\ \frac{4b^3}{9a^2}\end{matrix}\right)$ $[\operatorname{Re} s>0;\ \operatorname{Re}(3a\pm2b^{3/2})>0;\ \lvert\arg b\rvert<\pi/6]$
3	$e^{-2/3(ax)^{3/2}}\operatorname{Ai}(ax)$	$\frac{2^{(1-4s)/3}3^{(4s-7)/6}}{\sqrt{\pi}}a^{-s}\Gamma\left[\begin{matrix}\frac{2s}{3},\ \frac{2s+2}{3}\\ \frac{4s+5}{6}\end{matrix}\right]$ $[\operatorname{Re} s>0;\ \lvert\arg a\rvert<\pi/3]$
4	$e^{2/3(ax)^{3/2}}\operatorname{Ai}(ax)$	$\frac{2^{-(4s+2)/3}3^{(4s-7)/6}}{\pi^{3/2}}a^{-s}\Gamma\left(\frac{1-4s}{6}\right)\Gamma\left(\frac{2s}{3}\right)\Gamma\left(\frac{2s+2}{3}\right)$ $[0<\operatorname{Re} s<1/4;\ \lvert\arg a\rvert<\pi]$
5	$e^{-ax^3}\begin{Bmatrix}\operatorname{Ai}(bx)\\ \operatorname{Bi}(bx)\end{Bmatrix}$	$\frac{a^{-s/3}}{3^{(17\pm3)/12}}\Gamma\left[\begin{matrix}\frac{s}{3}\\ \frac{2}{3}\end{matrix}\right]{}_1F_1\left(\begin{matrix}\frac{s}{3}\\ \frac{2}{3};\ \frac{b^3}{9a}\end{matrix}\right)\mp\frac{a^{-(s+1)/3}b}{3^{(13\pm3)/12}}\Gamma\left[\begin{matrix}\frac{s+1}{3}\\ \frac{1}{3}\end{matrix}\right]{}_1F_1\left(\begin{matrix}\frac{s+1}{3}\\ \frac{4}{3};\ \frac{b^3}{9a}\end{matrix}\right)$ $[\operatorname{Re} a,\ \operatorname{Re} s>0]$
6	$e^{-2/3(ax)^{3/2}}\operatorname{Ai}'(ax)$	$-\frac{2^{-(4s+1)/3}3^{(4s-5)/6}}{\sqrt{\pi}}a^{-s}\Gamma\left[\begin{matrix}\frac{2s}{3},\ \frac{2s+4}{3}\\ \frac{4s+7}{6}\end{matrix}\right]$ $[\operatorname{Re} s>0;\ \lvert\arg a\rvert<\pi/3]$

3.18.3. Ai (bx) and trigonometric functions

Notation: $\delta=\begin{Bmatrix}1\\0\end{Bmatrix}$.

1	$\begin{Bmatrix}\sin(ax^{3/2})\\ \cos(ax^{3/2})\end{Bmatrix}\operatorname{Ai}(bx)$	$\frac{3^{(4s-7)/6+\delta}a^{\delta}}{2\pi\,b^{s+3\delta/2}}\Gamma\left(\frac{2s+3\delta}{6}\right)\Gamma\left(\frac{2s+3\delta+2}{6}\right){}_2F_1\left(\begin{matrix}\frac{2s+3\delta}{6},\ \frac{2s+3\delta+2}{6}\\ \frac{2\delta+1}{2};\ -\frac{9a^2}{4b^3}\end{matrix}\right)$ $[a>0;\ \operatorname{Re} s>-3\delta/2;\ \lvert\arg b\rvert<\pi/6]$

3.18.4. Ai (bx), Ai′ (bx), Bi (bx), and special functions

1	$\operatorname{Ei}(-ax^3)\begin{Bmatrix}\operatorname{Ai}(bx)\\ \operatorname{Bi}(bx)\end{Bmatrix}$	$\pm\frac{a^{-(s+1)/3}b}{3^{(25+3)/12}}\Gamma\left[\begin{matrix}\frac{s+1}{3},\ \frac{s+1}{3}\\ \frac{4}{3},\ \frac{s+4}{3}\end{matrix}\right]{}_2F_2\left(\begin{matrix}\frac{s+1}{3},\ \frac{s+1}{3}\\ \frac{4}{3},\ \frac{s+4}{3};\ \frac{b^3}{9a}\end{matrix}\right)$ $-\frac{a^{-s/3}}{3^{(17\pm3)/12}}\Gamma\left[\begin{matrix}\frac{s}{3},\ \frac{s}{3}\\ \frac{2}{3},\ \frac{s+3}{3}\end{matrix}\right]{}_2F_2\left(\begin{matrix}\frac{s}{3},\ \frac{s}{3}\\ \frac{2}{3},\ \frac{s+3}{3};\ \frac{b^3}{9a}\end{matrix}\right)$ $[a,\ \operatorname{Re} s>0]$

No.	$f(x)$	$F(s)$
2	$\operatorname{erfc}\left(ax^{3/2}\right)\begin{Bmatrix}\operatorname{Ai}(bx)\\ \operatorname{Bi}(bx)\end{Bmatrix}$	$\mp\frac{a^{-2(s+1)/3}b}{3^{(1\pm 3)/12}\sqrt{\pi}\,(s+1)}\Gamma\left[\begin{matrix}\frac{2s+5}{6}\\ \frac{1}{3}\end{matrix}\right]{}_2F_2\left(\begin{matrix}\frac{s+1}{3},\ \frac{2s+5}{6}\\ \frac{4}{3},\ \frac{s+4}{3};\ \frac{b^3}{9a^2}\end{matrix}\right)$ $+\frac{a^{-2s/3}}{3^{(5\pm 3)/12}\sqrt{\pi}\,s}\Gamma\left[\begin{matrix}\frac{2s+3}{6}\\ \frac{2}{3}\end{matrix}\right]{}_2F_2\left(\begin{matrix}\frac{s}{3},\ \frac{2s+3}{6}\\ \frac{2}{3},\ \frac{s+3}{3};\ \frac{b^3}{9a^2}\end{matrix}\right)$ $[\operatorname{Re}a,\ \operatorname{Re}s>0]$
3	$\Gamma\left(\nu,\ ax^3\right)\begin{Bmatrix}\operatorname{Ai}(bx)\\ \operatorname{Bi}(bx)\end{Bmatrix}$	$\mp\frac{a^{-(s+1)/3}b}{3^{(1\pm 3)/12}(s+1)}\Gamma\left[\begin{matrix}\frac{s+3\nu+1}{3}\\ \frac{1}{3}\end{matrix}\right]{}_2F_2\left(\begin{matrix}\frac{s+1}{3},\ \frac{s+3\nu+1}{3}\\ \frac{4}{3},\ \frac{s+4}{3};\ \frac{b^3}{9a}\end{matrix}\right)$ $+\frac{a^{-s/3}}{3^{(5\pm 3)/12}s}\Gamma\left[\begin{matrix}\frac{s+3\nu}{3}\\ \frac{2}{3}\end{matrix}\right]{}_2F_2\left(\begin{matrix}\frac{s}{3},\ \frac{s+3\nu}{3}\\ \frac{2}{3},\ \frac{s+3}{3};\ \frac{b^3}{9a}\end{matrix}\right)$ $[\operatorname{Re}a,\ \operatorname{Re}s,\ \operatorname{Re}(s+3\nu)>0]$
4	$J_\nu\left(ax^{3/2}\right)\operatorname{Ai}(bx)$	$\frac{3^{(4s-7)/6+\nu}a^\nu}{2^{\nu+1}\pi b^{s+3\nu/2}}\Gamma\left[\begin{matrix}\frac{2s+3\nu}{6},\ \frac{2s+3\nu+2}{6}\\ \nu+1\end{matrix}\right]{}_2F_1\left(\begin{matrix}\frac{2s+3\nu}{6},\ \frac{2s+3\nu+2}{6}\\ \nu+1;\ -\frac{9a^2}{4b^3}\end{matrix}\right)$ $[a,\ \operatorname{Re}(2s+3\nu)>0;\ \lvert\arg b\rvert<\pi/6]$
5	$I_\nu\left(\frac{2}{3}(ax)^{3/2}\right)\operatorname{Ai}(ax)$	$\frac{3^{(4s-7)/6}a^{-s}}{2\pi}\Gamma\left[\begin{matrix}\frac{2-2s}{3},\ \frac{2s+3\nu}{6},\ \frac{2s+3\nu+2}{6}\\ \frac{-2s+3\nu+4}{6},\ \frac{-2s+3\nu+6}{6}\end{matrix}\right]$ $[-3\operatorname{Re}\nu/2<\operatorname{Re}s<1;\ \lvert\arg a\rvert<\pi/3]$
6	$K_\nu\left(\frac{2}{3}(ax)^{3/2}\right)\operatorname{Ai}(ax)$	$\frac{3^{(4s-7)/6}a^{-s}}{4\pi}\Gamma\left[\begin{matrix}\frac{2s-3\nu}{6},\ \frac{2s+3\nu}{6},\ \frac{2s-3\nu+2}{6},\ \frac{2s+3\nu+2}{6}\\ \frac{2s+1}{3}\end{matrix}\right]$ $[\operatorname{Re}s>3\lvert\operatorname{Re}\nu\rvert/2;\ \lvert\arg a\rvert<\pi/3]$
7	$I_\nu\left(\frac{2}{3}(ax)^{3/2}\right)\operatorname{Ai}'(ax)$	$-\frac{3^{(4s-5)/6}a^{-s}}{\pi^{3/2}2^{(2s+5)/3}}\Gamma\left[\begin{matrix}\frac{1-2s}{6},\ \frac{2-s}{3},\ \frac{2s+3\nu}{6},\ \frac{2s+3\nu+4}{6}\\ \frac{-2s+3\nu+2}{6},\ \frac{-2s+3\nu+6}{6}\end{matrix}\right]$ $[-3\operatorname{Re}\nu/2<\operatorname{Re}s<1/2;\ \lvert\arg a\rvert<\pi/3]$
8	$K_\nu\left(\frac{2}{3}(ax)^{3/2}\right)\operatorname{Ai}'(ax)$	$\frac{3^{(4s-5)/6}a^{-s}}{4\sin(\nu\pi)}\Gamma\left(\frac{1-2s}{3}\right)\left(\Gamma\left[\begin{matrix}\frac{2s+3\nu}{6},\ \frac{2s+3\nu+4}{6}\\ \frac{-2s+3\nu+2}{6},\ \frac{-2s+3\nu+6}{6}\end{matrix}\right]-\Gamma\left[\begin{matrix}\frac{2s-3\nu}{6},\ \frac{2s-3\nu+4}{6}\\ -\frac{2s+3\nu-2}{6},\ -\frac{2s+3\nu-6}{6}\end{matrix}\right]\right)$ $[3\lvert\operatorname{Re}\nu\rvert/2<\operatorname{Re}s<1/2;\ \lvert\arg a\rvert<\pi/3]$

3.18.5. Products of Airy functions

1	$\operatorname{Ai}^2(ax)$	$\frac{2^{-2(s+1)/3}3^{-(2s+5)/6}}{\sqrt{\pi}}a^{-s}\Gamma\left[\begin{matrix}s\\ \frac{2s+5}{6}\end{matrix}\right]$ $[\operatorname{Re}s>0,\ \lvert\arg a\rvert<\pi/3]$

No.	$f(x)$	$F(s)$
2	$\mathrm{Ai}(ax)\,\mathrm{Bi}(ax)$	$\dfrac{2^{-(2s+5)/3}3^{2(s-2)/3}}{\pi^{3/2}}\,a^{-s}\,\Gamma\left[\begin{matrix}\frac{1-2s}{6},\ \frac{s}{3},\ \frac{s+2}{3}\\ \frac{2-s}{3}\end{matrix}\right]$ $[0<\mathrm{Re}\,s<1/2;\ \lvert\arg a\rvert<\pi/3]$
3	$\mathrm{Ai}(ax)\,\mathrm{Bi}(-ax)$	$\dfrac{12^{(s-5)/6}a^{-s}}{\sqrt{\pi}}\,\Gamma\left[\begin{matrix}\frac{s}{2},\ \frac{s+1}{6}\\ \frac{s+4}{6},\ \frac{2-s}{6}\end{matrix}\right]$ $[a,\ \mathrm{Re}\,s>0]$
4	$\mathrm{Ai}\left(ae^{i\pi/6}x\right)$ $\times\,\mathrm{Ai}\left(ae^{-i\pi/6}x\right)$	$\dfrac{2^{(s-8)/3}3^{(s-5)/6}}{\pi^{3/2}a^{s}}\,\Gamma\left(\dfrac{s}{2}\right)\Gamma\left(\dfrac{s+1}{6}\right)$ $[a,\ \mathrm{Re}\,s>0]$
5	$\mathrm{Ai}^2(-ax)+\mathrm{Bi}^2(-ax)$	$\dfrac{2^{(1-2s)/3}a^{-s}}{3^{(2s+5)/6}\pi^{3/2}}\,\Gamma(s)\,\Gamma\left(\dfrac{1-2s}{6}\right)$ $[a>0;\ 0<\mathrm{Re}\,s<1/2]$
6	$e^{-ax^3}\,\mathrm{Ai}^2(ax)$	$\dfrac{a^{-s/3}}{2^{2/3}3^{11/6}\sqrt{\pi}}\,\Gamma\left[\begin{matrix}\frac{s}{3}\\ \frac{5}{6}\end{matrix}\right]{}_2F_2\left(\begin{matrix}\frac{1}{6},\ \frac{s}{3}\\ \frac{1}{3},\ \frac{2}{3};\ \frac{4a^2}{9}\end{matrix}\right)-\dfrac{a^{(2-s)/3}}{3^{3/2}\pi}\,\Gamma\left(\dfrac{s+1}{3}\right)$ $\times\,{}_2F_2\left(\begin{matrix}\frac{1}{2},\ \frac{s+1}{3}\\ \frac{2}{3},\ \frac{4}{3};\ \frac{4a^2}{9}\end{matrix}\right)+\dfrac{a^{(4-s)/3}}{2^{1/3}3^{7/6}\sqrt{\pi}}\,\Gamma\left[\begin{matrix}\frac{s+2}{3}\\ \frac{1}{6}\end{matrix}\right]{}_2F_2\left(\begin{matrix}\frac{5}{6},\ \frac{s+2}{3}\\ \frac{4}{3},\ \frac{5}{3};\ \frac{4a^2}{9}\end{matrix}\right)$ $[a,\ \mathrm{Re}\,s>0]$
7	$\mathrm{Ai}(ax)\,\mathrm{Ai}'(ax)$	$-\dfrac{12^{-(2s+3)/6}}{\sqrt{\pi}}\,a^{-s}\,\Gamma\left[\begin{matrix}s\\ \frac{2s+3}{6}\end{matrix}\right]$ $[\mathrm{Re}\,s>0;\ \lvert\arg a\rvert<\pi/3]$
8	$\left(\mathrm{Ai}'(ax)\right)^2$	$\dfrac{2^{-(2s+7)/3}3^{(2s-2)/3}}{\pi^{3/2}}\,a^{-s}\,\Gamma\left[\begin{matrix}\frac{s}{3},\ \frac{s+2}{3},\ \frac{s+4}{3}\\ \frac{2s+7}{6}\end{matrix}\right]$ $[\mathrm{Re}\,s>0;\ \lvert\arg a\rvert<\pi/3]$
9	$\mathrm{Ai}'(ax)\,\mathrm{Bi}(ax)+\dfrac{1}{2\pi}$	$-\dfrac{12^{-(2s+3)/6}}{\pi^{3/2}}\,a^{-s}\sin\dfrac{s\pi}{3}\,\Gamma(s)\,\Gamma\left(\dfrac{3-2s}{6}\right)$ $[-1<\mathrm{Re}\,s<3/2;\ \lvert\arg a\rvert<\pi/3]$
10	$\mathrm{Ai}(ax)\,\mathrm{Bi}'(ax)-\dfrac{1}{2\pi}$	$-\dfrac{12^{-(2s+3)/6}}{\pi^{3/2}}\,a^{-s}\sin\dfrac{s\pi}{3}\,\Gamma(s)\,\Gamma\left(\dfrac{3-2s}{6}\right)$ $[-1<\mathrm{Re}\,s<3/2;\ \lvert\arg a\rvert<\pi/3]$
11	$J_\nu\left(ax^{3/2}\right)\mathrm{Ai}(bx)$ $\times\,\mathrm{Ai}'(bx)$	$-\dfrac{a^{\nu}b^{-s-3\nu/2}}{2^{2s/3+2\nu+1}3^{(2s+3\nu+3)/6}\sqrt{\pi}}\,\Gamma\left[\begin{matrix}\frac{2s+3\nu}{2}\\ \nu+1,\ \frac{2s+3\nu+3}{6}\end{matrix}\right]$ $\times\,{}_3F_2\left(\begin{matrix}\frac{2s+3\nu}{6},\ \frac{2s+3\nu+2}{6},\ \frac{2s+3\nu+4}{6}\\ \nu+1,\ \frac{2s+3\nu+3}{6};\ -\frac{9a^2}{16b^3}\end{matrix}\right)$ $[\mathrm{Re}\,a,\ \mathrm{Re}\,(s+3\nu/2)>0;\ \lvert\arg b\rvert<\pi/3]$

3.19. The Legendre Polynomials $P_n(z)$

More formulas can be obtained from the corresponding sections due to the relations

$$P_\nu(z) = P_\nu^0(z) = \mathrm{P}_\nu^0(z) = C_\nu^{1/2}(z) = P_\nu^{(0,0)}(z) = {}_2F_1\left(-\nu,\, \nu+1;\, 1;\, \frac{1-z}{2}\right).$$

3.19.1. $P_n(\varphi(x))$ and algebraic functions

Notation: $\varepsilon = 0$ or 1.

No.	$f(x)$	$F(s)$
1	$\theta(a-x)\, P_n\left(\frac{x}{a}\right)$	$\frac{a^s}{2}\,\Gamma\left[\begin{matrix}\frac{s}{2},\, \frac{s+1}{2}\\ \frac{s-n+1}{2},\, \frac{s+n+2}{2}\end{matrix}\right]$ $[a,\ \mathrm{Re}\, s > 0]$
2	$\theta(x-a)\, P_n\left(\frac{x}{a}\right)$	$\frac{a^s}{2^{s+1}\sqrt{\pi}}\,\Gamma\left[\begin{matrix}-\frac{s+n}{2},\, \frac{1-s+n}{2}\\ 1-s\end{matrix}\right]$ $[a>0;\ \mathrm{Re}\, s < -n]$
3	$(x^2-a^2)_+^{\alpha-1}\, P_n\left(\frac{x}{b}\right)$	$\frac{2^{n-1}a^{s+2\alpha+n-2}}{n!\, b^n}\left(\frac{1}{2}\right)_n \Gamma\left[\begin{matrix}\alpha,\, \frac{2-2\alpha-s-n}{2}\\ \frac{2-n-s}{2}\end{matrix}\right]$ $\times\, {}_3F_2\left(\begin{matrix}-\left[\frac{n}{2}\right],\, \frac{(-1)^n}{2}-\left[\frac{n}{2}\right],\, \frac{2-2\alpha-s-n}{2}\\ \frac{2-n-s}{2},\, \frac{1-2n}{2};\, \frac{b^2}{a^2}\end{matrix}\right)$ $[a>0;\ \mathrm{Re}\,\alpha > 0;\ \mathrm{Re}\,(s+2\alpha) < 2-n]$
4	$\frac{\theta(a-x)}{(b^2 \pm x^2)^\rho}\, P_{2n+\varepsilon}\left(\frac{x}{a}\right)$	$\frac{(-1)^n \left(\frac{1-s+\varepsilon}{2}\right)_n a^s}{2\left(\frac{s+\varepsilon}{2}\right)_{n+1} b^{2\rho}}\, {}_3F_2\left(\begin{matrix}\rho,\, \frac{s}{2},\, \frac{s+1}{2};\, \mp\frac{a^2}{b^2}\\ \frac{s+2n+\varepsilon+2}{2},\, \frac{s-2n-\varepsilon+1}{2}\end{matrix}\right)$ $\left[\mathrm{Re}\, s > -\varepsilon;\ \left\{\begin{matrix}a,\ \mathrm{Re}\, b > 0\\ b > a > 0\end{matrix}\right\}\right]$
5	$\frac{\theta(a-x)}{x^2-b^2}\, P_{2n+\varepsilon}\left(\frac{x}{a}\right)$	$\frac{(-1)^{n+1} a^s \left(\frac{1-s+\varepsilon}{2}\right)_n}{2b^2\left(\frac{s+\varepsilon}{2}\right)_{n+1}}\, {}_3F_2\left(\begin{matrix}1,\, \frac{s}{2},\, \frac{s+1}{2};\, \frac{a^2}{b^2}\\ \frac{s-2n-\varepsilon+1}{2},\, \frac{s+2n+\varepsilon+2}{2}\end{matrix}\right)$ $[b > a > 0;\ \mathrm{Re}\, s > -\varepsilon]$
6	$\frac{\theta(a-x)}{x^2-b^2}\, P_{2n+\varepsilon}\left(\frac{x}{a}\right)$	$\frac{(-1)^{\varepsilon+1}\pi b^{s-2}}{2}\tan^{2\varepsilon-1}\frac{s\pi}{2}\, P_{2n+\varepsilon}\left(\frac{b}{a}\right)$ $+\frac{(-1)^n a^{s-2}\left(\frac{3-s+\varepsilon}{2}\right)_n}{2\left(\frac{s+\varepsilon-2}{2}\right)_{n+1}}\, {}_3F_2\left(\begin{matrix}1,\, \frac{2-2n-s-\varepsilon}{2},\, \frac{2n-s+\varepsilon+3}{2}\\ \frac{3-s}{2},\, \frac{4-s}{2};\, \frac{b^2}{a^2}\end{matrix}\right)$ $[a > b > 0;\ \mathrm{Re}\, s > -\varepsilon]$
7	$\frac{\theta(x-a)}{(x^2 \pm b^2)^\rho}\, P_{2n+\varepsilon}\left(\frac{x}{a}\right)$	$\frac{(-1)^{n+1} a^{s-2\rho}\left(\frac{2\rho-s+\varepsilon+1}{2}\right)_n}{2\left(\frac{s-2\rho+\varepsilon}{2}\right)_{n+1}}\, {}_3F_2\left(\begin{matrix}\rho,\, \frac{2\rho+2n-s+\varepsilon+1}{2},\, \frac{2\rho-2n-s-\varepsilon}{2}\\ \frac{2\rho-s+1}{2},\, \frac{2\rho-s+2}{2};\, \mp\frac{b^2}{a^2}\end{matrix}\right)$ $\left[\mathrm{Re}\,(s-2\rho) < -2n-\varepsilon;\ \left\{\begin{matrix}a,\ \mathrm{Re}\, b > 0\\ a > b > 0\end{matrix}\right\}\right]$

No.	$f(x)$	$F(s)$
8	$\frac{\theta(x-a)}{x^2-b^2} P_{2n+\varepsilon}\left(\frac{x}{a}\right)$	$\frac{(-1)^n \left(\frac{1-s+\varepsilon}{2}\right)_n a^s}{2b^2 \left(\frac{s+\varepsilon}{2}\right)_{n+1}} {}_3F_2\left(\begin{matrix}1,\ \frac{s}{2},\ \frac{s+1}{2};\ \frac{a^2}{b^2}\\ \frac{s-2n-\varepsilon+1}{2},\ \frac{s+2n+\varepsilon+2}{2}\end{matrix}\right) + (-1)^{\varepsilon+1} \frac{\pi}{2} b^{s-2} \tan^{2\varepsilon-1} \frac{s\pi}{2} P_{2n+\varepsilon}\left(\frac{b}{a}\right)$ $[0 < a < b;\ \operatorname{Re} s < 2-2n-\varepsilon]$
9	$\frac{\theta(x-a)}{x^2-b^2} P_{2n+\varepsilon}\left(\frac{x}{a}\right)$	$(-1)^{n+1} \frac{\left(\frac{3-s+\varepsilon}{2}\right)_n a^{s-2}}{2\left(\frac{s-2+\varepsilon}{2}\right)_{n+1}} {}_3F_2\left(\begin{matrix}1,\ \frac{2n-s+\varepsilon+3}{2},\ \frac{2-2n-s-\varepsilon}{2}\\ \frac{3-s}{2},\ \frac{4-s}{2};\ \frac{b^2}{a^2}\end{matrix}\right)$ $[0 < b < a;\ \operatorname{Re} s < 2-2n-\varepsilon]$
10	$\frac{1}{(x+a)^\rho} P_n\left(\frac{2x}{b}+1\right)$	$a^{s-\rho} \mathrm{B}(s, \rho-s)\ {}_3F_2\left(\begin{matrix}-n,\ n+1,\ s\\ 1,\ s-\rho+1;\ \frac{a}{b}\end{matrix}\right)$ $[0 < \operatorname{Re} s < \operatorname{Re} \rho - n;\ \lvert\arg a\rvert < \pi]$
11	$(x-a)_+^{-1/2} P_{2n}\left(i\sqrt{\frac{x}{a}-1}\right)$	$\frac{(-1)^n a^{s-1/2}}{n!} \Gamma\left(\frac{2n+1}{2}\right) \Gamma\left[\begin{matrix}1-s+n,\ \frac{1-2s-2n}{2}\\ 1-s,\ 1-s\end{matrix}\right]$ $[a > 0;\ \operatorname{Re} s < 1/2 - n]$
12	$\theta(x-a) \times \left(\frac{x-a}{x}\right)^{(n-2[n/2]-1)/2} \times P_n\left(\sqrt{\frac{x-a}{x}}\right)$	$\frac{(-1)^{[n/2]} a^s}{[n/2]!} \Gamma\left[\begin{matrix}-s,\ -s,\ n-\left[\frac{n}{2}\right]+\frac{1}{2}\\ -s-\left[\frac{n}{2}\right],\ -s+n-\left[\frac{n}{2}\right]+\frac{1}{2}\end{matrix}\right]$ $[a > 0;\ \operatorname{Re} s < 0]$

3.19.2. $P_n(bx)$ and the exponential function

1	$\theta(x-a) e^{-bx} P_n\left(\frac{x}{a}\right)$	$\frac{2^n (1/2)_n e^{-ab}}{n!\, a^n b^{s+n}} \Gamma(s+n)\ {}_2F_2\left(\begin{matrix}-n,\ -n;\ 2ab\\ -2n,\ 1-s-n\end{matrix}\right)$ $[a,\ \operatorname{Re} b,\ \operatorname{Re} s > 0]$
2	$\theta(a-x) e^{-bx^2} P_n\left(\frac{x}{a}\right)$	$\sqrt{\pi}\left(\frac{a}{2}\right)^s \Gamma\left[\begin{matrix}s\\ \frac{s-n+1}{2},\ \frac{s+n+2}{2}\end{matrix}\right] {}_2F_2\left(\begin{matrix}\frac{s}{2},\ \frac{s+1}{2};\ -a^2b\\ \frac{s-n+1}{2},\ \frac{s+n+2}{2}\end{matrix}\right)$ $[a > 0;\ \operatorname{Re} s > ((-1)^n - 1)/2]$
3	$\theta(x-a) e^{b/x^2} P_n\left(\frac{x}{a}\right)$	$\frac{2^{-s-1} a^s}{\sqrt{\pi}} \Gamma\left[\begin{matrix}-\frac{s+n}{2},\ \frac{1-s+n}{2}\\ 1-s\end{matrix}\right] {}_2F_2\left(\begin{matrix}-\frac{s+n}{2},\ \frac{1-s+n}{2};\ \frac{b}{a^2}\\ \frac{1-s}{2},\ \frac{2-s}{2}\end{matrix}\right)$ $[a > 0;\ \operatorname{Re} b > 0]$

No.	$f(x)$	$F(s)$
4	$e^{-bx} P_n\left(\frac{2x}{a} \pm 1\right)$	$\frac{2^{2n}}{n!\, a^n b^{s+n}} \left(\frac{1}{2}\right)_n \Gamma(s+n)\; {}_2F_2\left(\begin{matrix} -n,\ -n;\ \pm ab \\ -2n,\ 1-s-n \end{matrix}\right)$ $[\operatorname{Re} b,\ \operatorname{Re} s > 0]$
5	$e^{-b/x} P_n\left(\frac{2x}{a} + 1\right)$	$b^s \Gamma(-s)\; {}_2F_2\left(\begin{matrix} -n,\ n+1 \\ 1,\ s+1;\ \frac{b}{a} \end{matrix}\right)$ $[\operatorname{Re} b > 0;\ \operatorname{Re} s < -n]$
6	$e^{-b\sqrt{x}} P_n\left(\frac{2x}{a} \pm 1\right)$	$\frac{2^{2n+1}}{n!\, a^n b^{2s+2n}} \left(\frac{1}{2}\right)_n \Gamma(2s+2n)\; {}_2F_3\left(\begin{matrix} -n,\ -n;\ \mp\frac{ab^2}{4} \\ -2n,\ \frac{1-2s-2n}{2},\ 1-s-n \end{matrix}\right)$ $[\operatorname{Re} b,\ \operatorname{Re} s > 0]$

3.19.3. $P_n(ax+b)$ and $\operatorname{Ei}(cx^r)$

Notation: $\varepsilon = 0$ or 1.

1	$\theta(a-x) \operatorname{Ei}(-bx^2)\, P_{2n+\varepsilon}\left(\frac{x}{a}\right)$	$\frac{(-1)^{n+1} \left(\frac{\varepsilon-s-1}{2}\right)_n a^{s+2}b}{2\left(\frac{s+\varepsilon+2}{2}\right)_{n+1}}\; {}_4F_4\left(\begin{matrix} 1,\ 1,\ \frac{s+2}{2},\ \frac{s+3}{2};\ -a^2b \\ 2,\ 2,\ \frac{s-2n-\varepsilon+3}{2},\ \frac{s+2n+\varepsilon+4}{2} \end{matrix}\right) + \frac{(-1)^n \left(\frac{\varepsilon-s+1}{2}\right)_n a^s}{2\left(\frac{s+\varepsilon}{2}\right)_{n+1}} \times \left[\mathbf{C} - \sum_{k=0}^{n} \frac{2}{2k+s+\varepsilon} - \sum_{k=0}^{n-1} \frac{2}{2k-s+\varepsilon+1} + \ln(a^2 b)\right]$ $[a > 0;\ \operatorname{Re} s > -\varepsilon;\ \lvert\arg b\rvert < \pi]$

3.19.4. $P_n(ax+b)$ and $\operatorname{si}(cx^r)$, $\operatorname{ci}(cx^r)$

1	$(a^2-x^2)_+^{1/2} \operatorname{si}(bx)\, P_n\left(\frac{x}{a}\right)$	$\sqrt{\pi}\left(\frac{a}{2}\right)^{s+1} b\,\Gamma\left[\begin{matrix} s+1 \\ \frac{s-n+2}{2},\ \frac{s+n+3}{2} \end{matrix}\right] {}_3F_4\left(\begin{matrix} \frac{1}{2},\ \frac{s+1}{2},\ \frac{s+2}{2};\ -\frac{a^2b^2}{4} \\ \frac{3}{2},\ \frac{3}{2},\ \frac{s-n+2}{2},\ \frac{s+n+3}{2} \end{matrix}\right) - \frac{\pi^{3/2} a^s}{2^{s+1}} \Gamma\left[\begin{matrix} s \\ \frac{s-n+1}{2},\ \frac{s+n+2}{2} \end{matrix}\right]$ $[a > 0;\ \operatorname{Re} s > ((-1)^n - 1)/2]$
2	$(a^2-x^2)_+^{1/2} \operatorname{ci}(bx)\, P_n\left(\frac{x}{a}\right)$	$-\frac{\sqrt{\pi}\, a^{s+2} b^2}{2^{s+4}} \Gamma\left[\begin{matrix} s+2 \\ \frac{s-n+3}{2},\ \frac{s+n+4}{2} \end{matrix}\right] {}_4F_5\left(\begin{matrix} 1,\ 1,\ \frac{s+2}{2},\ \frac{s+3}{2};\ -\frac{a^2b^2}{4} \\ \frac{3}{2},\ 2,\ 2,\ \frac{s-n+3}{2},\ \frac{s+n+4}{2} \end{matrix}\right) + \sqrt{\pi}\left(\frac{a}{2}\right)^s \Gamma\left[\begin{matrix} s \\ \frac{s-n+1}{2},\ \frac{s+n+2}{2} \end{matrix}\right] \left[\psi(s) - \frac{1}{2}\psi\left(\frac{s-n+1}{2}\right) - \frac{1}{2}\psi\left(\frac{s+n+2}{2}\right) + \ln\frac{ab}{2} + \mathbf{C}\right]$ $[a > 0;\ \operatorname{Re} s > ((-1)^n - 1)/2]$

3.19.5. $P_n(ax+b)$ and $\operatorname{erf}(cx^r)$, $\operatorname{erfc}(cx^r)$

1	$\theta(a-x)\operatorname{erfc}(bx)\,P_n\left(\frac{x}{a}\right)$	$-2^{-s}a^{s+1}b\,\Gamma\left[\begin{matrix}s+1\\ \frac{s-n+2}{2},\ \frac{s+n+3}{2}\end{matrix}\right]{}_3F_3\left(\begin{matrix}\frac{1}{2},\ \frac{s+1}{2},\ \frac{s+2}{2};\ -a^2b^2\\ \frac{3}{2},\ \frac{s-n+2}{2},\ \frac{s+n+3}{2}\end{matrix}\right)$ $+\sqrt{\pi}\left(\frac{a}{2}\right)^s\Gamma\left[\begin{matrix}s\\ \frac{s-n+1}{2},\ \frac{s+n+2}{2}\end{matrix}\right]$ $[a>0;\ \operatorname{Re}s>((-1)^n-1)/2]$

3.19.6. Products of $P_n(ax^r+b)$

Notation: $\varepsilon=0$ or 1.

1	$\theta(a-x)\,P_m\left(\frac{x}{a}\right)P_n\left(\frac{x}{a}\right)$	$\frac{\sqrt{\pi}\,(m+n)!\,a^s}{2^s m!\,n!}\Gamma\left[\begin{matrix}s\\ \frac{s-m-n+1}{2},\ \frac{s+m+n+2}{2}\end{matrix}\right]$ $\times{}_3F_2\left(\begin{matrix}-m,\ -n,\ \frac{s-m-n}{2};\ 1\\ -m-n,\ \frac{s-m-n+1}{2}\end{matrix}\right)$ $[a>0;\ \operatorname{Re}s>2[m/2]+2[n/2]-m-n]$
2	$\theta(x-a)\,P_m\left(\frac{2x}{a}-1\right)$ $\times P_n\left(\frac{2x}{b}\pm 1\right)$	$(-1)^{m+1}(\pm 1)^n a^s\frac{(1-s)_m}{(s)_{m+1}}\,{}_4F_3\left(\begin{matrix}-n,\ n+1,\ s,\ s;\ \mp\frac{a}{b}\\ 1,\ s-m,\ s+m+1\end{matrix}\right)$ $[a>0;\ \operatorname{Re}s<-m-n]$
3	$\theta(a-x)\,P_n\left(\sqrt{\frac{a}{x}}\right)$ $\times P_m\left(1-\frac{x}{b}\right)$	$a^s\,\Gamma\left[\begin{matrix}\frac{2s-n}{2},\ \frac{2s+n+1}{2}\\ \frac{2s+1}{2},\ s+1\end{matrix}\right]{}_4F_3\left(\begin{matrix}-m,\ m+1,\ \frac{2s-n}{2},\ \frac{2s+n+1}{2}\\ 1,\ \frac{2s+1}{2},\ s+1;\ \frac{a}{2b}\end{matrix}\right)$ $[a>0;\ \operatorname{Re}s>n/2]$
4	$\theta(a-x)\,(b-x)^m$ $\times P_n\left(\sqrt{\frac{a}{x}}\right)P_m\left(\frac{b+x}{b-x}\right)$	$a^s b^m\,\Gamma\left[\begin{matrix}\frac{2s-n}{2},\ \frac{2s+n+1}{2}\\ \frac{2s+1}{2},\ s+1\end{matrix}\right]{}_4F_3\left(\begin{matrix}-m,\ -m,\ \frac{2s-n}{2},\ \frac{2s+n+1}{2}\\ 1,\ \frac{2s+1}{2},\ s+1;\ \frac{a}{b}\end{matrix}\right)$ $[a>0;\ \operatorname{Re}s>n/2]$
5	$\theta(a-x)\,P_n\left(\frac{a}{x}\right)P_{2m+\varepsilon}(bx)$	$\frac{(-1)^m 2^{s+\varepsilon-1}a^{s+\varepsilon}b^{\varepsilon}}{\sqrt{\pi}\,m!}\left(\frac{2\varepsilon+1}{2}\right)_m\Gamma\left[\begin{matrix}\frac{s-n+\varepsilon}{2},\ \frac{s+n+\varepsilon+1}{2}\\ s+\varepsilon+1\end{matrix}\right]$ $\times{}_4F_3\left(\begin{matrix}-m,\ \frac{2m+2\varepsilon+1}{2},\ \frac{s-n+\varepsilon}{2},\ \frac{s+n+\varepsilon+1}{2}\\ \frac{2\varepsilon+1}{2},\ \frac{s+\varepsilon+1}{2},\ \frac{s+\varepsilon+2}{2};\ a^2b^2\end{matrix}\right)$ $[a>0;\ \operatorname{Re}s>n-\varepsilon]$
6	$\theta(a-x)\,P_m\left(1-\frac{2}{bx^2}\right)$ $\times P_n\left(\frac{x}{a}\right)$	$\frac{(-1)^m 2^{4m-s}a^{s-2m}b^{-m}}{m!}\Gamma\left[\begin{matrix}\frac{2m+1}{2},\ s-2m\\ \frac{s-2m-n+1}{2},\ \frac{s-2m+n+2}{2}\end{matrix}\right]$ $\times{}_4F_3\left(\begin{matrix}-m,\ -m,\ \frac{s-2m}{2},\ \frac{s-2m+1}{2};\ a^2b\\ -2m,\ \frac{s-2m-n+1}{2},\ \frac{s-2m+n+2}{2}\end{matrix}\right)$ $[a>0;\ \operatorname{Re}s>2m]$

3.20. The Chebyshev Polynomials $T_n(z)$

More formulas can be obtained from the corresponding sections due to the relations

$$T_n(z) = \frac{n}{2} \lim_{\lambda \to 0} \left[\frac{1}{\lambda} C_n^{\lambda}(z) \right], \quad T_{\nu}(z) = \frac{\Gamma(\nu+1)}{(1/2)_{\nu}} P_{\nu}^{(-1/2,\, -1/2)}(z),$$

$$T_{\nu}(z) = {}_2F_1\left(-\nu,\, \nu;\, \frac{1}{2};\, \frac{1-z}{2}\right).$$

3.20.1. $T_n(\varphi(x))$ and algebraic functions

Notation: $\varepsilon = 0$ or 1.

No.	$f(x)$	$F(s)$
1	$(a^2 - x^2)_+^{-1/2} T_n\left(\frac{x}{a}\right)$	$\frac{\pi a^{s-1}}{2^s} \Gamma\left[\begin{matrix} s \\ \frac{s+n+1}{2},\, \frac{s-n+1}{2} \end{matrix}\right]$ $[a,\ \mathrm{Re}\, s > 0]$
2	$\frac{(a^2 - x^2)_+^{-1/2}}{(b^2 \pm x^2)^{\rho}} T_n\left(\frac{x}{a}\right)$	$\frac{\pi a^{s-1}}{2^s b^{2\rho}} \Gamma\left[\begin{matrix} s \\ \frac{s-n+1}{2},\, \frac{s+n+1}{2} \end{matrix}\right] {}_3F_2\left(\begin{matrix} \rho,\, \frac{s}{2},\, \frac{s+1}{2};\, \mp\frac{a^2}{b^2} \\ \frac{s-n+1}{2},\, \frac{s+n+1}{2} \end{matrix}\right)$ $\left[\begin{Bmatrix} \mathrm{Re}\, b > 0 \\ b > a \end{Bmatrix};\ a,\ \mathrm{Re}\, s > 0\right]$
3	$(x^2 - a^2)_+^{-1/2} T_n\left(\frac{x}{a}\right)$	$\frac{a^{s-1}}{2^{s+1}} \Gamma\left[\begin{matrix} \frac{1-s+n}{2},\, \frac{1-s-n}{2} \\ 1-s \end{matrix}\right]$ $[a > 0;\ \mathrm{Re}\, s < 1-n]$
4	$(a-x)_+^{\alpha-1} T_{2n+\varepsilon}(bx)$	$\frac{(-1)^n (n+\varepsilon/2)\, a^{s+\alpha+\varepsilon-1} (2b)^{\varepsilon}}{n!} \mathrm{B}(\alpha,\, s+\varepsilon)$ $\times \Gamma(n+\varepsilon)\, {}_4F_3\left(\begin{matrix} -n,\, n+\varepsilon,\, \frac{s+\varepsilon}{2},\, \frac{s+\varepsilon+1}{2} \\ \frac{2\varepsilon+1}{2},\, \frac{s+\alpha+\varepsilon}{2},\, \frac{s+\alpha+\varepsilon+1}{2};\, a^2 b^2 \end{matrix}\right)$ $[a,\ \mathrm{Re}\, \alpha > 0;\ \mathrm{Re}\, s > -\varepsilon]$
5	$(x-a)_+^{\alpha-1} T_{2n+\varepsilon}(bx)$	$\frac{(-1)^n (n+\varepsilon/2)\, a^{s+\alpha+\varepsilon-1} (2b)^{\varepsilon}}{n!} \mathrm{B}(1-s-\alpha-\varepsilon,\, \alpha)$ $\times \Gamma(n+\varepsilon)\, {}_4F_3\left(\begin{matrix} -n,\, n+\varepsilon,\, \frac{s+\varepsilon}{2},\, \frac{s+\varepsilon+1}{2} \\ \frac{2\varepsilon+1}{2},\, \frac{s+\alpha+\varepsilon}{2},\, \frac{s+\alpha+\varepsilon+1}{2};\, a^2 b^2 \end{matrix}\right)$ $[a > 0;\ \mathrm{Re}\,(s+\alpha) < 1-2n-\varepsilon]$
6	$(a-x)_+^{-1/2} T_n\left(\frac{2x}{a} - 1\right)$	$\sqrt{\pi}\, a^{s-1/2} \Gamma\left[\begin{matrix} s,\, s+\frac{1}{2} \\ s-n+\frac{1}{2},\, s+n+\frac{1}{2} \end{matrix}\right]$ $[a,\ \mathrm{Re}\, s > 0]$
7	$(a-x)_+^{\alpha-1} T_n\left(\frac{2x}{a} - 1\right)$	$a^{s+\alpha-1}\, \mathrm{B}(\alpha,\, s)\, {}_3F_2\left(\begin{matrix} -n,\, n,\, \alpha \\ \frac{1}{2},\, s+\alpha;\, 1 \end{matrix}\right)$ $[a,\ \mathrm{Re}\, \alpha,\ \mathrm{Re}\, s > 0]$

No.	$f(x)$	$F(s)$
8	$\frac{(a-x)_+^{-1/2}}{(b\pm x)^\rho} T_n\left(\frac{2x}{a}-1\right)$	$(-1)^n \sqrt{\pi}\, a^{s-1/2} b^{-\rho} \left(\frac{1-2s}{2}\right)_n \Gamma\left[\begin{matrix} s \\ \frac{2s+2n+1}{2} \end{matrix}\right] \times {}_3F_2\left(\begin{matrix} \rho,\ s,\ \frac{2s+1}{2};\ \mp\frac{a}{b} \\ \frac{2s-2n+1}{2},\ \frac{2s+2n+1}{2} \end{matrix}\right)$ $\left[\left\{\begin{matrix} a>0 \\ b>a>0 \end{matrix}\right\};\ \mathrm{Re}\, s>0\right]$
9	$(x-a)_+^{-1/2} T_n\left(\frac{2x}{a}-1\right)$	$\sqrt{\pi}\, a^{s-1/2} \Gamma\left[\begin{matrix} \frac{1}{2}-s-n,\ \frac{1}{2}-s+n \\ 1-s,\ \frac{1}{2}-s \end{matrix}\right]$ $[a>0;\ \mathrm{Re}\, s<1/2-n]$
10	$\frac{1}{(x+a)^\rho} T_n\left(\frac{2x}{a}+1\right)$	$a^{s-\rho} \mathrm{B}(-s-n+\rho,\ s)\ {}_3F_2\left(\begin{matrix} -n,\ \frac{1-2n}{2},\ s \\ \frac{1}{2},\ \rho-n;\ 1 \end{matrix}\right)$ $[0<\mathrm{Re}\, s<\mathrm{Re}\, \rho-n;\ \lvert\arg a\rvert<\pi]$
11	$\frac{(x+a)^{-1/2}}{(x+b)^\rho} T_n\left(\frac{2x}{a}+1\right)$	$\frac{(-1)^n a^{s-\rho-1/2}}{\sqrt{\pi}} \left(\frac{1}{2}-s+\rho\right)_n \Gamma\left(\frac{1-2s-2n+2\rho}{2}\right) \times \Gamma(s-\rho)\ {}_3F_2\left(\begin{matrix} \rho,\ \frac{1-2s-2n+2\rho}{2},\ \frac{1-2s+2n+2\rho}{2} \\ 1-s+\rho,\ \frac{1-2s+2\rho}{2};\ \frac{b}{a} \end{matrix}\right) + a^{-1/2} b^{s-\rho} \mathrm{B}(s,\ \rho-s)\ {}_3F_2\left(\begin{matrix} \frac{-2n+1}{2},\ \frac{2n+1}{2},\ s \\ \frac{1}{2},\ s-\rho+1;\ \frac{b}{a} \end{matrix}\right)$ $[0<\mathrm{Re}\, s<1/2-n+\mathrm{Re}\, \rho;\ \lvert\arg a\rvert,\ \lvert\arg b\rvert<\pi]$
12	$(a-x)_+^{\alpha-1} T_n\left(\frac{2x}{b}\pm 1\right)$	$(\pm 1)^n a^{s+\alpha-1} \mathrm{B}(s,\ \alpha)\ {}_3F_2\left(\begin{matrix} -n,\ n,\ s \\ \frac{1}{2},\ s+\alpha;\ \mp\frac{a}{b} \end{matrix}\right)$ $[a,\ \mathrm{Re}\, \alpha,\ \mathrm{Re}\, s>0]$
13	$\frac{(a-x)_+^{\alpha-1}}{\sqrt{b\pm x}} T_n\left(\frac{2x}{b}\pm 1\right)$	$(\pm 1)^n a^{s+\alpha-1} b^{-1/2} \mathrm{B}(\alpha,\ s)\ {}_3F_2\left(\begin{matrix} \frac{-2n+1}{2},\ \frac{2n+1}{2},\ s \\ \frac{1}{2},\ s+\alpha;\ \mp\frac{a}{b} \end{matrix}\right)$ $\left[\left\{\begin{matrix} a>0 \\ b>a>0 \end{matrix}\right\};\ \mathrm{Re}\, \alpha,\ \mathrm{Re}\, s>0\right]$
14	$\frac{(x+a-b)^{-1/2}}{(x+a+b)^\rho} T_n\left(\frac{x+a}{b}\right)$	$\frac{2^{n-1}(\delta_{n,0}+1)(a+b)^{s+n-\rho-1/2}}{b^n} \mathrm{B}\left(\frac{1-2s-2n+2\rho}{2},\ s\right) \times {}_3F_2\left(\begin{matrix} \frac{1-2n}{2},\ 1-n,\ \frac{1-2s-2n+2\rho}{2} \\ \frac{1-2n+2\rho}{2},\ 1-2n;\ \frac{2b}{a+b} \end{matrix}\right)$ $[a>b>0;\ 0<\mathrm{Re}\, s<1/2-n+\mathrm{Re}\, \rho]$
15	$(a-x)_+^{\alpha-1} T_{2n+\varepsilon}(b(a-x))$	$\frac{(-1)^n (n+\varepsilon/2)\, a^{s+\alpha+\varepsilon-1} (2b)^\varepsilon}{n!} \mathrm{B}(\alpha+\varepsilon,\ s) \times \Gamma(n+\varepsilon)\ {}_4F_3\left(\begin{matrix} -n,\ n+\varepsilon,\ \frac{\alpha+\varepsilon}{2},\ \frac{\alpha+\varepsilon+1}{2} \\ \frac{2\varepsilon+1}{2},\ \frac{s+\alpha+\varepsilon}{2},\ \frac{s+\alpha+\varepsilon+1}{2};\ a^2b^2 \end{matrix}\right)$ $[a>0;\ \mathrm{Re}\, \alpha>-\varepsilon;\ \mathrm{Re}\, s>0]$

No.	$f(x)$	$F(s)$
16	$(a-x)_+^{-1/2}\,T_n\left(\frac{x+b}{a+b}\right)$	$n!\sqrt{\pi}\,a^{s-1/2}\,\Gamma\left[\begin{matrix}s\\ \frac{2s+2n+1}{2}\end{matrix}\right]P_n^{(s-1/2,\,-s-1/2)}\left(\frac{b}{a+b}\right)$ $[a,\ \operatorname{Re}s>0]$
17	$(a-x)_+^{\alpha-1}\,T_n\left(\frac{x+b}{a+b}\right)$	$a^{s+\alpha-1}\,\mathrm{B}\,(s,\,\alpha)\;{}_3F_2\left(\begin{matrix}-n,\,n,\,\alpha\\ \frac{1}{2},\,s+\alpha;\,\frac{a}{2(a+b)}\end{matrix}\right)$ $[a,\ \operatorname{Re}\alpha,\ \operatorname{Re}s>0]$
18	$\left(a^2-x^2\right)_+^{-1/2}\,T_n\left(\frac{a}{x}\right)$	$2^{s-2}a^{s-1}\,\Gamma\left[\begin{matrix}\frac{s-n}{2},\,\frac{s+n}{2}\\ s\end{matrix}\right]$ $[a,\ \operatorname{Re}s>0]$
19	$(a-x)_+^{-1/2}\,T_n\left(\frac{2a}{x}-1\right)$	$\sqrt{\pi}\,a^{s-1/2}\,\Gamma\left[\begin{matrix}s-n,\,s+n\\ s,\,s+\frac{1}{2}\end{matrix}\right]$ $[a>0;\ \operatorname{Re}s>n]$
20	$(x-a)_+^{-1/2}\,T_n\left(\frac{2a}{x}-1\right)$	$\sqrt{\pi}\,a^{s-1/2}\,\Gamma\left[\begin{matrix}\frac{1}{2}-s,\,1-s\\ 1-s-n,\,1-s+n\end{matrix}\right]$ $[a>0;\ \operatorname{Re}s<1/2]$
21	$\frac{1}{(x+a)^n}\,T_n\left(\frac{x-a}{x+a}\right)$	$\frac{(-1)^n\,2^{2n-1}a^{s-n}}{(2n-1)!}\,\Gamma\left[\begin{matrix}s,\,n-s,\,n-s+\frac{1}{2}\\ \frac{1}{2}-s\end{matrix}\right]$ $[0<\operatorname{Re}s<n;\ \lvert\arg a\rvert<\pi]$
22	$(a-x)_+^{-1/2}$ $\times\,T_n\left(\frac{x^2-8ax+8a^2}{x^2}\right)$	$\sqrt{\pi}a^{s-1/2}\,\Gamma\left[\begin{matrix}s-2n,\,s+2n\\ s,\,s+\frac{1}{2}\end{matrix}\right]$ $[a>0;\ \operatorname{Re}s>2n]$
23	$(x-a)_+^{-1/2}$ $\times\,T_n\left(\frac{x^2-8ax+8a^2}{a^2}\right)$	$\sqrt{\pi}a^{s-1/2}\,\Gamma\left[\begin{matrix}\frac{1}{2}-s-2n,\,\frac{1}{2}-s+2n\\ \frac{1}{2}-s,\,1-s\end{matrix}\right]$ $[a>0;\ \operatorname{Re}s<1/2-2n]$
24	$(a-x)_+^{\alpha-1}\,T_{2n+\varepsilon}\,(bx(a-x))$	$(-1)^n\,(2n+1)^{\varepsilon}\,a^{s+\alpha+2\varepsilon-1}b^{\varepsilon}\,\mathrm{B}\,(s+\varepsilon,\,\alpha+\varepsilon)$ $\times\,{}_6F_5\left(\begin{matrix}-n,\,n+\varepsilon,\,\Delta\,(2,\,s+\varepsilon)\,,\,\Delta\,(2,\,\alpha+\varepsilon)\\ \frac{2\varepsilon+1}{2},\,\Delta\,(4,\,s+\alpha+2\varepsilon)\,;\,\frac{a^4b^2}{16}\end{matrix}\right)$ $[a>0;\ \operatorname{Re}s,\ \operatorname{Re}\alpha>-\varepsilon]$
25	$(a-x)_+^{\alpha-1}\,T_n\left(\frac{b}{x\,(a-x)}\right)$	$2^{n-1}a^{s+\alpha-2n-1}b^n\,\mathrm{B}\,(s-n,\,\alpha-n)$ $\times\,{}_6F_5\left(\begin{matrix}\Delta\,(2,\,-n)\,,\,\Delta\,(2,\,s-n)\,,\,\Delta\,(2,\,\alpha-n)\\ 1-n,\,\Delta\,(4,\,s+\alpha-2n)\,;\,\frac{a^4}{16b^2}\end{matrix}\right)$ $[a>0;\ \operatorname{Re}s,\ \operatorname{Re}\alpha>n]$

No.	$f(x)$	$F(s)$
26	$(a-x)_+^{\alpha-1}$ $\times T_{2n+\varepsilon}\left(b\sqrt{x(a-x)}\right)$	$(-1)^n (2n+1)^{\varepsilon} a^{s+\alpha+\varepsilon-1} b^{\varepsilon}\, \mathrm{B}\left(\frac{2s+\varepsilon}{2}, \frac{2\alpha+\varepsilon}{2}\right)$ $\times {}_4F_3\left(\begin{matrix} -n,\, n+\varepsilon,\, \frac{2s+\varepsilon}{2},\, \frac{2\alpha+\varepsilon}{2} \\ \frac{2\varepsilon+1}{2},\, \frac{s+\alpha+\varepsilon}{2},\, \frac{s+\alpha+\varepsilon+1}{2};\, \frac{a^2b^2}{4} \end{matrix}\right)$ $[a>0;\ \mathrm{Re}\, s,\ \mathrm{Re}\,\alpha > -\varepsilon/2]$
27	$(a-x)_+^{\alpha-1} T_n\left(\frac{b}{\sqrt{x(a-x)}}\right)$	$2^{n-1} a^{s-n+\alpha-1} b^n\, \mathrm{B}\left(s-\frac{n}{2}, \alpha-\frac{n}{2}\right)$ $\times {}_4F_3\left(\begin{matrix} \frac{1-n}{2},\, -\frac{n}{2},\, s-\frac{n}{2},\, \alpha-\frac{n}{2} \\ 1-n,\, \frac{s-n+\alpha}{2},\, \frac{s-n+\alpha+1}{2};\, \frac{a^2}{4b^2} \end{matrix}\right)$ $[a>0;\ \mathrm{Re}\, s,\ \mathrm{Re}\,\alpha > n/2]$
28	$\theta(a-x)\, T_{2n+1}\left(\sqrt{1-\frac{x}{a}}\right)$	$\frac{2n+1}{2}\sqrt{\pi}\, a^s\, \Gamma\left[\begin{matrix} s,\, \frac{1}{2}-s+n \\ s+n+\frac{3}{2},\, \frac{1}{2}-s \end{matrix}\right]$ $[a,\ \mathrm{Re}\, s>0]$
29	$(a-x)_+^{-1/2}\, T_{2n}\left(\sqrt{1-\frac{x}{a}}\right)$	$\sqrt{\pi}\, a^{s-1/2}\, \Gamma\left[\begin{matrix} s,\, \frac{1}{2}-s+n \\ s+n+\frac{1}{2},\, \frac{1}{2}-s \end{matrix}\right]$ $[a,\ \mathrm{Re}\, s>0]$
30	$\theta(x-a)\, T_{2n+1}\left(i\sqrt{\frac{x}{a}-1}\right)$	$i\,\frac{2n+1}{2}\sqrt{\pi}\, a^s\, \Gamma\left[\begin{matrix} s+\frac{1}{2},\, -\frac{s+2n+1}{2} \\ s-n+\frac{1}{2},\, 1-s \end{matrix}\right]$ $[a>0;\ \mathrm{Re}\, s < -1/2-n]$
31	$(x-a)_+^{-1/2}\, T_{2n}\left(i\sqrt{\frac{x}{a}-1}\right)$	$\sqrt{\pi}\, a^{s-1/2}\, \Gamma\left[\begin{matrix} s+\frac{1}{2},\, \frac{1}{2}-s-n \\ s-n+\frac{1}{2},\, 1-s \end{matrix}\right]$ $[a>0;\ \mathrm{Re}\, s < 1/2-n]$
32	$\theta(a-x)\, T_{2n+1}\left(i\sqrt{\frac{a}{x}-1}\right)$	$i\,\frac{2n+1}{2}\sqrt{\pi}\, a^s\, \Gamma\left[\begin{matrix} s-n-\frac{1}{2},\, \frac{1}{2}-s \\ s+1,\, \frac{1}{2}-s-n \end{matrix}\right]$ $[a>0;\ \mathrm{Re}\, s > n+1/2]$
33	$(a-x)_+^{-1/2}\, T_{2n}\left(i\sqrt{\frac{a}{x}-1}\right)$	$\sqrt{\pi}\, a^{s-1/2}\, \Gamma\left[\begin{matrix} s-n,\, 1-s \\ s+\frac{1}{2},\, 1-s-n \end{matrix}\right]$ $[a>0;\ \mathrm{Re}\, s > n]$
34	$(x-a)_+^{-1/2}\, T_{2n}\left(\sqrt{1-\frac{a}{x}}\right)$	$\sqrt{\pi}\, a^{s-1/2}\, \Gamma\left[\begin{matrix} s+n,\, \frac{1}{2}-s \\ s,\, 1-s+n \end{matrix}\right]$ $[a>0;\ \mathrm{Re}\, s < 1/2]$
35	$\frac{1}{(x+a)^{n/2}}\, T_n\left(\sqrt{\frac{a}{x+a}}\right)$	$\frac{2^{n-1} a^{s-n/2}}{(n-1)!}\, \Gamma\left[\begin{matrix} s,\, \frac{n}{2}-s,\, \frac{n+1}{2}-s \\ \frac{1}{2}-s \end{matrix}\right]$ $[0<\mathrm{Re}\, s<n/2;\ n\geq 1;\ \lvert\arg a\rvert<\pi]$
36	$\frac{1}{(x+a)^{n/2}}\, T_n\left(\sqrt{\frac{x}{x+a}}\right)$	$\frac{2^{n-1} a^{s-n/2}}{(n-1)!}\, \Gamma\left[\begin{matrix} s,\, s+\frac{1}{2},\, \frac{n}{2}-s \\ s+\frac{1-n}{2} \end{matrix}\right]$ $[a>0;\ 0<\mathrm{Re}\, s<n/2;\ n\geq 1;\ \lvert\arg a\rvert<\pi]$

No.	$f(x)$	$F(s)$
37	$(a-x)_+^{-1/2} \dfrac{(bx+1)^{\alpha}}{\left[1-c(a-x)\right]^{\varepsilon/2}}$ $\times T_{2n+\varepsilon}\left(\sqrt{1+ac}-cx\right)$	$\sqrt{\pi}\, a^{s-1/2}\, \Gamma\left[\begin{matrix} s \\ \frac{2s+1}{2} \end{matrix}\right] F_3\left(-\alpha,\, -n,\, s,\, n+\varepsilon;\, \dfrac{2s+1}{2};\, -ab,\, -ac\right)$ $[a>0;\ \lvert\arg(1+ab)\rvert<\pi]$
38	$(a-x)_+^{(\varepsilon-1)/2}(b-x)^{n+\varepsilon/2}$ $\times T_{2n+\varepsilon}\left(c\sqrt{\dfrac{a-x}{b-x}}\right)$	$(-1)^n\left(n+\dfrac{1}{2}\right)^{\varepsilon}\sqrt{\pi}\, a^{s+\varepsilon-1/2}b^n c^{\varepsilon}\,\Gamma\left[\begin{matrix} s \\ \frac{2s+2\varepsilon+1}{2} \end{matrix}\right]$ $\times F_1\left(-n,\, s,\, n+\varepsilon;\, \dfrac{2s+2\varepsilon+1}{2};\, \dfrac{a}{b},\, \dfrac{ac^2}{b}\right)$ $[a,\ \operatorname{Re} s>0]$

3.20.2. $T_n(bx)$ and the exponential function

1	$\left(a^2-x^2\right)_+^{-1/2} e^{bx}\, T_n\left(\dfrac{x}{a}\right)$	$\dfrac{\pi}{2}\left(\dfrac{a}{2}\right)^{s-1}\Gamma\left[\begin{matrix} s \\ \frac{s-n+1}{2},\, \frac{s+n+1}{2} \end{matrix}\right] {}_2F_3\left(\begin{matrix} \frac{s}{2},\, \frac{s+1}{2};\, \frac{a^2b^2}{4} \\ \frac{1}{2},\, \frac{s-n+1}{2},\, \frac{s+n+1}{2} \end{matrix}\right)$ $+\dfrac{\pi}{2}\left(\dfrac{a}{2}\right)^{s} b\,\Gamma\left[\begin{matrix} s+1 \\ \frac{s-n+2}{2},\, \frac{s+n+2}{2} \end{matrix}\right] {}_2F_3\left(\begin{matrix} \frac{s+1}{2},\, \frac{s+2}{2};\, \frac{a^2b^2}{4} \\ \frac{3}{2},\, \frac{s-n+2}{2},\, \frac{s+n+2}{2} \end{matrix}\right)$ $[a>0;\ \operatorname{Re} s>\left((-1)^n-1\right)/2]$
2	$(a-x)_+^{-1/2} e^{bx}\, T_n\left(1-\dfrac{2x}{a}\right)$	$\sqrt{\pi}\, a^{s-1/2}\left(\dfrac{1}{2}-s\right)_n \Gamma\left[\begin{matrix} s \\ \frac{2s+2n+1}{2} \end{matrix}\right] {}_2F_2\left(\begin{matrix} s,\, \frac{2s+1}{2};\, ab \\ \frac{2s-2n+1}{2},\, \frac{2s+2n+1}{2} \end{matrix}\right)$ $[a,\ \operatorname{Re} s>0]$
3	$e^{-bx}\, T_n\left(\dfrac{2x}{a}\pm 1\right)$	$\dfrac{2^{2n-1}(\delta_{n,0}+1)}{a^n b^{s+n}}\Gamma(s+n)\; {}_2F_2\left(\begin{matrix} -n,\, \frac{1-2n}{2};\, \pm ab \\ 1-2n,\, 1-s-n \end{matrix}\right)$ $[\operatorname{Re} b,\ \operatorname{Re} s>0]$
4	$\dfrac{e^{-bx}}{\sqrt{x+a}}\, T_n\left(\dfrac{2x}{a}+1\right)$	$\dfrac{(-1)^n a^{s-1/2}}{\sqrt{\pi}}\left(\dfrac{1-2s}{2}\right)_n \Gamma(s)\,\Gamma\left(\dfrac{1-2s-2n}{2}\right)$ $\times {}_2F_2\left(\begin{matrix} s,\, \frac{2s+1}{2};\, ab \\ \frac{2s-2n+1}{2},\, \frac{2s+2n+1}{2} \end{matrix}\right)$ $+\dfrac{2^{2n-1}(\delta_{n,0}+1)}{a^n b^{s+n-1/2}}\Gamma\left(\dfrac{2s+2n-1}{2}\right)$ $\times {}_2F_2\left(\begin{matrix} \frac{1-2n}{2},\, 1-n;\, ab \\ 1-2n,\, \frac{3-2s-2n}{2} \end{matrix}\right)$ $[\operatorname{Re} b,\ \operatorname{Re} s>0;\ \lvert\arg a\rvert<\pi]$

No.	$f(x)$	$F(s)$
5	$(a-x)_+^{-1/2}\, e^{-b/x} \times T_n\left(\frac{2x}{a}-1\right)$	$(-1)^n \sqrt{\pi}\, a^{s-1/2} \left(\frac{1-2s}{2}\right)_n \Gamma\left[\begin{matrix} s \\ \frac{2s+2n+1}{2}\end{matrix}\right] \times {}_2F_2\left(\begin{matrix}\frac{-2s-2n+1}{2},\ \frac{-2s+2n+1}{2} \\ 1-s,\ \frac{1-2s}{2};\ -\frac{b}{a}\end{matrix}\right) + (-1)^n a^{-1/2} b^s\, \Gamma(-s)\, {}_2F_2\left(\begin{matrix}\frac{-2n+1}{2},\ \frac{2n+1}{2} \\ \frac{1}{2},\ s+1;\ -\frac{b}{a}\end{matrix}\right)$ $[a,\ \operatorname{Re} b>0]$
6	$\frac{e^{-b/x}}{\sqrt{x+a}}\, T_n\left(\frac{2x}{a}+1\right)$	$\frac{a^{s-1/2}}{\sqrt{\pi}} \left(\frac{1-2s}{2}\right)_n \Gamma\left(\frac{1-2s-2n}{2}\right) \Gamma(s) \times {}_2F_2\left(\begin{matrix}\frac{1-2s-2n}{2},\ \frac{1-2s+2n}{2} \\ 1-s,\ \frac{1-2s}{2};\ \frac{b}{a}\end{matrix}\right) + \frac{b^s}{\sqrt{a}}\, \Gamma(-s)\, {}_2F_2\left(\begin{matrix}\frac{-2n+1}{2},\ \frac{2n+1}{2} \\ \frac{1}{2},\ s+1;\ \frac{b}{a}\end{matrix}\right)$ $[\operatorname{Re} b>0;\ \operatorname{Re} s<1/2-n;\ \lvert\arg a\rvert<\pi]$
7	$(a-x)_+^{-1/2}\, e^{-b\sqrt{x}} \times T_n\left(1-\frac{2x}{a}\right)$	$\sqrt{\pi}\, a^{s-1/2} \left(\frac{1}{2}-s\right)_n \Gamma\left[\begin{matrix} s \\ \frac{2s+2n+1}{2}\end{matrix}\right] \times {}_2F_3\left(\begin{matrix}s,\ \frac{2s+1}{2};\ \frac{ab^2}{4} \\ \frac{1}{2},\ \frac{2s-2n+1}{2},\ \frac{2s+2n+1}{2}\end{matrix}\right) - \sqrt{\pi}\, a^s b\, (-s)_n \times \Gamma\left[\begin{matrix}\frac{2s+1}{2} \\ s+n+1\end{matrix}\right] {}_2F_3\left(\begin{matrix}\frac{2s+1}{2},\ s+1;\ \frac{ab^2}{4} \\ \frac{3}{2},\ s-n+1,\ s+n+1\end{matrix}\right)$ $[a,\ \operatorname{Re} s>0]$
8	$e^{-b\sqrt{x}}\, T_n\left(\frac{2x}{a}\pm 1\right)$	$\frac{2^{2n}(\delta_{n,0}+1)}{a^n b^{2s+2n}}\, \Gamma(2s+2n)\, {}_2F_3\left(\begin{matrix}-n,\ \frac{1-2n}{2};\ \mp\frac{ab^2}{4} \\ 1-2n,\ \frac{1-2s-2n}{2},\ 1-s-n\end{matrix}\right)$ $[\operatorname{Re} b,\ \operatorname{Re} s>0]$
9	$\frac{e^{-b\sqrt{x}}}{\sqrt{x+a}}\, T_n\left(\frac{2x}{a}+1\right)$	$\frac{(-1)^{n+1} a^s b}{\sqrt{\pi}} (-s)_n\, \Gamma(-s-n)\, \Gamma\left(\frac{2s+1}{2}\right) \times {}_2F_3\left(\begin{matrix}\frac{2s+1}{2},\ s+1;\ -\frac{ab^2}{4} \\ \frac{3}{2},\ s-n+1,\ s+n+1\end{matrix}\right) + \frac{(-1)^n a^{s-1/2}}{\sqrt{\pi}} \left(\frac{1}{2}-s\right)_n \Gamma(s)\, \Gamma\left(\frac{1-2s-2n}{2}\right) \times {}_2F_3\left(\begin{matrix}s,\ \frac{2s+1}{2};\ -\frac{ab^2}{4} \\ \frac{1}{2},\ \frac{2s-2n+1}{2},\ \frac{2s+2n+1}{2}\end{matrix}\right) + \frac{2^{2n}(\delta_{n,0}+1)}{a^n b^{2s+2n-1}} \times \Gamma(2s+2n-1)\, {}_2F_3\left(\begin{matrix}\frac{1-2n}{2},\ 1-n;\ -\frac{ab^2}{4} \\ 1-2n,\ 1-s-n,\ \frac{3-2s-2n}{2}\end{matrix}\right)$ $[\operatorname{Re} b,\ \operatorname{Re} s>0;\ \lvert\arg a\rvert<\pi]$

No.	$f(x)$	$F(s)$
10	$(a-x)_+^{-1/2}\, e^{-b/\sqrt{x}} \times T_n\left(\frac{2x}{a}-1\right)$	$(-1)^{n+1}\sqrt{\pi}\, a^{s-1} b\,(1-s)_n\, \Gamma\left[\begin{matrix}\frac{2s-1}{2}\\ s+n\end{matrix}\right] \times {}_2F_3\left(\begin{matrix}1-s-n,\ 1-s+n\\ \frac{3}{2},\ \frac{3-2s}{2},\ 1-s;\ \frac{b^2}{4a}\end{matrix}\right) + (-1)^n \sqrt{\pi}\, a^{s-1/2}\left(\frac{1-2s}{2}\right)_n \Gamma\left[\begin{matrix}s\\ \frac{2s+2n+1}{2}\end{matrix}\right] \times {}_2F_3\left(\begin{matrix}\frac{1-2s-2n}{2},\ \frac{1-2s+2n}{2}\\ \frac{1}{2},\ \frac{1-2s}{2},\ 1-s;\ \frac{b^2}{4a}\end{matrix}\right) + 2(-1)^n a^{-1/2} b^{2s}\Gamma(-2s)\, {}_2F_3\left(\begin{matrix}\frac{-2n+1}{2},\ \frac{2n+1}{2}\\ \frac{1}{2},\ \frac{2s+1}{2},\ s+1;\ \frac{b^2}{4a}\end{matrix}\right)$ $[a,\ \operatorname{Re} b>0]$
11	$e^{-b/\sqrt{x}}\, T_n\left(\frac{2x}{a}\pm 1\right)$	$2(\pm 1)^n b^{2s}\,\Gamma(-2s)\, {}_2F_3\left(\begin{matrix}-n,\ n;\ \mp\frac{b^2}{4a}\\ \frac{1}{2},\ \frac{2s+1}{2},\ s+1\end{matrix}\right)$ $[\operatorname{Re} b>0;\ \operatorname{Re} s<-n]$
12	$\frac{e^{-b/\sqrt{x}}}{\sqrt{x+a}}\, T_n\left(\frac{2x}{a}+1\right)$	$\frac{(-1)^n a^{s-1/2}}{\sqrt{\pi}}\left(\frac{1-2s}{2}\right)_n \Gamma\left(\frac{1-2s-2n}{2}\right) \times \Gamma(s)\, {}_2F_2\left(\begin{matrix}\frac{1-2s-2n}{2},\ \frac{1-2s+2n}{2}\\ \frac{1-2s}{2},\ 1-s;\ -\frac{b^2}{4a}\end{matrix}\right) + \frac{2b^{2s}}{\sqrt{a}}\Gamma(-2s)\, {}_2F_3\left(\begin{matrix}\frac{-2n+1}{2},\ \frac{2n+1}{2};\ -\frac{b^2}{4a}\\ \frac{1}{2},\ \frac{2s+1}{2},\ s+1\end{matrix}\right)$ $[\operatorname{Re} b>0;\ \operatorname{Re} s<1/2-n;\ \lvert\arg a\rvert<\pi]$
13	$\frac{e^{-bx}}{(x+a)^n}\, T_n\left(\frac{a-x}{a+x}\right)$	$\frac{a^{s-n}}{n!}\frac{\left(\frac{1}{2}-s\right)_n}{\left(\frac{1}{2}\right)_n}\Gamma(s)\,\Gamma(n-s)\, {}_2F_2\left(\begin{matrix}s,\ \frac{2s+1}{2};\ ab\\ s-n+1,\ \frac{2s-2n+1}{2}\end{matrix}\right) + (-1)^n b^{-s+n}\Gamma(s-n)\, {}_2F_2\left(\begin{matrix}n+\frac{1}{2},\ n;\ ab\\ \frac{1}{2},\ n-s+1\end{matrix}\right)$ $[\operatorname{Re} b,\ \operatorname{Re} s>0;\ \lvert\arg a\rvert<\pi]$
14	$(x+a)^n e^{-b/x} T_n\left(\frac{a-x}{a+x}\right)$	$a^n b^s\,\Gamma(-s)\, {}_2F_2\left(\begin{matrix}-n,\ \frac{1-2n}{2}\\ \frac{1}{2},\ s+1;\ \frac{b}{a}\end{matrix}\right)$ $[\operatorname{Re} b>0;\ \operatorname{Re} s<-n]$
15	$\frac{e^{-b/x}}{(x+a)^n}\, T_n\left(\frac{a-x}{a+x}\right)$	$\frac{na^{s-n}}{n!}\frac{\left(\frac{1-2s}{2}\right)_n}{\left(\frac{1}{2}\right)_n}\Gamma(n-s)\,\Gamma(s)\, {}_2F_2\left(\begin{matrix}n-s,\ \frac{1-2s+2n}{2}\\ 1-s,\ \frac{1-2s}{2};\ \frac{b}{a}\end{matrix}\right) + a^{-n} b^s\,\Gamma(-s)\, {}_2F_2\left(\begin{matrix}n,\ \frac{2n+1}{2}\\ \frac{1}{2},\ s+1;\ \frac{b}{a}\end{matrix}\right)$ $[\operatorname{Re} b>0;\ \operatorname{Re} s<n;\ \lvert\arg a\rvert<\pi]$

No.	$f(x)$	$F(s)$
16	$(x+a)^n e^{-b\sqrt{x}} T_n\left(\frac{a-x}{a+x}\right)$	$2(-1)^n b^{-2s-2n} \Gamma(2s+2n)\ {}_2F_3\left(\begin{matrix} -n,\ \frac{1-2n}{2};\ -\frac{ab^2}{4} \\ \frac{1}{2},\ \frac{1-2s-2n}{2},\ 1-s-n \end{matrix}\right)$ [$\operatorname{Re} b,\ \operatorname{Re} s > 0$]
17	$\frac{e^{-b\sqrt{x}}}{(x+a)^n} T_n\left(\frac{a-x}{a+x}\right)$	$-\frac{na^{s-n+1/2}b}{n!}\frac{(-s)_n}{\left(\frac{1}{2}\right)_n}\Gamma\left(\frac{2s+1}{2}\right)\Gamma\left(\frac{2n-2s-1}{2}\right) \times {}_2F_3\left(\begin{matrix} \frac{2s+1}{2},\ s+1;\ -\frac{ab^2}{4} \\ \frac{3}{2},\ s-n+1,\ \frac{2s-2n+3}{2} \end{matrix}\right) + \frac{na^{s-n}}{n!}\frac{\left(\frac{1-2s}{2}\right)_n}{\left(\frac{1}{2}\right)_n} \times \Gamma(n-s)\Gamma(s)\ {}_2F_3\left(\begin{matrix} s,\ \frac{2s+1}{2};\ -\frac{ab^2}{4} \\ \frac{1}{2},\ s-n+1,\ \frac{2s-2n+1}{2} \end{matrix}\right) + \frac{2(-1)^n}{b^{2(s-n)}}\Gamma(2s-2n)\ {}_2F_3\left(\begin{matrix} n,\ \frac{2n+1}{2};\ -\frac{ab^2}{4} \\ \frac{1}{2},\ \frac{1-2s+2n}{2},\ 1-s+n \end{matrix}\right)$ [$\operatorname{Re} b,\ \operatorname{Re} s > 0;\ \lvert\arg a\rvert < \pi$]
18	$(x+a)^n e^{-b/\sqrt{x}} T_n\left(\frac{a-x}{a+x}\right)$	$2a^n b^{2s}\Gamma(-2s)\ {}_2F_3\left(\begin{matrix} -n,\ \frac{-2n+1}{2};\ -\frac{b^2}{4a} \\ \frac{1}{2},\ \frac{2s+1}{2},\ s+1 \end{matrix}\right)$ [$\operatorname{Re} b > 0;\ \operatorname{Re} s < -n$]
19	$\frac{e^{-b/\sqrt{x}}}{(x+a)^n} T_n\left(\frac{a-x}{a+x}\right)$	$-\frac{na^{s-n-1/2}b}{n!}\frac{(1-s)_n}{\left(\frac{1}{2}\right)_n}\Gamma\left(\frac{1-2s+2n}{2}\right)\Gamma\left(\frac{2s-1}{2}\right) \times {}_2F_3\left(\begin{matrix} n-s+1,\ \frac{1-2s+2n}{2} \\ \frac{3}{2},\ \frac{3}{2}-s,\ 1-s;\ -\frac{b^2}{4a} \end{matrix}\right) + \frac{na^{s-n}}{n!}\frac{\left(\frac{1-2s}{2}\right)_n}{\left(\frac{1}{2}\right)_n} \times \Gamma(n-s)\Gamma(s)\ {}_2F_3\left(\begin{matrix} \frac{1-2s+2n}{2},\ n-s \\ \frac{1}{2},\ 1-s,\ \frac{1-2s}{2};\ -\frac{b^2}{4a} \end{matrix}\right) + 2a^{-n}b^{2s}\Gamma(-2s)\ {}_2F_3\left(\begin{matrix} n,\ \frac{2n+1}{2};\ -\frac{b^2}{4a} \\ \frac{1}{2},\ \frac{2s+1}{2},\ s+1 \end{matrix}\right)$ [$\operatorname{Re} b > 0;\ \operatorname{Re} s < n;\ \lvert\arg a\rvert < \pi$]

3.20.3. $T_n(bx)$ and hyperbolic functions

Notation: $\delta = \begin{Bmatrix} 1 \\ 0 \end{Bmatrix}$.

1	$(a^2-x^2)_+^{-1/2}\begin{Bmatrix} \sinh(bx) \\ \cosh(bx) \end{Bmatrix} \times T_n\left(\frac{x}{a}\right)$	$\frac{\pi a^{s+\delta-1}b^\delta}{2^{s+\delta}}\Gamma\left[\begin{matrix} s+\delta \\ \frac{s-n+\delta+1}{2},\ \frac{s+n+\delta+1}{2} \end{matrix}\right] \times {}_2F_3\left(\begin{matrix} \frac{s+\delta}{2},\ \frac{s+\delta+1}{2};\ \frac{a^2b^2}{4} \\ \frac{2\delta+1}{2},\ \frac{s-n+\delta+1}{2},\ \frac{s+n+\delta+1}{2} \end{matrix}\right)$ [$a > 0;\ \operatorname{Re} s > ((-1)^n - 2\delta - 1)/2$]

3.20.4. $T_n(ax+b)$ and trigonometric functions

Notation: $\delta = \begin{Bmatrix} 1 \\ 0 \end{Bmatrix}$.

1	$(a^2-x^2)_+^{-1/2} \begin{Bmatrix} \sin(bx) \\ \cos(bx) \end{Bmatrix} \times T_n\left(\frac{x}{a}\right)$	$\frac{\pi\, a^{s+\delta-1} b^\delta}{2^{s+\delta}} \Gamma\left[\begin{matrix} s+\delta \\ \frac{s-n+\delta+1}{2},\ \frac{s+n+\delta+1}{2} \end{matrix}\right] \times {}_2F_3\left(\begin{matrix} \frac{s+\delta}{2},\ \frac{s+\delta+1}{2};\ -\frac{a^2b^2}{4} \\ \frac{2\delta+1}{2},\ \frac{s-n+\delta+1}{2},\ \frac{s+n+\delta+1}{2} \end{matrix}\right)$ $[a>0;\ \operatorname{Re} s > ((-1)^n - 2\delta - 1)/2]$
2	$(a-x)_+^{-1/2} \begin{Bmatrix} \sin(b\sqrt{x}) \\ \cos(b\sqrt{x}) \end{Bmatrix} \times T_n\left(\frac{2x}{a}-1\right)$	$(-1)^n \sqrt{\pi}\, a^{s+(\delta-1)/2} b^\delta \left(\frac{1-2s-\delta}{2}\right)_n \times \Gamma\left[\begin{matrix} \frac{2s+\delta}{2} \\ \frac{2s+2n+\delta+1}{2} \end{matrix}\right] {}_2F_3\left(\begin{matrix} \frac{2s+\delta}{2},\ \frac{2s+\delta+1}{2};\ -\frac{ab^2}{4} \\ \frac{2\delta+1}{2},\ \frac{2s-2n+\delta+1}{2},\ \frac{2s+2n+\delta+1}{2} \end{matrix}\right)$ $[a>0;\ \operatorname{Re} s > -\delta/2]$

3.20.5. $T_n(ax+b)$ and the logarithmic function

Notation: $\varepsilon = 0$ or 1.

1	$(a-x)_+^{-1/2} \ln\frac{x}{a}\, T_n\left(\frac{x}{a}\right)$	$\sqrt{\pi}\, a^{s-1/2} \Gamma\left[\begin{matrix} s \\ \frac{2s+1}{2} \end{matrix}\right] \sum_{k=0}^{n} \frac{(-n)_k (n)_k}{2^k k! \left(\frac{2s+1}{2}\right)_k} \left[\psi(s) - \psi\left(\frac{2s+2k+1}{2}\right)\right]$ $[a>0;\ \operatorname{Re} s > ((-1)^n - 1)/2]$
2	$(a^2-x^2)_+^{-1/2} \ln(bx^2+1) \times T_n\left(\frac{x}{a}\right)$	$\frac{\pi}{2}\left(\frac{a}{2}\right)^{s+1} b\, \Gamma\left[\begin{matrix} s+2 \\ \frac{s-n+3}{2},\ \frac{s+n+3}{2} \end{matrix}\right] {}_4F_3\left(\begin{matrix} 1,\ 1,\ \frac{s+2}{2},\ \frac{s+3}{2};\ -a^2b \\ 2,\ \frac{s-n+3}{2},\ \frac{s+n+3}{2} \end{matrix}\right)$ $[a>0;\ \operatorname{Re} s > ((-1)^n - 5)/2;\ \lvert\arg b\rvert < \pi]$
3	$(a^2-x^2)_+^{-1/2} \ln\frac{b+x}{b-x} \times T_n\left(\frac{x}{a}\right)$	$\pi\left(\frac{a}{2}\right)^{s} b^{-1}\, \Gamma\left[\begin{matrix} s+1 \\ \frac{s-n+2}{2},\ \frac{s+n+2}{2} \end{matrix}\right] {}_4F_3\left(\begin{matrix} \frac{1}{2},\ 1,\ \frac{s+1}{2},\ \frac{s+2}{2};\ \frac{a^2}{b^2} \\ \frac{3}{2},\ \frac{s-n+2}{2},\ \frac{s+n+2}{2} \end{matrix}\right)$ $[a>0;\ \operatorname{Re} s > ((-1)^n - 1)/2;\ \lvert\arg b\rvert < \pi]$
4	$(a^2-x^2)_+^{-1/2} \times \ln\left(\sqrt{b^2x^2+1}+bx\right) \times T_n\left(\frac{x}{a}\right)$	$\frac{\pi}{2}\left(\frac{a}{2}\right)^{s} b\, \Gamma\left[\begin{matrix} s+1 \\ \frac{s-n+2}{2},\ \frac{s+n+2}{2} \end{matrix}\right] {}_4F_3\left(\begin{matrix} \frac{1}{2},\ \frac{1}{2},\ \frac{s+1}{2},\ \frac{s+2}{2};\ -a^2b^2 \\ \frac{3}{2},\ \frac{s-n+2}{2},\ \frac{s+n+2}{2} \end{matrix}\right)$ $[a,\ b>0;\ \operatorname{Re} s > ((-1)^n - 3)/2]$

No.	$f(x)$	$F(s)$
5	$\frac{(a^2-x^2)_+^{-1/2}}{\sqrt{b^2x^2+1}} T_n\left(\frac{x}{a}\right)$ $\times \ln\left(bx+\sqrt{b^2x^2+1}\right)$	$\frac{\pi}{2}\left(\frac{a}{2}\right)^s b\,\Gamma\left[\begin{matrix} s+1 \\ \frac{s-n+2}{2},\ \frac{s+n+2}{2} \end{matrix}\right] {}_4F_3\left(\begin{matrix} 1,\ 1,\ \frac{s+1}{2},\ \frac{s+2}{2};\ -a^2b^2 \\ \frac{3}{2},\ \frac{s-n+2}{2},\ \frac{s+n+2}{2} \end{matrix}\right)$ $[a,\ b>0;\ \operatorname{Re} s > ((-1)^n-3)/2]$
6	$(a^2-x^2)_+^{-1/2}$ $\times \ln^2\left(bx+\sqrt{b^2x^2+1}\right)$ $\times T_n\left(\frac{x}{a}\right)$	$\frac{\pi}{2}\left(\frac{a}{2}\right)^{s+1} b^2\,\Gamma\left[\begin{matrix} s+2 \\ \frac{s-n+3}{2},\ \frac{s+n+3}{2} \end{matrix}\right] {}_5F_4\left(\begin{matrix} 1,\ 1,\ 1,\ \frac{s+2}{2},\ \frac{s+3}{2};\ -a^2b^2 \\ \frac{3}{2},\ 2,\ \frac{s-n+3}{2},\ \frac{s+n+3}{2} \end{matrix}\right)$ $[a,\ b>0;\ \operatorname{Re} s > ((-1)^n-5)/2]$
7	$\theta(a-x)\ln\frac{\sqrt{a}+\sqrt{a-x}}{\sqrt{x}}$ $\times T_{2n+\varepsilon}(bx)$	$\frac{(-1)^n\sqrt{\pi}\,(2n+1)^{\varepsilon}a^{s+\varepsilon}b^{\varepsilon}}{2(s+\varepsilon)}\Gamma\left[\begin{matrix} s+\varepsilon \\ \frac{2s+2\varepsilon+1}{2} \end{matrix}\right]$ $\times {}_5F_4\left(\begin{matrix} -n,\ n+\varepsilon,\ \frac{s+\varepsilon}{2},\ \frac{s+\varepsilon}{2},\ \frac{s+\varepsilon+1}{2};\ a^2b^2 \\ \frac{2\varepsilon+1}{2},\ \frac{2s+2\varepsilon+1}{4},\ \frac{2s+2\varepsilon+3}{4},\ \frac{s+\varepsilon+2}{2} \end{matrix}\right)$ $[a>0;\ \operatorname{Re} s > -\varepsilon]$
8	$(a-x)_+^{-1/2}\ln\frac{x}{a}$ $\times T_n\left(\frac{2x}{a}-1\right)$	$(-1)^n\sqrt{\pi}\,a^{s-1/2}\left(\frac{1-2s}{2}\right)_n \Gamma\left[\begin{matrix} s \\ \frac{2s+2n+1}{2} \end{matrix}\right]\Big[\psi(s)$ $+\psi\left(\frac{1-2s}{2}\right)-\psi\left(\frac{2s+2n+1}{2}\right)-\psi\left(\frac{1-2s+2n}{2}\right)\Big]$ $[a,\ \operatorname{Re} s>0]$

3.20.6. $T_n(bx)$ and inverse trigonometric functions

Notation: $\varepsilon = 0$ or 1.

No.	$f(x)$	$F(s)$
1	$(a^2-x^2)_+^{-1/2}$ $\times \arcsin(bx)\,T_n\left(\frac{x}{a}\right)$	$\frac{\pi}{2}\left(\frac{a}{2}\right)^s b\,\Gamma\left[\begin{matrix} s+1 \\ \frac{s-n+2}{2},\ \frac{s+n+2}{2} \end{matrix}\right] {}_4F_3\left(\begin{matrix} \frac{1}{2},\ \frac{1}{2},\ \frac{s+1}{2},\ \frac{s+2}{2};\ a^2b^2 \\ \frac{3}{2},\ \frac{s-n+2}{2},\ \frac{s+n+2}{2} \end{matrix}\right)$ $[a>0;\ \operatorname{Re} s > ((-1)^n-3)/2]$
2	$\frac{(a^2-x^2)_+^{-1/2}}{\sqrt{1-b^2x^2}}$ $\times \arcsin(bx)\,T_n\left(\frac{x}{a}\right)$	$\frac{\pi}{2}\left(\frac{a}{2}\right)^s b\,\Gamma\left[\begin{matrix} s+1 \\ \frac{s-n+2}{2},\ \frac{s+n+2}{2} \end{matrix}\right] {}_4F_3\left(\begin{matrix} 1,\ 1,\ \frac{s+1}{2},\ \frac{s+2}{2};\ a^2b^2 \\ \frac{3}{2},\ \frac{s-n+2}{2},\ \frac{s+n+2}{2} \end{matrix}\right)$ $[a>0;\ \operatorname{Re} s > ((-1)^n-3)/2]$
3	$(a^2-x^2)_+^{-1/2}$ $\times \arctan(bx)\,T_n\left(\frac{x}{a}\right)$	$\frac{\pi}{2}\left(\frac{a}{2}\right)^s b\,\Gamma\left[\begin{matrix} s+1 \\ \frac{s-n+2}{2},\ \frac{s+n+2}{2} \end{matrix}\right] {}_4F_3\left(\begin{matrix} \frac{1}{2},\ 1,\ \frac{s+1}{2},\ \frac{s+2}{2};\ -a^2b^2 \\ \frac{3}{2},\ \frac{s-n+2}{2},\ \frac{s+n+2}{2} \end{matrix}\right)$ $[a>0;\ \operatorname{Re} s > ((-1)^n-3)/2]$

No.	$f(x)$	$F(s)$
4	$\left(a^2-x^2\right)_+^{-1/2}$ $\times \arcsin^2(bx)\,T_n\left(\frac{x}{a}\right)$	$\frac{\pi}{2}\left(\frac{a}{2}\right)^{s+1} b^2\,\Gamma\left[\begin{matrix} s+2 \\ \frac{s-n+3}{2},\ \frac{s+n+3}{2}\end{matrix}\right] {}_5F_4\left(\begin{matrix}1,\ 1,\ 1,\ \frac{s+2}{2},\ \frac{s+3}{2};\ a^2b^2 \\ \frac{3}{2},\ 2,\ \frac{s-n+3}{2},\ \frac{s+n+3}{2}\end{matrix}\right)$ $\left[a>0;\ \operatorname{Re} s>\left((-1)^n-5\right)/2\right]$
5	$\theta(a-x)\arccos\frac{x}{a}$ $\times T_{2n+\varepsilon}(bx)$	$\frac{(-1)^n\sqrt{\pi}\,(2n+1)^{\varepsilon}\,a^{s+\varepsilon}b^{\varepsilon}}{2(s+\varepsilon)}\,\Gamma\left[\begin{matrix}\frac{s+\varepsilon+1}{2} \\ \frac{s+\varepsilon+2}{2}\end{matrix}\right]$ $\times {}_4F_3\left(\begin{matrix}-n,\ n+\varepsilon,\ \frac{s+\varepsilon}{2},\ \frac{s+\varepsilon+1}{2};\ a^2b^2 \\ \frac{2\varepsilon+1}{2},\ \frac{s+\varepsilon+2}{2},\ \frac{s+\varepsilon+2}{2}\end{matrix}\right)$ $[a>0;\ \operatorname{Re} s>-\varepsilon]$

3.20.7. $T_n(ax+b)$ and $\operatorname{Ei}(cx^r)$

1	$\left(a^2-x^2\right)_+^{-1/2}\operatorname{Ei}(bx)$ $\times T_n\left(\frac{x}{a}\right)$	$\frac{\pi a^s b}{2^{s+1}}\,\Gamma\left[\begin{matrix}s+1 \\ \frac{s-n+2}{2},\ \frac{s+n+2}{2}\end{matrix}\right] {}_3F_4\left(\begin{matrix}\frac{1}{2},\ \frac{s+1}{2},\ \frac{s+2}{2};\ \frac{a^2b^2}{4} \\ \frac{3}{2},\ \frac{3}{2},\ \frac{s-n+2}{2},\ \frac{s+n+2}{2}\end{matrix}\right)$ $+\frac{\pi a^{s+1}b^2}{2^{s+4}}\,\Gamma\left[\begin{matrix}s+2 \\ \frac{s-n+3}{2},\ \frac{s+n+5}{2}\end{matrix}\right] {}_3F_4\left(\begin{matrix}1,\ 1,\ \frac{s+2}{2},\ \frac{s+3}{2};\ \frac{a^2b^2}{4} \\ \frac{3}{2},\ 2,\ 2,\ \frac{s-n+3}{2},\ \frac{s+n+3}{2}\end{matrix}\right)$ $+\frac{\pi a^{s-1}}{2^s}\,\Gamma\left[\begin{matrix}s \\ \frac{s-n+1}{2},\ \frac{s+n+1}{2}\end{matrix}\right]\left[\ln\frac{ab}{2}-\frac{1}{2}\psi\left(\frac{s+n+1}{2}\right)\right.$ $\left.-\frac{1}{2}\psi\left(\frac{s-n+1}{2}\right)+\psi(s)+\mathbf{C}\right]$ $\left[a>0;\ \operatorname{Re} s>\left((-1)^n-1\right)/2\right]$
2	$\left(a^2-x^2\right)_+^{-1/2}\operatorname{Ei}\left(bx^2\right)$ $\times T_n\left(\frac{x}{a}\right)$	$\frac{\pi a^{s+1}b}{2^{s+2}}\,\Gamma\left[\begin{matrix}s+2 \\ \frac{s-n+3}{2},\ \frac{s+n+3}{2}\end{matrix}\right] {}_4F_4\left(\begin{matrix}1,\ 1,\ \frac{s+2}{2},\ \frac{s+3}{2};\ a^2b \\ 2,\ 2,\ \frac{s-n+3}{2},\ \frac{s+n+3}{2}\end{matrix}\right)$ $+\frac{\pi a^{s-1}}{2^s}\,\Gamma\left[\begin{matrix}s \\ \frac{s-n+1}{2},\ \frac{s+n+1}{2}\end{matrix}\right]\left[\ln\frac{a^2b}{4}-\psi\left(\frac{s+n+1}{2}\right)\right.$ $\left.-\psi\left(\frac{s-n+1}{2}\right)+2\psi(s)+\mathbf{C}\right]$ $\left[a>0;\ \operatorname{Re} s>\left((-1)^n-1\right)/2\right]$
3	$(a-x)_+^{-1/2}\operatorname{Ei}(-bx)$ $\times T_n\left(\frac{2x}{a}-1\right)$	$(-1)^{n+1}\sqrt{\pi}\,a^{s+1/2}b\left(\frac{-2s-1}{2}\right)_n\Gamma\left[\begin{matrix}s+1 \\ \frac{2s+2n+3}{2}\end{matrix}\right]$ $\times {}_4F_4\left(\begin{matrix}1,\ 1,\ \frac{2s+3}{2},\ s+1;\ -ab \\ 2,\ 2,\ \frac{2s-2n+3}{2},\ \frac{2s+2n+3}{2}\end{matrix}\right)$ $+(-1)^n\sqrt{\pi}\,a^{s-1/2}\left(\frac{-2s+1}{2}\right)_n\Gamma\left[\begin{matrix}s \\ \frac{2s+2n+1}{2}\end{matrix}\right]$ $\times\left[\psi(s)-\psi\left(\frac{2s+2n+1}{2}\right)-\sum_{i=0}^{n-1}\frac{2}{2i-2s+1}+\ln(ab)+\mathbf{C}\right]$ $[a,\ \operatorname{Re} s>0]$

3.20.8. $T_n(ax+b)$ and $\operatorname{si}(cx^r)$, $\operatorname{ci}(cx^r)$

1	$(a-x)_+^{-1/2}\begin{Bmatrix}\operatorname{si}(b\sqrt{x})\\ \operatorname{ci}(b\sqrt{x})\end{Bmatrix} \times T_n\left(\frac{2x}{a}-1\right)$	$\frac{(-1)^{n+1}2^{\delta-2}\sqrt{\pi}\,a^{s+(\delta+1)/2}b^{\delta+2}}{3^{2\delta}}\left(\frac{-2s-\delta-1}{2}\right)_n\Gamma\left[\begin{matrix}\frac{2s+\delta+2}{2}\\ \frac{2s+2n+\delta+3}{2}\end{matrix}\right] \times {}_4F_5\left(\begin{matrix}1,\ \frac{\delta+2}{2},\ \frac{2s+\delta+2}{2},\ \frac{2s+\delta+3}{2};\ -\frac{ab^2}{4}\\ 2,\ \frac{\delta+4}{2},\ \frac{2\delta+3}{2},\ \frac{2s-2n+\delta+3}{2},\ \frac{2s+2n+\delta+3}{2}\end{matrix}\right) + (-1)^n\sqrt{\pi}\,a^{s+(\delta-1)/2}b^{\delta}\left(\frac{-2s-\delta+1}{2}\right)_n\Gamma\left[\begin{matrix}\frac{2s+\delta}{2}\\ \frac{2s+2n+\delta+1}{2}\end{matrix}\right] \times\left[\frac{1}{2}\psi(s)-\frac{1}{2}\psi\left(\frac{2s+2n+1}{2}\right)-\sum_{i=0}^{n-1}\frac{1}{2i-2s+1} + \frac{1}{2}\ln(ab^2)+\mathbf{C}\right]^{1-\delta}-\delta\frac{(-1)^n\pi^{3/2}a^{s-1/2}}{2} \times\left(\frac{1-2s}{2}\right)_n\Gamma\left[\begin{matrix}s\\ \frac{2s+2n+1}{2}\end{matrix}\right]$ $\left[a,\ \operatorname{Re}s>0;\ \delta=\begin{Bmatrix}1\\0\end{Bmatrix}\right]$

3.20.9. $T_n(ax+b)$ and $\operatorname{erf}(cx^r)$, $\operatorname{erfc}(cx^r)$

Notation: $\varepsilon = 0$ or 1.

1	$(a^2-x^2)_+^{-1/2} \times \operatorname{erf}(bx)\,T_n\left(\frac{x}{a}\right)$	$\frac{\sqrt{\pi}\,a^s b}{2^s}\Gamma\left[\begin{matrix}s+1\\ \frac{s-n+2}{2},\ \frac{s+n+2}{2}\end{matrix}\right]{}_3F_3\left(\begin{matrix}\frac{1}{2},\ \frac{s+1}{2},\ \frac{s+2}{2};\ -a^2b^2\\ \frac{3}{2},\ \frac{s-n+2}{2},\ \frac{s+n+2}{2}\end{matrix}\right)$ $[a>0;\ \operatorname{Re}s>((-1)^n-3)/2]$
2	$(a^2-x^2)_+^{-1/2}e^{b^2x^2} \times \operatorname{erf}(bx)\,T_n\left(\frac{x}{a}\right)$	$\frac{\sqrt{\pi}\,a^s b}{2^s}\Gamma\left[\begin{matrix}s+1\\ \frac{s-n+2}{2},\ \frac{s+n+2}{2}\end{matrix}\right]{}_3F_3\left(\begin{matrix}1,\ \frac{s+1}{2},\ \frac{s+2}{2};\ a^2b^2\\ \frac{3}{2},\ \frac{s-n+2}{2},\ \frac{s+n+2}{2}\end{matrix}\right)$ $[a>0;\ \operatorname{Re}s>((-1)^n-3)/2]$
3	$\operatorname{erfc}(ax)\,T_{2n+\varepsilon}(bx)$	$\frac{(-1)^n(2n+1)^{\varepsilon}b^{\varepsilon}}{\sqrt{\pi}\,(s+\varepsilon)\,a^{s+\varepsilon}}\Gamma\left(\frac{s+\varepsilon+1}{2}\right){}_3F_2\left(\begin{matrix}-n,\ n+\varepsilon,\ \frac{s+1}{2},\ \frac{s+2\varepsilon}{2}\\ \frac{2\varepsilon+1}{2},\ \frac{s+\varepsilon+2}{2};\ \frac{b^2}{a^2}\end{matrix}\right)$ $[\operatorname{Re}s>-\varepsilon;\ \lvert\arg a\rvert<\pi/4]$
4	$\operatorname{erfc}(bx)\,T_n\left(\frac{2x}{a}+1\right)$	$\frac{2n^2a^{-1}b^{-s-1}}{\sqrt{\pi}\,(s+1)}\Gamma\left(\frac{s+2}{2}\right){}_6F_4\left(\begin{matrix}\frac{1-n}{2},\ \frac{2-n}{2},\ \frac{n+1}{2},\ \frac{n+2}{2},\ \frac{s+1}{2},\ \frac{s+2}{2}\\ \frac{3}{2},\ \frac{3}{4},\ \frac{5}{4},\ \frac{s+3}{2};\ \frac{1}{a^2b^2}\end{matrix}\right) + \frac{b^{-s}}{\sqrt{\pi}\,s}\Gamma\left(\frac{s+1}{2}\right){}_6F_4\left(\begin{matrix}-\frac{n}{2},\ \frac{1-n}{2},\ \frac{n}{2},\ \frac{n+1}{2},\ \frac{s}{2},\ \frac{s+1}{2}\\ \frac{1}{2},\ \frac{1}{4},\ \frac{3}{4},\ \frac{s+2}{2};\ \frac{1}{a^2b^2}\end{matrix}\right)$ $[\operatorname{Re}s>0;\ \lvert\arg b\rvert<\pi/4]$
5	$\operatorname{erfc}(b\sqrt{x})\,T_n\left(\frac{2x}{a}\pm 1\right)$	$\frac{\delta_{n,0}+1}{2\sqrt{\pi}\,(s+n)}\left(\frac{4}{a}\right)^n b^{-2s-2n}\Gamma\left(s+n+\frac{1}{2}\right) \times {}_3F_3\left(\begin{matrix}-n,\ \frac{1}{2}-n,\ -s-n;\ \pm ab^2\\ 1-2n,\ \frac{1}{2}-s-n,\ 1-s-n\end{matrix}\right)$ $[\operatorname{Re}s>((-1)^n-1)/2;\ \lvert\arg b\rvert<\pi/4]$

No.	$f(x)$	$F(s)$
6	$(a-x)_+^{-1/2}$ $\times \begin{Bmatrix} \operatorname{erf}(b\sqrt{x}) \\ \operatorname{erfc}(b\sqrt{x}) \end{Bmatrix}$ $\times T_n\left(\frac{2x}{a}-1\right)$	$\pm 2(-1)^n a^s b(-s)_n \Gamma\left[\begin{matrix} \frac{2s+1}{2} \\ s+n+1 \end{matrix}\right] {}_3F_3\left(\begin{matrix} \frac{1}{2},\, \frac{2s+1}{2},\, s+1;\, -ab^2 \\ \frac{3}{2},\, s-n+1,\, s+n+1 \end{matrix}\right)$ $+\begin{Bmatrix} 0 \\ 1 \end{Bmatrix} (-1)^n \sqrt{\pi}\, a^{s-1/2} \left(\frac{1}{2}-s\right)_n \Gamma\left[\begin{matrix} s \\ s+n+\frac{1}{2} \end{matrix}\right]$ $[a>0;\ \operatorname{Re} s > -(1\pm 1)/4]$
7	$(a-x)_+^{-1/2} e^{b^2 x}$ $\times \operatorname{erf}(b\sqrt{x})$ $\times T_n\left(\frac{2x}{a}-1\right)$	$2(-1)^n a^s b(-s)_n \Gamma\left[\begin{matrix} \frac{2s+1}{2} \\ s+n+1 \end{matrix}\right] {}_3F_3\left(\begin{matrix} 1,\, \frac{2s+1}{2},\, s+1;\, ab^2 \\ \frac{3}{2},\, s-n+1,\, s+n+1 \end{matrix}\right)$ $[a>0;\ \operatorname{Re} s > -1/2]$

3.20.10. $T_n(bx)$ and $\Gamma(\nu, ax)$, $\gamma(\nu, ax)$

Notation: $\varepsilon = 0$ or 1.

1	$(a^2-x^2)_+^{-1/2}$ $\times \gamma(\nu, b^2x^2)\, T_n\left(\frac{x}{a}\right)$	$\frac{\pi}{2\nu}\left(\frac{a}{2}\right)^{s+2\nu-1} b^{2\nu}\, \Gamma\left[\begin{matrix} s+2\nu \\ \frac{s-n+2\nu+1}{2},\, \frac{s+n+2\nu+1}{2} \end{matrix}\right]$ $\times {}_3F_3\left(\begin{matrix} \nu,\, \frac{s+2\nu}{2},\, \frac{s+2\nu+1}{2};\, -a^2b^2 \\ \nu+1,\, \frac{s-n+2\nu+1}{2},\, \frac{s+n+2\nu+1}{2} \end{matrix}\right)$ $[a>0;\ \operatorname{Re}(s+2\nu) > ((-1)^n-1)/2]$
2	$(a^2-x^2)_+^{-1/2} e^{bx}$ $\times \gamma(\nu, bx)\, T_n\left(\frac{x}{a}\right)$	$\frac{\pi}{2\nu}\left(\frac{a}{2}\right)^{s+\nu-1} b^{\nu}\, \Gamma\left[\begin{matrix} s+\nu \\ \frac{s-n+\nu+1}{2},\, \frac{s+n+\nu+1}{2} \end{matrix}\right]$ $\times {}_3F_4\left(\begin{matrix} 1,\, \frac{s+\nu}{2},\, \frac{s+\nu+1}{2};\, \frac{a^2b^2}{4} \\ \frac{\nu+1}{2},\, \frac{\nu+2}{2},\, \frac{s-n+\nu+1}{2},\, \frac{s+n+\nu+1}{2} \end{matrix}\right)$ $+\frac{\pi}{2\nu(\nu+1)}\left(\frac{a}{2}\right)^{s+\nu} b^{\nu+1}\, \Gamma\left[\begin{matrix} s+\nu+1 \\ \frac{s-n+\nu+2}{2},\, \frac{s+n+\nu+2}{2} \end{matrix}\right]$ $\times {}_3F_4\left(\begin{matrix} 1,\, \frac{s+\nu+1}{2},\, \frac{s+\nu+2}{2};\, \frac{a^2b^2}{4} \\ \frac{\nu+2}{2},\, \frac{\nu+3}{2},\, \frac{s-n+\nu+2}{2},\, \frac{s+n+\nu+2}{2} \end{matrix}\right)$ $[a>0;\ \operatorname{Re}(s+\nu) > ((-1)^n-1)/2]$
3	$(a^2-x^2)_+^{-1/2} e^{b^2x^2}$ $\times \gamma(\nu, b^2x^2)\, T_n\left(\frac{x}{a}\right)$	$\frac{\pi}{2\nu}\left(\frac{a}{2}\right)^{s+2\nu-1} b^{2\nu}\, \Gamma\left[\begin{matrix} s+2\nu \\ \frac{s-n+2\nu+1}{2},\, \frac{s+n+2\nu+1}{2} \end{matrix}\right]$ $\times {}_3F_3\left(\begin{matrix} 1,\, \frac{s+2\nu}{2},\, \frac{s+2\nu+1}{2};\, a^2b^2 \\ \nu+1,\, \frac{s-n+2\nu+1}{2},\, \frac{s+n+2\nu+1}{2} \end{matrix}\right)$ $[a>0;\ \operatorname{Re}(s+2\nu) > ((-1)^n-1)/2]$

No.	$f(x)$	$F(s)$
4	$\Gamma(\nu,\, ax)\, T_{2n+\varepsilon}(bx)$	$\dfrac{(-1)^n (2n+1)^{\varepsilon} a^{-s-\varepsilon} b^{\varepsilon}}{s+\varepsilon} \Gamma(s+\nu+\varepsilon) \times {}_5F_2\left(\begin{matrix} -n,\, n+\varepsilon,\, \frac{s+\varepsilon}{2},\, \frac{s+\nu+\varepsilon}{2},\, \frac{s+\nu+\varepsilon+1}{2} \\ \frac{2\varepsilon+1}{2},\, \frac{s+\varepsilon+2}{2};\, \frac{4b^2}{a^2} \end{matrix}\right)$ $[\operatorname{Re} a > 0;\ \operatorname{Re} s,\ \operatorname{Re}(s+\nu) > -\varepsilon]$

3.20.11. $T_n(\varphi(x))$ and $J_\nu(cx^r)$, $I_\nu(cx)$

Notation: $\varepsilon = 0$ or 1.

1	$(a^2-x^2)_+^{-1/2} \left\{\begin{matrix} J_\nu(bx) \\ I_\nu(bx) \end{matrix}\right\} \times T_n\left(\dfrac{x}{a}\right)$	$\dfrac{\pi}{2}\left(\dfrac{a}{2}\right)^{s+\nu-1}\left(\dfrac{b}{2}\right)^{\nu} \Gamma\left[\begin{matrix} s+\nu \\ \nu+1,\, \frac{s-n+\nu+1}{2},\, \frac{s+n+\nu+1}{2} \end{matrix}\right] \times {}_2F_3\left(\begin{matrix} \frac{s+\nu}{2},\, \frac{s+\nu+1}{2};\, \frac{a^2b^2}{4} \\ \nu+1,\, \frac{s-n+\nu+1}{2},\, \frac{s+n+\nu+1}{2} \end{matrix}\right)$ $[a>0;\ \operatorname{Re}(s+\nu) > ((-1)^n-1)/2]$
2	$(a^2-x^2)_+^{-1/2} J_\nu\left(\dfrac{b}{x}\right) \times T_{2n+\varepsilon}\left(\dfrac{x}{a}\right)$	$\dfrac{(-1)^n \sqrt{\pi}\, a^{s-\nu-1} b^{\nu}}{2^{\nu+1}} \left(\dfrac{1-s+\nu+\varepsilon}{2}\right)_n \Gamma\left[\begin{matrix} \frac{s-\nu+\varepsilon}{2} \\ \nu+1,\, \frac{s+2n-\nu+\varepsilon+1}{2} \end{matrix}\right] \times {}_2F_3\left(\begin{matrix} \frac{1-s+2n+\nu+\varepsilon}{2},\, \frac{1-s-2n+\nu-\varepsilon}{2} \\ \nu+1,\, \frac{\nu-s+1}{2},\, \frac{\nu-s+2}{2};\, -\frac{b^2}{4a^2} \end{matrix}\right)$ $+ \dfrac{(-1)^n (2n+\varepsilon)\, n^{\varepsilon-1} a^{-\varepsilon-1} b^{s+\varepsilon}}{2^{s+2}} \Gamma\left[\begin{matrix} -\frac{s-\nu+\varepsilon}{2} \\ \frac{s+\nu+\varepsilon+2}{2} \end{matrix}\right] \times {}_2F_3\left(\begin{matrix} \frac{1-2n}{2},\, \frac{2n+2\varepsilon+1}{2};\, -\frac{b^2}{4a^2} \\ \frac{2\varepsilon+1}{2},\, \frac{s-\nu+\varepsilon+2}{2},\, \frac{s+\nu+\varepsilon+2}{2} \end{matrix}\right)$ $[a,\, b>0;\ \operatorname{Re} s > -\varepsilon-3/2]$
3	$(a-x)_+^{-1/2} \left\{\begin{matrix} J_\nu(b\sqrt{x}) \\ I_\nu(b\sqrt{x}) \end{matrix}\right\} \times T_n\left(\dfrac{2x}{a}-1\right)$	$\dfrac{(-1)^n \sqrt{\pi}\, a^{s+(\nu-1)/2} b^{\nu}}{2^{\nu}} \left(\dfrac{-2s-\nu+1}{2}\right)_n \times \Gamma\left[\begin{matrix} \frac{2s+\nu}{2} \\ \nu+1,\, \frac{2s+2n+\nu+1}{2} \end{matrix}\right] {}_2F_3\left(\begin{matrix} \frac{2s+\nu}{2},\, \frac{2s+\nu+1}{2};\, \mp\frac{ab^2}{4} \\ \nu+1,\, \frac{2s-2n+\nu+1}{2},\, \frac{2s+2n+\nu+1}{2} \end{matrix}\right)$ $[a>0;\ \operatorname{Re} s > -\operatorname{Re}\nu/2]$
4	$J_\nu(b\sqrt{x})\, T_n\left(\dfrac{2x}{a}\pm 1\right)$	$\dfrac{2^{2s+4n-1}(\delta_{n,0}+1)}{a^n b^{2s+2n}} \Gamma\left[\begin{matrix} \frac{2s+2n+\nu}{2} \\ \frac{2-2s-2n+\nu}{2} \end{matrix}\right] \times {}_2F_3\left(\begin{matrix} -n,\, \frac{1}{2}-n;\, \pm\frac{ab^2}{4} \\ 1-2n,\, \frac{2-2s-2n-\nu}{2},\, \frac{2-2s-2n+\nu}{2} \end{matrix}\right)$ $[b>0;\ -\operatorname{Re}\nu/2 < \operatorname{Re} s < 3/4-n]$

No.	$f(x)$	$F(s)$
5	$\dfrac{J_\nu(b\sqrt{x})}{\sqrt{x+a}}\,T_n\left(\dfrac{2x}{a}+1\right)$	$\dfrac{(-1)^n\,a^{s+(\nu-1)/2}b^\nu}{2^\nu\sqrt{\pi}}\left(\dfrac{1-2s-\nu}{2}\right)_n\Gamma\left[\begin{matrix}\frac{2s+\nu}{2},\ \frac{-2s-2n-\nu+1}{2}\\ \nu+1\end{matrix}\right]$ $\times\,{}_2F_3\left(\begin{matrix}\frac{2s+\nu}{2},\ \frac{2s+\nu+1}{2};\ \frac{ab^2}{4}\\ \nu+1,\ \frac{2s-2n+\nu+1}{2},\ \frac{2s+2n+\nu+1}{2}\end{matrix}\right)$ $+\,\dfrac{(\delta_{n,0}+1)\,2^{2s+4n-2}}{n!\,a^n b^{2s+2n-1}}\Gamma\left[\begin{matrix}\frac{2s+2n+\nu-1}{2}\\ \frac{-2s-2n+\nu+3}{2}\end{matrix}\right]$ $\times\,{}_2F_3\left(\begin{matrix}\frac{-2n+1}{2},\ -n+1;\ \frac{ab^2}{4}\\ -2n+1,\ \frac{-2s-2n-\nu+3}{2},\ \frac{-2s-2n+\nu+3}{2}\end{matrix}\right)$ $[b>0;\ -\operatorname{Re}\nu/2<\operatorname{Re}s<5/4-n;\ \lvert\arg a\rvert<\pi]$
6	$(a-x)_+^{-1/2}\,J_\nu\left(\dfrac{b}{\sqrt{x}}\right)$ $\times\,T_n\left(\dfrac{2x}{a}-1\right)$	$\dfrac{(-1)^n\,\sqrt{\pi}\,a^{s-(\nu+1)/2}b^\nu}{2^\nu}\left(\dfrac{1-2s+\nu}{2}\right)_n\Gamma\left[\begin{matrix}\frac{2s-\nu}{2}\\ \nu+1,\ \frac{2s+2n-\nu+1}{2}\end{matrix}\right]$ $\times\,{}_2F_3\left(\begin{matrix}\frac{1-2s-2n+\nu}{2},\ \frac{1-2s+2n+\nu}{2};\ -\frac{b^2}{4a}\\ \nu+1,\ \frac{1-2s+\nu}{2},\ \frac{2-2s+\nu}{2}\end{matrix}\right)$ $+\,\dfrac{(-1)^n\,a^{-1/2}b^{2s}}{2^{2s}}\Gamma\left[\begin{matrix}\frac{\nu-2s}{2}\\ \frac{2s+\nu+2}{2}\end{matrix}\right]{}_2F_3\left(\begin{matrix}\frac{-2n+1}{2},\ \frac{2n+1}{2};\ -\frac{b^2}{4a}\\ \frac{1}{2},\ \frac{2s-\nu+2}{2},\ \frac{2s+\nu+2}{2}\end{matrix}\right)$ $[a,\,b>0;\ \operatorname{Re}s>-3/4]$
7	$J_\nu\left(\dfrac{b}{\sqrt{x}}\right)T_n\left(\dfrac{2x}{a}\pm1\right)$	$(\pm1)^n\left(\dfrac{b}{2}\right)^{2s}\Gamma\left[\begin{matrix}\frac{\nu-2s}{2}\\ \frac{2s+\nu+2}{2}\end{matrix}\right]{}_2F_3\left(\begin{matrix}-n,\ n;\ \pm\frac{b^2}{4a}\\ \frac{1}{2},\ \frac{2s-\nu+2}{2},\ \frac{2s+\nu+2}{2}\end{matrix}\right)$ $[b>0;\ -3/4<\operatorname{Re}s<\operatorname{Re}\nu/2-n]$
8	$\dfrac{1}{\sqrt{x+a}}\,J_\nu\left(\dfrac{b}{\sqrt{x}}\right)$ $\times\,T_n\left(\dfrac{2x}{a}+1\right)$	$\dfrac{(-1)^n\,a^{s-(\nu+1)/2}b^\nu}{2^\nu\sqrt{\pi}}\left(\dfrac{1-2s+\nu}{2}\right)_n\Gamma\left[\begin{matrix}\frac{2s-\nu}{2},\ \frac{1-2s-2n+\nu}{2}\\ \nu+1\end{matrix}\right]$ $\times\,{}_2F_3\left(\begin{matrix}\frac{1-2s-2n+\nu}{2},\ \frac{1-2s+2n+\nu}{2}\\ \nu+1,\ \frac{1-2s+\nu}{2},\ \frac{2-2s+\nu}{2};\ \frac{b^2}{4a}\end{matrix}\right)$ $+\,\dfrac{a^{-1/2}b^{2s}}{2^{2s}}\Gamma\left[\begin{matrix}\frac{\nu-2s}{2}\\ \frac{2s+\nu+2}{2}\end{matrix}\right]{}_2F_3\left(\begin{matrix}\frac{-2n+1}{2},\ \frac{2n+1}{2};\ \frac{b^2}{4a}\\ \frac{1}{2},\ \frac{2s-\nu+2}{2},\ \frac{2s+\nu+2}{2}\end{matrix}\right)$ $[b>0;\ -3/4<\operatorname{Re}s<1/2-n+\operatorname{Re}\nu/2;\ \lvert\arg a\rvert<\pi]$
9	$(a-x)_+^{-1/2}$ $\times\begin{Bmatrix}J_\mu(b\sqrt{x})\,J_\nu(b\sqrt{x})\\ I_\mu(b\sqrt{x})\,I_\nu(b\sqrt{x})\end{Bmatrix}$ $\times\,T_n\left(1-\dfrac{2x}{a}\right)$	$\sqrt{\pi}\,a^{s+(\mu+\nu-1)/2}\,(b/2)^{\mu+\nu}\left(\dfrac{1-2s-\mu-\nu}{2}\right)_n$ $\times\,\Gamma\left[\begin{matrix}\frac{2s+\mu+\nu}{2}\\ \mu+1,\ \nu+1,\ \frac{2s+2n+\mu+\nu+1}{2}\end{matrix}\right]$ $\times\,{}_4F_5\left(\begin{matrix}\frac{\mu+\nu+1}{2},\ \frac{\mu+\nu+2}{2},\ \frac{2s+\mu+\nu}{2},\ \frac{2s+\mu+\nu+1}{2};\ \mp ab^2\\ \mu+1,\ \nu+1,\ \mu+\nu+1,\ \frac{2s-2n+\mu+\nu+1}{2},\ \frac{2s+2n+\mu+\nu+1}{2}\end{matrix}\right)$ $[a,\ \operatorname{Re}(2s+\mu+\nu)>0]$

No.	$f(x)$	$F(s)$
10	$(x+a)^n J_\nu(b\sqrt{x}) \times T_n\left(\frac{a-x}{a+x}\right)$	$(-1)^n \left(\frac{2}{b}\right)^{2s+2n} \Gamma\left[\begin{matrix}\frac{2s+2n+\nu}{2}\\ \frac{2-2s-2n+\nu}{2}\end{matrix}\right] {}_2F_3\left(\begin{matrix}-n,\ \frac{1-2n}{2};\ \frac{ab^2}{4}\\ \frac{1}{2},\ \frac{2-2s-2n-\nu}{2},\ \frac{2-2s-2n+\nu}{2}\end{matrix}\right)$ $[b>0;\ -\operatorname{Re}\nu/2<\operatorname{Re}s<3/4-n]$
11	$\frac{1}{(x+a)^n} J_\nu(b\sqrt{x}) \times T_n\left(\frac{a-x}{a+x}\right)$	$\frac{n a^{s-n+\nu/2} b^\nu}{2^\nu n!} \frac{\left(\frac{1-2s-\nu}{2}\right)_n}{\left(\frac{1}{2}\right)_n} \Gamma\left[\begin{matrix}\frac{2s+\nu}{2},\ \frac{2n-2s-\nu}{2}\\ \nu+1\end{matrix}\right]$ $\times {}_2F_3\left(\begin{matrix}\frac{2s+\nu}{2},\ \frac{2s+\nu+1}{2};\ \frac{ab^2}{4}\\ \nu+1,\ \frac{2s-2n+\nu+1}{2},\ \frac{2s-2n+\nu+2}{2}\end{matrix}\right) + (-1)^n \left(\frac{b}{2}\right)^{2(n-s)}$ $\times \Gamma\left[\begin{matrix}\frac{2s-2n+\nu}{2}\\ \frac{2-2s+2n+\nu}{2}\end{matrix}\right] {}_2F_3\left(\begin{matrix}n,\ \frac{2n+1}{2};\ \frac{ab^2}{4}\\ \frac{1}{2},\ \frac{2-2s+2n-\nu}{2},\ \frac{2-2s+2n+\nu}{2}\end{matrix}\right)$ $[b>0;\ -\operatorname{Re}\nu/2<\operatorname{Re}s<n+3/4;\ \lvert\arg a\rvert<\pi]$

3.20.12. $T_n(\varphi(x))$ and $K_\nu(cx^r)$

Notation: $\varepsilon = 0$ or 1.

1	$K_\nu(b\sqrt{x})\, T_n\left(\frac{2x}{a} \pm 1\right)$	$\frac{2^{2s+4n-2}(\delta_{n,0}+1)}{a^n b^{2s+2n}} \Gamma\left(s+n-\frac{\nu}{2}\right) \Gamma\left(s+n+\frac{\nu}{2}\right)$ $\times {}_2F_3\left(\begin{matrix}-n,\ \frac{1}{2}-n;\ \mp\frac{ab^2}{4}\\ 1-2n,\ \frac{2-2s-2n-\nu}{2},\ \frac{2-2s-2n+\nu}{2}\end{matrix}\right)$ $[b>0;\ \operatorname{Re}s>\lvert\operatorname{Re}\nu\rvert/2]$
2	$\frac{1}{\sqrt{x+a}} K_\nu(b\sqrt{x}) \times T_n\left(\frac{2x}{a}+1\right)$	$\frac{(-1)^n a^{s-(\nu+1)/2} b^{-\nu}}{2^{-\nu+1}\sqrt{\pi}} \left(\frac{1-2s+\nu}{2}\right)_n \Gamma\left(\frac{1-2n-2s+\nu}{2}\right)$ $\times \Gamma\left(\frac{2s-\nu}{2}\right) {}_2F_3\left(\begin{matrix}\frac{2s-\nu}{2},\ \frac{2s-\nu+1}{2};\ -\frac{ab^2}{4}\\ 1-\nu,\ \frac{s-2n-\nu+1}{2},\ \frac{s+2n-\nu+1}{2}\end{matrix}\right)$ $+ \frac{(-1)^n a^{s+(\nu-1)/2} b^\nu}{2^{\nu+1}\sqrt{\pi}} \left(\frac{1-2s-\nu}{2}\right)_n \Gamma\left(\frac{1-2s-2n-\nu}{2}\right)$ $\times \Gamma\left(\frac{2s+\nu}{2}\right) {}_2F_3\left(\begin{matrix}\frac{2s+\nu}{2},\ \frac{2s+\nu+1}{2};\ -\frac{ab^2}{4}\\ \nu+1,\ \frac{s-2n+\nu+1}{2},\ \frac{s+2n+\nu+1}{2}\end{matrix}\right)$ $+ \frac{(\delta_{n,0}+1)\, 2^{2s+4n-2}}{a^n b^{2s+2n-1}} \Gamma\left(\frac{2s+2n-\nu-1}{2}\right)$ $\times \Gamma\left(\frac{2s+2n+\nu-1}{2}\right) {}_2F_3\left(\begin{matrix}\frac{-2n+1}{2},\ -n+1;\ -\frac{ab^2}{4}\\ -2n+1,\ \frac{3-2s-2n-\nu}{2},\ \frac{3-2s-2n+\nu}{2}\end{matrix}\right)$ $[\operatorname{Re}b>0;\ \operatorname{Re}s>\lvert\operatorname{Re}\nu\rvert/2;\ \lvert\arg a\rvert<\pi]$

No.	$f(x)$	$F(s)$
3	$K_\nu\left(\frac{b}{\sqrt{x}}\right) T_n\left(\frac{2x}{a} \pm 1\right)$	$\frac{(\pm 1)^n}{2}\left(\frac{b}{2}\right)^{2s} \Gamma\left(\frac{\nu-2s}{2}\right)\Gamma\left(\frac{-\nu-2s}{2}\right) \times {}_2F_3\left(\begin{matrix}-n,\ n;\ \mp\frac{b^2}{4a}\\ \frac{1}{2},\ \frac{2s-\nu+2}{2},\ \frac{2s+\nu+2}{2}\end{matrix}\right)$ $[\operatorname{Re} b > 0;\ \operatorname{Re} s < -n - \lvert\operatorname{Re}\nu\rvert/2]$
4	$\frac{1}{\sqrt{x+a}} K_\nu\left(\frac{b}{\sqrt{x}}\right) \times T_n\left(\frac{2x}{a}+1\right)$	$\frac{b^{2s}}{2^{2s+1}\sqrt{a}} \Gamma\left(-\frac{\nu}{2}-s\right)\Gamma\left(\frac{\nu}{2}-s\right) {}_2F_3\left(\begin{matrix}\frac{-2n+1}{2},\ \frac{2n+1}{2};\ -\frac{b^2}{4a}\\ \frac{1}{2},\ \frac{2s-\nu+2}{2},\ \frac{2s+\nu+2}{2}\end{matrix}\right) + \frac{(-1)^n a^{s+(\nu-1)/2} b^{-\nu}}{2^{-\nu+1}\sqrt{\pi}}\left(\frac{1-2s-\nu}{2}\right)_n \times \Gamma\left[\nu,\ \frac{2s+\nu}{2},\ \frac{1-2s-2n-\nu}{2}\right] \times {}_2F_3\left(\begin{matrix}\frac{1-2s+2n-\nu}{2},\ \frac{1-2s-2n-\nu}{2}\\ 1-\nu,\ \frac{1-2s-\nu}{2},\ \frac{2-2s-\nu}{2};\ -\frac{b^2}{4a}\end{matrix}\right) + \frac{(-1)^n a^{s-(\nu+1)/2} b^{\nu}}{2^{\nu+1}\sqrt{\pi}}\left(\frac{1-2s+\nu}{2}\right)_n \times \Gamma\left[-\nu,\ \frac{2s-\nu}{2},\ \frac{1-2s-2n+\nu}{2}\right] \times {}_2F_3\left(\begin{matrix}\frac{1-2s+2n+\nu}{2},\ \frac{1-2s-2n+\nu}{2}\\ \nu+1,\ \frac{1-2s+\nu}{2},\ \frac{2-2s+\nu}{2};\ -\frac{b^2}{4a}\end{matrix}\right)$ $[\operatorname{Re} b > 0;\ \operatorname{Re} s < (1-2n-\lvert\operatorname{Re}\nu\rvert)/2;\ \lvert\arg a\rvert < \pi]$
5	$(x+a)^n K_\nu(b\sqrt{x}) \times T_n\left(\frac{a-x}{a+x}\right)$	$(-1)^n 2^{2s+2n-1} b^{-2s-2n} \Gamma\left(s+n-\frac{\nu}{2}\right)\Gamma\left(s+n+\frac{\nu}{2}\right) \times {}_2F_3\left(\begin{matrix}-n,\ \frac{1-2n}{2};\ -\frac{ab^2}{4}\\ \frac{1}{2},\ \frac{2-2s-2n-\nu}{2},\ \frac{2-2s-2n+\nu}{2}\end{matrix}\right)$ $[\operatorname{Re} b > 0;\ \operatorname{Re} s > \lvert\operatorname{Re}\nu\rvert/2]$

3.20.13. $T_n(bx)$ and $\mathbf{H}_\nu(ax)$, $\mathbf{L}_\nu(ax)$

1	$(a^2-x^2)_+^{-1/2} \begin{Bmatrix}\mathbf{H}_\nu(bx)\\ \mathbf{L}_\nu(bx)\end{Bmatrix} \times T_n\left(\frac{x}{a}\right)$	$\frac{\sqrt{\pi}}{2^{\nu+1}}\left(\frac{a}{2}\right)^{s+\nu} b^{\nu+1} \Gamma\left[\begin{matrix}s+\nu+1\\ \frac{2\nu+3}{2},\ \frac{s-n+\nu+2}{2},\ \frac{s+n+\nu+2}{2}\end{matrix}\right] \times {}_3F_3\left(\begin{matrix}1,\ \frac{s+\nu+1}{2},\ \frac{s+\nu+2}{2}\\ \frac{3}{2},\ \frac{s+n+\nu+2}{2},\ \frac{3s-3n+7\nu+6}{2};\ \mp\frac{a^2b^2}{4}\end{matrix}\right)$ $[a > 0;\ \operatorname{Re}(s+\nu) > ((-1)^n - 3)/2]$

3.20.14. $T_n(ax+b)$ and $P_m(\varphi(x))$

Notation: $\delta,\ \varepsilon = 0$ or 1.

1	$\theta(a-x)\, P_m\left(\frac{x}{a}\right) \times T_{2n+\varepsilon}(bx)$	$\frac{(-1)^n(2n+\varepsilon)\sqrt{\pi}}{2^{s+1}n!}\, a^{s+\varepsilon}b^{\varepsilon}\,\Gamma\left[\begin{matrix} n+\varepsilon,\ s+\varepsilon \\ \frac{s-m+\varepsilon+1}{2},\ \frac{s+m+\varepsilon+2}{2}\end{matrix}\right] \times {}_4F_3\left(\begin{matrix}-n,\ n+\varepsilon,\ \frac{s+\varepsilon}{2},\ \frac{s+\varepsilon+1}{2};\ a^2b^2 \\ \frac{2\varepsilon+1}{2},\ \frac{s-m+\varepsilon+1}{2},\ \frac{s+m+\varepsilon+2}{2}\end{matrix}\right)$ $[a>0;\ \operatorname{Re} s > ((-1)^m + (-1)^{\varepsilon})/2 - 1]$
2	$(x^2-a^2)_+^{-1/2} \times P_{2m+\varepsilon}\left(\frac{x}{b}\right) T_{2n+\delta}\left(\frac{x}{a}\right)$	$\frac{2^{2m+\varepsilon-1}\sqrt{\pi}\, a^{s+2m+\varepsilon-1}}{(2m+\varepsilon)!\, b^{2m+\varepsilon}} \left(\frac{1}{2}\right)_{2m+\varepsilon} \left(\frac{1-s-2m+\delta-\varepsilon}{2}\right)_n \times\Gamma\left[\begin{matrix}\frac{1-s-2m-2n-\delta-\varepsilon}{2} \\ \frac{2-s-2m-\delta-\varepsilon}{2}\end{matrix}\right] \times {}_4F_3\left(\begin{matrix}-m,\ \frac{1-2m-2\varepsilon}{2},\ \frac{1-s-2m-2n-\delta-\varepsilon}{2},\ \frac{1-s-2m+2n+\delta-\varepsilon}{2} \\ \frac{1-4m-2\varepsilon}{2},\ \frac{2-s-2m-2\varepsilon}{2},\ \frac{1-s-2m}{2};\ \frac{b^2}{a^2}\end{matrix}\right)$ $[a,\ b>0;\ \operatorname{Re} s < 1-2m-2n-\delta-\varepsilon]$
3	$\theta(x-a)\,(x^2-b^2)^{-1/2} \times P_{2m+\varepsilon}\left(\frac{x}{a}\right) T_{2n+\delta}\left(\frac{x}{b}\right)$	$\frac{(-1)^{m-1}\, 2^{2n+\delta-2} a^{s+2n+\delta-1}}{b^{2n+\delta}} \frac{\left(\frac{2-s-2n-\delta+\varepsilon}{2}\right)_m}{\left(\frac{s+2n+\delta+\varepsilon-1}{2}\right)_{m+1}} \times {}_4F_3\left(\begin{matrix}\frac{1-2n}{2},\ 1-n-\delta,\ \frac{2-s+2m-2n-\delta+\varepsilon}{2},\ \frac{1-s-2m-2n-\delta-\varepsilon}{2} \\ 1-2n-\delta,\ \frac{2-s-2n}{2},\ \frac{3-s-2n-2\delta}{2};\ \frac{b^2}{a^2}\end{matrix}\right)$ $[a>b>0;\ \operatorname{Re} s < 1-2m-2n-\delta-\varepsilon]$
4	$\theta(a-x)\, P_m\left(\frac{2x}{a}-1\right) \times T_n\left(\frac{2x}{a}-1\right)$	$\frac{(-1)^{m+n}(1-s)_m\, a^s}{(s)_{m+1}}\, {}_4F_3\left(\begin{matrix}-n,\ n,\ s,\ s;\ 1 \\ \frac{1}{2},\ s-m,\ s+m+1\end{matrix}\right)$ $[a,\ \operatorname{Re} s>0]$
5	$(a-x)_+^{-1/2}\, P_m\left(1-\frac{2x}{a}\right) \times T_n\left(1-\frac{2x}{a}\right)$	$\sqrt{\pi}\, a^{s-1/2}\left(\frac{1}{2}-s\right)_n \Gamma\left[\begin{matrix} s \\ \frac{2s+2n+1}{2}\end{matrix}\right] {}_4F_3\left(\begin{matrix}-m,\ m+1,\ s,\ \frac{2s+1}{2};\ 1 \\ 1,\ \frac{2s-2n+1}{2},\ \frac{2s+2n+1}{2}\end{matrix}\right)$ $[a,\ \operatorname{Re} s>0]$
6	$(a-x)_+^{-1/2}\, P_m(2bx-1) \times T_n\left(\frac{2x}{a}-1\right)$	$(-1)^{m+n}\sqrt{\pi}\, a^{s-1/2}\left(\frac{1-2s}{2}\right)_n \Gamma\left[\begin{matrix} s \\ \frac{2s+2n+1}{2}\end{matrix}\right] \times {}_4F_3\left(\begin{matrix}-m,\ m+1,\ s,\ \frac{2s+1}{2};\ ab \\ 1,\ \frac{2s-2n+1}{2},\ \frac{2s+2n+1}{2}\end{matrix}\right)$ $[a,\ \operatorname{Re} s>0]$
7	$\frac{\theta(a-x)}{\sqrt{b\pm x}}\, P_m\left(\frac{2x}{a}-1\right) \times T_n\left(\frac{2x}{b}\pm 1\right)$	$(-1)^m(\pm 1)^n a^s b^{-1/2}\frac{(1-s)_m}{(s)_{m+1}}\, {}_4F_3\left(\begin{matrix}\frac{-2n+1}{2},\ \frac{2n+1}{2},\ s,\ s;\ \mp\frac{a}{b} \\ \frac{1}{2},\ s-m,\ s+m+1\end{matrix}\right)$ $\left[\left\{\begin{matrix} a>0;\ \lvert\arg b\rvert<\pi \\ b>a>0\end{matrix}\right\};\ \operatorname{Re} s>0\right]$

No.	$f(x)$	$F(s)$
8	$(a-x)_+^{-1/2} P_{2m+\varepsilon}(b\sqrt{x})$ $\times T_n\left(\frac{2x}{a}-1\right)$	$\frac{(-1)^{m+n}\sqrt{\pi}\,a^{s+(\varepsilon-1)/2}(2b)^{\varepsilon}}{m!}\left(\frac{1}{2}\right)_{m+\varepsilon}\left(\frac{1-2s-\varepsilon}{2}\right)_n$ $\times\Gamma\left[\begin{matrix}\frac{2s+\varepsilon}{2}\\ \frac{2s+2n+\varepsilon+1}{2}\end{matrix}\right] {}_4F_3\left(\begin{matrix}-m,\ \frac{2m+2\varepsilon+1}{2},\ \frac{2s+\varepsilon}{2},\ \frac{2s+\varepsilon+1}{2};\ ab^2\\ \frac{2\varepsilon+1}{2},\ \frac{2s-2n+\varepsilon+1}{2},\ \frac{2s+2n+\varepsilon+1}{2}\end{matrix}\right)$ $[a,\,b>0;\ \operatorname{Re}s>-\varepsilon/2]$

3.20.15. Products of $T_n(\varphi(x))$

Notation: $\delta,\ \varepsilon=0$ or 1.

1	$(a^2-x^2)_+^{-1/2}$ $\times(b^2-x^2)^{-1/2}$ $\times T_{2m+\varepsilon}\left(\frac{x}{b}\right)T_{2n+\delta}\left(\frac{x}{a}\right)$	$\frac{(-1)^{m+n}\sqrt{\pi}\,(2m+1)^{\varepsilon}}{2}a^{s+\varepsilon-1}b^{-\varepsilon-1}\left(\frac{1-s+\delta-\varepsilon}{2}\right)_n$ $\times\Gamma\left[\begin{matrix}\frac{s+\delta+\varepsilon}{2}\\ \frac{s+2n+\delta+\varepsilon+1}{2}\end{matrix}\right] {}_4F_3\left(\begin{matrix}\frac{1-2m}{2},\ \frac{2m+2\varepsilon+1}{2},\ \frac{s+1}{2},\ \frac{s+2\varepsilon}{2}\\ \frac{2\varepsilon+1}{2},\ \frac{s-2n-\delta+\varepsilon+1}{2},\ \frac{s+2n+\delta+\varepsilon+1}{2};\ \frac{a^2}{b^2}\end{matrix}\right)$ $[b>a>0;\ \operatorname{Re}s>-\delta-\varepsilon]$
2	$(x^2-a^2)_+^{-1/2}$ $\times(x^2-b^2)^{-1/2}$ $\times T_{2m+\varepsilon}\left(\frac{x}{b}\right)T_{2n+\delta}\left(\frac{x}{a}\right)$	$2^{2m+\varepsilon-2}\sqrt{\pi}\,a^{s+2m+\varepsilon-2}b^{-2m-\varepsilon}\left(\frac{2-s-2m+\delta-\varepsilon}{2}\right)_n$ $\times\Gamma\left[\begin{matrix}\frac{2-s-2m-2n-\delta-\varepsilon}{2}\\ \frac{3-s-2m-\delta-\varepsilon}{2}\end{matrix}\right]$ $\times{}_4F_3\left(\begin{matrix}\frac{1-2m}{2},\ 1-m-\varepsilon,\ \frac{2-s-2m+2n+\delta-\varepsilon}{2},\ \frac{2-s-2m-2n-\delta-\varepsilon}{2}\\ \frac{3-s-2m-2\varepsilon}{2},\ 1-2m-\varepsilon,\ \frac{2-s-2m}{2};\ \frac{b^2}{a^2}\end{matrix}\right)$ $[a>b>0;\ \operatorname{Re}s<2-2m-2n-\delta-\varepsilon]$
3	$(x^2-a^2)_+^{-1/2}$ $\times(b^2-x^2)_+^{-1/2}$ $\times T_{2m+\varepsilon}\left(\frac{x}{b}\right)T_{2n+\delta}\left(\frac{x}{a}\right)$	$\frac{(-1)^m 2^{\varepsilon-2}(2m+\varepsilon)\sqrt{\pi}}{m!}a^{s+\varepsilon-1}b^{-\varepsilon-1}\left(\frac{1-s+\delta-\varepsilon}{2}\right)_n$ $\times\Gamma\left[\begin{matrix}m+\varepsilon,\ \frac{1-s-2n-\delta-\varepsilon}{2}\\ 1-\frac{s+\delta+\varepsilon}{2}\end{matrix}\right] {}_4F_3\left(\begin{matrix}\frac{1-2m}{2},\ \frac{2m+2\varepsilon+1}{2},\ \frac{s+1}{2},\ \frac{s+2\varepsilon}{2};\ \frac{a^2}{b^2}\\ \frac{2\varepsilon+1}{2},\ \frac{s-2n-\delta+\varepsilon+1}{2},\ \frac{s+2n+\delta+\varepsilon+1}{2}\end{matrix}\right)$ $+(-1)^m 2^{2n+\delta-2}\sqrt{\pi}\,a^{-2n-\delta}b^{s+2n+\delta-2}$ $\times\left(\frac{2-s-2n-\delta+\varepsilon}{2}\right)_m\Gamma\left[\begin{matrix}\frac{s+2n+\delta+\varepsilon-1}{2}\\ \frac{s+2m+2n+\delta+\varepsilon}{2}\end{matrix}\right]$ $\times{}_4F_3\left(\begin{matrix}\frac{1-2n}{2},\ 1-n-\delta,\ \frac{2-s-2m-2n-\delta-\varepsilon}{2},\ \frac{2-s+2m-2n-\delta+\varepsilon}{2}\\ -2n-\delta+1,\ \frac{-s-2n+2}{2},\ \frac{-s-2n-2\delta+3}{2};\ \frac{a^2}{b^2}\end{matrix}\right)$ $[b>a>0]$

No.	$f(x)$	$F(s)$
4	$(a-x)_+^{-1/2} T_m\left(1-\frac{2x}{a}\right) \times T_n\left(1-\frac{2x}{a}\right)$	$\sqrt{\pi}\, a^{s-1/2}\, \Gamma\left[\begin{matrix} s,\, n-s+\frac{1}{2} \\ \frac{1}{2}-s,\, s+n+\frac{1}{2}\end{matrix}\right] {}_4F_3\left(\begin{matrix} -m,\, m,\, s,\, s+\frac{1}{2};\, 1 \\ \frac{1}{2},\, s+n+\frac{1}{2},\, s-n+\frac{1}{2}\end{matrix}\right)$ $[a,\ \mathrm{Re}\, s>0]$
5	$(a-x)_+^{-1/2}(1-bx)^{-1/2} \times T_n\left(\frac{2x}{a}-1\right) \times T_m(2bx-1)$	$(-1)^{m+n}\sqrt{\pi}\, a^{s-1/2}\left(\frac{1}{2}-s\right)_n \Gamma\left[\begin{matrix} s \\ s+n+\frac{1}{2}\end{matrix}\right] \times {}_4F_3\left(\begin{matrix} -m+\frac{1}{2},\, m+\frac{1}{2},\, s,\, s+\frac{1}{2} \\ \frac{1}{2},\, s-n+\frac{1}{2},\, s+n+\frac{1}{2};\, ab\end{matrix}\right)$ $[a,\ \mathrm{Re}\, s>0;\ \lvert\arg(1-ab)\rvert<\pi]$
6	$(a-x)_+^{-1/2}(b\pm x)^{-1/2} \times T_m\left(\frac{2x}{a}-1\right) \times T_n\left(\frac{2x}{b}\pm 1\right)$	$(-1)^m(\pm 1)^n\sqrt{\pi}\, a^{s-1/2}b^{-1/2}\left(\frac{1}{2}-s\right)_m \times \Gamma\left[\begin{matrix} s \\ s+m+\frac{1}{2}\end{matrix}\right] {}_4F_3\left(\begin{matrix} -n+\frac{1}{2},\, n+\frac{1}{2},\, s,\, s+\frac{1}{2};\, \mp\frac{a}{b} \\ \frac{1}{2},\, s-m+\frac{1}{2},\, s+m+\frac{1}{2}\end{matrix}\right)$ $\left[\left\{\begin{matrix} a>0;\ \lvert\arg b\rvert<\pi \\ b>a>0\end{matrix}\right\};\ \mathrm{Re}\, s>0\right]$
7	$(a-x)_+^{-1/2}\left(1-b^2x\right)^{-1/2} \times T_{2m+\varepsilon}(b\sqrt{x}) \times T_n\left(\frac{2x}{a}-1\right)$	$(-1)^{m+n}(2m+1)^{\varepsilon}\sqrt{\pi}\, a^{s+(\varepsilon-1)/2}b^{\varepsilon} \times \left(\frac{1-2s-\varepsilon}{2}\right)_n \Gamma\left[\begin{matrix} \frac{2s+\varepsilon}{2} \\ \frac{2s+2n+\varepsilon+1}{2}\end{matrix}\right] \times {}_4F_3\left(\begin{matrix} \frac{-2m+1}{2},\, \frac{2m+2\varepsilon+1}{2},\, \frac{2s+\varepsilon}{2},\, \frac{2s+\varepsilon+1}{2} \\ \frac{2\varepsilon+1}{2},\, \frac{2s-2n+\varepsilon+1}{2},\, \frac{2s+2n+\varepsilon+1}{2};\, ab^2\end{matrix}\right)$ $[a>0;\ \mathrm{Re}\, s>-\varepsilon/2;\ \lvert\arg(1-ab^2)\rvert<\pi]$
8	$(a-x)_+^{-1/2} T_n\left(\sqrt{\frac{x}{a}}\right) \times T_m(\sqrt{1-bx})$	$\pi\left(\frac{\sqrt{a}}{2}\right)^{2s-1} \Gamma\left[\begin{matrix} 2s \\ \frac{2s-n+1}{2},\, \frac{2s+n+1}{2}\end{matrix}\right] {}_4F_3\left(\begin{matrix} -\frac{m}{2},\, \frac{m}{2},\, s,\, \frac{2s+1}{2} \\ \frac{1}{2},\, \frac{2s-n+1}{2},\, \frac{2s+n+1}{2};\, ab\end{matrix}\right)$ $[a>0;\ \mathrm{Re}\, s>0]$
9	$\left(a^2-x^2\right)_+^{-1/2} T_n\left(\frac{a}{x}\right) \times T_{2m+\varepsilon}(bx)$	$(-1)^m(2m+1)^{\varepsilon}\, 2^{s+\varepsilon-2}a^{s+\varepsilon-1}b^{\varepsilon}\, \Gamma\left[\begin{matrix} \frac{s-n+\varepsilon}{2},\, \frac{s+n+\varepsilon}{2} \\ s+\varepsilon\end{matrix}\right] \times {}_4F_3\left(\begin{matrix} -m,\, m+\varepsilon,\, \frac{s-n+\varepsilon}{2},\, \frac{s+n+\varepsilon}{2} \\ \frac{2\varepsilon+1}{2},\, \frac{s+\varepsilon}{2},\, \frac{s+\varepsilon+1}{2};\, a^2b^2\end{matrix}\right)$ $[a>0;\ \mathrm{Re}\, s>n-\varepsilon/2]$
10	$(a-x)_+^{-1/2} T_n\left(\sqrt{\frac{a}{x}}\right) \times T_m(bx+1)$	$(4a)^{s-1/2}\, \Gamma\left[\begin{matrix} \frac{2s-n}{2},\, \frac{2s+n}{2} \\ 2s\end{matrix}\right] {}_4F_3\left(\begin{matrix} -m,\, m,\, \frac{2s-n}{2},\, \frac{2s+n}{2} \\ \frac{1}{2},\, s,\, \frac{2s+1}{2};\, -\frac{ab}{2}\end{matrix}\right)$ $[a>0;\ \mathrm{Re}\, s>n/2]$

3.21. The Chebyshev Polynomials $U_n(z)$

More formulas can be obtained from the corresponding sections due to the relations

$$U_\nu(z) = \frac{1}{1-z^2}\left[T_\nu(z) - z\,T_{\nu+1}(z)\right], \quad U_\nu(z) = C_\nu^1(z),$$

$$U_\nu(z) = \frac{\Gamma(\nu+2)}{(3/2)_\nu}\,P_\nu^{(1/2,\,1/2)}(z), \quad U_\nu(z) = (\nu+1)\,{}_2F_1\left(-\nu,\,\nu+2;\,\frac{3}{2};\,\frac{1-z}{2}\right).$$

3.21.1. $U_n(\varphi(x))$ and algebraic functions

Notation: $\varepsilon = 0$ or 1.

No.	$f(x)$	$F(s)$
1	$(a^2-x^2)_+^{1/2}\,U_n\left(\frac{x}{a}\right)$	$\frac{n+1}{4}\sqrt{\pi}\,a^{s+1}\,\Gamma\left[\begin{matrix}\frac{s}{2},\,\frac{s+1}{2}\\\frac{s+n+3}{2},\,\frac{s-n+1}{2}\end{matrix}\right]$ $[a,\ \operatorname{Re} s > 0]$
2	$(x^2-a^2)_+^{1/2}\,U_n\left(\frac{x}{a}\right)$	$\frac{n+1}{4}\sqrt{\pi}\,a^{s+1}\,\Gamma\left[\begin{matrix}-\frac{s+n+1}{2},\,\frac{1-s+n}{2}\\\frac{1-s}{2},\,\frac{2-s}{2}\end{matrix}\right]$ $[a>0;\ \operatorname{Re} s < -(n+1)]$
3	$(a-x)_+^{\alpha-1}\,U_n\left(1-\frac{2x}{a}\right)$	$(n+1)\,a^{s+\alpha-1}\,\mathrm{B}(s,\,\alpha)\,{}_3F_2\left(\begin{matrix}-n,\,n+2,\,\alpha\\\frac{3}{2},\,s+\alpha;\,1\end{matrix}\right)$ $[a,\ \operatorname{Re}\alpha,\ \operatorname{Re} s > 0]$
4	$(a-x)_+^{\alpha-1}\,U_n\left(\frac{2x}{b}\pm 1\right)$	$(\pm 1)^n\,(n+1)\,a^{s+\alpha-1}\,\mathrm{B}(s,\,\alpha)\,{}_3F_2\left(\begin{matrix}-n,\,n+2,\,s\\\frac{3}{2},\,s+\alpha;\,\mp\frac{a}{b}\end{matrix}\right)$ $[a,\ \operatorname{Re}\alpha,\ \operatorname{Re} s > 0]$
5	$(x-a)_+^{-1/2}\,U_{2n}\left(i\sqrt{\frac{x}{a}-1}\right)$	$\sqrt{\pi}\,a^{s-1/2}\,\Gamma\left[\begin{matrix}s-\frac{1}{2},\,\frac{1}{2}-n-s\\s-n-\frac{1}{2},\,1-s\end{matrix}\right]$ $[\operatorname{Re} s < 1/2-n]$
6	$(a^2-x^2)_+\,U_n\left(\frac{x^2+a^2}{2ax}\right)$	$2\,(n+1)\,a^{s+2}\,\Gamma\left[\begin{matrix}s+n+2,\,s-n\\s+n+3,\,s-n+1\end{matrix}\right]$ $[a>0;\ \operatorname{Re} s > n]$
7	$(x^2-a^2)_+\,U_n\left(\frac{x^2+a^2}{2ax}\right)$	$2\,(n+1)\,a^{s+2}\,\Gamma\left[\begin{matrix}n-s,\,-n-s-2\\n-s+1,\,-n-s-1\end{matrix}\right]$ $[a>0;\ \operatorname{Re} s < -n-2]$
8	$(x-a)_+^{1/2}\,(2x-a)$ $\times U_n\left(\frac{8x^2-8ax+a^2}{a^2}\right)$	$\frac{n+1}{2}\sqrt{\pi}\,a^{s+3/2}\,\Gamma\left[\begin{matrix}2n-s+\frac{5}{2},\,-2n-s-\frac{3}{2}\\1-s,\,\frac{3}{2}-s\end{matrix}\right]$ $[a>0;\ \operatorname{Re} s < -2n-3/2]$
9	$(a-x)_+^{1/2}\,(2a-x)$ $\times U_n\left(\frac{x^2-8ax+8a^2}{x^2}\right)$	$\frac{n+1}{2}\sqrt{\pi}\,a^{s+3/2}\,\Gamma\left[\begin{matrix}s+2n+4,\,s-2n\\s+\frac{5}{2},\,s+3\end{matrix}\right]$ $[a>0;\ \operatorname{Re} s > 2n]$

No.	$f(x)$	$F(s)$
10	$\frac{x+2a}{(x+a)^{n+2}} \times U_n\left(\frac{x^2+2ax+2a^2}{2a(x+a)}\right)$	$\frac{a^{s-n-1}}{(2n+1)!}\Gamma\left[\begin{matrix} s,\ 1-s,\ 2n-s+3 \\ 2-s \end{matrix}\right]$ $[0<\operatorname{Re}s<1;\ \lvert\arg a\rvert<\pi]$
11	$\frac{2x+a}{(x+a)^{n+2}} \times U_n\left(\frac{2x^2+2ax+a^2}{2x(x+a)}\right)$	$\frac{a^{s-n-1}}{(2n+1)!}\Gamma\left[\begin{matrix} s-n,\ s+n+2,\ 1-s+n \\ s-n+1 \end{matrix}\right]$ $[n<\operatorname{Re}s<n+1;\ \lvert\arg a\rvert<\pi]$

3.21.2. Products of $U_n(\varphi(x))$

Notation: $\delta,\ \varepsilon = 0$ or 1.

No.	$f(x)$	$F(s)$
1	$(a^2-x^2)_+^{1/2}\sqrt{b^2-x^2} \times U_{2m+\varepsilon}\left(\frac{x}{b}\right)U_{2n+\delta}\left(\frac{x}{a}\right)$	$(-1)^{m+n}2^{\varepsilon-2}(m+1)^{\varepsilon}(2n+\delta+1)\sqrt{\pi}\,a^{s+\varepsilon+1}b^{-\varepsilon+1} \times \left(\frac{1-s+\delta-\varepsilon}{2}\right)_n\Gamma\left[\begin{matrix}\frac{s+\delta+\varepsilon}{2} \\ \frac{s+2n+\delta+\varepsilon+3}{2}\end{matrix}\right] \times {}_4F_3\left(\begin{matrix}-\frac{2m+1}{2},\ \frac{2m+2\varepsilon+1}{2},\ \frac{s+1}{2},\ \frac{s+2\varepsilon}{2} \\ \frac{2\varepsilon+1}{2},\ \frac{s-2n-\delta+\varepsilon+1}{2},\ \frac{s+2n+\delta+\varepsilon+3}{2};\ \frac{a^2}{b^2}\end{matrix}\right)$ $[b>a>0;\ \operatorname{Re}s>-\delta-\varepsilon]$
2	$(a^2-x^2)_+^{1/2}U_n\left(\frac{a}{x}\right) \times U_{2m+\varepsilon}(bx)$	$(-1)^m 2^{s+2\varepsilon-1}(m+1)^{\varepsilon}(n+1)\,a^{s+\varepsilon+1}b^{\varepsilon}\,\Gamma\left[\begin{matrix}\frac{s-n+\varepsilon}{2},\ \frac{s+n+\varepsilon+2}{2} \\ s+\varepsilon+2\end{matrix}\right] \times {}_4F_3\left(\begin{matrix}-m,\ m+\varepsilon+1,\ \frac{s-n+\varepsilon}{2},\ \frac{s+n+\varepsilon+2}{2} \\ \frac{2\varepsilon+1}{2},\ \frac{s+\varepsilon+2}{2},\ \frac{s+\varepsilon+3}{2};\ a^2b^2\end{matrix}\right)$ $[a>0;\ \operatorname{Re}s>n-\varepsilon]$
3	$(a-x)_+^{1/2}U_{2n}\left(\sqrt{\frac{x}{a}}\right) \times \left[U_m\left(\sqrt{1-bx}\right)\right]^2$	$(-1)^n(m+1)^2a^{s+1/2}\frac{\left(\frac{1-2s}{2}\right)_n}{(1/2)_n}\,\mathrm{B}\left(\frac{2n+3}{2},\ s\right) \times {}_5F_4\left(\begin{matrix}1,\ -m,\ m+2,\ \frac{2s+1}{2},\ s;\ ab \\ \frac{3}{2},\ 2,\ \frac{2s-2n+1}{2},\ \frac{2s+2n+3}{2}\end{matrix}\right)$ $[a,\ \operatorname{Re}s>0]$
4	$(a-x)_+^{\rho}P_n^{(\rho,\,\sigma)}\left(\frac{2x}{a}-1\right) \times \left[U_m\left(\sqrt{1-bx}\right)\right]^2$	$\frac{(-1)^n(m+1)^2a^{s+\rho}}{n!}(1-s+\sigma)_n\,\mathrm{B}(n+\rho+1,\ s) \times {}_5F_4\left(\begin{matrix}1,\ -m,\ m+2,\ s-\sigma,\ s;\ ab \\ \frac{3}{2},\ 2,\ s-n-\sigma,\ s+n+\rho+1\end{matrix}\right)$ $[a,\ \operatorname{Re}s>0;\ \operatorname{Re}\rho>-1]$

3.22. The Hermite Polynomials $H_n(z)$

More formulas can be obtained from the corresponding sections due to the relations

$$H_{2n+\varepsilon}(z) = (-1)^n 2^{2n+\varepsilon} n!\, z^{\varepsilon} L_n^{\varepsilon-1/2}(z^2), \quad \varepsilon = 0 \text{ or } 1;$$
$$H_n(z) = n! \lim_{\lambda\to\infty}\left[\lambda^{-n/2} C_n^{\lambda}\left(\frac{z}{\sqrt{\lambda}}\right)\right],$$
$$H_{2n+\varepsilon}(z) = (-1)^n \frac{(2n+\varepsilon)!}{n!}(2z)^{\varepsilon}\, {}_1F_1\left(-n;\, \varepsilon+\frac{1}{2};\, z^2\right), \quad \varepsilon = 0 \text{ or } 1;$$
$$H_{\nu}(z) = 2^{\nu}\sqrt{\pi}\left[\frac{1}{\Gamma\left(\frac{1-\nu}{2}\right)}\, {}_1F_1\left(-\frac{\nu}{2};\, \frac{1}{2};\, z^2\right) - \frac{2z}{\Gamma\left(-\frac{\nu}{2}\right)}\, {}_1F_1\left(\frac{1-\nu}{2};\, \frac{3}{2};\, z^2\right)\right],$$
$$H_n(z) = 2^n \Psi\left(-\frac{n}{2},\, \frac{1}{2};\, z^2\right) = 2^n \Psi\left(\frac{1-n}{2},\, \frac{3}{2};\, z^2\right),$$
$$H_{\nu}(z) = 2^{\nu/2} e^{z^2/2} D_{\nu}\left(\sqrt{2}\, z\right), \quad H_n(z) = 2^n e^{z^2} G_{12}^{20}\left(z^2 \,\middle|\, \begin{matrix}(1-n)/2\\ 0,\, 1/2\end{matrix}\right).$$

3.22.1. $H_n(bx)$ and algebraic functions

Notation: $\varepsilon = 0$ or 1.

No.	$f(x)$	$F(s)$
1	$(a-x)_+^{\alpha-1} H_{2n+\varepsilon}(bx)$	$\frac{(-1)^n (2n+\varepsilon)!}{n!} a^{s+\alpha+\varepsilon-1} (2b)^{\varepsilon}\, \mathrm{B}(s+\varepsilon,\, \alpha) \times {}_3F_3\left(\begin{matrix}-n,\, \frac{s+\varepsilon}{2},\, \frac{s+\varepsilon+1}{2};\, a^2b^2\\ \frac{2\varepsilon+1}{2},\, \frac{s+\alpha+\varepsilon}{2},\, \frac{s+\alpha+\varepsilon+1}{2}\end{matrix}\right)$ $[a,\ \mathrm{Re}\,\alpha > 0;\ \mathrm{Re}\, s > -\varepsilon]$
2	$(x-a)_+^{\alpha-1} H_{2n+\varepsilon}(bx)$	$\frac{(-1)^n (2n+\varepsilon)!}{n!} a^{s+\alpha+\varepsilon-1} (2b)^{\varepsilon}\, \mathrm{B}(1-s-\alpha-\varepsilon,\, \alpha) \times {}_3F_3\left(\begin{matrix}-n,\, \frac{s+\varepsilon}{2},\, \frac{s+\varepsilon+1}{2};\, a^2b^2\\ \frac{2\varepsilon+1}{2},\, \frac{s+\alpha+\varepsilon}{2},\, \frac{s+\alpha+\varepsilon+1}{2}\end{matrix}\right)$ $[a > 0;\ \mathrm{Re}(s+\alpha) < 1-2n-\varepsilon]$
3	$\left(a^2-x^2\right)_+^{\alpha-1} H_{2n+\varepsilon}(bx)$	$(-1)^n 2^{2n+\varepsilon-1} a^{s+2\alpha+\varepsilon-2} b^{\varepsilon} \left(\frac{2\varepsilon+1}{2}\right)_n \mathrm{B}\left(\alpha,\, \frac{s+\varepsilon}{2}\right) \times {}_2F_2\left(\begin{matrix}-n,\, \frac{s+\varepsilon}{2};\, a^2b^2\\ \frac{2\varepsilon+1}{2},\, \frac{s+2\alpha+\varepsilon}{2}\end{matrix}\right)$ $[a,\ \mathrm{Re}\,\alpha > 0;\ \mathrm{Re}\, s > -\varepsilon]$
4	$\left(x^2-a^2\right)_+^{\alpha-1} H_{2n+\varepsilon}(bx)$	$(-1)^n 2^{2n+\varepsilon-1} a^{s+2\alpha+\varepsilon-2} b^{\varepsilon} \left(\frac{2\varepsilon+1}{2}\right)_n \times \mathrm{B}\left(\alpha,\, \frac{2-2\alpha-s-\varepsilon}{2}\right) {}_2F_2\left(\begin{matrix}-n,\, \frac{s+\varepsilon}{2};\, a^2b^2\\ \frac{2\varepsilon+1}{2},\, \frac{s+2\alpha+\varepsilon}{2}\end{matrix}\right)$ $[a,\ \mathrm{Re}\,\alpha > 0;\ \mathrm{Re}(s+2\alpha) < 2-2n-\varepsilon]$

No.	$f(x)$	$F(s)$
5	$\frac{1}{(x^2+a^2)^\rho} H_{2n+\varepsilon}(bx)$	$(-1)^n 2^{2n+\varepsilon-1} a^{s-2\rho+\varepsilon} b^\varepsilon \left(\frac{2\varepsilon+1}{2}\right)_n \mathrm{B}\left(\frac{s+\varepsilon}{2}, \frac{2\rho-s-\varepsilon}{2}\right) \times {}_2F_2\left(\begin{matrix}-n, \frac{s+\varepsilon}{2}; -a^2b^2\\ \frac{2\varepsilon+1}{2}, \frac{s-2\rho+\varepsilon+2}{2}\end{matrix}\right)$ [$\mathrm{Re}\, a > 0;\ -\varepsilon < \mathrm{Re}\, s < 2\,\mathrm{Re}\,\rho - 2n - \varepsilon$]
6	$\frac{1}{(x+a)^\rho} \times H_{2n+\varepsilon}\left(\frac{bx}{x+a}\right)$	$(-1)^n 2^{2n+\varepsilon} \left(\frac{2\varepsilon+1}{2}\right)_n a^{s-\rho} b^\varepsilon \mathrm{B}(s+\varepsilon, \rho-s) \times {}_3F_3\left(\begin{matrix}-n, \frac{s+\varepsilon}{2}, \frac{s+\varepsilon+1}{2}\\ \frac{2\varepsilon+1}{2}, \frac{\rho+\varepsilon}{2}, \frac{\rho+\varepsilon+1}{2}; b^2\end{matrix}\right)$ [$-\varepsilon < \mathrm{Re}\, s < \mathrm{Re}\,\rho;\ \lvert\arg a\rvert < \pi$]
7	$\frac{1}{(x^2+a^2)^\rho} \times H_{2n+\varepsilon}\left(\frac{bx}{x^2+a^2}\right)$	$(-1)^n 2^{2n+\varepsilon-1} \left(\frac{2\varepsilon+1}{2}\right)_n a^{s-2\rho-\varepsilon} b^\varepsilon \times \mathrm{B}\left(\frac{s+\varepsilon}{2}, \frac{-s+2\rho+\varepsilon}{2}\right) {}_3F_3\left(\begin{matrix}-n, \frac{s+\varepsilon}{2}, \frac{-s+2\rho+\varepsilon}{2}\\ \frac{2\varepsilon+1}{2}, \frac{\rho+\varepsilon}{2}, \frac{\rho+\varepsilon+1}{2}; \frac{b^2}{4a^2}\end{matrix}\right)$ [$\mathrm{Re}\, a > 0;\ -\varepsilon < \mathrm{Re}\, s < 2\,\mathrm{Re}\,\rho + \varepsilon$]
8	$(a-x)_+^{\alpha-1} \times H_{2n+\varepsilon}(b\sqrt{x(a-x)})$	$\frac{(-1)^n (2n+\varepsilon)!}{n!} a^{s+\alpha+\varepsilon-1} (2b)^\varepsilon \mathrm{B}\left(\frac{2s+\varepsilon}{2}, \frac{2\alpha+\varepsilon}{2}\right) \times {}_3F_3\left(\begin{matrix}-n, \frac{2s+\varepsilon}{2}, \frac{2\alpha+\varepsilon}{2}; \frac{a^2b^2}{4}\\ \frac{2\varepsilon+1}{2}, \frac{s+\alpha+\varepsilon}{2}, \frac{s+\alpha+\varepsilon+1}{2}\end{matrix}\right)$ [$a > 0;\ \mathrm{Re}\, s,\ \mathrm{Re}\,\alpha > -\varepsilon/2$]

3.22.2. $H_n(bx)$ and the exponential function

Notation: $\varepsilon = 0$ or 1.

1	$e^{-ax} H_n(bx)$	$\frac{(2b)^n}{a^{s+n}} \Gamma(s+n)\, {}_2F_2\left(\begin{matrix}-\frac{n}{2}, \frac{1-n}{2}; -\frac{a^2}{4b^2}\\ \frac{1-s-n}{2}, \frac{2-s-n}{2}\end{matrix}\right)$ [$\mathrm{Re}\, a > 0;\ \mathrm{Re}\, s > 2[n/2] - n$]
2	$e^{-a^2x^2} H_n(ax)$	$\frac{2^{n-1}}{a^s} \Gamma\left[\begin{matrix}\frac{s}{2}, \frac{s+1}{2}\\ \frac{s-n+1}{2}\end{matrix}\right]$ [$\mathrm{Re}\, s > 0;\ \lvert\arg a\rvert < \pi/4$]
3	$e^{-ax^2} H_n(bx)$	$\frac{n!}{2} a^{-s/2} \Gamma\left(\frac{s-n}{2}\right) C_n^{((s-n)/2)}\left(\frac{b}{\sqrt{a}}\right)$ [$\mathrm{Re}\, a,\ \mathrm{Re}\, s > 0$]
4	$(a-x)_+^{\alpha-1} e^{-b^2x^2} \times H_{2n+\varepsilon}(bx)$	$(-1)^n 2^{2n+\varepsilon} a^{s+\alpha+\varepsilon-1} b^\varepsilon \left(\frac{2\varepsilon+1}{2}\right)_n \mathrm{B}(\alpha, s+\varepsilon) \times {}_3F_3\left(\begin{matrix}\frac{2n+2\varepsilon+1}{2}, \frac{s+1}{2}, \frac{s+2\varepsilon}{2}; -a^2b^2\\ \frac{2\varepsilon+1}{2}, \frac{s+\alpha+1}{2}, \frac{s+\alpha+2\varepsilon}{2}\end{matrix}\right)$ [$a,\ \mathrm{Re}\,\alpha > 0;\ \mathrm{Re}\, s > -\varepsilon$]

No.	$f(x)$	$F(s)$
5	$(x-a)_+^{\alpha-1}\, e^{-b^2x^2} \times H_{2n+\varepsilon}(bx)$	$(-1)^n\, 2^{2n+\varepsilon} a^{s+\alpha+\varepsilon-1} b^{\varepsilon} \left(\frac{2\varepsilon+1}{2}\right)_n \mathrm{B}(\alpha,\, 1-s-\alpha-\varepsilon) \times {}_3F_3\left(\begin{matrix}\frac{2n+2\varepsilon+1}{2},\, \frac{s+1}{2},\, \frac{s+2\varepsilon}{2};\, -a^2b^2\\ \frac{2\varepsilon+1}{2},\, \frac{s+\alpha+1}{2},\, \frac{s+\alpha+2\varepsilon}{2}\end{matrix}\right) + (-1)^n \frac{2^{2n+\varepsilon-1}}{b^{s+\alpha-1}} \left(\frac{2-s-\alpha+\varepsilon}{2}\right)_n \Gamma\left(\frac{s+\alpha+\varepsilon-1}{2}\right) \times {}_3F_3\left(\begin{matrix}\frac{1-\alpha}{2},\, \frac{2-\alpha}{2},\, \frac{2-s+2n-\alpha+\varepsilon}{2}\\ \frac{1}{2},\, \frac{2-s-\alpha}{2},\, \frac{3-s-\alpha}{2};\, -a^2b^2\end{matrix}\right) + (-1)^n (1-\alpha) \frac{2^{2n+\varepsilon-1}a}{b^{s+\alpha-2}} \left(\frac{3-s-\alpha+\varepsilon}{2}\right)_n \times \Gamma\left(\frac{s+\alpha+\varepsilon-2}{2}\right) {}_3F_3\left(\begin{matrix}\frac{2-\alpha}{2},\, \frac{3-\alpha}{2},\, \frac{3-s+2n-\alpha+\varepsilon}{2}\\ \frac{3}{2},\, \frac{3-s-\alpha}{2},\, \frac{4-s-\alpha}{2};\, -a^2b^2\end{matrix}\right)$ $[a,\ \mathrm{Re}\,\alpha>0;\ \lvert\arg b\rvert<\pi/4]$
6	$(a^2-x^2)_+^{\alpha-1}\, e^{-b^2x^2} \times H_{2n+\varepsilon}(bx)$	$(-1)^n\, 2^{2n+\varepsilon-1} a^{s+\varepsilon+2\alpha-1} b^{\varepsilon}\, \mathrm{B}\left(\alpha,\, \frac{s+\varepsilon}{2}\right) \left(\frac{2\varepsilon+1}{2}\right)_n \times {}_2F_2\left(\begin{matrix}\frac{2n+2\varepsilon+1}{2},\, \frac{s+\varepsilon}{2};\, -a^2b^2\\ \frac{2\varepsilon+1}{2},\, \frac{s+\varepsilon+2\alpha}{2}\end{matrix}\right)$ $[a,\ \mathrm{Re}\,\alpha>0;\ \mathrm{Re}\,s>-\varepsilon]$
7	$(x^2-a^2)_+^{\alpha-1}\, e^{-b^2x^2} \times H_{2n+\varepsilon}(bx)$	$(-1)^n\, 2^{2n+\varepsilon-1} a^{s+2\alpha+\varepsilon-2} b^{\varepsilon}\, \mathrm{B}\left(\alpha,\, \frac{2-s-2\alpha-\varepsilon}{2}\right) \times \left(\frac{2\varepsilon+1}{2}\right)_n {}_2F_2\left(\begin{matrix}\frac{2n+2\varepsilon+1}{2},\, \frac{s+\varepsilon}{2}\\ \frac{2\varepsilon+1}{2},\, \frac{s+2\alpha+\varepsilon}{2};\, -a^2b^2\end{matrix}\right) + (-1)^n\, 2^{2n+\varepsilon-1} b^{2-s-2\alpha} \left(\frac{3-s-2\alpha+\varepsilon}{2}\right)_n \times \Gamma\left(\frac{s+2\alpha+\varepsilon-2}{2}\right) {}_2F_2\left(\begin{matrix}1-\alpha,\, \frac{3-s+2n-2\alpha+\varepsilon}{2}\\ \frac{3-s-2\alpha}{2},\, \frac{4-s-2\alpha}{2};\, -a^2b^2\end{matrix}\right)$ $[a,\ \mathrm{Re}\,\alpha>0;\ \lvert\arg b\rvert<\pi/4]$
8	$(a^4-x^4)_+^{\alpha-1}\, e^{-b^2x^2} \times H_{2n+\varepsilon}(bx)$	$(-1)^n\, 2^{2n+\varepsilon-2} a^{s+4\alpha+\varepsilon-4} b^{\varepsilon}\, \mathrm{B}\left(\alpha,\, \frac{s+\varepsilon}{4}\right) \times \left(\frac{2\varepsilon+1}{2}\right)_n {}_3F_4\left(\begin{matrix}\frac{2n+3}{4},\, \frac{2n+4\varepsilon+1}{4},\, \frac{s+\varepsilon}{4};\, \frac{a^4b^4}{4}\\ \frac{1}{2},\, \frac{3}{4},\, \frac{4\varepsilon+1}{4},\, \frac{s+4\alpha+\varepsilon}{4}\end{matrix}\right) - (-1)^n\, 2^{2n+\varepsilon-2} a^{s+4\alpha+\varepsilon-2} b^{\varepsilon+2}\, \mathrm{B}\left(\alpha,\, \frac{s+\varepsilon+2}{4}\right) \times \left(\frac{2\varepsilon+3}{2}\right)_n {}_3F_4\left(\begin{matrix}\frac{2n+5}{4},\, \frac{2n+4\varepsilon+3}{4},\, \frac{s+\varepsilon+2}{4};\, \frac{a^4b^4}{4}\\ \frac{5}{4},\, \frac{3}{2},\, \frac{4\varepsilon+3}{4},\, \frac{s+4\alpha+\varepsilon+2}{4}\end{matrix}\right)$ $[a,\ \mathrm{Re}\,\alpha>0;\ \mathrm{Re}\,s>-\varepsilon]$

No.	$f(x)$	$F(s)$
9	$\frac{e^{-b^2x^2}}{(x+a)^\rho} H_{2n+\varepsilon}(bx)$	$(-1)^n 2^{2n+\varepsilon} a^{s-\rho+\varepsilon} b^\varepsilon \mathrm{B}(s+\varepsilon, \rho-s-\varepsilon) \times \left(\frac{2\varepsilon+1}{2}\right)_n {}_3F_3\left(\begin{matrix} \frac{2n+2\varepsilon+1}{2}, \frac{s+1}{2}, \frac{s+2\varepsilon}{2}; -a^2b^2 \\ \frac{2\varepsilon+1}{2}, \frac{s-\rho+2}{2}, \frac{s-\rho+2\varepsilon+1}{2} \end{matrix}\right) + (-1)^n 2^{2n+\varepsilon-1} b^{\rho-s} \left(\frac{1-s+\rho+\varepsilon}{2}\right)_n \Gamma\left(\frac{s-\rho+\varepsilon}{2}\right) \times {}_3F_3\left(\begin{matrix} \frac{\rho}{2}, \frac{\rho+1}{2}, \frac{1-s+2n+\rho+\varepsilon}{2} \\ \frac{1}{2}, \frac{1-s+\rho}{2}, \frac{2-s+\rho}{2}; -a^2b^2 \end{matrix}\right) - (-1)^n 2^{2n+\varepsilon-1} \rho a b^{1-s+\rho} \left(\frac{2-s+\rho+\varepsilon}{2}\right)_n \times \Gamma\left(\frac{s-\rho+\varepsilon-1}{2}\right) {}_3F_3\left(\begin{matrix} \frac{\rho+1}{2}, \frac{\rho+2}{2}, \frac{2-s+2n+\rho+\varepsilon}{2} \\ \frac{3}{2}, \frac{2-s+\rho}{2}, \frac{3-s+\rho}{2}; -a^2b^2 \end{matrix}\right)$ $[\mathrm{Re}\, s > -\varepsilon;\ \lvert\arg a\rvert, 4\lvert\arg b\rvert < \pi]$
10	$\frac{e^{-b^2x^2}}{(x^2+a^2)^\rho} H_{2n+\varepsilon}(bx)$	$(-1)^n 2^{2n+\varepsilon-1} a^{s-2\rho+\varepsilon} b^\varepsilon \left(\frac{2\varepsilon+1}{2}\right)_n \times \mathrm{B}\left(\frac{s+\varepsilon}{2}, \frac{2\rho-s-\varepsilon}{2}\right) {}_2F_2\left(\begin{matrix} \frac{s+\varepsilon}{2}, \frac{2n+2\varepsilon+1}{2}; a^2b^2 \\ \frac{2\varepsilon+1}{2}, \frac{s-2\rho+\varepsilon+2}{2} \end{matrix}\right) + (-1)^n 2^{2n+\varepsilon-1} b^{2\rho-s} \left(\frac{1-s+2\rho+\varepsilon}{2}\right)_n \Gamma\left(\frac{s-2\rho+\varepsilon}{2}\right) \times {}_2F_2\left(\begin{matrix} \rho, \frac{1-s+2n+2\rho+\varepsilon}{2}; a^2b^2 \\ \frac{1-s+2\rho}{2}, \frac{2-s+2\rho}{2} \end{matrix}\right)$ $[\mathrm{Re}\, a > 0;\ \mathrm{Re}\, s > -\varepsilon;\ \lvert\arg b\rvert < \pi/4]$
11	$\frac{e^{-b^2x^2}}{x-a} H_{2n+\varepsilon}(bx)$	$(-1)^{n+1} 2^{2n+\varepsilon} \pi \cot(s\pi)\, a^{s+\varepsilon-1} b^\varepsilon \left(\frac{2\varepsilon+1}{2}\right)_n \times {}_1F_1\left(\begin{matrix} \frac{2n+2\varepsilon+1}{2} \\ \frac{2\varepsilon+1}{2}; -a^2b^2 \end{matrix}\right) + \frac{(-1)^n 2^{2n+\varepsilon-1}}{b^{s-1}} \left(\frac{2-s+\varepsilon}{2}\right)_n \times \Gamma\left(\frac{s+\varepsilon-1}{2}\right) {}_2F_2\left(\begin{matrix} 1, \frac{2-s+2n+\varepsilon}{2} \\ \frac{2-s}{2}, \frac{3-s}{2}; -a^2b^2 \end{matrix}\right) + \frac{(-1)^n 2^{2n+\varepsilon-1} a}{b^{s-2}} \left(\frac{3-s+\varepsilon}{2}\right)_n \Gamma\left(\frac{s+\varepsilon-2}{2}\right) \times {}_2F_2\left(\begin{matrix} 1, \frac{3-s+2n+\varepsilon}{2} \\ \frac{3-s}{2}, \frac{4-s}{2}; -a^2b^2 \end{matrix}\right)$ $[a > 0;\ \mathrm{Re}\, s > -\varepsilon;\ \lvert\arg b\rvert < \pi/4]$

No.	$f(x)$	$F(s)$
12	$\dfrac{e^{-b^2x^2}}{x^2-a^2}\,H_{2n+\varepsilon}(bx)$	$(-1)^{n+1}\,2^{2n+\varepsilon-1}\pi\cot\dfrac{(s+\varepsilon)\pi}{2}\,a^{s+\varepsilon-2}b^{\varepsilon}\left(\dfrac{2\varepsilon+1}{2}\right)_n \times {}_1F_1\left(\begin{matrix}\frac{2n+2\varepsilon+1}{2}\\ \frac{2\varepsilon+1}{2};\ -a^2b^2\end{matrix}\right)+\dfrac{(-1)^n\,2^{2n+\varepsilon-1}}{b^{s-2}}\left(\dfrac{3-s+\varepsilon}{2}\right)_n \times\Gamma\left(\dfrac{s+\varepsilon-2}{2}\right){}_2F_2\left(\begin{matrix}1,\ \frac{3-s+2n+\varepsilon}{2}\\ \frac{3-s}{2},\ \frac{4-s}{2};\ -a^2b^2\end{matrix}\right)$ $[a>0;\ \operatorname{Re}s>-\varepsilon;\ \lvert\arg b\rvert<\pi/4]$
13	$e^{-a/x}H_{2n+\varepsilon}(bx)$	$(-1)^n\,2^{2n+\varepsilon}a^{s+\varepsilon}b^{\varepsilon}\left(\dfrac{2\varepsilon+1}{2}\right)_n\Gamma(-s-\varepsilon) \times{}_1F_3\left(\begin{matrix}-n;\ \frac{a^2b^2}{4}\\ \frac{2\varepsilon+1}{2},\ \frac{s+\varepsilon+1}{2},\ \frac{s+\varepsilon+2}{2}\end{matrix}\right)$ $[\operatorname{Re}a>0;\ \operatorname{Re}s<-2n-\varepsilon]$
14	$e^{-a/x^2}H_{2n+\varepsilon}(bx)$	$(-1)^n\,2^{2n+\varepsilon-1}a^{(s+\varepsilon)/2}b^{\varepsilon}\left(\dfrac{2\varepsilon+1}{2}\right)_n \times\Gamma\left(-\dfrac{s+\varepsilon}{2}\right){}_1F_2\left(\begin{matrix}-n;\ -ab^2\\ \frac{2\varepsilon+1}{2},\ \frac{s+\varepsilon+2}{2}\end{matrix}\right)$ $[\operatorname{Re}a>0;\ \operatorname{Re}s<-2n-\varepsilon]$
15	$e^{-a\sqrt{x}}H_{2n+\varepsilon}(bx)$	$(-1)^n\,2^{2n+2\varepsilon+1}a^{-2s-2\varepsilon}b^{\varepsilon}\left(\dfrac{1}{2}\right)_{n+\varepsilon}\Gamma(2s+2\varepsilon) \times{}_5F_1\left(\begin{matrix}-n,\ \Delta(4,\ 2s+2\varepsilon)\\ \frac{2\varepsilon+1}{2};\ \frac{256b^2}{a^4}\end{matrix}\right)$ $[\operatorname{Re}a>0;\ \operatorname{Re}s>-\varepsilon]$
16	$e^{-a/\sqrt{x}}H_{2n+\varepsilon}(bx)$	$(-1)^n\,2^{2n+2\varepsilon+1}a^{2s+2\varepsilon}b^{\varepsilon}\left(\dfrac{1}{2}\right)_{n+\varepsilon}\Gamma(-2s-2\varepsilon) \times{}_1F_5\left(\begin{matrix}-n;\ \frac{a^4b^2}{256}\\ \frac{2\varepsilon+1}{2},\ \Delta(4,\ 2s+2\varepsilon)\end{matrix}\right)$ $[\operatorname{Re}a>0;\ \operatorname{Re}s<-2n-\varepsilon]$
17	$e^{-ax-b^2x^2}H_{2n+\varepsilon}(bx)$	$\dfrac{(-1)^n\,2^{2n+\varepsilon-1}}{b^s}\left(\dfrac{1-s+\varepsilon}{2}\right)_n\Gamma\left(\dfrac{s+\varepsilon}{2}\right) \times{}_2F_2\left(\begin{matrix}\frac{s}{2},\ \frac{s+1}{2};\ \frac{a^2}{4b^2}\\ \frac{1}{2},\ \frac{s-2n-\varepsilon+1}{2}\end{matrix}\right)-\dfrac{(-1)^n\,2^{2n+\varepsilon-1}a}{b^{s+1}}\left(\dfrac{-s+\varepsilon}{2}\right)_n \times\Gamma\left(\dfrac{s+\varepsilon+1}{2}\right){}_2F_2\left(\begin{matrix}\frac{s+1}{2},\ \frac{s+2}{2};\ \frac{a^2}{4b^2}\\ \frac{3}{2},\ \frac{s-2n-\varepsilon+2}{2}\end{matrix}\right)$ $[\operatorname{Re}a>0;\ \operatorname{Re}s>-\varepsilon;\ \lvert\arg b\rvert<\pi/4]$

No.	$f(x)$	$F(s)$
18	$e^{-ax^4-b^2x^2}H_{2n+\varepsilon}(bx)$	$(-1)^n \frac{2^{2n+\varepsilon-2}b^\varepsilon}{a^{(s+\varepsilon)/4}}\left(\frac{2\varepsilon+1}{2}\right)_n \Gamma\left(\frac{s+\varepsilon}{4}\right) \times {}_3F_3\left(\begin{matrix}\frac{2n+2\varepsilon+1}{4}, \frac{2n+2\varepsilon+3}{4}, \frac{s+\varepsilon}{4} \\ \frac{1}{2}, \frac{3}{4}, \frac{4\varepsilon+1}{4}; \frac{b^4}{4a}\end{matrix}\right) - (-1)^n \frac{2^{2n+\varepsilon-2}b^{\varepsilon+2}}{a^{(s+\varepsilon+2)/4}}\left(\frac{2\varepsilon+3}{2}\right)_n \Gamma\left(\frac{s+\varepsilon+2}{4}\right) \times {}_3F_3\left(\begin{matrix}\frac{2n+2\varepsilon+3}{4}, \frac{2n+2\varepsilon+5}{4}, \frac{s+\varepsilon+2}{4} \\ \frac{5}{4}, \frac{3}{2}, \frac{4\varepsilon+3}{4}; \frac{b^4}{4a}\end{matrix}\right)$ [$\operatorname{Re} a > 0$; $\operatorname{Re} s > -\varepsilon$; $\lvert\arg b\rvert < \pi/4$]
19	$e^{-a/x-b^2x^2}H_{2n+\varepsilon}(bx)$	$(-1)^n 2^{2n+\varepsilon}a^{s+\varepsilon}b^\varepsilon \left(\frac{2\varepsilon+1}{2}\right)_n \Gamma(-s-\varepsilon) \times {}_1F_3\left(\begin{matrix}\frac{2n+2\varepsilon+1}{2}; -\frac{a^2b^2}{4} \\ \frac{2\varepsilon+1}{2}, \frac{s+\varepsilon+1}{2}, \frac{s+\varepsilon+2}{2}\end{matrix}\right) + \frac{(-1)^n 2^{2n+\varepsilon-1}}{b^s} \times \left(\frac{1-s+\varepsilon}{2}\right)_n \Gamma\left(\frac{s+\varepsilon}{2}\right) {}_1F_3\left(\begin{matrix}\frac{1-s+2n+\varepsilon}{2}; -\frac{a^2b^2}{4} \\ \frac{1}{2}, \frac{1-s}{2}, \frac{2-s}{2}\end{matrix}\right) - \frac{(-1)^n 2^{2n+\varepsilon-1}a}{b^{s-1}}\left(\frac{2-s+\varepsilon}{2}\right)_n \times \Gamma\left(\frac{s+\varepsilon-1}{2}\right) {}_1F_3\left(\begin{matrix}\frac{2-s+2n+\varepsilon}{2}; -\frac{a^2b^2}{4} \\ \frac{3}{2}, \frac{2-s}{2}, \frac{3-s}{2}\end{matrix}\right)$ [$\operatorname{Re} a > 0$; $\lvert\arg b\rvert < \pi/4$]
20	$e^{-a/x^2-b^2x^2}H_{2n+\varepsilon}(bx)$	$(-1)^n 2^{2n+\varepsilon-1}a^{(s+\varepsilon)/2}b^\varepsilon \left(\frac{2\varepsilon+1}{2}\right)_n \Gamma\left(-\frac{s+\varepsilon}{2}\right) \times {}_1F_2\left(\begin{matrix}\frac{2n+2\varepsilon+1}{2}; ab^2 \\ \frac{2\varepsilon+1}{2}, \frac{s+\varepsilon+2}{2}\end{matrix}\right) + \frac{(-1)^n 2^{2n+\varepsilon-1}}{b^s} \times \left(\frac{1-s+\varepsilon}{2}\right)_n \Gamma\left(\frac{s+\varepsilon}{2}\right) {}_1F_2\left(\begin{matrix}\frac{1-s+2n+\varepsilon}{2}; ab^2 \\ \frac{1-s}{2}, \frac{2-s}{2}\end{matrix}\right)$ [$\operatorname{Re} a > 0$; $\lvert\arg b\rvert < \pi/4$]
21	$e^{-a/x^4-b^2x^2}H_{2n+\varepsilon}(bx)$	$(-1)^n 2^{2n+\varepsilon-2}a^{(s+\varepsilon)/4}b^\varepsilon \left(\frac{2\varepsilon+1}{2}\right)_n \Gamma\left(-\frac{s+\varepsilon}{4}\right) \times {}_2F_4\left(\begin{matrix}\frac{2n+2\varepsilon+1}{4}, \frac{2n+2\varepsilon+3}{4}; -\frac{ab^4}{4} \\ \frac{1}{2}, \frac{3}{4}, \frac{4\varepsilon+1}{4}, \frac{s+\varepsilon+4}{4}\end{matrix}\right) - (-1)^n 2^{2n+\varepsilon-2}a^{(s+\varepsilon+2)/4}b^{\varepsilon+2} \times \left(\frac{2\varepsilon+3}{2}\right)_n \Gamma\left(-\frac{s+\varepsilon+2}{4}\right) {}_2F_4\left(\begin{matrix}\frac{2n+2\varepsilon+3}{4}, \frac{2n+2\varepsilon+5}{4}; -\frac{ab^4}{4} \\ \frac{5}{4}, \frac{3}{2}, \frac{4\varepsilon+3}{4}, \frac{s+\varepsilon+6}{4}\end{matrix}\right) + \frac{(-1)^n 2^{2n+\varepsilon-1}}{b^s}\left(\frac{1-s+\varepsilon}{2}\right)_n \Gamma\left(\frac{s+\varepsilon}{2}\right) \times {}_2F_4\left(\begin{matrix}\frac{1-s+2n+\varepsilon}{4}, \frac{3-s+2n+\varepsilon}{4}; -\frac{ab^4}{4} \\ \frac{1-s}{4}, \frac{2-s}{4}, \frac{3-s}{4}, \frac{4-s}{4}\end{matrix}\right)$ [$\operatorname{Re} a > 0$; $\lvert\arg b\rvert < \pi/4$]

No.	$f(x)$	$F(s)$
22	$(a-x)_+^{(\varepsilon-1)/2}\, e^{bx} \times H_{2n+\varepsilon}\left(c\sqrt{a-x}\right)$	$(-1)^n\, 2^{2n}\sqrt{\pi}\, a^{s+\varepsilon-1/2}c^{\varepsilon}\left(\frac{2\varepsilon+1}{2}\right)_n \Gamma\left[\begin{matrix} s \\ \frac{2s+2\varepsilon+1}{2}\end{matrix}\right] \times \Phi_2\left(s,\, -n;\, \frac{2s+2\varepsilon+1}{2};\, ab,\, ac^2\right)$ $[a,\ \operatorname{Re} s>0]$

3.22.3. $H_n(bx)$ and trigonometric functions

Notation: $\delta = \begin{Bmatrix} 1 \\ 0 \end{Bmatrix}$, $\varepsilon = 0$ or 1.

1	$e^{-b^2x^2}\begin{Bmatrix}\sin(ax)\\ \cos(ax)\end{Bmatrix} \times H_{2n+\varepsilon}(bx)$	$(-1)^n \frac{2^{2n+\varepsilon-1}a^{\delta}}{b^{s+\delta}}\left(\frac{1-s-\delta+\varepsilon}{2}\right)_n \Gamma\left(\frac{s+\delta+\varepsilon}{2}\right) \times {}_2F_2\left(\begin{matrix}\frac{s+1}{2},\, \frac{s+2\delta}{2};\, -\frac{a^2}{4b^2}\\ \frac{2\delta+1}{2},\, \frac{s-2n+\delta-\varepsilon+1}{2}\end{matrix}\right)$ $\left[\begin{matrix}a>0;\ \operatorname{Re} s>-\delta-\varepsilon;\\ \lvert\arg b\rvert<\pi/4\end{matrix}\right]$
2	$e^{-b^2x^2}\begin{Bmatrix}\sin(ax^2)\\ \cos(ax^2)\end{Bmatrix} \times H_{2n+\varepsilon}(bx)$	$(-1)^n \frac{2^{2n+\varepsilon-1}a^{\delta}}{b^{s+2\delta}}\left(\frac{1-s-2\delta+\varepsilon}{2}\right)_n \Gamma\left(\frac{s+2\delta+\varepsilon}{2}\right) \times {}_4F_3\left(\begin{matrix}\frac{s+2}{4},\, \frac{s+3}{4},\, \frac{s+4\delta}{4},\, \frac{s+4\delta+1}{4};\, -\frac{a^2}{b^4}\\ \frac{2\delta+1}{2},\, \frac{s-2n-\varepsilon+3}{4},\, \frac{s-2n+4\delta-\varepsilon+1}{4}\end{matrix}\right)$ $[a>0;\ \operatorname{Re} s>-2\delta-\varepsilon;\ \lvert\arg b\rvert<\pi/4]$
3	$e^{-b^2x^2}\begin{Bmatrix}\sin(a/x)\\ \cos(a/x)\end{Bmatrix} \times H_{2n+\varepsilon}(bx)$	$(-1)^n\, 2^{2n+\varepsilon-1}a^{\delta}b^{\delta-s}\left(\frac{1-s+\delta+\varepsilon}{2}\right)_n \Gamma\left(\frac{s-\delta+\varepsilon}{2}\right) \times {}_1F_3\left(\begin{matrix}\frac{1-s+2n+\delta+\varepsilon}{2};\, \frac{a^2b^2}{4}\\ \frac{2\delta+1}{2},\, \frac{2-s+\delta-\varepsilon}{2},\, \frac{1-s+\delta+\varepsilon}{2}\end{matrix}\right) \mp (-1)^n\, 2^{2n+\varepsilon}a^{s+\varepsilon}b^{\varepsilon} \times \left(\frac{2\varepsilon+1}{2}\right)_n \Gamma(-s-\varepsilon)\begin{Bmatrix}\sin[(s+\varepsilon)\pi/2]\\ \cos[(s+\varepsilon)\pi/2]\end{Bmatrix} \times {}_1F_3\left(\begin{matrix}\frac{2n+2\varepsilon+1}{2};\, \frac{a^2b^2}{4}\\ \frac{2\varepsilon+1}{2},\, \frac{s+\varepsilon+1}{2},\, \frac{s+\varepsilon+2}{2}\end{matrix}\right)$ $\left[\begin{matrix}a>0;\ \operatorname{Re}(s+\varepsilon)>-1;\\ \lvert\arg b\rvert<\pi/4\end{matrix}\right]$

3.22.4. $H_n(bx)$ and the logarithmic function

Notation: $\varepsilon = 0$ or 1.

1	$\theta(a-x)\ln\frac{\sqrt{a}+\sqrt{a-x}}{\sqrt{x}} \times H_{2n+\varepsilon}(bx)$	$\frac{(-1)^n\, 2^{\varepsilon-1}(2n+\varepsilon)!\sqrt{\pi}\, a^{s+\varepsilon}b^{\varepsilon}}{n!\,(s+\varepsilon)}\Gamma\left[\begin{matrix}s+\varepsilon\\ \frac{2s+2\varepsilon+1}{2}\end{matrix}\right] \times {}_4F_4\left(\begin{matrix}-n,\, \frac{s+\varepsilon}{2},\, \frac{s+\varepsilon}{2},\, \frac{s+\varepsilon+1}{2};\, a^2b^2\\ \frac{2\varepsilon+1}{2},\, \frac{2s+2\varepsilon+1}{4},\, \frac{2s+2\varepsilon+3}{4},\, \frac{s+\varepsilon+2}{2}\end{matrix}\right)$ $[a>0;\ \operatorname{Re} s>-\varepsilon]$

No.	$f(x)$	$F(s)$
2	$e^{-b^2x^2}\begin{Bmatrix}\ln(x+a)\\ \ln\|x-a\|\end{Bmatrix}$ $\times H_{2n+\varepsilon}(bx)$	$(-1)^n \frac{2^{2n+\varepsilon}\pi}{s+\varepsilon} a^{s+\varepsilon}b^{\varepsilon}\left(\frac{2\varepsilon+1}{2}\right)_n \begin{Bmatrix}\csc[(s+\varepsilon)\pi]\\ \cot[(s+\varepsilon)\pi]\end{Bmatrix} \times {}_2F_2\left(\begin{matrix}\frac{2n+2\varepsilon+1}{2}, \frac{s+\varepsilon}{2}; -a^2b^2\\ \frac{2\varepsilon+1}{2}, \frac{s+\varepsilon+2}{2}\end{matrix}\right) \pm (-1)^n \frac{2^{2n+\varepsilon-1}a}{b^{s-1}}\left(\frac{2-s+\varepsilon}{2}\right)_n \Gamma\left(\frac{s+\varepsilon-1}{2}\right) \times {}_3F_3\left(\begin{matrix}\frac{1}{2}, 1, \frac{2n-s+\varepsilon+2}{2}; -a^2b^2\\ \frac{3}{2}, \frac{2-s}{2}, \frac{3-s}{2}\end{matrix}\right) - (-1)^n \frac{2^{2n+\varepsilon-2}a^2}{b^{s-2}}\left(\frac{3-s+\varepsilon}{2}\right)_n \Gamma\left(\frac{s+\varepsilon-2}{2}\right) \times {}_3F_3\left(\begin{matrix}1, 1, \frac{2n-s+\varepsilon+3}{2}; -a^2b^2\\ 2, \frac{3-s}{2}, \frac{4-s}{2}\end{matrix}\right) + (-1)^n \frac{2^{2n+\varepsilon-2}}{b^s}\left(\frac{1-s+\varepsilon}{2}\right)_n \times \Gamma\left(\frac{s+\varepsilon}{2}\right)\left[\psi\left(\frac{s+\varepsilon}{2}\right) - \sum_{k=0}^{n-1}\frac{2}{2k-s+\varepsilon+1} - 2\ln b\right]$ $[\operatorname{Re} s > -\varepsilon;\ \|\arg a\|,\ 4\|\arg b\| < \pi]$
3	$e^{-b^2x^2}\begin{Bmatrix}\ln(x^2+a^2)\\ \ln\|x^2-a^2\|\end{Bmatrix}$ $\times H_{2n+\varepsilon}(bx)$	$(-1)^n \frac{2^{2n+\varepsilon}\pi}{s+\varepsilon} a^{s+\varepsilon}b^{\varepsilon}\left(\frac{2\varepsilon+1}{2}\right)_n \begin{Bmatrix}\csc[(s+\varepsilon)\pi/2]\\ \cot[(s+\varepsilon)\pi/2]\end{Bmatrix} \times {}_2F_2\left(\begin{matrix}\frac{2n+2\varepsilon+1}{2}, \frac{s+\varepsilon}{2}; \pm a^2b^2\\ \frac{2\varepsilon+1}{2}, \frac{s+\varepsilon+2}{2}\end{matrix}\right) \pm (-1)^n \frac{2^{2n+\varepsilon-1}a^2}{b^{s-2}}\left(\frac{3-s+\varepsilon}{2}\right)_n \Gamma\left(\frac{s+\varepsilon-2}{2}\right) \times {}_3F_3\left(\begin{matrix}1, 1, \frac{2n-s+\varepsilon+3}{2}; \pm a^2b^2\\ 2, \frac{3-s}{2}, \frac{4-s}{2}\end{matrix}\right) + (-1)^n \frac{2^{2n+\varepsilon-1}}{b^s}\left(\frac{1-s+\varepsilon}{2}\right)_n \Gamma\left(\frac{s+\varepsilon}{2}\right) \times \left[\psi\left(\frac{s+\varepsilon}{2}\right) - \sum_{k=0}^{n-1}\frac{2}{2k-s+\varepsilon+1} - 2\ln b\right]$ $[\operatorname{Re} a > 0;\ \operatorname{Re} s > -\varepsilon;\ \|\arg b\| < \pi/4]$

3.22.5. $H_n(bx)$ and inverse trigonometric functions

Notation: $\varepsilon = 0$ or 1.

1	$\theta(a-x)\arccos\frac{x}{a}$ $\times H_{2n+\varepsilon}(bx)$	$\frac{(-1)^n 2^{\varepsilon-1}(2n+\varepsilon)!\sqrt{\pi}\, a^{s+\varepsilon}b^{\varepsilon}}{n!\,(s+\varepsilon)} \Gamma\left[\begin{matrix}\frac{s+\varepsilon+1}{2}\\ \frac{s+\varepsilon+2}{2}\end{matrix}\right] \times {}_3F_3\left(\begin{matrix}-n, \frac{s+\varepsilon}{2}, \frac{s+\varepsilon+1}{2}; a^2b^2\\ \frac{2\varepsilon+1}{2}, \frac{s+\varepsilon+2}{2}, \frac{s+\varepsilon+2}{2}\end{matrix}\right)$ $[a > 0;\ \operatorname{Re} s > -\varepsilon]$

3.22.6. $H_n(bx)$ and $\mathrm{Ei}(ax^r)$

Notation: $\varepsilon = 0$ or 1.

1	$e^{-b^2x^2}\,\mathrm{Ei}(-ax)$ $\times H_{2n+\varepsilon}(bx)$	$(-1)^{n+1}\dfrac{2^{2n+\varepsilon-1}a}{b^{s+1}}\left(\dfrac{\varepsilon-s}{2}\right)_n\Gamma\left(\dfrac{s+\varepsilon+1}{2}\right)$ $\times {}_3F_3\left(\begin{matrix}\frac{1}{2},\ \frac{s+1}{2},\ \frac{s+2}{2};\ \frac{a^2}{4b^2}\\ \frac{3}{2},\ \frac{3}{2},\ \frac{s-2n-\varepsilon+2}{2}\end{matrix}\right)+(-1)^n\dfrac{2^{2n+\varepsilon-3}a^2}{b^{s+2}}$ $\times\left(\dfrac{\varepsilon-s-1}{2}\right)_n\Gamma\left(\dfrac{s+\varepsilon+2}{2}\right)$ $\times {}_4F_4\left(\begin{matrix}1,\ 1,\ \frac{s+2}{2},\ \frac{s+3}{2};\ \frac{a^2}{4b^2}\\ 2,\ 2,\ \frac{3}{2},\ \frac{s-2n-\varepsilon+3}{2}\end{matrix}\right)$ $+(-1)^n\dfrac{2^{2n+\varepsilon-1}}{b^s}\left(\dfrac{\varepsilon-s+1}{2}\right)_n\Gamma\left(\dfrac{s+\varepsilon}{2}\right)$ $\times\left[\mathbf{C}+\dfrac{1}{2}\psi\left(\dfrac{s+\varepsilon}{2}\right)-\sum_{k=0}^{n-1}\dfrac{1}{2k-s+\varepsilon+1}-\ln\dfrac{b}{a}\right]$ $[\mathrm{Re}\,s>-\varepsilon;\ \lvert\arg a\rvert,\ 4\lvert\arg b\rvert<\pi]$
2	$e^{-b^2x^2}\,\mathrm{Ei}(-ax^2)$ $\times H_{2n+\varepsilon}(bx)$	$(-1)^{n+1}\dfrac{2^{2n+\varepsilon}b^{\varepsilon}}{a^{(s+\varepsilon)/2}(s+\varepsilon)}\left(\dfrac{2\varepsilon+1}{2}\right)_n\Gamma\left(\dfrac{s+\varepsilon}{2}\right)$ $\times {}_3F_2\left(\begin{matrix}\frac{2n+2\varepsilon+1}{2},\ \frac{s+\varepsilon}{2},\ \frac{s+\varepsilon}{2}\\ \frac{2\varepsilon+1}{2},\ \frac{s+\varepsilon+2}{2};\ -\frac{b^2}{a}\end{matrix}\right)$ $[\mathrm{Re}\,(a+b^2)>0;\ \mathrm{Re}\,s>-\varepsilon]$
3	$e^{-(a+b^2)x^2}\,\mathrm{Ei}(ax^2)$ $\times H_{2n+\varepsilon}(bx)$	$\dfrac{2^{-s+2n+\varepsilon+2}\pi^{3/2}\csc(s\pi)}{ab^{s-2}\,\Gamma(3-s)\,\Gamma\left(\frac{s-2n-\varepsilon-1}{2}\right)}\,{}_3F_2\left(\begin{matrix}1,\ 1,\ \frac{3-s+2n+\varepsilon}{2}\\ \frac{3-s}{2},\ \frac{4-s}{2};\ -\frac{b^2}{a}\end{matrix}\right)$ $-\dfrac{2^{2n+2\varepsilon-1}\pi^{3/2}b^{\varepsilon}}{a^{(s+\varepsilon)/2}}\tan^{2\varepsilon-1}\dfrac{s\pi}{2}$ $\times\Gamma\left[\begin{matrix}\frac{s+\varepsilon}{2}\\ \frac{-2n-2\varepsilon+1}{2}\end{matrix}\right]{}_2F_1\left(\begin{matrix}\frac{2n+2\varepsilon+1}{2},\ \frac{s+\varepsilon}{2}\\ \frac{2\varepsilon+1}{2};\ -\frac{b^2}{a}\end{matrix}\right)$ $[a>0;\ \mathrm{Re}\,s>-\varepsilon;\ \lvert\arg b\rvert<\pi/4]$
4	$e^{-b^2x^2}\,\mathrm{Ei}(-ax^4)$ $\times H_{2n+\varepsilon}(bx)$	$(-1)^{n+1}\dfrac{2^{2n+\varepsilon}b^{\varepsilon}}{a^{(s+\varepsilon)/4}(s+\varepsilon)}\left(\dfrac{2\varepsilon+1}{2}\right)_n\Gamma\left(\dfrac{s+\varepsilon}{4}\right)$ $\times {}_4F_4\left(\begin{matrix}\frac{2n+3}{4},\ \frac{2n+4\varepsilon+1}{4},\ \frac{s+\varepsilon}{4},\ \frac{s+\varepsilon}{4}\\ \frac{1}{2},\ \frac{3}{4},\ \frac{4\varepsilon+1}{4},\ \frac{s+\varepsilon+4}{4};\ \frac{b^4}{4a}\end{matrix}\right)$ $+(-1)^n\dfrac{2^{2n+\varepsilon}b^{\varepsilon+2}}{a^{(s+\varepsilon+2)/4}(s+\varepsilon+2)}\left(\dfrac{2\varepsilon+3}{2}\right)_n$ $\times\Gamma\left(\dfrac{s+\varepsilon+2}{4}\right){}_4F_4\left(\begin{matrix}\frac{2n+5}{4},\ \frac{2n+2\varepsilon+3}{4},\ \frac{s+\varepsilon+2}{4},\ \frac{s+\varepsilon+2}{4}\\ \frac{5}{4},\ \frac{3}{2},\ \frac{4\varepsilon+3}{4},\ \frac{s+\varepsilon+6}{4};\ \frac{b^4}{4a}\end{matrix}\right)$ $[\mathrm{Re}\,a>0;\ \mathrm{Re}\,s>-\varepsilon]$

No.	$f(x)$	$F(s)$
5	$e^{\pm ax^4-b^2x^2}\operatorname{Ei}(\mp ax^4)$ $\times H_{2n+\varepsilon}(bx)$	$\mp\frac{2^{2n-s+\varepsilon+4}\sqrt{\pi}}{ab^{s-4}}\Gamma\left[\begin{matrix}s-4\\ \frac{s-2n-\varepsilon-3}{2}\end{matrix}\right]{}_4F_4\left(\begin{matrix}1,\ 1,\ \frac{2n-s+\varepsilon+5}{4},\ \frac{2n-s+\varepsilon+7}{4}\\ \frac{5-s}{4},\ \frac{6-s}{4},\ \frac{7-s}{4},\ \frac{8-s}{4};\ \mp\frac{b^4}{4a}\end{matrix}\right)$ $-\frac{2^{2n+\varepsilon-2}\pi^{3/2}}{a^{s/4}}\begin{Bmatrix}\csc\frac{s\pi}{4}\\ \cot\frac{s\pi}{4}\end{Bmatrix}\Gamma\left[\begin{matrix}\frac{s}{4}\\ -\frac{2n+\varepsilon-1}{2}\end{matrix}\right]$ $\times{}_3F_3\left(\begin{matrix}\frac{2n+\varepsilon+1}{4},\ \frac{2n+\varepsilon+3}{4},\ \frac{s}{4}\\ \frac{1}{4},\ \frac{1}{2},\ \frac{3}{4};\ \mp\frac{b^4}{4a}\end{matrix}\right)+\frac{2^{2n+\varepsilon-1}\pi^{3/2}b}{a^{(s+1)/4}}$ $\times\begin{Bmatrix}\csc\frac{(s+1)\pi}{4}\\ \cot\frac{(s+1)\pi}{4}\end{Bmatrix}\Gamma\left[\begin{matrix}\frac{s+1}{4}\\ -\frac{2n+\varepsilon}{2}\end{matrix}\right]{}_3F_3\left(\begin{matrix}\frac{2n+\varepsilon+2}{4},\ \frac{2n+\varepsilon+4}{4},\ \frac{s+1}{4}\\ \frac{1}{2},\ \frac{3}{4},\ \frac{5}{4};\ \mp\frac{b^4}{4a}\end{matrix}\right)$ $\mp\frac{2^{2n+\varepsilon-1}\pi^{3/2}b^2}{a^{(s+2)/4}}\begin{Bmatrix}\sec\frac{s\pi}{4}\\ \tan\frac{s\pi}{4}\end{Bmatrix}\Gamma\left[\begin{matrix}\frac{s+2}{4}\\ -\frac{2n+\varepsilon+1}{2}\end{matrix}\right]$ $\times{}_3F_3\left(\begin{matrix}\frac{2n+\varepsilon+3}{4},\ \frac{2n+\varepsilon+5}{4},\ \frac{s+2}{4}\\ \frac{3}{4},\ \frac{5}{4},\ \frac{3}{2};\ \mp\frac{b^4}{4a}\end{matrix}\right)\pm\frac{2^{2n+\varepsilon}\pi^{3/2}b^3}{3a^{(s+3)/4}}$ $\times\begin{Bmatrix}\sec\frac{(s+1)\pi}{4}\\ \tan\frac{(s+1)\pi}{4}\end{Bmatrix}\Gamma\left[\begin{matrix}\frac{s+3}{4}\\ -\frac{2n+\varepsilon+2}{2}\end{matrix}\right]{}_3F_3\left(\begin{matrix}\frac{2n+\varepsilon+4}{4},\ \frac{2n+\varepsilon+6}{4},\ \frac{s+3}{4}\\ \frac{5}{4},\ \frac{3}{2},\ \frac{7}{4};\ \mp\frac{b^4}{4a}\end{matrix}\right)$ $\left[\operatorname{Re}s>-\varepsilon;\ \lvert\arg b\rvert<\pi/4;\ \begin{Bmatrix}\lvert\arg a\rvert<\pi\\ a>0\end{Bmatrix}\right]$

3.22.7. $H_n(bx)$ and $\operatorname{si}(ax^r)$, $\operatorname{ci}(ax^r)$

Notation: $\varepsilon=0$ or 1.

1	$e^{-b^2x^2}\operatorname{si}(ax)\,H_{2n+\varepsilon}(bx)$	$2^{2n-s+\varepsilon-1}\sqrt{\pi}\,ab^{-s-1}\,\Gamma\left[\begin{matrix}s+1\\ \frac{s-2n-\varepsilon+2}{2}\end{matrix}\right]$ $\times{}_3F_3\left(\begin{matrix}\frac{1}{2},\ \frac{s+1}{2},\ \frac{s+2}{2};\ -\frac{a^2}{4b^2}\\ \frac{3}{2},\ \frac{3}{2},\ \frac{s-2n-\varepsilon+2}{2}\end{matrix}\right)-\frac{2^{2n-s+\varepsilon-1}\pi^{3/2}}{b^s}\Gamma\left[\begin{matrix}s\\ \frac{s-2n-\varepsilon+1}{2}\end{matrix}\right]$ $[a,\ \operatorname{Re}b>0;\ \operatorname{Re}s>-\varepsilon]$
2	$e^{-b^2x^2}\operatorname{ci}(ax)\,H_{2n+\varepsilon}(bx)$	$-2^{2n-s+\varepsilon-4}\sqrt{\pi}\,a^2b^{-s-2}\,\Gamma\left[\begin{matrix}s+2\\ \frac{s-2n-\varepsilon+3}{2}\end{matrix}\right]$ $\times{}_4F_4\left(\begin{matrix}1,\ 1,\ \frac{s+2}{2},\ \frac{s+3}{2};\ -\frac{a^2}{4b^2}\\ \frac{3}{2},\ 2,\ 2,\ \frac{s-2n-\varepsilon+3}{2}\end{matrix}\right)$ $+2^{2n-s+\varepsilon}\sqrt{\pi}\,b^{-s}\,\Gamma\left[\begin{matrix}s\\ \frac{s-2n-\varepsilon+1}{2}\end{matrix}\right]$ $\times\left[\psi(s)-\frac{1}{2}\psi\left(\frac{s-2n-\varepsilon+1}{2}\right)+\ln\frac{a}{2b}+\mathbf{C}\right]$ $[a,\ \operatorname{Re}b>0;\ \operatorname{Re}s>-\varepsilon]$

No.	$f(x)$	$F(s)$
3	$e^{-b^2x^2}\begin{Bmatrix}\operatorname{si}(ax^2)\\ \operatorname{ci}(ax^2)\end{Bmatrix} \times H_{2n+\varepsilon}(bx)$	$\frac{(-1)^{n+1}\,2^{2n+\varepsilon}b^{\varepsilon}}{a^{(s+\varepsilon)/2}(s+\varepsilon)}\begin{Bmatrix}\sin[(s+\varepsilon)\pi/4]\\ \cos[(s+\varepsilon)\pi/4]\end{Bmatrix}\left(\frac{2\varepsilon+1}{2}\right)_n\Gamma\left(\frac{s+\varepsilon}{2}\right) \times {}_5F_4\left(\begin{matrix}\frac{2n+3}{4},\ \frac{2n+4\varepsilon+1}{4},\ \frac{s+\varepsilon}{4},\ \frac{s+\varepsilon}{4},\ \frac{s+\varepsilon+2}{4}\\ \frac{1}{2},\ \frac{3}{4},\ \frac{4\varepsilon+1}{4},\ \frac{s+\varepsilon+4}{4};\ -\frac{b^4}{a^2}\end{matrix}\right) \pm\frac{(-1)^n\,2^{s+2n}b^{\varepsilon+2}}{a^{(s+\varepsilon+2)/2}(s+\varepsilon+2)}\begin{Bmatrix}\cos[(s+\varepsilon)\pi/4]\\ \sin[(s+\varepsilon)\pi/4]\end{Bmatrix} \times\left(\frac{2\varepsilon+3}{2}\right)_n\Gamma\left(\frac{s+\varepsilon+2}{2}\right) \times {}_5F_4\left(\begin{matrix}\frac{2n+5}{4},\ \frac{2n+4\varepsilon+3}{4},\ \frac{s+\varepsilon+2}{4},\ \frac{s+\varepsilon+2}{4},\ \frac{s+\varepsilon+4}{4}\\ \frac{5}{4},\ \frac{3}{2},\ \frac{4\varepsilon+3}{4},\ \frac{s+\varepsilon+6}{4};\ -\frac{b^4}{a^2}\end{matrix}\right)$ $[a>0;\ \operatorname{Re}s>-\varepsilon;\ \lvert\arg b\rvert<\pi/4]$

3.22.8. $H_n(bx)$ and $\operatorname{erf}(ax^r)$, $\operatorname{erfc}(ax^r)$

Notation: $\varepsilon = 0$ or 1.

1	$\operatorname{erfc}(ax)\,H_{2n+\varepsilon}(bx)$	$(-1)^n\frac{2^{2n+\varepsilon}a^{-s-\varepsilon}b^{\varepsilon}}{\sqrt{\pi}\,(s+\varepsilon)}\left(\frac{2\varepsilon+1}{2}\right)_n\Gamma\left(\frac{s+\varepsilon+1}{2}\right) \times {}_3F_2\left(\begin{matrix}-n,\ \frac{s+\varepsilon}{2},\ \frac{s+\varepsilon+1}{2};\ \frac{b^2}{a^2}\\ \frac{2\varepsilon+1}{2},\ \frac{s+\varepsilon+2}{2}\end{matrix}\right)$ $[\operatorname{Re}s>-\varepsilon;\ \lvert\arg a\rvert<\pi/4]$
2	$e^{-b^2x^2}\begin{Bmatrix}\operatorname{erf}(ax)\\ \operatorname{erfc}(ax)\end{Bmatrix} \times H_{2n}(bx)$	$\mp\frac{(-1)^n\,2^{2n}}{\sqrt{\pi}\,a^s s}\left(\frac{1}{2}\right)_n\Gamma\left(\frac{s+1}{2}\right){}_3F_2\left(\begin{matrix}\frac{2n+1}{2},\ \frac{s}{2},\ \frac{s+1}{2}\\ \frac{1}{2},\ \frac{s+2}{2};\ -\frac{b^2}{a^2}\end{matrix}\right) +\frac{(1\pm1)(-1)^n\,2^{2n-2}}{b^s}\left(\frac{1-s}{2}\right)_n\Gamma\left(\frac{s}{2}\right)$ $[\operatorname{Re}s>-(1\pm1)/2;\ \lvert\arg a\rvert,\ \lvert\arg b\rvert<\pi/4]$
3	$e^{-b^2x^2}\begin{Bmatrix}\operatorname{erf}(ax)\\ \operatorname{erfc}(ax)\end{Bmatrix} \times H_{2n+1}(bx)$	$\mp\frac{(-1)^n\,2^{2n+1}b}{\sqrt{\pi}a^{s+1}(s+1)}\left(\frac{3}{2}\right)_n\Gamma\left(\frac{s+2}{2}\right) \times {}_3F_2\left(\begin{matrix}\frac{2n+3}{2},\ \frac{s+1}{2},\ \frac{s+2}{2}\\ \frac{3}{2},\ \frac{s+3}{2};\ -\frac{b^2}{a^2}\end{matrix}\right) +\frac{(1\pm1)(-1)^n\,2^{2n-1}}{b^s}\left(\frac{2-s}{2}\right)_n\Gamma\left(\frac{s+1}{2}\right)$ $[\operatorname{Re}s>-1-(1\pm1)/2;\ \lvert\arg a\rvert,\ \lvert\arg b\rvert<\pi/4]$

No.	$f(x)$	$F(s)$
4	$e^{-b^2x^2}\left\{\begin{matrix}\operatorname{erf}(ax^2)\\ \operatorname{erfc}(ax^2)\end{matrix}\right\}$ $\times H_{2n+\varepsilon}(bx)$	$\mp\dfrac{(-1)^n 2^{2n+\varepsilon}b^{\varepsilon}}{\sqrt{\pi}a^{(s+\varepsilon)/2}(s+\varepsilon)}\left(\dfrac{2\varepsilon+1}{2}\right)_n\Gamma\left(\dfrac{s+\varepsilon+2}{4}\right)$ $\times\,{}_4F_4\left(\begin{matrix}\frac{2n+3}{4},\ \frac{2n+4\varepsilon+1}{4},\ \frac{s+\varepsilon}{4},\ \frac{s+\varepsilon+2}{4}\\ \frac{1}{2},\ \frac{3}{4},\ \frac{4\varepsilon+1}{4},\ \frac{s+\varepsilon+4}{4};\ \frac{b^4}{4a^2}\end{matrix}\right)$ $\pm\dfrac{(-1)^n 2^{2n+\varepsilon}b^{\varepsilon+2}}{\sqrt{\pi}a^{(s+\varepsilon)/2+1}(s+\varepsilon+2)}\left(\dfrac{2\varepsilon+3}{2}\right)_n$ $\times\Gamma\left(\dfrac{s+\varepsilon+4}{4}\right){}_4F_4\left(\begin{matrix}\frac{2n+5}{4},\ \frac{2n+4\varepsilon+3}{4},\ \frac{s+\varepsilon+2}{4},\ \frac{s+\varepsilon+4}{4}\\ \frac{5}{4},\ \frac{3}{2},\ \frac{4\varepsilon+3}{4},\ \frac{s+\varepsilon+6}{4};\ \frac{b^4}{4a^2}\end{matrix}\right)$ $+\dfrac{(-1)^n(1\pm1)2^{2n+\varepsilon-2}}{b^s}\left(\dfrac{1-s+\varepsilon}{2}\right)_n\Gamma\left(\dfrac{s+\varepsilon}{2}\right)$ $[\operatorname{Re}s>-\varepsilon-1\mp1;\ \lvert\arg a\rvert,\ \lvert\arg b\rvert<\pi/4]$
5	$e^{-b^2x^2}\left\{\begin{matrix}\operatorname{erf}(a\sqrt{x})\\ \operatorname{erfc}(a\sqrt{x})\end{matrix}\right\}$ $\times H_n(bx)$	$\pm\dfrac{2^{n-s+1/2}a}{b^{s+1/2}}\Gamma\left[\begin{matrix}\frac{2s+1}{2}\\ \frac{2s-2n+3}{4}\end{matrix}\right]{}_3F_3\left(\begin{matrix}\frac{1}{4},\ \frac{2s+1}{4},\ \frac{2s+3}{4};\ \frac{a^4}{4b^2}\\ \frac{1}{2},\ \frac{5}{4},\ \frac{2s-2n+3}{4}\end{matrix}\right)$ $\mp\dfrac{2^{n-s-1/2}a^3}{3b^{s+3/2}}\Gamma\left[\begin{matrix}\frac{2s+3}{2}\\ \frac{2s-2n+5}{4}\end{matrix}\right]{}_3F_3\left(\begin{matrix}\frac{3}{4},\ \frac{2s+3}{4},\ \frac{2s+5}{4};\ \frac{a^4}{4b^2}\\ \frac{3}{2},\ \frac{7}{4},\ \frac{2s-2n+5}{4}\end{matrix}\right)$ $+\dfrac{(1\mp1)2^{n-2}}{b^s}\Gamma\left[\begin{matrix}\frac{s}{2},\ \frac{s+1}{2}\\ \frac{s-n+1}{2}\end{matrix}\right]$ $[\operatorname{Re}s>(-1-(1\pm1)^n)/4;\ \lvert\arg a\rvert,\ \lvert\arg b\rvert<\pi/4]$

3.22.9. $H_n(bx)$ and $S(ax^r)$, $C(ax^r)$

Notation: $\delta=\left\{\begin{matrix}1\\0\end{matrix}\right\}$.

1	$e^{-b^2x^2}\left\{\begin{matrix}S(ax)\\ C(ax)\end{matrix}\right\}$ $\times H_{2n}(bx)$	$(-1)^n\dfrac{2^{2n-1/2}a^{\delta+1/2}}{3^{\delta}\sqrt{\pi}\,b^{s+\delta+1/2}}\left(\dfrac{1-2s-2\delta}{4}\right)_n\Gamma\left(\dfrac{2s+2\delta+1}{4}\right)$ $\times\,{}_3F_3\left(\begin{matrix}\frac{2\delta+1}{4},\ \frac{2s+3}{4},\ \frac{2s+4\delta+1}{4};\ -\frac{a^2}{4b^2}\\ \frac{2\delta+1}{2},\ \frac{2\delta+5}{4},\ \frac{2s-4n+2\delta+3}{4}\end{matrix}\right)$ $[a>0;\ \operatorname{Re}s>-(2\pm1)/2;\ \lvert\arg b\rvert<\pi/4]$
2	$e^{-b^2x^2}\left\{\begin{matrix}S(ax)\\ C(ax)\end{matrix}\right\}$ $\times H_{2n+1}(bx)$	$(-1)^n\dfrac{2^{2n+1/2}a^{\delta+1/2}}{3^{\delta}\sqrt{\pi}\,b^{s+\delta+1/2}}\left(\dfrac{3-2s-2\delta}{4}\right)_n\Gamma\left(\dfrac{2s+2\delta+3}{4}\right)$ $\times\,{}_3F_3\left(\begin{matrix}\frac{2\delta+1}{4},\ \frac{2s+3}{4},\ \frac{2s+4\delta+1}{4};\ -\frac{a^2}{4b^2}\\ \frac{2\delta+1}{2},\ \frac{2\delta+5}{4},\ \frac{2s-4n+2\delta+1}{4}\end{matrix}\right)$ $[a>0;\ \operatorname{Re}s>-1-(2\pm1)/2;\ \lvert\arg b\rvert<\pi/4]$

No.	$f(x)$	$F(s)$
3	$e^{-b^2x^2}\begin{Bmatrix} S(ax^2) \\ C(ax^2) \end{Bmatrix} \times H_{2n}(bx)$	$(-1)^n \dfrac{2^{2n-1/2}a^{\delta+1/2}}{3^\delta\sqrt{\pi}\,b^{s+2\delta+1}} \left(-\dfrac{s+2\delta}{2}\right)_n \Gamma\left(\dfrac{s+2\delta+1}{2}\right) \times {}_5F_4\left(\begin{matrix} \frac{2\delta+1}{4}, \frac{s+3}{4}, \frac{s+4}{4}, \frac{s+4\delta+1}{4}, \frac{s+4\delta+2}{4} \\ \frac{2\delta+1}{2}, \frac{2\delta+5}{4}, \frac{s-2n+4}{4}, \frac{s-2n+4\delta+2}{4}; -\frac{a^2}{b^4} \end{matrix}\right)$ $[a>0;\ \operatorname{Re} s > -2 \mp 1;\ \lvert\arg b\rvert < \pi/4]$
4	$e^{-b^2x^2}\begin{Bmatrix} S(ax^2) \\ C(ax^2) \end{Bmatrix} \times H_{2n+1}(bx)$	$(-1)^n \dfrac{2^{2n+1/2}a^{\delta+1/2}}{3^\delta\sqrt{\pi}\,b^{s+2\delta+1}} \left(\dfrac{1-s-2\delta}{2}\right)_n \Gamma\left(\dfrac{s+2\delta+2}{2}\right) \times {}_5F_4\left(\begin{matrix} \frac{2\delta+1}{4}, \frac{s+3}{4}, \frac{s+4}{4}, \frac{s+4\delta+1}{4}, \frac{s+4\delta+2}{4} \\ \frac{2\delta+1}{2}, \frac{2\delta+5}{4}, \frac{s-2n+3}{4}, \frac{s-2n+4\delta+1}{4}; -\frac{a^2}{b^4} \end{matrix}\right)$ $[a>0;\ \operatorname{Re} s > -3 \mp 1;\ \lvert\arg b\rvert < \pi/4]$

3.22.10. $H_n(bx)$ and $\gamma(\nu, ax^r)$, $\Gamma(\nu, ax^r)$

Notation: $\varepsilon = 0$ or 1.

No.	$f(x)$	$F(s)$
1	$\Gamma(\nu, ax)\,H_{2n+\varepsilon}(bx)$	$\dfrac{(-1)^n 2^{2n+2\varepsilon}a^{-s-\varepsilon}b^\varepsilon}{\sqrt{\pi}\,(s+\varepsilon)} \Gamma\left(n+\dfrac{2\varepsilon+1}{2}\right)\Gamma(s+\nu+\varepsilon) \times {}_4F_2\left(\begin{matrix} -n, \frac{s+\varepsilon}{2}, \frac{s+\nu+\varepsilon}{2}, \frac{s+\nu+\varepsilon+1}{2} \\ \frac{2\varepsilon+1}{2}, \frac{s-\varepsilon+4}{2}; \frac{4b^2}{a^2} \end{matrix}\right)$ $[\operatorname{Re} a > 0;\ \operatorname{Re}(s+\nu) > -\varepsilon]$
2	$e^{-b^2x^2}\begin{Bmatrix} \gamma(\nu, ax) \\ \Gamma(\nu, ax) \end{Bmatrix} \times H_{2n+\varepsilon}(bx)$	$\pm(-1)^n \dfrac{2^{2n+\varepsilon-1}a^\nu}{\nu\, b^{s+\nu}} \left(\dfrac{1-s-\nu+\varepsilon}{2}\right)_n \Gamma\left(\dfrac{s+\nu+\varepsilon}{2}\right) \times {}_3F_3\left(\begin{matrix} \frac{\nu}{2}, \frac{s+\nu}{2}, \frac{s+\nu+1}{2}; \frac{a^2}{4b^2} \\ \frac{1}{2}, \frac{\nu+2}{2}, \frac{s-2n+\nu-\varepsilon+1}{2} \end{matrix}\right) \mp (-1)^n \dfrac{2^{2n+\varepsilon-1}a^{\nu+1}}{(\nu+1)\,b^{s+\nu+1}} \left(\dfrac{-s-\nu+\varepsilon}{2}\right)_n \times \Gamma\left(\dfrac{s+\nu+\varepsilon+1}{2}\right) {}_3F_3\left(\begin{matrix} \frac{\nu+1}{2}, \frac{s+\nu+1}{2}, \frac{s+\nu+2}{2}; \frac{a^2}{4b^2} \\ \frac{3}{2}, \frac{\nu+3}{2}, \frac{s-2n+\nu-\varepsilon+2}{2} \end{matrix}\right) + (-1)^n \dfrac{(1\mp 1)\,2^{2n+\varepsilon-2}}{b^s} \left(\dfrac{1-s+\varepsilon}{2}\right)_n \times \Gamma(\nu)\,\Gamma\left(\dfrac{s+\varepsilon}{2}\right)$ $\left[\operatorname{Re} a > 0;\ \operatorname{Re}(s+\nu) > -\varepsilon;\ \begin{Bmatrix} \operatorname{Re}\nu > 0 \\ \operatorname{Re} s > -\varepsilon \end{Bmatrix},\ \lvert\arg b\rvert < \pi/4\right]$

No.	$f(x)$	$F(s)$
3	$e^{-b^2x^2}\begin{Bmatrix}\gamma(\nu,\,ax^2)\\ \Gamma(\nu,\,ax^2)\end{Bmatrix}$ $\times H_{2n+\varepsilon}(bx)$	$\mp(-1)^n\frac{2^{2n+\varepsilon}b^\varepsilon}{(s+\varepsilon)\,a^{(s+\varepsilon)/2}}\left(\frac{2\varepsilon+1}{2}\right)_n$ $\times\Gamma\left(\frac{s+2\nu+\varepsilon}{2}\right){}_3F_2\left(\begin{matrix}\frac{2n+2\varepsilon+1}{2},\,\frac{s+\varepsilon}{2},\,\frac{s+2\nu+\varepsilon}{2}\\ \frac{2\varepsilon+1}{2},\,\frac{s+\varepsilon+2}{2};\,-\frac{b^2}{a}\end{matrix}\right)$ $+(-1)^n\frac{(1\pm1)\,2^{2n+\varepsilon-2}}{b^s}\left(\frac{1-s+\varepsilon}{2}\right)_n\Gamma(\nu)\,\Gamma\left(\frac{s+\varepsilon}{2}\right)$ $\left[\operatorname{Re}a>0;\ \operatorname{Re}(s+2\nu)>-\varepsilon;\begin{Bmatrix}\operatorname{Re}\nu>0\\ \operatorname{Re}s>-\varepsilon\end{Bmatrix},\ \lvert\arg b\rvert<\pi/4\right]$

3.22.11. $H_n(bx)$ and $J_\nu(ax^r)$, $I_\nu(ax^r)$

Notation: $\varepsilon=0$ or 1.

1	$J_\nu(ax)\,H_n(bx)$	$\frac{2^{s+2n-1}b^n}{a^{n+s}}\Gamma\left[\begin{matrix}\frac{s+\nu+n}{2}\\ \frac{2-s-n+\nu}{2}\end{matrix}\right]{}_2F_2\left(\begin{matrix}-\frac{n}{2},\,\frac{1-n}{2};\,\frac{a^2}{4b^2}\\ \frac{2-s-n-\nu}{2},\,\frac{2-s-n+\nu}{2}\end{matrix}\right)$ $[a>0;\ 2\,[n/2]-n-\operatorname{Re}\nu<\operatorname{Re}s<3/2-n]$
2	$J_\nu\left(\frac{a}{x}\right)H_{2n+\varepsilon}(bx)$	$\frac{(-1)^n\,a^{s+\varepsilon}b^\varepsilon}{2^{s-2n+1}}\left(\frac{2\varepsilon+1}{2}\right)_n\Gamma\left[\begin{matrix}\frac{\nu-s-\varepsilon}{2}\\ \frac{s+\nu+\varepsilon+2}{2}\end{matrix}\right]$ $\times{}_1F_3\left(\begin{matrix}-n;\,-\frac{a^2b^2}{4}\\ \frac{2\varepsilon+1}{2},\,\frac{s+\nu+\varepsilon+2}{2},\,\frac{s-\nu+\varepsilon+2}{2}\end{matrix}\right)$ $[a>0;\ -\varepsilon-3/2<\operatorname{Re}s<\operatorname{Re}\nu-2n-\varepsilon]$
3	$e^{-b^2x^2}\begin{Bmatrix}J_\nu(ax)\\ I_\nu(ax)\end{Bmatrix}$ $\times H_{2n+\varepsilon}(bx)$	$(-1)^n\frac{2^{2n+\varepsilon-\nu-1}a^\nu}{b^{s+\nu}}\left(\frac{1-s-\nu+\varepsilon}{2}\right)_n\Gamma\left[\begin{matrix}\frac{s+\varepsilon+\nu}{2}\\ \nu+1\end{matrix}\right]$ $\times{}_2F_2\left(\begin{matrix}\frac{s+\nu}{2},\,\frac{s+\nu+1}{2};\,\mp\frac{a^2}{4b^2}\\ \nu+1,\,\frac{s-2n+\nu-\varepsilon+1}{2}\end{matrix}\right)$ $[\operatorname{Re}(s+\nu)>-\varepsilon;\ \lvert\arg a\rvert,\ 4\lvert\arg b\rvert<\pi]$
4	$e^{-b^2x^2}\begin{Bmatrix}J_\nu(ax^2)\\ I_\nu(ax^2)\end{Bmatrix}$ $\times H_{2n+\varepsilon}(bx)$	$(-1)^n\frac{2^{2n+\varepsilon-\nu-1}\,a^\nu}{b^{s+2\nu}}\left(\frac{1-s-2\nu+\varepsilon}{2}\right)_n\Gamma\left[\begin{matrix}\frac{s+\varepsilon+2\nu}{2}\\ \nu+1\end{matrix}\right]$ $\times{}_4F_3\left(\begin{matrix}\frac{s+2\nu}{4},\,\frac{s+2\nu+1}{4},\,\frac{s+2\nu+2}{4},\,\frac{s+2\nu+3}{4}\\ \nu+1,\,\frac{s-2n+2\nu-\varepsilon+1}{4},\,\frac{s-2n+2\nu-\varepsilon+3}{4};\,\mp\frac{a^2}{b^4}\end{matrix}\right)$ $\left[\begin{matrix}\operatorname{Re}(s+2\nu)>-\varepsilon;\ \lvert\arg b\rvert<\pi/4;\\ \begin{Bmatrix}a>0\\ \operatorname{Re}(b^2-a)>0;\ \lvert\arg a\rvert<\pi\end{Bmatrix}\end{matrix}\right]$

3.22.12. $H_n(bx)$ and $Y_\nu(ax^r)$, $K_\nu(ax^r)$

Notation: $\varepsilon = 0$ or 1.

1	$K_\nu(ax)\,H_n(bx)$	$\frac{2^{s+2n-2}b^n}{a^{s+n}}\Gamma\left(\frac{s+n-\nu}{2}\right)\Gamma\left(\frac{s+n+\nu}{2}\right) \times {}_2F_2\left(\begin{matrix}-\frac{n}{2},\ \frac{1-n}{2};\ -\frac{a^2}{4b^2}\\ \frac{2-s-n-\nu}{2},\ \frac{2-s-n+\nu}{2}\end{matrix}\right)$ $[\operatorname{Re} a > 0;\ \operatorname{Re} s > \lvert\operatorname{Re}\nu\rvert + 2[n/2] - n]$
2	$K_\nu\left(\frac{a}{x}\right)H_{2n+\varepsilon}(bx)$	$\frac{(-1)^n a^{s+\varepsilon}b^\varepsilon}{2^{s-2n+2}}\left(\frac{2\varepsilon+1}{2}\right)_n\Gamma\left(\frac{\nu-s-\varepsilon}{2}\right)\Gamma\left(-\frac{s+\nu+\varepsilon}{2}\right) \times {}_1F_3\left(\begin{matrix}-n;\ \frac{a^2b^2}{4}\\ \frac{2\varepsilon+1}{2},\ \frac{s+\nu+\varepsilon+2}{2},\ \frac{s-\nu+\varepsilon+2}{2}\end{matrix}\right)$ $[\operatorname{Re} a > 0;\ \operatorname{Re} s < -\lvert\operatorname{Re}\nu\rvert - 2n - \varepsilon]$
3	$e^{-b^2x^2}\left\{\begin{matrix}Y_\nu(ax)\\ K_\nu(ax)\end{matrix}\right\} \times H_{2n+\varepsilon}(bx)$	$\mp(-1)^n\frac{2^{2n+\nu+\varepsilon-2}}{\pi a^\nu b^{s-\nu}}\left(\frac{1-s+\nu+\varepsilon}{2}\right)_n\Gamma(\nu) \times \left\{\begin{matrix}2\\ \pi\end{matrix}\right\}\Gamma\left(\frac{s-\nu+\varepsilon}{2}\right){}_2F_2\left(\begin{matrix}\frac{s-\nu}{2},\ \frac{s-\nu+1}{2};\ \mp\frac{a^2}{4b^2}\\ 1-\nu,\ \frac{s-2n-\nu-\varepsilon+1}{2}\end{matrix}\right)$ $\mp(-1)^n\frac{2^{2n-\nu+\varepsilon-2}a^\nu}{\pi b^{s+\nu}}\left(\frac{1-s-\nu+\varepsilon}{2}\right)_n\Gamma(-\nu) \times \left\{\begin{matrix}2\cos(\nu\pi)\\ \pi\end{matrix}\right\}\Gamma\left(\frac{s+\varepsilon+\nu}{2}\right){}_2F_2\left(\begin{matrix}\frac{s+\nu}{2},\ \frac{s+\nu+1}{2};\ \mp\frac{a^2}{4b^2}\\ 1+\nu,\ \frac{s-2n+\nu-\varepsilon+1}{2}\end{matrix}\right)$ $[\operatorname{Re} s > \lvert\operatorname{Re}\nu\rvert - \varepsilon;\ \lvert\arg a\rvert,\ \lvert\arg b\rvert < \pi/4]$
4	$e^{-b^2x^2}\left\{\begin{matrix}Y_\nu(ax^2)\\ K_\nu(ax^2)\end{matrix}\right\} \times H_{2n+\varepsilon}(bx)$	$\mp\frac{(-1)^n 2^{-s+2n+3\nu+\varepsilon-(1\mp1)/2}\pi^{\mp1/2}}{a^\nu b^{s-2\nu}} \times \left(\frac{1-s+2\nu+\varepsilon}{2}\right)_n\Gamma\left[\begin{matrix}\nu,\ s-2\nu\\ \frac{s-2\nu-\varepsilon+1}{2}\end{matrix}\right] \times {}_4F_3\left(\begin{matrix}\frac{s-2\nu}{4},\ \frac{s-2\nu+1}{4},\ \frac{s-2\nu+2}{4},\ \frac{s-2\nu+3}{4};\ \mp\frac{a^2}{b^4}\\ 1-\nu,\ \frac{s-2n-2\nu-\varepsilon+1}{4},\ \frac{s-2n-2\nu-\varepsilon+3}{4}\end{matrix}\right)$ $\mp\frac{(-1)^n 2^{-s+2n-3\nu+\varepsilon-(1\mp1)/2}\pi^{\mp1/2}[\cos(\nu\pi)]^{(1\pm1)/2}}{a^{-\nu}b^{s+2\nu}} \times \left(\frac{1-s-2\nu+\varepsilon}{2}\right)_n\Gamma\left[\begin{matrix}-\nu,\ s+2\nu\\ \frac{s+2\nu-\varepsilon+1}{2}\end{matrix}\right] \times {}_4F_3\left(\begin{matrix}\frac{s+2\nu}{4},\ \frac{s+2\nu+1}{4},\ \frac{s+2\nu+2}{4},\ \frac{s+2\nu+3}{4};\ \mp\frac{a^2}{b^4}\\ 1+\nu,\ \frac{s-2n+2\nu-\varepsilon+1}{4},\ \frac{s-2n+2\nu-\varepsilon+3}{4}\end{matrix}\right)$ $\left[\left\{\begin{matrix}\operatorname{Re} b^2 > \lvert\operatorname{Im} a\rvert\\ \operatorname{Re}(a+b^2) > 0\end{matrix}\right\};\ \operatorname{Re} s > 2\lvert\operatorname{Re}\nu\rvert - \varepsilon;\ \lvert\arg a\rvert,\ 4\lvert\arg b\rvert < \pi\right]$

3.22.13. $H_n(bx)$ and $P_m(\varphi(x))$

Notation: $\varepsilon = 0$ or 1.

1	$\theta(a-x)\, P_m\left(\frac{x}{a}\right) \times H_{2n+\varepsilon}(bx)$	$(-1)^n 2^{2n-s+\varepsilon} a^{s+\varepsilon} b^{\varepsilon}\, \Gamma\left[\begin{matrix} \frac{2n+2\varepsilon+1}{2},\ s+\varepsilon \\ \frac{s-m+\varepsilon+1}{2},\ \frac{s+m+\varepsilon+2}{2}\end{matrix}\right] \times {}_3F_3\left(\begin{matrix} -n,\ \frac{s+\varepsilon}{2},\ \frac{s+\varepsilon+1}{2};\ a^2b^2 \\ \frac{2\varepsilon+1}{2},\ \frac{s-m+\varepsilon+1}{2},\ \frac{s+m+\varepsilon+2}{2}\end{matrix}\right)$ $[a>0;\ \operatorname{Re} s > ((-1)^m + (-1)^{\varepsilon})/2 - 1]$
2	$\theta(a-x)\, e^{-b^2x^2} \times P_n\left(\frac{2x^2}{a^2}-1\right) \times H_{2m+\varepsilon}(bx)$	$(-1)^{m+n} 2^{2m+\varepsilon-1} a^{s+\varepsilon} b^{\varepsilon} \left(\frac{2\varepsilon+1}{2}\right)_m \left(\frac{2-s-\varepsilon}{2}\right)_n \times \Gamma\left[\begin{matrix} \frac{s+\varepsilon}{2} \\ \frac{s+2n+\varepsilon+2}{2}\end{matrix}\right] {}_3F_3\left(\begin{matrix} \frac{2m+2\varepsilon+1}{2},\ \frac{s+\varepsilon}{2},\ \frac{s+\varepsilon}{2};\ -a^2b^2 \\ \frac{2\varepsilon+1}{2},\ \frac{s-2n+\varepsilon}{2},\ \frac{s+2n+\varepsilon+2}{2}\end{matrix}\right)$ $[a>0;\ \operatorname{Re} s > -\varepsilon]$
3	$\theta(x-a)\, P_m\left(\frac{a}{x}\right) \times H_{2n+\varepsilon}(bx)$	$\frac{(-1)^{n+1}}{\sqrt{\pi}} 2^{s+2n+2\varepsilon-1} a^{s+\varepsilon} b^{\varepsilon} \left(\frac{2\varepsilon+1}{2}\right)_n \times \Gamma\left[\begin{matrix} \frac{s-m+\varepsilon}{2},\ \frac{s+m+\varepsilon+1}{2} \\ s+\varepsilon+1\end{matrix}\right] {}_3F_3\left(\begin{matrix} -n,\ \frac{s-m+\varepsilon}{2},\ \frac{s+m+\varepsilon+1}{2} \\ \frac{2\varepsilon+1}{2},\ \frac{s+\varepsilon+1}{2},\ \frac{s+\varepsilon+2}{2};\ a^2b^2\end{matrix}\right)$ $[a>0;\ \operatorname{Re} s < (1-(-1)^m)/2 - 2n - \varepsilon]$

3.22.14. $H_n(bx)$ and $T_m(\varphi(x))$, $U_m(\varphi(x))$

Notation: $\varepsilon = 0$ or 1.

1	$(a^2-x^2)_+^{-1/2} e^{-b^2x^2} \times T_n\left(\frac{2x^2}{a^2}-1\right) \times H_{2m+\varepsilon}(bx)$	$(-1)^{m+n} 2^{2m+\varepsilon-1} \sqrt{\pi}\, a^{s+\varepsilon-1} b^{\varepsilon} \left(\frac{2\varepsilon+1}{2}\right)_m \left(\frac{1-s-\varepsilon}{2}\right)_n \times \Gamma\left[\begin{matrix} \frac{s+\varepsilon}{2} \\ \frac{s+2n+\varepsilon+1}{2}\end{matrix}\right] {}_3F_3\left(\begin{matrix} \frac{2m+2\varepsilon+1}{2},\ \frac{s+\varepsilon}{2},\ \frac{s+\varepsilon+1}{2};\ -a^2b^2 \\ \frac{2\varepsilon+1}{2},\ \frac{s-2n+\varepsilon+1}{2},\ \frac{s+2n+\varepsilon+1}{2}\end{matrix}\right)$ $[a>0;\ \operatorname{Re} s > -\varepsilon]$
2	$(a^2-x^2)_+^{1/2} e^{-b^2x^2} \times U_n\left(\frac{2x^2}{a^2}-1\right) \times H_{2m+\varepsilon}(bx)$	$(-1)^{m+n} 2^{2m+\varepsilon-2} (n+1) \sqrt{\pi}\, a^{s+\varepsilon+1} b^{\varepsilon} \left(\frac{2\varepsilon+1}{2}\right)_m \times \left(\frac{3-s-\varepsilon}{2}\right)_n \Gamma\left[\begin{matrix} \frac{s+\varepsilon}{2} \\ \frac{s+2n+\varepsilon+3}{2}\end{matrix}\right] \times {}_3F_3\left(\begin{matrix} \frac{2m+2\varepsilon+1}{2},\ \frac{s+\varepsilon-1}{2},\ \frac{s+\varepsilon}{2};\ -a^2b^2 \\ \frac{2\varepsilon+1}{2},\ \frac{s-2n+\varepsilon-1}{2},\ \frac{s+2n+\varepsilon+3}{2}\end{matrix}\right)$ $[a>0;\ \operatorname{Re} s > -\varepsilon]$

3.22.15. Products of $H_n(bx)$

Notation: δ, $\varepsilon = 0$ or 1.

1	$e^{-ax^2} H_{2m+\varepsilon}(bx) \times H_{2n+\varepsilon}(cx)$	$\frac{(-1)^{m+n}(2m+\varepsilon)!\,(2n+\varepsilon)!\,(bc)^{\varepsilon}}{2m!\,n!\,a^{s/2+\varepsilon}} \Gamma\left(\frac{s+2\varepsilon}{2}\right) \times F_2\left(\frac{s+2\varepsilon}{2}, -m, -n; \frac{2\varepsilon+1}{2}, \frac{2\varepsilon+1}{2}; \frac{b^2}{a}, \frac{c^2}{a}\right)$ $\left[\operatorname{Re} a > 0; \operatorname{Re} s > -2\varepsilon\right]$
2	$e^{-a^2x^2} H_m(ax) H_n(ax)$	$\frac{\sqrt{\pi}}{2^{s-m-n} a^s} \Gamma\left[\begin{matrix} s \\ \frac{s-m-n+1}{2}\end{matrix}\right] {}_2F_1\left(\begin{matrix}-m, -n; \\ \frac{s-m-n+1}{2}; \frac{1}{2}\end{matrix}\right)$ $[\operatorname{Re} a, \operatorname{Re} s > 0]$
3	$e^{-ax^2} H_{2m+\delta}(bx) \times H_{2n+\varepsilon}(cx)$	$\frac{2^{2n+2m+\delta+\varepsilon+1}\pi bc}{a^{(s+2)/2}} \Gamma\left[\begin{matrix} \frac{s+2}{2} \\ -\frac{2m+\delta}{2}, -\frac{2n+\varepsilon}{2}\end{matrix}\right] \times F_2\left(\frac{s+2}{2}, -\frac{2m+\delta-1}{2}, -\frac{2n+\varepsilon-1}{2}; \frac{3}{2}, \frac{3}{2}; \frac{b^2}{a}, \frac{c^2}{a}\right)$ $- \frac{2^{2n+2m+\delta+\varepsilon}\pi b}{a^{(s+1)/2}} \Gamma\left[\begin{matrix} \frac{s+1}{2} \\ -\frac{2m+\delta}{2}, -\frac{2n+\varepsilon-1}{2}\end{matrix}\right] \times F_2\left(\frac{s+1}{2}, -\frac{2m+\delta-1}{2}, -\frac{2n+\varepsilon}{2}; \frac{3}{2}, \frac{1}{2}; \frac{b^2}{a}, \frac{c^2}{a}\right)$ $- \frac{2^{2n+2m+\delta+\varepsilon}\pi c}{a^{(s+1)/2}} \Gamma\left[\begin{matrix} \frac{s+1}{2} \\ -\frac{2m+\delta-1}{2}, -\frac{2n+\varepsilon}{2}\end{matrix}\right] \times F_2\left(\frac{s+1}{2}, -\frac{2m+\delta}{2}, -\frac{2n+\varepsilon-1}{2}; \frac{1}{2}, \frac{3}{2}; \frac{b^2}{a}, \frac{c^2}{a}\right)$ $+ \frac{2^{2n+2m+\delta+\varepsilon-1}\pi}{a^{s/2}} \Gamma\left[\begin{matrix} \frac{s}{2} \\ -\frac{2m+\delta-1}{2}, -\frac{2n+\varepsilon-1}{2}\end{matrix}\right] \times F_2\left(\frac{s}{2}, -\frac{2m+\delta}{2}, -\frac{2n+\varepsilon}{2}; \frac{1}{2}, \frac{1}{2}; \frac{b^2}{a}, \frac{c^2}{a}\right)$ $\left[\operatorname{Re} a > 0; \operatorname{Re} s > -\delta-\varepsilon\right]$
4	$e^{-(a^2+b^2)x^2} H_{2m+\varepsilon}(ax) \times H_{2n+\delta}(bx)$	$\frac{(-1)^{m+n} 2^{2m+2n+\delta+\varepsilon-1} b^{\delta}}{a^{s+\delta}} \left(\frac{2\delta+1}{2}\right)_n \left(\frac{1-s-\delta+\varepsilon}{2}\right)_m \times \Gamma\left(\frac{s+\delta+\varepsilon}{2}\right) {}_3F_2\left(\begin{matrix}\frac{2n+2\delta+1}{2}, \frac{s+\delta}{2}, \frac{s+\delta+1}{2} \\ \frac{2\delta+1}{2}, \frac{s-2m+\delta-\varepsilon+1}{2}; -\frac{b^2}{a^2}\end{matrix}\right)$ $\left[\operatorname{Re}(a^2+b^2) > 0; \operatorname{Re} s > -\delta-\varepsilon\right]$
5	$e^{-a^2x^4-b^2x^2} H_{2m+\varepsilon}(ax^2) \times H_{2n+\delta}(bx)$	$\frac{(-1)^{m+n} 2^{2m+2n+\delta+\varepsilon-2} b^{\delta}}{a^{(s+\delta)/2}} \left(\frac{2\delta+1}{2}\right)_n \left(\frac{2-s-\delta+2\varepsilon}{4}\right)_m \times \Gamma\left(\frac{s+2\varepsilon+\delta}{4}\right) {}_4F_4\left(\begin{matrix}\frac{2n+3}{4}, \frac{2n+4\delta+1}{4}, \frac{s+\delta}{4}, \frac{s+\delta+2}{4} \\ \frac{1}{2}, \frac{3}{4}, \frac{4\delta+1}{4}, \frac{s-4m+\delta-2\varepsilon+2}{4}; \frac{b^4}{4b^2}\end{matrix}\right)$ $- \frac{(-1)^{m+n} 2^{2m+2n+\delta+\varepsilon-2} b^{\delta+2}}{a^{(s+\delta)/2+1}} \left(\frac{2\delta+3}{2}\right)_n \left(\frac{2\varepsilon-\delta-s}{4}\right)_m \times \Gamma\left(\frac{s+\delta+2\varepsilon+2}{4}\right) {}_4F_4\left(\begin{matrix}\frac{2n+5}{2}, \frac{2n+4\delta+3}{4}, \frac{s+\delta+2}{4}, \frac{s+\delta+4}{4} \\ \frac{3}{2}, \frac{5}{4}, \frac{4\delta+3}{4}, \frac{s-4m+\delta-2\varepsilon+4}{4}; \frac{b^4}{4a^2}\end{matrix}\right)$ $\left[\operatorname{Re} s > -\delta-2\varepsilon; \lvert\arg a\rvert, \lvert\arg b\rvert < \pi/4\right]$

3.23. The Laguerre Polynomials $L_n^\lambda(z)$

More formulas can be obtained from the corresponding sections due to the relations

$$L_n^{-1/2}(z) = \frac{(-1)^n}{n!\,2^{2n}} H_{2n}(\sqrt{z}), \quad L_n^{1/2}(z) = \frac{(-1)^n}{n!\,2^{2n+1}\sqrt{z}} H_{2n+1}(\sqrt{z}),$$

$$L_n^\lambda(z) = \lim_{\sigma\to\infty} P_n^{(\lambda,\,\sigma)}\left(1 - \frac{2z}{\sigma}\right),$$

$$L_\nu(z) = {}_1F_1(-\nu;\,1;\,z),$$

$$L_\nu^\lambda(z) = \frac{(\lambda+1)_\nu}{\Gamma(\nu+1)}\,{}_1F_1(-\nu;\,\lambda+1;\,z),$$

$$L_\nu^\lambda(z) = \frac{e^z}{\Gamma(\nu+1)} G_{12}^{11}\left(z\,\middle|\,\begin{matrix}-\lambda-\nu\\ 0,\,-\lambda\end{matrix}\right).$$

3.23.1. $L_n^\lambda(bx)$ and algebraic functions

No.	$f(x)$	$F(s)$
1	$(a-x)_+^{\alpha-1} L_n^\lambda(bx)$	$\frac{(\lambda+1)_n a^{s+\alpha-1}}{n!} \mathrm{B}(\alpha,\,s)\,{}_2F_2\left(\begin{matrix}-n,\,s;\,ab\\ \lambda+1,\,s+\alpha\end{matrix}\right)$ $[a,\ \mathrm{Re}\,\alpha,\ \mathrm{Re}\,s>0]$
2	$(x-a)_+^{\alpha-1} L_n^\lambda(bx)$	$\frac{(\lambda+1)_n a^{s+\alpha-1}}{n!} \mathrm{B}(\alpha,\,1-s-\alpha)\,{}_2F_2\left(\begin{matrix}-n,\,s;\,ab\\ \lambda+1,\,s+\alpha\end{matrix}\right)$ $[a,\ \mathrm{Re}\,\alpha>0;\ \mathrm{Re}(s+\alpha)<1-n]$
3	$\frac{1}{(x+a)^\rho} L_n^\lambda(bx)$	$\frac{(\lambda+1)_n a^{s-\rho}}{n!} \mathrm{B}(s,\,\rho-s)\,{}_2F_2\left(\begin{matrix}-n,\,s;\,-ab\\ \lambda+1,\,s-\rho+1\end{matrix}\right)$ $[\mathrm{Re}\,s>0;\ \mathrm{Re}(s-\rho)<-n;\ \lvert\arg a\rvert<\pi]$
4	$(a-x)_+^{\alpha-1}$ $\times L_n^\lambda(bx(a-x))$	$\frac{(\lambda+1)_n a^{s+\alpha-1}}{n!} \mathrm{B}(s,\,\alpha)\,{}_3F_3\left(\begin{matrix}-n,\,s,\,\alpha;\,\frac{a^2b}{4}\\ \lambda+1,\,\frac{s+\alpha}{2},\,\frac{s+\alpha+1}{2}\end{matrix}\right)$ $[a,\ \mathrm{Re}\,s,\ \mathrm{Re}\,\alpha>0]$
5	$\frac{1}{(x+a)^\rho} L_n^\lambda\left(\frac{b}{x+a}\right)$	$\frac{(\lambda+1)_n a^{s-\rho}}{n!} \mathrm{B}(s,\,\rho-s)\,{}_2F_2\left(\begin{matrix}-n,\,\rho-s\\ \lambda+1,\,\rho;\,\frac{b}{a}\end{matrix}\right)$ $[0<\mathrm{Re}\,s<\mathrm{Re}\,\rho;\ \lvert\arg a\rvert<\pi]$
6	$\frac{1}{(x+a)^\rho} L_n^\lambda\left(\frac{b^2x^2}{(x+a)^2}\right)$	$\frac{(\lambda+1)_n a^{s-\rho}}{n!} \mathrm{B}(s,\,\rho-s)\,{}_3F_3\left(\begin{matrix}-n,\,\frac{s}{2},\,\frac{s+1}{2}\\ \lambda+1,\,\frac{\rho}{2},\,\frac{\rho+1}{2};\,b^2\end{matrix}\right)$ $[0<\mathrm{Re}\,s<\mathrm{Re}\,\rho;\ \lvert\arg a\rvert<\pi]$

No.	$f(x)$	$F(s)$
7	$\dfrac{1}{(x^2+a^2)^\rho} \times L_n^\lambda\left(\dfrac{b^2x^2}{(x^2+a^2)^2}\right)$	$\dfrac{(\lambda+1)_n a^{s-2\rho}}{2n!} \mathrm{B}\left(\dfrac{s}{2}, \rho-\dfrac{s}{2}\right) {}_3F_3\left(\begin{matrix}-n, \frac{s}{2}, \frac{2\rho-s}{2}; \frac{b^2}{4a^2}\\ \lambda+1, \frac{\rho}{2}, \frac{\rho+1}{2}\end{matrix}\right)$ [$\mathrm{Re}\, a>0;\ 0<\mathrm{Re}\, s<2\,\mathrm{Re}\,\rho$]

3.23.2. $L_n^\lambda(bx)$ and the exponential function

1	$e^{-ax}L_n^\lambda(bx)$	$\dfrac{\Gamma(s)}{a^s} P_n^{(\lambda,\, s-n-\lambda-1)}\left(1-\dfrac{2b}{a}\right)$ [$\mathrm{Re}\, a,\ \mathrm{Re}\, s>0;\ \mathrm{Re}\,\lambda>-1$]
2	$e^{-ax}L_n^\lambda(ax)$	$\dfrac{(1-s+\lambda)_n}{n!\, a^s}\Gamma(s)$ [$\mathrm{Re}\, a,\ \mathrm{Re}\, s>0$]
3	$e^{-bx}L_n^\lambda(bx+ab)$	$\dfrac{\Gamma(s)}{b^s} L_n^{\lambda-s}(ab)$ [$a,\ \mathrm{Re}\, b,\ \mathrm{Re}\, s>0$]
4	$e^{-a/x}L_n^\lambda(bx)$	$\dfrac{(\lambda+1)_n a^s}{n!}\Gamma(-s)\, {}_1F_2\left(\begin{matrix}-n; -ab\\ \lambda+1, s+1\end{matrix}\right)$ [$\mathrm{Re}\, a>0;\ \mathrm{Re}\, s<-n$]
5	$e^{-a/\sqrt{x}}L_n^\lambda(bx)$	$\dfrac{2(\lambda+1)_n a^{2s}}{n!}\Gamma(-2s)\, {}_1F_3\left(\begin{matrix}-n; \frac{a^2b}{4}\\ \lambda+1, \frac{2s+1}{2}, s+1\end{matrix}\right)$ [$\mathrm{Re}\, a>0;\ \mathrm{Re}\, s<-n$]
6	$e^{-ax^2-bx}L_n^\lambda(bx)$	$\dfrac{(\lambda+1)_n}{2(n!)\, a^{s/2}}\Gamma\left(\dfrac{s}{2}\right) {}_3F_3\left(\begin{matrix}\frac{n+\lambda+1}{2}, \frac{n+\lambda+2}{2}, \frac{s}{2}\\ \frac{1}{2}, \frac{\lambda+1}{2}, \frac{\lambda+2}{2}; \frac{b^2}{4a}\end{matrix}\right) - \dfrac{(\lambda+2)_n b}{2(n!)\, a^{(s+1)/2}}\Gamma\left(\dfrac{s+1}{2}\right) {}_3F_3\left(\begin{matrix}\frac{n+\lambda+2}{2}, \frac{n+\lambda+3}{2}, \frac{s+1}{2}\\ \frac{3}{2}, \frac{\lambda+2}{2}, \frac{\lambda+3}{2}; \frac{b^2}{4a}\end{matrix}\right)$ [$\mathrm{Re}\, a,\ \mathrm{Re}\, s>0$]
7	$e^{-a\sqrt{x}-bx}L_n^\lambda(bx)$	$\dfrac{(1-s+\lambda)_n}{n!\, b^s}\Gamma(s)\, {}_2F_2\left(\begin{matrix}s, s-\lambda; \frac{a^2}{4b}\\ \frac{1}{2}, s-n-\lambda\end{matrix}\right) - \dfrac{a}{n!\, b^{s+1/2}} \times \left(\dfrac{1-2s+2\lambda}{2}\right)_n \Gamma\left(\dfrac{2s+1}{2}\right) {}_2F_2\left(\begin{matrix}\frac{2s+1}{2}, \frac{2s-2\lambda+1}{2}; \frac{a^2}{4b}\\ \frac{3}{2}, \frac{2s-2n-2\lambda+1}{2}\end{matrix}\right)$ [$\mathrm{Re}\, b,\ \mathrm{Re}\, s>0$]
8	$e^{-a/x-bx}L_n^\lambda(bx)$	$\dfrac{(\lambda+1)_n a^s}{n!}\Gamma(-s)\, {}_1F_2\left(\begin{matrix}n+\lambda+1; ab\\ \lambda+1, s+1\end{matrix}\right) + \dfrac{(1-s+\lambda)_n}{n!\, b^s}\Gamma(s)\, {}_1F_2\left(\begin{matrix}1-s+n+\lambda; ab\\ 1-s, 1-s+\lambda\end{matrix}\right)$ [$\mathrm{Re}\, a,\ \mathrm{Re}\, b>0$]

No.	$f(x)$	$F(s)$
9	$e^{-a/x^2-bx}L_n^\lambda(bx)$	$\frac{(\lambda+1)_n a^{s/2}}{2(n!)}\Gamma\left(-\frac{s}{2}\right){}_2F_4\left(\begin{matrix}\frac{n+\lambda+1}{2},\ \frac{n+\lambda+2}{2};\ -\frac{ab^2}{4}\\ \frac{1}{2},\ \frac{\lambda+1}{2},\ \frac{\lambda+2}{2},\ \frac{s+2}{2}\end{matrix}\right)$ $-\frac{(\lambda+2)_n a^{(s+1)/2}b}{2(n!)}\Gamma\left(-\frac{s+1}{2}\right)$ $\times{}_2F_4\left(\begin{matrix}\frac{n+\lambda+2}{2},\ \frac{n+\lambda+3}{2};\ -\frac{ab^2}{4}\\ \frac{3}{2},\ \frac{\lambda+2}{2},\ \frac{\lambda+3}{2},\ \frac{s+3}{2}\end{matrix}\right)$ $+\frac{(1-s+\lambda)_n b^{-s}}{n!}\Gamma(s)\,{}_2F_4\left(\begin{matrix}\frac{1-s+n+\lambda}{2},\ \frac{2-s+n+\lambda}{2};\ -\frac{ab^2}{4}\\ \frac{1-s}{2},\ \frac{2-s}{2},\ \frac{1-s+\lambda}{2},\ \frac{2-s+\lambda}{2}\end{matrix}\right)$ [$\operatorname{Re} a,\ \operatorname{Re} b>0$]
10	$e^{-a/\sqrt{x}-bx}L_n^\lambda(bx)$	$\frac{2(\lambda+1)_n a^{2s}}{n!}\Gamma(-2s)\,{}_1F_3\left(\begin{matrix}n+\lambda+1;\ -\frac{a^2b}{4}\\ \lambda+1,\ \frac{2s+1}{2},\ s+1\end{matrix}\right)$ $-\frac{(1-s+\lambda)_n}{n!\,b^s}\Gamma(s)\,{}_1F_3\left(\begin{matrix}1-s+n+\lambda;\ -\frac{a^2b}{4}\\ \frac{1}{2},\ 1-s,\ 1-s+\lambda\end{matrix}\right)$ $-\frac{a\Gamma\left(\frac{2s-1}{2}\right)}{n!\,b^{s-1/2}}\left(\frac{3-2s+2\lambda}{2}\right)_n{}_1F_3\left(\begin{matrix}\frac{3-2s+2n+2\lambda}{2};\ -\frac{a^2b}{4}\\ \frac{3}{2},\ \frac{3-2s}{2},\ \frac{3-2s+2\lambda}{2}\end{matrix}\right)$ [$\operatorname{Re} a,\ \operatorname{Re} b>0$]
11	$e^{-a\sqrt{x}}L_n^\lambda(bx)$	$\frac{(-1)^n 2b^n}{n!\,a^{2s+2n}}\Gamma(2s+2n)\,{}_2F_2\left(\begin{matrix}-n,\ -n-\lambda;\ -\frac{a^2}{4b}\\ \frac{1-2s-2n}{2},\ 1-s-n\end{matrix}\right)$ [$\operatorname{Re} a,\ \operatorname{Re} s>0$]
12	$(a-x)_+^{\alpha-1}e^{-bx}L_n^\lambda(bx)$	$\frac{(\lambda+1)_n a^{s+\alpha-1}}{n!}\mathrm{B}(\alpha,\ s)\,{}_2F_2\left(\begin{matrix}n+\lambda+1,\ s;\ -ab\\ \lambda+1,\ s+\alpha\end{matrix}\right)$ [$a,\ \operatorname{Re}\alpha,\ \operatorname{Re} s>0$]
13	$(x-a)_+^{\alpha-1}e^{-bx}L_n^\lambda(bx)$	$\frac{(\lambda+1)_n a^{s+\alpha-1}}{n!}\mathrm{B}(\alpha,\ 1-s-\alpha)\,{}_2F_2\left(\begin{matrix}n+\lambda+1,\ s;\ -ab\\ \lambda+1,\ s+\alpha\end{matrix}\right)$ $+\frac{b^{1-s-\alpha}}{n!}(2-s-\alpha+\lambda)_n\Gamma(s+\alpha-1)$ $\times{}_2F_2\left(\begin{matrix}1-\alpha,\ 2-s+n-\alpha+\lambda;\ -ab\\ 2-\alpha-s,\ 2-s-\alpha+\lambda\end{matrix}\right)$ [$a,\ \operatorname{Re}\alpha,\ \operatorname{Re} b>0$]
14	$(a^2-x^2)_+^{\alpha-1}e^{-bx}$ $\times L_n^\lambda(bx)$	$\frac{(\lambda+1)_n a^{s+2\alpha-2}}{2(n!)}\mathrm{B}\left(\alpha,\ \frac{s}{2}\right){}_3F_4\left(\begin{matrix}\frac{n+\lambda+1}{2},\ \frac{n+\lambda+2}{2},\ \frac{s}{2};\ \frac{a^2b^2}{4}\\ \frac{1}{2},\ \frac{\lambda+1}{2},\ \frac{\lambda+2}{2},\ \frac{s+2\alpha}{2}\end{matrix}\right)$ $-\frac{(\lambda+2)_n a^{s+2\alpha-1}b}{2(n!)}\mathrm{B}\left(\alpha,\ \frac{s+1}{2}\right)$ $\times{}_3F_4\left(\begin{matrix}\frac{n+\lambda+2}{2},\ \frac{n+\lambda+3}{2},\ \frac{s+1}{2};\ \frac{a^2b^2}{4}\\ \frac{3}{2},\ \frac{\lambda+2}{2},\ \frac{\lambda+3}{2},\ \frac{s+2\alpha+1}{2}\end{matrix}\right)$ [$a,\ \operatorname{Re}\alpha,\ \operatorname{Re} s>0$]

No.	$f(x)$	$F(s)$
15	$\left(x^2-a^2\right)_+^{\alpha-1} e^{-bx}$ $\times L_n^\lambda(bx)$	$\frac{(\lambda+1)_n a^{s+2\alpha-2}}{2(n!)} \mathrm{B}\left(\alpha, \frac{2-s-2\alpha}{2}\right) {}_3F_4\left(\begin{matrix}\frac{n+\lambda+1}{2}, \frac{n+\lambda+2}{2}, \frac{s}{2}; \frac{a^2b^2}{4} \\ \frac{1}{2}, \frac{\lambda+1}{2}, \frac{\lambda+2}{2}, \frac{s+2\alpha}{2}\end{matrix}\right)$ $-\frac{(\lambda+2)_n a^{s+2\alpha-1} b}{2(n!)} \mathrm{B}\left(\alpha, \frac{1-s-2\alpha}{2}\right)$ $\times {}_3F_4\left(\begin{matrix}\frac{n+\lambda+2}{2}, \frac{n+\lambda+3}{2}, \frac{s+1}{2}; \frac{a^2b^2}{4} \\ \frac{3}{2}, \frac{\lambda+2}{2}, \frac{\lambda+3}{2}, \frac{s+2\alpha+1}{2}\end{matrix}\right)$ $+\frac{b^{2-2\alpha-s}}{n!}(3-s-2\alpha+\lambda)_n \Gamma(s+2\alpha-2)$ $\times {}_3F_4\left(\begin{matrix}1-\alpha, \frac{3-s+n-2\alpha+\lambda}{2}, \frac{4-s+n-2\alpha+\lambda}{2}; \frac{a^2b^2}{4} \\ \frac{3-s-2\alpha+\lambda}{2}, \frac{4-s-2\alpha+\lambda}{2}, \frac{3-s-2\alpha}{2}, \frac{4-s-2\alpha}{2}\end{matrix}\right)$ $[a, \operatorname{Re}\alpha, \operatorname{Re}b>0]$
16	$(\sqrt{a}-\sqrt{x})_+^{\alpha-1} e^{-bx}$ $\times L_n^\lambda(bx)$	$\frac{2(\lambda+1)_n a^{s+(\alpha-1)/2}}{n!} \mathrm{B}(\alpha, 2s)\ {}_3F_3\left(\begin{matrix}n+\lambda+1, s, \frac{2s+1}{2}; -ab \\ \lambda+1, \frac{2s+\alpha}{2}, \frac{2s+\alpha+1}{2}\end{matrix}\right)$ $[a, \operatorname{Re}\alpha, \operatorname{Re}s>0]$
17	$(\sqrt{x}-\sqrt{a})_+^{\alpha-1} e^{-bx}$ $\times L_n^\lambda(bx)$	$\frac{2(\lambda+1)_n a^{s+(\alpha-1)/2}}{n!} \mathrm{B}(\alpha, 1-2s-\alpha)$ $\times {}_3F_3\left(\begin{matrix}n+\lambda+1, s, \frac{2s+1}{2}; -ab \\ \lambda+1, \frac{2s+\alpha}{2}, \frac{2s+\alpha+1}{2}\end{matrix}\right)$ $+\frac{b^{(1-\alpha)/2-s}}{n!}\left(\frac{3-2s-\alpha+2\lambda}{2}\right)_n \Gamma\left(\frac{2s+\alpha-1}{2}\right)$ $\times {}_3F_3\left(\begin{matrix}\frac{1-\alpha}{2}, \frac{2-\alpha}{2}, \frac{3-2s+n-\alpha+2\lambda}{2}; -ab \\ \frac{1}{2}, \frac{3-2s-\alpha}{2}, \frac{3-2s-\alpha+2\lambda}{2}\end{matrix}\right) + \frac{(1-\alpha)\sqrt{a}}{n!\, b^{s+\alpha/2-1}}$ $\times\left(\frac{4-2s-\alpha+2\lambda}{2}\right)_n \Gamma\left(\frac{2s+\alpha-2}{2}\right)$ $\times {}_3F_3\left(\begin{matrix}\frac{2-\alpha}{2}, \frac{3-\alpha}{2}, \frac{4-2s+2n-\alpha+2\lambda}{2}; -ab \\ \frac{3}{2}, \frac{4-2s-\alpha}{2}, \frac{4-2s-\alpha+2\lambda}{2}\end{matrix}\right)$ $[a, \operatorname{Re}\alpha, \operatorname{Re}b>0]$
18	$\frac{1}{(x+a)^\rho} e^{-bx} L_n^\lambda(bx)$	$\frac{(\lambda+1)_n a^{s-\rho}}{n!} \mathrm{B}(s, \rho-s)\ {}_2F_2\left(\begin{matrix}s, n+\lambda+1; ab \\ \lambda+1, s-\rho+1\end{matrix}\right)$ $+\frac{b^{\rho-s}}{n!}(1-s+\lambda+\rho)_n \Gamma(s-\rho)$ $\times {}_2F_2\left(\begin{matrix}\rho, 1-s+n+\lambda+\rho; ab \\ 1-s+\rho, 1-s+\lambda+\rho\end{matrix}\right)$ $[\operatorname{Re}b, \operatorname{Re}s>0; \lvert\arg a\rvert<\pi]$

No.	$f(x)$	$F(s)$
19	$\dfrac{1}{(x^2+a^2)^\rho} e^{-bx} L_n^\lambda(bx)$	$\dfrac{(\lambda+1)_n a^{s-2\rho}}{2(n!)} \mathrm{B}\left(\dfrac{s}{2}, \dfrac{2\rho-s}{2}\right) {}_3F_4\left(\begin{matrix} \frac{n+\lambda+1}{2}, \frac{n+\lambda+2}{2}, \frac{s}{2}; -\frac{a^2b^2}{4} \\ \frac{1}{2}, \frac{\lambda+1}{2}, \frac{\lambda+2}{2}, \frac{s-2\rho+2}{2} \end{matrix}\right)$ $- \dfrac{(\lambda+2)_n a^{s-2\rho+1} b}{2(n!)} \mathrm{B}\left(\dfrac{s+1}{2}, \dfrac{2\rho-s-1}{2}\right)$ $\times {}_3F_4\left(\begin{matrix} \frac{n+\lambda+2}{2}, \frac{n+\lambda+3}{2}, \frac{s+1}{2}; -\frac{a^2b^2}{4} \\ \frac{3}{2}, \frac{\lambda+2}{2}, \frac{\lambda+3}{2}, \frac{s-2\rho+3}{2} \end{matrix}\right)$ $+ \dfrac{b^{2\rho-s}}{n!} (1-s+\lambda+2\rho)_n \Gamma(s-2\rho)$ $\times {}_3F_4\left(\begin{matrix} \rho, \frac{1-s+n+\lambda+2\rho}{2}, \frac{2-s+n+\lambda+2\rho}{2}; -\frac{a^2b^2}{4} \\ \frac{1-s+2\rho}{2}, \frac{2-s+2\rho}{2}, \frac{1-s+\lambda+2\rho}{2}, \frac{2-s+\lambda+2\rho}{2} \end{matrix}\right)$ [Re a, Re b, Re $s > 0$]
20	$\dfrac{1}{(\sqrt{x}+\sqrt{a})^\rho} e^{-bx} L_n^\lambda(bx)$	$\dfrac{2(\lambda+1)_n a^{s-\rho/2}}{n!} \mathrm{B}(2s, \rho-2s) \, {}_3F_3\left(\begin{matrix} n+\lambda+1, s, \frac{2s+1}{2}; -ab \\ \lambda+1, \frac{2s-\rho+1}{2}, \frac{2s-\rho+2}{2} \end{matrix}\right)$ $+ \dfrac{b^{\rho/2-s}}{n!} \left(\dfrac{2-2s+2\lambda+\rho}{2}\right)_n \Gamma\left(\dfrac{2s-\rho}{2}\right)$ $\times {}_3F_3\left(\begin{matrix} \frac{\rho}{2}, \frac{\rho+1}{2}, \frac{2-2s+2n+2\lambda+\rho}{2}; -ab \\ \frac{1}{2}, \frac{2-2s+\rho}{2}, \frac{2-2s+2\lambda+\rho}{2} \end{matrix}\right)$ $- \dfrac{\sqrt{a}\, b^{(\rho+1)/2-s}}{n!} \rho \left(\dfrac{3-2s+2\lambda+\rho}{2}\right)_n \Gamma\left(\dfrac{2s-\rho-1}{2}\right)$ $\times {}_3F_3\left(\begin{matrix} \frac{\rho+1}{2}, \frac{\rho+2}{2}, \frac{3-2s+2n+2\lambda+\rho}{2} \\ \frac{3}{2}, \frac{3-2s+\rho}{2}, \frac{3-2s+2\lambda+\rho}{2}; -ab \end{matrix}\right)$ [Re b, Re $s > 0$; $\vert\arg a\vert < 2\pi$]
21	$(a-x)_+^\lambda e^{bx}$ $\times L_n^\lambda(c(a-x))$	$\dfrac{e^{ab}}{n!} a^{s+\lambda} \Gamma\left[\begin{matrix} s, n+\lambda+1 \\ s+\lambda+1 \end{matrix}\right] \Phi_2(s, -n; s+\lambda+1; ab, ac)$ [a, Re λ, Re $s > 0$]

3.23.3. $L_n^\lambda(bx)$ and trigonometric functions

Notation: $\delta = \begin{Bmatrix} 1 \\ 0 \end{Bmatrix}$.

1	$e^{-bx} \begin{Bmatrix} \sin(ax) \\ \cos(ax) \end{Bmatrix} L_n^\lambda(bx)$	$\dfrac{a^\delta (1-s+\lambda-\delta)_n}{n!\, b^{s+\delta}} \Gamma(s+\delta)$ $\times {}_4F_3\left(\begin{matrix} \frac{s+1}{2}, \frac{s+2\delta}{2}, \frac{s-\lambda+1}{2}, \frac{s-\lambda+2\delta}{2} \\ \frac{2\delta+1}{2}, \frac{s-n-\lambda+1}{2}, \frac{s-n-\lambda+2\delta}{2}; -\frac{a^2}{b^2} \end{matrix}\right)$ [a, Re $b > 0$; Re $s > -\delta$]

No.	$f(x)$	$F(s)$
2	$e^{-bx}\begin{Bmatrix}\sin(a\sqrt{x})\\ \cos(a\sqrt{x})\end{Bmatrix}$ $\times L_n^{\lambda}(bx)$	$\frac{a^{\delta}}{n!\,b^{s+\delta/2}}\left(\frac{2-2s+2\lambda-\delta}{2}\right)_n\Gamma\left(\frac{2s+\delta}{2}\right)$ $\times\,{}_2F_2\left(\begin{matrix}\frac{2s+\delta}{2},\ \frac{2s-2\lambda+\delta}{2};\ -\frac{a^2}{4b}\\ \frac{2\delta+1}{2},\ \frac{2s-2n-2\lambda+\delta}{2}\end{matrix}\right)$ $[a,\ \operatorname{Re} b>0;\ \operatorname{Re} s>-\delta/2]$
3	$e^{-bx}\begin{Bmatrix}\sin(a/\sqrt{x})\\ \cos(a/\sqrt{x})\end{Bmatrix}$ $\times L_n^{\lambda}(bx)$	$\frac{a^{\delta}b^{\delta/2-s}}{n!}\left(\frac{2-2s+2\lambda+\delta}{2}\right)_n\Gamma\left(\frac{2s-\delta}{2}\right)$ $\times\,{}_1F_3\left(\begin{matrix}\frac{2-2s+2n+2\lambda+\delta}{2};\ \frac{a^2b}{4}\\ \frac{2\delta+1}{2},\ \frac{2-2s+\delta}{2},\ \frac{2-2s+2\lambda+\delta}{2}\end{matrix}\right)$ $\mp\frac{2(\lambda+1)_n\,a^{2s}}{n!}\Gamma(-2s)\begin{Bmatrix}\sin(s\pi)\\ \cos(s\pi)\end{Bmatrix}$ $\times\,{}_1F_3\left(\begin{matrix}n+\lambda+1;\ \frac{a^2b}{4}\\ \lambda+1,\ \frac{2s+1}{2},\ s+1\end{matrix}\right)$ $[a,\ \operatorname{Re} b>0;\ \operatorname{Re} s>-1/2]$

3.23.4. $L_n^{\lambda}(bx)$ and the logarithmic function

1	$e^{-ax}\ln(ax)\,L_n^{\lambda}(ax)$	$\frac{(1-s+\lambda)_n}{n!\,a^s}\Gamma(s)\left[\psi(s)-\sum_{k=1}^{n}\frac{1}{k+\lambda-s}\right]$ $[\operatorname{Re} a,\ \operatorname{Re} s>0]$
2	$e^{-ax}\ln^2(ax)\,L_n^{\lambda}(ax)$	$\frac{(1-s+\lambda)_n}{n!\,a^s}\Gamma(s)\left\{\left[\psi(s)-\sum_{k=1}^{n}\frac{1}{k+\lambda-s}\right]^2+\psi'(s)-\sum_{k=1}^{n}\frac{1}{(k+\lambda-s)^2}\right\}$ $[\operatorname{Re} a,\ \operatorname{Re} s>0]$
3	$e^{-ax}\ln^m(ax)\,L_n^{\lambda}(ax)$	$\frac{a^{-s}}{n!}\frac{\partial^m[\Gamma(s)(1-s+\lambda)_n]}{\partial s^m}$ $[\operatorname{Re} a,\ \operatorname{Re} s>0]$
4	$e^{-bx}\begin{Bmatrix}\ln(x+a)\\ \ln\lvert x-a\rvert\end{Bmatrix}$ $\times L_n^{\lambda}(bx)$	$\frac{\pi(\lambda+1)_n\,a^s}{n!\,s}\begin{Bmatrix}\csc(s\pi)\\ \cot(s\pi)\end{Bmatrix}{}_2F_2\left(\begin{matrix}n+\lambda+1,\ s\\ \lambda+1,\ s+1;\ \pm ab\end{matrix}\right)$ $\pm\frac{ab^{1-s}}{n!}(2-s+\lambda)_n\Gamma(s-1)$ $\times\,{}_3F_3\left(\begin{matrix}1,\ 1,\ 2-s+n+\lambda\\ 2,\ 2-s,\ 2-s+\lambda;\ \pm ab\end{matrix}\right)$ $+\frac{(1-s+\lambda)_n}{n!\,b^s}\Gamma(s)\left[\psi(s)-\sum_{k=1}^{n}\frac{1}{k+\lambda-s}-\ln b\right]$ $\left[\operatorname{Re} b,\ \operatorname{Re} s>0,\ \begin{Bmatrix}\lvert\arg a\rvert<\pi\\ a>0\end{Bmatrix}\right]$

No.	$f(x)$	$F(s)$
5	$e^{-bx}\left\{\begin{matrix}\ln(x^2+a^2)\\ \ln\lvert x^2-a^2\rvert\end{matrix}\right\}$ $\times L_n^{\lambda}(bx)$	$\frac{\pi(\lambda+1)_n a^s}{n!\,s}\left\{\begin{matrix}\csc(s\pi/2)\\ \cot(s\pi/2)\end{matrix}\right\}$ $\times {}_3F_4\left(\begin{matrix}\frac{n+\lambda+1}{2},\ \frac{n+\lambda+2}{2},\ \frac{s}{2};\ \mp\frac{a^2b^2}{4}\\ \frac{1}{2},\ \frac{\lambda+1}{2},\ \frac{\lambda+2}{2},\ \frac{s+2}{2}\end{matrix}\right)\mp\frac{\pi a^{s+1}b(\lambda+2)_n}{n!\,(s+1)}$ $\times\left\{\begin{matrix}\sec(s\pi/2)\\ \tan(s\pi/2)\end{matrix}\right\}{}_3F_4\left(\begin{matrix}\frac{n+\lambda+2}{2},\ \frac{n+\lambda+3}{2},\ \frac{s+1}{2};\ \mp\frac{a^2b^2}{4}\\ \frac{3}{2},\ \frac{\lambda+2}{2},\ \frac{\lambda+3}{2},\ \frac{s+3}{2}\end{matrix}\right)$ $\pm\frac{a^2b^{2-s}}{n!}(3-s+\lambda)_n\Gamma(s-2)$ $\times{}_4F_5\left(\begin{matrix}1,\ 1,\ \frac{3-s+n+\lambda}{2},\ \frac{4-s+n+\lambda}{2};\ \mp\frac{a^2b^2}{4}\\ 2,\ \frac{3-s}{2},\ \frac{4-s}{2},\ \frac{3-s+\lambda}{2},\ \frac{4-s+\lambda}{2}\end{matrix}\right)$ $+\frac{2(1-s+\lambda)_n}{n!\,b^s}\Gamma(s)\left[\psi(s)-\sum_{k=1}^{n}\frac{1}{k+\lambda-s}-\ln b\right]$ $\left[\operatorname{Re}b,\ \operatorname{Re}s>0,\ \left\{\begin{matrix}\operatorname{Re}a>0\\ a>0\end{matrix}\right\}\right]$
6	$\theta(a-x)\ln\frac{\sqrt{a}+\sqrt{a-x}}{\sqrt{x}}$ $\times L_n^{\lambda}(bx)$	$\frac{\sqrt{\pi}(\lambda+1)_n a^s}{2(n!)\,s}\Gamma\left[\begin{matrix}s\\ s+\frac{1}{2}\end{matrix}\right]{}_3F_3\left(\begin{matrix}-n,\ s,\ s;\ ab\\ \lambda+1,\ s+\frac{1}{2},\ s+1\end{matrix}\right)$ $[a,\ \operatorname{Re}s>0]$

3.23.5. $L_m^{\lambda}(bx^r)$ and $\operatorname{Ei}(ax^r)$

1	$e^{-bx}\operatorname{Ei}(-ax)L_n^{\lambda}(bx)$	$-\frac{(\lambda+1)_n}{n!\,a^s s}\Gamma(s)\,{}_3F_2\left(\begin{matrix}n+\lambda+1,\ s,\ s\\ \lambda+1,\ s+1;\ -\frac{b}{a}\end{matrix}\right)$ $[\operatorname{Re}a,\ \operatorname{Re}(a+b),\ \operatorname{Re}s>0]$
2	$e^{(\pm a-b)x}\operatorname{Ei}(\mp ax)$ $\times L_n^{\lambda}(bx)$	$-\frac{\pi(\lambda+1)_n}{n!\,a^s}\Gamma(s)\left\{\begin{matrix}\csc(s\pi)\\ \cot(s\pi)\end{matrix}\right\}{}_2F_1\left(\begin{matrix}n+\lambda+1,\ s\\ \lambda+1;\ \pm\frac{b}{a}\end{matrix}\right)$ $\mp\frac{b^{1-s}}{n!\,a}(2-s+\lambda)_n\Gamma(s-1)\,{}_3F_2\left(\begin{matrix}1,\ 1,\ 2-s+n+\lambda\\ 2-s,\ 2-s+\lambda;\ \pm\frac{b}{a}\end{matrix}\right)$ $\left[\operatorname{Re}b,\ \operatorname{Re}s>0,\ \left\{\begin{matrix}\lvert\arg a\rvert<\pi\\ a>0\end{matrix}\right\}\right]$
3	$e^{-bx}\operatorname{Ei}(-ax^2)L_n^{\lambda}(bx)$	$-\frac{(\lambda+1)_n}{n!\,a^{s/2}s}\Gamma\left(\frac{s}{2}\right){}_4F_4\left(\begin{matrix}\frac{n+\lambda+1}{2},\ \frac{n+\lambda+2}{2},\ \frac{s}{2},\ \frac{s}{2}\\ \frac{1}{2},\ \frac{\lambda+1}{2},\ \frac{\lambda+2}{2},\ \frac{s+2}{2};\ \frac{b^2}{4a}\end{matrix}\right)$ $+\frac{(\lambda+2)_n b}{n!\,a^{(s+1)/2}(s+1)}\Gamma\left(\frac{s+1}{2}\right){}_4F_4\left(\begin{matrix}\frac{n+\lambda+2}{2},\ \frac{n+\lambda+3}{2},\ \frac{s+1}{2},\ \frac{s+1}{2}\\ \frac{3}{2},\ \frac{\lambda+2}{2},\ \frac{\lambda+3}{2},\ \frac{s+3}{2};\ \frac{b^2}{4a}\end{matrix}\right)$ $[\operatorname{Re}a,\ \operatorname{Re}s>0]$

No.	$f(x)$	$F(s)$
4	$e^{\pm ax^2-bx}\,\mathrm{Ei}\,(\mp ax^2)$ $\times L_n^\lambda(bx)$	$-\frac{\pi(\lambda+1)_n}{2(n!)\,a^{s/2}}\Gamma\left(\frac{s}{2}\right)\begin{Bmatrix}\csc(s\pi/2)\\ \cot(s\pi/2)\end{Bmatrix}{}_3F_3\left(\begin{matrix}\frac{n+\lambda+1}{2},\ \frac{n+\lambda+2}{2},\ \frac{s}{2}\\ \frac{1}{2},\ \frac{\lambda+1}{2},\ \frac{\lambda+2}{2};\ \mp\frac{b^2}{4a}\end{matrix}\right)$ $\pm\frac{\pi(\lambda+2)_n\,b}{2(n!)\,a^{(s+1)/2}}\Gamma\left(\frac{s+1}{2}\right)$ $\times\begin{Bmatrix}\sec(s\pi/2)\\ \tan(s\pi/2)\end{Bmatrix}{}_3F_3\left(\begin{matrix}\frac{n+\lambda+2}{2},\ \frac{n+\lambda+3}{2},\ \frac{s+1}{2}\\ \frac{3}{2},\ \frac{\lambda+2}{2},\ \frac{\lambda+3}{2};\ \mp\frac{b^2}{4a}\end{matrix}\right)$ $\mp\frac{b^{2-s}}{n!\,a}(3-s+\lambda)_n\,\Gamma(s-2)$ $\times{}_4F_4\left(\begin{matrix}1,\ 1,\ \frac{3-s+n+\lambda}{2},\ \frac{4-s+n+\lambda}{2};\ \mp\frac{b^2}{4a}\\ \frac{3-s}{2},\ \frac{4-s}{2},\ \frac{3-s+\lambda}{2},\ \frac{4-s+\lambda}{2}\end{matrix}\right)$ $\left[\mathrm{Re}\,b,\ \mathrm{Re}\,s>0,\ \begin{Bmatrix}\lvert\arg a\rvert<\pi\\ a>0\end{Bmatrix}\right]$

3.23.6. $L_n^\lambda(bx)$ and $\mathrm{si}\,(ax^r)$, $\mathrm{ci}\,(ax^r)$

1	$e^{-bx}\begin{Bmatrix}\mathrm{si}\,(ax)\\ \mathrm{ci}\,(ax)\end{Bmatrix}L_n^\lambda(bx)$	$-\frac{(\lambda+1)_n}{n!\,a^s s}\begin{Bmatrix}\sin(s\pi/2)\\ \cos(s\pi/2)\end{Bmatrix}\Gamma(s)\ {}_5F_4\left(\begin{matrix}\frac{n+\lambda+1}{2},\ \frac{n+\lambda+2}{2},\ \frac{s}{2},\ \frac{s}{2},\ \frac{s+1}{2}\\ \frac{1}{2},\ \frac{\lambda+1}{2},\ \frac{\lambda+2}{2},\ \frac{s+2}{2};\ -\frac{b^2}{a^2}\end{matrix}\right)$ $\pm\frac{(\lambda+2)_n\,b}{n!\,a^{s+1}(s+1)}\begin{Bmatrix}\cos(s\pi/2)\\ \sin(s\pi/2)\end{Bmatrix}\Gamma(s+1)$ $\times{}_5F_4\left(\begin{matrix}\frac{n+\lambda+2}{2},\ \frac{n+\lambda+3}{2},\ \frac{s+1}{2},\ \frac{s+1}{2},\ \frac{s+2}{2}\\ \frac{3}{2},\ \frac{\lambda+2}{2},\ \frac{\lambda+3}{2},\ \frac{s+3}{2};\ -\frac{b^2}{a^2}\end{matrix}\right)$ $[a,\ \mathrm{Re}\,b,\ \mathrm{Re}\,s>0]$
2	$e^{-bx}\,\mathrm{si}\,(a\sqrt{x})\,L_n^\lambda(bx)$	$-\frac{ab^{-(2s+1)/2}}{n!}\left(\frac{1-2s+2\lambda}{2}\right)_n\Gamma\left(\frac{2s+1}{2}\right)$ $\times{}_3F_3\left(\begin{matrix}\frac{1}{2},\ \frac{2s+1}{2},\ \frac{2s-2\lambda+1}{2};\ -\frac{a^2}{4b}\\ \frac{3}{2},\ \frac{3}{2},\ \frac{2s-2n-2\lambda+1}{2}\end{matrix}\right)-\frac{\pi(1-s+\lambda)_n}{2(n!)\,b^s}\Gamma(s)$ $[a,\ \mathrm{Re}\,b,\ \mathrm{Re}\,s>0]$
3	$e^{-bx}\,\mathrm{ci}\,(a\sqrt{x})\,L_n^\lambda(bx)$	$-\frac{a^2b^{-s-1}}{4(n!)}(\lambda-s)_n\,\Gamma(s+1)$ $\times{}_4F_4\left(\begin{matrix}1,\ 1,\ s+1,\ s-\lambda+1;\ -\frac{a^2}{4b}\\ \frac{3}{2},\ 2,\ 2,\ s-n-\lambda+1\end{matrix}\right)+\frac{b^{-s}}{n!}(1-s+\lambda)_n$ $\times\Gamma(s)\left[\frac{1}{2}\psi(s)-\frac{1}{2}\sum_{k=1}^{n}\frac{1}{k+\lambda-s}-\ln\frac{\sqrt{b}}{a}+\mathbf{C}\right]$ $[a,\ \mathrm{Re}\,b,\ \mathrm{Re}\,s>0]$

3.23.7. $L_n^\lambda(bx)$ and $\operatorname{erf}(ax^r)$, $\operatorname{erfc}(ax^r)$

1	$\operatorname{erfc}(a\sqrt{x})\, L_n^\lambda(bx)$	$\frac{(\lambda+1)_n}{n!\sqrt{\pi}\, a^{2s} s}\,\Gamma\left(\frac{2s+1}{2}\right)\, {}_3F_2\left(\begin{matrix}-n,\ s,\ \frac{2s+1}{2};\ \frac{b}{a^2}\\ \lambda+1,\ s+1\end{matrix}\right)$ $[\operatorname{Re} s>0;\ \lvert\arg a\rvert<\pi/4]$
2	$e^{-bx}\left\{\begin{matrix}\operatorname{erf}(ax)\\ \operatorname{erfc}(ax)\end{matrix}\right\} L_n^\lambda(bx)$	$\mp\frac{(\lambda+1)_n}{n!\sqrt{\pi}\, a^s s}\,\Gamma\left(\frac{s+1}{2}\right)\, {}_4F_4\left(\begin{matrix}\frac{n+\lambda+1}{2},\ \frac{n+\lambda+2}{2},\ \frac{s}{2},\ \frac{s+1}{2};\ \frac{b^2}{4a^2}\\ \frac{1}{2},\ \frac{\lambda+1}{2},\ \frac{\lambda+2}{2},\ \frac{s+2}{2}\end{matrix}\right)$ $\pm\frac{(\lambda+2)_n\, b}{n!\sqrt{\pi}\, a^{s+1}(s+1)}\,\Gamma\left(\frac{s+2}{2}\right)\, {}_4F_4\left(\begin{matrix}\frac{n+\lambda+2}{2},\ \frac{n+\lambda+3}{2},\ \frac{s+1}{2},\ \frac{s+2}{2}\\ \frac{3}{2},\ \frac{\lambda+2}{2},\ \frac{\lambda+3}{2},\ \frac{s+3}{2};\ \frac{b^2}{4a^2}\end{matrix}\right)$ $+\frac{1\pm1}{2}\,\frac{(1-s+\lambda)_n}{n!\, b^s}\,\Gamma(s)$ $[\operatorname{Re} a>0;\ \operatorname{Re} s>-\delta]$
3	$e^{-bx}\left\{\begin{matrix}\operatorname{erf}(a\sqrt{x})\\ \operatorname{erfc}(a\sqrt{x})\end{matrix}\right\}$ $\times L_n^\lambda(bx)$	$\mp\frac{(\lambda+1)_n}{n!\sqrt{\pi}\, a^{2s} s}\,\Gamma\left(\frac{2s+1}{2}\right)\, {}_3F_2\left(\begin{matrix}n+\lambda+1,\ s,\ \frac{2s+1}{2}\\ \lambda+1,\ s+1;\ -\frac{b}{a^2}\end{matrix}\right)$ $+\frac{1\pm1}{2}\,\frac{(1-s+\lambda)_n}{n!\, b^s}\,\Gamma(s)$ $[\operatorname{Re} b,\ \operatorname{Re}(a^2+b)>0;\ \operatorname{Re} s>-(1\pm1)/4]$
4	$e^{(a^2-b)x}\operatorname{erfc}(a\sqrt{x})$ $\times L_n^\lambda(bx)$	$\frac{(\lambda+1)_n\,\Gamma(s)}{n!\, a^{2s}\cos(s\pi)}\, {}_2F_1\left(\begin{matrix}n+\lambda+1,\ s\\ \lambda+1;\ \frac{b}{a^2}\end{matrix}\right)+\frac{\Gamma(s-1/2)}{n!\sqrt{\pi}\, ab^{s-1/2}}$ $\times\left(\frac{3-2s+2\lambda}{2}\right)_n\, {}_3F_2\left(\begin{matrix}\frac{1}{2},\ 1,\ \frac{3-2s+2n+2\lambda}{2}\\ \frac{3-2s}{2},\ \frac{3-2s+2\lambda}{2};\ \frac{b}{a^2}\end{matrix}\right)$ $[\operatorname{Re} b,\ \operatorname{Re}(b-a^2)>0;\ \operatorname{Re} s>0]$

3.23.8. $L_n^\lambda(bx)$ and $S(ax^r)$, $C(ax^r)$

Notation: $\delta=\left\{\begin{matrix}1\\0\end{matrix}\right\}$.

1	$e^{-bx}\left\{\begin{matrix}S(ax)\\ C(ax)\end{matrix}\right\} L_n^\lambda(bx)$	$\frac{\sqrt{2}\, a^{\delta+1/2}}{3^\delta n!\sqrt{\pi}\, b^{s+\delta+1/2}}\left(\frac{1-2s+2\lambda-2\delta}{2}\right)_n\Gamma\left(\frac{2s+2\delta+1}{2}\right)$ $\times\, {}_5F_4\left(\begin{matrix}\frac{2\delta+1}{4},\ \frac{2s+3}{4},\ \frac{2s+4\delta+1}{4},\ \frac{2s-2\lambda+3}{4},\ \frac{2s-2\lambda+4\delta+1}{4}\\ \frac{2\delta+1}{2},\ \frac{2\delta+5}{4},\ \frac{2s-2n-2\lambda+3}{4},\ \frac{2s-2n-2\lambda+4\delta+1}{4};\ -\frac{a^2}{b^2}\end{matrix}\right)$ $[a,\ \operatorname{Re} b>0;\ \operatorname{Re} s>-\delta-1/2]$
2	$e^{-bx}\left\{\begin{matrix}S(a\sqrt{x})\\ C(a\sqrt{x})\end{matrix}\right\}$ $\times L_n^\lambda(bx)$	$\frac{\sqrt{2}\, a^{\delta+1/2}}{3^\delta n!\sqrt{\pi}\, b^{s+(2\delta+1)/4}}\left(\frac{3-4s+4\lambda-2\delta}{4}\right)_n\Gamma\left(\frac{4s+2\delta+1}{4}\right)$ $\times\, {}_3F_3\left(\begin{matrix}\frac{2\delta+1}{4},\ \frac{4s+2\delta+1}{4},\ \frac{4s-4\lambda+2\delta+1}{4};\ -\frac{a^2}{4b}\\ \frac{2\delta+1}{2},\ \frac{2\delta+5}{4},\ \frac{4s-4n-4\lambda+2\delta+1}{4}\end{matrix}\right)$ $[a,\ \operatorname{Re} b>0;\ \operatorname{Re} s>-(2\delta+1)/4]$

3.23.9. $L_n^{\lambda}(bx)$ and $\gamma(\nu, ax^r)$, $\Gamma(\nu, ax^r)$

1	$e^{-bx}\begin{Bmatrix}\gamma(\mu, ax)\\ \Gamma(\mu, ax)\end{Bmatrix} \times L_n^{\lambda}(bx)$	$\mp\frac{(\lambda+1)_n}{n!\,a^s s}\Gamma(s+\mu)\,{}_3F_2\left(\begin{matrix}n+\lambda+1,\, s,\, s+\mu\\ \lambda+1,\, s+1;\, -\frac{b}{a}\end{matrix}\right) + \frac{1\pm1}{2}\frac{(1-s+\lambda)_n}{n!\,b^s}\Gamma(\mu)\Gamma(s)$ $\left[\operatorname{Re}a,\ \operatorname{Re}b,\ \operatorname{Re}(s+\mu)>0,\ \begin{Bmatrix}\operatorname{Re}\mu>0\\ \operatorname{Re}s>0\end{Bmatrix}\right]$
2	$e^{-bx}\begin{Bmatrix}\gamma(\mu, ax^2)\\ \Gamma(\mu, ax^2)\end{Bmatrix} \times L_n^{\lambda}(bx)$	$\mp\frac{(\lambda+1)_n}{n!\,a^{s/2}s}\Gamma\left(\frac{s+2\mu}{2}\right){}_4F_4\left(\begin{matrix}\frac{n+\lambda+1}{2},\, \frac{n+\lambda+2}{2},\, \frac{s}{2},\, \frac{s+2\mu}{2}\\ \frac{1}{2},\, \frac{\lambda+1}{2},\, \frac{\lambda+2}{2},\, \frac{s+2}{2};\, \frac{b^2}{4a}\end{matrix}\right) \pm\frac{(\lambda+2)_n b}{a^{(s+1)/2}(s+1)}\Gamma\left(\frac{s+2\mu+1}{2}\right) \times{}_4F_4\left(\begin{matrix}\frac{n+\lambda+2}{2},\, \frac{n+\lambda+3}{2},\, \frac{s+1}{2},\, \frac{s+2\mu+1}{2}\\ \frac{3}{2},\, \frac{\lambda+2}{2},\, \frac{\lambda+3}{2},\, \frac{s+3}{2};\, \frac{b^2}{4a}\end{matrix}\right) + \frac{1\pm1}{2}\frac{(1-s+\lambda)_n}{n!\,b^s}\Gamma(\mu)\Gamma(s)$ $\left[\operatorname{Re}a,\ \operatorname{Re}(s+2\mu)>0,\ \begin{Bmatrix}\operatorname{Re}\mu>0\\ \operatorname{Re}s>0\end{Bmatrix}\right]$
3	$e^{-bx}\begin{Bmatrix}\gamma(\mu, a\sqrt{x})\\ \Gamma(\mu, a\sqrt{x})\end{Bmatrix} \times L_n^{\lambda}(bx)$	$\pm\frac{a^{\mu}}{n!\,b^{s+\mu/2}\mu}\left(\frac{2-2s+2\lambda-\mu}{2}\right)_n\Gamma\left(\frac{2s+\mu}{2}\right) \times{}_3F_3\left(\begin{matrix}\frac{\mu}{2},\, \frac{2s+\mu}{2},\, \frac{2s-2\lambda+\mu}{2};\, \frac{a^2}{4b}\\ \frac{1}{2},\, \frac{\mu+2}{2},\, \frac{2s-2n-2\lambda+\mu}{2}\end{matrix}\right)\mp\frac{a^{\mu+1}}{n!\,b^{s+(\mu+1)/2}(\mu+1)} \times\Gamma\left(\frac{2s+\mu+1}{2}\right)\left(\frac{1-2s+2\lambda-\mu}{2}\right)_n \times{}_3F_3\left(\begin{matrix}\frac{\mu+1}{2},\, \frac{2s+\mu+1}{2},\, \frac{2s-2\lambda+\mu+1}{2};\, \frac{a^2}{4b}\\ \frac{3}{2},\, \frac{\mu+3}{2},\, \frac{2s-2n-2\lambda+\mu+1}{2}\end{matrix}\right) + \frac{1\pm1}{2}\frac{(1-s+\lambda)_n}{n!\,b^s}\Gamma(\mu)\Gamma(s)$ $\left[\operatorname{Re}b,\ \operatorname{Re}(2s+\mu)>0,\ \begin{Bmatrix}\operatorname{Re}\mu>0\\ \operatorname{Re}s>0\end{Bmatrix}\right]$

3.23.10. $L_n^{\lambda}(bx)$ and $J_{\mu}(ax^r)$, $I_{\mu}(ax^r)$

1	$J_{\mu}(a\sqrt{x})\,L_n^{\lambda}(bx)$	$\frac{1}{n!}\left(\frac{2}{a}\right)^{2s+2n}(-b)^n\,\Gamma\left[\begin{matrix}\frac{2s+\mu+2n}{2}\\ \frac{2-2s-2n+\mu}{2}\end{matrix}\right]{}_2F_2\left(\begin{matrix}-n,\, -n-\lambda;\, \frac{a^2}{4b}\\ \frac{2-2s-2n+\mu}{2},\, \frac{2-2s-2n-\mu}{2}\end{matrix}\right)$ $[a>0;\ -\operatorname{Re}\mu<2\operatorname{Re}s<3/2-2n]$

No.	$f(x)$	$F(s)$
2	$J_\mu\left(\frac{a}{\sqrt{x}}\right) L_n^\lambda(bx)$	$\frac{(\lambda+1)_n}{n!}\left(\frac{a}{2}\right)^{2s}\Gamma\left[\begin{matrix}\frac{\mu-2s}{2}\\ \frac{2s+\mu+2}{2}\end{matrix}\right] {}_1F_3\left(\begin{matrix}-n;\ -\frac{a^2b}{4}\\ \lambda+1,\ \frac{2s+\mu+2}{2},\ \frac{2s-\mu+2}{2}\end{matrix}\right)$ $[a>0;\ -3/2<2\operatorname{Re}s<\operatorname{Re}\mu-2n]$
3	$e^{-bx}\left\{\begin{matrix}J_\mu(ax)\\ I_\mu(ax)\end{matrix}\right\} L_n^\lambda(bx)$	$\frac{a^\mu(1-s+\lambda-\mu)_n}{2^\mu n!\,b^{s+\mu}}\Gamma\left[\begin{matrix}s+\mu\\ \mu+1\end{matrix}\right]$ $\times {}_4F_3\left(\begin{matrix}\frac{s+\mu}{2},\ \frac{s+\mu+1}{2},\ \frac{s-\lambda+\mu}{2},\ \frac{s-\lambda+\mu+1}{2}\\ \mu+1,\ \frac{s-n-\lambda+\mu}{2},\ \frac{s-n-\lambda+\mu+1}{2};\ \mp\frac{a^2}{b^2}\end{matrix}\right)$ $\left[\operatorname{Re}b,\ \operatorname{Re}(s+\mu)>0,\ \left\{\begin{matrix}a>0\\ \operatorname{Re}(b-a)>0\end{matrix}\right\}\right]$
4	$e^{-bx}\left\{\begin{matrix}J_\mu(a\sqrt{x})\\ I_\mu(a\sqrt{x})\end{matrix}\right\}$ $\times L_n^\lambda(bx)$	$\frac{a^\mu}{2^\mu n!\,b^{s+\mu/2}}\left(\frac{2-2s+2\lambda-\mu}{2}\right)_n\Gamma\left[\begin{matrix}\frac{s+\mu}{2}\\ \mu+1\end{matrix}\right]$ $\times {}_2F_2\left(\begin{matrix}\frac{2s+\mu}{2},\ \frac{2s-2\lambda+\mu}{2};\ \mp\frac{a^2}{4b^2}\\ \mu+1,\ \frac{2s-2n-2\lambda+\mu}{2}\end{matrix}\right)$ $[\operatorname{Re}b,\ \operatorname{Re}(2s+\mu)>0]$
5	$e^{-bx}J_\mu\left(\frac{a}{\sqrt{x}}\right) L_n^\lambda(bx)$	$\frac{(\lambda+1)_n}{n!}\left(\frac{a}{2}\right)^{2s}\Gamma\left[\begin{matrix}\frac{\mu-2s}{2}\\ \frac{2s+\mu+2}{2}\end{matrix}\right] {}_1F_3\left(\begin{matrix}n+\lambda+1;\ \frac{a^2b}{4}\\ \lambda+1,\ \frac{2s+\mu+2}{2},\ \frac{2s-\mu+2}{2}\end{matrix}\right)$ $+\frac{1}{n!}\left(\frac{a}{2}\right)^{\mu} b^{\mu/2-s}\left(\frac{2-2s+2\lambda+\mu}{2}\right)_n\Gamma\left[\begin{matrix}\frac{2s-\mu}{2}\\ \mu+1\end{matrix}\right]$ $\times {}_1F_3\left(\begin{matrix}\frac{2-2s+2n+2\lambda+\mu}{2};\ \frac{a^2b}{4}\\ \mu+1,\ \frac{2-2s+\mu}{2},\ \frac{2-2s+\mu+2\lambda}{2}\end{matrix}\right)$ $[\operatorname{Re}b>0;\ \operatorname{Re}s>-3/4]$

3.23.11. $L_n^\lambda(bx)$ and $Y_\mu(ax^r)$, $K_\mu(ax^r)$

1	$K_\mu(a\sqrt{x})\,L_n^\lambda(bx)$	$\frac{2^{2s+2n-1}(-b)^n}{n!\,a^{2s+2n}}\Gamma\left(\frac{2s+2n-\mu}{2}\right)\Gamma\left(\frac{2s+2n+\mu}{2}\right)$ $\times {}_2F_2\left(\begin{matrix}-n,\ -n-\lambda;\ -\frac{a^2}{4b}\\ \frac{2-2s-2n-\mu}{2},\ \frac{2-2s-2n+\mu}{2}\end{matrix}\right)$ $[\operatorname{Re}a>0;\ 2\operatorname{Re}s>\lvert\operatorname{Re}\mu\rvert]$
2	$K_\mu\left(\frac{a}{\sqrt{x}}\right) L_n^\lambda(bx)$	$\frac{(\lambda+1)_n\,a^{2s}}{2^{2s+1}n!}\Gamma\left(\frac{\mu-2s}{2}\right)\Gamma\left(-\frac{\mu+2s}{2}\right)$ $\times {}_1F_3\left(\begin{matrix}-n;\ \frac{a^2b}{4}\\ \lambda+1,\ \frac{2s+\mu+2}{2},\ \frac{2s-\mu+2}{2}\end{matrix}\right)$ $[\operatorname{Re}a>0;\ 2\operatorname{Re}s+\lvert\operatorname{Re}\mu\rvert<-2n]$

No.	$f(x)$	$F(s)$
3	$e^{-bx}\left\{\begin{matrix} Y_\mu(ax) \\ K_\mu(ax) \end{matrix}\right\} L_n^\lambda(bx)$	$\mp\pi^{(1\mp1)/2}\frac{(\lambda+1)_n\, 2^{s-2}}{n!\pi b^s}\Gamma\left(\frac{s+\mu}{2}\right)\Gamma\left(\frac{s-\mu}{2}\right)$ $\times\left(2\cos\frac{(s-\mu)\pi}{2}\right)^{(1\pm1)/2} {}_4F_3\left(\begin{matrix}\frac{n+\lambda+1}{2},\ \frac{n+\lambda+2}{2},\ \frac{s+\mu}{2},\ \frac{s-\mu}{2} \\ \frac{1}{2},\ \frac{\lambda+1}{2},\ \frac{\lambda+2}{2};\ \mp\frac{b^2}{a^2}\end{matrix}\right)$ $\mp\pi^{(1\mp1)/2}\frac{2^{s-1}b\,(\lambda+2)_n}{n!\pi a^{s+1}}\Gamma\left(\frac{s+\mu+1}{2}\right)\Gamma\left(\frac{s-\mu+1}{2}\right)$ $\times\left(2\sin\frac{(s-\mu)\pi}{2}\right)^{(1\pm1)/2}$ $\times\,{}_4F_3\left(\begin{matrix}\frac{n+\lambda+2}{2},\ \frac{n+\lambda+3}{2},\ \frac{s-\mu+1}{2},\ \frac{s+\mu+1}{2} \\ \frac{3}{2},\ \frac{\lambda+2}{2},\ \frac{\lambda+3}{2};\ \mp\frac{b^2}{a^2}\end{matrix}\right)$ $\left[\operatorname{Re} b>0;\ \operatorname{Re} s>\lvert\operatorname{Re}\mu\rvert,\ \left\{\begin{matrix} a>0 \\ \operatorname{Re}(a+b)>0\end{matrix}\right\}\right]$
4	$e^{-(a+b)x}K_\mu(ax)\,L_n^\lambda(bx)$	$\frac{\sqrt{\pi}\,(\lambda+1)_n}{n!\,(2a)^s}\Gamma\left[\begin{matrix} s+\mu,\ s-\mu \\ \frac{2s+1}{2}\end{matrix}\right] {}_3F_2\left(\begin{matrix} n+\lambda+1,\ s+\mu,\ s-\mu \\ \lambda+1,\ \frac{2s+1}{2};\ -\frac{b}{2a}\end{matrix}\right)$ $[\operatorname{Re}(2a+b)>0;\ \operatorname{Re} s>\lvert\operatorname{Re}\mu\rvert]$
5	$e^{-bx}\left\{\begin{matrix} Y_\mu(a\sqrt{x}) \\ K_\mu(a\sqrt{x}) \end{matrix}\right\}$ $\times\,L_n^\lambda(bx)$	$\mp\frac{2^{\mu-(1\mp1)/2}\pi^{(1\mp1)/2}}{n!\pi a^\mu b^{s-\mu/2}}\left(\frac{2-2s+2\lambda+\mu}{2}\right)_n\Gamma(\mu)$ $\times\,\Gamma\left(\frac{2s-\mu}{2}\right) {}_2F_2\left(\begin{matrix}\frac{2s-\mu}{2},\ \frac{2s-2\lambda-\mu}{2};\ \mp\frac{a^2}{4b} \\ 1-\mu,\ \frac{2s-2n-2\lambda-\mu}{2}\end{matrix}\right)$ $\mp\frac{\pi^{(1\mp1)/2}a^\mu}{2^{\mu+1}n!\pi b^{s+\mu/2}}(2\cos\mu\pi)^{(1\pm1)/2}\Gamma(-\mu)\Gamma\left(\frac{2s+\mu}{2}\right)$ $\times\left(\frac{2-2s+2\lambda-\mu}{2}\right)_n {}_2F_2\left(\begin{matrix}\frac{2s+\mu}{2},\ \frac{2s-2\lambda+\mu}{2};\ \mp\frac{a^2}{4b} \\ 1+\mu,\ \frac{2s-2n-2\lambda+\mu}{2}\end{matrix}\right)$ $[\operatorname{Re} b>0;\ 2\operatorname{Re} s>\lvert\operatorname{Re}\mu\rvert;\ \lvert\arg a\rvert<\pi]$
6	$e^{-bx}K_\mu\left(\frac{a}{\sqrt{x}}\right)L_n^\lambda(bx)$	$\frac{(\lambda+1)_n\, a^{2s}}{2^{2s+1}n!}\Gamma\left(\frac{\mu-2s}{2}\right)\Gamma\left(-\frac{\mu+2s}{2}\right)$ $\times\,{}_1F_3\left(\begin{matrix} n+\lambda+1;\ -\frac{a^2b}{4} \\ \lambda+1,\ \frac{2s-\mu+2}{2},\ \frac{2s+\mu+2}{2}\end{matrix}\right)+\frac{a^\mu}{2^{\mu+1}n!\,b^{s-\mu/2}}$ $\times\left(\frac{2-2s+2\lambda+\mu}{2}\right)_n\Gamma(-\mu)\Gamma\left(\frac{2s-\mu}{2}\right)$ $\times\,{}_1F_3\left(\begin{matrix}\frac{2-2s+n+2\lambda+\mu}{2};\ -\frac{a^2b}{4} \\ \mu+1,\ \frac{2-2s+\mu}{2},\ \frac{2-2s+2\lambda+\mu}{2}\end{matrix}\right)$ $+\frac{2^{\mu-1}}{n!\,a^\mu b^{s+\mu/2}}\left(\frac{2-2s+2\lambda-\mu}{2}\right)_n\Gamma(\mu)\Gamma\left(\frac{2s+\mu}{2}\right)$ $\times\,{}_1F_3\left(\begin{matrix}\frac{2-2s+n+2\lambda-\mu}{2};\ -\frac{a^2b}{4} \\ 1-\mu,\ \frac{2-2s-\mu}{2},\ \frac{2-2s+2\lambda-\mu}{2}\end{matrix}\right)$ $[\operatorname{Re} a,\ \operatorname{Re} b>0]$

3.23.12. $L_n^\lambda(bx^r)$ and $P_n(ax^p+c)$

1	$\theta(a-x)\,e^{-bx} \times P_n\left(\frac{2x}{a}-1\right) L_m^\lambda(bx)$	$\frac{(-1)^n(\lambda+1)_m a^s}{m!}\,\frac{(1-s)_n}{(s)_{n+1}}\,{}_3F_3\left(\begin{matrix} m+\lambda+1,\, s,\, s;\, -ab \\ \lambda+1,\, s-n,\, s+n+1 \end{matrix}\right)$ [a, $\operatorname{Re} s>0$]
2	$\theta(x-a)\,P_m\left(\frac{a}{x}\right) \times L_n^\lambda(b^2x^2)$	$\frac{(\lambda+1)_n 2^{s-1}a^s}{\pi^{3/2}\,n!}\sin(s\pi)\,\Gamma(-s)\,\Gamma\left(\frac{s-m}{2}\right) \times\Gamma\left(\frac{s+m+1}{2}\right){}_3F_3\left(\begin{matrix} -n,\, \frac{s-m}{2},\, \frac{s+m+1}{2} \\ \lambda+1,\, \frac{s+1}{2},\, \frac{s+2}{2};\, a^2b^2 \end{matrix}\right)$ [$a>0$; $\operatorname{Re} s<-2n$]

3.23.13. $L_n^\lambda(bx)$ and $T_n(ax+c)$, $U_n(ax+c)$

1	$(a-x)_+^{-1/2}\,e^{-bx} \times T_n\left(\frac{2x}{a}-1\right) L_m^\lambda(bx)$	$\frac{(-1)^n\sqrt{\pi}\,a^{s-1/2}}{m!}(\lambda+1)_m\left(\frac{1-2s}{2}\right)_n \times\Gamma\left[\begin{matrix} s \\ \frac{2s+2n+1}{2}\end{matrix}\right]{}_3F_3\left(\begin{matrix} m+\lambda+1,\, s,\, \frac{2s+1}{2};\, -ab \\ \lambda+1,\, \frac{2s-2n+1}{2},\, \frac{2s+2n+1}{2}\end{matrix}\right)$ [a, $\operatorname{Re} s>0$]
2	$(a-x)_+^{1/2}\,e^{-bx} \times U_n\left(\frac{2x}{a}-1\right) L_m^\lambda(bx)$	$\frac{(-1)^n(n+1)\sqrt{\pi}\,a^{s+1/2}}{2\,(m!)}(\lambda+1)_m\left(\frac{3-2s}{2}\right)_n \times\Gamma\left[\begin{matrix} s \\ \frac{2s+2n+3}{2}\end{matrix}\right]{}_3F_3\left(\begin{matrix} m+\lambda+1,\, \frac{2s-1}{2},\, s;\, -ab \\ \lambda+1,\, \frac{2s-2n-1}{2},\, \frac{2s+2n+3}{2}\end{matrix}\right)$ [a, $\operatorname{Re} s>0$]

3.23.14. $L_n^\lambda(bx^r)$ and $H_n(ax)$

Notation: $\varepsilon=0$ or 1.

1	$e^{-a^2x^2-bx}H_{2m+\varepsilon}(ax) \times L_n^\lambda(bx)$	$\frac{(-1)^m 2^{2m+\varepsilon-1}(\lambda+1)_n}{n!\,a^s}\Gamma\left(\frac{s+\varepsilon}{2}\right)\left(\frac{\varepsilon-s+1}{2}\right)_m \times{}_4F_4\left(\begin{matrix} \frac{n+\lambda+1}{2},\, \frac{n+\lambda+2}{2},\, \frac{s}{2},\, \frac{s+1}{2} \\ \frac{1}{2},\, \frac{\lambda+1}{2},\, \frac{\lambda+2}{2},\, \frac{s-2m-\varepsilon+1}{2};\, \frac{b^2}{4a^2}\end{matrix}\right)$ $-\frac{(-1)^m 2^{2m+\varepsilon-1}(\lambda+2)_n\,b}{n!\,a^{s+1}}\Gamma\left(\frac{s+\varepsilon+1}{2}\right)\left(\frac{\varepsilon-s}{2}\right)_m \times{}_4F_4\left(\begin{matrix} \frac{n+\lambda+2}{2},\, \frac{n+\lambda+3}{2},\, \frac{s+1}{2},\, \frac{s+2}{2} \\ \frac{3}{2},\, \frac{\lambda+2}{2},\, \frac{\lambda+3}{2},\, \frac{s-2m-\varepsilon+2}{2};\, \frac{b^2}{4a^2}\end{matrix}\right)$ [$\operatorname{Re} s>-\varepsilon$; $\lvert\arg a\rvert<\pi/4$]

No.	$f(x)$	$F(s)$
2	$e^{-(a^2+b)x^2} H_{2m+\varepsilon}(ax)$ $\times L_n^{\lambda}(bx^2)$	$\frac{(-1)^m 2^{2m+\varepsilon-1}(\lambda+1)_n}{n!\,a^s}\Gamma\left(\frac{s+\varepsilon}{2}\right)\left(\frac{\varepsilon-s+1}{2}\right)_m$ $\times {}_3F_2\left(\begin{matrix} n+\lambda+1,\ \frac{s}{2},\ \frac{s+1}{2} \\ \lambda+1,\ \frac{s-2m-\varepsilon+1}{2};\ -\frac{b}{a^2}\end{matrix}\right)$ $[\operatorname{Re}(a^2+b)>0;\ \operatorname{Re}s>-\varepsilon]$

3.23.15. Products of $L_n^{\lambda}(bx)$

1	$e^{-ax}L_m^{\lambda}(bx)\,L_n^{\lambda}(cx)$	$\frac{(\lambda+1)_m(\mu+1)_n}{m!\,n!\,a^s}F_2\left(s,\ -m,\ -n;\ \lambda+1,\ \mu+1;\ \frac{b}{a},\ \frac{c}{a}\right)$ $[\operatorname{Re}a,\ \operatorname{Re}s>0]$
2	$e^{-bx}L_m^{\lambda}(ax)\,L_n^{\mu}(bx)$	$\frac{(\lambda+1)_m(1-s+\mu)_n}{m!\,n!\,b^s}\Gamma(s)\,{}_3F_2\left(\begin{matrix}-m,\ s,\ s-\mu;\ \frac{a}{b} \\ \lambda+1,\ s-\mu-n\end{matrix}\right)$ $[\operatorname{Re}b,\ \operatorname{Re}s>0]$
3	$e^{-(a+b)x}L_m^{\lambda}(ax)\,L_n^{\mu}(bx)$	$\frac{(\mu+1)_n(1-s+\lambda)_m}{m!\,n!\,a^s}\Gamma(s)\,{}_3F_2\left(\begin{matrix} n+\mu+1,\ s,\ s-\lambda \\ \mu+1,\ s-m-\lambda;\ -\frac{b}{a}\end{matrix}\right)$ $[\operatorname{Re}(a+b),\ \operatorname{Re}s>0]$
4	$e^{-bx}L_m^{\lambda}(ax^2)\,L_n^{\mu}(bx)$	$\frac{(\lambda+1)_m(1-s+\mu)_n b^{-s}}{m!\,n!}\Gamma(s)$ $\times {}_5F_3\left(\begin{matrix}-m,\ \frac{s}{2},\ \frac{s+1}{2},\ \frac{s-\mu}{2},\ \frac{s-\mu+1}{2} \\ \lambda+1,\ \frac{s-n-\mu}{2},\ \frac{s-n-\mu+1}{2};\ \frac{4a}{b^2}\end{matrix}\right)$ $[\operatorname{Re}b,\ \operatorname{Re}s>0]$
5	$e^{-ax^2-bx}L_m^{\lambda}(ax^2)$ $\times L_n^{\mu}(bx)$	$\frac{(\mu+1)_n}{2(m!)\,n!\,a^{s/2}}\Gamma\left(\frac{s}{2}\right)\left(\frac{2-s+2\lambda}{2}\right)_m$ $\times {}_4F_4\left(\begin{matrix}\frac{n+\mu+1}{2},\ \frac{n+\mu+2}{2},\ \frac{s}{2},\ \frac{s-2\lambda}{2};\ \frac{b^2}{4a} \\ \frac{1}{2},\ \frac{\mu+1}{2},\ \frac{\mu+2}{2},\ \frac{s-2m-2\lambda}{2}\end{matrix}\right)$ $-\frac{(\mu+2)_n b}{2(m!)\,n!\,a^{(s+1)/2}}\left(\frac{1-s+2\lambda}{2}\right)_m\Gamma\left(\frac{s+1}{2}\right)$ $\times {}_4F_4\left(\begin{matrix}\frac{n+\mu+2}{2},\ \frac{n+\mu+3}{2},\ \frac{s+1}{2},\ \frac{s-2\lambda+1}{2};\ \frac{b^2}{4a} \\ \frac{3}{2},\ \frac{\mu+2}{2},\ \frac{\mu+3}{2},\ \frac{s-2m-2\lambda+1}{2}\end{matrix}\right)$ $[\operatorname{Re}a,\ \operatorname{Re}s>0]$
6	$e^{-ax}\prod_{k=1}^{n}L_{m_k}^{\lambda_k}(b_k x)$	$a^{-s}\prod_{k=1}^{n}\frac{(\lambda_k+1)_{m_k}}{m_k!}F_A^{(n)}\left(s,\ (-m_n);\ (\lambda_n)+1;\ \frac{(b_n)}{a}\right)$ $[\operatorname{Re}a,\ \operatorname{Re}s>0]$

3.24. The Gegenbauer Polynomials $C_n^\lambda(z)$

More formulas can be obtained from the corresponding sections due to the relations

$$\lim_{\lambda\to 0}\left[\frac{1}{\lambda}C_n^\lambda(z)\right]=\frac{2}{n}T_n(z),\quad C_n^{1/2}(z)=P_n(z)=P_n^{(0,0)}(z),$$
$$C_n^1(z)=U_n(z)=\frac{(n+1)!}{(3/2)_n}P_n^{(1/2,1/2)}(z)=\frac{1}{1-z^2}\left[zT_{n+1}(z)-T_{n+2}(z)\right],$$
$$C_\nu^\lambda(z)=\frac{(2\lambda)_\nu}{(\lambda+1/2)_n}P_\nu^{(\lambda-1/2,\,\lambda-1/2)}(z),$$
$$C_\nu^\lambda(z)=\frac{\Gamma(2\lambda+\nu)}{\Gamma(2\lambda)\Gamma(\nu+1)}\,{}_2F_1\left(-\nu,\,2\lambda+\nu;\,\lambda+\frac{1}{2};\,\frac{1-z}{2}\right).$$

3.24.1. $C_n^\lambda(\varphi(x))$ and algebraic functions

Notation: $\varepsilon=0$ or 1.

No.	$f(x)$	$F(s)$
1	$(a-x)_+^{\alpha-1}C_{2n+\varepsilon}^\lambda(bx)$	$\frac{(-1)^n a^{s+\alpha+\varepsilon-1}(2b)^\varepsilon}{n!}(\lambda)_{n+\varepsilon}\mathrm{B}(\alpha,\,s+\varepsilon)\times{}_4F_3\left(\begin{matrix}-n,\,n+\lambda+\varepsilon,\,\frac{s+\varepsilon}{2},\,\frac{s+\varepsilon+1}{2}\\ \frac{2\varepsilon+1}{2},\,\frac{s+\alpha+\varepsilon}{2},\,\frac{s+\alpha+\varepsilon+1}{2};\,a^2b^2\end{matrix}\right)$ $[a,\ \mathrm{Re}\,\alpha>0;\ \mathrm{Re}\,s>-\varepsilon]$
2	$(x-a)_+^{\alpha-1}C_{2n+\varepsilon}^\lambda(bx)$	$\frac{(-1)^n a^{s+\alpha+\varepsilon-1}(2b)^\varepsilon}{n!}(\lambda)_{n+\varepsilon}\mathrm{B}(1-s-\alpha-\varepsilon,\,\alpha)\times{}_4F_3\left(\begin{matrix}-n,\,n+\lambda+\varepsilon,\,\frac{s+\varepsilon}{2},\,\frac{s+\varepsilon+1}{2}\\ \frac{2\varepsilon+1}{2},\,\frac{s+\alpha+\varepsilon}{2},\,\frac{s+\alpha+\varepsilon+1}{2};\,a^2b^2\end{matrix}\right)$ $[a>0;\ \mathrm{Re}\,(s+\alpha)<1-2n-\varepsilon]$
3	$(a^2-x^2)_+^{\alpha-1}C_{2n+\varepsilon}^\lambda(bx)$	$\frac{(-1)^n a^{s+2\alpha+\varepsilon-2}b^\varepsilon}{2^{1-\varepsilon}n!}(\lambda)_{n+\varepsilon}\mathrm{B}\left(\alpha,\,\frac{s+\varepsilon}{2}\right){}_3F_2\left(\begin{matrix}-n,\,n+\lambda+\varepsilon,\,\frac{s+\varepsilon}{2}\\ \frac{2\varepsilon+1}{2},\,\frac{s+2\alpha+\varepsilon}{2};\,a^2b^2\end{matrix}\right)$ $[a,\ \mathrm{Re}\,\alpha>0;\ \mathrm{Re}\,s>-\varepsilon]$
4	$(a^2-x^2)_+^{\lambda-1/2}C_n^\lambda\left(\frac{x}{a}\right)$	$\frac{\pi(2\lambda)_n}{2^{s+2\lambda-1}n!}a^{s+2\lambda-1}\Gamma\left[\begin{matrix}n+2\lambda,\,s\\ \lambda,\,\frac{s-n+1}{2},\,\frac{s+n+2\lambda+1}{2}\end{matrix}\right]$ $[a>0;\ \mathrm{Re}\,\lambda>-1/2;\ \mathrm{Re}\,s>((-1)^n-1)/2]$
5	$(x^2-a^2)_+^{\lambda-1/2}C_n^\lambda\left(\frac{x}{a}\right)$	$\frac{(2\lambda)_n}{2^{s+1}n!\sqrt{\pi}}a^{s+2\lambda-1}\Gamma\left[\begin{matrix}\frac{2\lambda+1}{2},\,\frac{1-s+n}{2},\,\frac{1-s-n-2\lambda}{2}\\ 1-s\end{matrix}\right]$ $[a>0;\ \mathrm{Re}\,\lambda>-1/2;\ \mathrm{Re}\,(s+2\lambda)<1-n]$

No.	$f(x)$	$F(s)$
6	$\frac{1}{x^2-b^2}\left(a^2-x^2\right)_+^{\lambda-1/2} \times C_{2n+\varepsilon}^{\lambda}\left(\frac{x}{a}\right)$	$\frac{(-1)^{n+1}\sqrt{\pi}\,(2\lambda)_{2n+\varepsilon}\,a^{s+2\lambda-1}}{2^s\,(2n+\varepsilon)!\,b^2}\left(\frac{1-s+\varepsilon}{2}\right)_n \times\Gamma\left[\begin{matrix}\frac{2\lambda+1}{2},\ s\\ \frac{s-\varepsilon+1}{2},\ \frac{s+2n+2\lambda+\varepsilon+1}{2}\end{matrix}\right]{}_3F_2\left(\begin{matrix}1,\ \frac{s}{2},\ \frac{s+1}{2};\ \frac{a^2}{b^2}\\ \frac{s-2n-\varepsilon+1}{2},\ \frac{s+2n+2\lambda+\varepsilon+1}{2}\end{matrix}\right)$ $[0<a<b;\ \operatorname{Re}\lambda>-1/2;\ \operatorname{Re}s>-\varepsilon]$
7	$\frac{1}{x^2-b^2}\left(a^2-x^2\right)_+^{\lambda-1/2} \times C_n^{\lambda}\left(\frac{x}{a}\right)$	$\frac{\sqrt{\pi}\,(2\lambda)_n\,a^{s+2\lambda-3}}{2^{s-2}n!}\Gamma\left[\begin{matrix}\frac{2\lambda+1}{2},\ s-2\\ \frac{s-n-1}{2},\ \frac{s+n+2\lambda-1}{2}\end{matrix}\right] \times{}_3F_2\left(\begin{matrix}1,\ \frac{3-s+n}{2},\ -\frac{s+n+2\lambda-3}{2}\\ \frac{3-s}{2},\ \frac{4-s}{2};\ \frac{b^2}{a^2}\end{matrix}\right)$ $+\frac{2^{n-1}\pi^{3/2}a^{2\lambda-2}b^{s-3}}{(n+1)!}\tan\frac{s\pi}{2}\,\Gamma\left[\begin{matrix}\frac{n+2\lambda-1}{2}\\ \lambda,\ -\frac{n}{2}\end{matrix}\right]$ $\times\left[\left(a^2+2(\lambda-2)\,b^2\right){}_2F_1\left(\begin{matrix}\frac{n+2}{2},\ -\frac{n+2\lambda-2}{2}\\ \frac{1}{2};\ \frac{b^2}{a^2}\end{matrix}\right)\right.$ $\left.-\left(a^2-b^2\right){}_2F_1\left(\begin{matrix}\frac{n+2}{2},\ -\frac{n+2\lambda-2}{2}\\ -\frac{1}{2};\ \frac{b^2}{a^2}\end{matrix}\right)\right]$ $-\frac{\pi^{3/2}\,(2\lambda)_n\,a^{2\lambda-1}b^{s-2}}{2\,(n)!}\cot\frac{s\pi}{2}\,\Gamma\left[\begin{matrix}\frac{2\lambda+1}{2}\\ \frac{n+2\lambda+1}{2},\ \frac{1-n}{2}\end{matrix}\right] \times{}_2F_1\left(\begin{matrix}\frac{n+1}{2},\ -\frac{n+2\lambda-1}{2}\\ \frac{1}{2};\ \frac{b^2}{a^2}\end{matrix}\right)$ $[0<b<a;\ \operatorname{Re}\lambda>-1/2;\ \operatorname{Re}s>((-1)^n-1)/2]$
8	$\frac{1}{x^2-b^2}\left(x^2-a^2\right)_+^{\lambda-1/2} \times C_{2n+\varepsilon}^{\lambda}\left(\frac{x}{a}\right)$	$-\frac{(2\lambda)_{2n+\varepsilon}\,a^{s+2\lambda-1}}{2\,(2n+\varepsilon)!\,b^2}\left(\frac{1-s+\varepsilon}{2}\right)_n\Gamma\left[\begin{matrix}\frac{2\lambda+1}{2},\ \frac{1-s-2n-2\lambda-\varepsilon}{2}\\ \frac{2-s-\varepsilon}{2}\end{matrix}\right] \times{}_3F_2\left(\begin{matrix}1,\ \frac{s}{2},\ \frac{s+1}{2};\ \frac{a^2}{b^2}\\ \frac{s-2n-\varepsilon+1}{2},\ \frac{s+2n+2\lambda+\varepsilon+1}{2}\end{matrix}\right)$ $+\frac{2^{2n+\varepsilon-1}\pi^2 b^{s+2n+2\lambda+\varepsilon-3}}{(2n+\varepsilon)!\,a^{2n+\varepsilon}}\,\frac{\csc\left[(\lambda+\varepsilon)\,\pi\right]}{\Gamma(\lambda)\,\Gamma(1-2n-\lambda-\varepsilon)}$ $\times\tan\frac{(s+2\lambda+\varepsilon)\,\pi}{2}\,{}_2F_1\left(\begin{matrix}\frac{1-2n-2\lambda-\varepsilon}{2},\ \frac{2-2n-2\lambda-\varepsilon}{2}\\ 1-2n-\lambda-\varepsilon;\ \frac{a^2}{b^2}\end{matrix}\right)$ $[0<a<b;\ \operatorname{Re}\lambda>-1/2;\ \operatorname{Re}(s+2\lambda)<3-2n-\varepsilon]$
9	$\frac{1}{x^2-b^2}\left(x^2-a^2\right)_+^{\lambda-1/2} \times C_n^{\lambda}\left(\frac{x}{a}\right)$	$\frac{(2\lambda)_n\,a^{s+2\lambda-3}}{2^{s-1}\sqrt{\pi}\,n!}\Gamma\left[\begin{matrix}\frac{2\lambda+1}{2},\ \frac{3-s+n}{2},\ \frac{3-s-n-2\lambda}{2}\\ 3-s\end{matrix}\right] \times{}_3F_2\left(\begin{matrix}1,\ \frac{3-s+n}{2},\ \frac{3-s-n-2\lambda}{2}\\ \frac{3-s}{2},\ \frac{4-s}{2};\ \frac{b^2}{a^2}\end{matrix}\right)$ $[0<b<a;\ \operatorname{Re}\lambda>-1/2;\ \operatorname{Re}(s+2\lambda)<3-n]$

No.	$f(x)$	$F(s)$
10	$(x^2-a^2)_+^{\lambda-1/2}(x^2-b^2)^{\mu-1}$ $\times C_{2n+\varepsilon}^\lambda\left(\frac{x}{a}\right)$	$\frac{a^{s+2\mu+2\lambda-3}}{2(2n+\varepsilon)!}(2\lambda)_{2n+\varepsilon}\left(\frac{3-s-2\mu+\varepsilon}{2}\right)_n$ $\times\Gamma\left[\begin{matrix}\frac{2\lambda+1}{2}, -\frac{s+2n+2\lambda+2\mu+\varepsilon-3}{2}\\ -\frac{s+2\mu+\varepsilon-4}{2}\end{matrix}\right]$ $\times{}_3F_2\left(\begin{matrix}1-\mu, -\frac{s-2n+2\mu-\varepsilon-3}{2}, -\frac{s+2n+2\lambda+2\mu+\varepsilon-3}{2}\\ \frac{3-s-2\mu}{2}, \frac{4-s-2\mu}{2}; \frac{b^2}{a^2}\end{matrix}\right)$ $\left[\begin{matrix}a>b>0;\ \mathrm{Re}\,\lambda>-1/2;\\ \mathrm{Re}\,(s+2\mu+2\lambda)<3-2n-\varepsilon\end{matrix}\right]$
11	$(a^2-x^2)_+^{\lambda-1/2}\frac{1}{(b^2\pm x^2)^\rho}$ $\times C_n^\lambda\left(\frac{x}{a}\right)$	$\frac{\pi}{n!}\left(\frac{a}{2}\right)^{s+2\lambda-1}b^{-2\rho}\Gamma\left[\begin{matrix}s, n+2\lambda\\ \lambda, \frac{s-n+1}{2}, \frac{s+n+2\lambda+1}{2}\end{matrix}\right]$ $\times{}_3F_2\left(\begin{matrix}\rho, \frac{s}{2}, \frac{s+1}{2}; \mp\frac{a^2}{b^2}\\ \frac{s-n+1}{2}, \frac{s+n+2\lambda+1}{2}\end{matrix}\right)$ $\left[\left\{\begin{matrix}a,\ \mathrm{Re}\,b>0\\ b>a>0\end{matrix}\right\};\ \mathrm{Re}\,\lambda>-\frac{1}{2};\ \mathrm{Re}\,s>\frac{(-1)^n-1}{2}\right]$
12	$(x^2-a^2)_+^{\lambda-1/2}\frac{1}{(x^2+a^2)^\rho}$ $\times C_{2n+\varepsilon}^\lambda\left(\frac{x}{a}\right)$	$\frac{a^{s+2\lambda-2\rho-1}}{2(2n+\varepsilon)!}(2\lambda)_{2n+\varepsilon}\left(\frac{1-s+2\rho+\varepsilon}{2}\right)_n$ $\times\Gamma\left[\begin{matrix}\frac{2\lambda+1}{2}, \frac{1-s-2n-2\lambda+2\rho-\varepsilon}{2}\\ \frac{2-s+2\rho-\varepsilon}{2}\end{matrix}\right]$ $\times{}_3F_2\left(\begin{matrix}\rho, \frac{1-s+2n+2\rho+\varepsilon}{2}, \frac{1-s-2n-2\lambda+2\rho-\varepsilon}{2}\\ \frac{1-s+2\rho}{2}, \frac{2-s+2\rho}{2}; -\frac{b^2}{a^2}\end{matrix}\right)$ $\left[\begin{matrix}a,\ \mathrm{Re}\,b>0;\ \mathrm{Re}\,\lambda>-1/2;\\ \mathrm{Re}\,(s+2\lambda-2\rho)<1-2n-\varepsilon\end{matrix}\right]$
13	$(a^2-x^2)_+^{\mu-1}(b^2-x^2)^{\lambda-1/2}$ $\times C_n^\lambda\left(\frac{x}{b}\right)$	$\frac{\sqrt{\pi}\,a^{s+2\mu-2}b^{2\lambda-1}}{2(n)!}(2\lambda)_n\Gamma\left[\begin{matrix}\frac{2\lambda+1}{2}, \mu, \frac{s}{2}\\ -\frac{n-1}{2}, \frac{n+2\lambda+1}{2}, \frac{s+2\mu}{2}\end{matrix}\right]$ $\times{}_3F_2\left(\begin{matrix}\frac{n+1}{2}, -\frac{n+2\lambda-1}{2}, \frac{s}{2}\\ \frac{1}{2}, \frac{s+2\mu}{2}; \frac{a^2}{b^2}\end{matrix}\right)$ $-\frac{\sqrt{\pi}\,a^{s+2\mu-1}b^{2\lambda-2}}{n!}(2\lambda)_n\Gamma\left[\begin{matrix}\frac{2\lambda+1}{2}, \mu, \frac{s+1}{2}\\ -\frac{n}{2}, \frac{n+2\lambda}{2}, \frac{s+2\mu+1}{2}\end{matrix}\right]$ $\times{}_3F_2\left(\begin{matrix}\frac{n+2}{2}, -\frac{n+2\lambda-2}{2}, \frac{s+1}{2}\\ \frac{3}{2}, \frac{s+2\mu+1}{2}; \frac{a^2}{b^2}\end{matrix}\right)$ $\left[\begin{matrix}b>a>0;\ \mathrm{Re}\,\mu>0;\ \mathrm{Re}\,\lambda>-1/2;\\ \mathrm{Re}\,s>((-1)^n-1)/2\end{matrix}\right]$

No.	$f(x)$	$F(s)$
14	$(x^2-a^2)_+^{\mu-1}(x^2-b^2)^{\lambda-1/2} \times C_{2n+\varepsilon}^{\lambda}\left(\frac{x}{b}\right)$	$\frac{2^{2n+\varepsilon-1}a^{s+2n+2\lambda+2\mu+\varepsilon-3}}{(2n+\varepsilon)!\,b^{2n+\varepsilon}}(\lambda)_{2n+\varepsilon}\Gamma\left[\begin{matrix}\lambda+\varepsilon,\,1-\lambda-\varepsilon\\ \lambda,\,1-2n-\lambda-\varepsilon\end{matrix}\right] \times \mathrm{B}\left(\mu,\,-\frac{s+2n+2\lambda+2\mu+\varepsilon-3}{2}\right) \times {}_3F_2\left(\begin{matrix}\frac{1-2n-2\lambda-\varepsilon}{2},\,\frac{2-2n-2\lambda-\varepsilon}{2},\,-\frac{s+2n+2\lambda+2\mu+\varepsilon-3}{2}\\ 1-2n-\lambda-\varepsilon,\,-\frac{s+2n+2\lambda+\varepsilon-3}{2};\,\frac{b^2}{a^2}\end{matrix}\right)$ $\left[\begin{matrix}a>b>0;\ \mathrm{Re}\,\mu>0;\ \mathrm{Re}\,\lambda>-1/2\\ \mathrm{Re}\,(s+2\lambda+2\mu)<3-2n-\varepsilon\end{matrix}\right]$
15	$(a-x)_+^{\alpha-1}C_n^{\lambda}(1-bx)$	$\frac{(2\lambda)_n}{n!}a^{s+\alpha-1}\mathrm{B}(\alpha,\,s)\,{}_3F_2\left(\begin{matrix}-n,\,n+2\lambda,\,s\\ \frac{2\lambda+1}{2},\,s+\alpha;\,\frac{ab}{2}\end{matrix}\right)$ $[a,\ \mathrm{Re}\,\alpha,\ \mathrm{Re}\,s>0;\ \mathrm{Re}\,\lambda>-1/2]$
16	$(a-x)_+^{\lambda-1/2} \times C_n^{\lambda}(bx-ab+1)$	$(2\lambda)_n a^{s+\lambda-1/2}\Gamma\left[\begin{matrix}\frac{2\lambda+1}{2},\,s\\ \frac{2s+2\lambda+2n+1}{2}\end{matrix}\right]P_n^{(s+\lambda-1/2,\,\lambda-s-1/2)}(1-ab)$ $[a,\ \mathrm{Re}\,s>0;\ \mathrm{Re}\,\lambda>-1/2]$
17	$(a-x)_+^{\alpha-1}C_{2n+\varepsilon}^{\lambda}(b(a-x))$	$\frac{(-1)^n a^{s+\alpha+\varepsilon-1}(2b)^{\varepsilon}}{n!}(\lambda)_{n+\varepsilon}\mathrm{B}(\alpha+\varepsilon,\,s) \times {}_4F_3\left(\begin{matrix}-n,\,n+\lambda+\varepsilon,\,\frac{\alpha+\varepsilon}{2},\,\frac{\alpha+\varepsilon+1}{2}\\ \frac{2\varepsilon+1}{2},\,\frac{s+\alpha+\varepsilon}{2},\,\frac{s+\alpha+\varepsilon+1}{2};\,a^2b^2\end{matrix}\right)$ $[a,\ \mathrm{Re}\,s>0;\ \mathrm{Re}\,\alpha>-\varepsilon]$
18	$(a-x)_+^{\lambda-1/2}C_n^{\lambda}\left(\frac{x+b}{a+b}\right)$	$a^{s+\lambda-1/2}\frac{(2\lambda)_n}{\left(\frac{2s+2\lambda+1}{2}\right)_n}\mathrm{B}\left(\frac{2\lambda+1}{2},\,s\right) \times P_n^{(s+\lambda-1/2,\,-s+\lambda-1/2)}\left(\frac{b}{a+b}\right)$ $[a,\ \mathrm{Re}\,s>0;\ \mathrm{Re}\,\lambda>-1/2]$
19	$(a-x)_+^{\alpha-1}C_n^{\lambda}\left(\frac{x+b}{a+b}\right)$	$\frac{a^{s+\alpha-1}}{n!}(2\lambda)_n\mathrm{B}(s,\,\alpha)\,{}_3F_2\left(\begin{matrix}-n,\,n+2\lambda,\,\alpha\\ \frac{2\lambda+1}{2},\,s+\alpha;\,\frac{a}{2(a+b)}\end{matrix}\right)$ $[a,\ \mathrm{Re}\,\alpha,\ \mathrm{Re}\,s>0]$
20	$(a-x)_+^{\lambda-1/2}C_n^{\lambda}\left(\frac{2x}{a}-1\right)$	$\frac{(2\lambda)_n}{n!}a^{s+\lambda-1/2}\Gamma\left[\begin{matrix}\frac{2\lambda+1}{2},\,s,\,\frac{2s-2\lambda+1}{2}\\ \frac{2s-2n-2\lambda+1}{2},\,\frac{2s+2n+2\lambda+1}{2}\end{matrix}\right]$ $[a,\ \mathrm{Re}\,s>0;\ \mathrm{Re}\,\lambda>-1/2]$
21	$(x-a)_+^{\lambda-1/2}C_n^{\lambda}\left(\frac{2x}{a}-1\right)$	$\frac{(2\lambda)_n}{n!}a^{s+\lambda-1/2}\Gamma\left[\begin{matrix}\frac{2\lambda+1}{2},\,\frac{1-2s-2n-2\lambda}{2},\,\frac{1-2s+2n+2\lambda}{2}\\ 1-s,\,\frac{1-2s+2\lambda}{2}\end{matrix}\right]$ $[a>0;\ \mathrm{Re}\,\lambda>-1/2;\ \mathrm{Re}\,s<1/2-n-\mathrm{Re}\,\lambda]$

No.	$f(x)$	$F(s)$
22	$(x+a)^{\lambda-1/2}\, C_n^{\lambda}\left(\frac{2x}{a}+1\right)$	$\frac{(-1)^n\,(2\lambda)_n}{n!}\, a^{s+\lambda-1/2}\,\Gamma\begin{bmatrix} s,\ \frac{1-2s+2n+2\lambda}{2},\ \frac{1-2s-2n-2\lambda}{2} \\ \frac{1-2\lambda}{2},\ \frac{1-2s+2\lambda}{2} \end{bmatrix}$ $[0<\operatorname{Re} s<1/2-\operatorname{Re}\lambda-n;\ \lvert\arg a\rvert<\pi]$
23	$\frac{1}{(x+a)^{\rho}}\, C_n^{\lambda}\left(\frac{2x}{a}+1\right)$	$\frac{a^{s-\rho}}{n!}\,(2\lambda)_n\,\mathrm{B}\,(-s-n+\rho,\ s)\ {}_3F_2\left(\begin{matrix} -n,\ \frac{1-2n-2\lambda}{2},\ s \\ \frac{2\lambda+1}{2},\ \rho-n;\ 1 \end{matrix}\right)$ $[0<\operatorname{Re} s<\operatorname{Re}\rho-n;\ \lvert\arg a\rvert<\pi]$
24	$\frac{(a-x)_+^{\lambda-1/2}}{(b\pm x)^{\rho}}\, C_n^{\lambda}\left(\frac{2x}{a}-1\right)$	$\frac{(-1)^n\, a^{s+\lambda-1/2} b^{-\rho}}{n!}\,\frac{(2\lambda)_n\left(\frac{-2s+2\lambda+1}{2}\right)_n}{\left(\frac{2s+2\lambda+1}{2}\right)_n}\,\mathrm{B}\left(\frac{2\lambda+1}{2},\ s\right)$ $\times\ {}_3F_2\left(\begin{matrix} \rho,\ s,\ \frac{2s-2\lambda+1}{2};\ \mp\frac{a}{b} \\ \frac{2s-2n-2\lambda+1}{2},\ \frac{2s+2n+2\lambda+1}{2} \end{matrix}\right)$ $\left[\left\{\begin{matrix} a>0 \\ b>a>0 \end{matrix}\right\},\ \operatorname{Re}\lambda>-1/2;\ \operatorname{Re} s>0\right]$
25	$(a-x)_+^{\alpha-1}\,(b\pm x)^{\lambda-1/2}$ $\times\, C_n^{\lambda}\left(\frac{2x}{b}\pm 1\right)$	$\frac{(\pm 1)^n\, a^{s+\alpha-1} b^{\lambda-1/2}}{n!}\,(2\lambda)_n\,\mathrm{B}\,(\alpha,\ s)$ $\times\ {}_3F_2\left(\begin{matrix} \frac{-2n-2\lambda+1}{2},\ \frac{2n+2\lambda+1}{2},\ s \\ \frac{2\lambda+1}{2},\ s+\alpha;\ \mp\frac{a}{b} \end{matrix}\right)$ $[a,\ b,\ \operatorname{Re}\alpha,\ \operatorname{Re} s>0]$
26	$\frac{(x+a)^{\lambda-1/2}}{(x+b)^{\rho}}\, C_n^{\lambda}\left(\frac{2x}{a}+1\right)$	$\frac{(-1)^n\, a^{s+\lambda-\rho-1/2}}{n!}\,(2\lambda)_n\left(\frac{1}{2}-s+\lambda+\rho\right)_n$ $\times\,\Gamma\begin{bmatrix} s-\rho,\ \frac{1-2s-2n-2\lambda+2\rho}{2} \\ \frac{1-2\lambda}{2} \end{bmatrix}$ $\times\ {}_3F_2\left(\begin{matrix} \rho,\ \frac{1-2s-2n-2\lambda+2\rho}{2},\ \frac{1-2s+2n+2\lambda+2\rho}{2} \\ 1-s+\rho,\ \frac{1-2s+2\lambda+2\rho}{2};\ \frac{b}{a} \end{matrix}\right)$ $+\ \frac{a^{\lambda-1/2} b^{s-\rho}}{n!}\,(2\lambda)_n\,\mathrm{B}\,(s,\ \rho-s)\ {}_3F_2\left(\begin{matrix} \frac{-2n-2\lambda+1}{2},\ \frac{2n+2\lambda+1}{2},\ s \\ \frac{2\lambda+1}{2},\ s-\rho+1;\ \frac{b}{a} \end{matrix}\right)$ $[a>0;\ 0<\operatorname{Re} s<1/2-n-\operatorname{Re}\lambda+\operatorname{Re}\rho;\ \lvert\arg b\rvert<\pi]$
27	$\frac{(x+a-b)^{\lambda-1/2}}{(x+a+b)^{\rho}}$ $\times\, C_n^{\lambda}\left(\frac{x+a}{b}\right)$	$\frac{2^n\,(a+b)^{s+n+\lambda-\rho-1/2}}{n!\, b^n}\,(\lambda)_n\,\mathrm{B}\left(\frac{1-2s-2n-2\lambda+2\rho}{2},\ s\right)$ $\times\ {}_3F_2\left(\begin{matrix} \frac{1-2n-2\lambda}{2},\ 1-n-2\lambda,\ \frac{1-2s-2n-2\lambda+2\rho}{2} \\ \frac{1-2n-2\lambda+2\rho}{2},\ 1-2n-2\lambda;\ \frac{2b}{a+b} \end{matrix}\right)$ $[a>b>0;\ 0<\operatorname{Re} s<1/2-n-\operatorname{Re}\lambda+\operatorname{Re}\rho]$
28	$\left(a^2-x^2\right)_+^{\lambda-1/2}\, C_n^{\lambda}\left(\frac{a}{x}\right)$	$\frac{2^{s-1}}{n!}\, a^{s+2\lambda-1}\,\Gamma\begin{bmatrix} 2\lambda+n,\ \frac{s-n}{2},\ \frac{s+n+2\lambda}{2} \\ \lambda,\ s+2\lambda \end{bmatrix}$ $[a>0;\ \operatorname{Re}\lambda>-1/2;\ \operatorname{Re} s>n]$

No.	$f(x)$	$F(s)$
29	$\left(x^2-a^2\right)_+^{\lambda-1/2} C_n^{\lambda}\left(\frac{a}{x}\right)$	$\frac{\sqrt{\pi}\,(2\lambda)_n\,(2a)^{s+2\lambda-1}}{n!}\,\Gamma\left[\begin{matrix}\frac{2\lambda+1}{2},\ 1-s-2\lambda\\ \frac{2-s+n}{2},\ \frac{2-s-n-2\lambda}{2}\end{matrix}\right]$ $[a>0;\ \operatorname{Re}\lambda>-1/2;\ \operatorname{Re}(s+2\lambda)<1]$
30	$(x+a)^{\lambda-1/2} C_n^{\lambda}\left(\frac{2a}{x}+1\right)$	$\frac{(-1)^n\,(2\lambda)_n}{n!}\,a^{s+\lambda-1/2}\,\Gamma\left[\begin{matrix}s+n+2\lambda,\ s-n,\ \frac{1-2s-2\lambda}{2}\\ \frac{1-2\lambda}{2},\ s+2\lambda\end{matrix}\right]$ $[n<\operatorname{Re}s<1/2-\operatorname{Re}\lambda;\ \lvert\arg a\rvert<\pi]$
31	$(a-x)_+^{\lambda-1/2} C_n^{\lambda}\left(\frac{2a}{x}-1\right)$	$\frac{(2\lambda)_n}{n!}\,a^{s+\lambda-1/2}\,\Gamma\left[\begin{matrix}\frac{2\lambda+1}{2},\ s-n,\ s+n+2\lambda\\ \frac{2s+2\lambda+1}{2},\ s+2\lambda\end{matrix}\right]$ $[a>0;\ \operatorname{Re}\lambda>-1/2;\ \operatorname{Re}s>n]$
32	$\left(a^2-x^2\right)_+^{2\lambda-1}$ $\times\, C_n^{\lambda}\left(\frac{a}{2x}+\frac{x}{2a}\right)$	$\frac{a^{s+4\lambda-2}}{2\,(n!)}\,\Gamma\left[\begin{matrix}n+2\lambda,\ \frac{s-n}{2},\ \frac{s+n+2\lambda}{2}\\ \frac{s-n+2\lambda}{2},\ \frac{s+n+4\lambda}{2}\end{matrix}\right]$ $[a>0;\ \operatorname{Re}\lambda>-1/2;\ \operatorname{Re}s>n]$
33	$(x-a)_+^{\lambda-1/2} C_n^{\lambda}\left(\frac{2a}{x}-1\right)$	$\frac{(2\lambda)_n}{n!}\,a^{s+\lambda-1/2}\,\Gamma\left[\begin{matrix}\frac{2\lambda+1}{2},\ \frac{1-2s-2\lambda}{2},\ 1-s-2\lambda\\ 1-s+n,\ 1-s-n-2\lambda\end{matrix}\right]$ $[a>0;\ \operatorname{Re}\lambda>-1/2;\ \operatorname{Re}s<1/2-\operatorname{Re}\lambda]$
34	$(x+a)^{-n-2\lambda} C_n^{\lambda}\left(\frac{a-x}{a+x}\right)$	$\frac{a^{s-n-2\lambda}}{n!\left(\frac{2\lambda+1}{2}\right)_n}\,\Gamma\left[\begin{matrix}s,\ n-s+2\lambda,\ \frac{1-2s+2n+2\lambda}{2}\\ 2\lambda,\ \frac{1-2s+2\lambda}{2}\end{matrix}\right]$ $[0<\operatorname{Re}s<n+2\operatorname{Re}\lambda;\ \lvert\arg a\rvert<\pi]$
35	$(a-x)_+^{-n-2\lambda} C_n^{\lambda}\left(\frac{a+x}{a-x}\right)$	$\frac{a^{s-n-2\lambda}}{n!\left(\frac{2\lambda+1}{2}\right)_n}\,\Gamma\left[\begin{matrix}1-2\lambda,\ s,\ \frac{2s-2\lambda+1}{2}\\ s-n-2\lambda+1,\ \frac{2s-2n-2\lambda+1}{2}\end{matrix}\right]$ $[a,\ \operatorname{Re}s>0;\ \operatorname{Re}\lambda<1/2-n]$
36	$(x-a)_+^{-n-2\lambda} C_n^{\lambda}\left(\frac{x+a}{x-a}\right)$	$\frac{a^{s+n-2\lambda}}{n!\left(\frac{2\lambda+1}{2}\right)_n}\,\Gamma\left[\begin{matrix}1-2\lambda,\ n-s+2\lambda,\ \frac{1-2s+2n+2\lambda}{2}\\ 1-s,\ \frac{1-2s+2\lambda}{2}\end{matrix}\right]$ $[a>0;\ \operatorname{Re}\lambda<1/2-n;\ \operatorname{Re}s<n+2\operatorname{Re}\lambda]$
37	$(a-x)_+^{\alpha-1}$ $\times\, C_{2n+\varepsilon}^{\lambda}\left(bx\,(a-x)\right)$	$\frac{(-1)^n\,2^{\varepsilon}\,(\lambda)_{n+\varepsilon}}{n!}\,a^{s+\alpha+2\varepsilon-1}b^{\varepsilon}\,\mathrm{B}\,(s+\varepsilon,\ \alpha+\varepsilon)$ $\times\,{}_6F_5\left(\begin{matrix}-n,\ n+\lambda+\varepsilon,\ \Delta(2,\ \alpha+\varepsilon),\ \Delta(2,\ s+\varepsilon)\\ \frac{2\varepsilon+1}{2},\ \Delta(4,\ s+\alpha+2\varepsilon);\ \frac{a^4b^2}{16}\end{matrix}\right)$ $[a>0;\ \operatorname{Re}s,\ \operatorname{Re}\alpha>-\varepsilon]$

No.	$f(x)$	$F(s)$
38	$(a-x)_+^{\alpha-1} C_n^\lambda\left(\frac{b}{x(a-x)}\right)$	$\frac{(\lambda)_n}{n!} a^{s-2n+\alpha-1}(2b)^n \mathrm{B}(s-n, \alpha-n) \times {}_6F_5\left(\begin{matrix}\Delta(2, -n), \Delta(2, \alpha-n), \Delta(2, s-n) \\ 1-n-\lambda, \Delta(4, s-n+\alpha); \frac{a^4}{16b^2}\end{matrix}\right)$ $[a>0;\ \mathrm{Re}\, s,\ \mathrm{Re}\,\alpha > n]$
39	$(a-x)_+^{\alpha-1} \times C_{2n+\varepsilon}^\lambda\left(b\sqrt{x(a-x)}\right)$	$\frac{2^\varepsilon(-1)^n(\lambda)_{n+\varepsilon}}{n!} a^{s+\alpha+\varepsilon-1} b^\varepsilon \mathrm{B}\left(\frac{2s+\varepsilon}{2}, \frac{2\alpha+\varepsilon}{2}\right) \times {}_4F_3\left(\begin{matrix}-n, n+\lambda+\varepsilon, \frac{2\alpha+\varepsilon}{2}, \frac{2s+\varepsilon}{2} \\ \frac{2\varepsilon+1}{2}, \frac{s+\alpha+\varepsilon}{2}, \frac{s+\alpha+\varepsilon+1}{2}; \frac{a^2b^2}{4}\end{matrix}\right)$ $[a>0;\ \mathrm{Re}\,\alpha,\ \mathrm{Re}\, s > -\varepsilon/2]$
40	$(a-x)_+^{\alpha-1} C_n^\lambda\left(\frac{b}{\sqrt{x(a-x)}}\right)$	$\frac{2^n(\lambda)_n}{n!} a^{s-n+\alpha-1} b^n \mathrm{B}\left(\frac{2s-n}{2}, \frac{2\alpha-n}{2}\right) \times {}_4F_3\left(\begin{matrix}\frac{1-n}{2}, -\frac{n}{2}, s-\frac{n}{2}, \alpha-\frac{n}{2}; \frac{a^2}{4b^2} \\ -n-\lambda+1, \frac{s-n+\alpha}{2}, \frac{s-n+\alpha+1}{2}\end{matrix}\right)$ $[a>0;\ \mathrm{Re}\,\alpha,\ \mathrm{Re}\, s > n/2]$
41	$(a-x)_+^{-1/2} C_{2n}^\lambda\left(\sqrt{1-\frac{x}{a}}\right)$	$\frac{\sqrt{\pi}(\lambda)_n}{n!} a^{s-1/2} \Gamma\left[\begin{matrix}s, \frac{1-2s+2n+2\lambda}{2} \\ \frac{2s+2n+1}{2}, \frac{1-2s+2\lambda}{2}\end{matrix}\right]$ $[a,\ \mathrm{Re}\, s > 0]$
42	$\theta(a-x) C_{2n+1}^\lambda\left(\sqrt{1-\frac{x}{a}}\right)$	$\frac{\sqrt{\pi}(\lambda)_{n+1}}{n!} a^s \Gamma\left[\begin{matrix}s, \frac{1-2s+2n+2\lambda}{2} \\ \frac{2s+2n+3}{2}, \frac{1-2s+2\lambda}{2}\end{matrix}\right]$ $[a,\ \mathrm{Re}\, s > 0]$
43	$\theta(x-a) C_{2n+1}^\lambda\left(i\sqrt{\frac{x}{a}-1}\right)$	$\frac{i\sqrt{\pi}(\lambda)_{n+1}}{n!} a^s \Gamma\left[\begin{matrix}\frac{2s-2\lambda+1}{2}, -\frac{2s+2n+1}{2} \\ \frac{2s-2n-2\lambda+1}{2}, 1-s\end{matrix}\right]$ $[a>0;\ \mathrm{Re}\, s < -n-1/2]$
44	$(x-a)_+^{-1/2} C_{2n}^\lambda\left(\sqrt{1-\frac{a}{x}}\right)$	$\frac{\sqrt{\pi}(\lambda)_n}{n!} a^{s-1/2} \Gamma\left[\begin{matrix}s+n+\lambda, \frac{1-2s}{2} \\ s+\lambda, 1-s+n\end{matrix}\right]$ $[a>0;\ \mathrm{Re}\, s < 1/2]$
45	$\theta(x-a) C_{2n+1}^\lambda\left(\sqrt{1-\frac{a}{x}}\right)$	$\frac{\sqrt{\pi}(\lambda)_{n+1}}{n!} a^s \Gamma\left[\begin{matrix}\frac{2s+2\lambda+2n+1}{2}, -s \\ \frac{2s+2\lambda+1}{2}, \frac{3-2s+2n}{2}\end{matrix}\right]$ $[a>0;\ \mathrm{Re}\, s < 0]$
46	$(a-x)_+^{-1/2} C_{2n}^\lambda\left(i\sqrt{\frac{a}{x}-1}\right)$	$\frac{\sqrt{\pi}(\lambda)_n}{n!} a^{s-1/2} \Gamma\left[\begin{matrix}s-n, 1-s-\lambda \\ \frac{2s+1}{2}, 1-s-n-\lambda\end{matrix}\right]$ $[a>0;\ \mathrm{Re}\, s > n]$
47	$\theta(a-x) C_{2n+1}^\lambda\left(i\sqrt{\frac{a}{x}-1}\right)$	$\frac{i\sqrt{\pi}(\lambda)_{n+1}}{n!} a^s \Gamma\left[\begin{matrix}\frac{2s-2n-1}{2}, \frac{1-2s-2\lambda}{2} \\ s+1, \frac{1-2s-2n-2\lambda}{2}\end{matrix}\right]$ $[a>0;\ \mathrm{Re}\, s > n+1/2]$

No.	$f(x)$	$F(s)$
48	$(a-x)_+^{-n/2-\lambda}\, C_n^\lambda\left(\sqrt{\frac{a}{a-x}}\right)$	$\frac{(-2)^n}{n!}\, a^{s-n/2-\lambda}\, \Gamma\left[\begin{matrix} 1-\lambda,\ s,\ \frac{2s-2\lambda+1}{2} \\ \frac{2s-n-2\lambda+1}{2},\ \frac{2s-n-2\lambda+2}{2} \end{matrix}\right]$ $[a,\ \operatorname{Re} s>0;\ \operatorname{Re}\lambda<1-n]$
49	$(x-a)_+^{-n/2-\lambda}\, C_n^\lambda\left(\sqrt{\frac{x}{x-a}}\right)$	$\frac{(-2)^n}{n!}\, a^{s-n/2-\lambda}\, \Gamma\left[\begin{matrix} 1-\lambda,\ \frac{n-2s+2\lambda}{2},\ \frac{n-2s+1}{2} \\ \frac{1-2s}{2},\ 1-s \end{matrix}\right]$ $[a>0;\ \operatorname{Re}\lambda<1-n;\ \operatorname{Re} s<\operatorname{Re}\lambda+n/2]$
50	$(x+a)^{-n/2-\lambda}\, C_n^\lambda\left(\sqrt{\frac{a}{x+a}}\right)$	$\frac{2^n}{n!}\, a^{s-n/2-\lambda}\, \Gamma\left[\begin{matrix} s,\ \frac{n-2s+2\lambda}{2},\ \frac{n-2s+2\lambda+1}{2} \\ \lambda,\ \frac{1-2s+2\lambda}{2} \end{matrix}\right]$ $[0<\operatorname{Re} s<\operatorname{Re}\lambda+n/2;\ \lvert\arg a\rvert<\pi]$
51	$(x+a)^{-n/2-\lambda}\, C_n^\lambda\left(\sqrt{\frac{x}{x+a}}\right)$	$\frac{2^n}{n!}\, a^{s-n/2-\lambda}\, \Gamma\left[\begin{matrix} s,\ \frac{2s+1}{2},\ \frac{n-2s+2\lambda}{2} \\ \lambda,\ \frac{2s-n+1}{2} \end{matrix}\right]$ $[0<\operatorname{Re} s<\operatorname{Re}\lambda+n/2;\ \lvert\arg a\rvert<\pi]$
52	$(a-x)_+^{2\lambda-1}\, C_n^\lambda\left(\frac{x+a}{2\sqrt{ax}}\right)$	$\frac{a^{s+2\lambda-1}}{n!}\, \Gamma\left[\begin{matrix} n+2\lambda,\ \frac{2s-n}{2},\ \frac{2s+n+2\lambda}{2} \\ \frac{2s+n+4\lambda}{2},\ \frac{2s-n+2\lambda}{2} \end{matrix}\right]$ $[a,\ \operatorname{Re}\lambda>0;\ \operatorname{Re} s>n/2]$
53	$(x-a)_+^{2\lambda-1}\, C_n^\lambda\left(\frac{x+a}{2\sqrt{ax}}\right)$	$\frac{a^{s+2\lambda-1}}{n!}\, \Gamma\left[\begin{matrix} n+2\lambda,\ \frac{2-2s-n-4\lambda}{2},\ \frac{2-2s+n-2\lambda}{2} \\ \frac{2-2s+n}{2},\ \frac{2-2s-n-2\lambda}{2} \end{matrix}\right]$ $[a,\ \operatorname{Re}\lambda>0;\ \operatorname{Re} s<1-2\operatorname{Re}\lambda-n/2]$
54	$(a-x)_+^{-n/2-\lambda}$ $\times C_n^\lambda\left(\frac{2a-x}{2\sqrt{a(a-x)}}\right)$	$\frac{a^{s-n/2-\lambda}}{n!}\, \Gamma\left[\begin{matrix} 1-\lambda,\ s,\ n-s+2\lambda \\ s-\lambda+1,\ -s+2\lambda \end{matrix}\right]$ $[a,\ \operatorname{Re} s>0;\ \operatorname{Re}\lambda<1-n]$
55	$(x-a)_+^{-n/2-\lambda}$ $\times C_n^\lambda\left(\frac{2x-a}{2\sqrt{x(x-a)}}\right)$	$\frac{a^{s-n/2-\lambda}}{n!}\, \Gamma\left[\begin{matrix} 1-\lambda,\ \frac{2s+n+2\lambda}{2},\ \frac{-2s+n+2\lambda}{2} \\ \frac{2s-n+2\lambda}{2},\ \frac{2-2s+n}{2} \end{matrix}\right]$ $[a>0;\ \operatorname{Re}\lambda<1-n;\ \operatorname{Re} s<\operatorname{Re}\lambda+n/2]$
56	$(x+a)^{-n/2-\lambda}$ $\times C_n^\lambda\left(\frac{x+2a}{2\sqrt{a(x+a)}}\right)$	$\frac{a^{s-n/2-\lambda}}{n!}\, \Gamma\left[\begin{matrix} s,\ \lambda-s,\ -s+n+2\lambda \\ \lambda,\ 2\lambda-s \end{matrix}\right]$ $[0<\operatorname{Re} s<\operatorname{Re}\lambda;\ \lvert\arg a\rvert<\pi]$

No.	$f(x)$	$F(s)$
57	$(x+a)^{-n/2-\lambda} \times C_n^\lambda\left(\frac{2x+a}{2\sqrt{x(x+a)}}\right)$	$\frac{a^{s-n/2-\lambda}}{n!}\Gamma\left[\begin{matrix}\frac{2s+n+2\lambda}{2}, \frac{2s-n}{2}, \frac{-2s+n+2\lambda}{2}\\ \lambda, \frac{2s-n+2\lambda}{2}\end{matrix}\right]$ $[n/2<\operatorname{Re}s<\operatorname{Re}\lambda+n/2;\ \lvert\arg a\rvert<\pi]$
58	$(a-x)_+^{(\varepsilon-1)/2}(b-x)^{n+\varepsilon/2} \times C_{2n+\varepsilon}^\lambda\left(c\sqrt{\frac{a-x}{b-x}}\right)$	$\frac{(-1)^n\sqrt{\pi}(\lambda)_{n+\varepsilon}}{n!}a^{s+\varepsilon-1/2}b^n c^\varepsilon\,\Gamma\left[\begin{matrix}s\\ \frac{2s+2\varepsilon+1}{2}\end{matrix}\right] \times F_1\left(-n, s, n+\lambda+\varepsilon; \frac{2s+2\varepsilon+1}{2}; \frac{a}{b}, \frac{ac^2}{b}\right)$ $[a,\ \operatorname{Re}s>0]$
59	$(a-x)_+^{\lambda-1/2}\frac{(bx+1)^\alpha}{[1-c(a-x)]^{\varepsilon/2}} \times C_{2n+\varepsilon}^\lambda\left(\sqrt{1+ac-cx}\right)$	$\frac{(2\lambda)_{2n+\varepsilon}}{(2n+\varepsilon)!}a^{s+\lambda-1/2}\,\mathrm{B}\left(s, \frac{2\lambda+1}{2}\right) \times F_3\left(-\alpha, -n, s, n+\lambda+\varepsilon; s+\lambda+\frac{1}{2}; -ab, -ac\right)$ $[a,\ \operatorname{Re}s>0]$

3.24.2. $C_n^\lambda(bx)$ and the exponential function

Notation: $\varepsilon=0$ or 1.

1	$e^{-ax}C_{2n+\varepsilon}^\lambda(bx)$	$\frac{(-1)^n(\lambda)_{n+\varepsilon}(2b)^\varepsilon}{n!\,a^{s+\varepsilon}}\Gamma(s+\varepsilon) \times {}_4F_1\left(\begin{matrix}-n, n+\lambda+\varepsilon, \frac{s+\varepsilon}{2}, \frac{s+\varepsilon+1}{2}\\ \frac{2\varepsilon+1}{2}; \frac{4b^2}{a^2}\end{matrix}\right)$ $[\operatorname{Re}a>0;\ \operatorname{Re}s>-\varepsilon]$
2	$e^{-ax^2}C_{2n+\varepsilon}^\lambda(bx)$	$\frac{(-1)^n 2^{\varepsilon-1}(\lambda)_{n+\varepsilon}b^\varepsilon}{n!\,a^{(s+\varepsilon)/2}}\Gamma\left(\frac{s+\varepsilon}{2}\right){}_3F_1\left(\begin{matrix}-n, \lambda+n+\varepsilon, \frac{s+\varepsilon}{2}\\ \frac{2\varepsilon+1}{2}; \frac{b^2}{a}\end{matrix}\right)$ $[\operatorname{Re}a>0;\ \operatorname{Re}s>-\varepsilon]$
3	$(a^2-x^2)_+^{\lambda-1/2}e^{bx}C_n^\lambda\left(\frac{x}{a}\right)$	$\frac{\pi}{n!}\left(\frac{a}{2}\right)^{s+2\lambda-1}\Gamma\left[\begin{matrix}n+2\lambda, s\\ \lambda, \frac{s-n+1}{2}, \frac{s+n+2\lambda+1}{2}\end{matrix}\right] \times {}_2F_3\left(\begin{matrix}\frac{s}{2}, \frac{s+1}{2}; \frac{a^2b^2}{4}\\ \frac{1}{2}, \frac{s-n+1}{2}, \frac{s+n+2\lambda+1}{2}\end{matrix}\right)+\frac{\pi}{n!}\left(\frac{a}{2}\right)^{s+2\lambda}b \times \Gamma\left[\begin{matrix}n+2\lambda, s+1\\ \lambda, \frac{s-n+2}{2}, \frac{s+n+2\lambda+2}{2}\end{matrix}\right]{}_2F_3\left(\begin{matrix}\frac{s+1}{2}, \frac{s+2}{2}; \frac{a^2b^2}{4}\\ \frac{3}{2}, \frac{s-n+2}{2}, \frac{s+n+2\lambda+2}{2}\end{matrix}\right)$ $[a>0;\ \operatorname{Re}\lambda>-1/2;\ \operatorname{Re}s>((-1)^n-1)/2]$

No.	$f(x)$	$F(s)$
4	$\left(a^2-x^2\right)_+^{\lambda-1/2} e^{bx^2} C_n^\lambda\left(\frac{x}{a}\right)$	$\frac{\pi}{n!}\left(\frac{a}{2}\right)^{s+\lambda-1}\Gamma\left[\begin{matrix} n+2\lambda,\ s \\ \lambda,\ \frac{s-n+1}{2},\ \frac{s+n+2\lambda+1}{2}\end{matrix}\right] \times {}_2F_2\left(\begin{matrix}\frac{s}{2},\ \frac{s+1}{2};\ a^2b \\ \frac{s-n+1}{2},\ \frac{s+n+2\lambda+1}{2}\end{matrix}\right)$ $\left[a>0;\ \operatorname{Re}\lambda>-1/2;\ \operatorname{Re}s>\left((-1)^n-1\right)/2\right]$
5	$\left(x^2-a^2\right)_+^{\lambda-1/2} e^{-bx^2} \times C_{2n+\varepsilon}^\lambda\left(\frac{x}{a}\right)$	$\frac{(2\lambda)_{2n+\varepsilon}\, a^{s+2\lambda-1}}{2\,(2n+\varepsilon)!}\left(\frac{1-s+\varepsilon}{2}\right)_n \Gamma\left[\begin{matrix}\frac{2\lambda+1}{2},\ \frac{1-s-2n-2\lambda-\varepsilon}{2} \\ \frac{2-s-\varepsilon}{2}\end{matrix}\right] \times {}_2F_2\left(\begin{matrix}\frac{s}{2},\ \frac{s+1}{2};\ -a^2b \\ \frac{s-2n-\varepsilon+1}{2},\ \frac{s+2n+2\lambda+\varepsilon+1}{2}\end{matrix}\right) + \frac{2^{2n+\varepsilon-1}\,(\lambda)_{2n+\varepsilon}\, b^{-s/2-n-\lambda-\varepsilon/2+1/2}}{(2n+\varepsilon)!\, a^{2n+\varepsilon}} \times \Gamma\left(\frac{s+2n+2\lambda+\varepsilon-1}{2}\right) \times {}_2F_2\left(\begin{matrix}\frac{1-2n-2\lambda}{2},\ 1-n-\lambda-\varepsilon;\ -a^2b \\ 1-2n-\lambda-\varepsilon,\ \frac{3-s-2n-2\lambda-\varepsilon}{2}\end{matrix}\right)$ $\left[a,\ \operatorname{Re}b>0;\ \operatorname{Re}\lambda>-1/2\right]$
6	$\left(a^2-x^2\right)_+^{\lambda-1/2} e^{-b/x^2} \times C_{2n+\varepsilon}^\lambda\left(\frac{x}{a}\right)$	$\frac{(-1)^n\,(2\lambda)_{2n+\varepsilon}\, a^{s+2\lambda-1}}{2\,(2n+\varepsilon)!}\left(\frac{1-s+\varepsilon}{2}\right)_n \Gamma\left[\begin{matrix}\frac{2\lambda+1}{2},\ \frac{s+\varepsilon}{2} \\ \frac{s+2n+2\lambda+\varepsilon+1}{2}\end{matrix}\right] \times {}_2F_2\left(\begin{matrix}\frac{1-s-2n-2\lambda-\varepsilon}{2},\ \frac{1-s+2n+\varepsilon}{2} \\ \frac{1-s}{2},\ \frac{2-s}{2};\ -\frac{b}{a^2}\end{matrix}\right) + \frac{(-1)^n\, 2^{\varepsilon-1}\,(\lambda)_{n+\varepsilon}}{n!}\, a^{2\lambda-\varepsilon-1} b^{(s+\varepsilon)/2}\,\Gamma\left(-\frac{s+\varepsilon}{2}\right) \times {}_2F_2\left(\begin{matrix}\frac{2n+2\varepsilon+1}{2},\ \frac{1-2n-2\lambda}{2} \\ \frac{2\varepsilon+1}{2},\ \frac{s+\varepsilon+2}{2};\ -\frac{b}{a^2}\end{matrix}\right)$ $\left[a,\ \operatorname{Re}b>0;\ \operatorname{Re}\lambda>-1/2\right]$
7	$\left(x^2-a^2\right)_+^{\lambda-1/2} e^{-b/x^2} \times C_{2n+\varepsilon}^\lambda\left(\frac{x}{a}\right)$	$\frac{(2\lambda)_{2n+\varepsilon}\, a^{s+2\lambda-1}}{2\,(2n+\varepsilon)!}\left(\frac{1-s+\varepsilon}{2}\right)_n \Gamma\left[\begin{matrix}\frac{2\lambda+1}{2},\ \frac{1-s-2n-2\lambda-\varepsilon}{2} \\ \frac{2-\varepsilon-s}{2}\end{matrix}\right] \times {}_2F_2\left(\begin{matrix}\frac{1-s+2n+\varepsilon}{2},\ \frac{1-s-2n-2\lambda-\varepsilon}{2} \\ \frac{1-s}{2},\ \frac{2-s}{2};\ -\frac{b}{a^2}\end{matrix}\right)$ $\left[a>0;\ \operatorname{Re}\lambda>-1/2;\ \operatorname{Re}(s+2\lambda)<1-2n-\varepsilon\right]$
8	$e^{-bx}\, C_n^\lambda\left(\frac{x}{a}\pm 1\right)$	$\frac{2^n}{n!\, a^n b^{s+n}}\,(\lambda)_n\,\Gamma(s+n)\ {}_2F_2\left(\begin{matrix}-n,\ \frac{1-2n-2\lambda}{2};\ \pm 2ab \\ 1-2n-2\lambda,\ 1-s-n\end{matrix}\right)$ $\left[\operatorname{Re}b,\ \operatorname{Re}s>0\right]$

No.	$f(x)$	$F(s)$
9	$(a-x)_+^{\lambda-1/2}\, e^{bx}\, C_n^\lambda\left(1-\frac{2x}{a}\right)$	$\frac{a^{s+\lambda-1/2}}{n!}(2\lambda)_n\left(\frac{1}{2}-s+\lambda\right)_n\Gamma\left[\begin{matrix}\frac{2\lambda+1}{2},\, s\\ \frac{2s+2n+2\lambda+1}{2}\end{matrix}\right]$ $\times\,{}_2F_2\left(\begin{matrix}s,\, \frac{2s-2\lambda+1}{2};\, ab\\ \frac{2s-2n-2\lambda+1}{2},\, \frac{2s+2n+2\lambda+1}{2}\end{matrix}\right)$ $[a,\ \operatorname{Re} s>0;\ \operatorname{Re}\lambda>-1/2]$
10	$(x+a)^{\lambda-1/2}\, e^{-bx}$ $\times\, C_n^\lambda\left(\frac{2x}{a}+1\right)$	$\frac{(-1)^n a^{s+\lambda-1/2}}{n!}(2\lambda)_n\left(\frac{1-2s+2\lambda}{2}\right)_n\Gamma\left[\begin{matrix}s,\, \frac{1-2s-2n-2\lambda}{2}\\ \frac{1-2\lambda}{2}\end{matrix}\right]$ $\times\,{}_2F_2\left(\begin{matrix}s,\, \frac{2s-2\lambda+1}{2};\, ab\\ \frac{2s-2n-2\lambda+1}{2},\, \frac{2s+2n+2\lambda+1}{2}\end{matrix}\right)$ $+\,\frac{2^{2n}}{n!\, a^n b^{s+n+\lambda-1/2}}(\lambda)_n\,\Gamma\left(\frac{2s+2n+2\lambda-1}{2}\right)$ $\times\,{}_2F_2\left(\begin{matrix}\frac{1-2n-2\lambda}{2},\, 1-n-2\lambda;\, ab\\ 1-2n-2\lambda,\, \frac{3-2s-2n-2\lambda}{2}\end{matrix}\right)$ $[\operatorname{Re} b,\ \operatorname{Re} s>0;\ \lvert\arg a\rvert<\pi]$
11	$(a-x)_+^{\lambda-1/2}\, e^{-b/x}$ $\times\, C_n^\lambda\left(\frac{2x}{a}-1\right)$	$\frac{(-1)^n a^{s+\lambda-1/2}}{n!}(2\lambda)_n\left(\frac{1-2s+2\lambda}{2}\right)_n$ $\times\,\Gamma\left[\begin{matrix}\frac{2\lambda+1}{2},\, s\\ \frac{2s+2n+2\lambda+1}{2}\end{matrix}\right]{}_2F_2\left(\begin{matrix}\frac{1-2s-2n-2\lambda}{2},\, \frac{1-2s+2n+2\lambda}{2}\\ 1-s,\, \frac{1-2s+2\lambda}{2};\, -\frac{b}{a}\end{matrix}\right)$ $+\,\frac{(-1)^n a^{\lambda-1/2} b^s}{n!}(2\lambda)_n\,\Gamma(-s)\ {}_2F_2\left(\begin{matrix}\frac{-2n-2\lambda+1}{2},\, \frac{2n+2\lambda+1}{2}\\ \frac{2\lambda+1}{2},\, s+1;\, -\frac{b}{a}\end{matrix}\right)$ $[a,\ \operatorname{Re} b>0;\ \operatorname{Re}\lambda>-1/2]$
12	$(x+a)^{\lambda-1/2}\, e^{-b/x}$ $\times\, C_n^\lambda\left(\frac{2x}{a}+1\right)$	$\frac{a^{s+\lambda-1/2}}{n!}(2\lambda)_n\left(\frac{1-2s+2\lambda}{2}\right)_n\Gamma\left[\begin{matrix}\frac{1-2s-2n-2\lambda}{2},\, s\\ \frac{1-2\lambda}{2}\end{matrix}\right]$ $\times\,{}_2F_2\left(\begin{matrix}\frac{1-2s-2n-2\lambda}{2},\, \frac{1-2s+2n+2\lambda}{2}\\ 1-s,\, \frac{1-2s+2\lambda}{2};\, \frac{b}{a}\end{matrix}\right)$ $+\,\frac{a^{\lambda-1/2} b^s}{n!}(2\lambda)_n\,\Gamma(-s)\ {}_2F_2\left(\begin{matrix}\frac{-2n-2\lambda+1}{2},\, \frac{2n+2\lambda+1}{2}\\ \frac{2\lambda+1}{2},\, s+1;\, \frac{b}{a}\end{matrix}\right)$ $[\operatorname{Re} b>0;\ \operatorname{Re} s<1/2-n-\operatorname{Re}\lambda;\ \lvert\arg a\rvert<\pi]$
13	$e^{-b\sqrt{x}}\, C_n^\lambda\left(\frac{x}{a}\pm 1\right)$	$\frac{2^{n+1}}{n!\, a^n b^{2s+2n}}(\lambda)_n\,\Gamma(2s+2n)$ $\times\,{}_2F_3\left(\begin{matrix}-n,\, \frac{1-2n-2\lambda}{2};\, \mp\frac{ab^2}{2}\\ 1-2n-2\lambda,\, \frac{1-2s-2n}{2},\, 1-s-n\end{matrix}\right)$ $[\operatorname{Re} b,\ \operatorname{Re} s>0]$

No.	$f(x)$	$F(s)$
14	$(a-x)_+^{\lambda-1/2}\,e^{-b\sqrt{x}}$ $\times\,C_n^{\lambda}\left(1-\frac{2x}{a}\right)$	$\frac{a^{s+\lambda-1/2}}{n!}\,(2\lambda)_n\left(\frac{1}{2}-s+\lambda\right)_n\Gamma\left[\begin{matrix}\frac{2\lambda+1}{2},\ s\\ \frac{2s+2n+2\lambda+1}{2}\end{matrix}\right]$ $\times\,{}_2F_3\left(\begin{matrix}s,\ \frac{2s-2\lambda+1}{2};\ \frac{ab^2}{4}\\ \frac{1}{2},\ \frac{2s-2n-2\lambda+1}{2},\ \frac{2s+2n+2\lambda+1}{2}\end{matrix}\right)$ $-\frac{a^{s+\lambda}b}{n!}\,(2\lambda)_n\,(\lambda-s)_n\,\Gamma\left[\begin{matrix}\frac{2\lambda+1}{2},\ \frac{2s+1}{2}\\ s+n+\lambda+1\end{matrix}\right]$ $\times\,{}_2F_3\left(\begin{matrix}\frac{2s+1}{2},\ s-\lambda+1;\ \frac{ab^2}{4}\\ \frac{3}{2},\ s-n-\lambda+1,\ s+n+\lambda+1\end{matrix}\right)$ $[a,\ \operatorname{Re}s>0;\ \operatorname{Re}\lambda>-1/2]$
15	$(x+a)^{\lambda-1/2}\,e^{-b\sqrt{x}}$ $\times\,C_n^{\lambda}\left(\frac{2x}{a}+1\right)$	$-\frac{(-1)^n\,a^{s+\lambda}b}{n!}\,(2\lambda)_n\,(\lambda-s)_n\,\Gamma\left[\begin{matrix}-s-n-\lambda,\ \frac{2s+1}{2}\\ \frac{1-2\lambda}{2}\end{matrix}\right]$ $\times\,{}_2F_3\left(\begin{matrix}\frac{2s+1}{2},\ s-\lambda+1;\ -\frac{ab^2}{4}\\ \frac{3}{2},\ s-n-\lambda+1,\ s+n+\lambda+1\end{matrix}\right)$ $+\frac{(-1)^n\,a^{s+\lambda-1/2}}{n!}\,(2\lambda)_n\left(\frac{1}{2}-s+\lambda\right)_n\Gamma\left[\begin{matrix}s,\ \frac{1-2s-2n-2\lambda}{2}\\ \frac{1-2\lambda}{2}\end{matrix}\right]$ $\times\,{}_2F_3\left(\begin{matrix}s,\ s-\lambda+\frac{1}{2};\ -\frac{ab^2}{4}\\ \frac{1}{2},\ \frac{2s-2n-2\lambda+1}{2},\ \frac{2s+2n+2\lambda+1}{2}\end{matrix}\right)$ $+\frac{2^{2n+1}}{n!a^nb^{2s+2n+2\lambda-1}}\,(\lambda)_n\,\Gamma\left(2s+2n+2\lambda-1\right)$ $\times\,{}_2F_3\left(\begin{matrix}\frac{1-2n-2\lambda}{2},\ 1-n-2\lambda;\ -\frac{ab^2}{4}\\ 1-2n-2\lambda,\ \frac{3-2s-2n-2\lambda}{2},\ 1-s-n-\lambda\end{matrix}\right)$ $[\operatorname{Re}b,\ \operatorname{Re}s>0;\ \lvert\arg a\rvert<\pi]$
16	$e^{-b/\sqrt{x}}\,C_n^{\lambda}\left(\frac{x}{a}\pm1\right)$	$\frac{2\,(\pm1)^n\,b^{2s}}{n!}\,\Gamma\left(-2s\right)(2\lambda)_n\,{}_2F_3\left(\begin{matrix}-n,\ n+2\lambda;\ \mp\frac{b^2}{8a}\\ \frac{2\lambda+1}{2},\ \frac{2s+1}{2},\ s+1\end{matrix}\right)$ $[\operatorname{Re}b>0;\ \operatorname{Re}s<-n]$
17	$(a-x)_+^{\lambda-1/2}\,e^{-b/\sqrt{x}}$ $\times\,C_n^{\lambda}\left(\frac{2x}{a}-1\right)$	$\frac{(-1)^{n+1}\,a^{s+\lambda-1}b}{n!}\,(2\lambda)_n\,(1-s+\lambda)_n\,\Gamma\left[\begin{matrix}\frac{2\lambda+1}{2},\ \frac{2s-1}{2}\\ s+n+\lambda\end{matrix}\right]$ $\times\,{}_2F_3\left(\begin{matrix}1-s-n-\lambda,\ 1-s+n+\lambda\\ \frac{3}{2},\ \frac{3}{2}-s,\ 1-s+\lambda;\ \frac{b^2}{4a}\end{matrix}\right)$ $+\frac{(-1)^n\,a^{s+\lambda-1/2}}{n!}\,(2\lambda)_n\left(\frac{1-2s+2\lambda}{2}\right)_n\Gamma\left[\begin{matrix}\frac{2\lambda+1}{2},\ s\\ \frac{2s+2n+2\lambda+1}{2}\end{matrix}\right]$ $\times\,{}_2F_3\left(\begin{matrix}\frac{1-2s-2n-2\lambda}{2},\ \frac{1-2s+2n+2\lambda}{2}\\ \frac{1}{2},\ 1-s,\ \frac{1-2s+2\lambda}{2};\ \frac{b^2}{4a}\end{matrix}\right)+\frac{2\,(-1)^n\,a^{\lambda-1/2}b^{2s}}{n!}$ $\times\,(2\lambda)_n\,\Gamma\left(-2s\right)\,{}_2F_3\left(\begin{matrix}\frac{-2n-2\lambda+1}{2},\ \frac{2n+2\lambda+1}{2}\\ \frac{2\lambda+1}{2},\ \frac{2s+1}{2},\ s+1;\ \frac{b^2}{4a}\end{matrix}\right)$ $[a,\ \operatorname{Re}b>0;\ \operatorname{Re}\lambda>-1/2]$

No.	$f(x)$	$F(s)$
18	$(x+a)^{\lambda-1/2}e^{-b/\sqrt{x}}$ $\times C_n^\lambda\left(\frac{2x}{a}+1\right)$	$\frac{(-1)^n a^{s+\lambda-1/2}}{n!}(2\lambda)_n\left(\frac{1-2s+2\lambda}{2}\right)_n\Gamma\left[\begin{matrix}\frac{1-2s-2n-2\lambda}{2},\ s\\ \frac{1-2\lambda}{2}\end{matrix}\right]$ $\times {}_2F_2\left(\begin{matrix}\frac{1-2s-2n-2\lambda}{2},\ \frac{1-2s+2n+2\lambda}{2}\\ 1-s,\ \frac{1-2s+2\lambda}{2};\ -\frac{b^2}{4a}\end{matrix}\right)$ $+\frac{2a^{\lambda-1/2}b^{2s}}{n!}(2\lambda)_n\Gamma(-2s)\,{}_2F_3\left(\begin{matrix}\frac{-2n-2\lambda+1}{2},\ \frac{2n+2\lambda+1}{2}\\ \frac{2\lambda+1}{2},\ \frac{2s+1}{2},\ s+1;\ -\frac{b^2}{4a}\end{matrix}\right)$ $[\operatorname{Re} b>0;\ \operatorname{Re} s<1/2-n-\operatorname{Re}\lambda;\ \lvert\arg a\rvert<\pi]$
19	$(x+a)^{-n-2\lambda}e^{-bx}$ $\times C_n^\lambda\left(\frac{a-x}{a+x}\right)$	$\frac{a^{s-n-2\lambda}}{n!}\frac{\left(\frac{1}{2}-s+\lambda\right)_n}{\left(\lambda+\frac{1}{2}\right)_n}\Gamma\left[\begin{matrix}s,\ n-s+2\lambda\\ 2\lambda\end{matrix}\right]$ $\times {}_2F_2\left(\begin{matrix}s,\ \frac{2s-2\lambda+1}{2};\ ab\\ s-n-2\lambda+1,\ \frac{2s-2n-2\lambda+1}{2}\end{matrix}\right)$ $+\frac{(-1)^n b^{-s+n+2\lambda}}{n!}(2\lambda)_n\Gamma(s-n-2\lambda)$ $\times {}_2F_2\left(\begin{matrix}n+\lambda+\frac{1}{2},\ n+2\lambda;\ ab\\ \lambda+\frac{1}{2},\ 1-s+n+2\lambda\end{matrix}\right)$ $[\operatorname{Re} b,\ \operatorname{Re} s>0;\ \lvert\arg a\rvert<\pi]$
20	$(x+a)^n e^{-b/x}C_n^\lambda\left(\frac{a-x}{a+x}\right)$	$\frac{a^n b^s}{n!}(2\lambda)_n\Gamma(-s)\,{}_2F_2\left(\begin{matrix}-n,\ \frac{1-2n-2\lambda}{2}\\ \frac{2\lambda+1}{2},\ s+1;\ \frac{b}{a}\end{matrix}\right)$ $[\operatorname{Re} b>0;\ \operatorname{Re} s<-n]$
21	$(x+a)^{-n-2\lambda}e^{-b/x}$ $\times C_n^\lambda\left(\frac{a-x}{a+x}\right)$	$\frac{a^{s-n-2\lambda}}{n!}\frac{\left(\frac{1-2s+2\lambda}{2}\right)_n}{\left(\frac{2\lambda+1}{2}\right)_n}\Gamma\left[\begin{matrix}n-s+2\lambda,\ s\\ 2\lambda\end{matrix}\right]$ $\times {}_2F_2\left(\begin{matrix}\frac{2n-2s+2\lambda+1}{2},\ n-s+2\lambda\\ 1-s,\ \frac{1-2s+2\lambda}{2};\ \frac{b}{a}\end{matrix}\right)$ $+\frac{a^{-n-2\lambda}b^s}{n!}(2\lambda)_n\Gamma(-s)\,{}_2F_2\left(\begin{matrix}\frac{2n+2\lambda+1}{2},\ n+2\lambda\\ \frac{2\lambda+1}{2},\ s+1;\ \frac{b}{a}\end{matrix}\right)$ $[\operatorname{Re} b>0;\ \operatorname{Re} s<n+2\operatorname{Re}\lambda;\ \lvert\arg a\rvert<\pi]$
22	$(x+a)^n e^{-b\sqrt{x}}C_n^\lambda\left(\frac{a-x}{a+x}\right)$	$\frac{2(-1)^n b^{-2s-2n}}{n!}(2\lambda)_n\Gamma(2s+2n)$ $\times {}_2F_3\left(\begin{matrix}-n,\ \frac{1-2n-2\lambda}{2};\ -\frac{ab^2}{4}\\ \frac{2\lambda+1}{2},\ \frac{1-2s-2n}{2},\ 1-s-n\end{matrix}\right)$ $[\operatorname{Re} b,\ \operatorname{Re} s>0]$

No.	$f(x)$	$F(s)$
23	$(x+a)^{-n-2\lambda}\, e^{-b\sqrt{x}}$ $\times\, C_n^{\lambda}\left(\frac{a-x}{a+x}\right)$	$-\frac{a^{s-n-2\lambda+1/2}b}{n!}\frac{(\lambda-s)_n}{\left(\frac{2\lambda+1}{2}\right)_n}\Gamma\left[\begin{matrix}\frac{2s+1}{2},\ \frac{2n-2s+4\lambda-1}{2}\\ 2\lambda\end{matrix}\right]$ $\times\, {}_2F_3\left(\begin{matrix}\frac{2s+1}{2},\ s-\lambda+1;\ -\frac{ab^2}{4}\\ \frac{3}{2},\ s-n-\lambda+1,\ \frac{2s-2n-4\lambda+3}{2}\end{matrix}\right)$ $+\frac{a^{s-n-2\lambda}}{n!}\frac{\left(\frac{1-2s+2\lambda}{2}\right)_n}{\left(\frac{2\lambda+1}{2}\right)_n}\Gamma\left[\begin{matrix}n-s+2\lambda,\ s\\ 2\lambda\end{matrix}\right]$ $\times\, {}_2F_3\left(\begin{matrix}s,\ \frac{2s-2\lambda+1}{2};\ -\frac{ab^2}{4}\\ \frac{1}{2},\ s-n-2\lambda+1,\ \frac{2s-2n-2\lambda+1}{2}\end{matrix}\right)$ $+\frac{2(-1)^n b^{2(-s+n+2\lambda)}}{n!}(2\lambda)_n\Gamma(2s-2n-4\lambda)$ $\times\, {}_2F_3\left(\begin{matrix}n+2\lambda,\ \frac{2n+2\lambda+1}{2};\ -\frac{ab^2}{4}\\ \frac{2\lambda+1}{2},\ \frac{1-2s+2n+4\lambda}{2},\ 1-s+n+2\lambda\end{matrix}\right)$ $[\operatorname{Re} b,\ \operatorname{Re} s>0;\ \lvert\arg a\rvert<\pi]$
24	$(x+a)^n\, e^{-b/\sqrt{x}}\, C_n^{\lambda}\left(\frac{a-x}{a+x}\right)$	$\frac{2a^n b^{2s}}{n!}(2\lambda)_n\Gamma(-2s)\, {}_2F_3\left(\begin{matrix}-n,\ \frac{1-2n-2\lambda}{2};\ -\frac{b^2}{4a}\\ \frac{2\lambda+1}{2},\ \frac{2s+1}{2},\ s+1\end{matrix}\right)$ $[a,\ \operatorname{Re} b>0;\ \operatorname{Re} s<-n]$
25	$(x+a)^{-n-2\lambda}\, e^{-b/\sqrt{x}}$ $\times\, C_n^{\lambda}\left(\frac{a-x}{a+x}\right)$	$-\frac{a^{s-n-2\lambda-1/2}b}{n!}\frac{(1-s+\lambda)_n}{\left(\frac{2\lambda+1}{2}\right)_n}\Gamma\left[\begin{matrix}\frac{1-2s+2n+4\lambda}{2},\ \frac{2s-1}{2}\\ 2\lambda\end{matrix}\right]$ $\times\, {}_2F_3\left(\begin{matrix}1-s+n+\lambda,\ \frac{1-2s+2n+4\lambda}{2}\\ \frac{3}{2},\ \frac{3}{2}-s,\ 1-s+\lambda;\ -\frac{b^2}{4a}\end{matrix}\right)+\frac{a^{s-n-2\lambda}}{n!}\frac{\left(\frac{1-2s+2\lambda}{2}\right)_n}{\left(\frac{2\lambda+1}{2}\right)_n}$ $\times\,\Gamma\left[\begin{matrix}n-s+2\lambda,\ s\\ 2\lambda\end{matrix}\right]{}_2F_3\left(\begin{matrix}\frac{1-2s+2n+2\lambda}{2},\ n-s+2\lambda\\ \frac{1}{2},\ 1-s,\ \frac{1-2s+2\lambda}{2};\ -\frac{b^2}{4a}\end{matrix}\right)$ $+\frac{2a^{-n-2\lambda}b^{2s}}{n!}(2\lambda)_n\Gamma(-2s)\, {}_2F_3\left(\begin{matrix}n+2\lambda,\ \frac{2n+2\lambda+1}{2};\ -\frac{b^2}{4a}\\ \frac{2\lambda+1}{2},\ \frac{2s+1}{2},\ s+1\end{matrix}\right)$ $[a,\ \operatorname{Re} b>0;\ \operatorname{Re} s<n+2\operatorname{Re}\lambda]$

3.24.3. $C_n^{\lambda}(bx)$ and hyperbolic functions

Notation: $\delta=\left\{\begin{matrix}1\\0\end{matrix}\right\}$.

1	$(a^2-x^2)_+^{\lambda-1/2}\left\{\begin{matrix}\sinh(bx)\\ \cosh(bx)\end{matrix}\right\}$ $\times\, C_n^{\lambda}\left(\frac{x}{a}\right)$	$\frac{\pi}{n!}\left(\frac{a}{2}\right)^{s+2\lambda+\delta-1}b^{\delta}\,\Gamma\left[\begin{matrix}s+\delta,\ n+2\lambda\\ \lambda,\ \frac{s-n+\delta+1}{2},\ \frac{s+n+2\lambda+\delta+1}{2}\end{matrix}\right]$ $\times\, {}_2F_3\left(\begin{matrix}\frac{s+\delta}{2},\ \frac{s+\delta+1}{2};\ \frac{a^2b^2}{4}\\ \frac{2\delta+1}{2},\ \frac{s-n+\delta+1}{2},\ \frac{s+n+2\lambda+\delta+1}{2}\end{matrix}\right)$ $[a>0;\ \operatorname{Re}\lambda>-1/2;\ \operatorname{Re} s>-\delta]$

3.24.4. $C_n^\lambda(ax+b)$ and trigonometric functions

Notation: $\delta = \left\{\begin{matrix}1\\0\end{matrix}\right\}$, $\varepsilon = 0$ or 1.

1	$(a^2-x^2)_+^{\lambda-1/2} \times \left\{\begin{matrix}\sin(bx)\\ \cos(bx)\end{matrix}\right\} C_n^\lambda\left(\frac{x}{a}\right)$	$\frac{\pi}{n!}\left(\frac{a}{2}\right)^{s+2\lambda+\delta-1} b^\delta\,\Gamma\left[\begin{matrix}s+\delta,\ n+2\lambda\\ \lambda,\ \frac{s-n+\delta+1}{2},\ \frac{s+n+2\lambda+\delta+1}{2}\end{matrix}\right] \times {}_2F_3\left(\begin{matrix}\frac{s+\delta}{2},\ \frac{s+\delta+1}{2};\ -\frac{a^2b^2}{4}\\ \frac{2\delta+1}{2},\ \frac{s-n+\delta+1}{2},\ \frac{s+n+2\lambda+\delta+1}{2}\end{matrix}\right)$ $[a>0;\ \mathrm{Re}\,\lambda>-1/2;\ \mathrm{Re}\,s>((-1)^n-1)/2-\delta]$
2	$(x^2-a^2)_+^{\lambda-1/2} \times \left\{\begin{matrix}\sin(bx)\\ \cos(bx)\end{matrix}\right\} C_{2n+\varepsilon}^\lambda\left(\frac{x}{a}\right)$	$\frac{a^{s+2\lambda+\delta-1}b^\delta}{2(2n+\varepsilon)!}(2\lambda)_{2n+\varepsilon}\left(\frac{1-s-\delta+\varepsilon}{2}\right)_n \Gamma\left[\begin{matrix}\frac{2\lambda+1}{2},\ \frac{1-s-2n-2\lambda-\delta-\varepsilon}{2}\\ \frac{2-s-\delta-\varepsilon}{2}\end{matrix}\right] \times {}_2F_3\left(\begin{matrix}\frac{s+1}{2},\ \frac{s+2\delta}{2};\ -\frac{a^2b^2}{4}\\ \frac{2\delta+1}{2},\ \frac{s-2n+\delta-\varepsilon+1}{2},\ \frac{s+2n+2\lambda+\delta+\varepsilon+1}{2}\end{matrix}\right) \mp \frac{(-1)^n 2^{2n+\varepsilon} b^{1-s-2n-2\lambda-\varepsilon}}{(2n+\varepsilon)!\,a^{2n+\varepsilon}}(\lambda)_{2n+\varepsilon} \times \Gamma(s+2n+2\lambda+\varepsilon-1)\left\{\begin{matrix}\cos[(s+2\lambda+\varepsilon)\pi/2]\\ \sin[(s+2\lambda+\varepsilon)\pi/2]\end{matrix}\right\} \times {}_2F_3\left(\begin{matrix}\frac{1-2n-2\lambda}{2},\ 1-n-\lambda-\varepsilon;\ -\frac{a^2b^2}{4}\\ 1-2n-\lambda-\varepsilon,\ \frac{3-s-2n-2\lambda-2\varepsilon}{2},\ \frac{2-s-2n-2\lambda}{2}\end{matrix}\right)$ $[a,\ b>0;\ \mathrm{Re}\,\lambda>-1/2;\ \mathrm{Re}\,(s+2\lambda)<2-\varepsilon-2n]$
3	$(a-x)_+^{\lambda-1/2}\left\{\begin{matrix}\sin(b\sqrt{x})\\ \cos(b\sqrt{x})\end{matrix}\right\} \times C_n^\lambda\left(\frac{2x}{a}-1\right)$	$\frac{(-1)^n a^{s+\lambda+(\delta-1)/2}b^\delta}{n!}(2\lambda)_n\left(\frac{1-2s+2\lambda-\delta}{2}\right)_n \times \Gamma\left[\begin{matrix}\frac{2\lambda+1}{2},\ \frac{2s+\delta}{2}\\ \frac{2s+2n+2\lambda+\delta+1}{2}\end{matrix}\right] {}_2F_3\left(\begin{matrix}\frac{2s+\delta}{2},\ \frac{2s-2\lambda+\delta+1}{2};\ -\frac{ab^2}{4}\\ \frac{2\delta+1}{2},\ \frac{2s-2n-2\lambda+\delta+1}{2},\ \frac{2s+2n+2\lambda+\delta+1}{2}\end{matrix}\right)$ $[a>0;\ \mathrm{Re}\,\lambda>-1/2;\ \mathrm{Re}\,s>-\delta/2]$
4	$(x+a)^{\lambda-1/2} \times \left\{\begin{matrix}\sin(b/\sqrt{x})\\ \cos(b/\sqrt{x})\end{matrix}\right\} \times C_n^\lambda\left(\frac{2x}{a}+1\right)$	$\frac{(-1)^n a^{s+\lambda-(\delta+1)/2}b^\delta}{n!}(2\lambda)_n\left(\frac{1-2s+2\lambda+\delta}{2}\right)_n \times \Gamma\left[\begin{matrix}\frac{2s-\delta}{2},\ \frac{1-2s-2n-2\lambda+\delta}{2}\\ \frac{1-2\lambda}{2}\end{matrix}\right] \times {}_2F_3\left(\begin{matrix}\frac{1-2s-2n-2\lambda+\delta}{2},\ \frac{1-2s+2n+2\lambda+\delta}{2}\\ \frac{2\delta+1}{2},\ \frac{2-2s+\delta}{2},\ \frac{1-2s+2\lambda+\delta}{2};\ \frac{b^2}{4a}\end{matrix}\right) + \frac{2(-1)^\delta a^{\lambda-1/2}b^{2s}}{n!}(2\lambda)_n\,\Gamma(-2s)\left\{\begin{matrix}\sin(s\pi)\\ \cos(s\pi)\end{matrix}\right\} \times {}_2F_3\left(\begin{matrix}\frac{-2n-2\lambda+1}{2},\ \frac{2n+2\lambda+1}{2}\\ \frac{2\lambda+1}{2},\ \frac{2s+1}{2},\ s+1;\ \frac{b^2}{4a}\end{matrix}\right)$ $[b>0;\ -1/2<\mathrm{Re}\,s<(\delta+1)/2-n-\mathrm{Re}\,\lambda;\ \lvert\arg a\rvert<\pi]$

3.24.5. $C_n^\lambda(bx)$ and the logarithmic function

Notation: $\varepsilon = 0$ or 1.

1	$(a-x)_+^{\lambda-1/2}\ln\frac{x}{a} \times C_n^\lambda\left(\frac{2x-a}{a}\right)$	$\frac{(-1)^n a^{s+\lambda-1/2}}{n!}(2\lambda)_n\left(\frac{1-2s+2\lambda}{2}\right)_n\Gamma\left[\begin{matrix}\frac{2\lambda+1}{2},\ s\\ \frac{2s+2n+2\lambda+1}{2}\end{matrix}\right] \times\left[\psi(s)+\psi\left(\frac{1-2s+2\lambda}{2}\right)-\psi\left(\frac{2s+2n+2\lambda+1}{2}\right) - \psi\left(\frac{-2s+2n+2\lambda+1}{2}\right)\right]$ $[a,\ \operatorname{Re}s>0;\ \operatorname{Re}\lambda>-1/2]$
2	$(a^2-x^2)_+^{\lambda-1/2} \times\left\{\begin{matrix}\ln(x^2+b^2)\\ \ln\lvert x^2-b^2\rvert\end{matrix}\right\} \times C_n^\lambda\left(\frac{x}{a}\right)$	$\pm\frac{\sqrt{\pi}\,a^{s+2\lambda+1}}{2^{s+2}n!\,b^2}(2\lambda)_n\Gamma\left[\begin{matrix}\lambda+\frac{1}{2},\ s+2\\ \frac{s-n+3}{2},\ \frac{s+n+2\lambda+3}{2}\end{matrix}\right] \times{}_4F_3\left(\begin{matrix}1,\ 1,\ \frac{s+2}{2},\ \frac{s+3}{2};\ \mp\frac{a^2}{b^2}\\ 2,\ \frac{s-n+3}{2},\ \frac{s+n+2\lambda+3}{2}\end{matrix}\right) + \frac{\sqrt{\pi}\,a^{s+2\lambda-1}\ln b}{2^{s-1}n!}(2\lambda)_n\Gamma\left[\begin{matrix}\lambda+\frac{1}{2},\ s\\ \frac{s-n+1}{2},\ \frac{s+n+2\lambda+1}{2}\end{matrix}\right]$ $\left[\left\{\begin{matrix}a,\ \operatorname{Re}b>0\\ b>a>0\end{matrix}\right\};\ \operatorname{Re}\lambda>-1/2;\ \operatorname{Re}s>((-1)^n-1)/2\right]$
3	$(a^2-x^2)_+^{\lambda-1/2}\ln\lvert x^2-b^2\rvert \times C_n^\lambda\left(\frac{x}{a}\right)$	$-\frac{\pi(a/2)^{s+2\lambda-3}b^2}{n!}\Gamma\left[\begin{matrix}s-2,\ n+2\lambda\\ \lambda,\ \frac{s-n-1}{2},\ \frac{s+n+2\lambda-1}{2}\end{matrix}\right] \times{}_4F_3\left(\begin{matrix}1,\ 1,\ \frac{3-s+n}{2},\ -\frac{s+n+2\lambda-3}{2}\\ 2,\ \frac{3-s}{2},\ \frac{4-s}{2};\ \frac{b^2}{a^2}\end{matrix}\right) + \frac{2^{n+1}\pi^{3/2}a^{2\lambda-2}b^{s+1}}{n!\,(s+1)}\tan\frac{s\pi}{2}\Gamma\left[\begin{matrix}\frac{n+2\lambda+1}{2}\\ \lambda,\ -\frac{n}{2}\end{matrix}\right] \times{}_3F_2\left(\begin{matrix}\frac{n+2}{2},\ -\frac{n+2\lambda-2}{2},\ \frac{s+1}{2}\\ \frac{3}{2},\ \frac{s+3}{2};\ \frac{b^2}{a^2}\end{matrix}\right) + \frac{2^n\pi^{3/2}a^{2\lambda-1}b^s}{n!\,s}\cot\frac{s\pi}{2}\Gamma\left[\begin{matrix}\frac{n+2\lambda}{2}\\ \lambda,\ -\frac{n-1}{2}\end{matrix}\right] \times{}_3F_2\left(\begin{matrix}\frac{n+1}{2},\ -\frac{n+2\lambda-1}{2},\ \frac{s}{2}\\ \frac{1}{2},\ \frac{s+2}{2};\ \frac{b^2}{a^2}\end{matrix}\right) - \frac{\pi(a/2)^{s+2\lambda-1}}{n!}\Gamma\left[\begin{matrix}s,\ n+2\lambda\\ \lambda,\ \frac{s-n+1}{2},\ \frac{s+n+2\lambda+1}{2}\end{matrix}\right] \times\left[\psi\left(\frac{s+n+2\lambda+1}{2}\right)+\psi\left(\frac{s-n+1}{2}\right)-2\psi(s)+\ln\frac{4}{a^2}\right]$ $[a>0;\ a>b;\ \operatorname{Re}\lambda>-1/2;\ \operatorname{Re}s>((-1)^n-1)/2]$

No.	$f(x)$	$F(s)$
4	$\left(x^2-a^2\right)_+^{\lambda-1/2}$ $\times\left\{\begin{matrix}\ln\left(x^2+b^2\right)\\ \ln\lvert x^2-b^2\rvert\end{matrix}\right\}$ $\times C_n^\lambda\left(\frac{x}{a}\right)$	$\pm\frac{\pi^2(a/2)^{s+2\lambda+1}}{2b^2\,n!}\sec\frac{(s-n)\pi}{2}\sec\frac{(s+n+2\lambda)\pi}{2}$ $\times\Gamma\left[\begin{matrix}n+2\lambda\\ \lambda,\,-s-1,\,\frac{s-n+3}{2},\,\frac{s+n+2\lambda+3}{2}\end{matrix}\right]$ $\times{}_4F_3\left(\begin{matrix}1,\,1,\,\frac{s+2}{2},\,\frac{s+3}{2};\,\mp\frac{a^2}{b^2}\\ 2,\,\frac{s-n+3}{2},\,\frac{s+n+2\lambda+3}{2}\end{matrix}\right)-\frac{2^n\pi^2b^{s+n+2\lambda-1}}{(s+n+2\lambda-1)\,n!\,a^n}$ $\times\frac{\csc[(n+\lambda)\pi]}{\Gamma(\lambda)\,\Gamma(1-n-\lambda)}\left\{\begin{matrix}\sec[(s+n+2\lambda)\pi/2]\\ \tan[(s+n+2\lambda)\pi/2]\end{matrix}\right\}$ $\times{}_3F_2\left(\begin{matrix}\frac{1-n-2\lambda}{2},\,\frac{2-n-2\lambda}{2},\,\frac{1-s-n-2\lambda}{2}\\ 1-n-\lambda,\,\frac{3-s-n-2\lambda}{2};\,\mp\frac{a^2}{b^2}\end{matrix}\right)$ $+\frac{(a/2)^{s+2\lambda-1}\ln b}{n!}\Gamma\left[\begin{matrix}n+2\lambda,\,\frac{1-s+n}{2},\,\frac{1-s-n-2\lambda}{2}\\ \lambda,\,1-s\end{matrix}\right]$ $\left[\left\{\begin{matrix}a,\ \operatorname{Re}b>0\\ b>a>0\end{matrix}\right\};\ \operatorname{Re}\lambda>-1/2;\ \operatorname{Re}(s+2\lambda)<1-n\right]$
5	$\left(x^2-a^2\right)_+^{\lambda-1/2}$ $\times\ln\lvert x^2-b^2\rvert\,C_n^\lambda\left(\frac{x}{a}\right)$	$-\frac{(a/2)^{s+2\lambda-3}\,b^2}{2\,(n!)}\Gamma\left[\begin{matrix}n+2\lambda,\,\frac{3-s+n}{2},\,\frac{3-s-n-2\lambda}{2}\\ \lambda,\,3-s\end{matrix}\right]$ $\times{}_4F_3\left(\begin{matrix}1,\,1,\,\frac{3-s+n}{2},\,\frac{3-s-n-2\lambda}{2}\\ 2,\,\frac{3-s}{2},\,\frac{4-s}{2};\,\frac{b^2}{a^2}\end{matrix}\right)$ $-\frac{(a/2)^{s+2\lambda-1}}{2\,(n!)}\Gamma\left[\begin{matrix}n+2\lambda,\,\frac{1-s+n}{2},\,\frac{1-s-n-2\lambda}{2}\\ \lambda,\,1-s\end{matrix}\right]$ $\times\left[\psi\left(\frac{1-s-n-2\lambda}{2}\right)+\psi\left(\frac{1-s+n}{2}\right)-2\psi(1-s)+\ln\frac{4b^2}{a^2}\right]$ $-\frac{a^{s+2\lambda-1}\ln b}{n!}(2\lambda)_n\Gamma\left[\begin{matrix}\frac{2\lambda+1}{2},\,\frac{1-s+n}{2},\,\frac{1-s-n-2\lambda}{2}\\ \frac{1-s}{2},\,\frac{2-s}{2}\end{matrix}\right]$ $[a,\,b>0;\ \operatorname{Re}\lambda>-1/2;\ \operatorname{Re}(s+2\lambda)<1-n]$
6	$\left(a^2-x^2\right)_+^{\lambda-1/2}\ln\frac{1+bx}{1-bx}$ $\times C_n^\lambda\left(\frac{x}{a}\right)$	$\frac{\pi a^{s+2\lambda}b}{2^{s+2\lambda-1}n!}\Gamma\left[\begin{matrix}n+2\lambda,\,s+1\\ \lambda,\,\frac{s-n+2}{2},\,\frac{s+n+2\lambda+2}{2}\end{matrix}\right]{}_4F_3\left(\begin{matrix}\frac{1}{2},\,1,\,\frac{s+1}{2},\,\frac{s+2}{2};\,a^2b^2\\ \frac{3}{2},\,\frac{s-n+2}{2},\,\frac{s+n+2\lambda+2}{2}\end{matrix}\right)$ $[a>0;\ \operatorname{Re}\lambda>-1/2;\ \operatorname{Re}s>-1;\ \lvert\arg\left(1-a^2b^2\right)\rvert<\pi]$
7	$\left(a^2-x^2\right)_+^{\lambda-1/2}$ $\times\ln\left(bx+\sqrt{b^2x^2+1}\right)$ $\times C_n^\lambda\left(\frac{x}{a}\right)$	$\frac{\pi}{n!}\left(\frac{a}{2}\right)^{s+2\lambda}b\,\Gamma\left[\begin{matrix}n+2\lambda,\,s+1\\ \lambda,\,\frac{s-n+2}{2},\,\frac{s+n+2\lambda+2}{2}\end{matrix}\right]$ $\times{}_4F_3\left(\begin{matrix}\frac{1}{2},\,\frac{1}{2},\,\frac{s+1}{2},\,\frac{s+2}{2};\,-a^2b^2\\ \frac{3}{2},\,\frac{s-n+2}{2},\,\frac{s+n+2\lambda+2}{2}\end{matrix}\right)$ $[a>0;\ \operatorname{Re}\lambda>-1/2;\ \operatorname{Re}s>((-1)^n-1)/2;\ \lvert\arg\left(1+a^2b^2\right)\rvert<\pi]$

No.	$f(x)$	$F(s)$
8	$\dfrac{(a^2-x^2)_+^{\lambda-1/2}}{\sqrt{b^2x^2+1}}$ $\times \ln\left(bx+\sqrt{b^2x^2+1}\right)$ $\times C_n^{\lambda}\left(\dfrac{x}{a}\right)$	$\dfrac{\pi}{n!}\left(\dfrac{a}{2}\right)^{s+2\lambda} b\,\Gamma\left[\begin{matrix} n+2\lambda,\, s+1 \\ \lambda,\, \frac{s-n+2}{2},\, \frac{s+n+2\lambda+2}{2}\end{matrix}\right] \times {}_4F_3\left(\begin{matrix}1,\, 1,\, \frac{s+1}{2},\, \frac{s+2}{2};\, -a^2b^2 \\ \frac{3}{2},\, \frac{s-n+2}{2},\, \frac{s+n+2\lambda+2}{2}\end{matrix}\right)$ $[a>0;\ \operatorname{Re}\lambda>-1/2;\ \operatorname{Re}s>((-1)^n-3)/2;\ \lvert\arg(1+a^2b^2)\rvert<\pi]$
9	$\theta(a-x)\ln\dfrac{\sqrt{a}+\sqrt{a-x}}{\sqrt{x}}$ $\times C_{2n+\varepsilon}^{\lambda}(bx)$	$\dfrac{(-1)^n\sqrt{\pi}\,(\lambda)_{n+\varepsilon}}{2^{1-\varepsilon}n!\,(s+\varepsilon)}\, a^{s+\varepsilon}b^{\varepsilon}\,\Gamma\left[\begin{matrix} s+\varepsilon \\ \frac{2s+2\varepsilon+1}{2}\end{matrix}\right] \times {}_5F_4\left(\begin{matrix}-n,\, n+\lambda+\varepsilon,\, \frac{s+\varepsilon}{2},\, \frac{s+\varepsilon}{2},\, \frac{s+\varepsilon+1}{2} \\ \frac{2\varepsilon+1}{2},\, \frac{2s+2\varepsilon+1}{4},\, \frac{2s+2\varepsilon+3}{4},\, \frac{s+\varepsilon+2}{2};\, a^2b^2\end{matrix}\right)$ $[a>0;\ \operatorname{Re}\lambda>-1/2;\ \operatorname{Re}s>-\varepsilon]$
10	$(a^2-x^2)_+^{\lambda-1/2}$ $\times \ln^2\left(bx+\sqrt{b^2x^2+1}\right)$ $\times C_n^{\lambda}\left(\dfrac{x}{a}\right)$	$\dfrac{\pi}{n!}\left(\dfrac{a}{2}\right)^{s+2\lambda+1} b^2\,\Gamma\left[\begin{matrix} n+2\lambda,\, s+2 \\ \lambda,\, \frac{s-n+3}{2},\, \frac{s+n+2\lambda+3}{2}\end{matrix}\right] \times {}_5F_4\left(\begin{matrix}1,\, 1,\, 1,\, \frac{s+2}{2},\, \frac{s+3}{2};\, -a^2b^2 \\ \frac{3}{2},\, 2,\, \frac{s-n+3}{2},\, \frac{s+n+2\lambda+3}{2}\end{matrix}\right)$ $[a>0;\ \operatorname{Re}\lambda>-1/2;\ \operatorname{Re}s>((-1)^n-5)/2;\ \lvert\arg(1+a^2b^2)\rvert<\pi]$

3.24.6. $C_n^{\lambda}(bx)$ and inverse trigonometric functions

Notation: $\varepsilon=0$ or 1.

1	$\theta(a-x)\arccos\dfrac{x}{a}$ $\times C_{2n+\varepsilon}^{\lambda}(bx)$	$\dfrac{(-1)^n\sqrt{\pi}\,a^{s+\varepsilon}b^{\varepsilon}\,(\lambda)_{n+\varepsilon}}{2^{1-\varepsilon}n!\,(s+\varepsilon)}\,\Gamma\left[\begin{matrix} \frac{s+\varepsilon+1}{2} \\ \frac{s+\varepsilon+2}{2}\end{matrix}\right] \times {}_4F_3\left(\begin{matrix}-n,\, n+\lambda+\varepsilon,\, \frac{s+\varepsilon}{2},\, \frac{s+\varepsilon+1}{2} \\ \frac{2\varepsilon+1}{2},\, \frac{s+\varepsilon+2}{2},\, \frac{s+\varepsilon+2}{2};\, a^2b^2\end{matrix}\right)$ $[a>0;\ \operatorname{Re}\lambda>-1/2;\ \operatorname{Re}s>-\varepsilon]$
2	$(a^2-x^2)_+^{\lambda-1/2}$ $\times \arcsin(bx)\, C_n^{\lambda}\left(\dfrac{x}{a}\right)$	$\dfrac{\pi}{n!}\left(\dfrac{a}{2}\right)^{s+2\lambda} b\,\Gamma\left[\begin{matrix} n+2\lambda,\, s+1 \\ \lambda,\, \frac{s-n+2}{2},\, \frac{s+n+2\lambda+2}{2}\end{matrix}\right] \times {}_4F_3\left(\begin{matrix}\frac{1}{2},\, \frac{1}{2},\, \frac{s+1}{2},\, \frac{s+2}{2};\, a^2b^2 \\ \frac{3}{2},\, \frac{s-n+2}{2},\, \frac{s+n+2\lambda+2}{2};\, a^2b^2\end{matrix}\right)$ $[a>0;\ \operatorname{Re}\lambda>-1/2;\ \operatorname{Re}s>-1]$
3	$\dfrac{(a^2-x^2)_+^{\lambda-1/2}}{\sqrt{1-b^2x^2}}$ $\times \arcsin(bx)\, C_n^{\lambda}\left(\dfrac{x}{a}\right)$	$\dfrac{\pi}{n!}\left(\dfrac{a}{2}\right)^{s+2\lambda} b\,\Gamma\left[\begin{matrix} n+2\lambda,\, s+1 \\ \lambda,\, \frac{s-n+2}{2},\, \frac{s+n+2\lambda+2}{2}\end{matrix}\right] {}_4F_3\left(\begin{matrix}1,\, 1,\, \frac{s+1}{2},\, \frac{s+2}{2};\, a^2b^2 \\ \frac{3}{2},\, \frac{s-n+2}{2},\, \frac{s+n+2\lambda+2}{2}\end{matrix}\right)$ $[a>0;\ \operatorname{Re}\lambda>-1/2;\ \operatorname{Re}s>((-1)^n-3)/2]$

No.	$f(x)$	$F(s)$
4	$(a^2-x^2)_+^{\lambda-1/2}$ $\times \arcsin^2(bx)\, C_n^\lambda\left(\frac{x}{a}\right)$	$\frac{\pi}{n!}\left(\frac{a}{2}\right)^{s+2\lambda+1} b^2\,\Gamma\left[\begin{matrix} n+2\lambda,\ s+2 \\ \lambda,\ \frac{s-n+3}{2},\ \frac{s+n+2\lambda+3}{2}\end{matrix}\right]$ $\times {}_5F_4\left(\begin{matrix} 1,\ 1,\ 1,\ \frac{s+2}{2},\ \frac{s+3}{2};\ a^2b^2 \\ \frac{3}{2},\ 2,\ \frac{s-n+3}{2},\ \frac{s+n+2\lambda+3}{2}\end{matrix}\right)$ $[a>0;\ \operatorname{Re}\lambda>-1/2;\ \operatorname{Re}s>((-1)^n-5)/2]$
5	$(a^2-x^2)_+^{\lambda-1/2}$ $\times \arctan(bx)\, C_n^\lambda\left(\frac{x}{a}\right)$	$\frac{\pi}{n!}\left(\frac{a}{2}\right)^{s+2\lambda} b\,\Gamma\left[\begin{matrix} n+2\lambda,\ s+1 \\ \lambda,\ \frac{s-n+2}{2},\ \frac{s+n+2\lambda+2}{2}\end{matrix}\right]$ $\times {}_4F_3\left(\begin{matrix} \frac{1}{2},\ 1,\ \frac{s+1}{2},\ \frac{s+2}{2};\ -a^2b^2 \\ \frac{3}{2},\ \frac{s-n+2}{2},\ \frac{s+n+2\lambda+2}{2}\end{matrix}\right)$ $[a>0;\ \operatorname{Re}\lambda>-1/2;\ \operatorname{Re}s>((-1)^n-3)/2]$

3.24.7. $C_n^\lambda(ax+b)$ and $\operatorname{Ei}(ax^r)$

Notation: $\varepsilon=0$ or 1.

No.	$f(x)$	$F(s)$
1	$(a-x)_+^{\lambda-1/2}\operatorname{Ei}(-bx)$ $\times C_n^\lambda\left(\frac{2x}{a}-1\right)$	$\frac{(-1)^{n+1}a^{s+\lambda+1/2}b}{n!}(2\lambda)_n\left(\frac{-2s+2\lambda-1}{2}\right)_n$ $\times\Gamma\left[\begin{matrix}\frac{2\lambda+1}{2},\ s+1 \\ \frac{2s+2n+2\lambda+3}{2}\end{matrix}\right]{}_4F_4\left(\begin{matrix}1,\ 1,\ \frac{2s-2\lambda+3}{2},\ s+1;\ -ab \\ 2,\ 2,\ \frac{2s-2n-2\lambda+3}{2},\ \frac{2s+2n+2\lambda+3}{2}\end{matrix}\right)$ $+\frac{(-1)^n a^{s+\lambda-1/2}}{n!}(2\lambda)_n\left(\frac{-2s+2\lambda+1}{2}\right)_n$ $\times\Gamma\left[\begin{matrix}\frac{2\lambda+1}{2},\ s \\ \frac{2s+2n+2\lambda+1}{2}\end{matrix}\right]\left[\psi(s)-\psi\left(\frac{2s+2n+2\lambda+1}{2}\right)\right.$ $\left.-\sum_{i=0}^{n-1}\frac{2}{2i-2s+2\lambda+1}+\ln(ab)+\mathbf{C}\right]$ $\left[\begin{matrix}a,\ \operatorname{Re}s>0; \\ \operatorname{Re}\lambda>-1/2\end{matrix}\right]$
2	$(a^2-x^2)_+^{\lambda-1/2}\operatorname{Ei}(bx)$ $\times C_n^\lambda\left(\frac{x}{a}\right)$	$\frac{\pi}{n!}\left(\frac{a}{2}\right)^{s+2\lambda} b\,\Gamma\left[\begin{matrix} n+2\lambda,\ s+1 \\ \lambda,\ \frac{s-n+2}{2},\ \frac{s+n+2\lambda+2}{2}\end{matrix}\right]$ $\times {}_3F_4\left(\begin{matrix}\frac{1}{2},\ \frac{s+1}{2},\ \frac{s+2}{2};\ \frac{a^2b^2}{4} \\ \frac{3}{2},\ \frac{3}{2},\ \frac{s-n+2}{2},\ \frac{s+n+2\lambda+2}{2}\end{matrix}\right)+\frac{\pi a^{s+2\lambda+1}b^2}{2^{s+2\lambda+3}\,n!}$ $\times\Gamma\left[\begin{matrix} n+2\lambda,\ s+2 \\ \lambda,\ \frac{s-n+3}{2},\ \frac{s+n+2\lambda+3}{2}\end{matrix}\right]{}_3F_4\left(\begin{matrix}1,\ 1,\ \frac{s+2}{2},\ \frac{s+3}{2};\ \frac{a^2b^2}{4} \\ \frac{3}{2},\ 2,\ 2,\ \frac{s-n+3}{2},\ \frac{s+n+2\lambda+3}{2}\end{matrix}\right)$ $+\frac{\pi}{n!}\left(\frac{a}{2}\right)^{s+2\lambda-1}\Gamma\left[\begin{matrix} n+2\lambda,\ s \\ \lambda,\ \frac{s-n+1}{2},\ \frac{s+n+2\lambda+1}{2}\end{matrix}\right]\left[\psi(s)\right.$ $\left.-\frac{1}{2}\psi\left(\frac{s+n+2\lambda+1}{2}\right)-\frac{1}{2}\psi\left(\frac{s-n+1}{2}\right)+\ln\frac{ab}{2}+\mathbf{C}\right]$ $[a>0;\ \operatorname{Re}\lambda>-1/2;\ \operatorname{Re}s>((-1)^n-1)/2]$

No.	$f(x)$	$F(s)$
3	$\left(a^2-x^2\right)_+^{\lambda-1/2}\operatorname{Ei}\left(bx^2\right)$ $\times C_n^{\lambda}\left(\frac{x}{a}\right)$	$\frac{\pi}{n!}\left(\frac{a}{2}\right)^{2\lambda+s+1} b\,\Gamma\left[\begin{matrix} n+2\lambda,\ s+2 \\ \lambda,\ \frac{s-n+3}{2},\ \frac{s+n+2\lambda+3}{2}\end{matrix}\right]$ $\times {}_4F_4\left(\begin{matrix}1,\ 1,\ \frac{s+2}{2},\ \frac{s+3}{2};\ a^2b \\ 2,\ 2,\ \frac{s-n+3}{2},\ \frac{s+n+2\lambda+3}{2}\end{matrix}\right)$ $+\frac{\pi}{n!}\left(\frac{a}{2}\right)^{2\lambda+s+1}\Gamma\left[\begin{matrix} n+2\lambda,\ s \\ \lambda,\ \frac{s-n+1}{2},\ \frac{s+n+2\lambda+1}{2}\end{matrix}\right]$ $\times\left[2\psi(s)-\psi\left(\frac{s+n+2\lambda+1}{2}\right)-\psi\left(\frac{s-n+1}{2}\right)+\ln\frac{a^2b}{4}+\mathbf{C}\right]$ $\left[a>0;\ \operatorname{Re}\lambda>-1/2;\ \operatorname{Re}s>\left((-1)^n-1\right)/2\right]$
4	$\left(x^2-a^2\right)_+^{\lambda-1/2}$ $\times\operatorname{Ei}\left(-bx^2\right)C_n^{\lambda}\left(\frac{x}{a}\right)$	$-\frac{(a/2)^{s+2\lambda+1}b}{2(n!)}\Gamma\left[\begin{matrix} n+2\lambda,\ \frac{n-s-1}{2},\ -\frac{s+n+2\lambda+1}{2} \\ \lambda,\ -s-1\end{matrix}\right]$ $\times {}_4F_4\left(\begin{matrix}1,\ 1,\ \frac{s+2}{2},\ \frac{s+3}{2};\ -a^2b \\ 2,\ 2,\ \frac{s-n+3}{2},\ \frac{s+n+2\lambda+3}{2}\end{matrix}\right)$ $+\frac{\pi(2/a)^n b^{-(s+n+2\lambda-1)/2}}{(s+n+2\lambda-1)\,n!}\sec\frac{(s+n+2\lambda)\pi}{2}\frac{(\lambda)_n}{\Gamma\left(\frac{3-s-n-2\lambda}{2}\right)}$ $\times {}_3F_3\left(\begin{matrix}\frac{1-n-2\lambda}{2},\ \frac{2-n-2\lambda}{2},\ \frac{1-s-n-2\lambda}{2};\ -a^2b \\ 1-n-\lambda,\ \frac{3-s-n-2\lambda}{2},\ \frac{3-s-n-2\lambda}{2}\end{matrix}\right)$ $+\frac{(a/2)^{s+2\lambda-1}}{2(n!)}\Gamma\left[\begin{matrix} n+2\lambda,\ \frac{1-s+n}{2},\ \frac{1-s-n-2\lambda}{2} \\ \lambda,\ 1-s\end{matrix}\right]$ $\times\left[2\psi(1-s)-\psi\left(\frac{1-s-n-2\lambda}{2}\right)\right.$ $\left.-\psi\left(\frac{1-s+n}{2}\right)+\ln\frac{a^2b}{4}+\mathbf{C}\right]$ $\left[a,\ \operatorname{Re}b>0;\ \operatorname{Re}\lambda>-1/2\right]$

3.24.8. $C_n^{\lambda}(ax+b)$ and $\operatorname{si}(ax)$, $\operatorname{ci}(ax)$

Notation: $\delta=\left\{\begin{matrix}1\\0\end{matrix}\right\}$, $\varepsilon=0$ or 1.

1	$\left(a^2-x^2\right)_+^{\lambda-1/2}$ $\times\operatorname{si}(bx)\,C_n^{\lambda}\left(\frac{x}{a}\right)$	$\frac{\sqrt{\pi}\,(2\lambda)_n a^{s+2\lambda}b}{2^{s+1}n!}\Gamma\left[\begin{matrix}\frac{2\lambda+1}{2},\ s+1 \\ \frac{s-n+2}{2},\ \frac{s+n+2\lambda+2}{2}\end{matrix}\right]$ $\times {}_3F_4\left(\begin{matrix}\frac{1}{2},\ \frac{s+1}{2},\ \frac{s+2}{2};\ -\frac{a^2b^2}{4} \\ \frac{3}{2},\ \frac{3}{2},\ \frac{s-n+2}{2},\ \frac{s+n+2\lambda+2}{2}\end{matrix}\right)$ $-\frac{\pi^{3/2}(2\lambda)_n a^{s+2\lambda-1}}{2^{s+1}n!}\Gamma\left[\begin{matrix}\frac{2\lambda+1}{2},\ s \\ \frac{s-n+1}{2},\ \frac{s+n+2\lambda+1}{2}\end{matrix}\right]$ $\left[a>0;\ \operatorname{Re}\lambda>-1/2;\ \operatorname{Re}s>\left((-1)^n-1\right)/2\right]$

No.	$f(x)$	$F(s)$
2	$(a^2-x^2)_+^{\lambda-1/2}$ $\times \operatorname{ci}(bx)\, C_n^\lambda\left(\frac{x}{a}\right)$	$\frac{\pi}{n!}\left(\frac{a}{2}\right)^{s+2\lambda-1}\Gamma\left[\begin{matrix} n+2\lambda,\ s+2 \\ \lambda,\ \frac{s-n+1}{2},\ \frac{s+n+2\lambda+1}{2}\end{matrix}\right]$ $\times\left\{\frac{4}{s(s+1)}\left[\ln\frac{ab}{2}-\frac{1}{2}\psi\left(\frac{s+n+2\lambda+1}{2}\right)\right.\right.$ $\left.-\frac{1}{2}\psi\left(\frac{s-n+1}{2}\right)+\psi(s)+\mathbf{C}\right]$ $\left.-\frac{a^2b^2}{(s-n+1)(s+n+2\lambda+3)}\,{}_4F_5\left(\begin{matrix} 1,\ 1,\ \frac{s+2}{2},\ \frac{s+3}{2};\ -\frac{a^2b^2}{4} \\ \frac{3}{2},\ 2,\ 2,\ \frac{s-n+3}{2},\ \frac{s+n+2\lambda+3}{2}\end{matrix}\right)\right\}$ $[a>0;\ \operatorname{Re}\lambda>-1/2;\ \operatorname{Re}s>((-1)^n-1)/2]$
3	$(x^2-a^2)_+^{\lambda-1/2}\operatorname{si}(bx)$ $\times C_n^\lambda\left(\frac{x}{a}\right)$	$\frac{a^{s+2\lambda}b}{2^{s+2\lambda+1}n!}\Gamma\left[\begin{matrix} n+2\lambda,\ -\frac{s-n}{2},\ -\frac{s+n+2\lambda}{2} \\ \lambda,\ -s\end{matrix}\right]$ $\times\,{}_3F_4\left(\begin{matrix} \frac{1}{2},\ \frac{s+1}{2},\ \frac{s+2}{2};\ -\frac{a^2b^2}{4} \\ \frac{3}{2},\ \frac{3}{2},\ \frac{s-n+2}{2},\ \frac{s+n+2\lambda+2}{2}\end{matrix}\right)+\frac{2^n(s+n+2\lambda-1)^{-2}}{a^nb^{s+n+2\lambda-1}n!}$ $\times(\lambda)_n\cos\frac{(s+n+2\lambda)\pi}{2}\Gamma(s+n+2\lambda)$ $\times\,{}_3F_4\left(\begin{matrix} \frac{1-n-2\lambda}{2},\ \frac{2-n-2\lambda}{2},\ -\frac{s+n+2\lambda-1}{2};\ -\frac{a^2b^2}{4} \\ 1-n-\lambda,\ -\frac{s+n+2\lambda-2}{2},\ -\frac{s+n+2\lambda-3}{2},\ -\frac{s+n+2\lambda-3}{2}\end{matrix}\right)$ $-\frac{\pi a^{s+2\lambda-1}}{2^{s+2\lambda+1}n!}\Gamma\left[\begin{matrix} n+2\lambda,\ -\frac{s-n-1}{2},\ -\frac{s+n+2\lambda-1}{2} \\ \lambda,\ 1-s\end{matrix}\right]$ $[a,\ b>0;\ \operatorname{Re}\lambda>-1/2;\ \operatorname{Re}(s+2\lambda)<1-n]$
4	$(x^2-a^2)_+^{\lambda-1/2}\operatorname{ci}(bx)$ $\times C_n^\lambda\left(\frac{x}{a}\right)$	$-\frac{a^{s+2\lambda+1}b^2}{2^{s+2\lambda+4}n!}\Gamma\left[\begin{matrix} n+2\lambda,\ -\frac{s-n+1}{2},\ -\frac{s+n+2\lambda+1}{2} \\ \lambda,\ -s-1\end{matrix}\right]$ $\times\,{}_4F_5\left(\begin{matrix} 1,\ 1,\ \frac{s+2}{2},\ \frac{s+3}{2};\ -\frac{a^2b^2}{4} \\ \frac{3}{2},\ 2,\ 2,\ \frac{s-n+3}{2},\ \frac{s+n+2\lambda+3}{2}\end{matrix}\right)-\frac{2^n(s+n+2\lambda-1)^{-2}}{a^nb^{s+n+2\lambda-1}n!}$ $\times(\lambda)_n\sin\frac{(s+n+2\lambda)\pi}{2}\Gamma(s+n+2\lambda)$ $\times\,{}_3F_4\left(\begin{matrix} \frac{1-n-2\lambda}{2},\ \frac{2-n-2\lambda}{2},\ \frac{1-s-n-2\lambda}{2};\ -\frac{a^2b^2}{4} \\ 1-n-\lambda,\ \frac{2-s-n-2\lambda}{2},\ \frac{3-s-n-2\lambda}{2},\ \frac{3-n-s-2\lambda}{2}\end{matrix}\right)$ $+\frac{1}{2(n!)}\left(\frac{a}{2}\right)^{s+2\lambda-1}\Gamma\left[\begin{matrix} n+2\lambda,\ -\frac{s-n-1}{2},\ -\frac{s+n+2\lambda-1}{2} \\ \lambda,\ 1-s\end{matrix}\right]$ $\times\left[\psi(1-s)-\frac{1}{2}\psi\left(\frac{1-s-n-2\lambda}{2}\right)\right.$ $\left.-\frac{1}{2}\psi\left(\frac{1-s+n}{2}\right)+\ln\frac{ab}{2}+\mathbf{C}\right]$ $[a,\ b>0;\ \operatorname{Re}\lambda>-1/2;\ \operatorname{Re}(s+2\lambda)<1-n]$

No.	$f(x)$	$F(s)$
5	$(a-x)_+^{\lambda-1/2}$ $\times \begin{Bmatrix} \operatorname{si}(b\sqrt{x}) \\ \operatorname{ci}(b\sqrt{x}) \end{Bmatrix}$ $\times C_n^{\lambda}\left(\frac{2x}{a}-1\right)$	$\frac{(-1)^{n+1}\, 2^{\delta-2} a^{s+\lambda+(\delta+1)/2} b^{\delta+2}}{3^{2\delta} n!} (2\lambda)_n$ $\times \left(\frac{-2s+2\lambda-\delta-1}{2}\right)_n \Gamma\left[\begin{matrix} \frac{2\lambda+1}{2}, \frac{2s+\delta+2}{2} \\ \frac{2s+2n+2\lambda+\delta+3}{2} \end{matrix}\right]$ $\times {}_4F_5\left(\begin{matrix} 1, \frac{\delta+2}{2}, \frac{2s+\delta+2}{2}, \frac{2s-2\lambda+\delta+3}{2}; -\frac{ab^2}{4} \\ 2, \frac{\delta+4}{2}, \frac{2\delta+3}{2}, \frac{2s-2n-2\lambda+\delta+3}{2}, \frac{2s+2n+2\lambda+\delta+3}{2} \end{matrix}\right)$ $+ \frac{(-1)^n\, a^{s+\lambda+(\delta-1)/2} b^{\delta}}{n!} (2\lambda)_n \left(\frac{-2s+2\lambda-\delta+1}{2}\right)_n$ $\times \Gamma\left[\begin{matrix} \frac{2\lambda+1}{2}, \frac{2s+\delta}{2} \\ \frac{2s+2n+2\lambda+\delta+1}{2} \end{matrix}\right] \left[\frac{1}{2}\psi(s) - \frac{1}{2}\psi\left(\frac{2s+2n+2\lambda+1}{2}\right)\right.$ $\left. - \sum_{i=0}^{n-1} \frac{1}{2i-2s+2\lambda+1} + \frac{1}{2}\ln(ab^2) + \mathbf{C}\right]^{1-\delta}$ $- \delta \frac{(-1)^n\, \pi a^{s+\lambda-1/2}}{2(n!)} (2\lambda)_n \left(\frac{1-2s+2\lambda}{2}\right)_n \Gamma\left[\begin{matrix} \frac{2\lambda+1}{2}, s \\ \frac{2s+2n+2\lambda+1}{2} \end{matrix}\right]$ $[a,\ \operatorname{Re} s > 0;\ \operatorname{Re}\lambda > -1/2]$

3.24.9. $C_n^{\lambda}(ax+b)$ and $\operatorname{erf}(ax)$, $\operatorname{erfc}(ax)$

Notation: $\delta = \begin{Bmatrix} 1 \\ 0 \end{Bmatrix}$, $\varepsilon = 0$ or 1.

1	$\operatorname{erfc}(ax)\, C_{2n+\varepsilon}^{\lambda}(bx)$	$(-1)^n \frac{2^{\varepsilon} (\lambda)_{n+\varepsilon} a^{-s-\varepsilon} b^{\varepsilon}}{\sqrt{\pi}\,(s+\varepsilon)\, n!} \Gamma\left(\frac{s+\varepsilon+1}{2}\right)$ $\times {}_4F_2\left(\begin{matrix} -n, n+\lambda+\varepsilon, \frac{s+1}{2}, \frac{s+2\varepsilon}{2} \\ \frac{2\varepsilon+1}{2}, \frac{s+\varepsilon+2}{2}; \frac{b^2}{a^2} \end{matrix}\right)$ $[\operatorname{Re}\lambda > -1/2;\ \operatorname{Re} s > -\varepsilon;\ \lvert\arg a\rvert < \pi/4]$
2	$(a^2-x^2)_+^{\lambda-1/2} \operatorname{erf}(bx)$ $\times C_n^{\lambda}\left(\frac{x}{a}\right)$	$\frac{\sqrt{\pi} a^{s+2\lambda} b}{2^{s+2\lambda-1} n!} \Gamma\left[\begin{matrix} n+2\lambda, s+1 \\ \lambda, \frac{s-n+2}{2}, \frac{s+n+2\lambda+2}{2} \end{matrix}\right] {}_3F_3\left(\begin{matrix} \frac{1}{2}, \frac{s+1}{2}, \frac{s+2}{2}; -a^2b^2 \\ \frac{3}{2}, \frac{s-n+2}{2}, \frac{s+n+2\lambda+2}{2} \end{matrix}\right)$ $[a > 0;\ \operatorname{Re}\lambda > -1/2; \operatorname{Re} s > ((-1)^n - 3)/2]$
3	$(x^2-a^2)_+^{\lambda-1/2}$ $\times \begin{Bmatrix} \operatorname{erf}(bx) \\ \operatorname{erfc}(bx) \end{Bmatrix} C_n^{\lambda}\left(\frac{x}{a}\right)$	$\pm \frac{(a/2)^{s+2\lambda}\, b}{\sqrt{\pi}\, n!} \Gamma\left[\begin{matrix} n+2\lambda, \frac{n-s}{2}, -\frac{s+n+2\lambda}{2} \\ \lambda, -s \end{matrix}\right]$ $\times {}_3F_3\left(\begin{matrix} \frac{1}{2}, \frac{s+1}{2}, \frac{s+2}{2}; -a^2b^2 \\ \frac{3}{2}, \frac{s-n+2}{2}, \frac{s+n+2\lambda+2}{2} \end{matrix}\right) \mp \frac{a^{-n} b^{-s-n-2\lambda+1}}{2^{s+2\lambda-1} \pi\, n!} (\lambda)_n$ $\times \cos\frac{(s+n+2\lambda)\pi}{2} \Gamma\left(\frac{1-s-n-2\lambda}{2}\right) \Gamma(s+n+2\lambda-1) \times$

No.	$f(x)$	$F(s)$
		$\times\,{}_3F_3\left(\begin{matrix}\frac{1-n-2\lambda}{2},\ \frac{2-n-2\lambda}{2},\ \frac{1-s-n-2\lambda}{2};\ -a^2b^2\\ 1-n-\lambda,\ \frac{2-s-n-2\lambda}{2},\ \frac{3-s-n-2\lambda}{2}\end{matrix}\right)$ $+(1-\delta)\frac{(a/2)^{s+2\lambda-1}}{2(n!)}\Gamma\left[\begin{matrix}n+2\lambda,\ \frac{1-s+n}{2},\ -\frac{s+n+2\lambda-1}{2}\\ \lambda,\ 1-s\end{matrix}\right]$ $[a,\ \operatorname{Re} b>0;\ \operatorname{Re}\lambda>-1/2;\ \operatorname{Re}(s+2\lambda)<1-n \text{ for erf}]$
4	$(a^2-x^2)_+^{\lambda-1/2}e^{b^2x^2}$ $\times\operatorname{erf}(bx)\,C_n^\lambda\left(\frac{x}{a}\right)$	$\frac{\sqrt{\pi}a^{s+2\lambda}b}{2^{s+2\lambda-1}n!}\Gamma\left[\begin{matrix}n+2\lambda,\ s+1\\ \lambda,\ \frac{s-n+2}{2},\ \frac{s+n+2\lambda+2}{2}\end{matrix}\right]{}_3F_3\left(\begin{matrix}1,\ \frac{s+1}{2},\ \frac{s+2}{2};\ a^2b^2\\ \frac{3}{2},\ \frac{s-n+2}{2},\ \frac{s+n+2\lambda+2}{2}\end{matrix}\right)$ $[a>0;\ \operatorname{Re}\lambda>-1/2;\ \operatorname{Re}s>((-1)^n-3)/2]$
5	$\operatorname{erfc}(bx)\,C_n^\lambda\left(\frac{x}{a}+1\right)$	$\frac{n(n+2\lambda)a^{-1}b^{-s-1}}{(2\lambda+1)\sqrt{\pi}(s+1)n!}(2\lambda)_n\Gamma\left(\frac{s+2}{2}\right)$ $\times\,{}_6F_4\left(\begin{matrix}\frac{1-n}{2},\ \frac{2-n}{2},\ \frac{n+2\lambda+1}{2},\ \frac{n+2\lambda+2}{2},\ \frac{s+1}{2},\ \frac{s+2}{2}\\ \frac{3}{2},\ \frac{2\lambda+3}{4},\ \frac{2\lambda+5}{4},\ \frac{s+3}{2};\ \frac{1}{4a^2b^2}\end{matrix}\right)$ $+\frac{b^{-s}(2\lambda)_n}{\sqrt{\pi}\,s\,n!}\Gamma\left(\frac{s+1}{2}\right){}_6F_4\left(\begin{matrix}-\frac{n}{2},\ \frac{1-n}{2},\ \frac{n+2\lambda}{2},\ \frac{n+2\lambda+1}{2},\ \frac{s}{2},\ \frac{s+1}{2}\\ \frac{1}{2},\ \frac{2\lambda+1}{4},\ \frac{2\lambda+3}{4},\ \frac{s+2}{2};\ \frac{1}{4a^2b^2}\end{matrix}\right)$ $[\operatorname{Re}s>0;\ \lvert\arg b\rvert<\pi/4]$
6	$\operatorname{erfc}(b\sqrt{x})\,C_n^\lambda\left(\frac{x}{a}-1\right)$	$\frac{(2/a)^n b^{-2s-2n}}{\sqrt{\pi}(s+n)\,n!}(\lambda)_n\Gamma\left(s+n+\frac{1}{2}\right)$ $\times\,{}_3F_3\left(\begin{matrix}-n,\ \frac{1}{2}-n-\lambda,\ -s-n;\ -2ab^2\\ 1-2n-2\lambda,\ \frac{1}{2}-s-n,\ 1-s-n\end{matrix}\right)$ $[\operatorname{Re}\lambda>-1/2;\ \operatorname{Re}s>((-1)^n-1)/2;\ \lvert\arg b\rvert<\pi/4]$
7	$(a-x)_+^{\lambda-1/2}$ $\times\left\{\begin{matrix}\operatorname{erf}(b\sqrt{x})\\ \operatorname{erfc}(b\sqrt{x})\end{matrix}\right\}$ $\times C_n^\lambda\left(\frac{2x}{a}-1\right)$	$\pm\frac{2(-1)^n a^{s+\lambda}b}{\sqrt{\pi}\,n!}(2\lambda)_n(\lambda-s)_n\Gamma\left[\begin{matrix}\frac{2\lambda+1}{2},\ \frac{2s+1}{2}\\ s+n+\lambda+1\end{matrix}\right]$ $\times\,{}_3F_3\left(\begin{matrix}\frac{1}{2},\ \frac{2s+1}{2},\ s-\lambda+1;\ -ab^2\\ \frac{3}{2},\ s-n-\lambda+1,\ s+n+\lambda+1\end{matrix}\right)$ $+\left\{\begin{matrix}0\\ 1\end{matrix}\right\}\frac{(-1)^n a^{s+\lambda-1/2}}{n!}(2\lambda)_n\left(\lambda-s+\frac{1}{2}\right)_n\Gamma\left[\begin{matrix}\frac{2\lambda+1}{2},\ s\\ s+n+\lambda+\frac{1}{2}\end{matrix}\right]$ $[a>0;\ \operatorname{Re}\lambda>-1/2;\ \operatorname{Re}s>-(1\pm1)/4]$
8	$(a-x)_+^{\lambda-1/2}e^{b^2x}$ $\times\operatorname{erf}(b\sqrt{x})\,C_n^\lambda\left(\frac{2x}{a}-1\right)$	$\frac{2(-1)^n a^{s+\lambda}b}{\sqrt{\pi}\,n!}(2\lambda)_n(\lambda-s)_n\Gamma\left[\begin{matrix}\frac{2\lambda+1}{2},\ \frac{2s+1}{2}\\ s+n+\lambda+1\end{matrix}\right]$ $\times\,{}_3F_3\left(\begin{matrix}1,\ \frac{2s+1}{2},\ s-\lambda+1;\ ab^2\\ \frac{3}{2},\ s-n-\lambda+1,\ s+n+\lambda+1\end{matrix}\right)$ $[a>0;\ \operatorname{Re}\lambda>-1/2;\ \operatorname{Re}s>-1/2]$

3.24.10. $C_n^\lambda(bx)$ and $\Gamma(\nu, ax)$, $\gamma(\nu, ax)$

Notation: $\varepsilon = 0$ or 1.

1	$\Gamma(\nu, ax)\, C_{2n+\varepsilon}^\lambda(bx)$	$\dfrac{(-1)^n(\varepsilon+1)(\lambda)_{n+\varepsilon}\, a^{-s-\varepsilon} b^\varepsilon}{n!\,(s+\varepsilon)}\Gamma(s+\nu+\varepsilon) \times {}_5F_2\left(\begin{matrix}-n,\, n+\lambda+\varepsilon,\, \frac{s+\varepsilon}{2},\, \frac{s+\nu+\varepsilon}{2},\, \frac{s+\nu+\varepsilon+1}{2} \\ \frac{2\varepsilon+1}{2},\, \frac{s+\varepsilon+2}{2};\, \frac{4b^2}{a^2}\end{matrix}\right)$ $[\operatorname{Re} a > 0;\ \operatorname{Re} s > -\operatorname{Re}\nu - \varepsilon,\, 0]$
2	$(a^2-x^2)_+^{\lambda-1/2} e^{bx} \times \gamma(\nu, bx)\, C_n^\lambda\left(\dfrac{x}{a}\right)$	$\dfrac{\pi}{n!\,\nu}\left(\dfrac{a}{2}\right)^{s+2\lambda+\nu-1} b^\nu\, \Gamma\left[\begin{matrix}n+2\lambda,\, s+\nu \\ \lambda,\, \frac{s-n+\nu+1}{2},\, \frac{s+n+2\lambda+\nu+1}{2}\end{matrix}\right] \times {}_3F_4\left(\begin{matrix}1,\, \frac{s+\nu}{2},\, \frac{s+\nu+1}{2};\, \frac{a^2b^2}{4} \\ \frac{\nu+1}{2},\, \frac{\nu+2}{2},\, \frac{s-n+\nu+1}{2},\, \frac{s+n+2\lambda+\nu+1}{2}\end{matrix}\right)$ $+\dfrac{\pi}{n!\,\nu(\nu+1)}\left(\dfrac{a}{2}\right)^{s+2\lambda+\nu} b^{\nu+1}\, \Gamma\left[\begin{matrix}n+2\lambda,\, s+\nu+1 \\ \lambda,\, \frac{s-n+\nu+2}{2},\, \frac{s+n+2\lambda+\nu+2}{2}\end{matrix}\right] \times {}_3F_4\left(\begin{matrix}1,\, \frac{s+\nu+1}{2},\, \frac{s+\nu+2}{2};\, \frac{a^2b^2}{4} \\ \frac{\nu+2}{2},\, \frac{\nu+3}{2},\, \frac{s-n+\nu+2}{2},\, \frac{s+n+2\lambda+\nu+2}{2}\end{matrix}\right)$ $[a > 0;\ \operatorname{Re} s > -\operatorname{Re}\nu - \varepsilon]$
3	$(a^2-x^2)_+^{\lambda-1/2} \times \gamma(\nu, b^2x^2)\, C_n^\lambda\left(\dfrac{x}{a}\right)$	$\dfrac{\pi}{n!\,\nu}\left(\dfrac{a}{2}\right)^{s+2\lambda+2\nu-1} b^{2\nu}\, \Gamma\left[\begin{matrix}n+2\lambda,\, s+2\nu \\ \lambda,\, \frac{s-n+2\nu+1}{2},\, \frac{s+n+2\lambda+2\nu+1}{2}\end{matrix}\right] \times {}_3F_3\left(\begin{matrix}\nu,\, \frac{s+2\nu}{2},\, \frac{s+2\nu+1}{2};\, -a^2b^2 \\ \nu+1,\, \frac{s-n+2\nu+1}{2},\, \frac{s+n+2\lambda+2\nu+1}{2}\end{matrix}\right)$ $[a > 0;\ \operatorname{Re} s > -\operatorname{Re}\nu - \varepsilon]$
4	$(a^2-x^2)_+^{\lambda-1/2} e^{b^2x^2} \times \gamma(\nu, b^2x^2)\, C_n^\lambda\left(\dfrac{x}{a}\right)$	$\dfrac{\pi}{n!\,\nu}\left(\dfrac{a}{2}\right)^{s+2\lambda+2\nu-1} b^{2\nu}\, \Gamma\left[\begin{matrix}n+2\lambda,\, s+2\nu \\ \lambda,\, \frac{s-n+2\nu+1}{2},\, \frac{s+n+2\lambda+2\nu+1}{2}\end{matrix}\right] \times {}_3F_3\left(\begin{matrix}1,\, \frac{s+2\nu}{2},\, \frac{s+2\nu+1}{2};\, a^2b^2 \\ \nu+1,\, \frac{s-n+2\nu+1}{2},\, \frac{s+n+2\lambda+2\nu+1}{2}\end{matrix}\right)$ $[a > 0;\ \operatorname{Re} s > -\operatorname{Re}\nu - \varepsilon]$

3.24.11. $C_n^\lambda(bx)$ and Bessel functions

Notation: $\varepsilon = 0$ or 1.

1	$(a^2-x^2)_+^{\lambda-1/2} \times \begin{Bmatrix} J_\nu(bx) \\ I_\nu(bx)\end{Bmatrix} C_n^\lambda\left(\dfrac{x}{a}\right)$	$\dfrac{\pi}{n!}\left(\dfrac{a}{2}\right)^{s+2\lambda+\nu-1}\left(\dfrac{b}{2}\right)^\nu \Gamma\left[\begin{matrix}n+2\lambda,\, s+\nu \\ \lambda,\, \nu+1,\, \frac{s-n+\nu+1}{2},\, \frac{s+n+2\lambda+\nu+1}{2}\end{matrix}\right] \times {}_2F_3\left(\begin{matrix}\frac{s+\nu}{2},\, \frac{s+\nu+1}{2};\, \mp\frac{a^2b^2}{4} \\ \nu+1,\, \frac{s-n+\nu+1}{2},\, \frac{s+n+2\lambda+\nu+1}{2}\end{matrix}\right)$ $[a > 0;\ \operatorname{Re}\lambda > -1/2;\ \operatorname{Re}(s+\nu) > ((-1)^n - 1)/2]$

No.	$f(x)$	$F(s)$
2	$\left(x^2-a^2\right)_+^{\lambda-1/2} J_\nu(bx)$ $\times C_n^\lambda\left(\frac{x}{a}\right)$	$\frac{a^{s+2\lambda+\nu-1}b^\nu}{2^{s+2\lambda+2\nu}\,n!}\,\Gamma\left[\begin{matrix} n+2\lambda,\ \frac{1-s+n-\nu}{2},\ \frac{1-s-n-2\lambda-\nu}{2} \\ \lambda,\ \nu+1,\ 1-s-\nu \end{matrix}\right]$ $\times\,{}_2F_3\left(\begin{matrix} \frac{s+\nu}{2},\ \frac{s+\nu+1}{2};\ -\frac{a^2b^2}{4} \\ \nu+1,\ \frac{s-n+\nu+1}{2},\ \frac{s+n+2\lambda+\nu+1}{2} \end{matrix}\right)$ $+\,\frac{2^{s+2n+2\lambda-2}}{a^n b^{s+n+2\lambda-1}\,n!}\,(\lambda)_n\,\Gamma\left[\begin{matrix} \frac{s+n+2\lambda+\nu-1}{2} \\ \frac{3-s-n-2\lambda+\nu}{2} \end{matrix}\right]$ $\times\,{}_2F_3\left(\begin{matrix} \frac{1-n-2\lambda}{2},\ \frac{2-n-2\lambda}{2};\ -\frac{a^2b^2}{4} \\ 1-n-\lambda,\ \frac{3-s-n-2\lambda-\nu}{2},\ \frac{3-s-n-2\lambda+\nu}{2} \end{matrix}\right)$ $[a,\,b>0;\ \operatorname{Re}\lambda>-1/2;\ \operatorname{Re}(s+2\lambda)<3/2-n]$
3	$\left(a^2-x^2\right)_+^{\lambda-1/2} J_\nu\left(\frac{b}{x}\right)$ $\times C_{2n+\varepsilon}^\lambda\left(\frac{x}{a}\right)$	$\frac{(-1)^n\,a^{s+2\lambda-\nu-1}b^\nu}{2^{\nu+1}\,(2n+\varepsilon)!}\,(2\lambda)_{2n+\varepsilon}\left(\frac{1-s+\nu+\varepsilon}{2}\right)_n$ $\times\,\Gamma\left[\begin{matrix} \frac{2\lambda+1}{2},\ \frac{s-\nu+\varepsilon}{2} \\ \nu+1,\ \frac{s+2n+2\lambda-\nu+\varepsilon+1}{2} \end{matrix}\right]\,{}_2F_3\left(\begin{matrix} \frac{1-s+2n+\nu+\varepsilon}{2},\ \frac{1-s-2n-2\lambda+\nu-\varepsilon}{2} \\ \nu+1,\ \frac{1-s+\nu}{2},\ \frac{2-s+\nu}{2};\ -\frac{b^2}{4a^2} \end{matrix}\right)$ $+\,\frac{(-1)^n\,a^{2\lambda-\varepsilon-1}b^{s+\varepsilon}}{2^{s+1}\,n!}\,(\lambda)_{n+\varepsilon}\,\Gamma\left[\begin{matrix} -\frac{s-\nu+\varepsilon}{2} \\ \frac{s+\nu+\varepsilon+2}{2} \end{matrix}\right]$ $\times\,{}_2F_3\left(\begin{matrix} \frac{2n+2\varepsilon+1}{2},\ \frac{1-2\lambda-2n}{2};\ -\frac{b^2}{4a^2} \\ \frac{2\varepsilon+1}{2},\ \frac{s-\nu+\varepsilon+2}{2},\ \frac{s+\nu+\varepsilon+2}{2} \end{matrix}\right)$ $[a>0;\ \operatorname{Re}\lambda>-1/2;\ \operatorname{Re}s>-\varepsilon-3/2]$
4	$\left(x^2-a^2\right)_+^{\lambda-1/2} J_\nu\left(\frac{b}{x}\right)$ $\times C_{2n+\varepsilon}^\lambda\left(\frac{x}{a}\right)$	$\frac{a^{s+2\lambda-\nu-1}b^\nu}{2^{s+2\lambda}\,n!}\,\Gamma\left[\begin{matrix} n+2\lambda,\ \frac{1-s+n+\nu}{2},\ -\frac{s+n+2\lambda-\nu-1}{2} \\ \lambda,\ \nu+1,\ 1-s+\nu \end{matrix}\right]$ $\times\,{}_2F_3\left(\begin{matrix} \frac{1-s+n+\nu}{2},\ -\frac{s+n+2\lambda-\nu-1}{2} \\ \nu+1,\ \frac{1-s+\nu}{2},\ \frac{2-s+\nu}{2};\ -\frac{b^2}{4a^2} \end{matrix}\right)$ $[a,\,b>0;\ \operatorname{Re}\lambda>-1/2;\ \operatorname{Re}(s+2\lambda-\nu)<1-2n-\varepsilon]$
5	$J_\nu\left(b\sqrt{x}\right)C_n^\lambda\left(\frac{x\pm a}{a}\right)$	$\frac{2^{2s+3n}}{a^n b^{2s+2n}n!}\,(\lambda)_n\,\Gamma\left[\begin{matrix} \frac{2s+2n+\nu}{2} \\ \frac{2-2s-2n+\nu}{2} \end{matrix}\right]$ $\times\,{}_2F_3\left(\begin{matrix} -n,\ \frac{1}{2}-n-\lambda;\ \pm\frac{ab^2}{2} \\ 1-2n-2\lambda,\ \frac{2-2s-2n-\nu}{2},\ \frac{2-2s-2n+\nu}{2} \end{matrix}\right)$ $[b>0;\ \operatorname{Re}\lambda>-1/2;\ -\operatorname{Re}\nu/2<\operatorname{Re}s<3/4-n]$
6	$K_\nu\left(b\sqrt{x}\right)C_n^\lambda\left(\frac{x\pm a}{a}\right)$	$\frac{2^{2s+3n-1}}{a^n b^{2s+2n}n!}\,(\lambda)_n\,\Gamma\left(s+n-\frac{\nu}{2}\right)\Gamma\left(s+n+\frac{\nu}{2}\right)$ $\times\,{}_2F_3\left(\begin{matrix} -n,\ \frac{1}{2}-n-\lambda;\ \mp\frac{ab^2}{2} \\ 1-2n-2\lambda,\ \frac{2-2s-2n-\nu}{2},\ \frac{2-2s-2n+\nu}{2} \end{matrix}\right)$ $[b>0;\ \operatorname{Re}s>\lvert\operatorname{Re}\nu\rvert/2]$

No.	$f(x)$	$F(s)$
7	$J_\nu\left(\frac{b}{\sqrt{x}}\right) C_n^\lambda\left(\frac{x \pm a}{a}\right)$	$\frac{(\pm 1)^n}{n!}\left(\frac{b}{2}\right)^{2s}(2\lambda)_n \Gamma\left[\begin{matrix}\frac{\nu-2s}{2} \\ \frac{2s+\nu+2}{2}\end{matrix}\right] {}_2F_3\left(\begin{matrix}-n,\, n+2\lambda;\, \pm\frac{b^2}{8a} \\ \frac{2\lambda+1}{2},\, \frac{2s-\nu+2}{2},\, \frac{2s+\nu+2}{2}\end{matrix}\right)$ $[b>0;\ \operatorname{Re}\lambda > -1/2;\ -3/4 < \operatorname{Re}s < \operatorname{Re}\nu/2 - n]$
8	$K_\nu\left(\frac{b}{\sqrt{x}}\right) C_n^\lambda\left(\frac{x \pm a}{a}\right)$	$\frac{(\pm 1)^n}{2(n!)}\left(\frac{b}{2}\right)^{2s}(2\lambda)_n \Gamma\left(\frac{\nu-2s}{2}\right)\Gamma\left(\frac{-\nu-2s}{2}\right)$ $\times {}_2F_3\left(\begin{matrix}-n,\, n+2\lambda;\, \mp\frac{b^2}{8a} \\ \frac{2\lambda+1}{2},\, \frac{2s-\nu+2}{2},\, \frac{2s+\nu+2}{2}\end{matrix}\right)$ $[\operatorname{Re}b>0;\ \operatorname{Re}s < -n - \lvert\operatorname{Re}\nu\rvert/2]$
9	$(x+a)^n K_\nu(b\sqrt{x})$ $\times C_n^\lambda\left(\frac{a-x}{a+x}\right)$	$\frac{(-1)^n 2^{2s+2n-1} b^{-2s-2n}}{n!}(2\lambda)_n \Gamma\left(s+n-\frac{\nu}{2}\right)\Gamma\left(s+n+\frac{\nu}{2}\right)$ $\times {}_2F_3\left(\begin{matrix}-n,\, \frac{1-2n-2\lambda}{2};\, -\frac{ab^2}{4} \\ \frac{2\lambda+1}{2},\, \frac{2-2s-2n-\nu}{2},\, \frac{2-2s-2n+\nu}{2}\end{matrix}\right)$ $[\operatorname{Re}b>0;\ \operatorname{Re}s > \lvert\operatorname{Re}\nu\rvert/2]$
10	$(x+a)^{\lambda-1/2} K_\nu(b\sqrt{x})$ $\times C_n^\lambda\left(\frac{2x}{a}+1\right)$	$\frac{(-1)^n a^{s+\lambda-(\nu+1)/2} b^{-\nu}}{2^{-\nu+1} n!}(2\lambda)_n \left(\frac{1-2s+2\lambda+\nu}{2}\right)_n$ $\times \Gamma\left[\begin{matrix}\frac{2s-\nu}{2},\, \frac{1-2s-2n-2\lambda+\nu}{2} \\ \frac{1-2\lambda}{2}\end{matrix}\right]$ $\times {}_2F_3\left(\begin{matrix}\frac{2s-\nu}{2},\, \frac{2s-2\lambda-\nu+1}{2};\, -\frac{ab^2}{4} \\ 1-\nu,\, \frac{s-2n-2\lambda-\nu+1}{2},\, \frac{s+2n+2\lambda-\nu+1}{2}\end{matrix}\right)$ $+\frac{(-1)^n a^{s+\lambda+(\nu-1)/2} b^{\nu}}{2^{\nu+1} n!}(2\lambda)_n \left(\frac{1-2s+2\lambda-\nu}{2}\right)_n$ $\times \Gamma\left[\begin{matrix}\frac{2s+\nu}{2},\, \frac{1-2s-2n-2\lambda-\nu}{2} \\ \frac{1-2\lambda}{2}\end{matrix}\right]$ $\times {}_2F_3\left(\begin{matrix}\frac{2s+\nu}{2},\, \frac{2s-2\lambda+\nu+1}{2};\, -\frac{ab^2}{4} \\ \nu+1,\, \frac{s-2n-2\lambda+\nu+1}{2},\, \frac{s+2n+2\lambda+\nu+1}{2}\end{matrix}\right)$ $+\frac{2^{2s+4n+2\lambda-2}}{n!\, a^n b^{2s+2n+2\lambda-1}}(\lambda)_n$ $\times \Gamma\left(\frac{2s+2n+2\lambda-\nu-1}{2}\right)\Gamma\left(\frac{2s+2n+2\lambda+\nu-1}{2}\right)$ $\times {}_2F_3\left(\begin{matrix}\frac{-2n-2\lambda+1}{2},\, -n-2\lambda+1;\, -\frac{ab^2}{4} \\ 1-2n-2\lambda,\, \frac{3-2s-2n-2\lambda-\nu}{2},\, \frac{3-2s-2n-2\lambda+\nu}{2}\end{matrix}\right)$ $[\operatorname{Re}b>0;\ \operatorname{Re}s > \lvert\operatorname{Re}\nu\rvert/2;\ \lvert\arg a\rvert < \pi]$

No.	$f(x)$	$F(s)$
11	$(a-x)_+^{\lambda-1/2}K_\nu(b\sqrt{x})$ $\times C_n^\lambda\left(1-\frac{2x}{a}\right)$	$\frac{a^{s+\lambda-(\nu+1)/2}b^{-\nu}}{2^{-\nu+1}n!}(2\lambda)_n\left(\frac{1-2s+2\lambda+\nu}{2}\right)_n$ $\times\Gamma\left[\begin{matrix}\frac{2\lambda+1}{2},\ \nu,\ \frac{2s-\nu}{2}\\ \frac{2s+2n+2\lambda-\nu+1}{2}\end{matrix}\right]{}_2F_3\left(\begin{matrix}\frac{2s-\nu}{2},\ \frac{2s-2\lambda-\nu+1}{2};\ \frac{ab^2}{4}\\ 1-\nu,\ \frac{2s-2n-2\lambda-\nu+1}{2},\ \frac{2s+2n+2\lambda-\nu+1}{2}\end{matrix}\right)$ $+\frac{a^{s+\lambda+(\nu-1)/2}b^{\nu}}{2^{\nu+1}n!}(2\lambda)_n\left(\frac{1-2s+2\lambda-\nu}{2}\right)_n$ $\times\Gamma\left[\begin{matrix}\frac{2\lambda+1}{2},\ -\nu,\ \frac{2s+\nu}{2}\\ \frac{2s+2n+2\lambda+\nu+1}{2}\end{matrix}\right]{}_2F_3\left(\begin{matrix}\frac{2s+\nu}{2},\ \frac{2s-2\lambda+\nu+1}{2};\ \frac{ab^2}{4}\\ \nu+1,\ \frac{2s-2n-2\lambda+\nu+1}{2},\ \frac{2s+2n+2\lambda+\nu+1}{2}\end{matrix}\right)$ $[a>0;\ \operatorname{Re}\lambda>-1/2;\ \operatorname{Re}s>\lvert\operatorname{Re}\nu\rvert/2]$
12	$(x+a)^{\lambda-1/2}K_\nu\left(\frac{b}{\sqrt{x}}\right)$ $\times C_n^\lambda\left(\frac{2x}{a}+1\right)$	$\frac{a^{\lambda-1/2}b^{2s}}{2^{2s+1}n!}(2\lambda)_n\Gamma\left(-s-\frac{\nu}{2}\right)\Gamma\left(-s+\frac{\nu}{2}\right)$ $\times{}_2F_3\left(\begin{matrix}\frac{-2n-2\lambda+1}{2},\ \frac{2n+2\lambda+1}{2};\ -\frac{b^2}{4a}\\ \frac{2\lambda+1}{2},\ \frac{2s-\nu+2}{2},\ \frac{2s+\nu+2}{2}\end{matrix}\right)$ $+\frac{(-1)^n a^{s+\lambda+(\nu-1)/2}b^{-\nu}}{2^{-\nu+1}n!}(2\lambda)_n\left(\frac{1-2s+2\lambda-\nu}{2}\right)_n$ $\times\Gamma\left[\begin{matrix}\nu,\ \frac{2s+\nu}{2},\ \frac{1-2s-2n-2\lambda-\nu}{2}\\ \frac{1-2\lambda}{2}\end{matrix}\right]$ $\times{}_2F_3\left(\begin{matrix}\frac{1-2s+2n+2\lambda-\nu}{2},\ \frac{1-2s-2n-2\lambda-\nu}{2}\\ 1-\nu,\ \frac{2-2s-\nu}{2},\ \frac{1-2s+2\lambda-\nu}{2};\ -\frac{b^2}{4a}\end{matrix}\right)$ $+\frac{(-1)^n a^{s+\lambda-(\nu+1)/2}b^{\nu}}{2^{\nu+1}n!}(2\lambda)_n\left(\frac{1-2s+2\lambda+\nu}{2}\right)_n$ $\times\Gamma\left[\begin{matrix}-\nu,\ \frac{2s-\nu}{2},\ \frac{1-2s-2n-2\lambda+\nu}{2}\\ \frac{1-2\lambda}{2}\end{matrix}\right]$ $\times{}_2F_3\left(\begin{matrix}\frac{1-2s+2n+2\lambda+\nu}{2},\ \frac{1-2s-2n-2\lambda+\nu}{2}\\ \nu+1,\ \frac{2-2s+\nu}{2},\ \frac{1-2s+2\lambda+\nu}{2};\ -\frac{b^2}{4a}\end{matrix}\right)$ $[\operatorname{Re}b>0;\ \operatorname{Re}s<(1-2n-2\operatorname{Re}\lambda-\lvert\operatorname{Re}\nu\rvert)/2;\ \lvert\arg a\rvert<\pi]$
13	$(a-x)_+^{\lambda-1/2}$ $\times\left\{\begin{matrix}J_\nu(b\sqrt{x})\\ I_\nu(b\sqrt{x})\end{matrix}\right\}$ $\times C_n^\lambda\left(\frac{2x}{a}-1\right)$	$\frac{(-1)^n a^{s+\lambda+(\nu-1)/2}b^{\nu}}{2^{\nu}n!}(2\lambda)_n\left(\frac{1-2s+2\lambda-\nu}{2}\right)_n$ $\times\Gamma\left[\begin{matrix}\frac{2\lambda+1}{2},\ \frac{2s+\nu}{2}\\ \nu+1,\ \frac{2s+2n+2\lambda+\nu+1}{2}\end{matrix}\right]$ $\times{}_2F_3\left(\begin{matrix}\frac{2s+\nu}{2},\ \frac{2s-2\lambda+\nu+1}{2};\ \mp\frac{ab^2}{4}\\ \nu+1,\ \frac{2s-2n-2\lambda+\nu+1}{2},\ \frac{2s+2n+2\lambda+\nu+1}{2}\end{matrix}\right)$ $[a>0;\ \operatorname{Re}\lambda>-1/2;\ \operatorname{Re}s>-\operatorname{Re}\nu/2]$

No.	$f(x)$	$F(s)$
14	$(x+a)^{\lambda-1/2} J_\nu(b\sqrt{x})$ $\times C_n^\lambda\left(\frac{2x}{a}+1\right)$	$\frac{(-1)^n a^{s+\lambda+(\nu-1)/2} b^\nu}{2^\nu n!}(2\lambda)_n\left(\frac{1-2s+2\lambda-\nu}{2}\right)_n$ $\times\Gamma\left[\begin{matrix}\frac{2s+\nu}{2},\ \frac{1-2s-2n-2\lambda-\nu}{2}\\ \frac{1-2\lambda}{2},\ \nu+1\end{matrix}\right]$ $\times {}_2F_3\left(\begin{matrix}\frac{2s+\nu}{2},\ \frac{2s-2\lambda+\nu+1}{2};\ \frac{ab^2}{4}\\ \nu+1,\ \frac{2s-2n-2\lambda+\nu+1}{2},\ \frac{2s+2n+2\lambda+\nu+1}{2}\end{matrix}\right)$ $+\frac{2^{2s+4n+2\lambda-1}}{n!\,a^n b^{2s+2n+2\lambda-1}}(\lambda)_n\Gamma\left[\begin{matrix}\frac{2s+2n+2\lambda+\nu-1}{2}\\ \frac{3-2s-2n-2\lambda+\nu}{2}\end{matrix}\right]$ $\times {}_2F_3\left(\begin{matrix}\frac{1-2n-2\lambda}{2},\ 1-n-2\lambda;\ \frac{ab^2}{4}\\ 1-2n-2\lambda,\ \frac{3-2s-2n-2\lambda-\nu}{2},\ \frac{3-2s-2n-2\lambda+\nu}{2}\end{matrix}\right)$ $[b>0;\ -\operatorname{Re}\nu/2<\operatorname{Re}s<5/4-n-\operatorname{Re}\lambda;\ \lvert\arg a\rvert<\pi]$
15	$(a-x)_+^{\lambda-1/2} J_\nu\left(\frac{b}{\sqrt{x}}\right)$ $\times C_n^\lambda\left(\frac{2x}{a}-1\right)$	$\frac{(-1)^n a^{s+\lambda-(\nu+1)/2} b^\nu}{2^\nu n!}(2\lambda)_n\left(\frac{1-2s+2\lambda+\nu}{2}\right)_n$ $\times\Gamma\left[\begin{matrix}\frac{2\lambda+1}{2},\ \frac{2s-\nu}{2}\\ \nu+1,\ \frac{2s+2n+2\lambda-\nu+1}{2}\end{matrix}\right]{}_2F_3\left(\begin{matrix}\frac{1-2s-2n-2\lambda+\nu}{2},\ \frac{1-2s+2n+2\lambda+\nu}{2}\\ \nu+1,\ \frac{2-2s+\nu}{2},\ \frac{1-2s+2\lambda+\nu}{2};\ -\frac{b^2}{4a}\end{matrix}\right)$ $+\frac{(-1)^n a^{\lambda-1/2} b^{2s}}{2^{2s} n!}(2\lambda)_n\Gamma\left[\begin{matrix}\frac{\nu-2s}{2}\\ \frac{2s+\nu+2}{2}\end{matrix}\right]$ $\times {}_2F_3\left(\begin{matrix}\frac{-2n-2\lambda+1}{2},\ \frac{2n+2\lambda+1}{2};\ -\frac{b^2}{4a}\\ \frac{2\lambda+1}{2},\ \frac{2s-\nu+2}{2},\ \frac{2s+\nu+2}{2}\end{matrix}\right)$ $[a>0;\ \operatorname{Re}\lambda>-1/2;\ \operatorname{Re}s>-3/4]$
16	$(x+a)^{\lambda-1/2} J_\nu\left(\frac{b}{\sqrt{x}}\right)$ $\times C_n^\lambda\left(\frac{2x}{a}+1\right)$	$\frac{(-1)^n a^{s+\lambda-(\nu+1)/2} b^\nu}{2^\nu n!}(2\lambda)_n\left(\frac{1-2s+2\lambda+\nu}{2}\right)_n$ $\times\Gamma\left[\begin{matrix}\frac{2s-\nu}{2},\ \frac{1-2s-2n-2\lambda+\nu}{2}\\ \frac{1-2\lambda}{2},\ \nu+1\end{matrix}\right]$ $\times {}_2F_3\left(\begin{matrix}\frac{1-2s-2n-2\lambda+\nu}{2},\ \frac{1-2s+2n+2\lambda+\nu}{2}\\ \nu+1,\ \frac{2-2s+\nu}{2},\ \frac{1-2s+2\lambda+\nu}{2};\ \frac{b^2}{4a}\end{matrix}\right)+\frac{a^{\lambda-1/2} b^{2s}}{2^{2s} n!}$ $\times(2\lambda)_n\Gamma\left[\begin{matrix}\frac{\nu-2s}{2}\\ \frac{2s+\nu+2}{2}\end{matrix}\right]{}_2F_3\left(\begin{matrix}\frac{-2n-2\lambda+1}{2},\ \frac{2n+2\lambda+1}{2};\ \frac{b^2}{4a}\\ \frac{2\lambda+1}{2},\ \frac{2s-\nu+2}{2},\ \frac{2s+\nu+2}{2}\end{matrix}\right)$ $[b>0;\ -3/4<\operatorname{Re}s<1/2-n+\operatorname{Re}(\nu/2-\lambda);\ \lvert\arg a\rvert<\pi]$
17	$(x+a)^n J_\nu(b\sqrt{x})$ $\times C_n^\lambda\left(\frac{a-x}{a+x}\right)$	$\frac{(-1)^n}{n!}\left(\frac{2}{b}\right)^{2s+2n}(2\lambda)_n\Gamma\left[\begin{matrix}\frac{2s+2n+\nu}{2}\\ \frac{2-2s-2n+\nu}{2}\end{matrix}\right]$ $\times {}_2F_3\left(\begin{matrix}-n,\ \frac{-2n-2\lambda+1}{2};\ \frac{ab^2}{4}\\ \frac{2\lambda+1}{2},\ \frac{2-2s-2n-\nu}{2},\ \frac{2-2s-2n+\nu}{2}\end{matrix}\right)$ $[b>0;\ -\operatorname{Re}\nu/2<\operatorname{Re}s<3/4-n]$

No.	$f(x)$	$F(s)$
18	$(x+a)^{-n-2\lambda} J_\nu(b\sqrt{x})$ $\times C_n^\lambda\left(\frac{a-x}{a+x}\right)$	$\frac{a^{s-n-2\lambda+\nu/2}b^\nu}{2^\nu n!}\frac{\left(\frac{1-2s+2\lambda-\nu}{2}\right)_n}{\left(\frac{2\lambda+1}{2}\right)_n}\Gamma\left[\begin{matrix}\frac{2s+\nu}{2}, \frac{2n-2s+4\lambda-\nu}{2}\\ 2\lambda,\, \nu+1\end{matrix}\right]$ $\times {}_2F_3\left(\begin{matrix}\frac{2s-2\lambda+\nu+1}{2}, \frac{2s+\nu}{2}; \frac{ab^2}{4}\\ \nu+1, \frac{2s-2n-2\lambda+\nu+1}{2}, \frac{2s-2n-4\lambda+\nu+2}{2}\end{matrix}\right)$ $+\frac{(-1)^n}{n!}\left(\frac{b}{2}\right)^{2(n-s+2\lambda)}(2\lambda)_n\Gamma\left[\begin{matrix}\frac{2s-2n-4\lambda+\nu}{2}\\ \frac{2-2s+2n+4\lambda+\nu}{2}\end{matrix}\right]$ $\times {}_2F_3\left(\begin{matrix}n+2\lambda, \frac{2n+2\lambda+1}{2}; \frac{ab^2}{4}\\ \frac{2\lambda+1}{2}, \frac{2-2s+2n+4\lambda-\nu}{2}, \frac{2-2s+2n+4\lambda+\nu}{2}\end{matrix}\right)$ $[b>0;\ -\operatorname{Re}\nu/2<\operatorname{Re}s<n+2\operatorname{Re}\lambda+3/4]$
19	$(a-x)_+^{\lambda-1/2}$ $\times\left\{\begin{matrix}J_\mu(b\sqrt{x})\,J_\nu(b\sqrt{x})\\ I_\mu(b\sqrt{x})\,I_\nu(b\sqrt{x})\end{matrix}\right\}$ $\times C_n^\lambda\left(1-\frac{2x}{a}\right)$	$\frac{a^{s+(2\lambda+\mu+\nu-1)/2}(b/2)^{\mu+\nu}}{n!}(2\lambda)_n\left(\frac{1-2s+2\lambda-\mu-\nu}{2}\right)_n$ $\times\Gamma\left[\begin{matrix}\frac{2\lambda+1}{2}, \frac{2s+\mu+\nu}{2}\\ \mu+1, \nu+1, \frac{2s+2n+2\lambda+\mu+\nu+1}{2}\end{matrix}\right]{}_4F_5\left(\begin{matrix}\frac{\mu+\nu+1}{2}, \frac{\mu+\nu+2}{2}, \frac{2s+\mu+\nu}{2},\\ \mu+1, \nu+1, \mu+\nu+1,\end{matrix}\right.$ $\left.\begin{matrix}\frac{2s-2\lambda+\mu+\nu+1}{2}; \mp ab^2\\ \frac{2s-2n-2\lambda+\mu+\nu+1}{2}, \frac{2s+2n+2\lambda+\mu+\nu+1}{2}\end{matrix}\right)$ $[a,\ \operatorname{Re}(2s+\mu+\nu)>0;\ \operatorname{Re}\lambda>-1/2]$

3.24.12. $C_n^\lambda(bx)$ and $\mathbf{H}_\nu(ax)$, $\mathbf{L}_\nu(ax)$

1	$(a^2-x^2)_+^{\lambda-1/2}$ $\times\left\{\begin{matrix}\mathbf{H}_\nu(bx)\\ \mathbf{L}_\nu(bx)\end{matrix}\right\}C_n^\lambda\left(\frac{x}{a}\right)$	$\frac{\sqrt{\pi}}{2^\nu n!}\left(\frac{a}{2}\right)^{s+\nu+2\lambda}b^{\nu+1}\Gamma\left[\begin{matrix}s+\nu+1, n+2\lambda\\ \lambda, \frac{2\nu+3}{2}, \frac{s-n+\nu+2}{2}, \frac{s+n+\nu+2\lambda+2}{2}\end{matrix}\right]$ $\times {}_3F_3\left(\begin{matrix}1, \frac{s+\nu+1}{2}, \frac{s+\nu+2}{2}; \mp\frac{a^2b^2}{4}\\ \frac{3}{2}, \frac{s+n+\nu+2\lambda+2}{2}, \frac{3s-3n+7\nu+6}{2}\end{matrix}\right)$ $[a>0;\ \operatorname{Re}\lambda>-1/2;\ \operatorname{Re}(s+\nu)>-1]$

3.24.13. $C_n^\lambda(ax+b)$ and $P_m(cx^r+d)$

Notation: ε, $\delta=0$ or 1.

1	$\theta(a-x)\,P_m\left(\frac{x}{a}\right)$ $\times C_{2n+\varepsilon}^\lambda(bx)$	$\frac{(-1)^n\sqrt{\pi}}{n!}\left(\frac{a}{2}\right)^{s+\varepsilon}(2b)^\varepsilon(\lambda)_{n+\varepsilon}\Gamma\left[\begin{matrix}s+\varepsilon\\ \frac{s-m+\varepsilon+1}{2}, \frac{s+m+\varepsilon+2}{2}\end{matrix}\right]$ $\times {}_4F_3\left(\begin{matrix}-n, n+\lambda+\varepsilon, \frac{s+\varepsilon}{2}, \frac{s+\varepsilon+1}{2}; a^2b^2\\ \frac{2\varepsilon+1}{2}, \frac{s-m+\varepsilon+1}{2}, \frac{s+m+\varepsilon+2}{2}\end{matrix}\right)$ $[a>0;\ \operatorname{Re}s>((-1)^m-2\varepsilon-1)/2]$

No.	$f(x)$	$F(s)$
2	$(a^2-x^2)_+^{\lambda-1/2}$ $\times P_{2m+\varepsilon}\left(\frac{x}{b}\right) C_{2n+\delta}^{\lambda}\left(\frac{x}{a}\right)$	$\frac{(-1)^{m+n}\, 2^{\varepsilon-1} a^{s+2\lambda+\varepsilon-1}}{m!\,(2n+\delta)!\, b^{\varepsilon}} (2\lambda)_{2n+\delta} \left(\frac{1}{2}\right)_{m+\varepsilon} \left(\frac{1-s+\delta-\varepsilon}{2}\right)_n$ $\times \Gamma\left[\begin{matrix} \frac{2\lambda+1}{2}, \frac{s+\delta+\varepsilon}{2} \\ \frac{s+2n+2\lambda+\delta+\varepsilon+1}{2} \end{matrix}\right] {}_4F_3\left(\begin{matrix} -m, \frac{2m+2\varepsilon+1}{2}, \frac{s+1}{2}, \frac{s+2\varepsilon}{2};\, \frac{a^2}{b^2} \\ \frac{2\varepsilon+1}{2}, \frac{s-2n-\delta+\varepsilon+1}{2}, \frac{s+2n+2\lambda+\delta+\varepsilon+1}{2} \end{matrix}\right)$ $[b>a>0;\ \operatorname{Re}\lambda>-1/2;\ \operatorname{Re}s>-\delta-\varepsilon]$
3	$\theta(a-x)\,(b^2-x^2)^{\lambda-1/2}$ $\times P_{2m+\varepsilon}\left(\frac{x}{a}\right) C_{2n+\delta}^{\lambda}\left(\frac{x}{b}\right)$	$\frac{(-1)^{m+n}\, 2^{\delta-1} a^{s+\delta} b^{2\lambda-\delta-1}}{n!\,\left(\frac{s+\delta+\varepsilon}{2}\right)_{m+1}} (\lambda)_{n+\delta} \left(\frac{1-s-\delta+\varepsilon}{2}\right)_m$ $\times {}_4F_3\left(\begin{matrix} \frac{1-2n-2\lambda}{2}, \frac{2n+2\delta+1}{2}, \frac{s+1}{2}, \frac{s+2\delta}{2};\, \frac{a^2}{b^2} \\ \frac{2\delta+1}{2}, \frac{s-2m+\delta-\varepsilon+1}{2}, \frac{s+2m+\delta+\varepsilon+2}{2} \end{matrix}\right)$ $[b>a>0;\ \operatorname{Re}\lambda>-1/2;\ \operatorname{Re}s>-\delta-\varepsilon]$
4	$\theta(x-a)\,(x^2-b^2)^{\lambda-1/2}$ $\times P_{2m+\varepsilon}\left(\frac{x}{a}\right) C_{2n+\delta}^{\lambda}\left(\frac{x}{b}\right)$	$\frac{(-1)^{m-1}\, 2^{2n+\delta-1} a^{s+2n+2\lambda+\delta-1}}{(2n+\delta)!\, b^{2n+\delta}} \frac{(\lambda)_{2n+\delta} \left(\frac{2-s-2n-2\lambda-\delta+\varepsilon}{2}\right)_m}{\left(\frac{s+2n+2\lambda+\delta+\varepsilon-1}{2}\right)_{m+1}}$ $\times {}_4F_3\left(\begin{matrix} \frac{1-2n-2\lambda}{2}, 1-n-\lambda-\delta, \frac{2-s+2m-2n-2\lambda-\delta+\varepsilon}{2}, \frac{1-s-2m-2n-2\lambda-\delta-\varepsilon}{2} \\ 1-2n-\lambda-\delta, \frac{2-s-2n-2\lambda}{2}, \frac{3-s-2n-2\lambda-2\delta}{2};\, \frac{b^2}{a^2} \end{matrix}\right)$ $[a>b>0;\ \operatorname{Re}(s+2\lambda)<1-2m-2n-\delta-\varepsilon]$
5	$(x^2-a^2)_+^{\lambda-1/2}$ $\times P_{2m+\varepsilon}\left(\frac{x}{b}\right) C_{2n+\delta}^{\lambda}\left(\frac{x}{a}\right)$	$\frac{2^{2m+\varepsilon-1} a^{s+2m+2\lambda+\varepsilon-1}}{(2m+\varepsilon)!\,(2n+\delta)!\, b^{2m+\varepsilon}} (2\lambda)_{2n+\delta} \left(\frac{1}{2}\right)_{2m+\varepsilon}$ $\times \left(\frac{1-s-2m+\delta-\varepsilon}{2}\right)_n \Gamma\left[\begin{matrix} \frac{2\lambda+1}{2}, \frac{1-s-2m-2n-2\lambda-\delta-\varepsilon}{2} \\ \frac{2-s-2m-\delta-\varepsilon}{2} \end{matrix}\right]$ $\times {}_4F_3\left(\begin{matrix} -m, \frac{1-2m-2\varepsilon}{2}, \frac{1-s-2m+2n+\delta-\varepsilon}{2}, \frac{1-s-2m-2n-2\lambda-\delta-\varepsilon}{2} \\ \frac{1-4m-2\varepsilon}{2}, \frac{2-s-2m-2\varepsilon}{2}, \frac{1-s-2m}{2};\, \frac{b^2}{a^2} \end{matrix}\right)$ $\left[\begin{matrix} a>b>0;\ \operatorname{Re}\lambda>-1/2; \\ \operatorname{Re}(s+2\lambda)<1-2m-2n-\delta-\varepsilon \end{matrix}\right]$
6	$\theta(a-x)\, P_m\left(\frac{2x}{a}-1\right)$ $\times C_n^{\lambda}\left(\frac{2x}{a}-1\right)$	$\frac{(-1)^{m+n}\,(2\lambda)_n\,(1-s)_m\, a^s}{n!\,(s)_{m+1}} {}_4F_3\left(\begin{matrix} -n, n+2\lambda, s, s;\, 1 \\ \frac{2\lambda+1}{2}, s-m, s+m+1 \end{matrix}\right)$ $[a,\ \operatorname{Re}s>0]$
7	$(a-x)_+^{\lambda-1/2}\, P_m\left(1-\frac{2x}{a}\right)$ $\times C_n^{\lambda}\left(1-\frac{2x}{a}\right)$	$\frac{a^{s+\lambda-1/2}}{n!} (2\lambda)_n \left(\frac{1-2s+2\lambda}{2}\right)_n \Gamma\left[\begin{matrix} \frac{2\lambda+1}{2}, s \\ \frac{2s+2n+2\lambda+1}{2} \end{matrix}\right]$ $\times {}_4F_3\left(\begin{matrix} -m, m+1, \frac{2s-2\lambda+1}{2}, s;\, 1 \\ 1, \frac{2s-2n-2\lambda+1}{2}, \frac{2s+2n+2\lambda+1}{2} \end{matrix}\right)$ $[a,\ \operatorname{Re}s>0;\ \operatorname{Re}\lambda>-1/2]$

No.	$f(x)$	$F(s)$
8	$(a-x)_+^{\lambda-1/2} P_m(2bx-1) \times C_n^\lambda\left(\frac{2x}{a}-1\right)$	$\frac{(-1)^{m+n} a^{s+\lambda-1/2}}{n!}(2\lambda)_n\left(\frac{1-2s+2\lambda}{2}\right)_n \times \Gamma\left[\begin{matrix}\frac{2\lambda+1}{2},\, s\\ \frac{2s+2n+2\lambda+1}{2}\end{matrix}\right] {}_4F_3\left(\begin{matrix}-m,\, m+1,\, s,\, \frac{2s-2\lambda+1}{2};\, ab\\ 1,\, \frac{2s-2n-2\lambda+1}{2},\, \frac{2s+2n+2\lambda+1}{2}\end{matrix}\right)$ $[a,\ \operatorname{Re} s>0;\ \operatorname{Re}\lambda>-1/2]$
9	$\theta(a-x)(b\pm x)^{\lambda-1/2} \times P_m\left(\frac{2x}{a}-1\right) \times C_n^\lambda\left(\frac{2x}{b}\pm 1\right)$	$\frac{(-1)^m(\pm 1)^n a^s b^{\lambda-1/2}}{n!}\frac{(2\lambda)_n(1-s)_m}{(s)_{m+1}} \times {}_4F_3\left(\begin{matrix}\frac{-2n-2\lambda+1}{2},\, \frac{2n+2\lambda+1}{2},\, s,\, s\\ \frac{2\lambda+1}{2},\, s-m,\, s+m+1;\, \mp\frac{a}{b}\end{matrix}\right)$ $[a,\ b,\ \operatorname{Re} s>0]$
10	$(a-x)_+^{\lambda-1/2} P_{2m+\varepsilon}(b\sqrt{x}) \times C_n^\lambda\left(\frac{2x}{a}-1\right)$	$\frac{(-1)^{m+n} a^{s+\lambda+(\varepsilon-1)/2}(2b)^\varepsilon}{m!\,n!}\left(\frac{1}{2}\right)_{m+\varepsilon}(2\lambda)_n \times\left(\frac{1-2s+2\lambda-\varepsilon}{2}\right)_n \Gamma\left[\begin{matrix}\frac{2\lambda+1}{2},\, \frac{2s+\varepsilon}{2}\\ \frac{2s+2n+2\lambda+\varepsilon+1}{2}\end{matrix}\right] \times {}_4F_3\left(\begin{matrix}-m,\, \frac{2m+2\varepsilon+1}{2},\, \frac{2s+\varepsilon}{2},\, \frac{2s-2\lambda+\varepsilon+1}{2};\, ab^2\\ \frac{2\varepsilon+1}{2},\, \frac{2s-2n-2\lambda+\varepsilon+1}{2},\, \frac{2s+2n+2\lambda+\varepsilon+1}{2}\end{matrix}\right)$ $[a>0;\ \operatorname{Re}\lambda>-1/2;\ \operatorname{Re} s>-\varepsilon/2]$

3.24.14. $C_n^\lambda(bx)$ and $H_m(ax)$

Notation: $\delta,\ \varepsilon=0$ or 1.

1	$(a^2-x^2)_+^{\lambda-1/2} \times H_{2m+\varepsilon}(bx)\, C_n^\lambda\left(\frac{x}{a}\right)$	$\frac{(-1)^m \pi a^{s+2\lambda+\varepsilon-1} b^\varepsilon}{n!\, 2^{s+2\lambda-2m-1}}\left(\frac{2\varepsilon+1}{2}\right)_m \Gamma\left[\begin{matrix}s+\varepsilon,\, n+2\lambda\\ \lambda,\, \frac{s-n+\varepsilon+1}{2},\, \frac{s+2\lambda+n+\varepsilon+1}{2}\end{matrix}\right] \times {}_3F_3\left(\begin{matrix}-m,\, \frac{s+\varepsilon}{2},\, \frac{s+\varepsilon+1}{2};\, a^2b^2\\ \frac{2\varepsilon+1}{2},\, \frac{s-n+\varepsilon+1}{2},\, \frac{s+2\lambda+n+\varepsilon+1}{2}\end{matrix}\right)$ $[a>0;\ \operatorname{Re}\lambda>-1/2;\ \operatorname{Re} s>((-1)^n-2\varepsilon-1)/2]$
2	$(a^2-x^2)_+^{\lambda-1/2} e^{-b^2x^2} \times H_{2m+\varepsilon}(bx)\, C_n^\lambda\left(\frac{x}{a}\right)$	$\frac{(-1)^m \pi a^{s+2\lambda+\varepsilon-1} b^\varepsilon}{n!\, 2^{s+2\lambda-2m-1}}\left(\frac{2\varepsilon+1}{2}\right)_m \times\Gamma\left[\begin{matrix}s+\varepsilon,\, n+2\lambda\\ \lambda,\, \frac{s-n+\varepsilon+1}{2},\, \frac{s+2\lambda+n+\varepsilon+1}{2}\end{matrix}\right] \times {}_3F_3\left(\begin{matrix}\frac{2m+2\varepsilon+1}{2},\, \frac{s+\varepsilon}{2},\, \frac{s+\varepsilon+1}{2};\, -a^2b^2\\ \frac{2\varepsilon+1}{2},\, \frac{s-n+\varepsilon+1}{2},\, \frac{s+2\lambda+n+\varepsilon+1}{2}\end{matrix}\right)$ $[a>0;\ \operatorname{Re}\lambda>-1/2;\ \operatorname{Re} s>((-1)^n-2\varepsilon-1)/2]$

No.	$f(x)$	$F(s)$
3	$(x^2-a^2)_+^{\lambda-1/2}\, e^{-b^2x^2}$ $\times H_{2m+\varepsilon}(bx)$ $\times C_{2n+\delta}^{\lambda}\left(\frac{x}{a}\right)$	$(-1)^m \frac{2^{2m+2\varepsilon-1} a^{s+2\lambda+\varepsilon-1} b^{\varepsilon}}{(2n+\delta)!} \left(\frac{1}{2}\right)_{m+\varepsilon} (2\lambda)_{2n+\delta} \times \left(\frac{1-s+\delta-\varepsilon}{2}\right)_n \Gamma\left[\begin{matrix}\frac{2\lambda+1}{2},\ \frac{1-s-2n-2\lambda-\delta-\varepsilon}{2}\\ \frac{2-s-\delta-\varepsilon}{2}\end{matrix}\right] \times {}_3F_3\left(\begin{matrix}\frac{2\varepsilon+2m+1}{2},\ \frac{s+2\varepsilon}{2},\ \frac{s+1}{2};\ -a^2b^2\\ \frac{2\varepsilon+1}{2},\ \frac{s+2n+2\lambda+\delta+\varepsilon+1}{2},\ \frac{s-2n-\delta+\varepsilon+1}{2}\end{matrix}\right) + \frac{(-1)^m\, 2^{m+2n+\delta+\varepsilon-1}}{(2n+\delta)!\, a^{2n+\delta} b^{s+2n+2\lambda+\delta-1}} (\lambda)_{2n+\delta} \times \left(\frac{2-s-2n-2\lambda-\delta+\varepsilon}{2}\right)_m \Gamma\left(\frac{s+2n+2\lambda+\delta+\varepsilon-1}{2}\right) \times {}_3F_3\left(\begin{matrix}\frac{1-2n-2\lambda}{2},\ 1-n-\lambda-\delta,\ \frac{2-s+2m-2n+\varepsilon-2\lambda-\delta}{2}\\ 1-2n-\lambda-\delta,\ \frac{2-s-2n-2\lambda}{2},\ \frac{3-s-2n-2\lambda-2\delta}{2};\ -a^2b^2\end{matrix}\right)$ $[a,\ \mathrm{Re}\, b>0;\ \mathrm{Re}\,\lambda>-1/2]$
4	$(a-x)_+^{\lambda-1/2}\, e^{-b^2x}$ $\times H_{2m+\varepsilon}(b\sqrt{x})$ $\times C_n^{\lambda}\left(\frac{2x}{a}-1\right)$	$\frac{(-1)^{m+n}\, 2^{2m+\varepsilon}}{n!} a^{s+\lambda+(\varepsilon-1)/2} b^{\varepsilon} \left(\frac{2\varepsilon+1}{2}\right)_m (2\lambda)_n \times \left(\frac{1-2s+2\lambda-\varepsilon}{2}\right)_n \Gamma\left[\begin{matrix}\frac{2\lambda+1}{2},\ \frac{2s+\varepsilon}{2}\\ \frac{2s+2n+2\lambda+\varepsilon+1}{2}\end{matrix}\right] \times {}_3F_3\left(\begin{matrix}\frac{2m+2\varepsilon+1}{2},\ \frac{2s-2\lambda+\varepsilon+1}{2},\ \frac{2s+\varepsilon}{2};\ -ab^2\\ \frac{2\varepsilon+1}{2},\ \frac{2s-2n-2\lambda+\varepsilon+1}{2},\ \frac{2s+2n+2\lambda+\varepsilon+1}{2}\end{matrix}\right)$ $[a>0;\ \mathrm{Re}\,\lambda>-1/2;\ \mathrm{Re}\, s>-\varepsilon/2]$

3.24.15. $C_n^{\lambda}(bx)$ and $L_m^{\mu}(ax^r)$

Notation: $\delta = \begin{Bmatrix}1\\0\end{Bmatrix}$, $\varepsilon = 0$ or 1.

1	$(a^2-x^2)_+^{\lambda-1/2}$ $\times L_m^{\mu}(bx^2)\, C_n^{\lambda}\left(\frac{x}{a}\right)$	$\frac{\pi(\mu+1)_m}{m!\, n!} \left(\frac{a}{2}\right)^{s+2\lambda-1} \Gamma\left[\begin{matrix}s,\ n+2\lambda\\ \lambda,\ \frac{s-n+1}{2},\ \frac{s+n+2\lambda+1}{2}\end{matrix}\right] \times {}_3F_3\left(\begin{matrix}-m,\ \frac{s}{2},\ \frac{s+1}{2};\ a^2b\\ \mu+1,\ \frac{s-n+1}{2},\ \frac{s+n+2\lambda+1}{2}\end{matrix}\right)$ $[a>0;\ \mathrm{Re}\,\lambda>-1/2;\ \mathrm{Re}\, s>((-1)^n-1)/2]$
2	$(a^2-x^2)_+^{\lambda-1/2}\, e^{-bx^2}$ $\times L_m^{\mu}(bx^2)\, C_n^{\lambda}\left(\frac{x}{a}\right)$	$\frac{\pi(\mu+1)_m}{m!\, n!} \left(\frac{a}{2}\right)^{s+2\lambda-1} \Gamma\left[\begin{matrix}s,\ n+2\lambda\\ \lambda,\ \frac{s-n+1}{2},\ \frac{s+n+2\lambda+1}{2}\end{matrix}\right] \times {}_3F_3\left(\begin{matrix}m+\mu+1,\ \frac{s}{2},\ \frac{s+1}{2};\ -a^2b\\ \mu+1,\ \frac{s-n+1}{2},\ \frac{s+n+2\lambda+1}{2}\end{matrix}\right)$ $[a>0;\ \mathrm{Re}\,\lambda>-1/2;\ \mathrm{Re}\, s>((-1)^n-1)/2]$

No.	$f(x)$	$F(s)$
3	$(x^2-a^2)_+^{\lambda-1/2}e^{-bx^2}$ $\times L_m^\mu(bx^2)\,C_{2n+\varepsilon}^\lambda\left(\frac{x}{a}\right)$	$\frac{2^{2n+\varepsilon-1}b^{(1-s-\varepsilon)/2-n-\lambda}}{m!\,(2n+\varepsilon)!\,a^{2n+\varepsilon}}(\lambda)_{2n+\varepsilon}$ $\times\left(\frac{3-s-2n-2\lambda+2\mu-\varepsilon}{2}\right)_m\Gamma\left(\frac{s+2n+2\lambda+\varepsilon-1}{2}\right)$ $\times{}_3F_3\left(\begin{matrix}\frac{1-2n-2\lambda-\varepsilon}{2},\ \frac{2-2n-2\lambda-\varepsilon}{2},\ \frac{3-s+2m-2n-2\lambda+2\mu-\varepsilon}{2}\\ 1-2n-\lambda-\varepsilon,\ \frac{3-s-2n-2\lambda-\varepsilon}{2},\ \frac{3-s-2n-2\lambda+2\mu-\varepsilon}{2};\ -a^2b\end{matrix}\right)$ $+\frac{a^{s+2\lambda-1}(2\lambda)_{2n+\varepsilon}}{2(m!)(2n+\varepsilon)!}(\mu+1)_m\left(\frac{1-s+\varepsilon}{2}\right)_n$ $\times\Gamma\left[\begin{matrix}\frac{2\lambda+1}{2},\ \frac{1-s-2n-2\lambda-\varepsilon}{2}\\ \frac{2-s-\varepsilon}{2}\end{matrix}\right]{}_3F_3\left(\begin{matrix}m+\mu+1,\ \frac{s}{2},\ \frac{s+1}{2};\ -a^2b\\ \mu+1,\ \frac{s-2n-\varepsilon+1}{2},\ \frac{s+2n+2\lambda+\varepsilon+1}{2}\end{matrix}\right)$ $[a,\ \operatorname{Re}b>0;\ \operatorname{Re}\lambda>-1/2]$
4	$(a-x)_+^{\lambda-1/2}e^{-bx}$ $\times L_m^\mu(bx)\,C_n^\lambda\left(\frac{2x}{a}-1\right)$	$\frac{(-1)^n}{m!\,n!}a^{s+\lambda-1/2}(\mu+1)_m(2\lambda)_n\left(\frac{1-2s+2\lambda}{2}\right)_n$ $\times\Gamma\left[\begin{matrix}\frac{2\lambda+1}{2},\ s\\ \frac{2s+2n+2\lambda+1}{2}\end{matrix}\right]{}_3F_3\left(\begin{matrix}m+\mu+1,\ \frac{2s-2\lambda+1}{2},\ s;\ -ab\\ \mu+1,\ \frac{2s-2n-2\lambda+1}{2},\ \frac{2s+2n+2\lambda+1}{2}\end{matrix}\right)$ $[a,\ \operatorname{Re}s>0;\ \operatorname{Re}\lambda>-1/2]$

3.24.16. Products of $C_n^\lambda(bx)$

Notation: $\varepsilon,\ \delta=0$ or 1.

1	$(a^2-x^2)_+^{\lambda-1/2}$ $\times(b^2-x^2)^{\mu-1/2}$ $\times C_{2m+\varepsilon}^\mu\left(\frac{x}{b}\right)C_{2n+\delta}^\lambda\left(\frac{x}{a}\right)$	$\frac{(-1)^{m+n}2^{\varepsilon-1}(2\lambda)_{2n+\delta}(\mu)_{m+\varepsilon}}{m!\,(2n+\delta)!}a^{s+2\lambda+\varepsilon-1}b^{2\mu-\varepsilon-1}$ $\times\left(\frac{1-s+\delta-\varepsilon}{2}\right)_n\Gamma\left[\begin{matrix}\frac{2\lambda+1}{2},\ \frac{s+\delta+\varepsilon}{2}\\ \frac{s+2n+2\lambda+\delta+\varepsilon+1}{2}\end{matrix}\right]$ $\times{}_4F_3\left(\begin{matrix}\frac{1-2m-2\mu}{2},\ \frac{2m+2\varepsilon+1}{2},\ \frac{s+1}{2},\ \frac{s+2\varepsilon}{2}\\ \frac{2\varepsilon+1}{2},\ \frac{s-2n-\delta+\varepsilon+1}{2},\ \frac{s+2n+2\lambda+\delta+\varepsilon+1}{2};\ \frac{a^2}{b^2}\end{matrix}\right)$ $[b>a>0;\ \operatorname{Re}\lambda>-1/2;\ \operatorname{Re}s>-\delta-\varepsilon]$
2	$(x^2-a^2)_+^{\lambda-1/2}$ $\times(x^2-b^2)^{\mu-1/2}$ $\times C_{2m+\varepsilon}^\mu\left(\frac{x}{b}\right)C_{2n+\delta}^\lambda\left(\frac{x}{a}\right)$	$\frac{2^{2m+\varepsilon-1}(2\lambda)_{2n+\delta}(\mu)_{2m+\varepsilon}}{(2m+\varepsilon)!\,(2n+\delta)!}a^{s+2m+2\lambda+2\mu+\varepsilon-2}b^{-2m-\varepsilon}$ $\times\left(\frac{2-s-2m-2\mu+\delta-\varepsilon}{2}\right)_n$ $\times\Gamma\left[\begin{matrix}\frac{2\lambda+1}{2},\ \frac{2-s-2m-2n-2\lambda-2\mu-\delta-\varepsilon}{2}\\ \frac{3-s-2m-2\mu-\delta-\varepsilon}{2}\end{matrix}\right]\times$

No.	$f(x)$	$F(s)$
		$\times {}_4F_3\left(\begin{matrix}\frac{1-2m-2\mu}{2},\, 1-m-\mu-\varepsilon,\, \frac{2-s-2m+2n-2\mu+\delta-\varepsilon}{2},\, \frac{2-s-2m-2n-2\lambda-2\mu-\delta-\varepsilon}{2} \\ \frac{3-s-2m-2\mu-2\varepsilon}{2},\, 1-2m-\mu-\varepsilon,\, \frac{2-s-2m-2\mu}{2};\, \frac{b^2}{a^2}\end{matrix}\right)$ $\left[a>b>0;\ \operatorname{Re}\lambda>-1/2;\ \operatorname{Re}(s+2\lambda+2\mu)<2-2m-2n-\delta-\varepsilon\right]$
3	$(x^2-a^2)_+^{\lambda-1/2} \times (b^2-x^2)_+^{\mu-1/2} \times C_{2n+\delta}^{\lambda}\left(\frac{x}{a}\right) C_{2m+\varepsilon}^{\mu}\left(\frac{x}{b}\right)$	$\frac{(-1)^m 2^{\varepsilon-1} a^{s+2\lambda+\varepsilon-1} b^{2\mu-\varepsilon-1}}{m!\,(2n+\delta)!} (\mu)_{m+\varepsilon} (2\lambda)_{2n+\delta} \times \left(\frac{1-s+\delta-\varepsilon}{2}\right)_n \Gamma\left[\begin{matrix}\frac{2\lambda+1}{2},\, \frac{1-s-2n-2\lambda-\delta-\varepsilon}{2} \\ 1-\frac{s+\delta+\varepsilon}{2}\end{matrix}\right]$ $\times {}_4F_3\left(\begin{matrix}-m-\mu+\frac{1}{2},\, m+\varepsilon+\frac{1}{2},\, \frac{s+1}{2},\, \frac{s+2\varepsilon}{2} \\ \frac{2\varepsilon+1}{2},\, \frac{s-2n-\delta+\varepsilon+1}{2},\, \frac{s+2n+2\lambda+\delta+\varepsilon+1}{2};\, \frac{a^2}{b^2}\end{matrix}\right)$ $+\frac{(-1)^m 2^{2n+\delta-1} a^{-2n-\delta} b^{s+2n+2\mu+2\lambda+\delta-2}}{(2m+\varepsilon)!\,(2n+\delta)!}$ $\times (2\mu)_{2m+\varepsilon} (\lambda)_{2n+\delta} \left(\frac{2-s-2n-2\lambda-\delta+\varepsilon}{2}\right)_m$ $\times \Gamma\left[\begin{matrix}\frac{2\mu+1}{2},\, \frac{s+2n+2\lambda+\delta+\varepsilon-1}{2} \\ \frac{s+2m+2n+2\mu+2\lambda+\delta+\varepsilon}{2}\end{matrix}\right]$ $\times {}_4F_3\left(\begin{matrix}\frac{1-2n-2\lambda}{2},\, 1-n-\lambda-\delta,\, \frac{2-s-2m-2n-2\mu-2\lambda-\delta-\varepsilon}{2},\, \frac{2-s+2m-2n-2\lambda-\delta+\varepsilon}{2} \\ 1-2n-\lambda-\delta,\, \frac{2-s-2n-2\lambda}{2},\, \frac{3-s-2n-2\lambda-2\delta}{2};\, \frac{a^2}{b^2}\end{matrix}\right)$ $[b>a>0;\ \operatorname{Re}\lambda,\ \operatorname{Re}\mu>-1/2]$
4	$(a-x)_+^{\lambda-1/2} C_m^{\mu}\left(1-\frac{2x}{a}\right) \times C_n^{\lambda}\left(1-\frac{2x}{a}\right)$	$\frac{a^{s+\lambda-1/2}}{m!\,n!} \Gamma\left[\begin{matrix}\frac{2\lambda+1}{2},\, m+2\mu,\, n+2\lambda,\, s,\, \frac{1-2s+2n+2\lambda}{2} \\ 2\lambda,\, 2\mu,\, \frac{1-2s+2\lambda}{2},\, \frac{2s+2n+2\lambda+1}{2}\end{matrix}\right]$ $\times {}_4F_3\left(\begin{matrix}-m,\, m+2\mu,\, s,\, \frac{2s-2\lambda+1}{2};\, 1 \\ \frac{2\mu+1}{2},\, \frac{2s-2n-2\lambda+1}{2},\, \frac{2s+2n+2\lambda+1}{2}\end{matrix}\right)$ $[a,\ \operatorname{Re}s>0;\ \operatorname{Re}\lambda>-1/2]$
5	$(a-x)_+^{\lambda+\mu-1} C_m^{\mu}\left(1-\frac{2x}{a}\right) \times C_n^{\lambda}\left(1-\frac{2x}{a}\right)$	$\frac{a^{s+\lambda+\mu-1}}{m!\,n!} (2\mu)_m (2\lambda)_n \left(\frac{1-2s+2\lambda}{2}\right)_n \times \Gamma\left[\begin{matrix}\frac{2\lambda+1}{2},\, s \\ s+n+\lambda+\frac{1}{2}\end{matrix}\right]$ $\times {}_4F_3\left(\begin{matrix}-m-\mu+\frac{1}{2},\, m+\mu+\frac{1}{2},\, s-\lambda+\frac{1}{2},\, s \\ \frac{2\mu+1}{2},\, s-n-\lambda+\frac{1}{2},\, s+n+\lambda+\frac{1}{2};\, 1\end{matrix}\right)$ $[a,\ \operatorname{Re}s>0;\ \operatorname{Re}(\lambda+\mu)>0]$

No.	$f(x)$	$F(s)$
6	$(a-x)_+^{\lambda-1/2}(1-bx)^{\mu-1/2}$ $\times C_m^\mu(2bx-1)$ $\times C_n^\lambda\left(\frac{2x}{a}-1\right)$	$\frac{(-1)^{m+n}a^{s+\lambda-1/2}}{m!\,n!}(2\mu)_m(2\lambda)_n$ $\times\left(\frac{1-2s+2\lambda}{2}\right)_n\Gamma\left[\begin{matrix}\frac{2\lambda+1}{2},\ s\\ \frac{2s+2n+2\lambda+1}{2}\end{matrix}\right]$ $\times {}_4F_3\left(\begin{matrix}\frac{1-2m-2\mu}{2},\ \frac{2m+2\mu+1}{2},\ \frac{2s-2\lambda+1}{2},\ s\\ \frac{2\mu+1}{2},\ \frac{2s-2n-2\lambda+1}{2},\ \frac{2s+2n+2\lambda+1}{2};\ ab\end{matrix}\right)$ $[a,\ \operatorname{Re}s>0;\ \lvert\arg(1-ab)<\pi\rvert]$
7	$(a-x)_+^{\lambda-1/2}(b\pm x)^{\mu-1/2}$ $\times C_m^\lambda\left(\frac{2x}{a}-1\right)$ $\times C_n^\mu\left(\frac{2x}{b}\pm 1\right)$	$\frac{(-1)^m(\pm1)^n}{m!\,n!}a^{s+\lambda-1/2}b^{\mu-1/2}(2\lambda)_m(2\mu)_n$ $\times\left(\frac{1-2s+2\lambda}{2}\right)_m\Gamma\left[\begin{matrix}\frac{2\lambda+1}{2},\ s\\ \frac{2s+2m+2\lambda+1}{2}\end{matrix}\right]$ $\times {}_4F_3\left(\begin{matrix}\frac{1-2n-2\mu}{2},\ \frac{2n+2\mu+1}{2},\ \frac{2s-2\lambda+1}{2},\ s\\ \frac{2\mu+1}{2},\ \frac{2s-2m-2\lambda+1}{2},\ \frac{2s+2m+2\lambda+1}{2};\ \mp\frac{a}{b}\end{matrix}\right)$ $\left[a,\ \operatorname{Re}s>0;\ \left\{\begin{matrix}\lvert\arg b\rvert<\pi\\ b>a\end{matrix}\right\}\right]$
8	$(a-x)_+^{\lambda+\mu-1}C_{2m+\varepsilon}^\mu\left(\sqrt{\frac{x}{a}}\right)$ $\times C_n^\lambda\left(\frac{2x}{a}-1\right)$	$\frac{(-1)^{m+n}2^\varepsilon}{m!\,n!}a^{s+\lambda+\mu-1}(\mu)_{m+\varepsilon}(2\lambda)_n$ $\times\left(\frac{1-2s+2\lambda-\varepsilon}{2}\right)_n\Gamma\left[\begin{matrix}\frac{2\lambda+1}{2},\ \frac{2s+\varepsilon}{2}\\ \frac{2s+2n+2\lambda+\varepsilon+1}{2}\end{matrix}\right]$ $\times {}_4F_3\left(\begin{matrix}\frac{1-2m-2\mu}{2},\ \frac{2m+2\varepsilon+1}{2},\ \frac{2s-2\lambda+\varepsilon+1}{2},\ \frac{2s+\varepsilon}{2}\\ \frac{2\varepsilon+1}{2},\ \frac{2s-2n-2\lambda+\varepsilon+1}{2},\ \frac{2s+2n+2\lambda+\varepsilon+1}{2};\ 1\end{matrix}\right)$ $[a,\ \operatorname{Re}(\lambda+\mu)>0;\ \operatorname{Re}s>-\varepsilon/2]$
9	$(a-x)_+^{\lambda-1/2}(1-b^2x)^{\mu-1/2}$ $\times C_{2m+\varepsilon}^\mu(b\sqrt{x})$ $\times C_n^\lambda\left(\frac{2x}{a}-1\right)$	$\frac{(-1)^{m+n}}{m!\,n!}a^{s+\lambda+(\varepsilon-1)/2}(2b)^\varepsilon(\mu)_{m+\varepsilon}(2\lambda)_n$ $\times\left(\frac{1-2s+2\lambda-\varepsilon}{2}\right)_n\Gamma\left[\begin{matrix}\frac{2\lambda+1}{2},\ \frac{2s+\varepsilon}{2}\\ \frac{2s+2n+2\lambda+\varepsilon+1}{2}\end{matrix}\right]$ $\times {}_4F_3\left(\begin{matrix}\frac{1-2m-2\mu}{2},\ \frac{2m+2\varepsilon+1}{2},\ \frac{2s-2\lambda+\varepsilon+1}{2},\ \frac{2s+\varepsilon}{2}\\ \frac{2\varepsilon+1}{2},\ \frac{2s-2n-2\lambda+\varepsilon+1}{2},\ \frac{2s+2n+2\lambda+\varepsilon+1}{2};\ ab^2\end{matrix}\right)$ $[a>0;\ \operatorname{Re}\lambda>-1/2;\ \operatorname{Re}s>-\varepsilon/2;\ \lvert\arg(1-ab^2)<\pi\rvert]$
10	$(a^2-x^2)_+^{\lambda-1/2}C_m^\mu\left(\frac{b}{x}\right)$ $\times C_n^\lambda\left(\frac{x}{a}\right)$	$\frac{(-1)^m\pi}{m!\,n!}\left(\frac{a}{2}\right)^{s-m+2\lambda-1}(2b)^m(1-m-\mu)_m$ $\times\Gamma\left[\begin{matrix}n+2\lambda,\ s-m\\ \lambda,\ \frac{s-m-n+1}{2},\ \frac{s-m+n+2\lambda+1}{2}\end{matrix}\right]$ $\times {}_4F_3\left(\begin{matrix}-\frac{m}{2},\ \frac{1-m}{2},\ \frac{s-m}{2},\ \frac{s-m+1}{2};\ \frac{a^2}{b^2}\\ 1-m-\mu,\ \frac{s-m-n+1}{2},\ \frac{s-m+n+2\lambda+1}{2}\end{matrix}\right)$ $[a>0;\ \operatorname{Re}\lambda>-1/2;\ \operatorname{Re}s>m]$

3.25. The Jacobi Polynomials $P_n^{(\rho,\,\sigma)}(z)$

More formulas can be obtained from the corresponding section due to the relations

$$P_n^{(0,\,0)}(z) = P_n(z), \quad P_n^{(-1/2,\,-1/2)}(z) = \frac{(1/2)_n}{n!}\,T_n(z), \quad P_n^{(1/2,\,1/2)}(z) = \frac{(3/2)_n}{(n+1)!}\,U_n(z),$$

$$P_n^{(\lambda,\,\lambda)}(z) = \frac{(\lambda+1)_n}{(2\lambda+1)_n}\,C_n^{\lambda+1/2}(z),$$

$$P_\nu^{(\rho,\,\sigma)}(z) = \frac{\Gamma(\rho+\nu+1)}{\Gamma(\rho+1)\,\Gamma(\nu+1)}\,{}_2F_1\left(-\nu,\,\rho+\sigma+\nu+1;\,\rho+1;\,\frac{1-z}{2}\right).$$

3.25.1. $P_n^{(\rho,\,\sigma)}(\varphi(x))$ and algebraic functions

No.	$f(x)$	$F(s)$
1	$(2-x)_+^{\sigma}\,P_n^{(\rho,\,\sigma)}(1-x)$	$\frac{2^{s+\sigma}}{n!}\,\Gamma\begin{bmatrix} n+\sigma+1,\, s,\, 1-s+n+\rho \\ 1-s+\rho,\, s+n+\sigma+1 \end{bmatrix}$ [Re $\sigma > -1$; Re $s > 0$]
2	$(a-x)_+^{\alpha-1}\,P_n^{(\rho,\,\sigma)}(1-bx)$	$\frac{a^{s+\alpha-1}}{n!}\,\Gamma\begin{bmatrix} \alpha,\, n+\rho+1,\, s \\ \rho+1,\, s+\alpha \end{bmatrix}\,{}_3F_2\left(\begin{matrix} -n,\, n+\rho+\sigma+1,\, s \\ \rho+1,\, s+\alpha;\, \frac{ab}{2} \end{matrix}\right)$ [a, Re α, Re $s > 0$]
3	$(a-x)_+^{\alpha-1}\,(2-bx)^{\sigma}$ $\times\,P_n^{(\rho,\,\sigma)}(1-bx)$	$\frac{2^{\sigma}a^{s+\alpha-1}}{n!}\,\Gamma\begin{bmatrix} \alpha,\, n+\rho+1,\, s \\ \rho+1,\, s+\alpha \end{bmatrix}\,{}_3F_2\left(\begin{matrix} n+\rho+1,\, -n-\sigma,\, s \\ \rho+1,\, s+\alpha;\, \frac{ab}{2} \end{matrix}\right)$ [a, Re α, Re $s > 0$]
4	$(a-x)_+^{\rho}$ $\times\,P_n^{(\rho,\,\sigma)}(bx-ab+1)$	$a^{s+\rho}\,\Gamma\begin{bmatrix} n+\rho+1,\, s \\ s+n+\rho+1 \end{bmatrix}\,P_n^{(s+\rho,\,\sigma-s)}(1-ab)$ [a, Re $s > 0$; Re $\rho > -1$]
5	$(a-x)_+^{\rho}\,(bx+1)^{\alpha}$ $\times\,P_n^{(\rho,\,\sigma)}(1-cx+ac)$	$\frac{(\rho+1)_n}{n!}\,a^{s+\rho}\,\mathrm{B}(s,\,\rho+1)$ $\times\,F_3\left(-\alpha,\,-n,\,s,\,n+\rho+\sigma+1;\,s+\rho+1;\,-ab,\,-\frac{ac}{2}\right)$ [a, Re $s > 0$; Re $\rho > -1$; $\lvert\arg(ab+1)\rvert < \pi$]
6	$(a-x)_+^{\rho}\,P_n^{(\rho,\,\sigma)}\left(\frac{2x}{a}-1\right)$	$\frac{a^{s+\rho}}{n!}\,\Gamma\begin{bmatrix} s,\, s-\sigma,\, n+\rho+1 \\ s+n+\rho+1,\, s-n-\sigma \end{bmatrix}$ [a, Re $s > 0$; Re $\rho > -1$]
7	$(a-x)_+^{\alpha-1}$ $\times\,P_n^{(\rho,\,\sigma)}\left(\frac{2x}{b}\pm 1\right)$	$\frac{(\pm 1)^n\,a^{s+\alpha-1}}{n!}\,(\varphi+1)_n\,\mathrm{B}(\alpha,\,s)\,{}_3F_2\left(\begin{matrix} -n,\, n+\rho+\sigma+1,\, s \\ \varphi+1,\, s+\alpha;\, \mp\frac{a}{b} \end{matrix}\right)$ $\left[a,\ \mathrm{Re}\,\alpha,\ \mathrm{Re}\,s > 0;\ \varphi = \begin{Bmatrix} \rho \\ \sigma \end{Bmatrix}\right]$

No.	$f(x)$	$F(s)$
8	$(a-x)_+^{\sigma}$ $\times P_n^{(\rho,\sigma)}\left(\frac{a-b-2x}{a+b}\right)$	$a^{s+\sigma}\,\mathrm{B}(n+\sigma+1,\,s)\,P_n^{(\rho-s,\,s+\sigma)}\left(\frac{a-b}{a+b}\right)$ $[a>0;\ \rho>-1;\ \mathrm{Re}\,s>0;\ \mathrm{Re}\,\sigma>-1]$
9	$(x-a)_+^{\rho}\,P_n^{(\rho,\sigma)}\left(\frac{2x}{a}-1\right)$	$\frac{a^{s+\rho}}{n!}\,\Gamma\left[\begin{matrix}1-s+n+\sigma,\ -s-n-\rho,\ n+\rho+1\\ 1-s+\sigma,\ 1-s\end{matrix}\right]$ $[a>0;\ \mathrm{Re}\,\rho>-1;\ \mathrm{Re}\,s<-\mathrm{Re}\,\rho-n]$
10	$(x+a)^{\sigma}\,P_n^{(\rho,\sigma)}\left(\frac{2x}{a}+1\right)$	$\frac{a^{s+\sigma}}{n!}\,\Gamma\left[\begin{matrix}s,\ 1-s+n+\rho,\ -s-n-\sigma\\ 1-s+\rho,\ -n-\sigma\end{matrix}\right]$ $[0<\mathrm{Re}\,s<-\mathrm{Re}\,\sigma-n;\ \lvert\arg a\rvert<\pi]$
11	$\frac{(a-x)_+^{\rho}}{x-b}\,P_n^{(\rho,\sigma)}\left(\frac{2x}{a}-1\right)$	$\frac{(-1)^n\,a^{s+\rho-1}}{n!}\,(2-s+\sigma)_n\,\mathrm{B}(n+\rho+1,\,s-1)$ $\times\,{}_3F_2\left(\begin{matrix}1,\ 1-s-n-\rho,\ 2-s+n+\sigma\\ 2-s,\ 2-s+\sigma;\ \frac{b}{a}\end{matrix}\right)$ $-\pi\,(a-b)^{\rho}\,b^{s-1}\cot(s\pi)\,P_n^{(\rho,\sigma)}\left(\frac{2b-a}{a}\right)$ $[a>b]$
12		$=\frac{(-1)^{n+1}\,a^{s+\rho}}{n!\,b}\,(1-s+\sigma)_n\,\mathrm{B}(n+\rho+1,\,s)$ $\times\,{}_3F_2\left(\begin{matrix}1,\ s-\sigma,\ s;\ \frac{a}{b}\\ s-n-\sigma,\ s+n+\rho+1\end{matrix}\right)$ $[a<b]$ $[a,\ b,\ \mathrm{Re}\,s>0;\ \mathrm{Re}\,\rho>-1]$
13	$\frac{(x+a)^{\sigma}}{x-b}\,P_n^{(\rho,\sigma)}\left(\frac{2x}{a}+1\right)$	$\frac{a^{s+\sigma-1}}{n!}\,(2-s+\rho)_n\,\mathrm{B}(1-s-n-\sigma,\,s-1)$ $\times\,{}_3F_2\left(\begin{matrix}1,\ 1-s-n-\sigma,\ 2-s+n+\rho\\ 2-s,\ 2-s+\rho;\ -\frac{b}{a}\end{matrix}\right)$ $-\pi\,(a+b)^{\sigma}\,b^{s-1}\cot(s\pi)\,P_n^{(\rho,\sigma)}\left(\frac{a+2b}{a}\right)$ $[b>0;\ 0<\mathrm{Re}\,s<-\mathrm{Re}\,\sigma-n+1;\ \lvert\arg a\rvert<\pi]$
14	$(a-x)_+^{\alpha-1}\left\{\begin{matrix}(x+b)^{\sigma}\\ (b-x)^{\rho}\end{matrix}\right\}$ $\times P_n^{(\rho,\sigma)}\left(\frac{2x}{b}\pm1\right)$	$\frac{(\pm1)^n\,a^{s+\alpha-1}b^{\psi}}{n!}\,(\varphi+1)_n\,\mathrm{B}(\alpha,\,s)\,{}_3F_2\left(\begin{matrix}-n-\psi,\ n+\varphi+1,\ s\\ \varphi+1,\ s+\alpha;\ \mp\frac{a}{b}\end{matrix}\right)$ $\left[\left\{\begin{matrix}a>0;\ \lvert\arg b\rvert<\pi\\ b>a>0\end{matrix}\right\};\ \mathrm{Re}\,\alpha,\ \mathrm{Re}\,s>0;\ \varphi=\left\{\begin{matrix}\rho\\ \sigma\end{matrix}\right\},\ \psi=\left\{\begin{matrix}\sigma\\ \rho\end{matrix}\right\}\right]$

No.	$f(x)$	$F(s)$
15	$(a-x)_+^{\rho}(b\pm x)^{\tau} \times P_n^{(\rho,\sigma)}\left(\frac{2x}{a}-1\right)$	$\frac{(-1)^n a^{s+\rho}b^{\tau}}{n!}(1-s+\sigma)_n \mathrm{B}(n+\rho+1,\, s) \times {}_3F_2\left(\begin{matrix}-\tau,\, s-\sigma,\, s;\, \mp\frac{a}{b}\\ s-n-\sigma,\, s+n+\rho+1\end{matrix}\right)$ $\left[\left\{\begin{matrix}a>0;\ \lvert\arg b\rvert<\pi\\ b>a>0\end{matrix}\right\};\ \mathrm{Re}\,\rho>-1;\ \mathrm{Re}\,s>0\right]$
16	$(x+a)^{\sigma}(x+b)^{\tau} \times P_n^{(\rho,\sigma)}\left(\frac{2x}{a}+1\right)$	$\frac{a^{s+\sigma+\tau}}{n!}(1-s+\rho-\tau)_n \mathrm{B}(s+\tau,\, -s-n-\tau-\sigma) \times {}_3F_2\left(\begin{matrix}-\tau,\, -s-n-\sigma-\tau,\, 1-s+n+\rho-\tau\\ 1-s-\tau,\, 1-s+\rho-\tau;\, \frac{b}{a}\end{matrix}\right) + \frac{a^{\sigma}b^{s+\tau}}{n!}(\rho+1)_n \mathrm{B}(-s-\tau,\, s) \times {}_3F_2\left(\begin{matrix}-n-\sigma,\, n+\rho+1,\, s\\ \rho+1,\, s+\tau+1;\, \frac{b}{a}\end{matrix}\right)$ $[a>0;\ 0<\mathrm{Re}\,s<-\mathrm{Re}\,(\sigma+\tau)-n;\ \lvert\arg b\rvert<\pi]$
17	$(x+a-b)^{\rho}(x+a+b)^{\tau} \times P_n^{(\rho,\sigma)}\left(\frac{x+a}{b}\right)$	$\frac{(a+b)^{s+n+\rho+\tau}}{(2b)^n\, n!}(n+\rho+\sigma+1)_n \mathrm{B}(-s-n-\rho-\tau,\, s) \times {}_3F_2\left(\begin{matrix}-n-\rho-\sigma,\, -n-\rho,\, -s-n-\rho-\tau\\ -n-\rho-\tau,\, -2n-\rho-\sigma;\, \frac{2b}{a+b}\end{matrix}\right)$ $[a>b>0;\ 0<\mathrm{Re}\,s<-\mathrm{Re}\,(\tau+\rho)-n]$
18	$(a-x)_+^{\alpha-1} P_n^{(\rho,\sigma)}\left(\frac{x+b}{a+b}\right)$	$\frac{a^{s+\alpha-1}}{n!}(\rho+1)_n \mathrm{B}(\alpha,\, s)\ {}_3F_2\left(\begin{matrix}-n,\, n+\rho+\sigma+1,\, \alpha\\ \rho+1,\, s+\alpha;\, \frac{a}{2(a+b)}\end{matrix}\right)$ $[a,\ \mathrm{Re}\,\alpha,\ \mathrm{Re}\,s>0]$
19	$(a-x)_+^{\alpha-1}(x+a+2b)^{\sigma} \times P_n^{(\rho,\sigma)}\left(\frac{x+b}{a+b}\right)$	$\frac{2^{\sigma}(a+b)^{\sigma}a^{s+\alpha-1}}{n!}(\rho+1)_n \mathrm{B}(\alpha,\, s)\ {}_3F_2\left(\begin{matrix}-n-\sigma,\, n+\rho+1,\, \alpha\\ \rho+1,\, s+\alpha;\, \frac{a}{2(a+b)}\end{matrix}\right)$ $[a,\ b,\ \mathrm{Re}\,\alpha,\ \mathrm{Re}\,s>0]$
20	$(a-x)_+^{\rho} P_n^{(\rho,\sigma)}\left(\frac{2a}{x}-1\right)$	$\frac{a^{s+\rho}}{n!}\Gamma\left[\begin{matrix}s-n,\, s+\rho+\sigma+n+1,\, n+\rho+1\\ s+\rho+1,\, s+\rho+\sigma+1\end{matrix}\right]$ $[a>0;\ \mathrm{Re}\,\rho>-1;\ \mathrm{Re}\,s>n]$
21	$(x-a)_+^{\rho} P_n^{(\rho,\sigma)}\left(\frac{2a}{x}-1\right)$	$\frac{a^{s+\rho}}{n!}\Gamma\left[\begin{matrix}-s-\rho,\, -s-\rho-\sigma,\, n+\rho+1\\ 1-s+n,\, -s-n-\rho-\sigma\end{matrix}\right]$ $[a>0;\ \mathrm{Re}\,s<-\mathrm{Re}\,\rho<1]$

No.	$f(x)$	$F(s)$
22	$(x+a)^{\sigma} P_n^{(\rho,\sigma)}\left(\frac{2a}{x}+1\right)$	$\frac{a^{s+\sigma}}{n!}\Gamma\left[\begin{matrix} s-n,\, s+n+\rho+\sigma+1,\, -s-\sigma \\ s+\rho+\sigma+1,\, -n-\sigma \end{matrix}\right]$ $[n<\operatorname{Re}s<-\operatorname{Re}\sigma;\ \lvert\arg a\rvert<\pi]$
23	$(a-x)_+^{-(n+\rho+\sigma+1)}$ $\times P_n^{(\rho,\sigma)}\left(\frac{a+x}{a-x}\right)$	$\frac{a^{s-(n+\rho+\sigma+1)}}{n!}\Gamma\left[\begin{matrix} s,\, 1-s+n+\rho,\, -n-\rho-\sigma \\ s-n-\rho-\sigma,\, 1-s+\rho \end{matrix}\right]$ $[a,\ \operatorname{Re}s>0;\ \operatorname{Re}(\rho+\sigma)<-2n]$
24	$(x-a)_+^{-(n+\rho+\sigma+1)}$ $\times P_n^{(\rho,\sigma)}\left(\frac{x+a}{x-a}\right)$	$\frac{a^{s-(n+\rho+\sigma+1)}}{n!}\Gamma\left[\begin{matrix} s-\sigma,\, 1-s+n+\rho+\sigma,\, -n-\rho-\sigma \\ 1-s,\, s-n-\sigma \end{matrix}\right]$ $[a>0;\ \operatorname{Re}s<\operatorname{Re}(\rho+\sigma)+n+1<1-n]$
25	$(x+a)^{-(n+\rho+\sigma+1)}$ $\times P_n^{(\rho,\sigma)}\left(\frac{a-x}{a+x}\right)$	$\frac{a^{s-(n+\rho+\sigma+1)}}{n!}\Gamma\left[\begin{matrix} s,\, 1-s+n+\rho,\, 1-s+n+\rho+\sigma \\ 1-s+\rho,\, n+\rho+\sigma+1 \end{matrix}\right]$ $[0<\operatorname{Re}s<\operatorname{Re}(\rho+\sigma)+n+1;\ \lvert\arg a\rvert<\pi]$
26	$(x+a)^{-(n+\rho+\sigma+1)}$ $\times P_n^{(\rho,\sigma)}\left(\frac{b-x}{a+x}\right)$	$a^{s-(n+\rho+\sigma+1)}\,\mathrm{B}(1-s+n+\rho+\sigma,\, s)\,P_n^{(\rho-s,\sigma)}\left(\frac{b}{a}\right)$ $[0<\operatorname{Re}s<\operatorname{Re}(\rho+\sigma)+n+1;\ \lvert\arg a\rvert<\pi]$
27	$(a-x)_+^{\rho}$ $\times P_n^{(\rho,\sigma)}\left(\frac{a+b-x}{b}\right)$	$\frac{(2b)^{-n}(a+2b)^{s+n+\rho}}{n!}(n+\rho+\sigma+1)_n\,\mathrm{B}(n+\rho+1,\, s)$ $\times\, {}_2F_1\left(\begin{matrix} -n-\rho-\sigma,\, -s-n-\rho \\ -2n-\rho-\sigma;\ \frac{2b}{a+2b} \end{matrix}\right)$ $[a,\ \operatorname{Re}s>0;\ \operatorname{Re}\rho>-1]$

3.25.2. $P_n^{(\rho,\sigma)}(\varphi(x))$ and the exponential function

1	$e^{-bx}P_n^{(\rho,\sigma)}\left(\frac{2x}{a}\pm 1\right)$	$\frac{a^{-n}b^{-s-n}}{n!}(n+\rho+\sigma+1)_n\,\Gamma(s+n)$ $\times\, {}_2F_2\left(\begin{matrix} -n,\, -n-\varphi;\ ab \\ -2n-\rho-\sigma,\, 1-s-n \end{matrix}\right)$ $\left[\operatorname{Re}b,\ \operatorname{Re}s>0;\ \varphi=\left\{\begin{matrix}\rho\\ \sigma\end{matrix}\right\}\right]$
2	$(a-x)_+^{\sigma}\, e^{-bx}$ $\times P_n^{(\rho,\sigma)}\left(1-\frac{2x}{a}\right)$	$\frac{a^{s+\sigma}}{n!}(1-s+\rho)_n\,\mathrm{B}(n+\sigma+1,\, s)\ {}_2F_2\left(\begin{matrix} s-\rho,\, s;\ -ab \\ s-n-\rho,\, s+n+\sigma+1 \end{matrix}\right)$ $[a,\ \operatorname{Re}s>0;\ \operatorname{Re}\sigma>-1]$

No.	$f(x)$	$F(s)$
3	$(x+a)^{\sigma}e^{-bx}$ $\times P_n^{(\rho,\,\sigma)}\left(\frac{2x}{a}+1\right)$	$\frac{a^{-n}b^{-s-n-\sigma}}{n!}(n+\rho+\sigma+1)_n\,\Gamma(s+n+\sigma)$ $\times {}_2F_2\left(\begin{matrix}-n-\sigma,\,-n-\rho-\sigma;\,ab\\-2n-\rho-\sigma,\,1-s-n-\sigma\end{matrix}\right)$ $+\frac{a^{s+\sigma}}{n!}(1-s+\rho)_n\,\mathrm{B}(-s-n-\sigma,\,s)$ $\times {}_2F_2\left(\begin{matrix}s-\rho,\,s;\,ab\\s-n-\rho,\,s+n+\sigma+1\end{matrix}\right)$ $[\operatorname{Re}b,\ \operatorname{Re}s>0;\ \lvert\arg a\rvert<\pi]$
4	$e^{-b/x}P_n^{(\rho,\,\sigma)}\left(\frac{2x}{a}\pm1\right)$	$\frac{(\pm1)^n b^s}{n!}(\varphi+1)_n\,\Gamma(-s)\ {}_2F_2\left(\begin{matrix}-n,\,n+\rho+\sigma+1\\\varphi+1,\,s+1;\ \pm\frac{b}{a}\end{matrix}\right)$ $\left[\operatorname{Re}b>0;\ \operatorname{Re}s<-n;\ \varphi=\begin{Bmatrix}\rho\\\sigma\end{Bmatrix}\right]$
5	$(a-x)_+^{\rho}\,e^{-b/x}$ $\times P_n^{(\rho,\sigma)}\left(\frac{2x}{a}-1\right)$	$\frac{(-1)^n a^{\rho}b^s}{n!}(\sigma+1)_n\,\Gamma(-s)\ {}_2F_2\left(\begin{matrix}-n-\rho,\,n+\sigma+1\\\sigma+1,\,s+1;\,-\frac{b}{a}\end{matrix}\right)$ $+\frac{(-1)^n a^{s+\rho}}{n!}(1-s+\sigma)_n\,\mathrm{B}(n+\rho+1,\,s)$ $\times {}_2F_2\left(\begin{matrix}-s-n-\rho,\,1-s+n+\sigma\\1-s,\,1-s+\sigma;\,-\frac{b}{a}\end{matrix}\right)$ $[a,\ \operatorname{Re}b>0;\ \operatorname{Re}\rho>-1]$
6	$(x+a)^{\sigma}e^{-b/x}$ $\times P_n^{(\rho,\,\sigma)}\left(\frac{2x}{a}+1\right)$	$\frac{a^{s+\sigma}}{n!}(1-s+\rho)_n\,\mathrm{B}(-s-n-\sigma,\,s)$ $\times {}_2F_2\left(\begin{matrix}-s-n-\sigma,\,1-s+n+\rho\\1-s,\,1-s+\rho;\,\frac{b}{a}\end{matrix}\right)$ $+\frac{a^{\sigma}b^s}{n!}(\rho+1)_n\,\Gamma(-s)\ {}_2F_2\left(\begin{matrix}-n-\sigma,\,n+\rho+1\\\rho+1,\,s+1;\,\frac{b}{a}\end{matrix}\right)$ $[\operatorname{Re}b>0;\ \operatorname{Re}(s+\sigma)<-n;\ \lvert\arg a\rvert<\pi]$
7	$e^{-b\sqrt{x}}P_n^{(\rho,\,\sigma)}\left(\frac{2x}{a}\pm1\right)$	$\frac{2a^{-n}b^{-2s-2n}}{n!}(n+\rho+\sigma+1)_n\,\Gamma(2s+2n)$ $\times {}_2F_3\left(\begin{matrix}-n,\,-n-\varphi;\ \mp\frac{ab^2}{4}\\-2n-\rho-\sigma,\,1-s-n,\,\frac{1-2s-2n}{2}\end{matrix}\right)$ $\left[\operatorname{Re}b,\ \operatorname{Re}s>0;\ \varphi=\begin{Bmatrix}\rho\\\sigma\end{Bmatrix}\right]$

No.	$f(x)$	$F(s)$
8	$(a-x)_+^{\rho}\, e^{-b\sqrt{x}}$ $\times P_n^{(\rho,\sigma)}\left(\frac{2x}{a}-1\right)$	$\frac{(-1)^n a^{s+\rho}}{n!}(1-s+\sigma)_n \operatorname{B}(n+\rho+1,\, s) \times {}_2F_3\left(\begin{matrix} s-\sigma,\, s;\, \frac{ab^2}{4} \\ \frac{1}{2},\, s-n-\sigma,\, s+n+\rho+1 \end{matrix}\right) - \frac{(-1)^n a^{s+\rho+1/2} b}{n!}\left(\frac{1}{2}-s+\sigma\right)_n \operatorname{B}\left(n+\rho+1,\, s+\frac{1}{2}\right) \times {}_2F_3\left(\begin{matrix} s+\frac{1}{2},\, s-\sigma+\frac{1}{2};\, \frac{ab^2}{4} \\ \frac{3}{2},\, s-n-\sigma+\frac{1}{2},\, s+n+\rho+\frac{3}{2} \end{matrix}\right)$ $[a,\ \operatorname{Re} s>0;\ \operatorname{Re}\rho>-1]$
9	$(x+a)^{\sigma}\, e^{-b\sqrt{x}}$ $\times P_n^{(\rho,\sigma)}\left(\frac{2x}{a}+1\right)$	$-\frac{a^{s+\sigma+1/2} b}{n!}\left(\frac{1}{2}-s+\rho\right)_n \operatorname{B}\left(-s-n-\sigma-\frac{1}{2},\, s+\frac{1}{2}\right) \times {}_2F_3\left(\begin{matrix} \frac{2s+1}{2},\, \frac{2s-2\rho+1}{2};\, -\frac{ab^2}{4} \\ \frac{3}{2},\, \frac{2s-2n-2\rho+1}{2},\, \frac{2s+2n+2\sigma+3}{2} \end{matrix}\right) + \frac{a^{s+\sigma}}{n!}(1-s+\rho)_n \times \operatorname{B}(-s-n-\sigma,\, s)\; {}_2F_3\left(\begin{matrix} s-\rho,\, s;\, -\frac{ab^2}{4} \\ \frac{1}{2},\, s-n-\rho,\, s+n+\sigma+1 \end{matrix}\right) + \frac{2a^{-n} b^{-2s-2n-2\sigma}}{n!}(n+\rho+\sigma+1)_n \Gamma(2s+2n+2\sigma) \times {}_2F_3\left(\begin{matrix} -n-\sigma,\, -n-\rho-\sigma;\, -\frac{ab^2}{4} \\ -2n-\rho-\sigma,\, \frac{1}{2}-s-n-\sigma,\, 1-s-n-\sigma \end{matrix}\right)$ $[\operatorname{Re} b,\ \operatorname{Re} s>0;\ \|\arg a\|<\pi]$
10	$e^{-b/\sqrt{x}}\, P_n^{(\rho,\sigma)}\left(\frac{2x}{a}\pm 1\right)$	$\frac{2(\pm 1)^n b^{2s}}{n!}(\varphi+1)_n \Gamma(-2s)\; {}_2F_3\left(\begin{matrix} -n,\, n+\rho+\sigma+1 \\ \varphi+1,\, \frac{2s+1}{2},\, s+1;\, \mp\frac{b^2}{4a} \end{matrix}\right)$ $\left[\operatorname{Re} b>0;\ \operatorname{Re} s<-n;\ \varphi=\left\{\begin{matrix}\rho\\ \sigma\end{matrix}\right\}\right]$
11	$(x+a)^{\sigma}\, e^{-b/\sqrt{x}}$ $\times P_n^{(\rho,\sigma)}\left(\frac{2x}{a}+1\right)$	$\frac{a^{s+\sigma}}{n!}(1-s+\rho)_n \operatorname{B}(-s-n-\sigma,\, s) \times {}_2F_3\left(\begin{matrix} -s-n-\sigma,\, 1-s+n+\rho \\ \frac{1}{2},\, 1-s,\, 1-s+\rho;\, -\frac{b^2}{4a} \end{matrix}\right) + \frac{2a^{\sigma} b^{2s}}{n!}(\rho+1)_n \times \Gamma(-2s)\; {}_2F_3\left(\begin{matrix} -n-\sigma,\, n+\rho+1 \\ \rho+1,\, s+\frac{1}{2},\, s+1;\, -\frac{b^2}{4a} \end{matrix}\right)$ $[\operatorname{Re} b>0;\ \operatorname{Re}(s+\sigma)<-n;\ \|\arg a\|<\pi]$
12	$(x+a)^n\, e^{-bx}$ $\times P_n^{(\rho,\sigma)}\left(\frac{a-x}{a+x}\right)$	$\frac{(-1)^n b^{-s-n}}{n!}(\sigma+1)_n \Gamma(s+n)\; {}_2F_2\left(\begin{matrix} -n,\, -n-\rho;\, ab \\ \sigma+1,\, 1-s-n \end{matrix}\right)$ $[\operatorname{Re} b,\ \operatorname{Re} s>0]$

No.	$f(x)$	$F(s)$
13	$(x+a)^{-(n+\rho+\sigma+1)}\,e^{-bx}$ $\times P_n^{(\rho,\sigma)}\left(\frac{a-x}{a+x}\right)$	$\frac{a^{s-(n+\rho+\sigma+1)}}{n!}(-s+\rho+1)_n\,\mathrm{B}(s,\,-s+n+\rho+\sigma+1)$ $\times\,{}_2F_2\left(\begin{matrix}s,\,s-\rho;\,ab\\ s-n-\rho,\,s-n-\rho-\sigma\end{matrix}\right)$ $+\frac{(-1)^n b^{-s+n+\rho+\sigma+1}}{n!}(\sigma+1)_n\,\Gamma(s-n-\rho-\sigma-1)$ $\times\,{}_2F_2\left(\begin{matrix}n+\sigma+1,\,n+\rho+\sigma+1\\ \sigma+1,\,-s+n+\rho+\sigma+2;\,ab\end{matrix}\right)$ $[\operatorname{Re} b,\ \operatorname{Re} s>0;\ \lvert\arg a\rvert<\pi]$
14	$e^{-b/x}\,(x+a)^{-(n+\rho+\sigma+1)}$ $\times P_n^{(\rho,\sigma)}\left(\frac{a-x}{a+x}\right)$	$\frac{a^{-(n+\rho+\sigma+1)}b^s}{n!}(\rho+1)_n\,\Gamma(-s)\ {}_2F_2\left(\begin{matrix}n+\rho+1,\,n+\rho+\sigma+1\\ \rho+1,\,s+1;\,\frac{b}{a}\end{matrix}\right)$ $+\frac{a^{s-(n+\rho+\sigma+1)}}{n!}(1-s+\rho)_n\,\mathrm{B}(1-s+n+\rho+\sigma,\,s)$ $\times\,{}_2F_2\left(\begin{matrix}1-s+n+\rho,\,1-s+n+\rho+\sigma\\ 1-s,\,1-s+\rho;\,\frac{b}{a}\end{matrix}\right)$ $[\operatorname{Re} b>0;\ \operatorname{Re} s<\operatorname{Re}(\rho+\sigma)+n+1;\ \lvert\arg a\rvert<\pi]$
15	$e^{-b/x}\,(x+a)^n$ $\times P_n^{(\rho,\sigma)}\left(\frac{a-x}{a+x}\right)$	$\frac{a^n b^s}{n!}(\rho+1)_n\,\Gamma(-s)\ {}_2F_2\left(\begin{matrix}-n,\,-n-\sigma\\ \rho+1,\,s+1;\,\frac{b}{a}\end{matrix}\right)$ $[\operatorname{Re} b>0;\ \operatorname{Re} s<-n]$
16	$e^{-b\sqrt{x}}\,(x+a)^n$ $\times P_n^{(\rho,\sigma)}\left(\frac{a-x}{a+x}\right)$	$\frac{2(-1)^n b^{-2s-2n}}{n!}(\sigma+1)_n\,\Gamma(2s+2n)$ $\times\,{}_2F_3\left(\begin{matrix}-n,\,-n-\rho;\,-\frac{ab^2}{4}\\ \sigma+1,\,1-s-n,\,\frac{1-2s-2n}{2}\end{matrix}\right)$ $[\operatorname{Re} b,\ \operatorname{Re} s>0]$
17	$e^{-b\sqrt{x}}\,(x+a)^{-(n+\rho+\sigma+1)}$ $\times P_n^{(\rho,\sigma)}\left(\frac{a-x}{a+x}\right)$	$\frac{a^{s-(n+\rho+\sigma+1)}}{n!}(1-s+\rho)_n\,\mathrm{B}(1-s+n+\rho+\sigma,\,s)$ $\times\,{}_2F_3\left(\begin{matrix}s,\,s-\rho;\,-\frac{ab^2}{4}\\ \frac{1}{2},\,s-n-\rho-\sigma,\,s-n-\rho\end{matrix}\right)-\frac{a^{s-(n+\rho+\sigma+1/2)}b}{n!}$ $\times\left(\frac{1}{2}-s+\rho\right)_n\mathrm{B}\left(s+\frac{1}{2},\,\frac{1}{2}-s+n+\rho+\sigma\right)$ $\times\,{}_2F_3\left(\begin{matrix}\frac{2s+1}{2},\,\frac{2s-2\rho+1}{2};\,-\frac{ab^2}{4}\\ \frac{3}{2},\,\frac{2s-2n-2\rho+1}{2},\,\frac{2s-2n-2\rho-2\sigma+1}{2}\end{matrix}\right)+\frac{2(-1)^n}{n!}$ $\times\,b^{2(-s+n+\rho+\sigma+1)}(\sigma+1)_n\,\Gamma(2s-2n-2\rho-2\sigma-2)$ $\times\,{}_2F_3\left(\begin{matrix}n+\sigma+1,\,n+\rho+\sigma+1;\,-\frac{ab^2}{4}\\ \sigma+1,\,\frac{3-2s+2n+2\rho+2\sigma}{2},\,2-s+n+\rho+\sigma\end{matrix}\right)$ $[\operatorname{Re} b,\ \operatorname{Re} s>0;\ \lvert\arg a\rvert<\pi]$

No.	$f(x)$	$F(s)$
18	$e^{-b/\sqrt{x}}(x+a)^n \times P_n^{(\rho,\sigma)}\left(\frac{a-x}{a+x}\right)$	$\frac{2a^n b^{2s}}{n!}(\rho+1)_n\Gamma(-2s)\,{}_2F_3\left(\begin{matrix}-n,\,-n-\sigma;\,-\frac{b^2}{4a}\\ \rho+1,\,\frac{2s+1}{2},\,s+1\end{matrix}\right)$ [Re $b>0$; Re $s<-n$]
19	$\frac{e^{-b/\sqrt{x}}}{(x+a)^{n+\rho+\sigma+1}} \times P_n^{(\rho,\sigma)}\left(\frac{a-x}{a+x}\right)$	$\frac{2a^{-(n+\rho+\sigma+1)}b^{2s}}{n!}(\rho+1)_n\Gamma(-2s)\,{}_2F_3\left(\begin{matrix}n+\rho+1,\,n+\rho+\sigma+1\\ \rho+1,\,s+\frac{1}{2},\,s+1;\,-\frac{b^2}{4a}\end{matrix}\right) + \frac{a^{s-(n+\rho+\sigma+1)}}{n!}(1-s+\rho)_n\,\mathrm{B}(1-s+n+\rho+\sigma,\,s) \times {}_2F_3\left(\begin{matrix}1-s+n+\rho,\,1-s+n+\rho+\sigma\\ \frac{1}{2},\,1-s,\,1-s+\rho;\,-\frac{b^2}{4a}\end{matrix}\right) - \frac{a^{s-(n+\rho+\sigma+3/2)}b}{n!} \times \left(\frac{3}{2}-s+\rho\right)_n \mathrm{B}\left(\frac{3}{2}-s+n+\rho+\sigma,\,s-\frac{1}{2}\right) \times {}_2F_3\left(\begin{matrix}\frac{3}{2}-s+n+\rho,\,\frac{3}{2}-s+n+\rho+\sigma\\ \frac{3}{2},\,\frac{3}{2}-s,\,\frac{3}{2}-s+\rho;\,-\frac{b^2}{4a}\end{matrix}\right)$ [Re $b>0$; Re $s<$ Re$(\rho+\sigma)+n+1$; $\lvert\arg a\rvert<\pi$]

3.25.3. $P_n^{(\rho,\sigma)}(\varphi(x))$ and trigonometric functions

Notation: $\delta=\left\{\begin{matrix}1\\0\end{matrix}\right\}$.

1	$(a-x)_+^{\rho}\left\{\begin{matrix}\sin(b\sqrt{x})\\ \cos(b\sqrt{x})\end{matrix}\right\} \times P_n^{(\rho,\sigma)}\left(\frac{2x}{a}-1\right)$	$\frac{(-1)^n a^{s+\rho+\delta/2}b^{\delta}}{n!}\left(1-s+\sigma-\frac{\delta}{2}\right)_n \mathrm{B}\left(n+\rho+1,\,s+\frac{\delta}{2}\right) \times {}_2F_3\left(\begin{matrix}s-\sigma+\frac{\delta}{2},\,s+\frac{\delta}{2};\,-\frac{ab^2}{4}\\ \frac{2\delta+1}{2},\,\frac{2s-2n-2\sigma+\delta}{2},\,\frac{2s+2n+2\rho+\delta+2}{2}\end{matrix}\right)$ [a, Re $s>0$; Re $\rho>-1$]
2	$(x+a)^{\sigma}\left\{\begin{matrix}\sin(b\sqrt{x})\\ \cos(b\sqrt{x})\end{matrix}\right\} \times P_n^{(\rho,\sigma)}\left(\frac{2x}{a}+1\right)$	$\frac{a^{s+\sigma+\delta/2}b^{\delta}}{n!}\left(1-s+\rho-\frac{\delta}{2}\right)_n \mathrm{B}\left(-s-n-\sigma-\frac{\delta}{2},\,s+\frac{\delta}{2}\right) \times {}_2F_3\left(\begin{matrix}s-\rho+\frac{\delta}{2},\,s+\frac{\delta}{2};\,\frac{ab^2}{4}\\ \frac{2\delta+1}{2},\,\frac{2s-2n-2\rho+\delta}{2},\,\frac{2s+2n+2\sigma+\delta+2}{2}\end{matrix}\right) + \frac{2(-1)^n}{a^n b^{2s+2n+2\sigma}n!}\left\{\begin{matrix}\sin[(s+\sigma)\pi]\\ \cos[(s+\sigma)\pi]\end{matrix}\right\} \times (n+\rho+\sigma+1)_n\Gamma(2s+2n+2\sigma) \times {}_2F_3\left(\begin{matrix}-n-\sigma,\,-n-\rho-\sigma;\,\frac{ab^2}{4}\\ -2n-\rho-\sigma,\,\frac{1-2s-2n-2\sigma}{2},\,1-s-n-\sigma\end{matrix}\right)$ [$b>0$; $-\delta/2<$ Re $s<1/2-$ Re $\sigma-n$; $\lvert\arg a\rvert<\pi$]

No.	$f(x)$	$F(s)$
3	$(x+a)^{\sigma}\begin{Bmatrix}\sin(b/\sqrt{x})\\ \cos(b/\sqrt{x})\end{Bmatrix}$ $\times P_n^{(\rho,\sigma)}\left(\frac{2x}{a}+1\right)$	$\frac{a^{s+\sigma-\delta/2}b^{\delta}}{n!}\left(1-s+\rho+\frac{\delta}{2}\right)_n \mathrm{B}\left(-s-n-\sigma+\frac{\delta}{2},\, s-\frac{\delta}{2}\right)$ $\times {}_2F_3\left(\begin{matrix}1-s+n+\rho+\frac{\delta}{2},\; -s-n-\sigma+\frac{\delta}{2}\\ \delta+\frac{1}{2},\, 1-s+\frac{\delta}{2},\, 1-s+\rho+\frac{\delta}{2};\, \frac{b^2}{4a}\end{matrix}\right)$ $\mp\frac{2a^{\sigma}b^{2s}}{n!}(\rho+1)_n\Gamma(-2s)\begin{Bmatrix}\sin(s\pi)\\ \cos(s\pi)\end{Bmatrix}$ $\times {}_2F_3\left(\begin{matrix}-n-\sigma,\, n+\rho+1\\ \rho+1,\, \frac{2s+1}{2},\, s+1;\, \frac{b^2}{4a}\end{matrix}\right)$ $[b>0;\ -1/2<\operatorname{Re}s<\delta/2-\operatorname{Re}\sigma-n;\ \lvert\arg a\rvert<\pi]$
4	$(x+a)^{-(n+\rho+\sigma+1)}$ $\times\begin{Bmatrix}\sin(b\sqrt{x})\\ \cos(b\sqrt{x})\end{Bmatrix}$ $\times P_n^{(\rho,\sigma)}\left(\frac{a-x}{a+x}\right)$	$\frac{a^{s-(n+\rho+\sigma-\delta/2+1)}b^{\delta}}{n!}\left(1-s+\rho-\frac{\delta}{2}\right)_n$ $\times\mathrm{B}\left(s+\frac{\delta}{2},\, 1-s+n+\rho+\sigma-\frac{\delta}{2}\right)$ $\times {}_2F_3\left(\begin{matrix}s+\frac{\delta}{2}-\rho,\, s+\frac{\delta}{2};\, \frac{ab^2}{4}\\ \frac{2\delta+1}{2},\, \frac{2s-2n-2\rho+\delta}{2},\, \frac{2s-2n-2\rho-2\sigma+\delta}{2}\end{matrix}\right)$ $+\frac{2b^{2(-s+n+\rho+\sigma+1)}}{n!}(\sigma+1)_n$ $\times\Gamma(2s-2n-2\rho-2\sigma-2)\begin{Bmatrix}\sin[(\rho-s+\sigma)\pi]\\ \cos[(\rho-s+\sigma)\pi]\end{Bmatrix}$ $\times {}_2F_3\left(\begin{matrix}n+\sigma+1,\, n+\rho+\sigma+1;\, \frac{ab^2}{4}\\ \sigma+1,\, \frac{3-2s+2n+2\rho+2\sigma}{2},\, 2-s+n+\rho+\sigma\end{matrix}\right)$ $[b>0;\ -\delta/2<\operatorname{Re}s<\operatorname{Re}(\rho+\sigma)+n+3/2;\ \lvert\arg a\rvert<\pi]$

3.25.4. $P_n^{(\rho,\sigma)}(\varphi(x))$ and the logarithmic function

1	$(a-x)_+^{\rho}\ln\frac{x}{a}$ $\times P_n^{(\rho,\sigma)}\left(\frac{2x}{a}-1\right)$	$\frac{(-1)^n a^{s+\rho}}{n!}(1-s+\sigma)_n\mathrm{B}(n+\rho+1,\, s)\,[\psi(s)$ $+\psi(s-\sigma)-\psi(s+n+\rho+1)-\psi(s-n-\sigma)]$ $[a,\ \operatorname{Re}s>0;\ \operatorname{Re}\rho>-1]$

3.25.5. $P_n^{(\rho,\sigma)}(\varphi(x))$ and $\mathrm{Ei}(bx)$

1	$(a-x)_+^{\rho}\,\mathrm{Ei}(-bx) \times P_n^{(\rho,\sigma)}\left(\frac{2x}{a}-1\right)$	$\frac{(-1)^{n+1}a^{s+\rho+1}b}{n!}(\sigma-s)_n\,\mathrm{B}(n+\rho+1,\,s+1) \times {}_4F_4\left(\begin{matrix}1,\,1,\,s-\sigma+1,\,s+1;\,-ab\\ 2,\,2,\,s-n-\sigma+1,\,s+n+\rho+2\end{matrix}\right) + \frac{(-1)^n a^{s+\rho}}{n!}(\sigma-s+1)_n\,\mathrm{B}(n+\rho+1,\,s) \times \left[\psi(s)-\psi(s+n+\rho+1)-\sum_{j=0}^{n-1}\frac{1}{1-s+j+\sigma}+\ln(ab)+\mathbf{C}\right]$ $[a,\ \mathrm{Re}\,s>0;\ \mathrm{Re}\,\rho>-1]$

3.25.6. $P_n^{(\rho,\sigma)}(\varphi(x))$ and $\mathrm{si}(b\sqrt{x})$, $\mathrm{ci}(b\sqrt{x})$

Notation: $\delta=\left\{\begin{matrix}1\\0\end{matrix}\right\}$.

1	$(a-x)_+^{\rho}\left\{\begin{matrix}\mathrm{si}(b\sqrt{x})\\ \mathrm{ci}(b\sqrt{x})\end{matrix}\right\} \times P_n^{(\rho,\sigma)}\left(\frac{2x}{a}-1\right)$	$\frac{(-1)^{n+1}2^{\delta-2}a^{s+\rho+\delta/2+1}b^{\delta+2}}{3^{2\delta}n!}\left(\sigma-s-\frac{\delta}{2}\right)_n\mathrm{B}\left(n+\rho+1,\,s+\frac{\delta}{2}+1\right) \times {}_4F_5\left(\begin{matrix}1,\,\frac{\delta+2}{2},\,\frac{2s+\delta+2}{2},\,\frac{2s-2\sigma+\delta+2}{2};\,-\frac{ab^2}{4}\\ 2,\,\frac{\delta+4}{2},\,\frac{2\delta+3}{2},\,\frac{2s-2n-2\sigma+\delta+2}{2},\,\frac{2s+2n+2\rho+\delta+4}{2}\end{matrix}\right) + \frac{(-1)^n a^{s+\rho+\delta/2}b^{\delta}}{n!}\left(1-s+\sigma-\frac{\delta}{2}\right)_n\mathrm{B}\left(n+\rho+1,\,s+\frac{\delta}{2}\right) \times \left[\frac{1}{2}\psi(s)-\frac{1}{2}\psi(s+n+\rho+1)-\frac{1}{2}\sum_{i=0}^{n-1}\frac{1}{1-s+i+\sigma} + \frac{1}{2}\ln\left(ab^2\right)+\mathbf{C}\right]^{1-\delta} - \delta\frac{(-1)^n\pi a^{s+\rho}}{2(n!)}(1-s+\sigma)_n\,\mathrm{B}(n+\rho+1,\,s)$ $[a,\ \mathrm{Re}\,s>0;\ \mathrm{Re}\,\rho>-1]$

3.25.7. $P_n^{(\rho,\sigma)}(\varphi(x))$ and $\mathrm{erf}(bx^r)$, $\mathrm{erfc}(bx^r)$

1	$(a-x)_+^{\rho}\left\{\begin{matrix}\mathrm{erf}(bx)\\ \mathrm{erfc}(bx)\end{matrix}\right\} \times P_n^{(\rho,\sigma)}\left(\frac{2x}{a}-1\right)$	$\pm\frac{2(-1)^n a^{s+\rho+1}b}{\sqrt{\pi}\,n!}(\sigma-s)_n\,\mathrm{B}(n+\rho+1,\,s+1) \times {}_5F_5\left(\begin{matrix}\frac{1}{2},\,\frac{s+1}{2},\,\frac{s+2}{2},\,\frac{s-\sigma+1}{2},\,\frac{s-\sigma+2}{2};\,-a^2b^2\\ \frac{3}{2},\,\frac{s-n-\sigma+1}{2},\,\frac{s-n-\sigma+2}{2},\,\frac{s+n+\rho+2}{2},\,\frac{s+n+\rho+3}{2}\end{matrix}\right) + \left\{\begin{matrix}0\\1\end{matrix}\right\}\frac{(-1)^n a^{s+\rho}}{n!}(1-s+\sigma)_n\,\mathrm{B}(n+\rho+1,\,s)$ $[a>0;\ \mathrm{Re}\,\rho>-1;\ \mathrm{Re}\,s>-(1\pm1)/2]$

No.	$f(x)$	$F(s)$
2	$\operatorname{erfc}(b\sqrt{x})$ $\times P_n^{(\rho,\sigma)}\left(\frac{2x}{a}\pm 1\right)$	$\frac{(\pm 1)^n b^{-2s}}{\sqrt{\pi}\, n!\, s}(\varphi+1)_n \Gamma\left(\frac{2s+1}{2}\right) {}_4F_2\left(\begin{matrix}-n,\, n+\rho+\sigma+1,\, s,\, \frac{2s+1}{2}\\ \varphi+1,\, s+1;\, \mp\frac{1}{ab^2}\end{matrix}\right)$ $\left[\operatorname{Re} s>0;\ \lvert\arg b\rvert<\pi/4;\ \varphi=\begin{Bmatrix}\rho\\ \sigma\end{Bmatrix}\right]$
3	$(a-x)_+^{\rho}\begin{Bmatrix}\operatorname{erf}(b\sqrt{x})\\ \operatorname{erfc}(b\sqrt{x})\end{Bmatrix}$ $\times P_n^{(\rho,\sigma)}\left(\frac{2x}{a}-1\right)$	$\pm\frac{2(-1)^n a^{s+\rho+1/2} b}{\sqrt{\pi}\, n!}\left(\frac{1}{2}-s+\sigma\right)_n \mathrm{B}\left(n+\rho+1,\, \frac{2s+1}{2}\right)$ $\times {}_3F_3\left(\begin{matrix}\frac{1}{2},\, \frac{2s+1}{2}\, \frac{2s-2\sigma+1}{2};\, -ab^2\\ \frac{3}{2},\, \frac{2s-2n-2\sigma+1}{2},\, \frac{2s+2n+2\rho+3}{2}\end{matrix}\right)$ $+\begin{Bmatrix}0\\ 1\end{Bmatrix}\frac{(-1)^n a^{s+\rho}}{n!}(1-s+\sigma)_n \mathrm{B}(n+\rho+1,\, s)$ $[a>0;\ \operatorname{Re}\rho>-1;\ \operatorname{Re} s>-(1\pm 1)/4]$
4	$(a-x)_+^{\rho} e^{b^2x}\operatorname{erf}(b\sqrt{x})$ $\times P_n^{(\rho,\sigma)}\left(\frac{2x}{a}-1\right)$	$\frac{2(-1)^n a^{s+\rho+1/2} b}{\sqrt{\pi}\, n!}\left(\frac{1}{2}-s+\sigma\right)_n \mathrm{B}\left(n+\rho+1,\, s+\frac{1}{2}\right)$ $\times {}_3F_3\left(\begin{matrix}1,\, s+\frac{1}{2},\, s-\sigma+\frac{1}{2};\, ab^2\\ \frac{3}{2},\, s-n-\sigma+\frac{1}{2},\, s+n+\rho+\frac{3}{2}\end{matrix}\right)$ $[a>0,\ \operatorname{Re}\rho>-1;\ \operatorname{Re} s>-1/2]$
5	$(a-x)^n\operatorname{erfc}(b\sqrt{x})$ $\times P_n^{(\rho,\sigma)}\left(\frac{a+x}{a-x}\right)$	$\frac{(\rho+1)_n a^n}{n!\sqrt{\pi}\, b^{2s} s}\Gamma\left(\frac{2s+1}{2}\right) {}_4F_2\left(\begin{matrix}-n,\, -n-\sigma,\, s,\, \frac{2s+1}{2}\\ \rho+1,\, s+1;\, \frac{1}{ab^2}\end{matrix}\right)$ $[\operatorname{Re} s>0]$

3.25.8. $P_n^{(\rho,\sigma)}(\varphi(x))$ and $\gamma(\nu,\, bx)$

1	$(a-x)_+^{\rho}\gamma(\nu,\, bx)$ $\times P_n^{(\rho,\sigma)}\left(\frac{2x}{a}-1\right)$	$\frac{(-1)^n a^{s+\nu+\rho} b^{\nu}}{n!\,\nu}(1-s-\nu+\sigma)_n \Gamma\left[\begin{matrix}n+\rho+1,\, s+\nu\\ s+n+\nu+\rho+1\end{matrix}\right]$ $\times {}_3F_3\left(\begin{matrix}\nu,\, s+\nu,\, s+\nu-\sigma;\, -ab\\ \nu+1,\, s+n+\nu+\rho+1,\, s-n+\nu-\sigma\end{matrix}\right)$ $[a,\ \operatorname{Re}(s+\nu)>0;\ \operatorname{Re}\rho>-1]$

3.25.9. $P_n^{(\rho,\sigma)}(\varphi(x))$ and $I_{\nu}(bx^r)$, $J_{\nu}(bx^r)$

1	$J_{\nu}(b\sqrt{x})$ $\times P_n^{(\rho,\sigma)}\left(\frac{2x}{a}\pm 1\right)$	$\frac{2^{2s+2n}}{n!\, a^n b^{2s+2n}}(n+\rho+\sigma+1)_n \Gamma\left[\begin{matrix}\frac{2s+2n+\nu}{2}\\ \frac{2-2s-2n+\nu}{2}\end{matrix}\right]$ $\times {}_2F_3\left(\begin{matrix}-n,\, -n-\varphi;\, \pm\frac{ab^2}{4}\\ -2n-\rho-\sigma,\, \frac{2-2s-2n-\nu}{2},\, \frac{2-2s-2n+\nu}{2}\end{matrix}\right)$ $\left[b>0;\ -\operatorname{Re}\nu/2<\operatorname{Re} s<3/4-n;\ \varphi=\begin{Bmatrix}\rho\\ \sigma\end{Bmatrix}\right]$

No.	$f(x)$	$F(s)$
2	$(a-x)_+^{\rho}\left\{\begin{matrix}J_\nu(b\sqrt{x})\\ I_\nu(b\sqrt{x})\end{matrix}\right\}$ $\times P_n^{(\rho,\sigma)}\left(\dfrac{2x}{a}-1\right)$	$\dfrac{(-1)^n a^{s+\nu/2+\rho}b^\nu}{2^\nu n!\,\Gamma(\nu+1)}\left(1-s+\sigma-\dfrac{\nu}{2}\right)_n \mathrm{B}\left(n+\rho+1,\,s+\dfrac{\nu}{2}\right)$ $\times {}_2F_3\left(\begin{matrix}s+\frac{\nu}{2},\,s+\frac{\nu}{2}-\sigma;\,\mp\frac{ab^2}{4}\\ \nu+1,\,s-n+\frac{\nu}{2}-\sigma,\,s+n+\frac{\nu}{2}+\rho+1\end{matrix}\right)$ $[a,\ \operatorname{Re}(2s+\nu)>0;\ \operatorname{Re}\rho>-1]$
3	$(x+a)^{\sigma}J_\nu(b\sqrt{x})$ $\times P_n^{(\rho,\sigma)}\left(\dfrac{2x}{a}+1\right)$	$\dfrac{a^{s+\nu/2+\sigma}b^\nu}{2^\nu n!\,\Gamma(\nu+1)}\left(1-s+\rho-\dfrac{\nu}{2}\right)_n \mathrm{B}\left(-s-n-\sigma-\dfrac{\nu}{2},\,s+\dfrac{\nu}{2}\right)$ $\times {}_2F_3\left(\begin{matrix}s+\frac{\nu}{2},\,s+\frac{\nu}{2}-\rho;\,\frac{ab^2}{4}\\ \nu+1,\,\frac{2s-2n+\nu-2\rho}{2},\,\frac{2s+2n+\nu+2\sigma+2}{2}\end{matrix}\right)$ $+\dfrac{(2/b)^{2(s+n+\sigma)}}{n!\,a^n}(n+\rho+\sigma+1)_n\,\Gamma\left[\begin{matrix}\frac{2s+2n+\nu+2\sigma}{2}\\ \frac{2-2s-2n-2\sigma+\nu}{2}\end{matrix}\right]$ $\times {}_2F_3\left(\begin{matrix}-n-\sigma,\,-n-\rho-\sigma;\,\frac{ab^2}{4}\\ -2n-\rho-\sigma,\,\frac{2-2s-2n-2\sigma-\nu}{2},\,\frac{2-2s-2n-2\sigma+\nu}{2}\end{matrix}\right)$ $[b>0;\ -\operatorname{Re}\nu/2<\operatorname{Re}s<3/4-\operatorname{Re}\sigma-n;;\ \lvert\arg a\rvert<\pi]$
4	$(a-x)_+^{\rho}J_\nu\left(\dfrac{b}{\sqrt{x}}\right)$ $\times P_n^{(\rho,\sigma)}\left(\dfrac{2x}{a}-1\right)$	$\dfrac{(-1)^n a^{s+\rho-\nu/2}b^\nu}{2^\nu n!\,\Gamma(\nu+1)}\left(1-s+\dfrac{\nu}{2}+\sigma\right)_n \mathrm{B}\left(n+\rho+1,\,s-\dfrac{\nu}{2}\right)$ $\times {}_2F_3\left(\begin{matrix}-s+n+\frac{\nu}{2}+\sigma+1,\,-s-n+\frac{\nu}{2}-\rho\\ \nu+1,\,-s+\frac{\nu}{2}+1,\,-s+\frac{\nu}{2}+\sigma+1;\,-\frac{b^2}{4a}\end{matrix}\right)$ $+\dfrac{(-1)^n a^\rho b^{2s}}{2^{2s}n!}(\sigma+1)_n\,\Gamma\left[\begin{matrix}\frac{\nu}{2}-s\\ s+\frac{\nu}{2}+1\end{matrix}\right]$ $\times {}_2F_3\left(\begin{matrix}-n-\rho,\,n+\sigma+1;\,-\frac{b^2}{4a}\\ \sigma+1,\,\frac{2s-\nu+2}{2},\,\frac{2s+\nu+2}{2}\end{matrix}\right)$ $[a,\,b>0;\ \operatorname{Re}\rho>-1;\ \operatorname{Re}s>-3/4]$
5	$(x+a)^{\sigma}J_\nu\left(\dfrac{b}{\sqrt{x}}\right)$ $\times P_n^{(\rho,\sigma)}\left(\dfrac{2x}{a}+1\right)$	$\dfrac{a^{s-\nu/2+\sigma}b^\nu}{2^\nu n!\,\Gamma(\nu+1)}\left(1-s+\dfrac{\nu}{2}+\rho\right)_n \mathrm{B}\left(s-\dfrac{\nu}{2},\,\dfrac{\nu}{2}-s-n-\sigma\right)$ $\times {}_2F_3\left(\begin{matrix}1-s+n+\frac{\nu}{2}+\rho,\,-s-n+\frac{\nu}{2}-\sigma\\ \nu+1,\,1-s+\frac{\nu}{2},\,1-s+\frac{\nu}{2}+\rho;\,\frac{b^2}{4a}\end{matrix}\right)$ $+\dfrac{a^\sigma b^{2s}}{2^{2s}n!}(\rho+1)_n\,\Gamma\left[\begin{matrix}\frac{\nu}{2}-s\\ s+\frac{\nu}{2}+1\end{matrix}\right]$ $\times {}_2F_3\left(\begin{matrix}-n-\sigma,\,n+\rho+1\\ \rho+1,\,\frac{2s-\nu+2}{2},\,\frac{2s+\nu+2}{2};\,\frac{b^2}{4a}\end{matrix}\right)$ $[b>0;\ -3/4<\operatorname{Re}s<\operatorname{Re}(\nu/2-\sigma)-n;\ \lvert\arg a\rvert<\pi]$

No.	$f(x)$	$F(s)$
6	$(a-x)_+^{\sigma}$ $\times \left\{ \begin{matrix} J_\mu(b\sqrt{x})\, J_\nu(b\sqrt{x}) \\ I_\mu(b\sqrt{x})\, I_\nu(b\sqrt{x}) \end{matrix} \right\}$ $\times P_n^{(\rho,\sigma)}\left(1-\frac{2x}{a}\right)$	$\frac{a^{s+(\mu+\nu)/2+\sigma} b^{\mu+\nu}}{2^{\mu+\nu} n!} \left(1-s+\rho-\frac{\mu+\nu}{2}\right)_n$ $\times \Gamma\left[\begin{matrix} n+\sigma+1,\, \frac{2s+\mu+\nu}{2} \\ \mu+1,\, \nu+1,\, \frac{2s+2n+\mu+\nu+2\sigma+2}{2} \end{matrix}\right]$ $\times {}_4F_5\left(\begin{matrix} \frac{\mu+\nu+1}{2},\, \frac{\mu+\nu+2}{2},\, \frac{2s+\mu+\nu-2\rho}{2},\, \frac{2s+\mu+\nu}{2};\, \mp ab^2 \\ \mu+1,\, \nu+1,\, \mu+\nu+1,\, \frac{2s-2n+\mu+\nu-2\rho}{2},\, \frac{2s+2n+\mu+\nu+2\sigma+2}{2} \end{matrix}\right)$ $[a,\ \mathrm{Re}(2s+\mu+\nu)>0;\ \mathrm{Re}\,\sigma>-1]$
7	$(x+a)^n J_\nu(b\sqrt{x})$ $\times P_n^{(\rho,\sigma)}\left(\frac{a-x}{a+x}\right)$	$\frac{(-1)^n}{n!}\left(\frac{2}{b}\right)^{2s+2n} (\sigma+1)_n \Gamma\left[\begin{matrix} \frac{2s+2n+\nu}{2} \\ \frac{2-2s-2n+\nu}{2} \end{matrix}\right]$ $\times {}_2F_3\left(\begin{matrix} -n,\, -n-\rho;\, \frac{ab^2}{4} \\ \sigma+1,\, \frac{2-2s-2n-\nu}{2},\, \frac{2-2s-2n+\nu}{2} \end{matrix}\right)$ $[a,\ b>0;\ -\mathrm{Re}\,\nu/2<\mathrm{Re}\,s<3/4-n]$
8	$\frac{J_\nu(b\sqrt{x})}{(x+a)^{n+\rho+\sigma+1}}$ $\times P_n^{(\rho,\sigma)}\left(\frac{a-x}{a+x}\right)$	$\frac{a^{s-(n-\nu/2+\rho+\sigma+1)} (b/2)^\nu}{n!\,\Gamma(\nu+1)} \left(1-s+\rho-\frac{\nu}{2}\right)_n$ $\times \mathrm{B}\left(1-s+n-\frac{\nu}{2}+\rho+\sigma,\, s+\frac{\nu}{2}\right)$ $\times {}_2F_3\left(\begin{matrix} s+\frac{\nu}{2}-\rho,\, s+\frac{\nu}{2};\, \frac{ab^2}{4} \\ \nu+1,\, \frac{2s-2n+\nu-2\rho}{2},\, \frac{2s-2n+\nu-2\rho-2\sigma}{2} \end{matrix}\right)$ $+\frac{(-1)^n (b/2)^{2(-s+n+\rho+\sigma+1)}}{n!}$ $\times (\sigma+1)_n \Gamma\left[\begin{matrix} \frac{2s-2n+\nu-2\rho-2\sigma-2}{2} \\ \frac{4-2s+2n+\nu+2\rho+2\sigma}{2} \end{matrix}\right]$ $\times {}_2F_3\left(\begin{matrix} n+\sigma+1,\, n+\rho+\sigma+1;\, \frac{ab^2}{4} \\ \sigma+1,\, \frac{4-2s+2n-\nu+2\rho+2\sigma}{2},\, \frac{4-2s+2n+\nu+2\rho+2\sigma}{2} \end{matrix}\right)$ $[b>0;\ -\mathrm{Re}\,\nu/2<\mathrm{Re}\,s<\mathrm{Re}(\rho+\sigma)+n+7/4;\ \lvert\arg a\rvert<\pi]$

3.25.10. $P_n^{(\rho,\sigma)}(\varphi(x))$ and $K_\nu(bx^r)$

1	$K_\nu(b\sqrt{x})$ $\times P_n^{(\rho,\sigma)}\left(\frac{2x}{a}\pm 1\right)$	$\frac{2^{2s+2n-1}}{n!\, a^n b^{2s+2n}} (n+\rho+\sigma+1)_n \Gamma\left(s+n-\frac{\nu}{2}\right)\Gamma\left(s+n+\frac{\nu}{2}\right)$ $\times {}_2F_3\left(\begin{matrix} -n,\, -n-\varphi;\, \mp\frac{ab^2}{4} \\ -2n-\rho-\sigma,\, \frac{2-2s-2n-\nu}{2},\, \frac{2-2s-2n+\nu}{2} \end{matrix}\right)$ $\left[\mathrm{Re}\,b>0;\ \mathrm{Re}\,s>\lvert\mathrm{Re}\,\nu\rvert/2;\ \varphi=\left\{\begin{matrix}\rho\\ \sigma\end{matrix}\right\}\right]$

No.	$f(x)$	$F(s)$
2	$K_\nu\left(\frac{b}{\sqrt{x}}\right)$ $\times P_n^{(\rho,\sigma)}\left(\frac{2x}{a}\pm 1\right)$	$\frac{(\pm 1)^n b^{2s}}{2^{2s+1} n!}(\varphi+1)_n \Gamma\left(-s-\frac{\nu}{2}\right)\Gamma\left(-s+\frac{\nu}{2}\right)$ $\times {}_2F_3\left(\begin{matrix}-n,\, n+\rho+\sigma+1;\, \mp\frac{b^2}{4a}\\ \varphi+1,\, s-\frac{\nu}{2}+1,\, s+\frac{\nu}{2}+1\end{matrix}\right)$ $\left[\operatorname{Re} b>0;\ \operatorname{Re} s<-\lvert\operatorname{Re}\nu\rvert/2-n;\ \varphi=\left\{\begin{matrix}\rho\\ \sigma\end{matrix}\right\}\right]$
3	$(x+a)^n K_\nu(b\sqrt{x})$ $\times P_n^{(\rho,\sigma)}\left(\frac{a-x}{a+x}\right)$	$\frac{(-1)^n}{2(n!)}\left(\frac{2}{b}\right)^{2s+2n}(\sigma+1)_n \Gamma\left(s+n-\frac{\nu}{2}\right)\Gamma\left(s+n+\frac{\nu}{2}\right)$ $\times {}_2F_3\left(\begin{matrix}-n,\, -n-\rho;\, -\frac{ab^2}{4}\\ \sigma+1,\, \frac{2-2s-2n-\nu}{2},\, \frac{2-2s-2n+\nu}{2}\end{matrix}\right)$ $[\operatorname{Re} b>0;\ \operatorname{Re} s>\lvert\operatorname{Re}\nu\rvert/2]$

3.25.11. $P_n^{(\rho,\sigma)}(\varphi(x))$ and $P_m(\psi(x))$

Notation: $\varepsilon=0$ or 1.

No.	$f(x)$	$F(s)$
1	$(a-x)_+^\sigma P_m\left(1-\frac{2x}{a}\right)$ $\times P_n^{(\rho,\sigma)}\left(1-\frac{2x}{a}\right)$	$\frac{a^{s+\sigma}}{n!}(1-s+\rho)_n \mathrm{B}(n+\sigma+1,\, s)\ {}_4F_3\left(\begin{matrix}-m,\, m+1,\, s-\rho,\, s;\, 1\\ 1,\, s-n-\rho,\, s+n+\sigma+1\end{matrix}\right)$ $[a,\ \operatorname{Re} s>0;\ \operatorname{Re}\sigma>-1]$
2	$(a-x)_+^\rho P_{2m+\varepsilon}(bx)$ $\times P_n^{(\rho,\sigma)}\left(\frac{2x}{a}-1\right)$	$(-1)^{m+n}\frac{a^{s+\rho+\varepsilon}(2b)^\varepsilon}{m!\, n!}\left(\frac{1}{2}\right)_{m+\varepsilon}(1-s+\sigma-\varepsilon)_n \mathrm{B}(n+\rho+1,\, s+\varepsilon)$ $\times {}_6F_5\left(\begin{matrix}-m,\, m+\varepsilon+\frac{1}{2},\, \Delta(2,\, s+\varepsilon),\, \Delta(2,\, s-\sigma+\varepsilon);\, a^2b^2\\ \frac{2\varepsilon+1}{2},\, \Delta(2,\, s+n+\rho+\varepsilon+1),\, \Delta(2,\, s-n-\sigma+\varepsilon)\end{matrix}\right)$ $[a>0;\ \operatorname{Re}\rho>-1;\ \operatorname{Re} s>-\varepsilon]$
3	$(a-x)_+^\rho P_m(2bx-1)$ $\times P_n^{(\rho,\sigma)}\left(\frac{2x}{a}-1\right)$	$\frac{(-1)^{m+n} a^{s+\rho}}{n!}(1-s+\sigma)_n \mathrm{B}(n+\rho+1,\, s)$ $\times {}_4F_3\left(\begin{matrix}-m,\, m+1,\, s-\sigma,\, s;\, ab\\ 1,\, s-n-\sigma,\, s+n+\rho+1\end{matrix}\right)$ $[a,\ \operatorname{Re} s>0;\ \operatorname{Re}\rho>-1]$
4	$\theta(a-x)\left\{\begin{matrix}(x+b)^\sigma\\ (b-x)^\rho\end{matrix}\right\}$ $\times P_m\left(\frac{2x}{a}-1\right)$ $\times P_n^{(\rho,\sigma)}\left(\frac{2x}{b}\pm 1\right)$	$\frac{(-1)^m(\pm 1)^n b^\psi a^s}{n!}(\varphi+1)_n(1-s)_m$ $\times \Gamma\left[\begin{matrix}s\\ s+m+1\end{matrix}\right] {}_4F_3\left(\begin{matrix}-n-\psi,\, n+\varphi+1,\, s,\, s;\, \mp\frac{a}{b}\\ \varphi+1,\, s-m,\, s+m+1\end{matrix}\right)$ $\left[\left\{\begin{matrix}a>0;\ \lvert\arg b\rvert<\pi\\ b>a>0\end{matrix}\right\};\ \operatorname{Re} s>0;\ \varphi=\left\{\begin{matrix}\rho\\ \sigma\end{matrix}\right\},\ \psi=\left\{\begin{matrix}\sigma\\ \rho\end{matrix}\right\}\right]$

No.	$f(x)$	$F(s)$
5	$(a-x)_+^{\rho}$ $\times P_{2m+\varepsilon}\left(\sqrt{\frac{x}{a}}\right)$ $\times P_n^{(\rho,\,\sigma)}\left(\frac{2x}{a}-1\right)$	$\frac{2^{\varepsilon}(-1)^{m+n}a^{s+\rho}}{m!\,n!}\left(\frac{1}{2}\right)_{m+\varepsilon}\left(1-s+\sigma-\frac{\varepsilon}{2}\right)_n \mathrm{B}\left(n+\rho+1,\,s+\frac{\varepsilon}{2}\right)$ $\times {}_4F_3\left(\begin{matrix}-m,\ \frac{2m+2\varepsilon+1}{2},\ \frac{2s-2\sigma+\varepsilon}{2},\ \frac{2s+\varepsilon}{2};\ 1\\ \frac{2\varepsilon+1}{2},\ \frac{2s-2n-2\sigma+\varepsilon}{2},\ \frac{2s+2n+2\rho+\varepsilon+2}{2}\end{matrix}\right)$ $[a>0;\ \mathrm{Re}\,\rho>-1;\ \mathrm{Re}\,s>-\varepsilon/2]$
6	$(a-x)_+^{\rho}\,P_{2m+\varepsilon}(b\sqrt{x})$ $\times P_n^{(\rho,\,\sigma)}\left(\frac{2x}{a}-1\right)$	$\frac{(-1)^{m+n}a^{s+\rho+\varepsilon/2}(2b)^{\varepsilon}}{m!\,n!}\left(\frac{1}{2}\right)_{m+\varepsilon}\left(1-s+\sigma-\frac{\varepsilon}{2}\right)_n$ $\times \mathrm{B}\left(n+\rho+1,\,s+\frac{\varepsilon}{2}\right)$ $\times {}_4F_3\left(\begin{matrix}-m,\ \frac{2m+2\varepsilon+1}{2},\ \frac{2s-2\sigma+\varepsilon}{2},\ \frac{2s+\varepsilon}{2};\ ab^2\\ \frac{2\varepsilon+1}{2},\ \frac{2s-2n-2\sigma+\varepsilon}{2},\ \frac{2s+2n+2\rho+\varepsilon+2}{2}\end{matrix}\right)$ $[a>0;\ \mathrm{Re}\,\rho>-1;\ \mathrm{Re}\,s>-\varepsilon/2]$

3.25.12. $P_n^{(\rho,\,\sigma)}(\varphi(x))$ and $T_m(\psi(x))$

Notation: $\varepsilon=0$ or 1.

1	$(a-x)_+^{\sigma-1/2}$ $\times T_m\left(1-\frac{2x}{a}\right)$ $\times P_n^{(\rho,\,\sigma)}\left(1-\frac{2x}{a}\right)$	$\frac{a^{s+\sigma-1/2}}{n!}(1-s+\rho)_n\,\mathrm{B}(n+\sigma+1,\,s)$ $\times {}_4F_3\left(\begin{matrix}-m+\frac{1}{2},\ m+\frac{1}{2},\ s-\rho,\ s\\ \frac{1}{2},\ s-n-\rho,\ s+n+\sigma+1;\ 1\end{matrix}\right)$ $[a,\ \mathrm{Re}\,s>0;\ \mathrm{Re}\,\sigma>-1/2]$
2	$(a-x)_+^{\rho}\,T_{2m+\varepsilon}(bx)$ $\times P_n^{(\rho,\,\sigma)}\left(\frac{2x}{a}-1\right)$	$(-1)^{m+n}\frac{(m+\varepsilon/2)\,a^{s+\rho+\varepsilon}(2b)^{\varepsilon}}{m!\,n!}(1-s+\sigma-\varepsilon)_n$ $\times\Gamma(m+\varepsilon)\,\mathrm{B}(n+\rho+1,\,s+\varepsilon)$ $\times {}_6F_5\left(\begin{matrix}-m,\ m+\varepsilon,\ \Delta(2,\,s+\varepsilon),\ \Delta(2,\,s-\sigma+\varepsilon);\ a^2b^2\\ \frac{2\varepsilon+1}{2},\ \Delta(2,\,s+n+\rho+\varepsilon+1),\ \Delta(2,\,s-n-\sigma+\varepsilon)\end{matrix}\right)$ $[a>0;\ \mathrm{Re}\,\rho>-1;\ \mathrm{Re}\,s>-\varepsilon]$
3	$(a-x)_+^{\rho}(1-bx)^{-1/2}$ $\times T_m^{\lambda}(2bx-1)$ $\times P_n^{(\rho,\,\sigma)}\left(\frac{2x}{a}-1\right)$	$\frac{(-1)^{m+n}a^{s+\rho}}{n!}(1-s+\sigma)_n\,\mathrm{B}(n+\rho+1,\,s)$ $\times {}_4F_3\left(\begin{matrix}-m+\frac{1}{2},\ m+\frac{1}{2},\ s-\sigma,\ s\\ \frac{1}{2},\ s-n-\sigma,\ s+n+\rho+1;\ ab\end{matrix}\right)$ $[a,\ \mathrm{Re}\,s>0;\ \mathrm{Re}\,\rho>-1;\ \lvert\arg(1-ab)\rvert<\pi]$

No.	$f(x)$	$F(s)$
4	$(a-x)_+^{-1/2}\begin{Bmatrix}(x+b)^{\sigma}\\(b-x)^{\rho}\end{Bmatrix}$ $\times T_m\left(\frac{2x}{a}-1\right)$ $\times P_n^{(\rho,\sigma)}\left(\frac{2x}{b}\pm 1\right)$	$\frac{(-1)^m(\pm 1)^n\sqrt{\pi}\,a^{s-1/2}b^{\psi}}{n!}(\varphi+1)_n\left(\frac{1}{2}-s\right)_m$ $\times\Gamma\left[\begin{matrix}s\\\frac{2s+2m+1}{2}\end{matrix}\right]{}_4F_3\left(\begin{matrix}-n-\psi,\,n+\varphi+1,\,s,\,\frac{2s+1}{2}\\\varphi+1,\,\frac{2s-2m+1}{2},\,\frac{2s+2m+1}{2};\,\mp\frac{a}{b}\end{matrix}\right)$ $\left[\begin{Bmatrix}a>0;\,\lvert\arg b\rvert<\pi\\b>a>0\end{Bmatrix};\,\mathrm{Re}\,s>0;\,\varphi=\begin{Bmatrix}\rho\\\sigma\end{Bmatrix},\,\psi=\begin{Bmatrix}\sigma\\\rho\end{Bmatrix}\right]$
5	$(a-x)_+^{\rho-1/2}$ $\times T_{2m+\varepsilon}\left(\sqrt{\frac{x}{a}}\right)$ $\times P_n^{(\rho,\sigma)}\left(\frac{2x}{a}-1\right)$	$\frac{2^{\varepsilon}(-1)^{m+n}(2m+\varepsilon)\,a^{s+\rho-1/2}}{m!\,n!}\left(1-s+\sigma-\frac{\varepsilon}{2}\right)_n$ $\times\Gamma(m+\varepsilon)\,\mathrm{B}\left(n+\rho+1,\,s+\frac{\varepsilon}{2}\right)$ $\times{}_4F_3\left(\begin{matrix}\frac{-2m+1}{2},\,\frac{2m+2\varepsilon+1}{2},\,\frac{2s-2\sigma+\varepsilon}{2},\,\frac{2s+\varepsilon}{2}\\\frac{2\varepsilon+1}{2},\,\frac{2s-2n-2\sigma+\varepsilon}{2},\,\frac{2s+2n+2\rho+\varepsilon+2}{2};\,1\end{matrix}\right)$ $[a>0;\,\mathrm{Re}\,\rho>-1/2;\,\mathrm{Re}\,s>-\varepsilon/2]$
6	$(a-x)_+^{\rho}\left(1-b^2x\right)^{-1/2}$ $\times T_{2m+\varepsilon}(b\sqrt{x})$ $\times P_n^{(\rho,\sigma)}\left(\frac{2x}{a}-1\right)$	$\frac{(-1)^{m+n}(2m+\varepsilon)\,a^{s+\rho+\varepsilon/2}(2b)^{\varepsilon}}{2\,(m!\,n!)}\left(1-s+\sigma-\frac{\varepsilon}{2}\right)_n$ $\times\Gamma(m+\varepsilon)\,\mathrm{B}\left(n+\rho+1,\,s+\frac{\varepsilon}{2}\right)$ $\times{}_4F_3\left(\begin{matrix}\frac{-2m+1}{2},\,\frac{2m+2\varepsilon+1}{2},\,\frac{2s-2\sigma+\varepsilon}{2},\,\frac{2s+\varepsilon}{2}\\\frac{2\varepsilon+1}{2},\,\frac{2s-2n-2\sigma+\varepsilon}{2},\,\frac{2s+2n+2\rho+\varepsilon+2}{2};\,ab^2\end{matrix}\right)$ $\left[a>0;\,\mathrm{Re}\,\rho>-1;\,\mathrm{Re}\,s>-\varepsilon/2;;\,\left\lvert\arg\left(1-ab^2\right)\right\rvert<\pi\right]$

3.25.13. $P_n^{(\rho,\sigma)}(\varphi(x))$ and $U_m(\psi(x))$

Notation: $\varepsilon=0$ or 1.

No.	$f(x)$	$F(s)$
1	$(a-x)_+^{\sigma+1/2}$ $\times U_m\left(1-\frac{2x}{a}\right)$ $\times P_n^{(\rho,\sigma)}\left(1-\frac{2x}{a}\right)$	$\frac{(m+1)\,a^{s+\sigma+1/2}}{n!}(1-s+\rho)_n\,\mathrm{B}(n+\sigma+1,\,s)$ $\times{}_4F_3\left(\begin{matrix}-m-\frac{1}{2},\,m+\frac{3}{2},\,s-\rho,\,s;\,1\\\frac{3}{2},\,s-n-\rho,\,s+n+\sigma+1\end{matrix}\right)$ $[a,\ \mathrm{Re}\,s>0;\ \mathrm{Re}\,\sigma>-3/2]$
2	$(a-x)_+^{\rho}\,U_{2m+\varepsilon}(bx)$ $\times P_n^{(\rho,\sigma)}\left(\frac{2x}{a}-1\right)$	$(-1)^{m+n}\frac{(m+1)^{\varepsilon}\,a^{s+\rho+\varepsilon}(2b)^{\varepsilon}}{n!}(1-s+\sigma-\varepsilon)_n\,\mathrm{B}(n+\rho+1,\,s+\varepsilon)$ $\times{}_6F_5\left(\begin{matrix}-m,\,m+\varepsilon+1,\,\Delta(2,\,s+\varepsilon),\,\Delta(2,\,s-\sigma+\varepsilon);\,a^2b^2\\\frac{2\varepsilon+1}{2},\,\Delta(2,\,s+n+\rho+\varepsilon+1),\,\Delta(2,\,s-n-\sigma+\varepsilon)\end{matrix}\right)$ $[a>0;\ \mathrm{Re}\,\rho>-1;\ \mathrm{Re}\,s>-\varepsilon]$

No.	$f(x)$	$F(s)$
3	$(a-x)_+^{\rho}\sqrt{1-bx}$ $\times U_m(2bx-1)$ $\times P_n^{(\rho,\sigma)}\left(\frac{2x}{a}-1\right)$	$\frac{(-1)^{m+n}(m+1)a^{s+\rho}}{n!}(1-s+\sigma)_n\,\mathrm{B}(n+\rho+1,\,s)$ $\times {}_4F_3\left(\begin{matrix}-m-\frac{1}{2},\,m+\frac{3}{2},\,s-\sigma,\,s;\,ab\\ \frac{3}{2},\,s-n-\sigma,\,s+n+\rho+1\end{matrix}\right)$ $[a,\ \mathrm{Re}\,s>0;\ \mathrm{Re}\,\rho>-1;\ \lvert\arg(1-ab)<\pi\rvert]$
4	$(a-x)_+^{1/2}\left\{\begin{matrix}(x+b)^{\sigma}\\(b-x)^{\rho}\end{matrix}\right\}$ $\times U_m\left(\frac{2x}{a}-1\right)$ $\times P_n^{(\rho,\sigma)}\left(\frac{2x}{b}\pm1\right)$	$\frac{(-1)^m(\pm1)^n(m+1)\sqrt{\pi}\,a^{s+1/2}b^{\psi}}{2(n!)}(\varphi+1)_n\left(\frac{3}{2}-s\right)_m$ $\times\Gamma\left[\begin{matrix}s\\ \frac{2s+2m+3}{2}\end{matrix}\right]{}_4F_3\left(\begin{matrix}-n-\psi,\,n+\varphi+1,\,\frac{2s-1}{2},\,s\\ \varphi+1,\,\frac{2s-2m-1}{2},\,\frac{2s+2m+3}{2};\,\mp\frac{a}{b}\end{matrix}\right)$ $\left[\left\{\begin{matrix}a>0;\ \lvert\arg b\rvert<\pi\\ b>a>0\end{matrix}\right\};\ \mathrm{Re}\,s>0;\ \varphi=\left\{\begin{matrix}\rho\\ \sigma\end{matrix}\right\},\ \psi=\left\{\begin{matrix}\sigma\\ \rho\end{matrix}\right\}\right]$
5	$(a-x)_+^{\rho+1/2}$ $\times U_{2m+\varepsilon}\left(\sqrt{\frac{x}{a}}\right)$ $\times P_n^{(\rho,\sigma)}\left(\frac{2x}{a}-1\right)$	$\frac{2^{\varepsilon}(-1)^{m+n}(m+1)^{\varepsilon}a^{s+\rho+1/2}}{n!}\left(1-s+\sigma-\frac{\varepsilon}{2}\right)_n$ $\times\mathrm{B}\left(n+\rho+1,\,s+\frac{\varepsilon}{2}\right){}_4F_3\left(\begin{matrix}\frac{-2m-1}{2},\,\frac{2m+2\varepsilon+1}{2},\,\frac{2s-2\sigma+\varepsilon}{2},\,\frac{2s+\varepsilon}{2}\\ \frac{2\varepsilon+1}{2},\,\frac{2s-2n-2\sigma+\varepsilon}{2},\,\frac{2s+2n+2\rho+\varepsilon+2}{2};\,1\end{matrix}\right)$ $[a>0;\ \mathrm{Re}\,\rho>-3/2;\ \mathrm{Re}\,s>-\varepsilon/2]$
6	$(a-x)_+^{\rho}\sqrt{1-b^2x}$ $\times U_{2m+\varepsilon}(b\sqrt{x})$ $\times P_n^{(\rho,\sigma)}\left(\frac{2x}{a}-1\right)$	$\frac{(-1)^{m+n}(m+1)^{\varepsilon}a^{s+\rho+\varepsilon/2}(2b)^{\varepsilon}}{n!}\left(1-s+\sigma-\frac{\varepsilon}{2}\right)_n$ $\times\mathrm{B}\left(n+\rho+1,\,s+\frac{\varepsilon}{2}\right)$ $\times{}_4F_3\left(\begin{matrix}\frac{-2m-1}{2},\,\frac{2m+2\varepsilon+1}{2},\,\frac{2s-2\sigma+\varepsilon}{2},\,\frac{2s+\varepsilon}{2};\,ab^2\\ \frac{2\varepsilon+1}{2},\,\frac{2s-2n-2\sigma+\varepsilon}{2},\,\frac{2s+2n+2\rho+\varepsilon+2}{2}\end{matrix}\right)$ $[a>0;\ \mathrm{Re}\,\rho>-1;\ \mathrm{Re}\,s>-\varepsilon/2;\ \lvert\arg(1-ab^2)<\pi\rvert]$

3.25.14. $P_n^{(\rho,\sigma)}(\varphi(x))$ and $H_m(b\sqrt{x})$

Notation: $\varepsilon=0$ or 1.

1	$(a-x)_+^{\rho}e^{-b^2x}$ $\times H_{2m+\varepsilon}(b\sqrt{x})$ $\times P_n^{(\rho,\sigma)}\left(\frac{2x}{a}-1\right)$	$\frac{(-1)^{m+n}2^{2m+\varepsilon}a^{s+\rho+\varepsilon/2}b^{\varepsilon}}{n!}\left(\frac{2\varepsilon+1}{2}\right)_m$ $\times\left(1-s+\sigma-\frac{\varepsilon}{2}\right)_n\mathrm{B}\left(n+\rho+1,\,s+\frac{\varepsilon}{2}\right)$ $\times{}_3F_3\left(\begin{matrix}\frac{2m+2\varepsilon+1}{2},\,\frac{2s-2\sigma+\varepsilon}{2},\,\frac{2s+\varepsilon}{2};\,-ab^2\\ \frac{2\varepsilon+1}{2},\,\frac{2s+2n+2\rho+\varepsilon+2}{2},\,\frac{2s-2n-2\sigma+\varepsilon}{2}\end{matrix}\right)$ $[a>0;\ \mathrm{Re}\,\rho>-1;\ \mathrm{Re}\,s>-\varepsilon/2]$

3.25.15. $P_n^{(\rho,\sigma)}(\varphi(x))$ and $L_m^{\lambda}(bx)$

1	$(a-x)_+^{\rho}\, e^{-bx} L_m^{\lambda}(bx)$ $\times P_n^{(\rho,\sigma)}\left(\frac{2x}{a}-1\right)$	$\frac{(-1)^n a^{s+\rho}}{m!\, n!}(\lambda+1)_m (1-s+\sigma)_n \mathrm{B}(n+\rho+1,\, s)$ $\times {}_3F_3\left(\begin{matrix} m+\lambda+1,\, s-\sigma,\, s;\, -ab \\ \lambda+1,\, s-n-\sigma,\, s+n+\rho+1 \end{matrix}\right)$ $[a,\ \mathrm{Re}\, s>0;\ \mathrm{Re}\,\rho>-1]$

3.25.16. $P_n^{(\rho,\sigma)}(\varphi(x))$ and $C_m^{\lambda}(\psi(x))$

Notation: $\varepsilon=0$ or 1.

1	$(a-x)_+^{\lambda+\sigma-1/2}$ $\times C_m^{\lambda}\left(1-\frac{2x}{a}\right)$ $\times P_n^{(\rho,\sigma)}\left(1-\frac{2x}{a}\right)$	$\frac{a^{s+\lambda+\sigma-1/2}}{m!\, n!}(2\lambda)_m (1-s+\rho)_n \mathrm{B}(n+\sigma+1,\, s)$ $\times {}_4F_3\left(\begin{matrix} -m-\lambda+\frac{1}{2},\, m+\lambda+\frac{1}{2},\, s-\rho,\, s \\ \lambda+\frac{1}{2},\, s-n-\rho,\, s+n+\sigma+1;\, 1 \end{matrix}\right)$ $[a,\ \mathrm{Re}\, s>0;\ \mathrm{Re}\,(\lambda+\sigma)>-1/2]$
2	$(a-x)_+^{\rho}\, C_{2m+\varepsilon}^{\lambda}(bx)$ $\times P_n^{(\rho,\sigma)}\left(\frac{2x}{a}-1\right)$	$(-1)^{m+n}\frac{a^{s+\rho+\varepsilon}(2b)^{\varepsilon}}{m!\, n!}(\lambda)_{m+\varepsilon}$ $\times (1-s+\sigma-\varepsilon)_n \mathrm{B}(n+\rho+1,\, s+\varepsilon)$ $\times {}_6F_5\left(\begin{matrix} -m,\, m+\lambda+\varepsilon,\, \Delta(2,\, s+\varepsilon),\, \Delta(2,\, s-\sigma+\varepsilon);\, a^2b^2 \\ \frac{2\varepsilon+1}{2},\, \Delta(2,\, s+n+\rho+\varepsilon+1),\, \Delta(2,\, s-n-\sigma+\varepsilon) \end{matrix}\right)$ $[a>0;\ \mathrm{Re}\,\rho>-1;\ \mathrm{Re}\, s>-\varepsilon]$
3	$(a-x)_+^{\rho}(1-bx)^{\lambda-1/2}$ $\times C_m^{\lambda}(2bx-1)$ $\times P_n^{(\rho,\sigma)}\left(\frac{2x}{a}-1\right)$	$\frac{(-1)^{m+n} a^{s+\rho}}{m!\, n!}(2\lambda)_m (1-s+\sigma)_n \mathrm{B}(n+\rho+1,\, s)$ $\times {}_4F_3\left(\begin{matrix} -m-\lambda+\frac{1}{2},\, m+\lambda+\frac{1}{2},\, s-\sigma,\, s \\ \lambda+\frac{1}{2},\, s-n-\sigma,\, s+n+\rho+1;\, ab \end{matrix}\right)$ $[a,\ \mathrm{Re}\, s>0;\ \mathrm{Re}\,\rho>-1;\ \lvert\arg(1-ab)\rvert<\pi]$
4	$(a-x)_+^{\lambda-1/2}\left\{\begin{matrix}(x+b)^{\sigma} \\ (b-x)^{\rho}\end{matrix}\right\}$ $\times C_m^{\lambda}\left(\frac{2x}{a}-1\right)$ $\times P_n^{(\rho,\sigma)}\left(\frac{2x}{b}\pm 1\right)$	$\frac{(-1)^m (\pm 1)^n a^{s+\lambda-1/2} b^{\psi}}{m!\, n!}(2\lambda)_m (\varphi+1)_n$ $\times \left(\frac{1}{2}-s+\lambda\right)_m \Gamma\left[\begin{matrix} \frac{2\lambda+1}{2},\, s \\ \frac{2s+2m+2\lambda+1}{2} \end{matrix}\right]$ $\times {}_4F_3\left(\begin{matrix} -n-\psi,\, n+\varphi+1,\, s,\, \frac{2s-2\lambda+1}{2};\, \mp\frac{a}{b} \\ \varphi+1,\, \frac{2s-2m-2\lambda+1}{2},\, \frac{2s+2m+2\lambda+1}{2} \end{matrix}\right)$ $\left[\left\{\begin{matrix} a>0;\ \lvert\arg b\rvert<\pi \\ b>a>0 \end{matrix}\right\};\ \mathrm{Re}\, s>0;\ \varphi=\left\{\begin{matrix}\rho \\ \sigma\end{matrix}\right\},\ \psi=\left\{\begin{matrix}\sigma \\ \rho\end{matrix}\right\}\right]$

No.	$f(x)$	$F(s)$
5	$(a-x)_+^{\lambda+\rho-1/2}$ $\times C_{2m+\varepsilon}^{\lambda}\left(\sqrt{\frac{x}{a}}\right)$ $\times P_n^{(\rho,\sigma)}\left(\frac{2x}{a}-1\right)$	$\frac{2^{\varepsilon}(-1)^{m+n}a^{s+\lambda+\rho-1/2}}{m!\,n!}(\lambda)_{m+\varepsilon}\left(1-s+\sigma-\frac{\varepsilon}{2}\right)_n$ $\times \mathrm{B}\left(n+\rho+1,\,s+\frac{\varepsilon}{2}\right)$ $\times {}_4F_3\left(\begin{matrix}\frac{-2m-2\lambda+1}{2},\,\frac{2m+2\varepsilon+1}{2},\,\frac{2s+\varepsilon}{2},\,\frac{2s-2\sigma+\varepsilon}{2}\\ \frac{2\varepsilon+1}{2},\,\frac{2s-2n-2\sigma+\varepsilon}{2},\,\frac{2s+2n+2\rho+\varepsilon+2}{2};\,1\end{matrix}\right)$ $[a>0;\ \mathrm{Re}\,(\lambda+\rho)>-1/2;\ \mathrm{Re}\,s>-\varepsilon/2]$
6	$(a-x)_+^{\rho}\left(1-b^2x\right)^{\lambda-1/2}$ $\times C_{2m+\varepsilon}^{\lambda}(b\sqrt{x})$ $\times P_n^{(\rho,\sigma)}\left(\frac{2x}{a}-1\right)$	$\frac{(-1)^{m+n}a^{s+\rho+\varepsilon/2}(2b)^{\varepsilon}}{m!\,n!}(\lambda)_{m+\varepsilon}\left(1-s+\sigma-\frac{\varepsilon}{2}\right)_n$ $\times \mathrm{B}\left(n+\rho+1,\,s+\frac{\varepsilon}{2}\right)$ $\times {}_4F_3\left(\begin{matrix}\frac{-2m-2\lambda+1}{2},\,\frac{2m+2\varepsilon+1}{2},\,\frac{2s+\varepsilon}{2},\,\frac{2s-2\sigma+\varepsilon}{2}\\ \frac{2\varepsilon+1}{2},\,\frac{2s-2n-2\sigma+\varepsilon}{2},\,\frac{2s+2n+2\rho+\varepsilon+2}{2};\,ab^2\end{matrix}\right)$ $[a>0;\ \mathrm{Re}\,\rho>-1;\ \mathrm{Re}\,s>-\varepsilon/2;\ \lvert\arg\left(1-ab^2\right)\rvert<\pi]$
7	$(a-x)_+^{\rho}\left[C_m^{\lambda}\left(\sqrt{1-bx}\right)\right]^2$ $\times P_n^{(\rho,\sigma)}\left(\frac{2x}{a}-1\right)$	$\frac{(-1)^n a^{s+\rho}}{(m!)^2\,n!}\left[(2\lambda)_m\right]^2(1-s+\sigma)_n\,\mathrm{B}\,(n+\rho+1,\,s)$ $\times {}_5F_4\left(\begin{matrix}-m,\,\lambda,\,m+2\lambda,\,s-\sigma,\,s;\,ab\\ \frac{2\lambda+1}{2},\,2\lambda,\,s-n-\sigma,\,s+n+\rho+1\end{matrix}\right)$ $[a,\ \mathrm{Re}\,s>0;\ \mathrm{Re}\,\rho>-1]$

3.25.17. Products of $P_n^{(\rho,\sigma)}(ax+b)$

1	$(a-x)_+^{\nu+\sigma}$ $\times P_m^{(\lambda,\nu)}\left(1-\frac{2x}{a}\right)$ $\times P_n^{(\rho,\sigma)}\left(1-\frac{2x}{a}\right)$	$\frac{a^{s+\nu+\sigma}}{m!\,n!}(\lambda+1)_m(1-s+\rho)_n\,\mathrm{B}\,(n+\sigma+1,s)$ $\times {}_4F_3\left(\begin{matrix}-m-\nu,\,m+\lambda+1,\,s-\rho,\,s\\ \lambda+1,\,s-n-\rho,\,s+n+\sigma+1;\,1\end{matrix}\right)$ $[a,\ \mathrm{Re}\,s>0;\ \mathrm{Re}\,(\sigma+\nu)>-1]$
2	$(a-x)_+^{\lambda+\rho}$ $\times P_m^{(\lambda,\nu)}\left(\frac{2x}{a}-1\right)$ $\times P_n^{(\rho,\sigma)}\left(\frac{2x}{a}-1\right)$	$\frac{(-1)^{m+n}a^{s+\lambda+\rho}}{m!\,n!}(\nu+1)_m(1-s+\sigma)_n\,\mathrm{B}\,(n+\rho+1,\,s)$ $\times {}_4F_3\left(\begin{matrix}-m-\lambda,\,m+\nu+1,\,s-\sigma,\,s\\ \nu+1,\,s-n-\sigma,\,s+n+\rho+1;\,1\end{matrix}\right)$ $[a,\ \mathrm{Re}\,s>0;\ \mathrm{Re}\,(\lambda+\rho)>-1]$

No.	$f(x)$	$F(s)$
3	$(a-x)_+^{\rho}\, P_m^{(\lambda,\nu)}(bx\pm 1)$ $\times P_n^{(\rho,\sigma)}\left(\frac{2x}{a}-1\right)$	$\frac{(-1)^n (\pm 1)^m a^{s+\rho}}{m!\,n!} (\varphi+1)_m (1-s+\sigma)_n \; \mathrm{B}(n+\rho+1,\, s)$ $\times {}_4F_3\left(\begin{matrix}-m,\, m+\lambda+\nu+1,\, s-\sigma,\, s;\, \mp\frac{ab}{2}\\ \varphi+1,\, s-n-\sigma,\, s+n+\rho+1\end{matrix}\right)$ $\left[a,\ \mathrm{Re}\, s>0;\ \mathrm{Re}\,\rho>-1;\ \varphi=\begin{Bmatrix}\lambda\\ \nu\end{Bmatrix}\right]$
4	$(a-x)_+^{\rho}(1-bx)^{\lambda}$ $\times P_m^{(\lambda,\nu)}(2bx-1)$ $\times P_n^{(\rho,\sigma)}\left(\frac{2x}{a}-1\right)$	$\frac{(-1)^{m+n} a^{s+\rho}}{m!\,n!} (\nu+1)_m (1-s+\sigma)_n \mathrm{B}(n+\rho+1,\, s)$ $\times {}_4F_3\left(\begin{matrix}-m-\lambda,\, m+\nu+1,\, s-\sigma,\, s\\ \nu+1,\, s-n-\sigma,\, s+n+\rho+1;\, ab\end{matrix}\right)$ $[a,\ \mathrm{Re}\, s>0;\ \mathrm{Re}\,\rho>-1;\ \lvert\arg(1-ab)\rvert<\pi]$
5	$(a-x)_+^{\lambda}\begin{Bmatrix}(x+b)^{\sigma}\\ (b-x)^{\rho}\end{Bmatrix}$ $\times P_m^{(\lambda,\nu)}\left(\frac{2x}{a}-1\right)$ $\times P_n^{(\rho,\sigma)}\left(\frac{2x}{b}\pm 1\right)$	$\frac{(-1)^m (\pm 1)^n a^{s+\lambda} b^{\psi}}{m!\,n!} (1-s+\nu)_m \mathrm{B}(m+\lambda+1,\, s)$ $\times (\varphi+1)_n \, {}_4F_3\left(\begin{matrix}-n-\psi,\, n+\varphi+1,\, s-\nu,\, s;\, \mp\frac{a}{b}\\ \varphi+1,\, s-m-\nu,\, s+m+\lambda+1\end{matrix}\right)$ $\left[\varphi=\begin{Bmatrix}\rho\\ \sigma\end{Bmatrix},\ \psi=\begin{Bmatrix}\sigma\\ \rho\end{Bmatrix};\right.$ $\left.\begin{Bmatrix}a>0;\ \lvert\arg b\rvert<\pi\\ b>a>0\end{Bmatrix};\ \mathrm{Re}\,\lambda>-1;\ \mathrm{Re}\, s>0\right]$
6	$(a-x)_+^{\rho}\, P_m^{(\lambda,\nu)}\left(1-\frac{b}{x}\right)$ $\times P_n^{(\rho,\sigma)}\left(\frac{2x}{a}-1\right)$	$\frac{(-1)^n a^{s-m+\rho} (b/2)^m}{m!\,n!} (-2m-\lambda-\nu)_m$ $\times (m-s+\sigma+1)_n \Gamma\left[\begin{matrix}n+\rho+1, s-m\\ s-m+n+\rho+1\end{matrix}\right]$ $\times {}_4F_3\left(\begin{matrix}-m,\, -m-\lambda,\, s-m-\sigma,\, s-m;\, \frac{2a}{b}\\ -2m-\lambda-\nu,\, s-m-n-\sigma,\, s-m+n+\rho+1\end{matrix}\right)$ $[a>0;\ \mathrm{Re}\,\rho>-1;\ \mathrm{Re}\, s>m]$
7	$(a-x)_+^{\rho}(b-x)^m$ $\times P_m^{(\lambda,\nu)}\left(\frac{b+x}{b-x}\right)$ $\times P_n^{(\rho,\sigma)}\left(\frac{2x}{a}-1\right)$	$\frac{(-1)^n a^{s+\rho} b^m}{m!\,n!} (\lambda+1)_m (1-s+\sigma)_n \Gamma\left[\begin{matrix}n+\rho+1,\, s\\ s+n+\rho+1\end{matrix}\right]$ $\times {}_4F_3\left(\begin{matrix}-m,\, -m-\nu,\, s-\sigma,\, s;\, \frac{a}{b}\\ \lambda+1,\, s-n-\sigma,\, s+n+\rho+1\end{matrix}\right)$ $[a,\ \mathrm{Re}\, s>0;\ \mathrm{Re}\,\rho>-1]$

3.26. The Complete Elliptic Integrals $\mathbf{K}(z)$, $\mathbf{E}(z)$, and $\mathbf{D}(z)$

More formulas can be obtained from the corresponding section due to the relations

$$\mathbf{K}(z)=\frac{\pi}{2}\,{}_2F_1\left(\frac{1}{2},\,\frac{1}{2};\,1;\,z^2\right),\qquad \mathbf{E}(z)=\frac{\pi}{2}\,{}_2F_1\left(-\frac{1}{2},\,\frac{1}{2};\,1;\,z^2\right),$$

$$\mathbf{D}(z)=\frac{\pi}{4}\,{}_2F_1\left(\frac{1}{2},\,\frac{3}{2};\,2;\,z^2\right),\qquad \mathbf{K}(z)=\frac{1}{2}\,G_{22}^{12}\left(-z^2\,\middle|\,\begin{matrix}1/2,\,1/2\\0,\,0\end{matrix}\right),$$

$$\mathbf{E}(z)=-\frac{1}{4}\,G_{22}^{12}\left(-z^2\,\middle|\,\begin{matrix}1/2,\,3/2\\0,\,0\end{matrix}\right),\qquad \mathbf{D}(z)=\frac{1}{2}\,G_{22}^{12}\left(-z^2\,\middle|\,\begin{matrix}-1/2,\,1/2\\0,\,-1\end{matrix}\right).$$

3.26.1. $\mathbf{K}(\varphi(x))$

No.	$f(x)$	$F(s)$
1	$\mathbf{K}(iax)$	$\frac{a^{-s}}{4}\Gamma\left[\begin{matrix}\frac{s}{2},\,\frac{1-s}{2},\,\frac{1-s}{2}\\ \frac{2-s}{2}\end{matrix}\right]$ [Re $a>0$; $0<\operatorname{Re}s<1/2$]
2	$\mathbf{K}(iax)-\frac{\pi}{2}$	$\frac{a^{-s}}{4}\Gamma\left[\begin{matrix}\frac{s}{2},\,\frac{1-s}{2},\,\frac{1-s}{2}\\ \frac{2-s}{2}\end{matrix}\right]$ [Re $a>0$; $-1<\operatorname{Re}s<0$]
3	$\mathbf{K}(iax)$ $-\frac{\pi}{2}\sum_{k=0}^{n}\frac{(1/2)_k^2}{(k!)^2}\left(-a^2x^2\right)^k$	$\frac{a^{-s}}{4}\Gamma\left[\begin{matrix}\frac{s}{2},\,\frac{1-s}{2},\,\frac{1-s}{2}\\ \frac{2-s}{2}\end{matrix}\right]$ [Re $a>0$; $-n-1<\operatorname{Re}s<-n$]
4	$\mathbf{K}\left(\pm i\frac{a-x}{2\sqrt{ax}}\right)$	$\frac{a^s}{4\pi}\Gamma\left[\frac{2s+1}{4},\,\frac{2s+1}{4},\,\frac{1-2s}{4},\,\frac{1-2s}{4}\right]$ [$-1/2<\operatorname{Re}s<1/2$; $\lvert\arg a\rvert<\pi$]
5	$\mathbf{K}\left(\sqrt{\frac{a-x}{a}}\right)$	$\frac{a^s}{2\pi}\Gamma\left[s,\,s,\,\frac{1-2s}{2},\,\frac{1-2s}{2}\right]$ [$0<\operatorname{Re}s<1/2$]
6	$\mathbf{K}\left(\sqrt{\frac{a+x\operatorname{sgn}(x-a)}{2a}}\right)$	$\frac{a^s}{8\pi}\Gamma\left[\frac{s}{2},\,\frac{s+1}{2},\,\frac{1-2s}{4},\,\frac{1-2s}{4}\right]$ [$a>0$; $0<\operatorname{Re}s<1/2$]
7	$\mathbf{K}\left(\sqrt{\frac{\sqrt{a}-\sqrt{x+a}}{2\sqrt{a}}}\right)$	$\frac{\pi a^s}{2}\Gamma\left[\begin{matrix}s,\,\frac{1-4s}{4},\,\frac{1-4s}{4}\\ \frac{1}{4},\,\frac{1}{4},1-s\end{matrix}\right]$ [$0<\operatorname{Re}s<1/4$; $\lvert\arg a\rvert<\pi$]

3.26.2. $\mathbf{K}(\varphi(x))$ and algebraic functions

1	$\frac{1}{(x+a)^{\rho}}\mathbf{K}\left(\frac{b}{x+a}\right)$	$\frac{\pi a^{s-\rho}}{2}\mathrm{B}(s,\,\rho-s)\,{}_4F_3\left(\begin{matrix}\frac{1}{2},\,\frac{1}{2},\,\frac{\rho-s}{2},\,\frac{\rho-s+1}{2}\\ 1,\,\frac{\rho}{2},\,\frac{\rho+1}{2};\,\frac{b^2}{a^2}\end{matrix}\right)$ [$0<\operatorname{Re}s<\operatorname{Re}\rho$; $\lvert\arg a\rvert<\pi$]

No.	$f(x)$	$F(s)$
2	$\frac{1}{(x+a)^{\rho}}\mathbf{K}\left(\frac{bx}{x+a}\right)$	$\frac{\pi a^{s-\rho}}{2}\mathrm{B}(s,\,\rho-s)\;{}_4F_3\left(\begin{matrix}\frac{1}{2},\,\frac{1}{2},\,\frac{s}{2},\,\frac{s+1}{2}\\1,\,\frac{\rho}{2},\,\frac{\rho+1}{2};\,b^2\end{matrix}\right)$ $[0<\operatorname{Re}s<\operatorname{Re}\rho;\ \lvert\arg a\rvert<\pi]$
3	$\frac{1}{x+a}\mathbf{K}\left(\pm\frac{x-a}{x+a}\right)$	$\frac{a^{s-1}}{8\pi}\Gamma\left[\frac{s}{2},\,\frac{s}{2},\,\frac{1-s}{2},\,\frac{1-s}{2}\right]$ $[0<\operatorname{Re}s<1;\ \lvert\arg a\rvert<\pi]$
4	$\frac{1}{x+a}\mathbf{K}\left(\frac{\lvert x-a\rvert}{x+a}\right)$	$\frac{a^{s-1}}{8\pi}\Gamma\left[\frac{s}{2},\,\frac{s}{2},\,\frac{1-s}{2},\,\frac{1-s}{2}\right]$ $[\operatorname{Re}a>0;\ 0<\operatorname{Re}s<1]$
5	$\frac{1}{x+a}\mathbf{K}\left(\pm\frac{2\sqrt{ax}}{x+a}\right)$	$\frac{\pi a^{s-1}}{4}\Gamma\left[\begin{matrix}\frac{s}{2},\,\frac{1-s}{2}\\\frac{s+1}{2},\,\frac{2-s}{2}\end{matrix}\right]$ $[a>0;\ 0<\operatorname{Re}s<1]$
6	$\frac{1}{\lvert x-a\rvert}\mathbf{K}\left(\pm\frac{2i\sqrt{ax}}{x-a}\right)$	$\frac{\pi a^{s-1}}{4}\Gamma\left[\begin{matrix}\frac{s}{2},\,\frac{1-s}{2}\\\frac{s+1}{2},\,\frac{2-s}{2}\end{matrix}\right]$ $[a>0;\ 0<\operatorname{Re}s<1]$
7	$\frac{1}{(x+a)^{\rho}}\mathbf{K}\left(\frac{b}{\sqrt{x+a}}\right)$	$\frac{\pi a^{s-\rho}}{2}\mathrm{B}(s,\,\rho-s)\;{}_3F_2\left(\begin{matrix}\frac{1}{2},\,\frac{1}{2},\,\rho-s\\1,\,\rho;\,\frac{b^2}{a}\end{matrix}\right)$ $[0<\operatorname{Re}s<\operatorname{Re}\rho;\ \lvert\arg a\rvert<\pi]$
8	$\frac{1}{\sqrt{x+a}}\mathbf{K}\left(\sqrt{\frac{a}{x+a}}\right)$	$\frac{a^{s-1/2}}{2}\Gamma\left[\begin{matrix}s,\,s,\,\frac{1-2s}{2}\\\frac{2s+1}{2}\end{matrix}\right]$ $[0<\operatorname{Re}s<1/2;\ \lvert\arg a\rvert<\pi]$
9	$\frac{1}{\sqrt{x+a}}\mathbf{K}\left(\sqrt{\frac{x}{x+a}}\right)$	$\frac{a^{s-1/2}}{2}\Gamma\left[\begin{matrix}s,\,\frac{1-2s}{2},\,\frac{1-2s}{2}\\1-s\end{matrix}\right]$ $[0<\operatorname{Re}s<1/2;\ \lvert\arg a\rvert<\pi]$
10	$\frac{1}{\sqrt{x+a}}\mathbf{K}\left(\sqrt{\frac{\sqrt{a}-\sqrt{x+a}}{2\sqrt{a}}}\right)$	$\frac{\pi a^{s-1/2}}{2}\Gamma\left[\begin{matrix}s,\,\frac{3-4s}{4},\,\frac{3-4s}{4}\\\frac{3}{4},\,\frac{3}{4},\,1-s\end{matrix}\right]$ $[0<\operatorname{Re}s<3/4;\ \lvert\arg a\rvert<\pi]$
11	$\frac{1}{\sqrt{x+a}}\mathbf{K}\left(\frac{\lvert\sqrt{a}-\sqrt{x}\rvert}{\sqrt{2(x+a)}}\right)$	$\frac{a^{s-1/2}}{8\pi}\Gamma\left[\frac{s}{2},\,\frac{2s+1}{4},\,\frac{1-2s}{4},\,\frac{1-s}{2}\right]$ $[a>0;\ 0<\operatorname{Re}s<1/2]$
12	$\frac{1}{\sqrt[4]{x+a}}\mathbf{K}\left(\frac{\sqrt{\sqrt{x+a}-\sqrt{a}}}{\sqrt{2}\sqrt[4]{x+a}}\right)$	$2^{2s-1}\sqrt{\pi}\,a^{s-1/4}\Gamma\left[\begin{matrix}s,\,\frac{1-4s}{2}\\1-s\end{matrix}\right]$ $[0<\operatorname{Re}s<1/4;\ \lvert\arg a\rvert<\pi]$

No.	$f(x)$	$F(s)$
13	$\dfrac{1}{\sqrt[4]{x+a}}\,\mathbf{K}\left(\dfrac{\sqrt{\sqrt{x+a}-\sqrt{x}}}{\sqrt{2}\sqrt[4]{x+a}}\right)$	$2^{-2s-1/2}\sqrt{\pi}\,a^{s-1/4}\,\Gamma\left[\begin{matrix}2s,\ \frac{1-4s}{4}\\ \frac{4s+3}{4}\end{matrix}\right]$ $[0<\operatorname{Re}s<1/4;\ \lvert\arg a\rvert<\pi]$
14	$\dfrac{1}{\sqrt[4]{x+a}}\,\mathbf{K}\left(\pm i\dfrac{\sqrt{x+a}-\sqrt{a}}{2\sqrt[4]{a}\sqrt[4]{x+a}}\right)$	$\dfrac{a^{s-1/4}}{2}\,\Gamma\left[\begin{matrix}s,\ \frac{1-2s}{2},\ \frac{1-2s}{2}\\ 1-s\end{matrix}\right]$ $[0<\operatorname{Re}s<1/2;\ \lvert\arg a\rvert<\pi]$
15	$\dfrac{1}{\sqrt[4]{x+a}}\,\mathbf{K}\left(\pm i\dfrac{\sqrt{x+a}-\sqrt{x}}{2\sqrt[4]{x}\sqrt[4]{x+a}}\right)$	$\dfrac{a^{s-1/4}}{2}\,\Gamma\left[\begin{matrix}\frac{1-4s}{4},\ \frac{4s+1}{4},\ \frac{4s+1}{4}\\ \frac{4s+3}{4}\end{matrix}\right]$ $[-1/4<\operatorname{Re}s<1/4;\ \lvert\arg a\rvert<\pi]$
16	$\dfrac{1}{\sqrt{x+a}+\sqrt{a}}\,\mathbf{K}\left(\pm\dfrac{(\sqrt{x+a}-\sqrt{a})^2}{x}\right)$	$\dfrac{a^{s-1/2}}{4}\,\Gamma\left[\begin{matrix}s,\ \frac{1-2s}{2},\ \frac{1-2s}{2}\\ 1-s\end{matrix}\right]$ $[0<\operatorname{Re}s<1/2;\ \lvert\arg a\rvert<\pi]$
17	$\dfrac{1}{\sqrt{x+a}\pm\sqrt{a}}\,\mathbf{K}\left(\left\{\begin{matrix}1\\ \pm i\end{matrix}\right\}\dfrac{2\sqrt[4]{a}\sqrt[4]{x+a}}{\sqrt{x+a}\pm\sqrt{a}}\right)$	$\dfrac{a^{s-1/2}}{2}\,\Gamma\left[\begin{matrix}s,\ s,\ \frac{1-2s}{2}\\ \frac{2s+1}{2}\end{matrix}\right]$ $[0<\operatorname{Re}s<1/2;\ \lvert\arg a\rvert<\pi]$
18	$\dfrac{1}{\sqrt{x+a}+\sqrt{x}}\,\mathbf{K}\left(\dfrac{(\sqrt{x}-\sqrt{x+a})^2}{a}\right)$	$\dfrac{a^{s-1/2}}{4}\,\Gamma\left[\begin{matrix}s,\ s,\ \frac{1-2s}{2}\\ \frac{2s+1}{2}\end{matrix}\right]$ $[0<\operatorname{Re}s<1/2;\ \lvert\arg a\rvert<\pi]$
19	$\dfrac{1}{\sqrt{x+a}\pm\sqrt{x}}\,\mathbf{K}\left(\left\{\begin{matrix}1\\ \pm i\end{matrix}\right\}\dfrac{2\sqrt[4]{x}\sqrt[4]{x+a}}{\sqrt{x+a}\pm\sqrt{x}}\right)$	$\dfrac{a^{s-1/2}}{2}\,\Gamma\left[\begin{matrix}s,\ \frac{1-2s}{2},\ \frac{1-2s}{2}\\ 1-s\end{matrix}\right]$ $[0<\operatorname{Re}s<1/2;\ \lvert\arg a\rvert<\pi]$
20	$\dfrac{1}{\sqrt{\sqrt{x+a}+\sqrt{a}}}\,\mathbf{K}\left(\pm i\dfrac{\sqrt{x+a}-\sqrt{a}}{\sqrt{x}}\right)$	$2^{2s-3/2}\sqrt{\pi}\,a^{s-1/4}\,\Gamma\left[\begin{matrix}s,\ \frac{1-4s}{2}\\ 1-s\end{matrix}\right]$ $[0<\operatorname{Re}s<1/4;\ \lvert\arg a\rvert<\pi]$
21	$\dfrac{1}{\sqrt{\sqrt{x+a}+\sqrt{a}}}\,\mathbf{K}\left(\sqrt{\dfrac{\sqrt{a}-\sqrt{x+a}}{\sqrt{a}+\sqrt{x+a}}}\right)$	$2^{2s-3/2}\sqrt{\pi}\,a^{s-1/4}\,\Gamma\left[\begin{matrix}s,\ \frac{1-4s}{2}\\ 1-s\end{matrix}\right]$ $[0<\operatorname{Re}s<1/4;\ \lvert\arg a\rvert<\pi]$
22	$\dfrac{1}{\sqrt{\sqrt{x+a}+\sqrt{a}}}\,\mathbf{K}\left(\sqrt{\dfrac{\sqrt{x+a}-\sqrt{a}}{\sqrt{x+a}+\sqrt{a}}}\right)$	$\dfrac{a^{s-1/4}\pi}{2\sqrt{2}}\,\Gamma\left[\begin{matrix}s,\ \frac{1-4s}{4},\ \frac{1-4s}{4}\\ \frac{1}{4},\ \frac{1}{4},\ 1-s\end{matrix}\right]$ $[0<\operatorname{Re}s<1/4;\ \lvert\arg a\rvert<\pi]$

No.	$f(x)$	$F(s)$
23	$\dfrac{1}{\sqrt{\sqrt{x+a}+\sqrt{a}}} \times \mathbf{K}\left(\sqrt{\dfrac{2\sqrt{a}\sqrt{x+a}-x-2a}{x}}\right)$	$2^{2s-3/2}\sqrt{\pi}\,a^{s-1/4}\,\Gamma\left[\begin{matrix} s,\ \frac{1-4s}{2} \\ 1-s \end{matrix}\right]$ $[0<\operatorname{Re}s<1/4;\ \lvert\arg a\rvert<\pi]$
24	$\dfrac{1}{\sqrt{\sqrt{x+a}+\sqrt{a}}}\,\mathbf{K}\left(\pm\dfrac{\sqrt{x+a}-\sqrt{a}}{\sqrt{x}}\right)$	$\dfrac{a^{s-1/4}\pi}{2\sqrt{2}}\,\Gamma\left[\begin{matrix} s,\ \frac{1-4s}{4},\ \frac{1-4s}{4} \\ \frac{1}{4},\ \frac{1}{4},\ 1-s \end{matrix}\right]$ $[0<\operatorname{Re}s<1/4;\ \lvert\arg a\rvert<\pi]$
25	$\dfrac{1}{\sqrt{\sqrt{x+a}\pm\sqrt{a}}}\,\mathbf{K}\left(\sqrt{\dfrac{2\sqrt{a}}{\sqrt{a}\pm\sqrt{x+a}}}\right)$	$\dfrac{\pi a^{s-1/4}}{2}\,\Gamma\left[\begin{matrix} s,\ s,\ \frac{1-4s}{4} \\ \frac{1}{4},\ \frac{1}{4},\ \frac{4s+3}{4} \end{matrix}\right]$ $[0<\operatorname{Re}s<1/4;\ \lvert\arg a\rvert<\pi]$
26	$\dfrac{1}{\sqrt{\sqrt{x+a}\pm\sqrt{a}}} \times \mathbf{K}\left(\left\{\begin{matrix} 1 \\ i \end{matrix}\right\}\sqrt{\dfrac{2\sqrt{a}\left(\sqrt{x+a}\mp\sqrt{a}\right)}{x}}\right)$	$\dfrac{\pi a^{s-1/4}}{2}\,\Gamma\left[\begin{matrix} s,\ s,\ \frac{1-4s}{4} \\ \frac{1}{4},\ \frac{1}{4},\ \frac{4s+3}{4} \end{matrix}\right]$ $[0<\operatorname{Re}s<1/4;\ \lvert\arg a\rvert<\pi]$
27	$\dfrac{1}{\sqrt{\sqrt{x+a}+\sqrt{x}}}\,\mathbf{K}\left(\pm i\dfrac{\sqrt{x}-\sqrt{x+a}}{\sqrt{a}}\right)$	$\dfrac{\sqrt{\pi}\,a^{s-1/4}}{2^{2s+1}}\,\Gamma\left[\begin{matrix} \frac{1-4s}{4},\ 2s \\ \frac{4s+3}{4} \end{matrix}\right]$ $[0<\operatorname{Re}s<1/4;\ \lvert\arg a\rvert<\pi]$
28	$\dfrac{1}{\sqrt{\sqrt{x+a}+\sqrt{x}}}\,\mathbf{K}\left(\sqrt{\dfrac{\sqrt{x}-\sqrt{x+a}}{\sqrt{x}+\sqrt{x+a}}}\right)$	$\dfrac{\sqrt{\pi}\,a^{s-1/4}}{2^{2s+1}}\,\Gamma\left[\begin{matrix} \frac{1-4s}{4},\ 2s \\ \frac{4s+3}{4} \end{matrix}\right]$ $[0<\operatorname{Re}s<1/4;\ \lvert\arg a\rvert<\pi]$
29	$\dfrac{1}{\sqrt{\sqrt{x+a}+\sqrt{x}}}\,\mathbf{K}\left(\pm\dfrac{\sqrt{x}-\sqrt{x+a}}{\sqrt{a}}\right)$	$\dfrac{\pi a^{s-1/4}}{2\sqrt{2}}\,\Gamma\left[\begin{matrix} s,\ s,\ \frac{1-4s}{4} \\ \frac{1}{4},\ \frac{1}{4},\ \frac{4s+3}{4} \end{matrix}\right]$ $[0<\operatorname{Re}s<1/4;\ \lvert\arg a\rvert<\pi]$
30	$\dfrac{1}{\sqrt{\sqrt{x+a}+\sqrt{x}}}\,\mathbf{K}\left(\sqrt{\dfrac{\sqrt{x+a}-\sqrt{x}}{\sqrt{x+a}+\sqrt{x}}}\right)$	$\dfrac{\pi a^{s-1/4}}{2\sqrt{2}}\,\Gamma\left[\begin{matrix} s,\ s,\ \frac{1-4s}{4} \\ \frac{1}{4},\ \frac{1}{4},\ \frac{4s+3}{4} \end{matrix}\right]$ $[0<\operatorname{Re}s<1/4;\ \lvert\arg a\rvert<\pi]$
31	$\dfrac{1}{\sqrt{\sqrt{x+a}\pm\sqrt{x}}}\,\mathbf{K}\left(\sqrt{\dfrac{2\sqrt{x}}{\sqrt{x}\pm\sqrt{x+a}}}\right)$	$\dfrac{\pi a^{s-1/4}}{2}\,\Gamma\left[\begin{matrix} s,\ \frac{1-4s}{4},\ \frac{1-4s}{4} \\ \frac{1}{4},\ \frac{1}{4},\ 1-s \end{matrix}\right]$ $[0<\operatorname{Re}s<1/4;\ \lvert\arg a\rvert<\pi]$

No.	$f(x)$	$F(s)$
32	$(\sqrt{x+a}\pm\sqrt{a})$ $\times\mathbf{K}\left(\pm\begin{Bmatrix} i \\ 1 \end{Bmatrix}\frac{2\sqrt[4]{a}\sqrt[4]{x+a}}{\sqrt{x+a}\mp\sqrt{a}}\right)$	$\frac{a^{s+1/2}}{2}\Gamma\left[\begin{matrix} s+1,\ s+1,\ -\frac{2s+1}{2} \\ \frac{2s+3}{2} \end{matrix}\right]$ $[-1<\operatorname{Re}s<-1/2;\ \lvert\arg a\rvert<\pi]$
33	$(\sqrt{x+a}-\sqrt{a})\,\mathbf{K}\left(\pm\frac{(\sqrt{x+a}-\sqrt{a})^2}{x}\right)$	$\frac{a^{s+1/2}}{4}\Gamma\left[\begin{matrix} s+1,\ -\frac{2s+1}{2},\ -\frac{2s+1}{2} \\ -s \end{matrix}\right]$ $[-1<\operatorname{Re}s<-1/2;\ \lvert\arg a\rvert<\pi]$
34	$(\sqrt{x+a}\pm\sqrt{x})$ $\times\mathbf{K}\left(\pm\begin{Bmatrix} i \\ 1 \end{Bmatrix}\frac{2\sqrt[4]{x}\sqrt[4]{x+a}}{\sqrt{x+a}\mp\sqrt{x}}\right)$	$\frac{a^{s+1/2}}{2}\Gamma\left[\begin{matrix} s,\ \frac{1-2s}{2},\ \frac{1-2s}{2} \\ 1-s \end{matrix}\right]$ $[0<\operatorname{Re}s<1/2;\ \lvert\arg a\rvert<\pi]$
35	$(\sqrt{x+a}-\sqrt{x})\,\mathbf{K}\left(\pm\frac{(\sqrt{x}-\sqrt{x+a})^2}{a}\right)$	$\frac{a^{s+1/2}}{4}\Gamma\left[\begin{matrix} s,\ s,\ \frac{1-2s}{2} \\ \frac{2s+1}{2} \end{matrix}\right]$ $[0<\operatorname{Re}s<1/2;\ \lvert\arg a\rvert<\pi]$
36	$\sqrt{\sqrt{x+a}-\sqrt{a}}\,\mathbf{K}\left(\pm i\frac{\sqrt{x+a}-\sqrt{a}}{x}\right)$	$2^{2s-1/2}\sqrt{\pi}\,a^{s+1/4}\,\Gamma\left[\begin{matrix} \frac{2s+1}{2},\ -\frac{4s+1}{2} \\ \frac{1-2s}{2} \end{matrix}\right]$ $[-1/2<\operatorname{Re}s<-1/4;\ \lvert\arg a\rvert<\pi]$
37	$\sqrt{\sqrt{x+a}-\sqrt{a}}\,\mathbf{K}\left(\sqrt{\frac{\sqrt{a}-\sqrt{x+a}}{\sqrt{a}+\sqrt{x+a}}}\right)$	$2^{2s-1/2}\sqrt{\pi}\,a^{s+1/4}\,\Gamma\left[\begin{matrix} \frac{2s+1}{2},\ -\frac{4s+1}{2} \\ \frac{1-2s}{2} \end{matrix}\right]$ $[-1/2<\operatorname{Re}s<-1/4;\ \lvert\arg a\rvert<\pi]$
38	$\sqrt{\sqrt{x+a}-\sqrt{a}}\,\mathbf{K}\left(\pm\frac{\sqrt{x+a}-\sqrt{a}}{\sqrt{x}}\right)$	$\frac{\pi a^{s+1/4}}{2\sqrt{2}}\Gamma\left[\begin{matrix} \frac{2s+1}{2},\ -\frac{4s+1}{2},\ -\frac{4s+1}{2} \\ \frac{1}{4},\ \frac{1}{4},\ \frac{1-2s}{2} \end{matrix}\right]$ $[-1/2<\operatorname{Re}s<-1/4;\ \lvert\arg a\rvert<\pi]$
39	$\sqrt{\sqrt{x+a}-\sqrt{a}}\,\mathbf{K}\left(\sqrt{\frac{\sqrt{x+a}-\sqrt{a}}{\sqrt{x+a}+\sqrt{a}}}\right)$	$\frac{\pi a^{s+1/4}}{2\sqrt{2}}\Gamma\left[\begin{matrix} \frac{2s+1}{2},\ -\frac{4s+1}{2},\ -\frac{4s+1}{2} \\ \frac{1}{4},\ \frac{1}{4},\ \frac{1-2s}{2} \end{matrix}\right]$ $[-1/2<\operatorname{Re}s<-1/4;\ \lvert\arg a\rvert<\pi]$
40	$\sqrt{\sqrt{x+a}\pm\sqrt{a}}\,\mathbf{K}\left(\sqrt{\frac{2\sqrt{a}}{\sqrt{a}\mp\sqrt{x+a}}}\right)$	$\frac{\pi a^{s+1/4}}{2}\Gamma\left[\begin{matrix} \frac{2s+1}{2},\ \frac{2s+1}{2},\ -\frac{4s+1}{2} \\ \frac{1}{4},\ \frac{1}{4},\ \frac{4s+5}{4} \end{matrix}\right]$ $[-1/2<\operatorname{Re}s<-1/4;\ \lvert\arg a\rvert<\pi]$
41	$\sqrt{\sqrt{x+a}\pm\sqrt{a}}$ $\times\mathbf{K}\left(\begin{Bmatrix} i \\ 1 \end{Bmatrix}\sqrt{\frac{2\sqrt{a}\,(\sqrt{x+a}\pm\sqrt{a})}{x}}\right)$	$\frac{\pi a^{s+1/4}}{2}\Gamma\left[\begin{matrix} \frac{2s+1}{2},\ \frac{2s+1}{2},\ -\frac{4s+1}{2} \\ \frac{1}{4},\ \frac{1}{4},\ \frac{4s+5}{4} \end{matrix}\right]$ $[-1/2<\operatorname{Re}s<-1/4;\ \lvert\arg a\rvert<\pi]$

No.	$f(x)$	$F(s)$
42	$\sqrt{\sqrt{x+a}-\sqrt{x}}\,\mathbf{K}\left(\pm i\dfrac{\sqrt{x}-\sqrt{x+a}}{\sqrt{a}}\right)$	$\dfrac{\sqrt{\pi}\,a^{s+1/4}}{2^{2s+1}}\Gamma\left[\begin{matrix}\frac{1-4s}{4},\ 2s\\ \frac{4s+3}{4}\end{matrix}\right]$ $[0<\operatorname{Re}s<1/4;\ \lvert\arg a\rvert<\pi]$
43	$\sqrt{\sqrt{x+a}-\sqrt{x}}\,\mathbf{K}\left(\sqrt{\dfrac{\sqrt{x+a}-\sqrt{x}}{\sqrt{x+a}+\sqrt{x}}}\right)$	$\dfrac{\pi a^{s+1/4}}{2\sqrt{2}}\Gamma\left[\begin{matrix}s,\ s,\ \frac{1-4s}{4}\\ \frac{1}{4},\ \frac{1}{4},\ \frac{4s+3}{4}\end{matrix}\right]$ $[0<\operatorname{Re}s<1/4;\ \lvert\arg a\rvert<\pi]$
44	$\sqrt{\sqrt{x+a}\pm\sqrt{x}}\,\mathbf{K}\left(\sqrt{\dfrac{2\sqrt{x}}{\sqrt{x}\mp\sqrt{x+a}}}\right)$	$\dfrac{\pi a^{s+1/4}}{2}\Gamma\left[\begin{matrix}s,\ \frac{1-4s}{4},\ \frac{1-4s}{4}\\ \frac{1}{4},\ \frac{1}{4},\ 1-s\end{matrix}\right]$ $[0<\operatorname{Re}s<1/4;\ \lvert\arg a\rvert<\pi]$
45	$\dfrac{\sqrt{\sqrt{x+a}+\sqrt{a}}}{\sqrt{x+a}}\,\mathbf{K}\left(\sqrt{\dfrac{2\sqrt{a}}{\sqrt{a}-\sqrt{x+a}}}\right)$	$\dfrac{\pi a^{s-1/4}}{2}\Gamma\left[\begin{matrix}\frac{2s+1}{2},\ \frac{2s+1}{2},\ \frac{1-4s}{4}\\ \frac{3}{4},\ \frac{3}{4},\ \frac{4s+3}{2}\end{matrix}\right]$ $[-1/2<\operatorname{Re}s<1/4;\ \lvert\arg a\rvert<\pi]$
46	$\dfrac{\sqrt{\sqrt{x+a}-\sqrt{a}}}{\sqrt{x+a}}\,\mathbf{K}\left(\sqrt{\dfrac{\sqrt{x+a}-\sqrt{a}}{\sqrt{x+a}+\sqrt{a}}}\right)$	$\dfrac{\pi a^{s-1/4}}{2\sqrt{2}}\Gamma\left[\begin{matrix}\frac{2s+1}{2},\ \frac{1-4s}{4},\ \frac{1-4s}{4}\\ \frac{3}{4},\ \frac{3}{4},\ \frac{1-2s}{2}\end{matrix}\right]$ $[-1/2<\operatorname{Re}s<1/4;\ \lvert\arg a\rvert<\pi]$
47	$\dfrac{\sqrt{\sqrt{x+a}-\sqrt{a}}}{\sqrt{x+a}}\,\mathbf{K}\left(\pm\dfrac{\sqrt{x+a}-\sqrt{a}}{\sqrt{x}}\right)$	$\dfrac{\pi a^{s-1/4}}{2\sqrt{2}}\Gamma\left[\begin{matrix}\frac{2s+1}{2},\ \frac{1-4s}{4},\ \frac{1-4s}{4}\\ \frac{3}{4},\ \frac{3}{4},\ \frac{1-2s}{2}\end{matrix}\right]$ $[-1/2<\operatorname{Re}s<1/4;\ \lvert\arg a\rvert<\pi]$
48	$\dfrac{\sqrt{\sqrt{x+a}-\sqrt{a}}}{\sqrt{x+a}}\,\mathbf{K}\left(\sqrt{\dfrac{2\sqrt{a}}{\sqrt{x+a}+\sqrt{a}}}\right)$	$\dfrac{\pi a^{s-1/4}}{2}\Gamma\left[\begin{matrix}\frac{2s+1}{2},\ \frac{2s+1}{2},\ \frac{1-4s}{4}\\ \frac{3}{4},\ \frac{3}{4},\ \frac{4s+3}{4}\end{matrix}\right]$ $[-1/2<\operatorname{Re}s<1/4;\ \lvert\arg a\rvert<\pi]$
49	$\dfrac{\sqrt{\sqrt{x+a}-\sqrt{x}}}{\sqrt{x+a}}\,\mathbf{K}\left(\sqrt{\dfrac{\sqrt{x+a}-\sqrt{x}}{\sqrt{x+a}+\sqrt{x}}}\right)$	$\dfrac{\pi a^{s-1/4}}{2\sqrt{2}}\Gamma\left[\begin{matrix}s,\ s,\ \frac{3-4s}{4}\\ \frac{3}{4},\ \frac{3}{4},\ \frac{4s+1}{4}\end{matrix}\right]$ $[0<\operatorname{Re}s<3/4;\ \lvert\arg a\rvert<\pi]$
50	$\dfrac{\sqrt{\sqrt{x+a}-\sqrt{x}}}{\sqrt{x+a}}\,\mathbf{K}\left(\pm\dfrac{\sqrt{x}-\sqrt{x+a}}{\sqrt{a}}\right)$	$\dfrac{\pi a^{s-1/4}}{2\sqrt{2}}\Gamma\left[\begin{matrix}s,\ s,\ \frac{3-4s}{4}\\ \frac{3}{4},\ \frac{3}{4},\ \frac{4s+1}{4}\end{matrix}\right]$ $[0<\operatorname{Re}s<3/4;\ \lvert\arg a\rvert<\pi]$
51	$\dfrac{\sqrt{\sqrt{x+a}\mp\sqrt{x}}}{\sqrt{x+a}}\,\mathbf{K}\left(\sqrt{\dfrac{2\sqrt{x}}{\sqrt{x}\pm\sqrt{x+a}}}\right)$	$\dfrac{\pi a^{s-1/4}}{2}\Gamma\left[\begin{matrix}s,\ \frac{3-4s}{4},\ \frac{3-4s}{4}\\ \frac{3}{4},\ \frac{3}{4},\ 1-s\end{matrix}\right]$ $[0<\operatorname{Re}s<3/4;\ \lvert\arg a\rvert<\pi]$

3.26.3. $\theta(a-x)\,\mathbf{K}(\varphi(x))$ and algebraic functions

1	$(a-x)_+^{\alpha-1}\,\mathbf{K}(bx)$	$\dfrac{\pi a^{s+\alpha-1}}{2}\,\mathrm{B}(s,\,\alpha)\;{}_4F_3\left(\begin{matrix}\frac{1}{2},\,\frac{1}{2},\,\frac{s}{2},\,\frac{s+1}{2};\,a^2b^2\\ 1,\,\frac{s+\alpha}{2},\,\frac{s+\alpha+1}{2}\end{matrix}\right)$ $[a,\ \mathrm{Re}\,\alpha,\ \mathrm{Re}\,s>0]$
2	$\left(a^2-x^2\right)_+^{\alpha-1}\,\mathbf{K}(bx)$	$\dfrac{\pi a^{s+2\alpha-2}}{4}\,\mathrm{B}\left(\dfrac{s}{2},\,\alpha\right)\;{}_3F_2\left(\begin{matrix}\frac{1}{2},\,\frac{1}{2},\,\frac{s}{2};\,a^2b^2\\ 1,\,\frac{s+2\alpha}{2}\end{matrix}\right)$ $[a,\ \mathrm{Re}\,\alpha,\ \mathrm{Re}\,s>0]$
3	$(a-x)_+^{\alpha-1}\,\mathbf{K}(b(a-x))$	$\dfrac{\pi a^{s+\alpha-1}}{2}\,\mathrm{B}(s,\,\alpha)\;{}_4F_3\left(\begin{matrix}\frac{1}{2},\,\frac{1}{2},\,\frac{\alpha}{2},\,\frac{\alpha+1}{2};\,a^2b^2\\ \frac{3}{2},\,\frac{s+\alpha}{2},\,\frac{s+\alpha+1}{2}\end{matrix}\right)$ $[a,\ \mathrm{Re}\,\alpha,\ \mathrm{Re}\,s>0]$
4	$(a-x)_+^{\alpha-1}\,\mathbf{K}(bx(a-x))$	$\dfrac{\pi a^{s+\alpha-1}}{2}\,\mathrm{B}(s,\,\alpha)\;{}_6F_5\left(\begin{matrix}\frac{1}{2},\,\frac{1}{2},\,\Delta(2,\,\alpha),\,\Delta(2,\,s)\\ 1,\,\Delta(4,\,s+\alpha);\,\frac{a^4b^2}{16}\end{matrix}\right)$ $[a,\ \mathrm{Re}\,s,\ \mathrm{Re}\,\alpha>0]$
5	$\dfrac{\theta(a-x)}{x+a}\,\mathbf{K}\left(\pm\dfrac{a-x}{a+x}\right)$	$\dfrac{\pi a^{s-1}}{8}\,\Gamma\left[\begin{matrix}\frac{s}{2},\,\frac{s}{2}\\ \frac{s+1}{2},\,\frac{s+1}{2}\end{matrix}\right]$ $[a,\ \mathrm{Re}\,s>0]$
6	$(a-x)_+^{\alpha-1}\,\mathbf{K}\left(b\sqrt{a-x}\right)$	$\dfrac{\pi a^{s+\alpha-1}}{2}\,\mathrm{B}(s,\,\alpha)\;{}_3F_2\left(\begin{matrix}\frac{1}{2},\,\frac{1}{2},\,\alpha;\,ab^2\\ 1,\,s+\alpha\end{matrix}\right)$ $[a,\ \mathrm{Re}\,\alpha,\ \mathrm{Re}\,s>0]$
7	$\dfrac{\theta(a-x)}{(bx+1)^{\rho}}\,\mathbf{K}\left(c\sqrt{a-x}\right)$	$\dfrac{\pi a^s}{2s}\,F_3\left(\dfrac{1}{2},\,\rho,\,\dfrac{1}{2},\,s;\,s+1;\,ac^2,\,-ab\right)$ $[a,\ \mathrm{Re}\,s>0;\ \lvert\arg(1+ab)\rvert<\pi]$
8	$\theta(a-x)\,(x-b)_+^{\alpha-1}\,\mathbf{K}\left(c\sqrt{a-x}\right)$	$\dfrac{\pi(a-b)^{\alpha}\,b^{s-1}}{2\alpha}\,F_3\left(\dfrac{1}{2},\,1-s,\,\dfrac{1}{2},\,\alpha;\,\alpha+1;\right.$ $\left.c^2(a-b),\,\dfrac{b-a}{b}\right)$ $[a>b>0;\ \mathrm{Re}\,\alpha>0]$
9	$(a-x)_+^{\alpha-1}\,\mathbf{K}\left(b\sqrt{x(a-x)}\right)$	$\dfrac{\pi a^{s+\alpha-1}}{2}\,\mathrm{B}(s,\,\alpha)\;{}_4F_3\left(\begin{matrix}\frac{1}{2},\,\frac{1}{2},\,\alpha,\,s;\,\frac{a^2b^2}{4}\\ 1,\,\frac{s+\alpha}{2},\,\frac{s+\alpha+1}{2}\end{matrix}\right)$ $[a,\ \mathrm{Re}\,s,\ \mathrm{Re}\,\alpha>0]$
10	$\theta(a-x)\,\mathbf{K}\left(\pm i\dfrac{a-x}{2\sqrt{ax}}\right)$	$\dfrac{\pi a^s}{4}\,\Gamma\left[\begin{matrix}\frac{2s+1}{4},\,\frac{2s+1}{4}\\ \frac{2s+3}{4},\,\frac{2s+3}{4}\end{matrix}\right]$ $[a>0;\ \mathrm{Re}\,s>-1/2]$
11	$\theta(a-x)\,\mathbf{K}\left(\sqrt{\dfrac{a-x}{a}}\right)$	$\dfrac{\pi a^s}{2}\,\Gamma\left[\begin{matrix}s,\,s\\ \frac{2s+1}{2},\,\frac{2s+1}{2}\end{matrix}\right]$ $[a,\ \mathrm{Re}\,s>0]$

No.	$f(x)$	$F(s)$
12	$\theta(a-x)\,\mathbf{K}\left(\sqrt{\frac{a-x}{2a}}\right)$	$\frac{\pi^{3/2}a^s}{2^{s+1}}\,\Gamma\left[\begin{matrix} s \\ \frac{2s+3}{4},\ \frac{2s+3}{4}\end{matrix}\right]$ [a, $\operatorname{Re} s>0$]
13	$\frac{\theta(a-x)}{(bx+1)^{\rho}}\,\mathbf{K}\left(\sqrt{\frac{a-x}{a}}\right)$	$\frac{\pi a^s}{2}\,\Gamma\left[\begin{matrix} s,\ s \\ \frac{2s+1}{2},\ \frac{2s+1}{2}\end{matrix}\right]\,{}_3F_2\left(\begin{matrix}\rho,\ s,\ s;\ ab \\ \frac{2s+1}{2},\ \frac{2s+1}{2}\end{matrix}\right)$ [a, $\operatorname{Re} s>0$; $\lvert\arg(1-ab)\rvert<\pi$]
14	$\theta(a-x)\,\mathbf{K}\left(\sqrt{\frac{x-a}{x}}\right)$	$\frac{\pi a^s}{2}\,\Gamma\left[\begin{matrix}\frac{2s+1}{2},\ \frac{2s+1}{2} \\ s+1,\ s+1\end{matrix}\right]$ [$a>0$; $\operatorname{Re} s>-1/2$]
15	$\theta(a-x)\,\mathbf{K}\left(\sqrt{\frac{x-a}{2x}}\right)$	$2^{s-2}\sqrt{\pi}\,a^s\,\Gamma\left[\begin{matrix}\frac{2s+1}{4},\ \frac{2s+1}{4} \\ s+1\end{matrix}\right]$ [$a>0$; $\operatorname{Re} s>-1/2$]
16	$\frac{\theta(a-x)}{\sqrt{x+a}}\,\mathbf{K}\left(\sqrt{\frac{a-x}{a+x}}\right)$	$2^{s-3}\sqrt{\pi}\,a^{s-1/2}\,\Gamma\left[\begin{matrix}\frac{s}{2},\ \frac{s}{2} \\ \frac{2s+1}{2}\end{matrix}\right]$ [a, $\operatorname{Re} s>0$]
17	$\frac{\theta(a-x)}{\sqrt{x+a}}\,\mathbf{K}\left(\sqrt{\frac{x-a}{x+a}}\right)$	$\frac{\pi^{3/2}a^{s-1/2}}{2^{s+3/2}}\,\Gamma\left[\begin{matrix} s \\ \frac{2s+3}{4},\ \frac{2s+3}{4}\end{matrix}\right]$ [a, $\operatorname{Re} s>0$]
18	$\frac{\theta(a-x)}{\sqrt{b-x}}\,\mathbf{K}\left(c\sqrt{\frac{a-x}{b-x}}\right)$	$\frac{\pi a^s}{2s\sqrt{b}}\,F_1\left(\frac{1}{2},\ s,\ \frac{1}{2};\ s+1;\ \frac{a}{b},\ \frac{ac^2}{b}\right)$ [$b>a>0$; $\operatorname{Re} s>0$]
19	$\frac{\theta(a-x)}{\sqrt{a}\pm\sqrt{a-x}}\,\mathbf{K}\left(\pm\left\{\begin{matrix}1 \\ i\end{matrix}\right\}\frac{2\sqrt[4]{a}\sqrt[4]{a-x}}{\sqrt{a}\pm\sqrt{a-x}}\right)$	$\frac{\pi a^{s-1/2}}{2}\,\Gamma\left[\begin{matrix} s,\ s \\ \frac{2s+1}{2},\ \frac{2s+1}{2}\end{matrix}\right]$ [a, $\operatorname{Re} s>0$]
20	$\frac{\theta(a-x)}{\sqrt{\sqrt{a}\pm\sqrt{a-x}}}\,\mathbf{K}\left(\sqrt{\frac{2\sqrt{a-x}}{\sqrt{a-x}\pm\sqrt{a}}}\right)$	$2^{2s-3/2}\sqrt{\pi}\,a^{s-1/4}\,\Gamma\left[\begin{matrix} s,\ s \\ \frac{4s+1}{2}\end{matrix}\right]$ [a, $\operatorname{Re} s>0$]
21	$\frac{\theta(a-x)}{\sqrt{\sqrt{x}\pm\sqrt{x-a}}}\,\mathbf{K}\left(\sqrt{\frac{2\sqrt{x-a}}{\sqrt{x-a}\pm\sqrt{x}}}\right)$	$\frac{\pi^{3/2}a^{s-1/4}}{2^{2s}}\,\Gamma\left[\begin{matrix} 2s \\ \frac{4s+3}{4},\ \frac{4s+3}{4}\end{matrix}\right]$ [a, $\operatorname{Re} s>0$]
22	$\theta(a-x)\left(\sqrt{a}\pm\sqrt{a-x}\right)$ $\times\,\mathbf{K}\left(\left\{\begin{matrix} i \\ \pm 1\end{matrix}\right\}\frac{2\sqrt[4]{a}\sqrt[4]{a-x}}{\sqrt{a}\mp\sqrt{a-x}}\right)$	$\frac{\pi a^{s+1/2}}{2}\,\Gamma\left[\begin{matrix} s+1,\ s+1 \\ \frac{2s+3}{2},\ \frac{2s+3}{2}\end{matrix}\right]$ [$a>0$; $\operatorname{Re} s>-1$]
23	$\theta(a-x)\sqrt{\sqrt{a}\pm\sqrt{a-x}}$ $\times\,\mathbf{K}\left(\sqrt{\frac{2\sqrt{a-x}}{\sqrt{a-x}\mp\sqrt{a}}}\right)$	$2^{2s-1/2}\sqrt{\pi}\,a^{s+1/4}\,\Gamma\left[\begin{matrix}\frac{2s+1}{2},\ \frac{2s+1}{2} \\ \frac{4s+3}{2}\end{matrix}\right]$ [$a>0$; $\operatorname{Re} s>-1/2$]

No.	$f(x)$	$F(s)$
24	$\theta(a-x)\sqrt{\sqrt{x}\pm\sqrt{x-a}}$ $\times\mathbf{K}\left(\sqrt{\frac{2\sqrt{x-a}}{\sqrt{x-a}\mp\sqrt{x}}}\right)$	$\frac{\pi^{3/2}a^{s+1/4}}{2^{2s}}\Gamma\left[\begin{matrix}2s\\ \frac{4s+3}{4},\ \frac{4s+3}{4}\end{matrix}\right]$ [a, $\operatorname{Re}s>0$]

3.26.4. $\theta(x-a)\,\mathbf{K}(\varphi(x))$ and algebraic functions

1	$\frac{\theta(x-a)}{x+a}\mathbf{K}\left(\frac{x-a}{x+a}\right)$	$\frac{\pi a^{s-1}}{8}\Gamma\left[\begin{matrix}\frac{1-s}{2},\ \frac{1-s}{2}\\ \frac{2-s}{2},\ \frac{2-s}{2}\end{matrix}\right]$ [$a>0$; $\operatorname{Re}s<1$]
2	$\theta(x-a)\,\mathbf{K}\left(\sqrt{\frac{a-x}{a}}\right)$	$\frac{\pi a^{s}}{2}\Gamma\left[\begin{matrix}\frac{1-2s}{2},\ \frac{1-2s}{2}\\ 1-s,\ 1-s\end{matrix}\right]$ [$a>0$; $\operatorname{Re}s<1/2$]
3	$\theta(x-a)\,\mathbf{K}\left(\sqrt{\frac{a-x}{2a}}\right)$	$\frac{\sqrt{\pi}a^{s}}{2^{s+2}}\Gamma\left[\begin{matrix}\frac{1-2s}{4},\ \frac{1-2s}{4}\\ 1-s\end{matrix}\right]$ [$a>0$; $\operatorname{Re}s<1/2$]
4	$\theta(x-a)\,\mathbf{K}\left(\frac{i(x-a)}{2\sqrt{ax}}\right)$	$\frac{\pi a^{s}}{4}\Gamma\left[\begin{matrix}\frac{1-2s}{4},\ \frac{1-2s}{4}\\ \frac{3-2s}{4},\ \frac{3-2s}{4}\end{matrix}\right]$ [$a>0$; $\operatorname{Re}s<1/2$]
5	$\theta(x-a)\,\mathbf{K}\left(\sqrt{\frac{x-a}{x}}\right)$	$\frac{\pi a^{s}}{2}\Gamma\left[\begin{matrix}-s,\ -s\\ \frac{1-2s}{2},\ \frac{1-2s}{2}\end{matrix}\right]$ [$a>0$; $\operatorname{Re}s<0$]
6	$\theta(x-a)\,\mathbf{K}\left(\sqrt{\frac{x-a}{2x}}\right)$	$\frac{2^{s-1}\pi^{3/2}a^{s}}{2}\Gamma\left[\begin{matrix}-s\\ \frac{3-2s}{4},\ \frac{3-2s}{4}\end{matrix}\right]$ [$a>0$; $\operatorname{Re}s<0$]
7	$\frac{\theta(x-a)}{\sqrt{x+a}}\mathbf{K}\left(\sqrt{\frac{a-x}{a+x}}\right)$	$2^{s-2}\pi^{3/2}a^{s-1/2}\Gamma\left[\begin{matrix}\frac{1-2s}{2}\\ \frac{2-s}{2},\ \frac{2-s}{2}\end{matrix}\right]$ [$a>0$; $\operatorname{Re}s<1/2$]
8	$\frac{\theta(x-a)}{\sqrt{x+a}}\mathbf{K}\left(\sqrt{\frac{x-a}{x+a}}\right)$	$\frac{\sqrt{\pi}\,a^{s-1/2}}{2^{s+5/2}}\Gamma\left[\begin{matrix}\frac{1-2s}{4},\ \frac{1-2s}{4}\\ 1-s\end{matrix}\right]$ [$a>0$; $\operatorname{Re}s<1/2$]
9	$\frac{\theta(x-a)}{\sqrt{x}+\sqrt{x-a}}\mathbf{K}\left(\pm\frac{2\sqrt[4]{x}\sqrt[4]{x-a}}{\sqrt{x}+\sqrt{x-a}}\right)$	$\frac{\pi a^{s-1/2}}{2}\Gamma\left[\begin{matrix}\frac{1-2s}{4},\ \frac{1-2s}{4}\\ 1-s,\ 1-s\end{matrix}\right]$ [$a>0$; $\operatorname{Re}s<1/2$]
10	$\frac{\theta(x-a)}{\sqrt{x}-\sqrt{x-a}}\mathbf{K}\left(\pm\frac{2i\sqrt[4]{x}\sqrt[4]{x-a}}{\sqrt{x}+\sqrt{x-a}}\right)$	$\frac{\pi a^{s-1/2}}{2}\Gamma\left[\begin{matrix}\frac{1-2s}{4},\ \frac{1-2s}{4}\\ 1-s,\ 1-s\end{matrix}\right]$ [$a>0$; $\operatorname{Re}s<1/2$]
11	$\frac{\theta(x-a)}{\sqrt{\sqrt{a-x}\pm\sqrt{a}}}\mathbf{K}\left(\sqrt{\frac{2\sqrt{a-x}}{\sqrt{a-x}\pm\sqrt{a}}}\right)$	$2^{2s-1/2}\pi^{3/2}a^{s-1/4}\Gamma\left[\begin{matrix}\frac{1-4s}{2}\\ 1-s,\ 1-s\end{matrix}\right]$ [$a>0$; $\operatorname{Re}s<1/4$]

No.	$f(x)$	$F(s)$
12	$\dfrac{\theta(x-a)}{\sqrt{\sqrt{x}\pm\sqrt{x-a}}}\,\mathbf{K}\left(\sqrt{\dfrac{2\sqrt{x-a}}{\sqrt{x-a}\pm\sqrt{x}}}\right)$	$\dfrac{\sqrt{\pi}\,a^{s-1/4}}{2^{2s+1}}\,\Gamma\left[\begin{matrix}\frac{1-4s}{4},\ \frac{1-4s}{4}\\ 1-2s\end{matrix}\right]$ $[a>0;\ \operatorname{Re}s<1/4]$
13	$\theta(x-a)\left(\sqrt{x}+\sqrt{x-a}\right)$ $\times\mathbf{K}\left(\dfrac{2i\sqrt[4]{x}\sqrt[4]{x-a}}{\sqrt{x}-\sqrt{x-a}}\right)$	$\dfrac{\pi a^{s+1/2}}{2}\,\Gamma\left[\begin{matrix}\frac{1-2s}{2},\ \frac{1-2s}{2}\\ 1-s,\ 1-s\end{matrix}\right]$ $[a>0;\ \operatorname{Re}s<1/2]$
14	$\theta(x-a)\left(\sqrt{x}-\sqrt{x-a}\right)$ $\times\mathbf{K}\left(\dfrac{2\sqrt[4]{x}\sqrt[4]{x-a}}{\sqrt{x}+\sqrt{x-a}}\right)$	$\dfrac{\pi a^{s+1/2}}{2}\,\Gamma\left[\begin{matrix}\frac{1-2s}{2},\ \frac{1-2s}{2}\\ 1-s,\ 1-s\end{matrix}\right]$ $[a>0;\ \operatorname{Re}s<1/2]$
15	$\theta(x-a)\sqrt{\sqrt{a}\pm\sqrt{a-x}}$ $\times\mathbf{K}\left(\sqrt{\dfrac{2\sqrt{a-x}}{\sqrt{a-x}\mp\sqrt{a}}}\right)$	$\pi^{3/2}2^{2s+1/2}a^{s+1/4}\,\Gamma\left[\begin{matrix}-\frac{4s+1}{2}\\ \frac{1-2s}{2},\ \frac{1-2s}{2}\end{matrix}\right]$ $[a>0;\ \operatorname{Re}s<-1/4]$
16	$\theta(x-a)\sqrt{\sqrt{x}\pm\sqrt{x-a}}$ $\times\mathbf{K}\left(\sqrt{\dfrac{2\sqrt{x-a}}{\sqrt{x-a}\mp\sqrt{x}}}\right)$	$\dfrac{\pi a^{s+1/4}}{2}\,\Gamma\left[\begin{matrix}\frac{1-4s}{2},\ \frac{1-4s}{2}\\ \frac{1-2s}{2},\ 1-s\end{matrix}\right]$ $[a>0;\ \operatorname{Re}s<1/4]$

3.26.5. $\mathbf{E}(\varphi(x))$ and algebraic functions

Notation: $\varepsilon=0$ or 1.

1	$\mathbf{E}(iax)-\dfrac{\pi}{2}$	$-\dfrac{a^{-s}}{8}\,\Gamma\left[\begin{matrix}\frac{s}{2},\ \frac{1-s}{2},\ -\frac{s+1}{2}\\ \frac{2-s}{2}\end{matrix}\right]$ $[\operatorname{Re}a>0;\ -2<\operatorname{Re}s<1]$
2	$\mathbf{E}(iax)$ $-\dfrac{\pi}{2}\sum_{k=0}^{n}\dfrac{(-1/2)_k\,(1/2)_k}{(k!)^2}\left(-a^2x^2\right)^k$	$\dfrac{a^{-s}}{8}\,\Gamma\left[\begin{matrix}\frac{s}{2},\ \frac{1-s}{2},\ -\frac{s+1}{2}\\ \frac{2-s}{2}\end{matrix}\right]$ $[\operatorname{Re}a>0;\ -2n-2<\operatorname{Re}s<-1,-2n]$
3	$\dfrac{1}{x^2+a^2}\,\mathbf{E}\left(\dfrac{ix}{a}\right)$	$\dfrac{a^{s-2}}{2}\,\Gamma\left[\begin{matrix}\frac{s}{2},\ \frac{1-s}{2},\ \frac{3-s}{2}\\ \frac{2-s}{2}\end{matrix}\right]$ $[a>0;\ 0<\operatorname{Re}s<1]$
4	$\dfrac{1}{x^2+a^2}\,\mathbf{E}\left(\dfrac{ia}{x}\right)$	$\dfrac{a^{s-2}}{2}\,\Gamma\left[\begin{matrix}\frac{s+1}{2},\ \frac{s-1}{2},\ \frac{2-s}{2}\\ \frac{s}{2}\end{matrix}\right]$ $[a>0;\ 1<\operatorname{Re}s<2]$

No.	$f(x)$	$F(s)$
5	$\dfrac{1}{\left[(x+a)^2-b^2\right]^{\varepsilon}(x+a)^{\rho}}\,\mathbf{E}\left(\dfrac{b}{x+a}\right)$	$\dfrac{\pi a^{s-\rho-2\varepsilon}}{2}\,\mathrm{B}(s,\,\rho-s+2\varepsilon)\times{}_4F_3\left(\begin{matrix}\frac{2\varepsilon-1}{2},\,\frac{2\varepsilon+1}{2},\,\frac{\rho-s+2\varepsilon}{2},\,\frac{\rho-s+2\varepsilon+1}{2}\\1,\,\frac{\rho+2\varepsilon}{2},\,\frac{\rho+2\varepsilon+1}{2};\,\frac{b^2}{a^2}\end{matrix}\right)$ $[0<\operatorname{Re}s<\operatorname{Re}\rho+2\varepsilon;\ \lvert\arg a\rvert<\pi]$
6	$\dfrac{1}{\left[(x+a)^2-b^2x^2\right]^{\varepsilon}(x+a)^{\rho}}\,\mathbf{E}\left(\dfrac{bx}{x+a}\right)$	$\dfrac{\pi a^{s-\rho-2\varepsilon}}{2}\,\mathrm{B}(s,\,\rho-s+2\varepsilon)\times{}_4F_3\left(\begin{matrix}\frac{2\varepsilon-1}{2},\,\frac{2\varepsilon+1}{2},\,\frac{s}{2},\,\frac{s+1}{2}\\1,\,\frac{\rho+2\varepsilon}{2},\,\frac{\rho+2\varepsilon+1}{2};\,b^2\end{matrix}\right)$ $[0<\operatorname{Re}s<\operatorname{Re}\rho+2\varepsilon;\ \lvert\arg a\rvert<\pi]$
7	$\dfrac{1}{(x+a)^{\rho}}\,\mathbf{E}\left(\dfrac{b}{\sqrt{x+a}}\right)$	$\dfrac{\pi a^{s-\rho}}{2}\,\mathrm{B}(s,\,\rho-s)\,{}_3F_2\left(\begin{matrix}-\frac{1}{2},\,\frac{1}{2},\,\rho-s\\1,\,\rho;\,\frac{b^2}{a}\end{matrix}\right)$ $[0<\operatorname{Re}s<\operatorname{Re}\rho;\ \lvert\arg a\rvert<\pi]$
8	$\mathbf{E}\left(\sqrt{\dfrac{a-x}{a}}\right)$	$\dfrac{\pi a^{s}}{2}\,\Gamma\left[\begin{matrix}s,\,s+1\\\frac{2s+1}{2},\,\frac{2s+3}{2}\end{matrix}\right]$ $[a,\ \operatorname{Re}s>0]$
9	$\dfrac{1}{\sqrt{x+a}}\,\mathbf{E}\left(\sqrt{\dfrac{a}{x+a}}\right)$	$a^{s-1/2}\,\Gamma\left[\begin{matrix}\frac{1-2s}{2},\,s,\,s+1\\\frac{1+2s}{2}\end{matrix}\right]$ $[0<\operatorname{Re}s<1/2;\ \lvert\arg a\rvert<\pi]$
10	$\mathbf{E}\left(\sqrt{\dfrac{x-a}{x}}\right)$	$\dfrac{\pi a^{s}}{2}\,\Gamma\left[\begin{matrix}\frac{2s-1}{2},\,\frac{2s+1}{2}\\s,\,s+1\end{matrix}\right]$ $[a>0;\ \operatorname{Re}s>1/2]$
11	$\dfrac{1}{\sqrt{x+a}}\,\mathbf{E}\left(\sqrt{\dfrac{x}{x+a}}\right)$	$a^{s-1/2}\,\Gamma\left[\begin{matrix}\frac{1-2s}{2},\,\frac{3-2s}{2},\,s\\1-s\end{matrix}\right]$ $[0<\operatorname{Re}s<1/2;\ \lvert\arg a\rvert<\pi]$
12	$\dfrac{\sqrt{\sqrt{x+a}\pm\sqrt{a}}}{\sqrt{x+a}}\,\mathbf{E}\left(\sqrt{\dfrac{2\sqrt{a}}{\sqrt{a}\pm\sqrt{x+a}}}\right)$	$2\pi a^{s-1/4}\,\Gamma\left[\begin{matrix}s,\,s+1,\,\frac{1-4s}{4}\\\frac{1}{4},\,\frac{1}{4},\,\frac{4s+3}{4}\end{matrix}\right]$ $[0<\operatorname{Re}s<1/4;\ \lvert\arg a\rvert<\pi]$
13	$\dfrac{\sqrt{\sqrt{x+a}\pm\sqrt{x}}}{\sqrt{x+a}}\,\mathbf{E}\left(\sqrt{\dfrac{2\sqrt{x}}{\sqrt{x}\pm\sqrt{x+a}}}\right)$	$2\pi a^{s-1/4}\,\Gamma\left[\begin{matrix}s,\,\frac{1-4s}{4},\,\frac{5-4s}{4}\\\frac{1}{4},\,\frac{1}{4},\,1-s\end{matrix}\right]$ $[0<\operatorname{Re}s<1/4;\ \lvert\arg a\rvert<\pi]$

3.26.6. $\theta(a-x)\,\mathbf{E}(\varphi(x))$ and algebraic functions

Notation: $\varepsilon = 0$ or 1.

1	$\dfrac{(a-x)_+^{\alpha-1}}{(1-b^2x^2)^{\varepsilon}}\,\mathbf{E}(bx)$	$\dfrac{\pi a^{s+\alpha-1}}{2}\,\mathrm{B}(s,\alpha)\;{}_4F_3\left(\begin{matrix}\frac{2\varepsilon-1}{2},\,\frac{2\varepsilon+1}{2},\,\frac{s}{2},\,\frac{s+1}{2}\\ 1,\,\frac{s+\alpha}{2},\,\frac{s+\alpha+1}{2};\,a^2b^2\end{matrix}\right)$ [a, $\mathrm{Re}\,\alpha$, $\mathrm{Re}\,s>0$]
2	$\dfrac{(a^2-x^2)_+^{\alpha-1}}{(1-b^2x^2)^{\varepsilon}}\,\mathbf{E}(bx)$	$\dfrac{\pi a^{s+2\alpha-2}}{4}\,\mathrm{B}\left(\dfrac{s}{2},\,\alpha\right)\;{}_3F_2\left(\begin{matrix}\frac{2\varepsilon-1}{2},\,\frac{2\varepsilon+1}{2},\,\frac{s}{2}\\ 1,\,\frac{s+2\alpha}{2};\,a^2b^2\end{matrix}\right)$ [a, $\mathrm{Re}\,\alpha$, $\mathrm{Re}\,s>0$]
3	$\dfrac{(a-x)_+^{\alpha-1}}{\left[1-b^2(a-x)^2\right]^{\varepsilon}}\,\mathbf{E}(b(a-x))$	$\dfrac{\pi a^{s+\alpha-1}}{2}\,\mathrm{B}(s,\alpha)\;{}_4F_3\left(\begin{matrix}\frac{2\varepsilon-1}{2},\,\frac{2\varepsilon+1}{2},\,\frac{\alpha}{2},\,\frac{\alpha+1}{2}\\ 1,\,\frac{s+\alpha}{2},\,\frac{s+\alpha+1}{2};\,a^2b^2\end{matrix}\right)$ [a, $\mathrm{Re}\,\alpha$, $\mathrm{Re}\,s>0$]
4	$\dfrac{(a-x)_+^{\alpha-1}}{\left[1-b^2x^2(a-x)^2\right]^{\varepsilon}}\,\mathbf{E}(bx(a-x))$	$\dfrac{\pi a^{s+\alpha-1}}{2}\,\mathrm{B}(s,\alpha)$ $\times\,{}_6F_5\left(\begin{matrix}\frac{2\varepsilon-1}{2},\,\frac{2\varepsilon+1}{2},\,\Delta(2,\alpha),\,\Delta(2,s)\\ 1,\,\Delta(4,s+\alpha);\,\frac{a^4b^2}{16}\end{matrix}\right)$ [a, $\mathrm{Re}\,s$, $\mathrm{Re}\,\alpha>0$]
5	$\dfrac{(a-x)_+^{\alpha-1}}{\left[1-b^2(a-x)\right]^{\varepsilon}}\,\mathbf{E}\left(b\sqrt{a-x}\right)$	$\dfrac{\pi a^{s+\alpha-1}}{2}\,\mathrm{B}(s,\alpha)\;{}_3F_2\left(\begin{matrix}\frac{2\varepsilon-1}{2},\,\frac{2\varepsilon+1}{2},\,\alpha\\ 1,\,s+\alpha;\,ab^2\end{matrix}\right)$ [a, $\mathrm{Re}\,\alpha$, $\mathrm{Re}\,s>0$]
6	$\dfrac{\theta(a-x)}{\left[1-c^2(a-x)\right]^{\varepsilon}(bx+1)^{\rho}}\,\mathbf{E}\left(c\sqrt{a-x}\right)$	$\dfrac{\pi a^{s}}{2s}\,F_3\left(\dfrac{2\varepsilon-1}{2},\,\rho,\,\dfrac{2\varepsilon+1}{2},\,s;\,s+1;\,ac^2,\,-ab\right)$ [a, $\mathrm{Re}\,s>0$; $\lvert\arg(1+ab)\rvert<\pi$]
7	$\dfrac{\theta(a-x)(x-b)_+^{\alpha-1}}{\left[1-c^2(a-x)\right]^{\varepsilon}}\,\mathbf{E}\left(c\sqrt{a-x}\right)$	$\dfrac{\pi(a-b)^{\alpha}b^{s-1}}{2\alpha}\,F_3\left(\dfrac{2\varepsilon-1}{2},\,1-s,\,\dfrac{2\varepsilon+1}{2},\,\alpha;\,\alpha+1;\,c^2(a-b),\,\dfrac{b-a}{b}\right)$ [$a>b>0$; $\mathrm{Re}\,\alpha>0$]
8	$\dfrac{(a-x)_+^{\alpha-1}}{\left[1-b^2x(a-x)\right]^{\varepsilon}}\,\mathbf{E}\left(b\sqrt{x(a-x)}\right)$	$\dfrac{\pi a^{s+\alpha-1}}{2}\,\mathrm{B}(s,\alpha)\;{}_4F_3\left(\begin{matrix}\frac{2\varepsilon-1}{2},\,\frac{2\varepsilon+1}{2},\,\alpha,\,s\\ 1,\,\frac{s+\alpha}{2},\,\frac{s+\alpha+1}{2};\,\frac{a^2b^2}{4}\end{matrix}\right)$ [a, $\mathrm{Re}\,\alpha$, $\mathrm{Re}\,s>0$]
9	$\theta(a-x)\,\mathbf{E}\left(\sqrt{\dfrac{a-x}{a}}\right)$	$\dfrac{\pi a^{s}}{2}\,\Gamma\left[\begin{matrix}s,\,s+1\\ \frac{2s+1}{2},\,\frac{2s+3}{2}\end{matrix}\right]$ [a, $\mathrm{Re}\,s>0$]

No.	$f(x)$	$F(s)$
10	$\dfrac{\theta(a-x)}{(bx+1)^{\rho}}\,\mathbf{E}\left(\sqrt{\dfrac{a-x}{a}}\right)$	$\dfrac{\pi a^s}{2}\,\Gamma\left[\begin{matrix} s,\ s+1 \\ \frac{2s+1}{2},\ \frac{2s+3}{2}\end{matrix}\right]{}_3F_2\left(\begin{matrix}\rho,\ s,\ s+1;\ ab \\ \frac{2s+1}{2},\ \frac{2s+3}{2}\end{matrix}\right)$ $[a,\ \operatorname{Re} s>0;\ \lvert\arg(1-ab)\rvert<\pi]$
11	$\theta(a-x)\,\mathbf{E}\left(\sqrt{\dfrac{x-a}{x}}\right)$	$\dfrac{\pi a^s}{2}\,\Gamma\left[\begin{matrix}\frac{2s-1}{2},\ \frac{2s+1}{2} \\ s,\ s+1\end{matrix}\right]$ $[a>0;\ \operatorname{Re} s>1/2]$
12	$\dfrac{\theta(a-x)\,(b-x)^{\varepsilon-1/2}}{\left[c^2(x-a)+b-x\right]^{\varepsilon}}\,\mathbf{E}\left(c\sqrt{\dfrac{a-x}{b-x}}\right)$	$\dfrac{\pi a^s}{2s\sqrt{b}}\,F_1\left(\dfrac{1}{2},\ s,\ \dfrac{4\varepsilon-1}{2};\ s+1;\ \dfrac{a}{b},\ \dfrac{ac^2}{b}\right)$ $[b>a>0;\ \operatorname{Re} s>0]$
13	$\dfrac{\theta(a-x)}{\sqrt{a}\pm\sqrt{a-x}}\,\mathbf{E}\left(\pm\left\{\begin{matrix} i \\ 1\end{matrix}\right\}\dfrac{2\sqrt[4]{a}\sqrt[4]{a-x}}{\sqrt{a}\mp\sqrt{a-x}}\right)$	$\dfrac{\pi a^{s-1/2}}{2}\,\Gamma\left[\begin{matrix} s-1,\ s+1 \\ \frac{2s+1}{2},\ \frac{2s+1}{2}\end{matrix}\right]$ $[a>0;\ \operatorname{Re} s>1]$
14	$\dfrac{\theta(a-x)}{\sqrt{\sqrt{a}\pm\sqrt{a-x}}}\,\mathbf{E}\left(\sqrt{\dfrac{2\sqrt{a-x}}{\sqrt{a-x}\mp\sqrt{a}}}\right)$	$2^{2s-3/2}\sqrt{\pi}\,a^{s-1/4}\,\Gamma\left[\begin{matrix}\frac{2s-1}{2},\ \frac{2s+1}{2} \\ \frac{4s+1}{2}\end{matrix}\right]$ $[a>0;\ \operatorname{Re} s>1/2]$
15	$\theta(a-x)\left(\sqrt{a}-\sqrt{a-x}\right)$ $\times\,\mathbf{E}\left(\pm\dfrac{2i\sqrt[4]{a}\sqrt[4]{a-x}}{\sqrt{a}-\sqrt{a-x}}\right)$	$\dfrac{\pi a^{s+1/2}}{2}\,\Gamma\left[\begin{matrix} s,\ s+2 \\ \frac{2s+3}{2},\ \frac{2s+3}{2}\end{matrix}\right]$ $[a,\ \operatorname{Re} s>0]$
16	$\theta(a-x)\left(\sqrt{a-x}+\sqrt{a}\right)$ $\times\,\mathbf{E}\left(\pm\dfrac{2\sqrt[4]{a}\sqrt[4]{a-x}}{\sqrt{a}+\sqrt{a-x}}\right)$	$\dfrac{\pi a^{s+1/2}}{2}\,\Gamma\left[\begin{matrix} s,\ s+2 \\ \frac{2s+3}{2},\ \frac{2s+3}{2}\end{matrix}\right]$ $[a,\ \operatorname{Re} s>0]$
17	$\theta(a-x)\sqrt{\sqrt{a}\pm\sqrt{a-x}}$ $\times\,\mathbf{E}\left(\sqrt{\dfrac{2\sqrt{a-x}}{\sqrt{a-x}\pm\sqrt{a}}}\right)$	$2^{2s-1/2}\sqrt{\pi}\,a^{s+1/4}\,\Gamma\left[\begin{matrix} s,\ s+1 \\ \frac{4s+3}{2}\end{matrix}\right]$ $[a,\ \operatorname{Re} s>0]$
18	$\theta(a-x)\sqrt{\sqrt{x}\pm\sqrt{x-a}}$ $\times\,\mathbf{E}\left(\sqrt{\dfrac{2\sqrt{x-a}}{\sqrt{x-a}\pm\sqrt{x}}}\right)$	$\dfrac{\pi^{3/2}a^{s+1/4}}{2^{2s}}\,\Gamma\left[\begin{matrix} 2s \\ \frac{4s+1}{2},\ \frac{4s+5}{2}\end{matrix}\right]$ $[a,\ \operatorname{Re} s>0]$

3.26.7. $\theta(x-a)\,\mathbf{E}(\varphi(x))$ and algebraic functions

1	$\dfrac{\theta(x-a)}{\sqrt{x}-\sqrt{x-a}}\,\mathbf{E}\left(\dfrac{2\sqrt[4]{x(x-a)}}{\sqrt{x}+\sqrt{x-a}}\right)$	$\dfrac{\pi a^{s-1/2}}{2}\,\Gamma\left[\begin{matrix}\frac{3-2s}{2},\ -\frac{2s+1}{2}\\ 1-s,\ 1-s\end{matrix}\right]$ $[a>0;\ \operatorname{Re}s<-1/2]$
2	$\dfrac{\theta(x-a)}{\sqrt{x}+\sqrt{x-a}}\,\mathbf{E}\left(\pm\dfrac{2i\sqrt[4]{x(x-a)}}{\sqrt{x}-\sqrt{x-a}}\right)$	$\dfrac{\pi a^{s-1/2}}{2}\,\Gamma\left[\begin{matrix}\frac{3-2s}{2},\ -\frac{2s+1}{2}\\ 1-s,\ 1-s\end{matrix}\right]$ $[a>0;\ \operatorname{Re}s<-1/2]$
3	$\dfrac{\theta(x-a)}{\sqrt{\sqrt{x}\pm\sqrt{x-a}}}\,\mathbf{E}\left(\sqrt{\dfrac{2\sqrt{x-a}}{\sqrt{x-a}\mp\sqrt{x}}}\right)$	$2^{-2s-1}\sqrt{\pi}\,a^{s-1/4}\,\Gamma\left[\begin{matrix}\frac{3-4s}{4},\ -\frac{4s+1}{4}\\ 1-2s\end{matrix}\right]$ $[a>0;\ \operatorname{Re}s<-1/4]$
4	$\theta(x-a)\left(\sqrt{x-a}+\sqrt{x}\right)$ $\times\,\mathbf{E}\left(\pm\dfrac{2\sqrt[4]{x}\sqrt[4]{x-a}}{\sqrt{x}+\sqrt{x-a}}\right)$	$\dfrac{\pi a^{s+1/2}}{2}\,\Gamma\left[\begin{matrix}\frac{3-2s}{2},\ -\frac{2s+1}{2}\\ 1-s,\ 1-s\end{matrix}\right]$ $[a>0;\ \operatorname{Re}s<-1/2]$
5	$\theta(x-a)\left(\sqrt{x}-\sqrt{x-a}\right)$ $\times\,\mathbf{E}\left(\pm\dfrac{2i\sqrt[4]{x}\sqrt[4]{x-a}}{\sqrt{x}-\sqrt{x-a}}\right)$	$\dfrac{\pi a^{s+1/2}}{2}\,\Gamma\left[\begin{matrix}\frac{3-2s}{2},\ -\frac{2s+1}{2}\\ 1-s,\ 1-s\end{matrix}\right]$ $[a>0;\ \operatorname{Re}s<-1/2]$
6	$\theta(x-a)\sqrt{\sqrt{a}\pm\sqrt{a-x}}$ $\times\,\mathbf{E}\left(\sqrt{\dfrac{2\sqrt{a-x}}{\sqrt{a-x}\pm\sqrt{a}}}\right)$	$2^{2s+1/2}\pi^{3/2}a^{s+1/4}\,\Gamma\left[\begin{matrix}-\frac{4s+1}{2}\\ -s,\ 1-s\end{matrix}\right]$ $[a>0;\ \operatorname{Re}s<-1/4]$
7	$\theta(x-a)\sqrt{\sqrt{x}\pm\sqrt{x-a}}$ $\times\,\mathbf{E}\left(\sqrt{\dfrac{2\sqrt{x-a}}{\sqrt{x-a}\pm\sqrt{x}}}\right)$	$2^{-2s-1}\sqrt{\pi}\,a^{s+1/4}\,\Gamma\left[\begin{matrix}\frac{3-4s}{4},\ -\frac{4s+1}{4}\\ 1-2s\end{matrix}\right]$ $[a>0;\ \operatorname{Re}s<-1/4]$

3.26.8. $\mathbf{K}(\varphi(x))$, $\mathbf{E}(\varphi(x))$, and the exponential function

Notation: $\delta=\begin{Bmatrix}1\\0\end{Bmatrix}$.

1	$\theta(a-x)\,e^{bx}\begin{Bmatrix}\mathbf{K}\left(\sqrt{1-x/a}\right)\\ \mathbf{E}\left(\sqrt{1-x/a}\right)\end{Bmatrix}$	$\dfrac{\pi a^{s}}{2}\,\Gamma\left[\begin{matrix}s,\ s-\delta+1\\ \frac{2s+1}{2},\ \frac{2s-2\delta+3}{2}\end{matrix}\right]\,{}_2F_2\left(\begin{matrix}s,\ s-\delta+1;\ ab\\ \frac{2s+1}{2},\ \frac{2s-2\delta+3}{2}\end{matrix}\right)$ $[a,\ \operatorname{Re}s>0]$

No.	$f(x)$	$F(s)$
2	$\theta(a-x)\,e^{-b\sqrt{x}}\begin{Bmatrix}\mathbf{K}(1-x/a)\\ \mathbf{E}(1-x/a)\end{Bmatrix}$	$\frac{\pi a^s}{2}\Gamma\left[\begin{matrix}s,\ s-\delta+1\\ \frac{2s+1}{2},\ \frac{2s-2\delta+3}{2}\end{matrix}\right]{}_2F_3\left(\begin{matrix}s,\ s-\delta+1;\ \frac{ab^2}{4}\\ \frac{1}{2},\ \frac{2s+1}{2},\ \frac{2s-2\delta+3}{2}\end{matrix}\right) - \frac{\pi a^{s+1/2}b}{2}\Gamma\left[\begin{matrix}\frac{2s+1}{2},\ \frac{2s-2\delta+3}{2}\\ s+1,\ s-\delta+2\end{matrix}\right] \times {}_2F_3\left(\begin{matrix}\frac{2s+1}{2},\ \frac{2s-2\delta+3}{2};\ \frac{ab^2}{4}\\ \frac{3}{2},\ s+1,\ s-\delta+2\end{matrix}\right)$ $[a,\ \operatorname{Re} s>0]$

3.26.9. $\mathbf{K}(\varphi(x))$, $\mathbf{E}(\varphi(x))$, and hyperbolic or trigonometric functions

1	$\theta(a-x)\begin{Bmatrix}\sinh(b\sqrt{x})\\ \sin(b\sqrt{x})\end{Bmatrix}\mathbf{K}\left(\sqrt{1-\frac{x}{a}}\right)$	$\frac{\pi a^{s+1/2}b}{2}\Gamma\left[\begin{matrix}\frac{2s+1}{2},\ \frac{2s+1}{2}\\ s+1,\ s+1\end{matrix}\right]{}_2F_3\left(\begin{matrix}\frac{2s+1}{2},\ \frac{2s+1}{2};\ \pm\frac{ab^2}{4}\\ \frac{3}{2},\ s+1,\ s+1\end{matrix}\right)$ $[a>0;\ \operatorname{Re} s>-1/2]$
2	$\theta(a-x)\begin{Bmatrix}\sinh(b\sqrt{x})\\ \sin(b\sqrt{x})\end{Bmatrix}\mathbf{E}\left(\sqrt{1-\frac{x}{a}}\right)$	$\frac{\pi a^{s+1/2}b}{2}\Gamma\left[\begin{matrix}\frac{2s+1}{2},\ \frac{2s+3}{2}\\ s+1,\ s+2\end{matrix}\right]{}_2F_3\left(\begin{matrix}\frac{2s+1}{2},\ \frac{2s+3}{2};\ \pm\frac{ab^2}{4}\\ \frac{3}{2},\ s+1,\ s+2\end{matrix}\right)$ $[a>0;\ \operatorname{Re} s>-1/2]$
3	$\theta(a-x)\begin{Bmatrix}\cosh(b\sqrt{x})\\ \cos(b\sqrt{x})\end{Bmatrix}\mathbf{K}\left(\sqrt{1-\frac{x}{a}}\right)$	$\frac{\pi a^s}{2}\Gamma\left[\begin{matrix}s,\ s\\ \frac{2s+1}{2},\ \frac{2s+1}{2}\end{matrix}\right]{}_2F_3\left(\begin{matrix}s,\ s;\ \pm\frac{ab^2}{4}\\ \frac{1}{2},\ \frac{2s+1}{2},\ \frac{2s+1}{2}\end{matrix}\right)$ $[a,\ \operatorname{Re} s>0]$
4	$\theta(a-x)\begin{Bmatrix}\cosh(b\sqrt{x})\\ \cos(b\sqrt{x})\end{Bmatrix}\mathbf{E}\left(\sqrt{1-\frac{x}{a}}\right)$	$\frac{\pi a^s}{2}\Gamma\left[\begin{matrix}s,\ s+1\\ \frac{2s+1}{2},\ \frac{2s+3}{2}\end{matrix}\right]{}_2F_3\left(\begin{matrix}s,\ s+1;\ \pm\frac{ab^2}{4}\\ \frac{1}{2},\ \frac{2s+1}{2},\ \frac{2s+3}{2}\end{matrix}\right)$ $[a,\ \operatorname{Re} s>0]$

3.26.10. $\mathbf{K}(\varphi(x))$, $\mathbf{E}(\varphi(x))$, and the logarithmic function

Notation: $\delta=\begin{Bmatrix}1\\ 0\end{Bmatrix}$.

1	$\theta(1-x)\ln(ax+1)\begin{Bmatrix}\mathbf{K}(\sqrt{1-x})\\ \mathbf{E}(\sqrt{1-x})\end{Bmatrix}$	$\frac{\pi a}{2}\Gamma\left[\begin{matrix}s+1,\ s-\delta+2\\ \frac{2s+3}{2},\ \frac{2s-2\delta+5}{2}\end{matrix}\right]{}_4F_3\left(\begin{matrix}1,\ 1,\ s+1,\ s-\delta+2\\ 2,\ \frac{2s+3}{2},\ \frac{2s-2\delta+5}{2};\ -a\end{matrix}\right)$ $[\operatorname{Re} s>-1;\ \lvert\arg(1+a)\rvert<\pi]$
2	$\theta(a-x)\ln(bx^2+1) \times\begin{Bmatrix}\mathbf{K}(\sqrt{1-x/a})\\ \mathbf{E}(\sqrt{1-x/a})\end{Bmatrix}$	$\frac{\pi a^{s+2}b}{2}\Gamma\left[\begin{matrix}s+2,\ s-\delta+3\\ \frac{2s+5}{2},\ \frac{2s-2\delta+7}{2}\end{matrix}\right] \times {}_6F_5\left(\begin{matrix}1,\ 1,\ \frac{s+2}{2},\ \frac{s+3}{2},\ \frac{s-\delta+3}{2},\ \frac{s-\delta+4}{2};\ -a^2b\\ 2,\ \frac{2s+5}{4},\ \frac{2s+7}{4},\ \frac{2s-2\delta+7}{4},\ \frac{2s-2\delta+9}{4}\end{matrix}\right)$ $[a>0;\ \operatorname{Re} s>-2;\ \lvert\arg(1+a^2b)\rvert<\pi]$

No.	$f(x)$	$F(s)$
3	$\theta(a-x)\ln\frac{1+bx}{1-bx}\begin{Bmatrix}\mathbf{K}(\sqrt{1-x/a})\\ \mathbf{E}(\sqrt{1-x/a})\end{Bmatrix}$	$\pi a^{s+1}b\,\Gamma\begin{bmatrix}s+1,\ s-\delta+2\\ \frac{2s+3}{2},\ \frac{2s-2\delta+5}{2}\end{bmatrix}$ $\times\,{}_6F_5\left(\begin{matrix}\frac{1}{2},\ 1,\ \frac{s+1}{2},\ \frac{s+2}{2},\ \frac{s-\delta+2}{2},\ \frac{s-\delta+3}{2};\ a^2b^2\\ \frac{3}{2},\ \frac{2s+3}{4},\ \frac{2s+5}{4},\ \frac{2s-2\delta+5}{4},\ \frac{2s-2\delta+7}{4}\end{matrix}\right)$ $[a>0;\ \operatorname{Re}s>-1;\ \lvert\arg(1-a^2b^2)\rvert<\pi]$
4	$\theta(a-x)\ln\frac{1+b\sqrt{x}}{1-b\sqrt{x}}\begin{Bmatrix}\mathbf{K}(\sqrt{1-x/a})\\ \mathbf{E}(\sqrt{1-x/a})\end{Bmatrix}$	$\pi a^{s+1/2}b\,\Gamma\begin{bmatrix}\frac{2s+1}{2},\ \frac{2s-2\delta+3}{2}\\ s+1,\ s-\delta+2\end{bmatrix}$ $\times\,{}_4F_3\left(\begin{matrix}\frac{1}{2},\ 1,\ \frac{2s+1}{2},\ \frac{2s-2\delta+3}{2}\\ \frac{3}{2},\ s+1,\ s-\delta+2;\ ab^2\end{matrix}\right)$ $[a>0;\ \operatorname{Re}s>-1/2;\ \lvert\arg(1-ab^2)\rvert<\pi]$
5	$\theta(a-x)\ln\left(b\sqrt{x}+\sqrt{1+b^2x}\right)$ $\times\begin{Bmatrix}\mathbf{K}(\sqrt{1-x/a})\\ \mathbf{E}(\sqrt{1-x/a})\end{Bmatrix}$	$\frac{\pi a^{s+1/2}b}{2}\,\Gamma\begin{bmatrix}\frac{2s+1}{2},\ \frac{2s-2\delta+3}{2}\\ s+1,\ s-\delta+2\end{bmatrix}$ $\times\,{}_4F_3\left(\begin{matrix}\frac{1}{2},\ \frac{1}{2},\ \frac{2s+1}{2},\ \frac{2s-2\delta+3}{2}\\ \frac{3}{2},\ s+1,\ s-\delta+2;\ -ab^2\end{matrix}\right)$ $[a>0;\ \operatorname{Re}s>-1/2;\ \lvert\arg(1+ab^2)\rvert<\pi]$
6	$\theta(a-x)\ln\left(bx+\sqrt{b^2x^2+1}\right)$ $\times\begin{Bmatrix}\mathbf{K}(\sqrt{1-x/a})\\ \mathbf{E}(\sqrt{1-x/a})\end{Bmatrix}$	$\frac{\pi a^{s+1}b}{2}\,\Gamma\begin{bmatrix}s+1,\ s-\delta+2\\ \frac{2s+3}{2},\ \frac{2s-2\delta+5}{2}\end{bmatrix}$ $\times\,{}_6F_5\left(\begin{matrix}\frac{1}{2},\ \frac{1}{2},\ \frac{s+1}{2},\ \frac{s+2}{2},\ \frac{s-\delta+2}{2},\ \frac{s-\delta+3}{2};\ -a^2b^2\\ \frac{3}{2},\ \frac{2s+3}{4},\ \frac{2s+5}{4},\ \frac{2s-2\delta+5}{4},\ \frac{2s-2\delta+7}{4}\end{matrix}\right)$ $[a>0;\ \operatorname{Re}s>-1;\ \lvert\arg(1+a^2b^2)\rvert<\pi]$
7	$\frac{\theta(a-x)}{\sqrt{1+b^2x^2}}\ln\left(bx+\sqrt{b^2x^2+1}\right)$ $\times\begin{Bmatrix}\mathbf{K}(\sqrt{1-x/a})\\ \mathbf{E}(\sqrt{1-x/a})\end{Bmatrix}$	$\frac{\pi a^{s+1}b}{2}\,\Gamma\begin{bmatrix}s+1,\ s-\delta+2\\ \frac{2s+3}{2},\ \frac{2s-2\delta+5}{2}\end{bmatrix}$ $\times\,{}_6F_5\left(\begin{matrix}1,\ 1,\ \frac{s+1}{2},\ \frac{s+2}{2},\ \frac{s-\delta+2}{2},\ \frac{s-\delta+3}{2};\ -a^2b^2\\ \frac{3}{2},\ \frac{2s+3}{4},\ \frac{2s+5}{4},\ \frac{2s-2\delta+5}{4},\ \frac{2s-2\delta+7}{4}\end{matrix}\right)$ $[a>0;\ \operatorname{Re}s>-1;\ \lvert\arg(1+a^2b^2)\rvert<\pi]$
8	$\theta(a-x)\ln\frac{\sqrt{a-x}+\sqrt{a}}{\sqrt{x}}\begin{Bmatrix}\mathbf{K}(bx)\\ \mathbf{E}(bx)\end{Bmatrix}$	$\frac{\pi^{3/2}a^s}{4s}\,\Gamma\begin{bmatrix}s\\ \frac{2s+1}{2}\end{bmatrix}\,{}_5F_4\left(\begin{matrix}\pm\frac{1}{2},\ \frac{1}{2},\ \frac{s}{2},\ \frac{s}{2},\ \frac{s+1}{2};\ a^2b^2\\ 1,\ \frac{2s+1}{4},\ \frac{2s+3}{4},\ \frac{s+2}{2}\end{matrix}\right)$ $[a,\ \operatorname{Re}s>0]$
9	$\theta(a-x)\ln\frac{\sqrt{a^2-x^2}+a}{x}\begin{Bmatrix}\mathbf{K}(bx)\\ \mathbf{E}(bx)\end{Bmatrix}$	$\frac{\pi^{3/2}a^s}{4s}\,\Gamma\begin{bmatrix}\frac{s}{2}\\ \frac{s+1}{2}\end{bmatrix}\,{}_4F_3\left(\begin{matrix}\pm\frac{1}{2},\ \frac{1}{2},\ \frac{s}{2},\ \frac{s}{2}\\ 1,\ \frac{s+1}{2},\ \frac{s+2}{2};\ a^2b^2\end{matrix}\right)$ $[a,\ \operatorname{Re}s>0]$

No.	$f(x)$	$F(s)$
10	$\theta(a-x)\ln\frac{a+\sqrt{a^2-x^2}}{a-\sqrt{a^2-x^2}}\mathbf{K}(bx)$	$\frac{\pi^{3/2}a^s}{2s}\Gamma\left[\begin{matrix}\frac{s}{2}\\ \frac{s+1}{2}\end{matrix}\right]{}_4F_3\left(\begin{matrix}\frac{1}{2},\frac{1}{2},\frac{s}{2},\frac{s}{2};\ a^2b^2\\ 1,\frac{s+1}{2},\frac{s+2}{2}\end{matrix}\right)$ $[a,\ \operatorname{Re}s>0]$
11	$\frac{\theta(a-x)}{1-b^2x^2}\ln\frac{\sqrt{a-x}+\sqrt{a}}{\sqrt{x}}\mathbf{E}(bx)$	$\frac{\pi^{3/2}a^s}{4s}\Gamma\left[\begin{matrix}s\\ \frac{2s+1}{2}\end{matrix}\right]{}_5F_4\left(\begin{matrix}\frac{1}{2},\frac{3}{2},\frac{s}{2},\frac{s}{2},\frac{s+1}{2};\ a^2b^2\\ 1,\frac{2s+1}{4},\frac{2s+3}{4},\frac{s+2}{2}\end{matrix}\right)$ $[a,\ \operatorname{Re}s>0]$
12	$\frac{\theta(a-x)}{1-b^2x^2}\ln\frac{\sqrt{a^2-x^2}+a}{x}\mathbf{E}(bx)$	$\frac{\pi^{3/2}a^s}{4s}\Gamma\left[\begin{matrix}\frac{s}{2}\\ \frac{s+1}{2}\end{matrix}\right]{}_4F_3\left(\begin{matrix}\frac{1}{2},\frac{3}{2},\frac{s}{2},\frac{s}{2};\ a^2b^2\\ 1,\frac{s+1}{2},\frac{s+2}{2}\end{matrix}\right)$ $[a,\ \operatorname{Re}s>0]$
13	$\theta(a-x)\ln^2\left(b\sqrt{x}+\sqrt{b^2x+1}\right)$ $\times\begin{Bmatrix}\mathbf{K}\left(\sqrt{1-x/a}\right)\\ \mathbf{E}\left(\sqrt{1-x/a}\right)\end{Bmatrix}$	$\frac{\pi a^{s+1}b^2}{2}\Gamma\left[\begin{matrix}s+1,\ s-\delta+2\\ \frac{2s+3}{2},\frac{2s-2\delta+5}{2}\end{matrix}\right]$ $\times{}_5F_4\left(\begin{matrix}1,1,1,s+1,s-\delta+2\\ \frac{3}{2},2,\frac{2s+3}{2},\frac{2s-2\delta+5}{2};\ -a^2b\end{matrix}\right)$ $[a>0;\ \operatorname{Re}s>-1]$
14	$\theta(a-x)\ln^2\left(bx+\sqrt{b^2x^2+1}\right)$ $\times\begin{Bmatrix}\mathbf{K}\left(\sqrt{1-x/a}\right)\\ \mathbf{E}\left(\sqrt{1-x/a}\right)\end{Bmatrix}$	$\frac{\pi a^{s+2}b^2}{2}\Gamma\left[\begin{matrix}s+2,\ s-\delta+3\\ \frac{2s+5}{2},\frac{2s-2\delta+7}{2}\end{matrix}\right]$ $\times{}_7F_6\left(\begin{matrix}1,1,1,\frac{s+2}{2},\frac{s+3}{2},\frac{s-\delta+3}{2},\frac{s-\delta+4}{2};\ a^2b^2\\ \frac{3}{2},2,\frac{2s+5}{4},\frac{2s+7}{4},\frac{2s-2\delta+7}{4},\frac{2s-2\delta+9}{4}\end{matrix}\right)$ $[a>0;\ \operatorname{Re}s>-2]$

3.26.11. K $(\varphi(x))$, E $(\varphi(x))$, and inverse trigonometric functions

Notation: $\delta=\begin{Bmatrix}1\\0\end{Bmatrix}$.

1	$\theta(a-x)\arccos\frac{x}{a}\begin{Bmatrix}\mathbf{K}(bx)\\ \mathbf{E}(bx)\end{Bmatrix}$	$\frac{\pi^{3/2}a^s}{2s^2}\Gamma\left[\begin{matrix}\frac{s+1}{2}\\ \frac{s}{2}\end{matrix}\right]{}_4F_3\left(\begin{matrix}\pm\frac{1}{2},\frac{1}{2},\frac{s}{2},\frac{s+1}{2}\\ 1,\frac{s+2}{2},\frac{s+2}{2};\ a^2b^2\end{matrix}\right)$ $[a,\ \operatorname{Re}s>0]$
2	$\frac{\theta(a-x)}{1-b^2x^2}\arccos\frac{x}{a}\mathbf{E}(bx)$	$\frac{\pi^{3/2}a^s}{2s^2}\Gamma\left[\begin{matrix}\frac{s+1}{2}\\ \frac{s}{2}\end{matrix}\right]{}_4F_3\left(\begin{matrix}\frac{1}{2},\frac{3}{2},\frac{s}{2},\frac{s+1}{2}\\ 1,\frac{s+2}{2},\frac{s+2}{2};\ a^2b^2\end{matrix}\right)$ $[a,\ \operatorname{Re}s>0]$
3	$\theta(a-x)\arcsin(bx)\begin{Bmatrix}\mathbf{K}\left(\sqrt{1-x/a}\right)\\ \mathbf{E}\left(\sqrt{1-x/a}\right)\end{Bmatrix}$	$\frac{\pi a^{s+1}b}{2}\Gamma\left[\begin{matrix}s+1,\ s-\delta+2\\ \frac{2s+3}{2},\frac{2s-2\delta+5}{2}\end{matrix}\right]$ $\times{}_6F_5\left(\begin{matrix}\frac{1}{2},\frac{1}{2},\frac{s+1}{2},\frac{s+2}{2},\frac{s-\delta+2}{2},\frac{s-\delta+3}{2};\ a^2b^2\\ \frac{3}{2},\frac{2s+3}{4},\frac{2s+5}{4},\frac{2s-2\delta+5}{4},\frac{2s-2\delta+7}{4}\end{matrix}\right)$ $[a>0;\ \operatorname{Re}s>-1]$

No.	$f(x)$	$F(s)$
4	$\dfrac{\theta(a-x)}{\sqrt{1-b^2x^2}} \arcsin(bx) \times \begin{Bmatrix} \mathbf{K}\left(\sqrt{1-x/a}\right) \\ \mathbf{E}\left(\sqrt{1-x/a}\right) \end{Bmatrix}$	$\dfrac{\pi a^{s+1} b}{2} \Gamma\left[\begin{matrix} s+1,\ s-\delta+2 \\ \frac{2s+3}{2},\ \frac{2s-2\delta+5}{2} \end{matrix}\right] \times {}_6F_5\left(\begin{matrix} 1,\ 1,\ \frac{s+1}{2},\ \frac{s+2}{2},\ \frac{s-\delta+2}{2},\ \frac{s-\delta+3}{2};\ a^2b^2 \\ \frac{3}{2},\ \frac{2s+3}{4},\ \frac{2s+5}{4},\ \frac{2s-2\delta+5}{4},\ \frac{2s-2\delta+7}{4} \end{matrix}\right)$ $[a>0;\ \operatorname{Re} s>-1]$
5	$\theta(a-x) \arcsin(b\sqrt{x}) \times \begin{Bmatrix} \mathbf{K}\left(\sqrt{1-x/a}\right) \\ \mathbf{E}\left(\sqrt{1-x/a}\right) \end{Bmatrix}$	$\dfrac{\pi a^{s+1/2} b}{2} \Gamma\left[\begin{matrix} \frac{2s+1}{2},\ \frac{2s-2\delta+3}{2} \\ s+1,\ s-\delta+2 \end{matrix}\right] {}_4F_3\left(\begin{matrix} \frac{1}{2},\ \frac{1}{2},\ \frac{2s+1}{2},\ \frac{2s-2\delta+3}{2} \\ \frac{3}{2},\ s+1,\ s-\delta+2;\ ab^2 \end{matrix}\right)$ $[a>0;\ \operatorname{Re} s>-1/2]$
6	$\dfrac{\theta(a-x)}{\sqrt{1-b^2x}} \arcsin(b\sqrt{x}) \times \begin{Bmatrix} \mathbf{K}\left(\sqrt{1-x/a}\right) \\ \mathbf{E}\left(\sqrt{1-x/a}\right) \end{Bmatrix}$	$\dfrac{\pi a^{s+1/2} b}{2} \Gamma\left[\begin{matrix} \frac{2s+1}{2},\ \frac{2s-2\delta+3}{2} \\ s+1,\ s-\delta+2 \end{matrix}\right] {}_4F_3\left(\begin{matrix} 1,\ 1,\ \frac{2s+1}{2},\ \frac{2s-2\delta+3}{2} \\ \frac{3}{2},\ s+1,\ s-\delta+2;\ ab^2 \end{matrix}\right)$ $[a>0;\ \operatorname{Re} s>-1/2;\ \left\vert\arg\left(1-ab^2\right)\right\vert<\pi]$
7	$\theta(a-x) \arcsin^2(b\sqrt{x}) \times \begin{Bmatrix} \mathbf{K}\left(\sqrt{1-x/a}\right) \\ \mathbf{E}\left(\sqrt{1-x/a}\right) \end{Bmatrix}$	$\dfrac{\pi a^{s+1} b^2}{2} \Gamma\left[\begin{matrix} s+1,\ s-\delta+2 \\ \frac{2s+3}{2},\ \frac{2s-2\delta+5}{2} \end{matrix}\right] {}_5F_4\left(\begin{matrix} 1,\ 1,\ 1,\ s+1,\ s-\delta+2 \\ \frac{3}{2},\ 2,\ \frac{2s+3}{2},\ \frac{2s-2\delta+5}{2};\ ab^2 \end{matrix}\right)$ $[a>0;\ \operatorname{Re} s>-1]$
8	$\theta(a-x) \arcsin^2(bx) \times \begin{Bmatrix} \mathbf{K}\left(\sqrt{1-x/a}\right) \\ \mathbf{E}\left(\sqrt{1-x/a}\right) \end{Bmatrix}$	$\dfrac{\pi a^{s+2} b^2}{2} \Gamma\left[\begin{matrix} s+2,\ s-\delta+3 \\ \frac{2s+5}{2},\ \frac{2s-2\delta+7}{2} \end{matrix}\right] \times {}_7F_6\left(\begin{matrix} 1,\ 1,\ 1,\ \frac{s+2}{2},\ \frac{s+3}{2},\ \frac{s-\delta+3}{2},\ \frac{s-\delta+4}{2};\ a^2b^2 \\ \frac{3}{2},\ 2,\ \frac{2s+5}{4},\ \frac{2s+7}{4},\ \frac{2s-2\delta+7}{4},\ \frac{2s-2\delta+9}{4} \end{matrix}\right)$ $[a>0;\ \operatorname{Re} s>-2]$
9	$\theta(a-x) \arctan(bx) \times \begin{Bmatrix} \mathbf{K}\left(\sqrt{1-x/a}\right) \\ \mathbf{E}\left(\sqrt{1-x/a}\right) \end{Bmatrix}$	$\dfrac{\pi a^{s+1} b}{2} \Gamma\left[\begin{matrix} s+1,\ s-\delta+2 \\ \frac{2s+3}{2},\ \frac{2s-2\delta+5}{2} \end{matrix}\right] \times {}_6F_5\left(\begin{matrix} \frac{1}{2},\ 1,\ \frac{s+1}{2},\ \frac{s+2}{2},\ \frac{s-\delta+2}{2},\ \frac{s-\delta+3}{2};\ -a^2b^2 \\ \frac{3}{2},\ \frac{2s+3}{4},\ \frac{2s+5}{4},\ \frac{2s-2\delta+5}{4},\ \frac{2s-2\delta+7}{4} \end{matrix}\right)$ $[a>0;\ \operatorname{Re} s>-1]$
10	$\theta(a-x) \arctan(b\sqrt{x}) \times \begin{Bmatrix} \mathbf{K}\left(\sqrt{1-x/a}\right) \\ \mathbf{E}\left(\sqrt{1-x/a}\right) \end{Bmatrix}$	$\dfrac{\pi a^{s+1/2} b}{2} \Gamma\left[\begin{matrix} \frac{2s+1}{2},\ \frac{2s-2\delta+3}{2} \\ s+1,\ s-\delta+2 \end{matrix}\right] {}_4F_3\left(\begin{matrix} \frac{1}{2},\ 1,\ \frac{2s+1}{2},\ \frac{2s-2\delta+3}{2} \\ \frac{3}{2},\ s+1,\ s-\delta+2;\ -ab^2 \end{matrix}\right)$ $[a,\ \operatorname{Re} s>0]$

3.26.12. $\mathbf{K}(\varphi(x))$, $\mathbf{E}(\varphi(x))$, and $\mathrm{Li}_2(ax)$

Notation: $\delta = \left\{ \begin{matrix} 1 \\ 0 \end{matrix} \right\}$.

1	$\theta(a-x)\,\mathrm{Li}_2(bx) \times \left\{ \begin{matrix} \mathbf{K}(\sqrt{1-x/a}) \\ \mathbf{E}(\sqrt{1-x/a}) \end{matrix} \right\}$	$\frac{\pi a^{s+1} b}{2} \Gamma \left[\begin{matrix} s+1,\ s-\delta+2 \\ \frac{2s+3}{2},\ \frac{2s-2\delta+5}{2} \end{matrix} \right] {}_4F_3\left(\begin{matrix} 1,\ 1,\ 1,\ s+1,\ s-\delta+2 \\ 2,\ 2,\ \frac{2s+3}{2},\ \frac{2s-2\delta+5}{2};\ ab \end{matrix} \right)$ $[a>0,\ \mathrm{Re}\, s>-1;\ \lvert\arg(1-ab)\rvert<\pi]$

3.26.13. $\mathbf{K}(\varphi(x))$, $\mathbf{E}(\varphi(x))$, and $\mathrm{Si}(ax^r)$, $\mathrm{shi}(ax^r)$

1	$\theta(a-x) \left\{ \begin{matrix} \mathrm{shi}(bx) \\ \mathrm{Si}(bx) \end{matrix} \right\} \times \mathbf{K}\left(\sqrt{1-\frac{x}{a}}\right)$	$\frac{\pi a^{s+1} b^2}{2} \Gamma \left[\begin{matrix} s+1,\ s+1 \\ \frac{2s+3}{2},\ \frac{2s+3}{2} \end{matrix} \right] {}_5F_6\left(\begin{matrix} \frac{1}{2},\ \frac{s+1}{2},\ \frac{s+1}{2},\ \frac{s+2}{2},\ \frac{s+2}{2};\ \pm\frac{a^2b^2}{4} \\ \frac{3}{2},\ \frac{3}{2},\ \frac{2s+3}{4},\ \frac{2s+3}{4},\ \frac{2s+5}{4},\ \frac{2s+5}{4} \end{matrix} \right)$ $[a>0;\ \mathrm{Re}\, s>-1]$
2	$\theta(a-x) \left\{ \begin{matrix} \mathrm{shi}(bx) \\ \mathrm{Si}(bx) \end{matrix} \right\} \times \mathbf{E}\left(\sqrt{1-\frac{x}{a}}\right)$	$\frac{\pi a^{s+1} b^2}{2} \Gamma \left[\begin{matrix} s+1,\ s+2 \\ \frac{2s+3}{2},\ \frac{2s+5}{2} \end{matrix} \right] {}_5F_6\left(\begin{matrix} \frac{1}{2},\ \frac{s+1}{2},\ \frac{s+2}{2},\ \frac{s+2}{2},\ \frac{s+3}{2};\ \pm\frac{a^2b^2}{4} \\ \frac{3}{2},\ \frac{3}{2},\ \frac{2s+3}{4},\ \frac{2s+5}{4},\ \frac{2s+5}{4},\ \frac{2s+7}{4} \end{matrix} \right)$ $[a>0;\ \mathrm{Re}\, s>-1]$
3	$\theta(a-x) \left\{ \begin{matrix} \mathrm{shi}(b\sqrt{x}) \\ \mathrm{Si}(b\sqrt{x}) \end{matrix} \right\} \times \mathbf{K}\left(\sqrt{1-\frac{x}{a}}\right)$	$\frac{\pi a^{s+1/2} b}{2} \Gamma \left[\begin{matrix} \frac{2s+1}{2},\ \frac{2s+1}{2} \\ s+1,\ s+1 \end{matrix} \right] {}_3F_4\left(\begin{matrix} \frac{1}{2},\ \frac{2s+1}{2},\ \frac{2s+1}{2};\ \pm\frac{ab^2}{4} \\ \frac{3}{2},\ \frac{3}{2},\ s+1,\ s+1 \end{matrix} \right)$ $[a>0;\ \mathrm{Re}\, s>-1/2]$
4	$\theta(a-x) \left\{ \begin{matrix} \mathrm{shi}(a\sqrt{x}) \\ \mathrm{Si}(a\sqrt{x}) \end{matrix} \right\} \times \mathbf{E}\left(\sqrt{1-\frac{x}{a}}\right)$	$\frac{\pi a^{s+1/2} b}{2} \Gamma \left[\begin{matrix} \frac{2s+1}{2},\ \frac{2s+3}{2} \\ s+1,\ s+2 \end{matrix} \right] {}_3F_4\left(\begin{matrix} \frac{1}{2},\ \frac{2s+1}{2},\ \frac{2s+3}{2};\ \pm\frac{ab^2}{4} \\ \frac{3}{2},\ \frac{3}{2},\ s+1,\ s+2 \end{matrix} \right)$ $[a>0;\ \mathrm{Re}\, s>-1/2]$

3.26.14. $\mathbf{K}(\varphi(x))$, $\mathbf{E}(\varphi(x))$, and $\mathrm{ci}(ax)$, $\mathrm{chi}(ax)$

1	$\theta(a-x) \left\{ \begin{matrix} \mathrm{chi}(bx) \\ \mathrm{ci}(bx) \end{matrix} \right\} \times \mathbf{K}\left(\sqrt{1-\frac{x}{a}}\right)$	$\frac{\pi a^s}{2} \Gamma \left[\begin{matrix} s,\ s \\ \frac{2s+1}{2},\ \frac{2s+1}{2} \end{matrix} \right] \left[2\psi(s) - 2\psi\left(s+\frac{1}{2}\right) + \ln(ab) + \mathbf{C} \right] - \frac{\pi a^{s+2} b^2}{8} \Gamma \left[\begin{matrix} s+2,\ s+2 \\ \frac{2s+5}{2},\ \frac{2s+5}{2} \end{matrix} \right] \times {}_6F_7\left(\begin{matrix} 1,\ 1,\ \frac{s+2}{2},\ \frac{s+2}{2},\ \frac{s+3}{2},\ \frac{s+3}{2};\ \pm\frac{a^2b^2}{4} \\ \frac{3}{2},\ 2,\ 2,\ \frac{2s+5}{4},\ \frac{2s+5}{4},\ \frac{2s+7}{4},\ \frac{2s+7}{4} \end{matrix} \right)$ $[a,\ \mathrm{Re}\, s>0]$

No.	$f(x)$	$F(s)$
2	$\theta(a-x)\begin{Bmatrix}\operatorname{chi}(bx)\\ \operatorname{ci}(bx)\end{Bmatrix}$ $\times\mathbf{E}\left(\sqrt{1-\frac{x}{a}}\right)$	$\frac{\pi a^s}{2}\Gamma\left[\begin{matrix}s,\ s+1\\ \frac{2s+1}{2},\ \frac{2s+3}{2}\end{matrix}\right]\left[2\psi(s)-2\psi\left(s+\frac{1}{2}\right)\right.$ $\left.+\frac{1}{s(2s+1)}+\ln(ab)+\mathbf{C}\right]-\frac{\pi a^{s+2}b^2}{8}\Gamma\left[\begin{matrix}s+2,\ s+3\\ \frac{2s+5}{2},\ \frac{2s+7}{2}\end{matrix}\right]$ $\times{}_6F_7\left(\begin{matrix}1,\ 1,\ \frac{s+2}{2},\ \frac{s+3}{2},\ \frac{s+3}{2},\ \frac{s+4}{2};\ \pm\frac{a^2b^2}{4}\\ \frac{3}{2},\ 2,\ 2,\ \frac{2s+5}{4},\ \frac{2s+7}{4},\ \frac{2s+7}{4},\ \frac{2s+9}{4}\end{matrix}\right)$ $[a,\ \operatorname{Re}s>0]$

3.26.15. $\mathbf{K}(\varphi(x))$, $\mathbf{E}(\varphi(x))$, and $\operatorname{erf}(ax^r)$

Notation: $\delta=\begin{Bmatrix}1\\0\end{Bmatrix}$.

1	$\theta(a-x)\operatorname{erf}(bx)$ $\times\begin{Bmatrix}\mathbf{K}(\sqrt{1-x/a})\\ \mathbf{E}(\sqrt{1-x/a})\end{Bmatrix}$	$\frac{\pi a^{s+1}b}{2}\Gamma\left[\begin{matrix}s+1,\ s-\delta+2\\ \frac{2s+3}{2},\ \frac{2s-2\delta+5}{2}\end{matrix}\right]$ $\times{}_5F_5\left(\begin{matrix}\frac{1}{2},\ \frac{s+1}{2},\ \frac{s+2}{2},\ \frac{s-\delta+2}{2},\ \frac{s-\delta+3}{2};\ -a^2b^2\\ \frac{3}{2},\ \frac{2s+3}{4},\ \frac{2s+5}{4},\ \frac{2s-2\delta+5}{4},\ \frac{2s-2\delta+7}{4}\end{matrix}\right)$ $[a>0;\ \operatorname{Re}s>-1]$
2	$\theta(a-x)\,e^{b^2x^2}\operatorname{erf}(bx)$ $\times\begin{Bmatrix}\mathbf{K}(\sqrt{1-x/a})\\ \mathbf{E}(\sqrt{1-x/a})\end{Bmatrix}$	$\frac{\pi a^{s+1}b}{2}\Gamma\left[\begin{matrix}s+1,\ s-\delta+2\\ \frac{2s+3}{2},\ \frac{2s-2\delta+5}{2}\end{matrix}\right]$ $\times{}_5F_5\left(\begin{matrix}\frac{1}{2},\ \frac{s+1}{2},\ \frac{s+2}{2},\ \frac{s-\delta+2}{2},\ \frac{s-\delta+3}{2};\ a^2b^2\\ \frac{3}{2},\ \frac{2s+3}{4},\ \frac{2s+5}{4},\ \frac{2s-2\delta+5}{4},\ \frac{2s-2\delta+7}{4}\end{matrix}\right)$ $[a>0;\ \operatorname{Re}s>-1]$
3	$\theta(a-x)\operatorname{erf}(b\sqrt{x})$ $\times\begin{Bmatrix}\mathbf{K}(\sqrt{1-x/a})\\ \mathbf{E}(\sqrt{1-x/a})\end{Bmatrix}$	$\sqrt{\pi}\,a^{s+1/2}b\,\Gamma\left[\begin{matrix}\frac{2s+1}{2},\ \frac{2s-2\delta+3}{2}\\ s+1,\ s-\delta+2\end{matrix}\right]{}_3F_3\left(\begin{matrix}\frac{1}{2},\ \frac{2s+1}{2},\ \frac{2s-2\delta+3}{2};\ -ab^2\\ \frac{3}{2},\ s+1,\ s-\delta+2\end{matrix}\right)$ $[a>0;\ \operatorname{Re}s>-1/2]$
4	$\theta(a-x)\,e^{b^2x}\operatorname{erf}(b\sqrt{x})$ $\times\begin{Bmatrix}\mathbf{K}(\sqrt{1-x/a})\\ \mathbf{E}(\sqrt{1-x/a})\end{Bmatrix}$	$\sqrt{\pi}\,a^{s+1/2}b\,\Gamma\left[\begin{matrix}\frac{2s+1}{2},\ \frac{2s-2\delta+3}{2}\\ s+1,\ s-\delta+2\end{matrix}\right]{}_3F_3\left(\begin{matrix}1,\ \frac{2s+1}{2},\ \frac{2s-2\delta+3}{2};\ ab^2\\ \frac{3}{2},\ s+1,\ s-\delta+2\end{matrix}\right)$ $[a>0;\ \operatorname{Re}s>-1/2]$

3.26.16. $\mathbf{K}(\varphi(x))$, $\mathbf{E}(\varphi(x))$, and $S(a\sqrt{x})$, $C(a\sqrt{x})$

Notation: $\delta=\begin{Bmatrix}1\\0\end{Bmatrix}$.

1	$\theta(a-x)\,S(b\sqrt{x})$ $\times\begin{Bmatrix}\mathbf{K}(\sqrt{1-x/a})\\ \mathbf{E}(\sqrt{1-x/a})\end{Bmatrix}$	$\frac{a^{s+3/4}}{3}\sqrt{\frac{\pi b^3}{2}}\,\Gamma\left[\begin{matrix}\frac{4s+3}{4},\ \frac{4s-4\delta+7}{4}\\ \frac{4s+5}{4},\ \frac{4s-4\delta+9}{4}\end{matrix}\right]{}_3F_4\left(\begin{matrix}\frac{3}{4},\ \frac{4s+3}{4},\ \frac{4s-4\delta+7}{4};\ -\frac{ab^2}{4}\\ \frac{3}{2},\ \frac{7}{4},\ \frac{4s+5}{4},\ \frac{4s-4\delta+9}{4}\end{matrix}\right)$ $[a>0;\operatorname{Re}s>-3/4]$

No.	$f(x)$	$F(s)$
2	$\theta(a-x)\,C(b\sqrt{x})$ $\times \begin{Bmatrix} \mathbf{K}(\sqrt{1-x/a}) \\ \mathbf{E}(\sqrt{1-x/a}) \end{Bmatrix}$	$a^{s+1/4}\sqrt{\frac{\pi b}{2}}\,\Gamma\left[\begin{matrix} \frac{4s+1}{4},\ \frac{4s-4\delta+5}{4} \\ \frac{4s+3}{4},\ \frac{4s-4\delta+7}{4} \end{matrix}\right] {}_3F_4\left(\begin{matrix} \frac{1}{4},\ \frac{4s+1}{4},\ \frac{4s-4\delta+5}{4};\ -\frac{ab^2}{4} \\ \frac{1}{2},\ \frac{5}{4},\ \frac{4s+3}{4},\ \frac{4s-4\delta+7}{4} \end{matrix}\right)$ $[a>0;\ \operatorname{Re} s > -1/4]$

3.26.17. $\mathbf{K}(\varphi(x))$, $\mathbf{E}(\varphi(x))$, and $\gamma(\nu, ax)$

Notation: $\delta = \begin{Bmatrix} 1 \\ 0 \end{Bmatrix}$.

1	$\theta(a-x)\,\gamma(\nu, bx)$ $\times \begin{Bmatrix} \mathbf{K}(\sqrt{1-x/a}) \\ \mathbf{E}(\sqrt{1-x/a}) \end{Bmatrix}$	$\frac{\pi a^{s+\nu} b^{\nu}}{2\nu}\,\Gamma\left[\begin{matrix} s+\nu,\ s+\nu-\delta+1 \\ \frac{2s+2\nu+1}{2},\ \frac{2s+2\nu-2\delta+3}{2} \end{matrix}\right]$ $\times {}_3F_3\left(\begin{matrix} \nu,\ s+\nu,\ s+\nu-\delta+1;\ -ab \\ \nu+1,\ \frac{2s+2\nu+1}{2},\ \frac{2s+2\nu-2\delta+3}{2} \end{matrix}\right)$ $[a,\ \operatorname{Re}(s+\nu)>0]$
2	$\theta(a-x)\,e^{bx}\gamma(\nu, bx)$ $\times \begin{Bmatrix} \mathbf{K}(\sqrt{1-x/a}) \\ \mathbf{E}(\sqrt{1-x/a}) \end{Bmatrix}$	$\frac{\pi a^{s+\nu} b^{\nu}}{2\nu}\,\Gamma\left[\begin{matrix} s+\nu,\ s+\nu-\delta+1 \\ \frac{2s+2\nu+1}{2},\ \frac{2s+2\nu-2\delta+3}{2} \end{matrix}\right]$ $\times {}_3F_3\left(\begin{matrix} 1,\ s+\nu,\ s+\nu-\delta+1;\ ab \\ \nu+1,\ \frac{2s+2\nu+1}{2},\ \frac{2s+2\nu-2\delta+3}{2} \end{matrix}\right)$ $[a,\ \operatorname{Re}(s+\nu)>0]$

3.26.18. $\mathbf{K}(\varphi(x))$, $\mathbf{E}(\varphi(x))$, and $J_\nu(bx^r)$, $I_\nu(bx^r)$

Notation: $\delta = \begin{Bmatrix} 1 \\ 0 \end{Bmatrix}$.

1	$\theta(a-x)\begin{Bmatrix} J_\nu(bx) \\ I_\nu(bx) \end{Bmatrix}$ $\times \mathbf{K}\left(\sqrt{1-\frac{x}{a}}\right)$	$\frac{\pi a^{s+\nu} b^{\nu}}{2^{\nu+1}}\,\Gamma\left[\begin{matrix} s+\nu,\ s+\nu \\ \nu+1,\ \frac{2s+2\nu+1}{2},\ \frac{2s+2\nu+1}{2} \end{matrix}\right]$ $\times {}_4F_5\left(\begin{matrix} \frac{s+\nu}{2},\ \frac{s+\nu}{2},\ \frac{s+\nu+1}{2},\ \frac{s+\nu+1}{2};\ \mp\frac{a^2b^2}{4} \\ \nu+1,\ \frac{2s+2\nu+1}{4},\ \frac{2s+2\nu+1}{4},\ \frac{2s+2\nu+3}{4},\ \frac{2s+2\nu+3}{4} \end{matrix}\right)$ $[a,\ \operatorname{Re}(s+\nu)>0]$
2	$\theta(a-x)\begin{Bmatrix} J_\nu(bx) \\ I_\nu(bx) \end{Bmatrix}$ $\times \mathbf{E}\left(\sqrt{1-\frac{x}{a}}\right)$	$\frac{\pi a^{s+\nu} b^{\nu}}{2^{\nu+1}}\,\Gamma\left[\begin{matrix} s+\nu,\ s+\nu+1 \\ \nu+1,\ \frac{2s+2\nu+1}{2},\ \frac{2s+2\nu+3}{2} \end{matrix}\right]$ $\times {}_4F_5\left(\begin{matrix} \frac{s+\nu}{2},\ \frac{s+\nu+1}{2},\ \frac{s+\nu+1}{2},\ \frac{s+\nu+3}{2};\ \mp\frac{a^2b^2}{4} \\ \nu+1,\ \frac{2s+2\nu+1}{4},\ \frac{2s+2\nu+3}{4},\ \frac{2s+2\nu+3}{4},\ \frac{2s+2\nu+5}{4} \end{matrix}\right)$ $[a,\ \operatorname{Re}(s+\nu)>0]$
3	$\theta(a-x)\begin{Bmatrix} J_\nu(b\sqrt{x}) \\ I_\nu(b\sqrt{x}) \end{Bmatrix}$ $\times \mathbf{K}\left(\sqrt{1-\frac{x}{a}}\right)$	$\frac{\pi a^{s+\nu/2} b^{\nu}}{2^{\nu+1}}\,\frac{\Gamma^2\left(\frac{2s+\nu}{2}\right)}{\Gamma(\nu+1)\,\Gamma^2\left(\frac{2s+\nu+1}{2}\right)}\,{}_2F_3\left(\begin{matrix} \frac{2s+\nu}{2},\ \frac{2s+\nu}{2};\ \mp\frac{ab^2}{4} \\ \nu+1,\ \frac{2s+\nu+1}{2},\ \frac{2s+\nu+1}{2} \end{matrix}\right)$ $[a,\ \operatorname{Re}(s+\nu/2)>0]$

No.	$f(x)$	$F(s)$
4	$\theta(a-x)\begin{Bmatrix} J_\nu(b\sqrt{x}) \\ I_\nu(b\sqrt{x}) \end{Bmatrix} \times \mathbf{E}\left(\sqrt{1-\frac{x}{a}}\right)$	$\frac{\pi a^{s+\nu/2}b^\nu}{2^{\nu+1}}\Gamma\left[\begin{matrix} \frac{2s+\nu}{2}, \frac{2s+\nu+2}{2} \\ \nu+1, \frac{2s+\nu+1}{2}, \frac{2s+\nu+3}{2} \end{matrix}\right] \times {}_2F_3\left(\begin{matrix} \frac{2s+\nu}{2}, \frac{2s+\nu+2}{2}; \mp\frac{ab^2}{4} \\ \nu+1, \frac{2s+\nu+1}{2}, \frac{2s+\nu+3}{2} \end{matrix}\right)$ $[a,\ \operatorname{Re}(s+\nu/2)>0]$
5	$\theta(a-x)e^{bx}I_\nu(bx) \times \begin{Bmatrix} \mathbf{K}(\sqrt{1-x/a}) \\ \mathbf{E}(\sqrt{1-x/a}) \end{Bmatrix}$	$\frac{\pi a^{s+\nu}b^\nu}{2^{\nu+1}}\Gamma\left[\begin{matrix} s+\nu, s+\nu-\delta+1 \\ \nu+1, \frac{2s+2\nu+1}{2}, \frac{2s+2\nu-2\delta+3}{2} \end{matrix}\right] \times {}_3F_3\left(\begin{matrix} \frac{2\nu+1}{2}, s+\nu, s+\nu-\delta+1; 2ab \\ 2\nu+1, \frac{2s+2\nu+1}{2}, \frac{2s+2\nu-2\delta+3}{2} \end{matrix}\right)$ $[a,\ \operatorname{Re}(s+\nu)>0]$
6	$\theta(a-x) \times \begin{Bmatrix} J_\mu(b\sqrt{x})J_\nu(b\sqrt{x}) \\ I_\mu(b\sqrt{x})I_\nu(b\sqrt{x}) \end{Bmatrix} \times \mathbf{K}\left(\sqrt{1-\frac{x}{a}}\right)$	$\frac{\pi a^{s+(\mu+\nu)/2}b^{\mu+\nu}}{2^{\mu+\nu+1}}\Gamma\left[\begin{matrix} \frac{2s+\mu+\nu}{2}, \frac{2s+\mu+\nu}{2} \\ \mu+1, \nu+1, \frac{2s+\mu+\nu+1}{2}, \frac{2s+\mu+\nu+1}{2} \end{matrix}\right] \times {}_4F_5\left(\begin{matrix} \frac{\mu+\nu+1}{2}, \frac{\mu+\nu+2}{2}, \frac{2s+\mu+\nu}{2}, \frac{2s+\mu+\nu}{2}; \mp ab^2 \\ \mu+1, \nu+1, \mu+\nu+1, \frac{2s+\mu+\nu+1}{2}, \frac{2s+\mu+\nu+1}{2} \end{matrix}\right)$ $[a,\ \operatorname{Re}(2s+\mu+\nu)>0]$
7	$\theta(a-x) \times \begin{Bmatrix} J_\mu(b\sqrt{x})J_\nu(b\sqrt{x}) \\ I_\mu(b\sqrt{x})I_\nu(b\sqrt{x}) \end{Bmatrix} \times \mathbf{E}\left(\sqrt{1-\frac{x}{a}}\right)$	$\frac{\pi a^{s+(\mu+\nu)/2}b^{\mu+\nu}}{2^{\mu+\nu+1}}\Gamma\left[\begin{matrix} \frac{2s+\mu+\nu}{2}, \frac{2s+\mu+\nu+2}{2} \\ \mu+1, \nu+1, \frac{2s+\mu+\nu+1}{2}, \frac{2s+\mu+\nu+3}{2} \end{matrix}\right] \times {}_4F_5\left(\begin{matrix} \frac{\mu+\nu+1}{2}, \frac{\mu+\nu+2}{2}, \frac{2s+\mu+\nu}{2}, \frac{2s+\mu+\nu+2}{2}; \mp ab^2 \\ \mu+1, \nu+1, \mu+\nu+1, \frac{2s+\mu+\nu+1}{2}, \frac{2s+\mu+\nu+3}{2} \end{matrix}\right)$ $[a,\ \operatorname{Re}(2s+\mu+\nu)>0]$

3.26.19. $\mathbf{K}(\varphi(x))$, $\mathbf{E}(\varphi(x))$, and $\mathbf{H}_\nu(bx^r)$, $\mathbf{L}_\nu(bx^r)$

No.	$f(x)$	$F(s)$
1	$\theta(a-x)\begin{Bmatrix} \mathbf{H}_\nu(bx) \\ \mathbf{L}_\nu(bx) \end{Bmatrix} \times \mathbf{K}\left(\sqrt{1-\frac{x}{a}}\right)$	$\sqrt{\pi}\,a^{s+\nu+1}\left(\frac{b}{2}\right)^{\nu+1}\Gamma\left[\begin{matrix} s+\nu+1, s+\nu+1 \\ \frac{2\nu+3}{2}, \frac{2s+2\nu+3}{2}, \frac{2s+2\nu+3}{2} \end{matrix}\right] \times {}_4F_5\left(\begin{matrix} 1, \frac{s+\nu+1}{2}, \frac{s+\nu+1}{2}, \frac{s+\nu+2}{2}, \frac{s+\nu+2}{2}; \mp\frac{a^2b^2}{4} \\ \frac{3}{2}, \frac{2\nu+3}{2}, \frac{2s+2\nu+3}{4}, \frac{2s+2\nu+3}{4}, \frac{2s+2\nu+5}{4}, \frac{2s+2\nu+5}{4} \end{matrix}\right)$ $[a,\ \operatorname{Re}(s+\nu+1)>0]$
2	$\theta(a-x)\begin{Bmatrix} \mathbf{H}_\nu(bx)) \\ \mathbf{L}_\nu(bx) \end{Bmatrix} \times \mathbf{E}\left(\sqrt{1-\frac{x}{a}}\right)$	$\sqrt{\pi}\,a^{s+\nu+1}\left(\frac{b}{2}\right)^{\nu+1}\Gamma\left[\begin{matrix} s+\nu+1, s+\nu+2 \\ \frac{2\nu+3}{2}, \frac{2s+2\nu+3}{2}, \frac{2s+2\nu+5}{2} \end{matrix}\right] \times {}_5F_6\left(\begin{matrix} 1, \frac{s+\nu+1}{2}, \frac{s+\nu+2}{2}, \frac{s+\nu+2}{2}, \frac{s+\nu+3}{2}; \mp\frac{a^2b^2}{4} \\ \frac{3}{2}, \frac{2\nu+3}{2}, \frac{2s+2\nu+3}{4}, \frac{2s+2\nu+5}{4}, \frac{2s+2\nu+5}{4}, \frac{2s+2\nu+7}{4} \end{matrix}\right)$ $[a,\ \operatorname{Re}(s+\nu+1)>0]$

No.	$f(x)$	$F(s)$
3	$\theta(a-x)\left\{\begin{matrix}\mathbf{H}_\nu(b\sqrt{x})\\ \mathbf{L}_\nu(b\sqrt{x})\end{matrix}\right\}$ $\times\mathbf{K}\left(\sqrt{1-\frac{x}{a}}\right)$	$\sqrt{\pi}\,a^{s+(\nu+1)/2}\left(\frac{b}{2}\right)^{\nu+1}\Gamma\left[\begin{matrix}\frac{2s+\nu+1}{2},\ \frac{2s+\nu+1}{2}\\ \frac{2\nu+3}{2},\ \frac{2s+\nu+2}{2},\ \frac{2s+\nu+2}{2}\end{matrix}\right]$ $\times{}_3F_4\left(\begin{matrix}1,\ \frac{2s+\nu+1}{2},\ \frac{2s+\nu+1}{2};\ \mp\frac{ab^2}{4}\\ \frac{3}{2},\ \frac{2\nu+3}{2},\ \frac{2s+\nu+2}{2},\ \frac{2s+\nu+2}{2}\end{matrix}\right)$ $[a>0;\ \operatorname{Re}(2s+\nu)>-1]$
4	$\theta(a-x)\left\{\begin{matrix}\mathbf{H}_\nu(b\sqrt{x})\\ \mathbf{L}_\nu(b\sqrt{x})\end{matrix}\right\}$ $\times\mathbf{E}\left(\sqrt{1-\frac{x}{a}}\right)$	$\sqrt{\pi}a^{s+(\nu+1)/2}\left(\frac{b}{2}\right)^{\nu+1}\Gamma\left[\begin{matrix}\frac{2s+\nu+1}{2},\ \frac{2s+\nu+3}{2}\\ \frac{2\nu+3}{2},\ \frac{2s+\nu+2}{2},\ \frac{2s+\nu+4}{2}\end{matrix}\right]$ $\times{}_3F_4\left(\begin{matrix}1,\ \frac{2s+\nu+1}{2},\ \frac{2s+\nu+3}{2};\ \mp\frac{ab^2}{4}\\ \frac{3}{2},\ \frac{2\nu+3}{2},\ \frac{2s+\nu+2}{2},\ \frac{2s+\nu+4}{2}\end{matrix}\right)$ $[a>0;\ \operatorname{Re}(2s+\nu)>-1]$

3.26.20. $\mathbf{K}(bx)$, $\mathbf{E}(bx)$, and $T_n(ax)$

No.	$f(x)$	$F(s)$
1	$\left(a^2-x^2\right)_+^{-1/2}T_n\left(\frac{x}{a}\right)$ $\times\left\{\begin{matrix}\mathbf{K}(bx)\\ \mathbf{E}(bx)\end{matrix}\right\}$	$\frac{\pi^2}{4}\left(\frac{a}{2}\right)^{s-1}\Gamma\left[\begin{matrix}s\\ \frac{s-n+1}{2},\ \frac{s+n+1}{2}\end{matrix}\right]{}_4F_3\left(\begin{matrix}\pm\frac{1}{2},\ \frac{1}{2},\ \frac{s}{2},\ \frac{s+1}{2};\ a^2b^2\\ 1,\ \frac{s-n+1}{2},\ \frac{s+n+1}{2}\end{matrix}\right)$ $[a>0;\ \operatorname{Re}s>((-1)^n-1)/2]$
2	$\frac{\left(a^2-x^2\right)_+^{-1/2}}{1-b^2x^2}T_n\left(\frac{x}{a}\right)$ $\times\mathbf{E}(bx)$	$\frac{\pi^2}{4}\left(\frac{a}{2}\right)^{s-1}\Gamma\left[\begin{matrix}s\\ \frac{s-n+1}{2},\ \frac{s+n+1}{2}\end{matrix}\right]{}_4F_3\left(\begin{matrix}\frac{1}{2},\ \frac{3}{2},\ \frac{s}{2},\ \frac{s+1}{2};\ a^2b^2\\ 1,\ \frac{s-n+1}{2},\ \frac{s+n+1}{2}\end{matrix}\right)$ $[a>0;\ \operatorname{Re}s>((-1)^n-1)/2]$

3.26.21. $\mathbf{K}(\varphi(x))$, $\mathbf{E}(\varphi(x))$, and $L_n^\lambda(ax)$, $H_n(ax^r)$

Notation: $\delta=\left\{\begin{matrix}1\\ 0\end{matrix}\right\}$.

1	$\theta(a-x)\,L_n^\lambda(bx)$ $\times\left\{\begin{matrix}\mathbf{K}(\sqrt{1-x/a})\\ \mathbf{E}(\sqrt{1-x/a})\end{matrix}\right\}$	$\frac{\pi(\lambda+1)_n\,a^s}{2(n!)}\Gamma\left[\begin{matrix}s,\ s-\delta+1\\ \frac{2s+1}{2},\ \frac{2s-2\delta+3}{2}\end{matrix}\right]{}_3F_3\left(\begin{matrix}-n,\ s,\ s-\delta+1;\ ab\\ \lambda+1,\ \frac{2s+1}{2},\ \frac{2s-2\delta+3}{2}\end{matrix}\right)$ $[a,\ \operatorname{Re}s>0]$
2	$\theta(a-x)\,H_{2n}(bx)$ $\times\left\{\begin{matrix}\mathbf{K}(\sqrt{1-x/a})\\ \mathbf{E}(\sqrt{1-x/a})\end{matrix}\right\}$	$(-1)^n\frac{(2n)!}{n!}\frac{\pi a^s}{2}\Gamma\left[\begin{matrix}s+1,\ s-\delta+1\\ \frac{2s+1}{2},\ \frac{2s-2\delta+3}{2}\end{matrix}\right]$ $\times{}_5F_5\left(\begin{matrix}-n,\ \frac{s}{2},\ \frac{s+1}{2},\ \frac{s-\delta+1}{2},\ \frac{s-\delta+2}{2};\ -a^2b^2\\ \frac{1}{2},\ \frac{2s+1}{4},\ \frac{2s+3}{4},\ \frac{2s-2\delta+3}{4},\ \frac{2s-2\delta+5}{4}\end{matrix}\right)$ $[a,\ \operatorname{Re}s>0]$

No.	$f(x)$	$F(s)$
3	$\theta(a-x)\,H_{2n+1}(bx) \times \begin{Bmatrix} \mathbf{K}\left(\sqrt{1-x/a}\right) \\ \mathbf{E}\left(\sqrt{1-x/a}\right) \end{Bmatrix}$	$(-1)^n \frac{(2n+1)!}{n!} \pi a^{s+1} b\,\Gamma\left[\begin{matrix} s+1,\ s-\delta+2 \\ \frac{2s+3}{2},\ \frac{2s-2\delta+5}{2} \end{matrix}\right] \times {}_5F_5\left(\begin{matrix} -n,\ \frac{s+1}{2},\ \frac{s+2}{2},\ \frac{s-\delta+2}{2},\ \frac{s-\delta+3}{2};\ -a^2b^2 \\ \frac{3}{2},\ \frac{2s+3}{4},\ \frac{2s+5}{4},\ \frac{2s-2\delta+5}{4},\ \frac{2s-2\delta+7}{4} \end{matrix}\right)$ $[a>0;\ \operatorname{Re} s>-1]$
4	$\theta(a-x)\,H_{2n}(b\sqrt{x}) \times \begin{Bmatrix} \mathbf{K}\left(\sqrt{1-x/a}\right) \\ \mathbf{E}\left(\sqrt{1-x/a}\right) \end{Bmatrix}$	$(-1)^n 2^{2n-1} \pi a^s \left(\frac{1}{2}\right)_n \Gamma\left[\begin{matrix} s,\ s-\delta+1 \\ \frac{2s+1}{2},\ \frac{2s-2\delta+3}{2} \end{matrix}\right] \times {}_3F_3\left(\begin{matrix} -n,\ s,\ s-\delta+1;\ ab^2 \\ \frac{1}{2},\ \frac{2s+1}{2},\ \frac{2s-2\delta+3}{2} \end{matrix}\right)$ $[a,\ \operatorname{Re} s>0]$
5	$\theta(a-x)\,H_{2n+1}(b\sqrt{x}) \times \begin{Bmatrix} \mathbf{K}\left(\sqrt{1-x/a}\right) \\ \mathbf{E}\left(\sqrt{1-x/a}\right) \end{Bmatrix}$	$(-4)^n \pi a^{s+1/2} b \left(\frac{3}{2}\right)_n \Gamma\left[\begin{matrix} \frac{2s+1}{2},\ \frac{2s-2\delta+3}{2} \\ s+1,\ s-\delta+2 \end{matrix}\right] \times {}_3F_3\left(\begin{matrix} -n,\ \frac{2s+1}{2},\ \frac{2s-2\delta+3}{2};\ ab^2 \\ \frac{3}{2},\ s+1,\ s-\delta+2 \end{matrix}\right)$ $[a>0;\ \operatorname{Re} s>-1/2]$

3.26.22. $\mathbf{K}(bx)$, $\mathbf{E}(bx)$, and $C_n^\lambda(ax)$

1	$(a^2-x^2)_+^{\lambda-1/2} C_n^\lambda\left(\frac{x}{a}\right) \times \begin{Bmatrix} \mathbf{K}(bx) \\ \mathbf{E}(bx) \end{Bmatrix}$	$\frac{\pi^2}{2(n!)} \left(\frac{a}{2}\right)^{s+2\lambda-1} \Gamma\left[\begin{matrix} n+2\lambda,\ s \\ \lambda,\ \frac{s-n+1}{2},\ \frac{s+n+2\lambda+1}{2} \end{matrix}\right] \times {}_4F_3\left(\begin{matrix} \pm\frac{1}{2},\ \frac{1}{2},\ \frac{s}{2},\ \frac{s+1}{2};\ a^2b^2 \\ 1,\ \frac{s-n+1}{2},\ \frac{s+n+2\lambda+1}{2} \end{matrix}\right)$ $[a>0;\ \operatorname{Re}\lambda>-1/2;\ \operatorname{Re} s>((-1)^n-1)/2]$
2	$\frac{(a^2-x^2)_+^{\lambda-1/2}}{1-b^2x^2} C_n^\lambda\left(\frac{x}{a}\right) \times \mathbf{E}(bx)$	$\frac{\pi^2}{2(n!)} \left(\frac{a}{2}\right)^{s+2\lambda-1} \Gamma\left[\begin{matrix} n+2\lambda,\ s \\ \lambda,\ \frac{s-n+1}{2},\ \frac{s+n+2\lambda+1}{2} \end{matrix}\right] \times {}_4F_3\left(\begin{matrix} \frac{1}{2},\ \frac{3}{2},\ \frac{s}{2},\ \frac{s+1}{2};\ a^2b^2 \\ 1,\ \frac{s-n+1}{2},\ \frac{s+n+2\lambda+1}{2} \end{matrix}\right)$ $[a>0;\ \operatorname{Re}\lambda>-1/2;\ \operatorname{Re} s>((-1)^n-1)/2]$

3.26.23. $\mathbf{D}(\varphi(x))$ and various functions

1	$\theta(a-x)\ln\frac{a+\sqrt{a^2-x^2}}{x} \times \mathbf{D}(bx)$	$\frac{\pi^{3/2}a^s}{8s} \Gamma\left[\begin{matrix} \frac{s}{2} \\ \frac{s+1}{2} \end{matrix}\right] {}_4F_3\left(\begin{matrix} \frac{1}{2},\ \frac{3}{2},\ \frac{s}{2},\ \frac{s}{2};\ a^2b^2 \\ 2,\ \frac{s+1}{2},\ \frac{s+2}{2} \end{matrix}\right)$ $[a,\ \operatorname{Re} s>0]$
2	$\theta(a-x)\arccos\frac{x}{a}\ \mathbf{D}(bx)$	$\frac{\pi^{3/2}a^s}{4s^2} \Gamma\left[\begin{matrix} \frac{s+1}{2} \\ \frac{s}{2} \end{matrix}\right] {}_4F_3\left(\begin{matrix} \frac{1}{2},\ \frac{3}{2},\ \frac{s}{2},\ \frac{s+1}{2};\ a^2b^2 \\ 2,\ \frac{s+2}{2},\ \frac{s+2}{2} \end{matrix}\right)$ $[a,\ \operatorname{Re} s>0]$

No.	$f(x)$	$F(s)$
3	$(a-x)_+^{\alpha-1}\,\mathbf{D}\left(b\sqrt{x(a-x)}\right)$	$\dfrac{\pi a^{s+\alpha-1}}{4}\,\mathrm{B}(s,\,\alpha)\;{}_4F_3\left(\begin{matrix}\frac{1}{2},\,\frac{3}{2},\,\alpha,\,s;\,\frac{a^2b^2}{4}\\ 2,\,\frac{s+\alpha}{2},\,\frac{s+\alpha+1}{2}\end{matrix}\right)$ $[a,\ \mathrm{Re}\,s,\ \mathrm{Re}\,\alpha>0]$

3.26.24. Products of $\mathbf{K}(\varphi(x))$

1	$\theta(a-x)\,\mathbf{K}(bx)\,\mathbf{K}\left(\sqrt{1-\frac{x}{a}}\right)$	$\dfrac{\pi^2a^s}{4}\,\Gamma\left[\begin{matrix}s,\,s\\ \frac{2s+1}{2},\,\frac{2s+1}{2}\end{matrix}\right]\;{}_6F_5\left(\begin{matrix}\frac{1}{2},\,\frac{1}{2},\,\frac{s}{2},\,\frac{s}{2},\,\frac{s+1}{2},\,\frac{s+1}{2};\,a^2b^2\\ 1,\,\frac{2s+1}{4},\,\frac{2s+1}{4},\,\frac{2s+3}{4},\,\frac{2s+3}{4}\end{matrix}\right)$ $[a,\ \mathrm{Re}\,s>0]$
2	$\theta(a-x)\,\mathbf{K}(b\sqrt{x})\,\mathbf{K}\left(\sqrt{1-\frac{x}{a}}\right)$	$\dfrac{\pi^2a^s}{4}\,\Gamma\left[\begin{matrix}s,\,s\\ \frac{2s+1}{2},\,\frac{2s+1}{2}\end{matrix}\right]\;{}_4F_3\left(\begin{matrix}\frac{1}{2},\,\frac{1}{2},\,s,\,s;\,ab^2\\ 1,\,\frac{2s+1}{2},\,\frac{2s+1}{2}\end{matrix}\right)$ $[a,\ \mathrm{Re}\,s>0]$
3	$\dfrac{\theta(a-x)}{\sqrt{1+b^2x}}\,\mathbf{K}\left(\dfrac{b\sqrt{x}}{\sqrt{1+b^2x}}\right)\times\mathbf{K}\left(\sqrt{1-\frac{x}{a}}\right)$	$\dfrac{\pi^2a^s}{4}\,\Gamma\left[\begin{matrix}s,\,s\\ \frac{2s+1}{2},\,\frac{2s+1}{2}\end{matrix}\right]\;{}_4F_3\left(\begin{matrix}\frac{1}{2},\,\frac{1}{2},\,s,\,s;\,-ab^2\\ 1,\,\frac{2s+1}{2},\,\frac{2s+1}{2}\end{matrix}\right)$ $[a,\ \mathrm{Re}\,s>0;\ \lvert\arg(1+ab^2)\rvert<\pi]$
4	$\mathbf{K}^2\left(\sqrt{\dfrac{\sqrt{a}-\sqrt{x+a}}{2\sqrt{a}}}\right)$	$\dfrac{\sqrt{\pi}\,a^s}{4}\,\Gamma\left[\begin{matrix}s,\,\frac{1-2s}{4},\,\frac{1-2s}{4},\,\frac{1-2s}{4}\\ 1-s,\,1-s\end{matrix}\right]$ $[0<\mathrm{Re}\,s<1/2;\ \lvert\arg a\rvert<\pi]$
5	$\mathbf{K}^2\left(\sqrt{\dfrac{\sqrt{x}-\sqrt{x+a}}{2\sqrt{x}}}\right)$	$\dfrac{\sqrt{\pi}\,a^s}{4}\,\Gamma\left[\begin{matrix}-s,\,\frac{2s+1}{4},\,\frac{2s+1}{4},\,\frac{2s+1}{4}\\ s+1,\,s+1\end{matrix}\right]$ $[-1/2<\mathrm{Re}\,s<0;\ \lvert\arg a\rvert<\pi]$
6	$\dfrac{1}{\sqrt{x+a}+\sqrt{x}}\,\mathbf{K}^2\left(\pm\dfrac{\sqrt{x+a}-\sqrt{x}}{\sqrt{a}}\right)$	$\dfrac{\sqrt{\pi}\,a^{s-1/2}}{8}\,\Gamma\left[\begin{matrix}s,\,s,\,s,\,\frac{1-2s}{2}\\ \frac{2s+1}{2},\,\frac{2s+1}{2}\end{matrix}\right]$ $[0<\mathrm{Re}\,s<1/2;\ \lvert\arg a\rvert<\pi]$
7	$\dfrac{1}{\sqrt{x+a}+\sqrt{a}}\,\mathbf{K}^2\left(\pm\dfrac{\sqrt{x+a}-\sqrt{a}}{\sqrt{x}}\right)$	$\dfrac{\sqrt{\pi}\,a^{s-1/2}}{8}\,\Gamma\left[\begin{matrix}s,\,\frac{1-2s}{2},\,\frac{1-2s}{2},\,\frac{1-2s}{2}\\ 1-s,\,1-s\end{matrix}\right]$ $[0<\mathrm{Re}\,s<1/2;\ \lvert\arg a\rvert<\pi]$
8	$\mathbf{K}\left(i\sqrt{\dfrac{2\sqrt{x}\left(\sqrt{x}-\sqrt{x+a}\right)}{a}}\right)\times\mathbf{K}\left(i\sqrt{\dfrac{2\sqrt{x}\left(\sqrt{x}+\sqrt{x+a}\right)}{a}}\right)$	$\dfrac{\sqrt{\pi}\,a^s}{4}\,\Gamma\left[\begin{matrix}s,\,\frac{1-2s}{2},\,\frac{1-2s}{2},\,\frac{1-2s}{2}\\ 1-s,\,1-s\end{matrix}\right]$ $[0<\mathrm{Re}\,s<1/2;\ \lvert\arg a\rvert<\pi]$

No.	$f(x)$	$F(s)$
9	$\mathbf{K}\left(i\sqrt{\frac{2\sqrt{a}\,(\sqrt{a}-\sqrt{x+a})}{x}}\right) \times \mathbf{K}\left(i\sqrt{\frac{2\sqrt{a}\,(\sqrt{a}+\sqrt{x+a})}{x}}\right)$	$\frac{\sqrt{\pi}\,a^s}{4}\,\Gamma\left[\begin{matrix}-s,\ \frac{2s+1}{2},\ \frac{2s+1}{2},\ \frac{2s+1}{2}\\ s+1,\ s+1\end{matrix}\right]$ $[-1/2<\operatorname{Re}s<0;\ \lvert\arg a\rvert<\pi]$
10	$\mathbf{K}\left(\sqrt{1-\frac{(\sqrt{x+a}-\sqrt{a})^2}{x}}\right) \times \mathbf{K}\left(\sqrt{1-\frac{(\sqrt{x+a}+\sqrt{a})^2}{x}}\right)$	$\frac{\sqrt{\pi}\,a^s}{4}\,\Gamma\left[\begin{matrix}-s,\ \frac{2s+1}{2},\ \frac{2s+1}{2},\ \frac{2s+1}{2}\\ s+1,\ s+1\end{matrix}\right]$ $[-1/2<\operatorname{Re}s<0;\ \lvert\arg a\rvert<\pi]$
11	$\mathbf{K}\left(\sqrt{1-\frac{(\sqrt{x+a}-\sqrt{x})^2}{a}}\right) \times \mathbf{K}\left(\sqrt{1-\frac{(\sqrt{x+a}+\sqrt{x})^2}{a}}\right)$	$\frac{\sqrt{\pi}\,a^s}{4}\,\Gamma\left[\begin{matrix}s,\ \frac{1-2s}{2},\ \frac{1-2s}{2},\ \frac{1-2s}{2}\\ 1-s,\ 1-s\end{matrix}\right]$ $[0<\operatorname{Re}s<1/2;\ \lvert\arg a\rvert<\pi]$

3.26.25. Products of $\mathbf{K}(\varphi(x))$ and $\mathbf{E}(\varphi(x))$

Notation: $\varepsilon = 0$ or 1.

1	$\theta(a-x)\,\mathbf{E}\left(\sqrt{1-\frac{x}{a}}\right)\mathbf{K}(bx)$	$\frac{\pi^2 a^s}{4}\,\Gamma\left[\begin{matrix}s,\ s+1\\ \frac{2s+1}{2},\ \frac{2s+3}{2}\end{matrix}\right] \times {}_6F_5\left(\begin{matrix}\frac{1}{2},\ \frac{1}{2},\ \frac{s}{2},\ \frac{s+1}{2},\ \frac{s+1}{2},\ \frac{s+2}{2};\ a^2b^2\\ 1,\ \frac{2s+1}{4},\ \frac{2s+3}{4},\ \frac{2s+3}{4},\ \frac{2s+5}{4}\end{matrix}\right)$ $[a,\ \operatorname{Re}s>0]$
2	$\theta(a-x)\,\mathbf{E}\left(\sqrt{1-\frac{x}{a}}\right)\mathbf{K}(b\sqrt{x})$	$\frac{\pi^2 a^s}{4}\,\Gamma\left[\begin{matrix}s,\ s+1\\ \frac{2s+1}{2},\ \frac{2s+3}{2}\end{matrix}\right] {}_4F_3\left(\begin{matrix}\frac{1}{2},\ \frac{1}{2},\ s,\ s+1;\ ab^2\\ 1,\ \frac{2s+1}{2},\ \frac{2s+3}{2}\end{matrix}\right)$ $[a,\ \operatorname{Re}s>0]$
3	$\theta(a-x)\,\mathbf{E}(bx)\,\mathbf{K}\left(\sqrt{1-\frac{x}{a}}\right)$	$\frac{\pi^2 a^s}{4}\,\Gamma\left[\begin{matrix}s,\ s\\ \frac{2s+1}{2},\ \frac{2s+1}{2}\end{matrix}\right] \times {}_6F_5\left(\begin{matrix}-\frac{1}{2},\ \frac{1}{2},\ \frac{s}{2},\ \frac{s}{2},\ \frac{s+1}{2},\ \frac{s+1}{2};\ a^2b^2\\ 1,\ \frac{2s+1}{4},\ \frac{2s+1}{4},\ \frac{2s+3}{4},\ \frac{2s+3}{4}\end{matrix}\right)$ $[a,\ \operatorname{Re}s>0]$
4	$\frac{\theta(a-x)}{(1-b^2x)^{\varepsilon}}\,\mathbf{E}(b\sqrt{x})\,\mathbf{K}\left(\sqrt{1-\frac{x}{a}}\right)$	$\frac{\pi^2 a^s}{4}\,\Gamma\left[\begin{matrix}s,\ s\\ \frac{2s+1}{2},\ \frac{2s+1}{2}\end{matrix}\right] {}_4F_3\left(\begin{matrix}\frac{2\varepsilon-1}{2},\ \frac{2\varepsilon+1}{2},\ s,\ s;\ ab^2\\ 1,\ \frac{2s+1}{2},\ \frac{2s+1}{2}\end{matrix}\right)$ $[a,\ \operatorname{Re}s>0]$

No.	$f(x)$	$F(s)$
5	$\theta(a-x)\sqrt{1+b^2x}\,\mathbf{E}\left(\frac{b\sqrt{x}}{\sqrt{1+b^2x}}\right) \times \mathbf{K}\left(\sqrt{1-\frac{x}{a}}\right)$	$\frac{\pi^2 a^s}{4}\Gamma\left[\begin{matrix} s,\ s \\ \frac{2s+1}{2},\ \frac{2s+1}{2}\end{matrix}\right] {}_4F_3\left(\begin{matrix} -\frac{1}{2},\ \frac{1}{2},\ s,\ s;\ -ab^2 \\ 1,\ \frac{2s+1}{2},\ \frac{2s+1}{2}\end{matrix}\right)$ $[a,\ \operatorname{Re} s>0;\ \lvert\arg(1+ab^2)\rvert<\pi]$
6	$\frac{\theta(a-x)}{\sqrt{1+b^2x}}\mathbf{E}\left(\sqrt{1-\frac{x}{a}}\right) \times \mathbf{K}\left(\frac{b\sqrt{x}}{\sqrt{1+b^2x}}\right)$	$\frac{\pi^2 a^s}{4}\Gamma\left[\begin{matrix} s,\ s+1 \\ \frac{2s+1}{2},\ \frac{2s+3}{2}\end{matrix}\right] {}_4F_3\left(\begin{matrix} \frac{1}{2},\ \frac{1}{2},\ s,\ s+1;\ -ab^2 \\ 1,\ \frac{2s+1}{2},\ \frac{2s+3}{2}\end{matrix}\right)$ $[a,\ \operatorname{Re} s>0;\ \lvert\arg(1+ab^2)\rvert<\pi]$

3.26.26. Products of $\mathbf{E}(\varphi(x))$

Notation: $\varepsilon = 0$ or 1.

1	$\frac{\theta(a-x)}{(1-b^2x^2)^{\varepsilon}}\mathbf{E}(bx)\,\mathbf{E}\left(\sqrt{1-\frac{x}{a}}\right)$	$\frac{\pi^2 a^s}{4}\Gamma\left[\begin{matrix} s,\ s+1 \\ \frac{2s+1}{2},\ \frac{2s+3}{2}\end{matrix}\right] \times {}_6F_5\left(\begin{matrix} \frac{2\varepsilon-1}{2},\ \frac{2\varepsilon+1}{2},\ \frac{s}{2},\ \frac{s+1}{2},\ \frac{s+1}{2},\ \frac{s+2}{2};\ a^2b^2 \\ 1,\ \frac{2s+1}{4},\ \frac{2s+3}{4},\ \frac{2s+3}{4},\ \frac{2s+5}{4}\end{matrix}\right)$ $[a,\ \operatorname{Re} s>0]$
2	$\frac{\theta(a-x)}{(1-b^2x)^{\varepsilon}}\mathbf{E}(b\sqrt{x})\,\mathbf{E}\left(\sqrt{1-\frac{x}{a}}\right)$	$\frac{\pi^2 a^s}{4}\Gamma\left[\begin{matrix} s,\ s+1 \\ \frac{2s+1}{2},\ \frac{2s+3}{2}\end{matrix}\right] {}_4F_3\left(\begin{matrix} \frac{2\varepsilon-1}{2},\ \frac{2\varepsilon+1}{2},\ s,\ s+1;\ ab^2 \\ 1,\ \frac{2s+1}{2},\ \frac{2s+3}{2}\end{matrix}\right)$ $[a,\ \operatorname{Re} s>0]$
3	$\theta(a-x)\sqrt{1+b^2x}\,\mathbf{E}\left(\frac{b\sqrt{x}}{\sqrt{1+b^2x}}\right) \times \mathbf{E}\left(\sqrt{1-\frac{x}{a}}\right)$	$\frac{\pi^2 a^s}{4}\Gamma\left[\begin{matrix} s,\ s+1 \\ \frac{2s+1}{2},\ \frac{2s+3}{2}\end{matrix}\right] {}_4F_3\left(\begin{matrix} -\frac{1}{2},\ \frac{1}{2},\ s,\ s+1;\ -ab^2 \\ 1,\ \frac{2s+1}{2},\ \frac{2s+3}{2}\end{matrix}\right)$ $[a,\ \operatorname{Re} s>0;\ \lvert\arg(1+ab^2)\rvert<\pi]$

3.26.27. Products containing $\mathbf{D}(\varphi(x))$

1	$\theta(a-x)\,\mathbf{K}\left(\sqrt{1-\frac{x}{a}}\right)\mathbf{D}(b\sqrt{x})$	$\frac{\pi^2 a^s}{8}\Gamma\left[\begin{matrix} s,\ s \\ \frac{2s+1}{2},\ \frac{2s+1}{2}\end{matrix}\right] {}_4F_3\left(\begin{matrix} \frac{1}{2},\ \frac{3}{2},\ s,\ s;\ ab^2 \\ 2,\ \frac{2s+1}{2},\ \frac{2s+1}{2}\end{matrix}\right)$ $[a,\ \operatorname{Re} s>0]$
2	$\theta(a-x)\,\mathbf{E}\left(\sqrt{1-\frac{x}{a}}\right)\mathbf{D}(b\sqrt{x})$	$\frac{\pi^2 a^s}{8}\Gamma\left[\begin{matrix} s,\ s+1 \\ \frac{2s+1}{2},\ \frac{2s+3}{2}\end{matrix}\right] {}_4F_3\left(\begin{matrix} \frac{1}{2},\ \frac{3}{2},\ s,\ s+1;\ ab^2 \\ 2,\ \frac{2s+1}{2},\ \frac{2s+3}{2}\end{matrix}\right)$ $[a,\ \operatorname{Re} s>0]$

3.27. The Hypergeometric Function ${}_0F_1(b;\ z)$

More formulas can be obtained from the corresponding sections due to the relations

$$ {}_0F_1(b;\ \ z) = \Gamma(b)\, z^{(1-b)/2} J_{b-1}\left(2\sqrt{z}\right), \quad {}_0F_1(b;\ z) = \Gamma(b)\, z^{(1-b)/2} I_{b-1}\left(2\sqrt{z}\right), $$

$$ {}_0F_1(b;\ z) = \lim_{a\to\infty} {}_1F_1\left(a;\ b;\ \frac{z}{a}\right), \quad {}_0F_1(b;\ z) = e^{-2\sqrt{z}}\, {}_1F_1\left(b-\frac{1}{2};\ 2b-1;\ 4\sqrt{z}\right), $$

$$ {}_0F_1(b;\ z) = \Gamma(b)\, G_{02}^{10}\left(-z \middle| \begin{matrix} \cdot \\ 0,\ 1-b \end{matrix}\right). $$

3.27.1. ${}_0F_1(b;\ \omega x)$ and the exponential function

No.	$f(x)$	$F(s)$
1	$e^{2\sqrt{\omega x}}\, {}_0F_1(b;\ \omega x)$	$\frac{2^{2b-4s-1}\left(-\sqrt{\omega}\right)^{-2s}}{\sqrt{\pi}}\, \Gamma\left[\begin{matrix} b,\ \frac{2b-4s-1}{2},\ 2s \\ 2b-2s-1 \end{matrix}\right]$ $[\omega < 0;\ 0 < \operatorname{Re} s < (2\operatorname{Re} b - 1)/4]$
2	$e^{-2\sqrt{\omega x}}\, {}_0F_1(b;\ \omega x)$	$\frac{2^{2b-4s-1}\omega^{-s}}{\sqrt{\pi}}\, \Gamma\left[\begin{matrix} b,\ \frac{2b-4s-1}{2},\ 2s \\ 2b-2s-1 \end{matrix}\right]$ $[\omega < 0;\ 0 < \operatorname{Re} s < (2\operatorname{Re} b - 1)/4]$

3.27.2. ${}_0F_1(b;\ \omega x)$ and trigonometric functions

1	$\begin{Bmatrix} \sin\left(2\sqrt{\omega x}+\sigma\right) \\ \cos\left(2\sqrt{\omega x}+\sigma\right) \end{Bmatrix} \times {}_0F_1(b;\ -\omega x)$	$\frac{2^{2b-4s-1}\omega^{-s}}{\sqrt{\pi}} \begin{Bmatrix} \sin(s\pi+\sigma) \\ \cos(s\pi+\sigma) \end{Bmatrix} \Gamma\left[\begin{matrix} b,\ \frac{2b-4s-1}{2},\ 2s \\ 2b-2s-1 \end{matrix}\right]$ $[\omega > 0;\ 0 < \operatorname{Re} s < (2\operatorname{Re} b - 1)/4]$

3.27.3. ${}_0F_1(b;\ \omega x)$ and $\operatorname{sinc}\left(\sqrt{ax}\right)$

1	$\operatorname{sinc}\left(2\sqrt{\omega x}\right) {}_0F_1(b;\ -\omega x)$	$-\frac{2^{2b-4s}\omega^{-s}}{\sqrt{\pi}} \cos(s\pi)\, \Gamma\left[\begin{matrix} b,\ \frac{1-4s+2b}{2},\ 2s-1 \\ 2b-2s \end{matrix}\right]$ $[\omega > 0;\ 0 < \operatorname{Re} s < (2\operatorname{Re} b + 1)/4]$

3.27.4. ${}_0F_1(b;\ \omega x)$ and the Bessel functions

1	$J_\nu\left(2\sqrt{\omega x}\right) {}_0F_1(b;\ -\omega x)$	$\omega^{-s}\, \Gamma\left[\begin{matrix} b,\ b-2s,\ \frac{2s+\nu}{2} \\ \frac{\nu-2s+2}{2},\ \frac{2b-2s-\nu}{2},\ \frac{2b-2s+\nu}{2} \end{matrix}\right]$ $[\omega > 0;\ -\operatorname{Re}\nu/2 < \operatorname{Re} s < \operatorname{Re} b/2]$

No.	$f(x)$	$F(s)$
2	$J_{-b-n}\left(2\sqrt{\omega x}\right)\,{}_0F_1(b;\,-\omega x) - \frac{(-1)^{[(n+1)/2]}}{2^n\sqrt{\pi}}(\omega x)^{-(b+n)/2} \times \Gamma(b)\sum_{k=0}^{[n/2]}\frac{\left(\left[\frac{n}{2}\right]-k+1\right)_{n-[n/2]}}{k!} \times \Gamma\begin{bmatrix} k-n+\left[\frac{n}{2}\right]+\frac{1}{2} \\ b+k,\ k-n-b+1 \end{bmatrix}(-4\omega x)^k$	$\frac{(-1)^{n+1}}{\omega^s}\Gamma\begin{bmatrix} b,\ b-2s,\ \frac{2s-b+n+2}{2} \\ \frac{b+n-2s+2}{2},\ \frac{3b+n-2s}{2},\ \frac{2-b-n-2s}{2} \end{bmatrix}$ $[\omega>0;\ (\operatorname{Re} b-n)/2-1<\operatorname{Re} s<\operatorname{Re} b/2]$
3	$Y_\nu\left(2\sqrt{\omega x}\right)\,{}_0F_1(b;\,-\omega x)$	$-\frac{\omega^{-s}}{\pi}\cos\frac{(2s-\nu)\pi}{2}\,\Gamma\begin{bmatrix} b,\ b-2s,\ \frac{2s-\nu}{2},\ \frac{2s+\nu}{2} \\ \frac{2b-2s-\nu}{2},\ \frac{2b-2s+\nu}{2} \end{bmatrix}$ $[\omega>0;\ \lvert\operatorname{Re}\nu\rvert/2<\operatorname{Re} s<\operatorname{Re} b/2]$
4	$Y_{-b-1}\left(2\sqrt{\omega x}\right)\,{}_0F_1(b;\,-\omega x) + \frac{\cot(b\pi)}{\Gamma(-b)}(\omega x)^{-(b+1)/2}$	$\frac{2^{b-2s}\omega^{-s}}{\pi^{3/2}(b-2s+1)}\sin\frac{(2s+b)\pi}{2} \times\Gamma\begin{bmatrix} b,\ \frac{b-2s}{2},\ \frac{2s-b+3}{2},\ \frac{2s+b+1}{2} \\ \frac{3b-2s+1}{2} \end{bmatrix}$ $\begin{bmatrix} \omega>0;\ (\operatorname{Re} b-3)/2,\ -(\operatorname{Re} b+1)/2 \\ <\operatorname{Re} s<\operatorname{Re} b/2 \end{bmatrix}$
5	$Y_{-b}\left(2\sqrt{\omega x}\right)\,{}_0F_1(b;\,-\omega x) + \frac{\cos(b\pi)}{2\pi}\Gamma(b)(\omega x)^{-b/2}$	$-\frac{2^{b-2s-1}\omega^{-s}}{\pi^{3/2}}\cos\frac{(2s+b)\pi}{2}\,\Gamma\begin{bmatrix} b,\ \frac{b-2s+1}{2},\ \frac{2s-b}{2},\ \frac{2s+b}{2} \\ \frac{3b-2s}{2} \end{bmatrix}$ $[\omega>0;\ \lvert\operatorname{Re} b\rvert/2<\operatorname{Re} s<(\operatorname{Re} b+1)/2]$
6	$Y_b\left(2\sqrt{\omega x}\right)\,{}_0F_1(b;\,-\omega x) + \frac{\Gamma(b)}{\pi}(\omega x)^{-b/2}$	$-\frac{2^{b-2s-1}\omega^{-s}}{\sqrt{\pi}}\Gamma\begin{bmatrix} b,\ \frac{2s-b}{2},\ \frac{2s+b}{2} \\ \frac{3b-2s}{2},\ \frac{2s-b+1}{2} \end{bmatrix}$ $\begin{bmatrix} \omega>0;\ (\operatorname{Re} b-2)/2,\ -\operatorname{Re} b/2 \\ <\operatorname{Re} s<\operatorname{Re} b/2 \end{bmatrix}$
7	$Y_{b+1}\left(2\sqrt{\omega x}\right)\,{}_0F_1(b;\,-\omega x) + \frac{\Gamma(b+1)}{\pi}(\omega x)^{-(b+1)/2}$	$\frac{2^{b-2s}\omega^{-s}}{\sqrt{\pi}(b-2s+1)}\Gamma\begin{bmatrix} b,\ \frac{2s+b+1}{2},\ \frac{2s-b+3}{2} \\ \frac{2s-b+2}{2},\ \frac{3b-2s+1}{2} \end{bmatrix}$ $\begin{bmatrix} \omega>0;\ (\operatorname{Re} b-3)/2,\ -(\operatorname{Re} b+1)/2 \\ <\operatorname{Re} s<(\operatorname{Re} b+1)/2 \end{bmatrix}$
8	$Y_{\pm b\pm n}\left(2\sqrt{\omega x}\right)\,{}_0F_1(b;\,-\omega x) + \frac{(\mp 1)^n}{2^n\sqrt{\pi}}(\omega x)^{-(b+n)/2}\begin{Bmatrix} \csc(b\pi) \\ \cot(b\pi) \end{Bmatrix} \times\Gamma(b)\sum_{k=0}^{[n/2]}(-1)^{k+[(n+1)/2]} \times\frac{\left(\left[\frac{n}{2}\right]-k+1\right)_{n-[n/2]}}{k!} \times\Gamma\begin{bmatrix} k-n+\left[\frac{n}{2}\right]+\frac{1}{2} \\ b+k,\ k-n-b+1 \end{bmatrix}(4\omega x)^k$	$(\pm 1)^n\frac{\omega^{-s}}{\pi}\cos\frac{(2s\mp b+n)\pi}{2} \times\Gamma\begin{bmatrix} b,\ b-2s,\ \frac{2s-b+n+2}{2},\ \frac{2s+b+n}{2} \\ \frac{b+n-2s+2}{2},\ \frac{3b+n-2s}{2} \end{bmatrix}$ $\begin{bmatrix} \omega>0;\ (\operatorname{Re} b-n)/2-1,\ -(\operatorname{Re} b+n)/2 \\ <\operatorname{Re} s<\operatorname{Re} b/2 \end{bmatrix}$

No.	$f(x)$	$F(s)$
9	$\left\{\begin{matrix} H_\nu^{(1)}(2\sqrt{\omega x}) \\ H_\nu^{(2)}(2\sqrt{\omega x}) \end{matrix}\right\} {}_0F_1(b;\, -\omega x)$	$\dfrac{\omega^{-s}}{\pi}\Gamma\left[\begin{matrix} b,\, b-2s,\, \frac{2s+\nu}{2} \\ \frac{\nu-2s+2}{2},\, \frac{2b-2s-\nu}{2},\, \frac{2b-2s+\nu}{2} \end{matrix}\right] \times \left[\pi \mp i\cos\dfrac{(2s-\nu)\pi}{2}\Gamma\left(s-\dfrac{\nu}{2}\right)\Gamma\left(1-s+\dfrac{\nu}{2}\right)\right]$ $[\omega>0;\ \lvert\operatorname{Re}\nu\rvert/2<\operatorname{Re}s<\operatorname{Re}b/2]$
10	$K_{b-1}(2\sqrt{\omega x})\, {}_0F_1(b;\, -\omega x)$	$\dfrac{\omega^{-s}}{4}\Gamma\left[\begin{matrix} b,\, \frac{2s-b+1}{2},\, \frac{2s+b-1}{4} \\ \frac{3b-2s+1}{4} \end{matrix}\right]$ $[\operatorname{Re}s>\lvert 1-\operatorname{Re}b\rvert/2]$
11	$K_\nu(2\sqrt{\omega x})\, {}_0F_1(b;\, \omega x)$	$\dfrac{\omega^{-s}}{2}\Gamma\left[\begin{matrix} b,\, b-2s,\, \frac{2s-\nu}{2},\, \frac{2s+\nu}{2} \\ \frac{2b-2s-\nu}{2},\, \frac{2b-2s+\nu}{2} \end{matrix}\right]$ $[\lvert\operatorname{Re}\nu\rvert/2<\operatorname{Re}s<\operatorname{Re}b/2]$

3.27.5. ${}_0F_1(b;\, \omega x)$ and $\operatorname{ker}_\nu(\sqrt{ax})$, $\operatorname{kei}_\nu(\sqrt{ax})$

1	$\operatorname{ker}_{\pm b\mp 1}(2\sqrt{\omega x})\, {}_0F_1(b;\, i\omega x)$	$\pm\dfrac{i\omega^{-s}}{8}\left(\dfrac{2^{b-2s}}{\sqrt{\pi}}e^{i(\mp b-s)\pi/2}\Gamma\left[\begin{matrix} b,\, \frac{b-2s}{2},\, \frac{2s+b-1}{2},\, \frac{2s-b+1}{2} \\ \frac{3b-2s-1}{2} \end{matrix}\right] - e^{i(\pm b+s)\pi/2}\Gamma\left[\begin{matrix} b,\, \frac{2s+b-1}{4},\, \frac{2s-b+1}{2} \\ \frac{3b-2s+1}{4} \end{matrix}\right]\right)$ $[\omega>0;\ \lvert\operatorname{Re}b-1\rvert/2<\operatorname{Re}s<\operatorname{Re}b/2]$
2	$\operatorname{kei}_{\pm b\mp 1}(2\sqrt{\omega x})\, {}_0F_1(b;\, i\omega x)$	$\pm\dfrac{\omega^{-s}}{8}\left(\dfrac{2^{b-2s}}{\sqrt{\pi}}e^{i(\mp b-s)\pi/2}\Gamma\left[\begin{matrix} b,\, \frac{b-2s}{2},\, \frac{2s+b-1}{2},\, \frac{2s-b+1}{2} \\ \frac{3b-2s-1}{2} \end{matrix}\right] + e^{i(\pm b+s)\pi/2}\Gamma\left[\begin{matrix} b,\, \frac{2s+b-1}{4},\, \frac{2s-b+1}{2} \\ \frac{3b-2s+1}{4} \end{matrix}\right]\right)$ $[\omega>0;\ \lvert\operatorname{Re}b-1\rvert/2<\operatorname{Re}s<\operatorname{Re}b/2]$

3.27.6. ${}_0F_1(b;\, \omega x)$ and $\operatorname{Ai}(\sqrt[3]{ax})$, $\operatorname{Ai}'(\sqrt[3]{ax})$

1	$\operatorname{Ai}\left(3^{2/3}\sqrt[3]{\omega x}\right) {}_0F_1(b;\, \omega x)$	$\dfrac{3^{-1/6}\omega^{-s}}{2\pi}\Gamma\left[\begin{matrix} b,\, s,\, \frac{3s+1}{3},\, \frac{3b-6s-1}{3} \\ b-s,\, \frac{3b-3s-1}{3} \end{matrix}\right]$ $[0<\operatorname{Re}s<(3\operatorname{Re}b-1)/6;\ \lvert\arg\omega\rvert<\pi]$
2	$\operatorname{Ai}'\left(3^{2/3}\sqrt[3]{\omega x}\right) {}_0F_1(b;\, \omega x)$	$-\dfrac{3^{1/6}\omega^{-s}}{2\pi}\Gamma\left[\begin{matrix} b,\, s,\, \frac{3s+2}{3},\, \frac{3b-6s-2}{3} \\ b-s,\, \frac{3b-3s-2}{3} \end{matrix}\right]$ $[0<\operatorname{Re}s<(3\operatorname{Re}b-2)/6;\ \lvert\arg\omega\rvert<\pi]$

3.28. The Kummer Confluent Hypergeometric Function ${}_1F_1(a;\, b;\, z)$

More formulas can be obtained from the corresponding sections due to the relations

$$ {}_1F_1(a;\, b;\, z) = \lim_{\lambda\to\infty} {}_2F_1\left(a,\, \lambda;\, b;\, \frac{z}{\lambda}\right), \quad {}_1F_1(a;\, b;\, z) = \frac{\Gamma(b)}{\Gamma(a)} G_{12}^{11}\left(-z \,\middle|\, \begin{matrix} 1-a \\ 0,\, 1-b \end{matrix}\right). $$

3.28.1. ${}_1F_1(a;\, b;\, \omega x)$ and algebraic functions

No.	$f(x)$	$F(s)$
1	${}_1F_1\left(\begin{matrix} a;\, -\omega x \\ b \end{matrix}\right)$	$\omega^{-s}\Gamma\left[\begin{matrix} b,\, s,\, a-s \\ a,\, b-s \end{matrix}\right]$ $[0 < \operatorname{Re} s < \operatorname{Re} a;\ \operatorname{Re}\omega > 0]$
2	$(\sigma - x)_+^{\alpha-1}\, {}_1F_1\left(\begin{matrix} a;\, -\omega x \\ b \end{matrix}\right)$	$\sigma^{s+\alpha-1}\, \mathrm{B}(s, \alpha)\, {}_2F_2\left(\begin{matrix} a,\, s;\, -\sigma\omega \\ b,\, s+\alpha \end{matrix}\right)$ $[\sigma,\ \operatorname{Re}\alpha,\ \operatorname{Re} s > 0]$
3	$(x - \sigma)_+^{\alpha-1}\, {}_1F_1\left(\begin{matrix} a;\, -\omega x \\ b \end{matrix}\right)$	$\omega^{1-s-\alpha}\Gamma\left[\begin{matrix} b \\ b-s-\alpha+1 \end{matrix}\right] \mathrm{B}(a-s-\alpha+1,\, s+\alpha-1)$ $\times\, {}_2F_2\left(\begin{matrix} 1-\alpha,\, a-s-\alpha+1;\, -\sigma\omega \\ 1-s-\alpha+2,\, b-s-\alpha \end{matrix}\right)$ $+\, \sigma^{s+\alpha-1}\, \mathrm{B}(1-s-\alpha,\, \alpha)\, {}_2F_2\left(\begin{matrix} a,\, s;\, -\sigma\omega \\ b,\, s+\alpha \end{matrix}\right)$ $\left[\begin{matrix} \sigma,\ \operatorname{Re}\alpha > 0;\ (\operatorname{Re}\omega > 0;\ \operatorname{Re}(s-a+\alpha) < 1) \text{ or} \\ (\operatorname{Re}\omega = 0;\ \operatorname{Re} s < \operatorname{Re}(b-a-\alpha)+2) \end{matrix}\right]$
4	$\dfrac{1}{(x+\sigma)^{\rho}}\, {}_1F_1\left(\begin{matrix} a;\, -\omega x \\ b \end{matrix}\right)$	$\sigma^{s-\rho}\, \mathrm{B}(s,\, \rho-s)\, {}_2F_2\left(\begin{matrix} a,\, s;\, \sigma\omega \\ b,\, s-\rho+1 \end{matrix}\right)$ $+\, \omega^{\rho-s}\, \mathrm{B}(s-\rho,\, a-s+\rho)\, \Gamma\left[\begin{matrix} b \\ b-s+\rho \end{matrix}\right]$ $\times\, {}_2F_2\left(\begin{matrix} \rho,\, a-s+\rho;\, \sigma\omega \\ 1-s+\rho,\, b-s+\rho \end{matrix}\right)$ $\left[\begin{matrix} (\operatorname{Re}\omega > 0;\, 0 < \operatorname{Re} s < \operatorname{Re}(\rho+a)) \text{ or} \\ (\operatorname{Re}\omega = 0;\, 0 < \operatorname{Re} s < \operatorname{Re}(b-a+\rho)+1);\ \lvert\arg\sigma\rvert < \pi \end{matrix}\right]$
5	$\dfrac{1}{x-\sigma}\, {}_1F_1\left(\begin{matrix} a;\, -\omega x \\ b \end{matrix}\right)$	$\omega^{1-s}\, \mathrm{B}(a-s+1,\, s-1)\, \Gamma\left[\begin{matrix} b \\ b-s+1 \end{matrix}\right]$ $\times\, {}_2F_2\left(\begin{matrix} 1,\, a-s+1;\, -\sigma\omega \\ 2-s,\, b-s+1 \end{matrix}\right)$ $-\, \pi\sigma^{s-1}\cot(s\pi)\, {}_1F_1\left(\begin{matrix} a;\, -\sigma\omega \\ b \end{matrix}\right)$ $\left[\begin{matrix} \sigma > 0;\ (\operatorname{Re}\omega > 0;\, 0 < \operatorname{Re} s < \operatorname{Re} a + 1) \text{ or} \\ (\operatorname{Re}\omega = 0;\, 0 < \operatorname{Re} s < \operatorname{Re}(b-a)+2) \end{matrix}\right]$

No.	$f(x)$	$F(s)$
6	$\left(\sqrt{x}+\sqrt{x+\sigma}\right)^{\rho}$ $\times\, {}_1F_1\left(\begin{matrix} a;\, -\omega x \\ b \end{matrix}\right)$	$\frac{2^{\rho}}{\omega^{s+\rho/2}}\, \mathrm{B}\left(\frac{2b-2a-2s-\rho}{2},\, \frac{2s+\rho}{2}\right)$ $\times\, \Gamma\left[\begin{matrix} b \\ \frac{2b-2s-\rho}{2} \end{matrix}\right] {}_3F_3\left(\begin{matrix} -\frac{\rho}{2},\, \frac{1-\rho}{2},\, \frac{2a-2s-\rho}{2};\, \sigma\omega \\ 1-\rho,\, \frac{2-2s-\rho}{2},\, \frac{2b-2s-\rho}{2} \end{matrix}\right)$ $-\frac{\sigma^{s+\rho/2}\rho}{2^{2s}}\, \Gamma\left[\begin{matrix} 2s,\, \frac{-2s-\rho}{2} \\ \frac{2s-\rho+2}{2} \end{matrix}\right] {}_3F_3\left(\begin{matrix} a,\, s,\, \frac{2s+1}{2};\, \sigma\omega \\ b,\, \frac{2s-\rho+2}{2},\, \frac{2s+\rho+2}{2} \end{matrix}\right)$ $\left[\begin{matrix} (\mathrm{Re}\,\omega > 0;\, 0 < \mathrm{Re}\, s < \mathrm{Re}\,(a-\rho/2)) \text{ or} \\ (\mathrm{Re}\,\omega = 0;\, 0 < \mathrm{Re}\, s < \mathrm{Re}\,(b-a-\rho/2)+1);\, \lvert\arg\sigma\rvert < \pi \end{matrix}\right]$
7	$\frac{\left(\sqrt{x}+\sqrt{x+\sigma}\right)^{\rho}}{\sqrt{x+\sigma}}$ $\times\, {}_1F_1\left(\begin{matrix} a;\, -\omega x \\ b \end{matrix}\right)$	$\frac{2^{\rho}}{\omega^{s+(\rho-1)/2}}\, \mathrm{B}\left(\frac{2a-2s-\rho+1}{2},\, \frac{2s+\rho-1}{2}\right)$ $\times\, \Gamma\left[\begin{matrix} b \\ \frac{2b-2s-\rho+1}{2} \end{matrix}\right] {}_3F_3\left(\begin{matrix} \frac{1-\rho}{2},\, \frac{2-\rho}{2},\, \frac{2a-2s-\rho+1}{2};\, \sigma\omega \\ 1-\rho,\, \frac{3-2s-\rho}{2},\, \frac{2b-2s-\rho+1}{2} \end{matrix}\right)$ $+\frac{\sigma^{s+(\rho-1)/2}}{2^{2s-1}}\, \mathrm{B}\left(2s,\, \frac{1-2s-\rho}{2}\right) {}_3F_3\left(\begin{matrix} a,\, s,\, \frac{2s+1}{2};\, \sigma\omega \\ b,\, \frac{2s-\rho+1}{2},\, \frac{2s+\rho+1}{2} \end{matrix}\right)$ $\left[\begin{matrix} (\mathrm{Re}\,\omega > 0;\, 0 < \mathrm{Re}\, s < \mathrm{Re}\,(a+(1-\rho)/2)) \text{ or} \\ (\mathrm{Re}\,\omega = 0;\, 0 < \mathrm{Re}\, s < \mathrm{Re}\,(b-a+(3-\rho)/2));\, \lvert\arg\sigma\rvert < \pi \end{matrix}\right]$
8	$\theta(x-\sigma)(x-\tau)^{-a}$ $\times\, {}_1F_1\left(a;\, b;\, \frac{\omega x}{x-\tau}\right)$	$\frac{\sigma^{s-a}}{a-s}\, \Psi_1\left(a,\, a-s;\, a-s+1,\, b;\, \frac{\tau}{\sigma},\, \omega\right)$ $[\sigma > 0;\, \lvert\tau\rvert < \sigma;\, 0 < \mathrm{Re}\, s < \mathrm{Re}\, a]$
9	$(\sigma-x)_+^{b-1}(\tau-x)^{-a}$ $\times\, {}_1F_1\left(a;\, b;\, \frac{\omega(\sigma-x)}{\tau-x}\right)$	$\tau^{-a}\sigma^{s+b-1}\, \mathrm{B}(s,\, b)\, \Phi_1\left(a,\, s,\, s+b,\, \frac{\sigma}{\tau},\, \frac{\sigma\omega}{\tau}\right)$ $[\tau > \sigma > 0;\, \mathrm{Re}\, b,\, \mathrm{Re}\, s > 0]$

3.28.2. ${}_1F_1(a;\, b;\, \omega x)$ and the exponential function

1	$e^{-\sigma x}\, {}_1F_1\left(\begin{matrix} a;\, \omega x \\ b \end{matrix}\right)$	$\frac{\Gamma(s)}{\sigma^{s}}\, {}_2F_1\left(\begin{matrix} a,\, s \\ b;\, \frac{\omega}{\sigma} \end{matrix}\right)$ $\left[\begin{matrix} (\mathrm{Re}\,(\sigma-\omega) > 0,\, \mathrm{Re}\,\sigma > 0;\, \mathrm{Re}\, s > 0) \text{ or} \\ (\mathrm{Re}\,(\sigma-\omega) = 0,\, \mathrm{Re}\,\sigma > 0;\, 0 < \mathrm{Re}\, s < \mathrm{Re}\,(b-a)+1) \text{ or} \\ (\mathrm{Re}\,(\sigma-\omega) > 0,\, \mathrm{Re}\,\sigma = 0;\, 0 < \mathrm{Re}\, s < \mathrm{Re}\, a+1) \text{ or} \\ (\mathrm{Re}\,(\sigma-\omega) = 0,\, \mathrm{Re}\,\sigma = 0;\, 0 < \mathrm{Re}\, s < \mathrm{Re}\, a+1,\, \mathrm{Re}\,(b-a)+1) \end{matrix}\right]$
2	$e^{-\omega x}\, {}_1F_1\left(\begin{matrix} a;\, \omega x \\ b \end{matrix}\right)$	$\omega^{-s}\, \Gamma\left[\begin{matrix} s,\, b-a-s,\, b \\ b-s,\, b-a \end{matrix}\right]$ $\left[\begin{matrix} (\mathrm{Re}\,\omega > 0;\, 0 < \mathrm{Re}\, s < \mathrm{Re}\,(b-a)) \text{ or} \\ (\mathrm{Re}\,\omega = 0;\, 0 < \mathrm{Re}\, s < \mathrm{Re}\, a+1,\, \mathrm{Re}\,(b-a)) \end{matrix}\right]$

No.	$f(x)$	$F(s)$
3	$(\sigma-x)_+^{\alpha-1}\, e^{-\omega x} \times {}_1F_1\left(\begin{matrix} a;\ \omega x \\ b \end{matrix}\right)$	$\sigma^{s+\alpha-1}\,\mathrm{B}(s,\ \alpha)\ {}_2F_2\left(\begin{matrix} b-a,\ s;\ -\sigma\omega \\ b,\ s+\alpha \end{matrix}\right)$ $[\sigma,\ \operatorname{Re}\alpha,\ \operatorname{Re}s>0]$
4	$(x-\sigma)_+^{\alpha-1}\, e^{-\omega x} \times {}_1F_1\left(\begin{matrix} a;\ \omega x \\ b \end{matrix}\right)$	$\omega^{1-s-\alpha}\,\mathrm{B}(1-a+b-s-\alpha,\ s+\alpha-1)\,\Gamma\left[\begin{matrix} b \\ b-s-\alpha+1 \end{matrix}\right] \times {}_2F_2\left(\begin{matrix} 1-\alpha,\ 1-a+b-\alpha-s;\ -\sigma\omega \\ 2-s-\alpha,\ b-\alpha-s+1 \end{matrix}\right) + \sigma^{s+\alpha-1}\,\mathrm{B}(1-s-\alpha,\ \alpha)\ {}_2F_2\left(\begin{matrix} b-a,\ s;\ -\sigma\omega \\ b,\ s+\alpha \end{matrix}\right)$ $\left[\begin{matrix} \sigma>0;\ (\operatorname{Re}\omega>0;\ \operatorname{Re}s<\operatorname{Re}(b-a-\alpha)+1)\ \text{or} \\ (\operatorname{Re}\omega=0;\ \operatorname{Re}s<\operatorname{Re}(a-\alpha)+2,\ \operatorname{Re}(b-a-\alpha)+1) \end{matrix}\right]$
5	$\dfrac{e^{-\omega x}}{(x+\sigma)^{\rho}}\ {}_1F_1\left(\begin{matrix} a;\ \omega x \\ b \end{matrix}\right)$	$\sigma^{s-\rho}\,\mathrm{B}(s,\ \rho-s)\ {}_2F_2\left(\begin{matrix} b-a,\ s;\ \sigma\omega \\ b,\ s-\rho+1 \end{matrix}\right) + \omega^{\rho-s}\,\mathrm{B}(s-\rho,\ b-a+\rho-s)\,\Gamma\left[\begin{matrix} b \\ b+\rho-s \end{matrix}\right] \times {}_2F_2\left(\begin{matrix} \rho,\ b-a+\rho-s;\ \sigma\omega \\ \rho-s+1,\ b+\rho-s \end{matrix}\right)$ $\left[\begin{matrix} (\operatorname{Re}\omega>0;\ 0<\operatorname{Re}s<\operatorname{Re}(b-a+\rho))\ \text{or} \\ (\operatorname{Re}\omega=0;\ 0<\operatorname{Re}s<\operatorname{Re}(b-a+\rho),\ \operatorname{Re}(a+\rho)+1);\ \lvert\arg\sigma\rvert<\pi \end{matrix}\right]$
6	$\dfrac{e^{-\omega x}}{x-\sigma}\ {}_1F_1\left(\begin{matrix} a;\ \omega x \\ b \end{matrix}\right)$	$\omega^{1-s}\,\mathrm{B}(1-s-a+b,\ s-1)\,\Gamma\left[\begin{matrix} b \\ b-s+1 \end{matrix}\right] \times {}_2F_2\left(\begin{matrix} 1,\ 1-s-a+b \\ 2-s,\ 1-s+b;\ -\sigma\omega \end{matrix}\right) - \pi\sigma^{s-1}\cot(s\pi)\ {}_1F_1\left(\begin{matrix} b-a \\ b;\ -\sigma\omega \end{matrix}\right)$ $\left[\begin{matrix} \sigma>0;\ (\operatorname{Re}\omega>0;\ 0<\operatorname{Re}s<\operatorname{Re}(b-a)+1)\ \text{or} \\ (\operatorname{Re}\omega=0;\ 0<\operatorname{Re}s<\operatorname{Re}a+2,\ \operatorname{Re}(b-a)+1) \end{matrix}\right]$
7	$e^{-\sigma\sqrt{x}}\ {}_1F_1\left(\begin{matrix} a;\ -\omega x \\ b \end{matrix}\right)$	$\dfrac{2\sigma^{2(a-s)}}{\omega^{a}}\,\Gamma\left[\begin{matrix} b,\ 2s-2a \\ b-a \end{matrix}\right]\ {}_2F_2\left(\begin{matrix} a,\ a-b+1;\ \frac{\sigma^2}{4\omega} \\ \frac{2a-2s+1}{2},\ a-s+1 \end{matrix}\right) - \dfrac{\sigma}{\omega^{s+1/2}}\,\mathrm{B}\left(s+\dfrac{1}{2},\ a-s-\dfrac{1}{2}\right) \times\Gamma\left[\begin{matrix} b \\ b-s-\frac{1}{2} \end{matrix}\right]\ {}_2F_2\left(\begin{matrix} \frac{2s+1}{2},\ \frac{2s-2b+3}{2} \\ \frac{3}{2},\ \frac{2s-2a+3}{2};\ \frac{\sigma^2}{4\omega} \end{matrix}\right) + \omega^{-s}\,\mathrm{B}(s,\ a-s)\,\Gamma\left[\begin{matrix} b \\ b-s \end{matrix}\right]\ {}_2F_2\left(\begin{matrix} s,\ s-b+1;\ \frac{\sigma^2}{4\omega} \\ \frac{1}{2},\ s-a+1 \end{matrix}\right)$ $\left[\begin{matrix} (\operatorname{Re}\omega\ge 0;\ \operatorname{Re}\sigma,\ \operatorname{Re}s>0)\ \text{or} \\ (\operatorname{Re}\omega=\operatorname{Re}\sigma=0;\ 0<\operatorname{Re}s<\operatorname{Re}(b-a)+1/2,\ \operatorname{Re}a+1) \end{matrix}\right]$

No.	$f(x)$	$F(s)$
8	$e^{-\sigma\sqrt{x}-\omega x}\,{}_1F_1\left(\begin{matrix}a;\,\omega x\\ b\end{matrix}\right)$	$\frac{2\omega^{a-b}}{\sigma^{2(a-b+s)}}\Gamma\left[\begin{matrix}b,\,2a-2b+2s\\ a\end{matrix}\right]$ $\times\,{}_2F_2\left(\begin{matrix}1-a,\,b-a;\,\frac{\sigma^2}{4\omega}\\ b-a-s+\frac{1}{2},\,b-a-s+1\end{matrix}\right)$ $-\frac{\sigma}{\omega^{s+1/2}}\,\mathrm{B}\left(\frac{2b-2a-2s-1}{2},\,\frac{2s+1}{2}\right)$ $\times\,\Gamma\left[\begin{matrix}b\\ \frac{2b-2s-1}{2}\end{matrix}\right]{}_2F_2\left(\begin{matrix}\frac{2s+1}{2},\,\frac{2s-2b+3}{2}\\ \frac{3}{2},\,\frac{2s+2a-2b+3}{2};\,\frac{\sigma^2}{4\omega}\end{matrix}\right)$ $+\,\omega^{-s}\,\mathrm{B}(b-a-s,\,s)\,\Gamma\left[\begin{matrix}b\\ b-s\end{matrix}\right]{}_2F_2\left(\begin{matrix}s,\,s-b+1;\,\frac{\sigma^2}{4\omega}\\ \frac{1}{2},\,s+a-b+1\end{matrix}\right)$ $\left[\begin{matrix}(\mathrm{Re}\,\omega\geq 0;\ \mathrm{Re}\,\sigma,\ \mathrm{Re}\,s>0)\ \text{or}\\ (\mathrm{Re}\,\omega=\mathrm{Re}\,\sigma=0,\ 0<\mathrm{Re}\,s<\mathrm{Re}\,a+1/2,\ \mathrm{Re}\,(b-a)+1)\end{matrix}\right]$
9	$e^{-\sigma/x}\,{}_1F_1\left(\begin{matrix}a;\,-\omega x\\ b\end{matrix}\right)$	$\omega^{-s}\,\mathrm{B}(a-s,\,s)\,\Gamma\left[\begin{matrix}b\\ b-s\end{matrix}\right]{}_1F_2\left(\begin{matrix}a-s;\,\sigma\omega\\ 1-s,\,b-s\end{matrix}\right)$ $+\,\sigma^s\Gamma(-s)\ {}_1F_2\left(\begin{matrix}a;\,\sigma\omega\\ b,\,s+1\end{matrix}\right)$ $\left[\begin{matrix}(\mathrm{Re}\,\omega>0,\ \mathrm{Re}\,\sigma>0;\ \mathrm{Re}\,s<\mathrm{Re}\,a)\ \text{or}\\ (\mathrm{Re}\,\omega=0,\ \mathrm{Re}\,\sigma>0;\ \mathrm{Re}\,s<\mathrm{Re}\,a,\ \mathrm{Re}\,(b-a)+1)\ \text{or}\\ (\mathrm{Re}\,\omega>0,\ \mathrm{Re}\,\sigma=0;\ -1<\mathrm{Re}\,s<\mathrm{Re}\,a)\ \text{or}\\ (\mathrm{Re}\,\omega=0,\ \mathrm{Re}\,\sigma=0;\ -1<\mathrm{Re}\,s<\mathrm{Re}\,a,\ \mathrm{Re}\,(b-a)+1)\end{matrix}\right]$
10	$e^{-\omega x-\sigma/x}\,{}_1F_1\left(\begin{matrix}a;\,\omega x\\ b\end{matrix}\right)$	$\omega^{-s}\,\mathrm{B}(-a+b-s,\,s)\,\Gamma\left[\begin{matrix}b\\ b-s\end{matrix}\right]{}_1F_2\left(\begin{matrix}b-a-s;\,\sigma\omega\\ 1-s,\,b-s\end{matrix}\right)$ $+\,\sigma^s\Gamma(-s)\ {}_1F_2\left(\begin{matrix}b-a;\,\sigma\omega\\ b,\,s+1\end{matrix}\right)$ $\left[\begin{matrix}(\mathrm{Re}\,\omega>0,\ \mathrm{Re}\,\sigma>0;\ \mathrm{Re}\,s<\mathrm{Re}\,(b-a))\ \text{or}\\ (\mathrm{Re}\,\omega=0,\ \mathrm{Re}\,\sigma>0;\ \mathrm{Re}\,s<\mathrm{Re}\,a+1,\ \mathrm{Re}\,(b-a))\ \text{or}\\ (\mathrm{Re}\,\omega>0,\ \mathrm{Re}\,\sigma=0;\ -1<\mathrm{Re}\,s<\mathrm{Re}\,(b-a))\ \text{or}\\ (\mathrm{Re}\,\omega=0,\ \mathrm{Re}\,\sigma=0;\ -1<\mathrm{Re}\,s<\mathrm{Re}\,a+1,\ \mathrm{Re}\,(b-a))\end{matrix}\right]$
11	$\left(\sqrt{x}+\sqrt{x+\sigma}\right)^{\rho}e^{-\omega x}$ $\times\,{}_1F_1\left(\begin{matrix}a;\,\omega x\\ b\end{matrix}\right)$	$\frac{2^{\rho}}{\omega^{s+\rho/2}}\,\mathrm{B}\left(\frac{2b-2a-2s-\rho}{2},\,\frac{2s+\rho}{2}\right)$ $\times\,\Gamma\left[\begin{matrix}b\\ \frac{2b-2s-\rho}{2}\end{matrix}\right]{}_3F_3\left(\begin{matrix}-\frac{\rho}{2},\,\frac{1-\rho}{2},\,\frac{2b-2a-2s-\rho}{2};\,\sigma\omega\\ 1-\rho,\,\frac{2-2s-\rho}{2},\,\frac{2b-2s-\rho}{2}\end{matrix}\right)$ $-\,\frac{\sigma^{s+\rho/2}\rho}{2^{2s}}\,\Gamma\left[\begin{matrix}2s,\,\frac{-2s-\rho}{2}\\ \frac{2s-\rho+2}{2}\end{matrix}\right]{}_3F_3\left(\begin{matrix}b-a,\,s,\,\frac{2s+1}{2};\,\sigma\omega\\ b,\,\frac{2s-\rho+2}{2},\,\frac{2s+\rho+2}{2}\end{matrix}\right)$ $\left[\begin{matrix}(\mathrm{Re}\,\omega>0;\ 0<\mathrm{Re}\,s<\mathrm{Re}\,(b-a-\rho/2))\ \text{or}\\ (\mathrm{Re}\,\omega=0;\ 0<\mathrm{Re}\,s<\mathrm{Re}\,(a-\rho/2)+1,\ \mathrm{Re}\,(b-a-\rho/2));\ \lvert\arg\sigma\rvert<\pi\end{matrix}\right]$

No.	$f(x)$	$F(s)$
12	$\dfrac{\left(\sqrt{x}+\sqrt{x+\sigma}\right)^{\rho}}{\sqrt{x+\sigma}}\, e^{-\omega x} \times {}_1F_1\left(\begin{matrix} a;\, \omega x \\ b \end{matrix}\right)$	$\dfrac{2^{\rho}}{\omega^{s+(\rho-1)/2}}\, \mathrm{B}\left(\dfrac{2b-2a-2s-\rho+1}{2},\, \dfrac{2s+\rho-1}{2}\right) \times \Gamma\left[\begin{matrix} b \\ \frac{2b-2s-\rho+1}{2} \end{matrix}\right] {}_3F_3\left(\begin{matrix} \frac{1-\rho}{2},\, \frac{2-\rho}{2},\, \frac{2s-2a+2b-\rho+1}{2};\, \sigma\omega \\ 1-\rho,\, \frac{3-2s-\rho}{2},\, \frac{2b-2s-\rho+1}{2} \end{matrix}\right) + \dfrac{\sigma^{s+(\rho-1)/2}}{2^{2s-1}}\, \mathrm{B}\left(2s,\, \dfrac{1-2s-\rho}{2}\right) \times {}_3F_3\left(\begin{matrix} b-a,\, s,\, \frac{2s+1}{2};\, \sigma\omega \\ b,\, \frac{2s-\rho+1}{2},\, \frac{2s+\rho+1}{2} \end{matrix}\right)$
		$\left[\begin{matrix}(\mathrm{Re}\,\omega>0;\; 0<\mathrm{Re}\,s<\mathrm{Re}\,(b-a-\rho/2)+1/2)\;\text{or} \\ (\mathrm{Re}\,\omega=0;\; 0<\mathrm{Re}\,s<\mathrm{Re}\,(a-\rho/2)+3/2,\; \mathrm{Re}\,(b-a-\rho/2)+1/2);\; \lvert\arg\sigma\rvert<\pi\end{matrix}\right]$
13	$(\sigma-x)_+^{\alpha-1}\, e^{\tau x} \times {}_1F_1\left(\begin{matrix} a;\, \omega(\sigma-x) \\ b \end{matrix}\right)$	$\sigma^{s+\alpha-1}\, \mathrm{B}\,(s,\, \alpha)\, \Phi_2\,(s,\, a;\, s+b,\, \sigma\tau,\, \sigma\omega)$ $\quad [\sigma,\; \mathrm{Re}\,\alpha,\; \mathrm{Re}\,s>0]$

3.28.3. ${}_1F_1\,(a;\, b;\, \omega x)$ and trigonometric functions

1	$\sin(\sigma x)\; {}_1F_1\left(\begin{matrix} a;\, -\omega x \\ b \end{matrix}\right)$	$\sigma^{-s} \sin\dfrac{s\pi}{2}\, \Gamma(s)\; {}_4F_3\left(\begin{matrix} \frac{a}{2},\, \frac{a+1}{2},\, \frac{s}{2},\, \frac{s+1}{2} \\ \frac{1}{2},\, \frac{b}{2},\, \frac{b+1}{2} \end{matrix}\right) - \dfrac{a\sigma^{-s-1}\omega}{b} \cos\dfrac{s\pi}{2}\, \Gamma(s+1)\; {}_4F_3\left(\begin{matrix} \frac{a+1}{2},\, \frac{a+2}{2},\, \frac{s+1}{2},\, \frac{s+2}{2} \\ \frac{3}{2},\, \frac{b+1}{2},\, \frac{b+2}{2};\, -\frac{\omega^2}{\sigma^2} \end{matrix}\right)$ $\left[\begin{matrix}\sigma>0;\; (\mathrm{Re}\,\omega>0;\; -1<\mathrm{Re}\,s<\mathrm{Re}\,a+1)\;\text{or} \\ (\mathrm{Re}\,\omega=0;\; -1<\mathrm{Re}\,s<\mathrm{Re}\,a+1,\; \mathrm{Re}\,(b-a))\end{matrix}\right]$
2	$\cos(\sigma x)\; {}_1F_1\left(\begin{matrix} a;\, -\omega x \\ b \end{matrix}\right)$	$\dfrac{as\omega}{b\sigma^{s+1}} \sin\dfrac{s\pi}{2}\, \Gamma(s)\; {}_4F_3\left(\begin{matrix} \frac{a+1}{2},\, \frac{a+2}{2},\, \frac{s+1}{2},\, \frac{s+2}{2} \\ \frac{3}{2},\, \frac{b+1}{2},\, \frac{b+2}{2};\, -\frac{\omega^2}{\sigma^2} \end{matrix}\right) + \sigma^{-s} \cos\dfrac{s\pi}{2}\, \Gamma(s)\; {}_4F_3\left(\begin{matrix} \frac{a}{2},\, \frac{a+1}{2},\, \frac{s}{2},\, \frac{s+1}{2} \\ \frac{1}{2},\, \frac{b}{2},\, \frac{b+1}{2};\, -\frac{\omega^2}{\sigma^2} \end{matrix}\right)$ $\left[\begin{matrix}\sigma>0;\; (\mathrm{Re}\,\omega>0;\; 0<\mathrm{Re}\,s<\mathrm{Re}\,a+1)\;\text{or} \\ (\mathrm{Re}\,\omega=0;\; 0<\mathrm{Re}\,s<\mathrm{Re}\,a+1,\; \mathrm{Re}\,(b-a))\end{matrix}\right]$
3	$\sin(\sigma\sqrt{x})\; {}_1F_1\left(\begin{matrix} a;\, -\omega x \\ b \end{matrix}\right)$	$\dfrac{\sigma}{\omega^{s+1/2}}\, \mathrm{B}\left(\dfrac{2a-2s-1}{2},\, \dfrac{2s+1}{2}\right) \Gamma\left[\begin{matrix} b \\ \frac{2b-2s-1}{2} \end{matrix}\right] \times {}_2F_2\left(\begin{matrix} \frac{2s+1}{2},\, \frac{2s-2b+3}{2} \\ \frac{3}{2},\, \frac{2s-2a+3}{2};\, -\frac{\sigma^2}{4\omega} \end{matrix}\right) - \dfrac{2\sigma^{2a-2s}}{\omega^{a}} \sin[(a-s)\pi] \times \Gamma\left[\begin{matrix} b,\, 2s-2a \\ b-a \end{matrix}\right] {}_2F_2\left(\begin{matrix} a,\, a-b+1;\, -\frac{\sigma^2}{4\omega} \\ \frac{2a-2s+1}{2},\, a-s+1 \end{matrix}\right)$ $\left[\begin{matrix}\sigma>0;\; (\mathrm{Re}\,\omega>0;\; -1/2<\mathrm{Re}\,s<\mathrm{Re}\,a+1/2)\;\text{or} \\ (\mathrm{Re}\,\omega=0;\; -1/2<\mathrm{Re}\,s<\mathrm{Re}\,a+1/2,\; \mathrm{Re}\,(b-a)+1)\end{matrix}\right]$

No.	$f(x)$	$F(s)$
4	$\cos(\sigma\sqrt{x})\ {}_1F_1\left(\begin{matrix} a;\ -\omega x \\ b \end{matrix}\right)$	$\frac{2\sigma^{2a-2s}}{\omega a}\cos[(a-s)\pi]\,\Gamma\left[\begin{matrix} b,\ 2s-2a \\ b-a \end{matrix}\right] \times {}_2F_2\left(\begin{matrix} a,\ a-b+1;\ -\frac{\sigma^2}{4\omega} \\ \frac{2a-2s+1}{2},\ a-s+1 \end{matrix}\right) + \omega^{-s}\,\mathrm{B}(a-s,\ s)\,\Gamma\left[\begin{matrix} b \\ b-s \end{matrix}\right] {}_2F_2\left(\begin{matrix} s,\ s-b+1;\ -\frac{\sigma^2}{4\omega} \\ \frac{1}{2},\ s-a+1 \end{matrix}\right)$ $\left[\begin{matrix} \sigma>0;\ (\mathrm{Re}\,\omega>0;\ 0<\mathrm{Re}\,s<\mathrm{Re}\,a+1/2)\ \text{or} \\ (\mathrm{Re}\,\omega=0;\ 0<\mathrm{Re}\,s<\mathrm{Re}\,a+1/2,\ \mathrm{Re}\,(b-a)+1) \end{matrix}\right]$
5	$\sin\frac{\sigma}{\sqrt{x}}\ {}_1F_1\left(\begin{matrix} a;\ -\omega x \\ b \end{matrix}\right)$	$\frac{\pi\sigma^{2s}\sec(s\pi)}{\Gamma(2s+1)}\ {}_1F_3\left(\begin{matrix} a;\ \frac{\sigma^2\omega}{4} \\ b,\ \frac{2s+1}{2},\ s+1 \end{matrix}\right) - \frac{\pi\sigma\sec(s\pi)}{\omega^{s-1/2}}\,\Gamma\left[\begin{matrix} b,\ \frac{2a-2s+1}{2} \\ a,\ \frac{3}{2}-s,\ b-s+\frac{1}{2} \end{matrix}\right] \times {}_1F_3\left(\begin{matrix} \frac{2a-2s+1}{2};\ \frac{\sigma^2\omega}{4} \\ \frac{3}{2},\ \frac{3-2s}{2},\ \frac{2b-2s+1}{2} \end{matrix}\right)$ $\left[\begin{matrix} \sigma>0;\ (\mathrm{Re}\,\omega>0;\ -1/2<\mathrm{Re}\,s<\mathrm{Re}\,a+1/2)\ \text{or} \\ (\mathrm{Re}\,\omega=0;\ -1/2<\mathrm{Re}\,s<\mathrm{Re}\,a+1/2,\ \mathrm{Re}\,(b-a)+3/2) \end{matrix}\right]$
6	$\cos\frac{\sigma}{\sqrt{x}}\ {}_1F_1\left(\begin{matrix} a;\ -\omega x \\ b \end{matrix}\right)$	$\frac{\pi\csc(s\pi)}{\omega^s}\,\Gamma\left[\begin{matrix} b,\ a-s \\ a,\ 1-s,\ b-s \end{matrix}\right] {}_1F_3\left(\begin{matrix} a-s;\ \frac{\sigma^2\omega}{4} \\ \frac{1}{2},\ 1-s,\ b-s \end{matrix}\right) - \frac{\pi\sigma^{2s}\csc(s\pi)}{\Gamma(2s+1)}\ {}_1F_3\left(\begin{matrix} a;\ \frac{\sigma^2\omega}{4} \\ b,\ \frac{2s+1}{2},\ s+1 \end{matrix}\right)$ $\left[\begin{matrix} \sigma>0;\ (\mathrm{Re}\,\omega>0;\ -1/2<\mathrm{Re}\,s<\mathrm{Re}\,a)\ \text{or} \\ (\mathrm{Re}\,\omega=0;\ -1/2<\mathrm{Re}\,s<\mathrm{Re}\,a,\ \mathrm{Re}\,(b-a)+1) \end{matrix}\right]$
7	$e^{-\omega x}\sin(\sigma x)\ {}_1F_1\left(\begin{matrix} a;\ \omega x \\ b \end{matrix}\right)$	$\sigma^{-s}\sin\frac{s\pi}{2}\,\Gamma(s)\ {}_4F_3\left(\begin{matrix} \frac{b-a}{2},\ \frac{b-a+1}{2},\ \frac{s}{2},\ \frac{s+1}{2} \\ \frac{1}{2},\ \frac{b}{2},\ \frac{b+1}{2};\ -\frac{\omega^2}{\sigma^2} \end{matrix}\right) - \frac{a\sigma^{-s-1}\omega}{b}\cos\frac{s\pi}{2}\,\Gamma(s+1) \times {}_4F_3\left(\begin{matrix} \frac{b-a+1}{2},\ \frac{b-a+2}{2},\ \frac{s+1}{2},\ \frac{s+2}{2} \\ \frac{3}{2},\ \frac{b+1}{2},\ \frac{b+2}{2};\ -\frac{\omega^2}{\sigma^2} \end{matrix}\right)$ $\left[\begin{matrix} \sigma>0;\ (\mathrm{Re}\,\omega>0;\ -1<\mathrm{Re}\,s<\mathrm{Re}\,(b-a)+1)\ \text{or} \\ (\mathrm{Re}\,\omega=0;\ -1<\mathrm{Re}\,s<\mathrm{Re}\,a+1,\ \mathrm{Re}\,(b-a)+1) \end{matrix}\right]$
8	$e^{-\omega x}\cos(\sigma x)\ {}_1F_1\left(\begin{matrix} a;\ \omega x \\ b \end{matrix}\right)$	$\frac{(b-a)s\omega}{b\sigma^{s+1}}\sin\frac{s\pi}{2}\,\Gamma(s)\ {}_4F_3\left(\begin{matrix} \frac{b-a+1}{2},\ \frac{b-a+2}{2},\ \frac{s+1}{2},\ \frac{s+2}{2} \\ \frac{3}{2},\ \frac{b+1}{2},\ \frac{b+2}{2};\ -\frac{\omega^2}{\sigma^2} \end{matrix}\right) + \sigma^{-s}\cos\frac{s\pi}{2}\,\Gamma(s)\ {}_4F_3\left(\begin{matrix} \frac{b-a}{2},\ \frac{b-a+1}{2},\ \frac{s}{2},\ \frac{s+1}{2} \\ \frac{1}{2},\ \frac{b}{2},\ \frac{b+1}{2};\ -\frac{\omega^2}{\sigma^2} \end{matrix}\right)$ $\left[\begin{matrix} \sigma>0;\ (\mathrm{Re}\,\omega>0;\ 0<\mathrm{Re}\,s<\mathrm{Re}\,(b-a)+1)\ \text{or} \\ (\mathrm{Re}\,\omega=0;\ 0<\mathrm{Re}\,s<\mathrm{Re}\,a+1,\ \mathrm{Re}\,(b-a)+1) \end{matrix}\right]$

No.	$f(x)$	$F(s)$
9	$e^{-\omega x}\sin(\sigma\sqrt{x})$ $\times\, {}_1F_1\left(\begin{matrix}a;\ \omega x\\ b\end{matrix}\right)$	$\frac{2\omega^{a-b}}{\sigma^{2(s+a-b)}}\sin[(s+a-b)\pi]\,\Gamma\left[\begin{matrix}b,\ 2a-2b+2s\\ a\end{matrix}\right]$ $\times\, {}_2F_2\left(\begin{matrix}1-a,\ b-a;\ -\frac{\sigma^2}{4\omega}\\ \frac{2b-2a-2s+1}{2},\ b-a-s+1\end{matrix}\right)$ $+\frac{\sigma}{\omega^{s+1/2}}\,\mathrm{B}\left(\frac{2b-2a-2s-1}{2},\ \frac{2s+1}{2}\right)\Gamma\left[\begin{matrix}b\\ \frac{2b-2s-1}{2}\end{matrix}\right]$ $\times\, {}_2F_2\left(\begin{matrix}\frac{2s+1}{2},\ \frac{2s-2b+3}{2}\\ \frac{3}{2},\ \frac{2a-2b+2s+3}{2};\ -\frac{\sigma^2}{4\omega}\end{matrix}\right)$ $[\sigma>0;\ (\operatorname{Re}\omega>0;\ -1/2<\operatorname{Re}s<\operatorname{Re}(b-a)+1/2)$ or $(\operatorname{Re}\omega=0;\ -1/2<\operatorname{Re}s<\operatorname{Re}a+1,\ \operatorname{Re}(b-a)+1/2)]$
10	$e^{-\omega x}\cos(\sigma\sqrt{x})$ $\times\, {}_1F_1\left(\begin{matrix}a;\ \omega x\\ b\end{matrix}\right)$	$\frac{2\omega^{a-b}}{\sigma^{2(a-b+s)}}\cos[(s+a-b)\pi]\,\Gamma\left[\begin{matrix}b,\ 2s+2a-2b\\ a\end{matrix}\right]$ $\times\, {}_2F_2\left(\begin{matrix}1-a,\ b-a;\ -\frac{\sigma^2}{4\omega}\\ \frac{2b-2a-2s+1}{2},\ b-a-s+1\end{matrix}\right)$ $+\,\omega^{-s}\,\mathrm{B}(b-a-s,\ s)\,\Gamma\left[\begin{matrix}b\\ b-s\end{matrix}\right]{}_2F_2\left(\begin{matrix}s,\ s-b+1;\ -\frac{\sigma^2}{4\omega}\\ \frac{1}{2},\ s+a-b+1\end{matrix}\right)$ $[\sigma>0;\ (\operatorname{Re}\omega>0;\ 0<\operatorname{Re}s<\operatorname{Re}(b-a)+1/2)$ or $(\operatorname{Re}\omega=0;\ 0<\operatorname{Re}s<\operatorname{Re}a+1,\ \operatorname{Re}(b-a)+1/2)]$
11	$e^{-\omega x}\sin\frac{\sigma}{\sqrt{x}}\,{}_1F_1\left(\begin{matrix}a;\ \omega x\\ b\end{matrix}\right)$	$\frac{\pi\sigma^{2s}\sec(s\pi)}{\Gamma(2s+1)}\,{}_1F_3\left(\begin{matrix}b-a;\ \frac{\sigma^2\omega}{4}\\ b,\ \frac{2s+1}{2},\ s+1\end{matrix}\right)$ $-\frac{\pi\sigma\sec(s\pi)}{\omega^{s-1/2}}\,\Gamma\left[\begin{matrix}b,\ \frac{2b-2a-2s+1}{2}\\ b-a,\ \frac{3-2s}{2},\ \frac{2b-2s+1}{2}\end{matrix}\right]$ $\times\, {}_1F_3\left(\begin{matrix}\frac{2b-2a-2s+1}{2};\ \frac{\sigma^2\omega}{4}\\ \frac{3}{2},\ \frac{3-2s}{2},\ \frac{2b-2s+1}{2}\end{matrix}\right)$ $[\sigma>0;\ (\operatorname{Re}\omega>0;\ -1/2<\operatorname{Re}s<\operatorname{Re}(b-a)+1/2)$ or $(\operatorname{Re}\omega=0;\ -1/2<\operatorname{Re}s<\operatorname{Re}a+3/2,\ \operatorname{Re}(b-a)+1/2)]$
12	$e^{-\omega x}\cos\frac{\sigma}{\sqrt{x}}\,{}_1F_1\left(\begin{matrix}a;\ \omega x\\ b\end{matrix}\right)$	$\frac{\pi\csc(s\pi)}{\omega^{s}}\Gamma\left[\begin{matrix}b,\ b-a-s\\ b-a,\ 1-s,\ b-s\end{matrix}\right]{}_1F_3\left(\begin{matrix}b-a-s;\ \frac{\sigma^2\omega}{4}\\ \frac{1}{2},\ 1-s,\ b-s\end{matrix}\right)$ $-\frac{\pi\sigma^{2s}\csc(s\pi)}{\Gamma(2s+1)}\,{}_1F_3\left(\begin{matrix}b-a;\ \frac{\sigma^2\omega}{4}\\ b,\ \frac{2s+1}{2},\ s+1\end{matrix}\right)$ $[\sigma>0;\ (\operatorname{Re}\omega>0;\ -1/2<\operatorname{Re}s<\operatorname{Re}(b-a))$ or $(\operatorname{Re}\omega=0;\ -1/2<\operatorname{Re}s<\operatorname{Re}a+1,\ \operatorname{Re}(b-a))]$

3.28.4. ${}_1F_1\,(a;\ b;\ \omega x)$ and the logarithmic function

1	$\ln(\sigma x+1)\ {}_1F_1\left(\begin{matrix}a;\ -\omega x\\ b\end{matrix}\right)$	$\frac{\omega^{1-s}}{\sigma}\,\mathrm{B}\,(a-s+1,\ s-1)\ \Gamma\left[\begin{matrix}b\\ b-s+1\end{matrix}\right]\ {}_3F_3\left(\begin{matrix}1,\ 1,\ a-s+1;\ \frac{\omega}{\sigma}\\ 2,\ 2-s,\ b-s+1\end{matrix}\right)$ $+\,\omega^{-s}\,\mathrm{B}\,(s,\ a-s)\,\Gamma\left[\begin{matrix}b\\ b-s\end{matrix}\right]\left[\ln\sigma-\ln\omega-\psi\,(a-s)\right.$ $\left.+\,\psi\,(b-s)+\psi\,(s)\right]+\frac{\pi\csc(s\pi)}{\sigma^s s}\ {}_2F_2\left(\begin{matrix}a,\ s;\ \frac{\omega}{\sigma}\\ b,\ s+1\end{matrix}\right)$ $\left[\begin{matrix}(\mathrm{Re}\,\omega>0;\ -1<\mathrm{Re}\,s<\mathrm{Re}\,a)\ \text{or}\\ (\mathrm{Re}\,\omega=0;\ -1<\mathrm{Re}\,s<\mathrm{Re}\,a,\ \mathrm{Re}\,(b-a)+1)\,;\ \lvert\arg\sigma\rvert<\pi\end{matrix}\right]$
2	$\ln\lvert\sigma x-1\rvert\ {}_1F_1\left(\begin{matrix}a;\ -\omega x\\ b\end{matrix}\right)$	$-\frac{\omega^{1-s}}{\sigma}\,\mathrm{B}\,(a-s+1,\ s-1)$ $\times\,\Gamma\left[\begin{matrix}b\\ b-s+1\end{matrix}\right]\ {}_3F_3\left(\begin{matrix}1,\ 1,\ a-s+1;\ -\frac{\omega}{\sigma}\\ 2,\ 2-s,\ b-s+1\end{matrix}\right)$ $+\,\omega^{-s}\,\mathrm{B}\,(s,\ a-s)\,\Gamma\left[\begin{matrix}b\\ b-s\end{matrix}\right]\left[\ln\sigma-\ln\omega-\psi\,(a-s)\right.$ $\left.+\,\psi\,(b-s)+\psi\,(s)\right]+\frac{\pi\sigma^{-s}}{s}\cot(s\pi)\ {}_2F_2\left(\begin{matrix}a,\ s;\ -\frac{\omega}{\sigma}\\ b,\ s+1\end{matrix}\right)$ $\left[\begin{matrix}\sigma>0;\ (\mathrm{Re}\,\omega>0;\ -1<\mathrm{Re}\,s<\mathrm{Re}\,a)\ \text{or}\\ (\mathrm{Re}\,\omega=0;\ -1<\mathrm{Re}\,s<\mathrm{Re}\,a,\ \mathrm{Re}\,(b-a)+1)\end{matrix}\right]$
3	$e^{-\omega x}\ln(\sigma x+1)$ $\times\ {}_1F_1\left(\begin{matrix}a;\ \omega x\\ b\end{matrix}\right)$	$\frac{\omega^{1-s}}{\sigma}\,\mathrm{B}\,(b-a-s+1,\ s-1)\,\Gamma\left[\begin{matrix}b\\ b-s+1\end{matrix}\right]$ $\times\ {}_3F_3\left(\begin{matrix}1,\ 1,\ b-a-s+1;\ \frac{\omega}{\sigma}\\ 2,\ 2-s,\ b-s+1\end{matrix}\right)+\omega^{-s}\,\mathrm{B}\,(s,\ b-a-s)$ $\times\,\Gamma\left[\begin{matrix}b\\ b-s\end{matrix}\right]\left[\ln\sigma-\ln\omega-\psi\,(b-a-s)\right.$ $\left.+\,\psi\,(b-s)+\psi\,(s)\right]+\frac{\pi\csc(s\pi)}{\sigma^s s}\ {}_2F_2\left(\begin{matrix}b-a,\ s;\ \frac{\omega}{\sigma}\\ b,\ s+1\end{matrix}\right)$ $\left[\begin{matrix}(\mathrm{Re}\,\omega>0;\ -1<\mathrm{Re}\,s<\mathrm{Re}\,(b-a)+1)\ \text{or}\\ (\mathrm{Re}\,\omega=0;\ -1<\mathrm{Re}\,s<\mathrm{Re}\,a+1,\ \mathrm{Re}\,(b-a)+1)\,;\ \lvert\arg\sigma\rvert<\pi\end{matrix}\right]$
4	$e^{-\omega x}\ln\lvert\sigma x-1\rvert$ $\times\ {}_1F_1\left(\begin{matrix}a;\ \omega x\\ b\end{matrix}\right)$	$-\frac{\omega^{1-s}}{\sigma}\,\mathrm{B}\,(b-a-s+1,\ s-1)\,\Gamma\left[\begin{matrix}b\\ b-s+1\end{matrix}\right]$ $\times\ {}_3F_3\left(\begin{matrix}1,\ 1,\ b-a-s+1;\ -\frac{\omega}{\sigma}\\ 2,\ 2-s,\ b-s+1\end{matrix}\right)+\omega^{-s}\,\mathrm{B}\,(b-a-s,\ s)\,\Gamma\left[\begin{matrix}b\\ b-s\end{matrix}\right]$ $\times\,(\ln\sigma-\ln\omega)-\omega^{-s}\,\mathrm{B}\,(b-a-s,\ s)\,\Gamma\left[\begin{matrix}b\\ b-s\end{matrix}\right]\left[\psi\,(b-a-s)\right.$ $\left.-\,\psi\,(b-s)-\psi\,(s)\right]+\frac{\pi\sigma^{-s}}{s}\cot(s\pi)\ {}_2F_2\left(\begin{matrix}b-a,\ s;\ -\frac{\omega}{\sigma}\\ b,\ s+1\end{matrix}\right)$ $\left[\begin{matrix}\sigma>0;\ (\mathrm{Re}\,\omega>0;\ -1<\mathrm{Re}\,s<\mathrm{Re}\,a)\ \text{or}\\ (\mathrm{Re}\,\omega=0;\ -1<\mathrm{Re}\,s<\mathrm{Re}\,a,\ \mathrm{Re}\,(b-a)+1)\end{matrix}\right]$

3.28.5. ${}_1F_1\,(a;\, b;\, \omega x)$ and $\operatorname{erf}(\sigma\sqrt{x})$, $\operatorname{erfc}(\sigma\sqrt{x})$

1	$\operatorname{erf}(\sigma\sqrt{x})\ {}_1F_1\left(\begin{matrix} a;\ -\omega x \\ b \end{matrix}\right)$	$\omega^{-s}\,\mathrm{B}\,(a-s,\ s)\,\Gamma\left[\begin{matrix} b \\ b-s \end{matrix}\right] - \dfrac{\sigma^{-2s}}{\sqrt{\pi}s}\,\Gamma\left(\dfrac{2s+1}{2}\right)\,{}_3F_2\left(\begin{matrix} a,\ s,\ \frac{2s+1}{2} \\ b,\ s+1;\ -\frac{\omega}{\sigma^2} \end{matrix}\right)$ $\left[\begin{matrix} (\operatorname{Re}\omega>0;\ -1/2<\operatorname{Re}s<\operatorname{Re}a) \text{ or} \\ (\operatorname{Re}\omega=0;\ -1/2<\operatorname{Re}s<\operatorname{Re}a,\ \operatorname{Re}(b-a)+1);\ \lvert\arg\sigma\rvert<\pi/4 \end{matrix}\right]$
2	$\operatorname{erfc}(\sigma\sqrt{x})\ {}_1F_1\left(\begin{matrix} a;\ \omega x \\ b \end{matrix}\right)$	$\dfrac{\sigma^{-2s}}{\sqrt{\pi}s}\,\Gamma\left(\dfrac{2s+1}{2}\right)\,{}_3F_2\left(\begin{matrix} a,\ s,\ \frac{2s+1}{2} \\ b,\ s+1;\ \frac{\omega}{\sigma^2} \end{matrix}\right)$ $\left[\begin{matrix} (\operatorname{Re}(\sigma^2-\omega)>0;\ \operatorname{Re}s>0;\ \lvert\arg\sigma\rvert<\pi/4) \text{ or} \\ (\operatorname{Re}\omega<0;\ 0<\operatorname{Re}s<\operatorname{Re}a+3/2;\ \lvert\arg\sigma\rvert=\pi/4) \text{ or} \\ (\operatorname{Re}(\sigma^2-\omega)=0;\ 0<\operatorname{Re}s<\operatorname{Re}(b-a)+3/2;\ \lvert\arg\sigma\rvert<\pi/4) \text{ or} \\ (\operatorname{Re}(\sigma^2-\omega)=0;\ 0<\operatorname{Re}s<\operatorname{Re}a,\ \operatorname{Re}(b-a)+3/2;\ \lvert\arg\sigma\rvert=\pi/4) \end{matrix}\right]$
3	$e^{-\omega x}\operatorname{erf}(\sigma\sqrt{x}) \times {}_1F_1\left(\begin{matrix} a;\ \omega x \\ b \end{matrix}\right)$	$\omega^{-s}\,\mathrm{B}\,(b-a-s,\ s)\,\Gamma\left[\begin{matrix} b \\ b-s \end{matrix}\right] - \dfrac{\sigma^{-2s}}{\sqrt{\pi}s}\,\Gamma\left(\dfrac{2s+1}{2}\right)\,{}_3F_2\left(\begin{matrix} b-a,\ s,\ \frac{2s+1}{2} \\ b,\ s+1;\ -\frac{\omega}{\sigma^2} \end{matrix}\right)$ $\left[\begin{matrix} (\operatorname{Re}\omega>0;\ -1/2<\operatorname{Re}s<\operatorname{Re}(b-a)) \text{ or} \\ (\operatorname{Re}\omega=0;\ -1/2<\operatorname{Re}s<\operatorname{Re}a+1,\ \operatorname{Re}(b-a));\ \lvert\arg\sigma\rvert<\pi/4 \end{matrix}\right]$
4	$e^{-\omega x}\operatorname{erfc}(\sigma\sqrt{x}) \times {}_1F_1\left(\begin{matrix} a;\ \omega x \\ b \end{matrix}\right)$	$\dfrac{\sigma^{-2s}}{\sqrt{\pi}s}\,\Gamma\left(\dfrac{2s+1}{2}\right)\,{}_3F_2\left(\begin{matrix} b-a,\ s,\ \frac{2s+1}{2} \\ b,\ s+1;\ \frac{\omega}{\sigma^2} \end{matrix}\right)$ $\left[\begin{matrix} (\operatorname{Re}(\sigma^2+\omega)>0;\ \operatorname{Re}s>0;\ \lvert\arg\sigma\rvert<\pi/4) \text{ or} \\ (\operatorname{Re}\omega>0;\ 0<\operatorname{Re}s<\operatorname{Re}(b-a)+3/2;\ \lvert\arg\sigma\rvert=\pi/4) \text{ or} \\ (\operatorname{Re}(\sigma^2+\omega)=0;\ 0<\operatorname{Re}s<\operatorname{Re}a+3/2;\ \lvert\arg\sigma\rvert<\pi/4) \text{ or} \\ (\operatorname{Re}(\sigma^2+\omega)=0;\ 0<\operatorname{Re}s<\operatorname{Re}a,\ \operatorname{Re}(b-a)+3/2;\ \lvert\arg\sigma\rvert=\pi/4) \end{matrix}\right]$

3.28.6. ${}_1F_1\,(a;\, b;\, \omega x)$ and the Bessel functions

1	$J_\nu(\sigma x)\ {}_1F_1\left(\begin{matrix} a;\ -\omega x \\ b \end{matrix}\right)$	$2^{s-1}\omega^{-s}\left(\dfrac{\sigma^2}{\omega^2}\right)^{-s/2}\Gamma\left[\begin{matrix} \frac{s+\nu}{2} \\ \frac{2-s+\nu}{2} \end{matrix}\right]{}_4F_3\left(\begin{matrix} \frac{a}{2},\ \frac{a+1}{2},\ \frac{s-\nu}{2},\ \frac{s+\nu}{2} \\ \frac{1}{2},\ \frac{b}{2},\ \frac{b+1}{2};\ -\frac{\omega^2}{\sigma^2} \end{matrix}\right) - \dfrac{2^s a}{b\sigma^2}\,\omega^{2-s}\left(\dfrac{\sigma^2}{\omega^2}\right)^{(1-s)/2}\Gamma\left[\begin{matrix} \frac{s+\nu+1}{2} \\ \frac{1-s+\nu}{2} \end{matrix}\right] \times {}_4F_3\left(\begin{matrix} \frac{a+1}{2},\ \frac{a+2}{2},\ \frac{s-\nu+1}{2},\ \frac{s+\nu+1}{2} \\ \frac{3}{2},\ \frac{b+1}{2},\ \frac{b+2}{2};\ -\frac{\omega^2}{\sigma^2} \end{matrix}\right)$ $\left[\begin{matrix} \sigma>0;\ (\operatorname{Re}\omega>0;\ -\operatorname{Re}\nu<\operatorname{Re}s<\operatorname{Re}a+3/2) \text{ or} \\ (\operatorname{Re}\omega=0;\ -\operatorname{Re}\nu<\operatorname{Re}s<\operatorname{Re}a+3/2,\ \operatorname{Re}(b-a)+3/2) \end{matrix}\right]$

No.	$f\,(x)$	$F\,(s)$
2	$J_\nu\,(\sigma\sqrt{x})\;{}_1F_1\left(\begin{matrix}a;\,-\omega x\\ b\end{matrix}\right)$	$\frac{2^{2s-2a}}{\sigma^{2s}}\left(\frac{\omega}{\sigma^2}\right)^{-a}\Gamma\left[\begin{matrix}b,\ \frac{2s-2a+\nu}{2}\\ b-a,\ \frac{2a-2s+\nu+2}{2}\end{matrix}\right]{}_2F_2\left(\begin{matrix}a,\ a-b+1;\ -\frac{\sigma^2}{4\omega}\\ \frac{2a-2s-\nu+2}{2},\ \frac{2a-2s+\nu+2}{2}\end{matrix}\right)$ $+\frac{\sigma^{-2s}}{2^\nu}\left(\frac{\omega}{\sigma^2}\right)^{-s-\nu/2}\mathrm{B}\left(a-s-\frac{\nu}{2},\ s+\frac{\nu}{2}\right)$ $\times\Gamma\left[\begin{matrix}b\\ \nu+1,\ \frac{2b-2s-\nu}{2}\end{matrix}\right]{}_2F_2\left(\begin{matrix}\frac{2s+\nu}{2},\ \frac{2s-2b+\nu+2}{2}\\ \nu+1,\ \frac{2s-2a+\nu+2}{2};\ -\frac{\sigma^2}{4\omega}\end{matrix}\right)$ $\left[\begin{matrix}\sigma>0;\ (\mathrm{Re}\,\omega>0;\ -\mathrm{Re}\,\nu<\mathrm{Re}\,s<\mathrm{Re}\,a+3/4)\ \text{or}\\ (\mathrm{Re}\,\omega=0;\ -\mathrm{Re}\,\nu<\mathrm{Re}\,s<\mathrm{Re}\,a+3/4,\ \mathrm{Re}\,(b-a)+5/4)\end{matrix}\right]$
3	$J_\nu\left(\frac{\sigma}{\sqrt{x}}\right)\;{}_1F_1\left(\begin{matrix}a;\,-\omega x\\ b\end{matrix}\right)$	$\frac{\sigma^{2s}}{2^{2s}}\Gamma\left[\begin{matrix}\frac{\nu-2s}{2}\\ \frac{2s+\nu+2}{2}\end{matrix}\right]{}_1F_3\left(\begin{matrix}a;\ \frac{\sigma^2\omega}{4}\\ b,\ \frac{2s-\nu+2}{2},\ \frac{2s+\nu+2}{2}\end{matrix}\right)$ $+\frac{\omega^{-s}}{2^\nu}\left(\frac{1}{\sigma^2\omega}\right)^{-\nu/2}\mathrm{B}\left(\frac{2s-\nu}{2},\ \frac{2a-2s+\nu}{2}\right)$ $\times\Gamma\left[\begin{matrix}b\\ \nu+1,\ \frac{2b-2s+\nu}{2}\end{matrix}\right]{}_1F_3\left(\begin{matrix}\frac{2a-2s+\nu}{2};\ \frac{\sigma^2\omega}{4}\\ \nu+1,\ \frac{\nu-2s+2}{2},\ \frac{2b-2s+\nu}{2}\end{matrix}\right)$ $\left[\begin{matrix}\sigma>0;\ (\mathrm{Re}\,\omega>0;\ 3/4<\mathrm{Re}\,s<\mathrm{Re}\,(a+\nu/2)+1/4)\ \text{or}\\ (\mathrm{Re}\,\omega=0;\ 3/4<\mathrm{Re}\,s<\mathrm{Re}\,(a+\nu/2)+1/4,\ \mathrm{Re}\,(b-a-\nu/2)+5/4)\end{matrix}\right]$
4	$Y_\nu\,(\sigma x)\;{}_1F_1\left(\begin{matrix}a;\,-\omega x\\ b\end{matrix}\right)$	$-\frac{2^s a}{b\pi}\,\omega^{-s}\left(\frac{\sigma^2}{\omega^2}\right)^{-(s+1)/2}\sin\frac{(s-\nu)\,\pi}{2}\,\Gamma\left(\frac{s-\nu+1}{2}\right)$ $\times\Gamma\left(\frac{s+\nu+1}{2}\right){}_4F_3\left(\begin{matrix}\frac{a+1}{2},\ \frac{a+2}{2},\ \frac{s-\nu+1}{2},\ \frac{s+\nu+1}{2}\\ \frac{3}{2},\ \frac{b+1}{2},\ \frac{b+2}{2};\ -\frac{\omega^2}{\sigma^2}\end{matrix}\right)$ $-\frac{2^{s-1}}{\pi}\,\omega^{-s}\left(\frac{\sigma^2}{\omega^2}\right)^{-s/2}\cos\frac{(s-\nu)\,\pi}{2}\,\Gamma\left(\frac{s-\nu}{2}\right)$ $\times\Gamma\left(\frac{s+\nu}{2}\right){}_4F_3\left(\begin{matrix}\frac{a}{2},\ \frac{a+1}{2},\ \frac{s-\nu}{2},\ \frac{s+\nu}{2}\\ \frac{1}{2},\ \frac{b}{2},\ \frac{b+1}{2};\ -\frac{\omega^2}{\sigma^2}\end{matrix}\right)$ $\left[\begin{matrix}\sigma>0;\ (\mathrm{Re}\,\omega>0;\ \lvert\mathrm{Re}\,\nu\rvert<\mathrm{Re}\,s<\mathrm{Re}\,a+3/2)\ \text{or}\\ (\mathrm{Re}\,\omega=0;\ \lvert\mathrm{Re}\,\nu\rvert<\mathrm{Re}\,s<\mathrm{Re}\,a+3/2,\ \mathrm{Re}\,(b-a)+3/2)\end{matrix}\right]$
5	$Y_\nu\,(\sigma\sqrt{x})\;{}_1F_1\left(\begin{matrix}a;\,-\omega x\\ b\end{matrix}\right)$	$-\frac{2^{2s-2a}}{\pi}\,\sigma^{-2s}\left(\frac{\omega}{\sigma^2}\right)^{-a}\cos\frac{(2s-2a-\nu)\,\pi}{2}$ $\times\Gamma\left[\begin{matrix}b,\ \frac{2s-2a-\nu}{2},\ \frac{2s-2a+\nu}{2}\\ b-a\end{matrix}\right]{}_2F_2\left(\begin{matrix}a,\ a-b+1;\ -\frac{\sigma^2}{4\omega}\\ \frac{2a-2s-\nu+2}{2},\ \frac{2a-2s+\nu+2}{2}\end{matrix}\right)$ $-\frac{\sigma^{-2s}}{2^\nu\pi}\left(\frac{\omega}{\sigma^2}\right)^{-s-\nu/2}\cos\,(\pi\nu)\,\mathrm{B}\left(\frac{2a-2s-\nu}{2},\ \frac{2s+\nu}{2}\right)$ $\times\Gamma\left[\begin{matrix}b,\ -\nu\\ \frac{2b-2s-\nu}{2}\end{matrix}\right]{}_2F_2\left(\begin{matrix}\frac{2s+\nu}{2},\ \frac{2s-2b+\nu+2}{2}\\ \nu+1,\ \frac{2s+2a+\nu+2}{2};\ -\frac{\sigma^2}{4\omega}\end{matrix}\right)$ $-\,2^\nu\sigma^{-2s}\left(\frac{\omega}{\sigma^2}\right)^{\nu/2-s}\mathrm{B}\left(\frac{2a-2s+\nu}{2},\ \frac{2s-\nu}{2}\right)\times$

No.	$f(x)$	$F(s)$
		$\times \csc(\pi\nu)\,\Gamma\left[\begin{matrix} b \\ 1-\nu,\ \frac{2b-2s+\nu}{2}\end{matrix}\right] {}_2F_2\left(\begin{matrix}\frac{2s-\nu}{2},\ \frac{2s-2b-\nu+2}{2} \\ 1-\nu,\ \frac{2s-2a-\nu+2}{2};\ -\frac{\sigma^2}{4\omega}\end{matrix}\right)$ $\left[\begin{matrix}\sigma>0;\ (\operatorname{Re}\omega>0;\ \lvert\operatorname{Re}\nu\rvert<\operatorname{Re}s<\operatorname{Re}a+3/4)\ \text{or} \\ (\operatorname{Re}\omega=0;\ \lvert\operatorname{Re}\nu\rvert<\operatorname{Re}s<\operatorname{Re}a+3/4,\ \operatorname{Re}(b-a)+5/4)\end{matrix}\right]$
6	$K_\nu(\sigma x)\ {}_1F_1\left(\begin{matrix}a;\ -\omega x \\ b\end{matrix}\right)$	$\frac{2^{s-2}}{\omega^s}\left(\frac{\sigma^2}{\omega^2}\right)^{-s/2}\Gamma\left(\frac{s-\nu}{2}\right)\Gamma\left(\frac{s+\nu}{2}\right) {}_4F_3\left(\begin{matrix}\frac{a}{2},\ \frac{a+1}{2},\ \frac{s-\nu}{2},\ \frac{s+\nu}{2} \\ \frac{1}{2},\ \frac{b}{2},\ \frac{b+1}{2};\ \frac{\omega^2}{\sigma^2}\end{matrix}\right)$ $-\frac{2^{s-1}a}{b\omega^s}\left(\frac{\sigma^2}{\omega^2}\right)^{-(s+1)/2}\Gamma\left(\frac{s-\nu+1}{2}\right)\Gamma\left(\frac{s+\nu+1}{2}\right)$ $\times {}_4F_3\left(\begin{matrix}\frac{a+1}{2},\ \frac{a+2}{2},\ \frac{s-\nu+1}{2},\ \frac{s+\nu+1}{2} \\ \frac{3}{2},\ \frac{b+1}{2},\ \frac{b+2}{2};\ \frac{\omega^2}{\sigma^2}\end{matrix}\right)$
	$\left[\begin{matrix}(\operatorname{Re}\sigma>0,\ \operatorname{Re}(\sigma+\omega)>0;\ \lvert\operatorname{Re}\nu\rvert<\operatorname{Re}s)\ \text{or} \\ (\operatorname{Re}\sigma=0,\ \operatorname{Re}\omega>0;\ \lvert\operatorname{Re}\nu\rvert<\operatorname{Re}s<\operatorname{Re}a+3/2)\ \text{or} \\ (\operatorname{Re}\sigma=0,\ \operatorname{Re}\omega=0;\ \lvert\operatorname{Re}\nu\rvert<\operatorname{Re}s<\operatorname{Re}a+3/2,\ \operatorname{Re}(b-a)+3/2)\end{matrix}\right]$	
7	$K_\nu(\sigma\sqrt{x})\ {}_1F_1\left(\begin{matrix}a;\ -\omega x \\ b\end{matrix}\right)$	$2^{2s-2a-1}\sigma^{-2s}\left(\frac{\omega}{\sigma^2}\right)^{-a}\Gamma\left[\begin{matrix}b,\ \frac{2s-2a-\nu}{2},\ \frac{2s-2a+\nu}{2} \\ b-a\end{matrix}\right]$ $\times {}_2F_2\left(\begin{matrix}a,\ a-b+1;\ \frac{\sigma^2}{4\omega} \\ \frac{2a-2s-\nu+2}{2},\ \frac{2a-2s+\nu+2}{2}\end{matrix}\right)$ $+\frac{\sigma^{-2s}}{2^{\nu+1}}\left(\frac{\omega}{\sigma^2}\right)^{-s-\nu/2}\mathrm{B}\left(\frac{2a-2s-\nu}{2},\ \frac{2s+\nu}{2}\right)\Gamma\left[\begin{matrix}-\nu,\ b \\ \frac{2b-2s-\nu}{2}\end{matrix}\right]$ $\times {}_2F_2\left(\begin{matrix}\frac{2s+\nu}{2},\ \frac{2s-2b+\nu+2}{2};\ \frac{\sigma^2}{4\omega} \\ \nu+1,\ \frac{2s-2a+\nu+2}{2}\end{matrix}\right)$ $+\frac{2^{\nu-1}}{\sigma^{2s}}\left(\frac{\omega}{\sigma^2}\right)^{\nu/2-s}\mathrm{B}\left(\frac{2a-2s+\nu}{2},\ \frac{2s-\nu}{2}\right)\Gamma\left[\begin{matrix}\nu,\ b \\ \frac{2b-2s+\nu}{2}\end{matrix}\right]$ $\times {}_2F_2\left(\begin{matrix}\frac{2s-\nu}{2},\ \frac{2s-2b-\nu+2}{2};\ \frac{\sigma^2}{4\omega} \\ 1-\nu,\ \frac{2s-2a-\nu+2}{2}\end{matrix}\right)$
	$\left[\begin{matrix}(\operatorname{Re}\sigma>0,\ \operatorname{Re}\omega\geq 0;\ \lvert\operatorname{Re}\nu\rvert<\operatorname{Re}s)\ \text{or} \\ (\operatorname{Re}\sigma=0,\ \operatorname{Re}\omega>0;\ \lvert\operatorname{Re}\nu\rvert<\operatorname{Re}s<\operatorname{Re}a+1/2)\ \text{or} \\ (\operatorname{Re}\sigma=0,\ \operatorname{Re}\omega=0;\ \lvert\operatorname{Re}\nu\rvert<\operatorname{Re}s<\operatorname{Re}a+1/2,\ \operatorname{Re}(b-a)+1)\end{matrix}\right]$	
8	$e^{-\omega x}J_\nu(\sigma x)\ {}_1F_1\left(\begin{matrix}a;\ \omega x \\ b\end{matrix}\right)$	$\frac{2^s(a-b)}{b}\,\omega^{-s}\left(\frac{\sigma^2}{\omega^2}\right)^{-(s+1)/2}\Gamma\left[\begin{matrix}\frac{s+\nu+1}{2} \\ \frac{1-s+\nu}{2}\end{matrix}\right]$ $\times {}_4F_3\left(\begin{matrix}\frac{b-a+1}{2},\ \frac{b-a+2}{2},\ \frac{s-\nu+1}{2},\ \frac{s+\nu+1}{2} \\ \frac{3}{2},\ \frac{b+1}{2},\ \frac{b+2}{2};\ -\frac{\omega^2}{\sigma^2}\end{matrix}\right)$ $+\frac{2^{s-1}}{\omega^s}\left(\frac{\sigma^2}{\omega^2}\right)^{-s/2}\Gamma\left[\begin{matrix}\frac{s+\nu}{2} \\ \frac{2-s+\nu}{2}\end{matrix}\right] {}_4F_3\left(\begin{matrix}\frac{b-a}{2},\ \frac{b-a+1}{2},\ \frac{s-\nu}{2},\ \frac{s+\nu}{2} \\ \frac{1}{2},\ \frac{b}{2},\ \frac{b+1}{2};\ -\frac{\omega^2}{\sigma^2}\end{matrix}\right)$ $\left[\begin{matrix}\sigma>0;\ (\operatorname{Re}\omega>0;\ -\operatorname{Re}\nu<\operatorname{Re}s<\operatorname{Re}(b-a)+3/2)\ \text{or} \\ (\operatorname{Re}\omega=0;\ -\operatorname{Re}\nu<\operatorname{Re}s<\operatorname{Re}a+3/2,\ \operatorname{Re}(b-a)+3/2)\end{matrix}\right]$

No.	$f(x)$	$F(s)$
9	$e^{-\omega x} J_\nu(\sigma\sqrt{x})$ $\times\, {}_1F_1\left(\begin{matrix}a;\,\omega x\\ b\end{matrix}\right)$	$\frac{2^{2s+2a-2b}}{\sigma^{2s}}\left(\frac{\omega}{\sigma^2}\right)^{a-b}\Gamma\left[\begin{matrix}b,\ \frac{2s+2a-2b+\nu}{2}\\ a,\ \frac{2b-2a-2s+\nu+2}{2}\end{matrix}\right]$ $\times\, {}_2F_2\left(\begin{matrix}1-a,\ b-a;\ -\frac{\sigma^2}{4\omega}\\ b-a-s-\frac{\nu}{2}+1,\ b-a-s+\frac{\nu}{2}+1\end{matrix}\right)$ $+\frac{\sigma^{-2s}}{2^\nu}\left(\frac{\omega}{\sigma^2}\right)^{-s-\nu/2} \mathrm{B}\left(\frac{2b-2a-2s-\nu}{2},\ \frac{2s+\nu}{2}\right)$ $\times\,\Gamma\left[\begin{matrix}b\\ \nu+1,\ b-s-\frac{\nu}{2}\end{matrix}\right] {}_2F_2\left(\begin{matrix}\frac{2s+\nu}{2},\ \frac{2s-2b+\nu+2}{2};\ -\frac{\sigma^2}{4\omega}\\ \nu+1,\ \frac{2s+2a-2b+\nu+2}{2}\end{matrix}\right)$ $\left[\begin{matrix}\sigma>0;\ (\mathrm{Re}\,\omega>0;\ -\mathrm{Re}\,\nu<\mathrm{Re}\,s<\mathrm{Re}\,(b-a)+3/4)\ \text{or}\\ (\mathrm{Re}\,\omega=0;\ -\mathrm{Re}\,\nu<\mathrm{Re}\,s<\mathrm{Re}\,a+5/4,\ \mathrm{Re}\,(b-a)+3/4)\end{matrix}\right]$
10	$e^{-\omega x} J_\nu\left(\frac{\sigma}{\sqrt{x}}\right)$ $\times\, {}_1F_1\left(\begin{matrix}a;\,\omega x\\ b\end{matrix}\right)$	$\frac{\sigma^{2s}}{2^{2s}}\Gamma\left[\begin{matrix}\frac{\nu}{2}-s\\ s+\frac{\nu}{2}+1\end{matrix}\right] {}_1F_3\left(\begin{matrix}b-a;\ \frac{\sigma^2\omega}{4}\\ b,\ s-\frac{\nu}{2}+1,\ s+\frac{\nu}{2}+1\end{matrix}\right)$ $+\frac{\omega^{-s}}{2^\nu}\left(\frac{1}{\sigma^2\omega}\right)^{-\nu/2} \mathrm{B}\left(\frac{2s-\nu}{2},\ \frac{2b-2a-2s+\nu}{2}\right)$ $\times\,\Gamma\left[\begin{matrix}b\\ \nu+1,\ \frac{2b-2s+\nu}{2}\end{matrix}\right] {}_1F_3\left(\begin{matrix}\frac{2b-2a-2s+\nu}{2};\ \frac{\sigma^2\omega}{4}\\ \nu+1,\ \frac{\nu-2s+2}{2},\ \frac{2b-2s+\nu}{2}\end{matrix}\right)$ $\left[\begin{matrix}\sigma>0;\ (\mathrm{Re}\,\omega>0;\ 3/4<\mathrm{Re}\,s<\mathrm{Re}\,(b-a+\nu/2)+1/4)\ \text{or}\\ (\mathrm{Re}\,\omega=0;\ 3/4<\mathrm{Re}\,s<\mathrm{Re}\,(a-\nu/2)+5/4,\ \mathrm{Re}\,(b-a+\nu/2)+1/4)\end{matrix}\right]$
11	$e^{-\omega x} Y_\nu(\sigma x)\ {}_1F_1\left(\begin{matrix}a;\,\omega x\\ b\end{matrix}\right)$	$-\frac{2^s(b-a)}{b\pi\omega^s}\left(\frac{\sigma^2}{\omega^2}\right)^{-(s+1)/2}\sin\frac{(s-\nu)\pi}{2}\,\Gamma\left(\frac{s-\nu+1}{2}\right)$ $\times\,\Gamma\left(\frac{s+\nu+1}{2}\right) {}_4F_3\left(\begin{matrix}\frac{b-a+1}{2},\ \frac{b-a+2}{2},\ \frac{s-\nu+1}{2},\ \frac{s+\nu+1}{2}\\ \frac{3}{2},\ \frac{b+1}{2},\ \frac{b+2}{2};\ -\frac{\omega^2}{\sigma^2}\end{matrix}\right)$ $-\frac{2^{s-1}}{\pi\omega^s}\left(\frac{\sigma^2}{\omega^2}\right)^{-s/2}\cos\frac{(s-\nu)\pi}{2}\,\Gamma\left(\frac{s-\nu}{2}\right)$ $\times\,\Gamma\left(\frac{s+\nu}{2}\right) {}_4F_3\left(\begin{matrix}\frac{b-a}{2},\ \frac{b-a+1}{2},\ \frac{s-\nu}{2},\ \frac{s+\nu}{2}\\ \frac{1}{2},\ \frac{b+1}{2},\ \frac{b}{2};\ -\frac{\omega^2}{\sigma^2}\end{matrix}\right)$ $\left[\begin{matrix}\sigma>0;\ (\mathrm{Re}\,\omega>0;\ \lvert\mathrm{Re}\,\nu\rvert<\mathrm{Re}\,s<\mathrm{Re}\,(b-a)+3/2)\ \text{or}\\ (\mathrm{Re}\,\omega=0;\ \lvert\mathrm{Re}\,\nu\rvert<\mathrm{Re}\,s<\mathrm{Re}\,a+3/2,\ \mathrm{Re}\,(b-a)+3/2)\end{matrix}\right]$
12	$e^{-\omega x} Y_\nu(\sigma\sqrt{x})$ $\times\, {}_1F_1\left(\begin{matrix}a;\,\omega x\\ b\end{matrix}\right)$	$-\frac{2^{2(s+a-b)}}{\pi}\sigma^{-2s}\left(\frac{\omega}{\sigma^2}\right)^{a-b}\cos\frac{(2s+2a-2b-\nu)\pi}{2}$ $\times\,\Gamma\left[\begin{matrix}b,\ \frac{2s+2a-2b-\nu}{2},\ \frac{2s+2a-2b+\nu}{2}\\ a\end{matrix}\right]$ $\times\, {}_2F_2\left(\begin{matrix}1-a,\ b-a;\ -\frac{\sigma^2}{4\omega}\\ \frac{2b-2a-2s-\nu+2}{2},\ \frac{2b-2a-2s+\nu+2}{2}\end{matrix}\right)$ $-\frac{\sigma^{-2s}}{2^\nu\pi}\left(\frac{\omega}{\sigma^2}\right)^{-s-\nu/2}\cos(\nu\pi)\,\mathrm{B}\left(\frac{2b-2a-2s-\nu}{2},\ \frac{2s+\nu}{2}\right)\times$

No.	$f(x)$	$F(s)$
		$\times\,\Gamma\left[\begin{matrix}-\nu,\ b\\ \frac{2b-2s-\nu}{2}\end{matrix}\right]{}_2F_2\left(\begin{matrix}\frac{2s+\nu}{2},\ \frac{2s-2b+\nu+2}{2};\ -\frac{\sigma^2}{4\omega}\\ \nu+1,\ \frac{2s+2a-2b+\nu+2}{2}\end{matrix}\right)$ $-\frac{2^{\nu}}{\sigma^{2s}}\left(\frac{\omega}{\sigma^2}\right)^{\nu/2-s}\csc(\nu\pi)\,\mathrm{B}\left(\frac{2s-\nu}{2},\ \frac{2b-2a-2s+\nu}{2}\right)$ $\times\,\Gamma\left[\begin{matrix}b\\ 1-\nu,\ \frac{2b-2s+\nu}{2}\end{matrix}\right]{}_2F_2\left(\begin{matrix}\frac{2s-\nu}{2},\ \frac{2s-2b-\nu+2}{2};\ -\frac{\sigma^2}{4\omega}\\ 1-\nu,\ \frac{2s+2a-2b-\nu+2}{2}\end{matrix}\right)$ $\left[\begin{matrix}\sigma>0;\ (\mathrm{Re}\,\omega>0;\ \lvert\mathrm{Re}\,\nu\rvert<\mathrm{Re}\,s<\mathrm{Re}\,(b-a)+3/4)\ \text{or}\\ (\mathrm{Re}\,\omega=0;\ \lvert\mathrm{Re}\,\nu\rvert<\mathrm{Re}\,s<\mathrm{Re}\,a+5/4,\ \mathrm{Re}\,(b-a)+3/4)\end{matrix}\right]$
13	$e^{-\omega x}K_{\nu}(\sigma x)\ {}_1F_1\left(\begin{matrix}a;\ \omega x\\ b\end{matrix}\right)$	$\frac{2^{s-2}}{\omega^{s}}\left(\frac{\sigma^2}{\omega^2}\right)^{-s/2}\Gamma\left(\frac{s-\nu}{2}\right)\Gamma\left(\frac{s+\nu}{2}\right)$ $\times\,{}_4F_3\left(\begin{matrix}\frac{b-a}{2},\ \frac{b-a+1}{2},\ \frac{s-\nu}{2},\ \frac{s+\nu}{2}\\ \frac{1}{2},\ \frac{b}{2},\ \frac{b+1}{2};\ \frac{\omega^2}{\sigma^2}\end{matrix}\right)$ $+\frac{2^{s-1}(a-b)}{b\omega^{s}}\left(\frac{\sigma^2}{\omega^2}\right)^{-(s+1)/2}\Gamma\left(\frac{s-\nu+1}{2}\right)$ $\times\,\Gamma\left(\frac{s+\nu+1}{2}\right){}_4F_3\left(\begin{matrix}\frac{b-a+1}{2},\ \frac{b-a+2}{2},\ \frac{s-\nu+1}{2},\ \frac{s+\nu+1}{2}\\ \frac{3}{2},\ \frac{b+1}{2},\ \frac{b+2}{2};\ \frac{\omega^2}{\sigma^2}\end{matrix}\right)$
		$\left[\begin{matrix}(\mathrm{Re}\,\sigma>0,\ \mathrm{Re}\,(\sigma+\omega)>0;\ \lvert\mathrm{Re}\,\nu\rvert<\mathrm{Re}\,s)\ \text{or}\\ (\mathrm{Re}\,\sigma=0,\ \mathrm{Re}\,\omega>0;\ \lvert\mathrm{Re}\,\nu\rvert<\mathrm{Re}\,s<\mathrm{Re}\,(b-a)+3/2)\ \text{or}\\ (\mathrm{Re}\,\sigma=0,\ \mathrm{Re}\,\omega=0;\ \lvert\mathrm{Re}\,\nu\rvert<\mathrm{Re}\,s<\mathrm{Re}\,a+3/2,\ \mathrm{Re}\,(b-a)+3/2)\end{matrix}\right]$
14	$e^{-\omega x}K_{\nu}(\sigma\sqrt{x})$ $\times\,{}_1F_1\left(\begin{matrix}a;\ \omega x\\ b\end{matrix}\right)$	$2^{2(s+a-b)-1}\sigma^{-2s}\left(\frac{\omega}{\sigma^2}\right)^{a-b}\Gamma\left[\begin{matrix}b,\ \frac{2s+2a-2b-\nu}{2},\ \frac{2s+2a-2b+\nu}{2}\\ a\end{matrix}\right]$ $\times\,{}_2F_2\left(\begin{matrix}1-a,\ b-a;\ \frac{\sigma^2}{4\omega}\\ \frac{2b-2a-2s-\nu}{2},\ \frac{2b-2a-2s+\nu+2}{2}\end{matrix}\right)$ $+\frac{\sigma^{-2s}}{2^{\nu+1}}\left(\frac{\omega}{\sigma^2}\right)^{-s-\nu/2}\mathrm{B}\left(\frac{2b-2a-2s-\nu}{2},\ \frac{2s+\nu}{2}\right)$ $\times\,\Gamma\left[\begin{matrix}-\nu,\ b\\ \frac{2b-2s-\nu}{2}\end{matrix}\right]{}_2F_2\left(\begin{matrix}\frac{2s+\nu}{2},\ \frac{2s-2b+\nu+2}{2};\ \frac{\sigma^2}{4\omega}\\ \nu+1,\ \frac{2s+2a-2b+\nu+2}{2}\end{matrix}\right)$ $+2^{\nu-1}\sigma^{-2s}\left(\frac{\omega}{\sigma^2}\right)^{\nu/2-s}\mathrm{B}\left(\frac{2s-\nu}{2},\ \frac{2b-2a-2s+\nu}{2}\right)$ $\times\,\Gamma\left[\begin{matrix}\nu,\ b\\ \frac{2b-2s+\nu}{2}\end{matrix}\right]{}_2F_2\left(\begin{matrix}\frac{2s-\nu}{2},\ \frac{2s-2b-\nu+2}{2};\ \frac{\sigma^2}{4\omega}\\ 1-\nu,\ \frac{2s+2a-2b-\nu+2}{2}\end{matrix}\right)$
		$\left[\begin{matrix}(\mathrm{Re}\,\sigma>0,\ \mathrm{Re}\,\omega\geq 0;\ \lvert\mathrm{Re}\,\nu\rvert<\mathrm{Re}\,s)\ \text{or}\\ (\mathrm{Re}\,\sigma=0,\ \mathrm{Re}\,\omega>0;\ \lvert\mathrm{Re}\,\nu\rvert<\mathrm{Re}\,s<\mathrm{Re}\,(b-a)+1/2)\ \text{or}\\ (\mathrm{Re}\,\sigma=0,\ \mathrm{Re}\,\omega=0;\ \lvert\mathrm{Re}\,\nu\rvert<\mathrm{Re}\,s<\mathrm{Re}\,a+1,\ \mathrm{Re}\,(b-a)+1/2)\end{matrix}\right]$

3.28.7. ${}_1F_1(a;\, b;\, \omega x)$ and the Struve functions

1	$\mathbf{H}_\nu(\sigma\sqrt{x})\, {}_1F_1\left(\begin{matrix} a;\, \omega x \\ b \end{matrix}\right)$	$\frac{2^{2s-2a}\pi}{\omega^s}\left(\frac{\sigma^2}{\omega}\right)^{a-s}\Gamma\left[\begin{matrix} b \\ b-a,\, \frac{2a-2s-\nu+2}{2},\, \frac{2a-2s+\nu+2}{2} \end{matrix}\right]$ $\times \csc\frac{(2s-2a+\nu+1)\pi}{2}\, {}_2F_2\left(\begin{matrix} a,\, a-b+1;\, -\frac{\sigma^2}{4\omega} \\ a-s-\frac{\nu}{2}+1,\, a-s+\frac{\nu}{2}+1 \end{matrix}\right)$ $+\frac{\omega^{-s}}{2^\nu\sqrt{\pi}}\left(\frac{\sigma^2}{\omega}\right)^{(\nu+1)/2}\mathrm{B}\left(\frac{2a-2s-\nu-1}{2},\, \frac{2s+\nu+1}{2}\right)$ $\times\Gamma\left[\begin{matrix} b \\ \frac{2\nu+3}{2},\, \frac{2b-2s-\nu-1}{2} \end{matrix}\right]{}_3F_3\left(\begin{matrix} 1,\, \frac{2s+\nu+1}{2},\, \frac{2s-2b+\nu+3}{2} \\ \frac{3}{2},\, \frac{2\nu+3}{2},\, \frac{2s-2a+\nu+3}{2};\, -\frac{\sigma^2}{4\omega} \end{matrix}\right)$ $[\sigma>0;\ (\mathrm{Re}\,\omega>0;\ -\mathrm{Re}\,\nu-1<\mathrm{Re}\,s<\mathrm{Re}\,a+1/4,\ \mathrm{Re}\,(a-\nu/2)+1/2)$ or $(\mathrm{Re}\,\omega=0;\ -\mathrm{Re}\,\nu-1<\mathrm{Re}\,s<\mathrm{Re}\,(b-a)+5/4,\ \mathrm{Re}\,(a-\nu/2)+1/2,\ \mathrm{Re}\,(b-a-\nu/2)+3/2)]$
2	$e^{-\omega x}\,\mathbf{H}_\nu(\sigma\sqrt{x})$ $\times\, {}_1F_1\left(\begin{matrix} a;\, \omega x \\ b \end{matrix}\right)$	$2^{2(s+a-b)}\pi\omega^{-s}\left(\frac{\sigma^2}{\omega}\right)^{b-a-s}\csc\frac{(2s+2a-b+\nu+1)\pi}{2}$ $\times\Gamma\left[\begin{matrix} b \\ a,\, \frac{2b-2a-2s-\nu+2}{2},\, \frac{2b-2a-2s+\nu+2}{2} \end{matrix}\right]$ $\times\, {}_2F_2\left(\begin{matrix} 1-a,\, b-a;\, -\frac{\sigma^2}{4\omega} \\ \frac{2b-2a-2s-\nu+2}{2},\, \frac{2b-2a-2s+\nu+2}{2} \end{matrix}\right)$ $+\frac{\omega^{-s}}{2^\nu\sqrt{\pi}}\left(\frac{\sigma^2}{\omega}\right)^{(\nu+1)/2}\mathrm{B}\left(\frac{2b-2a-2s-\nu-1}{2},\, \frac{2s+\nu+1}{2}\right)$ $\times\Gamma\left[\begin{matrix} b \\ \frac{2\nu+3}{2},\, \frac{2b-2s-\nu-1}{2} \end{matrix}\right]{}_3F_3\left(\begin{matrix} 1,\, \frac{2s+\nu+1}{2},\, \frac{2s-2b+\nu+3}{2};\, -\frac{\sigma^2}{4\omega} \\ \frac{3}{2},\, \frac{2\nu+3}{2},\, \frac{2s+2a-2b+\nu+3}{2} \end{matrix}\right)$ $[\sigma>0;\ (\mathrm{Re}\,\omega>0;\ -\mathrm{Re}\,\nu-1<\mathrm{Re}\,s<\mathrm{Re}\,(b-a)+1/4,\ \mathrm{Re}\,(b-a-\nu/2)+1/2)$ or $(\mathrm{Re}\,\omega=0;\ -\mathrm{Re}\,\nu-1<\mathrm{Re}\,s<\mathrm{Re}\,a+5/4,\ \mathrm{Re}\,(a-\nu/2)+3/2,\ \mathrm{Re}\,(b-a-\nu/2)+1/2)]$

3.28.8. ${}_1F_1(a;\, b;\, \omega x)$ and $P_n(\varphi(x))$

1	$\theta(\sigma-x)\, P_n\left(\frac{2x}{\sigma}-1\right)$ $\times\, {}_1F_1\left(\begin{matrix} a;\, \omega x \\ b \end{matrix}\right)$	$\frac{(-1)^n(1-s)_n\,\sigma^s}{(s)_{n+1}}\, {}_3F_3\left(\begin{matrix} a,\, s,\, s;\, \sigma\omega \\ b,\, s-n,\, s+n+1 \end{matrix}\right)$ $[\sigma,\ \mathrm{Re}\,s>0]$
2	$\theta(x-\sigma)\, P_n\left(\frac{2x}{\sigma}-1\right)$ $\times\, {}_1F_1\left(\begin{matrix} a;\, -\omega x \\ b \end{matrix}\right)$	$\frac{(4/\sigma)^n}{n!}\left(\frac{1}{2}\right)_n \omega^{-s-n}\,\Gamma\left[\begin{matrix} b \\ b-n-s \end{matrix}\right]\mathrm{B}(a-n-s,\, s+n)$ $\times\, {}_3F_3\left(\begin{matrix} -n,\, -n,\, a-n-s;\, -\sigma\omega \\ -2n,\, 1-n-s,\, b-n-s \end{matrix}\right)$ $+\frac{(-1)^{n+1}\sigma^s(1-s)_n}{(s)_{n+1}}\, {}_3F_3\left(\begin{matrix} a,\, s,\, s;\, -\sigma\omega \\ b,\, s-n,\, s+n+1 \end{matrix}\right)$ $[\sigma>0;\ (\mathrm{Re}\,\omega>0;\ 0<\mathrm{Re}\,s<\mathrm{Re}\,a-n)$ or $(\mathrm{Re}\,\omega=0;\ 0<\mathrm{Re}\,s<\mathrm{Re}\,a-n,\ \mathrm{Re}\,(b-a)-n+1)]$

No.	$f(x)$	$F(s)$
3	$\theta(\sigma-x)\, P_n\left(\frac{2\sigma}{x}-1\right)$ $\times\, {}_1F_1\left(\begin{matrix} a;\ \omega x \\ b \end{matrix}\right)$	$(-1)^{n+1}\,\sigma^s \frac{(s+1)_n}{(-s)_{n+1}}\, {}_3F_3\left(\begin{matrix} a,\ s-n,\ s+n+1 \\ b,\ s+1,\ s+1;\ \sigma\omega \end{matrix}\right)$ $[\sigma>0;\ \operatorname{Re} s>n]$
4	$\theta(\sigma-x)\, P_n\left(\sqrt{\frac{x}{\sigma}}\right)$ $\times\, {}_1F_1\left(\begin{matrix} a;\ \omega x \\ b \end{matrix}\right)$	$\sigma^s\,\Gamma\left[\begin{matrix} s,\ \frac{2s+1}{2} \\ \frac{2s-n+1}{2},\ \frac{2s+n+2}{2} \end{matrix}\right] {}_3F_3\left(\begin{matrix} a,\ s,\ \frac{2s+1}{2};\ \sigma\omega \\ b,\ \frac{2s-n+1}{2},\ \frac{2s+n+2}{2} \end{matrix}\right)$ $[\sigma>0;\ \operatorname{Re} s>((-1)^n-1)/4]$
5	$\theta(x-\sigma)\, P_n\left(\sqrt{\frac{x}{\sigma}}\right)$ $\times\, {}_1F_1\left(\begin{matrix} a;\ -\omega x \\ b \end{matrix}\right)$	$\frac{2^n\omega^{-s-n/2}}{\sigma^{n/2}n!}\left(\frac{1}{2}\right)_n \mathrm{B}\left(\frac{2a-n-2s}{2},\ \frac{2s+n}{2}\right)$ $\times\,\Gamma\left[\begin{matrix} b \\ \frac{2b-n-2s}{2} \end{matrix}\right] {}_3F_3\left(\begin{matrix} -\frac{n}{2},\ \frac{1-n}{2},\ \frac{2a-n-2s}{2};\ -\sigma\omega \\ \frac{1-2n}{2},\ \frac{2-2s-n}{2},\ \frac{2b-n-2s}{2} \end{matrix}\right)$ $+\,\frac{(\sigma/4)^s}{\sqrt{\pi}}\,\Gamma\left[\begin{matrix} \frac{n-2s+1}{2},\ \frac{-2s-n}{2} \\ 1-2s \end{matrix}\right] {}_3F_3\left(\begin{matrix} a,\ s,\ \frac{2s+1}{2};\ -\sigma\omega \\ b,\ \frac{2s-n+1}{2},\ \frac{2s+n+2}{2} \end{matrix}\right)$ $\left[\begin{matrix} \sigma>0;\ (\operatorname{Re}\omega>0;\ \operatorname{Re} s<\operatorname{Re} a-n/2)\ \text{or} \\ (\operatorname{Re}\omega=0;\ \operatorname{Re} s<\operatorname{Re} a-n/2,\ \operatorname{Re}(b-a)-n/2+1) \end{matrix}\right]$
6	$\theta(\sigma-x)\, P_n\left(\sqrt{\frac{\sigma}{x}}\right)$ $\times\, {}_1F_1\left(\begin{matrix} a;\ \omega x \\ b \end{matrix}\right)$	$\frac{(4\sigma)^s}{\sqrt{\pi}}\,\Gamma\left[\begin{matrix} \frac{2s-n}{2},\ \frac{2s+n+1}{2} \\ 2s+1 \end{matrix}\right] {}_3F_3\left(\begin{matrix} a,\ \frac{2s-n}{2},\ \frac{2s+n+1}{2} \\ b,\ \frac{2s+1}{2},\ s+1;\ \sigma\omega \end{matrix}\right)$ $[\sigma>0;\ \operatorname{Re} s>n/2]$
7	$\theta(x-\sigma)\, P_n\left(\sqrt{\frac{\sigma}{x}}\right)$ $\times\, {}_1F_1\left(\begin{matrix} a;\ -\omega x \\ b \end{matrix}\right)$	$2^{2s+1}\sqrt{\pi}\,\sigma^s\,\Gamma\left[\begin{matrix} -2s \\ \frac{2-2s+n}{2},\ \frac{1-2s-n}{2} \end{matrix}\right] {}_3F_3\left(\begin{matrix} a,\ \frac{2s-n}{2},\ \frac{2s+n+1}{2};\ -\sigma\omega \\ b,\ s+\frac{1}{2},\ s+1 \end{matrix}\right)$ $+\,\frac{(1+(-1)^n)}{2\sqrt{\pi}\,\omega^s}\,\Gamma\left[\begin{matrix} b,\ \frac{n+1}{2} \\ b-s,\ \frac{n+2}{2} \end{matrix}\right] \mathrm{B}(a-s,\ s)$ $\times\, {}_3F_3\left(\begin{matrix} \frac{n+1}{2},\ -\frac{n}{2},\ a-s;\ -\sigma\omega \\ \frac{1}{2},\ 1-s,\ b-s \end{matrix}\right)$ $+\,\frac{((-1)^n-1)\sqrt{\sigma}}{\sqrt{\pi}\,\omega^{s-1/2}}\,\mathrm{B}\left(a-s+\frac{1}{2},\ s-\frac{1}{2}\right)$ $\times\,\Gamma\left[\begin{matrix} b,\ \frac{n+2}{2} \\ b-s+\frac{1}{2},\ \frac{n+1}{2} \end{matrix}\right] {}_3F_3\left(\begin{matrix} \frac{n+2}{2},\ \frac{1-n}{2},\ a-s+\frac{1}{2};\ -\sigma\omega \\ \frac{3}{2},\ \frac{3}{2}-s,\ b-s+\frac{1}{2} \end{matrix}\right)$ $\left[\begin{matrix} \sigma>0;\ (\operatorname{Re}\omega>0;\ \operatorname{Re} s<\operatorname{Re} a+(1-(-1)^n)/4)\ \text{or} \\ (\operatorname{Re}\omega=0;\ \operatorname{Re} s<\operatorname{Re} a+(1-(-1)^n)/4,\ \operatorname{Re}(b-a)+(5-(-1)^n)/4) \end{matrix}\right]$
8	$\theta(\sigma-x)\, e^{-\omega x} P_n\left(\frac{2x}{\sigma}-1\right)$ $\times\, {}_1F_1\left(\begin{matrix} a;\ \omega x \\ b \end{matrix}\right)$	$\frac{(-1)^n\,\sigma^s\,(1-s)_n}{(s)_{n+1}}\, {}_3F_3\left(\begin{matrix} b-a,\ s,\ s;\ -\sigma\omega \\ b,\ s-n,\ s+n+1 \end{matrix}\right)$ $[\sigma,\ \operatorname{Re} s>0]$

No.	$f(x)$	$F(s)$
9	$\theta(x-\sigma)\, e^{-\omega x} P_n\left(\frac{2x}{\sigma}-1\right) \times {}_1F_1\left(\begin{matrix} a;\, \omega x \\ b \end{matrix}\right)$	$\frac{(4/\sigma)^n}{n!}\left(\frac{1}{2}\right)_n \omega^{-s-n}\,\Gamma\left[\begin{matrix} b \\ b-s-n \end{matrix}\right] \mathrm{B}(b-a-s-n,\, s+n) \times {}_3F_3\left(\begin{matrix} -n,\, -n,\, b-a-n-s;\, -\sigma\omega \\ -2n,\, 1-s-n,\, b-n-s \end{matrix}\right) + \frac{(-1)^{n+1}\sigma^s (1-s)_n}{(s)_{n+1}}\, {}_3F_3\left(\begin{matrix} b-a,\, s,\, s;\, -\sigma\omega \\ b,\, s-n,\, s+n+1 \end{matrix}\right)$ $\left[\begin{matrix} \sigma>0;\ (\operatorname{Re}\omega>0;\ 0<\operatorname{Re}s<\operatorname{Re}(b-a)-n)\ \text{or} \\ (\operatorname{Re}\omega=0;\ 0<\operatorname{Re}s<\operatorname{Re}a-n+1,\ \operatorname{Re}(b-a)-n) \end{matrix}\right]$
10	$\theta(\sigma-x)\, e^{-\omega x} P_n\left(\frac{2\sigma}{x}-1\right) \times {}_1F_1\left(\begin{matrix} a;\, \omega x \\ b \end{matrix}\right)$	$(-1)^{n+1}\sigma^s \frac{(s+1)_n}{(-s)_{n+1}}\, {}_3F_3\left(\begin{matrix} b-a,\, s-n,\, s+n+1 \\ b,\, s+1,\, s+1;\, -\sigma\omega \end{matrix}\right)$ $[\sigma>0;\ \operatorname{Re}s>n]$
11	$\theta(\sigma-x)\, e^{-\omega x} P_n\left(\sqrt{\frac{x}{\sigma}}\right) \times {}_1F_1\left(\begin{matrix} a;\, \omega x \\ b \end{matrix}\right)$	$\frac{\sqrt{\pi}\,\sigma^s}{2^{2s-1}}\,\Gamma\left[\begin{matrix} 2s \\ \frac{2s-n+1}{2},\, \frac{2s+n+2}{2} \end{matrix}\right] {}_3F_3\left(\begin{matrix} b-a,\, s,\, \frac{2s+1}{2};\, -\sigma\omega \\ b,\, \frac{2s-n+1}{2},\, \frac{2s+n+2}{2} \end{matrix}\right)$ $[\sigma>0;\ \operatorname{Re}s>((-1)^n-1)/4]$
12	$\theta(x-\sigma)\, e^{-\omega x} P_n\left(\sqrt{\frac{x}{\sigma}}\right) \times {}_1F_1\left(\begin{matrix} a;\, \omega x \\ b \end{matrix}\right)$	$\frac{2^n \omega^{-s-n/2}}{n!\,\sigma^{n/2}}\left(\frac{1}{2}\right)_n \Gamma\left[\begin{matrix} b \\ \frac{2b-n-2s}{2} \end{matrix}\right] \mathrm{B}\left(\frac{2b-2a-n-2s}{2},\, \frac{2s+n}{2}\right) \times {}_3F_3\left(\begin{matrix} -\frac{n}{2},\, \frac{1-n}{2},\, \frac{2b-2a-n-2s}{2};\, -\sigma\omega \\ \frac{1-2n}{2},\, \frac{2-2s-n}{2},\, \frac{2b-n-2s}{2} \end{matrix}\right) + \frac{(\sigma/4)^s}{\sqrt{\pi}} \times \Gamma\left[\begin{matrix} \frac{1-2s+n}{2},\, \frac{-2s-n}{2} \\ 1-2s \end{matrix}\right] {}_3F_3\left(\begin{matrix} b-a,\, s,\, \frac{2s+1}{2};\, -\sigma\omega \\ b,\, \frac{2s-n+1}{2},\, \frac{2s+n+2}{2} \end{matrix}\right)$ $\left[\begin{matrix} \sigma>0;\ (\operatorname{Re}\omega>0;\ \operatorname{Re}s<\operatorname{Re}(b-a)-n/2)\ \text{or} \\ (\operatorname{Re}\omega=0;\ \operatorname{Re}s<\operatorname{Re}a-n/2+1,\ \operatorname{Re}(b-a)-n/2) \end{matrix}\right]$
13	$\theta(\sigma-x)\, e^{-\omega x} P_n\left(\sqrt{\frac{\sigma}{x}}\right) \times {}_1F_1\left(\begin{matrix} a;\, \omega x \\ b \end{matrix}\right)$	$\frac{(4\sigma)^s}{\sqrt{\pi}}\,\Gamma\left[\begin{matrix} \frac{2s-n}{2},\, \frac{2s+n+1}{2} \\ 2s+1 \end{matrix}\right] {}_3F_3\left(\begin{matrix} b-a,\, \frac{2s-n}{2},\, \frac{2s+n+1}{2} \\ b,\, \frac{2s+1}{2},\, s+1;\, -\sigma\omega \end{matrix}\right)$ $[\sigma>0;\ \operatorname{Re}s>n/2]$
14	$\theta(x-\sigma)\, e^{-\omega x} P_n\left(\sqrt{\frac{\sigma}{x}}\right) \times {}_1F_1\left(\begin{matrix} a;\, \omega x \\ b \end{matrix}\right)$	$2^{2s+1}\sqrt{\pi}\,\sigma^s\,\Gamma\left[\begin{matrix} -2s \\ \frac{2-2s+n}{2},\, \frac{1-2s-n}{2} \end{matrix}\right] {}_3F_3\left(\begin{matrix} b-a,\, \frac{2s-n}{2},\, \frac{2s+n+1}{2} \\ b,\, \frac{2s+1}{2},\, s+1;\, -\sigma\omega \end{matrix}\right) + \frac{(1+(-1)^n)}{2\sqrt{\pi}\,\omega^s}\,\Gamma\left[\begin{matrix} b,\, \frac{n+1}{2} \\ b-s,\, \frac{n+2}{2} \end{matrix}\right] {}_3F_3\left(\begin{matrix} -\frac{n}{2},\, \frac{n+1}{2},\, b-a-s \\ \frac{1}{2},\, 1-s,\, b-s;\, -\sigma\omega \end{matrix}\right) \times \mathrm{B}(b-a-s,\, s) + \frac{((-1)^n-1)\sqrt{\sigma}}{\sqrt{\pi}\,\omega^{s-1/2}}\,\mathrm{B}\left(b-a-s+\frac{1}{2},\, s-\frac{1}{2}\right) \times \Gamma\left[\begin{matrix} b,\, \frac{n+2}{2} \\ b-s+\frac{1}{2},\, \frac{n+1}{2} \end{matrix}\right] {}_3F_3\left(\begin{matrix} \frac{1-n}{2},\, \frac{n+2}{2},\, b-a-s+\frac{1}{2} \\ \frac{3}{2},\, \frac{3-2s}{2},\, b-s+\frac{1}{2};\, -\sigma\omega \end{matrix}\right)$ $\left[\begin{matrix} \sigma>0;\ (\operatorname{Re}\omega>0;\ \operatorname{Re}s<\operatorname{Re}(b-a)+(1-(-1)^n)/4)\ \text{or} \\ (\operatorname{Re}\omega=0;\ \operatorname{Re}s<\operatorname{Re}(b-a)+(1-(-1)^n)/4,\ \operatorname{Re}a+(5-(-1)^n)/4) \end{matrix}\right]$

3.28.9. ${}_1F_1(a;\, b;\, \omega x)$ and $T_n(\varphi(x))$

1	$(\sigma-x)_+^{-1/2}\, T_n\left(\frac{2x}{\sigma}-1\right) \times {}_1F_1\left(\begin{matrix} a;\ \omega x \\ b \end{matrix}\right)$	$(-1)^n \sqrt{\pi}\left(\frac{1}{2}-s\right)_n \sigma^{s-1/2}\,\Gamma\left[\begin{matrix} s \\ s+n+\frac{1}{2} \end{matrix}\right] \times {}_3F_3\left(\begin{matrix} a,\ s,\ s+\frac{1}{2};\ \sigma\omega \\ b,\ s-n+\frac{1}{2},\ s+n+\frac{1}{2} \end{matrix}\right)$ $[\sigma,\ \operatorname{Re} s>0]$
2	$(x-\sigma)_+^{-1/2}\, T_n\left(\frac{2x}{\sigma}-1\right) \times {}_1F_1\left(\begin{matrix} a;\ -\omega x \\ b \end{matrix}\right)$	$\frac{1}{2}\left(\frac{4}{\sigma}\right)^n \omega^{-s-n+1/2}\,\Gamma\left[\begin{matrix} b \\ \frac{2b-2n-2s+1}{2} \end{matrix}\right] \times \mathrm{B}\left(\frac{2a-2n-2s+1}{2},\ \frac{2s+2n-1}{2}\right) \times {}_3F_3\left(\begin{matrix} 1-n,\ \frac{1-2n}{2},\ \frac{2a-2n-2s+1}{2};\ -\sigma\omega \\ 1-2n,\ \frac{3-2n-2s}{2},\ \frac{2b-2n-2s+1}{2} \end{matrix}\right) + \sqrt{\pi}\,\sigma^{s-1/2} \times \left(\frac{1-2s}{2}\right)_n \Gamma\left[\begin{matrix} \frac{1-2n-2s}{2} \\ 1-s \end{matrix}\right] {}_3F_3\left(\begin{matrix} a,\ s,\ \frac{2s+1}{2};\ -\sigma\omega \\ b,\ \frac{2s-2n+1}{2},\ \frac{2s+2n+1}{2} \end{matrix}\right)$ $\left[\begin{matrix} \sigma>0;\ (\operatorname{Re}\omega>0;\ \operatorname{Re} s<\operatorname{Re} a-n+1/2)\ \text{or} \\ (\operatorname{Re}\omega=0;\ \operatorname{Re} s<\operatorname{Re} a-n+1/2,\ \operatorname{Re}(b-a)-n+1/2) \end{matrix}\right]$
3	$(\sigma-x)_+^{-1/2}\, T_n\left(\frac{2\sigma}{x}-1\right) \times {}_1F_1\left(\begin{matrix} a;\ \omega x \\ b \end{matrix}\right)$	$\sqrt{\pi}\,\sigma^{s-1/2}(s)_n\,\Gamma\left[\begin{matrix} s-n \\ \frac{2s+1}{2} \end{matrix}\right] {}_3F_3\left(\begin{matrix} a,\ s-n,\ s+n \\ b,\ s,\ \frac{2s+1}{2};\ \sigma\omega \end{matrix}\right)$ $[\sigma>0;\ \operatorname{Re} s>n]$
4	$(\sigma-x)_+^{-1/2}\, T_n\left(\sqrt{\frac{x}{\sigma}}\right) \times {}_1F_1\left(\begin{matrix} a;\ \omega x \\ b \end{matrix}\right)$	$\sqrt{\pi}\,\sigma^{s-1/2}\,\Gamma\left[\begin{matrix} s,\ \frac{2s+1}{2} \\ \frac{2s-n+1}{2},\ \frac{2s+n+1}{2} \end{matrix}\right] {}_3F_3\left(\begin{matrix} a,\ s,\ \frac{2s+1}{2};\ \sigma\omega \\ b,\ \frac{2s-n+1}{2},\ \frac{2s+n+1}{2} \end{matrix}\right)$ $[\sigma>0;\ \operatorname{Re} s>((-1)^n-1)/4]$
5	$(x-\sigma)_+^{-1/2}\, T_n\left(\sqrt{\frac{x}{\sigma}}\right) \times {}_1F_1\left(\begin{matrix} a;\ -\omega x \\ b \end{matrix}\right)$	$\frac{2^{n-1}\omega^{-s-n/2+1/2}}{\sigma^{n/2}}\,\Gamma\left[\begin{matrix} b \\ \frac{2b-n-2s+1}{2} \end{matrix}\right] \times \mathrm{B}\left(\frac{2a-n-2s+1}{2},\ \frac{2s+n-1}{2}\right) \times {}_3F_3\left(\begin{matrix} \frac{1-n}{2},\ \frac{2-n}{2},\ \frac{2a-n-2s+1}{2};\ -\sigma\omega \\ 1-n,\ \frac{3-2s-n}{2},\ \frac{2b-n-2s+1}{2} \end{matrix}\right) + \frac{1}{2}\left(\frac{\sigma}{4}\right)^{s-1/2} \Gamma\left[\begin{matrix} \frac{1-2s+n}{2},\ \frac{1-2s-n}{2} \\ 1-2s \end{matrix}\right] {}_3F_3\left(\begin{matrix} a,\ s,\ \frac{2s+1}{2};\ -\sigma\omega \\ b,\ \frac{2s-n+1}{2},\ \frac{2s+n+1}{2} \end{matrix}\right)$ $\left[\begin{matrix} \sigma>0;\ (\operatorname{Re}\omega>0;\ \operatorname{Re} s<\operatorname{Re} a-n/2+1/2)\ \text{or} \\ (\operatorname{Re}\omega=0;\ \operatorname{Re} s<\operatorname{Re} a-n/2+1/2,\ \operatorname{Re}(b-a)-n/2+3/2) \end{matrix}\right]$
6	$(\sigma-x)_+^{-1/2}\, T_n\left(\sqrt{\frac{\sigma}{x}}\right) \times {}_1F_1\left(\begin{matrix} a;\ \omega x \\ b \end{matrix}\right)$	$(4\sigma)^{s-1/2}\,\mathrm{B}\left(\frac{2s-n}{2},\ \frac{2s+n}{2}\right) {}_3F_3\left(\begin{matrix} a,\ \frac{2s-n}{2},\ \frac{2s+n}{2} \\ b,\ s,\ \frac{2s+1}{2};\ \sigma\omega \end{matrix}\right)$ $[\sigma>0;\ \operatorname{Re} s>n/2]$

No.	$f(x)$	$F(s)$
7	$(x-\sigma)_+^{-1/2}\, T_n\left(\sqrt{\frac{\sigma}{x}}\right) \times {}_1F_1\left(\begin{matrix} a;\, -\omega x \\ b \end{matrix}\right)$	$\frac{1+(-1)^n}{2\,\omega^{s-1/2}}\, \mathrm{B}\left(\frac{2a-2s+1}{2},\, \frac{2s-1}{2}\right) \times \Gamma\left[\begin{matrix} b \\ \frac{2b-2s+1}{2} \end{matrix}\right] {}_3F_3\left(\begin{matrix} \frac{1-n}{2},\, \frac{1+n}{2},\, \frac{2a-2s+1}{2} \\ \frac{1}{2},\, \frac{3-2s}{2},\, \frac{2b-2s+1}{2};\, -\sigma\omega \end{matrix}\right) + \frac{((-1)^n-1)\, n\sqrt{\sigma}}{2\,\omega^{s-1}}\, \mathrm{B}(a-s+1,\, s-1) \times \Gamma\left[\begin{matrix} b \\ b-s+1 \end{matrix}\right] {}_3F_3\left(\begin{matrix} \frac{2-n}{2},\, \frac{2+n}{2},\, a-s+1 \\ \frac{3}{2},\, 2-s,\, b-s+1;\, -\sigma\omega \end{matrix}\right) + 2^{2s}\pi\sigma^{s-1/2}\,\Gamma\left[\begin{matrix} 1-2s \\ \frac{2-2s-n}{2},\, \frac{2-2s+n}{2} \end{matrix}\right] {}_3F_3\left(\begin{matrix} a,\, \frac{2s-n}{2},\, \frac{2s+n}{2} \\ b,\, s,\, \frac{2s+1}{2};\, -\sigma\omega \end{matrix}\right)$ $\left[\begin{matrix} \sigma>0;\ (\operatorname{Re}\omega>0;\ \operatorname{Re}s<\operatorname{Re}a+(3-(-1)^n)/4) \text{ or} \\ (\operatorname{Re}\omega=0;\ \operatorname{Re}s<\operatorname{Re}a+(3-(-1)^n)/4,\ \operatorname{Re}(b-a)+(7-(-1)^n)/4) \end{matrix}\right]$
8	$(\sigma-x)_+^{-1/2}\, e^{-\omega x} \times T_n\left(\frac{2x}{\sigma}-1\right) \times {}_1F_1\left(\begin{matrix} a;\, \omega x \\ b \end{matrix}\right)$	$(-1)^n\sqrt{\pi}\,\sigma^{s-1/2}\left(\frac{1}{2}-s\right)_n \Gamma\left[\begin{matrix} s \\ s+n+\frac{1}{2} \end{matrix}\right] \times {}_3F_3\left(\begin{matrix} b-a,\, s,\, s+\frac{1}{2};\, -\sigma\omega \\ b,\, s-n+\frac{1}{2},\, s+n+\frac{1}{2} \end{matrix}\right)$ $[\sigma,\ \operatorname{Re}s>0]$
9	$(x-\sigma)_+^{-1/2}\, e^{-\omega x} \times T_n\left(\frac{2x}{\sigma}-1\right) \times {}_1F_1\left(\begin{matrix} a;\, \omega x \\ b \end{matrix}\right)$	$\frac{1}{2}\left(\frac{4}{\sigma}\right)^n \omega^{-s-n+1/2}\,\Gamma\left[\begin{matrix} b \\ \frac{2b-2s-2n+1}{2} \end{matrix}\right] \times \mathrm{B}\left(\frac{2b-2a-2s-2n+1}{2},\, \frac{2s+2n-1}{2}\right) \times {}_3F_3\left(\begin{matrix} 1-n,\, \frac{1-2n}{2},\, \frac{2b-2a-2s-2n+1}{2};\, -\sigma\omega \\ 1-2n,\, \frac{3-2s-2n}{2},\, \frac{2b-2s-2n+1}{2} \end{matrix}\right) + \sqrt{\pi}\,\sigma^{s-1/2}\left(\frac{1-2s}{2}\right)_n \Gamma\left[\begin{matrix} \frac{1-2s-2n}{2} \\ 1-s \end{matrix}\right] \times {}_3F_3\left(\begin{matrix} b-a,\, s,\, \frac{2s+1}{2};\, -\sigma\omega \\ b,\, \frac{2s-2n+1}{2},\, \frac{2s+2n+1}{2} \end{matrix}\right)$ $\left[\begin{matrix} \sigma>0;\ (\operatorname{Re}\omega>0;\ \operatorname{Re}s<\operatorname{Re}(b-a)-n+1/2) \text{ or} \\ (\operatorname{Re}\omega=0;\ \operatorname{Re}s<\operatorname{Re}a-n+3/2,\ \operatorname{Re}(b-a)-n+1/2) \end{matrix}\right]$
10	$(\sigma-x)_+^{-1/2}\, e^{-\omega x} \times T_n\left(\frac{2\sigma}{x}-1\right) \times {}_1F_1\left(\begin{matrix} a;\, \omega x \\ b \end{matrix}\right)$	$\sqrt{\pi}\,\sigma^{s-1/2}(s)_n\,\Gamma\left[\begin{matrix} s-n \\ \frac{2s+1}{2} \end{matrix}\right] {}_3F_3\left(\begin{matrix} b-a,\, s-n,\, s+n \\ b,\, s,\, \frac{2s+1}{2};\, -\sigma\omega \end{matrix}\right)$ $[\sigma>0;\ \operatorname{Re}s>n]$

No.	$f(x)$	$F(s)$
11	$(\sigma-x)_+^{-1/2}e^{-\omega x}$ $\times T_n\left(\sqrt{\frac{x}{\sigma}}\right){}_1F_1\left(\begin{matrix}a;\,\omega x\\ b\end{matrix}\right)$	$\frac{\pi\sigma^{s-1/2}}{2^{2s-1}}\Gamma\left[\begin{matrix}2s\\ \frac{2s-n+1}{2},\,\frac{2s+n+1}{2}\end{matrix}\right]{}_3F_3\left(\begin{matrix}b-a,\,s,\,\frac{2s+1}{2};\,-\sigma\omega\\ b,\,\frac{2s-n+1}{2},\,\frac{2s+n+1}{2}\end{matrix}\right)$ $[\sigma>0;\ \mathrm{Re}\,s>((-1)^n-1)/4]$
12	$(x-\sigma)_+^{-1/2}e^{-\omega x}$ $\times T_n\left(\sqrt{\frac{x}{\sigma}}\right)$ $\times{}_1F_1\left(\begin{matrix}a;\,\omega x\\ b\end{matrix}\right)$	$\frac{2^{n-1}\omega^{-s-n/2+1/2}}{\sigma^{n/2}}\Gamma\left[\begin{matrix}b\\ \frac{2b-n-2s+1}{2}\end{matrix}\right]$ $\times\mathrm{B}\left(\frac{2b-2a-n-2s+1}{2},\,\frac{2s+n-1}{2}\right)$ $\times{}_3F_3\left(\begin{matrix}\frac{1-n}{2},\,\frac{2-n}{2},\,\frac{2b-2a-n-2s+1}{2};\,-\sigma\omega\\ 1-n,\,\frac{3-2s-n}{2},\,\frac{2b-n-2s+1}{2}\end{matrix}\right)$ $+\frac{1}{2}\left(\frac{\sigma}{4}\right)^{s-1/2}\Gamma\left[\begin{matrix}\frac{1-2s+n}{2},\,\frac{1-2s-n}{2}\\ 1-2s\end{matrix}\right]{}_3F_3\left(\begin{matrix}b-a,\,s,\,\frac{2s+1}{2};\,-\sigma\omega\\ b,\,\frac{2s-n+1}{2},\,\frac{2s+n+1}{2}\end{matrix}\right)$ $\left[\begin{matrix}\sigma>0;\ (\mathrm{Re}\,\omega>0;\ \mathrm{Re}\,s<\mathrm{Re}\,(b-a)-n/2+1/2)\ \text{or}\\ (\mathrm{Re}\,\omega=0;\ \mathrm{Re}\,s<\mathrm{Re}\,a-n/2+3/2,\ \mathrm{Re}\,(b-a)-n/2+1/2)\end{matrix}\right]$
13	$(\sigma-x)_+^{-1/2}e^{-\omega x}$ $\times T_n\left(\sqrt{\frac{\sigma}{x}}\right){}_1F_1\left(\begin{matrix}a;\,\omega x\\ b\end{matrix}\right)$	$(4\sigma)^{s-1/2}\mathrm{B}\left(\frac{2s-n}{2},\,\frac{2s+n}{2}\right){}_3F_3\left(\begin{matrix}b-a,\,\frac{2s-n}{2},\,\frac{2s+n}{2}\\ b,\,s,\,\frac{2s+1}{2};\,-\sigma\omega\end{matrix}\right)$ $[\sigma>0;\ \mathrm{Re}\,s>n/2]$
14	$(x-\sigma)_+^{-1/2}e^{-\omega x}$ $\times T_n\left(\sqrt{\frac{\sigma}{x}}\right)$ $\times{}_1F_1\left(\begin{matrix}a;\,-\omega x\\ b\end{matrix}\right)$	$\frac{1+(-1)^n}{2\omega^{s-1/2}}\mathrm{B}\left(\frac{2b-2a-2s+1}{2},\,\frac{2s-1}{2}\right)$ $\times\Gamma\left[\begin{matrix}b\\ \frac{2b-2s+1}{2}\end{matrix}\right]{}_3F_3\left(\begin{matrix}\frac{1-n}{2},\,\frac{1+n}{2},\,\frac{2b-2a-2s+1}{2}\\ \frac{1}{2},\,\frac{3-2s}{2},\,\frac{2b-2s+1}{2};\,-\sigma\omega\end{matrix}\right)$ $+\frac{((-1)^n-1)\,n\sqrt{\sigma}}{2\omega^{s-1}}\mathrm{B}\,(b-a-s+1,\,s-1)$ $\times\Gamma\left[\begin{matrix}b\\ b-s+1\end{matrix}\right]{}_3F_3\left(\begin{matrix}\frac{2-n}{2},\,\frac{2+n}{2},\,b-a-s+1\\ \frac{3}{2},\,2-s,\,b-s+1;\,-\sigma\omega\end{matrix}\right)$ $+\frac{2^{2s}\pi}{\sigma^{1/2-s}}\Gamma\left[\begin{matrix}1-2s\\ \frac{2-2s-n}{2},\,\frac{2-2s+n}{2}\end{matrix}\right]{}_3F_3\left(\begin{matrix}b-a,\,\frac{2s-n}{2},\,\frac{2s+n}{2}\\ b,\,s,\,\frac{2s+1}{2};\,-\sigma\omega\end{matrix}\right)$ $\left[\begin{matrix}\sigma>0;\ (\mathrm{Re}\,\omega>0;\ \mathrm{Re}\,s<\mathrm{Re}\,(b-a)+(3-(-1)^n)/4)\ \text{or}\\ (\mathrm{Re}\,\omega=0;\ \mathrm{Re}\,s<\mathrm{Re}\,(b-a)+(3-(-1)^n)/4,\ \mathrm{Re}\,a+(7-(-1)^n)/4)\end{matrix}\right]$

3.28.10. ${}_1F_1(a;\,b;\,\omega x)$ and $U_n(\varphi(x))$

1	$(\sigma-x)_+^{1/2}U_n\left(\frac{2x}{\sigma}-1\right)$ $\times{}_1F_1\left(\begin{matrix}a;\,\omega x\\ b\end{matrix}\right)$	$\frac{(-1)^n(n+1)\sqrt{\pi}\,\sigma^{s+1/2}}{2}\left(\frac{3-2s}{2}\right)_n\Gamma\left[\begin{matrix}s\\ \frac{2s+2n+3}{2}\end{matrix}\right]$ $\times{}_3F_3\left(\begin{matrix}a,\,s,\,s-\frac{1}{2};\,\sigma\omega\\ b,\,\frac{2s-2n-1}{2},\,\frac{2s+2n+3}{2}\end{matrix}\right)$ $[\sigma,\ \mathrm{Re}\,s>0]$

No.	$f(x)$	$F(s)$
2	$(x-\sigma)_+^{1/2}\,U_n\left(\frac{2x}{\sigma}-1\right)$ $\times\,{}_1F_1\left(\begin{matrix}a;\,-\omega x\\ b\end{matrix}\right)$	$\frac{(n+1)\sqrt{\pi}\,\sigma^{s+1/2}}{2}\left(\frac{3-2s}{2}\right)_n\Gamma\left[\begin{matrix}-\frac{2s+2n+1}{2}\\ 1-s\end{matrix}\right]$ $\times\,{}_3F_3\left(\begin{matrix}a,\,s-\frac{1}{2},\,s;\,-\sigma\omega\\ b,\,\frac{2s-2n-1}{2},\,\frac{2s+2n+3}{2}\end{matrix}\right)$ $+\frac{(4/\sigma)^n}{\omega^{s+n+1/2}}\,\mathrm{B}\left(\frac{2a-2n-2s-1}{2},\,\frac{2s+2n+1}{2}\right)$ $\times\,\Gamma\left[\begin{matrix}b\\ \frac{2b-2n-2s-1}{2}\end{matrix}\right]$ $\times\,{}_3F_3\left(\begin{matrix}-n-1,\,-\frac{2n+1}{2},\,\frac{2a-2n-2s-1}{2};\,-\sigma\omega\\ -2n-1,\,\frac{1-2n-2s}{2},\,\frac{2b-2n-2s-1}{2}\end{matrix}\right)$ $\left[\begin{matrix}\sigma>0;\ (\mathrm{Re}\,\omega>0;\ \mathrm{Re}\,s<\mathrm{Re}\,a-n-1/2)\ \text{or}\\ (\mathrm{Re}\,\omega=0;\ \mathrm{Re}\,s<\mathrm{Re}\,a-n-1/2,\ \mathrm{Re}\,(b-a)-n+1/2)\end{matrix}\right]$
3	$(\sigma-x)_+^{1/2}\,U_n\left(\frac{2\sigma}{x}-1\right)$ $\times\,{}_1F_1\left(\begin{matrix}a;\,\omega x\\ b\end{matrix}\right)$	$\frac{(n+1)\sqrt{\pi}\,\sigma^{s+1/2}}{2}(s+2)_n\,\Gamma\left[\begin{matrix}s-n\\ \frac{2s+3}{2}\end{matrix}\right]{}_3F_3\left(\begin{matrix}a,\,s-n,\,s+n+2\\ b,\,s+\frac{3}{2},\,s+2;\,\sigma\omega\end{matrix}\right)$ $[\sigma>0;\ \mathrm{Re}\,s>n]$
4	$(\sigma-x)_+^{1/2}\,U_n\left(\sqrt{\frac{x}{\sigma}}\right)$ $\times\,{}_1F_1\left(\begin{matrix}a;\,\omega x\\ b\end{matrix}\right)$	$\frac{(n+1)\,\pi\sigma^{s+1/2}}{2^{2s}}\,\Gamma\left[\begin{matrix}2s\\ \frac{2s-n+1}{2},\,\frac{2s+n+3}{2}\end{matrix}\right]{}_3F_3\left(\begin{matrix}a,\,s,\,\frac{2s+1}{2};\,\sigma\omega\\ b,\,\frac{2s-n+1}{2},\,\frac{2s+n+3}{2}\end{matrix}\right)$ $[\sigma>0;\ \mathrm{Re}\,s>((-1)^n-1)/4]$
5	$(x-\sigma)_+^{1/2}\,U_n\left(\sqrt{\frac{x}{\sigma}}\right)$ $\times\,{}_1F_1\left(\begin{matrix}a;\,-\omega x\\ b\end{matrix}\right)$	$\frac{2^n\omega^{-s-n/2-1/2}}{\sigma^{n/2}}\,\Gamma\left[\begin{matrix}b\\ \frac{2b-n-2s-1}{2}\end{matrix}\right]$ $\times\,\mathrm{B}\left(\frac{2a-n-2s-1}{2},\,\frac{2s+n+1}{2}\right)$ $\times\,{}_3F_3\left(\begin{matrix}-\frac{n}{2},\,-\frac{n+1}{2},\,\frac{2a-n-2s-1}{2};\,-\sigma\omega\\ -n,\,\frac{1-2s-n}{2},\,\frac{2b-n-2s-1}{2}\end{matrix}\right)$ $+(n+1)\left(\frac{\sigma}{4}\right)^{s+1/2}\Gamma\left[\begin{matrix}\frac{n-2s+1}{2},\,-\frac{2s+n+1}{2}\\ 1-2s\end{matrix}\right]$ $\times\,{}_3F_3\left(\begin{matrix}a,\,s,\,\frac{2s+1}{2};\,-\sigma\omega\\ b,\,\frac{2s-n+1}{2},\,\frac{2s+n+3}{2}\end{matrix}\right)$ $\left[\begin{matrix}\sigma>0;\ (\mathrm{Re}\,\omega>0;\ \mathrm{Re}\,s<\mathrm{Re}\,a-n/2-1/2)\ \text{or}\\ (\mathrm{Re}\,\omega=0;\ \mathrm{Re}\,s<\mathrm{Re}\,a-n/2-1/2,\ \mathrm{Re}\,(b-a)-n/2+1/2)\end{matrix}\right]$
6	$(\sigma-x)_+^{1/2}\,U_n\left(\sqrt{\frac{\sigma}{x}}\right)$ $\times\,{}_1F_1\left(\begin{matrix}a;\,\omega x\\ b\end{matrix}\right)$	$2^{2s}(n+1)\,\sigma^{s+1/2}\,\Gamma\left[\begin{matrix}\frac{2s-n}{2},\,\frac{2s+n+2}{2}\\ 2s+2\end{matrix}\right]$ $\times\,{}_3F_3\left(\begin{matrix}a,\,\frac{2s-n}{2},\,\frac{2s+n+2}{2}\\ b,\,s+1,\,\frac{2s+3}{2};\,\sigma\omega\end{matrix}\right)$ $[\sigma>0;\ \mathrm{Re}\,s>n/2]$

No.	$f(x)$	$F(s)$
7	$(x-\sigma)_+^{1/2} U_n\left(\sqrt{\frac{\sigma}{x}}\right) \times {}_1F_1\left(\begin{matrix} a;\, -\omega x \\ b \end{matrix}\right)$	$(n+1)\,\pi\,(4\sigma)^{s+1/2}\,\Gamma\left[\begin{matrix} -2s-1 \\ \frac{n-2s+2}{2},\, -\frac{2s+n}{2} \end{matrix}\right] \times {}_3F_3\left(\begin{matrix} a,\, \frac{2s-n}{2},\, \frac{2s+n+2}{2};\, -\sigma\omega \\ b,\, s+1,\, s+\frac{3}{2} \end{matrix}\right) + \frac{1+(-1)^n}{2\,\omega^{s+1/2}} \times \Gamma\left[\begin{matrix} b \\ b-s-\frac{1}{2} \end{matrix}\right] \mathrm{B}\left(a-s-\frac{1}{2},\, s+\frac{1}{2}\right) \times {}_3F_3\left(\begin{matrix} -\frac{n+1}{2},\, \frac{n+1}{2},\, \frac{2a-2s-1}{2};\, -\sigma\omega \\ \frac{1}{2},\, \frac{1-2s}{2},\, \frac{2b-2s-1}{2} \end{matrix}\right) + \frac{(n+1)\left((-1)^n-1\right)\sqrt{\sigma}}{2\,\omega^s}\,\Gamma\left[\begin{matrix} b \\ b-s \end{matrix}\right] \mathrm{B}\,(a-s,\, s) \times {}_3F_3\left(\begin{matrix} -\frac{n}{2},\, \frac{n+2}{2},\, a-s;\, -\sigma\omega \\ \frac{3}{2},\, 1-s,\, b-s \end{matrix}\right)$
		$[\sigma > 0;\ (\mathrm{Re}\,\omega > 0;\ \mathrm{Re}\,s < \mathrm{Re}\,a - (1+(-1)^n)/4)$ or $(\mathrm{Re}\,\omega = 0;\ \mathrm{Re}\,s < \mathrm{Re}\,a - (1+(-1)^n)/4,\ \mathrm{Re}\,(b-a) + (3-(-1)^n)/4)]$
8	$(\sigma-x)_+^{1/2} e^{-\omega x} \times U_n\left(\frac{2x}{\sigma}-1\right) \times {}_1F_1\left(\begin{matrix} a;\, \omega x \\ b \end{matrix}\right)$	$\frac{(-1)^n\,(n+1)\sqrt{\pi}\,\sigma^{s+1/2}}{2}\left(\frac{3-2s}{2}\right)_n \Gamma\left[\begin{matrix} s \\ \frac{2s+2n+3}{2} \end{matrix}\right] \times {}_3F_3\left(\begin{matrix} b-a,\, s,\, s-\frac{1}{2};\, -\sigma\omega \\ b,\, \frac{2s-2n-1}{2},\, \frac{2s+2n+3}{2} \end{matrix}\right)$ $[\sigma,\ \mathrm{Re}\,s > 0]$
9	$(x-\sigma)_+^{1/2} e^{-\omega x} \times U_n\left(\frac{2x}{\sigma}-1\right) \times {}_1F_1\left(\begin{matrix} a;\, \omega x \\ b \end{matrix}\right)$	$\frac{(n+1)\sqrt{\pi}\,\sigma^{s+1/2}}{2}\left(\frac{3-2s}{2}\right)_n \times \Gamma\left[\begin{matrix} -\frac{2s+2n+1}{2} \\ 1-s \end{matrix}\right] {}_3F_3\left(\begin{matrix} b-a,\, s-\frac{1}{2},\, s;\, -\sigma\omega \\ b,\, \frac{2s-2n-1}{2},\, \frac{2s+2n+3}{2} \end{matrix}\right) + \frac{(4/\sigma)^n}{\omega^{s+n+1/2}}\,\mathrm{B}\left(\frac{2b-2a-2n-2s-1}{2},\, \frac{2s+2n+1}{2}\right) \times \Gamma\left[\begin{matrix} b \\ \frac{2b-2n-2s-1}{2} \end{matrix}\right] {}_3F_3\left(\begin{matrix} -n-1,\, -\frac{2n+1}{2},\, \frac{2b-2a-2n-2s-1}{2};\, -\sigma\omega \\ -2n-1,\, \frac{1-2n-2s}{2},\, \frac{2b-2n-2s-1}{2} \end{matrix}\right)$ $[\sigma > 0;\ (\mathrm{Re}\,\omega > 0;\ \mathrm{Re}\,s < \mathrm{Re}\,(b-a) - n - 1/2)$ or $(\mathrm{Re}\,\omega = 0;\ \mathrm{Re}\,s < \mathrm{Re}\,a - n + 1/2,\ \mathrm{Re}\,(b-a) - n - 1/2)]$
10	$(\sigma-x)_+^{1/2} e^{-\omega x} \times U_n\left(\frac{2\sigma}{x}-1\right) \times {}_1F_1\left(\begin{matrix} a;\, \omega x \\ b \end{matrix}\right)$	$\frac{(n+1)\sqrt{\pi}\,\sigma^{s+1/2}}{2}\,(s+2)_n\,\Gamma\left[\begin{matrix} s-n \\ \frac{2s+3}{2} \end{matrix}\right] \times {}_3F_3\left(\begin{matrix} b-a,\, s-n,\, s+n+2 \\ b,\, \frac{2s+3}{2},\, s+2;\, -\sigma\omega \end{matrix}\right)$ $[\sigma > 0;\ \mathrm{Re}\,s > n]$

No.	$f(x)$	$F(s)$
11	$(\sigma - x)_+^{1/2}\, e^{-\omega x}$ $\times U_n\left(\sqrt{\frac{x}{\sigma}}\right)$ $\times {}_1F_1\left(\begin{matrix} a;\, \omega x \\ b \end{matrix}\right)$	$\frac{(n+1)\,\pi\sigma^{s+1/2}}{2^{2s}}\,\Gamma\left[\begin{matrix} 2s \\ \frac{2s-n+1}{2},\, \frac{2s+n+3}{2} \end{matrix}\right] {}_3F_3\left(\begin{matrix} b-a,\, s,\, \frac{2s+1}{2};\, -\sigma\omega \\ b,\, \frac{2s-n+1}{2},\, \frac{2s+n+3}{2} \end{matrix}\right)$ $[\sigma > 0;\ \operatorname{Re} s > ((-1)^n - 1)/4]$
12	$(x - \sigma)_+^{1/2}\, e^{-\omega x}$ $\times U_n\left(\sqrt{\frac{x}{\sigma}}\right)$ $\times {}_1F_1\left(\begin{matrix} a;\, \omega x \\ b \end{matrix}\right)$	$\frac{2^n \omega^{-s-n/2-1/2}}{\sigma^{n/2}}\Gamma\left[\begin{matrix} b \\ \frac{2b-n-2s-1}{2} \end{matrix}\right]$ $\times \mathrm{B}\left(\frac{2b-2a-n-2s-1}{2},\, \frac{2s+n+1}{2}\right)$ $\times {}_3F_3\left(\begin{matrix} -\frac{n}{2},\, -\frac{n+1}{2},\, \frac{2b-2a-n-2s-1}{2};\, -\sigma\omega \\ -n,\, \frac{1-2s-n}{2},\, \frac{2b-n-2s-1}{2} \end{matrix}\right)$ $+ (n+1)\left(\frac{\sigma}{4}\right)^{s+1/2}\Gamma\left[\begin{matrix} \frac{n-2s+1}{2},\, -\frac{2s+n+1}{2} \\ 1-2s \end{matrix}\right]$ $\times {}_3F_3\left(\begin{matrix} b-a,\, s,\, \frac{2s+1}{2};\, -\sigma\omega \\ b,\, \frac{2s-n+1}{2},\, \frac{2s+n+3}{2} \end{matrix}\right)$ $\left[\begin{matrix} \sigma > 0;\ (\operatorname{Re}\omega > 0;\ \operatorname{Re} s < \operatorname{Re}(b-a) - n/2 - 1/2)\ \text{or} \\ (\operatorname{Re}\omega = 0;\ \operatorname{Re} s < \operatorname{Re} a - n/2 + 1/2,\ \operatorname{Re}(b-a) - n/2 - 1/2) \end{matrix}\right]$
13	$(\sigma - x)_+^{1/2}\, e^{-\omega x} U_n\left(\sqrt{\frac{\sigma}{x}}\right)$ $\times {}_1F_1\left(\begin{matrix} a;\, \omega x \\ b \end{matrix}\right)$	$2^{2s}(n+1)\,\sigma^{s+1/2}\,\Gamma\left[\begin{matrix} \frac{2s-n}{2},\, \frac{2s+n+2}{2} \\ 2s+2 \end{matrix}\right]$ $\times {}_3F_3\left(\begin{matrix} b-a,\, \frac{2s-n}{2},\, \frac{2s+n+2}{2} \\ b,\, s+1,\, \frac{2s+3}{2};\, -\sigma\omega \end{matrix}\right)$ $[\sigma > 0;\ \operatorname{Re} s > n/2]$
14	$(x - \sigma)_+^{1/2}\, e^{-\omega x}$ $\times U_n\left(\sqrt{\frac{\sigma}{x}}\right)$ $\times {}_1F_1\left(\begin{matrix} a;\, \omega x \\ b \end{matrix}\right)$	$(n+1)\,\pi\,(4\sigma)^{s+1/2}\,\Gamma\left[\begin{matrix} -2s-1 \\ \frac{n-2s+2}{2},\, \frac{-2s-n}{2} \end{matrix}\right]$ $\times {}_3F_3\left(\begin{matrix} b-a,\, \frac{2s-n}{2},\, \frac{2s+n+2}{2} \\ b,\, s+1,\, s+\frac{3}{2};\, -\sigma\omega \end{matrix}\right) + \frac{(1+(-1)^n)}{2\,\omega^{s+1/2}}$ $\times \Gamma\left[\begin{matrix} b \\ b-s-\frac{1}{2} \end{matrix}\right] \mathrm{B}\left(b-a-s-\frac{1}{2},\, s+\frac{1}{2}\right)$ $\times {}_3F_3\left(\begin{matrix} -\frac{n+1}{2},\, \frac{n+1}{2},\, b-a-s-\frac{1}{2} \\ \frac{1}{2},\, \frac{1-2s}{2},\, b-s-\frac{1}{2};\, -\sigma\omega \end{matrix}\right)$ $+ \frac{(n+1)((-1)^n - 1)\sqrt{\sigma}}{2\,\omega^s}\Gamma\left[\begin{matrix} b \\ b-s \end{matrix}\right]\mathrm{B}(b-a-s,\, s)$ $\times {}_3F_3\left(\begin{matrix} -\frac{n}{2},\, \frac{n+2}{2},\, b-a-s \\ \frac{3}{2},\, 1-s,\, b-s;\, -\sigma\omega \end{matrix}\right)$
		$\left[\begin{matrix} \sigma > 0;\ (\operatorname{Re}\omega > 0;\ \operatorname{Re} s < \operatorname{Re}(b-a) - (1+(-1)^n)/4)\ \text{or} \\ (\operatorname{Re}\omega = 0;\ \operatorname{Re} s < \operatorname{Re}(b-a) - (1+(-1)^n)/4,\ \operatorname{Re} a + (3-(-1)^n)/4) \end{matrix}\right]$

3.28.11. ${}_1F_1(a;\, b;\, \omega x)$ and $H_n(\sigma\sqrt{x})$

1	$e^{-\sigma^2 x} H_n(\sigma\sqrt{x}) \times {}_1F_1\left(\begin{matrix} a;\, -\omega x \\ b \end{matrix}\right)$	$\frac{2^{n-2s+1}\sqrt{\pi}}{\sigma^{2s}} \Gamma\left[\begin{matrix} 2s \\ \frac{2s-n+1}{2} \end{matrix}\right] {}_3F_2\left(\begin{matrix} a,\, s,\, \frac{2s+1}{2};\, -\frac{\omega}{\sigma^2} \\ b,\, \frac{2s-n+1}{2} \end{matrix}\right)$
	$\left[\begin{matrix} (\operatorname{Re}(\sigma^2+\omega) > 0;\ \operatorname{Re} s > [n/2] - n/2;\ \lvert\arg\sigma\rvert < \pi/4) \text{ or} \\ (\operatorname{Re}\omega > 0;\ [n/2] - n/2 < \operatorname{Re} s < \operatorname{Re} a - n/2 + 1;\ \lvert\arg\sigma\rvert = \pi/4) \text{ or} \\ (\operatorname{Re}\omega = 0;\ [n/2] - n/2 < \operatorname{Re} s < \operatorname{Re} a - n/2 + 1,\ \operatorname{Re}(b-a) - n/2 + 1;\ \lvert\arg\sigma\rvert = \pi/4) \end{matrix}\right]$	
2	$e^{-(\sigma^2+\omega)x} H_n(\sigma\sqrt{x}) \times {}_1F_1\left(\begin{matrix} a;\, \omega x \\ b \end{matrix}\right)$	$\frac{2^{n-2s+1}\sqrt{\pi}}{\sigma^{2s}} \Gamma\left[\begin{matrix} 2s \\ \frac{2s-n+1}{2} \end{matrix}\right] {}_3F_2\left(\begin{matrix} b-a,\, s,\, \frac{2s+1}{2};\, -\frac{\omega}{\sigma^2} \\ b,\, \frac{2s-n+1}{2} \end{matrix}\right)$
	$\left[\begin{matrix} (\operatorname{Re}(\sigma^2+\omega) > 0;\ \operatorname{Re} s > [n/2] - n/2;\ \lvert\arg\sigma\rvert < \pi/4) \text{ or} \\ (\operatorname{Re}\omega > 0;\ [n/2] - n/2 < \operatorname{Re} s < \operatorname{Re}(b-a) - n/2 + 1;\ \lvert\arg\sigma\rvert = \pi/4) \text{ or} \\ (\operatorname{Re}\omega = 0;\ [n/2] - n/2 < \operatorname{Re} s < \operatorname{Re} a - n/2 + 1,\ \operatorname{Re}(b-a) - n/2 + 1;\ \lvert\arg\sigma\rvert = \pi/4) \end{matrix}\right]$	

3.28.12. ${}_1F_1(a;\, b;\, \omega x)$ and $L_n^{\lambda}(\sigma x)$

1	$e^{-\sigma x} L_n^{\lambda}(\sigma x) \times {}_1F_1\left(\begin{matrix} a;\, -\omega x \\ b \end{matrix}\right)$	$\frac{\sigma^{a-s}}{n!\,\omega^a} (1-s+a+\lambda)_n \Gamma\left[\begin{matrix} b,\, s-a \\ b-a \end{matrix}\right] \times {}_3F_2\left(\begin{matrix} a,\, a-b+1,\, 1-s+a+n+\lambda \\ 1-s+a,\, 1-s+a+\lambda;\, -\frac{\sigma}{\omega} \end{matrix}\right) + \frac{(\lambda+1)_n}{n!\,\omega^s} \mathrm{B}(a-s,\, s) \Gamma\left[\begin{matrix} b \\ b-s \end{matrix}\right] {}_3F_2\left(\begin{matrix} s,\, s-b+1, n+\lambda+1 \\ s+a+1,\, \lambda+1;\, -\frac{\sigma}{\omega} \end{matrix}\right)$
		$\left[\begin{matrix} (\operatorname{Re}\sigma,\ \operatorname{Re}(\sigma+\omega),\ \operatorname{Re} s > 0) \text{ or} \\ (\operatorname{Re}\sigma = 0,\ \operatorname{Re}\omega > 0;\ 0 < \operatorname{Re} s < \operatorname{Re} a - n + 1) \text{ or} \\ (\operatorname{Re}\sigma = 0,\ \operatorname{Re}\omega = 0;\ 0 < \operatorname{Re} s < \operatorname{Re} a - n + 1,\ \operatorname{Re}(b-a) - n + 1) \end{matrix}\right]$
2	$e^{-(\sigma+\omega)x} L_n^{\lambda}(\sigma x) \times {}_1F_1\left(\begin{matrix} a;\, \omega x \\ b \end{matrix}\right)$	$\frac{\sigma^{b-a-s}}{n!\,\omega^{b-a}} (1-s+b-a+\lambda)_n \Gamma\left[\begin{matrix} b,\, s+a-b \\ a \end{matrix}\right] \times {}_3F_2\left(\begin{matrix} 1-a,\, b-a,\, 1-s+b-a+n+\lambda \\ 1-s+b-a,\, 1-s+b-a+\lambda;\, -\frac{\sigma}{\omega} \end{matrix}\right) + \frac{(\lambda+1)_n}{n!\,\omega^s} \mathrm{B}(b-a-s,\, s) \Gamma\left[\begin{matrix} b \\ b-s \end{matrix}\right] \times {}_3F_2\left(\begin{matrix} n+\lambda+1,\, s,\, s-b+1 \\ \lambda+1,\, s+a-b+1;\, -\frac{\sigma}{\omega} \end{matrix}\right)$
		$\left[\begin{matrix} (\operatorname{Re}\sigma,\ \operatorname{Re}(\sigma+\omega),\ \operatorname{Re} s > 0) \text{ or} \\ (\operatorname{Re}\sigma = 0,\ \operatorname{Re}\omega > 0;\ 0 < \operatorname{Re} s < \operatorname{Re}(b-a) - n + 1) \text{ or} \\ (\operatorname{Re}\sigma = 0,\ \operatorname{Re}\omega = 0;\ 0 < \operatorname{Re} s < \operatorname{Re} a - n + 1,\ \operatorname{Re}(b-a) - n + 1) \end{matrix}\right]$

3.28.13. ${}_1F_1(a;\, b;\, \omega x)$ and $C_n^\lambda(\varphi(x))$

1	$(\sigma - x)_+^{\lambda-1/2}\, C_n^\lambda\left(\frac{2x}{\sigma} - 1\right) \times {}_1F_1\left(\begin{matrix} a;\, \omega x \\ b \end{matrix}\right)$	$\frac{(-1)^n (2\lambda)_n \left(\frac{1}{2} - s + \lambda\right)_n \sigma^{s+\lambda-1/2}}{n!}\, \Gamma\left[\begin{matrix} s,\, \lambda + \frac{1}{2} \\ s + n + \lambda + \frac{1}{2} \end{matrix}\right] \times {}_3F_3\left(\begin{matrix} a,\, s,\, s - \lambda + \frac{1}{2};\, \sigma\omega \\ b,\, s - n - \lambda + \frac{1}{2},\, s + n + \lambda + \frac{1}{2} \end{matrix}\right)$ $\left[\sigma,\ \mathrm{Re}\, s > 0;\ \mathrm{Re}\,\lambda > -1/2\right]$
2	$(x - \sigma)_+^{\lambda-1/2}\, C_n^\lambda\left(\frac{2x}{\sigma} - 1\right) \times {}_1F_1\left(\begin{matrix} a;\, -\omega x \\ b \end{matrix}\right)$	$\frac{4^n \omega^{-n-s-\lambda+1/2}}{n!\, \sigma^n} (\lambda)_n\, \Gamma\left[\begin{matrix} b \\ \frac{2b-2n-2s-2\lambda+1}{2} \end{matrix}\right] \times \mathrm{B}\left(\frac{2a - 2n - 2s - 2\lambda + 1}{2},\, \frac{2s + 2n + 2\lambda - 1}{2}\right) \times {}_3F_3\left(\begin{matrix} 1 - n - 2\lambda,\, \frac{1-2n-2\lambda}{2},\, \frac{2a-2n-2s-2\lambda+1}{2};\, -\sigma\omega \\ 1 - 2n - 2\lambda,\, \frac{3-2n-2s-2\lambda}{2},\, \frac{2b-2n-2s-2\lambda+1}{2} \end{matrix}\right) + \frac{\sqrt{\pi}\, \sigma^{s+\lambda-1/2}}{2^{2\lambda-1} n!} \left(\frac{1 - 2s + 2\lambda}{2}\right)_n \times \Gamma\left[\begin{matrix} n + 2\lambda,\, \frac{1-2n-2s-2\lambda}{2} \\ \lambda,\, 1 - s \end{matrix}\right] {}_3F_3\left(\begin{matrix} a,\, s,\, \frac{2s-2\lambda+1}{2};\, -\sigma\omega \\ b,\, \frac{2s-2n-2\lambda+1}{2},\, \frac{2s+2n+2\lambda+1}{2} \end{matrix}\right)$ $\left[\sigma > 0;\ \mathrm{Re}\,\lambda > -1/2;\ (\mathrm{Re}\,\omega > 0;\ \mathrm{Re}\, s < \mathrm{Re}\,(a - \lambda) - n + 1/2)\ \text{or}\ (\mathrm{Re}\,\omega = 0;\ \mathrm{Re}\, s < \mathrm{Re}\,(a - \lambda) - n + 1/2,\ \mathrm{Re}\,(b - a - \lambda) - n + 1/2)\right]$
3	$(\sigma - x)_+^{\lambda-1/2}\, C_n^\lambda\left(\frac{2\sigma}{x} - 1\right) \times {}_1F_1\left(\begin{matrix} a;\, \omega x \\ b \end{matrix}\right)$	$\frac{2^{1-2\lambda} \sqrt{\pi}\, \sigma^{s+\lambda-1/2}}{n!} (s + 2\lambda)_n\, \Gamma\left[\begin{matrix} n + 2\lambda,\, s - n \\ \lambda,\, \frac{2s+2\lambda+1}{2} \end{matrix}\right] \times {}_3F_3\left(\begin{matrix} a,\, s - n,\, s + n + 2\lambda;\, \sigma\omega \\ b,\, \frac{2s+2\lambda+1}{2},\, s + 2\lambda \end{matrix}\right)$ $[\sigma > 0;\ \mathrm{Re}\,\lambda > -1/2;\ \mathrm{Re}\, s > n]$
4	$(\sigma - x)_+^{\lambda-1/2}\, C_n^\lambda\left(\sqrt{\frac{x}{\sigma}}\right) \times {}_1F_1\left(\begin{matrix} a;\, \omega x \\ b \end{matrix}\right)$	$\frac{(2\lambda)_n\, \sigma^{s+\lambda-1/2}}{n!}\, \Gamma\left[\begin{matrix} \frac{2\lambda+1}{2},\, s,\, \frac{2s+1}{2} \\ \frac{2s-n+1}{2},\, \frac{2s+2\lambda+n+1}{2} \end{matrix}\right] \times {}_3F_3\left(\begin{matrix} a,\, s,\, \frac{2s+1}{2};\, \sigma\omega \\ b,\, \frac{2s-n+1}{2},\, \frac{2s+2\lambda+n+1}{2} \end{matrix}\right)$ $[\sigma > 0;\ \mathrm{Re}\,\lambda > -1/2;\ \mathrm{Re}\, s > ((-1)^n - 1)/4]$
5	$(x - \sigma)_+^{\lambda-1/2}\, C_n^\lambda\left(\sqrt{\frac{x}{\sigma}}\right) \times {}_1F_1\left(\begin{matrix} a;\, -\omega x \\ b \end{matrix}\right)$	$\frac{2^n \omega^{-s-n/2-\lambda+1/2}}{n!\, \sigma^{n/2}} (\lambda)_n\, \Gamma\left[\begin{matrix} b \\ \frac{2b-n-2s-2\lambda+1}{2} \end{matrix}\right] \times \mathrm{B}\left(\frac{2a - n - 2s - 2\lambda + 1}{2},\, \frac{2s + n + 2\lambda - 1}{2}\right) \times {}_3F_3\left(\begin{matrix} \frac{1-n-2\lambda}{2},\, \frac{2-n-2\lambda}{2},\, \frac{2a-n-2s-2\lambda+1}{2};\, -\sigma\omega \\ 1 - n - \lambda,\, \frac{3-2s-n-2\lambda}{2},\, \frac{2b-n-2s-2\lambda+1}{2} \end{matrix}\right) + \frac{(\sigma/4)^{s+\lambda-1/2}}{n!} \times \Gamma\left[\begin{matrix} n + 2\lambda,\, \frac{n-2s+1}{2},\, \frac{1-n-2s-2\lambda}{2} \\ \lambda,\, 1 - 2s \end{matrix}\right] {}_3F_3\left(\begin{matrix} a,\, s,\, \frac{2s+1}{2};\, -\sigma\omega \\ b,\, \frac{2s-n+1}{2},\, \frac{2s+n+2\lambda+1}{2} \end{matrix}\right)$ $\left[\sigma > 0;\ \mathrm{Re}\,\lambda > -1/2;\ (\mathrm{Re}\,\omega > 0;\ \mathrm{Re}\, s < \mathrm{Re}\,(a - \lambda) - n/2 + 1/2)\ \text{or}\ (\mathrm{Re}\,\omega = 0;\ \mathrm{Re}\, s < \mathrm{Re}\,(a - \lambda) - n/2 + 1/2,\ \mathrm{Re}\,(b - a - \lambda) - n/2 + 3/2)\right]$

No.	$f(x)$	$F(s)$
6	$(\sigma - x)_+^{\lambda-1/2} C_n^{\lambda}\left(\sqrt{\frac{\sigma}{x}}\right) \times {}_1F_1\left(\begin{matrix} a;\ \omega x \\ b \end{matrix}\right)$	$\frac{2^{2s}\sigma^{s+\lambda-1/2}}{n!}\Gamma\left[\begin{matrix} n+2\lambda,\ \frac{2s-n}{2},\ \frac{2s+n+2\lambda}{2} \\ \lambda,\ 2s+2\lambda \end{matrix}\right] \times {}_3F_3\left(\begin{matrix} a,\ \frac{2s-n}{2},\ \frac{2s+n+2\lambda}{2} \\ b,\ s+\lambda,\ \frac{2s+2\lambda+1}{2};\ \sigma\omega \end{matrix}\right)$ $\left[\begin{matrix}\sigma>0,\ \operatorname{Re}\lambda>-1/2; \\ \operatorname{Re}s>n/2\end{matrix}\right]$
7	$(x - \sigma)_+^{\lambda-1/2} C_n^{\lambda}\left(\sqrt{\frac{\sigma}{x}}\right) \times {}_1F_1\left(\begin{matrix} a;\ -\omega x \\ b \end{matrix}\right)$	$\frac{2^{2s+1}\pi\sigma^{s+\lambda-1/2}}{n!}\Gamma\left[\begin{matrix} n+2\lambda,\ 1-2s-2\lambda \\ \lambda,\ \frac{n-2s+2}{2},\ \frac{2-n-2s-2\lambda}{2} \end{matrix}\right] \times {}_3F_3\left(\begin{matrix} a,\ \frac{2s-n}{2},\ \frac{2s+n+2\lambda}{2};\ -\sigma\omega \\ b,\ s+\lambda,\ s+\lambda+\frac{1}{2} \end{matrix}\right) + \frac{(1+(-1)^n)\,2^{n-1}}{n!\sqrt{\pi}\,\omega^{s+\lambda-1/2}} \times \Gamma\left[\begin{matrix} b,\ \frac{n+1}{2},\ \frac{n+2\lambda}{2} \\ \lambda,\ b-s-\lambda+\frac{1}{2} \end{matrix}\right] \mathrm{B}\left(a-s-\lambda+\frac{1}{2},\ s+\lambda-\frac{1}{2}\right) \times {}_3F_3\left(\begin{matrix} \frac{n+1}{2},\ \frac{1-n-2\lambda}{2},\ a-s-\lambda+\frac{1}{2};\ -\sigma\omega \\ \frac{1}{2},\ \frac{3}{2}-s-\lambda,\ b-s-\lambda+\frac{1}{2} \end{matrix}\right) + \frac{((-1)^n-1)\,2^n\sqrt{\sigma}}{n!\sqrt{\pi}\,\omega^{s+\lambda-1}}\,\mathrm{B}(a-s-\lambda+1,\ s+\lambda-1) \times \Gamma\left[\begin{matrix} b,\ \frac{n+2}{2},\ \frac{n+2\lambda+1}{2} \\ \lambda,\ b-s-\lambda+1 \end{matrix}\right] {}_3F_3\left(\begin{matrix} \frac{n+2}{2},\ \frac{2-n-2\lambda}{2},\ a-s-\lambda+1;\ -\sigma\omega \\ \frac{3}{2},\ 2-s-\lambda,\ b-s-\lambda+1 \end{matrix}\right)$ $\left[\begin{matrix}\sigma>0;\ \operatorname{Re}\lambda>-1/2;\ (\operatorname{Re}\omega>0;\ \operatorname{Re}s<\operatorname{Re}(a-\lambda)+(3-(-1)^n)/4)\ \text{or} \\ (\operatorname{Re}\omega=0;\ \operatorname{Re}s<\operatorname{Re}(a-\lambda)+(3-(-1)^n)/4,\ \operatorname{Re}(b-a-\lambda)+(7-(-1)^n)/4)\end{matrix}\right]$
8	$(\sigma - x)_+^{\lambda-1/2} e^{-\omega x} \times C_n^{\lambda}\left(\frac{2x}{\sigma}-1\right) \times {}_1F_1\left(\begin{matrix} a;\ \omega x \\ b \end{matrix}\right)$	$\frac{(-1)^n\sigma^{s+\lambda-1/2}(2\lambda)_n\left(\frac{1}{2}-s+\lambda\right)_n}{n!}\Gamma\left[\begin{matrix} s,\ \lambda+\frac{1}{2} \\ s+n+\lambda+\frac{1}{2} \end{matrix}\right] \times {}_3F_3\left(\begin{matrix} b-a,\ s,\ s-\lambda+\frac{1}{2};\ -\sigma\omega \\ b,\ s-n-\lambda+\frac{1}{2},\ s+n+\lambda+\frac{1}{2} \end{matrix}\right)$ $[\sigma,\ \operatorname{Re}s>0;\ \operatorname{Re}\lambda>-1/2]$
9	$(x - \sigma)_+^{\lambda-1/2} e^{-\omega x} \times C_n^{\lambda}\left(\frac{2x}{\sigma}-1\right) \times {}_1F_1\left(\begin{matrix} a;\ \omega x \\ b \end{matrix}\right)$	$\frac{4^n\omega^{-s-n-\lambda+1/2}}{n!\,\sigma^n}(\lambda)_n\Gamma\left[\begin{matrix} b \\ \frac{2b-2n-2s-2\lambda+1}{2} \end{matrix}\right] \times \mathrm{B}\left(\frac{2b-2a-2n-2s-2\lambda+1}{2},\ \frac{2s+2n+2\lambda-1}{2}\right) \times {}_3F_3\left(\begin{matrix} 1-n-2\lambda,\ \frac{1-2n-2\lambda}{2},\ \frac{2b-2a-2n-2s-2\lambda+1}{2};\ -\sigma\omega \\ 1-2n-2\lambda,\ \frac{3-2n-2s-2\lambda}{2},\ \frac{2b-2n-2s-2\lambda+1}{2} \end{matrix}\right) + \frac{\sqrt{\pi}\,\sigma^{s+\lambda-1/2}}{2^{2\lambda-1}n!}\left(\frac{1-2s+2\lambda}{2}\right)_n\Gamma\left[\begin{matrix} n+2\lambda,\ \frac{1-2n-2s-2\lambda}{2} \\ \lambda,\ 1-s \end{matrix}\right] \times {}_3F_3\left(\begin{matrix} b-a,\ s,\ \frac{2s-2\lambda+1}{2};\ -\sigma\omega \\ b,\ \frac{2s-2n-2\lambda+1}{2},\ \frac{2s+2n+2\lambda+1}{2} \end{matrix}\right)$ $\left[\begin{matrix}\sigma>0;\ \operatorname{Re}\lambda>-1/2;\ (\operatorname{Re}\omega>0;\ \operatorname{Re}s<\operatorname{Re}(b-a-\lambda)-n+1/2)\ \text{or} \\ (\operatorname{Re}\omega=0;\ \operatorname{Re}s<\operatorname{Re}(a-\lambda)-n+3/2,\ \operatorname{Re}(b-a-\lambda)-n+1/2)\end{matrix}\right]$

No.	$f(x)$	$F(s)$
10	$(\sigma - x)_+^{\lambda-1/2} e^{-\omega x} \times C_n^{\lambda}\left(\frac{2\sigma}{x} - 1\right) \times {}_1F_1\left(\begin{matrix} a;\, \omega x \\ b \end{matrix}\right)$	$\frac{2^{1-2\lambda}\sqrt{\pi}\,\sigma^{s+\lambda-1/2}}{n!} (s+2\lambda)_n\, \Gamma\left[\begin{matrix} n+2\lambda,\, s-n \\ \lambda,\, \frac{2s+2\lambda+1}{2} \end{matrix}\right] \times {}_3F_3\left(\begin{matrix} b-a,\, s-n,\, s+n+2\lambda \\ b,\, \frac{2s+2\lambda+1}{2},\, s+2\lambda;\, -\sigma\omega \end{matrix}\right)$ $[\sigma > 0;\ \mathrm{Re}\,\lambda > -1/2;\ \mathrm{Re}\,s > n]$
11	$(\sigma - x)_+^{\lambda-1/2} e^{-\omega x} \times C_n^{\lambda}\left(\sqrt{\frac{x}{\sigma}}\right) \times {}_1F_1\left(\begin{matrix} a;\, \omega x \\ b \end{matrix}\right)$	$\frac{\pi\sigma^{s+\lambda-1/2}}{2^{2s+2\lambda-2} n!} \Gamma\left[\begin{matrix} n+2\lambda,\, 2s \\ \lambda,\, \frac{2s-n+1}{2},\, \frac{2s+n+2\lambda+1}{2} \end{matrix}\right] \times {}_3F_3\left(\begin{matrix} b-a,\, s,\, \frac{2s+1}{2};\, -\sigma\omega \\ b,\, \frac{2s-n+1}{2},\, \frac{2s+n+2\lambda+1}{2} \end{matrix}\right)$ $[\sigma > 0;\ \mathrm{Re}\,\lambda > -1/2;\ \mathrm{Re}\,s > ((-1)^n - 1)/4]$
12	$(x - \sigma)_+^{\lambda-1/2} e^{-\omega x} \times C_n^{\lambda}\left(\sqrt{\frac{x}{\sigma}}\right) \times {}_1F_1\left(\begin{matrix} a;\, \omega x \\ b \end{matrix}\right)$	$\frac{2^n \omega^{-s-n/2-\lambda+1/2}}{n!\,\sigma^{n/2}} (\lambda)_n\, \Gamma\left[\begin{matrix} b \\ \frac{2b-n-2s-2\lambda+1}{2} \end{matrix}\right] \times \mathrm{B}\left(\frac{2b-2a-n-2s-2\lambda+1}{2},\, \frac{2s+n+2\lambda-1}{2}\right) \times {}_3F_3\left(\begin{matrix} \frac{1-n-2\lambda}{2},\, \frac{2-n-2\lambda}{2},\, \frac{2b-2a-n-2s-2\lambda+1}{2};\, -\sigma\omega \\ 1-n-\lambda,\, \frac{3-2s-n-2\lambda}{2},\, \frac{2b-n-2s-2\lambda+1}{2} \end{matrix}\right) + \frac{(\sigma/4)^{s+\lambda-1/2}}{n!} \Gamma\left[\begin{matrix} n+2\lambda,\, \frac{n-2s+1}{2},\, \frac{1-n-2s-2\lambda}{2} \\ \lambda,\, 1-2s \end{matrix}\right] \times {}_3F_3\left(\begin{matrix} b-a,\, s,\, \frac{2s+1}{2};\, -\sigma\omega \\ b,\, \frac{2s-n+1}{2},\, \frac{2s+n+2\lambda+1}{2} \end{matrix}\right)$
		$\left[\begin{matrix} \sigma > 0;\ \mathrm{Re}\,\lambda > -1/2;\ (\mathrm{Re}\,\omega > 0;\ \mathrm{Re}\,s < \mathrm{Re}\,(b-a-\lambda) - n/2 + 1/2)\ \text{or} \\ (\mathrm{Re}\,\omega = 0;\ \mathrm{Re}\,s < \mathrm{Re}\,(a-\lambda) - n/2 + 3/2,\ \mathrm{Re}\,(b-a-\lambda) - n/2 + 1/2) \end{matrix}\right]$
13	$(\sigma - x)_+^{\lambda-1/2} e^{-\omega x} \times C_n^{\lambda}\left(\sqrt{\frac{\sigma}{x}}\right) \times {}_1F_1\left(\begin{matrix} a;\, \omega x \\ b \end{matrix}\right)$	$\frac{2^{2s}\sigma^{s+\lambda-1/2}}{n!} \Gamma\left[\begin{matrix} n+2\lambda,\, \frac{2s-n}{2},\, \frac{2s+n+2\lambda}{2} \\ \lambda,\, 2s+2\lambda \end{matrix}\right] \times {}_3F_3\left(\begin{matrix} b-a,\, \frac{2s-n}{2},\, \frac{2s+n+2\lambda}{2} \\ b,\, s+\lambda,\, \frac{2s+2\lambda+1}{2};\, -\sigma\omega \end{matrix}\right)$ $[\sigma > 0,\ \mathrm{Re}\,\lambda > -1/2;\ \mathrm{Re}\,s > n/2]$
14	$(x - \sigma)_+^{\lambda-1/2} e^{-\omega x} \times C_n^{\lambda}\left(\sqrt{\frac{\sigma}{x}}\right) \times {}_1F_1\left(\begin{matrix} a;\, \omega x \\ b \end{matrix}\right)$	$\frac{2^{2s+1}\pi\sigma^{s+\lambda-1/2}}{n!} \Gamma\left[\begin{matrix} 1-2s-2\lambda,\, n+2\lambda \\ \lambda,\, \frac{n-2s+2}{2},\, \frac{2-n-2s-2\lambda}{2} \end{matrix}\right] \times {}_3F_3\left(\begin{matrix} b-a,\, \frac{2s-n}{2},\, \frac{2s+n+2\lambda}{2} \\ b,\, s+\lambda,\, s+\lambda+\frac{1}{2};\, -\sigma\omega \end{matrix}\right) + \frac{(1+(-1)^n)\, 2^{n-1}}{n!\sqrt{\pi}\,\omega^{s+\lambda-1/2}} \Gamma\left[\begin{matrix} b,\, \frac{n+1}{2},\, \frac{n+2\lambda}{2} \\ \lambda,\, b-s-\lambda+\frac{1}{2} \end{matrix}\right] \times$

No.	$f(x)$	$F(s)$
		$\times \mathrm{B}\left(b-a-s-\lambda+\frac{1}{2},\, s+\lambda-\frac{1}{2}\right) \times {}_3F_3\left(\begin{matrix}\frac{n+1}{2},\, \frac{1-n-2\lambda}{2},\, b-a-s-\lambda+\frac{1}{2}\\ \frac{1}{2},\, \frac{3}{2}-s-\lambda,\, b-s-\lambda+\frac{1}{2};\, -\sigma\omega\end{matrix}\right) + \frac{((-1)^n-1)\,2^n\sqrt{\sigma}}{n!\sqrt{\pi}\,\omega^{s+\lambda-1}}\, \mathrm{B}(b-a-s-\lambda+1,\, s+\lambda-1) \times \Gamma\left[\begin{matrix}b,\, \frac{n+2}{2},\, \frac{n+2\lambda+1}{2}\\ \lambda,\, b-s-\lambda+1\end{matrix}\right] {}_3F_3\left(\begin{matrix}\frac{n+2}{2},\, \frac{2-n-2\lambda}{2},\, b-a-s-\lambda+1\\ \frac{3}{2},\, 2-s-\lambda,\, b-s-\lambda+1;\, -\sigma\omega\end{matrix}\right)$
	$\left[\sigma>0;\ \mathrm{Re}\,\lambda>-1/2;\ (\mathrm{Re}\,\omega>0;\ \mathrm{Re}\,s<\mathrm{Re}\,(b-a-\lambda)+(3-(-1)^n)/4)\ \text{or}\ (\mathrm{Re}\,\omega=0;\ \mathrm{Re}\,s<\mathrm{Re}\,(b-a-\lambda)+(3-(-1)^n)/4,\ \mathrm{Re}\,(a-\lambda)+(7-(-1)^n)/4)\right]$	

3.28.14. ${}_1F_1(a;\, b;\, \omega x)$ and $P_n^{(\rho,\,\sigma)}(\varphi(x))$

1	$(\sigma-x)_+^{\mu}\, P_n^{(\mu,\,\nu)}\left(\frac{2x}{\sigma}-1\right) \times {}_1F_1\left(\begin{matrix}a;\, \omega x\\ b\end{matrix}\right)$	$\frac{\sigma^{s+\mu}}{n!}\,\Gamma\left[\begin{matrix}n+\mu+1,\, s,\, s-\nu\\ s+n+\mu+1,\, s-n-\nu\end{matrix}\right] \times {}_3F_3\left(\begin{matrix}a,\, s,\, s-\nu;\, \sigma\omega\\ b,\, s+n+\mu+1,\, s-n-\nu\end{matrix}\right)$ $\left[\sigma,\ \mathrm{Re}\,s>0;\ \mathrm{Re}\,\mu>-1\right]$
2	$(x-\sigma)_+^{\mu}\, P_n^{(\mu,\,\nu)}\left(\frac{2x}{\sigma}-1\right) \times {}_1F_1\left(\begin{matrix}a;\, -\omega x\\ b\end{matrix}\right)$	$\frac{\omega^{-s-n-\mu}}{n!\,\sigma^n}\,(n+\mu+\nu+1)_n\, \frac{\mathrm{B}(a-n-s-\mu,\, s+n+\mu)}{\Gamma(b-n-s-\mu)} \times \Gamma(b)\, {}_3F_3\left(\begin{matrix}-n-\mu,\, a-n-s-\mu,\, -n-\mu-\nu;\, -\sigma\omega\\ 1-n-s-\mu,\, b-n-s-\mu,\, -2n-\mu-\nu\end{matrix}\right) + \frac{\sigma^{s+\mu}}{n!}\,(\nu-s+1)_n\, \mathrm{B}(-s-n-\mu,\, n+\mu+1) \times {}_3F_3\left(\begin{matrix}a,\, s,\, s-\nu;\, -\sigma\omega\\ b,\, s+n+\mu+1,\, s-n-\nu\end{matrix}\right)$ $\left[\sigma>0;\ \mathrm{Re}\,\mu>-1;\ (\mathrm{Re}\,\omega>0;\ \mathrm{Re}\,s<\mathrm{Re}\,(a-\mu)-n)\ \text{or}\ (\mathrm{Re}\,\omega=0;\ \mathrm{Re}\,s<\mathrm{Re}\,(a-\mu)-n,\ \mathrm{Re}\,(b-a-\mu)-n+1)\right]$
3	$(\sigma-x)_+^{\mu}\, P_n^{(\mu,\,\nu)}\left(\frac{2\sigma}{x}-1\right) \times {}_1F_1\left(\begin{matrix}a;\, \omega x\\ b\end{matrix}\right)$	$\frac{\sigma^{s+\mu}}{n!}\,\Gamma\left[\begin{matrix}n+\mu+1,\, s-n,\, s+n+\mu+\nu+1\\ s+\mu+1,\, s+\mu+\nu+1\end{matrix}\right] \times {}_3F_3\left(\begin{matrix}a,\, s-n,\, s+n+\mu+\nu+1;\, \sigma\omega\\ b,\, s+\mu+1,\, s+\mu+\nu+1\end{matrix}\right)$ $\left[\sigma>0;\ \mathrm{Re}\,\mu>-1;\ \mathrm{Re}\,s>n\right]$
4	$(x-\sigma)_+^{\mu}\, P_n^{(\mu,\,\nu)}\left(\frac{2\sigma}{x}-1\right) \times {}_1F_1\left(\begin{matrix}a;\, -\omega x\\ b\end{matrix}\right)$	$\frac{(-1)^n\,(\nu+1)_n}{n!\,\omega^{s+\mu}}\,\mathrm{B}(a-s-\mu,\, s+\mu)\,\Gamma\left[\begin{matrix}b\\ b-s-\mu\end{matrix}\right] \times {}_3F_3\left(\begin{matrix}-n-\mu,\, n+\nu+1,\, a-s-\mu\\ \nu+1,\, 1-s-\mu,\, b-s-\mu;\, -\sigma\omega\end{matrix}\right) + \frac{(-1)^n\,\sigma^{s+\mu}}{n!}\,(s+\mu+\nu+1)_n\,\Gamma\left[\begin{matrix}n+\mu+1,\, -s-\mu\\ n-s+1\end{matrix}\right] \times$

No.	$f(x)$	$F(s)$
		$\times\, {}_3F_3\left(\begin{matrix} a,\, s-n,\, s+n+\mu+\nu+1 \\ b,\, s+\mu+1,\, s+\mu+\nu+1;\, -\sigma\omega \end{matrix}\right)$ $\left[\begin{matrix}\sigma>0;\ \operatorname{Re}\mu>-1;\ (\operatorname{Re}\omega>0;\ \operatorname{Re}s<\operatorname{Re}(a-\mu))\ \text{or} \\ (\operatorname{Re}\omega=0;\ \operatorname{Re}s<\operatorname{Re}(a-\mu),\ \operatorname{Re}(b-a-\mu)+1)\end{matrix}\right]$
5	$(\sigma-x)_+^{\mu}\, e^{-\omega x}$ $\times P_n^{(\mu,\,\nu)}\left(\frac{2x}{\sigma}-1\right)$ $\times\, {}_1F_1\left(\begin{matrix} a;\, \omega x \\ b \end{matrix}\right)$	$\frac{\sigma^{s+\mu}}{n!}\,\Gamma\left[\begin{matrix} n+\mu+1,\, s,\, s-\nu \\ s+n+\mu+1,\, s-n-\nu \end{matrix}\right]$ $\times\, {}_3F_3\left(\begin{matrix} b-a,\, s,\, s-\nu;\, -\sigma\omega \\ b,\, s+n+\mu+1,\, s-n-\nu \end{matrix}\right)$ $[\sigma,\ \operatorname{Re}s>0;\ \operatorname{Re}\mu>-1]$
6	$(x-\sigma)_+^{\mu}\, e^{-\omega x}$ $\times P_n^{(\mu,\,\nu)}\left(\frac{2x}{\sigma}-1\right)$ $\times\, {}_1F_1\left(\begin{matrix} a;\, \omega x \\ b \end{matrix}\right)$	$\frac{\omega^{-s-n-\mu}}{n!\,\sigma^{n}}\,(n+\mu+\nu+1)_n\,\mathrm{B}(b-a-n-s-\mu,\, s+n+\mu)$ $\times\,\Gamma\left[\begin{matrix} b \\ b-n-s-\mu \end{matrix}\right]$ $\times\, {}_3F_3\left(\begin{matrix} -n-\mu,\, b-a-n-s-\mu,\, -n-\mu-\nu;\, -\sigma\omega \\ 1-n-s-\mu,\, b-n-s-\mu,\, -2n-\mu-\nu \end{matrix}\right)$ $+\,\frac{\sigma^{s+\mu}}{n!}\,(\nu-s+1)_n\,\mathrm{B}(-s-n-\mu,\, n+\mu+1)$ $\times\, {}_3F_3\left(\begin{matrix} b-a,\, s,\, s-\nu;\, -\sigma\omega \\ b,\, s+n+\mu+1,\, s-n-\nu \end{matrix}\right)$ $\left[\begin{matrix}\sigma>0;\ \operatorname{Re}\mu>-1;\ (\operatorname{Re}\omega>0;\ \operatorname{Re}s<\operatorname{Re}(b-a-\mu)-n)\ \text{or} \\ (\operatorname{Re}\omega=0;\ \operatorname{Re}s<\operatorname{Re}(a-\mu)-n+1,\ \operatorname{Re}(b-a-\mu)-n)\end{matrix}\right]$
7	$(\sigma-x)_+^{\mu}\, e^{-\omega x}$ $\times P_n^{(\mu,\,\nu)}\left(\frac{2\sigma}{x}-1\right)$ $\times\, {}_1F_1\left(\begin{matrix} a;\, \omega x \\ b \end{matrix}\right)$	$\frac{\sigma^{s+\mu}}{n!}\,\Gamma\left[\begin{matrix} n+\mu+1,\, s-n,\, s+n+\mu+\nu+1 \\ s+\mu+1,\, s+\mu+\nu+1 \end{matrix}\right]$ $\times\, {}_3F_3\left(\begin{matrix} b-a,\, s-n,\, s+n+\mu+\nu+1;\, -\sigma\omega \\ b,\, s+\mu+1,\, s+\mu+\nu+1 \end{matrix}\right)$ $[\sigma>0;\ \operatorname{Re}\mu>-1;\ \operatorname{Re}s>n]$
8	$(x-\sigma)_+^{\mu}\, e^{-\omega x}$ $\times P_n^{(\mu,\,\nu)}\left(\frac{2\sigma}{x}-1\right)$ $\times\, {}_1F_1\left(\begin{matrix} a;\, \omega x \\ b \end{matrix}\right)$	$\frac{(-1)^n\,(\nu+1)_n}{n!\,\omega^{s+\mu}}\,\mathrm{B}(b-a-s-\mu,\, s+\mu)\,\Gamma\left[\begin{matrix} b \\ b-s-\mu \end{matrix}\right]$ $\times\, {}_3F_3\left(\begin{matrix} -n-\mu,\, b-a-s-\mu,\, n+\nu+1 \\ -s-\mu+1,\, b-s-\mu,\, \nu+1;\, -\sigma\omega \end{matrix}\right)$ $+\,\frac{(-1)^n\,\sigma^{s+\mu}}{n!}\,(s+\mu+\nu+1)_n\,\Gamma\left[\begin{matrix} n+\mu+1,\, -s-\mu \\ n-s+1 \end{matrix}\right]$ $\times\, {}_3F_3\left(\begin{matrix} b-a,\, s-n,\, s+n+\mu+\nu+1 \\ b,\, s+\mu+1,\, s+\mu+\nu+1;\, -\sigma\omega \end{matrix}\right)$ $\left[\begin{matrix}\sigma>0;\ \operatorname{Re}\mu>-1;\ (\operatorname{Re}\omega>0;\ \operatorname{Re}s<\operatorname{Re}(b-a-\mu))\ \text{or} \\ (\operatorname{Re}\omega=0;\ \operatorname{Re}s<\operatorname{Re}(a-\mu)+1,\ \operatorname{Re}(b-a-\mu))\end{matrix}\right]$

3.28.15. Products of ${}_1F_1\left(a;\, b;\, \omega x^r\right)$

1	${}_1F_1\left(\begin{matrix} a;\, -wx \\ b \end{matrix}\right) {}_1F_1\left(\begin{matrix} c;\, -\sigma x \\ d \end{matrix}\right)$	$\sigma^{a-s}\omega^{-a}\,\Gamma\left[\begin{matrix} b,\, d \\ b-a,\, a+d-s \end{matrix}\right] \mathrm{B}\left(a-s+c,\, s-a\right) \times {}_3F_2\left(\begin{matrix} a,\, a-b+1,\, a+c-s \\ a-s+1,\, a+d-s;\, -\frac{\sigma}{\omega} \end{matrix}\right) + \omega^{-s}\,\Gamma\left[\begin{matrix} b \\ b-s \end{matrix}\right] \mathrm{B}\left(a-s,\, s\right) {}_3F_2\left(\begin{matrix} c,\, s,\, s-b+1 \\ d,\, s-a+1;\, -\frac{\sigma}{\omega} \end{matrix}\right)$
		$\left[\begin{matrix} \left(\operatorname{Re}\sigma>0,\ \operatorname{Re}\omega>0;\ 0<\operatorname{Re}s<\operatorname{Re}(a+c)\right) \text{ or} \\ \left(\operatorname{Re}\sigma>0,\ \operatorname{Re}\omega=0;\ 0<\operatorname{Re}s<\operatorname{Re}(a+c),\ \operatorname{Re}(b+c-a)+1\right) \text{ or} \\ \left(\operatorname{Re}\sigma=0,\ \operatorname{Re}\omega>0;\ 0<\operatorname{Re}s<\operatorname{Re}(a+c),\ \operatorname{Re}(a+d-c)+1\right) \text{ or} \\ \left(\operatorname{Re}\sigma=0,\ \operatorname{Re}\omega=0;\ 0<\operatorname{Re}s<\operatorname{Re}(a+c),\ \operatorname{Re}(b+c-a)+1,\right. \\ \left.\operatorname{Re}(a+d-c)+1,\ \operatorname{Re}(b+d-a-c)+1\right) \end{matrix}\right]$
2	$\left(\sigma^2-x^2\right)_+^{\alpha-1}\, {}_1F_1\left(\begin{matrix} a;\, -\omega x \\ b \end{matrix}\right) \times {}_1F_1\left(\begin{matrix} a;\, \omega x \\ b \end{matrix}\right)$	$\frac{\sigma^{s+2\alpha-2}}{2}\,\mathrm{B}\left(\frac{s}{2},\, \alpha\right) {}_3F_4\left(\begin{matrix} a,\, b-a,\, \frac{s}{2};\, \frac{\sigma^2\omega^2}{4} \\ \frac{b}{2},\, \frac{b+1}{2},\, b,\, \frac{s+2\alpha}{2} \end{matrix}\right)$ $\left[\sigma,\ \operatorname{Re}\alpha,\ \operatorname{Re}s>0\right]$
3	${}_1F_1\left(\begin{matrix} a;\, -\omega x \\ b \end{matrix}\right) {}_1F_1\left(\begin{matrix} c;\, -\frac{\sigma}{x} \\ d \end{matrix}\right)$	$\Gamma\left[\begin{matrix} d \\ s+d \end{matrix}\right] \sigma^s\,\mathrm{B}\left(s+c,\, -s\right) {}_2F_3\left(\begin{matrix} a,\, s+c;\, \sigma\omega \\ b,\, s+1,\, s+d \end{matrix}\right) + \Gamma\left[\begin{matrix} b \\ b-s \end{matrix}\right] \omega^{-s}\,\mathrm{B}\left(a-s,\, s\right) {}_2F_3\left(\begin{matrix} c,\, a-s;\, \sigma\omega \\ d,\, 1-s,\, b-s \end{matrix}\right)$
		$\left[\begin{matrix} \left(\operatorname{Re}\sigma>0,\ \operatorname{Re}\omega>0;\ -\operatorname{Re}a<\operatorname{Re}s<\operatorname{Re}a\right) \text{ or} \\ \left(\operatorname{Re}\sigma>0,\ \operatorname{Re}\omega=0;\ -\operatorname{Re}a<\operatorname{Re}s<\operatorname{Re}a,\ \operatorname{Re}(b-a)+1\right) \text{ or} \\ \left(\operatorname{Re}\sigma=0,\ \operatorname{Re}\omega>0;\ -\operatorname{Re}a,\ \operatorname{Re}(c-d)-1<\operatorname{Re}s<\operatorname{Re}a\right) \text{ or} \\ \left(\operatorname{Re}\sigma=0,\ \operatorname{Re}\omega=0;\ -\operatorname{Re}a,\ \operatorname{Re}(c-d)-1<\operatorname{Re}s<\operatorname{Re}a,\ \operatorname{Re}(b-a)+1\right) \end{matrix}\right]$
4	$e^{-\omega x}\, {}_1F_1\left(\begin{matrix} a;\, -\sigma x \\ b \end{matrix}\right) \times {}_1F_1\left(\begin{matrix} c;\, \omega x \\ d \end{matrix}\right)$	$\sigma^{-a}\omega^{a-s}\,\Gamma\left[\begin{matrix} b,\, d \\ b-a,\, a+d-s \end{matrix}\right] \mathrm{B}\left(s-a,\, a-c+d-s\right) \times {}_3F_2\left(\begin{matrix} a,\, a-b+1,\, a-c+d-s \\ a-s+1,\, a+d-s;\, -\frac{\omega}{\sigma} \end{matrix}\right) + \sigma^{-s}\,\Gamma\left[\begin{matrix} b \\ b-s \end{matrix}\right] \mathrm{B}\left(a-s,\, s\right) {}_3F_2\left(\begin{matrix} d-c,\, s,\, s-b+1 \\ d,\, s-a+1;\, -\frac{\omega}{\sigma} \end{matrix}\right)$
		$\left[\begin{matrix} \left(\operatorname{Re}\sigma>0,\ \operatorname{Re}\omega>0;\ 0<\operatorname{Re}s<\operatorname{Re}(a-c+d)\right) \text{ or} \\ \left(\operatorname{Re}\sigma>0,\ \operatorname{Re}\omega=0;\ 0<\operatorname{Re}s<\operatorname{Re}(a-c+d),\ \operatorname{Re}(a+c)+1\right) \text{ or} \\ \left(\operatorname{Re}\sigma=0,\ \operatorname{Re}\omega>0;\ 0<\operatorname{Re}s<\operatorname{Re}(a-c+d),\ \operatorname{Re}(b+d-a-c)+1\right) \text{ or} \\ \left(\operatorname{Re}\sigma=0,\ \operatorname{Re}\omega=0;\ 0<\operatorname{Re}s<\operatorname{Re}(a-c+d),\ \operatorname{Re}(a+c)+1,\right. \\ \left.\operatorname{Re}(b+d-a-c)+1,\ \operatorname{Re}(b+c-a)+1\right) \end{matrix}\right]$

No.	$f(x)$	$F(s)$
5	$e^{-(\sigma+\omega)x}\,{}_1F_1\left(\begin{matrix}a;\,\sigma x\\ b\end{matrix}\right)\times{}_1F_1\left(\begin{matrix}c;\,\omega x\\ d\end{matrix}\right)$	$\sigma^{a-b}\omega^{-a+b-s}\,\Gamma\left[\begin{matrix}b,\,d\\ a,\,b-a+d-s\end{matrix}\right]\times\mathrm{B}(b-a-c+d-s,\,a-b+s)\times{}_3F_2\left(\begin{matrix}1-a,\,b-a,\,b-a-c+d-s\\ b-a-s+1,\,b-a+d-s;\,-\frac{\omega}{\sigma}\end{matrix}\right)+\sigma^{-s}\,\Gamma\left[\begin{matrix}b\\ b-s\end{matrix}\right]\mathrm{B}(b-a-s,\,s)\times{}_3F_2\left(\begin{matrix}d-c,\,s,\,s-b+1\\ d,\,s+a-b+1;\,-\frac{\omega}{\sigma}\end{matrix}\right)$
		$\left[\begin{matrix}(\mathrm{Re}\,\sigma>0,\ \mathrm{Re}\,\omega>0;\ 0<\mathrm{Re}\,s<\mathrm{Re}\,(b+d-a-c))\ \text{or}\\ (\mathrm{Re}\,\sigma>0,\ \mathrm{Re}\,\omega=0;\ 0<\mathrm{Re}\,s<\mathrm{Re}\,(b+d-a-c),\ \mathrm{Re}\,(b+c-a)+1)\ \text{or}\\ (\mathrm{Re}\,\sigma=0,\ \mathrm{Re}\,\omega>0;\ 0<\mathrm{Re}\,s<\mathrm{Re}\,(b+d-a-c),\ \mathrm{Re}\,(a+d-c)+1)\ \text{or}\\ (\mathrm{Re}\,\sigma=0,\ \mathrm{Re}\,\omega=0;\ 0<\mathrm{Re}\,s<\mathrm{Re}\,(b+d-a-c),\ \mathrm{Re}\,(b+c-a)+1,\\ \mathrm{Re}\,(a+d-c)+1,\ \mathrm{Re}\,(a+c)+1)\end{matrix}\right]$
6	$e^{-\sigma/x}\,{}_1F_1\left(\begin{matrix}a;\,-wx\\ b\end{matrix}\right)\times{}_1F_1\left(\begin{matrix}c;\,\frac{\sigma}{x}\\ d\end{matrix}\right)$	$\sigma^{s}\,\Gamma\left[\begin{matrix}d\\ s+d\end{matrix}\right]\mathrm{B}(s-c+d,\,-s)\,{}_2F_3\left(\begin{matrix}a,\,s-c+d;\,\sigma\omega\\ b,\,s+1,\,s+d\end{matrix}\right)+\omega^{-s}\,\Gamma\left[\begin{matrix}b\\ b-s\end{matrix}\right]\mathrm{B}(a-s,\,s)\,{}_2F_3\left(\begin{matrix}d-c,\,a-s;\,\sigma\omega\\ 1-s,\,b-s,\,d\end{matrix}\right)$
		$\left[\begin{matrix}(\mathrm{Re}\,\sigma>0,\ \mathrm{Re}\,\omega>0;\ \mathrm{Re}\,(c-d)<\mathrm{Re}\,s<\mathrm{Re}\,a)\ \text{or}\\ (\mathrm{Re}\,\sigma>0,\ \mathrm{Re}\,\omega=0;\ \mathrm{Re}\,(c-d)<\mathrm{Re}\,s<\mathrm{Re}\,a,\ \mathrm{Re}\,(b-a)+1)\ \text{or}\\ (\mathrm{Re}\,\sigma=0,\ \mathrm{Re}\,\omega>0;\ \mathrm{Re}\,(c-d),\ -\mathrm{Re}\,c-1<\mathrm{Re}\,s<\mathrm{Re}\,a)\ \text{or}\\ (\mathrm{Re}\,\sigma=0,\ \mathrm{Re}\,\omega=0;\ \mathrm{Re}\,(c-d),\ -\mathrm{Re}\,c-1<\mathrm{Re}\,s<\mathrm{Re}\,a,\ \mathrm{Re}\,(b-a)+1)\end{matrix}\right]$
7	$e^{-\omega x-\sigma/x}\,{}_1F_1\left(\begin{matrix}a;\,wx\\ b\end{matrix}\right)\times{}_1F_1\left(\begin{matrix}c;\,\frac{\sigma}{x}\\ d\end{matrix}\right)$	$\sigma^{s}\,\Gamma\left[\begin{matrix}d\\ s+d\end{matrix}\right]\mathrm{B}(s-c+d,\,-s)\,{}_2F_3\left(\begin{matrix}b-a,\,s-c+d\\ b,\,s+1,\,s+d;\,\sigma\omega\end{matrix}\right)+\omega^{-s}\,\Gamma\left[\begin{matrix}b\\ b-s\end{matrix}\right]\mathrm{B}(b-a-s,\,s)\,{}_2F_3\left(\begin{matrix}d-c,\,b-a-s\\ 1-s,\,b-s,\,d;\,\sigma\omega\end{matrix}\right)$
		$\left[\begin{matrix}(\mathrm{Re}\,\sigma>0,\ \mathrm{Re}\,\omega>0;\ \mathrm{Re}\,(c-d)<\mathrm{Re}\,s<\mathrm{Re}\,(b-a))\ \text{or}\\ (\mathrm{Re}\,\sigma>0,\ \mathrm{Re}\,\omega=0;\ \mathrm{Re}\,(c-d)<\mathrm{Re}\,s<\mathrm{Re}\,a+1,\ \mathrm{Re}\,(b-a))\ \text{or}\\ (\mathrm{Re}\,\sigma=0,\ \mathrm{Re}\,\omega>0;\ \mathrm{Re}\,(c-d),\ -\mathrm{Re}\,c-1<\mathrm{Re}\,s<\mathrm{Re}\,(b-a))\ \text{or}\\ (\mathrm{Re}\,\sigma=0,\ \mathrm{Re}\,\omega=0;\ \mathrm{Re}\,(c-d),\ -\mathrm{Re}\,c-1<\mathrm{Re}\,s<\mathrm{Re}\,a+1,\ \mathrm{Re}\,(b-a))\end{matrix}\right]$
8	$J_\nu(\sigma x)\,{}_1F_1\left(\begin{matrix}a;\,-\omega x\\ b\end{matrix}\right)\times{}_1F_1\left(\begin{matrix}a;\,\omega x\\ b\end{matrix}\right)$	$\frac{2^{s-1}}{\sigma^{s}}\,\Gamma\left[\begin{matrix}\frac{s+\nu}{2}\\ \frac{2-s+\nu}{2}\end{matrix}\right]{}_4F_3\left(\begin{matrix}a,\,b-a,\,\frac{s-\nu}{2},\,\frac{s+\nu}{2}\\ \frac{b}{2},\,\frac{b+1}{2},\,b;\,-\frac{\omega^2}{\sigma^2}\end{matrix}\right)$
		$[\sigma,\,\omega>0;\ -\mathrm{Re}\,\nu<\mathrm{Re}\,s<2\,\mathrm{Re}\,a+3/2,\ 2\,\mathrm{Re}\,(b-a)+3/2,\ \mathrm{Re}\,b+3/2]$

3.29. The Tricomi Confluent Hypergeometric Function $\Psi(a;\, b;\, z)$

In this section, we give some selected simple formulas. Many new transforms can be obtained from Section 3.28 due to the connection formula

$$\Psi\binom{a;\, z}{b} = \frac{\Gamma(b-1)}{\Gamma(a)}\, z^{1-b}\, {}_1F_1\binom{a-b+1}{2-b;\, z} + \frac{\Gamma(1-b)}{\Gamma(a-b+1)}\, {}_1F_1\binom{a;\, z}{b}.$$

More formulas can be obtained from the corresponding sections due to the relations

$$\Psi(a,\, b;\, z) = z^{-a} \lim_{c\to\infty} {}_2F_1\left(a,\, a-b+1;\, c;\, 1-\frac{c}{z}\right),$$

$$\Psi(a,\, b;\, z) = \frac{1}{\Gamma(a)\,\Gamma(a-b+1)}\, G_{12}^{21}\left(z \,\middle|\, \begin{matrix} 1-a \\ 0,\, 1-b \end{matrix}\right).$$

3.29.1. $\Psi(a;\, b;\, \omega x)$ and algebraic functions

No.	$f(x)$	$F(s)$
1	$\Psi\binom{a;\, \omega x}{b}$	$\omega^{-s}\,\Gamma\begin{bmatrix} s,\, s-b+1,\, a-s \\ a,\, a-b+1 \end{bmatrix}$ $[0,\ \operatorname{Re} b-1<\operatorname{Re} s<\operatorname{Re} a]$
2	$(\sigma-x)_+^{\mu-1}\,\Psi\binom{a;\, \omega x}{b}$	$\frac{\sigma^{s-b+\mu}}{\omega^{b-1}}\,\Gamma\begin{bmatrix} b-1 \\ a \end{bmatrix} \mathrm{B}(\mu,\, s-b+1)\, {}_2F_2\binom{a-b+1,\, s-b+1;\, \sigma\omega}{2-b,\, s-b+\mu+1}$ $+\,\sigma^{s+\mu-1}\,\Gamma\begin{bmatrix} 1-b \\ a-b+1 \end{bmatrix} \mathrm{B}(\mu,\, s)\, {}_2F_2\binom{a,\, s;\, \sigma\omega}{b,\, s+\mu}$ $[\sigma,\ \operatorname{Re}\mu>0;\ \operatorname{Re} s>0,\ \operatorname{Re} b-1]$
3	$(x-\sigma)_+^{\mu-1}\,\Psi\binom{a;\, \omega x}{b}$	$\frac{\sigma s^{-b+\mu}}{\omega^{b-1}}\,\Gamma\begin{bmatrix} b-1 \\ a \end{bmatrix} \mathrm{B}(\mu,\, b-s-\mu)\, {}_2F_2\binom{a-b+1,\, s-b+1; \sigma\omega}{2-b, s-b+\mu+1}$ $+\,\omega^{1-s-\mu}\,\Gamma\begin{bmatrix} s+\mu-1 \\ a \end{bmatrix} \mathrm{B}(a-s-\mu+1,\, s-b+\mu)$ $\times\, {}_2F_2\binom{1-\mu,\, a-s-\mu+1;\, \sigma\omega}{2-s-\mu,\, b-s-\mu+1}$ $+\,\sigma^{s+\mu-1}\,\Gamma\begin{bmatrix} 1-b \\ a-b+1 \end{bmatrix} \mathrm{B}(\mu,\, 1-s-\mu)\, {}_2F_2\binom{a,\, s;\, \sigma\omega}{b,\, s+\mu}$ $[\sigma,\ \operatorname{Re}\mu>0;\ \operatorname{Re}(s-a+\mu)<1;\ \lvert\arg\omega\rvert<\pi]$
4	$\frac{1}{x-\sigma}\,\Psi\binom{a;\, \omega x}{b}$	$\frac{\pi\sigma^{s-b}}{\omega^{b-1}} \cot[(b-s)\pi]\,\Gamma\begin{bmatrix} b-1 \\ a \end{bmatrix} {}_1F_1\binom{a-b+1}{2-b;\, \sigma\omega}$ $+\,\omega^{1-s}\,\Gamma\begin{bmatrix} s-b \\ a-b+1 \end{bmatrix} \mathrm{B}(s-1,\, a-s+1)\, {}_2F_2\binom{1,\, a-s+1;\, \sigma\omega}{2-s,\, b-s+1}$ $-\,\pi\sigma^{s-1}\cot(s\pi)\,\Gamma\begin{bmatrix} 1-b \\ a-b+1 \end{bmatrix} {}_1F_1\binom{a;\, \sigma\omega}{b}$ $[\sigma>0;\ 0,\ \operatorname{Re} b-1<\operatorname{Re} s<\operatorname{Re} a+1]$

No.	$f(x)$	$F(s)$
5	$\frac{1}{(x+\sigma)^{\rho}}\Psi\left(\begin{matrix}a;\, \omega x\\ b\end{matrix}\right)$	$\frac{\sigma^{s-b-\rho+1}}{\omega^{b-1}}\Gamma\left[\begin{matrix}b-1\\ a\end{matrix}\right]\mathrm{B}(s-b+1,\, b-s+\rho-1)$ $\times {}_2F_2\left(\begin{matrix}a-b+1,\, s-b+1\\ 2-b,\, s-b-\rho+2;\, -\sigma\omega\end{matrix}\right)+\omega^{\rho-s}\Gamma\left[\begin{matrix}s-b-\rho+1\\ a-b+1\end{matrix}\right]$ $\times \mathrm{B}(s-\rho,\, a-s+\rho)\, {}_2F_2\left(\begin{matrix}\rho,\, a-s+\rho;\, -\sigma\omega\\ \rho-s+1,\, \rho-s+b\end{matrix}\right)$ $+\sigma^{s-\rho}\Gamma\left[\begin{matrix}1-b\\ a-b+1\end{matrix}\right]\mathrm{B}(s,\, \rho)\, {}_2F_2\left(\begin{matrix}a,\, s;\, -\sigma\omega\\ b,\, s-\rho+1\end{matrix}\right)$ $[0,\ \mathrm{Re}\, b-1<\mathrm{Re}\, s<\mathrm{Re}\,(a+\rho);\ \lvert\arg\sigma\rvert,\ \lvert\arg\omega\rvert<\pi]$
6	$\left(\sqrt{x}+\sqrt{x+\sigma}\right)^{\nu}$ $\times\Psi\left(\begin{matrix}a;\, \omega x\\ b\end{matrix}\right)$	$-\frac{\nu\sigma^{s-b+\nu/2+1}}{2^{2s-2b+2}\omega^{b-1}}\Gamma\left[\begin{matrix}b-1,\, 2s-2b+2,\, \frac{2b-2s-\nu-2}{2}\\ a,\, \frac{2s-2b-\nu+4}{2}\end{matrix}\right]$ $\times {}_3F_3\left(\begin{matrix}a-b+1,\, s-b+1,\, \frac{2s-2b+3}{2};\, -\sigma\omega\\ 2-b,\, \frac{2s-2b-\nu+4}{2},\, \frac{2s-2b+\nu+4}{2}\end{matrix}\right)$ $+\frac{2^{\nu}}{\omega^{s+\nu/2}}\mathrm{B}\left(\frac{2a-2s-\nu}{2},\, \frac{2s+\nu}{2}\right)$ $\times\Gamma\left[\begin{matrix}\frac{2s-2b+\nu+2}{2}\\ a-b+1\end{matrix}\right]{}_3F_3\left(\begin{matrix}-\frac{\nu}{2},\, \frac{1-\nu}{2},\, \frac{2a-2s-\nu}{2};\, -\sigma\omega\\ 1-\nu,\, \frac{2-2s-\nu}{2},\, \frac{2b-2s-\nu}{2}\end{matrix}\right)$ $-\frac{\nu\sigma^{s+\nu/2}}{2^{2s}}\Gamma\left[\begin{matrix}1-b,\, -\frac{2s+\nu}{2},\, 2s\\ a-b+1,\, \frac{2s-\nu+2}{2}\end{matrix}\right]{}_3F_3\left(\begin{matrix}a,\, s,\, \frac{2s+1}{2};\, -\sigma\omega\\ b,\, \frac{2s-\nu+2}{2},\, \frac{2s+\nu+2}{2}\end{matrix}\right)$ $[0,\ \mathrm{Re}\, b-1<\mathrm{Re}\, s<\mathrm{Re}\,(a-\nu/2);\ \lvert\arg\sigma\rvert,\ \lvert\arg\omega\rvert<\pi]$
7	$\frac{\left(\sqrt{x}+\sqrt{x+\sigma}\right)^{\nu}}{\sqrt{x+\sigma}}$ $\times\Psi\left(\begin{matrix}a;\, \omega x\\ b\end{matrix}\right)$	$\frac{\sigma^{s-b+(\nu+1)/2}}{2^{2s-2b+1}\omega^{b-1}}\mathrm{B}\left(2s-2b+2,\, \frac{2b-2s-\nu-1}{2}\right)$ $\times\Gamma\left[\begin{matrix}b-1\\ a\end{matrix}\right]{}_3F_3\left(\begin{matrix}a-b+1,\, s-b+1,\, \frac{2s-2b+3}{2};\, -\sigma\omega\\ 2-b,\, \frac{2s-2b-\nu+3}{2},\, \frac{2s-2b+\nu+3}{2}\end{matrix}\right)$ $+\frac{2^{\nu}}{\omega^{s+(\nu-1)/2}}\mathrm{B}\left(\frac{2a-2s-\nu+1}{2},\, \frac{2s+\nu-1}{2}\right)$ $\times\Gamma\left[\begin{matrix}\frac{2s-2b+\nu+1}{2}\\ a-b+1\end{matrix}\right]{}_3F_3\left(\begin{matrix}\frac{1-\nu}{2},\, \frac{2-\nu}{2},\, \frac{2a-2s-\nu+1}{2};\, -\sigma\omega\\ 1-\nu,\, \frac{3-2s-\nu}{2},\, \frac{2b-2s-\nu+1}{2}\end{matrix}\right)$ $+\frac{\sigma^{s+(\nu-1)/2}}{2^{2s-1}}\mathrm{B}\left(2s,\, \frac{1-2s-\nu}{2}\right)$ $\times\Gamma\left[\begin{matrix}1-b\\ a-b+1\end{matrix}\right]{}_3F_3\left(\begin{matrix}a,\, s,\, \frac{2s+1}{2};\, -\sigma\omega\\ b,\, \frac{2s-\nu+1}{2},\, \frac{2s+\nu+1}{2}\end{matrix}\right)$ $[0,\ \mathrm{Re}\, b-1<\mathrm{Re}\, s<\mathrm{Re}\,(a-(\nu-1)/2);\ \lvert\arg\sigma\rvert,\ \lvert\arg\omega\rvert<\pi]$

3.29.2. $\Psi(a;\, b;\, \omega x)$ and the exponential function

1	$e^{-\omega x}\Psi\left(\begin{matrix} a;\ \omega x \\ b \end{matrix}\right)$	$\omega^{-s}\,\Gamma\left[\begin{matrix} s,\ s-b+1 \\ s+a-b+1 \end{matrix}\right]$ $\left[\begin{matrix} (\operatorname{Re}\omega>0;\ \operatorname{Re}s>0,\ \operatorname{Re}b-1)\ \text{or} \\ (\operatorname{Re}\omega=0;\ 0,\ \operatorname{Re}b-1<\operatorname{Re}s<\operatorname{Re}a+1) \end{matrix}\right]$
2	$e^{-\sigma x}\Psi\left(\begin{matrix} a;\ \omega x \\ b \end{matrix}\right)$	$\omega^{-s}\,\Gamma\left[\begin{matrix} s,\ s-b+1 \\ s+a-b+1 \end{matrix}\right] {}_2F_1\left(\begin{matrix} s,\ s-b+1;\ \frac{\omega-\sigma}{\omega} \\ s+a-b+1 \end{matrix}\right)$ $\left[\begin{matrix} (\operatorname{Re}\sigma>0;\ \operatorname{Re}s>0,\ \operatorname{Re}b-1)\ \text{or} \\ (\operatorname{Re}\sigma=0;\ 0,\ \operatorname{Re}b-1<\operatorname{Re}s<\operatorname{Re}a+1) \end{matrix}\right]$
3		$=\frac{\sigma^{b-s-1}}{\omega^{b-1}}\,\Gamma\left[\begin{matrix} b-1,\ s-b+1 \\ a \end{matrix}\right] {}_2F_1\left(\begin{matrix} a-b+1,\ s-b+1 \\ 2-b;\ \frac{\omega}{\sigma} \end{matrix}\right)$ $+\sigma^{-s}\,\Gamma\left[\begin{matrix} 1-b,\ s \\ a-b+1 \end{matrix}\right] {}_2F_1\left(\begin{matrix} a,\ s \\ b;\ \frac{\omega}{\sigma} \end{matrix}\right)$ $\left[\begin{matrix} (\operatorname{Re}\sigma>0;\ \operatorname{Re}s>0,\ \operatorname{Re}b-1)\ \text{or} \\ (\operatorname{Re}\sigma=0;\ 0,\ \operatorname{Re}b-1<\operatorname{Re}s<\operatorname{Re}a+1) \end{matrix}\right]$
4	$e^{-\sigma\sqrt{x}}\Psi\left(\begin{matrix} a;\ \omega x \\ b \end{matrix}\right)$	$\frac{2\sigma^{2a-2s}}{\omega^{a}}\,\Gamma(2s-2a)\ {}_2F_2\left(\begin{matrix} a,\ a-b+1;\ -\frac{\sigma^2}{4\omega} \\ \frac{2a-2s+1}{2},\ a-s+1 \end{matrix}\right)$ $-\frac{\sigma}{\omega^{s+1/2}}\,\mathrm{B}\left(\frac{2s+1}{2},\ \frac{2a-2s-1}{2}\right)$ $\times\Gamma\left[\begin{matrix} \frac{s-2b+3}{2} \\ a-b+1 \end{matrix}\right] {}_2F_2\left(\begin{matrix} \frac{2s+1}{2},\ \frac{2s-2b+3}{2} \\ \frac{3}{2},\ \frac{2s-2a+3}{2};\ -\frac{\sigma^2}{4\omega} \end{matrix}\right)$ $+\omega^{-s}\,\mathrm{B}(s,\ a-s)\,\Gamma\left[\begin{matrix} \frac{s-b+1}{2} \\ a-b+1 \end{matrix}\right] {}_2F_2\left(\begin{matrix} s,\ s-b+1 \\ \frac{1}{2},\ s-a+1;\ -\frac{\sigma^2}{4\omega} \end{matrix}\right)$ $\left[\begin{matrix} (\operatorname{Re}\sigma>0;\ \operatorname{Re}s>0,\ \operatorname{Re}b-1)\ \text{or} \\ (\operatorname{Re}\sigma=0;\ 0,\ \operatorname{Re}b-1<\operatorname{Re}s<\operatorname{Re}a+1/2) \end{matrix}\right]$
5	$e^{-\sigma\sqrt{x}-\omega x}\Psi\left(\begin{matrix} a;\ \omega x \\ b \end{matrix}\right)$	$\omega^{-s}\,\Gamma\left[\begin{matrix} s,\ s-b+1 \\ s+a-b+1 \end{matrix}\right] {}_2F_2\left(\begin{matrix} s,\ s-b+1;\ \frac{\sigma^2}{4\omega} \\ \frac{1}{2},\ s+a-b+1 \end{matrix}\right)$ $-\frac{\sigma}{\omega^{s+1/2}}\,\Gamma\left[\begin{matrix} \frac{2s+1}{2},\ \frac{2s-2b+3}{2} \\ \frac{2s+2a-2b+3}{2} \end{matrix}\right] {}_2F_2\left(\begin{matrix} \frac{2s+1}{2},\ \frac{2s-2b+3}{2};\ \frac{\sigma^2}{4\omega} \\ \frac{3}{2},\ \frac{2s+2a-2b+3}{2} \end{matrix}\right)$ $\left[\begin{matrix} (\operatorname{Re}\omega>0;\ \operatorname{Re}s>0,\ \operatorname{Re}b-1)\ \text{or} \\ (\operatorname{Re}\omega=0,\ \operatorname{Re}\sigma>0;\ \operatorname{Re}s>0,\ \operatorname{Re}b-1)\ \text{or} \\ (\operatorname{Re}\omega=0,\ \operatorname{Re}\sigma=0;\ \operatorname{Re}a+1>\operatorname{Re}s>0,\ \operatorname{Re}b-1) \end{matrix}\right]$
6	$e^{-\sigma/x}\Psi\left(\begin{matrix} a;\ \omega x \\ b \end{matrix}\right)$	$\frac{\sigma^{-b+s+1}}{\omega^{b-1}}\,\Gamma\left[\begin{matrix} b-1,\ b-s-1 \\ a \end{matrix}\right] {}_1F_2\left(\begin{matrix} a-b+1;\ -\sigma\omega \\ 2-b,\ s-b+2 \end{matrix}\right)$ $+\omega^{-s}\,\mathrm{B}(s,\ a-s)\,\Gamma\left[\begin{matrix} s-b+1 \\ a-b+1 \end{matrix}\right] {}_1F_2\left(\begin{matrix} a-s;\ -\sigma\omega \\ 1-s,\ b-s \end{matrix}\right)$ $+\sigma^{s}\,\Gamma\left[\begin{matrix} 1-b,\ -s \\ a-b+1 \end{matrix}\right] {}_1F_2\left(\begin{matrix} a;\ -\sigma\omega \\ b,\ s+1 \end{matrix}\right)$ $\left[\begin{matrix} (\operatorname{Re}\sigma>0;\ \operatorname{Re}s<\operatorname{Re}a)\ \text{or} \\ (\operatorname{Re}\sigma=0;\ -1,\ \operatorname{Re}b-2<\operatorname{Re}s<\operatorname{Re}a) \end{matrix}\right]$

No.	$f(x)$	$F(s)$
7	$e^{-\omega x-\sigma/x}\Psi\left(\begin{matrix}a;\,\omega x\\ b\end{matrix}\right)$	$\frac{\sigma^{s-b+1}}{\omega^{b-1}}\Gamma\left[\begin{matrix}b-1,\, b-s-1\\ a\end{matrix}\right]\,{}_1F_2\left(\begin{matrix}1-a;\,\sigma\omega\\ 2-b,\, s-b+2\end{matrix}\right)$ $+\omega^{-s}\,\Gamma\left[\begin{matrix}s,\, s-b+1\\ s+a-b+1\end{matrix}\right]\,{}_1F_2\left(\begin{matrix}s-a+b;\,\sigma\omega\\ 1-s,\, b-s\end{matrix}\right)$ $+\sigma^{s}\,\Gamma\left[\begin{matrix}1-b,\, -s\\ a-b+1\end{matrix}\right]\,{}_1F_2\left(\begin{matrix}b-a;\,\sigma\omega\\ b,\, s+1\end{matrix}\right)$ $\left[\begin{matrix}(\operatorname{Re}\sigma,\ \operatorname{Re}\omega>0)\ \text{or}\\ (\operatorname{Re}\sigma=0,\ \operatorname{Re}\omega>0;\ -1,\ \operatorname{Re}b-2<\operatorname{Re}s)\ \text{or}\\ (\operatorname{Re}\sigma>0,\ \operatorname{Re}\omega=0;\ \operatorname{Re}s<\operatorname{Re}a+1)\ \text{or}\\ (\operatorname{Re}\sigma=0,\ \operatorname{Re}\omega=0;\ -1,\ \operatorname{Re}b-2<\operatorname{Re}s<\operatorname{Re}a+1)\end{matrix}\right]$
8	$(\sigma-x)_+^{\mu-1}\,e^{-\omega x}$ $\times\Psi\left(\begin{matrix}a;\,\omega x\\ b\end{matrix}\right)$	$\frac{\sigma^{s-b+\mu}}{\omega^{b-1}}\,\Gamma\left[\begin{matrix}b-1\\ a\end{matrix}\right]\mathrm{B}(\mu,\, s-b+1)\,{}_2F_2\left(\begin{matrix}1-a,\, s-b+1;\, -\sigma\omega\\ 2-b,\, s-b+\mu+1\end{matrix}\right)$ $+\sigma^{s+\mu-1}\,\Gamma\left[\begin{matrix}1-b\\ a-b+1\end{matrix}\right]\mathrm{B}(\mu,\, s)\,{}_2F_2\left(\begin{matrix}b-a,\, s;\, -\sigma\omega\\ b,\, s+\mu\end{matrix}\right)$ $[\sigma,\ \operatorname{Re}\mu>0;\ \operatorname{Re}s>0,\ \operatorname{Re}b-1]$
9	$(x-\sigma)_+^{\mu-1}\,e^{-\omega x}$ $\times\Psi\left(\begin{matrix}a;\,\omega x\\ b\end{matrix}\right)$	$\frac{\sigma^{s-b+\mu}}{\omega^{b-1}}\,\Gamma\left[\begin{matrix}b-1\\ a\end{matrix}\right]\mathrm{B}(\mu,\, b-s-\mu)\,{}_2F_2\left(\begin{matrix}1-a,\, s-b+1;\, -\sigma\omega\\ 2-b,\, s-b+\mu+1\end{matrix}\right)$ $+\omega^{-s-\mu+1}\,\Gamma\left[\begin{matrix}s+\mu-1,\, s-b+\mu\\ a-b+s+\mu\end{matrix}\right]$ $\times\,{}_2F_2\left(\begin{matrix}1-\mu,\, b-a-s-\mu+1;\, -\sigma\omega\\ 2-s-\mu,\, b-s-\mu+1\end{matrix}\right)$ $+\sigma^{s+\mu-1}\,\Gamma\left[\begin{matrix}1-b\\ a-b+1\end{matrix}\right]\mathrm{B}(\mu,\, 1-s-\mu)\,{}_2F_2\left(\begin{matrix}b-a,\, s\\ b,\, s+\mu;\, -\sigma\omega\end{matrix}\right)$ $\left[\begin{matrix}\sigma,\ \operatorname{Re}\mu>0;\ \operatorname{Re}\omega>0\ \text{or}\\ (\operatorname{Re}\omega=0;\ \operatorname{Re}s<\operatorname{Re}(a-\mu)+2)\end{matrix}\right]$
10	$\frac{e^{-\omega x}}{(x+\sigma)^{\rho}}\,\Psi\left(\begin{matrix}a;\,\omega x\\ b\end{matrix}\right)$	$\frac{\sigma^{s-b-\rho+1}}{\omega^{b-1}}\,\Gamma\left[\begin{matrix}b-1\\ a\end{matrix}\right]\mathrm{B}(s-b+1,\, b-s+\rho-1)$ $\times\,{}_2F_2\left(\begin{matrix}1-a,\, s-b+1;\, \sigma\omega\\ 2-b,\, s-b-\rho+2\end{matrix}\right)$ $+\omega^{\rho-s}\,\Gamma\left[\begin{matrix}s-\rho,\, s-b-\rho+1\\ s+a-b-\rho+1\end{matrix}\right]\,{}_2F_2\left(\begin{matrix}\rho,\, b-a-s+\rho;\, \sigma\omega\\ \rho-s+1,\, \rho-s+b\end{matrix}\right)$ $+\sigma^{s-\rho}\,\Gamma\left[\begin{matrix}1-b\\ a-b+1\end{matrix}\right]\mathrm{B}(s,\, \rho-s)\,{}_2F_2\left(\begin{matrix}b-a,\, s;\, \sigma\omega\\ b,\, s-\rho+1\end{matrix}\right)$ $\left[\begin{matrix}(\operatorname{Re}\omega>0;\ \operatorname{Re}s>0,\ \operatorname{Re}b-1)\ \text{or}\\ (\operatorname{Re}\omega=0;\, 0,\ \operatorname{Re}b-1<\operatorname{Re}s<\operatorname{Re}\rho+1);\ \lvert\arg\sigma\rvert<\pi\end{matrix}\right]$

No.	$f(x)$	$F(s)$
11	$\frac{e^{-\omega x}}{x-\sigma}\,\Psi\left(\begin{matrix}a;\ \omega x\\ b\end{matrix}\right)$	$\frac{\pi\sigma^{s-b}}{\omega^{b-1}}\cot\left[(b-s)\pi\right]\Gamma\left[\begin{matrix}b-1\\ a\end{matrix}\right]{}_1F_1\left(\begin{matrix}1-a;\ -\sigma\omega\\ 2-b\end{matrix}\right)$ $+\omega^{1-s}\,\Gamma\left[\begin{matrix}s-1,\ s-b\\ s+a-b\end{matrix}\right]{}_2F_2\left(\begin{matrix}1,\ b-a-s+1;\ -\sigma\omega\\ 2-s,\ b-s+1\end{matrix}\right)$ $-\pi\sigma^{s-1}\cot(s\pi)\,\Gamma\left[\begin{matrix}1-b\\ a-b+1\end{matrix}\right]{}_1F_1\left(\begin{matrix}b-a\\ b;\ -\sigma\omega\end{matrix}\right)$ $\left[\begin{matrix}\sigma>0;\ (\operatorname{Re}\omega>0;\ \operatorname{Re}s>0,\ \operatorname{Re}b-1)\ \text{or}\\ (\operatorname{Re}\omega=0;\ 0,\ \operatorname{Re}b-1<\operatorname{Re}s<2)\end{matrix}\right]$
12	$\left(\sqrt{x}+\sqrt{\sigma+x}\right)^{\nu}e^{-\omega x}$ $\times\,\Psi\left(\begin{matrix}a;\ \omega x\\ b\end{matrix}\right)$	$\frac{\nu\sigma^{s-b+\nu/2+1}}{2^{2s-2b+2}\omega^{b-1}}\,\Gamma\left[\begin{matrix}b-1,\ 2s-2b+2,\ \frac{2b-2s-\nu-2}{2}\\ a,\ \frac{2s-2b-\nu+4}{2}\end{matrix}\right]$ $\times\,{}_3F_3\left(\begin{matrix}1-a,\ s-b+1,\ \frac{2s-2b+3}{2};\ \sigma\omega\\ 2-b,\ \frac{2s-2b-\nu+4}{2},\ \frac{2s-2b+\nu+4}{2}\end{matrix}\right)$ $+\frac{2^{\nu}}{\omega^{s+\nu/2}}\,\Gamma\left[\begin{matrix}s+\frac{\nu}{2},\ -b+s+\frac{\nu}{2}+1\\ a-b+s+\frac{\nu}{2}+1\end{matrix}\right]$ $\times\,{}_3F_3\left(\begin{matrix}-\frac{\nu}{2},\ \frac{1-\nu}{2},\ \frac{2b-2a-2s-\nu}{2};\ \sigma\omega\\ 1-\nu,\ \frac{2-2s-\nu}{2},\ \frac{2b-2s-\nu}{2}\end{matrix}\right)$ $-\frac{\nu\sigma^{s+\nu/2}}{2^{2s}}\,\Gamma\left[\begin{matrix}1-b,\ -\frac{2s+\nu}{2},\ 2s\\ a-b+1,\ \frac{2s-\nu+2}{2}\end{matrix}\right]{}_3F_3\left(\begin{matrix}b-a,\ s,\ \frac{2s+1}{2};\ \sigma\omega\\ b,\ \frac{2s-\nu+2}{2},\ \frac{2s+\nu+2}{2}\end{matrix}\right)$ $\left[\begin{matrix}(\operatorname{Re}\omega>0;\ \operatorname{Re}s>0,\ \operatorname{Re}b-1)\ \text{or}\\ (\operatorname{Re}\omega=0;\ 0,\ \operatorname{Re}b-1<\operatorname{Re}s<-\operatorname{Re}\nu/2+1);\ \lvert\arg\sigma\rvert<\pi\end{matrix}\right]$
13	$\frac{\left(\sqrt{x}+\sqrt{x+\sigma}\right)^{\nu}}{\sqrt{x+\sigma}}\,e^{-\omega x}$ $\times\,\Psi\left(\begin{matrix}a;\ \omega x\\ b\end{matrix}\right)$	$\frac{\sigma^{s-b+(\nu+1)/2}}{2^{2s-2b+1}\omega^{b-1}}\,\mathrm{B}\left(2s-2b+2,\ \frac{2b-2s-\nu-1}{2}\right)$ $\times\,\Gamma\left[\begin{matrix}b-1\\ a\end{matrix}\right]{}_3F_3\left(\begin{matrix}1-a,\ s-b+1,\ \frac{2s-2b+3}{2};\ \sigma\omega\\ 2-b,\ \frac{2s-2b-\nu+3}{2},\ \frac{2s-2b+\nu+3}{2}\end{matrix}\right)$ $+\frac{2^{\nu}}{\omega^{s+(\nu-1)/2}}\,\Gamma\left[\begin{matrix}\frac{2s+\nu-1}{2},\ \frac{2s-2b+\nu+1}{2}\\ \frac{2s+2a-2b+\nu+1}{2}\end{matrix}\right]$ $\times\,{}_3F_3\left(\begin{matrix}\frac{1-\nu}{2},\ \frac{2-\nu}{2},\ \frac{2b-2a-2s-\nu+1}{2};\ \sigma\omega\\ 1-\nu,\ \frac{3-2s-\nu}{2},\ \frac{2b-2s-\nu+1}{2}\end{matrix}\right)$ $+\frac{\sigma^{s+(\nu-1)/2}}{2^{2s-1}}\,\mathrm{B}\left(2s,\ \frac{1-2s-\nu}{2}\right)$ $\times\,\Gamma\left[\begin{matrix}1-b\\ a-b+1\end{matrix}\right]{}_3F_3\left(\begin{matrix}b-a,\ s,\ \frac{2s+1}{2};\ \sigma\omega\\ b,\ \frac{2s-\nu+1}{2},\ \frac{2s+\nu+1}{2}\end{matrix}\right)$ $\left[\begin{matrix}(\operatorname{Re}\omega>0;\ \operatorname{Re}s>0,\ \operatorname{Re}b-1)\ \text{or}\\ (\operatorname{Re}\omega=0;\ 0,\ \operatorname{Re}b-1<\operatorname{Re}s<-\operatorname{Re}\nu/2+3/2);\ \lvert\arg\sigma\rvert<\pi\end{matrix}\right]$

3.29.3. $\Psi(a;\, b;\, \omega x)$ and trigonometric functions

1	$\sin(\sigma x)\,\Psi\left(\begin{matrix} a;\ \omega x \\ b \end{matrix}\right)$	$\frac{a(a-b+1)\,\pi\sigma^{a-s+1}}{2\omega^{a+1}\Gamma(a-s+2)} \csc\frac{(a-s)\pi}{2} \times {}_4F_3\left(\begin{matrix} \Delta(2,\, a+1),\, \Delta(2,\, a-b+2) \\ \frac{3}{2},\, \Delta(2,\, a-s+2);\, -\frac{\sigma^2}{\omega^2} \end{matrix}\right) + \frac{\pi\sigma^{a-s}\omega^{-a}}{2\Gamma(a-s+1)} \times \sec\frac{(a-s)\pi}{2}\, {}_4F_3\left(\begin{matrix} \Delta(2,\, a),\, \Delta(2,\, a-b+1) \\ \frac{1}{2},\, \Delta(2,\, a-s+1);\, -\frac{\sigma^2}{\omega^2} \end{matrix}\right) + \frac{\sigma}{\omega^{s+1}} \mathrm{B}(a-s-1,\, s-b+2)\,\Gamma\left[\begin{matrix} s+1 \\ a \end{matrix}\right] \times {}_4F_3\left(\begin{matrix} \Delta(2,\, s+1),\, \Delta(2,\, s-b+2) \\ \frac{3}{2},\, \Delta(2,\, s-a+2);\, -\frac{\sigma^2}{\omega^2} \end{matrix}\right)$ $[\sigma > 0;\ -1,\ \mathrm{Re}\, b - 2 < \mathrm{Re}\, s < \mathrm{Re}\, a + 1;\ \lvert\arg\omega\rvert < \pi/2]$
2	$\cos(\sigma x)\,\Psi\left(\begin{matrix} a;\ \omega x \\ b \end{matrix}\right)$	$\omega^{-s}\,\mathrm{B}(a-s,\, s)\,\Gamma\left[\begin{matrix} s-b+1 \\ a-b+1 \end{matrix}\right] {}_4F_3\left(\begin{matrix} \Delta(2,\, s),\, \Delta(2,\, s-b+1) \\ \frac{1}{2},\, \Delta(2,\, s+a+2);\, -\frac{\sigma^2}{\omega^2} \end{matrix}\right) + \frac{a(a-b+1)\,\pi\sigma^{a-s+1}}{2\omega^{a+1}\Gamma(a-s+2)} \sec\frac{(a-s)\pi}{2} \times {}_4F_3\left(\begin{matrix} \Delta(2,\, a+1),\, \Delta(2,\, a-b+2) \\ \frac{3}{2},\, \Delta(2,\, a-s+2);\, -\frac{\sigma^2}{\omega^2} \end{matrix}\right) - \frac{\pi\sigma^{a-s}\omega^{-a}}{2\Gamma(a-s+1)} \times \csc\frac{(a-s)\pi}{2}\, {}_4F_3\left(\begin{matrix} \Delta(2,\, a),\, \Delta(2,\, a-b+1) \\ \frac{1}{2},\, \Delta(2,\, a-s+1);\, -\frac{\sigma^2}{\omega^2} \end{matrix}\right)$ $[\sigma > 0;\ 0,\ \mathrm{Re}\, b - 1 < \mathrm{Re}\, s < \mathrm{Re}\, a + 1;\ \lvert\arg\omega\rvert < \pi/2]$
3	$\sin(\sigma\sqrt{x})\,\Psi\left(\begin{matrix} a;\ \omega x \\ b \end{matrix}\right)$	$\frac{\sqrt{\pi}\,(2/\sigma)^{2(s-a)}}{\omega^a}\,\Gamma\left[\begin{matrix} \frac{2s-2a+1}{2} \\ a-s+1 \end{matrix}\right] {}_2F_2\left(\begin{matrix} a,\, a-b+1;\ \frac{\sigma^2}{4\omega} \\ a-s+\frac{1}{2},\, a-s+1 \end{matrix}\right) + \sigma\omega^{-s-1/2}\,\Gamma\left[\begin{matrix} \frac{2s+1}{2} \\ a,\ \frac{2s-2a+3}{2} \end{matrix}\right] \mathrm{B}\left(\frac{2a-2s-1}{2},\, \frac{2s-2b+3}{2}\right) \times {}_2F_2\left(\begin{matrix} \frac{2s+1}{2},\, \frac{2s-2b+3}{2} \\ \frac{3}{2},\, \frac{2s-2a+3}{2};\ \frac{\sigma^2}{4\omega} \end{matrix}\right)$ $[\sigma > 0;\ -1/2,\ \mathrm{Re}\, b - 3/2 < \mathrm{Re}\, s < \mathrm{Re}\, a + 1/2]$
4	$\cos(\sigma\sqrt{x})\,\Psi\left(\begin{matrix} a;\ \omega x \\ b \end{matrix}\right)$	$\frac{\sqrt{\pi}\,(2/\sigma)^{2s-2a}}{\omega^a}\,\Gamma\left[\begin{matrix} s-a \\ \frac{2a-2s+1}{2} \end{matrix}\right] {}_2F_2\left(\begin{matrix} a,\, a-b+1;\ \frac{\sigma^2}{4\omega} \\ \frac{2a-2s+1}{2},\, a-s+1 \end{matrix}\right) + \omega^{-s}\,\mathrm{B}(a-s,\, s)\,\Gamma\left[\begin{matrix} s-b+1 \\ a-b+1 \end{matrix}\right] {}_2F_2\left(\begin{matrix} s,\, s-b+1;\ \frac{\sigma^2}{4\omega} \\ \frac{1}{2},\, s-a+1 \end{matrix}\right)$ $[\sigma > 0;\ 0,\ \mathrm{Re}\, b - 1 < \mathrm{Re}\, s < \mathrm{Re}\, a + 1/2]$

No.	$f(x)$	$F(s)$
5	$e^{-\omega x}\sin(\sigma x)\,\Psi\binom{a;\,\omega x}{b}$	$\frac{\sigma}{\omega^{s+1}}\Gamma\left[\begin{matrix}s+1,\ s-b+2\\ a-b+s+2\end{matrix}\right]{}_4F_3\left(\begin{matrix}\Delta(2,\,s+1),\ \Delta(2,\,s-b+2)\\ \frac{3}{2},\ \Delta(2,\,s+a-b+2);\ -\frac{\sigma^2}{\omega^2}\end{matrix}\right)$ $\left[\begin{matrix}(\operatorname{Re}\omega>\lvert\operatorname{Im}\sigma\rvert;\ \operatorname{Re}s>-1,\ \operatorname{Re}b-2)\ \text{or}\\ (\operatorname{Re}^2\omega=\operatorname{Im}^2\sigma;\ -1,\ \operatorname{Re}b-2<\operatorname{Re}s<\operatorname{Re}a+1)\end{matrix}\right]$
6	$e^{-\omega x}\cos(\sigma x)\,\Psi\binom{a;\,\omega x}{b}$	$\omega^{-s}\Gamma\left[\begin{matrix}s,\ s-b+1\\ a-b+s+1\end{matrix}\right]{}_4F_3\left(\begin{matrix}\Delta(2,\,s),\ \Delta(2,\,s-b+1)\\ \frac{1}{2},\ \Delta(2,\,s+a-b+1);\ -\frac{\sigma^2}{\omega^2}\end{matrix}\right)$ $\left[\begin{matrix}(\operatorname{Re}\omega>\lvert\operatorname{Im}\sigma\rvert;\ \operatorname{Re}s>0,\ \operatorname{Re}b-1)\ \text{or}\\ (\operatorname{Re}^2\omega=\operatorname{Im}^2\sigma;\ 0,\ \operatorname{Re}b-1<\operatorname{Re}s<\operatorname{Re}a+1)\end{matrix}\right]$
7	$e^{-\omega x}\sin(\sigma\sqrt{x})$ $\times\Psi\binom{a;\,\omega x}{b}$	$\frac{\sigma}{\omega^{s+1/2}}\Gamma\left[\begin{matrix}\frac{2s+1}{2},\ \frac{2s-2b+3}{2}\\ \frac{2s+2a-2b+3}{2}\end{matrix}\right]{}_2F_2\left(\begin{matrix}\frac{2s+1}{2},\ \frac{2s-2b+3}{2}\\ \frac{3}{2},\ \frac{2s+2a-2b+3}{2};\ -\frac{\sigma^2}{4\omega}\end{matrix}\right)$ $\left[\begin{matrix}(\operatorname{Re}\omega>0;\ \operatorname{Re}s>-1/2,\ \operatorname{Re}b-3/2)\ \text{or}\\ (\sigma>0,\ \operatorname{Re}\omega=0;\ -1/2,\ \operatorname{Re}b-3/2<\operatorname{Re}s<\operatorname{Re}a+1/2)\end{matrix}\right]$
8	$e^{-\omega x}\cos(\sigma\sqrt{x})$ $\times\Psi\binom{a;\,\omega x}{b}$	$\omega^{-s}\Gamma\left[\begin{matrix}s,\ s-b+1\\ s+a-b+1\end{matrix}\right]{}_2F_2\left(\begin{matrix}s,\ s-b+1\\ 1,\ s+a-b+1;\ -\frac{\sigma^2}{4\omega}\end{matrix}\right)$ $\left[\begin{matrix}(\operatorname{Re}\omega>0;\ \operatorname{Re}s>0,\ \operatorname{Re}b-1)\ \text{or}\\ (\sigma>0,\ \operatorname{Re}\omega=0;\ 0,\ \operatorname{Re}b-1<\operatorname{Re}s<\operatorname{Re}a+1/2)\end{matrix}\right]$

3.29.4. $\Psi(a;\,b;\,\omega x)$ and the logarithmic function

1	$\ln(\sigma x+1)\,\Psi\binom{a;\,\omega x}{b}$	$\frac{\pi\sigma^{b-s-1}\omega^{1-b}}{s-b+1}\csc[(b-s)\pi]\,\Gamma\left[\begin{matrix}b-1\\ a\end{matrix}\right]$ $\times\ {}_2F_2\left(\begin{matrix}a-b+1,\ s-b+1\\ 2-b,\ s-b+2;\ -\frac{\omega}{\sigma}\end{matrix}\right)+\frac{\omega^{1-s}}{\sigma}\mathrm{B}(a-s+1,\ s-1)$ $\times\Gamma\left[\begin{matrix}s-b\\ a-b+1\end{matrix}\right]{}_3F_3\left(\begin{matrix}1,\ 1,\ a-s+1;\ -\frac{\omega}{\sigma}\\ 2,\ 2-s,\ b-s+1\end{matrix}\right)$ $+\,\omega^{-s}\mathrm{B}(a-s,\ s-b+1)\,\Gamma\left[\begin{matrix}s\\ a\end{matrix}\right]\ln\frac{\sigma}{\omega}$ $-\,\omega^{-s}\mathrm{B}(a-s,\ s)\,\Gamma\left[\begin{matrix}s-b+1\\ a-b+1\end{matrix}\right][\psi(a-s)-\psi(s-b+1)$ $-\,\psi(s)]+\frac{\pi\sigma^{-s}}{s}\csc(s\pi)\,\Gamma\left[\begin{matrix}1-b\\ a-b+1\end{matrix}\right]{}_2F_2\left(\begin{matrix}a,\ s;\ -\frac{\omega}{\sigma}\\ b,\ s+1\end{matrix}\right)$ $\left[\begin{matrix}(\operatorname{Re}\omega>0;\ \operatorname{Re}s>-1,\ \operatorname{Re}b-2)\ \text{or}\\ (\operatorname{Re}\omega=0;\ -1,\ \operatorname{Re}b-2<\operatorname{Re}s<\operatorname{Re}a);\ \lvert\arg\sigma\rvert<\pi\end{matrix}\right]$

No.	$f(x)$	$F(s)$
2	$e^{-\omega x}\ln(\sigma x+1)$ $\times\,\Psi\left(\begin{matrix}a;\ \omega x\\ b\end{matrix}\right)$	$\frac{\pi\sigma^{b-s-1}\omega^{1-b}}{s-b+1}\csc[(b-s)\pi]\,\Gamma\left[\begin{matrix}b-1\\ a\end{matrix}\right]{}_2F_2\left(\begin{matrix}1-a,\ s-b+1\\ 2-b,\ s-b+2;\ \frac{\omega}{\sigma}\end{matrix}\right)$ $-\frac{\pi\omega^{1-s}}{\sigma}\csc(s\pi)\,\Gamma\left[\begin{matrix}s-b\\ 2-s,\ a-b+s\end{matrix}\right]{}_3F_3\left(\begin{matrix}1,\ 1,\ 1-s-a+b;\ \frac{\omega}{\sigma}\\ 2,\ 2-s,\ 1-s+b\end{matrix}\right)$ $+\frac{\pi\sigma^{-s}}{s}\csc(s\pi)\,\Gamma\left[\begin{matrix}1-b\\ a-b+1\end{matrix}\right]{}_2F_2\left(\begin{matrix}b-a,\ s;\ \frac{\omega}{\sigma}\\ b,\ s+1\end{matrix}\right)$ $+\omega^{-s}\,\Gamma\left[\begin{matrix}s,\ s-b+1\\ s+a-b+1\end{matrix}\right]\left[\psi(s)+\psi(s-b)-\psi(s+a-b)+\ln\frac{\sigma}{\omega}\right]$ $\left[\begin{matrix}(\operatorname{Re}\omega>0;\ \operatorname{Re}s>-1,\ \operatorname{Re}b-2)\ \text{or}\\ (\operatorname{Re}\omega=0;\ -1,\ \operatorname{Re}b-2<\operatorname{Re}s<\operatorname{Re}a);\ \lvert\arg\sigma\rvert<\pi\end{matrix}\right]$

3.29.5. $\Psi(a;\,b;\,\omega x)$ and $\operatorname{Ei}(\sigma x)$

1	$\operatorname{Ei}(-\sigma x)\,\Psi\left(\begin{matrix}a;\ \omega x\\ b\end{matrix}\right)$	$\frac{\sigma^{b-s-1}\omega^{1-b}}{b-s-1}\Gamma\left[\begin{matrix}b-1,\ s-b+1\\ a\end{matrix}\right]{}_3F_2\left(\begin{matrix}a-b+1,\ s-b+1,\ s-b+1\\ 2-b,\ s-b+2;\ \frac{\omega}{\sigma}\end{matrix}\right)$ $-\frac{\sigma^{-s}}{s}\Gamma\left[\begin{matrix}1-b,\ s\\ a-b+1\end{matrix}\right]{}_3F_2\left(\begin{matrix}a,\ s,\ s\\ b,\ s+1;\ \frac{\omega}{\sigma}\end{matrix}\right)$ $\left[\begin{matrix}(\operatorname{Re}\sigma\ge 0;\ \operatorname{Im}\sigma\ne 0;\ 0,\ \operatorname{Re}b-1<\operatorname{Re}s<\operatorname{Re}a)\ \text{or}\\ (\sigma>0;\ \operatorname{Re}s>-1,\ \operatorname{Re}b-2)\end{matrix}\right]$
2	$e^{-\omega x}\operatorname{Ei}(-\sigma x)$ $\times\,\Psi\left(\begin{matrix}a;\ \omega x\\ b\end{matrix}\right)$	$\frac{\sigma^{b-s-1}\omega^{1-b}}{b-s-1}\Gamma\left[\begin{matrix}b-1,\ s-b+1\\ a\end{matrix}\right]{}_3F_2\left(\begin{matrix}1-a,\ s-b+1,\ s-b+1\\ 2-b,\ s-b+2;\ -\frac{\omega}{\sigma}\end{matrix}\right)$ $-\frac{\sigma^{-s}}{s}\Gamma\left[\begin{matrix}1-b,\ s\\ a-b+1\end{matrix}\right]{}_3F_2\left(\begin{matrix}b-a,\ s,\ s\\ b,\ s+1;\ -\frac{\omega}{\sigma}\end{matrix}\right)$ $\left[\begin{matrix}(\operatorname{Re}\omega,\ \operatorname{Re}(\sigma+\omega)>0;\ \operatorname{Im}\sigma\ne 0;\ \operatorname{Re}s>0,\ \operatorname{Re}b-1)\ \text{or}\\ (\operatorname{Re}(\sigma+\omega)>0;\ \operatorname{Im}\sigma=0;\ \operatorname{Re}s>0,\ \operatorname{Re}b-1)\ \text{or}\\ (\operatorname{Re}\sigma\ge 0;\ \operatorname{Re}\omega=0;\ \operatorname{Im}\sigma\ne 0;\ 0,\ \operatorname{Re}b-1<\operatorname{Re}s<\operatorname{Re}a+1)\ \text{or}\\ (\sigma>0;\ \operatorname{Re}\omega=0;\ \operatorname{Re}s>0,\ \operatorname{Re}b-1)\end{matrix}\right]$

3.29.6. $\Psi(a;\,b;\,\omega x)$ and $\operatorname{erf}(\sigma\sqrt{x})$, $\operatorname{erfc}(\sigma\sqrt{x})$

1	$\operatorname{erf}(\sigma\sqrt{x})\,\Psi\left(\begin{matrix}a;\ \omega x\\ b\end{matrix}\right)$	$\frac{\sigma^{2a-2s}\omega^{-a}}{\sqrt{\pi}\,(a-s)}\Gamma\left(\frac{2s-2a+1}{2}\right){}_3F_2\left(\begin{matrix}a,\ a-b+1,\ a-s;\ \frac{\sigma^2}{\omega}\\ \frac{2a-2s+1}{2},\ a-s+1\end{matrix}\right)$ $+\frac{2\sigma}{\sqrt{\pi}\,\omega^{s+1/2}}\mathrm{B}\left(\frac{2a-2s-1}{2},\ \frac{2s-2b+3}{2}\right)$ $\times\Gamma\left[\begin{matrix}\frac{2s+1}{2}\\ a\end{matrix}\right]{}_3F_2\left(\begin{matrix}\frac{1}{2},\ \frac{2s+1}{2},\ \frac{2s-2b+3}{2}\\ \frac{3}{2},\ \frac{2s-2a+3}{2};\ \frac{\sigma^2}{\omega}\end{matrix}\right)$ $[-1/2,\ \operatorname{Re}b-3/2<\operatorname{Re}s<\operatorname{Re}a;\ \lvert\arg\sigma\rvert<\pi/4]$

No.	$f(x)$	$F(s)$
2	$\operatorname{erfc}(\sigma\sqrt{x})\,\Psi\binom{a;\ \omega x}{b}$	$\dfrac{\sigma^{2b-2s-2}\omega^{1-b}}{\sqrt{\pi}\,(s-b+1)}\,\Gamma\left[\begin{matrix} b-1,\ \frac{2s-2b+3}{2} \\ a \end{matrix}\right] \times {}_3F_2\left(\begin{matrix} a-b+1,\ s-b+1,\ \frac{2s-2b+3}{2} \\ 2-b,\ s-b+2;\ \frac{\omega}{\sigma^2} \end{matrix}\right) + \dfrac{\sigma^{-2s}}{2^{2s-1}}\,\Gamma\left[\begin{matrix} 1-b,\ 2s \\ a-b+1,\ s \end{matrix}\right] {}_3F_2\left(\begin{matrix} a,\ s,\ \frac{2s+1}{2} \\ b,\ s+1;\ \frac{\omega}{\sigma^2} \end{matrix}\right)$ $\left[\begin{matrix} (\operatorname{Re} s > 0,\ \operatorname{Re} b - 1;\ \lvert\arg\sigma\rvert < \pi/4) \text{ or} \\ (0,\ \operatorname{Re} b - 1 < \operatorname{Re} s < \operatorname{Re} a + 3/2;\ \lvert\arg\sigma\rvert = \pi/4) \end{matrix}\right]$
3	$e^{-\omega x}\operatorname{erf}(\sigma\sqrt{x})$ $\times\,\Psi\binom{a;\ \omega x}{b}$	$\dfrac{2\sigma\omega^{-s-1/2}}{\sqrt{\pi}}\,\Gamma\left[\begin{matrix} \frac{2s+1}{2},\ \frac{2s-2b+3}{2} \\ \frac{2s+2a-2b+3}{2} \end{matrix}\right] {}_3F_2\left(\begin{matrix} \frac{1}{2},\ \frac{2s+1}{2},\ \frac{2s-2b+3}{2} \\ \frac{3}{2},\ \frac{2s+2a-2b+3}{2};\ -\frac{\sigma^2}{\omega} \end{matrix}\right)$ $\left[\begin{matrix} (\operatorname{Re}(\sigma^2+\omega) > 0;\ \operatorname{Re}\omega > 0;\ \operatorname{Re} s > -1/2,\ \operatorname{Re} b - 3/2) \text{ or} \\ (\operatorname{Re}(\sigma^2+\omega) = 0;\ \operatorname{Re}\omega > 0;\ \operatorname{Re} s > -1/2,\ \operatorname{Re} b - 3/2 < \operatorname{Re} s < 0,\ \operatorname{Re} a + 3/2) \text{ or} \\ (\operatorname{Re}\omega = 0;\ \operatorname{Re} s > -1/2,\ \operatorname{Re} b - 3/2 < \operatorname{Re} s < 0,\ \operatorname{Re} a + 1;\ \lvert\arg\sigma\rvert \le \pi/4) \end{matrix}\right]$
4	$e^{-\omega x}\operatorname{erfc}(\sigma\sqrt{x})$ $\times\,\Psi\binom{a;\ \omega x}{b}$	$\dfrac{\sigma^{2b-2s-2}\omega^{1-b}}{\sqrt{\pi}\,(s-b+1)}\,\Gamma\left[\begin{matrix} b-1,\ \frac{2s-2b+3}{2} \\ a \end{matrix}\right] {}_3F_2\left(\begin{matrix} 1-a,\ s-b+1,\ \frac{2s-2b+3}{2} \\ 2-b,\ s-b+2;\ -\frac{\omega}{\sigma^2} \end{matrix}\right) + \dfrac{\sigma^{-2s}}{\sqrt{\pi}\,s}\,\Gamma\left[\begin{matrix} 1-b,\ \frac{2s+1}{2} \\ a-b+1 \end{matrix}\right] {}_3F_2\left(\begin{matrix} b-a,\ s,\ \frac{2s+1}{2} \\ b,\ s+1;\ -\frac{\omega}{\sigma^2} \end{matrix}\right)$ $\left[\begin{matrix} (\operatorname{Re}(\sigma^2+\omega) > 0;\ \operatorname{Re} s > 0,\ \operatorname{Re} b - 1) \text{ or} \\ (\operatorname{Re}(\sigma^2+\omega) = 0;\ 0,\ \operatorname{Re} b - 1 < \operatorname{Re} s < \operatorname{Re} a + 3/2) \end{matrix}\right]$

3.29.7. $\Psi(a;\ b;\ \omega x)$ and the Bessel functions

No.	$f(x)$	$F(s)$
1	$J_\nu(\sigma x)\,\Psi\binom{a;\ \omega x}{b}$	$-\dfrac{2^{s-a-2}a\,(a-b+1)}{\sigma^{s-a-1}\omega^{a+1}}\,\Gamma\left[\begin{matrix} \frac{s-a+\nu-1}{2} \\ \frac{a-s+\nu+3}{2} \end{matrix}\right] \times {}_4F_3\left(\begin{matrix} \Delta(2,\ a+1),\ \Delta(2,\ a-b+2) \\ \frac{3}{2},\ \frac{a-s-\nu+3}{2},\ \frac{a-s+\nu+3}{2};\ -\frac{\sigma^2}{\omega^2} \end{matrix}\right) + \dfrac{2^{s-a-1}}{\sigma^{s-a}\omega^{a}}\,\Gamma\left[\begin{matrix} \frac{s-a+\nu}{2} \\ \frac{a-s+\nu+2}{2} \end{matrix}\right] {}_4F_3\left(\begin{matrix} \Delta(2,\ a),\ \Delta(2,\ a-b+1) \\ \frac{1}{2},\ \frac{a-s-\nu+2}{2},\ \frac{a-s+\nu+2}{2};\ -\frac{\sigma^2}{\omega^2} \end{matrix}\right) + \dfrac{(\sigma/2)^\nu}{\omega^{s+\nu}}\,\mathrm{B}(a-s-\nu,\ s+\nu)\,\Gamma\left[\begin{matrix} s-b+\nu+1 \\ \nu+1,\ a-b+1 \end{matrix}\right] \times {}_4F_3\left(\begin{matrix} \Delta(2,\ s+\nu),\ \Delta(2,\ s-b+\nu+1) \\ \nu+1,\ \Delta(2,\ s-a+\nu+1);\ -\frac{\sigma^2}{\omega^2} \end{matrix}\right)$ $\left[\begin{matrix} \sigma > 0;\ \operatorname{Re}(b-\nu) - 1,\ -\operatorname{Re}\nu < \operatorname{Re} s < \operatorname{Re} a + 3/2; \\ \lvert\arg\omega\rvert < \pi/2 \end{matrix}\right]$

No.	$f(x)$	$F(s)$
2	$J_\nu(\sigma\sqrt{x})\,\Psi\binom{a;\,\omega x}{b}$	$\frac{(\sigma/2)^{2a-2s}}{\omega^a}\Gamma\left[\begin{matrix}\frac{2s-2a+\nu}{2}\\ \frac{2a-2s+\nu+2}{2}\end{matrix}\right]{}_2F_2\left(\begin{matrix}a,\,a-b+1;\,\frac{\sigma^2}{4\omega}\\ \frac{2a-2s-\nu+2}{2},\,\frac{2a-2s+\nu+2}{2}\end{matrix}\right)$ $+\frac{(\sigma/2)^\nu}{\omega^{s+\nu/2}}\mathrm{B}\left(\frac{2a-2s-\nu}{2},\frac{2s+\nu}{2}\right)$ $\times\Gamma\left[\begin{matrix}\frac{2s-2b+\nu+2}{2}\\ \nu+1,\,a-b+1\end{matrix}\right]{}_2F_2\left(\begin{matrix}\frac{2s+\nu}{2},\,\frac{2s-2b+\nu+2}{2};\,\frac{\sigma^2}{4\omega}\\ \nu+1,\,\frac{2s-2a+\nu+2}{2}\end{matrix}\right)$ $[\sigma>0;\ \mathrm{Re}(b-\nu/2)-1,\ -\mathrm{Re}\,\nu/2<\mathrm{Re}\,s<\mathrm{Re}\,a+3/4]$
3	$e^{-\omega x}J_\nu(\sigma x)\,\Psi\binom{a;\,\omega x}{b}$	$\frac{(\sigma/2)^\nu}{\omega^{s+\nu}}\Gamma\left[\begin{matrix}s+\nu,\,s-b+\nu+1\\ \nu+1,\,s+a-b+\nu+1\end{matrix}\right]$ $\times{}_4F_3\left(\begin{matrix}\Delta(2,\,s+\nu),\,\Delta(2,\,s-b+\nu+1)\\ \Delta(2,\,s+a-b+\nu+1),\,\nu+1;\,-\frac{\sigma^2}{\omega^2}\end{matrix}\right)$ $\left[\begin{matrix}(\mathrm{Re}\,\omega>\lvert\mathrm{Im}\,\sigma\rvert;\ \mathrm{Re}\,s>\mathrm{Re}(b-\nu)-1,\ -\mathrm{Re}\,\nu)\ \text{or}\\ (\mathrm{Re}^2\,\omega=\mathrm{Im}^2\,\sigma;\ \mathrm{Re}(b-\nu)-1,\ -\mathrm{Re}\,\nu<\mathrm{Re}\,s<\mathrm{Re}\,a+3/2)\end{matrix}\right]$
4	$e^{-\omega x}J_\nu(\sigma\sqrt{x})$ $\times\Psi\binom{a;\,\omega x}{b}$	$\frac{(\sigma/2)^\nu}{\omega^{s+\nu/2}}\Gamma\left[\begin{matrix}\frac{2s+\nu}{2},\,\frac{2s-2b+\nu+2}{2}\\ \nu+1,\,\frac{2s+2a-2b+\nu+1}{2}\end{matrix}\right]{}_2F_2\left(\begin{matrix}\frac{2s+\nu}{2},\,\frac{2s-2b+\nu+2}{2};\,\frac{\sigma^2}{4\omega}\\ \nu+1,\,\frac{2s+2a-2b+\nu+2}{2}\end{matrix}\right)$ $\left[\begin{matrix}(\mathrm{Re}\,\omega>0;\ \mathrm{Re}\,s>\mathrm{Re}(b-\nu/2)-1,\ -\mathrm{Re}\,\nu/2)\ \text{or}\\ (\sigma>0;\ \mathrm{Re}\,\omega=0;\ \mathrm{Re}(b-\nu/2)-1,\ -\mathrm{Re}\,\nu/2<\mathrm{Re}\,s<\mathrm{Re}\,a+3/4)\end{matrix}\right]$
5	$e^{-\omega x}Y_\nu(\sigma\sqrt{x})$ $\times\Psi\binom{a;\,\omega x}{b}$	$\frac{(\sigma/2)^\nu}{\pi\omega^{s+\nu/2}}\cos(\pi\nu)\,\Gamma\left[\begin{matrix}\frac{2s+\nu}{2},\,\frac{2s-2b+\nu+2}{2}\\ -\nu,\,\frac{2s+2a-2b+\nu+2}{2}\end{matrix}\right]$ $\times{}_2F_2\left(\begin{matrix}\frac{2s+\nu}{2},\,\frac{2s-2b+\nu+2}{2};\,-\frac{\sigma^2}{4\omega}\\ \nu+1,\,\frac{2s+2a-2b+\nu+2}{2}\end{matrix}\right)-\frac{(2/\sigma)^\nu}{\pi\omega^{s-\nu/2}}$ $\times\Gamma\left[\begin{matrix}\nu,\,\frac{2s-\nu}{2},\,\frac{2s-2b-\nu+2}{2}\\ \frac{2s+2a-2b-\nu+2}{2}\end{matrix}\right]{}_2F_2\left(\begin{matrix}\frac{2s-\nu}{2},\,\frac{2s-2b-\nu+2}{2};\,-\frac{\sigma^2}{4\omega}\\ 1-\nu,\,\frac{2s+2a-2b-\nu+2}{2}\end{matrix}\right)$ $\left[\begin{matrix}(\mathrm{Re}\,\omega>0;\ \mathrm{Re}\,s>\mathrm{Re}\,b-\lvert\mathrm{Re}\,\nu\rvert/2-1,\ -\lvert\mathrm{Re}\,\nu\rvert/2)\ \text{or}\\ (\sigma>0;\ \mathrm{Re}\,\omega=0;\ \mathrm{Re}\,b-\lvert\mathrm{Re}\,\nu\rvert/2-1,\ -\lvert\mathrm{Re}\,\nu\rvert/2<\mathrm{Re}\,s<\mathrm{Re}\,a+3/4)\end{matrix}\right]$
6	$e^{-\omega x}I_\nu(\sigma x)\,\Psi\binom{a;\,\omega x}{b}$	$\frac{(\sigma/2)^\nu}{\omega^{s+\nu}}\Gamma\left[\begin{matrix}s+\nu,\,s-b+\nu+1\\ \nu+1,\,s+a-b+\nu+1\end{matrix}\right]$ $\times{}_4F_3\left(\begin{matrix}\Delta(2,\,s+\nu),\,\Delta(2,\,s-b+\nu+1)\\ \nu+1,\,\Delta(2,\,s+a-b+\nu+1);\,\frac{\sigma^2}{\omega^2}\end{matrix}\right)$ $\left[\begin{matrix}(\mathrm{Re}\,\omega>\lvert\mathrm{Re}\,\sigma\rvert;\ \mathrm{Re}\,s>\mathrm{Re}(b-\nu)-1,\ -\mathrm{Re}\,\nu)\ \text{or}\\ (\mathrm{Re}^2\,\omega=\mathrm{Re}^2\,\sigma;\ \mathrm{Re}(b-\nu)-1,\ -\mathrm{Re}\,\nu<\mathrm{Re}\,s<\mathrm{Re}\,a+3/2)\end{matrix}\right]$

No.	$f(x)$	$F(s)$
7	$e^{-\omega x} K_\nu(\sigma\sqrt{x}) \times \Psi\left(\begin{matrix} a;\ \omega x \\ b \end{matrix}\right)$	$\frac{2^{-\nu-1}\sigma^\nu}{\omega^{s+\nu/2}} \Gamma\left[\begin{matrix} -\nu,\ \frac{2s+\nu}{2},\ \frac{2s-2b+\nu+2}{2} \\ \frac{2s+2a-2b+\nu+2}{2} \end{matrix}\right] \times {}_2F_2\left(\begin{matrix} \frac{2s+\nu}{2},\ \frac{2s-2b-\nu+2}{2};\ \frac{\sigma^2}{4\omega} \\ \nu+1,\ \frac{2s+2a-2b+\nu+2}{2} \end{matrix}\right) + \frac{2^{\nu-1}\sigma^{-\nu}}{\omega^{s-\nu/2}} \times \Gamma\left[\begin{matrix} \nu,\ \frac{2s-\nu}{2},\ \frac{2s-2b-\nu+2}{2} \\ \frac{2s+2a-2b-\nu+2}{2} \end{matrix}\right] {}_2F_2\left(\begin{matrix} \frac{2s-\nu}{2},\ \frac{2s-2b-\nu+2}{2};\ \frac{\sigma^2}{4\omega} \\ 1-\nu,\ \frac{2s+2a-2b-\nu+2}{2} \end{matrix}\right)$
		$\left[\begin{matrix} (\operatorname{Re}\omega > 0;\ \operatorname{Re}s > \operatorname{Re}b - \lvert\operatorname{Re}\nu\rvert/2 - 1,\ -\lvert\operatorname{Re}\nu\rvert/2) \text{ or} \\ (\operatorname{Re}\sigma > 0;\ \operatorname{Re}\omega = 0;\ \operatorname{Re}s > \operatorname{Re}b - \lvert\operatorname{Re}\nu\rvert/2 - 1,\ -\lvert\operatorname{Re}\nu\rvert/2) \text{ or} \\ (\operatorname{Re}\sigma = 0;\ \operatorname{Re}\omega = 0;\ \operatorname{Re}b - \lvert\operatorname{Re}\nu\rvert/2 - 1,\ -\lvert\operatorname{Re}\nu\rvert/2 < \operatorname{Re}s < \operatorname{Re}a + 5/4) \end{matrix}\right]$

3.29.8. $\Psi(a;\ b;\ \omega x)$ and $P_n(\varphi(x))$

1	$\theta(\sigma - x) P_n\left(\frac{2x}{\sigma} - 1\right) \times \Psi\left(\begin{matrix} a;\ \omega x \\ b \end{matrix}\right)$	$\frac{\sigma^{s-b+1}}{\omega^{b-1}} \Gamma\left[\begin{matrix} b-1,\ s-b+1,\ s-b+1 \\ a,\ s-b-n+1,\ s-b+n+2 \end{matrix}\right] \times {}_3F_3\left(\begin{matrix} a-b+1,\ s-b+1,\ s-b+1;\ \sigma\omega \\ 2-b,\ s-b-n+1,\ s-b+n+2 \end{matrix}\right) + \sigma^s \Gamma\left[\begin{matrix} 1-b,\ s,\ s \\ a-b+1,\ s-n,\ s+n+1 \end{matrix}\right] {}_3F_3\left(\begin{matrix} a,\ s,\ s;\ \sigma\omega \\ b,\ s-n,\ s+n+1 \end{matrix}\right)$ $[\sigma > 0;\ \operatorname{Re}s > 0,\ \operatorname{Re}b - 1]$
2	$\theta(\sigma - x) P_n\left(\frac{2\sigma}{x} - 1\right) \times \Psi\left(\begin{matrix} a;\ \omega x \\ b \end{matrix}\right)$	$\frac{\sigma^{s-b+1}}{\omega^{b-1}} \Gamma\left[\begin{matrix} b-1,\ s-b-n+1,\ s-b+n+2 \\ a,\ s-b+2,\ s-b+2 \end{matrix}\right] \times {}_3F_3\left(\begin{matrix} a-b+1,\ s-b-n+1,\ s-b+n+2 \\ 2-b,\ s-b+2,\ s-b+2;\ \sigma\omega \end{matrix}\right) + \sigma^s \Gamma\left[\begin{matrix} 1-b,\ s-n,\ s+n+1 \\ a-b+1,\ s+1,\ s+1 \end{matrix}\right] {}_3F_3\left(\begin{matrix} a,\ s-n,\ s+n+1 \\ b,\ s+1,\ s+1;\ \sigma\omega \end{matrix}\right)$ $[\sigma > 0;\ \operatorname{Re}s > n,\ \operatorname{Re}b + n - 1]$
3	$\theta(\sigma - x) e^{-\omega x} \times P_n\left(\frac{2x}{\sigma} - 1\right) \times \Psi\left(\begin{matrix} a;\ \omega x \\ b \end{matrix}\right)$	$\frac{\sigma^{s-b+1}}{\omega^{b-1}} \Gamma\left[\begin{matrix} b-1,\ s-b+1,\ s-b+1 \\ a,\ s-b-n+1,\ s-b+n+2 \end{matrix}\right] \times {}_3F_3\left(\begin{matrix} 1-a,\ s-b+1,\ s-b+1;\ -\sigma\omega \\ 2-b,\ s-b-n+1,\ s-b+n+2 \end{matrix}\right) + \sigma^s \Gamma\left[\begin{matrix} 1-b,\ s,\ s \\ a-b+1,\ s-n,\ s+n+1 \end{matrix}\right] \times {}_3F_3\left(\begin{matrix} b-a,\ s,\ s;\ -\sigma\omega \\ b,\ s-n,\ s+n+1 \end{matrix}\right)$ $[\sigma > 0;\ \operatorname{Re}s > 0,\ \operatorname{Re}b - 1]$

No.	$f(x)$	$F(s)$
4	$\theta(\sigma-x)\, e^{-\omega x} \times P_n\left(\frac{2\sigma}{x}-1\right) \times \Psi\binom{a;\, \omega x}{b}$	$\frac{\sigma^{s-b+1}}{\omega^{b-1}}\Gamma\begin{bmatrix}b-1,\, s-b-n+1,\, s-b+n+2\\ a,\, s-b+2,\, s-b+2\end{bmatrix} \times {}_3F_3\left(\begin{matrix}1-a,\, s-b-n+1,\, s-b+n+2\\ 2-b,\, s-b+2,\, s-b+2;\, -\sigma\omega\end{matrix}\right) + \sigma^s\,\Gamma\begin{bmatrix}1-b,\, s-n,\, s+n+1\\ a-b+1,\, s+1,\, s+1\end{bmatrix} {}_3F_3\left(\begin{matrix}b-a,\, s-n,\, s+n+1\\ b,\, s+1,\, s+1;\, -\sigma\omega\end{matrix}\right)$ $[\sigma>0;\ \mathrm{Re}\, s>n,\ \mathrm{Re}\, b+n-1]$

3.29.9. $\Psi(a;\, b;\, \omega x)$ and $T_n(\varphi(x))$

1	$(\sigma-x)_+^{-1/2} \times T_n\left(\frac{2x}{\sigma}-1\right) \times \Psi\binom{a;\, \omega x}{b}$	$\frac{\sqrt{\pi}\,\sigma^{s-b+1/2}}{\omega^{b-1}}\Gamma\begin{bmatrix}b-1,\, s-b+1,\, \frac{2s-2b+3}{2}\\ a,\, \frac{2s-2b-2n+3}{2},\, \frac{2s-2b+2n+3}{2}\end{bmatrix} \times {}_3F_3\left(\begin{matrix}a-b+1,\, s-b+1,\, \frac{2s-2b+3}{2};\, \sigma\omega\\ 2-b,\, \frac{2s-2b-2n+3}{2},\, \frac{2s-2b+2n+3}{2}\end{matrix}\right) + \sqrt{\pi}\,\sigma^{s-1/2}\,\Gamma\begin{bmatrix}1-b,\, s,\, \frac{2s+1}{2}\\ a-b+1,\, \frac{2s-2n+1}{2},\, \frac{2s+2n+1}{2}\end{bmatrix} \times {}_3F_3\left(\begin{matrix}a,\, s,\, \frac{2s+1}{2};\, \sigma\omega\\ b,\, \frac{2s-2n+1}{2},\, \frac{2s+2n+1}{2}\end{matrix}\right)$ $[\sigma>0;\ \mathrm{Re}\, s>0,\ \mathrm{Re}\, b-1]$
2	$(\sigma-x)_+^{-1/2} \times T_n\left(\frac{2\sigma}{x}-1\right) \times \Psi\binom{a;\, \omega x}{b}$	$\frac{\sqrt{\pi}\,\sigma^{s-b+1/2}}{\omega^{b-1}}\Gamma\begin{bmatrix}b-1,\, s-b-n+1,\, s-b+n+1\\ a,\, \frac{2s-2b+3}{2},\, s-b+1\end{bmatrix} \times {}_3F_3\left(\begin{matrix}a-b+1,\, s-b-n+1,\, s-b+n+1\\ 2-b,\, \frac{2s-2b+3}{2},\, s-b+1;\, \sigma\omega\end{matrix}\right) + \sqrt{\pi}\,\sigma^{s-1/2}\,\Gamma\begin{bmatrix}1-b,\, s-n,\, s+n\\ a-b+1,\, \frac{2s+1}{2},\, s\end{bmatrix} {}_3F_3\left(\begin{matrix}a,\, s-n,\, s+n\\ b,\, \frac{2s+1}{2},\, s;\, \sigma\omega\end{matrix}\right)$ $[\sigma>0;\ \mathrm{Re}\, s>n,\ \mathrm{Re}\, b+n-1]$
3	$(\sigma-x)_+^{-1/2}\, e^{-\omega x} \times T_n\left(\frac{2x}{\sigma}-1\right) \times \Psi\binom{a;\, \omega x}{b}$	$\frac{\sqrt{\pi}\,\sigma^{s-b+1/2}}{\omega^{b-1}}\Gamma\begin{bmatrix}b-1,\, s-b+1,\, \frac{2s-2b+3}{2}\\ a,\, \frac{2s-2b-2n+3}{2},\, \frac{2s-2b+2n+3}{2}\end{bmatrix} \times {}_3F_3\left(\begin{matrix}1-a,\, s-b+1,\, \frac{2s-2b+3}{2};\, -\sigma\omega\\ 2-b,\, \frac{2s-2b-2n+3}{2},\, \frac{2s-2b+2n+3}{2}\end{matrix}\right) + \sqrt{\pi}\,\sigma^{s-1/2}\,\Gamma\begin{bmatrix}1-b,\, s,\, \frac{2s+1}{2}\\ a-b+1,\, \frac{2s-2n+1}{2},\, \frac{2s+2n+1}{2}\end{bmatrix} \times {}_3F_3\left(\begin{matrix}b-a,\, s,\, \frac{2s+1}{2};\, -\sigma\omega\\ b,\, \frac{2s-2n+1}{2},\, \frac{2s+2n+1}{2}\end{matrix}\right)$ $[\sigma>0;\ \mathrm{Re}\, s>0,\ \mathrm{Re}\, b-1]$

No.	$f(x)$	$F(s)$
4	$(\sigma-x)_+^{-1/2}\,e^{-\omega x}$ $\times T_n\left(\frac{2\sigma}{x}-1\right)$ $\times\Psi\binom{a;\,\omega x}{b}$	$\frac{\sqrt{\pi}\,\sigma^{s-b+1/2}}{\omega^{b-1}}\,\Gamma\left[\begin{matrix}b-1,\,s-b-n+1,\,s-b+n+1\\ a,\,\frac{2s-2b+3}{2},\,s-b+1\end{matrix}\right]$ $\times{}_3F_3\left(\begin{matrix}1-a,\,s-b-n+1,\,s-b+n+1\\ 2-b,\,\frac{2s-2b+3}{2},\,s-b+1;\,-\sigma\omega\end{matrix}\right)$ $+\sqrt{\pi}\,\sigma^{s-1/2}\,\Gamma\left[\begin{matrix}1-b,\,s-n,\,s+n\\ a-b+1,\,\frac{2s+1}{2},\,s\end{matrix}\right]{}_3F_3\left(\begin{matrix}b-a,\,s-n,\,s+n\\ b,\,\frac{2s+1}{2},\,s;\,-\sigma\omega\end{matrix}\right)$ $[\sigma>0;\ \operatorname{Re}s>n,\ \operatorname{Re}b+n-1]$

3.29.10. $\Psi(a;\,b;\,\omega x)$ and $U_n(\varphi(x))$

1	$(\sigma-x)_+^{1/2}$ $\times U_n\left(\frac{2x}{\sigma}-1\right)$ $\times\Psi\binom{a;\,\omega x}{b}$	$\frac{(n+1)\sqrt{\pi}\,\sigma^{s-b+3/2}}{2\,\omega^{b-1}}\,\Gamma\left[\begin{matrix}b-1,\,s-b+1,\,\frac{2s-2b+1}{2}\\ a,\,\frac{2s-2b-2n+1}{2},\,\frac{2s-2b+2n+5}{2}\end{matrix}\right]$ $\times{}_3F_3\left(\begin{matrix}a-b+1,\,s-b+1,\,\frac{2s-2b+1}{2};\,\sigma\omega\\ 2-b,\,\frac{2s-2b-2n+1}{2},\,\frac{2s-2b+2n+5}{2}\end{matrix}\right)$ $+\frac{(n+1)\sqrt{\pi}\,\sigma^{s+1/2}}{2}\,\Gamma\left[\begin{matrix}1-b,\,s,\,\frac{2s-1}{2}\\ a-b+1,\,\frac{2s-2n-1}{2},\,\frac{2s+2n+3}{2}\end{matrix}\right]$ $\times{}_3F_3\left(\begin{matrix}a,\,s,\,\frac{2s-1}{2};\,\sigma\omega\\ b,\,\frac{2s-2n-1}{2},\,\frac{2s+2n+3}{2}\end{matrix}\right)$ $[\sigma>0;\ \operatorname{Re}s>0,\ \operatorname{Re}b-1]$
2	$(\sigma-x)_+^{1/2}$ $\times U_n\left(\frac{2\sigma}{x}-1\right)$ $\times\Psi\binom{a;\,\omega x}{b}$	$\frac{(n+1)\sqrt{\pi}\,\sigma^{s-b+3/2}}{2\,\omega^{b-1}}\,\Gamma\left[\begin{matrix}b-1\\ a\end{matrix}\right]\Gamma\left[\begin{matrix}s-b-n+1,\,s-b+n+3\\ \frac{2s-2b+5}{2},\,s-b+3\end{matrix}\right]$ $\times{}_3F_3\left(\begin{matrix}a-b+1,\,s-b-n+1,\,s-b+n+3\\ 2-b,\,\frac{2s-2b+5}{2},\,s-b+3;\,\sigma\omega\end{matrix}\right)$ $+\frac{(n+1)\sqrt{\pi}\,\sigma^{s+1/2}}{2}\,\Gamma\left[\begin{matrix}1-b,\,s-n,\,s+n+2\\ a-b+1,\,\frac{2s+3}{2},\,s+2\end{matrix}\right]$ $\times{}_3F_3\left(\begin{matrix}a,\,s-n,\,s+n+2\\ b,\,\frac{2s+3}{2},\,s+2;\,\sigma\omega\end{matrix}\right)$ $[\sigma>0;\ \operatorname{Re}s>n,\ \operatorname{Re}b+n-1]$
3	$(\sigma-x)_+^{1/2}\,e^{-\omega x}$ $\times U_n\left(\frac{2x}{\sigma}-1\right)$ $\times\Psi\binom{a;\,\omega x}{b}$	$\frac{(n+1)\sqrt{\pi}\,\sigma^{s-b+3/2}}{2\,\omega^{b-1}}\,\Gamma\left[\begin{matrix}b-1,\,s-b+1,\,\frac{2s-2b+1}{2}\\ a,\,\frac{2s-2b-2n+1}{2},\,\frac{2s-2b+2n+5}{2}\end{matrix}\right]$ $\times{}_3F_3\left(\begin{matrix}1-a,\,s-b+1,\,\frac{2s-2b+1}{2};\,-\sigma\omega\\ 2-b,\,\frac{2s-2b-2n+1}{2},\,\frac{2s-2b+2n+5}{2}\end{matrix}\right)$ $+\frac{(n+1)\sqrt{\pi}\,\sigma^{s+1/2}}{2}\,\Gamma\left[\begin{matrix}1-b,\,s,\,\frac{2s-1}{2}\\ a-b+1,\,\frac{2s-2n-1}{2},\,\frac{2s+2n+3}{2}\end{matrix}\right]$ $\times{}_3F_3\left(\begin{matrix}b-a,\,s,\,\frac{2s-1}{2};\,-\sigma\omega\\ b,\,\frac{2s-2n-1}{2},\,\frac{2s+2n+3}{2}\end{matrix}\right)$ $[\sigma>0;\ \operatorname{Re}s>0,\ \operatorname{Re}b-1]$

No.	$f(x)$	$F(s)$
4	$(\sigma-x)_+^{1/2}\, e^{-\omega x} \times U_n\left(\frac{2\sigma}{x}-1\right) \times \Psi\left(\begin{matrix} a;\, \omega x \\ b \end{matrix}\right)$	$\frac{(n+1)\sqrt{\pi}\,\sigma^{s-b+3/2}}{2\,\omega^{b-1}}\Gamma\left[\begin{matrix} b-1 \\ a \end{matrix}\right]\Gamma\left[\begin{matrix} s-b-n+1,\, s-b+n+3 \\ \frac{2s-2b+5}{2},\, s-b+3 \end{matrix}\right] \times {}_3F_3\left(\begin{matrix} 1-a,\, s-b-n+1,\, s-b+n+3 \\ 2-b,\, \frac{2s-2b+5}{2},\, s-b+3;\, -\sigma\omega \end{matrix}\right) + \frac{(n+1)\sqrt{\pi}\,\sigma^{s+1/2}}{2}\Gamma\left[\begin{matrix} 1-b,\, s-n,\, s+n+2 \\ a-b+1,\, \frac{2s+3}{2},\, s+2 \end{matrix}\right] \times {}_3F_3\left(\begin{matrix} b-a,\, s-n,\, s+n+2 \\ b,\, \frac{2s+3}{2},\, s+2;\, -\sigma\omega \end{matrix}\right)$ $[\sigma>0;\ \mathrm{Re}\, s>n,\ \mathrm{Re}\, b+n-1]$

3.29.11. $\Psi(a;\, b;\, \omega x)$ and $H_n(\sigma\sqrt{x})$

No.	$f(x)$	$F(s)$
1	$e^{-\sigma^2 x} H_n(\sigma\sqrt{x}) \times \Psi\left(\begin{matrix} a;\, \omega x \\ b \end{matrix}\right)$	$\frac{\sqrt{\pi}\,\omega^{1-b}}{2^{2s-2b-n+1}\sigma^{2s-2b+2}}\Gamma\left[\begin{matrix} b-1,\, 2s-2b+2 \\ a,\, \frac{2s-2b-n+3}{2} \end{matrix}\right] \times {}_3F_2\left(\begin{matrix} a-b+1,\, s-b+1,\, \frac{2s-2b+3}{2} \\ 2-b,\, \frac{2s-2b-n+3}{2};\, \frac{\omega}{\sigma^2} \end{matrix}\right) + \frac{\sqrt{\pi}\,\sigma^{-2s}}{2^{2s-n-1}}\Gamma\left[\begin{matrix} 1-b,\, 2s \\ a-b+1,\, \frac{2s-n+1}{2} \end{matrix}\right] {}_3F_2\left(\begin{matrix} a,\, s,\, \frac{2s+1}{2} \\ b,\, \frac{2s-n+1}{2};\, \frac{\omega}{\sigma^2} \end{matrix}\right)$ $\left[\begin{matrix} (\mathrm{Re}\,\sigma^2>0;\ \mathrm{Re}\, s+(1-(-1)^n)/2>0,\ \mathrm{Re}\, b-1) \text{ or} \\ (\mathrm{Re}\,\sigma^2=0;\ 0,\ \mathrm{Re}\, b-1<\mathrm{Re}\, s-(1-(-1)^n)/2<\mathrm{Re}\, a+1) \end{matrix}\right]$
2	$e^{-(\sigma^2+\omega)x} H_n(\sigma\sqrt{x}) \times \Psi\left(\begin{matrix} a;\, \omega x \\ b \end{matrix}\right)$	$\frac{2^n\sqrt{\pi}}{\omega^s}\Gamma\left[\begin{matrix} s,\, s-b+1 \\ \frac{1-n}{2},\, s+a-b+1 \end{matrix}\right] {}_3F_2\left(\begin{matrix} \frac{n+1}{2},\, s,\, s-b+1 \\ \frac{1}{2},\, s+a-b+1;\, -\frac{\sigma^2}{\omega} \end{matrix}\right) - \frac{2^{n+1}\sqrt{\pi}\,\sigma}{\omega^{s+1/2}}\Gamma\left[\begin{matrix} \frac{2s+1}{2},\, \frac{2s-2b+3}{2} \\ -\frac{n}{2},\, \frac{2s+2a-2b+3}{2} \end{matrix}\right] {}_3F_2\left(\begin{matrix} \frac{n+2}{2},\, \frac{2s+1}{2},\, \frac{2s-2b+3}{2} \\ \frac{3}{2},\, \frac{2s+2a-2b+3}{2};\, -\frac{\sigma^2}{\omega} \end{matrix}\right)$ $\left[\begin{matrix} (\mathrm{Re}\,(\sigma^2+\omega)>0;\ \mathrm{Re}\, s+(1-(-1)^n)/2>0,\ \mathrm{Re}\, b-1) \text{ or} \\ (\mathrm{Re}\,(\sigma^2+\omega)=0;\ 0,\ \mathrm{Re}\, b-1<\mathrm{Re}\, s-(1-(-1)^n)/2<\mathrm{Re}\, a+1) \end{matrix}\right]$

3.29.12. $\Psi(a;\, b;\, \omega x)$ and $L_n^\lambda(\sigma x)$

No.	$f(x)$	$F(s)$
1	$e^{-\sigma x} L_n^\lambda(\sigma x)\, \Psi\left(\begin{matrix} a;\, \omega x \\ b \end{matrix}\right)$	$\frac{\omega^{-s}}{n!}\left(\frac{\sigma}{\omega}\right)^{a-s}\Gamma\left[\begin{matrix} s-a,\, 1-s+a+n+\lambda \\ 1-s+a+\lambda \end{matrix}\right] \times {}_3F_2\left(\begin{matrix} a,\, a-b+1,\, 1-s+a+n+\lambda \\ a-s+1,\, a-s+\lambda+1;\, \frac{\sigma}{\omega} \end{matrix}\right) + \frac{\omega^{-s}}{n!} \times \Gamma\left[\begin{matrix} n+\lambda+1,\, s,\, a-s,\, s-b+1 \\ a,\, a-b+1,\, \lambda+1 \end{matrix}\right] {}_3F_2\left(\begin{matrix} n+\lambda+1,\, s,\, s-b+1 \\ \lambda+1,\, s-a+1;\, \frac{\sigma}{\omega} \end{matrix}\right)$ $\left[\begin{matrix} (\mathrm{Re}\,\sigma>0;\ \mathrm{Re}\, s>0,\ \mathrm{Re}\, b-1) \text{ or} \\ (\mathrm{Re}\,\sigma=0;\ 0,\ \mathrm{Re}\, b-1<\mathrm{Re}\, s+n<\mathrm{Re}\, a+1) \end{matrix}\right]$

No.	$f(x)$	$F(s)$
2	$e^{-(\sigma+\omega)x} L_n^{\lambda}(\sigma x)$ $\times\, \Psi\left(\begin{matrix} a;\, \omega x \\ b \end{matrix}\right)$	$\frac{\omega^{-s}}{n!}\,\Gamma\left[\begin{matrix} n+\lambda+1,\, s,\, s-b+1 \\ \lambda+1,\, s+a-b+1 \end{matrix}\right] {}_3F_2\left(\begin{matrix} n+\lambda+1,\, s,\, s-b+1 \\ \lambda+1,\, s+a-b+1;\, -\frac{\sigma}{\omega} \end{matrix}\right)$ $\left[\begin{matrix} (\mathrm{Re}\,(\sigma+\omega)>0;\ \mathrm{Re}\, s>0,\ \mathrm{Re}\, b-1)\ \text{or} \\ (\mathrm{Re}\,(\sigma+\omega)=0;\ 0,\ \mathrm{Re}\, b-1<\mathrm{Re}\, s+n<\mathrm{Re}\, a+1) \end{matrix}\right]$

3.29.13. $\Psi(a;\, b;\, \omega x)$ and $C_n^{\lambda}(\varphi(x))$

1	$(\sigma-x)_+^{\lambda-1/2}$ $\times\, C_n^{\lambda}\left(\frac{2x}{\sigma}-1\right)$ $\times\, \Psi\left(\begin{matrix} a;\, \omega x \\ b \end{matrix}\right)$	$\frac{\sqrt{\pi}\,\sigma^{s-b+\lambda+1/2}}{2^{2\lambda-1} n!\,\omega^{b-1}}\,\Gamma\left[\begin{matrix} b-1,\, n+2\lambda,\, s-b+1,\, \frac{2s-2b-2\lambda+3}{2} \\ a,\, \lambda,\, \frac{2s-2b-2n-2\lambda+3}{2},\, \frac{2s-2b+2n+2\lambda+3}{2} \end{matrix}\right]$ $\times\, {}_3F_3\left(\begin{matrix} a-b+1,\, s-b+1,\, \frac{2s-2b-2\lambda+3}{2};\, \sigma\omega \\ 2-b,\, \frac{2s-2b-2n-2\lambda+3}{2},\, \frac{2s-2b+2n+2\lambda+3}{2} \end{matrix}\right)$ $+\, \frac{\sqrt{\pi}\,\sigma^{s+\lambda-1/2}}{2^{2\lambda-1} n!}\,\Gamma\left[\begin{matrix} 1-b,\, n+2\lambda,\, s,\, \frac{2s-2\lambda+1}{2} \\ \lambda,\, a-b+1,\, \frac{2s-2n-2\lambda+1}{2},\, \frac{2s+2n+2\lambda+1}{2} \end{matrix}\right]$ $\times\, {}_3F_3\left(\begin{matrix} a,\, s,\, \frac{2s-2\lambda+1}{2};\, \sigma\omega \\ b,\, \frac{2s-2n-2\lambda+1}{2},\, \frac{2s+2n+2\lambda+1}{2} \end{matrix}\right)$ $[\sigma>0;\ \mathrm{Re}\, s>0,\ \mathrm{Re}\, b-1]$
2	$(\sigma-x)_+^{\lambda-1/2}$ $\times\, C_n^{\lambda}\left(\frac{2\sigma}{x}-1\right)$ $\times\, \Psi\left(\begin{matrix} a;\, \omega x \\ b \end{matrix}\right)$	$\frac{\sqrt{\pi}\,\sigma^{s-b+\lambda+1/2}}{2^{2\lambda-1} n!\,\omega^{b-1}}\,\Gamma\left[\begin{matrix} b-1,\, n+2\lambda \\ a,\, \lambda \end{matrix}\right]$ $\times\, \Gamma\left[\begin{matrix} s-b-n+1,\, s-b+n+2\lambda+1 \\ \frac{2s-2b+2\lambda+3}{2},\, s-b+2\lambda+1 \end{matrix}\right]$ $\times\, {}_3F_3\left(\begin{matrix} a-b+1,\, s-b-n+1,\, s-b+n+2\lambda+1 \\ 2-b,\, \frac{2s-2b+2\lambda+3}{2},\, s-b+2\lambda+1;\, \sigma\omega \end{matrix}\right)$ $+\, \frac{\sqrt{\pi}\,\sigma^{s+\lambda-1/2}}{2^{2\lambda-1} n!}\,\Gamma\left[\begin{matrix} 1-b,\, n+2\lambda,\, s-n,\, s+n+2\lambda \\ \lambda,\, a-b+1,\, \frac{2s+2\lambda+1}{2},\, s+2\lambda \end{matrix}\right]$ $\times\, {}_3F_3\left(\begin{matrix} a,\, s-n,\, s+n+2\lambda \\ b,\, \frac{2s+2\lambda+1}{2},\, s+2\lambda;\, \sigma\omega \end{matrix}\right)$ $[\sigma>0;\ \mathrm{Re}\, s>n,\ \mathrm{Re}\, b+n-1]$
3	$(\sigma-x)_+^{\lambda-1/2}\, e^{-\omega x}$ $\times\, C_n^{\lambda}\left(\frac{2x}{\sigma}-1\right)$ $\times\, \Psi\left(\begin{matrix} a;\, \omega x \\ b \end{matrix}\right)$	$\frac{\sqrt{\pi}\,\sigma^{s-b+\lambda+1/2}}{2^{2\lambda-1} n!\,\omega^{b-1}}\,\Gamma\left[\begin{matrix} b-1,\, n+2\lambda,\, s-b+1,\, \frac{2s-2b-2\lambda+3}{2} \\ a,\, \lambda,\, \frac{2s-2b-2n-2\lambda+3}{2},\, \frac{2s-2b+2n+2\lambda+3}{2} \end{matrix}\right]$ $\times\, {}_3F_3\left(\begin{matrix} 1-a,\, s-b+1,\, \frac{2s-2b-2\lambda+3}{2};\, -\sigma\omega \\ 2-b,\, \frac{2s-2b-2n-2\lambda+3}{2},\, \frac{2s-2b+2n+2\lambda+3}{2} \end{matrix}\right)$ $+\, \frac{\sqrt{\pi}\,\sigma^{s+\lambda-1/2}}{2^{2\lambda-1} n!}\,\Gamma\left[\begin{matrix} 1-b,\, n+2\lambda,\, s,\, \frac{2s-2\lambda+1}{2} \\ \lambda,\, a-b+1,\, \frac{2s-2n-2\lambda+1}{2},\, \frac{2s+2n+2\lambda+1}{2} \end{matrix}\right]$ $\times\, {}_3F_3\left(\begin{matrix} b-a,\, s,\, \frac{2s-2\lambda+1}{2};\, -\sigma\omega \\ b,\, \frac{2s-2n-2\lambda+1}{2},\, \frac{2s+2n+2\lambda+1}{2} \end{matrix}\right)$ $[\sigma>0;\ \mathrm{Re}\, s>0,\ \mathrm{Re}\, b-1]$

No.	$f(x)$	$F(s)$
4	$(\sigma-x)_+^{\lambda-1/2}\,e^{-\omega x}$ $\times\, C_n^{\lambda}\left(\frac{2\sigma}{x}-1\right)$ $\times\,\Psi\left(\begin{matrix}a;\,\omega x\\ b\end{matrix}\right)$	$\frac{\sqrt{\pi}\,\sigma^{s-b+\lambda+1/2}}{2^{2\lambda-1}n!\,\omega^{b-1}}\,\Gamma\left[\begin{matrix}b-1,\,n+2\lambda\\ a,\,\lambda\end{matrix}\right]$ $\times\,\Gamma\left[\begin{matrix}s-b-n+1,\,s-b+n+2\lambda+1\\ \frac{2s-2b+2\lambda+3}{2},\,s-b+2\lambda+1\end{matrix}\right]$ $\times\,{}_3F_3\left(\begin{matrix}1-a,\,s-b-n+1,\,s-b+n+2\lambda+1\\ 2-b,\,\frac{2s-2b+2\lambda+3}{2},\,s-b+2\lambda+1;\,-\sigma\omega\end{matrix}\right)$ $+\,\frac{\sqrt{\pi}\,\sigma^{s+\lambda-1/2}}{2^{2\lambda-1}n!}\,\Gamma\left[\begin{matrix}1-b,\,n+2\lambda,\,s-n,\,s+n+2\lambda\\ \lambda,\,a-b+1,\,\frac{2s+2\lambda+1}{2},\,s+2\lambda\end{matrix}\right]$ $\times\,{}_3F_3\left(\begin{matrix}b-a,\,s-n,\,s+n+2\lambda\\ b,\,\frac{2s+2\lambda+1}{2},\,s+2\lambda;\,-\sigma\omega\end{matrix}\right)$ $[\sigma>0;\ \mathrm{Re}\,s>n,\ \mathrm{Re}\,b+n-1]$

3.29.14. $\Psi(a;\,b;\,\omega x)$ and $P_n^{(\mu,\,\nu)}(\varphi(x))$

No.	$f(x)$	$F(s)$
1	$(\sigma-x)_+^{\mu}$ $\times\,P_n^{(\mu,\,\nu)}\left(\frac{2\sigma}{x}-1\right)$ $\times\,\Psi\left(\begin{matrix}a;\,\omega x\\ b\end{matrix}\right)$	$-\frac{\pi\sigma^{s-b+\mu+1}}{n!\,\omega^{b-1}}\csc(b\pi)\,\Gamma\left[\begin{matrix}n+\mu+1\\ a,2-b\end{matrix}\right]$ $\times\,\Gamma\left[\begin{matrix}s-b-n+1,\,s-b+n+\mu+\nu+2\\ s-b+\mu+2,\,s-b+\mu+\nu+2\end{matrix}\right]$ $\times\,{}_3F_3\left(\begin{matrix}a-b+1,\,s-b-n+1,\,s-b+n+\mu+\nu+2\\ 2-b,\,s-b+\mu+2,\,s-b+\mu+\nu+2;\,\sigma\omega\end{matrix}\right)$ $+\,\frac{\pi\sigma^{s+\mu}}{n!}\csc[(1-b)\pi]\,\Gamma\left[\begin{matrix}n+\mu+1\\ b,\,a-b+1\end{matrix}\right]$ $\times\,\Gamma\left[\begin{matrix}s-n,\,s+n+\mu+\nu+1\\ s+\mu+1,\,s+\mu+\nu+1\end{matrix}\right]{}_3F_3\left(\begin{matrix}a,\,s-n,\,s+n+\mu+\nu+1\\ b,\,s+\mu+1,\,s+\mu+\nu+1;\,\sigma\omega\end{matrix}\right)$ $[\sigma>0;\ \mathrm{Re}\,s>n,\ \mathrm{Re}\,b+n-1]$
2	$(\sigma-x)_+^{\mu}\,e^{-\omega x}$ $\times\,P_n^{(\mu,\,\nu)}\left(\frac{2\sigma}{x}-1\right)$ $\times\,\Psi\left(\begin{matrix}a;\,\omega x\\ b\end{matrix}\right)$	$\frac{\pi\sigma^{s-b+\mu+1}}{n!\,\omega^{b-1}}\csc[(b-1)\pi]\,\Gamma\left[\begin{matrix}n+\mu+1\\ a,2-b\end{matrix}\right]$ $\times\,\Gamma\left[\begin{matrix}s-b-n+1,\,s-b+n+\mu+\nu+2\\ s-b+\mu+2,\,s-b+\mu+\nu+2\end{matrix}\right]$ $\times\,{}_3F_3\left(\begin{matrix}1-a,\,s-b-n+1,\,s-b+n+\mu+\nu+2\\ 2-b,\,s-b+\mu+2,\,s-b+\mu+\nu+2;\,-\sigma\omega\end{matrix}\right)$ $+\,\frac{\pi\sigma^{s+\mu}}{n!}\csc(b\pi)\,\Gamma\left[\begin{matrix}n+\mu+1,\,s-n,\,s+n+\mu+\nu+1\\ b,\,a-b+1,\,s+\mu+1,\,s+\mu+\nu+1\end{matrix}\right]$ $\times\,{}_3F_3\left(\begin{matrix}b-a,\,s-n,\,s+n+\mu+\nu+1\\ b,\,s+\mu+1,\,s+\mu+\nu+1;\,-\sigma\omega\end{matrix}\right)$ $[\sigma>0;\ \mathrm{Re}\,s>n,\ \mathrm{Re}\,b+n-1]$

3.29.15. $\Psi(a;\, b;\, \omega x)$ and $\mathbf{K}(\varphi(x))$, $\mathbf{E}(\varphi(x))$

Notation: $\delta = \begin{Bmatrix} 1 \\ 0 \end{Bmatrix}$.

1	$\theta(\sigma - x) \times \begin{Bmatrix} \mathbf{K}(\sqrt{1 - x/\sigma}) \\ \mathbf{E}(\sqrt{1 - x/\sigma}) \end{Bmatrix} \times \Psi\left(\begin{matrix} a;\, \omega x \\ b \end{matrix}\right)$	$\frac{\pi \sigma^{s-b+1}}{2\omega^{b-1}} \Gamma\left[\begin{matrix} b-1,\, s-b+1,\, s-b-\delta+2 \\ a,\, \frac{2s-2b+3}{2},\, \frac{2s-2b-2\delta+5}{2} \end{matrix}\right] \times {}_3F_3\left(\begin{matrix} a-b+1,\, s-b+1,\, s-b-\delta+2 \\ 2-b,\, \frac{2s-2b+3}{2},\, \frac{2s-2b-2\delta+5}{2};\, \sigma\omega \end{matrix}\right) + \frac{\pi\sigma^s}{2} \Gamma\left[\begin{matrix} 1-b,\, s,\, s-\delta+1 \\ a-b+1,\, \frac{2s+1}{2},\, \frac{2s-2\delta+3}{2} \end{matrix}\right] {}_3F_3\left(\begin{matrix} a,\, s,\, s-\delta+1;\, \sigma\omega \\ b,\, \frac{2s+1}{2},\, \frac{2s-2\delta+3}{2} \end{matrix}\right)$ $[\sigma > 0;\ \mathrm{Re}\, s > 0,\ \mathrm{Re}\, b - 1]$
2	$\theta(\sigma - x)\, e^{-\omega x} \times \begin{Bmatrix} \mathbf{K}(\sqrt{1 - x/\sigma}) \\ \mathbf{E}(\sqrt{1 - x/\sigma}) \end{Bmatrix} \times \Psi\left(\begin{matrix} a;\, \omega x \\ b \end{matrix}\right)$	$\frac{\pi \sigma^{s-b+1}}{2\omega^{b-1}} \Gamma\left[\begin{matrix} b-1,\, s-b+1,\, s-b-\delta+2 \\ a,\, \frac{2s-2b+3}{2},\, \frac{2s-2b-2\delta+5}{2} \end{matrix}\right] \times {}_3F_3\left(\begin{matrix} 1-a,\, s-b+1,\, s-b-\delta+2 \\ 2-b,\, \frac{2s-2b+3}{2},\, \frac{2s-2b-2\delta+5}{2};\, -\sigma\omega \end{matrix}\right) + \frac{\pi\sigma^s}{2} \Gamma\left[\begin{matrix} 1-b,\, s,\, s-\delta+1 \\ a-b+1,\, \frac{2s+1}{2},\, \frac{2s-2\delta+3}{2} \end{matrix}\right] \times {}_3F_3\left(\begin{matrix} b-a,\, s,\, s-\delta+1;\, -\sigma\omega \\ b,\, \frac{2s+1}{2},\, \frac{2s-2\delta+3}{2} \end{matrix}\right)$ $[\sigma > 0;\ \mathrm{Re}\, s > 0,\ \mathrm{Re}\, b - 1]$

3.29.16. $\Psi(a;\, b;\, \omega x)$ and ${}_1F_1(a;\, b;\, \sigma x)$

1	${}_1F_1\left(\begin{matrix} a;\, -\omega x \\ b \end{matrix}\right) \times \Psi\left(\begin{matrix} a;\, \omega x \\ b \end{matrix}\right)$	$\frac{2^{s-b-1}}{\sqrt{\pi}\, \omega^s} \Gamma\left[\begin{matrix} b,\, \frac{s}{2},\, a-\frac{s}{2},\, \frac{s-b+1}{2},\, \frac{s-b+2}{2} \\ a,\, b-\frac{s}{2},\, \frac{s}{2}+a-b+1 \end{matrix}\right]$ $\left[\begin{matrix} (\mathrm{Re}\,\omega > 0;\ 0,\ \mathrm{Re}\, b - 1 < \mathrm{Re}\, s < 2\,\mathrm{Re}\, a) \text{ or} \\ (\mathrm{Re}\,\omega = 0;\ 0,\ \mathrm{Re}\, b - 1 < \mathrm{Re}\, s < 2\,\mathrm{Re}\, a,\ \mathrm{Re}\, b + 1) \end{matrix}\right]$
2	$e^{-\omega x}\, {}_1F_1\left(\begin{matrix} b-a;\, \omega x \\ b \end{matrix}\right) \times \Psi\left(\begin{matrix} a;\, \omega x \\ b \end{matrix}\right)$	$\frac{2^{s-b-1}}{\sqrt{\pi}\, \omega^s} \Gamma\left[\begin{matrix} b,\, \frac{s}{2},\, a-\frac{s}{2},\, \frac{s-b+1}{2},\, \frac{s-b+2}{2} \\ a,\, b-\frac{s}{2},\, \frac{s}{2}+a-b+1 \end{matrix}\right]$ $\left[\begin{matrix} (\mathrm{Re}\,\omega > 0;\ 0,\ \mathrm{Re}\, b - 1 < \mathrm{Re}\, s < 2\,\mathrm{Re}\, a) \text{ or} \\ (\mathrm{Re}\,\omega = 0;\ 0,\ \mathrm{Re}\, b - 1 < \mathrm{Re}\, s < 2\,\mathrm{Re}\, a,\ \mathrm{Re}\, b + 1) \end{matrix}\right]$

3.29.17. Products of $\Psi(a;\, b;\, \omega x)$

1	$\Psi\left(\begin{matrix} a;\, -\omega x \\ b \end{matrix}\right) \Psi\left(\begin{matrix} a;\, \omega x \\ b \end{matrix}\right)$	$\frac{2^{s-b-1}}{\sqrt{\pi}\, (-\omega^2)^{s/2}} \Gamma\left[\begin{matrix} \frac{2a-s}{2},\, \frac{s}{2},\, \frac{s-b+1}{2},\, \frac{s-b+2}{2},\, \frac{s-2b+2}{2} \\ a,\, a-b+1,\, \frac{s+2a-2b+2}{2} \end{matrix}\right]$ $[\mathrm{Re}\,\omega > 0;\ 0,\ \mathrm{Re}\, b - 1,\ 2\,\mathrm{Re}\, b - 2 < \mathrm{Re}\, s < 2\,\mathrm{Re}\, a]$

No.	$f(x)$	$F(s)$
2	$\Psi\left(\begin{matrix}\mu;\,\sigma x\\ \nu\end{matrix}\right)\Psi\left(\begin{matrix}a;\,\omega x\\ b\end{matrix}\right)$	$\frac{\omega^{1-b}}{\sigma^{s-b+1}}\,\Gamma\left[\begin{matrix}b-1,\, s-b+1,\, b-s+\mu-1,\, s-b-\nu+2\\ a,\, \mu,\, \mu-\nu+1\end{matrix}\right]$ $\times\, {}_3F_2\left(\begin{matrix}a-b+1,\, s-b-\nu+2,\, s-b+1\\ 2-b,\, s-b-\mu+2;\, -\frac{\omega}{\sigma}\end{matrix}\right)$ $+\,\frac{\omega^{\mu-s}}{\sigma^{\mu}}\,\Gamma\left[\begin{matrix}a-s+\mu,\, s-\mu,\, s-b-\mu+1\\ a,\, a-b+1\end{matrix}\right]$ $\times\, {}_3F_2\left(\begin{matrix}\mu,\, \mu-\nu+1,\, a-s+\mu\\ 1-s+\mu,\, b-s+\mu;\, -\frac{\omega}{\sigma}\end{matrix}\right)$ $+\,\sigma^{-s}\,\Gamma\left[\begin{matrix}1-b,\, \mu-s,\, s,\, s-\nu+1\\ a-b+1,\, \mu,\, \mu-\nu+1\end{matrix}\right]{}_3F_2\left(\begin{matrix}a,\, s,\, s-\nu+1\\ b,\, s-\mu+1;\, -\frac{\omega}{\sigma}\end{matrix}\right)$ $[0,\ \operatorname{Re}\nu-1,\ \operatorname{Re}(b+\nu)-2,\ \operatorname{Re}b-1<\operatorname{Re}s<\operatorname{Re}(a+\mu)]$
3	$\Psi\left(\begin{matrix}\mu;\,\frac{\sigma}{x}\\ \nu\end{matrix}\right)\Psi\left(\begin{matrix}a;\,\omega x\\ b\end{matrix}\right)$	$\frac{\sigma^{s-b+1}}{\omega^{b-1}}\,\Gamma\left[\begin{matrix}b-1,\, b-s-1,\, b-s-\nu,\, s-b+\mu+1\\ a,\, \mu,\, \mu-\nu+1\end{matrix}\right]$ $\times\, {}_2F_3\left(\begin{matrix}a-b+1,\, s-b+\mu+1;\, \sigma\omega\\ 2-b,\, s-b+2,\, s-b+\nu+1\end{matrix}\right)$ $+\,\frac{\sigma^{1-\nu}}{\omega^{s+\nu-1}}\,\Gamma\left[\begin{matrix}\nu-1,\, 1-s+a-\nu,\, s+\nu-1,\, s-b+\nu\\ a,\, \mu,\, a-b+1\end{matrix}\right]$ $\times\, {}_2F_3\left(\begin{matrix}\mu-\nu+1,\, 1-s+a-\nu;\, \sigma\omega\\ 2-\nu,\, 2-s-\nu,\, 1-s+b-\nu\end{matrix}\right)$ $+\,\sigma^{s}\,\Gamma\left[\begin{matrix}1-b,\, -s,\, 1-s-\nu,\, s+\mu\\ a-b+1,\, \mu,\, \mu-\nu+1\end{matrix}\right]{}_2F_3\left(\begin{matrix}a,\, s+\mu;\, \sigma\omega\\ b,\, s+1,\, s+\nu\end{matrix}\right)$ $+\,\omega^{-s}\,\Gamma\left[\begin{matrix}1-\nu,\, s,\, a-s,\, s-b+1\\ a,\, a-b+1,\, \mu-\nu+1\end{matrix}\right]{}_2F_3\left(\begin{matrix}\mu,\, a-s;\, \sigma\omega\\ \nu,\, 1-s,\, b-s\end{matrix}\right)$ $[-\operatorname{Re}\mu,\ \operatorname{Re}(b-\mu)-1<\operatorname{Re}s<\operatorname{Re}a,\ \operatorname{Re}(a-\nu)+1]$
4	$e^{-\omega x}\Psi\left(\begin{matrix}\mu;\,\sigma x\\ \nu\end{matrix}\right)$ $\times\,\Psi\left(\begin{matrix}a;\,\omega x\\ b\end{matrix}\right)$	$\frac{\sigma^{1-\nu}}{\omega^{s-\nu+1}}\,\Gamma\left[\begin{matrix}\nu-1,\, s-\nu+1,\, s-b-\nu+2\\ \mu,\, s+a-b-\nu+2\end{matrix}\right]$ $\times\, {}_3F_2\left(\begin{matrix}\mu-\nu+1,\, s-\nu+1,\, s-b-\nu+2\\ 2-\nu,\, s+a-b-\nu+2;\, \frac{\sigma}{\omega}\end{matrix}\right)$ $+\,\omega^{-s}\,\Gamma\left[\begin{matrix}1-\nu,\, s,\, s-b+1\\ \mu-\nu+1,\, s+a-b+1\end{matrix}\right]{}_3F_2\left(\begin{matrix}\mu,\, s,\, s-b+1;\, \frac{\sigma}{\omega}\\ \nu,\, s+a-b+1\end{matrix}\right)$ $\left[\begin{matrix}(\operatorname{Re}\omega>0;\ \operatorname{Re}s>0,\ \operatorname{Re}b-1,\ \operatorname{Re}\nu-1,\ \operatorname{Re}(b+\nu)-2)\ \text{or}\\ (\operatorname{Re}\omega=0;\ 0,\ \operatorname{Re}b-1,\ \operatorname{Re}\nu-1,\ \operatorname{Re}(b+\nu)-2<\operatorname{Re}s<\operatorname{Re}(a+\mu)+1)\end{matrix}\right]$
5	$e^{-(\sigma+\omega)x}\Psi\left(\begin{matrix}\mu;\,\sigma x\\ \nu\end{matrix}\right)$ $\times\,\Psi\left(\begin{matrix}a;\,\omega x\\ b\end{matrix}\right)$	$\frac{\sigma^{1-\nu}}{\omega^{s-\nu+1}}\,\Gamma\left[\begin{matrix}\nu-1,\, s-\nu+1,\, s-b-\nu+2\\ \mu,\, s+a-b-\nu+2\end{matrix}\right]$ $\times\, {}_3F_2\left(\begin{matrix}1-\mu,\, s-\nu+1,\, s-b-\nu+2\\ 2-\nu,\, a-b+s-\nu+2;\, -\frac{\sigma}{\omega}\end{matrix}\right)+$

No.	$f(x)$	$F(s)$
		$+\,\omega^{-s}\Gamma\begin{bmatrix}1-\nu,\ s,\ s-b+1\\ \mu-\nu+1,\ a-b+s+1\end{bmatrix}{}_3F_2\left(\begin{matrix}\nu-\mu,\ s,\ s-b+1\\ \nu,\ a-b+s+1;\ -\frac{\sigma}{\omega}\end{matrix}\right)$ $\begin{bmatrix}(\operatorname{Re}(\sigma+\omega)>0;\ \operatorname{Re}s>0,\ \operatorname{Re}b-1,\ \operatorname{Re}\nu-1,\ \operatorname{Re}(b+\nu)-2)\text{ or}\\ (\operatorname{Re}(\sigma+\omega)=0;\ 0,\ \operatorname{Re}b-1,\ \operatorname{Re}\nu-1,\ \operatorname{Re}(b+\nu)-2<\operatorname{Re}s<\operatorname{Re}(a+\mu)+1)\end{bmatrix}$
6	$e^{-\sigma x}\Psi\left(\begin{matrix}\mu;\ \frac{\sigma}{x}\\ \nu\end{matrix}\right)\Psi\left(\begin{matrix}a;\ \omega x\\ b\end{matrix}\right)$	$\frac{\sigma^{s-b+1}}{\omega^{b-1}}\Gamma\begin{bmatrix}b-1,\ b-s-1,\ b-s-\nu,\ s-b+\mu+1\\ a,\ \mu,\ \mu-\nu+1\end{bmatrix}$ $\times\,{}_2F_3\left(\begin{matrix}1-a,\ s-b+\mu+1;\ -\sigma\omega\\ 2-b,\ s-b+2,\ s-b+\nu+1\end{matrix}\right)$ $+\,\frac{\sigma^{1-\nu}}{\omega^{s+\nu-1}}\Gamma\begin{bmatrix}\nu-1,\ s+\nu-1,\ s-b+\nu\\ \mu,\ s+a-b+\nu\end{bmatrix}$ $\times\,{}_2F_3\left(\begin{matrix}\mu-\nu+1,\ 1-s-a+b-\nu;\ -\sigma\omega\\ 2-\nu,\ 2-s-\nu,\ 1-s+b-\nu\end{matrix}\right)$ $-\,\sigma^s\csc(b\pi)\sin[(\mu-\nu)\pi]\,\Gamma\begin{bmatrix}\nu-\mu,\ -s,\ 1-s-\nu,\ s+\mu\\ b,\ a-b+1,\ \mu\end{bmatrix}$ $\times\,{}_2F_3\left(\begin{matrix}b-a,\ s+\mu;\ -\sigma\omega\\ b,\ s+1,\ s+\nu\end{matrix}\right)$ $+\,\omega^{-s}\Gamma\begin{bmatrix}1-\nu,\ s,\ s-b+1\\ \mu-\nu+1,\ s+a-b+1\end{bmatrix}{}_2F_3\left(\begin{matrix}\mu,\ b-a-s;\ -\sigma\omega\\ \nu,\ 1-s,\ b-s\end{matrix}\right)$ $\begin{bmatrix}(\operatorname{Re}\sigma>0;\ -\operatorname{Re}\mu,\ \operatorname{Re}(b-\mu)-1<\operatorname{Re}s)\text{ or}\\ (\operatorname{Re}\sigma=0;\ -\operatorname{Re}\mu,\ \operatorname{Re}(b-\mu)-1<\operatorname{Re}s\\ <\operatorname{Re}a+1,\operatorname{Re}(a-\nu)+2)\end{bmatrix}$
7	$J_\nu(\sigma x)\,\Psi\left(\begin{matrix}a;\ -i\omega x\\ b\end{matrix}\right)$ $\times\,\Psi\left(\begin{matrix}a;\ i\omega x\\ b\end{matrix}\right)$	$2^{s-2a-1}\left(\frac{\sigma^2}{\omega^2}\right)^{a-s/2}(\omega^2)^{-s/2}\,\Gamma\begin{bmatrix}\frac{s-2a+\nu}{2}\\ \frac{2a-s+\nu+2}{2}\end{bmatrix}$ $\times\,{}_4F_3\left(\begin{matrix}a,\ a-b+1,\ \frac{2a-b+1}{2},\ \frac{2a-b+2}{2}\\ 2a-b+1,\ \frac{2a-s-\nu+2}{2},\ \frac{2a-s+\nu+2}{2};\ \frac{\sigma^2}{\omega^2}\end{matrix}\right)+\frac{2^{-\nu-1}}{(\omega^2)^{s/2}}\left(\frac{\sigma^2}{\omega^2}\right)^{\nu/2}$ $\times\,\mathrm{B}\left(\frac{s+\nu}{2},\ \frac{2a-s-\nu}{2}\right)\Gamma\begin{bmatrix}\frac{s-2b+\nu+2}{2},\ s-b+\nu+1\\ \nu+1,\ a-b+1,\ \frac{s+2a-2b+\nu+2}{2}\end{bmatrix}$ $\times\,{}_4F_3\left(\begin{matrix}\frac{s+\nu}{2},\ \frac{s-2b+\nu+2}{2},\ \frac{s-b+\nu+1}{2},\ \frac{s-b+\nu+2}{2}\\ \nu+1,\ \frac{s-2a+\nu+2}{2},\ \frac{s+2a-2b+\nu+2}{2};\ \frac{\sigma^2}{\omega^2}\end{matrix}\right)$ $\begin{bmatrix}\sigma>0;\ \operatorname{Re}\omega\neq0;\\ -\operatorname{Re}\nu,\ \operatorname{Re}(2b-\nu)-2,\ \operatorname{Re}(b-\nu)-1<\operatorname{Re}s<2\operatorname{Re}a+3/2\end{bmatrix}$
8	$e^{-\omega x}J_\nu(\sigma x)\,\Psi\left(\begin{matrix}a;\ \omega x\\ b\end{matrix}\right)$ $\times\,\Psi\left(\begin{matrix}b-a;\ \omega x\\ b\end{matrix}\right)$	$\frac{2^{-\nu-1}\sigma^\nu}{\omega^{s+\nu}}\Gamma\begin{bmatrix}\frac{s+\nu}{2},\ \frac{s-2b+\nu+2}{2},\ s-b+\nu+1\\ \nu+1,\ \frac{s-2a+\nu+2}{2},\ \frac{s+2a-2b+\nu+2}{2}\end{bmatrix}$ $\times\,{}_4F_3\left(\begin{matrix}\frac{s+\nu}{2},\ \frac{s+\nu-2b+2}{2},\ \frac{s+\nu-b+1}{2},\ \frac{s+\nu-b+2}{2}\\ \nu+1,\ \frac{s+\nu-2a+2}{2},\ \frac{s+\nu+2a-2b+2}{2};\ -\frac{\sigma^2}{\omega^2}\end{matrix}\right)$ $\begin{bmatrix}(\operatorname{Re}\omega>\lvert\operatorname{Im}\sigma\rvert;\ \operatorname{Re}s>-\operatorname{Re}\nu,\ \operatorname{Re}(2b-\nu)-2,\ \operatorname{Re}(b-\nu)-1)\text{ or}\\ (\operatorname{Re}^2\omega=\operatorname{Im}^2\sigma;\ -\operatorname{Re}\nu,\ \operatorname{Re}(2b-\nu)-2,\ \operatorname{Re}(b-\nu)-1<\operatorname{Re}s<\operatorname{Re}b+3/2)\end{bmatrix}$

3.30. The Whittaker Functions $M_{\rho,\sigma}(z)$ and $W_{\rho,\sigma}(z)$

The Whittaker functions $M_{\rho,\sigma}(z)$ and $W_{\rho,\sigma}(z)$ are connected with the Kummer confluent hypergeometric function ${}_1F_1(a;\, b;\, z)$ and the Tricomi confluent hypergeometric function $\Psi(a;\, b;\, x)$ by the relations

$$M_{\rho,\sigma}(z) = z^{\sigma+1/2} e^{-z/2}\, {}_1F_1\left(\sigma - \rho + \frac{1}{2};\, 2\sigma + 1;\, z\right),$$

$$M_{\rho,\sigma}(z) = z^{\sigma+1/2} e^{-z/2} \left[z^{-2\sigma}\, \Gamma\left[\begin{matrix} 2\sigma \\ \frac{1}{2} + \sigma - \rho \end{matrix}\right] {}_1F_1\left(\begin{matrix} \frac{1}{2} - \sigma - \rho \\ 1 - 2\sigma;\, z \end{matrix}\right) + \Gamma\left[\begin{matrix} -2\sigma \\ \frac{1}{2} - \sigma - \rho \end{matrix}\right] {}_1F_1\left(\begin{matrix} \frac{1}{2} + \sigma - \rho \\ 1 + 2\sigma;\, z \end{matrix}\right) \right],$$

$$W_{\rho,\sigma}(z) = z^{\sigma+1/2} e^{-x/2}\, \Psi\left(\sigma - \rho + \frac{1}{2};\, 2\sigma + 1;\, z\right).$$

To evaluate the Mellin transform of functions containing $M_{\rho,\sigma}(z)$ and $W_{\rho,\sigma}(z)$, one can apply the above relations and the formulas of Sections 3.28 and 3.29. We present here only several of such formulas.

3.30.1. $W_{\rho,\sigma}(ax)$

No.	$f(x)$	$F(s)$
1	$W_{-1/2,0}(ax)$	$\frac{2^{s-1/2}}{a^s}\, \Gamma\left(\frac{2s+1}{2}\right) \left[\psi\left(\frac{2s+3}{4}\right) - \psi\left(\frac{2s+1}{4}\right)\right]$ $\left[\begin{matrix}(\operatorname{Re} a > 0;\ \operatorname{Re} s > -1/2) \text{ or} \\ (\operatorname{Re} a = 0;\ -1/2 < \operatorname{Re} s < 3/2)\end{matrix}\right]$
2	$W_{0,\sigma}(ax)$	$\frac{2^{2s-1}}{\sqrt{\pi}\, a^s}\, \Gamma\left(\frac{2s-2\sigma+1}{4}\right) \Gamma\left(\frac{2s+2\sigma+1}{4}\right)$ $\left[\begin{matrix}\sigma \neq 0;\ (\operatorname{Re} a > 0;\ \operatorname{Re} s > \lvert\operatorname{Re}\sigma\rvert - 1/2) \text{ or} \\ (\operatorname{Re} a = 0;\ \lvert\operatorname{Re}\sigma\rvert - 1/2 < \operatorname{Re} s < 1)\end{matrix}\right]$
3	$W_{\pm 1/2,\sigma}(ax)$	$\frac{2^{2s-1} a^{-s}}{\sqrt{\pi}\, \sigma^{(1\mp 1)/2}} \left[\Gamma\left(\frac{2s-2\sigma+1}{4}\right) \Gamma\left(\frac{2s+2\sigma+3}{4}\right) \pm \Gamma\left(\frac{2s-2\sigma+3}{4}\right) \Gamma\left(\frac{2s+2\sigma+1}{4}\right)\right]$ $\left[\begin{matrix}\sigma \neq 0;\ (\operatorname{Re} a > 0;\ \operatorname{Re} s > \lvert\operatorname{Re}\sigma\rvert - 1/2) \text{ or} \\ (\operatorname{Re} a = 0;\ \lvert\operatorname{Re}\sigma\rvert - 1/2 < \operatorname{Re} s < 1/2)\end{matrix}\right]$
4	$W_{\rho,\sigma}(ax)$	$a^{-s}\, \Gamma\left[\begin{matrix}\frac{2s-2\sigma+1}{2},\ \frac{2s+2\sigma+1}{2} \\ s - \rho + 1\end{matrix}\right] {}_2F_1\left(\begin{matrix}\frac{2s-2\sigma+1}{2},\ \frac{2s+2\sigma+1}{2} \\ s - \rho + 1;\ \frac{1}{2}\end{matrix}\right)$ $\left[\begin{matrix}\sigma \neq 0;\ (\operatorname{Re} a > 0;\ \operatorname{Re} s > \lvert\operatorname{Re}\sigma\rvert - 1/2) \text{ or} \\ (\operatorname{Re} a = 0;\ \lvert\operatorname{Re}\sigma\rvert - 1/2 < \operatorname{Re} s < -\operatorname{Re}\rho + 1)\end{matrix}\right]$

3.30.2. $M_{\rho,\sigma}(ax)$, $W_{\rho,\sigma}(bx)$, and the exponential function

1	$e^{-ax} M_{\rho,\sigma}(bx)$	$\frac{2^{s+\sigma+1/2} b^{\sigma+1/2}}{(2a-b)^{s+\sigma+1/2}} \Gamma\left(\frac{2s+2\sigma+1}{2}\right) {}_2F_1\left(\begin{matrix}\frac{2\rho+2\sigma+1}{2}, \frac{2s+2\sigma+1}{2} \\ 2\sigma+1;\ \frac{2b}{b-2a}\end{matrix}\right)$ $\left[\begin{matrix}(\operatorname{Re} a > \lvert\operatorname{Re} b\rvert/2;\ \operatorname{Re} s > -\operatorname{Re}\sigma - 1/2) \text{ or} \\ (\operatorname{Re} a = \operatorname{Re} b/2 > 0;\ -\operatorname{Re}\sigma - 1/2 < \operatorname{Re} s < \operatorname{Re}\rho + 1) \text{ or} \\ (\operatorname{Re} a = -\operatorname{Re} b/2 > 0;\ -\operatorname{Re}\sigma - 1/2 < \operatorname{Re} s < -\operatorname{Re}\rho + 1)\end{matrix}\right]$
2	$e^{-ax/2} M_{\rho,\sigma}(ax)$	$a^{-s}\Gamma\left[\begin{matrix}2\sigma+1,\ \rho-s,\ \frac{2s+2\sigma+1}{2} \\ \frac{2\rho+2\sigma+1}{2}, \frac{1-2s+2\sigma}{2}\end{matrix}\right]$ $\left[\begin{matrix}(\operatorname{Re} a > 0;\ -\operatorname{Re}\sigma - 1/2 < \operatorname{Re} s < \operatorname{Re}\rho) \text{ or} \\ (\operatorname{Re} a = 0;\ -\operatorname{Re}\sigma - 1/2 < \operatorname{Re} s < \operatorname{Re}\rho,\ 1-\operatorname{Re}\rho)\end{matrix}\right]$
3	$e^{-ax} W_{\rho,\sigma}(bx)$	$b^{-s}\Gamma\left[\begin{matrix}\frac{2s-2\sigma+1}{2}, \frac{2s+2\sigma+1}{2} \\ s-\rho+1\end{matrix}\right] {}_2F_1\left(\begin{matrix}\frac{2s-2\sigma+1}{2}, \frac{2s+2\sigma+1}{2} \\ s-\rho+1;\ \frac{b-2a}{2b}\end{matrix}\right)$ $\left[\begin{matrix}\sigma \neq 0;\ (\operatorname{Re}(2a+b) > 0;\ \operatorname{Re} s > \lvert\operatorname{Re}\sigma\rvert - 1/2) \text{ or} \\ (\operatorname{Re}(2a+b) = 0;\ \lvert\operatorname{Re}\sigma\rvert - 1/2 < \operatorname{Re} s < 1 - \operatorname{Re}\rho)\end{matrix}\right]$
4	$e^{-ax/2} W_{\rho,\sigma}(ax)$	$a^{-s}\Gamma\left[\begin{matrix}\frac{2s-2\sigma+1}{2}, \frac{2s+2\sigma+1}{2} \\ s-\rho+1\end{matrix}\right]$ $\left[\begin{matrix}\sigma \neq 0;\ (\operatorname{Re} a > 0;\ \operatorname{Re} s > \lvert\operatorname{Re}\sigma\rvert - 1/2) \text{ or} \\ (\operatorname{Re} a = 0;\ \lvert\operatorname{Re}\sigma\rvert - 1/2 < \operatorname{Re} s < 1 - \operatorname{Re}\rho)\end{matrix}\right]$
5	$e^{ax/2} W_{\rho,\sigma}(ax)$	$a^{-s}\Gamma\left[\begin{matrix}\frac{2s-2\sigma+1}{2}, \frac{2s+2\sigma+1}{2}, -s-\rho \\ \frac{1-2\rho-2\sigma}{2}, \frac{1-2\rho+2\sigma}{2}\end{matrix}\right]$ $[\lvert\operatorname{Re}\sigma\rvert - 1/2 < \operatorname{Re} s < -\operatorname{Re}\rho]$
6	$e^{-ax/2} W_{\rho,\sigma}(ax)$	$a^{-s}\Gamma\left[\begin{matrix}\frac{2s-2\sigma+1}{2}, \frac{2s+2\sigma+1}{2} \\ s-\rho+1\end{matrix}\right]$ $\left[\begin{matrix}(\operatorname{Re} s > \lvert\operatorname{Re}\sigma\rvert - 1/2;\ \lvert\arg a\rvert < \pi/2) \text{ or} \\ (\lvert\operatorname{Re}\sigma\rvert - 1/2 < \operatorname{Re} s < 1 - \operatorname{Re}\rho;\ \lvert\arg a\rvert = \pi/2)\end{matrix}\right]$

3.30.3. $W_{\rho,\sigma}(ax)$ and hyperbolic functions

1	$\left\{\begin{matrix}\sinh(ax/2) \\ \cosh(ax/2)\end{matrix}\right\}$ $\times W_{\rho,\sigma}(ax)$	$\frac{a^{-s}}{2}\Gamma\left[\begin{matrix}-s-\rho, \frac{2s-2\sigma+1}{2}, \frac{2s+2\sigma+1}{2} \\ \frac{1-2\rho-2\sigma}{2}, \frac{1-2\rho+2\sigma}{2}\end{matrix}\right] \mp \frac{a^{-s}}{2}\Gamma\left[\begin{matrix}\frac{2s-2\sigma+1}{2}, \frac{2s+2\sigma+1}{2} \\ s-\rho+1\end{matrix}\right]$ $[\lvert\operatorname{Re}\sigma\rvert - 1/2 < \operatorname{Re} s < -\operatorname{Re}\rho;\ \lvert\arg a\rvert \le \pi/2]$

3.30.4. $W_{\rho,\sigma}(ax)$ and $L_\rho^\sigma(bx)$

1	$e^{ax/2}L_{\rho-\sigma-1/2}^{2\sigma}(-ax) \times W_{\rho,\sigma}(ax)$	$\frac{a^{\sigma+1/2}\left(a^2\right)^{-(2s+2\sigma+1)/4}\cos[(\rho-\sigma)\pi]}{2\pi} \times \Gamma\left[\begin{matrix}\frac{2\rho+2\sigma+1}{2}, \frac{2s-2\sigma+1}{2}, \frac{2s+2\sigma+1}{4}, \frac{1-2s-4\rho+2\sigma}{4} \\ \frac{3-2s+6\sigma}{4}, \frac{2s-4\rho-2\sigma+3}{4}\end{matrix}\right]$ $\left[\begin{matrix}(\lvert\operatorname{Re}\sigma\rvert-1/2<\operatorname{Re}s<\operatorname{Re}(\sigma-2\rho)+1/2;\ \lvert\arg a\rvert<\pi/2) \text{ or} \\ (\lvert\operatorname{Re}\sigma\rvert-1/2<\operatorname{Re}s<\operatorname{Re}(\sigma-2\rho)+1/2,\ \operatorname{Re}\sigma+3/2;\ \lvert\arg a\rvert=\pi/2)\end{matrix}\right]$
2	$e^{-ax/2}L_{-\rho-\sigma-1/2}^{2\sigma}(ax) \times W_{\rho,\sigma}(ax)$	$\frac{a^{-s}}{2}\Gamma\left[\begin{matrix}\frac{2s-2\sigma+1}{2}, \frac{2s+2\sigma+1}{4}, \frac{1-2s-4\rho+2\sigma}{4} \\ \frac{1-2\rho-2\sigma}{2}, \frac{2s-4\rho-2\sigma+3}{4}, \frac{3-2s+6\sigma}{4}\end{matrix}\right]$ $\left[\begin{matrix}(\lvert\operatorname{Re}\sigma\rvert-1/2<\operatorname{Re}s<\operatorname{Re}(\sigma-2\rho)+1/2;\ \lvert\arg a\rvert<\pi/2) \text{ or} \\ (\lvert\operatorname{Re}\sigma\rvert-1/2<\operatorname{Re}s<\operatorname{Re}(\sigma-2\rho)+1/2,\ \operatorname{Re}\sigma+3/2;\ \lvert\arg a\rvert=\pi/2)\end{matrix}\right]$

3.30.5. $W_{\rho,\sigma}(ax)$ and ${}_1F_1(b;\,c;\,dx)$, $\Psi(b;\,c;\,dx)$

1	$e^{\pm ax/2}\,{}_1F_1\left(\begin{matrix}\frac{1\mp2\rho+2\sigma}{2} \\ 2\sigma+1;\ \mp ax\end{matrix}\right) \times W_{\rho,\sigma}(ax)$	$\frac{2^{s-\sigma-3/2}a^{-s}}{\sqrt{\pi}} \times \Gamma\left[\begin{matrix}2\sigma+1, \frac{2s-2\sigma+1}{4}, \frac{2s-2\sigma+3}{4}, \frac{2s+2\sigma+1}{4}, \frac{1-2s-4\rho+2\sigma}{4} \\ \frac{1-2\rho+2\sigma}{2}, \frac{3-2s+6\sigma}{4}, \frac{2s-4\rho-2\sigma+3}{4}\end{matrix}\right]$ $\left[\begin{matrix}(\lvert\operatorname{Re}\sigma\rvert-1/2<\operatorname{Re}s<1/2-\operatorname{Re}(2\rho-\sigma);\ \lvert\arg a\rvert<\pi/2) \text{ or} \\ (\lvert\operatorname{Re}\sigma\rvert-1/2<\operatorname{Re}s<1/2-\operatorname{Re}(2\rho-\sigma),\ \operatorname{Re}\sigma+3/2;\ \lvert\arg a\rvert=\pi/2)\end{matrix}\right]$
2	$e^{ax/2}\,\Psi\left(\begin{matrix}\frac{1-2\rho\pm2\sigma}{2} \\ 1\pm2\sigma;\ -ax\end{matrix}\right) \times W_{\rho,\sigma}(ax)$	$\frac{\sqrt{a}\,i^{\mp\sigma}2^{s\mp\sigma-3/2}\left(-a^2\right)^{-(2s+1)/4}}{\sqrt{\pi}} \times \Gamma\left[\begin{matrix}\frac{2s\mp6\sigma+1}{4}, \frac{2s-2\sigma+1}{4}, \frac{2s\mp2\sigma+3}{4}, \frac{2s+2\sigma+1}{4}, \frac{1-2s-4\rho\pm2\sigma}{4} \\ \frac{1-2\rho-2\sigma}{2}, \frac{1-2\rho+2\sigma}{2}, \frac{2s-4\rho\mp2\sigma+3}{4}\end{matrix}\right]$ $[\lvert\operatorname{Re}\sigma\rvert-1/2,\ \pm3\operatorname{Re}\sigma-1/2<\operatorname{Re}s<1/2-\operatorname{Re}(2\rho\mp\sigma)]$

3.30.6. Products of $M_{\mu,\nu}(ax)$ and $W_{\mu,\nu}(bx)$

1	$M_{\rho,\sigma}(-ax)\,W_{\rho,\sigma}(ax)$	$\frac{(-a)^{\sigma+1/2}}{2\,a^{s+\sigma+1/2}}\Gamma\left[\begin{matrix}2\sigma+1, -\frac{s+2\rho}{2}, s+1, \frac{s+2\sigma+1}{2} \\ \frac{1-2\rho+2\sigma}{2}, \frac{1-s+2\sigma}{2}, \frac{s-2\rho+2}{2}\end{matrix}\right]$ $\left[\begin{matrix}(-1,\ 2\operatorname{Re}\sigma-1<\operatorname{Re}s<-2\operatorname{Re}\rho;\ \lvert\arg a\rvert<\pi/2) \text{ or} \\ (-1,\ 2\operatorname{Re}\sigma-1<\operatorname{Re}s<1,\ -2\operatorname{Re}\rho;\ \lvert\arg a\rvert=\pi/2)\end{matrix}\right]$

No.	$f(x)$	$F(s)$
2	$M_{\rho,-\sigma}(-ax)\,W_{\rho,\sigma}(ax)$	$\frac{i(-1)^{-\sigma}}{2a^s}\Gamma\left[\begin{matrix}1-2\sigma,\ -\frac{s+2\rho}{2},\ s+1,\ \frac{s-2\sigma+1}{2}\\ \frac{1-2\rho-2\sigma}{2},\ \frac{1-s-2\sigma}{2},\ \frac{s-2\rho+2}{2}\end{matrix}\right]$ $\left[\begin{matrix}(-1,\ 2\operatorname{Re}\sigma-1<\operatorname{Re}s<-2\operatorname{Re}\rho;\ \lvert\arg a\rvert<\pi/2)\ \text{or}\\ (-1,\ 2\operatorname{Re}\sigma-1<\operatorname{Re}s<1,\ -2\operatorname{Re}\rho;\ \lvert\arg a\rvert=\pi/2)\end{matrix}\right]$
3	$M_{-\rho,\sigma}(ax)\,W_{\rho,\sigma}(ax)$	$\frac{a^{-s}}{2}\Gamma\left[\begin{matrix}2\sigma+1,\ s+1,\ \frac{s+2\sigma+1}{2},\ -\frac{s+2\rho}{2}\\ \frac{2\sigma-2\rho+1}{2},\ \frac{s-2\rho+2}{2},\ \frac{1-s+2\sigma}{2}\end{matrix}\right]$ $\left[\begin{matrix}(\operatorname{Re}a>0;\ -1,\ -2\operatorname{Re}\sigma-1<\operatorname{Re}s<-2\operatorname{Re}\rho)\ \text{or}\\ (\operatorname{Re}a=0;\ -1,\ -2\operatorname{Re}\sigma-1<\operatorname{Re}s<1,\ -2\operatorname{Re}\rho)\end{matrix}\right]$
4	$M_{-\rho,-\sigma}(ax)\,W_{\rho,\sigma}(ax)$	$\frac{a^{-s}}{2}\Gamma\left[\begin{matrix}1-2\sigma,\ -\frac{s+2\rho}{2},\ s+1,\ \frac{s-2\sigma+1}{2}\\ \frac{1-2\rho-2\sigma}{2},\ \frac{1-s-2\sigma}{2},\ \frac{s-2\rho+2}{2}\end{matrix}\right]$ $\left[\begin{matrix}(-1,\ 2\operatorname{Re}\sigma-1<\operatorname{Re}s<-2\operatorname{Re}\rho;\ \lvert\arg a\rvert<\pi/2)\ \text{or}\\ (-1,\ 2\operatorname{Re}\sigma-1<\operatorname{Re}s<1,\ -2\operatorname{Re}\rho;\ \lvert\arg a\rvert=\pi/2)\end{matrix}\right]$
5	$W_{\rho,\pm\sigma}(-ax)\,W_{\rho,\pm\sigma}(ax)$	$\frac{(-a^2)^{-s/2}}{2}\Gamma\left[\begin{matrix}-\frac{s+2\rho}{2},\ s+1,\ \frac{s-2\sigma+1}{2},\ \frac{s+2\sigma+1}{2}\\ \frac{1-2\rho-2\sigma}{2},\ \frac{1-2\rho+2\sigma}{2},\ \frac{s-2\rho+2}{2}\end{matrix}\right]$ $[2\lvert\operatorname{Re}\sigma\rvert-1<\operatorname{Re}s<-2\operatorname{Re}\rho]$
6	$W_{\rho,-\sigma}(\mp ax)$ $\times W_{\rho,\sigma}(\pm ax)$	$\frac{(-a^2)^{-s/2}}{2}\Gamma\left[\begin{matrix}-\frac{s+2\rho}{2},\ s+1,\ \frac{s-2\sigma+1}{2},\ \frac{s+2\sigma+1}{2}\\ \frac{1-2\rho-2\sigma}{2},\ \frac{1-2\rho+2\sigma}{2},\ \frac{s-2\rho+2}{2}\end{matrix}\right]$ $[2\lvert\operatorname{Re}\sigma\rvert-1<\operatorname{Re}s<-2\operatorname{Re}\rho]$
7	$W_{-\rho,\sigma}(ax)\,W_{\rho,\sigma}(ax)$	$\frac{a^{-s}}{2}\Gamma\left[\begin{matrix}s+1,\ \frac{s+2\sigma+1}{2},\ \frac{s-2\sigma+1}{2}\\ \frac{s-2\rho+2}{2},\ \frac{s+2\rho+2}{2}\end{matrix}\right]$ $\left[\begin{matrix}(\operatorname{Re}a>0;\ \operatorname{Re}s>-1,\ 2\lvert\operatorname{Re}\sigma\rvert-1)\ \text{or}\\ (\operatorname{Re}a=0;\ 2\lvert\operatorname{Re}\sigma\rvert-1<\operatorname{Re}s<1)\end{matrix}\right]$
8	$M_{\rho,\sigma}(-iax)\,M_{\rho,\sigma}(iax)$	$\frac{a^{-s}}{2}\Gamma\left[\begin{matrix}2\sigma+1,\ 2\sigma+1,\ \frac{s+2\sigma+1}{2},\ \frac{2\rho-s}{2},\ -\frac{s+2\rho}{2}\\ -s,\ \frac{1-2\rho+2\sigma}{2},\ \frac{1+2\rho+2\sigma}{2},\ \frac{2\sigma-s+1}{2}\end{matrix}\right]$ $[a>0;\ -2\operatorname{Re}\sigma-1<\operatorname{Re}s<1,\ -2\lvert\operatorname{Re}\rho\rvert]$
9	$W_{\rho,\sigma}(-iax)\,W_{\rho,\sigma}(iax)$	$\frac{a^{-s}}{2}\Gamma\left[\begin{matrix}s+1,\ \frac{s+2\sigma+1}{2},\ \frac{s-2\sigma+1}{2},\ -\frac{s+2\rho}{2}\\ \frac{1-2\rho+2\sigma}{2},\ \frac{1-2\rho-2\sigma}{2},\ \frac{s-2\rho+2}{2}\end{matrix}\right]$ $[2\lvert\operatorname{Re}\sigma\rvert-1<\operatorname{Re}s<-2\operatorname{Re}\rho;\ \lvert\arg a\rvert<\pi]$
10	$W_{\rho,-\sigma}(-iax)\,W_{\rho,\sigma}(iax)$	$\frac{a^{-s}}{2}\Gamma\left[\begin{matrix}s+1,\ \frac{s+2\sigma+1}{2},\ \frac{s-2\sigma+1}{2},\ -\frac{s+2\rho}{2}\\ \frac{1-2\rho+2\sigma}{2},\ \frac{1-2\rho-2\sigma}{2},\ \frac{s-2\rho+2}{2}\end{matrix}\right]$ $[2\lvert\operatorname{Re}\sigma\rvert-1<\operatorname{Re}s<-2\operatorname{Re}\rho;\ \lvert\arg a\rvert<\pi]$

3.31. The Gauss Hypergeometric Function ${}_2F_1(a,\, b;\, c;\, z)$

More formulas can be obtained from the corresponding sections due to the relations

$$ {}_2F_1\left(\begin{matrix}a,\, b\\ c;\, z\end{matrix}\right) = \Gamma\left[\begin{matrix}1-a,\, c\\ c-a\end{matrix}\right] P_{-a}^{(c-1,\, a+b-c)}(1-2z), $$

$$ {}_2F_1\left(\begin{matrix}a,\, b\\ c;\, z\end{matrix}\right) = \Gamma\left[\begin{matrix}c\\ a,\, b\end{matrix}\right] G_{22}^{12}\left(-z\,\middle|\,\begin{matrix}1-a,\, 1-b\\ 0,\, 1-c\end{matrix}\right), $$

$$ {}_2F_1\left(\begin{matrix}a,\, b\\ c;\, z\end{matrix}\right) = \Gamma\left[\begin{matrix}c\\ a,\, b,\, c-a,\, c-b\end{matrix}\right] G_{22}^{22}\left(1-z\,\middle|\,\begin{matrix}1-a,\, 1-b\\ 0,\, c-a-b\end{matrix}\right). $$

3.31.1. ${}_2F_1(a,\, b;\, c;\, \omega x)$ and algebraic functions

No.	$f(x)$	$F(s)$
1	${}_2F_1\left(\begin{matrix}a,\, b\\ c;\, -\omega x\end{matrix}\right)$	$\omega^{-s}\,\Gamma\left[\begin{matrix}c\\ a,\, b\end{matrix}\right]\Gamma\left[\begin{matrix}a-s,\, b-s,\, s\\ c-s\end{matrix}\right]$ $[0<\operatorname{Re} s<\operatorname{Re} a,\ \operatorname{Re} b;\ \lvert\arg\omega\rvert<\pi]$
2	${}_2F_1\left(\begin{matrix}a,\, b\\ c;\, -\omega x\end{matrix}\right)-1$	$-(-\omega)^{-s}\,\Gamma\left[\begin{matrix}c\\ a,\, b\end{matrix}\right]\Gamma\left[\begin{matrix}-s,\, a-s,\, b-s,\, s+1\\ 1-s,\, c-s\end{matrix}\right]$ $[-1<\operatorname{Re} s<0,\ \operatorname{Re} a,\ \operatorname{Re} b;\ \lvert\arg\omega\rvert<\pi]$
3	${}_2F_1\left(\begin{matrix}a,\, b\\ c;\, -\omega x\end{matrix}\right)-\sum_{k=0}^{n}\frac{(a)_k\,(b)_k}{k!\,(c)_k}(-\omega x)^k$	$(-1)^{n+1}\,\omega^{-s}\,\Gamma\left[\begin{matrix}c\\ a,\, b\end{matrix}\right]\Gamma\left[\begin{matrix}a-s,\, b-s,\, -n-s,\, s+n+1\\ 1-s,\, c-s\end{matrix}\right]$ $[-n-1<\operatorname{Re} s<-n,\ \operatorname{Re} a,\ \operatorname{Re} b;\ \lvert\arg\omega\rvert<\pi]$
4	$\frac{1}{x-\sigma}\,{}_2F_1\left(\begin{matrix}a,\, b\\ c;\, -\omega x\end{matrix}\right)$	$\omega^{1-s}\,\Gamma\left[\begin{matrix}c,\, a-s+1,\, b-s+1,\, s-1\\ a,\, b,\, c-s+1\end{matrix}\right]$ $\times\,{}_3F_2\left(\begin{matrix}1,\, a-s+1,\, b-s+1\\ 2-s,\, c-s+1;\, -\sigma\omega\end{matrix}\right)$ $-\,\pi\sigma^{s-1}\cot(s\pi)\,{}_2F_1\left(\begin{matrix}a,\, b\\ c;\, -\sigma\omega\end{matrix}\right)$ $[\sigma>0;\ 0<\operatorname{Re} s<\operatorname{Re} a+1,\ \operatorname{Re} b+1;\ \lvert\arg\omega\rvert<\pi]$
5	$(\sigma-x)_+^{\mu-1}\,{}_2F_1\left(\begin{matrix}a,\, b\\ c;\, -\omega x\end{matrix}\right)$	$\sigma^{s+\mu-1}\,\mathrm{B}(\mu,\, s)\,{}_3F_2\left(\begin{matrix}a,\, b,\, s\\ c,\, s+\mu;\, -\sigma\omega\end{matrix}\right)$ $[\sigma,\ \operatorname{Re}\mu,\ \operatorname{Re} s>0;\ \lvert\arg(1+\sigma\omega)\rvert<\pi]$

No.	$f(x)$	$F(s)$
6	$(x-\sigma)_+^{\mu-1}\ {}_2F_1\left(\begin{matrix}a,\,b\\c;\,-\omega x\end{matrix}\right)$	$\sigma^{s+\mu-1}\,\mathrm{B}(\mu,\,1-\mu-s)\ {}_3F_2\left(\begin{matrix}a,\,b,\,s;\,-\sigma\omega\\c,\,s+\mu\end{matrix}\right)$ $+\Gamma\left[\begin{matrix}c,\,s+\mu-1,\,a-\mu-s+1,\,b-\mu-s+1\\a,\,b,\,c-\mu-s+1\end{matrix}\right]$ $\times\,\omega^{1-s-\mu}\ {}_3F_2\left(\begin{matrix}1-\mu,\,a-\mu-s+1,\,b-\mu+1\\2-\mu-s,\,c-\mu-s+1;\,-\sigma\omega\end{matrix}\right)$ $[\sigma,\ \mathrm{Re}\,\mu>0;\ \mathrm{Re}\,s<\mathrm{Re}\,(a-\mu),\ \mathrm{Re}\,(b-\mu);\ \lvert\arg\omega\rvert<\pi]$
7	$\dfrac{1}{(x+\sigma)^{\rho}}\ {}_2F_1\left(\begin{matrix}a,\,b\\c;\,-\omega x\end{matrix}\right)$	$\sigma^{s-\rho}\,\mathrm{B}(s,\,\rho-s)\ {}_3F_2\left(\begin{matrix}a,\,b,\,s;\,\sigma\omega\\c,\,s-\rho+1\end{matrix}\right)$ $+\,\omega^{\rho-s}\,\Gamma\left[\begin{matrix}c,\,s-\rho,\,a+\rho-s,\,b+\rho-s\\a,\,b,\,c+\rho-s\end{matrix}\right]$ $\times\ {}_3F_2\left(\begin{matrix}\rho,\,a+\rho-s,\,b+\rho-s\\c+\rho-s,\,\rho-s+1;\,\sigma\omega\end{matrix}\right)$ $[0<\mathrm{Re}\,s<\mathrm{Re}\,(a+\rho),\ \mathrm{Re}\,(b+\rho);\ \lvert\arg\sigma\rvert,\ \lvert\arg\omega\rvert<\pi]$
8	${}_2F_1\left(\begin{matrix}a,\,b\\c;\,-i\omega x\end{matrix}\right)$ $+\ {}_2F_1\left(\begin{matrix}a,\,b\\c;\,i\omega x\end{matrix}\right)$	$2\omega^{-s}\cos\dfrac{s\pi}{2}\,\Gamma\left[\begin{matrix}c\\a,\,b\end{matrix}\right]\Gamma\left[\begin{matrix}a-s,\,b-s,\,s\\c-s\end{matrix}\right]$ $[0<\mathrm{Re}\,s<\mathrm{Re}\,a,\ \mathrm{Re}\,b;\ \lvert\arg\omega\rvert<\pi/2]$
9	${}_2F_1\left(\begin{matrix}a,\,b\\c;\,-i\omega x\end{matrix}\right)$ $-\ {}_2F_1\left(\begin{matrix}a,\,b\\c;\,i\omega x\end{matrix}\right)$	$-2i\,\omega^{-s}\sin\dfrac{s\pi}{2}\,\Gamma\left[\begin{matrix}c\\a,\,b\end{matrix}\right]\Gamma\left[\begin{matrix}a-s,\,b-s,\,s\\c-s\end{matrix}\right]$ $[-1<\mathrm{Re}\,s<\mathrm{Re}\,a,\ \mathrm{Re}\,b;\ \lvert\arg\omega\rvert<\pi/2]$
10	$(x+\omega)^{-b}$ $\times\ {}_2F_1\left(\begin{matrix}a,\,b;\,-\frac{x}{\omega}\\a+2b+1\end{matrix}\right)$	$b\,(2a+2b-s)\,\omega^{s-b}\,\Gamma\left[\begin{matrix}a+2b+1\\2b+1,\,a+b+1\end{matrix}\right]$ $\times\,\Gamma\left[\begin{matrix}s,\,2b-s,\,a+b-s\\a+2b-s+1\end{matrix}\right]$ $[0<\mathrm{Re}\,s<2\,\mathrm{Re}\,b,\ \mathrm{Re}\,(a+b),\ 2\,\mathrm{Re}\,(a+b)+1;\lvert\arg\omega\rvert<\pi]$
11	$(x+\omega)\ {}_2F_1\left(\begin{matrix}a,\,b;\,-\frac{x}{\omega}\\a-b+1\end{matrix}\right)$	$(a-2s-1)\,\omega^{s+1}\,\Gamma\left[\begin{matrix}a-b+1\\a,\,b-1\end{matrix}\right]\Gamma\left[\begin{matrix}a-s-1,\,b-s-1,\,s\\a-b-s+1\end{matrix}\right]$ $[0<\mathrm{Re}\,s<\mathrm{Re}\,a-1,\ \mathrm{Re}\,b-1,\ (\mathrm{Re}\,a+1)/2;\lvert\arg\omega\rvert<\pi]$
12	$(x+\omega)\ {}_2F_1\left(\begin{matrix}a,\,b;\,-\frac{x}{\omega}\\\frac{2a+b-1}{2}\end{matrix}\right)$	$\dfrac{2a-s-2}{2}\,\omega^{s+1}\,\Gamma\left[\begin{matrix}\frac{2a+b-1}{2}\\a,\,b-1\end{matrix}\right]\Gamma\left[\begin{matrix}a-s-1,\,b-s-1,\,s\\\frac{2a+b-2s-1}{2}\end{matrix}\right]$ $[0<\mathrm{Re}\,s<\mathrm{Re}\,a-1,\ \mathrm{Re}\,b-1;\ \lvert\arg\omega\rvert<\pi]$

No.	$f(x)$	$F(s)$
13	$(x+\omega)^{2b}\,{}_2F_1\left(\begin{matrix}a,\ b;\ -\frac{x}{\omega}\\ a-b+1\end{matrix}\right)$	$(a-2b-2s)\,\omega^{s+2b}\,\Gamma\left[\begin{matrix}a-b+1\\ -b,\ a-2b+1\end{matrix}\right]\Gamma\left[\begin{matrix}s,\ -b-s,\ a-2b-s\\ a-b-s+1\end{matrix}\right]$ $[0<\operatorname{Re}s<-\operatorname{Re}b,\ \operatorname{Re}(a-2b);\ \lvert\arg\omega\rvert<\pi]$
14	$(x+\omega)^{a+b-c}\,{}_2F_1\left(\begin{matrix}a,\ b\\ c;\ -\frac{x}{\omega}\end{matrix}\right)$	$\omega^{s+a+b-c}\,\Gamma\left[\begin{matrix}c\\ c-a,\ c-b\end{matrix}\right]\Gamma\left[\begin{matrix}c-a-s,\ c-b-s,\ s\\ c-s\end{matrix}\right]$ $[0<\operatorname{Re}s<\operatorname{Re}(c-a),\ \operatorname{Re}(c-b);\ \lvert\arg\omega\rvert<\pi]$
15	$\theta(\omega-x)\,{}_2F_1\left(\begin{matrix}a,\ c+n\\ c;\ \frac{x}{\omega}\end{matrix}\right)$	$\frac{\omega^s}{(c)_n}\Gamma\left[\begin{matrix}1-a,\ s,\ c-s+n\\ s-a+1,\ c-s\end{matrix}\right]$ $[\operatorname{Re}a<1-n;\ \omega,\ \operatorname{Re}s>0]$
16	$(\omega-x)_+^{b-c-n}\,{}_2F_1\left(\begin{matrix}-n,\ b\\ c;\ \frac{x}{\omega}\end{matrix}\right)$	$\frac{\omega^{s+b-c-n}}{(c)_n}\Gamma\left[\begin{matrix}b-c+1,\ s,\ c-s+n\\ s+b-c+1,\ c-s\end{matrix}\right]$ $[\omega,\ \operatorname{Re}(b-c-n+1),\ \operatorname{Re}s>0]$
17	$(x-\omega)_+^{b-c-n}\,{}_2F_1\left(\begin{matrix}-n,\ b\\ c;\ \frac{x}{\omega}\end{matrix}\right)$	$\frac{\omega^{s+b-c-n}}{(c)_n}\Gamma\left[\begin{matrix}b-c+1,\ s-c+1,\ c-b-s\\ s-c-n+1,\ 1-s\end{matrix}\right]$ $[\omega>0;\ \operatorname{Re}s<\operatorname{Re}(c-b)<1-n]$

3.31.2. ${}_2F_1\left(a,\ b;\ c;\ \frac{\omega}{x}\right)$ and algebraic functions

1	$(x+\omega)^{a+b-c}\,{}_2F_1\left(\begin{matrix}a,\ b\\ c;\ -\frac{\omega}{x}\end{matrix}\right)$	$\omega^{s+a+b-c}\,\Gamma\left[\begin{matrix}c\\ c-a,\ c-b\end{matrix}\right]\Gamma\left[\begin{matrix}c-a-b-s,\ s+a,\ s+b\\ s+a+b\end{matrix}\right]$ $[-\operatorname{Re}a,\ -\operatorname{Re}b<\operatorname{Re}s<\operatorname{Re}(c-a-b);\ \lvert\arg\omega\rvert<\pi]$
2	$(x-\omega)^{a+b-c}\,{}_2F_1\left(\begin{matrix}a,\ b\\ c;\ \frac{\omega}{x}\end{matrix}\right)$	$e^{i(-a+b+c)\pi}\omega^{s+a+b-c}\,\Gamma\left[\begin{matrix}c\\ c-a,\ c-b\end{matrix}\right]$ $\times\Gamma\left[\begin{matrix}c-a-b-s,\ s+a,\ s+b\\ s+a+b\end{matrix}\right]$ $[-\operatorname{Re}a,\ -\operatorname{Re}b<\operatorname{Re}s<\operatorname{Re}(c-a-b);\ 0<\arg\omega\le\pi]$
3	$(x-\omega)^{2b}\,{}_2F_1\left(\begin{matrix}a,\ b;\ \frac{\omega}{x}\\ a-b+1\end{matrix}\right)$	$e^{-i(s+2b)\pi}\,(2s+a+2b)\,\omega^{s+2b}\,\Gamma\left[\begin{matrix}a-b+1\\ -b,\ a-2b+1\end{matrix}\right]$ $\times\Gamma\left[\begin{matrix}-s-2b,\ s+a,\ s+b\\ s+a+b+1\end{matrix}\right]$ $\left[\begin{matrix}-\operatorname{Re}a,\ -\operatorname{Re}b,\ -\operatorname{Re}(a+2b)/2-1<\operatorname{Re}s<-2\operatorname{Re}b;\\ 0<\arg\omega\le\pi\end{matrix}\right]$

No.	$f(x)$	$F(s)$
4	$(x-\omega)^{-b} \times {}_2F_1\left(\begin{matrix} a,\ b;\ \frac{\omega}{x} \\ a+2b+1 \end{matrix}\right)$	$\frac{e^{-i(s-b)\pi}\omega^{s-b}}{2}\Gamma\left[\begin{matrix} a+2b+1 \\ 2b,\ a+b+1 \end{matrix}\right] \times\Gamma\left[\begin{matrix} b-s,\ s+a,\ s+b,\ s+2a+b+1 \\ s+a+b+1,\ s+2a+b \end{matrix}\right]$ $[-\operatorname{Re} a,\ -\operatorname{Re} b,\ -\operatorname{Re}(2a+b)-1<\operatorname{Re} s<\operatorname{Re} b;\ 0<\arg\omega\leq\pi]$
5	$(x+\omega)\,{}_2F_1\left(\begin{matrix} a,\ b;\ -\frac{\omega}{x} \\ \frac{2a+b-1}{2} \end{matrix}\right)$	$\frac{s+2a-1}{2}\,\omega^{s+1}\,\Gamma\left[\begin{matrix} \frac{2a+b-1}{2} \\ a,\ b-1 \end{matrix}\right]\Gamma\left[\begin{matrix} -s-1,\ s+a,\ s+b \\ \frac{2s+2a+b+1}{2} \end{matrix}\right]$ $[-\operatorname{Re} a,\ -\operatorname{Re} b<\operatorname{Re} s<-1;\ \lvert\arg\omega\rvert<\pi]$
6	$(x+\omega)\,{}_2F_1\left(\begin{matrix} a,\ b;\ -\frac{\omega}{x} \\ a-b+1 \end{matrix}\right)$	$2\,\omega^{s+1}\,\Gamma\left[\begin{matrix} a-b+1 \\ a,\ b-1 \end{matrix}\right]\Gamma\left[\begin{matrix} -s-1,\ s+a,\ s+b,\ \frac{2s+a+3}{2} \\ \frac{2s+a+1}{2},\ s+a-b+2 \end{matrix}\right]$ $[-\operatorname{Re} a,\ -\operatorname{Re} b<\operatorname{Re} s<-1;\ \lvert\arg\omega\rvert<\pi]$
7	$\theta(x-\omega)\ {}_2F_1\left(\begin{matrix} a,\ c+n \\ c;\ \frac{\omega}{x} \end{matrix}\right)$	$\frac{\omega^s}{(c)_n}\Gamma\left[\begin{matrix} 1-a,\ s+c+n,\ -s \\ s+c,\ 1-a-s \end{matrix}\right]$ $[\omega>0,\ \operatorname{Re} a<1-n;\ \operatorname{Re} s<0]$
8	$(x-\omega)_+^{b-c-n}\ {}_2F_1\left(\begin{matrix} -n,\ b \\ c;\ \frac{\omega}{x} \end{matrix}\right)$	$\frac{\omega^{s+b-c-n}}{(c)_n}\Gamma\left[\begin{matrix} b-c+1,\ c-b-s+n,\ s+b \\ n-s+1,\ s+b-n \end{matrix}\right]$ $[\omega>0,\ \operatorname{Re}(b-c)>n-1;\ \operatorname{Re} s<\operatorname{Re}(c-b)+n]$

3.31.3. ${}_2F_1(a,\ b;\ c;\ \omega x^r)$ and various functions

1	$\theta(1-x)\ {}_2F_1\left(\begin{matrix} a,\ b \\ c;\ x \end{matrix}\right) + \theta(x-1)\,x^{-a} \times\Gamma\left[\begin{matrix} 1-b,\ c \\ a-b+1,\ c-a \end{matrix}\right] \times {}_2F_1\left(\begin{matrix} a,\ a-c+1 \\ a-b+1;\ \frac{1}{x} \end{matrix}\right)$	$\Gamma\left[\begin{matrix} 1-b,\ c,\ s,\ a-s \\ a,\ s-b+1,\ c-s \end{matrix}\right]$ $\left[\begin{matrix} \operatorname{Re}(c-a-b)>-1;\ 0<\operatorname{Re} s<\operatorname{Re} a; \\ b\neq 1,2,\ldots;\ c\neq 0,-1,-2,\ldots \end{matrix}\right]$

3.31.4. ${}_2F_1\left(a,\ b;\ c;\ \dfrac{\omega-x}{\omega}\right)$ and algebraic functions

1	${}_2F_1\left(\begin{matrix} a,\ b \\ c;\ \frac{\omega-x}{\omega} \end{matrix}\right)$	$\omega^s\,\Gamma\left[\begin{matrix} c,\ a-s,\ b-s,\ s,\ s-a-b+c \\ a,\ b,\ c-a,\ c-b \end{matrix}\right]$ $[0,\ \operatorname{Re}(a+b-c)<\operatorname{Re} s<\operatorname{Re} a,\ \operatorname{Re} b;\ \lvert\arg\omega\rvert<\pi]$

No.	$f(x)$	$F(s)$
2	$(\sigma - x)_+^{\mu-1}\ {}_2F_1\left(\begin{matrix} a, b \\ c;\ \frac{\omega-x}{\omega} \end{matrix}\right)$	$\sigma^{s+\mu-1}\,\Gamma\begin{bmatrix} c,\ c-a-b,\ \mu,\ s \\ c-a,\ c-b,\ s+\mu \end{bmatrix} {}_3F_2\left(\begin{matrix} a,\ b,\ s;\ \frac{\sigma}{\omega} \\ a+b-c+1,\ s+\mu \end{matrix}\right)$ $+\dfrac{\sigma^{s-a-b+c+\mu-1}}{\omega^{c-a-b}}\,\Gamma\begin{bmatrix} c,\ a+b-c,\ \mu,\ s-a-b+c \\ a,\ b,\ s-a-b+c+\mu \end{bmatrix}$ $\times\ {}_3F_2\left(\begin{matrix} c-a,\ c-b,\ s-a-b+c;\ \frac{\sigma}{\omega} \\ c-a-b+1,\ s-a-b+c+\mu \end{matrix}\right)$ $[\sigma,\ \operatorname{Re}\mu > 0;\ \operatorname{Re}s > 0,\ \operatorname{Re}(a+b-c);\ \lvert\arg\omega\rvert < \pi]$
3	$(x - \sigma)_+^{\mu-1}\ {}_2F_1\left(\begin{matrix} a, b \\ c;\ \frac{\omega-x}{\omega} \end{matrix}\right)$	$\sigma^{s+\mu-a-1}\omega^{s}\,\Gamma\begin{bmatrix} c,\ b-a,\ \mu,\ a-\mu-s+1 \\ b,\ c-a,\ a-s+1 \end{bmatrix}$ $\times\ {}_3F_2\left(\begin{matrix} a,\ c-b,\ a-\mu-s+1 \\ a-b+1,\ a-s+1;\ \frac{\omega}{\sigma} \end{matrix}\right)$ $+\,\sigma^{s+\mu-b-1}\omega^{b}\Gamma\begin{bmatrix} a-b,\ c,\ \mu,\ b-\mu-s+1 \\ a,\ c-b,\ b-s+1 \end{bmatrix}$ $\times\ {}_3F_2\left(\begin{matrix} b,\ c-a,\ b-\mu-s+1 \\ b-a+1,\ b-s+1;\ \frac{\omega}{\sigma} \end{matrix}\right)$ $\begin{bmatrix} \sigma,\ \operatorname{Re}\mu > 0;\ \operatorname{Re}s < \operatorname{Re}(a-\mu+1), \\ \operatorname{Re}(b-\mu+1);\ \lvert\arg\omega\rvert < \pi \end{bmatrix}$
4	$\dfrac{1}{x-\sigma}\ {}_2F_1\left(\begin{matrix} a, b \\ c;\ \frac{\omega-x}{\omega} \end{matrix}\right)$	$\pi\sigma^{s-a-1}\omega^{a}\cot[(a-s)\pi]\,\Gamma\begin{bmatrix} b-a,\ c \\ b,\ c-a \end{bmatrix} {}_2F_1\left(\begin{matrix} a,\ c-b;\ \frac{\omega}{\sigma} \\ a-b+1 \end{matrix}\right)$ $+\,\pi\sigma^{s-b-1}\omega^{b}\cot[(b-s)\pi]\,\Gamma\begin{bmatrix} a-b,\ c \\ a,\ c-b \end{bmatrix} {}_2F_1\left(\begin{matrix} b,\ c-a;\ \frac{\omega}{\sigma} \\ b-a+1 \end{matrix}\right)$ $-\,\dfrac{\omega^{s}}{\sigma}\,\Gamma\begin{bmatrix} c,\ s,\ s-a-b+c,\ a-s,\ b-s \\ a,\ b,\ c-a,\ c-b \end{bmatrix}$ $\times\ {}_3F_2\left(\begin{matrix} 1,\ s,\ s-a-b+c;\ \frac{\omega}{\sigma} \\ s-a+1,\ s-b+1 \end{matrix}\right)$ $\begin{bmatrix} \sigma > 0;\ 0,\ \operatorname{Re}(a+b-c) < \operatorname{Re}s \\ < \operatorname{Re}a+1,\ \operatorname{Re}b+1;\ \lvert\arg\omega\rvert < \pi \end{bmatrix}$
5	$\dfrac{1}{(x+\sigma)^{\rho}}\ {}_2F_1\left(\begin{matrix} a, b \\ c;\ \frac{\omega-x}{\omega} \end{matrix}\right)$	$\sigma^{-\rho}\omega^{s}\,\Gamma\begin{bmatrix} c,\ s,\ a-s,\ b-s,\ s-a-b+c \\ a,\ b,\ c-a,\ c-b \end{bmatrix}$ $\times\ {}_3F_2\left(\begin{matrix} \rho,\ s,\ s-a-b+c;\ -\frac{\omega}{\sigma} \\ s-a+1,\ s-b+1 \end{matrix}\right) + \sigma^{s-\rho-a}\omega^{a}\,\Gamma\begin{bmatrix} b-a,\ c \\ b,\ c-a \end{bmatrix}$ $\times\,\Gamma\begin{bmatrix} s-a,\ a+\rho-s \\ \rho \end{bmatrix} {}_3F_2\left(\begin{matrix} a,\ c-b,\ a+\rho-s;\ -\frac{\omega}{\sigma} \\ a-s+1,\ a-b+1 \end{matrix}\right) + \sigma^{s-\rho-b}\omega^{b}$ $\times\,\Gamma\begin{bmatrix} a-b,\ c,\ s-b,\ b-s+\rho \\ a,\ c-b,\ \rho \end{bmatrix} {}_3F_2\left(\begin{matrix} b,\ c-a,\ b+\rho-s;\ -\frac{\omega}{\sigma} \\ b-a+1,\ b-s+1 \end{matrix}\right)$ $\begin{bmatrix} 0,\ \operatorname{Re}(a+b-c) < \operatorname{Re}s < \operatorname{Re}(a+\rho),\ \operatorname{Re}(b+\rho); \\ \lvert\arg\sigma\rvert,\ \lvert\arg\omega\rvert < \pi \end{bmatrix}$

No.	$f(x)$	$F(s)$
6	$(\omega-x)_+^{c-1}\,{}_2F_1\left(\begin{matrix}a,\,b\\ c;\,\frac{\omega-x}{\omega}\end{matrix}\right)$	$\omega^{s+c-1}\Gamma\left[\begin{matrix}c,\,s,\,s-a-b+c\\ s-a+c,\,s-b+c\end{matrix}\right]$ $[\omega,\ \operatorname{Re}c>0;\ \operatorname{Re}s>0,\ \operatorname{Re}(a+b-c)]$
7	$(\omega-x)^{c-1}\,{}_2F_1\left(\begin{matrix}a,\,b\\ c;\,\frac{\omega-x}{\omega}\end{matrix}\right)$	$\omega^{s+c-1}\Gamma\left[\begin{matrix}c,\,s,\,s-a-b+c\\ s-a+c,\,s-b+c\end{matrix}\right]-e^{ic\pi}\omega^{s+c-1}\Gamma\left[\begin{matrix}c,\,a-c-s+1,\,b-c-s+1\\ 1-s,\,a+b-c-s+1\end{matrix}\right]$ $\left[\begin{matrix}\operatorname{Re}c>0;\ 0,\ \operatorname{Re}(a+b-c)<\operatorname{Re}s\\ <\operatorname{Re}(a-c)+1,\ \operatorname{Re}(b-c)+1;\ \operatorname{Im}\omega>0\end{matrix}\right]$
8	$(x-\omega)_+^{c-1}\,{}_2F_1\left(\begin{matrix}a,\,b\\ c;\,\frac{\omega-x}{\omega}\end{matrix}\right)$	$\omega^{s+c-1}\Gamma\left[\begin{matrix}c,\,a-c-s+1,\,b-c-s+1\\ 1-s,\,a+b-c-s+1\end{matrix}\right]$ $[\omega,\ \operatorname{Re}c>0;\ \operatorname{Re}s<\operatorname{Re}(a-c)+1,\ \operatorname{Re}(b-c)+1]$
9	$(x-\omega)^{c-1}\,{}_2F_1\left(\begin{matrix}a,\,b\\ c;\,\frac{\omega-x}{\omega}\end{matrix}\right)$	$\omega^{s+c-1}\Gamma\left[\begin{matrix}c,\,a-c-s+1,\,b-c-s+1\\ 1-s,\,a+b-c-s+1\end{matrix}\right]-e^{ic\pi}\omega^{s+c-1}\Gamma\left[\begin{matrix}c,\,s-a-b+c,\,s\\ s-a+c,\,s-b+c\end{matrix}\right]$ $\left[\begin{matrix}\omega,\ \operatorname{Re}c>0;\ 0,\ \operatorname{Re}(a+b-c)<\operatorname{Re}s\\ <\operatorname{Re}(a-c)+1,\ \operatorname{Re}(b-c)+1\end{matrix}\right]$
10	$(\sigma-x)_+^{\mu-1}(\omega-x)^{c-1}\times{}_2F_1\left(\begin{matrix}a,\,b\\ c;\,\frac{\omega-x}{\omega}\end{matrix}\right)$	$\sigma^{s+\mu-1}\omega^{c-1}\Gamma\left[\begin{matrix}c,\,c-a-b,\,\mu,\,s\\ c-a,\,c-b,\,s+\mu\end{matrix}\right]\times{}_3F_2\left(\begin{matrix}a-c+1,\,b-c+1,\,s\\ a+b-c+1,\,s+\mu;\,\frac{\sigma}{\omega}\end{matrix}\right)+\frac{\sigma^{s-a-b+c+\mu-1}}{\omega^{1-a-b}}\Gamma\left[\begin{matrix}c,\,a+b-c,\,\mu,\,s-a-b+c\\ a,\,b,\,s+\mu-a-b+c\end{matrix}\right]\times{}_3F_2\left(\begin{matrix}1-a,\,1-b,\,s-a-b+c;\,\frac{\sigma}{\omega}\\ c-a-b+1,\,s+\mu-a-b+c\end{matrix}\right)$ $[0<\sigma<\omega;\ \operatorname{Re}\mu,\ \operatorname{Re}s,\ \operatorname{Re}(s-a-b+c)>0]$
11	$(\sigma-x)^{\mu-1}(\omega-x)_+^{c-1}\times{}_2F_1\left(\begin{matrix}a,\,b\\ c;\,\frac{\omega-x}{\omega}\end{matrix}\right)$	$\sigma^{\mu-1}\omega^{c+s-1}\Gamma\left[\begin{matrix}c,\,s,\,s-a-b+c\\ s-a+c,\,s-b+c\end{matrix}\right]\times{}_3F_2\left(\begin{matrix}1-\mu,\,s,\,s-a-b+c\\ s-a+c,\,s-b+c;\,\frac{\omega}{\sigma}\end{matrix}\right)$ $[0<\omega<\sigma;\ \operatorname{Re}c,\ \operatorname{Re}s,\ \operatorname{Re}(s-a-b+c)>0]$

No.	$f(x)$	$F(s)$
12	$(x-\sigma)_+^{\mu-1}(\omega-x)_+^{c-1}$ $\times\, {}_2F_1\left(\begin{matrix} a,\, b \\ c;\, \frac{\omega-x}{\omega} \end{matrix}\right)$	$\sigma^{s+\mu-1}\omega^{c-1}\,\Gamma\left[\begin{matrix} c,\, c-a-b,\, \mu,\, 1-s-\mu \\ c-a,\, c-b,\, 1-s \end{matrix}\right]$ $\times\, {}_3F_2\left(\begin{matrix} a-c+1,\, b-c+1,\, s \\ a+b-c+1,\, s+\mu;\, \frac{\sigma}{\omega} \end{matrix}\right) + \sigma^{s+\mu-a-b+c-1}\omega^{a+b-1}$ $\times\,\Gamma\left[\begin{matrix} c,\, a+b-c,\, \mu,\, a+b-c-\mu-s+1 \\ a,\, b,\, a+b-c-s+1 \end{matrix}\right]$ $\times\, {}_3F_2\left(\begin{matrix} 1-a,\, 1-b,\, s-a-b+c;\, \frac{\sigma}{\omega} \\ c-a-b+1,\, s+\mu-a-b+c \end{matrix}\right)$ $+\,\omega^{s+\mu+c-2}\,\Gamma\left[\begin{matrix} c,\, s+\mu-1,\, s-a-b+c+\mu-1 \\ s+\mu-a+c-1,\, s+\mu-b+c-1 \end{matrix}\right]$ $\times\, {}_3F_2\left(\begin{matrix} 1-\mu,\, a-c-\mu-s+2,\, b-c-\mu-s+2 \\ 2-\mu-s,\, a+b-c-\mu-s+2;\, \frac{\sigma}{\omega} \end{matrix}\right)$ $[0<\sigma<\omega;\ \operatorname{Re} c,\ \operatorname{Re}\mu>0]$
13	$\dfrac{(\omega-x)_+^{c-1}}{(x+\sigma)^{\rho}}\, {}_2F_1\left(\begin{matrix} a,\, b \\ c;\, \frac{\omega-x}{\omega} \end{matrix}\right)$	$\sigma^{-\rho}\omega^{s+c-1}\,\Gamma\left[\begin{matrix} c,\, s,\, s-a-b+c \\ s-a+c,\, s-b+c \end{matrix}\right]$ $\times\, {}_3F_2\left(\begin{matrix} \rho,\, s,\, s-a-b+c \\ s-a+c,\, s-b+c;\, -\frac{\omega}{\sigma} \end{matrix}\right)$ $[\omega,\ \operatorname{Re} c,\ \operatorname{Re} s,\ \operatorname{Re}(s-a-b+c)>0;\ \lvert\arg\sigma\rvert<\pi]$
14	$\dfrac{(\omega-x)_+^{c-1}}{x-\sigma}\, {}_2F_1\left(\begin{matrix} a,\, b \\ c;\, \frac{\omega-x}{\omega} \end{matrix}\right)$	$-\pi\sigma^{s-1}\omega^{c-1}\cot(s\pi)\,\Gamma\left[\begin{matrix} c,\, c-a-b \\ c-a,\, c-b \end{matrix}\right] {}_2F_1\left(\begin{matrix} a-c+1,\, b-c+1 \\ a+b-c+1;\, \frac{\sigma}{\omega} \end{matrix}\right)$ $+\,\pi\sigma^{s-a-b+c-1}\omega^{a+b-1}\cot[(a+b-c-s)\pi]$ $\times\,\Gamma\left[\begin{matrix} c,\, a+b-c \\ a,\, b \end{matrix}\right] {}_2F_1\left(\begin{matrix} 1-a,\, 1-b \\ c-a-b+1;\, \frac{\sigma}{\omega} \end{matrix}\right)$ $+\,\omega^{s+c-2}\,\Gamma\left[\begin{matrix} c,\, s-1,\, c-a-b+s-1 \\ s-a+c-1,\, s-b+c-1 \end{matrix}\right]$ $\times\, {}_3F_2\left(\begin{matrix} 1,\, a-c-s+2,\, b-c-s+2 \\ 2-s,\, a+b-c-s+2;\, \frac{\sigma}{\omega} \end{matrix}\right)$ $[0<\sigma<\omega;\ \operatorname{Re} c,\ \operatorname{Re} s,\ \operatorname{Re}(s-a-b+c)>0]$
15	$\dfrac{(\omega-x)_+^{c-1}}{x-\sigma}\, {}_2F_1\left(\begin{matrix} a,\, b \\ c;\, \frac{\omega-x}{\omega} \end{matrix}\right)$	$-\dfrac{\omega^{s+c-1}}{\sigma}\,\Gamma\left[\begin{matrix} c,\, s,\, s-a-b+c \\ s-a+c,\, s-b+c \end{matrix}\right] {}_3F_2\left(\begin{matrix} 1,\, s,\, s-a-b+c \\ s-a+c,\, s-b+c;\, \frac{\omega}{\sigma} \end{matrix}\right)$ $[0<\omega<\sigma;\ \operatorname{Re} c,\ \operatorname{Re} s,\ \operatorname{Re}(s-a-b+c)>0]$

3.31.5. ${}_2F_1\left(a,\, b;\, c;\, \frac{\omega}{x+\omega}\right)$ and algebraic functions

1	$(x+\omega)^{-a}\ {}_2F_1\left(\begin{matrix} a,\, b \\ c;\ \frac{\omega}{x+\omega}\end{matrix}\right)$	$\omega^{s-a}\,\Gamma\left[\begin{matrix} c,\, s,\, s-a-b+c,\, a-s \\ a,\, c-b,\, s-a+c\end{matrix}\right]$ $[0,\ \operatorname{Re}(a+b-c)<\operatorname{Re}s<\operatorname{Re}a;\ \lvert\arg\omega\rvert<\pi]$
2	$(x+\omega)^{a}\ {}_2F_1\left(\begin{matrix} a,\, 1-a \\ c;\ \frac{\omega}{x+\omega}\end{matrix}\right)$	$2\omega^{s+a}\,\Gamma\left[\begin{matrix} c,\, -s-a,\, s,\, s+c-1,\, \frac{2s+a+c+1}{2} \\ -a,\, c-a,\, s+a+c,\, \frac{2s+a+c-1}{2}\end{matrix}\right]$ $\left[\begin{matrix}0,\ 1-\operatorname{Re}c,\ -\operatorname{Re}(a+c+1)/2<\operatorname{Re}s<-\operatorname{Re}a; \\ \lvert\arg\omega\rvert<\pi\end{matrix}\right]$
3	$(x+\omega)^{-2a}$ $\times\ {}_2F_1\left(\begin{matrix} a,\, 2a+1 \\ c;\ \frac{\omega}{x+\omega}\end{matrix}\right)$	$\frac{\omega^{s-2a}}{2}\,\Gamma\left[\begin{matrix} c,\, 2a-s,\, s,\, s-3a+c-1,\, s-4a+2c-1 \\ 2a,\, c-a,\, s-2a+c,\, s-4a+2c-2\end{matrix}\right]$ $\left[\begin{matrix}0,\ \operatorname{Re}(4a-2c)+1,\ \operatorname{Re}(3a-c)+1<\operatorname{Re}s<2\operatorname{Re}a; \\ \lvert\arg\omega\rvert<\pi\end{matrix}\right]$
4	$(x+\omega)^{1-a}$ $\times\ {}_2F_1\left(\begin{matrix} a,\, b;\ \frac{\omega}{x+\omega} \\ \frac{a+b+1}{2}\end{matrix}\right)$	$2\omega^{s-a+1}\,\Gamma\left[\begin{matrix} \frac{a+b+1}{2},\, a-s-1,\, s,\, \frac{2s-a+3}{2},\, \frac{2s-a-b+1}{2} \\ a,\, \frac{a-b-1}{2},\, \frac{2s-a+1}{2},\, \frac{2s-a+b+3}{2}\end{matrix}\right]$ $\left[\begin{matrix}0,\ \operatorname{Re}(a-3)/2,\ \operatorname{Re}(a+b-1)/2<\operatorname{Re}s \\ <\operatorname{Re}a-1;\ \lvert\arg\omega\rvert<\pi\end{matrix}\right]$
5	$(x+\omega)^{1-b}$ $\times\ {}_2F_1\left(\begin{matrix} a,\, b;\ \frac{\omega}{x+\omega} \\ 2b-a-1\end{matrix}\right)$	$\frac{(s+b-1)\,\omega^{s-b+1}}{2}\,\Gamma\left[\begin{matrix} 2b-a-1 \\ b,\, 2b-2a-2\end{matrix}\right]$ $\times\,\Gamma\left[\begin{matrix} b-s-1,\, s,\, s-2a+b-1 \\ s-a+b\end{matrix}\right]$ $\left[\begin{matrix}0,\ -\operatorname{Re}b,\ \operatorname{Re}(2a-b)+1<\operatorname{Re}s<\operatorname{Re}b-1; \\ \lvert\arg\omega\rvert<\pi\end{matrix}\right]$

3.31.6. ${}_2F_1\left(a,\, b;\, c;\, \frac{x-\omega}{x}\right)$ and algebraic functions

1	$(\omega-x)_+^{c-1}\ {}_2F_1\left(\begin{matrix} a,\, b \\ c;\ \frac{x-\omega}{x}\end{matrix}\right)$	$\omega^{s+c-1}\,\Gamma\left[\begin{matrix} c,\, s+a,\, s+b \\ s+c,\, s+a+b\end{matrix}\right]$ $[\omega,\ \operatorname{Re}c>0;\ \operatorname{Re}s>-\operatorname{Re}a,\ -\operatorname{Re}b]$
2	$(\omega-x)^{c-1}\ {}_2F_1\left(\begin{matrix} a,\, b \\ c;\ \frac{x-\omega}{x}\end{matrix}\right)$	$\omega^{s+c-1}\,\Gamma\left[\begin{matrix} c,\, s+a,\, s+b \\ s+a+b,\, s+c\end{matrix}\right]$ $-\,e^{ic\pi}\omega^{s+c-1}\,\Gamma\left[\begin{matrix} c,\, 1-a-b-s,\, 1-c-s \\ 1-a-s,\, 1-b-s\end{matrix}\right]$ $\left[\begin{matrix}\operatorname{Re}c>0;\ -\operatorname{Re}a,\ -\operatorname{Re}b<\operatorname{Re}s \\ <\operatorname{Re}(1-a-b),\ \operatorname{Re}(1-c);\ 0<\arg\omega\le\pi\end{matrix}\right]$

No.	$f(x)$	$F(s)$
3	$(x-\omega)_+^{c-1}\, {}_2F_1\left(\begin{matrix} a,\, b \\ c;\ \frac{x-\omega}{x} \end{matrix}\right)$	$\omega^{s+c-1}\Gamma\left[\begin{matrix} c,\, 1-a-b-s,\, 1-c-s \\ 1-a-s,\, 1-b-s \end{matrix}\right]$ $[\omega,\ \operatorname{Re} c > 0;\ \operatorname{Re} s < 1-\operatorname{Re} c,\, 1-\operatorname{Re}(a+b)]$
4	$(x-\omega)^{c-1}\, {}_2F_1\left(\begin{matrix} a,\, b \\ c;\ \frac{x-\omega}{x} \end{matrix}\right)$	$\omega^{s+c-1}\Gamma\left[\begin{matrix} c,\, 1-a-b-s,\, 1-c-s \\ 1-a-s,\, 1-b-s \end{matrix}\right]$ $-\, e^{ic\pi}\omega^{s+c-1}\Gamma\left[\begin{matrix} c,\, s+a,\, s+b \\ s+a+b,\, s+c \end{matrix}\right]$ $\left[\begin{matrix} \operatorname{Re} c > 0;\ -\operatorname{Re} a,\ -\operatorname{Re} b < \operatorname{Re} s \\ < 1-\operatorname{Re} c,\, 1-\operatorname{Re}(a+b);\ \operatorname{Im}\omega < 0 \end{matrix}\right]$

3.31.7. ${}_2F_1\left(a,\, b;\, c;\, \dfrac{x}{x+\omega}\right)$ and algebraic functions

1	$(x+\omega)^{-a}\, {}_2F_1\left(\begin{matrix} a,\, b \\ c;\ \frac{x}{x+\omega} \end{matrix}\right)$	$\omega^{s-a}\Gamma\left[\begin{matrix} c,\, a-s,\, c-b-s,\, s \\ a,\, c-b,\, c-s \end{matrix}\right]$ $[0 < \operatorname{Re} s < \operatorname{Re} a,\ \operatorname{Re}(c-b);\ \lvert\arg\omega\rvert < \pi]$
2	$(x+\omega)^{a}\, {}_2F_1\left(\begin{matrix} a,\, 1-a \\ c;\ \frac{x}{x+\omega} \end{matrix}\right)$	$2\omega^{s+a}\Gamma\left[\begin{matrix} c \\ -a,\, c-a \end{matrix}\right]\Gamma\left[\begin{matrix} \frac{-2s-a+c+1}{2},\, -s-a,\ -s-a+c-1,\, s \\ \frac{-2s-a+c-1}{2},\, c-s \end{matrix}\right]$ $\left[\begin{matrix} 0 < \operatorname{Re} s < -\operatorname{Re} a,\ \operatorname{Re}(c-a)-1, \\ \operatorname{Re}(c-a+1)/2;\ \lvert\arg\omega\rvert < \pi \end{matrix}\right]$
3	$(x+\omega)^{1-a}$ $\times\, {}_2F_1\left(\begin{matrix} a,\, b;\ \frac{x}{x+\omega} \\ \frac{a+b+1}{2} \end{matrix}\right)$	$(a-2s-1)\,\omega^{s-a+1}\Gamma\left[\begin{matrix} \frac{a+b+1}{2} \\ a,\ \frac{a-b-1}{2} \end{matrix}\right]\Gamma\left[\begin{matrix} s,\, a-s-1,\ \frac{a-b-2s-1}{2} \\ \frac{a+b-2s+1}{2} \end{matrix}\right]$ $\left[\begin{matrix} 0 < \operatorname{Re} s < \operatorname{Re} a-1,\ \operatorname{Re}(a+1)/2, \\ \operatorname{Re}(a-b-1)/2;\ \lvert\arg\omega\rvert < \pi \end{matrix}\right]$
4	$(x+\omega)^{1-b}$ $\times\, {}_2F_1\left(\begin{matrix} a,\, b;\ \frac{x}{x+\omega} \\ 2b-a-1 \end{matrix}\right)$	$\dfrac{(2b-s-2)}{2}\,\omega^{s-b+1}\Gamma\left[\begin{matrix} 2b-a-1 \\ b,\, 2b-2a-2 \end{matrix}\right]$ $\times\,\Gamma\left[\begin{matrix} s,\, b-s-1,\, 2b-2a-s-2 \\ 2b-a-s-1 \end{matrix}\right]$ $[0 < \operatorname{Re} s < \operatorname{Re} b-1,\, 2\operatorname{Re}(b-a)-2;\ \lvert\arg\omega\rvert < \pi]$
5	$(x+\omega)^{-2a}$ $\times\, {}_2F_1\left(\begin{matrix} a,\, 2a+1 \\ c;\ \frac{x}{x+\omega} \end{matrix}\right)$	$\dfrac{\omega^{s-2a}}{2}\Gamma\left[\begin{matrix} c \\ 2a,\, c-a \end{matrix}\right]\Gamma\left[\begin{matrix} s,\, 2a-s,\, 2c-2a-s-1,\, c-a-s-1 \\ 2c-2a-s-2,\, c-s \end{matrix}\right]$ $\left[\begin{matrix} 0 < \operatorname{Re} s < 2\operatorname{Re} a,\ \operatorname{Re}(c-a)-1, \\ 2\operatorname{Re}(c-a)-1;\ \lvert\arg\omega\rvert < \pi \end{matrix}\right]$

3.31.8. ${}_2F_1\left(a,\, b;\, c;\, \frac{4\omega x}{(x+\omega)^2}\right)$ and algebraic functions

1	$(x+\omega)^{-2a} \times {}_2F_1\left(\begin{matrix} a,\, b \\ 2b;\, \frac{4\omega x}{(x+\omega)^2}\end{matrix}\right)$	$\frac{\omega^{s-2a}}{2}\Gamma\left[\begin{matrix}\frac{2b+1}{2},\, \frac{2b-2a+1}{2} \\ a\end{matrix}\right]\Gamma\left[\begin{matrix}\frac{s}{2},\, \frac{2a-s}{2} \\ \frac{s-2a+2b+1}{2},\, \frac{2b-s+1}{2}\end{matrix}\right]$ $[\operatorname{Re}\omega>0;\ 0<\operatorname{Re}s<2\operatorname{Re}a<2\operatorname{Re}b+1]$
2	$(x+\omega)^{-2a} \times {}_2F_1\left(\begin{matrix} a,\, \frac{2a+1}{2} \\ c;\, \frac{4\omega x}{(x+\omega)^2}\end{matrix}\right)$	$\omega^{s-2a}\,\Gamma\left[\begin{matrix}c,\, c-2a \\ 2a\end{matrix}\right]\Gamma\left[\begin{matrix}s,\, 2a-s \\ s-2a+c,\, c-s\end{matrix}\right]$ $[\operatorname{Re}\omega>0;\ 0<\operatorname{Re}s<2\operatorname{Re}a<\operatorname{Re}c]$
3	$(x+\omega)^{-2a} \times {}_2F_1\left(\begin{matrix} a,\, b \\ 2b;\, \frac{4\omega x}{(x+\omega)^2}\end{matrix}\right)$	$\frac{\omega^{s-2a}}{2}\Gamma\left[\begin{matrix}\frac{2b+1}{2},\, \frac{2b-2a+1}{2} \\ a\end{matrix}\right]\Gamma\left[\begin{matrix}\frac{s}{2},\, \frac{2a-s}{2} \\ \frac{2b-s+1}{2},\, \frac{s-2a+2b+1}{2}\end{matrix}\right]$ $[\omega>0;\ \operatorname{Re}(a-b)<1/2;\ 0<\operatorname{Re}s<2\operatorname{Re}a]$
4	$\frac{(x+\omega)^{-2b}}{\lvert x-\omega\rvert^{2b-2a}} \times {}_2F_1\left(\begin{matrix} a,\, b \\ 2b;\, \frac{4\omega x}{(x+\omega)^2}\end{matrix}\right)$	$\frac{\omega^{s+2a-4b}}{2}\Gamma\left[\begin{matrix}\frac{2b+1}{2},\, \frac{2a-2b+1}{2} \\ 2b-a\end{matrix}\right]\Gamma\left[\begin{matrix}\frac{s}{2},\, \frac{4b-2a-s}{2} \\ \frac{s+2a-2b+1}{2},\, \frac{2b-s+1}{2}\end{matrix}\right]$ $[\omega>0;\ \operatorname{Re}(b-a)<1/2;\ 0<\operatorname{Re}s<2\operatorname{Re}(2b-a)]$
5	$\frac{(x+\omega)^{-2a}}{\lvert x-\omega\rvert^{2c-4a-1}} \times {}_2F_1\left(\begin{matrix} a,\, \frac{2a+1}{2} \\ c;\, \frac{4\omega x}{(x+\omega)^2}\end{matrix}\right)$	$\omega^{s+2a-2c+1}\,\Gamma\left[\begin{matrix}c,\, 2a-c+1 \\ 2c-2a-1\end{matrix}\right]\Gamma\left[\begin{matrix}s,\, 2c-2a-s-1 \\ c-s,\, s+2a-c+1\end{matrix}\right]$ $[\omega>0;\ \operatorname{Re}(c-2a)<1;\ 0<\operatorname{Re}s<2\operatorname{Re}(c-a)-1]$

3.31.9. ${}_2F_1\left(a,\, b;\, c;\, -\frac{4\omega x}{(x-\omega)^2}\right)$ and algebraic functions

1	$\lvert x-\omega\rvert^{-2a} \times {}_2F_1\left(\begin{matrix} a,\, b \\ 2b;\, -\frac{4\omega x}{(x-\omega)^2}\end{matrix}\right)$	$\frac{\omega^{s-2a}}{2}\Gamma\left[\begin{matrix}\frac{2b+1}{2},\, \frac{2b-2a+1}{2} \\ a\end{matrix}\right]\Gamma\left[\begin{matrix}\frac{s}{2},\, \frac{2a-s}{2} \\ \frac{2b-s+1}{2},\, \frac{s-2a+2b+1}{2}\end{matrix}\right]$ $[\omega>0;\ \operatorname{Re}(a-b)<1/2;\ 0<\operatorname{Re}s<2\operatorname{Re}a]$
2	$\lvert x-\omega\rvert^{-2a} \times {}_2F_1\left(\begin{matrix} a,\, b;\, -\frac{4\omega x}{(x-\omega)^2} \\ a+b+\frac{1}{2}\end{matrix}\right)$	$\omega^{s-2a}\,\Gamma\left[\begin{matrix}\frac{2a+2b+1}{2},\, \frac{2b-2a+1}{2} \\ 2a\end{matrix}\right]\Gamma\left[\begin{matrix}s,\, 2a-s \\ \frac{2s-2a+2b+1}{2},\, \frac{2a+2b-2s+1}{2}\end{matrix}\right]$ $[\omega>0;\ 0<\operatorname{Re}s<2\operatorname{Re}a<2\operatorname{Re}b+1]$
3	$\frac{(x+\omega)^{2a-2b}}{\lvert x-\omega\rvert^{2b}} \times {}_2F_1\left(\begin{matrix} a,\, b \\ 2b;\, -\frac{4\omega x}{(x-\omega)^2}\end{matrix}\right)$	$\frac{\omega^{s+2a-4b}}{2}\Gamma\left[\begin{matrix}\frac{2b+1}{2},\, \frac{2a-2b+1}{2} \\ 2b-a\end{matrix}\right]\Gamma\left[\begin{matrix}\frac{s}{2},\, \frac{4b-2a-s}{2} \\ \frac{2b-s+1}{2},\, \frac{s+2a-2b+1}{2}\end{matrix}\right]$ $[\omega>0;\ \operatorname{Re}(b-a)<1/2;\ 0<\operatorname{Re}s<2\operatorname{Re}(2b-a)]$

No.	$f(x)$	$F(s)$
4	$\dfrac{x+1}{\lvert x-1\rvert^{2a}} \times {}_2F_1\left(\begin{matrix} a,\ b;\ -\frac{4x}{(x-1)^2} \\ \frac{2a+2b-1}{2} \end{matrix}\right)$	$\Gamma\left[\begin{matrix} \frac{2a+2b-1}{2},\ \frac{2b-2a+1}{2},\ s,\ 2a-s-1 \\ 2a-1,\ \frac{2s-2a+2b+1}{2},\ \frac{2a+2b-2s-1}{2} \end{matrix}\right]$ $[\operatorname{Re}(b-a) > -1/2;\ 0 < \operatorname{Re} s < 2\operatorname{Re} a - 1]$

3.31.10. ${}_2F_1\left(a, b; c; \dfrac{\alpha_1 x^3 + \beta_1 x^2 + \gamma_1 x + \delta_1}{\alpha_2 x^3 + \beta_2 x^2 + \gamma_2 x + \delta_2}\right)$ and algebraic functions

No.	$f(x)$	$F(s)$
1	$(x+\omega)^{-a} \times {}_2F_1\left(\begin{matrix} a,\ \frac{1-6a}{6} \\ \frac{1}{2};\ -\frac{x(8x+9\omega)^2}{27\omega^2(x+\omega)} \end{matrix}\right)$	$\sqrt{\pi}\,\omega^{s-a}\,\Gamma\left[\begin{matrix} s,\ 3a-s,\ \frac{1-3a-3s}{3} \\ 3a,\ \frac{1-3a}{3},\ \frac{1-2s}{2} \end{matrix}\right]$ $[\operatorname{Re}\omega \ge 0;\ 0 < \operatorname{Re} s < 1/3 - \operatorname{Re} a,\ 3\operatorname{Re} a]$
2	$(x+\omega)^{-a} \times {}_2F_1\left(\begin{matrix} a,\ \frac{1-6a}{6} \\ \frac{1}{2};\ -\frac{\omega(9x+8\omega)^2}{27x^2(x+\omega)} \end{matrix}\right)$	$\sqrt{\pi}\,\omega^{s-a}\,\Gamma\left[\begin{matrix} a-s,\ s+2a,\ \frac{3s-6a+1}{3} \\ 3a,\ \frac{1-3a}{3},\ \frac{2s-2a+1}{2} \end{matrix}\right]$ $[\operatorname{Re}\omega \ge 0;\ -2\operatorname{Re} a,\ 2\operatorname{Re} a - 1/3 < \operatorname{Re} s < \operatorname{Re} a]$
3	$(4x+\omega)^{-3a} \times {}_2F_1\left(\begin{matrix} a,\ \frac{3a+1}{3} \\ \frac{12a+5}{6};\ \frac{27\omega^2 x}{(4x+\omega)^3} \end{matrix}\right)$	$\dfrac{\omega^{s-3a}}{2^{6a}}\,\Gamma\left[\begin{matrix} \frac{12a+2}{3} \\ \frac{6a+1}{6},\ 3a \end{matrix}\right]\Gamma\left[\begin{matrix} s,\ 3a-s,\ \frac{6s-12a+1}{6} \\ \frac{3s+3a+2}{3} \end{matrix}\right]$ $[0,\ 2\operatorname{Re} a - 1/6 < \operatorname{Re} s < 3\operatorname{Re} a;\ \lvert\arg\omega\rvert < \pi]$
4	$(x+4\omega)^{-3a} \times {}_2F_1\left(\begin{matrix} a,\ \frac{3a+1}{3} \\ \frac{12a+5}{6};\ \frac{27\omega x^2}{(x+4\omega)^3} \end{matrix}\right)$	$\dfrac{\omega^{s-3a}}{2^{6a}}\,\Gamma\left[\begin{matrix} \frac{12a+2}{3} \\ \frac{6a+1}{6},\ 3a \end{matrix}\right]\Gamma\left[\begin{matrix} s,\ 3a-s,\ \frac{6a-6s+1}{6} \\ \frac{12a-3s+2}{3} \end{matrix}\right]$ $[0 < \operatorname{Re} s < 3\operatorname{Re} a,\ \operatorname{Re} a + 1/6;\ \lvert\arg\omega\rvert < \pi]$
5	$(3x+4\omega)^{-3a}(9x+8\omega) \times {}_2F_1\left(\begin{matrix} a,\ \frac{3a+1}{3} \\ \frac{3}{2};\ \frac{\omega(9x+8\omega)^2}{(3x+4\omega)^3} \end{matrix}\right)$	$\dfrac{3^{2-3a}\sqrt{\pi}\,\omega^{s-3a+1}}{2}\,\Gamma\left[\begin{matrix} 3a-s-1,\ \frac{3s-12a+7}{3},\ s \\ \frac{4-3a}{3},\ 3a-1,\ \frac{2s-6a+5}{2} \end{matrix}\right]$ $[\operatorname{Re} > 0;\ 4\operatorname{Re} a - 7/3 < \operatorname{Re} s < 3\operatorname{Re} a - 1]$
6	$(4x+3\omega)^{-3a}(8x+9\omega) \times {}_2F_1\left(\begin{matrix} a,\ \frac{3a+1}{3} \\ \frac{3}{2};\ \frac{x(8x+9\omega)^2}{(4x+3\omega)^3} \end{matrix}\right)$	$\dfrac{3^{2-3a}\sqrt{\pi}\,\omega^{s-3a+1}}{2}\,\Gamma\left[\begin{matrix} 3a-s-1,\ \frac{4-3a-3s}{3},\ s \\ \frac{4-3a}{3},\ 3a-1,\ \frac{3-2s}{2} \end{matrix}\right]$ $[0 < \operatorname{Re} s < 3\operatorname{Re} a - 1,\ 4/3 - \operatorname{Re} a;\ \lvert\arg\omega\rvert < \pi]$
7	$(3x-\omega)^{-3a}(9x+\omega) \times {}_2F_1\left(\begin{matrix} a,\ \frac{3a+1}{3} \\ \frac{3}{2};\ -\frac{\omega(9x+\omega)^2}{(3x-\omega)^3} \end{matrix}\right)$	$\dfrac{3^{2-3a}\sqrt{\pi}\,\omega^{s-3a+1}}{2}\,\Gamma\left[\begin{matrix} 3a-s-1,\ \frac{6s-12a+7}{6},\ s \\ \frac{6a+1}{6},\ 3a-1,\ \frac{2s-6a+5}{2} \end{matrix}\right]$ $[\operatorname{Re} \ge 0;\ 0,\ 2\operatorname{Re} a - 7/6 < \operatorname{Re} s < 3\operatorname{Re} a - 1]$

No.	$f(x)$	$F(s)$
8	$(3\omega-x)^{-3a}(x+9\omega)$ $\times\,{}_2F_1\left(\begin{matrix}a,\ \frac{3a+1}{3}\\ \frac{3}{2};\ \frac{x(x+9\omega)^2}{(x-3\omega)^3}\end{matrix}\right)$	$\frac{3^{2-3a}\sqrt{\pi}\,\omega^{s-3a+1}}{2}\Gamma\left[\begin{matrix}3a-s-1,\ \frac{6a-6s+1}{6},\ s\\ \frac{6a+1}{6},\ 3a-1,\ \frac{3-2s}{2}\end{matrix}\right]$ [$\mathrm{Re} \geq 0;\ 0,\ \mathrm{Re}\,a-1/6<\mathrm{Re}\,s<3\,\mathrm{Re}\,a-1$]

3.31.11. ${}_2F_1\left(a,\ b;\ c;\ \frac{\omega_1 x+\sigma_1}{\omega_2 x+\sigma_2}\right)$ and algebraic functions

1	$(\omega-x)_+^{-b}\,{}_2F_1\left(\begin{matrix}-n,\ b\\ c;\ \frac{\omega}{\omega-x}\end{matrix}\right)$	$(-1)^n\frac{\omega^{s-b}}{(c)_n}\Gamma\left[\begin{matrix}1-b,\ s,\ b-c-s+1\\ s-b+1,\ b-c-n-s+1\end{matrix}\right]$ [$\mathrm{Re}\,b<1-n;\ \omega,\ \mathrm{Re}\,s>0$]
2	$(x-\omega)_+^{-b}\,{}_2F_1\left(\begin{matrix}-n,\ b\\ c;\ \frac{\omega}{\omega-x}\end{matrix}\right)$	$\frac{\omega^{s-b}}{(c)_n}\Gamma\left[\begin{matrix}1-b,\ b-s,\ s-b+c+n\\ 1-s,\ s-b+c\end{matrix}\right]$ [$\omega>0;\ \mathrm{Re}\,s<\mathrm{Re}\,b<1-n$]
3	$(\omega-x)_+^{-b}\,{}_2F_1\left(\begin{matrix}-n,\ b\\ c;\ \frac{x}{x-\omega}\end{matrix}\right)$	$\frac{\omega^{s-b}}{(c)_n}\Gamma\left[\begin{matrix}1-b,\ s,\ c-s+n\\ s-b+1,\ c-s\end{matrix}\right]$ [$\mathrm{Re}\,b<1-n;\ \omega,\ \mathrm{Re}\,s>0$]
4	$(x-\omega)_+^{-b}\,{}_2F_1\left(\begin{matrix}-n,\ b\\ c;\ \frac{x}{x-\omega}\end{matrix}\right)$	$(-1)^n\frac{\omega^{s-b}}{(c)_n}\Gamma\left[\begin{matrix}1-b,\ s-c+1,\ b-s\\ s-c-n+1,\ 1-s\end{matrix}\right]$ [$\omega>0;\ \mathrm{Re}\,s<\mathrm{Re}\,b<1-n$]
5	$(\sigma-x)_+^{c-1}(\tau+x)^{\mu}$ $\times\,{}_2F_1\left(\begin{matrix}a,\ b\\ c;\ \omega(\sigma-x)\end{matrix}\right)$	$\sigma^{s+c-1}\tau^{\mu}\,\Gamma\left[\begin{matrix}c,\ s\\ s+c\end{matrix}\right]F_3\left(a,\ -\mu,\ b,\ s;\ s+c;\ \sigma\omega,\ -\frac{\sigma}{\tau}\right)$ [$\sigma,\ \mathrm{Re}\,c,\ \mathrm{Re}\,s>0$]
6	$\theta(x-\sigma)(x-\tau)^{-a}$ $\times\,{}_2F_1\left(\begin{matrix}a,\ b\\ c;\ \frac{\omega x}{x-\tau}\end{matrix}\right)$	$\frac{\sigma^{s-a}}{a-s}F_2\left(a,\ b,\ a-s;\ c,\ a-s+1;\ \omega,\ \frac{\tau}{\sigma}\right)$ [$\sigma>0;\ \lvert\omega\rvert+\lvert\tau/\sigma\rvert<1;\ \mathrm{Re}\,s<\mathrm{Re}\,a$]
7	$(\sigma-x)_+^{c-1}(\tau-x)^{-a}$ $\times\,{}_2F_1\left(\begin{matrix}a,\ b\\ c;\ \frac{\omega(\sigma-x)}{\tau-x}\end{matrix}\right)$	$\sigma^{s+c-1}\tau^{-a}\,\mathrm{B}(s,\ c)\,F_1\left(a,\ b,\ s;\ s+c;\ \frac{\sigma}{\tau},\ \frac{\sigma\omega}{\tau}\right)$ [$\sigma,\ \tau,\ \mathrm{Re}\,c,\ \mathrm{Re}\,s>0;\ \sigma<\tau,\ \lvert\omega\rvert<\tau$]
8	$(\sigma-x)_+^{c-1}\left(\frac{\sigma}{1-\sigma}-x\right)^{-a}$ $\times\,{}_2F_1\left(\begin{matrix}a,\ b\\ c;\ \frac{\omega(\sigma-x)}{(\sigma-1)x+\sigma}\end{matrix}\right)$	$(1-\sigma)^a\,\sigma^{s-a+c-1}\,\mathrm{B}(s,\ c)\,F_1(a;\ s,\ b;\ c+s;\ 1-\sigma,\ \omega)$ [$\sigma,\ \mathrm{Re}\,c,\ \mathrm{Re}\,s>0$]

3.31.12. ${}_2F_1\left(a,\, b;\, c;\, \frac{\sqrt{x}-\sqrt{x+\omega}}{2\sqrt{x}}\right)$ and algebraic functions

1	$\left(\sqrt{x+\omega}+\sqrt{x}\right)^{-a} \times {}_2F_1\left(\begin{matrix} a,\, b \\ a+1;\, \frac{\sqrt{x}-\sqrt{x+\omega}}{2\sqrt{x}} \end{matrix}\right)$	$\frac{2^{b-1} a\, \omega^{s-a/2}}{\sqrt{\pi}} \Gamma\left[\begin{matrix} \frac{-2s+a}{2},\, \frac{2s+a}{2},\, \frac{2s+b}{2},\, \frac{2s+b+1}{2} \\ \frac{2s+a+2}{2},\, \frac{2s+a+2b}{2} \end{matrix}\right]$ $\left[\begin{matrix} -\operatorname{Re} a/2,\, -\operatorname{Re} b/2 < \operatorname{Re} s < \operatorname{Re} a/2; \\ -\pi < \arg\omega \le \pi \end{matrix}\right]$
2	$\left(\sqrt{x+\omega}-\sqrt{x}\right)^{a} \times {}_2F_1\left(\begin{matrix} a,\, b \\ a+1;\, \frac{\sqrt{x}-\sqrt{x+\omega}}{2\sqrt{x}} \end{matrix}\right)$	$\frac{2^{b-1} a\, \omega^{s+a/2}}{\sqrt{\pi}} \Gamma\left[\begin{matrix} \frac{-2s+a}{2},\, \frac{2s+a}{2},\, \frac{2s+b}{2},\, \frac{2s+b+1}{2} \\ \frac{2s+a+2}{2},\, \frac{2s+a+2b}{2} \end{matrix}\right]$ $\left[\begin{matrix} -\operatorname{Re} a/2,\, -\operatorname{Re} b/2 < \operatorname{Re} s < \operatorname{Re} a/2; \\ \lvert\arg\omega\rvert < \pi \end{matrix}\right]$
3	$\left(\sqrt{x+\omega}+\sqrt{x}\right)^{b-2c+2} \times {}_2F_1\left(\begin{matrix} 1,\, b \\ c;\, \frac{\sqrt{x}-\sqrt{x+\omega}}{2\sqrt{x}} \end{matrix}\right)$	$\frac{(c-1)\, \omega^{s+b/2-c+1}}{\sqrt{\pi}} \Gamma\left[\begin{matrix} \frac{-2s-b+2c-2}{2},\, \frac{2s+1}{2},\, s+1,\, \frac{2s+b}{2} \\ \frac{2s-b+2c}{2},\, \frac{2s+b+2}{2} \end{matrix}\right]$ $\left[\begin{matrix} -1/2,\, -\operatorname{Re} b/2 < \operatorname{Re} s < \operatorname{Re}(2c-b-2)/2; \\ \lvert\arg\omega\rvert < \pi \end{matrix}\right]$
4	$\left(\sqrt{x+\omega}-\sqrt{x}\right)^{2c-b-2} \times {}_2F_1\left(\begin{matrix} 1,\, b \\ c;\, \frac{\sqrt{x}-\sqrt{x+\omega}}{2\sqrt{x}} \end{matrix}\right)$	$\frac{(c-1)\, \omega^{s-b/2+c-1}}{\sqrt{\pi}} \Gamma\left[\begin{matrix} \frac{-2s-b+2c-2}{2},\, \frac{2s+1}{2},\, s+1,\, \frac{2s+b}{2} \\ \frac{2s-b+2c}{2},\, \frac{2s+b+2}{2} \end{matrix}\right]$ $\left[\begin{matrix} -1/2,\, -\operatorname{Re} b/2 < \operatorname{Re} s < \operatorname{Re}(2c-b-2)/2; \\ -\pi < \arg\omega \le \pi \end{matrix}\right]$

3.31.13. ${}_2F_1\left(a,\, b;\, c;\, \frac{\sqrt{\omega}-\sqrt{x+\omega}}{2\sqrt{\omega}}\right)$ and algebraic functions

1	$\left(\sqrt{x+\omega}+\sqrt{\omega}\right)^{-a} \times {}_2F_1\left(\begin{matrix} a,\, b;\, \frac{\sqrt{\omega}-\sqrt{x+\omega}}{2\sqrt{\omega}} \\ a+1 \end{matrix}\right)$	$\frac{2^{2s-a} a\, \omega^{s-a/2}}{a-s} \Gamma\left[\begin{matrix} s,\, a+b-2s \\ a+b-s \end{matrix}\right]$ $[0 < \operatorname{Re} s < \operatorname{Re} a,\; \operatorname{Re}(a+b)/2;\; -\pi < \arg\omega \le \pi]$
2	$\left(\sqrt{x+\omega}+\sqrt{\omega}\right)^{b-2c+2} \times {}_2F_1\left(\begin{matrix} 1,\, b \\ c;\, \frac{\sqrt{\omega}-\sqrt{x+\omega}}{2\sqrt{\omega}} \end{matrix}\right)$	$\frac{(c-1)\, \omega^{s+b/2-c+1}}{\sqrt{\pi}} \Gamma\left[\begin{matrix} c-s-1 \\ c-s \end{matrix}\right] \Gamma\left[\begin{matrix} \frac{-2s-b+2c-1}{2},\, \frac{-2s-b+2c}{2},\, s \\ 2c-s-b-1 \end{matrix}\right]$ $\left[\begin{matrix} 0 < \operatorname{Re} s < \operatorname{Re} c-1,\; \operatorname{Re}(2c-b-1)/2; \\ -\pi < \arg\omega \le \pi \end{matrix}\right]$
3	$\left(\sqrt{x+\omega}-\sqrt{\omega}\right)^{2c-b-2} \times {}_2F_1\left(\begin{matrix} 1,\, b \\ c;\, \frac{\sqrt{\omega}-\sqrt{x+\omega}}{2\sqrt{\omega}} \end{matrix}\right)$	$\frac{(1-c)\, \omega^{s-b/2+c-1}}{\sqrt{\pi}} \Gamma\left[\begin{matrix} \frac{-2s+b-2c+3}{2},\, \frac{-2s+b-2c+4}{2} \\ 1-s \end{matrix}\right] \times \Gamma\left[\begin{matrix} b-c-s+1,\, s-b+2c-2 \\ b-c-s+2 \end{matrix}\right]$ $\left[\begin{matrix} \operatorname{Re}(b-2c)+2 < \operatorname{Re} s < \operatorname{Re}(b/2-c)+3/2, \\ \operatorname{Re}(b-c)+1;\; -\pi < \arg\omega \le \pi \end{matrix}\right]$

3.31.14. ${}_2F_1\left(a,\, b;\, c;\, \frac{\sqrt{x+\omega}-\sqrt{x}}{\sqrt{x+\omega}+\sqrt{x}}\right)$ and algebraic functions

1	$\left(\sqrt{x+\omega}+\sqrt{x}\right)^{-2a} \times {}_2F_1\left(\begin{matrix} a,\, b \\ a+1;\; \frac{\sqrt{x+\omega}-\sqrt{x}}{\sqrt{x+\omega}+\sqrt{x}}\end{matrix}\right)$	$\frac{a\,\omega^{s-a}}{2^b\sqrt{\pi}}\,\Gamma\left[\begin{matrix} a-s,\, \frac{2s-b+1}{2},\, \frac{2s-b+2}{2},\, s \\ s+1,\, s+a-b+1 \end{matrix}\right]$ $[0,\ \operatorname{Re}(b-1)/2<\operatorname{Re}s<\operatorname{Re}a;\ -\pi<\arg\omega\le\pi]$
2	$\left(\sqrt{x+\omega}+\sqrt{x}\right)^{1-b-c} \times {}_2F_1\left(\begin{matrix} 1,\, b \\ c;\; \frac{\sqrt{x+\omega}-\sqrt{x}}{\sqrt{x+\omega}+\sqrt{x}}\end{matrix}\right)$	$\frac{(c-1)\,\omega^{s+(1-b-c)/2}}{2\sqrt{\pi}}\,\Gamma\left[\begin{matrix} \frac{-2s+b+c-1}{2},\, \frac{2s-b+c-1}{2},\, \frac{2s+1}{2},\, s \\ \frac{2s-b+c+1}{2},\, \frac{2s+b+c-1}{2} \end{matrix}\right]$ $\left[\begin{matrix}0,\ \operatorname{Re}(b-c+1)/2<\operatorname{Re}s \\ <\operatorname{Re}(b+c-1)/2;\ -\pi<\arg\omega\le\pi\end{matrix}\right]$

3.31.15. ${}_2F_1\left(a,\, b;\, c;\, \frac{\sqrt{\pm x+\omega}-\sqrt{\omega}}{\sqrt{\pm x+\omega}+\sqrt{\omega}}\right)$ and algebraic functions

1	$\left(\sqrt{\omega-x}+\sqrt{\omega}\right)^{-2a} \times {}_2F_1\left(\begin{matrix} a,\, b \\ a+1;\; \frac{\sqrt{\omega-x}-\sqrt{\omega}}{\sqrt{\omega-x}+\sqrt{\omega}}\end{matrix}\right)$	$\frac{2^{2s-2a}a}{a-s}\,\omega^{-a}\left(-\frac{1}{\omega}\right)^{-s}\Gamma\left[\begin{matrix} 2a-b-2s+1,\, s \\ 2a-b-s+1 \end{matrix}\right]$ $\left[\begin{matrix}0<\operatorname{Re}s<\operatorname{Re}a,\ \operatorname{Re}(2a-b+1)/2; \\ -\pi<\arg\omega\le\pi\end{matrix}\right]$
2	$\left(\sqrt{x+\omega}+\sqrt{\omega}\right)^{-2a} \times {}_2F_1\left(\begin{matrix} a,\, b \\ a+1;\; \frac{\sqrt{x+\omega}-\sqrt{\omega}}{\sqrt{x+\omega}+\sqrt{\omega}}\end{matrix}\right)$	$\frac{a\,(4\omega)^{s-a}}{a-s}\,\Gamma\left[\begin{matrix} 2a-b-2s+1,\, s \\ 2a-b-s+1 \end{matrix}\right]$ $\left[\begin{matrix}0<\operatorname{Re}s<\operatorname{Re}a,\ \operatorname{Re}(2a-b+1)/2; \\ -\pi<\arg\omega\le\pi\end{matrix}\right]$
3	$\left(\sqrt{x+\omega}+\sqrt{\omega}\right)^{1-b-c} \times {}_2F_1\left(\begin{matrix} 1,\, b \\ c;\; \frac{\sqrt{x+\omega}-\sqrt{\omega}}{\sqrt{x+\omega}+\sqrt{\omega}}\end{matrix}\right)$	$\frac{(c-1)\,(4\omega)^{s+(1-b-c)/2}}{c-s-1}\,\Gamma\left[\begin{matrix} b+c-2s-1,\, s \\ b+c-s-1 \end{matrix}\right]$ $\left[\begin{matrix}0<\operatorname{Re}s<\operatorname{Re}c-1,\ \operatorname{Re}(b+c-1)/2; \\ -\pi<\arg\omega\le\pi\end{matrix}\right]$

3.31.16. ${}_2F_1\left(a,\, b;\, c;\, \frac{x-2\sqrt{\omega}\sqrt{x+\omega}+2\omega}{x}\right)$ and algebraic functions

1	$\left(\sqrt{x+\omega}-\sqrt{\omega}\right)^{2a} \times {}_2F_1\left(\begin{matrix} a,\, b;\; \frac{x-2\sqrt{\omega}\sqrt{x+\omega}+2\omega}{x} \\ a+1\end{matrix}\right)$	$\frac{a\,\omega^{s+a}}{2^b\sqrt{\pi}}\,\Gamma\left[\begin{matrix} -s-a,\, \frac{1-2s-2a-b}{2},\, \frac{2-2s-2a-b}{2},\, s+2a \\ 1-a-s,\, 1-b-s \end{matrix}\right]$ $\left[\begin{matrix}-2\operatorname{Re}a<\operatorname{Re}s<-\operatorname{Re}a,\ \operatorname{Re}(1-2a-b)/2; \\ -\pi<\arg\omega\le\pi\end{matrix}\right]$
2	$\left(\sqrt{x+\omega}-\sqrt{\omega}\right)^{b+c-1} \times {}_2F_1\left(\begin{matrix} 1,\, b \\ c;\; \frac{x-2\sqrt{\omega}\sqrt{x+\omega}+2\omega}{x}\end{matrix}\right)$	$\frac{(1-c)\,(4\omega)^{s+(b+c-1)/2}}{s+b}\,\Gamma\left[\begin{matrix} 1-b-c-2s,\, s+b+c-1 \\ -s \end{matrix}\right]$ $\left[\begin{matrix}\operatorname{Re}(1-b-c)<\operatorname{Re}s<-\operatorname{Re}b,\ \operatorname{Re}(1-b-c)/2; \\ -\pi<\arg\omega\le\pi\end{matrix}\right]$

3.31.17. ${}_2F_1\left(a,\, b;\, c;\, \dfrac{2x-2\sqrt{x}\sqrt{x+\omega}+\omega}{\omega}\right)$ and algebraic functions

1	$\left(\sqrt{x+\omega}-\sqrt{x}\right)^{2a} \times {}_2F_1\left(\begin{matrix} a,\, b;\; \frac{2x-2\sqrt{x}\sqrt{x+\omega}+\omega}{\omega} \\ a+1 \end{matrix}\right)$	$\dfrac{a\,\omega^{s+a}}{2^{2s}s}\Gamma\left[\begin{matrix} a-s,\, 2s-b+1 \\ s+a-b+1 \end{matrix}\right]$ $[0,\ \operatorname{Re}(b-1)/2<\operatorname{Re}s<\operatorname{Re}a;\ \lvert\arg\omega\rvert<\pi]$
2	$\left(\sqrt{x+\omega}-\sqrt{x}\right)^{b+c-1} \times {}_2F_1\left(\begin{matrix} 1,\, b \\ c;\; \frac{2x-2\sqrt{x}\sqrt{x+\omega}+\omega}{\omega} \end{matrix}\right)$	$\dfrac{(c-1)\,\omega^{s+(b+c-1)/2}}{2^{2s-1}(2s-b+c-1)}\Gamma\left[\begin{matrix} \frac{-2s+b+c-1}{2},\, 2s \\ \frac{2s+b+c-1}{2} \end{matrix}\right]$ $\left[\begin{matrix}0,\ \operatorname{Re}(b-c+1)/2<\operatorname{Re}s<\operatorname{Re}(b+c-1)/2; \\ \lvert\arg\omega\rvert<\pi\end{matrix}\right]$

3.31.18. ${}_2F_1\left(a,\, b;\, c;\, \dfrac{2x-2\sqrt{x}\sqrt{x+\omega}+\omega}{2\sqrt{x}\left(\sqrt{x}-\sqrt{x+\omega}\right)}\right)$ and algebraic functions

1	$\left(\sqrt{x+\omega}-\sqrt{x}\right)^{a} \times {}_2F_1\left(\begin{matrix} a,\, b;\; \frac{2x-2\sqrt{x}\sqrt{x+\omega}+\omega}{2\sqrt{x}\left(\sqrt{x}-\sqrt{x+\omega}\right)} \\ a+1 \end{matrix}\right)$	$\dfrac{2^{b-1}a\,\omega^{s+a/2}}{\sqrt{\pi}}\Gamma\left[\begin{matrix} \frac{-2s+a}{2},\, \frac{2s+a}{2},\, \frac{2s+b}{2},\, \frac{2s+b+1}{2} \\ \frac{2s+a+2}{2},\, \frac{2s+a+2b}{2} \end{matrix}\right]$ $\left[\begin{matrix}-\operatorname{Re}a/2,\ -\operatorname{Re}b/2<\operatorname{Re}s<\operatorname{Re}a/2; \\ \lvert\arg\omega\rvert<\pi\end{matrix}\right]$
2	$\left(\sqrt{x+\omega}-\sqrt{x}\right)^{2c-b-2} \times {}_2F_1\left(\begin{matrix} 1,\, b \\ c;\; \frac{2x-2\sqrt{x}\sqrt{x+\omega}+\omega}{2\sqrt{x}\left(\sqrt{x}-\sqrt{x+\omega}\right)} \end{matrix}\right)$	$\dfrac{(c-1)\,\omega^{s-b/2+c-1}}{\sqrt{\pi}}\Gamma\left[\begin{matrix} \frac{2c-b-2s-2}{2},\, \frac{2s+1}{2},\, s+1,\, \frac{2s+b}{2} \\ \frac{2s+2c-b}{2},\, \frac{2s+b+2}{2} \end{matrix}\right]$ $\left[\begin{matrix}-1/2,\ -\operatorname{Re}b/2<\operatorname{Re}s<\operatorname{Re}(c-b/2)-1; \\ -\pi<\arg\omega\le\pi\end{matrix}\right]$

3.31.19. ${}_2F_1\left(a,\, b;\, c;\, \dfrac{x-2\sqrt{\omega}\sqrt{x+\omega}+2\omega}{2\sqrt{\omega}\left(\sqrt{\omega}-\sqrt{x+\omega}\right)}\right)$ and algebraic functions

1	$\left(\sqrt{x+\omega}-\sqrt{\omega}\right)^{a} \times {}_2F_1\left(\begin{matrix} a,\, b;\; \frac{x-2\sqrt{\omega}\sqrt{x+\omega}+2\omega}{2\sqrt{\omega}\left(\sqrt{\omega}-\sqrt{x+\omega}\right)} \\ a+1 \end{matrix}\right)$	$\dfrac{2^{b-1}a\,\omega^{s+a/2}}{\sqrt{\pi}}\Gamma\left[\begin{matrix} -s,\, \frac{b-a-2s}{2},\, \frac{b-a-2s+1}{2},\, s+a \\ 1-s,\, b-s \end{matrix}\right]$ $[-\operatorname{Re}a<\operatorname{Re}s<0,\ \operatorname{Re}(b-a)/2;\ 0\le\arg\omega\le\pi]$
2	$\left(\sqrt{x+\omega}-\sqrt{\omega}\right)^{2c-b-2} \times {}_2F_1\left(\begin{matrix} 1,\, b;\; \frac{x-2\sqrt{\omega}\sqrt{x+\omega}+2\omega}{2\sqrt{\omega}\left(\sqrt{\omega}-\sqrt{x+\omega}\right)} \\ c \end{matrix}\right)$	$\dfrac{(c-1)\,\omega^{s-b/2+c-1}}{\sqrt{\pi}}\Gamma\left[\begin{matrix} \frac{b-2c-2s+3}{2},\, \frac{b-2c-2s+4}{2} \\ 1-s \end{matrix}\right]$ $\times\Gamma\left[\begin{matrix} b-c-s+1,\, s-b+2c-2 \\ b-c-s+2 \end{matrix}\right]$ $\left[\begin{matrix}\operatorname{Re}(b-2c)+2<\operatorname{Re}s<\operatorname{Re}(b-c)+1, \\ \operatorname{Re}(b-2c+3)/2;\ -\pi<\arg\omega\le\pi\end{matrix}\right]$

3.31.20. ${}_2F_1\left(a,\, b;\, c;\, \frac{x-\sqrt{x^2+\omega^2}}{2x}\right)$ and algebraic functions

1	$\left(\sqrt{x^2+\omega^2}+x\right)^{-a} \times {}_2F_1\left(\begin{matrix} a,\, b;\, \frac{x-\sqrt{x^2+\omega^2}}{2x} \\ a+1 \end{matrix}\right)$	$\frac{2^{b-2}a\,\omega^{s-a}}{\sqrt{\pi}}\,\Gamma\left[\begin{matrix} \frac{-s+a}{2},\, \frac{s+a}{2},\, \frac{s+b}{2},\, \frac{s+b+1}{2} \\ \frac{s+a+2}{2},\, \frac{s+a+2b}{2} \end{matrix}\right]$ $[-\operatorname{Re} a,\, -\operatorname{Re} b < \operatorname{Re} s < \operatorname{Re} a;\; -\pi/2 < \arg\omega \le \pi/2]$
2	$\left(\sqrt{x^2+\omega^2}+x\right)^{b-2c+2} \times {}_2F_1\left(\begin{matrix} 1,\, b;\, \frac{x-\sqrt{x^2+\omega^2}}{2x} \\ c \end{matrix}\right)$	$\frac{(c-1)\,\omega^{s+b-2c+2}}{2\sqrt{\pi}}\,\Gamma\left[\begin{matrix} \frac{2c-b-s-2}{2},\, \frac{s+1}{2},\, \frac{s+2}{2},\, \frac{s+b}{2} \\ \frac{s-b+2c}{2},\, \frac{s+b+2}{2} \end{matrix}\right]$ $\left[\begin{matrix} -1,\, -\operatorname{Re} b < \operatorname{Re} s < \operatorname{Re}(2c-b)-2; \\ -\pi/2 < \arg\omega \le \pi/2 \end{matrix}\right]$

3.31.21. ${}_2F_1\left(a,\, b;\, c;\, \frac{\omega-\sqrt{x^2+\omega^2}}{2\omega}\right)$ and algebraic functions

1	$\left(\sqrt{x^2+\omega^2}+\omega\right)^{-a} \times {}_2F_1\left(\begin{matrix} a,\, b;\, \frac{\omega-\sqrt{x^2+\omega^2}}{2\omega} \\ a+1 \end{matrix}\right)$	$\frac{2^{b-2}a\,\omega^{s-a}}{\sqrt{\pi}}\,\Gamma\left[\begin{matrix} \frac{2a-s}{2},\, \frac{a+b-s}{2},\, \frac{a+b-s+1}{2},\, \frac{s}{2} \\ \frac{2a-s+2}{2},\, \frac{2a+2b-s}{2} \end{matrix}\right]$ $\left[\begin{matrix} 0 < \operatorname{Re} s < 2\operatorname{Re} a,\; \operatorname{Re}(a+b); \\ -\pi/2 < \arg\omega \le \pi/2 \end{matrix}\right]$
2	$\left(\sqrt{x^2+\omega^2}+\omega\right)^{b-2c+2} \times {}_2F_1\left(\begin{matrix} 1,\, b;\, \frac{\omega-\sqrt{x^2+\omega^2}}{2\omega} \\ c \end{matrix}\right)$	$\frac{(c-1)\,\omega^{s+b-2c+2}}{2\sqrt{\pi}}\,\Gamma\left[\begin{matrix} \frac{2c-s-2}{2},\, \frac{2c-b-s-1}{2},\, \frac{2c-b-s}{2},\, \frac{s}{2} \\ \frac{2c-s}{2},\, \frac{4c-2b-s-2}{2} \end{matrix}\right]$ $\left[\begin{matrix} 0 < \operatorname{Re} s < 2\operatorname{Re} c-2,\; \operatorname{Re}(2c-b)-1; \\ -\pi/2 < \arg\omega \le \pi/2 \end{matrix}\right]$

3.31.22. ${}_2F_1\left(a,\, b;\, c;\, \frac{\sqrt{x^2+\omega^2}-x}{\sqrt{x^2+\omega^2}+x}\right)$ and algebraic functions

1	$\left(\sqrt{x^2+\omega^2}+x\right)^{-2a} \times {}_2F_1\left(\begin{matrix} a,\, b;\, \frac{\sqrt{x^2+\omega^2}-x}{\sqrt{x^2+\omega^2}+x} \\ a+1 \end{matrix}\right)$	$\frac{a\,\omega^{s-2a}}{2^{b+1}\sqrt{\pi}}\,\Gamma\left[\begin{matrix} \frac{2a-s}{2},\, \frac{s-b+1}{2},\, \frac{s-b+2}{2},\, \frac{s}{2} \\ \frac{s+2}{2},\, \frac{s+2a-2b+2}{2} \end{matrix}\right]$ $[0,\; \operatorname{Re} b-1 < \operatorname{Re} s < 2\operatorname{Re} a;\, -\pi/2 < \arg\omega \le \pi/2]$
2	$\left(\sqrt{x^2+\omega^2}+x\right)^{1-b-c} \times {}_2F_1\left(\begin{matrix} 1,\, b;\, \frac{\sqrt{x^2+\omega^2}-x}{\sqrt{x^2+\omega^2}+x} \\ c \end{matrix}\right)$	$\frac{(c-1)\,\omega^{s-b-c+1}}{4\sqrt{\pi}}\,\Gamma\left[\begin{matrix} \frac{b+c-s-1}{2},\, \frac{s}{2},\, \frac{s+1}{2},\, \frac{s-b+c-1}{2} \\ \frac{s-b+c+1}{2},\, \frac{s+b+c-1}{2} \end{matrix}\right]$ $\left[\begin{matrix} 0,\; \operatorname{Re}(b-c)+1 < \operatorname{Re} s < \operatorname{Re}(b+c)-1; \\ -\pi/2 < \arg\omega \le \pi/2 \end{matrix}\right]$

3.31.23. ${}_2F_1\left(a,\,b;\,c;\,\dfrac{\sqrt{x^2+\omega^2}-\omega}{\sqrt{x^2+\omega^2}+\omega}\right)$ and algebraic functions

1	$\left(\sqrt{x^2+\omega^2}+\omega\right)^{-2a} \times {}_2F_1\left(\begin{matrix}a,\,b;\,\frac{\sqrt{x^2+\omega^2}-\omega}{\sqrt{x^2+\omega^2}+\omega}\\ a+1\end{matrix}\right)$	$\dfrac{a\,\omega^{s-2a}}{2^{b+1}\sqrt{\pi}}\,\Gamma\left[\begin{matrix}\frac{2a-s}{2},\,\frac{2a-b-s+1}{2},\,\frac{2a-b-s+2}{2},\,\frac{s}{2}\\ \frac{2a-s+2}{2},\,\frac{4a-2b-s+2}{2}\end{matrix}\right]$ $\left[\begin{matrix}0<\operatorname{Re}s<2\operatorname{Re}a,\ \operatorname{Re}(2a-b)+1;\\ -\pi/2<\arg\omega\le\pi/2\end{matrix}\right]$
2	$\left(\sqrt{x^2+\omega^2}+\omega\right)^{1-b-c} \times {}_2F_1\left(\begin{matrix}1,\,b;\,\frac{\sqrt{x^2+\omega^2}-\omega}{\sqrt{x^2+\omega^2}+\omega}\\ c\end{matrix}\right)$	$\dfrac{(c-1)\,\omega^{s-b-c+1}}{4\sqrt{\pi}}\,\Gamma\left[\begin{matrix}\frac{b+c-s-1}{2},\,\frac{b+c-s}{2},\,\frac{2c-s-2}{2},\,\frac{s}{2}\\ \frac{2c-s}{2},\,\frac{2b+2c-s-2}{2}\end{matrix}\right]$ $\left[\begin{matrix}0<\operatorname{Re}s<2\operatorname{Re}c-2,\ \operatorname{Re}(b+c)-1;\\ -\pi/2<\arg\omega\le\pi/2\end{matrix}\right]$

3.31.24. ${}_2F_1\left(a,\,b;\,c;\,\dfrac{x^2-2\omega\sqrt{x^2+\omega^2}+2\omega^2}{x^2}\right)$ and algebraic functions

1	$\left(\sqrt{x^2+\omega^2}-\omega\right)^{2a} \times {}_2F_1\left(\begin{matrix}a,\,b;\,\frac{x^2-2\omega\sqrt{x^2+\omega^2}+2\omega^2}{x^2}\\ a+1\end{matrix}\right)$	$\dfrac{a\,\omega^{s+2a}}{2^{b+1}\sqrt{\pi}}\,\Gamma\left[\begin{matrix}\frac{1-s-2a-b}{2},\,\frac{2-s-2a-b}{2},\,\frac{-s-2a}{2},\,\frac{s+4a}{2}\\ \frac{2-2a-s}{2},\,\frac{2-2b-s}{2}\end{matrix}\right]$ $\left[\begin{matrix}-4\operatorname{Re}a<\operatorname{Re}s<-2\operatorname{Re}a,\ 1-\operatorname{Re}(2a+b);\\ \lvert\arg\omega\rvert<\pi/2\end{matrix}\right]$
2	$\left(\sqrt{x^2+\omega^2}-\omega\right)^{b+c-1} \times {}_2F_1\left(\begin{matrix}1,\,b;\,\frac{x^2-2\omega\sqrt{x^2+\omega^2}+2\omega^2}{x^2}\\ c\end{matrix}\right)$	$\dfrac{(c-1)\,\omega^{s+b+c-1}}{4\sqrt{\pi}}\,\Gamma\left[\begin{matrix}\frac{1-b-c-s}{2},\,\frac{2-b-c-s}{2},\,\frac{-s-2b}{2}\\ \frac{2-2b-s}{2}\end{matrix}\right] \times\Gamma\left[\begin{matrix}\frac{s+2b+2c-2}{2}\\ -\frac{s}{2}\end{matrix}\right]$ $\left[\begin{matrix}2-2\operatorname{Re}(b+c)<\operatorname{Re}s<-2\operatorname{Re}b,\\ 1-\operatorname{Re}(b+c);\ \lvert\arg\omega\rvert<\pi/2\end{matrix}\right]$

3.31.25. ${}_2F_1\left(a,\,b;\,c;\,\dfrac{2x^2-2x\sqrt{x^2+\omega^2}+\omega^2}{\omega^2}\right)$ and algebraic functions

1	$\left(\sqrt{x^2+\omega^2}-x\right)^{2a} \times {}_2F_1\left(\begin{matrix}a,\,b;\,\frac{2x^2-2x\sqrt{x^2+\omega^2}+\omega^2}{\omega^2}\\ a+1\end{matrix}\right)$	$\dfrac{a\,\omega^{s+2a}}{2^{b+1}\sqrt{\pi}}\,\Gamma\left[\begin{matrix}\frac{2a-s}{2},\,\frac{s-b+1}{2},\,\frac{s-b+2}{2},\,\frac{s}{2}\\ \frac{s+2}{2},\,\frac{s+2a-2b+2}{2}\end{matrix}\right]$ $[0,\ \operatorname{Re}b-1<\operatorname{Re}s<2\operatorname{Re}a;\ \lvert\arg\omega\rvert<\pi/2]$
2	$\left(\sqrt{x^2+\omega^2}-x\right)^{b+c-1} \times {}_2F_1\left(\begin{matrix}1,\,b;\,\frac{2x^2-2x\sqrt{x^2+\omega^2}+\omega^2}{\omega^2}\\ c\end{matrix}\right)$	$\dfrac{(c-1)\,\omega^{s+b+c-1}}{4\sqrt{\pi}}\,\Gamma\left[\begin{matrix}\frac{b+c-s-1}{2},\,\frac{s}{2},\,\frac{s+1}{2},\,\frac{s-b+c-1}{2}\\ \frac{s-b+c+1}{2},\,\frac{s+b+c-1}{2}\end{matrix}\right]$ $\left[\begin{matrix}0,\ \operatorname{Re}(b-c)+1<\operatorname{Re}s<\operatorname{Re}(b+c)-1;\\ \lvert\arg\omega\rvert<\pi/2\end{matrix}\right]$

3.31.26. ${}_2F_1\left(a,\, b;\, c;\, \dfrac{2x^2-2x\sqrt{x^2+\omega^2}+\omega^2}{2x\left(x-\sqrt{x^2+\omega^2}\right)}\right)$ and algebraic functions

1	$\left(\sqrt{x^2+\omega^2}-x\right)^a \times {}_2F_1\left(\begin{matrix} a,\, b;\, \frac{2x^2-2x\sqrt{x^2+\omega^2}+\omega^2}{2x\left(x-\sqrt{x^2+\omega^2}\right)} \\ a+1 \end{matrix}\right)$	$\dfrac{2^{b-2}a\,\omega^{s+a}}{\sqrt{\pi}}\,\Gamma\left[\begin{matrix} \frac{-s+a}{2},\, \frac{s+a}{2},\, \frac{s+b}{2},\, \frac{s+b+1}{2} \\ \frac{s+a+2}{2},\, \frac{s+a+2b}{2} \end{matrix}\right]$ $[\operatorname{Re}\omega>0;\, -\operatorname{Re}a,\, -\operatorname{Re}b<\operatorname{Re}s<\operatorname{Re}a]$
2	$\left(\sqrt{x^2+\omega^2}-x\right)^{2c-b-2} \times {}_2F_1\left(\begin{matrix} 1,\, b;\, \frac{2x^2-2x\sqrt{x^2+\omega^2}+\omega^2}{2x\left(x-\sqrt{x^2+\omega^2}\right)} \\ c \end{matrix}\right)$	$\dfrac{(c-1)\,\omega^{s-b+2c-2}}{2\sqrt{\pi}}\,\Gamma\left[\begin{matrix} \frac{2c-b-s-2}{2},\, \frac{s+1}{2},\, \frac{s+2}{2},\, \frac{s+b}{2} \\ \frac{s+b+2}{2},\, \frac{s-b+2c}{2} \end{matrix}\right]$ $\left[\begin{matrix} -1,\, -\operatorname{Re}b<\operatorname{Re}s<\operatorname{Re}(2c-b)-2; \\ -\pi/2<\arg\omega\le\pi/2 \end{matrix}\right]$

3.31.27. ${}_2F_1\left(a,\, b;\, c;\, \dfrac{x^2-2\omega\sqrt{x^2+\omega^2}+2\omega^2}{2\omega\left(\omega-\sqrt{x^2+\omega^2}\right)}\right)$ and algebraic functions

1	$\left(\sqrt{x^2+\omega^2}-\omega\right)^a \times {}_2F_1\left(\begin{matrix} a,\, b;\, \frac{x^2-2\omega\sqrt{x^2+\omega^2}+2\omega^2}{2\omega\left(\omega-\sqrt{x^2+\omega^2}\right)} \\ a+1 \end{matrix}\right)$	$\dfrac{2^{b-2}a\,\omega^{s+a}}{\sqrt{\pi}}\,\Gamma\left[\begin{matrix} -\frac{s}{2},\, \frac{b-a-s}{2},\, \frac{b-a-s+1}{2},\, \frac{s+2a}{2} \\ \frac{2-s}{2},\, \frac{2b-s}{2} \end{matrix}\right]$ $\left[\begin{matrix} -2\operatorname{Re}a<\operatorname{Re}s<0,\, \operatorname{Re}(b-a)\,; \\ -\pi/2<\arg\omega\le\pi/2 \end{matrix}\right]$
2	$\left(\sqrt{x^2+\omega^2}-\omega\right)^{2c-b-2} \times {}_2F_1\left(\begin{matrix} 1,\, b;\, \frac{x^2-2\omega\sqrt{x^2+\omega^2}+2\omega^2}{2\omega\left(\omega-\sqrt{x^2+\omega^2}\right)} \\ c \end{matrix}\right)$	$\dfrac{(c-1)\,\omega^{s-b+2c-2}}{2\sqrt{\pi}}\,\Gamma\left[\begin{matrix} \frac{b-2c-s+3}{2},\, \frac{b-2c-s+4}{2} \\ \frac{2-s}{2} \end{matrix}\right] \times\Gamma\left[\begin{matrix} \frac{2b-2c-s+2}{2},\, \frac{s-2b+4c-4}{2} \\ \frac{2b-2c-s+4}{2} \end{matrix}\right]$ $\left[\begin{matrix} 2\operatorname{Re}(b-2c)+4<\operatorname{Re}s<\operatorname{Re}(b-2c)+3, \\ 2\operatorname{Re}(b-c+1)\,;\, -\pi/2<\arg\omega\le\pi/2 \end{matrix}\right]$

3.31.28. ${}_2F_1(a,\, b;\, c;\, \varphi(x))$ and algebraic functions

No.	$f(x)$	$F(s)$
1	$\dfrac{\lvert 1\mp\sqrt{x}\rvert^{2a}}{\lvert 1-x\rvert^{2b}}\, {}_2F_1\left(\begin{matrix} a,\, b \\ 2b;\, \frac{\pm 4\sqrt{x}}{(1\pm\sqrt{x})^2} \end{matrix}\right)$	$\Gamma\left[\begin{matrix} \frac{2b+1}{2},\, \frac{2a-2b+1}{2},\, s,\, 2b-a-s \\ 2b-a,\, \frac{2s+2a-2b+1}{2},\, \frac{2b-2s+1}{2} \end{matrix}\right]$ $[\operatorname{Re}(a-b)>-1/2;\, 0<\operatorname{Re}s<\operatorname{Re}(2b-a)]$
2	$(\sigma-x)_+^{\mu-1}\, {}_2F_1\left(\begin{matrix} a,\, b \\ c;\, \omega x(\sigma-x) \end{matrix}\right)$	$\sigma^{s+\mu-1}\,\mathrm{B}(\mu,\, s)\, {}_4F_3\left(\begin{matrix} a,\, b,\, \mu,\, s;\, \frac{\sigma^2\omega}{4} \\ c,\, \frac{s+\mu}{2},\, \frac{s+\mu+1}{2} \end{matrix}\right)$ $[\sigma,\, \operatorname{Re}\mu,\, \operatorname{Re}s>0]$

3.31.29. ${}_2F_1(a,\ b;\ c;\ \varphi(x))$ and the exponential function

1	$e^{-\sigma x}\,{}_2F_1\left(\begin{matrix}a,\ b\\ c;\ -\omega x\end{matrix}\right)$	$\omega^{-s}\,\Gamma\left[\begin{matrix}c,\ s,\ a-s,\ b-s\\ a,\ b,\ c-s\end{matrix}\right]{}_2F_2\left(\begin{matrix}s,\ s-c+1;\ \frac{\sigma}{\omega}\\ s-a+1,\ s-b+1\end{matrix}\right)$ $+\frac{\sigma^{a-s}}{\omega^a}\Gamma\left[\begin{matrix}c,\ b-a,\ s-a\\ b,\ c-a\end{matrix}\right]{}_2F_2\left(\begin{matrix}a,\ a-c+1;\ \frac{\sigma}{\omega}\\ a-b+1,\ a-s+1\end{matrix}\right)$ $+\frac{\sigma^{b-s}}{\omega^b}\Gamma\left[\begin{matrix}c,\ a-b,\ s-b\\ a,\ c-b\end{matrix}\right]{}_2F_2\left(\begin{matrix}b,\ b-c+1;\ \frac{\sigma}{\omega}\\ b-a+1,\ b-s+1\end{matrix}\right)$ $[\operatorname{Re}\sigma,\ \operatorname{Re}s>0;\ \lvert\arg\omega\rvert<\pi]$
2	$e^{-\sigma x}\,{}_2F_1\left(\begin{matrix}a,\ b\\ c;\ 1-\omega x\end{matrix}\right)$	$\omega^{-s}\,\Gamma\left[\begin{matrix}c,\ s,\ a-s,\ b-s,\ s-a-b+c\\ a,\ b,\ c-a,\ c-b\end{matrix}\right]{}_2F_2\left(\begin{matrix}s,\ s-a-b+c;\ -\frac{\sigma}{\omega}\\ s-a+1,\ s-b+1\end{matrix}\right)$ $+\frac{\sigma^{a-s}}{\omega^a}\Gamma\left[\begin{matrix}b-a,\ c,\ s-a\\ b,\ c-a\end{matrix}\right]{}_2F_2\left(\begin{matrix}a,\ c-b;\ -\frac{\sigma}{\omega}\\ a-b+1,\ a-s+1\end{matrix}\right)$ $+\frac{\sigma^{b-s}}{\omega^b}\Gamma\left[\begin{matrix}a-b,\ c,\ s-b\\ a,\ c-b\end{matrix}\right]{}_2F_2\left(\begin{matrix}b,\ c-a;\ -\frac{\sigma}{\omega}\\ b-a+1,\ b-s+1\end{matrix}\right)$ $[\operatorname{Re}\sigma,\ \operatorname{Re}s,\ \operatorname{Re}(s-a-b+c)>0;\ \lvert\arg\omega\rvert<\pi]$
3	$(\sigma-x)_+^{c-1}\,e^{\tau x}$ $\times\,{}_2F_1\left(\begin{matrix}a,\ b\\ c;\ \omega(\sigma-x)\end{matrix}\right)$	$\sigma^{s+c-1}\,\mathrm{B}(s,\ c)\ \Xi_1(a,\ s,\ b;\ s+c;\ \sigma\omega,\ \sigma\tau)$ $[\sigma,\ \operatorname{Re}c,\ \operatorname{Re}s>0]$
4	$e^{-\sigma/x}\,{}_2F_1\left(\begin{matrix}a,\ b\\ c;\ -\omega x\end{matrix}\right)$	$\omega^{-s}\,\Gamma\left[\begin{matrix}c,\ s,\ a-s,\ b-s\\ a,\ b,\ c-s\end{matrix}\right]{}_2F_2\left(\begin{matrix}a-s,\ b-s;\ \sigma\omega\\ 1-s,\ c-s\end{matrix}\right)$ $+\sigma^s\,\Gamma(-s)\ {}_2F_2\left(\begin{matrix}a,\ b;\ \sigma\omega\\ c,\ s+1\end{matrix}\right)$ $[\operatorname{Re}\sigma>0;\ \operatorname{Re}s<\operatorname{Re}a,\ \operatorname{Re}b;\ \lvert\arg\omega\rvert<\pi]$
5	$e^{-\sigma\sqrt{x}}\,{}_2F_1\left(\begin{matrix}a,\ b\\ c;\ -\omega x\end{matrix}\right)$	$\frac{2\sigma^{2a-2s}}{\omega^a}\Gamma\left[\begin{matrix}c,\ b-a,\ 2s-2a\\ b,\ c-a\end{matrix}\right]{}_2F_3\left(\begin{matrix}a,\ a-c+1;\ -\frac{\sigma^2}{4\omega}\\ a-b+1,\ a-s+1,\ \frac{2a-2s-1}{2}\end{matrix}\right)$ $+\frac{2\sigma^{2b-2s}}{\omega^b}\Gamma\left[\begin{matrix}c,\ a-b,\ 2s-2b\\ a,\ c-b\end{matrix}\right]$ $\times\,{}_2F_3\left(\begin{matrix}b,\ b-c+1;\ -\frac{\sigma^2}{4\omega}\\ b-a+1,\ b-s+1,\ \frac{2b-2s-1}{2}\end{matrix}\right)$ $+\omega^{-s}\,\Gamma\left[\begin{matrix}c,\ s,\ a-s,\ b-s\\ a,\ b,\ c-s\end{matrix}\right]{}_2F_3\left(\begin{matrix}s,\ s-c+1;\ -\frac{\sigma^2}{4\omega}\\ \frac{1}{2},\ s-a+1,\ s-b+1\end{matrix}\right)$ $-\frac{\sigma}{\omega^{s+1/2}}\Gamma\left[\begin{matrix}c,\ \frac{2s+1}{2},\ \frac{2a-2s-1}{2},\ \frac{2b-2s-1}{2}\\ a,\ b,\ \frac{2c-2s-1}{2}\end{matrix}\right]{}_2F_3\left(\begin{matrix}\frac{2s+1}{2},\ \frac{2s-2c+3}{2};\ -\frac{\sigma^2}{4\omega}\\ \frac{3}{2},\ \frac{2s-2a+3}{2},\ \frac{2s-2b+3}{2}\end{matrix}\right)$ $[\operatorname{Re}\sigma,\ \operatorname{Re}s>0;\ \lvert\arg\omega\rvert<\pi]$

No.	$f(x)$	$F(s)$
6	$e^{-\sigma\sqrt{x}}\,{}_2F_1\left(\begin{matrix}a,\,b\\c;\,1-\omega x\end{matrix}\right)$	$\frac{2\sigma^{2a-2s}}{\omega^a}\Gamma\left[\begin{matrix}c,\,b-a,\,2s-2a\\b,\,c-a\end{matrix}\right]{}_2F_3\left(\begin{matrix}a,\,c-b;\,\frac{\sigma^2}{4\omega}\\a-b+1,\,a-s+1,\,\frac{2a-2s+1}{2}\end{matrix}\right)$ $+\frac{2\sigma^{2b-2s}}{\omega^b}\Gamma\left[\begin{matrix}c,\,a-b,\,2s-2b\\a,\,c-b\end{matrix}\right]$ $\times{}_2F_3\left(\begin{matrix}b,\,c-a;\,\frac{\sigma^2}{4\omega}\\b-a+1,\,b-s+1,\,\frac{2b-2s+1}{2}\end{matrix}\right)$ $+\omega^{-s}\Gamma\left[\begin{matrix}c,\,s,\,a-s,\,b-s,\,s-a-b+c\\a,\,b,\,c-a,\,c-b\end{matrix}\right]$ $\times{}_2F_3\left(\begin{matrix}s,\,s-a-b+c;\,\frac{\sigma^2}{4\omega}\\\frac{1}{2},\,s-a+1,\,s-b+1\end{matrix}\right)$ $-\frac{\sigma}{\omega^{s+1/2}}\Gamma\left[\begin{matrix}c,\,\frac{2s+1}{2},\,\frac{2a-2s-1}{2},\,\frac{2b-2s-1}{2},\,\frac{2s-2a-2b+2c+1}{2}\\a,\,b,\,c-a,\,c-b\end{matrix}\right]$ $\times{}_2F_3\left(\begin{matrix}\frac{2s+1}{2},\,\frac{2s-2a-2b+2c+1}{2};\,\frac{\sigma^2}{4\omega}\\\frac{3}{2},\,\frac{2s-2a+3}{2},\,\frac{2s-2b+3}{2}\end{matrix}\right)$ $[\operatorname{Re}\sigma,\ \operatorname{Re}s,\ \operatorname{Re}(s-a-b+c)>0;\ \lvert\arg\omega\rvert<\pi]$
7	$e^{-\sigma/\sqrt{x}}\,{}_2F_1\left(\begin{matrix}a,\,b\\c;\,-\omega x\end{matrix}\right)$	$\omega^{-s}\Gamma\left[\begin{matrix}c,\,s,\,a-s,\,b-s\\a,\,b,\,c-s\end{matrix}\right]{}_2F_3\left(\begin{matrix}a-s,\,b-s;\,-\frac{\sigma^2\omega}{4}\\\frac{1}{2},\,1-s,\,c-s\end{matrix}\right)$ $+\frac{\sigma^{2s}}{\sqrt{\pi}\,2^{2s}}\Gamma\left(\frac{1}{2}-s\right)\Gamma(-s)\,{}_2F_3\left(\begin{matrix}a,\,b;\,-\frac{\sigma^2\omega}{4}\\c,\,\frac{2s+1}{2},\,s+1\end{matrix}\right)$ $-\sigma\omega^{1/2-s}\Gamma\left[\begin{matrix}c,\,\frac{2s-1}{2},\,\frac{2a-2s+1}{2},\,\frac{2b-2s+1}{2}\\a,\,b,\,\frac{2c-2s+1}{2}\end{matrix}\right]$ $\times{}_2F_3\left(\begin{matrix}\frac{2a-2s+1}{2},\,\frac{2b-2s+1}{2};\,-\frac{\sigma^2\omega}{4}\\\frac{3}{2},\,\frac{3-2s}{2},\,\frac{2c-2s+1}{2}\end{matrix}\right)$ $[\operatorname{Re}\sigma>0;\ \operatorname{Re}s<\operatorname{Re}a,\ \operatorname{Re}b;\ -\pi<\arg\omega\le\pi]$
8	$(\omega-x)_+^{c-1}\,e^{-\sigma x}$ $\times{}_2F_1\left(\begin{matrix}a,\,b\\c;\,\frac{\omega-x}{\omega}\end{matrix}\right)$	$\omega^{s+c-1}\Gamma\left[\begin{matrix}c,\,s,\,s-a-b+c\\s-a+c,\,s-b+c\end{matrix}\right]{}_2F_2\left(\begin{matrix}s,\,s-a-b+c;\,-\sigma\omega\\s-a+c,\,s-b+c\end{matrix}\right)$ $[\omega,\ \operatorname{Re}c,\ \operatorname{Re}s,\ \operatorname{Re}(s-a-b+c)>0]$
9	$(\omega-x)_+^{c-1}\,e^{-\sigma\sqrt{x}}$ $\times{}_2F_1\left(\begin{matrix}a,\,b\\c;\,\frac{\omega-x}{\omega}\end{matrix}\right)$	$\omega^{s+c-1}\Gamma\left[\begin{matrix}c,\,s,\,s-a-b+c\\s-a+c,\,s-b+c\end{matrix}\right]{}_2F_3\left(\begin{matrix}s,\,s-a-b+c;\,\frac{\sigma^2\omega}{4}\\\frac{1}{2},\,s-a+c,\,s-b+c\end{matrix}\right)$ $-\sigma\omega^{s+c-1/2}\Gamma\left[\begin{matrix}c,\,\frac{2s+1}{2},\,\frac{2s-2a-2b+2c+1}{2}\\\frac{2s-2a+2c+1}{2},\,\frac{2s-2b+2c+1}{2}\end{matrix}\right]$ $\times{}_2F_3\left(\begin{matrix}\frac{2s+1}{2},\,\frac{2s-2a-2b+2c+1}{2};\,\frac{\sigma^2\omega}{4}\\\frac{3}{2},\,\frac{2s-2a+2c+1}{2},\,\frac{2s-2b+2c+1}{2}\end{matrix}\right)$ $[\omega,\ \operatorname{Re}c,\ \operatorname{Re}s,\ \operatorname{Re}(s-a-b+c)>0]$

No.	$f(x)$	$F(s)$
10	$(x-\omega)_+^{c-1}\,e^{-\sigma x}$ $\times\,{}_2F_1\left(\begin{matrix}a,\,b\\ c;\,\frac{\omega-x}{\omega}\end{matrix}\right)$	$\omega^{s+c-1}\,\Gamma\left[\begin{matrix}c,\,a-c-s+1,\,b-c-s+1\\ 1-s,\,a+b-c-s+1\end{matrix}\right]$ $\times\,{}_2F_2\left(\begin{matrix}s,\,s-a-b+c;\,-\sigma\omega\\ s-a+c,\,s-b+c\end{matrix}\right)$ $+\,\sigma^{a-c-s+1}\omega^a\,\Gamma\left[\begin{matrix}b-a,\,c,\,s-a+c-1\\ b,\,c-a\end{matrix}\right]$ $\times\,{}_2F_2\left(\begin{matrix}1-b,\,a-c+1;\,-\sigma\omega\\ a-b+1,\,a-c-s+2\end{matrix}\right)$ $+\,\sigma^{b-c-s+1}\omega^b\,\Gamma\left[\begin{matrix}a-b,\,c,\,s-b+c-1\\ a,\,c-b\end{matrix}\right]$ $\times\,{}_2F_2\left(\begin{matrix}1-a,\,b-c+1;\,-\sigma\omega\\ b-a+1,\,b-c-s+2\end{matrix}\right)$ $[\omega,\ \mathrm{Re}\,c,\ \mathrm{Re}\,\sigma>0]$
11	$(x-\omega)_+^{c-1}\,e^{-\sigma\sqrt{x}}$ $\times\,{}_2F_1\left(\begin{matrix}a,\,b\\ c;\,\frac{\omega-x}{\omega}\end{matrix}\right)$	$\omega^{c+s-1}\,\Gamma\left[\begin{matrix}c,\,a-c-s+1,\,b-c-s+1\\ 1-s,\,a+b-c-s+1\end{matrix}\right]$ $\times\,{}_2F_3\left(\begin{matrix}s,\,s-a-b+c;\,\frac{\sigma^2\omega}{4}\\ \frac{1}{2},\,s-a+c,\,s-b+c\end{matrix}\right)$ $+\,2\sigma^{2(a-c-s+1)}\omega^a\,\Gamma\left[\begin{matrix}b-a,\,c,\,2s-2a+2c-2\\ b,\,c-a\end{matrix}\right]$ $\times\,{}_2F_3\left(\begin{matrix}1-b,\,a-c+1;\,\frac{\sigma^2\omega}{4}\\ a-b+1,\,\frac{2a-2c-2s+3}{2},\,a-c-s+2\end{matrix}\right)$ $+\,2\sigma^{2(b-c-s+1)}\omega^b\,\Gamma\left[\begin{matrix}c,\,a-b,\,2s-2b+2c-2\\ a,\,c-b\end{matrix}\right]$ $\times\,{}_2F_3\left(\begin{matrix}1-a,\,b-c+1;\,\frac{\sigma^2\omega}{4}\\ b-a+1,\,\frac{2b-2c-2s+3}{2},\,\frac{2b-2c-2s+4}{2}\end{matrix}\right)$ $-\,\sigma\omega^{c+s-1/2}\,\Gamma\left[\begin{matrix}c,\,\frac{2a-2c-2s+1}{2},\,\frac{2b-2c-2s+1}{2}\\ \frac{1-2s}{2},\,\frac{2a+2b-2c-2s+1}{2}\end{matrix}\right]$ $\times\,{}_2F_3\left(\begin{matrix}\frac{2s+1}{2},\,\frac{2s-2a-2b+2c+1}{2};\,\frac{\sigma^2\omega}{4}\\ \frac{3}{2},\,\frac{2s-2a+2c+1}{2},\,\frac{2s-2b+2c+1}{2}\end{matrix}\right)$ $[\omega,\ \mathrm{Re}\,c,\ \mathrm{Re}\,\sigma>0]$

3.31.30. ${}_2F_1(a,\,b;\,c;\,\omega x+\sigma)$ and trigonometric functions

Notation: $\delta=\left\{\begin{matrix}1\\0\end{matrix}\right\}$.

1	$\left\{\begin{matrix}\sin(\sigma\sqrt{x})\\ \cos(\sigma\sqrt{x})\end{matrix}\right\}$ $\times\,{}_2F_1\left(\begin{matrix}a,\,b\\ c;\,-\omega x\end{matrix}\right)$	$\frac{2\sigma^{2a-2s}}{\omega^a}\,\Gamma\left[\begin{matrix}b-a,\,c,\,2s-2a\\ b,\,c-a\end{matrix}\right]\left\{\begin{matrix}\sin[(s-a)\pi]\\ \cos[(s-a)\pi]\end{matrix}\right\}$ $\times\,{}_2F_3\left(\begin{matrix}a,\,a-c+1;\,\frac{\sigma^2}{4\omega}\\ a-b+1,\,\frac{2a-2s+1}{2},\,a-s+1\end{matrix}\right)$ $+\,\frac{2\sigma^{2b-2s}}{\omega^b}\,\Gamma\left[\begin{matrix}a-b,\,c,\,2s-2b\\ a,\,c-b\end{matrix}\right]\left\{\begin{matrix}\sin[(s-b)\pi]\\ \cos[(s-b)\pi]\end{matrix}\right\}\times$

No.	$f(x)$	$F(s)$
		$\times\, {}_2F_3\left(\begin{matrix} b,\, b-c+1;\, \frac{\sigma^2}{4\omega} \\ b-a+1,\, \frac{2b-2s+1}{2},\, b-s+1 \end{matrix}\right)$ $+\,\frac{\sigma^\delta}{\omega^{s+\delta/2}}\,\Gamma\left[\begin{matrix} c,\, \frac{2s+\delta}{2},\, \frac{2a-2s-\delta}{2},\, \frac{2b-\delta-2s}{2} \\ a,\, b,\, \frac{2c-\delta-2s}{2} \end{matrix}\right]$ $\times\, {}_2F_3\left(\begin{matrix} s+\frac{\delta}{2},\, \frac{2s-2c+\delta+2}{2};\, \frac{\sigma^2}{4\omega} \\ \frac{2\delta+1}{2},\, \frac{2s-2a+\delta+2}{2},\, \frac{2s-2b+\delta+2}{2} \end{matrix}\right)$ $[\sigma>0;\ -\delta/2<\operatorname{Re}s<\operatorname{Re}a+1/2,\ \operatorname{Re}b+1/2;\ \lvert\arg\omega\rvert<\pi]$
2	$\begin{Bmatrix} \sin(\sigma/\sqrt{x}) \\ \cos(\sigma/\sqrt{x}) \end{Bmatrix}$ $\times\, {}_2F_1\left(\begin{matrix} a,\, b \\ c;\, -\omega x \end{matrix}\right)$	$\frac{\sigma^\delta}{\omega^{s-\delta/2}}\,\Gamma\left[\begin{matrix} c,\, \frac{2s-\delta}{2},\, \frac{2a+\delta-2s}{2},\, \frac{2b+\delta-2s}{2} \\ a,\, b,\, \frac{2c+\delta-2s}{2} \end{matrix}\right]$ $\times\, {}_2F_3\left(\begin{matrix} \frac{2a+\delta-2s}{2},\, \frac{2b+\delta-2s}{2};\, \frac{\sigma^2\omega}{4} \\ \frac{2\delta+1}{2},\, \frac{2c+\delta-2s}{2},\, \frac{\delta-2s+2}{2} \end{matrix}\right)$ $\mp\, 2\sigma^{2s}\,\Gamma(-2s)\begin{Bmatrix} \sin(s\pi) \\ \cos(s\pi) \end{Bmatrix}{}_2F_3\left(\begin{matrix} a,\, b;\, \frac{\sigma^2\omega}{4} \\ c,\, \frac{2s+1}{2},\, s+1 \end{matrix}\right)$ $[\sigma>0;\ -1/2<\operatorname{Re}s<\operatorname{Re}a+\delta/2,\ \operatorname{Re}b+\delta/2;\ \lvert\arg\omega\rvert<\pi]$
3	$\begin{Bmatrix} \sin(\sigma\sqrt{x}) \\ \cos(\sigma\sqrt{x}) \end{Bmatrix}$ $\times\, {}_2F_1\left(\begin{matrix} a,\, b \\ c;\, 1-\omega x \end{matrix}\right)$	$\frac{\sigma^\delta}{\omega^{s+\delta/2}}\,\Gamma\left[\begin{matrix} c,\, \frac{2s-2a-2b+2c+\delta}{2},\, \frac{2s+\delta}{2},\, \frac{2a-2s-\delta}{2},\, \frac{2b-2s-\delta}{2} \\ a,\, b,\, c-a,\, c-b \end{matrix}\right]$ $\times\, {}_2F_3\left(\begin{matrix} \frac{2s+\delta}{2},\, \frac{2s-2a-2b+2c+\delta}{2};\, -\frac{\sigma^2}{4\omega} \\ \frac{2\delta+1}{2},\, \frac{2s+\delta-2a+2}{2},\, \frac{2s+\delta-2b+2}{2} \end{matrix}\right)$ $+\,\frac{2\sigma^{2a-2s}}{\omega^a}\begin{Bmatrix} \sin[(s-a)\pi] \\ \cos[(s-a)\pi] \end{Bmatrix}\Gamma\left[\begin{matrix} b-a,\, c,\, 2s-2a \\ b,\, c-a \end{matrix}\right]$ $\times\, {}_2F_3\left(\begin{matrix} a,\, c-b;\, -\frac{\sigma^2}{4\omega} \\ a-b+1,\, \frac{2a-2s+1}{2},\, a-s+1 \end{matrix}\right)$ $+\,\frac{2\sigma^{2b-2s}}{\omega^b}\begin{Bmatrix} \sin[(s-b)\pi] \\ \cos[(s-b)\pi] \end{Bmatrix}\Gamma\left[\begin{matrix} a-b,\, c,\, 2s-2b \\ a,\, c-b \end{matrix}\right]$ $\times\, {}_2F_3\left(\begin{matrix} b,\, c-a;\, -\frac{\sigma^2}{4\omega} \\ b-a+1,\, \frac{2b-2s+1}{2},\, b-s+1 \end{matrix}\right)$ $\left[\begin{matrix} \sigma>0;\ \operatorname{Re}s,\ \operatorname{Re}(c-a-b+s)>-\delta/2; \\ \operatorname{Re}(s-a),\ \operatorname{Re}(s-b)<1/2;\ \lvert\arg\omega\rvert<\pi \end{matrix}\right]$
4	$(\omega-x)_+^{c-1}\begin{Bmatrix} \sin(\sigma\sqrt{x}) \\ \cos(\sigma\sqrt{x}) \end{Bmatrix}$ $\times\, {}_2F_1\left(\begin{matrix} a,\, b \\ c;\, \frac{\omega-x}{\omega} \end{matrix}\right)$	$\sigma^\delta\omega^{s+c+\delta/2-1}\,\Gamma\left[\begin{matrix} c,\, \frac{2s+\delta}{2},\, \frac{2s-2a-2b+2c+\delta}{2} \\ \frac{2s-2a+2c+\delta}{2},\, \frac{2s-2b+2c+\delta}{2} \end{matrix}\right]$ $\times\, {}_2F_3\left(\begin{matrix} \frac{2s+\delta}{2},\, \frac{2s-2a-2b+2c+\delta}{2};\, -\frac{\sigma^2\omega}{4} \\ \frac{2\delta+1}{2},\, \frac{2s-2a+2c+\delta}{2},\, \frac{2s-2b+2c+\delta}{2} \end{matrix}\right)$ $[\omega,\ \operatorname{Re}c,\ \operatorname{Re}s>0;\ \operatorname{Re}(s-a-b+c)>-\delta/2]$

No.	$f(x)$	$F(s)$
5	$(x-\omega)_+^{c-1}\left\{\begin{matrix}\sin(\sigma\sqrt{x})\\ \cos(\sigma\sqrt{x})\end{matrix}\right\} \times {}_2F_1\left(\begin{matrix}a,\, b\\ c;\, \frac{\omega-x}{\omega}\end{matrix}\right)$	$\sigma^{\delta}\omega^{c+s+\delta/2-1}\,\Gamma\left[\begin{matrix}c,\, \frac{2a-2c-\delta-2s+2}{2},\, \frac{2b-2c-\delta-2s+2}{2}\\ \frac{2-\delta-2s}{2},\, \frac{2a+2b-2c-\delta-2s+2}{2}\end{matrix}\right] \times {}_2F_3\left(\begin{matrix}s+\frac{\delta}{2},\, c-a-b+s+\frac{\delta}{2};\, -\frac{\sigma^2\omega}{4}\\ \frac{2\delta+1}{2},\, \frac{2s-2a+2c+\delta}{2},\, \frac{2s-2b+2c+\delta}{2}\end{matrix}\right) - 2\sigma^{2(a-c-s+1)}\omega^{a}\left\{\begin{matrix}\sin[(s-a+c)\pi]\\ \cos[(s-a+c)\pi]\end{matrix}\right\} \times \Gamma\left[\begin{matrix}b-a,\, c,\, 2s-2a+2c-2\\ b,\, c-a\end{matrix}\right] \times {}_2F_3\left(\begin{matrix}1-b,\, a-c+1;\, -\frac{\sigma^2\omega}{4}\\ a-b+1,\, a-c-s+\frac{3}{2},\, a-c-s+2\end{matrix}\right) - 2\sigma^{2(b-c-s+1)}\omega^{b}\left\{\begin{matrix}\sin[(s-b+c)\pi]\\ \cos[(s-b+c)\pi]\end{matrix}\right\} \times \Gamma\left[\begin{matrix}a-b,\, c,\, 2s-2b+2c-2\\ a,\, c-b\end{matrix}\right] \times {}_2F_3\left(\begin{matrix}1-a,\, b-c+1;\, -\frac{\sigma^2\omega}{4}\\ b-a+1,\, b-c-s+\frac{3}{2},\, b-c-s+2\end{matrix}\right)$ $[\sigma,\, \omega,\, \operatorname{Re} c>0;\ \operatorname{Re}(s-a+c),\, \operatorname{Re}(s-b+c)<3/2]$

3.31.31. ${}_2F_1(a,\, b;\, c;\, \varphi(x))$ and the Bessel functions

1	$J_\nu(\sigma\sqrt{x})\,{}_2F_1\left(\begin{matrix}a,\, b\\ c;\, -\omega x\end{matrix}\right)$	$\frac{(\sigma/2)^{\nu}}{\omega^{s+\nu/2}}\Gamma\left[\begin{matrix}c,\, \frac{2s+\nu}{2},\, \frac{2a-\nu-2s}{2},\, \frac{2b-\nu-2s}{2}\\ a,\, b,\, \nu+1,\, \frac{2c-\nu-2s}{2}\end{matrix}\right] \times {}_2F_3\left(\begin{matrix}\frac{2s+\nu}{2},\, \frac{2s+\nu-2c+2}{2};\, \frac{\sigma^2}{4\omega}\\ \nu+1,\, \frac{2s+\nu-2a+2}{2},\, \frac{2s+\nu-2b+2}{2}\end{matrix}\right) + \frac{(\sigma/2)^{2a-2s}}{\omega^{a}}\Gamma\left[\begin{matrix}c,\, b-a\\ b,\, c-a\end{matrix}\right] \times \Gamma\left[\begin{matrix}\frac{2s-2a+\nu}{2}\\ \frac{2-2s+2a+\nu}{2}\end{matrix}\right]{}_2F_3\left(\begin{matrix}a,\, a-c+1;\, \frac{\sigma^2}{4\omega}\\ a-b+1,\, \frac{2a+\nu-2s+2}{2},\, \frac{2a-\nu-2s+2}{2}\end{matrix}\right) + \frac{(\sigma/2)^{2b-2s}}{\omega^{b}}\Gamma\left[\begin{matrix}c,\, a-b,\, \frac{2s+\nu-2b}{2}\\ a,\, c-b,\, \frac{2b+\nu-2s+2}{2}\end{matrix}\right] \times {}_2F_3\left(\begin{matrix}b,\, b-c+1;\, \frac{\sigma^2}{4\omega}\\ b-a+1,\, \frac{2b-\nu-2s+2}{2},\, \frac{2b+\nu-2s+2}{2}\end{matrix}\right)$ $[\sigma,\, \operatorname{Re}(2s+\nu)>0;\ \operatorname{Re}(s-a),\, \operatorname{Re}(s-b)<3/4;\ \lvert\arg\omega\rvert<\pi]$
2	$J_\nu\left(\frac{\sigma}{\sqrt{x}}\right){}_2F_1\left(\begin{matrix}a,\, b\\ c;\, -\omega x\end{matrix}\right)$	$\left(\frac{\sigma}{2}\right)^{2s}\Gamma\left[\begin{matrix}\frac{\nu-2s}{2}\\ \frac{2s+\nu+2}{2}\end{matrix}\right]{}_2F_3\left(\begin{matrix}a,\, b;\, \frac{\sigma^2\omega}{4}\\ c,\, \frac{2s-\nu+2}{2},\, \frac{2s+\nu+2}{2}\end{matrix}\right) + \frac{(\sigma/2)^{\nu}}{\omega^{s-\nu/2}}\Gamma\left[\begin{matrix}c\\ a,\, b\end{matrix}\right] \times \Gamma\left[\begin{matrix}\frac{2s-\nu}{2},\, \frac{2a-2s+\nu}{2},\, \frac{2b-2s+\nu}{2}\\ \nu+1,\, \frac{2c-2s+\nu}{2}\end{matrix}\right]{}_2F_3\left(\begin{matrix}\frac{2a+\nu-2s}{2},\, \frac{2b+\nu-2s}{2};\, \frac{\sigma^2\omega}{4}\\ \nu+1,\, \frac{\nu-2s+2}{2},\, \frac{\nu+2c-2s}{2}\end{matrix}\right)$ $[\sigma>0;\ -3/4<\operatorname{Re} s<\operatorname{Re}(a+\nu/2),\, \operatorname{Re}(b+\nu/2);\ \lvert\arg\omega\rvert<\pi]$

No.	$f(x)$	$F(s)$
3	$J_\nu(\sigma\sqrt{x})$ $\times {}_2F_1\left(\begin{matrix} a, b \\ c; 1-\omega x\end{matrix}\right)$	$\frac{(\sigma/2)^{2a-2s}}{\omega^a}\Gamma\left[\begin{matrix} b-a, c, \frac{2s-2a+\nu}{2} \\ b, c-a, \frac{2-2s+2a+\nu}{2}\end{matrix}\right]$ $\times {}_2F_3\left(\begin{matrix} a, c-b; -\frac{\sigma^2}{4\omega} \\ a-b+1, \frac{2-2s+2a-\nu}{2}, \frac{2-2s+2a+\nu}{2}\end{matrix}\right) + \frac{(\sigma/2)^{2b-2s}}{\omega^b}\Gamma\left[\begin{matrix} a-b, c \\ a, c-b\end{matrix}\right]$ $\times \Gamma\left[\begin{matrix} \frac{2s-2b+\nu}{2} \\ \frac{2-2s+2b+\nu}{2}\end{matrix}\right] {}_2F_3\left(\begin{matrix} b, c-a; -\frac{\sigma^2}{4\omega} \\ 1-a+b, \frac{2-2s+2b-\nu}{2}, \frac{2-2s+2b+\nu}{2}\end{matrix}\right)$ $+ \frac{(\sigma/2)^{\nu}}{\omega^{s+\nu/2}}\Gamma\left[\begin{matrix} c, \frac{2s+\nu}{2}, \frac{2a-2s-\nu}{2}, \frac{2b-2s-\nu}{2}, \frac{2s-2a-2b+2c+\nu}{2} \\ a, b, c-a, c-b, \nu+1\end{matrix}\right]$ $\times {}_2F_3\left(\begin{matrix} \frac{2s+\nu}{2}, \frac{2s-2a-2b+2c+\nu}{2}; -\frac{\sigma^2}{4\omega} \\ \nu+1, \frac{2s-2a+\nu+2}{2}, \frac{2s-2b+\nu+2}{2}\end{matrix}\right)$ $\left[\begin{matrix} \sigma, \operatorname{Re}(s-a-b+c+\nu/2), \operatorname{Re}(2s+\nu) > 0; \\ \operatorname{Re}(s-a), \operatorname{Re}(s-b) < 3/4; \lvert\arg\omega\rvert < \pi\end{matrix}\right]$
4	$(\omega-x)_+^{c-1} J_\nu(\sigma\sqrt{x})$ $\times {}_2F_1\left(\begin{matrix} a, b \\ c; \frac{\omega-x}{\omega}\end{matrix}\right)$	$\left(\frac{\sigma}{2}\right)^{\nu}\omega^{s+c+\nu/2-1}\Gamma\left[\begin{matrix} c, \frac{2s+\nu}{2}, \frac{2s+2c-2a-2b+\nu}{2} \\ \nu+1, \frac{2s-2a+2c+\nu}{2}, \frac{2s-2b+2c+\nu}{2}\end{matrix}\right]$ $\times {}_2F_3\left(\begin{matrix} \frac{2s+\nu}{2}, \frac{2s-2a-2b+2c+\nu}{2}; -\frac{\sigma^2\omega}{4} \\ \nu+1, \frac{2s-2a+2c+\nu}{2}, \frac{2s-2b+2c+\nu}{2}\end{matrix}\right)$ $[\omega, \operatorname{Re} c, \operatorname{Re}(2s+\nu), \operatorname{Re}(c-a-b+s+\nu/2) > 0]$
5	$(x-\omega)_+^{c-1} J_\nu(\sigma\sqrt{x})$ $\times {}_2F_1\left(\begin{matrix} a, b \\ c; \frac{\omega-x}{\omega}\end{matrix}\right)$	$\left(\frac{\sigma}{2}\right)^{2(a-c-s+1)}\omega^{a}\Gamma\left[\begin{matrix} c, b-a, \frac{2s-2a+2c+\nu-2}{2} \\ c-a, b, \frac{2a-2c+\nu-2s+4}{2}\end{matrix}\right]$ $\times {}_2F_3\left(\begin{matrix} 1-b, a-c+1; -\frac{\sigma^2\omega}{4} \\ a-b+1, \frac{2a-2c+\nu-2s+4}{2}, \frac{2a-2c-\nu-2s+4}{2}\end{matrix}\right)$ $+ \left(\frac{\sigma}{2}\right)^{2(b-c-s+1)}\omega^{b}\Gamma\left[\begin{matrix} a-b, c, \frac{2s-2b+2c+\nu-2}{2} \\ a, c-b, \frac{2b-2c+\nu-2s+4}{2}\end{matrix}\right]$ $\times {}_2F_3\left(\begin{matrix} 1-a, b-c+1; -\frac{\sigma^2\omega}{4} \\ b-a+1, \frac{2b-2c-\nu-2s+4}{2}, \frac{2b-2c+\nu-2s+4}{2}\end{matrix}\right)$ $+ \left(\frac{\sigma}{2}\right)^{\nu}\omega^{s+c+\nu/2-1}\Gamma\left[\begin{matrix} c, \frac{2a-2c-\nu-2s+2}{2}, \frac{2b-2c-\nu-2s+2}{2} \\ \nu+1, \frac{2-\nu-2s}{2}, \frac{2a+2b-2c-\nu-2s+2}{2}\end{matrix}\right]$ $\times {}_2F_3\left(\begin{matrix} \frac{2s+\nu}{2}, \frac{2s-2a-2b+2c+\nu}{2}; -\frac{\sigma^2\omega}{4} \\ \nu+1, \frac{2s-2a+2c+\nu}{2}, \frac{2s-2b+2c+\nu}{2}\end{matrix}\right)$ $[\sigma, \omega > 0; \operatorname{Re}(s-a+c), \operatorname{Re}(s-b+c) < 7/4]$
6	$(\omega-x)_+^{c-1} I_\nu(\sigma\sqrt{x})$ $\times {}_2F_1\left(\begin{matrix} a, b \\ c; \frac{\omega-x}{\omega}\end{matrix}\right)$	$\left(\frac{\sigma}{2}\right)^{\nu}\omega^{s+c+\nu/2-1}\Gamma\left[\begin{matrix} c, \frac{2s+\nu}{2}, \frac{2s-2a-2b+2c+\nu}{2} \\ \nu+1, \frac{2s-2a+2c+\nu}{2}, \frac{2s-2b+2c+\nu}{2}\end{matrix}\right]$ $\times {}_2F_3\left(\begin{matrix} \frac{2s+\nu}{2}, \frac{2s-2a-2b+2c+\nu}{2}; \frac{\sigma^2\omega}{4} \\ \nu+1, \frac{2s-2a+2c+\nu}{2}, \frac{2s-2b+2c+\nu}{2}\end{matrix}\right)$ $[\omega, \operatorname{Re} c, \operatorname{Re}(2s+\nu), \operatorname{Re}(s-a-b+c+\nu/2) > 0]$

No.	$f(x)$	$F(s)$
7	$K_\nu(\sigma\sqrt{x}) \times {}_2F_1\left(\begin{matrix} a, b \\ c; -\omega x \end{matrix}\right)$	$\frac{(\sigma/2)^{2a-2s}}{2\omega^a}\Gamma\left[\begin{matrix} b-a, c, \frac{2s-2a+\nu}{2}, \frac{2s-2a-\nu}{2} \\ b, c-a \end{matrix}\right] \times {}_2F_3\left(\begin{matrix} a, a-c+1; -\frac{\sigma^2}{4\omega} \\ a-b+1, \frac{2a-2s+\nu+2}{2}, \frac{2a-2s-\nu+2}{2} \end{matrix}\right) + \frac{(\sigma/2)^{2b-2s}}{2\omega^b}\Gamma\left[\begin{matrix} a-b, c, \frac{2s-2b+\nu}{2}, \frac{2s-2b-\nu}{2} \\ a, c-b \end{matrix}\right] \times {}_2F_3\left(\begin{matrix} b, b-c+1; -\frac{\sigma^2}{4\omega} \\ b-a+1, \frac{2b-2s+\nu+2}{2}, \frac{2b-2s-\nu+2}{2} \end{matrix}\right) + \frac{(\sigma/2)^\nu}{2\omega^{s+\nu/2}}\Gamma\left[\begin{matrix} c, -\nu, \frac{2s+\nu}{2}, \frac{2a-2s-\nu}{2}, \frac{2b-2s-\nu}{2} \\ a, b, \frac{2c-2s-\nu}{2} \end{matrix}\right] \times {}_2F_3\left(\begin{matrix} \frac{2s+\nu}{2}, \frac{2s-2c+\nu+2}{2}; -\frac{\sigma^2}{4\omega} \\ \nu+1, \frac{2s-2a+\nu+2}{2}, \frac{2s-2b+\nu+2}{2} \end{matrix}\right) + \frac{(2/\sigma)^\nu}{2\omega^{s-\nu/2}}\Gamma\left[\begin{matrix} c, \nu, \frac{2s-\nu}{2}, \frac{2a-2s+\nu}{2}, \frac{2b-2s+\nu}{2} \\ a, b, \frac{2c-2s+\nu}{2} \end{matrix}\right] \times {}_2F_3\left(\begin{matrix} \frac{2s-\nu}{2}, \frac{2s-2c-\nu+2}{2}; -\frac{\sigma^2}{4\omega} \\ 1-\nu, \frac{2s-2a-\nu+2}{2}, \frac{2s-2b-\nu+2}{2} \end{matrix}\right)$ $[\operatorname{Re}\sigma > 0;\ \operatorname{Re}s > \lvert\operatorname{Re}\nu\rvert/2;\ \lvert\arg\omega\rvert < \pi]$
8	$K_\nu\left(\frac{\sigma}{\sqrt{x}}\right) \times {}_2F_1\left(\begin{matrix} a, b \\ c; -\omega x \end{matrix}\right)$	$\frac{(\sigma/2)^\nu}{2\omega^{s-\nu/2}}\Gamma\left[\begin{matrix} c, -\nu, \frac{2s-\nu}{2}, \frac{2a-2s+\nu}{2}, \frac{2b-2s+\nu}{2} \\ a, b, \frac{2c-2s+\nu}{2} \end{matrix}\right] \times {}_2F_3\left(\begin{matrix} \frac{2a-2s+\nu}{2}, \frac{2b-2s+\nu}{2}; -\frac{\sigma^2\omega}{4} \\ 1+\nu, \frac{2-2s+\nu}{2}, \frac{2c-2s+\nu}{2} \end{matrix}\right) + \frac{(\sigma/2)^{-\nu}}{2\omega^{s+\nu/2}}\Gamma\left[\begin{matrix} c, \nu, \frac{2s+\nu}{2}, \frac{2a-2s-\nu}{2}, \frac{2b-2s-\nu}{2} \\ a, b, \frac{2c-2s-\nu}{2} \end{matrix}\right] \times {}_2F_3\left(\begin{matrix} \frac{2a-2s-\nu}{2}, \frac{2b-2s-\nu}{2}; -\frac{\sigma^2\omega}{4} \\ 1-\nu, \frac{2-2s-\nu}{2}, \frac{2c-2s-\nu}{2} \end{matrix}\right) + \frac{(\sigma/2)^{2s}}{2}\Gamma\left(\frac{\nu-2s}{2}\right) \times \Gamma\left(-\frac{2s+\nu}{2}\right) {}_2F_3\left(\begin{matrix} a, b; -\frac{\sigma^2\omega}{4} \\ c, \frac{2s-\nu+2}{2}, \frac{2s+\nu+2}{2} \end{matrix}\right)$ $[\operatorname{Re}\sigma > 0;\ \operatorname{Re}s < \operatorname{Re}a - \lvert\operatorname{Re}\nu\rvert/2,\ \operatorname{Re}b - \lvert\operatorname{Re}\nu\rvert/2;\ \lvert\arg\omega\rvert < \pi]$
9	$(\omega - x)_+^{c-1} K_\nu(\sigma\sqrt{x}) \times {}_2F_1\left(\begin{matrix} a, b \\ c; \frac{\omega-x}{\omega} \end{matrix}\right)$	$\frac{2^{\nu-1}\omega^{s+c-\nu/2-1}}{\sigma^\nu}\Gamma\left[\begin{matrix} c, \nu, \frac{2s-\nu}{2}, \frac{2s-2a-2b+2c-\nu}{2} \\ \frac{2s-2a+2c-\nu}{2}, \frac{2s-2b+2c-\nu}{2} \end{matrix}\right] \times {}_2F_3\left(\begin{matrix} \frac{2s-\nu}{2}, \frac{2s-2a-2b+2c-\nu}{2}; \frac{\sigma^2\omega}{4} \\ 1-\nu, \frac{2s-2a+2c-\nu}{2}, \frac{2s-2b+2c-\nu}{2} \end{matrix}\right) + \frac{\sigma^\nu\omega^{s+c+\nu/2-1}}{2^{\nu+1}}\Gamma\left[\begin{matrix} c, -\nu, \frac{2s+\nu}{2}, \frac{2s-2a-2b+2c+\nu}{2} \\ \frac{2s-2a+2c+\nu}{2}, \frac{2s-2b+2c+\nu}{2} \end{matrix}\right] \times {}_2F_3\left(\begin{matrix} \frac{2s+\nu}{2}, \frac{2s-2a-2b+2c+\nu}{2}; \frac{\sigma^2\omega}{4} \\ 1+\nu, \frac{2s-2a+2c+\nu}{2}, \frac{2s-2b+2c+\nu}{2} \end{matrix}\right)$ $[\omega,\ \operatorname{Re}c > 0;\ \operatorname{Re}s,\ \operatorname{Re}(s-a-b+c) > \lvert\operatorname{Re}\nu\rvert/2]$

3.31.32. ${}_2F_1^2\,(a,\,b;\,c;\,\varphi\,(x))$

1	${}_2F_1^2\left(\begin{matrix}a,\,b;\,-\frac{x}{\omega}\\ \frac{2a+2b+1}{2}\end{matrix}\right)$	$\frac{2^{2a+2b-1}\omega^s}{\sqrt{\pi}}\,\Gamma\left[\begin{matrix}\frac{2a+2b+1}{2},\,\frac{2a+2b+1}{2}\\ 2a,\,2b\end{matrix}\right]\Gamma\left[\begin{matrix}s,\,2a-s,\,2b-s,\,a+b-s\\ 2a+2b-s,\,\frac{2a+2b-2s+1}{2}\end{matrix}\right]$ $[0<\operatorname{Re}s<2\operatorname{Re}a,\,2\operatorname{Re}b;\ \lvert\arg\omega\rvert<\pi]$
2	$(x+\omega)\,{}_2F_1^2\left(\begin{matrix}a,\,b;\,-\frac{x}{\omega}\\ \frac{2a+2b-1}{2}\end{matrix}\right)$	$\frac{2^{2a+2b-3}\omega^{s+1}}{\sqrt{\pi}}\,\Gamma\left[\begin{matrix}\frac{2a+2b-1}{2},\,\frac{2a+2b-1}{2}\\ 2a-1,\,2b-1\end{matrix}\right]$ $\times\,\Gamma\left[\begin{matrix}s,\,2a-s-1,\,2b-s-1,\,a+b-s-1\\ 2a+2b-s-2,\,\frac{2a+2b-2s-1}{2}\end{matrix}\right]$ $[0<\operatorname{Re}s<2\operatorname{Re}a-1,\,2\operatorname{Re}b-1;\ \lvert\arg\omega\rvert<\pi]$
3	${}_2F_1^2\left(\begin{matrix}a,\,b;\,-\frac{\omega}{x}\\ \frac{2a+2b+1}{2}\end{matrix}\right)$	$\frac{2^{2a+2b-1}\omega^s}{\sqrt{\pi}}\,\Gamma\left[\begin{matrix}\frac{2a+2b+1}{2},\,\frac{2a+2b+1}{2}\\ 2a,\,2b\end{matrix}\right]\Gamma\left[\begin{matrix}-s,\,s+2a,\,s+2b,\,s+a+b\\ s+2a+2b,\,\frac{2s+2a+2b+1}{2}\end{matrix}\right]$ $[-2\operatorname{Re}a,\,-2\operatorname{Re}b<\operatorname{Re}s<0;\ \lvert\arg\omega\rvert<\pi]$
4	$(x+\omega)\,{}_2F_1^2\left(\begin{matrix}a,\,b;\,-\frac{\omega}{x}\\ \frac{2a+2b-1}{2}\end{matrix}\right)$	$\frac{2^{2a+2b-3}\omega^{s+1}}{\sqrt{\pi}}\,\Gamma\left[\begin{matrix}\frac{2a+2b-1}{2},\,\frac{2a+2b-1}{2}\\ 2a-1,\,2b-1\end{matrix}\right]$ $\times\,\Gamma\left[\begin{matrix}-s-1,\,s+2a,\,s+2b,\,s+a+b\\ s+2a+2b-1,\,\frac{2s+2a+2b+1}{2}\end{matrix}\right]$ $[-2\operatorname{Re}a,\,-2\operatorname{Re}b<\operatorname{Re}s<-1;\ -\pi<\arg\omega\le\pi]$
5	$(x+\omega)^{-2a}$ $\times\,{}_2F_1^2\left(\begin{matrix}a,\,\frac{2a+1}{2}\\ c;\,\frac{\omega}{x+\omega}\end{matrix}\right)$	$\frac{4^{c-1}\omega^{s-2a}}{\sqrt{\pi}}\,\Gamma\left[\begin{matrix}c,\,c\\ 2a,\,2c-2a-1\end{matrix}\right]$ $\times\,\Gamma\left[\begin{matrix}s,\,2a-s,\,s-4a+2c-1,\,\frac{2s-4a+2c-1}{2}\\ s-2a+c,\,s-2a+2c-1\end{matrix}\right]$ $[0,\ \operatorname{Re}(2a-c)+1/2<\operatorname{Re}s<2\operatorname{Re}a;\ \lvert\arg\omega\rvert<\pi]$
6	$(x+\omega)^{-2a}$ $\times\,{}_2F_1^2\left(\begin{matrix}a,\,\frac{2a+1}{2}\\ c;\,\frac{x}{x+\omega}\end{matrix}\right)$	$\frac{4^{c-1}\omega^{s-2a}}{\sqrt{\pi}}\,\Gamma\left[\begin{matrix}c,\,c\\ 2a,\,2c-2a-1\end{matrix}\right]$ $\times\,\Gamma\left[\begin{matrix}s,\,2a-s,\,\frac{2c-2s-1}{2},\,2c-2a-s-1\\ c-s,\,2c-s-1\end{matrix}\right]$ $[0<\operatorname{Re}s<2\operatorname{Re}a,\,2\operatorname{Re}(c-a)-1;\ \lvert\arg\omega\rvert<\pi]$
7	${}_2F_1^2\left(\begin{matrix}a,\,b;\,\frac{\sqrt{\omega}-\sqrt{x+\omega}}{2\sqrt{\omega}}\\ \frac{a+b+1}{2}\end{matrix}\right)$	$\frac{2^{a+b-1}\omega^s}{\sqrt{\pi}}\,\Gamma\left[\begin{matrix}\frac{a+b+1}{2},\,\frac{a+b+1}{2}\\ a,\,b\end{matrix}\right]\Gamma\left[\begin{matrix}s,\,a-s,\,b-s,\,\frac{a+b-2s}{2}\\ a+b-s,\,\frac{a+b-2s+1}{2}\end{matrix}\right]$ $[0<\operatorname{Re}s<\operatorname{Re}a,\ \operatorname{Re}b;\ \lvert\arg\omega\rvert<\pi]$

No.	$f(x)$	$F(s)$
8	$\left(\sqrt{x+\omega}+\sqrt{\omega}\right)^{2-2c} \times {}_2F_1^2\left(\begin{matrix} a,\, 1-a \\ c;\, \frac{\sqrt{\omega}-\sqrt{x+\omega}}{2\sqrt{\omega}} \end{matrix}\right)$	$\frac{\omega^{s-c+1}}{\sqrt{\pi}}\Gamma\left[\begin{matrix} c,\, c \\ c-a,\, a+c-1 \end{matrix}\right] \times\Gamma\left[\begin{matrix} s,\, c-a-s,\, a+c-s-1,\, \frac{2c-2s-1}{2} \\ c-s,\, 2c-s-1 \end{matrix}\right]$ $[0<\operatorname{Re} s<\operatorname{Re}(c-a),\ \operatorname{Re}(a+c)-1;\ -\pi<\arg\omega\le\pi]$
9	$\left(\sqrt{x+\omega}-\sqrt{\omega}\right)^{2c-2} \times {}_2F_1^2\left(\begin{matrix} a,\, 1-a \\ c;\, \frac{\sqrt{\omega}-\sqrt{x+\omega}}{2\sqrt{\omega}} \end{matrix}\right)$	$\frac{\omega^{s+c-1}}{\sqrt{\pi}}\Gamma\left[\begin{matrix} c,\, c \\ c-a,\, a+c-1 \end{matrix}\right] \times\Gamma\left[\begin{matrix} \frac{3-2s-2c}{2},\, 2-s-a-c,\, 1-s+a-c,\, s+2c-2 \\ 1-s,\, 2-s-c \end{matrix}\right]$ $\left[\begin{matrix} 2-2\operatorname{Re} c<\operatorname{Re} s<2-\operatorname{Re}(a+c),\ 1+\operatorname{Re}(a-c); \\ -\pi<\arg\omega\le\pi \end{matrix}\right]$
10	${}_2F_1^2\left(\begin{matrix} a,\, b;\, \frac{\sqrt{x}-\sqrt{x+\omega}}{2\sqrt{x}} \\ \frac{a+b+1}{2} \end{matrix}\right)$	$\frac{2^{a+b-1}\omega^s}{\sqrt{\pi}}\Gamma\left[\begin{matrix} \frac{a+b+1}{2},\, \frac{a+b+1}{2} \\ a,\, b \end{matrix}\right]\Gamma\left[\begin{matrix} -s,\, s+a,\, s+b,\, \frac{2s+a+b}{2} \\ s+a+b,\, \frac{2s+a+b+1}{2} \end{matrix}\right]$ $[-\operatorname{Re} a,\ -\operatorname{Re} b<\operatorname{Re} s<0;\ \lvert\arg\omega\rvert<\pi]$
11	$\left(\sqrt{x+\omega}+\sqrt{x}\right)^{2-2c} \times {}_2F_1^2\left(\begin{matrix} a,\, 1-a \\ c;\, \frac{\sqrt{x}-\sqrt{x+\omega}}{2\sqrt{x}} \end{matrix}\right)$	$\frac{\omega^{s-c+1}}{\sqrt{\pi}}\Gamma\left[\begin{matrix} c,\, c \\ c-a,\, a+c-1 \end{matrix}\right]\Gamma\left[\begin{matrix} s+a,\, s-a+1,\, c-s-1,\, \frac{2s+1}{2} \\ s+1,\, s+c \end{matrix}\right]$ $[-1/2,\ -\operatorname{Re} a<\operatorname{Re} s<\operatorname{Re} c-1;\ -\pi<\arg\omega\le\pi]$
12	$\left(\sqrt{x+\omega}-\sqrt{x}\right)^{2c-2} \times {}_2F_1^2\left(\begin{matrix} a,\, 1-a \\ c;\, \frac{\sqrt{x}-\sqrt{x+\omega}}{2\sqrt{x}} \end{matrix}\right)$	$\frac{\omega^{s+c-1}}{\sqrt{\pi}}\Gamma\left[\begin{matrix} c,\, c \\ c-a,\, a+c-1 \end{matrix}\right]\Gamma\left[\begin{matrix} s+a,\, s-a+1,\, c-s-1,\, \frac{2s+1}{2} \\ s+1,\, s+c \end{matrix}\right]$ $\left[\begin{matrix} -1/2,\ -\operatorname{Re} a,\ \operatorname{Re} a-1<\operatorname{Re} s<\operatorname{Re} c-1; \\ -\pi<\arg\omega\le\pi \end{matrix}\right]$
13	$\left(\sqrt{x+\omega}+\sqrt{\omega}\right)^{-2a} \times {}_2F_1^2\left(\begin{matrix} a,\, b;\, \frac{\sqrt{x+\omega}-\sqrt{\omega}}{\sqrt{x+\omega}+\sqrt{\omega}} \\ a-b+1 \end{matrix}\right)$	$\frac{\omega^{s-a}}{4^b\sqrt{\pi}}\Gamma\left[\begin{matrix} a-b+1,\, a-b+1 \\ a,\, a-2b+1 \end{matrix}\right] \times\Gamma\left[\begin{matrix} s,\, a-s,\, a-2b-s+1,\, \frac{2a-2b-2s+1}{2} \\ a-b-s+1,\, 2a-2b-s+1 \end{matrix}\right]$ $[0<\operatorname{Re} s<\operatorname{Re} a,\ \operatorname{Re}(a-2b)+1;\ \lvert\arg\omega\rvert<\pi]$
14	$\left(\sqrt{x+\omega}-\sqrt{\omega}\right)^{2a} \times {}_2F_1^2\left(\begin{matrix} a,\, b;\, \frac{\sqrt{x+\omega}-\sqrt{\omega}}{\sqrt{x+\omega}+\sqrt{\omega}} \\ a-b+1 \end{matrix}\right)$	$\frac{\omega^{s+a}}{4^b\sqrt{\pi}}\Gamma\left[\begin{matrix} a-b+1,\, a-b+1 \\ a,\, a-2b+1 \end{matrix}\right] \times\Gamma\left[\begin{matrix} -s-a,\, s+2a,\, 1-s-a-2b,\, \frac{1-2s-2a-2b}{2} \\ 1-s-2b,\, 1-s-a-b \end{matrix}\right]$ $[-2\operatorname{Re} a<\operatorname{Re} s<-\operatorname{Re} a,\ 1-\operatorname{Re}(a+2b);\ \lvert\arg\omega\rvert<\pi]$

No.	$f(x)$	$F(s)$
15	$\left(\sqrt{x+\omega}+\sqrt{x}\right)^{-2a}$ $\times\, {}_2F_1^2\left(\begin{matrix} a,\, b;\; \frac{\sqrt{x+\omega}-\sqrt{x}}{\sqrt{x+\omega}+\sqrt{x}} \\ a-b+1 \end{matrix}\right)$	$\frac{\omega^{s-a}}{4^b\sqrt{\pi}}\,\Gamma\left[\begin{matrix} a-b+1,\, a-b+1 \\ a,\, a-2b+1 \end{matrix}\right]\Gamma\left[\begin{matrix} s,\, a-s,\, s-2b+1,\, \frac{2s-2b+1}{2} \\ s-b+1,\, s+a-2b+1 \end{matrix}\right]$ $[0,\, 2\operatorname{Re}b-1<\operatorname{Re}s<\operatorname{Re}a;\; \lvert\arg\omega\rvert<\pi]$
16	$\left(\sqrt{x+\omega}-\sqrt{x}\right)^{2a}$ $\times\, {}_2F_1^2\left(\begin{matrix} a,\, b;\; \frac{\sqrt{x+\omega}-\sqrt{x}}{\sqrt{x+\omega}+\sqrt{x}} \\ a-b+1 \end{matrix}\right)$	$\frac{\omega^{s+a}}{4^b\sqrt{\pi}}\,\Gamma\left[\begin{matrix} a-b+1,\, a-b+1 \\ a,\, a-2b+1 \end{matrix}\right]\Gamma\left[\begin{matrix} s,\, a-s,\, s-2b+1,\, \frac{2s-2b+1}{2} \\ s-b+1,\, s+a-2b+1 \end{matrix}\right]$ $[0,\, 2\operatorname{Re}b-1<\operatorname{Re}s<\operatorname{Re}a;\; \lvert\arg\omega\rvert<\pi]$
17	${}_2F_1^2\left(\begin{matrix} a,\, b;\; \frac{x-\sqrt{x^2+\omega^2}}{2x} \\ \frac{a+b+1}{2} \end{matrix}\right)$	$\frac{2^{a+b-2}\omega^s}{\sqrt{\pi}}\,\Gamma\left[\begin{matrix} \frac{a+b+1}{2},\, \frac{a+b+1}{2} \\ a,\, b \end{matrix}\right]\Gamma\left[\begin{matrix} -\frac{s}{2},\, \frac{s+2a}{2},\, \frac{s+2b}{2},\, \frac{s+a+b}{2} \\ \frac{s+a+b+1}{2},\, \frac{s+2a+2b}{2} \end{matrix}\right]$ $[-2\operatorname{Re}a,\, -2\operatorname{Re}b<\operatorname{Re}s<0;\; -\pi/2<\arg\omega\le\pi/2]$
18	$\left(\sqrt{x^2+\omega^2}+\omega\right)^{2-2c}$ $\times\, {}_2F_1^2\left(\begin{matrix} a,\, 1-a \\ c;\; \frac{\omega-\sqrt{x^2+\omega^2}}{2\omega} \end{matrix}\right)$	$\frac{\omega^{s-2c+2}}{2\sqrt{\pi}}\,\Gamma\left[\begin{matrix} c,\, c \\ c-a,\, a+c-1 \end{matrix}\right]\Gamma\left[\begin{matrix} \frac{s}{2},\, \frac{2c-s-1}{2},\, \frac{2c-s-2a}{2},\, \frac{2a+2c-s-2}{2} \\ \frac{2c-s}{2},\, \frac{4c-s-2}{2} \end{matrix}\right]$ $[0<\operatorname{Re}s<-2\operatorname{Re}(a-c),\, 2\operatorname{Re}c-1;\; -\pi/2<\arg\omega\le\pi/2]$
19	$\left(\sqrt{x^2+\omega^2}-\omega\right)^{2c-2}$ $\times\, {}_2F_1^2\left(\begin{matrix} a,\, 1-a \\ c;\; \frac{\omega-\sqrt{x^2+\omega^2}}{2\omega} \end{matrix}\right)$	$\frac{\omega^{s+2c-2}}{2\sqrt{\pi}}\,\Gamma\left[\begin{matrix} c,\, c \\ c-a,\, a+c-1 \end{matrix}\right]$ $\times\,\Gamma\left[\begin{matrix} \frac{3-s-2c}{2},\, \frac{4-s-2a-2c}{2},\, \frac{2-s+2a-2c}{2},\, \frac{s+4c-4}{2} \\ \frac{2-s}{2},\, \frac{4-s-2c}{2} \end{matrix}\right]$ $\left[\begin{matrix} 4-4\operatorname{Re}c<\operatorname{Re}s<2+2\operatorname{Re}(a-c),\, 3-2\operatorname{Re}c; \\ -\pi/2<\arg\omega\le\pi/2 \end{matrix}\right]$
20	$\left(\sqrt{x^2+\omega^2}+x\right)^{2-2c}$ $\times\, {}_2F_1^2\left(\begin{matrix} a,\, 1-a \\ c;\; \frac{x-\sqrt{x^2+\omega^2}}{2x} \end{matrix}\right)$	$\frac{\omega^{s-2c+2}}{\sqrt{\pi}\, s}\,\Gamma\left[\begin{matrix} c,\, c \\ c-a,\, a+c-1 \end{matrix}\right]\Gamma\left[\begin{matrix} \frac{2c-s-2}{2},\, \frac{s+1}{2},\, \frac{s-2a+2}{2},\, \frac{s+2a}{2} \\ \frac{s}{2},\, \frac{s+2c}{2} \end{matrix}\right]$ $[-1,\, -2\operatorname{Re}a<\operatorname{Re}s<2\operatorname{Re}c-2;\; -\pi/2<\arg\omega\le\pi/2]$
21	$\left(\sqrt{x^2+\omega^2}-x\right)^{2c-2}$ $\times\, {}_2F_1^2\left(\begin{matrix} a,\, 1-a \\ c;\; \frac{x-\sqrt{x^2+\omega^2}}{2x} \end{matrix}\right)$	$\frac{\omega^{s+2c-2}}{2\sqrt{\pi}}\,\Gamma\left[\begin{matrix} c,\, c \\ c-a,\, a+c-1 \end{matrix}\right]\Gamma\left[\begin{matrix} \frac{2c-s-2}{2},\, \frac{s+1}{2},\, \frac{s-2a+2}{2},\, \frac{s+2a}{2} \\ \frac{s+2}{2},\, \frac{s+2c}{2} \end{matrix}\right]$ $[-1,\, -2\operatorname{Re}a<\operatorname{Re}s<2\operatorname{Re}c-2;\; -\pi/2<\arg\omega\le\pi/2]$
22	$\left(\sqrt{x^2+\omega^2}+\omega\right)^{-2a}$ $\times\, {}_2F_1^2\left(\begin{matrix} a,\, b;\; \frac{\sqrt{x^2+\omega^2}-\omega}{\sqrt{x^2+\omega^2}+\omega} \\ a-b+1 \end{matrix}\right)$	$\frac{\omega^{s-2a}}{2^{2b+1}\sqrt{\pi}}\,\Gamma\left[\begin{matrix} a-b+1,\, a-b+1 \\ a,\, a-2b+1 \end{matrix}\right]$ $\times\,\Gamma\left[\begin{matrix} \frac{s}{2},\, \frac{2a-s}{2},\, \frac{1-s+2a-2b}{2},\, \frac{2-s+2a-4b}{2} \\ \frac{2-s+4a-4b}{2},\, \frac{2-s+2a-2b}{2} \end{matrix}\right]$ $[0<\operatorname{Re}s<2\operatorname{Re}a,\, 2\operatorname{Re}(a-b)+1;\; -\pi/2<\arg\omega\le\pi/2]$

No.	$f(x)$	$F(s)$
23	$\left(\sqrt{x^2+\omega^2}-\omega\right)^{2a} \times {}_2F_1^2\left(\begin{matrix} a,\, b;\, \frac{\sqrt{x^2+\omega^2}-\omega}{\sqrt{x^2+\omega^2}+\omega} \\ a-b+1 \end{matrix}\right)$	$\frac{\omega^{s+2a}}{2^{2b+1}\sqrt{\pi}}\Gamma\left[\begin{matrix} a-b+1,\, a-b+1 \\ a,\, a-2b+1 \end{matrix}\right] \times\Gamma\left[\begin{matrix} \frac{s+4a}{2},\, \frac{-2a-s}{2},\, \frac{1-s-2a-2b}{2},\, \frac{2-s-2a-4b}{2} \\ \frac{2-s-4b}{2},\, \frac{2-s-2a-2b}{2} \end{matrix}\right]$ $\left[-4\operatorname{Re}a<\operatorname{Re}s<-2\operatorname{Re}a,\, -2\operatorname{Re}(a+b)+1;\ \lvert\arg\omega\rvert<\pi/2\right]$
24	$\left(\sqrt{x^2+\omega^2}+x\right)^{-2a} \times {}_2F_1^2\left(\begin{matrix} a,\, b;\, \frac{\sqrt{x^2+\omega^2}-x}{\sqrt{x^2+\omega^2}+x} \\ a-b+1 \end{matrix}\right)$	$\frac{\omega^{s-2a}}{2^{2b+1}\sqrt{\pi}}\Gamma\left[\begin{matrix} a-b+1,\, a-b+1 \\ a,\, a-2b+1 \end{matrix}\right]\Gamma\left[\begin{matrix} \frac{s}{2},\, \frac{2a-s}{2},\, \frac{s-2b+1}{2},\, \frac{s-4b+2}{2} \\ \frac{s-2b+2}{2},\, \frac{s+2a-4b+2}{2} \end{matrix}\right]$ $[0,\, 2\operatorname{Re}b-1<\operatorname{Re}s<2\operatorname{Re}a;\ \lvert\arg\omega\rvert<\pi/2]$
25	$\left(\sqrt{x^2+\omega^2}-x\right)^{2a} \times {}_2F_1^2\left(\begin{matrix} a,\, b;\, \frac{\sqrt{x^2+\omega^2}-x}{\sqrt{x^2+\omega^2}+x} \\ a-b+1 \end{matrix}\right)$	$\frac{\omega^{s+2a}}{2^{2b+1}\sqrt{\pi}}\Gamma\left[\begin{matrix} a-b+1,\, a-b+1 \\ a,\, a-2b+1 \end{matrix}\right]\Gamma\left[\begin{matrix} \frac{s}{2},\, \frac{2a-s}{2},\, \frac{s-2b+1}{2},\, \frac{s-4b+2}{2} \\ \frac{s-2b+2}{2},\, \frac{s+2a-4b+2}{2} \end{matrix}\right]$ $[0,\, 2\operatorname{Re}b-1<\operatorname{Re}s<2\operatorname{Re}a;\ \lvert\arg\omega\rvert<\pi/2]$

3.31.33. ${}_2F_1\left(a_1, b_1; c_1; -\frac{x}{\omega}\right) {}_2F_1\left(a_2, b_2; c_2; -\frac{x}{\omega}\right)$ and algebraic functions

1	${}_2F_1\left(\begin{matrix} a,\, b;\, -\frac{x}{\omega} \\ \frac{2a+2b-1}{2} \end{matrix}\right) \times {}_2F_1\left(\begin{matrix} a,\, b;\, -\frac{x}{\omega} \\ \frac{2a+2b+1}{2} \end{matrix}\right)$	$\frac{2^{2a+2b-2}\omega^s}{\sqrt{\pi}}\Gamma\left[\begin{matrix} \frac{2a+2b-1}{2},\, \frac{2a+2b+1}{2} \\ 2a,\, 2b \end{matrix}\right] \times\Gamma\left[\begin{matrix} s,\, 2a-s,\, 2b-s,\, a+b-s \\ 2a+2b-s-1,\, \frac{2a+2b-2s+1}{2} \end{matrix}\right]$ $[0<\operatorname{Re}s<2\operatorname{Re}a,\, 2\operatorname{Re}b;\ \lvert\arg\omega\rvert<\pi]$
2	${}_2F_1\left(\begin{matrix} a,\, b;\, -\frac{x}{\omega} \\ \frac{2a+2b+1}{2} \end{matrix}\right) \times {}_2F_1\left(\begin{matrix} a,\, b+1;\, -\frac{x}{\omega} \\ \frac{2a+2b+1}{2} \end{matrix}\right)$	$\frac{2^{2a+2b-1}\omega^s}{\sqrt{\pi}}\Gamma\left[\begin{matrix} \frac{2a+2b+1}{2},\, \frac{2a+2b+1}{2} \\ 2a,\, 2b+1 \end{matrix}\right] \times\Gamma\left[\begin{matrix} s,\, 2a-s,\, 2b-s+1,\, a+b-s \\ 2a+2b-s,\, \frac{2a+2b-2s+1}{2} \end{matrix}\right]$ $\left[0<\operatorname{Re}s<2\operatorname{Re}a,\, 2\operatorname{Re}b+1,\ \operatorname{Re}(a+b);\ \lvert\arg\omega\rvert<\pi\right]$
3	${}_2F_1\left(\begin{matrix} a,\, b \\ \frac{2a+2b+1}{2};\, -\frac{x}{\omega} \end{matrix}\right) \times {}_2F_1\left(\begin{matrix} a+1,\, b \\ \frac{2a+2b+1}{2};\, -\frac{x}{\omega} \end{matrix}\right)$	$\frac{2^{2a+2b-1}\omega^s}{\sqrt{\pi}}\Gamma\left[\begin{matrix} \frac{2a+2b+1}{2},\, \frac{2a+2b+1}{2} \\ 2a+1,\, 2b \end{matrix}\right] \times\Gamma\left[\begin{matrix} s,\, 2a-s+1,\, 2b-s,\, a+b-s \\ 2a+2b-s,\, \frac{2a+2b-2s+1}{2} \end{matrix}\right]$ $\left[0<\operatorname{Re}s<2\operatorname{Re}a+1,\, 2\operatorname{Re}b,\ \operatorname{Re}(a+b);\ \lvert\arg\omega\rvert<\pi\right]$

No.	$f(x)$	$F(s)$
4	${}_2F_1\left(\begin{matrix} a,\, b;\, -\frac{x}{\omega} \\ \frac{2a+2b+1}{2} \end{matrix}\right) \times {}_2F_1\left(\begin{matrix} a+1,\, b+1;\, -\frac{x}{\omega} \\ \frac{2a+2b+3}{2} \end{matrix}\right)$	$\frac{2^{2a+2b}\omega^s}{\sqrt{\pi}}\Gamma\left[\begin{matrix} \frac{2a+2b+1}{2},\, \frac{2a+2b+3}{2} \\ 2a+1,\, 2b+1 \end{matrix}\right] \times \Gamma\left[\begin{matrix} s,\, 2a-s+1,\, 2b-s+1,\, a+b-s+1 \\ 2a+2b-s+1,\, \frac{2a+2b-2s+3}{2} \end{matrix}\right]$ $[0<\operatorname{Re} s<2\operatorname{Re} a+1,\, 2\operatorname{Re} b+1;\, \lvert\arg\omega\rvert<\pi]$
5	${}_2F_1\left(\begin{matrix} a,\, b \\ \frac{2a+2b+1}{2};\, -\frac{x}{\omega} \end{matrix}\right) \times {}_2F_1\left(\begin{matrix} \frac{1-2a}{2},\, \frac{1-2b}{2} \\ \frac{3-2a-2b}{2};\, -\frac{x}{\omega} \end{matrix}\right)$	$\frac{(1-2a-2b)\,\omega^s}{2\sqrt{\pi}}\,\frac{\cos[(a-b)\pi]}{\cos[(a+b)\pi]} \times \Gamma\left[\begin{matrix} s,\, \frac{1-2s}{2},\, \frac{2a-2b-2s+1}{2},\, \frac{2b-2a-2s+1}{2} \\ \frac{2a+2b-2s+1}{2},\, \frac{3-2a-2b-2s}{2} \end{matrix}\right]$ $[0<\operatorname{Re} s<1/2-\lvert\operatorname{Re}(a-b)\rvert;\, \lvert\arg\omega\rvert<\pi]$
6	$(x+\omega)\ {}_2F_1\left(\begin{matrix} \frac{3-2a}{2},\, \frac{3-2b}{2} \\ \frac{5-2a-2b}{2};\, -\frac{x}{\omega} \end{matrix}\right) \times {}_2F_1\left(\begin{matrix} a,\, b;\, -\frac{x}{\omega} \\ \frac{2a+2b-1}{2} \end{matrix}\right)$	$\frac{(2a+2b-3)\,\omega^{s+1}}{2\sqrt{\pi}}\,\frac{\cos[(a-b)\pi]}{\cos[(a+b)\pi]} \times \Gamma\left[\begin{matrix} s,\, \frac{1-2s}{2},\, \frac{2a-2b-2s+1}{2},\, \frac{2b-2a-2s+1}{2} \\ \frac{5-2a-2b-2s}{2},\, \frac{2a+2b-2s-1}{2} \end{matrix}\right]$ $[0<\operatorname{Re} s<1/2-\lvert\operatorname{Re}(a-b)\rvert;\, \lvert\arg\omega\rvert<\pi]$
7	$\sqrt{x+\omega}\,{}_2F_1\left(\begin{matrix} a,\, b;\, -\frac{x}{\omega} \\ \frac{2a+2b-1}{2} \end{matrix}\right) \times {}_2F_1\left(\begin{matrix} \frac{2a-1}{2},\, \frac{2b-1}{2} \\ \frac{2a+2b-1}{2};\, -\frac{x}{\omega} \end{matrix}\right)$	$\frac{2^{2a+2b-3}\omega^{s+1/2}}{\sqrt{\pi}}\Gamma\left[\begin{matrix} \frac{2a+2b-1}{2},\, \frac{2a+2b-1}{2} \\ 2a-1,\, 2b-1 \end{matrix}\right] \times \Gamma\left[\begin{matrix} s,\, 2a-s-1,\, 2b-s-1,\, a+b-s-1 \\ 2a+2b-s-2,\, \frac{2a+2b-2s-1}{2} \end{matrix}\right]$ $[0<\operatorname{Re} s<2\operatorname{Re} a-1,\, 2\operatorname{Re} b-1;\, \lvert\arg\omega\rvert<\pi]$
8	$\sqrt{x+\omega}\,{}_2F_1\left(\begin{matrix} a,\, b;\, -\frac{x}{\omega} \\ \frac{2a+2b-1}{2} \end{matrix}\right) \times {}_2F_1\left(\begin{matrix} \frac{2a-1}{2},\, \frac{2b+1}{2} \\ \frac{2a+2b-1}{2};\, -\frac{x}{\omega} \end{matrix}\right)$	$\frac{2^{2a+2b-3}\omega^{s+1/2}}{\sqrt{\pi}}\Gamma\left[\begin{matrix} \frac{2a+2b-1}{2},\, \frac{2a+2b-1}{2} \\ 2a-1,\, 2b \end{matrix}\right] \times \Gamma\left[\begin{matrix} s,\, 2a-s-1,\, 2b-s,\, a+b-s-1 \\ 2a+2b-s-2,\, \frac{2a+2b-2s-1}{2} \end{matrix}\right]$ $\left[\begin{matrix} 0<\operatorname{Re} s<2\operatorname{Re} a-1,\, 2\operatorname{Re} b,\, \operatorname{Re}(a+b)-1; \\ \lvert\arg\omega\rvert<\pi \end{matrix}\right]$
9	$\sqrt{x+\omega}\,{}_2F_1\left(\begin{matrix} a,\, b;\, -\frac{x}{\omega} \\ \frac{2a+2b+1}{2} \end{matrix}\right) \times {}_2F_1\left(\begin{matrix} \frac{2a+1}{2},\, \frac{2b+1}{2} \\ \frac{2a+2b+1}{2};\, -\frac{x}{\omega} \end{matrix}\right)$	$\frac{2^{2a+2b-1}\omega^{s+1/2}}{\sqrt{\pi}}\Gamma\left[\begin{matrix} \frac{2a+2b+1}{2},\, \frac{2a+2b+1}{2} \\ 2a,\, 2b \end{matrix}\right] \times \Gamma\left[\begin{matrix} s,\, 2a-s,\, 2b-s,\, a+b-s \\ 2a+2b-s,\, \frac{2a+2b-2s+1}{2} \end{matrix}\right]$ $[0<\operatorname{Re} s<2\operatorname{Re} a,\, 2\operatorname{Re} b;\, \lvert\arg\omega\rvert<\pi]$

No.	$f(x)$	$F(s)$
10	$\sqrt{x+\omega}\,{}_2F_1\left(\begin{matrix}a,\, b;\, -\frac{x}{\omega}\\ \frac{2a+2b-1}{2}\end{matrix}\right) \times {}_2F_1\left(\begin{matrix}1-a,\, 1-b\\ \frac{5-2a-2b}{2};\, -\frac{x}{\omega}\end{matrix}\right)$	$\frac{(2a+2b-3)\,\omega^{s+1/2}}{2\sqrt{\pi}}\frac{\cos[(a-b)\pi]}{\cos[(a+b)\pi]} \times \Gamma\left[\begin{matrix}s,\, \frac{1-2s}{2},\, \frac{2a-2b-2s+1}{2},\, \frac{2b-2a-2s+1}{2}\\ \frac{2a+2b-2s-1}{2},\, \frac{5-2a-2b-2s}{2}\end{matrix}\right]$ $[0<\operatorname{Re} s<1/2-\lvert\operatorname{Re}(a-b)\rvert;\ \lvert\arg\omega\rvert<\pi]$
11	$\sqrt{x+\omega}\,{}_2F_1\left(\begin{matrix}1-a,\, 1-b\\ \frac{3-2a-2b}{2};\, -\frac{x}{\omega}\end{matrix}\right) \times {}_2F_1\left(\begin{matrix}a,\, b;\, -\frac{x}{\omega}\\ \frac{2a+2b+1}{2}\end{matrix}\right)$	$\frac{(1-2a-2b)\,\omega^{s+1/2}}{2\sqrt{\pi}}\frac{\cos[(a-b)\pi]}{\cos[(a+b)\pi]} \times \Gamma\left[\begin{matrix}s,\, \frac{1-2s}{2},\, \frac{2a-2b-2s+1}{2},\, \frac{2b-2a-2s+1}{2}\\ \frac{2a+2b-2s+1}{2},\, \frac{3-2a-2b-2s}{2}\end{matrix}\right]$ $[0<\operatorname{Re} s<1/2-\lvert\operatorname{Re}(a-b)\rvert;\ \lvert\arg\omega\rvert<\pi]$
12	$\frac{1}{\sqrt{x+\omega}}\,{}_2F_1\left(\begin{matrix}a,\, b;\, -\frac{x}{\omega}\\ \frac{2a+2b+1}{2}\end{matrix}\right) \times {}_2F_1\left(\begin{matrix}\frac{2a+1}{2},\, \frac{2b+1}{2}\\ \frac{2a+2b+3}{2};\, -\frac{x}{\omega}\end{matrix}\right)$	$\frac{2^{2a+2b}\omega^{s-1/2}}{\sqrt{\pi}}\Gamma\left[\begin{matrix}\frac{2a+2b+1}{2},\, \frac{2a+2b+3}{2}\\ 2a+1,\, 2b+1\end{matrix}\right] \times \Gamma\left[\begin{matrix}s,\, 2a-s+1,\, 2b-s+1,\, a+b-s+1\\ 2a+2b-s+1,\, \frac{2a+2b-2s+3}{2}\end{matrix}\right]$ $[0<\operatorname{Re} s<2\operatorname{Re} a+1,\ 2\operatorname{Re} b+1;\ \lvert\arg\omega\rvert<\pi]$
13	$\frac{1}{\sqrt{x+\omega}}\,{}_2F_1\left(\begin{matrix}a,\, b;\, -\frac{x}{\omega}\\ \frac{2a+2b+1}{2}\end{matrix}\right) \times {}_2F_1\left(\begin{matrix}\frac{2a+1}{2},\, \frac{2b-1}{2};\, -\frac{x}{\omega}\\ \frac{2a+2b+1}{2}\end{matrix}\right)$	$\frac{2^{2a+2b-1}\omega^{s-1/2}}{\sqrt{\pi}}\Gamma\left[\begin{matrix}\frac{2a+2b+1}{2},\, \frac{2a+2b+1}{2}\\ 2a+1,\, 2b\end{matrix}\right] \times \Gamma\left[\begin{matrix}s,\, 2a-s+1,\, 2b-s,\, a+b-s\\ 2a+2b-s,\, \frac{2a+2b-2s+1}{2}\end{matrix}\right]$ $\left[\begin{matrix}0<\operatorname{Re} s<2\operatorname{Re} a+1, 2\operatorname{Re} b, \operatorname{Re}(a+b);\\ \lvert\arg\omega\rvert<\pi\end{matrix}\right]$
14	$\sqrt{x+\omega}\,{}_2F_1\left(\begin{matrix}\frac{2a-1}{2},\, \frac{2b-1}{2}\\ \frac{2a+2b-3}{2};\, -\frac{x}{\omega}\end{matrix}\right) \times {}_2F_1\left(\begin{matrix}a,\, b;\, -\frac{x}{\omega}\\ \frac{2a+2b-1}{2}\end{matrix}\right)$	$\frac{2^{2a+2b-4}\omega^{s+1/2}}{\sqrt{\pi}}\Gamma\left[\begin{matrix}\frac{2a+2b-3}{2},\, \frac{2a+2b-1}{2}\\ 2a-1,\, 2b-1\end{matrix}\right] \times \Gamma\left[\begin{matrix}s,\, 2a-s-1,\, 2b-s-1,\, a+b-s-1\\ \frac{2a+2b-2s-1}{2},\, 2a+2b-s-3\end{matrix}\right]$ $[0<\operatorname{Re} s<2\operatorname{Re} a-1,\ 2\operatorname{Re} b-1;\ \lvert\arg\omega\rvert<\pi]$
15	$\sqrt{x+\omega}\,{}_2F_1\left(\begin{matrix}a,\, b;\, -\frac{x}{\omega}\\ \frac{2a+2b-1}{2}\end{matrix}\right) \times {}_2F_1\left(\begin{matrix}\frac{2a+1}{2},\, \frac{2b+1}{2}\\ \frac{2a+2b+1}{2};\, -\frac{x}{\omega}\end{matrix}\right)$	$\frac{2^{2a+2b-2}\omega^{s+1/2}}{\sqrt{\pi}}\Gamma\left[\begin{matrix}\frac{2a+2b-1}{2},\, \frac{2a+2b+1}{2}\\ 2a,\, 2b\end{matrix}\right] \times \Gamma\left[\begin{matrix}s,\, 2a-s,\, 2b-s,\, a+b-s\\ \frac{2a+2b-2s+1}{2},\, 2a+2b-s-1\end{matrix}\right]$ $[0<\operatorname{Re} s<2\operatorname{Re} a,\ 2\operatorname{Re} b;\ \lvert\arg\omega\rvert<\pi]$

3.31.34. ${}_2F_1\left(\begin{matrix} a_1, b_1 \\ c_1;\ 1-\omega_1 x\end{matrix}\right) {}_2F_1\left(\begin{matrix} a_2, b_2 \\ c_2;\ 1-\omega_2 x\end{matrix}\right)$ and algebraic functions

1	$(\sigma-x)_+^{c-1}\ {}_2F_1\left(\begin{matrix} a, b \\ c;\ \frac{\sigma-x}{\sigma}\end{matrix}\right) \times {}_2F_1\left(\begin{matrix} a', b' \\ c';\ 1-\omega x\end{matrix}\right)$	$\sigma^{s+c-1}\Gamma\left[\begin{matrix} c, c', c'-a'-b', s, s-a-b+c \\ c'-a', c'-b', s-a+c, s-b+c\end{matrix}\right] \times {}_4F_3\left(\begin{matrix} a', b', s, s-a-b+c;\ \sigma\omega \\ a'+b'-c'+1, s-a+c, s-b+c\end{matrix}\right) + \frac{\sigma^{s-a'-b'+c+c'-1}}{\omega^{a'+b'-c'}}\Gamma\left[\begin{matrix} c, c', a'+b'-c \\ a', b', s-a-a'-b'+c+c'\end{matrix}\right] \times\Gamma\left[\begin{matrix} s-a'-b'+c', s-a-a'-b-b'+c+c' \\ s-a'-b-b'+c+c'\end{matrix}\right] \times {}_4F_3\left(\begin{matrix} c'-a', c'-b', s-a'-b'+c', s-a-a'-b-b'+c+c';\ \sigma\omega \\ c'-a'-b'+1, s-a-a'-b'+c+c', s-a'-b-b'+c+c'\end{matrix}\right)$ $\left[\begin{matrix} \omega,\ \mathrm{Re}\, c>0; \\ \mathrm{Re}\, s>0,\ \mathrm{Re}\,(a+b-c),\ \mathrm{Re}\,(a'+b'-c'), \\ \mathrm{Re}\,(a+a'+b+b'-c-c');\ \lvert\arg(1-\sigma\omega)\rvert<\pi\end{matrix}\right]$

3.31.35. ${}_2F_1\left(\begin{matrix} a_1, b_1 \\ c_1;\ \frac{\sqrt{\omega}-\sqrt{x+\omega}}{2\sqrt{\omega}}\end{matrix}\right) {}_2F_1\left(\begin{matrix} a_2, b_2 \\ c_2;\ \frac{\sqrt{\omega}-\sqrt{x+\omega}}{2\sqrt{\omega}}\end{matrix}\right)$ and algebraic functions

1	${}_2F_1\left(\begin{matrix} a, b;\ \frac{\sqrt{\omega}-\sqrt{x+\omega}}{2\sqrt{\omega}} \\ a+b-c+1\end{matrix}\right) \times {}_2F_1\left(\begin{matrix} a, b \\ c;\ \frac{\sqrt{\omega}-\sqrt{x+\omega}}{2\sqrt{\omega}}\end{matrix}\right)$	$(4\omega)^s\,\Gamma\left[\begin{matrix} c, a+b-c+1 \\ a, b\end{matrix}\right] \times\Gamma\left[\begin{matrix} s, a-s, b-s, a+b-2s \\ a+b-s, c-s, a+b-c-s+1\end{matrix}\right]$ $[0<\mathrm{Re}\, s<\mathrm{Re}\, a,\ \mathrm{Re}\, b;\ \lvert\arg\omega\rvert<\pi]$
2	${}_2F_1\left(\begin{matrix} a, 1-a \\ 2-c;\ \frac{\sqrt{\omega}-\sqrt{x+\omega}}{2\sqrt{\omega}}\end{matrix}\right) \times {}_2F_1\left(\begin{matrix} a, 1-a \\ c;\ \frac{\sqrt{\omega}-\sqrt{x+\omega}}{2\sqrt{\omega}}\end{matrix}\right)$	$\frac{(1-c)\,\omega^s}{\sqrt{\pi}}\,\frac{\sin(a\pi)}{\sin(c\pi)}\,\Gamma\left[\begin{matrix} s, \frac{1-2s}{2}, a-s, 1-a-s \\ c-s, 2-c-s\end{matrix}\right]$ $[0<\mathrm{Re}\, s<\mathrm{Re}\, a,\ 1-\mathrm{Re}\, a;\ \lvert\arg\omega\rvert<\pi]$
3	${}_2F_1\left(\begin{matrix} a, \frac{2a+1}{2};\ \frac{\sqrt{\omega}-\sqrt{x+\omega}}{2\sqrt{\omega}} \\ 2a-c+\frac{3}{2}\end{matrix}\right) \times {}_2F_1\left(\begin{matrix} a, \frac{2a+1}{2} \\ c;\ \frac{\sqrt{\omega}-\sqrt{x+\omega}}{2\sqrt{\omega}}\end{matrix}\right)$	$2^{8s-4a+1}\sqrt{\pi}\,\omega^s\,\Gamma\left[\begin{matrix} c \\ 2a\end{matrix}\right]\Gamma\left[\begin{matrix} s, 4a-4s, \frac{4a-2c+3}{2} \\ c-s, \frac{4a-2s+1}{2}, \frac{4a-2c-2s+3}{2}\end{matrix}\right]$ $[0<\mathrm{Re}\, s<\mathrm{Re}\, a;\ \lvert\arg\omega\rvert<\pi]$

No.	$f(x)$	$F(s)$
4	${}_2F_1\left(\begin{matrix}1-a,\, 1-b\\ \frac{3-a-b}{2};\ \frac{\sqrt{\omega}-\sqrt{x+\omega}}{2\sqrt{\omega}}\end{matrix}\right)$ $\times\, {}_2F_1\left(\begin{matrix}a,\, b;\ \frac{\sqrt{\omega}-\sqrt{x+\omega}}{2\sqrt{\omega}}\\ \frac{a+b+1}{2}\end{matrix}\right)$	$\frac{(1-a-b)\,\omega^s}{2\sqrt{\pi}}\,\frac{\cos[(a-b)\pi/2]}{\cos[(a+b)\pi/2]}$ $\times\,\Gamma\left[\begin{matrix}s,\ \frac{1-2s}{2},\ \frac{a-b-2s+1}{2},\ \frac{b-a-2s+1}{2}\\ \frac{a+b-2s+1}{2},\ \frac{3-a-b-2s}{2}\end{matrix}\right]$ $[0<\operatorname{Re}s<(1-\lvert\operatorname{Re}(a-b)\rvert)/2;\ \lvert\arg\omega\rvert<\pi]$
5	$\left(\sqrt{\omega}+\sqrt{x+\omega}\right)^{1-c}$ $\times\, {}_2F_1\left(\begin{matrix}a,\, b\\ c;\ \frac{\sqrt{\omega}-\sqrt{x+\omega}}{2\sqrt{\omega}}\end{matrix}\right)$ $\times\, {}_2F_1\left(\begin{matrix}a-c+1,\, b-c+1\\ a+b-c+1;\ \frac{\sqrt{\omega}-\sqrt{x+\omega}}{2\sqrt{\omega}}\end{matrix}\right)$	$(4\omega)^{s+(1-c)/2}\,\Gamma\left[\begin{matrix}c,\ a+b-c+1\\ a,\ b\end{matrix}\right]$ $\times\,\Gamma\left[\begin{matrix}s,\ a-s,\ b-s,\ a+b-2s\\ a+b-s,\ c-s,\ a+b-c-s+1\end{matrix}\right]$ $[0<\operatorname{Re}s<\operatorname{Re}a,\ \operatorname{Re}b;\ \lvert\arg\omega\rvert<\pi]$
6	$\left(\sqrt{\omega}+\sqrt{x+\omega}\right)^{a+b-c}$ $\times\, {}_2F_1\left(\begin{matrix}a,\, b\\ c;\ \frac{\sqrt{\omega}-\sqrt{x+\omega}}{2\sqrt{\omega}}\end{matrix}\right)$ $\times\, {}_2F_1\left(\begin{matrix}c-a,\, c-b;\ \frac{\sqrt{\omega}-\sqrt{x+\omega}}{2\sqrt{\omega}}\\ c-a-b+1\end{matrix}\right)$	$(4\omega)^{s+(a+b-c)/2}\,\Gamma\left[\begin{matrix}c,\ c-a-b+1\\ c-a,\ c-b\end{matrix}\right]$ $\times\,\Gamma\left[\begin{matrix}s,\ c-a-s,\ c-b-s,\ 2c-a-b-2s\\ c-s,\ c-a-b-s+1,\ 2c-a-b-s\end{matrix}\right]$ $[0<\operatorname{Re}s<\operatorname{Re}(c-a),\ \operatorname{Re}(c-b);\ \lvert\arg\omega\rvert<\pi]$
7	$\left(\sqrt{\omega}+\sqrt{x+\omega}\right)^{a+b-2c+1}$ $\times\, {}_2F_1\left(\begin{matrix}a,\, b\\ c;\ \frac{\sqrt{\omega}-\sqrt{x+\omega}}{2\sqrt{\omega}}\end{matrix}\right)$ $\times\, {}_2F_1\left(\begin{matrix}1-a,\, 1-b\\ c-a-b+1;\ \frac{\sqrt{\omega}-\sqrt{x+\omega}}{2\sqrt{\omega}}\end{matrix}\right)$	$(4\omega)^{s+(a+b-2c+1)/2}\,\Gamma\left[\begin{matrix}c,\ c-a-b+1\\ c-a,\ c-b\end{matrix}\right]$ $\times\,\Gamma\left[\begin{matrix}s,\ c-a-s,\ c-b-s,\ 2c-a-b-2s\\ c-s,\ c-a-b-s+1,\ 2c-a-b-s\end{matrix}\right]$ $[0<\operatorname{Re}s<\operatorname{Re}(c-a),\ \operatorname{Re}(c-b);\ \lvert\arg\omega\rvert<\pi]$
8	$\left(\sqrt{\omega}+\sqrt{x+\omega}\right)^{1-c}$ $\times\, {}_2F_1\left(\begin{matrix}a,\, 1-a\\ c;\ \frac{\sqrt{\omega}-\sqrt{x+\omega}}{2\sqrt{\omega}}\end{matrix}\right)$ $\times\, {}_2F_1\left(\begin{matrix}a-c+1,\, 2-a-c\\ 2-c;\ \frac{\sqrt{\omega}-\sqrt{x+\omega}}{2\sqrt{\omega}}\end{matrix}\right)$	$\frac{(1-c)\,\omega^{s+(1-c)/2}}{2^{c-1}\sqrt{\pi}}\,\frac{\sin(a\pi)}{\sin(c\pi)}\,\Gamma\left[\begin{matrix}s,\ \frac{1-2s}{2},\ a-s,\ 1-a-s\\ c-s,\ 2-c-s\end{matrix}\right]$ $[0<\operatorname{Re}s<\operatorname{Re}a,\ 1-\operatorname{Re}a;\ \lvert\arg\omega\rvert<\pi]$
9	$\left(\sqrt{\omega}+\sqrt{x+\omega}\right)^{1-c}$ $\times\, {}_2F_1\left(\begin{matrix}a,\ \frac{2a+1}{2}\\ c;\ \frac{\sqrt{\omega}-\sqrt{x+\omega}}{2\sqrt{\omega}}\end{matrix}\right)$ $\times\, {}_2F_1\left(\begin{matrix}a-c+1,\ \frac{2a-2c+3}{2}\\ \frac{4a-2c+3}{2};\ \frac{\sqrt{\omega}-\sqrt{x+\omega}}{2\sqrt{\omega}}\end{matrix}\right)$	$2^{8s-4a-c+2}\omega^{s+(1-c)/2}\,\Gamma\left[\begin{matrix}c,\ \frac{4a-2c+3}{2}\\ 2a\end{matrix}\right]$ $\times\,\Gamma\left[\begin{matrix}s,\ 4a-4s\\ c-s,\ \frac{4a-2s+1}{2},\ \frac{4a-2c-2s+3}{2}\end{matrix}\right]$ $[0<\operatorname{Re}s<\operatorname{Re}a;\ \lvert\arg\omega\rvert<\pi]$

No.	$f(x)$	$F(s)$
10	$\left(\sqrt{\omega}+\sqrt{x+\omega}\right)^{2a-2c+3/2}$ $\times {}_2F_1\left(\begin{matrix}\frac{1-2a}{2},\, 1-a\\ \frac{2c-4a+1}{2};\, \frac{\sqrt{\omega}-\sqrt{x+\omega}}{2\sqrt{\omega}}\end{matrix}\right)$ $\times {}_2F_1\left(\begin{matrix}a,\, \frac{2a+1}{2}\\ c;\, \frac{\sqrt{\omega}-\sqrt{x+\omega}}{2\sqrt{\omega}}\end{matrix}\right)$	$2^{8s+6a-6c+9/2}\sqrt{\pi}\,\omega^{a-c+s+3/4}\,\Gamma\left[\begin{matrix}c,\, \frac{2c-4a+1}{2}\\ 2c-2a-1\end{matrix}\right]$ $\times\Gamma\left[\begin{matrix}s,\, 4c-4a-4s-2\\ c-s,\, \frac{2c-4a-2s+1}{2},\, \frac{4c-4a-2s-1}{2}\end{matrix}\right]$ $[0<\operatorname{Re}s<\operatorname{Re}(c-a)-1/2;\ \lvert\arg\omega\rvert<\pi]$
11	$\left(\sqrt{\omega}+\sqrt{x+\omega}\right)^{(a+b-1)/2}$ $\times {}_2F_1\left(\begin{matrix}a,\, b;\, \frac{\sqrt{\omega}-\sqrt{x+\omega}}{2\sqrt{\omega}}\\ \frac{a+b+1}{2}\end{matrix}\right)$ $\times {}_2F_1\left(\begin{matrix}\frac{a-b+1}{2},\, \frac{b-a+1}{2}\\ \frac{3-a-b}{2};\, \frac{\sqrt{\omega}-\sqrt{x+\omega}}{2\sqrt{\omega}}\end{matrix}\right)$	$\frac{2^{(a+b-3)/2}\,(1-a-b)\,\omega^{s+(a+b-1)/4}}{\sqrt{\pi}}\cos\frac{(a-b)\pi}{2}$ $\times\sec\frac{(a+b)\pi}{2}\,\Gamma\left[\begin{matrix}s,\, \frac{1-2s}{2},\, \frac{a-b-2s+1}{2},\, \frac{b-a-2s+1}{2}\\ \frac{a+b-2s+1}{2},\, \frac{3-a-b-2s}{2}\end{matrix}\right]$ $[0<\operatorname{Re}s<1/2-\lvert\operatorname{Re}(a-b)\rvert;\ \lvert\arg\omega\rvert<\pi]$
12	$\left(\sqrt{\omega}+\sqrt{x+\omega}\right)^{(1-a-b)/2}$ $\times {}_2F_1\left(\begin{matrix}a,\, b;\, \frac{\sqrt{\omega}-\sqrt{x+\omega}}{2\sqrt{\omega}}\\ \frac{a+b+1}{2}\end{matrix}\right)$ $\times {}_2F_1\left(\begin{matrix}\frac{a-b+1}{2},\, \frac{b-a+1}{2}\\ \frac{a+b+1}{2};\, \frac{\sqrt{\omega}-\sqrt{x+\omega}}{2\sqrt{\omega}}\end{matrix}\right)$	$\frac{2^{(a+b-1)/2}\omega^{s+(1-a-b)/4}}{\sqrt{\pi}}\,\Gamma\left[\begin{matrix}\frac{a+b+1}{2},\, \frac{a+b+1}{2}\\ a,\, b\end{matrix}\right]$ $\times\Gamma\left[\begin{matrix}s,\, a-s,\, b-s,\, \frac{a+b-2s}{2}\\ \frac{a+b-2s+1}{2},\, a+b-s\end{matrix}\right]$ $[0<\operatorname{Re}s<\operatorname{Re}a,\ \operatorname{Re}b;\ \lvert\arg\omega\rvert<\pi]$
13	$\left(\sqrt{\omega}+\sqrt{x+\omega}\right)^{a-b}$ $\times {}_2F_1\left(\begin{matrix}a,\, \frac{2a+1}{2}\\ \frac{a+b+1}{2};\, \frac{\sqrt{\omega}-\sqrt{x+\omega}}{2\sqrt{\omega}}\end{matrix}\right)$ $\times {}_2F_1\left(\begin{matrix}b,\, \frac{2b+1}{2};\, \frac{\sqrt{\omega}-\sqrt{x+\omega}}{2\sqrt{\omega}}\\ b-a+1\end{matrix}\right)$	$2^{8s+a-5b+1}\sqrt{\pi}\omega^{s+(a-b)/2}\,\Gamma\left[\begin{matrix}\frac{2a+2b+1}{2},\, b-a+1\\ 2b\end{matrix}\right]$ $\times\Gamma\left[\begin{matrix}s,\, 4b-4s\\ \frac{4b-2s+1}{2},\, \frac{2a+2b-2s+1}{2},\, b-a-s+1\end{matrix}\right]$ $[0<\operatorname{Re}s<\operatorname{Re}b;\ \lvert\arg\omega\rvert<\pi]$
14	$\left(\sqrt{\omega}+\sqrt{x+\omega}\right)^{2a-c+1/2}$ $\times {}_2F_1\left(\begin{matrix}a,\, \frac{2a+1}{2}\\ c;\, \frac{\sqrt{\omega}-\sqrt{x+\omega}}{2\sqrt{\omega}}\end{matrix}\right)$ $\times {}_2F_1\left(\begin{matrix}c-a,\, \frac{2c-2a-1}{2}\\ \frac{2c-4a+1}{2};\, \frac{\sqrt{\omega}-\sqrt{x+\omega}}{2\sqrt{\omega}}\end{matrix}\right)$	$2^{8s+6a-5c+7/2}\sqrt{\pi}\omega^{s+(4a-2c+1)/4}\,\Gamma\left[\begin{matrix}c,\, \frac{2c-4a+1}{2}\\ 2c-2a-1\end{matrix}\right]$ $\times\Gamma\left[\begin{matrix}s,\, 4c-4a-4s-2\\ c-s,\, \frac{2c-4a-2s+1}{2},\, \frac{4c-4a-2s-1}{2}\end{matrix}\right]$ $[0<\operatorname{Re}s<\operatorname{Re}(c-a)-1/2;\ \lvert\arg\omega\rvert<\pi]$
15	$\left(\sqrt{\omega}+\sqrt{x+\omega}\right)^{1-c}$ $\times {}_2F_1\left(\begin{matrix}a,\, a-1\\ c;\, \frac{\sqrt{\omega}-\sqrt{x+\omega}}{2\sqrt{\omega}}\end{matrix}\right)$ $\times {}_2F_1\left(\begin{matrix}c-a,\, a+c-1\\ c;\, \frac{\sqrt{\omega}-\sqrt{x+\omega}}{2\sqrt{\omega}}\end{matrix}\right)$	$\frac{2^{c-1}\omega^{s+(1-c)/2}}{\sqrt{\pi}}\,\Gamma\left[\begin{matrix}c,\, c\\ c-a,\, c-a-1\end{matrix}\right]$ $\times\Gamma\left[\begin{matrix}s,\, c-a-s,\, a+c-s-1,\, \frac{2c-2s-1}{2}\\ c-s,\, 2c-s-1\end{matrix}\right]$ $[0<\operatorname{Re}s<\operatorname{Re}(c-a),\ \operatorname{Re}c-1/2;\ \lvert\arg\omega\rvert<\pi]$

3.31.36. ${}_2F_1\left(\begin{matrix} a_1, b_1 \\ c_1; \frac{\sqrt{x}-\sqrt{x+\omega}}{2\sqrt{x}} \end{matrix}\right) {}_2F_1\left(\begin{matrix} a_2, b_2 \\ c_2; \frac{\sqrt{x}-\sqrt{x+\omega}}{2\sqrt{x}} \end{matrix}\right)$ and algebraic functions

1	${}_2F_1\left(\begin{matrix} a, b; \frac{\sqrt{x}-\sqrt{x+\omega}}{2\sqrt{x}} \\ a+b-c+1 \end{matrix}\right) \times {}_2F_1\left(\begin{matrix} a, b \\ c; \frac{\sqrt{x}-\sqrt{x+\omega}}{2\sqrt{x}} \end{matrix}\right)$	$\left(\frac{\omega}{4}\right)^s \Gamma\left[\begin{matrix} c, a+b-c+1 \\ a, b \end{matrix}\right] \times \Gamma\left[\begin{matrix} -s, s+a, s+b, 2s+a+b \\ s+a+b, s+c, s+a+b-c+1 \end{matrix}\right]$ $[-\operatorname{Re} a, -\operatorname{Re} b < \operatorname{Re} s < 0; \lvert\arg\omega\rvert < \pi]$
2	${}_2F_1\left(\begin{matrix} 1-a, a \\ 2-c; \frac{\sqrt{x}-\sqrt{x+\omega}}{2\sqrt{x}} \end{matrix}\right) \times {}_2F_1\left(\begin{matrix} 1-a, a \\ c; \frac{\sqrt{x}-\sqrt{x+\omega}}{2\sqrt{x}} \end{matrix}\right)$	$\frac{(1-c)\,\omega^s}{\sqrt{\pi}} \frac{\sin(a\pi)}{\sin(c\pi)} \Gamma\left[\begin{matrix} -s, \frac{2s+1}{2}, s+a, s-a+1 \\ s+c, s-c+2 \end{matrix}\right]$ $[\operatorname{Re} a-1, -\operatorname{Re} a < \operatorname{Re} s < 0; \lvert\arg\omega\rvert < \pi]$
3	${}_2F_1\left(\begin{matrix} a, \frac{2a+1}{2}; \frac{\sqrt{x}-\sqrt{x+\omega}}{2\sqrt{x}} \\ \frac{4a-2c+3}{2} \end{matrix}\right) \times {}_2F_1\left(\begin{matrix} a, \frac{2a+1}{2} \\ c; \frac{\sqrt{x}-\sqrt{x+\omega}}{2\sqrt{x}} \end{matrix}\right)$	$\frac{\sqrt{\pi}\,\omega^s}{2^{8s+4a-1}} \Gamma\left[\begin{matrix} c, -s, 4s+4a, \frac{4a-2c+3}{2} \\ 2a, s+c, \frac{2s+4a+1}{2}, \frac{2s+4a-2c+3}{2} \end{matrix}\right]$ $[-\operatorname{Re} a < \operatorname{Re} s < 0; \lvert\arg\omega\rvert < \pi]$
4	${}_2F_1\left(\begin{matrix} 1-a, 1-b \\ \frac{3-a-b}{2}; \frac{\sqrt{x}-\sqrt{x+\omega}}{2\sqrt{x}} \end{matrix}\right) \times {}_2F_1\left(\begin{matrix} a, b; \frac{\sqrt{x}-\sqrt{x+\omega}}{2\sqrt{x}} \\ \frac{a+b+1}{2} \end{matrix}\right)$	$\frac{(1-a-b)\,\omega^s}{2\sqrt{\pi}} \frac{\cos[(a-b)\pi/2]}{\cos[(a+b)\pi/2]} \times \Gamma\left[\begin{matrix} -s, \frac{2s+1}{2}, \frac{2s+a-b+1}{2}, \frac{2s-a+b+1}{2} \\ \frac{2s-a-b+3}{2}, \frac{2s+a+b+1}{2} \end{matrix}\right]$ $[(\lvert\operatorname{Re}(a-b)\rvert-1)/2 < \operatorname{Re} s < 0; \lvert\arg\omega\rvert < \pi]$
5	$\left(\sqrt{x}+\sqrt{x+\omega}\right)^{1-c} \times {}_2F_1\left(\begin{matrix} a, b \\ c; \frac{\sqrt{x}-\sqrt{x+\omega}}{2\sqrt{x}} \end{matrix}\right) \times {}_2F_1\left(\begin{matrix} a-c+1, b-c+1 \\ a+b-c+1; \frac{\sqrt{x}-\sqrt{x+\omega}}{2\sqrt{x}} \end{matrix}\right)$	$4^{-s}\omega^{s+(1-c)/2} \Gamma\left[\begin{matrix} c, a+b-c+1 \\ a, b \end{matrix}\right] \times \Gamma\left[\begin{matrix} 2s+a+b-c+1, \frac{c-2s-1}{2} \\ \frac{2s+c+1}{2} \end{matrix}\right] \times \Gamma\left[\begin{matrix} \frac{2s+2a-c+1}{2}, \frac{2s+2b-c+1}{2} \\ \frac{2s+2a+2b-c+1}{2}, \frac{2s+2a+2b-3c+3}{2} \end{matrix}\right]$ $\left[\begin{matrix} \operatorname{Re}(c-2a-1)/2, \operatorname{Re}(c-2b-1)/2 \\ < \operatorname{Re} s < (\operatorname{Re} c-1)/2; \lvert\arg\omega\rvert < \pi \end{matrix}\right]$
6	$\left(\sqrt{x}+\sqrt{x+\omega}\right)^{a+b-c} \times {}_2F_1\left(\begin{matrix} a, b \\ c; \frac{\sqrt{x}-\sqrt{x+\omega}}{2\sqrt{x}} \end{matrix}\right) \times {}_2F_1\left(\begin{matrix} c-a, c-b; \frac{\sqrt{x}-\sqrt{x+\omega}}{2\sqrt{x}} \\ c-a-b+1 \end{matrix}\right)$	$4^{-s}\omega^{s+(a+b-c)/2} \Gamma\left[\begin{matrix} c, c-a-b+1 \\ c-a, c-b \end{matrix}\right] \times \Gamma\left[\begin{matrix} 2s+c, \frac{2s+a-b+c}{2}, \frac{2s-a+b+c}{2}, \frac{c-a-b-2s}{2} \\ \frac{2s+a+b+c}{2}, \frac{2s-a-b+c+2}{2}, \frac{2s+3c-a-b}{2} \end{matrix}\right]$ $\left[\begin{matrix} \operatorname{Re}(a-b-c)/2, \operatorname{Re}(b-a-c)/2, -\operatorname{Re} c/2 \\ < \operatorname{Re} s < \operatorname{Re}(c-a-b)/2; \lvert\arg\omega\rvert < \pi \end{matrix}\right]$

No.	$f(x)$	$F(s)$
7	$\left(\sqrt{x}+\sqrt{x+\omega}\right)^{a+b-2c+1}$ $\times {}_2F_1\left(\begin{matrix}a,\ b\\ c;\ \frac{\sqrt{x}-\sqrt{x+\omega}}{2\sqrt{x}}\end{matrix}\right)$ $\times {}_2F_1\left(\begin{matrix}1-a,\ 1-b;\ \frac{\sqrt{x}-\sqrt{x+\omega}}{2\sqrt{x}}\\ c-a-b+1\end{matrix}\right)$	$\frac{\omega^{s+(a+b-2c+1)/2}}{2^{2s}}\Gamma\left[\begin{matrix}c,\ c-a-b+1\\ c-a,\ c-b\end{matrix}\right]$ $\times\Gamma\left[\begin{matrix}2s+1,\ \frac{2s+a-b+1}{2},\ \frac{2s-a+b+1}{2},\ \frac{2c-a-b-2s-1}{2}\\ \frac{2s+a+b+1}{2},\ \frac{2s-a-b+3}{2},\ \frac{2s-a-b+2c+1}{2}\end{matrix}\right]$ $\left[(\lvert\mathrm{Re}\,(a-b)\rvert-1)/2<\mathrm{Re}\,s<\mathrm{Re}\,(2c-a-b-1)/2;\ \lvert\arg\omega\rvert<\pi\right]$
8	$\left(\sqrt{x}+\sqrt{x+\omega}\right)^{1-c}$ $\times {}_2F_1\left(\begin{matrix}a,\ 1-a\\ c;\ \frac{\sqrt{x}-\sqrt{x+\omega}}{2\sqrt{x}}\end{matrix}\right)$ $\times {}_2F_1\left(\begin{matrix}a-c+1,\ 2-a\\ 2-c;\ \frac{\sqrt{x}-\sqrt{x+\omega}}{2\sqrt{x}}\end{matrix}\right)$	$\frac{(1-c)\,\omega^{s+(1-c)/2}}{2^{c-1}\sqrt{\pi}}\,\frac{\sin(a\pi)}{\sin(c\pi)}$ $\times\Gamma\left[\begin{matrix}\frac{c-2s-1}{2},\ \frac{2s-c+2}{2},\ \frac{2s+2a-c+1}{2},\ \frac{2s-2a-c+3}{2}\\ \frac{2s+c+1}{2},\ \frac{2s-3c+5}{2}\end{matrix}\right]$ $\left[\mathrm{Re}\,(2a+c-3)/2,\ \mathrm{Re}\,(c-2a-1)/2<\mathrm{Re}\,s<(\mathrm{Re}\,c-1)/2;\ \lvert\arg\omega\rvert<\pi\right]$
9	$\left(\sqrt{x}+\sqrt{x+\omega}\right)^{1-c}$ $\times {}_2F_1\left(\begin{matrix}a,\ \frac{2a+1}{2}\\ c;\ \frac{\sqrt{x}-\sqrt{x+\omega}}{2\sqrt{x}}\end{matrix}\right)$ $\times {}_2F_1\left(\begin{matrix}a-c+1,\ \frac{2a-2c+3}{2}\\ \frac{4a-2c+3}{2};\ \frac{\sqrt{x}-\sqrt{x+\omega}}{2\sqrt{x}}\end{matrix}\right)$	$\frac{\sqrt{\pi}\,\omega^{s+(1-c)/2}}{2^{8s+4a-3c+2}}\Gamma\left[\begin{matrix}c,\ \frac{4a-2c+3}{2}\\ 2a\end{matrix}\right]$ $\times\Gamma\left[\begin{matrix}\frac{c-2s-1}{2},\ 4s+4a-2c+2\\ \frac{2s+c+1}{2},\ \frac{2s+4a-c+2}{2},\ \frac{2s+4a-3c+4}{2}\end{matrix}\right]$ $\left[\mathrm{Re}\,(c-2a-1)/2,\ \mathrm{Re}\,(2c-4a-3)/4<\mathrm{Re}\,s<(\mathrm{Re}\,c-1)/2;\ \lvert\arg\omega\rvert<\pi\right]$
10	$\left(\sqrt{x}+\sqrt{x+\omega}\right)^{2a-2c+3/2}$ $\times {}_2F_1\left(\begin{matrix}\frac{1-2a}{2},\ 1-a\\ \frac{2c-4a+1}{2};\ \frac{\sqrt{x}-\sqrt{x+\omega}}{2\sqrt{x}}\end{matrix}\right)$ $\times {}_2F_1\left(\begin{matrix}a,\ \frac{2a+1}{2}\\ c;\ \frac{\sqrt{x}-\sqrt{x+\omega}}{2\sqrt{x}}\end{matrix}\right)$	$\frac{\sqrt{\pi}\,\omega^{s+a-c+3/4}}{2^{8s+2a-2c+3/2}}\Gamma\left[\begin{matrix}c,\ \frac{2c-4a+1}{2}\\ 2c-2a-1\end{matrix}\right]$ $\times\Gamma\left[\begin{matrix}\frac{4c-4a-4s-3}{4},\ 4s+1\\ \frac{4s-4a+5}{4},\ \frac{4s+4a+3}{4},\ \frac{4s-4a+4c+1}{4}\end{matrix}\right]$ $\left[-1/4<\mathrm{Re}\,s<\mathrm{Re}\,(c-a)-3/4;\ \lvert\arg\omega\rvert<\pi\right]$

3.31.37. ${}_2F_1\left(\begin{matrix}a_1,\ b_1\\ c_1;\ -\frac{2\sqrt{x}\left(\sqrt{x}\pm\sqrt{x+\omega}\right)}{\omega}\end{matrix}\right){}_2F_1\left(\begin{matrix}a_2,\ b_2\\ c_2;\ -\frac{2\sqrt{x}\left(\sqrt{x}+\sqrt{x+\omega}\right)}{\omega}\end{matrix}\right)$ and algebraic functions

No.	$f(x)$	$F(s)$
1	${}_2F_1\left(\begin{matrix}a,\ b\\ c;\ -\frac{2\sqrt{x}\left(\sqrt{x}-\sqrt{x+\omega}\right)}{\omega}\end{matrix}\right)$ $\times {}_2F_1\left(\begin{matrix}a,\ b\\ c;\ -\frac{2\sqrt{x}\left(\sqrt{x}+\sqrt{x+\omega}\right)}{\omega}\end{matrix}\right)$	$\left(\frac{\omega}{4}\right)^{s}\Gamma\left[\begin{matrix}c,\ c\\ a,\ b,\ c-a,\ c-b\end{matrix}\right]$ $\times\Gamma\left[\begin{matrix}s,\ a-s,\ b-s,\ c-a-s,\ c-b-s\\ c-s,\ c-2s\end{matrix}\right]$ $\left[0<\mathrm{Re}\,s<\mathrm{Re}\,a,\ \mathrm{Re}\,b,\ \mathrm{Re}\,(c-a),\ \mathrm{Re}\,(c-b);\ \lvert\arg\omega\rvert<\pi\right]$

No.	$f(x)$	$F(s)$
2	$\left(\frac{2\sqrt{x}\left(\sqrt{x}+\sqrt{x+\omega}\right)}{\omega}+1\right)^a$ $\times\,{}_2F_1\left(\begin{matrix}a,\,b\\c;\,-\frac{2\sqrt{x}(\sqrt{x}+\sqrt{x+\omega})}{\omega}\end{matrix}\right)$ $\times\,{}_2F_1\left(\begin{matrix}a,\,c-b\\c;\,-\frac{2\sqrt{x}(\sqrt{x}+\sqrt{x+\omega})}{\omega}\end{matrix}\right)$	$\left(\frac{\omega}{4}\right)^s\Gamma\left[\begin{matrix}c,\,c\\a,\,b,\,c-a,\,c-b\end{matrix}\right]$ $\times\Gamma\left[\begin{matrix}s,\,a-s,\,b-s,\,c-a-s,\,c-b-s\\c-s,\,c-2s\end{matrix}\right]$ $\left[\begin{matrix}0<\operatorname{Re}s<\operatorname{Re}a,\,\operatorname{Re}b,\,\operatorname{Re}(c-a),\,\operatorname{Re}(c-b);\\ \lvert\arg\omega\rvert<\pi\end{matrix}\right]$
3	$\left(2x\pm2\sqrt{x}\sqrt{x+\omega}+\omega\right)^{-a-b+c}$ $\times\,{}_2F_1\left(\begin{matrix}a,\,b\\c;\,\pm\frac{2\sqrt{x}(\sqrt{x+\omega}\mp\sqrt{x})}{\omega}\end{matrix}\right)$ $\times\,{}_2F_1\left(\begin{matrix}c-a,\,c-b\\c;\,\mp\frac{2\sqrt{x}(\sqrt{x+\omega}\pm\sqrt{x})}{\omega}\end{matrix}\right)$	$4^{-s}\omega^{s-a-b+c}\,\Gamma\left[\begin{matrix}c,\,c,\,s,\,a-s,\,b-s\\a,\,b,\,c-a,\,c-b\end{matrix}\right]$ $\times\Gamma\left[\begin{matrix}c-a-s,\,c-b-s\\c-s,\,c-2s\end{matrix}\right]$ $\left[\left\{\begin{matrix}\lvert\arg\omega\rvert<\pi\\ \operatorname{Re}\omega\ge0\end{matrix}\right\};\;0<\operatorname{Re}s<\operatorname{Re}a,\,\operatorname{Re}b,\,\operatorname{Re}(c-a),\,\operatorname{Re}(c-b)\right]$

3.31.38. ${}_2F_1\left(\begin{matrix}a_1,\,b_1\\c_1;\,\frac{2\sqrt{\omega}(\sqrt{x+\omega}-\sqrt{\omega})}{x}\end{matrix}\right){}_2F_1\left(\begin{matrix}a_2,\,b_2\\c_2;\,-\frac{2\sqrt{\omega}(\sqrt{x+\omega}+\sqrt{\omega})}{x}\end{matrix}\right)$ and algebraic functions

1	${}_2F_1\left(\begin{matrix}a,\,b\\c;\,\frac{2\sqrt{\omega}(\sqrt{x+\omega}-\sqrt{\omega})}{x}\end{matrix}\right)$ $\times\,{}_2F_1\left(\begin{matrix}a,\,b\\c;\,-\frac{2\sqrt{\omega}(\sqrt{x+\omega}+\sqrt{\omega})}{x}\end{matrix}\right)$	$(4\omega)^s\,\Gamma\left[\begin{matrix}c,\,c\\a,\,b,\,c-a,\,c-b\end{matrix}\right]$ $\times\Gamma\left[\begin{matrix}-s,\,s+a,\,s+b,\,s-a+c,\,s-b+c\\s+c,\,2s+c\end{matrix}\right]$ $[-\operatorname{Re}a,\,-\operatorname{Re}b,\,\operatorname{Re}(a-c),\,\operatorname{Re}(b-c)<\operatorname{Re}s<0;\;\lvert\arg\omega\rvert<\pi]$
2	$\left(\sqrt{\omega}\pm\sqrt{x+\omega}\right)^{2(c-a-b)}$ $\times\,{}_2F_1\left(\begin{matrix}a,\,b\\c;\,\pm\frac{2\sqrt{\omega}(\sqrt{x+\omega}\mp\sqrt{\omega})}{x}\end{matrix}\right)$ $\times\,{}_2F_1\left(\begin{matrix}c-a,\,c-b\\c;\,\mp\frac{2\sqrt{\omega}(\sqrt{x+\omega}\pm\sqrt{\omega})}{x}\end{matrix}\right)$	$(4\omega)^{s-a-b+c}\,\Gamma\left[\begin{matrix}c,\,c,\,a+b-c-s,\,s-a+c\\a,\,b,\,c-a,\,c-b\end{matrix}\right]$ $\times\Gamma\left[\begin{matrix}s-b+c,\,s-a-2b+2c,\,s-2a-b+2c\\s-a-b+2c,\,2s-2a-2b+3c\end{matrix}\right]$ $\left[\begin{matrix}\operatorname{Re}(a-c),\,\operatorname{Re}(b-c),\,\operatorname{Re}(2a+b-2c),\\ \operatorname{Re}(a+2b-2c)<\operatorname{Re}s<\operatorname{Re}(a+b-c);\;\lvert\arg\omega\rvert<\pi\end{matrix}\right]$

3.31.39. ${}_2F_1\left(\begin{matrix}a_1,\,b_1\\c_1;\,\frac{2\sqrt{\omega}(\sqrt{\omega}+\sqrt{\omega-x})}{x}\end{matrix}\right){}_2F_1\left(\begin{matrix}a_2,\,b_2\\c_2;\,-\frac{2\sqrt{\omega}(\sqrt{\omega}+\sqrt{\omega-x})}{x}\end{matrix}\right)$ and algebraic functions

1	$\left(x-2\omega-2\sqrt{\omega}\sqrt{\omega-x}\right)^a$ $\times\,{}_2F_1\left(\begin{matrix}a,\,b\\c;\,\frac{2\sqrt{\omega}(\sqrt{\omega}+\sqrt{\omega-x})}{x}\end{matrix}\right)$ $\times\,{}_2F_1\left(\begin{matrix}a,\,c-b\\c;\,\frac{2\sqrt{\omega}(\sqrt{\omega}+\sqrt{\omega-x})}{x}\end{matrix}\right)$	$e^{-i(s+a)\pi}\,(4\omega)^{s+a}\,\Gamma\left[\begin{matrix}c,\,c\\a,\,b,\,c-a,\,c-b\end{matrix}\right]$ $\times\Gamma\left[\begin{matrix}s+2a,\,-s-a\\2s+2a+c\end{matrix}\right]\Gamma\left[\begin{matrix}s+a+b,\,s+c,\,s+a-b+c\\s+a+c\end{matrix}\right]$ $\left[\begin{matrix}-2\operatorname{Re}a,\,-\operatorname{Re}(a+b),\,-\operatorname{Re}c,\,\operatorname{Re}(b-a-c)\\<\operatorname{Re}s<-\operatorname{Re}a;\;0<\arg\omega\le\pi\end{matrix}\right]$

3.32. The Generalized Hypergeometric Function ${}_3F_2\left(\begin{matrix} a_1,\ a_2\ ,a_3 \\ b_1,\ b_2;\ z \end{matrix}\right)$

More formulas can be obtained from the corresponding sections due to the relations

$$ {}_3F_2\left(\begin{matrix} a_1,\ a_2,\ a_3 \\ b_1,\ b_2;\ z \end{matrix}\right) = \Gamma\left[\begin{matrix} b_1,\ b_2 \\ a_1,\ a_2,\ a_3 \end{matrix}\right] G_{33}^{13}\left(-z \,\middle|\, \begin{matrix} 1-a_1,\ 1-a_2,\ 1-a_3 \\ 0,\ 1-b_1,\ 1-b_2 \end{matrix}\right). $$

3.32.1. ${}_3F_2\left(\begin{matrix} a_1,\ a_2,\ a_3 \\ b_1,\ b_2;\ \varphi(x) \end{matrix}\right)$ and algebraic functions

No.	$f(x)$	$F(s)$
1	$(x+\sigma)^{2b+1}\ {}_3F_2\left(\begin{matrix} a,\ 2a-2,\ b;\ -\frac{x}{\sigma} \\ a-1,\ 2a-b-1 \end{matrix}\right)$	$(2a-2b-2s-3)\,\sigma^{s+2b+1} \times\Gamma\left[\begin{matrix} 2a-b-1 \\ -b-1,\ 2a-2b-2 \end{matrix}\right] \times\Gamma\left[\begin{matrix} s,\ -b-s-1,\ 2a-2b-s-3 \\ 2a-b-s-1 \end{matrix}\right]$ $\left[0<\operatorname{Re}s<-\operatorname{Re}b-1,\ \operatorname{Re}(a-b)-1/2,\ 2\operatorname{Re}(a-b)-3;\ \lvert\arg\sigma\rvert<\pi\right]$
2	$\left(\sqrt{x+\sigma}-\sqrt{x}\right)^a \times{}_3F_2\left(\begin{matrix} a,\ b,\ c;\ \frac{2\sqrt{x}\left(\sqrt{x}-\sqrt{x+\sigma}\right)}{\sigma}+1 \\ a-b+1,\ a-c+1 \end{matrix}\right)$	$\frac{\sigma^{s+a/2}}{2^{2s}}\Gamma\left[\begin{matrix} a-b+1,\ a-c+1 \\ a,\ a-b-c+1 \end{matrix}\right] \times\Gamma\left[\begin{matrix} 2s,\ \frac{a-2s}{2},\ \frac{2s+a-2b-2c+2}{2} \\ \frac{2s+a-2b+2}{2},\ \frac{2s+a-2c+2}{2} \end{matrix}\right]$ $\left[0,\ \operatorname{Re}(b-a/2+c)-1<\operatorname{Re}s<\operatorname{Re}a/2;\ \lvert\arg\sigma\rvert<\pi\right]$
3	$\left(\sqrt{x^2+\sigma^2}-x\right)^a \times{}_3F_2\left(\begin{matrix} a,\ b,\ c;\ 1-\frac{2x\left(\sqrt{x^2+\sigma^2}-x\right)}{\sigma^2} \\ a-b+1,\ a-c+1 \end{matrix}\right)$	$\frac{\sigma^{a+s}}{2^{s+1}}\Gamma\left[\begin{matrix} a-b+1,\ a-c+1 \\ a,\ a-b-c+1 \end{matrix}\right] \times\Gamma\left[\begin{matrix} s,\ \frac{a-s}{2},\ \frac{s+a-2b-2c+2}{2} \\ \frac{s+a-2b+2}{2},\ \frac{s+a-2c+2}{2} \end{matrix}\right]$ $\left[0,\ \operatorname{Re}(-a+2b+2c)-2<\operatorname{Re}s<\operatorname{Re}a;\ \lvert\arg\sigma\rvert<\pi/2\right]$
4	$\left(\sqrt{x+\sigma}-\sqrt{\sigma}\right)^a \times{}_3F_2\left(\begin{matrix} a,\ b,\ c;\ 1-\frac{2\sqrt{\sigma}\left(\sqrt{x+\sigma}-\sqrt{\sigma}\right)}{x} \\ a-b+1,\ a-c+1 \end{matrix}\right)$	$(4\sigma)^{s+a/2}\,\Gamma\left[\begin{matrix} a-b+1,\ a-c+1 \\ a,\ a-b-c+1 \end{matrix}\right] \times\Gamma\left[\begin{matrix} s+a,\ -2s-a,\ 1-b-c-s \\ 1-b-s,\ 1-c-s \end{matrix}\right]$ $\left[-\operatorname{Re}a<\operatorname{Re}s<-\operatorname{Re}a/2,\ 1-\operatorname{Re}(b+c);\ \lvert\arg\sigma\rvert<\pi\right]$

3.33. The Generalized Hypergeometric Functions ${}_pF_q\left((a_p);\ (b_q);\ z\right)$

More formulas can be obtained from the corresponding section due to the relation

$$ {}_pF_q\left(\begin{matrix} a_1,\, a_2, \ldots, a_p;\, z \\ b_1,\, b_2, \ldots, b_q \end{matrix}\right) = \Gamma\left[\begin{matrix} b_1,\, b_2, \ldots, b_q \\ a_1,\, a_2, \ldots, a_p \end{matrix}\right] G_{p,\,q+1}^{1,\,p}\left(-z \,\middle|\, \begin{matrix} 1-a_1,\, 1-a_2, \ldots, 1-a_p \\ 0,\, 1-b_1,\, 1-b_2, \ldots, 1-b_q \end{matrix}\right). $$

It is supposed that all hypergeometric functions in formulas exist. If at least one of the upper parameters of a hypergeometric function is a negative integer, then the corresponding function turns into a polynomial, and the conditions can be weakened.

Notation:

$$ \chi = \sum_{j=1}^{q} b_j - \sum_{i=1}^{p} a_i + \frac{p-q}{2} + 1. $$

The expression $\operatorname{Re} s < \operatorname{Re}(a_k + a)$ means that the inequality is valid for all $k = 1, 2, \ldots, p$.

3.33.1. ${}_pF_q\left((a_p);\ (b_q);\ \varphi(x)\right)$ and algebraic functions

No.	$f(x)$	$F(s)$
1	${}_pF_q\left(\begin{matrix}(a_p);\, -\omega x\\ (b_q)\end{matrix}\right)$	$\omega^{-s}\,\Gamma\left[\begin{matrix}(b_q),\, s,\, (a_p)-s\\ (a_p),\, (b_q)-s\end{matrix}\right]$ $\Big[[q=p-1;\ \lvert\arg\omega\rvert<\pi;\ 0<\operatorname{Re}s<\operatorname{Re}a_k]$ or $[q=p;\ (\operatorname{Re}\omega>0;\ 0<\operatorname{Re}s<\operatorname{Re}a_k)$ or $(\operatorname{Re}\omega=0;\ 0<\operatorname{Re}s<\operatorname{Re}a_k,\ 1-\operatorname{Re}\chi)]$ or $[q=p+1;\ \omega>0;\ 0<\operatorname{Re}s<\operatorname{Re}a_k,\ 1/2-\operatorname{Re}\chi] \Big]$
2	$(\sigma-x)_+^{\alpha-1}\ {}_pF_q\left(\begin{matrix}(a_p);\, -\omega x\\ (b_q)\end{matrix}\right)$	$\sigma^{s+\alpha-1}\,\mathrm{B}(s,\,\alpha)\ {}_{p+1}F_{q+1}\left(\begin{matrix}(a_p),\, s;\, -\sigma\omega\\ (b_q),\, s+\alpha\end{matrix}\right)$ $[\sigma,\ \operatorname{Re}\alpha,\ \operatorname{Re}s>0]$
3	$(x-\sigma)_+^{\alpha-1}$ $\times\ {}_pF_q\left(\begin{matrix}(a_p);\, -\omega x\\ (b_q)\end{matrix}\right)$	$\dfrac{\Gamma(s+\alpha-1)}{\omega^{s+\alpha-1}}\,\Gamma\left[\begin{matrix}(b_q),\, (a_p)-\alpha-s+1\\ (a_p),\, (b_q)-\alpha-s+1\end{matrix}\right]$ $\times\ {}_{p+1}F_{q+1}\left(\begin{matrix}1-\alpha,\, (a_p)-\alpha-s+1;\, -\sigma\omega\\ 2-\alpha-s,\, (b_q)-\alpha-s+1\end{matrix}\right)$ $+\,\sigma^{\alpha+s-1}\,\mathrm{B}(\alpha,\,1-\alpha-s)\ {}_{p+1}F_{q+1}\left(\begin{matrix}(a_p),\, s;\, -\sigma\omega\\ (b_q),\, s+\alpha\end{matrix}\right)$ $\Big[[q=p-1;\ \sigma,\ \operatorname{Re}\alpha>0;\ \lvert\arg\omega\rvert<\pi;\ \operatorname{Re}s<\operatorname{Re}(a_k-\alpha)+1]$ or $[q=p;\ \operatorname{Re}\alpha>0;\ (\sigma,\ \operatorname{Re}\omega>0;\ \operatorname{Re}s<\operatorname{Re}(a_k-\alpha)+1)$ or $(\sigma>0;\ \operatorname{Re}\omega=0;\ \operatorname{Re}s<\operatorname{Re}(a_k-\alpha)+1,\ 2-\operatorname{Re}(\alpha+\chi))]$ or $[q=p+1;\ \operatorname{Re}\alpha>0;\ \sigma,\ \omega>0;\ \operatorname{Re}s<\operatorname{Re}(a_k-\alpha)+1,\ 3/2-\operatorname{Re}(\alpha+\chi)] \Big]$

No.	$f(x)$	$F(s)$
4	$\frac{1}{\lvert x-\sigma\rvert^{\rho}}\,{}_pF_q\left(\begin{matrix}(a_p);\,-\omega x\\(b_q)\end{matrix}\right)$	$\frac{\Gamma(s-\rho)}{\omega^{s-\rho}}\,\Gamma\left[\begin{matrix}(b_q),\ (a_p)+\rho-s\\(a_p),\ (b_q)+\rho-s\end{matrix}\right]$ $\times\,{}_{p+1}F_{q+1}\left(\begin{matrix}\rho,\ (a_p)+\rho-s;\ -\sigma\omega\\\rho-s+1,\ (b_q)+\rho-s\end{matrix}\right)+\sigma^{s-\rho}\sec\frac{\rho\pi}{2}$ $\times\cos\frac{(2s-\rho)\pi}{2}\,\mathrm{B}(\rho-s,\ s)\,{}_{p+1}F_{q+1}\left(\begin{matrix}s,\ (a_p);\ -\sigma\omega\\s-\rho+1,\ (b_q)\end{matrix}\right)$ $[0<\rho<1]$ $\left[\begin{matrix}[q=p-1;\ \sigma>0;\ \lvert\arg\omega\rvert<\pi;\ 0<\operatorname{Re}s<\operatorname{Re}(a_k+\rho)]\ \text{or}\\ [q=p;\ (\sigma,\ \operatorname{Re}\omega>0;\ 0<\operatorname{Re}s<\operatorname{Re}(a_k+\rho))\ \text{or}\\ (\sigma>0;\ \operatorname{Re}\omega=0;\ 0<\operatorname{Re}s<\operatorname{Re}(a_k+\rho),\ \operatorname{Re}(\rho-\chi)+1)]\ \text{or}\\ [q=p+1;\ \sigma,\ \omega>0;\ 0<\operatorname{Re}s<\operatorname{Re}(a_k+\rho),\ \operatorname{Re}(\rho-\chi)+1/2]\end{matrix}\right]$
5	$\frac{1}{x-\sigma}\,{}_pF_q\left(\begin{matrix}(a_p);\,-\omega x\\(b_q)\end{matrix}\right)$	$\frac{\Gamma(s-1)}{\omega^{s-1}}\,\Gamma\left[\begin{matrix}(b_q),\ (a_p)-s+1\\(a_p),\ (b_q)-s+1\end{matrix}\right]$ $\times\,{}_{p+1}F_{q+1}\left(\begin{matrix}1,\ (a_p)-s+1;\ -\sigma\omega\\2-s,\ (b_q)-s+1\end{matrix}\right)$ $-\pi\sigma^{s-1}\cot(s\pi)\,{}_pF_q\left(\begin{matrix}(a_p);\ -\sigma\omega\\(b_q)\end{matrix}\right)$ $\left[\begin{matrix}[q=p-1;\ \sigma>0;\ \lvert\arg\omega\rvert<\pi;\ 0<\operatorname{Re}s<\operatorname{Re}a_k+1]\ \text{or}\\ [q=p;\ (\sigma,\ \operatorname{Re}\omega>0;\ 0<\operatorname{Re}s<\operatorname{Re}a_k+1)\ \text{or}\\ (\sigma>0;\ \operatorname{Re}\omega=0;\ 0<\operatorname{Re}s<\operatorname{Re}a_k+1,\ 2-\operatorname{Re}\chi)\ \text{or}\\ [q=p+1;\ \sigma,\ \omega>0;\ 0<\operatorname{Re}s<\operatorname{Re}a_k+1,\ 3/2-\operatorname{Re}\chi]\end{matrix}\right]$
6	$\frac{1}{(x+\sigma)^{\rho}}\,{}_pF_q\left(\begin{matrix}(a_p);\,-\omega x\\(b_q)\end{matrix}\right)$	$\sigma^{s-\rho}\,\mathrm{B}(\rho-s,\ s)\,{}_{p+1}F_{q+1}\left(\begin{matrix}(a_p),\ s;\ \sigma\omega\\(b_q),\ s-\rho+1\end{matrix}\right)$ $+\,\omega^{\rho-s}\,\Gamma\left[\begin{matrix}(b_q),\ s-\rho,\ (a_p)+\rho-s\\(a_p),\ (b_q)+\rho-s\end{matrix}\right]$ $\times\,{}_{p+1}F_{q+1}\left(\begin{matrix}\rho,\ (a_p)+\rho-s;\ \sigma\omega\\\rho-s+1,\ (b_q)+\rho-s\end{matrix}\right)$ $\left[\begin{matrix}[q=p-1;\ \lvert\arg\sigma\rvert,\ \lvert\arg\omega\rvert<\pi;\ 0<\operatorname{Re}s<\operatorname{Re}(a_k+\rho)]\ \text{or}\\ [q=p;\ (\lvert\arg\sigma\rvert<\pi;\ \operatorname{Re}\omega>0;\ 0<\operatorname{Re}s<\operatorname{Re}(a_k+\rho))\ \text{or}\\ (\lvert\arg\sigma\rvert<\pi;\ \operatorname{Re}\omega=0;\ 0<\operatorname{Re}s<\operatorname{Re}(a_k+\rho),\ \operatorname{Re}(\rho-\chi)+1)]\ \text{or}\\ [q=p+1;\ \lvert\arg\sigma\rvert<\pi;\ \omega>0;\ 0<\operatorname{Re}s<\operatorname{Re}(a_k+\rho),\ \operatorname{Re}(\rho-\chi)+1/2]\end{matrix}\right]$
7	$(\sigma^2-x^2)_+^{\alpha-1}$ $\,{}_pF_q\left(\begin{matrix}(a_p);\,-\omega x\\(b_q)\end{matrix}\right)$	$\prod_{i=1}^{p}a_i\prod_{j=1}^{q}b_j^{-1}\frac{\sigma^{s+2\alpha-1}\omega}{2}\,\mathrm{B}\left(\frac{s+1}{2},\ \alpha\right)$ $\times\,{}_{2p+1}F_{2q+2}\left(\begin{matrix}\frac{(a_p)+1}{2},\ \frac{(a_p)+2}{2},\ \frac{s+1}{2};\ \frac{\sigma^2\omega^2}{4}\\\frac{3}{2},\ \frac{(b_q)+1}{2},\ \frac{(b_q)+2}{2},\ \frac{s+2\alpha+1}{2}\end{matrix}\right)$ $+\,\frac{\sigma^{s+2\alpha-2}}{2}\,\mathrm{B}\left(\frac{s}{2},\ \alpha\right)\,{}_{2p+1}F_{2q+2}\left(\begin{matrix}\frac{(a_p)}{2},\ \frac{(a_p)+1}{2},\ \frac{s}{2};\ \frac{\sigma^2\omega^2}{4}\\\frac{1}{2},\ \frac{(b_q)}{2},\ \frac{(b_q)+1}{2},\ \frac{s+2\alpha}{2}\end{matrix}\right)$ $[\sigma,\ \operatorname{Re}\alpha,\ \operatorname{Re}s>0]$

No.	$f(x)$	$F(s)$
8	$\left(x^2-\sigma^2\right)_+^{\alpha-1} \times {}_pF_q\left(\begin{matrix}(a_p);\,-\omega x\\(b_q)\end{matrix}\right)$	$\frac{\Gamma(s+2\alpha-2)}{\omega^{s+2\alpha-2}}\,\Gamma\left[\begin{matrix}(b_q),\,(a_p)-2\alpha-s+2\\(a_p),\,(b_q)-2\alpha-s+2\end{matrix}\right] \times {}_{2p+1}F_{2q+2}\left(\begin{matrix}1-\alpha,\,\frac{(a_p)-2\alpha-s+2}{2},\,\frac{(a_p)-2\alpha-s+3}{2};\,\frac{\sigma^2\omega^2}{16}\\\frac{3-2\alpha-s}{2},\,\frac{4-2\alpha-s}{2},\,\frac{(b_q)-2\alpha-s+2}{2},\,\frac{(b_q)-2\alpha-s+3}{2}\end{matrix}\right) - \prod_{i=1}^{p}a_i\prod_{j=1}^{q}b_j^{-1}\,\frac{\sigma^{s+2\alpha-1}\omega}{2}\,\mathrm{B}\left(\alpha,\,\frac{1-2\alpha-s}{2}\right) \times {}_{2p+1}F_{2q+2}\left(\begin{matrix}\frac{(a_p)+1}{2},\,\frac{(a_p)+2}{2},\,\frac{s+1}{2};\,\frac{\sigma^2\omega^2}{16}\\\frac{3}{2},\,\frac{(b_q)+1}{2},\,\frac{(b_q)+2}{2},\,\frac{s+2\alpha+1}{2}\end{matrix}\right) + \frac{\sigma^{s+2\alpha-2}}{2}\,\mathrm{B}\left(\alpha,\,\frac{2-2\alpha-s}{2}\right) \times {}_{2p+1}F_{2q+2}\left(\begin{matrix}\frac{(a_p)}{2},\,\frac{(a_p)+1}{2},\,\frac{s}{2};\,\frac{\sigma^2\omega^2}{16}\\\frac{1}{2},\,\frac{(b_q)}{2},\,\frac{(b_q)+1}{2},\,\frac{s+2\alpha}{2}\end{matrix}\right)$
		$\left[\begin{matrix}[q=p-1;\,\sigma,\ \mathrm{Re}\,\alpha>0;\ \lvert\arg\omega\rvert<\pi;\ \mathrm{Re}\,s<\mathrm{Re}\,(a_k-2\alpha)+2]\ \text{or}\\ [q=p;\ \mathrm{Re}\,\alpha>0;\ (\sigma,\ \mathrm{Re}\,\omega>0;\ \mathrm{Re}\,s<\mathrm{Re}\,(a_k-2\alpha)+2)\ \text{or}\\ (\sigma>0;\ \mathrm{Re}\,\omega=0;\ \mathrm{Re}\,s<\mathrm{Re}\,(a_k-2\alpha)+2,\,3-\mathrm{Re}\,(2\alpha+\chi))]\ \text{or}\\ [q=p+1;\,\sigma,\,\omega,\ \mathrm{Re}\,\alpha>0;\ \mathrm{Re}\,s<\mathrm{Re}\,(a_k-2\alpha)+2,\,5/2-\mathrm{Re}\,(2\alpha+\chi)]\end{matrix}\right]$
9	$\frac{1}{\lvert x^2-\sigma^2\rvert^{\rho}}\,{}_pF_q\left(\begin{matrix}(a_p);\,-\omega x\\(b_q)\end{matrix}\right)$	$\left(\omega^2\right)^{\rho-s/2}\Gamma(s-2\rho)\,\Gamma\left[\begin{matrix}(b_q),\,(a_p)+2\rho-s\\(a_p),\,(b_q)+2\rho-s\end{matrix}\right] \times {}_{2p+1}F_{2q+2}\left(\begin{matrix}\rho,\,\frac{(a_p)+2\rho-s}{2},\,\frac{(a_p)+2\rho-s+1}{2};\,\frac{\sigma^2\omega^2}{16}\\\frac{2\rho-s+1}{2},\,\frac{2\rho-s+2}{2},\,\frac{(b_q)+2\rho-s}{2},\,\frac{(b_q)+2\rho-s+1}{2}\end{matrix}\right) + \prod_{i=1}^{p}a_i\prod_{j=1}^{q}b_j^{-1}\,\frac{\sigma^{s-2\rho+1}\sqrt{\omega^2}}{2} \times \sec\frac{\rho\pi}{2}\sin\frac{(s-\rho)\pi}{2}\,\mathrm{B}\left(\frac{s+1}{2},\,\frac{2\rho-s-1}{2}\right) \times {}_{2p+1}F_{2q+2}\left(\begin{matrix}\frac{s+1}{2},\,\frac{(a_p)+1}{2},\,\frac{(a_p)+2}{2};\,\frac{\sigma^2\omega^2}{16}\\\frac{3}{2},\,\frac{s-2\rho+3}{2},\,\frac{(b_q)+1}{2},\,\frac{(b_q)+2}{2}\end{matrix}\right) + \frac{\sigma^{s-2\rho}}{2}\sec\frac{\rho\pi}{2}\cos\frac{(s-\rho)\pi}{2}\,\mathrm{B}\left(\frac{s}{2},\,\frac{2\rho-s}{2}\right) \times {}_{2p+1}F_{2q+2}\left(\begin{matrix}\frac{s}{2},\,\frac{(a_p)}{2},\,\frac{(a_p)+1}{2};\,\frac{\sigma^2\omega^2}{16}\\\frac{1}{2},\,\frac{s-2\rho+2}{2},\,\frac{(b_q)}{2},\,\frac{(b_q)+1}{2}\end{matrix}\right)$ $[0<\rho<1/2]$
		$\left[\begin{matrix}[q=p-1;\,\sigma>0;\ \lvert\arg\omega\rvert<\pi;\ 0<\mathrm{Re}\,s<\mathrm{Re}\,(a_k+2\rho)]\ \text{or}\\ [q=p;\ (\sigma,\ \mathrm{Re}\,\omega>0;\ 0<\mathrm{Re}\,s<\mathrm{Re}\,(a_k+2\rho))\ \text{or}\\ (\sigma>0;\ \mathrm{Re}\,\omega=0;\ 0<\mathrm{Re}\,s<\mathrm{Re}\,(a_k+2\rho),\ \mathrm{Re}\,(2\rho-\chi)+1)]\ \text{or}\\ [q=p+1;\,\sigma,\,\omega>0;\ 0<\mathrm{Re}\,s<\mathrm{Re}\,(a_k+2\rho),\ \mathrm{Re}\,(2\rho-\chi)+1/2]\end{matrix}\right]$

No.	$f(x)$	$F(s)$
10	$\frac{1}{x^2-\sigma^2}\,{}_pF_q\left(\begin{matrix}(a_p);\,-\omega x\\ (b_q)\end{matrix}\right)$	$\frac{\Gamma(s-2)}{\omega^{s-2}}\,\Gamma\left[\begin{matrix}(b_q),\ (a_p)-s+2\\ (a_p),\ (b_q)-s+2\end{matrix}\right]$ $\times\,{}_{2p+1}F_{2q+2}\left(\begin{matrix}1,\ \frac{(a_p)-s+2}{2},\ \frac{(a_p)-s+3}{2};\ \frac{\sigma^2\omega^2}{4}\\ \frac{3-s}{2},\ \frac{4-s}{2},\ \frac{(b_q)-s+2}{2},\ \frac{(b_q)-s+3}{2}\end{matrix}\right)$ $-\prod_{i=1}^{p}a_i\prod_{j=1}^{q}b_j^{-1}\,\frac{\pi\omega\sigma^{s-1}}{2}\tan\frac{s\pi}{2}$ $\times\,{}_{2p}F_{2q+1}\left(\begin{matrix}\frac{(a_p)+1}{2},\ \frac{(a_p)+2}{2};\ \frac{\sigma^2\omega^2}{4}\\ \frac{3}{2},\ \frac{(b_q)+1}{2},\ \frac{(b_q)+2}{2}\end{matrix}\right)$ $-\frac{\pi\sigma^{s-2}}{2}\cot\frac{s\pi}{2}\,{}_{2p}F_{2q+1}\left(\begin{matrix}\frac{(a_p)}{2},\ \frac{(a_p)+1}{2};\ \frac{\sigma^2\omega^2}{4}\\ \frac{1}{2},\ \frac{(b_q)}{2},\ \frac{(b_q)+1}{2}\end{matrix}\right)$ $[q=p-1;\ \sigma>0;\ \lvert\arg\omega\rvert<\pi;\ 0<\operatorname{Re}s<\operatorname{Re}a_k+2]$ or $[q=p;\ (\sigma,\ \operatorname{Re}\omega>0;\ 0<\operatorname{Re}s<\operatorname{Re}a_k+2)$ or $(\sigma>0;\ \operatorname{Re}\omega=0;\ 0<\operatorname{Re}s<\operatorname{Re}a_k+2,\ 3-\operatorname{Re}\chi)$ or $[q=p+1;\ \sigma,\ \omega>0;\ 0<\operatorname{Re}s<\operatorname{Re}a_k+2,\ 5/2-\operatorname{Re}\chi]$
11	$\frac{1}{(x^2+\sigma^2)^{\rho}}\,{}_pF_q\left(\begin{matrix}(a_p);\,-\omega x\\ (b_q)\end{matrix}\right)$	$\omega^{2\rho-s}\,\Gamma(s-2\rho)\,\Gamma\left[\begin{matrix}(b_q),\ (a_p)+2\rho-s\\ (a_p),\ (b_q)+2\rho-s\end{matrix}\right]$ $\times\,{}_{2p+1}F_{2q+2}\left(\begin{matrix}\rho,\ \frac{(a_p)+2\rho-s}{2},\ \frac{(a_p)+2\rho-s+1}{2};\ -\frac{\sigma^2\omega^2}{16}\\ \frac{2\rho-s+1}{2},\ \frac{2\rho-s+2}{2},\ \frac{(b_q)+2\rho-s}{2},\ \frac{(b_q)+2\rho-s+1}{2}\end{matrix}\right)$ $-\prod_{i=1}^{p}a_i\prod_{j=1}^{q}b_j^{-1}\,\frac{\sigma^{s-2\rho+1}\omega}{2}\,\mathrm{B}\left(\frac{2\rho-s-1}{2},\ \frac{s+1}{2}\right)$ $\times\,{}_{2p+1}F_{2q+2}\left(\begin{matrix}\frac{s+1}{2},\ \frac{(a_p)+1}{2},\ \frac{(a_p)+2}{2};\ -\frac{\sigma^2\omega^2}{16}\\ \frac{3}{2},\ \frac{s-2\rho+3}{2},\ \frac{(b_q)+1}{2},\ \frac{(b_q)+2}{2}\end{matrix}\right)$ $+\frac{\sigma^{s-2\rho}}{2}\,\mathrm{B}\left(\frac{2\rho-s}{2},\ \frac{s}{2}\right)$ $\times\,{}_{2p+1}F_{2q+2}\left(\begin{matrix}\frac{s}{2},\ \frac{(a_p)}{2},\ \frac{(a_p)+1}{2};\ -\frac{\sigma^2\omega^2}{16}\\ \frac{1}{2},\ \frac{s-2\rho+2}{2},\ \frac{(b_q)}{2},\ \frac{(b_q)+1}{2}\end{matrix}\right)$ $[q=p-1;\ \lvert\arg\sigma\rvert<\pi/2;\ \lvert\arg\omega\rvert<\pi;\ 0<\operatorname{Re}s<\operatorname{Re}(a_k+2\rho)]$ or $[q=p;\ \lvert\arg\sigma\rvert<\pi/2;\ (\operatorname{Re}\omega>0;\ 0<\operatorname{Re}s<\operatorname{Re}(a_k+2\rho))$ or $(\operatorname{Re}\omega=0;\ 0<\operatorname{Re}s<\operatorname{Re}(a_k+2\rho),\ \operatorname{Re}(2\rho-\chi)+1)]$ or $[q=p+1;\ \lvert\arg\sigma\rvert<\pi/2;\ \omega>0;\ 0<\operatorname{Re}s<\operatorname{Re}(a_k+2\rho),\ \operatorname{Re}(2\rho-\chi)+1/2]$
12	$(\sigma-x)_+^{\alpha-1}$ $\times\,{}_pF_q\left(\begin{matrix}(a_p);\,-\omega x^2\\ (b_q)\end{matrix}\right)$	$\sigma^{s+\alpha-1}\,\mathrm{B}(\alpha,\,s)\,{}_{p+1}F_{q+1}\left(\begin{matrix}(a_p),\ \frac{s}{2},\ \frac{s+1}{2};\ -\sigma^2\omega\\ (b_q),\ \frac{s+\alpha}{2},\ \frac{s+\alpha+1}{2}\end{matrix}\right)$ $[\sigma,\ \operatorname{Re}\alpha,\ \operatorname{Re}s>0]$

No.	$f(x)$	$F(s)$
13	$(x-\sigma)_+^{\alpha-1}$ $\times\ {}_pF_q\left(\begin{matrix}(a_p);\ -\omega x^2\\ (b_q)\end{matrix}\right)$	$\frac{(1-\alpha)\,\sigma}{2\omega^{s+\alpha-2}}\,\Gamma\left(\frac{s+\alpha-2}{2}\right)\Gamma\left[\begin{matrix}(b_q),\ \frac{2(a_p)-\alpha-s+2}{2}\\ (a_p),\ \frac{(b_q)-\alpha-s+2}{2}\end{matrix}\right]$ $\times\ {}_{p+1}F_{q+1}\left(\begin{matrix}\frac{2-\alpha}{2},\ \frac{3-\alpha}{2},\ \frac{2(a_p)-\alpha-s+2}{2};\ -\sigma^2\omega\\ \frac{3}{2},\ \frac{4-\alpha-s}{2},\ \frac{2(b_q)-\alpha-s+2}{2}\end{matrix}\right)$ $+\ \frac{1}{2\omega^{s+\alpha-1}}\,\Gamma\left(\frac{s+\alpha-1}{2}\right)\Gamma\left[\begin{matrix}(b_q),\ \frac{2(a_p)-\alpha-s+1}{2}\\ (a_p),\ \frac{(b_q)-\alpha-s+1}{2}\end{matrix}\right]$ $\times\ {}_{p+1}F_{q+1}\left(\begin{matrix}\frac{1-\alpha}{2},\ \frac{2-\alpha}{2},\ \frac{2(a_p)-\alpha-s+1}{2};\ -\sigma^2\omega\\ \frac{1}{2},\ \frac{3-\alpha-s}{2},\ \frac{2(b_q)-\alpha-s+1}{2}\end{matrix}\right)$ $+\ \sigma^{s+\alpha-1}\,\mathrm{B}\,(\alpha,\ 1-\alpha-s)\ {}_{p+1}F_{q+1}\left(\begin{matrix}(a_p),\ \frac{s}{2},\ \frac{s+1}{2};\ -\sigma^2\omega\\ (b_q),\ \frac{s+\alpha}{2},\ \frac{s+\alpha+1}{2}\end{matrix}\right)$
		$\left[\begin{matrix}[q=p-1;\ \sigma,\ \mathrm{Re}\,\alpha>0;\ \lvert\arg\omega\rvert<\pi;\ \mathrm{Re}\,s<\mathrm{Re}\,(2a_k-\alpha)+1]\ \text{or}\\ [q=p;\ \mathrm{Re}\,\alpha>0;\ (\sigma,\ \mathrm{Re}\,\omega>0;\ \mathrm{Re}\,s<\mathrm{Re}\,(2a_k-\alpha)+1)\ \text{or}\\ (\sigma>0;\ \mathrm{Re}\,\omega=0;\ \mathrm{Re}\,s<\mathrm{Re}\,(2a_k-\alpha)+1,\ 3-\mathrm{Re}\,(\alpha+2\chi))]\ \text{or}\\ [q=p+1;\ \sigma,\ \omega,\ \mathrm{Re}\,\alpha>0;\ \mathrm{Re}\,s<\mathrm{Re}\,(2a_k-\alpha)+1,\ 2-\mathrm{Re}\,(\alpha+2\chi)]\end{matrix}\right]$
14	$\frac{1}{\lvert x-\sigma\rvert^{\rho}}\ {}_pF_q\left(\begin{matrix}(a_p);\ -\omega x^2\\ (b_q)\end{matrix}\right)$	$\frac{\rho\sigma\omega^{(\rho-s+1)/2}}{2}\Gamma\left(\frac{s-\rho+1}{2}\right)\Gamma\left[\begin{matrix}(b_q),\ \frac{2(a_p)+\rho-s+1}{2}\\ (a_p),\ \frac{2(b_q)+\rho-s+1}{2}\end{matrix}\right]$ $\times\ {}_{p+2}F_{q+2}\left(\begin{matrix}\frac{\rho+1}{2},\ \frac{\rho+2}{2},\ \frac{2(a_p)+\rho-s+1}{2};\ -\sigma^2\omega\\ \frac{3}{2},\ \frac{\rho-s+3}{2},\ \frac{2(b_q)+\rho-s+1}{2}\end{matrix}\right)$ $+\ \frac{\omega^{(\rho-s)/2}}{2}\Gamma\left(\frac{s-\rho}{2}\right)\Gamma\left[\begin{matrix}(b_q),\ \frac{2(a_p)+\rho-s}{2}\\ (a_p),\ \frac{2(b_q)+\rho-s}{2}\end{matrix}\right]$ $\times\ {}_{p+2}F_{q+2}\left(\begin{matrix}\frac{\rho}{2},\ \frac{\rho+1}{2},\ \frac{2(a_p)+\rho-s}{2};\ -\sigma^2\omega\\ \frac{1}{2},\ \frac{\rho-s+2}{2},\ \frac{2(b_q)+\rho-s}{2}\end{matrix}\right)$ $+\ \sigma^{s-\rho}\sec\frac{\rho\pi}{2}\cos\frac{(2s-\rho)\,\pi}{2}\,\mathrm{B}\,(s,\ \rho-s)$ $\times\ {}_{p+2}F_{q+2}\left(\begin{matrix}(a_p),\ \frac{s}{2},\ \frac{s+1}{2};\ -\sigma^2\omega\\ (b_q),\ \frac{s-\rho+1}{2},\ \frac{s-\rho+2}{2}\end{matrix}\right)$ $[0<\rho<1]$
		$\left[\begin{matrix}[q=p-1;\ \sigma>0;\ \lvert\arg\omega\rvert<\pi;\ 0<\mathrm{Re}\,s<\mathrm{Re}\,(2a_k+\rho)]\ \text{or}\\ [q=p;\ (\sigma,\ \mathrm{Re}\,\omega>0;\ 0<\mathrm{Re}\,s<\mathrm{Re}\,(2a_k+\rho))\ \text{or}\\ (\sigma>0;\ \mathrm{Re}\,\omega=0;\ 0<\mathrm{Re}\,s<\mathrm{Re}\,(2a_k+\rho),\ \mathrm{Re}\,(\rho-2\chi)+2)]\ \text{or}\\ [q=p+1;\ \sigma,\ \omega>0;\ 0<\mathrm{Re}\,s<\mathrm{Re}\,(2a_k+\rho),\ \mathrm{Re}\,(\rho-2\chi)+1]\end{matrix}\right]$
15	$\frac{1}{x-\sigma}\ {}_pF_q\left(\begin{matrix}(a_p);\ -\omega x^2\\ (b_q)\end{matrix}\right)$	$\frac{\sigma}{2\,\omega^{(s-2)/2}}\,\Gamma\left(\frac{s-2}{2}\right)\Gamma\left[\begin{matrix}(b_q),\ \frac{2(a_p)-s+2}{2}\\ (a_p),\ \frac{2(b_q)-s+2}{2}\end{matrix}\right]$ $\times\ {}_{p+1}F_{q+1}\left(\begin{matrix}1,\ \frac{2(a_p)-s+2}{2};\ -\sigma^2\omega\\ \frac{4-s}{2},\ \frac{2(b_q)-s+2}{2}\end{matrix}\right)$ $+\ \frac{1}{2\,\omega^{(s-1)/2}}\,\Gamma\left(\frac{s-1}{2}\right)\Gamma\left[\begin{matrix}(b_q),\ \frac{2(a_p)-s+1}{2}\\ (a_p),\ \frac{2(b_q)-s+1}{2}\end{matrix}\right]\ \times$

No.	$f(x)$	$F(s)$
		$\times\ {}_{p+1}F_{q+1}\left(\begin{matrix}1,\ \frac{2(a_p)-s+1}{2};\ -\sigma^2\omega\\ \frac{3-s}{2},\ \frac{2(b_q)-s+1}{2}\end{matrix}\right)$ $-\pi\sigma^{s-1}\cot(s\pi)\ {}_pF_q\left(\begin{matrix}(a_p);\ -\sigma^2\omega\\ (b_q)\end{matrix}\right)$
		$\left[\begin{matrix}[q=p-1;\ \sigma>0;\ \lvert\arg\omega\rvert<\pi;\ 0<\operatorname{Re}s<2\operatorname{Re}a_k+1]\ \text{or}\\ [q=p;\ (\sigma,\ \operatorname{Re}\omega>0;\ 0<\operatorname{Re}s<2\operatorname{Re}a_k+1)\ \text{or}\\ (\sigma>0;\ \operatorname{Re}\omega=0;\ 0<\operatorname{Re}s<2\operatorname{Re}a_k+1,\ 3-2\operatorname{Re}\chi)\ \text{or}\\ [q=p+1;\ \sigma,\ \omega>0;\ 0<\operatorname{Re}s<2\operatorname{Re}a_k+1,\ 2-2\operatorname{Re}\chi]\end{matrix}\right]$
16	$\frac{1}{(x+\sigma)^\rho}\ {}_pF_q\left(\begin{matrix}(a_p);\ -\omega x^2\\ (b_q)\end{matrix}\right)$	$\sigma^{s-\rho}\operatorname{B}(\rho-s,\ s)\ {}_{p+2}F_{q+2}\left(\begin{matrix}(a_p),\ \frac{s}{2},\ \frac{s+1}{2};\ -\sigma^2\omega\\ (b_q),\ \frac{s-\rho+1}{2},\ \frac{s-\rho+2}{2}\end{matrix}\right)$ $+\frac{\omega^{(\rho-s)/2}}{2}\Gamma\left[\begin{matrix}(b_q),\ \frac{s-\rho}{2},\ \frac{2(a_p)+\rho-s}{2}\\ (a_p),\ \frac{2(b_q)+\rho-s}{2}\end{matrix}\right]$ $\times\ {}_{p+2}F_{q+2}\left(\begin{matrix}\frac{\rho}{2},\ \frac{\rho+1}{2},\ \frac{2(a_p)+\rho-s}{2};\ -\sigma^2\omega\\ \frac{1}{2},\ \frac{\rho-s+2}{2},\ \frac{2(b_q)+\rho-s}{2}\end{matrix}\right)$ $-\frac{\rho\sigma\omega^{(\rho-s+1)/2}}{2}\Gamma\left[\begin{matrix}(b_q),\ \frac{s-\rho+1}{2},\ \frac{2(a_p)+\rho-s+1}{2}\\ (a_p),\ \frac{2(b_q)+\rho-s+1}{2}\end{matrix}\right]$ $\times\ {}_{p+2}F_{q+2}\left(\begin{matrix}\frac{\rho+1}{2},\ \frac{\rho+2}{2},\ \frac{2(a_p)+\rho-s+1}{2};\ -\sigma^2\omega\\ \frac{3}{2},\ \frac{\rho-s+3}{2},\ \frac{2(b_q)+\rho-s+1}{2}\end{matrix}\right)$
		$\left[\begin{matrix}[q=p-1;\ \lvert\arg\sigma\rvert,\ \lvert\arg\omega\rvert<\pi;\ 0<\operatorname{Re}s<\operatorname{Re}(2a_k+\rho)]\ \text{or}\\ [q=p;\ (\lvert\arg\sigma\rvert<\pi;\ \operatorname{Re}\omega>0;\ 0<\operatorname{Re}s<\operatorname{Re}(2a_k+\rho))\ \text{or}\\ (\lvert\arg\sigma\rvert<\pi;\ \operatorname{Re}\omega=0;\ 0<\operatorname{Re}s<\operatorname{Re}(2a_k+\rho),\ \operatorname{Re}(\rho-2\chi)+2)]\ \text{or}\\ [q=p+1;\ \lvert\arg\sigma\rvert<\pi;\ \omega>0;\ 0<\operatorname{Re}s<\operatorname{Re}(2a_k+\rho),\ \operatorname{Re}(\rho-2\chi)+1]\end{matrix}\right]$
17	$\frac{1}{(x+\sigma)^\rho}\ {}_pF_q\left(\begin{matrix}(a_p);\ \frac{b}{x+\sigma}\\ (b_q)\end{matrix}\right)$	$\sigma^{s-\rho}\operatorname{B}(s,\ \rho-s)\ {}_{p+1}F_{q+1}\left(\begin{matrix}(a_p),\ \rho-s\\ (b_q),\ \rho;\ \frac{b}{\sigma}\end{matrix}\right)$ $[0<\operatorname{Re}s<\operatorname{Re}\rho;\ \lvert\arg\sigma\rvert<\pi]$
18	$\frac{1}{(x+\sigma)^\rho}\ {}_pF_q\left(\begin{matrix}(a_p);\ \frac{bx}{x+\sigma}\\ (b_q)\end{matrix}\right)$	$\sigma^{s-\rho}\operatorname{B}(s,\ \rho-s)\ {}_{p+1}F_{q+1}\left(\begin{matrix}(a_p),\ s\\ (b_q),\ \rho;\ b\end{matrix}\right)$ $[0<\operatorname{Re}s<\operatorname{Re}\rho;\ \lvert\arg\sigma\rvert<\pi]$
19	$\frac{1}{(x+\sigma)^\rho}\ {}_pF_q\left(\begin{matrix}(a_p);\ \frac{b}{(x+\sigma)^2}\\ (b_q)\end{matrix}\right)$	$\sigma^{s-\rho}\operatorname{B}(s,\ \rho-s)\ {}_{p+1}F_{q+1}\left(\begin{matrix}(a_p),\ \frac{\rho-s}{2},\ \frac{\rho-s+1}{2}\\ (b_q),\ \frac{\rho}{2},\ \frac{\rho+1}{2};\ \frac{b}{\sigma^2}\end{matrix}\right)$ $[0<\operatorname{Re}s<\operatorname{Re}\rho;\ \lvert\arg\sigma\rvert<\pi]$
20	$\frac{1}{(x+\sigma)^\rho}\ {}_pF_q\left(\begin{matrix}(a_p);\ \frac{bx^2}{(x+\sigma)^2}\\ (b_q)\end{matrix}\right)$	$\sigma^{s-\rho}\operatorname{B}(s,\ \rho-s)\ {}_{p+2}F_{q+2}\left(\begin{matrix}(a_p),\ \frac{s}{2},\ \frac{s+1}{2}\\ (b_q),\ \frac{\rho}{2},\ \frac{\rho+1}{2};\ b\end{matrix}\right)$ $[0<\operatorname{Re}s<\operatorname{Re}\rho;\ \lvert\arg\sigma\rvert<\pi]$

3.33.2. ${}_pF_q\left((a_p);\,(b_q);\,\omega x^r\right)$ and the exponential function

Notation:

$$\mu=\sum_{i=1}^{p}a_i-\sum_{j=1}^{q}b_j+\frac{q-p+1}{2}.$$

1	$e^{-\sigma x}\,{}_pF_q\left(\begin{matrix}(a_p);\,-\omega x\\(b_q)\end{matrix}\right)$	$\dfrac{\Gamma(s)}{\sigma^s}\,{}_{p+1}F_q\left(\begin{matrix}(a_p),\,s\\(b_q);\,-\frac{\omega}{\sigma}\end{matrix}\right)$ $\Big[\big[q=p-1;\ \lvert\arg\omega\rvert<\pi;\ (\operatorname{Re}\sigma,\ \operatorname{Re}s>0)$ or $(\operatorname{Re}\sigma=0;\ 0<\operatorname{Re}s<\operatorname{Re}a_k+1)\big]$ or $\big[q=p;\ (\operatorname{Re}\sigma,\ \operatorname{Re}(\sigma+\omega),\ \operatorname{Re}s>0)$ or $(\operatorname{Re}\sigma>0;\ \operatorname{Re}(\sigma+\omega)=0;\ 0<\operatorname{Re}s<1-\operatorname{Re}\chi)$ or $(\operatorname{Re}\sigma=0;\ \operatorname{Re}\omega>0;\ 0<\operatorname{Re}s<\operatorname{Re}a_k+1)$ or $(\operatorname{Re}\sigma=\operatorname{Re}\omega=0;\ 0<\operatorname{Re}s<\operatorname{Re}a_k+1,\ 1-\operatorname{Re}\chi)\big]$ or $\big[q=p+1;\ (\operatorname{Re}\sigma,\ \operatorname{Re}s>0;\ \lvert\arg\omega\rvert<\pi)$ or $(\operatorname{Re}\sigma=0;\ \omega>0;\ 0<\operatorname{Re}s<\operatorname{Re}a_k+1,\ 1-\operatorname{Re}\chi)\big]$ or $\big[q\ge p+2;\ \operatorname{Re}\sigma,\ \operatorname{Re}s>0;\ \lvert\arg\omega\rvert<\pi\big]\Big]$
2	$e^{-\sigma x^k}\,{}_pF_q\left(\begin{matrix}(a_p);\,-\omega x^\ell\\(b_q)\end{matrix}\right)$	$\dfrac{k^{\mu-1}\ell^{s/k-1/2}\sigma^{-s/k}}{(2\pi)^{[(k-1)(p-q+1)+\ell-1]/2}}\,\Gamma\left[\begin{matrix}(b_q)\\(a_p)\end{matrix}\right]$ $\times\,G_{kp+\ell,\,kq+k}^{k,\,kp+\ell}\left(\dfrac{\ell^\ell\omega^k}{k^{k(q-p+1)}\sigma^\ell}\,\middle\vert\,\begin{matrix}\Delta(k,\,1-(a_p)),\Delta(\ell,\,s)\\\Delta(k,\,0),\Delta(k,\,1-(b_q))\end{matrix}\right)$ $[A=\min_{1\le i\le p}a_i]$ $\Big[\big[q=p-1;\ k>0;\ \lvert\arg\omega\rvert<\pi;\ (\operatorname{Re}\sigma,\ \operatorname{Re}s>0)$ or $(\operatorname{Re}\sigma=0;\ 0<\operatorname{Re}s<k+\ell A)\big]$ or $\big[q=p;\ \big(0<k<\ell;\ (\operatorname{Re}\sigma,\ \operatorname{Re}s>0;\ \operatorname{Re}\omega\ge0)$ or $(\operatorname{Re}\sigma=0;\ \operatorname{Re}\omega>0;\ 0<\operatorname{Re}s<k+\ell A)$ or $(\operatorname{Re}\sigma=\operatorname{Re}\omega=0;\ 0<\operatorname{Re}s<k+\ell A,\ \ell-\ell\operatorname{Re}\chi)\big)$ or $\big(k=\ell;\ (\operatorname{Re}\sigma,\ \operatorname{Re}(\sigma+\omega),\ \operatorname{Re}s>0)$ or $(\operatorname{Re}\sigma=0;\ \operatorname{Re}\omega>0;\ 0<\operatorname{Re}s<\ell+\ell A)$ or $(\operatorname{Re}\sigma>0;\ \operatorname{Re}(\sigma+\omega)=0;\ 0<\operatorname{Re}s<\ell-\ell\operatorname{Re}\chi)$ or $(\operatorname{Re}\sigma=\operatorname{Re}\omega=0;\ 0<\operatorname{Re}s<\ell+\ell A,\ \ell-\ell\operatorname{Re}\chi)\big)$ or $\big(k>\ell;\ (\operatorname{Re}\sigma,\ \operatorname{Re}s>0;\ \lvert\arg\omega\rvert<\pi)$ or $(\operatorname{Re}\sigma=0;\ \operatorname{Re}\omega>0;\ 0<\operatorname{Re}s<k+\ell A)$ or $(\operatorname{Re}\sigma=\operatorname{Re}\omega=0;\ 0<\operatorname{Re}s<k+\ell A,\ k-\ell\operatorname{Re}\chi)\big)\big]$ or $\big[q=p+1;\ \big(0<k<\ell/2;\ (\omega,\ \operatorname{Re}\sigma,\ \operatorname{Re}s>0)$ or $(\operatorname{Re}\sigma=0;\ \omega>0;\ 0<\operatorname{Re}s<k+\ell A,\ \ell/2-\ell\operatorname{Re}\chi)\big)$ or $\big(k=\ell/2;\ (2\lvert\operatorname{Im}\sqrt{\omega}\rvert<\operatorname{Re}\sigma;\ \operatorname{Re}s>0)$ or $(\operatorname{Re}\sigma=0;\ \omega>0;\ 0<\operatorname{Re}s<k+\ell A,\ \ell/2-\ell\operatorname{Re}\chi)\big)$ or $\big(k>\ell/2;\ (\operatorname{Re}\sigma,\ \operatorname{Re}s>0;\ \lvert\arg\omega\rvert<\pi)$ or $(\operatorname{Re}\sigma=0;\ \omega>0;\ 0<\operatorname{Re}s<k+\ell A,\ k-\ell\operatorname{Re}\chi)\big)\big]$ or $\big[q\ge p+2;\ \big(k=\ell/(q-p+1);\ (\operatorname{Re}\sigma,\ \operatorname{Re}s>0;\ \lvert\arg\omega\rvert<\pi)$ or $(\operatorname{Re}\sigma=0;\ \operatorname{Re}(-\omega)^{k/\ell}>0;\ 0<\operatorname{Re}s<\ell\operatorname{Re}a_k+k)\big)$ or $\big(k>\ell/(q-p+1);\ \operatorname{Re}\sigma,\ \operatorname{Re}s>0;\ \lvert\arg\omega\rvert<\pi\big)\big]\Big]$

No.	$f(x)$	$F(s)$
3	$e^{-\sigma x}\,{}_pF_q\left(\begin{matrix}(a_p);\,\omega x^{\ell}\\(b_q)\end{matrix}\right)$	$\frac{\Gamma(s)}{\sigma^s}\,{}_{p+\ell}F_q\left(\begin{matrix}(a_p),\,\Delta(\ell,\,s)\\(b_q);\,\left(\frac{\ell}{\sigma}\right)^{\ell}\omega\end{matrix}\right)$ $\left[\begin{matrix}p+\ell\le q+1;\ \operatorname{Re}s>0;\\ p+\ell<q;\ \operatorname{Re}\sigma>0;\\ p+\ell=q+1;\ \operatorname{Re}\left(\sigma+\ell\omega^{1/\ell}e^{2\pi ji/\ell}\right)>0\\ (j=0,\,1,\ldots,\ell-1)\end{matrix}\right]$
4	$e^{-\sigma x^2}\,{}_pF_q\left(\begin{matrix}(a_p);\,-\omega x\\(b_q)\end{matrix}\right)$	$-\prod_{i=1}^{p}a_i\prod_{j=1}^{q}b_j^{-1}\frac{\omega}{2\,\sigma^{(s+1)/2}}\Gamma\left(\frac{s+1}{2}\right)$ $\times\,{}_{2p+1}F_{2q+1}\left(\begin{matrix}\frac{(a_p)+1}{2},\,\frac{(a_p)+2}{2},\,\frac{s+1}{2}\\ \frac{3}{2},\,\frac{(b_q)+1}{2},\,\frac{(b_q)+2}{2};\,\frac{\omega^2}{4^{q-p+1}\sigma}\end{matrix}\right)$ $+\frac{\sigma^{-s/2}}{2}\Gamma\left(\frac{s}{2}\right)\,{}_{2p+1}F_{2q+1}\left(\begin{matrix}\frac{(a_p)}{2},\,\frac{(a_p)+1}{2},\,\frac{s}{2}\\ \frac{1}{2},\,\frac{(b_q)}{2},\,\frac{(b_q)+1}{2};\,\frac{\omega^2}{4^{q-p+1}\sigma}\end{matrix}\right)$ $\left[\begin{matrix}[q=p-1;\ \lvert\arg\omega\rvert<\pi;\ (\operatorname{Re}\sigma,\ \operatorname{Re}s>0)\ \text{or}\\ (\operatorname{Re}\sigma=0;\ 0<\operatorname{Re}s<\operatorname{Re}a_k+2)]\ \text{or}\\ [q=p;\ (\operatorname{Re}\sigma,\ \operatorname{Re}s>0;\ \lvert\arg\omega\rvert<\pi)\ \text{or}\\ (\operatorname{Re}\sigma=0;\ \operatorname{Re}\omega>0;\ 0<\operatorname{Re}s<\operatorname{Re}a_k+2)\ \text{or}\\ (\operatorname{Re}\sigma=\operatorname{Re}\omega=0;\ 0<\operatorname{Re}s<\operatorname{Re}a_k+2,\ 2-\operatorname{Re}\chi)]\ \text{or}\\ [q=p+1;\ (\operatorname{Re}\sigma,\ \operatorname{Re}s>0;\ \lvert\arg\omega\rvert<\pi)\ \text{or}\\ (\operatorname{Re}\sigma=0;\ \omega>0;\ 0<\operatorname{Re}s<\operatorname{Re}a_k+2,\ 2-\operatorname{Re}\chi)]\ \text{or}\\ [q\ge p+2;\ \operatorname{Re}\sigma,\ \operatorname{Re}s>0;\ \lvert\arg\omega\rvert<\pi]\end{matrix}\right]$
5	$e^{-\sigma/x}\,{}_pF_q\left(\begin{matrix}(a_p);\,-\omega x\\(b_q)\end{matrix}\right)$	$\omega^{-s}\Gamma\left[\begin{matrix}(b_q),\,s,\,(a_p)-s\\(a_p),\,(b_q)-s\end{matrix}\right]{}_pF_{q+1}\left(\begin{matrix}(a_p)-s;\,\sigma\omega\\(b_q)-s,\,1-s\end{matrix}\right)$ $+\sigma^s\Gamma(-s)\,{}_pF_{q+1}\left(\begin{matrix}(a_p);\,\sigma\omega\\(b_q),\,s+1\end{matrix}\right)$
	$\left[\begin{matrix}[q=p-1;\ \lvert\arg\omega\rvert<\pi;\ (\operatorname{Re}\sigma>0;\ \operatorname{Re}s<\operatorname{Re}a_k)\ \text{or}\ (\operatorname{Re}\sigma=0;\ \operatorname{Re}s>-1)]\ \text{or}\\ [q=p;\ (\operatorname{Re}\omega>0;\ \operatorname{Re}s<\operatorname{Re}a_k)\ \text{or}\ (\operatorname{Re}\omega=0;\ \operatorname{Re}s<\operatorname{Re}a_k,\ 1-\operatorname{Re}\chi)]\ \text{or}\\ [q=p+1;\ \omega>0;\ \operatorname{Re}s<\operatorname{Re}a_k,\ 1/2-\operatorname{Re}\chi]\end{matrix}\right]$	
6	$e^{-\sigma/x^2}\,{}_pF_q\left(\begin{matrix}(a_p);\,-\omega x\\(b_q)\end{matrix}\right)$	$\omega^{-s}\Gamma\left[\begin{matrix}(b_q),\,s,\,(a_p-s)\\(a_p),\,(b_q)-s\end{matrix}\right]{}_{2p}F_{2q+2}\left(\begin{matrix}\frac{(a_p)-s}{2},\,\frac{(a_p)-s+1}{2};\,-\frac{\sigma\omega^2}{4^{q-p+1}}\\ \frac{(b_q)-s}{2},\,\frac{(b_q)-s+1}{2},\,\frac{1-s}{2},\,\frac{2-s}{2}\end{matrix}\right)$ $-\prod_{i=1}^{p}a_i\prod_{j=1}^{q}b_j^{-1}\frac{\omega\sigma^{(s+1)/2}}{2}\Gamma\left(-\frac{s+1}{2}\right)$ $\times\,{}_{2p}F_{2q+2}\left(\begin{matrix}\frac{(a_p)+1}{2},\,\frac{(a_p)+2}{2};\,-\frac{\sigma\omega^2}{4^{q-p+1}}\\ \frac{3}{2},\,\frac{(b_q)+1}{2},\,\frac{(b_q)+2}{2},\,\frac{s+3}{2}\end{matrix}\right)$ $+\frac{\sigma^{s/2}}{2}\Gamma\left(-\frac{s}{2}\right)\,{}_{2p}F_{2q+2}\left(\begin{matrix}\frac{(a_p)}{2},\,\frac{(a_p)+1}{2};\,-\frac{\sigma\omega^2}{4^{q-p+1}}\\ \frac{1}{2},\,\frac{(b_q)}{2},\,\frac{(b_q)+1}{2},\,\frac{s+2}{2}\end{matrix}\right)$
	$\left[\begin{matrix}[q=p-1;\ \lvert\arg\omega\rvert<\pi;\ (\operatorname{Re}\sigma>0;\ \operatorname{Re}s<\operatorname{Re}a_k)\ \text{or}\ (\operatorname{Re}\sigma=0;\ -2<\operatorname{Re}s<\operatorname{Re}a_k)]\ \text{or}\\ [q=p;\ (\lvert\arg\omega\rvert<\pi/2;\ \operatorname{Re}s<\operatorname{Re}a_k)\ \text{or}\ (\lvert\arg\omega\rvert=\pi/2;\ \operatorname{Re}s<\operatorname{Re}a_k,\ 1-\operatorname{Re}\chi)]\ \text{or}\\ [q=p+1;\ \omega>0;\ \operatorname{Re}s<\operatorname{Re}a_k,\ 1/2-\operatorname{Re}\chi]\end{matrix}\right]$	

No.	$f(x)$	$F(s)$
7	$e^{-\sigma\sqrt{x}}\ {}_pF_q\left(\begin{matrix}(a_p);\ -\omega x\\ (b_q)\end{matrix}\right)$	$2\left(\sigma^2\right)^{-s}\Gamma(2s)\ {}_{p+2}F_q\left(\begin{matrix}(a_p),\ s,\ \frac{2s+1}{2}\\ (b_q);\ -\frac{4\omega}{\sigma^2}\end{matrix}\right)$ $\big[q=p-1;\ \lvert\arg\omega\rvert<\pi;\ (\operatorname{Re}\sigma,\ \operatorname{Re}s>0)$ or $(\operatorname{Re}\sigma=0;\ 0<\operatorname{Re}s<\operatorname{Re}a_k+1/2)\big]$ or $\big[q=p;\ (\operatorname{Re}\sigma,\ \operatorname{Re}s>0;\ \operatorname{Re}\omega\geq0)$ or $(\operatorname{Re}\sigma=0,\ \operatorname{Re}\omega>0;\ 0<\operatorname{Re}s<\operatorname{Re}a_k+1/2)$ or $(\operatorname{Re}\sigma=\operatorname{Re}\omega=0;\ 0<\operatorname{Re}s<\operatorname{Re}a_k+1/2,\ 1-\operatorname{Re}\chi)\big]$ or $\big[q=p+1;\ (2\lvert\operatorname{Im}\sqrt{\omega}\rvert<\operatorname{Re}\sigma;\ \operatorname{Re}s>0)$ or $(\operatorname{Re}\sigma=0;\ \omega>0;\ 0<\operatorname{Re}s<\operatorname{Re}a_k+1/2,\ 1/2-\operatorname{Re}\chi)$ or $(\operatorname{Re}\sigma>0;\ 2\lvert\operatorname{Im}\sqrt{\omega}\rvert+\operatorname{Re}\sigma=0;\ 0<\operatorname{Re}s<1/2-\operatorname{Re}\chi)\big]$ or $\big[q\geq p+2;\ \operatorname{Re}\sigma,\ \operatorname{Re}s>0;\ \lvert\arg\omega\rvert<\pi\big]$
8	$e^{-\sigma/\sqrt{x}}\ {}_pF_q\left(\begin{matrix}(a_p);\ -\omega x\\ (b_q)\end{matrix}\right)$	$-\left(\frac{1}{\sigma^2}\right)^{-s}\left(\sigma^2\omega\right)^{(1-2s)/2}\Gamma\left[\begin{matrix}\frac{2s-1}{2},\ (b_q),\ \frac{1-2s+2(a_p)}{2}\\ (a_p),\ \frac{1-2s+2(b_q)}{2}\end{matrix}\right]$ $\times\ {}_pF_{q+3}\left(\begin{matrix}\frac{1-2s+2(a_p)}{2};\ -\frac{\sigma^2\omega}{4}\\ \frac{3}{2},\ \frac{1-2s+2(b_q)}{2},\ \frac{3-2s}{2}\end{matrix}\right)$ $+\left(\frac{1}{\sigma^2}\right)^{-s}\left(\sigma^2\omega\right)^{-s}\Gamma\left[\begin{matrix}s,\ (b_q),\ (a_p)-s\\ (a_p),\ (b_q)-s\end{matrix}\right]$ $\times\ {}_pF_{q+3}\left(\begin{matrix}(a_p)-s;\ -\frac{\sigma^2\omega}{4}\\ \frac{1}{2},\ (b_q)-s,\ 1-s\end{matrix}\right)$ $+2\left(\frac{1}{\sigma^2}\right)^{-s}\Gamma(-2s)\ {}_pF_{q+3}\left(\begin{matrix}(a_p);\ -\frac{\sigma^2\omega}{4}\\ (b_q),\ \frac{2s+1}{2},\ s+1\end{matrix}\right)$ $\big[q=p-1;\ \lvert\arg\omega\rvert<\pi;\ (\operatorname{Re}\sigma>0;\ \operatorname{Re}s<\operatorname{Re}a_k)$ or $(\operatorname{Re}\sigma=0;\ \operatorname{Re}s>-1/2)\big]$ or $\big[q=p;\ (\operatorname{Re}\omega>0;\ \operatorname{Re}s<\operatorname{Re}a_k)$ or $(\operatorname{Re}\omega=0;\ \operatorname{Re}s<\operatorname{Re}a_k,\ 1-\operatorname{Re}\chi)\big]$ or $\big[q=p+1;\ \omega>0;\ \operatorname{Re}s<\operatorname{Re}a_k,\ 1/2-\operatorname{Re}\chi\big]$

3.33.3. ${}_pF_q\left((a_p);\ (b_q);\ \omega x^r\right)$ and the logarithmic function

1	$\theta(\sigma-x)\ln\frac{\sqrt{\sigma}+\sqrt{\sigma-x}}{\sqrt{x}}$ $\times\ {}_pF_q\left(\begin{matrix}(a_p)\\ (b_q);\ \omega x\end{matrix}\right)$	$\frac{\sqrt{\pi}\,\sigma^s}{2s}\Gamma\left[\begin{matrix}s\\ \frac{2s+1}{2}\end{matrix}\right]\ {}_{p+2}F_{q+2}\left(\begin{matrix}(a_p),\ s,\ s;\ \sigma\omega\\ (b_q),\ \frac{2s+1}{2},\ s+1\end{matrix}\right)$ $[\sigma,\ \operatorname{Re}s>0;\ \lvert\arg\omega\rvert<\pi]$
2	$\theta(\sigma-x)\ln\frac{\sqrt{\sigma}+\sqrt{\sigma-x}}{\sqrt{x}}$ $\times\ {}_pF_q\left(\begin{matrix}(a_p)\\ (b_q);\ \omega x^2\end{matrix}\right)$	$\frac{\sqrt{\pi}\,\sigma^s}{2s}\Gamma\left[\begin{matrix}s\\ \frac{2s+1}{2}\end{matrix}\right]\ {}_{p+3}F_{q+3}\left(\begin{matrix}(a_p),\ \frac{s}{2},\ \frac{s}{2},\ \frac{s+1}{2};\ \sigma^2\omega\\ (b_q),\ \frac{2s+1}{4},\ \frac{2s+3}{4},\ \frac{s+2}{2}\end{matrix}\right)$ $[\sigma,\ \operatorname{Re}s>0;\ \lvert\arg\omega\rvert<\pi]$

3.33.4. ${}_pF_q\left((a_p);\ (b_q);\ \omega x\right)$ and inverse trigonometric functions

1	$\theta(\sigma-x)\arccos\frac{x}{\sigma}$ $\times\ {}_pF_q\left(\begin{matrix}(a_p);\ \omega x\\(b_q)\end{matrix}\right)$	$\prod_{i=1}^{p} a_i \prod_{j=1}^{q} b_j^{-1}\,\frac{\sqrt{\pi}\,\omega\sigma^{s+1}}{2(s+1)}\,\Gamma\left[\begin{matrix}\frac{s+2}{2}\\\frac{s+3}{2}\end{matrix}\right]$ $\times\ {}_{2p+2}F_{2q+3}\left(\begin{matrix}\frac{(a_p)+1}{2},\ \frac{(a_p)+2}{2},\ \frac{s+1}{2},\ \frac{s+2}{2};\ \frac{\sigma^2\omega^2}{4^{q-p+1}}\\\frac{3}{2},\ \frac{(b_q)+1}{2},\ \frac{(b_q)+2}{2},\ \frac{s+3}{2},\ \frac{s+3}{2}\end{matrix}\right)$ $+\ \frac{\sqrt{\pi}\,\sigma^s}{2s}\,\Gamma\left[\begin{matrix}\frac{s+1}{2}\\\frac{s+2}{2}\end{matrix}\right]$ $\times\ {}_{2p+2}F_{2q+3}\left(\begin{matrix}\frac{(a_p)}{2},\ \frac{(a_p)+1}{2},\ \frac{s}{2},\ \frac{s+1}{2};\ \frac{\sigma^2\omega^2}{4^{q-p+1}}\\\frac{1}{2},\ \frac{(b_q)}{2},\ \frac{(b_q)+1}{2},\ \frac{s+2}{2},\ \frac{s+2}{2}\end{matrix}\right)$ $[\sigma,\ \operatorname{Re} s>0;\ \lvert\arg\omega\rvert<\pi]$
2	$\theta(\sigma-x)\arccos\sqrt{\frac{x}{\sigma}}$ $\times\ {}_pF_q\left(\begin{matrix}(a_p);\ \omega x\\(b_q)\end{matrix}\right)$	$\frac{\sqrt{\pi}\,\sigma^s}{2s}\,\Gamma\left[\begin{matrix}\frac{2s+1}{2}\\s+1\end{matrix}\right]\ {}_{p+2}F_{q+2}\left(\begin{matrix}(a_p),\ s,\ \frac{2s+1}{2};\ \sigma\omega\\(b_q),\ s+1,\ s+1\end{matrix}\right)$ $[\sigma,\ \operatorname{Re} s>0;\ \lvert\arg\omega\rvert<\pi]$

3.33.5. ${}_pF_q\left((a_p);\ (b_q);\ \omega x\right)$ and $\operatorname{Ei}(\sigma x^r)$

1	$\operatorname{Ei}(-\sigma x)\ {}_pF_q\left(\begin{matrix}(a_p);\ \omega x\\(b_q)\end{matrix}\right)$	$-\frac{\sigma^{-s}}{s}\,\Gamma(s)\ {}_{p+2}F_{q+1}\left(\begin{matrix}(a_p),\ s,\ s\\(b_q),\ s+1;\ \frac{\omega}{\sigma}\end{matrix}\right)$ $\left[\begin{matrix}[q=p-1;\ \sigma,\ \operatorname{Re} s>0;\ \lvert\arg\omega\rvert<\pi]\ \text{or}\\ [q=p;\ (\sigma,\ \sigma+\operatorname{Re}\omega,\ \operatorname{Re} s>0)\ \text{or}\\ (\sigma>0;\ \sigma+\operatorname{Re}\omega=0;\ 0<\operatorname{Re} s<2-\operatorname{Re}\chi)]\ \text{or}\\ [p=0;\ q=1;\ \operatorname{Im}\sigma\neq 0,\ \operatorname{Re}\sigma\geq 0;\ \omega>0;\\ 0<\operatorname{Re} s<(2\operatorname{Re} b_1+3)/4]\ \text{or}\\ [q\geq p+1;\ \sigma,\ \operatorname{Re} s>0;\ \lvert\arg\omega\rvert<\pi]\end{matrix}\right]$
2	$\operatorname{Ei}(-\sigma\sqrt{x})\ {}_pF_q\left(\begin{matrix}(a_p);\ \omega x\\(b_q)\end{matrix}\right)$	$\frac{\sigma^{-2s}}{s}\,\Gamma(2s)\ {}_{p+3}F_{q+1}\left(\begin{matrix}(a_p),\ s,\ s,\ \frac{2s+1}{2}\\(b_q),\ s+1;\ \frac{4\omega}{\sigma^2}\end{matrix}\right)$ $\left[\begin{matrix}[q=p-1;\ \sigma,\ \operatorname{Re} s>0;\ \lvert\arg\omega\rvert<\pi]\ \text{or}\\ [q=p;\ (\sigma,\ \operatorname{Re} s>0;\ \operatorname{Re}\omega\geq 0)\ \text{or}\\ (\sigma>0;\ \operatorname{Re}\omega=0;\ 0<\operatorname{Re} s<3/2-\operatorname{Re}\chi)]\ \text{or}\\ [p=0;\ q=1;\ \operatorname{Im}\sigma\neq 0;\ \omega>0;\\ 0<\operatorname{Re} s<(2\operatorname{Re} b_1+1)/4]\ \text{or}\\ [q=p+1;\ \lvert\operatorname{Im}\sqrt{\omega}\rvert<\sigma;\ \operatorname{Re} s>0]\ \text{or}\\ [q\geq p+2;\ \sigma,\ \operatorname{Re} s>0;\ \lvert\arg\omega\rvert<\pi]\end{matrix}\right]$

3.33.6. ${}_pF_q\left((a_p);\,(b_q);\,\omega x\right)$ and $\operatorname{erfc}(\sigma x^r)$

1	$\operatorname{erfc}(\sigma x)\,{}_pF_q\left(\begin{matrix}(a_p);\,-\omega x\\(b_q)\end{matrix}\right)$	$-\prod_{i=1}^{p}a_i\prod_{j=1}^{q}b_j^{-1}\frac{\sigma^{-s-1}\omega}{\sqrt{\pi}\,(s+1)}\Gamma\left(\frac{s+2}{2}\right)$ $\times\,{}_{2p+2}F_{2q+2}\left(\begin{matrix}\frac{(a_p)+1}{2},\,\frac{(a_p)+2}{2},\,\frac{s+1}{2},\,\frac{s+2}{2};\,\frac{\omega^2}{4^{q-p+1}\sigma^2}\\\frac{3}{2},\,\frac{(b_q)+1}{2},\,\frac{(b_q)+2}{2},\,\frac{s+3}{2}\end{matrix}\right)$ $+\frac{\sigma^{-s}}{\sqrt{\pi}\,s}\Gamma\left(\frac{s+1}{2}\right){}_{2p+2}F_{2q+2}\left(\begin{matrix}\frac{(a_p)}{2},\,\frac{(a_p)+1}{2},\,\frac{s}{2},\,\frac{s+1}{2};\,\frac{\omega^2}{4^{q-p+1}\sigma^2}\\\frac{1}{2},\,\frac{(b_q)}{2},\,\frac{(b_q)+1}{2},\,\frac{s+2}{2}\end{matrix}\right)$ $\left[\begin{matrix}[q=p-1;\ (\operatorname{Re}s>0;\ \lvert\arg\sigma\rvert<\pi/4)\text{ or }(0<\operatorname{Re}s<\operatorname{Re}a_k+3;\ \lvert\arg\sigma\rvert=\pi/4)]\text{ or}\\ [q=p;\ (\operatorname{Re}s>0;\ \lvert\arg\sigma\rvert<\pi/4)\text{ or }(\operatorname{Re}\omega>0;\ 0<\operatorname{Re}s<\operatorname{Re}a_k+3;\ \lvert\arg\sigma\rvert=\pi/4)\text{ or }(\operatorname{Re}\omega=0;\ \operatorname{Re}s<3-\operatorname{Re}\chi;\ \lvert\arg\sigma\rvert=\pi/4)]\text{ or}\\ [q=p+1;\ (\operatorname{Re}s>0;\ \lvert\arg\sigma\rvert<\pi/4)\text{ or }(0<\operatorname{Re}s<\operatorname{Re}a_k+3,\,3-\operatorname{Re}\chi;\ \lvert\arg\sigma\rvert=\pi/4)]\text{ or}\\ [q\ge p+2;\ \operatorname{Re}s>0;\ \lvert\arg\sigma\rvert<\pi/4]\end{matrix}\right]$
2	$\operatorname{erfc}(\sigma\sqrt{x})\,{}_pF_q\left(\begin{matrix}(a_p);\,-\omega x\\(b_q)\end{matrix}\right)$	$\frac{\sigma^{-2s}}{\sqrt{\pi}\,s}\Gamma\left(\frac{2s+1}{2}\right){}_pF_q\left(\begin{matrix}(a_p),\,s,\,\frac{2s+1}{2}\\(b_q),\,s+1;\,-\frac{\omega}{\sigma^2}\end{matrix}\right)$ $\left[\begin{matrix}[q=p-1;\ (\operatorname{Re}s>0;\ \lvert\arg\sigma\rvert<\pi/4)\text{ or }(0<\operatorname{Re}s<\operatorname{Re}a_k+3/2;\ \lvert\arg\sigma\rvert=\pi/4)]\text{ or}\\ [q=p;\ (\operatorname{Re}s>0;\ \lvert\arg\sigma\rvert<\pi/4)\text{ or }(\operatorname{Re}(\sigma^2+\omega)>0;\ 0<\operatorname{Re}s<\operatorname{Re}a_k+3/2;\ \lvert\arg\sigma\rvert=\pi/4)\text{ or }(\operatorname{Re}(\sigma^2+\omega)=0;\ \operatorname{Re}s<3/2-\operatorname{Re}\chi;\ \lvert\arg\sigma\rvert=\pi/4)]\text{ or}\\ [q=p+1;\ (\operatorname{Re}s>0;\ \lvert\arg\sigma\rvert<\pi/4)\text{ or }(0<\operatorname{Re}s<\operatorname{Re}a_k+3/2,\,3/2-\operatorname{Re}\chi;\ \lvert\arg\sigma\rvert=\pi/4)]\text{ or}\\ [q\ge p+2;\ \operatorname{Re}s>0;\ \lvert\arg\sigma\rvert<\pi/4]\end{matrix}\right]$

3.33.7. ${}_pF_q\left((a_p);\,(b_q);\,\omega x\right)$ and $\Gamma(\nu,\,\sigma x^r)$

1	$\Gamma(\nu,\,\sigma x)\,{}_pF_q\left(\begin{matrix}(a_p);\,-\omega x\\(b_q)\end{matrix}\right)$	$\frac{\sigma^{-s}}{s}\Gamma(s+\nu)\,{}_{p+2}F_{q+1}\left(\begin{matrix}(a_p),\,s,\,s+\nu\\(b_q),\,s+1;\,-\frac{\omega}{\sigma}\end{matrix}\right)$ $\left[\begin{matrix}[q=p-1;\ (\operatorname{Re}\sigma>0;\ \operatorname{Re}s>0,\,-\operatorname{Re}\nu)\text{ or }(\operatorname{Re}\sigma=0;\ 0,\,-\operatorname{Re}\nu<\operatorname{Re}s<2-\operatorname{Re}(\nu-a_k))]\text{ or}\\ [q=p;\ (\operatorname{Re}\sigma>0;\ \operatorname{Re}s>0,\,-\operatorname{Re}\nu)\text{ or }(\operatorname{Re}\sigma=0;\ 0,\,-\operatorname{Re}\nu<\operatorname{Re}s<2-\operatorname{Re}(\nu-a_k))\text{ or }(\operatorname{Re}(\sigma+\omega)>0;\ \operatorname{Re}s>0,\,-\operatorname{Re}\nu)\text{ or }(\operatorname{Re}(\sigma+\omega)=0;\ 0,\,-\operatorname{Re}\nu<\operatorname{Re}s<2-\operatorname{Re}(\nu+\chi))]\text{ or}\\ [q=p+1;\ (\operatorname{Re}\sigma>0;\ \operatorname{Re}s>0,\,-\operatorname{Re}\nu)\text{ or }(\operatorname{Re}\sigma=0;\ 0,\,-\operatorname{Re}\nu<\operatorname{Re}s<2-\operatorname{Re}(\nu+\chi),\,2-\operatorname{Re}(\nu-a_k))]\text{ or}\\ [q\ge p+2;\ \operatorname{Re}\sigma>0;\ \operatorname{Re}s>0,\,-\operatorname{Re}\nu]\end{matrix}\right]$

No.	$f(x)$	$F(s)$
2	$\Gamma(\nu,\, \sigma\sqrt{x})\; {}_pF_q\left(\begin{matrix}(a_p);\, -\omega x\\ (b_q)\end{matrix}\right)$	$\dfrac{\sigma^{-2s}}{s}\,\Gamma(2s+\nu)\; {}_{p+3}F_{q+1}\left(\begin{matrix}(a_p),\, s,\, \frac{2s+\nu}{2},\, \frac{2s+\nu+1}{2}\\ (b_q),\, s+1;\, -\frac{4\omega}{\sigma^2}\end{matrix}\right)$
		$\big[$ $[q=p-1;\ (\operatorname{Re}\sigma>0;\ \operatorname{Re}s>0,\, -\operatorname{Re}\nu/2)$ or $(\operatorname{Re}\sigma=0;\ 0,\, -\operatorname{Re}\nu/2<\operatorname{Re}s<1-\operatorname{Re}(\nu/2-a_k))]$ or $[q=p;\ (\operatorname{Re}\sigma>0;\ \operatorname{Re}\omega\geq 0;\ \operatorname{Re}s>0,\, -\operatorname{Re}\nu/2)$ or $(\operatorname{Re}\sigma=0;\ 0,\, -\operatorname{Re}\nu/2<\operatorname{Re}s<1-\operatorname{Re}(\nu/2-a_k))$ or $(\operatorname{Re}\omega>0;\ \operatorname{Re}s>0,\, -\operatorname{Re}\nu/2)$ or $(\operatorname{Re}\omega=0;\ 0,\, -\operatorname{Re}\nu/2<\operatorname{Re}s<3/2-\operatorname{Re}(\nu/2-\chi))]$ or $[q=p+1;\ (\operatorname{Re}\sigma>0;\ \operatorname{Re}s>0, -\operatorname{Re}\nu/2)$ or $(\operatorname{Re}\sigma=0;\ 0,\, -\operatorname{Re}\nu/2<\operatorname{Re}s<1-\operatorname{Re}(\nu/2+\chi))]$ or $[q\geq p+2;\ \operatorname{Re}\sigma>0;\ \operatorname{Re}s>0,\, -\operatorname{Re}\nu/2]$ $\big]$

3.33.8. ${}_pF_q((a_p);\, (b_q);\, \omega x^r)$ and $J_\nu(\sigma x)$, $Y_\nu(\sigma x)$

1	$J_\nu(\sigma x)\; {}_pF_q\left(\begin{matrix}(a_p);\, -\omega x\\ (b_q)\end{matrix}\right)$	$-\prod_{i=1}^{p} a_i \prod_{j=1}^{q} b_j^{-1}\, \dfrac{2^s}{\omega^s}\left(\dfrac{\omega^2}{\sigma^2}\right)^{(s+1)/2}\Gamma\left[\begin{matrix}\frac{s+\nu+1}{2}\\ \frac{1-s+\nu}{2}\end{matrix}\right]$ $\times\, {}_{2p+2}F_{2q+1}\left(\begin{matrix}\frac{(a_p)+1}{2},\, \frac{(a_p)+2}{2},\, \frac{s-\nu+1}{2},\, \frac{s+\nu+1}{2}\\ \frac{3}{2},\, \frac{(b_q)+1}{2},\, \frac{(b_q)+2}{2};\, -\frac{\omega^2}{4^{q-p}\sigma^2}\end{matrix}\right)$ $+\, \dfrac{2^{s-1}}{\omega^s}\left(\dfrac{\omega^2}{\sigma^2}\right)^{s/2}\Gamma\left[\begin{matrix}\frac{s+\nu}{2}\\ \frac{2-s+\nu}{2}\end{matrix}\right]$ $\times\, {}_{2p+2}F_{2q+1}\left(\begin{matrix}\frac{(a_p)}{2},\, \frac{(a_p)+1}{2},\, \frac{s-\nu}{2},\, \frac{s+\nu}{2}\\ \frac{1}{2},\, \frac{(b_q)}{2},\, \frac{(b_q)+1}{2};\, -\frac{\omega^2}{4^{q-p}\sigma^2}\end{matrix}\right)$
		$\big[$ $[q=p-1;\, \sigma>0;\, \lvert\arg\omega\rvert<\pi;\, -\operatorname{Re}\nu<\operatorname{Re}s<\operatorname{Re}a_k+3/2]$ or $[q=p;\ (\sigma,\ \operatorname{Re}\omega>0;\ -\operatorname{Re}\nu<\operatorname{Re}s<\operatorname{Re}a_k+3/2)$ or $(\sigma,\, \omega>0;\ -\operatorname{Re}\nu<\operatorname{Re}s<\operatorname{Re}a_k+3/2,\, 3/2-\operatorname{Re}\chi)]$ or $[q=p+1;\ (\sigma,\, \omega>0;\ -\operatorname{Re}\nu<\operatorname{Re}s<\operatorname{Re}a_k+3/2,\, 3/2-\operatorname{Re}\chi)]$ $\big]$
2	$Y_\nu(\sigma x)\; {}_pF_q\left(\begin{matrix}(a_p);\, -\omega x\\ (b_q)\end{matrix}\right)$	$-\prod_{i=1}^{p} a_i \prod_{j=1}^{q} b_j^{-1}\, \dfrac{2^s}{\pi\omega^s}\left(\dfrac{\omega^2}{\sigma^2}\right)^{(s+1)/2}\sin\dfrac{(s-\nu)\pi}{2}\,\Gamma\left(\dfrac{s-\nu+1}{2}\right)$ $\times\,\Gamma\left(\dfrac{s+\nu+1}{2}\right)\, {}_{2p+2}F_{2q+1}\left(\begin{matrix}\frac{(a_p)+1}{2},\, \frac{(a_p)+2}{2},\, \frac{s-\nu+1}{2},\, \frac{s+\nu+1}{2}\\ \frac{3}{2},\, \frac{(b_q)+1}{2},\, \frac{(b_q)+2}{2};\, -\frac{\omega^2}{4^{q-p}\sigma^2}\end{matrix}\right)$ $-\, \dfrac{2^{s-1}}{\pi\omega^s}\left(\dfrac{\omega^2}{\sigma^2}\right)^{s/2}\cos\dfrac{(s-\nu)\pi}{2}\,\Gamma\left(\dfrac{s-\nu}{2}\right)\Gamma\left(\dfrac{s+\nu}{2}\right)$ $\times\, {}_{2p+2}F_{2q+1}\left(\begin{matrix}\frac{(a_p)}{2},\, \frac{(a_p)+1}{2},\, \frac{s-\nu}{2},\, \frac{s+\nu}{2}\\ \frac{1}{2},\, \frac{(b_q)}{2},\, \frac{(b_q)+1}{2};\, -\frac{\omega^2}{4^{q-p}\sigma^2}\end{matrix}\right)$
		$\big[$ $[q=p-1;\, \sigma>0;\, \lvert\arg\omega\rvert<\pi;\, -\lvert\operatorname{Re}\nu\rvert<\operatorname{Re}s<\operatorname{Re}a_k+3/2]$ or $[q=p;\ (\sigma,\ \operatorname{Re}\omega>0;\ -\lvert\operatorname{Re}\nu\rvert<\operatorname{Re}s<\operatorname{Re}a_k+3/2)$ or $(\sigma,\, \omega>0;\ -\lvert\operatorname{Re}\nu\rvert<\operatorname{Re}s<\operatorname{Re}a_k+3/2,\, 3/2-\operatorname{Re}\chi)]$ or $[q=p+1;\ (\sigma,\, \omega>0;\ -\lvert\operatorname{Re}\nu\rvert<\operatorname{Re}s<\operatorname{Re}a_k+3/2,\, 3/2-\operatorname{Re}\chi)]$ $\big]$

3.33.9. ${}_pF_q\left((a_p);\,(b_q);\,\omega x\right)$ and $K_\nu\left(\sigma x^r\right)$

1	$K_\nu(\sigma x)\,{}_pF_q\left(\begin{matrix}(a_p);\,-\omega x\\(b_q)\end{matrix}\right)$	$-\frac{2^{s-1}\omega}{\sigma^{s+1}}\prod_{i=1}^{p}a_i\prod_{j=1}^{q}b_j^{-1}\,\Gamma\left(\frac{s-\nu+1}{2}\right)\Gamma\left(\frac{s+\nu+1}{2}\right)$ $\times\,{}_{2p+2}F_{2q+1}\left(\begin{matrix}\frac{(a_p)+1}{2},\,\frac{(a_p)+2}{2},\,\frac{s-\nu+1}{2},\,\frac{s+\nu+1}{2}\\\frac{3}{2},\,\frac{(b_q)+1}{2},\,\frac{(b_q)+2}{2};\,\frac{\omega^2}{4^{q-p}\sigma^2}\end{matrix}\right)+\frac{2^s}{4\sigma^s}\Gamma\left(\frac{s-\nu}{2}\right)$ $\times\,\Gamma\left(\frac{s+\nu}{2}\right)\,{}_{2p+2}F_{2q+1}\left(\begin{matrix}\frac{(a_p)}{2},\,\frac{(a_p)+1}{2},\,\frac{s-\nu}{2},\,\frac{s+\nu}{2}\\\frac{1}{2},\,\frac{(b_q)}{2},\,\frac{(b_q)+1}{2};\,\frac{\omega^2}{4^{q-p}\sigma^2}\end{matrix}\right)$ $\left[\left[q=p-1;\ \lvert\arg\omega\rvert<\pi;\ (\mathrm{Re}\,\sigma>0;\ \mathrm{Re}\,s>\lvert\mathrm{Re}\,\nu\rvert)\text{ or } (\sigma>0;\ \lvert\mathrm{Re}\,\nu\rvert<\mathrm{Re}\,s<\mathrm{Re}\,a_k+3/2)\right]\text{ or } \left[q=p;\ (\mathrm{Re}\,\sigma,\ \mathrm{Re}\,(\sigma+\omega)>0;\ \mathrm{Re}\,s>\lvert\mathrm{Re}\,\nu\rvert)\text{ or } (\sigma,\ \mathrm{Re}\,\omega>0;\ \lvert\mathrm{Re}\,\nu\rvert<\mathrm{Re}\,s<\mathrm{Re}\,a_k+3/2)\text{ or } (\mathrm{Re}\,\sigma>0;\ \mathrm{Re}\,(\sigma+\omega)=0;\ \lvert\mathrm{Re}\,\nu\rvert<\mathrm{Re}\,s<3/2-\mathrm{Re}\,\chi)\text{ or } (\sigma,\,\omega>0;\ \lvert\mathrm{Re}\,\nu\rvert<\mathrm{Re}\,s<3/2+\mathrm{Re}\,a_k,\,3/2-\mathrm{Re}\,\chi)\right]\text{ or } \left[q=p+1;\ (\mathrm{Re}\,\sigma>0;\ \mathrm{Re}\,s>\lvert\mathrm{Re}\,\nu\rvert)\text{ or } (\sigma,\,\omega>0;\ \lvert\mathrm{Re}\,\nu\rvert<\mathrm{Re}\,s<3/2+\mathrm{Re}\,a_k,\,3/2-\mathrm{Re}\,\chi)\right]\text{ or } \left[q\ge p+2;\ (\mathrm{Re}\,\sigma>0;\ \mathrm{Re}\,s>\lvert\mathrm{Re}\,\nu\rvert)\text{ or } (\sigma,\,\omega>0;\ \lvert\mathrm{Re}\,\nu\rvert<\mathrm{Re}\,s<3/2+\mathrm{Re}\,a_k)\right]\right]$
2	$K_\nu(\sigma\sqrt{x})\,{}_pF_q\left(\begin{matrix}(a_p);\,-\omega x\\(b_q)\end{matrix}\right)$	$\frac{2^{2s-1}}{\sigma^{2s}}\Gamma\left(\frac{2s-\nu}{2}\right)\Gamma\left(\frac{2s+\nu}{2}\right)\,{}_{p+2}F_q\left(\begin{matrix}(a_p),\,\frac{2s-\nu}{2},\,\frac{2s+\nu}{2}\\(b_q);\,-\frac{4\omega}{\sigma^2}\end{matrix}\right)$ $\left[\left[q=p-1;\ \lvert\arg\omega\rvert<\pi;\ (\mathrm{Re}\,\sigma>0;\ \mathrm{Re}\,s>\lvert\mathrm{Re}\,\nu\rvert/2)\text{ or } (\sigma>0;\ \lvert\mathrm{Re}\,\nu\rvert/2<\mathrm{Re}\,s<\mathrm{Re}\,a_k+3/4)\right]\text{ or } \left[q=p;\ (\mathrm{Re}\,\sigma>0;\ \mathrm{Re}\,\omega\ge 0;\ \mathrm{Re}\,s>\lvert\mathrm{Re}\,\nu\rvert/2)\text{ or } (\mathrm{Re}\,\sigma=0;\ \mathrm{Re}\,\omega>0;\ \lvert\mathrm{Re}\,\nu\rvert<\mathrm{Re}\,s<\mathrm{Re}\,a_k+3/4)\text{ or } (\mathrm{Re}\,\sigma=\mathrm{Re}\,\omega=0;\ \lvert\mathrm{Re}\,\nu\rvert/2<\mathrm{Re}\,s<3/4+\mathrm{Re}\,a_k,\,5/4-\mathrm{Re}\,\chi)\right]\text{ or } \left[q=p+1;\ (\mathrm{Re}\,\sigma>2\lvert\mathrm{Im}\,\sqrt{\omega}\rvert;\ \mathrm{Re}\,s>\lvert\mathrm{Re}\,\nu\rvert/2)\text{ or } (\sigma,\,\omega>0;\ \lvert\mathrm{Re}\,\nu\rvert/2<\mathrm{Re}\,s<3/4+\mathrm{Re}\,a_k,\,3/4-\mathrm{Re}\,\chi)\right]\text{ or } \left[q\ge p+2;\ (\mathrm{Re}\,\sigma>0;\ \mathrm{Re}\,s>\lvert\mathrm{Re}\,\nu\rvert/2)\right]\right]$
3	$e^{-\sigma x}K_\nu(\sigma x)$ $\times\,{}_pF_q\left(\begin{matrix}(a_p);\,-\omega x\\(b_q)\end{matrix}\right)$	$\sqrt{\pi}\,(2\sigma)^{-s}\,\Gamma\left[\begin{matrix}s-\nu,\,s+\nu\\\frac{2s+1}{2}\end{matrix}\right]\,{}_{p+2}F_{q+1}\left(\begin{matrix}(a_p),\,s-\nu,\,s+\nu\\(b_q),\,\frac{2s+1}{2};\,-\frac{\omega}{2\sigma}\end{matrix}\right)$ $\left[\left[q=p-1;\ \lvert\arg\omega\rvert<\pi;\ (\mathrm{Re}\,\sigma>0;\ \mathrm{Re}\,s>\lvert\mathrm{Re}\,\nu\rvert)\text{ or } (\sigma>0;\ \lvert\mathrm{Re}\,\nu\rvert<\mathrm{Re}\,s<\mathrm{Re}\,a_k+3/2)\right]\text{ or } \left[q=p;\ (\mathrm{Re}\,\sigma,\ \mathrm{Re}\,(2\sigma+\omega)>0;\ \mathrm{Re}\,s>\lvert\mathrm{Re}\,\nu\rvert)\text{ or } (\sigma,\ \mathrm{Re}\,\omega>0;\ \lvert\mathrm{Re}\,\nu\rvert<\mathrm{Re}\,s<\mathrm{Re}\,a_k+3/2)\text{ or } (\mathrm{Re}\,\sigma>0;\ \mathrm{Re}\,(2\sigma+\omega)=0;\ \lvert\mathrm{Re}\,\nu\rvert<\mathrm{Re}\,s<3/2-\mathrm{Re}\,\chi)\text{ or } (\sigma,\,\omega>0;\ \lvert\mathrm{Re}\,\nu\rvert<\mathrm{Re}\,s<3/2+\mathrm{Re}\,a_k,\,3/2-\mathrm{Re}\,\chi)\right]\text{ or } \left[q=p+1;\ (\mathrm{Re}\,\sigma>0;\ \mathrm{Re}\,s>\lvert\mathrm{Re}\,\nu\rvert)\text{ or } (\sigma,\,\omega>0;\lvert\mathrm{Re}\,\nu\rvert<\mathrm{Re}\,s<3/2+\mathrm{Re}\,a_k,\,3/2-\mathrm{Re}\,\chi)\right]\text{ or } \left[q\ge p+2;\ (\mathrm{Re}\,\sigma>0;\ \mathrm{Re}\,s>\lvert\mathrm{Re}\,\nu\rvert)\text{ or } (\sigma,\,\omega>0;\ \lvert\mathrm{Re}\,\nu\rvert<\mathrm{Re}\,s<\mathrm{Re}\,a_k+3/2)\right]\right]$

No.	$f(x)$	$F(s)$
4	$e^{-\sigma\sqrt{x}}K_\nu(\sigma\sqrt{x}) \times {}_pF_q\left(\begin{matrix}(a_p);\,-\omega x\\(b_q)\end{matrix}\right)$	$2\sqrt{\pi}\,(2\sigma)^{-2s}\,\Gamma\left[\begin{matrix}2s-\nu,\,2s+\nu\\\frac{4s+1}{2}\end{matrix}\right] \times {}_{p+4}F_{q+2}\left(\begin{matrix}(a_p),\,\frac{2s-\nu}{2},\,\frac{2s-\nu+1}{2},\,\frac{2s+\nu}{2},\,\frac{2s+\nu+1}{2}\\(b_q),\,\frac{4s+1}{4},\,\frac{4s+3}{4};\,-\frac{\omega}{\sigma^2}\end{matrix}\right)$
	$\left[\begin{matrix}[q=p-1;\ \lvert\arg\omega\rvert<\pi;\ (\operatorname{Re}\sigma>0;\ \operatorname{Re}s>\lvert\operatorname{Re}\nu\rvert/2)\ \text{or}\\ (\sigma>0;\ \lvert\operatorname{Re}\nu\rvert/2<\operatorname{Re}s<\operatorname{Re}a_k+3/4)]\ \text{or}\\ [q=p;\ (\operatorname{Re}\sigma>0;\ \operatorname{Re}\omega\ge 0;\ \operatorname{Re}s>\lvert\operatorname{Re}\nu\rvert/2)\ \text{or}\\ (\operatorname{Re}\sigma=0;\ \operatorname{Re}\omega>0;\ \lvert\operatorname{Re}\nu\rvert/2<\operatorname{Re}s<\operatorname{Re}a_k+3/4)\ \text{or}\\ (\operatorname{Re}\sigma=\operatorname{Re}\omega=0;\ \lvert\operatorname{Re}\nu\rvert/2<\operatorname{Re}s<3/4+\operatorname{Re}a_k,\ 5/4-\operatorname{Re}\chi)]\ \text{or}\\ [q=p+1;\ (\operatorname{Re}\sigma>\lvert\operatorname{Im}\sqrt{\omega}\rvert;\ \operatorname{Re}s>\lvert\operatorname{Re}\nu\rvert/2)\ \text{or}\\ (\sigma,\,\omega>0;\ \lvert\operatorname{Re}\nu\rvert/2<\operatorname{Re}s<3/4+\operatorname{Re}a_k,\ 3/4-\operatorname{Re}\chi)\ \text{or}\\ [q\ge p+2;\ (\operatorname{Re}\sigma>0;\ \operatorname{Re}s>\lvert\operatorname{Re}\nu\rvert/2)]\end{matrix}\right]$	
5	$K_\mu(\sigma x)\,K_\nu(\sigma x) \times {}_pF_q\left(\begin{matrix}(a_p);\,-\omega x\\(b_q)\end{matrix}\right)$	$-\prod_{i=1}^{p}a_i\prod_{j=1}^{q}b_j^{-1}\frac{2^{s-2}\omega}{\sigma^{s+1}} \times\Gamma\left[\begin{matrix}\frac{s-\mu-\nu+1}{2},\,\frac{s-\mu+\nu+1}{2},\,\frac{s+\mu-\nu+1}{2},\,\frac{s+\mu+\nu+1}{2}\\s+1\end{matrix}\right] \times {}_{2p+4}F_{2q+3}\left(\begin{matrix}\frac{(a_p)+1}{2},\,\frac{(a_p)+2}{2},\,\frac{s-\mu-\nu+1}{2},\,\frac{s-\mu+\nu+1}{2},\,\frac{s+\mu-\nu+1}{2},\,\frac{s+\mu+\nu+1}{2}\\\frac{3}{2},\,\frac{(b_q)+1}{2},\,\frac{(b_q)+2}{2},\,\frac{s+1}{2},\,\frac{s+2}{2};\,\frac{\omega^2}{4^{q-p+1}\sigma^2}\end{matrix}\right) + \frac{2^{s-3}}{\sigma^s}\,\Gamma\left[\begin{matrix}\frac{s-\mu-\nu}{2},\,\frac{s-\mu+\nu}{2},\,\frac{s+\mu-\nu}{2},\,\frac{s+\mu+\nu}{2}\\s\end{matrix}\right] \times {}_{2p+4}F_{2q+3}\left(\begin{matrix}\frac{(a_p)}{2},\,\frac{(a_p)+1}{2},\,\frac{s-\mu-\nu}{2},\,\frac{s-\mu+\nu}{2},\,\frac{s+\mu-\nu}{2},\,\frac{s+\mu+\nu}{2}\\\frac{1}{2},\,\frac{(b_q)}{2},\,\frac{(b_q)+1}{2},\,\frac{s}{2},\,\frac{s+1}{2};\,\frac{\omega^2}{4^{q-p+1}\sigma^2}\end{matrix}\right)$
	$\left[\begin{matrix}[q=p-1;\ \lvert\arg\omega\rvert<\pi;\ (\operatorname{Re}\sigma>0;\operatorname{Re}s>\lvert\operatorname{Re}\mu\rvert+\lvert\operatorname{Re}\nu\rvert)\ \text{or}\\ (\sigma>0;\ \lvert\operatorname{Re}\mu\rvert+\lvert\operatorname{Re}\nu\rvert<\operatorname{Re}s<\operatorname{Re}a_k+2)]\ \text{or}\\ [q=p;\ (\operatorname{Re}\sigma,\ \operatorname{Re}(2\sigma+\omega)>0;\ \operatorname{Re}s>\lvert\operatorname{Re}\mu\rvert+\lvert\operatorname{Re}\nu\rvert)\ \text{or}\\ (\operatorname{Re}\sigma=0;\ \operatorname{Re}\omega>0;\ \lvert\operatorname{Re}\mu\rvert+\lvert\operatorname{Re}\nu\rvert<\operatorname{Re}s<\operatorname{Re}a_k+2)\ \text{or}\\ (\operatorname{Re}\sigma>0,\ \operatorname{Re}(2\sigma+\omega)=0;\ \lvert\operatorname{Re}\mu\rvert+\lvert\operatorname{Re}\nu\rvert<\operatorname{Re}s<2-\operatorname{Re}\chi)\ \text{or}\\ (\operatorname{Re}\sigma=\operatorname{Re}\omega=0;\ \lvert\operatorname{Re}\mu\rvert+\lvert\operatorname{Re}\nu\rvert<\operatorname{Re}s<\operatorname{Re}a_k+2,\ 2-\operatorname{Re}\chi)]\ \text{or}\\ [q=p+1;\ (\operatorname{Re}\sigma>0;\ \operatorname{Re}s>\lvert\operatorname{Re}\mu\rvert+\lvert\operatorname{Re}\nu\rvert)\ \text{or}\\ (\sigma,\,\omega>0;\ \lvert\operatorname{Re}\mu\rvert+\lvert\operatorname{Re}\nu\rvert<\operatorname{Re}s<\operatorname{Re}a_k+2,\ 2-\operatorname{Re}\chi)]\ \text{or}\\ [q\ge p+2;\ \operatorname{Re}\sigma>0;\ \operatorname{Re}s>\lvert\operatorname{Re}\mu\rvert+\lvert\operatorname{Re}\nu\rvert]\end{matrix}\right]$	

No.	$f(x)$	$F(s)$
6	$K_\mu\left(\sigma\sqrt{x}\right)K_\nu\left(\sigma\sqrt{x}\right) \times {}_pF_q\left(\begin{matrix}(a_p);\,-\omega x\\(b_q)\end{matrix}\right)$	$\frac{2^{2s-2}}{\sigma^{2s}}\Gamma\left[\begin{matrix}\frac{2s-\mu-\nu}{2},\,\frac{2s-\mu+\nu}{2},\,\frac{2s+\mu-\nu}{2},\,\frac{2s+\mu+\nu}{2}\\2s\end{matrix}\right] \times {}_{p+4}F_{q+2}\left(\begin{matrix}(a_p),\,\frac{2s-\mu-\nu}{2},\,\frac{2s-\mu+\nu}{2},\,\frac{2s+\mu-\nu}{2},\,\frac{2s+\mu+\nu}{2}\\(b_q),\,s,\,\frac{2s+1}{2};\,-\frac{\omega}{\sigma^2}\end{matrix}\right)$
		$\left[\begin{matrix}[q=p-1;\ \lvert\arg\omega\rvert<\pi;\ (\operatorname{Re}\sigma>0;\operatorname{Re}s>(\lvert\operatorname{Re}\mu\rvert+\lvert\operatorname{Re}\nu\rvert)/2)\ \text{or}\\ (\sigma>0;\ (\lvert\operatorname{Re}\mu\rvert+\lvert\operatorname{Re}\nu\rvert)/2<\operatorname{Re}s<\operatorname{Re}a_k+1)]\ \text{or}\\ [q=p;\ (\operatorname{Re}\sigma>0,\ \operatorname{Re}\omega\ge0;\ \operatorname{Re}s>(\lvert\operatorname{Re}\mu\rvert+\lvert\operatorname{Re}\nu\rvert)/2)\ \text{or}\\ (\sigma>0;\ \operatorname{Re}\omega>0;\ (\lvert\operatorname{Re}\mu\rvert+\lvert\operatorname{Re}\nu\rvert)/2<\operatorname{Re}s<\operatorname{Re}a_k+1)\ \text{or}\\ (\sigma,\,\omega>0;\ (\lvert\operatorname{Re}\mu\rvert+\lvert\operatorname{Re}\nu\rvert)/2<\operatorname{Re}s<\operatorname{Re}a_k+1,\,3/2-\operatorname{Re}\chi)]\ \text{or}\\ [q=p+1;\ (\lvert\operatorname{Im}\sqrt{\omega}\rvert<\operatorname{Re}\sigma;\ \operatorname{Re}s>(\lvert\operatorname{Re}\mu\rvert+\lvert\operatorname{Re}\nu\rvert)/2)\ \text{or}\\ (\sigma,\,\omega>0;\ (\lvert\operatorname{Re}\mu\rvert+\lvert\operatorname{Re}\nu\rvert)/2<\operatorname{Re}s<\operatorname{Re}a_k+1,\,1-\operatorname{Re}\chi)]\ \text{or}\\ [q\ge p+2;\ \operatorname{Re}\sigma>0;\ \operatorname{Re}s>(\lvert\operatorname{Re}\mu\rvert+\lvert\operatorname{Re}\nu\rvert)/2]\end{matrix}\right]$

3.33.10. ${}_pF_q\left((a_p);\,(b_q);\,\omega x\right)$ and $\operatorname{Ai}(\sigma x^r)$

No.	$f(x)$	$F(s)$
1	$\operatorname{Ai}(\sigma x)\,{}_pF_q\left(\begin{matrix}(a_p);\,-\omega x\\(b_q)\end{matrix}\right)$	$\frac{3^{(4s-7)/6}}{2\pi\sigma^s}\Gamma\left(\frac{s}{3}\right)\Gamma\left(\frac{s+1}{3}\right) \times {}_{3p+2}F_{3q+2}\left(\begin{matrix}\Delta(3,\,(a_p)),\,\frac{s}{3},\,\frac{s+1}{3}\\\frac{1}{3},\,\frac{2}{3},\,\Delta(3,\,(b_q));\,-\frac{\omega^3}{3^{3(q-p)+1}\sigma^3}\end{matrix}\right) - \frac{3^{(4s-3)/6}\omega}{2\pi\sigma^{s+1}}\frac{\prod_{i=1}^p a_i}{\prod_{j=1}^q b_j}\Gamma\left(\frac{s+1}{3}\right)\Gamma\left(\frac{s+2}{3}\right) \times {}_{3p+2}F_{3q+2}\left(\begin{matrix}\Delta(3,\,(a_p)+1),\,\frac{s+1}{3},\,\frac{s+2}{3}\\\frac{2}{3},\,\frac{4}{3},\,\Delta(3,\,(b_q)+1);\,-\frac{\omega^3}{3^{3(q-p)+1}\sigma^3}\end{matrix}\right) + \frac{3^{(4s+1)/6}\omega^2}{4\pi\sigma^{s+2}}\frac{\prod_{i=1}^p a_i(a_i+1)}{\prod_{j=1}^q b_j(b_j+1)}\Gamma\left(\frac{s+2}{3}\right)\Gamma\left(\frac{s+3}{3}\right) \times {}_{3p+2}F_{3q+2}\left(\begin{matrix}\Delta(3,\,(a_p)+2),\,\frac{s+2}{3},\,\frac{s+3}{3}\\\frac{4}{3},\,\frac{5}{3},\,\Delta(3,\,(b_q)+2);\,-\frac{\omega^3}{3^{3(q-p)+1}\sigma^3}\end{matrix}\right)$
		$\left[\begin{matrix}[q=p-1;\ \lvert\arg\omega\rvert<\pi;\ (\lvert\arg\sigma\rvert<\pi/3;\ \operatorname{Re}s>0)\ \text{or}\\ (\lvert\arg\sigma\rvert=\pi/3;\ 0<\operatorname{Re}s<\operatorname{Re}a_k+7/4)]\\ [q=p;\ (\lvert\arg\sigma\rvert<\pi/3;\ \operatorname{Re}s>0)\ \text{or}\\ (\lvert\arg\sigma\rvert=\pi/3;\ \operatorname{Re}\omega>0;\ 0<\operatorname{Re}s<\operatorname{Re}a_k+7/4)\ \text{or}\\ (\lvert\arg\sigma\rvert=\pi/3;\ \operatorname{Re}\omega=0;\ 0<\operatorname{Re}s<\operatorname{Re}a_k+7/4,\,7/4-\operatorname{Re}\chi)]\\ [q=p+1;\ (\lvert\arg\sigma\rvert<\pi/3;\ \operatorname{Re}s>0)\ \text{or}\\ (\lvert\arg\sigma\rvert=\pi/3;\ \omega>0;\ 0<\operatorname{Re}s<\operatorname{Re}a_k+7/4,\,7/4-\operatorname{Re}\chi)]\\ [q\ge p+2;\ \lvert\arg\sigma\rvert<\pi/3;\ \operatorname{Re}s>0]\end{matrix}\right]$

No.	$f(x)$	$F(s)$
2	$\operatorname{Ai}^2(\sigma\sqrt[3]{x})\,{}_pF_q\left(\begin{matrix}(a_p);\,-\omega x\\(b_q)\end{matrix}\right)$	$\dfrac{2^{-2s-2/3}}{3^{s-1/6}\sqrt{\pi}\,\sigma^{3s}}\Gamma\left[\begin{matrix}3s\\\frac{6s+5}{6}\end{matrix}\right]{}_{p+3}F_{q+1}\left(\begin{matrix}(a_p),\,\Delta(3,\,3s)\\(b_q),\,\frac{6s+5}{6};\,-\frac{9\omega}{4\sigma^3}\end{matrix}\right)$
		$\left[\begin{array}{l}[q=p-1;\ \lvert\arg\omega\rvert<\pi;\ (\lvert\arg\sigma\rvert<\pi/3;\ \operatorname{Re}s>0)\ \text{or}\\ \quad(\lvert\arg\sigma\rvert=\pi/3;\ 0<\operatorname{Re}s<\operatorname{Re}a_k+2/3)]\\ [q=p;\ (\lvert\arg\sigma\rvert<\pi/3;\ \operatorname{Re}\omega\ge 0;\ \operatorname{Re}s>0)\ \text{or}\\ \quad(\lvert\arg\sigma\rvert=\pi/3;\ \operatorname{Re}\omega>0;\ 0<\operatorname{Re}s<\operatorname{Re}a_k+2/3)\ \text{or}\\ \quad(\lvert\arg\sigma\rvert=\pi/3;\ \operatorname{Re}\omega=0;\ 0<\operatorname{Re}s<\operatorname{Re}a_k+2/3,\ 7/6-\operatorname{Re}\chi)]\\ [q=p+1;\ \left(\lvert\operatorname{Im}\sqrt{\omega}\rvert<2\operatorname{Re}\sigma^{3/2}/3;\ \operatorname{Re}s>0\right)\ \text{or}\\ \quad\left(\lvert\arg\sigma\rvert=\pi/3;\ \lvert\operatorname{Im}\sqrt{\omega}\rvert<2\operatorname{Re}\sigma^{3/2}/3;\ 0<\operatorname{Re}s<\operatorname{Re}a_k+2/3\right)\ \text{or}\\ \quad(\lvert\arg\sigma\rvert=\pi/3;\ \omega>0;\ 0<\operatorname{Re}s<\operatorname{Re}a_k+2/3,\ 2/3-\operatorname{Re}\chi)]\\ [q\ge p+2;\ \lvert\arg\sigma\rvert<\pi/3;\ \operatorname{Re}s>0]\end{array}\right]$

3.33.11. ${}_pF_q((a_p);\,(b_q);\,\omega x^r)$ and $P_n(\varphi(x))$

Notation: $\varepsilon = 0$ or 1.

1	$\theta(\sigma-x)\,P_n\left(\dfrac{x}{\sigma}\right)$ $\times\,{}_pF_q\left(\begin{matrix}(a_p);\,\omega x\\(b_q)\end{matrix}\right)$	$\sqrt{\pi}\prod_{i=1}^{p}a_i\prod_{j=1}^{q}b_j^{-1}\left(\dfrac{\sigma}{2}\right)^{s+1}\omega\,\Gamma\left[\begin{matrix}s+1\\\frac{s-n+2}{2},\,\frac{s+n+3}{2}\end{matrix}\right]$ $\times\,{}_{2p+2}F_{2q+3}\left(\begin{matrix}\frac{(a_p)+1}{2},\,\frac{(a_p)+2}{2},\,\frac{s+1}{2},\,\frac{s+2}{2};\,\frac{\sigma^2\omega^2}{4^{q-p+1}}\\\frac{3}{2},\,\frac{(b_q)+1}{2},\,\frac{(b_q)+2}{2},\,\frac{s-n+2}{2},\,\frac{s+n+3}{2}\end{matrix}\right)$ $+\sqrt{\pi}\left(\dfrac{\sigma}{2}\right)^{s}\Gamma\left[\begin{matrix}s\\\frac{s-n+1}{2},\,\frac{s+n+2}{2}\end{matrix}\right]$ $\times\,{}_{2p+2}F_{2q+3}\left(\begin{matrix}\frac{(a_p)}{2},\,\frac{(a_p)+1}{2},\,\frac{s}{2},\,\frac{s+1}{2};\,\frac{\sigma^2\omega^2}{4^{q-p+1}}\\\frac{1}{2},\,\frac{(b_q)}{2},\,\frac{(b_q)+1}{2},\,\frac{s-n+1}{2},\,\frac{s+n+2}{2}\end{matrix}\right)$ $[\sigma>0;\ \operatorname{Re}s>2[n/2]-n]$
2	$\theta(\sigma-x)\,P_n\left(\dfrac{2x}{\sigma}-1\right)$ $\times\,{}_pF_q\left(\begin{matrix}(a_p);\,\omega x\\(b_q)\end{matrix}\right)$	$\sigma^s\,\Gamma\left[\begin{matrix}s,\,s\\s-n,\,s+n+1\end{matrix}\right]{}_{p+2}F_{q+2}\left(\begin{matrix}(a_p),\,s,\,s;\,\sigma\omega\\(b_q),\,s-n,\,s+n+1\end{matrix}\right)$ $[\sigma,\ \operatorname{Re}s>0]$
3	$\theta(\sigma-x)\,P_n\left(\dfrac{2x}{\sigma}-1\right)$ $\times\,{}_pF_q\left(\begin{matrix}(a_p);\,\omega x^2\\(b_q)\end{matrix}\right)$	$\sigma^s\,\Gamma\left[\begin{matrix}s,\,s\\s-n,\,s+n+1\end{matrix}\right]$ $\times\,{}_{p+4}F_{q+4}\left(\begin{matrix}(a_p),\,\frac{s}{2},\,\frac{s}{2},\,\frac{s+1}{2},\,\frac{s+1}{2};\,\sigma^2\omega\\(b_q),\,\frac{s-n}{2},\,\frac{s-n+1}{2},\,\frac{s+n+1}{2},\,\frac{s+n+2}{2}\end{matrix}\right)$ $[\sigma,\ \operatorname{Re}s>0]$

No.	$f(x)$	$F(s)$
4	$\theta(\sigma-x)\,P_n\left(\frac{\sigma}{x}\right) \times {}_pF_q\left(\begin{matrix}(a_p);\,\omega x\\(b_q)\end{matrix}\right)$	$\prod_{i=1}^{p} a_i \prod_{j=1}^{q} b_j^{-1} \frac{2^s\sigma^{s+1}\omega}{\sqrt{\pi}}\,\Gamma\left[\begin{matrix}\frac{s-n+1}{2},\ \frac{s+n+2}{2}\\ s+2\end{matrix}\right] \times {}_{2p+2}F_{2q+3}\left(\begin{matrix}\frac{(a_p)+1}{2},\ \frac{(a_p)+2}{2},\ \frac{s-n+1}{2},\ \frac{s+n+2}{2};\ \frac{\sigma^2\omega^2}{4^{q-p+1}}\\ \frac{3}{2},\ \frac{(b_q)+1}{2},\ \frac{(b_q)+2}{2},\ \frac{s+2}{2},\ \frac{s+3}{2}\end{matrix}\right) + \frac{2^{s-1}\sigma^s}{\sqrt{\pi}}\,\Gamma\left[\begin{matrix}\frac{s-n}{2},\ \frac{s+n+1}{2}\\ s+1\end{matrix}\right] \times {}_{2p+2}F_{2q+3}\left(\begin{matrix}\frac{(a_p)}{2},\ \frac{(a_p)+1}{2},\ \frac{s-n}{2},\ \frac{s+n+1}{2};\ \frac{\sigma^2\omega^2}{4^{q-p+1}}\\ \frac{1}{2},\ \frac{(b_q)}{2},\ \frac{(b_q)+1}{2},\ \frac{s+1}{2},\ \frac{s+2}{2}\end{matrix}\right)$ [$\sigma>0$; $\operatorname{Re} s>n$]
5	$\theta(\sigma-x)\,P_n\left(\frac{2\sigma}{x}-1\right) \times {}_pF_q\left(\begin{matrix}(a_p);\,\omega x\\(b_q)\end{matrix}\right)$	$\sigma^s\,\Gamma\left[\begin{matrix}s-n,\ s+n+1\\ s+1,\ s+1\end{matrix}\right] {}_{p+2}F_{q+2}\left(\begin{matrix}(a_p),\ s-n,\ s+n+1\\ (b_q),\ s+1,\ s+1;\ \sigma\omega\end{matrix}\right)$ [$\sigma>0$; $\operatorname{Re} s>n$]
6	$\theta(\sigma-x)\,P_n\left(\sqrt{\frac{x}{\sigma}}\right) \times {}_pF_q\left(\begin{matrix}(a_p);\,\omega x\\(b_q)\end{matrix}\right)$	$\frac{\sqrt{\pi}\,\sigma^s}{2^{2s-1}}\,\Gamma\left[\begin{matrix}2s\\ \frac{2s-n+1}{2},\ \frac{2s+n+2}{2}\end{matrix}\right] {}_{p+2}F_{q+2}\left(\begin{matrix}(a_p),\ s,\ \frac{2s+1}{2};\ \sigma\omega\\ (b_q),\ \frac{2s-n+1}{2},\ \frac{2s+n+2}{2}\end{matrix}\right)$ [$\sigma>0$; $\operatorname{Re} s>[n/2]-n/2$]
7	$\theta(\sigma-x)\,P_n\left(2\sqrt{\frac{x}{\sigma}}-1\right) \times {}_pF_q\left(\begin{matrix}(a_p);\,\omega x\\(b_q)\end{matrix}\right)$	$2\sigma^s\,\Gamma\left[\begin{matrix}2s,\ 2s\\ 2s-n,\ 2s+n+1\end{matrix}\right] \times {}_{p+4}F_{q+4}\left(\begin{matrix}(a_p),\ \Delta(2,\,2s),\ \Delta(2,\,2s);\ \sigma\omega\\ (b_q),\ \Delta(2,\,2s-n),\ \Delta(2,\,2s+n+1)\end{matrix}\right)$ [$\sigma>0$; $\operatorname{Re} s>0$]
8	$(\sigma-x)_+^{(\varepsilon-1)/2} \times P_{2n+\varepsilon}\left(\sqrt{1-\frac{x}{\sigma}}\right) \times {}_pF_q\left(\begin{matrix}(a_p);\,\omega x\\(b_q)\end{matrix}\right)$	$\frac{(-1)^n\sqrt{\pi}\,\sigma^{s+(\varepsilon-1)/2}}{n!}\left(\frac{1}{2}\right)_{n+\varepsilon}\Gamma\left[\begin{matrix}s,\ s\\ s-n,\ \frac{2s+2n+2\varepsilon+1}{2}\end{matrix}\right] \times {}_{p+2}F_{q+2}\left(\begin{matrix}(a_p),\ s,\ s;\ \sigma\omega\\ (b_q),\ s-n,\ \frac{2s+2n+2\varepsilon+1}{2}\end{matrix}\right)$ [σ, $\operatorname{Re} s>0$]
9	$\theta(\sigma-x)\,P_n\left(\sqrt{\frac{\sigma}{x}}\right) \times {}_pF_q\left(\begin{matrix}(a_p);\,\omega x\\(b_q)\end{matrix}\right)$	$\frac{(4\sigma)^s}{\sqrt{\pi}}\,\Gamma\left[\begin{matrix}\frac{2s-n}{2},\ \frac{s+2n+1}{2}\\ 2s+1\end{matrix}\right] {}_{p+2}F_{q+2}\left(\begin{matrix}(a_p),\ \frac{2s-n}{2},\ \frac{2s+n+1}{2}\\ (b_q),\ \frac{2s+1}{2},\ s+1;\ \sigma\omega\end{matrix}\right)$ [$\sigma>0$; $\operatorname{Re} s>n/2$]

No.	$f(x)$	$F(s)$
10	$\theta(\sigma-x)\,P_n\left(2\sqrt{\frac{\sigma}{x}}-1\right)$ $\times\,{}_pF_q\left(\begin{matrix}(a_p);\,\omega x\\(b_q)\end{matrix}\right)$	$2\,\sigma^s\,\Gamma\left[\begin{matrix}2s-n,\,2s+n+1\\2s+1,\,2s+1\end{matrix}\right]$ $\times\,{}_{p+4}F_{q+4}\left(\begin{matrix}(a_p),\,\frac{2s-n}{2},\,\frac{2s-n+1}{2},\,\frac{2s+n+1}{2},\,\frac{2s+n+2}{2}\\(b_q),\,\frac{2s+1}{2},\,\frac{2s+1}{2},\,s+1,\,s+1;\,\sigma\omega\end{matrix}\right)$ $[\sigma>0;\ \operatorname{Re}s>n/2]$

3.33.12. ${}_pF_q\left((a_p);\,(b_q);\,\omega x^r\right)$ and $T_n(\varphi(x))$

Notation: $\varepsilon=0$ or 1.

1	$\left(\sigma^2-x^2\right)_+^{-1/2}T_n\left(\frac{x}{\sigma}\right)$ $\times\,{}_pF_q\left(\begin{matrix}(a_p);\,\omega x\\(b_q)\end{matrix}\right)$	$\frac{\pi}{2}\prod_{i=1}^{p}a_i\prod_{j=1}^{q}b_j^{-1}\left(\frac{\sigma}{2}\right)^s\omega\,\Gamma\left[\begin{matrix}s+1\\\frac{s-n+2}{2},\,\frac{s+n+2}{2}\end{matrix}\right]$ $\times\,{}_{2p+2}F_{2q+3}\left(\begin{matrix}\frac{(a_p)+1}{2},\,\frac{(a_p)+2}{2},\,\frac{s+1}{2},\,\frac{s+2}{2};\,\frac{\sigma^2\omega^2}{4^{q-p+1}}\\\frac{3}{2},\,\frac{(b_q)+1}{2},\,\frac{(b_q)+2}{2},\,\frac{s-n+2}{2},\,\frac{s+n+2}{2}\end{matrix}\right)$ $+\,\frac{\pi}{2}\left(\frac{\sigma}{2}\right)^{s-1}\Gamma\left[\begin{matrix}s\\\frac{s-n+1}{2},\,\frac{s+n+1}{2}\end{matrix}\right]$ $\times\,{}_{2p+2}F_{2q+3}\left(\begin{matrix}\frac{(a_p)}{2},\,\frac{(a_p)+1}{2},\,\frac{s}{2},\,\frac{s+1}{2};\,\frac{\sigma^2\omega^2}{4^{q-p+1}}\\\frac{1}{2},\,\frac{(b_q)}{2},\,\frac{(b_q)+1}{2},\,\frac{s-n+1}{2},\,\frac{s+n+1}{2}\end{matrix}\right)$ $[\sigma>0;\ \operatorname{Re}s>2[n/2]-n]$
2	$(\sigma-x)_+^{-1/2}T_n\left(\frac{2x}{\sigma}-1\right)$ $\times\,{}_pF_q\left(\begin{matrix}(a_p);\,\omega x\\(b_q)\end{matrix}\right)$	$\frac{\pi\,\sigma^{s-1/2}}{2^{2s-1}}\,\Gamma\left[\begin{matrix}2s\\\frac{2s-2n+1}{2},\,\frac{2s+2n+1}{2}\end{matrix}\right]$ $\times\,{}_{p+2}F_{q+2}\left(\begin{matrix}(a_p),\,s,\,\frac{2s+1}{2};\,\sigma\omega\\(b_q),\,\frac{2s-2n+1}{2},\,\frac{2s+2n+1}{2}\end{matrix}\right)$ $[\sigma,\ \operatorname{Re}s>0]$
3	$(\sigma-x)_+^{-1/2}T_n\left(\frac{2x}{\sigma}-1\right)$ $\times\,{}_pF_q\left(\begin{matrix}(a_p);\,\omega x^2\\(b_q)\end{matrix}\right)$	$\frac{\pi\sigma^{s-1/2}}{2^{2s-1}}\,\Gamma\left[\begin{matrix}2s\\\frac{2s-2n+1}{2},\,\frac{2s+2n+1}{2}\end{matrix}\right]$ $\times\,{}_{p+4}F_{q+4}\left(\begin{matrix}(a_p),\,\Delta(4,\,2s);\,\sigma^2\omega\\(b_q),\,\Delta\left(2,\,\frac{2s-2n+1}{2}\right),\,\Delta\left(2,\,\frac{2s+2n+1}{2}\right)\end{matrix}\right)$ $[\sigma,\ \operatorname{Re}s>0]$
4	$(\sigma-x)_+^{-1/2}T_n\left(\frac{2\sigma}{x}-1\right)$ $\times\,{}_pF_q\left(\begin{matrix}(a_p);\,\omega x\\(b_q)\end{matrix}\right)$	$2^{2s-1}\sigma^{s-1/2}\,\mathrm{B}(s-n,\,s+n)\,{}_{p+2}F_{q+2}\left(\begin{matrix}(a_p),\,s-n,\,s+n\\(b_q),\,\frac{2s+1}{2},\,s;\,\sigma\omega\end{matrix}\right)$ $[\sigma>0;\ \operatorname{Re}s>n]$

No.	$f(x)$	$F(s)$
5	$(\sigma-x)_+^{-1/2}\,T_n\left(\frac{2\sigma}{x}-1\right)$ $\times\,{}_pF_q\left(\begin{matrix}(a_p);\,\omega x^2\\(b_q)\end{matrix}\right)$	$2^{2s-1}\,\sigma^{s-1/2}\,\mathrm{B}(s-n,\,s+n)$ $\times\,{}_{p+4}F_{q+4}\left(\begin{matrix}(a_p),\,\frac{s-n}{2},\,\frac{s-n+1}{2},\,\frac{s+n}{2},\,\frac{s+n+1}{2}\\(b_q),\,\frac{2s+1}{4},\,\frac{2s+3}{4},\,\frac{s}{2},\,\frac{s+1}{2};\,\sigma^2\omega\end{matrix}\right)$ $[\sigma>0;\ \operatorname{Re}s>n]$
6	$(\sigma-x)_+^{-1/2}\,T_n\left(\sqrt{\frac{x}{\sigma}}\right)$ $\times\,{}_pF_q\left(\begin{matrix}(a_p);\,\omega x\\(b_q)\end{matrix}\right)$	$\pi\left(\frac{\sqrt{\sigma}}{2}\right)^{2s-1}\Gamma\left[\begin{matrix}2s\\\frac{2s-n+1}{2},\,\frac{2s+n+1}{2}\end{matrix}\right]$ $\times\,{}_{p+2}F_{q+2}\left(\begin{matrix}(a_p),\,s,\,s+\frac{1}{2};\,\sigma\omega\\(b_q),\,\frac{2s-n+1}{2},\,\frac{2s+n+1}{2}\end{matrix}\right)$ $[\sigma>0;\ \operatorname{Re}s>[n/2]-n/2]$
7	$(\sigma-x)_+^{(\varepsilon-1)/2}$ $\times\,T_{2n+\varepsilon}\left(\sqrt{1-\frac{x}{\sigma}}\right)$ $\times\,{}_pF_q\left(\begin{matrix}(a_p);\,\omega x\\(b_q)\end{matrix}\right)$	$\frac{(-1)^n\,(2n+1)^{\varepsilon}\,\pi\,\sigma^{s+(\varepsilon-1)/2}}{2^{2s+\varepsilon-1}}\,\Gamma\left[\begin{matrix}2s\\\frac{2s-2n+1}{2},\,\frac{2s+2n+2\varepsilon+1}{2}\end{matrix}\right]$ $\times\,{}_{p+2}F_{q+2}\left(\begin{matrix}(a_p),\,s,\,\frac{2s+1}{2};\,\sigma\omega\\(b_q),\,\frac{2s-2n+1}{2},\,\frac{2s+2n+2\varepsilon+1}{2}\end{matrix}\right)$ $[\sigma,\ \operatorname{Re}s>0]$

3.33.13. ${}_pF_q\left((a_p);\,(b_q);\,\omega x^r\right)$ and $U_n(\varphi(x))$

Notation: $\varepsilon=0$ or 1.

No.	$f(x)$	$F(s)$
1	$(\sigma^2-x^2)_+^{1/2}\,U_n\left(\frac{x}{\sigma}\right)$ $\times\,{}_pF_q\left(\begin{matrix}(a_p);\,\omega x\\(b_q)\end{matrix}\right)$	$(n+1)\,\pi\prod_{i=1}^{p}a_i\prod_{j=1}^{q}b_j^{-1}\left(\frac{\sigma}{2}\right)^{s+2}\omega\,\Gamma\left[\begin{matrix}s+1\\\frac{s-n+2}{2},\,\frac{s+n+4}{2}\end{matrix}\right]$ $\times\,{}_{2p+2}F_{2q+3}\left(\begin{matrix}\frac{(a_p)+1}{2},\,\frac{(a_p)+2}{2},\,\frac{s+1}{2},\,\frac{s+2}{2};\,\frac{\sigma^2\omega^2}{4^{q-p+1}}\\\frac{3}{2},\,\frac{(b_q)+1}{2},\,\frac{(b_q)+2}{2},\,\frac{s-n+2}{2},\,\frac{s+n+4}{2}\end{matrix}\right)$ $+\,\pi\,(n+1)\left(\frac{\sigma}{2}\right)^{s+1}\Gamma\left[\begin{matrix}s\\\frac{s-n+1}{2},\,\frac{s+n+3}{2}\end{matrix}\right]$ $\times\,{}_{2p+2}F_{2q+3}\left(\begin{matrix}\frac{(a_p)}{2},\,\frac{(a_p)+1}{2},\,\frac{s}{2},\,\frac{s+1}{2};\,\frac{\sigma^2\omega^2}{4^{q-p+1}}\\\frac{1}{2},\,\frac{(b_q)}{2},\,\frac{(b_q)+1}{2},\,\frac{s-n+1}{2},\,\frac{s+n+3}{2}\end{matrix}\right)$ $[\sigma>0;\ \operatorname{Re}s>2\,[n/2]-n]$
2	$(\sigma-x)_+^{1/2}\,U_n\left(\frac{2x}{\sigma}-1\right)$ $\times\,{}_pF_q\left(\begin{matrix}(a_p);\,\omega x\\(b_q)\end{matrix}\right)$	$\frac{(n+1)\,\pi\,\sigma^{s+1/2}}{2^{2s-1}}\,\Gamma\left[\begin{matrix}2s-1\\\frac{2s-2n-1}{2},\,\frac{2s+2n+3}{2}\end{matrix}\right]$ $\times\,{}_{p+2}F_{q+2}\left(\begin{matrix}(a_p),\,s,\,\frac{2s-1}{2};\,\sigma\omega\\(b_q),\,\frac{2s-2n-1}{2},\,\frac{2s+2n+3}{2}\end{matrix}\right)$ $[\sigma,\ \operatorname{Re}s>0]$

No.	$f(x)$	$F(s)$
3	$(\sigma-x)_+^{1/2}\,U_n\left(\frac{2x}{\sigma}-1\right)$ $\times\,{}_pF_q\left(\begin{matrix}(a_p);\,\omega x^2\\(b_q)\end{matrix}\right)$	$\frac{(n+1)\,\pi\sigma^{s+1/2}}{2^{2s-1}}\,\Gamma\left[\begin{matrix}2s-1\\ \frac{2s-2n-1}{2},\ \frac{2s+2n+3}{2}\end{matrix}\right]$ $\times\,{}_{p+4}F_{q+4}\left(\begin{matrix}(a_p),\,\Delta(2,\,s),\,\Delta\left(2,\,\frac{2s-1}{2}\right);\,\sigma^2\omega\\(b_q),\,\Delta\left(2,\,\frac{2s-2n-1}{2}\right),\,\Delta\left(2,\,\frac{2s+2n+3}{2}\right)\end{matrix}\right)$ $[\sigma,\ \operatorname{Re}s>0]$
4	$(\sigma-x)_+^{1/2}\,U_n\left(\frac{2\sigma}{x}-1\right)$ $\times\,{}_pF_q\left(\begin{matrix}(a_p);\,\omega x\\(b_q)\end{matrix}\right)$	$2^{2s+1}\,(n+1)\,\sigma^{s+1/2}\,\Gamma\left[\begin{matrix}s-n,\,s+n+2\\2s+3\end{matrix}\right]$ $\times\,{}_{p+2}F_{q+2}\left(\begin{matrix}(a_p),\,s-n,\,s+n+2\\(b_q),\,\frac{2s+3}{2},\,s+2;\,\sigma\omega\end{matrix}\right)$ $[\sigma>0;\ \operatorname{Re}s>n]$
5	$(\sigma-x)_+^{1/2}$ $\times\,U_n\left(\frac{2\sigma}{x}-1\right)$ $\times\,{}_pF_q\left(\begin{matrix}(a_p);\,\omega x^2\\(b_q)\end{matrix}\right)$	$2^{2s+1}\,(n+1)\,\sigma^{s+1/2}\,\Gamma\left[\begin{matrix}s-n,\,s+n+2\\2s+3\end{matrix}\right]$ $\times\,{}_{p+4}F_{q+4}\left(\begin{matrix}(a_p),\,\frac{s-n}{2},\,\frac{s-n+1}{2},\,\frac{s+n+2}{2},\,\frac{s+n+3}{2}\\(b_q),\,\frac{2s+3}{4},\,\frac{2s+5}{4},\,\frac{2s+2}{2},\,\frac{2s+3}{2};\,\sigma\omega\end{matrix}\right)$ $[\sigma>0;\ \operatorname{Re}s>n]$
6	$(\sigma-x)_+^{1/2}\,U_n\left(\sqrt{\frac{x}{\sigma}}\right)$ $\times\,{}_pF_q\left(\begin{matrix}(a_p);\,\omega x\\(b_q)\end{matrix}\right)$	$\frac{(n+1)\,\pi\sigma^{s+1/2}}{2^{2s}}\,\Gamma\left[\begin{matrix}2s\\ \frac{2s-n+1}{2},\ \frac{2s+n+3}{2}\end{matrix}\right]$ $\times\,{}_{p+2}F_{q+2}\left(\begin{matrix}(a_p),\,s,\,\frac{2s+1}{2};\,\sigma\omega\\(b_q),\,\frac{2s-n+1}{2},\,\frac{2s+n+3}{2}\end{matrix}\right)$ $[\sigma>0;\ \operatorname{Re}s>[n/2]-n/2]$
7	$(\sigma-x)_+^{(\varepsilon-1)/2}$ $\times\,U_{2n+\varepsilon}\left(\sqrt{1-\frac{x}{\sigma}}\right)$ $\times\,{}_pF_q\left(\begin{matrix}(a_p);\,\omega x\\(b_q)\end{matrix}\right)$	$\frac{(-1)^n\,(n+1)^{\varepsilon}\,\pi\sigma^{s+(\varepsilon-1)/2}}{2^{2s-2}}\,\Gamma\left[\begin{matrix}2s-1\\ \frac{2s-2n-1}{2},\ \frac{2s+2n+2\varepsilon+1}{2}\end{matrix}\right]$ $\times\,{}_{p+2}F_{q+2}\left(\begin{matrix}(a_p),\,s,\,\frac{2s-1}{2};\,\sigma\omega\\(b_q),\,\frac{2s-2n-1}{2},\,\frac{2s+2n+2\varepsilon+1}{2}\end{matrix}\right)$ $[\sigma,\ \operatorname{Re}s>0]$

3.33.14. ${}_pF_q\left((a_p);\,(b_q);\,\omega x\right)$ and $H_n\left(\sigma x^r\right)$

1	$e^{-\sigma^2x^2}H_n(\sigma x)$ $\times\,{}_pF_q\left(\begin{matrix}(a_p);\,-\omega x\\(b_q)\end{matrix}\right)$	$-\prod_{i=1}^{p}a_i\prod_{j=1}^{q}b_j^{-1}\,\frac{2^{n-s-1}\sqrt{\pi}\,\omega}{\sigma^{s+1}}\,\Gamma\left[\begin{matrix}s+1\\ \frac{s-n+2}{2}\end{matrix}\right]$ $\times\,{}_{2p+2}F_{2q+2}\left(\begin{matrix}\frac{(a_p)+1}{2},\,\frac{(a_p)+2}{2},\,\frac{s+1}{2},\,\frac{s+2}{2};\,\frac{\omega^2}{4^{q-p+1}\sigma^2}\\ \frac{3}{2},\,\frac{(b_q)+1}{2},\,\frac{(b_q)+2}{2},\,\frac{s-n+2}{2}\end{matrix}\right)$ $+\,\frac{\sqrt{\pi}\,2^{n-s}}{\sigma^s}\,\Gamma\left[\begin{matrix}s\\ \frac{s-n+1}{2}\end{matrix}\right]\times$

No.	$f(x)$	$F(s)$
		$\times\, {}_{2p+2}F_{2q+2}\left(\begin{matrix}\frac{(a_p)}{2},\ \frac{(a_p)+1}{2},\ \frac{s}{2},\ \frac{s+1}{2};\ \frac{\omega^2}{4^{q-p+1}\sigma^2}\\ \frac{1}{2},\ \frac{(b_q)}{2},\ \frac{(b_q)+1}{2},\ \frac{s-n+1}{2}\end{matrix}\right)$
		$\left[\begin{matrix}\left[q=p-1;\ \lvert\arg\omega\rvert<\pi;\ \left(\operatorname{Re}\sigma^2>0;\ \operatorname{Re}s>2\left[n/2\right]-n\right)\ \text{or}\right.\\ \left.\left(\operatorname{Re}\sigma^2=0;\ 2\left[n/2\right]-n<\operatorname{Re}s<\operatorname{Re}a_k-n+2\right)\right]\ \text{or}\\ \left[q=p;\ \left(\operatorname{Re}\sigma^2>0;\ \lvert\arg\omega\rvert<\pi;\ \operatorname{Re}s>2\left[n/2\right]-n\right)\ \text{or}\right.\\ \left(\operatorname{Re}\sigma^2=0;\ \operatorname{Re}\omega>0;\ 2\left[n/2\right]-n<\operatorname{Re}s<\operatorname{Re}a_k-n+2\right)\ \text{or}\\ \left.\left(\operatorname{Re}\sigma^2=0;\ \operatorname{Re}\omega=0;\ 2\left[n/2\right]-n<\operatorname{Re}s<\operatorname{Re}a_k-n+2,\ 2-n-\operatorname{Re}\chi\right)\ \text{or}\right.\\ \left[q=p+1;\ \left(\operatorname{Re}\sigma^2>0;\ \lvert\arg\omega\rvert<\pi;\ \operatorname{Re}s>2\left[n/2\right]-n\right)\ \text{or}\right.\\ \left.\left(\operatorname{Re}\sigma^2=0;\ \omega>0;\ 2\left[n/2\right]-n<\operatorname{Re}s<\operatorname{Re}a_k-n+2,\ 2-n-\operatorname{Re}\chi\right)\right]\ \text{or}\\ \left[q\geq p+2;\ \operatorname{Re}\sigma^2>0;\ \lvert\arg\omega\rvert<\pi;\ \operatorname{Re}s>2\left[n/2\right]-n\right]\end{matrix}\right]$
2	$e^{-\sigma^2 x}H_n\left(\sigma\sqrt{x}\right)$ $\times\, {}_pF_q\left(\begin{matrix}(a_p);\ -\omega x\\ (b_q)\end{matrix}\right)$	$\frac{\sqrt{\pi}\,2^{1-2s+n}}{\sigma^{2s}}\,\Gamma\left[\begin{matrix}2s\\ \frac{2s-n+1}{2}\end{matrix}\right]\,{}_{p+2}F_{q+1}\left(\begin{matrix}(a_p),\ s,\ \frac{2s+1}{2}\\ (b_q),\ \frac{2s-n+1}{2};\ -\frac{\omega}{\sigma^2}\end{matrix}\right)$
		$\left[\begin{matrix}\left[q=p-1;\ \lvert\arg\omega\rvert<\pi;\ \left(\lvert\arg\sigma\rvert<\pi/4;\ \operatorname{Re}s>\left[n/2\right]-n/2\right)\ \text{or}\right.\\ \left.\left(\lvert\arg\sigma\rvert=\pi/4;\ \left[n/2\right]-n/2<\operatorname{Re}s<\operatorname{Re}a_k-n/2+1\right)\right]\ \text{or}\\ \left[q=p;\ \left(\lvert\arg\sigma\rvert<\pi/4;\ \operatorname{Re}\left(\sigma^2+\omega\right)>0;\ \operatorname{Re}s>\left[n/2\right]-n/2\right)\ \text{or}\right.\\ \left(\lvert\arg\sigma\rvert<\pi/4;\ \operatorname{Re}\left(\sigma^2+\omega\right)=0;\ \left[n/2\right]-n/2<\operatorname{Re}s<1-n/2-\operatorname{Re}\chi\right)\ \text{or}\\ \left(\lvert\arg\sigma\rvert=\pi/4;\ \operatorname{Re}\omega>0;\ \left[n/2\right]-n/2<\operatorname{Re}s<\operatorname{Re}a_k-n/2+1\right)\ \text{or}\\ \left.\left(\lvert\arg\sigma\rvert=\pi/4;\ \operatorname{Re}\omega=0;\ \left[n/2\right]-n/2<\operatorname{Re}s<\operatorname{Re}a_k-n/2+1,\ 1-n/2-\operatorname{Re}\chi\right)\right]\ \text{or}\\ \left[q=p+1;\ \left(\lvert\arg\sigma\rvert<\pi/4;\ \lvert\arg\omega\rvert<\pi;\ \operatorname{Re}s>\left[n/2\right]-n/2\right)\ \text{or}\right.\\ \left.\left(\lvert\arg\sigma\rvert=\pi/4;\ \omega>0;\ \left[n/2\right]-n/2<\operatorname{Re}s<\operatorname{Re}a_k-n/2+1,\ 1-n/2-\operatorname{Re}\chi\right)\right]\ \text{or}\\ \left[q\geq p+2;\ \lvert\arg\sigma\rvert<\pi/4;\ \lvert\arg\omega\rvert<\pi;\ \operatorname{Re}s>\left[n/2\right]-n/2\right]\end{matrix}\right]$

3.33.15. ${}_pF_q\left((a_p);\,(b_q);\,\omega x\right)$ and $L_n^\lambda\left(\sigma x^r\right)$

1	$e^{-\sigma x}L_n^\lambda(\sigma x)$ $\times\, {}_pF_q\left(\begin{matrix}(a_p);\ -\omega x\\ (b_q)\end{matrix}\right)$	$\frac{\sigma^{-s}}{n!}\,(1-s+\lambda)_n\,\Gamma(s)\ {}_{p+2}F_{q+1}\left(\begin{matrix}(a_p),\ s-\lambda,\ s\\ (b_q),\ s-n-\lambda;\ -\frac{\omega}{\sigma}\end{matrix}\right)$
		$\left[\begin{matrix}\left[q=p-1;\ \lvert\arg\omega\rvert<\pi;\ \left(\operatorname{Re}\sigma,\ \operatorname{Re}s>0\right)\ \text{or}\right.\\ \left.\left(\operatorname{Re}\sigma=0;\ 0<\operatorname{Re}s<\operatorname{Re}a_k-n+1\right)\right]\ \text{or}\\ \left[q=p;\ \left(\operatorname{Re}\sigma,\ \operatorname{Re}s>0;\ \operatorname{Re}\left(\sigma^2+\omega\right)>0\right)\ \text{or}\right.\\ \left(\operatorname{Re}\sigma>0;\ \operatorname{Re}\left(\sigma^2+\omega\right)=0;\ 0<\operatorname{Re}s<1-n-\operatorname{Re}\chi\right)\ \text{or}\\ \left(\operatorname{Re}\sigma=0;\ \operatorname{Re}\omega>0;\ 0<\operatorname{Re}s<\operatorname{Re}a_k-n+1\right)\ \text{or}\\ \left.\left(\operatorname{Re}\sigma=0;\ \operatorname{Re}\omega=0;\ 0<\operatorname{Re}s<\operatorname{Re}a_k-n+1,\ 1-n-\operatorname{Re}\chi\right)\right]\ \text{or}\\ \left[q=p+1;\ \left(\operatorname{Re}\sigma,\ \operatorname{Re}s>0;\ \lvert\arg\omega\rvert<\pi\right)\ \text{or}\right.\\ \left.\left(\operatorname{Re}\sigma=0;\ \omega>0;\ 0<\operatorname{Re}s<\operatorname{Re}a_k-n+1,\ 1-n-\operatorname{Re}\chi\right)\right]\ \text{or}\\ \left[q\geq p+2;\ \operatorname{Re}\sigma,\ \operatorname{Re}s>0;\ \lvert\arg\omega\rvert<\pi\right]\end{matrix}\right]$

No.	$f(x)$	$F(s)$
2	$e^{-\sigma\sqrt{x}} L_n^{\lambda}(\sigma\sqrt{x})$ $\times {}_pF_q\left(\begin{matrix}(a_p);\, -\omega x\\ (b_q)\end{matrix}\right)$	$\dfrac{2\sigma^{-2s}}{n!}(1-2s+\lambda)_n\,\Gamma(2s)$ $\times {}_{p+4}F_{q+2}\left(\begin{matrix}(a_p),\, \frac{2s-\lambda}{2},\, \frac{2s-\lambda+1}{2},\, s,\, \frac{2s+1}{2}\\ (b_q),\, \frac{2s-n-\lambda}{2},\, \frac{2s-n-\lambda+1}{2};\, -\frac{4\omega}{\sigma^2}\end{matrix}\right)$ $\left[\begin{matrix}[q=p-1;\ \lvert\arg\omega\rvert<\pi;\ (\mathrm{Re}\,\sigma,\ \mathrm{Re}\,s>0)\ \text{or}\\ (\mathrm{Re}\,\sigma=0;\ 0<\mathrm{Re}\,s<\mathrm{Re}\,a_k+(1-n)/2)]\ \text{or}\\ [q=p;\ (\mathrm{Re}\,\sigma,\ \mathrm{Re}\,s>0;\ \mathrm{Re}\,\omega\geq 0)\ \text{or}\\ (\mathrm{Re}\,\sigma=0;\ \mathrm{Re}\,\omega>0;\ 0<\mathrm{Re}\,s<\mathrm{Re}\,a_k+(1-n)/2)\ \text{or}\\ (\mathrm{Re}\,\sigma=\mathrm{Re}\,\omega=0;\ 0<\mathrm{Re}\,s<\mathrm{Re}\,a_k+(1-n)/2,\ 1-n/2-\mathrm{Re}\,\chi)]\ \text{or}\\ [q=p+1;\ (2\lvert\mathrm{Im}\,\sqrt{\omega}\rvert<\mathrm{Re}\,\sigma;\ \mathrm{Re}\,s>0)\ \text{or}\\ (\mathrm{Re}\,\sigma>0;\ 2\lvert\mathrm{Im}\,\sqrt{\omega}\rvert+\mathrm{Re}\,\sigma=0;\ 0<\mathrm{Re}\,s<(1-n)/2-\mathrm{Re}\,\chi)\ \text{or}\\ (\mathrm{Re}\,\sigma=0;\ \omega>0;\ \mathrm{Re}\,s<\mathrm{Re}\,a_k+(1-n)/2,\ (1-n)/2-\mathrm{Re}\,\chi)]\ \text{or}\\ [q\geq p+2;\ \mathrm{Re}\,\sigma,\ \mathrm{Re}\,s>0;\ \lvert\arg\omega\rvert<\pi]\end{matrix}\right]$

3.33.16. ${}_pF_q((a_p);\,(b_q);\,\omega x)$ and $C_n^{\lambda}(\varphi(x))$

Notation: $\varepsilon = 0$ or 1.

1	$(\sigma^2-x^2)_+^{\lambda-1/2}\, C_n^{\lambda}\left(\dfrac{x}{\sigma}\right)$ $\times {}_pF_q\left(\begin{matrix}(a_p);\, \omega x\\ (b_q)\end{matrix}\right)$	$\dfrac{\pi}{n!}\prod_{i=1}^{p} a_i \prod_{j=1}^{q} b_j^{-1}\left(\dfrac{\sigma}{2}\right)^{s+2\lambda}\omega\,\Gamma\left[\begin{matrix}n+2\lambda,\ s+1\\ \lambda,\ \frac{s-n+2}{2},\ \frac{s+n+2\lambda+2}{2}\end{matrix}\right]$ $\times {}_{2p+2}F_{2q+3}\left(\begin{matrix}\frac{(a_p)+1}{2},\, \frac{(a_p)+2}{2},\, \frac{s+1}{2},\, \frac{s+2}{2};\, \frac{\sigma^2\omega^2}{4^{q-p+1}}\\ \frac{3}{2},\, \frac{(b_q)+1}{2},\, \frac{(b_q)+2}{2},\, \frac{s-n+2}{2},\, \frac{s+n+2\lambda+2}{2}\end{matrix}\right)$ $+\dfrac{\pi}{n!}\left(\dfrac{\sigma}{2}\right)^{s+2\lambda-1}\Gamma\left[\begin{matrix}n+2\lambda,\ s\\ \lambda,\ \frac{s-n+1}{2},\ \frac{s+n+2\lambda+1}{2}\end{matrix}\right]$ $\times {}_{2p+2}F_{2q+3}\left(\begin{matrix}\frac{(a_p)}{2},\, \frac{(a_p)+1}{2},\, \frac{s}{2},\, \frac{s+1}{2};\, \frac{\sigma^2\omega^2}{4^{q-p+1}}\\ \frac{1}{2},\, \frac{(b_q)}{2},\, \frac{(b_q)+1}{2},\, \frac{s-n+1}{2},\, \frac{s+n+2\lambda+1}{2}\end{matrix}\right)$ $[\sigma>0;\ \mathrm{Re}\,\lambda>-1/2;\ \mathrm{Re}\,s>2[n/2]-n]$
2	$(\sigma-x)_+^{\lambda-1/2}$ $\times C_n^{\lambda}\left(\dfrac{2x}{\sigma}-1\right)$ $\times {}_pF_q\left(\begin{matrix}(a_p);\, \omega x\\ (b_q)\end{matrix}\right)$	$\dfrac{\sqrt{\pi}\,\sigma^{s+\lambda-1/2}}{2^{2\lambda-1}n!}\Gamma\left[\begin{matrix}n+2\lambda,\ s,\ \frac{2s-2\lambda+1}{2}\\ \lambda,\ \frac{2s-2n-2\lambda+1}{2},\ \frac{2s+2n+2\lambda+1}{2}\end{matrix}\right]$ $\times {}_{p+2}F_{q+2}\left(\begin{matrix}(a_p),\, s,\, \frac{2s-2\lambda+1}{2};\, \sigma\omega\\ (b_q),\, \frac{2s-2n-2\lambda+1}{2},\, \frac{2s+2n+2\lambda+1}{2}\end{matrix}\right)$ $[\sigma,\ \mathrm{Re}\,s>0;\ \mathrm{Re}\,\lambda>-1/2]$
3	$(\sigma-x)_+^{\lambda-1/2}$ $\times C_n^{\lambda}\left(\dfrac{2x}{\sigma}-1\right)$ $\times {}_pF_q\left(\begin{matrix}(a_p);\, \omega x^2\\ (b_q)\end{matrix}\right)$	$\dfrac{(-1)^n\sqrt{\pi}\,\sigma^{s+\lambda-1/2}}{2^{2\lambda-1}n!}\Gamma\left[\begin{matrix}n+2\lambda,\ s,\ \frac{1-2s+2n+2\lambda}{2}\\ \lambda,\ \frac{1-2s+2\lambda}{2},\ \frac{2s+2n+2\lambda+1}{2}\end{matrix}\right]$ $\times {}_pF_q\left(\begin{matrix}(a_p),\, \Delta(2,\, s),\, \Delta\left(2,\, \frac{2s-2\lambda+1}{2}\right);\, \sigma^2\omega\\ (b_q),\, \Delta\left(2,\, \frac{2s-2n-2\lambda+1}{2}\right),\, \Delta\left(2,\, \frac{2s+2n+2\lambda+1}{2}\right)\end{matrix}\right)$ $[\sigma,\ \mathrm{Re}\,s>0;\ \lambda>-1/2]$

No.	$f(x)$	$F(s)$
4	$(\sigma-x)_+^{\lambda-1/2} \times C_n^{\lambda}\left(\frac{2\sigma}{x}-1\right) \times {}_pF_q\left(\begin{matrix}(a_p);\ \omega x\\(b_q)\end{matrix}\right)$	$\frac{\sqrt{\pi}\,\sigma^{s+\lambda-1/2}}{2^{2\lambda-1}n!}\Gamma\left[\begin{matrix}n+2\lambda,\ s-n,\ s+n+2\lambda\\\lambda,\ \frac{2s+2\lambda+1}{2},\ s+2\lambda\end{matrix}\right] \times {}_{p+2}F_{q+2}\left(\begin{matrix}(a_p),\ s-n,\ s+n+2\lambda\\(b_q),\ \frac{2s+2\lambda+1}{2},\ s+2\lambda;\ \sigma\omega\end{matrix}\right)$ $[\sigma>0;\ \operatorname{Re}\lambda>-1/2;\ \operatorname{Re}s>n]$
5	$(\sigma-x)_+^{\lambda-1/2} \times C_n^{\lambda}\left(\frac{2\sigma}{x}-1\right) \times {}_pF_q\left(\begin{matrix}(a_p);\ \omega x^2\\(b_q)\end{matrix}\right)$	$\frac{\sqrt{\pi}\,\sigma^{s+\lambda-1/2}}{2^{2\lambda-1}n!}\Gamma\left[\begin{matrix}n+2\lambda,\ s-n,\ s+n+2\lambda\\\lambda,\ \frac{2s+2\lambda+1}{2},\ s+2\lambda\end{matrix}\right] \times {}_{p+4}F_{q+4}\left(\begin{matrix}(a_p),\ \Delta(2,\ s-n),\ \Delta(2,\ s+n+2\lambda)\\(b_q),\ \Delta\left(2,\ \frac{2s+2\lambda+1}{2}\right),\ \Delta(2,\ s+2\lambda);\ \sigma^2\omega\end{matrix}\right)$ $[\sigma>0;\ \operatorname{Re}\lambda>-1/2;\ \operatorname{Re}s>n]$
6	$(\sigma-x)^{-n-2\lambda} \times C_n^{\lambda}\left(\frac{\sigma+x}{\sigma-x}\right) \times {}_pF_q\left(\begin{matrix}(a_p);\ \omega x\\(b_q)\end{matrix}\right)$	$\frac{\sigma^{s-n-2\lambda}}{n!\left(\frac{2\lambda+1}{2}\right)_n}\Gamma\left[\begin{matrix}1-2\lambda,\ s,\ \frac{2s-2\lambda+1}{2}\\\frac{2s-2n-2\lambda+1}{2},\ s-n-2\lambda+1\end{matrix}\right] \times {}_{p+2}F_{q+2}\left(\begin{matrix}(a_p),\ s,\ \frac{2s-2\lambda+1}{2};\ \sigma\omega\\(b_q),\ s-n-2\lambda+1,\ \frac{2s-2n-2\lambda+1}{2}\end{matrix}\right)$ $[\sigma,\ \operatorname{Re}s>0;\ \operatorname{Re}\lambda<1/2-n]$
7	$(\sigma-x)_+^{\lambda-1/2}C_n^{\lambda}\left(\sqrt{\frac{x}{\sigma}}\right) \times {}_pF_q\left(\begin{matrix}(a_p);\ \omega x\\(b_q)\end{matrix}\right)$	$\frac{2\pi}{n!}\left(\frac{\sqrt{\sigma}}{2}\right)^{2s+2\lambda-1}\Gamma\left[\begin{matrix}n+2\lambda,\ 2s\\\lambda,\ \frac{2s-n+1}{2},\ \frac{2s+n+2\lambda+1}{2}\end{matrix}\right] \times {}_{p+2}F_{q+2}\left(\begin{matrix}(a_p),\ s,\ \frac{2s+1}{2};\ \sigma\omega\\(b_q),\ \frac{2s-n+1}{2},\ \frac{2s+n+2\lambda+1}{2}\end{matrix}\right)$ $[\sigma>0;\ \operatorname{Re}\lambda>-1/2;\ \operatorname{Re}s>[n/2]-n/2]$
8	$(\sigma-x)_+^{(\varepsilon-1)/2} \times C_{2n+\varepsilon}^{\lambda}\left(\sqrt{1-\frac{x}{\sigma}}\right) \times {}_pF_q\left(\begin{matrix}(a_p);\ \omega x\\(b_q)\end{matrix}\right)$	$\frac{(-1)^n\sqrt{\pi}\,\sigma^{s+(\varepsilon-1)/2}}{n!}(\lambda)_{n+\varepsilon}\Gamma\left[\begin{matrix}s,\ \frac{2s-2\lambda+1}{2}\\\frac{2s-2n-2\lambda+1}{2},\ \frac{2s+2n+2\varepsilon+1}{2}\end{matrix}\right] \times {}_{p+2}F_{q+2}\left(\begin{matrix}(a_p),\ s,\ \frac{2s-2\lambda+1}{2};\ \sigma\omega\\(b_q),\ \frac{2s-2n-2\lambda+1}{2},\ \frac{2s+2n+2\varepsilon+1}{2}\end{matrix}\right)$ $[\sigma,\ \operatorname{Re}s>0]$
9	$(\sigma-x)_+^{-n/2-\lambda} \times C_n^{\lambda}\left(\sqrt{\frac{\sigma}{\sigma-x}}\right) \times {}_pF_q\left(\begin{matrix}(a_p);\ \omega x\\(b_q)\end{matrix}\right)$	$\frac{\sigma^{s-n/2-\lambda}}{n!}(2\lambda-2s)_n\ \mathrm{B}(1-\lambda,\ s) \times {}_{p+2}F_{q+2}\left(\begin{matrix}(a_p),\ s,\ \frac{2s-2\lambda+1}{2};\ \sigma\omega\\(b_q),\ \frac{2s-n-2\lambda+1}{2},\ \frac{2s-n-2\lambda+2}{2}\end{matrix}\right)$ $[\sigma,\ \operatorname{Re}s>0;\ \operatorname{Re}\lambda<1-n]$

3.33.17. ${}_pF_q\left((a_p);\ (b_q);\ \omega x^r\right)$ and $P_n^{(\alpha,\,\beta)}\left(\varphi\left(x\right)\right)$

1	$(\sigma-x)_+^{\alpha}\,P_n^{(\alpha,\,\beta)}\left(\frac{2x}{\sigma}-1\right)\times{}_pF_q\left(\begin{matrix}(a_p);\,\omega x\\(b_q)\end{matrix}\right)$	$\frac{(-1)^n\,\sigma^{s+\alpha}}{n!}\,(1-s+\beta)_n\,\mathrm{B}\,(n+\alpha+1,\,s)\times{}_{p+2}F_{q+2}\left(\begin{matrix}(a_p),\,s,\,s-\beta;\,\sigma\omega\\(b_q),\,s-n-\beta,\,s+n+\alpha+1\end{matrix}\right)$ $[\sigma,\ \mathrm{Re}\,s>0;\ \mathrm{Re}\,\alpha>-1]$
2	$(\sigma-x)_+^{\alpha}\,P_n^{(\alpha,\,\beta)}\left(\frac{2x}{\sigma}-1\right)\times{}_pF_q\left(\begin{matrix}(a_p);\,\omega x^2\\(b_q)\end{matrix}\right)$	$\frac{(-1)^n\,\sigma^{s+\alpha}}{n!}\,(\beta-s+1)_n\,\mathrm{B}\,(n+\alpha+1,\,s)\times{}_{p+4}F_{q+4}\left(\begin{matrix}(a_p),\,\Delta\,(2,\,s),\,\Delta\,(2,\,s-\beta);\,\sigma^2\omega\\(b_q),\,\Delta\,(2,\,s-n-\beta),\,\Delta\,(2,\,s+n+\alpha+1)\end{matrix}\right)$ $[\sigma,\ \mathrm{Re}\,s>0;\ \mathrm{Re}\,\alpha>-1]$
3	$(\sigma-x)_+^{\alpha}\,P_n^{(\alpha,\,\beta)}\left(\frac{2\sigma}{x}-1\right)\times{}_pF_q\left(\begin{matrix}(a_p);\,\omega x\\(b_q)\end{matrix}\right)$	$\frac{\sigma^{s+\alpha}}{n!}\,(s+\alpha+\beta+1)_n\,\mathrm{B}\,(n+\alpha+1,\,s-n)\times{}_{p+2}F_{q+2}\left(\begin{matrix}(a_p),\,s-n,\,s+n+\alpha+\beta+1\\(b_q),\,s+\alpha+1,\,s+\alpha+\beta+1;\,\sigma\omega\end{matrix}\right)$ $[\sigma>0;\ \mathrm{Re}\,\alpha>-1;\ \mathrm{Re}\,s>n]$
4	$(\sigma-x)_+^{-n-\alpha-\beta-1}\times P_n^{(\alpha,\,\beta)}\left(\frac{\sigma+x}{\sigma-x}\right)\times{}_pF_q\left(\begin{matrix}(a_p);\,\omega x\\(b_q)\end{matrix}\right)$	$\frac{\sigma^{s-n-\alpha-\beta-1}}{n!}\,(1-s+\alpha)_n\,\mathrm{B}\,(-n-\alpha-\beta,\,s)\times{}_{p+2}F_{q+2}\left(\begin{matrix}(a_p),\,s,\,s-\alpha;\,\sigma\omega\\(b_q),\,s-n-\alpha,\,s-n-\alpha-\beta\end{matrix}\right)$ $[\sigma,\ \mathrm{Re}\,s>0;\ \mathrm{Re}\,(\alpha+\beta)<-n]$

3.33.18. ${}_pF_q\left((a_p);\ (b_q);\ \omega x^r\right)$ and $\mathbf{K}\left(\varphi\left(x\right)\right)$, $\mathbf{E}\left(\varphi\left(x\right)\right)$

Notation: $\delta=\left\{\begin{matrix}1\\0\end{matrix}\right\}$.

1	$\theta\,(\sigma-x)\left\{\begin{matrix}\mathbf{K}\big(\sqrt{1-x/\sigma}\big)\\\mathbf{E}\big(\sqrt{1-x/\sigma}\big)\end{matrix}\right\}\times{}_pF_q\left(\begin{matrix}(a_p);\,\omega x\\(b_q)\end{matrix}\right)$	$\frac{\pi\sigma^s}{2}\,\Gamma\left[\begin{matrix}s,\,s-\delta+1\\\frac{2s+1}{2},\,\frac{2s-2\delta+3}{2}\end{matrix}\right]\,{}_{p+2}F_{q+2}\left(\begin{matrix}(a_p),\,s,\,s-\delta+1;\,\sigma\omega\\(b_q),\,\frac{2s+1}{2},\,\frac{2s-2\delta+3}{2}\end{matrix}\right)$ $[\sigma,\ \mathrm{Re}\,s>0]$
2	$\theta\,(\sigma-x)\left\{\begin{matrix}\mathbf{K}\big(\sqrt{1-x/\sigma}\big)\\\mathbf{E}\big(\sqrt{1-x/\sigma}\big)\end{matrix}\right\}\times{}_pF_q\left(\begin{matrix}(a_p);\,\omega x^2\\(b_q)\end{matrix}\right)$	$\frac{\pi\sigma^s}{2}\,\Gamma\left[\begin{matrix}s,\,s-\delta+1\\\frac{2s+1}{2},\,\frac{2s-2\delta+3}{2}\end{matrix}\right]\times{}_{p+4}F_{q+4}\left(\begin{matrix}(a_p),\,\frac{s}{2},\,\frac{s+1}{2},\,\frac{s-\delta+1}{2},\,\frac{s-\delta+2}{2};\,\sigma^2\omega\\(b_q),\,\frac{2s+1}{4},\,\frac{2s+3}{4},\,\frac{2s-2\delta+3}{4},\,\frac{2s-2\delta+5}{4}\end{matrix}\right)$ $[\sigma,\ \mathrm{Re}\,s>0]$

3.33.19. ${}_pF_q\left((a_p);\,(b_q);\,\omega x^r\right)$ and $P_\nu^\mu\left(\varphi(x)\right)$, $\mathrm{P}_\nu^\mu\left(\varphi(x)\right)$

1	$(\sigma^2-x^2)_+^{-\mu/2}\,\mathrm{P}_\nu^\mu\left(\frac{x}{\sigma}\right) \times {}_pF_q\left(\begin{matrix}(a_p);\,\omega x\\(b_q)\end{matrix}\right)$	$\sqrt{\pi}\,\omega\left(\frac{\sigma}{2}\right)^{s-\mu+1}\prod_{i=1}^{p}a_i\prod_{j=1}^{q}b_j^{-1}\,\Gamma\left[\begin{matrix}s+1\\\frac{s-\mu-\nu+2}{2},\,\frac{s-\mu+\nu+3}{2}\end{matrix}\right] \times {}_{2p+2}F_{2q+3}\left(\begin{matrix}\frac{(a_p)+1}{2},\,\frac{(a_p)+2}{2},\,\frac{s+1}{2},\,\frac{s+2}{2};\,\frac{\sigma^2\omega^2}{4^{q-p+1}}\\\frac{3}{2},\,\frac{(b_q)+1}{2},\,\frac{(b_q)+2}{2},\,\frac{s-\mu-\nu+2}{2},\,\frac{s-\mu+\nu+3}{2}\end{matrix}\right) + \sqrt{\pi}\left(\frac{\sigma}{2}\right)^{s-\mu}\Gamma\left[\begin{matrix}s\\\frac{s-\mu-\nu+1}{2},\,\frac{s-\mu+\nu+2}{2}\end{matrix}\right] \times {}_{2p+2}F_{2q+3}\left(\begin{matrix}\frac{(a_p)}{2},\,\frac{(a_p)+1}{2},\,\frac{s}{2},\,\frac{s+1}{2};\,\frac{\sigma^2\omega^2}{4^{q-p+1}}\\\frac{1}{2},\,\frac{(b_q)}{2},\,\frac{(b_q)+1}{2},\,\frac{s-\mu-\nu+1}{2},\,\frac{s-\mu+\nu+2}{2}\end{matrix}\right)$ $[\sigma,\ \mathrm{Re}\,s>0;\ \mathrm{Re}\,\mu<1]$
2	$(\sigma-x)_+^{-\mu/2}\,\mathrm{P}_\nu^\mu\left(\frac{2x}{\sigma}-1\right) \times {}_pF_q\left(\begin{matrix}(a_p);\,\omega x\\(b_q)\end{matrix}\right)$	$\sigma^{s-\mu/2}\,\Gamma\left[\begin{matrix}\frac{2s-\mu}{2},\,\frac{2s+\mu}{2}\\\frac{2s-\mu-2\nu}{2},\,\frac{2s-\mu+2\nu+2}{2}\end{matrix}\right] \times {}_{p+2}F_{q+2}\left(\begin{matrix}(a_p),\,\frac{2s-\mu}{2},\,\frac{2s+\mu}{2};\,\sigma\omega\\(b_q),\,\frac{2s-\mu-2\nu}{2},\,\frac{2s-\mu+2\nu+2}{2}\end{matrix}\right)$ $[\sigma>0;\ \mathrm{Re}\,\mu<1;\ \mathrm{Re}\,s>\lvert\mathrm{Re}\,\mu\rvert/2]$
3	$(\sigma-x)_+^{-\mu/2}\,\mathrm{P}_\nu^\mu\left(\sqrt{\frac{x}{\sigma}}\right) \times {}_pF_q\left(\begin{matrix}(a_p);\,\omega x\\(b_q)\end{matrix}\right)$	$2^{1-2s+\mu}\sqrt{\pi}\,\sigma^{s-\mu/2}\,\Gamma\left[\begin{matrix}2s\\\frac{2s-\mu-\nu+1}{2},\,\frac{2s-\mu+\nu+2}{2}\end{matrix}\right] \times {}_{p+2}F_{q+2}\left(\begin{matrix}(a_p),\,s,\,\frac{2s+1}{2};\,\sigma\omega\\(b_q),\,\frac{2s-\mu-\nu+1}{2},\,\frac{2s-\mu+\nu+2}{2}\end{matrix}\right)$ $[\sigma,\ \mathrm{Re}\,s>0;\ \mathrm{Re}\,\mu<1]$
4	$(\sigma-x)_+^{-\mu/2}\,P_\nu^\mu\left(\frac{2\sigma}{x}-1\right) \times {}_pF_q\left(\begin{matrix}(a_p);\,\omega x\\(b_q)\end{matrix}\right)$	$\sigma^{s-\mu/2}\,\Gamma\left[\begin{matrix}s-\nu,\,s+\nu+1\\s+1,\,s-\mu+1\end{matrix}\right] \times {}_{p+2}F_{q+2}\left(\begin{matrix}(a_p),\,s-\nu,\,s+\nu+1;\,\sigma\omega\\(b_q),\,s+1,\,s-\mu+1\end{matrix}\right)$ $[\sigma>0;\ \mathrm{Re}\,\mu<1;\ \mathrm{Re}\,s>\mathrm{Re}\,\nu,\,-\mathrm{Re}\,\nu-1]$
5	$(\sigma^2-x^2)_+^{-\mu/2}\,P_\nu^\mu\left(\frac{\sigma}{x}\right) \times {}_pF_q\left(\begin{matrix}(a_p);\,\omega x\\(b_q)\end{matrix}\right)$	$\frac{2^s\omega\sigma^{s-\mu+1}}{\sqrt{\pi}}\prod_{i=1}^{p}a_i\prod_{j=1}^{q}b_j^{-1}\,\Gamma\left[\begin{matrix}\frac{s-\nu+1}{2},\,\frac{s+\nu+2}{2}\\s-\mu+2\end{matrix}\right] \times {}_{2p+2}F_{2q+3}\left(\begin{matrix}\frac{(a_p)+1}{2},\,\frac{(a_p)+2}{2},\,\frac{s-\nu+2}{2},\,\frac{s+\nu+3}{2};\,\frac{\sigma^2\omega^2}{4^{q-p+1}}\\\frac{3}{2},\,\frac{(b_q)+1}{2},\,\frac{(b_q)+2}{2},\,\frac{s-\mu+3}{2},\,\frac{s-\mu+4}{2}\end{matrix}\right) + \frac{2^{s-1}\sigma^{s-\mu}}{\sqrt{\pi}}\Gamma\left[\begin{matrix}\frac{s-\nu}{2},\,\frac{s+\nu+1}{2}\\s-\mu+1\end{matrix}\right] \times {}_{2p+2}F_{2q+3}\left(\begin{matrix}\frac{(a_p)}{2},\,\frac{(a_p)+1}{2},\,\frac{s-\nu}{2},\,\frac{s+\nu+1}{2};\,\frac{\sigma^2\omega^2}{4^{q-p+1}}\\\frac{1}{2},\,\frac{(b_q)}{2},\,\frac{(b_q)+1}{2},\,\frac{s-\mu+1}{2},\,\frac{s-\mu+2}{2}\end{matrix}\right)$ $[\sigma>0;\ \mathrm{Re}\,\mu<1;\ \mathrm{Re}\,s>\mathrm{Re}\,\nu,\,-\mathrm{Re}\,\nu-1]$

No.	$f(x)$	$F(s)$
6	$(\sigma-x)_+^{-\mu/2}\, P_\nu^\mu\left(\sqrt{\frac{\sigma}{x}}\right) \times {}_pF_q\left(\begin{matrix}(a_p);\, \omega x\\ (b_q)\end{matrix}\right)$	$\frac{2^{2s}\sigma^{s-\mu/2}}{\sqrt{\pi}}\,\Gamma\left[\begin{matrix}\frac{2s-\nu}{2},\, \frac{2s+\nu+1}{2}\\ 2s-\mu+1\end{matrix}\right] \times {}_{p+2}F_{q+2}\left(\begin{matrix}(a_p),\, \frac{2s-\nu}{2},\, \frac{2s+\nu+1}{2};\, \sigma\omega\\ (b_q),\, \frac{2s-\mu+1}{2},\, \frac{2s-\mu+2}{2}\end{matrix}\right)$ $[\sigma>0;\ \operatorname{Re}\mu<1;\ \operatorname{Re}s>\operatorname{Re}\nu,\, -\operatorname{Re}\nu-1]$
7	$(\sigma-x)_+^{-\mu/2}\, \mathrm{P}_\nu^\mu\left(\frac{2x}{\sigma}-1\right) \times {}_pF_q\left(\begin{matrix}(a_p);\, \omega x^2\\ (b_q)\end{matrix}\right)$	$\sigma^{s-\mu/2}\,\Gamma\left[\begin{matrix}\frac{2s-\mu}{2},\, \frac{2s+\mu}{2}\\ \frac{2s-\mu-2\nu}{2},\, \frac{2s-\mu+2\nu+2}{2}\end{matrix}\right] \times {}_{p+4}F_{q+4}\left(\begin{matrix}(a_p),\, \Delta\left(2,\, \frac{2s-\mu}{2}\right),\, \Delta\left(2,\, \frac{2s+\mu}{2}\right);\, \sigma^2\omega\\ (b_q),\, \Delta\left(2,\, \frac{2s-\mu-2\nu}{2}\right),\, \Delta\left(2,\, \frac{2s-\mu+2\nu+2}{2}\right)\end{matrix}\right)$ $[\sigma>0;\ \operatorname{Re}\mu<1;\ \operatorname{Re}s>\lvert\operatorname{Re}\mu\rvert/2]$
8	$(\sigma-x)_+^{-\mu/2}\, P_\nu^\mu\left(\frac{2\sigma}{x}-1\right) \times {}_pF_q\left(\begin{matrix}(a_p);\, \omega x^2\\ (b_q)\end{matrix}\right)$	$\sigma^{s-\mu/2}\,\Gamma\left[\begin{matrix}s-\nu,\, s+\nu+1\\ s+1,\, s-\mu+1\end{matrix}\right] \times {}_{p+4}F_{q+4}\left(\begin{matrix}(a_p),\, \frac{s-\nu}{2},\, \frac{s-\nu+1}{2},\, \frac{s+\nu+1}{2},\, \frac{s+\nu+2}{2}\\ (b_q),\, \frac{s+1}{2},\, \frac{s+2}{2},\, \frac{s-\mu+1}{2},\, \frac{s-\mu+2}{2};\, \sigma^2\omega\end{matrix}\right)$ $[\sigma>0;\ \operatorname{Re}\mu<1;\ \operatorname{Re}s>\operatorname{Re}\nu,\, -\operatorname{Re}\nu-1]$

3.33.20. ${}_pF_q((a_p);\, (b_q);\, \omega x^r)$ and $Q_\nu^\mu(\varphi(x))$

1	$(\sigma-x)_+^{\nu}\, Q_\nu^\mu\left(\frac{\sigma+x}{\sigma-x}\right) \times {}_pF_q\left(\begin{matrix}(a_p);\, \omega x\\ (b_q)\end{matrix}\right)$	$\frac{e^{i\mu\pi}\sigma^{s+\nu}}{2}\,\Gamma\left[\begin{matrix}\nu+1,\, \mu+\nu+1,\, \frac{2s-\mu}{2},\, \frac{2s+\mu}{2}\\ \frac{2s-\mu+2\nu+2}{2},\, \frac{2s+\mu+2\nu+2}{2}\end{matrix}\right] \times {}_{p+2}F_{q+2}\left(\begin{matrix}(a_p),\, \frac{2s-\mu}{2},\, \frac{2s+\mu}{2};\, \sigma\omega\\ (b_q),\, \frac{2s-\mu+2\nu+2}{2},\, \frac{2s+\mu+2\nu+2}{2}\end{matrix}\right)$ $[\sigma>0;\ \operatorname{Re}\nu>-1;\ \operatorname{Re}s>\lvert\operatorname{Re}\mu\rvert/2]$
2	$(\sigma-x)_+^{\nu}\, Q_\nu^\mu\left(\frac{\sigma+x}{\sigma-x}\right) \times {}_pF_q\left(\begin{matrix}(a_p);\, \omega x^2\\ (b_q)\end{matrix}\right)$	$\frac{e^{i\mu\pi}\sigma^{s+\nu}}{2}\,\Gamma\left[\begin{matrix}\nu+1,\, \mu+\nu+1,\, \frac{2s-\mu}{2},\, \frac{2s+\mu}{2}\\ \frac{2s-\mu+2\nu+2}{2},\, \frac{2s+\mu+2\nu+2}{2}\end{matrix}\right] \times {}_{p+4}F_{q+4}\left(\begin{matrix}(a_p),\, \Delta\left(2,\, \frac{2s-\mu}{2}\right),\, \Delta\left(2,\, \frac{2s+\mu}{2}\right);\, \sigma^2\omega\\ (b_q),\, \Delta\left(2,\, \frac{2s-\mu+2\nu+2}{2}\right),\, \Delta\left(2,\, \frac{2s+\mu+2\nu+2}{2}\right)\end{matrix}\right)$ $[\sigma>0;\ \operatorname{Re}\nu>-1;\ \operatorname{Re}s>\lvert\operatorname{Re}\mu\rvert/2]$

3.33.21. ${}_pF_q((a_p);\, (b_q);\, \omega x^r)$ and $\Psi(a,\, b;\, \sigma x)$

1	$e^{-\sigma x}\,\Psi(a,\, b;\, \sigma x)\times$	$\sigma^{-s}\,\Gamma\left[\begin{matrix}s,\, s-b+1\\ s+a-b+1\end{matrix}\right]\, {}_{p+2}F_{q+1}\left(\begin{matrix}(a_p),\, s,\, s-b+1\\ (b_q),\, s+a-b+1;\, -\frac{\omega}{\sigma}\end{matrix}\right)$

No.	$f(x)$	$F(s)$
	$\times\, {}_pF_q\left(\begin{matrix}(a_p);\,-\omega x\\(b_q)\end{matrix}\right)$	
	$\left[\begin{matrix}[q=p-1;\ \lvert\arg\sigma\rvert,\ \lvert\arg\omega\rvert<\pi;\ 0,\ \operatorname{Re}b-1<\operatorname{Re}s<\operatorname{Re}(a_k+a)]\ \text{or}\\ [q=p;\ \lvert\arg\sigma\rvert<\pi;\ (\operatorname{Re}\omega>0;\ 0,\ \operatorname{Re}b-1<\operatorname{Re}s<\operatorname{Re}(a_k+a))\ \text{or}\\ (\operatorname{Re}\omega=0;\ 0,\ \operatorname{Re}b-1<\operatorname{Re}s<\operatorname{Re}(a_k+a),\ \operatorname{Re}(a-\chi)+1)\ \text{or}\\ [q=p+1;\ \omega>0;\ 0,\ \operatorname{Re}b-1<\operatorname{Re}s<\operatorname{Re}(a_k+a),\ \operatorname{Re}(a-\chi)+1/2;\ \lvert\arg\sigma\rvert<\pi].\end{matrix}\right]$	
2	$e^{-\sigma x}\,\Psi(a,\,b;\,\sigma x)$ $\times\, {}_pF_q\left(\begin{matrix}(a_p);\,-\omega x^2\\(b_q)\end{matrix}\right)$	$\sigma^{-s}\,\Gamma\left[\begin{matrix}s,\ s-b+1\\ s+a-b+1\end{matrix}\right]\,{}_{p+4}F_{q+2}\left(\begin{matrix}(a_p),\ \frac{s}{2},\ \frac{s+1}{2},\ \frac{s-b+1}{2},\ \frac{s-b+2}{2}\\ (b_q),\ \frac{s+a-b+1}{2},\ \frac{s+a-b+2}{2};\ -\frac{4\omega}{\sigma^2}\end{matrix}\right)$
	$\left[\begin{matrix}[q=p-1;\ \lvert\arg\sigma\rvert,\ \lvert\arg\omega\rvert<\pi;\ 0,\ \operatorname{Re}b-1<\operatorname{Re}s<\operatorname{Re}(2a_k+a)]\ \text{or}\\ [q=p;\ \lvert\arg\sigma\rvert<\pi;\ (\operatorname{Re}\omega>0;\ 0,\ \operatorname{Re}b-1<\operatorname{Re}s<\operatorname{Re}(2a_k+a))\ \text{or}\\ (\operatorname{Re}\omega=0;\ 0,\ \operatorname{Re}b-1<\operatorname{Re}s<\operatorname{Re}(2a_k+a),\ \operatorname{Re}(a-2\chi)+2)\ \text{or}\\ [q=p+1;\ \omega>0;\ 0,\ \operatorname{Re}b-1<\operatorname{Re}s<\operatorname{Re}(2a_k+a),\ \operatorname{Re}(a-2\chi)+1;\ \lvert\arg\sigma\rvert<\pi].\end{matrix}\right]$	

3.33.22. ${}_pF_q((a_p);\,(b_q);\,\omega x^r)$ and ${}_2F_1(a,\,b;\,\varphi(x))$

1	$(\sigma-x)_+^{c-1}\,{}_2F_1\left(\begin{matrix}a,\,b\\ c;\,1-\frac{x}{\sigma}\end{matrix}\right)$ $\times\, {}_pF_q\left(\begin{matrix}(a_p);\,\omega x\\(b_q)\end{matrix}\right)$	$\sigma^{s+c-1}\,\Gamma\left[\begin{matrix}c,\ s,\ s-a-b+c\\ s-a+c,\ s-b+c\end{matrix}\right]$ $\times\,{}_{p+2}F_{q+2}\left(\begin{matrix}(a_p),\ s,\ s-a-b+c;\ \sigma\omega\\ (b_q),\ s-a+c,\ s-b+c\end{matrix}\right)$ $[\sigma,\ \operatorname{Re}c>0;\ \operatorname{Re}s>0,\ \operatorname{Re}(a+b-c)]$
2	$(\sigma-x)_+^{c-1}\,{}_2F_1\left(\begin{matrix}a,\,b\\ c;\,1-\frac{\sigma}{x}\end{matrix}\right)$ $\times\, {}_pF_q\left(\begin{matrix}(a_p);\,\omega x\\(b_q)\end{matrix}\right)$	$\sigma^{s+c-1}\,\Gamma\left[\begin{matrix}c,\ s+a,\ s+b\\ s+a+b,\ s+c\end{matrix}\right]\,{}_{p+2}F_{q+2}\left(\begin{matrix}(a_p),\ s+a,\ s+b;\ \sigma\omega\\ (b_q),\ s+a+b,\ s+c\end{matrix}\right)$ $[\sigma,\ \operatorname{Re}c>0;\ \operatorname{Re}s>-\operatorname{Re}a,\ -\operatorname{Re}b]$
3	$(\sigma-x)_+^{c-1}\,{}_2F_1\left(\begin{matrix}a,\,b\\ c;\,1-\frac{x}{\sigma}\end{matrix}\right)$ $\times\, {}_pF_q\left(\begin{matrix}(a_p);\,\omega x^2\\(b_q)\end{matrix}\right)$	$\sigma^{s+c-1}\,\Gamma\left[\begin{matrix}c,\ s,\ s-a-b+c\\ s-a+c,\ s-b+c\end{matrix}\right]$ $\times\,{}_{p+4}F_{q+4}\left(\begin{matrix}(a_p),\ \frac{s}{2},\ \frac{s+1}{2},\ \frac{s-a-b+c}{2},\ \frac{s-a-b+c+1}{2};\ \sigma^2\omega\\ (b_q),\ \frac{s-a+c}{2},\ \frac{s-a+c+1}{2},\ \frac{s-b+c}{2},\ \frac{s-b+c+1}{2}\end{matrix}\right)$ $[\sigma,\ \operatorname{Re}c>0;\ \operatorname{Re}s>0,\ \operatorname{Re}(a+b-c)]$
4	$(\sigma-x)_+^{c-1}\,{}_2F_1\left(\begin{matrix}a,\,b\\ c;\,1-\frac{\sigma}{x}\end{matrix}\right)$ $\times\, {}_pF_q\left(\begin{matrix}(a_p);\,\omega x^2\\(b_q)\end{matrix}\right)$	$\sigma^{s+c-1}\,\Gamma\left[\begin{matrix}c,\ s+a,\ s+b\\ s+a+b,\ s+c\end{matrix}\right]$ $\times\,{}_{p+4}F_{q+4}\left(\begin{matrix}(a_p),\ \frac{s+a}{2},\ \frac{s+a+1}{2},\ \frac{s+b}{2},\ \frac{s+b+1}{2};\ \sigma^2\omega\\ (b_q),\ \frac{s+a+b}{2},\ \frac{s+a+b+1}{2},\ \frac{s+c}{2},\ \frac{s+c+1}{2}\end{matrix}\right)$ $[\sigma,\ \operatorname{Re}c>0;\ \operatorname{Re}s>-\operatorname{Re}a,\ -\operatorname{Re}b]$

3.33.23. Products of ${}_pF_q\left((a_p);\;(b_q);\;\omega x^r\right)$

Notation:

$$g=\frac{(1-\ell)(m-n+1)+(1-k)(p-q+1)}{2};$$

$$\mu=\sum_{i=1}^{p}a_i-\sum_{j=1}^{q}b_j+\frac{q-p+1}{2};\;\rho=\sum_{i=1}^{m}c_i-\sum_{j=1}^{n}d_j+\frac{n-m+1}{2};$$

$$k,\,\ell,\,m,\,n,\,p,\,q=0,\,1,\,2,\ldots;\;k,\,\ell\neq 0;\;m\leq n+1;\;p\leq q+1.$$

1	$${}_mF_n\left(\begin{matrix}(c_m);\,-\sigma x\\(d_n)\end{matrix}\right)\times{}_pF_q\left(\begin{matrix}(a_p);\,-\omega x^{\ell/k}\\(b_q)\end{matrix}\right)$$	$$(2\pi)^g\,k^{\mu}\,\ell^{\rho+s(n-m+1)-1}\,\sigma^{-s}\,\Gamma\left[\begin{matrix}(b_q),\,(d_n)\\(a_p),\,(c_m)\end{matrix}\right]\times G^{k+\ell m,\,kp+\ell}_{kp+\ell n+\ell,\,kq+k+\ell m}\left(\frac{k^{k(p-q-1)}\omega^k}{\ell^{\ell(m-n-1)}\sigma^{\ell}}\,\middle\vert\,\begin{matrix}\Delta(\ell,\,1-s),\,\Delta(k,\,1-(a_p)),\,\Delta(\ell,\,(d_n)-s)\\\Delta(k,\,0),\,\Delta(\ell,\,(c_m)-s),\,\Delta(k,\,1-(b_q))\end{matrix}\right)$$
		and one of the following conditions hold (1) $mp\neq 0;\;m=n$ or $m=n+1;\;p=q$ or $p=q+1;$ $\vert\arg\sigma\vert<(m-n+1)\pi/2;\;\vert\arg\omega\vert<(p-q+1)\pi/2;\;\mathrm{Re}\,s>0;$ $\mathrm{Re}\left(s-c_j-a_i\ell/k\right)<0\;(j=1,2,\ldots,m;\;i=1,2,\ldots,p)$ (2) $m>0;\;m=n$ or $m=n+1;\;p=q-1$ or $p=q;$ $\vert\arg\sigma\vert<(m-n+1)\pi/2;\;\vert\arg\omega\vert=(p-q+1)\pi/2;\;\mathrm{Re}\,s>0;$ $\mathrm{Re}\left(s-c_j-a_i\ell/k\right)<0\;(j=1,2,\ldots,m;\;i=1,2,\ldots,p)$ $(p-q-1)\,\mathrm{Re}\,(s-c_j)-\ell\;\mathrm{Re}\,\mu/k>-3\ell/2k\;(j=1,2,\ldots,p)$ (3) $m=n-1;$ or $m=n;\;p>0;\;p=q$ or $p=q+1;$ $\vert\arg\sigma\vert=(m-n+1)\pi/2;\;\vert\arg\omega\vert<(p-q+1)\pi/2;\;\mathrm{Re}\,s>0;$ $\mathrm{Re}\left(s-c_j-a_i\ell/k\right)<0\;(j=1,2,\ldots,m;\;i=1,2,\ldots,p)$ $(m-n-1)\,\mathrm{Re}\left(s-a_i\ell/k\right)-\mathrm{Re}\,\rho>-3/2$
2	$${}_mF_n\left(\begin{matrix}(c_m);\,-\frac{\sigma}{x}\\(d_n)\end{matrix}\right)\times{}_pF_q\left(\begin{matrix}(a_p);\,-\omega x^{\ell/k}\\(b_q)\end{matrix}\right)$$	$$(2\pi)^g\,k^{\mu}\,\ell^{\rho+s+(m-n-1)-1}\,\sigma^{s}\,\Gamma\left[\begin{matrix}(b_q),\,(d_n)\\(a_p),\,(c_m)\end{matrix}\right]\times G^{k+\ell,\,kp+\ell m}_{kq+\ell m,\,kq+k+\ell n+\ell}\left(\frac{k^{k(p-q-1)}\omega^k}{\ell^{\ell(n-m+1)}\sigma^{-\ell}}\,\middle\vert\,\begin{matrix}\Delta(k,\,(a_p)),\,\Delta(\ell,\,1-s-(c_m))\\\Delta(\ell,\,-s),\,\Delta(k,\,0),\,\Delta(k,\,1-(b_q)),\,\Delta(\ell,\,1-s-(d_n))\end{matrix}\right)$$
		and one of the following conditions hold (1) $mp\neq 0;\;m=n$ or $m=n+1;\;p=q$ or $p=q+1;$ $\vert\arg\sigma\vert<(m-n+1)\pi/2;\;\vert\arg\omega\vert<(p-q+1)\pi/2;$ $\mathrm{Re}\,(a+c_j)>0\;(j=1,\ldots,m);\;\mathrm{Re}\left(s-a_j\ell/k\right)<0\;(j=1,2,\ldots,p)$ (2) $m=n$ or $m=n+1;\;p=q-1$ or $p=q;$ $\vert\arg\sigma\vert<(m-n+1)\pi/2;\;\vert\arg\omega\vert=(p-q+1)\pi/2;$ $\mathrm{Re}\,(a+c_j)>0\;(j=1,\ldots,m);\;\mathrm{Re}\left(s-a_j\ell/k\right)<0\;(j=1,2,\ldots,p)$ $\mathrm{Re}\left[(p-q-1)-\mu\ell/k\right]>-3\ell/(2k)$ (3) $m=n-1$ or $m=n;\;p=q$ or $p=q+1;$ $\vert\arg\sigma\vert=(m-n+1)\pi/2;\;\vert\arg\omega\vert<(p-q+1)\pi/2;$ $\mathrm{Re}\,(a+c_j)>0\;(j=1,\ldots,m);\;\mathrm{Re}\left(s-a_j\ell/k\right)<0\;(j=1,2,\ldots,p)$ $\mathrm{Re}\left[(n-m-1)\,s-\rho\right]>-3/2$

3.34. The Appell Functions

3.34.1. The Appell and algebraic functions

1	$(\sigma - x)_+^{c-1}$ $\times F_1(a, b, b'; c; w(\sigma - x), z(\sigma - x))$	$\sigma^{s+c-1} \mathrm{B}(s, c) F_1(a, b, b'; s+c; \sigma w, \sigma z)$ $[\sigma, \operatorname{Re} c, \operatorname{Re} s > 0]$
2	$\frac{1}{(x+\sigma)^b} F_1\left(a, b, b'; c; \frac{w}{x+\sigma}, z\right)$	$\sigma^{s-b} \mathrm{B}(s, b-s) F_1\left(a, b-s, b'; c; \frac{w}{\sigma}, z\right)$ $[0 < \operatorname{Re} s < \operatorname{Re} b;\ \lvert\arg \sigma\rvert < \pi]$
3	$\frac{1}{(x+\sigma)^b} F_1\left(a, b, b'; c; \frac{wx}{x+\sigma}, z\right)$	$\sigma^{s-b} \mathrm{B}(s, b-s) F_1(a, s, b'; c; w, z)$ $[0 < \operatorname{Re} s < \operatorname{Re} b;\ \lvert\arg \sigma\rvert < \pi]$
4	$\frac{1}{(x+\sigma)^a} F_1\left(a, b, b'; c; \frac{w}{x+\sigma}, \frac{z}{x+\sigma}\right)$	$\sigma^{s-a} \mathrm{B}(s, a-s) F_1\left(a-s, b, b'; c; \frac{w}{\sigma}, \frac{z}{\sigma}\right)$ $[0 < \operatorname{Re} s < \operatorname{Re} a;\ \lvert\arg \sigma\rvert < \pi]$
5	$\frac{1}{(x+\sigma)^a} F_1\left(a, b, b'; c; \frac{wx}{x+\sigma}, \frac{zx}{x+\sigma}\right)$	$\sigma^{s-a} \mathrm{B}(s, a-s) F_1(s, b, b'; c; w, z)$ $[0 < \operatorname{Re} s < \operatorname{Re} a;\ \lvert\arg \sigma\rvert < \pi]$
6	$(\sigma - x)_+^{c-1}$ $\times F_2(a, b, b'; c, c'; w(\sigma - x), z)$	$\sigma^{s+c-1} \mathrm{B}(s, c) F_2(a, b, b'; s+c, c'; \sigma w, z)$ $[\sigma, \operatorname{Re} c, \operatorname{Re} s > 0]$
7	$(x - \sigma)_+^{c-1}$ $\times F_2\left(a, b, b'; c, c'; \frac{w(x-\sigma)}{x}, z\right)$	$\sigma^{s+c-1} \mathrm{B}(c, 1-c-s) F_2(a, b, b'; 1-s, c'; w, z)$ $[\sigma, \operatorname{Re} c > 0;\ \operatorname{Re}(s+c) < 1]$
8	$\frac{1}{(x+\sigma)^b} F_2\left(a, b, b'; c, c'; \frac{w}{x+\sigma}, z\right)$	$\sigma^{s-b} \mathrm{B}(s, b-s) F_2\left(a, s, b'; c, c'; \frac{w}{\sigma}, z\right)$ $[0 < \operatorname{Re} s < \operatorname{Re} b;\ \lvert\arg \sigma\rvert < \pi]$
9	$\frac{1}{(x+\sigma)^b} F_2\left(a, b, b'; c, c'; \frac{wx}{x+\sigma}, z\right)$	$\sigma^{s-b} \mathrm{B}(s, b-s) F_2(a, s, b'; c, c'; w, z)$ $[0 < \operatorname{Re} s < \operatorname{Re} b;\ \lvert\arg \sigma\rvert < \pi]$
10	$\frac{1}{(x+\sigma)^a}$ $\times F_2\left(a, b, b'; c, c'; \frac{w}{x+\sigma}, \frac{z}{x+\sigma}\right)$	$\sigma^{s-a} \mathrm{B}(s, a-s) F_2\left(a-s, b, b'; c, c'; \frac{w}{\sigma}, \frac{z}{\sigma}\right)$ $[0 < \operatorname{Re} s < \operatorname{Re} a;\ \lvert\arg \sigma\rvert < \pi]$

No.	$f(x)$	$F(s)$
11	$\frac{1}{(x+\sigma)^a} \times F_2\left(a,\, b,\, b';\, c,\, c';\, \frac{wx}{x+\sigma},\, \frac{zx}{x+\sigma}\right)$	$\sigma^{s-a}\,\mathrm{B}(s,\, a-s)\, F_2(s,\, b,\, b';\, c,\, c';\, w,\, z)$ $[0<\operatorname{Re}s<\operatorname{Re}a;\ \lvert\arg\sigma\rvert<\pi]$
12	$(\sigma-x)_+^{c-1} \times F_3(a,\, a',\, b,\, b';\, c;\, w(\sigma-x),\, z(\sigma-x))$	$\sigma^{s+c-1}\,\mathrm{B}(s,\, c)\, F_3(a,\, a',\, b,\, b';\, s+c;\, \sigma w,\, \sigma z)$ $[\sigma,\ \operatorname{Re}c,\ \operatorname{Re}s>0]$
13	$\frac{1}{(x+\sigma)^a}\, F_3\left(a,\, a',\, b,\, b';\, c;\, \frac{w}{x+\sigma},\, z\right)$	$\sigma^{s-a}\,\mathrm{B}(s,\, a-s)\, F_3\left(a-s,\, a',\, b,\, b';\, c;\, \frac{w}{\sigma},\, z\right)$ $[0<\operatorname{Re}s<\operatorname{Re}a;\ \lvert\arg\sigma\rvert<\pi]$
14	$\frac{1}{(x+\sigma)^a}\, F_3\left(a,\, a',\, b,\, b';\, c;\, \frac{wx}{x+\sigma},\, z\right)$	$\sigma^{s-a}\,\mathrm{B}(s,\, a-s)\, F_3(s,\, a',\, b,\, b';\, c;\, w,\, z)$ $[0<\operatorname{Re}s<\operatorname{Re}a;\ \lvert\arg\sigma\rvert<\pi]$
15	$(1-x)_+^{c-1} \times F_3\left(a,\, a',\, b,\, b',\, c;\, 1-x,\, 1-\frac{1}{x}\right)$	$\Gamma(c)\,\Gamma\left[\begin{matrix} s+a',\, s+b',\, s+c-a-b \\ s+a'+b',\, s+c-a,\, s+c-b \end{matrix}\right]$ $\left[\begin{matrix}\operatorname{Re}c>0;\ \operatorname{Re}s>-\operatorname{Re}a',\ -\operatorname{Re}b'; \\ \operatorname{Re}(s-a-b+c)>0\end{matrix}\right]$
16	$(x-1)_+^{c-1} \times F_3\left(a,\, a',\, b,\, b',\, c; 1-x,\, 1-\frac{1}{x}\right)$	$\Gamma(c)\,\Gamma\left[\begin{matrix} 1-a'-b'-s,\, 1+a-c-s \\ 1-a'-s,\, 1-b'-s \end{matrix}\right] \times\Gamma\left[\begin{matrix} 1+b-c-s \\ 1+a+b-c-s \end{matrix}\right]$ $\left[\begin{matrix}\operatorname{Re}c>0;\ \operatorname{Re}s<1-\operatorname{Re}(a'+b'); \\ \operatorname{Re}s<1-\operatorname{Re}(a-c),\ 1-\operatorname{Re}(b-c)\end{matrix}\right]$
17	$(\sigma-x)_+^{c-1}\, F_4(a,\, b;\, c,\, c';\, w(\sigma-x),\, z)$	$\sigma^{s+c-1}\,\mathrm{B}(s,\, c)\, F_4(a,\, b;\, s+c,\, c';\, \sigma w,\, z)$ $[\sigma,\ \operatorname{Re}c,\ \operatorname{Re}s>0]$
18	$(x-\sigma)_+^{c-1}\, F_4\left(a,\, b;\, c,\, c';\, \frac{w(x-\sigma)}{x},\, z\right)$	$\sigma^{s+c-1}\,\mathrm{B}(c,\, 1-c-s)\, F_4(a,\, b;\, 1-s,\, c';\, w,\, z)$ $[\sigma,\ \operatorname{Re}c>0;\ \operatorname{Re}(s+c)<1]$
19	$\frac{1}{(x+\sigma)^a}\, F_4\left(a,\, b;\, c,\, c';\, \frac{w}{x+\sigma},\, \frac{z}{x+\sigma}\right)$	$\sigma^{s-a}\,\mathrm{B}(s,\, a-s)\, F_4\left(a-s,\, b;\, c,\, c';\, \frac{w}{\sigma},\, \frac{z}{\sigma}\right)$ $[0<\operatorname{Re}s<\operatorname{Re}a;\ \lvert\arg\sigma\rvert<\pi]$
20	$\frac{1}{(x+\sigma)^a}\, F_4\left(a,\, b;\, c,\, c';\, \frac{wx}{x+\sigma},\, \frac{zx}{x+\sigma}\right)$	$\sigma^{s-a}\,\mathrm{B}(s,\, a-s)\, F_4(s,\, b;\, c,\, c';\, w,\, z)$ $[0<\operatorname{Re}s<\operatorname{Re}a;\ \lvert\arg\sigma\rvert<\pi]$

3.35. The Humbert Functions

3.35.1. The Humbert and algebraic functions

1	$(\sigma - x)_+^{c-1}$ $\times \Phi_1(a, b; c; w(\sigma - x), z(\sigma - x))$	$\sigma^{s+c-1} \mathrm{B}(s, c) \Phi_1(a, b; s+c; \sigma w, \sigma z)$ $[\sigma, \operatorname{Re} c, \operatorname{Re} s > 0]$
2	$\frac{1}{(x+\sigma)^b} \Phi_1\left(a, b; c; \frac{w}{x+\sigma}, z\right)$	$\sigma^{s-b} \mathrm{B}(s, b-s) \Phi_1\left(a, b-s; c; \frac{w}{\sigma}, z\right)$ $[0 < \operatorname{Re} s < \operatorname{Re} b;\ \lvert\arg \sigma\rvert < \pi]$
3	$\frac{1}{(x+\sigma)^b} \Phi_1\left(a, b; c; \frac{wx}{x+\sigma}, z\right)$	$\sigma^{s-b} \mathrm{B}(s, b-s) \Phi_1(a, s; c; w, z)$ $[0 < \operatorname{Re} s < \operatorname{Re} b;\ \lvert\arg \sigma\rvert < \pi]$
4	$\frac{1}{(x+\sigma)^a} \Phi_1\left(a, b, c; \frac{w}{x+\sigma}, \frac{z}{x+\sigma}\right)$	$\sigma^{s-a} \mathrm{B}(s, a-s) \Phi_1\left(a-s, b; c; \frac{w}{\sigma}, \frac{z}{\sigma}\right)$ $[0 < \operatorname{Re} s < \operatorname{Re} a;\ \lvert\arg \sigma\rvert < \pi]$,
5	$\frac{1}{(x+\sigma)^a} \Phi_1\left(a, b, b'; c; \frac{wx}{x+\sigma}, \frac{zx}{x+\sigma}\right)$	$\sigma^{s-a} \mathrm{B}(s, a-s) \Phi_1(a-s, b, b'; c; w, z)$ $[0 < \operatorname{Re} s < \operatorname{Re} a;\ \lvert\arg \sigma\rvert < \pi]$
6	$(\sigma - x)_+^{c-1}$ $\times \Phi_2(b, b'; c; w(\sigma - x), z(\sigma - x))$	$\sigma^{s+c-1} \mathrm{B}(s, c) \Phi_2(b, b'; s+c; \sigma w, \sigma z)$ $[\sigma, \operatorname{Re} c, \operatorname{Re} s > 0]$
7	$\frac{1}{(x+\sigma)^b} \Phi_2\left(b, b'; c; \frac{w}{x+\sigma}, z\right)$	$\sigma^{s-b} \mathrm{B}(s, b-s) \Phi_2\left(b-s, b'; c; \frac{w}{\sigma}, z\right)$ $[0 < \operatorname{Re} s < \operatorname{Re} b;\ \lvert\arg \sigma\rvert < \pi]$
8	$\frac{1}{(x+\sigma)^b} \Phi_2\left(b, b'; c; \frac{wx}{x+\sigma}, z\right)$	$\sigma^{s-b} \mathrm{B}(s, b-s) \Phi_2(s, b'; c; w, z)$ $[0 < \operatorname{Re} s < \operatorname{Re} b;\ \lvert\arg \sigma\rvert < \pi]$
9	$(\sigma - x)_+^{c-1} \Phi_3(a; c; w(\sigma - x), z(\sigma - x))$	$\sigma^{s+c-1} \mathrm{B}(s, c) \Phi_3(a; s+c; \sigma w, \sigma z)$ $[\sigma, \operatorname{Re} c, \operatorname{Re} s > 0]$

No.	$f(x)$	$F(s)$
10	$\frac{1}{(x+\sigma)^b}\Phi_3\left(b;\, c;\, \frac{w}{x+\sigma},\, z\right)$	$\sigma^{s-b}\,\mathrm{B}(s,\, b-s)\,\Phi_3\left(b-s;\, c;\, \frac{w}{\sigma},\, z\right)$ $[0<\operatorname{Re}s<\operatorname{Re}b;\ \lvert\arg\sigma\rvert<\pi]$
11	$\frac{1}{(x+\sigma)^b}\Phi_3\left(b;\, c;\, \frac{wx}{x+\sigma},\, z\right)$	$\sigma^{s-b}\,\mathrm{B}(s,\, b-s)\,\Phi_3(s;\, c;\, w,\, z)$ $[0<\operatorname{Re}s<\operatorname{Re}b;\ \lvert\arg\sigma\rvert<\pi]$
12	$(\sigma-x)_+^{c-1}$ $\times\,\Xi_1(a,\, a',\, b;\, c;\, w(\sigma-x),\, z(\sigma-x))$	$\sigma^{s+c-1}\,\mathrm{B}(s,\, c)\,\Xi_1(a,\, a',\, b;\, s+c;\, \sigma w, \sigma z)$ $[\sigma,\ \operatorname{Re}c,\ \operatorname{Re}s>0]$
13	$\frac{1}{(x+\sigma)^a}\Xi_1\left(a,\, a',\, b;\, c;\, \frac{w}{x+\sigma},\, z\right)$	$\sigma^{s-a}\,\mathrm{B}(s,\, a-s)\,\Xi_1\left(a-s,\, a',\, b;\, c;\, \frac{w}{\sigma},\, z\right)$ $[0<\operatorname{Re}s<\operatorname{Re}a;\ \lvert\arg\sigma\rvert<\pi]$
14	$\frac{1}{(x+\sigma)^a}\Xi_1\left(a,\, a',\, b;\, c;\, \frac{wx}{x+\sigma},\, z\right)$	$\sigma^{s-a}\,\mathrm{B}(s,\, a-s)\,\Xi_1(s,\, a',\, b;\, c;\, w,\, z)$ $[0<\operatorname{Re}s<\operatorname{Re}a;\ \lvert\arg\sigma\rvert<\pi]$,
15	$(\sigma-x)_+^{c-1}$ $\times\,\Xi_2(a,\, b;\, c;\, w(\sigma-x),\, z(\sigma-x))$	$\sigma^{s+c-1}\,\mathrm{B}(s,\, c)\,\Xi_2(a,\, b;\, s+c;\, \sigma w,\, \sigma z)$ $[\sigma,\ \operatorname{Re}c,\ \operatorname{Re}s>0]$
16	$\frac{1}{(x+\sigma)^a}\Xi_2\left(a,\, b;\, c;\, \frac{w}{x+\sigma},\, z\right)$	$\sigma^{s-a}\,\mathrm{B}(s,\, a-s)\,\Xi_2\left(a-s,\, b;\, c;\, \frac{w}{\sigma},\, z\right)$ $[0<\operatorname{Re}s<\operatorname{Re}a;\ \lvert\arg\sigma\rvert<\pi]$
17	$\frac{1}{(x+\sigma)^a}\Xi_2\left(a,\, b;\, c;\, \frac{wx}{x+\sigma},\, z\right)$	$\sigma^{s-a}\,\mathrm{B}(s,\, a-s)\,\Xi_2(s,\, b;\, c;\, w,\, z)$ $[0<\operatorname{Re}s<\operatorname{Re}a;\ \lvert\arg\sigma\rvert<\pi]$
18	$(\sigma-x)_+^{c-1}\,\Psi_1(a,\, b;\, c,\, c';\, w(\sigma-x),\, z)$	$\sigma^{s+c-1}\,\mathrm{B}(s,\, c)\,\Psi_1(a,\, b;\, s+c,\, c';\, \sigma w,\, z)$ $[\sigma,\ \operatorname{Re}c,\ \operatorname{Re}s>0]$

No.	$f(x)$	$F(s)$
19	$(x-\sigma)_+^{c-1}\,\Psi_1\left(a,\,b;\,c,\,c';\,\frac{w\,(x-\sigma)}{x},\,z\right)$	$\sigma^{s+c-1}\,\mathrm{B}\,(c,\,1-c-s)\,\Psi_1\,(a,\,b;\,1-s,\,c';\,w,\,z)$ $[\sigma,\ \mathrm{Re}\,c>0;\ \mathrm{Re}\,(s+c)<1]$
20	$\frac{1}{(x+\sigma)^b}\,\Psi_1\left(a,\,b;\,c,\,c';\,\frac{w}{x+\sigma},\,z\right)$	$\sigma^{s-b}\,\mathrm{B}\,(s,\,b-s)\,\Psi_1\left(a,\,s;\,c,\,c';\,\frac{w}{\sigma},\,z\right)$ $[0<\mathrm{Re}\,s<\mathrm{Re}\,b;\ \lvert\arg\sigma\rvert<\pi]$
21	$\frac{1}{(x+\sigma)^b}\,\Psi_1\left(a,\,b;\,c,\,c';\,\frac{wx}{x+\sigma},\,z\right)$	$\sigma^{s-b}\,\mathrm{B}\,(s,\,b-s)\,\Psi_1\,(a,\,s;\,c,\,c';\,w,\,z)$ $[0<\mathrm{Re}\,s<\mathrm{Re}\,b;\ \lvert\arg\sigma\rvert<\pi]$
22	$\frac{1}{(x+\sigma)^a}\,\Psi_1\left(a,\,b;\,c,\,c';\,\frac{w}{x+\sigma},\,\frac{z}{x+\sigma}\right)$	$\sigma^{s-a}\,\mathrm{B}\,(s,\,a-s)\,\Psi_1\left(a-s,\,b;\,c,\,c';\,\frac{w}{\sigma},\,\frac{z}{\sigma}\right)$ $[0<\mathrm{Re}\,s<\mathrm{Re}\,a;\ \lvert\arg\sigma\rvert<\pi]$
23	$\frac{1}{(x+\sigma)^a}\,\Psi_1\left(a,\,b;\,c,\,c';\,\frac{wx}{x+\sigma},\,\frac{zx}{x+\sigma}\right)$	$\sigma^{s-a}\,\mathrm{B}\,(s,\,a-s)\,\Psi_1\,(s,\,b;\,c,\,c';\,w,\,z)$ $[0<\mathrm{Re}\,s<\mathrm{Re}\,a;\ \lvert\arg\sigma\rvert<\pi]$
24	$(\sigma-x)_+^{c-1}\,\Psi_2\,(a;\,c,\,c';\,w\,(\sigma-x)\,,\,z)$	$\sigma^{s+c-1}\,\mathrm{B}\,(s,\,c)\,\Psi_2\,(a;\,s+c,\,c';\,\sigma w,\,z)$ $[\sigma,\ \mathrm{Re}\,c,\ \mathrm{Re}\,s>0]$
25	$(x-\sigma)_+^{c-1}\,\Psi_2\left(a;\,c,\,c';\,\frac{w\,(x-\sigma)}{x},\,z\right)$	$\sigma^{s+c-1}\,\mathrm{B}\,(c,\,1-c-s)\,\Psi_2\,(a;\,1-s,\,c';\,w,\,z)$ $[\sigma,\ \mathrm{Re}\,c>0;\ \mathrm{Re}\,(s+c)<1]$
26	$\frac{1}{(x+\sigma)^a}\,\Psi_2\left(a;\,c,\,c';\,\frac{w}{x+\sigma},\,\frac{z}{x+\sigma}\right)$	$\sigma^{s-a}\,\mathrm{B}\,(s,\,a-s)\,\Psi_2\left(a-s;\,c,\,c';\,\frac{w}{\sigma},\,\frac{z}{\sigma}\right)$ $[0<\mathrm{Re}\,s<\mathrm{Re}\,a;\ \lvert\arg\sigma\rvert<\pi]$
27	$\frac{1}{(x+\sigma)^a}\,\Psi_2\left(a;\,c,\,c';\,\frac{wx}{x+\sigma},\,\frac{zx}{x+\sigma}\right)$	$\sigma^{s-a}\,\mathrm{B}\,(s,\,a-s)\,\Psi_2\,(s;\,c,\,c';\,w,\,z)$ $[0<\mathrm{Re}\,s<\mathrm{Re}\,a;\ \lvert\arg\sigma\rvert<\pi]$

3.35.2. The Humbert and the exponential functions

1	$e^{-px}\Phi_1(a,\ b;\ w,\ zx)$	$\dfrac{\Gamma(s)}{p^s} F_1\left(a,\ b,\ s;\ c;\ w,\ \dfrac{z}{p}\right)$ [$\operatorname{Re} p > 0,\ \operatorname{Re} z;\ \operatorname{Re} s > 0$]
2	$e^{-px}\Phi_2(b,\ b';\ c;\ wx,\ z)$	$\dfrac{\Gamma(s)}{p^s} \Xi_1\left(b,\ b';\ s;\ c;\ \dfrac{w}{p},\ z\right)$ [$\operatorname{Re} p > 0,\ \operatorname{Re} w;\ \operatorname{Re} s > 0$]
3	$e^{-px}\Phi_2(b,\ b';\ c;\ wx,\ zx)$	$\dfrac{\Gamma(s)}{p^s} F_1\left(s,\ b,\ b';\ c;\ \dfrac{w}{p},\ \dfrac{z}{p}\right)$ [$\operatorname{Re} p > 0,\ \operatorname{Re} z,\ \operatorname{Re} w;\ \operatorname{Re} s > 0$]
4	$e^{-px}\Phi_3(b;\ c;\ w,\ zx)$	$\dfrac{\Gamma(s)}{p^s} \Phi_2\left(b,\ s;\ c;\ w,\ \dfrac{z}{p}\right)$ [$\operatorname{Re} p > 0,\ \operatorname{Re} z;\ \operatorname{Re} s > 0$]
5	$e^{-px}\Phi_3(b;\ c;\ wx,\ z)$	$\dfrac{\Gamma(s)}{p^s} \Xi_2\left(s,\ b;\ c;\ \dfrac{w}{p},\ z\right)$ [$\operatorname{Re} p > 0,\ \operatorname{Re} w;\ \operatorname{Re} s > 0$]
6	$e^{-px}\Phi_3(b;\ c;\ wx,\ zx)$	$\dfrac{\Gamma(s)}{p^s} \Phi_1\left(s,\ b;\ c;\ \dfrac{w}{p},\ \dfrac{z}{p}\right)$ [$\operatorname{Re} p > 0,\ \operatorname{Re} w;\ \operatorname{Re} s > 0$]
7	$e^{-p\sqrt{x}}\Phi_3(b;\ c;\ w,\ zx)$	$\dfrac{2\Gamma(2s)}{p^{2s}} \Xi_1\left(s,\ b,\ \dfrac{2s+1}{2};\ c;\ \dfrac{4z}{p^2},\ w\right)$ [$\operatorname{Re} p > 2\lvert\operatorname{Re}(\sqrt{z})\rvert;\ \operatorname{Re} s > 0$]
8	$e^{-px}\Psi_1(a,\ b;\ c,\ c';\ w,\ zx)$	$\dfrac{\Gamma(s)}{p^s} F_2\left(a,\ b,\ s;\ c,\ c';\ w,\ \dfrac{z}{p}\right)$ [$\operatorname{Re} p > 0,\ \operatorname{Re} z;\ \operatorname{Re} s > 0$]
9	$e^{-px}\Psi_2(a;\ c,\ c';\ wx,\ z)$	$\dfrac{\Gamma(s)}{p^s} \Psi_1\left(a,\ s;\ c,\ c';\ \dfrac{w}{p}, z\right)$ [$\operatorname{Re} p > 0,\ \operatorname{Re} w;\ \operatorname{Re} s > 0$]
10	$e^{-px}\Psi_2(a;\ c,\ c';\ wx,\ zx)$	$\dfrac{\Gamma(s)}{p^s} F_4\left(s,\ a;\ c,\ c';\ \dfrac{w}{p},\ \dfrac{z}{p}\right)$ [$\operatorname{Re} p > 0,\ \operatorname{Re} w,\ \operatorname{Re} z;\ \operatorname{Re} s > 0$]
11	$e^{-px}\Xi_1(a,\ a',\ b;\ c;\ w,\ zx)$	$\dfrac{\Gamma(s)}{p^s} F_3\left(a,\ a',\ b,\ s;\ c;\ w,\ \dfrac{z}{p}\right)$ [$\operatorname{Re} p > 0,\ \operatorname{Re} z;\ \operatorname{Re} s > 0$]
12	$e^{-px}\Xi_2(a,\ b;\ c;\ w,\ zx)$	$\dfrac{\Gamma(s)}{p^s} \Xi_1\left(a,\ s,\ b;\ c;\ w,\ \dfrac{z}{p}\right)$ [$\operatorname{Re} p > 0,\ \operatorname{Re} z;\ \operatorname{Re} s > 0$]
13	$e^{-p\sqrt{x}}\Xi_2(a,\ b;\ c;\ w,\ zx)$	$\dfrac{2\Gamma(2s)}{p^{2s}} F_3\left(a,\ s,\ b,\ s+\dfrac{1}{2};\ c;\ w,\ \dfrac{4z}{p^2}\right)$ [$\operatorname{Re} p > 2\lvert\operatorname{Re}(\sqrt{z})\rvert;\ \operatorname{Re} s > 0$]

3.36. The Meijer G-Function

More formulas can be obtained from the corresponding section due to the relations

$$G_{pq}^{mn}\left(z\,\middle|\,\begin{matrix}(a_p)\\(b_q)\end{matrix}\right)=\sum_{k=1}^{m}\Gamma\left[\begin{matrix}(b_m)'-b_k,\ b_k-(a_n)+1\\a_{n+1}-b_k,\ldots,a_p-b_k,\ b_k-b_{m+1}+1,\ldots,b_k-b_q+1\end{matrix}\right]$$
$$\times\, z^{b_k}\,{}_pF_q\left(\begin{matrix}b_k-(a_p)+1;\ (-1)^{p-m-n}\,z\\b_k-(b_q)'+1\end{matrix}\right).$$

The notations $(b_m)'-b_k$ and $b_k-(b_q)'+1$ mean that the term with b_k-b_k is absent.

$$\left[\begin{matrix}(p<q)\ \text{or}\ (p=q,\ m+n>p)\ \text{or}\\(p=q,\ m+n=p;\ |z|<1);\\b_j-b_k\neq 0,\ \pm1,\ \pm2,\ldots;\ j\neq k;\ j,\ k=1,\ 2,\ldots,\ m.\end{matrix}\right]$$

$$G_{pq}^{mn}\left(z\,\middle|\,\begin{matrix}(a_p)\\(b_q)\end{matrix}\right)=\sum_{k=1}^{n}\Gamma\left[\begin{matrix}a_k-(a_n)',\ (b_m)-a_k+1\\a_k-b_{m+1},\ldots,a_k-b_q,\ a_{n+1}-a_k+1,\ldots,a_p-a_k+1\end{matrix}\right]$$
$$\times\, z^{a_k-1}\,{}_pF_q\left(\begin{matrix}(b_q)-a_k+1;\ \frac{(-1)^{q-m-n}}{z}\\(a_p)'-a_k+1\end{matrix}\right).$$

The notations $a_k-(a_n)'$ and $(a_p)'-a_k+1$ mean that the term with a_k-a_k is absent.

$$\left[\begin{matrix}(p>q)\ \text{or}\ (p=q,\ m+n=p+1;\ z\notin(-1,\ 0))\ \text{or}\\(p=q,\ m+n>p+1)\ \text{or}\ (p=q,\ m+n=p;\ |z|>1);\\a_j-a_k\neq 0,\ \pm1,\ \pm2,\ldots;\ j\neq k;\ j,\ k=1,\ 2,\ldots,\ n.\end{matrix}\right].$$

Notation:

$$m,\ n,\ p,\ q,\ r,\ t,\ u,\ v=0,\ 1,\ 2,\ldots;\ \sigma,\ \omega\in\mathbb{C};\ \sigma\neq 0;\ \omega\neq 0;$$
$$0\le m\le q;\ 0\le n\le p;\ 0\le r\le v;\ 0\le t\le u;$$
$$b^*=r+t-\frac{u+v}{2},\quad c^*=m+n-\frac{p+q}{2};$$
$$\mu=\sum_{j=1}^{q}b_j-\sum_{i=1}^{p}a_i+\frac{p-q}{2}+1,\quad \rho=\sum_{h=1}^{v}d_h-\sum_{g=1}^{u}c_g+\frac{u-v}{2}+1;$$
$$k,\ \ell=1,\ 2,\ldots;\ \Delta(k,a)=\frac{a}{k},\ \frac{a+1}{k},\ldots,\ \frac{a+k-1}{k};$$
$$\Delta(k,(a_p))=\Delta(k,\ a_1),\ \Delta(k,\ a_2),\ldots,\Delta(k,\ a_p);$$
$$\varphi=q-p-\frac{\ell}{k}(v-u);\ \eta=1-s(v-u)-\mu-\rho.$$

Conditions A:

1° $a_i-b_j\neq 1,2,\ldots\quad (i=1,\ldots,n;\ j=1,\ldots,m);$

$c_g-d_h\neq 1,2,\ldots\quad (g=1,\ldots,t;\ h=1,\ldots,r);$

2° $\operatorname{Re}\left(s+d_h+\frac{\ell}{k}b_j\right)>0\quad (j=1,\ldots,m;\ h=1,\ldots,r);$

3° $\operatorname{Re}\left(s+c_g+\frac{\ell}{k}\,a_i\right)<\frac{\ell}{k}+1 \quad (i=1,\ldots,n;\ g=1,\ldots,t)$;

4° $(p-q)\operatorname{Re}(s+c_g-1)-\frac{\ell}{k}\operatorname{Re}\mu>-\frac{3\ell}{2k} \quad (g=1,\ldots,t)$;

5° $(p-q)\operatorname{Re}(s+d_h)-\frac{\ell}{k}\operatorname{Re}\mu>-\frac{3\ell}{2k} \quad (h=1,\ldots,r)$;

6° $(u-v)\operatorname{Re}\left(s+\frac{\ell}{k}\,a_i-\frac{\ell}{k}\right)-\operatorname{Re}\rho>-\frac{3}{2} \quad (i=1,\ldots,n)$;

7° $(u-v)\operatorname{Re}\left(s+\frac{\ell}{k}\,b_j\right)-\operatorname{Re}\rho>-\frac{3}{2} \quad (j=1,\ldots,m)$;

8° $|\varphi|+2\operatorname{Re}\left[(q-p)(v-u)\,s+\frac{\ell}{k}\,(v-u)(\mu-1)+(q-p)(\rho-1)\right]>0$;

9° $|\varphi|-2\operatorname{Re}\left[(q-p)(v-u)\,s+\frac{\ell}{k}\,(v-u)(\mu-1)+(q-p)(\rho-1)\right]>0$;

10° $\varphi=0;\ c^*+r\,(b^*-1)\le 0;\ \left|\arg\left(1-z_0\sigma^{-\ell}\omega^k\right)\right|<\pi$;

$$z_0=\left(\frac{\ell}{k}\right)^{l(v-u)}\exp\left[-\left(\ell b^*+kc^*\right)\pi i\right]$$

and $z_0=\sigma^\ell\omega^{-k}$ provided that $\operatorname{Re}\left[(v-u)\,s+\mu+\rho\right]<1$.

11° One of the following conditions holds:

$\lambda_c>0$ or $\lambda_c=0,\ \lambda_r\neq 0,\ \operatorname{Re}\eta>-1$ or $\lambda_c=\lambda_r=0,\ \operatorname{Re}\eta>0$.

$$\lambda_c=(q-p)\,|\omega|^{1/(q-p)}\cos\widetilde{\psi}+(v-u)\,|\sigma|^{1/(v-u)}\cos\theta,$$

$$\widetilde{\psi}=\frac{1}{q-p}\left[|\arg\omega|+(q-m-n)\,\pi\right],\quad \theta=\frac{1}{v-u}\left[|\arg\sigma|+(v-r-t)\,\pi\right];$$

$$\lambda_r=(q-p)\,|\omega|^{1/(q-p)}\operatorname{sgn}(\arg\omega)\sin\widetilde{\psi}$$

$$+(v-u)\,|\sigma|^{1/(v-u)}\operatorname{sgn}(\arg\sigma)\sin\theta \qquad \text{for } \arg\omega\arg\sigma\neq 0;$$

$$\lambda_r=\lambda_r^+\lambda_r^-,\qquad \lambda_r^\pm=\lim_{\arg\sigma\to\pm 0}\lambda_r \qquad \text{for } \arg\sigma=0 \text{ and } \arg\omega\neq 0;$$

$$\lambda_r=\widetilde{\lambda}_r^+\widetilde{\lambda}_r^-,\qquad \widetilde{\lambda}_r^\pm=\lim_{\arg\omega\to\pm 0}\lambda_r \qquad \text{for } \arg\sigma\neq 0 \text{ and } \arg\omega=0;$$

$$\lambda_r=\bar{\lambda}_r^+\bar{\lambda}_r^-,\qquad \bar{\lambda}_r^\pm=\lim_{\substack{\arg\omega\to 0\\ \arg\sigma\to\pm 0}}\lambda_r \qquad \text{for } \arg\sigma=\arg\omega=0.$$

$$G_{pq}^{mn}\left(z\,\middle|\,\begin{matrix}(a_p)\\(b_q)\end{matrix}\right)=(2\pi)^{(1-k)c^*}\,k^\mu\,G_{kp,\,kq}^{km,\,kn}\left(\frac{z^k}{k^{k(q-p)}}\,\middle|\,\begin{matrix}\Delta\left(k,\ (a_p)\right)\\ \Delta\left(k,\ (b_q)\right)\end{matrix}\right) \qquad [k=1,\,2,\ldots].$$

3.36.1. $G_{pq}^{mn}\left(\omega x \middle| \begin{matrix}(a_p)\\(b_q)\end{matrix}\right)$

No.	$f(x)$	$F(s)$
1	$G_{pq}^{mn}\left(\omega x \middle| \begin{matrix}(a_p)\\(b_q)\end{matrix}\right)$	$\omega^{-s}\Gamma\begin{bmatrix}1-(a_n)-s,\ s+(b_m)\\ s+a_{n+1},\dots,s+a_p,\ 1-b_{m+1}-s,\dots,1-b_q-s\end{bmatrix}$ $\Big[[q=p-2;\ c^*\geq 0;\ (-1)^{q-m-n}\,\omega<0;\ -\operatorname{Re}b_k,\ -\operatorname{Re}\chi-1/2<\operatorname{Re}s<1-\operatorname{Re}a_k]$ or $[q=p-1;\ c^*\geq 0;\ ((-1)^{q-m-n}\operatorname{Re}\omega<0;\ -\operatorname{Re}b_k<\operatorname{Re}s<1-\operatorname{Re}a_k)$ or $(\operatorname{Re}\omega=0;\ -\operatorname{Re}b_k,\ -\operatorname{Re}\chi-1<\operatorname{Re}s<1-\operatorname{Re}a_k)]$ or $[q=p;\ c^*>0;\ (\lvert\arg\omega\rvert<(2m+2n-p-q)\pi/2;\ -\operatorname{Re}b_k<\operatorname{Re}s<1-\operatorname{Re}a_k)$ or $(\omega>0;\ c^*=0;\ \sum_{k=1}^{p}\operatorname{Re}(a_k-b_k)>0;\ -\operatorname{Re}b_k<\operatorname{Re}s<1-\operatorname{Re}a_k)]$ or $[q=p+1;\ c^*\geq 0;\ ((-1)^{p-m-n}\operatorname{Re}\omega<0;\ -\operatorname{Re}b_k<\operatorname{Re}s<1-\operatorname{Re}a_k)$ or $(\operatorname{Re}\omega=0;\ -\operatorname{Re}b_k<\operatorname{Re}s<1-\operatorname{Re}a_k,\ 1-\operatorname{Re}\chi)]$ or $[q=p+2;\ c^*\geq 0;\ (-1)^{p-m-n}\,\omega<0;\ -\operatorname{Re}b_k<\operatorname{Re}s<1-\operatorname{Re}a_k,\ 1/2-\operatorname{Re}\chi]\Big]$
2	$G_{pq}^{mn}\left(x \middle| \begin{matrix}(a_p)\\(b_q)\end{matrix}\right)$	$\Gamma\begin{bmatrix}1-(a_n)-s,\ s+(b_m)\\ s+a_{n+1},\dots,s+a_p,\ 1-b_{m+1}-s,\dots,1-b_q-s\end{bmatrix}$ $\Big[-\min_{1\leq j\leq m}\operatorname{Re}b_j<\operatorname{Re}s<1-\max_{1\leq k\leq n}\operatorname{Re}a_k$ and either $[0\leq n\leq p;\ 0\leq m\leq q;\ 2(m+n)>p+q]$ or $[0\leq n\leq p\leq q-2$ (or $0\leq m\leq q\leq p-2$); $2(m+n)=p+q;\ (q-p)\operatorname{Re}s<\frac{q-p+1}{2}+\operatorname{Re}\left(\sum_{k=1}^{p}a_k-\sum_{j=1}^{q}b_j\right)]$ or $[p=q\geq 1;\ m+n=p;\ \sum_{j=1}^{p}\operatorname{Re}(a_j-b_j)>0]\Big]$

3.36.2. $G_{pq}^{mn}\left(\omega x \middle| \begin{matrix}(a_p)\\(b_q)\end{matrix}\right)$ and algebraic functions

1	$(a-x)_+^{\alpha-1}$ $\times G_{pq}^{mn}\left(\omega x^{\ell/k} \middle| \begin{matrix}(a_p)\\(b_q)\end{matrix}\right)$	$\frac{k^{\mu}\,a^{s+\alpha-1}\Gamma(\alpha)}{\ell^{\alpha}\,(2\pi)^{c^*(k-1)}}\,G_{kp+\ell,\,kq+\ell}^{km,\,kn+\ell}\left(\frac{\omega^k a^{\ell}}{k^{k(q-p)}} \middle| \begin{matrix}\Delta(\ell,1-s),\ \Delta(k,(a_p))\\ \Delta(k,(b_q)),\ \Delta(\ell,1-s-\alpha)\end{matrix}\right)$ [see Conditions A with $\sigma=1/a;\ r=u=v=1;\ t=d_1=0;\ c_1=\alpha$]

No.	$f(x)$	$F(s)$
2	$(x-a)_+^{\alpha-1}$ $\times G_{pq}^{mn}\left(\omega x^{\ell/k}\,\middle\vert\,\begin{matrix}(a_p)\\(b_q)\end{matrix}\right)$	$\dfrac{k^{\mu}\,\ell^{-\alpha}}{(2\pi)^{c^*(k-1)}\,a^{1-s-\alpha}}\,\Gamma(\alpha)$ $\times G_{kp+\ell,\,kq+\ell}^{km+\ell,\,kn}\left(\dfrac{\omega^k a^\ell}{k^{k(q-p)}}\,\middle\vert\,\begin{matrix}\Delta(k,\,(a_p)),\,\Delta(k,\,1-s)\\\Delta(k,\,1-s-\alpha),\,\Delta(k,\,(b_q))\end{matrix}\right)$ [see Conditions A with $\sigma=1/a;\ r=d_1=0;\ t=u=v=1;\ c_1=\alpha$]
3	$\dfrac{1}{(x+a)^{\beta}}$ $\times G_{pq}^{mn}\left(\omega x^{\ell/k}\,\middle\vert\,\begin{matrix}(a_p)\\(b_q)\end{matrix}\right)$	$\dfrac{k^{\mu}\,\ell^{\beta-1}\,a^{s-\beta}}{(2\pi)^{c^*(k-1)+\ell-1}\,\Gamma(\beta)}$ $\times G_{kp+\ell,\,kq+\ell}^{km+\ell,\,kn+\ell}\left(\dfrac{\omega^k a^\ell}{k^{k(q-p)}}\,\middle\vert\,\begin{matrix}\Delta(\ell,\,1-s),\,\Delta(k,(a_p))\\\Delta(\ell,\,\beta-s),\,\Delta(k,(b_q))\end{matrix}\right)$ [see Conditions A with $\sigma=1/a;\ r=t=u=v=1;\ c_1=1-\beta;\ d_1=0$]
4	$\dfrac{1}{x-a}\,G_{pq}^{mn}\left(\omega x^{\ell/k}\,\middle\vert\,\begin{matrix}(a_p)\\(b_q)\end{matrix}\right)$	$-\dfrac{\pi\,k^{\mu}\,a^{s-1}}{(2\pi)^{c^*(k-1)}}\,G_{kp+2\ell,\,kq+2\ell}^{km+\ell,\,kn+\ell}\left(\dfrac{\omega^k a^\ell}{k^{k(q-p)}}\,\middle\vert\,\begin{matrix}\Delta(\ell,\,1-s),\,\Delta(k,(a_p)),\,\Delta\left(\ell,\,\frac{1-2s}{2}\right)\\\Delta(\ell,\,1-s),\,\Delta(k,(b_q)),\,\Delta\left(\ell,\,\frac{1-2s}{2}\right)\end{matrix}\right)$ [see Conditions A with $\sigma=1/a;\ r=t=1;\ u=v=2;\ c_1=d_1=0;\ c_2=d_2=1/2$]
5	$\dfrac{1}{(x+a)^{\beta}}$ $\times G_{pq}^{mn}\left(\dfrac{\omega(x+a)^\ell}{x^k}\,\middle\vert\,\begin{matrix}(a_p)\\(b_q)\end{matrix}\right)$	$\dfrac{\sqrt{2\pi}\,k^{s-1/2}\,\ell^{1/2-\beta}\,a^{s-\beta}}{(\ell-k)^{s-\beta+1/2}}\,G_{p+\ell,\,q+\ell}^{m+\ell,\,n}\left(\dfrac{\omega\ell^\ell}{k^k}\left(\dfrac{a}{\ell-k}\right)^{\ell-k}\,\middle\vert\,\begin{matrix}(a_p),\,\Delta(\ell,\,\beta)\\\Delta(k,\,s),\,\Delta(\ell-k,\,\beta-s),\,(b_q)\end{matrix}\right)$ [$0<k<\ell;\ c^*>0;\ \left\vert\arg\left(\omega a^{\ell-k}\right)\right\vert<c^*\pi;$ $-k+k\max\limits_{1\le j\le n}\operatorname{Re}a_j<\operatorname{Re}s$ $<\operatorname{Re}\beta+(\ell-k)\left[1-\max\limits_{1\le j\le n}\operatorname{Re}a_j\right]$]
6	$\dfrac{1}{(x+a)^{\beta}}$ $\times G_{pq}^{mn}\left(\dfrac{\omega(x+a)^\ell}{x^k}\,\middle\vert\,\begin{matrix}(a_p)\\(b_q)\end{matrix}\right)$	$\dfrac{\sqrt{2\pi}\,k^{s-1/2}\,\ell^{1/2-\beta}\,a^{s-\beta}}{(k-\ell)^{1/2+s-\beta}}$ $\times G_{p+k,\,q+k}^{m+k,\,n+k-\ell}\left(\dfrac{\omega\ell^\ell}{k^k}\left(\dfrac{a}{k-\ell}\right)^{\ell-k}\,\middle\vert\,\begin{matrix}\Delta(\ell-k,\,s-\beta+1),\,(a_p),\,\Delta(\ell,\,\beta)\\\Delta(k,\,s),\,(b_q)\end{matrix}\right)$ [$0<\ell<k;\ c^*>0;\ \left\vert\arg\left(\omega a^{\ell-k}\right)\right\vert<c^*\pi;$ $-k+k\max\limits_{1\le j\le n}\operatorname{Re}a_j<\operatorname{Re}s$ $<\operatorname{Re}\beta+(k-\ell)\min\limits_{1\le j\le m}\operatorname{Re}b_j$]

No.	$f(x)$	$F(s)$
7	$\dfrac{1}{(x+a)^{\beta}} \times G_{pq}^{mn}\left(\dfrac{\omega(x+a)^{\ell}}{x^{k}} \,\middle\vert\, \begin{matrix}(a_p)\\(b_q)\end{matrix}\right)$	$\sqrt{2\pi}\,(ka)^{s-\beta}\,\Gamma(\beta-s)\,G_{p+k,\,q+k}^{m+k,\,n}\left(\omega \,\middle\vert\, \begin{matrix}(a_p),\ \Delta(k,\,\beta)\\ \Delta(k,\,s),\ (b_q)\end{matrix}\right)$ $\left[\begin{matrix}\ell=k>0;\ c^{*}>0;\ \lvert\arg\omega\rvert<c^{*}\pi;\\ -k+k\max\limits_{1\le j\le n}\operatorname{Re}a_j<\operatorname{Re}s<\operatorname{Re}\beta\end{matrix}\right]$
8	$\dfrac{1}{(x+1)^{\beta}} \times G_{pq}^{mn}\left(\dfrac{\omega x}{x+1} \,\middle\vert\, \begin{matrix}(a_p)\\(b_q)\end{matrix}\right)$	$\Gamma(\beta-s)\,G_{p+1,\,q+1}^{m,\,n+1}\left(\omega \,\middle\vert\, \begin{matrix}1-s,\ (a_p),\ \Delta(k,\,\beta)\\ (b_q),\ 1-\beta\end{matrix}\right)$ $\left[\begin{matrix}c^{*}>0;\ \lvert\arg\omega\rvert<c^{*}\pi;\\ -\min\limits_{1\le j\le m}\operatorname{Re}b_j<\operatorname{Re}s<\operatorname{Re}\beta\end{matrix}\right]$

3.36.3. $G_{pq}^{mn}\left(\omega x^{\sigma} \,\middle\vert\, \begin{matrix}(a_p)\\(b_q)\end{matrix}\right)$ and the exponential function

1	$e^{-\sigma x}\,G_{pq}^{mn}\left(\omega x^{\ell/k} \,\middle\vert\, \begin{matrix}(a_p)\\(b_q)\end{matrix}\right)$	$\dfrac{k^{\mu}\,\ell^{s-1/2}\,\sigma^{-s}}{(2\pi)^{(\ell-1)/2+(k-1)c^{*}}}\,G_{kp+\ell,\,kq}^{km,\,kn+\ell}\left(\dfrac{\omega^{k}\ell^{\ell}}{\sigma^{\ell}k^{k(q-p)}} \,\middle\vert\, \begin{matrix}\Delta(\ell,\,1-s),\ \Delta(k,\,(a_p))\\ \Delta(k,\,(b_q))\end{matrix}\right)$ $\left[\begin{matrix}\text{see Conditions A with}\\ r=v=1;\ t=u=d_1=0\end{matrix}\right]$
2	$e^{-\sigma x}\,G_{pq}^{mn}\left(\omega x \,\middle\vert\, \begin{matrix}(a_p)\\(b_q)\end{matrix}\right)$	$\sigma^{-s}\,G_{p+1,\,q}^{m,\,n+1}\left(\dfrac{\omega}{\sigma} \,\middle\vert\, \begin{matrix}1-s,\ (a_p)\\ (b_q)\end{matrix}\right)$

$\Big[$ $[q=p-2;\ (-1)^{q-m-n}\,\omega<0;\ (\operatorname{Re}\sigma>0;\ -\operatorname{Re}b_k,\ -\operatorname{Re}\chi-1/2<\operatorname{Re}s)$ or
$(\operatorname{Re}\sigma=0;\ -\operatorname{Re}b_k,\ -\operatorname{Re}\chi-1/2<\operatorname{Re}s<2-\operatorname{Re}a_k)]$
$[q=p-1;\ (\operatorname{Re}\sigma>0;\ (-1)^{q-m-n}\,\omega<0;\ -\operatorname{Re}b_k<\operatorname{Re}s)$ or
$(\operatorname{Re}\sigma>0;\ \operatorname{Re}\omega=0;\ -\operatorname{Re}b_k,\ -\operatorname{Re}\chi-1<\operatorname{Re}s)$ or
$(\operatorname{Re}\sigma=0;\ (-1)^{q-m-n}\,\omega<0;\ -\operatorname{Re}b_k<\operatorname{Re}s<2-\operatorname{Re}a_k)$ or
$(\operatorname{Re}\sigma=\operatorname{Re}\omega=0;\ -\operatorname{Re}b_k,\ -\operatorname{Re}\chi-1<\operatorname{Re}s<2-\operatorname{Re}a_k)]$ or
$[q=p;\ (\operatorname{Re}\sigma>0;\ \operatorname{Re}s>-\operatorname{Re}b_k)$ or
$(\operatorname{Re}\sigma=0;\ -\operatorname{Re}b_k<\operatorname{Re}s<2-\operatorname{Re}a_k);$
$((m+n>p;\ \lvert\arg\omega\rvert<(m+n-p)\,\pi)$ or
$(m+n=p;\ \omega>0;\ \sum_{k=1}^{p}\operatorname{Re}(a_k-b_k)>0))]$ or
$[q=p+1;\ (\operatorname{Re}\sigma,\ \operatorname{Re}(\sigma-(-1)^{p-m-n}\,\omega)>0;\ \operatorname{Re}s>-\operatorname{Re}b_k)$ or
$(\operatorname{Re}\sigma>0;\ \operatorname{Re}(\sigma-(-1)^{p-m-n}\,\omega)=0;\ -\operatorname{Re}b_k<\operatorname{Re}s<1-\operatorname{Re}\chi)$ or
$(\operatorname{Re}\sigma=0;\ (-1)^{p-m-n}\operatorname{Re}\omega<0;\ -\operatorname{Re}b_k<\operatorname{Re}s<2-\operatorname{Re}a_k)$ or
$(\operatorname{Re}\sigma=\operatorname{Re}\omega=0;\ -\operatorname{Re}b_k<\operatorname{Re}s<2-\operatorname{Re}a_k,\ 1-\operatorname{Re}\chi)]$ or
$[q=p+2;\ (\operatorname{Re}\sigma>0;\ (-1)^{p-m-n}\,\omega<0;\ -\operatorname{Re}b_k<\operatorname{Re}s)$ or
$(\operatorname{Re}\sigma=0;\ \operatorname{Re}s<2-\operatorname{Re}a_k,\ 1-\operatorname{Re}\chi)]$ or
$[q\ge p+3;\ \operatorname{Re}\sigma>0;\ \operatorname{Re}s>-\operatorname{Re}b_k]$ $\Big]$

3.36.4. $G_{pq}^{mn}\left(\omega x^{\sigma}\,\middle|\,\begin{matrix}(a_p)\\(b_q)\end{matrix}\right)$ and trigonometric functions

1	$\sin(bx)$ $\times G_{pq}^{mn}\left(\omega x^{2\ell/k}\,\middle|\,\begin{matrix}(a_p)\\(b_q)\end{matrix}\right)$	$\frac{k^{\mu}\,(2\ell)^{s-1/2}\,b^{-s}}{2\,(2\pi)^{(k-1)c^*-1/2}}\,G_{kp+2\ell,\,kq}^{km,\,kn+\ell}\left(\frac{\omega^k\,(2\ell)^{2\ell}}{b^{2\ell}k^{k(q-p)}}\,\middle|\,\begin{matrix}\Delta\left(\ell,\,\frac{1-s}{2}\right),\,\Delta(k,\,(a_p)),\,\Delta\left(\ell,\,\frac{2-s}{2}\right)\\ \Delta(k,\,(b_q))\end{matrix}\right)$ [see Conditions A with s being replaced by $s/2$ and with $\sigma=b^2/4;\ r=1;\ t=u=0;\ v=2;\ d_1=1/2;\ d_2=0$]
2	$\sin(bx)\,G_{pq}^{mn}\left(\omega x^{2}\,\middle|\,\begin{matrix}(a_p)\\(b_q)\end{matrix}\right)$	$\frac{\sqrt{\pi}}{b}\,G_{p+2,\,q}^{m,\,n+1}\left(\frac{4\omega}{b^2}\,\middle|\,\begin{matrix}0,\,(a_p),\,\frac{1}{2}\\(b_q)\end{matrix}\right)$ [$c^*>0;\ b>0;\ \operatorname{Re}b_j>-1\ (j=1,\ldots,m);$ $\operatorname{Re}a_i<1/2\ (i=1,\ldots,n);\ \lvert\arg\omega\rvert<c^*\pi$]
3	$\cos(bx)$ $\times G_{pq}^{mn}\left(\omega x^{2\ell/k}\,\middle|\,\begin{matrix}(a_p)\\(b_q)\end{matrix}\right)$	$\frac{k^{\mu}\,(2\ell)^{s-1/2}\,b^{-s}}{2\,(2\pi)^{c^*(k-1)-1/2}}$ $\times\,G_{kp+2\ell,\,kq}^{km,\,kn+\ell}\left(\frac{\omega^k\,(2\ell)^{2\ell}}{b^{2\ell}k^{k(q-p)}}\,\middle|\,\begin{matrix}\Delta\left(\ell,\,\frac{2-s}{2}\right),\,\Delta(k,\,(a_p)),\,\Delta\left(\ell,\,\frac{1-s}{2}\right)\\ \Delta(k,\,(b_q))\end{matrix}\right)$ [see Conditions A with s being replaced by $s/2$ and with $\sigma=b^2/4;\ r=1;\ t=u=0;\ v=2;\ d_1=d_2=1/2$]
4	$\cos(bx)\,G_{pq}^{mn}\left(\omega x^{2}\,\middle|\,\begin{matrix}(a_p)\\(b_q)\end{matrix}\right)$	$\frac{\sqrt{\pi}}{b}\,G_{p+2,\,q}^{m,\,n+1}\left(\frac{4\omega}{b^2}\,\middle|\,\begin{matrix}\frac{1}{2},\,(a_p),\,0\\(b_q)\end{matrix}\right)$ [$c^*;\ b>0;\ \operatorname{Re}b_j>-1/2\ (j=1,\ldots,m);$ $\operatorname{Re}a_i<1/2\ (i=1,\ldots,n);\ \lvert\arg\omega\rvert<c^*\pi$]

3.36.5. $G_{pq}^{mn}\left(\omega x^{\sigma}\,\middle|\,\begin{matrix}(a_p)\\(b_q)\end{matrix}\right)$ and the Bessel functions

1	$J_{\nu}(bx)$ $\times G_{pq}^{mn}\left(\omega x^{2\ell/k}\,\middle|\,\begin{matrix}(a_p)\\(b_q)\end{matrix}\right)$	$\frac{k^{\mu}\,(2\ell)^{s-1}}{(2\pi)^{(k-1)c^*}\,b^{s}}\,G_{kp+2\ell,\,kq}^{km,\,kn+\ell}\left(\frac{\omega^k\,(2\ell)^{2\ell}}{b^{2\ell}k^{k(q-p)}}\,\middle|\,\begin{matrix}\Delta\left(\ell,\,\frac{2-s-\nu}{2}\right),\,\Delta(k,\,(a_p)),\,\Delta\left(\ell,\,\frac{2-s+\nu}{2}\right)\\ \Delta(k,\,(b_q))\end{matrix}\right)$ [see Conditions A with s being replaced by $s/2$ and with $\sigma=b^2/4;\ r=1;\ t=u=0;\ v=2;\ d_1=\nu/4;\ d_2=-\nu/4$]

No.	$f(x)$	$F(s)$
2	$J_\nu(bx)\, G_{pq}^{mn}\left(\omega x^2 \,\middle\vert\, \begin{matrix}(a_p)\\(b_q)\end{matrix}\right)$	$\dfrac{2^{s-1}}{b^s} G_{p+2,\,q}^{m,\,n+1}\left(\dfrac{4\omega}{b^2} \,\middle\vert\, \begin{matrix}\frac{2-s-\nu}{2},\ (a_p),\ \frac{2-s+\nu}{2}\\(b_q)\end{matrix}\right)$ $\left[\begin{matrix}c^*>0;\ b>0;\ \lvert\arg\omega\rvert<c^*\pi;\\ \operatorname{Re}(b_j+(s+\nu)/2)>0\ (j=1,\ldots,m),\\ \operatorname{Re}(a_i+s/2)<5/4\ (i=1,\ldots,n)\end{matrix}\right]$
3	$J_\nu(a\sqrt{x})$ $\times\, G_{pq}^{mn}\left(\omega x \,\middle\vert\, \begin{matrix}(a_p)\\(b_q)\end{matrix}\right)$	$\left(\dfrac{2}{a}\right)^{2s} G_{p+2,\,q}^{m,\,n+1}\left(\dfrac{4\omega}{a^2} \,\middle\vert\, \begin{matrix}\frac{2-2s-\nu}{2},\ (a_p),\ \frac{2-2s+\nu}{2}\\(b_q)\end{matrix}\right)$ $\left[\begin{matrix}c^*>0;\ a>0;\ \lvert\arg\omega\rvert<c^*\pi;\\ -\operatorname{Re}\nu/2-\min\limits_{1\le j\le m}\operatorname{Re}b_j<\operatorname{Re}s<7/4-\max\limits_{1\le i\le n}\operatorname{Re}a_i\end{matrix}\right]$
4	$Y_\nu(bx)$ $\times\, G_{pq}^{mn}\left(\omega x^{2\ell/k} \,\middle\vert\, \begin{matrix}(a_p)\\(b_q)\end{matrix}\right)$	$\dfrac{k^\mu (2\ell)^{s-1}}{(2\pi)^{(k-1)c^*}\, b^s} G_{kp+3\ell,\,kq+\ell}^{km,\,kn+2\ell}\left(\dfrac{\omega^k (2\ell)^{2\ell}}{b^{2\ell}k^{k(q-p)}} \,\middle\vert\, \begin{matrix}\Delta\left(\ell,\frac{2-s-\nu}{2}\right),\ \Delta\left(\ell,\frac{2-s+\nu}{2}\right),\ \Delta(k,(a_p)),\ \Delta\left(\ell,\frac{3-s+\nu}{2}\right)\\ \Delta(k,(b_q)),\ \Delta\left(\ell,\frac{3-s+\nu}{2}\right)\end{matrix}\right)$ [see Conditions A with s being replaced by $s/2$ and with $\sigma=b^2/4;\ r=2;\ t=0;\ u=1;\ v=3;$ $c_1=d_3=(1-\nu)/2;\ d_1=-\nu/2;\ d_2=\nu/2$]
5	$Y_\nu(a\sqrt{x})$ $\times\, G_{pq}^{mn}\left(\omega x \,\middle\vert\, \begin{matrix}(a_p)\\(b_q)\end{matrix}\right)$	$\left(\dfrac{2}{a}\right)^{2s} G_{p+3,\,q+1}^{m,\,n+2}\left(\dfrac{4\omega}{a^2} \,\middle\vert\, \begin{matrix}\frac{2-2s-2\nu}{2},\ \frac{2-2s+2\nu}{2},\ (a_p),\ \frac{3-2s+\nu}{2}\\(b_q),\ \frac{3-2s+\nu}{2}\end{matrix}\right)$ $\left[\begin{matrix}c^*>0;\ a>0;\ \lvert\arg\omega\rvert<c^*\pi\\ -\min\limits_{1\le j\le m}\operatorname{Re}b_j-\operatorname{Re}\nu/2<\operatorname{Re}s<7/4-\max\limits_{1\le i\le n}\operatorname{Re}a_i\end{matrix}\right]$
6	$K_\nu(bx)$ $\times\, G_{pq}^{mn}\left(\omega x^{2\ell/k} \,\middle\vert\, \begin{matrix}(a_p)\\(b_q)\end{matrix}\right)$	$\dfrac{\pi k^\mu (2\ell)^{s-1}}{(2\pi)^{(k-1)c^*+\ell}\, b^s} G_{kp+2\ell,\,kq}^{km,\,kn+2\ell}\left(\dfrac{\omega^k (2\ell)^{2\ell}}{b^{2\ell}k^{k(q-p)}} \,\middle\vert\, \begin{matrix}\Delta\left(\ell,\frac{2-s-\nu}{2}\right),\ \Delta\left(\ell,\frac{2-s+\nu}{2}\right),\ \Delta(k,(a_p))\\ \Delta(k,(b_q))\end{matrix}\right)$ [see Conditions A with s being replaced by $s/2$ and with $\sigma=b^2/4;\ r=v=2;\ t=u=0;\ d_1=-\nu/2;\ d_2=\nu/2$]
7	$K_\nu(a\sqrt{x})$ $\times\, G_{pq}^{mn}\left(\omega x \,\middle\vert\, \begin{matrix}(a_p)\\(b_q)\end{matrix}\right)$	$\dfrac{1}{2}\left(\dfrac{2}{a}\right)^{2s} G_{pq}^{mn}\left(\dfrac{4\omega}{a^2} \,\middle\vert\, \begin{matrix}\frac{2-2s-\nu}{2},\ \frac{2-2s+\nu}{2},\ (a_p)\\(b_q)\end{matrix}\right)$ $\left[\begin{matrix}c^*>0;\ a>0;\ \lvert\arg\omega\rvert<c^*\pi;\\ \operatorname{Re}s>\operatorname{Re}\nu/2-\min\limits_{1\le j\le m}\operatorname{Re}b_j\end{matrix}\right]$

3.36.6. $G_{pq}^{mn}\left(\omega x^{\sigma}\,\middle|\,\begin{matrix}(a_p)\\(b_q)\end{matrix}\right)$ and orthogonal polynomials

1	$\left(a^2-x^2\right)_+^{\lambda-1/2} C_r^{\lambda}\left(\frac{x}{a}\right)$ $\times G_{pq}^{mn}\left(\omega x^{2\ell/k}\,\middle|\,\begin{matrix}(a_p)\\(b_q)\end{matrix}\right)$	$\frac{k^{\mu}\,a^{s+2\lambda-1}}{2\,r!\,(2\pi)^{(k-1)c^*}\,\ell^{\lambda+1/2}}\,(2\lambda)_r\,\Gamma\left(\frac{2\lambda+1}{2}\right)$ $\times G_{kp+2\ell,\,kq+2\ell}^{km,\,kn+2\ell}\left(\frac{\omega^k k^{k(p-q)}}{\sigma^{2\ell}}\,\middle|\,\begin{matrix}\Delta(2\ell,\,1-s),\ \Delta(k,\,(a_p))\\ \Delta(k,\,(b_q)),\ \Delta\left(\ell,\,\frac{1-s-r-2\lambda}{2}\right),\ \Delta\left(\ell,\,\frac{1-s+r}{2}\right)\end{matrix}\right)$ [see Conditions A with s being replaced by $s/2$ and with $\sigma=a^{-2}$; $t=0$; $r=u=v=2$; $c_1=(r+2\lambda+1)/2$; $c_2=(1-r)/2$; $d_1=0$; $d_2=1/2$; $r=0,\,1,\,2,\ldots$]
2	$(a-x)_+^{\alpha} P_r^{(\alpha,\,\beta)}\left(\frac{2x}{a}-1\right)$ $\times G_{pq}^{mn}\left(\omega x^{\ell/k}\,\middle|\,\begin{matrix}(a_p)\\(b_q)\end{matrix}\right)$	$\frac{k^{\mu}\,a^{s+\alpha}\,\Gamma(\alpha+r+1)}{(2\pi)^{(k-1)c^*}\,\ell^{\alpha+1}r!}\,G_{kp+2\ell,\,kq+2\ell}^{km,\,kn+2\ell}\left(\frac{\omega^k a^{\ell}}{k^{k(q-p)}}\,\middle|\,\begin{matrix}\Delta(\ell,\,1-s),\ \Delta(\ell,\,1-s+\beta),\ \Delta(k,\,(a_p))\\ \Delta(k,\,(b_q)),\ \Delta(\ell,\,1-s+r+\beta),\ \Delta(\ell,\,-s-r-\alpha)\end{matrix}\right)$ [see Conditions A with $\sigma=1/a$; $t=0$; $r=u=v=2$; $c_1=\alpha+r+1$; $c_2=-v-r$; $d_1=0$; $d_2=-v$; $r=0,\,1,\,2,\ldots$]

3.36.7. $G_{pq}^{mn}\left(\omega x^{\sigma}\,\middle|\,\begin{matrix}(a_p)\\(b_q)\end{matrix}\right)$ and the Legendre function

1	$\left(a^2-x^2\right)_+^{-\lambda/2} P_{\nu}^{\lambda}\left(\frac{x}{a}\right)$ $\times G_{pq}^{mn}\left(\omega x^{2\ell/k}\,\middle|\,\begin{matrix}(a_p)\\(b_q)\end{matrix}\right)$	$\frac{k^{\mu}\,(2\ell)^{\lambda-1}\,a^{s-\lambda}}{(2\pi)^{(k-1)c^*}}\,G_{kp+2\ell,\,kq+2\ell}^{km,\,kn+2\ell}\left(\frac{a^{2\ell}\omega^k}{k^{k(q-p)}}\,\middle|\,\begin{matrix}\Delta(2\ell,\,1-s),\ \Delta(k,\,(a_p))\\ \Delta(k,\,(b_q)),\ \Delta\left(\ell,\,\frac{\lambda-s-\nu}{2}\right),\ \Delta\left(\ell,\,\frac{1-s+\lambda+\nu}{2}\right)\end{matrix}\right)$ [see Conditions A with s being replaced by $s/2$ and with $\sigma=1/a^2$; $t=0$; $r=u=v=2$; $c_1=(1-\lambda-\nu)/2$; $c_2=(2-\lambda+\nu)/2$; $d_1=0$; $d_2=1/2$]
2	$\left(x^2-a^2\right)_+^{-\lambda/2} P_{\nu}^{\lambda}\left(\frac{x}{a}\right)$ $\times G_{pq}^{mn}\left(\omega x^{2\ell/k}\,\middle|\,\begin{matrix}(a_p)\\(b_q)\end{matrix}\right)$	$\frac{k^{\mu}\,(2\ell)^{\lambda-1}\,a^{s-\lambda}}{(2\pi)^{(k-1)c^*}}\,G_{kp+2\ell,\,kq+2\ell}^{km+2\ell,\,kn}\left(\frac{a^{2\ell}\omega^k}{k^{k(q-p)}}\,\middle|\,\begin{matrix}\Delta(k,\,(a_p)),\ \Delta(2\ell,\,-s)\\ \Delta\left(\ell,\,\frac{\lambda-s-\nu}{2}\right),\ \Delta\left(\ell,\,\frac{1-s+\lambda+\nu}{2}\right),\ \Delta(k,\,(b_q))\end{matrix}\right)$ [see Conditions A with s being replaced by $s/2$ and with $\sigma=1/a^2$; $r=0$; $t=u=v=2$; $c_1=(1-\lambda-\nu)/2$; $c_2=(2-\lambda+\nu)/2$; $d_1=0$; $d_2=1/2$]

3.36.8. $G_{pq}^{mn}\left(\omega x^{\sigma}\,\middle|\,\begin{matrix}(a_p)\\(b_q)\end{matrix}\right)$ and the Struve function

1	$\mathbf{H}_{\nu}\left(2\sqrt{x}\right) \times G_{pq}^{mn}\left(\omega x\,\middle|\,\begin{matrix}(a_p)\\(b_q)\end{matrix}\right)$	$G_{p+3,\,q+1}^{m+1,\,n+1}\left(\omega\,\middle|\,\begin{matrix}\frac{1-2s-\nu}{2},\ (a_p),\ \frac{2-2s+\nu}{2},\ \frac{2-2s-\nu}{2}\\ \frac{1-2s-\nu}{2},\ (b_q)\end{matrix}\right)$ $\left[c^{*}>0;\ \lvert\arg\omega\rvert<c^{*}\pi;\ \operatorname{Re}s>-(1+\operatorname{Re}\nu)/2-\min_{1\le j\le m}\operatorname{Re}b_j;\ \operatorname{Re}s<1-\max_{1\le i\le n}\operatorname{Re}a_i-\max\left[-3/4,\ \operatorname{Re}(\nu-1)/2\right]\right]$

3.36.9. $G_{pq}^{mn}\left(\omega x^{\sigma}\,\middle|\,\begin{matrix}(a_p)\\(b_q)\end{matrix}\right)$ and the Whittaker functions

1	$e^{-\sigma x/2}\,W_{\mu,\,\nu}(\sigma x) \times G_{pq}^{mn}\left(\omega x^{\ell/k}\,\middle|\,\begin{matrix}(a_p)\\(b_q)\end{matrix}\right)$	$\frac{k^{\mu}\,\ell^{s+\mu-1/2}\,\sigma^{-s}}{(2\pi)^{(\ell-1)/2+(k-1)c^{*}}}\,\Gamma\left[\begin{matrix}2\nu+1\\ \frac{2\mu+2\nu+1}{2}\end{matrix}\right] \times G_{kp+2\ell,\,kq+\ell}^{km+\ell,\,kn+\ell}\left(\frac{\omega^{k}\ell^{\ell}}{\sigma^{\ell}k^{k(q-p)}}\,\middle|\,\begin{matrix}\Delta\left(\ell,\ \frac{1-2s-2\nu}{2}\right),\ \Delta(k,\ (a_p)),\ \Delta\left(\ell,\ \frac{1-2s+2\nu}{2}\right)\\ \Delta(\ell,\ \mu-s),\ \Delta(k,\ (b_q))\end{matrix}\right)$ [see Conditions A with $v=2$; $r=t=u=1$; $c_1=1-\mu$; $d_1=1/2+\nu$; $d_2=1/2-\nu$]
2	$e^{-\sigma x/2}\,W_{\mu,\,\nu}(\sigma x) \times G_{pq}^{mn}\left(\omega x^{\ell/k}\,\middle|\,\begin{matrix}(a_p)\\(b_q)\end{matrix}\right)$	$\frac{k^{\mu}\,\ell^{s+\mu-1/2}\,\sigma^{-s}}{(2\pi)^{(\ell-1)/2+(k-1)c^{*}}} \times G_{kp+2\ell,\,kq+\ell}^{km,\,kn+2\ell}\left(\frac{\omega^{k}\ell^{\ell}}{\sigma^{\ell}k^{k(q-p)}}\,\middle|\,\begin{matrix}\Delta\left(\ell,\ \frac{1-2s-2\nu}{2}\right),\ \Delta\left(\ell,\ \frac{1-2s+2\nu}{2}\right),\ \Delta(k,\ (a_p))\\ \Delta(k,\ (b_q)),\ \Delta(\ell,\ \mu-s)\end{matrix}\right)$ [see Conditions A with $t=0$; $r=v=2$; $u=1$; $c_1=1-\mu$; $d_1=1/2+\nu$; $d_2=1/2-\nu$]
3	$e^{\sigma x/2}\,W_{\mu,\,\nu}(\sigma x) \times G_{pq}^{mn}\left(\omega x^{\ell/k}\,\middle|\,\begin{matrix}(a_p)\\(b_q)\end{matrix}\right)$	$(2\pi)^{3(1-\ell)/2+(1-k)c^{*}}\,\frac{k^{\mu}\,\ell^{s-\mu-1/2}\,\sigma^{-s}}{\Gamma\left(\frac{1-2\mu-2\nu}{2}\right)\Gamma\left(\frac{1-2\mu+2\nu}{2}\right)} \times G_{kp+2\ell,\,kq+\ell}^{km+\ell,\,kn+2\ell}\left(\frac{\omega^{k}\ell^{\ell}}{\sigma^{\ell}k^{k(q-p)}}\,\middle|\,\begin{matrix}\Delta\left(\ell,\ \frac{1-2s-2\nu}{2}\right),\ \Delta\left(\ell,\ \frac{1-2s+2\nu}{2}\right),\ \Delta(k,\ (a_p))\\ \Delta(\ell,\ -s-\mu),\ \Delta(k,\ (b_q))\end{matrix}\right)$ [see Conditions A with $r=v=2$; $t=u=1$; $c_1=\mu+1$; $d_1=1/2+\nu$; $d_2=1/2-\nu$]

3.36.10. $G_{pq}^{mn}\left(\omega x^{\sigma}\,\middle|\,\begin{matrix}(a_p)\\(b_q)\end{matrix}\right)$ and hypergeometric functions

1	${}_2F_1\left(\begin{matrix}a,\,b\\c;\,1-\sigma x\end{matrix}\right)$ $\times G_{pq}^{mn}\left(\omega x^{\ell/k}\,\middle|\,\begin{matrix}(a_p)\\(b_q)\end{matrix}\right)$	$\dfrac{k^{\mu}\,\ell^{c-2}\,\sigma^{-\rho}}{(2\pi)^{2(\ell-1)+(k-1)c^*}}\,\Gamma\left[\begin{matrix}c\\a,\,b,\,c-a,\,c-b\end{matrix}\right]$ $\times G_{kp+2\ell,\,kq+2\ell}^{km+2\ell,\,kn+2\ell}\left(\dfrac{\omega^k\sigma^{-\ell}}{k^{k(q-p)}}\,\middle|\,\begin{matrix}\Delta(\ell,\,1-s),\,\Delta(\ell,\,1-s+a+b-c),\,\Delta(k,\,(a_p))\\\Delta(\ell,\,a-s),\,\Delta(\ell,\,b-s),\,\Delta(k,\,(b_q))\end{matrix}\right)$ [see Conditions A with $r=t=u=v=2$; $c_1=1-a$; $c_2=1-b$; $d_1=0$; $d_2=c-a-b$]
2	$(d-x)_+^{c-1}\;{}_2F_1\left(\begin{matrix}a,\,b\\c;\,\frac{d-x}{d}\end{matrix}\right)$ $\times G_{pq}^{mn}\left(\omega x^{\ell/k}\,\middle|\,\begin{matrix}(a_p)\\(b_q)\end{matrix}\right)$	$\dfrac{k^{\mu}\,\ell^{-c}\,\Gamma(c)}{(2\pi)^{(k-1)c^*}\,d^{1-s-c}}\,G_{kp+2\ell,\,kq+2\ell}^{km,\,kn+2\ell}\left(\dfrac{\omega^k d^{\ell}}{k^{k(q-p)}}\,\middle|\,\begin{matrix}\Delta(\ell,\,1-s),\,\Delta(\ell,\,1-s+a+b-c),\,\Delta(k,\,(a_p))\\\Delta(k,\,(b_q)),\,\Delta(\ell,\,1-s+a-c),\,\Delta(\ell,\,1-s+b-c)\end{matrix}\right)$ [see Conditions A with $\sigma=1/d$; $r=u=v=2$; $t=0$; $c_1=c-a$; $c_2=c-b$; $d_1=0$; $d_2=c-a-b$]
3	$(x-d)_+^{c-1}\;{}_2F_1\left(\begin{matrix}a,\,b\\c;\,\frac{d-x}{d}\end{matrix}\right)$ $\times G_{pq}^{mn}\left(\omega x^{\ell/k}\,\middle|\,\begin{matrix}(a_p)\\(b_q)\end{matrix}\right)$	$\dfrac{k^{\mu}\,\ell^{-c}\,\Gamma(c)}{(2\pi)^{(k-1)c^*}\,d^{1-s-c}}\,G_{kp+2\ell,\,kq+2\ell}^{km+2\ell,\,kn}\left(\dfrac{\omega^k d^{\ell}}{k^{k(q-p)}}\,\middle|\,\begin{matrix}\Delta(k,\,(a_p)),\,\Delta(\ell,\,1-s),\,\Delta(\ell,\,1-s+a+b-c)\\\Delta(\ell,\,1-s+a-c),\,\Delta(\ell,\,1-s+b-c),\,\Delta(k,\,(b_q))\end{matrix}\right)$ [see Conditions A with $\sigma=1/d$; $r=u=v=2$; $s=0$; $c_1=c-a$; $c_2=c-b$; $d_1=0$; $d_2=c-a-b$]
4	${}_2F_1\left(\begin{matrix}a,\,b\\c;\,-\sigma x\end{matrix}\right)$ $\times G_{pq}^{mn}\left(\omega x^{\ell/k}\,\middle|\,\begin{matrix}(a_p)\\(b_q)\end{matrix}\right)$	$\dfrac{k^{\mu}\,\ell^{a+b-c-1}\,\sigma^{-s}}{(2\pi)^{\ell-1+(k-1)c^*}}\,\Gamma\left[\begin{matrix}c\\a,\,b\end{matrix}\right]\,G_{kp+2\ell,\,kq+2\ell}^{km+2\ell,\,kn+\ell}\left(\dfrac{\sigma^{-l}\omega^k}{k^{k(q-p)}}\,\middle|\,\begin{matrix}\Delta(\ell,\,1-s),\,\Delta(k,\,(a_p)),\,\Delta(\ell,\,c-s)\\\Delta(\ell,\,a-s),\,\Delta(\ell,\,b-s),\,\Delta(k,\,(b_q))\end{matrix}\right)$ [see Conditions A with $r=1$; $t=u=v=2$; $c_1=1-a$; $c_2=1-b$; $d_1=0$; $d_2=1-c$]
5	${}_rF_t\left(\begin{matrix}(c_r);\,-\sigma x\\(d_t)\end{matrix}\right)$ $\times G_{pq}^{mn}\left(\omega x^{\ell/k}\,\middle|\,\begin{matrix}(a_p)\\(b_q)\end{matrix}\right)$	$\dfrac{k^{\mu}\,\ell^{\eta}\,\sigma^{-s}}{(2\pi)^{(1+r-t)(\ell-1)/2+(k-1)c^*}}\,\Gamma\left[\begin{matrix}(d_t)\\(c_r)\end{matrix}\right]$ $\times G_{kp+t\ell+\ell,\,kq+r\ell}^{km+r\ell,\,kn+\ell}\left(\dfrac{\omega^k k^{k(p-q)}}{\sigma^{\ell}\ell^{\ell(r-t-1)}}\,\middle|\,\begin{matrix}\Delta(\ell,\,1-s),\,\Delta(k,\,(a_p)),\,\Delta(\ell,\,(d_t)-s)\\\Delta(\ell,\,(c_r)-s),\,\Delta(k,\,(b_q))\end{matrix}\right)$ [see Conditions A with $t=u=r$; $r=1$; $,v=l+1$; $(c_u)=1-(c_r)$; $(d_v)=0,\,1-(d_t)$]

3.36.11. Products of two Meijer's G-functions

Notation:

$$\psi=\frac{1}{p-q}\left(\sum_{j=1}^{q}b_j-\sum_{i=1}^{p}a_i+\frac{p-q+1}{2}\right),\quad \chi=\frac{1}{v-u}\left(\sum_{j=1}^{v}d_j-\sum_{i=1}^{u}c_i+\frac{u-v+1}{2}\right).$$

$$\sum\nolimits_L=\sum_{k=1}^{m}\frac{\prod\limits_{j=1;\,j\neq k}^{m}\Gamma(b_j-b_k)\prod\limits_{j=1}^{n}\Gamma(b_k-a_j+1)\prod\limits_{j=1}^{r}\Gamma(s+b_k+d_j)\prod\limits_{j=1}^{t}\Gamma(1-b_k-c_j-s)}{\prod\limits_{j=n+1}^{p}\Gamma(a_j-b_k)\prod\limits_{j=m+1}^{q}\Gamma(b_k-b_j+1)\prod\limits_{j=t+1}^{u}\Gamma(s+b_k+c_j)\prod\limits_{j=r+1}^{v}\Gamma(1-b_k-d_j-s)}$$

$$\times\left(\frac{\omega}{\sigma}\right)^{b_k}{}_{p+v}F_{q+u-1}\left(\begin{matrix}b_k-(a_p)+1,\ s+b_k+(d_v);\ (-1)^{p+v-m-n-r-t}\frac{\omega}{\sigma}\\ b_k-(b_q)'+1,\ s+b_k+(c_u)\end{matrix}\right)$$

$$+\sum_{k=1}^{t}\frac{\prod\limits_{j=1;\,j\neq k}^{t}\Gamma(c_k-c_j)\prod\limits_{j=1}^{r}\Gamma(d_j-c_k+1)\prod\limits_{j=1}^{n}\Gamma(2-a_j-c_k-s)\prod\limits_{j=1}^{m}\Gamma(s+b_j+c_k-1)}{\prod\limits_{j=r+1}^{v}\Gamma(c_k-d_j)\prod\limits_{j=t+1}^{u}\Gamma(c_j-c_k+1)\prod\limits_{j=m+1}^{q}\Gamma(2-b_j-c_k-s)\prod\limits_{j=n+1}^{p}\Gamma(s+a_j+c_k-1)}$$

$$\times\left(\frac{\omega}{\sigma}\right)^{1-s-c_k}{}_{p+v}F_{q+u-1}\left(\begin{matrix}(d_v)-c_k+1,\ 2-(a_p)-c_k-s;\ (-1)^{p+v-m-n-r-t}\frac{\omega}{\sigma}\\ (c_u)'-c_k+1,\ 2-(b_q)-c_k-s\end{matrix}\right),$$

$$\sum\nolimits_R=\sum_{k=1}^{n}\frac{\prod\limits_{j=1;\,j\neq k}^{n}\Gamma(a_k-a_j)\prod\limits_{j=1}^{m}\Gamma(b_j-a_k+1)\prod\limits_{j=1}^{t}\Gamma(2-a_k-c_j-s)\prod\limits_{j=1}^{r}\Gamma(s+a_k+d_j-1)}{\prod\limits_{j=n+1}^{p}\Gamma(a_j-a_k+1)\prod\limits_{j=m+1}^{q}\Gamma(a_k-b_j)\prod\limits_{j=r+1}^{v}\Gamma(2-a_k-d_j-s)\prod\limits_{j=t+1}^{u}\Gamma(s+a_k+c_j-1)}$$

$$\times\left(\frac{\omega}{\sigma}\right)^{a_k-1}{}_{q+u}F_{p+v-1}\left(\begin{matrix}(b_q)-a_k+1,\ 2-a_k-(c_u)-s;\ (-1)^{q+u-m-n-r-t}\frac{\sigma}{\omega}\\ (a_p)'-a_k+1,\ 2-a_k-(d_v)-s\end{matrix}\right)$$

$$+\sum_{k=1}^{r}\frac{\prod\limits_{j=1;\,j\neq k}^{r}\Gamma(d_j-d_k)\prod\limits_{j=1}^{t}\Gamma(d_k-c_j+1)\prod\limits_{j=1}^{m}\Gamma(s+b_j+d_k)\prod\limits_{j=1}^{n}\Gamma(1-a_j-d_k-s)}{\prod\limits_{j=r+1}^{v}\Gamma(d_k-d_j+1)\prod\limits_{j=n+1}^{p}\Gamma(s+a_j+d_k)\prod\limits_{j=t+1}^{u}\Gamma(c_j-d_k)\prod\limits_{j=m+1}^{q}\Gamma(1-s-b_j-d_k)}$$

$$\times\left(\frac{\omega}{\sigma}\right)^{-s-d_k}{}_{q+u}F_{p+v-1}\left(\begin{matrix}s+(b_q)+d_k,\ d_k-(c_u)+1;\ (-1)^{q+u-m-n-r-t}\frac{\sigma}{\omega}\\ d_k-(d_v)'+1,\ s+(a_p)+d_k\end{matrix}\right).$$

Conditions B:

B1 $(v\le u-3)$:

B1.1 $(q=p-1)$

$$\left[\begin{matrix}m=0;\ n\ge q+1;\ 2n-2q+2r+2t-u-v\ge 1;\\ (-1)^{q-n}\operatorname{Re}\omega<0;\ \operatorname{Re}s<2-\operatorname{Re}(a_i+c_g)\end{matrix}\right]$$

B2 $(v = u - 2)$:

B2.1 $(q = p - 2)$

$$\left[\begin{array}{l} m+n \geq q+1;\ r+t \geq u-1;\ (-1)^{v-r-t}\sigma < 0;\ (-1)^{q-m-n}\omega < 0; \\ -\operatorname{Re}(\chi+\psi) - \left(1-\delta_{0,\,\omega-(-1)^{q-m-n-v+r+t}\sigma}\right)/2,\ -\operatorname{Re}(b_j+\psi) - 1/2, \\ -\operatorname{Re}(\chi+d_h) - 1/2,\ -\operatorname{Re}(b_j+d_h) < \operatorname{Re}s < 2 - \operatorname{Re}(a_i+c_g) \end{array}\right]$$

or

B2.2 $(q = p - 1)$

$$\left[\begin{array}{l} m+n \geq q+1;\ r+t \geq u-1;\ (-1)^{v-r-t}\sigma < 0; \\ -\operatorname{Re}(b_j+\psi) - 1/2,\ -\operatorname{Re}(b_j+d_h) < \operatorname{Re}s < 2 - \operatorname{Re}(a_i+c_g); \\ (-1)^{q-m-n}\operatorname{Re}\omega < 0 \text{ or } (\operatorname{Re}\omega = 0;\ -\operatorname{Re}(\chi+\psi) - 1,\ -\operatorname{Re}(\chi+d_h) - 1 < \operatorname{Re}s) \end{array}\right]$$

or

B2.3 $(q = p)$

$$\left[\begin{array}{l} r+t \geq u-1;\ (-1)^{v-r-t}\sigma < 0; \\ -\operatorname{Re}(b_j+\psi) - 1/2,\ -\operatorname{Re}(b_j+d_h) < \operatorname{Re}s < 2 - \operatorname{Re}(a_i+c_g); \\ \left(m+n > p;\ |\arg\omega| < (m+n-p)\pi\right) \text{ or} \\ \left(m+n = p;\ \omega > 0;\ \sum_{k=1}^{p}\operatorname{Re}(a_k-b_k) > 0\right) \end{array}\right]$$

or

B2.4 $(q = p + 1)$

$$\left[\begin{array}{l} m+n \geq p+1;\ r+t \geq u-1;\ (-1)^{v-r-t}\sigma < 0; \\ -\operatorname{Re}(b_j+\psi) - 1/2,\ -\operatorname{Re}(b_j+d_h) < \operatorname{Re}s < 2 - \operatorname{Re}(a_i+c_g); \\ (-1)^{p-m-n}\operatorname{Re}\omega < 0 \text{ or } \left(\operatorname{Re}\omega = 0;\ \operatorname{Re}s < 2 - \operatorname{Re}(c_g+\chi)\right) \end{array}\right]$$

or

B2.5 $(q = p + 2)$

$$\left[\begin{array}{l} m+n \geq p+1;\ r+t \geq u-1; \\ (-1)^{v-r-t}\sigma < 0;\ (-1)^{p-m-n}\omega < 0; \\ -\operatorname{Re}(b_j+\psi) - 1/2,\ -\operatorname{Re}(b_j+d_h) < \operatorname{Re}s \\ < 2 - \operatorname{Re}(a_i+c_g),\ 3/2 - \operatorname{Re}(c_g+\chi) \end{array}\right]$$

B3 $(v = u - 1)$:

B3.1 $(q = p - 2)$

$$\left[\begin{array}{l} m+n \geq q+1;\ r+t \geq u;\ (-1)^{q-m-n}\omega < 0; \\ -\operatorname{Re}(d_h+\chi) - 1/2,\ -\operatorname{Re}(b_j+d_h) < \operatorname{Re}s \\ \qquad < 2 - \operatorname{Re}(a_i+c_g);\ (-1)^{v-r-t}\operatorname{Re}\sigma < 0 \text{ or} \\ (\operatorname{Re}\sigma = 0;\ -\operatorname{Re}(b_j+\psi) - 1,\ -\operatorname{Re}(\chi+\psi) - 1 < \operatorname{Re}s) \end{array}\right]$$

or

B3.2 $(q = p - 1)$

$$\left[\begin{array}{l} m = 0;\ n \geq q+1;\ r+t \geq u; \\ (-1)^{q-n}\operatorname{Re}\omega < 0; \operatorname{Re}s < 2 - \operatorname{Re}(a_i+c_g); \\ \left(\operatorname{Re}((-1)^{q-n}/\omega + (-1)^{v-r-t}/\sigma) < 0 \text{ or}\right. \\ \left(\operatorname{Re}((-1)^{q-n}\sigma + (-1)^{v-r-t}\omega) = 0;\ -\operatorname{Re}(\chi+\psi) - 1 < \operatorname{Re}s\right), \end{array}\right]$$

$$\left[\begin{array}{l} m+n\geq q+1;\ r+t\geq u;\ -\operatorname{Re}(b_j+d_h)<\operatorname{Re}s \\ <2-\operatorname{Re}(a_i+c_g);\ \big(\big((-1)^{v-r-t}\operatorname{Re}\sigma<0;\ \big((-1)^{q-m-n}\operatorname{Re}\omega<0\text{ or} \\ (\operatorname{Re}\omega=0;\ -\operatorname{Re}(d_h+\chi)-1<\operatorname{Re}s)\big)\big)\text{ or} \\ \big(\operatorname{Re}\sigma=0;\ -\operatorname{Re}(b_j+\psi)-1<\operatorname{Re}s; \\ \big((-1)^{q-m-n}\operatorname{Re}\omega<0\text{ or} \\ \big(\operatorname{Re}\omega=0;\ -\operatorname{Re}(d_h+\chi)-1,\ -\operatorname{Re}(\chi+\psi)-1<\operatorname{Re}s\big)\big)\big)\big) \end{array}\right]$$

or

B3.3 $(q=p)$

$$\left[\begin{array}{l} r+t\geq u;\ -\operatorname{Re}(b_j+d_h)<\operatorname{Re}s<2-\operatorname{Re}(a_i+c_g); \\ (-1)^{v-r-t}\operatorname{Re}\sigma<0\text{ or }\ (\operatorname{Re}\sigma=0;\ -\operatorname{Re}(b_j+\psi)-1<\operatorname{Re}s); \\ \big(m+n>p;\ |\arg\omega|<(m+n-p)\pi\big)\text{ or} \\ \big(m+n=p;\ \omega>0;\ \sum_{k=1}^{p}\operatorname{Re}(a_k-b_k)>0\big) \end{array}\right]$$

or

B3.4 $(q=p+1)$

$$\left[\begin{array}{l} m+n\geq p+1;\ r+t\geq u;\ -\operatorname{Re}(b_j+d_h)<\operatorname{Re}s<2-\operatorname{Re}(a_i+c_g); \\ \big((-1)^{v-r-t}\operatorname{Re}\sigma<0\text{ or }\ (\operatorname{Re}\sigma=0;\ -\operatorname{Re}(b_j+\psi)-1<\operatorname{Re}s)\big); \\ (-1)^{p-m-n}\operatorname{Re}\omega<0\text{ or }\ (\operatorname{Re}\omega=0;\ \operatorname{Re}s<2-\operatorname{Re}(c_g+\chi)) \end{array}\right]$$

or

B3.5 $(q=p+2)$

$$\left[\begin{array}{l} m+n\geq p+1;\ r+t\geq u;\ (-1)^{p-m-n}\omega<0; \\ -\operatorname{Re}(b_j+d_h)<\operatorname{Re}s<2-\operatorname{Re}(a_i+c_g),\ 3/2-\operatorname{Re}(c_g+\chi); \\ \big((-1)^{v-r-t}\operatorname{Re}\sigma<0\text{ or }\big(\operatorname{Re}\sigma=0;\ -\operatorname{Re}(b_j+\psi)-1<\operatorname{Re}s\big) \end{array}\right]$$

B4 $(v=u)$:

B4.1 $(q=p-2)$

$$\left[\begin{array}{l} m+n\geq q+1;\ (-1)^{q-m-n}\omega<0;\ -\operatorname{Re}(b_j+d_h),\ -\operatorname{Re}(d_h+\chi)-1/2< \\ <\operatorname{Re}s<2-\operatorname{Re}(a_i+c_g);\ \big(\big(r+t>v;\ |\arg\sigma|<(r+t-v)\pi\big)\text{ or} \\ \big(r+t=v;\ \sigma>0;\ \sum_{j=1}^{u}\operatorname{Re}(c_j-d_j)>0\big)\big) \end{array}\right]$$

or

B4.2 $(q=p-1)$

$$\left[\begin{array}{l} m+n\geq q+1;\ -\operatorname{Re}(b_j+d_h)<\operatorname{Re}s<2-\operatorname{Re}(a_i+c_g); \\ \big(\big(r+t>v;\ |\arg\sigma|<(r+t-v)\pi\big)\text{ or} \\ \big(r+t=v;\ \sigma>0;\ \sum_{j=1}^{u}\operatorname{Re}(c_j-d_j)>0\big)\big); \\ (-1)^{q-m-n}\operatorname{Re}\omega<0\text{ or }\ (\operatorname{Re}\omega=0;\ -\operatorname{Re}(d_h+\chi)-1<\operatorname{Re}s) \end{array}\right]$$

or

B4.3 $(q=p)$

$$\left[\begin{array}{l} r+t>u;\ |\arg\sigma|<(r+t-u)\pi;\ -\operatorname{Re}(b_j+d_h)<\operatorname{Re}s<2-\operatorname{Re}(a_i+c_g); \\ \big(\big(m+n>p;\ |\arg\omega|<(m+n-p)\pi\big)\text{ or} \\ \big(m+n=p;\ \omega>0;\ \sum_{k=1}^{p}\operatorname{Re}(a_k-b_k)>0\big)\big), \end{array}\right]$$

$$\left[\begin{array}{l} r+t=u;\ \sigma>0;\ -\operatorname{Re}(b_j+d_h)<\operatorname{Re}s<2-\operatorname{Re}(a_i+c_g);\\ \left(\left(\sum_{j=1}^{u}\operatorname{Re}(c_j-d_j)>0;\right.\right.\\ \big((m+n>p;\ |\arg\omega|<(m+n-p)\pi\big)\ \text{or}\\ \big(m+n=p;\ \omega>0;\ \omega\neq\sigma;\ \sum_{k=1}^{p}\operatorname{Re}(a_k-b_k)>0)\big)\big)\ \text{or}\\ \big(m+n=p;\ \omega=\sigma;\ \sum_{k=1}^{p}\operatorname{Re}(a_k-b_k)+\sum_{j=1}^{u}\operatorname{Re}(c_j-d_j)>1)\big) \end{array}\right]$$

or

B4.4 $(q=p+1)$

$$\left[\begin{array}{l} m+n\geq p+1;\ -\operatorname{Re}(b_j+d_h)<\operatorname{Re}s<2-\operatorname{Re}(a_i+c_g);\\ \big((r+t>u;\ |\arg\sigma|<(r+t-u)\pi\big)\ \text{or}\\ \big(r+t=u;\ \sigma>0;\ \sum_{j=1}^{u}\operatorname{Re}(c_j-d_j)>0)\big);\\ \big((-1)^{p-m-n}\operatorname{Re}\omega<0\ \text{or}\ (\operatorname{Re}\omega=0;\ \operatorname{Re}s<2-\operatorname{Re}(c_g+\chi))\big) \end{array}\right]$$

or

B4.5 $(q=p+2)$

$$\left[\begin{array}{l} m+n\geq p+1;\ (-1)^{p-m-n}\,\omega<0;\\ -\operatorname{Re}(b_j+d_h)<\operatorname{Re}s<2-\operatorname{Re}(a_i+c_g),\ 3/2-\operatorname{Re}(c_g+\chi);\\ \big(r+t>u;\ |\arg\sigma|<(r+t-u)\pi\big)\ \text{or}\\ \big(r+t=u;\ \sigma>0;\ \sum_{j=1}^{u}\operatorname{Re}(c_j-d_j)>0\big) \end{array}\right]$$

B5 $(v=u+1)$:

B5.1 $(q=p-2)$

$$\left[\begin{array}{l} m+n\geq q+1;\ r+t\geq v;\ (-1)^{q-m-n}\,\omega<0;\\ -\operatorname{Re}(b_j+d_h),\ -\operatorname{Re}(d_h+\chi)-1/2<\operatorname{Re}s<2-\operatorname{Re}(a_i+c_g);\\ (-1)^{u-r-t}\operatorname{Re}\sigma<0\ \text{or}\ (\operatorname{Re}\sigma=0;\ \operatorname{Re}s<2-\operatorname{Re}(a_i+\psi)) \end{array}\right]$$

or

B5.2 $(q=p-1)$

$$\left[\begin{array}{l} m+n\geq q+1;\ r+t\geq v;\ -\operatorname{Re}(b_j+d_h)<\operatorname{Re}s<2-\operatorname{Re}(a_i+c_g);\\ (-1)^{u-r-t}\operatorname{Re}\sigma<0\ \text{or}\ (\operatorname{Re}\sigma=0;\ \operatorname{Re}s<2-\operatorname{Re}(a_i+\psi));\\ (-1)^{q-m-n}\operatorname{Re}\omega<0\ \text{or}\ (\operatorname{Re}\omega=0;\ -\operatorname{Re}(d_h+\chi)-1<\operatorname{Re}s) \end{array}\right]$$

or

B5.3 $(q=p)$

$$\left[\begin{array}{l} r+t\geq u+1;\ -\operatorname{Re}(b_j+d_h)<\operatorname{Re}s<2-\operatorname{Re}(a_i+c_g);\\ (-1)^{u-r-t}\operatorname{Re}\sigma<0\ \text{or}\ (\operatorname{Re}\sigma=0;\ \operatorname{Re}s<2-\operatorname{Re}(a_i+\psi));\\ \big(m+n>p;\ |\arg\omega|<(m+n-p)\pi\big)\ \text{or}\\ \big(m+n=p;\ \omega>0;\ \sum_{k=1}^{p}\operatorname{Re}(a_k-b_k)>0\big) \end{array}\right]$$

or

B5.4 $(q=p+1)$

$$\left[\begin{array}{l} nt=0;\ v=u+1;\ m+n\geq p+1;\ r+t\geq u+1;\\ (-1)^{p-m}\operatorname{Re}\omega<0\ \text{for}\ n=0;\ (-1)^{u-r}\operatorname{Re}\sigma<0\ \text{for}\ t=0;\\ -\operatorname{Re}(b_j+d_h)<\operatorname{Re}s;\ (-1)^{u-r-t}\operatorname{Re}\sigma+(-1)^{p-m-n}\operatorname{Re}\omega<0\ \text{or}\\ \big((-1)^{u-r-t}\operatorname{Re}\sigma+(-1)^{p-m-n}\operatorname{Re}\omega=0;\ \operatorname{Re}s<1-\operatorname{Re}(\chi+\psi)\big) \end{array}\right]$$

$$\left[\begin{array}{l} m+n \geq p+1;\, r+t \geq u+1;\; -\operatorname{Re}(b_j+d_h) < \operatorname{Re} s < 2-\operatorname{Re}(a_i+c_g)\,; \\ \big(\big((-1)^{u-r-t}\operatorname{Re}\sigma < 0;\; \big((-1)^{p-m-n}\operatorname{Re}\omega < 0 \text{ or} \\ \big(\operatorname{Re}\omega = 0;\; \operatorname{Re} s < 2-\operatorname{Re}(c_g+\chi)\big)\big)\big) \text{ or} \\ \big(\operatorname{Re}\sigma = 0;\; \operatorname{Re} s < 2-\operatorname{Re}(a_i+\psi)\,;\; \big((-1)^{p-m-n}\operatorname{Re}\omega < 0 \text{ or} \\ \big(\operatorname{Re}\omega = 0;\; \operatorname{Re} s < 1-\operatorname{Re}(\chi+\psi)\,,\, 2-\operatorname{Re}(c_g+\chi)\big)\big)\big)\big) \end{array}\right]$$

or

B5.5 $(q = p+2)$

$$\left[\begin{array}{l} m+n \geq p+1;\, r+t \geq u+1;\; (-1)^{p-m-n}\,\omega < 0; \\ -\operatorname{Re}(b_j+d_h) < \operatorname{Re} s < 2-\operatorname{Re}(a_i+c_g)\,,\, 3/2-\operatorname{Re}(c_g+\chi)\,; \\ (-1)^{u-r-t}\operatorname{Re}\sigma < 0 \text{ or } \big(\operatorname{Re}\sigma = 0;\; \operatorname{Re} s < 1-\operatorname{Re}(\chi+\psi)\,,\, 2-\operatorname{Re}(a_i+\psi)\big) \end{array}\right]$$

B6 $(v = u+2)$:

B6.1 $(q = p-2)$

$$\left[\begin{array}{l} m+n \geq q+1;\, r+t \geq v-1;\; (-1)^{u-r-t}\,\sigma < 0;\; (-1)^{q-m-n}\,\omega < 0; \\ -\operatorname{Re}(b_j+d_h)\,,\, -\operatorname{Re}(d_h+\chi)-1/2 < \\ < \operatorname{Re} s < 2-\operatorname{Re}(a_i+c_g)\,,\, 3/2-\operatorname{Re}(a_i+\psi) \end{array}\right]$$

or

B6.2 $(q = p-1)$

$$\left[\begin{array}{l} m+n \geq q+1;\, r+t \geq v-1;\; (-1)^{u-r-t}\,\sigma < 0; \\ -\operatorname{Re}(b_j+d_h) < \operatorname{Re} s < 2-\operatorname{Re}(a_i+c_g)\,,\, 3/2-\operatorname{Re}(a_i+\psi)\,; \\ (-1)^{q-m-n}\operatorname{Re}\omega < 0 \text{ or } \big(\operatorname{Re}\omega = 0;\; -\operatorname{Re}(d_h+\chi)-1 < \operatorname{Re} s\big) \end{array}\right]$$

or

B6.3 $(q = p)$

$$\left[\begin{array}{l} r+t \geq u+1;\; (-1)^{u-r-t}\,\sigma < 0; \\ -\operatorname{Re}(b_j+d_h) < \operatorname{Re} s < 2-\operatorname{Re}(a_i+c_g)\,,\, 3/2-\operatorname{Re}(a_i+\psi)\,; \\ \big(\big(m+n > p;\; |\arg\omega| < (m+n-p)\,\pi\big) \text{ or} \\ \big(m+n = p;\; \omega > 0;\; \sum_{k=1}^{p}\operatorname{Re}(a_k-b_k) > 0\big)\big) \end{array}\right]$$

or

B6.4 $(q = p+1)$

$$\left[\begin{array}{l} m+n \geq p+1;\, r+t \geq u+1;\; (-1)^{u-r-t}\,\sigma < 0; \\ -\operatorname{Re}(b_j+d_h) < \operatorname{Re} s < 2-\operatorname{Re}(a_i+c_g)\,,\, 3/2-\operatorname{Re}(a_i+\psi)\,; \\ (-1)^{p-m-n}\operatorname{Re}\omega < 0 \text{ or } (\operatorname{Re}\omega = 0;\; \operatorname{Re} s < 2-\operatorname{Re}(c_g+\chi)\,,\, 1-\operatorname{Re}(\chi+\psi)) \end{array}\right]$$

or

B6.5 $(q = p+2)$

$$\left[\begin{array}{l} m+n \geq p+1;\, r+t \geq u+1; \\ (-1)^{u-r-t}\,\sigma < 0; (-1)^{p-m-n}\,\omega < 0;\; -\operatorname{Re}(b_j+d_h) < \operatorname{Re} s < 2-\operatorname{Re}(a_i+c_g)\,, \\ 3/2-\operatorname{Re}(c_g+\chi)\,,\, 3/2-\operatorname{Re}(a_i+\psi)\,,\; \big(1-\delta_{0,\,\sigma-(-1)^{-m-n+p+r+t-u}\omega}\big)/2-\operatorname{Re}(\chi+\psi) \end{array}\right]$$

B7 $(v \geq u+3)$:

B7.1 $(q = p+1)$

$$\left[\; n = 0;\; m \geq p+1;\; 2m-2p+2r+2t-u-v \geq 1;\; (-1)^{p-m}\operatorname{Re}\omega < 0;\; -\operatorname{Re}(b_j+d_h) < \operatorname{Re} s \;\right]$$

No.	$f(x)$	$F(s)$
1	$G_{pq}^{mn}\left(\omega x \middle\vert \begin{matrix}(a_p)\\(b_q)\end{matrix}\right)$ $\times G_{uv}^{rt}\left(\sigma x \middle\vert \begin{matrix}(c_u)\\(d_v)\end{matrix}\right)$	$\sigma^{-s} G_{p+v,\,q+u}^{m+t,\,n+r}\left(\frac{\omega}{\sigma} \middle\vert \begin{matrix}(a_n),\,1-s-(d_v),\,a_{n+1},\ldots,a_p\\(b_m),\,1-s-(c_u),\,b_{m+1},\ldots,b_q\end{matrix}\right)$ $\left[\begin{matrix}0\le m\le q;\ 0\le n\le p;\ 0\le r\le v;\ 0\le t\le u;\ 0\le q-p\le 2;\\ -2\le v-u;\ \sigma\in\mathbb{C};\ \omega\in\mathbb{C};\ \sigma\ne 0;\ \omega\ne 0;\ a_i-b_j\ne 1,2,\ldots;\\ i=1,\ldots,n;\ j=1,\ldots,m;\ c_g-d_h\ne 1,2,\ldots;\\ g=1,\ldots,t;\ h=1,\ldots,r;\\ \text{see Conditions B1–B7}\end{matrix}\right]$
2	$G_{pq}^{mn}\left(\omega x \middle\vert \begin{matrix}(a_p)\\(b_q)\end{matrix}\right)$ $\times G_{uv}^{rt}\left(\sigma x \middle\vert \begin{matrix}(c_u)\\(d_v)\end{matrix}\right)$	$\frac{\sigma^{-s}}{2\pi i}\int_{\gamma-i\infty}^{\gamma+i\infty}\frac{\prod_{j=1}^{m}\Gamma(b_j+\tau)\prod_{g=1}^{t}\Gamma(1-s-c_g+\tau)}{\prod_{k=n+1}^{p}\Gamma(a_k+\tau)\prod_{k=r+1}^{v}\Gamma(1-s-d_k+\tau)}$ $\times\frac{\prod_{i=1}^{n}\Gamma(1-a_i-\tau)\prod_{h=1}^{r}\Gamma(s+d_h-\tau)}{\prod_{k=m+1}^{q}\Gamma(1-b_k-\tau)\prod_{k=t+1}^{u}\Gamma(s+c_k-\tau)}\left(\frac{\omega}{\sigma}\right)^{-\tau}d\tau$ $\left[\begin{matrix}-\operatorname{Re}b_k,\ \operatorname{Re}(s+c_k)-1<\gamma=\operatorname{Re}\tau<1-\operatorname{Re}a_k,\ \operatorname{Re}(s+d_k);\\ \text{see Conditions B1–B7}\end{matrix}\right]$
3	$G_{pq}^{mn}\left(\omega x \middle\vert \begin{matrix}(a_p)\\(b_q)\end{matrix}\right)$ $\times G_{uv}^{rt}\left(\sigma x \middle\vert \begin{matrix}(c_u)\\(d_v)\end{matrix}\right)$	$\sigma^{-s}\sum\nolimits_L$ $\left[\begin{matrix}(q+u>p+v;\ \lvert\omega/\sigma\rvert<\infty)\ \text{or}\ (q+u=p+v;\ \lvert\omega/\sigma\rvert<1);\\ a_i-a_k,\ d_h-d_f,\ s+a_i+d_h\ne 0,\ \pm1,\ \pm2,\ldots\ \text{for}\\ 1\le i\le n,\ 1\le k\le n,\ 1\le h\le r,\ 1\le f\le r,\ j\ne k,\ h\ne f;\\ \text{see Conditions B1–B7}\end{matrix}\right]$
4		$\sigma^{-s}\sum\nolimits_R$ $\left[\begin{matrix}(q+u<p+v;\ \lvert\omega/\sigma\rvert<\infty)\ \text{or}\ (q+u=p+v;\ \lvert\omega/\sigma\rvert>1);\\ b_j-b_k,\ c_g-c_f,\ s+b_j+c_g\ne 0,\ \pm1,\ \pm2,\ldots\ \text{for}\\ 1\le j\le m,\ 1\le k\le m,\ 1\le g\le t,\ 1\le f\le t,\ j\ne k,\ g\ne f;\\ \text{see Conditions B1–B7}\end{matrix}\right]$
5		$\sigma^{-s}\sum\nolimits_L=\sigma^{-s}\sum\nolimits_R$ $\left[\begin{matrix}q+u=p+v;\ m+n+r+t-(p+q+u+v)/2>0;\\ \text{see Conditions B1–B7}\end{matrix}\right]$

The following formula is valid if the integers k and ℓ are mutually prime. If this is not the case and M is the greatest common divisor of k and ℓ, one should make the change of variable of integration $x\to x^{1/M}$:

6	$G_{pq}^{mn}\left(\omega x^{\ell} \middle\vert \begin{matrix}(a_p)\\(b_q)\end{matrix}\right)$ $\times G_{uv}^{rt}\left(\sigma x^{k} \middle\vert \begin{matrix}(c_u)\\(d_v)\end{matrix}\right)$	$\frac{k^{\mu-1}\ell^{\rho+s(v-u)/k-1}\sigma^{-s/k}}{(2\pi)^R}G_{kp+\ell v,\,kq+\ell u}^{km+\ell t,\,kn+\ell r}\left(\frac{k^{k(p-q)}\omega^k}{\ell^{\ell(u-v)}\sigma^{\ell}} \middle\vert \begin{matrix}\Delta(k,\,(a_n)),\,\Delta\left(\ell,\,1-(d_v)-\frac{s}{k}\right),\,\Delta(k,\,(a_{n+1})),\ldots,\Delta(k,\,(a_p))\\ \Delta(k,\,(b_m)),\,\Delta\left(\ell,\,1-(c_u)-\frac{s}{k}\right),\,\Delta(k,\,(b_{m+1})),\ldots,\Delta(k,\,(b_q))\end{matrix}\right)$

No.	$f(x)$	$F(s)$
		$\left[\begin{array}{l} R=(k-1)\,c^*+(\ell-1)\,b^*; \\ k,\,\ell=1,\,2,...;\ 0\le m\le q;\ 0\le n\le p;\ 0\le r\le v; \\ 0\le t\le u;\ 0\le q-p\le 2;\ -2\le v-u;\ \sigma\ne 0;\ \omega\ne 0; \\ a_i-b_j\ne 1,\,2,\ldots;\ i=1,\,2,...,n;\ j=1,\,2,\ldots,m; \\ c_g-d_h\ne 1,\,2,\ldots;\ g=1,\ldots,t;\ h=1,\ldots,r; \\ \text{see Conditions B1–B7 with the substitution} \\ m\to km,\ n\to kn,\ p\to kp,\ q\to kq, \\ r\to \ell r,\ t\to \ell t,\ u\to \ell u,\ v\to \ell v, \\ a_p\to\Delta(k,\ (a_p)),\,b_q\to\Delta(k,\ (b_q)), \\ c_u\to\Delta(k,\ (c_u)),\ d_v\to\Delta(k,\ (d_v)), \\ \sigma\to\sigma^\ell\ell^{-\ell(v-u)},\ \omega\to\omega^k k^{-k(q-p)},\ s\to s/(k\ell) \end{array}\right]$
7	$G_{pq}^{mn}\left(\omega x^{\ell/k}\,\middle\vert\,\begin{matrix}(a_p)\\(b_q)\end{matrix}\right)$ $\times\, G_{uv}^{rt}\left(\sigma x\,\middle\vert\,\begin{matrix}(c_u)\\(d_v)\end{matrix}\right)$	$\frac{k^\mu\,\ell^{\rho+s(v-u)-1}\,\sigma^{-s}}{(2\pi)^{(\ell-1)b^*+(k-1)c^*}}G_{kp+\ell v,\,kq+\ell u}^{km+\ell t,\,kn+\ell r}\left(\frac{k^{k(p-q)}\omega^k}{\ell^{\ell(u-v)}\sigma^\ell}\,\middle\vert\,\begin{matrix}\Delta(k,(a_n)),\ \Delta(\ell,\,1-(d_v)-s),\ \Delta(k,\,a_{n+1}),\ldots,\Delta(k,\,a_p)\\ \Delta(k,(b_m)),\ \Delta(\ell,\,1-(c_u)-s),\ \Delta(k,\,b_{m+1}),\ldots,\Delta(k,\,b_q)\end{matrix}\right)$

[One of the following conditions holds (if $mr=0$ or $nt=0$, the 2° and 3° are omitted, respectively):

1) $mnrt\ne 0;\ b^*,\ c^*>0;\ |\arg\sigma|<b^*\pi;\ |\arg\omega|<c^*\pi;\ 1^\circ–3^\circ;$

2) $u=v;\ b^*=0;\ c^*,\ \sigma>0;\ |\arg\omega|<c^*\pi;\ |\mathrm{Re}\,\rho|<1;\ 1^\circ–3^\circ;$

3) $p=q;\ b^*,\ \omega>0;\ c^*=0;\ |\arg\sigma|<b^*\pi;\ |\mathrm{Re}\,\mu|<1;\ 1^\circ–3^\circ;$

4) $p=q;\ u=v;\ b^*=c^*=0;\ \sigma,\ \omega>0;\ \mathrm{Re}\,\mu,\ \mathrm{Re}\,\rho<1;\ \sigma^l\ne\omega^k;\ 1^\circ–3^\circ;$

5) $p=q;\ u=v;\ b^*=c^*=0;\ \sigma,\ \omega>0;\ \mathrm{Re}\,(\mu+\rho)<2;\ \sigma^l=\omega^k;\ 1^\circ–3^\circ;$

6) $p>q;\ r>0;\ b^*>0;\ c^*\ge 0;\ |\arg\sigma|<b^*\pi;\ |\arg\omega|=c^*\pi;\ 1^\circ–3^\circ;\ 5^\circ;$

7) $p<q;\ t>0;\ b^*>0;\ c^*\ge 0;\ |\arg\sigma|<b^*\pi;\ |\arg\omega|=c^*\pi;\ 1^\circ–4^\circ;$

8) $m>0;\ u>v;\ b^*\ge 0;\ c^*>0;\ |\arg\sigma|=b^*\pi;\ |\arg\omega|<c^*\pi;\ 1^\circ–3^\circ;\ 7^\circ;$

9) $n>0;\ u<v;\ b^*\ge 0;\ c^*>0;\ |\arg\sigma|<b^*\pi;\ |\arg\omega|=c^*\pi;\ 1^\circ–3^\circ;\ 6^\circ;$

10) $p>q;\ u=v;\ b^*=0;\ c^*\ge 0;\ \sigma>0;\ |\arg\omega|=c^*\pi;\ \mathrm{Re}\,\rho<1;\ 1^\circ–3^\circ;\ 5^\circ;$

11) $p<q;\ u=v;\ b^*=0;\ c^*\ge 0;\ \sigma>0;\ |\arg\omega|=c^*\pi;\ \mathrm{Re}\,\rho<1;\ 1^\circ–4^\circ;$

12) $p=q;\ u>v;\ b^*\ge 0;\ c^*=0;\ |\arg\sigma|=b^*\pi;\ \omega>0;\ \mathrm{Re}\,\mu<1;\ 1^\circ–3^\circ;\ 7^\circ;$

13) $p=q;\ u<v;\ b^*\ge 0;\ c^*=0;\ |\arg\sigma|=b^*\pi;\ \omega>0;\ \mathrm{Re}\,\mu<1;\ 1^\circ–3^\circ;\ 6^\circ;$

14) $p<q;\ u>v;\ b^*,\ c^*\ge 0;\ |\arg\sigma|=b^*\pi;\ |\arg\omega|=c^*\pi;\ 1^\circ–4^\circ;\ 7^\circ;$

15) $p>q;\ u<v;\ b^*,\ c^*\ge 0;\ |\arg\sigma|=b^*\pi;\ |\arg\omega|=c^*\pi;\ 1^\circ–3^\circ;\ 5^\circ;\ 6^\circ;$

16) $p>q;\ u>v;\ b^*,\ c^*\ge 0;\ |\arg\sigma|=b^*\pi;\ |\arg\omega|=c^*\pi;\ 1^\circ–3^\circ;\ 5^\circ;\ 7^\circ;\ 8^\circ;\ 10^\circ;$

17) $p<q;\ u<v;\ b^*,\ c^*\ge 0;\ |\arg\sigma|=b^*\pi;\ |\arg\omega|=c^*\pi;\ 1^\circ–4^\circ;\ 6^\circ;\ 9^\circ;\ 10^\circ;$

18) $t=0;\ r,\ b^*,\ \varphi>0;\ |\arg\sigma|<b^*\pi;\ 1^\circ–2^\circ;$

19) $t>0;\ r=0;\ b^*>0;\ \varphi<0;\ |\arg\sigma|<b^*\pi;\ 1^\circ;\ 3^\circ;$

20) $m>0;\ n=0;\ c^*>0;\ \varphi<0;\ |\arg\omega|<c^*\pi;\ 1^\circ–2^\circ;$]

No.	$f(x)$	$F(s)$
		21) $m=0;\ n>0;\ c^*,\ \varphi>0;\ \lvert\arg\omega\rvert<c^*\pi;\ 1^\circ;\ 3^\circ;$ 22) $rt=0;\ b^*,\ c^*>0;\ \lvert\arg\sigma\rvert<b^*\pi;\ \lvert\arg\omega\rvert<c^*\pi;\ 1^\circ\text{–}3^\circ;$ 23) $mn=0;\ b^*,\ c^*>0;\ \lvert\arg\sigma\rvert<b^*\pi;\ \lvert\arg\omega\rvert<c^*\pi;\ 1^\circ\text{–}3^\circ;$ 24) $m+n>p;\ t=\varphi=0;\ r,\ b^*>0;\ c^*<0;\ \lvert\arg\sigma\rvert<b^*\pi;$ $\lvert\arg\omega\rvert<(m+n-p+1)\pi;\ 1^\circ;\ 2^\circ;\ 10^\circ;\ 11^\circ;$ 25) $m+n>q;\ r=\varphi=0;\ t,\ b^*>0;\ c^*<0;\ \lvert\arg\sigma\rvert<b^*\pi;$ $\lvert\arg\omega\rvert<(m+n-q+1)\pi;\ 1^\circ;\ 3^\circ;\ 10^\circ;\ 11^\circ;$ 26) $p=q-1;\ t=\varphi=0;\ r>0;\ b^*>0;\ c^*\ge 0;\ \lvert\arg\sigma\rvert<b^*\pi;$ $c^*\pi<\lvert\arg\omega\rvert<(c^*+1)\pi;\ 1^\circ;\ 2^\circ;\ 10^\circ;\ 11^\circ;$ 27) $p=q+1;\ r=\varphi=0;\ t>0;\ b^*>0;\ c^*\ge 0;\ \lvert\arg\sigma\rvert<b^*\pi;$ $c^*\pi<\lvert\arg\omega\rvert<(c^*+1)\pi;\ 1^\circ;\ 3^\circ;\ 10^\circ;\ 11^\circ;$ 28) $p<q-1;\ t=\varphi=0;\ r>0;\ b^*>0;\ c^*\ge 0;\ \lvert\arg\sigma\rvert<b^*\pi;$ $c^*\pi<\lvert\arg\omega\rvert<(m+n-p+1)\pi;\ 1^\circ;\ 2^\circ;\ 10^\circ;\ 11^\circ;$ 29) $p>q-1;\ r=\varphi=0;\ t>0;\ b^*>0;\ c^*\ge 0;\ \lvert\arg\sigma\rvert<b^*\pi;$ $c^*\pi<\lvert\arg\omega\rvert<(m+n-q+1)\pi;\ 1^\circ;\ 3^\circ;\ 10^\circ;\ 11^\circ;$ 30) $n=\varphi=0;\ r+t>u;\ m>0;\ b^*<0;\ c^*>0;\ \lvert\arg\sigma\rvert<(r+t-u+1)\pi;$ $\lvert\arg\omega\rvert<c^*\pi;\ 1^\circ;\ 2^\circ;\ 10^\circ;\ 11^\circ;$ 31) $m=\varphi=0;\ r+t>v;\ n>0;\ b^*<0;\ c^*>0;\ \lvert\arg\sigma\rvert<(r+t-v+1)\pi;$ $\lvert\arg\omega\rvert<c^*\pi;\ 1^\circ;\ 3^\circ;\ 10^\circ;\ 11^\circ;$ 32) $n=\varphi=0;\ u=v-1;\ m>0;\ b^*\ge 0;\ c^*>0;\ b^*\pi<\lvert\arg\sigma\rvert<(b^*+1)\pi;$ $\lvert\arg\omega\rvert<c^*\pi;\ 1^\circ;\ 2^\circ;\ 10^\circ;\ 11^\circ;$ 33) $m=\varphi=0;\ u=v+1;\ n>0;\ b^*\ge 0;\ c^*>0;\ b^*\pi<\lvert\arg\sigma\rvert<(b^*+1)\pi;$ $\lvert\arg\omega\rvert<c^*\pi;\ 1^\circ;\ 3^\circ;\ 10^\circ;\ 11^\circ;$ 34) $n=\varphi=0;\ u<v-1;\ m>0;\ b^*\ge 0;\ c^*>0;$ $b^*\pi<\lvert\arg\sigma\rvert<(r+t-u+1)\pi;\ \lvert\arg\omega\rvert<c^*\pi;\ 1^\circ;\ 2^\circ;\ 10^\circ;\ 11^\circ;$ 35) $m=\varphi=0;\ u>v+1;\ n>0;\ b^*\ge 0;\ c^*>0;$ $b^*\pi<\lvert\arg\sigma\rvert<(r+t-v+1)\pi;\ \lvert\arg\omega\rvert<c^*\pi;\ 1^\circ;\ 3^\circ;\ 10^\circ;\ 11^\circ.$

$$b^*=r+t-\frac{u+v}{2},\qquad c^*=m+n-\frac{p+q}{2}.$$

8	$G_{uv}^{rt}\left(x+\sigma\,\middle\vert\,\begin{matrix}(c_u)\\(d_v)\end{matrix}\right)\times G_{pq}^{mn}\left(\omega x\,\middle\vert\,\begin{matrix}(a_p)\\(b_q)\end{matrix}\right)$	$\sum_{k=0}^{\infty}\frac{(-\sigma)^k}{k!}\,G_{p+v+1,\,q+u+1}^{m+t,\,n+r+1}\left(\omega\,\middle\vert\,\begin{matrix}1-s,\ k-s-(d_v)+1,\ a_{n+1},\dots,a_p\\(b_m),\ k-s-(c_u)+1,\ k-s+1,\ b_{m+1},\dots,b_q\end{matrix}\right)$ $\left[b^*,\ c^*>0;\ \lvert\arg\sigma\rvert<\pi;\ \lvert\arg\omega\rvert<c^*\pi;\ -\min_{1\le j\le m}\operatorname{Re}b_j<\operatorname{Re}s<2-\max_{1\le i\le n}\operatorname{Re}a_i-\max_{1\le k\le t}\operatorname{Re}c_k\right]$

3.37. Various Special Functions

3.37.1. The exponential integral $E_\nu(z)$

More formulas can be obtained from the corresponding sections due to the relations

$$E_\nu(z) = z^{\nu-1}\Gamma(1-\nu, z), \quad E_\nu(z) = z^{\nu-1}\Gamma(1-\nu) - \frac{1}{1-\nu}\,{}_1F_1\left(\begin{matrix}1-\nu\\ 2-\nu; -z\end{matrix}\right),$$

$$E_\nu(z) = z^{\nu-1}e^{-z}\Psi(\nu, \nu; z), \quad E_\nu(z) = G_{12}^{20}\left(z\,\middle|\,\begin{matrix}\nu\\ \nu-1, 0\end{matrix}\right).$$

No.	$f(x)$	$F(s)$
1	$E_\nu(ax)$	$\frac{a^{-s}}{s+\nu-1}\Gamma(s)$ $\left[\begin{matrix}(\operatorname{Re} a > 0;\ \operatorname{Re} s > 1-\operatorname{Re}\nu, 0) \text{ or}\\ (\operatorname{Re} a = 0;\ 0, 1-\operatorname{Re}\nu < \operatorname{Re} s < 2)\end{matrix}\right]$
2	$E_\nu(ax) - \Gamma(1-\nu)(ax)^{\nu-1} + \frac{1}{1-\nu}$	$\frac{a^{-s}}{s+\nu-1}\Gamma(s)$ $[\operatorname{Re} a \geq 0;\ -1 < \operatorname{Re} s < 1-\operatorname{Re}\nu, 0]$
3	$E_\nu(ax) - \Gamma(1-\nu)(ax)^{\nu-1} + \sum_{k=0}^{n}\frac{(-ax)^k}{k!\,(k-\nu+1)}$	$\frac{a^{-s}}{s+\nu-1}\Gamma(s)$ $[\operatorname{Re} a \geq 0;\ -n-1 < \operatorname{Re} s < -n, 1-\operatorname{Re}\nu]$
4	$e^{ax}E_\nu(ax)$	$\pi a^{-s}\csc(s\pi)\,\Gamma\left[\begin{matrix}s+\nu-1\\ \nu\end{matrix}\right]$ $[0, 1-\operatorname{Re}\nu < \operatorname{Re} s < 1]$
5	$\Gamma(1-\nu, -ax)\,E_\nu(ax)$	$\frac{\pi}{s+\nu-1}\,a^{(\nu-s-1)/2}(-a)^{(1-\nu-s)/2}\sec\frac{(s-\nu)\pi}{2}\,\Gamma\left[\begin{matrix}s\\ \nu\end{matrix}\right]$ $[\lvert 1-\operatorname{Re}\nu\rvert < \operatorname{Re} s < \operatorname{Re}\nu + 1]$
6	$E_\nu(-ax)\,E_\nu(ax)$	$\frac{\pi}{s+2\nu-2}\,a^{-s/2}(-a)^{-s/2}\csc\frac{s\pi}{2}\,\Gamma\left[\begin{matrix}s+\nu-1\\ \nu\end{matrix}\right]$ $[0, 1-\operatorname{Re}\nu, 2-2\operatorname{Re}\nu < \operatorname{Re} s < 2]$
7	$E_\nu(-i\sqrt{ax})\,E_\nu(i\sqrt{ax})$	$\frac{\pi a^{-s}}{s+\nu-1}\csc(s\pi)\,\Gamma\left[\begin{matrix}2s+\nu-1\\ \nu\end{matrix}\right]$ $[0, 1-\operatorname{Re}\nu,\ (1-\operatorname{Re}\nu)/2 < \operatorname{Re} s < 1]$

3.37.2. The theta functions $\theta_j(b,\, ax)$

No.	$f(x)$	$F(s)$
1	$\theta(a-x)\begin{Bmatrix}\theta_1(b,\, x/a)\\ \theta_2(b,\, x/a)\end{Bmatrix}$	$\dfrac{\pi a^s s^{-1/2}}{\cosh(\sqrt{s}\,\pi)}\begin{Bmatrix}\sinh(2b\sqrt{s})\\ \sinh[(\pi-2b)\sqrt{s}]\end{Bmatrix}$ $[a,\ \operatorname{Re} s>0;\ -(1\pm1)\pi\le b\le(3\mp1)\pi/4]$
2	$\theta(a-x)\begin{Bmatrix}\theta_3(b,\, x/a)\\ \theta_4(b,\, x/a)\end{Bmatrix}$	$\dfrac{\pi a^s s^{-1/2}}{\sinh(\sqrt{s}\,\pi)}\begin{Bmatrix}\cosh[(\pi-2b)\sqrt{s}]\\ \cosh(2b\sqrt{s})\end{Bmatrix}$ $[a,\ \operatorname{Re} s>0;\ -(1\mp1)\pi\le b\le(3\pm1)\pi/4]$
3	$\begin{Bmatrix}\theta_1(b,\, e^{-x})\\ \theta_2(b,\, e^{-x})\end{Bmatrix}$	$2^{2s-1}\pi^{2s-1/2}\,\Gamma\left(\dfrac{1-2s}{2}\right)\left[\zeta\left(1-2s,\, \dfrac{(3\pm3)\pi+4b}{8\pi}\right)+\zeta\left(1-2s,\, \dfrac{(5\mp3)\pi-4b}{8\pi}\right)-\zeta\left(1-2s,\, \dfrac{(3\mp1)\pi+4b}{8\pi}\right)-\zeta\left(1-2s,\, \dfrac{(5\pm1)\pi-4b}{8\pi}\right)\right]$ $[\operatorname{Re} s>0;\ -(1\pm1)\pi/4\le b\le(3\mp1)\pi/4]$
4	$\begin{Bmatrix}\theta_3(b,\, e^{-x})-1\\ \theta_4(b,\, e^{-x})-1\end{Bmatrix}$	$\Gamma(s)\left[\operatorname{Li}_{2s}\left(\pm e^{-2ib}\right)+\operatorname{Li}_{2s}\left(\pm e^{2ib}\right)\right]$ $[\operatorname{Re} s>1/2;\ -(1\mp1)\pi/4\le b\le(3\pm1)\pi/4]$
5	$\begin{Bmatrix}\theta_1(\pi/2,\, e^{-x})\\ \theta_2(0,\, e^{-x})\end{Bmatrix}$	$2\Gamma(s)\,\zeta\left(2s,\, \dfrac{1}{2}\right)$ $[\operatorname{Re} s>1/2]$
6	$\begin{Bmatrix}\theta_3(0,\, e^{-x})-1\\ \theta_4(\pi/2,\, e^{-x})-1\end{Bmatrix}$	$2\Gamma(s)\,\zeta(2s,\, 0)$ $[\operatorname{Re} s>1/2]$
7	$\begin{Bmatrix}\theta_3(\pi/2,\, e^{-x})-1\\ \theta_4(0,\, e^{-x})-1\end{Bmatrix}$	$2\left(2^{1-2s}-1\right)\Gamma(s)\,\zeta(2s,\, 0)$ $[\operatorname{Re} s>1/2]$
8	$-\theta_2(0,\, e^{-x})+\theta_3(0,\, e^{-x})-\theta_4(0,\, e^{-x})$	$2\left(2^{2s}-1\right)\left(2^{1-2s}-1\right)\Gamma(s)\,\zeta(2s,\, 0)$ $[\operatorname{Re} s>1/2]$

3.37.3. The generalized Fresnel integrals $S(z, \nu)$ and $C(z, \nu)$

More formulas can be obtained from the corresponding sections due to the relations

$$\begin{Bmatrix} S(z, \nu) \\ C(z, \nu) \end{Bmatrix} = \begin{Bmatrix} \sin(\nu\pi/2) \\ \cos(\nu\pi/2) \end{Bmatrix} \Gamma(\nu) - \frac{z^{\nu+\delta}}{\nu+\delta} \, {}_1F_2\left(\begin{matrix} \frac{\nu+\delta}{2}; -\frac{z^2}{4} \\ \frac{2\delta+1}{2}, \frac{\nu+\delta+2}{2} \end{matrix} \right), \quad \delta = \begin{Bmatrix} 1 \\ 0 \end{Bmatrix}.$$

1	$C(ax, \nu)$	$\frac{2^{s+\nu-1}\sqrt{\pi}}{s\, a^s} \Gamma\left[\begin{matrix} \frac{s+\nu}{2} \\ \frac{1-s-\nu}{2} \end{matrix}\right]$	$[a > 0;\ 0,\ -\operatorname{Re}\nu < \operatorname{Re}s < 2 - \operatorname{Re}\nu]$
2	$S(ax, \nu)$	$\frac{2^{s+\nu-1}\sqrt{\pi}}{s\, a^s} \Gamma\left[\begin{matrix} \frac{s+\nu+1}{2} \\ \frac{2-s-\nu}{2} \end{matrix}\right]$	$[a > 0;\ 0,\ -\operatorname{Re}\nu - 1 < \operatorname{Re}s < 2 - \operatorname{Re}\nu]$

3.37.4. The integral Bessel functions

More formulas can be obtained from the corresponding sections due to the relations

$$Ji_\nu(z) = -\frac{z^\nu}{2^\nu \nu^2 \Gamma(\nu)} \, {}_1F_2\left(\begin{matrix} \frac{\nu}{2}; -\frac{z^2}{4} \\ \frac{\nu+2}{2}, \nu+1 \end{matrix} \right) + \frac{1}{\nu};$$

$$\begin{Bmatrix} Yi(z) \\ Ki(z) \end{Bmatrix} = \pm\frac{1}{2\nu} \begin{Bmatrix} \cot(\nu\pi/2) \\ \pi\csc(\nu\pi/2) \end{Bmatrix} \pm \frac{\Gamma(-\nu)}{2\nu\pi} \begin{Bmatrix} \cos\nu\pi \\ \pi \end{Bmatrix} \left(\frac{z}{2}\right)^\nu {}_1F_2\left(\begin{matrix} \frac{\nu}{2}; \mp\frac{z^2}{4} \\ \frac{\nu+2}{2}, \nu+1 \end{matrix} \right)$$

$$\mp \frac{\Gamma(\nu)}{2\nu\pi} \begin{Bmatrix} 1 \\ \pi \end{Bmatrix} \left(\frac{z}{2}\right)^{-\nu} {}_1F_2\left(\begin{matrix} -\frac{\nu}{2}; \mp\frac{z^2}{4} \\ \frac{2-\nu}{2}, 1-\nu \end{matrix} \right), \quad \nu \neq \pm n.$$

1	$Ji_\nu(ax)$	$\frac{2^{s-1}}{s\, a^s} \Gamma\left[\begin{matrix} \frac{s+\nu}{2} \\ \frac{2-s+\nu}{2} \end{matrix}\right]$	$[a > 0;\ -\operatorname{Re}\nu,\ 0 < \operatorname{Re}s < 2]$
2	$Ki_\nu(ax)$	$\frac{2^{s-2}}{s\, a^s} \Gamma\left(\frac{s-\nu}{2}\right) \Gamma\left(\frac{s+\nu}{2}\right)$	$[a > 0;\ \operatorname{Re}s > \lvert\operatorname{Re}\nu\rvert]$
3	$Yi_\nu(ax)$	$\frac{2^{s-1}}{s\, a^s} \Gamma\left[\begin{matrix} \frac{s-\nu}{2}, \frac{s+\nu}{2} \\ \frac{3-s+\nu}{2}, \frac{s-\nu-1}{2} \end{matrix}\right]$	$[a > 0;\ \lvert\operatorname{Re}\nu\rvert < \operatorname{Re}s < 2]$

3.37.5. The Lommel functions

1	$s_{\mu,\nu}(ax)$	$\frac{2^{s+\mu-2}}{a^s} \Gamma\left[\begin{matrix} \frac{\mu-\nu+1}{2}, \frac{\mu+\nu+1}{2}, \frac{-s-\mu+1}{2}, \frac{s+\mu+1}{2} \\ \frac{2-s-\nu}{2}, \frac{2-s+\nu}{2} \end{matrix}\right]$ $[a > 0;\ \lvert\operatorname{Re}(s+\mu)\rvert < 1,\ \operatorname{Re}s < 3/2]$
2	$S_{\mu,\nu}(ax)$	$\frac{2^{s+\mu-2}}{a^s} \Gamma\left[\begin{matrix} \frac{s-\nu}{2}, \frac{s+\nu}{2}, \frac{-s-\mu+1}{2}, \frac{s+\mu+1}{2} \\ \frac{1-\mu-\nu}{2}, \frac{1-\mu+\nu}{2} \end{matrix}\right]$ $[a > 0;\ \lvert\operatorname{Re}(s+\mu)\rvert < 1,\ \operatorname{Re}s < 3/2]$

3.37.6. The Owen and $\mathcal{H}$-functions

No.	$f(x)$	$F(s)$
1	$T(ax,\, b)$	$\frac{2^{s/2-2}b}{\pi a^s}\Gamma\left(\frac{s}{2}\right){}_2F_1\left(\begin{matrix}\frac{1}{2},\, \frac{s+2}{2}\\ \frac{3}{2};\, -b^2\end{matrix}\right)$ $\left[\operatorname{Re} a^2,\ \operatorname{Re}\left(a^2+a^2b^2\right)<0;\ \operatorname{Re} s>0\right]$
2	$e^{-cx^2}T(ax,\, b)$	$\frac{2^{s/2-2}b}{\pi\left(a^2+2c\right)^{s/2}}\Gamma\left(\frac{s}{2}\right)F_1\left(\frac{1}{2},\, 1,\, \frac{s}{2};\, \frac{3}{2};\, -b^2,\, -\frac{a^2b^2}{a^2+2c}\right)$ $\left[\operatorname{Re} a^2,\ \operatorname{Re}\left(a^2+a^2b^2\right)<2\operatorname{Re} c;\ \operatorname{Re} s>0\right]$
3	$\mathcal{H}_\nu(x,\, a,\, b)$	$\frac{2^{s/2-3}\left(1-a^2\right)^\nu a^{s+1}}{\pi}\left[\sqrt{\pi}\,\Gamma\left[\begin{matrix}\frac{2\nu+1}{2},\, \frac{s}{2}\\ \nu+1\end{matrix}\right]{}_2F_1\left(\begin{matrix}\frac{2\nu+1}{2},\, \frac{s+2}{2}\\ \nu+1;\, 1-a^2\end{matrix}\right)\right.$ $-\frac{2}{(2\nu+1)\left(1+a^2b^2\right)^{\nu+1/2}}\Gamma\left(\frac{s}{2}\right)$ $\left.\times F_1\left(\nu+\frac{1}{2};\, \frac{1}{2},\, \frac{s}{2}+1;\, \nu+\frac{3}{2};\, \frac{1}{1+a^2b^2},\, \frac{1-a^2}{1+a^2b^2}\right)\right]$ $[0<a\leq 1;\ \operatorname{Re} s>0]$

3.37.7. The Bessel–Maitland and generalized Bessel–Maitland functions

1	$J_\nu^\mu(ax)$	$a^{-s}\Gamma\left[\begin{matrix}s\\ 1-\mu s+\nu\end{matrix}\right]$ $\left[\begin{matrix}(a>0,\ \operatorname{Re}\mu<1;\ \operatorname{Re} s>0)\ \text{or}\\ (a>0,\ \mu=1;\ 0<\operatorname{Re} s<(2\operatorname{Re}\nu+3)/4)\end{matrix}\right]$
2	$J_{\nu,\lambda}^\mu(ax)$	$\frac{2^{s-1}}{a^s}\Gamma\left[\begin{matrix}-\frac{s+2\lambda+\nu-2}{2},\, \frac{s+2\lambda+\nu}{2}\\ \frac{2-s-\nu}{2},\, \frac{2-\mu s-(2\lambda+\nu)\mu+2\lambda+2\nu}{2}\end{matrix}\right]$ $\left[\begin{matrix}(a>0,\ \operatorname{Re}\mu<1;\ -\operatorname{Re}(2\lambda+\nu)<\operatorname{Re} s<2-\operatorname{Re}(2\lambda+\nu))\ \text{or}\\ (a>0,\ \mu=1;\ -\operatorname{Re}(2\lambda+\nu)<\operatorname{Re} s<3/2,\, 2-\operatorname{Re}(2\lambda+\nu))\end{matrix}\right]$

3.37.8. Other functions

1	$E_\rho(-x;\, \mu)$	$\Gamma\left[\begin{matrix}s,\, 1-s\\ \mu-\frac{s}{\rho}\end{matrix}\right]$ $\left[\begin{matrix}(\rho>1/2;\ 0<\operatorname{Re} s<1)\ \text{or}\\ (\rho=1/2;\ 0<\operatorname{Re} s<1,\ \operatorname{Re}\mu/2)\end{matrix}\right]$
2	$\mu(ae^{-x},\, 1)$	$\frac{\pi(1-s)}{\sin(s\pi)}\mu(a,\, 1-s)$
3	$\theta(1-x)\,\mu(-\ln x,\, \lambda)$	$\frac{1}{s\ln^{\lambda+1}s}$ $[\operatorname{Re}\lambda>-1;\ \operatorname{Re} s>1]$

No.	$f(x)$	$F(s)$	
4	$\mu(z, \lambda, x+\rho)$	$\Gamma(s)\,\mu(z, s+\lambda, \rho)$	$[\operatorname{Re}\lambda, \operatorname{Re}\rho > -1;\ \operatorname{Re}s > 0]$
5	$e^{\rho x}\mu(ae^{-x}, 1, \rho)$	$\Gamma(s)\,\Gamma(2-s)\,\mu(a, 1-s, \rho)$	
6	$\theta(1-x)\,\mu(-\ln x, \lambda, \rho)$	$\dfrac{1}{s^{\rho+1}\ln^{\lambda+1}s}$	$[\operatorname{Re}\lambda, \operatorname{Re}\rho > -1;\ \operatorname{Re}s > 1]$
7	$\dfrac{\theta(1-x)}{\sqrt{-\ln x}}\,\mu\left(a\sqrt{-\ln x}, \lambda, \rho\right)$	$\dfrac{2^{\lambda+1}\sqrt{\pi}}{\sqrt{s}}\,\mu\left(\dfrac{a^2}{4s}, \lambda, \dfrac{\rho}{2}\right)$	$[a > 0;\ \operatorname{Re}\lambda > -1;\ \operatorname{Re}\rho > -2]$
8	$\theta(1-x)\,\mu\left(a\sqrt{-\ln x}, \lambda, \rho\right)$	$\dfrac{2^{\lambda}\sqrt{\pi}\,a}{s^{3/2}}\,\mu\left(\dfrac{a^2}{4s}, \lambda, \dfrac{\rho-1}{2}\right)$	$[a > 0;\ \operatorname{Re}\lambda, \operatorname{Re}\rho > -1]$
9	$\theta(a-x)\,\nu\left(\dfrac{x}{a}\right)$	$a^s\displaystyle\int_0^\infty \dfrac{dt}{(t+s)\,\Gamma(t+1)}$	$[a, \operatorname{Re}s > 0]$
10	$\nu(e^{-x})$	$\dfrac{\pi}{\sin(s\pi)}\,\mu(1, -s)$	$[\operatorname{Re}s > 0]$
11	$\nu\left(ae^{-bx}\right)$	$\dfrac{\pi}{b^s\sin(s\pi)}\,\mu(a, -s)$	$[a, \operatorname{Re}b > 0;\ 0 < \operatorname{Re}s < 1]$
12	$\theta(1-x)\,\nu(-\ln x)$	$\dfrac{1}{s\ln s}$	$[\operatorname{Re}s > 1]$
13	$\dfrac{\theta(1-x)\,\nu\left(a\sqrt{-\ln x}\right)}{\sqrt{-\ln x}}$	$\dfrac{2\sqrt{\pi}}{\sqrt{s}}\,\nu\left(\dfrac{a^2}{4s}\right)$	$[a, \operatorname{Re}s > 0]$
14	$\nu(a, x+\rho)$	$\Gamma(s)\,\mu(a, s, \rho)$	$[\operatorname{Re}\rho > -1;\ a, \operatorname{Re}s > 0]$
15	$\theta(1-x)\,\nu(-\ln x, \rho)$	$\dfrac{1}{s^{\rho+1}\ln s}$	$[\operatorname{Re}\rho > -1;\ \operatorname{Re}s > 1]$
16	$\dfrac{\theta(1-x)\,\nu\left(a\sqrt{-\ln x}, \rho\right)}{\sqrt{-\ln x}}$	$\dfrac{2\sqrt{\pi}}{\sqrt{s}}\,\nu\left(\dfrac{a^2}{4s}, \dfrac{\rho}{2}\right)$	$[a, \operatorname{Re}s > 0;\ \operatorname{Re}\rho > -2]$
17	$\theta(1-x)\,\nu\left(a\sqrt{-\ln x}, \rho\right)$	$\dfrac{\sqrt{\pi}\,a}{s^{3/2}}\,\nu\left(\dfrac{a^2}{4s}, \dfrac{\rho-1}{2}\right)$	$[a, \operatorname{Re}s > 0;\ \operatorname{Re}\rho > -1]$

Appendix I

Some Properties of the Mellin Transforms

The integral

$$F(s) = \int_0^\infty x^{s-1} f(x)\, dx \tag{I.1}$$

is called the Mellin transform of the function $f(x)$.

The notations $\mathfrak{M}[f(x)](s)$ and $\mathfrak{M}[f(x);\, s]$ are used as well. Here $f(x)$ denotes a function of the real variable x, $0 \le x < \infty$, which is Lebesgue integrable over any interval $(0,\, A)$, $A > 0$, and $s = \sigma + i\tau$ is a complex number.

The Mellin transform is closely connected with the Fourier and Laplace transforms. The substitution $x = e^{-t}$ transforms (I.1) into the two-sided Laplace transform,

$$F(s) = \int_{-\infty}^{\infty} e^{-ts} f\left(e^{-t}\right) dt.$$

Change of variables $x = e^y$, $f(x) = g(y)$ in (I.1) yields

$$F(s) = \int_{-\infty}^{\infty} e^{sy} g(y)\, dy = (Fg)(is),$$

where

$$(Fg)(\xi) = \int_{-\infty}^{\infty} e^{-iy\xi} g(y)\, dy$$

is the Fourier transform of the function $g(y)$. Below, relations are given between the Mellin transform and some other integral transforms [15, 22].

1. **The Fourier cosine transform**:

$$\mathfrak{F}_c[f(t);\, x] = \sqrt{\frac{2}{\pi}} \int_0^\infty \cos(xt)\, f(t)\, dt,$$

$$\mathfrak{M}[\mathfrak{F}_c[f(t);\, s]] = \sqrt{\frac{2}{\pi}} \cos\frac{s\pi}{2}\, \Gamma(s)\, \mathfrak{M}[f(x);\, 1-s].$$

2. **The Fourier sine transform**:

$$\mathfrak{F}_s[f(t);\, x] = \sqrt{\frac{2}{\pi}} \int_0^\infty \sin(xt)\, f(t)\, dt,$$

$$\mathfrak{M}[\mathfrak{F}_s[f(t);\, x];\, z] = \sqrt{\frac{2}{\pi}} \sin\frac{z\pi}{2}\, \Gamma(z)\, \mathfrak{M}[f(x);\, 1-z].$$

3. **The Laplace transform:**

$$\mathfrak{L}\left[f(t);\, x\right] = \int_0^\infty e^{-xt} f(t)\, dt,$$
$$\mathfrak{M}\left[\mathfrak{L}\left[f(t);\, x\right];\, s\right] = \Gamma(s)\, \mathfrak{M}\left[f(x);\, 1-s\right].$$

4. **The Hankel transform:**

$$H_\nu\left[f(t);\, x\right] = \int_0^\infty \sqrt{xt}\, J_\nu(xt)\, f(t)\, dt,$$
$$\mathfrak{M}\left[H_\nu\left[f(t);\, x\right];\, s\right] = 2^{s-1/2}\, \Gamma\left[\begin{matrix}\frac{2s+2\nu+1}{4}\\ \frac{3-2s+2\nu}{4}\end{matrix}\right] \mathfrak{M}\left[f(x);\, 1-s\right].$$

5. **The Meijer transform:**

$$K_\nu\left[f(t);\, x\right] = \int_0^\infty \sqrt{xt}\, K_\nu(xt)\, f(t)\, dt,$$
$$\mathfrak{M}\left[K_\nu\left[f(t);\, x\right];\, s\right] = 2^{s-3/2}\, \Gamma\left(\frac{2s+2\nu+1}{4}\right) \Gamma\left(\frac{2s-2\nu+1}{4}\right) \mathfrak{M}\left[f(x);\, 1-s\right].$$

6. **The Y_ν–Bessel transform:**

$$Y_\nu\left[f(t);\, x\right] = \int_0^\infty \sqrt{xt}\, Y_\nu(xt)\, f(t)\, dt,$$
$$\mathfrak{M}\left[Y_\nu\left[f(t);\, x\right];\, s\right] = \frac{2^{s-1/2}}{\pi} \sin\frac{(2\nu-2s-3)\,\pi}{4}$$
$$\times\, \Gamma\left(\frac{2s-2\nu+1}{4}\right) \Gamma\left(\frac{2s+2\nu+1}{4}\right) \mathfrak{M}\left[f(x);\, 1-s\right].$$

7. **The $\mathbf{H}_\nu$–Struve transform:**

$$\mathbf{H}_\nu\left[f(t);\, x\right] = \int_0^\infty \sqrt{xt}\, \mathbf{H}_\nu(xt)\, f(t)\, dt,$$
$$\mathfrak{M}\left[\mathbf{H}_\nu\left[f(t);\, x\right];\, s\right] = 2^{s-1/2} \tan\frac{(2s+2\nu+1)\,\pi}{4}\, \Gamma\left[\begin{matrix}\frac{2s+2\nu+1}{2}\\ \frac{3-2s+2\nu}{4}\end{matrix}\right] \mathfrak{M}\left[f(x);\, 1-s\right].$$

8. **The Hilbert transform:**

$$\mathcal{H}\left[f(t);\, x\right] = \int_0^\infty \frac{f(t)}{t-x}\, dt,$$
$$\mathfrak{M}\left[\mathcal{H}\left[f(t);\, x\right];\, s\right] = \frac{\Gamma(s)\,\Gamma(1-s)}{\Gamma\left(s+\frac{1}{2}\right)\Gamma\left(\frac{1}{2}-s\right)}\, \mathfrak{M}\left[f(x);\, s\right] = \cos(s\pi)\, \mathfrak{M}\left[f(x);\, s\right].$$

9. **The generalized Stieltjes transform:**

$$\mathcal{S}_\nu\left[f(t);\, x\right] = \int_0^\infty \frac{f(t)}{(x+t)^\nu}\, dt,$$
$$\mathfrak{M}\left[\mathcal{S}_\nu\left[f(t);\, x\right];\, s\right] = \mathrm{B}(s,\, \nu-s)\, \mathfrak{M}\left[f(x);\, s-\nu+1\right].$$

10. **The Liouville fractional integrals** [24]:

$$I_{0+}^{\nu}[f(t);x] = \frac{1}{\Gamma(\nu)}\int_0^x (x-t)^{\nu-1} f(t)\,dt,$$

$$\mathfrak{M}\left[I_{0+}^{\nu}[f(t);x];s\right] = \frac{\Gamma(1-s-\nu)}{\Gamma(1-s)}\mathfrak{M}[f(x);s+\nu].$$

$$I_{-}^{\nu}[f(t);x] = \frac{1}{\Gamma(\nu)}\int_x^\infty (t-x)^{\nu-1} f(t)\,dt,$$

$$\mathfrak{M}\left[I_{-}^{\nu}[f(t);x];s\right] = \frac{\Gamma(s)}{\Gamma(s+\nu)}\mathfrak{M}[f(x);s+\nu].$$

The inverse formula. The Mellin transform can be inverted under some conditions. For example, if $f(x)$ is analytic on $0 < x < \infty$ and satisfies the asymptotic conditions

$$f(x) = O\left(x^{-\alpha}\right), \quad x \to 0,$$
$$f(x) = O\left(x^{-\beta}\right), \quad x \to \infty,$$

where $\alpha < \beta$, then the function $F(s)$, defined by (I.1), is analytic in the strip $\alpha < \operatorname{Re} s < \beta$, and

$$f(x) = \frac{1}{2\pi i}\int_{\sigma-i\infty}^{\sigma+i\infty} x^{-s}F(s)\,ds = \mathfrak{M}^{-1}[F(s)], \qquad \alpha < \sigma < \beta. \tag{I.2}$$

The simplest sufficient condition for the validity of the formulae (I.1) and (I.2) is provided by the continuity of $f(x)$ on $0 < x < \infty$ and the existence of the integral

$$\int_0^\infty x^{\sigma-1}|f(x)|\,dx < \infty. \tag{I.3}$$

Let us note two important properties of the Mellin transform [11]:

The convolution formula. If $F(s)$ and $G(s)$ are the Mellin transforms of $f(x)$ and $g(x)$, then

$$\mathfrak{M}\left[\int_0^\infty f(\xi)\, g\left(\frac{x}{\xi}\right)\frac{d\xi}{\xi};s\right] = F(s)\,G(s). \tag{I.4}$$

The commutation formula. We have

$$\mathfrak{M}\left[x\,\frac{df(x)}{dx};s\right] = -s\,\mathfrak{M}[f(x);s] \tag{I.5}$$

provided that $f(x)$ and $xf'(x)$ satisfy the condition (I.3), and

$$\lim_{x\to 0} x^s f(x) = \lim_{x\to\infty} x^s f(x) = 0.$$

One more important formula: If

$$F(s) = \int_0^\infty x^{s-1} f(x)\,dx, \qquad G(s) = \int_0^\infty x^{s-1} g(x)\,dx,$$

and

$$h(t) = \int_0^\infty f(x)\,g(xt)\,dx,$$

then

$$\int_0^\infty t^{s-1} h(t)\,dt = F(1-s)\,G(s).$$

In conclusion, we mention one more version of the Mellin transform that is useful in the theory of Dirichlet series [10]. Let

$$\Phi(s) = \sum_{n=1}^{\infty} a_n n^{-s}, \qquad \operatorname{Re} s > \alpha,$$

and

$$\varphi(x) = \sum_{n=1}^{\infty} a_n e^{-nx}, \qquad x > 0.$$

Then we have

$$\Phi(s) = \frac{1}{\Gamma(s)} \int_0^{\infty} x^{s-1} \varphi(x)\, dx \tag{I.6}$$

and

$$\varphi(x) = \frac{1}{2\pi i} \int_{\sigma - i\infty}^{\sigma + i\infty} x^{-s}\, \Gamma(s)\, \Phi(s)\, ds, \qquad \sigma > \alpha. \tag{I.7}$$

Putting $\Phi(s) = 1$ in (I.6), we obtain the integral representation of the gamma function:

$$\Gamma(s) = \int_0^{\infty} e^{-x} x^{s-1}\, dx, \quad \operatorname{Re} s > 0.$$

For $\Phi(s) = \zeta(s)$ in (I.6), we get the integral representation of the Riemann zeta function:

$$\zeta(s) = \frac{1}{\Gamma(s)} \int_0^{\infty} \frac{x^{s-1}}{e^x - 1}\, dx, \quad \operatorname{Re} s > 1.$$

Putting $\Phi(s) = 1$ in (I.7), we obtain the integral representation of the exponential function:

$$e^{-x} = \frac{1}{2\pi i} \int_{\sigma - i\infty}^{\sigma + i\infty} x^{-s}\, \Gamma(s)\, ds, \qquad \sigma > 0,\ |\arg x| < \pi/2.$$

For $\Phi(s) = \Gamma(s)$ in (I.7), we arrive at the integral representation of the Macdonald's function:

$$2K_0\left(2\sqrt{x}\right) = \frac{1}{2\pi i} \int_{\sigma - i\infty}^{\sigma + i\infty} x^{-s}\, \Gamma^2(s)\, ds, \qquad x,\, \sigma > 0.$$

Evaluation of integrals. We illustrate the Mellin transformation method in evaluation of integrals by some examples.

Example I.1. Let us derive the relation

$$\int_0^{\infty} t^{\alpha-1} e^{-t-t/x}\, dt = \Gamma(\alpha) \left(1 + \frac{1}{x}\right)^{-\alpha}, \tag{I.8}$$

where $\operatorname{Re}\alpha$, $\operatorname{Re}(1 + 1/x) > 0$. The integral has the form of the Mellin convolution of the functions

$$f(t) = t^{\alpha} e^{-t}, \qquad g(t) = e^{-1/t}.$$

Their Mellin transforms are

$$F(s) = \Gamma(s + \alpha), \qquad \operatorname{Re}(s + \alpha) > 0,$$

and

$$G(s) = \Gamma(-s), \qquad \operatorname{Re} s < 0.$$

Denoting the integral by $I(x, \alpha)$ we obtain its Mellin transform in the form

$$\mathfrak{M}\left[I(x, \alpha); s\right] = \Gamma(s + \alpha)\Gamma(-s), \qquad -\operatorname{Re}\alpha < \operatorname{Re} s < 0.$$

From the formula 2.1.2.3 we have

$$\mathfrak{M}\left[(1+x)^{-\alpha};\, s\right] = \frac{1}{\Gamma(\alpha)}\Gamma(s)\,\Gamma(\alpha-s), \qquad 0<\operatorname{Re}s<\operatorname{Re}\alpha,$$

whence, due to the relation 1.1.2.3, we get

$$\mathfrak{M}\left[\left(1+\frac{1}{x}\right)^{-\alpha};\, s\right] = \frac{1}{\Gamma(\alpha)}\Gamma(-s)\,\Gamma(s+\alpha), \qquad -\operatorname{Re}\alpha<\operatorname{Re}s<0,$$

and, finally,

$$I(x,\alpha) = \Gamma(\alpha)\left(1+\frac{1}{x}\right)^{-\alpha}, \qquad \operatorname{Re}\alpha,\ \operatorname{Re}(1+1/x)>0.$$

Example I.2. Let us evaluate the integral

$$I(a,b,\alpha,\mu,\nu) = \int_0^\infty t^{\alpha-1}\,K_\mu(at)\,I_\nu(bt)\,dt. \tag{I.9}$$

Making use of the formula $I_\nu(z) = (-i)^\nu J_\nu(iz)$, we transform the function I_ν into J_ν, for which the Mellin transform exists. Then the integral (I.9) takes the form

$$I(a,b,\alpha,\mu,\nu) = (-i)^\nu \int_0^\infty t^{\alpha-1}K_\mu(at)\,J_\nu(ibt)\,dt.$$

After substitutions

$$b\to -ic, \quad t\to\frac{2\sqrt{\tau}}{c}, \quad c\to\sqrt{x}$$

and

$$f(\eta) = K_\mu\left(\frac{2}{\sqrt{\eta}}\right)\eta^{-\alpha/2}, \qquad g(\tau) = J_\nu\left(2\sqrt{\tau}\right)$$

we obtain a relation of the form (I.4):

$$I(a,b,\alpha,\mu,\nu) = (-i)^\nu\,2^{\alpha-1}a^{-\alpha}\,\mathfrak{M}^{-1}\left[F(s)\,G(s)\right].$$

The images of the corresponding functions can be found by making use of formulae (1.1.5.2), (3.14.1.3), and (3.10.1.2):

$$\begin{aligned} F(s) &= \int_0^\infty t^{s-1}K_\mu\left(\frac{2}{\sqrt{t}}\right)t^{-\alpha/2}\,dt \\ &= \frac{1}{2}\Gamma\left[\frac{\alpha+\mu}{2}-s,\ \frac{\alpha-\mu}{2}-s\right], \quad \operatorname{Re}s < -\frac{|\operatorname{Re}(\alpha\pm\mu)|}{2}; \\ G(s) &= \int_0^\infty t^{s-1}J_\nu(2\sqrt{t})\,dt \\ &= \Gamma\left[\begin{matrix} s+\frac{\nu}{2} \\ 1-s+\frac{\nu}{2}\end{matrix}\right], \quad -\frac{\operatorname{Re}\nu}{2} < \operatorname{Re}s < \frac{3}{4}. \end{aligned}$$

Multiplying them, we obtain

$$F(s)\,G(s) = \frac{1}{2}\Gamma\left[\begin{matrix}\frac{\alpha+\mu}{2}-s,\ \frac{\alpha-\mu}{2}-s,\ s+\frac{\nu}{2} \\ 1-s+\frac{\nu}{2}\end{matrix}\right].$$

Now, with the aid of formulae 8.4.49.13 from [20], we find $\mathfrak{M}^{-1}\left[F(s)\,G(s)\right]$, and thereby the value of the integral (I.9):

$$\begin{aligned} I(a,b,\alpha,\mu,\nu) &= 2^{\alpha-2}a^{-\alpha-\nu}b^\nu\,\Gamma\left[\frac{\alpha+\mu+\nu}{2},\ \frac{\alpha-\mu+\nu}{2},\ \nu+1\right] \\ &\quad\times {}_2F_1\left(\frac{\alpha+\mu+\nu}{2},\ \frac{\alpha-\mu+\nu}{2};\ \nu+1;\ \frac{b^2}{a^2}\right), \\ &\qquad \operatorname{Re}(\alpha+\nu\pm\mu),\ \operatorname{Re}(a\pm b)>0. \end{aligned}$$

Example I.3. Consider the integral equation

$$y(x) + \int_0^\infty y(\xi)\, f\left(\frac{x}{\xi}\right) \frac{d\xi}{\xi} = g(x), \tag{I.10}$$

where f and g are known functions. Applying the Mellin transform (I.1) and the relation (I.4), we obtain the equality

$$Y(s) + F(s)\, Y(s) = G(s),$$

where Y, F, and G are the Mellin transforms of y, f, and g, respectively, and hence

$$Y(s) = \frac{G(s)}{1 + F(s)}.$$

Applying the inversion formula (I.2), we find the required solution

$$y(x) = \mathfrak{M}^{-1}\left[\frac{G(s)}{1 + F(s)}\right].$$

Example I.4. Consider the Laplace equation in polar coordinates [11]

$$\Delta u = \left(\frac{\partial^2}{\partial r^2} + \frac{1}{r}\frac{\partial}{\partial r} + \frac{1}{r^2}\frac{\partial^2}{\partial \varphi^2}\right) u = 0 \tag{I.11}$$

in the sector $0 < \varphi < \varphi_0 < 2\pi$, $0 < r < \infty$ with Dirichlet boundary conditions

$$u(r, \varphi)|_{\varphi=0} = u_0(r), \qquad u(r, \varphi)|_{\varphi=\varphi_0} = u_1(r). \tag{I.12}$$

We suppose that the solution is bounded at infinity and the so-called "Meixner condition on the edge" $\lim_{r\to 0} \sqrt{r}\, \frac{\partial u}{\partial r} = 0$ is satisfied. These conditions guarantee the uniqueness of the solution. Applying the Mellin transform with respect to the variable r, we get

$$U(s, \varphi) = \int_0^\infty r^{s-1} u(r, \varphi)\, dr.$$

The functions $U_0(s)$ and $U_1(s)$ are defined similarly. Now, by taking the commutation relation (I.5) into account, (I.11) and (I.12) are reduced to the ordinary differential equation for $U(s, \varphi)$

$$\left(U''_{\varphi\varphi}(s, \varphi)\right)^2 + U(s, \varphi) = 0 \tag{I.13}$$

with the boundary conditions

$$U(s, 0) = U_0(s), \qquad U(s, \varphi_0) = U_1(s). \tag{I.14}$$

Solving this boundary value problem and applying the inversion formula (I.2), we find the solution $u(r, \varphi)$.

Some other applications can be found, for example, in [2].

Appendix II

Conditions of Convergence

Exploring conditions of convergence of integrals at a point, we often can replace integrands with simpler asymptotic expressions containing only power, exponential, and trigonometric functions and providing the same conditions. For example, instead of behavior of the functions e^{ax-x^2} and $\sin\left(x^2+2x+a\right)$, when $x \to \infty$, we can consider behavior of e^{ax} and $\sin x^2$, respectively, and get the same conditions of convergence of the corresponding integral at infinity.

Below, we give some model integrals, their conditions of convergence, and a list of asymptotic analogues of elementary and special functions. Conditions for the majority of other integrals can be obtained by replacing integrands with their asymptotic analogues and comparing them with the formulas 1–9. Note that some integrals can require deeper investigation of asymptotics.

I. Convergence at $x = 0$:

1. $\displaystyle\int_0^1 x^\alpha\,dx \qquad [\operatorname{Re}\alpha > -1].$

2. $\displaystyle\int_0^1 x^\alpha e^{ax^\beta}\,dx \qquad \begin{bmatrix} (\operatorname{Re} a,\ \beta > 0;\ \operatorname{Re}\alpha > -1) \text{ or} \\ (\operatorname{Re} a < 0;\ \beta < 0) \text{ or} \\ (\operatorname{Re} a < 0;\ \beta \ge 0;\ \operatorname{Re}\alpha > -1) \text{ or} \\ (\operatorname{Re} a = 0;\ \operatorname{Re}\alpha > -1) \end{bmatrix}.$

II. Convergence at $x = \infty$:

3. $\displaystyle\int_1^\infty x^\alpha\,dx \qquad [\operatorname{Re}\alpha < -1].$

4. $\displaystyle\int_1^\infty x^\alpha e^{-ax}\,dx \qquad \begin{bmatrix} (\operatorname{Re} a > 0) \text{ or} \\ (\operatorname{Re} a = 0;\ \operatorname{Im} a \ne 0;\ \operatorname{Re}\alpha < 0) \end{bmatrix}.$

5. $\displaystyle\int_1^\infty x^\alpha e^{-ax^\beta}\,dx \qquad \begin{bmatrix} (\operatorname{Re} a > 0;\ \beta > 0) \text{ or} \\ (\operatorname{Re} a > 0;\ \beta > 0;\ \operatorname{Re}\alpha < -1) \text{ or} \\ (\operatorname{Re} a < 0;\ \beta < 0;\ \operatorname{Re}\alpha < -1) \text{ or} \\ (\operatorname{Re} a = 0;\ \operatorname{Re}\alpha < \beta - 1) \end{bmatrix}.$

6. $\displaystyle\int_1^\infty x^\alpha \begin{Bmatrix} \sin(ax) \\ \cos(ax) \end{Bmatrix} dx \qquad [\operatorname{Im} a = 0;\ \operatorname{Re}\alpha < 0].$

7. $\displaystyle\int_1^{\infty} x^{\alpha} \begin{Bmatrix} \sin(ax^{\beta}) \\ \cos(ax^{\beta}) \end{Bmatrix} dx$ $[\operatorname{Im} a = 0;\ \beta > 0;\ \operatorname{Re}\alpha < \beta - 1]$.

8. $\displaystyle\int_1^{\infty} x^{\alpha} e^{ax^{\beta}} \begin{Bmatrix} \sin(bx^{\gamma}) \\ \cos(bx^{\gamma}) \end{Bmatrix} dx$ $\left[\begin{array}{l} (\operatorname{Re} a > 0;\ \operatorname{Im} b = 0;\ \alpha > 0;\ \beta < 0;\ \operatorname{Re}\alpha < \gamma - 1) \text{ or} \\ (\operatorname{Re} a < 0;\ \operatorname{Im} b = 0;\ \beta,\ \gamma > 0) \text{ or} \\ (\operatorname{Re} a = \operatorname{Im} b = 0;\ \alpha,\ \beta > 0;\ \operatorname{Re}\alpha < \beta + 1,\ \gamma + 1) \end{array}\right]$.

The Cauchy principal value of the integral

$$\int_a^b f(x)\, dx$$

with a singular point $x = c \in (a,\, b)$ is defined as

$$\lim_{\varepsilon \to 0}\left(\int_a^{c-\varepsilon} f(x)\, dx + \int_{c+\varepsilon}^b f(x)\, dx\right);$$

for example,

$$\int_a^b \frac{1}{x - c}\, dx = \ln\frac{b - c}{c - a}; \quad 0 < a < c < b.$$

III. Convergence at $x = c$ (Cauchy principal value):

9. $\displaystyle\int_a^b \frac{1}{x^r - c^r}\, dx$ $[0 < a < c < b]$.

Asymptotic analogues of elementary and special functions

Definition. A set of functions $\{f_1(z),\, f_2(z),\, \ldots,\, f_n(z)\}$, such that the integral

$$\int_a^b f(x)\, g(x)\, dx$$

converges or diverges at a point $x = c$ simultaneously with all integrals

$$\int_a^b f_i(x)\, g(x)\, dx,$$

is called asymptotic analogue of the function $f(x)$ at the point $x = c$. Note that asymptotic analogue is not the main term of asymptotics, though in some cases it can coincide with it. We use the notation

$$f(z) \Longrightarrow \{f_1(z),\, f_2(z),\, \ldots,\, f_n(z)\}.$$

For $z \to \infty$, the functions $\sin z$ and $\sinh z$ in asymptotic analogues can be replaced with $\cos z$ and $\cosh z$, respectively.

For example, for the error function $\operatorname{erf}(z)$ that has asymptotic behaviour of the form

$$\operatorname{erf}(z) \sim \frac{2z}{\sqrt{\pi}} - \frac{2z^3}{3\sqrt{\pi}} + \ldots, \qquad z \to 0,$$

$$\operatorname{erf}(z) \sim 1 + e^{-z^2}\left(-\frac{1}{\sqrt{\pi} z} + \frac{1}{2\sqrt{\pi} z^3} + \ldots\right), \qquad z \to \infty,$$

we write

$$\operatorname{erf}(z) \Longrightarrow \left[\begin{array}{ll} z, & z \to 0, \\ \left\{1,\, \dfrac{e^{-z^2}}{z}\right\}, & z \to \infty. \end{array}\right.$$

More examples:

$$\sin(az) \Longrightarrow \left[\begin{array}{ll} z, & z \to 0, \\ \sin(az), & |z| \to \infty;\ \operatorname{Im}(az) = 0, \\ e^{|\operatorname{Im}(az)|}, & |z| \to \infty;\ \operatorname{Im}(az) \neq 0. \end{array}\right.$$

$$J_\nu(az) \Longrightarrow \left[\begin{array}{ll} z^\nu, & z \to 0, \\ \dfrac{\sin(az)}{\sqrt{z}}, & |z| \to \infty;\ \operatorname{Im}(az) = 0, \\ \dfrac{e^{|\operatorname{Im}(az)|}}{\sqrt{z}}, & |z| \to \infty;\ \operatorname{Im}(az) \neq 0. \end{array}\right.$$

Table of asymptotic analogues

$$(a+bz^r)^s \Longrightarrow \left[\begin{array}{ll} 1, & z \to 0;\ r > 0, \\ z^{rs}, & z \to 0;\ r < 0, \\ \left\{\left[a + \left(\left(-\frac{a}{b}\right)^{1/r}\right)^r b\right]^s,\ \left[z - \left(-\frac{a}{b}\right)^{1/r}\right]^s\right\}, & z \to \left(-\frac{a}{b}\right)^{1/r}, \\ z^{rs}, & |z| \to \infty;\ r > 0, \\ 1, & |z| \to \infty;\ r < 0. \end{array}\right.$$

$$\sqrt{bz^r} \Longrightarrow \left[\begin{array}{ll} z^{r/2}, & z \to 0, \\ z^{r/2}, & |z| \to \infty. \end{array}\right.$$

$$\operatorname{Ai}(z) \Longrightarrow \left[\begin{array}{ll} 1, & z \to 0, \\ \dfrac{e^{-2z^{3/2}/3}}{\sqrt[4]{z}}, & |z| \to \infty;\ -\dfrac{2\pi}{3} < \arg z \leq \dfrac{2\pi}{3}, \\ \left\{\dfrac{e^{-2z^{3/2}/3}}{\sqrt[4]{z}},\ \dfrac{e^{2z^{3/2}/3}}{\sqrt[4]{z}}\right\}, & |z| \to \infty;\ \text{otherwise}. \end{array}\right.$$

$$\operatorname{Ai}'(z) \Longrightarrow \left[\begin{array}{ll} 1, & z \to 0, \\ \sqrt[4]{z}e^{-2z^{3/2}/3}, & |z| \to \infty;\ -\dfrac{2\pi}{3} < \arg z \leq \dfrac{2\pi}{3}, \\ \left\{\sqrt[4]{z}\, e^{-2z^{3/2}/3},\ \sqrt[4]{z}\, e^{2z^{3/2}/3}\right\}, & |z| \to \infty;\ \text{otherwise}. \end{array}\right.$$

$$\arccos z \Longrightarrow \left[\begin{array}{ll} 1, & z \to 0, \\ \sqrt{1-z}, & z \to 1, \\ 1, & z \to -1, \\ \{1,\ \ln z\}, & |z| \to \infty. \end{array}\right.$$

$$\operatorname{arccosh} z \Longrightarrow \left[\begin{array}{ll} 1, & z \to 0, \\ \sqrt{z-1}, & z \to 1, \\ 1, & z \to -1, \\ \{1,\ \ln z\}, & |z| \to \infty. \end{array}\right.$$

$$\operatorname{arccot} z \Longrightarrow \left[\begin{array}{ll} 1, & z \to 0, \\ \{1,\ \ln(z-i)\}, & z \to i, \\ \{1,\ \ln(z+i)\}, & z \to -i, \\ \dfrac{1}{z}, & |z| \to \infty. \end{array}\right.$$

$$\operatorname{arccoth} z \Longrightarrow \left[\begin{array}{ll} 1, & z \to 0, \\ \{1,\ \ln(z-1)\}, & z \to 1, \\ \{1,\ \ln(z+1)\}, & z \to -1, \\ \dfrac{1}{z}, & |z| \to \infty. \end{array}\right.$$

$$\operatorname{arccsc} z \Longrightarrow \left[\begin{array}{ll} \{1,\ \ln z\}, & z \to 0, \\ 1, & z \to 1, \\ 1, & z \to -1, \\ \dfrac{1}{z}, & |z| \to \infty. \end{array}\right.$$

$$\operatorname{arccsch} z \Longrightarrow \left[\begin{array}{ll} \{1,\ \ln z\}, & z \to 0, \\ 1, & z \to i, \\ 1, & z \to -i, \\ \dfrac{1}{z}, & |z| \to \infty. \end{array}\right.$$

$$\operatorname{arcsec} z \Longrightarrow \left[\begin{array}{ll} \{1,\ \ln z\}, & z \to 0, \\ \sqrt{z-1}, & z \to 1, \\ 1, & z \to -1, \\ 1, & |z| \to \infty. \end{array}\right.$$

$$\operatorname{arcsech} z \Longrightarrow \left[\begin{array}{ll} \{1,\ \ln z\}, & z \to 0, \\ \sqrt{1-z}, & z \to 1, \\ 1, & z \to -1, \\ 1, & |z| \to \infty. \end{array}\right.$$

$$\arcsin z \Longrightarrow \left[\begin{array}{ll} z, & z \to 0, \\ 1, & z \to 1, \\ 1, & z \to -1, \\ \{1,\ \ln z\}, & |z| \to \infty. \end{array}\right.$$

$$\operatorname{arcsinh} z \Longrightarrow \begin{bmatrix} z, & z \to 0, \\ 1, & z \to i, \\ 1, & z \to -i, \\ \{1,\ \ln z\}, & |z| \to \infty. \end{bmatrix}$$

$$\arctan z \Longrightarrow \begin{bmatrix} z, & z \to 0, \\ \{1,\ \ln(z-i)\}, & z \to i, \\ \{1,\ \ln(z+i)\}, & z \to -i, \\ 1, & |z| \to \infty. \end{bmatrix}$$

$$\operatorname{arctanh} z \Longrightarrow \begin{bmatrix} z, & z \to 0, \\ \{1,\ \ln(1-z)\}, & z \to 1, \\ \{1,\ \ln(1+z)\}, & z \to -1, \\ 1, & |z| \to \infty. \end{bmatrix}$$

$$\operatorname{bei}_\nu(z) \Longrightarrow \begin{bmatrix} \left\{\nu z^\nu,\ z^{\nu+2}\right\}, & z \to 0, \\ \left\{\dfrac{e^{(-1)^{1/4}z}}{\sqrt{z}},\ \dfrac{e^{(-1)^{3/4}z}}{\sqrt{z}},\ \dfrac{e^{-(-1)^{1/4}z}}{\sqrt{z}},\ \dfrac{e^{-(-1)^{3/4}z}}{\sqrt{z}}\right\}, & |z| \to \infty. \end{bmatrix}$$

$$\operatorname{ber}_\nu(z) \Longrightarrow \begin{bmatrix} z^\nu, & z \to 0, \\ \left\{\dfrac{e^{(-1)^{1/4}z}}{\sqrt{z}},\ \dfrac{e^{(-1)^{3/4}z}}{\sqrt{z}},\ \dfrac{e^{-(-1)^{1/4}z}}{\sqrt{z}},\ \dfrac{e^{-(-1)^{3/4}z}}{\sqrt{z}}\right\}, & |z| \to \infty. \end{bmatrix}$$

$$\operatorname{Bi}(z) \Longrightarrow \begin{bmatrix} 1, & z \to 0, \\ \left\{\dfrac{e^{-2z^{3/2}/3}}{\sqrt[4]{z}},\ \dfrac{e^{2z^{3/2}/3}}{\sqrt[4]{z}}\right\}, & |z| \to \infty. \end{bmatrix}$$

$$\operatorname{Bi}'(z) \Longrightarrow \begin{bmatrix} 1, & z \to 0, \\ \left\{\sqrt[4]{z}\,e^{-2z^{3/2}/3},\ \sqrt[4]{z}\,e^{2z^{3/2}/3}\right\}, & |z| \to \infty. \end{bmatrix}$$

$$C(z) \Longrightarrow \begin{bmatrix} \sqrt{z}, & z \to 0, \\ \left\{1,\ \dfrac{\sin z}{\sqrt{z}}\right\}, & |z| \to \infty. \end{bmatrix}$$

$$C_n^\lambda(z) \Longrightarrow \begin{bmatrix} z^{n-2[n/2]}, & z \to 0, \\ z^n, & |z| \to \infty. \end{bmatrix}$$

$$\operatorname{chi}(z) \Longrightarrow \begin{bmatrix} \{1,\ \ln z\}, & z \to 0, \\ \left\{1,\ \dfrac{\sinh z}{z}\right\}, & |z| \to \infty;\ \arg z = \pi/2, \\ \dfrac{\sinh z}{z}, & |z| \to \infty;\ \arg z \neq \pi/2. \end{bmatrix}$$

$$\operatorname{ci}(z) \Longrightarrow \left[\begin{array}{ll} \{1,\ \ln z\}, & z \to 0, \\ \left\{1,\ \dfrac{\sin z}{z}\right\}, & z \to -\infty, \\ \dfrac{\sin z}{z}, & |z| \to \infty;\ \arg z \neq \pi. \end{array}\right.$$

$$\cos z \Longrightarrow \left[\begin{array}{ll} 1, & z \to 0, \\ \cos z, & z \to \infty;\ \operatorname{Im} z = 0, \\ e^{|\operatorname{Im} z|}, & |z| \to \infty;\ \operatorname{Im} z \neq 0. \end{array}\right.$$

$$\cosh z \Longrightarrow \left[\begin{array}{ll} 1, & z \to 0, \\ \cosh z, & z \to \infty;\ \operatorname{Re} z = 0, \\ e^{|\operatorname{Re} z|}, & |z| \to \infty;\ \operatorname{Re} z \neq 0. \end{array}\right.$$

$$\cot z \Longrightarrow \left[\begin{array}{ll} \dfrac{1}{z}, & z \to 0, \\ \cot z, & z \to \infty;\ \operatorname{Im} z = 0, \\ \dfrac{1}{z - n\pi} & z \to n\pi;\ n = 0,\ \pm 1,\ \pm 2, \ldots \\ 1, & |z| \to \infty;\ \operatorname{Im} z \neq 0. \end{array}\right.$$

$$\coth z \Longrightarrow \left[\begin{array}{ll} \dfrac{1}{z}, & z \to 0, \\ \coth z, & z \to \infty;\ \operatorname{Re} z = 0, \\ \dfrac{1}{z - n\pi i}, & z \to n\pi i;\ n = 0,\ \pm 1,\ \pm 2, \ldots, \\ 1, & |z| \to \infty;\ \operatorname{Re} z \neq 0. \end{array}\right.$$

$$\csc z \Longrightarrow \left[\begin{array}{ll} \dfrac{1}{z}, & z \to 0, \\ \dfrac{1}{z - n\pi}, & z \to n\pi;\ n = 0,\ \pm 1,\ \pm 2, \ldots, \\ \csc z, & z \to \infty;\ \operatorname{Im} z = 0, \\ e^{-|\operatorname{Im} z|}, & |z| \to \infty;\ \operatorname{Im} z \neq 0. \end{array}\right.$$

$$\operatorname{csch} z \Longrightarrow \left[\begin{array}{ll} \dfrac{1}{z}, & z \to 0, \\ \dfrac{1}{z - n\pi i}, & z \to n\pi i,\ n = 0,\ \pm 1,\ \pm 2, \ldots, \\ \operatorname{csch} z, & z \to \infty;\ \operatorname{Re} z = 0, \\ e^{-|\operatorname{Re} z|}, & |z| \to \infty;\ \operatorname{Re} z \neq 0. \end{array}\right.$$

$$\mathbf{D}(z) \Longrightarrow \left[\begin{array}{ll} 1, & z \to 0, \\ \{1,\ \ln(1 - z)\}, & z \to 1, \\ \{1,\ \ln(1 + z)\}, & z \to -1, \\ \dfrac{1}{z}, & |z| \to \infty. \end{array}\right.$$

$$D_\nu(z) \Longrightarrow \begin{bmatrix} 1, & z \to 0, \\ z^\nu e^{-z^2/4}, & |z| \to \infty;\ -\dfrac{\pi}{2} < \arg z \le \dfrac{\pi}{2}, \\ \left\{ z^\nu e^{-z^2/4},\ \dfrac{z^{-\nu-2} e^{z^2/4}}{\Gamma(-\nu)} \right\}, & |z| \to \infty\ . \end{bmatrix}$$

$$\mathbf{E}_\nu(z) \Longrightarrow \begin{bmatrix} 1, & z \to 0, \\ \left\{ \dfrac{1}{z},\ \dfrac{\cos z}{\sqrt{z}} \right\}, & |z| \to \infty;\ \operatorname{Im} z = 0;\ \nu = \pm\dfrac{1}{2},\ \pm\dfrac{3}{2}, \ldots, \\ \left\{ \dfrac{1}{z},\ \dfrac{e^{|\operatorname{Im} z|}}{\sqrt{z}} \right\} & |z| \to \infty;\ \operatorname{Im} z \neq 0;\ \nu \neq \pm\dfrac{1}{2},\ \pm\dfrac{3}{2}, \ldots \end{bmatrix}$$

$$\mathbf{E}(z) \Longrightarrow \begin{bmatrix} 1, & z \to 0, \\ 1, & z \to 1, \\ 1, & z \to -1, \\ z, & |z| \to \infty. \end{bmatrix}$$

$$\operatorname{Ei}(z) \Longrightarrow \begin{bmatrix} \{1,\ \ln z\}, & z \to 0, \\ \dfrac{e^z}{z}, & |z| \to \infty;\ \operatorname{Im} z = 0, \\ \left\{ 1,\ \dfrac{e^z}{z} \right\}, & |z| \to \infty;\ \operatorname{Im} z \neq 0, \ldots \end{bmatrix}$$

$$e^z \Longrightarrow \begin{bmatrix} 1, & z \to 0, \\ \cos(\operatorname{Im} z), & |z| \to \infty;\ \operatorname{Re} z = 0, \\ e^z, & |z| \to \infty;\ \operatorname{Re} z \neq 0. \end{bmatrix}$$

$$\operatorname{erf}(z) \Longrightarrow \begin{bmatrix} z, & z \to 0, \\ \left\{ 1,\ \dfrac{e^{-z^2}}{z} \right\}, & |z| \to \infty. \end{bmatrix}$$

$$\operatorname{erfc}(z) \Longrightarrow \begin{bmatrix} 1, & z \to 0, \\ \dfrac{e^{-z^2}}{z}, & |z| \to \infty;\ -\dfrac{\pi}{2} < \arg z \le \dfrac{\pi}{2}, \\ \left\{ 1,\ \dfrac{e^{-z^2}}{z} \right\} & |z| \to \infty;\ \text{otherwise}. \end{bmatrix}$$

$$\operatorname{erfi}(z) \Longrightarrow \begin{bmatrix} z, & z \to 0, \\ \left\{ 1,\ \dfrac{e^{z^2}}{z} \right\}, & |z| \to \infty. \end{bmatrix}$$

$${}_0F_1(b;\ z) \Longrightarrow \begin{bmatrix} 1, & z \to 0, \\ z^{(1-2b)/4} \cos\left(2\sqrt{-z}\right), & z \to -\infty, \\ z^{(1-2b)/4} e^{2|\operatorname{Im}(\sqrt{-z})|}, & |z| \to \infty;\ \arg z \neq \pi. \end{bmatrix}$$

$$
{}_1F_1\left(\begin{matrix} a;\, z \\ b \end{matrix}\right) \Longrightarrow \left[\begin{array}{ll} 1, & z \to 0, \\ \left\{z^{-a},\, z^{a-b}e^{z}\right\}, & |z| \to \infty. \end{array}\right.
$$

$$
{}_2F_1\left(\begin{matrix} a,\, b \\ c;\, z \end{matrix}\right) \Longrightarrow \left[\begin{array}{ll} 1, & z \to 0, \\ \left\{1,\, (1-z)^{c-a-b}\right\}, & |z| \to 1;\, c-a-b \neq 0, \\ \{1,\, \ln(1-z)\}, & |z| \to 1;\, c-a-b = 0, \\ \left\{z^{-a},\, z^{-b}\right\}, & |z| \to \infty. \end{array}\right.
$$

$$
{}_3F_2\left(\begin{matrix} a_1,\, a_2,\, a_3 \\ b_1,\, b_2;\, z \end{matrix}\right) \Longrightarrow \left[\begin{array}{ll} 1, & z \to 0, \\ \left\{1,\, (1-z)^{b_1+b_2-a_1-a_2-a_3}\right\}, & z \to 1;\, b_1+b_2-a_1-a_2-a_3 \neq 0, \\ \{1,\, \ln(1-z)\}, & z \to 1;\, b_1+b_2-a_1-a_2-a_3 = 0, \\ \left\{z^{-a_1},\, z^{-a_2},\, z^{-a_3}\right\}, & |z| \to \infty. \end{array}\right.
$$

$$
{}_1F_2\left(\begin{matrix} a_1;\, z \\ b_1,\, b_2 \end{matrix}\right) \Longrightarrow \left[\begin{array}{ll} 1, & z \to 0, \\ \left\{z^{-a_1},\, z^{(2a_1-2b_1-2b_2+1)/4}\cos\left(2\sqrt{-z}\right)\right\}, & z \to -\infty, \\ \left\{z^{-a_1},\, z^{(2a_1-2b_1-2b_2+1)/4}e^{2|\mathrm{Im}(\sqrt{-z})|}\right\}, & |z| \to \infty;\, \arg z \neq \pi. \end{array}\right.
$$

$$
{}_pF_q\left(\begin{matrix} (a_p)\,;\, z \\ (b_q) \end{matrix}\right) \Longrightarrow \left[\begin{array}{ll} 1, & z \to 0, \\ \left\{1,\, (1-z)^{\sum_{j=1}^{q} b_j - \sum_{i=1}^{q+1} a_i}\right\}, & z \to 1;\, p = q+1; \\ & \sum\limits_{j=1}^{q} b_j - \sum\limits_{i=1}^{q+1} a_i \neq 0, \\ \{1,\, \ln(1-z)\}, & z \to 1;\, p = q+1; \\ & \sum\limits_{j=1}^{q} b_j - \sum\limits_{i=1}^{q+1} a_i = 0, \\ \left\{z^{-a_1},\, z^{-a_2},\, \ldots,\, z^{-a_p}\right\}, & |z| \to \infty;\, p = q+1, \\ \left\{z^{-a_1},\, z^{-a_2},\, \ldots,\, z^{-a_p},\, z^{\chi}e^{z}\right\}, & |z| \to \infty;\, p = q, \\ \left\{z^{-a_1},\, z^{-a_2},\, \ldots,\, z^{-a_p},\, z^{\chi}\cos\left(2\sqrt{-z}\right)\right\} & |z| \to \infty;\, p = q-1, \\ \left\{z^{-a_1},\, z^{-a_2},\, \ldots,\, z^{-a_p},\right. & |z| \to \infty;\, p < q-1, \\ \quad \left. z^{\chi}\exp\left[(q-p+1)\, z^{1/(q-p+1)}\right]\right\}, & \\ \quad \chi = \dfrac{1}{q-p+1}\left(\dfrac{q-p}{2} + \sum\limits_{i=1}^{p} a_i - \sum\limits_{j=1}^{q} b_j\right). & \end{array}\right.
$$

$$G^{m,n}_{p,q}\left(z\,\middle|\,\begin{matrix}(a_p)\\(b_q)\end{matrix}\right)\Longrightarrow\left[\begin{array}{ll}\{z^{b_1},\,z^{b_2},\ldots,z^{b_m}\}, & z\to0;\ p=q,\\ \{z^{b_1},\,z^{b_2},\ldots,z^{b_m},\ z^{\chi}\exp\left[(-1)^{q-m-n}z^{-1}\right]\}, & z\to0;\ p=q+1,\\ \{z^{b_1},\,z^{b_2},\ldots,z^{b_m}, & z\to0;\ p=q+2,\\ \qquad z^{\chi}\cos\left(2\sqrt{(-1)^{q-m-n-1}z^{-1}}\right)\}, & \\ \Big\{z^{b_1},\,z^{b_2},\ldots,z^{b_m}, & \\ \qquad z^{\chi}\exp\left[(p-q)\,(-z)^{1/(q-p)}\right]\Big\}, & z\to0;\ p>q+2,\\ \Big\{1,\ (1-(-1)^{p-m-n}z)^{\sum_{i=1}^{p}(a_i-b_i)-1}\Big\}, & z\to(-1)^{m+n-p};\\ & p=q;\\ & \sum\limits_{i=1}^{p}(a_i-b_i)\neq1,\\ \{1,\ \ln(1-(-1)^{p-m-n}z)\}, & z\to(-1)^{m+n-p};\\ & p=q;\\ & \sum_{i=1}^{p}(a_i-b_i)=1,\\ 1, & z\to(-1)^{m+n-p};\\ & p\neq q,\\ \{z^{a_1-1},\,z^{a_2-1},\ldots,z^{a_n-1}\}, & |z|\to\infty;\ p=q,\\ \{z^{a_1-1},\,z^{a_2-1},\ldots,z^{a_n-1}, & |z|\to\infty;\ p=q-1,\\ \qquad z^{\chi}\exp\left[(-1)^{p-m-n}z\right]\}, & \\ \{z^{a_1-1},\,z^{a_2-1},\ldots,z^{a_n-1}, & |z|\to\infty;\ p=q-2,\\ \qquad z^{\chi}\cos\left(2\sqrt{(-1)^{p-m-n-1}z}\right)\}, & \\ \Big\{z^{a_1-1},\,z^{a_2-1},\ldots,z^{a_n-1}, & |z|\to\infty;\ p<q-2,\\ \qquad z^{\chi}\exp\left[(q-p)\,(-z)^{1/(q-p)}\right]\Big\}, & \\ \quad\chi=\dfrac{1}{q-p}\left(\dfrac{p-q+1}{2}+\sum\limits_{j=1}^{q}b_j-\sum\limits_{i=1}^{p}a_i\right) & \end{array}\right.$$

$$\mathbf{H}_{\nu}(z)\Longrightarrow\left[\begin{array}{ll}z^{\nu+1}, & z\to0,\\ \left\{z^{\nu-1},\ \dfrac{\cos z}{\sqrt{z}}\right\}, & |z|\to\infty;\ \operatorname{Im}z=0,\\ \left\{z^{\nu-1},\ \dfrac{e^{|\operatorname{Im}z|}}{\sqrt{z}}\right\} & |z|\to\infty;\ \operatorname{Im}z\neq0.\end{array}\right.$$

$$H^{(1)}_{\nu}(z)\Longrightarrow\left[\begin{array}{ll}\{z^{\nu},\,z^{-\nu}\}, & z\to0,\\ \dfrac{\cos z}{\sqrt{z}}, & |z|\to\infty;\ \operatorname{Im}z=0,\\ \dfrac{e^{|\operatorname{Im}z|}}{\sqrt{z}}, & |z|\to\infty;\ \operatorname{Im}z\neq0.\end{array}\right.$$

$$H_\nu^{(2)}(z) \Longrightarrow \left[\begin{array}{ll} \{z^\nu,\, z^{-\nu}\}, & z \to 0, \\ \dfrac{\cos z}{\sqrt{z}}, & |z| \to \infty;\ \operatorname{Im} z = 0, \\ \dfrac{e^{|\operatorname{Im} z|}}{\sqrt{z}} & |z| \to \infty;\ \operatorname{Im} z \neq 0. \end{array}\right.$$

$$H_n(z) \Longrightarrow \left[\begin{array}{ll} z^{n-2[n/2]}, & z \to 0, \\ z^n, & |z| \to \infty. \end{array}\right.$$

$$I_\nu(z) \Longrightarrow \left[\begin{array}{ll} z^\nu, & z \to 0, \\ \dfrac{\cosh z}{\sqrt{z}}, & |z| \to \infty;\ \operatorname{Re} z = 0, \\ \dfrac{e^{|\operatorname{Re} z|}}{\sqrt{z}}, & |z| \to \infty;\ \operatorname{Re} z \neq 0. \end{array}\right.$$

$$J_\nu(z) \Longrightarrow \left[\begin{array}{ll} z^\nu, & z \to 0, \\ \dfrac{\cos z}{\sqrt{z}}, & |z| \to \infty;\ \operatorname{Im} z = 0, \\ \dfrac{e^{|\operatorname{Im} z|}}{\sqrt{z}}, & |z| \to \infty;\ \operatorname{Im} z \neq 0. \end{array}\right.$$

$$\mathbf{J}_\nu(z) \Longrightarrow \left[\begin{array}{ll} 1, & z \to 0, \\ \left\{\dfrac{1}{z},\, \dfrac{\cos z}{\sqrt{z}}\right\}, & |z| \to \infty;\ \operatorname{Im} z = 0, \\ \left\{\dfrac{1}{z},\, \dfrac{e^{|\operatorname{Im} z|}}{\sqrt{z}}\right\}, & |z| \to \infty;\ \operatorname{Im} z \neq 0. \end{array}\right.$$

$$\mathbf{K}(z) \Longrightarrow \left[\begin{array}{ll} 1, & z \to 0, \\ \{1,\ \ln(1-z)\}, & z \to 1, \\ \{1,\ \ln(1+z)\} & z \to -1, \\ \dfrac{\ln z}{z} & |z| \to \infty. \end{array}\right.$$

$$K_\nu(z) \Longrightarrow \left[\begin{array}{ll} \{z^\nu,\, z^{-\nu}\}, & z \to 0, \\ \dfrac{e^{-z}}{\sqrt{z}}, & |z| \to \infty. \end{array}\right.$$

$$\operatorname{kei}_\nu(z) \Longrightarrow \left[\begin{array}{ll} \{z^\nu,\, z^{-\nu}\}, & z \to 0, \\ \left\{\dfrac{e^{(-1)^{1/4} z}}{\sqrt{z}},\, \dfrac{e^{(-1)^{3/4} z}}{\sqrt{z}},\, \dfrac{e^{-(-1)^{1/4} z}}{\sqrt{z}},\, \dfrac{e^{-(-1)^{3/4} z}}{\sqrt{z}}\right\}, & |z| \to \infty. \end{array}\right.$$

$$\operatorname{ker}_\nu(z) \Longrightarrow \left[\begin{array}{ll} \{z^\nu,\, z^{-\nu}\}, & z \to 0, \\ \left\{\dfrac{e^{(-1)^{1/4} z}}{\sqrt{z}},\, \dfrac{e^{(-1)^{3/4} z}}{\sqrt{z}},\, \dfrac{e^{-(-1)^{1/4} z}}{\sqrt{z}},\, \dfrac{e^{-(-1)^{3/4} z}}{\sqrt{z}}\right\}, & |z| \to \infty. \end{array}\right.$$

$$\mathbf{L}_\nu(z) \Longrightarrow \left[\begin{array}{ll} z^{\nu+1}, & z \to 0, \\ \left\{z^{\nu-1},\, \dfrac{\cosh z}{\sqrt{z}}\right\}, & |z| \to \infty;\ \operatorname{Re} z = 0, \\ \left\{z^{\nu-1},\, \dfrac{e^{|\operatorname{Re} z|}}{\sqrt{z}}\right\}, & |z| \to \infty;\ \operatorname{Re} z \neq 0. \end{array}\right.$$

$$L_n^\lambda(z) \Longrightarrow \begin{bmatrix} 1, & z \to 0, \\ z^n, & |z| \to \infty. \end{bmatrix}$$

$$\mathrm{Li}_\nu(z) \Longrightarrow \begin{bmatrix} z, & z \to 0, \\ \left\{1,\ (z-1)^{\nu-1}\right\}, & z \to 1, \\ \{1,\ \ln^\nu z\}, & |z| \to \infty. \end{bmatrix}$$

$$\mathrm{Li}_2(z) \Longrightarrow \begin{bmatrix} z, & z \to 0, \\ 1, & z \to 1, \\ \left\{1,\ \ln^2 z\right\}, & |z| \to \infty. \end{bmatrix}$$

$$\ln(1+az^r) \Longrightarrow \begin{bmatrix} z^r, & z \to 0;\ r > 0, \\ \ln z, & z \to 0;\ r < 0, \\ \ln z, & |z| \to \infty;\ r > 0, \\ z^r, & |z| \to \infty;\ r < 0. \end{bmatrix}$$

$$M_{\rho,\sigma}(z) \Longrightarrow \begin{bmatrix} z^{\sigma+1/2}, & z \to 0, \\ \left\{z^\rho e^{-z/2},\ z^{-\rho} e^{z/2}\right\}, & |z| \to \infty. \end{bmatrix}$$

$$P_n(z) \Longrightarrow \begin{bmatrix} z^{n-2[n/2]}, & z \to 0, \\ z^n, & |z| \to \infty. \end{bmatrix}$$

$$P_\nu(z) \Longrightarrow \begin{bmatrix} \left\{\dfrac{1}{\Gamma\left(\frac{1-\nu}{2}\right)\Gamma\left(\frac{2+\nu}{2}\right)},\ \dfrac{z}{\Gamma\left(-\frac{\nu}{2}\right)\Gamma\left(\frac{1+\nu}{2}\right)}\right\}, & z \to 0, \\ 1, & z \to 1, \\ \{1,\ \ln(z+1)\} & z \to -1, \\ \left\{z^\nu,\ \dfrac{z^{-\nu-1}}{\Gamma(-\nu)}\right\}, & |z| \to \infty. \end{bmatrix}$$

$$P_\nu^\mu(z) \Longrightarrow \begin{bmatrix} 1, & z \to 0, \\ (1-z)^{-\mu/2}, & z \to 1, \\ \left\{(z+1)^{\mu/2},\ (z+1)^{-\mu/2}\right\}, & z \to -1, \\ \left\{z^\nu,\ \dfrac{z^{-\nu-1}}{\Gamma(-\nu)}\right\}, & |z| \to \infty. \end{bmatrix}$$

$$\mathrm{P}_\nu^\mu(z) \Longrightarrow \begin{bmatrix} 1, & z \to 0, \\ (z-1)^{-\mu/2}, & z \to 1, \\ \left\{(z+1)^{\mu/2},\ (z+1)^{-\mu/2}\right\}, & z \to -1, \\ \left\{z^\nu,\ \dfrac{z^{-\nu-1}}{\Gamma(-\nu)}\right\}, & |z| \to \infty. \end{bmatrix}$$

$$P_n^{(\rho,\sigma)}(z) \Longrightarrow \begin{bmatrix} 1, & z \to 0, \\ z^n, & |z| \to \infty. \end{bmatrix}$$

$$Q_\nu(z) \Longrightarrow \left[\begin{array}{ll} 1, & z \to 0, \\ \{1,\ \ln(1-z)\}, & z \to 1, \\ \{1,\ \ln(z+1)\}, & z \to -1, \\ \left\{z^\nu,\ z^{-\nu-1}\right\}, & |z| \to \infty. \end{array}\right.$$

$$Q_\nu^\mu(z) \Longrightarrow \left[\begin{array}{ll} 1, & z \to 0, \\ \left\{(1-z)^{\mu/2},\ (1-z)^{-\mu/2}\right\}, & z \to 1, \\ \left\{(z+1)^{\mu/2},\ (z+1)^{-\mu/2}\right\}, & z \to -1, \\ \left\{z^\nu,\ z^{-\nu-1}\right\}, & |z| \to \infty. \end{array}\right.$$

$$\mathrm{Q}_\nu^\mu(z) \Longrightarrow \left[\begin{array}{ll} 1, & z \to 0, \\ \left\{(z-1)^{\mu/2},\ (z-1)^{-\mu/2}\right\}, & z \to 1, \\ \left\{(z+1)^{\mu/2},\ (z+1)^{-\mu/2}\right\}, & z \to -1, \\ z^{-\nu-1}, & |z| \to \infty. \end{array}\right.$$

$$S(z) \Longrightarrow \left[\begin{array}{ll} z^{3/2}, & z \to 0, \\ \left\{1,\ \dfrac{\cos z}{\sqrt{z}}\right\}, & |z| \to \infty. \end{array}\right.$$

$$S_{\mu,\nu}(z) \Longrightarrow \left[\begin{array}{ll} z^{\mu+1}, & z \to 0, \\ \dfrac{\cos z}{\sqrt{z}}, & |z| \to \infty;\ \operatorname{Im} z = 0, \\ \dfrac{e^{|\operatorname{Im} z|}}{\sqrt{z}}, & |z| \to \infty;\ \operatorname{Im} z \neq 0. \end{array}\right.$$

$$s_{\mu,\nu}(z) \Longrightarrow \left[\begin{array}{ll} z^{\mu+1}, & z \to 0, \\ \dfrac{\cos z}{\sqrt{z}}, & |z| \to \infty;\ \operatorname{Im} z = 0, \\ \dfrac{e^{|\operatorname{Im} z|}}{\sqrt{z}}, & |z| \to \infty;\ \operatorname{Im} z \neq 0. \end{array}\right.$$

$$\sec z \Longrightarrow \left[\begin{array}{ll} 1, & z \to 0, \\ \dfrac{1}{z - \left(n+\frac{1}{2}\right)\pi}, & z \to \left(n+\dfrac{1}{2}\right)\pi;\ n = 0,\ \pm 1,\ \pm 2, \ldots, \\ \sec z, & |z| \to \infty;\ \operatorname{Im} z = 0, \\ e^{-|\operatorname{Im} z|}, & |z| \to \infty;\ \operatorname{Im} z \neq 0. \end{array}\right.$$

$$\operatorname{sech} z \Longrightarrow \left[\begin{array}{ll} 1, & z \to 0, \\ \dfrac{1}{z - \left(n+\frac{1}{2}\right)\pi i}, & z \to \left(n+\dfrac{1}{2}\right)\pi i;\ n = 0,\ \pm 1,\ \pm 2, \ldots, \\ \operatorname{sech} z, & |z| \to \infty;\ \operatorname{Im} z = 0, \\ e^{-|\operatorname{Re} z|}, & |z| \to \infty;\ \operatorname{Im} z \neq 0. \end{array}\right.$$

$$\sin z \Longrightarrow \left[\begin{array}{ll} z, & z \to 0, \\ \sin z, & |z| \to \infty;\ \operatorname{Im} z = 0, \\ e^{|\operatorname{Im} z|}, & |z| \to \infty;\ \operatorname{Im} z \neq 0. \end{array}\right.$$

$$\sinh z \Longrightarrow \left[\begin{array}{ll} z, & z \to 0, \\ \sinh z, & |z| \to \infty;\ \operatorname{Re} z = 0, \\ e^{|\operatorname{Re} z|}, & |z| \to \infty;\ \operatorname{Re} z \neq 0. \end{array}\right.$$

$$\operatorname{sinc}(z) \Longrightarrow \left[\begin{array}{ll} 1, & z \to 0, \\ \dfrac{\sin z}{z}, & |z| \to \infty;\ \operatorname{Im} z = 0, \\ \dfrac{e^{|\operatorname{Im} z|}}{z}, & |z| \to \infty;\ \operatorname{Im} z \neq 0. \end{array}\right.$$

$$\operatorname{shi}(z) \Longrightarrow \left[\begin{array}{ll} z, & z \to 0, \\ \left\{1,\ \dfrac{\cosh z}{z}\right\}, & |z| \to \infty \end{array}\right.$$

$$\operatorname{Si}(z) \Longrightarrow \left[\begin{array}{ll} z, & z \to 0, \\ \left\{1,\ \dfrac{\cos z}{z}\right\}, & |z| \to \infty \end{array}\right.$$

$$T(z, a) \Longrightarrow \left[\begin{array}{ll} 1, & z \to 0. \end{array}\right.$$

$$T_n(z) \Longrightarrow \left[\begin{array}{ll} z^{n-2[n/2]}, & z \to 0, \\ z^n, & |z| \to \infty. \end{array}\right.$$

$$\tan z \Longrightarrow \left[\begin{array}{ll} z, & z \to 0, \\ \dfrac{1}{z - \left(n + \frac{1}{2}\right)\pi}, & z \to \left(n + \dfrac{1}{2}\right)\pi;\ n = 0,\ \pm 1,\ \pm 2, \ldots, \\ \tan z, & |z| \to \infty;\ \operatorname{Im} z = 0, \\ 1, & |z| \to \infty;\ \operatorname{Im} z \neq 0. \end{array}\right.$$

$$\tanh z \Longrightarrow \left[\begin{array}{ll} z, & z \to 0, \\ \dfrac{1}{z - \left(n + \frac{1}{2}\right)\pi i}, & z \to \left(n + \dfrac{1}{2}\right)\pi i;\ n = 0,\ \pm 1,\ \pm 2, \ldots, \\ \tanh z, & |z| \to \infty;\ \operatorname{Re} z = 0, \\ 1, & |z| \to \infty;\ \operatorname{Re} z \neq 0. \end{array}\right.$$

$$U_n(z) \Longrightarrow \left[\begin{array}{ll} z^{n-2[n/2]}, & z \to 0, \\ z^n, & |z| \to \infty. \end{array}\right.$$

$$W_{\rho,\sigma}(z) \Longrightarrow \left[\begin{array}{ll} \left\{z^{\sigma+1/2},\ z^{1/2-\sigma}\right\}, & z \to 0, \\ z^{\rho} e^{-z/2}, & |z| \to \infty. \end{array}\right.$$

$$Y_\nu(z) \Longrightarrow \left[\begin{array}{ll} \{z^{\nu},\ z^{-\nu}\}, & z \to 0, \\ \dfrac{\cos z}{\sqrt{z}}, & |z| \to \infty;\ \operatorname{Im} z = 0, \\ \dfrac{e^{|\operatorname{Im} z|}}{\sqrt{z}}, & |z| \to \infty;\ \operatorname{Im} z \neq 0. \end{array}\right.$$

$$\mathrm{B}(z,\,\beta) \Longrightarrow \left[\begin{array}{ll} \dfrac{1}{z}, & z\to 0;\ \beta\neq 0,-1,-2,\ldots, \\ \dfrac{1}{z-k}, & z\to k;\ k=0,-1,-2;\ k+\beta\neq 0,-1,-2,\ldots, \\ z^{-\beta}, & |z|\to\infty. \end{array}\right.$$

$$\Gamma(z) \Longrightarrow \left[\begin{array}{ll} \dfrac{1}{z}, & z\to 0, \\ \dfrac{1}{z-n}, & z\to n;\ n=0,-1,-2,\ldots, \\ \dfrac{z^{z}}{\sqrt{z}\,e^{z}}, & |z|\to\infty. \end{array}\right.$$

$$\Gamma(\nu,\,z) \Longrightarrow \left[\begin{array}{ll} \{1,\,z^{\nu}\}, & z\to 0, \\ z^{\nu-1}e^{-z}, & |z|\to\infty. \end{array}\right.$$

$$\gamma(\nu,\,z) \Longrightarrow \left[\begin{array}{ll} z^{\nu}, & z\to 0, \\ \left\{1,\,z^{\nu-1}e^{-z}\right\}, & |z|\to\infty. \end{array}\right.$$

$$\zeta(z) \Longrightarrow \left[\begin{array}{ll} 1, & z\to 0, \\ \zeta(z), & |z|\to\infty. \end{array}\right.$$

$$\zeta(z,\,v) \Longrightarrow \left[\begin{array}{ll} 1, & z\to 0. \end{array}\right.$$

$$\theta_j(z,\,q) \Longrightarrow \left[\begin{array}{ll} z, & z\to 0;\ j=1, \\ 1, & z\to 0;\ j=2,\,3,\,4. \end{array}\right.$$

$$\Phi(z,\,s,\,v) \Longrightarrow \left[\begin{array}{ll} 1, & z\to 0, \\ \left\{\dfrac{1}{z},\,z^{-v}\ln^{s-1}z\right\}, & |z|\to\infty;\ \operatorname{Re}v,\ \operatorname{Re}s>0. \end{array}\right.$$

$$\Psi(a;\,b;\,z) \Longrightarrow \left[\begin{array}{ll} \left\{1,\,z^{1-b}\right\}, & z\to 0, \\ z^{-a}, & |z|\to\infty. \end{array}\right.$$

$$\psi(z) \Longrightarrow \left[\begin{array}{ll} \dfrac{1}{z}, & z\to 0, \\ \dfrac{1}{z+k}, & z\to -k;\ k=0,1,2,\ldots, \\ \ln z, & |z|\to\infty. \end{array}\right.$$

$$\psi^{(n)}(z) \Longrightarrow \left[\begin{array}{ll} z^{-n-1}, & z\to 0, \\ (z+k)^{-n-1}, & z\to -k;\ k,\,n=0,1,2,\ldots, \\ z^{-n}, & |z|\to\infty;\ n\neq 0. \end{array}\right.$$

Bibliography

[1] Bateman H., Erdélyi A., Magnus W., Oberhettinger F., Tricomi F. G. Tables of Integral Transforms, Vols. 1–2. McGraw–Hill, New York, 1954.

[2] Bertrand J., Bertrand P., Ovarlez J. The Transforms and Applications Handbook. Chapman & Hall/CRC, Boca Raton, 2000.

[3] Brychkov Yu. A. Handbook of Special Functions: Derivatives, Integrals, Series and Other Formulas. Chapman & Hall/CRC, Boca Raton, 2008.

[4] Brychkov Yu. A., Glaeske H.-J., Marichev O. I. Factorization of Integral Transforms of Convolution Type. Journal of Soviet Mathematics, 1985, 30:3, 2071–2094.

[5] Brychkov Yu. A., Glaeske H.-J., Prudnikov A. P., Vu Kim Tuan. Multidimensional Integral Transformations. Gordon and Breach, New York–London, 1992.

[6] Brychkov Yu. A., Prudnikov A. P. Integral Transformations of Generalized Functions. Gordon and Breach, New York–London, 1989.

[7] Colombo S. Les Transformations de Mellin et de Hankel. Centre National de la Recherche Scientifique, Paris, 1959.

[8] Colombo S., Lavoine J. Transformation de Laplace et de Mellin. Formulaires. Mode d'Utilisation. Gauthier–Villars, Paris, 1972.

[9] Debnath L., Bhatta D. Integral Transforms and Their Applications. Chapman & Hall/CRC, Boca Raton, 2015.

[10] Ditkin V. A., Prudnikov A. P. Integral Transforms and Operational Calculus. Pergamon Press, New York, 1965.

[11] Fedoryuk M. V. Integral Transforms, Analysis-1, Itogi Nauki i Tekhniki. Ser. Sovrem. Probl. Mat. Fund. Napr., 13, VINITI, Moscow, 1986, 211–253.

[12] Fikioris G. Integral Transforms and Their Applications. Morgan & Claypool, San Rafael, 2007.

[13] Gradshteyn I. S., Ryzhik I. M. Table of Integrals, Series and Products. Academic Press, New York, 2014.

[14] Marichev O. I. Handbook of Integral Transforms of Higher Transcendental Functions: Theory and Algorithmic Tables. Chichester, Ellis Horwood, 1983.

[15] Oberhettinger F. Tables of Mellin Transforms. Springer, Berlin, 1974.

[16] Paris R. B., Kaminski D. Asymptotics and Mellin–Barnes Integrals. Cambridge University Press, Cambridge, 2001.

[17] Poularikas A. D. Transforms and Applications Handbook. Chapman & Hall/CRC, Boca Raton, 2010.

[18] Prudnikov A. P., Brychkov Yu. A., Marichev O. I. Integrals and Series. Vol. 1: Elementary Functions. Gordon and Breach, New York, 1986.

[19] Prudnikov A. P., Brychkov, Yu. A., Marichev O. I. Integrals and Series. Vol. 2: Special Functions. Gordon and Breach, New York, 1986.

[20] Prudnikov A. P., Brychkov Yu. A., Marichev O. I. Integrals and Series. Vol. 3: More Special Functions. Gordon and Breach, New York, 1990.

[21] Prudnikov A. P., Brychkov Yu. A., Marichev O. I. Evaluation of Integrals and Mellin Transform, Journal of Soviet Mathematics, 1991, 54:6, pp.1239–1341.

[22] Prudnikov A. P., Brychkov Yu. A., Marichev O. I. Integrals and Series, Vol. 4: Laplace Transforms. Gordon and Breach, New York, 1992.

[23] Prudnikov A. P., Brychkov Yu. A., Marichev O. I. Integral and Series, Vol. 5: Inverse Laplace Transforms. Gordon and Breach, New York, 1992.

[24] Samko S. G., Kilbas A. A., Marichev O. I. Fractional Integrals and Derivatives. Theory and Applications. Gordon and Breach, New York, 1993.

[25] Sasiela R. J. Electromagnetic Wave Propagation in Turbulence: Evaluation and Application of Mellin Transforms. Springer, Berlin–Heidelsberg, 2012.

[26] Savischenko N. V. Special Integral Functions Used in Wireless Communications Theory. World Scientific, Singapore, 2014.

[27] Sneddon I. N. The Use of Integral Transform. McGraw–Hill, New York, 1972.

[28] Zemanian A. H. Generalized Integral Transformations, Dover Publications, New York, 1987.

Index of Notations for Functions and Constants

$\mathrm{Ai}\,(z) = \dfrac{1}{\pi}\sqrt{\dfrac{z}{3}}\,K_{1/3}\left(\dfrac{2}{3}\,z^{3/2}\right)$ is the Airy function

$\arccos z$, $\operatorname{arccot} z = \arctan\dfrac{1}{z}$, $\operatorname{arccsc} z = \arcsin\dfrac{1}{z}$, $\operatorname{arcsec} z = \arccos\dfrac{1}{z}$, $\arcsin z$, $\arctan z$ are inverse trigonometric functions

$\operatorname{arccosh} z$, $\operatorname{arcsinh} z$, $\operatorname{arctanh} z$, $\operatorname{arccsch} z = \operatorname{arcsinh}\dfrac{1}{z}$, $\operatorname{arcsech} z = \operatorname{arccosh}\dfrac{1}{z}$, $\operatorname{arccoth} z = \operatorname{arctanh}\dfrac{1}{z}$ are inverse hyperbolic functions

$\arg z$ is the argument of the complex number z, $z = |z|e^{i\arg z}$

B_n are the Bernoulli numbers

$B_n\,(z)$ are the Bernoulli polynomials

$\mathrm{bei}_\nu\,(z)$, $\mathrm{ber}_\nu\,(z)$ are the Kelvin functions

$$\mathrm{ber}_\nu\,x + i\,\mathrm{bei}_\nu\,x = J_\nu\left(e^{3\pi i/4}x\right) = e^{\nu\pi i}J_\nu\left(e^{-\pi i/4}x\right) = e^{\nu\pi i/2}I_\nu\left(e^{\pi i/4}x\right) = e^{3\nu\pi i/2}I_\nu\left(e^{-3\pi i/4}x\right)$$

$\mathrm{Bi}\,(z) = \sqrt{\dfrac{z}{3}}\left[I_{-1/3}\left(\dfrac{2}{3}\,z^{3/2}\right) + I_{1/3}\left(\dfrac{2}{3}\,z^{3/2}\right)\right]$ is the Airy function

$\mathbf{C} = -\psi\,(1) = 0{,}577\,215\,664\,9\ldots$ is the Euler constant

$C\,(z) = \dfrac{1}{\sqrt{2\pi}}\displaystyle\int_0^z \dfrac{\cos t}{\sqrt{t}}\,dt$ is the Fresnel cosine integral

$C\,(z,\,\nu) = \displaystyle\int_z^\infty t^{\nu-1}\cos t\,dt \quad [\mathrm{Re}\,\nu < 1]$ is the generalized Fresnel cosine integral

$C_n^\lambda\,(z) = \dfrac{(2\lambda)_n}{n!}\,{}_2F_1\left(-n,\,n+2\lambda;\,\lambda+\dfrac{1}{2};\,\dfrac{1-z}{2}\right)$ are the Gegenbauer polynomials

$\mathrm{chi}\,(z) = \mathbf{C} + \ln z + \displaystyle\int_0^z \dfrac{\cosh t - 1}{t}\,dt$ is the hyperbolic cosine integral

$\mathrm{ci}\,(z) = -\displaystyle\int_z^\infty \dfrac{\cos t}{t}\,dt$ is the cosine integral

$\cos z = \dfrac{e^{iz} + e^{-iz}}{2}$

$\cosh z = \dfrac{e^z + e^{-z}}{2}$

$\coth z = \dfrac{\cosh z}{\sinh z}$

$\cot z = \dfrac{\cos z}{\sin z}$

$\csc z = \dfrac{1}{\sin z}$

$\operatorname{csch} z = \dfrac{1}{\sinh z}$

$D = \dfrac{d}{dz}$, $D_a = \dfrac{d}{da}$

$\mathbf{D}(k) = \displaystyle\int_0^{\pi/2} \frac{\sin^2 t\, dt}{\sqrt{1 - k^2 \sin^2 t}}$ is the complete elliptic integral

$D(\varphi, k) = \displaystyle\int_0^{\varphi} \frac{\sin^2 t\, dt}{\sqrt{1 - k^2 \sin^2 t}}$ is the elliptic integral

$D_\nu(z) = 2^{\nu/2} e^{-z^2/4}\, \Psi\left(-\dfrac{\nu}{2}, \dfrac{1}{2}; \dfrac{z^2}{2}\right)$ is the parabolic cylinder function

$\mathbf{E}(k) = \displaystyle\int_0^{\pi/2} \sqrt{1 - k^2 \sin^2 t}\, dt$ is the complete elliptic integral of the second kind

E_n are the Euler numbers

$E_n(z)$ are the Euler polynomials

$E_\nu(z) = \displaystyle\int_1^{\infty} \frac{e^{-zt}}{t^\nu}\, dt$ $\quad [\operatorname{Re} z > 0]$ is the exponential E-integral

$\mathbf{E}_\nu(z) = \dfrac{1}{\pi} \displaystyle\int_0^{\pi} \sin(\nu t - z \sin t)\, dt$ is the Weber function

$E_\rho(z; \mu) = \displaystyle\sum_{k=0}^{\infty} \frac{z^k}{\Gamma(\mu + \rho^{-1} k)}$ $\quad [\rho > 0]$ is the Mittag–Leffler function

$\operatorname{Ei}(z) = \displaystyle\int_{-\infty}^{z} \frac{e^t}{t}\, dt$ is the exponential integral

$\operatorname{erf}(z) = \dfrac{2}{\sqrt{\pi}} \displaystyle\int_0^{z} e^{-t^2}\, dt$ is the error function

$\operatorname{erfc}(z) = 1 - \operatorname{erf}(z) = \dfrac{2}{\sqrt{\pi}} \displaystyle\int_z^{\infty} e^{-t^2}\, dt$ is the complementary error function

$\operatorname{erfi}(z) = \dfrac{2}{\sqrt{\pi}} \displaystyle\int_0^{z} e^{t^2}\, dt$ is the error function of imaginary argument

$${}_1F_1\left(\begin{matrix} a;\, z \\ b \end{matrix}\right) \equiv {}_1F_1\left(\begin{matrix} a \\ b;\, z \end{matrix}\right) \equiv {}_1F_1(a;\, b;\, z) = \sum_{k=0}^{\infty} \frac{(a)_k\, z^k}{(b)_k\, k!}$$

is the Kummer confluent hypergeometric function

$${}_2F_1\left(\begin{matrix} a,\, b;\, z \\ c \end{matrix}\right) \equiv {}_2F_1\left(\begin{matrix} a,\, b \\ c;\, z \end{matrix}\right) \equiv {}_2F_1(a,\, b;\, c;\, z) = \sum_{k=0}^{\infty} \frac{(a)_k\, (b)_k}{(c)_k} \frac{z^k}{k!} \qquad [|z| < 1],$$

$$= \frac{\Gamma(c)}{\Gamma(a)\,\Gamma(c - b)} \int_0^1 t^{b-1} (1 - t)^{c-b-1} (1 - tz)^{-a}\, dt$$

$$[\operatorname{Re} c > \operatorname{Re} b > 0;\ |\arg(1 - z)| < \pi]$$

is the Gauss hypergeometric function

$${}_pF_q\left(\begin{matrix} (a_p);\, z \\ (b_q) \end{matrix}\right) \equiv {}_pF_q\left(\begin{matrix} (a_p) \\ (b_q);\, z \end{matrix}\right) \equiv {}_pF_q((a_p);\, (b_q);\, z)$$

$$\equiv {}_pF_q(a_1, \ldots, a_p;\ b_1, \ldots, b_q;\ z) = \sum_{k=0}^{\infty} \frac{(a_1)_k\, (a_2)_k \cdots (a_p)_k}{(b_1)_k\, (b_2)_k \cdots (b_q)_k} \frac{z^k}{k!}$$

is the generalized hypergeometric function

$F_j^{(n)}(\ldots;\ \ldots;\ z_1,\ldots,z_n)$ $\quad [j = A, B, C, D]$ are the Lauricella functions:

$$F_A^{(n)}(a,\ b_1,\ldots,b_n;\ c_1,\ldots,c_n;\ z_1,\ldots,z_n) = \\ = \sum_{k_1,\ldots,k_n=0}^{\infty} \frac{(a)_{k_1+\ldots+k_n}(b_1)_{k_1}\ldots(b_n)_{k_n}}{(c_1)_{k_1}\ldots(c_n)_{k_n}} \frac{z_1^{k_1}\ldots z_n^{k_n}}{k_1!\ldots k_n!}, \qquad \left[\sum_{j=1}^{n}|z_j| < 1\right]$$

$$F_B^{(n)}(a_1,\ldots,a_n,\ b_1,\ldots,b_n;\ c;\ z_1,\ldots,z_n) = \\ = \sum_{k_1,\ldots,k_n=0}^{\infty} \frac{(a_1)_{k_1}\ldots(a_n)_{k_n}(b_1)_{k_1}\ldots(b_n)_{k_n}}{(c)_{k_1+\ldots+k_n}} \frac{z_1^{k_1}\ldots z_n^{k_n}}{k_1!\ldots k_n!}, \qquad [|z_j| < 1,\ j = 1, 2, \ldots, n]$$

$$F_C^{(n)}(a,\ b;\ c_1,\ldots,c_n;\ z_1,\ldots,z_n) = \sum_{k_1,\ldots,k_n=0}^{\infty} \frac{(a)_{k_1+\ldots+k_n}(b)_{k_1+\ldots+k_n}}{(c_1)_{k_1}\ldots(c_n)_{k_n}} \frac{z_1^{k_1}\ldots z_n^{k_n}}{k_1!\ldots k_n!}, \qquad \left[\sum_{j=1}^{n}\sqrt{|z_j|} < 1\right]$$

$$F_D^{(n)}(a,\ b_1,\ldots,b_n;\ c;\ z_1,\ldots,z_n) = \sum_{k_1,\ldots,k_n=0}^{\infty} \frac{(a)_{k_1+\ldots+k_n}(b_1)_{k_1}\ldots(b_n)_{k_n}}{(c)_{k_1+\ldots+k_n}} \frac{z_1^{k_1}\ldots z_n^{k_n}}{k_1!\ldots k_n!}, \qquad [|z_j| < 1,\ j = 1, 2, \ldots, n]$$

$F_j(\ldots;\ w,\ z)$ $\quad [j = 1, 2, 3, 4]$ are the Appell functions:

$$F_1(a,\ b,\ b';\ c;\ w,\ z) = \sum_{k,\ell=0}^{\infty} \frac{(a)_{k+\ell}(b)_k(b')_\ell}{(c)_{k+\ell}} \frac{w^k z^\ell}{k!\,\ell!} \qquad [|w|,\ |z| < 1],$$

$$F_2(a,\ b,\ b';\ c,\ c';\ w,\ z) = \sum_{k,\ell=0}^{\infty} \frac{(a)_{k+\ell}(b)_k(b')_\ell}{(c)_k(c')_\ell} \frac{w^k z^\ell}{k!\,\ell!} \qquad [|w| + |z| < 1],$$

$$F_3(a,\ a',\ b,\ b';\ c;\ w,\ z) = \sum_{k,\ell=0}^{\infty} \frac{(a)_k(a')_\ell(b)_k(b')_\ell}{(c)_{k+\ell}} \frac{w^k z^\ell}{k!\,\ell!} \qquad [|w|,\ |z| < 1],$$

$$F_4(a,\ b;\ c,\ c';\ w,\ z) = \sum_{k,\ell=0}^{\infty} \frac{(a)_{k+\ell}(b)_{k+\ell}}{(c)_k(c')_\ell} \frac{w^k z^\ell}{k!\,\ell!} \qquad \left[\sqrt{|w|} + \sqrt{|z|} < 1\right]$$

$\mathbf{G} = \sum_{k=0}^{\infty} \frac{(-1)^k}{(2k+1)^2} = 0{,}915\,965\,594\,2\ldots$ is the Catalan constant

$$G_{pq}^{mn}\left(z \,\middle|\, \begin{matrix}(a_p)\\(b_q)\end{matrix}\right) \equiv G_{pq}^{mn}\left(z \,\middle|\, \begin{matrix}a_1,\ldots,a_p\\b_1,\ldots,b_q\end{matrix}\right) \\ = \frac{1}{2\pi i}\int_L \frac{\Gamma(b_1+s)\ldots\Gamma(b_m+s)\,\Gamma(1-a_1-s)\ldots\Gamma(1-a_n-s)}{\Gamma(a_{n+1}+s)\ldots\Gamma(a_p+s)\,\Gamma(1-b_{m+1}-s)\ldots\Gamma(1-b_q-s)}\, z^{-s}\,ds,$$

$L = L_{\pm\infty},\ L_{i\infty}$ is the Meijer G-function

$\mathbf{H}_\nu(z) = \frac{2}{\sqrt{\pi}}\left(\frac{z}{2}\right)^{\nu+1} \frac{1}{\Gamma\left(\nu+\frac{3}{2}\right)}\, {}_1F_2\left(\begin{matrix}1;\ -\frac{z^2}{4}\\ \frac{3}{2},\ \nu+\frac{3}{2}\end{matrix}\right)$ is the Struve function

$H_\nu^{(1)}(z)$, $H_\nu^{(2)}(z)$ are the Hankel functions of the first and second kind (the Bessel functions of the third kind $H_\nu^{(1)}(z) = J_\nu(z) + i\,Y_\nu(z)$, $H_\nu^{(2)}(z) = J_\nu(z) - i\,Y_\nu(z)$)

$H_n(z) = (-1)^n e^{z^2} \frac{d^n}{dz^n} e^{-z^2}$ are the Hermite polynomials

$\mathcal{H}_\nu(z, a, b) = \frac{(1-a^2)^\nu}{2\pi} \int_0^b \frac{e^{-z^2(t^2+1)/[2(a^2t^2+1)]}}{(t^2+1)(a^2t^2+1)^\nu} \, dt$

$I_\nu(z) = \frac{1}{\Gamma(\nu+1)} \left(\frac{z}{2}\right)^\nu {}_0F_1\left(\nu+1; \frac{z^2}{4}\right) = e^{-\nu\pi i/2} J_\nu\left(e^{\pi i/2} z\right)$ is the modified Bessel function of the first kind

$J_\nu(z) = \frac{1}{\Gamma(\nu+1)} \left(\frac{z}{2}\right)^\nu {}_0F_1\left(\nu+1; -\frac{z^2}{4}\right)$ is the Bessel function of the first kind

$\mathbf{J}_\nu(z) = \frac{1}{\pi} \int_0^\pi \cos(\nu t - z \sin t) \, dt$ is the Anger function

$J_\nu^\mu(z) = \sum_{k=0}^\infty \frac{(-z)^k}{k!\,\Gamma(k\mu+\nu+1)}$ $[\mu > -1]$ is the Bessel–Maitland function

$J_{\nu,\lambda}^\mu(z) = \sum_{k=0}^\infty \frac{(-1)^k (z/2)^{2k+2\lambda+\nu}}{\Gamma(k+\lambda+1)\,\Gamma(k\mu+\nu+\lambda+1)}$ $[\mu > 0]$ is the generalized Bessel–Maitland function

$Ji_\nu(z) = \int_z^\infty \frac{J_\nu(t)}{t} \, dt$ is the integral Bessel function of the first kind

$\mathbf{K}(k) = \int_0^{\pi/2} \frac{dt}{\sqrt{1 - k^2 \sin^2 t}}$ is the complete elliptic integral of the first kind

$K_\nu(z) = \frac{\pi[I_{-\nu}(z) - I_\nu(z)]}{2 \sin \nu\pi}$ $[\nu \neq n]$, $K_n(z) = \lim_{\nu \to n} K_\nu(z)$ $[n = 0, \pm 1, \pm 2, \ldots]$

is the Macdonald function (the modified Bessel function of the third kind)

$\mathrm{kei}_\nu(z)$, $\mathrm{ker}_\nu(z)$ are the Kelvin functions

$$\mathrm{ker}_\nu x + i\,\mathrm{kei}_\nu x = e^{-\nu\pi i/2} K_\nu\left(e^{\pi i/4} x\right) = \frac{1}{2}\pi i H_\nu^{(1)}\left(e^{3\pi i/4} x\right) = -\frac{1}{2}\pi i e^{-\nu\pi i} H_\nu^{(2)}\left(e^{-\pi i/4} x\right)$$

$Ki_\nu(z) = \int_z^\infty \frac{K_\nu(t)}{t} \, dt$ is the modified integral Bessel function

$\mathbf{L}_\nu(z) = e^{-(\nu+1)\pi i/2} \mathbf{H}_\nu\left(e^{\pi i/2} z\right)$ is the modified Struve function

$L_n(z) = L_n^0(z)$ are the Laguerre polynomials

$L_n^\lambda(z) = \frac{z^{-\lambda} e^z}{n!} \frac{d^n}{dz^n} \left(z^{n+\lambda} e^{-z}\right)$ are the generalized Laguerre polynomials

$$\mathrm{Li}_\nu(z) = \sum_{k=1}^\infty \frac{z^k}{k^\nu} \qquad [|z| < 1]$$

$$= \frac{z}{\Gamma(\nu)} \int_0^\infty \frac{t^{\nu-1}\, dt}{e^t - z} \qquad [\mathrm{Re}\,\nu > 0;\ |\arg(1-z)| < \pi]$$

is the polylogarithm of the order ν

$\mathrm{Li}_2(z)$ is the Euler dilogarithm

$M_{\varkappa,\mu}(z) = z^{\mu+1/2} e^{-z/2} {}_1F_1\left(\begin{matrix}\mu - \varkappa + \frac{1}{2} \\ 2\mu + 1;\ z\end{matrix}\right)$ is the Whittaker confluent hypergeometric function

$P_n(z) = \frac{2^{-n}}{n!} \frac{d^n}{dz^n} (z^2 - 1)^n$ are the Legendre polynomials

$$P_\nu(z) \equiv P_\nu^0(z) = {}_2F_1\left(\begin{matrix} -\nu,\ 1+\nu \\ 1;\ \frac{1-z}{2} \end{matrix}\right) \qquad [|\arg(1+z)| < \pi]$$

is the Legendre function of the first kind

$$P_\nu^\mu(z) = \frac{1}{\Gamma(1-\mu)} \left(\frac{z+1}{z-1}\right)^{\mu/2} {}_2F_1\left(\begin{matrix} -\nu,\ \nu+1 \\ 1-\mu;\ \frac{1-z}{2} \end{matrix}\right) \qquad [|\arg(z \pm 1)| < \pi;\ \mu \neq m;\ m = 1, 2, \ldots]$$

$$P_\nu^m(z) = (z^2 - 1)^{m/2} \left(\frac{d}{dz}\right)^m P_\nu(z) \qquad [|\arg(z-1)| < \pi;\ m = 1, 2, \ldots]$$

$$\mathrm{P}_\nu^\mu(x) = \frac{1}{\Gamma(1-\mu)} \left(\frac{1+x}{1-x}\right)^{\mu/2} {}_2F_1\left(\begin{matrix} -\nu,\ \nu+1 \\ 1-\mu;\ \frac{1-x}{2} \end{matrix}\right) \qquad [-1 < x < 1;\ \mu \neq m;\ m = 1, 2, \ldots]$$

$$\mathrm{P}_\nu^m(x) = (-1)^m (1 - x^2)^{m/2} \left(\frac{d}{dx}\right)^m P_\nu(x) \qquad [-1 < x < 1;\ m = 1, 2, \ldots]$$

is the associated Legendre function of the first kind

$$P_n^{(\rho,\sigma)}(z) = \frac{(-1)^n}{2^n n!} (1-z)^{-\rho} (1+z)^{-\sigma} \frac{d^n}{dz^n} \left[(1-z)^{\rho+n} (1+z)^{\sigma+n}\right]$$

$$= \frac{(\rho+1)_n}{n!} {}_2F_1\left(\begin{matrix} -n,\ \rho+\sigma+n+1 \\ \rho+1;\ \frac{1-z}{2} \end{matrix}\right) \text{ are the Jacobi polynomials}$$

$Q_\nu(z) \equiv Q_\nu^0(z)$ is the Legendre function of the second kind

$$\mathrm{Q}_\nu(z) \equiv \mathrm{Q}_\nu^0(z) = Q_\nu(z) + \frac{1}{2} [\ln(z-1) - \ln(1-z)] P_\nu(z)$$

$$Q_\nu^\mu(z) = \frac{e^{i\mu\pi}\sqrt{\pi}}{2^{\nu+1}} \Gamma\left[\begin{matrix} \mu+\nu+1 \\ \nu+3/2 \end{matrix}\right] z^{-\mu-\nu-1} (z^2-1)^{\mu/2} {}_2F_1\left(\begin{matrix} \frac{\mu+\nu+1}{2},\ \frac{\mu+\nu+2}{2} \\ \nu+\frac{3}{2};\ \frac{1}{z^2} \end{matrix}\right)$$

$$[|\arg z|,\ |\arg(z \pm 1)| < \pi;\ \nu + 1/2,\ \mu + \nu \neq -1, -2, -3, \ldots]$$

$$Q_{-n-3/2}^\mu(z) = \frac{e^{i\mu\pi}\sqrt{\pi}\,\Gamma(\mu+n+3/2)}{2^{n+3/2}(n+1)!} z^{-\mu-n-3/2} (z^2-1)^{\mu/2} {}_2F_1\left(\begin{matrix} \frac{2\mu+2n+3}{4},\ \frac{2\mu+2n+5}{4} \\ n+2;\ \frac{1}{z^2} \end{matrix}\right)$$

$$[|\arg z|,\ |\arg(z \pm 1)| < \pi;\ \mu + \nu \neq -1, -2, -3, \ldots]$$

$$\mathrm{Q}_\nu^\mu(x) = \frac{e^{-i\mu\pi}}{2} \left[e^{-\mu\pi/2} Q_\nu^\mu(x + i0) + e^{i\mu\pi/2} Q_\nu^\mu(x - i0)\right]$$

$$= \frac{\pi}{2\sin\mu\pi} \left[\mathrm{P}_\nu^\mu(x)\cos\mu\pi - \Gamma\left[\begin{matrix} \nu+\mu+1 \\ \nu-\mu+1 \end{matrix}\right] \mathrm{P}_\nu^{-\mu}(x)\right]$$

$$[-1 < x < 1;\ \mu \neq \pm m;\ \mu + \nu \neq -1, -2, -3, \ldots],$$

$$= (-1)^m (1-x^2)^{m/2} \left(\frac{d}{dx}\right)^m \mathrm{Q}_\nu(x) \qquad [\mu = m;\ \nu \neq -m-1, -m-2, \ldots],$$

$$= (-1)^m \Gamma\left[\begin{matrix} \nu-m+1 \\ \mu+m+1 \end{matrix}\right] \mathrm{Q}_\nu^m(x) \qquad [\mu = -m;\ \nu \neq -m-1, -m-2, \ldots]$$

is the associated Legendre function of the second kind

$S(z) = \frac{1}{\sqrt{2\pi}} \int_0^z \frac{\sin t}{\sqrt{t}}\, dt$ is the Fresnel cosine integral

$S(z, \nu) = \int_z^\infty t^{\nu-1} \sin t\, dt \quad [\operatorname{Re}\nu < 1]$ is the generalized Fresnel sine integral

$$S_{\mu,\,\nu}(z) = s_{\mu,\nu}(z) + 2^{\mu-1}\Gamma\begin{bmatrix}\nu,\ (\nu+\mu+1)/2\\ (\nu-\mu+1)/2\end{bmatrix}\left(\frac{z}{2}\right)^{-\nu}{}_0F_1\left(1-\nu;\ -\frac{z^2}{4}\right)+$$

$$+\,2^{\mu-1}\Gamma\begin{bmatrix}-\nu,\ (1+\mu-\nu)/2\\ (1-\mu-\nu)/2\end{bmatrix}\left(\frac{z}{2}\right)^{\nu}{}_0F_1\left(1+\nu;\ -\frac{z^2}{4}\right)$$ is the Lommel function

$$s_{\mu,\,\nu}(z) = \frac{z^{\mu+1}}{(\mu+1)^2-\nu^2}\,{}_1F_2\left(1;\ \frac{\mu+\nu+3}{2},\ \frac{\mu-\nu+3}{2};\ -\frac{z^2}{4}\right)$$ is the Lommel function

$$\operatorname{sgn} x = \begin{cases}1, & x>0,\\ 0, & x=0,\\ -1, & x<0\end{cases}$$

$$\sec z = \frac{1}{\cos z}$$

$$\operatorname{sech} z = \frac{1}{\cosh z}$$

$$\operatorname{shi}(z) = \int_0^z \frac{\sinh t}{t}\,dt = -i\operatorname{Si}(iz)$$ is the hyperbolic sine integral

$$\operatorname{Si}(z) = \int_0^z \frac{\sin t}{t}\,dt$$ is the sine integral

$$\operatorname{si}(z) = \operatorname{Si}(z) - \frac{\pi}{2} = -\int_z^\infty \frac{\sin t}{t}\,dt$$ is the sine integral

$$\sin z = \frac{e^{iz}-e^{-iz}}{2i},$$

$$\operatorname{sinc} z = \frac{\sin z}{z}$$

$$\sinh z = \frac{e^{z}-e^{-z}}{2},$$

$$T(z,\,a) = \frac{1}{2\pi}\int_0^a \frac{e^{-(1+t^2)z^2/2}}{1+t^2}\,dt \quad [|\arg a|<\pi]$$ is the Owen function

$$T_n(z) = \cos(n\arccos z) = {}_2F_1\left(\begin{matrix}-n,\ n\\ \frac{1}{2};\ \frac{1-z}{2}\end{matrix}\right)$$ are the Chebyshev polynomials of the first kind

$$\tanh z = \frac{\sinh z}{\cosh z},$$

$$U_n(z) = \frac{\sin[(n+1)\arccos z]}{\sqrt{1-z^2}} = (n+1)\,{}_2F_1\left(\begin{matrix}-n,\ n+2\\ \frac{3}{2};\ \frac{1-z}{2}\end{matrix}\right)$$ are the Chebyshev polynomials of the second kind

$$W_{\varkappa,\,\mu}(z) = z^{\mu+1/2}e^{-z/2}\,\Psi\left(\begin{matrix}\mu-\varkappa+\frac{1}{2}\\ 2\mu+1;\ z\end{matrix}\right)$$ is the Whittaker confluent hypergeometric function

$$Y_\nu(z) = \frac{\cos\nu\pi J_\nu(z) - J_{-\nu}(z)}{\sin\nu\pi} \quad [\nu\neq n], \qquad Y_n(z) = \lim_{\nu\to n} Y_\nu(z) \quad [n=0,\pm1,\pm2,\ldots]$$

is the Neumann function (the Bessel function of the second kind)

$$Yi_\nu(z) = \int_z^\infty \frac{Y_\nu(t)}{t}\,dt$$ is the integral Bessel function of the second kind

$\mathrm{B}(\alpha,\beta)=\dfrac{\Gamma(\alpha)\Gamma(\beta)}{\Gamma(\alpha+\beta)}$ is the beta function

$\mathrm{B}_z(\alpha,\beta)=\int_0^z t^{\alpha-1}(1-t)^{\beta-1}\,dt$ $\quad[\operatorname{Re}\alpha>1;\ z<1]$ is the incomplete beta function

$\Gamma(z)=\int_0^\infty t^{z-1}e^{-t}\,dt$ $\quad[\operatorname{Re}z>0]$ is the gamma function

$\Gamma(\nu,z)=\int_z^\infty t^{\nu-1}e^{-t}\,dt$ is the complementary incomplete gamma function

$\gamma(\nu,z)=\Gamma(\nu)-\Gamma(\nu,z)=\int_0^z t^{\nu-1}e^{-t}\,dt$ $\quad[\operatorname{Re}\nu>0]$ is the incomplete gamma function

$$\Gamma\begin{bmatrix}(a_p)\\(b_q)\end{bmatrix}\equiv\Gamma\begin{bmatrix}a_1,\ldots,a_p\\b_1,\ldots,b_q\end{bmatrix}\equiv\frac{\prod_{k=1}^{p}\Gamma(a_k)}{\prod_{\ell=1}^{q}\Gamma(b_\ell)}$$

$$\Gamma[(a_p)]\equiv\Gamma[a_1,\ldots,a_p]\equiv\prod_{k=1}^{p}\Gamma(a_k)$$

$$\Delta(k,a)=\frac{a}{k},\frac{a+1}{k},\ldots,\frac{a+k-1}{k}$$

$$\Delta(k,(a_p))=\frac{(a_p)}{k},\frac{(a_p)+1}{k},\ldots,\frac{(a_p)+k-1}{k}$$

$\delta_{m,n}=\begin{cases}0, & m\neq n,\\1, & m=n\end{cases}$ is the Kronecker symbol

$\zeta(z)=\sum_{k=1}^{\infty}\dfrac{1}{k^z}$ $\quad[\operatorname{Re}z>1]$ is the Riemann zeta function

$\zeta(z,v)=\sum_{k=0}^{\infty}\dfrac{1}{(v+k)^z}$ $\quad[\operatorname{Re}z>1;\ v\neq0,-1,-2,\ldots]$ is the Hurwitz zeta function

$\theta_j(z,q)$ $\quad[j=1,2,3,4]$ are the theta functions:

$$\theta_1(z,q)=2\sum_{k=0}^{\infty}(-1)^k q^{(k+1/2)^2}\sin(2k+1)z,$$

$$\theta_2(z,q)=2\sum_{k=0}^{\infty}q^{(k+1/2)^2}\cos(2k+1)z,$$

$$\theta_3(z,q)=1+2\sum_{k=1}^{\infty}q^{k^2}\cos(2kz),$$

$$\theta_4(z,q)=1+2\sum_{k=1}^{\infty}(-1)^k q^{k^2}\cos(2kz)$$

$\theta(x)=\begin{cases}1, & x\geq0,\\0, & x<0\end{cases}$ is the Heaviside function

$$\lambda(z,a)=\int_0^a z^{-t}\,\Gamma(t+1)\,dt$$

$$\mu(z,\lambda) = \int_0^\infty \frac{t^\lambda z^t}{\Gamma(\lambda+1)\Gamma(t+1)}\,dt \qquad [\operatorname{Re}\lambda > -1]$$

$$\mu(z,\lambda,\rho) = \int_0^\infty \frac{t^\lambda z^{t+\rho}}{\Gamma(\lambda+1)\Gamma(t+\rho+1)}\,dt \qquad [\operatorname{Re}\lambda > -1]$$

$$\nu(z) = \int_0^\infty \frac{z^t}{\Gamma(t+1)}\,dt$$

$$\nu(z,\rho) = \int_0^\infty \frac{z^{t+\rho}}{\Gamma(t+\rho+1)}\,dt$$

$\Xi_j(\ldots;\, w,\, z)$ $[j = 1, 2]$ are the Humbert functions:

$$\Xi_1(a, a', b; c; w, z) = \sum_{k,\,\ell=0}^{\infty} \frac{(a)_k (a')_\ell (b)_k}{(c)_{k+\ell}} \frac{w^k z^\ell}{k!\,\ell!} \qquad [|w| < 1]$$

$$\Xi_2(a, b; c; w, z) = \sum_{k,\,\ell=0}^{\infty} \frac{(a)_k (b)_k}{(c)_{k+\ell}} \frac{w^k z^\ell}{k!\,\ell!} \qquad [|w| < 1]$$

$$\Phi(z, s, v) = \sum_{k=0}^{\infty} \frac{z^k}{(v+k)^s} \qquad [|z| < 1;\ v \neq 0, -1, -2, \ldots]$$

$\Phi_j(\ldots;\, w,\, z)$ $[j = 1, 2, 3]$ are the Humbert functions:

$$\Phi_1(a, b; c; w, z) = \sum_{k,\,\ell=0}^{\infty} \frac{(a)_{k+\ell} (b)_k}{(c)_{k+\ell}} \frac{w^k z^\ell}{k!\,\ell!} \qquad [|w| < 1]$$

$$\Phi_2(b, b'; c; w, z) = \sum_{k,\,\ell=0}^{\infty} \frac{(b)_k (b')_\ell}{(c)_{k+\ell}} \frac{w^k z^\ell}{k!\,\ell!}$$

$$\Phi_3(b; c; w, z) = \sum_{k,\,\ell=0}^{\infty} \frac{(b)_k}{(c)_{k+\ell}} \frac{w^k z^\ell}{k!\,\ell!}$$

$$\Psi\binom{a;\ z}{b} \equiv \Psi\binom{a}{b;\ z} \equiv \Psi(a; b; z) = \frac{\Gamma(b-1)}{\Gamma(a)} z^{1-b}\, {}_1F_1\binom{1+a-b}{2-b;\ z} + \frac{\Gamma(1-b)}{\Gamma(1+a-b)}\, {}_1F_1\binom{a;\ z}{b} \qquad [b \neq 0, \pm 1, \pm 2, \ldots]$$

$$\Psi(a; n; z) = \lim_{b\to n} \Psi(a; b; z) \qquad [n = 0, \pm 1, \pm 2, \ldots]$$

is the Tricomi confluent hypergeometric function

$\Psi_j(\ldots;\, w,\, z)$ $[j = 1, 2]$ are the Humbert functions:

$$\Psi_1(a, b; c, c'; w, z) = \sum_{k,\,\ell=0}^{\infty} \frac{(a)_{k+\ell} (b)_k}{(c)_k (c')_\ell} \frac{w^k z^\ell}{k!\,\ell!} \qquad [|w| < 1]$$

$$\Psi_2(a; c, c'; w, z) = \sum_{k,\,\ell=0}^{\infty} \frac{(a)_{k+\ell}}{(c)_k (c')_\ell} \frac{w^k z^\ell}{k!\,\ell!}$$

$\psi(z) = [\ln\Gamma(z)]' = \dfrac{\Gamma'(z)}{\Gamma(z)}$ is the psi function (digamma function)

$\psi^{(n)}(z) = \dfrac{d^n}{dz^n}\psi(z)$ is the polygamma function

Index of Notations for Symbols

$(a_p) = a_1,\, a_2, \ldots, a_p$

$(a_p) + b = a_1 + b,\, a_2 + b, \ldots, a_p + b$

$(a_p)/b = a_1/b,\, a_2/b, \ldots, a_p/b$

$(a_p)' - a_j = a_1 - a_j, \ldots, a_{j-1} - a_j,\, a_{j+1} - a_j, \ldots, a_p - a_j \qquad [1 \le j \le p]$

$(a)_k = a\,(a+1)\ldots(a+k-1) = \Gamma\,(a+k)\,/\Gamma\,(a) \quad [k = 1, 2, 3, \ldots],\ (a)_0 = 1$

is the Pochhammer symbol

$\Delta\,(k,\, a) = \frac{a}{k},\, \frac{a+1}{k}, \ldots, \frac{a+k-1}{k}$

$\Delta\,(k,\, (a_p)) = \frac{(a_p)}{k},\, \frac{(a_p)+1}{k}, \ldots, \frac{(a_p)+k-1}{k}$

$n! = 1 \cdot 2 \cdot 3 \ldots (n-1)\, n = (1)_n, \quad 0! = 1! = (-1)! = 1$

$(2n)!! = 2 \cdot 4 \cdot 6 \ldots (2n-2)\, 2n = 2^n n!, \quad 0!! = (-1)!! = 1$

$(2n+1)!! = 1 \cdot 3 \cdot 5 \ldots (2n+1) = \frac{2^{n+1}}{\sqrt{\pi}}\,\Gamma\left(n + \frac{3}{2}\right) = \left(\frac{3}{2}\right)_n 2^n$

$\binom{n}{k} = \frac{n\,(n-1)\ldots(n-k+1)}{k!} = \frac{n!}{k!\,(n-k)!} = \frac{(-1)^k\,(-n)_k}{k!}, \quad \binom{n}{0} = 1$

$\operatorname{Re} a,\ \operatorname{Re} b > c$ means $\operatorname{Re} a > c$ and $\operatorname{Re} b > c$

$[x] = n \quad [n \le x < n+1,\ n = 0, \pm 1, \pm 2, \ldots]$ is the integer part of x

$$x_+^\lambda = \begin{cases} x^\lambda, & x > 0, \\ 0, & x < 0 \end{cases}$$

$$\prod (a_p)_k = \prod_{j=1}^{p} (a_j)_k, \quad \prod ((a_p) + b)_k = \prod_{j=1}^{p} (a_j + b)_k$$

$$\prod_{k=m}^{n} a_k = a_m a_{m+1} \ldots a_n \quad [n \ge m], \qquad = 1 \quad [n < m]$$

$$\prod_{k=1}^{\infty} a_k\,(z) = \lim_{n \to \infty} \prod_{k=1}^{n} a_k\,(z)$$

$$\sum_{k=m}^{n} a_k = a_m + a_{m+1} + \ldots + a_n \quad [n \ge m], \qquad = 0 \quad [n < m]$$

$$\sum_{k=1}^{\infty} a_k\,(z) = \lim_{n \to \infty} \sum_{k=1}^{n} a_k\,(z)$$